Fundamentals of Mathematics

Co-ordinate Geometry

Second Edition

Sanjay Mishra

B.Tech.
Indian Institute of Technology, Varanasi

Title : Fundamentals of Mathematics Co-ordinate Geometry

Language : English

Author : Sanjay Mishra

Typeset & Published by :

Career Launcher Infrastructure (P) Ltd.

A-45, Mohan Cooperative Industrial Area, Near Mohan Estate Metro Station, New Delhi - 110044

Marketed by :

G.K. Publications (P) Ltd.

Plot No. 63, Sector-27A, Near Sector - 28 Metro Station, Faridabad, Haryana-121003

ISBN : **978-81-93975-89-3**

Printer's Details : Printed in India, New Delhi.

For product information :
Visit ***www.gkpublications.com*** or email to ***gkp@gkpublications.com***

Preface

Man always has been curious about everything around him, like various shapes of hills, rocks, caves, trees, etc. He also divided lands for cultivation and other purposes. He observed the various-shaped objects and gradually started to draw figures of those objects. In this process, he needed to learn about the dimensions of the things around. This need gave rise to a new branch of mathematics, 'Synthetic Geometry'. The word 'Geometry' is a combination of two Greek words, *geo* (earth) and *metry* (measure).

Synthetic geometry can be regarded as a game played by axiomatic rules created by the ancient Indians and the Greeks. The noted Greek mathematician Euclid and his predecessors as well as his disciples were quite convinced about certain rules. They thought those these rules adequately represented the laws of the real world around them. From this period to the 17th century, only geometric logics and arguments were applied in the study of geometry, which is now familiar in the name of 'synthetic geometry'. It is a study of various theorems, their proofs, solving problems with the help of these theorems and different construction techniques. Although this concept was noble and quite useful in solving problems related to straight lines, triangles, circles and their properties, but was not equally convenient for the study and derivation of the properties of many other curves like, ellipse, parabola, hyperbola, and so on.

In the 17th century-geometry was associated with algebra and many problems of synthetic geometry were conveniently solved applying the concepts of algebra. The study of geometry with the help of algebra is called 'analytical geometry'. The systematic use of algebra in geometry was first carried out by a famous French philosopher René Descartes (1596–1650) in his book *La Géométrie* (published in 1637).

'Co-ordinate Geometry' is defined as the branch of mathematics which includes the study of different curves and figures by representing points in a plane by ordered pairs of real numbers called 'Cartesian Co-ordinates' and it also includes representation of lines and other specific types of curves by algebraic equations in the variables x and y or x, y and z. The notion of co-ordinates lead to the idea of locus and consequently, laid open the path for mathematical analysis, conceptualization of functions, their limits, continuity, differentiability, and ultimately culminated into the advent of the modern form of calculus.

Co-ordinate geometry has a major share in the syllabus of IIT-JEE and other competitive examinations, so its in-depth analysis is important. During my high school days, as an IIT-JEE aspirant, and later, as a mathematics tutor, Mathematics for last fifteen years, I always realized the need of a comprehensive textbook for this subject. I, therefore, always had an insatiable desire to write one.

This book has been written with the objective of providing a textbook as well as an exercise book, focusing on problem solving. I feel, this will not only fulfil the need of a beginner, pre-college student (i.e., students of XI and XII standards), but also meet the requirements of the advanced level students who are preparing for various entrance examinations like IIT-JEE, AIEEE, BIT-SAT, and other state engineering entrance examinations. This book, *Fundamentals of Mathematics—Co-ordinate Geometry*, develops a deep insights into topics, such as Points and Cartesian System, Straight Lines and Family of Straight Lines, Circles and Family of Circle, Parabola, Ellipse and Hyperbola. I personally experienced in my teaching career that the last three conic sections are the most scary from a student's point of view, but highly scoring topics of mathematics as far as competitive exams are concerned. One of the reasons for the phobia in the students' minds against these topics is non-familiarity with these curves in basic classes and lack of good books that lay down these concepts in a student-friendly and lucid manner.

The well-arranged content list will help students and teachers to conveniently access the chapters and sub-topics of their interest. Each chapter is divided into several topics. Each topic contains theory and sometimes sub-topics with sufficient number of worked-out illustrative problems. Students can develop applicative ability of the concepts learned. This is followed by a textual exercise of both objective and subjective type problems, as per the requirements. At the end of the theory of each chapter, a large set of solved examples of both objective and subjective types is given. This will involve application of all the concepts learnt in the chapter so that students can develop mastery over the chapter. The tutorial exercise given at the end contains a large number of multiple-choice problems with single and multiple correct answers, comprehension passages, column matching problems, numerical integer type questions to facilitate the students to do thorough revision of the entire chapter and to enhance their level of understanding of the topics. For teachers, this text book will be quite helpful as it will provide a set of well-graded problems and well-arranged topics that can be used to give home assignments to their students.

All suggestions for improvement are welcome and shall be gratefully acknowledged.

—Sanjay Mishra

Acknowledgements

I am grateful to GK Publications for keeping faith in me and providing me with an opportunity to transform my yearning, my vast experience of years of teaching, and my knowledge comprehensively into the present textbook, *Fundamentals of Mathematics—Co-ordinate Geometry*. I would like to thank all my friends and teachers, for their valuable criticism, support and advice that was helped me to carve out this work. I am pleased to award special acknowledgements to all my pupils. During my interactions with my students, I received much of the inspiration of writing this textbook. I feel that by interacting with them, I have learnt much more than I could have ever taught them. I wish to thank my parents and all my family members, for their patience and support in bringing out this book and contributing their valuable time for this cause. I extend my special thanks to my team, especially to my assisting teachers, managers, and computer opera for their hard work and dedication in completing this task. I extend my special thanks to my team, Rakesh Gupta, Sanket Sinha, and the DTP Operators of MIIT (Edu.) Services Pvt. Ltd., for their hard work and dedication in comthis task.

—Sanjay Mishra

Contents

CHAPTER 1

Point and Cartesian System

■ INTRODUCTION

The word Geometry is a combination of two Greek words *Geo* (earth) and *Metry* (measure). It can be defined as a branch of mathematics which developed to facilitate the study and measurement of various land form. A famous Greek mathematician Euclid (300 BC) in his first systematic treaty on geometry mentioned some axioms and postulates and regarded them as the rules that adequately represent the laws of real world around them. The set of these laws and their applications are together called as Synthetic Geometry. Some of the postluates of Euclidean Geometry are mentioned below.

■ POSTULATES OF EUCLIDEAN GEOMETRY

1. **Notion of point:** Point is a dimensionless hypothetical object. It is a geometric construction with no dimension.

   However, it is noted that howsoever small dot is placed on a piece of paper, it is theoretically not a point but a combination of infinitely many points. But practically small dots are considered to be points as their all dimensions are reasonably small enough to be ignored and taken as zero. Moreover, we study these ideas in relative sense not in absolute sense, e.g., a dot with area 10^{-7} square cm can be considered as a point in comparison to a circular field with area few square meters but a circle with area 10 cm^2 cannot be treated so, whereas the same circle with 10 cm^2 area can be regarded as a point with respect to plane fields expanding over square miles.

2. **Line:** It is a locus of a point defined as one dimensional curve, it has only length, no breadth and thickness.

   Lines can be classified in two ways: (i) curved line; (ii) straight line.

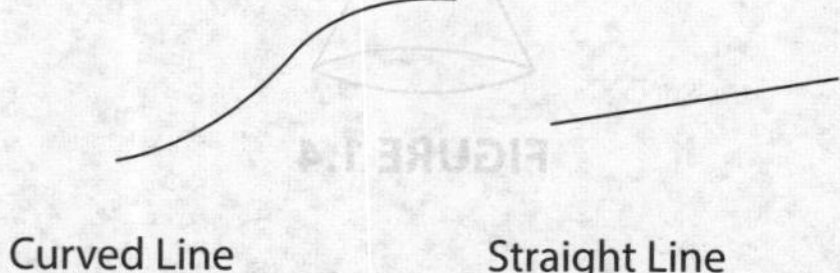

FIGURE 1.1

3. **Idea of intermediacy:** Between any two distinct points on a straight line there always lies another point howsoever close they are.

5. **Line segment and rays:** If a line is truncated by two points (say A and B), then it is known as '*line segment*'. Clearly, a line segment has an initial point (A) and a final point (B). Therefore, has a fixed finite length. When the final point B lies at infinity, then the semi-infinite length line segments obtained are called 'rays'.

6. **Surface:** It is a two-dimensional construction, can be termed as a locus of a line, i.e., having length and breadth, but no thickness. Practically, it has thickness as small as compaired to other two dimensions that it can be ignored.

   Surfaces may be 'plane' as well as 'curved'. If we choose two arbitrary points (A and B) on a surface and joint them by a straight line segment (AB), and every point on the line segment is contained in the surface, then it is regarded as Euclidean plane surface, otherwise known as curved surface.

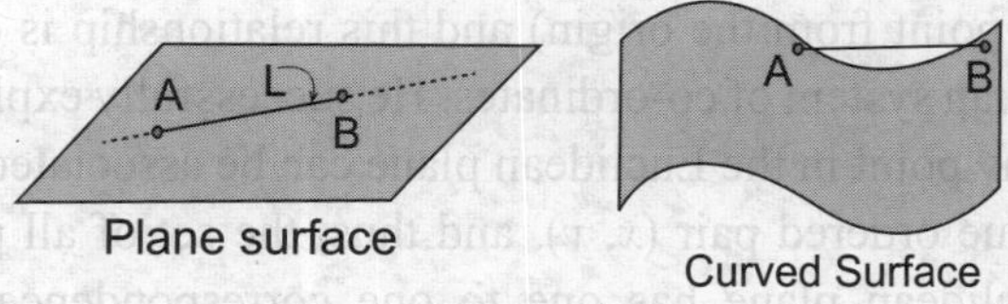

FIGURE 1.2

A straight line L divides an Euclidean plane (P) into two disjoint half planes P_1 and P_2 such that the plane (P) is union of three disjoint set of points P_1, P_2 and L, where L is the set of points on the line.

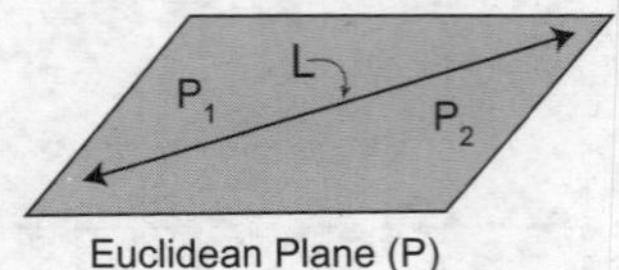

FIGURE 1.3

7. **Solid:** A geometrical construction having three dimensions, obtained by translation or rotation of surfaces is called '*solid*'.

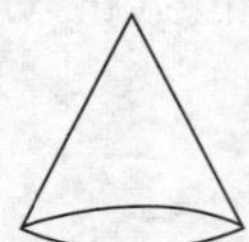

FIGURE 1.4

Frame of Reference

Several methods have been developed by mathematicians to uniquely locate the position of a point in the space. So that various useful observations can be conveniently made with the help of set of fixed points/lines/surfaces.

A set of fixed points/lines/surfaces with respect to which observations are made is called frame of reference.

■ CO-ORDINATE SYSTEMS

In the 17th centuary *AD*, the synthetic geometry was associated with algebra and the study of geometry with the systematic application of algebra was first carried out by a famous french Philospher Rene Descartes (1596-1650) in his book *La Geometrie* (published in 1637). The study of geometry with help of algebraic equations is called analytical geometry/Cartesian co-ordinate geometry.

Co-ordinate geometry begins with the study of the *Concept of Point.* Descartes established a relationship between the basic geometric concept of point and the ordered pair of real number (x, y) (where x represents the horizontal distance of the point from origin and y represents the vertical distance of the point from the origin) and this relationship is called Cartesian system of co-ordinates. He successfully explained that any point in the Euclidean plane can be associated with a unique ordered pair (x, y), and thus, the set of all points in Euclidean plane has one to one correspondence with the set of ordered pairs represented by Cartesian product $X \times Y$ ($\mathbb{R} \times \mathbb{R}$). This is the reason why Euclidean plane is also called Cartesian plane or $\mathbb{R} \times \mathbb{R}$ plane or $\mathbb{R}^2$ plane.

The most commonly used frame of references (co-ordinate systems) are of the following three types:

(1) Rectangular co-ordinate system

(2) Oblique co-ordinate system

(3) Polar co-ordinate system

Rectangular Co-ordinate System

It consists of two mutually perpendicular lines intersecting at O called origin as shown below.

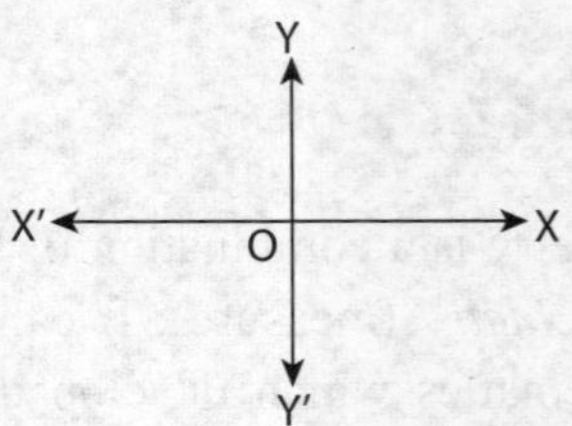

FIGURE 1.5

- Origin is the point from where the observations are made with the help of two axes and a suitably chosen sign convention.
- Horizontal line $X'OX$ is called x-axis (*abscissa axis*).
- Vertical line $Y'OY$ is called y-axis (*ordinate axis*).

Sign Convention

All the distances measured towards OX and OY directions are cosidered to be positive whereas towards OX' and OY' directions are termed negative and any distance measured from origin to any other direction than the above four is taken always positive. The angle measured in clockwise direction is considered negative, whereas that measured in counterclockwise sense is taken as positive angle.

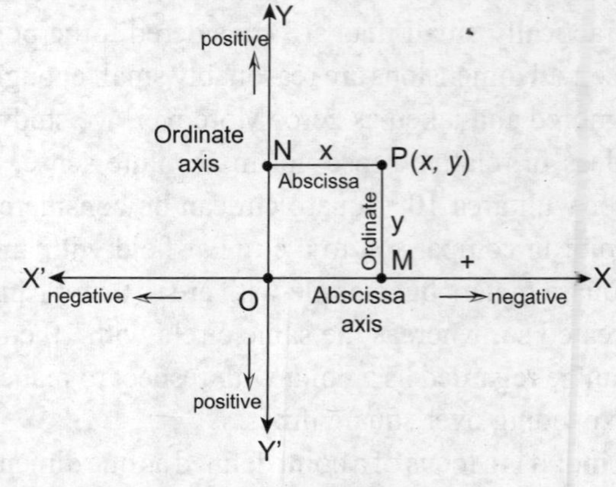

FIGURE 1.6

- **Representation of point:** Any point P in x-y plane can be represented by unique ordered pair of two real numbers x and y as (x, y) and it is defined as co-ordinates of the point P.
- Here x is *abscissa* of point (OM or PN).
- y is *ordinate* of point (ON or PM).
- Therefore, the x-y plane (Cartesian plane) is algebraically represented as Cartesian product of two sets of real numbers, so it is called $\mathbb{R} \times \mathbb{R}$ plane or $(\mathbb{R})^2$ plane. where $\mathbb{R}\times\mathbb{R} = \{(x, y) : x \in \mathbb{R} \text{ and } y \in \mathbb{R}\}$.

$\mathbb{R}^+ \times \mathbb{R}^+$ = set of all points in the 1st quadrant

$\mathbb{R}^+ \times \mathbb{R}^-$ = set of all points in the 4th quadrant

$\mathbb{R}^- \times \mathbb{R}^+$ = set of all points in the 2nd quadrant

$\mathbb{R}^- \times \mathbb{R}^-$ = set of all points in the 3rd quadrant

REMARKS

(i) *The ordinate of every point on x-axis is 0, that is why equation of x-axis is $y = 0$.*

(ii) *The absicissa of every point on y-axis is 0, that is why equation of y-axis is $x = 0$.*

(iii) *The abscissa* and *ordinate of the origin O are both zero, i.e., (0, 0).*

(iv) *The abscissa* and *ordinate of a point is its algebraic length of perpendicular distance from y-axis* and *x-axis, respectively.*

(v)

Quadrants	**XOY (I)**	**X'OY (II)**	**X'OY'(III)**	**(XOY'(IV)**
Sign of x co-ordinates	+	–	–	+
Sign of y co-ordinates	+	+	–	–
Sign of (x, y)	(+, +)	(–, +)	(–, –)	(+, –)

ILLUSTRATION 1: Locate (1, 2), (–1, 2), (–1, –2) and (1, –2) in rectangular co-ordinate system and then find the area of figure obtained by joining them with straight line segments in the given order.

SOLUTION: Clearly, the figure obtained is rectangle ABCD with area 8 square unit.

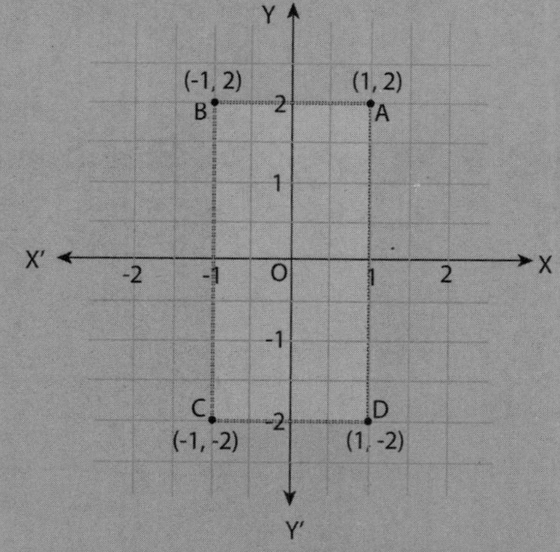

FIGURE 1.7

ILLUSTRATION 2: Locate the position of following points in the correct quadrant.

(a) $(a\alpha^2 + b\alpha + c,\ b^2 - 4ac)\ \forall \alpha \in \mathbb{R}$ when $a > 0$ and $b^2 < 4ac$

(b) $(\cos\theta + \sin\theta,\ \cos\theta - \sin\theta)$ when $\theta \in \left(\frac{\pi}{4}, \frac{3\pi}{4}\right)$

SOLUTION: (a) Given, $b^2 < 4ac \Rightarrow b^2 - 4ac < 0$

Therefore, expression $a\alpha^2 + b\alpha + c$ has same sign as its leading co-efficient.

$\Rightarrow$ Abscissa of point, i.e., $x = a\alpha^2 + b\alpha + c > 0 \ \forall \alpha \in \mathbb{R} \quad a > 0$

and ordinate of point i.e., $y = b^2 - 4ac < 0$. Clearly, the point lies in IVth quadrant.

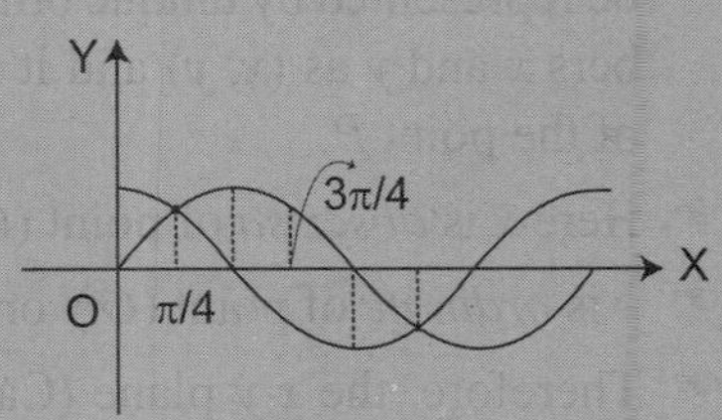

FIGURE 1.8

(b) Consider the co-ordinate of the point $x = \cos\theta + \sin\theta$;

$y = \cos\theta - \sin\theta$

As $\forall \theta \in \left(\frac{\pi}{4}, \frac{3\pi}{4}\right)$; $\sin\theta > \cos\theta \Rightarrow \cos\theta - \sin\theta < 0 \Rightarrow y < 0$

whereas $\cos\theta + \sin\theta$ is positive if $\theta \in \left(\frac{\pi}{4}, \frac{3\pi}{4}\right)$.

$\Rightarrow$ x > 0. Hence, point lies in the IVth quadrant.

ILLUSTRATION 3: Find the quadrant where the following points are located.

(a) $\left(\cos 10° - \cos 13°, \tan\frac{\pi}{9} - \tan\frac{\pi}{7}\right)$ (b) $(\cos 18° - \sin 15°, \cos 105° - \tan 15°)$

(c) $(\sec^2 9° - \tan^2 9°, \cos 17° - 1)$ (d) $\left(\log_2\left(\frac{x}{x^2+1}\right), \log_{1/2}(x^2+x+1)\right) \forall x \in \mathbb{R}^+$

SOLUTION: (a) Since $\cos x$ decreases with increase of $x \ \forall \ x \in (0, \pi)$ thus $\cos 10° > \cos 13°$

$\Rightarrow x = \cos 10° - \cos 13° > 0$

and as $\tan x$ increases with x in first quadrant and

$\because \frac{\pi}{9} < \frac{\pi}{7} \Rightarrow \tan\frac{\pi}{9} < \tan\frac{\pi}{7} \Rightarrow y = \tan\frac{\pi}{9} - \tan\frac{\pi}{7} < 0$

Thus, $x > 0$ and $y < 0$, so the point lies in 4th quadrant.

(b) Since $\sin\theta$ is an increasing function in the interval $[0, \pi/2]$ and $\cos 18° = \sin 72°$

$\Rightarrow \sin 72° > \sin 15° \Rightarrow \cos 18° - \sin 15° > 0$

Now, $\cos 105° - \tan 15° = \cos(90° + 15°) - \tan 15°$

$= -\sin 15° - \tan 15° < 0$ ($\sin 15°, \tan 15°$ are both positive)

$\therefore$ Abscissa of point is positive and ordinate negative, consequently the point lies in the IVth quadrant.

(c) Since $\sec^2 9° - \tan^2 9° = 1$ (as $\sec^2\theta - \tan^2\theta = 1$) and $\cos 17° - 1 < 0$ as $\cos\theta \le 1 \ \forall \ \vartheta \in \mathbb{R}$

$\therefore$ Abscissa of point is positive, whereas the ordinate is negative.

$\Rightarrow$ point lies in IVth quadrant.

(d) Let $y = \frac{x}{x^2+1}$; $x \in \mathbb{R}^+ \Rightarrow x^2y - x + y = 0$, but $x \in \mathbb{R}^+ \Rightarrow D \ge 0 \Rightarrow 1 - 4y^2 \ge 0$

$\Rightarrow \frac{-1}{2} \le y \le \frac{1}{2}$. But $x \in \mathbb{R}^+$, therefore y should be positive. $\Rightarrow \frac{x}{x^2+1} \in \left(0, \frac{1}{2}\right]$

Alternatively, $y = \frac{x}{x^2+1} = \frac{1}{x + \frac{1}{x}}$; $x \in \mathbb{R}^+$

by AM ≥ GM

$$\Rightarrow \frac{x+\frac{1}{x}}{2} \geq 1 \Rightarrow \frac{1}{x+\frac{1}{x}} \in \left(0, \frac{1}{2}\right]$$

$\Rightarrow \log_2\left(\frac{x}{x^2+1}\right) < 0$ as $\log_b a < 0$ for a and b on opposite side of 1. Thus, abscissa is negative.

Now, $x^2 + x + 1 = x^2 + x + 1/4 + 3/4$

$$= \left(x+\frac{1}{2}\right)^2 + \frac{3}{4} \geq \frac{3}{4} \Rightarrow x^2 + x + 1 \in \left[\frac{3}{4}, \infty\right)$$

But, for $x \in \mathbb{R}^+$, $x^2 + x + 1$ is an increasing function and $x^2 + x + 1 \in (1, \infty)$.

$\Rightarrow \log_{1/2}(x^2 + x + 1) < 0 \Rightarrow$ ordinate is also negative $\Rightarrow$ Point lies in the IIIrd quadrant.

■ OBLIQUE CO-ORDINATE SYSTEM

It consists of two axes which are not perpendicular, i.e., inclined at certain angle ω (ω ≠ 90°) as shown in the figure. That is why the system is known as oblique co-ordinate system. The concept of assigning co-ordinates and sign convention is the same as the rectangular co-ordinate system. The point P with co-ordinates (x', y') means $OM = x'$ and $PM = y'$.

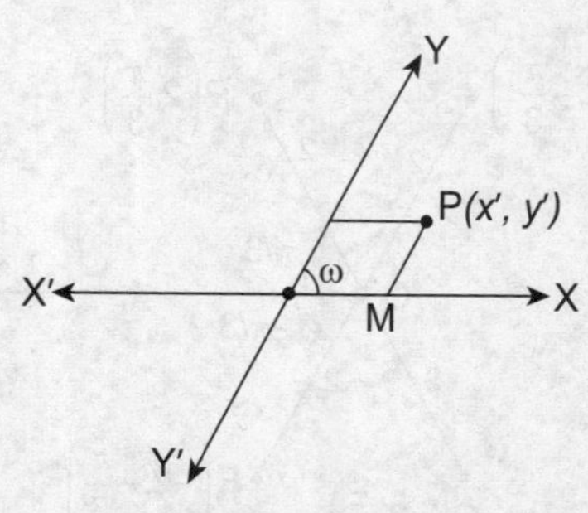

FIGURE 1.9

Relation between rectangular co-ordinate system and oblique co-ordinate system: Consider that a point P having its rectangular co-ordinates (x, y) and oblique co-ordinates (x', y') as shown in the figure below.

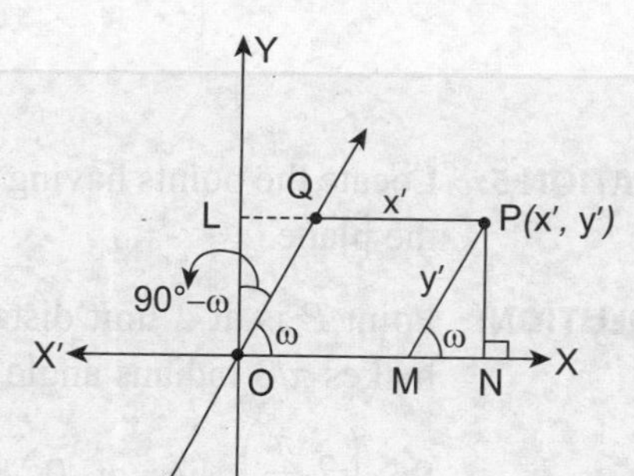

FIGURE 1.10

Now, $x = LP = LQ + QP = y' \cos\omega + x'$ and $y = PN = y' \sin\omega$, Therefore, $x = x' + y' \cos\omega$ and $y = y' \sin\omega$.

ILLUSTRATION 4: Let $P(2,3)$ in a rectangular co-ordinate system. Then, obtain this point with respect to oblique co-ordinate system, where angle between two axes is 60°.

SOLUTION: $x = x' + y' \cos\omega$

$\Rightarrow 2 = x' + y' \cos 60°$ .... (i)

and $y = y' \sin\omega$

$\Rightarrow 3 = y' \sin 60° \Rightarrow y' = 2\sqrt{3}$

$\therefore$ From (i), we get $x' = 2 - 2\sqrt{3}\,(1/2) = 2 - \sqrt{3}$

$\therefore$ Oblique co-ordinates of point P are $(2 - \sqrt{3}, 2\sqrt{3})$.

Polar Co-ordinate System

This system has its frame of reference that consists of a point O (called as *Pole*) and a Ray (OX) originating from pole known as *initial line.*

Pole Initial Line
O ———— X.

FIGURE 1.11

The line joining any point P to pole (O), i.e., OP ($OP = r$) is called '*radius vector*' and the angle $\angle XOP = \theta$ that radius vector subtends with initial line in anti-clock-wise sense is called *vectorial angle*. Position of any point P lying in the plane containing O and initial line OX, can be located uniquely by an ordered pair (r, θ) which is also called the *polar co-ordinates* of point P.

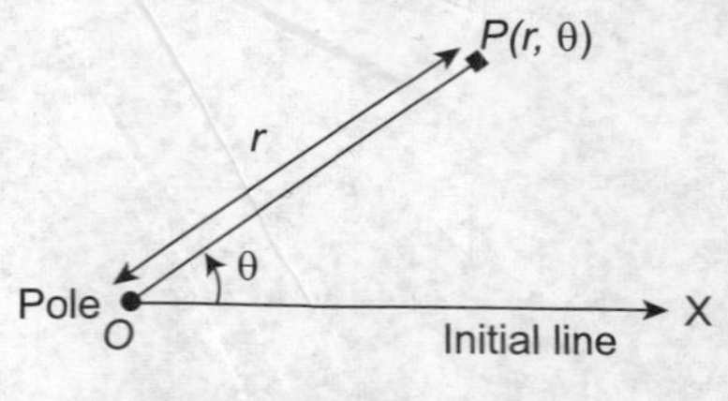

FIGURE 1.12

REMARKS

(i) *Replacing r by –r, the position of point (r, θ) gets reflected in pole (origin). This process is called 'plunging of points'. So (r, θ) and (– r, θ) are mirror images of each other in pole.*

(ii) *Adding 2π or 360° (or any integral multiple of 2π or 360°) to the vectorial angle does not alter the final position of revolving line so that (r, θ) is always the same point as $(r, \theta + 2n\pi$ or $n \times (360°)$, where $n \in \mathbb{Z}$.*

(iii) *Adding π or 180° or any odd multiple of π to the vectorial angle* and *changing the sign of radius vector gives the same point as original. Thus the point (r, θ) is same as $(-r, \theta + \pi)$ or $(-r, \theta + (2n + 1)\pi)$.*

ILLUSTRATION 5: Locate the points having polar co-ordinates $P\left(2, \frac{\pi}{3}\right), Q\left(-2, \frac{\pi}{3}\right), R\left(-2, -\frac{\pi}{3}\right)$ and $S\left(2, -\frac{\pi}{3}\right)$ on the plane.

SOLUTION: Point P is at 2 unit distance from O and OP makes $\pi/3$ radians angle with OX.

So $\left(2, \frac{\pi}{3}\right)$ lies at P. Similarly, Q denotes $\left(-2, \frac{\pi}{3}\right)$, R denotes $\left(-2, -\frac{\pi}{3}\right)$ and S denotes its image in pole, i.e., $\left(2, -\frac{\pi}{3}\right)$.

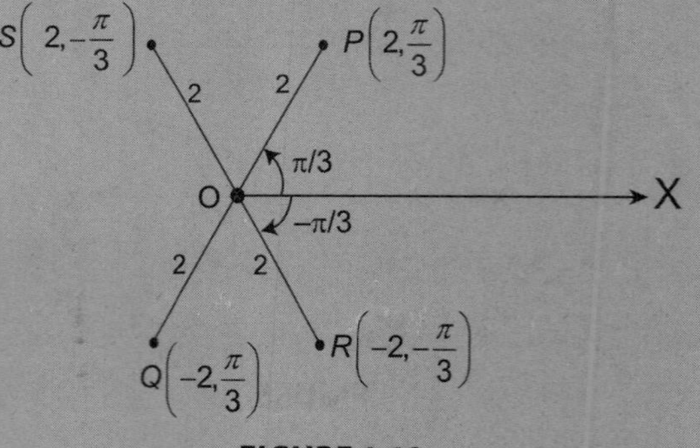

FIGURE 1.13

Relation Between the Polar and Cartesian Co-ordinates

Let $P(x, y)$ be the cartesian co-ordinates with respect to OX and OY and (r, θ) be its polar co-ordinates with respect to pole O and initial line OX. It is clear from from the that

$$OM = x = r\cos\theta \quad \text{.... (1)}$$

and

$$MP = y = r\sin\theta \quad \text{.... (2)}$$

Squaring and adding (1) and (2), we get:

$$x^2 + y^2 = r^2 \quad \text{or} \quad r = \sqrt{(x^2 + y^2)}$$

Y, P(x,y), r, y, θ, O, M, X

FIGURE 1.14

Dividing (2) by (1), we get $\tan\theta = y/x$.

Now, we have the following cases:

Case I: If P is in the first quadrant, then $\tan\theta = y/x$

$\Rightarrow \theta = \tan^{-1} y/x$

Case II: If P is in the second quadrant, then $\tan\theta = y/x$

$\Rightarrow \theta = \pi - \tan^{-1} |y/x|$

Case III: If P is in the third quadrant, then $\tan\theta = y/x$

$\Rightarrow \theta = \pi + \tan^{-1} |y/x| \text{ or } -\pi + \tan^{-1} |y/x|$

Case IV: If P is in the fourth quadrant, then $\tan\theta = y/x$

$\Rightarrow \theta = -\tan^{-1} |y/x| \text{ or } 2\pi - \tan^{-1} |y/x|$

Therefore, by using $r = \sqrt{x^2 + y^2}$ and the above four cases, we can find the polar co-ordinates of P, when its rectangular co-ordinates are known.

DISTANCE BETWEEN TWO POINTS LYING IN A PLANE

1. When Co-ordinates of Two Points are Given in Rectangular Form

Let, $P(x_1, y_1)$ and $Q(x_2, y_2)$ be two given points, then the distance PQ between them is given by.

$$PQ = d = \sqrt{(x_2 - x_1)^2 + (y_2 - y_1)^2}$$

Proof: From the figure:

$QM = QN - MN = y_2 - y_1$

$PM = ON - OL = x_2 - x_1$

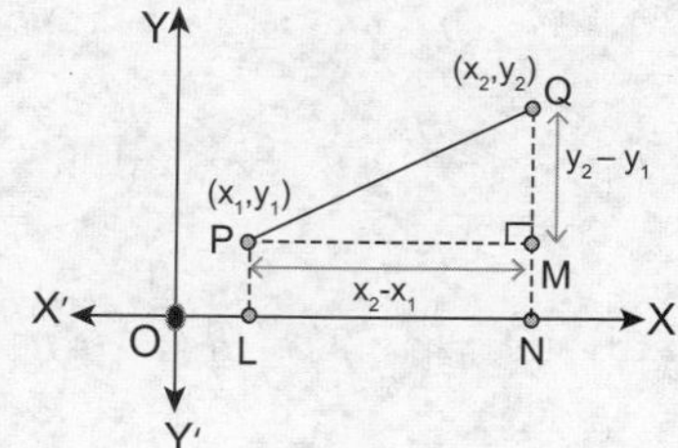

FIGURE 1.15

Now, applying Pythagorous theorem, we have:

$PQ^2 = PM^2 + QM^2$

$$\Rightarrow PQ = \sqrt{PM^2 + QM^2}$$

$$\Rightarrow d = \sqrt{(x_2 - x_1)^2 + (y_2 - y_1)^2}$$

REMARKS

(i) *When the line PQ is parallel to the y-axis, the abscissa of points P and Q will be equal, i.e.,* $x_1 = x_2$. *Therefore,* $PQ = |y_2 - y_1|$.

(ii) *When the segment PQ is parallel to the x-axis, the ordinates of the points P and Q will be equal, i.e.,* $y_1 = y_2$. *Therefore,* $PQ = |x_2 - x_1|$.

(iii) *The above result holds good even when P and Q lies in the different quadrants.*

ILLUSTRATION 6: Find the distance between points

(a) $P(-2,1)$ and $Q(1, -3)$ (b) $P(5\tan\theta, 5)$ and origin $(0,0)$.

SOLUTION: (a) $PQ = \sqrt{(-2-1)^2 + (1-(-3))^2} = \sqrt{(-3)^2 + (1+3)^2} = \sqrt{9+16} = \sqrt{25} = 5.$

(b) $PQ = \sqrt{(5\tan\theta - 0)^2 + (5-0)^2} = \sqrt{25\tan^2\theta + 25} = 5\sqrt{\tan^2\theta + 1} = 5|\sec\theta|$

ILLUSTRATION 7: Show that the triangle whose vertices are $A(-3, -4)$, $B(2, 6)$ and $C(-6, 10)$ is right angled.

SOLUTION: We observe that

$AB^2 = (-3-2)^2 + (-4-6)^2 = 125$; $BC^2 = (2-(-6))^2 + (6-10)^2 = 80$

$CA^2 = (-6-(-3))^2 + (10-(-4))^2 = 205$

Thus, $CA^2 = 205 = 125 + 80 = AB^2 + BC^2$ and hence, ABC is right, angled at B.

ILLUSTRATION 8: Prove that the points $A(3, -5)$, $B(-5, -4)$, $C(7, 10)$ and $D(15, 9)$ taken in order are the vertices of a parallelogram.

SOLUTION: Since $AB^2 = (3-(-5))^2 + (-5-(-4))^2 = 65$; $BC^2 = (-5-7)^2 + (-4-10)^2 = 340$

$CD^2 = (7-15)^2 + (10-9)^2 = 65$; $DA^2 = (15-3)^2 + (9-(-5))^2 = 340$

Clearly, opposite sides are equal. So, $ABCD$ is a parallelogram.

ILLUSTRATION 9: If P and Q are two points whose co-ordinates are $(at^2, 2at)$ and $\left(\frac{a}{t^2}, -\frac{2a}{t}\right)$, respectively, and S is the point $(a,0)$. Show that $\frac{1}{SP}+\frac{1}{SQ}$ is independent of t.

SOLUTION: We have, $SP = \sqrt{(at^2-a)^2+(2at-0)^2} = |a|\sqrt{(t^2-1)^2+4t^2}$

$$= |a|\sqrt{(t^2+1)^2} = \left|a(t^2+1)\right| = a(t^2+1)$$

and, $$SQ = \sqrt{\left(\frac{a}{t^2}-a\right)^2+\left(-\frac{2a}{t}-0\right)^2} = \sqrt{a^2\frac{(1-t^2)^2}{t^4}+\frac{4a^2}{t^2}}$$

$$= \frac{|a|}{t^2}\sqrt{(1-t^2)^2+4t^2} = \frac{|a|}{t^2}\sqrt{(1+t^2)^2} = \frac{|a|}{t^2}(1+t^2)$$

$$SP = \sqrt{(at^2-a)^2+(2at-0)^2} = |a|\sqrt{(t^2-1)^2+4t^2} = |a|(1+t^2)$$

$$\therefore \quad \frac{1}{SP}+\frac{1}{SQ} = \frac{1}{|a|(t^2+1)}+\frac{t^2}{|a|(t^2+1)} = \frac{1+t^2}{|a|(t^2+1)} = \frac{1}{|a|},$$ which is independent of t.

ILLUSTRATION 10: Find the centre and radius of the circle passing through the vertices of a triangle ABC where the co-ordinates of vertices are given as: $A(0, 1)$, $B(1, 0)$, $C(2, 3)$.

SOLUTION: Let O' be the centre of the circle circumscribing ΔABC.

$(O'A)^2 = (O'B)^2 \Rightarrow (x-0)^2 + (y-1)^2 = (x-1)^2 + (y-0)^2$

$\Rightarrow -2y = -2x \Rightarrow y = x$ ...(i)

$(O'B)^2 = (O'C)^2 \Rightarrow (x-1)^2 + (y-0)^2 = (x-2)^2 + (y-3)^2$

$\Rightarrow 2x + 6y = 12 \Rightarrow x + 3y = 6$ ...(ii)

Solving the equations (i) and (ii), we get $4x = 6 \Rightarrow x = 3/2$

so $y = 3/2$ from equation (i) $\because x = 3/2$

so the centre is $(3/2, 3/2)$

Therefore, the radius of circumcircle $R = \sqrt{(3/2-1)^2+(3/2-0)^2}$

$= \sqrt{1/4+9/4} = \frac{\sqrt{10}}{2}$.

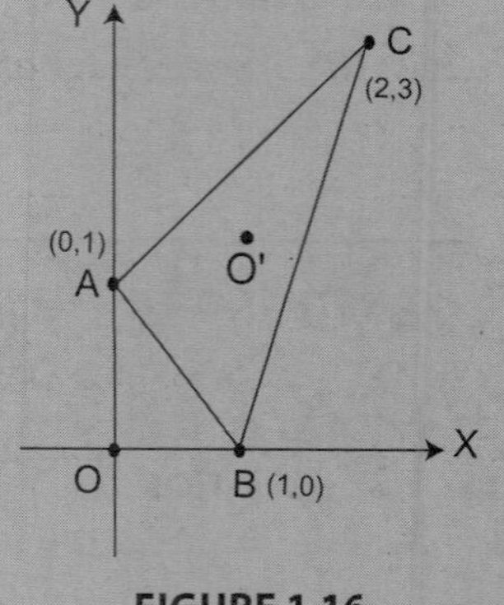

FIGURE 1.16

ILLUSTRATION 11: In any triangle ABC, prove that $AB^2 + AC^2 = 2(AD^2 + DC^2)$, where D is the middle point of BC.

SOLUTION: Take B as origin, BC as the axis of x and a line through B perpendicular to BC as the axis of y. Let $BC = a$, so that C is the point $(a, 0)$ and let A be the point (x_1, y_1). Then D is the point $(a/2, 0)$.

Hencem $AD^2 = \left(x_1 - \frac{a}{2}\right)^2 + y_1^2$ and $DC^2 = \left(\frac{a}{2}\right)^2$

Hencem $2(AD^2 + DC^2) = 2\left[x_1^2 + y_1^2 - ax_1 + a^2/4 + a^2/4\right]$

$= 2x_1^2 + 2y_1^2 - 2ax_1 + a^2.$

Also, $AC^2 = (x_1 - a)^2 + y_1^2$ and $AB^2 = x_1^2 + y_1^2$

Therefore, $AB^2 + AC^2 = 2x_1^2 + 2y_1^2 - 2ax_1 + a^2$

Hence, $AB^2 + AC^2 = 2(AD^2 + DC^2)$.

FIGURE 1.17

2. When the Co-ordinates of Points are Given in Oblique System

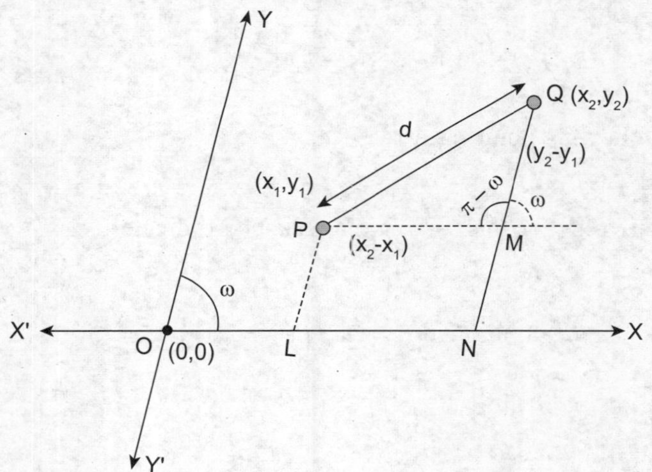

FIGURE 1.18

If the co-ordinate system is oblique, i.e., if the co-ordinate axes are inclined at an angle ω. In this case, the distance between two points P and Q will be given by

$$PQ = d = \sqrt{(x_2 - x_1)^2 + (y_2 - y_1)^2 + 2(x_2 - x_1)(y_2 - y_1)\cos\omega}$$

Hint: (a) To derive the above expression, students are advised to apply cosine formula for $\angle QPM$ in the triangle. PMQ,

i.e., $\cos(\pi - \omega) = \dfrac{(x_2 - x_1)^2 + (y_2 - y_1)^2 - d^2}{2(x_2 - x_1)(y_2 - y_1)}$

(b) If the distance between two points is zero, then both the points have same respective co-ordinates and vice versa.

ILLUSTRATION 12: Let, $A(4,5)$ and $B(2,3)$ be two points in an oblique co-ordinate system, in which the axes are inclined at an angle of $30°$. Then, find the distance AB.

SOLUTION: $AB = \sqrt{(4-2)^2 + (5-3)^2 + 2(4-2)(5-3)\cos 30°}$;

$\Rightarrow AB = \sqrt{2^2 + 2^2 + (2)(2)(2) \times \frac{\sqrt{3}}{2}}$

$= \sqrt{4 + 4 + 4\sqrt{3}}$

$\Rightarrow AB = 2\sqrt{2 + \sqrt{3}}$

3. When the Co-ordinates of Points are Given in Polar Form

Let, $P(r_1, \theta_1)$ and $Q(r_2, \theta_2)$ be two given points in polar co-ordinate system, then the distance PQ between them shall be given as:

$$PQ = d = \sqrt{r_1^2 + r_2^2 - 2r_1r_2\cos(\theta_1 - \theta_2)}$$

Proof: In ΔPOQ, applying cosine rule for the $\angle POQ$,

$$\cos(\theta_1 - \theta_2) = \frac{(OP)^2 + (OQ)^2 - (PQ)^2}{2OP.OQ}$$

$$OP = r_1 \text{ and } OQ = r_2$$

$$\therefore \quad (PQ)^2 = r_1^2 + r_2^2 - 2r_1r_2\cos(\theta_1 - \theta_2)$$

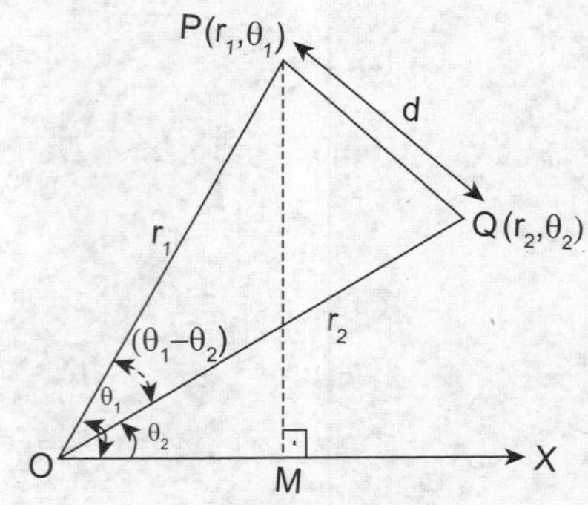

FIGURE 1.19

$$\therefore \quad PQ = \sqrt{r_1^2 + r_2^2 - 2r_1r_2\cos(\theta_1 - \theta_2)}$$

ILLUSTRATION 13: Find the distance between points $P\left(3, -\dfrac{\pi}{6}\right)$ and $Q\left(2, \dfrac{\pi}{6}\right)$.

SOLUTION: $PQ = \sqrt{3^2 + 2^2 - 2\times3\times2\times\cos\left(-\dfrac{\pi}{6} - \dfrac{\pi}{6}\right)} \Rightarrow PQ = \sqrt{13 - 12\times\dfrac{1}{2}} = \sqrt{7}$

ILLUSTRATION 14: (a) Find the polar co-ordinates of the following points (x, y):

(i) $(1, \sqrt{3})$ (ii) $(\sqrt{2}, 1)$ (iii) $(-2, -2)$.

(b) If polar co-ordinates of any points are $(2, \pi/3)$, then find its Cartesian co-ordinates.

SOLUTION: (a) Let the polar co-ordinate of the point be (r, θ), so

(i) $1 = r\cos\theta$ and $\sqrt{3} = r\sin\theta \Rightarrow r = \sqrt{1+3} = 2$ and $\theta = \tan^{-1}\left(\dfrac{y}{x}\right) = \tan^{-1}\sqrt{3} = \dfrac{\pi}{3}$

Hence, the polar co-ordinates of the point are $\left(2, \dfrac{\pi}{3}\right)$.

(ii) $\sqrt{2} = r\cos\theta$ and $1 = r\sin\theta \Rightarrow r = \sqrt{1+2} = \sqrt{3}$ and $\theta = \tan^{-1}\left(\dfrac{1}{\sqrt{2}}\right)$,

hence the polar co-ordinates are $\left(\sqrt{3}, \tan^{-1}\dfrac{1}{\sqrt{2}}\right)$

(iii) Since the cartesian co-ordinate of P are $(-2, -2)$, therefore

$x = -2 = r\cos\theta$; $y = -2 = r\sin\theta \Rightarrow r = \sqrt{4+4} = 2\sqrt{2}$

since the point lies in third quadrant, $\theta = \pi + \tan^{-1}|y/x| = \pi + \tan^{-1}1$

$\Rightarrow \theta = \pi + \dfrac{\pi}{4} = \dfrac{5\pi}{4}$

$\therefore$ The co-ordinates of P are $\left(2\sqrt{2}, \dfrac{5\pi}{4}\right)$.

(b) Since polar co-ordinates are $(2, \pi/3) \Rightarrow$ the point lies in first quadrant.

$\Rightarrow x = 2\cos\pi/3 = 1$ and $y = 2\sin\pi/3 = \sqrt{3}$

Hence, the cartesian co-ordinates of the point P are $(1, \sqrt{3})$.

ILLUSTRATION 15: Transform the polar equation of the curve $r = a\cos 2\theta$ into Cartesian form.

SOLUTION: Since $x = r\cos\theta, y = r\sin\theta \therefore x^2 + y^2 = r^2$ ... (i)

and $\cos 2\theta = \cos^2\theta - \sin^2\theta = \left(\frac{x^2}{r^2} - \frac{y^2}{r^2}\right)$

$\Rightarrow a\cos 2\theta = \frac{a}{r^2}(x^2 - y^2)$ ...(ii)

$\because r = a\cos 2\theta \Rightarrow r = \frac{a}{r^2}(x^2 - y^2) \Rightarrow r^3 = a(x^2 - y^2) \Rightarrow (x^2 + y^2)^{3/2} = a(x^2 - y^2)$

This is the required equation in the Cartesian form.

TEXTUAL EXERCISE–1 (SUBJECTIVE)

1. Find the distance between the points:
(a) $R(a + b, a - b)$ and $S(a - b, -a - b)$
(b) $A(at_1^2, 2at_1)$ and $B(at_2^2, 2at_2)$
(c) $L(a\cos\alpha, a\sin\alpha)$ and $M(a\cos\beta, a\sin\beta)$
(d) $(a\cos\theta, b\sin\theta)$ and $(a\cos\phi, b\sin\phi)$
(e) $(ct_1, c/t_1)$ and $(ct_2, c/t_2)$

2. Find the rectangular co-ordinates of the points whose polar co-ordinates are
(i) $\left(5, \pi - \tan^{-1}\left(\frac{4}{3}\right)\right)$
(ii) $\left(5\sqrt{2}, \frac{\pi}{4}\right)$

3. If the point (x, y) is equidistant from the points $(a + b, b - a)$ and $(a - b, a + b)$ prove that $bx - ay = 0$.

4. Prove that the points $(2a, 4a)$, $(2a, 6a)$ and $(2a + \sqrt{3}a, 5a)$ are the vertices of an equilateral triangle whose side is $|2a|$.

5. The base of an equilateral triangle with side $2a$ lies along the y-axis such that the mid-point of the base is at the origin. Find the vertices of the triangle.

6. Prove that the distance of the point $(b + a\cos\alpha, c + a\sin\alpha)$ from the point (b, c) is independent of α.

7. Find the distance between the points $(a + r\cos\alpha, b + r\sin\alpha)$ and $(a + r\cos\beta, b + r\sin\beta)$ where r is positive.

8. In which quadrant, do following points lie?
(a) $\left(a^2 + b^2 - ab, 2ab - a^2 - b^2\right): a, b \in \mathbb{R}$ and $a \neq b$
(b) $\left(a + b + c, a^3 + b^3 + c^3 - 3abc\right)$
where a, b, c are different real numbers of same sign.

9. If $P(\cos\theta + \sin\theta, \cos\theta - \sin\theta)$ for $\theta \in [0, \pi/4]$; then find in which quadrant this point P lies.

10. A line segment AB is of length 10 unit, given co-ordinates of $A(2, 3)$ and abscissa of B be 10, then prove that ordinate of B is either –3 or 9.

Answer Keys

1. (a) $2\sqrt{a^2 + b^2}$ (b) $a|(t_2 - t_1)|\sqrt{(t_2 + t_1)^2 + 4}$ (c) $2|a|\sin\left|\left(\frac{\alpha - \beta}{2}\right)\right|$

(d) $\sqrt{a^2(\cos\phi - \cos\theta)^2 + b^2(\sin\phi - \sin\theta)^2}$ (e) $\frac{c(t_2 - t_1)}{t_1 t_2}\sqrt{1 + (t_1 t_2)^2}$

2. (i) (–3, 4) (ii) (5, 5)

5. $(\sqrt{3}a, 0), (0, a), (0, -a)$ or $(-\sqrt{3}a, 0), (0, a), (0, -a)$

7. $2r\sin\left(\frac{\alpha - \beta}{2}\right)$

8. (a) Fourth quadrant (b) 1st or 3rd quadrant **9.** First quadrant or on x-axis when $\theta = \pi/4$

TEXTUAL EXERCISE–1 (OBJECTIVE)

1. The polar co-ordinates of the point whose cartesian co-ordinates are (–1, –1) is

(a) $\left(\sqrt{2}, \frac{\pi}{4}\right)$ (b) $\left(\sqrt{2}, \frac{5\pi}{4}\right)$

(c) $\left(\sqrt{2}, -\frac{\pi}{4}\right)$ (d) $\left(\sqrt{2}, \frac{3\pi}{4}\right)$

2. The cartesian co-ordinates of the point whose polar co-ordinates are $\left(2\sqrt{2}, -\frac{3\pi}{4}\right)$.

(a) (–2, 2) (b) (2, –2)

(c) (–2, –2) (d) (2, 4)

3. Cartesian equation of the polar equation $r = a \sin\theta$ is

(a) $x^2 + y^2 = ay$ (b) $x^2 + y^2 = -ay$

(c) $x^2 - y^2 = ay$ (d) None of these

4. Cartesian equation is given by $(x - a)^2 + y^2 = a^2$. Then its polar co-ordinate equation is given by

(a) $x = a(1 + \cos\theta)$, $y = a \sin\theta$

(b) $x = a(1 - \cos\theta)$, $y = a \cos\theta$

(c) $x = a(1 + \sin\theta)$, $y = a \cos\theta$

(d) $x = a(1 - \sin\theta)$, $y = a\cos\theta$

5. Given P' is reflection of P in x-axis. Then the polar co-ordinates of P' in the figure is

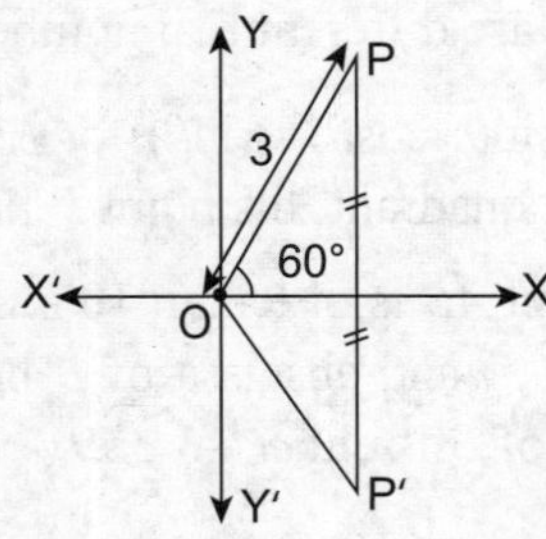

FIGURE 1.20

(a) $\left(3, \frac{\pi}{3}\right)$ (b) $\left(3, \frac{-\pi}{3}\right)$

(c) $\left(-3, \frac{-\pi}{3}\right)$ (d) $\left(-3, \frac{\pi}{3}\right)$

6. The Cartesian co-ordinates of the point Q in the figure are

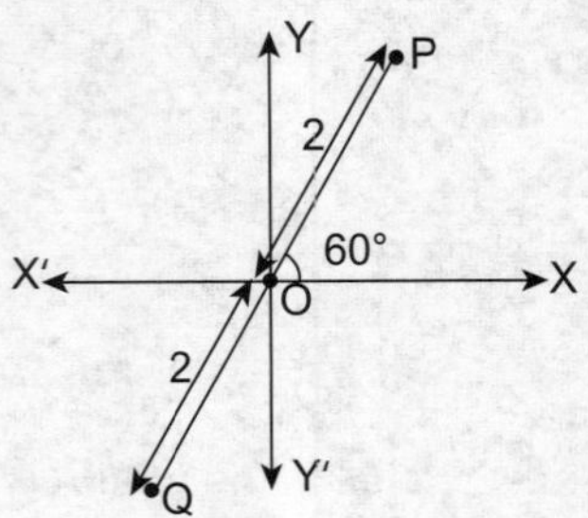

FIGURE 1.21

(a) $\left(1, \sqrt{3}\right)$ (b) $\left(-1, \sqrt{3}\right)$

(c) $\left(1, -\sqrt{3}\right)$ (d) $\left(-1, -\sqrt{3}\right)$

7. The distance between $A(2, 15°)$ and $B(1, 75°)$ is

(a) $\sqrt{6}$ (b) 6

(c) $\sqrt{3}$ (d) 3

8. Transformation of polar equation $r = a$ to cartesian equation is

(a) $x^2 - y^2 = a$

(b) $x^2 - y^2 = ax$

(c) $x^2 + y^2 = a^2$

(d) None of these

9. Three given points which satisfy the given condition $4(PQ)^2 + (PR)^2 = (QR)^2$

(a) $P(1,2)$, $Q(4,3)$ and $R(2,5)$

(b) $P(2,2)$, Q $(5,2)$ and $R(3,4)$

(c) $P(4,2)$, $Q(3,1)$ and $R(4,5)$

(d) $P(-5,1)$, $Q(1,5)$ and $R(4,7)$

10. The common property of points lying on x-axis, is

(a) $x = 0$ (b) $y = 0$

(c) $x = 0, y = 0$ (d) None of these

11. Common property of the bisector of Ist and IIIrd quadrant is

(a) $y = x$ (b) $y = -x$

(c) $x^2 = y^2$ (d) None of these

Answer Keys

1. (b) **2.** (c) **3.** (a) **4.** (a) **5.** (b) **6.** (d) **7.** (c) **8.** (c) **9.** (c) **10.** (b)
11. (a)

APPLICATION OF DISTANCE FORMULA

Let A, B, C are three points lying on a plane, then while analysing their relative positions with respect to each other, two conditions arise:

(i) Either they are collinear, or
(ii) They form a triangle

Distance formula can be used to determine whether a set of three points is collinear or forms a triangle as discussed below.

1. Collinearity of Three Given Points

The three given points A, B, C are collinear, i.e., lie on the same straight line, iff the sum of the distances between two points of two pairs is equal to distance between the points of third pair.

i.e., $AB + BC = AC$; A —— B —— C

or $AC + BC = AB$; A —— C —— B

or $BA + AC = BC$; B —— A —— C

2. Formation of Different Triangles

Three points A, B, C form triangles of following types:

(a) **Equilateral triangle:** If $AB = BC = CA$.

(b) **Isosceles triangle:** If any two sides are equal and their sum is not equal to third side. i.e., their sum is greater than third side.
i.e., $AB = BC$ and $AB + BC > AC$
or $AC = CB$ and $AC + CB > AB$
or $BA = AC$ and $BA + AC > BC$

(c) **Scalene triangle:** If all the distances AB, BC, CA are different
i.e., $AB \neq BC \neq AC$, $AC \neq AB$ and A, B and C are non-collinear.

(d) **Right angled triangle:** If the sum of squares of any two distances is equal to the square of third distance.
i.e., $AB^2 + BC^2 = AC^2$ or $AB^2 + AC^2 = BC^2$ or $AC^2 + BC^2 = AB^2$

3. Formation of Quadrilaterals

Similarly, four points A, B, C, D lying on a plane can form quadrilateral of different types, provided no three of them are collinear.

Quadrilaterals are classified in the table ahead, alongwith their elementary properties.

S. No.	Quadrilateral	Side/Angle	Diagonals and Angle Between Diagonals	Diagram
1	Parallelogram, but not rectangle.	Opposite sides are equal, no angle is right angle.	Diagonals are unequal, but bisect each other.	
2	Rectangle.	A parallelogram having all angles right angles.	Diagonals are equal and bisect each other (*not necessarily at right angle*).	
3	Rhombus, but not square.	All four sides are equal, no angle is right angle.	Diagonals are unequal, but bisect each other at right angle.	
4	Square.	Rhombus all angles having right angles.	Diagonals are equal bisecting each other at right angle.	

REMARKS

(i) *It should be kept in mind that four points A, B, C, D would form a quadrilateral iff no three of the given four points are collinear.*

(ii) *If the co-ordinates of points A and B are* (x_1, y_1) and (x_2, y_2), *then the co-ordinates of mid-point of line segment AB would be* $\left(\frac{x_1 + x_2}{2}, \frac{y_1 + y_2}{2}\right)$. *Students are very much familiar with this formula known as* ***mid-point formula*** *as they learnt it in their junior classes, and it would be exaplained in the article section formula of this chapter.*

ILLUSTRATION 16: Using distance formula, show that the points $P(1, 5)$; $Q(2, 4)$; $R(3, 3)$ are collinear.

SOLUTION: $PQ = \sqrt{(2-1)^2+(4-5)^2} = \sqrt{2}$; $QR = \sqrt{(1)^2+(1)^2} = \sqrt{2}$; $PR = \sqrt{(2)^2+(2)^2} = 2\sqrt{2}$

$\therefore$ $PQ + QR = PR$. **Hence, Proved.**

ILLUSTRATION 17: Prove that the points (2, –2), (–3,8) and (–1,4) are collinear.

SOLUTION: Let the given points be A (2, –2), B(–3, 8) and C(–1, 4), then

$$AB = \sqrt{(2-(-3))^2+(-2-8)^2} = \sqrt{25+100} = \sqrt{125} = 5\sqrt{5}$$

$$AC = \sqrt{(2-(-1))^2+(-2-4)^2} = \sqrt{9+36} = \sqrt{45} = 3\sqrt{5}$$

$$BC = \sqrt{(-3-(-1))^2+(8-4)^2} = \sqrt{4+16} = \sqrt{20} = 2\sqrt{5}$$

Hence, $AC + BC = AB$. So A, B, C are collinear.

ILLUSTRATION 18: There are four points (0,–1), (6,7), (–2,3) and (8,3). Find out which kind of quadrilateral these points would form.

SOLUTION: Let A(0, –1), B(6, 7), C(–2, 3) and D(8, 3) be the given points. Then,

$$AD = \sqrt{(8-0)^2+(3+1)^2} = \sqrt{64+16} = 4\sqrt{5}; \; BC = \sqrt{(6+2)^2+(7-3)^2} = \sqrt{64+16} = 4\sqrt{5}$$

$$AC = \sqrt{(-2-0)^2+(3+1)^2} = \sqrt{4+16} = 2\sqrt{5} \text{ and } BD = \sqrt{(8-6)^2+(7-3)^2} = \sqrt{4+16} = 2\sqrt{5}$$

$\therefore$ $AD = BC$ and $AC = BD$.

If we take the points as shown in the figure, then we observe that opposite sides are equal, so the figure is a parallelogram. Moreover,

$$AB = \sqrt{(6-0)^2+(7+1)^2} = \sqrt{36+64} = 10$$

$$CD = \sqrt{(8+2)^2+(3-3)^2} = 10.$$

Clearly, $AB^2 = AD^2 + DB^2$ and $CD^2 = CB^2 + BD^2$.

$\Rightarrow$ $\angle ADB = DBC = 90°$. $\Rightarrow$ $ADBC$ is a rectangle.

FIGURE 1.22

ILLUSTRATION 19: The triangle OAB is right angled with right angle at O, where points O, A, B are (0, 0), ($\cos\theta$, $\sin\theta$) and ($\cos\phi$, $\sin\phi$), respectively, then θ and ϕ are connected by the relation.

(a) $\sin\left(\frac{\theta-\phi}{2}\right) = \pm\frac{1}{\sqrt{2}}$

(b) $\cos\left(\frac{\theta-\phi}{2}\right) = \pm\frac{1}{\sqrt{2}}$

(c) $\cos(\theta - \phi) = 0$

(d) None of these

SOLUTION: Clearly, $OA = OB = 1$ as $\cos^2 x + \sin^2 x = 1 \;\; \forall x \in \mathbb{R}$

$\therefore$ $AB^2 = OA^2 + OB^2 = 2$

or $(\cos\theta - \cos\phi)^2 + (\sin\theta - \sin\phi)^2 = 2$ or $1 + 1 - 2(\cos\theta \,.\, \cos\phi + \sin\theta \,.\, \sin\phi) = 2$

or $\cos(\theta - \phi) = 0$ or $2\cos^2\left(\frac{\theta-\phi}{2}\right) - 1 = 0$ or $1 - 2\sin^2\left(\frac{\theta-\phi}{2}\right) = 0$

Option (a), (b), (c) are correct.

ILLUSTRATION 20: Find the nature of triangle formed, having vertices (–2, 2), (8, –2), (–4, –3).

SOLUTION: Let A(–2, 2), B(8, –2) and C(–4, –3), then $AB = \sqrt{10^2+4^2} = 2\sqrt{29}$

$BC = \sqrt{12^2+1} = \sqrt{145}$; $CA = \sqrt{4+25} = \sqrt{29}$

Clearly, $AB^2 + AC^2 = BC^2$ (Pythagoras theorem) $\Rightarrow$ Δ is right angled triangle.

ILLUSTRATION 21: Find the number of integer points on x-axis whose distance (p) from the point (2, 3) is such that $p \in (3, 5]$. (Point (x, y) is called integer point if $x, y \in \mathbb{Z}$.)

SOLUTION: Let P be (2, 3) and M is foot of perpendicular from P to x-axis.

$\Rightarrow$ $PM = 3$, Q be geometrically any arbitrary point on x-axis 5 units away from P.

$\therefore$ By Pythagoras theorem $MQ = \sqrt{5^2 - 3^2} = 4$.

$\Rightarrow$ All integer points between (–2, 0) to (6, 0) on x-axis are desired points except for (2, 0).

$\Rightarrow$ number of points = 8

Aliter: Let Q is $(x, 0)$

$\Rightarrow \; 3 < PQ \le 5 \Rightarrow 3 < \sqrt{(x-2)^2 + 9} \le 5$

$\Rightarrow 9 < (x-2)^2 + 9 \le 25 \;\; \Rightarrow 0 < (x-2)^2 \le 16$

$\Rightarrow 0 < |x - 2| \le 4 \Rightarrow x \in [-2, 6]$, but $x \ne 2$

$\Rightarrow$ $x = -2, -1, 0, 1, 3, 4, 5, 6$ number of points is clearly 8.

FIGURE 1.23

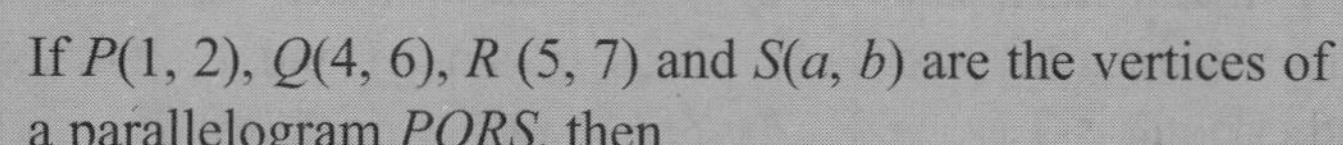

ILLUSTRATION 22: If $P(1, 2)$, $Q(4, 6)$, R (5, 7) and $S(a, b)$ are the vertices of a parallelogram $PQRS$, then

(a) $a = 2, b = 4$ (b) $a = 3, b = 4$

(c) $a = 2, b = 3$ (d) $a = 3, b = 5$

SOLUTION: $PQRS$ will represent a paralleogram if and only if the mid-point of PR is same as that of QS.

That is, if and only if $\dfrac{1+5}{2} = \dfrac{4+a}{2}$ and $\dfrac{2+7}{2} = \dfrac{6+b}{2} \Rightarrow a = 2$ and $b = 3$.

ILLUSTRATION 23: The points (– 2, 3), (3, 8) and (4,1) are the vertices of

(a) an isosceles triangle (b) an equilateral triangle

(c) right angled triangle (d) None of these

SOLUTION: From figure, it is clear that points are non collinear and hence, form a triangle. In $\Delta\, ABC$,

$AB = \sqrt{(3+2)^2 + (8-3)^2} = \sqrt{50}$;

$BC = \sqrt{(4+2)^2 + (1-3)^2} = \sqrt{40}$

$AC = \sqrt{(3-4)^2 + (8-1)^2} = \sqrt{50}$

$\therefore$ $\Delta\, ABC$ is isosceles

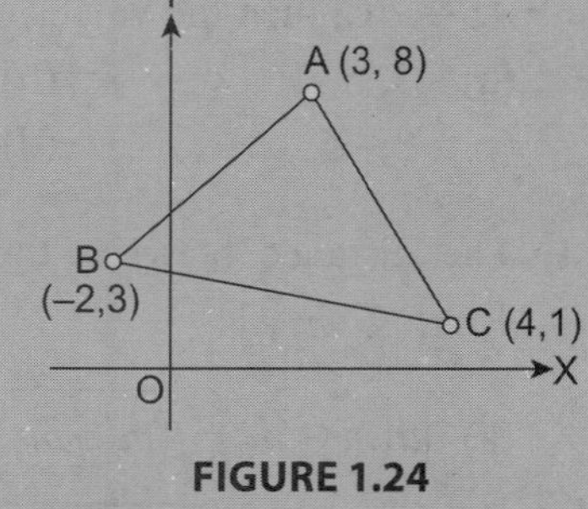

FIGURE 1.24

TEXTUAL EXERCISE–2 (SUBJECTIVE)

1. (a) Prove that the points (1, 1), (–2, 7) and (3, –3) are collinear.
   (b) Check whether the points (0, 8/3), (1, 3) and (82, 30) are the vertices of an isosceles triangle?

2. Show that the point A (0, –1), B (2, 1), C (0, 3) and D (–2, 1) are the vertices of a square.

3. Show that the points (a, a), $(-a, -a)$ and $(-a\sqrt{3}, a\sqrt{3})$ are vertices of an equilateral triangle.

4. If $A \equiv (3, 4)$ and B is a variable point on the lines $|x| = 6$. If $AB \le 4$, then find the number of positions of B with integral co-ordinates.

5. Find the number of integral points on co-ordinate axes which are at a distance c units ($c \le 3$) from (2, 3).

6. If O is the origin and the co-ordinates of any two points Q_1 and Q_2 are (x_1, y_1) and (x_2, y_2) respectively, then prove that $OQ_1 \cdot OQ_2 \cos \angle Q_1 OQ_2 = x_1x_2 + y_1y_2$

7. Show that the points $A(3, 3)$, $B(h, 0)$ and $C(0, k)$ are collinear if $\frac{1}{h}+\frac{1}{k}=\frac{1}{3}$.

8. Find the distance between the points $(at_1^2, 2at_1)$ and $(at_2^2, 2at_2)$, where t_1 and t_2 are the roots of the equation $x^2 - 2\sqrt{3}x + 2 = 0$ and $a > 0$.

Answer Keys

1. (b) no 4. 5 points 5. 6 such points exist. 8. $4a(|t_1 - t_2|)$ or $8a$ as $|t_1 - t_2| = 2$

TEXTUAL EXERCISE–2 (OBJECTIVE)

1. The co-ordinates of a point are (0, 1) and the ordinate of another point is –3. If the distance between the two points is 5, then the abscissa of another point is
(a) 3 (b) –3
(c) ±3 (d) 1

2. One of the vertices of a square is origin and adjacent sides of the square are coincident with positive axes. If length of side is 5 then which will not be its one of the vertices?
(a) (0, 5) (b) (5, 0)
(c) (–5, –5) (d) (0, 0)

3. If the points $A(6,-1)$, $B(1,3)$ and $C(x,8)$ be such that $AB = BC$, then the values of x is
(a) –3, 5 (b) 3, –5
(c) –3, –5 (d) 3, 5

4. The distance between the points $\left(am_1^2, 2am_1\right)$ and $\left(am_2^2, 2am_2\right)$ is
(a) $|a(m_1 - m_2)|\sqrt{(m_1+m_2)^2+4}$
(b) $(m_1 - m_2)\sqrt{(m_1+m_2)^2+4}$
(c) $a(m_1 - m_2)\sqrt{(m_1+m_2)^2-4}$
(d) $(m_1 - m_2)\sqrt{(m_1+m_2)^2-4}$

5. If the points (0,0), $\left(2, 2\sqrt{3}\right)$ and (a, b) be the vertices of an equilateral triangle, then the ordered pair (a, b) given as
(a) $(-2, 2\sqrt{3})$ (b) $(2, -2\sqrt{3})$
(c) (4, 0) (d) (–4,0)

6. If the co-ordinates of vertices of ΔOAB are (0,0); $(\cos\alpha, \sin\alpha)$ and $(-\sin\alpha, \cos\alpha)$ respectively, then $OA^2 + OB^2$ is equal to
(a) 0 (b) 1
(c) 2 (d) 3

7. Given points $A(5,2)$, $B(2,3)$, $C(6,5)$ are vertices of which of the following geometrical figures?
(a) equilateral triangle
(b) right angled isosceles triangle
(c) scalene triangle
(d) None of these

8. Given points $P(0,0)$, $Q(3, \pi/2)$ and $R(3, \pi/6)$ are the vertices of
(a) isosceles triangle but not equilateral
(b) right angled isoscles triangle
(c) equilateral triangle
(d) None of these

9. If $A(1,4)$ and $B(4,8)$ are two points, P is a point on AB such that $AP = AB + BP$. If $AP = 10$, then the co-ordinates of P are
(a) (12, 7) (b) (7, 12)
(c) (2, –3) (d) (–3, 2)

10. The points $A(a, b)$, $B(a_1, b_1)$ and $C(2a_1 - a, 2b_1 - b)$ are
(a) vertices of an scalene triangle
(b) vertices of right angled triangle
(c) collinear
(d) vertices of an isosceles triangle

11. The point whose abscissa is equal to its ordinate and which is equidistant from the points (1,0) and (0,3) is
(a) (1,1) (b) (2,2)
(c) (3,3) (d) (4,4)

ILLUSTRATION 21: Find the number of integer points on x-axis whose distance (p) from the point (2, 3) is such that $p \in (3, 5]$. (Point (x, y) is called integer point if $x, y \in \mathbb{Z}$.)

SOLUTION: Let P be (2, 3) and M is foot of perpendicular from P to x-axis.

$\Rightarrow$ $PM = 3$, Q be geometrically any arbitrary point on x-axis 5 units away from P.

$\therefore$ By Pythagoras theorem $MQ = \sqrt{5^2 - 3^2} = 4$.

$\Rightarrow$ All integer points between (–2, 0) to (6, 0) on x-axis are desired points except for (2, 0).

$\Rightarrow$ number of points = 8

Aliter: Let Q is (x, 0)

$\Rightarrow 3 < PQ \le 5 \Rightarrow 3 < \sqrt{(x-2)^2 + 9} \le 5$

$\Rightarrow 9 < (x-2)^2 + 9 \le 25 \Rightarrow 0 < (x-2)^2 \le 16$

$\Rightarrow 0 < |x - 2| \le 4 \Rightarrow x \in [-2, 6]$, but $x \ne 2$

$\Rightarrow x = -2, -1, 0, 1, 3, 4, 5, 6$ number of points is clearly 8.

FIGURE 1.23

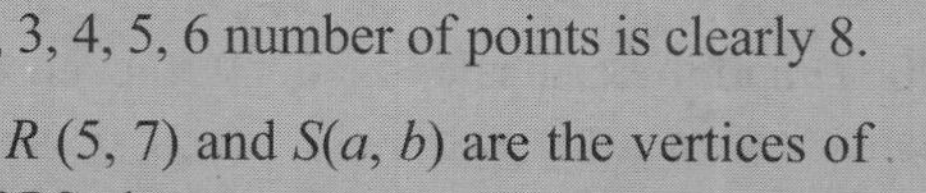

ILLUSTRATION 22: If $P(1, 2)$, $Q(4, 6)$, R (5, 7) and $S(a, b)$ are the vertices of a parallelogram $PQRS$, then

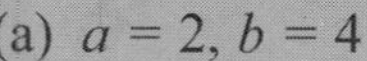

(a) $a = 2, b = 4$ (b) $a = 3, b = 4$

(c) $a = 2, b = 3$ (d) $a = 3, b = 5$

SOLUTION: $PQRS$ will represent a paralleogram if and only if the mid-point of PR is same as that of QS.

That is, if and only if $\frac{1+5}{2} = \frac{4+a}{2}$ and $\frac{2+7}{2} = \frac{6+b}{2} \Rightarrow a = 2$ and $b = 3$.

ILLUSTRATION 23: The points (– 2, 3), (3, 8) and (4,1) are the vertices of

(a) an isosceles triangle (b) an equilateral triangle

(c) right angled triangle (d) None of these

SOLUTION: From figure, it is clear that points are non collinear and hence, form a triangle. In $\Delta\ ABC$,

$AB = \sqrt{(3+2)^2 + (8-3)^2} = \sqrt{50}$;

$BC = \sqrt{(4+2)^2 + (1-3)^2} = \sqrt{40}$

$AC = \sqrt{(3-4)^2 + (8-1)^2} = \sqrt{50}$

$\therefore$ $\Delta\ ABC$ is isosceles

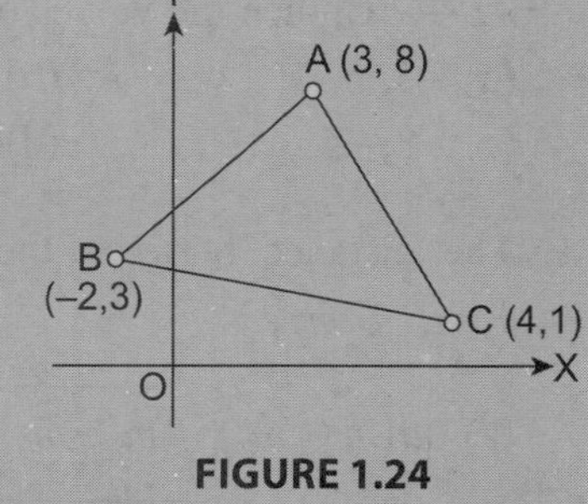

FIGURE 1.24

TEXTUAL EXERCISE–2 (SUBJECTIVE)

1. (a) Prove that the points (1, 1), (–2, 7) and (3, –3) are collinear.
   (b) Check whether the points (0, 8/3), (1, 3) and (82, 30) are the vertices of an isosceles triangle?

2. Show that the point A (0, –1), B (2, 1), C (0, 3) and D (–2, 1) are the vertices of a square.

3. Show that the points (a, a), ($-a$, $-a$) and ($-a\sqrt{3}$, $a\sqrt{3}$) are vertices of an equilateral triangle.

4. If $A \equiv (3, 4)$ and B is a variable point on the lines $|x| = 6$. If $AB \le 4$, then find the number of positions of B with integral co-ordinates.

5. Find the number of integral points on co-ordinate axes which are at a distance c units ($c \le 3$) from (2, 3).

6. If O is the origin and the co-ordinates of any two points Q_1 and Q_2 are (x_1, y_1) and (x_2, y_2) respectively, then prove that $OQ_1 \cdot OQ_2 \cos \angle Q_1 OQ_2 = x_1x_2 + y_1y_2$

7. Show that the points $A(3, 3)$, $B(h, 0)$ and $C(0, k)$ are collinear if $\frac{1}{h}+\frac{1}{k}=\frac{1}{3}$.

8. Find the distance between the points $(at_1^2, 2at_1)$ and $(at_2^2, 2at_2)$, where t_1 and t_2 are the roots of the equation $x^2 - 2\sqrt{3}x + 2 = 0$ and $a > 0$.

Answer Keys

1. (b) no **4.** 5 points **5.** 6 such points exist. **8.** $4a(|t_1 - t_2|)$ or $8a$ as $|t_1 - t_2| = 2$

TEXTUAL EXERCISE–2 (OBJECTIVE)

1. The co-ordinates of a point are (0, 1) and the ordinate of another point is –3. If the distance between the two points is 5, then the abscissa of another point is

(a) 3 (b) –3
(c) ±3 (d) 1

2. One of the vertices of a square is origin and adjacent sides of the square are coincident with positive axes. If length of side is 5 then which will not be its one of the vertices?

(a) (0, 5) (b) (5, 0)
(c) (–5, –5) (d) (0, 0)

3. If the points $A(6,–1)$, $B(1,3)$ and $C(x,8)$ be such that $AB = BC$, then the values of x is

(a) –3, 5 (b) 3, –5
(c) –3, –5 (d) 3, 5

4. The distance between the points $\left(am_1^2, 2am_1\right)$ and $\left(am_2^2, 2am_2\right)$ is

(a) $\left|a\left(m_1 - m_2\right)\right|\sqrt{\left(m_1 + m_2\right)^2 + 4}$

(b) $\left(m_1 - m_2\right)\sqrt{\left(m_1 + m_2\right)^2 + 4}$

(c) $a\left(m_1 - m_2\right)\sqrt{\left(m_1 + m_2\right)^2 - 4}$

(d) $\left(m_1 - m_2\right)\sqrt{\left(m_1 + m_2\right)^2 - 4}$

5. If the points (0,0), $\left(2, 2\sqrt{3}\right)$ and (a, b) be the vertices of an equilateral triangle, then the ordered pair (a, b) given as

(a) $(–2, 2\sqrt{3})$ (b) $(2, –2\sqrt{3})$
(c) (4, 0) (d) (–4,0)

6. If the co-ordinates of vertices of ΔOAB are (0,0); $(\cos \alpha, \sin \alpha)$ and $(–\sin\alpha, \cos\alpha)$ respectively, then $OA^2 + OB^2$ is equal to

(a) 0 (b) 1
(c) 2 (d) 3

7. Given points $A(5,2)$, $B(2,3)$, $C(6,5)$ are vertices of which of the following geometrical figures?

(a) equilateral triangle
(b) right angled isosceles triangle
(c) scalene triangle
(d) None of these

8. Given points $P(0,0)$, $Q(3, \pi/2)$ and $R(3, \pi/6)$ are the vertices of

(a) isosceles triangle but not equilateral
(b) right angled isoscles triangle
(c) equilateral triangle
(d) None of these

9. If $A(1,4)$ and $B(4,8)$ are two points, P is a point on AB such that $AP = AB + BP$. If $AP = 10$, then the co-ordinates of P are

(a) (12, 7) (b) (7, 12)
(c) (2, –3) (d) (–3, 2)

10. The points $A(a, b)$, $B(a_1, b_1)$ and $C(2a_1 – a, 2b_1 – b)$ are

(a) vertices of an scalene triangle
(b) vertices of right angled triangle
(c) collinear
(d) vertices of an isosceles triangle

11. The point whose abscissa is equal to its ordinate and which is equidistant from the points (1,0) and (0,3) is

(a) (1,1) (b) (2,2)
(c) (3,3) (d) (4,4)

12. If $A(9, -9)$, $B(1, 3)$ are the ends a right angled isosceles triangle, then the third vertex is
(a) (8, –1) (b) (–8, 2)
(c) (8, –8) (d) None of these

13. A triangle ABC right angled at A, has points A and B as (2, 3) and (0, –1) respectively. If $BC = 5$ units, then the point C is
(a) (4, 2) (b) (–4, 2)
(c) (0, 4) (d) (3, –3)

14. If the point $(x, -1),(3, y),(-2,3)$ and $(-3,-2)$ be the vertices of a parallelogram, then
(a) $x = 2, y = 4$ (b) $x = 1, y = 2$
(c) $x = 4, y = 2$ (d) None of these

15. If the three vertices of a rectangle taken in order are the points (2, –2), (8, 4) and (5, 7). The co-ordinates of the fourth vertex is
(a) (1,1) (b) (1,–1)
(c) (–1,1) (d) None of these

16. If $P(1, 2)$, $Q(4, 6)R(5, 7)$ and $S(a, b)$ are the vertices of a parallelogram $PQRS$, then
(a) $a = 2, b = 4$ (b) $a = 3, b = 4$
(c) $a = 2, b = 3$ (d) $a = 3, b = 5$

Answer Keys

1. (c)	**2.** (c)	**3.** (a)	**4.** (a)	**5.** (a), (c)	**6.** (c)	**7.** (b)	**8.** (c)	**9.** (b)	**10.** (c)
11. (b)	**12.** (a)	**13.** (a), (c)	**14.** (a)	**15.** (c)	**16.** (c)				

SECTION FORMULA (DIVISION OF A LINE SEGMENT BY A POINT)

If a point P lies on the line segment AB such that $PA : PB = m : n \in \mathbb{R}$, then it is said that P divides AB in the ratio $m : n$. If the ratio is positive we say that P divides line segment AB internally in the ratio $m : n$. i.e., P lies between A and B, on the line segment AB.

If the ratio is negative, then we say that P divides line segments AB externally in the ratio $|m|: |n|$ i.e., P lies outside the line segment AB, either on AB produced or on BA produced.

Depending upon the location of point P i.e., in between AB or out side AB section formula is known as section formula for internal division formula or section formula for external division.

(a) Internal Division: Co-ordinates of a point which divides the line segment joining two points $A(x_1, y_1)$ and $B(x_2, y_2)$ in the ratio $m: n$, internally are

$$\left(\frac{mx_2 + nx_1}{m+n}, \frac{my_2 + ny_1}{m+n}\right)$$

Proof: From the similar triangles ΔAHP and ΔPKB,

we have $\dfrac{AP}{PB} = \dfrac{AH}{PK} = \dfrac{PH}{BK}$

$$\Rightarrow \frac{m}{n} = \frac{x - x_1}{x_2 - x} = \frac{y - y_1}{y_2 - y}.$$

$$\Rightarrow x = \frac{mx_2 + nx_1}{m+n}, \quad y = \frac{my_2 + ny_1}{m+n}$$

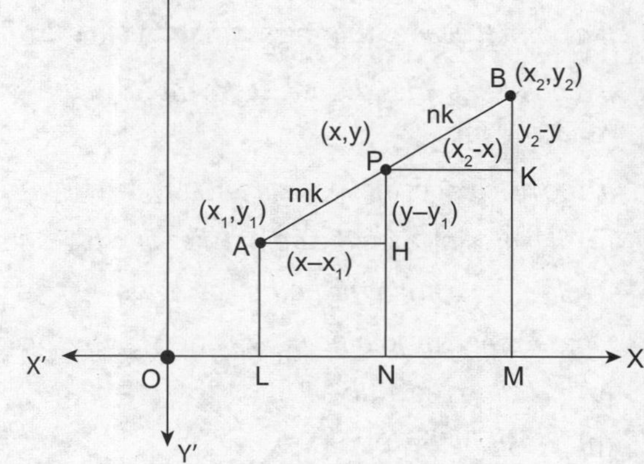

FIGURE 1.25

REMARKS

(i) *The diagram given below helps in memorising the section formula.*

m n
A(x1,y1) P B(x2,y2)

FIGURE 1.26

(ii) *If P is the mid-point of AB, then it divides AB in the ratio 1: 1, so its co-ordinates are* $\left(\dfrac{x_1 + x_2}{2}, \dfrac{y_1 + y_2}{2}\right)$.

(b) External Division: Co-ordinates of a point P which divides the line segment joining two points $A(x_1,y_1)$ and $B(x_2,y_2)$ in the ratio $m: n$, externally are $\left(\frac{mx_2-nx_1}{m-n}, \frac{my_2-ny_1}{m-n}\right)$

Proof: Let the point $P(x,y)$ lies on AB produced such that $PA : PB$ is $m : n$ as shown in figure below.

In the given figure, consider the similar triangles ΔPAH and ΔPBK,

$$\because \frac{AP}{BP}=\frac{AH}{BK}=\frac{PH}{PK}$$

$$\Rightarrow \frac{m}{n}=\frac{x-x_1}{x-x_2}=\frac{y-y_1}{y-y_2}$$

$$\Rightarrow x=\frac{mx_2-nx_1}{m-n},\ y=\frac{my_2-ny_1}{m-n}$$

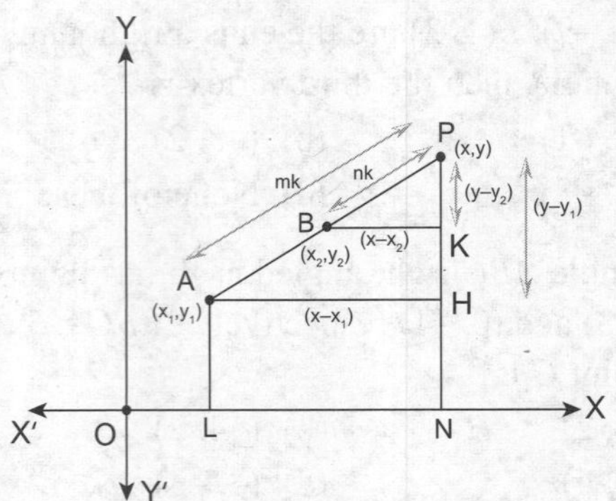

FIGURE 1.27

CONCLUSION: From both the results, we may conclude that

$$x=\frac{\left(\pm\frac{m}{n}\right)x_2+x_1}{\left(\pm\frac{m}{n}\right)+1}\ \&\ y=\frac{\left(\pm\frac{m}{n}\right)y_2+y_1}{\left(\pm\frac{m}{n}\right)+1}$$

; where (m/n) represents ratio for internal division and $(-m/n)$ stands for external division.

REMARKS

(i) *Co-ordinates of any point on the line segment joining two points* $A(x_1, y_1)$ and $B(x_2, y_2)$ and *dividing it in the ratio* λ*: 1 is given by* $\left(\frac{x_1+\lambda x_2}{\lambda+1}, \frac{y_1+\lambda y_2}{\lambda+1}\right)$, $(\lambda \neq -1)$; *here* $\lambda = m/n > 0$ *(for internal division) and* $\lambda = -m/n < 0$ *(for external division)*

(ii) *If a point P divides the line segment AB in the ratio* $\lambda : 1$ and $\lambda = m/n > 0$, *then point P divides AB internally.*

(iii) *If a point P divides the line segment AB in the ratio* $\lambda : 1$ and $\lambda = m/n < 0$, *then point P divides AB externally in the ratio* $|m| : |n|$.

(iv) *If a point P divides the line segment AB externally in the ratio m : n, then the following statements hold good:*

(a) *If* $|m| > |n|$, *then the point P lies on AB produced.*
(b) *If* $|m| < |n|$, *then the point P lies on BA produced.*
(c) *If* $|m| = 0$, *then the point P coincides with A.*
(d) *If* $|n| = 0$, *then the point P coincides with B.*

(v) *Lines formed by joining* (x_1, y_1) and (x_2, y_2) *is divided by*

(a) *x-axis in the ratio* $-y_1/y_2$
(b) *y-axis in the ratio* $-x_1/x_2$

If the ratio is positive, the axis divides it internally and *if it is negative, axis divides externally.*

(vi) *If mid-points of the sides of a triangle ABC are D, E, F respectively of BC, CA, AB as shown in the figure, then* $A(x_E+x_F-x_D, y_E+y_F-y_D)$, $B(x_D+x_F-x_E, y_D+y_F-y_E)$, *and* $C(x_D+x_E-x_F, y_D+y_E-y_F)$ *and the area of each of the four triangles formed by joining the mid-points of the sides are equal.*

$\therefore$ *Area of* $\Delta ABC = 4 \times$ *Area of* ΔDEF

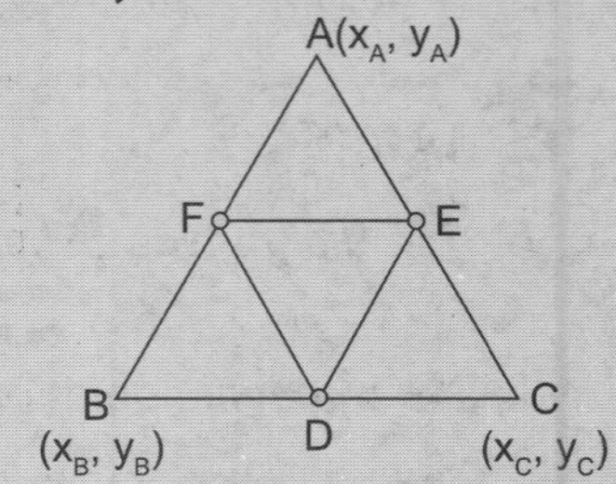

FIGURE 1.28

ILLUSTRATION 24: Find the co-ordinates of the point which divides the line segment joining the points $A(-1,3)$ and $(4, -7)$ in the ratio of 3: 2 internally.

3 2
A(-1,3) C(a,b) B(4,-7)

FIGURE 1.29

SOLUTION: Let 'C' be point of division of AB in the ratio 3: 2 (internally)

$$\Rightarrow a = \frac{mx_2 + nx_1}{m+n}, b = \frac{my_2 + ny_1}{m+n} \quad \therefore a = \frac{3.4 + 2(-1)}{3+2} = 2 \text{ and } b = \frac{3(-7)+2(3)}{3+2} = -3$$

$\Rightarrow C(2,-3)$ **Ans.**

ILLUSTRATION 25: Find the co-ordinates of the point which divides the line segment joining the points $A(-1, 3)$ and $(4, -7)$ in the ratio of 3: 2 externally.

A B C
(-1,3) (4,-7) (a,b)

FIGURE 1.30

SOLUTION: Let 'C' be point of division of AB in the ratio 3: 2 (externally)

$$\Rightarrow \frac{AC}{BC} = \frac{3}{2}; \text{ Now, } a = \frac{mx_2 - nx_1}{m-n}, b = \frac{my_2 - ny_1}{m-n}$$

$$\Rightarrow a = \frac{3(4)-(2)(-1)}{3-2} = 14; b = \frac{3(-7)-2(3)}{3-2} = -27$$

$\therefore$ $C(14,-27)$ **Ans.**

ILLUSTRATION 26: Find the co-ordinates of the point which divides the line segment joining the points $A(6,3)$ and $B(-4, 5)$ in the ratio 3 : 2

(a) internally and (b) externally.

SOLUTION: Let $P(x, y)$ be the required point.

(a) For internal division, we have (as shown in Figure 1.31)

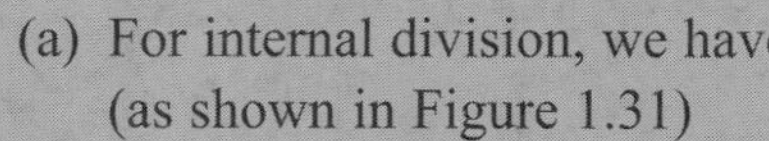

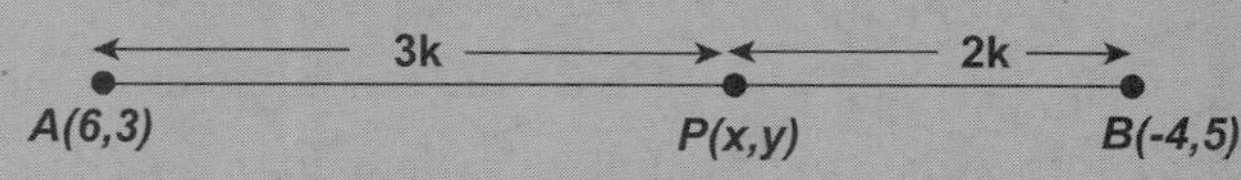

FIGURE 1.31

$$x = \frac{3\times(-4)+2\times 6}{3+2} \text{ and } y = \frac{3\times 5+2\times 3}{3+2}$$

$\Rightarrow x = 0$ and $y = 21/5$. So the co-ordinates of P are $(0, 21/5)$

(b) For external division (refer to Figure 1.32) we have

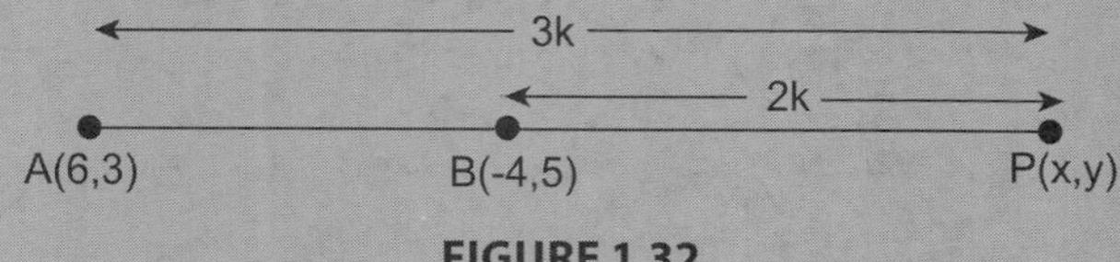

FIGURE 1.32

$$x = \frac{3\times -4 - 2\times 6}{3-2} \text{ and}$$

$$y = \frac{3\times 5 - 2\times 3}{3-2} \Rightarrow x = -24 \text{ and } y = 9$$

So the co-ordinates of P are $(-24, 9)$.

ILLUSTRATION 27: Determine the ratio in which the line $3x + y - 9 = 0$ divides the line segment joining the points (1,3) and (2,7).

SOLUTION: Let the line $3x + y - 9 = 0$ divides the line segment joining $A(1, 3)$ and $B(2, 7)$ in the ratio $k : 1$ at point C, then the co-ordinates of C are $\left(\frac{2k+1}{k+1}, \frac{7k+3}{k+1}\right)$

$\because$ C lies on line $3x + y - 9 = 0$ so it must satisfy its equation, thus

$$3\left(\frac{2k+1}{k+1}\right) + \left(\frac{7k+3}{k+1}\right) - 9 = 0 \Rightarrow k = 3/4 \text{ (which is } \textit{positive}\text{)}$$

Consequently, the required ratio of division is 3 : 4 internally.

ILLUSTRATION 28: The co-ordinates of three consecutive vertices of a parallelogram are (1,3); (–1,2) and (2,5). Then find the co-ordinates of the fourth vertex.

SOLUTION: Let fourth vertex be (α, β). There are three possibilities for fourth vertex (α, β). i.e., D_1, D_2, D_3 as been shown in figure below.

Since $ABCD_i$ is a parallelogram (not in same order) for each $i = 1, 2, 3$. The diagonals bisect each other. Therefore,

FIGURE 1.33

For parallelogram $ABCD_1$:

Mid-point of BD_1 = mid-point of AC

$$\Rightarrow \left(\frac{\alpha-1}{2}, \frac{\beta+2}{2}\right) = \left(\frac{1+2}{2}, \frac{3+5}{2}\right)$$

$$\text{or} \quad \left(\frac{\alpha-1}{2}, \frac{\beta+2}{2}\right) = \left(\frac{3}{2}, 4\right)$$

On equating abscissa and ordinates, we get

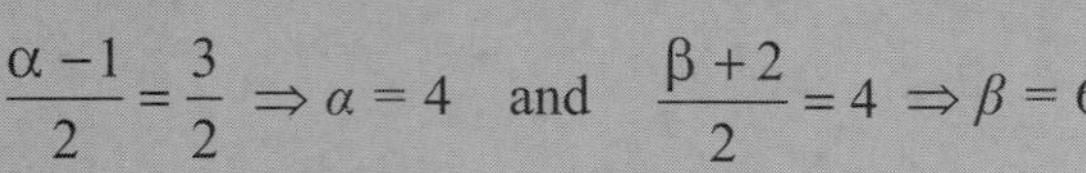

$$\frac{\alpha-1}{2} = \frac{3}{2} \Rightarrow \alpha = 4 \quad \text{and} \quad \frac{\beta+2}{2} = 4 \Rightarrow \beta = 6$$

Hence, the co-ordinates of the fourth vertex $D_1(\alpha, \beta)$ are (4,6)

For parallelogram $ABCD_2$: Mid-point of AD_2 = mid-point of BC

$$\Rightarrow \left(\frac{\alpha+1}{2}, \frac{\beta+3}{2}\right) = \left(\frac{-1+2}{2}, \frac{5+2}{2}\right) \quad \text{or} \quad \left(\frac{\alpha+1}{2}, \frac{\beta+3}{2}\right) = \left(\frac{1}{2}, \frac{7}{2}\right)$$

On equating abscissae and ordinates, we get $\alpha = 0, \beta = 4$.

Hence the co-ordinates of the fourth vertex $D_2(\alpha, \beta)$ are (0,4).

For parallelogram $ABCD_3$: Mid-point of CD_3 = mid-point of AB

$$\Rightarrow \left(\frac{\alpha+2}{2}, \frac{\beta+5}{2}\right) = \left(\frac{1-1}{2}, \frac{3+2}{2}\right) \quad \text{or} \quad \left(\frac{\alpha+2}{2}, \frac{\beta+5}{2}\right) = \left(0, \frac{5}{2}\right)$$

On equating abscissae and ordinates, we get $\alpha = -2, \beta = 0$

Hence, the co-ordinates of the fourth vertex $D_3(\alpha, \beta)$ are (–2,0).

ILLUSTRATION 29: In what ratio does the x-axis, divide the line segment joining the points (2,–3) and (5,6)?

SOLUTION: Let the required ratio be $\lambda : 1$. Then the co-ordinates of the point of division are $\left(\frac{5\lambda+2}{\lambda+1}, \frac{6\lambda-3}{\lambda+1}\right)$.

Since this point lies on x axis, so y co-ordiante is zero. $\frac{6\lambda-3}{\lambda+1} = 0 \Rightarrow \lambda = 1/2$

thus the ratio of division is 1 : 2 (i.e., in ratio $-y_1 : y_2$).

ILLUSTRATION 30: If ABC is a triangle and D, E and F are middle points of side BC, CA and AB respectively and $D(1, 2)$, $E(0, -1)$, $F(2, -1)$, then find the co-ordinates of A, B and C.

SOLUTION: Let A, B and C the their co-ordinates (x_1, y_1); (x_2, y_2); (x_3, y_3) respectively.

$$\therefore \quad \frac{x_2+x_3}{2} = 1, \text{ and } \frac{y_2+y_3}{2} = 2 \Rightarrow x_2 + x_3 = 2 \text{ and } y_2 + y_3 = 4 \qquad \text{...(i)}$$

Similarly, E and F are the mid-points of CA and AB respectively.

$$\therefore \quad \frac{x_1+x_3}{2} = 0 \text{ and } \frac{y_1+y_3}{2} = -1 \Rightarrow x_1 + x_3 = 0 \text{ and } y_1 + y_3 = -2 \qquad \text{...(ii)}$$

and $\frac{x_1+x_2}{2}=2$ and $\frac{y_1+y_2}{2}=-1 \Rightarrow x_1+x_2=4$ and $y_1+y_2=-2$ ...(iii)

Adding (i), (ii) and (iii), we get $x_1+x_2+x_3=3$ and $y_1+y_2+y_3=0$ ...(iv)

From (i) and (iv), we get $x_1=1$ and $y_1=-4$. So, the co-ordinates of A are (1, –4)

From (ii) and (iv), we get $x_2+0=3$ and $y_2-2=0$

$\Rightarrow x_2=3$ and $y_2=2$. So, co-ordinates of B are (3, 2)

From (iii) and (iv), we get $x_3+4=3$ and $y_3-2=0$

$\Rightarrow x_3=-1$ and $y_3=2$. So, co-ordinates of C are (–1, 2)

Hence, the vertices of the triangle ABC are A(1, –4), B(3, 2) and C(–1, 2).

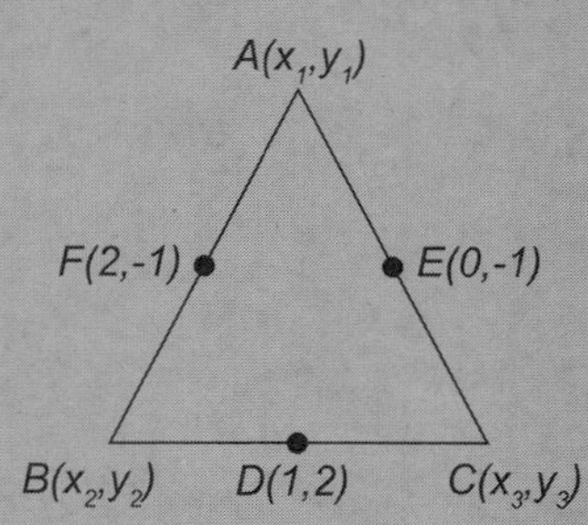

FIGURE 1.34

ILLUSTRATION 31: ABC is a triangle and D, E and F are the middle points of the sides BC, CA and AB; prove that the point which divides AD internally in the ratio 2 : 1 also divides the lines BE and CF in the same ratio. Hence prove that the medians of a triangle meet in a point.

SOLUTION: Let the co-ordinates of the vertices A, B and C be (x_1, y_1), (x_2, y_2) and (x_3, y_3) respectively and

D, E and F be middle points of the sides BC, AC and AB, respectively. Thus, AD, BE and CF are three medians of the triangle ABC. By mid-point formula, the co-ordinates of points D, E and F are respectively.

$$D:\left(\frac{x_2+x_3}{2},\frac{y_2+y_3}{2}\right), E:\left(\frac{x_1+x_3}{2},\frac{y_1+y_3}{2}\right) \text{ and } F:\left(\frac{x_1+x_2}{2},\frac{y_1+y_2}{2}\right)$$

Now consider a point G_1, which divides the segment AD, internally in the ratio 2: 1. By the section formula, we can obtain the co-ordinates of G_1 as

$$\left(\frac{2\left(\frac{x_2+x_3}{2}\right)+1(x_1)}{2+1},\frac{2\left(\frac{y_2+y_3}{2}\right)+1(y_1)}{2+1}\right) \text{ or } \left(\frac{x_1+x_2+x_3}{3},\frac{y_1+y_2+y_3}{3}\right)$$

In the figure where $D\equiv\left(\frac{x_2+x_3}{2},\frac{y_2+y_3}{2}\right)$,

$F\equiv\left(\frac{x_1+x_2}{2},\frac{y_1+y_2}{2}\right)$ and $E\equiv\left(\frac{x_1+x_3}{2},\frac{y_1+y_3}{2}\right)$

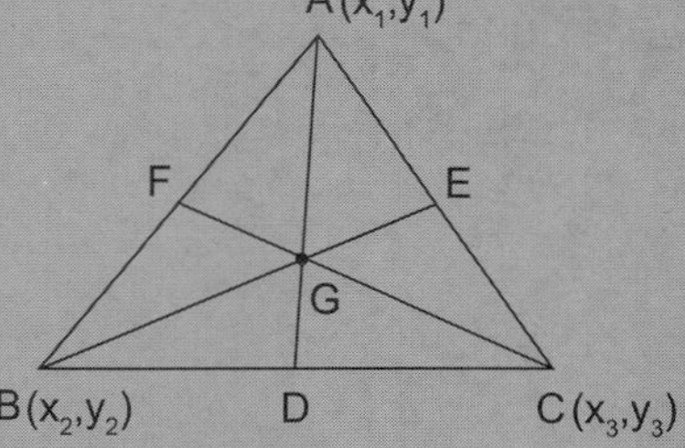

FIGURE 1.35

Also, consider another point G_2 on the median BE, which divides it internally in ratio 2: 1.

Then by section formula again the co-ordinates of G_2 are $\left(\frac{x_1+x_2+x_3}{3},\frac{y_1+y_2+y_3}{3}\right)$

Similarly, the co-ordinates of the point G_3 on the median CF which divides it internally in the ratio 2 : 1 are $\left(\frac{x_1+x_2+x_3}{3},\frac{y_1+y_2+y_3}{3}\right)$. Thus all the three points G_1, G_2, G_3 coincide with each other. This is clear now that the point $\left(\frac{x_1+x_2+x_3}{3},\frac{y_1+y_2+y_3}{3}\right)$ is common point to all the medians. Therefore, the medians of the triangle ABC are concurrent and the point of concurrency is called centroid. Hence co-ordinates of centroid are $\left(\frac{x_1+x_2+x_3}{3}, \frac{y_1+y_2+y_3}{3}\right)$.

Note: The result for the co-ordinates of the centroid of the triangle can be used as a formula.

ILLUSTRATION 32: The co-ordinates of the centroid of a triangle are $(\sqrt{3}, 2)$ and two of its vertices are $(2\sqrt{3}, -1)$ and $(2\sqrt{3}, 5)$. find the third vertex of the triangle.

SOLUTION: Let the third vertex of the triangle be $P(x, y)$. Thus the centroid of the triangle is

$$\left(\frac{x+2\sqrt{3}+2\sqrt{3}}{3}, \frac{y-1+5}{3}\right) \text{ or } \left(\frac{x+4\sqrt{3}}{3}, \frac{y+4}{3}\right)$$

But it is given that the centroid has the co-ordinates $(\sqrt{3}, 2)$.

Therefore $\frac{x+4\sqrt{3}}{3} = \sqrt{3}$ or $x = -\sqrt{3}$ and $\frac{y+4}{3} = 2$

or $y = 2$

Thus the co-ordinates of the third vertex are $(-\sqrt{3}, 2)$.

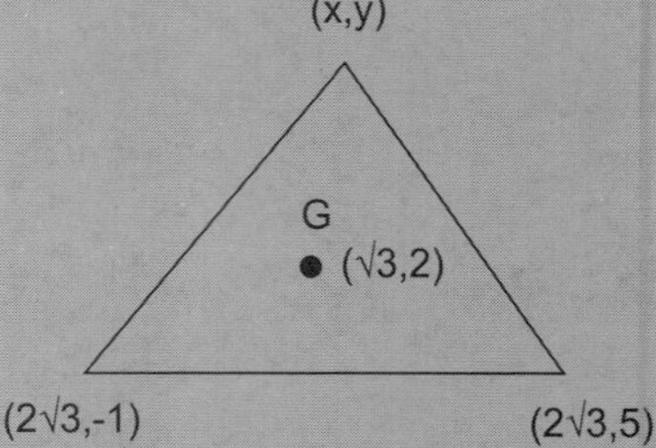

FIGURE 1.36

POINTS OF TRI-SECTION

These are the points that divide a line segment AB, in to three equal parts.

$\therefore$ If P and Q are points of trisection of AB then P divides AB in ratio 1:2 and Q divides AB, in ratio 2:1.

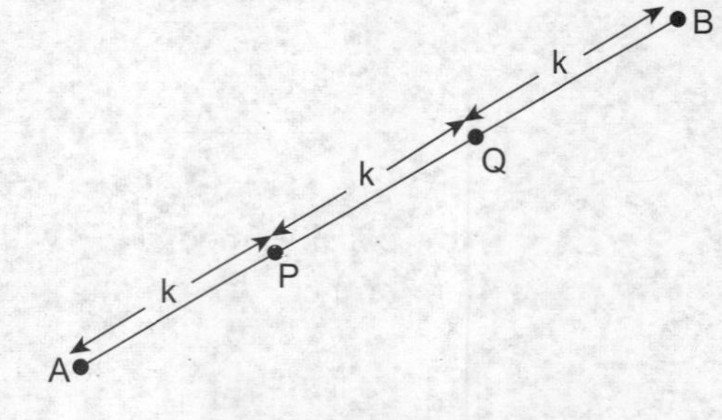

FIGURE 1.37

ILLUSTRATION 33: Find the co-ordinates of the points which trisect the line segment joining $(1, -2)$ and $(-3, 4)$.

SOLUTION: Let the points which trisect the line segment are $C(p, q)$ and $D(r, s)$

$\Rightarrow AC = CD = DB$

$\because \frac{AC}{CB} = \frac{1}{2}$ and $\frac{AD}{DB} = \frac{2}{1}$

$\because$ $C(p, q)$ divides AB in 1:2 ratio

$$\therefore \quad p = \frac{1(-3)+2.1}{1+2} = -\frac{1}{3} \text{ and } q = \frac{1.4+2(-2)}{1+2} = \frac{0}{3} = 0 \;\therefore\; C\left(-\frac{1}{3}, 0\right)$$

A(1,-2) C(p,q) D(r,s) B(-3,4)

FIGURE 1.38

Now, since $D(r, s)$ divides AB in ratio 2: 1

$$\therefore \quad r = \frac{2(-3)+1(1)}{2+1} = \frac{-6+1}{3} = \frac{-5}{3};\; s = \frac{2.(4)+1(-2)}{2+1} = \frac{6}{3} = 2 \Rightarrow D\left(\frac{-5}{3}, 2\right)$$

APPLICATIONS OF SECTION FORMULA

Application-1: If a line segment AB, where $A(x_1, y_1)$ and $B(x_2, y_2)$ is divided by a straight line $L \equiv ax + by + c = 0$ in ratio λ: 1, then $\lambda = -\dfrac{ax_1 + by_1 + c}{ax_2 + by_2 + c} = -\dfrac{L_1}{L_2}$.

Proof: Let the point $P(x_p, y_p)$ of line $L = ax + by + c = 0$ divides AB in the ratio $\lambda : 1$,

then $x_p = \dfrac{\lambda x_2 + x_1}{\lambda + 1}$ and $y_p = \dfrac{\lambda y_2 + y_1}{\lambda + 1}$

and as the point P lies on the line $L = 0$, therefore its co-ordinates satisfy the equation of the line.

$$\Rightarrow ax_p + by_p + c = 0$$

$$\Rightarrow a\left(\frac{\lambda x_2 + x_1}{\lambda + 1}\right) + b\left(\frac{\lambda y_2 + y_1}{\lambda + 1}\right) + c = 0$$

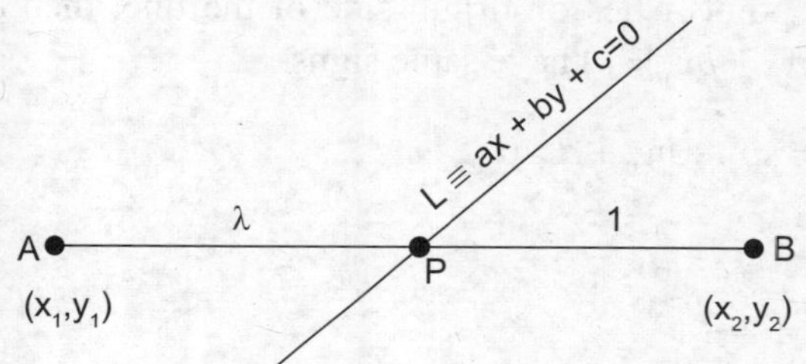

FIGURE 1.39

$$\Rightarrow a(\lambda x_2 + x_1) + b(\lambda y_2 + y_1) + c(\lambda + 1) = 0$$

$$\Rightarrow \lambda(ax_2 + by_2 + c) + (ax_1 + by_1 + c) = 0$$

$$\Rightarrow \lambda = -\left(\frac{ax_1 + by_1 + c}{ax_2 + by_2 + c}\right) = -\left(\frac{L_1}{L_2}\right).$$

here L_1 is the value of expression of line replacing (x_1, y_1) in place of (x, y). Similarly, L_2 is the value of expression of line replacing (x_2, y_2) in place of (x, y).

Application-2: Relative position of two points with respect to a line.

Case I: If $ax_1 + by_1 + c_1$ and $ax_2 + by_2 + c_2$ have opposite signs, then $\lambda > 0$ and line divides internally the line segment joining $A(x_1, y_1)$ and $B(x_2, y_2)$, so both points lie on opposite sides of line $ax + by + c = 0$.

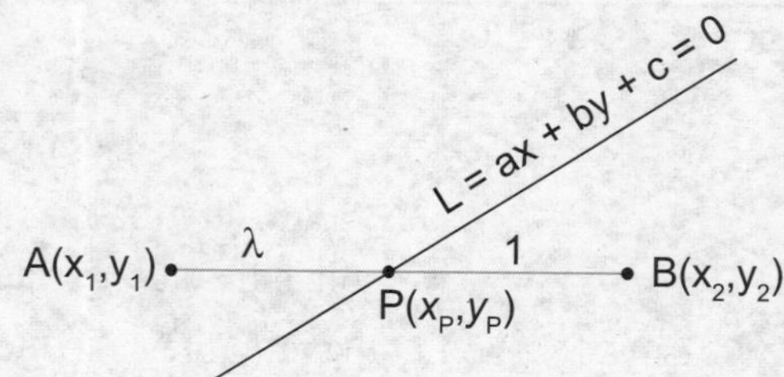

FIGURE 1.40

Proof: If A and B lie on opposite sides of line L: $ax + by + c = 0$, then division is internal

$$\Rightarrow \lambda > 0$$

$$\Rightarrow -\left(\frac{ax_1 + by_1 + c}{ax_2 + by_2 + c}\right) = -\left(\frac{L_1}{L_2}\right) > 0$$

$\Rightarrow$ L_1 and L_2 have opposite signs.

Case II: If $ax_1 + by_1 + c_1$ and $ax_2 + by_2 + c_2$ have same signs, then $\lambda < 0$ and line divides externally the line segment joining $A(x_1, y_1)$ and $B(x_2, y_2)$, so both points lie on same side of the line $ax + by + c = 0$.

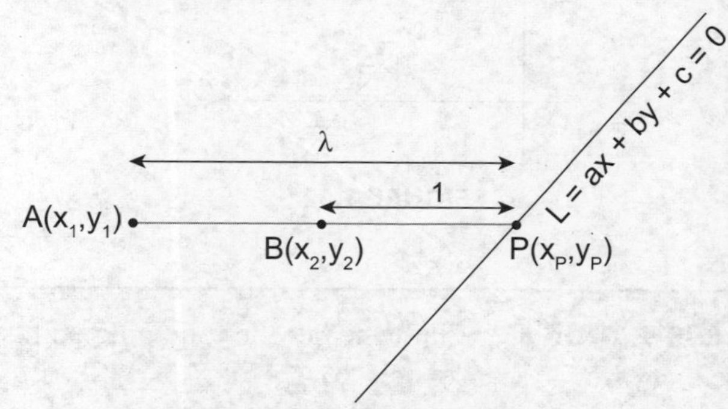

FIGURE 1.41

Proof: If $A(x_1, y_1)$ and $B(x_2, y_2)$ lie on the same side of line $L = ax + by + c = 0$, then division is external.

$$\Rightarrow \lambda < 0$$

$$\Rightarrow -\left(\frac{ax_1 + by_1 + c}{ax_2 + by_2 + c}\right)$$

$$= -\left(\frac{L_1}{L_2}\right) < 0$$

$\Rightarrow$ L_1 and L_2 have same signs.

ILLUSTRATION 34: Find the ratio in which the line segment AB is divided by the straight line $L = 0$ and hence conclude about the relative position of the points A and B w.r.t. the line $L = 0$.

(a) $A(3,1)$; $B(7,4)$; $L = 3x + 2y - 12 = 0$ (b) $A(1,2)$; $B(0,1)$; $L = 2x - y - 3 = 0$

SOLUTION: (a) Let the ratio in which the point P on line $3x + 2y - 12 = 0$ divides the line segment AB be $\lambda : 1$.

$$\therefore\ \lambda = -\frac{L_1}{L_2} = -\frac{3(3)+2(1)-12}{3(7)+2(4)-12} = \frac{1}{17} > 0$$

i.e., the required ratio is 1 : 17.(internally)

$\Rightarrow$ points lie on opposite side of line $3x + 2y - 12 = 0$

(b) Let the ratio in which the point P on line $2x - y - 3 = 0$ divides the line segment AB be λ: 1.

$$\therefore\ \lambda = -\frac{L_1}{L_2} = -\frac{2(1)-(2)-3}{2(0)-(1)-3} = -\frac{3}{4} < 0;$$ i.e., the required ratio is 3: 4 (externally)

$\Rightarrow$ points lie on same side of line $2x - y - 3 = 0$.

Application-3: Position of point $A(x_0, y_0)$ and origin w.r.t. a line $L \equiv ax + by + c = 0$: where $c > 0$.

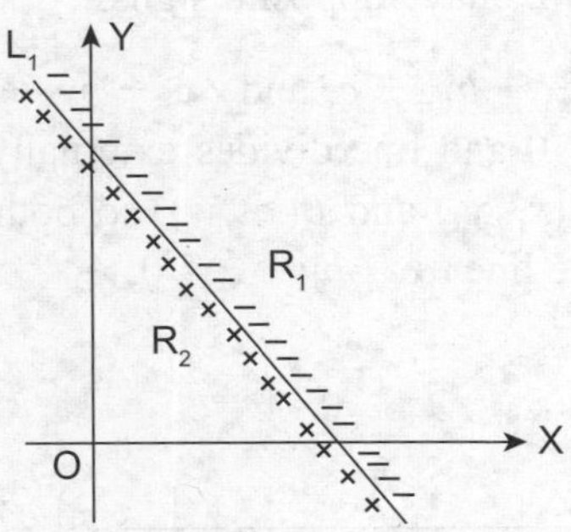

FIGURE 1.42

(a) If point A lies on non-origin side of the line, then c and $(ax_0 + by_0 + c)$ have opposite signs.

$$\Rightarrow (ax_0 + by_0 + c) < 0$$

(b) If point A lies on origin side of the line, then c and $(ax_0 + by_0 + c)$ have same signs.

$$\Rightarrow ax_0 + by_0 + c > 0$$

ILLUSTRATION 35: Find whether the following points lie on origin side/non origin side of the given line;

(a) (–1, 2) w.r.t. $x + y + 2 = 0$

(b) (2, 3) w.r.t. $2x + 3y - 4 = 0$

SOLUTION: (a) Consider $(x_0, y_0) \equiv (-1, 2)$ and the given line is $x + y + 2 = 0$; $c = 2 > 0$

Now, $x_0 + y_0 + 2 = -1 + 2 + 2 = 3 > 0$

$\therefore$ (–1, 2) lies on origin side.

(b) Consider $(x_0, y_0) \equiv (2, 3)$ and the given line is $2x + 3y - 4 = 0$; $c = -4 < 0$;

making c positive, we have equation of line as $-2x - 3y + 4 = 0$; $c = 4 > 0$

Now, $-2x_0 - 3y_0 + 4 = -4 - 9 + 4 = -9 < 0$

$\therefore$ (2, 3) lies on non-origin side.

Application-4: Position of a point A (x, y) w.r.t to two lines $L_1 = a_1x + b_1y + c_1$ such that $c_1 > 0$

and $L_2 = a_2x + b_2y + c_2$ such that $c_2 > 0$

Since for the points on origin side of both the lines will have positive sign of expression for both the lines and non origin sides have negative sign of expression. Therefore marking the origin sign of lines with positive sign and non-origin side with negative sign, a sign scheme is evolved for the four regions of xy plane divided by the pair of lines $L_1 = 0$ and $L_2 = 0$ (as shown in the figure).

$$(x, y) \in R_1 \Leftrightarrow \begin{cases} L_1 = a_1x + b_1y + c_1 < 0 \\ L_2 = a_2x + b_2y + c_2 > 0 \end{cases}$$

$$(x, y) \in R_2 \Leftrightarrow \begin{cases} L_1 = a_1x + b_1y + c_1 > 0 \\ L_2 = a_2x + b_2y + c_2 < 0 \end{cases}$$

$$(x,y)\in R_3 \Leftrightarrow \begin{cases} L_1 = a_1x+b_1y+c_1<0 \\ L_2 = a_2x+b_2y+c_2<0 \end{cases}$$

$$(x,y)\in R_4 \Leftrightarrow \begin{cases} L_1 = a_1x+b_1y+c_1>0 \\ L_2 = a_2x+b_2y+c_2>0 \end{cases}$$

So we can conclude following three facts:

F$_1$: Point lies in the angle containing origin if $L_1 > 0$ and $L_2 > 0$

F$_2$: Point lies in vertically opposite angle to that containing origin if $L_1 < 0$ and $L_2 < 0$

F$_3$: Point lies in the adjacent angle to that containing origin if L_1 and L_2 have opposite sign. i.e., $L_1 . L_2 < 0$.

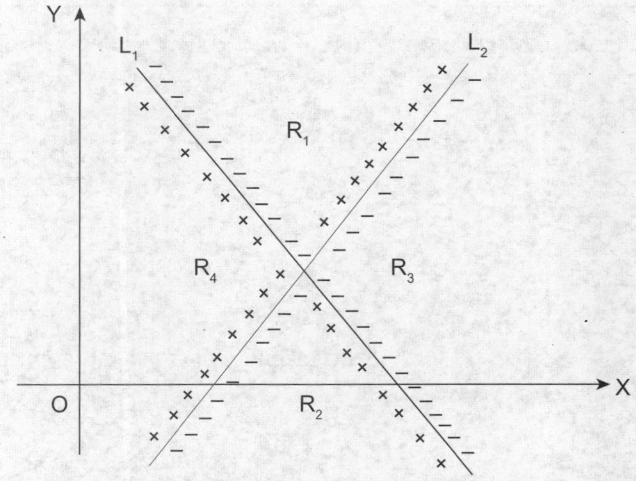

FIGURE 1.43

ILLUSTRATION 36: Find the set of values α for which, with respect to two straight lines $3x + 2y - 6 = 0$ and $x - 3y + 3 = 0$ the point $(3\alpha, 2\alpha - 1)$ lies in.

(a) The angle containing the origin

(b) Vertically opposite angle to that containing origin

(c) Adjacent angle to that containing origin; (given $\alpha \notin \left\{2, \frac{8}{13}\right\}$)

SOLUTION: (a) If $(3\alpha, 2\alpha - 1)$ lies on same angle as that of origin. i.e., in region R_1 then after making c_1 and c_2 positve i.e., $L_1 = x - 3y + 3$; $L_2 = -3x - 2y + 6$

$\Rightarrow L_2 > 0; L_1 > 0$

$\Rightarrow 3\alpha - 3(2\alpha - 1) + 3 > 0; - 3(3\alpha) - 2(2\alpha - 1) + 6 > 0$

$\Rightarrow \alpha < 2; -13\alpha > -8 \Rightarrow \alpha < 8/13 \Rightarrow \alpha, 2; \alpha < 8/13$

$\Rightarrow \alpha \in \left(-\infty, \frac{8}{13}\right)$

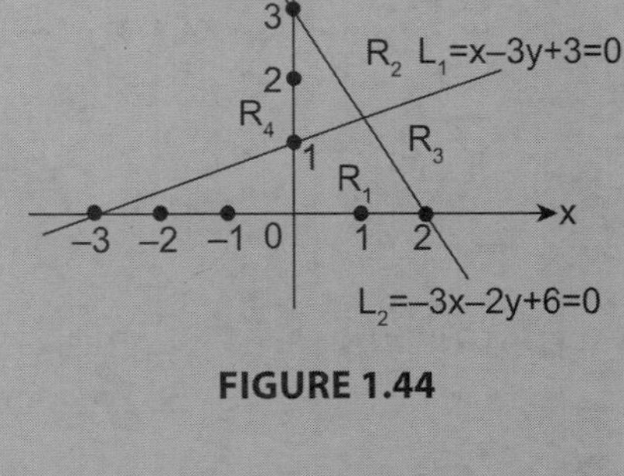

FIGURE 1.44

(b) $(3\alpha, 2\alpha - 1)$ lies in vertically opposite angle to that containing origin. i.e., in region R_2 if $L_1 < 0$ and $L_2 < 0$

$\Rightarrow 3\alpha - 3(2\alpha - 1) + 3 < 0$ and $-3(3\alpha) - 2(2\alpha - 1) + 6 < 0$

$\Rightarrow -3\alpha + 6 < 0$ and $-13\alpha + 8 < 0$

$\Rightarrow \alpha > 2$ and $\alpha > 8/13 \Rightarrow \alpha \in (2, \infty)$

(c) $(3\alpha, 2\alpha - 1)$ lies in adjacent angle to that containing origin if $L_1L_2 < 0$

$\Rightarrow [(3\alpha) - 3(2\alpha - 1) + 3] \times [-3(3\alpha) - 2(2\alpha - 1) + 6] < 0.$

$\Rightarrow (-3\alpha + 6)(-13\alpha + 8) < 0 \Rightarrow (3\alpha - 6)(13\alpha - 8) < 0 \Rightarrow \alpha \in \left(\frac{8}{13}, 2\right)$.

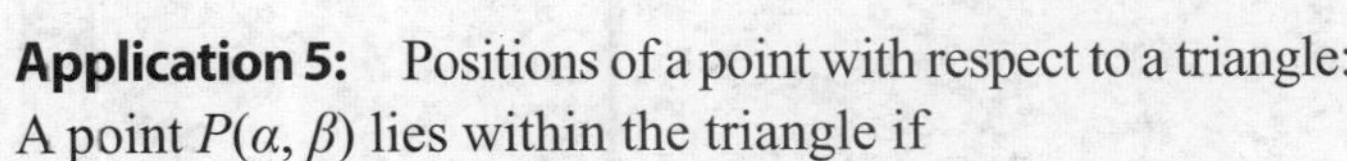

Application 5: Positions of a point with respect to a triangle: A point $P(\alpha, \beta)$ lies within the triangle if

(i) P and A lie on same side of BC

(ii) P and B lie on same side of AC

(iii) P and C lie on same side of AB

the required condition can be obtianed by taking the intersection of above said three conditions.

A(x$_1$,y$_1$), L$_1$, L$_3$, P(α,β), B (x$_2$,y$_2$), L$_2$, C (x$_3$,y$_3$)

FIGURE 1.45

ILLUSTRATION 37: Find α, if $(\alpha, 3\alpha)$ lies inside the triangle having sides along the lines $2x + 3y = 1$, $x + 2y - 3 = 0$, $6y = 5x - 1$.

SOLUTION: Given lines are $AB : 2x + 3y - 1 = 0$ ....(i)

$AC : x + 2y - 3 = 0$ .....(ii)

$BC : 5x - 6y - 1 = 0$ ....(iii)

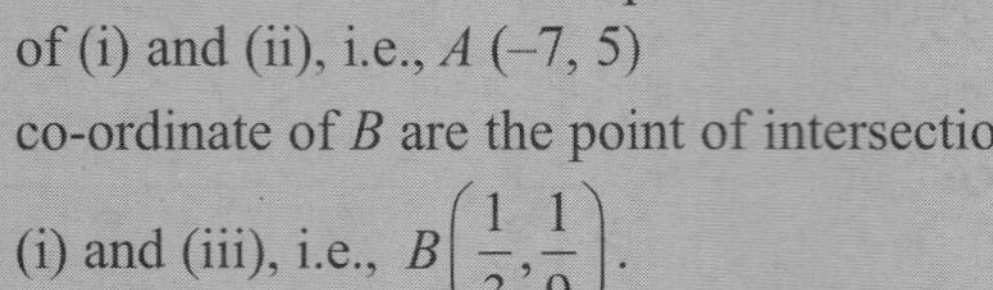

$\therefore$ Co-ordinates of A are the point of intersection of (i) and (ii), i.e., $A(-7, 5)$

co-ordinate of B are the point of intersection of (i) and (iii), i.e., $B\left(\frac{1}{3}, \frac{1}{9}\right)$.

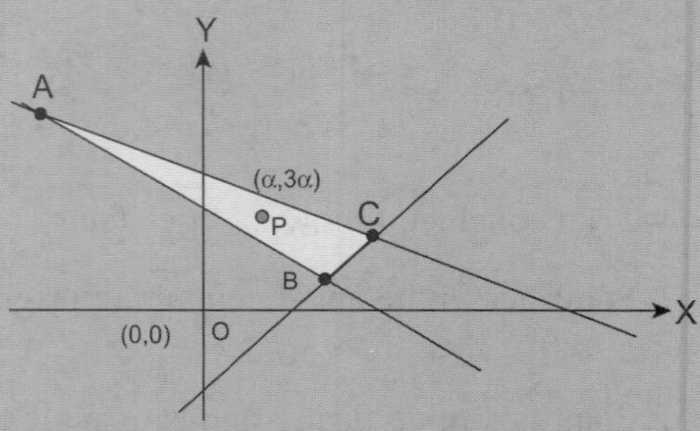

FIGURE 1.46

Co-ordinate of C are the point of intersection of (ii) and (iii), i.e., $C\left(\frac{5}{4}, \frac{7}{8}\right)$

Now, $P(\alpha, 3\alpha)$ and $C\left(\frac{5}{4}, \frac{7}{8}\right)$ lie on same side of AB; $(2\alpha + 9\alpha - 1)\left(\frac{5}{2} + \frac{21}{8} - 1\right) > 0$

$\Rightarrow (11\alpha - 1)\left(\frac{20 + 21 - 8}{8}\right) > 0 \Rightarrow \alpha > \frac{1}{11}$ ...(iv)

Also, $P(\alpha, 3\alpha)$ and $B\left(\frac{1}{3}, \frac{1}{9}\right)$ lies on same side of AC.

$\Rightarrow (\alpha + 6\alpha - 3)\left(\frac{1}{3} + \frac{2}{9} - 3\right) > 0 \Rightarrow (7\alpha - 3)\left(\frac{3 + 2 - 27}{9}\right) > 0 \Rightarrow 7\alpha - 3 < 0 \Rightarrow \alpha < 3/7$ ....(v)

Also, $P(\alpha, 3\alpha)$ and A lies on same side of BC; $(5\alpha - 18\alpha - 1)(-35 - 30 - 1) > 0$.

$\Rightarrow (-13\alpha - 1) < 0 \Rightarrow 13\alpha + 1 > 0 \Rightarrow \alpha > -1/13$ ....(vi)

Taking intersection of the solution sets of inqualities (iv), (v) and (vi)

$\Rightarrow \alpha \in \left(\frac{1}{11}, \frac{3}{7}\right)$.

ILLUSTRATION 38: Given that $A(1, 1)$ and $B(2, -3)$ are two points and D is a point on AB produced such that $AD = 5AB$. Find the co-ordinates of D.

SOLUTION: We have, $AD = 5AB$. Therefore, $BD = 4AB$.

Thus, D divides AB externally in the ratio $AD : BD = 5 : 4$.

K 3K
A(1,1) B(2,-3) D

FIGURE 1.47

Hence, the co-ordinate of D are $\left(\frac{5(2) - 4(1)}{5 - 4}, \frac{5(-3) - 4(1)}{5 - 4}\right) = (6, -19)$.

ILLUSTRATION 39: If A divides OP internally in the ratio $k_1 : k_2$ and B, externally, in the ratio $k_1 : k_2$, then prove that OP is the harmonic mean of OA and OB.

SOLUTION: We have, $\frac{1}{OA} = \frac{k_1 + k_2}{k_1 OP}$ and $\frac{1}{OB} = \frac{k_1 - k_2}{k_1 OP}$

$\Rightarrow \frac{1}{OA} + \frac{1}{OB} = \frac{2}{OP}$

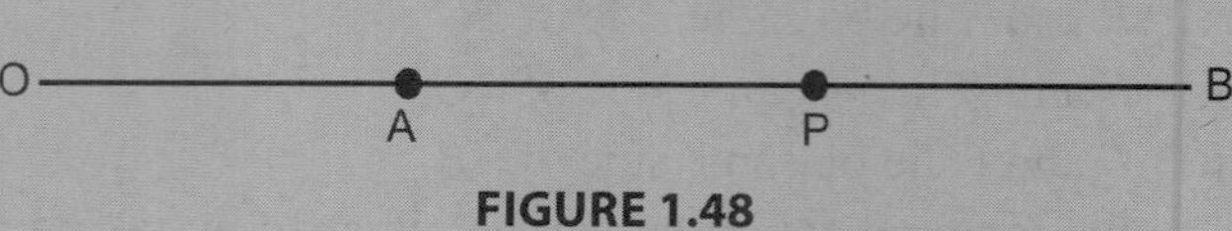

FIGURE 1.48

$\Rightarrow OA$, OP and OB are in H.P.

Application-6: To test co-linearity of three points.

If three points A, B, C are collinear then B divides AC in ratio $\lambda : 1$.

$$\Rightarrow x_B = \frac{\lambda x_c + x_A}{\lambda + 1} \quad ...(i)$$

$$\text{and } y_B = \frac{\lambda y_c + y_A}{\lambda + 1} \quad ...(ii)$$

Solving these two equations if we get same value of λ ($\lambda \neq -1$) from both equations, then points A, B, C will be collinear. Otherwise, they will form a triangle.

Also Since $x_B = \frac{\lambda}{\lambda+1} x_C + \frac{1}{\lambda+1} x_A$

If $x_B = mx_C + nx_A$ and $y_B = my_C + ny_A$ such that $m + n = 1$. Then A, B, C must be collinear.

Application-7: Menelau's Theorem

If R, Q, P are three points on the sides AB, AC and BC produced respectively of a triangle ABC such that the points R, Q, P are collinear, then $\left(\frac{PB}{PC}\right) \cdot \left(\frac{QC}{QA}\right) \cdot \left(\frac{RA}{RB}\right) = -1$.

Here $\left(\frac{PB}{PC}\right)$ denotes ratio in which P divides BC.

Similarly $\left(\frac{QC}{QA}\right)$ and $\left(\frac{RA}{RB}\right)$ are ratios

Ceva's Theorem:

If D, E, F are points on sides BC, CA, and AB, respectively, of triangle ABC, such that AD, BE and CF are concurrent, then $\left(\frac{BD}{DC}\right) \cdot \left(\frac{CE}{AE}\right) \cdot \left(\frac{AF}{BF}\right) = 1$. Here, $\left(\frac{BD}{DC}\right)$ is ratio in which D divides BC. Similarly, $\left(\frac{CE}{AE}\right)$ and $\left(\frac{AF}{BF}\right)$

ILLUSTRATION 40: Test whether the points $A(2, -2)$; $B(-3, 8)$ and $C(-1, 4)$ are collinear or not.

SOLUTION : Suppose that B divides AC internally in the ratio $\lambda : 1$.

$$\Rightarrow -3 = \frac{\lambda(-1) + 1(2)}{\lambda + 1}; \; 8 = \frac{\lambda(4) + 1(-2)}{\lambda + 1}.$$

$$\Rightarrow -3\lambda - 3 = -\lambda + 2; \; 8\lambda + 8 = 4\lambda - 2$$

$$\Rightarrow 2\lambda = -5 \Rightarrow \lambda = -5/2.$$

$\Rightarrow$ point B divides AC externally in the ratio $5 : 2$.

ILLUSTRATION 41: In which ratio the line joining (2,3) and (4,1) divides the line segment joining (1,2) and (4,3)?

(a) 1 : 2 (b) 2 : 1

(c) 1 : 1 (d) 2 : 3

SOLUTION : Let the line divides the line segment in the ratio of k: 1.

Now, co-ordinate of point is $P\left(\frac{4k+1}{k+1}, \frac{3k+2}{k+1}\right)$

Now, locus of any point line on lying joining (2,3) and (4,1) is $\frac{3-1}{2-4} = \frac{y-1}{x-4}$.

$$\Rightarrow -\frac{2}{2} = \frac{y-1}{x-4} \Rightarrow y-1 = -(x-4)$$

$$\Rightarrow y-1 = -x+4 \Rightarrow y+x = 5$$

So, point P will also satisfy $\frac{3k+2}{k+1} + \frac{4k+1}{k+1} = 5 \Rightarrow k = 1$

$\therefore$ Required ratio is k: 1 = 1: 1.

ILLUSTRATION 42: The range of values of α in the interval $(0, \pi)$ such that the points (3, 5) and $(\sin\alpha, \cos\alpha)$ lies on the same side of the line $x + y - 1 = 0$

SOLUTION: The value of expression $L = x + y - 1$ for (3, 5) is 7, i.e., + *ve* and hence L for $(\sin \alpha, \cos \alpha)$ to lie on same side as that of (3, 5); $(\sin \alpha + \cos \alpha - 1) > 0$

$$\Rightarrow \frac{1}{\sqrt{2}}\sin\alpha + \frac{1}{\sqrt{2}}\cos\alpha > \frac{1}{\sqrt{2}}$$

$\Rightarrow \sin(\pi/4 + \alpha) > \sin(\pi/4)$ or $(3\pi/4)$

$$\therefore \quad \frac{\pi}{4} < \frac{\pi}{4} + \alpha < \frac{3\pi}{4} \text{ or } 0 < \alpha < \frac{\pi}{2} \quad \therefore \ \alpha \in \left(0, \frac{\pi}{2}\right).$$

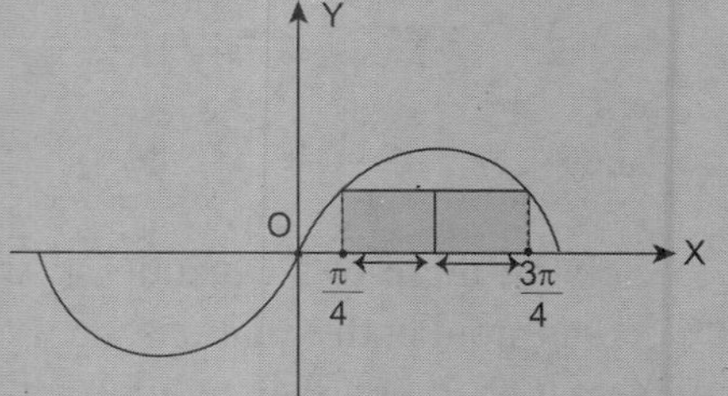

FIGURE 1.49

ILLUSTRATION 43: Find the set of values of 'b' for which the origin and the point (1,1) lie on the straight line $a^2x + aby + 1 = 0 \ \forall \ a \in \mathbb{R}, b > 0$.

SOLUTION: Condition for (x_1, y_1) and (x_2, y_2) lying on the same side w.r.t. $ax + by + c = 0$

$$\frac{ax_1 + by_1 + c}{ax_2 + by_2 + c} > 0 \Rightarrow \frac{1}{a^2 + ab + 1} > 0$$

$a^2 + ab + 1 > 0$

It is quadratic in 'a' (so to be positive $\forall a \in \mathbb{R}$, $D < 0$ and the leading co-efficient must be positive), so $b^2 - 4 < 0 \Rightarrow b \in (-2, 0) \cup (0, 2)$ but $b > 0 \Rightarrow b \in (0, 2)$.

ILLUSTRATION 44: Determine the range of values of $\theta \in [0, 2\pi]$ for which the point $(\cos\theta, \sin\theta)$ lies inside the triangle formed by the lines $x + y = 2$; $x - y = 1$ and $6x + 2y - \sqrt{10} = 0$.

SOLUTION: The point $P(\cos\theta, \sin\theta)$ and (0, 0) lies on the same side of line $x + y = 2$

$\Rightarrow (\cos\theta + \sin\theta - 2)(0 - 0 - 2) > 0$

$\cos\theta + \sin\theta - 2 < 0$ which is always true as $|\cos\theta + \sin\theta| \le \sqrt{2}$

the piont $P(\cos\theta, \sin\theta)$ and (0, 0) lies on the same side of line $x - y = 1$

$\Rightarrow (\cos\theta - \sin\theta - 1)(0 - 0 - 1) > 0$

$\Rightarrow \cos\theta - \sin\theta < 1$

$$\Rightarrow \cos\left(\theta + \frac{\pi}{4}\right) < \frac{1}{\sqrt{2}} \Rightarrow \theta \in \left(0, \frac{3\pi}{2}\right)$$

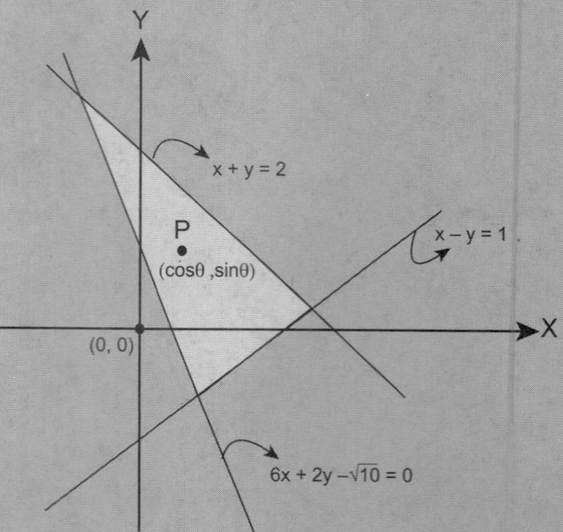

FIGURE 1.50

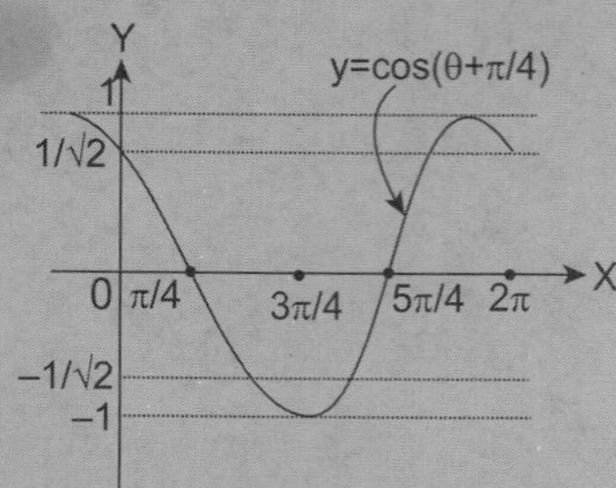

FIGURE 1.51

and $P(\cos\theta, \sin\theta)$ and (0, 0) lies on opposite sides of line $6x + 2y - \sqrt{10} = 0$

$\Rightarrow (6\cos\theta + 2\sin\theta - \sqrt{10})(0 + 0 - \sqrt{10}) < 0$

$\Rightarrow \frac{6}{\sqrt{40}}\cos\theta + \frac{2}{\sqrt{40}}\sin\theta > \frac{1}{2} \Rightarrow \frac{3}{\sqrt{10}}\cos\theta + \frac{1}{\sqrt{10}}\sin\theta > \frac{1}{2}$

$\Rightarrow \sin(\theta + \alpha) > \frac{1}{2} \Rightarrow \theta + \alpha \in \left(\frac{\pi}{6}, \frac{5\pi}{6}\right)$; as $\theta \in \left(0, \frac{3\pi}{2}\right)$ only

where $\sin\alpha = \frac{3}{\sqrt{10}}, \cos\alpha = \frac{1}{\sqrt{10}} \Rightarrow \tan\alpha = 3 \Rightarrow \alpha = \tan^{-1} 3$.

$\Rightarrow \theta \in \left(\frac{\pi}{6} - \alpha, \frac{5\pi}{6} - \alpha\right)$

$\Rightarrow \theta \in \left(\frac{\pi}{6} - \tan^{-1} 3, \frac{5\pi}{6} - \tan^{-1} 3\right)$ for $\theta \in [0, 2\pi]$

$\therefore \theta \in \left(0, \frac{5\pi}{6} - \tan^{-1} 3\right) \because \left(\frac{\pi}{6} - \tan^{-1}(3)\right)$ is negative angle.

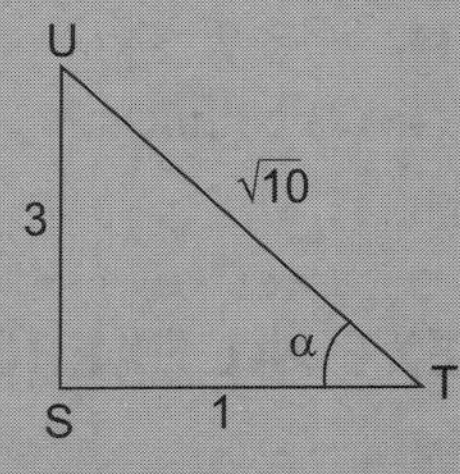

FIGURE 1.52

TEXTUAL EXERCISE–3 (SUBJECTIVE)

1. (a) Find the co-ordinates of the points which externally divide the line segment AB in the given ratios:
   (i) $A(0, -4)$; $B(8, 0)$; Ratio 4: 3
   (ii) $A(1, 2)$; $B(-4, -3)$; Ratio 2: 3.
   (b) Find the co-ordinates of the points which divide, internally and externally, the line joining the point $(a + b, a - b)$ to the point $(a - b, a + b)$ in the ratio $a : b$.

2. If the mid-point of join of $(x, y + 1)$ and $(x + 1, y + 2)$ is (3/2, 5/2); then find the mid-point of join of $(x - 1, y + 1)$ and $(x + 1, y - 1)$.

3. Prove that the lines joining the midpoints of the opposite sides of a quadrilateral bisect one another.

4. The points (3, – 4) and (–6, 2) are the extremities of a diagonal of a parallelogram. If the third vertex is (–1, –3), find the co-ordinates of the fourth vertex.

5. Find the third vertex
   (i) of a triangle if two of its vertices are at (–2, 4), (7, –3) and the centroid at (3, 2).
   (ii) if two vertices of an equilateral triangle are (0, 0) and $(3, \sqrt{3})$.

6. Prove that the points (–2, –2), (1, 0), (4, 4) and (1, 2) are the vertices of a parallelogram.

7. (a) In what ratio the join of the points (7, 3) and (–4,5) divided by the y-axis?
   (b) Find the point P if $Q = (-2, 4)$ is the point on the line segment OP such that $OQ = 1/3\ OP$, O being the origin.

8. If $0 < t < 1$, the straight line joining the points (x_1, y_1) and (x_2, y_2) is divided by the point $P(x_1 + t(x_2 - x_1), y_1 + t(y_2 - y_1))$ in the ratio $t : 1 - t$ internally, prove it.

9. Find the range of $a \in (0, 2\pi)$ such that the point (3, –1) and (sin$a$, –cos$a$) lie on the opposite sides of the line $x - y + 1 = 0$.

10. Show that the origin is within the triangle formed by the lines $4x + 7y + 19 = 0$, $4x + y - 11 = 0$ and $4x - 5y + 7 = 0$.

11. Derive the conditions to be imposed on β so that $(0, \beta)$ should lie on or inside the triangle having sides

    $y + 3x + 2 = 0$, $3y - 2x - 5 = 0$ and $4y + x - 14 = 0$

12. A triangle ABC is formed by the lines $2x - 3y - 6 = 0$ $3x - y + 3 = 0$ and $3x + 4y - 12 = 0$. If the points $P(\alpha, 0)$ and $Q(0, \beta)$ always lie on or inside the $\Delta\ ABC$, then find α and β.

13. If the points (5, 6)and (cos θ, sin θ), where $0 < \theta < \pi$, lie on the opposite sides of equation $x + y - 1 = 0$; then find the set of values of θ.

14. Test whether the given set of point are collinear or not.
(a) (–2, 1) (0, 5) and (5, 15)
(b) (3, 1) (6, 4) and (4, 5)

Answer Keys

1. (a) (i) (32, 12) (ii) (11, 12) (b) $\left(\frac{a^2+b^2}{a+b}, \frac{a^2-b^2+2ab}{a+b}\right)$ 2. (1, 1) 4. (–2, 1)

5. (i) (4, 5), (ii) $(2\sqrt{3}, 0), (3, -\sqrt{3})$ 7. (a) 7: 4 (b) (–6, 12) 9. $(\pi, 3\pi/2)$

11. $5/3 < \beta < 7/2$ 12. $\alpha \in [-1, 3]$ and $\beta \in [-2, 3]$ 13. $\theta \in (\pi/2, \pi)$

14. (a) collinear (b) non-collinear

TEXTUAL EXERCISE–3 (OBJECTIVE)

1. The distance of the middle point of the line joining the points $(a\sin\theta, 0)$ and $(0, a\cos\theta)$ from the origin is
(a) $\frac{a}{2}$ (b) $\frac{1}{2}a(\sin\theta + \cos\theta)$
(b) $a(\sin\theta + \cos\theta)$ (d) a

2. The ratio in which x-axis divides the join of the points (2, –3) and (5, 6) is
(a) 2 : 1 (b) 1 : 2
(c) 2 : –1 (d) None of these

3. P and Q are points on the line joining $A(-2, 5)$ and $B(3,1)$ such that $AP = PQ = QB$. Then the mid-point of PQ is
(a) $\left(\frac{1}{2}, 3\right)$ (b) $\left(-\frac{1}{2}, 4\right)$
(c) (2,3) (d) (1,4)

4. If the middle point of the line segment joining the points (5, a) and (b, 7) be (3, 5), then $(a, b) =$
(a) (3, 1) (b) (1, 3)
(c) (–2, –2) (d) (–3, –1)

5. If the three vertices of a rectangle taken in order are the points (2, –2), (8, –4) and (5, 7). The co-ordinates of the fourth vertex is
(a) (1, 1) (b) (1, –1)
(c) (–1, 1) (d) None

6. Co-ordinates of point of trisection as line joining $A(3, -1)$ and $B(-6,5)$ are
(a) (0, 1) and (–3, 3) (b) (1,0) and (3,–3)
(c) (1, 1) and (2, 4) (d) None of these

7. All points lying inside the triangle formed by the points (1, 3), (5, 0) and (–1, 2) satisfy
(a) $2x + y \geq 0$
(b) $3x + 4y - 15 \geq 0$
(c) $x + 3y - 10 \leq 0$
(d) $-2x - y \geq 0$

8. The position of the point (1, 1) w.r.t. the lines $L_1 = 2x + 3y + 7 = 0$ and $L_2 = 4x + 6y - 19 = 0$ is such that
(a) it is below L_2
(b) it is above L_1
(c) it is between L_1 and L_2
(d) it is above both L_1 and L_2

9. If (a, a^2) falls inside the angle made by the line $y = x/2, x > 0$ and $y = 3x, x > 0$ then a belong to:
(a) (3, 0) (b) $\left(\frac{1}{2}, 3\right)$
(c) $\left(-3, -\frac{1}{2}\right)$ (d) $\left(0, \frac{1}{2}\right)$

10. ($\sin\theta$, $\cos\theta$) and (3, 2) lie on the same side of the line $x + y = 1$, then θ lies in
(a) $(0, \pi/2)$ (b) $(0, \pi)$
(c) $\left(\frac{\pi}{4}, \frac{\pi}{2}\right)$ (d) $\left(0, \frac{\pi}{4}\right)$.

11. The three vertices of a parallelogram taken in order are (–1, 0), (3, 1) and (2, 2), respectively. Find the co-ordinates of the fourth vertex.
(a) (–2, 1) (b) (–2, –1)
(c) (6, 3) (d) None of these

Answer Keys

1. (a) **2.** (b) **3.** (a) **4.** (a) **5.** (d) **6.** (a) **7.** (a, c) **8.** (a, b, c) **9.** (b) **10.** (a) **11.** (a, c)

■ AREA OF GEOMETRICAL FIGURES

Surface is an object having two dimensions. It can be plane or curved. So, they possess some, that is, plane surfaces like triangle, quadrilateral and curved surfaces like sphere, cone, etc. In the following articles, we will study the area of some regular plane surfaces.

1. Area of Triangle

Consider a triangle ABC, whose vertices have co-ordiantes $A(x_1, y_1)$, $B(x_2, y_2)$ and $C(x_3, y_3)$. Then the area of triangle ABC is given by

$$\Delta = \frac{1}{2}[x_1(y_2 - y_3) + x_2(y_3 - y_1) + x_3(y_1 - y_2)].$$

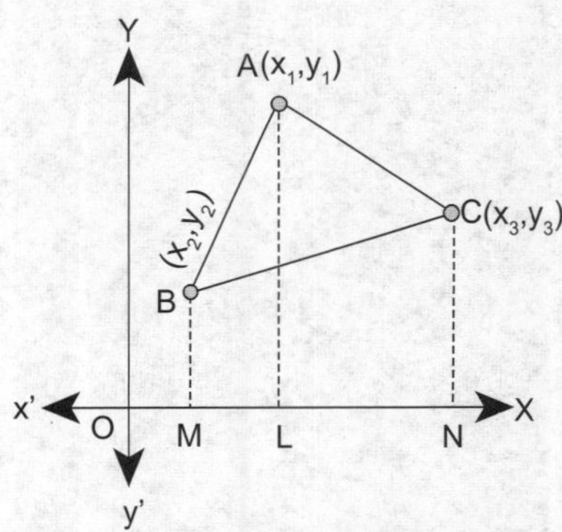

FIGURE 1.53

Proof: We know that, Area of trapezium = 1/2 (Sum of parallel sides) (Distance between them). Therefore AL, BM and CN are the perpendiculars drawn from A, B, C on x-axis (as shown in the figure). Now express the area of triangle ABC as sum and difference of areas of three trapeziums as decribed here.

Area of ΔABC = Area of trapezium $ABML$ + Area of trapezium $ALNC$ – Area of trapezium $BMNC$

$\therefore$ Area of ΔABC

$$= \frac{1}{2}(BM + AL)(ML) + \frac{1}{2}(AL + CN)(LN) - \frac{1}{2}(BM + CN)(MN)$$

$$= \frac{1}{2}(y_2 + y_1)(x_1 - x_2) + \frac{1}{2}(y_1 + y_3)(x_3 - x_1) - \frac{1}{2}(y_2 + y_3)(x_3 - x_2)$$

$$= \frac{1}{2}\begin{bmatrix} x_1(y_2 + y_1 - y_1 - y_3) + \\ x_2(-y_2 - y_1 + y_2 + y_3) + \\ x_3(y_1 + y_3 - y_2 - y_3) \end{bmatrix}$$

$$= \frac{1}{2}[x_1(y_2 - y_3) + x_2(y_3 - y_1) + x_3(y_1 - y_2)]$$

Sign Convention: The area calculated by the above formula comes out to be positive, if and only if points $A(x_1, y_1)$, $B(x_2, y_2)$ and $C(x_3, y_3)$ form a close loop in anti-clockwise order, if the order is reversed (i.e., clockwise) The above expression generates a value of same magnitude but with negative sign, consequently depending upon the location of points in the quadrants and their relative cyclic order the value of above expression can be positive or negative, so to deal with this problem we define the area as the modulus of above expression, since area is a positive scalar quantity.

REMARKS

(i) *Area of triangle ABC* $= \frac{1}{2}| x_1(y_2 - y_3) + x_2(y_3 - y_1) + x_3(y_1 - y_2)|$

(ii) *Remember the cyclic order of the subscript* 123; 231; 312 *in above expression (see diagram)*

(iii) *This expression for the area can also be written in the determinant form as follows:*

$$\Delta = \frac{1}{2}\begin{vmatrix} x_1 & y_1 & 1 \\ x_2 & y_2 & 1 \\ x_3 & y_3 & 1 \end{vmatrix}$$

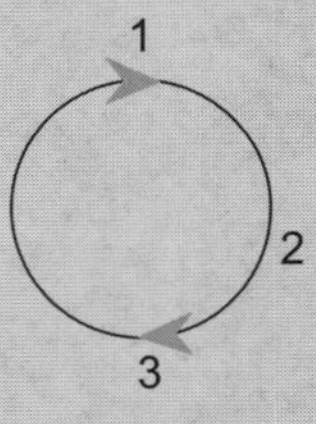

FIGURE 1.54

(iv) *If area of Δ ABC is zero, then the points are collinear. Hence the essential condition for three points to be collinear is given as :*

$$\text{Area of } \Delta ABC = 0 \Rightarrow \begin{vmatrix} x_1 & y_1 & 1 \\ x_2 & y_2 & 1 \\ x_3 & y_3 & 1 \end{vmatrix} = 0$$

(vi) *If the co-ordinates of vertices of Δ ABC are given in polar form $(r_1, \theta_1), (r_2, \theta_2), (r_3, \theta_3)$, then the area of Δ will be given by*

$$\Delta = \frac{1}{2}\left|r_2 r_3 \sin(\theta_3 - \theta_2) + r_3 r_1 \sin(\theta_1 - \theta_3) + r_1 r_2 \sin(\theta_2 - \theta_1)\right|.$$

ILLUSTRATION 45: If the area of triangle with vertices (4, 0), (1, 1), (3, a) be 2 units, then find the value of a.

SOLUTION: According to the question, the area of triangle (say Δ) is given as

$$\Delta = \frac{1}{2}\left\|\begin{matrix} 4 & 0 & 1 \\ 1 & 1 & 1 \\ 3 & a & 1 \end{matrix}\right\| = 2$$

$$\Rightarrow \frac{1}{2}|4(1-a) + 1(a-3)| = 2$$

$$\Rightarrow |4 - 4a + a - 3| = 4 \Rightarrow |1 - 3a| = 4$$

$$\Rightarrow 1 - 3a = \pm 4 \Rightarrow a = -1 \text{ or } a = 5/3.$$

ILLUSTRATION 46: Find the area of a triangle whose vertices are $A(3,2)$; $B(11,8)$ and $C(8,12)$.

SOLUTION: Let $A \equiv (x_1, y_1) \equiv (3, 2)$, $B \equiv (x_2, y_2) \equiv (11, 8)$ and $C \equiv (x_3, y_3) \equiv (8, 12)$. Then,

$$\text{Area of } \Delta ABC = \frac{1}{2}\begin{vmatrix} x_1 & y_1 & 1 \\ x_2 & y_2 & 1 \\ x_3 & y_3 & 1 \end{vmatrix} = \frac{1}{2}\begin{vmatrix} 3 & 2 & 1 \\ 11 & 8 & 1 \\ 8 & 12 & 1 \end{vmatrix}$$

$$= \frac{1}{2}|\{3(8-12) + 11(12-2) + 8(2-8)\}| = \frac{1}{2}|\{-12 + 110 - 48\}|$$

$= 25$ sq. units

If we change the order of points to ACB for (x_1, y_1), (x_2, y_2), (x_3, y_3), then second row (R_2) and third row (R_3) get interchanged so making the values negative.

ILLUSTRATION 47: Find the area of a triangle whose vertices are $A(1, \pi/8)$, $B(1, 5\pi/8)$ and $C(\sqrt{2}, 3\pi/8)$.

SOLUTION: Area of Δ having vertices given in polar form is given by

$$\Delta = \frac{1}{2}\left|r_1 r_2 \sin(\theta_2 - \theta_1) + r_2 r_3 \sin(\theta_3 - \theta_1) + r_3 r_1 \sin(\theta_1 - \theta_3)\right|$$

$$\Rightarrow \Delta = \frac{1}{2}\left|1 \sin(5\pi/8 - \pi/8) + \sqrt{2}\sin(3\pi/8 - 5\pi/8) + \sqrt{2}\sin(\pi/8 - 3\pi/8)\right|$$

$$\Rightarrow \Delta = \frac{1}{2}\left|\sin\frac{\pi}{2} - \sqrt{2}\sin\frac{\pi}{4} - \sqrt{2}\sin\frac{\pi}{4}\right| \quad \Delta = \left|\frac{1}{2}[1-1-1]\right| = \frac{1}{2}. \text{ sq. units.}$$

2. Stair Method to Find the Area of *n-sided* Polygon

To find the area of polygon whose vertices are (x_1, y_1), (x_2, y_2), (x_3, y_3), ..., (x_n, y_n) follow the given steps:

Step (1). Put the co-ordinates of first vertex along first row and second vertex along second row and so on. Finally, repeat the co-ordinates of first vertex along the last row of a $(n + 1) \times 2$ order matrix formed and multiply by 1/2 as shown below and consider the magnitude of the expression.

$$\frac{1}{2}\left\|\begin{matrix} x_1 & y_1 \\ x_2 & y_2 \\ x_3 & y_3 \\ \vdots & \vdots \\ \vdots & \vdots \\ x_n & y_n \\ x_1 & y_1 \end{matrix}\right\| = 1/2\,|\{(x_1 y_2 - x_2 y_1) + (x_2 y_3 - x_3 y_2) + ..+ (x_n y_1 - x_1 y_n)\}|$$

Step (2). Now, to calculate the area of polygon go on multiplying the left element of each row to the right element of next row and add them (say the sum obtained is S_1).

Step (3). Now multiplying left element of each row starting from last row by right element of upper row to get the sum (S_2). The area of polygon would be given by $(1/2)\,|S_1 - S_2|$. e.g., Area of ΔABC with vertices (x_1, y_1); (x_2, y_2); (x_3, y_3) is given by

$$\Delta = \frac{1}{2}\left\|\begin{matrix} x_1 & y_1 \\ x_2 & y_2 \\ x_3 & y_3 \\ x_1 & y_1 \end{matrix}\right\| = \frac{1}{2}\left|\begin{matrix}(x_1 y_2 + x_2 y_3 + x_3 y_1) - \\ (x_1 y_3 + x_3 y_2 + x_2 y_1)\end{matrix}\right|$$

3. Area of General Quadrilateral

If $A(x_1, y_1)$, $B(x_2, y_2)$, $C(x_3, y_3)$ and $D(x_4, y_4)$ are vertices of a quadrilateral, then its area will be given by

$$\frac{1}{2}\left\|\begin{matrix} x_1 & y_1 \\ x_2 & y_2 \\ x_3 & y_3 \\ x_4 & y_4 \\ x_1 & y_1 \end{matrix}\right\| = \frac{1}{2}\left\|\begin{bmatrix} (x_1 y_2 - x_2 y_1) \\ +(x_2 y_3 - x_3 y_2) \\ +(x_3 y_4 - x_4 y_3) \\ +(x_4 y_1 - x_1 y_4) \end{bmatrix}\right\|$$

Proof: Area of quadrilateral $ABCD$ = area of trapezium $ALND$ + area of trapezium $DNRC$ + area of trapezium $CRMB$ – area of trapezium $ALMB$

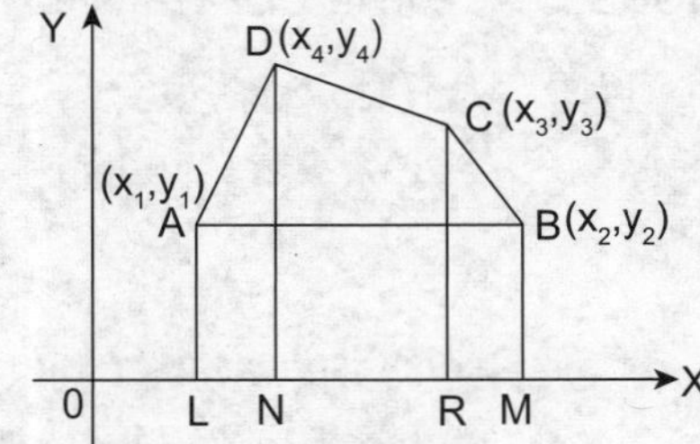

FIGURE 1.55

Substituting the value of area of trapezium, we get area of quadrilateral $ABCD$.

$$= \frac{1}{2}|[(x_4 - x_1)(y_1 + y_4) + (x_3 - x_4)(y_3 + y_4) + (x_2 - x_3)(y_2 + y_3) - (x_2 - x_1)(y_2 + y_1)]|$$

$$= \frac{1}{2}\left\|\begin{bmatrix}(x_1 y_2 - x_2 y_1) + (x_2 y_3 - x_3 y_2) + \\ (x_3 y_4 - x_4 y_3) + (x_4 y_1 - x_1 y_4)\end{bmatrix}\right\| = \frac{1}{2}\left\|\begin{matrix} x_1 & y_1 \\ x_2 & y_2 \\ x_3 & y_3 \\ x_4 & y_4 \\ x_1 & y_1 \end{matrix}\right\|$$

REMARKS

(i) If the *area of a quadrilateral joining four points is zero, then four points are collinear.*

(ii) *If vertices of ΔABC are taken in counter clockwise order, then the above formula generates the area of ΔABC. But, if vertices are chosen in clockwise order, then this formula generates negative of the value of area.*

(iii) *If the area of Δ is given* and *co-ordinate of some vertex has to be obtained, always use the modulus sign in formula.*

(iv) *If points $A_1, A_2, A_3, ..., A_n$ are collinear, then area of polygon obtained by above formula comes out to be zero.*

(v) *If $a_1x + b_1y + c_1 = 0$; $a_2x + b_2y + c_2 = 0$* and *$a_3x + b_3y + c_3 = 0$ are the sides of a triangle, then the area of the triangle is given by (without solving the vertices)*

$$\Delta = \frac{1}{2C_1C_2C_3}\begin{vmatrix} a_1 & b_1 & c_1 \\ a_2 & b_2 & c_2 \\ a_3 & b_3 & c_3 \end{vmatrix}$$ *where C_1, C_2, C_3 are the co-factors of c_1, c_2, c_3 in the determinant*

i.e., $C_1 = a_2b_3 - a_3b_2$, $C_2 = a_3b_1 - a_1b_3$ and $C_3 = a_1b_2 - a_2b_1$.

ILLUSTRATION 48: Find the area of quadrilateral whose vertices are $A(1,1)$, $B(3,4)$, $C(5,-2)$ and $D(4,-7)$.

SOLUTION: Area of quadrilateral

$$=\frac{1}{2}[(1\times 4)-(3\times 1)+(3\times(-2))-(5\times 4)+5(-7)-(4\times(-2))+(4\times 1)-(1\times(-7))] = \frac{41}{2} \text{ sq. unit.}$$

ILLUSTRATION 49: If $A(6,3)$, $B(-3,5)$, $C(4,-2)$ and $D(x,3x)$ are four points such that $\Delta DBC/\Delta ABC = 1:2$, then what is the value of x?

SOLUTION: Since ratio of area of triangle is given as

$$\frac{\Delta DBC}{\Delta ABC}=\pm\frac{1}{2} \Rightarrow 2\begin{vmatrix} x & 3x & 1 \\ -3 & 5 & 1 \\ 4 & -2 & 1 \end{vmatrix} = \pm\begin{vmatrix} 6 & 3 & 1 \\ -3 & 5 & 1 \\ 4 & -2 & 1 \end{vmatrix} \Rightarrow x = 11/8; -3/8.$$

ILLUSTRATION 50: Prove that an equilateral triangle cannot have all three vertices with rational co-ordinates.

SOLUTION: Let $P(x_1, y_1)$, $Q(x_2, y_2)$ and $R(x_3, y_3)$ be the vertices of a triangle PQR, where (x_i, y_i): $i = 1, 2, 3$ are rationals. Then, area of ΔPQR is given by

$$\Delta = \frac{1}{2}[x_1(y_2-y_3)+x_2(y_3-y_1)+x_3(y_1-y_2)]$$

which is clearly a rational number [$\because x_i, y_i$ are rationals]

let the triangle PQR be an equilateral triangle,

$$\Delta = \frac{\sqrt{3}}{4}(\text{side})^2 = \frac{\sqrt{3}}{4}(AB)^2 \qquad [\because PQ = QR = RP]$$

$$= \frac{\sqrt{3}}{4} \ (a \text{ positive rational}) \qquad \left[\begin{array}{l}\because \text{ vertices are rational} \\ \therefore AB^2 = (x_1-x_2)^2+(y_1-y_2)^2 \text{ is a positive rational}\end{array}\right]$$

which is an irrational number

This is a contradiction to the fact that the area is a rational number.

Hence, the triangle cannot be equilateral.

ILLUSTRATION 51: Prove that the points $(a, b+c)$, $(b, c+a)$ and $(c, a+b)$ are collinear.

SOLUTION: Let $A = (x_1, y_1) = (a, b+c)$, $B = (x_2, y_2) = (b, c+a)$ and $C = (x_3, y_3) = (c, a+b)$ be three points. Then,

$$2(\text{Area of } \Delta ABC) = x_1(y_2-y_3)+x_2(y_3-y_1)+x_3(y_1-y_2) = \sum x_1(y_2-y_3) = \sum a(c-b) = 0$$

$$= a\{(c+a)-(a+b)\}+b\{(a+b)-(b+c)\}+c\{(b+c)-(c+a)\}$$

$$= a(c-b)+b(a-c)+c(b-a) = 0 \Rightarrow \text{Area of } \Delta ABC = 0$$

Hence, the given points are collinear.

$$\text{Aliter: } \Delta = \begin{vmatrix} a & b+c & 1 \\ b & c+a & 1 \\ c & a+b & 1 \end{vmatrix} \xrightarrow{C_1 \to C_1 + C_2} \begin{vmatrix} a+b+c & b+c & 1 \\ a+b+c & c+a & 1 \\ a+b+c & a+b & 1 \end{vmatrix}$$

(taking $a+b+c$ common from column 1)

$$= (a+b+c)\begin{vmatrix} 1 & b+c & 1 \\ 1 & c+a & 1 \\ 1 & a+b & 1 \end{vmatrix} = 0$$

ILLUSTRATION 52: Find the area of the quadrilateral whose vertices are $(-3, 2)$, $(7, -6)$, $(-5, -4)$ and $(5, 4)$.

SOLUTION: First of all, we plot the points. The vertices of the quadrilateral taken in order are $A(5, 4)$, $B(-3, 2)$, $C(-5, -4)$, $D(7, -6)$.

Now area of quadrilateral $ABCD = \frac{1}{2}\begin{Vmatrix} -5 & -4 \\ 7 & -6 \\ 5 & 4 \\ -3 & 2 \\ -5 & -4 \end{Vmatrix}$.

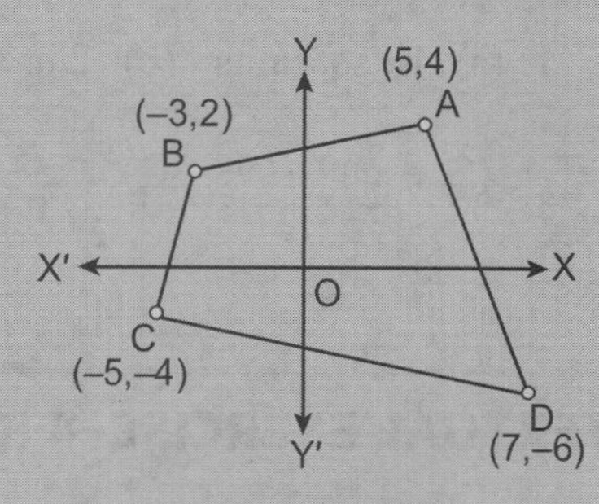

FIGURE 1.56

Applying formula, $= \frac{1}{2}|(30+28+10+12)-(-28-30-12-10)|$

$= \frac{1}{2}|80+80| = 80$ sq. units

ILLUSTRATION 53: The vertices of a triangle ABC are $(\lambda, 2-2\lambda)$ $(-\lambda + 1, 2\lambda)$ and $(-4, -\lambda, 6 - 2\lambda)$. If its area be 70 units then find number of integral values of λ.

SOLUTION: $\Delta = \frac{1}{2}\begin{vmatrix} \lambda & 2-2\lambda & 1 \\ -\lambda+1 & 2\lambda & 1 \\ -4-\lambda & 6-2\lambda & 1 \end{vmatrix} = \pm 70$. Now, apply new transformations $R_2 \to R_2 - R_1, R_3 \to R_3 - R_1$ to make two zero.

$\therefore \begin{vmatrix} \lambda & 2(1-\lambda) & 1 \\ 1-2\lambda & 2(2\lambda-1) & 0 \\ -2(\lambda+2) & 4 & 0 \end{vmatrix} = \pm 140$ or $4(1 - 2\lambda) + 4(2\lambda^2 + 3\lambda - 2) = \pm 140$

or $2\lambda^2 + \lambda - 1 = \pm 35$ or $2\lambda^2 + \lambda - 36 = 0$ or $2\lambda^2 + \lambda + 34 = 0$

(roots of 2nd equation are non real as $D < 0$)

$\Rightarrow (\lambda - 4)(2\lambda + 9) = 0 \therefore \lambda = 4, -9/2$. Hence, there is only one integral value of λ.

TEXTUAL EXERCISE–4 (SUBJECTIVE)

1. Find the area of triangles whose co-ordinates are given by

(a) $(1, 3), (-7, 6), (5, -1)$

(b) $(a \cos \phi_1, b \sin \phi_1), (a \cos \phi_2, b \sin \phi_2), (a \cos \phi_3, b \sin \phi_3)$

2. Find the area of triangle having vertices at $(at_1^2, 2at_1), (at_2^2, 2at_2)$ and $(at_3^2, 2at_3)$.

3. If two opposite vertex of a square are (x_1, y_1) and (x_2, y_2), then what will be its area?

4. Prove that the area of the triangle with vertices $(t, t-2)$, $(t + 2, t + 2)$ and $(t + 3, t)$ is independent of t.

5. Using the area of the triangle or otherwise, show that the points $(\text{cosec}^2 \theta, 0)$, $(0, \sec^2 \theta)$, $(1, 1)$ are collinear.

6. Evaluate

(a) the area of the quadrilateral whose vertices are $A(0, 0)$, $B(1, 3)$, $C(2, 5)$ and $D(-1, 4)$.

(b) the area of the triangle, the co-ordinates of mid-points of whose sides are $(2, -1)$, $(3, 2)$ and $(5, 9)$.

7. The vertices of ΔABC are $A(3, 0)$, $B(0, 6)$ and $C(6, 9)$. A straight line DE divides AB and AC in the ratio 1: 2 at D and E respectively, prove that $\frac{\Delta ABC}{\Delta ADE} = 9$

8. The co-ordinates of points A, B, C and P are $(6, 3)$, $(-3, 5)$, $(4, -2)$ and (x, y) respectively, prove that $\frac{\Delta PBC}{\Delta ABC} = \frac{|x+y-2|}{7}$

9. If $(1, 4)$ be the centroid, of a triangle and the co-ordinates of its any two vertices be $(4, -8)$ and $(-9, 7)$, find the area of the triangle.

Answer Keys

1. (a) 10 sq. units (b) $2ab\sin\frac{\phi_2-\phi_3}{2}.\sin\frac{\phi_3-\phi_1}{2}.\sin\frac{\phi_1-\phi_2}{2}$ **2.** Area of Δ is given as $a^2|(t_1-t_2)(t_2-t_3)(t_3-t_1)|$

3. $\frac{(x_2-x_1)^2+(y_2-y_1)^2}{2}$ **6.** (a) 6 sq. units (b) 2 sq. units. **9.** $\frac{333}{2}$ sq. units.

TEXTUAL EXERCISE–4 (OBJECTIVE)

1. If (– 4, 0) and (1, –1) are two vertices of a triangle of area 4 square units, then its third vertex lies on

(a) $y = x$ (b) $5x + y + 2 = 0$

(c) $x + 5y - 4 = 0$ (d) None of these

2. The centroid of a triangle is (1, 4) and the co-ordinates of its two vertices are (4,–3) and (–9, 7). Then the area of the triangle is

(a) $\frac{183}{2}$ (b) $-\frac{183}{2}$

(c) 183 (d) None of these

3. $A(-5, 0)$ and $B(3, 0)$ are two of the vertices of a triangle ABC. Its area is 20 square cms. The vertex C lies on the line $x - y = 2$. The co-ordinates of C are

(a) (–7,– 5) or (3, 5)

(b) (– 3, –5) or (–5, 7)

(c) (7, 5) or (3, 5)

(d) (– 3, – 5) or (7, 5)

4. If $\begin{vmatrix} x_1 & y_1 & 1 \\ x_2 & y_2 & 1 \\ x_3 & y_3 & 1 \end{vmatrix} = \begin{vmatrix} a_1 & b_1 & 1 \\ a_2 & b_2 & 1 \\ a_3 & b_3 & 1 \end{vmatrix}$, then the two triangle with vertices $(x_1, y_1),(x_2, y_2), (x_3, y_3)$ and $(a_1b_1), (a_2, b_2)$ (a_3, b_3) must be

(a) similar (b) congruent

(c) never congruent (d) None of these

5. If $A(6,3)$; $B(-3,5)$; $C(4,-2)$ and $D(x,3x)$ are four points If the ratio of area of ΔDBC and ΔABC is 1: 2 then the value of x, will be

(a) 53/57 (b) 8/11

(c) 3 (d) None of these

6. The vertices of the triangle ABC are (2,1), (4,3) and (2,5). D, E, F are the mid-points of the sides. The area of the triangle DEF is

(a) 1 (b) 1.5

(c) 3 (d) 4

Answer Keys

1. (c) **2.** (a) **3.** (d) **4.** (d) **5.** (a) **6.** (a)

■ SLOPE OF LINE SEGMENT

Slope of line segment AB joining two points $A(x_1, y_1)$ and $B(x_2, y_2)$ is measure of steepness of the line segment, with respect to horizontal. It is defined as change in y-co-ordinate for unit change in x-co-ordinate, thus tell us about the rise along Y-axis for unit change in x (i.e., $\frac{\Delta y}{\Delta x}$). Therefore initially it used to be represented by n : 1. Trigonometrically, slope can be defined as "*The trigonometrical tangent of the angle that a line segment makes with the positive direction of the x-axis in anti-clockwise sense is called the slope or gradient of the line*". It is denoted by m.

Case I: $m = \tan\theta = \frac{y_2 - y_1}{x_2 - x_1}$ (see given Figure 1.57).

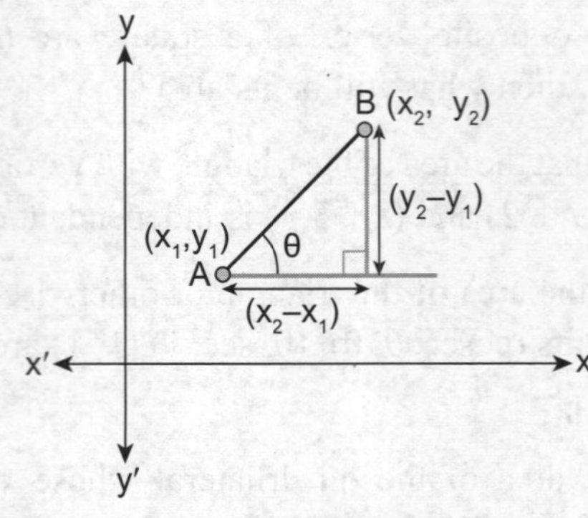

FIGURE 1.57

Case II: $m = -\tan(\pi - \theta)$

$$= -\frac{y_2 - y_1}{x_1 - x_2}$$

$$= \frac{y_2 - y_1}{x_2 - x_1}$$

(see given Figure 1.58)

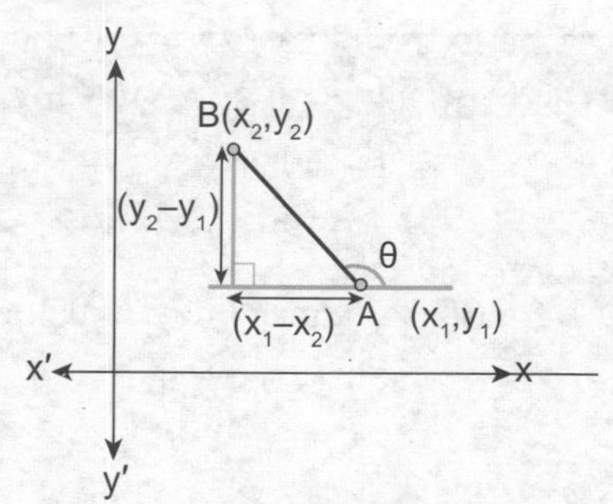

FIGURE 1.58

REMARKS

(i) *Slope of a line segment remains invariant for interchange of terminal points. i.e., Slope of AB = slope of BA*

(ii) *Slope is positive when θ lies in $(2n\pi, 2n\pi + \frac{\pi}{2})$* and *negative if θ lies in $(2n\pi + \frac{\pi}{2}, (2n+1)\pi)$*

(iii) *Slope of the line segment parallel to x-axis is zero.*

(iv) *Slope of the line segment parallel to y-axis (i.e., perpendicular to x- axis) is infinite. (see Figure 1.59)*

(v) *If the points A* and *B coincide. ⇒ Slope is indeterminate. (see Figure 1.60)*

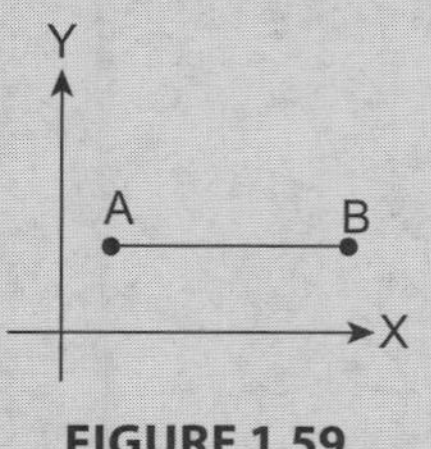

FIGURE 1.59

FIGURE 1.60

ILLUSTRATION 54: Find the slope of a line which passes through points (3,2) and (–1,5).

SOLUTION: We know that the slope of a line passing through two points (x_1, y_1) and (x_2, y_2) is given by $m = \frac{y_2 - y_1}{x_2 - x_1}$.

Here, the line passes through (3, 2) and (–1, 5). So, its slope is given by $m = \frac{5-2}{-1-3} = -\frac{3}{4}$.

ILLUSTRATION 55: Find the slope of a line whose inclination to the positive direction of x-axis in anti-clockwise sense is

(a) 60°　　(b) 0°

(c) 150°　　(d) 120°

SOLUTION: Let m denote the slope of line

(a) $m = \tan 60° = \sqrt{3}$

(b) $m = \tan 0° = 0.$

(c) $m = \tan 150° = -\cot 60° = -\frac{1}{\sqrt{3}}$

(d) $m = \tan 120° = -\cot 30° = -\sqrt{3}.$

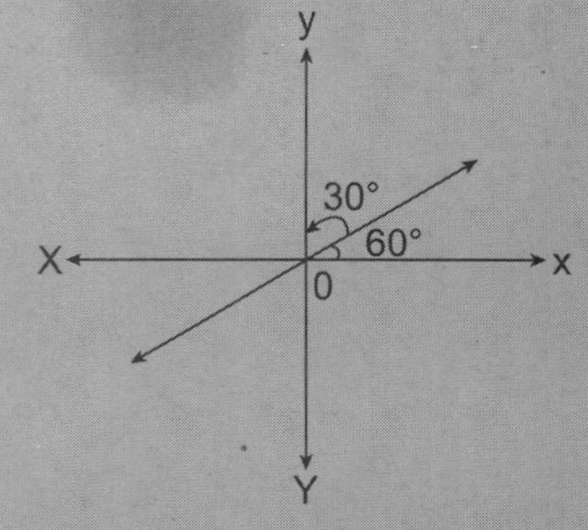

FIGURE 1.61

ILLUSTRATION 56: Slope of a line making following angles with positive direction of axis of y.

(i) 30° (ii) 150°

(iii) 120° (iv) $-\pi/4$

SOLUTION: (i) Since line AB makes 30° with y-axis hence the inclination of line with positive x-axis is 120°

$\therefore\ m = \tan(90° + 30°) = -\cot 30° = -\sqrt{3}$

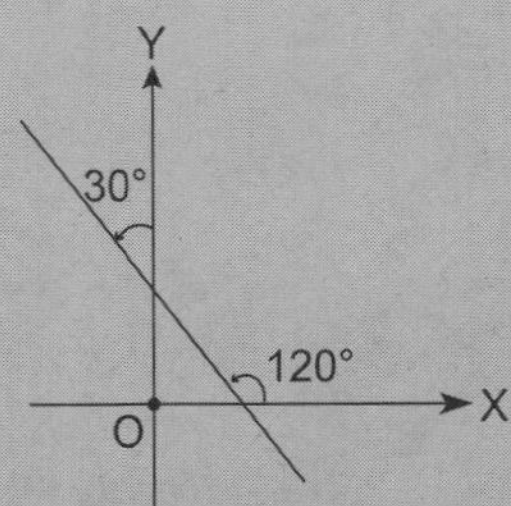

FIGURE 1.62

(ii) Clearly, angle with positive, x axis is 60°

$\Rightarrow$ Slope (m) = tan 60° = $\sqrt{3}$

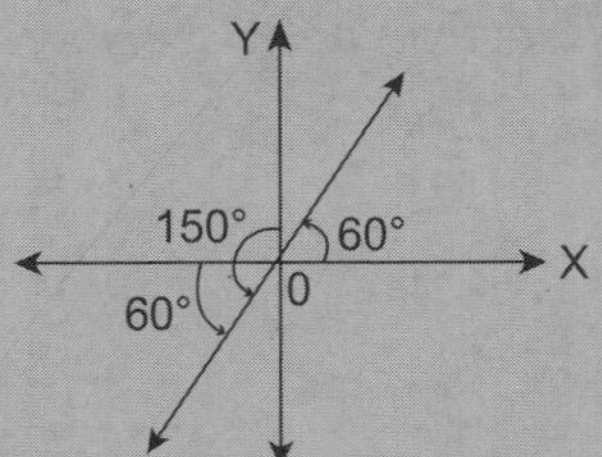

FIGURE 1.63

(iii) Clearly, the angle with positive x axis = 30°

$\Rightarrow$ Slop (m) = tan 30° = $\frac{1}{\sqrt{3}}$

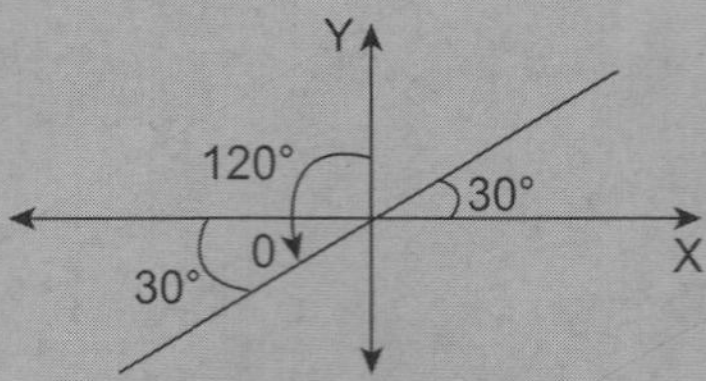

FIGURE 1.64

(iv) Clearly, the angle with positive x axis = 45°

$\Rightarrow m = \tan(\pi/4) = 1$

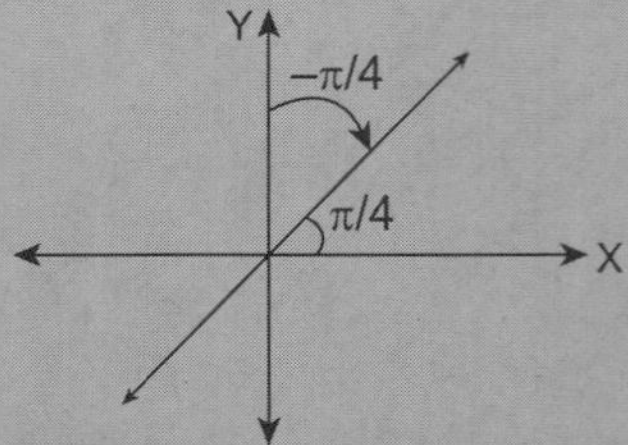

FIGURE 1.65

ANGLE BETWEEN TWO LINE SEGMENTS

The angle θ between the line segments having slopes m_1 and m_2 is given by $\tan\theta = \pm\frac{m_2 - m_1}{1+m_1m_2}$.

Proof: $m_1 = \tan\theta_1$ and $m_2 = \tan\theta_2$

Let θ be the angle between the given line segments.

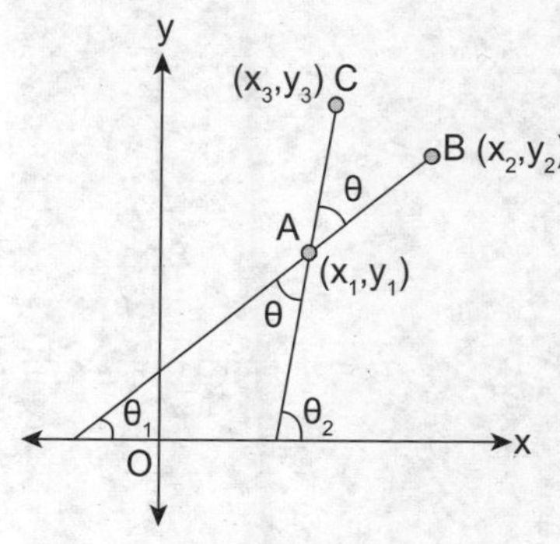

FIGURE 1.66

Since sum of the two internal angles of any triangle is equal to the opposite external angle, therefore $\theta + \theta_1 = \theta_2$

$\Rightarrow\ \theta = \theta_2 - \theta_1 \Rightarrow \tan\theta = \tan(\theta_2 - \theta_1)$

$$\Rightarrow\ \tan\theta = \frac{\tan\theta_2 - \tan\theta_1}{1+\tan\theta_2\tan\theta_1}$$

$$\left[\text{Using}: \tan(A-B) = \frac{\tan A - \tan B}{1+\tan A\tan B}\right]$$

$$\Rightarrow\ \tan\theta = \frac{m_2 - m_1}{1+m_1m_2} \quad \text{...(i)}$$

$(\pi - \theta)$ is also an angle between AB and AC.

$$\text{So, } \tan(\pi-\theta) = -\tan\theta = -\frac{m_2 - m_1}{1+m_1m_2} \quad \text{... (ii)}$$

From (i) and (ii), we find that the angle between two line segments of slopes

m_1 and m_2 is given by $\tan\theta = \pm\left(\frac{m_2 - m_1}{1+m_1.m_2}\right)$.

REMARKS

(i) *The acute angle θ between the line segments is given by* $\tan\theta = \left|\frac{m_2 - m_1}{1+m_1m_2}\right|$

i.e., $\theta = \tan^{-1}\left|\frac{m_2 - m_1}{1+m_1m_2}\right|$

(ii) *The obtuse angle θ between the line segments is given by* $\tan\theta = -\left|\frac{m_2 - m_1}{1+m_1m_2}\right|$

i.e., $\theta = \pi - \tan^{-1}\left|\frac{m_2 - m_1}{1+m_1m_2}\right|$

(iii) *Two line segments are parallel when $\theta = 0$*

$\Rightarrow \frac{m_2 - m_1}{1+m_1m_2} = 0 \Rightarrow m_2 = m_1.$

slopes of parallel line segments are equal.

(iv) *Two line segments are perpendicular when $\theta = 90°$*

$\Rightarrow \frac{m_2 - m_1}{1+m_1m_2} = \infty \Rightarrow m_1m_2 = -1$

The product of slopes of two perpendicular lines is –1.

Hence, if m is the slope of the line, then slope of line perpendicular to it is –(1/m).

(i.e., their slopes are negative reciprocal of each other)

(v) *If slope of AB and CD are such that their product is equal to 1.*

$\Rightarrow$ *both the lines will make the same angle with the lines* $y = x + c$ *and* $y = -x + c$

If AB forms angle α *with x-axis, then line CD forms angle* $90° - \alpha$ *with x-axis.*

i.e., α *angle with y axis.*

Since $m_1 m_2 = 1$

$\Rightarrow \tan\alpha \cdot \tan\psi = 1.$

$\Rightarrow \tan\psi = \cot\alpha$

$\Rightarrow \tan(90° - \alpha) = \tan\psi$

$\Rightarrow \psi = 90° - \alpha.$

(vi) *Three points A, B, C will be collinear if slope of AB = slope of BC or slope of AC = Slope of BC or slope of AB = slope of AC. This is the another condition of collinearity of three points.*

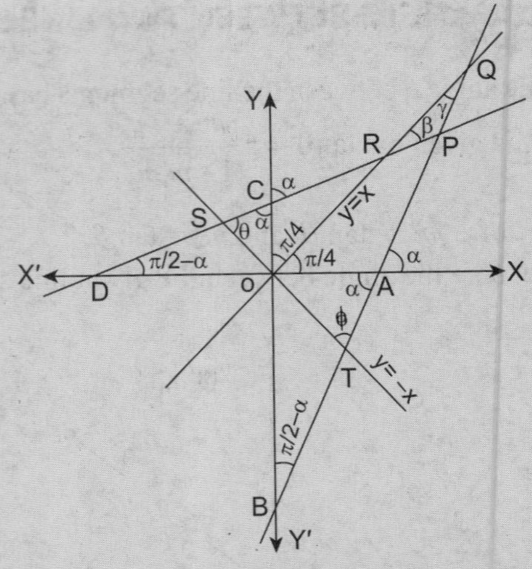

FIGURE 1.67

ILLUSTRATION 57: Let $A(6,4)$ and $B(2,12)$ be two given points. Find the slope of a line perpendicular to AB.

SOLUTION: Let m be the slope of AB. Then, $m = \dfrac{12-4}{2-6} = \dfrac{8}{-4} = -2$.

So, the slope of a line perpendicular to $AB = -\dfrac{1}{m} = \dfrac{1}{2}$.

ILLUSTRATION 58: Without using pythagoras theorem, show that $A(4,4)$, $B(3,5)$ and $C(-1,-1)$ are the vertices of a right angled triangle.

SOLUTION: In ΔABC, we have m_1 = slope of $AB = \dfrac{4-5}{4-3} = -1$; m_2 = slope of $AC = \dfrac{4-(-1)}{4-(-1)} = 1$.

Clearly, $m_1 m_2 = -1$. This shows that $AB \perp AC$, *i.e.*, $\angle CAB = \pi/2$. Hence, the given points are the vertices of a right angled triangle.

ILLUSTRATION 59: If $A(-2, 1)$, $B(2,3)$ and $C(-2, -4)$ are three points, find the angle between BA and BC.

SOLUTION: Let m_1 and m_2 be the slopes of BA and BC, respectively. Then,

$$m_1 = \frac{3-1}{2-(-2)} = \frac{1}{2} \text{ and } m_2 = \frac{-4-3}{-2-2} = \frac{7}{4}$$

Let θ be the acute angle between BA and BC. Then

$$\tan\theta = \left|\frac{m_2 - m_1}{1 + m_1 m_2}\right| = \left|\frac{(7/4 - 1/2)}{(1 + 7/4 \times 1/2)}\right| = \left|\frac{10/8}{15/8}\right| = \frac{2}{3}$$

$\Rightarrow$ Acute angle between the two line segments $= \tan^{-1}\left(\dfrac{2}{3}\right)$.

Thus, obtuse angle between BA and $BC = \pi - \tan^{-1}\left(\dfrac{2}{3}\right) \Rightarrow \theta = \tan^{-1}\left(\dfrac{2}{3}\right)$ or $\pi - \tan^{-1}\left(\dfrac{2}{3}\right)$.

ILLUSTRATION 60: If $A(4,-1)$, $B(2,-3)$ are the ends of the hypotenuse of a right angled isosceles triangle, then find the co-ordinate of its third vertex.

SOLUTION: Let the co-ordinate of third vertex P be (h, k) given $PA = PB$.

Slope of $AB = \frac{-3+1}{2-4} = 1$

$\Rightarrow$ Slope of $AP = 0$ and slope of $BP = \infty$

$\Rightarrow \frac{k+1}{h-4} = 0$ and $\frac{-3-k}{2-h} = \infty \Rightarrow k = -1$ and $h = 2$ or

Slope of $AP = \infty$ and slope of $BP = 0$.

$\Rightarrow h = 4$ and $k = -3$

$\therefore$ (2,–1) or (4,–3)

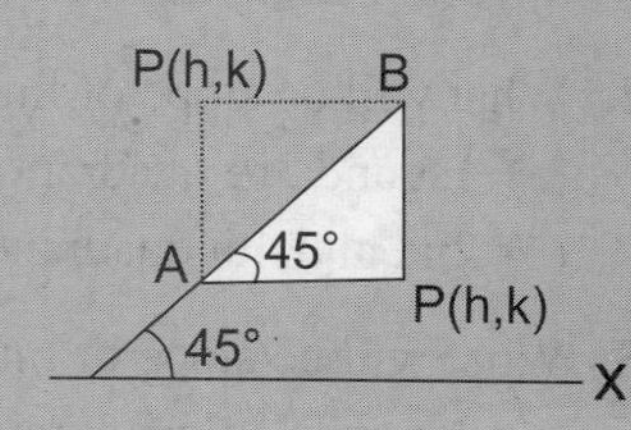

FIGURE 1.68

ILLUSTRATION 61: A triangle ABC right triangled at A, has points A and B as (2, 3) and (0,–1), respectively. If $BC = 5$ units, then find point C.

SOLUTION: Let the point C be (x, y); Since given that $BC = 5$

$\Rightarrow x^2 + (y + 1)^2 = 25$ ....(i)

Also $AB \perp AC \Rightarrow m_1 m_2 = -1$

$\Rightarrow \left(\frac{y-3}{x-2}\right)(2) = -1 \Rightarrow 2y - 6 = 2 - x$

$\Rightarrow x + 2y = 8$ ... (ii)

substituting $x = 2(4 - y)$ in equation (i)

$\therefore 4(y-4)^2 + (y + 1)^2 = 25$

$\Rightarrow y^2 - 6y + 8 = 0$

$\therefore y = 4,2$; for $y = 4$, we get $x = 0$

and for $y = 2 \Rightarrow x = 4$

$\therefore$ Point C is (0,4) or (4,2).

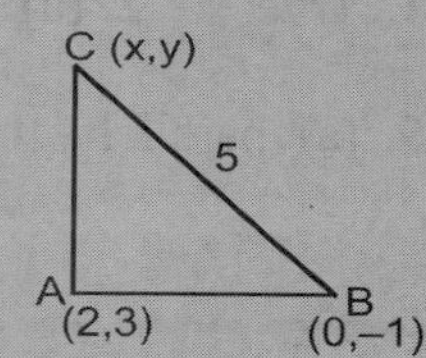

FIGURE 1.69

ILLUSTRATION 62: Without applying distance formula prove that the triangle having the vertices A (0, 1), B (2, 1) and $C(1, \sqrt{3} + 1)$ is equilateral.

SOLUTION: Obtaining the slopes of the three sides of triangle we get

Slope of $AB = m_1 = \frac{1-1}{2-0} = 0$, Slope of $BC = m_2 = \frac{\sqrt{3}}{1-2} = -\sqrt{3}$

Slope of $AC = m_3 = \frac{\sqrt{3}+1-1}{1-0} = \sqrt{3}$

So, angle $\angle CAB = \tan^{-1}\left(\frac{m_3 - m_1}{1 + m_1 m_3}\right) = \tan^{-1}\sqrt{3} = \frac{\pi}{3}$

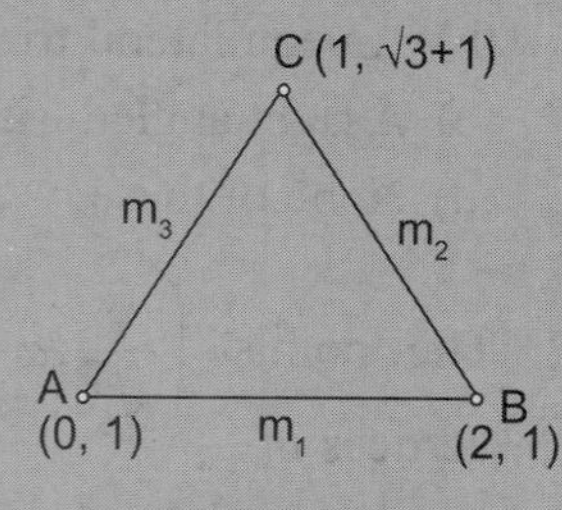

FIGURE 1.70

Similarly, the angle $ABC = \tan^{-1}\left(\frac{m_1 - m_2}{1 + m_1 m_2}\right) = \tan^{-1}\sqrt{3} = \frac{\pi}{3}$. Since two angles of a triangle are equal to $\frac{\neq}{3}$, so the third angle will also be same. Hence the triangle is equilateral triangle.

TEXTUAL EXERCISE–5 (SUBJECTIVE)

1. Slope of a given line is independent of its sense of direction along the line. (True/False)
2. What will be slope of line bisecting the
   (a) 1st and 3rd quadrant
   (b) 2nd and 4th quadrant
3. What will be slope of lines trisecting the
   (a) 1st and 3rd quadrant
   (b) 2nd and 4th quadrant
4. Determine x so that the line passing through (3, 4) and (x, 5) makes 135° angle with the positive direction of x-axis.
5. A quadrilateral has the vertices at the points (–4, 2), (2, 6), (8, 5) and (9, –7). Show that the mid-points of the sides of this quadrilateral are the vertices of a parallelogram.
6. Prove that A (4, 3), B (6, 4), C (5, 6) and D (3, 5) are the angular points of a square.
7. If a, b, c are distinct real numbers, show that the points (a, a^2), (b, b^2) and (c, c^2) are not collinear.
8. Applying and without applying section formula prove that (1, 3), (2, 5), (4, 9) are collinear.
9. State whether the two lines in each of the problems are parallel, perpendicular or neither
   (a) through (5, 6) and (2, 3), through (9, –2) and (6, –5).
   (b) through (8, 2) and (–5, 3), through (6, 16) and (5, 3)
   (c) through (2, –5) and (–5, 2), through (6, 3) and (1, 1)
10. Without using Pythagora theorem, show that (4, 4), (3, 5) and (– 1, –1) are the vertices of a right angled triangle.

Answer Keys

1. True **2.** (a) 1 (b) –1 **3.** (a) $\frac{1}{\sqrt{3}}, \sqrt{3}$ (b) $-\frac{1}{\sqrt{3}}, -\sqrt{3}$ **4.** 2

9. (a) parallel (b) perpendicular (c) neither

TEXTUAL EXERCISE–5 (OBJECTIVE)

1. The three points (– 2, 2), (8, –2) and (–4, –3) are the vertices of
   (a) An isosceles triangle
   (b) An equilateral triangle
   (c) A right angled triangle
   (d) None of these
2. The points $\left(\frac{a}{\sqrt{3}}, a\right), \left(\frac{2a}{\sqrt{3}}, 2a\right), \left(\frac{a}{\sqrt{3}}, 3a\right)$ are the vertices of
   (a) An equilateral triangle
   (b) An isosceles triangle
   (c) A right angled triangle
   (d) None of these
3. The point (a, b), (c, d) and $\left(\frac{kc+la}{k+l}, \frac{kd+lb}{k+l}\right)$ are
   (a) Vertices of an equilateral triangle
   (b) Vertices of an isosceles triangle
   (c) Vertices of a right angled triangle
   (d) Collinear
4. The points (0,8/3), (1,3) and (82,30) are the vertices of
   (a) An equilateral triangle
   (b) An isosceles triangle
   (c) A right angled triangle
   (d) None of these
5. The points $(-a, -b)$,(a, b), (a^2, ab) are
   (a) Vertices of an equilateral triangle
   (b) Vertices of a right angled triangle
   (c) Vertices of an isosceles triangle
   (d) Collinear
6. The points $(-a, -b)$, (0,0), (a, b) and (a^2, ab) are
   (a) Collinear
   (b) Vertices of a rectangle
   (c) Vertices of a parallelogram
   (d) None of these

7. The points (0,0), $(a,0)$ and $\left(\frac{a}{2}, \frac{a\sqrt{3}}{2}\right)$ are vertices of

(a) Isosceles triangle (b) Equilateral triangle
(c) Scalene triangle (d) None of these

8. A triangle with vertices (4,0), (–1, – 1),(3,5) is
(a) Isosceles and right angled
(b) Isosceles but not right angled
(c) Right angled but not isosceles
(d) Neither right angled nor isosceles

9. The points (1,1), $(0, \sec^2\theta)$, $(\text{cosec}^2\theta, 0)$ are collinear for

(a) $\theta = \frac{n\pi}{2}$ (b) $\theta \neq \frac{n\pi}{2}$
(c) $\theta = n\pi$ (d) None of these

Answer Keys

1. (c) **2.** (b) **3.** (d) **4.** (d) **5.** (d) **6.** (a) **7.** (a),(b) **8.** (a) **9.** (b)

STANDARD POINTS OF A TRIANGLE

The triangle is a magical geometric figure having many cyclic and symmetric properties, it is a surprise to know that the medians, altitudes, perpendicular bisectors of sides, internal angle bisector, two external and one internal angle bisectors are all concurrent and their point of concurrency are, respectively, known as centroid(G), Orthocentre (O or H), Circumcentre (C), Incentre (I), and Excentres (I_1, I_2, I_3) respectively. Let $A(x_1, y_1)$, $B(x_2, y_2)$ and $C(x_3, y_3)$ are vertices of any triangle ABC, then the definition of the above points and their properties are mentioned as follows:

1. Centroid

The centroid of a triangle is the point of intersection of its medians. It is also called as median point/centre of mass. It is denoted by G.

Co-ordinates of centroid G are given by:

$$\left(\frac{x_1+x_2+x_3}{3}, \frac{y_1+y_2+y_3}{3}\right)$$

FIGURE 1.71

Proof: Very clearly, Area of ΔBGD = Area of ΔCDG

$\Rightarrow \Delta_1 = \Delta_2$

$EF \parallel BC \Rightarrow$ Area of ΔBCF = Area of ΔBEC

$\Rightarrow \Delta_1 + \Delta_6 + \Delta_2 = \Delta_1 + \Delta_2 + \Delta_3 \Rightarrow \Delta_6 = \Delta_3$

Similarly, we can prove $\Delta_1 = \Delta_4$ and $\Delta_2 = \Delta_5$

$\Rightarrow \Delta_1 = \Delta_2 = \Delta_3 = \Delta_4 = \Delta_5 = \Delta_6 = \Delta$ (say)

Now $\frac{\text{area of } \Delta GDC}{\text{area of } \Delta ADC} = \frac{\frac{1}{2} \times GD \times h}{\frac{1}{2} AD \times h}$

$$= \frac{GD}{AD} = \frac{\Delta_2}{\Delta_2 + \Delta_3 + \Delta_4} = \frac{\Delta}{3\Delta} = \frac{1}{3}$$

$$\Rightarrow \frac{GD}{AG+GD} = \frac{1}{3} \Rightarrow \frac{AG}{GD} = \frac{2}{1}$$

Thus, centroid divides medians in the ratio 2:1

Further, co-ordinates of D are $\left(\frac{x_2+x_3}{2}, \frac{y_2+y_3}{2}\right)$

$\because$ D is the mid-point of the BC

and centroid G divides AD in the ratio 2: 1, internally. Hence, applying section formula, co-ordinates of G are given by

$$\left(\frac{1.x_1 + 2\left(\frac{(x_2+x_3)}{2}\right)}{1+2}, \frac{1.y_1 + 2\left(\frac{(y_2+y_3)}{2}\right)}{1+2}\right)$$

$$= \left(\frac{x_1+x_2+x_3}{3}, \frac{y_1+y_2+y_3}{3}\right)$$

Properties of Centroid of Triangle

1. The centroid (G) divides the medians in the ratio 2 : 1. That is $AG : GD :: BG : GE :: CG : GF :: 2 : 1$.
2. Co-ordinates of centroid are symmetrical with respect to the co-ordinates of the vertices.

3. Centroid always lies inside the triangle and also called as mean centre of vertices.
4. All the three medians of a triangle are concurrent, i.e., if we join a vertex to the point of intersection of medians from the other two vertices and extend it will meet the opposite side in its mid-point.
5. Line segments joining centroid to vertices divide the triangle into three triangles of equal area. i.e., Area (ΔGAB) = Area (ΔGBC) = Area (ΔGAC).
6. For equilateral triangle centroid coincides with other points of triangle i.e., orthocentre (O), circumcentre (C) and incentre (I).
7. Medians divide the triangle into six triangles of equal area, i.e.,

   Area (ΔGDB) = Area(ΔGDC) = Area (ΔGEC) = Area (ΔGEA)

   = Area (ΔGAF) = Area (ΔGFB)

   i.e., $\Delta_1 = \Delta_2 = \Delta_3 = \Delta_4 = \Delta_5 = \Delta_6$.
8. If mid-points of the sides of a triangle ABC are D, E, F respectively of BC, CA, AB as shown in the figure, then $A(x_E + x_F - x_D, y_E + y_F - y_D)$, $B(x_D + x_F - x_E, y_D + y_F - y_E)$, and $C(x_D + x_E - x_F, y_D + y_E - y_F)$ and the area of each of the four triangles formed by joining the mid-points of the sides are equal

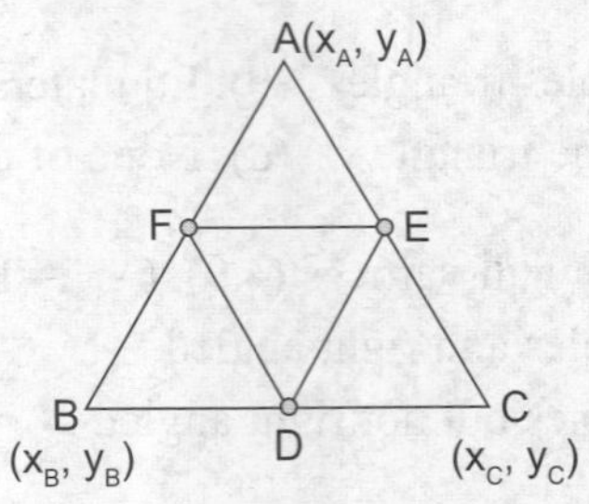

FIGURE 1.72

$BDEF$ is parallelogram and diagonals bisect each

So, $\frac{x_B + x_E}{2}$ $\frac{x_D + x_F}{2}$ $\Rightarrow x_B = x_D + x_F - x_E$

Similarly,

$\therefore$ *Area of* $\Delta ABC = 4 \times$ *Area of* ΔDEF

i.e., Area of a triangle is four times the area of the triangle formed by joining the mid-points of its sides.

9. ΔABC and $\Delta\, DEF$ are similar.

$$\Rightarrow \frac{\text{Area of } \Delta ABC}{\text{Area of } \Delta DEF} = \frac{(BC)^2}{(EF)^2} = 4\,.$$

ILLUSTRATION 63: Two vertices of a triangle are (3, –5) and (–7, 4). If its centroid is (2, –1), find the third vertex.

SOLUTION: Let the co-ordinates of the third vertex be (x, y). Then

$$\frac{x+3-7}{3} = 2 \quad \text{and} \quad \frac{y-5+4}{3} = -1$$

$$\Rightarrow x - 4 = 6 \quad \text{and} \quad y - 1 = -3 \Rightarrow x = 10 \quad \text{and} \quad y = -2$$

Thus, the co-ordinates of the third vertex are (10, –2).

ILLUSTRATION 64: If α, β, γ satisfy the equation $3(\tan^3\theta) - 3a\tan^2\theta + 3b\tan\theta - 1 = 0$, then find the centroid of the triangle with vertices ($\tan\alpha$, $\cot\alpha$), ($\tan\beta$, $\cot\beta$) and ($\tan\gamma$, $\cot\gamma$)

SOLUTION: Since $\tan\alpha$, $\tan\beta$, $\tan\gamma$ are the roots of, $3(\tan^3\theta) - 3a\tan^2\theta + 3b\tan\theta - 1 = 0$.

$\Rightarrow \tan\alpha + \tan\beta + \tan\gamma = -(-3a)/3 = a$ ... (i)

and $\tan\alpha.\tan\beta + \tan\beta.\tan\gamma + \tan\alpha.\tan\gamma = 3b/3 = b$ ... (ii)

and $\tan\alpha.\tan\beta.\tan\gamma = 1/3$ .. (iii)

Let the centroid of the triangle be (x_G, y_G), therefore; $x_G = \frac{\tan\alpha + \tan\beta + \tan\gamma}{3} = \frac{a}{3}$

$$y_G = \frac{\cot\alpha + \cot\beta + \cot\gamma}{3} = \frac{\tan\alpha.\tan\beta + \tan\beta.\tan\gamma + \tan\alpha.\tan\gamma}{3\tan\alpha\tan\beta\tan\gamma} = \frac{b}{3.1/3} = b$$

Consequently, the centroid is (a/3, b)

ILLUSTRATION 65: The area of a triangle is 3 square units two of its vertices are A(3,1), B(1, –3) and the centroid of the triangle lies on x-axis. Find the co-ordiantes of the third vertex C.

SOLUTION: Let the third vertex be (x, y) so $\dfrac{1+(-3)+y}{3} = 0 \Rightarrow y = 2$

So the third vertex can be taken as $(x, 2)$, given that area of triangle is 3 sq. unit

$$\frac{1}{2}\left\|\begin{matrix} 3 & 1 & 1 \\ 1 & -3 & 1 \\ x & 2 & 1 \end{matrix}\right\| = 3 \;(\because R_1 \to R_1 - R_2) \Rightarrow \frac{1}{2}\left\|\begin{matrix} 2 & 4 & 0 \\ 1 & -3 & 1 \\ (x-1) & 5 & 0 \end{matrix}\right\| = 3 \;(\because R_3 \to R_3 - R_2)$$

$$\Rightarrow \frac{1}{2}|-(10 - 4(x-1))| = 3 \Rightarrow \frac{1}{2}|\,14 - 4x\,| = 3$$

$$\Rightarrow 4x = 14 + 6 = 20 \text{ or } 4x = 14 - 6 = 8 \Rightarrow x = 5 \text{ or } 2.$$

Thus the co-ordinate of third vertex are (5, 2) or (2, 2).

2. Circumcircle and Circumcentre

The perpendicular bisectors of sides of a triangle are concurrent and their point of intersection is called circumcentre. It is denoted by $C(C_0)$ and it is equidistant from all three vertices.

(i.e., $C_0A = C_0B = C_0C = R$; R is circumradius)

Taking R as radius and C_0 as centre the circle drawn circumscribes the ΔABC and called as circumcircle of ΔABC. The co-ordinates of circumcentre are given by:

$$\left(\frac{x_1 \sin 2A + x_2 \sin 2B + x_3 \sin 2C}{\sin 2A + \sin 2B + \sin 2C}, \frac{y_1 \sin 2A + y_2 \sin 2B + y_3 \sin 2C}{\sin 2A + \sin 2B + \sin 2C}\right)$$

Proof: In the given diagram, C_0 is the circumcentre and the angle subtended at the centre by any chord is twice that of the angle subtended at any point on the circumference.

Therefore, $\angle BC_0C = 2A$.; $\angle BC_0A = 2C$; $\angle CC_0A = 2B$ also $AC_0 = BC_0 = CC_0 = R$.

Applying sine formula, we get:

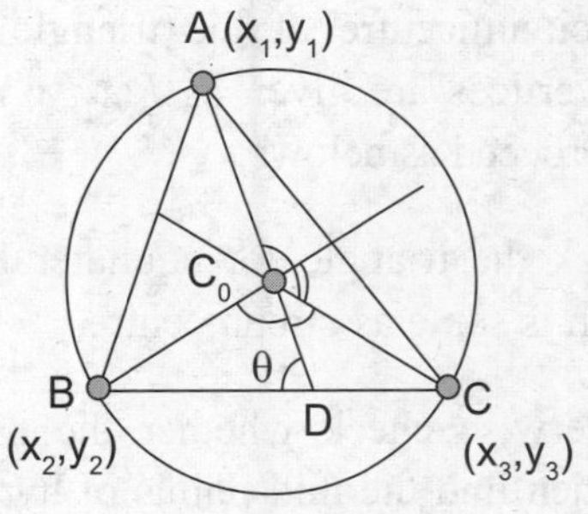

FIGURE 1.73

In triangle BC_0D, $\dfrac{R}{\sin\theta} = \dfrac{BD}{\sin(\pi - 2C)}$

and in triangle CC_0D, $\dfrac{R}{\sin(\pi - \theta)} = \dfrac{CD}{\sin(\pi - 2B)}$

Dividing the above two relations, $\dfrac{BD}{CD} = \dfrac{\sin 2C}{\sin 2B}$ ....(i)

Alternatively

Let P be circumcentre of ΔABC and AP intersect BC at D

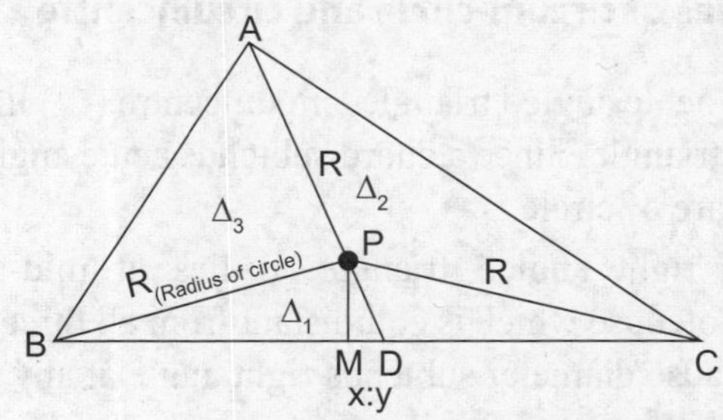

FIGURE 1.74

$\angle APB = 2C$; $\angle BPC = 2A$; $\angle CPA = 2B$

$$\frac{x}{y} = \frac{BD}{CD} = \frac{\text{Area}\,\Delta\,ABD}{\text{Area}\,\Delta\,ADC} = \frac{\text{Area}\,\Delta PBD}{\text{Area}\,CPD}$$

$$= \frac{\text{Area}\,\Delta ABD - \text{Area}\,\Delta PBD}{\text{Area}\,\Delta ADC - \text{Area}\,\Delta CPD} = \frac{\Delta_3}{\Delta_2}$$

$$\Rightarrow \frac{x}{y} = \frac{\frac{R^2 \sin 2C}{2}}{\frac{R^2 \sin 2B}{2}} = \frac{\text{Sin}2C}{\text{Sin}2B} \quad .....(i)$$

Also $\dfrac{PA}{PD}=\dfrac{\text{Area }\Delta APB}{\text{Area }\Delta BPD}=\dfrac{\text{Area }\Delta ACP}{\text{Area }\Delta CPD}$

$$=\frac{\text{Area }(\Delta APB+\Delta ACP)}{\text{Area }\Delta BPD+\text{Area }\Delta CPD}=\frac{\Delta_3+\Delta_2}{\Delta_1}$$

$$=\frac{\sin 2C+\sin 2B}{\sin 2A}$$

$$\Rightarrow \frac{AC_0}{C_0D}=\frac{AP}{PD}=\frac{\sin 2B+\sin 2C}{\sin 2A}$$

Therefore applying section formula, co-ordinates of D are given as:

$$(x_D,y_D)\equiv\left(\frac{x_2\sin 2B+x_3\sin 2C}{\sin 2B+\sin 2C},\frac{y_2\sin 2B+y_3\sin 2C}{\sin 2B+\sin 2C}\right)$$

And applying section formula for AC_0D, the co-ordinates of Circumcentre C_0 are given by:

$$=\left(\frac{(\sin 2B+\sin 2C)x_D+x_A\sin 2A}{\sin 2A+\sin 2B+\sin 2C},\right.$$
$$\left.\frac{(\sin 2B+\sin 2C)y_D+y_A\sin 2A}{\sin 2A+\sin 2B+\sin 2C}\right)$$

$$=\left(\frac{x_1\sin 2A+x_2\sin 2B+x_3\sin 2C}{\sin 2A+\sin 2B+\sin 2C},\right.$$
$$\left.\frac{y_1\sin 2A+y_2\sin 2B+y_3\sin 2C}{\sin 2A+\sin 2B+\sin 2C}\right).$$

Properties of circum-circle and circumcentre

1. For acute angled triangle, circumcentre (C_0) lies inside the triangle. Since a chord subtends acute angle in major arc of circle.
2. For right angled triangle, it lies at mid-point of hypotenuse which is equidistant from all three vertices because diameter subtends right angle at any point on circumference.
3. For obtuse angled triangle, it lies outside the triangle, since chord subtends obtuse angle on miror arc of circle.
4. For equilateral triangle it coincides with centroid and other points of triangle i.e., incentre and orthocentre.
5. For isosceles triangle centroid, orthocentre, circumcentre, and incentre of triangle are collinear.
6. Centroid, orthocentre and circumcentre are collinear and Centroid (G) divides the line joining orthocentre (O) and circumcentre (C_0) in the ratio 2: 1 (OG: GC_0)

Proof: Let, here O is orthocenter and C_0 is circum centre of $DABC$. We claim that G is centroid of $DABC$

Clearly $AM = c\,CosA$ (By ΔABM)

Now, in ΔOAM, $\dfrac{OA}{AM}=\sec\Delta OAM$

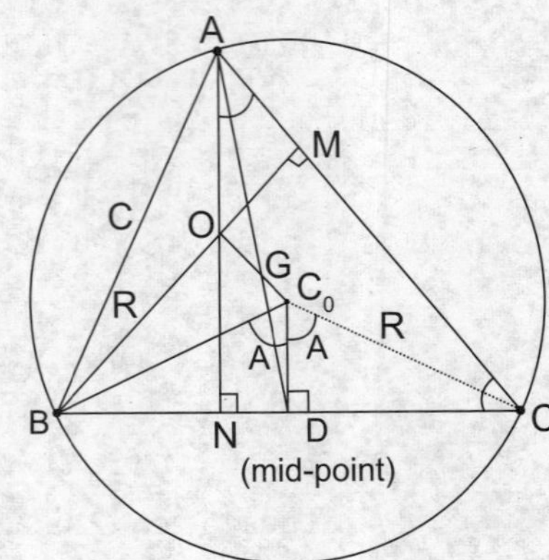

FIGURE 1.75

$$=\sec\left(\frac{\pi}{2}-C\right)=\text{cosecC}$$

$\Rightarrow OA = \text{AMcosec}C = C\cos\text{A}\ \text{cosecC}$

$=\dfrac{c}{\sin C}\cos A = 2R\cos A$ (By since formula)

Also in ΔBC_0D, $C_0D = R\cos\text{A}$

$\therefore \dfrac{OA}{C\,D}=\dfrac{2R\cos A}{R\cos A}=\dfrac{2}{1}$

Now $\Delta OAG \sim \Delta C_oDG$ (By AAA similarly)

$\Rightarrow \dfrac{AG}{GD}=\dfrac{OA}{C_oD}=\dfrac{2}{1}$

$\Rightarrow$ G is centroid of ΔABC

This O, G and C_o are collinear. Also by similarty of ΔOAG and C_oDG

$OG : GC_o = AG : GD = 2:1$

Hence centroid divides the line segment joining orthocenter and circumcentre in ratio 2 :1

Method to find out circumcentre

To find the circumcentre of the triangle when the co-ordinates of vertices are given as $P(x_1, y_1)$, $Q(x_2, y_2)$ and $R(x_3, y_3)$, we proceed as below:

Step I: Check the triangle for equilateral if so find the centroid which is same as circumcentre.

Step II: Otherwise check whether the triangle is right angled if so then find the mid-points of hypotenuse which is same as circumcentre.

Step III: If the triangle PQR is scalene, as shown in the figure with circumcentre (x_c, y_c) we have two methods to find the co-ordiantes of circumcentre.

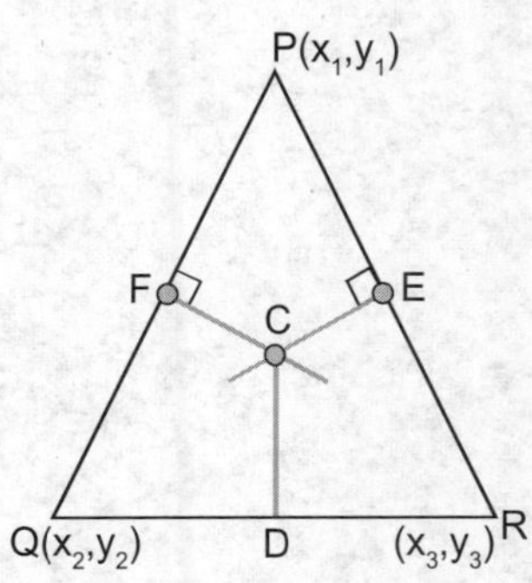

FIGURE 1.76

Method-1: In ΔPQR, since $(CP)^2 = (CQ)^2 = (CR)^2$

$$\Rightarrow (x_c - x_1)^2 + (y_c - y_1)^2$$
$$= (x_c - x_2)^2 + (y_c - y_2)^2$$
$$= (x_c - x_3)^2 + (y_c - y_3)^2$$

Solving any two of the above linear equations, we can find the co-ordinates of circum-centre and the obtained co-ordinates would always satisfy the third equation and that indicates the concurrency of perpendicular bisectors.

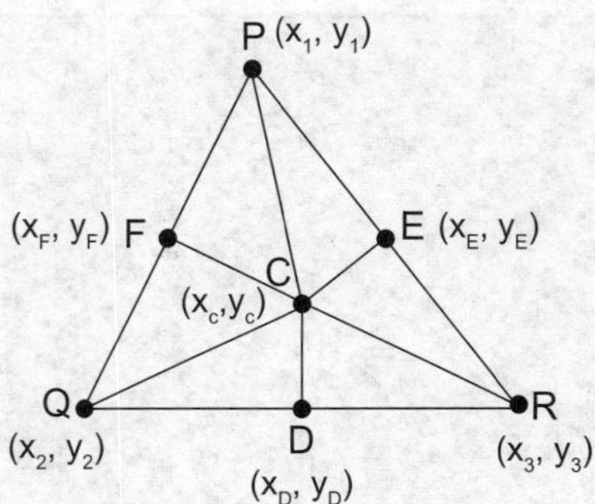

FIGURE 1.77

Method-2: The line CD is perpendicular to QR, CE is perpendicular to PR and CF is perpendicular to PQ. So

$$m_{CD}.m_{QR} = \frac{y_C - y_D}{x_C - x_D}.\frac{y_3 - y_2}{x_3 - x_2} = -1 \quad ....(i)$$

$$m_{CE}.m_{PR} = \frac{y_C - y_E}{x_C - x_E}.\frac{y_3 - y_1}{x_3 - x_1} = -1 \quad ....(ii)$$

$$m_{FC}.m_{PQ} = \frac{y_C - y_F}{x_C - x_F}.\frac{y_1 - y_2}{x_1 - x_2} = -1 \quad ....(iii)$$

Solving any two of the above the co-ordinates of the circumcentre (x_C, y_C) can be obtained. By substituting these co-ordinates in the third equation establishing the equality, we can show that perpendicular bisectors of sides of triangle are concurrent.

ILLUSTRATION 66: The vertices of a triangle are $A(4, -3)$, $B(-2, 1)$ and $C(2, 3)$. Find the co-ordinate of the circumcentre of the triangle.

SOLUTION: Since sides are $BC = 2\sqrt{5}$; $CA = 2\sqrt{10}$; $AB = 2\sqrt{13}$

So, the triangle is scalene. Let the co-ordinate of circumcentre be $O(x, y)$

$\therefore \; OA^2 = OB^2 \Rightarrow (x - 4)^2 + (y + 3)^2 = (x + 2)^2 + (y - 1)^2$

$\Rightarrow 20 - 8x + 6y = 4x - 2y \Rightarrow 12x - 8y = 20 \Rightarrow 3x - 2y = 5$ ...(i)

and $OB^2 = OC^2 \Rightarrow (x + 2)^2 + (y - 1)^2 = (x - 2)^2 + (y - 3)^2$

$\Rightarrow 8x + 4y = 8$ ...(ii)

Solving both linear equations (i) and (ii) we get

circumcentre as $\left(\frac{9}{7}, -\frac{4}{7}\right)$.

ILLUSTRATION 67: Find the centre and radius of the circumcircle of the triangle whose vertices are $A(-2,3)$, $B(2, -1)$ and $C(4,0)$.

SOLUTION: Let the circumcentre of Δ be $P(\alpha, \beta)$

$\therefore \; PA = PB = PC$ $\quad\quad \because \; PA = PB$

$\Rightarrow PA^2 = PB^2 \Rightarrow (\alpha + 2)^2 + (\beta - 3)^2 = (\alpha - 2)^2 + (\beta + 1)^2$

$\Rightarrow \alpha^2 + 4 + 4\alpha + \beta^2 + 9 - 6\beta = \alpha^2 + 4 - 4\alpha + \beta^2 + 1 + 2\beta$

$\Rightarrow 4\alpha - 6\beta + 13 = 2\beta - 4\alpha + 5 \Rightarrow 8\alpha - 8\beta + 8 = 0$

$\Rightarrow \alpha - \beta = -1$ ...(i)

$\because PB = PC$ $\therefore PB^2 = PC^2$

$\Rightarrow (\alpha-2)^2 + (\beta+1)^2 = (\alpha-4)^2 + (\beta-0)^2 \Rightarrow \alpha^2 + 4 - 4\alpha + \beta^2 + 1 + 2\beta = \alpha^2 + 16 - 8\alpha + \beta^2$

$\Rightarrow 4\alpha + 2\beta + 5 - 16 = 0 \Rightarrow 4\alpha + 2\beta = 11$ ....(ii)

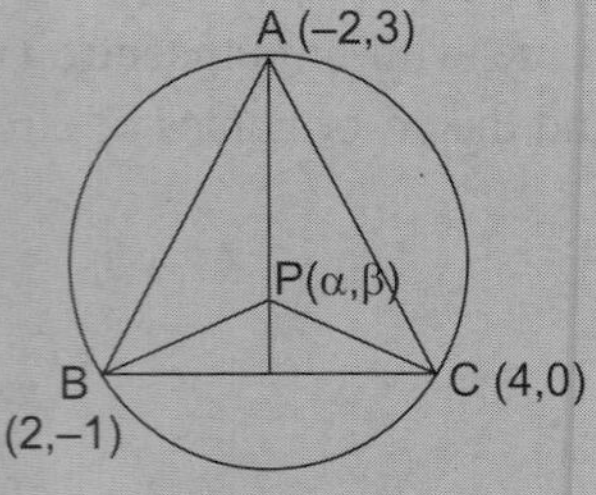

FIGURE 1.78

Now solving (i) and (ii); $\alpha = \frac{3}{2}, \beta = \frac{5}{2}$ $\therefore$ Centre of circumcircle is $\left(\frac{3}{2}, \frac{5}{2}\right)$

$\therefore$ Radius $= PA = \sqrt{\left(\frac{3}{2}+2\right)^2 + \left(\frac{5}{2}-3\right)^2} = \frac{5\sqrt{2}}{2}$

ILLUSTRATION 68: The line $ax + by + c = 0$ cuts the x-axis at P and y-axis at Q. Find the co-ordinates of the orthocentre, centroid and circumcentre of the ΔOPQ.

SOLUTION: ΔOPC is a right angled triangle with vertices $O(0, 0)$, $P(-c/a, 0)$, $Q(0, -c/b)$. Orthocentre is right angular vertex i.e., origin $(0, a)$

Centroid is $\left(\left(\frac{0+0+(-c/a)}{3}\right), \frac{0+0-c/b}{3}\right) \equiv \left(\frac{-c}{3a}, \frac{-c}{3b}\right)$

Circumcentre "C" is mid-point of $PQ = \left(\frac{-c}{2a}, \frac{-c}{2b}\right)$

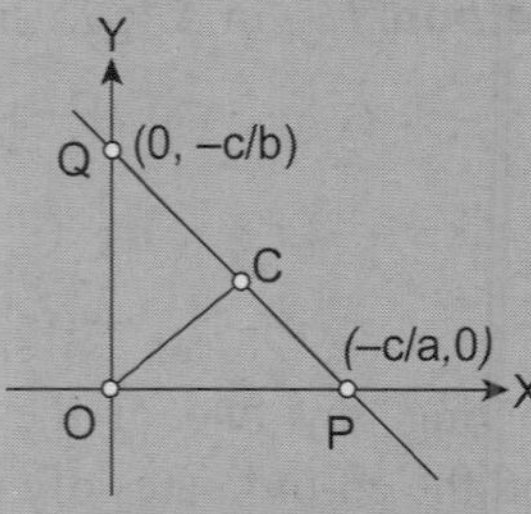

FIGURE 1.79

ILLUSTRATION 69: The equations of the three sides of a triangle are $x = 2$, $y + 1 = 0$ and $x + 2y = 4$. Find the co-ordinates of the circumcentre of the triangle.

SOLUTION: One side of the triangle is parallel to the y-axis and another side is paralllel to the x-axis. So, the triangle is a right angled triangle. Hence, the middle point of the hypotenuse is the circumcentre. Solving $AC(x = 2)$, $BC(x + 2y = 4)$ we get one end of the hypotenuse and solving $AB(y + 1 = 0)$, $x + 2y = 4$ we get the other end. Their co-ordinates are $C(2, 1)$ and $B(6, -1)$

$\therefore$ Circumcentre is $\left(\frac{2+6}{2}, \frac{1-1}{2}\right) = (4, 0)$

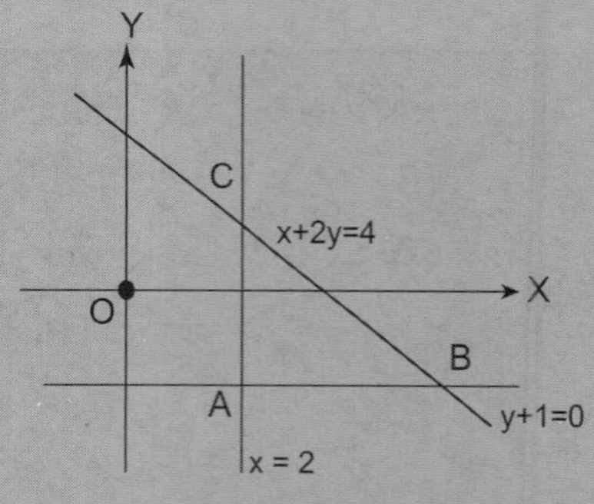

FIGURE 1.80

3. Orthocentre

The altitudes of triangle are concurrent and the point of intersection of altitudes is known as orthocentre. It is denoted by O or H and its co-ordinates are given by

$$\left(\frac{x_1 \tan A + x_2 \tan B + x_3 \tan C}{\tan A + \tan B + \tan C}, \frac{y_1 \tan A + y_2 \tan B + y_3 \tan C}{\tan A + \tan B + \tan C}\right)$$

Proof: Consider a ΔABC with AD, BE and CF as altitudes as shown in the figure.

$\frac{AD}{BD} = \tan B$ and $\frac{AD}{CD} = \tan C \Rightarrow \frac{BD}{CD} = \frac{\tan C}{\tan B}$

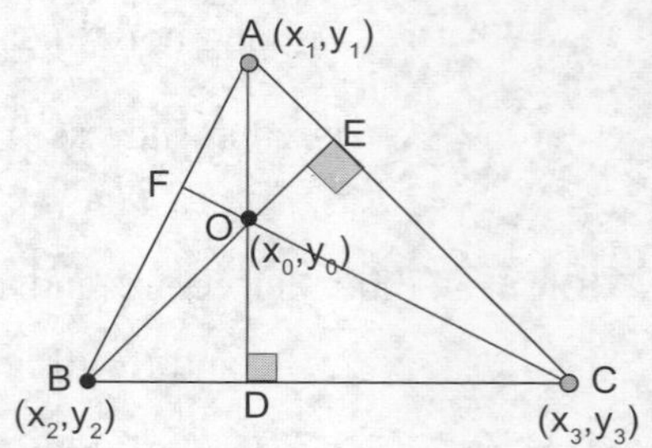

FIGURE 1.81

Applying section formula for BC we can get the co-ordinates of D as:

$$x_D = \frac{x_3 \tan C + x_2 \tan B}{\tan B + \tan C} \text{ and } y_D = \frac{y_3 \tan C + y_2 \tan B}{\tan B + \tan C}$$

In ΔBAE

$$AE = c \cos A \quad \text{.... (i)}$$

and in ΔOEA : $OA = AE \sec\left(\frac{\pi}{2} - C\right)$

$$\Rightarrow OA = c \cos A \operatorname{cosec} C = \frac{c \cos A}{\sin C} \quad \text{.... (ii)}$$

Also, in ΔABD; $BD = c \cos B$ and

$$OD = \text{BD} \tan\left(\frac{\pi}{2} - C\right) \quad \text{......(iii)}$$

From equation (ii) and (iii), we get

$$\frac{OA}{OD} = \frac{\cos A}{\cos B \cos C} = \frac{\sin A}{\cos B \cos C\left(\frac{\sin A}{\cos A}\right)}$$

$$= \frac{\sin B \cos C + \cos B \sin C}{(\cos B \cos C)\tan A} = \frac{\tan B + \tan C}{\tan A}$$

$$\Rightarrow \frac{AO}{OD} = \frac{\tan B + \tan C}{\tan A}$$

And again applying section formula for AD we can get the co-ordinates of O as:

$$x_0 = \frac{(\tan B + \tan C)\left(\frac{x_3 \tan C + x_2 \tan B}{\tan B + \tan C}\right) + x_1 \tan A}{\tan A + \tan B + \tan C} = \frac{x_1 \tan A + x_2 \tan B + x_3 \tan C}{\tan A + \tan B + \tan C}$$

$$y_0 = \frac{(\tan B + \tan C)\left(\frac{y_3 \tan C + y_2 \tan B}{\tan B + \tan C}\right) + y_1 \tan A}{\tan A + \tan B + \tan C} = \frac{y_1 \tan A + y_2 \tan B + y_3 \tan C}{\tan A + \tan B + \tan C}.$$

Properties of orthocentre of triangle

1. For acute angled triangle it lies inside the triangle.
2. For right angled triangle it is at right angled vertex.
3. For abtuse angled triangle it lies outside the triangle. (for above three properties refer to diagram)
4. For any triangle orthocentre (O), circumcentre(C) and centroid (G) are always collinear. $OG : CG :: 2 : 1$
5. For isosceles triangle the incentre (I) is also collinear to above three points.
6. For isosceles triangle the orthocenter lies at the median through vertex (point of intersection of equal sides).
7. For equilateral triangle it coincides with centroid.

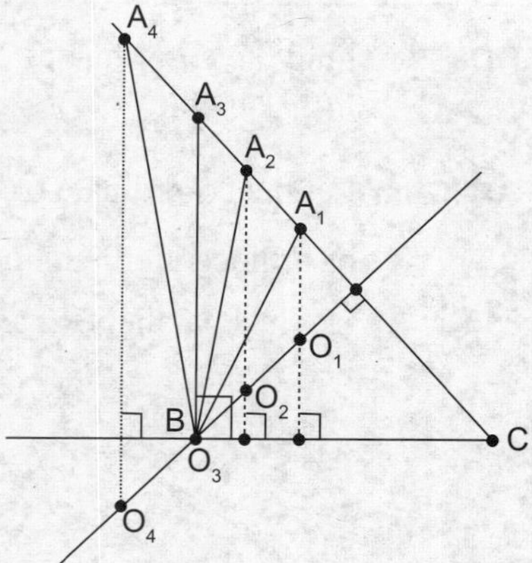

FIGURE 1.82

Method to find out orthocentre

To find the orthocentre of the triangle with co-ordinates of vertices given as (x_1, y_1), (x_2, y_2) and (x_3, y_3), we proceed as below:

Step I: Check whether the triangle is equilateral if so, find centroid which is same as orthocentre.

Step II: Check whether the triangle is right angled if so, then find the co-ordinate of right angular vertex .

Step III: If the triangle is scalene triangle (say ABC) as shown in the figure with orthocentre (x_0, y_0), the line OA is perpendicular to BC, OB is perpendicular to AC and OC is perpendicular to AB. So

$$m_{OA}.m_{BC} = \frac{y_o - y_1}{x_0 - x_1}.\frac{y_3 - y_2}{x_3 - x_2} = -1 \quad \text{....(i)}$$

$$m_{OB}.m_{AC} = \frac{y_o - y_2}{x_0 - x_2}.\frac{y_3 - y_1}{x_3 - x_1} = -1 \quad \text{....(ii)}$$

$$m_{OC}.m_{AB} = \frac{y_o - y_3}{x_0 - x_3}.\frac{y_1 - y_2}{x_1 - x_2} = -1 \quad \text{....(iii)}$$

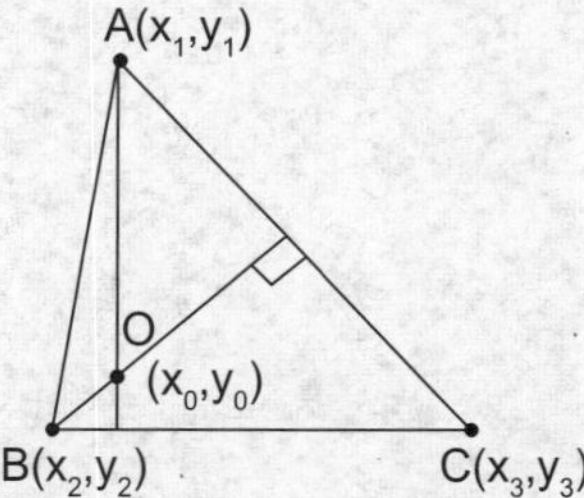

FIGURE 1.83

Solving any two of the above equations the co-ordinates of the orthocentre (x_0, y_0) can be obtained. Since they always satisfy the third equation, which shows that altitudes of triangle are concurrent.

ILLUSTRATION 70: Find the co-ordinates of the orthocentre of the triangle whose angular points are A (1, 2), B(2, 3) and C(4, 3).

SOLUTION: Let the co-ordinate of orthocentre be (x_0, y_0).

or orthocentre lie on pependicular on BC from A

i.e., line parallel to y-axis so x co-ordinate of O is 1

i.e., $x_o = 1$

Also, $m_{OB} . m_{AC} = -1 \Rightarrow \dfrac{y_0 - 3}{x_0 - 2} . \dfrac{3-2}{4-1} = -1$

$\Rightarrow y_0 - 3 = 3 \Rightarrow y_0 = 6$. Hence, the orthocentre is (1, 6)

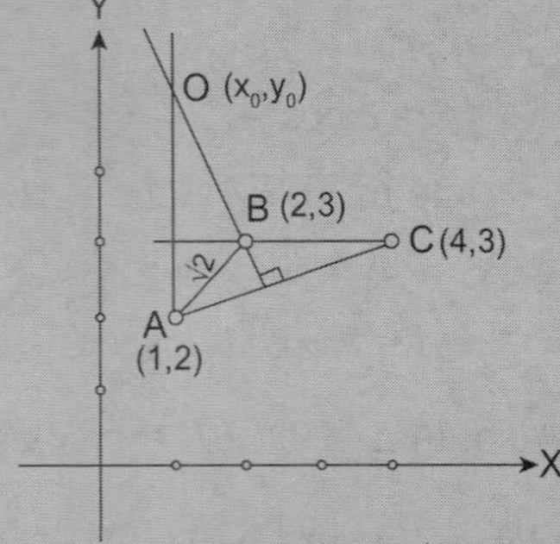

FIGURE 1.84

ILLUSTRATION 71: ΔABC has vertices $A(a\cos\alpha, a\sin\alpha)$, $B(a\cos\beta, a\sin\beta)$ and $C(a\cos\gamma, a\sin\gamma)$. Prove that its orthocentre is given by $H\,(a(\cos\alpha + \cos\beta + \cos\gamma), a(\sin\alpha + \sin\beta + \sin\gamma))$.

SOLUTION: Centroid is $G\left(\dfrac{a(\cos\alpha + \cos\beta + \cos\gamma)}{3}, \dfrac{a(\sin\alpha + \sin\beta + \sin\gamma)}{3}\right)$

Circumcentre (C) is clearly origin (0, 0)

$\because$ Centroid (G) divides line segment joining O (orthocentre) and circumcentre (C) in the ratio 2 : 1 internally.

Apply section formulae. So, co-ordinate of orthocentre: $\dfrac{x_0}{3} = x_G$ and $\dfrac{y_0}{3} = y_G$

$\Rightarrow x_0 = 3x_G$ and $y_0 = 3y_G \Rightarrow x_0 = a(\cos\alpha + \cos\beta + \cos\gamma)$ and $y_0 = a(\sin\alpha + \sin\beta + \sin\gamma)$

ILLUSTRATION 72: Two vertices of ΔABC are A(2,–1) and B(–1, –3). If the orthocentre of the triangle is at H(1,2) then find the co-ordinates of third vertex.

SOLUTION: Consider the altitude AD of ΔABC (as shown in figure)

$\Rightarrow$ Slope of $AD = \dfrac{3}{-1} = -3$ and Slope of $BC = \dfrac{k+3}{h+1}$

$\therefore \; BC \perp AD \Rightarrow m_1 m_2 = -1$; we get $\left(\dfrac{k+3}{h+1}\right)(-3) = -1$

$\Rightarrow 3k + 9 = h + 1 \Rightarrow h - 3k = 8$ ....(1)

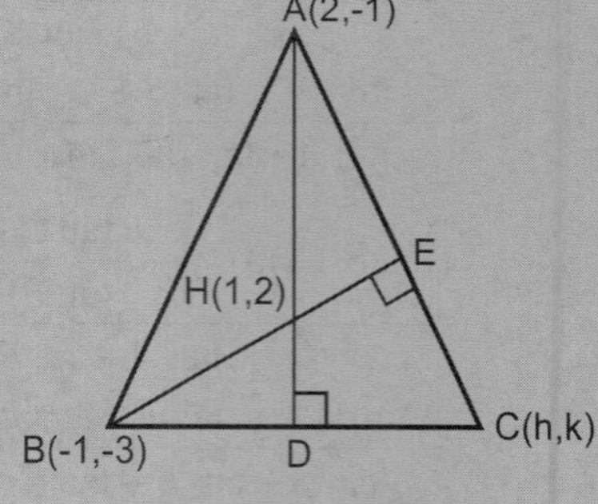

FIGURE 1.85

Similiary, Slope of BE = Slope of $BH = \dfrac{5}{2}$

Slope of $AC = \dfrac{k+1}{h-2}$

As we know $BE \perp AC$

$\Rightarrow \left(\dfrac{5}{2}\right)\left(\dfrac{k+1}{h-2}\right) = -1 \Rightarrow 5k + 5 = -2h + 4 \Rightarrow 2h + 5k + 1 = 0$ ....(2)

Solving the equations (1) and (2) we get $11k = -17$

$\Rightarrow k = -\dfrac{17}{11}$; substituting in equation (1)

$\Rightarrow h + 3 \times \dfrac{17}{11} = 8 \qquad \Rightarrow h = 8 - \left(\dfrac{51}{11}\right) = \dfrac{88-51}{11} \qquad \Rightarrow h = \dfrac{37}{11}$

Therefore, the co-ordinates of third vertex C are given as $(h, k) = \left(\dfrac{37}{11}, \dfrac{-17}{11}\right)$.

ILLUSTRATION 73: The vertices of a triangle are (2, a), (3, b) and $\left(\sqrt{c^4}, -1\right)$

(i) Prove that the centroid of this triangle can never lie on y-axis.

(ii) Obtain the condition under which the centroid lie on x-axis.

SOLUTION: Centroid of the triangle is G, i.e., $\left(\dfrac{2+3+\sqrt{c^4}}{3}, \dfrac{a+b-1}{3}\right)$ i.e., $\left(\dfrac{5+\sqrt{c^4}}{3}, \dfrac{a+b-1}{3}\right)$

(i) Let us asume that G lies on y-axis,

So, $\dfrac{5+c^2}{3} = 0 \Rightarrow c = \pm i\sqrt{5}$

$\therefore$ No real values of c exists. So centroid can never lie on y-axis.

(ii) If the centroid lie on x-axis, then y-co-ordinate must be zero.

$\therefore \quad \dfrac{a+b-1}{3} = 0 \Rightarrow a + b - 1 = 0$

$\Rightarrow a + b = 1$, which is the required condition.

ILLUSTRATION 74: If a vertex of an equilateral triangle is the origin and the side opposite to it has the equation $x + y = 1$, then find the orthocentre of the triangle.

SOLUTION: Consider ΔOPQ as shown in the figure. Since, ΔOPQ is the equilateral triangle, thus median $OC \perp PQ$. As we know that in an equilateral triangle the orthocentre and the centroid are the same. Consequently, the orthocentre H divides OC internally in the ratio 2 : 1.

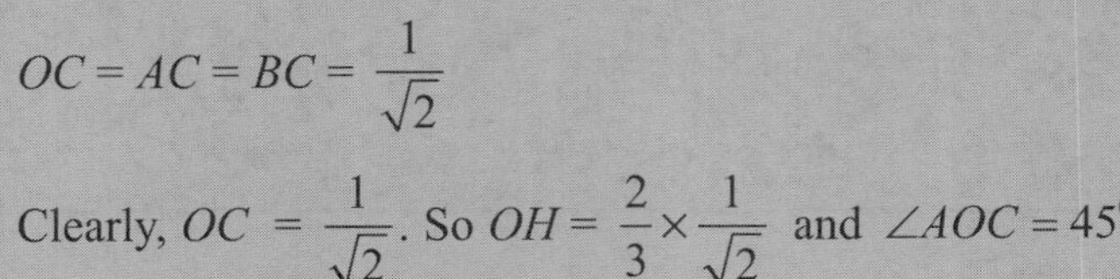

$$OC = AC = BC = \frac{1}{\sqrt{2}}$$

Clearly, $OC = \dfrac{1}{\sqrt{2}}$. So $OH = \dfrac{2}{3} \times \dfrac{1}{\sqrt{2}}$ and $\angle AOC = 45^0$

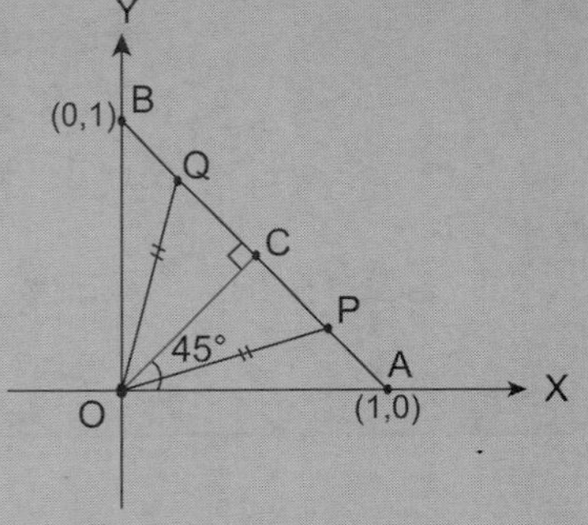

FIGURE 1.86

$$\therefore \quad H \equiv \left(\frac{2}{3\sqrt{2}}\cos 45^\circ, \frac{2}{3\sqrt{2}}\sin 45^\circ\right) \equiv \left(\frac{1}{3}, \frac{1}{3}\right)$$

ILLUSTRATION 75: Two vertices of a triangle are (5, –1) and (–2, 3). If the orthocentre of the triangle is the origin, then find the co-ordinates of the third vertex:

SOLUTION: Let the vertex A be (p, q). O is orthocentre (0, 0) which is the point of intersection of altitudes through A and B.

Now, slope of $BC = \dfrac{3-(-1)}{-2-5} = \dfrac{4}{7}$

slope of $AD = \dfrac{q}{p}$ but $m_1 m_2 = -1$

$\therefore \quad \left(-\dfrac{4}{7}\right) \times \dfrac{q}{p} = -1$ or $7p = 4q$ ....(i)

slope of $CA = \dfrac{q-3}{p+2}$

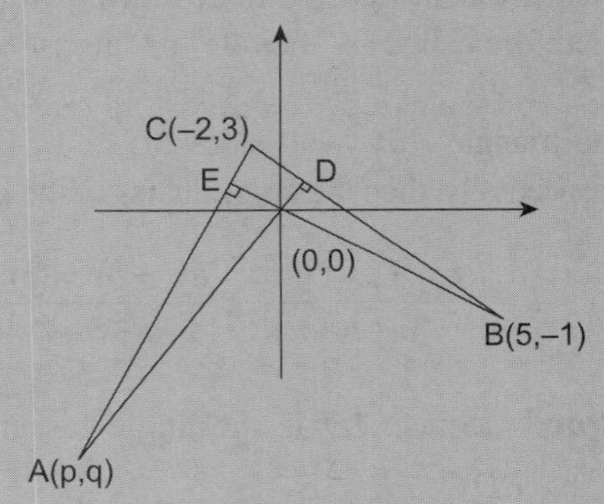

FIGURE 1.87

slope of BE = slope of $BO = \frac{-1}{5}$; But $m_1 m_2 = -1$

$\therefore \left(\frac{q-3}{p+2}\right)\left(-\frac{1}{5}\right) = -1$ or $5p - q + 13 = 0$ ......(ii)

On solving equation (i) and (ii), we get $p = -4$, $q = -7$. Hence, the third vertex A is the point (–4, –7).

ILLUSTRATION 76: If a triangle has its orthocentre at (1, 1) and circumcentre at (3/2, 3/4), then the co-ordinates of the centroid of the triangle are

(a) $\left(\frac{4}{3}, -\frac{5}{6}\right)$ (b) $\left(\frac{4}{3}, \frac{5}{6}\right)$

(c) $\left(-\frac{4}{3}, \frac{5}{6}\right)$ (d) $\left(-\frac{4}{3}, -\frac{5}{6}\right)$

SOLUTION: Since the centroid divides the line joining the orthocentre and circumcentre in the ratio 2:1 internally, therefore, if the centroid is (x, y), then $x = \frac{2.\frac{3}{2}+1.1}{2+1} = \frac{4}{3}$ and $y = \frac{2.\frac{3}{4}+1.1}{2+1} = \frac{5}{6}$.

$\therefore$ Co-ordinates of centroid are (4/3, 5/6)

ILLUSTRATION 77: If the centroid and circumcentre of a triangle are (3, 3) and (6, 2), respectively, then the orthocentre is

(a) (–3, 5) (b) (–3, 1)

(c) (3, –1) (d) (9, 5)

SOLUTION: Let orthocentre be (α, β); then $3 = \frac{2(6)+1.\alpha}{2+1}$ and $3 = \frac{2(2)+1.\beta}{2+1}$

$\Rightarrow \alpha = -3$ and $\beta = 5$ $\therefore$ orthocentre is (–3, 5)

4. Incircle and Incentre

The internal angle bisectors of the triangle are concurrent and their point of intersection is called incentre (I). Consequently incentre is a point equidistant from all the three sides of a triangle and this distance is called the inradius (r) of the triangle. Taking I as a centre and r as radius if a circle is drawn, then it touches all the three sides of triangle internally and therefore called as “incircle” or “inscribed circle” of triangle.

If $A(x_1, y_1)$, $B(x_2, y_2)$ and $C(x_3, y_3)$ are the vertices of the triangle ABC with sides BC, CA, AB of lengths a, b, c respectively, then the co-ordinates of the incentre are given by

$$I = \left(\frac{ax_1 + bx_2 + cx_3}{a+b+c}, \frac{ay_1 + by_2 + cy_3}{a+b+c}\right)$$

Proof: Since AD is the angle bisector of angle A, we have, $\frac{BD}{DC} = \frac{AB}{AC} = \frac{c}{b}$

$$\Rightarrow \frac{DC}{BD} = \frac{b}{c} \Rightarrow \frac{DC}{BD} + 1 = \frac{b}{c} + 1 \Rightarrow BD = \frac{ac}{b+c}$$

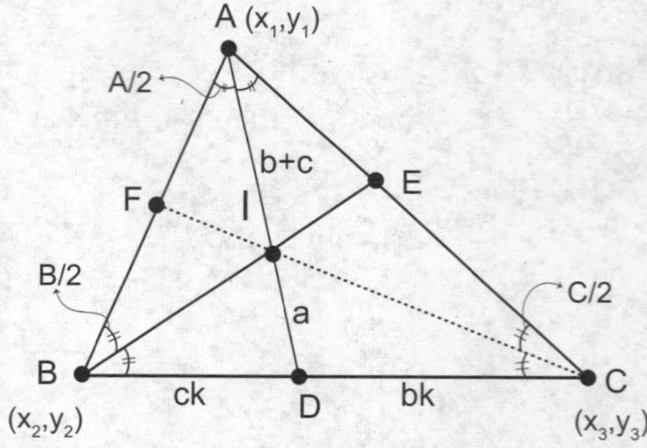

FIGURE 1.88

Since BI is the bisector of $\angle B$, so it divides AD in the ratio $AB : BD$.

$$\therefore \frac{AI}{ID} = \frac{AB}{BD} = \frac{c}{\frac{ac}{b+c}} = \frac{b+c}{a} \Rightarrow AI : ID = b + c : a$$

Since D, divides BC in the ratio $c : b$.

$\therefore$ Co-ordinates of D are $\left(\dfrac{bx_2+cx_3}{b+c}, \dfrac{by_2+cy_3}{b+c}\right)$.

Since I divides AD in the ratio $b + c : a$. Hence, co-ordinates of I are

$$\left(\frac{ax_1+(b+c)\left(\frac{bx_2+cx_3}{b+c}\right)}{a+b+c}, \frac{ay_1+(b+c)\left(\frac{by_2+cy_3}{b+c}\right)}{a+b+c}\right)$$

or $\left(\dfrac{ax_1+bx_2+cx_3}{a+b+c}, \dfrac{ay_1+by_2+cy_3}{a+b+c}\right)$.

Properties of incircle and incentre

1. Incentre lies always inside the triangle.
2. For equilateral triangle it coincides with centroid.
3. For isosceles triangle it is collinear with orthocentre, centroid and circumcentre.
4. $r = \dfrac{\Delta}{s}$, i.e., in radius is the ratio of area of triangle to its semi-perimeter.

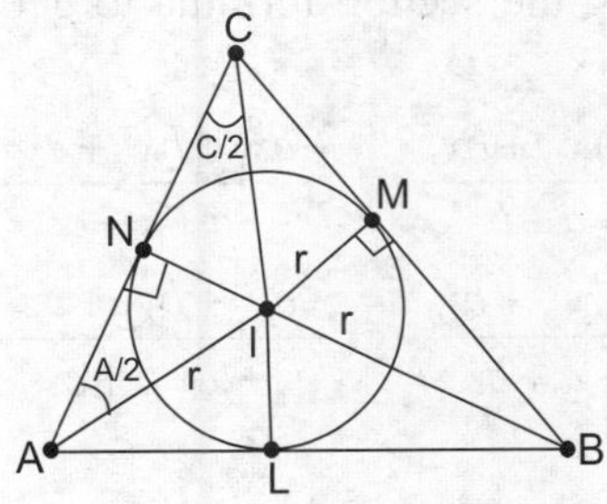

FIGURE 1.89

Proof: Area of ΔABC = Area of ΔBIA + Area of ΔCIA + Area of ΔBIC

$$\Rightarrow \Delta = \frac{1}{2}r(c)+\frac{1}{2}r(b)+\frac{1}{2}r(a) \Rightarrow \Delta = \frac{r}{2}(a+b+c)$$

$$\Rightarrow \Delta = r.s \Rightarrow r = \frac{\Delta}{s}.$$

5. $r = (s-a)\tan\dfrac{A}{2} = (s-b)\tan\dfrac{B}{2} = (s-c)\tan\dfrac{C}{2}$

Proof: $AL = AN = x$, $BL = BM = y$, $CM = CN = z$ (say)

$\Rightarrow a = z + y$, $b = x + z$ and $c = x + y$

$\Rightarrow a + b + c = 2(x + y + z) = 2s$

$$\Rightarrow x+y+z = s \Rightarrow \begin{cases} x = s-a \\ y = s-b \\ z = s-c \end{cases}$$

$$\Rightarrow \text{in } \Delta NIA : \tan\frac{A}{2} = \frac{r}{x} \Rightarrow r = x\tan\frac{A}{2}.$$

$$\Rightarrow r = (s-a)\tan\frac{A}{2}$$

Similarly, $r = (s-b)\tan\dfrac{B}{2} = (s-c)\tan\dfrac{C}{2}$.

6. $r = 4R\sin\dfrac{A}{2}\sin\dfrac{B}{2}\sin\dfrac{C}{2}$

Proof: $\because c = AL + BL$

$$\Rightarrow c = r\cot\frac{A}{2}+r\cot\frac{B}{2}$$

$$\Rightarrow c = r\left(\frac{\cos\frac{A}{2}\sin\frac{B}{2}+\cos\frac{B}{2}\sin\frac{A}{2}}{\sin\frac{A}{2}\sin\frac{B}{2}}\right)$$

$$\Rightarrow c = \frac{r\sin\left(\frac{A+B}{2}\right)}{\sin\frac{A}{2}\sin\frac{B}{2}}$$

$$\Rightarrow 2R(\sin C) = \frac{r\cos\frac{C}{2}}{\sin\frac{A}{2}\sin\frac{B}{2}}$$

$$\Rightarrow r = \frac{4R\sin\frac{C}{2}\cos\frac{C}{2}.\sin\frac{A}{2}\sin\frac{A}{2}}{\cos\frac{C}{2}}$$

$$\Rightarrow r = 4R\sin\frac{A}{2}\sin\frac{B}{2}\sin\frac{C}{2}.$$

ILLUSTRATION 78: Find the co-ordinates of the incentre of the triangle having its co-ordinates $A(4,2)$, $B(3,5)$ and $C(6,5)$.

SOLUTION: $a = BC = \sqrt{(6-3)^2+(5-5)^2} = 3$

$b = AC = \sqrt{(6-4)^2+(5-2)^2} = \sqrt{2^2+3^2} = \sqrt{4+9} = \sqrt{13}$

and $c = AB = \sqrt{(3-4)^2+(5-2)^2} = \sqrt{1^2+3^2} = \sqrt{9+1} = \sqrt{10}$

Let (p, q) be co-ordinates of incentre of ΔABC, then by using formula of incentre, we have:

$$p = \frac{ax_1 + bx_2 + cx_3}{a+b+c} = \frac{12 + 3\sqrt{13} + 6\sqrt{10}}{3 + \sqrt{13} + \sqrt{10}} \text{ and}$$

$$q = \frac{ay_1 + by_2 + cy_3}{a+b+c} = \frac{(3)(2) + (\sqrt{13})(5) + (\sqrt{10})(\sqrt{5})}{3 + \sqrt{13} + \sqrt{10}} = \frac{6 + 5\sqrt{13} + 5\sqrt{10}}{3 + \sqrt{13} + \sqrt{10}}.$$

5. Ex-circles and Ex-centres

The circle which touches the side BC and two sides AB and AC produced is called **excribed circle** opposite to the angle A. The bisectors of the external angles B and C^* meet at a point I_A which is the centre of the excribed circle opposite to the vertex A.

The co-ordinates of I_A are given by

$$\left(\frac{-ax_1 + bx_2 + cx_3}{-a+b+c}, \frac{-ay_1 + by_2 + cy_3}{-a+b+c}\right)$$

The co-ordinates of I_B and I_C (centres of escribed circles opposite to the vertex B and C, respectively) are given by $I_B\left(\frac{ax_1 - bx_2 + cx_3}{a-b+c}, \frac{ay_1 - by_2 + cy_3}{a-b+c}\right)$ and $I_C\left(\frac{ax_1 + bx_2 - cx_3}{a+b-c}, \frac{ay_1 + by_2 - cy_3}{a+b-c}\right)$, respectively.

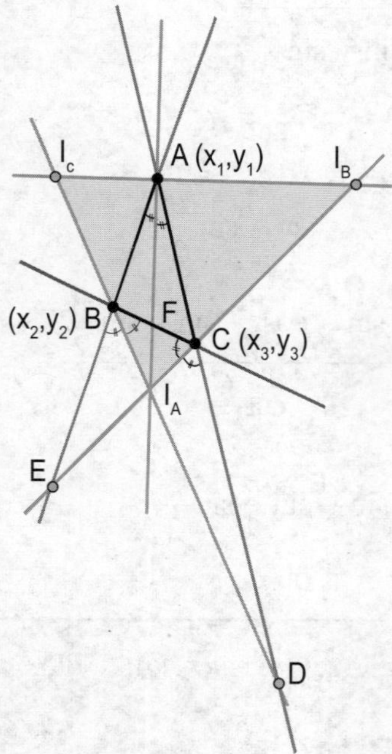

FIGURE 1.90

Proof: Since external bisector of angle B divides the opposite side externally in the ratio of sides containing angle B.

$$\therefore \quad \frac{AD}{DC} = \frac{BA}{BC} \Rightarrow \frac{AD}{DC} = \frac{c}{a}$$

So, the co-ordinate of D are given as:

$$x_D = \frac{-ax_1 + cx_3}{-a+c};\ y_D = \frac{-ay_1 + cy_3}{-a+c} \qquad \text{...(i)}$$

and $\frac{AD}{CD} = \frac{c}{a} \Rightarrow$

and $AD - CD = b \Rightarrow CD.\left(\frac{c}{a} - 1\right) = b \Rightarrow CD = \frac{ab}{c-a}$

I_A divides BD in the ratio $= \frac{BI_A}{DI_A} = \frac{BC}{CD} = \frac{c-a}{b}$.

So applying the section formula to get co-ordinate of I_A, we get:

$$x_A = \frac{(c-a)x_D + bx_B}{c-a+b} = \frac{-ax_1 + bx_2 + cx_3}{-a+b+c}.$$

$$y_A = \frac{(c-a)y_D + by_B}{c-a+b} = \frac{-ay_1 + by_2 + cy_3}{-a+b+c}.$$

Properties of ex-circles and ex-centres

1. Distance of each ex-centre is same from all three sides of triangle. Called ex-radii, r_A, r_B, r_C.
2. Considering I_A as centre and r_A as ex-radius, if a circle is drawn, the circle touches the sides of ΔABC, BC externally and AC and AB produced internally. It is called as **ex-circle** opposite to A. There are three such circles as shown below.

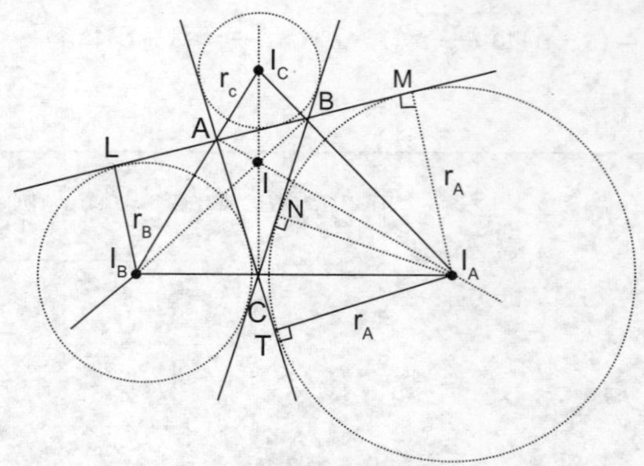

FIGURE 1.91

- The ex-radii, r_A, r_B, r_C are as given below.

$$r_a = \frac{\Delta}{s-a} = s\tan\frac{A}{2} = 4R\sin\frac{A}{2}\cos\frac{B}{2}\cos\frac{C}{2}$$

$$r_a = \frac{\Delta}{s-b} = s\tan\frac{B}{2} = 4R\sin\frac{B}{2}\cos\frac{C}{2}\cos\frac{A}{2}$$

$$r_C = \frac{\Delta}{s-c} = s\tan\frac{C}{2} = 4R\sin\frac{C}{2}\cos\frac{A}{2}\cos\frac{B}{2}.$$

Proof: ar ΔABC = arΔ ACI_A + ar ΔABI_A – arΔBCI_A

$\Rightarrow$ $\Delta = 1/2(AC).r_A + 1/2(AB).r_A - 1/2(BC).r_A$

$\Rightarrow$ $\Delta = 1/2\ (b + c - a).r_A$

$\Rightarrow$ $\Delta = 1/2\ (a + b + c - 2a).r_A$

$\Rightarrow$ $\Delta = 1/2(2s - 2a).r_A$

$\Rightarrow$ $\Delta = (s - a).r_A$

$\Rightarrow$ $r_A = \Delta/(s - a)$; similarly $r_B = \Delta/(s - b)$ and $r_c = \Delta/(s - c)$

Next, in $\Delta\ AI_AM$,

$r_A = AM\tan A/2$ ....(i)

Now, $AM = (AB + BM)$

$= (AB + BN)$

$= (AB + CB - CN)$

$= [AB + CB - (CT)]$

$= [AB + CB - (AT - AC)]$

$= (AB + BC + AC - AT)$

$= (AB + BC + CA - AM)$

$\Rightarrow$ $2AM = c + a + b$

$\Rightarrow$ $AM = S$ ...(ii)

$\therefore$ From (i) & (ii) $r_A = s\tan A/2$

Similarly $r_B = s\tan B/2$ and $r_c = s\tan C/2$

Further

$a = BC = BN + NC$

$$= r_A\cot\left(\frac{\pi}{2} - \frac{B}{2}\right) + r_A\cot\left(\frac{\pi}{2} - \frac{C}{2}\right)$$

$$= r_A\tan\frac{B}{2} + r_A\tan\frac{C}{2} \Rightarrow r_A = 4R\sin\frac{A}{2}\cos\frac{B}{2}\cos\frac{C}{2}$$

(using the sine formula)

- $\Delta\ ABC$ is pedal Δ of $\Delta\ I_A I_B I_c$

  $AI_A \perp I_B I_C$ and $BI_B \perp I_A I_C$ and $CI_C \perp I_A I_B$

  And $\Delta\ I_A I_B I_c$ is called ex-central triangle of ΔABC.
- Incentre of ΔABC (Pedal Δ) is same as orthocenter of ex-central triangle.

ILLUSTRATION 79: If the co-ordinates of the vertices of the triangle are $A(2, 2)$, $B(3, -4)$ and $C(1, 2)$ then find the excentre opposite to the vertex A.

SOLUTION: $a = |BC| = \sqrt{(1-3)^2 + (2+4)^2} = \sqrt{4+36} = \sqrt{40}$,

$b = |AC| = \sqrt{(1-2)^2 + (2-2)^2} = 1$, and $c = |AB| = \sqrt{(3-2)^2 + (-4-2)^2} = \sqrt{37}$

$$x = \frac{-ax_1 + bx_2 + cx_3}{-a+b+c} = \frac{-(\sqrt{40})(2) + (1)(1) + (\sqrt{37})(1)}{-\sqrt{40} + 1 + \sqrt{37}}$$

$$y = \frac{-ay_1 + by_2 + cy_3}{-a+b+c} = \frac{-(\sqrt{40})(2) + (\sqrt{37})(2) + (1)(-4)}{-\sqrt{40} + \sqrt{37} + 1}$$

$\Rightarrow$ $x = \dfrac{-2\sqrt{40} + \sqrt{37} + 1}{-\sqrt{40} + \sqrt{37} + 1}$; $y = \dfrac{-2\sqrt{40} + 2\sqrt{37} - 4}{-\sqrt{40} + \sqrt{37} + 1}$ are the co-ordinates of excentre opposite to vertex A.

■ NINE POINT CIRCLE

The nine point circle also called **Euler's circle** or the **Feuerbach circle** is a circle passing through the feet of altitudes (i.e., D, E, F), midpoints of sides BC, CA, AB, respectively, (i.e., H, I, J) and the mid-points of the line joining the orthocentre O to the angular points A, B, C (i.e., K, L, M). Thus the nine points D, E, F, H, I, J, K, L, M all lie on a circle. This circle is known as **nine point circle** and its centre is called the **nine point centre**.

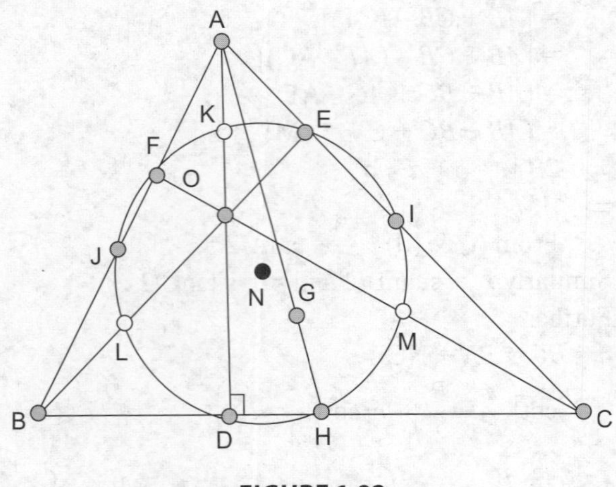

FIGURE 1.92

The nine point centre of a triangle is collinear with the circumcentre and the orthocentre and bisects the segment joining them radius of nine-point circle of a triangle is half the radius of the circumcircle.

Properties of nine point circle

1. The orthocentre, the nine-point centre, the centroid and the circumcentre all lie on a straight line
2. If O is orthocentre, N is nine point centre, G is centroid and C is circumcentre, then to memorize, remember. ONGC (i.e., Oil and Natural Gas Corporation).

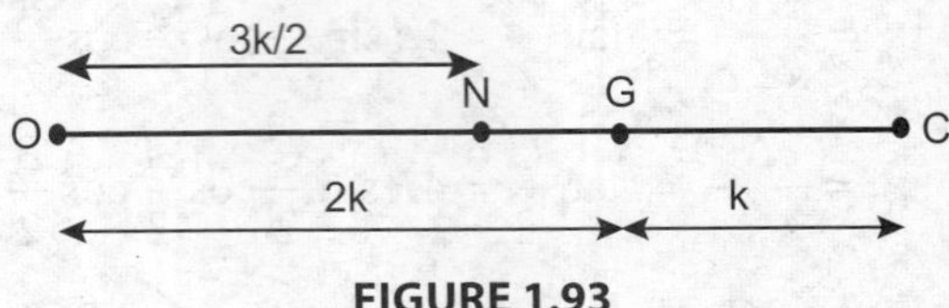

FIGURE 1.93

Alphabets are initials of 4 respective points two alphabet in left of G and 1 to its right signifies the ratio 2 : 1 in which G divides OC. But, N is mid-point of OC.

3. Radius of nine point circle = $\frac{1}{2}$ × Radius of circumcircle.
4. Nine point centre is the circumcentre of pedal triangle.
5. If circumcentres of triangle be origin and centroid has co-ordinate (x, y), then
   - Co-ordinate of orthocentre = $(3x, 3y)$
   - Co-ordinate of nine point centre = $\left(\frac{3x}{2}, \frac{3y}{2}\right)$.

ILLUSTRATION 80: The co-ordinates of vertices P, Q, R of ΔPQR are (0, 2); (4, 0) and (2, 4), respectively, then find the co-ordinates of (i) centroid (G) (ii) circumcentre (iii) orthocenter and (iv) nine-point centre Also find the nine point radius, circumcentre and circumradius of Pedal Δ of ΔPQR.

SOLUTION: (i) Co-ordinates centroid (G) are given by $G \equiv \left(\frac{x+x_2+x_3}{3}, \frac{y_1+y_2+y_3}{3}\right)$

$$\equiv \left(\frac{0+4+2}{3}, \frac{2+0+4}{3}\right) \equiv (2,2)$$

(ii) Let C be circumcentre of ΔPQR and Let its co-ordinates be (x,y)

$\Rightarrow PC = QC = RC$

$$\Rightarrow \sqrt{(x-0)^2+(y-2)^2} = \sqrt{(x-4)^2+(y-0)^2} = \sqrt{(x-2)^2+(y-4)^2}$$

$$\Rightarrow \sqrt{x^2+(y-2)^2} = \sqrt{(x-4)^2+y^2} = \sqrt{(x-2)^2+(y-4)^2}$$

$\Rightarrow x^2 + y^2 - 4y + 4 = x^2 + y^2 - 8x + 16 = x^2 + y^2 - 4x - 8y + 20$

$\Rightarrow -4y + 4 = -8x + 16 = -4x - 8y + 20 \Rightarrow 8x - 4y = 12$ and $4x - 8y = -4$

Or equivalently $2x - y = 3$ and $x - 2y = -1$

$\Rightarrow x = 7/3, y = 5/3$

$$\therefore \; C \equiv \left(\frac{7}{3}, \frac{5}{3}\right)$$

(iii) Let the co-ordinates of orthocentre be (u,v), i.e., $O \equiv (u,v)$, We know that centroid divides the line joining orthocentre and circumcentre in the ratio 2 : 1

$\therefore$ The co-ordinates of orthocentre are given by $2 = \dfrac{2\left(\frac{7}{3}\right)+1(u)}{2+1}, 2 = \dfrac{2\left(\frac{5}{3}\right)+1(v)}{2+1}$

$$\underset{O(u,v)\quad G(2,2)\quad C\left(\frac{7}{3},\frac{5}{3}\right)}{\overline{\qquad 2:1 \qquad}}$$

$\Rightarrow u = 6 - \dfrac{14}{3}; v = 6 - \dfrac{10}{3}$

$\Rightarrow u = 4/3; v = 8/3$ $\quad\therefore\quad O \equiv \left(\dfrac{4}{3},\dfrac{8}{3}\right)$.

(iv) we know that nine point centre is the mid-point of line segment joining the orthocentre and circumcentre i.e., mid-point of $O\left(\dfrac{4}{3},\dfrac{8}{3}\right)$ and $C\left(\dfrac{7}{3},\dfrac{5}{3}\right)$

i.e., $N \equiv \left(\dfrac{\frac{4}{3}+\frac{7}{3}}{2}, \dfrac{\frac{8}{3}+\frac{5}{3}}{2}\right) = \left(\dfrac{11}{6},\dfrac{13}{6}\right)$. Now, nine-point radius $= R_9 = 1/2(\text{R})$

$$= \frac{1}{2}PC = \frac{1}{2}\sqrt{\left(\frac{7}{3}-0\right)^2+\left(\frac{5}{3}-2\right)^2} = \frac{1}{2}\sqrt{\frac{49}{9}+\frac{1}{9}} = \frac{1}{2}\sqrt{\frac{50}{9}} = \frac{1}{2}\frac{(5)}{3}\sqrt{2} = \frac{5}{6}\sqrt{2}$$

As we know that nine point circle is nothing, but circumcircle of pedal triangle. Therefore, circumcentre of pedal Δ is the same as nine point centre $N\left(\dfrac{11}{6},\dfrac{13}{6}\right)$ and circumradius of pedal triangle is nine point radius $R_9 = \dfrac{5}{6}\sqrt{2}$.

ILLUSTRATION 81: If in ΔABC, $BC = 6$, $CA = 3$ and $AB = 4$ and D and E trisect BC and $\angle CAE = \theta$, then find $\tan\theta$

SOLUTION: D and E trisect $BC \Rightarrow BD = DE = EC = 6/3 = 2$

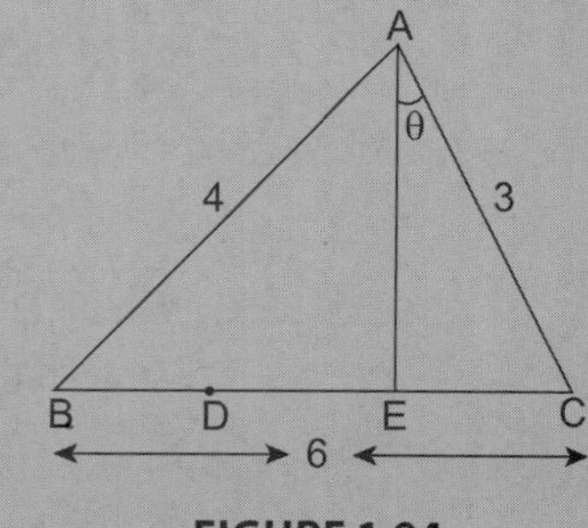

FIGURE 1.94

Now, in ΔACE, by cosine formula $\cos C = \dfrac{(AC)^2+(CE)^2-(AE)^2}{2(AC)(CE)}$

$\Rightarrow$ Cos $C = \dfrac{9+4-(AE)^2}{2(3)(2)}$ Also, $\cos C = \dfrac{(6)^2+(3)^2-(4)^2}{2(6)(3)}$

Using cosine formula in ΔABC $\dfrac{(6)^2+(3)^2-(4)^2}{36} = \dfrac{13-(AE)^2}{12}$

$\Rightarrow \dfrac{29}{3} = 13-(AE)^2 \Rightarrow (AE)^2 = 13 - \dfrac{29}{3} = \dfrac{10}{3} \Rightarrow AE = \sqrt{\dfrac{10}{3}}$

From ΔACE, by the cosine formula $\cos\theta = \dfrac{(AC)^2+(AE)^2-(EC)^2}{2(AC)(AE)} = \dfrac{9+\left(\frac{10}{3}\right)-(4)}{2(3)\left(\sqrt{\frac{10}{3}}\right)}$

$\Rightarrow \cos\theta = \dfrac{\frac{25}{3}}{2\sqrt{30}} = \dfrac{25}{6\sqrt{30}} \Rightarrow \sec\theta = \dfrac{6\sqrt{30}}{25}$

$\therefore \tan\theta = \sqrt{\sec^2\theta - 1} = \sqrt{\dfrac{36\times 30}{625}-1} = \sqrt{\dfrac{1080-625}{625}} = \sqrt{\dfrac{455}{625}} = \sqrt{\dfrac{91}{125}}$

ILLUSTRATION 82: The foot of altitude through vertex P of a ΔPQR is (2,3) and mid-points of sides PQ and PR are respectively, (–1,3) and (5,4), then find the Nine point centre and Nine point radius. Hence, find the circumradius of ΔPQR.

SOLUTION: We know that nine point circle passes through 9 points, i.e., 3 feet of altitudes, 3 mid-points of line segments joining vertices to orthocentre of given triangle and 3 mid-points of sides of triangle and any 3 points of these 9 points are sufficient to draw the Nine point circle let it be $x^2 + y^2 + 2gx + 2fy + c = 0$

$\Rightarrow 4 + 9 + 4g + 6f + c = 0$ ((2, 3) foot of altitude)

and $1 + 9 - 2g + 6f + c = 0$; { (–1, 3) and (5, 4) lie on circle}

and $25 + 16 + 10g + 8f + c = 0$

$\Rightarrow 4g + 6f + c = -13$ ....(i)

$-2g + 6f + c = -10$ ...(ii)

and $10g + 8f + c = -41$ ... (iii)

$\Rightarrow 6g = -3 \Rightarrow g = -1/2$ and $12g + 2f = -31$

$\Rightarrow -6 + 2f = -31 \Rightarrow 2f = -25 \Rightarrow f = -25/2$

and $c = -10 + 2g - 6f = -10 - 1 + 75 = 64$

$\therefore$ The equation of nine point circle is $x^2 + y^2 - x - 25y + 64 = 0$.

$\therefore$ Its centre i.e., nine point centre is $\left(\frac{1}{2}, \frac{25}{2}\right)$ and its radius i.e., nine point radius is given by

$$= \sqrt{\left(\frac{1}{2}\right)^2 + \left(\frac{25}{2}\right)^2 - 64} = \sqrt{\frac{1}{4} + \frac{625}{4} - \frac{256}{4}} = \frac{1}{2}\sqrt{370} = R_9$$

$\therefore$ Circum radius of $\Delta PQR = 2R_9 = \sqrt{370}$

ILLUSTRATION 83: If a triangle has its orthocentre at (1, 1) and circumcentre at (3/2, 3/4), then find the centroid and nine point centre.

SOLUTION: Since centroid divides the line segment joining the orthocentre and circumcentre in the ratio 2:1 (internally) and if centroid is (x, y) then

$$\therefore \quad x = \frac{2.\frac{3}{2} + 1.1}{2+1} = \frac{4}{3} \text{ and } y = \frac{2.\frac{3}{4} + 1.1}{2+1} = \frac{5}{6}$$

$\therefore$ centroid is (4/3, 5/6)

and since nine point centre is mid-point of orthocentre and circumcentre

thus nine point centre is $\left(\frac{1+3/2}{2}, \frac{1+3/4}{2}\right)$ i.e., $\left(\frac{5}{4}, \frac{7}{8}\right)$

ILLUSTRATION 84: If the co-ordinates of vertices of a triangle are (0,6), (8,12) and (8,0), then find its (a) centroid (b) in-centre (c) orthocentre

SOLUTION: (a) Centroid is $\left(\frac{0+8+6}{3}, \frac{6+12+0}{3}\right)$ or $\left(\frac{16}{3}, 6\right)$

(b) $a = BC = \sqrt{(8-8)^2+(12-0)^2} = 12,$

$b = CA = \sqrt{(0-8)^2+(6-0)^2} = 10,$

$c = AB = \sqrt{(0-8)^2+(6-12)^2} = 10$

The co-ordinates of the in-centre are

$$\left(\frac{12\times0+12\times8+10\times8}{12+10+10}, \frac{12\times6+10\times12+10\times0}{12+10+10}\right)$$

or $\left(\frac{160}{32}, \frac{192}{32}\right) = (5, 6)$

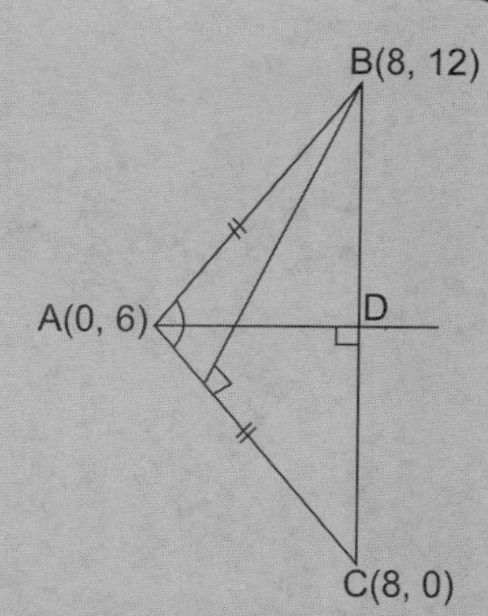

FIGURE 1.95

(c) As the triangle is isosceles Δ

Equation of altitude AD will be $y = 6$ ...(i)

Now equation of altitude BE will be $(y-12) = -\left(\frac{0-8}{6-0}\right)(x-8)$

$\Rightarrow y - 12 = 4/3(x - 8)$ ...(ii)

Now, orthocenter is given by the point of intersection of (i) and (ii) i.e., $\left(\frac{7}{2}, 6\right)$

ILLUSTRATION 85: A triangle OAB is formed by the lines $y = 2x$, $x + y = 1$ and y-axis then which of the following points lie inside the triangle?

(a) centroid (b) circum-centre

(c) orthocentre (d) incentre

SOLUTION: Clearly, $\angle AC_o = 45°$ and $\tan\phi = 2 > 1 = \tan 45°$ $\angle\, \phi > 45°$

$\therefore$ $\angle AC_o + \phi > 90°$ $\angle\, \theta < 90°$ $\angle\, \psi > 90°$

Thus the given triangle is an obtuse angled triangle, therefore, both the points circumcentre and orthocentre lie outside of the triangle. So the point lying inside the triangle are centroid and incentre.

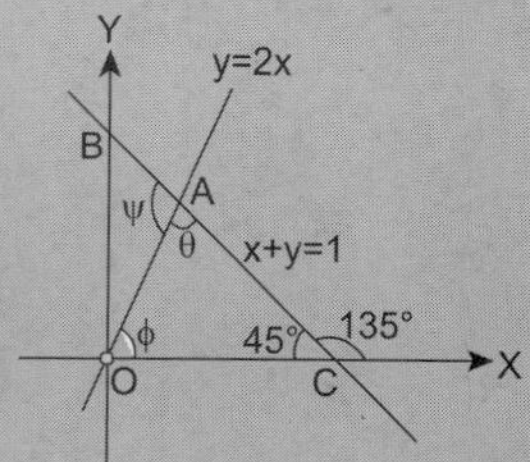

FIGURE 1.96

ILLUSTRATION 86: Prove that the incentre is a point of concurency for Internal angle bisectors of a triangle.

SOLUTION: Given $A \equiv (x_1, y_1)$, $B(x_2, y_2)$, $C \equiv (x_3, y_3)$ be the vertices of ΔABC and $BC = a$, $CA = b$ and $AB = c$. Let AD be the bisector of A. We know that the bisector of an angle of a triangle divides the opposite side in the ratio of the sides containing the angle.

$\therefore \quad \frac{BD}{DC} = \frac{AB}{AC} = \frac{c}{b}$; Hence D divides BC in the ratio c: b

$\therefore$ Co-ordinate of D are $\left(\frac{cx_3+bx_2}{c+b}, \frac{cy_3+by_2}{c+b}\right)$

From (1), $\frac{DC}{BD} = \frac{b}{c}$ or $\frac{DC}{BD} + 1 = \frac{b}{c} + 1$ or $\frac{DC+BD}{BD} = \left(\frac{b+c}{c}\right)$

or $\frac{a}{BD} = \left(\frac{b+c}{c}\right)$

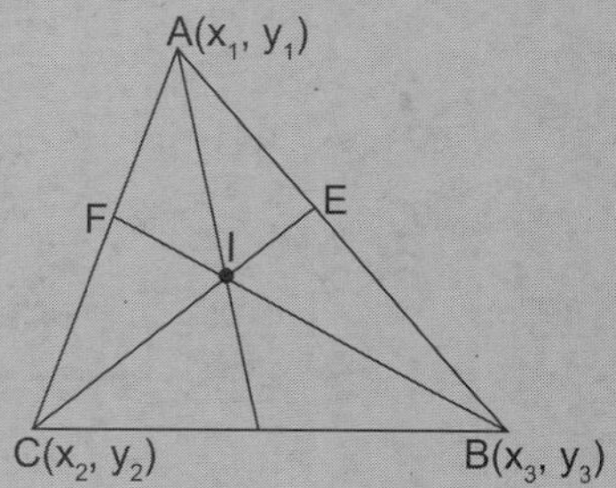

FIGURE 1.97

$\therefore \quad BD = \frac{a.c}{(b+c)}$. Also in ΔABD, BI is the bisector of B

Then $\dfrac{AI}{ID}=\dfrac{AB}{BD}=\dfrac{c}{\left(\dfrac{ac}{b+c}\right)}=\dfrac{b+c}{a}$

$\therefore$ I divides AD in the ratio $b+c:a$

$\therefore$ Co-ordinates of I are $\left(\dfrac{(b+c)\left(\dfrac{cx_3+bx_2}{c+b}\right)+a.x_1}{b+c+a},\dfrac{(b+c).\left(\dfrac{cy_3+by_2}{c+b}\right)+a.y_1}{b+c+a}\right)$

i.e., $\left(\dfrac{ax_1+bx_2+cx_3}{a+b+c},\dfrac{ay_1+by_2+cy_3}{a+b+c}\right)$

Similarly, we can show that the co-ordinates of the point which divides BE internally in the $c+a:b$ and the co-ordinates of the point which divides CF internally in the ratio $a+b:c$ will be each

$\left(\dfrac{ax_1+bx_2+cx_3}{a+b+c},\dfrac{ay_1+by_2+cy_3}{a+b+c}\right)$ and $CE=\dfrac{ab}{(c+a)}$, $AE=\dfrac{bc}{(c+a)}$; $AF=\dfrac{bc}{a+b}$, $BF=\dfrac{ac}{a+b}$

Thus, the three internal bisectors of the angles of a triangle meet in a point I

$$I=\left(\frac{ax_1+bx_2+cx_3}{a+b+c},\frac{ay_1+by_2+cy_3}{a+b+c}\right)$$

ILLUSTRATION 87: If $D(-2, 3)$, $E(4, -3)$ and $F(4, 5)$ are the mid-points of the sides BC, CA and AB of triangle ABC, then find $\sqrt{\left(|AG|^2+|BG|^2+|CG|^2\right)}$ where G is the centroid of ΔABC. Find its incentre also.

SOLUTION: Consider a ΔABC with vertices $A(x_1, y_1)$, $B(x_2, y_2)$ and $C(x_3, y_3)$ as shown in figure. Co-ordinates of $A: x_1=x_F+x_E-x_D=4+4-(-2)=10$

$y_1=y_F+y_E-y_D=5+(-3)-(3)=-1$

Co-ordinates of $B: x_2=x_F+x_D-x_E=4+(-2)-4=-2$

$y_2=y_F+y_D-y_E=5+3-(-3)=11$

Similarly, $x_3=4-2-4=-2$ and $y_3=3+(-3)-5=-5$

$\Rightarrow A(10,-1), B(-2,11), C(-2,-5)$

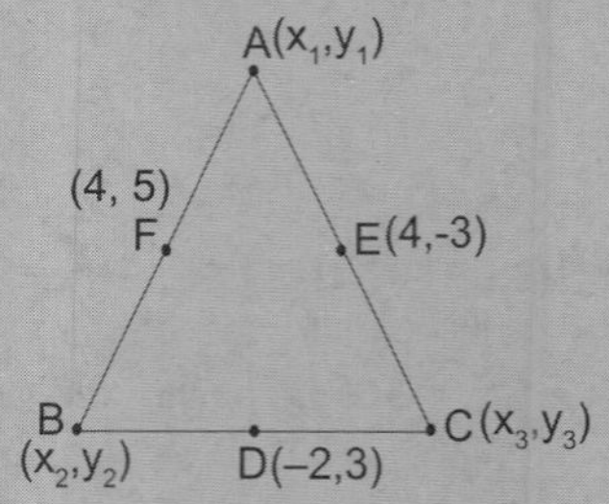

FIGURE 1.98

Now, co-ordinates of centroid are given by $x_G=\dfrac{10-4}{3}\Rightarrow\dfrac{6}{3}=2$;

$y_G=\dfrac{11-6}{3}=\dfrac{5}{3}$

$\Rightarrow AG^2=(10-2)^2+\left(-1-\dfrac{5}{3}\right)^2=8^2+\left(-\dfrac{8}{3}\right)^2=64+\dfrac{64}{9}=\dfrac{640}{9}$;

$BG^2=(-2-2)^2+\left(11-\dfrac{5}{3}\right)^2=4^2+\left(\dfrac{28}{3}\right)^2=16+\dfrac{784}{9}=\dfrac{928}{9}$;

and $CG^2=(-2-2)^2+\left(-5-\dfrac{5}{3}\right)^2=\dfrac{544}{9}$

$\therefore\ \sqrt{\left(|AG|^2+|BG|^2+|CG|^2\right)}=\sqrt{\dfrac{640}{9}+\dfrac{928}{9}+\dfrac{544}{9}}\Rightarrow\dfrac{1}{3}\times\sqrt{2112}=\dfrac{8\sqrt{33}}{3}=8\left(\sqrt{\dfrac{11}{3}}\right)$

TEXTUAL EXERCISE–6 (SUBJECTIVE)

1. If (α, β), (x, y), (p, q) are the co-ordinates of the circumcentre, centroid, orthocentre of a triangle, then prove that $3x = 2\alpha + p$ and $3y = 2\beta + q$.
2. Find the co-ordinates of the centroid and circumcentre of the triangle whose vertices are (2, 3), (3, 4) and (6, 8) and hence find the co-ordinates of its orthocentre and nine-point centre.
3. Find the co-ordinates of the incentre of the triangle whose vertices are (4, 1), (1, 5) and (–2, 1).
4. Find the incentre of the triangle whose vertices (7, –36), (7, 20) and (–8, 0).
5. If (0, 1/2), (1/2, 1/2) and (1/2, 0) be the middle points of the sides of a triangle, find the co-ordinates of its incentre.
6. Two vertices of a triangle are (4, –3) and (–2, 5). If the co-ordinates of the orthocentre of triangle are (1, 2), find the co-ordinates of third vertex.
7. A and B are two fixed points whose co-ordinates are respectively (3,2) and (5, 1). ABP is an equilateral triangle on AB situated on the side opposite to that of origin. Find the co-ordinates of P and those of the orthocentre of triangle ABP.
8. The circumcentre of a triangle with vertices $A(a, a \tan \alpha)$ $B(b, b \tan \beta)$, $C(c, c \tan \gamma)$ lies at the origin, where $\alpha + \beta + \gamma = \pi$; ($a, b, c > 0$ and α, β, γ being acute angles). Show that its orthocentre lies on the line $4(x \cos \alpha/2 \cos \beta/2 \cos \gamma/2 - y \sin \alpha/2 \sin \beta/2 \sin \gamma/2) = y$. Also find the orthocentre.

Answer Keys

2. G(11/3, 5), Q(38, –24), N(125/6, –19/2) (–27/2, 39/2) **3.** (1, 5/2) **4.** (0, –1) **5.** $(1-1/\sqrt{2}, 1-1/\sqrt{2})$
6. (33, 26) **7.** $[(4 + \sqrt{3}/2, 3/2 + \sqrt{3}), (4 + \sqrt{3}/2, 3/2 + \sqrt{3}/2)]$
8. $[R(1 + 4\sin \alpha/2 \sin \beta/2 \sin \gamma/2), 4R \cos \alpha/2 \cos \beta/2 \cos \gamma/2]$

TEXTUAL EXERCISE–6 (OBJECTIVE)

1. If the vertices of a triangle be $(a, b - c)$, $(b, c - a)$ and $(c, a - b)$, then the centroid of the triangle lies
   (a) At origin (b) on x-axis
   (c) on y-axis (d) None of these
2. If the vertices of a triangle be $(a,1)$,$(b,3)$ and $(4,c)$ then the centroid of the triangle will lie on x-axis, if
   (a) $a + c = -4$ (b) $a + b = -4$
   (c) $c = -4$ (d) $b + c = -4$
3. Two vertices of a triangle are (5,4) and (– 2, 4). If its centroid is (5,6) then the third vertex has the co-ordinates
   (a) (12,10) (b) (10,12)
   (c) (–10, 12) (d) (12, –10)
4. The incentre of the triangle formed by (0,0), (5,12), (16,12) is
   (a) (7, 9) (b) (9, 7)
   (c) (– 9, 7) (d) (–7, 9)
5. The equation of the sides of a triangle are $x + y - 5 = 0$; $x - y + 1 = 0$ and $y - 1 = 0$, then the co-ordinates of the circumcentre are
   (a) (2, 1) (b) (1, 2)
   (c) (2, –2) (d) (1, –2)
6. If two vertices of a triangle are (6, 4), (2, 6) and its centroid is (4, 6), then the third vertex is
   (a) (4,8) (b) (8,4)
   (c) (6,4) (d) None of these
7. The incentre of the triangle with vertices $(1, \sqrt{3})$,(0, 0) and (2, 0) is
   (a) $\left(1, \frac{\sqrt{3}}{2}\right)$ (b) $\left(\frac{2}{3}, \frac{1}{\sqrt{3}}\right)$
   (c) $\left(\frac{2}{3}, \frac{\sqrt{3}}{2}\right)$ (d) $\left(1, \frac{1}{\sqrt{3}}\right)$
8. If $A(x_1, y_1)$, $B(x_2, y_2)$ and $C(x_3, y_3)$ are the vertices of a triangle, then the excentre with respect to B is
   (a) $\left(\frac{ax_1 - bx_2 + cx_3}{a-b+c}, \frac{ay_1 - by_2 + cy_3}{a-b+c}\right)$
   (b) $\left(\frac{ax_1 + bx_2 - cx_3}{a+b-c}, \frac{ay_1 + by_2 - cy_3}{a+b-c}\right)$

(c) $\left(\dfrac{ax_1-bx_2-cx_3}{a-b-c}, \dfrac{ay_1-by_2-cy_3}{a-b-c}\right)$

(d) None of these

9. If a vertex of a triangle is (1,1) and the mid-points of two sides through this vertex are (–1, 2) and (3,2), then the centroid of the triangle is

(a) $\left(1, \dfrac{7}{3}\right)$ (b) $\left(\dfrac{1}{3}, \dfrac{7}{3}\right)$

(c) $\left(-1, \dfrac{7}{3}\right)$ (d) $\left(\dfrac{-1}{3}, \dfrac{7}{3}\right)$

10. Orthocentre of triangle with vertices (0, 0), (3, 4) and (4, 0) is

(a) $\left(3, \dfrac{5}{4}\right)$ (b) (3, 12)

(c) $\left(3, \dfrac{3}{4}\right)$ (d) (3, 9)

11. The incentre of the triangle formed by the axes and the line $x/a + y/b = 1$ is

(a) $\left(\dfrac{a}{2}, \dfrac{b}{2}\right)$

(b) $\left(\dfrac{a}{3}, \dfrac{b}{3}\right)$

(c) $\left(\dfrac{ab}{a+b+\sqrt{ab}}, \dfrac{ab}{a+b+\sqrt{ab}}\right)$

(d) $\left(\dfrac{ab}{a+b+\sqrt{a^2+b^2}}, \dfrac{ab}{a+b+\sqrt{a^2+b^2}}\right)$

12. In the ΔABC, the co-ordinates of B are (0, 0) $AB = 2$, $\angle ABC = \pi/3$ and the middle point of BC has the co-ordinates (2, 0). The centroid of the triangle is

(a) $\left(\dfrac{1}{2}, \dfrac{\sqrt{3}}{2}\right)$ (b) $\left(\dfrac{5}{3}, \dfrac{1}{\sqrt{3}}\right)$

(c) $\left(\dfrac{4+\sqrt{3}}{3}, \dfrac{1}{3}\right)$ (d) None of these

13. The vertices of a triangle are $[at_1t_2, a(t_1+t_2)]$, $[at_2t_3, a(t_2+t_3)]$, $[at_3t_1, a(t_3+t_1)]$, then the co-ordinates of its orthocentre are

(a) $[a, a(t_1+t_2+t_3+t_1t_2t_3)]$

(b) $[-a, a(t_1+t_2+t_3+t_1t_2t_3)]$

(c) $[-a(t_1+t_2+t_3+t_1t_2t_3), a]$

(d) None of these

14. The orthocentre of the triangle with vertices $\left(2, \dfrac{\sqrt{3}-1}{2}\right)$; $\left(\dfrac{1}{2}, -\dfrac{1}{2}\right)$ and $\left(2, -\dfrac{1}{2}\right)$?

(a) $\left(\dfrac{3}{2}, \dfrac{\sqrt{3}-3}{6}\right)$ (b) $\left(2, -\dfrac{1}{2}\right)$

(c) $\left(\dfrac{4}{5}, \dfrac{\sqrt{3}-2}{4}\right)$ (d) $\left(\dfrac{1}{2}, -\dfrac{1}{2}\right)$

15. Point of intersection of the diagonals of a square is at origin and co-ordinate axes are drawn along the diagonals. If each side is of length a, then one which is not the vertex of square is

(a) $\left(a\sqrt{2}, 0\right)$ (b) $\left(0, \dfrac{a}{\sqrt{2}}\right)$

(c) $\left(\dfrac{a}{\sqrt{2}}, 0\right)$ (d) $\left(-\dfrac{a}{\sqrt{2}}, 0\right)$

16. ABC is an isosceles triangle of non-zero area. If the co-ordinates of the base are B (1, 3) and C (–2, 7), then the co-ordinates of vertex A can be

(a) (1,6) (b) $\left(-\dfrac{1}{2}, 5\right)$

(c) $\left(\dfrac{5}{6}, 6\right)$ (d) None of these

17. If the vertices P, Q, R, of a triangle PQR are rational points, which of the following points of the triangle PQR is (are) always rational points (s)?

(a) Centroid (b) Incentre

(c) Circumcentre (d) Orthocentre

18. If $\tan\alpha$, $\tan\beta$, $\tan\gamma$ are the roots of equation $x^3 - 3ax^2 + 3bx - 1 = 0$, the centroid of the triangle whose vertices are $(\tan\alpha, \cot\alpha)$, $(\tan\beta, \cot\beta)$ and $(\tan\gamma, \cot\gamma)$, is

(a) $(1/a, b)$ (b) $(a, 1/b)$

(c) (a, b) (d) None of these

19. If a triangle has its orthocentre at (1, 1) and circumcentre at (3/2, 3/4) then the centroid and nine point centre respectivey are

(a) (4/3, 5/6); (7/6, 11/12)

(b) (1/3, 5/6); (5/4, 1/8)

(c) (2/3, 5/6); (5/4, 3/8)

(d) None of these

Answer Keys

1. (b)	**2.** (c)	**3.** (a)	**4.** (a)	**5.** (a)	**6.** (a)	**7.** (d)	**8.** (a)	**9.** (a)	**10.** (c)
11. (d)	**12.** (b)	**13.** (b)	**14.** (b)	**15.** (a)	**16.** (c)	**17.** (a,c,d)	**18.** (c)	**19.** (a)	

■ SELECTION OF AXES

Sometimes by a suitable selection of axes, the length of solutions of the problem becomes short, therefore such selections converts conveniently the difficult problem to easier one. But one should always take care that in the process of such selection of axes, the generality of problem remains preserved.

Here generality of the problem preserved means, by suitable choice of origin and consideration of axis, any most general case can be converted to chosen case.

Let us take some simple cases to understand and master the above art where by proper choice of co-ordinate axes the difficult problems can be solved in easier ways.

Case 1. **To choose two general points:** Select them to be $P(-\lambda, 0)$ and $Q(\lambda, 0)$ because any two general points can be converted to above two points by shifting the origin to mid-point of PQ and rotating the axes by suitable angle θ. (as shown in Figure 1.99)

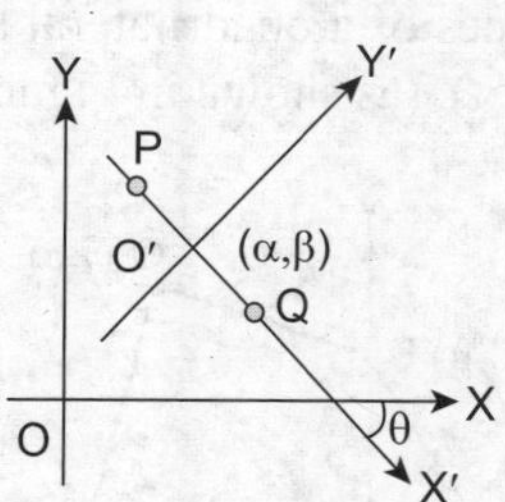

FIGURE 1.99

Case 2. **In order to select two perpendicular lines:** Take them as co-ordinate axes, because by shifting the origin to point of intersection (α, β) of given two lines and rotating the axes by a suitable angle θ, any two given general lines can be converted to co-ordinate axes. (as shown in Figure 1.100)

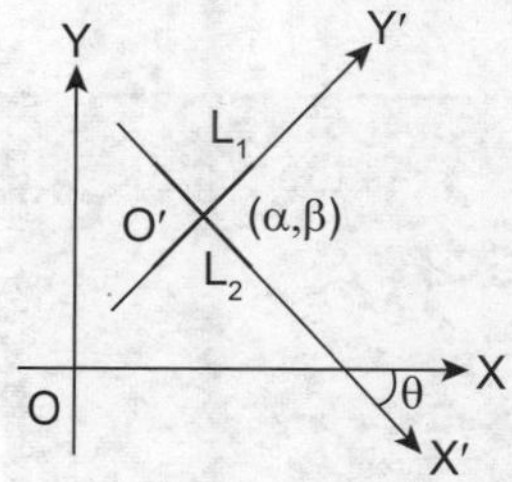

FIGURE 1.100

Case 3. **When problem is related to triangles**

(a) **Right angled triangle:** A general right angled triangle can be considered (as shown in the Figure 1.101). So that vertices are $(0, 0)$, $(\alpha, 0)$ and $(0, \beta)$; α, β are independent parameters.

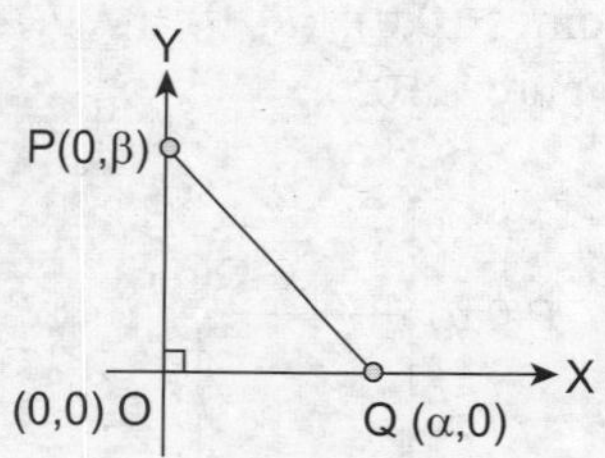

FIGURE 1.101

(b) **Isosceles triangle:** For an isosceles triangle, consider the terminal points of base as $(-\alpha, 0)$, $(\alpha, 0)$ and the vertex $(0, \beta)$ on y-axis. (as shown in Figure 1.102). Here α, β are independent parameters.

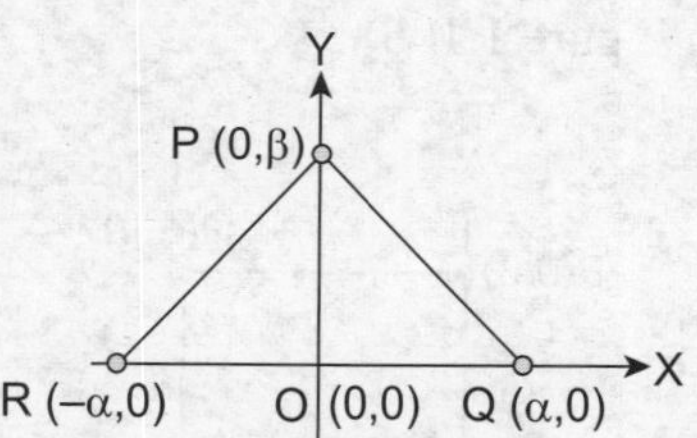

FIGURE 1.102

(c) **Equilateral triangle:** For an equilateral triangle, consider the terminal points of base as $(-\alpha, 0)$ and $(\alpha, 0)$ and the vertex $(0, \sqrt{3}\alpha)$ on y-axis (as shown in figure 1.103).

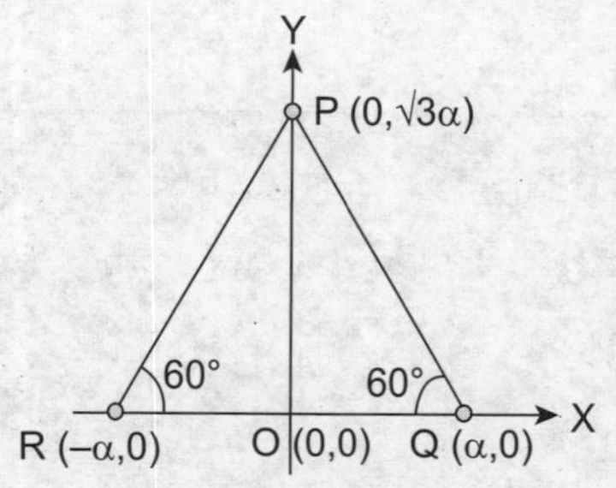

FIGURE 1.103

(d) **Any scalene triangle:** For a scalene triangle, consider the terminal points of base as $(-\alpha, 0)$, $(\alpha, 0)$ and the vertex a general point (β, γ) (as shown in Figure 1.104)

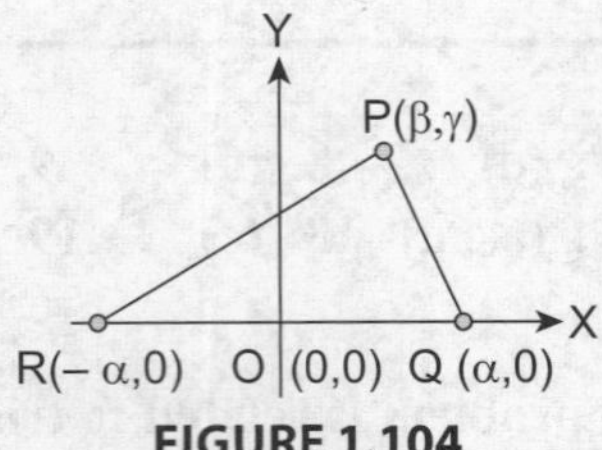

FIGURE 1.104

Case 4. **When problem is related to quadrilaterals**

(a) **Square:** To choose the general square, the axes should be oriented in such a way so that the vertices of a square becomes (0,0), $(\alpha, 0)$, (α, α) and $(0, \alpha)$, (as shown in Figure 1.105).

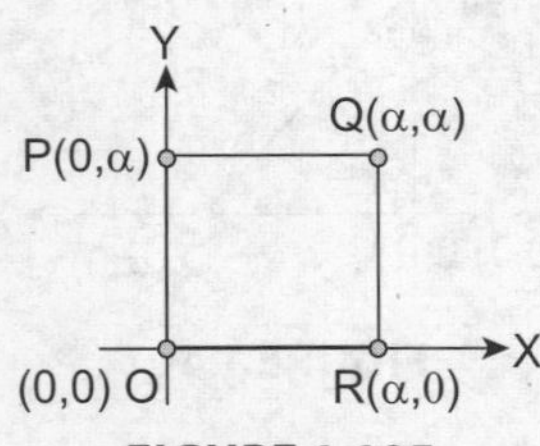

FIGURE 1.105

(b) **Rectangle:** To choose the general rectangle, the axes should be oriented in such a way so that the vertices of a rectangle becomes (0, 0), $(\alpha, 0)$, (α, β) and $(0, \beta)$, (as shown in Figure 1.106).

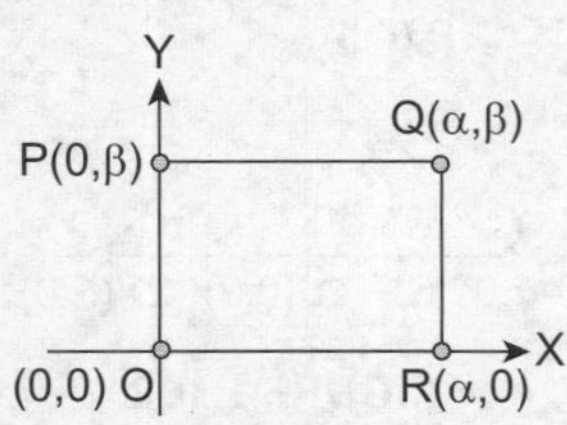

FIGURE 1.106

(c) **Rhombus:** To choose the general rhombus, the axes should be oriented in such a way so that the vertices of a rhombus becomes $(-\alpha, 0)$, $(0, -\beta)$, $(\alpha, 0)$, and $(0, \beta)$ (as shown in Figure 1.107).

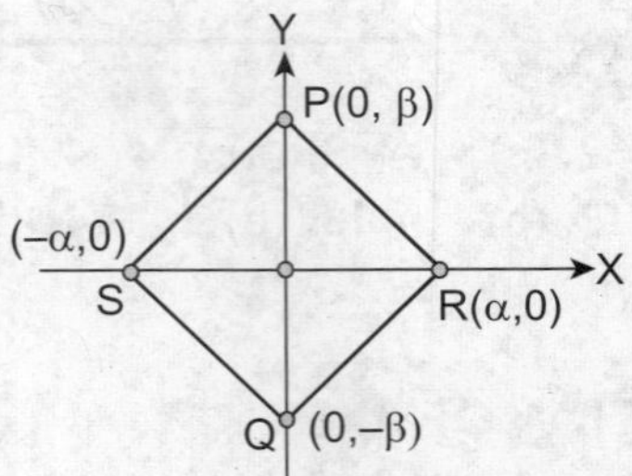

FIGURE 1.107

(d) **Parallelogram:** To choose the general parallelogram, the axes should be oriented in such a manner that the vertices of a parallelogram becomes (as given in the Figure 1.108).

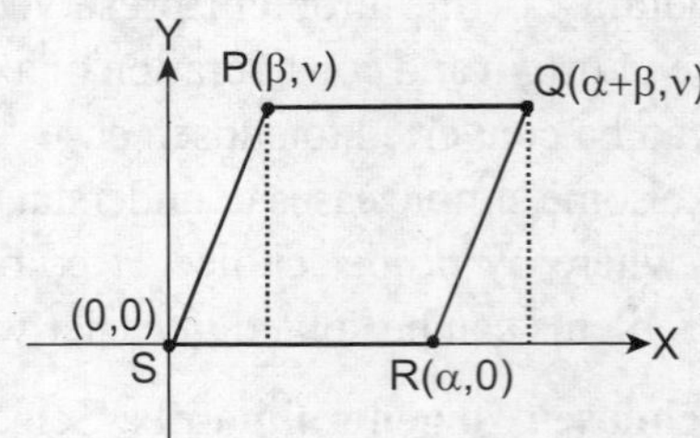

FIGURE 1.108

(e) **General quadrilateral:** To choose a general quadrilateral, the axes should be oriented in such a manner that the vertices of a quadrilateral becomes $(\alpha, 0)$, $(\beta, 0)$, (γ, δ), $(0, \mu)$. (as shown in Figure 1.109)

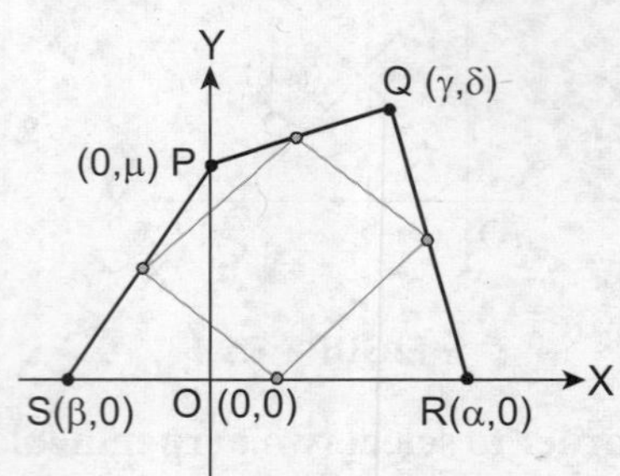

FIGURE 1.109

ILLUSTRATION 88: In any ΔABC prove that $AB^2 + AC^2 = 2(AO^2 + CO^2)$,where O is the middle point of side BC.

SOLUTION: We take O as the origin and OX and OY as the x and y axes respectively. Let $BC = 2\alpha$, then $B \equiv (\alpha, 0)$, $C \equiv (-\alpha, 0)$, $A \equiv (\beta, \gamma)$. Considering the ΔABC as shown in given diagram,

we get $AB^2 + AC^2 = (\beta - \alpha)^2 + \gamma^2 + (\beta + \alpha)^2 + \gamma^2$

$= 2(\alpha^2 + \beta^2 + \gamma^2 + \alpha\beta - \alpha\beta) = 2(\alpha^2 + \beta^2 + \gamma^2)$ ....(i)

and $AO^2 = \beta^2 + \gamma^2$ and $OC^2 = \alpha^2$

$\Rightarrow AO^2 + CO^2 = \alpha^2 + \beta^2 + \gamma^2$ .....(ii)

$\therefore$ From (i) and (ii), $AB^2 + AC^2 = 2(AO^2 + OC^2)$

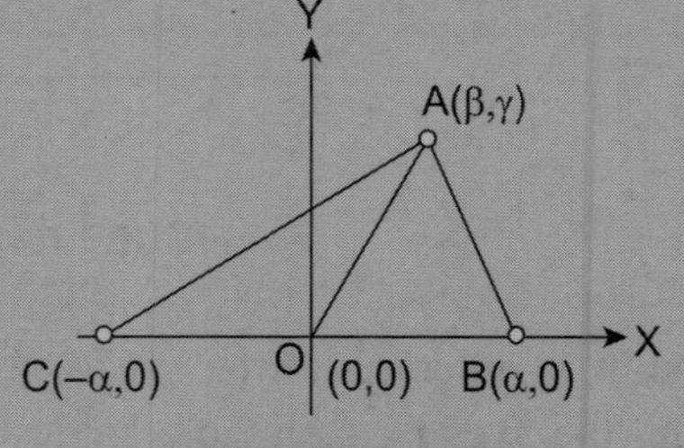

FIGURE 1.110

ILLUSTRATION 89: Prove that the mid-point of the hypotenuse of a right angled triangle is equidistant from all three vertices of the triangle.

SOLUTION: Consider the right angled ΔOAB, where $O(0, 0)$ as shown in figure. Let M be the mid-point of hypotenuse with terminal points $A(\alpha, 0)$, $B(0, \beta)$.

So co-ordinates of M are $(\alpha/2, \beta/2)$

$\Rightarrow MA = MB = \frac{1}{2}\sqrt{\alpha^2+\beta^2}$

and $OM = \sqrt{(\alpha/2-0)^2+(\beta/2-0)^2} = \frac{1}{2}\sqrt{\alpha^2+\beta^2}$

Therefore, $MO = MA = MB$

Aliter: $\angle AOB = 90°$

$\Rightarrow$ Circle drawn on AB as diameter must passes through O, consequently, $OM = MA = MB$ as these are all radii of that circle.

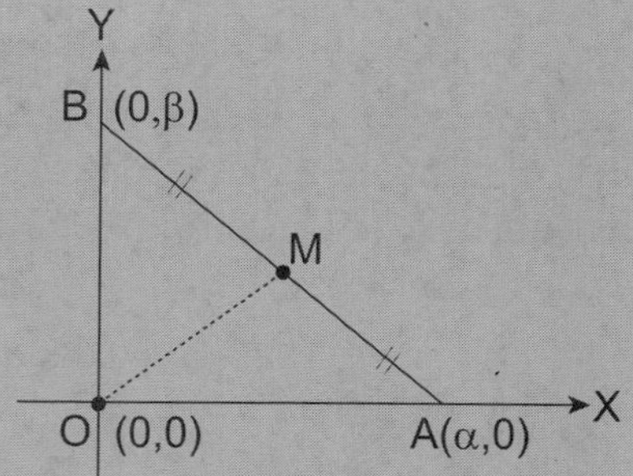

FIGURE 1.111

ILLUSTRATION 90: Show that the triangle, the co-ordinates of whose vertices are given by integers, can never be an equilateral triangle.

SOLUTION: Let $A \equiv (0, 0)$, $B \equiv (a, 0)$ and $C \equiv (b, c)$ be the vertices of an equilateral triangle ABC where a, b, c are positive integers then, $|AB| = |BC| = |CA|$

$\Rightarrow (AB)^2 = (BC)^2 = (CA)^2 \Rightarrow a^2 = (a-b)^2 + c^2 = b^2 + c^2$

From first two members, we get $b^2 + c^2 = 2ab$ ...(i)

and taking first and third members, we get $b^2 + c^2 = a^2$ ... (ii)

From (i) and (ii), we get $a = 2b$ ($\because a \neq 0$)

From (ii), $b^2 + c^2 = (2b)^2$ or $c^2 = 3b^2$ or $c = \pm b\sqrt{3}$

which is impossible, since b and c are positive integers.

Aliter: $\tan A = \frac{c}{a/2} \Rightarrow \tan 60^0 = \frac{2c}{a} \Rightarrow 2c = a\sqrt{3} \Rightarrow$ which is a contradiction, as a and c are positive integers.

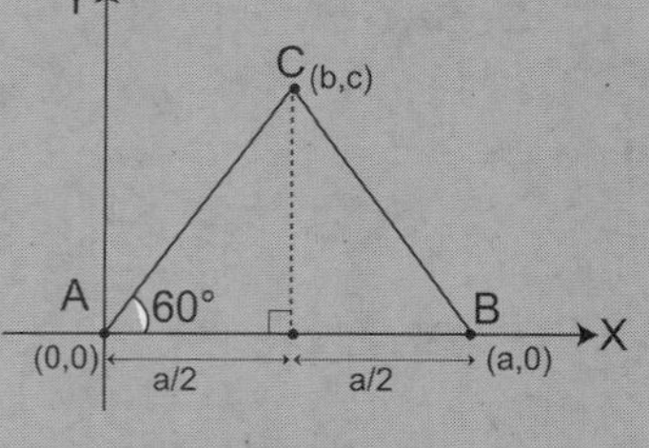

FIGURE 1.113

ILLUSTRATION 91: Show that the line segment joining the mid-points of any two sides of a triangle is half of the third side.

SOLUTION: We take O as the origin and OX and OY as the x and y axis respectively.

Let $BC = 2a$, then $B \equiv (-a, 0)$, $C \equiv (a, 0)$

Let $A \equiv (b, c)$ if E and F are the mid-points of sides AC and AB respectively,

then $E \equiv \left(\frac{a+b}{2}, \frac{c}{2}\right)$ and $F \equiv \left(\frac{b-a}{2}, \frac{c}{2}\right)$

Now $FE = \sqrt{\left(\frac{a+b}{2}-\frac{b-a}{2}\right)^2+\left(\frac{c}{2}-\frac{c}{2}\right)^2} = a$

$= \frac{1}{2}(2a) = \frac{1}{2}(BC)$

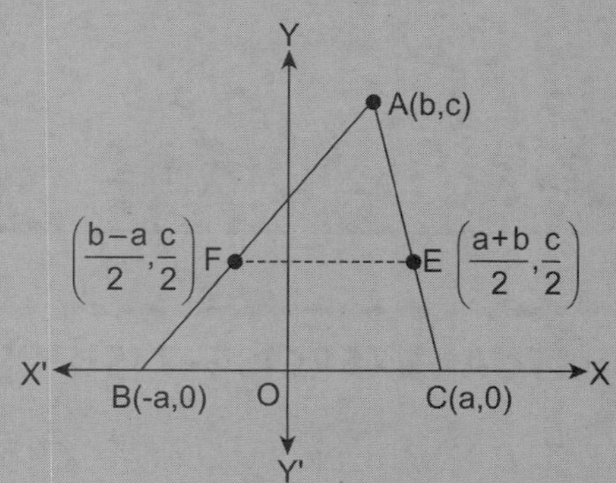

FIGURE 1.112

Hence the line segment joining the mid-points of any two sides of a triangle is half of the third side.

ILLUSTRATION 92: Prove that the area of a triangle is four times the area of the triangle formed by joining the mid-points of its sides.

SOLUTION: Let ABC be a triangle. Let O be the middle point of BC and $BC = 4\alpha$. We take O as the origin and OC along positive x-axis. Then $B \equiv (-2\alpha, 0)$, $C \equiv (2\alpha, 0)$, $A \equiv (2\beta, 2\gamma)$.

Let E and F be the middle points of AC and AB, respectively.

Then $O \equiv (0, 0)$, $E \equiv (\beta + \alpha, \gamma)$, $F \equiv (\beta - \alpha, \gamma)$

Now area of

$$\Delta ABC = \frac{1}{2}(\text{base} \times \text{altitude}) = \frac{1}{2}|8\alpha\gamma| = 4|\alpha\gamma| \quad ...(i)$$

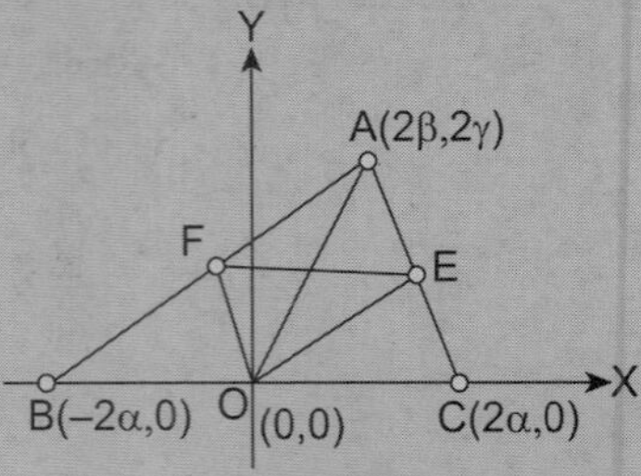

FIGURE 1.113

$$\text{Area of } \Delta OEF = \text{mod of } \frac{1}{2}\begin{vmatrix} 0 & 0 & 1 \\ \beta+\alpha & \gamma & 1 \\ \beta-\alpha & \gamma & 1 \end{vmatrix} = \left|\frac{1}{2}[(\beta+\alpha)\gamma - (\beta-\alpha)\gamma]\right| = |\alpha\gamma| \quad ..(ii)$$

From (i) and (ii) it follows that area of $\Delta ABC = 4$ area of $\Delta\ OEF$.

Aliter: *FE* is mid parallel to base *BC*.

$\Rightarrow FE = |2\alpha|$; Altitude $= |\gamma|$

$$\Rightarrow \text{Area } \Delta OEF = \frac{1}{2}|2\alpha\gamma| = |\alpha\gamma| = \frac{1}{4}|4\alpha\gamma| = \frac{1}{4}(\text{Area } \Delta ABC)$$

ILLUSTRATION 93: If G be the centroid of the ΔABC and O be any other point in the plane of the triangle ABC, then prove that $OA^2 + OB^2 + OC^2 = GA^2 + GB^2 + GC^2 + 3GO^2$.

SOLUTION: Let G be the origin and GO be along positive x-axis. Let $O \equiv (\alpha, 0)$, $A \equiv (x_1, y_1)$ $B \equiv (x_2, y_2)$ and $C \equiv (x_3, y_3)$

Now, L.H.S. $= OA^2 + OB^2 + OC^2$

$$= (x_1 - \alpha)^2 + y_1^2 + (x_2 - \alpha)^2 + y_2^2 + (x_3 - \alpha)^2 + y_3^2$$

$$= x_1^2 + x_2^2 + x_3^2 + y_1^2 + y_2^2 + y_3^2 + 3\alpha^2 - 2\alpha(x_1 + x_2 + x_3)$$

$$= x_1^2 + x_2^2 + x_3^2 + y_1^2 + y_2^2 + y_3^2 + 3\alpha^2$$

$$\because \frac{x_1 + x_2 + x_3}{3} = 0 \text{ (}x\text{-co-ordinate of } G\text{)}$$

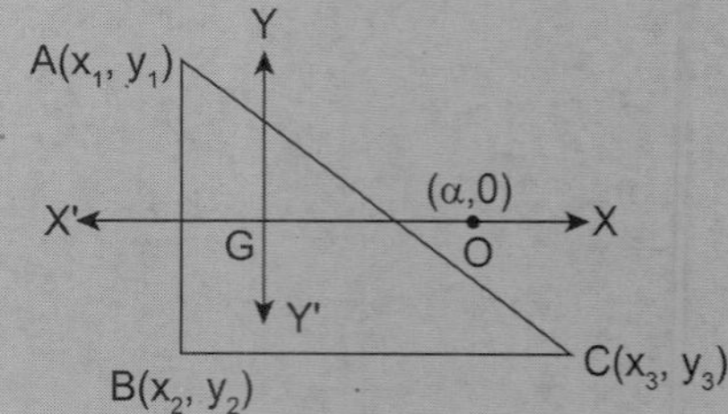

FIGURE 1.114

and RHS $= GA^2 + GB^2 + GC^2 + 3GO^2 = x_1^2 + y_1^2 + x_2^2 + y_2^2 + x_3^2 + y_3^2 + 3\alpha^2$

$\therefore$ L.H.S = R.H.S

TEXTUAL EXERCISE–7 (SUBJECTIVE)

1. If P, Q, R divide the sides BC, CA and AB of ΔABC in the same ratio, prove that the centroid of the triangles ABC and PQR coincide.
2. Prove that in any triangle four times the sum of the squares of the medians is equal to three times the sum of the squares of the sides.
3. Prove that the medians of triangles are concurrent and the point of concurrency divides the median in the ratio 2:1.
4. Prove that if all the medians of a triangle are equal,then the triangle is equilateral.

5. Prove that the line segments joining the mid points of any quadrilateral always construct a parallelogram.

6. Prove that the perpendicular bisectors of sides of a triangle are concurrent.

7. ABC is a given triangle in which $AB = AC$. The sides AB and AC are produced to P and Q respectively such that $BP \cdot CQ = AB^2$. Prove that the line PQ always passes through a fixed point.

8. Use co-ordinate geometry to prove that altitudes in a triangle are concurrent.

■ GEOMETRICAL TRANSFORMATIONS

Geometrical transformation is any geometric operation undergoing through which the co-ordinate of the point changes It is of two types:

(i) Linear transformation
(ii) Non-linear transformation.

(i) Linear transformation: A transformation in which the origin of reference frame doesn't change and the new co-ordinate obtained is linear function of old co-ordinate. i.e., $x' = ax + by$ and $y' = cx + dy$ is called linear transformation. In such stated transformation, the straight line remains straight.

(ii) Non-linear transformation: The remaining transformations are called non-linear transformations.

T$_1$: Reflection of point in x-axis $(x, y) \xrightarrow{T_1} (x, -y)$

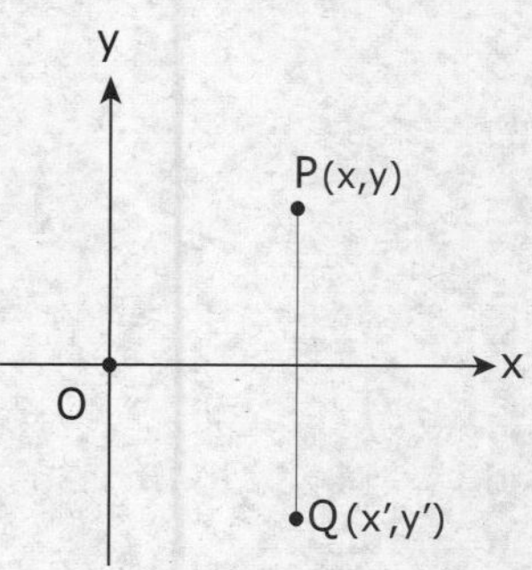

FIGURE 1.115

T$_2$: Reflection of point in y-axis $(x, y) \xrightarrow{T_2} (-x, y)$

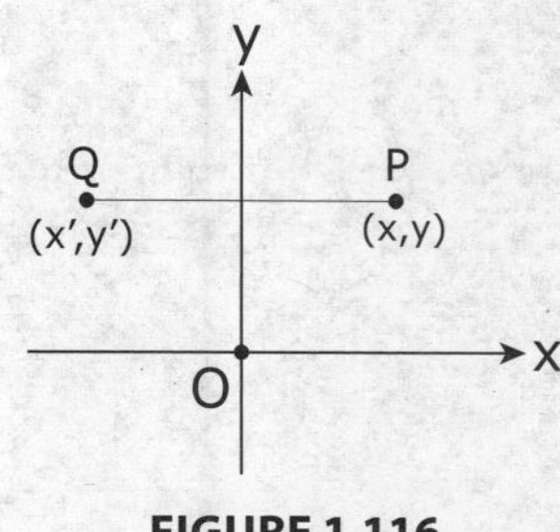

FIGURE 1.116

T$_3$: Reflection of point in origin $(x, y) \xrightarrow{T_3} (-x, -y)$

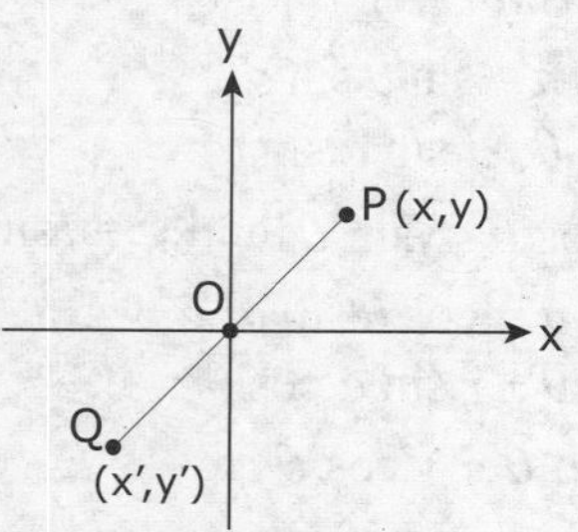

FIGURE 1.117

T$_4$: Reflection of point in the line $y = x$, $(x, y) \xrightarrow{T_4} (y, x)$

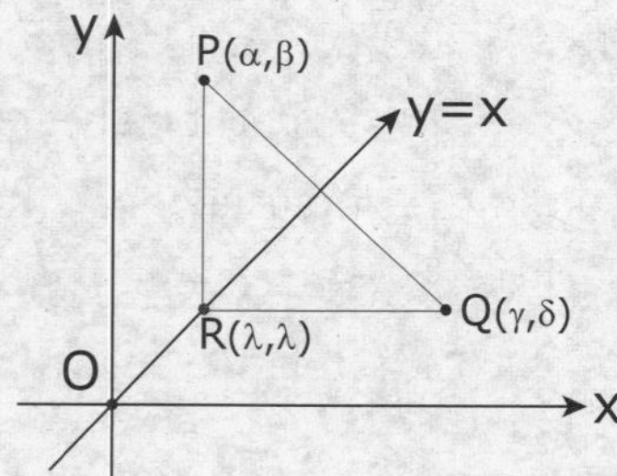

FIGURE 1.118

Proof. Mid-point of PQ lies on $y = x$

$\Rightarrow \alpha + \gamma = \beta + \delta$ ...(i)

$\therefore$ $PQ \perp$ to line $y = x \Rightarrow \dfrac{\beta - \delta}{\alpha - \gamma} = -1$.

$\Rightarrow \beta - \delta = \gamma - \alpha \Rightarrow \alpha + \beta = \gamma + \delta$ ...(ii)

Adding equation (i) and (ii), we get $2\alpha = 2\delta \Rightarrow \alpha = \delta$

Subtracting equation (i) from (ii), we get $\beta - \gamma = \gamma - \beta$

$\Rightarrow \beta = \gamma$. Thus $(\alpha, \beta) \xrightarrow{T_4} (\beta, \alpha)$

T$_5$: Rotation of point about origin by angle θ

$(x, y) \longrightarrow (x', y') \equiv (x\cos\theta - y\sin\theta, x\sin\theta + y\cos\theta)$

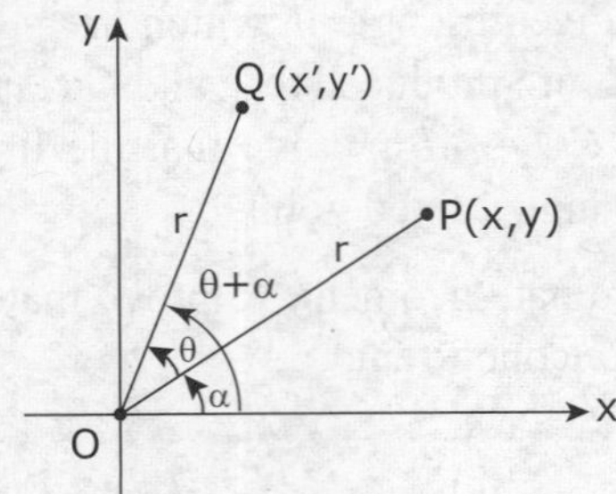

FIGURE 1.119

Proof: $\because x = r\cos\alpha$ and $y = r\sin\alpha$

Also $x' = r\cos(\theta + \alpha)$

$= r[\cos\theta \cos\alpha - \sin\theta \sin\alpha]$

$= r.\cos\alpha.\cos\theta - r.\sin\alpha.\sin\theta$

$\Rightarrow$ $x' = x\cos\theta - y\sin\theta$ and $y' = r\sin(\theta + \alpha)$

$= r[\sin\theta \cos\alpha + \cos\theta \sin\alpha]$

$= r.\cos\alpha.\sin\theta + r.\sin\alpha.\cos\theta$

$\Rightarrow$ $y' = x\sin\theta + y\cos\theta$.

T_6 : Reflection of point in the line $y = x\tan\theta$

$(x, y) \xrightarrow{T_6} (x', y') \equiv (x\cos 2\theta + y\sin 2\theta, x\sin 2\theta - y\cos 2\theta)$

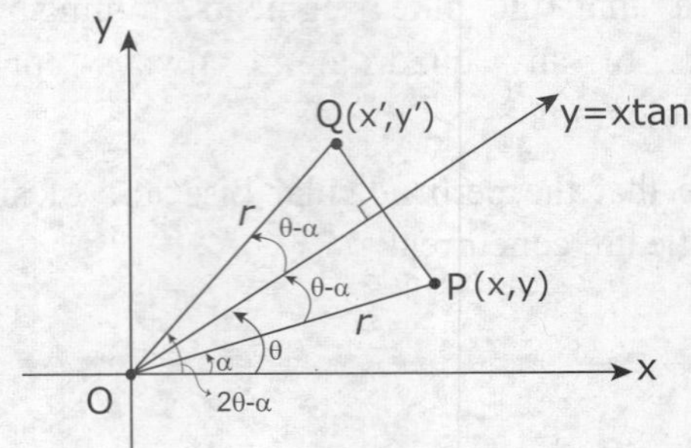

FIGURE 1.120

Proof: $\because$ x = r cos α & y = r sin β

$\Rightarrow$ $x' = r\cos(2\theta - \alpha)$

$= r(\cos\alpha \cos 2\theta + \sin\alpha \sin 2\theta)$

$\Rightarrow$ $x' = r\cos\alpha \cos 2\theta + r\sin\alpha \sin 2\theta$

$= x\cos 2\theta + y\sin 2\theta$ and $y' = r\sin(2\theta - \alpha)$

$= r(\cos\alpha \sin 2\theta - \sin\alpha \cos 2\theta)$

$\Rightarrow y' = x\sin 2\theta - y\cos 2\theta$

ILLUSTRATION 94: If a point $P(3, 2)$ undergoes following transformations in succession.

T_1 : Reflection about origin.

T_2 : Translation by 4 unit along +ve x-axis.

T_3 : Reflection in x-axis.

T_4 : Rotation about origin by 30°. Find the co-ordinates of new point after each transformations.

SOLUTION: After T_1, (3, 2) transforms to (–3, –2)

After T_2, (–3, –2) transforms to (1, –2)

After T_3, (1, –2) transforms to (1, 2)

After T_4, (1, 2) transforms to $x' = 1\cos 30° - 2\sin 30°$ and $y' = 1\sin 30° + 2\cos 30°$

i.e., $((\sqrt{3} - 2)/2, (2\sqrt{3} + 1)/2)$

ILLUSTRATION 95: Find the co-ordinates of new point if (2, 6) is reflected about the straight line $y = \sqrt{3}x$.

SOLUTION: We know that if (x, y) is reflected about the line $y = x\tan\theta$, then the co-ordinates of transformed points are given by (x', y');

where $x' = x\cos 2\theta + y\sin 2\theta$ and $y' = x\sin 2\theta - y\cos 2\theta$.

$\therefore$ (2, 6) when reflected about the line $y = \sqrt{3}x = (\tan 60°)x$, the point transforms to

$x' = 2\cos 120° + 6\sin 120°$

and $y' = 2\sin 120° - 6\cos 120°$

i.e., $((-1 + 3\sqrt{3}), (\sqrt{3} + 3))$

TRANSFORMATION OF AXES

There are two ways of transformation of co-ordinate axes as given below:

(a) By shifting of origin to some other point and keeping the co-ordinate axes parallel to original axes.
(b) Rotation of axes by desirable angle keeping the origin fixed i.e., without changing the origin. Its detailed explanations are as given below:

(a) By shifting of origin to some other point and keeping the co-ordinate axes parallel to original axes

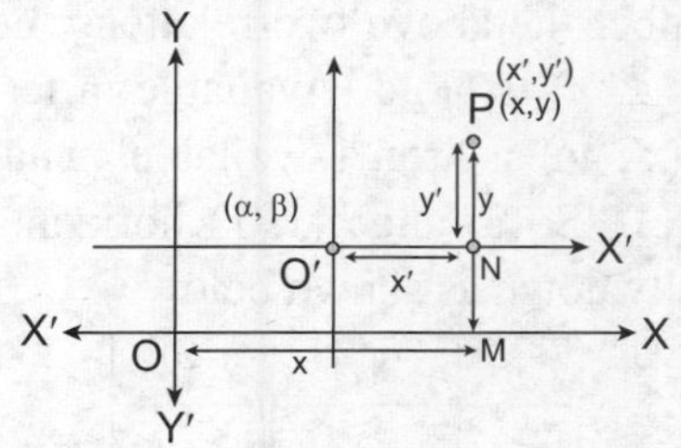

FIGURE 1.121

Let origin O (0, 0) be shifted to a point (α, β) by moving the x-axis and y-axis parallel to themselves. If the co-ordinate of point P with reference to old axes are (x, y) and with respect to new axes be (x', y'). Then from the figure, $O'N = OM - \alpha$ and $PN = PM - MN$.

$$\Rightarrow \begin{cases} x' = x - \alpha \\ y' = y - \beta \end{cases} \Rightarrow P(x', y') = P(x-\alpha,\ y-\beta)$$

And the old co-ordinates of P in terms of new are given by $x = x' + \alpha$ and $y = y' + \beta$

Conclusions:

(i) New co-ordinates of point P in terms of old $x' = x - h$ and $y' = y - k$.

(ii) Old co-ordinates of point P in terms of new $x = x' + h$ and $y = y' + k$.

(iii) The transformation equation of a locus $f(x, y) = 0$ is obtained by replacing x by $x + h$ and y by $y + k$

$$f(x, y) = 0 \xrightarrow[y \to y+k]{x \to x+h} f(x+h,\ y+k) = 0$$

(iv) Given the transformation equation of a locus $g(x, y) = 0$ the original (old) equation is obtained by replacing x by $x - h$ and y by $y - k$.

$$g(x-h,\ y-k) = 0 \xleftarrow[y \to y-k]{x \to x-h} g(x, y) = 0$$

(v) By shifting origin to a suitable point, any two of the following three achieved at a time:

Removal of constant term from equation of locus.

Removal of term containing linear in x.

Removal of term containing linear in y.

ILLUSTRATION 96: If the axes are translated parallel so that the origin is shifted to the point (–2,1), then what will be the new co-ordinates of (4, –5).

SOLUTION: The new co-ordinates: $x' = x - \alpha$ and $y' = y - \beta$;

where (α, β) is the new origin and (x,y) are old co-ordinates of point

$\Rightarrow x' = 4 - (-2) = 6$ and $y' = (-5 - 1) = -6$. So the new co-ordinates are (6, –6).

ILLUSTRATION 97: Transform the following equation of curve $2x^2 + 4xy + 5y^2 - 4x + 22y + 7 = 0$, by shifting the origin to (–2, 3) and translating axes parallel to the original axes.

SOLUTION: We substitute $x = x' - 2$ and $y = y' + 3$, and the equation becomes

$$2(x-2)^2 + 4(x-2)(y+3) + 5(y+3)^2 - 4(x-2) + 22(y+3) + 7 = 0,$$

We get the transformed equation $2x^2 + 4xy + 5y^2 + 44y + 110 = 0$.

ILLUSTRATION 98: Find the transformed equation of the curve $2x^2 + y^2 - 3x + 5y - 8 = 0$, when the origin is transferred to the point (–1, 2) without changing the direction of axes.

SOLUTION: If $P(x, y)$ be any point on the curve and (x', y') be the co-ordinates of P with respect to new axes, then $x = x' - 1$, $y = y' + 2$.

Hence the transformed equation will be $2(x' - 1)^2 + (y' + 2)^2 - 3(x' - 1) + 5(y' + 2) - 8 = 0$

or $2(x'^2 - 2x' + 1) + y'^2 + 4y' + 4 - 3x' + 3 + 5y' + 10 - 8 = 0$ or $2x'^2 + y'^2 - 7x' + 9y' + 11 = 0$.

Hence the transformed equation of curve is $2x^2 + y^2 - 7x + 9y + 11 = 0$.

ILLUSTRATION 99: Shift the origin to a suitable point so that the equation $y^2 + 4y + 8x - 2 = 0$ after transformation will not has terms containing y and the constant.

SOLUTION: Let the origin be shifted to the point (h, k) and let $P(x, y)$ be any point on the curve and (x', y') be the co-ordinate of P with respect to new axes, then $x = x' + h$ and $y = y' + k$. Hence, new equation will be

$(y' + k)^2 + 4(y' + k) + 8(x' + h) - 2 = 0 \Rightarrow y'^2 + (2k + 4)y' + 8x + (k^2 + 4k + 8h - 2) = 0$

Thus new equation of curve will be $y^2 + (2k + 4)y + 8x + (k^2 + 4k + 8h - 2) = 0$

Since this equation is required to be free from the term containing 'y' and the constant, we have to take $k = -2$ and $h = 3/4$. Hence the point to which the origin be shifted is $(3/4, -2)$

(b) Rotation of axes by desirable angle keeping the origin fixed i.e., without changing the origin

Let OX and OY be the old axes and OX' and OY' be the new axes obtained by rotating the old axes through an angle θ in anti-clockwise sense about origin; then the old co-ordinates of $P(x, y)$ with respect to new co-ordinate axes can be found as follows.

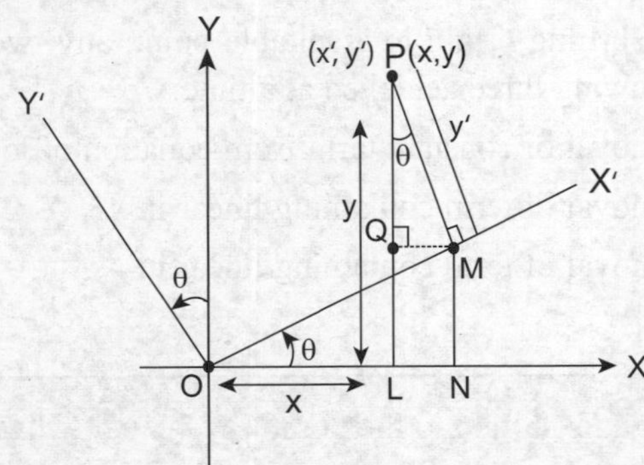

FIGURE 1.122

$x = OL = ON - NL,\ y = PL = PQ + QL$

$$\Rightarrow \left.\begin{aligned} x &= x'\cos\theta - y'\sin\theta \\ y &= y'\cos\theta + x'\sin\theta \end{aligned}\right\} \quad .....(i)$$

Similarly, the co-ordinates of point with respect to new axes in terms of old co-ordinates x, y, can be obtained by solving (i) for x', $y'$and given by $\left.\begin{aligned} x' &= x\cos\theta + y\sin\theta \\ y' &= -x\sin\theta + y\cos\theta \end{aligned}\right\}$ ..(ii)

Aid to memory

(i) Light-heavy method (An interesting way to remember above transformation)

To remember the above two relations. Let us take the case as: "Old (x, y) is Gold (Available in less quantity so light)", new (x', y') is iron (Available in larger quantity so heavy metal). So arrange (x, y) horizontally (up) and (x', y') vertically down as shown below:

	x↓	y↓
x' →	cos θ	sin θ
y' →	−sin θ	cos θ

FIGURE 1.123

(To remember the table, take first row as Civil Services i.e., C for cos θ and S for sin θ and IInd row can be obtained as derivative of Ist row with respect to θ.)

To find x and y in terms of x' and y', look in the columns of x and y:

	↓ x	↓ y
x'	cos θ ↓	sin θ ↓
y'	−sin θ	cos θ

FIGURE 1.124

$\therefore \quad x = x'\cos\theta - y'\sin\theta,\ y = x'\sin\theta + y'\cos\theta$

For finding x and y in terms of x and y, look in the row of x', y'; $x' = x\cos\theta + y\sin\theta,\ y' = -x\sin\theta + y\cos\theta$

	x	y
x'→	cos θ →	sin θ
y'→	−sin θ →	cos θ

FIGURE 1.125

REMARKS

1. *x', y' are new co-ordinates where as x, y are old co-ordinates.*
2. *For the rotation of axes in anti-clockwise, θ is positive* and *if clockwise, then θ is negative.*

ILLUSTRATION 100: If the axes are turned through 135° anti-clockwise, then find the transformed equation of the curve $4x^2 + 4y^2 + 3xy = 3$. As the given equation is a relation between old co-ordinates (x, y) of the point in original set of axes. To obtain the relation in new co-ordinates after rotation of axes by 135° anti-clockwise, we have to replace old co-ordinates by, a function of new co-ordinates in the given equation.

SOLUTION: Here $\theta = 135°$ so $\sin\theta = \frac{1}{\sqrt{2}}$; $\cos\theta = -\frac{1}{\sqrt{2}}$

Replacing the old co-ordinates in term of new (x, y) by $(x \cos\theta - y \sin\theta;\ x\sin\theta + y\cos\theta)$, that is

$$= \left(-\frac{x}{\sqrt{2}} - \frac{y}{\sqrt{2}}, \frac{x}{\sqrt{2}} - \frac{y}{\sqrt{2}}\right) = \left(\frac{-1}{\sqrt{2}}(x+y), \frac{1}{\sqrt{2}}(x-y)\right)$$

$\therefore$ Transformed equation will be $4.\frac{1}{2}(x+y)^2 + 4.\frac{1}{2}(x-y)^2 - \frac{3}{2}(x^2 - y^2) = 3$

$\Rightarrow$ $4(x+y)^2 + 4(x-y)^2 - 3(x^2 - y^2) = 6$

$\Rightarrow$ $4\left[2x^2 + 2y^2\right] - 3x^2 + 3y^2 = 6$

$\Rightarrow$ $5x^2 + 11y^2 = 6$ which is an equation of standard ellipse with centre at origin and major axis and minor axis of length $\frac{2\sqrt{6}}{\sqrt{5}}, \frac{2\sqrt{6}}{\sqrt{11}}$, respectively.

(ii) Complex number method As any point P having co-ordinate (x, y) is regarded as a complex number $z(x + iy)$ treating x-y cartesian plane as Argand plane, therefore the transformed co-ordinates (x', y') of the point $P(x, y)$ can also be obtained by applying the rotation theorem of complex numbers, which states that "if a complex number $x + iy$ (refer as point (x, y)) is multiplied by another complex number $\cos\theta + i \sin\theta$, then the resulting complex number $x' + iy'$ (refer as point (x', y')) is obtained by geometrically rotating $x + iy$ about origin by an angle θ in anti-clockwise sense". (For proof look into the chapter complex number.).

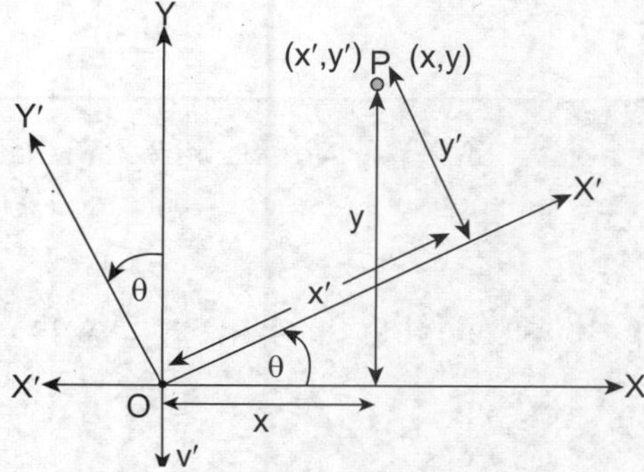

FIGURE 1.126

Rotating the axes by angle θ in anti-clockwise sense has same effect over the co-ordinates of the point, as keeping the axes fixed and rotating the point about origin by an angle θ in clockwise sense. So the co-ordinate of new point can be obtained by rotating the point (x, y) about origin in clockwise sense by angle θ. (i.e., $-\theta$).

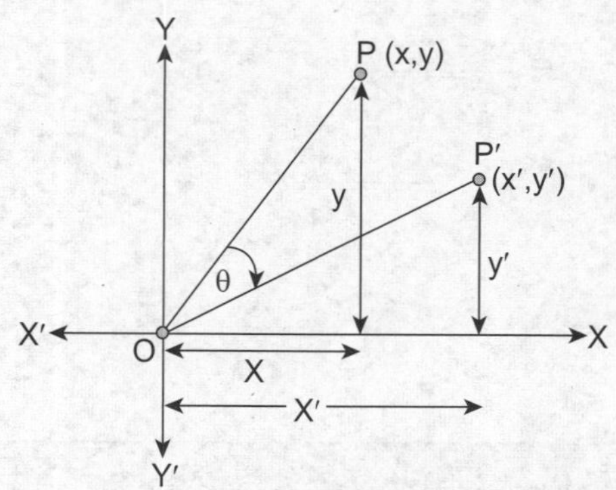

FIGURE 1.127

$\Rightarrow$ $(x' + iy') = (x + iy)(\cos\theta - i \sin\theta)$ ....(i)

$\Rightarrow$ $(x' + iy') = (x \cos\theta + y \sin\theta) + i(-x \sin\theta + y\cos\theta)$.

On comparing real and imaginary parts, we get the new co-ordinates in terms of old as below.

$$\begin{cases} x' = x\cos\theta + y\sin\theta \\ y' = -x\sin\theta + y\cos\theta \end{cases}$$

from equation (i) we can also write

$(x + iy) = (x' + iy')\, e^{i\theta}$

$\Rightarrow (x + iy) = (x' + iy')\,(\cos\theta + i\sin\theta)$

$= (x'\cos\theta - y'\sin\theta) + i\,(x'\sin\theta + y'\cos\theta)$

On comparing real and imaginary parts, we get the new co-ordinates in terms of new as below.

$$\begin{cases} x = x'\cos\theta - y'\sin\theta \\ y = x'\sin\theta + y'\cos\theta \end{cases}$$

(c) Rotation of axes by desirble angle and shifting the origin to some other point

Given a point $P(x, y)$ in x-y plane. If origin is shifted to new point (α, β) by translating the axes parallel to original axes and then the axes are rotated about the new origin (α, β) by an angle ϕ in the anti-clockwise sense or first the axes rotated by some angle ϕ and then translated parallel to itself to the point (α, β), then relation between (x, y) and (x', y'), are obtained as given here.

Let the co-odinates of $P(x, y)$ in the new co-ordinate system be $(x'y')$, then $x = \alpha + x'\cos\phi - y'\sin\phi$;

$y = \beta + x'\sin\phi + y'\cos\phi$;

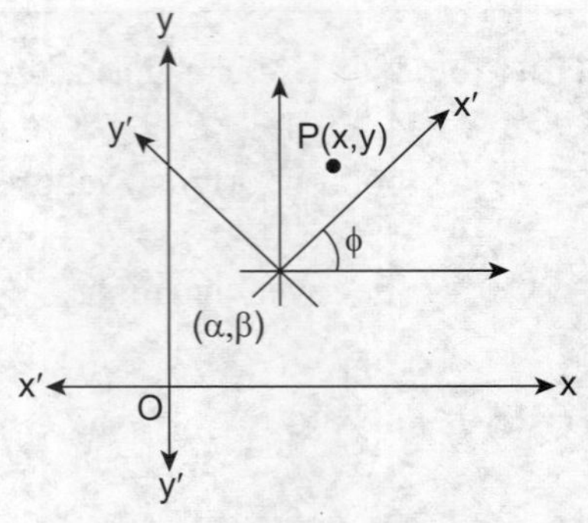

FIGURE 1.128

$$(x, y) \xrightarrow[\text{to } (\alpha,\beta)]{\text{shifting origin}} (x_1, y_1) \xrightarrow{\text{Axes rotated by } \phi} (x', y')$$

$$\Rightarrow \begin{cases} x_1 = x - \alpha \\ y_1 = y - \beta \end{cases} \quad \text{...(i)}$$

$$\Rightarrow \begin{cases} x_1 = x'\cos\phi - y'\sin\phi \\ y_1 = x'\sin\phi + y'\cos\phi \end{cases} \quad \text{...(ii)}$$

Using (i) and (ii) set of equations together, we get

$x = \alpha + x'\cos\phi - y'\sin\phi$

$y = \beta + x'\sin\phi + y'\cos\phi$

NOTE

If two mutually perpendicular lines $ax + by + c = 0$ and $bx - ay + d = 0$ are taken as new axes, then new co-ordinates of $P(x, y)$ are (x', y'), given by

$$x' = \frac{bx - ay + d}{\sqrt{a^2 + b^2}}$$

and $$y' = \frac{ax + by + c}{\sqrt{a^2 + b^2}}$$

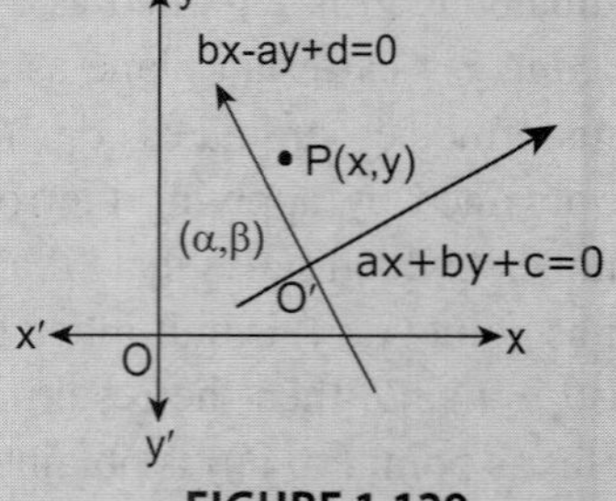

FIGURE 1.129

ILLUSTRATION 101: If (x, y) and (X, Y) be the co-ordinates of the same point referred to two sets of rectangular axes with the same origin and if $ux + vy$, where u and v are independent of x and y become $VX + UY$, show that $u^2 + v^2 = U^2 + V^2$.

SOLUTION: Let the axes be rotated through an angle θ anti-clockwise and if (x, y) be the point with respect to old axes and (X, Y) be the co-ordinates with respect to new axes, then $x + iy = (X + iY)\,e^{i\theta} = (X + iY)(\cos\theta + i\sin\theta)$

then we get $x = X\cos\theta - Y\sin\theta$; $y = X\sin\theta + Y\cos\theta$,

then $ux + vy = u\,(X\cos\theta - Y\sin\theta) + v\,(X\sin\theta + Y\cos\theta)$

$= (u\cos\theta + v\sin\theta)\,X + (-u\sin\theta + v\cos\theta)\,Y = VX + UY$ (given)

On comparing the co-efficient of X and Y, we get $u\cos\theta + v\sin\theta = V$ ..(i)

and $-u\sin\theta + v\cos\theta = U$ ..(ii)

Squaring and adding (i) and (ii), we get $u^2 + v^2 = U^2 + V^2$

ILLUSTRATION 102: Find the transformed equation for the curve $4xy - 3x^2 = a^2$, when the axes are rotated by an angle $\tan^{-1} 2$.

SOLUTION: Given that $\theta = \tan^{-1}2 \Rightarrow \tan\theta = 2$. So, $\cos\theta = \frac{1}{\sqrt{5}}$ and $\sin\theta = \frac{2}{\sqrt{5}}$

$$\Rightarrow x = x'\cos\theta - y'\sin\theta = \frac{x' - 2y'}{\sqrt{5}} \text{ and } y = x'\sin\theta + y'\cos\theta = \frac{2x' + y'}{\sqrt{5}}$$

putting this in given equation, we get $4\left(\frac{x'-2y'}{\sqrt{5}} \cdot \frac{2x'+y'}{\sqrt{5}}\right) - 3\left(\frac{x'-2y'}{\sqrt{5}}\right)^2 = a^2$

$$\Rightarrow 8x'^2 - 12x'y' - 8y'^2 - 3x'^2 - 12y'^2 + 12x'y' = 5a^2$$

$$\Rightarrow 5x'^2 - 20y'^2 = 5a^2 \qquad \Rightarrow x^2 - 4y^2 = a^2$$

ILLUSTRATION 103: Through what angle should the axes be rotated so that the equation $9x^2 - 2\sqrt{3}\,xy + 7y^2 = 10$ may be changed to $3x^2 + 5y^2 = 5$?

SOLUTION: Let the required angle be θ, then replacing (x, y) by $(x\cos\theta - y\sin\theta, x\sin\theta + y\cos\theta)$

$9x^2 - 2\sqrt{3}\,xy + 7y^2 = 10$ transform to

$9(x\cos\theta - y\sin\theta)^2 - 2\sqrt{3}(x\cos\theta - y\sin\theta)(x\sin\theta + y\cos\theta) + 7(x\sin\theta + y\cos\theta)^2 = 10$

$\Rightarrow x^2(9\cos^2\theta - 2\sqrt{3}\sin\theta\cos\theta + 7\sin^2\theta) + 2xy(-9\sin\theta\cos\theta - \sqrt{3}\cos 2\theta + 7\sin\theta\cos\theta) + y^2(9\sin^2\theta + 2\sqrt{3}\sin\theta\cos\theta + 7\cos^2\theta) = 10$

On comparing with $3x^2 + 5y^2 = 5$ (co-efficient of $xy = 0$)

We get $-9\sin\theta\cos\theta - \sqrt{3}\cos 2\theta + 7\sin\theta\cos\theta = 0$

or $\sin 2\theta = -\sqrt{3}\cos 2\theta$ or $\tan 2\theta = -\sqrt{3} = \tan(180° - 60°)$

or $2\theta = 120°$ $\therefore \theta = 60°$.

ILLUSTRATION 104: Prove that if the axes be turned through $\frac{3\pi}{4}$, the equation $x^2 - y^2 = 4$ is transformed to the form $xy = \lambda$. Then find the value of λ.

SOLUTION: Replacing (x,y) by $(x\cos\theta - y\sin\theta, x\sin\theta + y\cos\theta)$;

$\sin\theta = 1/\sqrt{2}$, $\cos\theta = -1/\sqrt{2}$ i.e., $\left(\frac{-1}{\sqrt{2}}(x+y), \frac{1}{\sqrt{2}}(x-y)\right)$

$\therefore x^2 - y^2 = 4 \Rightarrow \frac{1}{2}(x+y)^2 - \frac{1}{2}(x-y)^2 = 4 \Rightarrow (2x)(2y) = 8 \Rightarrow xy = 2 \Rightarrow \lambda = 2$

TEXTUAL EXERCISE–8 (SUBJECTIVE)

1. At what point the origin be shifted, if the co-ordinates of point (4, 5) become (–3, 9) ?

2. What does the equation $(a-b)(x^2+y^2) - 2abx = 0$, become if the origin be moved to the point $\left(\frac{ab}{a-b}, 0\right)$?

3. On shifting the origin to the point (1, –1), the axes remaining parallel to the original axes, the equation of a curve becomes $x^2 + y^2 + 3x - 4y + 2 = 0$. Find its original equation.

4. Find the co-ordinates of the point to which the origin should be shifited so that the transformed equation of the equation $x^2 + 3xy + 4y^2 - 4x - 6y + 5 = 0$ doesn't contain linear terms in x, y. Also find the new equation.

5. Transform the equations w.r.t. axes inclined at 45° to the original axes:
(a) $x^2 - y^2 = a^2$
(b) $17x^2 - 16xy + 17y^2 = 225$
(c) $y^4 + x^4 + 6x^2 y^2 = 2$
(d) $x^2 + 2xy \tan 2\alpha - y^2 = a^2$

6. Find the angle through which the axes may be turned so that the equation $Ax + By + C = 0$ may be reduced to the form x = constant, and determine the value of this constant.

7. State which of the following statements is true:
(a) Constant terms in the equation can be removed by shifting the origin to a suitable point.
(b) The terms containing xy from the equation of any curve can be removed by rotating the axes through a suitable angle without shifting the origin.

8. If (x, y) be the co-ordinates of a point referred to axes OX, OY and (x', y') be the co-ordinates of the same point referred to axes OX', OY' and if $ax^2 + 2hxy + by^2$ in which a, b, h are independent of x, and y, become $a'x'^2 + 2h'x'y' + b'y'^2$, then prove that $a + b = a' + b'$ and $ab - h^2 = a'b' - h'^2$.

9. Find the angle through which the axes be rotated so that the expression $ax^2 + 2hxy + by^2$ may become of the form $a'x'^2 + b'y'^2$.

10. Transform the equation $x^2 + 4xy + y^2 - 2x + 2y + 4 = 0$ in to the form $Y^2/b^2 - X^2/a^2 = 1$.

Answer Keys

1. (7, –4) 2. $(a - b)^2 (x'^2 + y'^2) = a^2b^2$. 3. $x^2 + y^2 + x - 2y - 3 = 0$ 4. (2,0); $x^2 + 4y^2 + 3xy + 1 = 0$
5. (a) $2x'y' + a^2 = 0$ (b) $9x'^2 + 25y'^2 = 225$ (c) $x'^4 + y'^4 = 1$ (d) $(x^2 - y^2) \tan 2x - 2xy = a^2$
6. $\tan^{-1}\left(\frac{B}{A}\right); \frac{-C}{\sqrt{A^2 + B^2}}$ 7. (a) true (b) true 9. $\frac{1}{2}\cot^{-1}\left(\frac{a-b}{2h}\right)$ 10. $\frac{Y'^2}{6} - \frac{X'^2}{2} = 1$.

TEXTUAL EXERCISE-7 (OBJECTIVE)

1. If the origin is shifted to the point (1, –2) without rotation of axes, then equation
(i) $2x^2 + y^2 - 4x + 4y = 0$ becomes
(a) $x^2 + y^2 = 6$ (b) $2x^2 + y^2 = 6$
(c) $x^2 + 2y^2 = 6$ (d) None of these
(ii) $y^2 - 4x + 4y + 8 = 0$ becomes
(a) $y^2 = 4x$ (b) $y = 4x$
(c) $y^2 = -4x$ (d) None of these

2. In order to make the equation $y^2 + 4y + 8x - 2 = 0$ free from constant term and term of y the origin of the co-ordinate system should be shifted to
(a) (3/4, – 2) (b) (1/4, – 2)
(c) (3/4, – 1) (d) None of these

3. The equation $2x^2 + 4xy - 5y^2 + 20x - 22y - 14 = 0$ when refered to rectangular axes through the point (– 2, – 3) and new axes being inclined at an angle of 45° with the old, transforms to
(a) $x^2 - 14xy - 7y^2 - 2 = 0$
(b) $x^2 + 14xy - 7y^2 - 2 = 0$
(c) $x^2 - 14xy + 7y^2 + 2 = 0$
(d) None of these

4. Without changing the direction of co-ordinate axes, origin is transferred to (h, k), so that the linear (one degree) terms in the equation $x^2 + y^2 - 4x + 6y - 7 = 0$ are eliminated. Then the point (h, k) is
(a) (3, 2) (b) (–3, 2)
(c) (2, – 3) (d) None of these

5. If the expression $x^2 + 4xy + y^2$ transformed to $Ax^2 + By^2$ by rotation of axes through an angle $\theta \left(0 \le \theta \le \frac{\pi}{2}\right)$, then θ is equal to
(a) $\theta = \pi/6$ (b) $\theta = \pi/4$
(c) $\theta = \pi/3$ (d) None of these

6. The point (4,1) undergoes the following three transformations successively:
   I. reflection about the line $y = x$
   II. translation through a distance 2 units along the positive direction of x – axis.
   III. rotation through an angle $\pi/4$ about the origin in the anti-clockwise direction.

   Then the final position of the point is given by the co-ordinates.
   (a) $(1/\sqrt{2}, 7/\sqrt{2})$
   (b) $(-\sqrt{2}, 7\sqrt{2})$
   (c) $(-1/\sqrt{2}, 7/\sqrt{2})$
   (d) $(\sqrt{2}, 7\sqrt{2})$

7. The point (4,1) undergoes the following two sucessive transformations
   (i) Reflection about the lline $y = x$
   (ii) Translation through a distance 2 units along the positive x-axis.
   (iii) Then the final co-ordinates of the point are
   (a) (4, 3) (b) (3, 4)
   (c) (1, 4) (d) (7/2, 7/2)

Answer Keys

1. (i) (b) (ii) (a) **2.** (a) **3.** (a) **4.** (c) **5.** (b) **6.** (c) **7.** (b)

LOCUS

Locus of a moving point is defined as the path traced out by that point under the influence of one or more than one given geometrical conditions and if a point moves satisfying the given condition(*s*), then it describes a definite curve, and equation that relates the co-ordinates x and y of the moving points is called the equation of locus/equation of such curve.

Thus we conclude as, 'The equation to a curve is a mathematical relation between the x and y co-ordinates of any point on the curve and which holds for every other point on the curve'. That is:

(i) If a point moves in a plane under the geometrical condition that its distance from a fixed point O in the same plane is always equal to a constant quantity a, then the curve traced out by the moving point will be a circle with centre at O and radius a. Thus locus of the point is a circle with centre O and radius a. Thus the equation of locus is $x^2 + y^2 = a^2$.

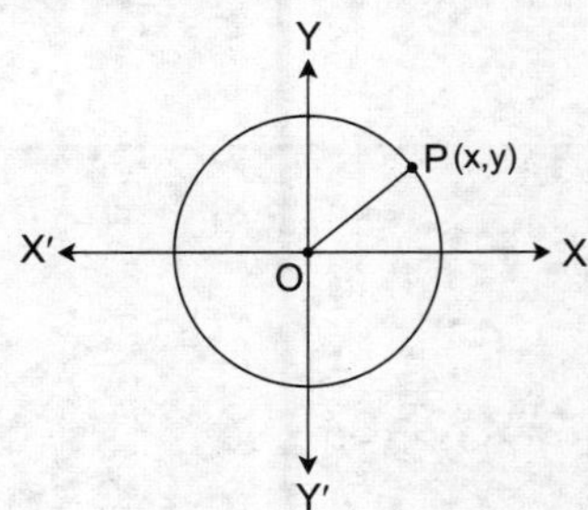

FIGURE 1.130

(ii) The perpendicular bisector of a line segment AB (where $A(a, 0)$ and $B(0, b)$) is the locus of a point moving on the plane OAB, which is equidistant from the end points A and B.

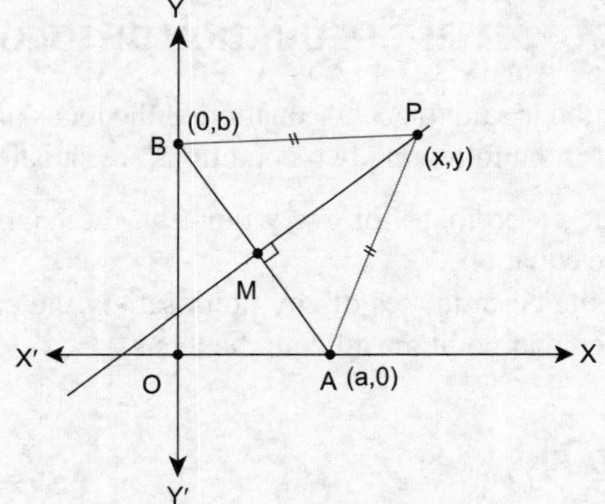

FIGURE 1.131

Equation of locus of $P(x, y)$ is given by $(PA)^2 = (PB)^2$

$\Rightarrow (x - a)^2 + y^2 = x^2 + (y - b)^2$

$\Rightarrow -2ax + a^2 = -2by + b^2$

$\Rightarrow ax - by = \dfrac{a^2 - b^2}{2}$

(iii) Angle bisector of an angle is the locus of a point moving on the plane of angle which is equidistant from two arms of angle. In the chapter straight lines, we will learn how to find the equation of angle bisectors of any given pair of straight lines.

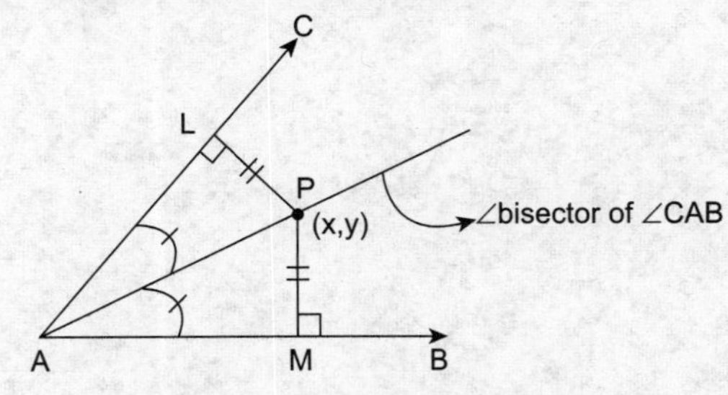

FIGURE 1.132

(iv) Parabola is the locus of a point moving on a plane so that its distance from a fixed point (focus) is always equal to its distance from a fixed straight line (directrix) in the same plane.

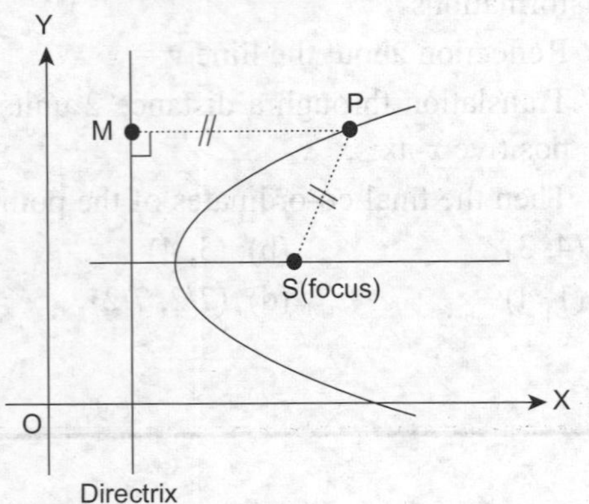

FIGURE 1.133

PROPERTIES OF EQUATION OF LOCUS

An equation is said to be the equation of the locus of a moving point if the following two conditions are satisfied.

(i) The co-ordinates of every point on the locus satisfy the equation.
(ii) If the co-ordinates of any point satisfy the equation, then that point must lie on the locus.

That is, if $x^2 + y^2 = a^2$ is the equation of a locus and (x_1, y_1), (x_2, y_2), (x_3, y_3) are the points on the locus, then $x_1^2 + y_1^2 = a^2$, $x_2^2 + y_2^2 = a^2$ and $x_3^3 + y_3^3 = a^2$, and if $\alpha^2 + \beta^2 = a^2$, then point (α, β) must lie on the locus. Consequently, no point of the path should be left out (i.e., excluded from the eqaution) and equation shall never include a point that actually does not lie on the path of moving point.

METHOD TO FIND OUT THE LOCUS

In order to find out the equation of locus, we follow the steps as given below:

Step I: Let $(h\ k)$ be the co-ordinates of the moving point say P.

Step II: Now apply the geometrical conditions on h, k. This gives a relation/relations between h and k including some parameters.

Step III: Eliminate the parameters to get a relation between h and k.

Step IV: Now replace h by x and k by y in the eliminant and resulting equation thus obtained would be the equation of the locus.

REMARKS

1. *Sometimes the co-ordinates of the moving point itself is taken as (x, y). But this should be done only when no equation is given in question* and *co-ordinate of no point is given as (x, y). In this case, the relation in x* and *y can be directly obtained by eliminating the variable.*
2. *If x* and *y co-ordinates of the moving point are given in terms of a third variable (called the parameter, e.g., t, θ, ϕ, α, β etc.), eliminate the parameter to obtain the relation in x* and *y and simplify this relation. This will give the required locus.*
3. *Make suitable choice of the origin* and *the axes if co-ordinates of no point* and *equation of no curve is given in the question.*

ILLUSTRATION 105: Find the locus of a point which moves such that its distance from the point (0, 0) is twice its distance from the y-axis.

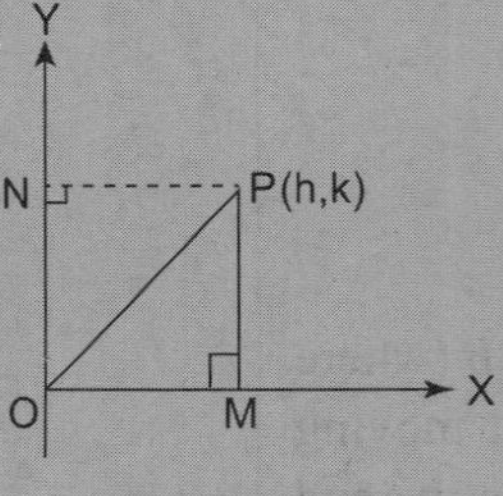

FIGURE 1.134

SOLUTION: Let $P(h, k)$ be the co-ordinates of the point whose locus is required. By hypothesis,

$|OP| = 2|PN| \Rightarrow \sqrt{h^2 + k^2} = 2|h|$

$\Rightarrow 3h^2 - k^2 = 0$

Replacing (h, k) by (x, y), we get $3x^2 - y^2 = 0$, which is the required locus of P.

ILLUSTRATION 106: The locus of the point P equidistant from the points (x_1, y_1) and (x_2, y_2) is $(x_1 - x_2)x + (y_1 - y_2)y + c = 0$, then find the value of c.

SOLUTION: Let the point P has co-ordinates (x, y) and A and B have (x_1, y_1) and (x_2, y_2) respectively, then

$PA = PB \Rightarrow PA^2 = PB^2$

$\Rightarrow (x - x_1)^2 + (y - y_1)^2 = (x - x_2)^2 + (y - y_2)^2$

$\Rightarrow 2x(x_1 - x_2) + 2y(y_1 - y_2) + (x_2^2 + y_2^2 - x_1^2 - y_1^2) = 0$

$\Rightarrow x(x_1 - x_2) + y(y_1 - y_2) + \frac{1}{2}(x_2^2 + y_2^2 - x_1^2 - y_1^2) = 0$ ...(i)

Comparing with given equation of locus, we get $c = 1/2\left(x_2^2 + y_2^2 - x_1^2 - y_1^2\right)$

ILLUSTRATION 107: (a) A point moves so that the algebraic sum of its distances from two given perpendicular axes is equal to a constant quantity a; find the equation of its locus.

(b) The sum of the squares of the distances of a moving point from the two fixed points $(a, 0)$ and $(-a, 0)$ is equal to a constant quantity $2c^2$. Find the equation of its locus when $c^2 > a^2$.

SOLUTION: (a) Take the two straight lines as the axes of co-ordinates. Let (x, y) be any point satisfying the given condition. Then we have $x + y = a$. Being the relation connecting the co-ordinates of any point on the locus, is the equation of the locus. Being a first degree equation in x and y represents a straight line.

It will be found in the next chapter that this equation represents a straight line.

(b) Let (x, y) be any position of the moving point. Then the condition of the question gives $\{(x - a)^2 + y^2\} + \{(x + a)^2 + y^2\} = 2c^2$ i.e., $x^2 + y^2 = c^2 - a^2$

This being the relation between the co-ordinates of any and every point that satisfies the given condition is the equation of the required locus.

This equation tells us that the square of the distance of the point (x, y) from the origin is constant and equal to $c^2 - a^2$ and therefore the locus of the point is a circle whose centre is the origin and radius $\sqrt{c^2 - a^2}$

ILLUSTRATION 108: If the co-ordinates of a variable point P be $(a \cos\theta, b \sin\theta)$, where θ is a variable quantity, find the locus of P.

SOLUTION: Let $P \equiv (x, y)$

According to question, $x = a\cos\theta$ ...(i) $y = b\sin\theta$ ...(ii)

From (i), $\cos^2\theta = \frac{x^2}{a^2}$ and from (ii) $\sin^2\theta = \frac{y^2}{b^2}$

Now to eliminate the parameter θ, adding the above equations, we get

$\frac{x^2}{a^2} + \frac{y^2}{b^2} = \cos^2\theta + \sin^2\theta$, therefore $\frac{x^2}{a^2} + \frac{y^2}{b^2} = 1$ is the required equation of locus.

ILLUSTRATION 109: A stick of length λ rests against the floor and the wall of a room. If the rod begins to slide on the floor, find the locus of a point which divides the rod in the ratio 2 : 1

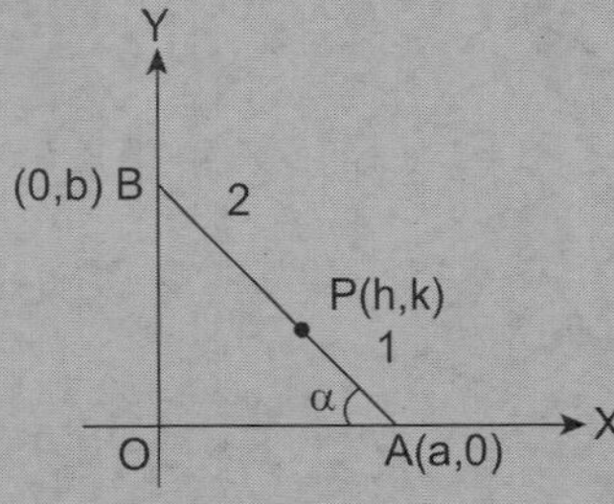

FIGURE 1.135

SOLUTION: Let the cross-section of the floor and wall be taken as the co-ordinate axes and AB be one of the position of the stick as shown in figure. Let the point P be (h, k) and the co-ordinates of A and B are $(a, 0)$ and $(0, b)$, respectively.

But given $|AB| = \lambda$

$\Rightarrow (AB)^2 = \lambda^2 \qquad \Rightarrow a^2 + b^2 = \lambda^2$

Also, $h = \dfrac{2a+0}{3} \qquad \Rightarrow a = \dfrac{3h}{2}$ and $k = \dfrac{2(0)+1.b}{3} \Rightarrow \qquad b = 3k$

$\because a^2 + b^2 = \lambda^2 \Rightarrow \left(\dfrac{3h}{2}\right)^2 + (3k)^2 = \lambda^2 \qquad \Rightarrow h^2 + 4k^2 = \left(\dfrac{2\lambda}{3}\right)^2$

Replacing h by x and k by y, we get $x^2 + 4y^2 = \left(\dfrac{2\lambda}{3}\right)^2$ as the equation of required locus.

Aliter: Let the angle OAB be α, therefore $OA = \lambda \cos\alpha$ and $OB = \lambda \sin\alpha$

Applying section formula, the co-ordinates of P can be written as $h = \dfrac{2\lambda\cos\alpha}{3}$ and $k = \dfrac{\lambda\sin\alpha}{3}$.

$\Rightarrow \sin\alpha = \dfrac{3k}{\lambda}$ and $\cos\alpha = \dfrac{3h}{2\lambda}$

But we know that $\sin^2\alpha + \cos^2\alpha = 1$, therefore $\dfrac{9k^2}{\lambda^2} + \dfrac{9h^2}{4\lambda^2} = 1 \Rightarrow h^2 + 4k^2 = \left(\dfrac{2\lambda}{3}\right)^2$

Replacing h by x and k by y, we get $x^2 + 4y^2 = \left(\dfrac{2\lambda}{3}\right)^2$.

REMARK

Point (a,b) will lie on the curve f(x, y) = 0 if and only if f(a, b) = 0. If two equations represent the same locus (curve), then the corresponding co-efficients in the two equations are proportional.

ILLUSTRATION 110: If the equations $ax^2 + 2hxy + by^2 = 0$ and $y^2 - (m_1 + m_2)\, xy + m_1 m_2 x^2 = 0$ represent the same curve, find $m_1 + m_2$ and $m_1 m_2$.

SOLUTION: Given equations are $ax^2 + 2hxy + by^2 = 0$ ..(i)

and $m_1 m_2 x^2 - (m_1 + m_2)\, xy + y^2 = 0$ ...(ii)

Since equations (i) and (ii) represent the same curve, therefore equations (i) and (ii) must be identical.

$$\therefore \quad \frac{m_1 m_2}{a} = -\frac{(m_1 + m_2)}{2h} = \frac{1}{b} \qquad .....(iii)$$

From 1st and 3rd member, we have $m_1 m_2 = a/b$

From 2nd and 3rd member, we have $m_1 + m_2 = -\dfrac{2h}{b}$

ILLUSTRATION 111: A variable line through (p, q) cuts the axes of co-ordinates at A and B respectively. Lines are drawn through A parallel to y-axis and through B parallel to x-axis. If they meet at P, then find locus of P.

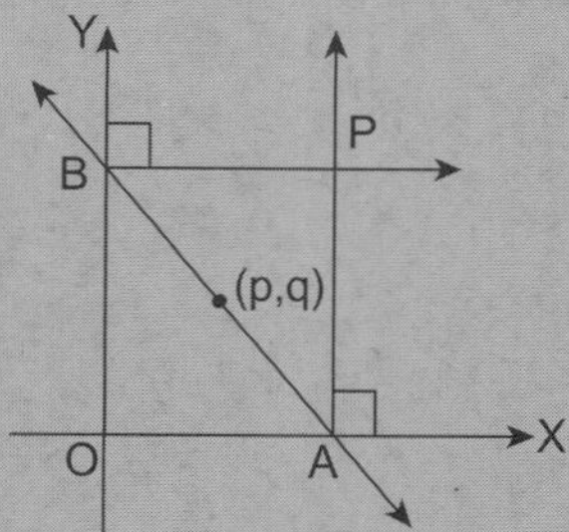

FIGURE 1.136

SOLUTION: Let the line be $\dfrac{x}{a} + \dfrac{y}{b} = 1$; $(a, b$ variable)

It passes through (p, q)

$$\Rightarrow \frac{p}{a} + \frac{q}{b} = 1 \qquad .... (i)$$

$A(a, 0)$, $B(0, b)$. The other lines are $x = a$ and $y = b$, which meet at $P(x, y) \equiv P(a,b)$.

Eliminating the variables a, b, the required locus is $\dfrac{p}{x} + \dfrac{q}{y} = 1$.

ILLUSTRATION 112: A variable line cuts the axes of co-ordinates in points A and B such that $OA + OB = c$. Then find the locus of foot of perpendicular from origin to the line.

SOLUTION: Line is $\dfrac{x}{a} + \dfrac{y}{b} = 1$, where $a + b = c$ ..(i)

If (h, k) be the foot of perpendicular from (0, 0), then

$$\frac{h-0}{\frac{1}{a}} = \frac{k-0}{\frac{1}{b}} = -\frac{(0+0-1)}{\frac{1}{a^2} + \frac{1}{b^2}} = \frac{a^2b^2}{a^2 + b^2}$$ (by formula of foot of perpendicular from a point)

$$\Rightarrow h = \frac{ab^2}{a^2 + b^2} \qquad .... (ii)$$

$$\text{and } k = \frac{a^2 b}{a^2 + b^2} \qquad ..... (iii)$$

In order to find the locus, we have to eliminate a, b between (i), (ii) and (iii) as given below:

$$h^2 + k^2 = \frac{a^2b^2}{(a^2 + b^2)^2}(a^2 + b^2) = \frac{a^2b^2}{a^2 + b^2}$$ (squaring and adding (i) and (ii)) .... (iv)

from (ii) and (iii) $\frac{1}{h}+\frac{1}{k}=\frac{a^2+b^2}{ab}\left[\frac{1}{a}+\frac{1}{b}\right]$

$$=\frac{a^2+b^2}{a^2b^2}(a+b)=c\ .\ \frac{a^2+b^2}{a^2b^2} \quad \text{... (v)}$$

$\therefore\ (h^2+k^2)\left(\frac{1}{h}+\frac{1}{k}\right)=c$ [by (iv) and (v)]

$\therefore$ locus of point is $(x^2+y^2)(y+x)=cxy$

Aliter method: Let $A(a, 0)$ and $B(0, b)$, then equation of variable line is $\frac{x}{a}+\frac{y}{b}=1$ ......(i);

where $c = a + b$. Now let (h, k) be the foot of $\perp r$ drawn from origin to AB.

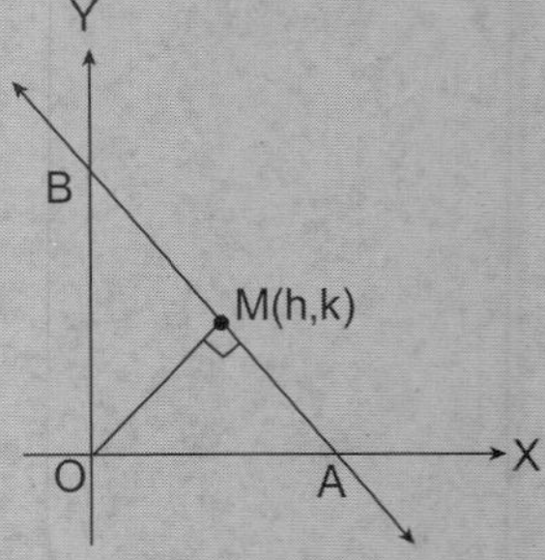

FIGURE 1.137

$\therefore$ (Slope of OM) $=\frac{-1}{\text{slope of } AB}$

$\Rightarrow\ \frac{k}{h}=\frac{-1}{(-1/a)/(1/b)}$ $\Rightarrow\ \frac{k}{h}=\frac{a}{b}$

$\Rightarrow\ kb = ah$ ....(ii)

$\because\ (h, k)$ lies on line $\frac{x}{a}+\frac{y}{b}=1$

$\Rightarrow\ \frac{h}{a}+\frac{k}{b}=1$ $\Rightarrow\ \frac{h}{a}+\frac{k^2}{ah}=1$ [from (ii)]

$\Rightarrow\ a=h+\frac{k^2}{h}$......(iii) Also $a + b = c$

$\Rightarrow\ a+\frac{ah}{k}=c$ [from (iii)] $\Rightarrow\ a\left(1+\frac{h}{k}\right)=c$

$\Rightarrow\ \left(h+\frac{k^2}{h}\right)\left(1+\frac{h}{k}\right)=c$ $\Rightarrow\ \frac{(h^2+k^2)}{h}\frac{(k+h)}{k}=c$

or $(h^2+k^2)(k+h)=ckh \Rightarrow$ The required locus is $(x^2+y^2)(x+y)=cxy$

ILLUSTRATION 113: Through the point (5, 12), a straight line is drawn to meet the axes in points A and B. If the rectangle $OACB$ is completed, then find locus of the vertex C.

SOLUTION: Any line through (5, 12) is $y - 12 = m(x - 5)$; where m is variable

Putting $y = 0$ and $x = 0$ the points A and B are

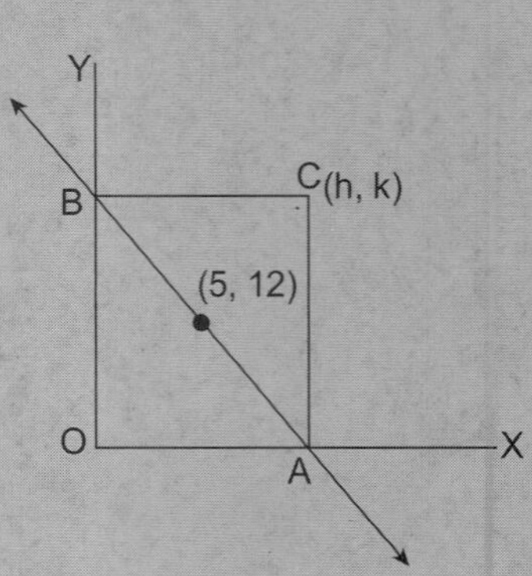

FIGURE 1.138

$A\equiv\left(5-\frac{12}{m}, 0\right)$; $B \equiv (0, 12 - 5m)$

If C be the point (h, k), then $h=5-\frac{12}{m}$, $k = 12 - 5m$

In order to find the locus, eliminate the variable m

$(h-5)=-\frac{12}{m}$ and $(k - 12) = - 5m$

or Multiplying, we get $(h-5)(k-12) = 60$

or $hk - 5k - 12h = 0$

or $\frac{5}{h} + \frac{12}{k} = 1$ [Divided by hk]

$\therefore$ $\frac{5}{x} + \frac{12}{y} = 1$ is the required locus.

ILLUSTRATION 114: Let a given line L_1 intersect the X and Y axes at P and Q respectively. Let another line L_2 perpendicular to L_1 cut the X-Y axes at R and S respectively. Find the locus of point of intersection of PS and QR.

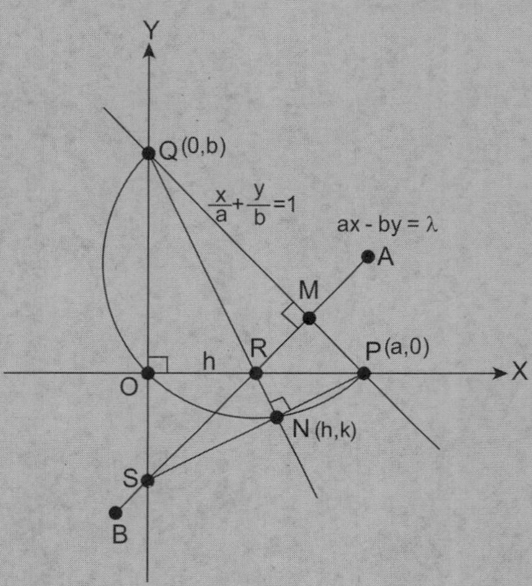

FIGURE 1.139

SOLUTION: Let $PQ(L_1)$ be a fixed line such that $P(a, 0)$ and $Q(0, b)$ and AB a variable line which slides such that it remains perpendicular to PQ, as shown in the figure. AB cuts PQ at M, X-axis at R and Y- axis at S. If QR produced cuts PS at $N(h, k)$. Consider the ΔPSQ.

$\therefore$ $OP \perp SQ$ and $SM \perp PQ$

$\Rightarrow$ R is orthocenter therefore QN is an altitude of ΔPSQ

And thus QN is perpendicular to SP

$\Rightarrow m_{QN} . m_{SP} = -1 \Rightarrow \left(\frac{k-b}{h}\right)\left(\frac{k}{h-a}\right) = -1$

$\Rightarrow (y-b)y = -x(x-a)$ replacing h, k by x, y respectively, we get

$\Rightarrow x(x-a) + y(y-b) = 0 \Rightarrow x^2 + y^2 - ax - by = 0$

$\Rightarrow \left(x - \frac{a}{2}\right)^2 + \left(y - \frac{b}{2}\right)^2 = \frac{a^2+b^2}{4}$ which is the required of a circle having centre at $\left(\frac{a}{2}, \frac{b}{2}\right)$ and radius $r = \frac{\sqrt{a^2+b^2}}{2}$

Concluding that the locus of the point N is a circle with extremities of diameter as P,Q.

ILLUSTRATION 115: Ends A, B of a straight line segment of constant length 'c' slides upon the fixed rectangular axes OX, OY respectively. If the rectangle $OAPB$ is completed, then find the locus of foot of perpendicular, drawn from P to AB.

SOLUTION: Consider a line segment AB of constant length 'c' that slides upon the fixed rectangular axes OX, OY respectively. At an instant let AB makes angle θ with x-axis as shown in the figure, then the end points have co-ordinates $A(c\cos\theta, 0)$ and $B(0, c\sin\theta)$. Now the rectangle $OAPB$ is completed and a perpendicular PM from P to AB is drawn as shown in figure below.

In ΔMAP : $\dfrac{AM}{AP} = \sin\theta$

$\Rightarrow AM = AP\sin\theta = c\sin^2\theta$ .... (i)

In ΔAMN : $\dfrac{AN}{AM} = \cos\theta$ and $\dfrac{MN}{AM} = \sin\theta$

$\Rightarrow AN = c\sin^2\theta\cos\theta$ .... (ii)

and $MN = c\sin^3\theta$ .... (iii)

$\Rightarrow k = c\sin^3\theta$

Since $h = ON = OA - AN$

$\Rightarrow h = c\cos\theta - c\sin^2\theta\cos\theta$

$\Rightarrow h = c\cos\theta\left(1-\sin^2\theta\right) = c\cos^3\theta$

$\because \sin^2\theta + \cos^2\theta = 1$

$\Rightarrow \left(\dfrac{k}{c}\right)^{\frac{2}{3}} + \left(\dfrac{h}{c}\right)^{\frac{2}{3}} = 1$ replacing h, k by x, y respectively we get $x^{\frac{2}{3}} + y^{\frac{2}{3}} = c^{\frac{2}{3}}$ as the equation of desired locus.

FIGURE 1.140

ILLUSTRATION 116: Two straight lines rotate about two fixed points (a, 0) and (–a, 0) respectively . If they start from their position of coincidence such that one rotates at the rate double that of the other, then find the locus of their point of intersection.

SOLUTION: Let both lines are initially along x axis at $t = 0$ and start rotating in anti-clockwise sense and at an instant 't' they intersect at $P(h, k)$. As AP rotates with double rate than that of BP, therefore

If $\angle PBX = \theta$

$\Rightarrow \angle PAX = 2\theta$

Consequently, $\angle APB = \theta$

ΔPAB is isosceles triangle with $PA = AB$

$\Rightarrow \sqrt{(h-a)^2 + k^2} = |2a|$

$\Rightarrow h^2 + a^2 - 2ah + k^2 = 4a^2$

Replacing h, k by x, y respectively we get

$x^2 + y^2 - 2ax - 3a^2 = 0$, which is desired locus of point of intersection of rotating lines.

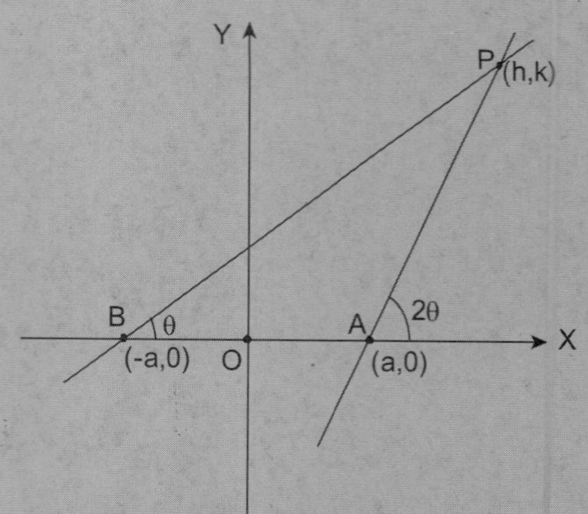

Figure 1.141

INTERSECTION OF LOCI

Let $C_1 = 0, C_2 = 0, ..., C_n = 0$ be n given equations of curves representing the loci $L_1, L_2, ..., L_n$ respectively. Then the locus of point satisfying all the equations $C_1 = 0, C_2 = 0, ..., C_n = 0$ is called the intersection of loci. If (x, y) satisfy all the above equations, then point $(x, y) \in L_i$ for all $i = 1,2,...n \Rightarrow$ the point $(x, y) \in L_1 \cap L_2 \cap ..... \cap L_n$. Hence, the required locus is $L_1 \cap L_2 \cap L_3 ..... \cap L_n$.

ILLUSTRATION 117: Find the intersection of the loci whose equations are $2x - y - 3 = 0$ and $x + 2y - 4 = 0$.

SOLUTION: Solving the two given equations simultaneously, we get $x = 2, y = 1$

$\Rightarrow L_1 \cap L_2 = \{2, 1\}$

ILLUSTRATION 118: Find the points of intersection of the loci $x^2 + y^2 = 25$ and $x = 6$.

SOLUTION: Putting $x = 6$ in $x^2 + y^2 = 25$

then we get $y^2 = -11$, which has no real roots. Thus the loci do not have any point in common.

$\therefore L_1 \cap L_2 = \phi$ (null set)

LOCUS REPRESENTED BY COMBINED EQUATIONS

(a) The locus represented by the equation $C_1.C_2, ..., C_n = 0$

$C_1 .C_2, ..., C_n = 0$

$\Rightarrow$ either $C_1 = 0$ or $C_2 = 0$ or ....

or $C_n = 0$ and let L be the locus of a point (x,y) satisfying $C_1.C_2, ..., C_n = 0$ then $(x, y) \in L$ iff $(x, y) \in L_1$ or L_2. ..

or L_n; where $L_1, L_2, ..., L_n$ are the loci represented by the curves $C_1 = 0, C_2 = 0, ..., C_n = 0$ respectively

$\Rightarrow L = L_1 \cup L_2, ..., \cup L_n$

Thus to obtain the union of loci whose equations are known, first of all write their equation such that right hand side is zero e.g. $C_1 = 0, C_2 = 0, ..., C_n = 0$,

then combined locus: $C_1 . C_2 . C_3, ..., C_n = 0$.

ILLUSTRATION 119: What locus is represented by $xy - 3x - y + 3 = 0$?

SOLUTION: Since $xy - 3x - y + 3 = 0$

$\Rightarrow (x - 1)(y - 3) = 0$

Let $C_1: (x - 1) = 0$ and $C_2: (y - 3) = 0$

So if L_1 and L_2 be the loci represented by equation C_1 and C_2, respectively.

$\therefore L_1 = \{ (1, y): y \in \mathbb{R}\}$ = straight line || to y – axis and $L_2 = \{ (x, 3): x \in \mathbb{R}\}$ = straight line || to x - axis

Let L be the locus represented by $xy - 3x - y + 3 = 0$.

$\therefore L = L_1 \cup L_2$ = pair of perpendicular lines ($x = 1, y = 3$)

ILLUSTRATION 120: What locus is represented by equation $(ax + by - 1)(bx - ay + 1) = 0$?

SOLUTION: Let $C_1 = ax + by - 1 = 0$ and $C_2 = bx - ay + 1 = 0$ and L_1 and L_2 be the loci represented by them, respectively.

Further, let L be the locus represented by $(ax + by - 1)(bx - ay + 1) = 0$

then $L_1 = ((x, y): ax + by - 1 = 0)$
and $L_2 = ((x, y): bx - ay + 1 = 0)$
as $(ax + by - 1)(bx - ay + 1) = 0$
$\Rightarrow ax + by - 1 = 0$ or $bx - ay + 1 = 0$
we get $L = L_1 \cup L_2$ = Pair of lines $ax + by - 1 = 0$ and $bx - ay + 1 = 0$

ILLUSTRATION 121: Identify the locus represented by the equation $(x + 1)(x^2 + y^2 - 1)(y^2 - 4(x - 1)) = 0$.

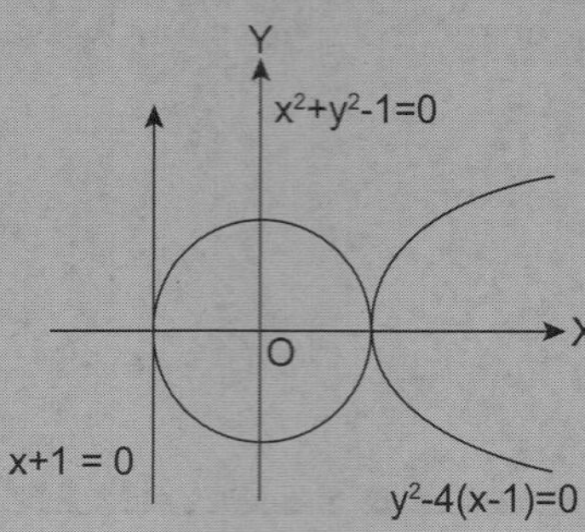

FIGURE 1.142

SOLUTION: $(x + 1)(x^2 + y^2 - 1)(y^2 - 4(x - 1)) = 0$ representes the union of loci L_1, L_2, L_3, respectively. where L_1 represents locus which is a straight line parallel to y-axis, i.e., $x = -1$.

L_2: represents locus which is a circle represented by $x^2 + y^2 = 1$

L_3: represents locus which is parabola represented by $y^2 = 4(x - 1)$

$\therefore$ $L = L_1 \cup L_2 \cup L_3$ is a locus of point which can move on any of these curves.

(b) Intersection of Loci :

Intersection of loci $S = 0$ and $S' = 0$ is defined as set of those points which lie on both the curves $S = 0$ and $S' = 0$.
That is, set of common points.

$\Rightarrow S_1 \cap S_2 = \{(x, y): f(x, y) = 0 \text{ and } g(x, y) = 0\}$

And its equation is given as

$\Rightarrow |f(x,y)| + |g(x,y)| = 0 \text{ or } |S| + |S'| = 0$

or $\sqrt{S} + \sqrt{S'} = 0$

or $S^2 + S'^2 = 0$

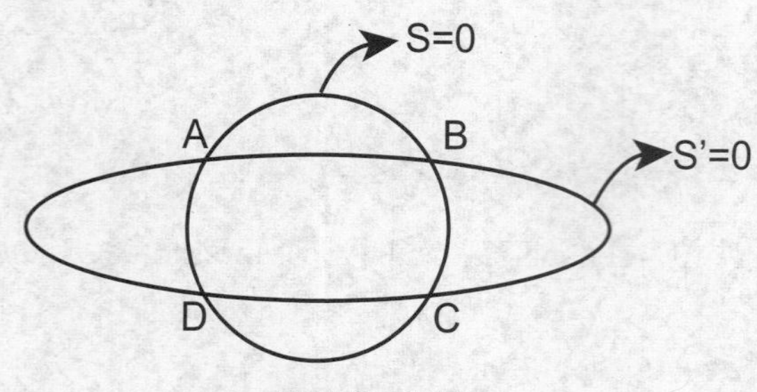

FIGURE 1.143

(c) The locus represented by the equation $C_1 + kC_2 = 0$

Let L_1 and L_2 be the loci represented by the equations $C_1 = 0$ and $C_2 = 0$ respectively and k be any real number, and let L be the locus represented by the equation $C_1 + kC_2 = 0$.

If $(x, y) \in L_1$ and $(x, y) \in L_2$.

$\Rightarrow$ (x, y) satisfies the equations $C_1 = 0$ and $C_2 = 0$

$\Rightarrow$ (x, y) also satisfies the equations $C_1 + kC_2 = 0$

$\Rightarrow$ $(x, y) \in L$ but converse is not necessarily true (think why)

Thus, $L_1 \cap L_2 \in L$, hence the locus contains all the points which satisfy both $C_1 = 0$ and $C_2 = 0$. But, L can include infinitely many other points as well thus L represented by equation $C_1 + kC_2 = 0$ is an equation of locus passing through intersection of $C_1 = 0$ and $C_2 = 0$.

ILLUSTRATION 122: What locus is represented by the equation $(x + y - 2) + k(x - y - 4) = 0$, $k \in \mathbb{R}$

SOLUTION: Let $C_1 = x + y - 2 = 0$ and $C_2 = x - y - 4 = 0$. Solving $C_1 = 0$ and $C_2 = 0$ simultaneously we get $x = 3, y = -1$ and let L_1 and L_2 be the loci represented by $C_1 = 0$ and $C_2 = 0$ respectively.
Further let L be the locus represented by $(x + y - 2) + k(x - y - 4) = 0$
Then $(3, -1) \in L_1$, $(3, -1) \in L_2$ hence $(3, -1) \in L$
i.e., the locus represented by the equation $(x + y - 2) + h(x - y - 4) = 0$ is a family of lines passing through $(3, -1)$; i.e., the point of intersection of lines $x + y - 2 = 0$ and $x - y - 4 = 0$. Hence we get infinitely many different lines for different values of k.

ILLUSTRATION 123: (a) If $A(\cos\theta, \sin\theta)$, $B(\sin\theta, \cos\theta)$, $C(1, 0)$ are the vertices of a ΔABC. Find the locus of its centroid as well as orthocentre if θ varies.

(b) If $A(\cos\alpha, \sin\alpha)$, $B(\sin\alpha, -\cos\alpha)$, $C(0, 1)$ are the vertices of triangle ABC where α varies, then find the locus of centroid and orthocentre of triangle ABC.

SOLUTION: (a) Let Centroid be (h, k) i.e., $3h - 1 = \sin\theta + \cos\theta = 3k \Rightarrow 3h - 3k = 1 \Rightarrow h - k = 1/3$

By replacing h, k by x, y respectively, we get $x - y = 1/3 \Rightarrow 3y = 3x - 1$,
thus the locus of centroid is straight line $3x - 3y - 1 = 0$.
A, B, C lie on a unit circle with centre at origin
$\therefore$ circumcentre of ΔABC is C' i.e., $(0, 0)$. As we know that centroid divides OC' in ratio 2: 1. where $O(x_0, y_0)$ is orthocentre.

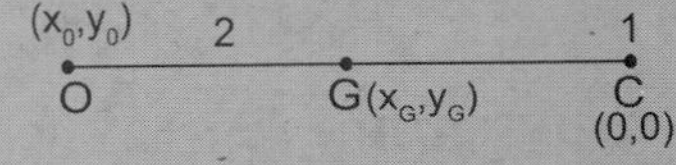

FIGURE 1.144

$$\therefore \quad x_G = \frac{\cos\theta + \sin\theta + 1}{3} = \frac{x_0}{3} \Rightarrow \cos\theta + \sin\theta = x_0 - 1$$

$$\text{Also } y_G = \frac{\cos\theta + \sin\theta}{3} = \frac{y_0}{3} \Rightarrow \cos\theta + \sin\theta = y_0$$

$\Rightarrow x_0 - 1 = y_0$
replacing x_0, y_0 by x, y respectively, we get $x - y = 1$
Thus locus of orthocentre is clearly a straight line $x - y - 1 = 0$.

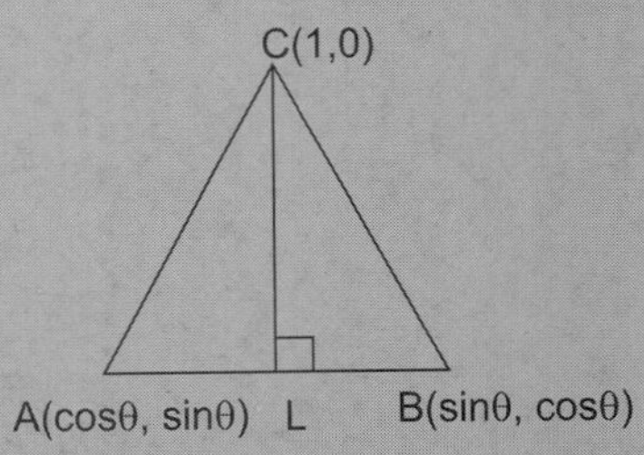

FIGURE 1.145

Aliter: Slope of $AB = -1$ (independent of θ)
$\therefore$ Equation of CL is $y = x - 1$.
$\therefore$ For every value of θ orthocentre lies on line CL i.e $y = x - 1$ is the required locus of orthocentre.

(b) Let the co-ordinates of centroid be (x_0, y_0). Therefore

$$x_0 = \frac{\cos\alpha + \sin\alpha}{3} \text{ and } y_0 = \frac{\sin\alpha - \cos\alpha + 1}{3}$$

$\Rightarrow \cos\alpha + \sin\alpha = 3x_0$ .....(i)
and $\sin\alpha - \cos\alpha = 3y_0 - 1$ ...(ii)
Eliminating α by squaring and adding (i) and (ii), we get
$(3x_0)^2 + (3y_0 - 1)^2 = 2$
Replacing x_0 and y_0 by x and y respectively, we get
$9x^2 + 9y^2 - 6y - 1 = 0$

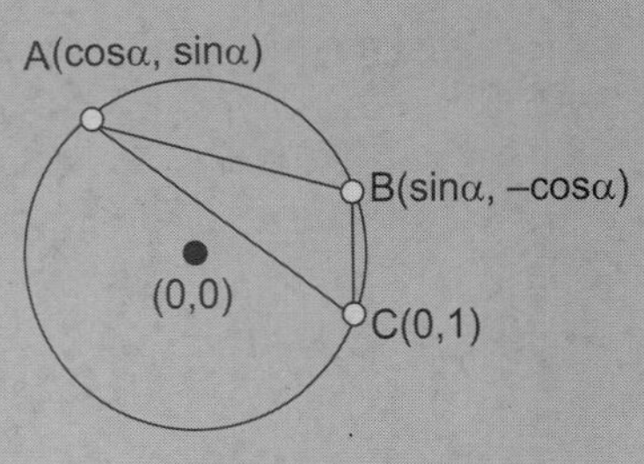

FIGURE 1.146

Let the co-ordinates of orthocentre be (h, k) and since G divides OC in the ratio 2:1; where C is the circumcentre of triangle, here it is (0,0)

$$\Rightarrow x_0 = \frac{h + 2(0)}{3} \Rightarrow h = 3x_0 \text{ and } y_0 = \frac{k + 2(0)}{3} \Rightarrow k = 3y_0$$

$\Rightarrow$ cosα + $\sin\alpha$ = h and sinα – cosα = $k - 1$

Squaring and adding, we get $h^2 + (k-1)^2 = 2$

Replacing h and k by x and y respectively, we get $x^2 + y^2 - 2y - 1 = 0$

ILLUSTRATION 124: Find the equation of locus of a point which moves so that the difference of its distances from points (4,0) and (–4,0) is 4 units.

SOLUTION: Let $P(h,k)$ be any point moving so that given condition is satisfied

Let $A(4,0)$ and B (–4,0)

$\Rightarrow |PA - PB| = 4 \Rightarrow PA - PB = \pm 4 \Rightarrow \left(\sqrt{(h-4)^2 + (k-0)^2}\right) = \pm 4 + \left(\sqrt{(h+4)^2 + (k-0)^2}\right)$

$\Rightarrow (h-4)^2 + (k)^2 = 16 + (h+4)^2 + k^2 \pm 8\left(\sqrt{(h+4)^2 + k^2}\right)$

$\Rightarrow (h-4)^2 - (h+4)^2 = 16 \pm 8\left(\sqrt{(h+4)^2 + k^2}\right)$

$\Rightarrow (h-4-h-4)(h-4+h+4) = 16 \pm 8\left(\sqrt{(h+4)^2 + k^2}\right) \Rightarrow (-8)(2h) = 16 \pm 8\left(\sqrt{(h+4)^2 + k^2}\right)$

$\Rightarrow -16h - 16 = \pm 8\left(\sqrt{(h+4)^2 + k^2}\right) \Rightarrow -16(h+1) = \pm 8\left(\sqrt{(h+4)^2 + k^2}\right)$

Squaring and simplifying, we get $3h^2 = k^2 + 12$

$\therefore$ The required locus is $3x^2 - y^2 = 12$

ILLUSTRATION 125: Find the locus of a point whose co-ordinates are given by $x = a\cos\theta$ and $y = a\sin 2\theta$

SOLUTION: Here locus can be formed by eliminating θ

$\cos\theta = \frac{x}{a} \Rightarrow \cos^2\theta = \frac{x^2}{a^2}$

$\sin^2\theta = 1 - \frac{x^2}{a^2} = \frac{a^2 - x^2}{a^2} \quad \Rightarrow \sin\theta = \frac{\pm\sqrt{a^2 - x^2}}{a}$

$\Rightarrow y = a\,(2\sin\theta\cos\theta) = a\left(2\left(\pm\frac{\sqrt{a^2 - x^2}}{a}\right) \times \frac{(x)}{a}\right)$

$ya^2 = \pm 2x\sqrt{a^2 - x^2} \Rightarrow y^2a^4 = 4x^2\left(a^2 - x^2\right) \Rightarrow y^2a^4 = 4x^2a^2 - 4x^4$

ILLUSTRATION 126: Find the locus of the point whose position vector is given by $(a + ib)^5 + (b + ia)^5$ (where a, b are real parameters)

SOLUTION: Since to find locus we have to eliminate constants from equation of position vector Let $a = \gamma\cos\theta$, $b = \gamma\sin\theta$. Then, $(a + ib)^5 + (b + ia)^5$

$= \gamma^5[(\cos\theta + i\sin\theta)^5 + (\sin\theta + i\cos\theta)^5] = \gamma 5\left[(\cos 5\theta + i\sin 5\theta) + \left\{\cos\left(\frac{\pi}{2} - \theta\right) + i\sin\left(\frac{\pi}{2} - \theta\right)\right\}^5\right]$

$= \gamma 5\left[(\cos 5\theta + i\sin 5\theta) + \left\{\cos 5\left(\frac{\pi}{2} - \theta\right) + i\sin 5\left(\frac{\pi}{2} - \theta\right)\right\}\right]$

$= \gamma^5[\cos 5\theta + i\sin 5\theta + \sin 5\theta + i\cos 5\theta]$

This is a complex number whose real and imaginary parts are equal. So, locus of such a point will be '$y = x$'

TEXTUAL EXERCISE–9 (SUBJECTIVE)

1. If the co-ordinates of a variable point P be $(t + 1/t, t - 1/t)$, where t is a variable quantity, then find the locus of P.

2. Find the locus of a point such that the line segment having end points (2,0) and (–2,0) subtend a right angle at that point.

3. What loci are represented by the following equations?
(a) $xy - ay = 0$ (b) $x^3 - x^2 - x + 1 = 0$
(c) $x^3 - xy^2 = 0$ (d) $x^2 - y^2 = 0$
(e) $x^2 - xy = 0$ (f) $x^3 + y^3 = 0$
(g) $x^2 + y^2 = 0$ (h) $x^2y = 0$
(i) $(x^2 - 1)(y^2 - 4) = 0$ (j) $(x - a)^2 + y^2 = 0$
(k) $(x^2 - 1)^2 + (y^2 - 4)^2 = 0$

4. Find the locus of a point which moves such that its distance from the point (0, 0) is twice its distance from the y-axis.

5. The position of a moving point in the xy-plane at time t is $(u\cos\alpha.t,\ u\sin\alpha.\ t - kt^2)$, where u, α, k are constants. Find the locus of the moving point.

6. A (2, 3) is a fixed point and $Q(3\cos\theta, 2\sin\theta)$ a variable point. If P divides AQ internally in the ratio 3 : 1, find the locus of P.

7. P is the point (–1, 2). A variable line through P cuts the co-ordinate axes at A and B respectively. Q is a point on AB such that PA, PQ, PB are in H.P. Show that the locus of Q is the line $y = 2x$. (A, B lie on the same side of P) Also show that in general the locus of Q is the rhombus whose sides are $y = 2x$, $y = -2x + 4$, $y = -2x - 4$ and $y = 2x + 8$ excluding the vertices.

8. If $A(\cos t, \sin t)$, $B(-\sin t, \cos t)$, $C(1, 2)$ are the vertices of a ΔABC find the locus of its centroid if t varies.

Answer Keys

1. $x^2 - y^2 = 4$ 2. $x^2 + y^2 = 4$.

3. (a) Pair of perpendicular straight lines passing through $(a, 0)$.
(b) Pair of parallel straight lines $x = \pm 1$.
(c) Set of three straight lines $y = x$, $y = -x$ and y-axis.
(d) Angle bisectors of quadrants.
(e) y-axis and angle bisector of Ist, IIIrd quadrant.
(f) Angle bisector of IInd and IVth quadrant.
(g) Origin.
(h) Co-ordinate axes.
(i) Set of four lines forming a rectangle $x = \pm 1$ and $y = \pm 2$.
(j) $(a, 0)$
(k) Set of four points (1, –2); (1, 2); (–1, 2); (–1, –2).

4. $3x^2 - y^2 = 0$ 5. $y = x\tan\alpha - \dfrac{kx^2}{u^2\cos^2\alpha}$ 6. $\left(\dfrac{4x-2}{9}\right)^2 + \left(\dfrac{4y-3}{6}\right)^2 = 1$

8. $3(x^2 + y^2) - 2x - 4y + 1 = 0$.

TEXTUAL EXERCISE–8 (OBJECTIVE)

1. The equation of the locus of a point whose distance from (a, 0) is equal to its distance from y-axis, is
(a) $y^2 - 2ax = a^2$ (b) $y^2 - 2ax + a^2 = 0$
(c) $y^2 + 2ax + a^2 = 0$ (d) $y^2 + 2ax = a^2$

2. The locus of a point P which moves in such a way that the segment OP, where O is the origin, has slope $\sqrt{3}$ is
(a) $x - \sqrt{3}y = 0$ (b) $x + \sqrt{3}y = 0$
(c) $\sqrt{3}x + y = 0$ (d) $\sqrt{3}x - y = 0$

3. If $P \equiv (1, 0)$, $Q \equiv (-1, 0)$ and $R \equiv (2, 0)$ are three given points, then the locus of a point S satisfying the relation $SQ^2 + SR^2 = 2\,SP^2$ is
(a) A straight line parallel to x-axis
(b) A circle through origin
(c) A circle with centre at the origin
(d) A straight line parallel to y-axis

4. The co-ordinates of the points O, A and B are (0, 0), (0, 4) and (6, 0) respectively. If a point P moves such that the area of ΔPOA is always twice the area of

ΔPOB, then the equation to both parts of the locus of P is

(a) $(x-3y)(x+3y)=0$ (b) $(x-3y)(x+y)=0$

(c) $(3x-y)(3x+y)=0$ (d) None of these

5. A points P moves so that its distance from the point (a, 0) is always equal to its distance from the line $x+a=0$. The locus of the point is

(a) $y^2=4ax$ (b) $x^2=4ay$

(c) $y^2+4ax=0$ (d) $x^2+4ay=0$

6. A point move so that its distance from the point (–1, 0) is always three times its distance from the point (0, 2). The locus of the points is

(a) A line (b) A circle

(c) A parabola (d) An ellipse

7. The locus of a point which moves so that it is always equidistant from the point A $(a, 0)$ and B $(-a, 0)$ is

(a) A circle

(b) Perpendicular bisector of the line segment AB

(c) A line parallei to x - axis ‘

(d) None of these

8. If the co-ordinates of a point be given by the equations $x=b\sec\phi$, $y=a\tan\phi$, then its locus is

(a) A straight line (b) A circle

(c) An ellipse (d) A hyperbola

9. The co-ordinates of the point A and B are $(ak, 0)$ and $\left(\frac{a}{k},0\right);(k\neq\pm1)$. If a point P moves so that $PA=kPB$, then the equation to the locus of P is

(a) $k^2(x^2+y^2)-a^2=0$

(b) $x^2+y^2-k^2a^2=0$

(c) $x^2+y^2+a^2=0$

(d) $x^2+y^2-a^2=0$

10. The locus of the mid-point of the distance between the axes for the variable line $x\cos\alpha+y\sin\alpha=p$, where p is constant, is

(a) $x^2+y^2=4p^2$ (b) $\frac{1}{x^2}+\frac{1}{y^2}=\frac{4}{p^2}$

(c) $x^2+y^2=\frac{4}{p^2}$ (d) $\frac{1}{x^2}+\frac{1}{y^2}=\frac{2}{p^2}$

11. The locus of a point whose distance from the point $(-g,-f)$ is always ‘a’ will be, (where $k=g^2+f^2-a^2$)

(a) $x^2+y^2+2gx+2fy+k=0$

(b) $x^2-y^2+2gx+2fy+k=0$

(c) $x^2+y^2+2xy+2gx+2fy+k=0$

(d) None of these

12. A point moves such that the sum of its distance from two fixed points $(ae, 0)$ and $(-ae, 0)$ is always $2a$. Then equation of its locus is

(a) $\frac{x^2}{a^2}+\frac{y^2}{a^2(1-e^2)}=1$ (b) $\frac{x^2}{a^2}-\frac{y^2}{a^2(1-e^2)}=1$

(c) $\frac{x^2}{a^2(1-e^2)}-\frac{y^2}{a^2}=1$ (d) None of these

13. A point moves in such a way that its distance from origin is always 4 . Then the locus of the point is

(a) $x^2+y^2=4$ (b) $x^2+y^2=16$

(c) $x^2+y^2=2$ (d) None of these

14. If A $(-a,0)$ and $B(a, 0)$ are two fixed points, then the locus of the point on which the line AB subtends a right angle, is

(a) $x^2+y^2=2a^2$ (b) $x^2-y^2=0$

(c) $x^2+y^2+a^2=0$ (d) $x^2+y^2=a^2$

15. If A and B are two fixed points and P is a variable point such that $PA+PB=4$, then the locus of P (when $AB<4$) is *a/an*

(a) Parabola (b) Ellipse

(c) Hyperbola (d) None of these

16. If A and B are two points in a plane, so that $|PA-PB|$ = constant ($<AB$), then the locus of P is

(a) Hyperbola (b) Circle

(c) Parabola (d) Ellipse

17. The locus of P such that the area of ΔPAB = 12sq. units, where A (2,3) and B (– 4, 5) is

(a) $(x+3y-1)(x+3y-23)=0$

(b) $(x+3y+1)(x+3y-23)=0$

(c) $(3x+y-1)(3x+y-23)=0$

(d) $(3x+y+1)(3x+y+23)=0$

18. The position of *a* moving point in the XY – plane at time t is given by $\left((u\cos\alpha)t,(u\sin\alpha)t-\frac{1}{2}gt^2\right)$, where u,α,g are constants. The locus of the moving point is

(a) A circle (b) A parabola

(c) An ellipse (d) None of these

19. If the equation of the locus of a point equidistant from the points (a_1,b_1) and (a_2,b_2) is $(a_1-a_2)x+(b_1-b_2)y+c=0$, then the value of c is

(a) $a_1^2-a_2^2+b_1^2-b_2^2$ (b) $\sqrt{a_1^2+b_1^2-a_2^2-b_2^2}$

(c) $\frac{1}{2}(a_1^2+a_2^2+b_1^2+b_2^2)$ (d) $\frac{1}{2}(a_2^2+b_2^2-a_1^2-b_1^2)$

20. If sum of distances of a point from the origin and lines $x = 2$ is 4, then its locus is

(a) $x^2 - 12y = 36$

(b) $y^2 + 12x + 4 = 0$

(c) $y^2 - 12x = 36$

(d) $x^2 + 12y = 36$

21. Locus of centroid of the triangle whose vertices are (a cos$t$, $a$sin$t$), ($b$sin t, $-b$cos$t$) and (1, 0), where t is a parameter; is

(a) $(3x - 1)^2 + (3y)^2 = a^2 - b^2$

(b) $(3x - 1)^2 + (3y)^2 = a^2 + b^2$

(c) $(3x + 1)^2 + (3y)^2 = a^2 + b^2$

(d) $(3x + 1)^2 + (3y)^2 = a^2 - b^2$

22. Let P be the point (1,0) and Q be a point of the locus $y^2 = 8x$. The locus of mid-point of PQ is

(a) $x^2 + 4y + 2 = 0$ (b) $x^2 - 4y + 2 = 0$

(c) $y^2 - 4x + 2 = 0$ (d) $y^2 + 4x + 2 = 0$

23. The ends of a rod of length ℓ move on two mutually perpendicular lines . The locus of the point on the rod which divides it in the ratio 1: 2 is

(a) $36x^2 + 9y^2 = 4\ell^2$ (b) $36 x^2 + 9y^2 = \ell^2$

(c) $9x^2 + 36 y^2 = 4\ell^2$ (d) None of these

Answer Keys

1. (b)	**2.** (d)	**3.** (d)	**4.** (a)	**5.** (a)	**6.** (b)	**7.** (b)	**8.** (d)	**9.** (d)	**10.** (b)
11. (a)	**12.** (a)	**13.** (b)	**14.** (d)	**15.** (b)	**16.** (a)	**17.** (b)	**18.** (b)	**19.** (d)	**20.** (b)
21. (b)	**22.** (c)	**23.** (a)							

MULTIPLE-CHOICE QUESTIONS

SECTION–I

OBJECTIVE-TYPE SOLVED EXAMPLES

1. x-co-ordinates of two points P and Q are the roots of equation $x^2 + 4x + 3 = 0$ and their y-coordinates are the roots of equation $x^2 - x - 6 = 0$. If x-coordinate of P is less than x-co-ordinate of Q and y-coordinate of P is greater than the y-co-ordinate of Q and co-ordinates of a third point R be (3, –5), then the length of the bisector of the interior angle R is

(a) $\frac{7\sqrt{2}}{3}$ (b) $\frac{14\sqrt{2}}{3}$

(c) $\frac{5\sqrt{2}}{3}$ (d) None of these

Solution: (b) Roots of the equation $x^2 + 4x + 3 = 0$ are –1, –3 and those of equation $x^2 - x - 6 = 0$ are –2, 3. Since x and y co-ordinates of P are respectively less than and greater than the corresponding co-ordinates of Q.

$\therefore$ $P(-3, 3)$ and $Q(-1, -2)$; Given $R(3, -5)$

Now, $PR = \sqrt{(3+3)^2 + (-5-3)^2} = 10$

and $RQ = \sqrt{(3+1)^2 + (-5+2)^2} = 5$

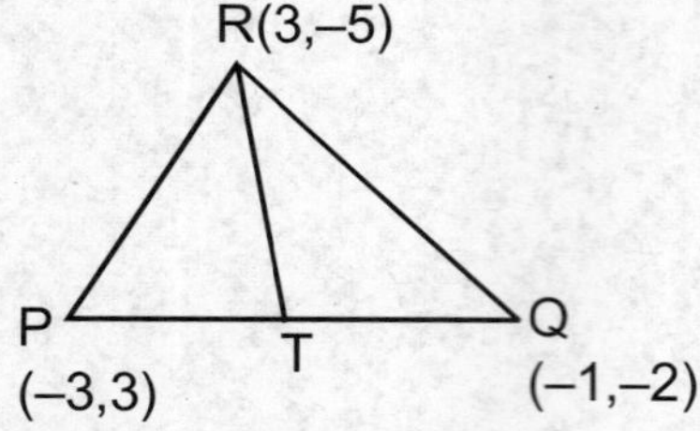

Let RT be the bisector of $\angle PRQ$,

then $\frac{PT}{TQ} = \frac{RP}{RQ} = \frac{10}{5} = \frac{2}{1}$

Thus, T divides PQ internally in the ratio 2:1.

$\therefore$ By section formula, co-ordinates of T will be given by

$$\left(\frac{2(-1)+1(-3)}{2+1}, \frac{2(-2)+1\times 3}{2+1}\right), \text{ i.e., } \left(-\frac{5}{3}, -\frac{1}{3}\right)$$

$\therefore$ $RT = \sqrt{(3+5/3)^2 + (-5+1/3)^2} = \frac{14\sqrt{2}}{3}$ units

2. If the point A is symmetric to the point $B(4, -1)$ with respect to the bisector of the first quadrant, then the length of AB is

(a) 5 (b) $5\sqrt{2}$

(c) $3\sqrt{2}$ (d) 3

Solution: (b) Let $A \equiv (a, b)$, the co-ordinates of the mid-point C of AB are $\left(\frac{a+4}{2}, \frac{b-1}{2}\right)$

Since it lies on the line $y = x$
(the bisector of first quadrant)

$\therefore$ $\frac{b-1}{2} = \frac{a+4}{2} \Rightarrow a - b = -5$ ...(i)

Since AB is perpendicular to the line $y = x$

$\therefore$ Product of slope of AB and slope of the line $y = x$ is equal to –1.

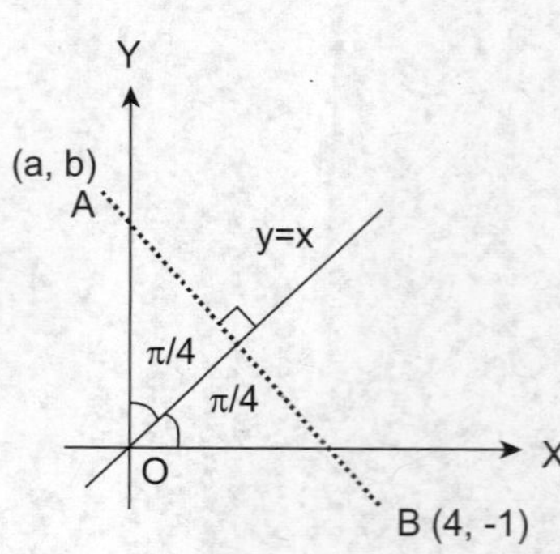

$\Rightarrow$ $\frac{b+1}{a-4} \times 1 = -1$

$\Rightarrow$ $b + 1 = -a + 4$

$\Rightarrow$ $a + b = 3$ ...(ii)

Solving (i) and (ii). we get $a = -1$, $b = 4$.

$\therefore$ Coordinates of the point A are $(-1, 4)$

$\therefore$ $AB = \sqrt{(-1-4)^2 + (4+1)^2} = 5\sqrt{2}$

3. The distance between two parallel lines is unity. A point P lies between the lines at a distance 'a' from one of them.

The length of a side of an equilateral triangle PQR, vertex Q of which lies on one of the parallel lines and vertex R lies on the other line is

(a) $\frac{2}{\sqrt{3}}\sqrt{a^2+a+1}$ (b) $\frac{2}{\sqrt{3}}\sqrt{a^2-a+1}$

(c) $\frac{1}{\sqrt{3}}\sqrt{a^2+a+1}$ (d) $\frac{1}{\sqrt{3}}\sqrt{a^2-a+1}$

Solution: (b) Let one of the parallel lines be x-axis and other be $y = 1$, with Q at origin as shown below.

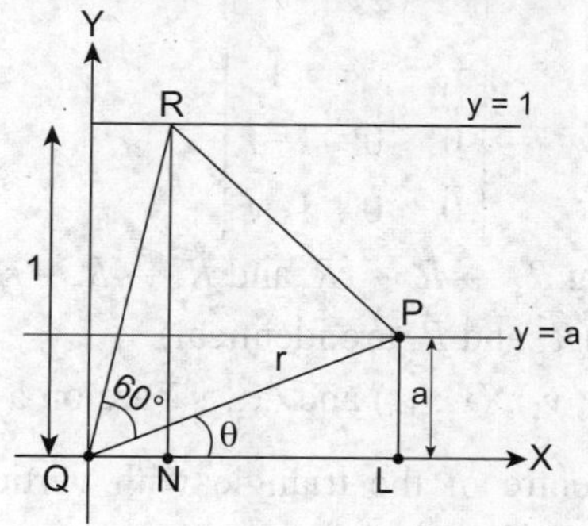

Let $PQ = QR = RP = r$

and $\angle PQX = \theta$, then $\angle RQX = \frac{\pi}{3}+\theta$

Given, $PL = a$, $RN = 1$

Now, $a = PL = r\sin\theta$

and $1 = RN = r\sin\left(\frac{\pi}{3}+\theta\right)$

$\Rightarrow 1 = r\left(\sin\frac{\pi}{3}\cos\theta+\cos\frac{\pi}{3}\sin\theta\right)$

$\Rightarrow r\left(\frac{\sqrt{3}}{2}\cos\theta+\frac{1}{2}\sin\theta\right)=1$

$\Rightarrow r\left(\frac{\sqrt{3}}{2}\sqrt{1-\frac{a^2}{r^2}}+\frac{1}{2}.\frac{a}{r}\right)=1 \quad \left(\because \sin\theta=\frac{a}{r}\right)$

$\Rightarrow \frac{\sqrt{3}}{2}\sqrt{r^2-a^2}=1-\frac{a}{2}$

$\Rightarrow r^2-a^2=\frac{(2-a)^2}{3}=\frac{4+a^2-4a}{3}$

$\Rightarrow r^2=\frac{4+a^2-4a}{3}+a^2=\frac{4+4a^2-4a}{3}$

$\therefore \quad r=\frac{2}{\sqrt{3}}\sqrt{a^2-a+1}$.

4. The co-ordinates of the orthocentre of the triangle having sides $3x - 2y = 6$, $3x + 4y + 12 = 0$ and $3x - 8y + 12 = 0$ are

(a) $\left(-\frac{1}{6},\frac{23}{9}\right)$ (b) $\left(-\frac{1}{6},-\frac{23}{9}\right)$

(c) $\left(\frac{1}{6},\frac{23}{9}\right)$ (d) None of these

Solution: (b) The three sides of the triangle are

$3x - 2y = 6$ ...(i)

$3x + 4y = -12$ ...(ii)

$3x - 8y = -12$ ...(iii)

Solving equation (i) and (iii), we get $A \equiv (4, 3)$

Solving equations (i) and (ii), we get $B \equiv (0, -3)$

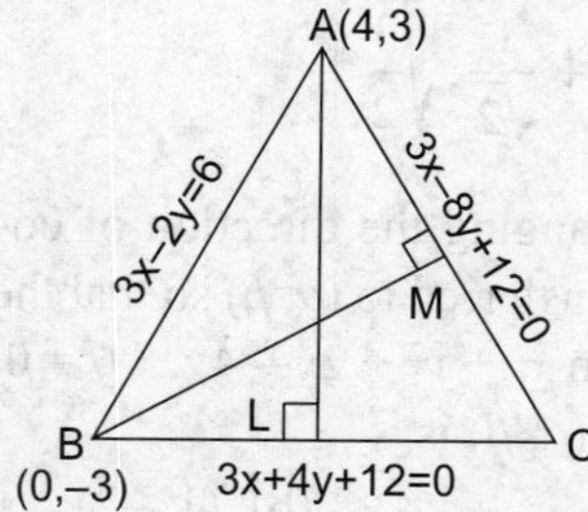

Since $AL \perp BC$ and it passes through A (4, 3)

$\therefore$ equation of line AL is $4x - 3y - 7 = 0$ ..(iv)

Since $BM \perp AC$ and it passes through B (0, –3)

$\therefore$ equation of line BM is $8x + 3y + 9 = 0$ ...(v)

Solving (iv) and (v), we get $x = -1/6$, $y = -23/9$

Hence the co-ordinates of the orthocentre are $\left(-\frac{1}{6},-\frac{23}{9}\right)$.

5. A line joining two points $A(2,0)$ and $B(3, 1)$ is rotated about A in anti-clockwise direction through an angle 15°. If B goes to C in the new position, then the co-ordinates of C are

(a) $\left(2,\sqrt{\frac{3}{2}}\right)$ (b) $\left(2,-\sqrt{\frac{3}{2}}\right)$

(c) $\left(2+\frac{1}{\sqrt{2}},\sqrt{\frac{3}{2}}\right)$ (d) None of these

Solution: (c) Slope of line $AB = \frac{0-1}{2-3} = 1 = \tan 45^\circ$

$\therefore \quad \angle BAX = 45°$

Given $\angle CAB = 15°$

$\therefore \quad \angle CAX = 60°$

$\therefore$ Slope of line $AC = \tan 60° = \sqrt{3}$

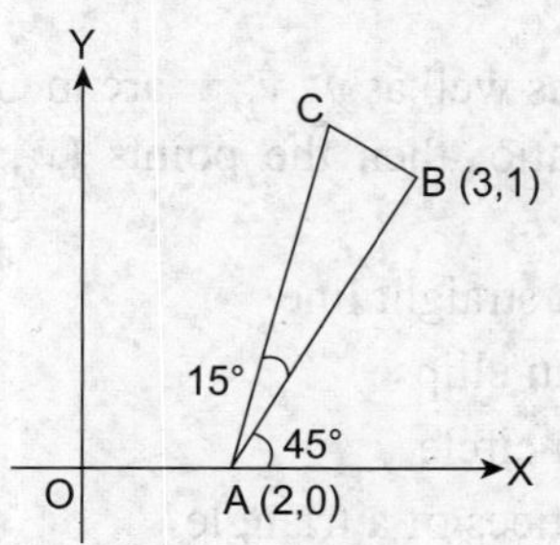

Now line AC makes an angle of 60° with positive direction of x-axis and

$AC = AB = \sqrt{(3-2)^2+(1-0)^2} = \sqrt{2}$

∴ Co-ordinates of C are

$(2+\sqrt{2}\cos 60°, 0+\sqrt{2}\sin 60°)$

i.e., $\left(2+\frac{1}{\sqrt{2}}, \sqrt{\frac{3}{2}}\right)$

6. Without changing the direction of co-ordinates axes, origin is transferred to (α, β) so that the linear terms in the equation $x^2 + y^2 + 2x - 4y + 6 = 0$ are eliminated. The points (α, β) is

(a) (–1, 2) (b) (1, –2)
(c) (1, 2) (d) (–1, –2)

Solution: (a) The given equation is $x^2 + y^2 + 2x - 4y + 6 = 0$ ......(i)

Putting $x = x' + \alpha$, $y = y' + \beta$ in (i), we get

$$x'^2 + y'^2 + x'(2\alpha + 2) + y'(2\beta - 4) + (\alpha^2 + \beta^2 + 2\alpha - 4\beta + 6) = 0$$

To eliminate linear terms, we should have

$2\alpha + 2 = 0$ and $2\beta - 4 = 0$

$\Rightarrow \alpha = -1$ and $\beta = 2$

$\therefore (\alpha, \beta) = (-1, 2)$

7. The number of integeral values of m for which the x-co-ordinate of the point of intersection of the lines $3x + 4y = 9$ and $y = mx + 1$ is also an integer is

(a) 2 (b) 0
(c) 4 (d) 1

Solution: (a) The given lines are

$3x + 4y = 9$ ....(i)

and $y = mx + 1$ ...(ii)

Solving (i) and (ii), we get the x-co-ordinate of the point of intersection as $x = \frac{5}{4m+3}$

Since x-co-ordinate is an integer

$\therefore 4m + 3 = \pm 5$ or $4m + 3 = \pm 1$

Solving these, only integeral values of m are –1 and –2

$\therefore m = -1, -2$

8. If x_1, x_2, x_3 as well as y_1, y_2, y_3 are in G.P. with the same common ratio, then the points (x_1, y_1), (x_2, y_2) and (x_3, y_3)

(a) lie on a straight line
(b) lie on an ellipse
(c) lie on a circle
(d) are vertices of a triangle

Solution: (a) Let $\frac{x_2}{x_1} = \frac{x_3}{x_2} = r$ and $\frac{y_2}{y_1} = \frac{y_3}{y_2} = r$

$\Rightarrow x_2 = x_1 r$, $x_3 = x_1 r^2$, $y_2 = y_1 r$ and $y_3 = y_1 r^2$

We have $\Delta = \begin{vmatrix} x_1 & y_1 & 1 \\ x_2 & y_2 & 1 \\ x_3 & y_3 & 1 \end{vmatrix} = \begin{vmatrix} x_1 & y_1 & 1 \\ x_1 r & y_1 r & 1 \\ x_1 r^2 & y_1 r^2 & 1 \end{vmatrix}$

$= \begin{vmatrix} x_1 & y_1 & 1 \\ 0 & 0 & 1-r \\ 0 & 0 & 1-r \end{vmatrix}$

(applying $R_3 \to R_3 - rR_2$ and $R_2 \to R_2 - rR_1$)

$= 0$ (R_2 and R_3 are identical)

Thus (x_1, y_1), (x_2, y_2) and (x_3, y_3) are on a straight line.

9. The incentre of the triangle with vertices $(1, \sqrt{3})$, (0, 0), and (2, 0) is

(a) $\left(1, \frac{\sqrt{3}}{2}\right)$ (b) $\left(\frac{2}{3}, \frac{1}{\sqrt{3}}\right)$
(c) $\left(\frac{2}{3}, \frac{\sqrt{3}}{2}\right)$ (d) $\left(1, \frac{1}{\sqrt{3}}\right)$

Solution: (d) Let $A \equiv (1, \sqrt{3})$, $B \equiv (0, 0)$ and $C \equiv (2, 0)$

Then $AB = \sqrt{1+3} = 2$, $BC = 2$ and

$CA = \sqrt{(1-2)^2 + (\sqrt{3}-0)^2} = 2$

Thus $AB = BC = CA$

As $\Delta\ ABC$ is an equilateral triangle, the incentre coincides with the centroid of the triangle which is

$I\left(\frac{0+1+2}{3}, \frac{0+0+\sqrt{3}}{3}\right)$, i.e., $I\left(1, \frac{1}{\sqrt{3}}\right)$

10. If the vertices P, Q, R of a ΔPQR are rational points, then which of the following points of the ΔPQR is (are) always rational point(s)?

(a) centroid (b) incentre
(c) circumcentre (d) orthocentre

Solution: (a, c, d) Let $P \equiv (x_1, y_1)$, $Q \equiv (x_2, y_2)$; $R \equiv (x_3, y_3)$ where; x_i, y_i ($i = 1, 2, 3$) are rational numbers.

Now, the centroid of ΔPQR is

$\left(\frac{x_1+x_2+x_3}{3}, \frac{y_1+y_2+y_3}{3}\right)$

which is rational point. Incentre depends on sides of the triangle which may not be rational even if vertices are so. Thus co-ordinates of incentre need not to be rational points through the equations of perpendicular bisectors and altitude would have rational co-efficients and hence the points of intersection of perpendicular bisectors i.e., circumcentre and that of altitude i.e., orthocentre would have rational co-ordinates.

11. The co-ordinates of the middle points of the sides of a triangle are (4, 2), (3, 3) and (2, 2), the co-ordinates of its centroid are

(a) (3, 7/3) (b) (3,3)
(c) (4, 3) (d) None of these

Solution: (a) Let D, E, F are mid-points of BC, AC, AB respectively.

∴ FD is parallel to AC and DE parallel to AB i.e., $FDEA$ is a parallelogram and hence diagonals AD and FE bisect each other. Thus we can say that AD is a median of both ΔABC and ΔDEF. By similar arguments BE is a median of both ΔABC and ΔDEF and hence centroid G of ΔABC and ΔDEF are the same.

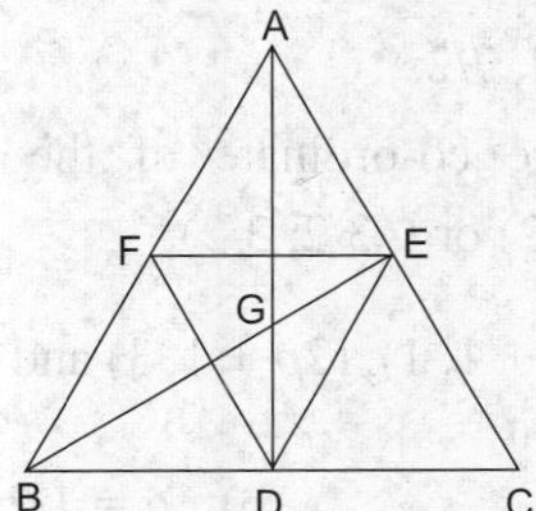

So if mid-point of triangle are given to be (4, 2), (3, 3), (2, 2), then the centroid of the triangle itself is $\left(\frac{4+3+2}{3}, \frac{2+3+2}{3}\right)$ i.e., (3, 7/3).

12. A straight line L with negative slope passes through the point (8, 2) and cuts the positive co-ordinate axes at points P and Q. AS L varies, the absolute minimum value of $OP + OQ$ is (O is origin)

(a) 10 (b) 18
(c) 16 (d) 12

Solution: (b) The equation of the line L be $y - 2 = m(x - 8)$, m < 0. Co-ordinates of P and Q are $P\left(8-\frac{2}{m}, 0\right)$ and $Q(0, 2 - 8m)$.

So, $OP + OQ = 8 - \frac{2}{m} + 2 - 8m = 10 + \frac{2}{(-m)} + 8(-m)$

$\geq 10 + 2\sqrt{\frac{2}{(-m)} \times 8(-m)} = 18$

($AM \geq GM$ for positive real numbers)

So, absolute minimum value of $OP + OQ = 18$.

13. If the distance of any point (x, y) from origin is defined as $d(x, y) = |x| + |y|$, then the locus $d(x, y) = 1$ is a

(a) circle of area 2 sq units
(b) square of area 2 sq units
(c) square of area 1 sq units
(d) None of the above

Solution: (b) $d(x, y) = 1 \Rightarrow |x| + |y| = 1$ The graph of which is shown in the figure.

The graph is a square $ABCD$ and $AB = BC = CD = DA = \sqrt{2}$ Where, $OA = 1 = $ OB, $\angle ABC = 90°$

∴ Area $= AB \cdot AD = \sqrt{2} \times \sqrt{2} = 2$ sq unit

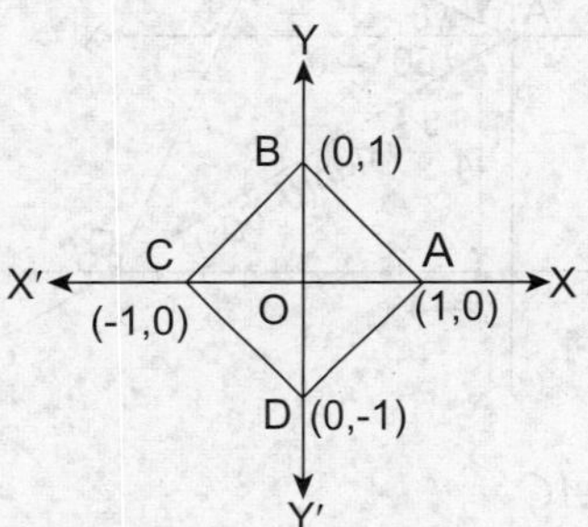

14. If the point (cos θ, sin θ) does not fall in the region of $|y| = |x - 1|$ in which the origin lies, then θ belongs to

(a) $\left(\frac{\pi}{2}, \frac{3\pi}{2}\right)$ b) $\left(-\frac{\pi}{2}, \frac{\pi}{2}\right)$
(c) $(0, \pi)$ (d) None of these

Solution: (b) The point (cos θ, sin θ) lies on a circle $x^2 + y^2 = 1$

$|y| = |x - 1| \quad \Rightarrow y = \pm|x - 1|$

∴ $y = \pm(x - 1)$ and $y = \pm(1 - x)$ for $x \geq 1$ and $x < 1$ resppectively

From the figure, θ can vary from $-\frac{\pi}{2}$ to $\frac{\pi}{2}$

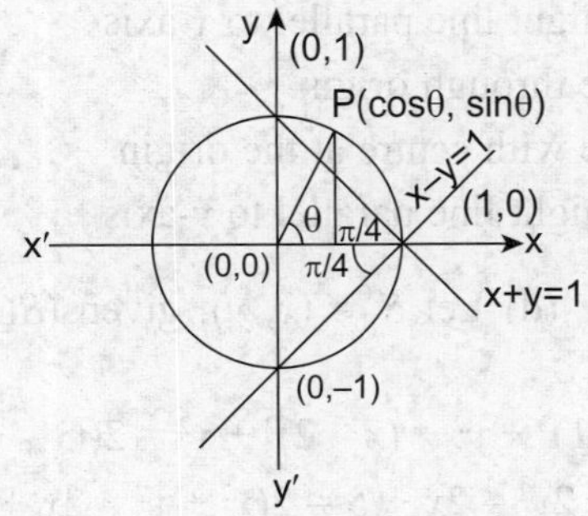

15. ABC is an equilateral triangle such that the vertices B and C lie on two parallel lines 6 units apart from each other. If A lies between the parallel lines at a distance 4 from one of them, then the length of a side of the equilateral triangle is

(a) 8 (b) $\sqrt{\dfrac{88}{3}}$

(c) $\dfrac{4\sqrt{7}}{\sqrt{3}}$ (d) None of these

Solution: (c) Let, $A(0, 0)$, and co-ordinate of B are $(2 \cot \theta, 2)$ and co-ordinate of C are $(4 \cot (60° – \theta), – 4)$

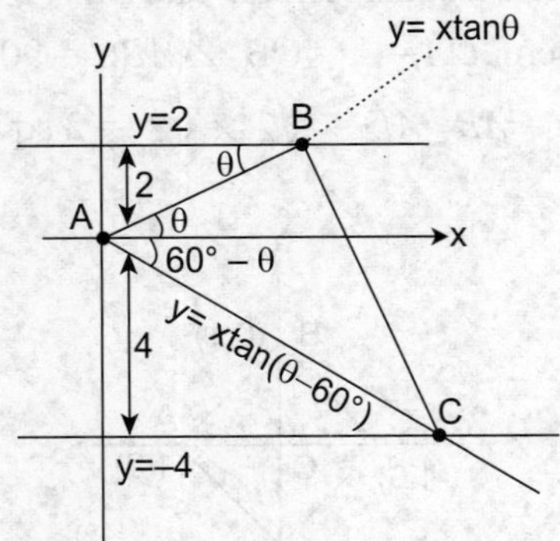

$\therefore\ AB = AC$

$\Rightarrow (AB)^2 = (AC)^2$

$\Rightarrow 4 \cot^2 \theta + 4 = 16 \cot^2 (60° – \theta) + 16$

$4 \operatorname{cosec}^2 \theta = 16 \operatorname{cosec}^2 (60° – \theta)$

$\Rightarrow \sin (60° – \theta) = 2 \sin \theta$

$\Rightarrow \left(\dfrac{\sqrt{3}}{2}\cos\theta - \dfrac{1}{2}\sin\theta\right) = 2\sin\theta$

$\therefore\ \tan\theta = \dfrac{\sqrt{3}}{5}$

$\therefore$ The required length $= AB = \sqrt{(4\cot^2\theta + 4)} = \dfrac{4\sqrt{7}}{\sqrt{3}}$.

16. If $P(1, 0)$, $Q(–1, 0)$ and $R(2, 0)$ are three given points, then the locus of point S satisfying the relation $(SQ)^2 + (SR)^2 = 2(SP)^2$ is

(a) a straight line parallel to x-axis
(b) circle through origin
(c) circle with centre at the origin
(d) a straight line parallel to y-axis

Solution: (d) Let $S = (x, y)$, given $(SQ)^2 + (SR)^2 = 2(SP)^2$

$\Rightarrow (x + 1)^2 + y^2 + (x – 2)^2 + y^2 = 2((x – 1)^2 + y^2)$

$\Rightarrow 2x^2 + 2y^2 – 2x + 5 = 2(x^2 + y^2 – 2x + 1)$

$\Rightarrow 2x + 3 = 0 \Rightarrow x = -\dfrac{3}{2}$

A straight line parallel to y-axis.

17. The area of a triangle is 5 sq units. Two of its vertices are (2, 1) and (3, –2). The third vertex lies on $y = x + 3$. The co-ordinates of the third vertex can be

(a) (–3/2, 3/2) (b) (3/4, –3/2)
(c) (7/2, 13/2) (d) (–1/4, 11/4)

Solution: (a, c) As the third vertex lies on the line $y = x + 3$, its co-ordinates are of the form $(x, x + 3)$. The area of the triangle with vertices (2, 1), (3, –2) and $(x, x + 3)$ is given by

$$\frac{1}{2}\left\|\begin{matrix} x & x+3 & 1 \\ 2 & 1 & 1 \\ 3 & -2 & 1\end{matrix}\right\| = 5 \text{ (given)}$$

$\Rightarrow |2x – 2| = 5$

$\therefore\ 2x – 2 = \pm 5$

$\Rightarrow x = – 3/2, 7/2.$

Thus, the co-ordinate of the third vertex are (7/2, 13/2) or (–3/2, 3/2).

18. The point $(p + 1, 1)$; $(2p + 1, 3)$ and $(2p + 2, 2p)$ are collinear, then

(a) $p = – 1$ (b) $p = 1/2$
(c) $p = 2$ (d) $p = –1/2$

Solution: (c, d) Let $A \equiv (p + 1, 1)$, $B \equiv (2p + 1, 3)$, and $C \equiv (2p + 2, 2p)$; As A, B, C are collinear.

$\therefore$ Slope of AB = Slope of AC

$\Rightarrow \dfrac{3-1}{2p+1-p-1} = \dfrac{2p-1}{2p+2-p-1}$

$\Rightarrow p = 2$ or $–1/2$.

19. The co-ordinates of the points A and B are respectively (–3, 2) and (2, 3). P and Q are points on the line joining A and B such that $AP = PQ = QB$. A square $PQRS$ is constructed on PQ as one side, the co-ordinates of R can be

(a) $\left(-\dfrac{4}{3}, \dfrac{7}{3}\right)$ (b) $\left(0, \dfrac{13}{3}\right)$

(c) $\left(\dfrac{1}{3}, \dfrac{8}{3}\right)$ (d) $\left(\dfrac{2}{3}, 1\right)$

Solution: (b, d) P, Q divides AB in the ratio of 1: 2 and 2: 1 respectively and hence the co-ordinates of P and Q are

$$\left[\frac{1\times 2+2(-3)}{3}, \frac{1\times 3+2\times 2}{3}\right]$$

and $\left[\frac{2\times 2+1(-3)}{3}, \frac{2\times 3+1\times 2}{3}\right]$

$\Rightarrow P\left(-\frac{4}{3}, \frac{7}{3}\right)$ and $Q\left(\frac{1}{3}, \frac{8}{3}\right)$ then $PQ = \frac{\sqrt{26}}{3}$

Let the co-ordinates of R be (x, y)

Then, $QR = PQ \Rightarrow \left(x-\frac{1}{3}\right)^2+\left(y-\frac{8}{3}\right)^2 = \frac{26}{9}$ ..(i)

and QR is perpendicular to PQ

$$\Rightarrow \frac{y-\frac{8}{3}}{x-\frac{1}{3}} \times \frac{\frac{8}{3}-\frac{7}{3}}{\frac{1}{3}+\frac{4}{3}} = -1$$

$$\Rightarrow \left(y-\frac{8}{3}\right) = -5\left(x-\frac{1}{3}\right) \quad .....(2)$$

From equation, (i) and (ii), we get

$$26\left(x-\frac{1}{3}\right)^2 = 26/9 \quad \Rightarrow \quad x-\frac{1}{3} = \pm\frac{1}{3}$$

$\Rightarrow x = 0$ or $x = 2/3$

From equation (ii), when $x = 0$, $y = 13/3$ and when $x = 2/3$, $y = 1$

So, the required co-ordinates of R are $\left(0, \frac{13}{3}\right)$ or $\left(\frac{2}{3}, 1\right)$

20. A line passing through the point (2,2) encloses with axes an algebraic area λ. The intercepts on the axes made by the line are given by the roots of equation.

(a) $x^2 - |\lambda|x + |\lambda| = 0$ (b) $x^2 - |\lambda|x + 2|\lambda| = 0$
(c) $x^2 + |\lambda|x + 2|\lambda| = 0$ (d) None of these

Solution: (b) Let the line cut intercepts 'a' and 'b' respectively from the axes

$\therefore$ Area $= \frac{1}{2}(ab) = |\lambda| \quad \Rightarrow ab = |2\lambda|$ ..... (i)

Also equation of line is $\frac{x}{a}+\frac{y}{b} = 1$

As (2, 2) lies on it, $\frac{2}{a}+\frac{2}{b} = 1$

$\Rightarrow 2(a + b) = ab$ .... (ii)

$\therefore$ using (i) in (ii), we get $2(a + b) = 2|\lambda|$

$\Rightarrow a+\frac{2|\lambda|}{a} = |\lambda| \Rightarrow a^2 - a|\lambda| + 2|\lambda| = 0$

Hence intercepts on axes mades by the line are given by the roots of $x^2 - |\lambda|x + 2|\lambda| = 0$.

21. A ray of light is sent along the line which passes through the point (2,3).The ray is reflected from the point on x-axis. If the reflected ray passes through the point (6,4).The co-ordinates of P are

(a) $\left(\frac{26}{7}, 0\right)$ (b) $\left(-\frac{26}{7}, 0\right)$
(c) $\left(\frac{13}{7}, 0\right)$ (d) $\left(\frac{-13}{7}, 0\right)$

Solution: (a) Let the reflected ray make an angle θ with + *ve* direction of x axis, then the incident ray makes angle $(\pi - \theta)$ with positive direction of x-axis. Now the slope of the incident ray is

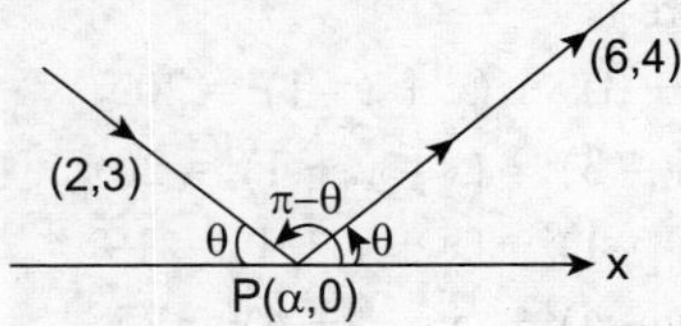

$$\Rightarrow \frac{0-3}{\alpha-2} = \tan(\pi - \theta)$$

$$\Rightarrow \tan(\pi - \theta) = \frac{0-3}{\alpha-2} \quad .....(1)$$

Slope of reflected ray is

$$\frac{4-0}{6-\alpha} = \tan\theta \quad .....(2)$$

from (1) and (2) we get, $\alpha = \frac{26}{7}$.

$\Rightarrow$ The co-ordinates of point P are $\left(\frac{26}{7}, 0\right)$.

22. $OPQR$ is a square and M, N are the mid-points of the sides PQ and QR respectively. If the ratio of the area of the square and ΔOMN is λ: 6, then the value of $\frac{\lambda}{4}$ is equal to

(a) 2 (b) 1
(c) 4 (d) None of these

Solution: (c) We can assume that OP and OR are along x-axis and y-axis respectively.

Let $OP = a$, then area (square $OPQR$) = a^2

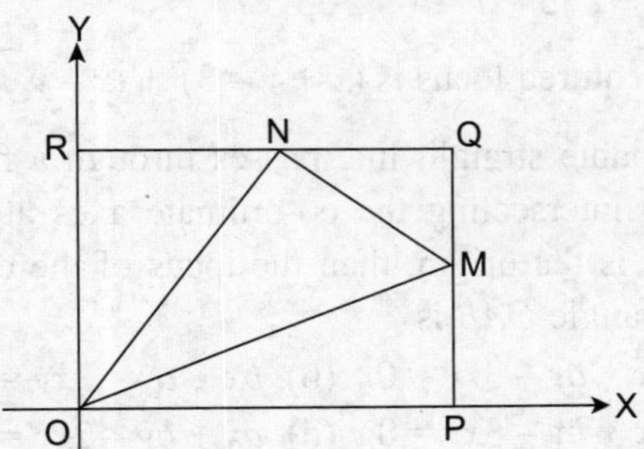

Co-ordinates of M and N are $\left(a, \frac{a}{2}\right)$ and $\left(\frac{a}{2}, a\right)$ respectively

$\therefore$ area $\Delta OMN = \frac{1}{2}\begin{vmatrix} a & a/2 \\ a/2 & a \end{vmatrix} = \frac{3a^2}{8}$.

$\therefore \frac{8}{3} = \frac{\lambda}{6}$ $\quad\therefore\quad \lambda = 16 \Rightarrow \frac{\lambda}{4} = \frac{16}{4} = 4$.

23. Vertices of a triangle are (1, 2); $\left(\sqrt{5}\cos\theta, \sqrt{5}\sin\theta\right)$ and $\left(\sqrt{5}\sin\theta, -\sqrt{5}\cos\theta\right)$. Then locus of its orthocentre is

(a) $(x - y + 3)^2 + (x + y - 1)^2 = 20$
(b) $(x + y - 3)^2 + (x - y + 1)^2 = 20$
(c) $(x + y - 1)^2 + (x + y + 1)^2 = 19$
(d) $(x + y - 4)^2 + (x - y + 4)^2 = 25$

Solution: (b) Let $A \equiv (1,2)$, $B \equiv \left(\sqrt{5}\cos\theta, \sqrt{5}\sin\theta\right)$ and $C \equiv \left(\sqrt{5}\sin\theta, -\sqrt{5}\cos\theta\right)$. Distance of A, B, C from (0,0) is $\sqrt{5}$ units

$\Rightarrow$ circumcentre of ΔABC is origin (0, 0)

Let $G(h, k)$ be the centroid of triangle

$\Rightarrow 3h = 1 + \sqrt{5}(\cos\theta + \sin\theta)$

and $3k = 2 + \sqrt{5}(\sin\theta - \cos\theta)$

If 'O' is orthocentre of triangle having co-ordinates (a, b), then $OG : GC = 2:1$

$\Rightarrow a = 3h, b = 3k$

$\Rightarrow \cos\theta + \sin\theta = \frac{a-1}{\sqrt{5}}$ ....(i)

and $\sin\theta - \cos\theta = \frac{b-2}{\sqrt{5}}$

$\sin\theta = \frac{a+b-3}{2\sqrt{5}}, \cos\theta = \frac{a-b+1}{2\sqrt{5}}$

$\Rightarrow \left(\frac{a+b-3}{2\sqrt{5}}\right)^2 + \left(\frac{a-b+1}{2\sqrt{5}}\right)^2 = 1$

$\therefore$ Required locus is $(x + y - 3)^2 + (x - y + 1)^2 = 20$.

24. A variable straight line passes through a fixed point (a, b) intersecting the co-ordinate axes at A and B. If 'O' is the origin, then the locus of the centroid of the triangle OAB is

(a) $bx + ay - 3xy = 0$ (b) $bx + ay - 2xy = 0$
(c) $ax + by - 3xy = 0$ (d) $ax + by - 2xy = 0$

Solution: (a) Equation of line AB is $y - b = m(x - a)$

$\Rightarrow A \equiv (a - b/m, 0)$ and $B \equiv (0, b - am)$

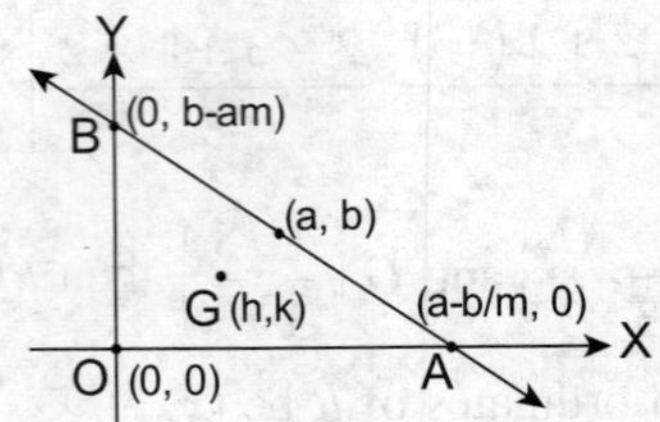

$\therefore$ Centroid of ΔOAB is given by

$\therefore G\left(\frac{a - \frac{b}{m}}{3}, \frac{b - am}{3}\right)$

$\Rightarrow h = \frac{a - \frac{b}{m}}{3}, k = \frac{b - am}{3}$

On eliminating 'm', we get required locus $bh + ak - 3hk = 0$, replacing (h, k) by (x, y), we get $bx + ay - 3xy = 0$.

25. Consider three points $P = (-\sin(\beta - \alpha), -\cos\beta)$, $Q = (\cos(\beta - \alpha), \sin\beta)$ and $R = (\cos(\beta - \alpha + \theta), \sin(\beta - \theta))$, where $0 < \alpha, \beta, \theta < \frac{\pi}{4}$. Then

(a) P lies on the line segment RQ
(b) Q lies on the line segment PR
(c) R lies on the line segment QP
(d) P, Q, R are non-collinear.

Solution: (d) $P \equiv (-\sin(\beta - \alpha), -\cos\beta)$,
$Q \equiv (\cos(\beta - \alpha), \sin\beta)$
and $R \equiv (\cos(\beta - \alpha + \theta)), \sin(\beta - \theta)$; where
$0 < \alpha, \beta, \theta < \frac{\pi}{4}$

$x_R = \cos(\beta - \alpha)\cos\theta - \sin(\beta - \alpha)\sin\theta$

$\Rightarrow x_R = x_Q \cdot \cos\theta + x_P \cdot \sin\theta$

and $y_R = \sin\beta\cos\theta - \cos\beta\sin\theta$

$\Rightarrow y_R = y_Q \cdot \cos\theta + y_P \cdot \sin\theta$

For P, Q, R to be collinear $\sin\theta + \cos\theta = 1$ (by area of triangle $PQR = 0$)

$\Rightarrow \sin\left(\theta + \frac{\pi}{4}\right) = \frac{1}{\sqrt{2}}$

$\Rightarrow$ not possible for the given interval $\theta \in \left(0, \frac{\pi}{4}\right)$

$\Rightarrow$ non-collinear.

SECTION-II

SUBJECTIVE-TYPE SOLVED EXAMPLES

1. Show that the equation of the locus of a point which moves so that the sum of its distances from two given points $(k,0)$ and $(-k,0)$ is equal to $2a$ is $\frac{x^2}{a^2}+\frac{y^2}{a^2-k^2}=1$; where $a > |k|$.

Solution: Let $P(\alpha, \beta)$ be a point which is moving according to given conditions, then $PA + PB = 2a$, where $A \equiv (k,0)$ and $B \equiv (-k,0)$

$$\Rightarrow \sqrt{(\alpha-k)^2+(\beta-0)^2}+\sqrt{(\alpha+k)^2+(\beta-0)^2}=2a$$

$$\Rightarrow \sqrt{(\alpha-k)^2+\beta^2}=2a-\sqrt{(\alpha+k)^2+\beta^2}$$

$$\Rightarrow (\alpha-k)^2+\beta^2=4a^2+(\alpha+k)^2+\beta^2-4a\sqrt{(\alpha+k)^2+\beta^2}$$

(On squaring on both sides)

$$\Rightarrow 4a\sqrt{(\alpha+k)^2+\beta^2}=4\alpha k+4a^2$$

$$\Rightarrow a\sqrt{(\alpha+k)^2+\beta^2}=a^2+\alpha k$$

$$\Rightarrow a^2\{\alpha^2+k^2+2\alpha k+\beta^2\}=a^4+2a^2\alpha k+\alpha^2k^2$$

$$\Rightarrow \frac{\alpha^2}{a^2}+\frac{\beta^2}{a^2-k^2}=1$$

$\therefore \frac{x^2}{a^2}+\frac{y^2}{a^2-k^2}=1$ is the required locus of point P.

2. The vertices of a triangle are $A(x_1, x_1 \tan\alpha)$, $B(x_2, x_2 \tan\beta)$ and $C(x_3, x_3 \tan\gamma)$. If the circumcentre of ΔABC coincides with the origin and $H(a, b)$ be its orthocentre, then find $a : b$.

Solution: Let R be the radius of the circumcircle and O be the origin, then, $AO = \sqrt{(x_1^2 + x_1^2\tan^2\alpha)}$

$\Rightarrow R = x_1 \sec\alpha \Rightarrow x_1 = R\cos\alpha$

Similarly, $x_2 = R\cos\beta$ and $x_3 = R\cos\gamma$

So, the co-ordinates of vertices are $A(R\cos\alpha, R\sin\alpha)$, $B(R\cos\beta, R\sin\beta)$, $C(R\cos\gamma, R\sin\gamma)$

Hence, the co-ordinates of centroid G are

$$\left(\frac{\Sigma R\cos\alpha}{3}, \frac{\Sigma R\sin\alpha}{3}\right)$$

Since, the orthocentre $H(a, b)$, circumcentre $O(0, 0)$ and the centroid G are collinear, therefore slope of OH = slope of OG

$$\Rightarrow \frac{b}{a}=\frac{R(\sin\alpha+\sin\beta+\sin\gamma)}{R(\cos\alpha+\cos\beta+\cos\gamma)}$$

$$\therefore \frac{a}{b}=\frac{\cos\alpha+\cos\beta+\cos\gamma}{\sin\alpha+\sin\beta+\sin\gamma}.$$

3. Find the distance between the circumcentre and orthocenter of the triangle whose vertices are $(0, 0)$, $(6,8)$ and $(-4, 3)$.

Solution: Let $O(0, 0)$; $B(6, 8)$; $C(-4, 3)$ be the vertices of ΔOBC as shown below:

Slope of $OB = \frac{4}{3}$; Slope of $OC = -\frac{3}{4}$

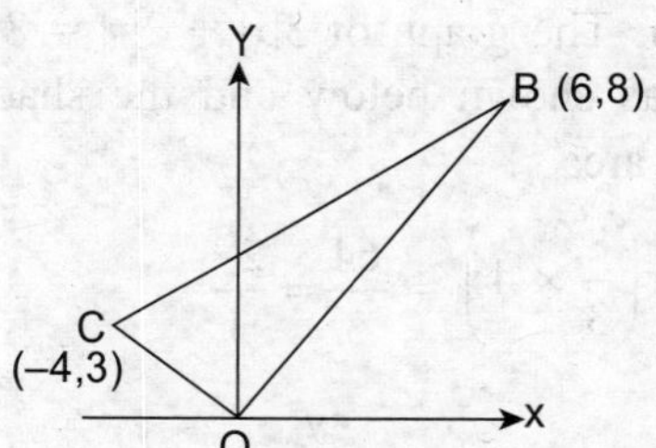

$\therefore \angle BOC = \pi/2$

ΔOBC is right angled at O

Circumcentre ≡ mid-point of hypotenuse $= \left(1, \frac{11}{2}\right)$.

Orthocenter ≡ vertex $O(0, 0)$ (as ΔOBC is right angled triangled)

Distance between circumcentre and orthocentre

$$= \sqrt{\left(1+\frac{121}{4}\right)} = \frac{5\sqrt{5}}{2} \text{ units.}$$

4. A man starts from the point $P(-3, 4)$ and reaches point $Q(0, 1)$ after touching x axis at R, such that $PR + RQ$ is minimum, then find the co-ordinates of R.

Solution: Let $R = (\alpha, 0)$;

For $PR + RQ$ to be minimum, it should be the path of light and thus we have

$$\Delta APR \sim \Delta BQR \Rightarrow \frac{AR}{RB}=\frac{PA}{QB} \Rightarrow \frac{|\alpha+3|}{|\alpha|}=\frac{4}{1}$$

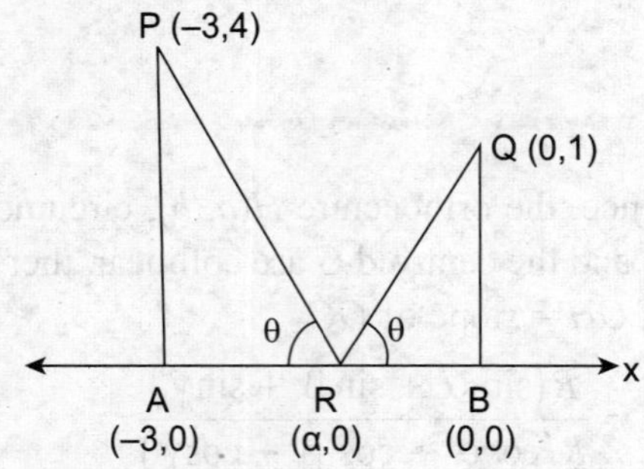

$\Rightarrow$ $|\alpha + 3| = 4|\alpha|$

Clearly, $\alpha > -3 \Rightarrow \alpha + 3 > 0 \Rightarrow \alpha + 3 = 4|\alpha|$,

But, $\alpha < 0 \Rightarrow \alpha + 3 = -4\alpha$

$\Rightarrow$ $5\alpha = -3$

$\Rightarrow$ $\alpha = -3/5$

$\therefore$ Co-ordinates of R are $\left(\frac{-3}{5}, 0\right)$.

5. If $f(x + y) = f(x) \cdot f(y)$; $x, y \in \mathbb{R}$ and $f(1) = 2$, then area enclosed by $3\,|x| + 2\,|y| \le 8$ in terms of image $f(a)$ for some a.

Solution: The graph of $3|x| + 2|y| = 8$ is a rhombus ABCD as shown below and the shaded portion is required area

$$= 4 \times \frac{1}{2}\left(\frac{8}{3} \times 4\right) = \frac{64}{3} = \frac{2^6}{3}.$$

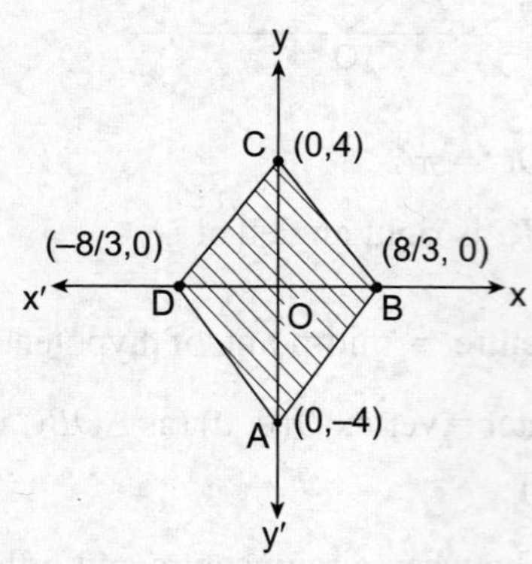

Also, $f(x + y) = f(x)\,f(y)$

$\therefore f(2) = f(1)\,f(1) = 2^2$ (taking $x = y = 1$)

$f(3) = f(1 + 2) = f(1)\,f(2) = 2^3$ (taking $x = 1$, $y = 2$) or $x = 2$, $y = 1$

and so on

$\therefore$ $f(n) = 2^n$

Hence, area $= \dfrac{2^6}{3} = \dfrac{f(6)}{3}$ sq. units.

6. Let AB be a line segment of length 4 units with the point A on the line $y = 2x$ and B on the line $y = x$. Then, find locus of middle point of all such line segments.

Solution: Let $B = (a, a)$ and middle point of AB is (h, k). Then, $A \equiv (2h - a, 2k - a)$

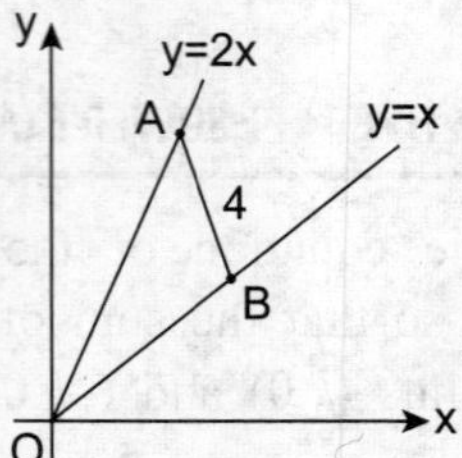

Which lies on $y = 2x$, so, $(2k - a) = 2(2h - a)$

$\therefore$ $a = 4h - 2k$ .....(i)

Also $|AB| = 4$ (given)

$\Rightarrow$ $\sqrt{(2h - 2a)^2 + (2k - 2a)^2} = 4$

or $(h - a)^2 + (k - a)^2 = 4$

or $[h - (4h - 2k)]^2 + [k - (4h - 2k)]^2 = 4$

$\Rightarrow$ $25h^2 + 13k^2 - 36hk = 4$

$\Rightarrow$ Required locus is $25x^2 + 13y^2 - 36xy - 4 = 0$;

Here, $h^2 < ab$ and $\Delta \ne 0$

$\Rightarrow$ The required locus in an ellipse.

7. If the area of the triangle whose vertices are (b, c), (c, a) and (a, b) is Δ, and the area of triangle whose vertices are $(ac - b^2, ab - c^2)$, $(ba - c^2, bc - a^2)$ and $(cb - a^2, ca - b^2)$ is Δ', then find the relation between Δ and Δ'.

Solution: The area of the triangle formed by the first set of vertices is $\Delta = \frac{1}{2}\left\|\begin{matrix} b & c & 1 \\ c & a & 1 \\ a & b & 1 \end{matrix}\right\|$

$= 1/2\,(a^2 + b^2 + c^2 - bc - ca - ab)$ ....(i)

and the area of the triangle formed by the second set of vertices is

$$\Delta = \frac{1}{2}\left\|\begin{matrix} ac - b^2 & ab - c^2 & 1 \\ ba - c^2 & bc - a^2 & 1 \\ cb - a^2 & ca - b^2 & 1 \end{matrix}\right\|$$

Applying $R_2 \to R_2 - R_1$ and $R_3 \to R_3 - R_1$, we get

$$\Delta' = \frac{1}{2}\left\|\begin{matrix} ac - b^2 & ab - c^2 & 1 \\ (b - c)(a + b + c) & (c - a)(a + b + c) & 0 \\ (b - a)(a + b + c) & (c - b)(a + b + c) & 0 \end{matrix}\right\|$$

$$\frac{1}{2}(a+b+c)^2\begin{Vmatrix} b-c & c-a \\ b-a & c-b \end{Vmatrix}$$

$= (1/2)\ (a + b + c)^2 . [a^2 + b^2 + c^2 - ab - bc - ca]$

$\therefore\ \Delta' = (a + b + c)^2\ \Delta$ [from Equation (i)]

8. Given the base and the product of the tangents of the halves of the base angles of a triangle, show that the locus of the vertex of the triangle is an ellipse.

Solution: Let ABC be the triangle, where BC and $\tan B/2\ \tan C/2$ are fixed quantities. To show that the locus of A is an ellipse. From an elementary trigonometric property of a triangle, we have

$$\tan\frac{B}{2} = \sqrt{\frac{(s-c)(s-a)}{s(s-b)}}$$

and $\tan\dfrac{C}{2} = \sqrt{\dfrac{(s-a)(s-b)}{s(s-c)}}$; where $2s = a + b + c$

Hence $\tan\dfrac{B}{2}.\tan\dfrac{C}{2}$ = constant (k, say)

implies $\sqrt{\dfrac{(s-c)(s-a)}{s(s-b)}}\cdot\sqrt{\dfrac{(s-a)(s-b)}{s(s-c)}} = k$

i.e., $(s - a) = ks$. Since $BC = a$(given) it follows that

$s = \dfrac{a}{1-k}$ = constant i.e., $\dfrac{a+b+c}{2}$ = constant

or $b + c$ = (constant) i.e., $BA + CA$ = constant.

Since B and C are fixed points and A moves such that the sum of its distances from B and C is fixed, so it follows that A describes an ellipse with B and C as its foci.

9. If t_1, t_2 and t_3 are distinct, show that the points $(t_1, 2at_1 + at_1^3)$; $(t_2, 2at_2 + at_2^3)$ and $(t_3, 2at_3 + at_3^3)$ are collinear if $t_1 + t_2 + t_3 = 0$. Given that $a \neq 0$.

Solution: The given points are collinear if

$$\begin{vmatrix} t_1 & 2at_1+at_1^3 & 1 \\ t_2 & 2at_2+at_2^3 & 1 \\ t_3 & 2at_3+at_3^3 & 1 \end{vmatrix} = 0 \Rightarrow a\begin{vmatrix} t_1 & 2t_1+t_1^3 & 1 \\ t_2 & 2t_2+t_2^3 & 1 \\ t_3 & 2t_3+t_3^3 & 1 \end{vmatrix} = 0$$

Applying $R_2 \to R_2 - R_1$, $R_3 \to R_3 - R_1$, we get

$$\begin{vmatrix} t_1 & 2t_1+t_1^3 & 1 \\ t_2-t_1 & 2(t_2-t_1)+(t_2^3-t_1^3) & 0 \\ t_3-t_1 & 2(t_3-t_1)+(t_3^3-t_1^3) & 0 \end{vmatrix} = 0$$

$$\Rightarrow (t_2-t_1)(t_3-t_1)\begin{vmatrix} t_1 & 2t_1+t_1^3 & 1 \\ 1 & 2+t_2^2+t_1^2+t_2t_1 & 0 \\ 1 & 2+t_3^2+t_1^2+t_3t_1 & 0 \end{vmatrix} = 0$$

$$\Rightarrow (t_2-t_1)(t_3-t_1)(t_3-t_2)(t_3+t_2+t_1) = 0$$

$$\Rightarrow t_1+t_2+t_3 = 0 \qquad [\because t_1 \neq t_2 \neq t_3 \neq t_1].$$

10. Prove that the points $P(k, 2 - 2k)$, $Q(-k + 1, 2k)$ and $R(-4 - k, 6 - 2k)$ are collinear for $k = -1$ and $k = 1/2$.

Solution: Slope method: $m_{PQ} = m_{PR}$

$$\Rightarrow \frac{1-2k}{4k-2} = \frac{-4-k-k}{6-2k-2+2k}$$

$$\Rightarrow \frac{1-2k}{-2(1-2k)} = \frac{-2(2+k)}{4}$$

when $k \neq 1/2$ $\Rightarrow -\dfrac{1}{2} = \dfrac{-1}{2}(k+2)$

$\Rightarrow k + 2 = 1$ $\Rightarrow k = -1$

when $k = 1/2$; points P and Q become coincident at (1/2, 1)

$\therefore$ P, Q and R will be collinear

Area method: The given points are collinear if area of triangle PQR is zero.

$$\Rightarrow \begin{vmatrix} k & 2-2k & 1 \\ -k+1 & 2k & 1 \\ -4-k & 6-2k & 1 \end{vmatrix} = 0$$

$$\Rightarrow \begin{vmatrix} k & 2-2k & 1 \\ -2k+1 & 4k-2 & 0 \\ -4-2k & 4 & 0 \end{vmatrix} = 0$$

(operating $R_2 \to R_2 - R_1$ and $R_3 \to R_3 - R_1$)

$\Rightarrow 4(-2k + 1) - (-4 - 2k)(4k - 2) = 0$

$\Rightarrow (1 - 2k)(4 - 8 - 4k) = 0$

$\Rightarrow (1 - 2k)(k + 1) = 0$

$\Rightarrow k = -1$ or $k = 1/2$.

11. If $A = (at^2, 2at)$, $B = \left(\dfrac{a}{t^2}, \dfrac{2a}{t}\right)$ and $S(a, 0)$, then prove that $2a$ is the harmonic mean of SA and SB $(a > 0)$.

Solution: Here $SA^2 = (a - at^2)^2 + (0 - 2at)^2 = a^2\{(1 - t^2)^2 + 4t^2\}$

$\therefore\ SA^2 = a^2(1 + t^2)^2$

$\Rightarrow SA = a(1 + t^2)$

and $SB^2 = \left(a - \dfrac{a}{t^2}\right)^2 + \left(0 - \dfrac{2a}{t}\right)^2$

$$= a^2\left\{\left(1-\frac{1}{t^2}\right)^2+\frac{4}{t^2}\right\}$$

$$\therefore \quad SB^2 = a^2\left(1+\frac{1}{t^2}\right)^2; \quad \therefore \quad SB = a\left(1+\frac{1}{t^2}\right)$$

$$\therefore \quad \frac{1}{SA}+\frac{1}{SB}=\frac{1}{a(1+t^2)}+\frac{1}{a\left(1+\frac{1}{t^2}\right)}$$

$$=\frac{1}{a}\left\{\frac{1}{1+t^2}+\frac{t^2}{1+t^2}\right\}=\frac{1}{a}\cdot\frac{1+t^2}{1+t^2}=\frac{1}{a}=\frac{2}{2a}$$

$\therefore$ $2a$ is the harmonic mean of SA and SB.

12. The point A divides the join of $P \equiv (-5, 1)$ and $Q = (3, 5)$ in the ratio k: 1. Find the two values of k for which the area of ΔABC, where $B \equiv (1, 5)$, $C \equiv (7, -2)$ is equal to 2 square units.

Solution: Co-ordinates of A, dividing the join of $P \equiv (-5, 1)$ and $Q \equiv (3, 5)$ in the ratio k: 1 are given by

$$\left(\frac{3k-5}{k+1}, \frac{5k+1}{k+1}\right).$$

Also, area of the ΔABC is given by

$$\Delta = \left|\frac{1}{2}\sum x_1(y_2-y_3)\right|$$

$$=\frac{1}{2}\left|\left[x_1(y_2-y_3)+x_2(y_3-y_1)+x_3(y_1-y_2)\right]\right|$$

$$=\left|\frac{1}{2}\left\{\frac{3k-5}{k+1}(7)+\left(-2-\frac{5k+1}{k+1}\right)+7\left(\frac{5k+1}{k+1}-5\right)\right\}\right|=2$$

$$\Rightarrow (1/2)\left\{\frac{3k-5}{k+1}(7)+\left(-2-\frac{5k+1}{k+1}\right)+7\left(\frac{5k+1}{k+1}-5\right)\right\}=\pm 2$$

$\Rightarrow 14k - 66 = 4k + 4, \Rightarrow 10k = 70, \Rightarrow k = 7$

or $14k - 66 = -4k - 4 \Rightarrow 18k = 62,$

$\Rightarrow k = (31/9).$

Therefore value of $k = 7, 31/9$.

13. Two vertices of a triangle are B (3, –1) and C (–2, 3) and its orthocentre is origin, find the co-ordinates of the third vertex.

Solution: Let the required vertex be A (h, k). Point O (0, 0) will be the orthocentre iff

(a) AOK be perpendicular to BC

(b) CON be perpendicular to AB as shown below

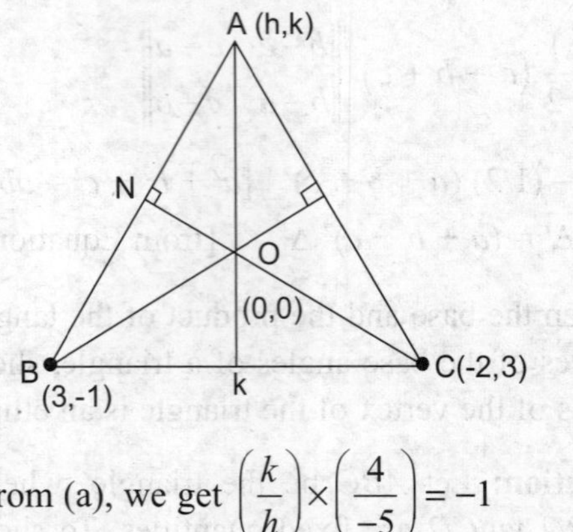

From (a), we get $\left(\frac{k}{h}\right)\times\left(\frac{4}{-5}\right)=-1$

i.e., $4k = 5h$ ... (i)

From (b) we get $\left(\frac{3}{-2}\right)\times\left(\frac{k+1}{h-3}\right)=-1$

i.e., $3k = 2h - 9$ ... (ii)

Solving (i) and (ii), we get $h=\frac{-36}{7}$ and $k=\frac{-45}{7}$.

$$\therefore \quad A \equiv \left(\frac{-36}{7}, \frac{-45}{7}\right).$$

14. A line L intersects three sides BC, CA and AB of a triangle in P, Q, R respectively, then $\frac{BP}{PC}\cdot\frac{CQ}{QA}\cdot\frac{AR}{RB}=-1$ where $\frac{BP}{PC},\frac{CQ}{QA},\frac{AR}{RB}$ represents ratios in which P, Q and R divides BC, CA, and AR respectively internally.

Solution: Let the straight line $L : ax + by + c = 0$ divides BC externally in ratio $|\lambda| : 1$ i.e., $BP : CP = \lambda : 1$; $\lambda < 0$, AC in the ratio $\mu : 1$ internally i.e., $CQ : QA = \mu : 1$; $\mu > 0$; AB in the ratio $\nu : 1$ internally i.e., $AR : RB = \nu : 1$ as shown below.

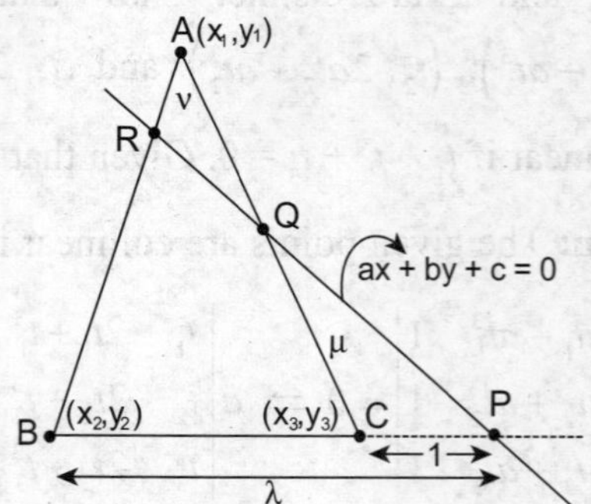

$$\Rightarrow \lambda = -\left(\frac{L_2}{L_3}\right) = -\left(\frac{ax_2+by_2+c}{ax_3+by_3+c}\right)=\frac{BP}{PC}$$

$$\Rightarrow \mu = -\left(\frac{L_3}{L_1}\right) = -\left(\frac{ax_3+by_3+c}{ax_1+by_1+c}\right)=\frac{CQ}{QA}$$

$$\Rightarrow \nu = -\left(\frac{L_1}{L_2}\right) = -\left(\frac{ax_1 + by_1 + c}{ax_2 + by_2 + c}\right) = \frac{AR}{RB}$$

$$\Rightarrow \lambda.\mu.\nu = -\left(\frac{L_2}{L_3}.\frac{L_3}{L_1}.\frac{L_1}{L_2}\right) = -1$$

$$\Rightarrow \left(\frac{PB}{PC}\right)\left(\frac{CQ}{AQ}\right)\left(\frac{AR}{BR}\right) = -1$$

15. Find the orthocentre of the triangle whose vertices are $(at_1t_2, a(t_1 + t_2))$, $(at_2t_3, a(t_2 + t_3))$ and $(at_1t_3, a(t_1 + t_3))$.

Solution: Let ABC be a triangle whose vertices are $A(at_1t_2, a(t_1 + t_2))$; $B(at_2t_3, a(t_2 + t_3))$ and $C(at_1t_3, a(t_1 + t_3))$. Then

$$\text{Slope of } BC = \frac{a(t_2 + t_3) - a(t_1 + t_3)}{at_2t_3 - at_1t_3} = \frac{1}{t_3}$$

$$\text{Slope of } AC = \frac{a(t_1 + t_3) - a(t_1 + t_2)}{at_1t_3 - at_1t_2} = \frac{1}{t_1}$$

So, the equation of a line through A perpendicular to BC is

$$y - a(t_1 + t_2) = -t_3(x - at_1t_2) \quad \text{... (i)}$$

and the equation of a line through B perpendicular to AC is

$$y - a(t_2 + t_3) = -t_1(x - at_2t_3) \quad \text{... (ii)}$$

The point of intersection of (i) and (ii) is the orthocentre.

Subtracting (ii) from (i), we get $x = -a$.

Putting $x = -a$ in (i), we get $y = a(t_1 + t_2 + t_3 + t_1t_2t_3)$

Hence, the co-ordinates of the orthocentre are

$$(-a, a(t_1 + t_2 + t_3 + t_1t_2t_3)).$$

Remark: The above example shows that the orthocentre of triangle formed by the tangents at any three points on parabola lies on its directrix.

16. Find the co-ordinates of the vertices of a square inscribed in the triangle with vertices $A(0, 0)$, $B(2, 1)$, $C(3, 0)$; given that two of its vertices are on the side AC and one on each of the sides AB and BC.

Solution: Let $PQRS$ be the square inscribed in the triangle ABC with length of each side equal to a. Let the co-ordinates of P be $(p, 0)$.

Then the co-ordinates of Q, R, S will be $(p + a, 0)$, $(p + a, a)$ and (p, a) respectively.

Now equation of AB is $y = (1/2)x$ and $S(p, a)$ lies on it.

Therefore, $a = p/2$ or $p = 2a$. Also equation of BC is $x + y = 3$ and $R(p + a, a)$ lies on it.

Therefore, $p + a + a = 3 \Rightarrow 4a = 3$

$\Rightarrow a = 3/4$ and $p = 3/2$.

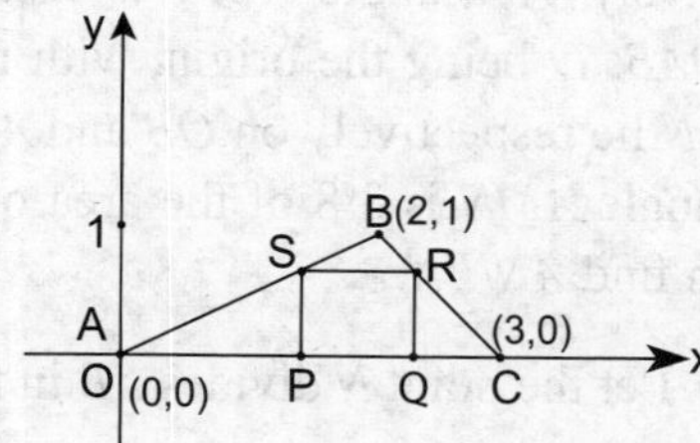

Hence, the required co-ordinates are (3/2, 0); (9/4, 0); (9/4, 3/4) and (3/2, 3/4).

17. Two points P and Q are given, R is a variable point on one side of line PQ such that $\angle RPQ - \angle RQP$ is positive constant 2α. Find the locus of point R.

Solution: Without loss of generality let us assume that origin is the mid-point of line segment PQ, where the co-ordinates of P and Q are (-a,0) and (a, 0) respectively. Let at an instant the co-ordinates of variable point R be (h, k) as shown below.

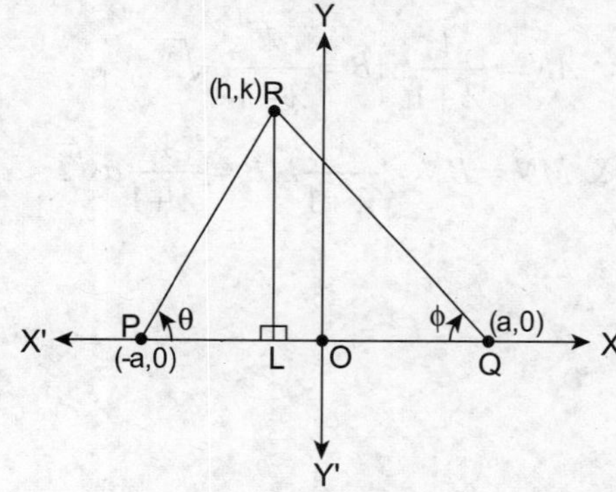

$A.T.Q$, $\angle RPQ - \angle RQP = 2\alpha$; $\alpha > 0$

$\Rightarrow \theta - \phi = 2\alpha > 0$ (constant angle) ....(i)

$$\text{and } \tan\theta = \frac{k}{h+a} \Rightarrow \tan\phi = \frac{k}{a-h} \quad \text{........(ii)}$$

$$\Rightarrow \frac{\tan\theta - \tan\phi}{1 + \tan\theta\tan\phi} = \tan 2\alpha \text{ (using (i) and (ii))}$$

$$\Rightarrow \frac{\frac{k}{a+h} - \frac{k}{a-h}}{1 + \frac{k^2}{a^2 - h^2}} = \tan 2\alpha$$

$$\Rightarrow \frac{2kh}{(h^2 - k^2 - a^2)} = \tan 2\alpha$$

$$\Rightarrow (h^2 - k^2 - a^2)\tan 2\alpha - 2hk = 0 \text{ or}$$

$\left(x^2-y^2-a^2\right)\tan 2\alpha - 2xy = 0$ is the required locus of point R.

18. The line $x + y = a$ meets the axis of x and y at A and B respectively. A triangle AMN is inscribed in the triangle OAB, O being the origin, with right angle at N. M and N lie respectively on OB and AB. If the area of the triangle AMN is 3/8 of the area of the triangle OAB, then find AN/BN.

Solution: Let the point N divides AB in the ratio λ : 1 i.e., $AN : BN$ = λ : 1

Now according to the question,

$$\text{Area of } \Delta AMN = \frac{3}{8}(ar\,\Delta OAB)$$

$$\Rightarrow \frac{1}{2}(AN)(MN) = \frac{3}{8}\left(\frac{1}{2}a^2\right)$$

$$\Rightarrow \frac{1}{2}AN \times MN = \frac{3}{16}a^2 \quad \text{.... (i)}$$

$$\Rightarrow \frac{1}{2}\left(\frac{\lambda}{\lambda+1}\right)a\sqrt{2}\left(\frac{1}{\lambda+1}\right)a\sqrt{2} = \frac{3}{16}a^2$$

$$\Rightarrow \begin{cases} \because AN = \dfrac{\lambda}{\lambda+1}.AB = \dfrac{\lambda}{\lambda+1}a\sqrt{2} \\ \& \; MN = BN = \dfrac{1}{\lambda+1}.AB = \dfrac{1}{\lambda+1}a\sqrt{2} \end{cases}$$

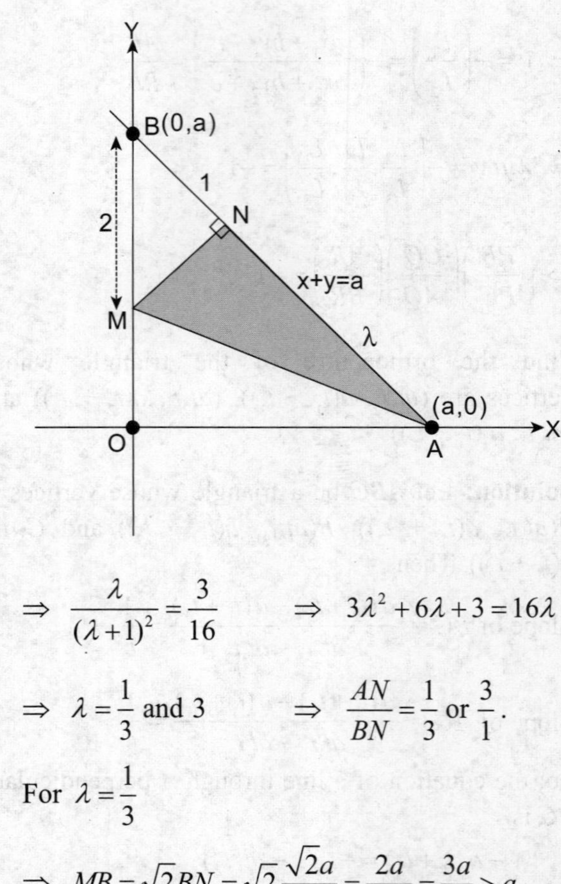

$$\Rightarrow \frac{\lambda}{(\lambda+1)^2} = \frac{3}{16} \quad \Rightarrow 3\lambda^2 + 6\lambda + 3 = 16\lambda$$

$$\Rightarrow \lambda = \frac{1}{3} \text{ and } 3 \quad \Rightarrow \frac{AN}{BN} = \frac{1}{3} \text{ or } \frac{3}{1}.$$

For $\lambda = \frac{1}{3}$

$$\Rightarrow MB = \sqrt{2}BN = \sqrt{2}\frac{\sqrt{2}a}{\lambda+1} = \frac{2a}{4/3} = \frac{3a}{2} > a$$

(a contradiction) $\therefore \quad \frac{AN}{BN} = \lambda = 3:1$

TUTORIAL EXERCISE

SECTION–III

OBJECTIVE ASSIGNMENT (ONLY ONE CORRECT ANSWER)

1. If A and B are the points (–3, 4) and (3, –4) respectivley, then the co-ordinates of the points C on AB produced such that $AC = 3BC$ are
(a) (0, 0) (b) (–6, 8)
(c) (6, –8) (d) (–6, –8)

2. The points $(3a, 0)$, $(0, 3b)$ and $(a, 2b)$ are
(a) vertices of an equilateal triangle
(b) vertices of an isosceles triangle
(c) vertices of a right angled isosceles Δ
(d) collinear

3. If the line segment joining (2,3) and (–1, 2) is divided internally in the ratio 3: 4 by the line $x + 2y = k$, then k is
(a) $\frac{41}{7}$ (b) $\frac{5}{7}$
(c) $\frac{36}{7}$ (d) $\frac{31}{7}$

4. The opposite angular points of a square are (3,4) and (1,–1). Then the co-ordinates of other two vertices are
(a) $D(1/2, 9/2)$, $B(-1/2, 5/2)$
(b) $D(-1/2, 9/2)$, $B(1/2, 5/2)$
(c) $D(9/2, 1/2)$, $B(-1/2, 5/2)$
(d) None of these

5. The vertices of the triangle ABC are $A(1,2)$, $B(3,4)$ and $C(2,3)$, then the greatest angle of the triangle is
(a) 75° (b) 105°
(c) 120° (d) None of these

6. $P(3, 1)$, $Q(6, 5)$ and $R(x, y)$ are three points such that the angle PRQ is a right angle and the area of $\Delta RPQ = 7$. Then the number of such points R is
(a) 0 (b) 1
(c) 2 (d) 4

8. If three vertices of a rhombus taken in order are (2, –1), (3, 4) and (– 2, 3) then the fourth vertex is
(a) (–3, –2) (b) (3, 2)
(c) (2, 3) (d) (1, 2)

9. If the vertices of triangle be $(a, 1)$ $(b, 3)$ and $(4, c)$ then the centroid of the triangle will lie on x–axis if
(a) $a + c = -4$ (b) $a + b = -4$
(c) $c = -4$ (d) $b + c = -4$

10. Co-ordinates of the orthocentre of the triangle whose sides are $x = 3$, $y = 4$ and $3x + 4y = 6$, will be
(a) (0, 0) (b) (3, 0)
(c) (0, 4) (d) (3, 4)

11. If the vertices A and B of a triangle ABC are given by (2, 5) and (4, –11) and C moves along the line $9x + 7y + 4 = 0$, the locus of the centroid of the triangle ABC is a straight line parallel to
(a) AB (b) BC
(c) CA (d) $9x + 7y + 4 = 0$

12. The locus of point P which divides the line joining (1, 0) and $(2\cos\theta, 2\sin\theta)$ internally in the ratio 2:3 for all θ, is a
(a) straight line (b) circle
(c) pair of straight lines (d) parabola

15. Let $S_1, S_2, \ldots$ be squares such that for each $n \geq 1$, the length of side of S_n equals the length of a diagonal of S_{n+1}. If the length of a side of S_1 is 10 cm, then which of the following values of n is not the area of S_n less than 1 cm^2?
(a) 7 (b) 8
(c) 9 (d) 10

16. If the axes are rotated through an angle of 30° in the clockwise direction, the point $(4, -2\sqrt{3})$ in the new system was formerly
(a) $(2, \sqrt{3})$ (b) $(\sqrt{3}, -5)$
(c) $(\sqrt{3}, 2)$ (d) (2, 3)

17. Let $A = (1, 0)$ and $B = (2,1)$. The line AB turns about A through an angle $\pi/6$ in the clockwise sense and the new position of B is B'. Then B' has the co-ordinates
(a) $\left(\frac{3+\sqrt{3}}{2}, \frac{\sqrt{3}-1}{2}\right)$ (b) $\left(\frac{3-\sqrt{3}}{2}, \frac{\sqrt{3}+1}{2}\right)$
(c) $\left(\frac{1-\sqrt{3}}{2}, \frac{1+\sqrt{3}}{2}\right)$ (d) None of these

18. If (–1, 2); (2, –1) and (3, –1) are three vertices of a parallelogram, then the fourth vertex (a, b) will be
(a) $a = 0, b = 2$ (b) $a = 2, b = 0$
(c) $a = -2, b = 0$ (d) None of these

19. The co-ordinates of the point of trisection of the join of the points (–2,3), (3,–1) nearer to (–2, 3) is
(a) (–1/3, 5/3) (b) (4/3, 1/3)
(c) (–1/3, 2) (d) (1/3, 5/3)

20. The three vertices of a parallelogram are $(a + b, a - b)$, $(2a + b, 2a - b)$ and $(a - b, a + b)$, the fourth vertex is
(a) $(-b, b)$ (b) $(-b, -b)$
(c) (b, a) (d) None of these

21. If the sum of the distances of the point from two perpendicular lines in a plane is 1, then its locus is
(a) Square (b) Circle
(c) Straight line (d) Two intersecting lines

22. ABC is an isosceles triangle. If the co-ordinates of the base are $B(a + b, b - a)$ and $C(a - b, a + b)$ then the co-ordinates of the vertex A can be
(a) $(-a, b)$ (b) (b, a)
(c) $(a/b, b/a)$ (d) $(1, b/a)$

23. ABC is an isosceles triangle. If the co-ordinates of the base are $B(1, 3)$ and $C(-2, 7)$, the co-ordinates of vertex A (not bying on BC) can be
(a) (1, 6) (b) (–1/2, 5)
(c) (5/6, 6) (d) (7, 1/8)

24. The abscissa of two points P,Q are the roots of the equation $x^2 + 2ax - b^2 = 0$ and their ordinates are the roots of $x^2 + 2cx - d^2 = 0$, then the distance PQ in terms of a, b, c, d is
(a) $\sqrt{a^2+b^2+c^2+d^2}$ (b) $\sqrt{a^2+b^2+2c^2d^2}$
(c) $2\sqrt{a^2+b^2+c^2+d^2}$ (d) $\frac{1}{2}\sqrt{a^2+b^2+c^2+d^2}$

25. The point A divides the join of the points (–5, 1) and (3, 5) in the ratio k: 1 and the co-ordinates of the point B and C are (1,5) and (7,–2) respectively. If the area of the ΔABC be 5 square units, then k is
(a) 6, 7 (b) $\frac{31}{9}$, 9
(c) $\frac{7}{3}$, 19 (d) 7,9

26. Area of a triangle with vertices (a, b), (x_1, y_1) and (x_2, y_2) where a, x_1 and x_2 are in G.P. with common ratio r and b, y_1 and y_2 are in G.P. with common ratio s, is given by
(a) $ab(r - 1)(s - 1)(s - r)$
(b) $ab(r + 1)(s + 1)(s + r)$
(c) $\frac{1}{2}\ ab(r - 1)(s - 1)(s - r)$
(d) None of these

27. The area of the pentagon whose vertices are (4, 1), (3, 6), (– 5, 1) (–3, –3) and (–3, 0) is
(a) 30 $unit^2$ (b) 60 $unit^2$
(c) 120 $unit^2$ (d) None of these

28. The area of the triangle formed by the lines $y = m_1x + c_1$, $y = m_2x + c_2$ and $x = 0$ is
(a) $\frac{1}{2}\frac{(c_1+c_2)^2}{|m_1-m_2|}$ (b) $\frac{1}{2}\frac{(c_1-c_2)^2}{|m_1+m_2|}$
(c) $\frac{1}{2}\frac{(c_1-c_2)^2}{|m_1-m_2|}$ (d) $\frac{(c_1-c_2)^2}{|m_1-m_2|}$

30. Given the points $A(0, a)$ and $B(0, -a)$, the equation of the locus of point $P(x, y)$ scuh that $|AP–BP| = 6$ is
(a) $\frac{x^2}{a^2-9}-\frac{y^2}{9}=1$ (b) $\frac{x^2}{a^2-9}-\frac{y^2}{9}+1=0$
(c) $\frac{y^2}{9}-\frac{x^2}{9-a^2}=1$ (d) $\frac{x^2}{9}-\frac{y^2}{a^2-9}=1$

31. Two points A and B move on the x-axis and the y-axis respectively such that the distance between the two points is always the same. The locus of the middle point of AB is
(a) a straight line (b) a pair of straight lines
(c) a circle (d) None of these

32. Let $A = (1,2)$, $B = (3, 4)$ and let $C = (x,y)$ be a point such that $(x-1)(x - 3) + (y - 2)(y - 4) = 0$. If ar($\Delta ABC$) = 1, then maximum number of positions of C in the x–y plane is
(a) 2 (b) 4
(c) 8 (d) None of these

33. The area of a parallelogram formed by the lines $ax \pm by \pm c = 0$, is
(a) $\frac{c^2}{ab}$ (b) $\frac{4c^2}{a^2+b^2}$
(c) $\frac{c^2}{2ab}$ (d) None of these

34. The line $\frac{x}{3}+\frac{y}{4}=1$ meets the axis of y and axis of x at A and B respectively. A square is $ABCD$ is constructed on the line segment AB away from the origin, the co-ordinates of the vertex of the square farthest from the origin are

(a) (7, 3) (b) (4, 7)
(c) (6, 4) (d) (3, 8)

35. If $P(2, 1)$, $Q(4, -1)$, R (3, 2) are the vertices of a triangle and if through P and R lines parallel to the opposite sides are drawn to intersect at S, then the area of $PQRS$ in square units is

(a) 4 (b) 8
(c) 12 (d) 16

36. Let $P = (1, 1)$ and $Q = (3, 2)$. The point R on the x-axis such that $PR + RQ$ is the minimum is

(a) (5/3, 0) (b) (1/3, 0)
(c) (3, 0) (d) None of these

37. The vertex O of an isosceles ΔOAB lies at the origin and the equation of the base AB is $x–y + 1 = 0$. If $OA = OB = 6$, then area of the triangle OAB is

(a) $\frac{\sqrt{71}}{2}$ sq. units (b) $\frac{\sqrt{142}}{2}$ sq. units

(c) $\frac{2}{\sqrt{71}}$ sq. units (d) $\sqrt{142}$ sq. units

38. A point moves in the x-y plane such that the sum of its distances from two mutually perpendicular lines is always equal to 3. The area enclosed by the locus of the point is

(a) 18 unit² (b) 9/2 unit²
(c) 9 unit² (d) None of these

39. The co-ordinates of a point on the line $x + y = 3$ such that the point is at equal distances from the lines $|x| = |y|$ are

(a) (3, 0) (b) (0, 6)
(c) (–3, 0) (d) (0, – 3)

40. A point moves such that its distance from the point (4, 0) is half that of its distance from the line $x = 16$. The locus of the point is

(a) $3x^2 + 4y^2 = 192$ (b) $4x^2 + 3y^2 = 192$
(c) $x^2 + y^2 = 192$ (d) None of these

41. If $bx + cy = a$, where a, b, c are of the same sign, be a line such that the area enclosed by the line and the axes of reference is 1/8 unit², then

(a) b, a, c are in G.P. (b) b, $2a$, $-c$ are in G.P.
(c) b, $a/2$, c are in A.P. (d) b, $\pm 2a$, c are in G.P

42. If the extremities of the base of an isosceles triangle are the points $(2a,0)$ and $(0,a)$ and the equation of one of the sides is $x = 2a$, then the area of the triangle is

(a) $5a^2$ sq. units (b) $\frac{5}{2}a^2$ sq. units

(c) $\frac{25}{2}a^2$ sq. units (d) None of these

43. Three vertices of a quadrilateral in order are (6,1), (7,2) and (–1,0). If the area of the quadrilateral is 4 unit², then the locus of the fourth vertex has the equation

(a) $x - 7y = 1$
(b) $x - 7y + 15 = 0$
(c) $(x - 7y)^2 + 14(x - 7y) - 15 = 0$
(d) None of these

44. If O is the origin and co-ordinates of A and B are (a, b) and (b, a) respectively, then the locus of the point P which moves such that the area of the ΔPAB is one third of the area of the ΔOAB is

(a) $4(x + y) = 3(a + b)$ (b) $3(x + y) = 4(a + b)$
(c) $3(x + y) = (a + b)$ (d) $2(x + y) = 3(a + b)$

45. A line $2x + 3y–1 = 0$ intersects the three sides BC, CA and AB of the ΔABC (where co-ordinates of A, B, C are $A(x_1, y_1)$; $B(x_2, y_2)$ and $C(x_3, y_3)$) in P,Q,R respectively, then

(a) $\frac{BP}{PC} = \frac{2x_2 + 3y_2 + 1}{2x_3 + 3y_3 + 1}$

(b) $\frac{CQ}{QA} = \frac{2x_3 + 3y_3 + 1}{2x_1 + 3y_1 + 1}$

(c) $\frac{AR}{RB} = \frac{2x_1 + 3y_1 + 1}{2x_2 + 3y_2 - 1}$

(d) $\frac{BP}{PC} \times \frac{CQ}{QA} \times \frac{AR}{RB} = -1$

46. If one vertex of an equilateral triangle of side a lies at the origin and the other lies on the line $x–\sqrt{3}\,y = 0$, which one can't be third vertex?

(a) $(0, a)$ (b) $\left(\frac{\sqrt{3}a}{2}, -\frac{a}{2}\right)$

(c) $(0, -a)$ (d) None of these

47. If $A = \left(\sqrt{1-t^2} + t, 0\right)$ and $B = \left(\sqrt{1-t^2} - t, 2t\right)$ are two variable points, where $\boldsymbol{t}$ is a parameter. The locus of the middle point of AB is

(a) a straight line (b) a pair of lines
(c) a circle (d) None of these

48. The line $\frac{x}{a} + \frac{y}{b} = 1$ meets the axis of x and y at A and B respectively and the line $y = x$ at C so that area of the ΔAOC is twice the area of the ΔBOC, O being the origin, then the co-ordinates of C are:

(a) $\left(\frac{a}{3}, \frac{a}{3}\right)$ (b) $\left(\frac{2a}{3}, \frac{2a}{3}\right)$

(c) $\left(\frac{b}{3}, \frac{b}{3}\right)$ (d) None of these

49. P, Q, R are three collinear points. If the co-ordinates of P and R are (3, 4) and (11, 10) respectively and PQ is equal to 2.5 units, the co-ordinates of Q are

(a) $\left(5, \frac{11}{2}\right)$ (b) $\left(1, \frac{11}{2}\right)$

(c) $\left(5, \frac{5}{2}\right)$ (d) None of these

50. Two consecutive vertices of a rectangle of area 10 *unit*^{2} are (1, 3) and (–2, –1). Other two vertices are equal to

(a) $\left(+\frac{3}{5}, \frac{21}{5}\right), \left(+\frac{18}{5}, \frac{1}{5}\right)$

(b) $\left(-\frac{3}{5}, \frac{21}{5}\right), \left(-\frac{2}{5}, -\frac{11}{5}\right)$

(c) $\left(-\frac{2}{5}, -\frac{11}{5}\right), \left(\frac{13}{5}, \frac{9}{5}\right)$

(d) $\left(\frac{13}{5}, \frac{9}{5}\right), \left(-\frac{18}{5}, \frac{1}{5}\right)$

51. The altitudes of a ΔABC are respectively AD, BE, CF. If the points A, D, E and F have the co-ordinates (–4,5), $\left(\frac{16}{5}, -\frac{23}{5}\right)$, (4,1) and (–1,–4) respectively, the other vertices are.

(a) (1, –7), (8, 2) (b) (0, –7), (8,–1)

(c) (–7, 1), (2, 8) (b) (–7, 0), (–1,8)

52. The position of a moving point in the x - y plane at time t is given by $(u \cos\alpha.t,\ u \sin\alpha.t - pt^2)$ where u, α, p are constants. The equation of the locus of the moving point is

(a) $y = x\cos\alpha - \dfrac{px^2}{u^2 \tan^2\alpha}$

(b) $y = x\cos\alpha - \dfrac{ux^2}{p^2 \tan^2\alpha}$

(c) $y = x\tan\alpha - \dfrac{px^2}{u^2 \cos^2\alpha}$

(d) None of these

53. A variable striaght line of slope 4 intersects the hyperbola $xy = 1$ at two points. The locus of the point which divides the line segment between these two points in the ratio 1: 2 is

(a) $16x^2 + 10xy + y^2 - 2 = 0$

(b) $14x^2 + 7xy + y^2 - 2 = 0$

(c) $12x^2 + 14xy + 2y^2 - 2 = 0$

(d) None of these

54. The x-co-ordiantes of the vertices of a square of unit area are the roots of the equation $x^2 - 3\,|\,x\,| + 2 = 0$ and the y-co-ordinates of the vertices are the roots of the equation $y^2 – 3y + 2 = 0$. The vertices of the square are the sets of points are

(a) (1,1), (2,1), (2,2), (1,2) or (–2,1), (–1,1), (–1,2), (–2,2)

(b) (1,1), (1,2), (2,2), (2,1) or (–2,1), (–1,1), (–1,2), (–2,–2)

(c) (1,1), (2,1), (2,2), (1,2) or (–1,2), (–1,1), (–2,1), (–1, –1)

(d) None of these

55. ABC is a variable triangle with a fixed centroid (5, 5). The side $BC = 13$ and B and C move on the x and y axis respectively. The equation of the locus of the vertex A is

(a) $x^2 + y^2 - 15x - 15y + 235 = 0$

(b) $x^2 + y^2 + 32x - 32y - 281 = 0$

(c) $x^2 + y^2 - 30x - 30y + 281 = 0$

(d) None of these

56. One end of a thin straight elastic string is fixed at A (4, –1) and the other end B is at (1, 2) in the unstretched conditon. If the string is stretched to tripple its length, then the new position of B is

(a) (– 5, 8) (b) (–8, 5)

(c) (–2, 3) (d) (–3, –2)

57. The locus of a point at which two given portions of the same straight line subtend equal angles is

(a) an ellipse (b) a striaght line

(c) a parabola (d) a circle

58. In a triangle ABC, co-ordinates of A are (1, 2) and the equations of the medians through B and C are $x + y = 5$ and $x = 4$ respectively. The co-ordinates of C and B are respectively.

(a) (3, 3), (7, 2) (b) (–4, 3), (7, –2)

(c) (–4, –3), (7, –2) (d) (4, 3), (7, –2)

60. The distance of the point ($a \cos\alpha$, $a \sin\alpha$) and ($a \cos \beta$, $a \sin \beta$)

(a) $|a|$ (b) $|\,a\sin(\alpha-\beta)\,|$

(c) $2\,|\,a\,|\left|\sin\dfrac{(\alpha-\beta)}{2}\right|$ (d) None of these.

61. S be a square of unit area, any quadrilateral which has one vertex on each side of square. If a, b, c, d denote length of sides of the quadrilateral then

(a) $3 \le a^2 + b^2 + c^2 + d^2 \le 4$

(b) $2 \le a^2 + b^2 + c^2 + d^2 \le 4$

(c) $0 \le a^2 + b^2 + c^2 + d^2 \le 2$

(d) None of these.

62. The line joining A (b cosθ, b sin θ) and B (a cos ϕ, a sin ϕ) is produced to the point M (x, y), so that AM and BM are in the ratio $b : a$ then

(a) $x + y\tan\left(\frac{\theta-\phi}{2}\right) = 0$

(b) $x + y\tan\left(\frac{\theta+\phi}{2}\right) = 0$

(c) $y + x\tan\left(\frac{\theta-\phi}{2}\right) = 0$

(d) None of these.

63. $P(x_1, y_1)$, $Q(x_2, y_2)$, $R(x_3, y_3)$ are the vertices of ΔPQR. The side PQ is divided by the point S in the ratio $\alpha : \beta$ and then the line segment SR is divided by the point T in the ratio $\gamma : \alpha + \beta$. the co-ordinates of T is

(a) $\left(\frac{-\beta x_1 + \alpha x_2 + \gamma x_3}{\alpha - \beta + \gamma}, \frac{-\beta y_1 + \alpha y_2 + \gamma y_3}{\alpha - \beta + \gamma}\right)$

(b) $\left(\frac{\beta x_1 - \alpha x_2 + \gamma x_3}{-\alpha + \beta + \gamma}, \frac{\beta y_1 - \alpha y_2 + \gamma y_3}{-\alpha + \beta + \gamma}\right)$

(c) $\left(\frac{\beta x_1 + \alpha x_2 + \gamma x_3}{\alpha + \beta + \gamma}, \frac{\beta y_1 + \alpha y_2 + \gamma y_3}{\alpha + \beta + \gamma}\right)$

(d) None of these.

64. If the lines $a_1x + b_1y + c_1 = 0$ and $a_2x + b_2y + c_2 = 0$ cut the co-ordinate axes in concyclic points, then

(a) $a_1b_1 = a_2b_2$ (b) $\frac{a_1}{a_2} = \frac{b_1}{b_2}$

(c) $a_1 + a_2 = b_1 + b_2$ (d) $a_1a_2 = b_1b_2$

65. If $\alpha_1, \alpha_2, \alpha_3, \beta_1, \beta_2, \beta_3$ are the values of n for which $\sum_{r=0}^{n-1} x^{2r}$ is divisible by $\sum_{r=0}^{n-1} x^r$ then the triangle having vertices $(\alpha_1, \beta_1), (\alpha_2, \beta_2), (\alpha_3, \beta_3)$ cannot be

(a) an isosceles triangle

(b) a right angled isosceles triangle

(c) a right angled triangle

(d) an equilateral triangle

66. The equation of a pair of straight lines is $ax^2 + 2hxy + by^2 = 0$. The angle by which the axes be rotated so that the terms containing xy in the equation may be removed is

(a) $\frac{1}{2}\tan^{-1}\frac{h}{a-b}$ (b) $\frac{1}{2}\tan^{-1}\frac{2h}{a-b}$

(c) $\frac{1}{2}\tan^{-1}\frac{2h}{a+b}$ (d) None of these

67. The graph of the function

$y = \cos x\cos(x+2) - \cos^2(x+1)$ is

(a) A straight line passing through (0, -sin²1) with slope 2

(b) A straight line passing through (0,0)

(c) a parabola with vertex (1, -sin²1)

(d) a straight line passing through the point $\left(\frac{\pi}{2}, -\sin^2 1\right)$ and parallel to the x-axis.

71. Through the centroid of an equilateral triangle a line parallel to the base is drawn. On this line, an arbitary point P is taken inside the triangle. Let h denote the distance of P from the base of the triangle. Let h_1 and h_2 be the distances of P from the other two sides of the triangle, then

(a) h is the H.M.of h_1, h_2

(b) h is the G.M.of h_1, h_2

(c) h is the A.M.of h_1, h_2

(d) None of these

SECTION–IV

OBJECTIVE-TYPE (MORE THAN ONE COREECT ANSWER)

1. If $A(a, a)$; $B(-a, -a)$ are two vertices of an equilateral triangle, then third vertex is

(a) $\left(\frac{a\sqrt{3}}{2}, -\frac{a\sqrt{3}}{2}\right)$ (b) $(-a\sqrt{3}, a\sqrt{3})$

(c) $(a\sqrt{3}, -a\sqrt{3})$ (d) $(-a\sqrt{3}, -a\sqrt{3})$

2. Let S_1, S_2, ... be squares such that for each $n \ge 1$, the length of side of S_n equals the length of a diagonal of S_{n+1}. If the length of a side of S_1 is 10 cm, then which of the following values of n is the area of S_n less than 1 cm^2?

(a) 7 (b) 8

(c) 9 (d) 10

3. The points (a, b), (b, c) and (c, a) form the vertices of a right angled triangle, if

(a) $a = b$
(b) $b = c$
(c) $c = a$
(d) a, b, c are three consecutive integers

4. The co-ordinates of a point on the line $x + y = 3$ such that the point is at equal distances from the lines $|x| = |y|$ are
(a) (3, 0) (b) (0, 3)
(c) (–3, 0) (d) (0, – 3)

5. If O is the origin and co-ordinates of A and B are (a, b) and (b, a) respectively, then the locus of the point P which moves such that the area of the ΔPAB is one third of the area of the ΔOAB is
(a) $4(x + y) = 3(a + b)$ (b) $3(x + y) = 4(a + b)$
(c) $3(x + y) = 2(a + b)$ (d) $2(x + y) = 3(a + b)$

6. If one vertex of an equilateral triangle of side a lies at the origin and the other lies on the line $x - \sqrt{3}y = 0$, the co-ordinates of the third vertex are
(a) $(0, a)$ (b) $\left(\frac{\sqrt{3}a}{2}, -\frac{a}{2}\right)$
(c) $(0, -a)$ (d) $\left(-\frac{\sqrt{3}a}{2}, \frac{a}{2}\right)$

7. The line $\frac{x}{a} + \frac{y}{b} = 1$ meets the axis of x and y at A and B respectively and the line $y = x$ at C so that area of the triangle AOC is twice the area of the triangle BOC, O being the origin, the co-ordinate of C are
(a) $\left(\frac{a}{3}, \frac{a}{3}\right)$ (b) $\left(\frac{2a}{3}, \frac{2a}{3}\right)$
(c) $\left(\frac{b}{3}, \frac{b}{3}\right)$ (d) $\left(\frac{2b}{3}, \frac{2b}{3}\right)$

8. Given a Δ ABC with co-ordinates of its vertices as $A(6, 8)$; $B(2, -4)$ and $C(-6, 4)$. The angle between the side AB and the median drawn from the vertex A is:
(a) $\frac{\pi}{2} - \tan^{-1}2$ (b) $\sin^{-1}\frac{2}{\sqrt{5}}$
(c) $\cos^{-1}\frac{1}{\sqrt{5}}$ (d) $\frac{\pi}{4} - \tan^{-1}3$

9. The vertices of a Δ ABC are $A(-5,-2)$; $B(7,6)$ and $C(5, -4)$. Then:
(a) Measure of angle B is $\frac{\pi}{4}$.
(b) Equation of the altitude drawn from the vertex C has the equation, $3x + 2y - 7 = 0$.
(c) Orthocentre of the triangle does not lie inside the Δ ABC.
(d) Distance between centroid and circumcentre of the Δ ABC is $\frac{4\sqrt{13}}{3}$.

10. Let (x_1, y_1); $(x_2; y_2)$ and $(x_3; y_3)$ be respectively the vertices of a Δ ABC, then:
(a) $(x_1 - x_2)(x_1 - x_3) + (y_1 - y_2)(y_1 - y_3)$ is positive, therefore angle A is acute
(b) $(x_1 - x_2)(x_1 - x_3) + (y_1 - y_2)(y_1 - y_3)$ is negative, therefore angle A is obtuse
(c) $(x_1 - x_2)(x_1 - x_3) + (y_1 - y_2)(y_1 - y_3)$ is positive, therefore A is inside the circle of least radius passing through (x_2, y_2) and (x_3, y_3)
(d) $(x_1 - x_2)(x_1 - x_3) + (y_1 - y_2)(y_1 - y_3)$ is negative, therefore A is outside the circle of least radius passing through (x_2, y_2) and (x_3, y_3)

11. The vertices $(-\lambda, \mu)$; $(0,0)$; (λ, μ) and $(\lambda^{2-}, \lambda\mu)$ can be
(a) collinear
(b) used to form a parallelogram
(c) used to form a rectangle
(d) concylic

12. The abscissae of vertices of a square of unit area are roots of equation $x^2 - 3|x| + 2 = 0$ and the ordinates are roots of equation $y^2 - 3y + 2 = 0$. Then which of the following is/are the vertex (vertices) of square
(a) (-2,1) (b) (1,2)
(c) (-1,2) (d) (-2,2)

SECTION–V

ASSERTION AND REASON-TYPE

(a) Both A and R are individually true and R is the correct explanation of A.
(b) Both A and R are individually true but R not the correct explanation of A.
(c) A is true but R is false.
(d) A is false but R is true.

1. A: The points (2,1) and (– 3, 5) lie on opposite side of the line $3x - 2y + 1 = 0$.
R: The algebraic perpendicular distance from the given point to the line have opposite sign.

2. A: The points A, B and C are collinear then area of $\Delta ABC = 0$

R: $AB + BC = AC$

3. A: If centroid and circumcentre of a triangle are known, then its orthocentre and nine point centre can be found.

R: Orthocentre, nine point centre, centroid and circumcentre are collinear.

4. A: The quadrilateral whose vertices (in order) are A (1, 0), B (0, 3), C (– 2, 0) and D (0, 2) can not be convex.

R: A quadrilateral $ABCD$ (in order) is convex if and only if any diagonal is taken, then remaining vertices must be on the opposite sides of it.

5. A line segment AB is divided internally and externally in the same ratio at P and Q respectively and M is the mid-point of AB.

A M P B Q

A: MP, MB, MQ are in G.P.

R: AP, AB and AQ are in H.P.

6. Let $A(x_1, y_1)$, $B(x_2, y_2)$ and $C(x_3, y_3)$ are the vertices of a triangle ABC.

A: If angle C is obtuse then then quantity $(x_3 - x_1)(x_3 - x_2) + (y_3 - y_1)(y_3 - y_2)$ is negative.

R: Diameter of a circle subtends obtuse angle at any point lying inside the semi-circle.

7. A: The co-ordinates of vertices of ΔABC are $A(400,205)$; $B(500, 300)$ and $C(600,55)$ and ΔABC has area 'k' square units undergoes successive transformations; $f_1(x, y) \to (y, x)$; $f_2(x,y) \to (x + k, y)$; $f_3(x, y) \to (x, y + k)$ has area 'k' square unit

R: ΔABC and ΔA'B'C' are reflection of each other on line $y = x$ and the further two transformations just translate the Δ by k units horizontally and k units vertically successively.

8. A: The orthocentre, circumcentre, incentre, centroid of Δ formed by the vertices (1,1); (1,a-1) and (a-1,1) lie on the line $y = x$.

R: (1,a-1) and (a-1,1) are reflections of each other on line $y = x$.

9. A: If three points are given, fourth point can be chosen to form three parallelogram of unequal areas.

R : Each side of a non-zero area triangle formed by three vertices can be used as a diagonal of a parallelgram.

10. A: Le PQ be a line segment of length 6cm. A point R moves on the plane such that its distance from point P is always 4cm. and $\pi/2 \geq \angle RPQ \geq \angle RQP$, then the area swept out by PR is equal to 4π - 8 $\cos^{-1}(3/4)$ or $8\sin^{-1}3/4$.

R: PR sweeps out 1/6 portion of circle on the left of perpendicular bisector of line segment PQ.

SECTION–VI

LINKED COMPREHENSION-TYPE

A: A (x_1, y_1), $B(x_2, y_2)$, $C(x_3, y_3)$ are the vertices of a triangle ABC. $lx + my + n = 0$ is an equation of the line L.

1. If L intersect the sides BC, CA and AB of the triangle ABC at P, Q, R respectively, then $\frac{BP}{PC} \times \frac{CQ}{QA} \times \frac{AR}{RB}$ is equal to

(a) –1 (b) –1/2
(c) 1/2 (d) 1

2. If the centroid of the triangle ABC is the origin and algebraic sum of the length of the perpendiculars from the vertices of the triangle ABC on the line L is equal to 1 then sum of the squares of the reciprocal of intercepts made by L on the co-ordinate axes is equal to

(a) 0 (b) 4
(c) 9 (d) 16

3. If P divides BC in the ratio 2: 1 and Q divides CA in the ratio 1: 3, then R divides AB in the ratio

(a) 2: 3 internally (b) 2: 3 enternally
(c) 3: 2 internally (d) 3: 2 externally

B: A (0, 3), B (– 2, 0) and C(6, 1) be the vertices of a triangle and M $(\beta, \beta + 1)$ be a moving point then

4. M lies on the curve

(a) $y = x + 1$ (b) $y = x^2$
(c) $x = y + 1$ (d) None of these

5. If M and A lie on same side of BC then

(a) $\beta > 2$ (b) $\beta < 2$

(c) $\beta > -\frac{6}{7}$ (d) $\beta < \frac{3}{4}$

6. M lies within $\Delta\, ABC$ if

(a) $-\frac{6}{7} < \beta < 4$ (b) $-4 < \beta < -\frac{6}{7}$

(c) $-\frac{6}{7} < \beta < \frac{3}{2}$ (d) None of these

C: Let A (0, β), B (– 2, 0) and C (1, 1) be the vertices of a triangle then

7. Angle A of the triangle ABC will be obtuse if β lies in

(a) (–1, 2) (b) $\left(2, \frac{5}{2}\right)$

(c) $\left(-1, \frac{2}{3}\right) \cup \left(\frac{2}{3}, 2\right)$ (d) None of these

8. If I_1 is the interval of values of β for which A is obtuse and I_2 be the interval of values of β for which A is largest angle of ΔABC, then

(a) $I_1 = I_2$ (b) I_1 is a subset of I_2

(c) I_2 is a subset of I_1 (d) None of these.

9. All the values of β for which angle A of the triangle ABC is largest lie in interval.

(a) (– 2, 1)

(b) $\left(-2, \frac{2}{3}\right) \cup \left(\frac{2}{3}, 1\right)$

(c) $\left(-2, \frac{2}{3}\right) \cup \left(\frac{2}{3}, \sqrt{6}\right)$

(d) None of these

D: The locus of a moving point is the path traced out by that point under one or more given conditions. Technically, a locus represents the 'set of all points' which lies on it.

A relation $f(x, y) = 0$ between x and y which is satisfied by each point on the locus and such that each point satisfying the equation is on the locus is called the equation of the locus.

On the basis of above information, answer the following questions:

10. Locus of centroid of the triangle whose vertices are (a cos$t$, a sin t), (b sin t, – b cos t) and (1, 0) where t is a parameter, is

(a) $(3x-1)^2 + (3y)^2 = a^2 - b^2$

(b) $(3x-1)^2 + (3y)^2 = a^2 + b^2$

(c) $(3x+1)^2 + (3y)^2 = a^2 + b^2$

(d) $(3x+1)^2 + (3y)^2 = a^2 - b^2$

11. Let A(2, – 3) and B(– 2, 1) be vertices of a ΔABC. If the centroid of this triangle moves on the line $2x + 3y = 1$, then the locus of the vertex C is the line

(a) $2x + 3y = 9$ (b) $2x - 3y = 7$

(c) $3x + 2y = 5$ (d) $3x - 2y = 3$

12. If the equation of the locus of a point equidistant from the points (a_1, b_1) and (a_2, b_2) is $(a_1 - a_2)x + (b_1 - b_2)y + c = 0$, then the value of 'c' is

(a) $\frac{1}{2}(a_2^2 + b_2^2 - a_1^2 - b_1^2)$

(b) $a_1^2 - a_2^2 + b_1^2 - b_2^2$

(c) $\frac{1}{2}(a_1^2 + a_2^2 + b_1^2 + b_2^2)$

(d) $\sqrt{(a_1^2 + b_1^2 - a_2^2 - b_2^2)}$

13. The area of the triangle formed by the intersection of a line parallel to x-axis and passing through $P(h, k)$ with the lines $y = x$ and $x + y = 2$ is $4h^2$. Then the locus of the point P is

(a) $y = 3x - 2$ or $y = - 3x - 2$

(b) $y = 2x - 1$ or $y = - 2x - 1$

(c) $y = 3x + 2$ or $y = - 3x + 2$

(d) $y = 2x + 1$ or $y = - 2x + 1$.

E: For points $P \equiv (x_1, y_1)$ and $Q \equiv (x_2, y_2)$ of the co-ordinate plane, a new distance d (P, Q) is defined by $d(P,Q) = |x_1 - x_2| + |y_1 - y_2|$. Let $O \equiv (0, 0)$; $A \equiv (1, 2)$, $B \equiv (2, 3)$ and $C \equiv (4, 3)$ are four fixed points on $x - y$ plane. On the basis of above information, answer the following questions:

14. Let R (x, y) such that R is equidistant from the point O and A with respect to new distance and if $0 \le x < 1$ and $0 \le y < 2$, then R lie on a line segment whose equation is

(a) $x + y = 3$ (b) $x + 2y = 3$

(c) $2x + y = 3$ (d) $2x + 2y = 3$

15. Let S (x, y) such that S is equidistant from points O and B with respect to new distance and if $x \ge 2$ and $0 \le y < 3$, then locus of S is

(a) a line segment of finite length

(b) a line of infinite length

(c) a ray of finite length

(d) a ray of infinite length

16. Let T (x, y) such that T is equidistant from point O and C with respect to new distance and if T lie in first quadrant, then T consists of the union of a line segment of finite length and an infinite ray whose labelled diagram is

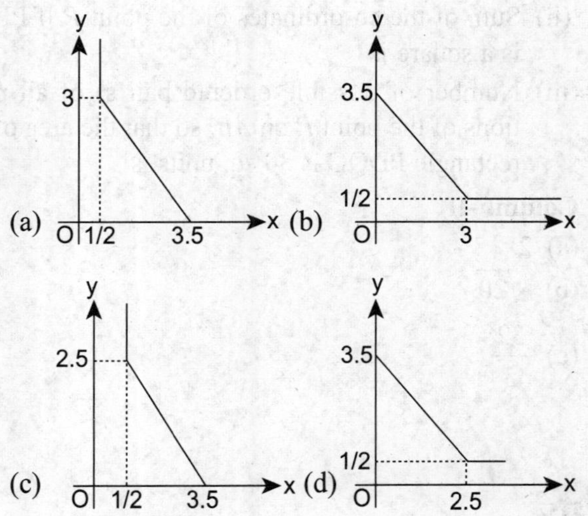

17. Distance between the circumcentre and orthocentre of ΔABC is

(a) 5 (b) $\sqrt{29}$

(c) 7 (d) $\sqrt{37}$

18. Nine point centre of the ΔABC is

(a) (1, 6) (b) (1, 5/2)

(c) (3, 1) (d) (2, 7/2)

SECTION–VII

MATRIX-MATCH TYPE

1. If Δ denotes the area of the triangle with vertices $(p + 1, 1)$, $(2p + 1, 3)$ and $(2p + 2, 2p)$

Column-I	Column-II
(i) $p = 0$	A. $\Delta = 7/2$
(ii) $p = \pm 1$	B. $\Delta = 25/2$
(iii) $p = 3$	C. $\Delta = 3/2$
(iv) $p = -3$	D. $\Delta = 1$

2. If A $(2a, 4a)$ and B $(2a, 6a)$ are two vertices of a triangle ABC and the vertex C is given by

Column-I	Column-II
(i) $(4a, 5a)$	A. equilateral
(ii) $((2+\sqrt{3})a, 5a)$	B. right angled
(iii) $(6a, 4a)$	C. Obtuse angled
(iv) $(a, 3a)$	D. isosceles

3. The two intersecting lines divides the X-Y plane in regions as shown in the figure. If point (α^2, α) lying in X-Y plane, then match the region mentioned in column I with the set containing all possible values of α given in column II.

Column-I	Column-II
(i) R_1	(a) $\alpha \in (-2, 0)$
(ii) R_2	(b) $\alpha \in (-\infty, -2) \cup (1, \infty)$
(iii) R_3	(c) $\alpha \in (1, \infty)$
(iv) R_4	(d) $\alpha \in \{\}$
(v) R_5	(e) $a > 0$

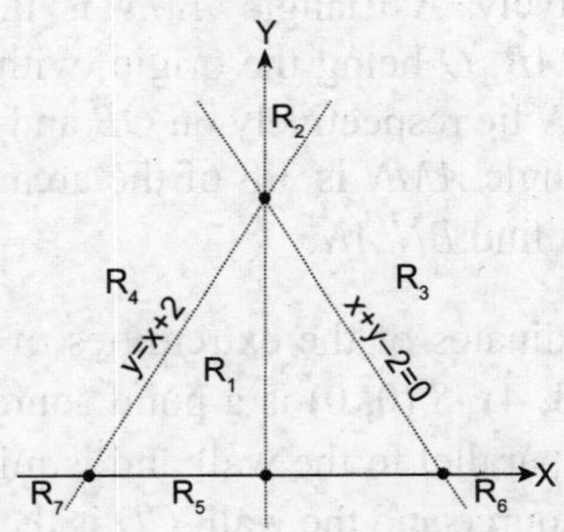

4. Column-I

(i) The lines $y = 0$, $y = 1$; $x - 6y + 4 = 0$ and $x + 6y - 9 = 0$ constitutes a figure which is

(ii) The points $A(a, 0)$, $B(0, b)$, $C(c, 0)$ and $D(0, d)$ are such that $ac = bd$ and a, b, c, d are all non-zero. The points A, B, C and D always consitute

(iii) The figure formed by the four lines $ax \pm by \pm c = 0$ $(a \neq b)$, is

(iv) The line pairs $x^2 - 8x + 12 = 0$ and $y^2 - 14y + 45 = 0$ constitute a figure which is

Column-II

(a) a cyclic quadrilateral

(b) a rhombus

(c) a square

(d) a trapezium

5. In the diagram, a line is drawn through the points $A(0, 16)$ and $B(8, 0)$. Point P is chosen in the first quadrant on the line through A and B. Points C and D are chosen on the x and y axis respectively, so that PDOC is a rectangle. Then match the following columns:

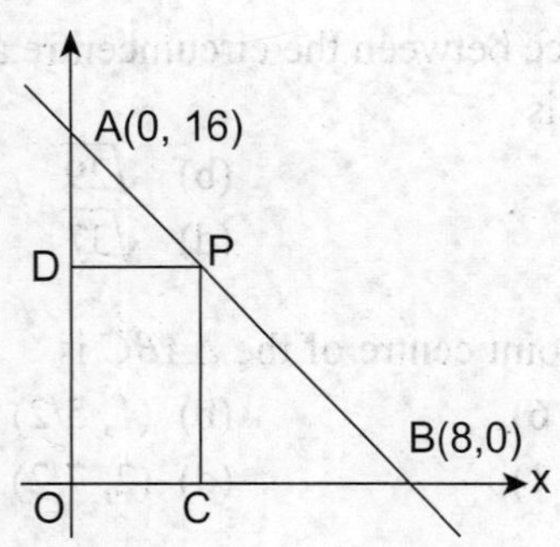

Column-I

(i) Perpendicular distance of the line AB from point (2,2) is

(ii) Sum of the co-ordinates of the point P if PDOC is a square is

(iii) Number of possible ordered pair(s) of all positions of the point P on AB, so that the area of the rectangle PDOC is 30 sq. units is

Column-II

(a) 2

(b) $\sqrt{20}$

(c) $\frac{32}{3}$

SECTION–VIII

INTEGER-TYPE

1. The line $x + y = a$ meets the axis of x and y at A and B respectively. A triangle AMN is inscribed in the triangle OAB, O being the origin, with right angle at N. M and N lie respectively on OB and AB. If the area of the triangle AMN is 3/8 of the area of the triangle OAB, then find BN/AN.

2. The co-ordinates of the extremities of a rod are $A(1, 2)$ and $B(3, 4)$. S (0, 0) is a point source of light The rod AB is parallel to the wall and is midway between the point source and the wall. CD is the shadow of AB on the wall. S, AB and CD are in the same horizontal plane. Find the sum of absiccissae of ends C, D of the shadow.

3. Two vertices of an equilateral triangle are (0, 0) and (0, 2√3). If the third vertex is $(x_3\ y_3)$ in first quadrant, then evaluate $x_3 + y_3^2$.

4. A rod AB = 8 slides along OX and rod CD = 6 slides along OY. If A, B, C, D always lie on a circle and the locus of the centre of the circle is $4\ (x^2 - y^2) = k$, find the value of k.

5. If the points P,Q,R lie on the curve having equation $xy = 16$, and (h,k) are the co-ordinates of orthocentre of ΔPQR, then evaluate (hk).

6. If the line segment joining the points (0,–1) and (15,2) is divide by the line segment joining the points (-1,2) and (4,-5) internally in the ratio $k_1 : k_2$, then evaluate $(k_1)^{k_2} + (k_2)^{k_1}$; g.c.d $(k_1, k_2) = 1$.

7. If the line joining the points $A(3\cos\alpha, 3\sin\alpha)$ and B $(2\cos\beta, 2\sin\beta)$ is produced to a point $C(h,k)$ such that AC :BC is 3 :2, then evaluate $h + k \tan\left(\frac{\alpha+\beta}{2}\right) + 4$.

8. If $\tan\alpha$, $\tan\beta$, $\tan\gamma$ are the roots of the equation $t^3 - 12t^2 + 15t - 1 = 0$; then the centroid of triangle having vertices $(\tan\alpha, \cot\alpha)$; $(\tan\beta, \cot\beta)$; $(\tan\gamma, \cot\gamma)$ is given by $G(h, k)$; then evaluate $(h + k)/(k - h)$.

9. If the area of the triangle formed by joining the mid-points of sides of a ΔABC is 3 square units, then find the area of figure obtained (i.e., new addition) when three triangle ABC′, BCA′,ACB′, similar to ΔABC are drawn in the exterior region of ΔABC.

10. If $A(\alpha,\beta)$; $B(2,1)$; $C(-2,1)$ are the vertices of a triangle, where $\alpha = \lim_{n\to\infty}\lim_{k\to\infty}\cos^{2n}(2k!\pi/3)$ and $\beta = \lim_{n\to\infty}\lim_{k\to\infty}\cos^{2n}\left(\frac{\sqrt{3}}{2}k!\pi\right)$; (where n, k are positive intger), then evaluate the area of ΔABC in square units.

11. The orthocentre and circumcentre of Δ formed by the straight line $2x + 3y = 6a$ with co-ordinate axis lie on the line $y = kx$, then evaluate 6k.

12. The vertices of Δ are (1,2); (3,5); and (4,8).The vertices undergo the following true successive transformations:

(i) $f_1\ (x,y) \to (y, x)$

(ii) $f_2\ (x,y) \to \left(\frac{x-y}{3}, \frac{x+y}{3}\right)$

(iii) $f_3\ (x,y) \to (x + 2y, 2y)$; then find 3A, where A is the area of final figure joining the new transformed vertices.

13. If the system of following equations in x & y $2x + 3y + 4 = 0$; $3x + 5y + 6 = 0$; $2x^2 + 6xy + 5y^2 + 8x + 12y + 1 = t$, consists of a solution. Then find the numeric value of t^4.

14. Family of lines $x(a - b) + y = 1$ where a and b are the roots of the equation $x^3 - 3x^2 + x + 1 = 0$ and $[a + b] = 1$ (where [.] denotes the greatest integer function), such that it intercepts a triangle of area A with co-ordinate axes. Find the maximum value of 2A.

15. A line 'L' is drawn from (4, 3) to meet the lines $L_1 : 3x + 4y + 5 = 0$ and $L_2 : 3x + 4y + 15 = 0$ at points A and B respectively. From 'A' a line, perpendicular to L is drawn meeting the line L_2 at A_1. Similarly, from point 'B' a line, perpendicular to L is drawn meeting the line L_1 at B_1. Thus parallelogram AA_1BB_1 is formed. Find the least value of area of parallelogram AA_1BB_1.

Answer Keys

SECTION–III

1. (c) 2. (d) 3. (a) 4. (c) 5. (d) 6. (a) 7. (a) 8. (c) 9. (d) 10. (d)
11. (b) 12. (a) 13. (b) 14. (a) 15. (a) 16. (a) 17. (a) 18. (a) 19. (d) 20. (c)
21. (c) 22. (c) 23. (c) 24. (a) 25. (c) 26. (b) 27. (c) 28. (b) 29. (b) 30. (b)
31. (a) 32. (a) 33. (a) 34. (a) 35. (a) 36. (a) 37. (d) 38. (b) 39. (c) 40. (b)
41. (d) 42. (d) 43. (c) 44. (a) 45. (a) 46. (c) 47. (b) 48. (c) 49. (a) 50. (a)
51. (c) 52. (a) 53. (d) 54. (d) 55. (c) 56. (b) 57. (a) 58. (c) 59. (d) 60. (d)
61. (b) 62. (d) 63. (c)

SECTION–IV

1. (b, c) 2. (b, c, d) 3. (a, b, c,) 4. (a, b) 5. (b, c) 6.(a, b, c, d) 7. (a, d) 8. (a, d)
9. (a, b, c) 10. (a, b) 11. (a, b, c, d) 12. (a, b, c, d)

SECTION–V

1. (a) 2. (b) 3. (b) 4. (a) 5. (a) 6. (a) 7. (a) 8. (b) 9. (d) 10. (c)

SECTION–VI

1. (a) 2. (c) 3. (d) 4. (a) 5. (c) 6. (c) 7. (c) 8. (b) 9. (c) 10. (b)
11. (a) 12. (a) 13. (d) 14. (d) 15. (d) 16. (a) 17. (b) 18. (d)

SECTION–VII

1. (i) → (d) (ii) → (c) (iii) → (a) (iv) → (b)
2. (i) → (d) (ii) → (a),(d) (iii) → (b) (iv) → (c)
3. (i) → (d) (ii) → (d) (iii) → (b) (iv) → (d) (v) → (a)
4. (i) → (a), (d) (ii) → (a) (iii) → (b), (d) (iv) → (a), (b), (c), (d)
5. (i) → (b) (ii) → (c) (iii) → (a)

SECTION–VIII

1. 3 2. 8 3. 6 4. 28 5. 16 6. 17 7. 4 8. 9 9. 36 10. 2
11. 4 12. 2 13. 2401 14. 1 15. 8

HINTS AND SOLUTIONS

TEXTUAL EXERCISE-1 (SUBJECTIVE)

1. (a) $R(a+b, a-b)$, $S(a-b, -a-b)$

Distance $RS = \sqrt{(2b)^2 + (2a)^2} = 2\sqrt{a^2+b^2}$

(b) $A(at_1^2, 2at_1), B(at_2^2, 2at_2)$

Distance $AB = \sqrt{a^2(t_1^2 - t_2^2)^2 + (2a)^2(t_1 - t_2)^2}$

$= a|t_1 - t_2| \sqrt{t_1^2 + t_2^2 + 2t_1t_2 + 4} = a|t_1 - t_2| \sqrt{(t_1+t_2)^2 + 4}$

(c) L (a cosα, a sinα), M (a cosβ, a sinβ)

Distance $LM = \sqrt{a^2(\cos\alpha - \cos\beta)^2 + a^2(\sin\alpha - \sin\beta)^2}$

$= |a|\sqrt{2 - 2\cos(\alpha - \beta)} = 2|a|\left|\sin\left(\frac{\alpha - \beta}{2}\right)\right|$

(d) Let $P(a\cos\theta, b\sin\theta)$, $Q(a\cos\phi, b\sin\phi)$

Distance $PQ = \sqrt{a^2(\cos\theta - \cos\phi)^2 + b^2(\sin\theta - \sin\phi)^2}$

(e) Let $A(ct_1, c/t_1)$ and $B(ct_2, c/t_2)$

Distance $AB = \sqrt{c^2(t_1 - t_2)^2 + c^2\{t_2 - t_1\}^2 / t_1^2t_2^2}$

$= |c||t_1 - t_2|\frac{1}{t_1t_2}\sqrt{1 + t_1^2t_2^2}$

2. (i) $A\left(5, \pi - \tan^{-1}\left(\frac{4}{3}\right)\right)$

Now $r = 5$, $\theta = \pi - \tan^{-1}(4/3)$ means 2nd quadrant

So $x = r\cos\theta = 5.\left(-\frac{3}{5}\right) = -3$ and $y = r\sin\theta = 5.\left(\frac{4}{5}\right) = 4$

$\Rightarrow A = (-3, 4)$

(ii) $B\left(5\sqrt{2}, \frac{\pi}{4}\right)$

Now $r = 5\sqrt{2}$ and $\theta = \frac{\pi}{4}$ means Ist quadrant

So $x = 5\sqrt{2}\cos\frac{\pi}{4} = 5$ and $y = 5\sqrt{2}\sin\frac{\pi}{4} = 5$

$\Rightarrow B = (5, 5)$

3. According to the given equation $\{x-(a+b)\}^2 + \{y-(b-a)\}^2 = \{x-(a-b)\}^2 + \{y-(a+b)\}^2$

$\Rightarrow -2x(a+b) - 2y(b-a) = -2x(a-b) - 2y(a+b)$

Or $2bx + 2ax + 2yb - 2ay + 2bx - 2ax - 2ay - 2by = 0$

Hence $4(bx - ay) = 0$; i.e., $bx - ay = 0$

4. Let A(2a, 4a), $B(2a, 6a)$ and $C(2a + \sqrt{3}a, 5a)$

Observe that $AB = \sqrt{(2a-2a)^2 + (ba-4a)^2} = |2a|$,

$BC = \sqrt{\left(a\sqrt{3}\right)^2 + (-a)^2} = |2a|$ and $AC = \sqrt{(a\sqrt{3})^2 + a^2} = |2a|$

Hence ABC is an equilateral triangle with side $|2a|$.

5. Let the two vertices on y-axis be $(0, a)$ & $(0, -a)$, so that mid point is (0, 0) obviously the third vertex is say $(x, 0)$, then $x^2 + (-a)^2 = x^2 + a^2 = (2a)^2$

$\Rightarrow x = \pm a\sqrt{3}$. Hence the vertices are $(0, a)$, $(0, -a)$ and $(\pm a\sqrt{3}, 0)$ as two such triangles are possible.

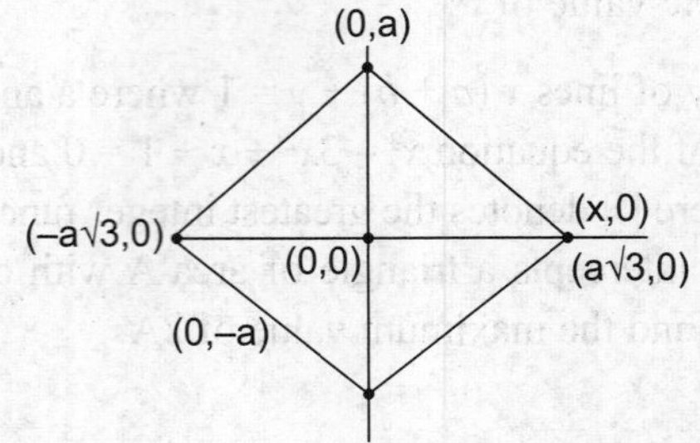

6. Distance $d = \sqrt{(b + a\cos\alpha - b)^2 + (c + a\sin\alpha - c)^2}$

$= \sqrt{a^2\cos^2\alpha + a^2\sin^2\alpha} = |a|$ which is independent of α.

7. Distance

$d = \sqrt{(a + r\cos\alpha - a - r\cos\beta)^2 + (b + r\sin\alpha - b - r\sin\beta)^2}$

$= r\sqrt{(\cos\alpha - \cos\beta)^2 + (\sin\alpha - \sin\beta)^2}$

$= r\sqrt{2 - 2\cos(\alpha - \beta)} = 2r\left|\sin\left(\frac{\alpha - \beta}{2}\right)\right|$

8. (a) Given point $P(a^2 + b^2 - ab, 2ab - a^2 - b^2)$ where $a, b \in \mathbb{R}$ and $a \neq b$.

Now observe that $(a-b)^2 > 0$ (as $a \neq b$)

So $a^2 + b^2 - 2ab > 0 \Rightarrow \frac{a^2 + b^2}{2} - ab > 0$

$\therefore \frac{a^2+b^2}{2} + \frac{a^2+b^2}{2} - ab > \frac{a^2+b^2}{2} > 0$

So $a^2 + b^2 - ab >$ (i.e., +ve)

$\Rightarrow -(a^2 + b^2 - 2ab) = -(a-b)^2 < 0$ (i.e., –ve)

so x is +ve and y is –ve

$\Rightarrow$ Fourth quadrant

(b) Given point $Q(a + b + c, a^3 + b^3 + c^3 - 3abc)$ where a, b, c are of same sign but distinct real numbers.

Now, observe that $a^3 + b^3 + c^3 - 3ab = (a + b + c)\{a^2 + b^2 + c^2 - (ab + bc + ca)\}$

$= \frac{1}{2}(a+b+c)\{(a-b)^2 + (b-c)^2 + (c-a)^2\}$ as a, b, c are distinct

So $(a-b)^2 + (b-c)^2 + (c-a)^2 > 0$

Hence $x = (a + b + c)$ and $y = (a^3 + b^3 + c^3 - 3abc)$ are of the same sign (both +ve) or (both –ve)

$\Rightarrow$ First or Third quadrant

9. $P(\cos\theta + \sin\theta, \cos\theta - \sin\theta)$ where $\theta \in \left[0, \frac{\pi}{4}\right]$

Now $x = \cos\theta + \sin\theta = \sqrt{2}\sin\left(\frac{\pi}{4} + \theta\right) > 0$ and

$y = \cos\theta - \sin\theta = \sqrt{2}\sin\left(\frac{\pi}{4} - \theta\right) \geq 0$ as $\theta \in \left[0, \frac{\pi}{4}\right]$

$\therefore$ P lies in First quadrant (Or on x-axis when $\theta = \pi/4$)

10. Given $AB = 10$ units, $A(2, 3)$.

Let $B = (10, y)$, so $8^2 + (y-3)^2 = 100$

i.e., $(y-3)^2 = 36 = (\pm 6)^2 \Rightarrow y = -3, 9$

TEXTUAL EXERCISE–1 (OBJECTIVE)

1. (b) $A = (-1, -1) \Rightarrow r = \sqrt{2}$ and $\theta = \pm\pi + \tan^{-1}(1)$

So $A = \left(\sqrt{2}, -\frac{3\pi}{4}\right)$ or $\left(\sqrt{2}, \frac{5\pi}{4}\right)$

2. (c) $A\left(2\sqrt{2}, -\frac{3\pi}{4}\right)$

$\Rightarrow r = 2\sqrt{2}$ and $\theta = -\frac{3\pi}{4}$ means third quadrant

So $x = 2\sqrt{2}\cos\frac{3\pi}{4} = -2$ and $y = 2\sqrt{2}\sin\frac{3\pi}{4} = -2$

$\Rightarrow A(-2, -2)$

3. (a) $r = a\sin\theta$

Since $x = r\cos\theta = a\sin\theta\cos\theta$ & $y = r\sin\theta = a\sin^2\theta$

$\Rightarrow x^2 + y^2 = a^2\sin^2\theta = a.a\sin^2\theta = ay$

So $x^2 + y^2 = ay$

4. (a) $x - a = a\cos\theta \Rightarrow x = a(1 + \cos\theta)$ and $y = a\sin\theta$

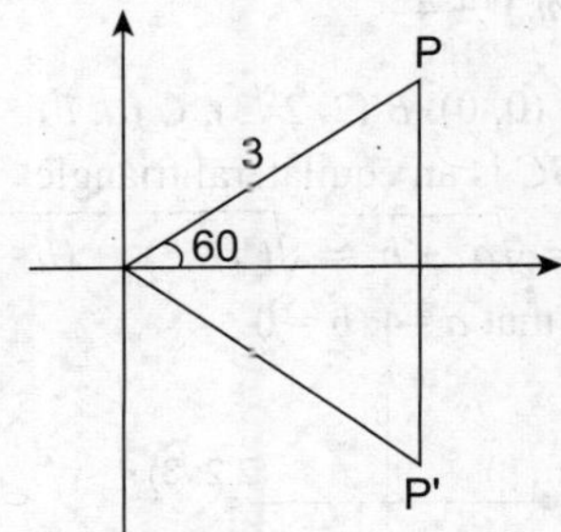

5. (b) Given $P\left(3, \frac{\pi}{3}\right)$; so $P'\left(3 - \frac{\pi}{3}\right)$

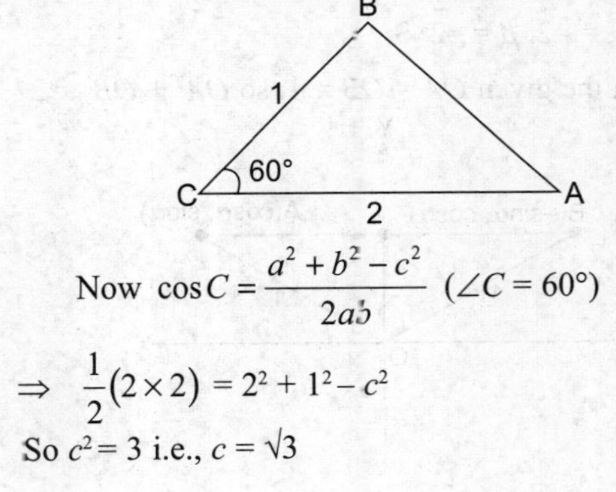

6. (d) $P = (2\cos 60°, 2\sin 60°)$

$Q = \left\{2\cos\left(-\frac{2\pi}{3}\right), 2\sin\left(-\frac{2\pi}{3}\right)\right\} = \left(-1, -\sqrt{3}\right)$

7. (c) Consider a ΔABC as shown below

Now $\cos C = \frac{a^2 + b^2 - c^2}{2ab}$ $(\angle C = 60°)$

$\Rightarrow \frac{1}{2}(2 \times 2) = 2^2 + 1^2 - c^2$

So $c^2 = 3$ i.e., $c = \sqrt{3}$

8. (c) $r = a$; So $r^2 = a^2$

$\Rightarrow x^2 + y^2 = a^2$

9. (c) By checking we can conclude that points in (c) option satisfy the given condition of $4(PQ)^2 + (PR)^2 = (QR)^2$

10. (b)

11. (a)

TEXTUAL EXERCISE–2 (SUBJECTIVE)

1. (a) Let $A(1, 1)$, $B(-2, 7)$ and $C(3, -3)$

$\Rightarrow AB = \sqrt{3^2 + 6^2} = \sqrt{45} = 3\sqrt{5}$;

$BC = \sqrt{5^2 + 10^2} = \sqrt{125} = 5\sqrt{5}$;

$AC = \sqrt{2^2 + 4^2} = \sqrt{20} = 2\sqrt{5}$

Observe that $AB + AC = AB$. Hence Points are collinear.

(b) Let $A\left(0, \frac{8}{3}\right); B(1,3); C(82,30)$

$AB = \sqrt{1^2 + \left(\frac{1}{3}\right)^2} = \frac{1}{3}\sqrt{10}$;

$AC = 82\sqrt{1 + \left(\frac{1}{3}\right)^2} = \frac{82}{3}\sqrt{10}$;

$BC = \sqrt{(81)^2 + (27)^2} = 27\sqrt{10}$

Observe that $AB + BC = AC$, so these points are collinear.

Observe that no two sides are equal, so these points do not form an isosceles triangle.

2. Given: $A(0, -1)$, $B(2, 1)$, $C(0, 3)$, $D(-2, 1)$

$AB = \sqrt{8} = BC = CD = AD$ .This gives a rhombus or squares, also $AC = 4 \Rightarrow AB^2 + BC^2 = AC^2$

$\therefore$ $ABCD$ is a square.

3. Let $A(a, a)$, $B(-a, -a)$, $C(-a\sqrt{3}, a\sqrt{3})$

$\Rightarrow AB = \sqrt{8a^2} = 2a\sqrt{2}$; BC

$= \sqrt{a^2\left(\sqrt{3}-1\right)^2 + a^2\left(\sqrt{3}+1\right)^2} = \sqrt{8a^2} = 2a\sqrt{2}$

And $AC = \sqrt{a^2\left(\sqrt{3}+1\right)^2 + a^2\left(\sqrt{3}-1\right)^2} = 2a\sqrt{2}$

Observe that $AB = BC = AC$

$\Rightarrow \Delta ABC$ is equilateral

4. $A(3, 4)$; B is variable point on $|x| = 6$

$\Rightarrow B(6, y)$ or $B(-6, y)$ where $y \in \mathbb{R}$, $AB \le 4$

Let $AM \perp L_1$ where L_1 is $x = 6$

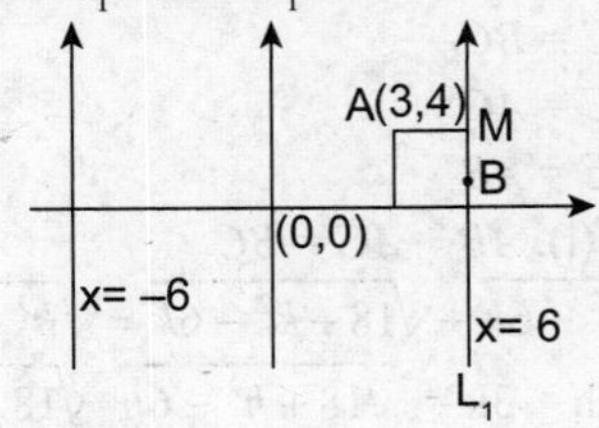

Now consider variable points (6, y)

So $AB = \sqrt{3^2 + (4-y)^2} \le 4$ or $9 + (4-y)^2 \le 16$

$\Rightarrow \; -\sqrt{7} \le \sqrt{(4-y)^2} \le \sqrt{7}$. Since B is an integral point.

Hence we get 5 such points.

$\Rightarrow y = \{2, 3, 4, 5, 6\}$

5. Let A (2, 3) on x-axis (2, 0) is at 3 units on y-axis.
Let B (0, y) be a variable point, then $\sqrt{2^2 + (3-y)^2} \le 3$ gives $|y - 3| \le \sqrt{5}$ (y is an integral point)
Hence $y = \{1, 2, 3, 4, 5\}$
$\therefore$ In all we get six (6) such points

6. Let $\angle Q_1 O Q_2 = \theta$
Now $\theta = |\theta_1 - \theta_2|$

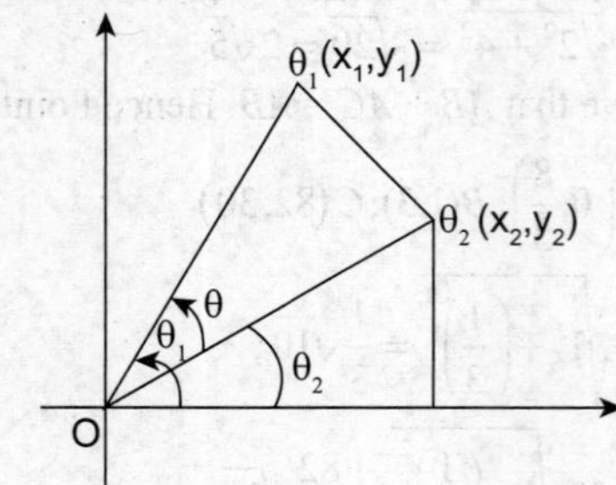

$\Rightarrow \cos\theta = \dfrac{(x_2^2 + y_2^2) + (x_1^2 + y_1^2) - (x_2 - x_1)^2 - (y_2 - y_1)^2}{2\sqrt{x_2^2 + y_2^2}\sqrt{x_1^2 + y_1^2}}$

So, $O\theta_1 O\theta_2 \cos\theta = \dfrac{2x_1x_2 + 2y_1y_2}{2} = x_1x_2 + y_1y_2$

7. $AB = \sqrt{(3-h)^2 + (3-0)^2} = \sqrt{18 + h^2 - 6h}$

$BC = \sqrt{h^2 + k^2} = \sqrt{h^2 + k^2}$

$AC = \sqrt{(3-0)^2(3-k)^2} = \sqrt{18 + k^2 - 6k}$

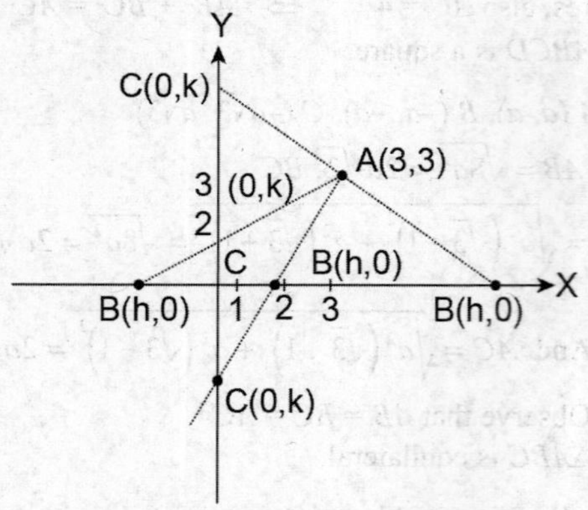

From the given figure given above clearly, we have following possibilities

$AB + AC = BC$ ...(i)

$AB + BC = AC$ ...(ii)

and $AC + BC = AB$ ..(iii)

Considering (i), $AB + AC = BC$

$\Rightarrow \sqrt{18 + h^2 - 6h} + \sqrt{18 + k^2 - 6k} = \sqrt{h^2 + k^2}$

$\Rightarrow$ – 18 + 3h + 3k = $\sqrt{18 + h^2 - 6h}\ \sqrt{18 + k^2 - 6h}$

Squaring again and solving, we get $0 = 9h^2 + 9k^2 + 18hk + h^2k^2 - 6h^2k - 6hk^2$

$\Rightarrow (3h + 3k - hk)^2 = 0 \quad \Rightarrow \quad 3h + 3k = hk$

$\Rightarrow \dfrac{1}{h} + \dfrac{1}{k} = \dfrac{1}{3}$

Similarly same result can be obtained using (ii) and (iii)

8. Let $A(a_1^2, 2at_1)$ & $B(at_2^2, 2at_2)$

$AB = \sqrt{a^2(t_2^2 - t_1^2)^2 + 4a^2(t_2 - t_1)^2}$

$= a|t_2 - t_1|\sqrt{t_2^2 + t_1^2 + 2t_1t_2 + 4}$ (Given a > 0)

Now $t_1 + t_2$ are the roots of equation $x^2 - 2\sqrt{3}\,x + 2 = 0$, so $t_1 t_2 = 2$ and $t_2 + t_1 = 2\sqrt{3}$

Hence $AB = a\sqrt{(t_2 + t_1)^2 - 4t_1t_2}\sqrt{(2\sqrt{3})^2 + 4}$

= a(2) (4) = 8a or 4a $|t_2 - t_1|$

TEXTUAL EXERCISE–2 (OBJECTIVE)

1. **(c)** Let A (0, 1) and B (x, –3), so $AB = \sqrt{x^2 + 4^2} = 5$
$\Rightarrow x^2 + 16 = 25$, so $x \pm 3$

2. **(c)** Since the sides are along positive axes, so (–5, –5) can not be vertex.

3. **(a)** Given A (6, –1), B (1, 3) and C (x, 8)
From $AB = BC$, we get $5^2 + 4^2 = (x - 1)^2 + 5^2$
$\Rightarrow x - 1 = \pm 4$; So $x = 5$ or -3

4. **(a)** Let $A(am_1^2, 2am_1)$ and $B(am_2^2, 2am_2)$
So $AB = \sqrt{a^2(m_2^2 - m_1^2)^2 + 4a^2(m_2 - m_1)^2} = |a|\,|m_2 - m_1|$
$\sqrt{(m_2 + m_1)^2 + 4}$

5. **(a), (c)** Let A (0, 0), B (2, $2\sqrt{3}$), C (a, b)
Since ABC is an equilateral triangle
$\therefore \quad AB = 4 = \sqrt{a^2 + b^2} = \sqrt{(a-2)^2 + (b - 2\sqrt{3})^2}$
Observe that $a = 4$, $b = 0$

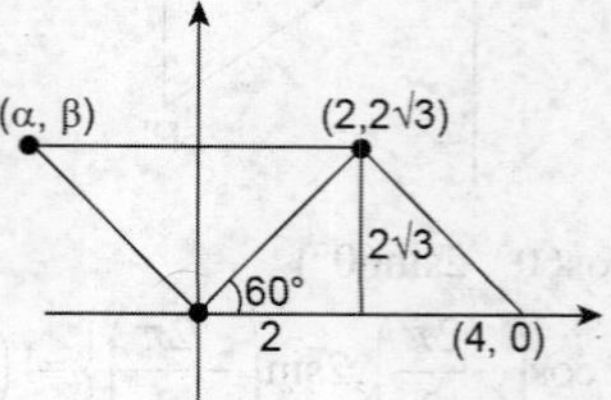

Since $\dfrac{\alpha + 4}{2} = \dfrac{0 + 2}{2}$; $\dfrac{\beta + 0}{2} = \dfrac{0 + 2\sqrt{3}}{2}$
$\Rightarrow \alpha = -2$; $\beta = -2\sqrt{3}$

6. **(c)** From the given $OA = OB = 1$, so $OA^2 + OB^2 = 2$

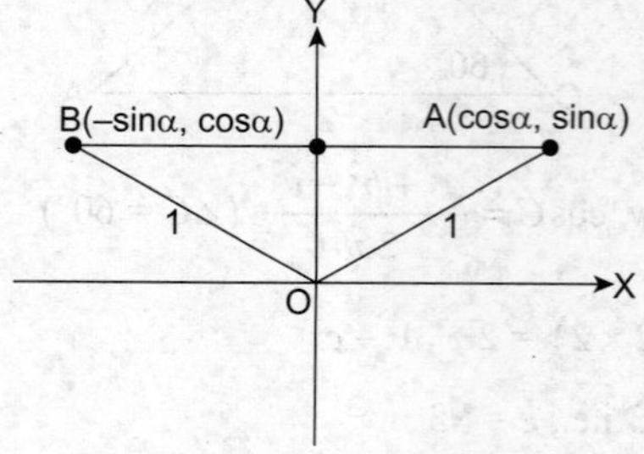

Observe that $\angle AOB = 90°$
Since $\cos(90° + \alpha) = -\sin\alpha$ and $\sin(90 + \alpha) = \cos\alpha$

7. **(b)** Given: $A(5, 2)$, $B(2, 3)$, $C(6, 5)$
$\Rightarrow AB = \sqrt{10}, BC = 2\sqrt{5} = \sqrt{20}, AC = \sqrt{10}$
Observe that $AB^2 - AC^2 = BC^2$, also $AB = AC$
So it is right angled isosceles triangle

8. **(c)** Given: $P(0, 0)$, $Q\left(3, \frac{\pi}{2}\right), R\left(3, \frac{\pi}{6}\right)$

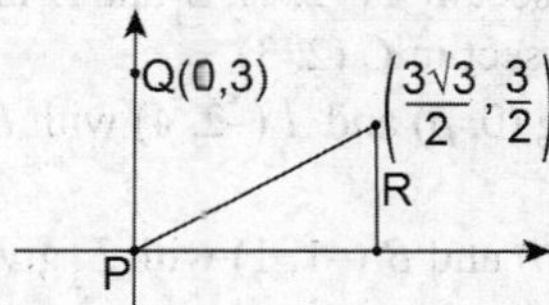

From the given it is very clear that $PQ = PR = 3$, also $\angle QPR = 60°$
$\Rightarrow \Delta PQR$ is equilateral triangle

9. **(b)** Given: $A(1, 4)$ and $B(4, 8)$, also $AP = AB + BP$
Now $AB = \sqrt{3^2 + 4^2} = 5$, so $P(7, 12)$ or $(12, 7)$
On checking $P(7, 12)$ gives $AP = 10$ and $BP = 5$
$\Rightarrow$ P (7, 12)

10. **(c)** Let $A(a, b)$, $B(a_1, b_1)$, $C(2a_1 - a, 2b_1 - b)$
$AB = \sqrt{(a - a_1)^2 + (b - b_1)^2}$
$BC = \sqrt{(a_1 - a)^2 + (b_1 - b)^2}$
$AC = \sqrt{(2a_1 - 2a)^2 + (2b_1 - 2b)^2}$
$\Rightarrow AC = 2\sqrt{(a_1 - a)^2 - (b_1 - b)^2}$
$\Rightarrow AC = AB + BC$
$\Rightarrow A, B, C$ are collinear

11. **(b)** Let (a, a) be point, then $(a - 1)^2 + a^2 = a^2 + (a - 3)^2$
$\Rightarrow a - 1 = a - 3$ (not possible) or $a - 1 = 3 - a$
$\Rightarrow a = 2$
So (2, 2) is the point

12. **(a)** $A(9, -9)$ and $B(1, 3)$
$\Rightarrow$ Slope of $AB = -3/2$
Now, mid point is $(5, -3)$, so equation of right bisector of AB is $(y + 3) = \frac{2}{3}(x - 5)$ or $3y + 9 = 2x - 10$ i.e., $2x - 3y - 19 = 0$.
Observe that $(8, -1)$ satisfies .

13. **(a), (c)** $A(2, 3)$, $B(0, -1)$
$\Rightarrow$ Slope of $AB = 2$ and slope of $CA = -1/2$
Equation of AC $(y - 3) = -\frac{1}{2}(x - 2)$ or $2y - 6 = -x + 2$
i.e., $x + 2y - 8 = 0$
Observe that (4, 2) and (0, 4) satisfies.
Now BC = 5 units which is satisfied by both these points.

14. **(a)** Given $A(x, -1)$, $B(3, y)$, $C(-2, 3)$, $D(-3, -2)$
So $x - 2 = 0$ $\Rightarrow x = 2$ and $-1 + 3 = y - 2$
$\Rightarrow y = 4$

15. **(c)** Given $A(2, -2)$, $B(8, 4)$ and $C(5, 7)$
Let $D(x, y)$, then $7 = x + 18$
$\Rightarrow x = -1$ and $5 = 4 + y \Rightarrow y = 1$
$\Rightarrow$ D (–1, 1)

16. **(c)** P (1, 2), $Q(4, 6)$, $R(5, 7)$, S (a, b)
So $4 + a = 5 + 1$ $\Rightarrow a = 2$ and $6 + b = 2 + 7$
$\Rightarrow$ b = 3
Hence S (2, 3)

TEXTUAL EXERCISE–3 (SUBJECTIVE)

1.(a) **(i)** $A(0, -4)$, $B(8, 0)$, external division in 4 : 3
$\Rightarrow x = \frac{4(8) - 3(0)}{4 - 3} = 32$ and $y = \frac{4(0) - 3(-4)}{4 - 3} = 12$
$\Rightarrow$ (32, 12)

(ii) $A(1, 2)$, $B(-4, -3)$, external division in the ratio 2 : 3
$\Rightarrow x = \frac{2(-4) - 3(1)}{2 - 3} = 11$ and $y = \frac{2(-3) - 3(2)}{2 - 3} = 12$
$\Rightarrow$ (11, 12)

(b) Let $P(a + b, a - b)$ and $Q(a - b, a + b)$ ratio is $a : b$, $(a + b \neq 0)$
$\Rightarrow x = \frac{a^2 - ab + b^2 + ab}{a + b} = \frac{a^2 + b^2}{a + b}$ and
$y = \frac{a^2 + ab + ab - b^2}{a + b} = \frac{a^2 + 2ab - b^2}{a + b}$
$\Rightarrow \left(\frac{a^2 + b^2}{a + b}, \frac{a^2 - b^2 + 2ab}{a + b}\right)$

2. Mid point of $A(x, y + 1)$ and $B(x + 1, y + 2)$ is $P\left(x + \frac{1}{2}, y + \frac{3}{2}\right) = \left(\frac{3}{2}, \frac{5}{2}\right)$, so $x = y = 1$
Hence mid point of $(x - 1, y + 1)$ and $(x + 1, y - 1)$ is (x, y) = (1, 1)

3. Let $A(x_1, y_1)$, $B(x_2, y_2)$, $C(x_3, y_3)$ and $D(x_4, y_4)$, taken in order form a quadrilateral mid point of AD is
$E = \left(\frac{x_1 + x_4}{2}, \frac{y_1 + y_4}{2}\right)$
Mid point of BC is $F = \left(\frac{x_2 + x_3}{2}, \frac{y_2 + y_3}{2}\right)$

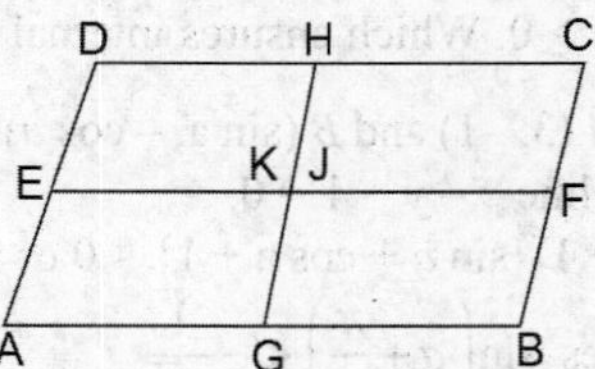

$\Rightarrow$ Mid point of EF is
$J = \left(\frac{x_1 + x_4 + x_2 + x_3}{4}, \frac{y_1 + y_2 + y_3 + y_4}{4}\right)$.

Similarly mid point of CD is $H = \left(\frac{x_3 + x_4}{2}, \frac{y_3 + y_4}{2}\right)$ and mid point of AB is $G = \left(\frac{x_1 + x_2}{2}, \frac{y_1 + y_2}{2}\right)$

$\Rightarrow$ Mid point of GH is

$k = \dfrac{x_1+x_2+x_3+x_4}{4}, \dfrac{y_1+y_2+y_3+y_4}{4}$

$\Rightarrow$ Observe that J & K coincide.

4. Let A (3, –4) and C (–6, 2) be the extremities of a diagonal. Third vertex is B (–1, –3).
Now, let D (x, y) be the required vertex, then $x - 1 = 3 = 6$
$\Rightarrow$ $x = -2$ and
$y - 3 = 2 - 4$ $\qquad \Rightarrow$ $y = 1$
So D (–2, 1)

5. **(i)** Given A (–2, 4), B (7, –3) and centroid G (3, 2)
$\Rightarrow$ $C_x = \{3 \times 3 - (-2 + 7)\} = 4$ and $C_y = 2 \times 3 - (4 - 3) = 5$
Using $\left(\dfrac{x_1+x_2+x_3}{3}, \dfrac{y_1+y_2+y_3}{3}\right) = G$
$\therefore$ C (4, 5)
(ii) Two vertices of an equilateral triangle are A (0, 0), B (3, √3).
Observe that slope of AB = 1/√3 (i.e., π/6 or 30°)
$\Rightarrow$ Third vertex will be (0, 2√3), or (3, –√3) {from slope the angle θ = π/2 or π/6}

6. Let A (–2, –2), B (1, 0), C (4, 4) and D (1, 2)
$AB = \sqrt{13}$, BC = 5, AC = 6√2, AD = 5, CD = √13.
Observe that AB = CD and BC = AD. Also mid point of AC and BD coincide at (1, 1).
Hence it is a parallelogram.

7. **(a)** A (7, 3) and B (–4, 5).
The join will be divided by y – axis in $\lambda = \dfrac{-7}{(-4)} = 7:4$ (internally)
(b) Given Q (–2, 4), OQ = $\dfrac{1}{3}$ OP, OP = 3OQ
$\Rightarrow$ P (–6, 12)

8. Given A (x_1, y_1) and B (x_2, y_2). A point that divides the line segment AB in the ratio t : 1 – t internally will be
$M\left(\dfrac{tx_2+(1-t)x_1}{t+1-t}, \dfrac{ty_2+(1-t)y_1}{t+1-t}\right) = x_1 + t\,(x_2 - x_1),\ y_1 + t\,(y_2 - y_1)$
Which happens to be the given point. Since 0 < t < 1
$\Rightarrow$ 0 < 1 – t < 1
So (t : 1 – t) > 0. Which ensures internal division.

9. Given that A (3, –1) and B (sin a, – cos a) lie on the opposite sides of the line $x - y + 1 = 0$.
$\Rightarrow$ (3 + 1 + 1){sin a + cos a + 1} < 0 or 1 + √2 sin(a + π/4) < 0 gives $\sin\left(a+\dfrac{\pi}{4}\right) < -\dfrac{1}{\sqrt{2}}$
So $\dfrac{5\pi}{4} < a+\dfrac{\pi}{4} < \dfrac{7\pi}{4}$ or $\pi < a < \dfrac{3\pi}{2}$

10 Let $L_1 : 4x + 7y + 19 = 0$
$L_2 : 4x + y - 11 = 0$ or $-4x - y + 11 = 0$
$L_3 : 4x - 5y + 7 = 0$
L_1 and L_2 intersect at A (4, –5); L_1 & L_3 intersect in B(–3, –1) and L_2 & L_3 intersect at C (2, 3)
Now easily we can observe (or check) that A (4, –5) and (0, 0) lie on the same side of L_3.
Similarly B (–3, –1) and (0, 0) with L_2 and C(2, 3) and (0, 0) with L_1
$\Rightarrow$ (0, 0) lies inside the ΔABC.

11. Let L_1: $3x + y + 2 = 0$
$L_2 : 2x - 3y + 5 = 0$
$L_3 : -x - 4y + 14 = 0$
L_1 and L_3 intersect in A (–2, 4); L_1 and L_2 intersect in B (–1, 1);
L_2 and L_3 intersect in C (2, 3)
Now checking (0, β) and A (–2, 4) with L_2 gives $-3\beta + 5 \le 0$
$\Rightarrow$ $\beta \ge 5/3$ ……(i)
Similarly (0, β) and B (–1, 1) with L_3 gives $-4\beta + 14 \ge 0$
$\Rightarrow$ $\beta \le 7/2$ ……(ii)
Further (0, β) and C (2, 3) with L_1 gives $\beta + 2 > 0$
$\Rightarrow$ $\beta \ge -2$ ……(iii)
From (i), (ii) & (iii), we get $5/3 \le \beta \le 7/2$

12. Let $L_1 : 2x - 3y - 6 = 0$ or $2x + 3y + 6 = 0$
L_2: $3x - y + 3 = 0$
L_3: $3x + 4y - 12 = 0$ or $-3x + 4y + 12 = 0$
Point of intersection of L_1 and L_2 is $A\left(\dfrac{-15}{7}, \dfrac{-24}{7}\right)$; L_2 & L_3 in B (0, 3)
L_1 and L_3 in $C\left(\dfrac{60}{17}, \dfrac{6}{17}\right)$
Now P (α, 0) and Q (0, β) will line inside the ΔABC
On checking, we get
(i) 3α – 12 ≤ 0 $\qquad \Rightarrow$ α ≤ 4 and $4\beta - 12 \le 0$
$\qquad \Rightarrow$ $\beta \le 3$
(ii) 2α – 6 ≤ 0 $\qquad \Rightarrow$ α ≤ 3 and $-3\beta - 6 \le 0$
$\qquad \Rightarrow$ $\beta \ge -2$
(iii) 3α + 3 ≥ 0 $\qquad \Rightarrow$ α ≥ –1 and $-\beta + 3 \ge 0$
$\qquad \Rightarrow$ $\beta \le 3$
From (i), (ii) and (iii), we get –1 ≤ α ≤ 3 and $-2 \le \beta \le 3$ i.e., α ∈ [–1, 3] and β [–2, 3]

13. Given: A (5, 6) and B (cosθ, sinθ) where $\theta \in (0, \pi)$ lie on the opposite sides of $x + y - 1 = 0$
$\Rightarrow$ $\dfrac{L(x_1, y_1)}{L(x_2, y_2)} < 0$
$\Rightarrow$ cosθ + sinθ – 1 < 0 i.e., $\sqrt{2}\left\{\sin\left(\dfrac{\pi}{4}+\theta\right)\right\} < 1$ or $\sin\left(\dfrac{\pi}{4}+\theta\right) < \dfrac{1}{\sqrt{2}}$
Since $\theta \in (0, \pi)$ given so $\left(\theta+\dfrac{\pi}{4}\right) \in \left(\dfrac{\pi}{4}, \dfrac{5\pi}{4}\right)$.
Now $\sin\left(\dfrac{\pi}{4}+\theta\right) < \dfrac{1}{\sqrt{2}}$ gives $\theta \in \left(\dfrac{\pi}{2}, \pi\right)$

14. **(a)** A (–2, 1), B (0, 5), C (5, 15)
Let B divide AC in the ratio λ : 1, then
$B = \left(\dfrac{5\lambda-2}{\lambda+1}, \dfrac{15\lambda+1}{\lambda+1}\right) = (0,5)$

$\frac{5\lambda-2}{\lambda+1}=0$ gives λ = 2/5, putting in $\frac{15\lambda+1}{\lambda+1}$ gives $\frac{6+1}{1+\frac{2}{5}}=5$ which is true

⇒ Points are collinear

(b) A (3, 1), B (6, 4) and C (4, 5). Let B divide AC in the ratio λ : 1

⇒ $\left(\frac{4\lambda+3}{\lambda+1},\frac{5\lambda+1}{\lambda+1}\right)=(6,4)$

$\frac{4\lambda+3}{\lambda+1}=6$ gives 4λ + 3 = 6λ + 6

⇒ λ = –3/2 and $\frac{5\lambda+1}{\lambda+1}=4$ gives 5λ + 1 = 4λ + 4 ⇒ λ = 3

Values are different so these points are not collinear.

TEXTUAL EXERCISE–3 (OBJECTIVE)

1. (a) Mid point of (a sinθ, 0) and (0, a cosθ) is $\left(\frac{a}{2}\sin\theta,\frac{a}{2}\cos\theta\right)$ and its distance form the origin is $a/2$

2. (b) x – axis will divide the join of A (2, –3) and B (5, 6) in the ratio $\lambda=\frac{-(-3)}{6}=1:2$

3. (a) A (–2, 5), B (3, 1). Now $AP = PQ =$ QB
⇒ Mid point of PQ is same as the mid point of AB which is $\left(\frac{1}{2},3\right)$

4. (a) Mid point of (5, a) and (b, 7) is (3, 5)
⇒ 5 + b = 6, so b = 1 and a + 7 = 10 so a = 3
Hence (a, b) = (3, 1)

5. (d) Let A (2, –2), B (8, –4), C (5, 7) and D(x, y) be the vertices of a rectangle, then 2 + 5 = 8 + x
⇒ x = –1 and y – 4 = 7 – 2
⇒ y = 9
Hence D = (–1, 9)

6. (a) A (3, –1); B (–6, 5).
Let P & Q trisect it then $P=\left(\frac{-6+6}{3},\frac{5-2}{3}\right)=(0,1)$ and $Q=\left(\frac{-9}{3}+\frac{9}{3}\right)=(-3,3)$

7. (a), (c) Let A (1, 3), B (5, 0), C (–1, 2)
Equation of
(i) AB is L_1 : 3x + 4y – 15 = 0
(ii) BC is L_2 : x + 3y – 5 = 0
(iii) AC is L_3 : x – 2y + 5 = 0
On checking A, B, C respectively with L_2, L_3, L_1; we get x + 3y – 5 ≥ 0; x – 2y + 5 ≥ 0, 3x + 4y – 15 ≤ 0
Now, x + 3y – 5 ≥ 0 and x – 2y + 5 ≥ 0
⇒ (x + 3y – 5) + (x – 2y + 5) ≥ 0
⇒ 2x + y ≥ 0 ∴ option (a)
Also 3x + 4y – 15 ≤ 0 and x – 2y + 5 ≥ 0 or –x + 2y – 5 ≤ 0
⇒ (3x + 4y – 15) + (–x + 2y – 5) ≤ 0
⇒ 2x + 6y – 20 ≤ 0 ⇒ x + 3y – 10 ≤ 0
∴ option (c)

8. (a), (b), (c) L_1 : 2x + 3y + 7 = 0 & L_2 : 4x + 6y – 19 = 0
For point P (1, 1)
⇒ L_1 (1, 1) = 12 and (b_1 = 3) and L_2 (1, 1) = – 9 (b_2 = 6)
So P (1, 1) lies above L_1 but below L_2
Now for C_1 = 7 > 0 and C_2 = 19 > 0; we get L_1 (1, 1) > 0 & L_2 (1, 1) > 0
⇒ (1, 1) lies between L_1 & L_2

9. (b) Let L_1 : x – 2y = 0 (for x > 0) and L_2: 3x – y = 0 (for x > 0)
⇒ The point (a, a^2) should be such that it is above L_1 but below L_2
i.e., $\frac{L_1(a,a^2)}{b_1}>0$ and $\frac{L_2(a,a^2)}{b_2}<0$

Hence $\frac{a-2a^2}{-2}>0$ or $2a^2 - a > 0$, i.e., a(2a – 1) > 0
So a ∈ (–∞, 0) ∪ (1/2, ∞) ...(i)

Similarly $\frac{3a-a^2}{-1}<0$ ⇒ a^2 – 3a < 0 or a (a – 3) < 0
⇒ a ∈ (0, 3) ...(ii)
From (i) and (ii) $a\in\left(\frac{1}{2},3\right)$

10. (a) (sinθ, cosθ) and (3, 2) lie on the same side of $x + y - 1 = 0$
⇒ $\frac{L(\sin\theta,\cos\theta)}{L(3,2)}>0$ i.e., $\frac{\sin\theta+\cos\theta-1}{3+2-1}>0$
⇒ $\sqrt{2}\sin\left(\frac{\pi}{4}+\theta\right)>1$, hence $\left(\theta+\frac{\pi}{4}\right)\in\left(\frac{\pi}{4},\frac{3\pi}{4}\right)$
So $\theta\in\left(0,\frac{\pi}{2}\right)$

11. (a), (c) Let A (–1, 0), B (3, 1), C (2, 2) and D (x, y) be the vertices of a ||gm taken in order
So x + 3 = 2 – 1 ⇒ x = –2
and y + 1 = 2 + 0 ⇒ y = 1
Hence D_1 (–2, 1)
Similarly x – 1 = 3 + 2
⇒ x = 6
and y – 0 = 1 + 2 ⇒ y = 3, so D_2 (6, 3)
On the same line x + 2 = 3 – 1
⇒ x = 0 and y + 2 = 1 + 0
⇒ y = – 1, so D_3 (0, –1)

TEXTUAL EXERCISE–4 (SUBJECTIVE)

1. (a) Let A (1, 3), B (–7, 6), C (5, –1)
Area of $\Delta ABC=\frac{1}{2}\left\|\begin{matrix}1&3&1\\-7&6&1\\5&-1&1\end{matrix}\right\|=\frac{1}{2}(20)=10$ square units.

(b) Let $A(a\cos\phi_1, b\sin\phi_1), B(a\cos\phi_2, b\sin\phi_2), C(a\cos\phi_3, b\sin\phi_3)$

$$\text{Area of } \Delta ABC = 1/2 \left\|\begin{matrix} a\cos\phi_1 & b\sin\phi_1 & 1 \\ a\cos\phi_2 & b\sin\phi_2 & 1 \\ a\cos\phi_3 & b\sin\phi_3 & 1 \end{matrix}\right\|$$

$$= \frac{1}{2}\left\|\begin{matrix} a\cos\phi_1 & b\sin\phi_1 & 1 \\ a(\cos\phi_2 - \cos\phi_1) & b(\sin\phi_2 - \sin\phi_1) & 0 \\ a(\cos\phi_3 - \cos\phi_1) & b(\sin\phi_3 - \sin\phi_1) & 0 \end{matrix}\right\|$$

$$= \frac{ab}{2}\left\{\begin{matrix}(\cos\phi_2 - \cos\phi_1)(\sin\phi_3 - \sin\phi_1) - \\ (\cos\phi_3 - \cos\phi_1)(\sin\phi_2 - \sin\phi_1)\end{matrix}\right\}$$

By proper grouping, we get

$$\Delta ABC = \frac{ab}{2}\sin\left(\frac{\phi_2 - \phi_3}{2}\right).\sin\left(\frac{\phi_3 - \phi_1}{2}\right)\sin\left(\frac{\phi_1 - \phi_2}{2}\right)$$

2. Let $A(at_1^2, 2at_1), B(at_2^2, 2at_2), C(at_3^2, 2at_3)$

$$\Delta ABC = \frac{1}{2}\left\|\begin{matrix} at_1^2 & 2at_1 & 1 \\ at_2^2 & 2at_2 & 1 \\ at_3^2 & 2at_3 & 1 \end{matrix}\right\| = \frac{1}{2}\left\|\begin{matrix} at_1^2 & 2at_1 & 1 \\ a(t_2^2 - t_1^2) & 2a(t_2 - t_1) & 0 \\ a(t_3^2 - t_1^2) & 2a(t_3 - t_1) & 0 \end{matrix}\right\|$$

$$= \frac{2a^2}{2}\{(t_2 - t_1)(t_2 + t_1)(t_3 - t_1) - (t_3 - t_1)(t_3 + t_1)(t_2 - t_1)\}$$

$$= a^2|(t_3 - t_1)(t_2 - t_1)||\{t_2 + t_1 - t_3 - t_1\}|$$

$$= a^2|(t_2 - t_1)(t_2 - t_3)(t_3 - t_4)|$$

3. Two opposite vertex of a square are $A(x_1, y_1)$ and $C(x_2, y_2)$

Now area of square $ABCD = \frac{Ac^2}{2} = \frac{(x_2 - x_1)^2 + (y_2 - y_1)^2}{2}$

($\theta \neq n\pi/2$ where $n \in \mathbb{Z}$}

4. Let $A(t, t - 2), B(t + 2, t + 2), C(t + 3, t)$

$$\text{Area } \Delta ABC = \frac{1}{2}\left\|\begin{matrix} t & t-2 & 1 \\ t+2 & t+2 & 1 \\ t+3 & t & 1 \end{matrix}\right\|$$

$$= \frac{1}{2}\left\|\begin{matrix} t & t-2 & 1 \\ 2 & 4 & 0 \\ 3 & 2 & 0 \end{matrix}\right\| = \frac{8}{2} = 4 \text{ square units, which is independent}$$

of t.

5. Let $A(\cos^2\theta, 0), B(0, \sec^2\theta), C(1, 1)$

$$\text{Area of } \Delta ABC = \frac{1}{2}\left\|\begin{matrix} \text{cosec}^2\theta & 0 & 1 \\ 0 & \sec^2\theta & 1 \\ 1 & 1 & 1 \end{matrix}\right\|$$

$$= \frac{1}{2}\left|\begin{matrix} \text{cosec}^2\theta & 0 & 1 \\ -\text{cosec}^2\theta & \sec^2\theta & 0 \\ 1 & 1 - \sec^2\theta & 0 \end{matrix}\right| = \frac{1}{2}$$

$\{\sec^2\theta . \text{cosec}^2\theta - \text{cosec}^2\theta - \sec^2\theta\}$

$$= \frac{1}{2}\left\{\frac{1}{\sin^2\theta.\cos^2\theta} - \frac{\sin^2\theta + cos^2\theta}{\sin^2\theta\, cos^2}\right\} = 0.$$

Hence these points are collinear.

6 (a) Given $A(0, 0), B(1, 3), C(2, 5)$ & $D(-1, 4)$ is a quadrilateral.

So area of quadrilateral

$$ABCD = \frac{1}{2}\left|\begin{matrix} x_1 & y_1 \\ x_2 & y_2 \\ x_3 & y_3 \\ x_4 & y_4 \\ x_1 & y_1 \end{matrix}\right| = \frac{1}{2}\left|\begin{matrix} 0 & 0 \\ 1 & 3 \\ 2 & 5 \\ -1 & 4 \\ 0 & 0 \end{matrix}\right| = \frac{1}{2}\{(5-6) + (8+5) + 0\}$$

= 6 square units.

(b) Let ABC be the triangle and let mid-points of sides be $D(2, -1)$, E (3, 2), F (5, 9), then $\Delta ABC = 4\ \Delta DEF$

$$\text{So } \Delta ABC = 2\left|\begin{matrix} 2 & -1 & 1 \\ 3 & 2 & 1 \\ 5 & 9 & 1 \end{matrix}\right| = 2\left|\begin{matrix} 2 & -1 & 1 \\ 1 & 3 & 0 \\ 2 & 7 & 0 \end{matrix}\right| = 2(7 - 6) = 2$$

square units

7. Given $A(3, 0), B(0, 6)$ and $C(6, 9)$, DE divides AB and AC at D & E respectively.

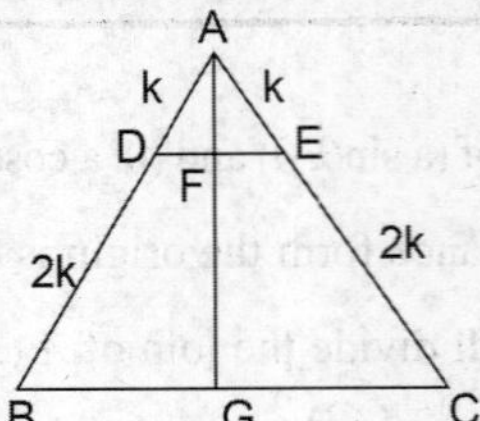

Since ΔABC & ΔADE are similar $AD : AB = AE : AC$

Now, $DE: BC = 1 : 3$, $(DE \parallel BC)$.

Let $AG \perp BC$ & $AG \perp DE$. Hence AF $= \frac{1}{3}$ AG

$\Rightarrow$ Area of $\Delta ABC = \frac{1}{2} BC.AG = \frac{9}{2} DE.$AF and area of $\Delta ADE = \frac{1}{2} DE.$AF

Hence $\frac{\text{area of } \Delta ABC}{\text{area of } \Delta ADE} = 9$

8. Given: $A(6, 3), B(-3, 5), C(4, -2)$ and $P(x, y)$

$$\text{Area of } \Delta PBC = \frac{1}{2}\left\|\begin{matrix} x & y & 1 \\ -3 & 5 & 1 \\ 4 & -2 & 1 \end{matrix}\right\| = \frac{1}{2}\left|\begin{matrix} (x+3) & (y-5) & 0 \\ -7 & 7 & 0 \\ 4 & -2 & 1 \end{matrix}\right|$$

$$= \frac{1}{2}|7(x + 3 + y - 5)| = \frac{7}{2}|x + y - 2| \text{ square units}$$

$$\text{Similarly Area of } \Delta ABC = \frac{1}{2}\left\|\begin{matrix} 6 & 3 & 1 \\ -3 & 5 & 1 \\ 4 & -2 & 1 \end{matrix}\right\| = \frac{1}{2}\left\|\begin{matrix} 0 & 0 & 1 \\ -9 & 2 & 1 \\ -2 & -5 & 1 \end{matrix}\right\|$$

$$= \frac{49}{2} \text{ square units}$$

$$\Rightarrow \frac{\text{Area of } \Delta PBC}{\text{Area of } \Delta ABC} = \frac{|x + y - 2|}{7} \text{ Hence proved}$$

9. Let $C(x, y)$ be the third coordinate then $x = 3 - (4 - 9) = 8$ and $y = 12(-8 + 7) = 13$

Hence area of $\Delta ABC = \frac{1}{2}\begin{Vmatrix} 8 & 13 & 1 \\ 4 & -8 & 1 \\ -9 & 7 & 1 \end{Vmatrix}$

$= \frac{1}{2}\begin{Vmatrix} 0 & 0 & 1 \\ -4 & -21 & 1 \\ -17 & -6 & 1 \end{Vmatrix} = \frac{1}{2}|\{24-357\}| = \frac{333}{2}$ Square Units

TEXTUAL EXERCISE–4 (OBJECTIVE)

1. (c) Given $A(-4, 0)$, $B(1, -1)$, $C(p, q)$ and $\Delta ABC = 4$ square units

$AB = \sqrt{5^2 + 1^2} = \sqrt{26}$

Equation of AB is $x + 5y + 4 = 0$, point C is $\frac{8}{\sqrt{26}}$ units form AB which is possible when the point $C(p, q)$ lies on $x + 5y - 4 = 0$ as distance between these parallel lines is $\frac{8}{\sqrt{26}}$ units

2. (a) From the given data the third vertex is $C(8, 8)$, when $A(-9, 7)$, $B(4, -3)$ and G (1, 4)

$\Delta ABC = \frac{1}{2}\begin{Vmatrix} -9 & 7 & 1 \\ 4 & -3 & 1 \\ 8 & 8 & 1 \end{Vmatrix} = \frac{1}{2}\begin{Vmatrix} 0 & 0 & 1 \\ 13 & -10 & 1 \\ 17 & 1 & 1 \end{Vmatrix}$

$= \frac{1}{2}\{13 + 170\} = \frac{183}{2}$ square units

3. (d) The vertices of ΔABC are $A(-5, 0)$, $B(3, 0)$ and C lies on $x - y = 2$

Area of $\Delta ABC = 20$ units. Equation of AB is $y = 0$

$AB = 8$ units

$\Rightarrow$ Point C is 5 units form AB. Let (x_1, y_1) be on C

$\Rightarrow y_1 = \pm 5$

So $x = 7, -3$ respectively. Hence $C(7, 5)$ or $(-3, -5)$

4. (d) Given: two Δ's with equal area, then the Δ's may or may not be similar or congruent

5. (a) Given $A(6, 3)$, $B(-3, 5)$, $C(4, -2)$ & $D(x, 3x)$

Area of $\Delta DBC = \frac{1}{2}\begin{vmatrix} -3 & 5 & 1 \\ 4 & -2 & 1 \\ x & 3x & 1 \end{vmatrix} = \frac{1}{2}\begin{vmatrix} 0 & 0 & 1 \\ 7 & -7 & 1 \\ x+3 & 3x-5 & 1 \end{vmatrix}$

$= \frac{7}{2}|4x - 2| = 7|(2x-1)|$ and area of ΔABC

$= \frac{1}{2}\begin{vmatrix} x & 3x \\ 6 & 3 \\ -3 & 5 \\ 4 & -2 \\ x & 3x \end{vmatrix} = \frac{1}{2}|\{25 - x\}|$

From the given (2)7| (2x – 1) = 1/2 |(25 – x)| or 56x – 28 = 25 – x

$\Rightarrow 57x = 53$ gives $x = 53/57$

6. (a) The vertices of ΔABC are given as $A(2, 1)$, $B(4, 3)$, $C(2, 5)$ and the mid point are D, E, F

Now $\Delta ABC = \frac{1}{2}\begin{Vmatrix} 2 & 1 & 1 \\ 4 & 3 & 1 \\ 2 & 5 & 1 \end{Vmatrix} = \frac{1}{2}\begin{Vmatrix} 0 & 1 & 1 \\ 2 & 3 & 1 \\ 0 & 5 & 1 \end{Vmatrix}$

$= \begin{Vmatrix} 1 & 1 \\ 5 & 1 \end{Vmatrix} = 4$ square units

Hence area of $\Delta DEF = 1$ square units

TEXTUAL EXERCISE–5 (SUBJECTIVE)

1. True, slope of AB = slope of BA

2. (a) $m = 1$

(b) $m = -1$

3. (a) $m_1 = \frac{1}{\sqrt{3}}, m_2 = \sqrt{3}$

(b) $m_1 = \frac{-1}{\sqrt{3}}, m_2 = -\sqrt{3}$

4. Slope $= \tan 135^0 = -1$; So $\frac{5-4}{x-3} = -1$

So $x = 2$

5. $A(-4, 2)$, $B(2, 6)$, $C(8, 5)$, $D(9, -7)$.

Let E, F, G, H be the respective mid points, then E (–1, 4), $F\left(5, \frac{11}{2}\right), G\left(\frac{17}{2}, -1\right), H\left(\frac{5}{2}, \frac{-5}{2}\right)$.

$\Rightarrow EF = \sqrt{6^2 + \frac{9}{4}} = \sqrt{\frac{153}{4}} = \frac{\sqrt{153}}{2}$

$\Rightarrow GH = \sqrt{6^2 + \frac{9}{4}} = \frac{\sqrt{153}}{2}$ and

$FG = \sqrt{\frac{49}{4} + \frac{169}{4}} = \sqrt{\frac{218}{4}} = \frac{\sqrt{218}}{2}$

$\Rightarrow EH = \sqrt{\frac{49}{4} + \frac{169}{4}} = \frac{\sqrt{218}}{2}$

As $EF = GH$ and $FG = EH$, so ||gm is formed.

6. Given $A(4, 3)$, $B(6, 4)$, $C(5, 6)$, $D(3, 5)$

$\Rightarrow AB = \sqrt{5}, BC = \sqrt{5}, CD = \sqrt{5}, AD = \sqrt{5}$

Slope of $BA = \frac{1}{2}$, slope of $BC = \frac{2}{-1} = -2$

As $BA \perp BC$ and $AB = BC = CD = AD$

$\Rightarrow$ A square is formed

7. Let $A(a, a^2)$, $B(b, b^2)$ and $C(c, c^2)$

Now slope of $AB = \frac{b^2 \quad a^2}{b \quad a} = b + a$

(as a, b, c are distinct)

Similarly slope of $BC = \frac{c^2 - b^2}{c - b} = c + b$

Since $a + b \neq b + c$ (otherwise $a = c$)

So A, B, C are not collinear

8. Let A (1, 3), B (2, 5), C (4, 9)
Slope of $AB = 2/1 = 2$
Slope of $BC = 4/2 = 2$. Observe that slopes are equal hence these points are collinear.

9. **(a)** Slope of line L_1 through (5, 6) and (2, 3) in $m_1 = 1$
Slope of line L_2 through (9, –2) & (6, –5) is $m_2 = 1$
$\Rightarrow$ Lines are parallel
(b) Slope of lines L_1 through (8, 2) and (5, 3) in $m_1 = -1/13$
Slope of line L_2 through (6, 16) and (5, 3) is $m_2 = 13$
$\Rightarrow$ Lines are perpendicular
(c) Slope of line L_1 through (2, –5) and (–5, 2) is $m_1 = -1$
Slope of line L_2 through (6, 3) and (1, 1) is $m_2 = 2/5$
$\Rightarrow$ Lines are neither parallel not perpendicular

10. Let A (4, 4), B (3, 5), C (–1, –1)
Slope of AB is $m_1 = -1$, slope of AC is $m_2 = 1$
As $m_1 \perp m_2$, so a right angled triangle is formed.

TEXTUAL EXERCISE–5 (OBJECTIVE)

1. **(c)** Let A (–2, 2), B (8, –2), C (–4, –3)
Now, $AB = \sqrt{10^2 + 4^2} = \sqrt{116}$
$BC = \sqrt{12^2 + 1^2} = \sqrt{145}$
$AC = \sqrt{4 + 25} = \sqrt{29}$
$\Rightarrow AB^2 + AC^2 = BC^2 \quad \Rightarrow$ right angled triangle

2. **(b)** Let $A\left(\frac{a}{\sqrt{3}}, a\right), B\left(\frac{2a}{\sqrt{3}}, 2a\right), C\left(\frac{a}{\sqrt{3}}, 3a\right)$

Now $AB = \sqrt{\frac{a^2}{3} + a^2} = \frac{2|a|}{\sqrt{3}}$

$BC = \sqrt{\frac{a^2}{3} + a^2} = \frac{2|a|}{\sqrt{3}}$

$AC = 2|a|. \quad \Rightarrow AB = BC \neq AC$
$\Rightarrow$ It is an isosceles triangle

3. **(d)** Let P(a, b), Q(c, d) & $R\left(\frac{kc + \ell a}{k + \ell}, \frac{kd + \ell b}{k + \ell}\right)$
Since R lies on the line joining P & Q
$\Rightarrow$ PQR are collinear

4. **(d)** Let $A\left(0, \frac{8}{3}\right)$, B (1, 3), C (82, 30)

Now, $AB = \sqrt{1^2 + \frac{1}{9}} = \frac{\sqrt{10}}{3}$

$BC = \sqrt{81^2 + 27^2} = 27\sqrt{10}$

$AC = \sqrt{82 + \left(\frac{82}{3}\right)^2} = \frac{82}{3}\sqrt{10}$

$\Rightarrow AB + BC = AC$
$\Rightarrow$ These points are collinear

5. **(d)** Let A (–a, –b), B (a, b) and C (a^2, ab)
We can easily observe that mid point of AB is O (0, 0), so A, O, B are collinear.
Also A (OB) = OC and slope of OB = slope of OC i.e., O, B, C are collinear hence A, B, C are collinear

6. **(a)** Let A (–a, –b), B (0, 0), C (a, b) and D (a^2, ab), we can easily observe that A, B, C are collinear.
Also $BC = \frac{1}{a}$ BD and slope of BC = slope of BD
$\Rightarrow$ Points are collinear

7. **(a), (b)** Let A (0, 0), B (a, 0) and $C\left(\frac{a}{2}, \frac{a\sqrt{3}}{2}\right)$

Now, $AB = a$, $AC = \frac{a}{2}\sqrt{1+3} = a$, $BC = \frac{a}{2}\sqrt{1+3} = a$
$\Rightarrow \Delta ABC$ is equilateral
Since an equilateral triangle is also isosceles.

8. **(a)** Let A (4, 0), B (–1, –1) and C (3, 5)
Now $AB = \sqrt{5^2 + 1^2} = \sqrt{26}$, $AC = \sqrt{1^2 + 5^2} = \sqrt{26}$,
$BC = \sqrt{4^2 + 6^2} = \sqrt{52}$
$\Rightarrow AB = AC$, also $AB^2 + AC^2 = BC^2$
Hence it is a right angled isosceles.

9. **(b)** Let A (1, 1), B (0, $\sec^2\theta$), C ($\operatorname{cosec}^2\theta$, 0)
Area of ΔABC

$$= \frac{1}{2}\left\|\begin{matrix} 1 & 1 & 1 \\ 0 & \sec^2\theta & 1 \\ \operatorname{cosec}^2\theta & 0 & 1 \end{matrix}\right\| = \frac{1}{2}\left\|\begin{matrix} 1 & 1-\sec^2\theta & 0 \\ -\operatorname{cosec}^2\theta & \sec^2\theta & 0 \\ \operatorname{cosec}^2\theta & 0 & 1 \end{matrix}\right\|$$

$$= \frac{1}{2}\{\sec^2\theta + \operatorname{cosec}^2\theta - \sec^2\theta \operatorname{cosec}^2\theta\}$$

$$= \frac{1}{2}\left\{\frac{\sin^2\theta + \cos^2\theta}{\sin^2\theta.\cos^2\theta} - \frac{1}{\sin^2.\cos^2\theta}\right\} = 0$$

So these points will be collinear, when $\sin\theta \neq 0, \cos\theta \neq 0$
i.e., $\theta \neq \frac{n\pi}{2}; n \in \mathbb{Z}$

TEXTUAL EXERCISE–6 (SUBJECTIVE)

1. Given: circum centre C (α, β), centroid G (x, y) and ortho-center O (p, q), then from ONGC (rule)
We know that G divides OC in the ratio 2 : 1
$\Rightarrow x = \frac{2\alpha + p}{3}$ and $y = \frac{2\beta + q}{3}$
Hence $3x = 2\alpha + p$ and $3y = 2\beta + q$.

2. Let the vertices of the triangle be P (2, 3), Q (3, 4), R (6, 8)
$\Rightarrow$ Centroid $G = \left(\frac{11}{3}, 5\right)$;

using $G = \left(\frac{x_1 + x_2 + x_3}{3}, \frac{y_1 + y_2 + y_3}{3}\right)$
Let circum-centre C be at (x, y), then $(x - 2)^2 + (y - 3)^2 = (x - 3)^2 + (y - 4)^2 = (x - 6)^2 + (y - 8)^2$
From the first two equations $2x + 2y - 12 = 0$ or $x + y - 6 = 0$

Similarly last two equations give $6x + 8y - 75 = 0$

So $x = -\frac{27}{2}, y = \frac{39}{2} \Rightarrow G = \left(\frac{11}{3}, 5\right) \& C\left(\frac{-27}{2}, \frac{39}{2}\right)$

From ONGC, we get $\frac{x + 2\left(-\frac{27}{2}\right)}{3} = \frac{11}{3} \Rightarrow x = 38$ and

$\frac{y + 2\left(\frac{39}{2}\right)}{3} = 5 \Rightarrow y = -24$

Hence orthocenter O (or H) is (38, –24)

$\Rightarrow N = \left(\frac{125}{6}, -\frac{19}{2}\right)$

3. Let the vertices be P (4, 1), Q (1, 5) and R (–2, 1)

$PQ = \sqrt{3^2 + 4^2} = 5$. $PR = 6$, $QR = 5$

Hence it is an isosceles triangle

$\Rightarrow$ Incentre will lie on QS

Now in radius $r = \frac{\Delta}{s} = \frac{\frac{1}{2} \times 6 \times 4}{8} = \frac{12}{8} = \frac{3}{2}$

$\therefore$ Incentre = (1, 5/2)

Aliter: $I = \left(\frac{6(1) + 5(4) - (-2)(5)}{16}, \frac{6(5) + 1(5) + 1(5)}{16}\right)$

$= \left(\frac{16}{16}, \frac{40}{16}\right) = \left(1, \frac{5}{2}\right)$

4. Let the vertices be P (7, –36), Q (7, 20) and R (–8, 0)

$\Rightarrow PQ = 56$, $PR = \sqrt{15^2 + 36^2} = 39$, $QR = \sqrt{15^2 + 20^2} = 25$

Now $I = \left(\frac{ax_1 + bx_2 + cx_3}{a + b + c}, \frac{ay_1 + by_2 + cy_3}{a + b + c}\right)$

So $I = \left(\frac{25(7) + 7(39) + 56(-8)}{25 + 39 + 55}, \frac{(-36)(27) + (20)(39) + 0}{25 + 39 + 56}\right)$

$I = \left(0, -\frac{120}{120}\right) = (0, -1)$

5. Mid points are at $D\left(0, \frac{1}{2}\right); E\left(\frac{1}{2}, 0\right)$ and $F\left(\frac{1}{2}, \frac{1}{2}\right)$

$\Rightarrow$ The triangle has vertices at (0, 0), (0, 1) and (1, 0) which is right angled isosceles

$I = \left(\frac{1}{2 + \sqrt{2}}, \frac{1}{2 + \sqrt{2}}\right) = \left(1 - \frac{1}{\sqrt{2}}, 1 - \frac{1}{\sqrt{2}}\right)$

6. Given P (4, –3); Q (–2, 5) and orthocenter O (1, 2). Let the third vertex be R (h, k)

Slope of $PQ = m_1 = \frac{8}{-6} = \frac{-4}{3}$

Slope of $RO = m_2 = \frac{k - 2}{h - 1} = \frac{3}{4}$ as $m_1 m_2 = -1$

Similarly slope of OQ $= m_1 = \frac{3}{-3} = -1$

So slope of RP is $m_2 = \frac{k + 3}{h - 4} = 1$

Now $4k - 8 = 3h - 3$ …(i)

And $k + 3 = h - 4$ ….(ii)

Putting $k = h - 7$, we get $4h - 28 - 8 = 3h - 3$

So h = 33, $k = 26$

7 ABP is an equilateral triangle A (3, 2), B (5, 1), $AB = \sqrt{5}$, mid point of AB is D (4, 3/2)

Slope of $AB = -1/2$

Equation of PD is $\left(y - \frac{3}{2}\right) = (2)(x - 4)$ or $2x - y - 13/2 = 0$

$\left(\text{or } y = 2x - \frac{13}{2}\right)$; $PD = \frac{\sqrt{5}\sqrt{3}}{2} = \frac{\sqrt{15}}{2}$

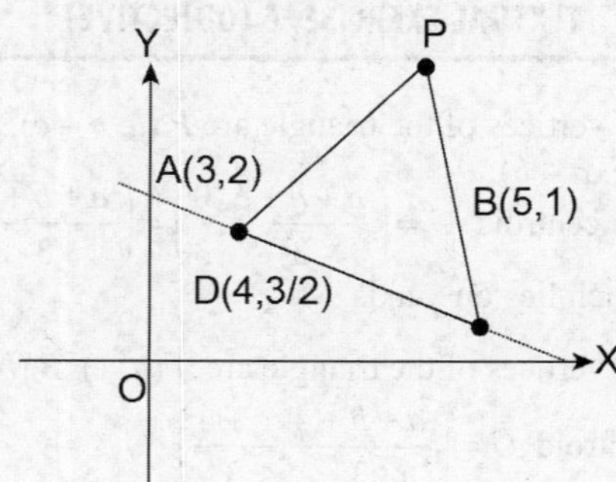

From $(x - 4)^2 + \left(2x - \frac{13}{2} - \frac{3}{2}\right)^2 = \frac{15}{4}$, We get $5(x - 4)^2 = \frac{15}{4}$

$\Rightarrow x - 4 = \pm\frac{\sqrt{3}}{2}$ or $x = \left(4 + \frac{\sqrt{3}}{2}\right), \left(4 - \frac{\sqrt{3}}{2}\right)$

From the given considerations $x = 4 + \sqrt{3}/2$ and

$y = 8 - \frac{13}{2} + \frac{\sqrt{3}}{2} = \frac{3}{2} + \frac{\sqrt{3}}{2}$

$\Rightarrow P\left(4 + \frac{\sqrt{3}}{2}, \frac{3}{2} + \frac{\sqrt{3}}{2}\right)$

8. Circum-centre $C \equiv (0,0)$

Centroid $G \equiv \left(\frac{a + b + c}{3}, \frac{a\tan\alpha + b\tan\beta + c\tan\gamma}{3}\right)$

Let orthocenter be (x, y); then by ONGC; i.e., centroid divides line segment joining orthocenter and circum-centre in the ratio 2 : 1

$\Rightarrow x = a + b + c$, $y = a \tan\alpha + b \tan\beta + c \tan\gamma$

But R = circum-radius of ΔABC

$= |a|\ |\sec\alpha| = |b|\ |\sec\beta| = |c|\ |\sec\gamma|$

$= a \sec\alpha = b \sec\beta = c \sec\gamma$ ($a, b, c > 0$, α, β, γ are acute angles)

$\Rightarrow a = R\cos\alpha$, $b = R\cos\beta$, $c\ R\cos\gamma$

$\Rightarrow$ Orthocenter $\equiv$ (R ($\cos\alpha + \cos\beta + \cos\gamma$), R ($\sin\alpha + \sin\beta + \sin\gamma$))

Further, $\alpha + \beta + \gamma = \pi$ (given)

$\Rightarrow \cos\alpha + \cos\beta + \cos\gamma = 4\sin\frac{\alpha}{2}\sin\frac{\beta}{2}\sin\frac{\gamma}{1} + 1$ and $\sin\alpha + \sin\beta + \sin\gamma = 4\cos\frac{\alpha}{2}\cos\frac{\beta}{2}\cos\frac{\gamma}{2}$

⇒ Orthocenter

$$\equiv \left(R\left(4\sin\frac{\alpha}{2}\sin\frac{\beta}{2}\sin\frac{\gamma}{2}+1\right), 4R\cos\frac{\alpha}{2}\cos\frac{\beta}{2}\cos\frac{\gamma}{2}\right)$$

Now, substituting

$$x = R\left(4\sin\frac{\alpha}{2}\sin\frac{\beta}{2}\sin\frac{\gamma}{2}+1\right) \text{ and } y = 4R\cos\frac{\alpha}{2}\cos\frac{\beta}{2}\cos\frac{\gamma}{2}$$

In $4\left(x\cos\frac{\alpha}{2}\cos\frac{\beta}{2}\cos\frac{\gamma}{2} - y\sin\frac{\alpha}{2}\sin\frac{\beta}{2}\sin\frac{\gamma}{2}\right) = y$,

we see that it is satisfied.

Thus orthocenter lies on the given line.

TEXTUAL EXERCISE–6 (OBJECTIVE)

1. (b) The vertices of the triangle are $P(a, b-c)$, $Q(b, c-a)$, $R(c, a-b)$

The centroid $G = \left(\frac{a+b+c}{3}, \frac{0}{3}\right) = \left(\frac{a+b+c}{3}, 0\right)$,

Which lies on x-axis.

2. (c) The vertices of the triangle are $A(a, 1)$, $B(b, 3)$, $C(4, c)$

⇒ Centroid $G = \left(\frac{a+b+4}{3}, \frac{c+4}{3}\right)$

The centroid will lie on x – axis if $\frac{c+4}{3} = 0$

⇒ $c = -4$

3. (a) Given vertex $A(5, 4)$, $B(-2, 4)$ and centroid G (5, 6)

⇒ Third vertex C (15 – 3, 18 – 8) = (12, 10)

4. (a) The vertices of the triangle are $O(0,0)$, $P(5,12)$, $Q(16,12)$

⇒ OP = 13, OQ = 20, PQ = 11

So in centre $I = \left(\frac{5\times 20+16\times 13}{13+20+11}, \frac{12\times 20+12\times 13}{13+20+11}\right)$

$= \left(\frac{308}{44}, \frac{396}{44}\right) = (7, 9)$

5. (a) Let $L_1: x + y - 5 = 0$

$L_2: x - y + 1 = 0$

$L_3: y = 1$

The points of intersection are $P(0, 1)$, $Q(2, 3)$ and $R(4, 1)$

The triangle formed is right angled isosceles

⇒ The circumcentre is the mid –point of hypogenous i.e., mid point of PR

⇒ (2, 1)

6. (a) The vertices (6, 4), $Q(2, 6)$ and centroid $G(4, 6)$ (given)

⇒ The third vertex R = (12 – 8, 18 – 10) = (4, 8)

7. (d) The vertices are $P(1, \sqrt{3})$, $O(0, 0)$, $R(2, 0)$ observe that OP = 2, OR = 2, PR = 2 (given)

⇒ OPR is an equilateral triangle so incentre is the same as centroid i.e., $\left(1, \frac{1}{\sqrt{3}}\right)$

8. (a) Vertices are $A(x_1, y_1)$, $B(x_2, y_2)$ and $C(x_3, y_3)$ (given)

⇒ Ex–centre w.r.t. to B is

$$I_2 = \left(\frac{ax_1 - bx_2 + cx_3}{a-b+c}, \frac{ay_1 - by_2 + cy_3}{a-b+c}\right)$$

9. (a) vertex $A(1, 1)$ and two mid points $D(-1, 2)$, $F(3, 2)$ (given)

⇒ B (–3, 3) and $C(5, 3)$

Hence the centroid G is $\left(1, \frac{7}{3}\right)$

10. (c) Given $O(0, 0)$, $P(3, 4)$, $R(4, 0)$, as OR is horizontal

⇒ Altitude through vertex P is $x = 3$

⇒ Slope of $PR = -4$

⇒ Equation of altitude through vertex O (0, 0) is

$y = \frac{x}{4} \Rightarrow x = 3, y = \frac{3}{4}$

Hence orthocenter is at $\left(3, \frac{3}{4}\right)$

11. (d) Let the vertices be $O(0, 0)$, $P(a, 0)$, $R(0, b)$

⇒ In centre $I = \left(\frac{ab}{a+b+\sqrt{a^2+b^2}}, \frac{ab}{a+b+\sqrt{a^2+b^2}}\right)$

12. (b) From the given $A(1, \pm\sqrt{3})$, $B(0, 0)$, $C(4, 0)$ (as shown)

⇒ Centroid $G = \left(\frac{5}{3}, \frac{1}{\sqrt{3}}\right)$

13. (b) The vertices of triangle are

$P(at_1t_2, a(t_1+t_2)), Q(at_2t_3, a(t_2+t_3))$, $R(at_1t_3, at(t_1+t_3))$

⇒ Slope $PQ = \frac{1}{t_2}$

⇒ Equation of altitude through R is $y - a(t_1 + t_3) = -t_2$ $(x - at_1t_3)$ gives $t_2x + y = a(t_1 + t_3 + t_1t_2t_3)$

Similarly $t_1x + y = a(t_2 + t_3 + t_1t_2t_3)$

$t_3x + y = a(t_1 + t_2 + t_1t_2t_3)$

⇒ $(t_2 - t_1)x = a(t_1 - t_2)$ gives $x = -a$ and $y = a(t_1 + t_2 + t_3 + t_1t_2t_3)$, so the orthocenter is at $(-a, a(t_1 + t_2 + t_3 + t_1t_2t_3))$

14. (b) Let the vertices be $P\left(2, \frac{\sqrt{3}-1}{2}\right), Q\left(\frac{1}{2}, -\frac{1}{2}\right), R\left(2, -\frac{1}{2}\right)$

These will form a right angled triangle

⇒ Orthocenter is at the vertex with 90° angle i.e., (2, –1/2)

15. (a) From the given information $(a\sqrt{2}, 0)$ is not a vertex

16. (c) Given $B(1, 3)$ and $C(-2, 7)$, so BC = 5 units and slope of BC is $m = -4/3$

Now equation of right bisector of BC is $3x - 4y + \frac{43}{2} = 0$

Which is satisfied by $\left(-\frac{1}{2}, 5\right)$ as well as $\left(\frac{5}{6}, 6\right)$ but for $\left(-\frac{1}{2}, 5\right)$ the area of triangle becomes zero.

17. (a), (c), (d) When vertices are rational points then all the mid-points slopes etc will be rational (excepting the infinite /undefined slope if any), so orthocenter as well as circum-centre will be rational. The centroid G is obviously rational. There is no guarantee for the incentre.

18. (c) Vertices of the triangle are P(tan α, cot α), Q(tan β, cot β), R(tan γ, cot γ) (given)

$\Rightarrow$ Centroid of the triangle is

$$G\left(\frac{\tan\alpha+\tan\beta+\tan\gamma}{3},\frac{\cot\alpha+\cot\beta+\cot\gamma}{3}\right)$$

Now $\tan\alpha$, $\tan\beta$, $\tan\gamma$ are the roots of $x^3-3ax^2+3bx-1=0$

$\Rightarrow$ $\tan\alpha+\tan\beta+\tan\gamma=3a$ and $\tan\alpha\tan\beta+\tan\beta\tan\gamma+\tan\alpha\tan\gamma=3b$ dividing by $\tan\alpha\tan\beta\tan\gamma$, we get

$$\cot\alpha+\cot\beta+\cot\gamma=\frac{3b}{\tan\alpha\tan\beta\tan\gamma}$$

Since $\tan\alpha\tan\beta\tan\gamma=1$ from the equation so

$$G=\left(\frac{3a}{3},\frac{3b}{3(1)}\right)=(a,b)$$

19. **(a)** Given orthocenter O (1, 1) & circum-centre $C\left(\frac{3}{2},\frac{3}{4}\right)$ from ONGC rule

$$\Rightarrow\ N=\left(\frac{2+\frac{3}{2}}{3},\frac{2+\frac{3}{4}}{3}\right)=\left(\frac{7}{6},\frac{11}{12}\right)\text{ and}$$

$$G=\left(\frac{3+1}{3},\frac{\frac{3}{2}+1}{3}\right)=\left(\frac{4}{3},\frac{5}{6}\right)$$

TEXTUAL EXERCISE-7 (SUBJECTIVE)

1. Let $A(\beta,\gamma)$, $B(-\alpha,0)$, $C(\alpha,0)$ be the vertices of ΔABC then its centroid $G_1=\left(\frac{\beta}{3},\frac{\gamma}{3}\right)$

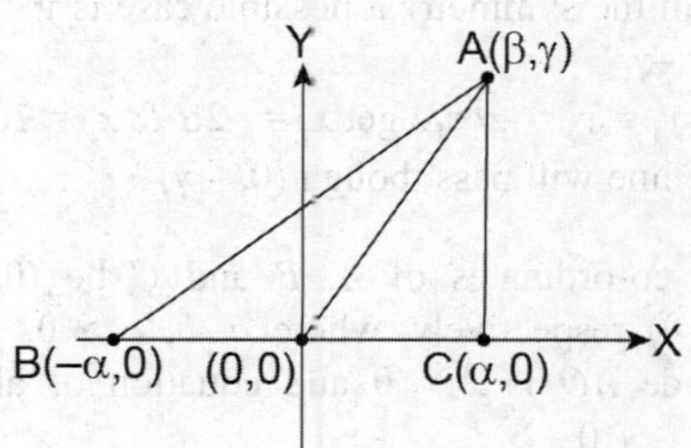

Let P, Q, R divide sides BC, CA and AB in the ratio $\lambda:1$ then $P=\left(\frac{\lambda\alpha-\alpha}{1+\lambda},0\right)$;

$$Q=\left(\frac{\lambda\beta+\alpha}{\lambda+1},\frac{\lambda\gamma}{\lambda+1}\right)\text{ and }R=\left(\frac{-\lambda\alpha+\beta}{1+\lambda},\frac{\gamma}{\lambda+1}\right)$$

Now the centroid of ΔPQR will be

$$G_2=\left(\frac{\beta\lambda+\beta}{3(\lambda+1)},\frac{\gamma\lambda+\gamma}{3(\lambda+1)}\right)=\left(\frac{\beta}{3},\frac{\gamma}{3}\right)$$

So $G_1=G_2$

2. Let the vertices be $A(\beta,\gamma)$, $B(-\alpha,0)$, $C(\alpha,0)$

So, $AB^2+BC^2+CA^2=(\beta+\alpha)^2+\gamma^2+(\beta-\alpha)^2+\gamma^2+4\alpha^2$

$=2\beta^2+6\alpha^2+2\gamma^2$ ....(i)

Now consider the medians

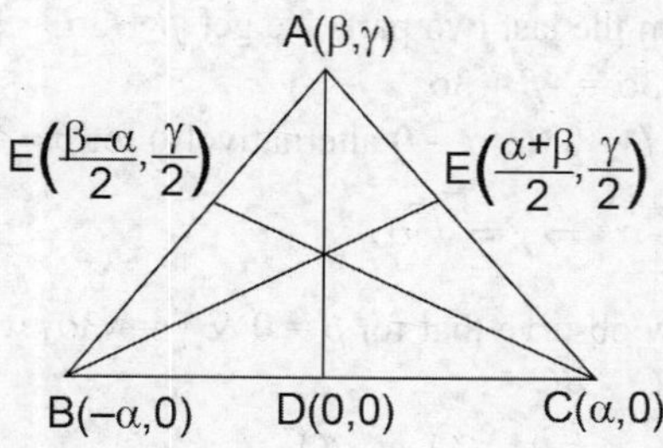

$AD^2+BE^2+CF^2=\beta^2+\gamma^2+$

$$\left(\frac{\beta-\alpha-2\alpha}{2}\right)^2+\left(\frac{\gamma}{2}\right)^2+\left(\frac{\alpha+\beta+2\alpha}{2}\right)^2+\left(\frac{\gamma}{2}\right)^2$$

$$=\frac{3}{2}\beta^2+\frac{9}{2}\alpha^2+\frac{3}{2}\gamma^2 \qquad \text{......(ii)}$$

$\Rightarrow$ 4 (ii) = 3(i)

3. Let the vertices be $A(\beta,\gamma)$, $B(-\alpha,0)$ and $C(\alpha,0)$, so the median AD is $\gamma x-\beta y=0$,

Median BE is $y=\frac{\gamma}{\beta+3\alpha}(x+\alpha)$ i.e., $x\gamma-(\beta+3\alpha)y+\gamma\alpha=0$

Median CF is $y=\frac{\gamma}{\beta-3\alpha}(x-\alpha)$ i.e., $x\gamma-(\beta-3\alpha)y-\gamma\alpha=0$

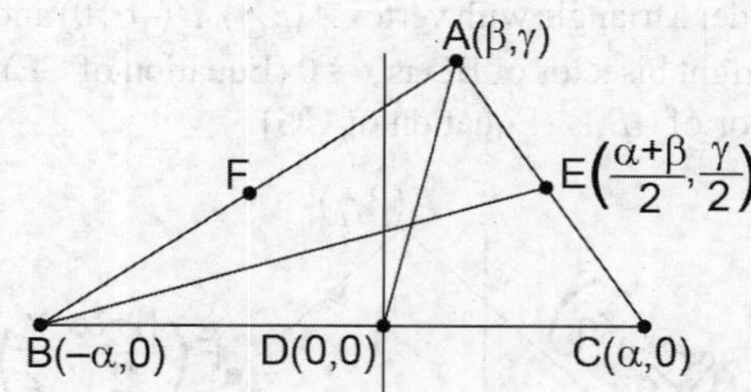

Solving the equations for BE and CF, we get $6\alpha y=3\alpha\gamma$ so $y=\frac{\gamma}{3}$ and $\gamma x=(\beta-3\alpha)\gamma/3+\gamma\alpha$

So $x=\beta/3$ $\Rightarrow$ centroid $G=\left(\frac{\beta}{3},\frac{\gamma}{3}\right)$

Which also satisfies $\gamma x-\beta y=0$, also it divides AD in the ratio 2 : 1 similarly it divides BE & CF in 2 : 1

4. Consider a triangle with $A(\beta,\gamma)$; $B(-\alpha,0)$ and $C(\alpha,0)$ as vertices.

Since medians are equal $\therefore$ $AD^2=BE^2=CF^2$

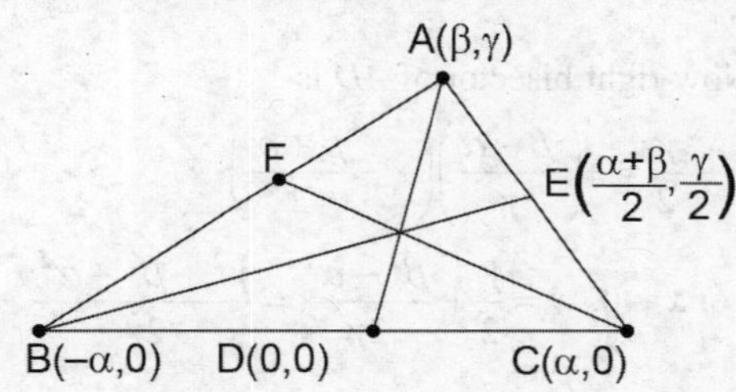

$$\Rightarrow\ \beta^2+\gamma^2=\left(\frac{\beta+3\alpha}{2}\right)^2+\left(\frac{\gamma}{2}\right)^2=\left(\frac{\beta-3\alpha}{2}\right)^2+\left(\frac{\gamma}{2}\right)^2$$

From the last two parts we get $|\beta + 3\alpha| = |\beta - 3\alpha|$

$\Rightarrow \beta + 3\alpha = -\beta + 3\alpha$

i.e., $\beta = 0$ (or $\alpha = 0$ alternatively) for $\beta = 0$,

$\frac{3}{4}\gamma^2 = \frac{9}{4}\alpha^2 \Rightarrow \gamma = \sqrt{3}\alpha$

Now observe that for $\beta = 0$ & $\gamma = \sqrt{3}\alpha$, we will get $AB = AC = BC$

5. Now $EH^2 = \left(\frac{p}{2} - \frac{\alpha}{2}\right)^2 + \left(\frac{q}{2}\right)^2$

D(p,q) C(β,γ)
$\left(\frac{p}{2},\frac{q}{2}\right)$ H
$\left(\frac{p+\beta}{2},\frac{q+\gamma}{2}\right)$
F$\left(\frac{\alpha+\beta}{2},\frac{\gamma}{2}\right)$
A(0,0) E(α/2,0) B(α,0)

$GF^2 = \left(\frac{p-\alpha}{2}\right)^2 + \left(\frac{q}{2}\right)^2$, so $EH = GF$

Similarly $HG = EF = \left(\frac{\beta}{2}\right)^2 + \left(\frac{\gamma}{2}\right)^2$

Hence a ||gm is formed

6. Consider a triangle with vertex $A(\beta, \gamma)$, $B(-\alpha, 0)$ and $D(\alpha, 0)$. Now right bisector of BD is $x = 0$ (Equation of CE) and right bisector of AB is (Equation of CG)

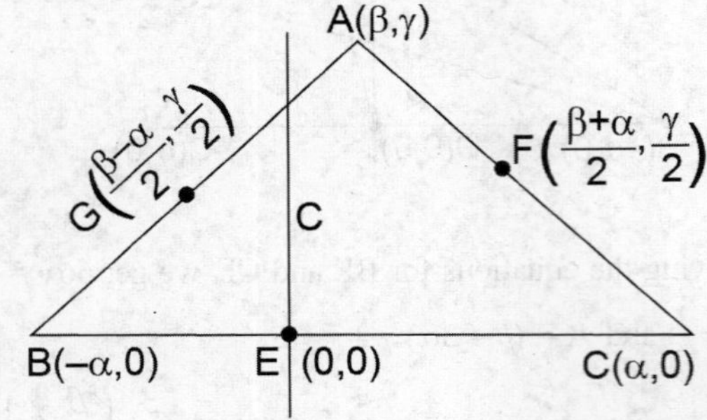

$$\Rightarrow y - \frac{\gamma}{2} = \frac{(-\beta+\alpha)}{\gamma}\left\{x + \frac{\alpha - \beta}{2}\right\}$$

For $x = 0$, $y = \frac{\gamma}{2} + \frac{\beta^2 - \alpha^2}{2\gamma} = \frac{\gamma^2 + \beta^2 - \alpha^2}{2\gamma}$

So the point of intersection is $C\left(0, \frac{\gamma^2 + \beta^2 - \alpha^2}{2\gamma}\right)$

Now right bisector of AD is

$$y - \frac{\gamma}{2} = -\left(\frac{\beta - \alpha}{\gamma}\right)\left(x - \frac{\beta + \alpha}{2}\right)$$

For $x = 0$, $y = \frac{\gamma}{2} + \frac{\beta^2 - \alpha^2}{2\gamma} = \frac{\gamma^2 + \beta^2 - \alpha^2}{2\gamma}$

The point $\left(0, \frac{\gamma^2 + \beta^2 - \alpha^2}{2\gamma}\right)$ is C

So All right bisector intersect in C hence concurrent.

7. Let $A(0, \gamma)$, $B(-\alpha, 0)$ and $C(\alpha, 0)$. Equation of AB is $y = \frac{\gamma}{\alpha}(x + \alpha)$ and equation of AC is $y = \frac{-\gamma}{\alpha}(x - \alpha)$

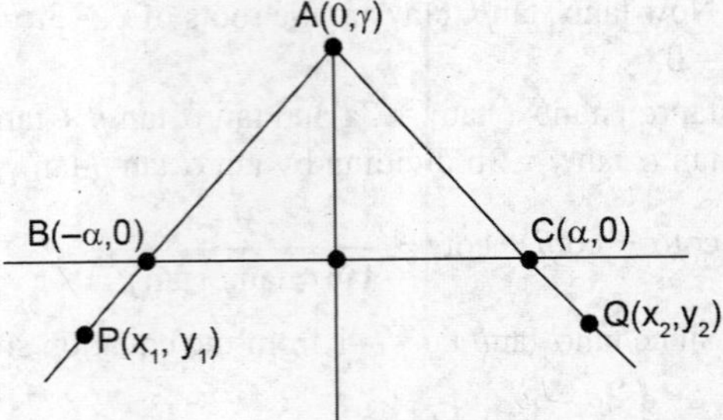

Since BP. CQ = AB^2 OR $BP^2.CQ^2 = AB^4$

So let $P(x_1, y_1)$ and $Q(x_2, y_2)$ be the points (respectively on AB & AC)

$\Rightarrow y_1 = \frac{\gamma}{\alpha}(x_1 + \alpha)$ and $y_2 = \frac{-\gamma}{\alpha}(x_2 - \alpha)$

Now $BP^2.CQ^2 = AB^4$ gives

$$\{(x_1+\alpha)^2 + y_1^2\}\{(x_2-\alpha)^2 + y_2^2\} = (\alpha^2 + \gamma^2)$$

Putting the values we get

$$\left\{\left(\frac{\alpha}{\gamma}y_1\right)^2 + y_1^2\right\}.\left\{\left(\frac{-\alpha}{\gamma}y_2\right)^2 + y_2^2\right\} = (\alpha^2 + \gamma^2)^2 \text{ gives}$$

$$(y_1^2.y_2^2)\left\{1 + \frac{\alpha^2}{\gamma^2}\right\}^2 = (\alpha^2 + \gamma^2)^2$$

So $(y_1y_2)^2 = (\gamma^2)^2$ or $y_1y_2 = \pm\gamma^2$

We consider only the positive rule, so $y_1y_2 = \gamma^2$

From the symmetry a possible case is $y_1 = y_2$ then $y_1 = y_2 = -\gamma$

For $y_1 = y_2 = -\gamma$ we get $x_1 = -2\alpha$ & $x_2 = 2\alpha$

The line will pass though $(0, -\gamma)$

8. Let the co-ordinates of A, B and C be $(0, a)$, $(-b, 0)$ and $(c, 0)$ respectively, where $a, b, c > 0$. The equation of altitude AO is $x = 0$ and equation of altitude BL is

$(y - 0) = -\left(\frac{0 - c}{a - 0}\right)(x + b)$ i.e., $y = \frac{c}{a}(x + b)$

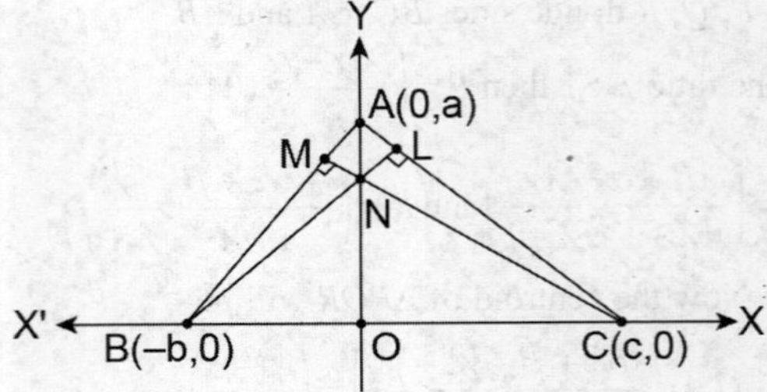

$\therefore$ AO and BL intersect at $N\left(0, \frac{bc}{a}\right)$. Now equation of altitude CM is $(y - 0) = -\left(\frac{b}{a - 0}\right)(x - c)$

$\Rightarrow y = -\frac{b}{a}(x - c)$

Clearly, $N\left(0,\dfrac{bc}{a}\right)$ lies on $y = \dfrac{-b}{a}(x-c)$ i.e., altitudes AO, BL and CM are concurrent.

Note that $N\left(0,\dfrac{bc}{a}\right)$ is the orthocenter of ΔABC.

TEXTUAL EXERCISE–8 (SUBJECTIVE)

1. $P(x, y) = (4, 5)$. Let the origin be shifted to $O'(h, k)$
$\Rightarrow x' = x - h$ & $y' = y - k$.
Now $P(x', y') = (-3, 9) \Rightarrow -3 = 4 - h$
$\Rightarrow h = 7$ and $9 = 5 - k \Rightarrow k = -4$
$\Rightarrow O' = (7, -4)$

2. When origin is shifted to $\left(\dfrac{ab}{a-b},0\right)$, then $x = \left(x' + \dfrac{ab}{a-b}\right)$ & $y' = y$ so the equation becomes

$$(a-b)\left\{x'^2 + \frac{a^2b^2}{(a-b)^2} + \frac{2abx}{(a-b)} + y'^2\right\} - 2ab\left(x' + \frac{ab}{a-b}\right) = 0$$

or $(a-b)^2\{x'^2 + y'^2\} + a^2b^2 + 2ab(a-b)x - 2ab(a-b)x - 2a^2b^2 = 0$ gives $(a-b)^2(x'^2 + y'^2) = a^2b^2$

3. After shifting the origin to $(1, -1)$ the new equation is $x^2 + y^2 + 3x - 4y + 2 = 0$ or $\left(x+\dfrac{3}{2}\right)^2 + (y-2)^2 - \dfrac{17}{4} = 0$ to get the original equation replace x by $(x-1)$ and y by $(y+1)$ hence the new equation will be $\left(x+\dfrac{1}{2}\right)^2 + (y-1)^2 - \dfrac{17}{4} = 0$

$\Rightarrow x^2 + x + y^2 - 2y - 3 = 0$

4. The equation of the curve is $x^2 + 3xy + 4y^2 - 4x - 6y + 5 = 0$

Let the origin be shifted to (h, k) and (x_1, y_1) be the new co-ordinates then $x = x_1 + h$ and $y = y_1 + k$
So, $(x_1 + h)^2 + 3(x_1 + h)(y_1 + k) + 4(y_1 + k)^2 - 4(x_1 + h) - 6(y_1 + k) + 5 = 0$
$\Rightarrow x_1^2 + 4y_1^2 + 3x_1y_1 + x_1(2h + 3k - 4) + y_1(3h + 8k - 6) + (h^2 + 4k^2 + 3hk - 4h - 6k + 5) = 0$
Solving $2h + 3k - 4 = 0$ and $3h + 8k - 6 = 0$, We get $h = 2$ and $k = 0$, so $h^2 + 4k^2 + 3hk - 4h - 6k + 5 = 1$
Hence the equation becomes $x_1^2 + 4y_1^2 + 3x_1y_1 + 1 = 0$, when origin is shifted to $(2, 0)$

5. (a) Given equation is $x^2 - y^2 = a^2$. Let new system by x' and y'

$x = x'\cos\theta - y'\sin\theta = \left(\dfrac{x'}{\sqrt{2}} - \dfrac{y'}{\sqrt{2}}\right)$ and $y = x'\sin\theta + y'\cos\theta = \left(\dfrac{x'}{\sqrt{2}} + \dfrac{y'}{\sqrt{2}}\right)$

So the equation becomes

$\left(\dfrac{x'}{\sqrt{2}} - \dfrac{y'}{\sqrt{2}}\right)^2 - \left(\dfrac{x'}{\sqrt{2}} + \dfrac{y'}{\sqrt{2}}\right) = a^2$ gives $-2x'y' = a^2$ or $2x'y' + a^2 = 0$

(b) $17x^2 + 17y^2 - 16xy = 225$

$$\Rightarrow 17\left\{\frac{x'}{\sqrt{2}} - \frac{y'}{\sqrt{2}}\right\}^2 + 17\left\{\frac{x'}{\sqrt{2}} + \frac{y'}{\sqrt{2}}\right\}^2 - 16\left\{\frac{x'^2}{2} - \frac{y'^2}{2}\right\} = 225$$

$\Rightarrow 9x'^2 + 25y'^2 = 225$

(c) $x^4 + y^4 + 6x^2y^2 = 2$ becomes $(x^2 + y^2)^2 + 4x^2y^2 = 2$, we get

$$\left\{\frac{x'^2}{2} + \frac{y'^2}{2} - x'y' + \frac{x'^2}{2} + \frac{y'^2}{2} + x'y'\right\}^2 + 4\left\{\frac{x'^2}{2} - \frac{y'^2}{2}\right\}^2 = 2$$

or $x'^4 + y'^4 + 2x'^2y'^2 + x'^4 + y'^4 - 2x'^2y'^2 = 2$

$\Rightarrow x'^4 + y'^4 = 1$

(d) $x^2 - y^2 + 2xy\tan 2\alpha = a^2$

$$\Rightarrow \left(\frac{x'}{\sqrt{2}} - \frac{y'}{\sqrt{2}}\right)^2 - \left(\frac{x'}{\sqrt{2}} + \frac{y'}{\sqrt{2}}\right)^2 + 2\left(\frac{x'^2}{2} - \frac{y'^2}{2}\right)\tan 2\alpha = a^2$$

$\Rightarrow (x'^2 - y'^2)\tan 2\alpha - 2x'y' = a^2$

6. The given equation is $Ax + By + C = 0$. Let the axes be rotated anticlockwise by θ and X, Y be the new coordinate axes $x = X\cos\theta - y\sin\theta$ & $y = X\sin\theta + y\cos\theta$
$\Rightarrow A(X\cos\theta - Y\sin\theta) + B(X\sin\theta + Y\cos\theta) + C = 0$
$\Rightarrow (A\cos\theta + B\sin\theta)X + Y(B\cos\theta - A\sin\theta) + C = 0$
Since it is to match with $x = C_2$ (constant).
So $B\cos\theta = A\sin\theta \Rightarrow \tan\theta = B/A$ or $\theta = \tan^{-1}(B/A)$

i.e., $\cos\theta = \dfrac{A}{\sqrt{A^2+B^2}}$ and $\sin\theta = \dfrac{B}{\sqrt{A^2+B^2}}$

Now, the equation becomes

$\sqrt{A^2+B^2}\,X = -C$ or $X = -\dfrac{C}{\sqrt{A^2+B^2}}$

$\Rightarrow \theta = \tan^{-1}\left(\dfrac{B}{A}\right)$ and constant $C_2 = \dfrac{-C}{\sqrt{A^2+B^2}}$

7. (a) True
(b) True

8. Let the axes be rotated by θ in anticlockwise direction when (x, y) is the point in old system and it becomes (x', y') in the new system. So $x = x'\cos\theta - y'\sin\theta$ and $y = x'\sin\theta + y'\cos\theta$
Hence $ax^2 + by^2 + 2hxy$ expression will become $a(x'\cos\theta - y'\sin\theta)^2 + b(x'\sin\theta + y'\cos\theta)^2 + 2h(x'\cos\theta - y'\sin\theta)(x'\sin\theta + y'\cos\theta)$
$\Rightarrow \{ax'^2\cos^2\theta + bx'^2\sin^2\theta\} + 2hx'^2\sin\theta\cos\theta + \{ay'^2\sin^2\theta + by'^2\cos^2\theta\} - 2hy'^2\sin\theta\cos\theta - 2a\sin\theta\cos\theta\, x'y' + 2b\sin\theta\cos\theta\, x'y' + 2hx'y'(\cos^2\theta - \sin^2\theta)$
Or $x'^2\{a\cos^2\theta + b\sin^2\theta + 2h\sin\theta\cos\theta\} + y'^2\{a\sin^2\theta + b\cos^2\theta - 2h\sin\theta\cos\theta\} + x'y'\{(b-a)\sin 2\theta + 2h\cos 2\theta\}$
Comparing with $a'x'^2 + b'y'^2 + 2h'x'y'$, we get $a + b = a' + b'$ and similarly we will get $ab - h^2 = a'b' - h'^2$

9. Let $P(x, y)$ be a point in original system by rotating the axes in anticlockwise direction by θ (angle), $P(x', y')$ are the new coordinates, so $x = x'\cos\theta - y'\sin\theta$ & $y = x'\sin\theta + y'\cos\theta$

Putting in $ax^2 + by^2 + 2hxy$ the expression becomes $a(x'\cos\theta - y'\sin\theta)^2 + b(x'\sin\theta + y'\cos\theta)^2 + 2h(x'\cos\theta - y'\sin\theta)(x'\sin\theta + y'\cos\theta)$

$\Rightarrow x'^2$ {$a\cos^2\theta + b\sin^2\theta + 2h\cos\theta\sin\theta$} $+ y'^2$ {$a\sin^2\theta + b\cos^2\theta - 2h\sin\theta\cos\theta$} $+ x'y'$ {$2b\sin\theta\cos\theta - 2a\cos\theta\sin\theta + 2h(\cos^2\theta - \sin^2\theta)$}

Comparison gives $a'x'^2 + b'y'^2$, so $(b - a)\sin 2\theta = -2h\cos 2\theta$

Hence $\tan 2\theta = \dfrac{h}{a-b} \Rightarrow \theta = \dfrac{1}{2}\tan^{-1}\left(\dfrac{2h}{a-b}\right)$ or

$\theta = \dfrac{1}{2}\cot^{-1}\left(\dfrac{a-b}{2h}\right)$

10. Let the origin be shifted to (h, k), so we get $(x + h)^2 + (y + k)^2 + 4(x + h)(y + k) - 2(x + h) + 2(y + k) + 4 = 0$
or $x^2 + y^2 + 4 \times y + (2h + 4k - 2)x + (4h + 2k + 2)y + (h^2 + k^2 + 4hk - 2h + 2k + 4) = 0$

We first eliminate x & y so we get $h = -1$ & $k = 1$ and the equation becomes $x^2 + y^2 + 4 \times y + 6 = 0$

Now consider rotation (in anticlock direction) of axes by angle $\theta \Rightarrow X = x'\cos\theta - y'\sin\theta$ and $y = x'\sin\theta + y'\cos\theta$
Now the equation becomes x'^2 {$1 + 2\sin 2\theta$} $+ y'^2 (1 - 2\sin 2\theta) + 4x'y'(\cos 2\theta) + 6 = 0$ for matching we need to eliminate $x'y'$, so $\cos 2\theta = 0$
$\Rightarrow \sin 2\theta = 1$ (for $\theta = \pi/4$)
The equation becomes $3x'^2 + y'^2(-1) = -6$

$\Rightarrow \dfrac{y'^2}{6} - \dfrac{x'^2}{2} = 1$

TEXTUAL EXERCISE-7 (OBJECTIVE)

1. (i) (b) Origin is shifted to $(1, -2)$, so $x = X + 1$, $y = Y - 2$ and $2(x - 1)^2 + (y + 2)^2 - 6 = 0$
$\Rightarrow 2x^2 + y^2 = 6$
(ii) (a) $(y + 2)^2 - (x - 1) = 0 \Rightarrow$
$y^2 - 4x = 0$ or $y^2 = 4x$

2. (a) The equation is $(y + 2)^2 + 8\left(x - \dfrac{3}{4}\right) = 0$

So by shifting origin to $\left(-,\ 2\right)$ the equation $Y^2 + 8X = 0$

The new origin is at $\left(\dfrac{3}{4}, -2\right)$

3. (a) When origin is shifted to $(-2, -3)$ and axes are rotated by 45°, we get $x = -2 + \dfrac{X}{\sqrt{2}} - \dfrac{Y}{\sqrt{2}}$ and

$y = -3 + \dfrac{X}{\sqrt{2}} + \dfrac{Y}{\sqrt{2}}$

Now, $2x^2 + 20x + 4xy - 5y^2 - 22y - 14 = 0$
$\Rightarrow 2(x + 2)^2 - 5(y + 3)^2 + 4(x + 2)(y + 3) - 1 = 0$
So, we get

$$2\left(\frac{X}{\sqrt{2}} - \frac{Y}{\sqrt{2}}\right)^2 - 5\left(\frac{X}{\sqrt{2}} + \frac{Y}{\sqrt{2}}\right)^2 + 4\left(\frac{X}{\sqrt{2}} - \frac{Y}{\sqrt{2}}\right)\left(\frac{X}{\sqrt{2}} + \frac{Y}{\sqrt{2}}\right) = 1$$

$\Rightarrow \dfrac{X^2}{2} - \dfrac{7Y^2}{2} - 7XY = 1$ or $x^2 - 14xy - 7y^2 - 2 = 0$

4. (c) The given equation is $(x - 2)^2 + (y + 3)^2 = 20$
When origin is shifted to $(2, -3)$ the equation becomes $X^2 + Y^2 = 20$, which is free from first degree terms.

5. (b) On rotation by θ the expression $x^2 + y^2 + 4xy$ becomes $(X\cos\theta - Y\sin\theta) + (X\sin\theta + Y\cos\theta)^2 + 4(X\cos\theta - Y\sin\theta)(X\sin\theta + Y\cos\theta)$
$\Rightarrow X^2(\cos^2\theta + \sin^2\theta + 4\sin\theta\cos\theta) + Y^2$ {$\sin^2\theta + \cos^2\theta - 4\sin\theta\cos\theta$} $+ 4$ {$\cos^2\theta - \sin^2\theta$} XY
Comparing with $AX^2 + BY^2$, we get $\cos 2\theta = 0$, so $\theta = \pi/4$

6. (c) On reflection in $y = x$ point $(4, 1)$ becomes $(1, 4)$ on translation by 2 units along positive x-axis the point is $(3, 4)$

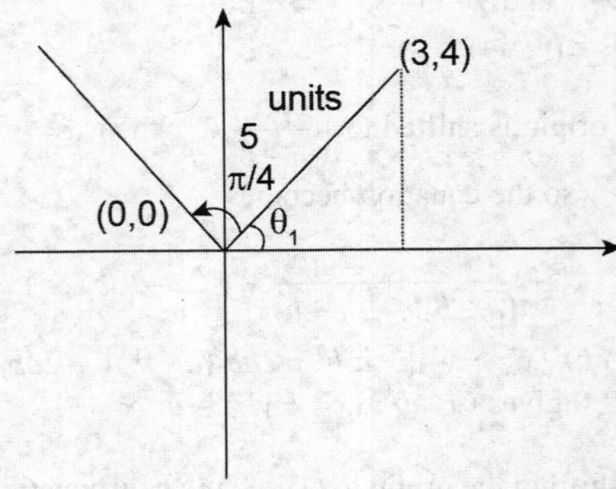

When rotation about origin in anticlock wise sense by $\pi/4$ is given, we get $\tan\theta_1 = 4/3$, $\tan \pi/4 = 1$ now at final position $\theta = \theta_1 + \pi/4$

$\Rightarrow \tan\theta = \dfrac{\tan\theta_1 + 1}{1 - (1)\tan\theta_1} = \dfrac{1 + \frac{4}{3}}{1 - \frac{4}{3}} = -7$

Hence $\sin\theta = \dfrac{7}{5\sqrt{2}}$ and $\cos\theta = \dfrac{-1}{5\sqrt{2}}$ and $r = 5$ units

The point is $(r\cos\theta, r\sin\theta) = \left(-\dfrac{1}{\sqrt{2}}, \dfrac{7}{\sqrt{2}}\right)$

7. (b) As shown above the final position is at $(3, 4)$

TEXTUAL EXERCISE-9 (SUBJECTIVE)

1. Given $P\left(t + \dfrac{1}{t}, t - \dfrac{1}{t}\right)$, so $x = t + \dfrac{1}{t}$ and $y = t - \dfrac{1}{t}$

$\Rightarrow x^2 = t^2 + \dfrac{1}{t^2} + 2$ and $y^2 = t^2 + \dfrac{1}{t^2} - 2$

Hence $x^2 - y^2 = 4$ or $\dfrac{x^2}{4} - \dfrac{y^2}{4} = 1$. Which is a rectangular hyperbola.

2. Let $P(x, y)$ be a point on the locus. As $PA \perp PB$

so, $\left(\dfrac{y-0}{x+2}\right)\left(\dfrac{y-0}{x-2}\right) = -1$

$\Rightarrow y^2 = -(x^2 - 4)$ or $x^2 + y^2 = 4$

3. (a) $y(x-a)=0$, gives $y=0$ or $x=a$; $y=0$ is x-axis and $x=a$ is a vertical line through $(a, 0)$

$\Rightarrow$ A horizontal and a vertical and passing through $(a, 0)$

(b) $x^3-x^2-x+1=0$ gives $x^2(x-1)-(x-1)=0$

So, $(x^2-1)(x-1)=(x+1)(x-1)^2=0$

The locus gives two parallel vertical lines passing through $(-1, 0)$ & $(1, 0)$

(c) $x(x^2-y^2)=0$ or $x(x+y)(x-y)=0$

$x=0$ is the y-axis and $y=x$, $y=-x$ are the other lines

$\Rightarrow$ Set of three straight lines $y=\pm x$ & y-axis

(d) $x^2-y^2=0 \quad \Rightarrow \quad (x+y)(x-y)=0$

$\Rightarrow$ Set of two straight lines $y=\pm x$ i.e., Angle bisectors of quadrants.

(e) $x(x-y)=0$ gives set of two straight lines $y=x$ & y-axis i.e., y-axis & angle bisector of Ist & IIIrd quadrant

(f) $x^3+y^3=0$ gives $(x+y)\{x^2+y^2-xy\}=0$

$\Rightarrow$ $x^2+y^2-xy=0$ has only one solutions viz $(0, 0)$. This can be proved as $y^2-xy+x=0$

$$\Rightarrow \quad y=\frac{x\pm\sqrt{-3x^2}}{2}=\frac{x\pm|x|\sqrt{-3}}{2}$$

$\Rightarrow$ Solution set is line $y=-x$ {which also includes point $(0,0)$} i.e., Angle bisector of 2nd &4th quadrant

(g) $x^2+y^2=0$ only solution is point $x=0$ and $y=0$, i.e., origin

(h) $x^2y=0 \Rightarrow x=0$ or $y=0$, i.e., co-ordinate axes.

(i) $(x^2-1)(y^2-4)=0$, gives $(x+1)(x-1)(y-2)(y+2)=0$

Which gives $x=\pm 1$, $y=\pm 2$

We get a set of four lines $x=\pm 1$, $y=\pm 2$, which will also form a rectangle.

(j) $(x-a)^2+y^2=0$, which is a point circle centered at $(a, 0)$

$\Rightarrow$ Point $(a, 0)$

(k) $(x^2-1)^2+(y^2-4)^2=0$ is possible only when $x^2-1=0$ and $y^2-4=0$. i.e., $x=\pm 1$ and $y=\pm 2$ so we get set of four point $(1, 2)$; $(-1, 2)$; $(1, -2)$; $(-1, -2)$

4. Let $P(x, y)$ be the point then according to the given $\sqrt{x^2+y^2}=2|x|$ or $x^2+y^2=4x^2$ which gives $3x^2-y^2=0$.

5. Position of a moving point at time t is $x(t)=\mu t\cos\alpha + y(t)=\mu t\sin\alpha - kt^2$

So $\dfrac{y+kt^2}{x}=\tan\alpha \quad \Rightarrow y=x\tan\alpha-kt^2$

Putting $t=\dfrac{x}{u\cos\alpha}$ we get $y=x\tan\alpha-\dfrac{k\,x^2}{u^2\cos^2\alpha}$

6. Given $A(2, 3)$ and $Q(3\cos\theta, 2\sin\theta)$; P divides AQ internally in the ratio $3:1$

$$\Rightarrow \quad P(x,y)=\left(\frac{9\cos\theta+2}{4},\frac{6\sin\theta+3}{4}\right) \text{ gives}$$

$$\frac{4x-2}{9}=\cos\theta \text{ and } \frac{4y-3}{6}=\sin\theta$$

Squaring and adding, we get $\left(\dfrac{4x-2}{9}\right)^2+\left(\dfrac{4y-3}{6}\right)^2=1$

or $\left\{\dfrac{\left(x-\frac{1}{2}\right)}{\left(\frac{9}{4}\right)}\right\}^2+\left\{\dfrac{\left(y-\frac{3}{4}\right)}{\left(\frac{6}{4}\right)}\right\}^2=1$, which is an ellipse.

7. A line with real non-zero slope m ($m<0$) will have equation $y-2=m(x+1)$ (m is a variable).

It will intersect axes at $A\left(-\dfrac{m+2}{m},0\right)$ and $B(0, m+2)$

Now $PA=\sqrt{(a+1)^2+2^2}=\sqrt{\dfrac{4}{m^2}(m^2+1)}=\dfrac{2}{|m|}\sqrt{1+m^2}$;

$PB=\sqrt{1+m^2}$

Let $Q(h, k)$, then $PQ=\sqrt{(h+1)^2+(k-2)^2}$, also $(k-2)=m(h+1)$, so $PQ=|(h+1)|\sqrt{1+m^2}$

Now PA, PQ, PB are in HP

$\Rightarrow \quad \dfrac{1}{PA},\dfrac{1}{PQ},\dfrac{1}{PB}$ are A.P.

Hence $\dfrac{2}{PQ}=\dfrac{|m|}{2\sqrt{1+m^2}}+\dfrac{1}{\sqrt{m^2+1}}=\dfrac{2}{(h+1)\sqrt{m^2+1}}$

So, we get $\{|m|+2\}(h+1)=4$ i.e., $(h+1)=\dfrac{4}{|m|+2}=\dfrac{4}{2-m}$

(as $m<0$), we will get $h=\dfrac{2+m}{2-m}$ and $k=\dfrac{2m+4}{2-m}=2h$

$\Rightarrow \quad y=2x$

The other equations can be shown similarly from other possibilities take $m>0$ and when p lies in between A & B.

8. The vertices of a triangle ABC are $A(\cos t, \sin t)$, $B(-\sin t, \cos t)$, $C(1, 2)$

$$\text{Centroid } G=\left(\frac{1+\cos t-\sin t}{3},\frac{2+\sin t+\cos t}{3}\right)=G(x,y)$$

$\Rightarrow$ $3x-1=\cos t-\sin t$ and $3y-2=\cos t+\sin t$

So, $(3x-1)^2=1-2\sin t\cos t$ and $(3y-2)^2=1+2\sin t\cos t$

On adding, we get $(3x-1)^2+(3y-2)^2=2$

$\Rightarrow$ $3(x^2+y^2)-2x-4y+1=0$

TEXTUAL EXERCISE-8 (OBJECTIVE)

1. (b) $\sqrt{(x-a)^2+y^2}=|x|$ so, $x^2+a^2-2ax+y^2=x^2$ or $y^2+a^2=2ax$ or $y^2+a^2=2ax$

2. (d) Slope of line segment OP is $m=\sqrt{3}$

$\Rightarrow \quad \dfrac{y}{x}=\sqrt{3} \qquad \Rightarrow y=\sqrt{3}x$ or $\sqrt{3}x-y=0$

3. (d) $SQ^2=(x+1)^2+y^2$, $SR^2=(x-2)^2+y^2$

$SP^2=(x-1)^2+y^2$. Consider $SQ^2+SR^2=2SP^2$

According to the given $2x^2+2y^2-2x+5=2x^2+2y^2-4x+2$

$\Rightarrow$ $2x+3=0$, so $x=-3/2$

This is a straight line parallel to y-axis

4. (a) Let $P(x, y)$, according to the given $\Delta POA = 2\ \Delta POB$. Observe that AO = 4 units and BO = 6 units, considering AO as base the altitude of $\Delta POA = |x|$ and considering BO as base the altitude of $\Delta POB = |y|$

So $\frac{1}{2} \times 4 \times |x| = 2\left(\frac{1}{2} \times 6 \times |y|\right)$

$\Rightarrow |2x| = |6y|$

So P(x, y) has a locus $y = \frac{x}{3}$ or $y = -\frac{x}{3}$

$\Rightarrow (x + 3y)(x - 3y) = 0$

5. (a) From the given $(x - a)^2 + y^2 = (x + a)^2$

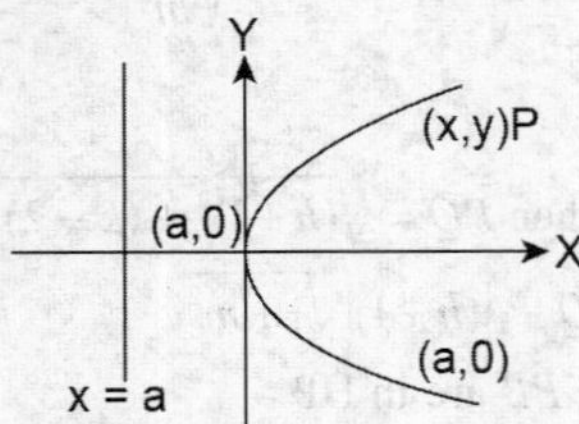

$\Rightarrow y^2 = 4ax$

6. (b) $(x + 1)^2 + y^2 = 9\{x^2 + (y - 2)^2\}$ (given)

$\Rightarrow 8(x^2 + y^2) - 2x - 36y + 35 = 0$; which is a circle

7. (b) $A(a, 0)$, $B(-a, 0)$. Let $P(x, y)$ be the point equidistant from A & B

$\Rightarrow (x - a)^2 + y^2 = (x + a)^2 + y^2$

$\Rightarrow x = 0$ which is the right bisector of line segment AB.

8. (d) $\frac{x}{b} = \sec\phi$ and $\frac{y}{a} \equiv \tan\phi$ (given)

So $\frac{x^2}{b^2} = \sec^2\phi = \tan^2\phi + 1 = \left(\frac{y}{a}\right)^2 + 1$

$\Rightarrow \frac{x^2}{b^2} - \frac{y^2}{a^2} = 1$, which is a hyperbola.

9. (d) $A(ak, 0)$, $B\left(\frac{a}{k}, 0\right)$. (Given)

Let $P(x, y)$ be a point, so $(x - ak)^2 + y^2 = k^2\left\{\left(x - -\right) + y^2\right\}$

or $x^2 + a^2k^2 - 2akx + y^2 = k^2x^2 + a^2 - 2akx + k^2y^2$

$\Rightarrow (k^2 - 1)\{x^2 + y^2 - a^2\} = 0$. Since $k \neq \pm 1$, so $k^2 - 1 \neq 0$

Hence $x^2 + y^2 - a^2 = 0$

10. (b) $x \cos\alpha + y \sin\alpha = p$ gives two points $A(p \sec\alpha, 0)$ and $B(0, p \operatorname{cosec}\alpha)$

The mid point of AB is $\left(\frac{p}{2}\sec\alpha, \frac{p}{2}\operatorname{cosec}\alpha\right)$

Now $x = (p/2)\sec\alpha$, so $\cos\alpha = p/2x$ and $y = P/2 \operatorname{cosec}\alpha$, so $\sin\alpha = p/2y$

Squaring and adding, we get $\frac{p^2}{4}\left\{\frac{1}{x^2} + \frac{1}{y^2}\right\} = 1$

$\Rightarrow \frac{1}{x^2} + \frac{1}{y^2} = \frac{4}{p^2}$

11. (a) Let $P(x, y)$ be a point, so $(x + g)^2 + (y + f)^2 = a^2$

$\Rightarrow x^2 + y^2 + 2gx + 2fy + g^2 + f^2 - a^2 = 0$, when $k = g^2 + f^2 - a^2$ then the equation is $x^2 + y^2 + 2gx + 2fy + k = 0$

12. (a) $\sqrt{(x - ae)^2 + y^2} + \sqrt{(x + ae)^2 + y^2} = 2a$ (given)

By squaring and suitable rearrangement, we will get $\frac{x^2}{a^2} + \frac{y^2}{a^2(1 - e^2)} = 1$

13. (b) $(x - 0)^2 + (y - 0)^2 = 16$ (given)

$\Rightarrow x^2 + y^2 = 16$

14. (d) Let $P(x, y)$ be a point then $\frac{y - a}{x} \cdot \frac{y + a}{x} = -1$

$\Rightarrow y^2 - a^2 = -x^2$ or $x^2 + y^2 = a^2$

15. (b) $AB < 4$ (given), then $PA + PB = 4$ gives an ellipse

16. (a) Given equation is $|PA - PB|$ = constant ($< AB$) this will represent a hyperbola.

17. (b) Given $A(2, 3)$, $B(-4, 5)$ (given)

$\Rightarrow AB = \sqrt{6^2 + 2^2} = 2\sqrt{10}$ units

Equation of AB: $x + 3y - 11 = 0$

For an area of 12 square units, we get $\frac{1}{2} \times 2\sqrt{10} \times h = 12$

$\Rightarrow h = \frac{12}{\sqrt{10}}$ units.

$\Rightarrow$ The locus of P(x, y) will be at a distance of $\frac{12}{\sqrt{10}}$ units from AB

$\Rightarrow$ Locus is $x + 3y + 1 = 0$ or $x + 3y - 23 = 0$ combined locus is $(x + 3y + 1)(x + 3y - 23) = 0$

18. (b) $x(t) = (u\cos\alpha)t$ and $y = (u\sin\alpha)t - \frac{1}{2}gt^2$

$\Rightarrow y + \frac{gt^2}{2} = x\tan\alpha$, so $y = x\tan\alpha - \frac{g\,x^2}{2\,u^2}\sec^2\alpha$

This can be rewritten as $\frac{g\sec^2\alpha}{2u^2}x^2 - x\tan\alpha - y = 0$, which is a parabola.

19. (d) Given $A(a_1, b_1)$ and $B(a_2, b_2)$. The locus of P is the right bisector of line segment AB

$\Rightarrow (x - a_1)^2 + (y - b_1)^2 = (x - a_2)^2 + (y - b_2)^2$

$\Rightarrow 2(a_1 - a_2)x + 2(b_1 - b_2)y + (a_2^2 - a_1^2) + (b_2^2 - b_1^2) = 0$

$\Rightarrow c = \frac{(a_2^2 - a_1^2) + (b_2^2 - b_1^2)}{2}$

20. (b) Let $P(x, y)$ be the point, then $|x - 2| + \sqrt{x^2 + y^2} = 4$ or $\sqrt{x^2 + y^2} = 4 - |x - 2|$.

On squaring, we get $x^2 + y^2 = 16 + (x^2 + 4 - 4x) - 8|x - 2|$

$\Rightarrow y^2 + 4x + 20 + 8(x - 2) = 0$ if $x \geq 2$

$\Rightarrow y^2 + 12x + 4 = 0$ (for $x \geq 2$)

$\Rightarrow y^2 - 4x + 36 = 0$ for $x < 2$

21. $A(a\cos t, a\sin t)$, $B(b\sin t, -b\cos t)$, $C(1, 0)$ (given)

Then centroid $G = \left(\frac{1 + a\cos t + b\sin t}{3}, \frac{a\sin t - b\cos t}{3}\right)$, So

$3x - 1 = a\cos t + b\sin t$ and $3y = a\sin t - b\cos t$

Squaring and adding, we get $(3x - 1)^2 + 3y^2 = a^2(\cos^2 t + \sin^2 t) + b^2(\sin^2 t + \cos^2 t)$

$\Rightarrow (3x - 1)^2 + 3y^2 = a^2 + b^2$

22. (c) $P(1, 0)$ and $Q(h, k)$ lies on $y^2 = 8x$. Mid point of PQ will be $M\left(\frac{h+1}{2}, \frac{k}{2}\right)$

Now $k^2 = 8h$, putting $\frac{h+1}{2} = x$ and $\frac{k}{2} = y$, We get $4y^2 = 8(2x - 1)$

$\Rightarrow y^2 - 4x + 2 = 0$

23. (a) The equation of line (rod) is $\frac{x}{a} + \frac{y}{b} = 1$ and $a^2 + b^2 = \ell^2$

Putting $\frac{a}{3} = x$ and $\frac{2b}{3} = y$

$\Rightarrow 9x^2 + \frac{9}{4}y^2 = \ell^2$ or $36x^2 + 9y^2 = 4\ell^2$

TEXTUAL EXERCISE SECTION–III (OBJECTIVE)

1. (c) $A(-3, 4)$, $B(3, -4)$ (given). Observe that AB passes through origin $AC = 3BC$

$\Rightarrow OC = 2OB \Rightarrow$ $C(6, -8)$

2. (d) Let $A(3a, 0)$, $B(0, 3b)$ and $C(a, 2b)$, observe that C divides AB in the ratio 2 : 1

As $C = \left(\frac{3a}{3}, \frac{6b}{3}\right) = (a, 2b)$, so A, B, C are collinear.

3. (a) Let $A(2, 3)$ and $B(-1, 2)$. Now L: $x + 2y - k = 0$ divides line segment in the ratio 3: 4 internally

$\Rightarrow \frac{L(2,3)}{L(-1,2)} = -\frac{3}{4}$; so $\frac{2+6-k}{-1+4-k} = -\frac{3}{4}$

$\Rightarrow 7k = 41$, so $k = \frac{41}{7}$

4. (c) $A(3,4), C(1,-1)$ (given). Equation of AC is $5x - 2y - 7 = 0$

Equation of right bisector is $4x + 10y - 23 = 0$

$\therefore$ $D\left(\frac{9}{2}, \frac{1}{2}\right)$ and $B\left(-\frac{1}{2}, \frac{5}{2}\right)$ lie on this equation.

$\Rightarrow \frac{\left|\frac{45}{2} - 8\right|}{\sqrt{29}} = \frac{\left|-\frac{5}{2} - \frac{10}{2} - 7\right|}{\sqrt{29}} = \frac{\sqrt{29}}{2}$

5. (d) $A(1, 2)$, $B(3, 4)$, $C(2, 3)$ (given)

$\Rightarrow AB = 2\sqrt{2}, BC = \sqrt{2}$ and $AC = \sqrt{2}$, as $AB = BC + AC$, so these points are collinear.

$\Rightarrow$ Greatest angle is 180°

6. (a) Given $P(3, 1)$, $Q(6, 5)$ and $R(x, y)$.

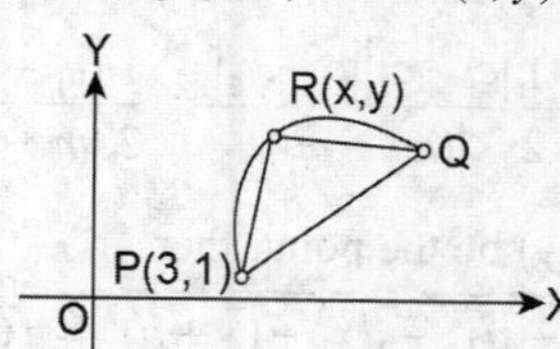

ΔPRQ will be right angled. As area of $\Delta PRQ = 7$ square units, distance of R from line PQ is $h = \frac{14}{5}$ units but hypo-tenuous $PQ = 5$ units, as $2h > PQ$, so no such point is possible.

7. (a) The three vertices of a rhombus taken in order are $A(2, -1)$, $B(3, 4)$, $C(-2, 3)$, then the fourth vertex $D(x, y)$ is $x + 3 = 0$

$\Rightarrow x = -3$ and $y + 4 = 3 - 1 = 2$, so $y = -2$ i.e., $D = (-3, 2)$

8. (c) Given: $A(a, 1)$, $B(b, 3)$, $C(4, c)$, the centroid G will lie on x-axis when $\frac{1+3+C}{3} = 0$ i.e., $C = -4$.

9. (d) The sides of a triangle are $x = 3$, $y = 4$ and $3x + 4y - 6 = 0$, as $x = 3 + y = 4$ are at right angle to each other

$\therefore$ Orthocenter will lie at that vertex i.e., (3, 4)

10. (d) Given: $A(2, 5)$, $B(4, -11)$. Equation of AB is $8x + y - 21 = 0$

Now C moves along $9x + 7y + 4 = 0$

Let $C(x_1, y_1)$ be a point then $C = \left(x_1, -\frac{(9x_1 + 4)}{7}\right)$

So, centroid G of ΔABC is

$$G = \left(\frac{2+4+x_1}{3}, \frac{5 - 11 - \frac{9}{7}x_1 - \frac{4}{7}}{3}\right) \text{ i.e., } x = 2 + \frac{1}{3}x_1$$

and $3y = \frac{-46}{7} - \frac{9}{7}x_1$

Eliminating x_1, we get $3x - 6 = \frac{-46}{9} - \frac{7}{3}y$ i.e., $7y + 9x - 8/3 = 0$ which is parallel to $9x + 7y + 4 = 0$

11. (b) Given $A(1, 0)$, $B(2\cos\theta, 2\sin\theta)$, P divides AB internally in the ratio 2 : 3

So $P = \left(\frac{3 + 4\cos\theta}{5}, \frac{4\sin\theta}{5}\right)$

$\Rightarrow 5x - 3 = 4\cos\theta$ and $5y = 4\sin\theta$.

On squaring and adding, we get $(5x - 3)^2 + (5y)^2 = 16$ i.e., $25x^2 + 25y^2 + 9 - 30x - 16 = 0$

$\Rightarrow 25(x^2 + y^2) - 30x - 7 = 0$, which is a circle

12. (a) From the given it is possible to conclude that area $S_n = \frac{S_{n+1}}{2}$

Now $S_1 = 10 \times 10 = 100$ square cm

Let S_n be the area less than 1 cm^2, so $2^{n-1} > 100$

$\Rightarrow n - 1 \geq 7$ i.e., $n \geq 8$, so $n = 7$ will not give area less than $1 cm^2$

13. (b) In the new system (x', y') is $(4, -2\sqrt{3})$, here $\theta = -30°$, so

$x = x'\cos\theta - y'\sin\theta = \frac{4\sqrt{3}}{2} - \frac{2\sqrt{3}}{2} = \sqrt{3}$

And $y = x'\sin\theta + y'\cos\theta = -\frac{4}{2} - 2\sqrt{3}\frac{\sqrt{3}}{2} = -5$

So old position is $(\sqrt{3}, -5)$

14. (a) As shown in the figure, AB is inclined to x-axis at 45°. When AB is rotated clockwise by 30° then

$B = (1 + \sqrt{2}\cos 15°, 0 + \sqrt{2}\sin 15°)$

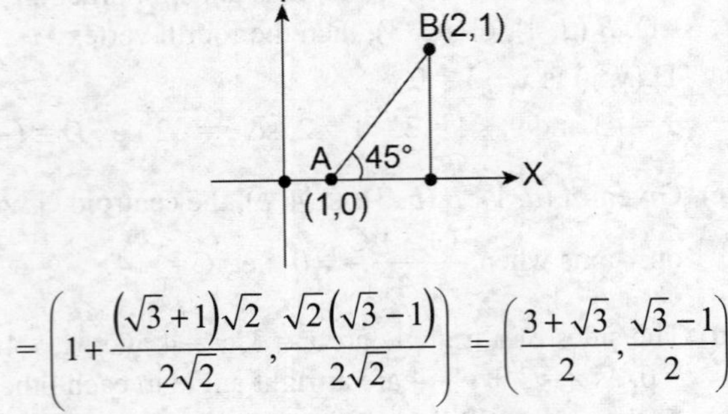

$$= \left(1+\frac{(\sqrt{3}+1)\sqrt{2}}{2\sqrt{2}}, \frac{\sqrt{2}(\sqrt{3}-1)}{2\sqrt{2}}\right) = \left(\frac{3+\sqrt{3}}{2}, \frac{\sqrt{3}-1}{2}\right)$$

15. (a) $A(-1, 2)$, $B(2, -1)$ and $C(3, -1)$ (given). The fourth vertex of ||gm may be any one of $D_1 = (0, 2)$, $D_2 = (6, -4)$, $D_3(-2, 2)$

16. (a) $A(-2, 3)$, $B(3, -1)$ (given). The point of trisection nearer to $A(-2, 3)$ is $P\left(-\frac{1}{3}, \frac{5}{3}\right)$

17. (a) The three vertices of a ||gm are $A(a+b, a-b)$, $B(2a+b, 2a-b)$, $C(a-b, a+b)$, the fourth vertex will be $D_1 = (-b, b)$

18. (a) From the given $|x| + |y| = 1$. This will form a square.

19. (d) B $(a+b, b-a)$, $C(a-b, a+b)$, $BC = 2\sqrt{a^2+b^2}$. The equation of right bisector of BC is $bx - ay = 0$.

Observe that $\left(1, \frac{b}{a}\right)$ satisfies this equation

20. (c) ΔABC is an isosceles (given)

Where $B(1, 3)$, $C(-2, 7)$, so mid point of BC is $\left(-\frac{1}{2}, 5\right)$

$\Rightarrow BC = 5$ and equation of right bisector of BC is $6x - 8y + 43 = 0$.

Observe that $\left(-\frac{1}{2}, 5\right)$ is the mid point and $\left(\frac{5}{6}, 6\right)$ satisfies the equation.

21. (c) Roots of $x^2 + 2ax - b^2 = 0$ are $x = \frac{-2a \pm 2\sqrt{a^2+b^2}}{2}$

$= \left(-a-\sqrt{a^2+b^2}\right), \sqrt{a^2+b^2} - a\right)$

Similarly roots of $x^2 + 2cx - d^2 = 0$ are $x = -c \pm \sqrt{c^2+d^2}$

There are two possible sets of P, Q but the distance for both the cases will be

$PQ = \sqrt{4(a^2+b^2) + 4(c^2+d^2)} = 2\sqrt{a^2+b^2+c^2+d^2}$

22. (c) Let $A(x, y)$, then $A = (x, y) = \left(\frac{3k-5}{k+1}, \frac{5k+1}{k+1}\right)$, $B(1, 5)$, $C(7, -2)$

Now $\Delta ABC = 5$ square units

$$\Rightarrow \left\|\begin{matrix} x & y & 1 \\ 7 & -2 & 1 \\ 1 & 5 & 1 \end{matrix}\right\| = 10$$

$$\Rightarrow \left|\begin{matrix} (x-7) & (y+2) & 0 \\ 6 & -7 & 0 \\ 1 & 5 & 1 \end{matrix}\right| = |49 - 7x - 6y - 12| = 10$$

$\Rightarrow 37 - 7x - 6y = \pm 10$

$\Rightarrow 7x + 6y - 27 = 0$ or $7x + 6y - 47 = 0$

Solving in terms of k, we get $21k - 35 + 30k + 6 = 27k + 27$

$\Rightarrow 24k = 56$, so $k = 7/3$ and

$21k - 35 + 30k + 6 = 47k + 47$

$\Rightarrow 4k = 76$, so $k = 19$. Hence $k = 7/3, 19$

Aliter: B $(1, 5)$, $C(7, -2)$, $BC = \sqrt{6^2+7^2} = \sqrt{85}$;

Equation of BC is $7x + 6y - 37 = 0$; area of $\Delta ABC = 5$ units

Distance of A form BC is $d = \frac{20}{\sqrt{85}}$ units

Since $A(x, y)$ divides $(-5, 1)$ and $(3, 5)$ in the ratio k: 1

$\Rightarrow x = \frac{3k-5}{k+1}$ and $y = \frac{5k+1}{k+1}$

$\Rightarrow \frac{|21k - 35 + 30k + 6 - 37|}{(k+1)\sqrt{85}} = \frac{10}{\sqrt{85}}$; on solving, we get $k = \frac{7}{3}, 19$

23. (c) The vertices of ΔABC are $A(a, b)$, B (ar, bs), $C(ar^2, bs^2)$ (given)

$$\Rightarrow \Delta ABC = \frac{1}{2}\left\|\begin{matrix} a & b & 1 \\ ar & bs & 1 \\ ar^2 & bs^2 & 1 \end{matrix}\right\|$$

$$= \frac{ab}{2}\left\|\begin{matrix} 1 & 1 & 1 \\ r & s & 1 \\ r^2 & s^2 & 1 \end{matrix}\right\| = \frac{ab}{2}\left\|\begin{matrix} 0 & 0 & 1 \\ r-s & s-1 & 1 \\ (r-s)(r+s) & s^2-1 & 1 \end{matrix}\right\|$$

$$= \frac{ab}{2}(r-s)(s-1)\left|\begin{matrix} 1 & 1 \\ r+s & s+1 \end{matrix}\right| = \pm\frac{ab}{2}(r-s)(s-1)(r-1)$$

24. (a) The vertices of the pentagon are $A(4, 1)$, $B(3, 6)$, $C(-5, 1)$, $D(-3, -3)$, $E(-3, 0)$

$$\text{Area} = \frac{1}{2}\left|\begin{matrix} 4 & 1 \\ 3 & 6 \\ -5 & 1 \\ -3 & -3 \\ -3 & 0 \\ 4 & 1 \end{matrix}\right| = \frac{1}{2}|(24-3) + (3+30) + (15+3)$$

$+ (0 - 9) + (-3 - 0)| = 30$ square units.

25. (c) The point of intersection of $y = m_1x + c_1$ and $y = m_2x + c_2$ has x-coordinate as $x = \frac{c_1 - c_2}{m_2 - m_1}$, so area of

$$\Delta ABC = \frac{1}{2}\frac{|c_1-c_2||c_1-c_2|}{|(m_2-m_1)|} = \frac{1}{2}\frac{(c_1-c_2)^2}{|m_1-m_2|}$$

26. (b) Let $P(x, y)$ be the point, then

$\sqrt{x^2+(y-a)^2} - \sqrt{x^2+(y+a)^2} = \pm 6$

Solving we will get $\frac{y^2}{9} - \frac{x^2}{a^2-9} = 1$

or $\frac{x^2}{a^2-9} - \frac{y^2}{9} + 1 = 0$

27. **(c)** Let at a given time the position of A is $(a, 0)$ and that of B is $(0, b)$

$\Rightarrow$ $a^2 + b^2 = d^2$. Now the mid point is $\left(\frac{a}{2}, \frac{b}{2}\right)$, so $2x$ = a and $2y$ = b

$\Rightarrow$ $4x^2 + 4y^2 = d^2$ or $\left(x^2 + y^2 = \frac{d^2}{4}\right)$, which gives a circle.

28. **(b)** Given vertices A (1, 2), B = (3, 4)

$\Rightarrow$ $AB = 2\sqrt{2}$

Now, ΔABC = 1 square unit, so distance of C from AB is $1/\sqrt{2} < \sqrt{2}$. So, we get 4 possible points

29. **(b)** The ||gm is formed by ax ± by ± c = 0, distance between the set of parallel lines is $\frac{2|c|}{\sqrt{a^2+b^2}}$

$\Rightarrow$ Area of the ||gm = $\frac{4c^2}{a^2+b^2}$

30. **(b)** The given line is $L: \frac{x}{3} + \frac{y}{4} = 1$

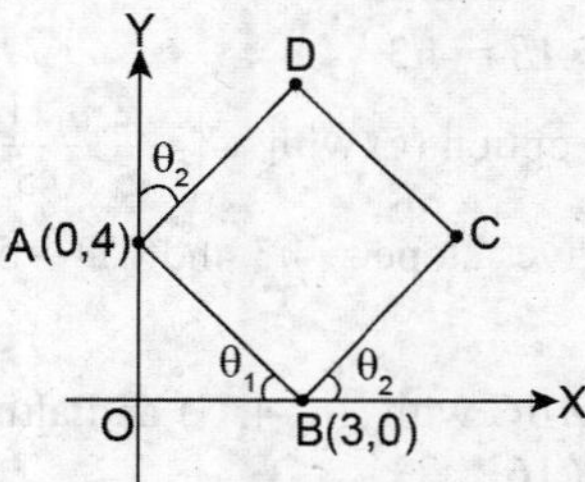

So, A (0, 4) and B (3, 0). Observe that AB = 5 units, $\sin\theta_1 = \cos\theta_2$ = 4/5 and $\cos\theta_1 = \sin\theta_2$ = 3/5

So, as shown C (7, 3) and D (4, 7), obviously (4, 7) is farthest from the origin.

31. **(a)** Given vertices of a triangle is P (2, 1), Q (4, –1) and R (3, 2) .

Under the given conditions a ||gm will be formed, so area of ||gm $PQRS$ = |2 area of ΔPQR

$$= \begin{vmatrix} 2 & 1 & 1 \\ 4 & -1 & 1 \\ 3 & 2 & 1 \end{vmatrix} = \begin{vmatrix} 2 & 1 & 1 \\ 2 & -2 & 0 \\ 1 & 1 & 0 \end{vmatrix} = 4 \text{ square units}$$

32. **(a)** Given point is P (1, 1) and Q (3, 2), consider the image of Q (3, 2) in x-axis, so Q' = (3, –2).

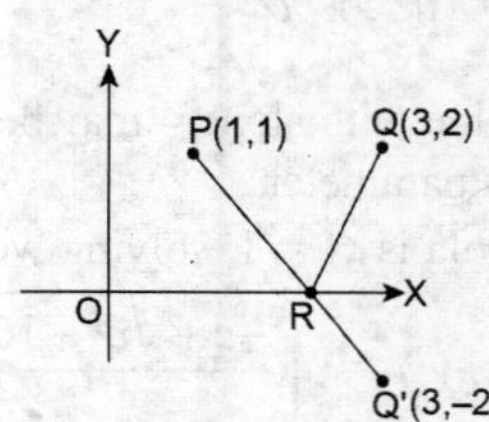

Line joining PQ' will intersect the x-axis at 5/3. Hence $\left(\frac{5}{3}, 0\right)$ is the point.

33. **(a)** Given OA = OB = 6 square units and point O (0, 0). Now A & B lie on $x - y + 1 = 0$

Distance of O from $AB = \frac{1}{\sqrt{2}}$

$AB = 2\sqrt{6^2 - \frac{1}{2}} = 2\sqrt{\frac{71}{2}}$ and area ΔOAB

$= \frac{1}{2} \times \frac{1}{\sqrt{2}} \times 2\sqrt{\frac{71}{2}} = \frac{\sqrt{71}}{2}$ square units.

34. **(a)** Let x & y axes be the perpendicular lines, then a rhombus square is formed, so the area enclosed = 18 square units

35. **(a)** The line is $x + y = 3$. Observe that (0, 3) and (3, 0) are equidistant from $|x| = |y|$.

36. **(a)** Let $P(x, y)$ be the point, then $\sqrt{(x-4)^2 + y^2} = \frac{1}{2}|16 - x|$

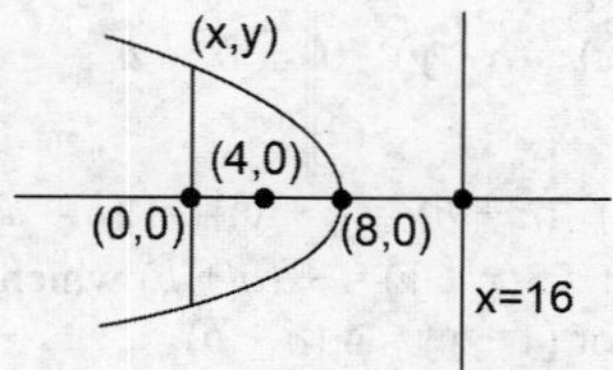

On squaring, we get $4\{x^2 + 16 - 8x + y^2\} = 256 + x^2 - 32x$

$\Rightarrow$ $3x^2 + 4y^2 = 192$

37. **(d)** Given equation is $bx + cy$ = a, where a, b, c are of the same sign. So the line intersects the axes at

$A\left(\frac{a}{b}, 0\right)$ and $B\left(o, \frac{a}{c}\right)$

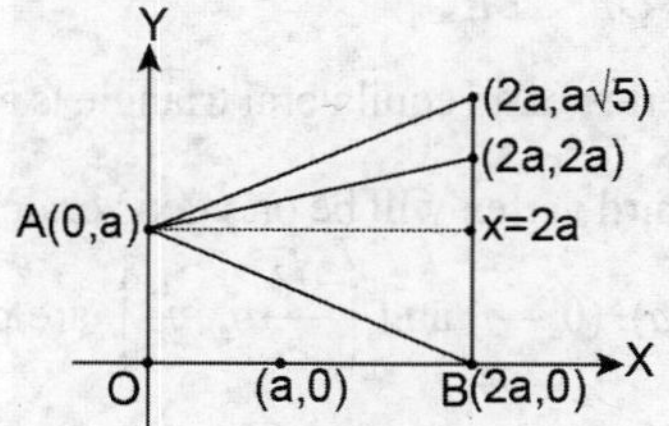

$\Rightarrow$ Area enclosed between the axes and the line

$= \frac{1}{2} \cdot \frac{a}{b} \cdot \frac{a}{c} = \frac{a^2}{2bc} = \frac{1}{8}$

$\Rightarrow$ $4a^2$ = bc, so b, $\pm 2a$, c are in G.P.

38. **(b)** Let C $(2a, y)$, A $(0, a)$, B $(2a, 0)$

$\Rightarrow$ $(2a^2) + (y - a)^2 = y^2$ gives $4a^2 + a^2 = 2ay$, so $y = \frac{5a}{2}$

Now Area $\Delta ABC = \frac{1}{2} \times \frac{5a}{2} \times 2a = \frac{5}{2}a^2$ square units.

39. (c) Given vertex are A (6, 1), B (7, 2), C (–1, 0).

Let the fourth vertex be (x, y), then area of quadrilateral

$$ABCD = \frac{1}{2}\left\|\begin{matrix} 6 & 1 \\ 7 & 2 \\ -1 & 0 \\ x & y \\ 6 & 1 \end{matrix}\right\| = 4$$

$\Rightarrow$ $(12 - 7) + (0 + 2) + (6 - y) + (x - 6y) = \pm 8$

$\Rightarrow$ $x - 7y = 1$ or $x - 7y = -15$

Let $x - 7y = t$, then $t = 1$, $t = -15$

$\Rightarrow$ $t^2 + 14t - 15 = 0$

So the locus of $D(x, y)$ is $(x - 7y)^2 + 14(x - 7y) - 15 = 0$

40. (b) From the given area of $\Delta PAB = (1/3)\ \Delta OAB$

$$\Rightarrow \frac{1}{2}\left\|\begin{matrix} b & a & 1 \\ a & b & 1 \\ x & y & 1 \end{matrix}\right\| = \frac{1}{6}\left|\begin{matrix} b & a & 1 \\ a & b & 1 \\ 0 & 0 & 1 \end{matrix}\right|$$

$$\Rightarrow 3\left\|\begin{matrix} (b-a) & (a-b) & 0 \\ (a-x) & (b-y) & 0 \\ x & y & 1 \end{matrix}\right\| = |b^2 - a^2|$$

$\Rightarrow$ $3|(b - a)\ \{(a + b) - (x + y)\}| = |(b - a)\ (b + a)|$

$\Rightarrow$ $3(a + b) - 3(x + y) = \pm (a + b)$ which gives $(x + y) = 2(a + b)$ or $(x + y) = 4(a + b)$

41. (d) Given vertex are $A(x_1, y_1)$, $B(x_2, y_2)$ and $C(x_3, y_3)$

L: $2x + 3y - 1 = 0$

$$\Rightarrow \frac{BP}{PC} = \frac{-L(x_2, y_2)}{L(x_3, y_3)},\ \frac{CQ}{QA} = \frac{-L(x_3, y_3)}{L(x_1, y_1)},$$

$$\frac{AR}{RB} = \frac{-L(x_1, y_1)}{L(x_2, y_2)}$$

$$\Rightarrow \frac{BP}{PC} \times \frac{CQ}{QA} \times \frac{AR}{BR} = -1$$

42. (d) One vertex of an equilateral triangle is at (0, 0). The possible third vertex will be on y-axis or on line $y = -\frac{1}{\sqrt{3}}x$, so $(0, a)$, $(0, -a)$ and $\left(\frac{\sqrt{3}}{2}a, -\frac{a}{2}\right)$ are possible

43. (c) Given $A\left(\sqrt{1-t^2} + t, 0\right)$ and $B\left(\sqrt{1-t^2} - t, 2t\right)$

So mid point of AB is $M\left(\sqrt{1-t^2}, t\right)$

$\Rightarrow$ $x = \sqrt{1-t^2}$ and $y = t$, on squaring and adding, we get $x^2 + y^2 = 1$, which gives a circle.

44. (a) Area $\Delta AOC = 2$area ΔBOC

Now $\frac{x}{a} + \frac{y}{b} = 1$ and $y = x$ intersect at $C\left(\frac{ab}{a+b}, \frac{ab}{a+b}\right)$

$$\Rightarrow 2 \times \frac{1}{2} b\left(\frac{ab}{a+b}\right) = \frac{1}{2} a\left(\frac{ab}{a+b}\right);\ \text{so } a = 2b$$

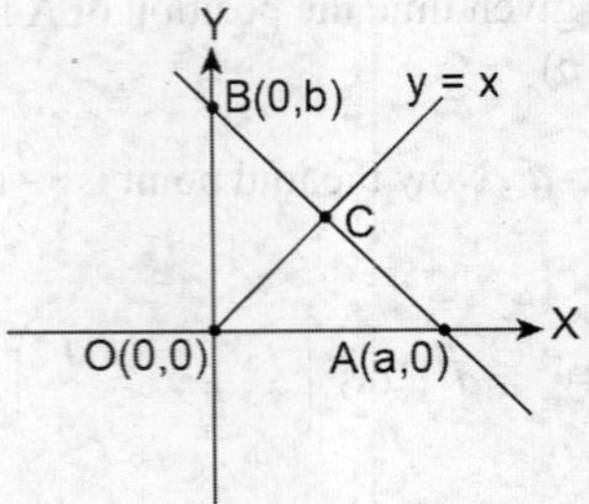

Hence $C = \left(\frac{a}{3}, \frac{a}{3}\right)$

45. (a) Given vertex are P (3, 4), $Q(x_1, y_1)$ and R (11, 10) are collinear and $PQ = 2.5$ units

Equation of PR is $3x - 4y + 7 = 0$

Now (5, 11/2) satisfies, also distance

$$d = \sqrt{2^2 + \left(\frac{3}{2}\right)^2} = 2.5 \text{ units}$$

46. (c) Given vertex are A (1, 3) and B (–2, –1), so $AB = 5$units.

Since area = 10square units

$\therefore$ $BC = AD = 2$ units

Slope of $AB = 4/3$

Observe option (c) with $C\left(-\frac{2}{5}, -\frac{11}{5}\right)$ & $D\left(\frac{13}{5}, \frac{9}{5}\right)$

Which gives slope = 4/3 and $CD = 5$ units. Also AD is 2 units

47. (b) Given ΔABC with A (–4, 5) and altitudes AD, BE, CF where $D\left(\frac{16}{5}, -\frac{23}{5}\right)$, E (4, 1), F (–1, –4)

$$\text{Slope of } AD = \frac{48}{5\left(-\frac{36}{5}\right)} = -\frac{4}{3}$$

$\Rightarrow$ slope of $BC = \frac{3}{4}$

Observe that points under option (b) are suitable as B (0, –7), C (8, –1) gives slope of $m = \frac{6}{8} = \frac{3}{4}$

Also equation of AB is $3x + y + 7 = 0$ and (–1, –4) lies on it

48. (c) Given $x(t) = \mu t\cos\alpha$ and $y(t) + pt^2 = \mu t \sin t\alpha$

$$\Rightarrow y = x\tan\alpha - \frac{px^2}{u^2\cos^2\alpha}$$

49. (a) A variable line with slope 4 may be taken as $y = 4x + c$ where c is a parameter.

The hyperbola is $xy = 1$, solving we get $4x^2 + cx - 1 = 0$

Gives $M(x_1, y_1) = \left(\frac{-c + \sqrt{c^2+16}}{8}, \frac{c + \sqrt{c^2+16}}{2}\right)$ and

$N(x_2, y_2) = \left(\frac{-c - \sqrt{c^2+16}}{8}, \frac{c - \sqrt{c^2+16}}{2}\right)$

Let (h, k) be the point which divides MN in ratio 1 : 2

$\Rightarrow\ (h,k) = \left(\dfrac{-3c+\sqrt{c^2+16}}{24}, \dfrac{3c+\sqrt{c^2+16}}{6}\right)$

$\Rightarrow$ Product $hk = \dfrac{16-8c^2}{24\times 6} = \dfrac{2-c^2}{18}$

Hence $24x = \sqrt{18-18xy} - 3c$ and $6y = \sqrt{18-18xy} + 3c$

On adding, we get $24x + 6y = 2\sqrt{18-18xy}$

$\Rightarrow\ 4x + y = \sqrt{2-2xy}$

On squaring both sides, we get $16x^2 + y^2 + 8xy = 2 - 2xy$

$\Rightarrow\ 16x^2 + 10xy + y^2 - 2 = 0$

50. **(a)** $|x|^2 - 3|x| + 2 = 0 \Rightarrow (|x| - 1)(|x| - 2) = 0$

So $x = \pm 1, \pm 2$

Similarly $y^2 - 3y + 2 = 0$ gives $y = 1, 2$

Since the area is 1 square units

We have two possibilities:

(i) (1, 1), (1, 2), (2, 1), (2, 2) or

(ii) (–1, 1), (–1, 2), (–2, 1), (–2, 2)

51. **(c)** Let $A(x, y)$ and at any time $B(b, 0)$ and $C(0, c)$ where $b^2 + c^2 = 169$ ($BC = 13$ units)

Centroid $G = \left(\dfrac{x+b}{3}, \dfrac{y+c}{3}\right) = (5,5)$

$\Rightarrow\ x + b = 15, y + c = 15$

Putting in $b^2 + c^2 = 169$, we get $(15 - x)^2 + (15 - y)^2 = 169$ i.e., $x^2 + y^2 - 30x - 30y + 281 = 0$

52. **(a)** Given vertex are $A(4, -1)$ and $B(1, 2)$.

Let B' (x, y) be the new position of B in the stretched condition

$\Rightarrow$ B divides AB' in the ratio 1 : 2.

So $\left(\dfrac{8+x}{3}, \dfrac{-2+y}{3}\right) = (1,2)$, so $B(-5, 8)$

53. **(d)** From the given the locus is a circle.

54. **(d)** Given coordinates are $A(1, 2)$. Equation of medians through B and C are $x + y = 5$, $x = 4$ respectively point of intersection of medians is centroid G = (4, 1). Now observe that $C(4, y_1)$ and let $B(x_2, y_2)$, so $\dfrac{1+4+x_2}{3} = 4$

$\Rightarrow\ x_2 = 7$

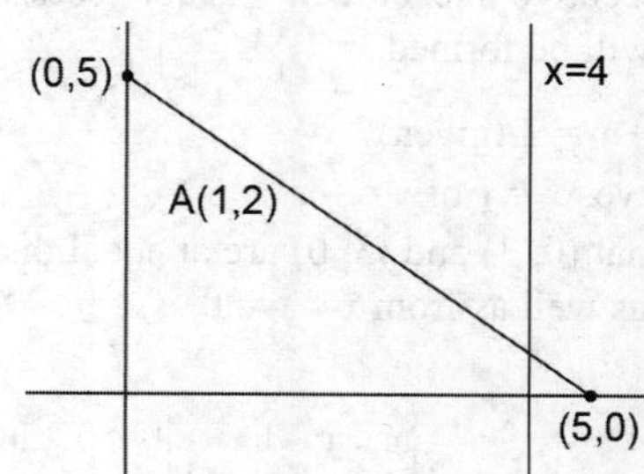

Also $x_2 + y_2 = 5$ (as B lies on $x + y = 5$)

$\Rightarrow\ y_2 = -2$

So, $B(7, -2)$ and $\dfrac{2+y_1-2}{3} = 1$

$\Rightarrow\ y_1 = 3$

We get $B(7, -2)$, $C(4, 3)$

55. **(c)** Given A($a\cos\alpha$, $a\sin\alpha$) and $B(a\cos\beta, a\sin\beta)$

$\Rightarrow\ AB = a\sqrt{(\cos\alpha - \cos\beta)^2 + (\sin\alpha - \sin\beta)^2}$

$= |a|\sqrt{2\{1-\cos(\alpha-\beta)\}} = \left|2a\sin\left(\dfrac{\alpha-\beta}{2}\right)\right|$

56. **(b)** From the given $EB^2 + BF^2 = a^2 = (x_2 - x_1)^2 + (y_2 - y_1)^2$

$CF^2 + GC^2 = b^2 = (x_3 - x_2)^2 + (y_3 - y_2)^2$

$GD^2 + HD^2 = c^2 = (x_4 - x_3)^2 + (y_4 - y_3)^2$

$AH^2 + AE^2 = d^2 = (x_4 - x_1)^2 + (y_4 - y_1)^2$

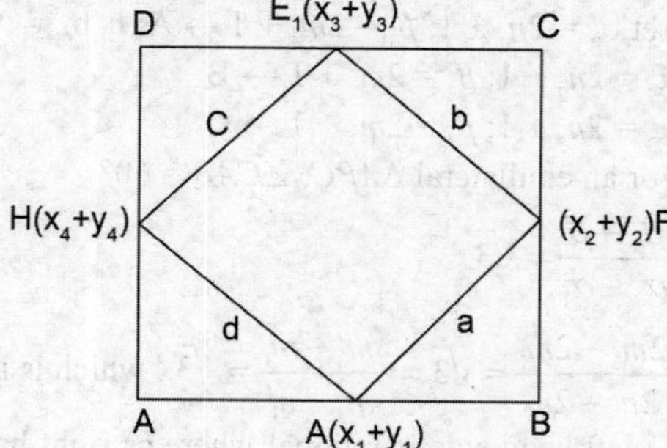

$\Rightarrow\ AE^2 + EB^2 \le AB^2 = 1$, so $a^2 + b^2 + c^2 + d^2 \le 4$

Also in ΔEFG $a^2 + b^2 \ge EG^2 \ge BC^2$. Similarly for other combination

$\therefore\ a^2 + b^2 + c^2 + d^2 \ge 2$

Hence $2 \le a^2 + b^2 + c^2 + d^2 \le 4$

57. **(a)** $A(b\cos\theta, b\sin\theta)$, $B(a\cos\phi, a\sin\phi)$

M (x, y) and AM : BM = b : a or AM : MB = b : $-a$

$\text{M}(x, y) = \left\{\dfrac{ab(\cos\phi - \cos\theta)}{b-a}, \dfrac{ab(\sin\phi - \sin\theta)}{b-a}\right\}$ or

$\dfrac{x}{\cos\theta - \cos\phi} = \dfrac{y}{\sin\theta - \sin\phi} = \dfrac{ab}{a-b}$

$\Rightarrow\ \dfrac{x}{(-2)\sin\left(\dfrac{\theta+\phi}{2}\right)\sin\left(\dfrac{\theta-\phi}{2}\right)} = \dfrac{y}{2\sin\left(\dfrac{\theta+\phi}{2}\right)\cos\left(\dfrac{\theta-\phi}{2}\right)}$

$\Rightarrow\ -x = y\tan\left(\dfrac{\theta-\phi}{2}\right)$ or $x + y\tan\left(\dfrac{\theta-\phi}{2}\right) = 0$

58. **(c)** Co-ordinates of S are (x_4, y_4) as S

$(x_4, y_4) = \left(\dfrac{\alpha x_2 + \beta x_1}{\alpha+\beta}, \dfrac{\alpha y_2 + \beta y_1}{\alpha+\beta}\right)$

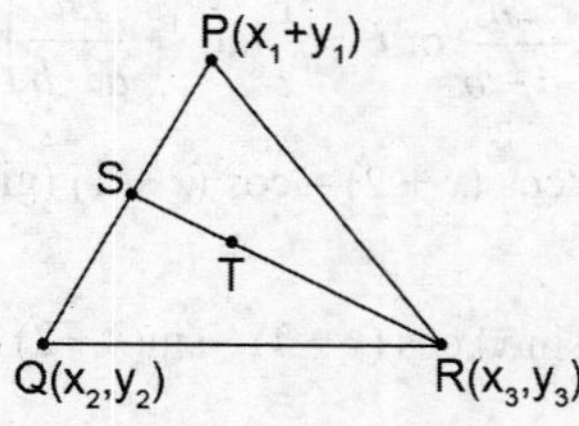

Now T divides SR in the ratio $\gamma : (\alpha + \beta)$

$\Rightarrow\ T(x_5, y_5) = \left\{\dfrac{(\alpha+\beta)x_4 + \gamma x_3}{(\alpha+\beta+\gamma)}, \dfrac{(\alpha+\beta)y_4 + \gamma y_3}{\alpha+\beta+\gamma}\right\}$

$= \left(\dfrac{\beta x_1 + \alpha x_2 + \gamma x_3}{\alpha+\beta+\gamma}, \dfrac{\beta y_1 + \alpha y_2 + \gamma y_3}{\alpha+\beta+\gamma}\right)$

59. (d) Lines L_1: $a_1x + b_1y + c_1 = 0$ and L_2: $a_2x + b_2y + c_2 = 0$ intersect in concyclic points then $a_1a_2 = b_1b_2$

60. (d) $\dfrac{\sum_{r=0}^{n-1} x^{2r}}{\sum_{r=0}^{n-1} x^{r}} \in \mathbb{Z} \Rightarrow 1\dfrac{\left(1-(x^2)^n\right)}{1-x^2} \times \dfrac{(1-x)}{1(1-x^n)} \in \mathbb{Z}$

$\Rightarrow \dfrac{1-x^{2n}}{(1+x)(1-x^n)} \in \mathbb{Z} = \dfrac{(1+x^n)}{(1+x)} \in \mathbb{Z}$

$\Rightarrow$ n must be an odd positive integer

Let $\alpha_1 = 2n_1 + 1; \beta_1 = 2m_1 + 1 \rightarrow A$; $n_i\ m_i \in W$

$\alpha_2 = 2n_2 + 1; \beta_2 = 2m_2 + 1 \rightarrow B$

$\alpha_3 = 2n_3 + 1; \beta_3 = 2m_3 + 1 \rightarrow C$

For an equilateral ΔABC, $\angle CAB = 60°$

$\Rightarrow \dfrac{\beta_3 - \beta_1}{\alpha_3 - \alpha_1} = \sqrt{3}$

$\Rightarrow \dfrac{2m_3 - 2m_1}{2n_3 - 2n_1} = \sqrt{3} \Rightarrow \dfrac{m_3 - m_1}{n_3 - n_1} = \sqrt{3}$, which is impossible as left hand side is rational where as right hand side is irrational.

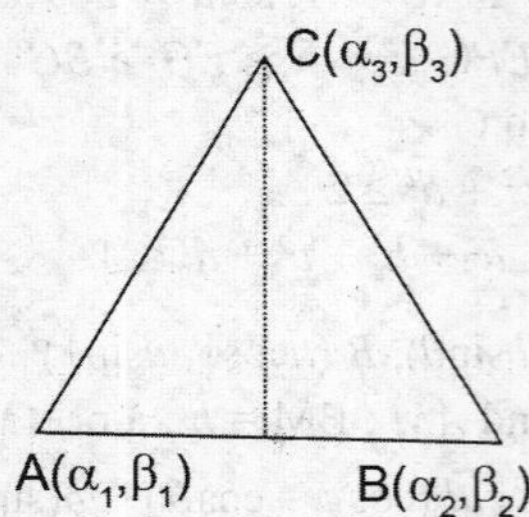

Now for $A \equiv (1, 1)$; $B \equiv (3, 1)$ and $C \equiv (1, 3)$

ΔABC is isosceles as well as right angled

61. (b) As axes are rotated by angle θ anticlockwise $P(x, y)$ becomes $P(x', y')$ where $x = x'\cos\theta - y'\sin\theta$ and $y = x'\sin\theta + y'\cos\theta$

Putting in $ax^2 + 2hxy + by^2 = 0$, we get $a(x'\cos\theta - y'\sin\theta)^2 + 2h(x'\cos\theta - y'\sin\theta)(x'\sin\theta + y'\sin\theta) + b(x'\sin\theta + y'\cos\theta)^2 = 0$

To eliminate xy' terms, we put $2b\sin\theta\cos\theta - 2a\sin\theta\cos\theta + 2h(\cos^2\theta - \sin^2\theta) = 0$.

$\Rightarrow (b - a)\sin 2\theta = -2h\cos 2\theta$, i.e.,

$\tan 2\theta = \dfrac{2h}{a-b}$ or $\theta = \dfrac{1}{2}\tan^{-1}\left(\dfrac{2h}{a-b}\right)$

62. (d) $y = \cos x \cos(x + 2) - \cos^2(x + 1)$ (given)

$\Rightarrow \dfrac{dy}{dx} = (-\sin x)\cos(x+2) - \sin(x+2)\cos x + 2\cos(x+1)\sin(x+1)$

$\Rightarrow \dfrac{dy}{dx} = \sin(2x + 2) - \sin(2x + 2) = 0$

So the graph is parallel to x–axis, when $x = \pi/2$, $y = -\sin^2 1$, so the graph passes through $\left(\dfrac{\pi}{2}, -\sin^2 1\right)$ and it is parallel to x-axis

63. (c) Since the ΔABC is equilateral as shown $h = \alpha/\sqrt{3}$.

Let $DP = k$, then $PE = \left(\dfrac{4\alpha}{3} - k\right)$

$\Rightarrow h_1 = DP\sin B = k\dfrac{\sqrt{3}}{2}$ and $h_2 = \left(\dfrac{4\alpha}{3} - k\right)\dfrac{\sqrt{3}}{2}$

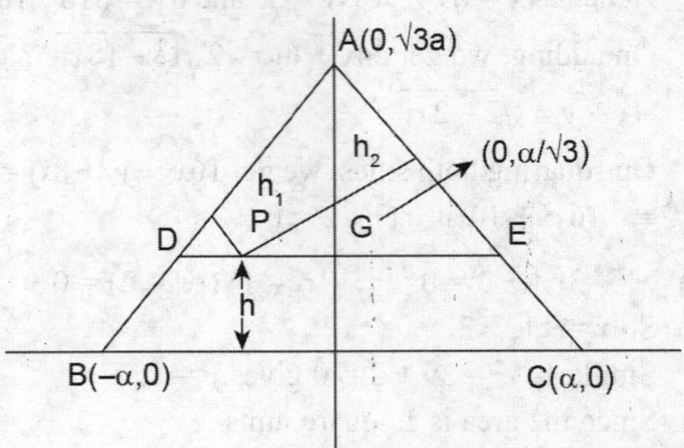

As $h_2 = PE\sin C$. Eliminating k, we get $h_2 = \dfrac{2\alpha}{\sqrt{3}} - h_1$ or $h_2 = 2h - h_1$; hence $h_1 + h_2 = 2h$.

So, h is AM between h_1 & h_2

SECTION–IV (MORE THAN ONE ARE CORRECT)

1. (b), (c) Given vertex are $A(a, a)$, $B(-a, -a)$

So, $AB = 2\sqrt{2}a$ vertex C will lie on $y = -x$ $(OC = \sqrt{6}a)$

So, $C'(a\sqrt{3}, -a\sqrt{3})$ and $C(-a\sqrt{3}, a\sqrt{3})$

2. (b), (c), (d) Given $S_1 = 10 \times 10 = 100$ square units

From the given area $S_{n+1} = \dfrac{100}{2^n}$ or $S_n = \dfrac{100}{2^{n-1}}$

When $S_n < 1$ square unit, then $2^{n-1} > 100$

$\Rightarrow n - 1 \geq 7$ or $n \geq 8$

3. (a), (b), (c) Let $A(a, b)$, $B(b, c)$ and $C(c, a)$

If a right angled triangle is formed at B then $AB^2 + BC^2 = AC^2$

$\Rightarrow \{(a - b)^2 + (b - c)^2\} + \{(b - c)^2 + (c - a)^2\} = (c - a)^2 + (b - a)^2$

$\Rightarrow 2(b - c)^2 = 0$, so $b = c$

Similarly, for other points having right angle, if a, b, c are three consecutive integer then it is not necessary that a right angled Δ will be formed.

4. (a), (b) $x + y = 3$ (given)

$|x| = |y|$ gives $x = y$ or $y = -x$

Observe that (0, 3) and (3, 0) are at equal distance (s) form $x + y = 0$ as well as from $x - y = 0$

5. (b), (c) From given $\dfrac{3}{2}\begin{vmatrix} x & y & 1 \\ b & a & 1 \\ a & b & 1 \end{vmatrix} = \dfrac{1}{2}\begin{vmatrix} 0 & 0 & 1 \\ b & a & 1 \\ a & b & 1 \end{vmatrix}$

$\Rightarrow 3\begin{vmatrix} x-b & y-a & 0 \\ b-a & a-b & 0 \\ a & b & 1 \end{vmatrix} = \left|(b^2 - a^2)\right|$

$\Rightarrow$ $3\{a - b)(x - b) + (a - b)(y - a)\} = |b^2 - a^2|$ or $3\{(x + y) - (b + a)\} = \pm(a + b)$

$\Rightarrow$ $3(x + y) = 4(a + b)$ or $3(x + y) = 2(a + b)$

6. (a), (b), (c), (d) From the given the third vertex will be either on y-axis at $(0, a)$, $(0, -a)$ or on line $y = -\frac{x}{\sqrt{3}}$ at $\left(\frac{\sqrt{3}}{2}a, \frac{-a}{2}\right), \left(\frac{-\sqrt{3}a}{2}, \frac{a}{2}\right)$

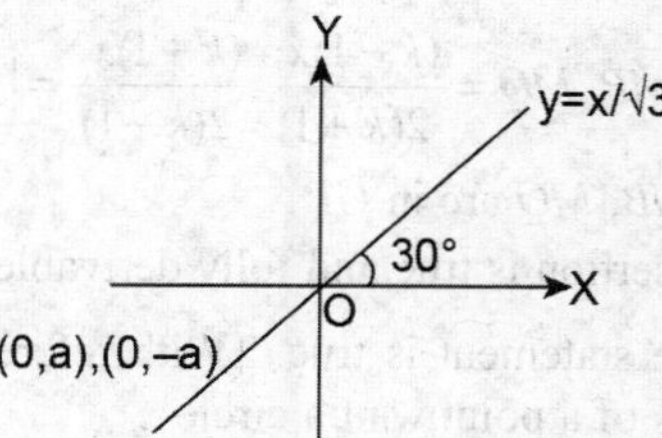

7. (a), (d) $A(a, 0)$, $B(0, b)$, $C\left(\frac{ab}{a+b}, \frac{ab}{a+b}\right)$

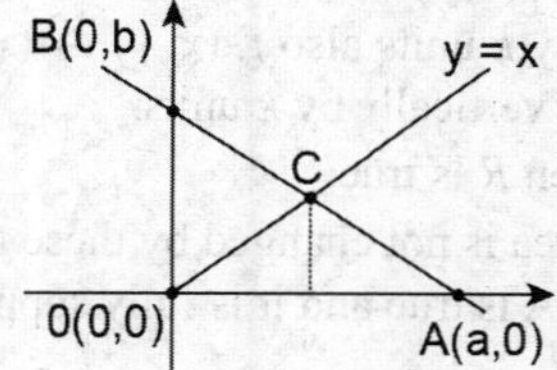

Area ΔAOC = 2area ΔBOC, so $a^2b = \frac{2b^2}{2(a+b)}$ gives $a = 2b$, so $C\left(\frac{2b}{3}, \frac{2b}{3}\right) = \left(\frac{a}{3}, \frac{a}{3}\right)$

8. (a), (d) Slope of AB, $m_1 = 3$; Slope of AD, $m_2 = 1$

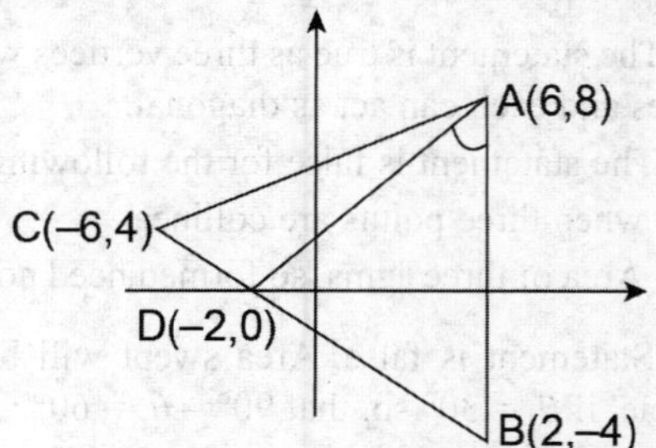

$\tan\theta = \left|\frac{m_1 - m_2}{1 + m_1m_2}\right| = \frac{2}{1+3} = \frac{1}{2} \Rightarrow \theta = \tan^{-1} 0.5 = \cot^{-1} 2$

$= \frac{\pi}{2} - \tan^{-1} 2$

Also $\sin\theta = \frac{1}{\sqrt{5}}$ & $\cos\theta = \frac{2}{\sqrt{5}}$; Also $\theta = -\tan^{-1} 0.5 = \frac{\pi}{4} - \tan^{-1} 3$

9. (a), (b), (c) Given $A(-5, -2)$, $B(7, 6)$ and $C(5, -4)$

Slope of AB is $m_1 = \frac{8}{12} = \frac{2}{3}$; Slope of $BC = \frac{10}{2} = 5$; $\tan B = \frac{5 - 2/3}{1 + 10/3} = \frac{13}{13}$

$\Rightarrow$ $\angle B = 45°$

Equation of altitude through $C(5, -4)$ is $y + 4 = \left(\frac{-3}{2}\right)(x - 5)$ or $3x + 2y - 7 = 0$

Observe that $AB^2 = 208$, $BC^2 = 104$ and $AC^2 = 104$, so $\angle C = 90°$.

$\therefore$ Orthocenter will lie at C which is not inside the ΔABC

As $\angle C = 90°$

$\Rightarrow$ Circum-center C' is at the mid point of AB i.e., $C'(1, 2)$ and centroid $G = \left(\frac{7}{3}, 0\right)$

So $GC' = \sqrt{\frac{16}{9} + 4} = \sqrt{\frac{52}{9}} = \frac{2}{3}\sqrt{13} \neq \frac{4}{3}\sqrt{13}$

10. (a), (b) Consider the concept of power of a point w.r.to a circle $S(x, y) = 0$.

If $S(x_1, y_1) > 0$, the point $P(x_1, y_1)$ will lie outside the circle and the angle at P formed by a chord of the circle will be acute.

Observe that $(x - x_2)(x - x_3) + (y - y_2)(y - y_3) = 0$ is the equation of a circle with $B(x_2, y_2)$ and $C(x_3, y_3)$ as its diameter.

11. (a), (b), (c), (d) Let $A(-\lambda, \mu)$, $B(\lambda, \mu)$, $C(\lambda^2, \lambda\mu)$, $O(0, 0)$. These points can be collinear if $\mu = 0$

As AO = BO so under certain conditions a ||gm or a rectangle can be formed.

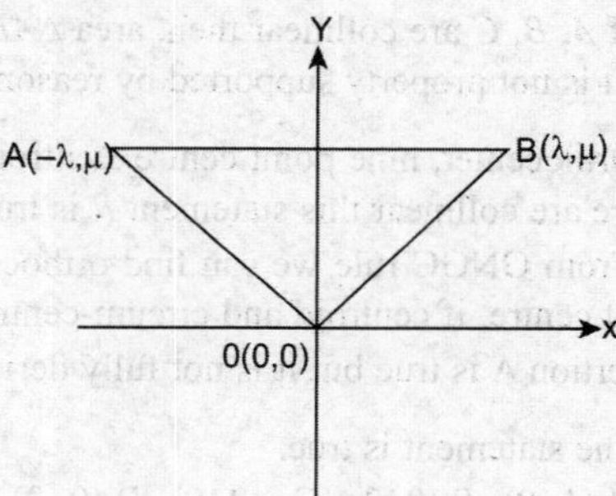

For OAB the centre of a circle will be at $\left(0, \frac{\lambda^2 + \mu^2}{2\mu}\right)$ and its radius will be $r = \left(\frac{\lambda^2 + \mu^2}{2\mu}\right)$

Distance of $(\lambda^2, \lambda\mu)$ from $\left(0, \frac{\lambda^2 + \mu^2}{2\mu}\right)$ will be

$$\sqrt{\lambda^4 + \left(\frac{\lambda^2 + \mu^2 - 2\lambda\mu^2}{2\mu}\right)^2} = \frac{\lambda^2 + \mu^2}{2\mu}$$

On Squaring, we get

$$\lambda^4 + \frac{(\lambda^2 + \mu^2 - 2\lambda\mu^2)^2}{4\mu^2} = \frac{\lambda^4 + \mu^4 + 2\lambda^2\mu^2}{4\mu^2}$$

$\Rightarrow$ $(\lambda^2)^2 = \frac{4\lambda\mu^2\{\lambda^2 + \mu^2 - \lambda\mu^2\}}{4\mu^2}$

$\Rightarrow$ $\lambda^3 = \lambda^2 + \mu^2 - \lambda\mu^2$ or $\lambda^3 - \lambda^2 = \mu^2(1 - \lambda)$ i.e., $\lambda^2(\lambda - 1) = \mu^2(1 - \lambda)$

$\Rightarrow$ $\lambda = 1$

i.e., when $\lambda = 1$, we get $(-1, \mu)$, $(0, 0)$, $(1, \mu)$, $(1, \mu)$, which are three non-collinear points for $\mu \neq 0$.

So these will be concylic

12. **(a), (b), (c), (d)** $x^2 - 3|x| + 2 = 0$ gives $x = \pm 1, \pm 2$ and $y^2 - 3y + 2 = 0$ gives $y = 1, 2$

As shown we get two squares each in Ist and IInd quadrant

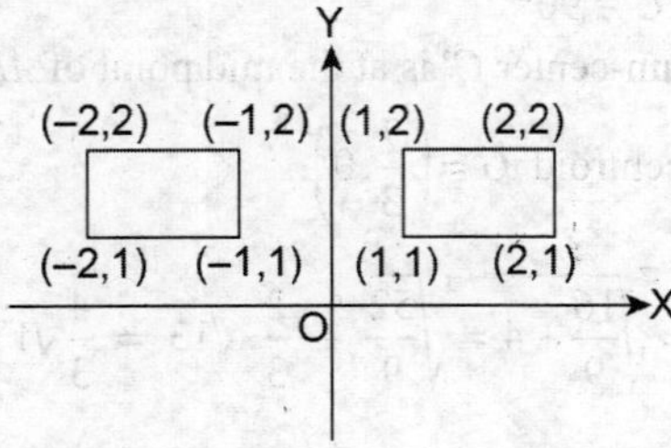

SECTION-V (ASSERTION AND REASON-TYPE)

1. **(a)** **R**: *R* is true

A: L: $3x - 2y + 1 = 0$; A (2, 1) and B (−3, 5)

L (2, 1) = 6 − 2 + 1 = 5

L (−3, 5) = −9 − 10 + 1 = −18

⇒ These points A and *B* lie on the opposite side of L.

So assertion A is true and fully supported by R.

2. **(b)** **R**: $AB + BC = AC$ is true only if A, B, C are collinear not otherwise.

A: If A, B, C are collinear then, area $\Delta ABC = 0$, is true but it is not property supported by reason.

3. **(b)** **R**: Orthocenter, nine point centre, centroid and circum-centre are collinear this statement *R* is true.

A: From ONGC rule we can find orthocenter and nine point centre, if centroid and circum-centre are known.

∴ Assertion A is true but it is not fully derivable from R.

4. **(a)** **R**: The statement is true.

A: A (1, 0), B (0, 3), C (−2, 0), D (0, 2) (given) is true and it is fully supported by reason

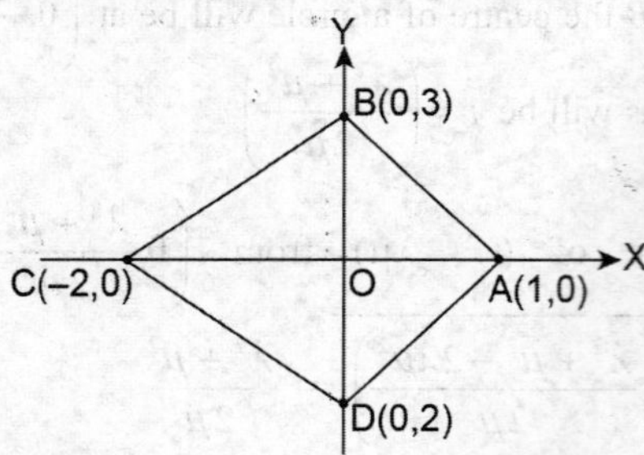

5. **(a)** **R**: For the sake of simplification let origin be at A and *x*-axis be along *AB*, so A (0, 0) and B (x_3, 0)

So, $M = \left(\frac{X_3}{2}, 0\right)$. Let *P* divide *AB* internally and *Q* externally in the ratio $k : 1$

So $P = \left(\frac{kx_3}{k+1}, 0\right)$ and $Q = \left(\frac{kx_3}{k-1}, 0\right)$

$\Rightarrow AP = \frac{kx_3}{k+1}, AB = x_3, AQ = \frac{kx_3}{k-1}$

So AP, *AB* and *AQ* will be in H.P.

If $\frac{2}{x_3} = \frac{k+1}{kx_3} + \frac{k-1}{kx_3} = \frac{2k}{kx_3} = \frac{2}{x_3}$, Which is true so, *AP*, *AB* and *AQ* are in HP, so *R* is true

A: $MP = \frac{kx_3}{k+1} - \frac{x_3}{2} = \frac{2kx_3 - kx_3 - x_3}{2(k+1)} = \frac{(k-1)x_3}{2(k+1)}$

$MB = \frac{x_3}{2}, MQ = \frac{kx_3}{k-1} - \frac{x_3}{2} = \frac{(k+1)x_3}{2(k-1)}$

Now, $MP.\ MQ = \frac{(k-1)x_3}{2(k+1)} \cdot \frac{(k+1)x_3}{2(k-1)} = \left(\frac{x_3}{2}\right)^2 = MB^2$

⇒ *MP*, *MB*, *MQ* are in *GP*

So Assertion is true and fully derivable form reason R.

6. **(a)** **R**: The statement is true. {With reference to the topic powers of a point w.r.t a circle}.

A: The statement is true and it is fully supported by R.

7. **(a)** **R**: $f_1(x, y) \to (y, x)$ will give the image of any figure in $y = x$ and $f_2(x, y) \to (x + k, y)$ will move the figure horizontally by *k* units also $f_3(x, y) \to (x, y + k)$ will move the figure vertically by *k* units.

The reason *R* is true

A: The area is not changed by these transformations, so assertion A is true and it is fully supported by reason R.

8. **(b)** **R**: $(1, a-1)$ and $(a-1, 1)$ are reflection of each other in $y = x$. The statement *R* is true.

A: (1, 1), $(1, a-1)$, $(a-1, 1)$. This is isosceles right angled triangle, so O (1, 1) and $C\left(\frac{a}{2}, \frac{a}{2}\right)$, which are only $y = x$, so *N* and G will be on same line. A is true but it is not fully supported by R.

9. **(d)** **R**: The statement is true as three vertices will form three sides and each can act as diagonal.

A: The statement is false for the following reasons:

(i) when three points are collinear

(ii) Area of three ||gms, so formed need not be different.

10. **(c)** **R**: Statement is false. Area swept will be 1/6th of the circle, if $\theta_1 = 30°$ so, that $90° - \theta_1 = 60°$

(Area above *x*-axis will be considered)

RMR' is the right bisector where $\theta_1 = \theta_2$

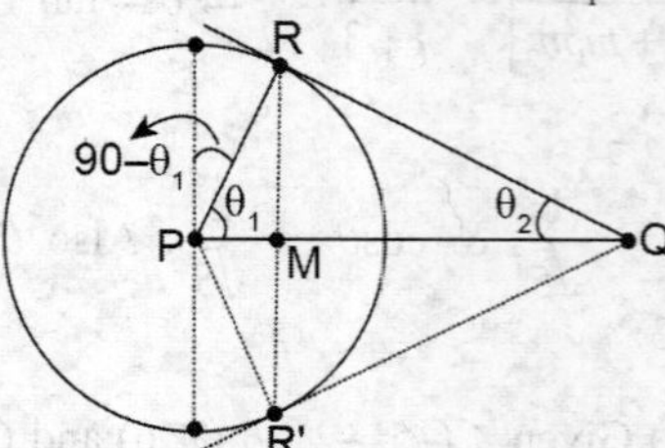

A: $\cos\theta_1 = \frac{3}{4} \Rightarrow \theta_1 = \cos^{-1}\frac{3}{4}$, so $\left(\frac{\pi}{2} - \cos^{-1}\frac{3}{4}\right)$ is the angle

∴ Area swept = $\frac{16\pi}{2\pi}\left(\frac{\pi}{2} - \cos^{-1}\frac{3}{4}\right) = 4\pi - 8\cos^{-1}\frac{3}{4}$ or

$\frac{16\pi}{2\pi}\left(\sin^{-1}\frac{3}{4}\right) = 8\sin^{-1}\left(\frac{3}{4}\right)$ ⇒ Assertion is true

SECTION–VI (LINKED COMPREHENSION-TYPE)

Comprehension A:

1. **(a)** L: $x + my + n = 0$, as shown

$\frac{BP}{PC} = \frac{-L(x_2, y_2)}{L(x_3, y_3)}, \frac{CQ}{QA} = \frac{-L(x_3, y_3)}{L(x_1, y_1)}$ and

$\frac{AR}{RB} = \frac{-L(x_1, y_1)}{L(x_2, y_2)} \Rightarrow \frac{BP}{PC} \times \frac{CQ}{QA} \times \frac{AR}{RB} = -1$

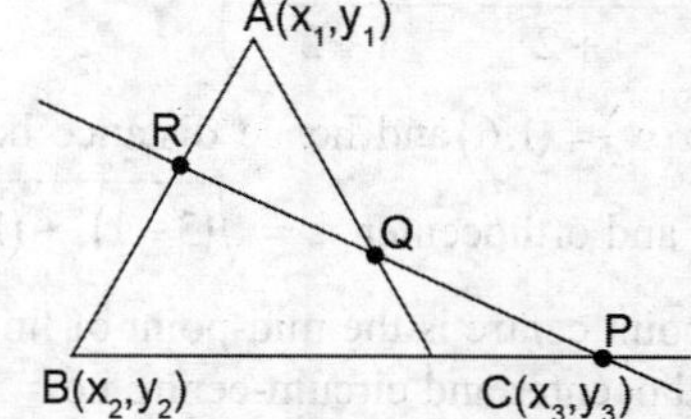

2. **(c)** Since centroid is at origin, so $x_1 + x_2 + x_3 = 0$ and $y_1 + y_2 + y_3 = 0$

From the given $\frac{\ell x_1 + my_1 + \ell x_2 + my_2 + \ell x_3 + my_3 + 3n}{\sqrt{\ell^2 + m^2}} = 1$

$\Rightarrow (x_1 + x_2 + x_3) \quad + (y_1 + y_2 + y_3)\, m + 3n = \sqrt{\ell^2 + m^2}$

$\therefore \quad {}^2 + m^2 = 9n^2$

To find : $\left(\frac{\ell}{-n}\right)^2 + \left(\frac{m}{-n}\right)^2 = \frac{\ell^2 + m^2}{n^2} = 9$

3. **(d)** From **1** we know that $\frac{BP}{PC} \times \frac{CQ}{QA} \times \frac{AR}{RB} = -1$

$\Rightarrow \quad \frac{AR}{RB} = (-1) \times \frac{1}{2} \times \frac{3}{1} = \frac{-3}{2}$

So, R divides AB in the ratio 3 : 2 externally

Comprehension B:

4. **(a)** $A\ (0, 3), B\ (-2, 0), C\ (6, 1)$ and $M\ (\beta, \beta + 1)$

$\Rightarrow x = \beta$ and $y = \beta + 1$, so $y = x + 1$

5. **(c)** Equation of BC is L_1: $x - 8y + 2 = 0$

Now A and M lie on the side of BC, if L_1: $(\beta, \beta + 1)$. L_1 (A) > 0

$\Rightarrow L_1 (0, 3) \times L_1 (\beta, \beta + 1) > 0$ i.e., $(-2, 2)\ (\beta - 8\beta - 8 + 2) > 0$, so $7\beta + 6 > 0$ or $\beta > -6/7$

6. **(c)** M will lie inside ΔABC, when $L_1 (\beta, \beta + 1) L_1 (A) > 0$, $L_2 (\beta, \beta + 1)\ L_2 (\beta) > 0$ and $L_3 (\beta, \beta + 1).L_3 (c) > 0$

As in Q # 5, $L_1 (\beta, \beta + 1)$. L_1 (A) > 0 gives $\beta > -\frac{6}{7}$ ...(i)

Similarly $L_2 (\beta, \beta + 1).L_2$ (B) > 0 gives

$(-11)\ (\beta + 3\beta + 3 - 9) > 0$ i.e., $\beta < 3/2$ ...(ii)

{AC is L_2: $x + 3y - 9 = 0$}

$L_3 (\beta, \beta + 1).L_3$ (C) > 0 gives

$(3\beta - 2\beta - 2 + 6)(22) > 0$, i.e., $\beta > -4$ ....(iii)

(AB is L_3: $3x - 2y + 6 = 0$)

From (i), (ii), (iii), we get $\frac{-6}{7} < \beta < \frac{3}{2}$

Comprehension C:

7. **(c)** Given vertex is $A\ (0, \beta), B\ (-2, 0), C\ (1, 1)$, angle A will be obtuse if $BC^2 > AB^2 + AC^2$

i.e., $10 > 4 + \beta^2 + 1 + \beta^2 + 1 - 2\beta$ or $2(\beta^2 - \beta - 2) < 0$

$\Rightarrow (\beta + 1)\ (\beta - 2) < 0$, so $\beta \in (-1, 2)$

Also for $\beta = 2/3$; A, B, C will be collinear and hence ΔABC is impossible. $\therefore\ \beta \in \left(-1, \frac{2}{3}\right) \cup \left(\frac{2}{3}, 2\right) \to I_1$

8. **(b)** A will be the largest angle of ΔABC, when $BC > AB$ and $BC > AC$, so $10 > 4 + \beta^2$

$\Rightarrow \beta^2 < 6$ i.e., $\beta \in \left(-\sqrt{6}, \sqrt{6}\right)$ ....(i)

And $10 > 1 + \beta^2 - 2\beta + 1$

$\Rightarrow \beta^2 - 2\beta - 8 < 0$ or $(\beta - 4)\ (\beta + 2) < 0$, so $\beta \in (-2, 4)$ ...(ii)

From (i) and (ii), we get $\beta \in \left(-2, \sqrt{6}\right)$; $\beta \neq \frac{2}{3}$

$\Rightarrow \quad \beta \in \left(-2, \frac{2}{3}\right) \cup \left(\frac{2}{3}, \sqrt{6}\right) = I_2$

Now, $I_1 = \left(-1, \frac{2}{3}\right) \cup \left(\frac{2}{3}, 2\right)$ and $I_2 = \left(-2, \frac{2}{3}\right) \cup \left(\frac{2}{3}, \sqrt{6}\right)$

Clearly, I_1 is a subset of I_2

9. **(c)** As solved above $\beta \in \left(-2, \frac{2}{3}\right) \cup \left(\frac{2}{3}, \sqrt{6}\right) = I_2$, when A is the largest angle.

Comprehension D:

10. **(b)** A (a cos t, a sin t), B (b sin t, $-b$ cos t), C (1, 0)

Centroid $G(x, y) = \left(\frac{a\cos t + b\sin t + 1}{3}, \frac{a\sin t - b\cos t}{3}\right)$

So $3x - 1 = a\cos t + b\sin t$ and $3y = a\sin t - b\cos t$

On squaring and adding, we get $(3x - 1)^2 + (3y)^2 = a^2 + b^2$

11. **(a)** Given vertex are $A\ (2, -3), B\ (-2, 1), C\ (x_3, y_3)$

Now G centroid lies on $2x + 3y = 1$

$\Rightarrow \quad \frac{x_3}{3}, \frac{y_3 - 2}{3}$ satisfies it so $2x_3 + 3y_3 - 6 = 3$ i.e., $2x + 3y = 9$

12. **(a)** $P\ (a_1, b_1), Q\ (a_2, b_2)$, now R is equidistant from P and Q

So, $\left(\frac{a_1 + a_2}{2}, \frac{b_1 + b_2}{2}\right)$ will satisfy the equation $(a_1 - a_2)$

$x + (b_1 - b_2)\, y + c = 0$

$\Rightarrow \quad \frac{a_1^2 - a_2^2}{2} + \frac{b_1^2 - b_2^2}{2} = -c \Rightarrow c = \frac{a_2^2 - a_1^2 + b_2^2 - b_1^2}{2}$

13. **(d)** As shown in the figure the vertices are (1, 1), (k, k), (2–k, k)

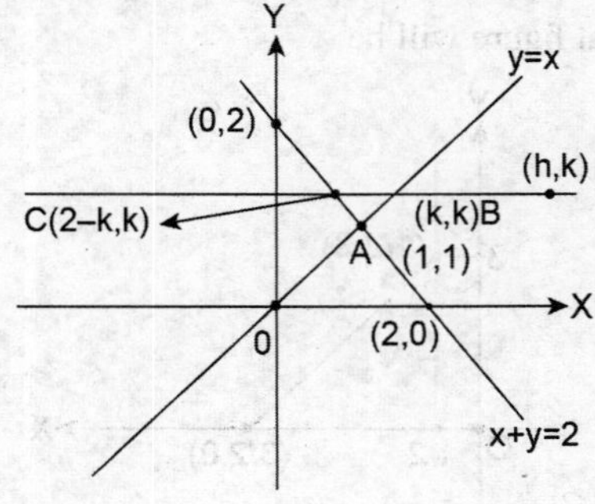

So area of $\Delta ABC = \frac{1}{2}|(2k-2)(k-1)| = 4h^2$

$\Rightarrow$ $P(h, k)$ satisfies $(k-1)^2 = (2h)^2$

So either $2x = y - 1$ or $y - 1 = -2x$ i.e., $y = 2x + 1$ or $y = -2x + 1$

Comprehension E:

14. (d) New distance d is given by $d = |x_2 - x_1| + |y_2 - y_1|$

Now $O(0, 0)$, $A(1, 2)$, $B(2, 3)$, $C(4, 3)$

For $R(x, y)$ is equidistance from O and A

$\Rightarrow |x| + |y| = |x-1| + |y-2|$

Since $0 \le x < 1$ and $0 \le y < 2$, so we get $x + y = 1 - x + 2 - y$ or $2x + 2y = 3$

15. (d) For S (x, y) is equidistant from O and B

$\Rightarrow |x| + |y| = |x-2| + |y-3|$

Since $x \ge 2$ and $0 \le y < 3$, so we get $x + y = x - 2 + 3 - y$ or $2y = 1$

Now, $y = \frac{1}{2}$ lies parallel to x-axis (also it satisfies $0 \le y < 3$)

As $x \ge 2$

$\Rightarrow$ A ray starting from $(2, 1/2)$ and going parallel to positive x-axis is formed which will have infinite length.

16. (a) T (x, y) is equidistant from O and C

$\Rightarrow |x| + |y| = |x-4| + |y-3|$. Since T lies in I[st] quadrant

$\therefore$ $x \ge 0, y \ge 0$

Case (i): $0 \le x < 4$ and $0 \le y < 3$, then $x + y = 4 - x + 3 - y$ gives $2x + 2y = 7$ but $y < 3$

$\Rightarrow$ L (1, 3), M (7/2, 0)

Case (ii): $0 \le x < 4$; $y \ge 3$

$\Rightarrow x + y = -x + 4 + y - 3$

$\Rightarrow 2x = 1$

$\Rightarrow x = 1/2$; $y \ge 3$

$\Rightarrow$ It is a ray starting from point $\left(\frac{1}{2}, 3\right)$ and going along positive y-axis

Case (iii): $x \ge 4$; $0 \le y < 3$

$\Rightarrow x + y = x - 4 - y + 3$

$\Rightarrow 2y = -1$

$\Rightarrow y = -1/2 \in [0,3)$

Case (iv): $x \ge 4$; $y \ge 3$

$\Rightarrow x + y = x - 4 + y - 3$

$\Rightarrow 0 = -7$, which is impossible.

$\therefore$ Final figure will be

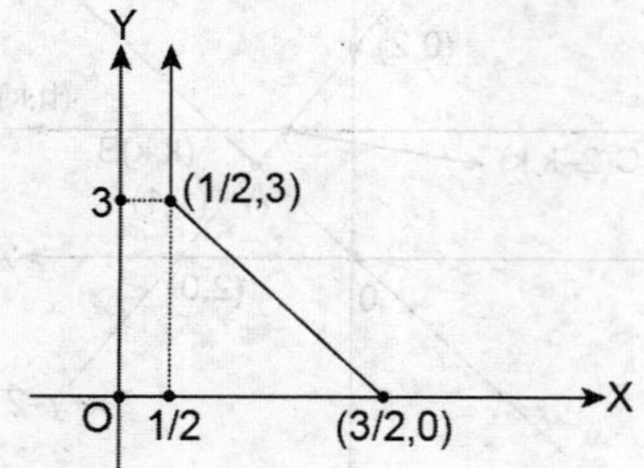

17. (b) Let $C'(h, k)$ be the circum-centre of ΔABC, then $C'A = C'B = C'C$

$\Rightarrow (h-1)^2 + (k-2)^2 = (h-2)^2 + (k-3)^2 = (h-4)^2 + (k-3)^2$

$\Rightarrow -2h - 4k + 5 = -4h - 6k + 13 = -8h - 6k + 25$

$\Rightarrow 2h + 2k = 8$ and $4h = 12 \Rightarrow h = 3, k = 1$

$\Rightarrow$ C' $(h, k) \equiv (3, 1)$ and (centroid) is

$G\left(\frac{1+2+4}{3}, \frac{2+3+3}{3}\right) \equiv \left(\frac{7}{3}, \frac{8}{3}\right)$

Now, by ONGC Rule, Orthocenter is given by equation

$\left(\frac{7}{3}, \frac{8}{3}\right) = \left(\frac{1.x_1 + 2(3)}{1+2}, \frac{1.y_1 + 2(1)}{1+2}\right)$

$\Rightarrow O \equiv (x_1 y_1) = (1,6)$ and hence distance between circum-centre and orthocenter = $= \sqrt{(3-1)^2 + (1-6)^2} = \sqrt{29}$

18. (d) Nine point centre is the mid-point of line segment joining orthocenter and circum-centre i.e..

$N = \left(\frac{1+3}{2}, \frac{6+1}{2}\right) \equiv \left(2, \frac{7}{2}\right)$

SECTION–VIII (MATRIX-MATCH TYPE)

1. (i) → D; (ii) → C; (iii) → A, (iv) → B

Given $A(p+1, 1)$, $B(2p+1, 3)$ and $C(2p+2, 2p)$

Area $\Delta ABC = \frac{1}{2}\begin{Vmatrix} p+1 & 1 & 1 \\ 2p+1 & 3 & 1 \\ 2p+2 & 2p & 1 \end{Vmatrix} = \frac{1}{2}\begin{Vmatrix} p+1 & 1 & 1 \\ p & 2 & 0 \\ p & (2p-3) & 0 \end{Vmatrix}$

$= \left|\frac{1}{2}(2p^2 - 3p - 2)\right|$

(i) When $p = 0$, $\Delta = 1$ $\Rightarrow$ (i) → D

(ii) When $p = \pm 1$, $\Delta = 3/2$

$\Rightarrow$ (ii) → C

(iii) When p = 3, $\Delta = 7/2$

$\Rightarrow$ (iii) → A

(iv) When $p = -3$, $\Delta = 25/2$

$\Rightarrow$ (iv) → B

2. (i) → D; (ii) → A,D; (iii) → B; (iv) → C

Given vertex are $A(2a, 4a)$; $B(2a, 6a)$

(i) When $C(4a, 5a)$, then $AC^2 = 5a^2$, $BC^2 = 5a^2$, $BC^2 = 5a^2$ and $AB^2 = 4a^2$.

$\Rightarrow$ ΔABC is isosceles. $\Rightarrow$ (i) → D

(ii) When $C((2+\sqrt{3})a, 5a)$, then $AC^2 = 4a^2$

$BC^2 = 4a^2$ and $AB^2 = 4a^2$

$\Rightarrow$ ΔABC is equilateral and isosceles also

$\Rightarrow$ (ii) → A, D

(iii) When $C(6a, 4a)$, then $AC^2 = 16a^2$, $BC = 20a^2$ and $AB^2 = 4a^2$

$\Rightarrow$ ΔABC is right angled

$\Rightarrow$ (iii) → B

(iv) When $C(a, 3a)$, then $AC^2 = 2a^2$, $BC^2 = 10a^2$ and $AB^2 = 4a^2$

Observe that $BC^2 > AC^2 + AB^2$

$\Rightarrow$ $\angle A$ is obtuse $\Rightarrow$ (iv) → (c)

3. (i) → (d); (ii) → (d); (iii) → (b); (iv) → (d); (v) → (a)

Given $L_1: x - y + 2 = 0$ and $L_2: x + y - 2 = 0$

(i) In R_1, $x < 0$

$\Rightarrow$ $\alpha^2 < 0$, which is impossible for any real values of α

$\Rightarrow$ (i) → (d)

(ii) R_2: $L_1(\alpha^2, \alpha) < 0$ and $L_2(\alpha^2, \alpha) > 0$

$\Rightarrow$ $\alpha^2 - \alpha + 2 < 0$, which is never true.

$\Rightarrow$ (ii) → (d)

(iii) R_3: $L_1(\alpha^2, \alpha) > 0$ and $L_2(\alpha^2, \alpha) > 0$, so $\alpha^2 - \alpha + 2 > 0$ which is always true and $\alpha^2 + \alpha - 2 > 0$ i.e., $(\alpha + 2)(\alpha - 1) > 0$ gives $\alpha \in (-\infty, -2) \cup (1, \infty)$, overall solution is $\alpha \in (-\infty, -2) \cup (1, \infty)$

$\Rightarrow$ (iii) → (b)

(iv) R_4: $L_1(\alpha^2, \alpha) < 0$ and $L_2(\alpha^2, \alpha) < 0$ gives $\alpha^2 - \alpha + 2 < 0$, which is never true

$\Rightarrow$ (iv) → (d)

(v) R_5: $L_1(\alpha^2, \alpha) > 0$ and $L_2(\alpha^2, \alpha) < 0$ and $y < 0$ i.e., $\alpha < 0$

So over all solution will be $\alpha \in (-2, 0)$

$\Rightarrow$ (v) → (a)

4. (i) → (a), (d); (ii) → (a); (iii) → (b) and (d); (iv) → (a), (b), (c), (d)

(i) Observe that a trapezium is formed which has a line of symmetry $x = 5/2$. so these points will also be concyclic.

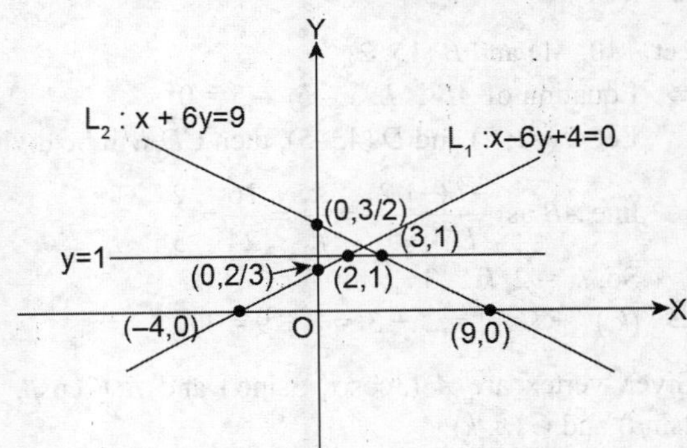

Centre of circle at $\left(\frac{5}{2}, \frac{-31}{2}\right)$ $\Rightarrow$ (i) → (a), (d)

(ii) $ac = bd$

Now, $OA.OB = ab$ and $OC.OD = c.d$

$\Rightarrow$ $OA.OB = OC.OD$ ($ac = bd$)

$\Rightarrow$ $ABCD$ is a cyclic quadrilateral

$\Rightarrow$ (ii) → (c)

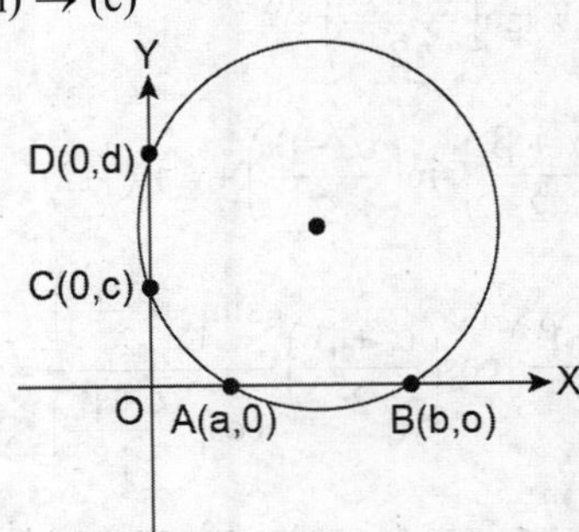

(iii) $ax \pm by \pm c = 0$ will form two sets of lines of || lines, so parallelogram is formed (as $a \neq b$). A ||gm is also a trapezium, since in this case the distance between the set of parallel lines is equal $\left(= \frac{|c|}{\sqrt{a^2 + b^2}}\right)$, so a rhombus is formed.

$\Rightarrow$ (iii) → (b), (d)

(iv) The line pair equation $x^2 - 8x + 12 = 0$, gives $(x - 6)(x - 2) = 0$

$\Rightarrow$ $x = 2, 6$

Similarly, $y^2 - 14y + 45 = 0$, gives $(y - 9)(y - 5) = 0$

$\Rightarrow$ $y = 5, 9$

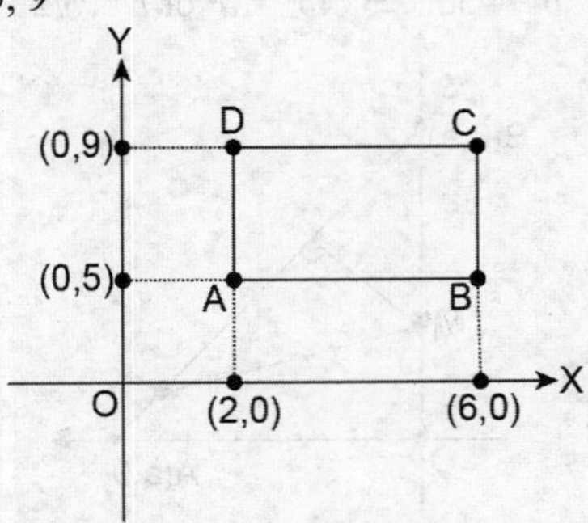

So a rectangle is formed which is also a cyclic quadrilateral as well as a trapezium.

Also $AB = AD = 4$ units

$\Rightarrow$ $ABCD$ is a square and hence also a rhombus

$\Rightarrow$ (iv) → (a), (b), (c), (d)

5. $A(0, 16)$, $B(8, 0)$, $y = -2(x - 8)$ gives $2x + y - 16 = 0$

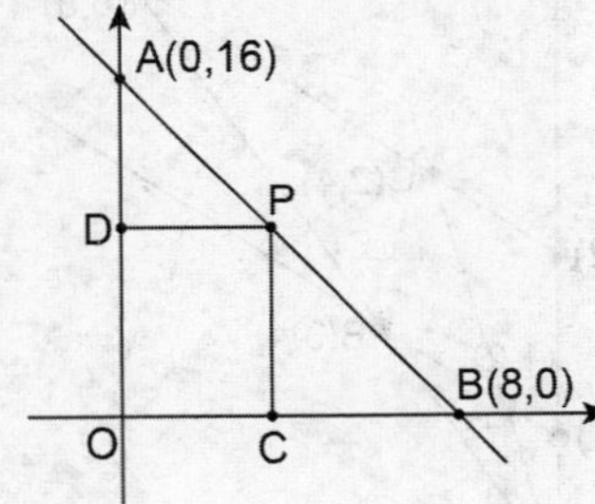

(i) Distance of (2, 2) from line AB is $\frac{10}{\sqrt{5}} = 2\sqrt{5} = \sqrt{20}$

$\Rightarrow$ (i) → (b)

(ii) $PDOC$ will be a square when $P(x, y)$ is at $y = x$ i.e., $P\left(\frac{16}{3}, \frac{16}{3}\right)$

So sum of the coordinates $= \frac{32}{3}$

$\Rightarrow$ (ii) → (c)

(iii) When area of $PODC = 30$ sq units, then for $P(x, y)$, we get $2x + y = 16$ and $xy = 30$.

Solving for x, we get $(x - 5)(x - 3) = 0$ for $x = 3$, $y = 10$ and for $x = 5$, $y = 6$

$\Rightarrow$ $P(3, 10)$ and $P(5, 6)$ are the two possible sets

$\Rightarrow$ (iii) → (a)

SECTION–VIII (INTEGER-TYPE)

1. $x + y = a$ will give $A(a, 0)$, $B(0, a)$, so ΔOAB area $= \frac{a^2}{2}$

Since $\angle N = 90°$

$\Rightarrow$ $MN \perp AB$, i.e., equation of MN is $x - y$ = constant = b (say)

Hence, we get M $(0, -b)$ and $N\left(\frac{a+b}{2}, \frac{a-b}{2}\right)$

Area ΔAMN = (3/8) $\Delta OAB = \frac{3a^2}{16} = \frac{1}{2}$

$\left|\frac{(a+b)}{\sqrt{2}}(AN)\right| = \left|\frac{1}{2}\frac{(a+b)(a-b)}{2}\right|$

So, $4|a^2 - b^2| = 3a^2 \Rightarrow 4b^2 = a^2$ or $b = a/2$

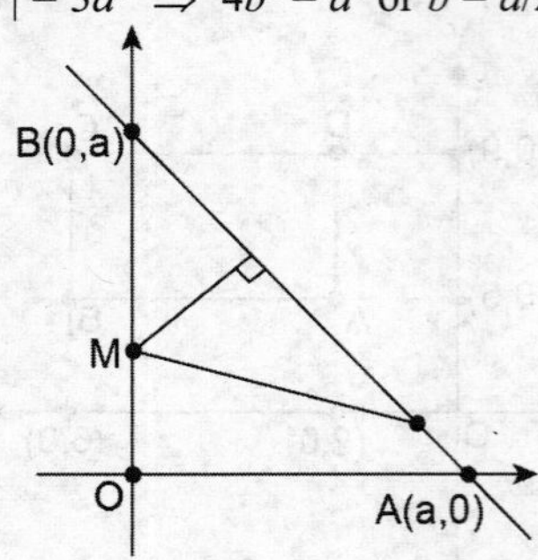

$AN = \frac{a}{2\sqrt{2}}$, hence $\frac{AN}{BN} = \frac{a}{2\sqrt{2}\left\{a\sqrt{2} - \frac{a}{2\sqrt{2}}\right\}} = \frac{1}{3}$ so $\frac{BN}{AN} = 3$

2. Given vertex are A (1, 2), B (3, 4)

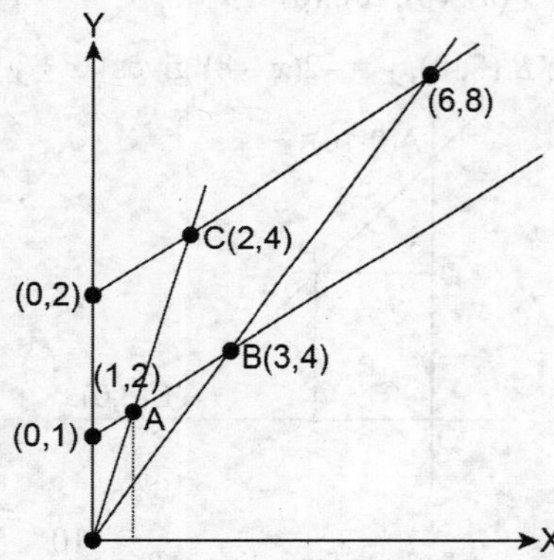

AB is mid-way between source S (0, 0) and CD, ΔSAB and ΔSCD are similar.

$\Rightarrow$ C (2, 4) and D (6, 8), hence required sum = 2 + 6 = 8.

3. From the given vertex $C(x, \sqrt{3})$, so $x^2 + (\sqrt{3})^2 = 12$

$\Rightarrow$ $x = \pm\sqrt{3}$

Hence, $C = (3, \sqrt{3})$

$\therefore$ $x_3 + y_3^2 = 3 + 3 = 6$

4. Let the centre of the circle be at C (h, k)

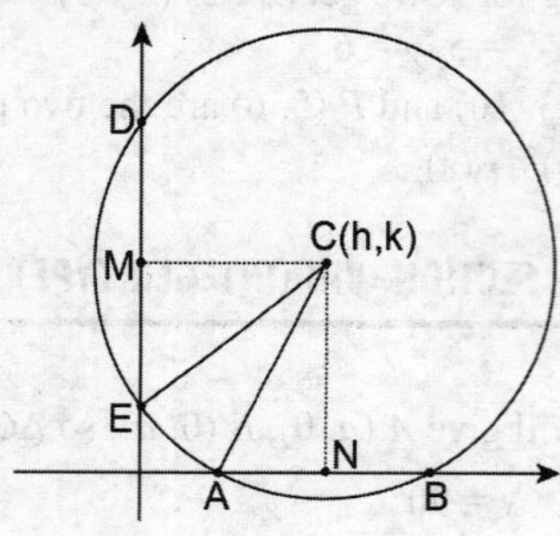

$\Rightarrow$ $CE^2 = CM^2 + ME^2$

$= h^2 + 3^2 = 9 + h^2$

Similarly, $AC^2 = CN^2 + AN^2 = k^2 + 16$.

Since CE = AC = Radius

$\therefore$ $h^2 + 9 = k^2 + 16$ $\Rightarrow h^2 - k^2 = 7$

$\therefore$ $4(x^2 - y^2) = 7 \times 4 = 28$

5. Given $xy = 16$ is a curve and let A, B, C be the three points on this curve as $A\left(t_1, \frac{16}{t_1}\right), B\left(t_2, \frac{16}{t_2}\right)$ and $C\left(t_3, \frac{16}{t_3}\right)$

Slope of $AB = -\frac{16}{t_1t_2}$

$\Rightarrow$ Equation of altitude through C (i.e., CF) is $y - \frac{16}{t_3} = \frac{t_1t_2}{16}(x - t_3)$

$\Rightarrow$ $16(t_3y - 16) = t_1t_2t_3(x - t_3)$

Similarly, the altitude through A (i.e., AD) is $16(t_1y - 16) = t_1t_2t_3(x - t_1)$

Point of intersection will be $y = -\frac{t_1t_2t_3}{16}$ and $x = \frac{-256}{t_1t_2t_3}$

So, orthocenter $(h, k) = \left(\frac{-256}{t_1t_2t_3}, \frac{-t_1t_2t_3}{16}\right)$

$\Rightarrow$ $hk = 16$

6. Let A (0, –1) and B (15, 2)

$\Rightarrow$ Equation of AB is $L: x - 5y - 5 = 0$

Let C (–1, 2) and D (4, –5), then CD will be divided by line AB as $\frac{-L(-1,2)}{L(4,-5)} = \frac{k_1}{k_2} = \frac{16}{24} = \frac{2}{3}$

So, $k_1 = 2, k_2 = 3$

$\Rightarrow$ $(k_1)^{k2} + (k_2)^{k1} = 2^3 + 3^2 = 8 + 9 = 17$

7. Given vertex are A (3cosα, 3sinα) and B (2cosβ, 2cosβ, 2sinβ) and C (h, k)

$\Rightarrow$ AC : CB = 3 : –2 (given)

$\Rightarrow$ $(h, k) = (-6\cos\alpha + 6\cos\beta, -6\sin\alpha + 6\sin\beta)$

$\Rightarrow$ $h = (-6)\left(-2\sin\frac{\alpha+\beta}{2}\sin\frac{\alpha-\beta}{2}\right)$ and

$k = (-12)\sin\frac{\alpha-\beta}{2}\cos\frac{\alpha+\beta}{2}$

Hence $h + k\tan\left(\frac{\alpha+\beta}{2}\right) + 4$

$= 12\sin\left(\frac{\alpha+\beta}{2}\right)\sin\left(\frac{\alpha-\beta}{2}\right) + 4 + (-12)$

$\sin\left(\frac{\alpha-\beta}{2}\right)\cos\left(\frac{\alpha+\beta}{2}\right) \cdot \frac{\sin\left(\frac{\alpha+\beta}{2}\right)}{\cos\left(\frac{\alpha+\beta}{2}\right)} = 4$

8. tanα, tanβ, tanγ are the roots of equation $t^3 - 12t^2 + 15t - 1 = 0$ (given).

So, tanα + tanβ + tanγ = 12 and tanα tanβ tanγ = 1. Also Σtanα tanβ = 15

When A (tanα, cotα), B (tanβ, cotβ) and C (tanγ, cotγ)
The centroid G (h, k)

$$=\left(\frac{\tan\alpha+\tan\beta+\tan\gamma}{3},\frac{\cot\alpha+\cot\beta+\cot\gamma}{3}\right)$$

$$=\left(\frac{12}{3},\frac{\Sigma\tan\alpha.\tan\beta}{3(\tan\alpha\tan\beta\tan\gamma)}\right)=(4,5)\Rightarrow\ \frac{k+h}{k-h}=\frac{5+4}{5-4}=9$$

9. Given that area of ΔDEF = 3 square units
$\Rightarrow$ Area of ΔABC = 12 square units

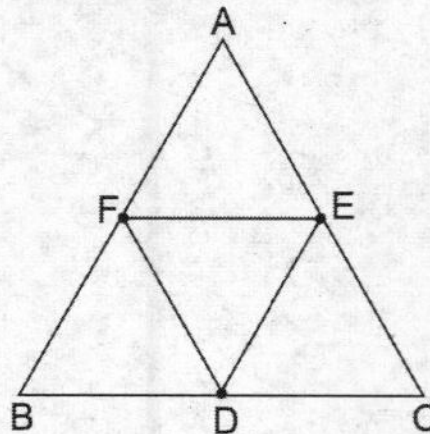

When three more figures. Similar to ΔABC are added then total Area = 4 × ΔABC = 48 square units
Area of new additions = 36 sq units.

10. Given $\alpha=\lim\limits_{n\to\infty}.\lim\limits_{k\to\infty}\cos^{2n}\left(2k!\frac{\pi}{3}\right)$

Observe that $2k!\ \frac{\pi}{3}$ = 2π m (i.e., integral multiple of 2π), where m is a positive integer

$\Rightarrow\ \cos\left(2k!\frac{\pi}{3}\right)=1$, so α = 1. Now, $\beta=\lim\limits_{n\to\infty}\ \lim\limits_{k\to\infty}$

$\cos^{2n}\left(\frac{\sqrt{3}}{2}k!\pi\right)$

Observe that $\frac{\sqrt{3}}{2}k!\pi=\sqrt{3}m\pi$, which is not an integral

multiple of 2π $\Rightarrow\ \cos\left(\frac{\sqrt{3}}{2}k!\,\pi\right)\neq\pm1$, so $\beta=0$

Now area of ΔABC where A (α, β), B (2, 1), C (–2, 1) is

$$\Delta ABC=\frac{1}{2}\left\|\begin{matrix}1&0&1\\2&1&1\\-2&1&1\end{matrix}\right\|=\frac{1}{2}\{1+3\}=2\text{ square units.}$$

11. $2x + 3y$ = 6a will meet the co-ordinate axes at A $(3a, 0)$ and B $(0, 2a)$.
Now, orthocenter is at (0, 0) and circum-centre is at $C\left(\frac{3a}{2},a\right)$

So, OC line has equation $y=\frac{2}{3}x=kx$. Hence $6k = 4$

12. Given, A (1, 2), B (2, 5), C (4, 8). Now f_1 $(x, y) \to (y, x)$.
So, A_1 (2, 1), B_1 (5, 3), C_1 (8, 4) and f_2 (x, y) $\to$ $\left(\frac{x-y}{3},\frac{x+y}{3}\right)$. So $A_2\left(\frac{1}{3},1\right),B_2\left(\frac{2}{3},\frac{8}{3}\right),C_2\left(\frac{4}{3},4\right)$.
And f_3 $(x, y) \to (x + 2y, 2y)$

So, $A_3\left(\frac{7}{3},2\right);B_3\left(6,\frac{16}{3}\right),C_3\left(\frac{28}{3},8\right)$

$$\text{Now, }\Delta A_3B_3C_3=\frac{1}{2}\left\|\begin{matrix}\frac{7}{3}&2&1\\6&\frac{16}{3}&1\\\frac{28}{3}&8&1\end{matrix}\right\|$$

$$=\frac{1}{2}\left\|\begin{matrix}\frac{7}{3}&2&1\\\frac{11}{3}&\frac{10}{3}&0\\\frac{10}{3}&\frac{8}{3}&0\end{matrix}\right\|=\frac{1}{2}\times\frac{12}{9}=\frac{2}{3}\text{ square units}$$

Hecne, $3DA_3B_3C_3$ = 2square units.

13. $2x^2 + 6xy + 5y^2 + 8x + 12y + 1 = t$
$\Rightarrow$ $x(2x + 3y + 4) + y(3x + 5y + 6) + 4x + 6y+1–t = 0$
So, we have equation as
$2x + 3y + 4 = 0$ ... (i)
$3x + 5y + 6 = 0$ ... (ii)
and $4x + 6y + 1 – t = 0$ ...(iii)

For existence of solution $\begin{vmatrix}2&3&4\\3&5&6\\4&6&1-t\end{vmatrix}=0$

$\Rightarrow$ t = –7 $\Rightarrow$ $t^4 = 2401$

14. $a + b + c = 3$; $a + b = 3 – c$; $[a + b] = [3 – c] = 1$
$\Rightarrow 1 < c \le 2$
Area $A=\frac{1}{2}\frac{1}{3-c}$
For A maximum $3 – c$ should be minimum $A_{max}=\frac{1}{2}$ $2A = 1$.

15. L_1 and L_2 are parallel lines.
Line L divides this parallelogram in two triangles of equal area. Altitudes of these triangle's is fixed

$=h=\frac{|15-5|}{5}=2$, base length of each triangle is = h tanθ + h cotθ

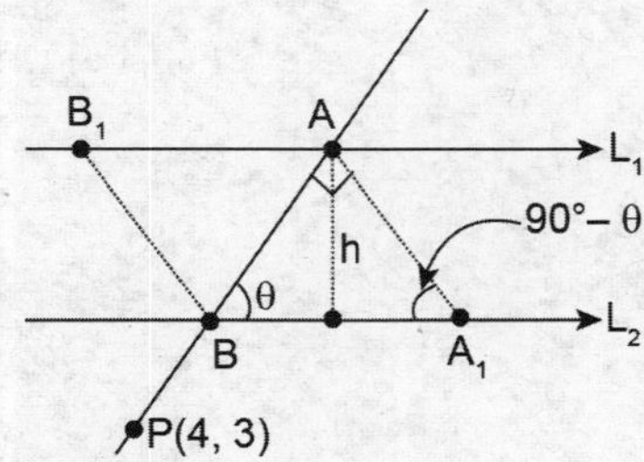

CHAPTER 2

Straight Lines

INTRODUCTION

In co-ordinate geometry, we study the geometry of points, straight lines, curves and surfaces using the representation of points by their co-ordinates which are ordered pairs of real numbers and applying their algebraic and other properties to the maximum extent possible. Since we have seen in the previous chapter that the point in cartesian plane have one to one correspondence with the ordered pair of real numbers. When a point moves in accordance with certain given conditions, its path is called its locus and the equation of locus is an equation of the form $f(x, y) = 0$ which is satisfied by the each and every point on the locus and by no other point.

Straight line can also be defined as a locus, over which if two arbitrary points taken, the line segment joining them completely lies on the curve i.e., each point on the line segment (chord) is common with the points of curve.

A straight line represents shortest route between two points in the space and therefore there can be only one straight line between any two given points and any linear equation of the type $ax + by + c = 0$ represents a straight line in two dimensional x-y plane.

When we state a line passing through two given points we are imposing two conditions on the line. The condition can also be imposed in the several other manners e.g., line passing through one point and having a given fixed slope, line cutting given intercept on both axes etc. These two conditions are sufficient to uniquely define a line.

In this chapter, we will also learn various methods to impose conditions and finding the equation of line under those conditions.

Having defined a line uniquely in space, we will learn to divide it in a given ratio, distance of a point from line, angle between lines, transformation of equation of line with the change of co-ordinate axes, combined equation of pair of straight lines and various other applications of above said studies.

DEFINITION OF A STRAIGHT LINE

A straight line is the locus of a point which moves such that the slope of the line segment joining its any two positions (during the journey) remains constant i.e., fixed, therefore it is a curve such that every point on the line segment joining any two points of it lies on the curve.

General Equation of a Straight Line

Every first degree equation in x, y i.e., $ax + by + c = 0$ *represents a straight line.* Thus a line is also defined as the locus of a point satisfying the condition $ax + by + c = 0$, where a, b, c are constants.

Proof: Let (x_1, y_1) and (x_2, y_2) be two arbitrary points lying on the curve $ax + by + c = 0$, therefore (x_1, y_1) and (x_2, y_2) will satisfy the equation $ax + by + c = 0$.

$$\Rightarrow \quad ax_1 + by_1 + c = 0 \qquad \text{......(i)}$$

$$\text{and } ax_2 + by_2 + c = 0 \qquad \text{...(ii)}$$

Subtracting (i) from (ii), we get $a(x_2 - x_1) + b(y_2 - y_1) = 0$

$$\Rightarrow \quad \frac{y_2 - y_1}{x_2 - x_1} = -\frac{a}{b} \text{ which is constant}$$

So the equation $ax + by + c = 0$ represents the most general form of a straight line.

Dividing the above equation by 'c'; we get:

$$\frac{a}{c}x + \frac{b}{c}y + 1 = 0 \qquad \text{...(iii)}$$

Substituting $l = a/c =$ and $m = b/c$; equation (iii) becomes $lx + my + 1 = 0$. Now this equation has two independent variables l and m.

ILLUSTRATION 1: Write the equation of straight line passing through (1, 0) and (3, 2).

SOLUTION: Let the equation of line be $lx + my + 1 = 0$ ....(i)

(1, 0) lies on the line

$\therefore$ (1, 0) will satisfy the equation (i),

$\Rightarrow l(1) + m(0) + 1 = 0$ $\qquad \Rightarrow l = -1$ and similarly for (3, 2)

$l(3) + m(2) + 1 = 0 \Rightarrow m = 1$

$\therefore$ equation of line is $-x + y + 1 = 0$

ILLUSTRATION 2: Write the equation of a straight lines passing through (1, 0).

SOLUTION: Let the equation of the line be $lx + my + 1 = 0$

$\Rightarrow l(1) + m(0) + 1 = 0 \Rightarrow l = -1$

$\therefore$ Equation becomes $-x + my + 1 = 0$.

This means that m can takes any real value

$\therefore$ $-x + my + 1 = 0$ represents a family of straight lines passing through (1, 0) except horizontal line $y = 0$, i.e., x-axis.

Equation of Straight Line Parallel to Axes

(i) Equation of a straight line which is parallel to x-axis and at a distance b units from it is given by $y = b$, $b >$ or < 0 according as it is above or below x-axis.

$\therefore$ equation of x-axis is $y = 0$.

(ii) Similarly, for any line parallel to y axis and at a distance a units from it, is given by $x = a$, $a >$ or < 0 according as the line lies on right or left side of y-axis. Equation of y-axis is $x = 0$.

NOTE

The combined equation of the co-ordinate axis is xy = 0

ILLUSTRATION 3: Find the equation of a line passing through (0, 2) and parallel to x-axis.

SOLUTION: line passes through the point (0,2) and is parallel to x-axis, its equation will be $y = b$, where $b = 2$ i.e., $y = 2$

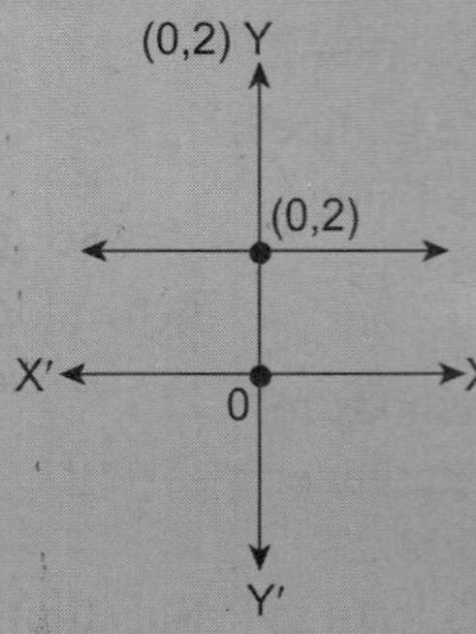

FIGURE 2.1

■ SLOPE OF A LINE

Slope of a line describes the inclination of the line with the positive direction of x-axis and it is defined as the trigonometrical tangent of the angle that a line makes with the positive direction of the x-axis in anti-clockwise sense.

Let $P(x_1, y_1)$ and $Q(x_2, y_2)$ be two points on a line making an angle θ with positive direction of x-axis as shown below.

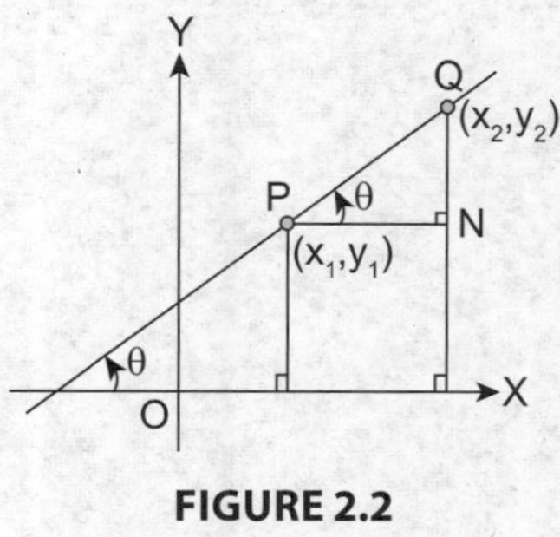

FIGURE 2.2

In ΔPQN, $\tan\theta = \dfrac{QN}{PN} = \dfrac{y_2 - y_1}{x_2 - x_1}$;

$$\Rightarrow \quad m = \tan\theta = \frac{y_2 - y_1}{x_2 - x_1}$$

Thus slope [gradient] of the line

PQ is $m = \tan\theta$

$$= \frac{y_2 - y_1}{x_2 - x_1} = \frac{\text{Difference of ordinates}}{\text{Difference of abscissae}}$$

REMARKS

1. *Slope of a line does not change if the direction of the line is reversed.*
2. *Slopes of parallel lines are same.*
3. *Slope of lines perpendicular to each other are negative reciprocal to each other.*

ILLUSTRATION 4: Find the slope of line such that

(i) line subtends an angle of 15° with +ve direction of x-axis, measured anti-clockwise.
(ii) line subtends an angle of 45° with –ve direction of x-axis, measured anti-clockwise.
(iii) line subtends an angle 15° with +ve y-axis, measured clockwise.
(iv) line subtends an angle of 30° with –ve y-axis measured anti-closkwise.
(v) line equally inclined with the axes.
(vi) line remains at a constant distance from y-axis.
(vii) line remains at a constant distance from x-axis.

SOLUTION: (i) Slope = m = $\tan\theta$; where θ is the angle subtended by straight line with the positive direction of x-axis.

$$\Rightarrow \; m = \tan 15^o = \tan(45^o - 30^o) = \frac{1 - 1/\sqrt{3}}{1 + 1/\sqrt{3}} = \frac{\sqrt{3} - 1}{\sqrt{3} + 1} = 2 - \sqrt{3}$$

(ii) Here the straight line subtends an angle of 45° with the –ve x-axis as shown in Figure 2.3.

$\Rightarrow \theta = \alpha = 45°$

$\therefore$ slope = m = tan 45° = 1

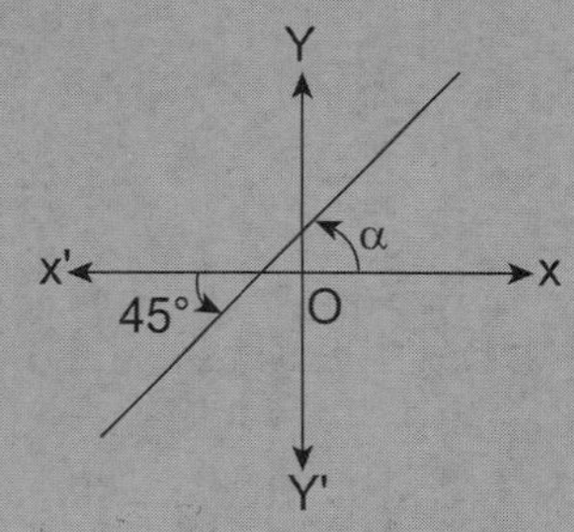

FIGURE 2.3

(iii) Here the straight line subtends an angle of 15° with +ve y–axis measured clockwise as shown in Figure 2.4.

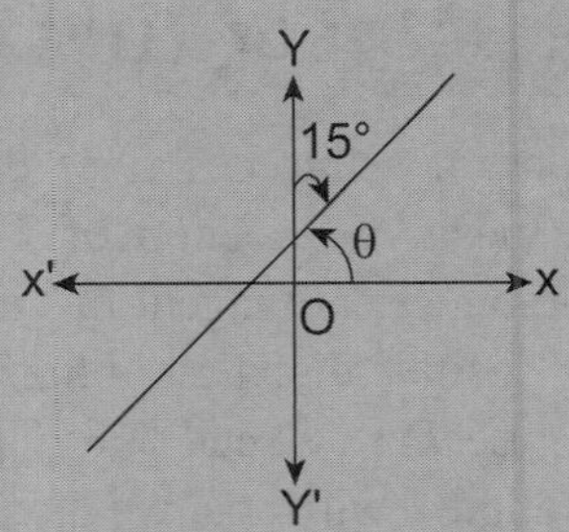

FIGURE 2.4

$\therefore \quad \theta = 90° - 15° = 75°$

$\Rightarrow$ slope (m) = $\tan\theta = \tan 75° = \tan(45° + 30°)$

$$= \frac{1+1/\sqrt{3}}{1-1/\sqrt{3}} = \frac{\sqrt{3}+1}{\sqrt{3}-1} = 2+\sqrt{3}$$

(iv) Line is subtending 30° with –ve y-axis measured anti-clockwise as shown in Figure 2.5.

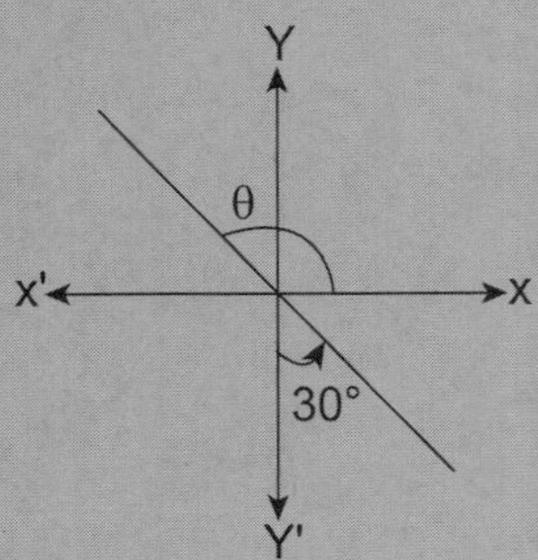

FIGURE 2.5

$\therefore$ Slope (m) = $\tan\theta = \tan(90° + 30°) = \tan 120° = -\tan 60° = -\sqrt{3}$

(v) Line is equally inclined with axes implies the line will be in the position as shown in Figure 2.6.

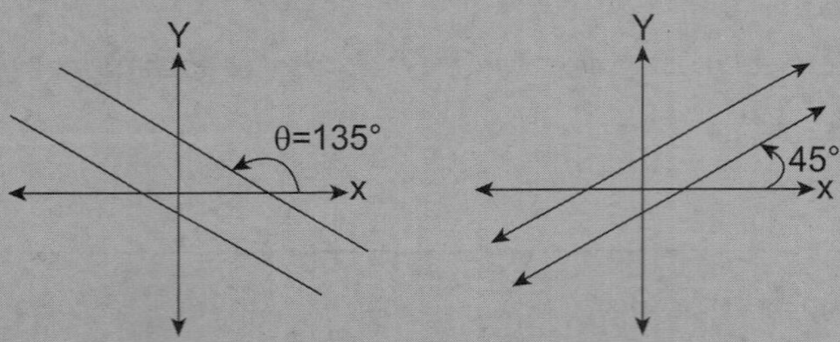

FIGURE 2.6

$\Rightarrow$ either $\theta = 135^o$ or $\theta = 45^o \Rightarrow$ slope (m) = $\tan 135^0$ or $\tan 45^0 = -1$ or 1

(vi) Line is at a constant distance from y-axis $\Rightarrow$ line is parallel to y-axis $\Rightarrow \theta = 90^o$

$\Rightarrow$ slope = $\tan\theta = \infty$ i.e., infinite

(vii) Line is at a constant distance from x-axis

$\Rightarrow$ Line is parallel to x-axis $\Rightarrow \theta = 0^o$ or $180^o \Rightarrow$ Slope = $\tan\theta = 0$

ILLUSTRATION 5: Find the slope of line parallel to line joining the points (2,4) and (5,7).

SOLUTION: As the slopes of parallel lines are same

$\therefore$ slope of required line $= \dfrac{y_2 - y_1}{x_2 - x_1} = \dfrac{7-4}{5-2} = 1$

ILLUSTRATION 6: Find the slope of line ⊥r to line joining the points (3,7) and (6,13)

SOLUTION : Slope of given line (m) $= \dfrac{y_2 - y_1}{x_2 - x_1} = \dfrac{13-7}{6-3} = \dfrac{6}{3} = 2 \therefore$ slope of required line = m′ = $\dfrac{-1}{m} = -1/2$

■ DIFFERENT FORMS OF THE EQUATION OF STRAIGHT LINE

Slope Intercept Form

The equation of a line with slope m making an intercept c on y-axis is $y = mx + c$.

Proof: Let (h, k) and $(0, c)$ be two points on the cartesian plane

$$\Rightarrow \frac{k-c}{h-0} = m \Rightarrow k = mh + c$$

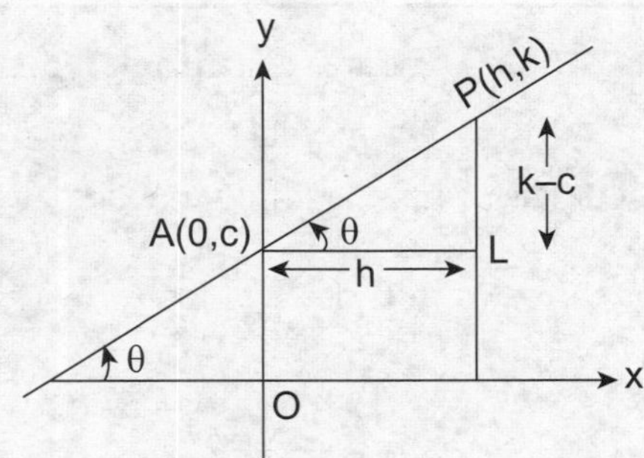

FIGURE 2.7

Replacing k by y and h by x, we get the equation of line as: $y = mx + c$

REMARKS

(i) *Equation of any line parallel to y-axis cannot be expressed in this form.*

(ii) *$c < 0$ if intercept is on negative y-axis otherwise; $c > 0$ if intercept is on positive y-axis.*

(iii) *If the line passes through the origin, then $c = 0$. Thus the equation of line with slope m and passing through the origin is $y = mx$.*

ILLUSTRATION 7: Find the equation of a straight line passing through the point $(0, -4)$ and subtending an angle 30° with the +ve y-axis measured clockwise.

SOLUTION: As shown in Figure 2.8, the line is subtending an angle of 30° measured clockwise with the +ve y-axis, that means line is subtending an angle 60° with +ve x-axis anticlock wise.

$\Rightarrow \theta = 60°$

$\Rightarrow$ slope of required line $m = \tan \theta = \tan 60° = \sqrt{3}$

Thus the equation of line by slope intercept form is given by $y = \sqrt{3}\ x - 4$; as $c = OB = -4$

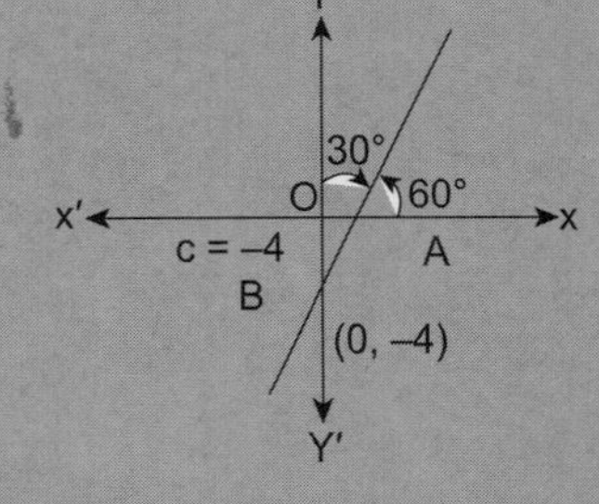

FIGURE 2.8

ILLUSTRATION 8: (a) Find the equation of the straight line which makes an intercept of 2 units on positive y-axis and having inclination of 45° with positive x-axis.

(b) Find the equation of line which cuts off an intercept *of* –2 units on y-axis having slope = 1/2.

(c) Find the equations of straight line cutting off an intercept *of* – 5 on the y-axis and is equally inclined to the axes.

SOLUTION: (a) Since the line is inclined at 45° with the x-axis $\therefore m = \tan 45° = 1$

The length of intercept on y-axis $= 2 \Rightarrow c = 2$

$\therefore$ The equation of the line is given by $y = x + 2$

(b) $m = 1/2$ and $c = -2$ (given)

$\therefore$ The equation of line is given by $y = x/2 - 2 \Rightarrow 2y = x - 4$

(c) Since a line is equally inclined to the axes therefore it will make either 45° or 135° with positive direction of x-axis so its slope may be 1 or –1.

Thus, the equation of line can be given as $y = x - 5$ or $y = -x - 5$

ILLUSTRATION 9: The area of a triangle is 5, two of its vertices are (2, 1) and (3, –2). The third vertex lies on $y = x + 3$. Find the third vertex.

SOLUTION: Let the third vertex be given by $(h, h + 3)$ as shown in in Figure 2.9.

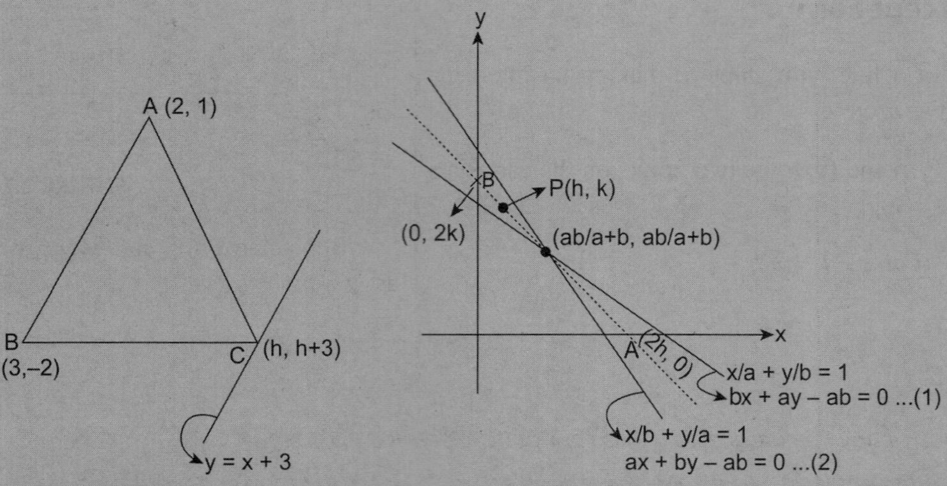

FIGURE 2.9 **FIGURE 2.10**

Now the area of triangle $= \frac{1}{2}\begin{vmatrix} 2 & 1 & 1 \\ 3 & -2 & 1 \\ h & h+3 & 1 \end{vmatrix} = \pm 5$

$\Rightarrow 2(-2 - h - 3) - 1\,(3 - h) + 1\,(3h + 9 + 2h) = \pm 10$

$\Rightarrow 4h - 4 = \pm 10$

For $4h - 4 = 10 \Rightarrow h = 7/2$

For $4h - 4 = -10 \Rightarrow h = -3/2$

Thus $C\left(\frac{7}{2}, \frac{7}{2} + 3\right)$ or $C\left(-\frac{3}{2}, -\frac{3}{2} + 3\right)$

i.e., either $\left(\frac{7}{2}, \frac{13}{2}\right)$ or $\left(-\frac{3}{2}, \frac{3}{2}\right)$ are the possible co-ordinates of third vertex C.

ILLUSTRATION 10: Find the equation of line inclined at an angle α with y-axis and cuts an intercept of length 'a' unit on x-axis.

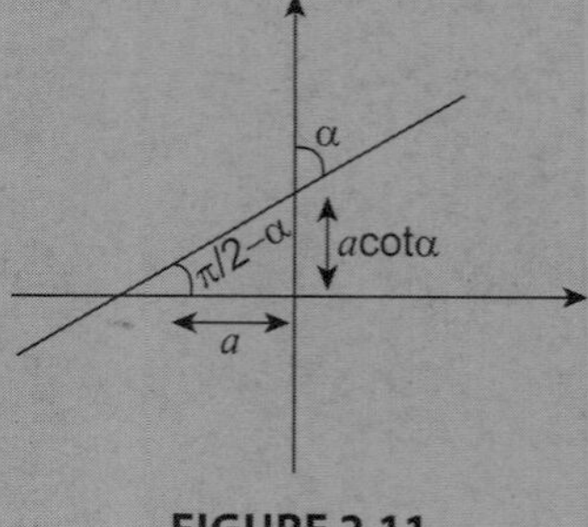

FIGURE 2.11

SOLUTION: Inclination of line with x-axis is $\pi/2 - \alpha$

Therefore the slope of line is $\tan(\pi/2 - \alpha) = \cot \alpha$

and length of y-intercept $= a \cot \alpha$

So the equation of line will be given by $y = (x + a) \cot \alpha$.

ILLUSTRATION 11: Find the equation of the straight line cutting off an intercept of 4 units on positive direction of y-axis and inclined at an angle $\sin^{-1}\left(\frac{3}{5}\right)$ to positive y-axis measured clockwise.

SOLUTION: Here, $\theta = \sin^{-1}\left(\dfrac{3}{5}\right)$

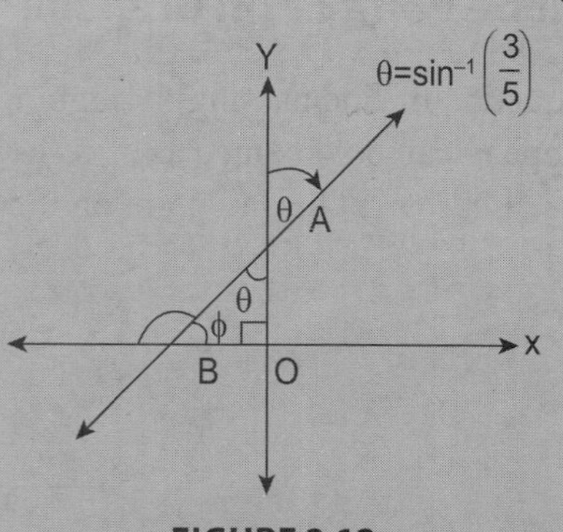

FIGURE 2.12

$\angle BAO = \theta$ (vertically opposite angles)

$\angle BAO + \angle ABO = 90°$ (due to sum of angles of triangle)

$\Rightarrow \angle ABO = 90° - \angle BAO$

$\Rightarrow \tan(\angle ABO) = \tan(90° - \angle BAO)$

$= \cot(\angle BAO) = \cot\left(\sin^{-1}\dfrac{3}{5}\right) = \dfrac{4}{3}$

$\Rightarrow m = \dfrac{4}{3}$; Also given $c = 4$

$\Rightarrow y = \dfrac{4}{3}x + 4$ is the required equation of line.

Reduction of General Form into Slope-Intercept Form

Let $Ax + By + C = 0$ be general equation of straight line.

Now $Ax + By + C = 0 \Rightarrow y = \left(\dfrac{-A}{B}\right)x + \left(\dfrac{-C}{B}\right)$

which is of the form of $y = mx + c$,

where, slope $= m = \dfrac{-A}{B} = \dfrac{\text{-coefficient of } x}{\text{coefficient of } y}$

and intercept on y-axis $= \dfrac{-C}{B} = -\dfrac{\text{constant term}}{\text{coefficient of } y} = c$

ILLUSTRATION 12: Convert the equation $8x - 15y + 51 = 0$ in the slope intercept form and hence find its slope and its intercept on the y-axis.

SOLUTION: Equation of given line is $8x - 15y + 51 = 0$

$\Rightarrow 15y = 8x + 51 \Rightarrow y = \left(\dfrac{8}{15}\right)x + \left(\dfrac{51}{15}\right)$ which is in slope intercept form $y = mx + c$.

$\Rightarrow m = 8/15$ (slope) and $c = 51/15$

ILLUSTRATION 13: Find the slopes of line

(i) parallel to line $2x - 3y + 7 = 0$

(ii) perpendicular to line $3x - 4y + 10 = 0$

SOLUTION: (i) slope of line $2x - 3y + 7 = 0$ is given by (m) $= \dfrac{-\text{coefficient of x}}{\text{coefficient of y}} = -\left(\dfrac{2}{-3}\right) = \dfrac{2}{3}$

As slopes of parallel lines are same

$\Rightarrow$ slope of required line = 2/3

(ii) slope of line $3x - 4y + 10 = 0$ is (m) $= -\left(\dfrac{3}{-4}\right) = \dfrac{3}{4}$

As slopes of lines perpendicular to each other is –ve reciprocal of each other, so the slope of required line will be $m' = \dfrac{-1}{m} = -4/3$

Slope Point Form of a Line

Equation of line passing through point $A(x_1, y_1)$ and having slope m can be obtained as as follows.

Let $P(x, y)$ be any point on the line as shown in Figure 2.13.

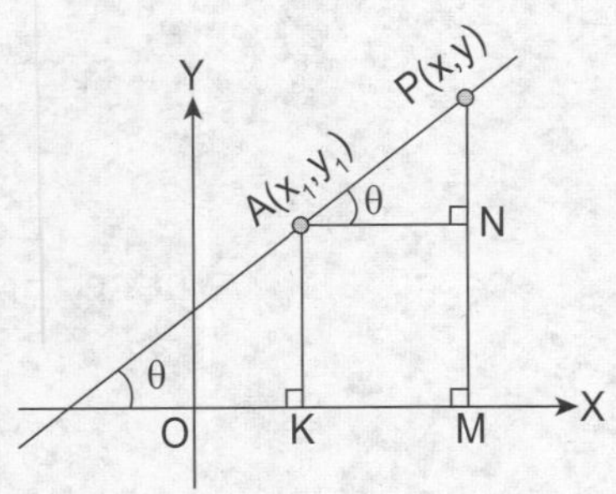

FIGURE 2.13

Then, slope of the line is:

$$m = \tan\theta = \frac{PN}{AN} = \frac{PM - NM}{AN}$$

$$= \frac{PM - AK}{OM - OK} = \frac{y - y_1}{x - x_1}$$

$$\therefore \quad y - y_1 = m(x - x_1)$$

Thus the equation of line is given by $y - y_1 = m(x - x_1)$

ILLUSTRATION 14: Find the equation of a line passing through (2, –3) and inclined at an angle of 135° with the + *ve* direction of *x*-axis.

SOLUTION: m = slope of the line = $\tan 135° = -1$, $x_1 = 2$, $y_1 = -3$

So the equation of the line is $(y - y_1) = m(x - x_1)$

or $y - (-3) = -1(x - 2) \Rightarrow y + 3 = -x + 2 \Rightarrow x + y + 1 = 0$.

ILLUSTRATION 15: Determine the equation of line through the point (–4, –3) and parallel to *x*-axis.

SOLUTION: $m = 0$, $x_1 = -4$, $y_1 = -3$

So the equation of the line is $y - y_1 = m(x - x_1) \Rightarrow y + 3 = 0(x + 4) \Rightarrow y = -3$

ILLUSTRATION 16: If the co-ordinates of the vertex A of a ΔABC are (1, 2) and the equation of the perpendicular bisectors of AB and AC are $3x + 4y - 1 = 0$ and $4x + 3y - 5 = 0$, then find the area of ΔABC.

SOLUTION: Equation of line AB will be $(y - 2) = (4/3)(x - 1)$

or $4x - 3y + 2 = 0$ ..... (i)

$\therefore$ Mid-point of AB is given by the point of intersection of line (i) and $3x + 4y - 1 = 0$ is L $\left(\frac{-1}{5}, \frac{2}{5}\right)$

Similarly, mid-point of AC will be $M\left(\frac{1}{5}, \frac{7}{5}\right)$

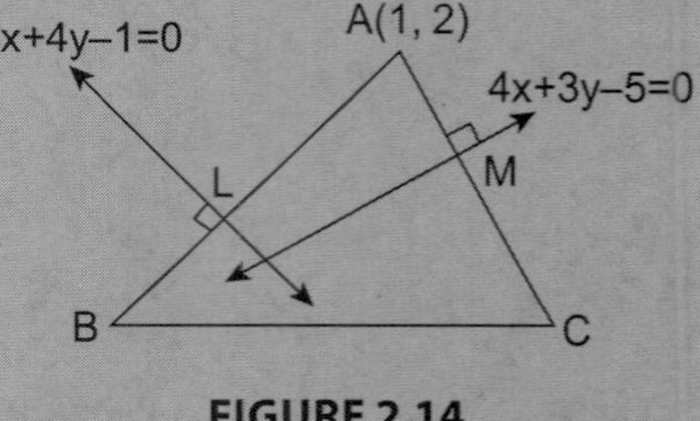

FIGURE 2.14

$\therefore$ co-ordinate of B and C will be $B\left(\frac{-7}{5}, -\frac{6}{5}\right)$ and $C\left(\frac{-3}{5}, \frac{4}{5}\right)$

$$\therefore \text{ Area of } \Delta ABC = \frac{1}{2}\begin{vmatrix} 1 & 2 \\ -\frac{7}{5} & -\frac{6}{5} \\ -\frac{3}{5} & \frac{4}{5} \\ 1 & 2 \end{vmatrix} = \frac{1}{2}\left|\frac{8}{5} - \frac{46}{25} - \frac{10}{5}\right| = \frac{1}{2}\left|\frac{-46}{25} - \frac{2}{5}\right| = \frac{1}{2}\left(\frac{56}{25}\right) = \frac{28}{25} \text{ sq. units.}$$

ILLUSTRATION 17: If the circumcentre of a triangle lies at the origin and the centroid is the middle point of the line joining the points $(a^2 + 1, a^2 + 1)$ and $(2a, -2a)$; find the equation of line on which the orthocentre lies.

SOLUTION: We know that orthocentre (O), centroid (G) and circumcentre (C) of a Δ are collinear

$\therefore$ orthocentre lies on the line joining the circumcentre (0, 0) and centroid which is the middle point of line joining the points $(a^2 + 1, a^2 + 1)$ and $(2a - 2a)$; i.e.,

$$G\left(\frac{a^2+1+2a}{2}, \frac{a^2+1-2a}{2}\right) = \left(\frac{(a+1)^2}{2}, \frac{(a-1)^2}{2}\right)$$

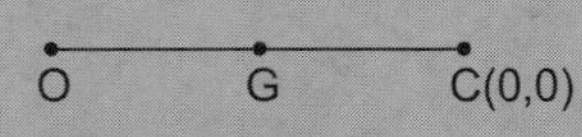

FIGURE 2.15

$\therefore$ Orthocentre lies on the line

$$(y-0) = \frac{\frac{(a-1)^2}{2}-0}{\frac{(a+1)^2}{2}-0}(x-0) \text{ or } (a+1)^2 y = (a-1)^2 x \text{ or } (a-1)^2 x - (a+1)^2 y = 0.$$

Two Point Form Equation of a Line

To find the equation of line passing through two points $A(x_1, y_1)$ and $B(x_2, y_2)$ as shown in Figure 2.16.

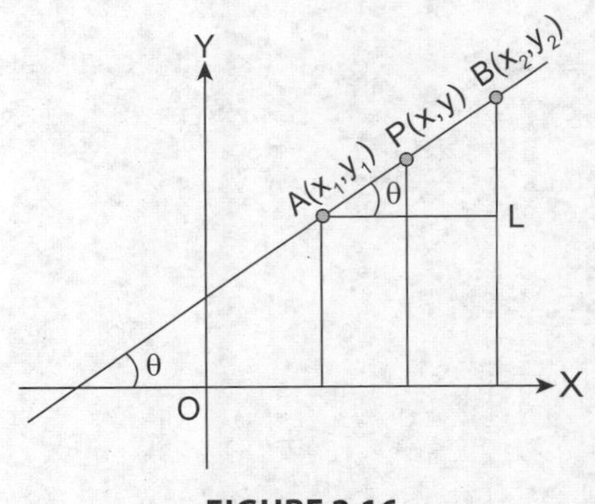

FIGURE 2.16

Let m be the slope of such line.

Then $m = \tan\theta = \frac{BL}{AL} = \frac{y_2 - y_1}{x_2 - x_1}$, so the equation of the line is $y - y_1 = m(x - x_1)$ ....(i)

Putting the value of m in equation (i)

we get, $(y - y_1) = \frac{y_2 - y_1}{x_2 - x_1}(x - x_1)$

This is the required equation of the line.

ILLUSTRATION 18: Find the equation of the line joining the following pairs of points.

(a) (–1, 3) and (4, –2) (b) (2, 5) and (1, 2)

SOLUTION: (a) The two points are (–1, 3) and (4, –2), so the equation of the line in two-point form is

$$y - 3 = \frac{-2-3}{4+1}(x+1) \Rightarrow y - 3 = -x - 1 \Rightarrow x + y - 2 = 0$$

(b) The equation of the line passing through points $P(2, 5)$ and $Q(1, 2)$ is given by

$$\Rightarrow (y - y_1) = \frac{y_2 - y_1}{x_2 - x_1}(x - x_1) \Rightarrow (y - 5) = \frac{2-5}{1-2}(x-2)$$

$$\Rightarrow (y-5) = \frac{-3}{-1}(x-2) \Rightarrow (y-5) = 3(x-2) \Rightarrow y - 5 - 3x + 6 = 0 \Rightarrow y - 3x + 1 = 0$$

ILLUSTRATION 19: Find the equation of the line which passes through (0, 1) and bisects the line joining the points (–2, 2) and (3, 4).

SOLUTION: Mid-points of segment joining (–2, 2) and (3, 4) is (1/2, 3).

Therefore we require a line passing through (0, 1) and(1/2, 3)

which is given by $y-1=\frac{3-1}{1/2}(x-0) \Rightarrow y-4x=1$

ILLUSTRATION 20: If the real numbers a_1, a_2, a_3 as well as b_1, b_2, b_3 are in geometric progression with same common ratio, then prove that the points (a_1, b_1), (a_2, b_2) and (a_3, b_3) lie on a straight line.

SOLUTION: Given $\frac{a_2}{a_1}=\frac{a_3}{a_2}=\lambda$ and $\frac{b_2}{b_1}=\frac{b_3}{b_2}=\lambda \Rightarrow a_2=\lambda a_1, a_3=\lambda a_2$ and $b_2=b_1\lambda; b_3=b_2\lambda$

$\Rightarrow$ It will be sufficient to prove that the area subtended by three points = 0

Thus, $\Delta=\frac{1}{2}\begin{vmatrix} a_1 & b_1 & 1 \\ a_2 & b_2 & 1 \\ a_3 & b_3 & 1 \end{vmatrix} \Rightarrow \Delta=\frac{1}{2}\begin{vmatrix} a_1 & b_1 & 1 \\ \lambda a_1 & b_1\lambda & 1 \\ \lambda^2 a_1 & b_1\lambda^2 & 1 \end{vmatrix}$

$=\frac{1}{2}a_1 b_1\begin{vmatrix} 1 & 1 & 1 \\ \lambda & \lambda & 1 \\ \lambda^2 & \lambda^2 & 1 \end{vmatrix}=0$ (C_1 and C_2 are identical)

$\therefore$ They lie on a straight line.

ILLUSTRATION 21: Show that the points (1, 4), (3, –2) and (–3, 16) are collinear. Also find the equations of this straight line.

SOLUTION: The equation of line passing through points (1, 4) and (3, –2) is given by

$(y-4)=\left(\frac{-2-4}{3-1}\right)(x-1) \Rightarrow y-4=-3(x-1)$

$\Rightarrow y+3x-7=0$ ....(i)

Then, put (– 3, 16) in this equation i.e., $3(-3)+16-7=0 \Rightarrow 0=0$

$\Rightarrow C(-3, 16)$ satisfies the equation (i)

$\therefore$ All three points lie on a straight line $y+3x-7=0$

ILLUSTRATION 22: Find the equation of the straight line which passes through the points (3, 2) and (4, 5), then find the co-ordinates of the points on the line that are 5 units away from the point (3, 2).

SOLUTION: Equation of required line is given by $(y-2)=\frac{5-2}{4-3}(x-3) \Rightarrow (y-2)=3(x-3)$

$\Rightarrow 3x-y-7=0$ ......(i)

Let the required points be represented by (x, y).

Now, as given distance = 5 unit from the point (3, 2)

$\Rightarrow (x-3)^2+(y-2)^2=25$, and as the points (x, y) lies on line (i)

$\Rightarrow \left((y-2)\frac{1}{3}\right)^2+(y-2)^2=25 \Rightarrow \frac{10}{9}(y-2)^2=25$

$\Rightarrow (y-2)^2=\frac{25\times 9}{10}=\frac{45}{2} \Rightarrow y-2=\pm 3\sqrt{\frac{5}{2}} \Rightarrow y=2\pm 3\sqrt{\frac{5}{2}}$

$$\Rightarrow\ x = \frac{y+7}{3} = \frac{9 \pm 3\sqrt{\frac{5}{2}}}{3} \Rightarrow x = 3 \pm \sqrt{\frac{5}{2}}$$

$\therefore$ The required points are $\left(3+\sqrt{\frac{5}{2}}, 2+3\sqrt{\frac{5}{2}}\right)$ and $\left(3-\sqrt{\frac{5}{2}}, 2-3\sqrt{\frac{5}{2}}\right)$.

ILLUSTRATION 23: A square of side 'a' above the x-axis has one vertex at the origin. The side passing through the origin makes an angle $\alpha(0 < \alpha < \pi/4)$ with the positive direction of x-axis. Find the equation of its diagonal not passing through the origin.

SOLUTION: We are to find the equation of diagonal AC of square $OABC$ as shown in Figure 2.17.

Now, co-ordinates of A are $(a \cos \alpha, a \sin \alpha)$ and that of C are

$$\left[a\cos\left(\frac{\pi}{2}+\alpha\right), a\sin\left(\frac{\pi}{2}+\alpha\right)\right] \equiv [-a\sin\alpha, a\cos\alpha]$$

$\therefore$ By two point form equation of AC will be

$$[y - a\sin\alpha] = \frac{a\cos\alpha - a\sin\alpha}{-a\sin\alpha - a\cos\alpha}(x - a\cos\alpha)$$

$\Rightarrow y \sin \alpha - a \sin^2\alpha + y\cos \alpha - a \sin \alpha \cos \alpha.$

$= x \sin \alpha - a \sin\alpha \cos \alpha - x \cos \alpha + a \cos^2 \alpha$

$\Rightarrow x \sin\alpha - x \cos \alpha - y \sin \alpha - y\cos \alpha + a = 0$

or $y(\sin\alpha + \cos \alpha) + (\cos\alpha - \sin \alpha)x = a$

or $y(\cos \alpha + \sin \alpha) - x(\sin\alpha - \cos \alpha) = a.$

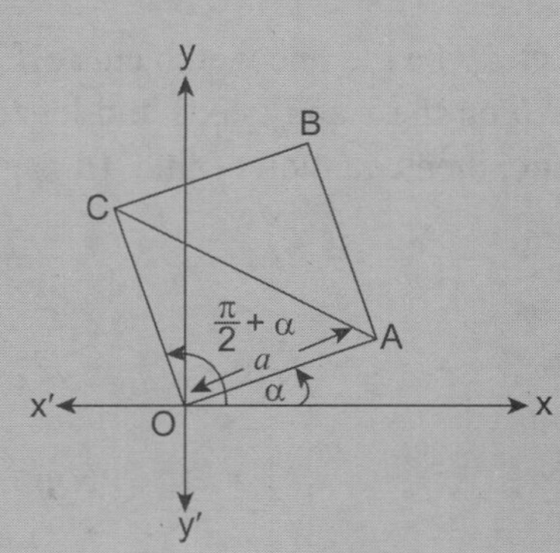

FIGURE 2.17

ILLUSTRATION 24: Two consecutive side of parallelogram are $4x + 5y = 0$ and $7x + 2y = 0$. If the equation to one diagonal is $11x + 7y = 9$, find the equation to the other diagonal.

SOLUTION: $4x + 5y = 0$ ...(1)

$11x + 7y = 9$ ...(2)

$7x + 2y = 0$ ...(3)

for $D \Rightarrow$

using (1) and (2)

$44x + 55y = 0$

$44x + 28y = 36$

$\underline{-\quad -\quad -}$

$27y = -36 \Rightarrow y = -\frac{4}{3}$

put in (1), we get $4x = \frac{20}{3} \Rightarrow x = \frac{5}{3}$

for $B \Rightarrow$ using (2) and (3)

$22x + 14y = 18$

$49x + 14y = 0$

$\underline{-\quad -\quad -}$

$-27x = 18 \Rightarrow x = -\frac{2}{3}$

put in (2)

$$\Rightarrow 7y = 9 + \frac{22}{3} \Rightarrow y = \frac{27+22}{21} = \frac{49}{21} \Rightarrow y = \frac{7}{3}$$

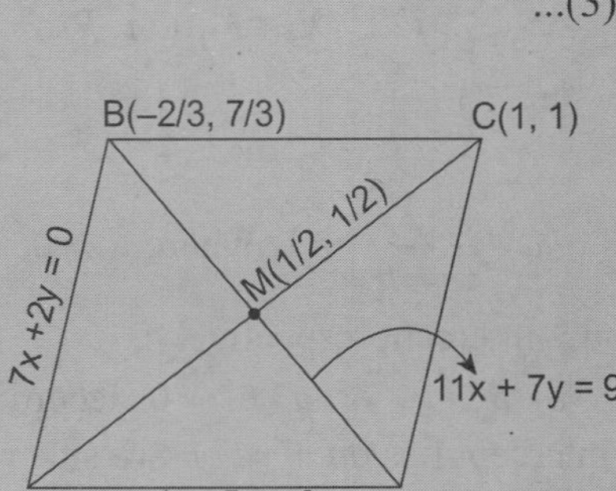

FIGURE 2.18

Using co-ordinates of B and D; we can find the co-ordinates of M (mid-point of BD)

$$\Rightarrow \ M = \left(\frac{1}{2}, \frac{1}{2}\right)$$

$\therefore$ Co-ordinates of $C = (1, 1)$

using two-point from equation of AC is $(y-0) = \left(\frac{1-0}{1-0}\right)(x-0) \Rightarrow y = x$

Intercept Form of a Line and Concept of Line at Infinity

Let AB be the line which cuts off intercepts $OA = a$ and $OB = b$ on the x and y-axis and let $P(x, y)$ be any point on the line; then equation of line AB is given by $x/a + y/b = 1$.

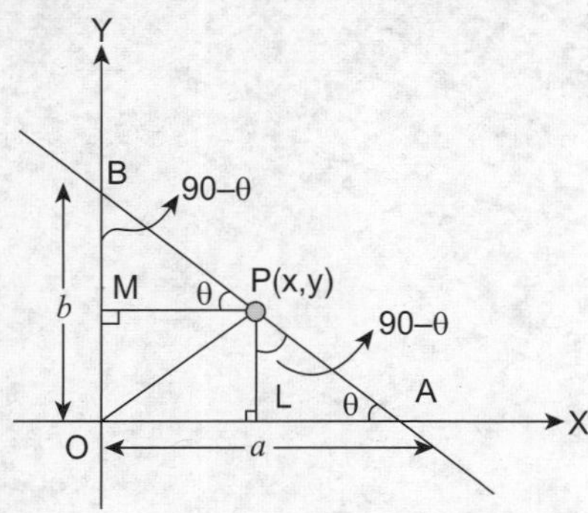

FIGURE 2.19

Proof I: Area of ΔOAB = Area of ΔOPA + Area of ΔOPB

$$\Rightarrow \ \frac{1}{2}\ OA.\ OB = \frac{1}{2}\ OA.\ PL + \frac{1}{2}\ OB.PM$$

$$\Rightarrow \ \frac{1}{2}\ ab = \frac{1}{2}\ ay + \frac{1}{2}\ bx \Rightarrow \ ab = ay + bx$$

$\therefore \ \frac{x}{a} + \frac{y}{b} = 1$, where $a \neq 0$, $b \neq 0$ are x – intercept and y-intercept respectively.

a, $b > 0$ or a, $b < 0$ accordingly as x-intercept or y-intercept lies *on + ve or –ve* side respectively.

Proof II: In similar triangles BMP and PLA,

$$\frac{BM}{PL} = \frac{MP}{LA} \Rightarrow \frac{b-y}{y} = \frac{x}{a-x}$$

$(\ \ \angle PMB = \angle PLA = 90^0$ and $\angle PAL = \angle MPB = \theta)$

$$\Rightarrow \ ab - ay - bx + xy = xy \Rightarrow bx + ay = ab \Rightarrow \frac{x}{a} + \frac{y}{b} = 1$$

Proof III: Since $OA = a$ and $OB = b$, therefore $A\ (a, 0)$ and $B(0, b)$. Since points $P(x, y)$, $A(a, 0)$ and $B(0, b)$ are collinear therefore $\begin{vmatrix} x & y & 1 \\ a & 0 & 1 \\ 0 & b & 1 \end{vmatrix} = 0$

or $x(0 - b) - y\ (a - 0) + 1\ (ab - 0) = 0$ or $bx + ay = ab$

$$\Rightarrow \ \frac{x}{a} + \frac{y}{b} = 1$$

Line at infinity

Let us consider a line $ax + by + c = 0$. Converting this line to intercept form, we get $\frac{ax}{-c} + \frac{by}{-c} = 1$

$$\Rightarrow \ \frac{x}{-c/a} + \frac{y}{-c/b} = 1$$

So x-intercept of line $= -\frac{c}{a}$ units

and y-intercept of line $= -\frac{c}{b}$ units

Case I: When $a \to 0$; magnitude of x intercept goes to infinity so line tends to become parallel to x-axis. $(y = -c/b)$

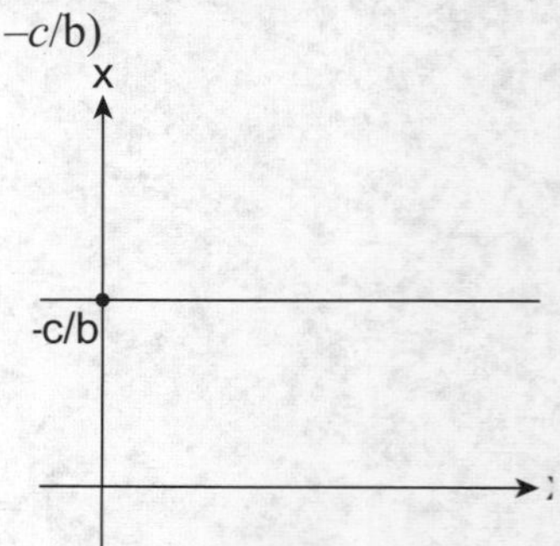

FIGURE 2.20

Case II: When $b \to 0$; magnitude of y intercept goes to infinity so line tends to become parallel to y-axis. $(x = -c/a)$

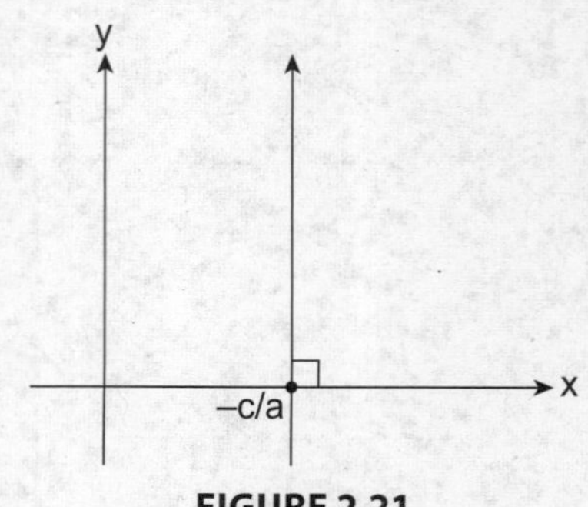

FIGURE 2.21

ILLUSTRATION 25: Find the equation of line

(i) having x-intercept = 4 and y-intercept = –3.

(ii) passing through the points (–4,0) and (0,5).

(iii) making equal intercepts from positive axes and passing through the point (4,5).

(iv) making equal intercepts from axes but opposite in sign and passing through the point where the line $2x - 3y = -2$ meets the line $y - x = 0$.

SOLUTION: (i) Here x-intercept = a = 4;

y-intercept = b = –3;

∴ By intercept form, equation of straight line will be $\frac{x}{a}+\frac{y}{b}=1$

i.e., $\frac{x}{4}+\frac{y}{-3}=1$

or $3x - 4y = 12$

(ii) As the line is passing through the points $A(-4,0)$ and B (0,5) as shown in Figure 2.22.

Thus $a = -4$; $b = 5$

∴ equation of straight line is given by

Intercept form $\frac{x}{a}+\frac{y}{b}=1$

i.e., $\frac{x}{-4}+\frac{y}{5}=1$

or $5x - 4y + 20 = 0$

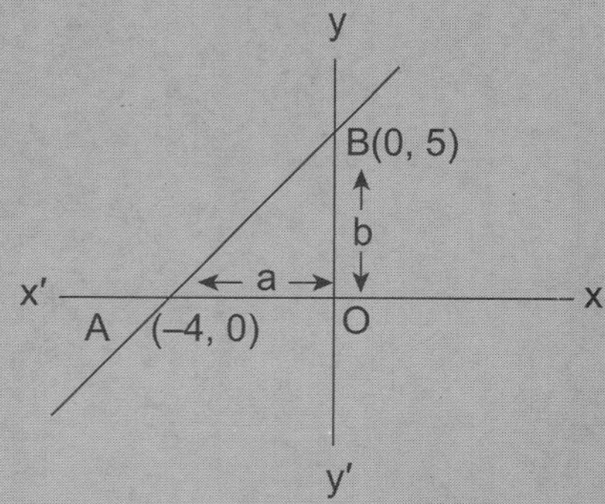

FIGURE 2.22

(iii) Let the intercepts made by straight line from positive axes be k; where $k > 0$

Thus the equation of required straight line is given by intercept form $\frac{x}{a}+\frac{y}{b}=1; a=b=k$

i.e., $\frac{x}{k}+\frac{y}{k}=1$

or $x + y = k$ ...(i)

Now (i) passes through the point (4,5) ⇒ $k = 9$

∴ required equation of straight line is $x + y = 9$

(iv) The straight line is making equal intercepts from axes which are of opposite signs, so let x-intercept = $a = k$ and y – intercept = $b = -k$

Thus the equation of straight line, by intercept form will be given by $\frac{x}{k}+\frac{y}{k}=$

i.e., $x - y = k$ ...(i)

Now (i) passes through the point where the straight line $2x - 3y = -2$ meets the line $y - x = 0$

⇒ At that point $2x - 3x = -2$

⇒ $x = 2 \Rightarrow y = 2$

Thus the line passes through the point (2,2).

∴ (2,2) would satisfy (i)

⇒ k = 2–2 =0, resubstituting the value of k in (i), we get the equation of reqd. straight line i.e., $x - y = 0$ or $x = y$

ILLUSTRATION 26: Find the equation of straight line parallel to line $2x + 3y = 6$ and at a distance of 2 units from it.

SOLUTION: Given line is $2x + 3y = 6$

or $\frac{x}{3}+\frac{y}{2}=1$ .....(i)

$\Rightarrow$ x-intercept of line (i) $a = 3$ and y-intercepts of line (i) $= b = 2$

And slope of (i) $= -\frac{\text{coefficient of x}}{\text{coefficient of y}} = \frac{-2}{3} = \tan\theta$

Let AB and CD be two straight lines parallel to line (i) and at a distance of 2 units from it as shown in the diagram.

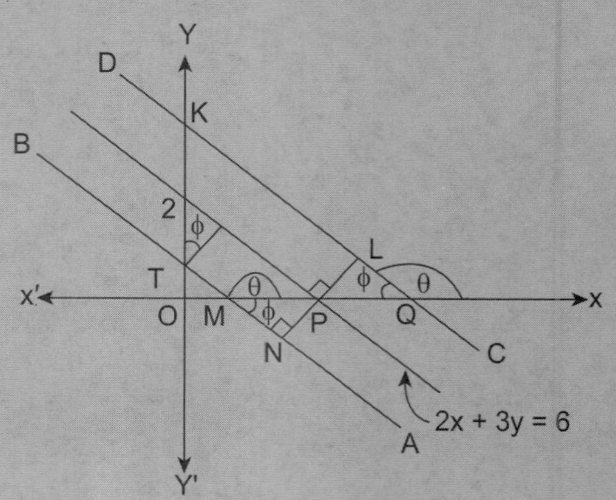

FIGURE 2.23

$\Rightarrow PL = PN = 2$

In ΔPQL, $\phi = \pi - \theta$

$\Rightarrow \tan\phi = -\tan\theta = -\left(\frac{-2}{3}\right) = \frac{2}{3}$

And $\tan\phi = \frac{PL}{QL} = \frac{2}{QL} = \frac{2}{3} \Rightarrow QL = 3$

$\therefore$ $PQ = \sqrt{QL^2 + PL^2} = \sqrt{9+4} = \sqrt{13}$

Similarly, in Δ PMN; $PN = 2$, $MN = 3$

$\therefore$ $PM = \sqrt{MN^2 + PN^2} = \sqrt{9+4} = \sqrt{13}$

$\Rightarrow OM = OP - PM = 3 - \sqrt{13}$; $OQ = OP + PQ = 3 + \sqrt{13}$

Similarly, $OT = 2 - \frac{2\sqrt{13}}{3}$ and $OK = 2+\frac{2\sqrt{13}}{3}$

$\therefore$ Equation of required straight lines are $\frac{x}{OQ}+\frac{y}{OK}=1$ and $\frac{x}{OM}+\frac{y}{OT}=1$

$$\Rightarrow \frac{x}{3+\sqrt{13}}+\frac{y}{\left(2+\frac{2\sqrt{13}}{3}\right)}=1 \text{ and } \frac{x}{3-\sqrt{13}}+\frac{y}{\left(2-\frac{2\sqrt{13}}{3}\right)}=1$$

$$\Rightarrow \frac{x}{3+\sqrt{13}}+\frac{3}{2\left(3+\sqrt{13}\right)}y=1 \; and \; \frac{x}{3-\sqrt{13}}+\frac{3y}{2\left(3-\sqrt{13}\right)}=1$$

i.e., $2x + 3y = 6+2\sqrt{13}$ and $2x + 3y = 6-2\sqrt{13}$

NOTE: Although after discussing the distance between parallel line, this can be done with less effort.

ILLUSTRATION 27: A straight line moves so that the sum of the reciprocals of its intercepts on the co-ordinate axes is constant. Show that it passes through a fixed point.

SOLUTION: Let the variable line be $\frac{x}{a}+\frac{y}{b}=1$.

Now we are given that $\frac{1}{a}+\frac{1}{b}=k$ (where k is any constant)

$\Rightarrow \frac{1}{k}.\frac{1}{a}+\frac{1}{k}.\frac{1}{b}=1 \Rightarrow \left(\frac{1}{k},\frac{1}{k}\right)$ satisfies the equation of the line

Therefore, the variable line takes the form $\frac{x}{a}+\left(k-\frac{1}{a}\right)y-1=0$ or $\frac{1}{a}(x-y)+(ky-1)=0$. This represents a straight line through the intersection of $x - y = 0$ and $ky - 1 = 0$. They intersect at $(1/k, 1/k)$ and hence $\frac{x}{a}+\frac{y}{b}=1$ always passes through the fixed point $\left(\frac{1}{k},\frac{1}{k}\right)$.

ILLUSTRATION 28: Find the equation of the straight line when the portion of it intercepts between the axes is divided by the point (3, 1) in the ratio 1 : 3.

SOLUTION: Let the required straight line meet the x-axis at $A(a, 0)$ and meet the y-axis at $B(0, b)$. It is given that the point $C(3, 1)$ divides AB or BA in the ratio 1 : 3. Therefore by the section formula, C must be the point $\left(\frac{1.0+3.a}{4},\frac{1.b+3.0}{4}\right)=\left(\frac{3a}{4},\frac{b}{4}\right)$

if $\frac{AC}{CB}=\frac{1}{3}$ or the point $\left(\frac{1.a+3.0}{4},\frac{1.0+3b}{4}\right)=\left(\frac{a}{4},\frac{3b}{4}\right)$ if $\frac{BC}{CA}=\frac{1}{3}$

$\therefore\ C(3, 1)=\left(\frac{3a}{4},\frac{b}{4}\right)$ gives a = 4, $b = 4$ and $C(3, 1)=\left(\frac{a}{4},\frac{3b}{4}\right)$

gives a = 12, $b = 4/3$. Hence the required line is either $\frac{x}{4}+\frac{y}{4}=1$ or $\frac{x}{12}+\frac{y}{4/3}=1$

i.e., either $x + y = 4$ or $x + 9y = 12$.

ILLUSTRATION 29: Find the locus of the mid-points of the portion of the lines $x\cos\theta + y\sin\theta = p$ (p is constant) which is intercepted between the axes.

SOLUTION: Let the family of lines intersect the x-axis at A $(a, 0)$ and y-axis at $B(0, b)$.

Let $M(h, k)$ be the mid-point of AB

$\Rightarrow h = a/2, k = b/2$ ......(i)

Now a and b are to be eliminated, because these are extra variables.

Using the given equation, $x\cos\theta + y\sin\theta = p$

For point A: $y = 0, \Rightarrow x = \frac{p}{\cos\theta} = p\sec\theta \Rightarrow a = p\sec\theta$

For point B: $x = 0 \Rightarrow y = p\,\text{cosec}\theta \Rightarrow b = p\,\text{cosec}\theta$

Substituting in (i), we get, $2h = p\sec\theta$ and $2k = p\,\text{cosec}\theta$

$\Rightarrow \cos\theta = p/2h$ and $\sin\theta = p/2k$

Eliminating extra variable θ.

$$\Rightarrow \left(\frac{p}{2h}\right)^2+\left(\frac{p}{2k}\right)^2=1\ (\quad \sin^2\theta+\cos^2\theta=1)$$

$$\therefore \text{ locus of } P(h, k) \text{ is } \frac{1}{x^2}+\frac{1}{y^2}=\frac{4}{p^2}$$

ILLUSTRATION 30: A line having slope of $-1/5$ meets the co-ordinate axes at A and B. If the area of $\Delta\ OAB$ is 10 *sq*. units where 'O' is the oigin. then find equation of drawn line.

SOLUTION: Let the equation of line be $\frac{x}{a}+\frac{y}{b}=1$.

$\Rightarrow$ $(b/a) . 5 = 1 \Rightarrow 5b = a$

Area of $\Delta_{OAB} = 1/2\ |ab|$

$\Rightarrow$ $10 = 1/2\ |5b^2| \Rightarrow b^2 = 4$

$\Rightarrow$ $b = \pm 2,\ a = \pm\ 10$

The line can be $x/10 + y/2 = 1$ or $x/10 + y/2 = -1$

ILLUSTRATION 31: A variable line, drawn through the point of intersection of the straight line $\frac{x}{a}+\frac{y}{b}=1$ and $\frac{x}{b}+\frac{y}{a}=1$ meets the cordinate axes in A and B. Show that the locus of the mid-point of AB is the curve $2xy(a + b) = ab(x + y)$.

SOLUTION: Let the co-ordinates of mid-point P of AB be (h, k)

$\therefore$ AB intersects x-axis at $(2h, 0)$ and y-axis at $(0, 2k)$

$\therefore$ Equation of AB is $\frac{x}{2h}+\frac{y}{2k}=1$

point of intersection of given two lines i.e., $bx + ay - ab = 0$ and $ax + by - ab = 0$ are given by

$\Rightarrow$ $x=\frac{ab}{a+b}$ and $y=\frac{ab}{a+b}$

If equation AB passes through $\left(\frac{ab}{a+b}, \frac{ab}{a+b}\right)$, then $\frac{ab}{(a+b)2h}+\frac{ab}{(a+b)2k}=1$

(Since line passes through this point) therefore locus is $ab\ (x + y) = 2xy\ (a + b)$

ILLUSTRATION 32: Find the equations of the straight lines which passes through the point (3, 2) and makes intercepts a, b respectively on the x-axis and y-axis so that $a - b = 2$.

SOLUTION: Let the equation of line be $\frac{x}{a}+\frac{y}{b}=1$

$\Rightarrow$ $bx + ay = ab$ ....(i)

this line passes through (3, 2)

$\Rightarrow$ $3b + 2a = ab$ ... (ii)

Also $a - b = 2$ (given)

Substituting $a = 2 + b$ in equation (ii) we get

$\Rightarrow$ $3b + 2(2 + b) = (2 + b)b$

$\Rightarrow$ $3b + 4 + 2b = 2b + b^2$

$\Rightarrow$ $b^2 - 3b - 4 = 0 \Rightarrow b^2 - 4b + b - 4 = 0$

$\Rightarrow$ $b(b - 4) + 1(b - 4) = 0 \Rightarrow b = -1, 4$

$\Rightarrow$ $a = 1, 6$

thus $a = 1, b = -1; a = 6, b = 4$

Substituting these values of a and b in equation (i), we get equation of the required line as $x - y - 1 = 0$ or $2x + 3y = 12$

Reduction of General Form into Intercept Form (x/a + y/b = 1)

Let $Ax + By + C = 0$ be the general equation of straight line

$\therefore \quad Ax + By + C = 0 \Rightarrow Ax + By = -C$

$$\Rightarrow \quad \frac{x}{-C/A} + \frac{y}{-C/B} = 1 \Rightarrow \frac{x}{a} + \frac{y}{b} = 1;$$

Where intercept on x-axis

$$= a = \frac{-C}{A} = -\frac{\text{constant term}}{\text{coefficient of x}}, \text{ intercept on}$$

$$y\text{-axis} = b = \frac{-C}{B} = -\frac{\text{constant term}}{\text{coefficient of } y}$$

ILLUSTRATION 33: Convert the general equation $8x - 15y + 51 = 0$ into its intercept form and hence find its intercepts on x and y-axis.

SOLUTION: Equation of given line is $8x - 15y + 51 = 0$

$$\Rightarrow 8x - 15y = -51 \Rightarrow \frac{8x}{-51} - \frac{15y}{-51} = 1 \Rightarrow \frac{x}{(-51/8)} + \frac{y}{(51/15)} = 1$$

This equation is of the form $\frac{x}{a} + \frac{y}{b} = 1$

$\Rightarrow$ on x-axis : $a = \frac{-51}{8}$

and on y-axis : $b = \frac{51}{15} = \frac{17}{5}$

TEXTUAL EXERCISE–1 (SUBJECTIVE)

1. Find the equation of line making an angle of 150° with the x-axis and cutting off an intercept 2 from y-axis.
2. Find the equation of the line cutting off an intercept 3 from the negative direction of the axis of y and inclined at 120° to the axis of x.
3. Find the area of the triangle formed by the lines $y = m_1x + c_1$, $y = m_2x + c_2$ and $x = 0$.
4. Find the equation of the lines through the point of intersection of $2x + 3y - 7 = 0$ and $x + 3y - 5 = 0$ which cuts the co-ordinate axes at points which are equidistant from origin.
5. Find the equation of a straight line passing through the point (5, 2) and the intersection of the lines

   $3x + y - 2 = 0$, $x + 5y - 7 = 0$
6. Find the equation of the line through the intersection of $2x + 4y + 7 = 0$, $x - y + 2 = 0$ and with slope 3.
7. Find the equation of the line through the intersection of the lines $x + 2y = 3$, $4x - y + 7 = 0$ and which is parallel to $x - 3y + 2 = 0$.
8. Find the equation of line through the point of intersection of $2x + 3y = 4$, $x - 5y + 7 = 0$ and perpendicular to the line $x = 0$.
9. A line is such that its segment between the straight lines $5x - y - 4 = 0$ and $3x + 4y - 4 = 0$ is bisected at the point (1, 5), then find its equation.
10. Find the equation to the straight line cutting an intercept *of* –4 unit on x and 2 unit on y-axis.
11. Find the equation of the straight line which passes through (1, –2) and cuts off equal intercepts on the axes.
12. Find the equation of line passing through (x_1, y_1) if this point bisects the segment of the line between the axes and write the equation if point (x_1, y_1) is (5, 2).
13. Find the equation of a line (lines) passing through the point (2, 3) and cuts off intercepts on the axes which are equal in magnitude but opposite in sign.
14. Find the equation of the lines, which cut off intercepts on the axes whose sum and product are 1 and – 6 respectively.
15. Find the equation of a straight line through the point (2, 2) such that the sum of its intercept on the axes is 9.

Answer Keys

1. $x + \sqrt{3}y = 2\sqrt{3}$ **2.** $y + x\sqrt{3} + 3 = 0$ **3.** $\frac{1}{2}\frac{(c_1 - c_2)^2}{|m_1 - m_2|}$

4. $x + y - 3 = 0$; $x - y - 1 = 0$; $x = 2y$;

5. $9x - 67y + 89 = 0$ **6.** $3x - y + 7 = 0$ **7.** $9x - 27y + 68 = 0$ **8.** $y = 18/13$

9. $83x - 35y + 92 = 0$ **10.** $x/-4 + y/2 = 1$ **11.** $x + y + 1 = 0$ **12.** $\frac{x}{x_1} + \frac{y}{y_1} = 2$, $2x + 5y = 20$

13. $x - y + 1 = 0$, $3x - 2y = 0$ **14.** $2x - 3y = 6$ or $2y - 3x = 6$ **15.** $x + 2y = 6$, $2x + y = 6$

TEXTUAL EXERCISE–1 (OBJECTIVE)

1. If the co-ordinates of the points A and B be (3, 3) and (7, 6), then the length of portion of the line AB intercepted between the axes is

(a) $\frac{5}{4}$ (b) $\frac{\sqrt{10}}{4}$

(c) $\frac{\sqrt{13}}{3}$ (d) None of these

2. If the transversal $y = m_r x$; $r = 1, 2, 3$ cuts off equal intercepts on the transversal $x + y = 1$, then $1 + m_1$, $1 + m_2$, $1 + m_3$ are in

(a) A.P. (b) G.P.

(c) H.P. (d) None of these

3. The angle between the lines whose intercepts on the axes are (1, –2) and (2, –1) respectively is

(a) $\tan^{-1}\left(\frac{4}{3}\right)$ (b) $\tan^{-1}\left(\frac{3}{4}\right)$

(c) $\tan^{-1}\left(\frac{2}{3}\right)$ (d) $\tan^{-1}\left(\frac{3}{2}\right)$

4. The equation to the line bisecting the join of (1, –2) and (2, –5) and having its intercepts on the x-axis and y-axis in the ratio of 2 : 3 is

(a) $3x + 4y + 5/2 = 0$ (b) $4x + 3y + 5/2 = 0$

(c) $3x + 2y + 5/2 = 0$ (d) None of these

5. The equation of the straight line which is passing through the point (1, 2) and makes an intercept of 3 units on the x-axis is

(a) $y = 2x + 4$ (b) $x + y = 3$

(c) $2y + x - 4 = 0$ (d) $3x + 2y + 4 = 0$

6. The equation of a line passing through (1, –1) and making an angle 45° with x-axis is

(a) $x - y = 2$ (b) $2x - y = 2$

(c) $x - 2y = 2$ (d) $x - y = 1$

7. The equation of a line passing through (2, 3) and making an angle of 135° with the positive direction of x-axis is

(a) $2x + y - 5 = 0$ (b) $x + y - 5 = 0$

(c) $x + 2y - 5 = 0$ (d) $2x + y - 6 = 0$

8. The equation of a straight line which passes through the point (–2, 3) and makes an angle of 60° with the positive direction of x-axis is

(a) $y + \sqrt{3}x - \sqrt{3}(\sqrt{3} + 2) = 0$

(b) $y - \sqrt{3}x + \sqrt{3}(\sqrt{3} + 2) = 0$

(c) $y - \sqrt{3}x - \sqrt{3}(\sqrt{3} + 2) = 0$

(d) None of these

9. The equation of line/s passing through (2, 1) and equally inclined to the axes is

(a) $y - x + 1 = 0$ and $y + x = 3$

(b) $2y - x + 1 = 0$ and $2y + x = 3$

(c) $y + x - 5 = 0$ and $y - x = 6$

(d) None of these

10. The co-ordinates of the points A, B, C, D, be (a, b), (a', b'), $(-a, b)$ and $(a', -b')$ respectively, then the equation of the line bisecting the line segments AB and CD is.

(a) $2a'y - 2bx = ab - a'b'$

(b) $2ay - 2b'x = ab - a'b'$

(c) $2ay - 2b'x = a'b - ab'$

(d) None of these

11. If the co-ordinates of the vertices of a triangle ABC be (– 1, 6); (– 3, –9); and (5, –8) respectively, then the equation of the median through C is

(a) $13x - 14y - 47 = 0$ (b) $13x - 14y + 47 = 0$
(c) $13x + 14y + 47 = 0$ (c) $13x + 14y - 47 = 0$

12. A line passing through the point (2, 2) and the axes enclose an area λ. The intercepts on the axes made by the line are given by the two roots of
(a) $x^2 - 2\,|\,\lambda\,|\,x + |\,\lambda\,| = 0$
(b) $x^2 + |\,\lambda\,|\,x + 2|\,\lambda\,| = 0$
(c) $x^2 - |\,\lambda\,|\,x + 2\,|\,\lambda\,| = 0$
(d) None of these

13. If distance formula between two points (x_1, y_1) and (x_2, y_2) be redefined as $|\,x_1 - x_2\,| + |\,y_1 - y_2\,|$, then the locus of a point which is at a constant distance of 5 units from (3, 5) (w.r.t. new formula) is
(a) circle (b) line segment
(c) quadrilateral (d) square

14. A variable straight line passes through the points of intersection of the lines, $x + 2y = 1$ and $2x - y = 1$ and meets the co-ordinate axes in A and B. The locus of the middle point of AB is
(a) $x + 3y - 10\,xy = 0$
(b) $x - 3y + 10\,xy = 0$
(c) $x + 3y + 10\,xy = 0$
(d) None of these

Answer Keys

1. (a)	**2.** (c)	**3.** (b)	**4.** (c)	**5.** (b)	**6.** (a)	**7.** (b)	**8.** (c)	**9.** (a)	**10.** (b)
11. (c)	**12.** (c)	**13.** (d)	**14.** (a)						

Normal/Perpendicular Form of a Line

Equation of a straight line whose perpendicular distance from origin (OL) is p and the directed distance OL makes an angle α (measured in anticlockswise sense) with positive direction of x-axis, is $x \cos \alpha + y \sin \alpha = p$, where $p > 0$ and $0 \leq \alpha < 2\pi$.

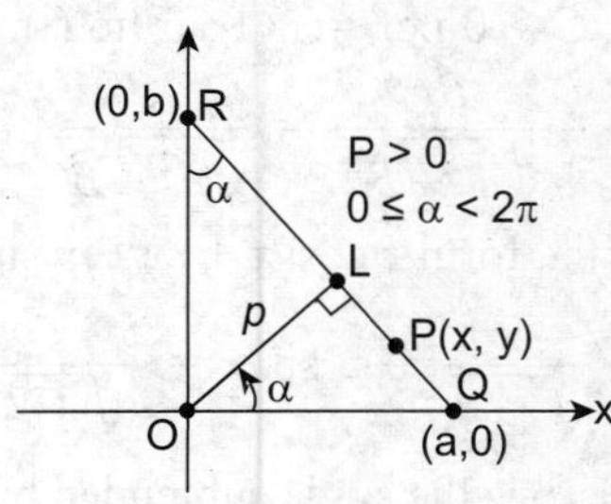

FIGURE 2.24

Proof I: Let x and y intercept of the above line are a and b respectively, then

$$\sec\alpha = \frac{a}{p} \Rightarrow a = p\sec\alpha$$

$$\text{and } \operatorname{cosec}\alpha = \frac{b}{p} \Rightarrow b = p\operatorname{cosec}\alpha$$

So the equation of the line in intercept form can be written as

$$\frac{x}{p\sec\alpha} + \frac{y}{p\operatorname{cosec}\alpha} = 1$$

or $x \cos \alpha + y \sin \alpha = p$

Proof II: $OQ = p \sec\alpha$ and since $\angle ORL = \alpha$

$\therefore \quad OR = p \operatorname{cosec} \alpha$

Hence $Q \equiv (p \sec\alpha, 0)$; $R \equiv (0, p\operatorname{cosec} \alpha)$

Now the points $P(x, y)$, Q $(p \sec\alpha, 0)$ and R $(0, p \operatorname{cosec} \alpha)$ are collinear, therefore

$$\begin{vmatrix} x & y & 1 \\ p\sec\alpha & 0 & 1 \\ 0 & p\operatorname{cosec}\alpha & 1 \end{vmatrix} = 0 \text{ or}$$

$$x(-p\operatorname{cosec}\alpha) - yp\sec\alpha + 1.p^2\sec\alpha\operatorname{cosec}\alpha = 0$$

$$\Rightarrow p\sec\alpha\operatorname{cosec}\alpha = x\operatorname{cosec}\alpha + y\sec\alpha$$

$$\Rightarrow p = \frac{x}{\sec\alpha} + \frac{y}{\operatorname{cosec}\alpha}$$

$$\Rightarrow x\cos\alpha + y\sin\alpha = p$$

NOTE

In normal form of equation of a straight line, p is always taken as positive and *α is measured from positive direction of x-axis in anti-clockwise direction between* 0 and 2π.

ILLUSTRATION 34: Find the equation of the line which is at a distance of 3 units from the origin and the perpendicular from the origin to the line makes an angle of 30° with the positive direction of the x-axis.

SOLUTION: Given that $p = 3$, $\alpha = 30°$

$\therefore$ equation of the line in the normal form is, $x \cos 30° + y \sin 30° = 3 \Rightarrow \sqrt{3}x + y = 6$

ILLUSTRATION 35: Find the equation of the straight line upon which the length of the perpendicular from the origin is 2 and the perpendicular makes an angle of 30° with the positive direction of y-axis in clockwise direction.

SOLUTION: Since, line which is drawn from the origin to the straight line makes angle of 30° with the y-axis.

$\Rightarrow$ angle which it makes with x-axis in anti-clockwise direction is 60°.

$\Rightarrow x \cos\alpha + y \sin\alpha = p$ where $\alpha = 60°$, $p = 2$

$\Rightarrow x \cos 60° + y \sin 60° = 2 \Rightarrow x(1/2) + y(\sqrt{3}/2) = 2 \Rightarrow x + \sqrt{3}y = 4.$

ILLUSTRATION 36: Find the equation of the straight line upon which the length of the perpendicular from the origin in first quadrant is 6 and the gradient of this perpendicular is 3/4.

SOLUTION: Let α be the angle subtended by perpendicular from origin upon the required line.

$\therefore \tan\alpha = 3/4$, $p = 6 \Rightarrow \cos\alpha = 4/5$ and $\sin\alpha = 3/5$

$\therefore x \cos\alpha + y \sin\alpha = p \Rightarrow x(4/5) + y(3/5) = 6 \Rightarrow 4x + 3y = 30$

Reduction of general form $Ax + By + C = 0$ to normal form ($x \cos\alpha + y \sin\alpha = p$)

First take C to right hand side and make it positive, then divide the whole equation by $\sqrt{A^2+B^2}$ then,

Since p has to be always positive, so we can discuss above procedure in the following two cases:

Case I: When $C < 0$ i.e., $-C > 0$ dividing both sides of equation (i) by $\sqrt{A^2+B^2}$, we have,

$$\frac{Ax}{\sqrt{A^2+B^2}} + \frac{By}{\sqrt{A^2+B^2}} = -\frac{C}{\sqrt{A^2+B^2}}$$

which is of the form $x \cos\alpha + y \sin\alpha = p$

where $\cos\alpha = \dfrac{A}{\sqrt{A^2+B^2}}$,

$\sin\alpha = \dfrac{B}{\sqrt{A^2+B^2}}$ and $p = -\dfrac{C}{\sqrt{A^2+B^2}}$

Case II: When $C > 0$ i.e., $-C < 0$; from (i), $-Ax - By = C$

$$\frac{-Ax}{\sqrt{A^2+B^2}} - \frac{By}{\sqrt{A^2+B^2}} = \frac{C}{\sqrt{A^2+B^2}}$$

which is of the form $x \cos(\pi + \alpha) + y \sin(\pi + \alpha) = p$

where $\cos\alpha = \dfrac{A}{\sqrt{A^2+B^2}}$, $\sin\alpha = \dfrac{B}{\sqrt{A^2+B^2}}$ and $p = \dfrac{C}{\sqrt{A^2+B^2}}$; $(\pi + \alpha)$ is the angle subtended by perpendicular drawn from origin upon the given straight line.

ILLUSTRATION 37: For the straight line $8x - 15y + 51 = 0$, find the length of the perpendicular from the origin to this line and the inclination of the perpendicular with the positive x-axis.

SOLUTION: Equation of line $8x - 15y + 51 = 0$

$\Rightarrow$ divide both sides by $\sqrt{8^2+(-15)^2}$

$$\Rightarrow \frac{8x}{\sqrt{8^2+(-15)^2}} - \frac{15y}{\sqrt{8^2+(-15)^2}} = \frac{-51}{\sqrt{8^2+(-15)^2}} \Rightarrow \frac{8x}{17} - \frac{15y}{17} = \frac{-51}{17} = -3$$

$$\therefore \frac{-8x}{17} + \frac{15y}{17} = 3$$

this equation is of the form $x \cos\alpha + y \sin\alpha = p$

$$\Rightarrow \cos\alpha = \frac{-8}{17} \text{ and } \sin\alpha = \frac{15}{17} \text{ and } p = 3$$

$$\Rightarrow \tan\alpha = \frac{15/17}{-8/17} = \frac{15}{-8}. \Rightarrow \alpha = \tan^{-1}\left(\frac{-15}{8}\right)$$

■ PARAMETRIC/SYMMETRIC OR DISTANCE FORM OF LINE

Let a line meet x-axis at A and y-axis at B and passes through the point $Q(x_1, y_1)$ and make an angle θ with the positive direction of x-axis. Let $P(x, y)$ be any point on the line at a distance r from $Q(x_1, y_1)$.

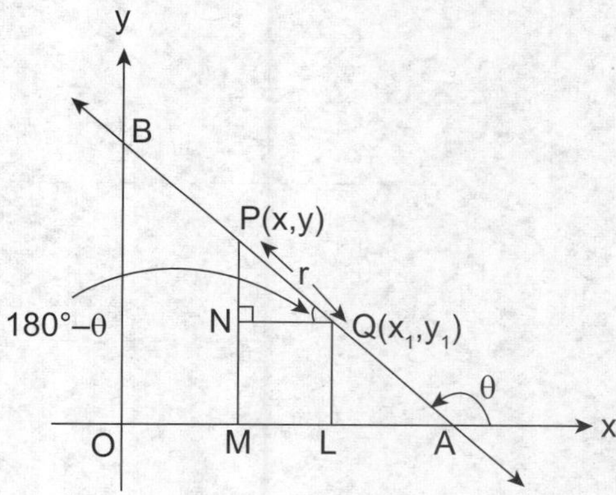

FIGURE 2.25

Then $QN = ML = OL - OM = x_1 - x$

and $PN = PM - NM = y - y_1$

From ΔPQN, $\cos(180° - \theta) = \dfrac{QN}{PQ} = \dfrac{x_1 - x}{r}$

$$\Rightarrow -\cos\theta = \frac{x_1 - x}{r}$$

$$\Rightarrow \cos\theta = \frac{x - x_1}{r} \quad .....(i)$$

$$\text{Again } \sin\theta = \frac{PN}{PQ} \Rightarrow \sin\theta = \frac{y - y_1}{r} \quad .....(ii),$$

From (i) and (ii) we get, $\dfrac{x - x_1}{\cos\theta} = \dfrac{y - y_1}{\sin\theta} = r$

This is the equation of the line in the distance form. Also known as parametric or symmetric form.

REMARKS

(i) *The equation of the line is* $\dfrac{x - x_1}{\cos\theta} = \dfrac{y - y_1}{\sin\theta} = \pm r$

$\therefore\ x - x_1 = \pm r\cos\theta$ and $y - y_1 = \pm r\sin\theta \Rightarrow x = x_1 \pm r\cos\theta$ and $y = y_1 \pm r\sin\theta$

Thus the co-ordinates of any point on the line at a distance r from the given point (x_1, y_1) are $(x_1 \pm r\cos\theta, y_1 \pm r\sin\theta)$

(ii) *If θ is variable* and x_1, y_1, r *are fixed, then the same equation will represent a circle whose centre is (x_1, y_1)* and *radius is r.*

Proof: $\dfrac{x - x_1}{r} = \cos\theta$ and $\dfrac{y - y_1}{r} = \sin\theta$

Now we know that $\sin^2\theta + \cos^2\theta = 1 \Rightarrow (x - x_1)^2 + (y - y_1)^2 = r^2$

Which is equation of a circle with centre at (x_1, y_1) and *radius = r*

(iii) *This form is selected where the portion of a line or line segment is to be discussed or the co-ordinates of a point on the line at certain distance from a fixed point of the line is to be obtained.*

ILLUSTRATION 38: Find the equation of the line passing through $P(4, 5)$ and making an angle of 30° with positive x-axis measured anticlockwise. Also find the co-ordinates of points which are at a distance of 4 units on either side of P.

SOLUTION: Since the equation of the line passing through $P(4,. 5)$ and making an angle 30° is given by

$$\frac{x-4}{\cos 30°} = \frac{y-5}{\sin 30°} \Rightarrow \frac{2(x-4)}{\sqrt{3}} = \frac{2(y-5)}{1} \Rightarrow x - \sqrt{3}\,y - 4 + 5\sqrt{3} = 0$$

Point on the line at a distance 4 from (4, 5) is given by $\dfrac{x-4}{\cos 30°} = \dfrac{y-5}{\sin 30°} = \pm 4$

$\Rightarrow x = 4 \pm 4\cos 30°$ and $y = 5 \pm 4\sin 30°$

$\Rightarrow x = 4 \pm 2\sqrt{3}\,;\, y = 5 \pm 2 \Rightarrow (4 + 2\sqrt{3}, 7)$ and $(4 - 2\sqrt{3}, 3)$ are the required points.

ILLUSTRATION 39: A line is such that its segment between the straight line $5x - y - 4 = 0$ and $3x + 4y - 4 = 0$ is bisected at the point (1, 5), obtain the equation of line.

SOLUTION: Let the angle that the required line makes with the positive x-axis be θ

$\therefore$ Co-ordinate of points A, B (at a distance "r") from (1, 5) be given by $(1 + r\cos\theta, 5 + r\sin\theta)$ and $(1 - r\cos\theta, 5 - r\sin\theta)$ respectively.

Point A lies on $5x - y - 4 = 0$,

then $5(1 + r\cos\theta) - (5 + r\sin\theta) - 4 = 0$

$$\Rightarrow r = \frac{4}{5\cos\theta - \sin\theta} \qquad ...(1)$$

point B lies on $3x + 4y - 4 = 0$, then $3(1 - r\cos\theta) + 4(5 - r\sin\theta) - 4 = 0$

$$\Rightarrow r = \frac{19}{3\cos\theta + 4\sin\theta} \qquad ...(2)$$

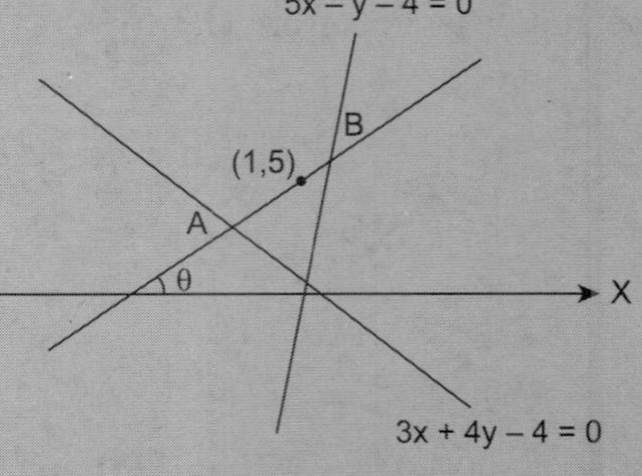

FIGURE 2.26

equating (1) and (2), we get

$$\Rightarrow \frac{4}{5\cos\theta - \sin\theta} = \frac{19}{3\cos\theta + 4\sin\theta} \Rightarrow 4(3 + 4\tan\theta) = 19(5 - \tan\theta)$$

$$\Rightarrow 12 + 16\tan\theta = 95 - 19\tan\theta \Rightarrow \tan\theta = \frac{83}{35} = m$$

$\therefore$ equation of line AB is $(y - 5) = \dfrac{83}{35}(x - 1)$

$83x - 35y + 92 = 0$ **Ans**

ILLUSTRATION 40: A line through (1, 2) makes an angle of $3\pi/4$ with negative direction of x-axis measured clockwise. Find the length of the line segment intercepted between (1, 2) and the line $x + y - 4 = 0$.

SOLUTION: Line makes an angle of $3\pi/4$ with the negative direction of x-axis. So it makes an angle of $\pi/4$ in the positive direction of x-axis.

$\therefore$ equation of line passing through (1, 2) in parametric form is $\dfrac{x-1}{\cos(\pi/4)} = \dfrac{y-2}{\sin(\pi/4)} = r$

$$\Rightarrow x = 1 + \frac{r}{\sqrt{2}}\,;\, y = 2 + \frac{r}{\sqrt{2}}.$$

Now, this point lies on the line $x + y - 4 = 0$

$$\Rightarrow 1 + \frac{r}{\sqrt{2}} + 2 + \frac{r}{\sqrt{2}} - 4 = 0 \Rightarrow \sqrt{2}r = 1$$

$$\Rightarrow r = \frac{1}{\sqrt{2}} \quad \therefore AP = \frac{1}{\sqrt{2}}.$$

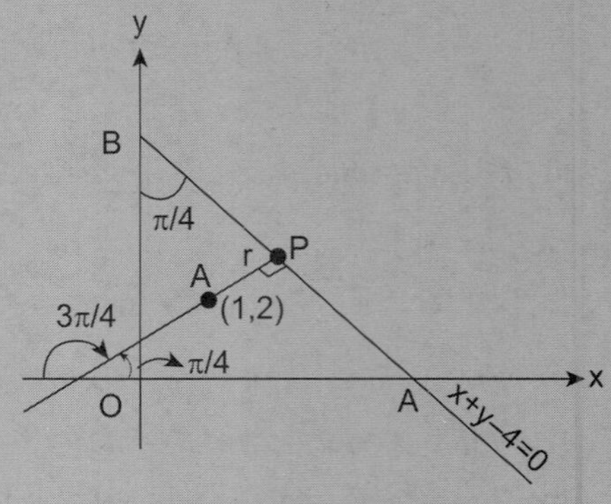

FIGURE 2.27

ILLUSTRATION 41: Prove that the length of perpendicular drawn from the point (x_1, y_1) to a given line $ax + by + c = 0$ (where b/a is $\tan\theta$) is equal to $\left|\dfrac{ax_1 + by_1 + c}{a\cos\theta + b\sin\theta}\right|$.

SOLUTION: Let Q be the foot of perpendicular from P to the line $ax + by + c = 0$ and $PQ = r$

Therefore the slope of the line $PQ = b/a$. If PQ makes angle θ with +ve x-axis, then

$\tan\theta = b/a$; $\sin\theta = \dfrac{b}{\sqrt{a^2+b^2}}$; $\cos\theta = \dfrac{a}{\sqrt{a^2+b^2}}$

Co-ordinate of $Q \equiv (x_1 + r\cos\theta,\ y_1 + r\sin\theta)$

$\Rightarrow$ Q lies on $ax + by + c = 0$

$\Rightarrow\ ax_1 + ar\cos\theta + by_1 + br\sin\theta + c = 0$

$\Rightarrow\ ax_1 + by_1 + c = -(ar\cos\theta + br\sin\theta)$

$\Rightarrow\ r = \left|\dfrac{ax_1 + by_1 + c}{a\cos\theta + b\sin\theta}\right|$.

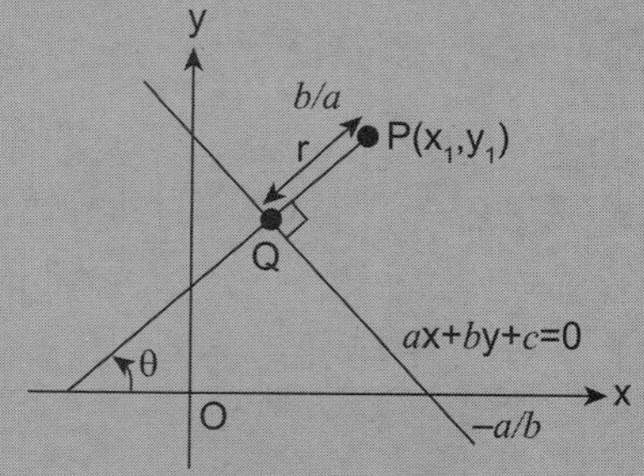

FIGURE 2.28

ILLUSTRATION 42: Find the distance of a line $2x - y + 3 = 0$ from the point (2, 1) measured parallel to the line $x + y = 1$.

SOLUTION: ***Method I:*** PQ is a line through (2, 1) parallel to $x + y = 1$

$\Rightarrow y - 1 = -1\,(x - 2) \Rightarrow y - 1 = -x + 2 \Rightarrow y + x = 3 \Rightarrow \theta \equiv (0, 3)$

$\Rightarrow r = \sqrt{(2)^2 + (-2)^2} = 2\sqrt{2}$

Method II: Equation of PQ is $\dfrac{x-2}{-1/\sqrt{2}} = \dfrac{y-1}{1/\sqrt{2}} = r$

$$Q \equiv \left(2 - \frac{r}{\sqrt{2}},\ 1 + \frac{r}{\sqrt{2}}\right) \Rightarrow 2\left(2 - \frac{r}{\sqrt{2}}\right) - \left(1 + \frac{r}{\sqrt{2}}\right) + 3 = 0$$

$$\Rightarrow\ 4 - r\sqrt{2} - 1 - \frac{r}{\sqrt{2}} + 3 = 0 \Rightarrow \frac{-r(2+1)}{\sqrt{2}} = -6$$

$$\Rightarrow\ r = \frac{6\sqrt{2}}{3} = 2\sqrt{2} \Rightarrow |r| = 2\sqrt{2}.$$

P(2,1)

r

x+y=1

Q

FIGURE 2.29

ILLUSTRATION 43: Find the equation of the straight line through the origin whose intercepts between the straight lines $2x + 3y = 12$ and $2x + 3y = 15$ is 3 units.

SOLUTION: Assume the equation of the required straight line to be $y = mx$. Here m is to be determined.

The intersection of this with $2x + 3y = 12$ is $P = \left(\dfrac{12}{2+3m}, \dfrac{12m}{2+3m}\right)$

and with $2x + 3y = 15$ is $Q = \left(\dfrac{15}{2+3m}, \dfrac{15m}{2+3m}\right)$

The requirement is that the distance $PQ = 3$. This means

$$9 = \left(\frac{3}{2+3m}\right)^2 + \left(\frac{3m}{2+3m}\right)^2 \text{ which leads to } m = \frac{-3 \pm \sqrt{3}}{4}.$$

Thus the required lines are $y = \dfrac{-3+\sqrt{3}}{4}x$ and $y = \dfrac{-3-\sqrt{3}}{4}x$.

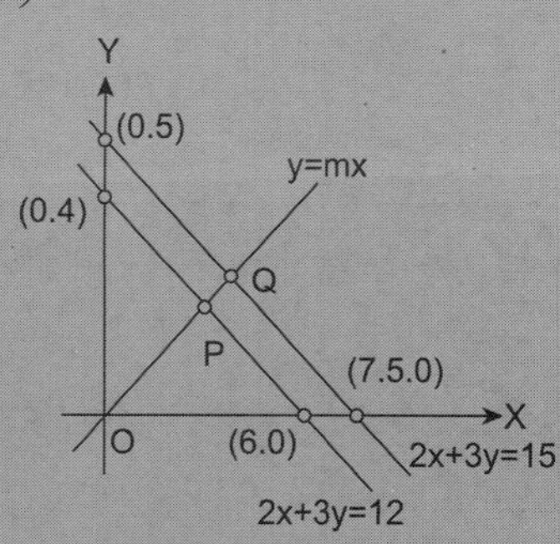

FIGURE 2.30

ILLUSTRATION 44: A line joining two points $A(2, 0)$ and $B(3, 1)$ is rotated about A in anti-clockwise direction through an angle 15°.

(a) Find the equation of the line in the new position.

(b) If B goes to C in the new position, what will be the co-ordinates of C?

SOLUTION: Given A ≡ (2, 0) and $B \equiv (3, 1)$

Slope of line $AB = \dfrac{0-1}{2-3} = 1 = \tan 45^\circ$

$\therefore \quad \angle BAX = 45°$

Given $\angle CAB = 15° \therefore \angle CAX = 60°$

(i) Slope of line $AC = \tan 60° = \sqrt{3}$

$\therefore$ equation of line AC will be

$y - 0 = \sqrt{3}(x-2)$ or $\sqrt{3}x - y - 2\sqrt{3} = 0$

(ii) Line AC makes an angle of 60° with the positive direction of x-axis and

$AC = AB = \sqrt{(3-2)^2 + (1-0)^2} = \sqrt{2} \;\therefore\; C \equiv (2+\sqrt{2}\cos 60°, 0+\sqrt{2}\sin 60°)$

or $C \equiv \left(2+\dfrac{1}{\sqrt{2}}, \sqrt{\dfrac{3}{2}}\right)$

FIGURE 2.31

■ OBLIQUE DISTANCE OF A POINT FROM A LINE

Using this method, we can evaluate the distance of a point $P(x_1, y_1)$ from a line

$ax + by + c = 0$ along $y = mx + c$:

Given $L_1 : ax + by + c = 0$ ...(i)

Let line parallel to $y = mx + c$ through

P cuts $ax + by + c = 0$ at $Q(x_0, y_0)$ as shown in Figure 2.32.

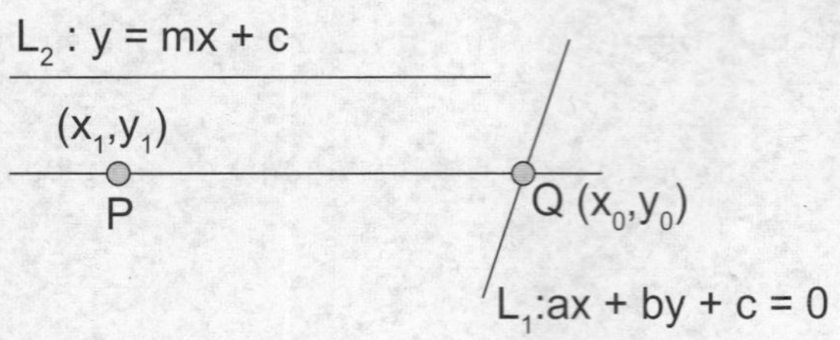

FIGURE 2.32

Thus we are to find distance PQ

Method I: Equation of $PQ : y - y_1 = m(x - x_1)$ ...(ii)

Solving (i) and (ii), we get co-ordinates of $Q \equiv (x_0, y_0)$ and applying distance formula.

$$d(P, Q) = \sqrt{(x_1 - x_0)^2 + (y_1 - y_0)^2}$$

Method II: Let $m = \tan\theta$

Equation of PQ in symmetric form is $\dfrac{x - x_1}{\cos\theta} = \dfrac{y - y_1}{\sin\theta} = r$

For $Q : (x_1 + r\cos\theta, y_1 + r\sin\theta)$ must satisfy $L_1 : ax + by + c = 0$

$\Rightarrow a(x_1 + r\cos\theta) + b(y_1 + r\sin\theta) + c = 0$

$\Rightarrow r = -\left(\dfrac{(ax_1 + by_1 + c)}{a\cos\theta + b\sin\theta}\right)$

The sign of r indicates the position of line w.r.t. point and $|r|$ is the required distance.

ILLUSTRATION 45: Find the distance of the line $3x - y = 0$ from the point (4, 1) measured along a line making an angle of 135° with x-axis.

SOLUTION: The straight line L through, A (4, 1) making an angle of 135° with x-axis is

$$\frac{x-4}{\cos 135°} = \frac{y-1}{\sin 135°} = r, \text{ i.e., } \frac{x-4}{-1/\sqrt{2}} = \frac{y-1}{1/\sqrt{2}} = r$$

Any point P on this straight line is of the form $x = 4 - \frac{r}{\sqrt{2}}$, $y = 1 + \frac{r}{\sqrt{2}}$. Where r is the algebraic distance AP. If this point $\left(4-\frac{r}{\sqrt{2}}, 1+\frac{r}{\sqrt{2}}\right)$ lies on $3x - y = 0$, then

$$3\left(4-\frac{r}{\sqrt{2}}\right)-\left(1+\frac{r}{\sqrt{2}}\right)=0, \Rightarrow r=\frac{11\sqrt{2}}{4}$$

Therefore $r = \frac{11\sqrt{2}}{4}$ units. Thus the distance of $3x - y = 0$ from (4, 1) measured along L is $\frac{11\sqrt{2}}{4}$ units.

ILLUSTRATION 46: Find the direction in which a straight line must be drawn through the point (1, 2) so that its point of intersection with the line $x + y = 4$ may be at a distance $\sqrt{\frac{2}{3}}$ from the point (1, 2).

SOLUTION: Let $P \equiv (1, 2)$

Let AB be the given line $x + y = 4$ ....(i)

Let the line through P making an angle θ with x-axis cuts the line AB at Q at a distance $\sqrt{\frac{2}{3}}$ from P, then

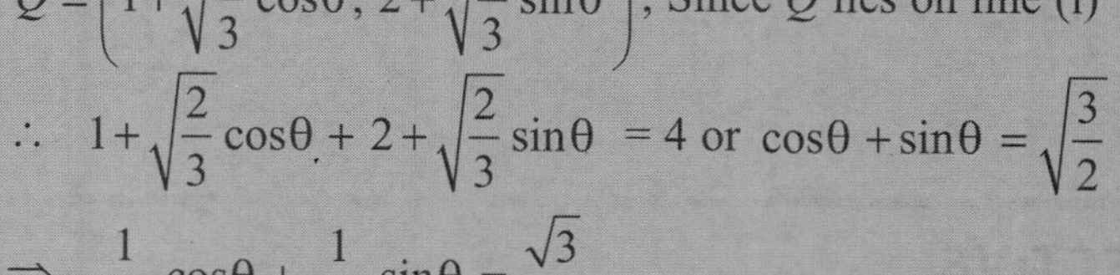

$Q \equiv \left(1+\sqrt{\frac{2}{3}}\cos\theta,\ 2+\sqrt{\frac{2}{3}}\sin\theta\right)$; Since Q lies on line (i)

$$\therefore\ 1+\sqrt{\frac{2}{3}}\cos\theta + 2+\sqrt{\frac{2}{3}}\sin\theta = 4 \text{ or } \cos\theta+\sin\theta=\sqrt{\frac{3}{2}}$$

FIGURE 2.33

$$\Rightarrow \frac{1}{\sqrt{2}}\cos\theta+\frac{1}{\sqrt{2}}\sin\theta=\frac{\sqrt{3}}{2}$$

$$\Rightarrow \cos 45^\circ.\cos\theta+\sin 45^\circ.\sin\theta=\frac{\sqrt{3}}{2}$$

$$\Rightarrow \cos(\theta-45^\circ)=\cos 30^\circ \qquad \therefore\ \theta-45^\circ = 2n\pi \pm 30^\circ,\ n\in\mathbb{Z}$$

or $\theta = 15^\circ, 75^\circ$ (taking only those values of θ which lie between 0° and 180°)

TEXTUAL EXERCISE–2 (SUBJECTIVE)

1. Find the equation of a line for which
(a) $p = 4$, $\alpha = 150°$ (b) $p = 8$, $\alpha = 300°$
(c) $p = 2$, $\cos\alpha = 3/5$

2. Find the equation of the straight line on which the length of the perpendicular from the origin is 2 and the perpendicular makes an angle α with x-axis such that $\sin\alpha = 1/3$.

3. A line makes 45° angle with x-axis and at a distance of $\sqrt{2}$ from the origin, find its equation.

4. The perpendicular distance of a line from the origin is 5unit and its slope *is* –1, find the equation of the line.

5. Find the equation of line
(a) whose x-intercept is 7 and distance from origin is 2.
(b) whose nearest point to the origin is (3, –4)

6. Write the following straight lines in the parametric form $\frac{x-x_1}{\cos\theta}=\frac{y-y_1}{\sin\theta}=r.$
(i) through (2, 3) with slope 2
(ii) through (1, 4) with slope $-1/\sqrt{3}$
(iii) through (1, 3) and (4, 2)

7. Find the points on the straight line $3x - 2y = 2$ at a distance of 3 units from the straight line $3x + 4y = 8$.

8. If the straight line drawn through the point $P(3, 2)$ and making an angle $\pi/6$ with the x-axis meets the line $3x - 4y + 8 = 0$ at Q. Find the length of PQ.

9. Find the co-ordinates of the points at a distance $4\sqrt{2}$ units from the point $(-2, 3)$ along the line making an angle of 45° with the positive direction of x-axis.

10. The extremities of the diagonal of a square are $(1, 1)$, $(-2, -1)$. Obtain the other two vertices and the equation of the other diagonal.

11. Find the equation to the straight line which passes through the point $(-4, 3)$ and is such that the portion of it between the axes is divided by the point in the ratio $5 : 3$.

12. $ABCD$ is a square having vertices A and B on positive x and y axes respectively if $C(12, 17)$. Find co-ordinates of points A, B, D.

13. Let 'P' be a point on the line $8y = 15x$ and Q lies on $10y = 33x$ and the mid-point of PQ is $(8, 6)$. Find the co-ordinates of P and Q.

Answer Keys

1. (a) $-\sqrt{3}x + y = 8$ (b) $x - \sqrt{3}y = 16$ (c) $3x \pm 4y = 10$ **2.** $\pm 2\sqrt{2}x + y = 6$ **3.** $y - x = 2, x - y = 2$

4. $x + y = \pm 5\sqrt{2}$ **5.** (a) $2x \pm 3\sqrt{5}y = 14$ (b) $3x - 4y = 25$

6. (i) $\frac{x-2}{1/\sqrt{5}} = \frac{y-3}{2/\sqrt{5}} = r$ (ii) $\frac{x-1}{-\sqrt{3}/2} = \frac{y-4}{1/2} = r$ (iii) $\frac{x-1}{3/\sqrt{10}} = \frac{y-3}{-1/\sqrt{10}} = r$

7. $(3, 7/2)$ and $(-1/3, -3/2)$ **8.** $\frac{18}{11}\left(3\sqrt{3}+4\right)$ **9.** $(2, 7), (-6, -1)$ **10** $(-3/2, 3/2)$ and $(1/2, -3/2)$

11. $9x - 20y + 96 = 0$ **12.** $A(5, 0)$; $B(0, 12)$; $D(17, 5)$ or $A(-29, 0)$, $B(0, -12)$, $D(-17, 29)$

13. $P \equiv \left(\frac{544}{19}, \frac{1020}{19}\right); Q \equiv \left(-\frac{240}{19}, \frac{792}{19}\right)$

TEXTUAL EXERCISE–2 (OBJECTIVE)

1. A line has intercepts 'a' and 'b' on the co-ordinate axes. If keeping the origin fixed, the co-ordinate axes are rotated through 90° (in anti-clockwise direction), same line has intercepts p and q, then
   (a) $p = a, q = b$ (b) $p = b, q = -a$
   (c) $b^2 + q^2 = a^2 + p^2$ (d) $\frac{1}{a^2} + \frac{1}{q^2} = \frac{1}{b^2} + \frac{1}{p^2}$

2. The line L has intercepts a and b on the co-ordinate axes. Keeping the origin fixed, the co-ordinate axes are rotated through a fixed angle. The line L has now intercepts p and q on the rotated axes. Then
   (a) $a^2 + b^2 = p^2 + q^2$ (b) $1/a^2 + 1/b^2 = 1/p^2 + 1/q^2$
   (c) $a^2 + p^2 = b^2 + q^2$ (d) $1/a^2 + 1/p^2 = 1/b^2 + 1/q^2$

3. The equation $\sqrt{3}x + y + 2 = 0$ in the normal form is
   (a) $x\cos\frac{7\pi}{6} + y\sin\frac{7\pi}{6} = 2$
   (b) $x\cos\frac{7\pi}{6} + y\sin\frac{7\pi}{6} = 1$
   (c) $x\cos\frac{5\pi}{6} + y\sin\frac{5\pi}{6} = 8$
   (d) None of these

4. The equation $3x + 4y - 12 = 0$ in its slope intercept form and the intercept it makes on y-axis are respectively
   (a) $y = \frac{4}{3}x + 2$, y-intercept = 1
   (b) $y = \frac{4}{3}x - 2$, y-intercept = 1
   (c) $y = \frac{1}{2}x + 2$, y-intercept = 5
   (d) $y = -\frac{3}{4}x + 3$, y-intercept = 3

5. The normal form of the line $4x - 3y + 5 = 0$ is
   (a) $\frac{4x}{5} - \frac{3y}{5} = 1$ (b) $\frac{4x}{5} + \frac{3y}{5} = 1$
   (c) $-\frac{4x}{5} + \frac{3y}{5} = 1$ (d) $-\frac{3x}{5} - \frac{4y}{5} = 1$

6. The lines $x \cos \alpha + y \sin \alpha = p_1$ and $x \cos\beta + y \sin\beta = p_2$ will be perpendicular if
   (a) $\alpha = \beta$ (b) $|\alpha - \beta| = \pi/2$
   (c) $\alpha = \pi/2$ (d) $\alpha \pm \beta = \pi/2$

7. Angles made with the x-axis by two lines drawn through the point (1, 2) and cutting the line $x + y = 4$ at a distance $1/3\sqrt{6}$ from the point (1, 2) are

(a) $\frac{\neq}{6}$ and $\frac{\neq}{3}$ (b) $\frac{\neq}{8}$ and $\frac{3\neq}{8}$

(c) $\frac{\neq}{12}$ and $\frac{5\neq}{12}$ (d) None of these

8. If p_1 and p_2 are the perpendiculars from the origin to the lines $x \sec\alpha + y \operatorname{cosec}\alpha = a$ and $x \cos\alpha - y \sin\alpha = a \cos 2\alpha$. Then $4p_1^2 + p_2^2$ equals

(a) $2a^2$ (b) a^2

(c) $3a^2$ (d) None of these

9. Straight line through point P (0, 3) making an angle 30° with positive X-axis meets line $x + y = 6$ at Q. The length PQ and if distance of a point R along the line from (0, 3) is 6, then the co-ordinates of R are respectively.

(a) Length: $\frac{6}{\sqrt{3}-1}$; co-ordinates: $(3\sqrt{3}, 6)$; $(-3\sqrt{3}, 0)$

(b) Length: $\frac{6}{\sqrt{3}-1}$; co-ordinates: $(3\sqrt{3}, 6)$; $(3\sqrt{3}, 0)$

(c) Length: $\frac{6}{\sqrt{3}+1}$; co-ordinates: $(3\sqrt{3}, 6)$; $(3\sqrt{3}, 0)$

(d) Length: $\frac{6}{\sqrt{3}-1}$; co-ordinates: $(3\sqrt{3}, 6)$; $(3\sqrt{3}, 0)$

Answer Keys

1. (b, c) 2. (b) 3. (b) 4. (d) 5. (c) 6. (b) 7. (c) 8. (b) 9. (a)

■ POSITION OF POINT WITH RESPECT TO LINE

Two points $P(x_1, y_1)$ and $Q(x_2, y_2)$ lie on the same side or on the opposite side of the line $ax + by + c = 0$ according as $ax_1 + by_1 + c$ and $ax_2 + by_2 + c$ are of the same sign or opposite signs respectively.

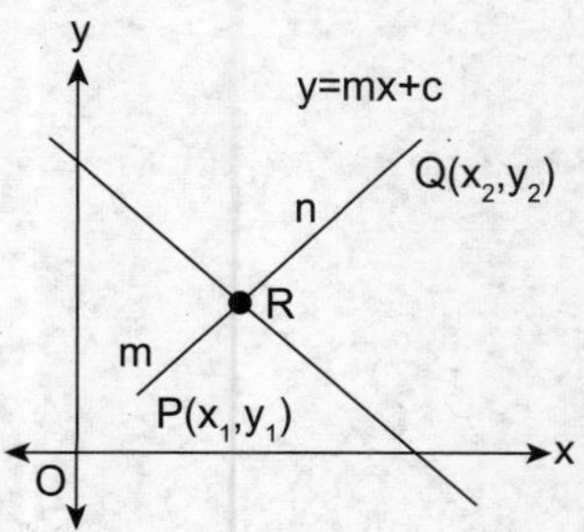

FIGURE 2.34

Proof: The co-ordinates of the point R which divides the line joining P and Q in the ratio $m : n$ are

$$\left(\frac{mx_2 + nx_1}{m+n}, \frac{my_2 + ny_1}{m+n}\right)$$

If this point lie on $ax + by + c = 0$, then

$$a\left(\frac{mx_2 + nx_1}{m+n}\right) + b\left(\frac{my_2 + ny_1}{m+n}\right) + c = 0$$

$$\Rightarrow m(ax_2 + by_2 + c) + n(ax_1 + by_1 + c) = 0$$

$$\Rightarrow \frac{m}{n} = -\left(\frac{ax_1 + by_1 + c}{ax_2 + by_2 + c}\right)$$

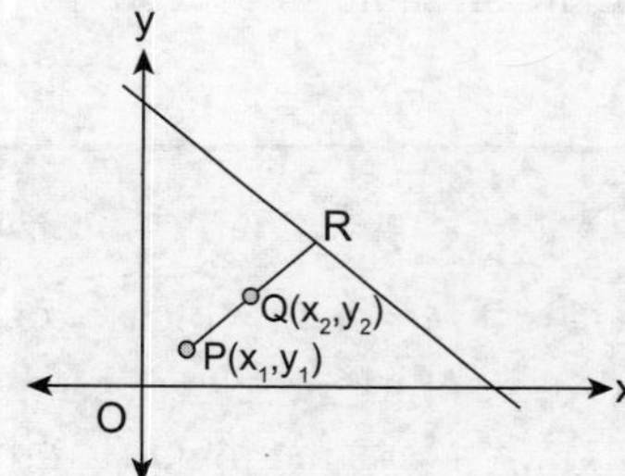

FIGURE 2.35

❑ If the point R is between the points P and Q, then the ratio $m : n$ is positive. So from the above equation, we get

$$-\left(\frac{ax_1 + by_1 + c}{ax_2 + by_2 + c}\right) > 0 \Rightarrow \frac{ax_1 + by_1 + c}{ax_2 + by_2 + c} < 0$$

$\therefore$ $ax_1 + by_1 + c$ and $ax_2 + by_2 + c$ are of opposite signs

❑ If point R is not between P and Q, then the ratio $m : n$ is negative

$$\Rightarrow -\left(\frac{ax_1 + by_1 + c}{ax_2 + by_2 + c}\right) < 0 \Rightarrow \frac{ax_1 + by_1 + c}{ax_2 + by_2 + c} > 0$$

$\Rightarrow$ $ax_1 + by_1 + c$ and $ax_2 + by_2 + c$ are of same sign.

NOTE

If the location of single point is to be defined, then the other point taken as origin and w.r.t. origin, the location of the point w.r.t. the line is defined.

Aid to memory: Two points $P(x_1, y_1)$ and $Q(x_2, y_2)$ will be located at the same side of the line. if they give the same sign. of the expression $ax_i + by_i + c \; \forall i \in \{1, 2\}$. Otherwise they will lie on the opposite side of the line.

ILLUSTRATION 47: Find out whether the points (2, 3); (–4, 5) are on the same or opposite side of the line $3x - 4y = 8$.

SOLUTION: The essential condition for the points to be on the same side or opposite side of line is that, $ax_1 + by_1 + c$ and $ax_2 + by_2 + c$ are of same sign or opposite signs respectively.

Let $c = 3x - 4y - 8$ then value of c_1 at (2, 3) is

$3 \times 2 - 4 \times 3 - 8 \Rightarrow c_1 = 6 - 12 - 8 \Rightarrow c_1 = -14$

c_2 at (–4, 5) is given by $c_2 = 3 \times -4 - 4 \times 5 - 8 \Rightarrow c_2 = -40$

Since c_1 and c_2 are of same sign, therefore the two points are on the same side of the given line.

Position of a Point with Respect to a Triangle

A point $P(\alpha, \beta)$ lies within the triangle iff

(i) P and A lie on same side of BC
(ii) P and B lie on same side of AC
(iii) P and C lie on same side of AB

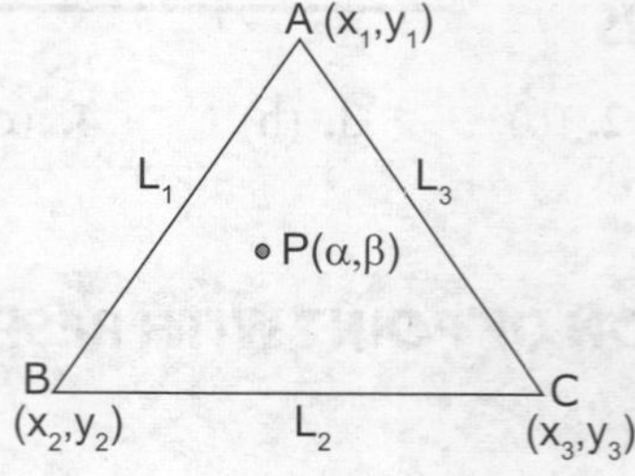

FIGURE 2.36

ILLUSTRATION 48: Find 'a' if $(\alpha, 3\alpha)$ lies inside the triangle having sides $x + y - 1 = 0$; $2x + y - 5 = 0$ and $x - y + 3 = 0$

SOLUTION: ***Method1:*** (Please refer to chapter of point and cartesian system)

Method 2: Let us consider lines $L_1 : x + y - 1 = 0$; $L_2 : 2x + y - 5 = 0$ and $L_3 : x - y + 3 = 0$

Now, the point $(\alpha, 3\alpha)$ will be on the line $y = 3x$

$\therefore$ Lets consider $L : y - 3x = 0$

Point of intersection of line L with

$$L_1 \Rightarrow P \equiv \left(\frac{1}{4}, \frac{3}{4}\right)$$

$$L_2 \Rightarrow Q \equiv (1, 3)$$

$$L_3 \Rightarrow R \equiv \left(\frac{3}{2}, \frac{9}{2}\right)$$

Clearly, P point lies on the line segment AB.

Similarly, Q point lies on the line segment AC.

But R point doesn't lie on the line segment BC as abscissa of C < abscissa of R

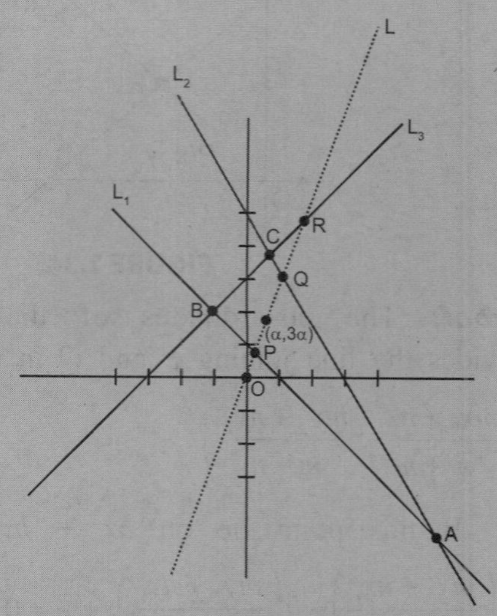

FIGURE 2.37

$\therefore$ P and Q points will determine the boundaries of $(\alpha, 3\alpha)$

Comparing the abscissae of P, $(\alpha, 3\alpha)$ and Q:

we get $\frac{1}{4} < \alpha < 1$

ANGLE BETWEEN TWO STRAIGHT LINES

Consider two straight lines $y = m_1x + c_1$ and $y = m_2x + c_2$. as shown in the figure, where β and α are the angles made by the given lines with the positive x-axis. Therefore we have $\tan\alpha = m_1$ and $\tan\beta = m_2$ and the angles between the straight lines θ. Then the two angles θ and $\pi - \theta$ between these straight lines are supplementary angles as shown in Figure 2.38..

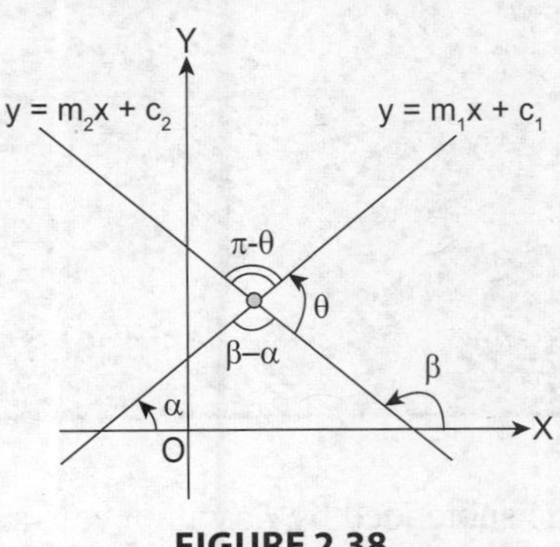

FIGURE 2.38

From the figure, it is clear that the two angles between the two straight lines are $\theta = \pi - \beta + \alpha$ and the other angle is $(\beta - \alpha)$. So angle θ can be given by the equation

$$\tan\theta = \pm\tan(\alpha - \beta) = \pm\frac{\tan\alpha - \tan\beta}{1 + \tan\alpha\,\tan\beta} = \pm\frac{m_1 - m_2}{1 + m_1m_2}$$

If $\frac{m_1 - m_2}{1 + m_1m_2} > 0$, then $\tan\theta = \frac{m_1 - m_2}{1 + m_1m_2}$ gives the acute angle between the two straight lines and

If $\frac{m_1 - m_2}{1 + m_1m_2} < 0$, then $\tan\theta = \frac{m_1 - m_2}{1 + m_1m_2}$ gives the obtuse angle between the straight lines.

ILLUSTRATION 49: The angle between two lines passing through a point (2, 3) is 45°. If the slope of one of the lines is 2, find the slope of the other.

SOLUTION: Let m be the slope of other line

$\Rightarrow \tan 45° = \left|\frac{m - 2}{1 + 2m}\right|$

$\Rightarrow |1 + 2m| = |m - 2|$

$\Rightarrow m = -3$ or $1/3$

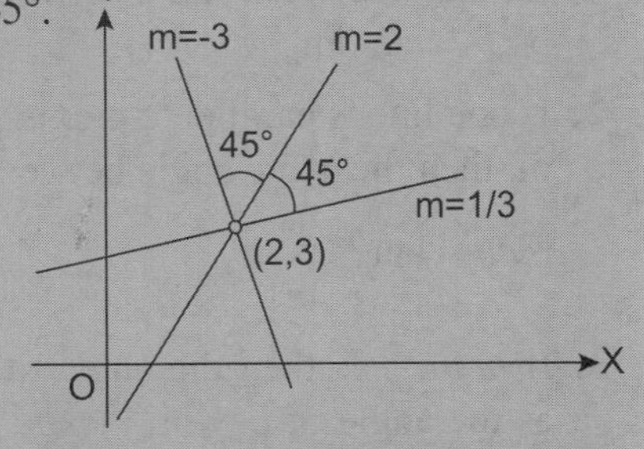

FIGURE 2.39

ILLUSTRATION 50: Two equal sides of an isosceles triangle are given by the equations $7x - y + 3 = 0$ and $x + y - 3 = 0$ and its third side passes through the point $(1, -10)$. Determine the equation of the third side.

SOLUTION: Let AB and AC be equal sides of isosceles triangle ABC as shown Figure 2.40.

Equation of AB is $7x - y + 3 = 0$, then the slope of $AB = 7$

Equation of AC is $x + y - 3 = 0$, then the slope of $AC = -1$

Let the slope of BC be m

$\therefore \left|\frac{m - 7}{1 + 7m}\right| = \left|\frac{-1 - m}{1 + (-1)m}\right| \Rightarrow \frac{m - 7}{1 + 7m} = \pm\left(\frac{-1 - m}{1 - m}\right)$

$\Rightarrow 6m^2 + 16m - 6 = 0$ for positive sign and $m^2 + 1 = 0$ for negative sign

$\Rightarrow 3m^2 + 8m - 3 = 0$ as $m^2 + 1 = 0$ is impossible for real 'm'

$\Rightarrow 3m^2 + 9m - m - 3 = 0$

$\Rightarrow (3m - 1)(m + 3) = 0 \Rightarrow m = \frac{1}{3}, -3$

$\therefore$ Required equation of possible lines are given by

$y + 10 = \frac{1}{3}(x - 1)$ and $y + 10 = -3(x - 1)$ i.e., $x - 3y = 31$ and $3x + y + 7 = 0$

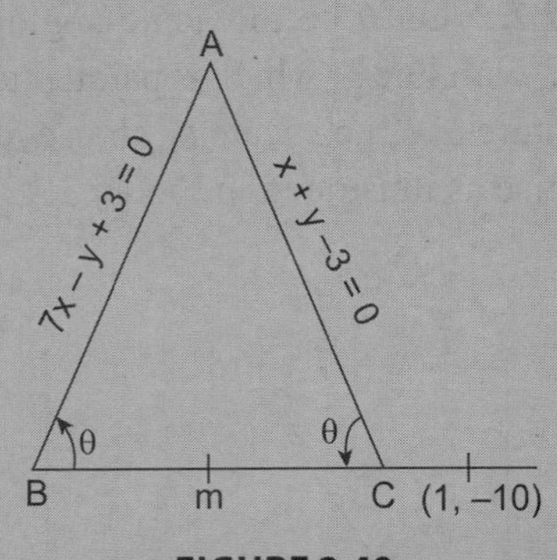

FIGURE 2.40

ILLUSTRATION 51: Two sides of a rhombous $ABCD$ are parallel to the lines $y = x + 2$ and $y = 7x + 3$. If the diagonals of the rhombous intersect at the point (1, 2) and the vertex A is on the y-axis, find the possible co-ordinates of A.

SOLUTION: Let the co-ordinates of point A be $(0, \lambda)$ as shown Figure 2.41.

Slope of $AP = \dfrac{\lambda-2}{0-1} = 2-\lambda$

$$\Rightarrow \tan\theta = \frac{(2-\lambda)-1}{1+(2-\lambda)} = \frac{7-(2-\lambda)}{1+7(2-\lambda)} \Rightarrow \frac{1-\lambda}{3-\lambda} = \frac{5+\lambda}{15-7\lambda}$$

$\Rightarrow (\lambda-1)(7\lambda-15) = (5+\lambda)(3-\lambda)$

$\Rightarrow 7\lambda^2 - 22\lambda + 15 + \lambda^2 + 2\lambda - 15 = 0$

$\Rightarrow 8\lambda^2 - 20\lambda = 0 \therefore \lambda = 0, 5/2$

$\Rightarrow A = (0, 0)$ and $(0, 5/2)$.

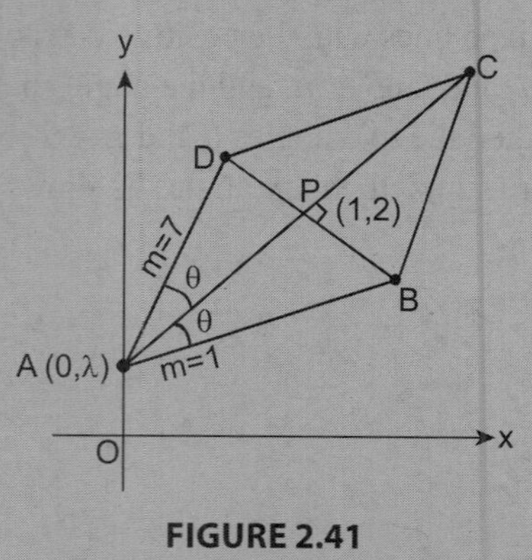

FIGURE 2.41

Some Important Results

1. If one line is parallel to x-axis and slope of the other is m, then the acute angle between the two lines is given by $\theta = \tan^{-1}|m|$.
2. If one line is parallel to y-axis and slope of the other is m, then the acute angle between the two lines is given by $\theta = \tan^{-1}\dfrac{1}{|m|}$.
3. If $m_1m_2 = 1$; then the angle of L_1 with x-axis is same as the angle of L_2 with y-axis. Hence both lines make same angle with the lines $y = x + k$ and $y = -x + k$ (where k is any constant)
4. If $m_1 + m_2 = 0$, then L_1 and L_2 form an isosceles triangle with x-axis as well as with y-axis.

Proof 1: Let θ be the acute angle subtended by straight line L_2 with line L_1 which is parallel to x-axis. Depending on the nature of θ, i.e., acute or obtuse following two possibilities arise as illustrated in Figure 2.42.

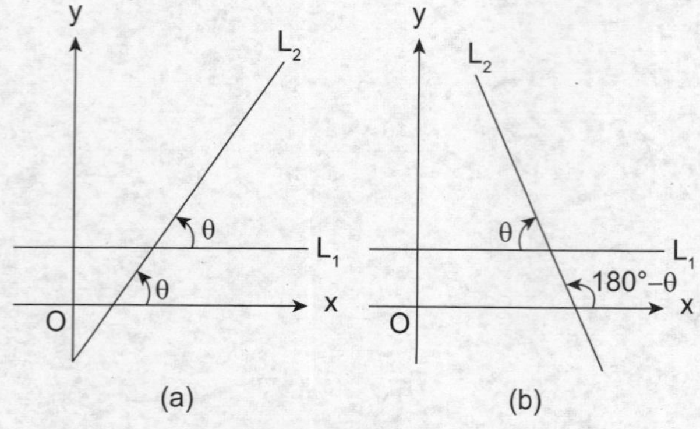

FIGURE 2.42

when the angle subtended by L_2 with +ve x-axis is acute as shown in figure (a), now L_2 being a transversal of parallel lines L_1 and x-axis subtend an acute angle θ with L_1 as well as with +ve x-axis and hence slope of $L_2 = \tan\theta = m = |m|$ (as $m > 0$) $\Rightarrow \theta = \tan^{-1}|m|$ i.e., the acute angle between two lines is $\tan^{-1}|m|$

When angle subtended by L_2 with positive direction of x-axis is obtuse as shown in figure (b), then slope of $L_2 = \tan(180° - \theta) = m = -|m|$ (as $m < 0$) $\Rightarrow -\tan\theta = -|m|$ $\Rightarrow \theta = \tan^{-1}|m|$

Proof 2: Let ϕ be the acute angle between L_1 and L_2. Let L_1 be parallel to y-axis. Now depending upon the nature of angle (θ) subtended by L_2 the +ve direction of x-axis i.e., acute or obtuse there arise two possibilities as illustrated in Figure 2.43.

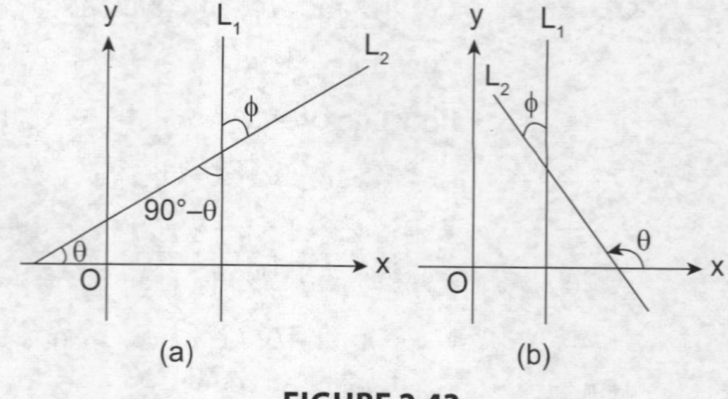

FIGURE 2.43

When the angle subtended by line L_2 with the +ve direction of x-axis is acute i.e., $\theta < 90°$ as illustrated in figure (a), then slope of $L_2 = \tan\theta = m = |m|$ (as $m > 0$). Now ϕ is the angle between L_1 and L_2, then $\phi = 90° - \theta \Rightarrow \tan\phi = \cot\theta \Rightarrow \tan\phi = 1/\tan\theta = 1/|m| \Rightarrow \phi = \tan^{-1}(1/|m|)$.

When the angle subtended by line L_2 with the +ve direction of x-axis is obtuse i.e., $\theta > 90°$ as illustrated in figure (b), then slope of $L_2 = \tan\theta = m = -|m|$ (as $m < 0$). Now ϕ is the angle between L_1 and L_2, then $\phi = \theta - 90° \Rightarrow \tan\phi = -\cot\theta = -1/\tan\theta = 1/|m| \Rightarrow \phi = \tan^{-1}(1/|m|)$

Proof 3: Let θ be the angle subtended by lines L_1 and L_2 with the positive direction of x-axis and y-axis respectively as shown in Figure 2.44.

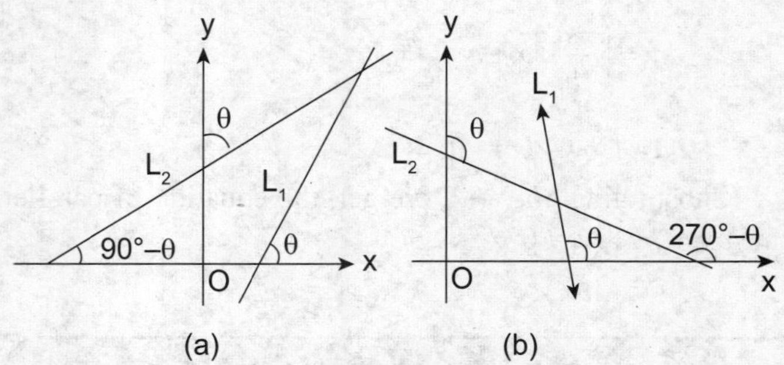

FIGURE 2.44

If θ is acute as shown in figure (a), then slope of L_1 $(m_1) = \tan\theta$ and slope of L_2 $(m_2) = \tan(90° - \theta) = \cot\theta \Rightarrow m_1 m_2 = 1$.

If θ is obtuse as shown in figure (b), then slope of L_1 $(m_1) = \tan\theta$ and slope of L_2 $(m_2) = \tan(270° - \theta) = \cot\theta \Rightarrow m_1 m_2 = 1$.

Let us prove next part by taking the line $y = x$ and $y = -x$ as shown in Figure 2.45.

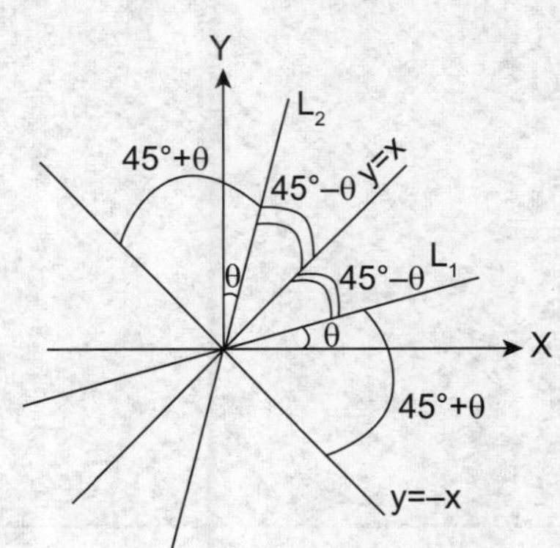

FIGURE 2.45

If L_1 makes an angle θ with + *ve* x-axis.

$\Rightarrow$ $m_1 = \tan\theta$ $\Rightarrow$ $m_2 = \tan(90° - \theta)$

$\Rightarrow$ $m_2 = \cot\theta$ $\Rightarrow$ $m_1 m_2 = \tan\theta \cdot \cot\theta = 1$

$\therefore$ L_2 makes an angle of $(90° - \theta)$ with (+) *ve* x-axis

As is evident from the diagram; the lines L_1 and L_2 makes angle $45° - \theta$ with the line $y = x$ and similarly L_1 and L_2 makes an angle of $45° + \theta$ with the line $y = -x$

Now; we know; that the slope of the line does not change by addition of a constant on either side of equation of a straight line.

$\Rightarrow$ Slope of $y = x$ is the same as the slope of $y = x + k$ and similarly slope of $y = -x$ is the same as that of $y = -x + k$

$\therefore$ the angles that L_1 and L_2 make with $y = x$ and $y = -x$ will be same with the lines $y = x + k$ and $y = -x + k$ respectively

Proof 4: Let $m_1 + m_2 = 0 \Rightarrow m_2 = -m_1 \Rightarrow$ if θ is the angle subtended by L_2 with the positive direction of x-axis, then $180° - \theta$ is the angle subtended by L_1 with the positive direction of x-axis as shown in Figure 2.46.

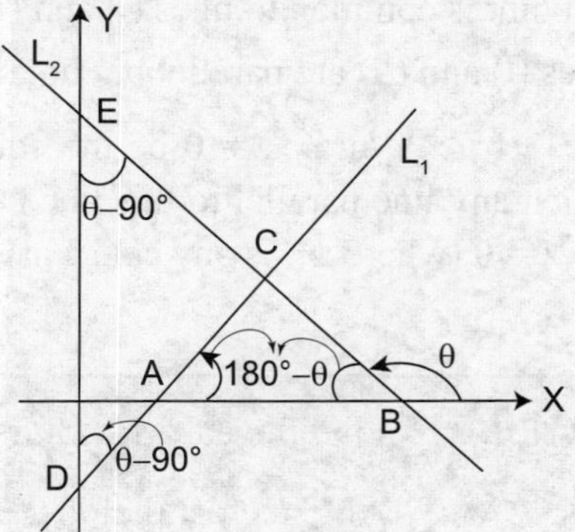

FIGURE 2.46

Clearly, $\angle CAB = \angle CBA = 180° - \theta$ and $\angle CED = \angle CDE = \theta - 90°$.

$\Rightarrow$ ΔABC and ΔCED are isosceles triangles.

Conditions for two Lines to be Parallel/Coincident/Perpendicular

Let $a_1x + b_1y + c_1 = 0$ .....(i)

and $a_2x + b_2y + c_2 = 0$ ......(ii)

be to given lines.

1. *Condition for lines to be parallel (not coincident)*

If $\frac{a_1}{a_2} = \frac{b_1}{b_2} \neq \frac{c_1}{c_2}$, then the two lines are parallel (but not coincident)

Proof: $\frac{a_1}{a_2} = \frac{b_1}{b_2} \Rightarrow \frac{a_1}{b_1} = \frac{a_2}{b_2}$

$\Rightarrow$ $-\frac{a_1}{b_1} = \frac{-a_2}{b_2} \Rightarrow m_1 = m_2$

$\Rightarrow$ $\tan\alpha = \frac{m_1 - m_2}{1 + m_1 m_2} = 0 \Rightarrow \alpha = 0$

$\therefore$ Lines are parallel

In other words, we can say; that if two lines (with slopes m_1 and m_2) are parallel, then $m_1 = m_2$

Now let $\frac{a_1}{a_2} = \frac{b_1}{b_2} = \lambda \neq \frac{c_1}{c_2}$, then (x', y') lying on (i)

$\Rightarrow a_1x' + b_1y' = -c_1$ ......(iii)

Now $a_2x' + b_2y' = \frac{a_1}{\lambda}x' + \frac{b_1y'}{\lambda}$

$= \frac{1}{\lambda}(a_1x' + b_1y') = -\frac{c_1}{\lambda} \neq -c_2$ using (iii)

Thus $a_2x' + b_2y' \neq -c_2$

$\Rightarrow$ (x', y') does not lie on (ii)

$\Rightarrow$ No point is common to lines (i) and (ii)

$\Rightarrow$ Lines (i) and (ii) are parallel but not coincident

Application: If $ax + by + c = 0$ be the equation of the given line, then any line parallel to this line can be written as $ax + by + \lambda = 0$ (where 'λ' is any constant)

Proof: For $ax + by + c = 0$, slope $= m = \frac{-a}{b}$

Any line parallel to $ax + by + c = 0$ will also have slope = 'm'

$\therefore$ Equation of parallel line will be given by

$$y = \left(\frac{-a}{b}\right)x + c$$

$\Rightarrow ax + by - bc = 0$

Substituting $-bc = \lambda$, we get, the equation of parallel line $ax + by + \lambda = 0$

ILLUSTRATION 52: Find the equation of a line passing through (4, 7) and parallel to $2x + 3y = 7$.

SOLUTION: Any line parallel to $2x + 3y = 7$ is $2x + 3y = \lambda$

This line passes through the point (4, 7) $\therefore 2(4) + 3(7) = \lambda \Rightarrow \lambda = 29$

therefore equation of required line is $2x + 3y = 29$

ILLUSTRATION 53: Find the equation of a straight line parallel to $2x + 3y + 11 = 0$ and which is such that the sum of its intercepts on the axes is 15.

SOLUTION: The equation of the line parallel to the line $2x + 3y + 11 = 0$ is

$2x + 3y + \lambda = 0$ ..(i)

where λ is a constant

The intercept by the line on x-axis is given by $= -\lambda/2$

and on the y-axis $= -\lambda/3$

It is given that the sum of the intercepts on the axes is 15.

$\therefore -(\lambda/2) + (-\lambda/3) = 15 \Rightarrow -5\lambda/6 = 15 \Rightarrow \lambda = -18$

$\therefore$ the equation of required line is $2x + 3y - 18 = 0$

2. *Condition for lines to be coincident*

If $\frac{a_1}{a_2} = \frac{b_1}{b_2} = \frac{c_1}{c_2}$, then the two lines will be coincident.

Proof: In condition (1) we already proved that, if $\frac{a_1}{a_2} = \frac{b_1}{b_2}$, then the lines $a_1x + b_1y + c_1 = 0$ and $a_2x + b_2y + c_2 = 0$ have same slope i.e., are parallel.

Now let $\frac{a_1}{a_2} = \frac{b_1}{b_2} = \frac{c_1}{c_2} = \lambda$

$\Rightarrow a_1 = a_2\lambda,\ b_1 = b_2\lambda,\ c_1 = c_2\lambda$

Now if (x', y') lies on line (i),

then $a_1x' + b_1y' + c_1 = 0$

$\Rightarrow a_2\lambda x' + b_2\lambda y' + c_2\lambda = 0$

$\Rightarrow \lambda(a_2x' + b_2y' + c_2) = 0$

$\Rightarrow a_2x' + b_2y' + c_2 = 0$

(x', y') also lies on line (ii), thus every point lying on line (i) also lies on line (ii) and hence the two lines are coincident.

3. *Condition for perpendicularity of two lines*

Two straight lines $a_1x + b_1y + c_1 = 0$ and $a_2x + b_2y + c_2 = 0$ will be perpendicular to each other if $a_1a_2 + b_1b_2 = 0$

or $\frac{a_1}{b_1} = -\frac{b_2}{a_2}$

$\Rightarrow -m_1 = 1/m_2$ or $m_2 = -1/m_1$

or $m_1m_2 = -1$ i.e., the product of slopes of two straight lines is -1

Proof: Let θ be the angle between the two straight lines and m_1, m_2 be their slopes, then $\tan\theta = \dfrac{m_1 - m_2}{1 + m_1m_2}$

Now if the two straight lines are perpendicular to each other, then $\theta = 90° \Rightarrow \tan\theta = \infty$

$$\Rightarrow \frac{m_1 - m_2}{1 + m_1m_2} = \infty$$

$$\Rightarrow 1 + m_1m_2 = 0$$

$$\Rightarrow m_1m_2 = -1$$

$$\Rightarrow m_1 = -1/m_2$$

$$\Rightarrow \frac{-a_1}{b_1} = \frac{b_2}{a_2}$$

$$\Rightarrow a_1a_2 + b_1b_2 = 0$$

Application: If equation of given line be $ax + by + c = 0$; then any line perpendicular to it will be of the form $bx - ay + \lambda = 0$

Proof: Slope of given line $m_1 = \dfrac{-a}{b}$

Let the slope of the required line be m_2

We know, that $m_1m_2 = -1$ for perpendicularity

$$\Rightarrow m_2 \times \frac{-a}{b} = -1 \Rightarrow m_2 = \frac{b}{a}$$

$\therefore$ Requied line is $y = m_2x + c \Rightarrow y = \dfrac{b}{a}x + c$

$$\Rightarrow ay - bx - ac = 0$$

Substituting $-ac = \lambda$ (a constant); we get

$$ay - bx = \lambda$$

i.e., interchange co-efficients of x and y and change the sign of any one of them and replace the constant by λ.

ILLUSTRATION 54: Find the magnitudes of the angles of the triangle the equations of whose sides are $x + y - 2 = 0$, $x - y + 2 = 0$ and $y + 2 = 0$.

SOLUTION: Let $L_1 = x + y - 2 = 0$ ....(AB)

$L_2 = x - y + 2 = 0$.......... (BC)

$L_3 = y + 2 = 0$ .... (AC)

Clearly, $m_{AB} = -1$; $m_{BC} = 1$; $m_{AC} = 0$

$\Rightarrow (m_{AB})(m_{BC}) = -1$

$\Rightarrow \angle B = 90°$ and $\angle A = \angle C = 45°$

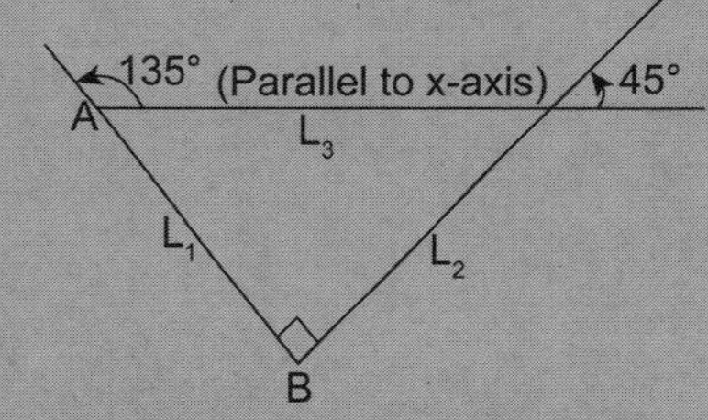

FIGURE 2.47

ILLUSTRATION 55: One side of a rectangle lies along the line $4x + 7y + 5 = 0$. Two of its vertices are $(-3, 1)$ and $(1, 1)$. Find the equations of the other three sides.

SOLUTION: Let the equation of AD be $4x + 7y + 5 = 0$; As $BC \parallel AD$, therefore the equation of BC be $4x + 7y + \lambda = 0$

As it passes through $(1, 1)$ we get; $4 + 7 + \lambda = 0$

$\Rightarrow \lambda = -11. \Rightarrow$ Equation of BC is $4x + 7y - 11 = 0$

Equation of AB is $y - 1 = \dfrac{7}{4}(x + 3)$ or $7x - 4y + 25 = 0$

Let equation of CD be; $7x - 4y + k = 0$

As it passes through $(1, 1)$ we get; $7 - 4 + k = 0 \Rightarrow k = -3$

$\therefore$ equation of CD is $7x - 4y - 3 = 0$

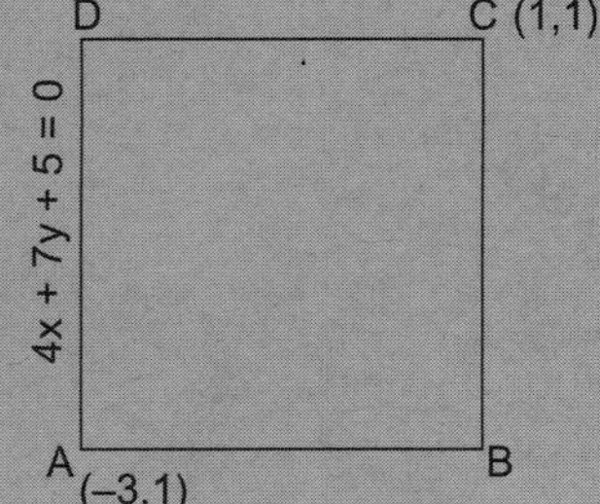

FIGURE 2.48

ILLUSTRATION 56: The line $3x + 2y = 24$ meets the y-axis at A and the x-axis at B. The perpendicular bisector of AB meets a line through $(0, -1)$ parallel to x-axis at C. Find the area of the triangle ABC.

SOLUTION: Let the equation of CD be $2x - 3y = \lambda$ ...(1)

As it passes through $(4, 6)$, $\Rightarrow \boxed{\lambda = -10}$ (As CD is perpendicular bisector of AB, it passes through mid-point of AB)

$\therefore$ equation of $CD : 2x - 3y + 10 = 0$ ...(2)

co-ordinate of $C\left(-\frac{13}{2}, -1\right)$. Now the area of

$$\text{triangle } \Delta ABC = \frac{1}{2}\left\|\begin{matrix} 0 & 12 & 1 \\ 8 & 0 & 1 \\ -\frac{13}{2} & -1 & 1 \end{matrix}\right\|$$

$$= \frac{1}{2}\left|0 - 12\left(8 + \frac{13}{2}\right) + 1(-8)\right| = \frac{1}{2}\left|-6(29) - 8\right|$$

$= 91$ sq. units

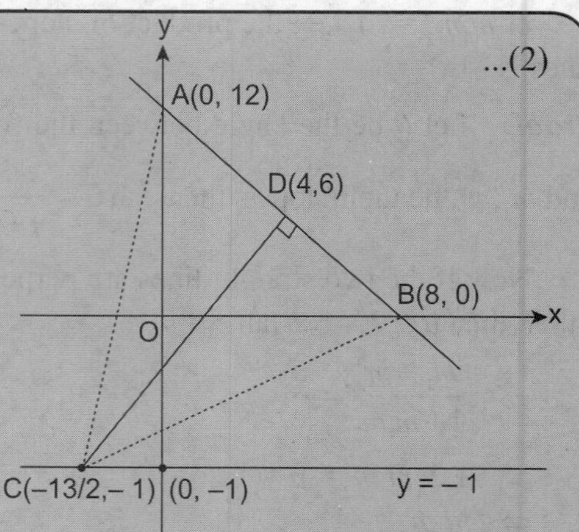

FIGURE 2.49

ILLUSTRATION 57: An equilateral triangle has its centroid at the origin and one side is $x + y = 1$. Find the other sides of the triangle.

SOLUTION: Let the vertex not on the line $x + y = 1$, be $A(x_1, y_1)$. Let the other two vertices be $B(b, 1 - b)$ and $C(c, 1 - c)$. Note that we have here used the fact that B and C lie on $x + y = 1$. We are given that G is $(0, 0)$. Let AG meet $x + y = 1$ at D. Since G is the centroid, we have

$$b + c + x_1 = 0 = 1 - b + 1 - c + y_1 \quad \text{... (1)}$$

But $AD \perp BC$ (the triangle is equilateral).

$\therefore$ 'm' of $AD \times$ 'm' of $BC = -1$.

$\frac{y_1}{x_1} \times (-1) = -1$ so that $y_1 = x_1$.

Substituting this in (1) we get $b + c = 1$. This gives $c = 1 - b$. Thus the three vertices of the triangle are $A(x_1, x_1)$, $B(b, 1 - b)$ and $C(1 - b, b)$. Again equation (1) gives

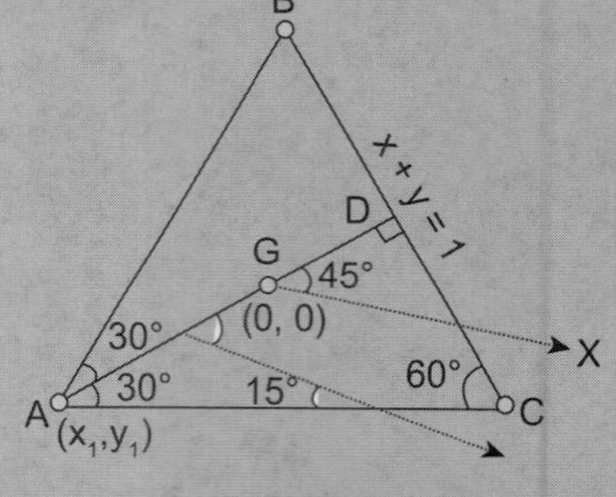

FIGURE 2.50

$x_1 + b + 1 - b = 0$, which means $x_1 = -1$. Hence A is $(-1, -1)$.

Now equation to AC is $y + 1 = \frac{b+1}{2-b}(x+1)$

which reduces to $y = \frac{b+1}{2-b}x + \frac{2b-1}{2-b}$ ... (2)

Similarly, equation to AB is $y = \frac{2-b}{b+1}x + \frac{1-2b}{b+1}$ ... (3)

Now we have to find only b. Now angle between AC and BC is 60°.

$$\Rightarrow \sqrt{3} = \left|\frac{\frac{b+1}{2-b} + 1}{1 + \left(\frac{b+1}{2-b}\right)(-1)}\right| \Rightarrow \frac{3}{1-2b} = \pm\sqrt{3} \Rightarrow b = \frac{1 \pm \sqrt{3}}{2} \quad \text{... (4)}$$

$\Rightarrow$ B and C have co-ordinates $\left(\frac{1+\sqrt{3}}{2}, \frac{1-\sqrt{3}}{2}\right)$ and $\left(\frac{1-\sqrt{3}}{2}, \frac{1+\sqrt{3}}{2}\right)$ respectively, or $\left(\frac{1-\sqrt{3}}{2}, \frac{1+\sqrt{3}}{2}\right)$ and $\left(\frac{1+\sqrt{3}}{2}, \frac{1-\sqrt{3}}{2}\right)$

$\therefore$ Equation of AC and AB will be among $(y + 1) = (2 - \sqrt{3})(x + 1)$ and $(y + 1) = (2 + \sqrt{3})(x + 1)$

Aliter: Let the vertex not in the line $x + y = 1$ be $A(x_1, y_1)$

We know that AD ⊥r BC

$\Rightarrow m_{AD} = 1 \Rightarrow \angle XGD = 45°$

and $\angle DAB = 30°$

inclination of AB with +ve x-axis is 75°

also $\angle DAC = 30°$

inclination of AC with x-axis is 15°

$$\therefore \quad m_{AB} = \tan 75° = \frac{\sqrt{3}+1}{\sqrt{3}-1} = 2+\sqrt{3}$$

$$m_{AC} = \tan 15° = \frac{\sqrt{3}-1}{\sqrt{3}+1} = 2-\sqrt{3}$$

Equation of line AB:

$(y + 1) = (2 + \sqrt{3})(x + 1)$ and equation of line AC will be $y + 1 = (2 - \sqrt{3})(x + 1)$

TEXTUAL EXERCISE–3 (SUBJECTIVE)

1. Find whether the following pairs of lines are coincident, parallel (not coincident), intersecting or perpendicular:

(a) $2x + 3y - 5 = 0$, $4x + 6y - 10 = 0$
(b) $3x - 4y = 5$, $3x + 4y = 7$
(c) $x - 2y = 7$, $x - 2y = 5$

2. Show that line joning (2, –3) and (–5, 1) is parallel to the line joining (7, –1) and (0, 3).

3. Prove that the points (–4, –1); (–2, –4); (4, 0) and (2, 3) are the vertices of a rectangle.

4. Find the angle between the lines $x/a + y/b = 1$ and $x/a - y/b = 1$.

5. Find the obtuse angle between the lines $x - 2y + 3 = 0$ and $3x + y - 1 = 0$.

6. Find the value of k for which the line $3x + 5y + 6 + k(2x - 3y + 1) = 0$ is perpendicular to the line $7x + 5y - 4 = 0$.

7. A vertex of an equilateral triangle is at (2, 3) and the opposite side is $x + y = 2$. Find the equation to the other sides of the triangle.

Answer Keys

1. (i) coincident (ii) intersecting (iii) parallel **4.** $\tan^{-1}\dfrac{2ab}{a^2 - b^2}$

5. $\pi - \tan^{-1}(7)$ **6.** 46 **7.** $y - 3 = (2 \pm \sqrt{3})(x - 2)$

TEXTUAL EXERCISE–3 (OBJECTIVE)

1. The co-ordinates of the orthocentre of the triangle whose angular points are (1, 2); (2, 3) and (4, 3) is

(a) (1, 7) (b) (1, 8)
(c) (1, 6) (d) (1, 9)

2. The vertices of a triangle are (4, –3); (– 2, 1) and (2, 3). The co-ordinates of the circumcentre of the Δ are

(a) (9/7, 4/7) (b) (3/7, –4/7)
(c) (9/7, 3/7) (d) (9/7, –4/7)

3. The inradius of the triangle formed by the lines $x = 0$, $y = 0$ and $x/3 + y/4 = 1$ is
(a) 2 (b) 1
(c) 8 (d) 9

4. The line $7x - 9y - 19 = 0$ is $\perp$ to the line through the points $(\alpha, 3)$ and $(4, 1)$. The value of α is
(a) $\alpha = 22/9$ (b) $\alpha = 9/22$
(c) $\alpha = -22/9$ (d) $\alpha = -9/22$

5. The equation of the line that has x-intercept -3 and is perpendicular to the line $3x + 5y = 4$ is
(a) $x - y + 15 = 0$
(b) $5x + 3y + 15 = 0$
(c) $5x - 3y - 10 = 0$
(d) $5x - 3y + 15 = 0$

6. The triangle formed by the lines $\sqrt{3}x + y - 2 = 0$, $\sqrt{3}x - y + 1 = 0$ and $y = 0$ is
(a) an isoceles triangle
(b) a right -angled triangle
(c) an equilateral triangle
(d) None of these

7. The centroid of a triangle formed by the points $(0, 0)$; $(\cos\theta, \sin\theta)$ and $(\sin\theta, -\cos\theta)$ lies on the line $y = 2x$. Then, θ is
(a) $\tan^{-1} 2$ (b) $\tan^{-1} 1/3$
(c) $\tan^{-1} 1/2$ (d) $\tan^{-1}(-3)$

8. The triangle formed by $x^2 - 3y^2 = 0$ and $x = 4$ is
(a) isosceles (b) equilateral
(c) right angled (d) None of these

9. The equation(s) of the line/lines through $(1, 1)$ and making an angle of 45° with the line $x + y = 0$
(a) $x - 1 = 0$ (b) $x - y = 0, y - 1 = 0$
(c) $x + y - 2 = 0$ (d) $x - 1 = 0, y - 1 = 0$

Answer Keys

1. (c) 2. (d) 3. (b) 4. (a) 5. (d) 6. (a), (c) 7. (d) 8. (a), (b) 9. (d)

■ STRAIGHT LINE THROUGH A GIVEN POINT (x_1, y_1) MAKING AN ANGLE α WITH A GIVEN STRAIGHT LINE $y = mx + c$

Let $P(x_1, y_1)$ be the given point and let the given line be LMN, making an angle θ with the axis of x. Then $\tan\theta = m$.

Let PMR and PNS be two such lines which make angle α with the given line. Let these lines meet the axis of x at R and S. Let PMR and PNS make angles θ_1 and θ_2 with the positive direction of x-axis. Then the equation of the two required lines are

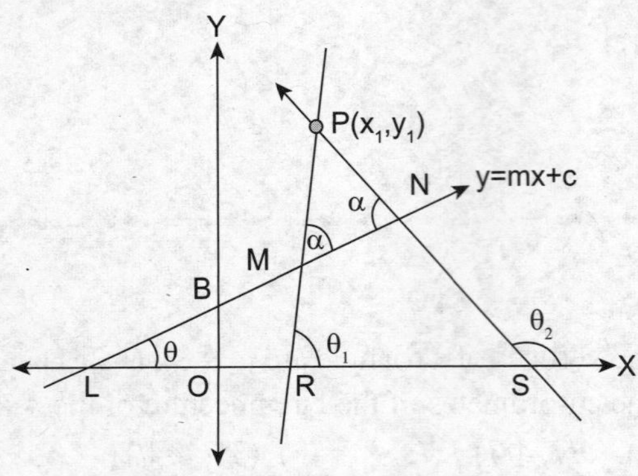

FIGURE 2.51

$$(y - y_1) = \tan\theta_1 (x - x_1) \quad \text{...(i)}$$

$$\text{and } (y - y_1) = \tan\theta_2 (x - x_1) \quad \text{..(ii)}$$

In ΔLMR, we have $\theta_1 = \theta + \alpha$ and in ΔLNS, we have $\theta_2 = \theta + 180° - \alpha$

$$\Rightarrow \tan\theta_1 = \tan(\theta + \alpha) = \frac{\tan\theta + \tan\alpha}{1 - \tan\theta\tan\alpha}$$

$$= \frac{m + \tan\alpha}{1 - m\tan\alpha} \text{ and } \theta_2 = \theta + 180° - \alpha$$

$$\Rightarrow \tan\theta_2 = \tan(180° + \theta - \alpha)$$

$$= \tan(\theta - \alpha) = \frac{\tan\theta - \tan\alpha}{1 + \tan\theta\tan\alpha}$$

$$= \frac{m - \tan\alpha}{1 + m\tan\alpha}$$

On substituting the values of $\tan\theta_1$ and $\tan\theta_2$ in (i) and (ii), we get $(y - y_1) = \dfrac{m + \tan\alpha}{1 - m\tan\alpha}(x - x_1)$ and

$$(y - y_1) = \frac{m - \tan\alpha}{1 + m\tan\alpha}(x - x_1)$$

These are the equations of the two required lines.

ILLUSTRATION 58: Find the equations of the two straight lines through (7, 9) and making an angle of 60° with the line $x - \sqrt{3}y - 2\sqrt{3} = 0$

SOLUTION: We know that the equation of two straight lines which pass through a point (x_1, y_1) and subtend a given angle

α with given straight line $y = mx + c$ are $y - y_1 = \dfrac{m \pm \tan\alpha}{1 \mp m\tan\alpha}(x - x_1)$

Here $x_1 = 7, y_1 = 9$; $\alpha = 60°$ and m = slope of the line $x - \sqrt{3}y - 2\sqrt{3} = 0$ i.e., $m = 1/\sqrt{3}$

So the equation of the required lines are $y - 9 = \dfrac{(1/\sqrt{3}) + \tan 60°}{1 - (1/\sqrt{3})\tan 60°}(x - 7)$ and

$$y - 9 = \frac{(1/\sqrt{3}) - \tan 60°}{1 + (1/\sqrt{3})\tan 60°}(x - 7)$$

or $(y - 9)\left(1 - \dfrac{\tan 60°}{\sqrt{3}}\right) = \left(\dfrac{1}{\sqrt{3}} + \tan 60°\right)(x - 7)$ and

$$(y - 9)\left(1 + \frac{1}{\sqrt{3}}\tan 60°\right) = \left(\frac{1}{\sqrt{3}} - \tan 60°\right)(x - 7)$$

or $0 = \left(\dfrac{1}{\sqrt{3}} + \sqrt{3}\right)(x - 7) \Rightarrow x - 7 = 0$

and $(y - 9)(2) = \left(\dfrac{1}{\sqrt{3}} - \sqrt{3}\right)(x - 7) \Rightarrow x + \sqrt{3}y = 7 + 9\sqrt{3}$

Hence the required lines are $x = 7$ and $x + \sqrt{3}y = 7 + 9\sqrt{3}$.

ILLUSTRATION 59: A pair of straight lines drawn through the origin form with the line $2x + 3y = 6$ an isosceles right angled triangle, then find the sides and the area of the triangle thus formed.

SOLUTION: Let the straight lines be $y = mx$

It makes an angle of 45° with $2x + 3y = 6\left(\text{slope} - \dfrac{2}{3}\right)$

$\therefore \quad \tan 45° = \left|\dfrac{m - (-2/3)}{1 + m(-2/3)}\right| \Rightarrow \dfrac{3m + 2}{3 - 2m} = \pm 1$

$\Rightarrow \; 3m + 2 = \pm(3 - 2m)$

$\therefore \quad m = 1/5, -5$

Hence the sides are $x - 5y = 0$, $5x + y = 0$ and $2x + 3y = 6$

If $p = OD$ be perpendicular from O to BC, as shown Figure:

$\therefore \quad \text{Area} = \dfrac{1}{2}\sqrt{2}p \times \sqrt{2}p = p^2 = \dfrac{36}{13}$

$\left(\because \dfrac{2x}{\sqrt{13}} + \dfrac{3y}{\sqrt{13}} = \dfrac{6}{\sqrt{13}}\text{; which is in normal form, implies } p = \dfrac{6}{\sqrt{13}}\right)$

FIGURE 2.52

ILLUSTRATION 60: If the straight line drawn through the point $P(\sqrt{3}, 2)$ and inclined at an angle $\frac{\neq}{6}$ with the x-axis meets the line $\sqrt{3}x - 4y + 8 = 0$ at Q, then find the length PQ.

SOLUTION: Point Q is on $\sqrt{3}x-4y+8$, so it will satisfy this equation

$$\Rightarrow \sqrt{3}\left(\sqrt{3}\pm\frac{\sqrt{3}r}{2}\right)-4\left(2\pm\frac{r}{2}\right)+8=0$$

$\Rightarrow$ $r = \pm 6$, thus length of $PQ = 6$

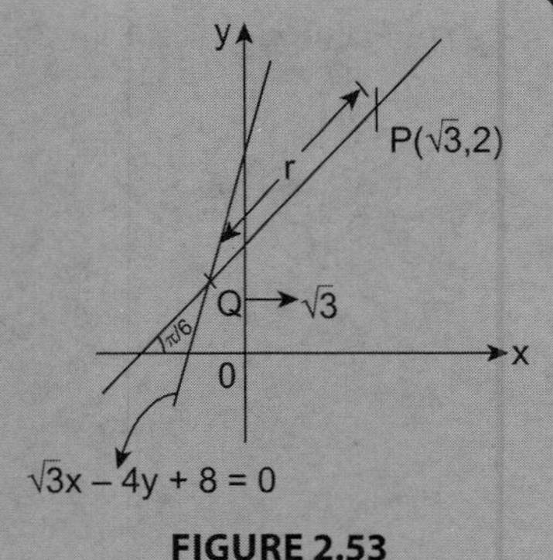

FIGURE 2.53

DISTANCE BETWEEN TWO PARALLEL LINES

Perpendicular distance between two parallel lines (d) is the difference of the algebraic distances p_1 and p_2 of these lines from the origin (if the origin lies between two lines, then d becomes sum of the perpendicular distances). i.e., $d=|p_1-p_2|$

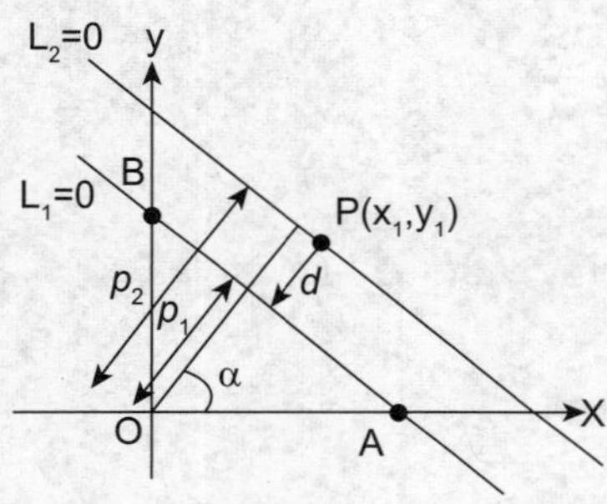

FIGURE 2.54

Let $ax + by + c = 0$ and $ax + by + c' = 0$ be two parallel straight lines, then converting these lines into normal form we get,

$$L_1: \frac{ax}{\sqrt{a^2+b^2}}+\frac{by}{\sqrt{a^2+b^2}}$$

$$=-\frac{c}{\sqrt{a^2+b^2}}$$

$$\Rightarrow p_1=\left|\frac{-c}{\sqrt{a^2+b^2}}\right|$$

(perpendicular distance from origin to L_1)

$$L_2: \frac{ax}{\sqrt{a^2+b^2}}+\frac{by}{\sqrt{a^2+b^2}}=-\frac{c'}{\sqrt{a^2+b^2}}$$

$$\Rightarrow p_2=\left|\frac{-c'}{\sqrt{a^2+b^2}}\right|$$

(perpendicular distance from origin to L_2)

Therefore the distance between parallel lines L_1 and L_2 is given by $|p_1-p_2|=\left|\frac{c-c'}{\sqrt{a^2+b^2}}\right|$.

ILLUSTRATION 61: Find the distance between the parallel lines $3x - 4y + 9 = 0$ and $6x - 8y - 15 = 0$.

SOLUTION: Given straight lines are $3x - 4y + 9 = 0$ .... (i)

and $6x - 8y - 15 = 0$ .... (ii)

or $3x - 4y + 9 = 0$ .... (iii)

and $3x - 4y - 15/2 = 0$ .... (iv)

$\Rightarrow$ $a = 3, b = -4, c = 9, c' = -15/2$

$$\Rightarrow d=\left|\frac{c-c'}{\sqrt{9+16}}\right|=\left|\frac{9+\frac{15}{2}}{5}\right|=\left|\frac{33}{10}\right|=\frac{33}{10}$$

■ DISTANCE OF A POINT FROM A LINE

The perpendicular distance (d) of any point $P(x_1, y_1)$ from the given line $ax + by + c = 0$ is given by $d = \frac{|ax_1 + by_1 + c|}{\sqrt{a^2 + b^2}}$

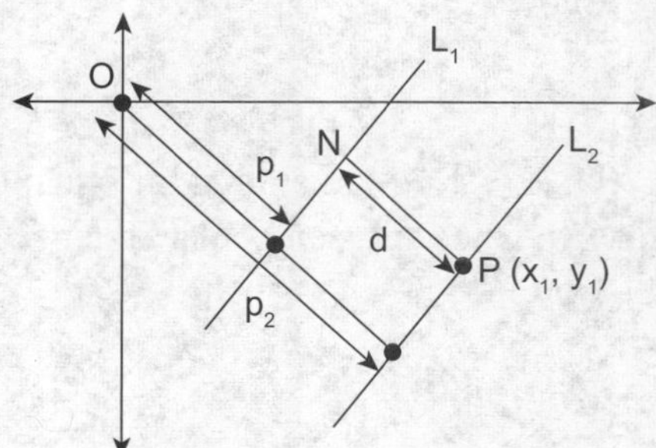

FIGURE 2.55

Proof: Let the given line be L_1: $ax + by + c = 0$ .....(i)
Equation of line passing through (x_1, y_1) parallel to above line is, $ax + by = ax_1 + by_1$ ...(ii)

Thus the distance between these two parallel lines is given by $d = |p_1 - p_2| = \frac{|ax_1 + by_1 + c|}{\sqrt{a^2 + b^2}}$

Aliter: $\perp$r distance of a point from a line

Proof: Slope of line $L_1 = \tan\theta = m = -a/b$

$\therefore$ Slope of $\perp r$ line through (x_1, y_1)

$= \tan\phi$

$= \frac{-1}{\tan\theta} = \frac{-1}{m} = \frac{b}{a}$

$$\Rightarrow \quad \sin\theta = \frac{b}{\sqrt{a^2 + b^2}} \text{ and } \cos\theta = \frac{a}{\sqrt{a^2 + b^2}}$$

And we have already established the formulae for oblique distance of a point from a line

i.e., $r = -\left(\frac{ax_1 + by_1 + c}{a\cos\theta + b\sin\theta}\right)$

$$\Rightarrow \quad |r| = \left|\frac{ax_1 + by_1 + c}{\frac{a^2}{\sqrt{a^2 + b^2}} + \frac{b^2}{\sqrt{a^2 + b^2}}}\right| = \left|\frac{ax_1 + by_1 + c}{\sqrt{a^2 + b^2}}\right|$$

REMARK

The distance of a line $ax + by + c = 0$ from origin is $\frac{|c|}{\sqrt{a^2 + b^2}}$.

ILLUSTRATION 62: Find the distance of point (–1, 1) from the line $12x + 5y + 9 = 0$

SOLUTION: The required distance $= \left|\frac{-12 + 5 + 9}{\sqrt{(12)^2 + (5)^2}}\right| = \frac{2}{13}$. $\left(\because \frac{|ax_1 + by_1 + c|}{\sqrt{a^2 + b^2}}\right)$

ILLUSTRATION 63: Find the co-ordinates of a point on $x + y + 3 = 0$ whose distance from $x + 2y + 2 = 0$ is $\sqrt{5}$.

SOLUTION: Let the required point be (x_1, y_1). Since it lies on $x + y + 3 = 0$

$\therefore \quad x_1 + y_1 + 3 = 0$ ....(i)

Now, length of the perpendicular from (x_1, y_1) to $x + 2y + 2 = 0$ is $\sqrt{5}$

$$\Rightarrow \quad \left|\frac{x_1 + 2y_1 + 2}{\sqrt{1^2 + 2^2}}\right| = \sqrt{5} \qquad \text{...(ii)}$$

Solving (i) and (ii) we get $x_1 = -9$ and $y_1 = 6$. Hence the required point is (– 9, 6)

ILLUSTRATION 64: If p is the length of the perpendicular from origin to the line $x/a + y/b = 1$, then prove that $1/p^2 = 1/a^2 + 1/b^2$.

SOLUTION: The given line is $bx + ay - ab = 0$ ..(i)

It is given that p = length of the perpendicular from the origin to (i)

$$\Rightarrow p = \frac{|b(0) + a(0) - ab|}{\sqrt{a^2 + b^2}} = \frac{ab}{\sqrt{a^2 + b^2}}$$

$$\Rightarrow p^2 = \frac{a^2b^2}{a^2 + b^2} \Rightarrow \frac{1}{p^2} = \frac{a^2 + b^2}{a^2b^2} \Rightarrow \frac{1}{p^2} = \frac{1}{a^2} + \frac{1}{b^2}$$

ILLUSTRATION 65: The vertices of a triangle OBC are O (0, 0); B(– 3, –1); C (– 1, –3), then find the equation of the line parallel to BC and intersecting the sides OB and OC, whose perpendicular distance from the point (0, 0) is 1/2.

SOLUTION: Slope of $BC = m = \frac{-3+1}{-1+3} \Rightarrow \boxed{m = -1}$

So, let the equation of line be $x + y + c = 0$

$\perp$ distance of a point (x_1, y_1) upon the line $ax + by + c = 0$ is

$$OP = \left|\frac{ax_1 + by_1 + c}{\sqrt{a^2 + b^2}}\right|$$

$$\Rightarrow \left|\frac{0 + 0 - c}{\sqrt{1+1}}\right| = \frac{1}{2} \Rightarrow c = \pm\frac{1}{\sqrt{2}}$$

$\therefore$ y intercept = '$-c$' should be negative

$\therefore$ $c > 0 \Rightarrow x + y + \frac{1}{\sqrt{2}} = 0$ is the required line.

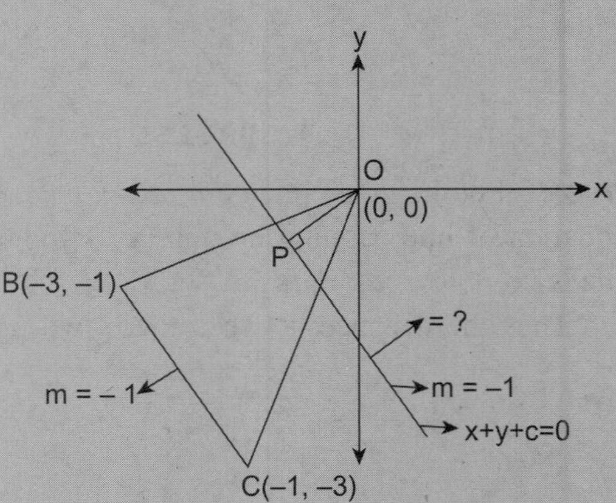

FIGURE 2.56

ILLUSTRATION 66: Find the equation of the straight lines passing through (– 2, –7) and having an intercept of length 3 between the straight lines $4x + 3y = 12$, $4x + 3y = 3$.

SOLUTION: Let B and C be any two points on line $4x + 3y = 12$ and $4x + 3y = 3$ respectively and A be a point on $4x + 3y = 3$ such that $AB = 3$, and having slope (m) = tan θ

$$\therefore \; BC = \frac{|12 - 3|}{\sqrt{4^2 + 3^2}} = \frac{9}{5};$$

and $AB = 3$, $AC = \frac{12}{5}$

$$\Rightarrow \tan\theta = \frac{BC}{AC} = \frac{9/5}{12/5} = \frac{3}{4}$$

θ is the angle between line AB and parallel lines

$$\Rightarrow \tan\theta = \left|\frac{m + \frac{4}{3}}{1 - \frac{4m}{3}}\right| \Rightarrow \left|\frac{3m + 4}{4m - 3}\right| = \frac{3}{4}$$

$$\Rightarrow \frac{3m - 4}{4m - 3} = \frac{3}{4} \text{ or } \frac{3m + 4}{4m - 3} = -\frac{3}{4}$$

$$\Rightarrow 12m - 16 = 12m - 9 \text{ or } 12m + 16 = -12m + 9$$

$$\Rightarrow 12 - \frac{16}{m} = 12 - \frac{9}{m} \text{ or } 24m + 7 = 0$$

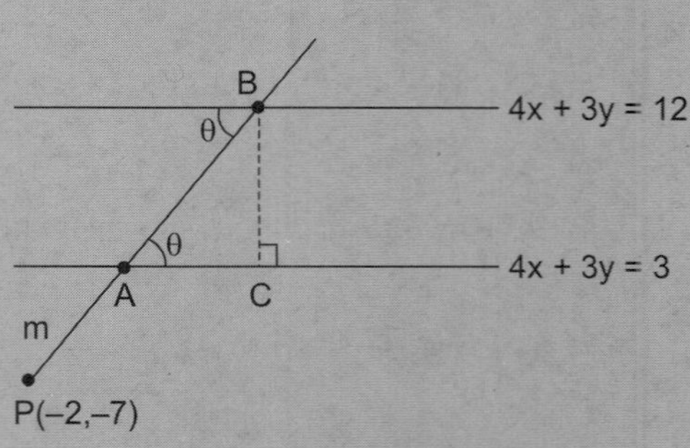

FIGURE 2.57

$$\Rightarrow m \to \infty \text{ or } m = -\frac{7}{24}$$

$$\therefore \quad x = -2 \text{ or } y + 7 = -\frac{7}{24}(x+2) \text{ or } 7x + 24y + 182 = 0.$$

■ IMAGE OF A POINT IN A LINE

If the image of the point $P(x_1, y_1)$ with respect to the line $ax + by + c = 0$ is the point $Q(x_2, y_2)$, then

$$\frac{x_2 - x_1}{a} = \frac{y_2 - y_1}{b} = -2\frac{(ax_1 + by_1 + c)}{a^2 + b^2}$$

Proof 1: As line segment PQ is ⊥r to given line $ax + by + c = 0$

$$\Rightarrow \text{(slope of } PQ) = \frac{-1}{\text{slope of given line}}$$

$$\Rightarrow \left(\frac{y_2 - y_1}{x_2 - x_1}\right)\left(-\frac{a}{b}\right) = -1 \Rightarrow \frac{y_2 - y_1}{b} = \frac{x_2 - x_1}{a}$$

$$\Rightarrow \frac{b(y_2 - y_1)}{b^2} = \frac{a(x_2 - x_1)}{a^2} = \frac{a(x_2 - x_1) + b(y_2 - y_1)}{a^2 + b^2}$$

$$= \frac{(ax_2 + by_2 + c) - (ax_1 + by_1 + c)}{a^2 + b^2}$$

(using ratio and proportion)

$$\text{Thus, } \frac{y_2 - y_1}{b} = \frac{x_2 - x_1}{a}$$

$$= \frac{(ax_2 + by_2 + c) - (ax_1 + by_1 + c)}{a^2 + b^2} \quad \text{..... (i)}$$

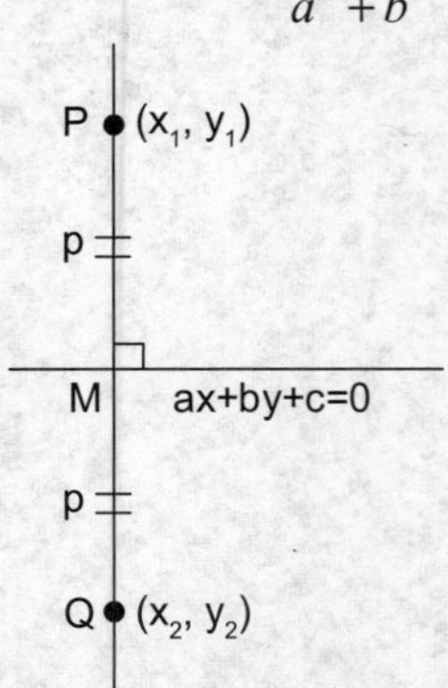

FIGURE 2.58

M(mid-point of PQ) lies on line $ax + by + c = 0$

$$\Rightarrow a\left(\frac{x_1 + x_2}{2}\right) + b\left(\frac{y_1 + y_2}{2}\right) + c = 0$$

$$\Rightarrow (ax_1 + by_1 + c) + (ax_2 + by_2 + c) = 0$$

$$\Rightarrow ax_2 + by_2 + c = -(ax_1 + by_1 + c) \quad \text{..... (ii)}$$

∴ using (ii) in (i), we get

$$\Rightarrow \frac{y_2 - y_1}{b} = \frac{x_2 - x_1}{a} = \frac{-2(ax_1 + by_1 + c)}{a^2 + b^2}.$$

Proof 2: As $Q(x_2, y_2)$ is the image of $P(x_1, y_1)$ with respect to the line $ax + by + c = 0$ we must have PQ perpendicular to $ax + by + c = 0$ and $PM = MQ$ (see figure); where M is the point of intersection of PQ with the straight line $ax + by + c = 0$.

∴ Slope PQ × Slope of $(ax + by + c = 0) = -1$

$$\text{i.e., } \frac{y_2 - y_1}{x_2 - x_1} \times \frac{-a}{b} = -1$$

$$\text{This implies that } \frac{y_2 - y_1}{b} = \frac{x_2 - x_1}{a} = \lambda(say)$$

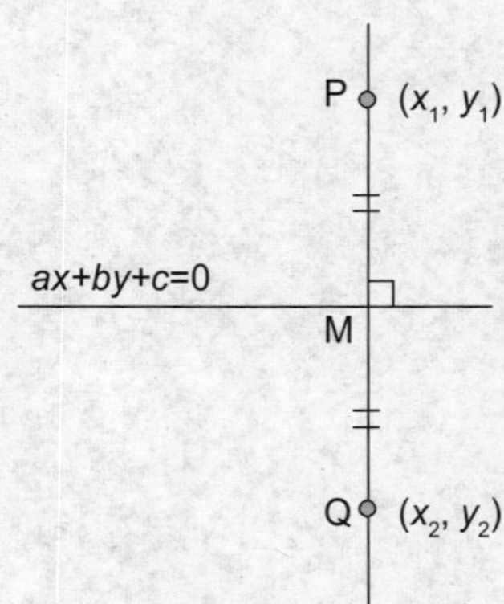

FIGURE 2.59

Then $x_2 = x_1 + a\lambda$ and $y_2 = y_1 + b\lambda$. This gives the mid-point M of PQ as

$$\left(\frac{x_1 + x_1 + a\lambda}{2}, \frac{y_1 + y_1 + b\lambda}{2}\right)$$

$$\therefore \quad M \text{ is the point } \left(\frac{2x_1 + a\lambda}{2}, \frac{2y_1 + b\lambda}{2}\right)$$

$$\Rightarrow a\left(\frac{2x_1 + a\lambda}{2}\right) + b\left(\frac{2y_1 + b\lambda}{2}\right) + c = 0$$

$$\therefore \quad (a^2 + b^2)\lambda = -2(ax_1 + by_1 + c)$$

$$\text{or} \quad \lambda = -2\frac{(ax_1 + by_1 + c)}{a^2 + b^2}.$$

$$\text{Thus } \frac{x_2 - x_1}{a} = \frac{y_2 - y_1}{b} = \lambda = -2\frac{(ax_1 + by_1 + c)}{a^2 + b^2}.$$

ILLUSTRATION 67: Show that the image of the point (3, 8) in the line $x + 3y = 7$ is (–1, –4).

SOLUTION: Let the co-ordinate of image be (x_0, y_0)

$$\Rightarrow \frac{x_0-3}{1}=\frac{y_0-8}{3}=\frac{-2[3+3(8)-7]}{1+9} \Rightarrow \frac{x_0-3}{1}=\frac{y_0-8}{3}=-4$$

$\Rightarrow x_0 = -1$ and $y_0 = -4$, therefore image is (–1, –4)

Aliter: Equation of line through (3, 8) perpendicular to $x + 3y - 7 = 0$ is $3x - y - 1 = 0$

Point of intersection of $x + 3y - 7 = 0$ and $3x - y - 1 = 0$ is (1, 2).

So applying section formula, we get $\frac{x_0+3}{2}=1$ and $\frac{y_0+8}{2}=2$

$\Rightarrow x_0 = -1$ and $y_0 = -4$

ILLUSTRATION 68: Find the equation of sides of a triangle having (4, – 1) as a vertex, if the lines $x - 1 = 0$ and $x - y - 1 = 0$ are equations of two internal bisectors of its angles.

SOLUTION: Let CF and BE be the angle bisectors of $\angle C$ and $\angle B$ respectively. Let the image of point A in the line FC be $P_2 \equiv (x_2, y_2)$ and the image of point A on the line BE be $P_1 \equiv (x_1, y_1)$

Now, $M_1 \equiv \left(\frac{x_1+4}{2}, \frac{y_1-1}{2}\right)$ and M_1 lies on $x - 1 = 0$

$\Rightarrow \frac{x_1+4}{2}-1=0 \Rightarrow x_1=-2$ and slope of

$AP_1 = \frac{y_1+1}{x_1-4}=0$

(AP_1 is $\perp r$ to BE and slope of $BE = \infty$)

$\Rightarrow y_1 = -1 \Rightarrow P_1 \equiv (-2, -1)$

Also, $M_2 \equiv \left(\frac{x_2+4}{2}, \frac{y_2-1}{2}\right)$ and M_2 lies on $x - y - 1 = 0$

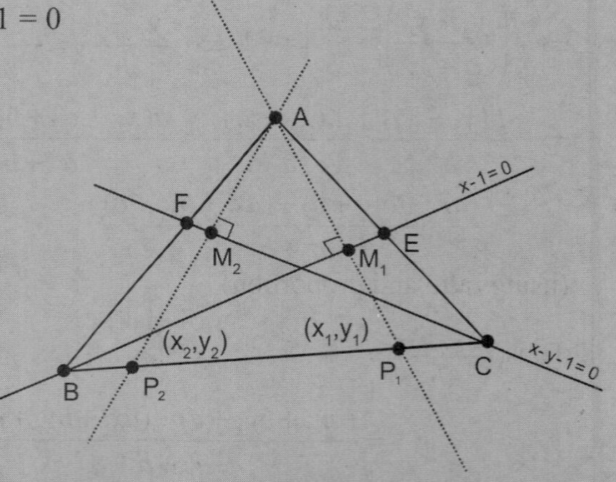

FIGURE 2.60

$\Rightarrow \frac{x_2+4}{2}-\frac{y_2-1}{2}-1=0 \Rightarrow x_2+4-y_2+1-2=0 \Rightarrow x_2-y_2+3=0$ ...(1)

and slope of $AP_2 = \frac{y_2+1}{x_2-4}=-1$ ($AP_2 \perp CF$ and slope of $CF = 1$)

$\Rightarrow y_2 + 1 = -x_2 + 4 \Rightarrow x_2 + y_2 = 3$ ....(2)

Solving equation (1) and (2); we get $P_2 \equiv (0, 3)$

Now equation of line P_1P_2 is $y - 3 = \left(\frac{3-(-1)}{0-(-2)}\right)(x-0)$

$\Rightarrow y-3=\frac{4}{2}(x-0) \Rightarrow y-2x-3=0 \Rightarrow$ Equation of $BC \equiv y-2x-3=0$

Now co-ordinates of point B can be obtained by the intersection of BE and BC

$\Rightarrow B \equiv (1, 5)$

Similarly co-ordinates of point C can be obtained by the intersection of CF and BC

$\Rightarrow C \equiv (-4, -5)$

$\therefore$ We can get the equations of lines of triangle as follows:

$$AB \equiv (y+1) = \left(\frac{5-(-1)}{1-4}\right)(x-4)$$

$$\Rightarrow\ y+1 = \frac{6}{-3}(x-4)\ \Rightarrow 2x + y - 7 = 0 \text{ and}$$

$$AC \equiv (y+1) = \left(\frac{-5-(-1)}{-4-4}\right)(x-4) \Rightarrow y+1 = \left(\frac{-4}{-8}\right)(x-4)$$

$\Rightarrow\ 2y + 2 = x - 4 \Rightarrow x - 2y - 6 = 0$ and BC $\equiv y - 2x - 3 = 0$ (already derived).

TEXTUAL EXERCISE-4 (SUBJECTIVE)

1. Find the distance between the lines $7x + 24y + 3 = 0$ and $7x + 24y + 28 = 0$.
2. Find the distance between the parallel straight lines $y = mx + c$ and $y = mx + d$.
3. The line $x\cos\theta + y\sin\theta = p$ meets the axes of co-ordinates at A and B respectively. Through A and B lines are drawn parallel to axes so as to meet the perpendicular drawn from origin to given line in P and Q respectively; then show that $|PQ| = \dfrac{4p|\cos 2\theta|}{\sin^2 2\theta}$.
4. (a) Find the equation to the sides of an isosceles right-angled triangle, the equation of whose hypotenuse is $3x + 4y = 4$ and the opposite vertex is the point (2, 2).

   (b) A line $x - 2y = 1$ through the point $A(1, 0)$ meets the line BC whose equation is $x - y + 1 = 0$ at the point B. Find the equation to the line AC so that $AB = AC$.
5. Find the perpendicular distance between following pair of lines

   (a) $3x + 4y - 5 = 0$ and $6x + 8y = 5$.

   (b) $x\cos(\pi + \alpha) + y\sin(\pi + \alpha) = p_1$ and $x\cos\alpha + y\sin\alpha = p_2$ (where p_1 and p_2 are positive real numbers)
6. (a) Find the integer values of m, for which the x-co-ordinate of the point of intersection of the lines $3x + 4y = 9$ and $y = mx + 1$ is also an integer.

   (b) If the lines $y = 3x + 1$ and $2y = x + 3$ are equally inclined to the line $y = mx + 4$, then find the value of m.

Answer Keys

1. 1 unit **2.** $\left|\dfrac{c-d}{\sqrt{1+m^2}}\right|$ **4.** (a) $7y - x - 12 =$; $7x + y = 16$ (b) $y = 2x - 2$

5. (a) 1/2 (b) $p_1 + p_2$ **6.** (a) $-1, -2$ (b) $\dfrac{+5\sqrt{2}}{7}$

TEXTUAL EXERCISE-4 (OBJECTIVE)

1. The distance between the parallel lines

   $y = 2x + 4$ and $6x = 3y + 5$ is

   (a) $\dfrac{17}{\sqrt{3}}$ (b) 1

   (c) $\dfrac{3}{\sqrt{5}}$ (d) $\dfrac{17\sqrt{5}}{15}$
2. The vertex of an equilateral triangle is (2, 3) and the opposite side is $x + y = 2$. The area of the triangle is

   (a) $\dfrac{3\sqrt{6}}{8}$ (b) $\dfrac{3\sqrt{3}}{2}$

   (c) $\dfrac{3\sqrt{3}}{4}$ (d) None of these

3. Let the algebraic sum of the perpendicular distances from the points (2, 0); (0, 2) and (1, 1) to a variable straight line be zero, then the line passes through a fixed point whose co-ordinates are:
(a) (1, 2) (b) (2, 1)
(c) (1, 1) (d) (2, 2)

4. A line parallel to the straight line, $3x - 4y - 2 = 0$ and at a distance of 4 units from it is:
(a) $3x - 4y + 20 = 0$ (b) $4x - 3y + 12 = 0$
(c) $3x - 4y + 18 = 0$ (d) $3x - 4y - 22 = 0$

5. If the sum of the distances of a point from two perpendicular lines in a plane is 1, then its locus is
(a) square (b) circle
(c) a straight line (d) two intersecting lines

6. Two particles start from the same point (2, –1), one moving 2 units along the line $x + y = 1$ and the other 5 units along the line $x - 2y = 4$. If the particles move towards increasing y, then the distance between their new positions is
(a) $\sqrt{29+2\sqrt{10}}$ (b) $\sqrt{-29+2\sqrt{10}}$
(c) $\sqrt{-29+4\sqrt{10}}$ (d) None of these

7. Straight line L is perpendicular to the line $5x - y = 1$. The area of the triangle formed by the line L and co-ordinate axes is 5, then the equation of the line is/are
(a) $x + 5y = 5\sqrt{2}$ (b) $x + 5y = -5\sqrt{2}$
(c) $x + 5y = 6\sqrt{2}$ (d) $x + 5y = -6\sqrt{2}$

8. The co-ordinates of point P on the line $y = x$ for which $PA + PB$ is minimum, where A ≡ (1 3) and B ≡ (3, 2) are
(a) (7/3, 7/3) (b) (2/3, 2/3)
(c) (–7/3, –7/3) (d) None of these

Answer Keys

1. (d) 2. (b) 3. (c) 4. (c, d) 5. (a) 6. (a) 7. (a, b) 8. (a)

FAMILY OF STRAIGHT LINES

Definition of Family of Lines

A set of lines (infinite in number) is called a family of lines if each of them have a common characteristic like set of lines having fixed slope, a fixed intercept, passing through a fixed point e.g.,

1. $y = mx + 2$; where m is a parameter representing family of lines passing through (0, 2) for different real values of m, one can get different lines of this family by taking different values of 'm'.

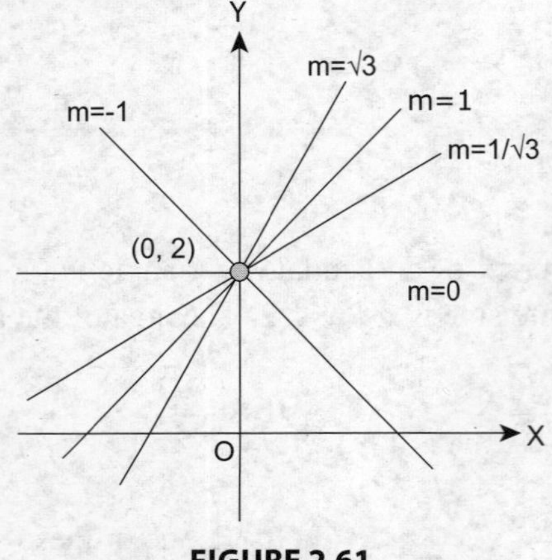

FIGURE 2.61

2. Similarly, the equation $y = 2x + c$ represents family of parallel lines (with slope 2) having variable intercepts

3. Family of straight lines parallel to the line $ax + by + c = 0$ is given by $ax + by + k = 0$, where k is a parameter.

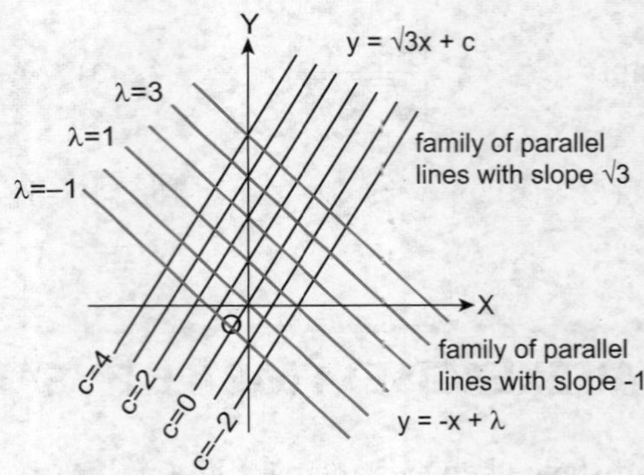

FIGURE 2.62

4. Family of straight lines perpendicular to the line $ax + by + c = 0$ is given by $bx - ay + k = 0$, where k is a parameter.

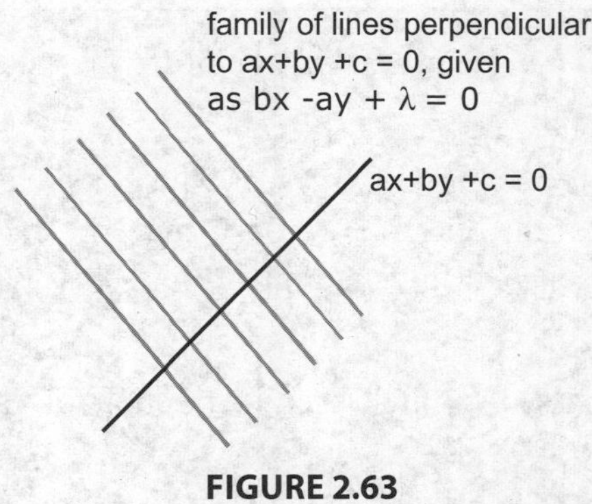

FIGURE 2.63

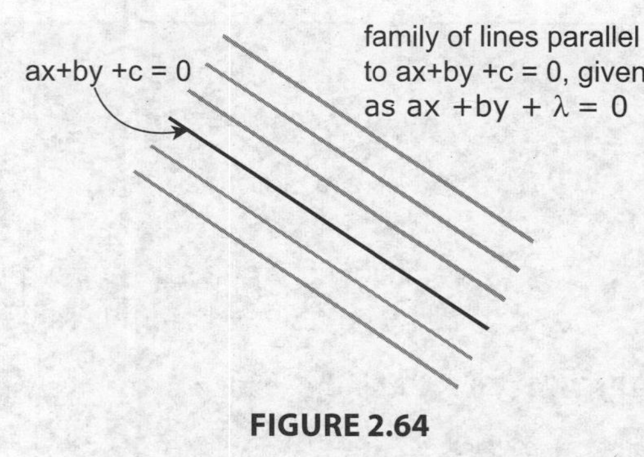

FIGURE 2.64

NOTE

General equation of line $ax + by + c = 0$ contains three parameters a, b, c but the number of parameters can be reduced to 2 without loosing the generality of line. e.g., dividing the above equation by -c and substituting $-a/c = l$ and $-b/c = m$. The above equation reduces to $lx + my = 1$; where l and m are two parameters can be called as effective parameters. This is the reason why two conditions are neceessary as well as sufficient to uniquely determine a line.

If only one condition is given out of the two, then the equation contains a parameter and therefore it represents a family of lines. And as per the known parameters, all the lines must have a common characteristic. And unknown parameters present in the equation represents the individual (variable) characteristics of members of the family.

Family of Lines Passing Through Intersection of Two Lines

If $L_1 \equiv a_1x + b_1y + c_1 = 0$ and $L_2 \equiv a_2x + b_2y + c_2 = 0$ are two straight lines (not parallel), then $L_1 + \lambda L_2 \equiv a_1x + b_1y + c_1 + \lambda (a_2x + b_2y + c_2) = 0$ represents family of lines passing through the point of intersection of $L_1 = 0$ and $L_2 = 0$. (Here λ is a parameter).

Proof: Consider two straight lines $L_1 \equiv a_1x + b_1y + c_1 = 0$, $L_2 \equiv a_2x + b_2y + c_2 = 0$ intersecting at a point $P\ (h, k)$. Now $L_1 + \lambda L_2 \equiv (a_1x + b_1y + c_1) + \lambda(a_2x + b_2y + c_2) = 0$ (*) is again a linear equation, namely $(a_1 + \lambda a_2)x + (b_1 + \lambda b_2)y + (c_1 + \lambda c_2) = 0$ and hence is also a straight line. Further, the point (h, k) satisfies $(a_1h + b_1k + c_1) + \lambda(a_2h + b_2k + c_2) = 0 + \lambda.0 = 0$. Therefore, (*) is a straight line passing through the point of intersection of L_1 and L_2.

Conversely, suppose L is any straight line passing through the point of intersection of L_1 and L_2. Let $px + qy + r = 0$ be the straight line L. Then L can be written as $y - k = m(x - h)$; where m is its slope. Solving $a_1h + b_1k + c_1 = 0$, $a_2h + b_2k + c_2 = 0$; we get $h = \dfrac{b_1c_2 - b_2c_1}{a_1b_2 - a_2b_1}$ and $k = \dfrac{a_2c_1 - a_1c_2}{a_1b_2 - a_2b_1}$

$$\therefore \quad y - \frac{a_2c_1 - a_1c_2}{a_1b_2 - a_2b_1} = m\left(x - \frac{b_1c_2 - b_2c_1}{a_1b_2 - a_2b_1}\right)$$

is the equation of L.

Simplifying we get, $m(a_1b_2 - a_2b_1)x - y(a_1b_2 - a_2b_1) + (a_2c_1 - a_1c_2) - m(b_1c_2 - b_2c_1) = 0$,

i.e., $(a_2 + mb_2)(a_1x + b_1y + c_2) - (a_1 + mb_1)(a_2x + b_2y + c_2) = 0$

which is of the form $a_1x + b_1y + c_1 + \lambda(a_2x + b_2y + c_2) = 0$ where $\lambda = -\dfrac{a_1 + mb_1}{a_2 + mb_2}$

NOTE

If λ is a parameter, $L_1 + \lambda L_2 = 0$ gives us the family of straight lines passing through the intersection of $L_1 = 0$ and $L_2 = 0$ (because it is a linear equation in x and y having a parameter λ which is always satisfied by the point of intersection of $L_1 = 0$ and $L_2 = 0$).

ILLUSTRATION 69: The family of lines $x(a + 2b) + y(a + 3b) = a + b$ passes through a point for all values of a and b. Find the point.

SOLUTION: The given equation can be written as $a(x + y - 1) + b(2x + 3y - 1) = 0$ which is the equation of a line passing through the point of intersection of the lines $x + y - 1 = 0$ and $2x + 3y - 1 = 0$. The point of intersection of these lines is $(2, -1)$. Hence the given family of lines passes through the point $(2, -1)$ for all values of a and b.

ILLUSTRATION 70: Find the equation of the straight line which passes through the intersection of the lines $x - y - 1 = 0$ and $2x - 3y + 1 = 0$ and is parallel to

(a) x-axis (b) y-axis (c) $3x + 4y = 14$

SOLUTION: The equation of any line through the intersection of the lines $x - y - 1 = 0$ and $2x - 3y + 1 = 0$ is $(x - y - 1) + \lambda(2x - 3y + 1) = 0$ or $(2\lambda + 1)x - y(3\lambda + 1) + \lambda - 1 = 0$ ...(i)

(a) The line in (i) will be parallel to x-axis if it is of the form y = constant. Therefore co-efficient of x in (i) = 0. i.e., $2\lambda + 1 = 0 \Rightarrow \lambda = -1/2$

Putting $\lambda = -1/2$ in (i), we get $y = 3$.

This is the equation of the required line.

(b) The line in (i) will be parallel to y-axis if it is of the form $x = \mu$(constant), so the co-efficient of y in

(i) = 0 i.e., $3\lambda + 1 = 0 \Rightarrow \lambda = -1/3$

Putting $\lambda = -1/3$ in (i), we get $x = 4$. This is the equation of the required line.

(c) The line (i) is parallel to the line $3x + 4y - 14 = 0$. Therefore their slopes are equal

i.e., $\dfrac{2\lambda+1}{3\lambda+1} = -\dfrac{3}{4} \Rightarrow \lambda = -\dfrac{7}{17}$

Putting this value of λ in (i), we get the equation of the required line as

$$\left(-\frac{14}{17}+1\right)x - \left(-\frac{21}{17}+1\right)y - \frac{7}{17} - 1 = 0 \Rightarrow 3x + 4y = 24$$

ILLUSTRATION 71: Find the equation of the straight line which passes through the pcint of intersection of the straight lines $x + 2y = 5$ and $3x + 7y = 17$ and is perpendicular to the straight line $3x + 4y = 10$.

SOLUTION: The equation of any line through the intersection of the lines $x + 2y - 5 = 0$ and $3x + 7y - 17 = 0$ is $(x + 2y - 5) + \lambda(3x + 7y - 17) = 0$

or $x(3\lambda + 1) + y(7\lambda + 2) - (17\lambda + 5) = 0$ ...(i)

$\therefore$ The slope of this line is $m_1 = -\left(\dfrac{3\lambda+1}{7\lambda+2}\right)$

The slope of the line $3x + 4y - 10 = 0$ is $m_2 = -3/4$

As the lines are perpendicular $m_1 m_2 = -1$

$$\Rightarrow -\left(\frac{3\lambda+1}{7\lambda+2}\right)\left(-\frac{3}{4}\right) = -1 \Rightarrow \lambda = -\frac{11}{37}$$

Putting this value of λ in (i), the equation of the required line is

$$x\left(-\frac{33}{37}+1\right) + y\left(-\frac{77}{37}+2\right) - \left(-\frac{187}{37}+5\right) = 0 \Rightarrow 4x - 3y + 2 = 0$$

ILLUSTRATION 72: The equation of the sides of a triangle are $x + 2y = 0$, $4x + 3y = 5$ and $3x + y = 0$. Find the orthocentre of the triangle.

SOLUTION: Let AB be $x + 2y = 0$, BC be $4x + 3y = 5$ and CA be $3x + y = 0$. The slope of BC is $-4/3$ and therefore the slope of AD is 3/4. Hence the equation to AD is $3x - 4y = 0$. Similarly one finds that the altitude BE must be of the form $x + 2y + \lambda(4x + 3y - 5) = 0$ or $(1 + 4\lambda)x + (2 + 3\lambda)y - 5\lambda = 0$.

Now, -1 = (Slope of BE) (Slope of CA) = $-\left(\frac{1+4\lambda}{2+3\lambda}\right)(-3)$ or

$3(1 + 4\lambda) = -(2 + 3\lambda)$

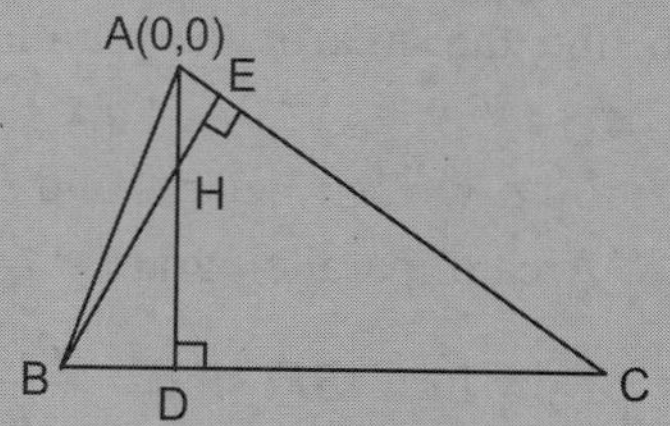

FIGURE 2.65

This gives $\lambda = -1/3 \Rightarrow$ Equation of BE is given by $x - 3y - 5 = 0$.

Thus the orthocentre H of $\Delta\, ABC$ can be obtained by solving $3x - 4y = 0$, $x - 3y = 5$.

We get H as $(-4, -3)$.

ILLUSTRATION 73: Prove that the diagonals of the parallelogram formed by the lines $ax + by + c = 0$, $ax + by + c' = 0$, $a'x + b'y + c = 0$ and $a'x + b'y + c' = 0$ will be at right angles if $a^2 + b^2 = a'^2 + b'^2$

SOLUTION: Now the diagonal AC is of the form $ax + by + c + \lambda(a'x + b'y + c) = 0$ as it passes through the intersection of $ax + by + c = 0$ and $a'x + b'y + c = 0$

Slope of $AC = -\left(\frac{a+\lambda a'}{b+\lambda b'}\right)$

Similarly, the diagonal AC is also of the form $(ax + by + c') + \mu(a'x + b'y + c') = 0$

Thus AC is given by $(a + \lambda a')x + (b + \lambda b')y + (1 + \lambda)c = 0$ and $(a + \mu a')x + (b + \mu b')y + (1 + \mu)c' = 0$.

$\therefore$ We must have $\frac{a+\lambda a'}{a+\mu a'} = \frac{b+\lambda b'}{b+\mu b'} = \left(\frac{1+\lambda}{1+\mu}\right)\frac{c}{c'}$

$\Rightarrow \lambda(a'b - ab') = \mu(a'b - ab')$

Now $\frac{a}{b} \neq \frac{a'}{b'}$ (Why ?) and hence $\lambda = \mu$. This means that $1 = \left(\frac{1+\lambda}{1+\lambda}\right)\frac{c}{c'}$

But $c \neq c'$ (why ?) and hence $\lambda = -1$.

Thus $\lambda = \mu = -1$.

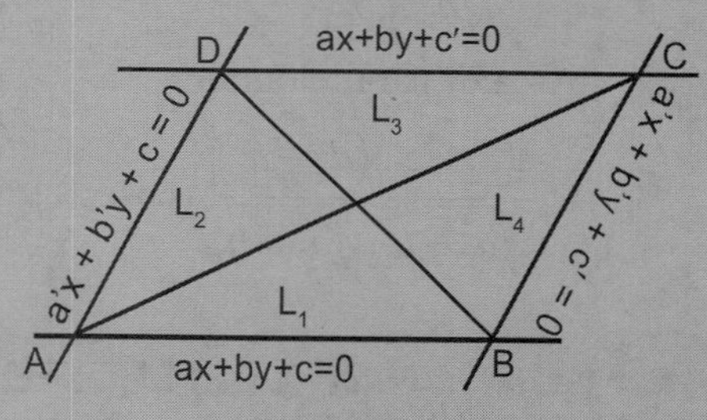

FIGURE 2.66

Therefore equation to AC is $(a - a')x + (b - b')y = 0$.

Similarly, equation to BD is $(a + a')x + (b + b')y + c + c' = 0$.

(It is easily observed that $(ax + by + c) - (a'x + b'y + c) = 0$ passes through A and C; $(ax + by + c) + (a'x + b'y + c') = 0$ passes through B and D). Now AC is perpendicular to BD if and only if the product of their slopes is -1; which happens iff $\left(-\frac{a-a'}{b-b'}\right).\left(-\frac{a+a'}{b+b'}\right) = -1$ which on simplification gives $a^2 + b^2 = a'^2 + b'^2$.

REMARK

If the sides of parallelogram are $L_1 \equiv ax + by + c = 0$, $L_2 \equiv a'x + b'y + c' = 0$ and $L_3 \equiv ax + by + c' = 0$ and $L_4 = a'x + b'y + c = 0$, then the diagonals are given by $L_1 - L_2 = 0$ and $L_1 + L_4 = 0$.

AREA OF PARALLELOGRAM

To find the area of the parallelogram formed by the line L_1: $a_1x + b_1y + c_1 = 0$, L_2: $a_1x + b_1y + c_2 = 0$,

L_3: $a_2x + b_2y + d_1 = 0$ and L_4: $a_2x + b_2y + d_2 = 0$

Area of parallelogram = $(AB)\ (AD)\ \sin\theta$

Now In ΔABF: $\sin\theta = \frac{p_2}{AB} \Rightarrow p_2 = AB\sin\theta$

$\Rightarrow AB = p_2 \operatorname{cosec}\theta$

Similarly, in ΔAED: $AD = p_1 \operatorname{cosec}\theta$

$\therefore$ Area of parallelogram $= p_1p_2 \operatorname{cosec}\theta$

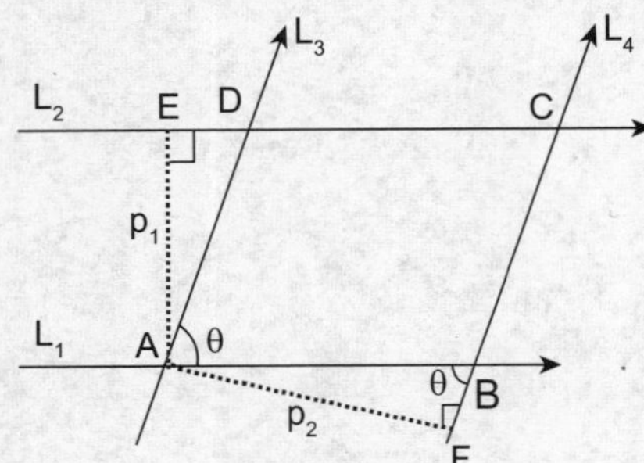

FIGURE 2.67

Now p_1 is the perpendicular distance beween L_1 and L_2

$$\Rightarrow p_1 = \frac{|c_2 - c_1|}{\sqrt{a_1^2 + b_1^2}}; \text{ Similarly, } p_2 = \frac{|d_2 - d_1|}{\sqrt{a_2^2 + b_2^2}}$$

$\therefore$ Area of parallelogram

$$= \frac{|(c_2 - c_1)(d_2 - d_1)|}{\sqrt{a_1^2 + b_1^2}\sqrt{a_2^2 + b_2^2}}(\operatorname{cosec}\theta)$$

EQUATION OF A REFLECTED RAY IN A MIRROR

Given a line mirror $L_M \equiv ax + by + c = 0$ and a ray is incident along the line $L_I \equiv a_1x + b_1y + c_1 = 0$.

To find the equation of reflected ray L_R.

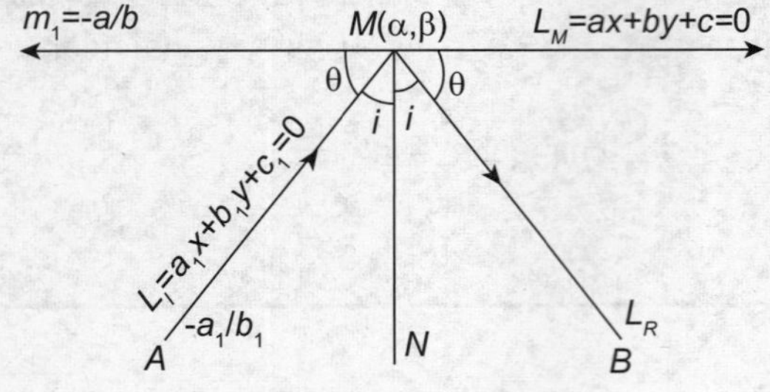

FIGURE 2.68

Method 1

Step I: Set slope of reflected ray = m and the angle between mirror and incident ray be θ.

$$\Rightarrow \tan\theta = \left|\frac{m + a/b}{1 - m(a/b)}\right| = \left|\frac{\frac{a}{b} - \frac{a_1}{b_1}}{1 + \frac{aa_1}{bb_1}}\right| = \left|\frac{ab_1 - a_1b}{aa_1 + bb_1}\right|$$

Upon solving, we get two values of m

$m = -a/b$ or $m = m_0$

But $m = -a/b$ is the slope of the incident ray

$\therefore$ slope of the reflected ray is the other value of m, i.e., m_0

Step II: Solve: $ax + by + c = 0$ and $a_1x + b_1y + c_1 = 0$ to get the co-ordinates of $M(\alpha, \beta)$

Now the equation of the reflected ray is

$L_R = (y - \beta) - m_0(x - \alpha) = 0$

Method 2

Choose a point $P(p, q)$ on the incident ray and get its image $Q(r, s)$ in line mirror L_M.

$$\Rightarrow \frac{r - p}{a} = \frac{s - q}{b} = \frac{-2(ap + bq + c)}{a^2 + b^2}$$

We already know the point of intersection of L_i and L_M i.e., $M(\alpha, \beta)$

$\therefore$ Equation of reflected ray is $y - \beta = \frac{s - \beta}{r - \alpha}(x - \alpha)$

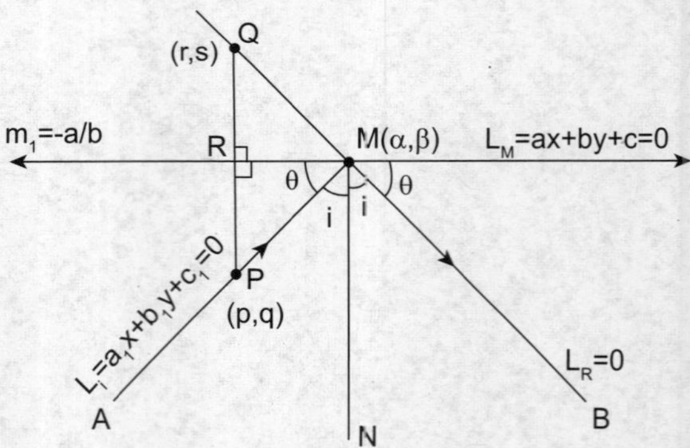

FIGURE 2.69

Method 3 The equation of the reflected ray is given by $L_I + \lambda L_M = 0$ (Since the reflected ray passes through the point of intersection of L_I and L_M)

i.e., $L_R \equiv (a_1x + b_1y + c_1) + \lambda(ax + by + c) = 0$

Now, $R(\gamma, \delta)$ is a point on L_M.

Also $\angle PRM = \angle QRM = 90°$

and $\angle QMR = \angle PMR = \theta$ and RM is common.

Therefore by ASA congurency $\Delta PRM \cong \Delta QRM$;

$\Rightarrow$ ⊥r distance of R from QM and PM will be same

$$\Rightarrow \frac{|(a_1\gamma + b_1\delta + c_1) + \lambda(a\gamma + b\delta + c)|}{\sqrt{(a_1 + a\lambda)^2 + (b_1 + b\lambda)^2}} = \frac{|a_1\gamma + b_1\delta + c_1|}{\sqrt{a_1^2 + b_1^2}}$$

$$\Rightarrow \frac{|a_1\gamma + b_1\delta + c_1|}{\sqrt{(a_1 + a\lambda)^2 + (b_1 + b\lambda)^2}} = \frac{|a_1\gamma + b_1\delta + c_1|}{\sqrt{a_1^2 + b_1^2}}$$

(Since $a\gamma + b\delta + c = 0$)

$$\Rightarrow a_1^2 + a^2\lambda^2 + 2a\,a_1\lambda + b_1^2 + b^2\lambda^2 + 2bb_1\lambda = a_1^2 + b_1^2$$

$$\Rightarrow \lambda\left(\left(a^2 + b^2\right)\lambda + 2\left(a_1 a + bb_1\right)\right) = 0$$

$$\Rightarrow \begin{cases} \lambda = 0\left(\text{incident ray}\right) \text{ or} \\ \lambda = \dfrac{-2\left(aa_1 + bb_1\right)}{a^2 + b^2}\left(\text{reflected ray}\right) \end{cases}$$

$\Rightarrow$ Equation of reflected ray is

$$\Rightarrow L_I + \frac{-2\left(aa_1 + bb_1\right)}{a^2 + b^2} L_M = 0$$

ILLUSTRATION 74: Prove that all lines represented by the equation $(2\cos\theta + 3\sin\theta)x + (3\cos\theta - 5\sin\theta)y - 5\cos\theta + 2\sin\theta = 0$ pass through fixed point for all values of θ. Find the co-ordinates of this points and its reflection in the line $x + y = \sqrt{2}$.

SOLUTION: On rearranging the given equation $\sin\theta\,[3x - 5y + 2] + \cos\theta\,[2x + 3y - 5] = 0$

or $(3x - 5y + 2) + \cot\theta\,(2x + 3y - 5) = 0$

This represents a family of lines passing through the intersection of lines $3x - 5y + 2 = 0$ and $2x + 3y - 5 = 0$ which is fixed point ($\sin\theta = 0$ is excluded from this equation).

But if we arrange equation as $(2x + 3y - 5) + \tan\theta\,(3x - 5y + 2) = 0$, it still passes through the same fixed point which includes $\sin\theta = 0$ and excludes $\cos\theta = 0$

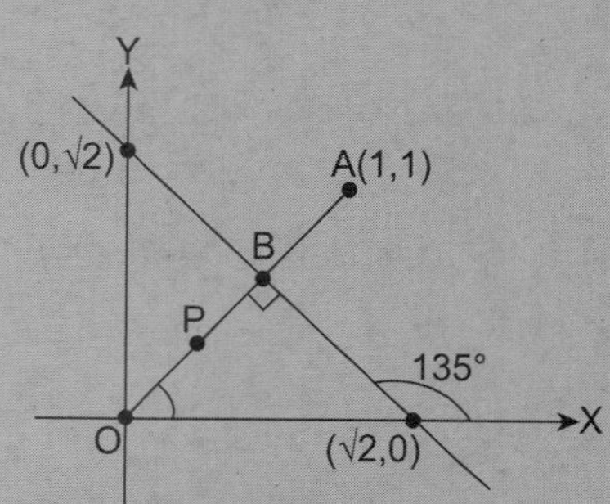

FIGURE 2.70

We get fixed point by solving $3x - 5y + 2 = 0$ and

$2x + 3y - 5 = 0$. we get $\dfrac{x}{25-6} = \dfrac{y}{4+15} = \dfrac{1}{9+10}$

$\Rightarrow x = 1, y = 1$ fixed point is (1, 1); also it can be verified that for $\cos\theta = 0$ and $\sin\theta = 0$, (1, 1) also satisfy the given equation of line.

Let reflection of $A(1, 1)$ in $x + y - \sqrt{2} = 0$ be point P. Then $|PB| = |AB|$; where B is foot of perpendicular from $A(1, 1)$ on line $x + y - \sqrt{2} = 0$

Now, $|AB| = \dfrac{|1+1-\sqrt{2}|}{\sqrt{2}} = \sqrt{2} - 1 = r$

P is on line AB and at a distance $2r$ from A $\Rightarrow$ equation of AP is $\dfrac{x-1}{\cos\theta} = \dfrac{y-1}{\sin\theta}; \theta = 45^\circ$

$\therefore$ co-ordinates of point P are given by

$$\frac{x_p - 1}{1} = \frac{y_p - 1}{1} = \frac{-2\left(2-\sqrt{2}\right)}{2} = \sqrt{2} - 2 \left(\because \frac{x - x_1}{a} = \frac{y - y_1}{b} = \frac{-2(ax_1 + by_1 + c)}{a^2 + b^2}\right)$$

$$\Rightarrow x_p = y_p = \sqrt{2} - 1$$

ILLUSTRATION 75: Starting from the origin, a beam of light hits a mirror (in the form of a line) at the point $A(4, 8)$ and is reflected towards the point $B(8, 12)$. Compute the slope of the mirror.

SOLUTION: Let m be slope of mirror, m_1 and m_2 be that of incident ray and reflected ray respectively. Now angle between normal and incident ray is equal to the angle between normal and reflected ray = θ (say) as shown here.

$$\Rightarrow \tan\theta = \left|\frac{2 + \frac{1}{m}}{1 + (2)\left(-\frac{1}{m}\right)}\right| = \left|\frac{1 + \frac{1}{m}}{1 + (1)\left(-\frac{1}{m}\right)}\right| \Rightarrow \left|\frac{2m+1}{m-2}\right| = \left|\frac{m+1}{m-1}\right| \Rightarrow \frac{2m+1}{m-2} = \pm\left(\frac{m+1}{m-1}\right)$$

$$\Rightarrow \frac{2m+1}{m-2} = \frac{m+1}{m-1} \quad \text{or} \quad \frac{2m+1}{m-2} = -\left(\frac{m+1}{m-1}\right)$$

$\Rightarrow 2m^2 - 2m + m - 1 = m^2 + m - 2m - 2$

or $2m^2 - 2m + m - 1 = -m^2 + m + 2$

$\Rightarrow m^2 = -1$ or $3m^2 - 2m - 3 = 0$

$$\Rightarrow m = \frac{2 \pm \sqrt{4 + 4(3)(3)}}{2 \times 3}$$

$$\Rightarrow m = \frac{1+\sqrt{10}}{3}, \text{ or } m = \left(\frac{1-\sqrt{10}}{3}\right) (\text{rejected})$$

since $m \in (1, 2)$. Therefore $m = \dfrac{1+\sqrt{10}}{3}$

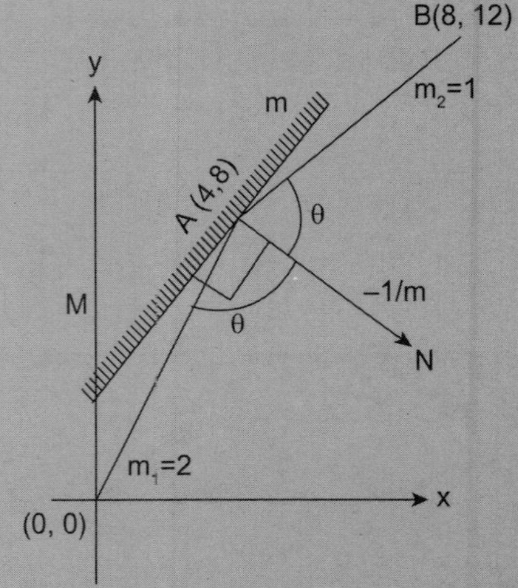

FIGURE 2.71

ILLUSTRATION 76: A ray of light is sent along the line $x - 2y - 3 = 0$. Upon reaching the line $3x - 2y - 5 = 0$, the ray is reflected from it. Find the equation of line containing the reflected ray.

SOLUTION: ***Method I:*** Given equation of *LM* is $3x - 2y - 5 = 0$ ....(i)

and the equation of *PA* is $x - 2y - 3 = 0$ ...(ii)

Solving (i) and (ii), we get $x = 1, y = -1 \therefore A(1, -1)$

Let slope of AQ = m, slope of $LM = \dfrac{3}{2}$,

slope of $PA = \dfrac{1}{2}$ Let $\angle LAP = \alpha$, then $\angle QAM = \alpha$

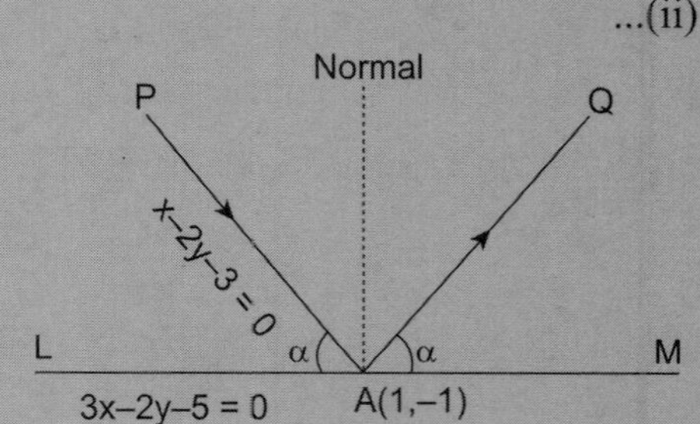

FIGURE 2.72

Now $\angle LAP = \alpha \quad \therefore \tan\alpha = \left|\dfrac{\frac{3}{2} - \frac{1}{2}}{1 + \frac{3}{2}.\frac{1}{2}}\right| = \dfrac{4}{7}$ ....(iii)

again $\angle QAM = \alpha \quad \therefore \tan\alpha = \left|\dfrac{m - \frac{3}{2}}{1 + m.\frac{3}{2}}\right| = \left|\dfrac{2m-3}{2+3m}\right|$ ...(iv)

From (iii) and (iv), we have $\left|\dfrac{2m-3}{2+3m}\right| = \dfrac{4}{7}$

$\Rightarrow \dfrac{2m-3}{2+3m} = \pm\dfrac{4}{7} \quad \therefore m = \dfrac{1}{2}, \dfrac{29}{2}$; But slope of $AP = \dfrac{1}{2} \quad \therefore$ slope of $AQ = \dfrac{29}{2}$

Now equation of AQ will be $(y+1) = \dfrac{29}{2}(x-1)$ or $29x - 2y - 31 = 0$

Method II: Marking the direction of arrow in anti-clockwise direction in *PAL* and *QAM* and equating the two values of tanα we get, $\dfrac{\frac{1}{2} - \frac{3}{2}}{1 + \frac{1}{2}.\frac{3}{2}} = \dfrac{\frac{3}{2} - m}{1 + \frac{3}{2}.m} = -\dfrac{4}{7} = \dfrac{3-2m}{2+3m}$

$\Rightarrow m = \dfrac{29}{2} \therefore$ equation of AQ will be $29x - 2y - 31 = 0$

TEXTUAL EXERCISE–5 (SUBJECTIVE)

1. Find the co-ordinates of the foot of the perpendicular from a point $(-1, 3)$ to the line $3x - 4y - 16 = 0$.

2. Find the image of the point $(-8, 12)$ with respect to the mirror $4x + 7y + 13 = 0$.

3. If the image of the point $(2, 1)$ with respect to the line mirror be $(5, 2)$, then the equation of the mirror.

4. Find the reflection of the point $(4, -13)$ in the line $5x + y + 6 = 0$.

5. Find the foot of perpendicular and mirror image of the following point P, with respect to the mentioned equation of the mirror
(i) L_m: $P(1, 2)$ and L_m: $x + 2y = 10$
(ii) L_m: $P(2, 3)$ and L_m : $2x - 3y + 18 = 0$

6. In a ΔABC, if the equations of the sides AB, BC and CA are $2x - y + 4 = 0$, $x - 2y - 1 = 0$ and $x + 3y - 3 = 0$ respectively, then find the image of point B w.r.t. the side CA.

Answer Keys

1. (68/25, –49/25) **2.** (–16, –2) **3.** $3x + y = 12$ **4.** (–1, –14)

5. (i) foot of perpendicular $M(2, 4)$; mirror image $Q(3, 6)$ (ii) Foot of perpendicular $M(0, 6)$; mirror image $Q(-2, 9)$

6. $\left(-\frac{3}{5}, \frac{26}{5}\right)$.

TEXTUAL EXERCISE–5 (OBJECTIVE)

1. The image of the point $A(1, 2)$ by the line mirror $y = x$ is the point B and the image of B by the line mirror $y = 0$ is the point (α, β); where
(a) $\alpha = 3, \beta = 1$ (b) $\alpha = 2, \beta = -1$
(c) $\alpha = -2, \beta = 1$ (d) $\alpha = -2, \beta = -1$

2. A straight line passes throughn a fixed point (h, k). The locus of the foot of the perpendicular on it drawn from the origin is
(a) $2x^2 + 2y^2 - hx - ky = 0$
(b) $x^2 - y^2 - hx - ky = 0$
(c) $x - y - hx - ky = 0$
(d) $x^2 + y^2 - hx - ky = 0$

3. The equation of the line mirror L_M in which the mirror image of the point $P(3, 2)$ be $Q(7, 4)$ is
(a) $x - y = 13$ (b) $2x + y = 13$
(c) $x + y = 13$ (d) $2x + 2y = 13$

4. A light ray through point $(2, 3)$ falls on a mirror with equation $2x - y + 7 = 0$ such that the reflected ray passes through the point $(6, 11)$. At the point of incidence, the equation of normal to mirror is
(a) $2x - 2y = 18$ (b) $x - 2y = 18$
(c) $x + 2y = 18$ (d) $2x + 2y = 18$

5. A ray is incident along $3x - y = 5$ to the line mirror $x = 2y$, the equation of reflected ray will be
(a) $x + 3y = 5$ (b) $x + 3y = 8$
(c) $x - 3y = 5$ (d) None of these

6. The equation of mirror and reflected ray if equation of normal and incident ray is given as $2x - 3y - 5 = 0$ and $2x - y = 3$ respectively are
(a) $x + y = 1$, $2x - 29y - 31 = 0$
(b) $3x - 2y = 1$, $2x + 29y + 31 = 0$
(c) $3x + 2y = 1$, $2x - 29y - 31 = 0$
(d) $x - y = 1$, $x - y - 31 = 0$

7. The equation of reflected ray and normal at the point of incidence, if the ray is incident along $2x - y = 4$ and gets reflected from the mirror $x - y = 1$ respectively are
(a) $x + y = 5$, $x - 2y + 1 = 0$
(b) $x - y = 5$, $x + 2y + 3 = 0$
(c) $2x + 2y = 5$, $x - 2y + 4 = 0$
(d) None of these

Answer Keys

1. (b) **2.** (d) **3.** (b) **4.** (c) **5.** (a) **6.** (c) **7.** (a)

EQUATION OF THE BISECTORS OF THE ANGLES BETWEEN LINES

Let $L_1 \equiv a_1x + b_1y + c_1 = 0$ .....(i)

and $L_2 \equiv a_2x + b_2y + c_2 = 0$ ......(ii)

be two intersecting lines, then the equations of the lines bisecting the angles between L_1 and L_2 are given by

$$\frac{a_1x+b_1y+c_1}{\sqrt{a_1^2+b_1^2}} = \pm\frac{a_2x+b_2y+c_2}{\sqrt{a_2^2+b_2^2}}$$

Proof: The bisectors of the angles between two straight lines are the locus of a point which is equidistant from the two lines. So let $P(h, k)$ be a point equidistant from the lines (i) and (ii). Then $PL = PM$

$$\Rightarrow \left|\frac{a_1h+b_1k+c_1}{\sqrt{a_1^2+b_1^2}}\right| = \left|\frac{a_2h+b_2k+c_2}{\sqrt{a_2^2+b_2^2}}\right|$$

$\therefore$ locus of (h, k) i.e., the equation of the bisectors are

$$\frac{a_1x+b_1y+c_1}{\sqrt{a_1^2+b_1^2}} = \pm\frac{a_2x+b_2y+c_2}{\sqrt{a_2^2+b_2^2}}$$

$(\because |a| = |b| \Leftrightarrow a = \pm b)$

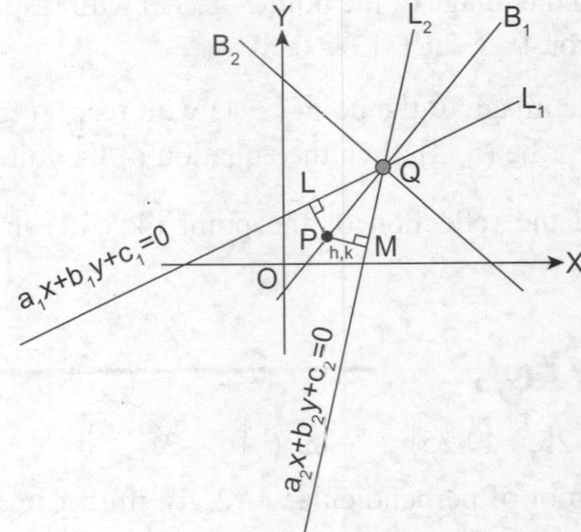

FIGURE 2.73

ILLUSTRATION 77: Two sides of a rhombus $ABCD$ are parallel to the lines $y = x + 2$ and $y = 7x + 3$. If the diagonals of the rhombus intersect at the point (1, 2) and the vertex A is on y-axis. Find the equations of diagonals and hence find the possible co-ordinates of A.

SOLUTION: Let the co-ordinates of A be $(0, a)$ as shown in Figure 2.74.

As the sides of the rhombus are parallel to line $y = x + 2$ and $y = 7x + 3$, the diagonals of the rhombus are parallel to the bisectors of the angle between given lines.

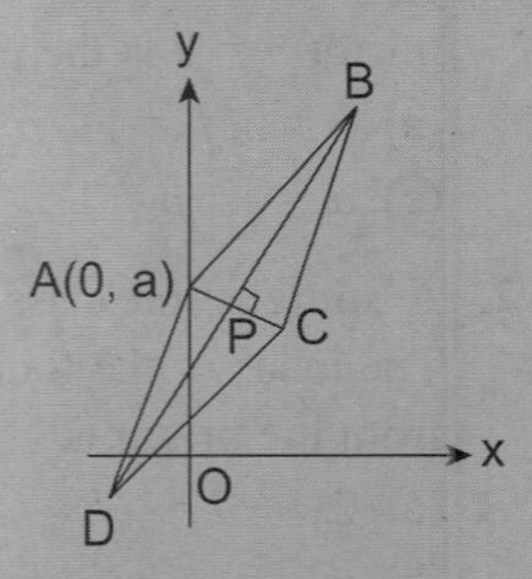

FIGURE 2.74

Equation of the bisectors of the angles between the lines

$y = x + 2$ and $y = 7x + 3$ are $\dfrac{y-x-2}{\sqrt{1+1}} = \pm\dfrac{y-7x-3}{\sqrt{1+49}}$

or $5(y - x - 2) = \pm(y - 7x - 3)$

Thus the diagonals are parallel to lines $2x + 4y - 7 = 0$ and $12x - 6y + 13 = 0$

Let P be the point of intersection of diagonals

$\Rightarrow P \equiv (1, 2)$

$\Rightarrow$ Equation of bisectors will be $(y-2) = \frac{1}{2}(x-1)$ and $(y - 2) = 2(x - 1)$

i.e., $x + 2y - 5 = 0$ and $2x - y = 0$

Therefore, slope of AP is either 2 or $-1/2$

$\therefore \dfrac{a-2}{0-1} = 2, -\dfrac{1}{2} \Rightarrow a = 0, 5/2$

Hence the possible co-ordinates of A are (0, 0) and (0, 5/2)

■ BISECTOR OF ANGLE CONTAINING THE ORIGIN

Let the origin be contained in angle α between the straight lines L_1 and L_2 and hence we are to find the equation of bisector of angle α. Let $P(x, y)$ be any point on this angle bisector. Then origin and $P(x, y)$ are on the same side of L_1 as well L_2 as shown in Figure 2.75.

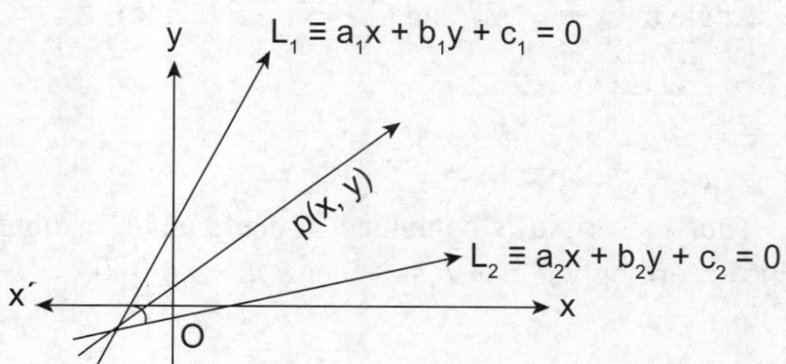

FIGURE 2.75

Without loss of generality let $c_1, c_2 > 0$

Then $a_1x + b_1y + c_1 > 0$ and $a_2x + b_2y + c_2 > 0$

$\Rightarrow |a_1x + b_1y + c_1| = a_1x + b_1y + c_1$

and $|a_2x + b_2y + c_2| = a_2x + b_2y + c_2$

∴ Equation of bisector

$$\left|\frac{a_1x+b_1y+c_1}{\sqrt{a_1^2+b_1^2}}\right| = \left|\frac{a_2x+b_2y+c_2}{\sqrt{a_2^2+b_2^2}}\right|$$

becomes $\dfrac{a_1x+b_1y+c_1}{\sqrt{a_1^2+b_1^2}} = \dfrac{a_2x+b_2y+c_2}{\sqrt{a_2^2+b_2^2}}$

which is the required equation of angle bisector of angle α containing the origin

Algorithm

Let the equations of the two lines be $a_1x + b_1y + c_1 = 0$ and $a_2x + b_2y + c_2 = 0$. To find the bisectors of the angle containing the origin, following steps are taken:

Step 1: See whether the constant terms c_1 and c_2 in the equations of two lines are positive or not. If not, then multiply both the sides of the equation *by* –1 to make the constant terms positive.

Step 2: Now obtain the bisector corresponding to the positive sign i.e.,

$\dfrac{a_1x+b_1y+c_1}{\sqrt{a_1^2+b_1^2}} = \dfrac{a_2x+b_2y+c_2}{\sqrt{a_2^2+b_2^2}}$. It is the required bisector of the angle containing the origin.

ILLUSTRATION 78: Prove that the internal bisectors of the angles of a triangle meet in a point.

SOLUTION: Let the equations of the sides BC, CA and AB of a ΔABC be respectively.

$a_1x + b_1y + c_1 = 0;$

$a_2x + b_2y + c_2 = 0;$

and $a_3x + b_3y + c_3 = 0$

and where c_1, c_2 and c_3 are positive

Let the origin O be within the triangle (without loss of generality)

Then the equation of the internal bisectors of the angle BAC is

$$\frac{(a_2x+b_2y+c_2)}{\sqrt{(a_2^2+b_2^2)}} = \frac{(a_3x+b_3y+c_3)}{\sqrt{(a_3^2+b_3^2)}} \quad \text{... (i)}$$

Similarly, the equations of the internal bisectors ABC and BCA are respectively.

$$\frac{(a_3x+b_3y+c_3)}{\sqrt{(a_3^2+b_3^2)}} = \frac{(a_1x+b_1y+c_1)}{\sqrt{(a_1^2+b_1^2)}} \quad \text{.... (ii)}$$

and $$\frac{(a_1x+b_1y+c_1)}{\sqrt{(a_1^2+b_1^2)}} = \frac{(a_2x+b_2y+c_2)}{\sqrt{(a_2^2+b_2^2)}} \quad \text{.... (iii)}$$

Adding the equations (i) and (ii), we get (iii)

Hence internal bisectors of a triangle are concurrent.

■ CONDITION FOR ORIGIN TO LIE THE OBTUSE OR ACUTE ANGLE

Let $a_1x + b_1y + c_1 = 0$ ....(i)

and $a_2x + b_2y + c_2 = 0$ ...(ii)

be two given straight lines L_1 and L_2 respectively, then origin would lie in acute angle if $a_1a_2 + b_1b_2 < 0$.

And origin would lie in obtuse angle if $a_1a_2 + b_1b_2 > 0$; provided $c_1, c_2 > 0$.

Proof: Let (x_1, y_1) and (x_2, y_2) be the feet of perpendiculars from origin O on line L_1 and L_2 as shown.

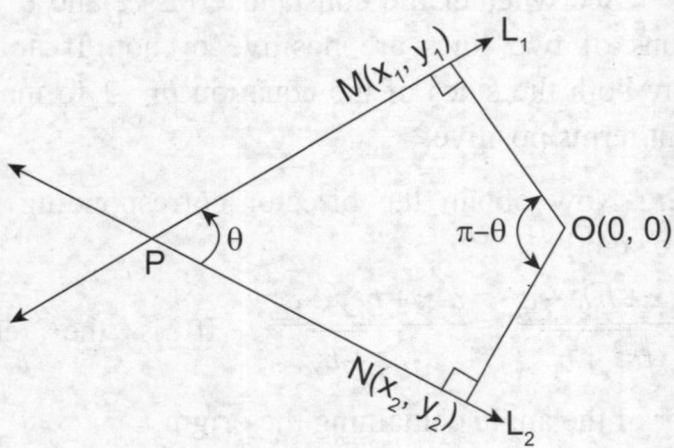

FIGURE 2.76

Let θ be the angle between the lines L_1 and L_2. Thus $(\pi - \theta)$ will be the angle between the perpendicular OM and ON. Thus feet of ⊥ rs $M(x_1, y_1)$ and $N(x_2, y_2)$ are given by

$$\frac{x_1 - 0}{a_1} = \frac{y_1 - 0}{b_1} = \frac{-(a_1(0) + b_1(0) + c_1)}{(a_1^2 + b_1^2)}$$ [By foot of perpendicular formula]

and $$\frac{x_2 - 0}{a_2} = \frac{y_2 - 0}{b_2} = \frac{-(a_2(0) + b_2(0) + c_2)}{(a_2^2 + b_2^2)}$$

$$\Rightarrow (x_1, y_1) \equiv \left(\frac{-a_1c_1}{a_1^2 + b_1^2}, \frac{-b_1c_1}{a_1^2 + b_1^2}\right) \equiv (-a_1\alpha, -b_1\alpha)$$

and $$(x_2, y_2) \equiv \left(\frac{-a_2c_2}{a_2^2 + b_2^2}, \frac{-b_2c_2}{a_2^2 + b_2^2}\right) \equiv (-a_2\beta, -b_2\beta)$$

where $\alpha = \frac{c_1}{a_1^2 + b_1^2}$ and $\beta = \frac{c_2}{a_2^2 + b_2^2}$

Here $\alpha, \beta > 0$ provided $c_1, c_2 > 0$ (if not so, we can make them so)

$\therefore \quad \overrightarrow{OM} = -a_1\alpha\hat{i} - b_1\alpha\hat{j}$ and $\overrightarrow{ON} = -a_2\beta\hat{i} - b_2\beta\hat{j}$

Case i: when θ is acute

$\therefore$ Angle between $\overrightarrow{OM}$ and $\overrightarrow{ON} = \pi - \theta$ (obtuse)

$\Rightarrow \overrightarrow{OM}.\overrightarrow{ON} < 0 \left(\because \overrightarrow{OM}.\overrightarrow{ON} = |\overrightarrow{OM}||\overrightarrow{ON}|.\cos(\pi - \theta)\right)$

$\Rightarrow a_1a_2\,\alpha\beta + b_1b_2\,\alpha\beta < 0$

$\Rightarrow \alpha\beta\,(a_1a_2 + b_1b_2) < 0$

$\Rightarrow a_1a_2 + b_1b_2 < 0$ as $\alpha\beta > 0$

Case ii: when θ is obtuse:

Angle between $\overrightarrow{OM}$ and $\overrightarrow{ON} = \pi - \theta$ (acute)

$\Rightarrow \overrightarrow{OM}.\overrightarrow{ON} > 0$

$\Rightarrow \alpha\beta\,(a_1a_2 + b_1b_2) > 0$

Now as $a\beta > 0$

Thus the origin is contained in acute angle or obtuse angle accordingly $a_1a_2 + b_1b_2 < 0$ or $a_1a_2 + b_1b_2 > 0$.

Algorithm

Let the equations of the two lines be $a_1x + b_1y + c_1 = 0$ and $a_2x + b_2y + c_2 = 0$. To determine whether the origin lies in the acute angle or obtuse angle between the lines we proceed as follows:

Step 1: See whether the constant terms c_1 and c_2 in the equations of two lines are positive or not if not, then multiply both the sides of the equations *by* –1 to make the constant terms positive.

Step 2: Determine the sign of the expression $a_1a_2 + b_1b_2$.

Step 3: If $a_1a_2 + b_1b_2 > 0$, then the origin lies in the obtuse angle and if $a_1a_2 + b_1b_2 < 0$, then the origin lies in the acute angle. If $a_1a_2 + b_1b_2 = 0$, then the lines L_1 and L_2 are perpendicular to each other and the origin lies in one of the four right angles.

■ BISECTOR OF ACUTE AND OBTUSE ANGLE

Let the equations of the two lines be $a_1x + b_1y + c_1 = 0$ and $a_2x + b_2y + c_2 = 0$. To separate the bisectors of the obtuse and acute angles between the lines we proceed as follows:

Step 1: See whether the constant terms c_1 and c_2 in the equations of two lines are positive or not. If not, then multiply both the sides of the equations *by* –1 to make the constant terms positive.

Step 2: Determine the sign of the expression $a_1a_2 + b_1b_2$.

Step 3: If $a_1a_2 + b_1b_2 > 0$, then the origin lies in obtuse angle and the bisector corresponding to " + " sign gives the equation of bisector of angle contaning the origin i.e., of obtuse angle bisector. And hence the bisector corresponding to " – " sign will be the bisector of the acute angle between the lines i.e.,

$$\frac{a_1x+b_1y+c_1}{\sqrt{a_1^2+b_1^2}}=\frac{a_2x+b_2y+c_2}{\sqrt{a_2^2+b_2^2}} \text{ and}$$

$$\frac{a_1x+b_1y+c_1}{\sqrt{a_1^2+b_1^2}}=-\frac{a_2x+b_2y+c_2}{\sqrt{a_2^2+b_2^2}}$$

are the bisectors of obtuse and acute angles respectively.

If $a_1a_2 + b_1b_2 < 0$, then the origin lies in acute angle and hence the bisector corresponding to " + " sign gives the equation of bisector of angle containing the origin and hence of acute angle and hence " – " sign gives the obtuse angle bisectors respectively i.e.,

$$\frac{a_1x+b_1y+c_1}{\sqrt{a_1^2+b_1^2}}=+\frac{a_2x+b_2y+c_2}{\sqrt{a_2^2+b_2^2}}$$

and $$\frac{a_1x+b_1y+c_1}{\sqrt{a_1^2+b_1^2}}=-\frac{a_2x+b_2y+c_2}{\sqrt{a_2^2+b_2^2}}$$

are the bisectors of acute and obtuse angles respectively.

Aliter: Let L_3 and L_4 be the two bisectors. If $a_1a_2 + b_1b_2 = 0$, then the given lines are perpendicular to each other else they will contain acute or obtuse angle. i.e., $a_1a_2 + b_1b_2 \neq 0$. Let θ be the angle between L_1 and L_2 which is bisected by one of the bisectors (say L_3). Then angle between L_1 and L_3 is $\theta/2$. Now find tan $(\theta/2)$.

Two cases arise:

Case i: If $0 < \tan\frac{\theta}{2} < 1$, then $0 < \theta < \frac{\pi}{2}$

Thus L_3 will be bisecting the acute angles between L_1 and L_2 and L_4 will be that of obtuse angle.

Case ii: If $\tan\frac{\theta}{2} > 1$, then $\frac{\pi}{2} < \theta < \pi$

Thus L_3 will be bisecting the obtuse angle between L_1 and L_2 and L_4 will be bisecting acute angle.

ILLUSTRATION 79: Find the equation of the obtuse angle bisector of lines $12x - 5y + 7 = 0$ and $3y - 4x - 1 = 0$

SOLUTION: First make the constant terms (c_1 and c_2) positive, the two equations become $12x - 5y + 7 = 0$ and $4x - 3y + 1 = 0$

Now check $a_1 a_2 + b_1 b_2 = 12 \times 4 + (-5) \times (-3) = 63$ which is positive

Hence + sign give the obtuse angle bisector. The obtuse angle bisector is

$$\Rightarrow \frac{12x-5y+7}{\sqrt{(12)^2+(-5)^2}}=+\frac{4x-3y+1}{\sqrt{(4)^2+(-3)^2}}$$

$\Rightarrow$ $5(12x - 5y + 7) = 13(4x - 3y + 1)$

$\Rightarrow$ $4x + 7y + 11 = 0$ is the required obtuse angle bisector.

ILLUSTRATION 80: Find the bisector between the line $3x - 4y + 7 = 0$ and $12x + 5y - 2 = 0$ and identify the bisector.

(i) containing origin

(ii) not containing origin

(iii) acute angle bisector

(iv) obtuse angle bisector

SOLUTION: Let the equations of lines be: $3x - 4y + 7 \equiv a_1x + b_1y + c_1 = 0$

and $-12x - 5y + 2 \equiv a_2x + b_2y + c_2 = 0$

Here, $c_1 > 0$ and $c_2 > 0$

Now calculating $a_1a_2 + b_1b_2 \equiv (3) \times (-12) + (-4) \times (-5) = -16 < 0$; Equation of bisectors are

$$\frac{(3x-4y+7)}{\sqrt{3^2+4^2}}=\frac{(-12x-5y+2)}{\sqrt{12^2+5^2}} \text{ i.e., } 11x-3y+9=0$$

and $$\frac{(3x-4y+7)}{\sqrt{3^2+4^2}}=-\frac{(-12x-5y+2)}{\sqrt{12^2+5^2}} \text{ i.e., } 21x+77y-101=0$$

(i) Bisector of angle containing origin: $11x - 3y + 9 = 0$ (+ve sign bisector)
(ii) Bisector of angle not containing origin: $21x + 77y - 101 = 0$ (other bisector)
(iii) Acute angle bisector: $11x - 3y + 9 = 0$ ($a_1a_2 + b_1b_2 < 0 \Rightarrow$ +ve sign bisector will be of acute angle)
(iv) Obtuse angle bisector: $21x + 77y - 101 = 0$ (other bisector)

EQUATION OF BISECTOR OF ANGLE BETWEEN TWO LINES CONTAINING A GIVEN POINT A(α, β)

Let $a_1x + b_1y + c_1 = 0$ ...(i)
and $a_2x + b_2y + c_2 = 0$ ...(ii)
be two given lines and α be the angle containing the point $A(\alpha,\beta)$ as shown in Figure 2.77.

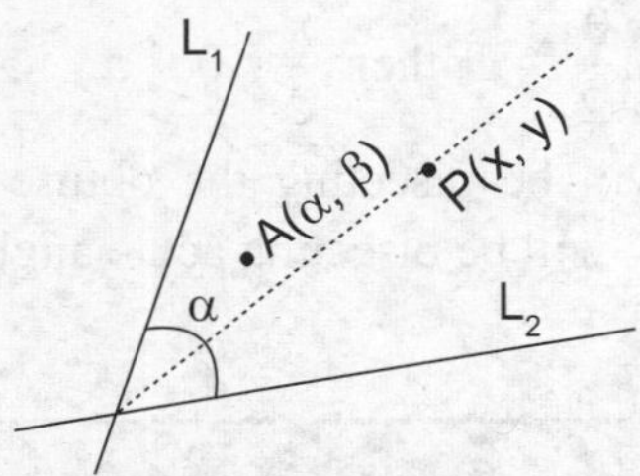

FIGURE 2.77

Let $P(x, y)$ be any point on the required bisector. Then A and P are on same side of lines L_1 and L_2.

$$\Rightarrow \frac{a_1x+b_1y+c_1}{a_1\alpha+b_1\beta+c_1} > 0 \text{ and } \frac{a_2x+b_2y+c_2}{a_2\alpha+b_2\beta+c_2} > 0$$

$\Rightarrow$ If $a_1\alpha + b_1\beta + c_1$ and $a_2\alpha + b_2\beta + c_2$ are of same sign, then $a_1x + b_1y + c_1$ and $a_2x + b_2y + c_2$ will also be of same sign.

$\therefore$ Equations of bisectors

$$\frac{|a_1x+b_1y+c_1|}{\sqrt{a_1^2+b_1^2}} = \frac{|a_2x+b_2y+c_2|}{\sqrt{a_2^2+b_2^2}}$$

reduces to $\dfrac{a_1x+b_1y+c_1}{\sqrt{a_1^2+b_1^2}} = \dfrac{a_2x+b_2y+c_2}{\sqrt{a_2^2+b_2^2}}$

and if $a_1\alpha + b_1\beta + c_1$ and $a_2\alpha + b_2\beta + c_2$ are of opposite sign, then $a_1x + b_1y + c_1$ and $a_2x + b_2y + c_2$ will also be of opposite signs,

$\therefore$ equations of bisectors

$$\frac{|a_1x+b_1y+c_1|}{\sqrt{a_1^2+b_1^2}} = \frac{|a_2x+b_2y+c_2|}{\sqrt{a_2^2+b_2^2}}$$

Reduces to

$$\frac{a_1x+b_1y+c_1}{\sqrt{a_1^2+b_1^2}} = \frac{-(a_2x+b_2y+c_2)}{\sqrt{a_2^2+b_2^2}}$$

Algorithm

Step I: Find the sign of expression $\dfrac{a_1\alpha+b_1\beta+c_1}{a_2\alpha+b_2\beta+c_2}$.

Step II: If it *is* +ve, then sele*ct* +ve sign bisector

i.e., $\dfrac{a_1x+b_1y+c_1}{\sqrt{a_1^2+b_1^2}} = \dfrac{(a_2x+b_2y+c_2)}{\sqrt{a_2^2+b_2^2}}$ will be the required bisector.

and if it is –ve, then select –ve sign bisector

i.e., $\dfrac{a_1x+b_1y+c_1}{\sqrt{a_1^2+b_1^2}} = \dfrac{-(a_2x+b_2y+c_2)}{\sqrt{a_2^2+b_2^2}}$ will be the required bisector.

ILLUSTRATION 81: Find the bisector of the angle between the lines $2x + y - 6 = 0$ and $2x - 4y + 7 = 0$ which contains the point (1, 2).

SOLUTION: Value of $2x + y - 6$ at (1, 2) *is* – 2 (negative)

and value of $2x - 4y + 7$ at (1, 2) is 1 (positive) i.e., opposite sign $\Rightarrow \dfrac{-2}{1} < 0$

$\Rightarrow$ Equation of bisector containing the point (1, 2) is $\dfrac{(2x+y-6)}{\sqrt{(2^2+1^2)}} = \dfrac{-(2x-4y+7)}{\sqrt{(2)^2+(-4)^2}}$

$\Rightarrow 2(2x + y - 6) + (2x - 4y + 7) = 0$ or $6x - 2y - 5 = 0$.

ILLUSTRATION 82: Find the equation of the bisector of the angle between two lines $3x - 4y + 12 = 0$ and $12x - 5y + 7 = 0$ which contains the points $(-1, 4)$.

SOLUTION: Let the equations of the two lines be

$$a_1x + b_1y + c_1 = 0 \quad \text{.... (i)}$$

and $$a_2x + b_2y + c_2 = 0 \quad \text{.... (ii)}$$

The equation of the bisector of the angle between the two lines containing the points (h, k) will be $\frac{(a_1x+b_1y+c_1)}{\sqrt{(a_1^2+b_1^2)}} = \frac{(a_2x+b_2y+c_2)}{\sqrt{a_2^2+b_2^2}}$ or $\frac{(a_1x+b_1y+c_1)}{\sqrt{(a_1^2+b_1^2)}} = -\frac{(a_2x+b_2y+c_2)}{\sqrt{(a_2^2+b_2^2)}}$

according as $a_1h + b_1k + c_1$ and $a_2h + b_2k + c_2$ are of the same sign or opposite sign at $(-1, 4)$

Now, $3x - 4y + 12 < 0$ at $(-1, 4)$ and $12x - 5y + 7 < 0$ at $(-1, 4)$

$\Rightarrow \frac{3x-4y+12}{12x-5y+7} > 0$ at $(-1, 4)$ $\therefore$ we have to take the bisector with + *ve* sign

i.e., $\left(\frac{3x-4y+12}{5}\right) = \left(\frac{12x-5y+7}{13}\right) \Rightarrow 21x + 27y - 121 = 0.$

Given line L_1: $4x + 3y - 6$ and the equation of the pair of bisectors of the angle that L_1 forms with same other line L_2 is $B_1 := 9x - 7y - 41$ and $B_2 : 7x - 9y - 3 = 0$, which of these is the bisector of obtuse angle between L_1 and L_2.

Solution : Let us take one point on L_1: (0, 2)

Distance of B_1 from (0, 2) $= \frac{|-55|}{\sqrt{9^2+7^2}} = d_1$

Distance of B_2 form (0, 2) $= \frac{|-21|}{\sqrt{9^2+7^2}} = d_2$

Clearly $d_1 > d_2$

$\Rightarrow$ B_1 is the obtuse angle bisector

$\Rightarrow$ Obtuse angle bisector is $9x - 7y - 41 = 0$

■ TO MAKE OUT WHICH OF THE GIVEN IS THE ACUTE (OR OBTUSE) ANGLE BISECTOR

Given one line of a pair of lines L_1 and L_2 and their corresponding pair of bisector B_1 and B_2. The notion is: from any point (except point of intersection) on line L_1 (or L_2), the acute angle bisector will be nearer than the obtuse angle bisectors.

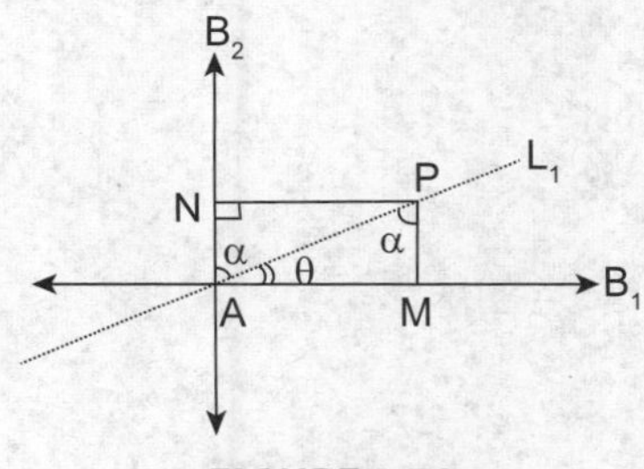

FIGURE 2.78

P is a point on line L_1 (without loss of generality of L_1 or L_2)

$\therefore \quad \theta < 45°$ and $\alpha > 45°$

In the right ΔAMP, clearly $AM > PM$

$\Rightarrow$ PM < PN

$\Rightarrow$ Acute angle bisector is nearer to P.

■ TO DETERMINE IF P(α, β) LIES IN ACUTE/OBTUSE ANGLE BETWEEN THE TWO STRAIGHT LINES L_1 AND L_2

Let $a_1x + b_1y + c_1 = 0$ and $a_2x + b_2y + c_2 = 0$ be two straight lines

First of all, let us find whether the origin lies in acute or obtuse angle. For it we make $c_1, c_2 > 0$ and find the sign of expression $a_1a_2 + b_1b_2$, if $a_1a_2 + b_1b_2 > 0$, then origin lies in obtuse angle and if $a_1a_2 + b_1b_2 < 0$, then origin lies in acute angle as shown in Figure 2.79.

Acute angle, Obtuse angle, O, L_1, L_2 — $(a_1a_2 + b_1b_2 < 0;\ c_1, c_2 > 0)$ Acute angle, Obtuse angle, O, L_1, L_2 — $(a_1a_2 + b_1b_2 > 0;\ c_1, c_2 > 0)$

FIGURE 2.79

Next we find the location of point P w.r.t, origin i.e., point P lies in same angle, opposite angle or adjacent angle to that in which origin lies.

Case I: Point $P(\alpha, \beta)$ and origin lies in same angle as shown ahead:

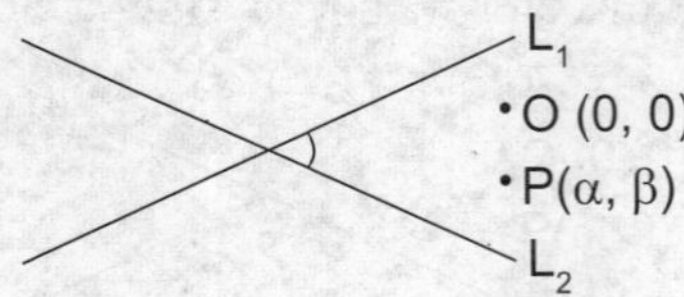

FIGURE 2.80

Then c_1 and $a_1\alpha + b_1\beta + c_1$ have same sign and c_2 and $a_2\alpha + b_2\beta + c_2$ both are of same sign.

$\Rightarrow$ $a_1\alpha + b_1\beta + c_1$ and $a_2\alpha + b_2\beta + c_2$ both are +ve

$\Rightarrow$ $P(\alpha,\beta)$ lies on same angle acute/obtuse in which origin lies.

Case II: Point $P(\alpha,\beta)$ and origin lie in opposite angles.

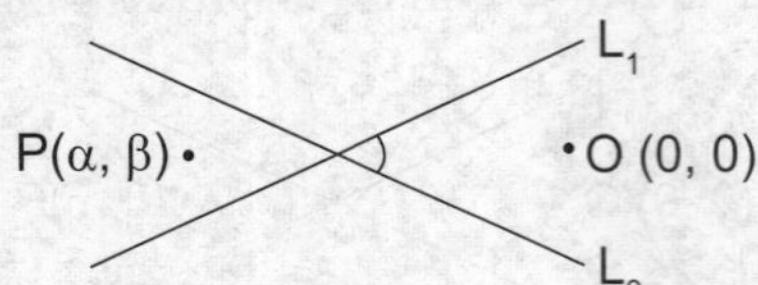

FIGURE 2.81

$\Rightarrow$ c_1 and $a_1\alpha + b_1\beta + c_1$ have opposite signs i.e., $a_1\alpha + b_1\beta + c_1 < 0$

Also c_2 and $a_2\alpha + b_2\beta + c_2$ have opposite signs

$\Rightarrow$ $a_2\alpha + b_2\beta + c_2 < 0$

Case III: Point $P(\alpha,\beta)$ and origin lies in adjacent angles.

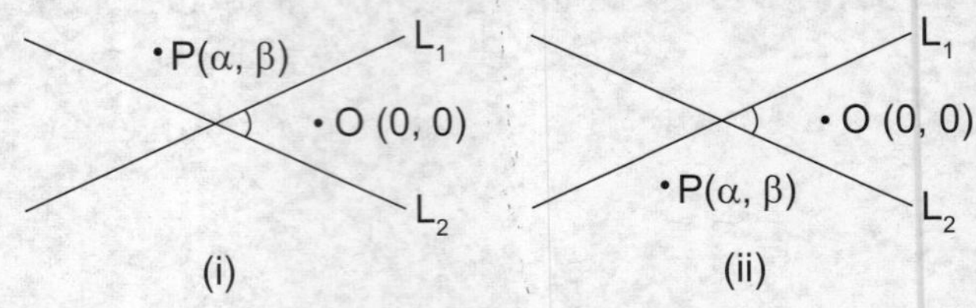

FIGURE 2.82

In figure (i) c_1 and $a_1\alpha + b_1\beta + c_1$ have opposite sign.

$\Rightarrow$ $a_1\alpha + b_1\beta + c_1 < 0$

Also c_2 and $a_2\alpha + b_2\beta + c_2$ have same sign.

$\Rightarrow$ $a_2\alpha + b_2\beta + c_2 > 0$

In figure (ii) c_1 and $a_1\alpha + b_1\beta + c_1$ have same sign.

$\Rightarrow$ $a_1\alpha + b_1\beta + c_1 > 0$

Also, and c_2 and $a_2\alpha + b_2\beta + c_2$ have opposite sign.

$\Rightarrow$ $a_2\alpha + b_2\beta + c_2 < 0$

Algorithm

Step 1: Locate the sign of c_1 and c_2. If they *are* +ve then its ok, otherwise make them +ve by multiplying with –1.

Step 2: If $a_1a_2 + b_1b_2 > 0$, then origin lies in obtuse angle and if $a_1a_2 + b_1b_2 < 0$, then origin lies in acute angle.

Step 3: Find the sign of $\dfrac{a_1\alpha + b_1\beta + c_1}{a_2\alpha + b_2\beta + c_2}$.

If $\dfrac{a_1\alpha + b_1\beta + c_1}{a_2\alpha + b_2\beta + c_2} > 0$, then point $P(\alpha, \beta)$ lies on same natured angle acute or obtuse as that of origin containing angle and

if $\dfrac{a_1\alpha + b_1\beta + c_1}{a_2\alpha + b_2\beta + c_2} < 0$, then $P(\alpha, \beta)$ lies on opposite natured angle as that of origin containing angle.

ILLUSTRATION 83: Check whether $P \equiv (3, 7)$ lies in the actue/obtuse angle formed by the line L_1: $4x + y - 3 = 0$ and $L_2: x - 2y + 4 = 0$.

SOLUTION: ***Setp 1:*** Let : $L_1 : -4x - y + 3 \equiv a_1x + b_1y + c_1 = 0$ and $L_2 : x - 2y + 4 \equiv a_2x + b_2y + c_2 = 0$

Now, since $c_1 > 0$ and $c_2 > 0$

Therefore we check $a_1a_2 + b_1b_2$

$a_1a_2 + b_1b_2 = (-4 \times 1) + (-1) \times (-2) = -2$

$a_1a_2 + b_1b_2 < 0$

$\therefore$ Origin lies in acute angle.

Step 2: To check whether point $P(\alpha, \beta)$ and origin lies in same opposite or adjacent angle formed by $L_1 = 0$ and $L_2 = 0$.

For $P \equiv (3,7)$; $a_1\alpha + b_1\beta + c_1 = -4 \times 3 - 1 \times 7 + 3 = -12 - 7 + 3 = -16$ and $a_2\alpha + b_2\beta + c_2 = (3) - 2 \times (7) + 4 = -7$. Therefore, $a_1\alpha + b_1\beta + c_1$ and $a_2\alpha + b_2\beta + c_2$ are of same sign

⇒ Nature of angle in which point $P(3, 7)$ lies is same as that of angle in which origin lies, i.e., in acute angle also, since value of expressions $a_1\alpha + b_1\beta + c_1 < 0$ and $a_2\alpha + b_2\beta + c_2 < 0$.

⇒ Point $P \equiv (\alpha, \beta)$ lies in the opposite angle formed by lines $L_1 = 0$ and $L_2 = 0$ to that of origin

∴ Point $P \equiv (\alpha, \beta)$ lies in acute angle but vertically opposite to that containing origin.

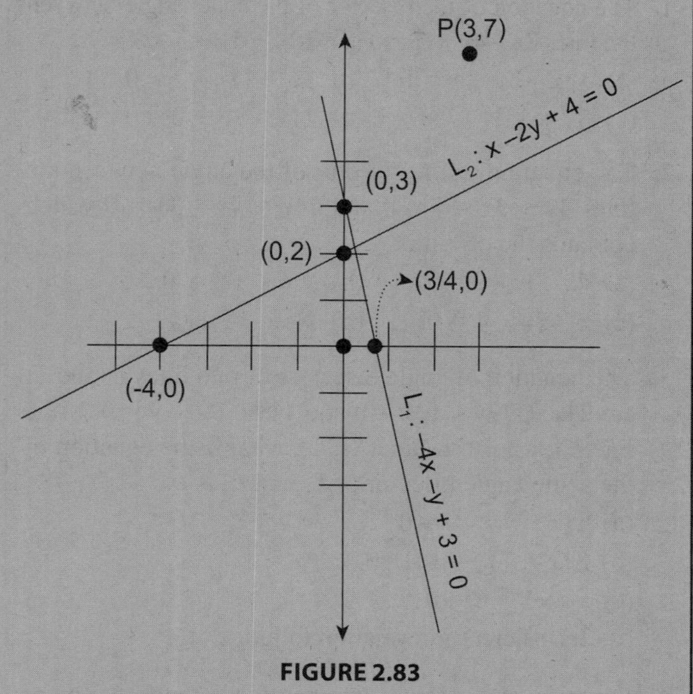

FIGURE 2.83

TEXTUAL EXERCISE–6 (SUBJECTIVE)

1. For the straight lines $4x + 3y - 6 = 0$ and $5x + 12y + 9 = 0$, find the equation of the
   (i) bisector of the obtuse angle between them
   (ii) bisector of the acute angle between them
   (iii) bisector of the angle which contains (1, 2).
2. Find the equation of obtuse angle bisector of lines $12x - 5y + 7 = 0$ and $3y - 4x - 1 = 0$.
3. Find the equation of the bisectors of the angle between the lines $y - b = \frac{2m}{1-m^2}(x-a)$ and $y - b = \frac{2M}{1-M^2}(x-a)$.
4. (a) A ray travelling along $x = 1$ gets reflected in $x + y = 1$, then find the equation of the reflected ray.
   (b) A ray of light is sent along the line $x - 2y - 3 = 0$. Upon reaching the line $3x - 2y - 5 = 0$, ray is reflected from it. Find the equation of line containing the reflected ray.
   (c) Find the image of the line $2x + 3y = 5$ in both the co-ordinate axes respectively and also in the origin.
5. If A (x_1, y_1); $B(x_2, y_2)$ and $C(x_3, y_3)$ are the vertices of a triangle ABC. Show that the equation of the internal bisector of angle A, is $b\begin{vmatrix} x & y & 1 \\ x_1 & y_1 & 1 \\ x_2 & y_2 & 1 \end{vmatrix} + c\begin{vmatrix} x & y & 1 \\ x_1 & y_1 & 1 \\ x_3 & y_3 & 1 \end{vmatrix} = 0$. (Where $b = AC$ and $c = AB$).

Answer Keys

1. (i) $7x + 9y - 3 = 0$ (ii) $9x - 7y - 41 = 0$ (iii) $9x - 7y - 41 = 0$ 2. $4x + 7y + 11 = 0$

3. $y - b = \left(\frac{m+M}{1-mM}\right)(x-a) y - b = \left(\frac{mM-1}{m+M}\right)(x-a)$

4. (a) $y = 0$ i.e., x-axis (b) $29x - 2y - 31 = 0$ (c) $2x - 3y = 5$; $2x - 3y + 5 = 0$

TEXTUAL EXERCISE–6 (OBJECTIVE)

1. The equation of the bisector of the acute angle between the lines $3x - 4y + 7 = 0$ and $12x + 5y - 2 = 0$ is
   (a) $11x - 3y + 9 = 0$ (b) $3x + 11y - 3 = 0$
   (c) $3x + 11y - 3 = 0$ (d) $11x - 3y + 2 = 0$

2. The equation of the bisector of the angle between the lines $4x + 3y - 7 = 0$ and $24x + 7y - 31 = 0$ which contains the origin is
   (a) $4x - y + 9 = 0$ (b) $x + y + 1 = 0$
   (c) $x - 3y + 1 = 0$ (d) None of these

3. The equation of angle bisectors of two lines L_1 and L_2 are $21x + 77y - 101 = 0$ and $11x - 3y + 9 = 0$. If the line L_1 passes through $(-1, 1)$, what is the equation of the acute angle bisector of L_1 and L_2 is
   (a) $11x - 3y + 9 = 0$
   (b) $21x + 77y - 101 = 0$
   (c) $x - 3y + 1 = 0$
   (d) Insufficient information to judge.

4. Equation of a straight line passing through the point (4, 5) and equally inclined to the lines, $3x = 4y + 7$ and $5y = 12x + 6$ is
   (a) $9x - 7y = 1$ (b) $9x + 7y = 71$
   (c) $7x + 9y = 73$ (d) None of these

5. Let $P \equiv (-1, 0)$; $Q \equiv (0, 0)$ and $R \equiv (3, 3\sqrt{3})$ be three points. Equation of the bisector of the $\angle PQR$ is
   (a) $x + \sqrt{3}y = 0$
   (b) $x + \sqrt{3}/2y = 0$
   (c) $\sqrt{3}/2\, x + y = 0$
   (d) $\sqrt{3}x + y = 0$

6. If one of the lines of $my^2 + (1 - m^2)xy - mx^2 = 0$ is a bisector of the angle between the lines $xy = 0$, then m is
   (a) $-1/2$ (b) -2
   (c) ± 1 (d) 2

7. A straight line through the point $A(3, 4)$ is such that its intercept between the axes is bisected at A. Its equation is
   (a) $3x - 4y + 7 = 0$
   (b) $4x + 3y = 24$
   (c) $3x + 4y = 25$
   (d) $x + y = 7$

Answer Keys

1. (a) **2.** (c) **3.** (a) **4.** (a, c) **5.** (d) **6.** (c) **7.** (b)

■ INTERSECTION OF TWO LINES AND CONDITION FOR CONCURRENCY OF THREE LINES

Intersection of Two Lines

Let the equations of two lines be $a_1x + b_1y + c_1 = 0$ ...(i)
and $a_2x + b_2y + c_2 = 0$ ...(ii)

Let these two lines intersect at a point $P(x_1, y_1)$. Then (x_1, y_1) satisfies each of the given equations.

$\therefore \quad a_1x_1 + b_1y_1 + c_1 = 0$ and $a_2x_1 + b_2y_1 + c_2 = 0$

Solving these two equations, we get

$$\frac{x_1}{b_1c_2 - b_2c_1} = \frac{y_1}{c_1a_2 - c_2a_1} = \frac{1}{a_1b_2 - a_2b_1}$$

Hence, the co-ordinates of the point of intersection of (i) and (ii) are $\left(\dfrac{b_1c_2 - b_2c_1}{a_1b_2 - a_2b_1}, \dfrac{c_1a_2 - c_2a_1}{a_1b_2 - a_2b_1}\right)$

ILLUSTRATION 84: Find the co-ordinates of the point of intersection of the lines $2x - y + 3 = 0$ and $x + 2y - 4 = 0$

SOLUTION: Solving the two equations

$$\frac{x}{4-6} = \frac{y}{3+8} = \frac{1}{4+1} \Rightarrow \frac{x}{-2} = \frac{y}{11} = \frac{1}{5} \Rightarrow x = \frac{-2}{5}, y = \frac{11}{5}$$

CONDITION FOR CONCURRENCY OF THREE LINES

Three lines are said to be concurrent if they pass through a common point. Thus, if three lines are concurrent the point of intersection of two lines lies on the third line.

Let $a_1x + b_1y + c_1 = 0$; $a_2x + b_2y + c_2 = 0$

and $a_3x + b_3y + c_3 = 0$

be three concurrent lines. Then the point of intersection of (i) and (ii) must lie on the third. The co-ordinates of the point of intersection of (i) and (ii) are $\left(\frac{b_1c_2 - b_2c_1}{a_1b_2 - a_2b_1}, \frac{c_1a_2 - c_2a_1}{a_1b_2 - a_2b_1}\right)$

This point lies on (iii), therefore $a_3\left(\frac{b_1c_2 - b_2c_1}{a_1b_2 - a_2b_1}\right) + b_3\left(\frac{c_1a_2 - c_2a_1}{a_1b_2 - a_2b_1}\right) + c_3 = 0$

$$\Rightarrow a_3(b_1c_2 - b_2c_1) + b_3(c_1a_2 - c_2a_1) + c_3(a_1b_2 - a_2b_1) = 0$$

$$\Rightarrow \Delta = \begin{vmatrix} a_1 & b_1 & c_1 \\ a_2 & b_2 & c_2 \\ a_3 & b_3 & c_3 \end{vmatrix} = 0$$

This is the required condition for concurrency of lines.

NOTE

$\Delta = 0 \Rightarrow$ lines are concurrent

❑ **Theorem:** *Another condition of concurrency of three lines; $L_1 : a_1x + b_1y + c_1 = 0$; $L_2 : a_2x + b_2y + c_2 = 0$ and L_3: $a_3x + b_3y` + c_3 = 0$ are concurrent iff there exists three constants $\lambda_1, \lambda_2, \lambda_3$ not all zeros such that $\lambda_1L_1 + \lambda_2 L_2 + \lambda_3 L_3 = 0$*

i.e., $\lambda_1(a_1x + b_1y + c_1) + \lambda_2(a_2x + b_2y + c_2) + \lambda_3(a_3x + b_3y + c_3) = 0$

Proof: Suppose there exist $\lambda_1, \lambda_2, \lambda_3$ all non-zeros such that

$\lambda_1(a_1x + b_1y + c_1) + \lambda_2(a_2x + b_2y + c_2) + \lambda_3(a_3x + b_3y + c_3) = 0.$

We may assume that $\lambda_3 \neq 0$. Then the above condition gives

$$\frac{\lambda_1}{\lambda_3}(a_1x + b_1y + c_1) + \frac{\lambda_2}{\lambda_3}(a_2x + b_2y + c_2) + (a_3x + b_3y + c_3 = 0) \quad \text{... (1)}$$

If (h, k) is the point of intersection of $a_1x + b_1y + c_1 = 0$ and $a_2x + b_2y + c_2 = 0$,

substituting in (1) we see that $a_3h + b_3k + c_3 = 0$.

Hence (h, k) lies on the third line $a_3x + b_3y + c_3 = 0$.

In other words, the three lines are concurrent.

Conversely, if the three lines are concurrent, we can write $a_3x + b_3y + c_3 = 0$ in the form

$(a_1x + b_1y + c_1) + \lambda(a_2x + b_2y + c_2) = 0$ for some constant λ.

This means that $a_3 = k(a_1 + \lambda a_2)$,

$b_3 = k(b_1 + \lambda b_2)$; $c_3 = k(c_1 + \lambda c_2)$ for some constant k.

Therefore, we get

$$a_1x + b_1y + c_1 + \lambda(a_2x + b_2y + c_2) - \frac{1}{k}(a_3x + b_3y + c_3) \equiv 0$$

The proof is now complete.

ILLUSTRATION 85: Show that the straight lines $2x + 7y + 27 = 0$, $5x + 13y - 17 = 0$ and $12x + 33y - 7 = 0$ are concurrent.

SOLUTION: We note that $(2x + 7y + 27) + 2(5x + 13y - 17) - (12x + 33y - 7) \equiv 0$.

Hence the three straight lines are concurrent.

ILLUSTRATION 86: Find the equations of the medians of a triangle formed by the lines $x + y - 6 = 0$, $x - 3y - 2 = 0$ and $5x - 3y + 2 = 0$.

SOLUTION: The given equations are: $x + y - 6 = 0$ ..(i)

$x - 3y - 2 = 0$ ...(ii)

and $5x - 3y + 2 = 0$ ..(iii)

Suppose equations (i), (ii) and (iii) represent the sides AB, BC and CA respectively of ΔABC.

Solving (i) and (ii), we get $x = 5$ and $y = 1$. Thus AB and BC intersect at $B(5, 1)$.

Solving (ii) and (iii), we get $x = -1$ and $y = -1$. Thus BC and CA intersect at $C(-1, -1)$.

Solving (i) and (iii), we get $x = 2$ and $y = 4$. Thus AB and CA intersect at $A(2, 4)$.

Thus the co-ordinates of the vertices A, B and C of triangle ABC are (2, 4), (5, 1) and (–1, –1) respectively.

Let D, E and F be the mid-points of sides BC, CA and AB respectively. Then the co-ordinates of D, E and F are $\left(\frac{5-1}{2}, \frac{1-1}{2}\right) \equiv (2, 0)$; $\left(\frac{2-1}{2}, \frac{4-1}{2}\right) \equiv (1/2, 3/2)$ and $\left(\frac{2+5}{2}, \frac{4+1}{2}\right)$ $= (7/2, 5/2)$ respectively.

Now, equation of AD is $y - 4 = \frac{0-4}{2-2}(x-2)$

i.e., $x - 2 = \frac{2-2}{0-4}(y-4)$

$\Rightarrow \; x - 2 = 0 \Rightarrow x = 2$

Similarly, the equation of line BE is $x + 9y - 14 = 0$

Similarly, the equation of line CF is $7x - 9y - 2 = 0$

Hence the equations of the medians of the triangle are

$x = 2$, $x + 9y - 14 = 0$ and $7x - 9y - 2 = 0$.

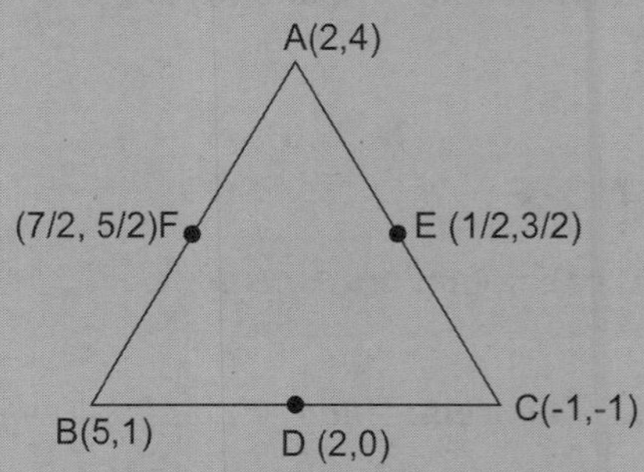

FIGURE 2.84

ILLUSTRATION 87: Find the value of λ, if the lines $3x - 4y - 13 = 0$, $8x - 11y - 33 = 0$ and $2x - 3y + \lambda = 0$ are concurrent.

SOLUTION: The given lines are concurrent if $\begin{vmatrix} 3 & -4 & -13 \\ 8 & -11 & -33 \\ 2 & -3 & \lambda \end{vmatrix} = 0$

$\Rightarrow \; 3(-11\lambda - 99) + 4(8\lambda + 66) - 13(-24 + 22) = 0$

$\Rightarrow \; -\lambda - 7 = 0 \Rightarrow \lambda = -7$

ILLUSTRATION 88: Show that the lines $2x + 3y - 8 = 0$, $x - 5y + 9 = 0$ and $3x + 4y - 11 = 0$ are concurrent.

SOLUTION: ***Method 1:*** Solving the first two equations, we see that their point of intersection is (1, 2) which satisfies the third equation. i.e., $3 \times 1 + 4 \times 2 - 11 = 0$. Hence the given lines are concurrent.

Method 2: We have $\begin{vmatrix} a_1 & b_1 & c_1 \\ a_2 & b_2 & c_2 \\ a_3 & b_3 & c_3 \end{vmatrix} = \begin{vmatrix} 2 & 3 & -8 \\ 1 & -5 & 9 \\ 3 & 4 & -11 \end{vmatrix}$ Applying $C_3 \to C_3 + C_1 + 2C_2$; we get

$= \begin{vmatrix} 2 & 3 & 0 \\ 1 & -5 & 0 \\ 3 & 4 & 0 \end{vmatrix} = 0$. Hence the given lines are concurrent.

ILLUSTRATION 89: Find the value of a for which the three lines $2x + y - 1 = 0$, $ax + 3y - 3 = 0$, and $3x + 2y - 2 = 0$ are concurrent.

SOLUTION: Solving $2x + y - 1 = 0$ and $3x + 2y - 2 = 0$, we get the point (0, 1) as the point of intersection. Now, whatever be 'a', (0, 1) always lies on $ax + 3y - 3 = 0$.

Hence the three lines are concurrent for all values of a.

ILLUSTRATION 90: Area of parallelogram formed by the lines $y = mx$, $y = mx + 1$, $y = nx$ and $y = nx + 1$

SOLUTION: Co-ordinates of point B:

point of intersection of AB and OB

$\Rightarrow B \equiv \left(\frac{1}{m-n}, \frac{m}{m-n}\right)$

$\Rightarrow |BD| = \left|\frac{1}{m-n}\right|$ and $|OA| = 1$ as co-ordinates of A are (0, 1)

$\Rightarrow$ Area of $\Delta AOB = \frac{1}{2} \times |BD| \times |OA| = \frac{1}{2}\left|\frac{1}{m-n}\right|$

$\Rightarrow$ Area of parallelogram $= \frac{1}{|m-n|}$

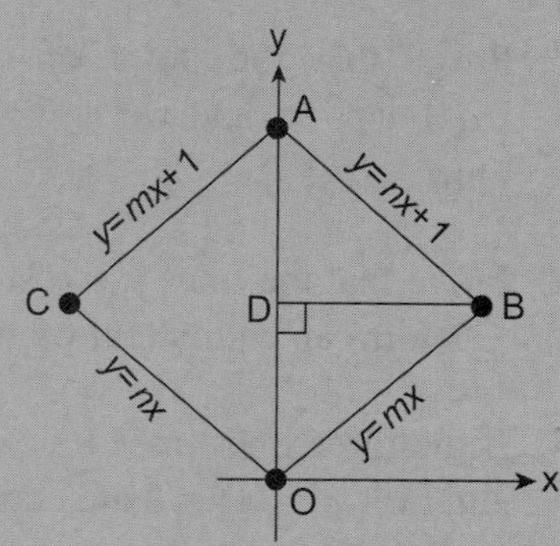

FIGURE 2.85

ILLUSTRATION 91: Find the number of integer values of m, for which the x co-ordinate of the point of intersection of the lines $3x + 4y = 9$ and $y = mx + 1$ is also an integer.

SOLUTION: $3x + 4y = 9$ ...(1)

$y = mx + 1$ ...(2)

using (1) and (2) $x = \frac{5}{3+4m}$

for x to be an integer $3 + 4m = \pm 5$ or ± 1

therefore four values of 'm' are possible and given by $m = -2, -1, -1/2, 1/2$

ILLUSTRATION 92: If $L_1 : x + 3y - 5 = 0$;

$L_2 : 3x - ky - 1 = 0$;

$L_3 : 5x + 2y - 12 = 0$; are three straight lines,

Then answer the questions that follows.

(a) (L_1, L_2, L_3 are concurrent), then find value of k.

(b) one of L_1, L_2, L_3 is parallel to at least one of the other two, then find the value of k.

(c) L_1, L_2, L_3 form a triangle, then find the value of k.

(d) L_1, L_2, L_3 do not form a triangle, then find the value of k.

SOLUTION: (a) Point of intersection of L_1 and L_3 are given by $\frac{x}{-36+10} = \frac{y}{-25+12} = \frac{1}{2-15}$

$x = 2, y = 1$

L_1, L_2, L_3 are concurrent if point (2, 1) lies on L_2

$\Rightarrow 6 - k - 1 = 0 \Rightarrow k = 5$

(b) Either $L_1 \parallel L_2$ or $L_3 \parallel L_2$, then $\frac{1}{3} = \frac{3}{-k}$ or $\frac{3}{5} = -\frac{k}{2}$

$\Rightarrow k = -9$ or $k = -\frac{6}{5}$

(c) L_1, L_2, L_3 form a triangle, if they are not concurrent or not parallel.

$\therefore\ k \neq 5, -9, -\frac{6}{5}$. i.e., $k \in \mathbb{R} \sim \{5, -9, -6/5\}$

(d) $k = 5, -9, -\frac{6}{5}$

TEXTUAL EXERCISE–7 (SUBJECTIVE)

1. Prove that the point of intersection of the lines $x/a + y/b = 1$ and $x/b + y/a = 1$ lies on the line $(x + y)(a + b) = 2ab$.

2. Prove that the lines $y = \sqrt{3}x + 1$, $y = 4$ and $y = -\sqrt{3}x + 2$ form an equilateral triangle.

3. If the three lines $ax + a^2y + 1 = 0$, $bx + b^2y + 1 = 0$ and $cx + c^2y + 1 = 0$ are concurrent, show that at least two of three constants a, b, c are equal.

4. Find the value of m, so that the straight lines $y = x + 1$, $y = 2(x + 1)$ and $y = mx + 3$ are concurrent.

5. Find the value of m, so that the lines $3x + y + 2 = 0$, $2x - y + 3 = 0$ and $x + my - 3 = 0$ may be concurrent.

6. Find the value of m for which the two lines $mx + (2m + 3)y + m + 6 = 0$ and $(2m + 1)x + (m - 1)y + (m - 9) = 0$ intersect at a point on the y-axis.

7. Find the value of m so that lines $y = x + 1$, $2x + y = 16$ and $y = mx - 4$ may be concurrent.

Answer Keys

4. 3 **5.** 4 **6.** –1 or 21 **7.** 2.

TEXTUAL EXERCISE–7 (OBJECTIVE)

1. A line passing through the point (2, 2) and the axes enclose an area λ. The intercepts on the axes made by the line are given by the two roots of
   (a) $x^2 - 2|\lambda|x + |\lambda| = 0$
   (b) $x^2 + |\lambda|x + 2|\lambda| = 0$
   (c) $x^2 - |\lambda|x + 2|\lambda| = 0$
   (d) None of these

2. Let a, b, c be distinct non-negative numbers. If the lines $ax + ay + c = 0$, $x + 1 = 0$ and $cx + cy + b = 0$ pass through the same point, then c is
   (a) the A.M. of a and b (b) the G.M. of a and b
   (c) the H.M. of a and b (d) equal to zero.

3. The set of lines $ax + by + c = 0$, where $3a + 2b + 4c = 0$, is concurrent at the point:
   (a) $\left(\frac{3}{4}, \frac{3}{4}\right)$ (b) $\left(\frac{1}{2}, \frac{1}{2}\right)$
   (c) $\left(\frac{3}{4}, \frac{1}{2}\right)$ (d) (1, 1)

4. If $a^2 + b^2 - c^2 - 2ab = 0$, then the family of straight lines $ax + by + c = 0$ is concurrent at
   (a) (–1, 1) only (b) (1, –1) only
   (c) (–1, 1) or (1, –1) (d) None of these

5. If a, b, c are in harmonical progression, then the line, $bcx + cay + ab = 0$ passes through a fixed point whose co-ordinates are
   (a) (1, 2) (b) (–1, 2)
   (c) (–1, –2) (d) (1, –2)

Answer Keys

1. (c) **2.** (b) **3.** (c) **4.** (c) **5.** (d)

■ PAIR OF STRAIGHT LINES

In the previous chapters, we have learnt

(i) Intersection of two loci $L_1 = 0$ and $L_2 = 0$ can be given by $L_1^2 + L_2^2 = 0$ or $|L_1| + |L_2| = 0$.

(ii) Family of locus passing through the intersection of two given loci $L_1 = 0$ and $L_2 = 0$ is given by $L_1 + \lambda L_2 = 0$, where λ is a real number.

In this chapter, we shall study the union of two loci.

Union of two straight lines is called Pair of straight lines and the equation of pair of straight lines $y = m_1x + c_1$ and $y = m_2x + c_2$ is given by: $(y - m_1x - c_1)(y - m_2x - c_2) = 0$ and is called joint equation of pair of lines i.e, $y^2 + m_1m_2x^2 - (m_1 + m_2)xy + (m_1c_2 + m_2c_1)x - (c_1 + c_2)y + c_1c_2 = 0$ ....(i)

Which is of the form $ax^2 + 2hxy + by^2 + 2gx + 2fy + c = 0$ ......(ii).

This equation is known as the general equation of second degree in x and y.

ILLUSTRATION 93: Find the joint equation of lines $y = 2x$ and $y = -3x$.

SOLUTION: Given equations can be rewritten as $y - 2x = 0$ and $y + 3x = 0$

$\therefore$ joint equation of lines is $(y - 2x)(y + 3x) = 0 \Rightarrow y^2 + xy - 6x^2 = 0$

REMARKS

(i) *The general equation of* pair of straight lines *is represented by the most general equation of second degree in* x and *y, but any equation in* x and *y in degree two does not always represent* pair of straight lines.

(ii) *The equation (ii) is called the general equation of conic because by this equation, we can represent all the conic like circle, parabola, ellipse, hyperbola, also it will represent pair of straight lines if it can be resolved into two linear factors.*

(iii) *$(y - m_1x - c_1)(y - m_2x - c_2)(y - m_3x - c_3) \ldots (y - m_nx - c_n) = 0$ represents the joint equation of n-straight lines, $y = m_ix + c_i$ where $i \in \{1, 2, 3, \ldots, n\}$.*

CONDITION FOR THE GENERAL EQUATION TO REPRESENT A PAIR OF STRAIGHT LINES AND METHOD TO SEPERATE THEM

Let $ax^2 + by^2 + 2hxy + 2gx + 2fy + c = 0$ be the given general equation of second degree in x and y.

Considering equation as a quadratic equation in y, we have, $by^2 + 2(hx + f)y + ax^2 + 2gx + c = 0$

$$\Rightarrow \quad y = \frac{-2(hx+f) \pm 2\sqrt{(hx+f)^2 - b(ax^2+2gx+c)}}{2b}$$

$$\Rightarrow \quad by = -(hx+f) \pm \sqrt{(hx+f)^2 - b(ax^2+2gx+c)}$$

$$\Rightarrow \quad by + hx + f = \pm\sqrt{(h^2-ab)x^2 + 2(hf-bg)x + f^2 - bc} \quad \ldots\text{(i)}$$

$$\Rightarrow \quad hx + by + f = \pm\sqrt{(h^2-ab)}\sqrt{(x-\alpha)(x-\beta)}$$

where α and β are roots of quadratic equation

$(h^2 - ab)x^2 + 2(hf - bg)x + f^2 - bc = 0$

Thus equation (i) represents pair of straight lines (which can be easly seperated) if $\alpha = \beta$, i.e.,

Discriminant = 0

$\Rightarrow \Delta = 4(hf - bg)^2 - 4(h^2 - ab)(f^2 - bc) = 0$

$\Rightarrow b^2g^2 - 2hfgb + h^2bc + abf^2 - ab^2c = 0$

Dividing the above equation by 'b' ($b \neq 0$); we get $\Delta = abc + 2fgh - af^2 - bg^2 - ch^2 = 0$

$$\Rightarrow \Delta = \begin{vmatrix} a & h & g \\ h & b & f \\ g & f & c \end{vmatrix} = 0$$

Aliter:

Consider equation as a quadratic in x

$\Rightarrow \quad ax^2 + 2(hy + g)x + by^2 + 2fy + c = 0$

$$\Rightarrow \quad x = \frac{-2(hy+g) \pm 2\sqrt{(hy+g)^2 - a(by^2+2fy+c)}}{2a}$$

$$\Rightarrow \quad ax = -(hy+g) \pm \sqrt{(h^2-ab)y^2 + 2(gh-af)y + (g^2-ac)}$$

$$\Rightarrow \quad ax + hy + g = \pm\sqrt{h^2-ab} \times \sqrt{(y-\gamma)(y-\delta)} \quad \ldots\ldots\text{(iv)}$$

where γ and δ are the roots of the quadratic $(h^2 - ab)y^2 + 2(gh - af)y + (g^2 - ac) = 0$

The equation (iv) represents pair of straight lines if $\gamma = \delta$; i.e., Discriminant = 0

$\Rightarrow \quad \Delta = 4(gh - af)^2 - 4(h^2 - ab)(g^2 - ac) = 0$

$\Rightarrow \quad g^2h^2 + a^2f^2 - 2afgh - h^2g^2 + h^2ac + g^2ab - a^2bc = 0$

$\Rightarrow \quad a^2f^2 - 2afgh + h^2ac + g^2ab - a^2bc = 0$

Dividing the above equation by 'a' ($a \neq 0$); we get, $\Delta = abc + 2fgh - af^2 - ch^2 - bg^2 = 0$

$$\Rightarrow \quad \Delta = \begin{vmatrix} a & h & g \\ h & b & f \\ g & f & c \end{vmatrix} = 0$$

ILLUSTRATION 94: Show that the equation $3x^2 + 7xy + 2y^2 + 5x + 5y + 2 = 0$ represents a pair of straight lines.

SOLUTION: The given equation is of the form $ax^2 + 2hxy + by^2 + 2gx + 2fy + c = 0$, comparing the given equation with $ax^2 + 2hxy + by^2 + 2gx + 2fy + c = 0$, we obtain $a = 3$, $h = 7/2$, $b = 2$, $g = 5/2$, $f = 5/2$ and $c = 2$. Now $abc + 2fgh - af^2 - bg^2 - ch^2$

$$= 12 + 2 \times 5/2 \times 5/2 \times 7/2 - 3(5/2)^2 - 2\,(5/2)^2 - 2\,(7/2)^2 = 0$$

Hence, the given equation represents a pair of straight lines.

ILLUSTRATION 95: Find the value of k so that the following equation may represent pairs of straight line $kxy - 8x + 9y - 12 = 0$.

SOLUTION: Given equation is $kxy - 8x + 9y - 12 = 0$ or $2kxy - 16x + 18y - 24 = 0$.

Here $a = 0$, $b = 0$, $c = -24$, $f = 9$, $g = -8$, $h = k$.

Now $\Delta = abc + 2fgh - af^2 - bg^2 - ch^2$

$= 0.0(-24) + 2.9.(-8)\,(k) - 0\,(9)^2 - 0\,(-8)^2 - (-24)\,k^2 = 0$

or $24\,k^2 - 144\,k = 0$

Solving, we get $k = 0$ or $k = 6$.

If $k = 0$, it will give an equation of 1st degree. Hence $k = 6$

ILLUSTRATION 96: What relations must hold between the co-efficients of the equation

(i) $ax^2 + by^2 + cx + cy = 0$

(ii) $ay^2 + bxy + dy + cx = 0$

so that each of them may represent a pair of straight lines?

SOLUTION: (i) $ax^2 + by^2 + cx + cy = 0$

If it represents 2 straight lines, Δ must be zero.

Hence $\Delta = a.b.0 + 2.\frac{c}{2}.\frac{c}{2}.0 - a.\left(\frac{c}{2}\right)^2 - b.\left(\frac{c}{2}\right)^2 - 0 = 0$

or $-\frac{ac^2}{4} - \frac{bc^2}{4} = 0$ or $c^2\,(a + b) = 0$.

So either $c = 0$ or $(a + b) = 0$

(ii) $\Delta = 0.a.0 + 2.\frac{d}{2}.\frac{c}{2}.\frac{b}{2} - 0\left(\frac{d}{2}\right)^2 - a\left(\frac{c}{2}\right)^2 - 0\left(\frac{b}{2}\right)^2 = 0$

or $\frac{bdc}{4} - \frac{ac^2}{4} = 0$ or $c\,(bd - ac) = 0$

Hence, either $c = 0$ or $bd - ac = 0$

ILLUSTRATION 97: Find the separate equation of lines represented by equation $x^2 - 5xy + 6y^2 = 0$

SOLUTION: $x^2 - 3xy - 2xy + 6y^2 = 0$

$\Rightarrow x\,(x - 3y) - 2y\,(x - 3y) = 0$

$\Rightarrow (x - 2y)\,(x - 3y) = 0$

$\Rightarrow x - 2y = 0$ and $x - 3y = 0$

Aliter: $x = \dfrac{5y \pm \sqrt{(-5y)^2 - 4(6y^2)}}{2}$

$\Rightarrow x = \dfrac{5y \pm \sqrt{25y^2 - 24y^2}}{2} \Rightarrow x = \dfrac{5y \pm y}{2} \Rightarrow x - 3y = 0$ and $x - 2y = 0$

ILLUSTRATION 98: Prove that the equation $y^3 - x^3 + 3xy(y - x) = 0$, represents three straight lines equally inclined to one another.

SOLUTION: Given equation is $y^3 - x^3 + 3xy(y - x) = 0$

or $(y - x)(y^2 + xy + x^2) + 3xy(y - x) = 0$

or $(y - x)(y^2 + xy + x^2 + 3xy) = 0$ or

or $(y - x)[y + (2 - \sqrt{(3)}x][y + (2 + \sqrt{3})x] = 0$

Hence, $y - x = 0 \Rightarrow y = (\tan 45°)x$ ... (1)

or $y + (2 - \sqrt{3})x = 0 \Rightarrow y = (\tan 165°)x$ ... (2)

and $y + (2 + \sqrt{3})x = 0 \Rightarrow y = (\tan 105°)x$ ... (3)

which represent three straight lines equally inclined to each other.

ILLUSTRATION 99: Find the separate equations of two straight lines whose joint equation is $ab(x^2 - y^2) + (a^2 - b^2)xy = 0$.

SOLUTION: $abx^2 + (a^2 - b^2)xy - aby^2 = 0$

$\Rightarrow abx^2 + a^2xy - b^2xy - aby^2 = 0$

$\Rightarrow ax(bx + ay) - by(bx + ay) = 0$

$\Rightarrow (ax - by)(bx + ay) = 0$

$\Rightarrow ax - by = 0$ and $bx + ay = 0$

ILLUSTRATION 100: Find what straight lines are represented by the equation $x^3 - 6x^2 + 11x - 6 = 0$. Also determine the angle between them.

SOLUTION: Given equation is $x^3 - 6x^2 + 11x - 6 = 0$

By remainder theorem, the factors are $(x - 1)(x - 2)(x - 3) = 0$

Hence the lines represented are $x - 1 = 0$, $x - 2 = 0$ and $x - 3 = 0$ and clearly, all the lines are parallel to y-axis. Therefore angles between the lines is 0.

TEXTUAL EXERCISE–8 (SUBJECTIVE)

1. Show that following equations represent pair of straight lines and find the lines:
   (a) $x^2 - 7xy + 12y^2 = 0$ (b) $4x^2 - 24xy + 11y^2 = 0$
   (c) $y^2 = 0$ (d) $xy - ay = 0$
   (e) $x^3 - x^2 - x + 1 = 0$
2. If the following equations, represent a pair of straight lines, then find the values of λ.
   (i) $x^2 - 3xy + \lambda y^2 + 3x - 5y + 2 = 0$
   (ii) $\lambda x^2 - 10xy + 12y^2 + 5x - 16y - 3 = 0$
3. What straight lines are represented by equation $y^3 - xy^2 - 14x^2y + 24x^3 = 0$
4. Find the value of k so that the following equation may represent a pair of straight lines $2x^2 + xy - y^2 + kx + 6y - 9 = 0$.
5. (a) Find the values of λ for which the following equations represents a pair of straight lines $3x^2 - 10xy + 7y^2 + 2\lambda x - 14y - 42 = 0$.
   (b) Check whether that the equation $x^2 + 3xy + 2y^2 - x - 4y - 6 = 0$ represents a pair of straight lines?

Answer Keys

1. (a) $x = 3y$; $x = 4y$ (b) $y = 2x$; $2x = 11y$ (c) $y = 0, y = 0$ (d) $y = 0, x = a$
   (e) $(x - 1)^2(x + 1) = 0$
2. (i) $\lambda = 2$ (ii) $\lambda = 2$
3. $y = 2x$, $y = 3x$, $y = -4x$.
4. $k = -3$
5. (a) $\lambda = 5 \pm 2\sqrt{7}$, (b) Yes

TEXTUAL EXERCISE–8 (OBJECTIVE)

1. If the slope of one of the lines represented by $ax^2 + 2hxy + by^2 = 0$ be the square of the other, then
   (a) $a^2b + ab^2 - 6abh + 8h^3 = 0$
   (b) $a^2b + ab^2 + 6abh + 8h^3 = 0$
   (c) $a^2b + ab^2 - 3abh + 8h^3 = 0$
   (d) $a^2b + ab^2 - 6abh - 8h^3 = 0$

2. The value of h for which the equation $3x^2 + 2hxy - 3y^2 - 40x + 30y - 75 = 0$ represents a pair of straight lines, are
   (a) 4, 4 (b) 4, 6
   (c) 4, –4 (d) 0, 4

3. The equation of lines passing through the origin and parallel to the lines $y = m_1x + c_1$ and $y = m_2x + c_2$ is
   (a) $m_1m_2x^2 - (m_1 + m_2)xy + y^2 = 0$
   (b) $m_1m_2x^2 + (m_1 + m_2)xy + y^2 = 0$
   (c) $m_1m_2y^2 - (m_1 + m_2)xy + x^2 = 0$
   (d) $m_1m_2y^2 + (m_1 + m_2)xy + x^2 = 0$

4. If the equation $Ax^2 + 2Bxy + Cy^2 + Dx + Ey + F = 0$ represents a pair of straight lines, then $B^2 - AC$ is
   (a) < 0 (b) $= 0$
   (c) > 0 (d) None of these

5. If the slope of one of the lines represented by the equation $ax^2 + 2hxy + by^2 = 0$ be λ times that of the other, then
   (a) $4\lambda h = ab(1 + \lambda)$ (b) $\lambda h = ab(1 + \lambda)^2$
   (c) $4\lambda h^2 = ab(1 + \lambda)^2$ (d) None of these

6. The equation $4x^2 + 12xy + 9y^2 + 2gx + 2fy + c = 0$ will represent two real parallel straight lines, if
   (a) $g = 4, f = 9, c = 0$ (b) $g = 2, f = 3, c = 1$
   (c) $g = 2, f = 3, c > 25$ (d) $g = 4, f = 9, c > 1$

7. The equations of the lines represented by the equation $ax^2 + (a + b)xy + by^2 + x + y = 0$ are
   (a) $ax + by + 1 = 0, x + y = 0$
   (b) $ax + by - 1, x + y = 0$
   (c) $ax + by + 1 = 0, x - y = 0$
   (d) None of these

8. The value of λ for which the equation $x^2 - \lambda xy + 2y^2 + 3x - 5y + 2 = 0$ may represent a pair of straight lines is
   (a) 2 (b) 3
   (c) 4 (d) 1

9. For what value of 'p', $y^2 + xy + px^2 - x - 2y = 0$ represents two straight lines?
   (a) 2 (b) 1/3
   (c) 1/4 (d) 1/2

10. If one of the lines given by $6x^2 - xy + 4cy^2 = 0$ is $3x + 4y = 0$, then c equals
    (a) –3 (b) –1
    (c) 3 (d) 1

Answer Keys

1. (a) **2.** (a) **3.** (a) **4.** (d) **5.** (c) **6.** (b) **7.** (a) **8.** (b) **9.** (c) **10.** (a)

POINT OF INTERSECTION OF PAIR OF STRAIGHT LINES

Method I Given equation of second degree $ax^2 + 2hxy + by^2 + 2gx + 2fy + c = 0$.

Dividing the equation by b we get,

$$y^2 + \frac{a}{b}x^2 + \frac{2h}{b}xy + \frac{2g}{b}x + \frac{2f}{b}y + \frac{c}{b} = 0 \qquad ...(i)$$

Let the lines represented by equation is $y = m_1x + c_1$ and $y = m_2x + c_2$.

$$\Rightarrow \quad (y - m_1x - c_1)(y - m_2x - c_2) = 0$$

$$\Rightarrow \quad (y - m_1x)(y - m_2x) - c_2(y - m_1x) - c_1(y - m_2x) + c_1c_2 = 0$$

$$\Rightarrow \quad y^2 + m_1m_2x^2 - (m_1 + m_2)xy + (m_1c_2 + c_1m_2)x - (c_1 + c_2)y + c_1c_2 = 0 \qquad ...(ii)$$

Comparing the co-efficients of equation (i) and (ii), we get $m_1m_2 = a/b$; $m_1 + m_2 = -2h/b$; $m_1c_2 + c_1m_2 = 2g/b$; $c_1 + c_2 = -2f/b$; $c_1 c_2 = c/b$

Point of intersection:

The point of intersection is $\left(-\frac{c_1 - c_2}{m_1 - m_2}, \frac{c_1m_2 - c_2m_1}{m_2 - m_1}\right)$

but $c_1 - c_2 = \pm\sqrt{(c_1 + c_2)^2 - 4c_1c_2} = \pm\frac{2\sqrt{f^2 - bc}}{b}$

and $m_1 - m_2 = \pm\sqrt{(m_1 + m_2)^2 - 4m_1m_2} = \pm\frac{2\sqrt{h^2 - ab}}{b}$

Also $c_1m_2 - c_2m_1 = \pm\sqrt{(c_1m_2 + c_2m_1)^2 - 4c_1c_2m_1m_2}$

$$= \pm\sqrt{\frac{4g^2}{b^2} - 4\frac{ac}{b^2}} = \pm\frac{2\sqrt{g^2 - ac}}{b}$$

So the point of intersection is given by

$$\left(\pm\sqrt{\frac{f^2 - bc}{h^2 - ab}}, \pm\sqrt{\frac{g^2 - ac}{h^2 - ab}}\right)$$

Out of above four possible points, only one point gives us the point of intersection of lines of given pair and can be obtained by substituting in the given equation of pair of straight lines.

Method II

Given pair of straight line is $ax^2 + 2hxy + by^2 + 2gx + 2fy + c = 0$. Let the point of intersection of the lines represented by this equation be (α, β), then by shifting the origin to (α, β) it should become homogenous because these lines will become lines passing through origin and since axes are not rotated so the slopes will be same.

Replacing 'x' by $x' + \alpha$ and y by $y' + \beta$ in given equation, we get

$a(x'+\alpha)^2 + 2h(x'+\alpha)(y'+\beta) + b(y'+\beta)^2 + 2g(x'+\alpha) + 2f(y'+\beta) + c = 0$ has,

Co-efficient of $x' = 0 \Rightarrow 2a\alpha + 2h\beta + 2g = 0$ ...(i)

and co-efficient of $y' = 0 \Rightarrow 2b\beta + 2h\alpha + 2f = 0$ ...(ii)

and constant term = 0

$\Rightarrow a\alpha^2 + b\beta^2 + 2hg\alpha\beta + 2g\alpha + 2f\beta + c = 0$ ...(iii)

which is ever satisfied as (α, β) lies on straight line.

From the equation (i) and (ii), $a\alpha + h\beta + g = 0$; $h\alpha + b\beta + f = 0$ and solving we get

$$\frac{\alpha}{\begin{vmatrix} -g & h \\ -f & b \end{vmatrix}} = \frac{\beta}{\begin{vmatrix} a & -g \\ h & -f \end{vmatrix}} = \frac{1}{\begin{vmatrix} a & h \\ h & b \end{vmatrix}}$$

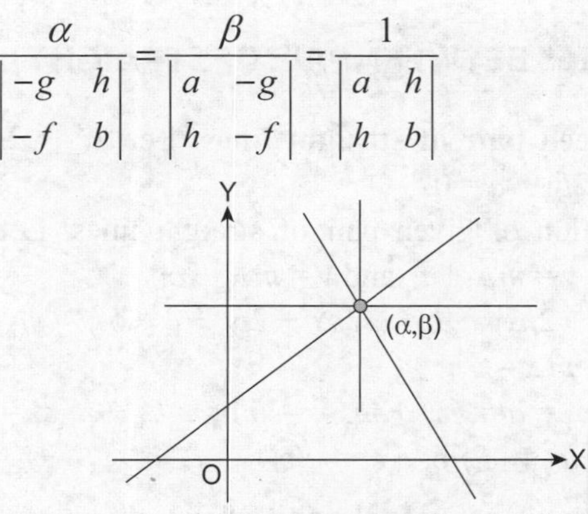

FIGURE 2.86

$$\Rightarrow \alpha = -\frac{hf - bg}{h^2 - ab};\ \beta = -\frac{gh - af}{h^2 - ab}$$

So the co-ordinates of point of intersection is also given by $\left(\dfrac{bg - hf}{h^2 - ab}, \dfrac{af - gh}{h^2 - ab}\right)$

Aid to Memory: Partially differentiating the expression with respect to x and y and equating to zero and solving them for x and y as given here:

Let $f = ax^2 + 2hxy + by^2 + 2gx + 2fy + c = 0$

Differentiating f partially w.r.t x and y, we get

$$\frac{\partial f}{\partial x} = 2ax + 2hy + 2g = 0;$$

and $\dfrac{\partial f}{\partial y} = 2hx + 2by + 2f = 0$, on solving we get

$$\Rightarrow \frac{x}{fh - bg} = \frac{y}{gh - af} = \frac{1}{ab - h^2}$$

$$\Rightarrow (x, y) \equiv \left(\frac{bg - hf}{h^2 - ab}, \frac{af - gh}{h^2 - ab}\right)$$

ILLUSTRATION 101: Find the point of intersection of pair of straight lines $x^2 - xy + 2y^2 - y - 5x - 6 = 0$.

SOLUTION: $S: x^2 - xy + 2y^2 - y - 5x - 6 = 0$. Let (α, β) be the point of intersection of both lines represented by $S = 0$

Now, shifting the origin to (α, β), then equation $S = 0$ must transform to homogenous form

$\therefore$ $(x + \alpha)^2 + 2(y + \beta)^2 + (-1)(x + \alpha)(y + \beta) - (y + \beta) - 5(x + \alpha) - 6 = 0$ must be homogeneous in x and y.

$\Rightarrow$ Co-efficient of $x = 0 \Rightarrow \alpha + \left(\dfrac{-1}{2}\right)\beta + \left(\dfrac{-5}{2}\right) = 0$

Similarly, co-efficient of $y = 0 \Rightarrow \left(\dfrac{-1}{2}\right)\alpha + 2\beta + \left(\dfrac{-1}{2}\right) = 0$

On solving the above equations, we get $(\alpha, \beta) \equiv (3, 1)$

Aliter: $\left(\dfrac{\partial S}{\partial x}\right) = 0 \Rightarrow 2x - y - 5 = 0$; $\left(\dfrac{\partial S}{\partial y}\right) = 0 \Rightarrow -x + 4y - 1 = 0$

Solving; we get point of intersection $\equiv (3, 1)$

■ ANGLES BETWEEN PAIR OF STRAIGHT LINES

Angle between pair of straight lines: Let $ax^2 + 2hxy + by^2 + 2gx + 2fy + c = 0$ ......(i)
be the equation of given pair of straight lines. Let their equations be $y = m_1x + c_1$ and $y = m_2x + c_2$

$\therefore\ ax^2 + 2hxy + by^2 + 2gx + 2fy + c \equiv b(y - m_1x - c_1)(y - m_2x - c_2)$

$\Rightarrow\ m_1m_2 = a/b;\ m_1 + m_2 = -2h/b;\ m_1c_2 + m_2c_1 = 2g/b;\ c_1 + c_2 = -2f/b;\ c_1c_2 = c/b$

Let θ be the angle between the pair of straight lines, then

$$\tan\theta = \frac{m_1 - m_2}{1 + m_1m_2} \Rightarrow \tan\theta = \frac{\pm\sqrt{(m_1+m_2)^2 - 4m_1m_2}}{1+m_1m_2},$$

$$\Rightarrow\ \tan\theta = \pm\frac{\sqrt{\left(\frac{-2h}{b}\right)^2 - 4\left(\frac{a}{b}\right)}}{1+\frac{a}{b}} = \frac{\pm\frac{2}{b}\sqrt{h^2-ab}}{\frac{a+b}{b}}$$

$$= \frac{\pm 2\sqrt{h^2-ab}}{a+b}$$

$\therefore$ Angle of intersection

$$\theta = \tan^{-1}\left(\frac{2\sqrt{h^2-ab}}{|a+b|}\right) \text{ for } \theta \text{ to be acute}$$

$$\text{and } \pi - \tan^{-1}\left(\frac{2\sqrt{h^2-ab}}{|a+b|}\right) \text{ for } \theta \text{ to be obtuse}$$

■ SUFFICIENT CONDITIONS FOR PAIR OF STRAIGHT LINES $ax^2 + 2hxy + by^2 + 2gx + 2fy + c = 0$ TO REPRESENT INTERSECTING/ PARALLEL/COINCIDENT LINES

Let $ax^2 + 2hxy + by^2 + 2gx + 2fy + c = 0$ be given pair of straight lines,

$$\text{then } x = \frac{-(2hy+2g) \pm \sqrt{4(hy+g)^2 - 4a(by^2+2fy+c)}}{2a}$$

$$\Rightarrow x = \frac{-(hy+g) \pm \sqrt{h^2y^2 + g^2 + 2hgy - aby^2 - 2afy - ac}}{a}$$

$$\Rightarrow x = \frac{-(hy+g) \pm \sqrt{(h^2-ab)y^2 + 2(hg-af)y + (g^2-ac)}}{a}$$

$$\Rightarrow ax + hy + g = \pm\sqrt{(h^2-ab)y^2 + 2(hg-af)y + (g^2-ac)}$$ .....(i)

Now clearly, (i) would represent a pair of coincident straight lines if $h^2 - ab = hg - af = g^2 - ac = 0$ The acute angle between the two straight lines is given by $\theta = \tan^{-1}\left(\frac{2\sqrt{h^2-ab}}{|a+b|}\right) = 0$.

For coincident lines $h^2 - ab = 0$

The point of intersection between the pair of straight lines is given by

$$\left(\frac{bg-hf}{h^2-ab}, \frac{af-gh}{h^2-ab}\right) \text{ or } \left(\pm\sqrt{\frac{f^2-bc}{h^2-ab}}, \pm\sqrt{\frac{g^2-ac}{h^2-ab}}\right)$$

For coincident lines, the abscissa and ordinates should be indeterminate and hence, $h^2 - ab = 0 \Rightarrow bg - hf = f^2 - bc = af - gh = g^2 - ac = 0$

Hence $h^2 - ab = 0$ ensures for straight lines to be parallel or coincident and $af - gh = bg - hf = f^2 - bc = g^2 - ac$ guarantees for coincidence of straight lines.

Hence the sufficient conditions for straight lines represented by $ax^2 + 2hxy + bg^2 + 2gx + 2fy + c = 0$ to be coincident are

$h^2 - ab = af - gh = bg - hf = f^2 - bc = g^2 - ac = 0$

Now $h^2 - ab = 0$ for parallel lines, Equation (i) reduces to $ax + by + g = \pm\sqrt{2(hg-af)y + g^2 - ac}$

$\therefore$ for straight lines we must have $hg - af = 0$ and for parallel but not coincident straight lines we must have $g^2 - ac > 0$.

Thus the sufficient conditions for pair of straight lines $ax^2 + 2hxy + bg^2 + 2gx + 2fy + c = 0$ to represent parallel but not coincident lines is $h^2 - ab = 0$; $g^2 - ac > 0$.

For intersecting lines $\tan\theta \neq 0 \Rightarrow h^2 - ab \geq 0$, which is the sufficient conditions for pair of straight lines to be intersecting

Conclusions

(a) If $h^2 - ab > 0$
$\Rightarrow$ two real and distinct intersecting straight lines.

(b) If $h^2 - ab < 0$
$\Rightarrow$ two imaginary lines as in this case θ is imaginary.

(c) If $h^2 - ab = 0$
$\Rightarrow$ two parallel lines if $g^2 - ac \neq 0$ as in this case point of intersection lies at infinity and the lines have same slope i.e., lines are parallel but not intersecting.

(d) If $h^2 - ab = 0$
$\Rightarrow$ two coincident lines if $bg - hf = af - gh = f^2 - bc = g^2 - ac = 0$.

(e) If $a + b = 0$
$\Rightarrow$ both lines are perpendicular as in this case $\tan\theta$ tends to infinity.

(f) If $h = 0$
$\Rightarrow$ The two lines makes an isosceles triangle with x-axis and y-axis.

REMARKS

1. *Only homogeneous part of equation of* pair of straight lines *governs the slope of the lines represented.*
2. *As we know that* pair of straight lines *is a conic section obtained when a plane perpendicular to base of cone cuts it passing through the vertex. When the straight lines are parallel, the cone is of infinite height* and *the angle between the opposite generaters tends to become zero.*

ILLUSTRATION 102: Find the angle between the pair of straight lines $x^2 - 5xy + 6y^2 = 0$.

SOLUTION: $m_1 + m_2 = \frac{-2h}{b} = \frac{5}{6}$; $m_1 m_2 = a/b = 1/6$

Therefore acute angle θ between the lines is given by $\tan\theta = \left|\frac{m_1 - m_2}{1 + m_1 m_2}\right| = \frac{\sqrt{(m_1+m_2)^2 - 4m_1m_2}}{|1+m_1m_2|}$

$$\Rightarrow \tan\theta = \frac{\sqrt{\left(\frac{5}{6}\right)^2 - \frac{4}{6}}}{1+\frac{1}{6}} = \frac{\sqrt{\frac{25}{36} - \frac{4}{6}}}{\left|1+\frac{1}{6}\right|} = \frac{\sqrt{\frac{25-24}{36}}}{\frac{7}{6}} = 1/7 \Rightarrow \theta = \tan^{-1}\left(\frac{1}{7}\right)$$

Aliter: Acute angle θ is given by $\theta = \tan^{-1} 2\frac{\sqrt{h^2 - ab}}{|a+b|} = \tan^{-1} 2\frac{\sqrt{\left(\frac{-5}{2}\right)^2 - (1)(6)}}{|1+6|} = \tan^{-1}\frac{1}{7}$

ILLUSTRATION 103: Test whether the equation $x^2 + xy - 2y^2 + 3y - 1 = 0$ represent a pair of straight line or not, if yes, then find both equation of lines and their point of intersection as well as angle of intersection.

SOLUTION: As we know that any equation will represent pair of straight lines if, it satisfies given condition $\Delta = abc + 2fgh - af^2 - bg^2 - ch^2 = 0$

Now, comparing the given equation $x^2 + xy - 2y^2 + 3y - 1 = 0$ with $ax^2 + 2hxy + by^2 + 2gx + 2fy + c = 0$ we have, $a = 1$; $2h = 1 \Rightarrow h = 1/2$; $b = -2$; $2g = 0$; $2f = 3$

$\Rightarrow f = 3/2$; $c = -1$

$\therefore$ $\Delta = (1 \times -2 \times -1) + 2(0) - (1)(3/2)^2 - 0 - (-1 \times 1/4) = 0$

$\therefore$ It represents pair of straight lines and the lines represented are given by: $hx + by + f = \pm\sqrt{h^2 - ab}\,(x - \alpha)$; where α is a root of equation $(h^2 - ab)x^2 + 2(hf - bg)x + (f^2 - bc) = 0$; substituting the values; we have $\frac{9}{4}x^2 + \frac{3}{2}x + \frac{1}{4} = 0$

$\Rightarrow$ $9x^2 + 6x + 1 = 0 \Rightarrow (3x + 1)^2 = 0 \Rightarrow x = -1/3$

$\Rightarrow$ lines are $\frac{1}{2}x - 2y + \frac{3}{2} = \pm\sqrt{\frac{9}{4}}\left(x + \frac{1}{3}\right) \Rightarrow \frac{1}{2}x - 2y + \frac{3}{2} = \pm\frac{3}{2}\left(x + \frac{1}{3}\right)$

or $3x - 12y + 9 = \pm 3(3x + 1) \Rightarrow 3x - 12y + 9 = (9x + 3)$ or $-9x - 3$

$\Rightarrow$ $6x + 12y - 6 = 0$ or $12x - 12y + 12 = 0 \Rightarrow x + 2y - 1 = 0$ or $x - y + 1 = 0$

$\therefore$ The point of intersection are given by $\left(-\frac{1}{3}, \frac{2}{3}\right)$.

Aliter: $x^2 + xy - 2y^2 + 3y - 1 = 0$

Treating the above equation as quadratic in 'x', we can find the roots as

$2x = -y \pm \sqrt{y^2 - 4(-2y^2 + 3y - 1)}$

$\Rightarrow 2x + y = \pm\sqrt{9y^2 - 12y + 4} \Rightarrow 2x + y = \pm (3y - 2)$

$\Rightarrow 2x + y + 3y - 2 = 0$ or $2x + y - 3y + 2 = 0 \Rightarrow x + 2y - 1 = 0$ or $x - y + 1 = 0$

$\therefore$ The above equation represents a pair of straight lines

$\Rightarrow$ Their point of intersection on simultaneously solving; we get $(x, y) = \left(\frac{-1}{3}, \frac{2}{3}\right)$

Acute angle between them: $\theta = \tan^{-1}\left|\frac{m - m_2}{1 + m_1 m_2}\right| = \tan^{-1}\left|\frac{1-(-1/2)}{1+1(-1/2)}\right| = \tan^{-1}\left|\frac{3/2}{1/2}\right| = \tan^{-1} 3$

Aliter: $m_1 + m_2 = \frac{-2h}{b} = \frac{-2(1/2)}{-2} = \frac{1}{2}$ ......(1)

$m_1\, m_2 = \frac{a}{b} = \frac{-1}{2}$ ....(2)

$c_1 + c_2 = -\frac{2f}{b} = \frac{-2\times(3/2)}{-2} = \frac{3}{2}$ ......(3)

$c_1.c_2 = \frac{c}{b} = \frac{-1}{-2} = \frac{1}{2}$ ....(4)

and $m_1 c_2 + m_2 c_1 = \frac{2g}{b} = \frac{2\times 0}{-2} = 0$ .....(5)

Solving (1) and (2)

$\Rightarrow m_1 = \frac{1}{2} - m_2; \Rightarrow \left(\frac{1}{2} - m_2\right)(m_2) = \frac{-1}{2} \Rightarrow m_2 - 2m_2^2 = -1$

$\Rightarrow 2m_2^2 - m_2 - 1 = 0 \Rightarrow 2m_2^2 - 2m_2 + m_2 - 1 = 0$

$\Rightarrow (2m_2 + 1)(m_2 - 1) = 0 \Rightarrow m_2 = \frac{-1}{2}$ or $m_2 = 1$

Let $m_2 = \frac{-1}{2}$, $\therefore m_1 = 1$ ....(6)

Solving (3) and (4)

$c_1 + c_2 = 3/2$ and $c_1 c_2 = 1/2$

and either $c_1 = 1/2$ and $c_2 = 1$ ......(7)

or $c_1 = 1$ and $c_2 = 1/2$ ....(8)

Putting values from (6) and (7) in (5); we get

$m_1 c_2 + m_2 c_1 = 1 \times 1 + \frac{-1}{2} \times \frac{1}{2} = \frac{3}{4}$

But, we already know that $m_1 c_2 + m_2 c_1 = 0$

$\therefore$ our supposition (7) is false

Now, puting values from (6) and (8) in (5); we get

$m_1 c_2 + m_2 c_1 = 1 \times \frac{1}{2} + 1 \times \left(\frac{-1}{2}\right) = 0$

$\therefore$ Required lines are $y = m_1 x + c_1 \Rightarrow y = x + 1$

and $y = m_2 x + c_2 \Rightarrow y = \frac{-1}{2}x + \frac{1}{2} \Rightarrow 2y + x = 1$

ILLUSTRATION 104: Prove that triangle formed by the lines $ax^2 + 2hxy + by^2 = 0$ and $lx + my - 1 = 0$ is isosceles if the condition $h(l^2 - m^2) = lm(a - b)$ is satisfied.

SOLUTION: Let the slopes of the two lines be m_1 and m_2

terms containing f and g are not there so, values of $c_1 + c_2$ and c_1c_2 is 0

$\therefore$ c_1 and $c_2 = 0$

Then equation of two lines of given pair of straight lines are $y = m_1x$ ...... (i)

and $y = m_2x$ .....(ii)

and the third line is $lx + my - 1 = 0$ ...(iii)

with slope $M = -l/m$, therefore for isosceles triangle angle between (i) and (iii) must be same as that between (ii) and (iii)

$$\Rightarrow \left|\frac{m_1 - M}{1 + m_1M}\right| = \left|\frac{M - m_2}{1 + m_2M}\right| \Rightarrow \frac{m_1 - M}{1 + m_1M} = \pm\frac{M - m_2}{1 + m_2M}$$

$$\Rightarrow (m_2 + m_1) - 2M + 2m_1m_2M - (m_2 + m_1)M^2 = 0 \text{ or } M^2 = -1 \text{ (impossible)}$$

$$\Rightarrow \left(\frac{-2h}{b}\right) + 2\times\left(\frac{l}{m}\right) + 2\times\frac{a}{b}\times\left(\frac{-l}{m}\right) + \frac{2h}{b}\times\frac{l^2}{m^2} = 0$$

$$\Rightarrow hm^2 - lbm + alm - hl^2 = 0 \Rightarrow h(l^2 - m^2) = lm(a - b)$$

ILLUSTRATION 105: Find the equation of lines represented by pair of straight lines $y^2 - 5xy + 6x^2 = 0$ and area of triangle formed by these lines with the line $x = 6$.

SOLUTION: $y^2 - 5xy + 6x^2 = 0$

represents a homogenous pair of straight lines, passing through origin

Let the lines represented by the above equation

be $y - m_1x = 0$ and $y - m_2x = 0$

$\therefore$ $(y - m_1x)(y - m_2x) = y^2 - 5xy + 6x^2$

upon solving, we get $\Rightarrow m_1 = 3$ and $m_2 = 2$

$\therefore$ Equations are $y = 3x$ and $y = 2x$

Now area of triangle ΔABC

$$= \frac{1}{2}\left|x_1(y_2 - y_3) + x_2(y_3 - y_1) + x_3(y_1 - y_2)\right|$$

$$= \frac{1}{2}\left|0 + 6(18 - 0) + 6(0 - 12)\right|$$

$$= \frac{1}{2}\left|6\times18 - 12\times6\right| = \frac{36}{2} = 18 \text{ square units.}$$

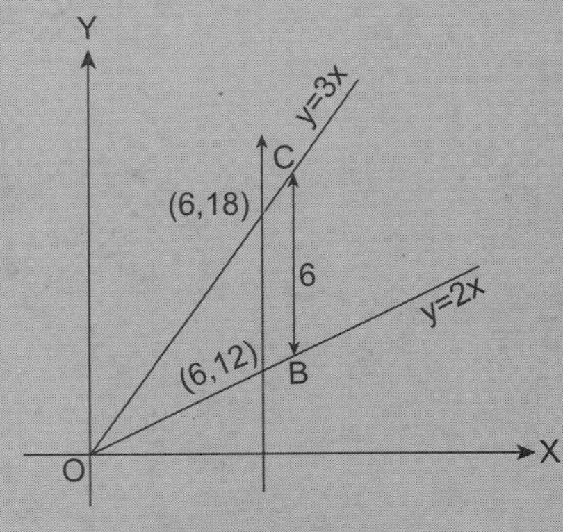

FIGURE 2.87

ILLUSTRATION 106: Find the co-ordinates of centroid of the triangle whose sides are $12x^2 - 20xy + 7y^2 = 0$ and $2x - 3y + 4 = 0$.

SOLUTION: $12x^2 - 20xy + 7y^2 = 0$; comparing it with $ax^2 + 2hxy + by^2 + 2gx + 2fy + c = 0$; we get

$a = 12, 2h = -20, b = 7; g = 0, f = 0, c = 0$

Now $m_1 + m_2 = \dfrac{-2h}{b}$ and $m_1m_2 = \dfrac{a}{b}$

$$\Rightarrow m_1 + m_2 = \frac{-(-20)}{7} \text{ and } m_1m_2 = \frac{12}{7}$$

$\Rightarrow m_1 = \frac{20}{7} - m_2$ and $m_2\left(\frac{20}{7} - m_2\right) = \frac{12}{7}$

$\Rightarrow \frac{20}{7}m_2 - m_2^2 = \frac{12}{7}$

$\Rightarrow 20m_2 - 7m_2^2 = 12$

$\Rightarrow 7m_2^2 - 20m_2 + 12 = 0 \Rightarrow 7m_2^2 - 14m_2 - 6m_2 + 12 = 0$

$\Rightarrow 7m_2(m_2 - 2) - 6(m_2 - 2) = 0 \Rightarrow (7m_2 - 6)(m_2 - 2) = 0$

$\Rightarrow m_2 = 2$ or $m_2 = \frac{6}{7}$ when $m_2 = 2$; $m_1 = \frac{6}{7}$ and when $m_2 = \frac{6}{7}; m_1 = 2$

$\therefore$ The equatios of straight lines are $y = 2x$ and $y = \frac{6}{7}x$ (say l_1 and l_2 respectively) and $2x - 3y + 4 = 0$ be l_3

$\therefore$ Point of intersection of l_1 and l_3 are $(1, 2) \equiv A$(say)

$\therefore$ Similarly, point of intersection of l_2 and l_3 be 'B' $\equiv (7, 6)$

Now vertices of triangle ABC are given by $A(1, 2)$; $B(7, 6)$; $C(0, 0)$

$\therefore$ the co-ordinates of centroid are given by $G\left(\frac{8}{3}, \frac{8}{3}\right)$.

ILLUSTRATION 107: Prove that the following equation represents two straight lines. Also find their point of intersection and the angle between them $x^2 - 5xy + 4y^2 + x + 2y - 2 = 0$.

SOLUTION: Given equation is $x^2 - 5xy + 4y^2 + x + 2y - 2 = 0$.

Here a = 1, $b = 4$, $c = -2$, $h = -5/2$, $g = 1/2$, $f = 1$

$\therefore \quad \Delta = (1)(4)(-2) + (2)(1)\left(\frac{1}{2}\right)\left(-\frac{5}{2}\right) - 1(1)^2 - 4\left(\frac{1}{2}\right)^2 - (-2)\left(-\frac{5}{2}\right)^2 = 0$

Hence it represents two straight lines

Now $x^2 - 5xy + 4y^2 = (x - 4y)(x - y)$.

So let $x^2 - 5xy + 4y^2 + x + 2y - 2 \equiv (x - 4y + A)(x - y + B)$ ... (1)

Comparing the co-efficient of x and y, we get $A + B = 1$ and $-A - 4B = 2$.

Solving, we get A = 2, $B = -1$.

Putting in (1), the required equations are $x - 4y + 2 = 0$ ... (2)

and $x - y - 1 = 0$ ... (3)

Solving (2) and (3), the point of intersection is (2, 1).

if θ is the acute angle between them, then $\tan\theta = \frac{2\left\{\sqrt{\left(-\frac{5}{2}\right)^2 - 1(4)}\right\}}{|1+4|} = \frac{3}{5}$

$\Rightarrow \theta = \tan^{-1}\left(\frac{3}{5}\right)$

ILLUSTRATION 108: Prove that the following equation represents two straight lines; also find their point of intersection and the angle between them $3y^2 - 8xy - 3x^2 - 29x + 3y - 18 = 0$.

SOLUTION: Given equation is $3y^2 - 8xy - 3x^2 - 29x + 3y - 18 = 0$.

Here $a = -3$, $b = 3$, $c = -18$, $f = 3/2$, $g = -29/2$, $h = -4$

Hence $\Delta = (-3)(3)(-18) + 2\left(\frac{3}{2}\right)\left(-\frac{29}{2}\right)(-4) - (-3)\left(\frac{3}{2}\right)^2 - (3)\left(-\frac{29}{2}\right)^2 - (-18)(-4)^2 = 0$

Hence two straight lines are represented by given second degree equation.

Now $3y^2 - 8xy - 3x^2 = (3y + x)(y - 3x)$,

Hence let $3y^2 - 8xy - 3x^2 - 29x + 3y - 18 \equiv (3y + x + A)(y - 3x + B)$ ... (1)

Equating the co-efficients of x and y, we get

$-3A + B = -29$ and $A + 3B = 3$.

Solving, we get $A = 9$ and $B = -2$

Substituting in (1), and equating each to zero, the equations are $3y + x + 9 = 0$ ... (2)

and $y - 3x - 2 = 0$ ... (3)

Solving (2) and (3), we get the point of intersection as $\left(-\frac{3}{2}, -\frac{5}{2}\right)$

If θ is the acute angle between them, then we get $\tan\theta = \frac{2\sqrt{(-4)^2 - (-3)(3)}}{|-3+3|}$;

$\Rightarrow \theta = 90°$

ILLUSTRATION 109: Prove that the following equation represents two straight lines; find also their point of intersection and the angle between them $y^2 + xy - 2x^2 - 5x - y - 2 = 0$.

SOLUTION: $y^2 + xy - 2x^2 - 5x - y - 2 = 0$

Here $a = -2$, $b = 1$, $c = -2$, $f = -1/2$, $g = -5/2$, $h = 1/2$

Hence $\Delta = (-2)(1)(-2) + 2\left(-\frac{1}{2}\right)\left(-\frac{5}{2}\right)\left(\frac{1}{2}\right) - (-2)\left(-\frac{1}{2}\right)^2 - 1.\left(-\frac{5}{2}\right)^2 - (-2).\left(\frac{1}{2}\right)^2 = 0$

So the equation represents two straight lines

Now $y^2 + xy - 2x^2 = (y - x)(y + 2x)$.

Hence let $y^2 + xy - 2x^2 - 5x - y - 2 \equiv (y - x + A)(y + 2x + B)$ ....(i)

Comparing the co-efficients of y and x, we get $A + B = -1$ and $2A - B = -5$

Solving, we get $A = -2$, $B = 1$

Substituting in (i) and putting each factor equal to zero, the equations are $y - x - 2 = 0$ ..(ii)

and $y + 2x + 1 = 0$ ...(iii)

Solving (ii) and (iii), the point of intersection is $(-1, 1)$

If θ be the acute angle between them, then we get $\tan\theta = \frac{2\sqrt{(1/2)^2 - (-2)(1)}}{|-2-1|} = 1 \Rightarrow \theta = \pi/4$

ILLUSTRATION 110: Show that lines $x^2 - 4xy + y^2 = 0$ and $x + y - 1 = 0$ form an equilateral triangle

SOLUTION: $x^2 - 4xy + y^2 = 0$

Angle between the two lines $= \theta = \tan^{-1}\left|\frac{2\sqrt{h^2 - ab}}{a+b}\right| = \tan^{-1}\left(\frac{2\sqrt{2^2 - 1}}{1+1}\right) = 60^0$

Also $m_1m_2 = a/b = 1$

Now; we know that if two lines $y = m_1x + c_1$ and $y = m_2x + c_2$ are such that $m_1m_2 = 1$; then the two lines makes same angle with the line $y = \pm x + c$ and hence with $y = -x + 1 \Rightarrow \Delta$ formed is isosceles with one angle 60°.

Also $x^2 - 4xy + y^2 = 0$ and $x + y = 1$ intersect where $x^2 - 4x(1 - x) + (1 - x)^2 = 0$

$$\Rightarrow\ x = \frac{1}{2}\left(1 \pm \frac{1}{\sqrt{3}}\right) \in (0,1)$$

$\Rightarrow$ Both points of intersection lie in first quadrant as shown in Figure 2.88.:

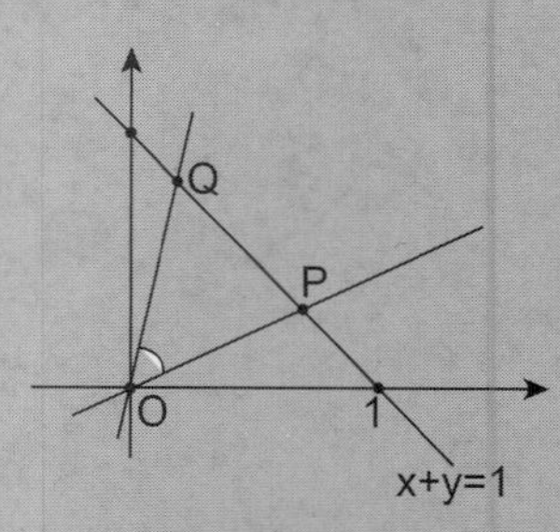

FIGURE 2.88

$\angle POQ$ is acute i.e., 60° and $\angle OPQ = \angle OQP$

$\Rightarrow$ ΔOPQ is equilateral triangle.

Note that it $\angle QOP$ was 120° i.e., obtuse, then the Δ will be isosceles but not equilateral.

ILLUSTRATION 111: Find the condition that the slope of one of lines represeented by $ax^2 + 2hxy + by^2 = 0$ should be 2 times the slope of the other.

SOLUTION: Let the lines represented be $y = m_1x$ and $y = m_2x$

$$\therefore\ m_1 + m_2 = \frac{-2h}{b} \text{ and } m_1m_2 = \frac{a}{b};\ m_1 = 2m_2 \text{ (given)}$$

$$\Rightarrow 3m_2 = -\frac{2h}{b} \text{ and } 2m_2^2 = \frac{a}{b}$$

$$\Rightarrow m_2 = \frac{-2h}{3b} \text{ and } 2\left(\frac{-2h}{3b}\right)^2 = \frac{a}{b}.$$

$$\Rightarrow 2\times\frac{4h^2}{9b^2} = \frac{a}{b} \Rightarrow 8h^2 = 9ab.$$

ILLUSTRATION 112: Find the condition that the slope of one of lines represented by $ax^2 + 2hxy + by^2 = 0$ be the 2nd power of the other.

SOLUTION: Let m and m^2 be the slopes of the lines represented by $ax^2 + 2hxy + by^2 = 0$

$$\Rightarrow m + m^2 = \frac{-2h}{b} \text{ and } m \times m^2 = \frac{a}{b} \Rightarrow m = \left(\frac{a}{b}\right)^{1/3}$$

$$\Rightarrow \left(\frac{a}{b}\right)^{1/3} + \left(\frac{a}{b}\right)^{2/3} = \frac{-2h}{b}$$

ILLUSTRATION 113: Obtain the condition that one of the lines given by $ax^2 + 2hxy + by^2 = 0$ may be perpendicular to one of the lines given by $a'x^2 + 2h'xy + b'y^2 = 0$.

SOLUTION: Since both pair are passing through origin, let $y = mx$ be one of the lines represented by $ax^2 + 2hxy + by^2 = 0$

then $ax^2 + 2hx(mx) + b(mx)^2 = 0$

$\Rightarrow bm^2 + 2hm + a = 0$ .....(1)

then $y = -\dfrac{1}{m}x$ be of the line represented by

$a'x^2 + 2h'xy + b'y^2 = 0$

then $a'x^2 + 2h'x\left(-\dfrac{1}{m}x\right) + b'\left(-\dfrac{1}{m}x\right)^2 = 0$

$\Rightarrow a'm^2 - 2h'm + b' = 0$ ..(2)

Solving (1) and (2), by cross multiplication rule, we get the following equation:

$$\Rightarrow \quad \frac{m^2}{2hb'+2h'a} = \frac{m}{aa'-bb'} = \frac{1}{-2h'b-2a'h}$$

$$\therefore \quad m^2 = -\frac{(hb'+h'a)}{(h'b+a'h)};\ m = \frac{(bb'-aa')}{2(a'h+h'b)}$$

Eliminating m, we obtain

$4\,(ha' + h'b)\,(h'a + hb') + (bb' - aa')^2\,(h'b + a'h) = 0.$

ILLUSTRATION 114: Show that angle between the pair of line $\beta\,(a + 2h\cos\beta + b\cos^2\beta)x^2 + 2\,\{(b-a)\cos\beta - (\cos^2\beta - 1)h\}xy + (a\cos^2\beta - 2h\cos\beta + b)y^2 = 0$ is independent of value of β; where $\beta \in (0, \pi)$

SOLUTION: Comparing the given equation with the general equation $Ax^2 + 2Hxy + By^2 = 0$

we have, $A = a + 2h\cos\beta + b\cos^2\beta$

$H = \{(b-a)\cos\beta - (\cos^2\beta - 1)h\}$

$B = (a\cos^2\beta - 2h\cos\beta + b)$

Now, acute angle θ is given by $\tan\theta = \dfrac{2\sqrt{H^2 - AB}}{|A+B|}$

$$\frac{2\sqrt{\{(b-a)\cos\beta - (\cos^2\beta - 1)h\}^2 - (a + 2h\cos\beta + b\cos^2\beta)(a\cos^2\beta + b - 2h\cos\beta)}}{|a + 2h\cos\beta + b\cos^2\beta + a\cos^2\beta - 2h\cos\beta + b|}$$

$$= \frac{2\sqrt{(\cos^2\beta + 1)^2\,(h^2 - ab)}}{|(a+b)(\cos^2\beta + 1)|} \quad \text{(On simplification)}$$

$$= \frac{2\sqrt{h^2 - ab}}{|a+b|} \quad \text{Hence the value of } \theta \text{ is independent of '}\beta\text{'}$$

ILLUSTRATION 115: Find the straight lines represented by the equation $x^2 + 2xy\sec\theta + y^2 = 0$ and determine the angle between them.

SOLUTION: $x^2 + 2xy\sec\theta + y^2 = 0$

$\Rightarrow \quad m^2 + 2m\sec\theta + 1 = 0$ (dividing by x^2 and putting $y/x = m$)

$$\Rightarrow \quad m = \frac{-2\sec\theta \pm \sqrt{(4\sec^2\theta - 4)}}{2} = -\sec\theta \pm \tan\theta = \frac{-1 \pm \sin\theta}{\cos\theta}$$

$$\Rightarrow \quad \frac{y}{x} = \frac{-1 \pm \sin\theta}{\cos\theta}$$

Hence the equations are $y\cos\theta + x(1 + \sin\theta) = 0$ and $y\cos\theta + x(1 - \sin\theta) = 0$

If ϕ is the acute angle between them, then $\tan\phi = \dfrac{2\sqrt{\sec \quad 1}}{|1 \quad 1|} = |\tan\theta| \Rightarrow \phi = \tan^{-1}|\tan\theta|$.

ILLUSTRATION 116: Show that the two straight lines represented by equation $x^2\,(\tan^2\theta + \cos^2\theta) - 2xy\tan\theta + y^2\sin^2\theta = 0$, subtend with the axis of x angles such that the difference of their tangents is 2.

SOLUTION: The given equation is $x^2\,(\tan^2\theta + \cos^2\theta) - 2xy\tan\theta + y^2\sin^2\theta = 0.$ ... (1)

Let the equation (1) represent two straight lines $y = m_1x$ and $y = m_2x$; where m_1 and m_2 are their slopes since (1) represents two straight lines passing through origin.

The combined equation will be $(y - m_1x)(y - m_2x) = 0$

or $y^2 - (m_1 + m_2)xy + m_1m_2x^2 = 0$ ... (2)

As (1) and (2) represent same straight lines, the co-efficients of x^2, y^2 and xy must be proportional i.e., $\dfrac{1}{\sin^2\theta} = \dfrac{-(m_1+m_2)}{-2\tan\theta} = \dfrac{\tan^2\theta+\cos^2\theta}{m_1m_2}$

Hence $m_1 + m_2 = \dfrac{2\tan\theta}{\sin^2\theta} = \dfrac{2}{\sin\theta\cos\theta}$

and $m_1m_2 = \dfrac{\tan^2\theta+\cos^2\theta}{\sin^2\theta} = \dfrac{\sin^2\theta+\cos^4\theta}{\sin^2\theta\cos^2\theta}$

Therefore $(m_1 - m_2)^2 = (m_1 + m_2)^2 - 4\,m_1m_2$

$$= \frac{4}{\sin^2\theta\cos^2\theta} - \frac{4(\sin^2\theta+\cos^4\theta)}{\sin^2\theta\cos^2\theta} = 4\left[\frac{1-\sin^2\theta-\cos^4\theta}{\sin^2\theta\cos^2\theta}\right]$$

$$= 4\left[\frac{\cos^2\theta-\cos^4\theta}{\sin^2\theta\cos^2\theta}\right] = \frac{4\cos^2\theta(1-\cos^2\theta)}{\cos^2\theta\sin^2\theta} = \frac{4\cos^2\theta\sin^2\theta}{\cos^2\theta\sin^2\theta} = 4$$

Hence $(m_1 - m_2)^2 = 4$ or $m_1 - m_2 = \pm 2$ or $\tan\theta_1 - \tan\theta_2 = \pm 2$,

$\Rightarrow$ $|\tan\theta_1 - \tan\theta_2| = 2$; where θ_1 and θ_2 are the inclinations of the lines represented by (1) with the x-axis.

TEXTUAL EXERCISE–9 (SUBJECTIVE)

1. For what value of λ, does the equation $12x^2 - 10xy + 2y^2 + 11x - 5y + \lambda = 0$ represent two straight lines? Also find the angle between them.

2. Determine the angles between the pair of straight lines $x^2 + 2xy\cot\theta + y^2 = 0$.

3. Find the straight lines represented by the equation $y^3 - xy^2 - 14x^2y + 24x^3 = 0$ and determine the angle between them.

4. Pair of straight lines $7x^2 - y^2 + 6xy - 18x + 4y - 9 = 0$ represents two sides of an isosceles triangle and third side passes through (1, –2) Then find the equation of the third side.

Answer Keys

1. $\lambda = 2$; $\tan^{-1}(1/7)$. **2.** $\varphi = \tan^{-1}[\operatorname{cosec}\theta\sqrt{\cos 2\theta}\,]$

3. $y - 2x = 0$; $y - 3x = 0$; $y + 4x = 0$; $\theta_1 = \tan^{-1}(1/7)$; $\theta_2 = \tan^{-1}(-6/7)$ and $\theta_3 = \tan^{-1}(-7/11)$

4. $x - 3y - 7 = 0$ or $3x + y - 1 = 0$

TEXTUAL EXERCISE–9 (OBJECTIVE)

1. The lines $a^2x^2 + bcy^2 = a(b + c)xy$ will be coincident, if
   (a) $a = 0$ or $b = c$ (b) $a = b$ or $a = c$
   (c) $c = 0$ or $a = b$ (d) $a = b + c$

2. The equation $2x^2 + 4xy - py^2 + 4x + qy + 1 = 0$ will represent two mutually perpendicular straight lines, if
   (a) $p = 1$ and $q = 2$ or 6
   (b) $p = 2$ and $q = 0$ or 6
   (c) $p = 2$ and $q = 0$ or 8
   (d) $p = -2$ and $q = -2$ or 8

3. The angle between the lines represented by the equation $ax^2 + xy + by^2 = 0$ will be 45°, if

(a) $a = 1, b = 6$
(b) $a = 1, b = -6$
(c) $a = 6, b = 1$
(d) None of these

4. If the lines represented by the equation $2x^2 - 3xy + y^2 = 0$ subtend angles α and β with x-axis, then $cot^2 \alpha + cot^2 \beta =$

(a) 0 (b) 3/2
(c) 7/4 (d) 5/4

5. The lines $(lx + my)^2 - 3(mx - ly)^2 = 0$ and $lx + my + n = 0$ form

(a) An isosceles triangle
(b) A right angled triangle
(c) An equilateral triangle
(d) None of these

6. The angle between the lines represented by the equation $\lambda x^2 + (1 - \lambda)^2 xy - \lambda y^2 = 0$, is

(a) 30^0 (b) 45^0
(c) 60^0 (d) 90^0

7. If the sum of the slopes of the lines represented by the equation $x^2 - 2xy \tan A - y^2 = 0$ be 4, then $\angle A =$

(a) 0^0 (b) 45^0
(c) 60^0 (d) $\tan^{-1}(-2)$

8. The co-ordinates of the centroid of the triangle whose sides are $12x^2 - 20xy + 7y^2 = 0$ and $2x - 3y + 4 = 0$ are

(a) $\left(\frac{8}{3}, \frac{8}{3}\right)$ (b) $\left(\frac{4}{3}, \frac{4}{3}\right)$
(c) $\left(\frac{2}{3}, \frac{3}{4}\right)$ (d) None

9. The area of the parallelogram formed by the pair of lines $6x^2 + 6y^2 + ax + ay - 13xy - a^2 = 0$ and $6x^2 + 6y^2 - 13xy + 5ax - 6a^2 = 0$ is:

(a) $\frac{4a^2}{7}$ (b) $\frac{2}{5}a^2$
(c) $\frac{3}{2}a^2$ (d) None of these

10. The orthocentre of the triangle formed by the pair of lines $2x^2 + 6y^2 + 7xy - 3x - 5y + 1 = 0$ and $ax + by - 1 = 0$ is origin, then (a, b) is given by

(a) (2, –4) (b) (–8, 8)
(c) (4, 4) (d) None of these

Answer Keys

1. (a) **2.** (c) **3.** (b) **4.** (d) **5.** (a), (c) **6.** (d) **7.** (d) **8.** (a) **9.** (b) **10.** (b)

HOMOGENOUS EQUATION IN TWO VARIABLES x AND y

The equation $a_0 y^n + a_1 y^{n-1} x + ... + a_n x^n = 0$ is called homogeneous equation of n degree in x and y.

$$\Rightarrow \quad a_0\left(\frac{y}{x}\right)^n + a_1\left(\frac{y}{x}\right)^{n-1} + ... + a_n = 0$$

is n^{th} degree polynomial equation and let $m_1, m_2, m_3, ..., m_n$, be its roots then above equation will be identical with

$$a_0\left(\frac{y}{x} - m_1\right)\left(\frac{y}{x} - m_2\right)\left(\frac{y}{x} - m_3\right)......\left(\frac{y}{x} - m_n\right) = 0$$

$$\Rightarrow \quad a_0\left(y - m_1 x\right)\left(y - m_2 x\right)......\left(y - m_n x\right) = 0$$

which represents n straight lines $y - m_i x = 0$ (where i = 1, 2, 3, ..., n), all of which clearly pass through origin.

PAIR OF STRAIGHT LINES THROUGH THE ORIGIN

The homogenous equation of second degree $ax^2 + 2hxy + by^2 = 0$ always represent a pair of straight lines through origin. Let these lines be $y = m_1 x$ and $y = m_2 x$

$$\Rightarrow \quad b\,(y/x)^2 + 2h\,(y/x) + a = 0$$

$$\Rightarrow \quad y/x = \frac{-2h \pm \sqrt{4h^2 - 4ab}}{2b}$$ solving the quadratic equation in (y/x)

$$\Rightarrow \quad y = \left(\frac{-h \pm \sqrt{h^2 - ab}}{b}\right)x \Rightarrow y = m_1 x \text{ or } y = m_2 x;$$

where $m_1 = \frac{-h + \sqrt{h^2 - ab}}{b}$ and $m_2 = \frac{-h - \sqrt{h^2 - ab}}{b}$

Since $h^2 \geq ab$, values of m_1 and m_2 are real.

Clearly $y = m_1 x$ and $y = m_2 x$ are straight lines passing through the origin. Hence $ax^2 + 2hxy + by^2 = 0$ represents a pair of straight lines passing through origin.

REMARKS

1. *According to the value of m_1 and m_2. The lines are*
   (i) *Real and distinct, if $h^2 > ab$;*
   (ii) *coincident, if $h^2 = ab$;*
   (iii) *imaginary if $h^2 < ab$*

2. $m_1 = \dfrac{-h+\sqrt{h^2-ab}}{b}$ and $m_2 = \dfrac{-h-\sqrt{h^2-ab}}{b}$

   $$m_1 + m_2 = \frac{-h+\sqrt{h^2-ab}}{b} + \frac{-h-\sqrt{h^2-ab}}{b} = \frac{-2h}{b}$$

   $$\text{and } m_1 m_2 = \left(\frac{-h+\sqrt{h^2-ab}}{b}\right)\left(\frac{-h-\sqrt{h^2-ab}}{b}\right) = \frac{a}{b}$$

   Thus, if $y = m_1x$ and $y = m_2x$ are lines represented by the equation

   $ax^2 + 2hxy + by^2 = 0$, then $m_1 + m_2 = \dfrac{-2h}{b}$ and $m_1 m_2 = \dfrac{a}{b}$

ILLUSTRATION 117: Show that the equation $6x^2 - 5xy + y^2 = 0$ represents a pair of distinct straight lines, each passing through the origin. Find the separate equations of these lines.

SOLUTION: The given equation is a homogeneous equation of second degree. So, it represents a pair of straight lines passing through the origin. Comparing the given equation with $ax^2 + 2hxy + by^2 = 0$, we obtain $a = 6$, $b = 1$ and $2h = -5$.

$\therefore\ h^2 - ab = \dfrac{25}{4} - 6 = \dfrac{1}{4} > 0 \Rightarrow h^2 > ab$

Hence the given equation represents a pair of distinct lines passing through the origin.

Now, $6x^2 - 5xy + y^2 = 0 \Rightarrow (y/x)^2 - 5(y/x) + 6 = 0$ (dividing by x^2)

$\Rightarrow (y/x)^2 - 3(y/x) - 2(y/x) + 6 = 0 \Rightarrow (y/x - 3)(y/x - 2) = 0$

$\Rightarrow y/x - 3 = 0$ or $y/x - 2 = 0 \Rightarrow y - 3x = 0$ or $y - 2x = 0$

ILLUSTRATION 118: Find the equation of pair of straight lines through origin and passing through (0, 1) and

(a) parallel to lines of $x^2 - xy - 2y^2 = 0$

(b) perpendicular to lines of $x^2 - xy - 2y^2 = 0$

SOLUTION: Let $L_1 = 0$ and $L_2 = 0$ be the lines represented by $x^2 - xy - 2y^2 = 0$

$\Rightarrow L_1.L_2 \equiv (x - 2y)(x + y) = 0$

$\therefore$ Slopes of L_1 and L_2 are -1 and $1/2$

(a) Now, we need to find lines with slopes -1 and $1/2$ passing through (0, 1)

$\therefore$ equations are $y - 1 = (-1)x \Rightarrow y + x - 1 = 0$

and $y - 1 = (1/2)x \Rightarrow 2y - x - 1 = 0$

(b) Slopes of lines $\perp$ to $L_1 = 0$ and $L_2 = 0$ are -2 and 1

$\therefore$ equations of required lines are

$y - 1 = (1)x \Rightarrow y - x - 1 = 0$

and $y - 1 = (-2)x \Rightarrow y + 2x - 1 = 0$

ILLUSTRATION 119: Prove that $3x^2 - 8xy - 3y^2 = 0$ and $x + 2y = 3$ enclose a right angled isosceles triangle.

SOLUTION: $3x^2 - 8xy - 3y^2 = 0$ and $x + 2y = 3$

Now, $3y^2 - 3x^2 + 8xy = 0$

$\Rightarrow 3\left(\frac{y}{x}\right)^2 - 3 + 8\left(\frac{y}{x}\right) = 0$

$\Rightarrow 3m^2 + 8m - 3 = 0$; where $y/x = m$

$\Rightarrow 3m^2 + 9m - m - 3 = 0$

$\Rightarrow 3m(m + 3) - 1\,(m + 3) = 0$

$\Rightarrow (3m - 1)\,(m + 3) = 0$

$\Rightarrow m = 1/3$ and -3

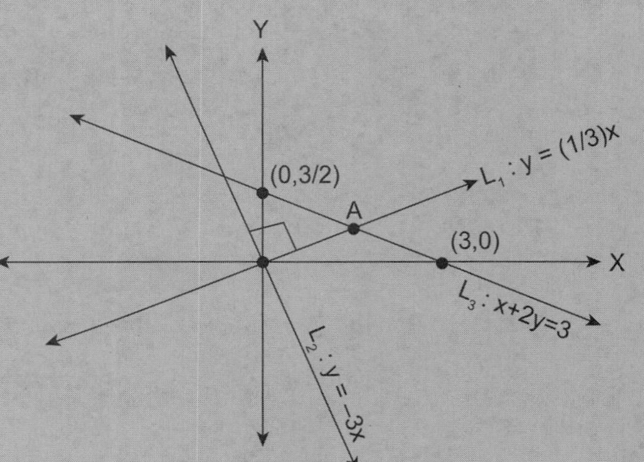

FIGURE 2.89

$\therefore\ L_1\!: y = \frac{1}{3}x$ and $L_2 : y = -3x$ and L_3:

$x + 2y - 3 = 0$ are three given lines

$\Rightarrow$ Acute angle θ_1 between L_1 and L_3 is given by $\tan\theta_1 = \left|\frac{\frac{1}{3} - \left(\frac{-1}{2}\right)}{1 + \left(\frac{1}{3} \times \frac{-1}{2}\right)}\right| = \left|\frac{\frac{2+3}{6}}{1 - \frac{1}{6}}\right| = 1$

$\Rightarrow \theta_1 = \pi/4$

and acute angle θ_2 between L_2 and L_3 is given by $\tan\theta_2 = \left|\frac{-3 + \frac{1}{2}}{1 + \left(3 \times \frac{1}{2}\right)}\right| = \left|\frac{\frac{-5}{2}}{\frac{5}{2}}\right| = 1$

$\Rightarrow \theta_2 = \pi/4$

Also $m_1 m_2 = -1 \Rightarrow L_1 \perp\text{r}\ L_2$

$\therefore\ L_1, L_2, L_3$ form a right angled isosceles triangle OAB.

ILLUSTRATION 120: Find the equations of both diagonals of the parallelogram formed by the pairs of straight lines $x^2 - 4xy + y^2 = 0$ and $x^2 - 4xy + y^2 - 10x - 3 = 0$.

SOLUTION: Let $S_1\!: x^2 - 4xy + y^2 = 0$ (i.e., OA and OC)

and $S_2\!: x^2 - 4xy + y^2 - 10x - 3 = 0$ (i.e., AB and BC)

be pairs of straight lines representing opposite pairs of parallel sides of parallelogram as shown in Figure 2.90.

To find *OB*

We know the co-ordinates of O i.e., (0, 0) (The point of intersection of the lines representened by $S_1 = 0$)

Now, we can find the co-ordinates of point B (The point of intersection of $S_2 = 0$)

by $\left(\frac{\partial S_2}{\partial x}\right) = 0 \Rightarrow x - 2y - 5 = 0$

and $\left(\frac{\partial S_2}{\partial y}\right) = 0 \Rightarrow 2y = 4x \Rightarrow y = 2x$

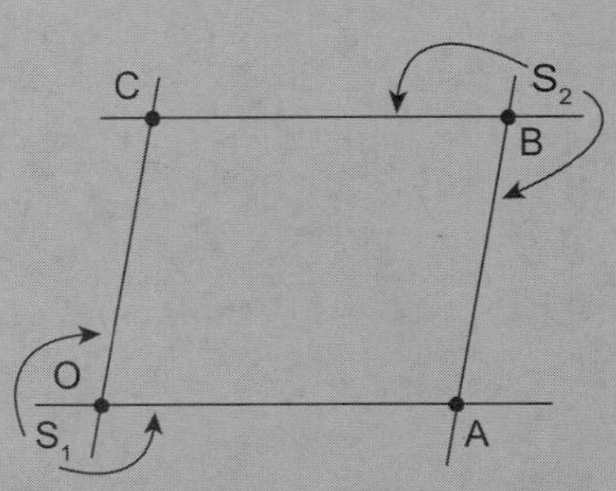

FIGURE 2.90

Solving the above two equations, we get $B \equiv \left(\frac{-5}{3}, \frac{-10}{3}\right)$

$\therefore$ The line OB is given by $(y - 0) = \left(\frac{10/3}{5/3}\right)(x - 0) \Rightarrow y = 2x$

To find AC:

Point A lies on $S_1 = 0$ and also on $S_2 = 0$ similarly, for point C.

$\therefore$ AC is a line passing through the point of intersection of two loci $S_1 = 0$ and $S_2 = 0$

$\therefore$ equation of AC is given by $S_1 - S_2 = 0$

$\Rightarrow$ $10x + 3 = 0$

ILLUSTRATION 121: Find the pair of straight lines whose lines are perpendicular to the lines of pair of straight lines $ax^2 + by^2 + 2hxy = 0$.

SOLUTION: Let the slopes of lines which are perpendicular to given line be m_1' and m_2'

Now, $m_1 + m_2 = \dfrac{-2h}{b}$ and $m_1m_2 = \dfrac{a}{b}$

$\Rightarrow$ $m_1' + m_2' = \dfrac{-1}{m_1} + \dfrac{-1}{m_2} = \dfrac{-m_1 - m_2}{m_1 m_2} = \dfrac{2h}{a}$

and $m_1'm_2' = \dfrac{-1}{m_1} \times \dfrac{-1}{m_2} = \dfrac{b}{a}$

$\Rightarrow$ equation of lines are $bx^2 + ay^2 - 2hxy = 0$

ILLUSTRATION 122: Find the condition in parameters a, b, c, d so that the set of three lines represented by $ax^3 + bx^2 y + cxy^2 + dy^3 = 0$ are such that

(a) two of the lines are perpendicular.

(b) two lines make same angle with $y = x$

(c) For two of these three lines, sum of slopes = product of slopes

(d) two of them make an isosceles triangle with a line parallel to x-axis

SOLUTION: $ax^3 + bx^2y + cxy^2 + dy^3 = 0$

$\Rightarrow$ $d\left(\dfrac{y}{x}\right)^3 + c\left(\dfrac{y}{x}\right)^2 + b\left(\dfrac{y}{x}\right) + a = 0$

$\Rightarrow$ $dm^3 + cm^2 + bm + a = 0$; where $m = y/x$ ......(1)

$\Rightarrow$ $d(y - m_1x)(y - m_2x)(y - m_3x) = 0$

$\Rightarrow$ $m_1 + m_2 + m_3 = \dfrac{-c}{d}$; ....(2)

$m_1m_2 + m_2m_3 + m_3m_1 = \dfrac{b}{d}$; $m_1m_2m_3 = \dfrac{-a}{d}$

(a) If two lines are perpendicular, then

let $m_1m_2 = -1$ (without loss of generality)

$\therefore$ From (2) $m_3 = \dfrac{a}{d}$ $\Rightarrow$ $m_3 = \dfrac{a}{d}$

Now, since m_3 satisfies equation (1)

$\Rightarrow$ $a + bm_3 + cm_3^2 + dm_3^3 = 0$

$\Rightarrow$ $a + b\left(\dfrac{a}{d}\right) + c\left(\dfrac{a}{d}\right)^2 + d\left(\dfrac{a}{d}\right)^3 = 0$

$\Rightarrow$ $ad^2 + abd + a^2c + a^3 = 0$

$\Rightarrow$ $a^2 + d^2 + ac + bd = 0$ is the required conditions.

(b) We know lines make equal angles with line $y = x$; then the product of their slopes = 1

$\Rightarrow m_1m_2 = 1$ (let us assume without loss of generality)

$\Rightarrow m_3 = \dfrac{-a}{d}$ (using $m_1m_2 = 1$ in (2))

Now since m_3 satisfies equation (1)

$\Rightarrow \left(a + bm_3 + cm_3^2 + dm_3^3\right) = 0$

$$\Rightarrow a + b\left(\frac{-a}{d}\right) + c\left(\frac{-a}{d}\right)^2 + d\left(\frac{-a}{d}\right)^3 = 0$$

$\Rightarrow ad^3 + b(-a)d^2 + c(-a)^2 d + d(-a)^3 = 0$

$\Rightarrow d^3 - bd^2 + ca^2d - da^3 = 0$

$\Rightarrow d^2 - a^3 + ca^2 - bd = 0$ is the required condition.

(c) For two of these three lines, sum of slopes = product of slopes $\Rightarrow m_1 + m_2 = m_1m_2$

$\therefore$ Substituting this value of $m_1 m_2$ in equation (2); we get

$\Rightarrow (m_1 + m_2)m_3 = \dfrac{-a}{d} \Rightarrow m_1m_3 + m_2m_3 = \dfrac{-a}{d}$

Substituting this value of $m_1m_3 + m_2m_3$ in equation (3); we get

$\Rightarrow m_1m_2 = \dfrac{a+b}{d} \Rightarrow m_1 + m_2 = \dfrac{a+b}{d}$

Again substituting this value of $m_1 + m_2$ in (2); we get

$\Rightarrow m_3 = -\dfrac{(a+b+c)}{d}$; m_3 satisfies the equation (1), we get

$\Rightarrow a + bm_3 + cm_3^2 + dm_3^3 = 0$

$$\Rightarrow a + b\left(\frac{-(a+b+c)}{d}\right) + c\left(\frac{-(a+b+c)}{d}\right)^2 + d\left(\frac{-(a+b+c)}{d}\right)^3 = 0$$

$\Rightarrow ad^3 - b(a+b+c)d^2 + cd(a+b+c)^2 - d(a+b+c)^3 = 0$

$\Rightarrow ad^2 - b(a+b+c)d + c(a+b+c)^2 - (a+b+c)^3 = 0$ which is the required condition.

(d) Now two lines make isosceles triangle with line parallel to x-axis $\angle OAB = \angle ABO$

$\Rightarrow \tan(\angle OAB) = \tan(\angle OBA) \Rightarrow \left|\dfrac{m_1 - 0}{1+0}\right| = \left|\dfrac{0 - m_2}{1+0}\right|$

$\therefore |m_1| = |m_2| \Rightarrow m_1 \pm m_2 = 0$; but for isosceles triangle with one side parallel to x-axis m_1 and m_2 must be of opposite signs. Therefore $m_1 + m_2 = 0$

$\Rightarrow m_3 = \dfrac{-c}{d}$

Substituting this value of m_3 in equation (1); we get

$a + bm_3 + cm_3^2 + dm_3^3 = 0$

$$\Rightarrow a + b\left(\frac{-c}{d}\right) + c\left(\frac{-c}{d}\right)^2 + d\left(\frac{-c}{d}\right)^3 = 0$$

$\Rightarrow ad^3 + b(-c)d^2 + c(c)^2 d + d(-c)^3 = 0$

$\Rightarrow ad^2 - bcd = 0 \Rightarrow ad - bc = 0$

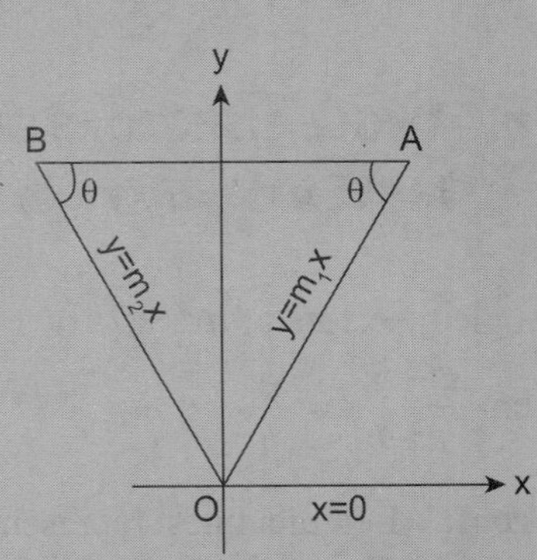

FIGURE 2.91

TEXTUAL EXERCISE–10 (OBJECTIVE)

1. Area of triangle formed by pair of lines $6x^2 + 6y^2 + 13xy + 22x + 23y + 20 = 0$ and $x + 3y + 1 = 0$ is
(a) $\frac{32}{105}$ (b) $\frac{32}{91}$
(c) $\frac{31}{92}$ (d) None

2. Equation $ax^3 - 10x^2y - xy^2 + 4y^3 = 0$ represented three straight lines, out of these three, two lines makes equal angle with $y = x$ and $a > 0$ .Then value of a is
(a) 6 (b) 7
(c) 8 (d) 4

3. The value of λ, for which the pair of lines representing $-2x^2 - 3xy + \lambda y^2 = 0$ and $x + y - 1 = 0$ encloses a right angled triangle is
(a) 2 (b) 3
(c) –1 (d) None of these

4. The values of '$\lambda > 0$' for which the pair of lines $x^2 + 2\lambda xy + y^2 = 0$, represents two real and distinct lines are given by
(a) $\lambda \in (-\infty, 1)$ (b) $\lambda \in (1, \infty)$
(c) $\lambda \in (1,5)$ (d) None of these

5. The pair of lines represented by homogenous set of the equation $30x^2 + 5xy + (a^2 - 2)y^2 = 0$ are at right angles to each other for
(a) for all a
(b) two values of a
(c) for one value of a
(d) for no value of a

6. Area enclosed by curves $y^2 + 6x^2 - 5xy + 3x - y = 0$ and $y^2 + 6x^2 - 5xy + 2x - y = 0$ is
(a) 2 (b) 1/2
(c) 1 (d) 1/3

7. Let ABC be a right angled isosceles triangle and right angled at $A(4,1)$. If the equation of BC is $2x + 4y = 5$, the equation representing the pair of lines AB and AC is
(a) $3x^2 - 3y^2 - 8xy - 38y - 16x + 13 = 0$
(b) $3x^2 - 4y^2 + 8xy - 11y - 17x + 22 = 0$
(c) $3x^2 - 3y^2 + 6xy - 21y - 42x + 42 = 0$
(d) None of these

8. If the point $(1 + \cos\theta, \sin\theta)$ lies between the region corresponding to the acute angle between the pair of lines $x^2 - 9xy + 18y^2 = 0$, then
(a) $\theta \in R - \frac{n\pi}{2}, n \in R$
(b) $\theta \in R - \frac{n\pi}{2}, n \in W$
(c) $\theta \in R - \frac{n\pi}{3}, n \in R$
(d) None of these

Answer Keys

1. (a) 2. (b) 3. (a) 4. (b) 5. (d) 6. (c) 7. (a) 8. (d)

■ ANGLE BISECTORS OF PAIR OF STRAIGHT LINES $ax^2 + 2hxy + by^2 = 0$

Angle bisectors of $ax^2 + 2hxy + by^2 = 0$ are given by equation $\frac{x^2 - y^2}{a - b} = \frac{xy}{h}$

Proof: Let the lines represented by the above pair of straight lines be $y = m_1x$ and $y = m_2x$ and B_1 and B_2 be two bisectors of angles between the lines as shown in Figure 2.92.

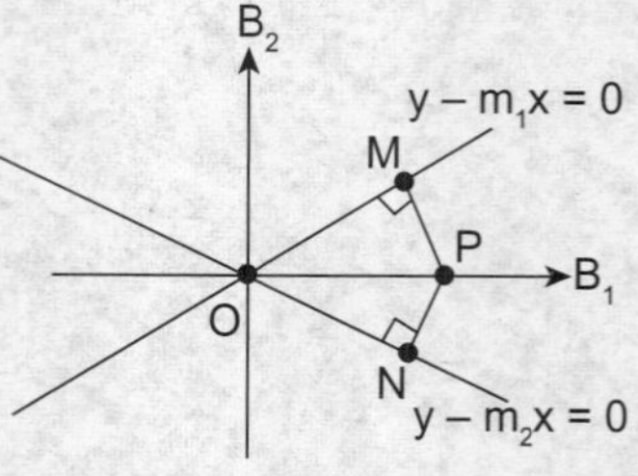

FIGURE 2.92

$$\Rightarrow m_1 + m_2 = -\frac{2h}{b} \text{ and } m_1m_2 = \frac{a}{b}$$

Let $P(x, y)$ be any point on bisector B_1 (or B_2)

$$\Rightarrow PM = PN \Rightarrow \frac{y - m_1x}{\sqrt{1+m_1^2}} = \pm\frac{y-m_2x}{\sqrt{1+m_2^2}}$$

Squaring both sides, we get $\frac{y^2+m_1^2x^2-2ym_1x}{1+m_1^2}$

$$= \frac{y^2+m_2^2x^2-2m_2xy}{1+m_2^2}$$

$$\Rightarrow (1+m_2^2)(y^2+m_1^2x^2-2ym_1x) = (1+m_1^2)(y^2+m_2^2x^2-2m_2xy)$$

$$\Rightarrow (m_2^2-m_1^2)y^2 + x^2\left(m_1^2(1+m_2^2)-m_2^2(1+m_1^2)\right) - 2xy\left(m_1.(1+m_2^2)-m_2(1+m_1^2)\right) = 0$$

$$\Rightarrow (m_2^2 - m_1^2)y^2 + (m_1^2 - m_2^2)x^2 - 2xy[(m_1 - m_2) + m_1m_2(m_2 - m_1)] = 0$$

$$\Rightarrow (m_2+m_1)y^2 - x^2(m_1+m_2) + 2xy(1-m_1m_2) = 0$$

Substituting, $m_1+m_2 = -\frac{2h}{b}$ and $m_1m_2 = \frac{a}{b}$

in the above equation, we get

$$\Rightarrow \left(-\frac{2h}{b}\right)(y^2-x^2) + 2xy\left(1-\frac{a}{b}\right) = 0$$

$$\Rightarrow (-2h)(y^2-x^2) + 2xy(b-a) = 0$$

$$\Rightarrow h(x^2-y^2) + xy(b-a) = 0 \Rightarrow \frac{x^2-y^2}{a-b} = \frac{xy}{h}$$

$\therefore$ $\frac{x^2-y^2}{a-b} = \frac{xy}{h}$ are the equations of angle bisectors of $ax^2 + 2hxy + by^2 = 0$

■ ANGLE BISECTORS OF PAIR OF STRAIGHT LINES $ax^2 + 2hxy + by^2 + 2gx + 2fy + c = 0$

The pair of bisectors of the lines represented by $ax^2 + 2hxy + by^2 + 2gx + 2fy + c = 0$ are given by $\frac{(x-\alpha)^2-(y-\beta)^2}{(a-b)} = \frac{(x-\alpha)(y-\beta)}{h}$;

where (α, β) is the point of intersection of lines represented by given pair of straight lines.

Proof: Because (α,β) is the point of intersection of the lines represented by

$$ax^2 + 2hxy + by^2 + 2gx + 2fy + c = 0 \quad ....(1)$$

Shifting the origin to points $P(\alpha, \beta)$ without rotating the co-ordinate axes, i.e., replacing x by $X + \alpha$ and y by $Y + \beta$, the equation (1) becomes

$a(X + \alpha)^2 + 2h(X + \alpha)(Y + \beta) + b(Y + \beta)^2 + 2g(X + \alpha) + 2f(Y + \beta) + c = 0$

$\Rightarrow (aX^2 + 2hXY + bY^2) + 2X(a\alpha + h\beta + g) + 2Y(h\alpha + b\beta + f) + a\alpha^2 + 2h\alpha\beta + b\beta^2 + 2g\alpha + 2f\beta + c = 0$ ...(2)

The above equation represents a pair of straight lines passing through the new origin. i.e., $P(\alpha, \beta)$.

So it must be homogeneous equation of second degree in X and Y.

$\Rightarrow a\alpha + h\beta + g = 0$ ....(3)

and $h\alpha + b\beta + f = 0$ ....(4)

and $a\alpha^2 + 2h\alpha\beta + b\beta^2 + 2g\alpha + 2f\beta + c = 0$ (which is always true as (α, β) lies on given equation) ...(5)

Now (2) becomes, $aX^2 + 2hXY + bY^2 = 0$ ...(6)

Therefore the equations of the bisectors of the angles between the lines given by (6) are

$$\frac{X^2-Y^2}{a-b} = \frac{XY}{h} \text{ (w.r.t. new origin)} \quad ...(7)$$

Replacing X by $x - \alpha$ and Y by $y - \beta$ in (7), we get

$$\frac{(x-\alpha)^2-(y-\beta)^2}{(a-b)} = \frac{(x-\alpha)(y-\beta)}{h} \text{ (w.r.t. old origin)}$$

Which is the required equations of the bisectors of the angle between the lines represented by given pair of lines (1).

ILLUSTRATION 123: Find the nature of triangle formed by the pair of lines $3x^2 - 8xy - 3y^2 = 0$ and the straight line $x + 2y = 3$

SOLUTION: Given equations are $3x^2 - 8xy - 3y^2 = 0$ and $x + 2y = 3$

Let l_1, l_2 and l_3 be the lines represented by the above equation

$\therefore$ l_1: $y = \frac{1}{3}x$; l_2: $y = -3x$; l_3: $x + 2y = 3$

Clearly, l_1 and l_2 are $\perp$ to each other, so, l_1 and l_2 enclose right angle with each other.

angle bisector of l_1 and l_2 are given by $\dfrac{x^2-y^2}{a-b}=\dfrac{xy}{h} \Rightarrow \dfrac{x^2-y^2}{3+3}=\dfrac{xy}{-4}$

$\Rightarrow \dfrac{x^2-y^2}{6}=\dfrac{xy}{-4} \Rightarrow -4\left(x^2-y^2\right)=6xy$

$\Rightarrow (4x-2y)(x+2y)=0$

$\Rightarrow 2x=y$ and $y=\dfrac{-1}{2}x$

Slope of line $l_3=-1/2$ and slope of $2x=y$ is 2

$\therefore$ we observe that $2x=y$ is $\perp$ to $l_3=0$,

$\Rightarrow$ Thus the given Δ is a right angled isosceles triangle

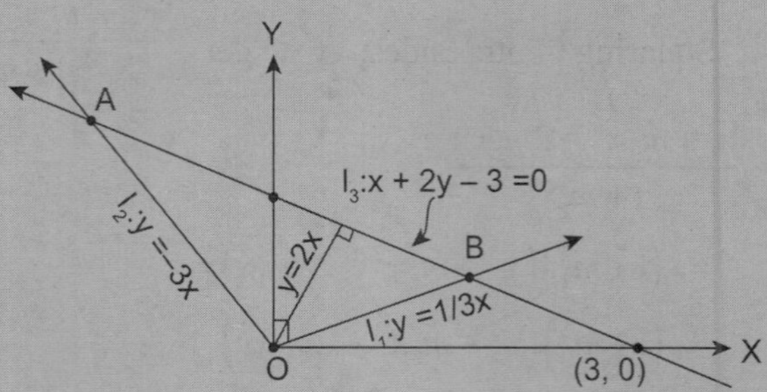

FIGURE 2.93

ILLUSTRATION 124: If pairs of straight lines $x^2-2pxy-y^2=0$ and $x^2-2qxy-y^2=0$ be such that each pair bisects the angle between the other pair, prove that $pq=-1$

SOLUTION: According to the question, the equation of the bisectors of the angle between the lines

$x^2-2pxy-y^2=0$ ...(1)

is $x^2-2qxy-y^2=0$ ....(2)

But the equation of bisectors of the angle between the lines represented by (1) is

$$\frac{x^2-y^2}{1-(-1)}=\frac{xy}{-p}$$

i.e., $-px^2-2xy+py^2=0$ ......(3)

Since (2) and (3) are identical, $\therefore \dfrac{1}{-p}=\dfrac{-2q}{-2}=\dfrac{-1}{p} \Rightarrow pq=-1$

ILLUSTRATION 125: If one of the lines represented by the line pair $ax^2+2hxy+by^2=0$ bisects the angle between co-ordinate axes, then prove that $(a+b)^2=4h^2$.

SOLUTION: Let $ax^2+2hxy+by^2=0$ represent lines with slopes m_1 and m_2

$\Rightarrow m_1+m_2=\dfrac{-2h}{b}$ and $m_1m_2=\dfrac{a}{b}$

It is given that one line bisects the co-ordinate axes.

Case I: Let $m_2=\tan 45^0=1$

$\Rightarrow m_1+1=\dfrac{-2h}{b}$ and $m_1=\dfrac{a}{b}$

$\Rightarrow \dfrac{a}{b}+=\dfrac{h}{b}$

$\Rightarrow (a+b)=(-2h)$ ....(1)

Case II: Let $m_2=\tan 135^0=-1$

$m_1-1=\dfrac{-2h}{b}$ and $m_1\times(-1)=\dfrac{a}{b} \Rightarrow \dfrac{-a}{b}-1=\dfrac{-2h}{b}$

$\Rightarrow (a+b)=(2h)$ ....(2)

Therefore from (1) and (2); we get $(a+b)^2=4h^2$

■ ISOINCLINED PAIR OF STRAIGHT LINES

If two pairs of straight lines have same angle bisectors then they are called isoinclined pairs of straight lines. The angle between two straight lines (taking one from each pair) equals the angle between the remaining two.

Let S_1: $a_1x^2 + 2h_1xy + b_1y^2 + 2g_1x + 2f_1y + c_1 = 0$

represents union of $L_1 = 0$ and $L_2 = 0$

and S_2: $a_2x^2 + 2h_2xy + b_2y^2 + 2g_2x + 2f_2y + c_2 = 0$

represents union of $L'_1 = 0$ and $L'_2 = 0$. Let they be isoinclined pairs having common bisectors B_1 and B_2 as shown in Figure 2.94.

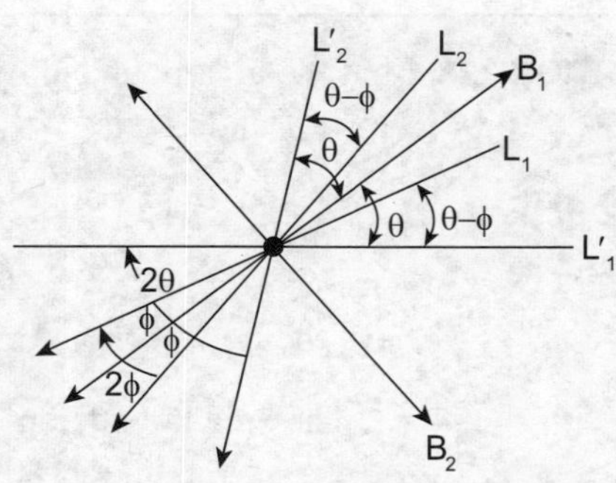

FIGURE 2.94

If the angle between L_1 and L_2 be 2ϕ and angle between L'_1 and L'_2 be 2θ, then from the diagram it is clear that the angles between L_1 and L'_1 is equal to $\theta - \phi$ which is same as the angle between L_2 and L'_2. Similarly, the angle between L_1 and L'_2 is $\theta + \phi$ which is same as the angle between L_2 and L'_1.

REMARK

Isoinclined pair of straight lines must have the same point of intersection.

ILLUSTRATION 126: Prove that the angle between one of the lines given by $ax^2 + 2hxy + by^2 = 0$ and one of the lines $ax^2 + 2hxy + by^2 + \lambda(x^2 + y^2) = 0$ is equal to the angle between the other two lines of the system.

SOLUTION: Bisectors of $ax^2 + 2hxy + by^2 + \lambda(x^2 + y^2) = 0$ are given by

$$\frac{x^2 - y^2}{(a+\lambda)-(b+\lambda)} = \frac{xy}{h}$$

or $\dfrac{x^2 - y^2}{(a-b)} = \dfrac{xy}{h}$

Which is same as bisector of $ax^2 + 2hxy + by^2 = 0$

ILLUSTRATION 127: Show that the pair of lines given by

$a^2x^2 + 2h(a + b)\,xy + b^2y^2 = 0$ is equally inclined to the pair given by $ax^2 + 2hxy + by^2 = 0$

SOLUTION: Given pair of lines are

$S_1 \equiv a^2x^2 + 2h\,(a + b)xy + b^2y^2 = 0$

and $S_2 \equiv ax^2 + 2hxy + by^2 = 0$

Equation of bisectors of first pair (S_1) is $\dfrac{x^2 - y^2}{a^2 - b^2} = \dfrac{xy}{h(a+b)}$

$\Rightarrow \dfrac{x^2 - y^2}{a-b} = \dfrac{xy}{h}$

And equation of bisector of second pair (S_2) is $\dfrac{x^2 - y^2}{a-b} = \dfrac{xy}{h}$

$\Rightarrow$ S_1 and S_2 have same bisector

$\Rightarrow$ S_1 and S_2 are isoinclined

ILLUSTRATION 128: Find the area bounded by the line $x + y = 3$ angle bisectors of the pair of straight lines $x^2 - y^2 + 2y = 1$ and x-axis.

SOLUTION: $x^2 - y^2 + 2y = 1$

$x = \pm(y - 1)$ i.e., $y = x + 1$ and $y = -x + 1$, which have their angle bisectors $x = 0$ and $y = 1$.

$\therefore$ The area bounded by straight line $x = 0$, $y = 1$, $y = -x + 3$ and $y = 0$

= area of trapezium $OBCD = 1/2\ (OD + BC) \times OB$

$= \dfrac{1}{2}(3+2)\times 1 = \dfrac{5}{2}$ square units.

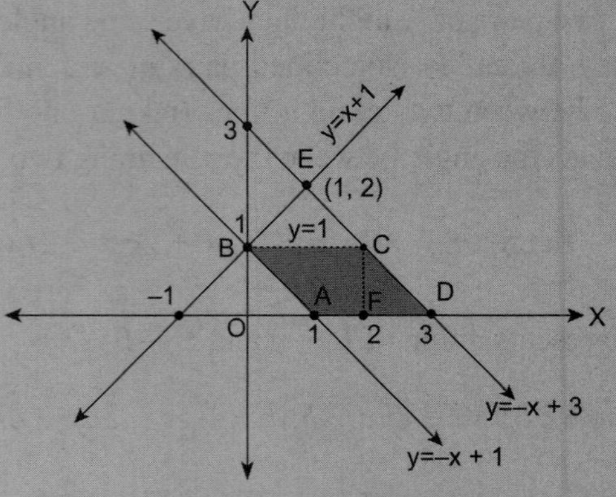

FIGURE 2.95

ILLUSTRATION 129: The lines represented by $x^2 + 2\lambda xy + 2y^2 = 0$ and $(\lambda + 1)\, x^2 - 8xy + y^2 = 0$ are equally inclined, then find the value of λ.

SOLUTION: $x^2 + 2\lambda xy + 2y^2 = 0$ ....(i)

and $(\lambda + 1)x^2 - 8xy + y^2 = 0$ ....(ii)

Equations of angle bisectors of (i) and (ii) are equals

$\Rightarrow \dfrac{x^2 - y^2}{1-2} = \dfrac{xy}{\lambda}$ and $\dfrac{x^2 - y^2}{(\lambda+1)-1} = \dfrac{xy}{-4}$ must be same

$\Rightarrow \dfrac{\lambda}{-1} = \dfrac{-4}{\lambda} \Rightarrow \lambda^2 = 4 \Rightarrow \lambda = \pm 2$

TEXTUAL EXERCISE–10 (SUBJECTIVE)

1. The bisectors of two lines L_1 and L_2 are given by $3x^2 - 8xy - 3y^2 + 10x + 20y - 25 = 0$. If the line L_1 passes through origin. Find the equation of L_2.
2. Prove that the angle between one of the lines represented by $(a + \lambda)x^2 + 2hxy + (\lambda + b)y^2 = 0$ and one of the lines, represented by $(a + \mu)\, x^2 + 2hxy + (b + \mu)y^2 = 0$ is equal to angle between the other two lines of the system.
3. One of the bisectors of the angle between the lines, $a(x - 1)^2 + 2h\,(x - 1)(y - 2) + b\,(y - 2)^2 = 0$ is $x + 2y - 5 = 0$, then find the other bisector and the relation between a, b and h.
4. The base of a triangle passes through a fixed point (λ, μ) and its sides are respectively bisected at right angles by the lines $y^2 - 8xy - 9x^2 = 0$. Determine the locus of its vertex.
5. Show that the lines $2x^2 + 6xy + y^2 = 0$ are equally inclined to the lines $4x^2 + 18xy + y^2 = 0$.
6. Prove that the equation $(x^3 - 3xy^2) + y^3 - 3x^2y = 0$ represents three straight lines equally inclined to each other.

Answer Keys

1. $x + 2y - 5 = 0$ **3.** $2x - y = 0$, $2a + 3h = 2b$ **4.** $4(x^2 + y^2) + (4\mu + 5\lambda)x + (4\lambda - 5\mu)y = 0$

TEXTUAL EXERCISE–11 (OBJECTIVE)

1. The equation of the bisector of the acute angle between the lines $36x^2 - 20y^2 - 33xy + 78x + 43y - 14 = 0$ is:
(a) $23x - 3y + 9 = 0$ (b) $11x - 3y + 9 = 0$
(c) $11x - 9y + 3 = 0$ (d) None of these

2. If the pairs $a_1x^2 + 2h_1xy + b_1y^2 = 0$ and $a_2x^2 + 2h_2xy + b_2y^2 = 0$ have one line common, then
$$\frac{(a_1b_2 - a_2b_1)^2}{(b_1h_2 - b_2h_1)(h_1a_2 - a_1h_2)} =$$
(a) 1 (b) 2
(c) 4 (d) None

3. The pairs of straight lines $ax^2 + 2hxy - ay^2 = 0$ and $hx^2 - 2axy - hy^2 = 0$ are such that
(a) One pair bisects the angle between the other pair.
(b) The lines of one pair are perpendicular to the lines of the other pair.
(c) They constitute a square in the $x - y$ plane.
(d) None of these

4. If the line $y = mx$ bisects the angle between the lines $ax^2 + 2hxy + by^2 = 0$, then m is a root of the quadratic equation
(a) $hx^2 + (a - b)x - h = 0$
(b) $x^2 + h(a - b)x - 1 = 0$
(c) $(a - b)x^2 + hx - (a - b) = 0$
(d) $(a - b)x^2 - hx - (a - b) = 0$

Answer Keys

1. (b) **2.** (c) **3.** (a) **4.** (a)

■ EQUATION OF PAIR OF STRAIGHT LINES JOINING ORIGIN TO THE POINT OF INTERSECTION OF A CURVE AND A STRAIGHT LINE

Given a straight line $lx + my = n$ .... (i) and a conic

$S \equiv ax^2 + 2hxy + by^2 + 2gx + 2fy + c = 0$ .... (ii)

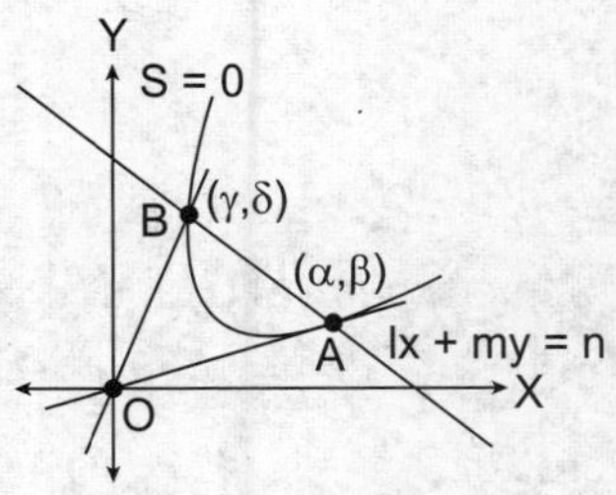

FIGURE 2.96

We are to find a pair of straight line passing through origin which is satisfied by point of intersection $A(\alpha, \beta)$ and $B(\gamma, \delta)$ of line $lx + my = n$ and the given curve (ii), that is, we require a homogeneous equation of degree two that is satisfied by the co-ordinates of $A(\alpha, \beta)$ and $B(\gamma, \delta)$.

Since $l\alpha + m\beta = n$ and $S_{(\alpha,\beta)} = a\alpha^2 + 2h\alpha\beta + b\beta^2 + 2g\alpha + 2f\beta + c = 0$

Therefore, homogenising $S = 0$ with the help of given line we get the required combined equation of pair of straight lines, OA and OB as given below:

$$\frac{lx + my}{n} = 1$$

Substituting $1 = \dfrac{lx + my}{n}$ in $S = 0$, we get:

$$ax^2 + 2hxy + by^2 + 2gx(1) + 2fy(1) + c(1)^2 = 0$$

$$\Rightarrow ax^2 + 2hxy + by^2 + 2gx\left(\frac{lx + my}{n}\right) + 2fy\left(\frac{lx + my}{n}\right) + c\left(\frac{lx + my}{n}\right)^2 = 0$$

$$\Rightarrow \underbrace{ax^2 + 2hxy + by^2}_{\text{Homogeneous}} + \underbrace{2\underbrace{(gx + fy)}_{\text{Linear}}\left(\frac{lx + my}{n}\right)}_{\text{Homogeneous}} + \underbrace{c\left(\frac{lx + my}{n}\right)^2}_{\text{Homogeneous}} = 0$$

ILLUSTRATION 130: Prove that the acute angle between the lines joining the origin to the points of intersection of the line $y = 3x + 2$ with the curve $x^2 + 2xy + 3y^2 + 4x + 8y = 11$ is $\tan^{-1}\left(\frac{2\sqrt{2}}{3}\right)$.

SOLUTION: The equation of the given straight line is $y = 3x + 2$

$$\Rightarrow \quad \frac{y-3x}{2} = 1 \qquad \text{..(i)}$$

The equation of the given curve is $x^2 + 2xy + 3y^2 + 4x + 8y - 11 = 0$ ..(ii)

The joint equation of the straight lines joining the origin to the points of intersection of (i) and (ii) is a homogeneous equation of second degree obtained with the help of (i) and (ii).

Making the equation (ii) homogenous of the second degree in x and y with the help of (i),

we get $x^2 + 2xy + 3y^2 + 4x\left(\frac{y-3x}{2}\right) + 8y\left(\frac{y-3x}{2}\right) - 11\left(\frac{y-3x}{2}\right)^2 = 0$

$\Rightarrow \quad 4x^2 + 8xy + 12y^2 + 2(8y^2 - 12x^2 - 20xy) - 11(y^2 - 6xy + 9x^2) = 0$

$\Rightarrow \quad 7x^2 - 2xy - y^2 = 0$

This is the required equation. Comparing this equation with $ax^2 + 2h\,xy + y^2 = 0$, we obtain $a = 7$, $b = -1$ and $h = -1$

Let θ be the required acute angle. Then $\tan\theta = \frac{2\sqrt{h^2 - ab}}{a+b}$

$$\Rightarrow \quad \tan\theta = \frac{2\sqrt{1+7}}{7-1} = \frac{2\sqrt{2}}{3} \Rightarrow \theta = \tan^{-1}\left(\frac{2\sqrt{2}}{3}\right).$$

ILLUSTRATION 131: Find the equation to the pair of straight lines joining the origin to the intersection of the straight line $y = mx + c$ and the curve $x^2 + y^2 = a^2$. Prove that they are at right angles if $2c^2 = a^2(1 + m^2)$.

SOLUTION: Straight line is given by $y = mx + c$ or $\frac{y}{c} - \frac{mx}{c} = 1$ .....(i)

The curve is given by $x^2 + y^2 = a^2$ or $x^2 + y^2 - a^2 = 0$ .....(ii)

Making (ii) homogenous with the help of (i), we get

$$x^2 + y^2 - a^2\left(\frac{y}{c} - \frac{mx}{c}\right)^2 = 0 \text{ or } x^2 + y^2 - a^2\left(\frac{y^2}{c^2} + \frac{m^2x^2}{c^2} - \frac{2myx}{c^2}\right) = 0$$

or $c^2x^2 + c^2y^2 - a^2y^2 - a^2m^2x^2 + 2a^2myx = 0$

or $x^2(c^2 - a^2m^2) + y^2(c^2 - a^2) + 2a^2myx = 0$ .....(iii)

This equation represents the two lines joining the points of intersection of (i) and (ii) with the origin

If the lines represented by (iii) are at right angles, then $(c^2 - a^2m^2) + (c^2 - a^2) = 0$

(as the sum of the co-efficients of x^2 and y^2 must be zero) or $2c^2 = a^2(1 + m^2)$

ILLUSTRATION 132: Prove that the straight lines joining the origin to the points of intersection of the straight line $kx + hy = 2hk$ with the curve $(x - h)^2 + (y - k)^2 = c^2$ are at right angles if $h^2 + k^2 = c^2$.

SOLUTION: The given line is $kx + hy = 2hk$ or $\frac{x}{2h} + \frac{y}{2k} = 1$ ...(i)

The given curve is $(x-h)^2 + (y-k)^2 = c^2$

or $x^2 + y^2 - 2hx - 2ky + h^2 + k^2 - c^2 = 0$ ....(ii)

Making (ii) homogenous with the help of (i), we get

$$x^2+y^2-2hx\left(\frac{x}{2h}+\frac{y}{2k}\right)-2ky\left(\frac{x}{2h}+\frac{y}{2k}\right)+(h^2+k^2-c^2)\left(\frac{x}{2h}+\frac{y}{2k}\right)^2=0 \quad \text{...(iii)}$$

Co-efficient of x^2 in (iii) is $\frac{(h^2+k^2-c^2)}{4h^2}$; Co-efficient of y^2 in (iii) is $\frac{(h^2+k^2-c^2)}{4k^2}$

If the lines represented by (iii) are perpendicular to each other, then

$$\frac{(h^2+k^2-c^2)}{4h^2}+\frac{(h^2+k^2-c^2)}{4k^2}=0 \text{ or } (h^2+k^2-c^2)\left(\frac{1}{4h^2}+\frac{1}{4k^2}\right)=0$$

As $\left(\frac{1}{4h^2}+\frac{1}{4k^2}\right)$ can't be zero, being sum of the reciprocals of two squares, hence

$(h^2+k^2-c^2)=0$ or $h^2+k^2=c^2$

ILLUSTRATION 133: Show that the straight lines joining the origin to the other two points of intersection of the curves whose equations are $ax^2+2hxy+by^2+2gx=0$ and $a'x^2+2h'xy+b'y^2+2g'x=0$ will be at right angles if $g(a'+b')-g'(a+b)=0$.

SOLUTION: The two curves are given by $ax^2+2hxy+by^2+2gx=0$ ...(i)

and $a'x^2+2h'xy+b'y^2+2g'x=0$ ...(ii)

To get the joint equation of the straight lines joining the point of intersection of (i) and (ii) to the origin, we make (ii) homogenous with the help of (i) i.e., we eliminate the terms in first degree from (i) and (ii). Hence multiplying (i) by g' and (ii) by g and subtracting, we get

$(g'ax^2+2g'hxy+g'by^2+2gg'x)-(ga'x^2+2h'gxy+gb'y^2+2gg'x)=0$

or $x^2(g'a-ga')+2(g'h-gh')xy+(g'b-gb')y^2=0$ ...(iii)

which is the required equation.

If the lines represented by (iii) be at right angles, then $(g'a-ga')+(g'b-gb')=0$

or $g'(a+b)-g(a'+b')=0$. or $g(a'+b')-g'(a+b)=0$

ILLUSTRATION 134: Find the angle subtended by chord of the ellipse $2x^2+3y^2=5$ on the line $3x+4y=5$ at origin.

SOLUTION: Equation of given ellipse is $\frac{2x^2}{5}+\frac{3y^2}{5}=1$ ....(1)

and that of line is: $3x+4y=5$ ....(2)

making (1) homogenous with the helop of (2), we get $\frac{2x^2}{5}+\frac{3y^2}{5}=\left(\frac{3x+4y}{5}\right)^2$

$$\Rightarrow 5^2\left(\frac{2x^2}{5}+\frac{3y^2}{5}\right)=(3x)^2+(4y)^2+2(3)(4)(xy)$$

$$\Rightarrow \left(\frac{5^2}{5/2}-3^2\right)x^2+\left(\frac{5^2}{5/3}-4^2\right)y^2-2(3)(4)(xy)=0 \Rightarrow x^2-y^2-24xy=0$$

which is of the form $ax^2+2hxy+by^2=0 \Rightarrow a=1, b=-1, h=-12$

Now, if θ is the required angle, then $\tan\theta=\frac{2\sqrt{h^2-ab}}{|a+b|}$ but, $a+b=0 \Rightarrow \theta=90°$

ILLUSTRATION 135: Show that if all the chords of the curve $3x^2-y^2-2x+4y=0$, intercepted by a family of straight lines subtend right angles at the origin, then the family of straight lines passes through a fixed point. Also find the co-ordinates of the fixed point.

SOLUTION: Let the line be $ax + by = 1$ ....(1)

and the cuvre is $3x^2 - y^2 - 2x + 4y = 0$ ....(ii)

Homogenising (ii) with (i) we get

$3x^2 - y^2 - 2x\,(ax + by) + 4y\,(ax + by) = 0$

$\Rightarrow\ 3x^2 - y^2 - 2ax^2 - 2bxy + 4axy + 4by^2 = 0 \Rightarrow (3 - 2a)x^2 + (4b - 1)y^2 + (4a - 2b)xy = 0$

Now, since above two lines are perpendicular to each other

$\therefore$ cooff of x^2 + coeff of $y^2 = 0$

$\Rightarrow\ 3 - 2a + 4b - 1 = 0 \Rightarrow 2a - 4b = 2$ or $a - 2b = 1$

$\therefore$ comparing this with (1), we get (1, –2) is the required fixed point.

ILLUSTRATION 136: A line L passing through the point (2, 1) intersects the curve $4x^2 + y^2 - x + 4y - 2 = 0$ at the points A, B. If the lines joining origin and the points A, B are such that the co-ordinate axis are the bisectors between them, then find the equation of line L.

SOLUTION: Line passing through (2, 1), having slope m is given by $(y - 1) = m\,(x - 2)$

$\Rightarrow\ y - mx = 1 - 2m$

$\Rightarrow\ \dfrac{y - mx}{1 - 2m} = 1$ ....(i)

Homogenizing the given second degree curve with (i) we get,

$$4x^2 + y^2 - x\left(\frac{y - mx}{1 - 2m}\right) + 4y\left(\frac{y - mx}{1 - 2m}\right) - 2\left(\frac{y - mx}{1 - 2m}\right)^2 = 0$$

$$\Rightarrow\ 4x^2 + y^2 - \frac{xy}{(1-2m)} + \frac{mx^2}{(1-2m)} + \frac{4y^2}{(1-2m)} - \frac{4x\,ym}{(1-2m)} - 2\left(\frac{y\quad mx}{1-2m}\right) = 0$$

$$\Rightarrow\ \left(4 + \frac{m}{(1-2m)}\right)x^2 + \left(1 + \frac{4}{1-2m}\right)y^2 - \left(\frac{4xym + xy}{1-2m}\right) - 2\left(\frac{y^2 + m^2x^2 - 2mxy}{(1-2m)^2}\right) = 0$$

$$\Rightarrow\ \left(4 + \frac{m}{1-2m} - \frac{2m^2}{(1-2m)^2}\right)x^2 + \left(1 + \frac{4}{1-2m} - \frac{2}{(1-2m)^2}\right)y^2 + \frac{4mxy}{(1-2m)^2} - \left(\frac{4m+1}{1-2m}\right)xy = 0$$

$$\Rightarrow\ \left(4 + \frac{m}{1-2m} - \frac{2m^2}{(1-2m)^2}\right)x^2 + \left(1 + \frac{4}{1-2m} - \frac{2}{(1-2m)^2}\right)y^2 + \left(\frac{4m}{(1-2m)^2} - \frac{(4m+1)}{(1-2m)}\right)xy = 0$$

Now equation of angle bisectors is given by

$$\frac{x^2 - y^2}{4 + \dfrac{m}{1-2m} - \dfrac{2m^2}{(1-2m)^2} - 1 + \dfrac{-4}{(1-2m)} + \dfrac{2}{(1-2m)^2}} = \frac{xy}{\dfrac{1}{2}\left(\dfrac{4m}{(1-2m)^2} - \dfrac{(4m+1)}{1-2m}\right)}$$ ......(ii)

Now since it is given that the equation of bisectors are co-ordinate axes $x = 0$ and $y = 0$

i.e., $xy = 0$ ......(iii)

Thus (ii) and (iii) are identical

$\Rightarrow\ \dfrac{4m}{(1-2m)^2} = \dfrac{4m+1}{(1-2m)}$

$\Rightarrow\ 4m = (4m + 1)(1 - 2m)\quad \Rightarrow 8m^2 + 4m - 2m - 1 = 0$

$\Rightarrow\ 4m(2m + 1) - 1\,(2m + 1) = 0 \Rightarrow m = \dfrac{1}{4}$ or $m = \dfrac{-1}{2}$

$\therefore$ Possible equation of line $L \equiv 2y + x = 4$ and $4y - x = 2$

TEXTUAL EXERCISE–11 (SUBJECTIVE)

1. Prove that the straight lines joining the origin to the points of intersection of the line $7x - y + 2 = 0$ and the curve $2x^2 + y^2 + x + y = 0$ are at right angles to one another.

2. Show that the straight lines joining the origin to the points of intersection of the curves $x^2 + y^2 = a^2$ and $x^2 + y^2 + 2(gx + fy) = 0$ are given by $a^2 (x^2 + y^2) - 4 (gx + fy)^2 = 0$.

3. Find the value of 'm' if the lines joining the origin to the points common to $x^2 + y^2 + x - 2y - m = 0$ and $x + y = 1$ are at right angles.

4. Find the value of 'm', if the lines joining the origin a nd the´ points of intersection of $y = nx + 1$ and $x^2 + y^2 = 1$ perpendicular to one another.

Answer Keys

3. 1/2 **4.** ±1

TEXTUAL EXERCISE–12 (OBJECTIVE)

1. A pair of perpendicular straight lines is drawn through the origin forming with the line $2x + 3y = 6$ an isosceles triangle right angled at the origin. The equation to the line pair is
   (a) $5x^2 - 24xy - 5y^2 = 0$
   (b) $5x^2 - 26xy - 5y^2 = 0$
   (c) $5x^2 + 24xy - 5y^2 = 0$
   (d) $5x^2 + 26xy - 5y^2 = 0$

2. If a pair of perpendicular straight lines drawn through the origin forms an isosceles triangle with the line $2x + 3y = 6$, then area of the triangle so formed is
   (a) 36/13 (b) 12/17
   (c) 13/5 (d) 17/13

3. The lines joining the origin to the point of intersection of $3x^2 + mxy - 4x + 1 = 0$ and $2x + y - 1 = 0$ are at right angles. Then which of the following is/are possible value of m?
   (a) –4 (b) 4
   (c) 7 (d) 3

4. If the lines joining the origin to the intersection of the line $y = mx + 2$ and the curve $x^2 + y^2 = 1$ are at right angles, then
   (a) $m^2 = 1$ (b) $m^2 = 3$
   (c) $m^2 = 7$ (d) $2m^2 = 1$

5. The lines joining the origin to the points of intersection of $x^2 + y^2 + 2gx + c = 0$ and $x^2 + y^2 + 2fy - c = 0$ are at right angles, if
   (a) $g^2 + f^2 = c$ (b) $g^2 - f^2 = c$
   (c) $g^2 - f^2 = 2c$ (d) $g^2 + f^2 = c^2$

Answer Keys

1. (a) **2.** (a) **3.** (a,b,c,d) **4.** (c) **5.** (c)

MULTIPLE-CHOICE QUESTIONS

SECTION–I

OBJECTIVE-TYPE SOLVED EXAMPLES

1. The base BC of ΔABC is bisected at (p, q) and equation of sides AB and AC are $px + qy = 1$ and $qx + py = 1$. Then equation of median through A is

(a) $(2q - 1)(px + qy = pq)$

(b) $(2pq - 1)(px + qy - 1) = (p^2 + q^2 - 1)(qx + py - 1)$

(c) $(px + qy - 1)(qx + py - 1) = 0$

(d) None of these

Solution: (b) Any line through A is given by $(px + qy - 1) + \lambda(qx + py - 1) = 0$ which is passing through (p, q). Hence $\lambda = \dfrac{-(p^2+q^2-1)}{(2pq-1)}$

Thus the required line is

$$(px + qy - 1) - \frac{p^2+q^2-1}{2pq-1}(qx + py - 1) = 0$$

2. The equation of a straight line equally inclined to the axis and equidistant from the points $(1, -2)$ and $(3, 4)$ is

(a) $x + y + 1 = 0$ (b) $x + y + 2 = 0$

(c) $x - y - 2 = 0$ (d) $x - y - 1 = 0$

Solution: (d) All the lines given in four options are equally inclined to the axis. Middle point of the line joining points $(1, -2)$ and $(3, 4)$ is $(2, 1)$ which lies on line $x - y - 1 = 0$.

3. The orthocentre of the triangle formed by $(0, 0)$, $(8, 0)$, $(4, 6)$ is

(a) $(4, 8/3)$ (b) $(3, 4)$

(c) $(4, 3)$ (d) $(-3, 4)$

Solution: (a) Let ABC be the given triangle and the vertices be $A(0, 0)$, $B(8, 0)$ and $C(4, 6)$.

Slope of $BC = (6 - 0)/(4 - 8) = -3/2$.

$\therefore$ Equation of line through A and $\perp$ to BC is $y - 0 = (2/3)(x - 0)$

i.e., $2x - 3y = 0$ ....(1)

And slope of $CA = (6 - 0)/(4 - 0) = 3/2$

$\therefore$ Equation of line through B and $\perp$ to CA is $y - 0 = (-2/3)(x - 8)$

i.e., $2x + 3y = 16$. ......(2)

Solving (1) and (2), the orthocentre is $(4, 8/3)$, which is given in (a).

4. The point on the line $3x - 4y = 25$ which is nearest to the origin is

(a) $(-4, 5)$ (b) $(3, -4)$

(c) $(3, 4)$ (d) $(3, 5)$

Solution: (b) The point on the line $3x - 4y = 25$ that is nearest from origin is the foot of perpendicular from origin upon the line itself which is given by:

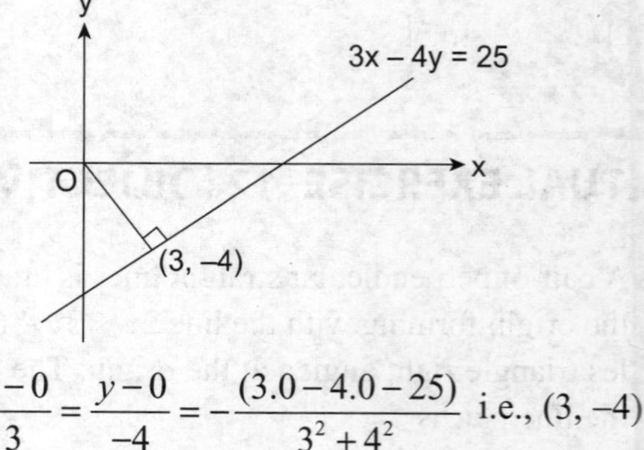

$$\frac{x-0}{3} = \frac{y-0}{-4} = -\frac{(3.0-4.0-25)}{3^2+4^2} \text{ i.e., } (3, -4)$$

5. The straight lines $x + y - 4 = 0$, $3x + y + 7 = 0$ and $x + 3y - 10 = 0$ forms a triangle which is

(a) isosceles (b) right angled

(c) equilateral (d) obtuse triangle

Solution: (a, d) One method to discuss the nature of ΔABC is to determine the points of intersection of lines which gives the co-ordinates of vertices A, B and C and then find the length AB, BC and AC. Another method is to find the internal angles of ΔABC. Slopes of sides AB, BC and CA are $-\frac{1}{3}, -1, -3$ respectively which is indicated in figure. Therefore

$$\tan A = \left(\frac{-3+1/3}{1+1}\right) = -\frac{4}{3}\text{[obtuse angle]}$$

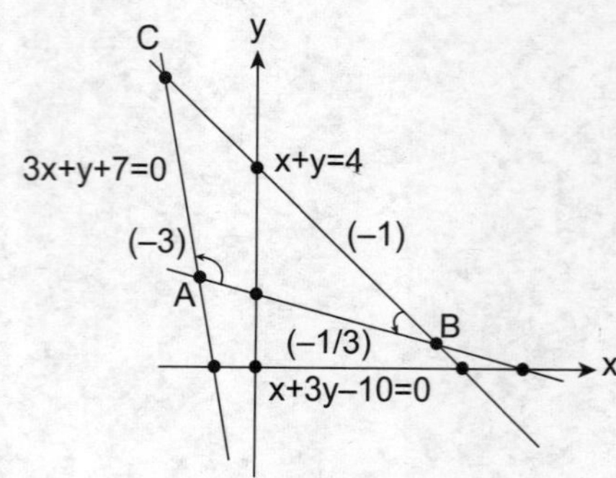

$\tan B = \left(\frac{-1/3+1}{1+1/3}\right) = \frac{1}{2}$; $\tan C\left(\frac{-1+3}{1+3}\right) = \frac{1}{2}$

$\Rightarrow$ ΔABC is both isosceles and obtuse.

6. The equation $x^2y^2 - 2xy^2 - 3y^2 - 4x^2y + 8xy + 12y = 0$ represents

(a) a pair of straight lines

(b) a pair of straight lines and a circle

(c) a pair of straight lines and a parabola

(d) a set of four lines forming a square

Solution: (d) Given equation is

$x^2y^2 - 2xy^2 - 3y^2 - 4x^2y + 8xy + 12y = 0$

$\Rightarrow y^2(x^2 - 2x - 3) - 4y(x^2 - 2x - 3) = 0$

$\Rightarrow y(y - 4)(x - 3)(x + 1) = 0$

$\Rightarrow y = 0, y = 4, x = 3, x = -1$

Hence the equation represents four straight lines which evidently form a square.

7. If one of the diagonals of a square is along the line $x = 2y$ and one of its vertices is (3,0), then its sides through this vertex are given by the equations

(a) $y - 3x + 9 = 0, 3y + x - 3 = 0$

(b) $y - 3x + 9 = 0, 3y + x - 3 = 0$

(c) $y - 3x + 9 = 0, 3y - x + 3 = 0$

(d) $y - 3x + 3 = 0, 3y + x + 9 = 0$

Solution: (a) Diagonal of the square is along

$x - 2y = 0$ ...(1)

The point (3,0) does not lie on (1)

Let the side through this vertex be $y - 0 = m(x - 3)$ .....(2)

Acute angle between side (2) and diagonal (1) is 45^0.

$\Rightarrow \left|\frac{m-1/2}{1+m(1/2)}\right| = \tan 45^o$

$\Rightarrow \frac{2m-1}{2+m} = \pm 1$ $\quad\Rightarrow m = 3, -1/3$

$\therefore$ from (2), the required sides are $y - 3x + 9 = 0$ and $3y + x - 3 = 0$

8. The equation of the line with gradient $-3/2$, which is concurrent with the lines $4x + 3y - 7 = 0$ and $8x + 5y - 1 = 0$ is

(a) $3x + 2y - 2 = 0$ (b) $3x + 2y - 63 = 0$

(c) $2y - 3x - 2 = 0$ (d) None of these

Solution: (a) Equation of any line through the point of intersection of $4x + 3y - 7 = 0$ and $8x + 5y - 1 = 0$ is given by $4x + 3y - 7 + \lambda(8x + 5y - 1) = 0$

or, $(4 + 8\lambda)x + (3 + 5\lambda)y - 7 - \lambda = 0$ .....(i)

Slope of the line $= \frac{-(4+8\lambda)}{3+5\lambda}$ and it is given to be $-\frac{3}{2}$

$\therefore \frac{-(4+8\lambda)}{3+5\lambda} = -\frac{3}{2}$

$\Rightarrow \lambda = 1$

By putting this value of λ in (i), we get equation of required line i.e., $3x + 2y - 2 = 0$.

9. The algebraic sum of the perpendicular distances from the points (2, 0), (0, 2) and (1, 1) to a variable straight line is zero. The line passes through a fixed point whose co-ordinates are

(a) (1, 2) (b) (2, 1)

(c) (1, 1) (d) (2, 2)

Solution: (c) Let the line by $lx + my + 1 = 0$ ..... (1)

According to the given condition,

$\frac{2l+1+2m+1+l+m+1}{\sqrt{l^2+m^2}} = 0$

$\Rightarrow l + m + 1 = 0$

$\Rightarrow l(1) + m(1) + 1 = 0$

Comparing it with (1), we find that line (1) is passing through (1,1)

10. The straight line passing through the point of intersection of the straight lines $x - 3y + 1 = 0$ and $2x + 5y - 9 = 0$ and having infinite slope and at a distance 2 units from the origin has the equation

(a) $x = 2$ (b) $x = -2$

(c) $y = 1$ (d) None of these

Solution: (a) Any line through the intersection of given lines is $(x - 3y + 1) + \lambda(2x + 5y - 9) = 0$

Now its slope will be infinite if it is perpendicular to the x-axis.

Hence the co-efficient of y will be zero.

$\Rightarrow -3 + 5\lambda = 0$ $\quad\Rightarrow \lambda = 3/5$

Putting λ, the line is $x - 2 = 0$. Clearly, its distance from (0, 0) is 2.

11. The line L has intercepts a and b on the co-ordinate axes. Keeping the origin fixed, the co-ordinate axes are rotated through a fixed angle. The line L has now intercepts p and q on the rotated axes. Then

(a) $a^2 + b^2 = p^2 + q^2$ (b) $1/a^2 + 1/b^2 = 1/p^2 + 1/q^2$

(c) $a^2 + p^2 = b^2 + q^2$ (d) $1/a^2 + 1/p^2 = 1/b^2 + 1/q^2$

Solution: (b) The equation of the line L is $x/a + y/b = 1$ ... (1)

After the rotation of axes, the line L has intercepts p and q on axes.

In this system, equation of the line is $X/p + Y/q = 1$
Since the origin and the line, both are fixed, the distance between them remains the same.

$$\Rightarrow \left|\frac{-1}{\sqrt{1/a^2+1/b^2}}\right| = \left|\frac{-1}{\sqrt{1/p^2+1/q^2}}\right|$$

$$\Rightarrow \frac{1}{a^2}+\frac{1}{b^2}=\frac{1}{p^2}+\frac{1}{q^2}$$

12. If a, b, c are in A.P. then $ax + by + c = 0$ represents
(a) a single line
(b) a family of concurrent lines
(c) Family of parallel lines
(d) None of these

Solution: (b) a, b, c are in *A.P.*
$\Rightarrow 2b = a + c$ or $c = 2b - a$; substituting the value of c in $ax + by + c = 0$,
we get $a(x - 1) + b(y + 2) = 0$.
This denotes a family of lines all passing through the point of intersection of lines
$x - 1 = 0$ and $y + 2 = 0$ i.e., family of concurrent lines all passing through point $(1, -2)$

13. The equation of the line bisecting the obtuse angle between $y - x = 2$ and $\sqrt{3}y + x = 5$ is
(a) $\frac{y-x-2}{\sqrt{2}}=\frac{\sqrt{3}y+x-5}{2}$
(b) $\frac{y+x-2}{\sqrt{2}}=\frac{\sqrt{3}y+x-5}{2}$
(c) $\frac{-y+x+2}{\sqrt{2}}=\frac{\sqrt{3}y-x-5}{2}$
(d) None of these

Solution: (a) Bisectors of given lines are

$$\frac{y-x-2}{\sqrt{2}}=\pm\left(\frac{\sqrt{3}y+x-5}{2}\right)$$

$\Rightarrow -(\sqrt{2}+1)x + (\sqrt{2}-\sqrt{3})y = 2\sqrt{2} - 5$ ... (1)
and $(1-\sqrt{2})x + (\sqrt{2}+\sqrt{3})y = 2\sqrt{2} + 5$ ... (2)
slope of line $y - x = 2$ is $(m_1 = 1)$ and slope of first bisector $\frac{\sqrt{2}+1}{\sqrt{2}-\sqrt{3}} = m_2$

Now $\tan\theta = \frac{m_2 - m_1}{1+m_1m_2} = \frac{\sqrt{3}+1}{1+2\sqrt{2}-\sqrt{3}} > 1.$

Therefore $\theta > 45°$
So the bisector of obtuse angle is

$$\frac{y-x-2}{\sqrt{2}}=\frac{\sqrt{3}y+x-5}{2}$$

14. A point starts from (1, 2) and its projections on the x and the y axes are moving with velocity of 3 m/s and 2m/s respectively. Its locus is
(a) $2x - 3y + 4 = 0$ (b) $3x - 2y + 1 = 0$
(c) $3y - 2x + 4 = 0$ (d) $2y - 3x + 1 = 0$

Solution: (a) Here $\frac{dx}{dt} = 3$

$\Rightarrow x = 3t + A$ and $\frac{dy}{dt} = 2 \Rightarrow y = 2t + B$

$\therefore$ $2x - 3y = 2A - 3B = C$(say) it passes through $P(1, 2)$.
So, $C = -4$
$\Rightarrow$ required locus is $2x - 3y + 4 = 0$

15. A (1, 3) and B (7, 5) are two opposite vertices of a square. The equation of side through A are
(a) $x + 2y - 7 = 0$ (b) $x - 2y + 5 = 0$
(c) $2x + y - 5 = 0$ (d) $2x - y + 1 = 0$

Solution: (a), (d) Let m be slope of the line through A

Slope of $AB = \frac{5-3}{7-1} = \frac{1}{3}$ (diagonal)

$$\Rightarrow \tan 45° = \left|\frac{m-1/3}{1+m/3}\right|$$

$$\Rightarrow m = 2 \text{ or } -\frac{1}{2}$$

Required equations are $x + 2y - 7 = 0$
or $2x - y + 1 = 0$

16. The image of the pair of lines represented by $ax^2 + 2hxy + by^2 = 0$ by the line mirror $y = 0$ is
(a) $ax^2 - 2hxy - by^2 = 0$
(b) $bx^2 - 2hxy + ay^2 = 0$
(c) $bx^2 + 2hxy + ay^2 = 0$
(d) $ax^2 - 2hxy + by^2 = 0$

Solution: (d) Let $y = m_1x$ and $y = m_2x$ be the lines represented by $ax^2 + 2hxy + by^2 = 0$
Then their images in $y = 0$ are $y = -m_1x$ and $y = -m_2x$ and so their combined equation is $y^2 + m_1m_2x^2 + xy(m_1 + m_2) = 0$

$$\Rightarrow y^2 + \frac{a}{b}x^2 + xy\left(\frac{-2h}{b}\right) = 0$$

$$\Rightarrow ax^2 - 2hxy + by^2 = 0$$

17. If the lines represented by $x^2 - 2pxy - y^2 = 0$ are rotated about the origin through an angle θ, one in clockwise direction and other in anti-clockwise direction, then the equation of the bisectors of the angle between the lines in the new positions is

(a) $px^2 + 2xy - py^2 = 0$ (b) $px^2 + 2xy + py^2 = 0$
(c) $x^2 - 2pxy - y^2 = 0$ (d) None of these

Solution: (a) The bisector of the angles between the lines in new position are same as the bisectors of the angles between their original position.

Therefore, the required equation is $\frac{x^2 - y^2}{1-(-1)} = \frac{xy}{-p}$

$\Rightarrow px^2 + 2xy - py^2 = 0$

18. The straight line $x - y = 2$ cuts the co-ordinate axes in A and B. On AB a square is constructed away from the origin, 'p' denotes the perpendicular distance from (0,0) to a side of the square, then
(a) maximum value of p is $3\sqrt{2}$
(b) Area of square is 8(square units)
(c) $x + y = 2$ is the equation to one of the sides of the square
(d) None of the above

Solution: (a, b, c)

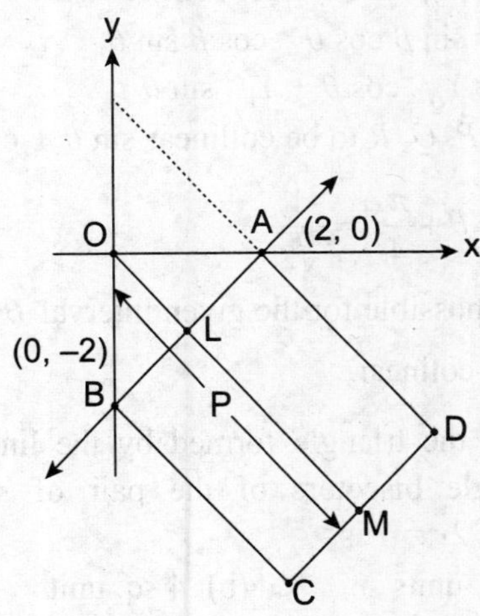

Area of square = $(2\sqrt{2})^2 = 8$ square units

$p_{max} = OM = OL + LM$

$= \frac{|-2|}{\sqrt{2}} + 2\sqrt{2} = 3\sqrt{2}$

Clearly, equation of side $AD = x + y = 2$
So (a), (b), (c) all three options are correct.

19. If the two lines represented by $x^2(\tan^2\theta + \cos^2\theta) - 2xy\tan\theta + y^2\sin^2\theta = 0$ make angles α, β with x-axis, then
(a) $\tan\alpha + \tan\beta = 4\ \text{cosec}\ 2\theta$
(b) $\tan\alpha\tan\beta = \sec^2\theta + \tan^2\theta$
(c) $\tan\alpha.\tan\beta = \sec^2\theta + \sin^2\theta$
(d) None of these

Solution: (a), (c) Let the slopes of lines be represented by m_1 and m_2

$\Rightarrow m_1 = \tan\alpha,\ m_2 = \tan\beta$

$m_1 + m_2 = \frac{-2h}{b} = \frac{2\tan\theta}{\sin^2\theta}$

$\Rightarrow \tan\alpha + \tan\beta = \frac{4}{2\sin\theta\cos\theta} = 4\text{cosec}\ 2\theta$

(a) option is correct.

Also $m_1 m_2 = \frac{a}{b} = \frac{\tan^2\theta + \cos^2\theta}{\sin^2\theta} = \sec^2\theta + \cot^2\theta$

$\Rightarrow \tan\alpha.\ \tan\beta = \sec^2\theta + \cot^2\theta$
Therefore (b) option is false and (c) is true.

20. Point $P(2,4)$ is translated through a distance $3\sqrt{2}$ units measured parallel to the line $y - x - 1 = 0$, in the direction of decreasing ordinates, to reach at Q. If 'R' is the image of Q with respect to the line $y - x - 1 = 0$, then coordinates of R are given by
(a) (–1, 1) (b) (5, 7)
(c) (6, 6) (d) (0, 0)

Solution: (d) Let Q be (x_q, y_q)

$\Rightarrow \frac{x_q - 2}{1/\sqrt{2}} = \frac{y_q - 4}{1/\sqrt{2}} = -3\sqrt{2}$

$\Rightarrow (x_q, y_q) \equiv (-1, 1)$; Let R be (x_r, y_r)

$\Rightarrow \frac{x_r + 1}{-1} = \frac{y_r - 1}{1} = \frac{-2(1+1-1)}{1^2 + 1^2}$

$\Rightarrow (x_r, y_r) = (0, 0)$

21. The equation $x^3 + y^3 = 0$ represents
(a) three real straight lines
(b) three points
(c) the combined equation of a straight line and a circle
(d) None of these

Solution: (d) Here $(x + y)(x^2 - xy + y^2) = 0$.
Clearly, $x + y = 0$ is a real straight line.
But $x^2 - xy + y^2 = 0$ represents two imaginary straight lines because $D = (-1)^2 - 4.1.1 < 0$

22. The acute angle between the pair of lines $y^2 - 2xy\ \text{cosec}\theta + x^2 = 0$, $0 \le \theta \le \pi/2$ is
(a) $\pi/2$ (b) θ
(c) $\pi/2 - \theta$ (d) None of these

Solution: (c) Acute angle between pair of straight lines $ax^2 + 2hxy + by^2 = 0$ is $\tan^{-1}\left(\frac{2\sqrt{h^2 - ab}}{|a+b|}\right)$

$= \tan^{-1}\left(\frac{2\sqrt{\text{cosec}^2\theta - 1}}{2}\right) = \tan^{-1}(|\cot\theta|)$

$= \tan^{-1}(\tan(\pi/2 - \theta)) = (\pi/2 - \theta)$

$\left[\begin{array}{l} \because -\frac{\pi}{2} \le -\theta < 0 \\ 0 \le \frac{\pi}{2} - \theta < \frac{\pi}{2} \end{array}\right]$

23. The triangle formed by the straight lines whose combined equation is $(y^2 - 4xy - x^2)(x + y - 1) = 0$ is

(a) equilateral (b) right angled
(c) Acute angled (d) obtuse angled

Solution: (b) The pair $y^2 - 4xy - x^2 = 0$ contains lines which are at right angles because $a + b = 0$.

24. The lines represented by $x^2 + 2\lambda\, xy + 2y^2 = 0$ and the lines represented by $(1 + \lambda)\, x^2 - 8xy + y^2 = 0$ are equally inclined, then λ is

(a) any real number (b) greater than 2
(c) ± 2 (d) less than -2

Solution: (c) Using the fact that the two pairs have the same bisectors of angles.

$\Rightarrow \frac{x^2 - y^2}{1 - 2} = \frac{xy}{\lambda} \approx \frac{x^2 - y^2}{1 + \lambda - 1} = \frac{xy}{-4}$

$\Rightarrow \lambda(x^2 - y^2) + xy = 0$ and $4(x^2 - y^2) + \lambda xy = 0$ are same equations

$\Rightarrow \frac{\lambda}{4} = \frac{1}{\lambda} \qquad \Rightarrow \lambda^2 = 4$

$\Rightarrow \lambda = \pm 2$

25. The equation $x^3 + x^2y - xy^2 = y^3$ represents

(a) three real straight lines
(b) lines in which two of them are perpendicular to each other
(c) lines in which two of them are coincident
(d) None of these

Solution: (a, b, c) The equation is $x^2(x + y) - y^2(x + y) = 0$ or $(x + y)^2(x - y) = 0$

It represents the lines $x + y = 0$, $x + y = 0$, $x - y = 0$

26. If one of the lines of $my^2 + (-m^2 + 1)\, xy - mx^2 = 0$ is a bisector of the angle between the lines $xy = 0$, then m is

(a) 1 (b) 2
(c) $-1/2$ (d) -1

Solution: (a, d) Here $my(y - mx) + x(y - mx) = 0$

i.e., $(y - mx)(my + x) = 0$

So the lines are $y = mx$ or $y = \frac{-1}{m}x$

$\therefore$ Bisectors between the lines $xy = 0$ are $y = x$ or $y = -x$

$\therefore\ m = 1$ or -1

27. Consider three points $P = (-\sin(\beta - \alpha), -\cos\beta)$, $Q = (\cos(\beta - \alpha), \sin\beta)$ and $R = (\cos(\beta - \alpha + \theta), \sin(\beta - \theta))$, where $0 < \alpha, \beta, \theta < \frac{\pi}{4}$. Then,

(a) P lies on the line segment RQ
(b) Q lies on the line segment PR
(c) R lies on the line segment QP
(d) P, Q, R are non-collinear

Solution: (d) $P \equiv (-\sin(\beta - \alpha), -\cos\beta)$

$Q \equiv (\cos(\beta - \alpha), \sin\beta)$

$R \equiv (\cos(\beta - \alpha + \theta), \sin(\beta - \theta))$

Also $0 < \alpha, \beta, \theta < \frac{\pi}{4}$ (given)

$\Rightarrow X_R = \cos(\beta - \alpha)\cos\theta - \sin(\beta - \alpha)\sin\theta$

$\Rightarrow X_R = X_Q \cdot \cos\theta + X_P \cdot \sin\theta;$

$Y_R = \sin\beta\cos\theta - \cos\beta\sin\theta;$

$\Rightarrow Y_R = Y_Q \cdot \cos\theta + Y_P \cdot \sin\theta$

For P, Q, R to be collinear $\sin\theta + \cos\theta = 1$

$\Rightarrow \sin\left(\theta + \frac{\pi}{4}\right) = \frac{1}{\sqrt{2}}$

$\Rightarrow$ not possible for the given interval $\theta \in \left(0, \frac{\pi}{4}\right)$

$\Rightarrow$ non-colinear.

28. Area of the triangle formed by the line $x + y = 3$ and angle bisectors of the pair of straight lines $x^2 - y^2 + 2y = 1$ is

(a) 2 sq. units (b) 4 sq. units
(c) 6 sq. units (d) 8 sq. units

Solution: (a)

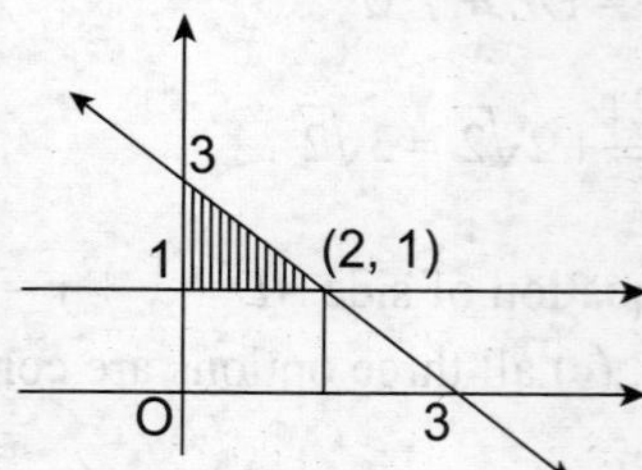

$\Rightarrow x^2 - y^2 + 2y = 1$

$\Rightarrow x = \pm(y - 1)$

Bisector of above lines are $y = 1$ and $x = 0$.

So area between, $x = 0$, $y = 1$ and $x + y = 3$ is given by

Area $= \frac{1}{2} \times 2 \times 2 = 2$ sq. units

29. A straight line through the origin O meets the parallel lines $4x + 2y = 9$ and $2x + y + 6 = 0$ at points P and Q respectively. Then the point O divides the segment PQ in the ratio

(a) 1: 2 (b) 3: 4
(c) 2: 1 (d) 4: 3

Solution: (b)

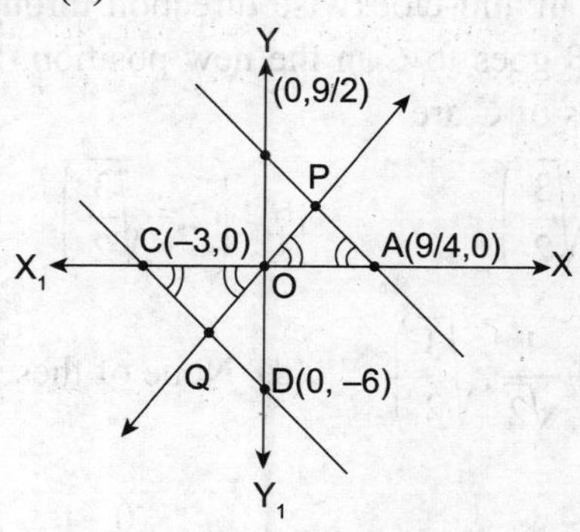

as $\Delta OPA \sim \Delta OQC$ ∴ $\dfrac{OP}{OQ} = \dfrac{OA}{OC} = \dfrac{9/4}{3} = \dfrac{3}{4}$

30. A straight line L with negative slope passes through the point (8, 2) and cuts the positive co-ordinate axes at points P and Q. Find the absolute minimum value of $OP + OQ$, as L varies, where O is the origin.

(a) 8 (b) 12
(c) 18 (d) None of these

Solution: (c) Equation of line L is $y - 2 = m(x - 8)$; where $m < 0$

$\Rightarrow P \equiv \left(8 - \dfrac{2}{m}, 0\right)$ and Q $(0, 2 - 8m)$

Now, $OP + OQ = \left|8 - \dfrac{2}{m}\right| + |2 - 8m|$

$= 10 + \dfrac{2}{(-m)} - 8 \geq 10 + 2\sqrt{\dfrac{2}{-m} \times 8(-m)} = 18$

Thus $OP + PQ \geq 18$

31. The number of integer values of m, for which the x-co-ordinate of the point of intersection of the lines, $3x + 4y = 9$ and $y = mx + 1$ is also an integer is

(a) 2 (b) 0
(c) 4 (d) 1

Solution: Intersection of $3x + 4y = 9$ and $y = mx + 1$ for x-co-ordinate $3x + 4(mx + 1) = 9 \Rightarrow (3 + 4m)\,x = 5$

$\Rightarrow x = \dfrac{5}{3 + 4m}$

For x being an integer $3 + 4m$ should be divisor of 5 i.e., 1, –1, 5 *or* –5

For $3 + 4m = 1$ $\Rightarrow$ $m = -\dfrac{1}{2}$(Not integer)

For $4m + 3 = -1$ $\Rightarrow$ $m = -1$ (integer)

For $3 + 4m = 5$ $\Rightarrow$ –(Not an integer)

For $3 + 4m = -5$ $\Rightarrow$ $m = -2$ (integer)

∴ There are two integral values of m

32. Let PQR be a right angled isosceles triangle, right angled at $P(2, 1)$. If the equations of the line QR is $2x + y = 3$, then the equation representing the pair of lines PQ and PR is

(a) $3x^2 - 3y^2 + 8xy + 2x + 10y + 25 = 0$
(b) $3x^2 - 3y^2 + 8xy - 20x - 10y + 25 = 0$
(c) $3x^2 - 3y^2 + 8xy + 10x + 15y + 20 = 0$
(d) $3x^2 - 3y^2 - 8xy - 10x - 15y - 20 = 0$

Solution: (b) Let m be the slope of PQ then

$\tan 45° = \left|\dfrac{m - (-2)}{1 + m(-2)}\right|$

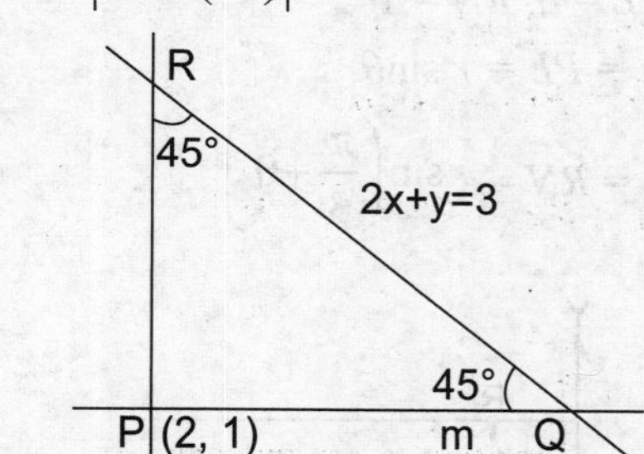

$\Rightarrow 1 = \left|\dfrac{m + 2}{1 - 2m}\right|$ $\Rightarrow \pm 1 = \dfrac{m + 2}{1 - 2m}$

$\Rightarrow m + 2 = 1 - 2m$ *or* $-1 + 2m = m + 2$

$m = -\dfrac{1}{3}$ or $m = 3$

PR makes 45° with PQ

equation of PQ $y - 1 = -\dfrac{1}{3}(x - 2)$

$\Rightarrow x + 3y - 5 = 0$

equation of PR is $y - 1 = 3(x - 2)$

$\Rightarrow 3x - y - 5 = 0$

∴ combined equation of PQ and PR is $(x + 3y - 5)(3x - y - 5) = 0$

$\Rightarrow 3x^2 - 3y^2 + 8xy - 20x - 10y + 25 = 0$

33. The diagonals of a parallelogram $PQRS$ are along the lines $x + 3y = 4$ and $6x - 2y = 7$, then $PQRS$ must be a

(a) rectangle (b) square
(c) cyclic quadrilateral (d) rhombus

Solution: (d) Slope of $x + 3y = 4$ is $-\frac{1}{3}$

slope of $6x - 2y = 7$ is 3 two lines are perpendiculars

$\Rightarrow$ diagonals are perpendicular.

$\Rightarrow$ *PQRS* is a Rhombus.

34. The distance between two parallel lines is unity. A point *P* lies between the lines at a distance '*a*' from one of them. The length of a side of an equilateral triangle *PQR*, vertex *Q* of which lies on one of the parallel lines and vertex *R* lies on the other line is

(a) $\frac{2}{\sqrt{3}}\sqrt{a^2+a+1}$ (b) $\frac{2}{\sqrt{3}}\sqrt{a^2-a+1}$

(c) $\frac{1}{\sqrt{3}}\sqrt{a^2+a+1}$ (d) $\frac{1}{\sqrt{3}}\sqrt{a^2-a+1}$

Solution: (b) Let $PQ = QR = RP = r$

and $\angle PQX = \theta$ then $\angle RQX = \frac{\pi}{3}+\theta$

Given $PL = a$, $RN = 1$

Now $a = PL = r\sin\theta$

and $1 = RN = r\sin\left(\frac{\pi}{3}+\theta\right)$

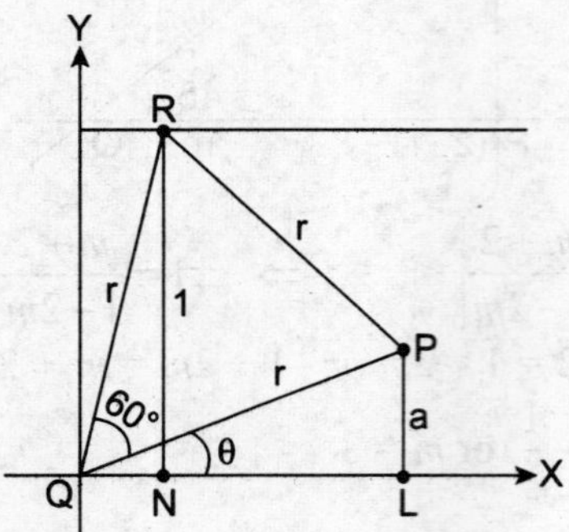

$\Rightarrow 1 = r\left(\sin\frac{\pi}{3}\cos\theta + \cos\frac{\pi}{3}\sin\theta\right)$

$\Rightarrow r\left(\frac{\sqrt{3}}{2}\cos\theta + \frac{1}{2}\sin\theta\right) = 1$

$\Rightarrow r\left(\frac{\sqrt{3}}{2}\sqrt{1-\frac{a^2}{r^2}} + \frac{1}{2}\cdot\frac{a}{r}\right) = 1 \quad \left(\because \sin\theta = \frac{a}{r}\right)$

$\Rightarrow \frac{\sqrt{3}}{2}\sqrt{r^2-a^2} = 1-\frac{a}{2}$ or

$\Rightarrow r^2 - a^2 = \frac{(2-a)^2}{3} = \frac{4+a^2-4a}{3}$

$\Rightarrow r^2 = \frac{4+a^2-4a}{3} + a^2 = \frac{4+4a^2-4a}{3}$

$\therefore \quad r = \frac{2}{\sqrt{3}}\sqrt{a^2-a+1}$.

35. A line joining two points $A(2,0)$ and $B(3, 1)$ is rotated about *A* in anti-clockwise direction through an angle 15°. If *B* goes to *C* in the new position, then the co-ordinates of *C* are

(a) $\left(2, \sqrt{\frac{3}{2}}\right)$ (b) $\left(2, -\sqrt{\frac{3}{2}}\right)$

(c) $\left(2+\frac{1}{\sqrt{2}}, \sqrt{\frac{3}{2}}\right)$ (d) None of these

Solution: (c) Slope of line $AB = \frac{0-1}{2-3} = 1 = \tan 45^\circ$

$\therefore \quad \angle BAX = 45°$

Given $\angle CAB = 15°$

$\therefore \quad \angle CAX = 60°$

$\therefore$ Slope of line $AC = \tan 60° = \sqrt{3}$

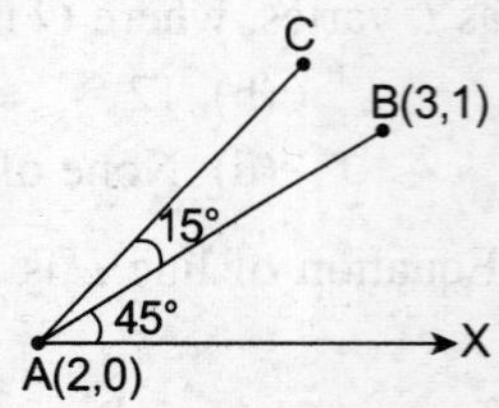

Now line *AC* makes an angle of 60° with positive direction of *x*-axis and

$AC = AB = \sqrt{(3-2)^2+(1-0)^2} = \sqrt{2}$

$\therefore$ Co-ordinates of *C* are

$(2+\sqrt{2}\cos 60^\circ, 0+\sqrt{2}\sin 60^\circ)$

i.e., $\left(2+\frac{1}{\sqrt{2}}, \sqrt{\frac{3}{2}}\right)$

36. If the distance of any point (x, y) from origin is defined as $d(x, y) = |x| + |y|$, then the locus $d(x, y) = 1$ is a

(a) circle of area 2 sq unit

(b) square of area 2 sq unit

(c) square of area 1 sq unit

(d) None of the above

Solution: (b) $d(x, y) = 1 \Rightarrow |x| + |y| = 1$ The graph of which is shown in the figure

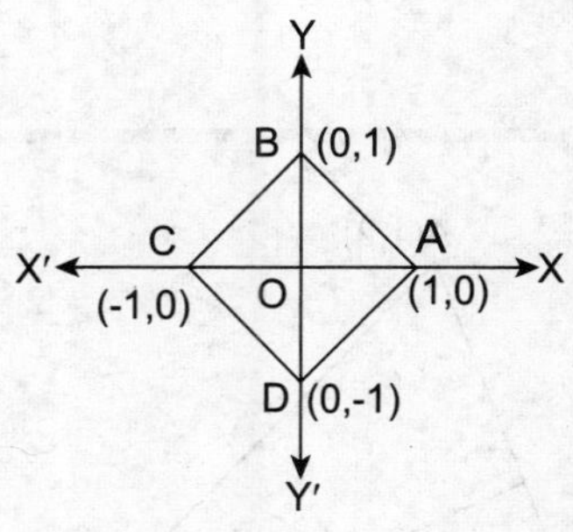

The graph is a square $AB = BC = CD = DA = \sqrt{2}$
Where, $OA = 1 = OB$, $\angle ABC = 90°$

$\therefore$ Area $= AB \,.\, AD = \sqrt{2} \times \sqrt{2} = 2$ sq unit

37. A system of lines is given as $y = m_i x + c_i$, where m_i can take any value out of 0, 1, – 1 and when m_i is positive, then c_i can be 1 or – 1; when m_i equal 0, c_i can be 0 or 1 and when m_i equals to – 1, c_i can take 0 or 2. Then the area enclosed by all these straight lines is

(a) $\frac{3}{\sqrt{2}}(\sqrt{2} - 1)$ sq unit

(b) $\frac{3}{\sqrt{2}}$ sq unit

(c) $\frac{3}{2}$ sq unit

(d) None of these

Solution: (c) Lines are $y = 1$, $y = 0$

$y = -x$, $y = -x + 2$

$y = x + 1$, $y = x - 1$

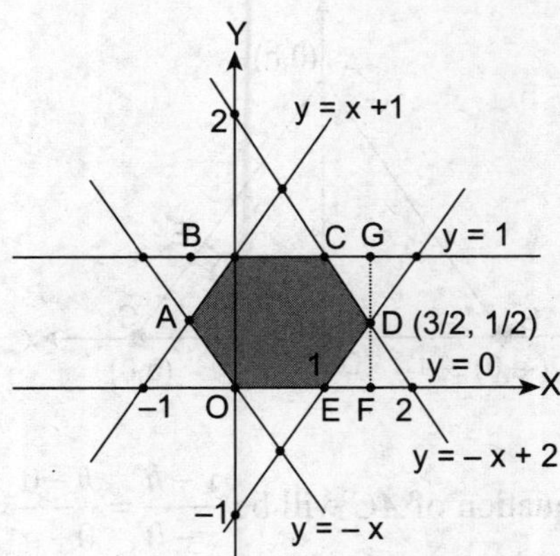

Area of $OABCDE$ = area of $OBGF$

$$= \frac{3}{2} \times 1 = \frac{3}{2} \text{ sq unit}$$

38. The equation of the perpendicular bisectors of the sides AB and AC of triangle ABC are $x - y + 5 = 0$ and $x + 2y = 0$ respectively. If the point A is $(1, -2)$, the equation of the line BC is.

(a) $14x + 23y = 40$ (b) $14x - 23y = 40$

(c) $14x + 23y + 40 = 0$ (d) None of these

Solution: (a) Given equation of line LF is $x - y + 5 = 0$ ...(i)

and equation of line ME is $x + 2y = 0$ ...(ii)

Since AB is perpendicular to LF and passes through $A(1, -2)$ therefore equation of AB is $x + y - (1 - 2) = 0$

or $x + y = -1$ ...(iii)

Similarly, equation of line AC is $2x - y - (2.1 + 2) = 0$

$\Rightarrow 2x - y = 4$ ...(iv)

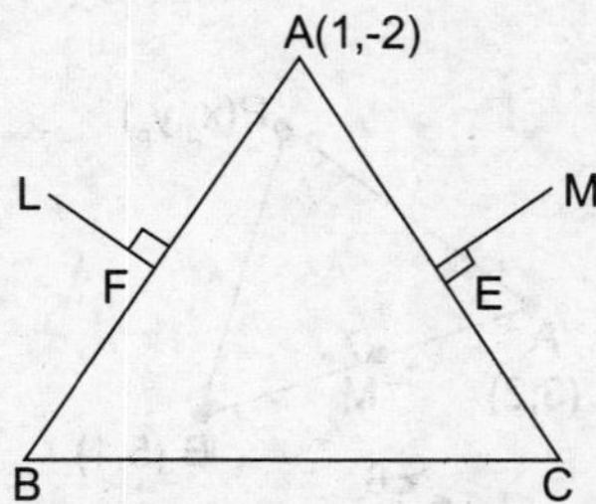

Solving (i) and (iii), we get $x = -3$, $y = 2$

$\therefore F \equiv (-3, 2)$

Solving (ii) and (iv) we get $x = 8/5$, $y = -4/5$

$\therefore E \equiv (8/5, -4/5)$

Let $B \equiv (x_1, y_1)$ and $C \equiv (x_2, y_2)$

Since F is the middle point of AB

$$\therefore \frac{x_1 + 1}{2} = -3 \Rightarrow x_1 = -7$$

$$\text{and } \frac{y_1 - 2}{2} = 2 \Rightarrow y_1 = 6$$

Hence $B \equiv (-7, 6)$. Again E is the middle point of AC

$$\therefore \frac{8}{5} = \frac{x_2 + 1}{2}$$

$$\Rightarrow x_2 = \frac{11}{5} \text{ and } -\frac{4}{5} = \frac{y_2 - 2}{2}$$

$$\Rightarrow y_2 = \frac{2}{5}$$

Hence $C \equiv \left(\frac{11}{5}, \frac{2}{5}\right)$

$\Rightarrow$ equation of BC is $14x + 23y = 40$

SECTION–II

SUBJECTIVE-TYPE SOLVED EXAMPLES

1. A and B are two fixed points having co-ordinates (3, 2) and (5, 1). PAB is an equilateral triangle on the side of AB opposite to origin. Find co-ordinates of P.

Solution: Co-ordinates of mid-point of AB are (4, 3/2). Slope of AB is −1/2 and hence slope of ⊥r bisector of AB is 2.

Point P will lie on this perpendicular bisector, so that MP = (side of equilateral triangle) × sin 60°

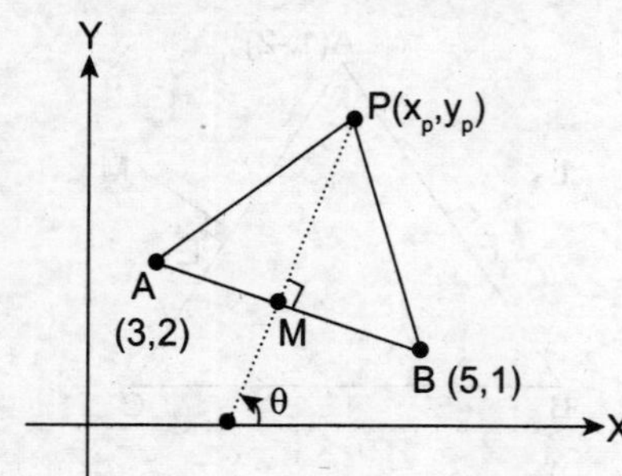

$$\therefore\ MP = \frac{\sqrt{15}}{2} \text{ (where } M \text{ is mid-point of } AB).$$

$$\Rightarrow \frac{x_p - 4}{\cos\theta} = \frac{y_p - 3/2}{\sin\theta} = \pm r$$

where (x_p, y_p) are co-ordinate of point P and θ is the angle that MP makes with x-axis.

Since, $\tan\theta = 2 \Rightarrow \cos\theta = \frac{1}{\sqrt{5}}$, $\sin\theta = \frac{2}{\sqrt{5}}$

And for point away from the origin, we have

$$\frac{x_p - 4}{1/\sqrt{5}} = \frac{y_p - 3/2}{2/\sqrt{5}} = +\frac{\sqrt{15}}{2}$$

$$\Rightarrow \left(4 + \frac{\sqrt{3}}{2}, \frac{3}{2} + \sqrt{3}\right) \equiv (x_p, y_p)$$

2. In a triangle ABC, co-ordinates of A are (1, 2) and the equations to the medians through B and C are $x + y = 5$ and $x = 4$ respectively. Find the co-ordinates of B and C.

Solution: Let the co-ordinates of B be (x_1, y_1) and that of C be $(4, y)$. Since the medians through B and C intersect at the point (4, 1), the co-ordinates of the centroid G of the triangle ABC are (4, 1).

$$\Rightarrow \frac{x_1 + 4 + 1}{3} = 4 \Rightarrow x_1 = 7$$

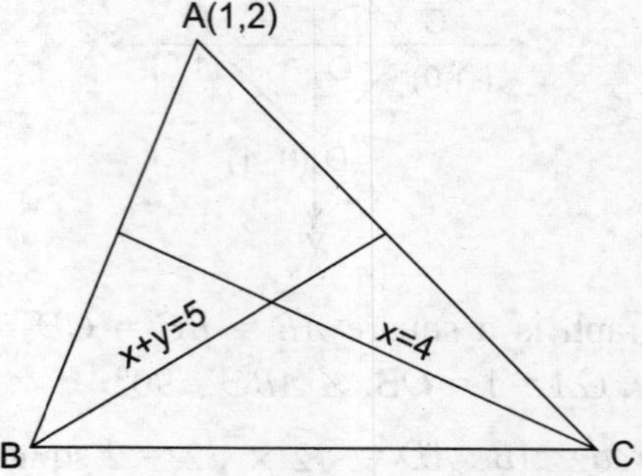

Since $B(x_1, y_1)$ lies on $x + y = 5$

$\Rightarrow y_1 = 5 - x_1 = 5 - 7 = -2$

So that the co-ordinates of B are (7, −2)

Also $\frac{y_1 + y + 2}{3} = 1$

$\Rightarrow y = 3 - 2 - y_1 = 3 - 2 + 2 = 3$

So that the co-ordinates of C are (4, 3).

3. Let ABC be a triangle with $AB = AC$. If D is the mid-point of BC. E is the foot of the ⊥ drawn from D to AC and F the mid-point of DE. Find the co-ordinates of point F.

Solution: Since $AB = AC$, D is mid-point of BC. Let us select mid-point of BC, i.e., D as origin and BC along x-axis and A lying on y-axis as shown in the figure, below. Let the co-ordinate of A be $(0, h)$ and that of B and C be respectively $(-a, 0)$ and $(a, 0)$. Co-ordinates of the points A, B, C are as shown in the figure.

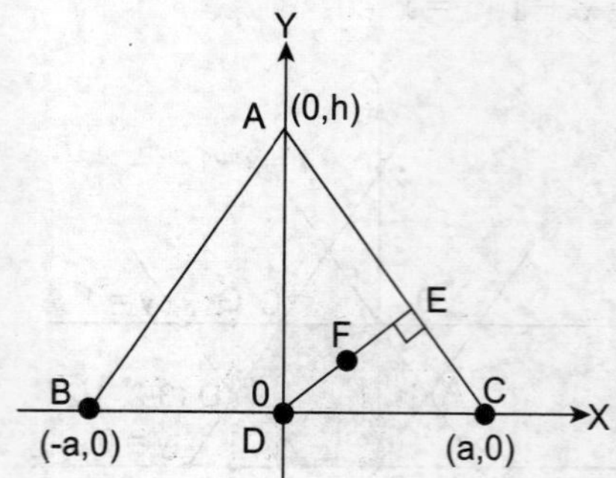

Now equation of AC will be $\frac{y-h}{x-0} = \frac{h-0}{0-a} = -\frac{h}{a}$;

where (x, y) is any arbitray point on AC

$$\Rightarrow y - h = -\frac{h}{a}(x) \Rightarrow ay + hx = ah \quad \text{... (i)}$$

since $DE \perp AC$, slope of $DE = a/h$

Equation of DE is $y - 0 = \frac{a}{h}(x - 0)$

$\Rightarrow hy - ax = 0$ .... (ii)

solving (i) and (ii), we get the co-ordinates of the point E. as $x_E = \dfrac{ah^2}{a^2+h^2}$ and $y_E = \dfrac{a^2h}{a^2+h^2}$

Since, F is the mid-point of DE.

$x_F = \dfrac{ah^2}{2(a^2+h^2)}$ and $y_F = \dfrac{a^2h}{2(a^2+h^2)}$.

4. From the origin two lines are drawn such that they divide the segment of the line $3x + y - 12 = 0$ in 3 equal parts between co-ordinates axes. Find area of quadrilateral formed between these two lines and lines $3x + y - 12 = 0$ and $6x + 2y - 50 = 0$.

Solution: Let the given line $3x + y = 12$ ...(i)

intersects x-axis and y-axis respectively at A and B

$\therefore$ $A \equiv (4, 0)$ and $B \equiv (0, 12)$

L_1, L_2 are two lines drawn from O (0, 0) such that these lines divide $3x + y = 12$ in 3 equal parts between co-ordinate axis i.e., $AC = CD = DB$

Now, C is dividing AB in ratio 1: 2 while D is dividing AB in 2: 1.

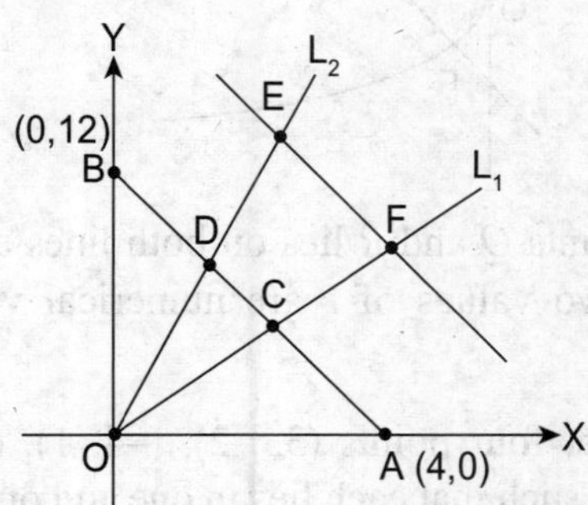

Let, co-ordinates of be $C(x_2, y_2)$ and that of D be (x_3, y_3)

$$x_2 = \frac{1\times 0 + 2\times 4}{1+2} = \frac{8}{3};$$

$$y_2 = \frac{1\times 12 + 2\times 0}{1+2} = 4, \text{ so, } C \equiv \left(\frac{8}{3}, 4\right)$$

$$\text{and } x_3 = \frac{2\times 0 + 1\times 4}{2+1} = \frac{4}{3};$$

$$y_3 = \frac{2\times 12 + 1\times 0}{2+1} = 8 \text{ so } D \equiv (4/3, 8).$$

Now, slope of line L_1 is $m_1 = \dfrac{4-0}{8/3-0} = \dfrac{3}{2}$.

Therefore equation of line L_1 is $y - 0 = \dfrac{3}{2}(x - 0)$

$\Rightarrow y = \dfrac{3}{2}x$

Similarly, slope of line L_2 is $m_2 = \dfrac{8-0}{4/3-0} = 6$

$\therefore$ equation of L_2 is $y - 0 = 6(x - 0) \Rightarrow y = 6x$

Quadrilateral formed between L_1, L_2, $3x + y = 12$ and $6x + 2y = 50$ is $CDEF$. As slope of $EF = -\dfrac{6}{2} = -\dfrac{1}{3}$ that is same as that of given line AB.

So, $CDEF$ is a trapezium

$\therefore$ Area of trapezium $= \dfrac{1}{2}$ (sum of parallel sides) $\times$ (perpendicular distance between the parallel sides)

$= \dfrac{1}{2}$ $(CD + EF) \times$ (perpendicular distance between CD and EF); point of intersection of L_1 and $6x + 2y = 50$ gives F, while that of L_2 and $6x + 2y = 50$ gives E.

$F \equiv (50/9, 25/3)$, $E \equiv (25/9, 50/3)$

$$CD = \sqrt{(8/3-4/3)^2 + (4-8)^2} = \frac{4}{3}\sqrt{10}$$

$$EF = \sqrt{(50/9-25/9)^2 + (25/3-50/3)^2} = \frac{25}{9}\sqrt{10}$$

Perpendicular distance between CD and EF:

$$\frac{50/2-12}{\sqrt{3^2+1^2}} = \frac{13}{\sqrt{10}}.$$

$$\text{Area of trapezium} = \frac{1}{2}\left(\frac{4}{3}\sqrt{10} + \frac{25}{9}\sqrt{10}\right)\times\frac{13}{\sqrt{10}} = \frac{481}{18}.$$

5. The sides of a triangle are $U_r \equiv x \cos \alpha_r + y \sin\alpha_r - p_r = 0$ for $r = 1, 2, 3$. Show that its orthocentre is given by $U_1 \cos(\alpha_2 - \alpha_3) = U_2 \cos(\alpha_3 - \alpha_1) = U_3 \cos(\alpha_1 - \alpha_2)$.

Solution: Let BC be $U_1 \equiv x \cos\alpha_1 + y \sin\alpha_1 - p_1 = 0$

CA be $U_2 = 0$ and AB be $U_3 = 0$

Then the altitude AD through A is of the form

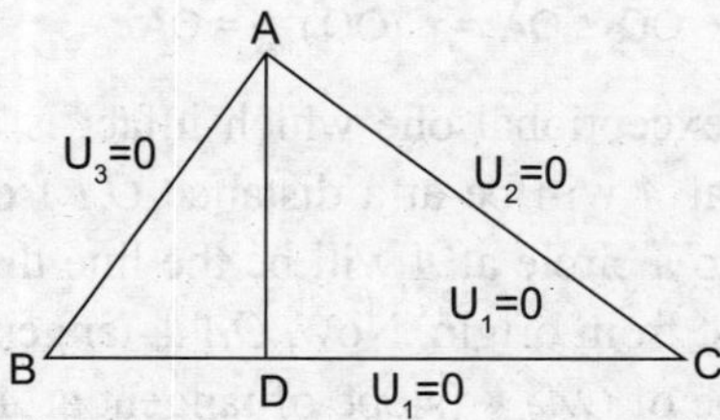

$U_2 + \lambda U_3 = (x \cos\alpha_2 + y \sin\alpha_2 - p_2) + \lambda[(x \cos\alpha_3 + y \sin\alpha_3 - p_3] = 0$,

i.e., AD is given by $(\cos\alpha_2 + \lambda\cos\alpha_3)x + (\sin\alpha_2 + \lambda\sin\alpha_3)y - (p_2 + \lambda p_3) = 0$.

Slope of $AD = -\dfrac{\cos\alpha_2 + \lambda\cos\alpha_3}{\sin\alpha_2 + \lambda\sin\alpha_3}$. Now $AD \perp BC$ gives

$$\left(-\frac{\cos\alpha_2 + \lambda\cos\alpha_3}{\sin\alpha_2 + \lambda\sin\alpha_3}\right)(-\cot\alpha_1) = -1,$$

or $\cos\alpha_1(\cos\alpha_2 + \lambda\cos\alpha_3) + \sin\alpha_1(\sin\alpha_2 + \lambda\sin\alpha_3) = 0$,
This gives $(\cos\alpha_1\cos\alpha_2 + \sin\alpha_1\sin\alpha_2) + \lambda(\cos\alpha_1\cos\alpha_3 + \sin\alpha_1\sin\alpha_3) = 0$

$$\Rightarrow \lambda = -\frac{\cos(\alpha_1 - \alpha_2)}{\cos(\alpha_3 - \alpha_1)}.$$

$\therefore$ Equation to AD is $U_2 - \dfrac{\cos(\alpha_1 - \alpha_2)}{\cos(\alpha_3 - \alpha_1)}U_3 = 0$

or equivalently $U_2\cos(\alpha_3 - \alpha_1) = U_3\cos(\alpha_1 - \alpha_2)$.
Similarly, the altitude BE is $U_3\cos(\alpha_1 - \alpha_2) = U_1\cos(\alpha_2 - \alpha_3)$

$\therefore$ The orthocentre is given by
$U_1\cos(\alpha_2 - \alpha_3) = U_2\cos(\alpha_3 - \alpha_1) = U_3\cos(\alpha_1 - \alpha_2)$.

6. Find the equation of the line, through the intersection of $2x + 3y - 7 = 0$ and $x + 3y - 5 = 0$ and having distance from origin as large as possible.

Solution: Point of intersection of two lines is A (2, 1). Now, with OA as radius and O itself as centre draw a circle. There will be infinitely many lines through A and each except one of them produces a chord of circle and hence their distance from origin i.e., centre of circle is less than OA i.e., radius of circle.

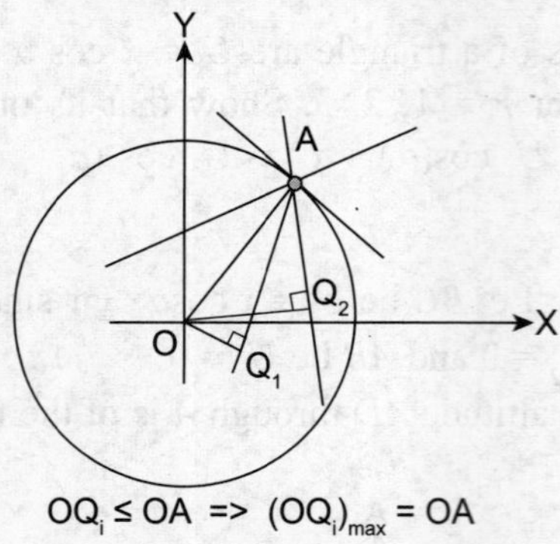

$OQ_i \le OA \Rightarrow (OQ_i)_{max} = OA$

But the exceptional one which infact is a tangent to a circle at A will be at a distance OA from O. Thus, tangent to a circle at A will be the line through A and is farthest from origin. Now, $OA \perp$ tangent at A

$\therefore$ (slope of OA) × (slope of tangent at A) = −1.

or, $\dfrac{1-0}{2-0}$ × (slope of tangent at A) = −1

$\Rightarrow$ Slope of tangen at A = −2.

$\therefore$ equation of required line is $(y - 1) = -2(x - 2)$ or $2x + y - 5 = 0$

7. A line which makes an acute angle θ with the positive direction of the x axis is drawn through the point $P(3, 4)$ to cut the curve $y^2 = 4x$ at Q and R. Show that the lengths of the segments PQ and PR are numerical values of the roots of the equation $r^2\sin^2\theta + 4r(2\sin\theta - \cos\theta) + 4 = 0$.

Solution: Equation of line passing through point (3, 4) and making angle θ with +ve x-axis is

$$\left.\begin{aligned} \frac{x-3}{\cos\theta} = \frac{y-4}{\sin\theta} = r \Rightarrow x = 3 + r\cos\theta \\ \text{and } y = 4 + r\sin\theta \end{aligned}\right\} \quad \text{....(i)}$$

r is the distance of general point (x, y) from point P. For intersection of line (i) and curve $y^2 = 4x$; $(4 + r\sin\theta)^2 = 4(3 + r\cos\theta)$

$\Rightarrow 16 + r^2\sin^2\theta + 8r\sin\theta = 12 + 4r\cos\theta$
$\Rightarrow r^2\sin^2\theta + 4r(2\sin\theta - \cos\theta) + 4 = 0$

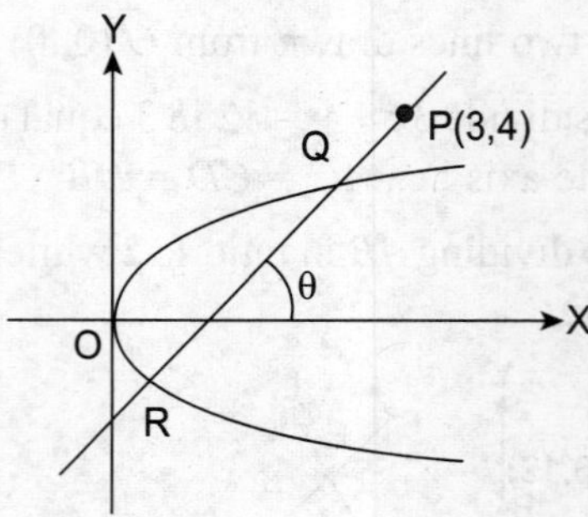

Since points Q and R lies on both lines as well as on curve, two values of r are numerical values of PQ and PR.

8. Show that four points (3, −2), (−4, 1), (−1, −4) and (1, 3) are such that each lies in one and only one of the 4 regions into which the straight lines $x - 2y = 2$ and $3x + 2y + 6 = 0$ divide the plane.

Solution: Let points $A \equiv (3, -2)$, $B \equiv (-4, 1)$, $C \equiv (-1, -4)$, $D \equiv (1, 3)$
$L_1 \equiv x - 2y - 2 = 0$,
$L_2 \equiv 3x + 2y + 6 = 0$
On substituting the points A, B, C, D on lines L_1 and L_2 we have
$L_1(A) \equiv 3 - 2(-2) - 2 = 5 > 0$;
$L_1(B) \equiv -4 - 2 - 2 = -8 < 0$;
$L_2(A) \equiv 3 \times 3 + 2(-2) + 6 = 11 > 0$;
$L_2(B) \equiv 3(-4) + 2(1) + 6 = -4 < 0$;
$L_1(C) = -1 - 2(-4) - 2 = 5 > 0$;
$L_1(D) = 1 - 2(3) - 2 = -7 < 0$;
$L_2(C) = 3(-1) + 2(-4) + 6 = -5 < 0$;
$L_2(D) \equiv 3(1) + 2(3) + 6 = 15 > 0$

Without loss of generality let A lies at position as shown in the Figure, below.

From the given equations, we observe the following:

(i) A and C are on the same side of line L_1 while at opposite sides of line L_2.

(ii) B and D lie on same side of L_1 where as on opposite side on L_2.

(iii) A and D are on opposite side of L_1 where as on same side of L_2.

(iv) B and C are on opposite sides of L_1 where as on same side of L_2.

Thus we see that given 4 points are such that each lie in one and only one of the 4 regions into which given 2 straight lines divide the plane as shown in the figure, below.

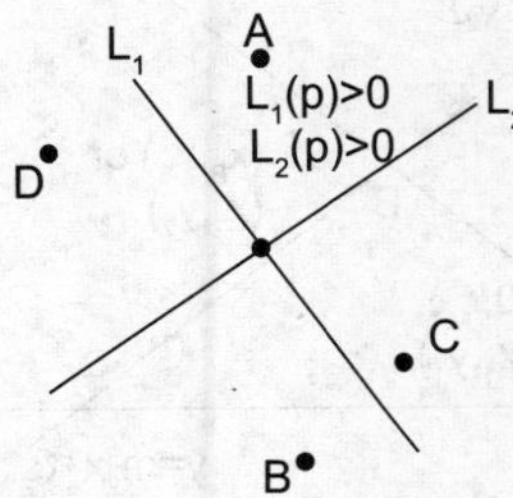

9. *RA* and *RB* are perpendiculars from *R* to the lines $y = 3x$ and $y = 2x$ respectively. If *R* moves on line $2x + 3y = 6$, then find the locus of the mid-point of *AB*.

Solution: Let $\left(\alpha, \frac{6-2\alpha}{3}\right)$, be the co-ordinates of *R*

Since RA is perpendicular to line $y = 3x$.

we have: $3\times\dfrac{\frac{6-2\alpha}{3}-3x_1}{\alpha - x_1} = -1$; where $A \equiv (x_1, 3x_1)$.

$\Rightarrow \dfrac{6-2\alpha-9x_1}{\alpha - x_1} = -1$

$\Rightarrow 6 - 2a - 9x_1 = -\alpha + x_1$

$\Rightarrow 10x_1 = 6 - \alpha \Rightarrow x_1 = \dfrac{6-\alpha}{10}$ ....(i)

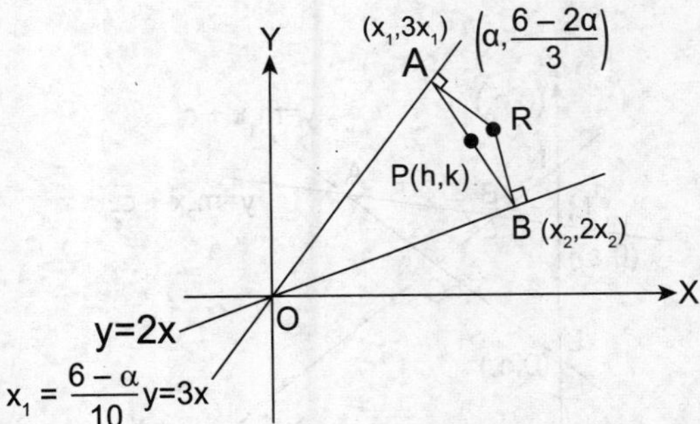

Since *RB* is perpendicular to line $y = 2x$.

We have $2 \times \dfrac{\frac{6-2\alpha}{3}-2x_2}{\alpha - x_2} = -1$

$\Rightarrow x_2 = \dfrac{12-\alpha}{15}$ ....(ii)

Let $P(h, k)$ be the mid-point of AB

$\Rightarrow h = \dfrac{x_1 + x_2}{2}$ and $k = \dfrac{3x_1 + 2x_2}{2}$ ....(iii)

Substituting the values of x_1 and x_2 in equation (iii),

we get $h = \dfrac{42-5\alpha}{60}$ and $k = \dfrac{102-13\alpha}{60}$

Eliminating α from these two equations, we get $65h = 25k + 3$

i.e., the locus of $P(h, k)$ is $65x - 25y - 3 = 0$.

10. The altitudes AD, *BE* and *CF* of a triangle *ABC* are $x + y = 0$, $x = 4y$ and $2x = y$, respectively. The co-ordinates of *A* are $(a, -a)$. If '*a*' varies, then find the locus of centroid of ΔABC.

Solution: Since line *AC* is perpendicular to *BE*.,

let equation of *AC* be $y = -4x + \lambda$

But the point $A(a, -a)$ lies on *AC*

$\Rightarrow -a = -4a + \lambda \Rightarrow \lambda = 3a$

$\Rightarrow$ Equation of *AC* is $y = -4x + 3a$

Co-ordinates of point *C* are $(x_3, 3a - 4x_3)$

Since point *C* lies on $y = 2x$ as well, we have $3a - 4x_3 = 2x_3 \Rightarrow x_3 = a/2$

$\therefore$ co-ordinates of point *C* are $(a/2, a)$

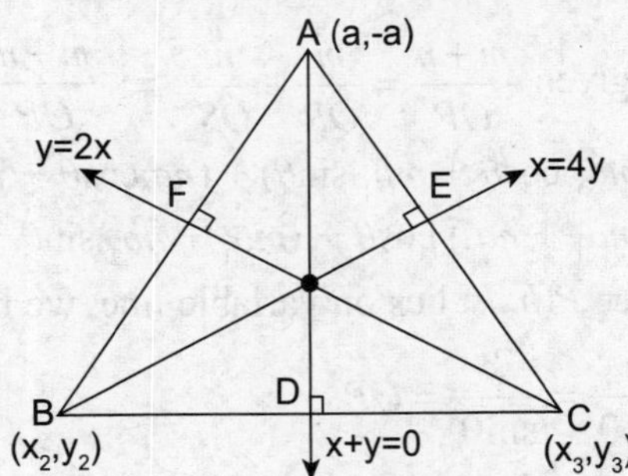

Similarly, co-ordinates of the point *B* are $\left(-\frac{2}{3}a, -\frac{a}{6}\right)$.

Let co-ordinates of centroid be (h, k).

$\Rightarrow h = \dfrac{a + \frac{a}{2} - \frac{2}{3}a}{3} = \dfrac{5a}{18}$ and $k = \dfrac{-a + a - \frac{a}{6}}{3} = \dfrac{-a}{18}$

Eliminating a, we get $x + 5y = 0$ which is the required locus.

11. A variable line is drawn through a point O to cut two fixed straight lines, L_1 and L_2 in R and S. A point P is chosen on the variable line such that

$\frac{m+n}{OP} = \frac{m}{OR} + \frac{n}{OS}$. Show that the locus of P is a straight line passing through the point of intersection of L_1 and L_2; m and n are variables.

Solution: Let two fixed straight lines L_1 and L_2 be

$L_1 \equiv a_1x + b_1y - 1 = 0$ ..(i)

and $L_2 \equiv a_2x + b_2y - 1 = 0$ ...(ii)

Let the variable line through origin be $\frac{x-0}{\cos\theta} = \frac{y-0}{\sin\theta} = r$

for point R: $r = OR$

$x_R = OR\cos\theta$, $y_R = OR\sin\theta$

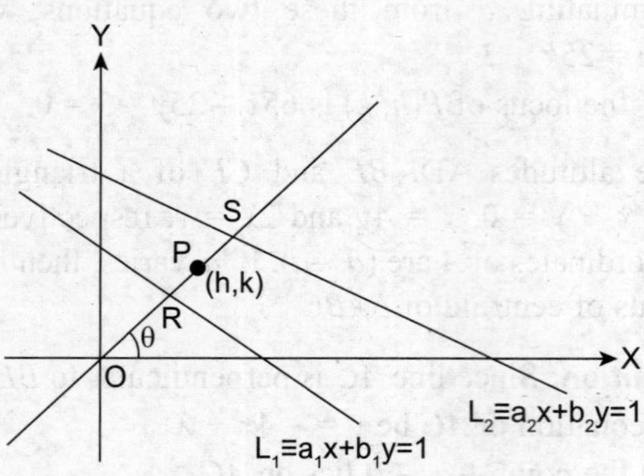

since point R lies on L_1, we have $a_1\, OR\cos\theta + b_1\, OR \sin\theta = 1$

$\Rightarrow \frac{1}{OR} = a_1\cos\theta + b_1\sin\theta,$

Similarly, $\frac{1}{OS} = a_2\cos\theta + b_2\sin\theta$

we are given $\frac{m+n}{OP} = \frac{m}{OR} + \frac{n}{OS}. \Rightarrow \frac{m+n}{OP}$

$= (ma_1\cos\theta + mb_1\sin\theta) + (na_2\cos\theta + nb_2\sin\theta)$

$= (ma_1 + na_2)\cos\theta + (mb_1 + nb_2)\sin\theta$ ....(iv)

Since $P(h, k)$ lies on variable line, we have,

$$\frac{h}{\cos\theta} = \frac{k}{\sin\theta} = OP$$

$\Rightarrow \cos\theta = \frac{h}{OP}$ and $\sin\theta = \frac{k}{OP}$ ....(v)

Substituting values of $\cos\theta$ and $\sin\theta$ from equation (v) to equation (iv)

we get, $\frac{m+n}{OP} = (ma_1 + na_2)\frac{h}{OP} + (mb_1 + nb_2)\frac{k}{OP}$

$\Rightarrow m(a_1h + b_1k - 1) + n(a_2h + b_2k - 1) = 0$

Therefore, locus of the point $P(h, k)$ is

$$(a_1x + b_1y - 1) + \frac{n}{m}(a_2x + b_2y - 1) = 0$$

which is a straight line passing through the intersection of L_1 and L_2.

12. The equations of the sides of a triangle are $y = m_1x + c_1$, $y = m_2x + c_2$ and $x = 0$. Prove that the area of the triangle is $\frac{1(c_1 - c_2)^2}{2|m_1 - m_2|}$ and hence show that the area of the triangle formed by the lines $y = m_rx + c_r (r = 1, 2, 3)$ is given by $\frac{(c_1 - c_2)^2}{2|m_1 - m_2|} + \frac{(c_2 - c_3)^2}{2|m_2 - m_3|} - \frac{(c_1 - c_3)^2}{2|m_1 - m_3|}$.

Solution: The sides of the triangle are

$y = m_1x + c_1$ ...(i)

$y = m_2x + c_2$ ...(ii)

and $x = 0$ ..(iii)

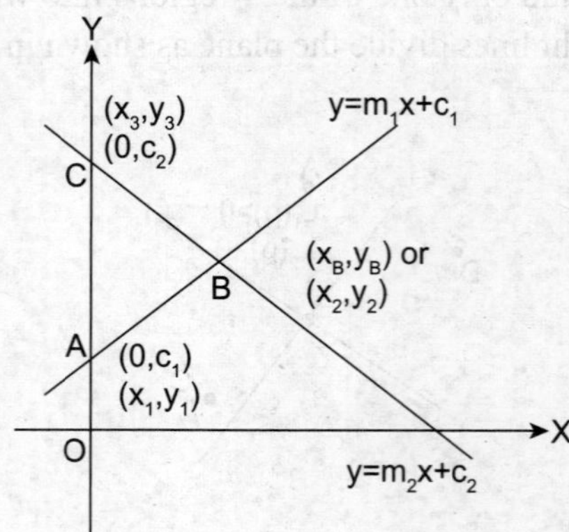

On solving (i) and (ii), we get one vertex as $B\left(\frac{c_1 - c_2}{m_2 - m_1}, \frac{c_1m_2 - m_1c_2}{m_2 - m_1}\right)$

Similarly, other vertices are $(0, c_1)$ and $(0, c_2)$

Area of triangle

$$= \left|\frac{1}{2}\begin{vmatrix} 0 & c_1 & 1 \\ 0 & c_2 & 1 \\ \frac{c_1 - c_2}{m_2 - m_1} & \frac{c_1m_2 - m_1c_2}{m_2 - m_1} & 1 \end{vmatrix}\right| = \frac{1}{2}\frac{(c_1 - c_2)^2}{|m_1 - m_2|}$$

$$= \frac{1}{2}\frac{(c_2 - c_2)^2}{|m_1 - m_2|}$$

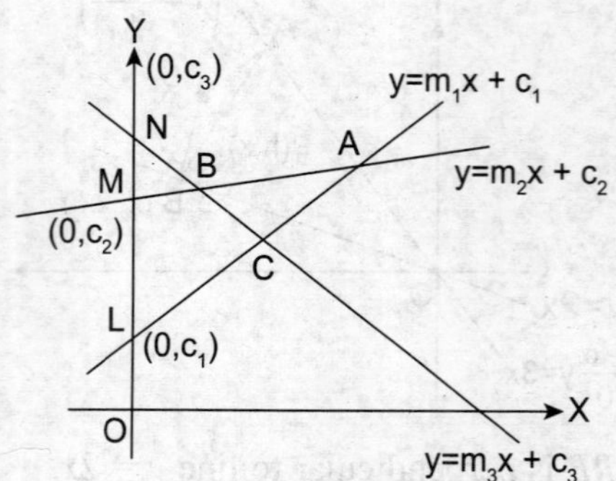

Now considering the diagram (ii),
Area of ΔABC = Area of ΔAML + *Area* of ΔBNM – *Area* of ΔCNL

$$= \frac{(c_1-c_2)^2}{2|m_1-m_2|} + \frac{(c_2-c_3)^2}{2|m_2-m_3|} - \frac{(c_1-c_3)^2}{2|m_1-m_3|}$$

(By using the above result)

13. Show that the parallelogram $ax + by + c = 0$ and $ax + by + d = 0$; $a'x + b'y + c' = 0$ and $a'x + b'y + d' = 0$ will be rhombus, if $(a^2 + b^2)(c' - d')^2 = (a'^2 + b'^2)(c - d)^2$ and area of the parallelogram is $\dfrac{|(c-d)||(c'-d')||ab'-a'b|}{(a^2+b^2)(a'^2-b'^2)}$.

Solution: At first strike, we think to obtain the co-ordinates of vertices of the quadrilateral and try to find the length of sides of the quadrilateral and try to equal the length. But it will be complicated, so it would be better to use the fact that in rhombus the distance between the parallel sides are equal. i.e., $\dfrac{|(c-d)|}{\sqrt{a^2+b^2}} = \dfrac{|(c'-d')|}{\sqrt{a'^2+b'^2}}$

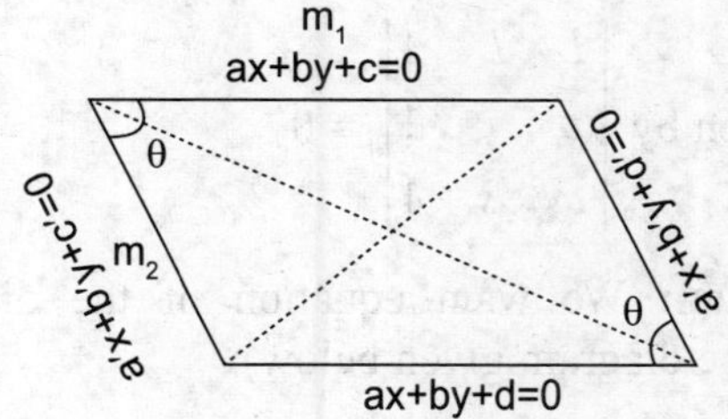

By squaring, we get $(c - d)^2 (a'^2 + b'^2) = (c' - d')^2 (a^2 + b^2)$.

Area of parallelogram $= p_1p_2 \sin\theta$ and required

Area $= \dfrac{|(c-d)|}{\sqrt{a^2+b^2}} \times \dfrac{|(c'-d')|}{\sqrt{a'^2+b'^2}} \times \sin\theta$; p_1, p_2 are distance between the parallel sides.

To find $\sin\theta$, we first calculate $\tan\theta$; $m_1 = -a/b$;

$$m_2 = -a'/b',\ \tan\theta_{acute} = \frac{\left|\frac{a}{b}-\frac{a'}{b'}\right|}{\left|1+\frac{a}{b}\cdot\frac{a'}{b'}\right|} = \left|\frac{ab'-a'b}{aa'+bb'}\right|$$

$$\Rightarrow \sin\theta = \sqrt{\frac{1}{1+\cot^2\theta}} = \sqrt{\frac{\tan^2\theta}{1+\tan^2\theta}}$$

Substituting the values for (i)

$$= \frac{\sqrt{(ab'-a'b)^2}}{\sqrt{(aa'+bb')^2+(ab'-a'b)^2}}$$

$$= \frac{|ab'-a'b|}{\sqrt{(aa')^2+(bb')^2+(ab')^2+(a'b)^2}}$$

$$= \frac{|ab'-a'b|}{\sqrt{(a^2+b^2)(a'^2+b'^2)}}$$

$\Rightarrow$ Area of parallelogram:

$$\frac{|(c-d)|}{\sqrt{a^2+b^2}} \times \frac{|(c'-d')|}{\sqrt{a'^2+b'^2}} \times \sin\theta$$

$$= \frac{|c-d||c'-d'|}{\sqrt{a^2+b^2}\sqrt{a'^2+b'^2}} \frac{|ab'-a'b|}{\sqrt{(a^2+b^2)(a'^2+b'^2)}}$$

$$= \frac{|(c-d)||(c'-d')||ab'-a'b|}{(a^2+b^2)(a'^2+b'^2)}$$

14. If through the angular points of a triangle, straight lines are drawn parallel to the sides and if the intersection of these lines be joined to the opposite angular points of the triangle, show that the joining lines so obtained will meet at a point.

Solution:

(i) We can find a line through the point *A*, which is parallel to *BC*. Proceed to solve the problem.

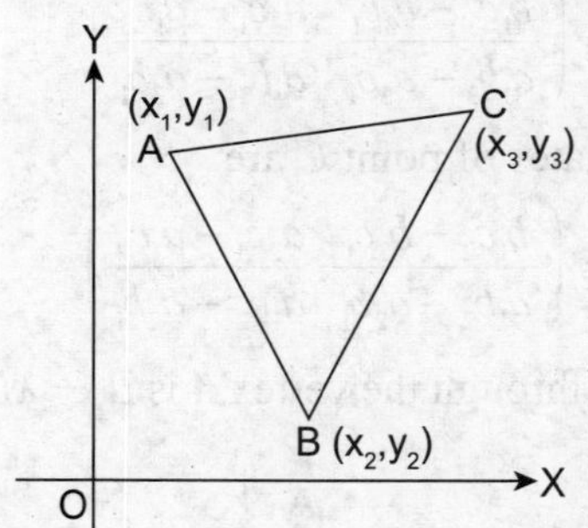

(ii) Another way is to take our axes such that the mid-point of *BC* is taken as origin and the *x*-axis is taken along the side *BC*. this is, *B* (–α, 0) and *C* (α, 0) and now you can proceed to solve the problem further (see figure shown below).

(iii) Draw the lines through vertices and parallel to opposite base.

Observe that quadrilateral *PBCA* is a parallelogram. Quadrilateral *BARC* is also a parallelogram.

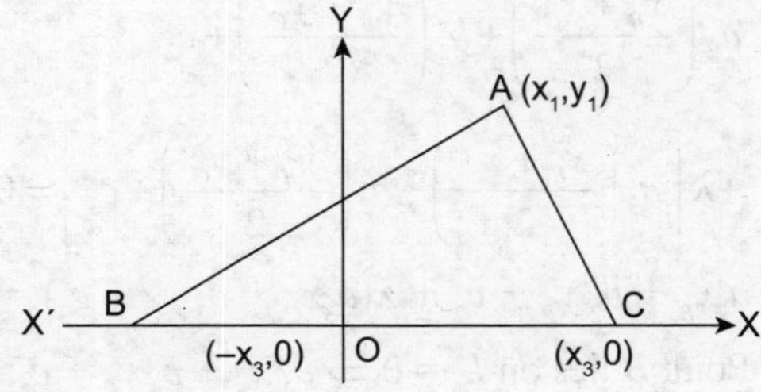

$\Rightarrow$ $PA = BC$ and $AR = BC \Rightarrow PA = AR$

$\Rightarrow$ AQ is the median of the triangle PQR

Similarly, BR and CP are also the medians. We know that medians are concurrent and meet at a point G(centroid).

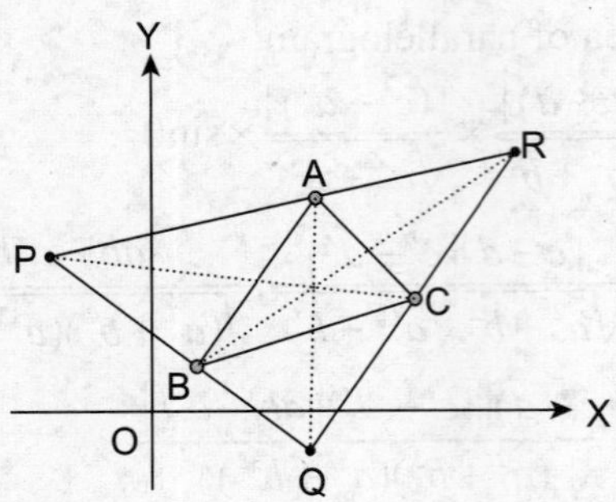

15. The sides of a Δ ABC are BC: $L_1 = 0$, CA: $L_2 = 0$, AB: $L_3 = 0$ where $L_n \equiv a_n x + b_n y + c_n$; $n = 1, 2, 3$. Prove that the median from A is given by

$$\frac{L_2}{\begin{vmatrix} a_1 & a_2 \\ b_1 & b_2 \end{vmatrix}} + \frac{L_3}{\begin{vmatrix} a_1 & a_3 \\ b_1 & b_3 \end{vmatrix}} = 0.$$

Solution: Co-ordinates of point B are

$$(x_B, y_B) \equiv \left(\frac{b_1c_3 - b_3c_1}{a_1b_3 - a_3b_1}, \frac{a_3c_1 - a_1c_3}{a_1b_3 - a_3b_1}\right)$$

Co-ordinates of point C are

$$(x_C, y_C) \equiv \left(\frac{b_1c_2 - b_2c_1}{a_1b_2 - a_2b_1}, \frac{a_2c_1 - a_1c_2}{a_1b_2 - a_2b_1}\right)$$

Any line through the vertex A is $L_2 + \lambda L_3 = 0$ as shown below.

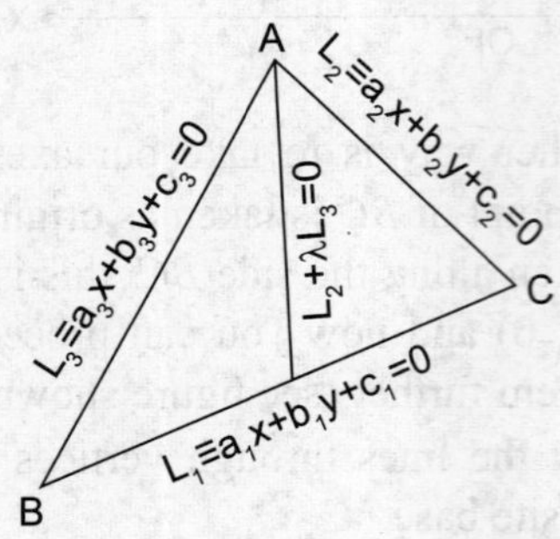

$\Rightarrow$ $(a_2x + b_2y + c_2) + \lambda\,(a_3x + b_3y + c_3) = 0$ .....(i)

Equation (i) must pass through the mid-point of BC

$$\Rightarrow a_2\left(\frac{x_B + x_C}{2}\right) + b_2\left(\frac{y_B + y_C}{2}\right) + c_2$$

$$+\lambda\left[a_3\left(\frac{x_B + x_C}{2}\right) + b_3\left(\frac{y_B + y_C}{2}\right) + c_3\right] = 0$$

$\Rightarrow$ $a_2x_B + b_2y_B + c_2 + \lambda(a_3x_C + b_3y_C + c_3) = 0$ ....(ii)

(Point B lies on $L_3 = 0 \Rightarrow a_3x_B + b_3y_B + c_3 = 0$,

and point C lies on $L_2 = 0 \Rightarrow a_2x_C + b_2y_C + c_2 = 0$))

$$\Rightarrow a_2\left(\frac{b_1c_3 - b_3c_1}{a_1b_3 - a_3b_1}\right) + b_2\left(\frac{a_3c_1 - a_1c_3}{a_1b_3 - a_3b_1}\right) + c_2 + \lambda$$

$$\left(a_3\left(\frac{b_1c_2 - b_2c_1}{a_1b_2 - a_2b_1}\right) + b_3\left(\frac{a_2c_1 - a_1c_2}{a_1b_2 - a_2b_1}\right) + c_3\right) = 0$$

$$\Rightarrow \frac{1}{a_1b_3 - a_3b_1} - \frac{\lambda}{a_1b_2 - a_2b_1} = 0$$

$$\Rightarrow \lambda = \frac{a_1b_2 - a_2b_1}{a_1b_3 - a_3b_1}$$

$\Rightarrow$ Equation of median through point A is

$$L_2 + \frac{a_1b_2 - a_2b_1}{a_1b_3 - a_3b_1}L_3 = 0$$

$$\Rightarrow \frac{L_2}{\begin{vmatrix} a_1 & a_2 \\ b_1 & b_2 \end{vmatrix}} + \frac{L_3}{\begin{vmatrix} a_1 & a_3 \\ b_1 & b_3 \end{vmatrix}} = 0$$

16. Show that the diagonal of a parallelogram whose sides are $L = p$, $L = q$, $L' = r$ and $L' = s$; where $L = ax + by + c$ and $L' = a'x + b'y + c'$, passing through the point of intersection of $L = p$ and $L' = r$,

is given by $\begin{vmatrix} L & L' & 1 \\ p & r & 1 \\ q & s & 1 \end{vmatrix} = 0$

Solution: We want equation of the diagonal BD (see the diagram given below)

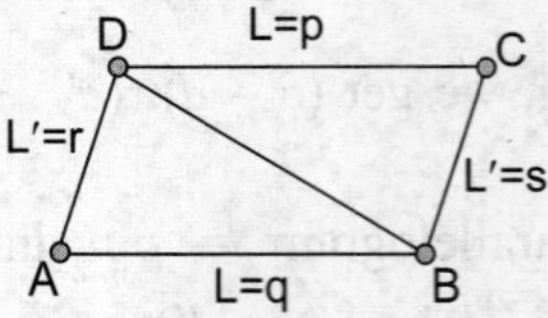

Any line through the point D is $(L - p) + \lambda(L' - r) = 0$

or $x\,(a + \lambda a') + y(b + \lambda b') + c + \lambda c' - p - r\lambda = 0$ ....(i)

Any line through the point B is $(L - q) + \beta\,(L' - s) = 0$

or $x\,(a + \beta\, a') + (b + \beta b')y + c + \beta c' - q - s\beta = 0$ ....(ii)

Since, Equations (i) and (ii) represent the diagonal BD of the parallelogram.

$$\Rightarrow \frac{a + \lambda a'}{a + \beta a'} = \frac{b + \lambda b'}{b + \beta b'} = \frac{c + \lambda c' - p - r\lambda}{c + \beta c' - q - s\beta} \quad \text{....(iii)}$$

i.e., $\dfrac{a + \lambda a'}{a + \beta a'} = \dfrac{b + \lambda b'}{b + \beta b'} \Rightarrow \lambda = \beta$

$$\Rightarrow 1 = \frac{c + \lambda c' - p - r\lambda}{c + \beta c' - q - s\beta} \Rightarrow \lambda = \frac{q - p}{r - s} \quad (\because \lambda = \beta)$$

∴ The equation of the required diagonal is $(L-p)+\left(\frac{q-p}{r-s}\right)(L'-r)=0 \Rightarrow \begin{vmatrix} L & L' & 1 \\ p & r & 1 \\ q & s & 1 \end{vmatrix}=0$

Aliter: Given that diagonal BD passing through the points of intersection of $L=p$, $L'=r$, and $L=q$, $L'=s$, the equation of the line BD can be written as $(L-p)+\lambda(L'-r)=0$ or $(L-q)+\mu(L'-s)=0$ eliminating λ and μ from these equations, we have

$$\begin{vmatrix} L-p & L'-r \\ L-q & L'-s \end{vmatrix}=0$$

$$\Rightarrow \begin{vmatrix} L & L' & 1 \\ p-L & r-L' & 0 \\ q-L & s-L' & 0 \end{vmatrix}=0 \Rightarrow \begin{vmatrix} L & L' & 1 \\ p & r & 1 \\ q & s & 1 \end{vmatrix}=0$$

17. $A(3, 0)$ and $B(6, 0)$are two fixed points and $U(x_1, y_1)$ is a variable point of the plane. AU and BU meet the y-axis at C and D, respectively, and AD meets OU at V. Prove that CV passes through (2, 0) for any position of U in the plane.

Solution: Let the co-ordinate of U be (x_1, y_1)

∴ equation of AU is $y-y_1=\frac{0-y_1}{3-x_1}(x-x_1)$

so that the co-ordinates of C are $\left(0, \frac{3y_1}{3-x_1}\right)$.

Similarly, the co-ordinates of D are $\left(0, \frac{6y_1}{6-x_1}\right)$

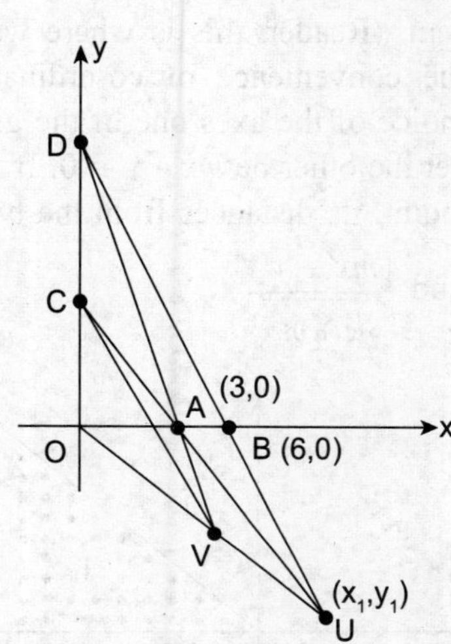

Now, equation of AD is $\frac{x}{3}+\frac{y(6-x_1)}{6y_1}=1$ ...(i)

and equation of OU is $y_1x=x_1y$ ...(ii)

Solving (i) and (ii), we get $\frac{x_1y}{3y_1}+\frac{y(6-x_1)}{6y_1}=1$

$$\Rightarrow y(2x_1+6-x_1)=6y_1 \Rightarrow y=\frac{6y_1}{6+x_1} \Rightarrow x=\frac{6x_1}{6+x_1}$$

Hence, the co-ordinates of V are $\left(\frac{6x_1}{6+x_1}, \frac{6y_1}{6+x_1}\right)$

Therefore, equation of CV is

$$y-\frac{3y_1}{3-x_1}=\frac{\frac{6y_1}{6+x_1}-\frac{3y_1}{3-x_1}}{\frac{6x_1}{6+x_1}-0}(x-0)$$

$$\Rightarrow y=\frac{3y_1}{3-x_1}-\frac{9x_1y_1}{6x_1(3-x_1)}x=\frac{3y_1}{3-x_1}\left(1-\frac{x}{2}\right).$$

Which clearly passes through (2, 0) irrespective of the positon of $U(x_1, y_1)$.

18. One diagonal of a square is the portion of the line $\frac{x}{a}+\frac{y}{b}=1$ intercepted between the axes. Show that the extremities of the other diagonal are $\left(\frac{a+b}{2}, \frac{a+b}{2}\right)$ and $\left(\frac{a-b}{2}, \frac{b-a}{2}\right)$.

Solution: Let BD be the line whose equation is $x/a + y/b = 1$. Then the co-ordinates of B are $(a, 0)$ and those of D are $(0, b)$. Since the other diagonal, AC, passes through the mid-point E of BD and is perpendicular to it, the equation of AC is

$$y-\frac{b}{2}=\frac{a}{b}\left(x-\frac{a}{2}\right) \quad ...(i)$$

Let the co-ordinates of C be (x_1, y_1).

Then (x_1, y_1) lies on (i) and $(DC)^2 = BD^2/2$

$$\Rightarrow 2[x_1^2+(y_1-b)^2]=a^2+b^2$$

$$\Rightarrow 2\left[x_1^2+\left\{\frac{a}{b}\left(x_1-\frac{a}{2}\right)-\frac{b}{2}\right\}^2\right]=a^2+b^2$$

$$\Rightarrow 2\left[\left(1+\frac{a^2}{b^2}\right)x_1^2-\frac{2a}{b}x_1+\left(\frac{a^2+b^2}{2b}\right)+\frac{(a^2+b^2)^2}{4b^2}\right]=a^2+b^2$$

$$\Rightarrow 4x_1^2-4ax_1+a^2+b^2=2b^2$$

$$\Rightarrow (2x_1-a)^2=b^2$$

$$\Rightarrow x_1=(a\pm b)/2$$

Taking the positive sign, we get $\left(\frac{a+b}{2}, \frac{a+b}{2}\right)$.

and taking the negative sign, we get $\left(\frac{a-b}{2}, \frac{b-a}{2}\right)$.

as the co-ordinates of the extremities of the diagonal AC.

19. A variable straight line of slope 4 intersects the hyperbola $xy = 1$ at two points. Find the locus of the point which divides the line segment between these two points in the ratio 1: 2.

Solution: Let the equation of the line be $y = 4x + c$ where c is a parameter. It intersects the hyperbola $xy = 1$ at two points, for which $x(4x + c) = 1$.

$\Rightarrow 4x^2 + cx - 1 = 0$

Let x_1 and x_2 be the roots of this equation. then $x_1 + x_2 = -c/4$ and $x_1x_2 = -1/4$. If A and B are the points of intersection of the line and the hyperbola, then the co-ordinates of A are $(x_1, 1/x_1)$ and that of B are $(x_2, 1/x_2)$

Let $R(h, k)$ be the point which divides AB in ratio 1: 2, then

$$h = \frac{2x_1 + x_2}{3} \text{ and } k = \frac{2/x_1 + 1/x_2}{3} = \frac{2x_2 + x_1}{3x_1x_2}$$

$\Rightarrow 2x_1 + x_2 = 3h$ ....(i)

and $x_1 + 2x_2 = 3(-1/4)k = (-3/4)k$ ...(ii)

Adding (i) and (ii), we get $3(x_1 + x_2) = 3[h - k/4]$

$\Rightarrow 3(-c/4) = 3(h - k/4) \Rightarrow h - k/4 = -c/4$ ....(iii)

Subtracting (ii) from (i), we get $x_1 - x_2 = 3(h + k/4)$

$\Rightarrow (x_1 - x_2)^2 = 9(h + k/4)^2 \Rightarrow c^2/16 + 1 = 9(h + k/4)^2$

$\Rightarrow (h - k/4)^2 + 1 = 9(h + k/4)^2$

$$\Rightarrow h^2 - \frac{1}{2}hk + \frac{k^2}{16} + 1 = 9\left(h^2 + \frac{1}{2}hk + \frac{k^2}{16}\right)$$

$\Rightarrow 16h^2 + 10hk + k^2 - 2 = 0$

So that the locus of $R(h, k)$ is $16x^2 + 10xy + y^2 - 2 = 0$

20. Given n straight lines and a fixed point O, a straight line is drawn through O meeting lines in the points $R_1, R_2, R_3, \ldots, R_n$ and on it a point R is taken such that

$\frac{n}{OR} = \frac{1}{OR_1} + \frac{1}{OR_2} + \ldots + \frac{1}{OR_n}$ Show that the locus of R is a straight line.

Solution: Let the equation of the given lines be $a_ix + b_iy + c_i = 0$, $i = 1, 2, \ldots, n$ and the point O be the origin (0, 0). Then equation of line through O can be written

as $\frac{x}{\cos\theta} = \frac{y}{\sin\theta} = r$;

where θ is the angle made by the line with the positive direction of x-axis and r is the distance of any point on the line from the origin O.

Let $r, r_1, r_2, \ldots, r_n$ be the distances of the points. $R, R_1, R_2, \ldots R_n$ from O.

$\Rightarrow OR = r$ and $OR_i = r_i$ $(i = 1, 2, \ldots, n)$

Then co-ordinates of R are $(r\cos\theta, r\sin\theta)$ and of R_i are $(r_i\cos\theta, r_i\sin\theta)$; $i = 1, \ldots, n$

Since R_i lies on $a_ix + b_iy + c_i = 0$

$\Rightarrow a_ir_i\cos\theta + b_ir_i\sin\theta + c_i = 0$ for $i = 1, 2, \ldots n$.

$$\Rightarrow -\frac{a_i}{c_i}\cos\theta - \frac{b_i}{c_i}\sin\theta = \frac{1}{r_i}, \; i = 1, 2, \ldots, n$$

$$\Rightarrow \sum_{i=1}^{n}\frac{1}{r_i} = -\left[\sum_{i=1}^{n}\frac{a_i}{c_i}\right]\cos\theta - \left[\sum_{i=1}^{n}\frac{b_i}{c_i}\right]\sin\theta$$

$$\Rightarrow \frac{n}{r} = -\left[\sum_{i=1}^{n}\frac{a_i}{c_i}\right]\cos\theta - \left[\sum_{i=1}^{n}\frac{b_i}{c_i}\right]\sin\theta$$

$$\Rightarrow \left(\sum_{i=1}^{n}\frac{a_i}{c_i}\right)r\cos\theta + \left(\sum_{i=1}^{n}\frac{b_i}{c_i}\right)r\sin\theta + n = 0$$

Hence the locus of R is $\left(\sum_{i=1}^{n}\frac{a_i}{c_i}\right)x + \left(\sum_{i=1}^{n}\frac{b_i}{c_i}\right)y + n = 0$

which is a straight line.

21. Find the locus of a point which moves such that the sum of the perpendicular distances from it on two given straight lines is a constant.

Solution: We may take one of the two given straight lines to be our x-axis and the point of intersection of the given lines as our origin. Observe that we have the freedom of choosing our axes, depending on the problem. (Reader, this is where we intelligently exploit the convenience of co-ordinate geometry). By our choice of the axes one of the given lines is $y = 0$ and let the other be $mx - y = 0$. If $P(x', y')$ is the variable point, its distances from the two given lines are $|y'|$ and $\frac{|mx' - y'|}{\sqrt{1+m^2}}$.

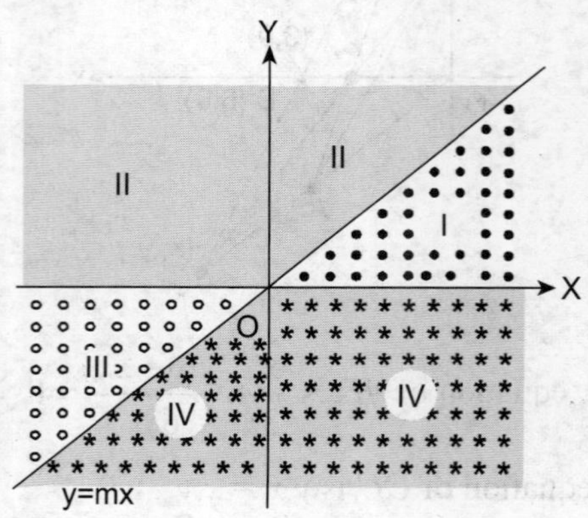

We are given that P moves such that $|y'| + \dfrac{|mx' - y'|}{\sqrt{1+m^2}} = k$ = constant.

$\therefore$ The locus of P is $\sqrt{1+m^2}\ |y| + |mx - y| = k\sqrt{1+m^2}$.

In region I where $y > 0$, $mx > y$ the locus is the straight line $mx + (\sqrt{1+m^2} - 1)\,y - k\sqrt{1+m^2} = 0$.

In region II where $y > 0$, $mx < y$ the locus is the straight line $(1+\sqrt{1+m^2})\,y - mx = k\sqrt{1+m^2}$.

In region III, where $y < 0$, $mx < y$ the locus is the straight line $(1-\sqrt{1+m^2})\,y - mx = k\sqrt{1+m^2}$.

In region IV, the locus is

$mx - (1+\sqrt{1+m^2})\,y = k\sqrt{1+m^2}$.

22. Find the incentre of the triangle whose sides have the equations $x + y - 7 = 0$, $x - y + 1 = 0$ and $x - 3y + 5 = 0$.

Solution: Let $I\ (x_1, y_1)$ be the incentre of ΔABC whose sides have the given equations. Then the perpendicular distance of I from BC is $= \dfrac{|x_1 + y_1 - 7|}{\sqrt{2}}$.

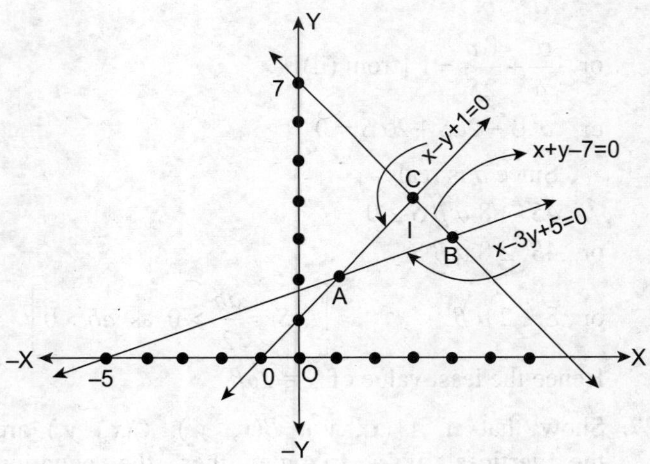

From the figure, it is clear that (0, 0) and I are on the same sides of BC.

Hence $x_1 + y_1 - 7 < 0$

$\therefore\ r = \dfrac{-(x_1 + y_1 - 7)}{\sqrt{2}}$. Similarly, the distance of I

from $CA = \dfrac{|x_1 - y_1 + 1|}{\sqrt{2}} = \dfrac{(x_1 - y_1 + 1)}{\sqrt{2}}$

(Since O and I are on the same sides of $x - y + 1 = 0$.)

The distance of I from AB is

$$r = \frac{|x_1 - 3y_1 + 5|}{\sqrt{1+9}} = -\frac{(x_1 - 3y_1 + 5)}{\sqrt{10}}$$

Since O and I are on the opposite side of AB

Thus,

$$r = \frac{-(x_1 + y_1 - 7)}{\sqrt{2}} = \frac{(x_1 - y_1 + 1)}{\sqrt{2}} = \frac{-(x_1 - 3y_1 + 5)}{\sqrt{10}}$$

which leads to $x_1 = 3$, $y_1 = 1 + \sqrt{5}$.

$\therefore$ The incentre is $(3, 1 + \sqrt{5})$.

23. The straight line $lx + my + n = 0$ bisects an angle between a pair of lines of which $px + qy + r = 0$ is one. Find the other line.

Solution: The required line is of the form $(px + qy + r) + k(lx + my + n) = 0$.

For any point $P(x', y')$ on the bisector $lx + my + n = 0$, we must have $PA = PB$

$$\therefore\ \frac{|px' + qy' + r|}{\sqrt{p^2 + q^2}} = \frac{|px' + qy' + r|}{\sqrt{(p + kl)^2 + (q + km)^2}}$$

(Since $lx' + my' + n = 0$)

$\Rightarrow p^2 + q^2 = (p + kl)^2 + (q + km)^2$

$\therefore\ (l^2 + m^2)\,k^2 + 2(pl + qm)\,k = 0$

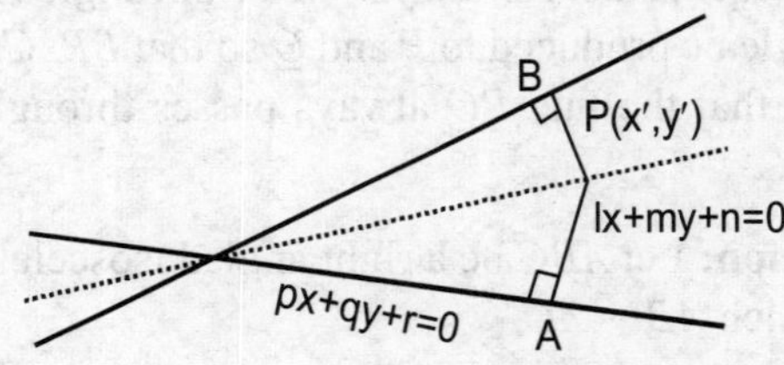

Hence $k = 0$ or $k = -2\dfrac{(pl + qm)}{l^2 + m^2}$

Here $k \neq 0$ (Why ?) and hence $k = -2\dfrac{(pl + qm)}{l^2 + m^2}$.

This gives the required line as

$$px + qy + r - 2\frac{(pl + qm)}{l^2 + m^2}(lx + my + n) = 0$$

or $(l^2 + m^2)(px + qy + r) - 2(pl + qm)(lx + my + n) = 0$

24. $ABCD$ is a variable rectangle having its sides parallel to fixed directions. The vertices B and D lie on $x = a$ and $x = -a$ and A lies on the line $y = 0$. Find the locus of C.

Solution: Let A be $(x_1, 0)$, B be (a, y_2) and D be $(-a, y_4)$. We are given AB and AD have fixed directions and hence their slopes are constants, say m_1 and m_2 respectively.

$$\therefore\ \frac{y_2}{a - x_1} = m_1 \text{ and } \frac{y_4}{-a - x_1} = m_2$$

Further, $m_1 m_2 = -1$

$$\Rightarrow \frac{y_2}{a - x_1} = m_1 \text{ and } \frac{y_4}{-a - x_1} = -\frac{1}{m_1}$$

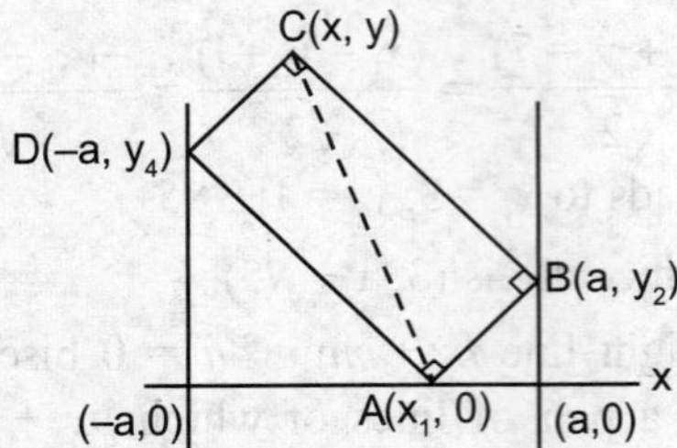

The mid-point of BD is $\left(0, \frac{y_2 + y_4}{2}\right) = \left(\frac{x_1 + x}{2}, \frac{y}{2}\right)$

= mid-point of AC (where C is taken to be (x, y))

This gives $x = -x_1$ and $y = y_2 + y_4$. so C is $(-x_1, y_2 + y_4)$.

Also $\frac{y_2}{a - x_1} = m_1$ and $\frac{y_4}{a + x_1} = +\frac{1}{m_1}$ gives the locus of C as:

$(m_1^2 - 1)x + m_1 y = (m_1^2 + 1)a$; m_1 is slope of AB

25. If the equal sides AB and AC of a right angled isosceles triangle be produced to P and Q so that $BP.\ CQ = AB^2$, show that the line PQ always passes through a fixed point.

Solution: Let ABC be a right angled isosceles triangle in which $AB = AC$.

We take A as the origin and AB and AC as x and y axis respectively.

Let $AB = AC = a$ and $AP = h$, $AQ = k$

Now equation of line PQ will be $\frac{x}{h} + \frac{y}{k} = 1$ ...(i)

Given $BP.CQ = AB^2$ or $(h - a)(k - a) = a^2$

or $hk - ak - ah = 0$ or $ak + ha = hk$

or $\frac{a}{h} + \frac{a}{k} = 1$ ...(ii)

From (ii) it follows that line (i) i.e., line PQ passes through the fixed point (a, a).

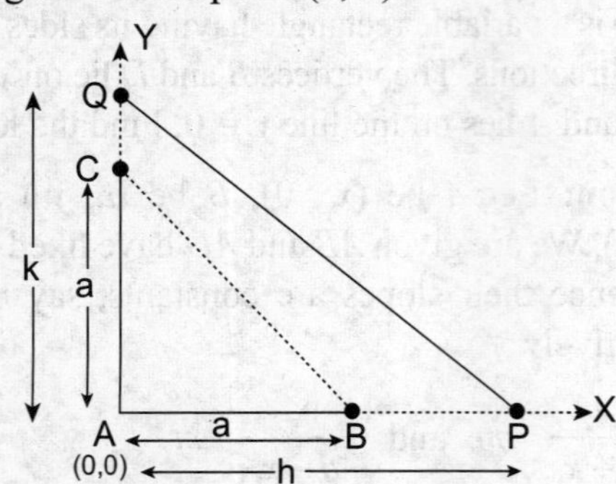

26. Through the point $P(\alpha, \beta)$ where $\alpha\beta > 0$ the straight line $\frac{x}{a} + \frac{y}{b} = 1$ is drawn so as to form with co-ordinate axes a triangle of area S. If $ab > 0$, find the least value of S.

Solution: Given $P \equiv (\alpha, \beta)$ and given line is $\frac{x}{a} + \frac{y}{b} = 1$ ...(i)

If line (i) cuts x and y axes at A and B respectively, then $A \equiv (a, 0)$ and $B \equiv (0, b)$

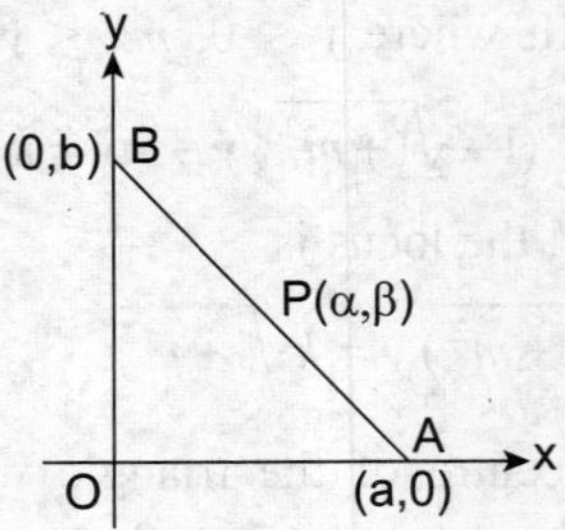

Given area of $\Delta OAB = S$

$\therefore \left|\frac{1}{2}ab\right| = S$ or $ab = 2S$ [$ab > 0$] ...(ii)

Since line (i) passes through $P(\alpha, \beta)$

$\therefore \frac{\alpha}{a} + \frac{\beta}{b} = 1$

or $\frac{\alpha}{a} + \frac{\beta a}{2S} = 1$ [from (ii)]

or $a^2\beta - 2aS + 2\alpha S = 0$

Since a is real

$\therefore 4S^2 - 8\alpha\beta S \geq 0$

or $4S^2 \geq 8\alpha\beta S$

or $S \geq 2\alpha\beta$ $\left[\because S = \frac{ab}{2} > 0 \text{ as } ab > 0\right]$

Hence the least value of $S = 2\alpha\beta$

27. Show that if $A(x_1, y_1)$; $B(x_2, y_2)$; $C(x_3, y_3)$ are the vertices of a triangle, then the equation of the internal bisector of angle A is given by

$$b\begin{vmatrix} x & y & 1 \\ x_1 & y_1 & 1 \\ x_2 & y_2 & 1 \end{vmatrix} + c\begin{vmatrix} x & y & 1 \\ x_1 & y_1 & 1 \\ x_3 & y_3 & 1 \end{vmatrix} = 0 \text{ where } b = AC$$

and $c = AB$.

Solution: Let AD be the internal bisector of $\angle BAC$,

then $\frac{BD}{DC} = \frac{AB}{AC} = \frac{c}{b}$

Thus D divides BC internally in the ratio $c : b$, therefore

$$D \equiv \left(\frac{bx_2 + cx_3}{b+c}, \frac{by_2 + cy_3}{b+c}\right)$$

Let the equation of line AD be $lx + my + n = 0$ ...(i)

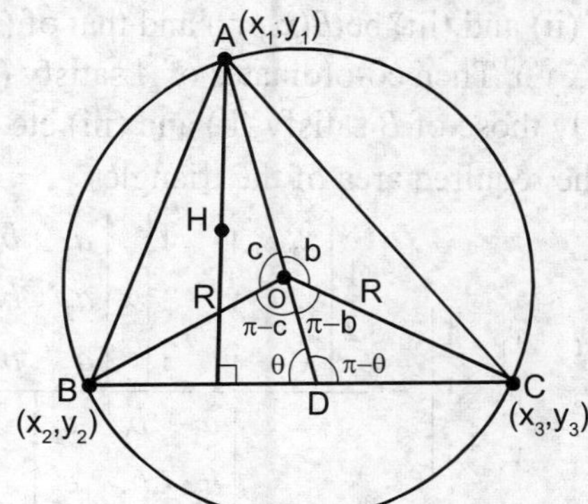

Since A and D lie on line AD $\therefore$ $lx_1 + my_1 + n = 0$ ....(ii)

and $l\left(\frac{bx_2 + cx_3}{b+c}\right) + m\left(\frac{by_2 + cy_3}{b+c}\right) + n = 0$

or $l(bx_2 + cx_3) + m(by_2 + cy_3) + n(b + c) = 0$ ...(iii)

Eliminating l, m, n from equation (i), (ii) and (iii)

we get $\begin{vmatrix} x & y & 1 \\ x_1 & y_1 & 1 \\ bx_2 + cx_3 & by_2 + cy_3 & b+c \end{vmatrix} = 0$

$$\Rightarrow \begin{vmatrix} x & y & 1 \\ x_1 & y_1 & 1 \\ bx_2 & by_2 & b \end{vmatrix} + \begin{vmatrix} x & y & 1 \\ x_1 & y_1 & 1 \\ cx_3 & cy_3 & c \end{vmatrix} = 0$$

$$\Rightarrow b\begin{vmatrix} x & y & 1 \\ x_1 & y_1 & 1 \\ x_2 & y_2 & 1 \end{vmatrix} + c\begin{vmatrix} x & y & 1 \\ x_1 & y_1 & 1 \\ x_3 & y_3 & 1 \end{vmatrix} = 0$$

28. The base of a triangle passes through a fixed point [(f, g) and its sides are respectively bisected at right angles by the lines $y^2 - 8xy - 9x^2 = 0$. Determine the locus of its vertex.

Solution: Given equation of lines is $y^2 - 8xy - 9x^2 = 0$

$\Rightarrow y^2 - 9xy + xy - 9x^2 = 0$

$\Rightarrow (x + y)(y - 9x) = 0$

$\Rightarrow y + x = 0$ ... (i)

and $\Rightarrow y - 9x = 0$ ... (ii)

Let ABC be the triangle with vertex $A(\alpha, \beta)$ and $B(x_1, y_1)$, $C(x_2, y_2)$.

Locus of point $A(\alpha, \beta)$ is required where given conditions are

(i) Line $y + x = 0$ is the perpendicular bisector of AB.

(ii) Line $y - 9x = 0$ is the perpendicular bisector of AC.

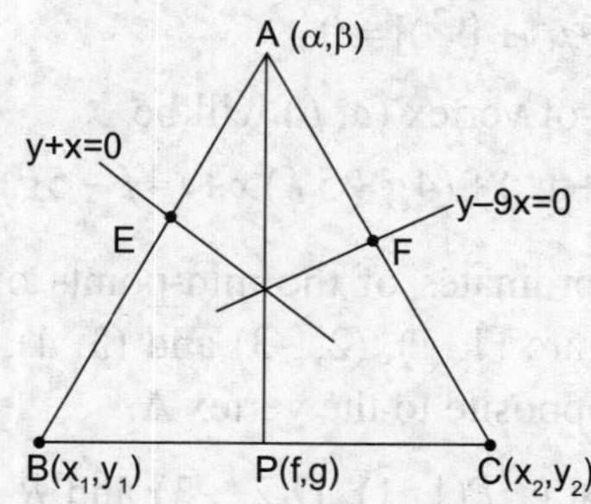

Using (i), co-ordinates of mid-point $E\left(\frac{\alpha + x_1}{2}, \frac{\beta + y_1}{2}\right)$ will satisfy line (i),

$\Rightarrow x_1 + y_1 + (\alpha + \beta) = 0$ ... (iii)

and $\left(\frac{\beta - y_1}{\alpha - x_1}\right)(-1) = -1$

$\Rightarrow x_1 - y_1 + (\beta - \alpha) = 0$ ... (iv)

solving (iii) and (iv) we get $x_1 = -\beta$, $y_1 = -\alpha$

Next using (ii), co-ordinates of mid-point $F\left(\frac{\alpha + x_2}{2}, \frac{\beta + y_2}{2}\right)$ will satisfy line (ii)

$\Rightarrow \frac{\beta + y_2}{2} - 9\left(\frac{\alpha + x_2}{2}\right) = 0$

$\Rightarrow 9x_2 - y_2 = \beta - 9\alpha$ ... (v)

and $\left(\frac{\beta - y_2}{\alpha - x_2}\right)(9) = -1$

$\Rightarrow x_2 + 9y_2 = 9\beta + \alpha$ ... (vi)

solving (v) and (vi), we get

$$x_2 = \frac{9\beta - 40\alpha}{41},\ y_2 = \frac{9\alpha + 40\beta}{41}.$$

To get a relation in α, β, we need to eliminate x_1, y_1, x_2, y_2 from above results; and so we need a relation in all these six elements.

Since points P, B, C are collinear, therefore,

$$\begin{vmatrix} f & g & 1 \\ x_1 & y_1 & 1 \\ x_2 & y_2 & 1 \end{vmatrix} = 0$$

i.e., $f(y_1 - y_2) - g(x_1 - x_2) + (x_1y_2 - x_2y_1) = 0$

$$\Rightarrow f\left[-\alpha - \frac{9\alpha + 40\beta}{41}\right] - g\left[-\beta - \frac{9\beta - 40\alpha}{41}\right]$$

$$-\left[\beta\left(\frac{9\alpha + 40\beta}{41}\right) - \alpha\left(\frac{9\beta - 40\alpha}{41}\right)\right] = 0$$

$$\Rightarrow f[-50\alpha - 40\beta] - g[-50\beta + 40\alpha]$$
$$-[40(\alpha^2 + \beta^2)] = 0.$$

$\therefore$ Locus of vertex (α, β) will be
$$4(x^2 + y^2) + (4g + 5f)x + (4f - 5g)y = 0$$

29. If the co-ordinates of the mid-points of the sides of a triangle are (1, 1), (2, –3) and (3, 4), then find the excentre opposite to the vertex A.

Solution: Let $P(1, 1)$, $Q(2, -3)$ and R (3, 4) are the mid-points of the sides BC, CA and AB respectively. Let A (x_1, y_1), $B(x_2, y_2)$ and $C(x_3, y_3)$

Dignoals of parallalogram biseet each other

$x_1 = 3 + 2 - 1 = 4$

$y_1 = 4 - 3 - 1 = 0$

So co-ordinates of A are (4, 0)

Similarty co-ordinates of B and C are veopertively (2, 8) and (0, –6)

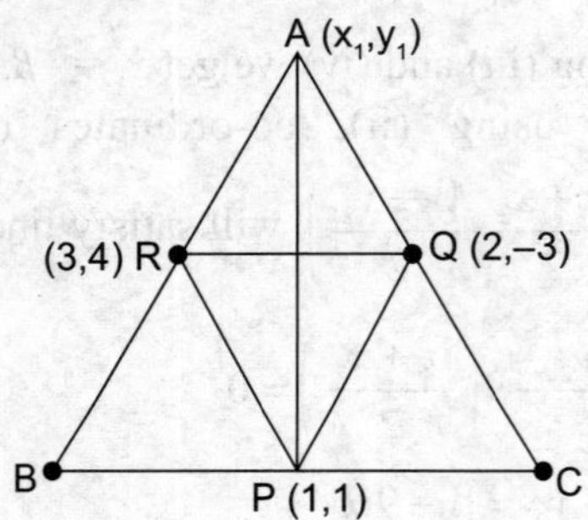

So co-ordinates of C are (0, – 6)

$$a = BC = \sqrt{(2-0)^2 + (8+6)^2} = 10\sqrt{2}\,;$$

$$b = CA = \sqrt{(0-4)^2 + (-6-0)^2} = 2\sqrt{13} \text{ and}$$

$$c = AB = \sqrt{(4-2)^2 + (0-8)^2} = 2\sqrt{17}$$

The co-ordinates of the excentre opposite to A are

$$\left(\frac{-ax_1 + bx_2 + cx_3}{-a+b+c}, \frac{-ay_1 + by_2 + cy_3}{-a+b+c}\right)$$

$$\equiv \left(\frac{-20\sqrt{2} + 2\sqrt{13}}{-5\sqrt{2} + \sqrt{13} + \sqrt{17}}, \frac{8\sqrt{13} - 6\sqrt{17}}{-5\sqrt{2} + \sqrt{13} + \sqrt{17}}\right)$$

30. Show that the area of the triangle whose sides are $a_r x + b_r y + c_r = 0$, $r = 1, 2, 3$ is $\dfrac{\varnothing^2}{2|C_1C_2C_3|}$, where C_1, C_2 and C_3 are the co-factors of c_1, c_2 and c_3 respectively in the determinant $\begin{vmatrix} a_1 & b_1 & c_1 \\ a_2 & b_2 & c_2 \\ a_3 & b_3 & c_3 \end{vmatrix}$

Solution: Given lines are $a_1x + b_1y + c_1 = 0$ ....(i)

$a_2x + b_2y + c_2 = 0$ ...(ii)

and $a_3x + b_3y + c_3 = 0$ ...(iii)

Let the point of intersections of (i) and (ii) be $A(x_1, y_1)$, that of (ii) and (iii) be $B(x_2, y_2)$ and that of (iii) and (i) be $C(x_3, y_3)$. Then co-ordinates of A satisfy (i) and (ii). similarly those of B satisfy (ii) and (iii) etc.

Now the required area of the triangle

$$= \frac{1}{2}\begin{vmatrix} x_1 & y_1 & 1 \\ x_2 & y_2 & 1 \\ x_3 & y_3 & 1 \end{vmatrix} = \frac{\frac{1}{2}\begin{vmatrix} x_1 & y_1 & 1 \\ x_2 & y_2 & 1 \\ x_3 & y_3 & 1 \end{vmatrix} \times \begin{vmatrix} a_1 & b_1 & c_1 \\ a_2 & b_2 & c_2 \\ a_3 & b_3 & c_3 \end{vmatrix}}{\begin{vmatrix} a_1 & b_1 & c_1 \\ a_2 & b_2 & c_2 \\ a_3 & b_3 & c_3 \end{vmatrix}}$$

[Multiplying the numerator and denominator by the same determinant]

$$= \frac{1}{2}\begin{vmatrix} a_1x_1 + b_1y_1 + c_1 & a_2x_1 + b_2y_1 + c_2 & a_3x_1 + b_3y_1 + c_3 \\ a_1x_2 + b_1y_2 + c_1 & a_2x_2 + b_2y_2 + c_2 & a_3x_2 + b_3y_2 + c_3 \\ a_1x_3 + b_1y_3 + c_1 & a_2x_3 + b_2y_3 + c_2 & a_3x_3 + b_3y_3 + c_3 \end{vmatrix} \div \Delta$$

where Δ is the determinant in the denominator of previous step.

$$= \frac{1}{2}\begin{vmatrix} 0 & 0 & a_3x_1 + b_3y_1 + c_3 \\ a_1x_2 + b_1y_2 + c_1 & 0 & 0 \\ 0 & a_2x_3 + b_2y_3 + c_2 & 0 \end{vmatrix} \div \Delta$$

Since $A(x_1, y_1)$ satisfies (i) and (ii); B (x_2, y_2) satisfies (ii) and (iii) etc.

$$|\,[\frac{1}{2}(a_1x_2 + b_1y_2 + c_1)(a_2x_3 + b_2y_3 + c_2)$$
$$(a_3x_1 + b_3y_1 + c_3) \div \Delta\,| \quad .... \text{(A)}$$

Now (x_1, y_1) satisfies (i) and (ii) so we have

$a_1x_1 + b_1y_1 + c_1 = 0$ ....(iv)

$a_2x_1 + b_2y_1 + c_2 = 0$ .....(v)

and $a_3x_1 + b_3y_1 + c_3 = \lambda_1$

i.e., $a_3x_1 + b_3y_1 + (c_3 - \lambda_1) = 0$.... (vi)

Eliminating x_1, y_1 from (iv), (v) and (vi) we get

$$\begin{vmatrix} a_1 & b_1 & c_1 \\ a_2 & b_2 & c_2 \\ a_3 & b_3 & c_3 - \lambda_1 \end{vmatrix} = 0$$

$$\Rightarrow \begin{vmatrix} a_1 & b_1 & c_1 \\ a_2 & b_2 & c_2 \\ a_3 & b_3 & c_3 \end{vmatrix} + \begin{vmatrix} a_1 & b_1 & 0 \\ a_2 & b_2 & 0 \\ a_3 & b_3 & -\lambda_1 \end{vmatrix} = 0$$

$$\Rightarrow \Delta - \lambda_1 \begin{vmatrix} a_1 & b_1 \\ a_2 & b_2 \end{vmatrix} = 0$$

$\Rightarrow \Delta - \lambda_1 C_3 = 0$

$\Rightarrow \lambda_1 = \Delta/C_3$; where C_3 is the co-factor of c_3 in Δ

Similarly, taking $a_1x_2 + b_1y_2 + c_1 = \lambda_2$

and $a_2x_3 + b_2y_3 + c_2 = \lambda_3$

we can prove as above that $\lambda_2 = \Delta/C_1$ and $\lambda_3 = \Delta/C_2$

From (A) the required area of the $\Delta = \left|\frac{1}{2}(\lambda_1\lambda_2\lambda_3) \div \Delta\right|$

$$= \left|\frac{1}{2}\left[\frac{\Delta}{C_1}\cdot\frac{\Delta}{C_2}\cdot\frac{\Delta}{C_3}\right] \div \Delta\right| = \frac{\varnothing^2}{2|C_1C_2C_3|}$$

Hence area of the triangle $= \dfrac{\varnothing^2}{2|C_1C_2C_3|}$

31. Let ABC be a triangle with $AB = AC$. If D is the mid-point of BC, E the foot of the perpendicular drawn from D to AC and F the mid-point of DE, prove that AF is perpendicular to BE.

Solution: Given $AB = AC$ and $BD = DC$; where $\angle ADC = 90°$

We take D as the origin and DC and DA as x and y axes respectively

Let $BC = 2\alpha$ and $AD = \beta$, then

$B \equiv (-\alpha, 0)$, $C \equiv (\alpha, 0)$ and $A \equiv (0, \beta)$.

Now equation of AC is $\dfrac{x}{\alpha} + \dfrac{y}{\beta} = 1$ ....(i)

Since $DE \perp AC$ and it passes through (0, 0)

$\therefore$ equation of DE is $\dfrac{x}{\beta} - \dfrac{y}{\alpha} = 0$ ....(ii)

Solving (i) and (ii), we get $E \equiv \left(\dfrac{\alpha\beta^2}{\alpha^2+\beta^2}, \dfrac{\alpha^2\beta}{\alpha^2+\beta^2}\right)$

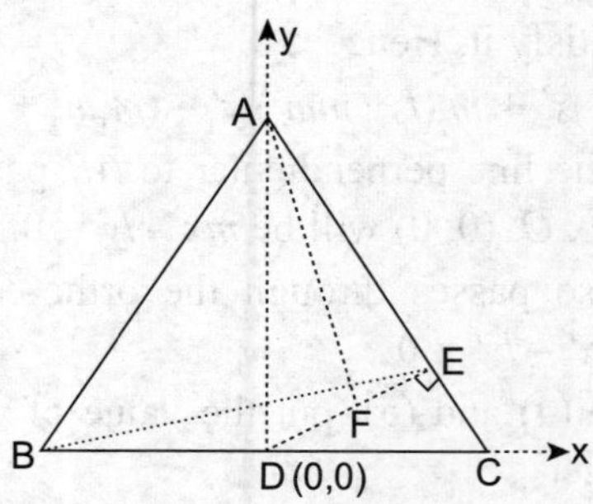

Since F is the mid-point of DE

$$\therefore\ F \equiv \left(\frac{\alpha\beta^2}{2(\alpha^2+\beta^2)}, \frac{\alpha^2\beta}{2(\alpha^2+\beta^2)}\right)$$

$$\text{Slope of } AF = \frac{\dfrac{\alpha^2\beta}{2(\alpha^2+\beta^2)} - \beta}{\dfrac{\alpha\beta^2}{2(\alpha^2+\beta^2)} - 0}$$

$$= \frac{-(2\beta^2+\alpha^2)}{\alpha\beta} = m_1 \text{ (say)}$$

$$\text{Slope of } BE = \frac{\dfrac{\alpha^2\beta}{\alpha^2+\beta^2} - 0}{\dfrac{\alpha\beta^2}{(\alpha^2+\beta^2)} + \alpha}$$

$$= \frac{\alpha\beta}{2\beta^2+\alpha^2} = m_2 \text{ (say)}$$

Clearly $m_1 m_2 = -1$; Hence $AF \perp BE$.

32. A variable line cuts the axes of co-ordinates in point A and B such that $OA + OB =$ c. Find the locus of foot of perpendicular from origin to the line.

Solution: Let the co-ordinates of A and B be $(a, 0)$ and $(0, b)$ respectively. Therefore the equation of variable line is given by $\dfrac{x}{a} + \dfrac{y}{b} = 1$; where $a + b = c$ ... (i)

If (h, k) be the foot of perpendicular from (0, 0), as shown ahead:

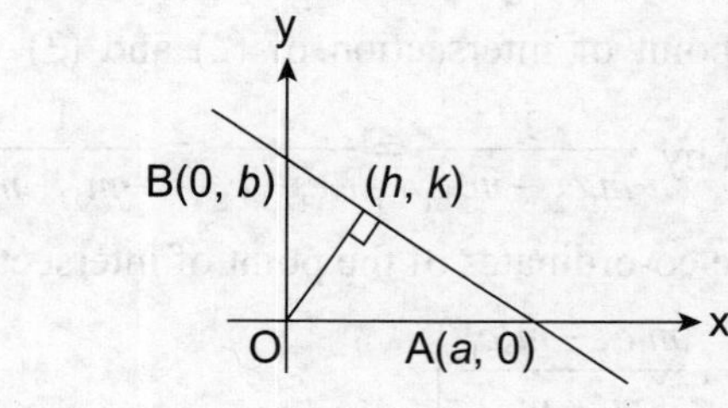

$$\Rightarrow \frac{h-0}{1/a} = \frac{k-0}{1/b} = \frac{-(-1)}{\left(\dfrac{1}{a^2}+\dfrac{1}{b^2}\right)} = \frac{a^2b^2}{a^2+b^2}$$

$$\therefore\ h = \frac{ab^2}{a^2+b^2};\ k = \frac{a^2b}{a^2+b^2} \quad \text{.... (ii)}$$

In order to find the locus, we have to eliminate a, b between (i), (ii)

From (ii)

$$h^2 + k^2 = \frac{(ab^2)^2}{(a^2+b^2)^2} + \frac{(a^2b)^2}{(a^2+b^2)^2} = \frac{a^2b^2}{(a^2+b^2)} \quad \text{....(iii)}$$

Also

$$\frac{1}{h} + \frac{1}{k} = \frac{a^2+b^2}{ab}\left[\frac{1}{a}+\frac{1}{b}\right] = \frac{a^2+b^2}{a^2b^2}(a+b) = c.\frac{a^2+b^2}{a^2b^2}$$

$$\Rightarrow \frac{a^2b^2}{(a^2+b^2)}\left(\frac{1}{h}+\frac{1}{k}\right) = c \quad \text{..... (iv)}$$

$$\Rightarrow (h^2+k^2)\left(\frac{1}{h}+\frac{1}{k}\right) = c \text{ by (iii) and (iv).}$$

$\therefore$ Required locus is $(x^2+y^2)\left(\dfrac{1}{x}+\dfrac{1}{y}\right) = c$

33. If the equation $ax^2 + 2hxy + by^2 + 2gx + 2fy + c = 0$, represent two straight lines, prove that the square of the distance of their point of intersection from the origin is $\dfrac{c(a+b)-f^2-g^2}{ab-h^2}$.

Solution: Let the lines represented by $ax^2 + 2hxy + by^2 + 2gx + 2fy + c = 0$ ... (1)

be $y - m_1x - c_1 = 0$ ... (2)

and $y - m_2x - c_2 = 0$ ... (3)

The combined equation must be similar to (1), so

$ax^2 + 2hxy + by^2 + 2gx + 2fy + c = b(y - m_1x - c_1)(y - m_2x - c_2)$

(We multiply R.H.S. by b so as to make the co-efficient of term say y^2 on R.H.S equal to co-efficient of y^2 on L.H.S., so the co-efficients must also be equal). So equating other co-efficients,

we get $m_1 + m_2 = -\dfrac{2h}{b}$; $m_1m_2 = \dfrac{a}{b}$

$\Rightarrow c_1m_2 + c_2m_1 = \dfrac{2g}{b}$; $c_1 + c_2 = \dfrac{-2f}{b}$ and $c_1c_2 = \dfrac{c}{b}$.

The point of intersection of (2) and (3) will be given by $\dfrac{y}{m_1c_2 - m_2c_1} = \dfrac{x}{-c_1+c_2} = \dfrac{1}{-m_2+m_1}$

Hence the co-ordinates of the point of intersection are

$$\left(\frac{c_2-c_1}{m_1-m_2}, \frac{m_1c_2-m_2c_1}{m_1-m_2}\right)$$

The square of the distance of this point from the origin (0, 0) is

$$\left[\frac{c_2-c_1}{m_1-m_2}\right]^2 + \left[\frac{m_1c_2-m_2c_1}{m_1-m_2}\right]^2$$

$$= \frac{\left[(c_1+c_2)^2 - 4c_1c_2\right] + (m_1c_2+m_2c_1)^2 - 4m_1m_2c_1c_2}{(m_1+m_2)^2 - 4m_1m_2}$$

$$= \frac{\left[\frac{4f^2}{b^2} - \frac{4c}{b}\right] + \left[\frac{4g^2}{b^2} - \frac{4ac}{b^2}\right]}{\frac{4h^2}{b^2} - \frac{4a}{b}}$$

$$= \frac{f^2 - bc + g^2 - ac}{h^2 - ab} = \frac{f^2+g^2-c(b+a)}{h^2-ab}$$

$$= \frac{c(a+b)-f^2-g^2}{ab-h^2}.$$

34. Show that the orthocentre of the triangle formed by the straight lines $ax^2 + 2hxy + by^2 = 0$ and $lx + my = 1$ is a point (x', y') such that $\dfrac{x'}{l} = \dfrac{y'}{l} = \dfrac{a+b}{l^2b - 2hlm + am^2}$.

Solution: Let the equation $ax^2 + 2hxy + by^2 = 0$ represent OA and OB respectively as $y = m_1x$ ... (1)

and $y = m_2x$ ... (2), Then $m_1 + m_2 = -\dfrac{2h}{b}$

and $m_1m_2 = \dfrac{a}{b}$

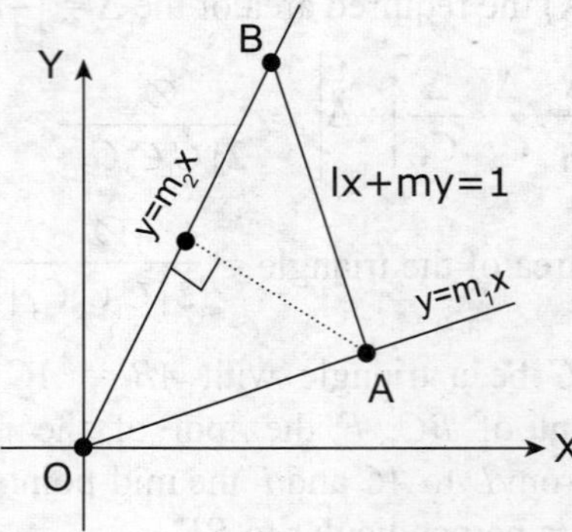

Let the line AB is given by $lx + my = 1$ ... (3)

Solving (1) and (3), we get $x = \dfrac{1}{l+mm_1}$ $y = \dfrac{m_1}{l+mm_1}$

Hence the co-ordiantes of A are $\left(\dfrac{1}{l+mm_1}, \dfrac{m_1}{l+mm_1}\right)$

and $B \equiv \left(\dfrac{1}{l+mm_2}, \dfrac{m_2}{l+mm_2}\right)$

The equation of the line perpendicular to OB passing through opposite vertex A, will be

$$y - \frac{m_1}{l+mm_1} = -\frac{1}{m_2}\left(x - \frac{1}{l+mm_1}\right)$$

or $(l + mm_1)x + m_2(l + mm_1)y - (m_1m_2 + 1) = 0$.

As the line passes through the orthocentre (x', y'), so it will satisfy it. Hence

$(l + mm_1)x' + m_2(l + mm_1)y' - (m_1m_2 + 1) = 0$ ... (4)

Again, the line perpendicular to AB passing through the vertex O. (0, 0) will be $mx - ly = 0$.

As it also passes through the ortho-centre (x', y'), hence $mx' - ly' = 0$ ... (5)

To solve (4) and (5), put the value of y' from (5) in (4); we get

$$(l+mm_1)x' + m_2(l+mm_1)\frac{mx'}{l} - (m_1m_2+1) = 0$$

or $\dfrac{x'}{l}\left[l^2 + lmm_1 + mm_2l + m^2m_1m_2\right] = 1 + m_1m_2$

or $\dfrac{x'}{l}\left[l^2 + lm(m_1+m_2) + m^2m_1m_2\right] = 1 + m_1m_2$

Putting the values of $m_1 + m_2$ and m_1m_2 *as* $-\dfrac{2h}{b}$

and $\frac{a}{b}$ respectively, $\frac{x'}{l}\left[l^2 - lm\frac{2h}{b} + m^2\frac{a}{b}\right] = 1 + \frac{a}{b}$

or $\frac{x'}{l} = \frac{a+b}{l^2 b - 2hlm + am^2}$

Similarly, putting the value of x' from (5) in (4), we get $\frac{y'}{l} = \frac{a+b}{l^2 b - 2hlm + am^2}$

Hence $\frac{x'}{l} = \frac{y'}{l} = \frac{a+b}{l^2 b - 2hlm + am^2}$

35. Find the locus of the orthocentre of a triangle of which two sides are given in position and whose third side goes through a fixed point.

Solution: Let the line $lx + my = 1$ given in last question passes through some fixed point $P \equiv (\alpha, \beta)$. Then $l\alpha + m\beta = 1$ ... (1)

If (x', y') be the orthocentre, then

$$\frac{x'}{l} = \frac{y'}{m} = \frac{a+b}{am^2 - 2hlm + bl^2} = \frac{x'+y'}{l+m}. \quad ....(2)$$

As $\frac{x'}{l} = \frac{y'}{m} = \frac{\alpha x'}{\alpha l} = \frac{\beta y'}{\beta m} = \frac{\alpha x' + \beta y'}{\alpha l + \beta m} = \alpha x' + \beta y'$

[as $\alpha l + \beta m = 1$ by (1)]

$\therefore \quad \frac{x'}{l} = \frac{y'}{m} = \alpha x' + \beta y'$.

$$l = \frac{x'}{\alpha x' + \beta y'}, \; m = \frac{y'}{\alpha x' + \beta y'}$$

Putting these values in (2), we get

$$\frac{a+b}{a\left(\frac{y'}{\alpha x' + \beta y'}\right)^2 - 2h\frac{x'}{\alpha x' + \beta y'}\frac{y'}{\alpha x' + \beta y'} + b\left(\frac{x'}{\alpha x' + \beta y'}\right)^2}.$$

$$= \frac{x'+y'}{\frac{x'}{\alpha x' + \beta y'} + \frac{y'}{\alpha x' + \beta y'}}$$

$$\Rightarrow \frac{(a+b)(\alpha x' + \beta y')^2}{a y'^2 - 2hx'y' + bx'^2} = \frac{(x'+y')(\alpha x' + \beta y')}{(x'+y')}$$

Simplifying and generalising for x' and y', we get the required locus as $(a + b)(\alpha x + \beta y) = ay^2 - 2hxy + bx^2$.

36. Show that two of the straight lines represented by the equation $ay^4 + bxy^3 + cx^2y^2 + dx^3y + ex^4 = 0$ will be at right angles if $(b + d)(ad + be) + (e - a)^2 (a + c + e) = 0$.

Solution: The given equation is $ay^4 + bxy^3 + cx^2y^2 + dx^3y + ex^4 = 0$ .... (1)

Which is homogeneous of degree form.

If two of the lines represented by (1) be at right angles, their equation must be of the form $x^2 + pxy - y^2 = 0$ as the sum of the co-efficients of x^2 and y^2 is zero. Hence let

$ay^4 + bxy^3 + cx^2y^2 + dx^3y + ex^4 \equiv (x^2 + pxy - y^2)(ex^2 + qxy - ay^2)$

As the co-efficient of x^4 on both sides is same, other co-efficients will also be equal.

Now compare different co-efficients.

Comparing xy^3, we have $-pa - q = b$ ... (2)

Comparing x^2y^2, we have $-a + pq - e = c$. ... (3)

Comparing x^3y, we have $q + ep = d$ ......(4)

Now eliminate p and q from (2), (3) and (4).

From (2) and (4); $p = \frac{b+d}{e-a}$ and $q = -\frac{(da+eb)}{e-a}$

Substituting these values in (3),

we get $-a - \frac{b+d}{e-a}.\frac{da+eb}{e-a} - e = c$

or $(a + e + c)(e - a)^2 + (b + d)(ad + be) = 0$

37. Prove that two of the lines represented by the equation $ax^4 + bx^3y + cx^2y^2 + dxy^3 + ay^4 = 0$ will bisect the angles between the other two if $c + 6a = 0$ and $b + d = 0$.

Solution: The equation is given by

$ax^4 + bx^3y + cx^2y^2 + dxy^3 + ay^4 = 0$ ... (1)

The equation being homogeneous equation of fourth degree will represent 4 straight lines through the origin.

Let one pair of these be $(px^2 + 2qxy + ry^2) = 0$ ... (2)

The equations of the bisectors of the angles between the lines given by (2) will be

$\frac{x^2 - y^2}{p - r} = \frac{xy}{q}$ or $q(x^2 - y^2) - (p - r)xy = 0$

$ax^4 + bx^3y + cx^2y^2 + dxy^3 + ay^4 \equiv (px^2 + 2qxy + ry^2)[qx^2 - qy^2 - (p - r)xy] = 0$

Comparing the co-efficient of different terms in order,

we get $\frac{a}{pq} = \frac{b}{-p^2 + pr + 2q^2} = \frac{c}{3q(r-p)}$

$$= \frac{d}{-2q^2 - pr + r^2} = -\frac{a}{qr}$$

By first and fifth members, we have $r = -p$. Putting in second and fourth ratios.

$$\frac{b}{-p^2-p^2+2q^2}=\frac{d}{-2q^2+p^2+p^2}$$

or $b = -d$ or $b + d = 0$.

Again by first and third ratios

$$\frac{a}{pq}=\frac{c}{3q.(-p-p)} \quad \text{or} \quad a=-\frac{c}{6} \quad \text{or } 6a+c=0$$

38. If the lines L_1 and L_2 and their angle bisectors are represented by the equation $ay^4 + bxy^3 + cx^2y^2 + dx^3y + ex^4 = 0$ and slope of L_1 is 1, show that slope of L_2 is $-e/a$ and hence show that $ea^4 - dea^3 + ce^2a^2 + a(e^4 - be^3) = 0$

Solution: Given equation is $ay^4 + bxy^3 + cx^2y^2 + dx^3y + ex^4 = 0$ ....(i)

(i) is a homogenous equation of degree four. It represents four lines passing through the origin i.e., of the form $y = mx$

From (i), we have

$$a(y/x)^4+b(y/x)^3+c(y/x)^2+d(y/x)+e=0$$

$\Rightarrow am^4 + bm^3 + cm^2 + dm + e = 0$ ...(ii)

$\Rightarrow m_1 m_2 m_3 m_4 = e/a$, where m_1, m_2, m_3 and m_4 are four roots of (ii)

Also let m_1, m_2 be the slopes of L_1 and L_2 respectively and m_3, m_4 be the slopes of angle bisectors.

Clearly, $m_3 m_4 = -1$. Given $m_1 = 1$, we get $m_2 = -e/a$

$\Rightarrow$ slope of $L_2 = -e/a$

Putting $m = -e/a$ in (ii), we have

$$a(-e/a)^4+b(-e/a)^3+c(-e/a)^2+d(-e/a)+e=0$$

$\Rightarrow ea^4 - dea^3 + ce^2a^2 + a(e^4 - be^3) = 0.$

39. Let a, b, c be real numbers with $a^2 + b^2 + c^2 = 1$. Show that the equation

$$\begin{vmatrix} ax-by-c & bx+ay & cx+a \\ bx+ay & -ax+by-c & cy+b \\ cx+a & cy+b & -ax-by+c \end{vmatrix}=0$$

represents a straight line.

Solution: $C_1 \to aC_1$

$$\Delta=\frac{1}{a}\begin{vmatrix} a^2x-aby-ac & bx+ay & cx+a \\ abx+a^2y & -ax+by-c & cy+b \\ acx+a^2 & cy+b & -ax-by+c \end{vmatrix}$$

$C_1 \to C_1 + bC_2 + CC_3$ gives

$$\Delta=\frac{1}{a}\begin{vmatrix} (a^2+b^2+c^2)x & ay+bx & cx+a \\ (a^2+b^2+c^2)y & by-c-ax & cy+b \\ (a^2+b^2+c^2) & b+cy & -ax-by+c \end{vmatrix}$$

$$=\frac{1}{a}\begin{vmatrix} x & ay+bx & cx+a \\ y & by-c-ax & b+cy \\ 1 & b+cy & c-ax-by \end{vmatrix}$$

($a^2 + b^2 + c^2 = 1$) given

$$\Rightarrow \Delta=\frac{1}{a}\begin{vmatrix} x & ay & a \\ y & -c-ax & b \\ 1 & cy & -ax-by \end{vmatrix} \quad (C_3 \to C_3 - cC_1)$$

$$=\frac{1}{ax}\begin{vmatrix} x^2 & axy & ax \\ y & -c-ax & b \\ 1 & cy & -ax-by \end{vmatrix}$$

$R_1 \to R_1 + yR_2 + R_3$, gives

$$\Delta=\frac{1}{ax}\begin{vmatrix} x^2+y^2+1 & 0 & 0 \\ y & -c-ax & b \\ 1 & cy & -ax-by \end{vmatrix}$$

$\Rightarrow \Delta = (x^2 + y^2 + 1)(ax + by + c)$

Given $\Delta = 0 \Rightarrow ax + by + c = 0$ which represent a straight line

40. Using co-ordinate geometry, prove that three altitudes of any triangle are concurrent.

Solution: $A(x_1, y_1)$, $B(x_2, y_2)$ and $C(x_3, y_3)$ be the vertices of ΔABC. Equation of altitude AD is

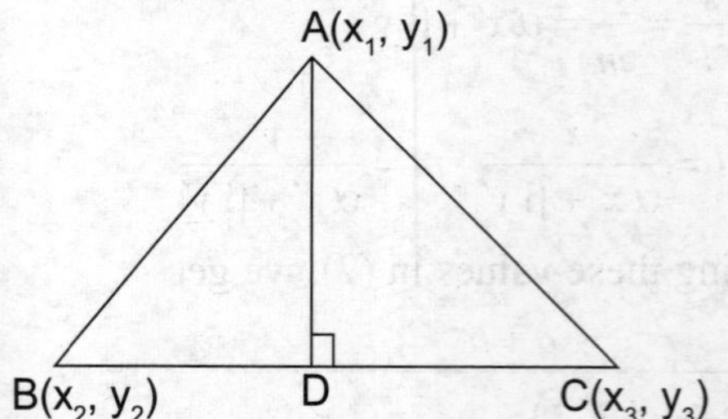

$$y-y_1=-\left[\frac{x_2-x_3}{y_2-y_3}\right](x-x_1)$$

$\Rightarrow (x - x_1)(x_2 - x_3) + (y - y_1)(y_2 - y_3) = 0$ ....(i)

Similarly, equation of other two altitudes are

$(x - x_2)(x_3 - x_1) + (y - y_2)(y_3 - y_1) = 0$ ......(ii)

and $(x - x_3)(x_1 - x_2) + (y - y_3)(y_1 - y_2) = 0$ ....(iii)

altitude will be concurrent iff

$$\begin{vmatrix} x_2-x_3 & y_2-y_3 & -x_1(x_2-x_3)-y_1(y_2-y_3) \\ x_3-x_1 & y_3-y_1 & -x_2(x_3-x_1)-y_2(y_3-y_1) \\ x_1-x_2 & y_1-y_2 & -x_3(x_1-x_2)-y_3(y_1-y_2) \end{vmatrix}=0$$

on opearating $R_1 \to R_1 + R_2 + R_3$, we observed that R_1 becomes zero and hence the altitude are concurrent

Assertion and Reason Type

(a) If both assertion and reason are correct and reason is the correct explanation of the assertion.
(b) If both assertion and reason are correct but reason is not correct explanation of the assertion.
(c) If assertion is correct, but reason is incorrect
(d) If assertion is incorrect, but reason is correct

41. Lines L_1: $y - x = 0$ and L_2: $2x + y = 0$ intersect the line L_3: $y + 2 = 0$ at P and Q respectively. The bisector of the acute angle between L_1 and L_2 intersects L_3 at R.

A: The ratio $PR: RQ$ equal $2\sqrt{2}: \sqrt{5}$.

R: In any triangle, bisector of an angle divides the triangle into two similar triangles.

Ans. (c)

Solution: Points of intersection of L_1 and L_2, i.e., A is (0, 0). Points of intersection L_2 and L_3 is $Q(1, -2)$ and that of L_1 and L_3 is $P(-2, -2)$.

Now, $AP = \sqrt{(-2-0)^2 + (-2-0)^2} = 2\sqrt{2}$

and $AQ = \sqrt{(1-0)^2 + (-2-0)^2} = \sqrt{5}$

By angle bisector theorem, $RP:RQ = AP:AQ = 2\sqrt{2}:\sqrt{5}$

∴ Assertion is correct but reason is incorrect.

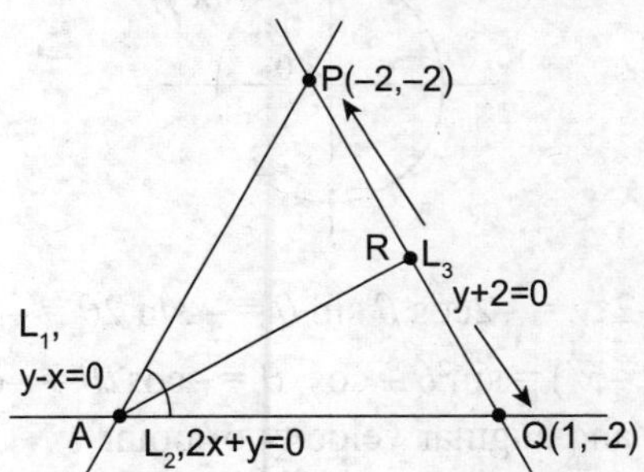

42. **A:** The equation $bx^3 + 12x^2y - 9xy^2 + 2y^3 = 0$ has at least two coincident lines. If $b = -4$ *or* -5.

R: If $x = \alpha$ is a repeated root of $f(x) = 0$, then $f(\alpha) = f'(\alpha) = 0$.

Ans. (a)

Solution: For $b = -4$, $-4x^3 + 12x^2y - 9xy^2 + 2y^3 = 0$ .... (i)

$y = 2x$ satisfies, (i) or $x = y/2$. On differentiating (i) w.r.t. x, treating y as constant we get

$-12x^2 + 24xy - 9y^2 = 0$ .... (ii)

Again $x = y/2$ satisfies (ii)

∴ $(y - 2x)$ is repeated factor of (i)

∴ Equation (i) represents at least two coincident lines

Similarly, we can show that for $b = -5$, $(y - x)$ is a repeated factor of (i)

∴ Assertion is correct

Also if $f(x) = (x - \alpha)^2 g(x)$, then $f(\alpha) = 0$ and

$f'(x) = 2(x - \alpha)\, g(x) + (x - \alpha)^2 g'(x)$

$\Rightarrow f'(\alpha) = 0$

Thus if $x = \alpha$ is a repeated root of $f(x)$, then $f(\alpha) = f'(\alpha) = 0$

∴ Reason is also correct.

43. **A:** If $\frac{x}{2} + \frac{y}{3} = 1$ and $\frac{x}{4} + \frac{y}{5} = -1$ cut x and y-axes at four concyclic points P, Q; R and S respectively, then the orthocentre of ΔPQR is $\left(0, \frac{8}{3}\right)$.

R: If chords PR and QS of a circle intersect at the origin, then $OP.\, OR = OQ.OS$.

Ans. (b)

Solution: The orthocentre of ΔPQR be $H(0, a)$

$\Rightarrow m_{RH} \cdot m_{QP} = -1$

$\Rightarrow \left(\frac{a-0}{4}\right) \cdot \left(\frac{3}{2}\right) = 1$

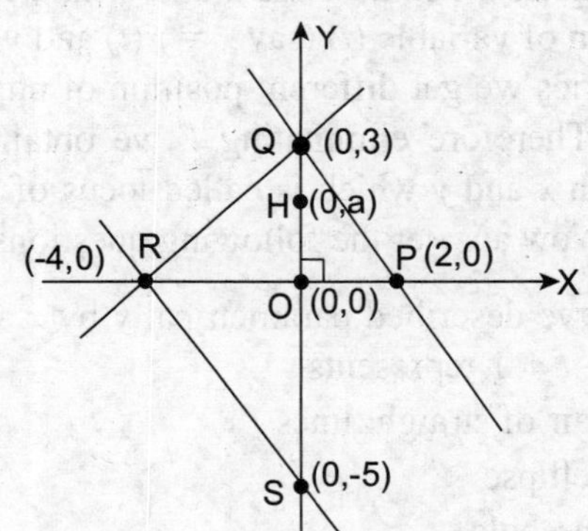

$\Rightarrow 3a = 8 \Rightarrow a = 8/3$.

∴ orthocentre of ΔPQR is $H\left(0, \frac{8}{3}\right)$.

Also, Reason is true but not the correct explanation of Assertion.

44. **A:** There are two mutually perpendicular lines intersecting the co-ordinate axis, then the angle which one line makes with positive x-axis anti-clockwise is equal to the angle which other line makes with positive y-axis anti-clockwise.

R: All quadrilaterals which are formed by point of intersection of pair of lines with co-ordinate axes are concyclic.

Ans. (c)

Solution: Assertion is true, as we can take any generalized case, we can see $\angle 1 = \angle 2$

Reason is false as this is true only in case, when all the lines are mutually perpendicular to each other.

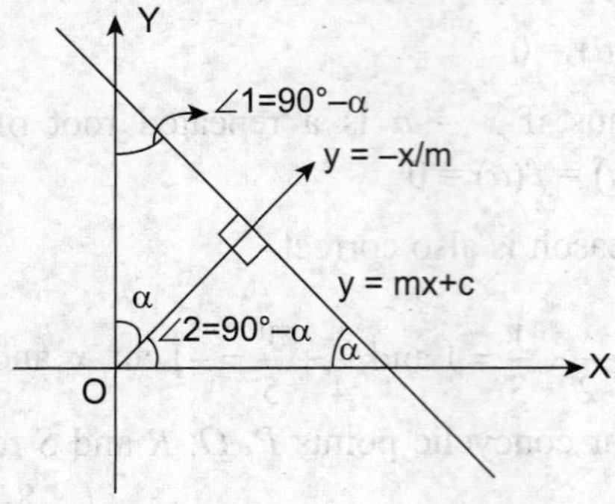

Comprehension Type Questions

A: "Path traced by a moving point is called its locus" In co-ordinate geometry, obtaining locus of a point means establishing a relation between its abscissa and ordinate in all positions of its motion. Idea of locus can be effectively defined by means of a parameter which is called arbitrary constant in co-ordinate geometry. When a moving point is taking various positions such that its abscissa (x) and ordinate (y) are depending on a parameter t, i.e., abscissa x and ordinate y are some function of variable (t) (say $x = f(t)$ and $y = g(t)$) and as t varies we get different position of point (x, y) on locus. Therefore eliminating t, we obtain a relation between x and y which is called locus of the moving point. Now answer the following questions.

45. The curve described parametrically by $x = t^2 + t + 1$, $y = t^2 - t + 1$ represents

(a) a pair of straight lines
(b) an ellipse
(c) a parabola
(d) a hyperbola

Ans. (c)

Solution: $x = t^2 + t + 1$ .... (i)

$y = t^2 - t + 1$ .... (ii)

on adding, $\dfrac{x+y}{2} = t^2 + 1$ .... (iii)

on subtraction $\dfrac{x-y}{2} = t$ .... (iv)

$\therefore$ from (iii) and (iv) we get

$$\frac{x+y}{2} = \left(\frac{x-y}{2}\right)^2 + 1$$

$\Rightarrow x^2 + y^2 - 2xy - 2x - 2y + 4 = 0$ .... (v)

Comparing (v) with equation $ax^2 + by^2 + 2hxy + 2gx + 2fy + c = 0$

we have, $a = 1, b = 1, h = -1, g = -1, f = -1, c = 4$

$$\Delta = \begin{vmatrix} a & h & g \\ h & b & f \\ g & f & c \end{vmatrix} = \begin{vmatrix} 1 & -1 & -1 \\ -1 & 1 & -1 \\ -1 & -1 & 4 \end{vmatrix}$$

$= 1(3) + 1(-5) - 1(2) = -4 \neq 0$

Also $h^2 - ab = (-1)^2 - (1)(1) = 0$

$\therefore$ $\Delta \neq 0$ and $h^2 - ab = 0$

$\Rightarrow$ equation (5) represents a parabola.

46. If the point (x, y) is moving counterclock wise along the unit circle at constant angular speed ω, how is the point $(-2xy, y^2 - x^2)$ moving?

(a) clockwise along the unit circle at angular speed ω.
(b) counterclockwise along the unit circle at angular speed ω.
(c) clockwise along the unit circle at angular speed 2ω.
(d) counterclockwise along the unit circle at angular speed 2ω.

Ans. (c)

Solution: Let $x = \cos\theta$; $\omega = \dfrac{d\theta}{dt}$ ....(i)

and $y = \sin\theta$

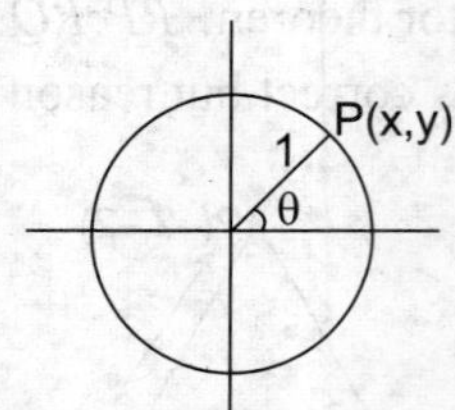

Now, $-2xy = -2\cos\theta\sin\theta = -\sin 2\theta = -\sin\alpha$

and $(y^2 - x^2) = \sin^2\theta - \cos^2\theta = -\cos 2\theta = -\cos\alpha$; where $2\theta = \alpha$ and angular velocity of point

$(-2xy, y^2 - x^2)$ is $\dfrac{d\alpha}{dt} = \dfrac{d}{dt}(2\theta) = 2\dfrac{d\theta}{dt} = 2\omega$

$\therefore$ The point $(-2xy, y^2 - x^2)$ is moving on a unit circle clockwise with angular velocity 2ω.

47. A line AB is drawn from $A(-a, 0)$ $(a > 0)$ (B lies on y-axis). From the point B segments BM and BN of a length b (b = constant) are laid off in either directions along the ray. As the ray revolves about the point A the points M and N describes a curve called the conchoid. Then its equation in Cartesian system of co-ordinates is

(a) $x^2y^2 + (x + a)(x^2 - b^2) = 0$
(b) $xy + (x + a)(x - b) = 0$
(c) $x^2y^2 + (x + a)^2(x^2 - b^2) = 0$
(d) None of these

Ans. (c)

Solution: We are to find the locus of point M or N when the ray AB revolves about A.

Now, co-ordinate of M are given by
$(b\cos\theta, k + b\sin\theta) = (b\cos\theta, a\tan\theta + b\sin\theta)$
Let $x = b\cos\theta$, $a\tan\theta + b\sin\theta = y$

$\Rightarrow \cos\theta = x/b \therefore a\left[\dfrac{\sqrt{b^2-x^2}}{x}\right]+b\left[\dfrac{\sqrt{b^2-x^2}}{b}\right] = y$

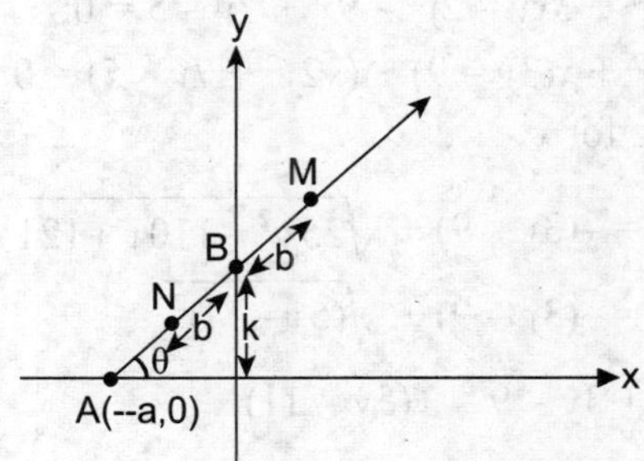

$\Rightarrow \dfrac{a}{x}\sqrt{b^2-x^2}+\sqrt{b^2-x^2} = y$

$\Rightarrow \sqrt{b^2-x^2}\left[\dfrac{a}{x}+1\right] = y$

$\Rightarrow (b^2-x^2)\left(\dfrac{a+x}{x}\right)^2 = y^2$

$\Rightarrow (b^2 - x^2)(a + x)^2 = x^2y^2$
$\Rightarrow (x^2 - b^2)(a + x)^2 + x^2y^2 = 0$

B: Given two straight lines AB and AC whose equations are $3x + 4y = 5$ and $4x - 3y = 15$ respectively. Then the possible equation of line BC through (1, 2), such that ΔABC is isosceles, is $L_1 = 0$ or $L_2 = 0$, then answer the following questions:

48. If $L_1 \equiv ax + by + c = 0$ and $L_2 \equiv dx + ey + f = 0$ where $a, b, c, d, e, f \in \mathbb{Z}$, and $a, d > 0$, then $c + f =$
(a) 1 (b) 2
(c) 3 (d) 4

49. A straight line through $P(2, c + f - 1)$. inclined at an angle of 60° with positive y-axis. The co-ordinates of one of the points on it at a distance $(c + f)$ units from point P is (c, f) (obtained from previous question) given by
(a) $(2+2\sqrt{3},5)$ (b) $(3+2\sqrt{3},3)$
(c) $(2+3\sqrt{2},4)$ (d) $(2+3\sqrt{2},3)$

50. If (a, b) are the co-ordinates of the point obtained in previous question, then find the equation of line which is at the distance $|b - 2a - 1|$ units from origin and make equal intercept on co-ordinate axis in first quadrant.
(a) $z+y+4\sqrt{6}=0$ (b) $x+y+2\sqrt{6}=6$
(c) $x+y-4\sqrt{6}=0$ (d) $x+y-2\sqrt{6}=0$

Solution: Slopes of the lines
$3x + 4y = 5$ is $m_1 = -\dfrac{3}{4}$

and $4x - 3y = 15$ is $m_2 = \dfrac{4}{3}$ $\quad m_1m_2 = -1$

$\therefore$ given lines are perpendicular and $\angle A = \dfrac{\pi}{2}$

Now required equation of BC is

$$(y-2)=\frac{\pm 1-\frac{3}{4}}{1\pm\frac{3}{4}}(x-1) \qquad \text{.......(i)}$$

$\therefore$ equation of BC is (on solving (i))
$x - 7y + 13 = 0$ and $7x + y - 9 = 0$
$L_1 \equiv x - 7y + 13 = 0$
$L_2 \equiv 7x + y - 9 = 0$

51. **(d)** $c + f = 4$

52. **(a)** Equation of a straight line through (2, 3) and inclined at an angle of (π/3) with y-axis ≡ ((π/6) with x-axis) is $\dfrac{x-2}{\cos(\pi/6)} = \dfrac{y-3}{\sin(\pi/6)}$

$\Rightarrow x-\sqrt{3}y = 2-3\sqrt{3}$

Points at a distance $c + f = 4$ units from point P are
$(2 + 4\cos(\pi/6), 3 + 4\sin(\pi/6)) \equiv (2+2\sqrt{3},5)$
and $(2 - 4\cos(\pi/6), 3 - 4\sin(\pi/6)) \equiv (2-2\sqrt{3},1)$
only (a) is true out of given options.

53. **(c)** Let required line be $x + y = k$
which is at $|b - 2a - 1| = |5-4-4\sqrt{3}-1| = 4\sqrt{3}$ units from origin
$\therefore$ Required line is $x+y-4\sqrt{6}=0$ (since intercepts are on positive axes only)

Solved Column Matching

54. **Column-I**

(i) Nearest point on the line $3x - 4y = 25$ from origin is
(ii) The circumcentre of Δ formed by $(xy - 4x - 3y + 12)(x + y - 6) = 0$ is
(iii) The orthocentre of the Δ formed by the lines $2x^2 + 3xy - 2y^2 - 9x + 7y - 5 = 0$ and $4x + 5y - 3 = 0$ is
(iv) If line $(p + 2q)x + (p - 3q)y = p - q$ for different values of p and q passes through a fixed point, then that point is

Column-II

(a) $\left(\dfrac{5}{2},\dfrac{7}{2}\right)$

(b) (3,–4)

(c) $\left(\dfrac{3}{5},\dfrac{11}{5}\right)$

(d) $\left(\dfrac{2}{5},\dfrac{3}{5}\right)$

(e) (2, –1)

Solution: (i) Nearest point on the line $3x - 4y = 25$ from orgin will be the foot of ⊥r drawn from the origin on the straight line $3x - 4y = 25$ ...(i)

Equation of OL will be $y = -\dfrac{4}{3}x$ .... (ii)

point of intersection of (i) and (ii) will be $3x - 4\left(-\dfrac{4}{3}x\right) = 25$

$\Rightarrow \dfrac{9x+16x}{3} = 25 \Rightarrow x = 3$ and $y = -4$

∴ Nearest point is (3, –4) ∴ (i) → (b)

(ii) $(xy - 4x - 3y + 12)(x + y - 6) = 0$

$\Rightarrow [x(y - 4) - 3(y - 4)](x + y - 6) = 0$

$\Rightarrow (x - 3)(y - 4)(x + y - 6) = 0$

∴ $x - 3 = 0, y - 4 = 0, x + y - 6 = 0$ are the sides of Δ.

ΔABC is rt. ∠d at C

∴ Circumcentre will be the mid-point of AB where $A(3, 3)$ and $B(2, 4)$

∴ Circumcentre will be $\left(\dfrac{5}{2},\dfrac{7}{2}\right)$ ∴ (ii) → (a)

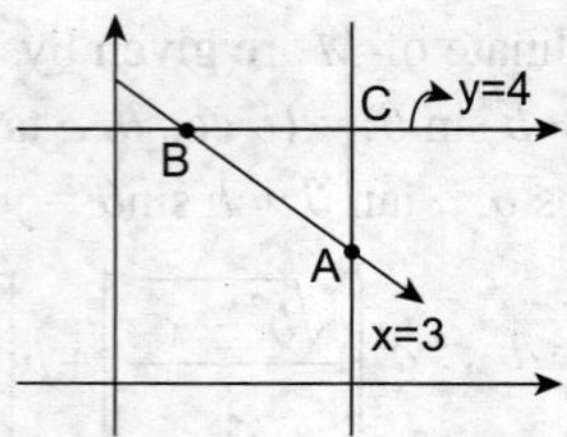

(iii) $2x^2 + 3xy - 2y^2 - 9x + 7y - 5 = 0;$

$2x^2 + x(3y - 9) + (-2y^2 + 7y - 5) = 0$

Solving for x

$\Rightarrow 4x = -(3y - 9) \pm \sqrt{25y^2 - 110y + 121}$

$\Rightarrow 4x = -(3y - 9) \pm \sqrt{(5y - 11)^2}$

$\Rightarrow 4x + 3y - 9 = \pm(5y - 11)$

$\Rightarrow 4x - 2y + 2 = 0$ or $4x + 8y - 20 = 0$

$\Rightarrow 2x - y + 1 = 0$ .... (i)

and $x + 2y - 5 = 0$ .... (ii)

and $4x + 5y - 3 = 0$ .... (iii)

We are to find the orthocentre of Δ formed by sides (i), (ii) and (iii). But (i) and (ii) are ⊥r to each other.

∴ their point of intersection will be the orthocentre of Δ

Now, point of intersection of (i) and (ii) will be $\left(\dfrac{3}{5},\dfrac{11}{5}\right)$ ∴ (iii) → (c).

(iv) $(p + 2q)x + (p - 3q)y = p - q$

$\Rightarrow \left(\dfrac{p}{q}+2\right)x + \left(\dfrac{p}{q}-3\right)y = \dfrac{p}{q}-1$

$\Rightarrow (2x - 3y + 1) + \dfrac{p}{q}(x + y - 1) = 0$

⇒ Required point is point of intersection of $2x - 3y + 1 = 0$ and $x + y - 1 = 0$ i.e., $\left(\dfrac{2}{5},\dfrac{3}{5}\right)$

∴ (IV)→(D)

TUTORIAL EXERCISE

SECTION–III

OBJECTIVE ASSIGNMENT (ONLY ONE CORRECT ANSWER)

1. If $a_1x + b_1y + 1 = 0$, $a_2x + b_2y + 1 = 0$ and $a_3x + b_3y + 1 = 0$ are concurrent lines, then the points (a_1, b_1), (a_2, b_2) and (a_3, b_3)
(a) constitute the vertices of an equilateral triangle.
(b) constitute the vertices of a right Δ.
(c) lie on the same straight line.
(d) constitute the vertices of a right isosceles triangle.

2. A variable straight line passes through a fixed point (a, b) intersecting the co-ordinates axes at A and B. If 'O' is the origin then the locus of the centroid of the triangle OAB is:
(a) $bx + ay - 3xy = 0$ (b) $bx + ay - 2xy = 0$
(c) $ax + by - 3xy = 0$ (d) None of the above

3. If $\frac{x}{c}+\frac{y}{d}=1$ be any line through the intersection of lines $\frac{x}{a}+\frac{y}{b}=1$ and $\frac{x}{b}+\frac{y}{a}=1$ then
(a) $\frac{1}{a}+\frac{1}{d}=\frac{1}{b}+\frac{1}{c}$ (b) $\frac{1}{b}+\frac{1}{d}=\frac{1}{c}+\frac{1}{a}$
(c) $\frac{1}{c}+\frac{1}{d}=\frac{1}{a}+\frac{1}{b}$ (d) None of these

4. If $U = x + 2y - 3 = 0$ and $V = 4x - y + 7 = 0$ represent two straight lines, then $U + KV = 0$ represent a straight line parallel to $5x+4y=0$ if k equals
(a) 6/11 (b) –6/11
(c) 2/7 (d) –2/7

5. If $P = (1, 0)$, $Q = (-1, 0)$ and $R = (2, 0)$ are three given points, then the locus of point S satisfying the relation $SQ^2 + SR^2 = 2SP^2$ is
(a) a straight line parallel to x-axis
(b) circle through origin
(c) circle with centre at the origin
(d) a straight line parallel to y-axis

6. The product of perpendiculars drawn from the point (1, 2) to the pair of lines $x^2 + 4xy + y^2 = 0$ is equal to
(a) 13/4 (b) 3/4
(c) 9/16 (d) None of these

7. The equation of the image of the pair of rays $y = |x|$ by the line $x = 1$ is
(a) $|y| = x + 2$ (b) $|y| + 2 = x$
(c) $y = |x - 2|$ (d) None of these

8. The equation $(x + y + 1)^2 + k(x^2 + y^2) = 0$ will represent two straight lines if k is
(a) 0 only (b) 3 only
(c) 0 or –3 (d) None of these

9. If the equation $12x^2 - 10xy + 2y^2 + 14x - 5y - c = 0$ represents two straight lines, then c, is
(a) 2 (b) 0
(c) 1 (d) –2

10. The straight line, $ax + by = 1$ makes with the curve $px^2 + 2axy + qy^2 = r$, a chord which subtends a right angle at the origin is
(a) $r(a^2 + b^2) = p + q$ (b) $r(a^2 + p^2) = q + b$
(c) $r(b^2 + q^2) = p + a$ (d) None of these

11. If a circle of radius 4 is touching the lines $x^2 + 4xy + y^2 = 0$ in acute angle part, then length of chord of contact to this circle is
(a) $4(3)^{1/4}$ (b) $8\sqrt{3}$
(c) $4\sqrt{3}$ (d) None of these

12. The straight lines, $12x - 5y - 17 = 0$ and $24x - 10y + 44 = 0$ are tangents to the same circle. Then the radius of the circle is:
(a) 1 (b) 3/2
(c) 2 (d) None of these

13. The line $x + 3y - 2 = 0$ bisects the angle between a pair of straight lines of which one has equation $x - 7y + 5 = 0$. The equation of the other line is:
(a) $3x + 3y - 1 = 0$ (b) $x - 3y + 2 = 0$
(c) $5x + 5y - 3 = 0$ (d) None of these

14. A variable rectangle $PQRS$ has its sides parallel to fixed directions. Q and S lie respectively on the lines $x = a$, $x = -a$ and P lies on the x-axis. Then the locus of R is:
(a) a straight line (b) a circle
(c) a parabola (d) None of these

15. Points A (2, 1), B(3, –7), C (lying on line $3x - 2y = 1$) and D form a llgm $ABCD$. The locus of D is
(a) $3x + 2y + 18 = 0$ (b) $3x + 2y = 12$
(c) $3x = 2y + 14$ (d) $2y = 3x + 18$

16. P is a point on either of the two line $y - \sqrt{3}|x| = 2$ at a distance of 5 units from their point of intersection. Then co-ordinates of the foot of the ⊥ from P on the acute angle bisector of these lines are

(a) $\left(0, \frac{4+5\sqrt{3}}{2}\right)$ or $\left(0, \frac{4-5\sqrt{3}}{2}\right)$ depending upon a choice of point P.

(b) $\left(0, \frac{4+5\sqrt{3}}{2}\right)$

(c) $\left(0, \frac{4-5\sqrt{3}}{2}\right)$

(d) $\left(5, \frac{5\sqrt{3}}{2}\right)$

17. If distance formula between two points (x_1, y_1) and (x_2, y_2) be redefined as $|x_1 - x_2| + |y_1 - y_2|$, then the locus of a point which is at a constant distance 5 units from (3, 5) (w.r.t. new formula) is
(a) circle (b) line segment
(c) quadrilateral (d) square

18. If $ax^3 + by^3 + cx^2y + dxy^2 = 0$ represents three distinct straight lines, so that each line is angle bisector of the other two, then
(a) $b + c = 0$ (b) $b + 3c = 0$
(c) $c + 3b = 0$ (d) $a + b = 0$

19. If a pair of straight line represented by $ax^2 + 2hxy + by^2 = 0$ is such that the slope of one line is double of other, then $ab: h^2$ is equal to
(a) 8: 9 (b) 9: 8
(c) 4: 9 (d) 9: 4

20. If $x^2 + \lambda y^2 + 16x - 16y + 55 = 0$ represent a pair of straight lines (where λ is any real number) and θ be angle between the lines then $\sin^2\theta$ is
(a) $\frac{3025}{5329}$ (b) $\frac{2304}{3025}$
(c) $\frac{2304}{5329}$ (d) None of these

21. The set of values of α, if origin lies in the bisector of acute angle of the lines $\alpha x + 2y - 3 = 0$ and $x + y + 7 = 0$ is
(a) $\alpha < -2$ (b) $\alpha > -2$
(c) $\alpha > 2$ (d) $\alpha < 2$

22. Distance between the two lines represented by the line pair $x^2 - 4xy + 4y^2 + x - 2y - 6 = 0$ is
(a) $2/\sqrt{5}$ (b) $\sqrt{5}$
(c) $4/\sqrt{5}$ (d) None of these

23. If the equation, $2x^2 + kxy - 3y^2 - x - 4y - 1 = 0$ represents a pair of lines, then the positive integer value of k is.
(a) 1 (b) 5
(c) –1 (d) –5

24. If $9a^2 + 16b^2 - c^2 - 24ab = 0$, then the family of straight lines $ax + by + c = 0$ are concurrent. The points of concurrency are
(a) (3, –4), (–3,4) (b) (5, –2), (–2,5)
(c) (1, –2), (–5,7) (d) None of these

25. If A and B are two points on the line $3x + 4y + 15 = 0$, such that $OA = OB = 9$ units, where O is the origin, then area of the ΔOAB is
(a) $18\sqrt{2}$ (b) $14\sqrt{2}$
(c) $12\sqrt{2}$ (d) $16\sqrt{2}$

26. On the portion of the straight line $x + y = 2$ which is intercepted between the axes, a square is constructed, away from the origin, with this portion as one of its side. If p denotes the perpendicular distance of a side of this square from the origin, then the maximum value of p
(a) $2\sqrt{3}$ (b) $\sqrt{9}$
(c) $3\sqrt{2}$ (d) $\sqrt{5}$

27. The line $x + y = 1$ meets x-axis at A and y-axis at B. P is the mid-point of AB. P_1 is the foot of the perpendicular from P to OA: M_1 is that of P_1 from OP; P_2 is that of M_1 from OA; M_2 is that of P_2 from OP; P_3 is that of M_2 from OA and so on. If P_n denotes the nth foot of the perpendicular on OA from M_{n-1}, then OP_n equal
(a) $1/3^n$ (b) $1/2^n$
(c) $1/4^n$ (d) $1/7^n$

28. If pair of straight lines $ax^2 + 2hxy - ay^2 = 0$ and $bx^2 + 2gxy - by^2 = 0$ be such that each bisects the angle between the other, value of then the $a^2b + 2gh$ is
(a) 1 (b) 2
(c) 0 (d) 4

29. Consider the triangle OAB, where O is the origin. If $B \equiv (3, 4)$ and the orthocentre of the triangle is $P \equiv (1, 4)$, then the co-ordinates of A are
(a) (0, 13/2) (b) (0, 17/2)
(c) (0, 24/5) (d) (0, 19/4)

30. The angle between the lines belonging to the family of lines $(x+3y-7)+\lambda(x+4y-3)=0$ and $(3x+y-7)+\mu(x+y-3)=0$ which are farthest from origin (where λ and μ are real parameters) is
(a) $\sec^{-1}(2/9)$ (b) $\tan^{-1}(27/34)$
(c) $\sin^{-1}(3/9)$ (d) $\cos^{-1}(1/7)$

31. Let $2x+3y=6$ be a line meeting the co-ordinate axes at A and B respectively. A variable line $\frac{x}{a}+\frac{y}{b}=1$ meets

the axes at P and Q respectively in such a way that the lines BP and AQ always meet at right angle at R. Then the locus of the orthocentre of the ΔARB is

(a) $x^2 + y^2 + 8x + 5y = 0$
(b) $2x^2 + y^2 - x - y = 0$
(c) $x^2 + y^2 - 3x - 2y = 0$
(d) $x^2 - y^2 + 3x + 2y = 0$

32. $ABCD$ is a square of side length a. Its side AB slides between x and y-axes in first quadrant. Then the locus of the foot of perpendicular dropped from the point E on the diagonal AC, where E is the mid-point of the side AD is.

(a) $(y - x)^2 + (x - 3y)^2 = a^2/4$
(b) $(y + x)^2 - (3x + y)^2 = a^2/7$
(c) $(y - 2x)^2 + (x - 2y)^2 = a^2/4$
(d) $(y + x)^2 + (3x + y)^2 = a^2/5$.

33. A line is such that its segment between the straight lines $5x - y - 4 = 0$ and $3x + 4y - 4 = 0$ is bisected at the point (1, 5), then its equation is

(a) $83x - 35y + 92 = 0$ (b) $35x - 83y + 92 = 0$
(c) $35x + 35y + 92 = 0$ (d) None of these

34. The angle between the lines whose intercepts on the axes are a, $-b$ and b, $-a$ respectively, is

(a) $\tan^{-1}\left|\dfrac{a^2 - b^2}{ab}\right|$ (b) $\tan^{-1}\left|\dfrac{b^2 - a^2}{2}\right|$
(c) $\tan^{-1}\left|\dfrac{b^2 - a^2}{2ab}\right|$ (d) None of these

35. A straight line passes through a fixed point (h, k), The locus of the of the foot of perpendicular on it drawn from the origin is

(a) $x^2 + y^2 - hx - ky = 0$
(b) $x^2 + y^2 + hx + ky = 0$
(c) $3x^2 + 3y^2 + hk - ky = 0$
(d) None of these

36. One diagonal of a square is along the line $8x - 15y = 0$ and one of its vertex is (1, 2). then the equation of the sides of the square passing through this vertex, are

(a) $23x + 7y = 9$, $7x + 23y = 53$
(b) $23x - 7y + 9 = 0$, $7x + 23y + 53 = 0$
(c) $23x - 7y - 9 = 0$, $7x + 23y - 53 = 0$
(d) None of these

37. The ends of the base of an isosceles triangle are at $(2a, 0)$ and $(0, a)$. The equation of one side is $x = 2a$. the equation of the other side is

(a) $x + 2y - a = 0$ (b) $x + 2y = 2a$
(c) $3x + 4y - 4a = 0$ (d) $3x - 4y + 4a = 0$

38. If one of the diagonals of a square is along the line $x = 2y$ and one of its vertices is (3, 0), then its sides through this vertex are given by the equations

(a) $y - 3x + 9 = 0$, $3y + x - 3 = 0$
(b) $y - 3x + 9 = 0$, $3y + x - 3 = 0$
(c) $y - 3x + 9 = 0$, $3y - x + 3 = 0$
(d) $y - 3x + 3 = 0$, $3y + x + 9 = 0$

39. The line L has intercepts a and b on the co-ordinate axes. Keeping the origin fixed, the co-ordinate axes are rotated through a fixed angle. The line L has now intercepts p and q on the rotated axes. Then

(a) $a^2 + b^2 = p^2 + q^2$ (b) $1/a^2 + 1/b^2 = 1/p^2 + 1/q^2$
(c) $a^2 + p^2 = b^2 + q^2$ (d) $1/a^2 + 1/p^2 = 1/b^2 + 1/q^2$

40. A point starts from (1, 2) and its projections on the x and the y axes are moving with velocity of $3m/s$ and $2m/s$ respectively (along with the positive direction). Its locus is

(a) $2x - 3y + 4 = 0$ (b) $3x - 2y + 1 = 0$
(c) $3y - 2x + 4 = 0$ (d) $2y - 3x + 1 = 0$

41. If the lines represented by $x^2 - 2pxy - y^2 = 0$ are rotated about the origin through an angle θ, one in clockwise direction and other in anti-clockwise direction, then the equation of the bisectors of the angle between the lines in the new position is

(a) $px^2 + 2xy - py^2 = 0$ (b) $px^2 + 2xy + py^2 = 0$
(c) $x^2 - 2pxy - y^2 = 0$ (d) None of these

42. If the two lines represented by $x^2(\tan^2\theta + \cos^2\theta) - 2xy \tan\theta + y^2 \sin^2\theta = 0$ make angles α, β with x-axis, then

(a) $\tan\alpha + \tan\beta = 4 \operatorname{cosec} 2\theta$
(b) $\tan\alpha \tan\beta = \sec^2\theta + \tan^2\theta$
(c) $\dfrac{\tan\alpha}{\tan\beta} = \dfrac{2 + \sin 2\theta}{2 - \sin 2\theta}$
(d) None of these

43. If the roots of the equation $ax^2 + bx + 10 = 0$ are not real and distinct where $a, b \in R$ and α and β are the values of a and b for which $5a + b$ is minimum, then the family of lines $\alpha(4x + 2y + 3) + \beta(x - y - 1) = 0$ are concurrent at

(a) (1,–1) (b) (1, 1)
(c) $\left(-\dfrac{1}{6}, -\dfrac{7}{6}\right)$ (d) $\left(\dfrac{1}{6}, \dfrac{7}{6}\right)$

44. The family of lines is given as : $a(3x + 4y + 6) + b(x + y + 2) = 0$. The line of the family situated at the greatest distance from the point $P(2, 3)$ has the equation

(a) $4x + 3y + 8 = 0$ (b) $5x + 3y + 10 = 0$
(c) $15x + 8y + 30 = 0$ (d) None of these

45. Which of the following inequations can never be satisfied by any interior point of triangle formed by vertices (1, 3); (5, 0); (–1, 2)?

(a) $x + 3y \geq 5$ (b) $3x + 4y - 15 \leq 0$
(c) $x - 2y + 5 \geq 0$ (d) None of thesse

46. If the slope of one line is double the slope of another line and the combined equation of the pair of lines $x^2/a + 2xy/h + y^2/b = 0$ then $ab : h^2$ is equal to

(a) 9 : 8 (b) 3 : 2
(c) 8 : 3 (d) None of these

47. The line $x + 2y = 4$ is translated parallel to itself by 3 units in the sense of increasing x and then rotated by 30° in the anti-clockwise direction about the point where the shifted line cuts the x-axis. The equation of the line in the new position is (θ is the angle subtendent by given line with x-axis in anti-clockwise direction

(a) $y = \tan(\theta + 30°)(x - 4 - 3\sqrt{5})$
(b) $y = \tan(30° - \theta)(x - 4 - 3\sqrt{5})$
(c) $y = \tan(\theta + 30°)(x + 4 + 3\sqrt{5})$
(d) $y = \tan(30° - \theta)(x + 4 + 3\sqrt{5})$

48. Drawn from the origin are two mutually perpendicular straight lines forming an isosceles triangle together with the straight line, $2x + y = a$. Then the area of the triangle is

(a) $\frac{a^2}{2}$ (b) $\frac{a^2}{3}$
(c) $\frac{a^2}{5}$ (d) None of these

49. The equations of three lines, AB, CD and EF are, $(b - c)x + (c - a)y + (a - b) = 0$, $(c - a)x + (a - b)y + (b - c) = 0$ and $(a - b)x + (b - c)y + (c - a) = 0$. Which one of the following inferences is correct?

(a) the lines are parallel to each other
(b) AB and BC are perpendicular to EF
(c) all the lines are coincident
(d) the lines are concurrent

50. The product of perpendiculars from the points $\{\pm\sqrt{a^2 - b^2}, 0\}$ on the straight line $\frac{x}{a}\cos\phi + \frac{y}{b}\sin\phi = 1$ is

(a) a^2 (b) b^2
(c) ab (d) $a^2 b^2$

51. If $a^2 + b^2 - c^2 - 2ab = 0$ then the family of straight lines $ax + by + c = 0$ is concurrent at

(a) (–1, 1) only (b) (1, –1) only
(c) (–1, 1) or (1, –1) (d) None of these

52. If a, b, c are in harmonical progression then the line, $bcx + cay + ab = 0$ passes through a fixed point whose co-ordinates are

(a) (1, 2) (b) (–1, 2)
(c) (–1, –2) (d) (1, –2)

53. The orthocentre of the triangle ABC is 'B' and the circumcentre is 'S' (a, b). If A is the origin then the co-ordinates of C are

(a) $(2a, 2b)$ (b) $\left(\frac{a}{2}, \frac{b}{2}\right)$
(c) $(\sqrt{a^2 + b^2}, 0)$ (d) None of these

54. The lines $3x + 4y = 9$ and $4x - 3y + 12 = 0$ intersect at P. The first line intersects x-axis at A and the second line intersects y-axis at B Then the circumradius of the triangle PAB is

(a) 3/2 (b) 5/2
(c) 10 (d) None of these

55. A variable straight line passes through the points of intersection of the lines, $x + 2y = 1$ and $2x - y = 1$ and meets the co-ordinate axes in A and B. The locus of the middle point of AB is

(a) $x + 3y - 10xy = 0$ (b) $x - 3y + 10xy = 0$
(c) $x + 3y + 10xy = 0$ (d) None of these

56. If two sides of a triangle are represented by $x^2 - 7xy + 6y^2 = 0$ and the centroid is (1, 0) then the equation of third side is

(a) $2x + 7y + 3 = 0$ (b) $2x - 7y + 3 = 0$
(c) $2x + 7y - 3 = 0$ (d) $2x - 7y - 3 = 0$

57. Let $O(0, 0)$, $P(3, 4)$, $Q(6, 0)$ be the vertices of the triangle OPQ. The point R inside the triangle OPQ is such that the triangles OPR, PQR, OQR are of equal area. The co-ordinates of R are

(a) $\left(\frac{4}{3}, 3\right)$ (b) $\left(3, \frac{2}{3}\right)$
(c) $\left(3, \frac{4}{3}\right)$ (d) $\left(\frac{4}{3}, \frac{2}{3}\right)$

58. The number of integral points (integral point means both the co-ordinates should be integer) exactly in the interior of the triangle with vertices (0, 0), (0, 21) and (21, 0), is

(a) 133 (b) 190
(c) 233 (d) 105

59. Orthocentre of triangle with vertices (0, 0), (3, 4) and (4, 0) is

(a) $\left(3, \frac{5}{4}\right)$ (b) (3, 12)
(c) $\left(3, \frac{3}{4}\right)$ (d) (3, 9)

60. The area bounded by the curves $y = |x| - 1$ and $y = -|x| + 1$ is

(a) 1 (b) 2

(c) $2\sqrt{2}$ (d) 4

61. Let $0 < \alpha < \frac{\pi}{2}$ be fixed angle. If $P = (\cos\theta, \sin\theta)$ and $Q = (\cos(\alpha - \theta), \sin(\alpha - \theta))$, then Q is obtained from P by

(a) colockwise rotation around origin through an angle α

(b) anti-clockwise rotation around origin through an angle α

(c) reflection in the line through origin with slope $\tan\alpha$

(d) reflection in the line through origin with slope angle $(\alpha/2)$

62. Area of the parallelogram formed by the lines $y = mx$, $y = mx + 1$, $y = nx$ and $y = nx + 1$ equals

(a) $\frac{|m+n|}{(m-n)^2}$ (b) $\frac{2}{|m+n|}$

(c) $\frac{1}{|m+n|}$ (d) $\frac{1}{|m-n|}$

63. The incentre of the triangle with vertices $(1, \sqrt{3}), (0,0)$ and $(2, 0)$ is

(a) $\left(1, \frac{\sqrt{3}}{2}\right)$ (b) $\left(\frac{2}{3}, \frac{1}{\sqrt{3}}\right)$

(c) $\left(\frac{2}{3}, \frac{\sqrt{3}}{2}\right)$ (d) $\left(1, \frac{1}{\sqrt{3}}\right)$

64. If x_1, x_2, x_3 as well as y_1, y_2, y_3 are in G.P. with the same common ratio, then the points $(x_1, y_1), (x_2, y_2)$ and (x_3, y_3)

(a) lie on a straight line

(b) lie on the ellipse

(c) lie on a circle

(d) are vertices of a triangle

SECTION–IV

MORE THAN ONE CORRECT ANSWER

1. If the given lines $y = m_1x + c_1, y = m_2x + c_2$ and $y = m_3x + c_3$ be concurrent, then

(a) $m_1(c_2 - c_3) + m_2(c_3 - c_1) + m_3(c_1 - c_2) = 0$

(b) $m_1(c_2 - c_1) + m_2(c_3 - c_2) + m_3(c_1 - c_3) = 0$

(c) $c_1(m_2 - m_3) + c_2(m_3 - m_1) + c_3(m_1 - m_2) = 0$

(d) None of these

2. Two roads are represented by the equations $y - x = 6$ and $x + y = 8$. An inspection bungalow has to be so constructed that it is at a distance of 100 from each of the roads. Possible location of the bungalow is given by

(a) $(100\sqrt{2}+1, 7)$ (b) $(1 - 100\sqrt{2}, 7)$

(c) $(1, 7 + 100\sqrt{2})$ (d) $(1, 7 - 100\sqrt{2})$

3. The straight line $x - y = 2$ cuts the co-ordinate axes in A and B. On AB a square is constructed away from the origin, p denotes the $\perp$ distance from $(0, 0)$ to a side of the square, then

(a) maximum value of p is $3\sqrt{2}$

(b) Area of square is 8(square units)

(c) $x + y = 2$ is the equation to one of the sides of the square.

(d) None of these

4. The equation $x^3 + x^2y - xy^2 = y^3$ represents

(a) three real straight lines

(b) lines in which two of them are perpendicular to each other

(c) lines in which two of them are coincident

(d) None of these

5. If one of the lines of $my^2 + (1 - m^2)xy - mx^2 = 0$ is a bisector of the angle between the lines $xy = 0$, then m is

(a) 1 (b) 2

(c) –1/2 (d) –1

6. The value of 'c' for which the lines joining the origin to the points of intersection of the line $y = x\sqrt{3} + c$ and the curve $x^2 + y^2 = 2$ are perpendicular to each other are

(a) 1 (b) 0

(c) –2 (d) 2

7. If $(\alpha, -\alpha)$ be an end of diagonal of square and the other diagonal has the equation $x - y = \alpha$, then another vertex of the square can be

(a) $(\alpha - \beta, \alpha)$ (b) $(\alpha, 0)$

(c) $(0, -\alpha)$ (d) $(\alpha + \beta, \beta)$

8. One side of the square is inclined to the x-axis at an angle α and one of its extremities is at the origin. If the side of the square is 4, the equations of the diagonals of the square are

(a) $x(\cos\alpha + \sin\alpha) - y(\cos\alpha - \sin\alpha) = 0$,

(b) $x(\cos\alpha - \sin\alpha) + y(\cos\alpha + \sin\alpha) = 4$

(c) $x(\cos\alpha + \sin\alpha) + y(\cos\alpha - \sin\alpha) = 0$

(d) None of these

9. A canal is 41/2 km from a place and the shrotest route from this place to canal is exactly north east. A village

P is 3 km north and 4 km east from the place. And another village Q is 3 km east and 4 km north from the place, then

(a) P lies on the canal
(b) Q does not lies on the canal
(c) P and Q are both on the same side of canal
(d) P and Q lies on that side of the canal on which given place lies.

10. A line is drawn through $A(4, -1)$ parallel to the line $3x - 4y + 1 = 0$. The co-ordinates of a point on this line which is at a distance of 5 units from A are

(a) (8, 2) (b) (0, –4)
(c) (2, 8) (d) None of these

11. The co-ordinates of the extremities of one diagonal of a square are (1, 1) and (–2,–1). Then

(a) One of the other vertices is $\left(\frac{1}{2}, \frac{-3}{2}\right)$
(b) One of the other vertices is $\left(\frac{-3}{2}, \frac{3}{2}\right)$
(c) Equation of the other diagonal is $6x + 4y + 3 = 0$.
(d) None of these

12. One diagonal of a sqaure lies along the line $x - 2y + 2 = 0$ and one vertex, of the square is (1, 4). Then

(a) Equation of one of the sides is $x + 3y - 13 = 0$
(b) Equation of one of the sides is $3x - y + 1 = 0$
(c) Equation of the other diagonal is $x + 3y - 3 = 0$
(d) None of these

13. The equations of lines joining the origin to the point of intersection of circle $x^2 + y^2 = 3$ and the line $x + y = 2$ is

(a) $y - (3 + 2\sqrt{2})\, x = 0$ (b) $x - (3 + 2\sqrt{2})\, y = 0$
(c) $y - (3 - 2\sqrt{2})\, x = 0$ (d) None of these

14. Three lines $px + qy + r = 0$, $qx + ry + p = 0$ and $rx + py + q = 0$ are concurrent if

(a) $p + q + r = 0$ (b) $p^2 + q^2 + r^2 = pq + qr + rp$
(c) $p^3 + q^3 + r^3 = 3pqr$ (d) None of these

15. The lines $(lx + my)^2 - 3(mx - ly)^2 = 0$ and $lx + my + n = 0$ form

(a) An isosceles triangle
(b) A right angled triangle
(c) An equilateral triangle
(d) None of these

SECTION–V

ASSERTION AND REASON-TYPE

The questions given below consist of an assertion (A) and the reason (R). Use the following keys to choose the appropriate answer.

(a) If both assertion and reason are correct and reason is the correct explanation of the assertion.
(b) If both assertion and reason are correct but reason is not correct explanation of the assertion.
(c) If assertion is correct, but reason is incorrect
(d) If assertion is incorrect, but reason is correct

Now consider the following statements:

1. A: The line $2x + y + 6 = 0$ is perpendicular to the line $x - 2y + 5 = 0$ and second line passes through (1, 3).
R: Product of the slopes of the perpendicular lines is equal to – 1.

2. A: If the diagonals of the quadrilateral formed by the lines $px + qy + r = 0$; $p'x + q'y + r = 0$; $px + qy + r' = 0$; $p'x + q'y + r' = 0$; are at right angles, then $p^2 + q^2 = p'^2 + q'^2$.
R: Diagonals of a rhombus are bisected and perpendicular to each other.

3. A: If point of intersection of the lines $4x + 3y = \lambda$ and $3x - 4y = \mu$ $\forall\, \lambda, \mu \in \mathbb{R}$ is (x_1, y_1) then the locus of (λ, μ) is $x + 7y = 0$, $\forall\, x_1 = y_1$.
R: If $4\lambda + 3\mu > 0$ and $3\lambda - 4\mu > 0$, then (x_1, y_1) is in first quadrant.

4. A: Let the lines $2x + 3y + 19 = 0$ and $9x + 6y - 17 = 0$ cut the x-axis in A, B and y axis in C, D. Then points A, B, C, D are concylic.
R: $OA. OB = OC. OD$, where O is origin
$\therefore$ A, B, C, D points are concyclic

5. A: All chords of the curve $4x^2 + y^2 - x + 4y = 0$, which subtends right angle at the origin passes through the point $\left(\frac{1}{5}, -\frac{4}{5}\right)$.
R: Chords of any curve, subtending right angle at origin passes through a fixed point.

6. A: Let the vertices of a ΔABC be $A(-5, -2)$: $B(7,6)$ and $C(5, -4)$, then co-ordinates of circumcentre are (1, 2).
R: For a right angle triangle, mid-point of hypotenuous is the circumcentre of the triangle.

7. A: If $f - 2h = a + b$, then one line of the pair of lines $ax^2 + 2hxy + by^2 = 0$ bisects the angle between co-ordinate axes in positive quadrant.

R: If $ax + y(2h + a) = 0$ is a factor of $ax^2 + 2hxy + by^2 = 0$ then $b + 2h + a = 0$.

8. A: Two of the straight lines represented by the equation $ax^3 + bx^2y + cxy^2 + dy^3 = 0$ will be right angled if $a^2 + ac + bc + d^2 = 0$.

R: Product of the slopes of two perpendicular lines (with real shapes) is – 1.

9. A: The equation $2x^2 + 3xy - 2y^2 + 5x - 5y + 3 = 0$ represents a pair of ⊥ straight lines.

R: A pair of lines given by $ax^2 + 2hxy + by^2 + 2gx + 2fy + c = 0$ are perpendicular, if $a + b = 0$.

10. A: The joint equation of lines $y = x$ and $y = -x$ is $y^2 = -x^2$ i.e., $x^2 + y^2 = 0$

R: The joint equation of lines $ax + by = 0$ and $cx + dy = 0$ is $(ax + by)(cx + dy) = 0$; where a, b, c, d are constants.

SECTION–VI

LINKED COMPREHENSION-TYPE

A: In a triangle ABC, if the equations of sides AB, BC and CA are $2x - y + 4 = 0$, $x - 2y - 1 = 0$ and $x + 3y - 3 = 0$ respectively, then

1. Tangent of internal angle A is equal to
(a) 1/2 (b) –3
(c) 7 (d) –7

2. The equation of external bisector of angle B is;
(a) $x - y + 1 = 0$ (b) $x + y + 5 = 0$
(c) $x - y - 1 = 0$ (d) $x + y - 5 = 0$

3. The image of point B w.r.t. the side CA is
(a) $\left(-\frac{3}{5}, \frac{26}{5}\right)$ (b) $\left(-\frac{3}{5}, -\frac{26}{5}\right)$
(c) $\left(\frac{3}{5}, -\frac{26}{5}\right)$ (d) $\left(\frac{3}{5}, \frac{26}{5}\right)$

B: $A(1, 3)$ and $C\left(-\frac{2}{5}, -\frac{2}{5}\right)$ are the vertices of a triangle ABC and the equation of the angle bisector of $\angle ABC$ is $x + y = 2$

4. Equation of side BC is
(a) $7x + 3y - 4 = 0$ (b) $7x + 3y + 4 = 0$
(c) $7x - 3y + 4 = 0$ (d) $7x - 3y - 4 = 0$

5. Co-ordinates of vertex B are
(a) $\left(\frac{3}{10}, \frac{17}{10}\right)$ (b) $\left(\frac{17}{10}, \frac{3}{10}\right)$
(c) $\left(-\frac{5}{2}, \frac{9}{2}\right)$ (d) (1,1)

6. Equation of side AB is
(a) $3x + 7y = 24$ (b) $3x + 7y + 24 = 0$
(c) $13x + 7y + 8 = 0$ (d) $13x - 7y + 8 = 0$

C: If co-ordinate axes of system is rotated by angle θ without shifting the origin the co-ordinate (x, y) of a point P changes to (X, Y) So the equation of each locus gets transformed and transformed equation can be obtained by replacing $x = X\cos\theta + Y\sin\theta$, $y = X\sin\theta - Y\cos\theta$ in the equation of locus $f(x, y) = 0$. So if after transformation $f(x,y) = x^2 + 2xy + y^2$ transforms to $aX^2 + bY^2$, then

7. Value of 'a' is equal to
(a) –1 (b) 3
(c) 2 (d) 0

8. Which one of the following is correct?
(a) $a + 2b = 2$ (b) $2a + b = 2$
(c) $3a + b = 0$ (d) $3a - b = 0$

9. Which one of the following is correct?
(a) $\cos 2\theta = 1$ (b) $\theta = \pi/2$
(c) $\sin 2\theta = 0$ (d) $\sec 2\theta$ is not defined.

D: The equation of the straight line passing through (x_1, y_1) and making an angle θ with the positive direction of the x-axis is $\frac{x - x_1}{\cos\theta} = \frac{y - y_1}{\sin\theta} = r$, where r is the directed distance between the points (x, y) and (x_1, y_1).

10. The angle made with x-axis of a straight line drawn through (2, 3) so that it intersects the line $x + y - 7 = 0$ at a distance $\sqrt{2}$ from (2, 3) is
(a) $\frac{\neq}{4}$ (b) $\frac{3\neq}{4}$
(c) $\frac{\neq}{6}$ (d) $\frac{\neq}{3}$.

11. Which of the following points lie at a distance of 4 units from the point $P(2, 3)$ on the line through P whose inclination with negative direction of x-axis is 30°?
(a) $\left(2 - 2\sqrt{3}, 1\right)$ (b) $\left(2 + 2\sqrt{3}, 0\right)$
(c) $\left(2 - 2\sqrt{3}, 0\right)$ (d) $\left(2\sqrt{3}, 1\right)$.

12. The sides AB and AC of a ΔABC are given by the equations $3x + y = 4$ and $x + 3y = 4$ respectively

and mid-point of BC is (2, 2). Then the equation of BC is
(a) $x - y + 4 = 0$ (b) $x + 3y + 4 = 0$
(c) $x + y = 4$ (d) $x + y + 4 = 0$.

E: Solving problems in co-ordinate geometry with given set of conditions causes difficulties at times. One such case is equation of family of lines. The equation $y-1=m(x-2)$ represents a family of lines passing through (2, 1). But this does not include the line $x - 2 = 0$. Since the slope of the line $x - 2 = 0$ is infinity. Therefore in case there are two solutions to a problem (known by geometrical considerations) and only one solution is being obtained algebraically we have to re-examine. For example, if line $x - 2 = 0$ is a solution of a problem which leads to a degenerated quadratic equaiton, we will take the other root of the equation as ∞.

13. A line $y = mx$ through origin cuts the parallel lines $x + y = 3$ and $x + y = 5$ at A and B respectively. The distance between the two points of intersection is d. If $AB = 2$, then the quadratic equation formed is
(a) $(4 - d^2)\, m^2 + 2md^2 + 4 - d^2 = 0$
(b) $(4 - d^2)\, m^2 + 2md^2 + d^2 - 4 = 0$
(c) $(4 - d^2)\, m^2 + 2md^2 + 4 - d^2 = 0$
(d) None of these

14. In Q13, if $d = 2$ then
(a) There is only one line (x-axis) cutting intercept 2 between the parallel lines
(b) There is only one line (y-axis) cutting intercept 2 between the parallel lines
(c) There are two lines (inclined at acute angle to each other) through origin cutting intercepts of 2 units between the parallel lines
(d) None of these.

F: Three lines are said to be concurrent, if they pass through a common point. Thus, if three lines are concurrent the point of intersection of two lines lies on the third line. Let $a_1x + b_1y + c_1 = 0$; $a_2x + b_2y + c_2 = 0$ and $a_3x + b_3y + c_3 = 0$ be three concurrent lines. Then the point of intersection of (i) and (ii) must lie on the third. The co-ordinates of the point of intersection of (i) and (ii) are $\left(\dfrac{b_1c_2 - b_2c_1}{a_1b_2 - a_2b_1}, \dfrac{c_1a_2 - c_2a_1}{a_1b_2 - a_2b_1}\right)$

This point lies on (iii), therefore $a_3\left(\dfrac{b_1c_2 - b_2c_1}{a_1b_2 - a_2b_1}\right) + b_3\left(\dfrac{c_1a_2 - c_2a_1}{a_1b_2 - a_2b_1}\right) + c_3 = 0$

$\Rightarrow a_3(b_1c_2 - b_2c_1) + b_3(c_1 a_2 - c_2 a_1) + c_3 (a_1b_2 - a_2 b_1) = 0$

$\Rightarrow \Delta = \begin{vmatrix} a_1 & b_1 & c_1 \\ a_2 & b_2 & c_2 \\ a_3 & b_3 & c_3 \end{vmatrix} = 0$; it is

the required condition for concurrency of lines.

15. If lines $y = m_1x + c_1, y = m_2x + c_2$ and $y = m_3x + c_3$ are concurrent then
(a) $-\dfrac{m_2 - m_1}{m_3 - m_2} = \dfrac{c_2 - c_1}{c_3 - c_1}$
(b) $\dfrac{m_2 - m_1}{m_3 - m_2} = \dfrac{c_2 - c_1}{c_3 - c_2}$
(c) $\dfrac{m_1 - m_2}{m_3 - m_2} = \dfrac{c_2 - c_1}{c_3 - c_2}$
(d) None

16. If $al + bm + cn = 0$, then family of lines $lx + my + n = 0$ is always passing through a fixed point (where a, b, c are constants) then fixed point is
(a) $\left(\dfrac{c}{a}, \dfrac{c}{b}\right)$ (b) $\left(-\dfrac{a}{c}, \dfrac{b}{c}\right)$
(c) $\left(\dfrac{a}{c}, \dfrac{b}{c}\right)$ (d) None

17. The lines $ax + by + c = 0$, $bx + cy + a = 0$, $cx + ay + b = 0$ are concurrent where a, b, c are sides of $\Delta\, ABC$ in usual notations then
(a) $\sin^3 A + \sin^3 B + \sin^3 C = 3 \sin A \sin B \sin C$
(b) $\cos^3 A + \cos^3 B + \cos^3 C = 3 \sin A \sin B \sin C$
(c) $\sin^3 A + \sin^3 B + \sin^3 C = 3 \cos A \cos B \cos C$
(d) None of these

G: If the lines represented by $2x^2 - 5xy + 2y^2 = 0$ be the two sides of a parallelogram and the line $5x + 2y = 1$ be one of its diagonal. On the basis of above information, answer the following questions:

18. The equation of the other diagonal is
(a) $10x - 11y = 0$ (b) $11x - 10y = 0$
(c) $3x - 2y = 0$ (d) $2x - 3y = 0$

19. The centroid of the parallelogram is
(a) $\left(\dfrac{5}{72}, \dfrac{11}{36}\right)$ (b) $\left(\dfrac{11}{72}, \dfrac{5}{36}\right)$
(c) $\left(\dfrac{5}{36}, \dfrac{11}{72}\right)$ (d) $\left(\dfrac{11}{36}, \dfrac{5}{72}\right)$

20. The area of the parallelogram is

(a) $\frac{1}{36}$ sq units (b) $\frac{1}{18}$ sq units

(c) $\frac{1}{9}$ sq units (d) None of these

21. The ratio of the longer side to smaller side is

(a) 6: 5 (b) 7: 6

(c) 5: 4 (d) 4: 3

H: Let ABC be an acute angled triangle and AD, BE and CF are its medians, where E and F are the points E(3, 4) and F(1, 2) respectively and centroid of ΔABC is G(3, 2), then answer the following questions:

22. The equation of side AB is

(a) $2x + y = 4$

(b) $x + y - 3 = 0$

(c) $4x - 2y = 0$

(d) None of these

23. Co-ordinates of (third) mid point D are

(a) (7, –4) (b) (5, 0)

(c) (7, 4) (d) (–3, 0)

24. Height of altitude drawn from point A is (in units)

(a) $4\sqrt{2}$ (b) $3\sqrt{2}$

(c) $6\sqrt{2}$ (d) $2\sqrt{3}$

SECTION–VII

MATRIX-MATCH TYPE

1. Reflection of the line $x + y + 1 = 0$ in the line

Column-1	Column-II
(i) $2x + y + 1 = 0$ is	(a) $x + 7y - 11 = 0$
(ii) $x - 2y + 1 = 0$ is	(b) $7x + y + 1 = 0$
(iii) $x + 2y - 1 = 0$ is	(c) $7x + y - 11 = 0$
(iv) $2x + y - 1 = 0$ is	(d) $7x + y + 7 = 0$

2. Column-1

(i) Lines $x - 2y - 6 = 0, 3x+y-4=0$ and $\lambda x + 4y + \lambda^2 =0$, are concurrent, then value of λ is

(ii) The points$(\lambda + 1, 1)$; $(2\lambda + 1, 3)$ and $(2\lambda + 2, 2\lambda)$ are collinear, then the value of λ is

(iii) If line $x + y - 1 - \lambda = 0$, passing through the intersection of $x - y + 1 = 0$ and $3x + y - 5 = 0$ is perpendicular to one of them, then the value of λ is

(iv) If line $y - x - 1 + \lambda = 0$ is equally inclined to axes and equidistant from the points (1, –2) and (3, 4), then λ is

Column-II

(a) 2

(b) 4

(c) –1/2

(d) –4

3. Column-I

(i) The number of lintegral values of 'a' for which the point $P(a, a^2)$ lies completely inside the triangle formed by the lines $x = 0$, $y = 0$ and $x + 2y = 3$.

(ii) ΔABC with $AB = 13$, $BC = 5$ and $AC = 12$ slides on the co-ordinate axes with A and B on the positive x-axis and +ve y-axis respectively, the locus of vertex C is a line $12x + ky = 0$,then the value of k is

(iii) The reflection of the point $(t - 1, 2t + 2)$ in a line is $(2t + 1, t)$ then the equation of the line has slope equals to

(iv) In a ΔABC, the bisector of angles B (internal) and C (external) lie along the lines $x = y$ and $y = 0$. If A is (1, 2) then $\sqrt{10}\ d(A, BC)$ where $d(A, BC)$ represents distance of point A from side BC.

Column-II

(a) 1

(b) 4

(c) 0

(d) 5

4. Column-I

(i) If the slope of one of the lines represented by $ax^2 - 6xy + y^2 = 0$ is square of the other, then a is

(ii) If two of the lines given by the equation $ax^3 - 9x^2y - xy^2 + 4y^3 = 0$ are perpendicular, then a is

(iii) If two of the lines represented by $x^4 + x^3y + cx^2y^2 - xy^3 + y^4 = 0$ bisect the angle between the other two having sum of their slopes as 2, then the value of c is

Column-II

(a) –4

(b) –6

(c) 8

(d) 5

(e) –27

SECTION–VIII

INTEGER-TYPE

1. If the straight lines joining the origin to points of intersection of the straight line $4x + 3y = 24$ and the curve $(x - 3)^2 + (y - 4)^2 = c^2$, are at right angles, then find the value of $|c|$.
2. The pair of lines joining the origin to the point of intersection of the curve $\frac{x^2}{16}+\frac{y^2}{9}=1$ by the line $2x + 2y + k = 0$ are coincident. Then find value of $|k|$.
3. Two sides of a triangle have the combined equation $x^2 - 3y^2 - 2xy + 8y - 4 = 0$. The third side, which is variable always passes through the point $(-5, 1)$. If the range of values of the slope of the third line so that the origin is an interior point of the interval (a, b), then find the value of $\left(a+\frac{1}{b^2}\right)$.
4. Consider the set of all triangles OPQ where 'O' is the origin and P and Q are distinct points in the plane with non negative integral co-ordinates (x, y) such that $5x + y = 99$. Find the number of such distinct triangles whose area is a positive integer.
5. Consider two points $A \equiv (1, 2)$ and $B \equiv (3, -1)$. Let M be a point on the straight line $L \equiv x + y = 0$. If M be a point on the line $L = 0$ such that $|AM - BM|$ is maximum, then find the distance of M from $N \equiv (1, 1)$.
6. Line $\frac{x}{6}+\frac{y}{8}=1$, intersects the x and y axes at M and N respectively. If the co-ordinates of the point P lying inside the triangle OMN (where 'O' is origin) are (a, b) such that the areas of the triangle POM, PON and PMN are equal. Find the radius of the circle escribed opposite to the angle N.
7. The point A divides the join of $P(-5,1)$, $Q(3,5)$ in the ratio k: 1, find the integer value of k for which the area of ΔABC where B is $(1,5)$ and C is $(7, -2)$ is equal to 2 units in magnitude.
8. Triangle ABC lies in the cartesian plane and has an area of 70 sq. units. The co-ordinates of B and C are $(12, 19)$ and $(23,20)$ respectively and the co-ordinates of A are (p, q). The line contianing the median to the side BC has slope -5. Find the largest possible value of $(p + q)$.
9. The two line pairs $y^2 - 4y + 3 = 0$ and $x^2 + 4xy + 4y^2 - 5x - 10y + 4 = 0$ enclose a 4 sided convex polygon, then find the area of polygon.
10. Find the value of 'a' s.t. the portion of the line $ax + 3y - 1 = 0$, intercepted between the lines $ax + y + 1 = 0$ and $x + 3y = 0$ subtend a right angle at origin.
11. Find the value of $|m|$ if the lines joining the origin and the points of intersection of the line $y = mx + 1$ and $x^2 + y^2 = 1$ perpendicular to one another.

Answer Keys

SECTION–III

1. (c)	**2.** (a)	**3.** (c)	**4.** (c)	**5.** (d)	**6.** (a)	**7.** (c)	**8.** (a)	**9.** (d)	**10.** (a)
11. (c)	**12.** (b)	**13.** (c)	**14.** (a)	**15.** (d)	**16.** (b)	**17.** (d)	**18.** (c)	**19.** (a)	**20.** (c)
21. (b)	**22.** (b)	**23.** (a)	**24.** (a)	**25.** (a)	**26.** (c)	**27.** (b)	**28.** (c)	**29.** (d)	**30.** (b)
31. (c)	**32.** (a)	**33.** (a)	**34.** (c)	**35.** (a)	**36.** (c)	**37.** (d)	**38.** (a)	**39.** (b)	**40.** (a)
41. (a)	**42.** (a)	**43.** (c)	**44.** (a)	**45.** (d)	**46.** (a)	**47.** (a)	**48.** (c)	**49.** (d)	**50.** (b)
51. (c)	**52.** (d)	**53.** (a)	**54.** (b)	**55.** (a)	**56.** (d)	**57.** (c)	**58.** (b)	**59.** (c)	**60.** (b)
61. (d)	**62.** (d)	**63.** (d)	**64.** (a)						

SECTION–IV

1. (a, c)	**2.** (a, b, c, d)	**3.** (a, b, c)	**4.** (a, b, c)	**5.** (a, d)	**6.** (c, d)	**7.** (b, c)
8. (a, b)	**9.** (b, c, d)	**10.** (a, b)	**11.** (a, b, c)	**12.** (a,b)	**13.** (a, c)	**14.** (a, b, c)
15. (a,c)						

SECTION–V

1. (a) **2.** (a) **3.** (b) **4.** (a) **5.** (c) **6.** (a) **7.** (b) **8.** (a) **9.** (d) **10.** (d)

SECTION–VI

1. (d) **2.** (b) **3.** (a) **4.** (b) **5.** (c) **6.** (a) **7.** (c) **8.** (d) **9.** (b) **10.** (a)
11. (a) **12.** (c) **13.** (c) **14.** (d) **15.** (b) **16.** (c) **17.** (a) **18.** (b) **19.** (c) **20.** (a)
21. (a) **22.** (a) **23.** (b) **24.** (c)

SECTION–VII

1. (i) → (b) (ii) → (d) (iii) → (a) (iv) → (c)
2. (i) → (a, d) (ii) → (a, c) (iii) → (a) (iv) → (a)
3. (i) → (c) (ii) → (d) (iii) → (a) (iv) → (b)
4. (i) → (c, e) (ii) → (a, d) (iii) → (a)

SECTION–VIII

1. 5 **2.** 10 **3.** 24 **4.** 90 **5.** 10 **6.** 4 **7.** 7 **8.** 47 **9.** 6 **10.** –6
11. 1

HINTS AND SOLUTIONS

TEXTUAL EXERCISE-1 (SUBJECTIVE)

1. Given $m = \tan 150° = -1/\sqrt{3}$, $c = 2$. The required equation is $y = -\frac{x}{\sqrt{3}} + 2$ or $x + y\sqrt{3} = 2\sqrt{3}$

2. Given $m = \tan 120° = -\sqrt{3}$ and $c = -3$. The required equation is $y = -\sqrt{3}\,x - 3$ or $y + x\sqrt{3} + 3 = 0$

3. $x = 0$ is y axis, $y = m_1 x + c_1$ and $y = m_2 x + c_2$, x-coordinate of point of intersection will be $x = \frac{c_2 - c_1}{m_1 - m_2}$.

   Hence the area of Δ formed. $= \frac{1}{2}\frac{(c_2 - c_1)^2}{|m_1 - m_2|}$

4. $L_1: 2x + 3y = 7$ and $L_2: x + 3y = 5$. The point f intersection is $P(2, 1)$.

   A line through (2, 1) with slope m will be $y - 1 = m(x - 2)$ or $mx - y = 2m - 1$

   $\Rightarrow$ x-intercept $= \frac{2m-1}{m}$ and y-intercept $= 1 - 2m$

   Since intercepts are equal in magnitude $\left|\frac{2m-1}{m}\right| = |1 - 2m|$

   $\Rightarrow \frac{2m-1}{m} = \pm(1 - 2m)$

   $\Rightarrow \frac{2m-1}{m} - (1 - 2m) = 0$ or $\frac{2m-1}{m} - (2m - 1) = 0$

   $\Rightarrow (2m-1)\left(\frac{1}{m} + 1\right) = 0$ or $(2m-1)\left(\frac{1}{m} - 1\right) = 0$

   $\Rightarrow m = \frac{1}{2}, -1, 1$

   $\therefore$ Lines are $(y-1) = \frac{1}{2}(x-2) \Rightarrow x - 2y = 0; m = \frac{1}{2}$

   $(y-1) = -1(x-2) \Rightarrow x + y - 3 = 0; m = -1$

   $(y-1) = 1(x-2) \Rightarrow x - y - 1 = 0; m = 1$

5. $L_1: 3x + y = 2$ and $L_2: x + 5y = 7$, so $15x + 5y = 10$

   $\Rightarrow x = \frac{3}{14}, y = \frac{19}{14}$

   Line passing through (5, 2) and $\left(\frac{3}{14}, \frac{19}{14}\right)$ is $y - 2 = \frac{9}{67}(x - 5)$ i.e., $9x - 67y + 89 = 0$

6. $L_1: 2x + 4y = -7$ and $L_2: x - y = -2$ or $2x - 2y = -4$ gives $6y = -3$, so $y = -1/2$ and $x = -5/2$

   The point of intersection is $P(-5/2, -1/2)$. The line through P with slope $m = 3$ will be $y + 1/2 = 3(x + 5/2)$ or $3x - y + 7 = 0$

7. $L_1: x + 2y = 3$ and $L_2: 4x - y = -7$ or $8x - 2y = -14$, so $9x = -11$

   $\therefore\ x = -\frac{11}{9}, y = \frac{19}{9}$

   Any line parallel to $x - 3y + 2 = 0$ is $x - 3y + c = 0$. Since $P\left(\frac{-11}{9}, \frac{19}{9}\right)$ satisfies so $c = \frac{68}{9}$ and the equation will be $9x - 27y + 68 = 0$

8. $L_1: 2x + 3y = 4$ and $L_2: x - 5y = -7$ or $2x - 10y = -14$, so $y = 18/13$ and $x = -1/13$

   Point of intersection $P\left(\frac{-1}{13}, \frac{18}{13}\right)$.

   A line perpendicular to $x = 0$ is $y =$ constant

   $\Rightarrow y = \frac{18}{13}$

9. Let (x_1, y_1) be on L_1, so $L_1(x_1, y_1)$: $5x_1 - y_1 = 4$ and let (x_2, y_2) be on L_2, so $L_2(x_2, y_2)$: $3x_2 + 4y_2 = 4$

   Now, from the mid-point $x_1 + x_2 = 2$ and $y_1 + y_2 = 10$, so $x_1 = 2 - x_2$ and $y_1 = 10 - y_2$

   Putting in L_1 we get $5x_2 - y_2 = -4$ or $20x_2 - 4y_2 = -16$.

   Now $3x_2 + 4y_2 = 4$ (from L_2) gives $x_2 = -\frac{12}{23}$ and $y_2 = \frac{32}{33}$.

   $\left\{\text{No need to find } x_1 \text{ and } y_1 \left(x_1 = \frac{58}{23}, y_1 = \frac{198}{23}\right)\right\}$

   The straight line passing through (1, 5) and $\left(-\frac{12}{23}, \frac{32}{23}\right)$ is $y - 5 = \frac{83}{35}(\quad 1)$ or $83x - 35y + 92 = 0$

10. Given: x-intercept $a = -4$ units and y-intercept $b = 2$ units.

    The equation of the straight line will be $\frac{x}{(-4)} + \frac{y}{2} = 1$ or $x - 2y + 4 = 0$

11. Equation of straight line through (1, –2) will be $y + 2 = m(x - 1)$ or $mx - y = (m + 2)$

    The intercepts will be $a = \frac{m+2}{m}$ and $b = -(m + 2)$ from the given $\frac{m+2}{m} = -(m+2) \Rightarrow m = -2, -1$

    $m = -1$ will give intercepts of same sign, where as $m = -2$ will give $a = b = 0$ i.e., the straight line will pass through origin

    $\therefore$ Line is either $x + y + 1 = 0$ or $2x + y = 0$

12. As in 9. Let $(a, 0)$ and $(0, b)$ be the points on x-axis and y-axis respectively, so $2x_1 = a$ and $2y_1 = b$ and equation of line is $\frac{x}{a} + \frac{y}{b} = 1$ Putting the values of a and b, we get $\frac{x}{x_1} + \frac{y}{y_1} = 2$, when $(x_1, y_1) = (5, 2)$, we get $\frac{x}{5} + \frac{y}{2} = 2$ or $2x + 5y = 20$

13. A straight line passing through (2, 3) will be $y - 3 = m(x - 2)$ or $mx - y = (2m - 3)$ intercepts are $a = \frac{2m-3}{m}$ and $b = 3 - 2m$.

Since intercepts are equal but of opposite signs

$\Rightarrow \quad \dfrac{2m-3}{m} = 2m-3$, so $m = 1$ or $m = 3/2$.

When $m = 1$ the line is $x - y + 1 = 0$ and when $m = 3/2$ the line is $3x - 2y = 0$

14. Let the line be $y = mx + c$

$\Rightarrow \; a = -c/m, b = c$

According to the given $ab = -6$ and $a + b = 1$, so

$-\dfrac{c}{m} + c = 1$ or $c\left\{\dfrac{m-1}{m}\right\} = 1$ gives $-\dfrac{c^2}{m} = -6$

$\Rightarrow \; c^2 = 6m$ (so $m > 0$)

Putting in ——, we get $\dfrac{m^2}{(m-1)} = 6m$

$m \neq 0, 1$, we get $m = 6\{m^2 + 1 - 2m\}$ gives $6m^2 - 13m + 6$

$= 0$, so $m = \dfrac{13 \pm \sqrt{169-144}}{12} = \dfrac{13 \pm 5}{12} = \dfrac{3}{2}, \dfrac{2}{3}$ and $c = \dfrac{m}{m-1}$

gives $c = 3$ for

$m_1 = 3/2$ and $c = -2$ for $m_2 = 2/3$

The straight lines are $y = \dfrac{3}{2}x + 3$ or $3x - 2y + 6 = 0$

Similarly $y = \dfrac{2}{3}x - 2$ or $2x - 3y - 6 = 0$

15. A straight line through (2, 2) will be $y - 2 = m(x - 2)$ or $mx - y = 2(m - 1)$ intercepts are $a = \dfrac{2(m-1)}{m}$ and $b = 2(1 - m)$

According to the given $a + b = 9$ or $\dfrac{2(m-1)}{m} + 2(1-m) = 9$

gives $2(m - 1)\{m - 1\} = -9m$ or $2m^2 + 5m + 2 = 0$ gives $m = -1/2, -2$ and the straight lines are $2x + y = 6$ or $x + 2y = 6$.

TEXTUAL EXERCISE–1 (OBJECTIVE)

1. (a) Straight line passing through $A(3, 3)$, $B(7, 6)$ is $y - 3 = \dfrac{3}{4}(x - 3)$ or $3x - 4y = -3$

So the straight line intersects axes at $C(-1, 0)$ and $D(0, 3/4)$

$\Rightarrow \quad CD = \sqrt{1 + \dfrac{9}{16}} = \dfrac{5}{4}$ Units

2. (c) Let $y = m_r x$ intersect with $x + y = 1$ at A, B, C respectively for $r = 1, 2, 3$

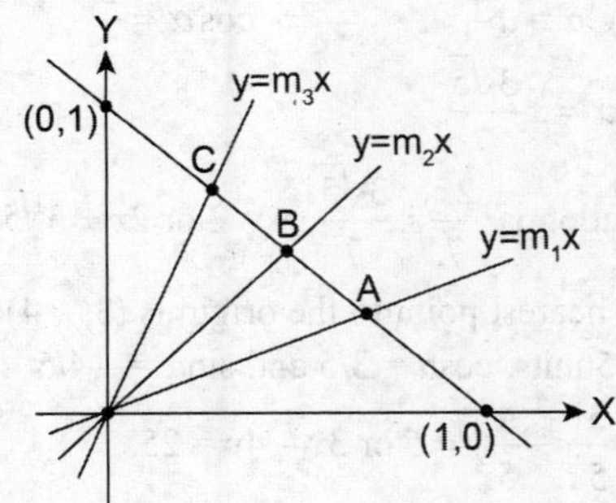

$\Rightarrow \quad A\left(\dfrac{1}{m_1+1}, \dfrac{m_1}{m_1+1}\right)$, $B\left(\dfrac{1}{m_2+1}, \dfrac{m_2}{m_2+1}\right)$ and $C\left(\dfrac{1}{m_3+1}, \dfrac{m_3}{m_3+1}\right)$

As $AB = BC$

So $\dfrac{2}{m_2+1} = \dfrac{1}{m_1+1} + \dfrac{1}{m_3+1}$ and $\dfrac{2m_2}{m_2+1} = \dfrac{m_1}{m_1+1} + \dfrac{m_3}{m_3+1}$

or $2 - \dfrac{2}{m_2+1} = 1 - \dfrac{1}{m_1+1} + 1 - \dfrac{1}{m_3+1}$

Again we get $\dfrac{2}{m_2+1} = \dfrac{1}{m_1+1} + \dfrac{1}{m_3+1}$, so $(1 + m_1), (1 + m_2), (1 + m_3)$ are in H.P.

3. (b) Line L_1 has intercepts on the axes are as $(1, -2)$

$\Rightarrow$ Slope of L_1 is $m_1 = 2$

Intercepts of L_2 are $(2, -1)$

$\Rightarrow$ Slope of L_2 is $m_2 = \dfrac{1}{2}$

Hence $\tan\theta = \dfrac{\left|2 - \dfrac{1}{2}\right|}{1 + 2\left(\dfrac{1}{2}\right)} = \dfrac{3}{4} \Rightarrow \theta = \tan^{-1}(3/4)$

4. (c) The required straight line pass through the mid points $A(1, -2)$ and $B(2, -5)$ i.e., $\left(\dfrac{3}{2}, -\dfrac{7}{2}\right)$

Hence the equation of straight line is $y + \dfrac{3}{2} = m\left(x - \dfrac{3}{2}\right)$

or $mx - y = \dfrac{7}{2} + \dfrac{3m}{2}$

So, intercept $a = \dfrac{7+3m}{2m}$ and $b = \dfrac{-(3m+7)}{2}$

According to the given $\dfrac{a}{b} = \dfrac{2}{3}$

$\Rightarrow \quad \dfrac{21+9m}{2m} = \dfrac{(-2)(3m+7)}{2}$ or $\dfrac{3}{m} = -2$ (For $3m + 7 \neq 0$)

and the equation is $6x + 4y + 5 = 0$

5. (b) From the given it can be concluded that the line passes through (1, 2) and (3, 0), so the equation is $x + y = 3$

6. (a) From the given, we get $m = 1 = \tan 45°$, point $P(1, -1)$

$\Rightarrow$ The equation is $y + 1 = x - 1$ or $x - y = 2$

7. (b) From the given, we get $m = \tan 135° = -1$ and point $P(2, 3)$

$\Rightarrow$ The equation is $y - 3 = (-1)(x - 2)$ or $x + y = 5$

8. (c) From the given, we get $m = \tan 60° = \sqrt{3}$ and points $P(-2, 3)$

$\Rightarrow$ The equation is $y - 3 = \sqrt{3}(x + 2)$ or $x\sqrt{3} - y + (3 + 2\sqrt{3}) = 0$

9. (a) From the given, $m = \pm 1$ and point $P(2, 1)$

$\Rightarrow$ The equations are (i) $y - 1 = (x - 2)$ i.e., $x - y = 1$ and (ii) $y - 1 = (-1)(x - 2)$ i.e., $x + y = 3$

10. **(b)** Given vertex are $A(a, b), B(a', b'), C(-a, b), D(a', -b')$

$\Rightarrow$ Mid point of AB is $P\left(\frac{a+a'}{2}, \frac{b+b'}{2}\right)$ and mid point of CD is $Q\left(\frac{a'-a}{2}, \frac{b-b'}{2}\right)$

Slope of PQ is $m = \frac{\left(\frac{b+b'}{2} - \frac{b-b'}{2}\right)}{\left(\frac{a+a'}{2} - \frac{a'-a}{2}\right)} = \frac{b'}{a}$ and the equation of PQ will be $y - \frac{b+b'}{2} = \frac{b'}{a}\left\{x - \frac{a'-a}{2}\right\}$

$\Rightarrow 2ay - a(b + b') = 2b'x - (a' - a)b'$ or $2ay - 2b'x = ab - a'b'$

11. **(c)** Given points $A(-1, 6), B(-3, -9), C(5, -8)$, mid point of AB is $F = (-2, -3/2)$

$\Rightarrow$ Equation of CF is $13x + 14y + 47 = 0$

12. **(c)** Line passing through $(a, 0)$ and $(0, b)$ is $\frac{x}{a} + \frac{y}{b} = 1$.

Since (2, 2) satisfies

$\Rightarrow \frac{2(a+b)}{ab} = 1$, so $2(a + b) = ab$, also $\frac{ab}{2} = |\lambda|$

$\Rightarrow 2(a + b) = 2|\lambda|$

i.e., sum of roots $(a + b) = |\lambda|$ and product of roots $(ab) = 2|\lambda|$

so, the equation will be $x^2 - |\lambda|x + |2\lambda| = 0$

13. **(d)** Let P (x, y) be the variable point, so $|x - 3| + |y - 5| = 5$

Shifting origin to (3, 5), we observe $|X| + |Y| = 5$.

Which will form square with centre at (3, 5)

14. **(a)** Line through the intersection point $P\left(\frac{3}{5}, \frac{1}{5}\right)$ will be

$y - \frac{1}{5} = m\left(x - \frac{3}{5}\right)$ or $mx - y = \frac{3m-1}{5}$

Points of intersection with axes are

$A\left(\frac{3m-1}{5m}, 0\right), B\left(0, \frac{1-3m}{5}\right)$; mid point is

$M\left(\frac{3m-1}{10m}, \frac{1-3m}{10}\right) \Rightarrow m = \frac{1}{3-10x} = \frac{1-10y}{3}$ and we get

$(10x - 3)(10y - 1) = 3$ or $x + 3y - 10xy = 0$

TEXTUAL EXERCISE–2 (SUBJECTIVE)

1. **(a)** Given $p = 4$ and $\alpha = 150°$

$\Rightarrow \cos\alpha = -\frac{\sqrt{3}}{2}$, $\sin\alpha = 1/2$.

The equation of straight line in perpendicular form is $x\cos\alpha + y\sin\alpha = p$ gives $-\frac{\sqrt{3}}{2}x + \frac{1}{2}y = 4$ or $x\sqrt{3} - y + 8 = 0$

(b) Given $p = 8$, $\alpha = 300°$

$\Rightarrow \cos\alpha = 1/2$, $\sin\alpha = -\frac{\sqrt{3}}{2}$

Hence the equation is $x\cos\alpha + y\sin\alpha = p$ gives $\frac{x}{2} - \frac{\sqrt{3}}{2}y = 8$ or $x - y\sqrt{3} = 16$

(c) Given $p = 2$, $\cos\alpha = 3/5$

$\Rightarrow \sin\alpha\ \alpha = \pm 4/5$ (1st & 4th quadrant)

The equation will be $3/5\,x \pm 4/5y = 2$ or $3x \pm 4y = 10$

2. Given $p = 2$ and $\sin\alpha = 1/3$

$\Rightarrow \cos\alpha = \pm\frac{2\sqrt{2}}{3}$.

The equation will be $x\left(\pm\frac{2\sqrt{2}}{3}\right) + \frac{1}{3}y = 2$ or $(\pm 2\sqrt{2})x + y = 6$

3. Given $p = \sqrt{2}$ and $\alpha = 135°, -\pi/4$

Case (i): $\cos\alpha = \frac{}{\sqrt{}}$ and $\sin\alpha\ \frac{1}{\sqrt{2}}$

$\Rightarrow x\left(\frac{-1}{\sqrt{2}}\right) + \frac{y}{\sqrt{2}} = \sqrt{2}$ gives $x - y + 2 = 0$

Case (ii): $\cos\alpha = \frac{1}{\sqrt{2}}$ and $\sin\alpha = \frac{-1}{\sqrt{2}}$ and the equation $\frac{x}{\sqrt{2}} - \frac{y}{\sqrt{2}} = \sqrt{2}$ gives $x - y = 2$

4. **Case (i):** $p = 5$ and $\alpha = 45°$

The equations is $\frac{x}{\sqrt{2}} + \frac{y}{\sqrt{2}} = 5$ or $x + y = 5\sqrt{2}$

Case (ii): $p = 5$ and $\alpha = -135°$

$\Rightarrow \sin\alpha = -\frac{1}{\sqrt{2}} = \cos\alpha$

The equation will be $\frac{-x}{\sqrt{2}} + \frac{-y}{\sqrt{2}} = 5$ gives $x + y + 5\sqrt{2} = 0$

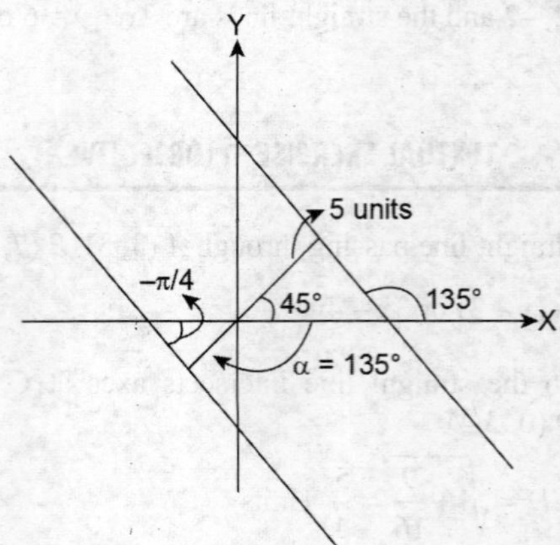

5. **(a)** x-intercepts; $a = 7$ units and $p = 2$

$\Rightarrow p\sec\alpha = a \qquad \Rightarrow \cos\alpha = \frac{2}{7}$

$\therefore \sin\alpha = \pm\frac{3\sqrt{5}}{7}$

The equation is $\frac{2x}{7} \pm \frac{3\sqrt{5}}{7}y = 2$ or $2x \pm 3\sqrt{5}y = 14$

(b) The nearest point to the origin is $(3, -4)$

$\Rightarrow p = 5$units, $\cos\alpha = 3/5$ and $\sin\alpha = -4/5$

$\Rightarrow \frac{3}{5}x - \frac{4}{5}y = 5$ or $3x - 4y = 25$.

6. (i) Line through $P(2, 3)$ with slope $m = 2$ gives $\tan\theta = 2/1$
$\Rightarrow$ $\sin\theta = 2/\sqrt{5}$ and $\cos\theta = 1/\sqrt{5}$.
The parameter form is $\dfrac{x-2}{\left(\frac{1}{\sqrt{5}}\right)} = \dfrac{y-3}{\left(\frac{2}{\sqrt{5}}\right)} = r$

(ii) Line through (1, 4) with slope $m = -1/\sqrt{3}$, so $\theta = 150°$, we get $\cos\theta = -\dfrac{\sqrt{3}}{2}$ and $\sin\theta = \dfrac{1}{2}$

The equation will be $\dfrac{x-1}{\left(-\frac{\sqrt{3}}{2}\right)} = \dfrac{y-4}{\left(\frac{1}{2}\right)} = r$ or

$\dfrac{x-1}{\frac{\sqrt{3}}{2}} = \dfrac{y-4}{\left(-\frac{1}{2}\right)} = r$

(iii) Straight line through (1, 3) and (4, 2), slope $m = \tan\theta = -1/3$, either $\cos\theta = \dfrac{3}{\sqrt{10}}, \sin\theta = \dfrac{-1}{\sqrt{10}}$ or $\cos\theta = -\dfrac{3}{\sqrt{10}}, \sin\theta = \dfrac{1}{\sqrt{10}}$

The equation is $\dfrac{x-1}{\left(\frac{3}{\sqrt{10}}\right)} = \dfrac{y-3}{\left(-1/\sqrt{10}\right)} = r$

7. A line at a distance of 3 units from $3x + 4y = 8$ is $3x + 4y = 23$ or $3x + 4y + 7 = 0$.
Intersection of these lines with $3x - 2y = 2$ gives $P_1(3, 7/2)$ and $P_2(-1/3, -3/2)$

8. Given P (3, 2), slope $m = \tan\pi/6 = \dfrac{1}{\sqrt{3}}$. So $\sin\theta = \dfrac{1}{2}$ and $\cos\theta = \dfrac{\sqrt{3}}{2}$

Hence the equation of straight line is $\dfrac{x-3}{\left(\frac{\sqrt{3}}{2}\right)} = \dfrac{y-2}{\left(\frac{1}{2}\right)} = r$

$\Rightarrow$ $x = 3 + \dfrac{\sqrt{3}}{2}r$ and $y = 2 + \dfrac{r}{2}$

The line will intersect the line $3x - 4y + 8 = 0$ at Q, so

$3\left(3+\dfrac{\sqrt{3}r}{2}\right) - 4\left(2+\dfrac{r}{2}\right) + 8 = 0$ gives $\left(\dfrac{3\sqrt{3}-4}{2}\right)r = -9$

$\Rightarrow$ $|r| = \dfrac{18}{11}\left(3\sqrt{3}+4\right)$ units.

9. From the given $x = -2 + \dfrac{r}{\sqrt{2}}$, $y = 3 + \dfrac{r}{\sqrt{2}}$ for $r = \pm 4\sqrt{2}$, we get $x = -2 \pm 4 = 2, 6$ and $y = 3 \pm 4 = 7, -1$
The point P (2, 7); (–6, –1)

10. Let A (1, 1) and C (–2, –1) $\Rightarrow AC = \sqrt{13}$

$\Rightarrow$ Mid point of AC is M $\left(\dfrac{-1}{2}, 0\right)$ and slope of $m = \dfrac{2}{3}$, so

slope of $BD = -\dfrac{3}{2}$

$\Rightarrow$ $\cos\theta = \dfrac{2}{\sqrt{3}}$, $\sin = -\dfrac{3}{\sqrt{13}}$

Hence $x = \dfrac{-1}{2} \pm \dfrac{2}{\sqrt{3}}, \dfrac{\sqrt{13}}{2} = \dfrac{1}{2}, \dfrac{-3}{2}$ and

$y = 0 \pm \left(\dfrac{-3}{\sqrt{13}}\right)\dfrac{\sqrt{13}}{2} = -\dfrac{3}{2}, \dfrac{3}{2}$ respectively.

Hence $B\left(\dfrac{1}{2}, -\dfrac{3}{2}\right)$ and $D\left(-\dfrac{3}{2}, \dfrac{3}{2}\right)$

11. Let A (a, 0) and B (0, b) be the points of intersect on the respective axes where P (–4, 3) divides AB in the ratio 5 : 3

$\Rightarrow$ $P\left(\dfrac{3a}{8}, \dfrac{5b}{8}\right) = (-4, 3)$

$\Rightarrow$ $a = -\dfrac{32}{3}, b = \dfrac{24}{5}$

So the equation of straight line is $\dfrac{3x}{-32} + \dfrac{5y}{24} = 1$ or $9x - 20y + 96 = 0$

12. Let A (a, 0), B (0, b) and D (x_4, y_4) where $ABCD$ is square.

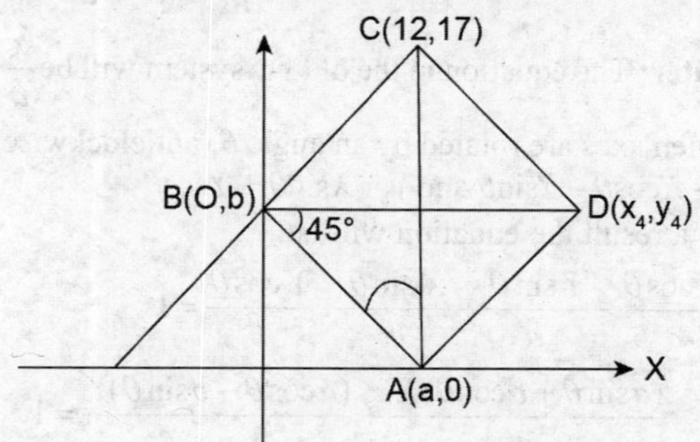

Now slope of $AB \times$ slope of $BC = -1$ gives $\left(\dfrac{b}{-a}\right)\left(\dfrac{17-b}{12}\right) = -1$,
so $b(17 - b) = 12a$ ...(i)
From $AB^2 = BC^2$ we get, $a^2 + b^2 = 144 + (17 - b)^2$

$\Rightarrow$ $a^2 + b^2 = 144 + \left(\dfrac{12a}{b}\right)^2 = \dfrac{144(a^2+b^2)}{b^2}$

$\Rightarrow$ $b^2 = 144$ i.e., $b = \pm 12$
Case (i): $b = 12$, then $a = 5 \Rightarrow x_4 = 17$ and $y_4 = 5$
Hence A (5, 0), B (0, 12) and (17, 5)
Case (ii): When $b = -12$, then $a = -29$
$\Rightarrow$ $x_4 = -27$ and $y_4 = 29$
So, A (–29, 0), B (0, –12) and D (–17, 29)

13. Point P lies on $8y = 15x$, so let $P\left(x_1, \dfrac{15}{8}x_1\right)$ and Q lies on $10y = 33x$, so let $Q\left(x_2, \dfrac{33}{10}x_2\right)$

Mid point of P and Q is M (8, 6)

$\Rightarrow$ $x_1 + x_2 = 16$ and $\dfrac{15}{8}x_1 + \dfrac{33}{10}x_2 = 12$ or $25x_1 + 44x_2 = 160$

On solving, we get $19x_1 = 16 \times 34$, so

$x_1 = \dfrac{544}{19}, y_1 = \dfrac{15x_1}{8} = \dfrac{1020}{19}$

and $x_2 = -\dfrac{240}{19}, y_2 = \dfrac{33}{10}x_2 = \dfrac{-792}{19}$

$\therefore$ Coordinates of $P \equiv \left(\dfrac{544}{19}, \dfrac{1020}{19}\right)$ and that of

$Q \equiv \left(-\dfrac{240}{19}, -\dfrac{792}{19}\right)$

TEXTUAL EXERCISE-2 (OBJECTIVE)

1. **(b), (c)** When the needful is done, then $p^2 = b^2$ and $p^2 = a^2$

$\Rightarrow\ b^2 + q^2 = p^2 + a^2$

2. **(b)** Rotate the axes through an angle θ, so that $OM \perp AB$ (Now $OY \parallel AB$).

Let $OM = p$ (perpendicular distance)

From area of ΔOAB, $\dfrac{AB \times OM}{2} = \dfrac{ab}{2}$, we get

$|p| = \dfrac{|ab|}{\sqrt{a^2+b^2}} \Rightarrow \dfrac{1}{p^2} = \dfrac{a^2+b^2}{a^2b^2}$

Obviously $q = \infty$ (**as** $AB \parallel OY$)

Easily we can conclude that $\dfrac{1}{p^2} + \dfrac{1}{q^2} = \dfrac{1}{a^2} + \dfrac{1}{b^2}$

Aliter: The equation in the old *x-y* system will be $\dfrac{x}{a} + \dfrac{y}{b} = 1$.

When axes are rotated by an angle θ (anticlockwise), we get $x = X\cos\theta - Y\sin\theta$ and $y = X\sin\theta + Y\cos\theta$

As a result the equation will be

$$\frac{X\cos\theta}{a} - \frac{Y\sin\theta}{a} + \frac{X\sin\theta}{b} + \frac{Y\cos\theta}{b} = 1$$

$$\Rightarrow \frac{(a\sin\theta + b\cos\theta)X}{ab} + \frac{(a\cos\theta - b\sin\theta)Y}{ab} = 1$$

$$\Rightarrow \frac{1}{p} = \frac{a\sin\theta + b\cos\theta}{ab} \text{ and } \frac{1}{q} = \frac{a\cos\theta = b\sin\theta}{ab}$$

Squaring and adding, we get $\dfrac{1}{p^2} + \dfrac{1}{q^2} = \dfrac{a^2+b^2}{a^2+b^2} = \dfrac{1}{a^2} + \dfrac{1}{b^2}$

3. **(b)** The given equation is $x\sqrt{3} + y = -2$

So $x\dfrac{\sqrt{3}}{2} + \dfrac{1}{2}y = -1$ from $x\cos\theta + y\sin\theta = p$ or

$-\dfrac{\sqrt{3}}{2}x - \dfrac{1}{2}y = 1 \Rightarrow \theta = \dfrac{7\pi}{2}$ (Anticlockwise)

So equation will be $x\cos\dfrac{7\pi}{6} + y\sin\dfrac{7\pi}{6} = 1$

4. **(d)** $4y = -3x + 12$ or $y = -\dfrac{3}{4}x + 3$

So, *y*-intercept = 3

5. **(c)** $4x - 3y = -5$ gives $\left(-\dfrac{4}{5}\right)x + \dfrac{3}{5}y = 1$

6. **(b)** The lines $x\cos\alpha + y\sin\alpha = p_1$ and $x\cos\beta + y\sin\beta = p_2$ will be perpendicular when $|\alpha - \beta| = \pi/2$

7. **(c)** A general point on the line through (1, 2) with θ angle with horizontal is $(1 + r\cos\theta, 2 + r\sin\theta)$

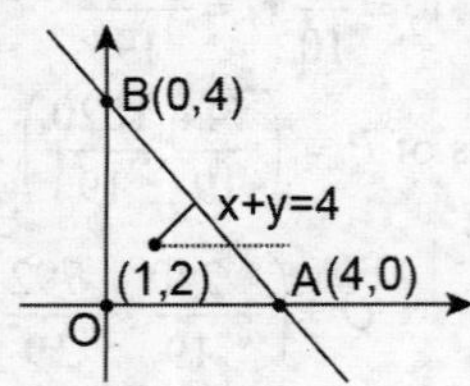

For $|r| = \dfrac{\sqrt{6}}{3}$ the point will satisfy $x + y = 4$

$\Rightarrow\ 1 + \dfrac{\sqrt{6}}{3}\cos\theta + 2 + \dfrac{\sqrt{6}}{3}\sin\theta = 4$ gives $\sqrt{6}$ {$\sin\theta$ + $\cos\theta$} = 3

or $\sin\left(\theta + \dfrac{\pi}{4}\right) = \dfrac{\sqrt{3}}{2}$ gives

$\theta + \dfrac{\pi}{4} = \dfrac{\pi}{6}, \dfrac{2\pi}{6} \Rightarrow \theta = \dfrac{\pi}{12}, \dfrac{5\pi}{12}$

8. **(b)** $\dfrac{x}{\cos\alpha} + \dfrac{y}{\sin\alpha} = a$ or $\dfrac{x}{\sin\left(\dfrac{\pi}{2} - \alpha\right)} + \dfrac{y}{\cos\left(\dfrac{\pi}{2} - \alpha\right)} = a$

$\Rightarrow\ x\cos\left(\dfrac{\pi}{2} - \alpha\right) + y\sin\left(\dfrac{\pi}{2} - \alpha\right) = a\sin\left(\dfrac{\pi}{2} - \alpha\right).\cos\left(\dfrac{\pi}{2} - \alpha\right)$

or $x\cos\left(\dfrac{\pi}{2} - \alpha\right) + y\sin\left(\dfrac{\pi}{2} - \alpha\right) = \dfrac{a}{2}\sin(\pi - 2\alpha)$

$\Rightarrow\ p_1 = \dfrac{a}{2}\sin(\pi - 2\alpha) \Rightarrow 2p_1 = a\sin 2\alpha$

Similarly $x\cos(-\alpha) + y\sin(-\alpha) = a\cos 2\alpha$

$\Rightarrow\ p_2 = a\cos^2\alpha$

Hence $4p_1^2 + p_2^2 = a^2\{\sin^2 2\alpha + \cos^2 2\alpha\} = a^2$

9. **(a)** $x+y=6$. A general point along PQ is $\left(r\cos\dfrac{\pi}{6}, 3 + r\sin\dfrac{\pi}{6}\right)$

If Q lies on $x + y = 6$, then $r\left\{\cos\dfrac{\pi}{6} + \sin\dfrac{\pi}{6}\right\} = 3$

$\Rightarrow\ r = \dfrac{6\{\sqrt{3}-1\}}{2} = \dfrac{6}{\sqrt{3}+1}$

When $|r| = 6$, then $R = (3\sqrt{3}, 6)$ or $(-3\sqrt{3}, 0)$

TEXTUAL EXERCISE-3 (SUBJECTIVE)

1. **(a)** $L_1: 2x + 3y - 5 = 0$ and $L_2: 4x + 6y - 10 = 0$. Observe that $\dfrac{a_1}{a_2} = \dfrac{b_1}{b_2} = \dfrac{c_1}{c_2} = \dfrac{1}{2}$

$\Rightarrow$ Lines are coincidental

(b) $L_1: 3x - 4y = 5$; $L_2: 3x + 4y = 7$, so $m_1 = \dfrac{3}{4}$ and $m_2 = -\dfrac{3}{4}$

As $m_1 \neq m_2$ and $m_1 m_2 \neq -1$

$\Rightarrow$ Lines are intersecting

(c) $L_1: x - 2y = 7$ and $L_2: x - 2y = 5$ as $\dfrac{a_1}{a_2} = \dfrac{b_1}{b_2} \neq \dfrac{c_1}{c_2}$

$\Rightarrow$ Lines are parallel.

2. Straight line joining (2, –3) and (–5, 1) is $L_1: 4x + 7y + 13 = 0$ and Straight line joining (7, –1) and (0, 3) is $L_2: 4x + 7y - 21 = 0$,

We observe that $\dfrac{a_1}{a_2} = \dfrac{b_1}{b_2} \neq \dfrac{c_1}{c_2}$

$\Rightarrow$ Lies are parallel but not coincident.

3. Let $A(-4, -1)$; $B(-2, -4)$; $C(4, 0)$ and $D(2, 3)$

Slope of AB is $m_{AB} = \frac{-3}{2}$ and slope of AD is $m_{AD} = \frac{4}{6} = \frac{2}{3}$

Now $m_{AB} \times m_{AD} = -1 \Rightarrow$ Lines are perpendicular.

Similarly slope of BC is $m_{BC} = 2/3$, so $AB \perp BC$.

Slope of CD is $m_{CD} = \frac{-3}{2}$, so $CD \perp BC$

$\Rightarrow$ A rectangle will be formed.

4. Given L_1: $\left(\frac{1}{a}\right)x + \left(\frac{1}{b}\right)y = 1 \Rightarrow$ slope $m_1 = \frac{-b}{a}$

L_2: $\left(\frac{1}{a}\right)x - \left(\frac{1}{b}\right)y = 1 \Rightarrow$ slope $m_2 = \frac{b}{a}$

Since $m_1 = -m_2$, so lines are equally inclined with x-axis as well as y-axis.

$$\tan\theta = \left|\frac{2\frac{b}{a}}{1+\left(-\frac{b^2}{a^2}\right)}\right| = \left|\frac{2ba^2}{a(a^2-b^2)}\right| \left|\frac{2ab}{a^2-b^2}\right|$$

When $a \neq b$, then $\theta = \tan^{-1}\frac{2ab}{a^2-b^2}$ (when $a = b$, then $\theta = 90°$).

5. Given L_1: $x - 2y + 3 = 0 \Rightarrow m_1 = 1/2$ and

L_2: $3x + y - 1 = 0 \Rightarrow m_2 = -3$

So $\tan\theta = \left|\frac{\frac{1}{2}+3}{1-\frac{3}{2}}\right| = \left|\frac{7}{2\left(-\frac{1}{2}\right)}\right| = |(-7)|$

For obtuse angle, we get $\theta = \pi + \tan^{-1}(-7)$ or $\theta = \pi - \tan^{-1}(7)$.

6. Let L_1: $7x + 5y - 4 = 0 \Rightarrow m_1 = -7/5$ and L_2: $(3x + 5y + 6) + k(2x - 3y + 1) = 0 \Rightarrow m_2 = \frac{2k+3}{3k-5}$ as $L_1 \perp L_2$

$\therefore \quad m_2 = \frac{2k+3}{3k-5} = \frac{5}{7}$ gives $14k + 21 = 15k - 25$

$\Rightarrow k = 46$

7. Given $A(2, 3)$, now points B and C lie on L_1: $x + y = 2$

$\Rightarrow m_{BC} = -1$

Let m_{AB} and m_{AC} be the slope of AB and AC, respectively.

$\Rightarrow \frac{m_{AB}+1}{1-m_{AB}} = \sqrt{3} \qquad \Rightarrow \quad m_{AB} = \frac{-1+\sqrt{3}}{\sqrt{3}+1} = (2-\sqrt{3})$

So equation of AB is $y - 3 = (2 - \sqrt{3})(x - 2)$

Similarly, $\frac{-1-m_{AC}}{1-m_{AC}} = \sqrt{3} \Rightarrow m_{AC} = \frac{\sqrt{3}+1}{\sqrt{3}-1} = (2+\sqrt{3})$

So equation of AC is $y - 3 = (2 + \sqrt{3})(x - 2)$

Hence the required equations are $y - 3 = (2 \pm \sqrt{3})(x - 2)$

TEXTUAL EXERCISE–3 (OBJECTIVE)

1. (c) Let $A(1, 2)$, $B(2, 3)$ and $C(4, 3)$, slope of $AB = 1$, slope of $BC = 0$

$\Rightarrow$ AD has equation $x = 1$, now $CF \perp AB$

$\Rightarrow$ Equation of CF is $y - 3 = (-1)(x - 4)$

$\Rightarrow$ $x + y = 7$, putting $x = 1$ we get, $y = 6$. So orthocenter is at $(1, 6)$

2. (d) The vertices are $A(4, -3)$, $B(-2, 1)$, $C(2, 3)$, mid point of AB is $(1, -1)$

Slope of AB is $m_{AB} = -\frac{4}{6} = -\frac{2}{3}$

Slope of BC is $m_{BC} = \frac{1}{2}$, mid point of BC is $(0, 2)$

Slope of AC is $m_{AC} = -3$

Equation of right bisector of AB is $y + 1 = \frac{3}{2}(x - 1)$ or $3x - 2y = 5$

Similarly right bisector of BC is $y - 2 = (-2)x$ or $2x + y = 2$

$\Rightarrow$ Circum centre is $\left(\frac{9}{7}, -\frac{4}{9}\right)$

3. (b) The straight lines are $x = 0$, $y = 0$ and $\frac{x}{3} + \frac{y}{4} = 1$

$\Rightarrow$ Area of $\Delta ABC = \frac{3 \times 4}{2} = 6$ square units and semi-perimeter $s = \frac{3+4+5}{2} = 6$ units

So, $r = \frac{\Delta}{s} = 1$ unit.

4. (a) Slope of straight line L_1: $7x - 9y - 19 = 0$ is $m_1 = 7/9$ as line L_2 through $A(\alpha, 3)$, $B(4, 1)$ has slope $m_2 = \frac{2}{\alpha - 4}$

so, when $L_1 \perp L_2$

Then, $m_1 m_2 = -1$, so $\frac{2}{\alpha-4} = -\frac{9}{7}$

$\Rightarrow \alpha = \frac{22}{9}$

5. (d) The required line perpendicular to $3x + 5y = 4$ has slope $m_2 = 5/3$

Since the x-intercept is -3, so $m_2 x - y + 5 = 0$ is the equation i.e., $5x - 3y + 15 = 0$

6. (a), (c) L_1: $\sqrt{3}x + y - 2 = 0$

$\Rightarrow m_1 = -\sqrt{3}$ and L_2: $\sqrt{3}x - y + 1 = 0$

$\Rightarrow m_2 = \sqrt{3}$

L_3: $y = 0$, As $m_1 + m_2 = 0$ and L_3 is horizontal (x-axis)

$\Rightarrow \Delta$ is isosceles

Further $\tan(\pm 60°) = \pm\sqrt{3}$

$\Rightarrow \Delta$ is equilateral

7. (d) Given: The vertices are $A(0, 0)$, $B(\cos\theta, \sin\theta)$, $C(\sin\theta, -\cos\theta)$

$\Rightarrow$ Centroid $G = \left(\frac{\cos\theta+\sin\theta}{3}, \frac{\sin\theta-\cos\theta}{3}\right)$; since it lies on $y = 2x$

$\Rightarrow \left(\frac{\sin\theta-\cos\theta}{3}\right) = 2\left(\frac{\cos\theta+\sin\theta}{3}\right)$ or $3\cos\theta = -\sin\theta$

$\Rightarrow \tan\theta = -3 \Rightarrow \theta = \tan^{-1}(-3)$

8. **(a), (b)** $x = 4$ is vertical line through (4, 0)

Now $3y^2 = x^2 \Rightarrow y = \pm\frac{x}{\sqrt{3}}$

So an equilateral Δ is formed which is also isosceles .

9. **(d)** Slope of line $y = -x$ is $m = -1$. So $\theta = 135°$ or $-45°$

$\Rightarrow$ The required lines will be horizontal and vertical passing through (1, 1)

$\therefore$ $x = 1, y = 1$ are the lines

TEXTUAL EXERCISE–4 (SUBJECTIVE)

1. Distance between L_1: $7x + 24y + 3 = 0$ and L_2: $7x + 24y + 28 = 0$ is $d = \frac{|28-3|}{\sqrt{7^2+24^2}} = 1$ unit

2. Distance between L_1: $mx - y + c = 0$ and L_2: $mx - y + d = 0$ is $p = \frac{|c-d|}{\sqrt{1+m^2}}$

3. L_1: $x\cos\theta + y\sin\theta = p$ (Here $\alpha = \theta$)

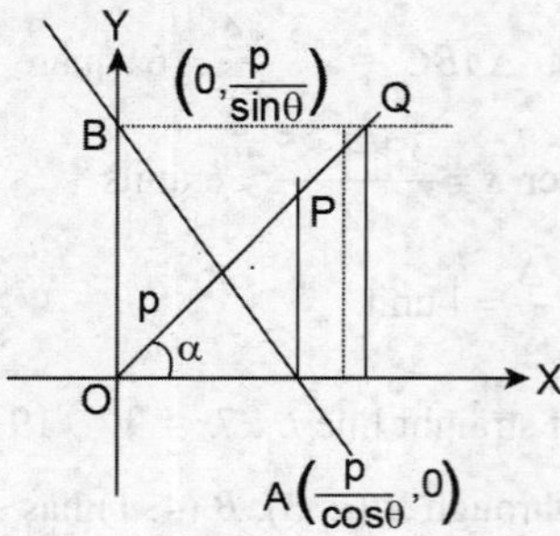

From the given equation of AP is $x = \frac{p}{\cos\theta}$ and equation BQ is $y = \frac{p}{\sin\theta}$

As OP (or OQ) is perpendicular to the line L_1

$\therefore$ Equation of OP is $x\sin\theta - y\cos\theta = 0$

$\Rightarrow OP = OA\sec\alpha = \frac{p}{\cos\alpha}.\sec\alpha = \frac{p}{\cos^2\alpha}$ and $OQ = OB$

$\text{cosec}\alpha = \frac{p}{\sin\alpha}.\text{cosec}\alpha = \frac{p}{\sin^2\alpha}$

$\Rightarrow PQ = \left|\frac{p}{\sin^2\alpha} - \frac{p}{\cos^2\alpha}\right| = \frac{|p(\cos^2\alpha - \sin^2\alpha)|}{\sin^2\alpha.\cos^2\alpha}$

$= \frac{4p\cos 2\alpha}{\sin^2 2\alpha}$ or $\frac{4p|\cos 2\theta|}{\sin^2 2\theta}$

4. **(a)** Given vertex $A = (2, 2)$ = hypotenuous BC lies on $3x + 4y = 4$

Slope of L_1: $m_1 = -\frac{3}{4} = \tan\theta$

$\Rightarrow$ Slope of L_2 is $m_2 = \tan\left(\theta + \frac{\pi}{4}\right) = \frac{m_1+1}{1-m_1}$

$\Rightarrow m_2 = \frac{1-3/4}{1-\left(-\frac{3}{4}\right)} = \frac{1}{7}$

$\Rightarrow$ Slope of L_3 is $m_3 = \tan\left(\theta - \frac{\pi}{4}\right) = \frac{m_1 - 1}{1+m_1}$

$\Rightarrow m_3 = \frac{\left(-\frac{3}{4}\right)-1}{1+\left(-\frac{3}{4}\right)} = -7$

The line L_2 through (2, 2) is $x - 7y + 12 = 0$

The line L_3 through (2, 2) is $7x + y = 16$

(b) Line L_1: $x - 2y = 1$ passes though A (1, 0)

$\Rightarrow m_1 = 1/2$

Equation of BC is L_2: $x - y + 1 = 0 \Rightarrow m_2 = 1$

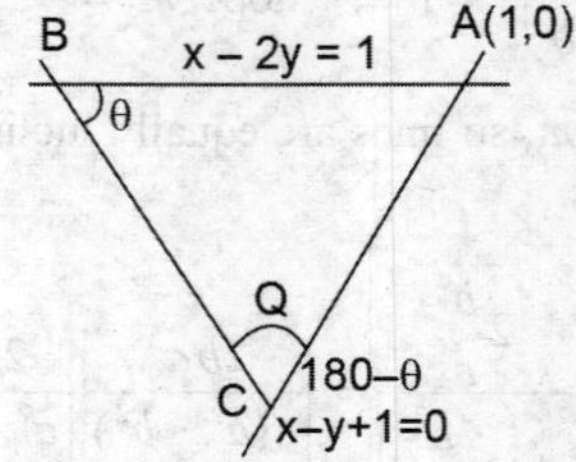

As $AB = AC$

$\Rightarrow \tan\theta = \frac{1-\frac{1}{2}}{1+\left(\frac{1}{2}\right)(1)} = \frac{1}{3}$

Slope of AC is $m_3 = \frac{m_2 + \tan\theta}{1 - m_2\tan\theta} = \frac{1+\frac{1}{3}}{1-\left(\frac{1}{3}\right)(1)} = 2$

Equation of AC is $y = 2x - 2$ or $2x - y = 2$

5. **(a)** Given L_1: $3x + 4y - 5 = 0$ and L_2: $3x + 4y - 5/2 = 0$

$d = \frac{\left|-5+\frac{5}{2}\right|}{\sqrt{3^2+4^2}} = \frac{1}{2}$ Unit

(b) Given L_1: $x\cos(\pi + x) + y\sin(\pi + \alpha) - p_1 = 0$ or $x\cos\alpha + y\sin\alpha + p_1 = 0$

L_2: $x\cos\alpha + y\sin\alpha - p_2 = 0$

$\Rightarrow$ Distance $d = \frac{|p_1 + p_2|}{1}$ (**as** $p_1 > 0, p_2 > 0$)

$\Rightarrow d = p_1 + p_2$

6. **(a)** Given L_1: $3x + 4y = 9$ and L_2: $mx - y = -1$ or $4mx - 4y = -4$

On adding, we get $(3 + 4m)x = 5 \Rightarrow x = \frac{5}{3+4m}$

Now m and x are both integers so we have $3 + 4m \in \{-5, -1, 1, 5\}$

Now, $3 + 4m = 5 \quad \Rightarrow m = -2$ and $3 + 4m = -1$

$\Rightarrow 4m = -4 \quad \Rightarrow m = -1$

For $3 + 4m = 1 \quad \Rightarrow 4m = -2$

$\Rightarrow m = \frac{1}{2} \notin \mathbb{Z}$

For $3 + 4m = 5 \quad \Rightarrow m = \frac{1}{2} \notin \mathbb{Z}$

$\therefore \quad m = -1, -2$

(b) Given that L_1: $y = 3x + 1$

$\Rightarrow m_1 = 3$ and L_2: $y = \frac{x}{2} + \frac{3}{2} \Rightarrow m_2 = \frac{1}{2}$

Further L_1 and L_2 are equally inclined to the line $y = mx + 4$

$\Rightarrow \quad \frac{m - m_2}{1 + mm_2} = \frac{m_1 - m}{1 + mm_1} \quad \Rightarrow \quad \frac{m - \frac{1}{2}}{1 + \frac{m}{2}} = \frac{3 - m}{1 + 3m}$

i.e., $6m^2 - m - 1 = -m^2 + m + 6$ or $7m^2 - 2m - 7 = 0$

$\Rightarrow \quad m = \frac{1 \pm 5\sqrt{2}}{7}$.

The acute angle bisector is with $m = \frac{1 + 5\sqrt{2}}{7}$ (both m_1 and m_2 are positive)

TEXTUAL EXERCISE-4 (OBJECTIVE)

1. (d) Given L_1: $2x - y + 4 = 0$ and L_2: $6x - 3y - 5 = 0$ or $2x - y - \frac{5}{3} = 0$

Distance $d = \frac{|4 - (-5/3)|}{\sqrt{2^2 + 1^2}} = \frac{17}{3\sqrt{5}} = \frac{17}{15}\sqrt{5}$

2. (b) Vertex of equilateral ΔABC is at A (2, 3) and B, C points are on line $x + y - 2 = 0$

Now $d = \frac{\sqrt{3}}{2} a$ (where a is the side), so $a = \frac{2}{\sqrt{3}} d$ and

d (perpendicular distance) $= \frac{|2 + 3 - 2|}{\sqrt{2}} = \frac{3}{\sqrt{2}}$

So, ΔABC area $= \frac{\sqrt{3} a^2}{4} = \frac{\sqrt{3}}{4}\left(\frac{2}{\sqrt{3}} d\right)^2 = \frac{\sqrt{3}}{4} \cdot \frac{4}{3} \cdot \frac{9}{2} = \frac{3}{2}\sqrt{3}$ square units

3. (c) Let the line be L : $mx - y + c = 0$

From the given: $\frac{2m + c}{\sqrt{1 + m^2}} + \frac{c - 2}{\sqrt{1 + m^2}} + \frac{m - 1 + c}{\sqrt{1 + m^2}} = 0$

$\Rightarrow$ $3m + 3c - 3 = 0$ or $m - 1 + c = 0$

$\Rightarrow$ which means line always passes through (1, 1) as L (1, 1) $= m - 1 + c = 0$

4. (c), (d) A line parallel to L: $3x - 4y - 2 = 0$ at a distance of 4 units from L will be $3x - 4y - 2 \pm 20 = 0$

$\Rightarrow$ $3x - 4y - 22 = 0$ or $3x - 4y + 18 = 0$

5. (a) Let the two perpendicular lines be x-axis and y-axis, so $|x| + |y| = 1$, which will give a square bounded by $x + y = 1$, $x + y = -1$, $x - y = 1$, $x - y = -1$

6. (a) Given A (2, –1) and line L_1: $x + y = 1$

A general point on L_1 is $\left(2 + r_1 \cos\frac{3\pi}{4}, -1 + r_1 \sin\frac{3\pi}{4}\right)$

When moving $r_1 = 2$ units (so that y-increases).

The position of particle will be

$B\left(2 - \frac{2}{\sqrt{2}}, -1 + \frac{2}{\sqrt{2}}\right) = \left(2 - \sqrt{2}, \sqrt{2} - 1\right)$

Similarly a general point on L_2: $x - 2y = 4$ will be $(2 + r_2 \cos\theta, -1 + r_2 \sin\theta)$ (where $\tan\theta = 1/2$)

When the particle has moved (then) $r_2 = 5$ units (so that y-increases)

The position of particle will be $C\left(2 + 2\sqrt{5}, -1 + \sqrt{5}\right)$

So distance BC is $\sqrt{\left(2\sqrt{5} + \sqrt{2}\right)^2 + \left(\sqrt{5} - \sqrt{2}\right)^2} = \sqrt{29 + 2\sqrt{10}}$ units.

7. (a), (b) Straight line L is perpendicular to $5x - y = 1$, so let L be $x + 5y + c = 0$

$\Rightarrow$ x-intercept 'a' $= -c$ and y-intercept 'b' $= -\frac{c}{5}$

ΔOAB are $a = \frac{c^2}{10} = 5$ units (given)

So $c = \pm 5\sqrt{2}$ and the line will be $x + 5y \pm 5\sqrt{2} = 0$

8. (a) Obviously $PA + PB$ will be minimum when A, P, B are in a straight line.

So $\frac{3\lambda + 1}{\lambda + 1} = \frac{2\lambda + 3}{\lambda + 1}$ (as P is on $y = x$)

gives $\lambda = 2$. Hence $P\left(\frac{7}{3}, \frac{7}{3}\right)$

TEXTUAL EXERCISE-5 (SUBJECTIVE)

1. Given L: $3x - 4y - 16 = 0$ and P (–1, 3). Let M (h, k) be the foot of perpendicular then $\frac{-1 - h}{3} = \frac{3 - k}{-4} = \frac{-31}{25}$

$\Rightarrow \quad h = \frac{68}{25}$ and $k = -\frac{49}{25} \quad \Rightarrow \quad M\left(\frac{68}{25}, -\frac{49}{25}\right)$

2. Given L: $4x + 7y + 13 = 0$ and P (–8, 12).

Let Q (h, k) be the image of P, then $\frac{-8 - h}{4} = \frac{12 - k}{7} = \frac{2(65)}{65} \quad \Rightarrow \quad h = -16, k = -2$

$\Rightarrow$ Q (–16, –2)

3. Image of the point P (2, 1) is Q (5, 2).

Mid point of PQ is $M\left(\frac{7}{2}, \frac{3}{2}\right)$ and slope of $PQ = \frac{1}{3}$

Equation of line mirror is $y - \frac{3}{2} = (-3)\left(x - \frac{7}{2}\right)$ or $3x + y - 12 = 0$

4. Given L: $5x + y + 6 = 0$ and P (4, –13). Let Q (h, k) be the image of P in L, then $\frac{4 - h}{5} = \frac{-13 - k}{1} = \frac{2(13)}{26}$

$\Rightarrow \quad h = -1, k = -14 \Rightarrow Q$ (–1, –14)

5. (i) P (1, 2), L_M: $x + 2y - 10 = 0$.

Let M (h, k) be foot of perpendicular and Q (, m) be the image, then $\frac{1 - h}{1} = \frac{2 - k}{2} = \frac{-5}{5} \Rightarrow h = 2, k = 4$

$\Rightarrow$ Foot is M (2, 4) $\quad \Rightarrow Q$ (3, 6)

(ii) P (2, 3) and $L_M = 2x - 3y + 18 = 0$.

Let M (h, k) be the foot of perpendicular and Q (, m) be the image, then $\frac{2 - h}{2} = \frac{3 - k}{-3} = \frac{13}{13}$

$\Rightarrow$ $h = 0, k = 6$

$\Rightarrow$ Foot is $M(0, 6)$ $\qquad \Rightarrow Q(-2, 9)$

6. Equation of AB is $2x - y + 4 = 0$. Equation of BC is $x - 2y - 1 = 0 \Rightarrow$ Point B is $(-3, -2)$

Now AC line acts as mirror, so L_M: $x + 3y - 3 = 0$

$\Rightarrow$ Image of P is Q (h, k) which is given by $\frac{-3-h}{1} = \frac{-2-k}{3} = \frac{2(-12)}{10}$

$\Rightarrow$ $h = -3/5$ and $k = 26/5$

$\Rightarrow$ $Q\left(-\frac{3}{5}, \frac{26}{5}\right)$

TEXTUAL EXERCISE-5 (OBJECTIVE)

1. **(b)** Given A (1, 2) and L_m: $y = x \Rightarrow B$ (2, 1)

Image of B in $y = 0$, (i.e. x-axis) is $(\alpha, \beta) = (2, -1)$

$\Rightarrow$ $\alpha = 2, \beta = -1$

2. **(d)** Let the slope of line be m and the line is $L_1 : y - k = m(x - h)$ or $mx - y + (k - mh) = 0$

Now the equation of foot of perpendicular from origin will be $y = -\frac{1}{m}x$ or $x + my = 0$

If (p, q) is the foot of perpendicular, then $q = \frac{k - mh}{1 + m^2}$

and $p = \frac{m(mh - k)}{1 + m^2} = -mq$

Putting $m = \frac{y - k}{(x - h)}$ and replacing p by 'x' and q by 'y',

we get $x = \frac{-(y - k)}{(x - h)}y$ or $x^2 - hx + y^2 - ky = 0$

3. **(b)** Image of P (3, 2) is Q (7, 4). Slope of $PQ = \frac{1}{2}$ and mid point of PQ is M foot = (5, 3)

$\Rightarrow$ Equation of Line mirror L_m is $y - 3 = (-2)(x - 5)$ or $2x + y = 13$

4. **(c)** Let P (2, 3) and L_m: $2x - y + 7 = 0$. Image P is $Q(h, k)$ given by $\frac{2-h}{2} = \frac{3-k}{-1} = \frac{2(8)}{5}$

$\Rightarrow$ $h = \frac{-22}{5}, k = \frac{31}{5}$

Given point is R (6, 11), so equation of RQ is $6x - 13y + 107 = 0$ which intersects L_m at $\left(\frac{4}{5}, \frac{43}{5}\right)$

$\therefore$ The equation of normal is $x + 2y = 18$

5. **(a)** Equation of L_m: $2y = x$ and equation of line of incident ray is $3x - y = 5 \Rightarrow m_1 = 3$

Point of intersection is (2, 1)

Slope of normal is $m_2 = -2$

Let m_3 be the slope of reflected ray, then

$\frac{m_3 - m_2}{1 + m_2 m_3} = \frac{m_2 - m_1}{1 + m_1 m_2}$

$\Rightarrow$ $\frac{m_3 + 2}{1 - 2m_3} = \frac{-5}{-5} = 1 \Rightarrow m_3 = -1/3$

$\therefore$ The equation of reflected ray is $y - 1 = \left(-\frac{1}{3}\right)(x - 2)$

$\Rightarrow$ $x + 3y = 5$

6. **(c)** Equation of normal L_N: $2x - 3y - 5 = 0 \Rightarrow m_2 = 2/3$

Equation of incident ray L_I : $2x - y = 3$

$\Rightarrow$ $m_1 = 2$, so the point of incident is $(1, -1)$

Hence equation line mirror $L_M : y + 1 = (-3/2)(x - 1)$ or $3x + 2y = 1$

Let slope of reflected ray be m_3 then $\frac{m_3 - m_2}{1 + m_2 m_3} = \frac{m_2 - m_1}{1 + m_1 m_2}$

$\Rightarrow$ $\frac{m_3 - 2/3}{1 + \frac{2}{3}m_3} = \frac{\frac{2}{3} - 2}{1 + \frac{4}{3}} \Rightarrow 29m_3 = 2$ or $m_3 = \frac{2}{29}$

$\therefore$ The equation of reflected ray is L_R: $y + 1 = (2/29)(x - 1)$ or $2x - 29y = 31$

7. **(a)** $L_I : 2x - y = 4 \quad \Rightarrow m_1 = 2$ and $L_M : x - y = 1$

$\Rightarrow$ Point of incident (3, 2)

$\Rightarrow$ Equation of normal L_N: $x + y = 5$

$\Rightarrow$ $m_2 = -1$

$\Rightarrow$ Slope of reflected ray $m_3 = \frac{1}{2}\left(\text{From } \frac{m_3 + 1}{1 - m_3} = \frac{-3}{-1}\right)$

$\Rightarrow$ Equation of reflected ray L_R: $x - 2y + 1 = 0$

TEXTUAL EXERCISE-6 (SUBJECTIVE)

1. Given L_1: $-4x - 3y + 6 = 0$ and $L_2 : 5x + 12y + 9 = 0$

As $c_1 > 0$, $c_2 > 0$ and $a_1a_2 + b_1b_2 = -20 - 36 = -56 < 0$

So origin lies in acute angle, and the positive sign taken gives bisector of angle containing the origin.

(i) Obtuse angle bisector is $\frac{5x + 12y + 9}{13} = \frac{-4x - 3y + 6}{5}$

i.e., $77x + 99y - 33 = 0$ or $7x + 9y = 3$

(ii) Acute angle bisector is $\frac{5x + 12y + 9}{13} = \frac{4x + 3y - 6}{5}$ i.e.,

$(25 - 52)x + (60 - 39)y + (45 + 78) = 0$ or $9x - 7y - 41 = 0$

(iii) $L_1(1, 2) = -4 < 0$ and $c_1 = 6 > 0$

$\Rightarrow$ (1, 2) and origin lie on opposite side of L_1 and L_2 (1, 2) $= 38 > 0$ and $C_2 = 9 > 0 \Rightarrow$ (1, 2) and origin on the same side of L_2

This means the point P(1,2) lies in the obtuse angle which has bisector as $9x - 7y = 41$

2. Given L_1: $12x - 5y + 7 = 0$ and L_2: $4x - 3y + 1 = 0$, as $c_1 > 0$, $c_2 > 0$ and $a_1a_2 + b_1b_2 = 48 + 15 = 63 > 0$

$\Rightarrow$ Origin lies in obtuse angle, so the bisector is

$\frac{12x - 5y + 7}{13} = \frac{4x - 3y + 1}{5}$

$\Rightarrow$ $4x + 7y + 11 = 0$

3. Given L_1: $y - b = \frac{2m}{1 - m^2}(x - a)$

$\Rightarrow$ Slope $= \dfrac{2m}{1-m^2}$ and $\theta_1 = 2\tan^{-1} m$

$L_2: y - b = \dfrac{2M}{1-M^2}(x-a) \Rightarrow \dfrac{2M}{1-M^2}$ and $\theta_2 = 2\tan^{-1} M$

Equation of bisectors will be

$$\frac{2mx + (m^2-1)y + (b - bm^2 - 2am)}{(m^2+1)} = \pm \frac{2Mx + (M^2-1)y + b - bM^2 - 2aM}{(M^2+1)}$$

$\Rightarrow \; y - b = \dfrac{m+M}{1-mM}(x-a)$ or $y - b = \dfrac{mM-1}{m+M}(x-a)$

4. (a) The incident ray travels along $x = 1$

$L_M: x + y = 1 \Rightarrow$ Point of incidence is (1, 0)

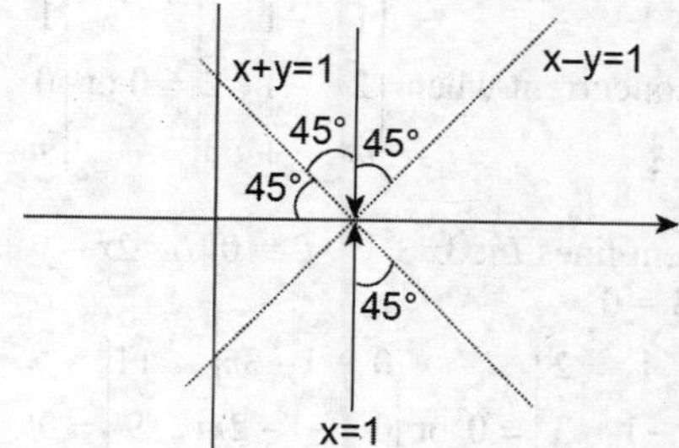

Equation of normal is $y = x - 1$

Equation of reflected ray is $y = 0$

(b) Line of incident $L_I: x - 2y - 3 = 0$ and $L_M = 3x - 2y - 5 = 0$

$\Rightarrow$ Point of incident is (1, –1). Now image of P (3, 0) lying on L_1 will be in L_2 as Q (h, k)

Hence $\dfrac{3-h}{3} = \dfrac{0-k}{-2} = \dfrac{2(9-5)}{13}$

$\Rightarrow \; h = \dfrac{15}{13}, k = \dfrac{16}{13}$

So the equation of reflected ray is $y + 1 = \dfrac{29}{2}(x-1)$ or $29x - 2y = 31$

(c) $2x + 3y = 5$. The ray will meet x-axis at $\left(\dfrac{5}{2}, 0\right)$ and y-axis at (0, 5/3), image of (0, 5/3) in x-axis is (0, –5/3). Hence the equation of reflected ray from x-axis is

$y = \dfrac{-5}{3\left(-\dfrac{5}{2}\right)}\left(x - \dfrac{5}{2}\right) \Rightarrow 3y = 2x - 5$ or $2x - 3y = 5$

Similarly image of (5/2, 0) is y-axis is (–5/2, 0), so the equation of reflection ray from y-axis is $y = 2/3\,(x + 5/2)$

$\Rightarrow \; 2x - 3y + 5 = 0$

5. Given $A = (x_1, y_1)$, $B = (x_2, y_2)$ and $C\,(x_3, y_3)$

Equation of AB is $\begin{vmatrix} x & y & 1 \\ x_1 & y_1 & 1 \\ x_2 & y_2 & 1 \end{vmatrix} = 0$ or $(x_2 - x_1)(y - y_1) - (y_2 - y_1)(x - x_1) = 0$

Similarly equation of line AC is $\begin{vmatrix} x & y & 1 \\ x_1 & y_1 & 1 \\ x_2 & y_2 & 1 \end{vmatrix} = 0$

Without any loss of generality let the origin be within the ΔABC

Equation of bisector of $\angle A$ is

$$\frac{(y_2 - y_1)x - (x_2 - x_1)y + (x_2y_1 - x_1y_2)}{\sqrt{(x_2-x_1)^2 + (y_2-y_1)^2}} = \frac{-\{(y_3 - y_1)x - (x_3 - x_1)y + (x_3y_1 - x_1y_3)\}}{\sqrt{(x_3-x_1)^2 + (y_3-y_1)^2}}$$

Now $\sqrt{(x_2-x_1)^2 + (y_2-y_1)^2} = AB = c$ and $\sqrt{(x_3-x_1)^2 (y_3-y_1)^2} = AC = b$

{Note: negative sign is taken on the basis of anticlockwise rotation}

$$\Rightarrow \; b\begin{vmatrix} x & y & 1 \\ x_1 & y_1 & 1 \\ x_2 & y_2 & 1 \end{vmatrix} + c\begin{vmatrix} x & y & 1 \\ x_1 & y_1 & 1 \\ x_3 & y_3 & 1 \end{vmatrix} = 0$$

TEXTUAL EXERCISE–6 (OBJECTIVE)

1. (a) Given L_1: $3x - 4y + 7 = 0$ and L_2: $-12x - 5y + 2 = 0$, As $c_1, c_2 > 0$

Now $a_1a_2 + b_1b_2 = -36 + 20 = -16 < 0$

So origin lies in acute angle which will have bisector as $\dfrac{3x - 4y + 7}{5} = \dfrac{-12x - 5y + 2}{13}$ or $(39 + 60)\,x - (52 - 25)$

$y + (91 - 10) = 0$

$\Rightarrow \; 11x - 3y + 9\} = 0$

2. (c) Given L_1: $-4x - 3y + 7 = 0$ and L_2: $-24x - 7y + 31 = 0$, as $c_1, c_2 > 0$

Now $a_1a_2 + b_1b_2 > 0$

So origin lies in obtuse angle which has equation of bisector as $\dfrac{(-1)\{4x + 3y - 7\}}{5} = \dfrac{(-1)\{24x + 7y - 31\}}{25}$

$\Rightarrow \; 20x - 40y + 20 = 0$ or $x - 2y + 1 = 0$

3. (a) Give (–1, 1) is a point that lies on L_1.

$\therefore$ From any point (other than the point of intersection) on line L_1 the acute angle is nearer than the obtuse angle.

$\Rightarrow d_1$: Distance of (–1, 1) from $21x + 77y - 101$ is $\dfrac{|-45|}{\sqrt{21^2 + 77^2}}$ and d_2: distance of (–1, 1) from $11x - 3y + 9 = 0$ is $\dfrac{|-5|}{\sqrt{11^2 + 3^2}}$ Clearly $d_2 < d_1$

$\Rightarrow$ The acute angle bisector is $11x - 3y + 9 = 0$

4. (a), (c) Given L_1: $-3x + 4y + 7 = 0$ and L_2: $12x - 5y + 6 = 0$

Equation of bisectors is $\dfrac{12x - 5y + 6}{13} = \pm\dfrac{(-3x + 4y + 7)}{5}$

$\Rightarrow$ $99x - 77y - 61 = 0$ or $21x + 27y + 121 = 0$

Lines parallel to these through (4, 5) are $99x - 77y - 11 = 0$

$\Rightarrow$ $9x - 7y - 1 = 0$ or $21x + 27y - 219 = 0$

$\Rightarrow$ $7x + 9y = 73$

5. (d) Given $P(-1, 0)$, $Q(0, 0)$ and $R(3, 3\sqrt{3})$.

Observe that $\angle RQX = 60°$ so, $\angle PQR = 120°$ (**as** PQ is along x-axis)

$\Rightarrow$ Equation of bisector of $\angle PQR$ is $y = -\sqrt{3}\,x$ or $x\sqrt{3} + y = 0$

6. (c) Given: one of the line in $my^2 + (1 - m^2)xy - mx^2 = 0$ is a bisector of angle between $xy = 0$

As $x = 0 \Rightarrow$ y-axis and $y = 0$

$\Rightarrow$ x-axis which will have $x \pm y = 0$ as the bisectors.

Hence $m = 1$ or $m = -1$ will satisfy the above equation.

7. Let $B(0, b)$ and $C(c, 0)$ and the mid point of BC is $A(3, 4)$

$\Rightarrow$ $\left(\frac{c}{2}, \frac{b}{2}\right) = (3, 4)$, so $c = 6$, $b = 8$ and the equation of BC is $\frac{x}{6} + \frac{y}{8} = 1$ or $4x + 3y = 24$

TEXTUAL EXERCISE–7 (SUBJECTIVE)

1. To prove: L_1, L_2 and L_3 are concurrent

Given: L_1: $\frac{x}{a} + \frac{y}{b} - 1 = 0$; L_2: $\frac{x}{b} + \frac{y}{a} - 1 = 0$ and L_3: $(x + y)(a + b) - 2ab = 0$

Now,

$$\begin{vmatrix} \frac{1}{a} & \frac{1}{b} & -1 \\ \frac{1}{b} & \frac{1}{a} & -1 \\ (a+b) & (a+b) & -2ab \end{vmatrix} = \begin{vmatrix} \left(\frac{1}{a} - \frac{1}{b}\right) & \frac{1}{b} & -1 \\ -\left(\frac{1}{a} - \frac{1}{b}\right) & \frac{1}{a} & -1 \\ 0 & (a+b) & -2ab \end{vmatrix}$$

$$= \left(\frac{1}{a} - \frac{1}{b}\right)\begin{vmatrix} 1 & \frac{1}{b} & -1 \\ -1 & \frac{1}{a} & -1 \\ 0 & (a+b) & -2ab \end{vmatrix}$$

$$= \left(\frac{1}{a} - \frac{1}{b}\right)\begin{vmatrix} 1 & \frac{1}{b} & -1 \\ 0 & \left(\frac{1}{a} + \frac{1}{b}\right) & -2 \\ 0 & (a+b) & -2ab \end{vmatrix}$$

$$= \left(\frac{1}{a} - \frac{1}{b}\right)\{-2b - 2a + 2a + 2b\} = 0$$

2. Given L_1: $y = \sqrt{3}\,x + 1$; L_2: $y = 4$; L_3: $-\sqrt{3}\,x + 2$

Observe that respective slopes are $m_1 = \sqrt{3}$, $m_2 = 0$, $m_3 = -\sqrt{3}$

$\Rightarrow$ m_1 and m_3 are at 60° to m_2. Hence an equilateral Δ will be formed.

3. Given that L_1, L_2, L_3 are concurrent where L_1: $ax + a^2y + 1 = 0$; L_2: $bx + b^2y + 1 = 0$ and L_3: $cx + c^2y + 1 = 0$.

Since lines are concurrent

$$\therefore \begin{vmatrix} a & a^2 & 1 \\ b & b^2 & 1 \\ c & c^2 & 1 \end{vmatrix} = 0 \text{ i.e., } \begin{vmatrix} (a-b) & (a-b)(a+b) & 0 \\ (b-c) & (b-c)(b+c) & 0 \\ c & c^2 & 1 \end{vmatrix}$$

$$= (a-b)(b-c)\begin{vmatrix} 1 & a+b \\ 1 & b+c \end{vmatrix} = (a-b)(b-c)(c-a) = 0$$

So either $a = b$ or $b = c$ or $c = a$, i.e., at least two of the three constants a, b, c are equal.

4. L_1: $x - y + 1 = 0$; L_2: $2x - y + 2 = 0$ and L_3: $mx - y + 3 = 0$ will be concurrent when $\begin{vmatrix} 1 & -1 & 1 \\ 2 & -1 & 2 \\ m & -1 & 3 \end{vmatrix} = 0$ or $\begin{vmatrix} 1 & -1 & 1 \\ 0 & 1 & 0 \\ m & -1 & 3 \end{vmatrix} = 0$

$\Rightarrow$ $m = 3$

5. Concurrent lines L_1: $3x + y + 2 = 0$; L_2: $2x - y + 3 = 0$; L_3: $x + my - 3 = 0$

$$\Rightarrow \begin{vmatrix} 3 & 1 & 2 \\ 2 & -1 & 3 \\ 1 & m & -3 \end{vmatrix} = 0 \text{ or } \begin{vmatrix} 0 & 1-3m & 11 \\ 0 & -1-2m & 9 \\ 1 & m & -3 \end{vmatrix} = 0$$

$\Rightarrow$ $5m = 20$ or $m = 4$

6. Take L_1 and L_2 as given and L_3: $x = 0$. The lines will be concurrent when $\begin{vmatrix} (2m+1) & (m-1) & (m-9) \\ m & (2m+3) & (m+6) \\ 1 & 0 & 0 \end{vmatrix} = 0$

$\Rightarrow$ $(m^2 + 5m - 6) - (2m^2 - 15m - 27) = 0$ or $m^2 - 20m - 21 = 0$

$\Rightarrow$ $(m - 21)(m + 1) = 0$ $\Rightarrow$ $m = -1, 21$

Aliter: The lines L_1: $mx + (2m + 3)y + (m + 6) = 0$ and L_2: $(2m + 1)x + (m - 1)y + (m - 9) = 0$, will intersect (at a point) on y-axis when L_1 and L_2 give $x = 0$ as its solution.

$\Rightarrow$ $m(m - 1)x + (m - 1)(m + 6) - (2m + 1)(2m + 3)x - (2m + 3)(m - 9) = 0$

Or $\{(m^2 - m) - (4m^2 + 8m + 3)\}x = 2m^2 - 15m - 27 - m^2 - 5m + 6$ gives $(-x)\{3m^2 + 7m + 3\} = m^2 - 20m - 21$

Putting $m^2 - 20m - 21 = 0$, we get $m = \frac{20 \pm \sqrt{400 + 84}}{2}$;

$m = \frac{20 \pm 22}{2} = 21, -1$

Observe that for $m = -1$, $3m^2 + 7m + 3 = -1 \neq 0$ and for $m = 21$; $3m^2 + 7m + 3 = 1473 \neq 0$

So for $m = 21, -1$ the lines will intersect on y-axis

$\therefore$ $m = -1$ or 21

7. Given L_1: $x - y + 1 = 0$; L_2: $2x + y - 16 = 0$; and L_3: $mx - y - 4 = 0$

Lines will be concurrent when $\begin{vmatrix} 1 & -1 & 1 \\ 2 & 1 & -16 \\ m & -1 & -4 \end{vmatrix} = 0$ or

$$\begin{vmatrix} 1 & 0 & 0 \\ 2 & 3 & -18 \\ m & (m-1) & (-4-m) \end{vmatrix} = 0$$

$\Rightarrow -12 - 3m + 18m - 18 = 0 \Rightarrow 15m = 30 \Rightarrow m = 2$

TEXTUAL EXERCISE-7 (OBJECTIVE)

1. (c) The line is $\frac{x}{a}+\frac{y}{b}=1$. Since (2, 2) lies on it

$\therefore \quad \frac{1}{a}+\frac{1}{b}=\frac{1}{2}$ or $2(a+b)=ab$

$\Rightarrow$ Area $\Delta OAB = \frac{ab}{2} = |\lambda|$ and $(a+b)=|\lambda|$ and $ab = 2|\lambda|$

$\Rightarrow$ a and b are the roots of $x^2 - |\lambda|x + 2|\lambda| = 0$

2. (b) Given a, b, c are distinct and non-negative.

L_1: $ax + ay + c = 0$; L_2: $x + 1 = 0$; L_3: $cx + cy + b = 0$ are concurrent

$$\Rightarrow \begin{vmatrix} a & a & c \\ 1 & 0 & 1 \\ c & c & b \end{vmatrix} = 0 \text{ or } \begin{vmatrix} a-c & a & c \\ 0 & 0 & 1 \\ c-b & c & b \end{vmatrix} = 0$$

$\Rightarrow c^2 - ac + ac - ab = 0$ or $c^2 = ab$

$\Rightarrow$ a, c, b are in G.P. i.e., c is G.M. of a and b

3. (c) $ax + by + c = 0$ and $3a + 2b + 4c = 0$

$\Rightarrow \quad \frac{3}{4}a+\frac{1}{2}b+c=0$

$\therefore \quad x=\frac{3}{4}, y=\frac{1}{2}$ satisfies the equation irrespective of the individual values of a, b, c subject to the given condition of $\frac{3}{4}a+\frac{b}{2}+c=0$. The lines will always pass though $\left(\frac{3}{4},\frac{1}{2}\right)$

4. (c) Given $a^2 + b^2 - 2ab = c^2$, so $(a - b) = \pm c$

Either $a - b + c = 0$ or $a - b - c = 0$

Now $ax + by + c = 0$, becomes $a - b + c = 0$, when $x = 1$, $y = -1$

Similarly $ax + by + c = 0$, becomes $-a + b + c = 0$, when $x = -1$, $y = 1$

$\Rightarrow$ Point is $(-1, 1)$ or $(1, -1)$

5. (d) Since a, b, c are in H.P.

$\Rightarrow ab + bc = 2ac$ or $bc - 2ac + ab = 0$

Now, (for $x = 1$, $y = -2$), $bcx + cay + ab = 0$, gives $bc - 2ac + ab = 0$

$\Rightarrow$ Point is $(1, -2)$

TEXTUAL EXERCISE-8 (SUBJECTIVE)

1. (a) $x^2 - 7xy + 12y^2 = 0$

$\Rightarrow (x - 3y)(x - 4y) = 0$

$\Rightarrow y=\frac{x}{3}$ and $y=\frac{x}{4}$ are the two straight lines.

(b) $4x^2 - 24xy + 11y^2 = 0$

$\Rightarrow (2x - 11y)(2x - y) = 0$

$\Rightarrow y=\frac{2}{11}x$ and $y = 2x$

(c) $y^2 = 0$, gives $y = 0$, $y = 0$ (i.e., x-axis) coincident line.

(d) $xy - ay = 0 \qquad \Rightarrow y(x - a) = 0$

$\Rightarrow y = 0$ and $x = a$ are the straight line.

(e) $x^3 - x^2 - x + 1 = 0 \qquad \Rightarrow (x^2 - 1)(x - 1) = 0$

$\Rightarrow (x - 1)(x - 1)(x + 1) = 0$

$\Rightarrow x = 1, x = 1, x = -1$ are the straight lines

2. (i) $x^2 - 3xy + \lambda y^2 + 3x - 5y + 2 = 0$

Comparing with $ax^2 + 2hxy + by^2 + 2gx + 2fy + c = 0$,

we get $a = 1$, $h = -3/2$, $b = \lambda$, $g=\frac{3}{2}$, $f = -5/2$, $c = 2$

The equation will represent a pair straight lines when

$$\Delta = \begin{vmatrix} a & h & g \\ h & b & f \\ g & f & c \end{vmatrix} = 0 \text{ i.e., } \Delta = \begin{vmatrix} 1 & -\frac{3}{2} & \frac{3}{2} \\ -\frac{3}{2} & \lambda & -\frac{5}{2} \\ \frac{3}{2} & -\frac{5}{2} & 2 \end{vmatrix} = 0$$

$\Rightarrow \quad -\frac{25}{4}+2\lambda+\frac{27}{4}-\frac{9\lambda}{4}=0$

$\Rightarrow \lambda = 2$

(ii) $\lambda x^2 - 10xy + 12y^2 + 5x - 16y - 3 = 0$

Comparing with $ax^2 + 2hxy + by^2 + 2gx + 2fy + c = 0$, we get $a = \lambda$, $b = 12$, $h = -5$, $g = 5/2$, $f = -8$, $c = -3$

Now, $\Delta = \begin{vmatrix} a & h & g \\ h & b & f \\ g & f & c \end{vmatrix} = 0$

$$\Rightarrow \begin{vmatrix} \lambda & -5 & \frac{5}{2} \\ -5 & 12 & -8 \\ \frac{5}{2} & -8 & -3 \end{vmatrix} = 0 \text{ or } \lambda(12 - 112) + (5/2)(70 + 10) = 0$$

$\Rightarrow 100\lambda = 200 \Rightarrow \lambda = 2$

3. Given $y^3 - xy^2 - 14x^2y + 24x^3 = 0$

Which is homogeneous and cubic in x and y. Let $y = \alpha x$, βx, γx be its roots, then $(\alpha + \beta + \gamma)x = x$ i.e., $\alpha + \beta + \gamma = 1$,

Similarly $(\alpha\beta + \beta\gamma + \gamma\alpha)x^2 = -14x^2$ i.e., $\alpha\beta + \beta\gamma + \gamma\alpha = -14$ and $(\alpha\beta\gamma)x^3 = -24x^3 \Rightarrow \alpha\beta\gamma = -24$

We can also form $\alpha^2 + \beta^2 + \gamma^2 = 29$, as $\alpha\beta\gamma = -24$, so either all of α, β, γ are negative or only one is negative.

Since $\alpha + \beta + \gamma = 1$, so only one will be negative.

Consider integral values of α, β, γ. We can assume $\alpha = 2$, $\beta = 3$ and $\gamma = -4$, then $\alpha^2 + \beta^2 + \gamma^2 = 29$ is satisfied.

Now $\alpha\beta + \beta\gamma + \gamma\alpha = 6 - 12 - 8 = -14$ is also true, so $y = 2x$, $y = 3x$ and $y = -4x$

4. Given $2x^2 + xy - y^2 + kx + 6y - 9 = 0$

Comparing with $ax^2 + by^2 + 2hxy + 2gx + 2fy + c = 0$

We get $a = 2$, $h = 1/2$, $b = -1$, $g = k/2$, $f = 3$, $c = -9$

Now, $\Delta = \begin{vmatrix} a & h & g \\ h & b & f \\ g & f & c \end{vmatrix} = 0 \Rightarrow \begin{vmatrix} 2 & \frac{1}{2} & \frac{k}{2} \\ \frac{1}{2} & -1 & 3 \\ \frac{k}{2} & 3 & -9 \end{vmatrix} = 0$

$\Rightarrow \left(-\frac{1}{2}\right)\left\{-\frac{81}{2}+36-\frac{3k}{2}\right\}+\frac{k}{2}\left(\frac{27}{2}+\frac{k}{2}-12\right)=0$

$\Rightarrow \frac{k^2+6k+9}{4}=0 \quad \Rightarrow k=-3$

5. **(a)** $3x^2 - 10xy + 7y^2 + 2\lambda x - 14y - 42 = 0$

Comparing with $ax^2 + by^2 + 2hxy + 2gx + 2fy + c = 0$

We get $a = 3, b = 7, h = -5, g = \lambda, f = -7, c = -42$

Now $\Delta = \begin{vmatrix} a & h & g \\ h & b & f \\ g & f & c \end{vmatrix} = 0 \Rightarrow \begin{vmatrix} 3 & -5 & \lambda \\ -5 & 7 & -7 \\ \lambda & -7 & -42 \end{vmatrix} = 0$

$\Rightarrow \lambda = 5 \pm 2\sqrt{7}$

(b) Given $x^2 + 3xy + 2y^2 - x - 4y - 6 = 0$

Comparing with $ax^2 + by^2 + 2hxy + 2gx + 2fy + c = 0$

We get $a = 1, b = 2, h = 3/2, g = -1/2, f = -2, c = -6$

Now, $\Delta = \begin{vmatrix} a & h & g \\ h & b & f \\ g & f & c \end{vmatrix} = \begin{vmatrix} 1 & \frac{3}{2} & -\frac{1}{2} \\ \frac{3}{2} & 2 & -2 \\ -\frac{1}{2} & -2 & -6 \end{vmatrix}$

$= \begin{vmatrix} 0 & 0 & -\frac{1}{2} \\ -\frac{5}{2} & -4 & -2 \\ -\frac{25}{2} & -20 & -6 \end{vmatrix} = \left(-\frac{1}{2}\right)\{50-50\} = 0$

So it will represent a pair of straight lines.

TEXTUAL EXERCISE–8 (OBJECTIVE)

1. **(a)** Given: $y^2 + \frac{2hx}{b}y + \frac{a}{b}x^2 = 0$

Let $y = mx$ and $y = m^2x$ be the solution, then $x\,(m + m^2)$

$= -\frac{2hx}{b}$ and $m^3x^2 = \frac{a}{b}x^2$

$\Rightarrow m^2 + m = -\frac{2h}{b}$ and $m^3 = \frac{a}{b}$

Cubing $m\,(m+1) = \frac{-2h}{b}$, we get $m^3\,\{m^3+1+3\,(m^2+m)\}$

$= \frac{-8h^3}{b^3}$

$\Rightarrow \frac{a}{b}\left\{\frac{a}{b}+1+3\left(\frac{-2h}{b}\right)\right\} = \frac{-8h^3}{b^3}$, will give $ab\,\{a + b - 6h\}$

$= -8h^3$ or $a^2b + ab^2 - 6abh + 8h^3 = 0$

Aliter: $ax^2 + 2hxy + by^2 = 0$, which is homogenous

$\Rightarrow y = \frac{-2hx + \sqrt{4h^2x^2 - 4ax^2}}{2b} = \frac{x\left\{-h \pm \sqrt{h^2 - ab}\right\}}{b}$

Now, let $m_2 = \frac{-h + \sqrt{h^2 - ab}}{b}$ be the square of

$m_1 = \frac{-h - \sqrt{h^2 - ab}}{b}$

$\Rightarrow \frac{h^2 + (h^2 - ab) + 2h\sqrt{h^2 - ab}}{b^2} = \frac{-h + \sqrt{h^2 - ab}}{b}$ i.e., $2h^2$

$- ab + 2h\,\sqrt{h^2 - ab} = -hb + b\sqrt{h^2 - ab}$

i.e., $(2h - b)\,\sqrt{h^2 - ab} = (ab - 2h^2 - hb)$

Squaring again, we get $(4h^2 + b^2 - 4hb)\,(h^2 - ab) = a^2$

$b^2 + (4h^4 + h^2b^2 + 4h^3b) - 2ab\,(2h^2 - hb)$

$\Rightarrow a^2b + ab^2 - 6abh + 8h^3 = 0$

2. **(a)** $3x^2 + 2hxy - 3y^2 - 40x + 30y - 75 = 0$

Comparing with $ax^2 + by^2 + 2hxy + 2gx + 2fy + c = 0$

We get $a = 3, b = -3, h = h, g = -20, f = 15, c = -75$

Now, $\Delta = \begin{vmatrix} a & h & g \\ h & b & f \\ g & f & c \end{vmatrix} = 0$

$\Rightarrow \begin{vmatrix} 3 & h & -20 \\ h & -3 & 15 \\ -20 & 15 & -75 \end{vmatrix} = 0$ or $\begin{vmatrix} 3 & h & (5h-20) \\ h & -3 & 0 \\ -20 & 15 & 0 \end{vmatrix} = 0$

$\Rightarrow (5h - 20)\,\{15h - 60\} = 0 \Rightarrow h = 4, 4$

3. **(a)** Lines through the origin and parallel to the given lines will be $y = m_1x$ and $y = m_2x$ and the combined equation will be

$y^2 - (m_1 + m_2)\,xy + m_1m_2x^2 = 0$

4. **(d)** $Ax^2 + 2Bxy + Cy^2 + Dx + E + F = 0$, will represent a pair of straight lines then $Bx + Cy + E/2 = 1$

$= \pm\sqrt{\left(B^2 - AC\right)x^2 + 2\left(\frac{BE}{2} - \frac{CD}{2}\right)x + \frac{E^2}{4} - CF}$

$B^2 - AC$ may be positive or zero depending upon other constants

5. **(c)** As worked for Question no. 1

Similarly let $y = mx$ and $y = m\lambda x$ be the solution

$\Rightarrow y^2 + \frac{2hx}{b}y + \frac{a}{b}x^2 = 0$

$\Rightarrow m\,(1 + \lambda) = -\frac{2h}{b}$ and $m^2\lambda = a/b$

Squaring, we get $m^2\,(\lambda + 1)^2 = \left(-\frac{2h}{b}\right)^2 = \frac{4h^2}{b^2}$

or $\left(\frac{a}{b\lambda}\right)(\lambda+1)^2 = \frac{4h^2}{b^2} \quad \Rightarrow ab(\lambda + 1)^2 = 4h^2\lambda$

6. **(b)** $9y^2 + 12xy + 4x^2 + 2gx + 2fy + c = 0$ will represents two real parallel straight lines, as $(3y + 2x + c_1)$ $(3y + 2x + c_2) = 0$

$\Rightarrow$ $2gx = 2(c_1 + c_2)x$ and $3(c_1 + c_2)y = 2fy$ and $c_1c_2 = c$

$\Rightarrow$ $g = c_1 + c_2, f = (3/2)(c_1 + c_2)$ and $c = c_1c_2$

$\Rightarrow$ Option (b) is suitable when $c_1 = c_2 = 1$

7. **(a)** $ax^2 + by^2 + (a + b)xy + x + y = 0$

We can rewrite as $(ax + by + 1)(x + y) = 0$

Aliter: Splitting as $ax^2 + by^2 + axy + bxy + (x + y) = 0$

$\Rightarrow$ $ax(x + y) + by(y + x) + (x + y) = 0$

$\Rightarrow$ $(x + y)\{ax + by + 1\} = 0$

8. **(b)** $x^2 - \lambda xy + 2y^2 + 3x - 5y + 2 = 0$

$\Rightarrow$ $a = 1, b = 2, h = -\lambda/2, g = 3/2, f = -5/2, c = 2$

$$\text{Now, } \Delta = \begin{vmatrix} a & h & g \\ h & b & f \\ g & f & c \end{vmatrix} = \begin{vmatrix} 1 & -\frac{\lambda}{2} & \frac{3}{2} \\ -\frac{\lambda}{2} & 2 & -\frac{5}{2} \\ \frac{3}{2} & -\frac{5}{2} & 2 \end{vmatrix} = 0$$

$$\Rightarrow \left(-\frac{9}{4}\right) + \frac{\lambda}{2}\left(-\lambda + \frac{15}{4}\right) + \frac{3}{2}\left(\frac{5\lambda}{4} - 3\right) = 0$$

$$\Rightarrow -\frac{\lambda^2}{2} + \frac{15\lambda}{4} - \frac{27}{4} = 0 \text{ or } 2\lambda^2 - 15\lambda + 27 = 0$$

$$\Rightarrow \lambda = \frac{15 \pm 3}{4} = 3, \frac{9}{2}$$

9. **(c)** $px^2 + y^2 + xy - x - 2y = 0$, gives $a = p, b = 1, h = 1/2, g = -1/2, f = -1, c = 0$

$$\text{Now, } \Delta = \begin{vmatrix} a & h & g \\ h & b & f \\ g & f & c \end{vmatrix} = 0$$

$$\Rightarrow \begin{vmatrix} p & \frac{1}{2} & -\frac{1}{2} \\ \frac{1}{2} & 1 & -1 \\ -\frac{1}{2} & -1 & 0 \end{vmatrix} = 0 \text{ or } \begin{vmatrix} p & \frac{1}{2} & 0 \\ \frac{1}{2} & 1 & 0 \\ -\frac{1}{2} & -1 & -1 \end{vmatrix} = 0$$

$$\Rightarrow (-1)\left\{p - \frac{1}{4}\right\} = 0 \quad \Rightarrow p = 1/4$$

10. **(a)** $6x^2 - xy + 4cy^2 = 0$ is homogenous and one line is $3x + 4y = 0$

So let $(3x + 4y)(2x + py) = 6x^2 - xy + 4cy^2$

$\Rightarrow$ $6x^2 + 4py^2 + (8 + 3p)xy = 0$

Hence $8 + 3p = -1 \Rightarrow p = -3$

$\Rightarrow$ $4cy^2 = 4py^2 \Rightarrow c = p = -3$

TEXTUAL EXERCISE-9 (SUBJECTIVE)

1. $12x^2 - 10xy + 2y^2 + 11x - 5y + \lambda = 0$, gives $a = 12, b = 2, h = -5, g = 11/2, f = -5/2, c = \lambda$

$$\text{Now, } \Delta = \begin{vmatrix} a & h & g \\ h & b & f \\ g & f & c \end{vmatrix} = 0 \Rightarrow \begin{vmatrix} 12 & -5 & \frac{11}{2} \\ -5 & 2 & -\frac{5}{2} \\ \frac{11}{2} & -\frac{5}{2} & \lambda \end{vmatrix} = 0$$

$\Rightarrow$ $-\lambda + 2 = 0 \quad \Rightarrow \lambda = 2$

$$\text{Now, } \tan\theta = \frac{2\sqrt{h^2 - ab}}{|a + b|} = \frac{2\sqrt{25 - 24}}{14} = \frac{1}{7}$$

Hence $\theta = \tan^{-1}(1/7)$

2. Given $x^2 + 2xy\cot\theta + y^2 = 0$, gives $a = 1, b = 1, h = \cot\theta$

$$\text{Now, } \tan\theta = \frac{2\sqrt{h^2 - ab}}{|a + b|} = \frac{2\sqrt{\cot^2\theta - 1}}{2}$$

$$= \frac{\sqrt{\cos 2\theta}}{|\sin\theta|} \Rightarrow \theta = \tan^{-1}\left\{\operatorname{cosec}\theta\sqrt{\cos 2\theta}\right\}$$

3. Given $y^3 - xy^2 - 14x^2y + 24x^3 = 0$ is a cubic and homogenous in x and y.

$\Rightarrow$ $y = 2x, y = 3x, y = -4x$ are the solutions as worked in 3. {Textual exercise –8 (subjective)}

Let $m_1 = 2, m_2 = 3, m_3 = -4$

$$\Rightarrow \tan\theta_1 = \frac{|m_2 - m_1|}{1 + m_1m_2} = \frac{|3 - 2|}{1 + 6} = \frac{1}{7}$$

$\Rightarrow$ GWH28-DGCMP-P6RC4-6J4MT-3HFDY

$$\Rightarrow \tan\theta_2 = \frac{m_2 - m_3}{1 + m_2m_3} = \frac{3 + 4}{1 - 12} = \frac{-7}{11}$$

$$\Rightarrow \theta_2 = \tan^{-1}\left(\frac{-7}{11}\right)$$

$$\Rightarrow \tan\theta_3 = \frac{m_1 - m_3}{1 + m_1m_3} = \frac{2 + 4}{1 - 8} = \frac{-6}{7}$$

$\Rightarrow$ = *tan* (--)

Note: positive or negative values may arise on account of clockwise or anticlockwise direction.

4. Given: $S = 7x^2 - y^2 + 6xy - 18x + 4y - 9 = 0$

The point of intersection is given $\dfrac{\partial S}{\partial x} = 14x + 6y - 18 = 0$

and $\dfrac{\partial S}{\partial y} = -2y + 6x + 4 = 0$

Solving $\dfrac{\partial S}{\partial x} = \dfrac{\partial S}{\partial y} = 0$, we get $x = \dfrac{3}{16}, y = \dfrac{25}{16}$

Now $7X^2 - Y^2 + 6XY = 0 \Rightarrow \quad (7X - Y)(X + Y) = 0$

Slopes are $m_1 = -1, m_2 = 7$

Let m be the slope of the third line that forms an isosceles triangle

$$\text{So, } \left(\frac{m_1 - M}{1 + m_1M}\right) = -\left(\frac{m_2 - M}{1 + m_2M}\right)$$

$$\Rightarrow \frac{-1 - M}{1 - M} = \frac{7 - M}{1 + 7M} \quad \Rightarrow M = -3, 1/3$$

$$\text{Now } \frac{-1 - M}{1 - M} = -\frac{7 - M}{1 + 7M}$$

$\Rightarrow$ $8m^2 = -8$ i.e., $M^2 = -1$ (not acceptable)

When $M = -1/3$, then we get $3(y + 2) = x - 1$ or $x - 3y - 7 = 0$

When $M = -3$, then we get $y + 2 = (-3)(x - 1)$ or $3x + y - 1 = 0$

TEXTUAL EXERCISE-9 (OBJECTIVE)

1. **(a)** The lines are $a^2x^2 + bcy^2 - a(b + c)xy = 0$

Which is homogenous, so the lines will be coincident if

$$\left(ax - \sqrt{bc}\,y\right)^2 = a^2x^2 + bcy^2 - a(b + c)xy$$

i.e., $-2a\sqrt{bc} = -(a)(b + c)$

$\Rightarrow$ (either $a = 0$) or $4bc = b^2 + c^2 + 2bc$

$\Rightarrow$ $(b - c)^2 = 0$ i.e., $b = c$. Hence either $a = 0$ or $b = c$

2. **(c)** $2x^2 + 4xy - py^2 + 4x + qy + 1 = 0$ will represent a pair of straight lines if $\Delta = 0$ where $a = 2, h = 2, b = -p, g = 2, f_2 = 9/2, c = 1$

$$\Rightarrow \begin{vmatrix} 2 & 2 & 2 \\ 2 & -p & \frac{q}{2} \\ 2 & \frac{q}{2} & 1 \end{vmatrix} = 0 \text{ or } \begin{vmatrix} 2 & 2 & 2 \\ 0 & (-p-2) & \left(\frac{q}{2} - 2\right) \\ 0 & \left(\frac{q}{2} - 2\right) & (-1) \end{vmatrix} = 0$$

$$\Rightarrow 2\left\{(p+2) - \left(\frac{q}{2} - 2\right)^2\right\}$$

$$\Rightarrow \left(\frac{q}{2} - 2\right)^2 = (p + 2)$$

Further lines will be perpendicular, when $a + b = 0$

$\Rightarrow 2 - p = 0 \Rightarrow p = 2$ then $\frac{q}{2} - 2 = \pm 2$

$\Rightarrow q = 0, 8$

Hence $p = 2$ and $q = 0$ or 8

3. **(b)** Given: $ax^2 + xy + by^2 = 0 \quad \Rightarrow h = 1/2$

$$\text{Now, } \tan\theta = \left|\frac{2\sqrt{h^2 - ab}}{a + b}\right| = 1$$

$$\Rightarrow \frac{2\sqrt{\frac{1}{4} - ab}}{|a + b|} = 1 \text{ or } a^2 + b^2 + 2ab = 4\left(\frac{1}{4} - ab\right)$$

$$\Rightarrow a^2 + b^2 + 6ab = 1 \left\{\text{also } \frac{1}{4} - ab > 0, \text{ so } ab < \frac{1}{4}\right\}$$

Hence $a = 1, b = -6$ is a suitable choice

4. **(d)** Given: $2x^2 - 3xy + y^2 = 0 \quad \Rightarrow (2x - y)(x - y) = 0$

$\Rightarrow \tan\alpha = 1, \tan\beta = 2$

$$\text{Hence } \cot^2\alpha + \cot^2\beta = (1)^2 + \frac{1}{4} = \frac{5}{4}$$

5. **(a), (c)** Given $(\ell x + my)^2 - 3(mx - \ell y)^2 = 0$

$$\Rightarrow \left(\ell x + my + mx\sqrt{3} - \sqrt{3}\ell y\right)\left(\ell x + my + \sqrt{3}\ell y - \sqrt{3}mx\right) = 0$$

$$\Rightarrow y = \frac{\left(\ell + m\sqrt{3}\right)}{\left(\sqrt{3}\ell - m\right)}x, \; y = \frac{\sqrt{3}m - \ell}{(\sqrt{3}\ell + m)}x$$

$$\Rightarrow m_1 = \frac{\left(\ell + \sqrt{3}m\right)}{\left(\sqrt{3}\ell - m\right)}, m_2 = \frac{\left(\sqrt{3}m - \ell\right)}{\left(\sqrt{3}\ell + m\right)} \text{ and } m_3 = \frac{-\ell}{m}$$

The given can be written as ($^2 - 3m^2)x^2 + (m^2 - 3$ $^2)y^2 + 8$ $mxy = 0$

$$\Rightarrow \tan\theta_1 = \frac{2\sqrt{16\ell^2m^2 + 3\ell^4 + 3m^4 - 10\ell^2m^{-2}}}{(-2)\left(\ell^2 + m^2\right)}$$

$$\Rightarrow \tan\theta_1 = \left|\frac{2\sqrt{3}(\ell^2 + m^2)}{-2(\ell^2 + m^2)}\right| = \sqrt{3} \Rightarrow \theta = 60°$$

Now m_1 and m_3 are at angle θ_2, then

$$\tan\theta_2 = \left|\frac{m_1 - m_3}{1 + m_1m_3}\right| = \frac{\left|\sqrt{3}(m^2 + \ell^2)\right|}{-\left(\ell^2 + m^2\right)} = \sqrt{3}$$

$\Rightarrow \theta_2 = 60°$

Hence an equilateral triangle is formed which is also an isosceles one

6. **(d)** Given $\lambda x^2 + (1 - \lambda)^2 xy - \lambda y^2 = 0$

$\Rightarrow a = \lambda, b = -\lambda$ and $h = (1/2)(1 - \lambda)^2$

$$\text{Now, } \tan\theta = \left|\frac{2\sqrt{h^2 - ab}}{a + b}\right|, \text{ as } a + b = 0$$

$\Rightarrow \theta = 90°$

7. **(d)** Given $x^2 - 2xy\tan A - y^2 = 0 \Rightarrow 2\tan A = 4 \Rightarrow A = \tan^{-1}(2)$

8. **(a)** Given $12x^2 - 20xy + 7y^2 = 0$

$\Rightarrow (2x - y)(6x - 7y) = 0 \Rightarrow y = 2x$ and $y = \frac{6x}{7}$ are the two lines passing through origin $O(0, 0)$. The points of intersection with $2x - 3y + 4 = 0$ are respectively $A(1, 2)$ and $B(7, 6)$

Hence the centroid is at $G\left(\frac{8}{3}, \frac{8}{3}\right)$

9. **(b)** Given set is $(3x - 2y - a)(2x - 3y + a) = 0$

Similarly other set is $(3x - 2y - 2a)(2x - 3y + 3a) = 0$

distance between $3x - 2y - a = 0$ and $3x - 2y - 2a = 0$ is $d = \frac{a}{\sqrt{13}}$ units.

Now point of intersection of $3x - 2y - a = 0$ with $2x - 3y + a = 0$ is $A(a, a)$ and point of intersection $3x - 2y - a = 0$ and $2x - 3y + 3a = 0$ is $B\left(\frac{9}{5}a, \frac{11}{5}a\right)$

$$\Rightarrow \text{Distance } AB = \sqrt{\left(\frac{4a}{5}\right)^2 + \left(\frac{6}{5}a\right)^2} = \frac{2a}{5}\sqrt{13} \text{ units}$$

$$\text{Hence area enclosed } AB \times d = \frac{a}{\sqrt{13}} \times \frac{2a\sqrt{13}}{5} = \frac{2a^2}{5}$$

10. Given $2x^2 + 6y^2 + 7xy - 3x - 5y + 1 = 0$

$$\text{The point of intersection is } \frac{\partial S}{\partial x} = \frac{\partial S}{\partial y} = 0$$

So $4x + 7y - 3 = 0$ and $12y + 7x - 5 = 0$, gives $x = -1, y = 1$

$\Rightarrow$ Lines joining $(-1, 1)$ and origin $O(0, 0)$ is perpendicular to $ax + by - 1 = 0$

$\therefore$ Product of slopes $m_1m_2 = -1$ i.e., $(-1)\left(-\dfrac{a}{b}\right) = -1$

$\Rightarrow a = -b$

Hence $(-8, 8)$ satisfies the requirements

TEXTUAL EXERCISE–10 (OBJECTIVE)

1. Given: $6x^2 + 6y^2 + 13xy + 22x + 23y + 20 = 0$ easily we can get the lines as $(2x + 3y + 4)(3x + 2y + 5) = 0$

Let: $L_1: 2x + 3y + 4 = 0$; $L_2: 3x + 2y + 5 = 0$; $L_3: x + 3y + 1 = 0$

The points of intersection are

$A\left(-\dfrac{7}{5}, \dfrac{-2}{5}\right), B\left(\dfrac{-13}{7}, \dfrac{2}{7}\right), C\left(-3, \dfrac{2}{3}\right)$

$$\text{Area } \Delta ABC = \frac{1}{2}\left\|\begin{matrix} -\frac{7}{5} & -\frac{2}{5} & 1 \\ -\frac{13}{7} & \frac{2}{7} & 1 \\ -3 & \frac{2}{3} & 1 \end{matrix}\right\| = \frac{1}{2}\left\|\begin{matrix} -\frac{7}{5} & -\frac{2}{5} & 1 \\ -\frac{16}{35} & \frac{24}{35} & 0 \\ -\frac{8}{5} & \frac{16}{15} & 0 \end{matrix}\right\|$$

$$= \frac{1}{2}\left\{-\frac{256}{35\times15} + \frac{192}{35\times5}\right\} = \frac{(288-128)}{35\times15} = \frac{160}{35\times15} = \frac{32}{105}$$

square units.

2. (b) Given $4y^3 - xy^2 - 10x^2y + ax^3 = 0$

Let $y = m_1x$, $y = m_2x$ and $y = m_3x$ be the solution, then

$m_1 + m_2 + m_3 = \dfrac{1}{4}$ and $m_1m_2 + m_1m_3 + m_2m_3 = -\dfrac{10}{4}$ and $m_1m_2m_3 = -\dfrac{9}{4}$

Since two lines are equally inclined with $y = x$, so let $m_1m_2 = 1$ then $m_3 = -\dfrac{a}{4} \Rightarrow 1 + m_1\left(-\dfrac{a}{4}\right) + m_2\left(-\dfrac{a}{4}\right) = \dfrac{-10}{4}$

i.e., $(m_1 + m_2)\left(-\dfrac{a}{4}\right) = -\dfrac{14}{4}$ and $m_1 + m_2 = \dfrac{1}{4} + \dfrac{a}{4} = \dfrac{a+1}{4}$

Hence $\dfrac{(-a)(a+1)}{16} = -\dfrac{14}{4}$

$\therefore a^2 + a - 56 = 0 \quad \Rightarrow (a + 8)(a - 7) = 0$

$a > 0$

$\Rightarrow a = 7$

3. (a) Given equation is $2x^2 - \lambda y^2 + 3xy = 0$ will represent perpendicular lines when $a + b = 0$ i.e., $\lambda = 2$ and $h^2 - ab = \dfrac{9}{4} + 4 > 0$

$\Rightarrow$ The lines will be at 90°

4. (b) Given equation is $x^2 + 2\lambda xy + y^2 = 0$ will represent two real and distinct lines, when $h^2 - ab > 0$ i.e., $\lambda^2 - 1 > 0$

i.e., $\lambda^2 > 1$

Since $\lambda > 0 \quad \Rightarrow \lambda \in (1, \infty)$

5. (d) Given equation is $30x^2 + 5xy + (a^2 - 2)y^2 = 0$. The lines will be perpendicular to each other when $a + b = 0$ i.e., $30 + a^2 - 2 = 0$ which does not give any real value of a.

6. (c) $6x^2 + y^2 - 5xy + 3x - y = 0$ can be written as $(3x - y)(2x - y + 1) = 0$

Similarly $6x^2 + y^2 - 5xy + 2x - y = 0$ can be written as $(3x - y + 1)(2x - y) = 0$

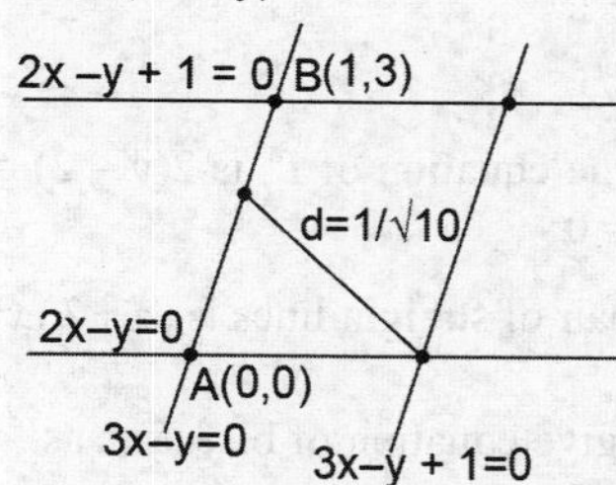

As shown point of intersection of $2x - y = 0$ and $3x - y = 0$ is $A(0, 0)$ and point of intersection of $3x - y = 0$ and $2x - y + 1 = 0$ is $B(1, 3)$, so $AB = \sqrt{10}$ units.

Now, distance between $3x - y = 0$ and $3x - y + 1 = 0$ is $d = \dfrac{1}{\sqrt{10}}$ units.

Hence the area of $\|gm$ is 1 square unit.

7. (a) Given BC is $L_3: 2x + 4y - 5 = 0$ and slope of $BC = -1/2$

Let m be the slope of straight line at 45° to BC, then

$\tan\dfrac{\pi}{4} = \left|\dfrac{m + \frac{1}{2}}{1 - \frac{m}{2}}\right| = 1$, gives $\dfrac{2m+1}{2-m} = \pm 1$

$\Rightarrow m = 1/3, -3$

Hence the equation of AB and AC will be

(i) $3(y - 1) = x - 4$ or $x - 3y - 1 = 0$

(ii) $y - 1 = (-3)(x - 4)$ or $3x + y - 13 = 0$

Hence the combined equation is $(3x + y - 13)(x - 3y - 1) = 0$ or $3x^2 - 3y^2 - 8xy - 16x + 38y + 13 = 0$

8. Given $x^2 - 9xy + 18y^2 = 0$, which is homogenous in x and y and it can be written as $(x - 6y)(x - 3y) = 0$

So $y = \dfrac{x}{3}, y = \dfrac{x}{6}$ are the straight lines which are at an acute angle in Ist & 3rd quadrant.

The point $P(1 + \cos\theta, \sin\theta)$ will lie in acute angle when

$\dfrac{1}{6} < \dfrac{\sin\theta}{1+\cos\theta} < \dfrac{1}{3}$ { $1 + \cos\theta \geq 0$, so the point P can not lie in 2nd or 3rd quadrant}.

This will give $\dfrac{1}{6} < \tan\dfrac{\theta}{2} < \dfrac{1}{3}$ in the Ist quadrant only

TEXTUAL EXERCISE–10 (SUBJECTIVE)

1. Given the combined equation of bisectors B_1 and B_2 is $3x^2 - 8xy - 3y^2 + 10x + 20y - 25 = 0$

Which can be rewritten as $(3x + y - 5)(x - 3y + 5) = 0$

$\Rightarrow$ Point of intersection is $P(1, 2)$

Now L_1 passes through the origin.

$\Rightarrow$ $L_1: y = 2x$ and $m_1 = 2$

Slope of the bisector $L_3: x - 3y + 5 = 0$ is $m_3 = 1/3$

Slope of L_2 is m_2 given by $\dfrac{m_1 - m_3}{1 + m_1m_3} = \dfrac{m_3 - m_2}{1 + m_2m_3}$

$\Rightarrow \dfrac{2 - \frac{1}{3}}{1 + \frac{2}{3}} = \dfrac{\frac{1}{3} - m_2}{1 + \frac{m_2}{3}} \quad \Rightarrow \quad 1 - 3m_2 = 3 + m_2$

$\Rightarrow$ $m_2 = -1/2$

Hence the equation of L_2 is $2(y - 2) = -(x - 1)$ or $x + 2y - 5 = 0$

2. First set of pair of straight lines is $(a + \lambda)x^2 + 2hxy + (\lambda + b)y^2 = 0$

Which will give equation of bisectors as

$$\frac{x^2 - y^2}{(a+\lambda) - (b+\lambda)} = \frac{xy}{h} \text{ or } \frac{x^2 - y^2}{a-b} = \frac{xy}{h}$$

The second pair is $(a + \mu)x^2 + 2hxy + (b + \mu)y^2 = 0$

The equation of bisector s is

$$\frac{x^2 - y^2}{(a+\mu) - (b+\mu)} = \frac{xy}{h} \text{ or } \frac{x^2 - y^2}{a-b} = \frac{xy}{h}$$

Which is the same as for the other set so the lines are isoinclined. Hence the angle will be the same.

3. Given equation is $a(x - 1)^2 + 2h(x - 1)(y - 2) + b(y - 2)^2 = 0$

The combined equation of bisectors will be $\dfrac{(x-1)^2 - (y-2)^2}{a-b} = \dfrac{(x-1)(y-2)}{h}$

Putting $x - 1 = X$ and $y - 2 = Y$ we get $hX^2 - hY^2 - (a - b)XY = 0$ and $x + 2y - 5 = 0$ becomes $X + 2Y = 0$

Now, $(X + 2Y)(hX - \frac{h}{2}Y) = hX^2 - hY^2 - (a - b)XY$

Easily we can deduce that $(2X - Y) = 0$ is the other line subject to the condition that $2(a - b) = -3h$

Now, $2(x - 1) - (y - 2) = 0$

$\Rightarrow$ $2x - y = 0$

$\therefore$ Other line is $2x - y = 0$ and the relation is $2a + 3h = 2b$

4. Given pair of lines (perpendicular bisectors of AB and AC) is $y^2 - 8xy - 9x^2 = 0$

$\Rightarrow$ $(y - 9x)(y + x) = 0$

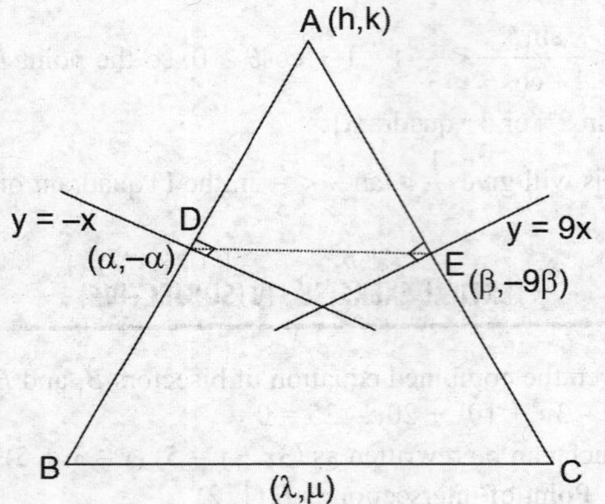

Point is will be the image of $A(h, k)$ on line $x + y = 0$, given by $\dfrac{x-h}{1} = \dfrac{y-k}{1} = \dfrac{-2(h+k)}{2}$

$\Rightarrow x = -h - k + h = -k$ and $y = -h$

Similarly co-ordinates of C will be the image of $A(h, k)$ on line $9x - y = 0$

Given by $\dfrac{x-h}{9} = \dfrac{y-k}{-1} = \dfrac{-2(9h-k)}{82}$

$\Rightarrow \quad x = \dfrac{-81h + 9k}{41} + h = \dfrac{9k - 40h}{41}$ and $y = \dfrac{9h - k}{41} + k$

$\Rightarrow \quad y = \dfrac{9h + 40k}{41}$

Now equation of BC is given by $(y + h) = \dfrac{\frac{9h+40k}{41} + h}{\frac{9k-40h}{41} + k}(x + k)$

$\Rightarrow \quad y + h = \dfrac{50h + 40k}{50k - 40h}(x + k)$

Since it passes through $(\lambda, -\mu)$

$\Rightarrow$ $(5k - 4h)(\mu + h) = (5h + 4k)(\lambda + k)$

$\Rightarrow$ Locus is $(5y - 4x)(u + x) = (5x + 4y)(\lambda + y)$

$\Rightarrow$ $4(x^2 + y^2) + (5\lambda + 4\mu)x + (4\lambda - 5\mu)y = 0$

5. Given a pair of straight lines by $2x^2 + 6xy + y^2 = 0$.

Now the equation of angle bisectors will be $\dfrac{x^2 - y^2}{2-1} = \dfrac{xy}{3}$

or $3(x^2 - y^2) = xy$

The other pair of straight lines is $4x^2 + 18xy + y^2 = 0$, which will have the equation of angle bisectors as $\dfrac{x^2 - y^2}{4-1} = \dfrac{xy}{9}$

or $3(x^2 - y^2) = xy$

As bisectors have the same equation

$\therefore$ Lines are inclined.

6. $1 - \dfrac{3y^2}{x^2} + \dfrac{y^3}{x^3} - \dfrac{3y}{x} = 0$

Put $\dfrac{y}{x} = m \quad \Rightarrow \quad m^3 - 3m^2 - 3m + 1 = 0$

Clearly $m = -1$ is root

$\Rightarrow$ $(m + 1)(m^2 - 4m + 1) = 0$

$\Rightarrow \quad m = -1, m = \dfrac{4 \pm \sqrt{12}}{2}$

$\Rightarrow \quad m = -1, \; m = 2 \pm \sqrt{3}$

Let $m_1 = -1, \; m_2 = 2 + \sqrt{3}, \; m_3 = 2 - \sqrt{3}$;

Let $\tan\alpha = \dfrac{3+\sqrt{3}}{1 - 2 - \sqrt{3}} = \dfrac{3+\sqrt{3}}{-1-\sqrt{3}} = \dfrac{\sqrt{3}(\sqrt{3}+1)}{-(\sqrt{3}+1)} = -\sqrt{3}$

$\Rightarrow$ $\alpha = 120°$.

Similarly $\tan\beta = \dfrac{2 - \sqrt{3} - 2 - \sqrt{3}}{1 + (1)} = -\sqrt{3} \Rightarrow B = 120°$

Clearly $\gamma = 120°$

TEXTUAL EXERCISE–11 (OBJECTIVE)

1. **(b)** Given pair of lines is $36x^2 - 20y^2 - 33xy + 78x + 43y - 14 = 0$

$\Rightarrow$ $(3x - 4y + 7)(12x + 5y - 2) = 0$

Now, $3x - 4y + 7 = 0$ and $-12x - 5y + 2 = 0$

$\Rightarrow$ $a_1a_2 + b_1b_2 = -36 + 20 = -16 < 0$

$\Rightarrow$ Origin lies in acute angle

$\Rightarrow$ Equation of bisector of acute angle will be given by $\frac{3x-4y+7}{5} = \frac{(-12x+5y+2)}{13}$

$\Rightarrow$ $39x - 52y + 91 = -60x - 25y + 10$

$\Rightarrow$ $99x - 27y + 81 = 0$

$\Rightarrow$ $11x - 3y + 9 = 0$

2. **(c)** Let $y = mx$ be the common line, then $x^2\{b_1 m^2 + 2h_1 m + a_1\} = 0$ and $x^2\{b_2 m^2 + 2h_2 m + a_2\} = 0$

$$\Rightarrow \frac{m^2}{2(h_1a_2 - h_2a_1)} = \frac{m}{(a_1b_2 - a_2b_1)} = \frac{1}{2(b_1h_2 - b_2h_1)}$$

$$\Rightarrow m^2 = \frac{4(h_1a_2 - h_2a_1)}{4(b_1h_2 - b_2h_1)} = \frac{(a_1b_2 - a_2b_1)^2}{4(b_1h_2 - b_2h_1)^2}$$

Hence $\frac{(a_1b_2 - a_2b_1)^2}{(h_1a_2 - h_2a_1)(b_1h_2 - b_2h_1)} = 4$

3. **(a)** $ax^2 + 2hxy - ay^2 = 0$, the lines are perpendicular to each other as $A - B = a - a = 0$

$\Rightarrow$ Equation of angle bisector will be $\frac{x^2 - y^2}{2a} = \frac{xy}{h}$ or $hx^2 - hy^2 - 2axy = 0$

This equation is the same as the given equation for the second set of lines

$\Rightarrow$ Each set of lines is the angles bisectors of the other

4. **(a)** Equation of angle bisectors for the lines $ax^2 + 2hxy + by^2 = 0$ is $\frac{x^2 - y^2}{a-b} = \frac{xy}{h}$ or $hx^2 - (a-b)xy - hy^2 = 0$ or $hy^2 + (a-b)xy - hx^2$

Since $y = mx$ is one of these bisectors

$\Rightarrow$ $x^2\{hm^2 + (a-b)m - h\} = 0$

$\Rightarrow$ $hm^2 + (a-b)m - h = 0$, which can be written as $hx^2 + (a-b)x - h = 0$

TEXTUAL EXERCISE–11 (SUBJECTIVE)

1. The line is $\frac{y-7x}{2} = 1$ and the curve is $2x^2 + y^2 + (x + y)(1) = 0$

On Homogenizing, we get $2x^2 + y^2 + \frac{(x+y)(y-7x)}{2} = 0$

$\Rightarrow$ $(-1)\{3x^2 + 6xy - 3y^2\} = 0$ or $x^2 + 2xy - y^2 = 0$

Since $a + b = 0 \Rightarrow$ lines are at right angles

2. Give equation is $x^2 + y^2 = a^2$.......(i) and $x^2 + y^2 + 2(gx + fy) = 0$ ......(ii)

Operating (ii) – (i), we get $2(gx + fy) = -a^2$

To homogenize square both sides and get $4(gx + fy)^2 = a^4 = a^2.a^2 = a^2(x^2 + y^2)$ or $a^2(x^2 + y^2) - 4(gx + fy)^2 = 0$

3. Given equation are $x + y = 1$ ....(i) and $x^2 + y^2 + (x - 2y) - m = 0$ ....(ii)

Homogenizing (ii) using (i), we get $x^2 + y^2 + (x - 2y)(x + y) - m(x + y)^2 = 0$ or $(m - 2)x^2 + (m + 1)y^2 + (2m + 1)xy = 0$

Since the lines are perpendicular to each other

$\therefore$ $a + b = 0$

$\Rightarrow$ $(m - 2) + (m + 1) = 0 \Rightarrow m = 1/2$

4. Given equation are: $y - mx = 1$...(i) and $x^2 + y^2 = 1$...(ii)

Homogenizing (ii) with (i), we get $x^2 + y^2 = (y - mx)^2$

$\Rightarrow$ $(m^2 - 1)x^2 + 0y^2 - 2mxy = 0$

The lines will be perpendicular to each other when $a + b = 0$ i.e., $m^2 - 1 + 0 = 0$

$\Rightarrow$ $m = \pm 1$

TEXTUAL EXERCISE–12 (OBJECTIVE)

1. **(a)** From the given we can conclude that $2x + 3y = 0$ and $3x - 2y = 0$ are the bisectors of the pair of lines

i.e., $\frac{x^2 - y^2}{a-b} = \frac{xy}{h}$

Also the lines are angle bisectors of $(2x + 3y)(3x - 2y) = 0$

$\Rightarrow$ $6x^2 - 6y^2 + 5xy = 0$

$\Rightarrow$ The lines are $\frac{x^2 - y^2}{6+6} = \frac{xy}{(5/2)}$

$\Rightarrow$ $5x^2 - 5y^2 - 24xy = 0$

2. **(a)** Distance of origin from the line $2x + 3y - 6 = 0$ is $p = \frac{|-6|}{\sqrt{13}}$

Since an isosceles ΔOAB is formed.

$\Rightarrow$ Area of $\Delta OAB = p^2 = 36/13$

3. **(a), (b), (c), (d)** $2x + y = 1$ ....(i) and $3x^2 + mxy - 4x + 1 = 0$ ....(ii)

Homogenizing (ii) with (i), we get $3x^2 + mxy - 4x(2x + y) + (2x + y)^2 = 0$

$\Rightarrow$ $(-x^2) + y^2 + mxy = 0$

The line area at right angle as $a + b = 0$, when $h^2 - ab > 0$

i.e., $\frac{m^2}{4} + 1 > 0$

Which is always true.

4. **(c)** Given $\frac{y-3x}{2} = 1$ ....(i) and

$x^2 + y^2 = 1$....(ii)

Homogenizing (ii) with (i), we get $4(x^2 + y^2) = y^2 + m^2x^2 - 2mxy$ or $(m^2 - 4)x^2 - 3y^2 - 2mxy = 0$

The lines will be at right angles when $a + b = m^2 - 4 - 3 = 0$

$\Rightarrow$ $m^2 = 7$

5. **(c)** Given: $x^2 + y^2 + 2gx + c = 0$...(i) and $x^2 + y^2 + 2fy - c = 0$....(ii)

Operating (i) – (ii), we get $2(gx - fy) = -2c$ or $fy - gx = c$

Also $2(x^2 + y^2 + gx + fy) = 0$

$\Rightarrow x^2 + y^2 + (gx + fy)\dfrac{(fy - gx)}{c} = 0$

Hence $f^2y^2 - g^2x^2 + cx^2 + cy^2 = 0$ or $(c - g^2)x^2 + (f^2 + c)y^2 = 0$

The lines will be at right angles, when $c - g^2 + f^2 + c = 0$

$\Rightarrow f^2 - g^2 = -2c$ or $g^2 - f^2 = 2c$

TEXTUAL EXERCISE SECTION–III (OBJECTIVE)

1. (c) Given $L_1: a_1x + b_1y + 1 = 0$; $L_2: a_2x + b_2x + 1 = 0$ and $L_3: a_3x + b_3y + 1 = 0$ lines are concurrent.

So $\begin{vmatrix} a_1 & b_1 & 1 \\ a_2 & b_2 & 1 \\ a_3 & b_3 & 1 \end{vmatrix} = 0 \Rightarrow \begin{vmatrix} a_1 & b_1 & 1 \\ a_2 - a_1 & b_2 - b_1 & 0 \\ a_3 - a_2 & b_3 - b_2 & 0 \end{vmatrix} = 0$

$\Rightarrow (a_2 - a_1)(b_3 - b_2) = (a_3 - a_2)(b_2 - b_1)$ or

$\dfrac{b_3 - b_2}{a_3 - a_2} = \dfrac{b_2 - b_1}{a_2 - a_1} = m$(say), so same slope

$\Rightarrow (a_1, b_1), (a_2, b_2), (a_3, b_3)$ are lying on the same straight line.

2. (a) Let m be the slope of line through $P(a, b)$, then $y - b = m(x - a)$

$\Rightarrow A\left(a - \dfrac{b}{m}, 0\right)$, $B(0, b - am)$ and the centroid G

$\left(\dfrac{a}{3} - \dfrac{b}{3m}, \dfrac{b - a_m}{3}\right)$

$\Rightarrow 3x = a - \dfrac{b}{m}$ or $m = \dfrac{b}{a - 3x}$ and $b - am = 3y$ or

$m = \dfrac{b - 3y}{a} \qquad \Rightarrow \dfrac{b}{a - 3x} = \dfrac{b - 3y}{a}$

$\Rightarrow$ ab $= ab - 3bx - 3ay + 9xy$ or $bx + ay - 3xy = 0$

3. (c) Lines is concurrent.

So $\begin{vmatrix} \frac{1}{a} & \frac{1}{b} & 1 \\ \frac{1}{b} & \frac{1}{a} & 1 \\ \frac{1}{c} & \frac{1}{d} & 1 \end{vmatrix} = 0$

$\Rightarrow \left(\dfrac{1}{b} - \dfrac{1}{a}\right)\left(\dfrac{1}{d} - \dfrac{1}{a}\right) = \left(\dfrac{1}{a} - \dfrac{1}{b}\right)\left(\dfrac{1}{c} - \dfrac{1}{b}\right)$

$\Rightarrow \left(\dfrac{1}{a} - \dfrac{1}{d}\right)\left(\dfrac{1}{a} - \dfrac{1}{b}\right) = \left(\dfrac{1}{a} - \dfrac{1}{b}\right)\left(\dfrac{1}{c} - \dfrac{1}{b}\right)$

$\Rightarrow \dfrac{1}{a} + \dfrac{1}{b} = \dfrac{1}{c} + \dfrac{1}{d}$

4. (c) Given $L_1: U = x + 2y - 3 = 0$ and $L_2: V = 4x - y + 7 = 0$

Now, $U + kV = 0$ will represent (a line parallel to) $5x + 4y = 0$, when slope $m = -5/4$

$\Rightarrow m = \dfrac{-(1 + 4k)}{(2 - k)} = -\dfrac{5}{4} \Rightarrow 16k + 4 = 10 - 5k$

$\Rightarrow k = \dfrac{6}{21} = \dfrac{2}{7}$

5. (d) Given $P = (1, 0)$, $Q(-1, 0)$ and $R = (2, 0)$, now $S(x, y)$ satisfies $SQ^2 + SR^2 = 2SP^2$

i.e., $(x + 1)^2 + y^2 + (x - 2)^2 + y^2 = 2\{(x - 1)^2 + y^2\}$

$\Rightarrow x^2 + 1 + 2x + y^2 + x^2 + 4 - 4x + y^2 = 2x^2 + 2 - 4x + 2y^2$

$\Rightarrow 2x = -3 \Rightarrow x = -3/2$, which is a straight parallel to y-axis

6. (a) Product of perpendicular from (x_1, y_1) to $ax^2 + by^2 + 2hxy = 0$ is given by (product) $d_1d_2 = \dfrac{|ax_1^2 + by_1^2 + 2hx_1y_1|}{\sqrt{(a - b)^2 + 4h^2}}$

$\Rightarrow$ For P (1, 2) and line pair $x^2 + y^2 + 4xy = 0$, we get $d_1d_2 = \dfrac{|1 + 4 + 8|}{\sqrt{0 + 16}} = \dfrac{13}{4}$

Aliter: The pair of lines is $y^2 + 4xy + x^2 = 0$

$\Rightarrow y = (-2 \pm \sqrt{3})x$ or $\{y + (2 - \sqrt{3})x\}\{y + (2 + \sqrt{3})x\} = 0$

Product of distances from (1, 2) will be

$d_1d_2 = \dfrac{|2 - \sqrt{3} + 2||2 + \sqrt{3} + 2|}{\sqrt{(7 + 1 - 4\sqrt{3})}\sqrt{(7 + 1 + 4\sqrt{3})}} = \dfrac{13}{\sqrt{64 - 48}} = \dfrac{13}{4}$

7. (c) Let $L_1: y = x$

Image of $y = x$ in $x = 1$ will be $y - 1 = -(x - 1)$ gives $y = 2 - x$.

Similarly image of $L_2: y = -x$ in $x = 1$ will be $y + 1 = x - 1$ or $y = x - 2$, so combined equation is $y = |x - 2|$

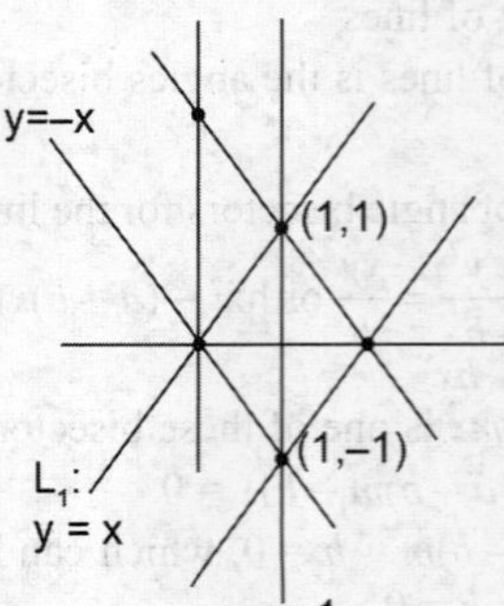

8. (a) $(x + y + 1) + k(x^2 + y^2) = 0$ will represent two straight when $\Delta = 0$

Now $(k + 1)x^2 + (k + 1)y^2 + 2xy + 2x + 2y + 1 = 0$

So $\Delta = \begin{vmatrix} (k+1) & 1 & 1 \\ 1 & (k+1) & 1 \\ 1 & 1 & 1 \end{vmatrix} = 0$ or $\begin{vmatrix} k & 0 & 0 \\ 0 & k & 0 \\ 1 & 1 & 1 \end{vmatrix} = 0$

$\Rightarrow k^2 = 0 \Rightarrow k = 0$

9. (d) Given equation is $12x^2 - 10xy + 2y^2 + 14x - 5y - c = 0$ will represent two straight lines when $\Delta = \begin{vmatrix} a & h & g \\ h & b & f \\ g & f & c \end{vmatrix} = 0$

i.e., $\begin{vmatrix} 12 & -5 & -7 \\ -5 & 2 & -\frac{5}{2} \\ 7 & -\frac{5}{2} & -c \end{vmatrix} = 0$

$\Rightarrow$ $25c - 24c - 75 - 98 + \dfrac{175+175}{2} = 0 \Rightarrow c = -2$

10. **(a)** Given L: $ax + by = 1$ …..(i) and curve C: $px^2 + 2axy + qy^2 - r = 0$

Homogenizing (ii) with (i), we get $px^2 + 2axy + qy^2 - r(ax + by)^2 = 0$

$\Rightarrow$ $(p - ra^2)x^2 + (q - rb^2)y^2 + (2a - 2arb)xy = 0$, which will give a right angle at origin when $(p - ra^2) + (q - rb^2) = 0$

i.e., $r(a^2 + b^2) = p + q$

11. **(c)** Given equation is $x^2 + y^2 + 4xy = 0$

$\Rightarrow$ $\tan\theta = \dfrac{2\sqrt{4-1}}{2} = \sqrt{3}$, gives $\theta = 60°$

As shown $CM = 4$ units and $\theta/2 = 30°$, so $OC = 8$ units and $OM = 4\sqrt{3}$ units

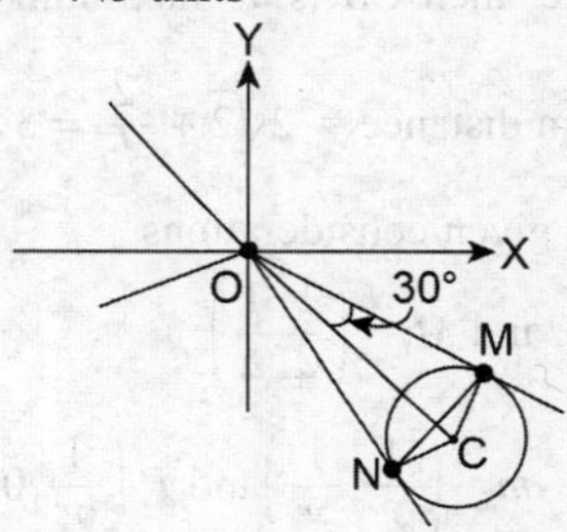

$\therefore$ Length of chord of contact $= 2\times\dfrac{4\sqrt{3}}{2} = 4\sqrt{3}$ units

12. **(b)** Given equation L_1 : $12x - 5y - 17 = 0$ and L_2: $24x - 10y + 44 = 0$ or $12x - 5y + 22 = 0$ are tangents to a circle.

Observe that $L_1 \parallel L_2$ and distance between the lines will be diameter of the circle.

$\Rightarrow$ $d = \dfrac{|22+17|}{13} = 3$ $\Rightarrow$ r = 3/2 units

13. **(c)** B_1: $x + 3y - 2 = 0 \Rightarrow m_3 = -1/3$ and L_1: $x - 7y + 5 = 0$

$\Rightarrow$ $m_1 = 1/7$

Intersect a $\left(-\dfrac{1}{10}, \dfrac{7}{10}\right)$ only $5x + 5y - 3 = 0$ passes through it.

Now, $\tan\theta = \dfrac{m_1 - m_3}{1 + m_1m_3} = \dfrac{m_3 - m_2}{1 + m_2m_3}$

$\Rightarrow$ $\tan\theta = \dfrac{\frac{1}{7}+\frac{1}{3}}{1-\frac{1}{21}} = \dfrac{1}{2} = \dfrac{-\frac{1}{3} - m_2}{1 - \frac{m_2}{3}}$

$\Rightarrow$ $3 - m_2 = -2 - 6m_2$

$\Rightarrow$ $m_2 = -1$, which is same as the slope of $5x + 5y - 3 = 0$

14. **(a)** Since $PQ \perp PS$

$m_1m_2 = \left(\dfrac{y_1}{a - x_1}\right)\left(\dfrac{y_2}{-a - x_1}\right) = -1$

$\Rightarrow$ $a^2 - x_1^2 = y_1y_2$ or $y_1y_2 + x_1^2 = a^2$

Also mid points of QS is $\left(0, \dfrac{y_1 + y_2}{2}\right)$ which is the mid point PR

$\Rightarrow$ $R\left(-x_1, \dfrac{y_1 + y_2}{2}\right)$

Now, $\dfrac{y_1 + y_2}{2} = \dfrac{m_1^2(a - x_1) + (a + x_1)}{2m_1}$. Since m_1 is also a fixed constant

$\Rightarrow$ $R = \left(-x_1, \dfrac{(m_1^2 + 1)a + (1 - m_1^2)x_1}{2m_1}\right)$

Hence $y = \left(\dfrac{m_1^2 - 1}{2m_1}\right)(-x_1) + \dfrac{(m_1^2 + 1)a}{2m_1}$

Which gives a straight line as m_1 and 'a' are constants

15. **(d)** A (2, 1), B (3, –7), C are lying on $3x - 2y = 1$ equation AB is L_1: $8x + y - 17 = 0$

Let $C(x_1, y_1)$, then $3x_1 - 2y_1 = 1$

Let D (x_2, y_2), then $x_1 + 2 = x_2 + 3$

$\Rightarrow$ $x_1 = x_2 + 1$ and $y_1 + 1 = y_2 - 7$

$\Rightarrow$ $y_1 = y_2 = -8$

Hence $3(x_2 + 1) - 2(y_2 - 8) = 1$

$\Rightarrow$ $3x_2 - 2y_2 + 18 = 0$ or $3x + 18 = 2y$

16. **(b)** L_1: $y - \sqrt{3}x = 2$ where $x \ge 0$ and L_2: $y + \sqrt{3}x = 2$ where $x \le 0$

Observe that the acute angle bisector is $x = 0$ and the foot of perpendicular from P will be at $x = 0$, $y = 2 + 5\cos 30°$ i.e., $\left(0, \dfrac{4 + 5\sqrt{3}}{2}\right)$

17. **(d)** According to the new formula $d = |x_1 - x_2| + |y_1 - y_2|$

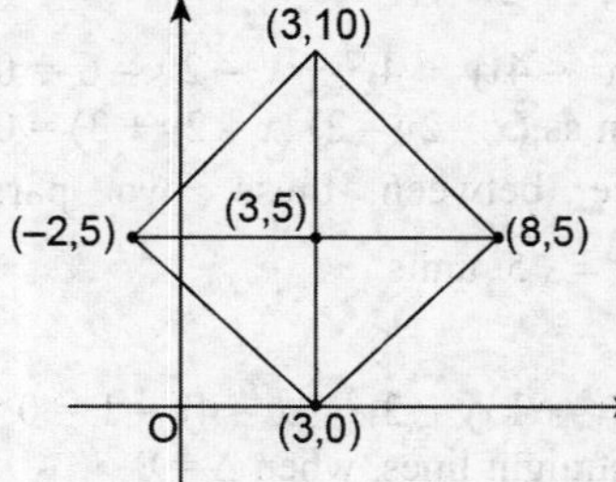

Case 1: For $x \ge 3, y \ge 5$ $\Rightarrow$ $x + y = 13$, represents AB

Case 2: For $x \le 3, y \ge 5$ $\Rightarrow$ $y - x = 7$, represents BC

Case 3: For $x \le 3, y \le 5$ $\Rightarrow$ $x + y = 3$, represents CD

Case 4: For $x \ge 3, y \le 5$ $\Rightarrow$ $x - y = 3$, represents DA

$\Rightarrow$ Thus square $ABCD$ is obtained.

18. **(c)** Given $ax^3 + by^3 + cx^2y + dxy^2 = 0$

Which is homogenous in x and y (degree = 3)

According to the given this will represent three distinct lines as $y = \pm\sqrt{3}x$ and $y = 0$ (without any loss of generality)

$\Rightarrow$ $\left(\sqrt{3}x - y\right)(y)\left(\sqrt{3}x + y\right) = 0$

$\Rightarrow$ $(3x^2 - y^2)y = 0$ or $0x^3 - y^3 + 3x^2y + 0xy^2 = 0$

Comparison gives $c + 3b = 0$

19. (a) Given equation is $ax^2 + 2hxy + by^2 = 0$

$y = mx$ and $y = 2mx$ is the solution, then $\frac{2h}{b} = -3m$ and $\frac{a}{b} = 2m^2$

$\Rightarrow \frac{a}{2b} = \frac{4h^2}{9b^2} \Rightarrow \frac{ab}{h^2} = \frac{8}{9}$

20. (c) Given equation is $x^2 + \lambda y^2 + 16x - 16y + 55 = 0$

This will represent a pair of straight when $\Delta = 0$

$\Rightarrow \begin{vmatrix} 1 & 0 & 8 \\ 0 & \lambda & -8 \\ 8 & -8 & 55 \end{vmatrix} = 0 \quad \Rightarrow (55\lambda - 64) + 8(-8\lambda) = 0$

$\Rightarrow \lambda = -\frac{64}{9}$

Now, For $\lambda = -\frac{64}{9}$, $\tan\theta = \left|\frac{2\sqrt{h^2 - ab}}{a+b}\right|$

$\Rightarrow \tan\theta = \left|\frac{2\sqrt{0 + \frac{64}{9}}}{1 + \left(-\frac{64}{9}\right)}\right| = \frac{16(9)}{3(55)} = \frac{48}{55}$

Now, $\sin^2\theta = \frac{\tan^2\theta}{1 + \tan^2\theta} = \frac{\left(\frac{48}{55}\right)^2}{1 + \left(\frac{48}{55}\right)^2} = \frac{(48)^2}{(55)^2 + (48)^2}$

$= \frac{2304}{5329}$.

21. (b) Given, L_1: $-\alpha x - 2y + 3 = 0$ and L_2: $x + y + 7 = 0$

Here, $c_1, c_2 > 0$ and $a_1a_2 + b_1b_2 = -\alpha - 2$.

Now, the origin will be in the acute angle, if $-(\alpha + 2) < 0$, i.e., $\alpha > -2$

22. (b) Given, $x^2 - 4xy + 4y^2 + x - 2y - 6 = 0$, which can be rewritten as $(x - 2y - 2)(x - 2y + 3) = 0$.

$\Rightarrow$ Distance between these two parallel lines is $d = \frac{5}{\sqrt{5}} = \sqrt{5}$ units.

23. (a) Given $2x^2 + kxy - 3y^2 - x - 4y - 1 = 0$ will represent a pair of straight lines, when $\Delta = 0$

$\Rightarrow \begin{vmatrix} 2 & \frac{k}{2} & -\frac{1}{2} \\ \frac{k}{2} & -3 & -2 \\ -\frac{1}{2} & -2 & -1 \end{vmatrix} = 0$

$\Rightarrow \begin{vmatrix} 2 & \left(\frac{k}{2} - 8\right) & -\frac{9}{2} \\ \frac{k}{2} & (-2k - 3) & (-2 - k) \\ -\frac{1}{2} & 0 & 0 \end{vmatrix} = 0$

Hence, $\left(-\frac{1}{2}\right)\left\{16 + 7k - \frac{k^2}{2} - 9k - \frac{27}{2}\right\} = 0$ or

$\left(\frac{1}{4}\right)\left\{k^2 + 4k - 5\right\} = 0 \quad \Rightarrow k = 1, -5.$

24. (a) $9a^2 + 16b^2 - 24ab = c^2$ can be rewritten as $(3a - 4y^2)^2 = (\pm c)^2$

Since, family of straight lines $ax + by + c = 0$ are concurrent. $3a - 4y + c = 0$ or $3a - 4y - c = 0$. From, $ax + by + c = 0$, we get $a(3) + b(-4) + c = 0$

$\Rightarrow$ (3, –4) is the point .

Or $3a - 4y - c = 0$ can be written as $a(-3) + b(4) + c = 0$ i.e., $3a - 4b - c = 0$

$\Rightarrow$ (–3, 4) is the point .

Hence point of concurrency is (3, –4) or (–3, 4)

25. (a) Given $OA = OB = 9$ units. Where A and B lie on $3x + 4y + 15 = 0$

Now, distance of origin from the line is $p = \frac{15}{5} = 3$ units

$\Rightarrow$ area $\Delta OAB = 3\sqrt{9^2 - 3^2} = 9 \times 2\sqrt{2} = 18\sqrt{2}$ square units

26. (c) Side of the square = $2\sqrt{2}$units. As shown let $ABCD$ be the square, then CB is at maximum distance from the origin.

Maximum distance = $2\sqrt{2} + \frac{2}{\sqrt{2}} = 3\sqrt{2}$

27. (b) From the given considerations

$P_1\left(\frac{1}{2}, 0\right)$ and $M_1 = \left(\frac{1}{4}, \frac{1}{4}\right)$

$\Rightarrow P_2\left(\frac{1}{4}, 0\right)$; $m_{n-1}\left(\frac{1}{2^n}, \frac{1}{2^n}\right)$ and $P_n\left(\frac{1}{2^n}, 0\right)$

$\Rightarrow OP_n = \frac{1}{2^n}$

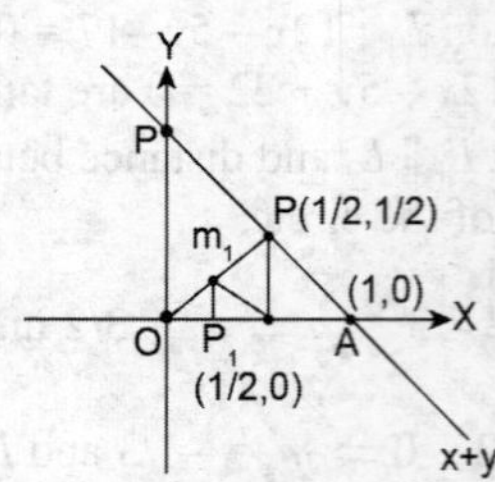

28. (c) Given $ax^2 + 2hxy - ay^2 = 0$.....(i) and $bx^2 + 2hxy - by^2 = 0$ ....(ii) are bisectors of each other (Both sets are having lines at 90°).

Now, equation of bisectors is $\frac{x^2 - y^2}{ab} = \frac{xy}{h}$.

So, for (i) the equation will be $h(x^2 - y^2) = a(-a)xy$, or $hx^2 + a^2xy - hy^2 = 0$

$\Rightarrow bx^2 + 2gxy - by^2 = 0$

Comparison with (ii), we get $\frac{h}{b} = \frac{a^2}{2g} \Rightarrow a^2b = 2gh$.

Hence, $a^2b - 2gh = 0$

29. (d) Since altitude from B is horizontal.

$\therefore$ OA is vertical, so let $A(0, y_1)$. Now, altitude from A will be $(y - 4) = (-3/4)(x - 1)$.

So, $y_1 = \frac{19}{4} \Rightarrow A(0, 19/4)$.

30. **(b)** The family lines $(x + 3y - 7) + \lambda (x + 4y - 3) = 0$, will pass through P_1 (19, –4) $\Rightarrow$ slope of $OP_1 = \frac{-4}{19} = m_1$.

Similarly, family lines $(3x + y - 7) + \mu (x + y - 3) = 0$, will pass through P_2 (2, 1).

$\Rightarrow$ slope of $OP_2 = 1/2 = m_2$

Slope of line passing though P_1 and perpendicular to OP_1 is $m_3 = \frac{19}{4}$ and line perpendicular to OP_2 is with slope $m_4 = -2$

$$\Rightarrow \tan\theta = \left|\frac{m_3 - m_4}{1 + m_3 m_4}\right| = \left|\frac{\frac{19}{4}+2}{1-\frac{19}{2}}\right| = \left|\frac{27}{2(17)}\right|$$

$$\Rightarrow \theta = \tan^{-1}\left(\frac{27}{34}\right)$$

31. **(c)** A (3, 0), B (0, 2), P (a, 0) and Q (0, b)

AQ perpendicular BP and the point of intersection is at R, so ΔARB is right angled at R.

$\Rightarrow$ Orthocenter of ΔARB is at R.

Also from the given considerations $\frac{x}{a}+\frac{y}{b}=1$ is at right angle to $2x + 3y = 6$.

$\Rightarrow$ Slope of $\frac{x}{a}+\frac{y}{b}=1$ is $m = \frac{-b}{a} = \frac{(-1)}{(-2/3)}$

$\Rightarrow b = -\frac{3}{2}a \Rightarrow P (a, 0)$ and $Q\left(0, -\frac{3}{2}a\right)$

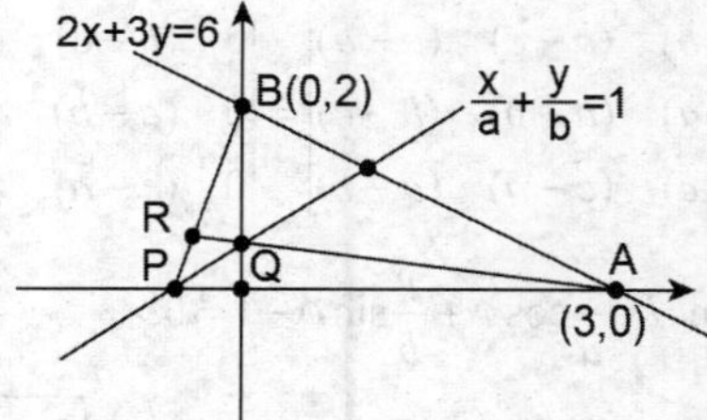

AQ: $y = \frac{a}{2}(x-3)$ and BP: $y = \frac{-2x}{a} + 2$ and point of intersection $R = \left(\frac{4-9a^2}{3(4-a^2)}, \frac{2a(a-3)}{(4-a^2)}\right)$.

Eliminating 'a' we get $a = \frac{2y}{x-3} = \frac{-2x}{y-2}$ gives the locus of R as $y^2 - 2y = 3x - x^2$ or $x^2 + y^2 - 3x - 2y = 0$

32. **(a)** Let A (x_1, 0) and B (0, y_1). As shown $x_1^2 + y_1^2 = a^2$.

Let P (h, k) be the foot of perpendicular from E as shown and N be the foot from D on x-axis.

Since ΔOAB and ΔADN are congruent

$\therefore$ $AN = OB = y_1$ and $DN = OA = x_1$

$\Rightarrow D (x_1 + y_1, x_1)$

$\Rightarrow$ E being mid-point of AD, will be $E = \left(\frac{2x_1 + y_1}{2}, \frac{x_1}{2}\right)$

Also mid-point of AB is $F\left(\frac{x_1}{2}, \frac{y_1}{2}\right)$. Observe that P (h, k) is the mid pt of E and F.

$\Rightarrow P (h, k) = \left(\frac{3x_1 + y_1}{4}, \frac{x_1 + y_1}{4}\right)$

Hence, $4h = 3x_1 + y_1$ and $4k = x_1 + y_1$ gives $4h - 4k = 2x_1$.

$\Rightarrow x_1 = 2h - 2k$

$\therefore$ $y_1 = 4k - x_1 = 6k - 2h$

From $x_1^2 + y_1^2 = a^2$ we get $4 (h - k)^2 + 4 (3k - h)^2 = a^2$ or $(k - h)^2 + (h - 3k)^2 = a^2/4$ i.e., $(y - x)^2 + (x - 3y)^2 = a^2/4$

33. **(a)** Let (x_1, y_1) be the point on L_1, so $5x_1 - y_1 = 4$ and (x_2, y_2) be the point on L_2, so $3x_2 + 4y_2 = 4$ as (1, 5) is the mid point.

$\Rightarrow$ $x_2 = 2 - x_1$ and $y_2 = 10 - y_1$.

Hence, $3(2 - x_1) + 4 (10 - y_1) = 4$.

$\Rightarrow$ $3x_1 + 4y_1 = 42$ solving with $20x_1 - 4y_1 = 16$.

We get, $x_1 = \frac{58}{23}$ and $y_1 = \frac{198}{23}$.

Hence, the equation $y - 5 = (83/35) (x - 1)$.

$\Rightarrow$ $83x - 35y + 92 = 0$.

34. **(c)** Line L_1 intercepts a, $-b$ will have slope $m_1 = -\frac{(-b)}{a} = \frac{b}{a}$

Similarly the line L_2 with intercepts b, $-a$ will have slope $m_2 = \frac{a}{b}$.

$$\Rightarrow \tan\theta = \left|\frac{m_1 - m_2}{1 + m_1 m_2}\right| = \left|\frac{a^2 - b^2}{ab(2)}\right| \Rightarrow \theta = \tan^{-1}\left|\left(\frac{b^2 - a^2}{2ab}\right)\right|$$

35. **(a)** Let m be the slope of any (non-vertical) line through (h, k), then L: $y - k = m (x - h)$.

Now the equation of the line that is perpendicular from the origin is $y = -\frac{x}{m}$

Hence $m = \frac{y-k}{x-h} = \frac{-x}{y} \Rightarrow y^2 - ky + x^2 - xh = 0$

36. **(c)** Given diagonal is $8x = 15y$. The other diagonal through (1, 2) will have slope $m_2 = -\frac{15}{8}$

Hence the slope of sides $m_3 = \frac{\frac{-15}{8} - 1}{1 + \left(-\frac{15}{8}\right)} = \frac{23}{7}$

$\Rightarrow$ L_3: $23x - 7y = 9$

And $m_4 = \frac{\frac{-15}{8}+1}{1 + \frac{15}{8}} = -\frac{7}{23} \Rightarrow L_4$: $7x + 23y = 53$

37. **(d)** Let C $(2a, y)$, then $y^2 = (2a)^2 + (y - a)^2$

$\Rightarrow$ $y = 5a/2$

$\therefore$ Line BC is $3x - 4y + 4a = 0$

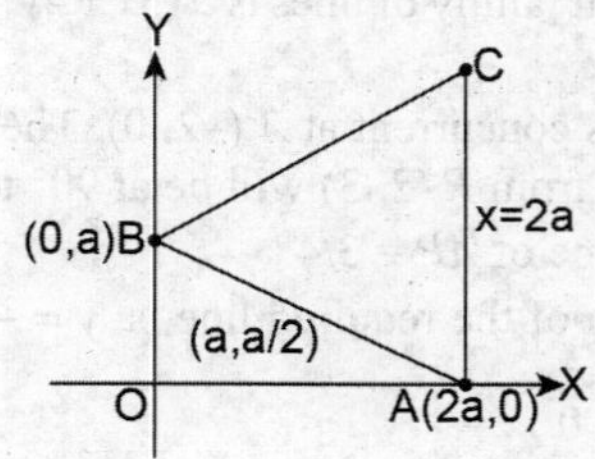

38. **(a)** One of the diagonals is along $x = 2y$, so the slope of other diagonal through (3, 0) will be $m_2 = -2$

$\Rightarrow$ Slope of side through (3, 0) is

$$\tan\theta = \left|\frac{m_2 - 1}{1 + m_2(1)}\right| = \left|\frac{3}{1-2}\right| = 3$$

Hence the sides through (3, 0) are $L_1: y = 3x - 9$ and $L_2: x + 3y = 3$

39. **(b)** Original equation of line is $\frac{x}{a} + \frac{y}{b} - 1 = 0$ after rotation the equation of line is $\frac{x}{p} + \frac{y}{q} - 1 = 0$

Since there is no translation (no change is origin)

$\therefore$ Distance of origin remains the same

Hence $\left|\frac{a^2b^2}{a^2+b^2}\right| = \left|\frac{p^2q^2}{p^2+q^2}\right| \Rightarrow \frac{a^2+b^2}{a^2b^2} = \frac{p^2+q^2}{p^2q^2}$

or $\frac{1}{a^2} + \frac{1}{b^2} = \frac{1}{p^2} + \frac{1}{q^2}$

40. **(a)** From the given we can conclude that after 1sec the point will be at B (4, 4) starting from (1, 2)

Hence $y - 2 = (2/3)(x - 1)$ or $2x - 3y + 4 = 0$

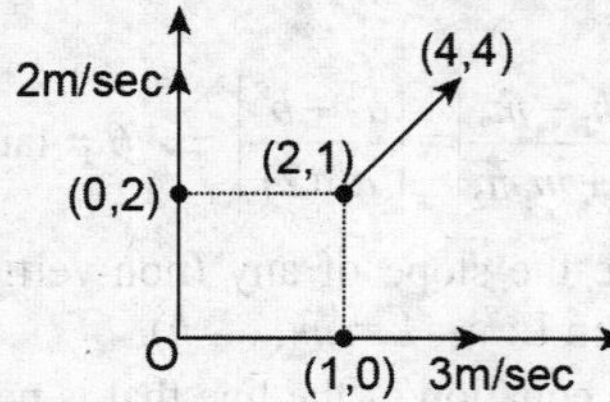

41. **(a)** When the lines are rotated in the given manner the angle bisectors remain the same as these are isoinclined pairs.

For $x^2 - 2pxy - y^2 = 0$ the angle bisectors are $\frac{x^2-y^2}{a-b} = \frac{xy}{h}$

$\Rightarrow$ The required lines are $\frac{x^2-y^2}{2} = \frac{xy}{-p}$

$\Rightarrow px^2 + 2xy - py^2 = 0$

42. **(a)** The lines are $x^2(\tan^2\theta + \cos^2\theta) - 2xy\tan\theta + y^2\sin^2\theta = 0$ or $y^2\sin^2\theta - 2xy\tan\theta + x^2(\tan^2\theta + \cos^2\theta) = 0$

Since α, β are angles with x-axis

$\therefore$ $\tan\alpha + \tan\beta$

$$= \frac{2\tan\theta}{\sin^2\theta} = \frac{2\sin\theta}{\cos\theta\sin^2\theta} = \frac{4}{2\sin\theta\cos\theta} = 4\operatorname{cosec}2\theta$$

43. **(c)** Observe that $\alpha(4x + 2y + 3) + \beta(x - y - 1) = 0$ are concurrent at $\left(-\frac{1}{6}, \frac{-7}{6}\right)$. (For all values of α and β)

44. **(a)** The given family of lines is $a(3x + 4y + 6) + b(x + y + 2) = 0$

Which is concurrent at A (–2, 0). The line at maximum distance from P (2, 3) will be at 90° to AP

Now slope of $AP = 3/4$

$\Rightarrow$ Equation of the required line is $y = -\frac{4}{3}(x+2)$ or $4x + 3y + 8 = 0$

45. **(d)** Points of triangle are A (1, 3), B (5, 0), C (–1, 2) and in equations are $E_1: (x + 3y - 5 \geq 0)$; $E_2: (3x + 4y - 15 \leq 0)$ and

$E_3: (x - 2y + 5 \geq 0)$

$E_1(A)$: True; $E_1(B)$: True; $E_1(C)$: True;

$E_2(A)$: True; $E_2(B)$: True; $E_2(C)$: True;

$E_3(A)$: True; $E_3(B)$: True; $E_3(C)$: True;

46. **(a)** Given $\frac{x^2}{a} + \frac{2xy}{h} + \frac{y^2}{b} = 0$; where $m_1 = m$ and $m_2 = 2m$

$\Rightarrow m_1 + m_2 = \frac{-2a}{h} = 3m$ and $m_1m_2 = \frac{a}{b} = 2m^2$

$\Rightarrow m^2 = \left(\frac{-2a}{3h}\right)^2 = \frac{a}{2b}$. Hence $\frac{ab}{h^2} = \frac{9}{8}$

47. **(a)** $x + 2y - 4 = 0$ (given)

When translated under given condition will be $x + 2y - 4 - 3\sqrt{5} = 0$

When rotated by 30° anticlockwise about $(4 + 3\sqrt{5}, 0)$, we get the equation as $y = \tan(\theta + 30°)(x - 4 - 3\sqrt{5})$.

48. **(c)** The given line is $2x + y - a = 0$, which is at distance of $\frac{a}{\sqrt{5}}$ units from the origin.

As an isosceles right triangle formed

$\Rightarrow$ Area of $\Delta OAB = \frac{1}{2} \times 2 \times \frac{a^2}{5} = \frac{a^2}{5}$ square units

49. **(d)** Observe that the lines will be concurrent since

$$\begin{vmatrix} (a-b) & (b-c) & (c-a) \\ (c-a) & (a-b) & (b-c) \\ (b-c) & (c-a) & (a-b) \end{vmatrix} = \begin{vmatrix} 0 & (b-c) & (c-a) \\ 0 & (a-b) & (b-c) \\ 0 & (c-a) & (a-b) \end{vmatrix} = 0$$

50. **(b)** Given $L: \frac{x}{a}\cos\phi + \frac{y}{b}\sin\phi - 1 = 0$

Product of perpendiculars from $\left(\pm\sqrt{a^2-b^2}, 0\right)$ on the line is $d_1d_2 = \frac{\left|\frac{\sqrt{a^2-b^2}\cos\phi}{a} - 1\right|\left|\frac{-\sqrt{a^2-b^2}\cos\phi}{a} - 1\right|}{\frac{\cos^2\phi}{a^2} + \frac{\sin^2\phi}{b^2}}$

$$= \frac{(a^2\sin^2\phi + b^2\cos^2\phi)a^2b^2}{a^2\{a^2\sin^2\phi + b^2\cos^2\phi\}} = b^2$$

51. **(c)** Given $a^2 + b^2 - 2ab = c^2$ or $(a - b) = \pm c$, so either $a - b - c = 0$ or $a - b + c = 0$

Now $ax + by + c = 0$ becomes $a - b - c = 0$, when $x = -1, y = 1$

$\Rightarrow$ Concurrency at (–1, 1)

Similarly for $x = 1, y = -1$, we get $a - b + c = 0$

$\therefore$ Intersection at (1, –1)

52. **(d)** Given a, b, c are in HP $\Rightarrow ab + bc = 2ac$

Now $bcx + acy + ab = 0$ becomes $bc - 2ac + ab = 0$, when $x = 1$ and $y = -2$

$\Rightarrow$ Concurrent at (1, –2)

53. **(a)** From the given B is 90°. Since circum-centre S (a, b) which is the mid point of AC

As A (0, 0) $\Rightarrow C = (2a, ab)$

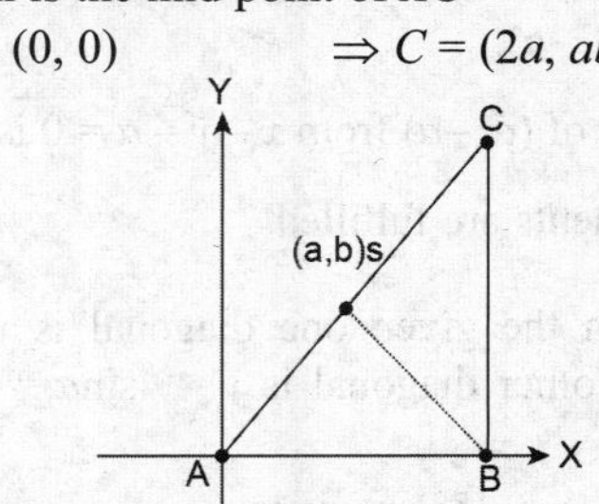

54. **(b)** Observe that the given lines L_1: $3x + 4y - 9 = 0$ and L_2: $4x - 3y + 12 = 0$ are at 90° to each other from the given A (3, 0) and B (0, 4) and the circum radius $r = 5/2$ units.

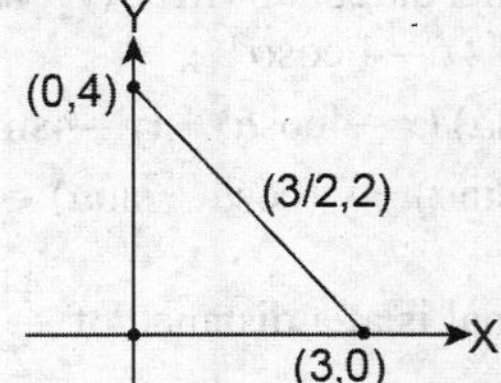

55. **(a)** The point of intersection of L_1: $x + 2y = 1$ and L_2: $2x - y = 1$ is $\left(\frac{3}{5}, \frac{1}{5}\right)$

Any non vertical with slope m will be $y - \frac{1}{5} = m\left(x - \frac{3}{5}\right)$

$\Rightarrow A\left(\frac{3m-1}{5m}, 0\right)$ and $B\left(0, \frac{1-3m}{5}\right)$

And mid point of AB is $P\left(\frac{3m-1}{10m}, \frac{1-3m}{10}\right)$

$\Rightarrow \left(x - \frac{3}{10}\right) = -\frac{1}{10m} \Rightarrow \dfrac{\quad}{3\quad 10}$

Similarly $10y - 1 = -3m \Rightarrow m = \frac{1-10y}{3}$

Hence $\frac{1}{3-10x} = \frac{1-10y}{3} \Rightarrow x + 3y - 10xy = 0$

56. **(d)** The triangle has two sides given by $6y^2 - 7xy + x^2 = 0$ or $(6y - x)(y - x) = 0$

One vertex is O (0, 0) and centroid is (1, 0)

$\Rightarrow$ Mid point of AB is (3/2, 0)

Let A (a, a) be the point on $y = x$, then B $(3 - a, -a)$

$\Rightarrow 6(-a) - (3 - a) = 0 \Rightarrow a = -\frac{3}{5}$

Hence $y - 0 = \frac{2}{7}\left(x - \frac{3}{2}\right) \Rightarrow 2x - 7y - 3 = 0$

57. **(c)** Given O (0, 0), P (3, 4) and Q (6, 0), we know that centroid divides the area equally

$\Rightarrow R\left(3, \frac{4}{3}\right)$

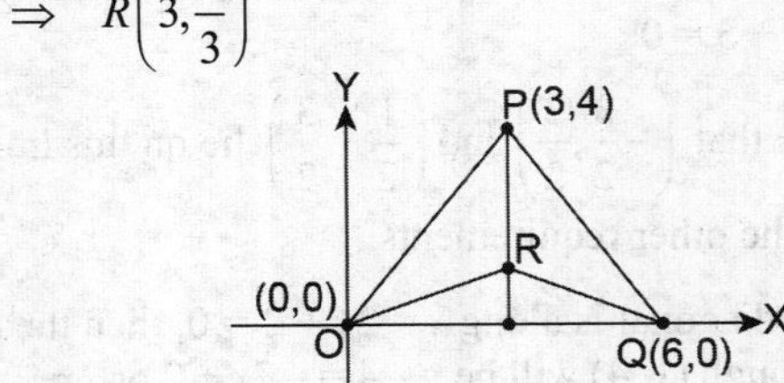

58. **(b)** According to the given, the number of integral points = 19 + 18 + 17 + ... + 1 = 190

59. **(c)** Let O (0, 0), A (3, 4), B (4, 0) be the vertices then altitude through A is $x = 3$.

Now slope of OA is $m = \frac{4}{3}$, so altitude through B is $y - 0 = -\frac{3}{4}(x - 4)$

$\Rightarrow$ $3x + 4y = 12$ for $x = 3$, $y = 3/4$

$\Rightarrow$ Orthocenter is (3, 3/4)

60. **(b)** $y = |x| - 1$ gives $y = x - 1$ for $x \geq 0$ and $y = -x - 1$ for $x \leq 0$.

Similarly $y = -|x| + 1$ gives $y = -x + 1$ for $x \leq 0$ and $y = 1 + x$ for $x \leq 0$. Area locked is $A = 2$ square units

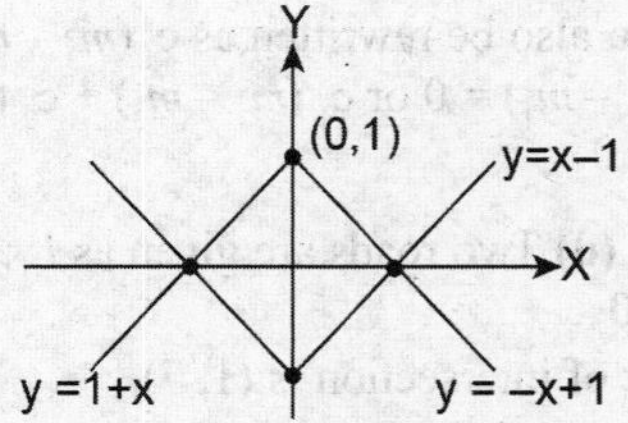

61. **(d)** Given $0 < \alpha < \pi/2$; $P = (\cos\theta, \sin\theta)$ and Q $(\cos(\alpha - \theta), \sin(\alpha - \theta))$

As shown the angle of Q is $\frac{\alpha}{2} - \theta + \frac{\alpha}{2} - \theta + \theta = \alpha - \theta$

$\Rightarrow Q = (\cos(\alpha - \theta), \sin(\alpha - \theta))$

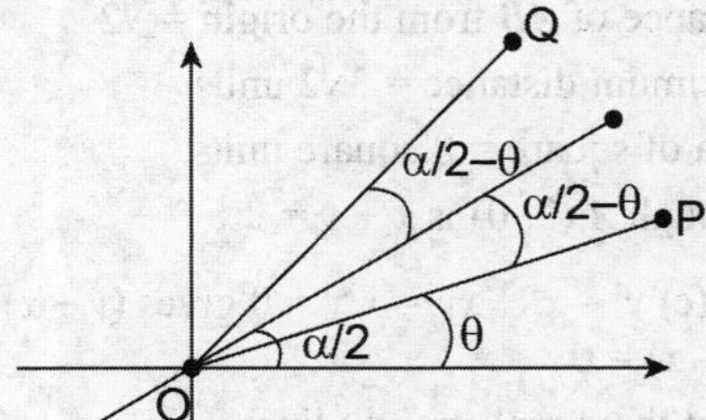

62. **(d)** Observe that $y = mx$ and $y = mx + 1$ are at a distance $d = \frac{1}{\sqrt{m^2+1}}$

Now $y = nx + 1$ and $y = mx$ intersect at

$P\left(\frac{1}{m-n}, \frac{m}{m-n}\right) \Rightarrow OP = \sqrt{\frac{m^2+1}{(m-n)^2}}$

So area of || gm $OPQR = OP \times d = \frac{1}{|m-n|}$.

63. **(d)** As we can observe the Δ formed by (0, 0), $(1, \sqrt{3})$ and (2, 0) is equilateral

So in centre merges with centroid G at $\left(1, \frac{1}{\sqrt{3}}\right)$

64. **(a)** Given $x_3 = rx_2$, $x_2 = rx_1$, also $y_3 = r\,y_2$, $y_2 = ry_1$

$\Rightarrow \frac{y_3 - y_2}{x_3 - x_2} = \frac{y_2 - y_1}{x_2 - x_1}$ i.e., slopes are equal and (x_2, y_2) is a common point

$\Rightarrow$ Points lie on a straight line

SECTION–IV (ONE OR MORE THAN IS/ARE CORRECT)

1. **(a), (c)** L_1: $m_1 x - y + c_1 = 0$, L_2: $m_2 x - y + c_2 = 0$ and L_3: $m_3 x - y + c_3 = 0$ are concurrent

$$\Rightarrow \Delta = 0 \text{ i.e., } \begin{vmatrix} m_1 & -1 & c_1 \\ m_2 & -1 & c_2 \\ m_3 & -1 & c_3 \end{vmatrix} = 0$$

$$\text{or } \begin{vmatrix} m_1 & -1 & c_1 \\ (m_2 - m_1) & 0 & (c_2 - c_1) \\ (m_3 - m_2) & 0 & (c_3 - c_2) \end{vmatrix} = 0$$

$\Rightarrow$ $(m_2 - m_1)(c_3 - c_2) = (m_3 - m_2)(c_2 - c_1)$ or $m_1 (c_2 - c_3) + m_2 (c_3 - c_1) + m_3 (c_1 - c_2) = 0$

This can also be rewritten as $c_1 (m_3 - m_2) + c_2 (m_1 - m_3) + c_3 (m_2 - m_1) = 0$ or $c_1 (m_2 - m_3) + c_2 (m_3 - m_1) + c_3 (m_1 - m_2) = 0$

2. **(a), (b), (c), (d)** Two roads are given as $L_1 : y - x - 6 = 0$ and $x + y - 8 = 0$

So the point of intersection is (1, 7).

The bungalow will be on the bisectors given by $y = 7$ and $x = 1$.

Hence possible locations are $(1 \pm 100\sqrt{2}, 7)$ or $(1, 7 \pm 100\sqrt{2})$

3. **(a), (b), (c)** From the given A (2, 0), B (0, –2)

$\Rightarrow$ $AB = 2\sqrt{2}$

Distance of AB from the origin = $\sqrt{2}$

Maximum distance = $3\sqrt{2}$ units

Area of square = 8 square units.

Side through A (2, 0) is $x + y = 2$

4. **(a), (b), (c)** $y^3 - x^3 + xy^2 - x^2y = 0$ gives $(y + x)(y^2 - x^2)$ or $(y + x)^2 (y - x) = 0$

So we get three real straight lines

$y + x = 0$, $y + x = 0$ are coincident and $y + x = 0$ and $y - x = 0$ are at 90°

5. **(a), (d)** $my^2 + (1 - m^2) xy - mx^2 = 0$ or $y^2 - mxy - x^2 = 0$

Represents two lines at 90° to each other through origin.

Since $y = x$ or $y = - x$ is one of the possibility, so automatically both are the angle bisectors of $xy = 0$

Hence $(y + x)(y - x) = 0$ and $my^2 + (1 - m^2)xy - mx^2 = 0$ are identical

$\Rightarrow$ $1 - m^2 = 0$

$\Rightarrow$ $m = \pm 1$ which is true and satisfies all the equations

6. **(c), (d)** Given $L : \sqrt{3}x - y + c = 0$ ....(i)

and $x^2 + y^2 - 2\ 0$ ...(ii)

Homogenizing (ii) with (i), we get $x^2 + y^2 - 2\left(\dfrac{y - \sqrt{3}\,x}{c}\right)^2 = 0$

$\Rightarrow$ $c^2x^2 + c^2y^2 - 2\,\{y^2 + 3x^2 - 2\sqrt{3}\,xy\} = 0$ or $(c^2 - 6)x^2 + (c^2 - 2)y^2 + 4\sqrt{3}\,xy = 0$

These lines will be at right angle when $c^2 - 6 + c^2 - 2 = 0$, i.e., $c = \pm 2$

7. **(b), (c)** One diagonal is $x - y - \alpha = 0$

One vertex is $(\alpha, -\alpha)$. Now $(0, -\alpha)$ and $(\alpha, 0)$ are at $|\alpha|$ distance from $(\alpha, -\alpha)$.

Also distance of $(\alpha, -\alpha)$ from $x - y - \alpha = 0$ is $\dfrac{|\alpha|}{\sqrt{2}}$, so necessary requirements are fulfilled.

8. **(a), (b)** From the given one diagonal is $y = \tan(45° + \alpha)\,x$ and the other diagonal is $y - 4\sin\alpha = \tan(135° + \alpha)(x - 4\cos\alpha)$.

Now $y = \tan(\alpha + 45°)\,x$, gives

$$y\left\{\frac{\cos\alpha}{\sqrt{2}} - \frac{\sin\alpha}{\sqrt{2}}\right\} = x\left\{\frac{\sin\alpha}{\sqrt{2}} + \frac{\cos\alpha}{\sqrt{2}}\right\}$$

or $x(\sin\alpha + \cos\alpha) - y\,\{\cos\alpha - \sin\alpha)\} = 0$

Similarly the other diagonal will be $(y - 4\sin\alpha)\,\{-\sin(45° + \alpha)\} = \cos(45° + \alpha)\,\{x - 4\cos\alpha\}$

$\Rightarrow$ $(\cos\alpha - \sin\alpha)(x - 4\cos\alpha) + (y - 4\sin\alpha)(\cos\alpha + \sin\alpha) = 0$

$\Rightarrow$ $x(\cos\alpha - \sin\alpha) + y(\cos\alpha + \sin\alpha) = 4\sin^2\alpha + 4\cos^2\alpha = 4$

9. **(b), (c), (d)** Canal is at a distance of $\dfrac{41}{2}$ km from the place (say at origin)

P and Q are at a distance of 5 units from the place

$\Rightarrow$ P and Q are on the same side of canal on which the given place lies.

Villages are not the bank of the canal

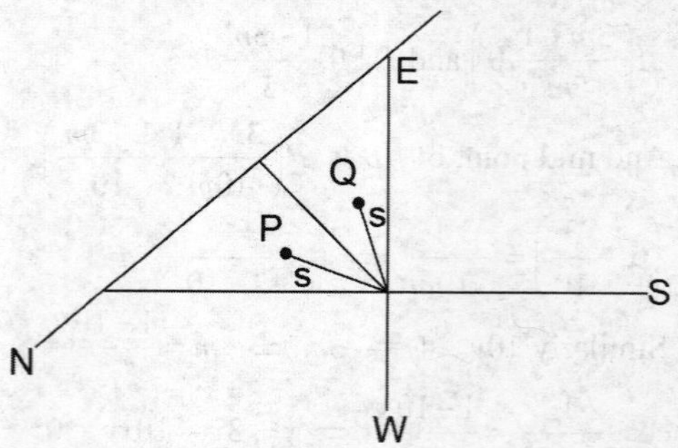

10. **(a), (b)** Line through A (4, –1) that is ||el to $3x - 4y + 1 = 0$ is $3x - 4y = 16$

Now $\cos\alpha = \dfrac{4}{5}$, $\sin\alpha = \dfrac{3}{5}$

Points on this line at a distance of 5 units will be (8, 2) and (0, –4)

11. **(a), (b), (c)** Coordinates of exterminates of one diagonal is A (1, 1) and C (–2, –1)

$\Rightarrow$ Mid points is (–1/2, 0) and its slope is $m = 2/3$

$\Rightarrow$ Equation of the other diagonal is $y = \left(\dfrac{-3}{2}\right)\left(x + \dfrac{1}{2}\right)$ or $6x + 4y + 3 = 0$

Observe that $\left(-\dfrac{3}{2}, \dfrac{3}{2}\right)$ and $\left(\dfrac{1}{2}, -\dfrac{3}{2}\right)$ lie on this line and satisfy the other requirements

12. **(a), (b)** One diagonal is along $x - 2y + 2 = 0$, then the other diagonal through (1, 4) will be $y - 4 = -2x + 2$ or $2x + y = 6$.

Further (1, 4) satisfies $x + 3y - 13 = 0$ and $3x - y + 1 = 0$ angle of $x + 3y - 13 = 0$ with $x - 2y + 2 = 0$ is

$$\tan\theta = \frac{\left|-2+\frac{1}{3}\right|}{1+\left(\frac{2}{3}\right)} = \frac{5}{3\left(\frac{5}{3}\right)} = 1 \Rightarrow \theta = 45°$$

⇒ lines are the sides

13. **(a), (c)** Given $L: \frac{x+y}{2} = 1$ ……(i)

and C: $x^2 + y^2 - 3 = 0$ …..(ii)

Homogenizing (ii) with (i), we get $x^2 + y^2 - 3\left(\frac{x+y}{2}\right)^2 = 0$

i.e., $4x^2 + 4y^2 - 3x^2 - 3y^2 - 6xy = 0$

⇒ $y^2 - 6xy + x^2 = 0$ ⇒ $y = \frac{6 \pm 4\sqrt{2}}{2}x$

Hence $(3 + 2\sqrt{2})x - y = 0$ and $y - (3 - 2\sqrt{2})x = 0$ are the lines

14. **(a), (b), (c)** Lines $L_1: px + qy + r = 0$; $L_2: qx + ry + p = 0$ and $L_3: rx + py + q = 0$ are concurrent

⇒ $\Delta = 0$

$$\Rightarrow \begin{vmatrix} p & q & r \\ q & r & p \\ r & p & q \end{vmatrix} = 3pqr - (p^3 + q^3 + r^3) = 0$$

⇒ $(p + q + r)\{p^2 + q^2 + r^2 - (pq + qr + pr)\} = 0$

Either $p + q + r = 0$ or $p^2 + q^2 + r^2 = pq + qr + pr$

Also $p^3 + q^3 + r^3 = 3pqr$

15. **(a), (c)** $(\ \ x + my)^2 - 3(mx - \ \ y)^2 = 0$

⇒ $(\ \ ^2 - 3m^2)x^2 + (m^2 - 3\ \ ^2)y^2 + 8m\ \ xy = 0$

⇒ $\{(\ \ + \sqrt{3}\,m)x + (m - \sqrt{3}\ \ )y\}\{(\ \ - \sqrt{3}\,m)x + (m + \sqrt{3}\ \ )y\} = 0$

Let slope of $(\ \ + \sqrt{3}\,m)x + (m - \sqrt{3}\ \ )y = 0$ be M_1 and slope of $(\ \ - \sqrt{3}\,m)x + (m + \sqrt{3}\ \ )y = 0$ be M_2

Slope of $\ \ x + my + n = 0$ is $M_3 = -\ \ /m$

$$\text{Now } \tan\theta_1 = \left|\frac{M_3 - M_1}{1 + M_1M_3}\right| = \left|\frac{-\frac{\ell}{m} + \frac{\ell + \sqrt{3}m}{m - \sqrt{3}\ell}}{1 + \frac{\ell}{m}\cdot\frac{\ell + \sqrt{3}m}{m - \sqrt{3}\ell}}\right|$$

$$= \left|\frac{\sqrt{3}\ell^2 + \sqrt{3}m^2}{\ell^2 + m^2}\right| = \sqrt{3} \Rightarrow \theta_1 = 60°$$

Similarly, $\tan\theta_2 = \left|\frac{M_2 - M_1}{1 + M_2M_1}\right| = \sqrt{3} \Rightarrow \theta_2 = 60°$. Hence an equilateral triangle is formed which is also an isosceles one.

SECTION–V (ASSERTION AND REASON)

1. **(a)** **R:** The statement is true if $m_1m_2 = -1$, then lines are perpendicular

A: The lines $2x + y + 6 = 0$ and $x - 2y + 5 = 0$ have slopes $m_1 = -2$ and $m_2 = 1/2$, respectively.

⇒ The product $m_1m_2 = -1$ also line $x - 2y + 5 = 0$ passes through (1, 3)

⇒ Which is true and fully supported by *R*.

2. **(a)** **R:** statement is true by definition

A: We get two sets of || lines. || gm's formed since diagonals are given to be at right angles, so automatically they are also bisected

⇒ $p^2 + q^2 = (p')^2 + (q')^2$ and it is fully supported by R.

3. **(b)** $4x + 3y = \lambda$ and $3x - 4y = \mu$

⇒ $25x = 4\lambda + 3\mu$ i.e., $x = \frac{4\lambda + 3\mu}{25}$ and $y = \frac{3\lambda - 4\mu}{25}$

Now $x_1 = \frac{4\lambda + 3\mu}{25}$ and $y_1 = \frac{3\lambda - 4\mu}{25}$

For $x_1 = y_1$, we get $\lambda = -7\mu$

⇒ $\lambda + 7\mu = 0$ or λ, μ satisfy $x + 7y = 0$

⇒ Assertion is true

R: statement that (x_1, y_1) will be in first quadrant (if $4\lambda + 3\mu > 0$ and $3\lambda - 4\mu > 0$) is true but in no way we can drive $x + 7y = 0$

(or $\lambda + 7\mu = 0$) from this statement.

4. **(a)** **R:** Statement is true. *OA.OB* = *OC.OD* means the points are concylic (Power of a point w.r.t. a circle)

Here the point is origin.

A: Line $2x + 3y + 19 = 0$ intersects the axes at $A\left(-\frac{19}{2}, 0\right)$ and $C\left(0, -\frac{19}{3}\right)$

And line $9x + 6y = 17$ intersects the axes at $B\left(\frac{17}{9}, 0\right)$ and (0,17/6)

Now $OA.OB = \frac{19}{2} \times \frac{17}{9}$ and $OC.OD = \frac{19}{3} \times \frac{17}{6}$

Which are equal

⇒ Statement is true and derivable from *R*.

5. **(c)** **R:** Reason is false, as shown consider *AB* and CD as two chords associated with coordinate axes for a circle $x^2 + y^2 = r^2$

$OA = OB = OC = OD = r$. The chords do not pass through any fixed point.

A: $4x^2 + y^2 - x + 4y = 0$

Let the chord be along a line $\frac{\ell x + my}{n} = 1$

Homogenizing, we get $4x^2 + y^2 + (4y - x)\frac{(\ell x + my)}{n} = 0$

⇒ $4nx^2 + ny^2 + 4my^2 - \ \ x^2 + (4\ \ - m)xy = 0$

For right angle purpose $a + b = 0$

⇒ $(4n - \ \ ) + (n + 4m) = 0$ i.e., $\ \ - 4m + 5n = 0$ or

$\frac{\ell}{5} - \frac{4}{5}m + n = 0$, which can be written as $\ \ x + my + n = 0$

where $x = \frac{1}{5}$ and $y = -\frac{4}{5}$

Hence the chords lie on the lines passing through $\left(\frac{1}{5}, -\frac{4}{5}\right)$

⇒ *A* is true

6. (a) **R**: Statement is true

A: The given points are $A(-5, -2)$, $B(7, 6)$ and $C(5, -4)$.

Now $AB^2 = 208$, $AC^2 = 104$, $BC^2 = 104$

$\Rightarrow$ The ΔABC is right angled isosceles.

Hence circum-centre is the mid point of AB i.e., (1, 2)

Statement (A) is true and supported by **R**

7. (b) **R:** statement is true

$ax + y(2h + a) = 0$ is a factor of $ax^2 + 2hxy + by^2 = 0$

$$\Rightarrow \quad m_1 = -\frac{a}{2h+a},\ m_1 + m_2 = -\frac{2h}{b},\ m_1m_2 = \frac{a}{b}$$

$$\text{Hence } m_2 = \frac{a}{2h+a} - \frac{2h}{b} = \frac{ab - 4h^2 - 2ah}{b(2h+a)}$$

$$\text{Also } m_2 = \frac{a\{2h+a\}}{b(-a)} = \frac{-(2h+a)}{b}$$

$$\Rightarrow \quad \frac{ab - 2h(2h+a)}{b(2h+a)} = \frac{-(2h+a)}{b}$$

$\Rightarrow$ $ab = (2h + a)\{2h - 2h - a\} = -a(2h + a)$

$\Rightarrow$ $b + 2h + a = 0$

A: The statement is true

If $a + b = f - 2h$ (where $f = 0$), then the lines $by^2 + 2hxy + ax^2 = 0$

$$\Rightarrow \quad m_1 + m_2 = \frac{-2h}{b} \text{ and } m_1m_2 = a/b$$

Since $a + b = -2h$

$$\Rightarrow \quad \frac{-2h}{b} = 1 + \frac{a}{b} \qquad \Rightarrow \quad m_1 + m_2 = 1 + m_1m_2$$

Hence one of the slopes is 1 which means angle bisector of 1st quadrant .

$\Rightarrow$ A is true but is not where derivable from **R**

8. (a) **R:** The statement is true

A: The given three lines are $dy^3 + cxy^2 + ax^3 = 0$

Let $y = m_1x$, $y = m_2x$ and $y = m_3x$ be the three lines

$\Rightarrow$ $m_1 + m_2 + m_3 = -c/d$; $m_1m_2 + m_1m_3 + m_2m_3 = b/d$ and $m_1m_2m_3 = -a/d$

If two lines are at right angles (say m_1 and m_2), then

$$m_3 = \frac{a}{d} \text{ and } -1 + m_3(m_1 + m_2) = b/d$$

$$\Rightarrow \quad \frac{a}{d}(m_1 + m_2) = \frac{b+d}{d}$$

$$\Rightarrow m_1 + m_2 = \frac{b+d}{a}$$

$$\therefore \quad m_1 + m_2 + m_3 = \frac{a}{d} + \frac{b+d}{a} = -\frac{c}{d}$$

$\Rightarrow$ $a^2 + b^2 + d^2 = -ac$ or $a^2 + ac + b^2 + d^2 = 0$

Hence the statement A is true and it is support by **R.**

9. (d) **R:** The pair of lines is given by $ax^2 + 2hxy + by^2 + 2gx + 2fy + c = 0$ as $a + b = 0$

$\Rightarrow$ Lines are perpendicular to each other

$\Rightarrow$ R is true

A: The equation $2x^2 + 3xy - 2y^2 + 5x - 5y + 3 = 0$ has $a + b = 0$

$$\text{Now } \Delta = \begin{vmatrix} 2 & \frac{3}{2} & \frac{5}{2} \\ \frac{3}{2} & -2 & -\frac{5}{2} \\ \frac{5}{2} & -\frac{5}{2} & 3 \end{vmatrix} = \begin{vmatrix} \frac{7}{2} & -\frac{1}{2} & 0 \\ \frac{3}{2} & -2 & -\frac{5}{2} \\ \frac{5}{2} & -\frac{5}{2} & 3 \end{vmatrix} = \frac{-343}{8} + \frac{43}{8}$$

$$\Rightarrow \quad \Delta = \frac{-300}{8} = \frac{-75}{2}$$

$\Rightarrow$ The equation does not represent two straight lines A is false

10. (d) **R:** The statement is true

A: The joint equation of $y = x$ and $y = -x$ is $(y - x)(y + x) = 0$ i.e., $y^2 - x^2 = 0$ or $x^2 - y^2 = 0$ is $(y - x)(y + x) = 0$ i.e., $y^2 - x^2 = 0$ or $x^2 - y^2 = 0 \Rightarrow$ A is false

SECTION–VI (COMPREHENSION-TYPE)

COMPREHENSION A

1. (d) Given the side of a triangle along AB: $2x - y + 4 = 0$; BC: $x - 2y - 1 = 0$; AC: $x + 3y - 3 = 0$

Angle A is formed by AB and AC having slopes $m_1 = 2$, $m_2 = -1/3$ receptively {since $0 > m_1m_2 = -2/3 > -1$}

$$\Rightarrow \quad \tan A = \frac{-\left(2 + \frac{1}{3}\right)}{1 + 2\left(-\frac{1}{3}\right)} = \frac{-7}{3\left(\frac{1}{3}\right)} = -7$$

2. (b) Angle B is formed by AB and BC (B is an acute angle)

AB: $2x - y + 4 = 0$ and BC: $-x + 2y + 1 = 0$, as $c_1, c_2 > 0$

$\Rightarrow$ $a_1a_2 + b_1b_2 = -2 - 2 = -4 < 0 \Rightarrow$ Origin lines in acute angle

Hence, the external angle bisector will be

$$\frac{2x - y + 4}{\sqrt{5}} = \frac{-(-x + 2y + 1)}{\sqrt{5}},\ \text{i.e., } x + y + 5 = 0.$$

3. (a) Point B is $(-3, -2)$ and its image Q in AC will be give by $\frac{-3 - x_1}{1} = \frac{-2 - y_1}{3} = \frac{2(-12)}{10}$ i.e., $\left(-\frac{3}{5}, \frac{26}{5}\right)$

COMPREHENSION B

Two vertices of a triangle ΔABC are $A(1, 3)$, $C\left(-\frac{2}{5}, -\frac{2}{5}\right)$ and the equation of angle bisector of $\angle ABC$ is $x + y = 2$ and its slope $m_4 = -1$. Let $B(x_1, 2 - x_1)$, then slope of AB is

$$m_1 = \frac{1 + x_1}{1 - x_1} \text{ and slope of } BC \text{ is } m_2 = \frac{x_1 - \frac{12}{5}}{-\frac{2}{5} - x_1}$$

$$\text{Now } \frac{m_4 - m_1}{1 + m_1m_4} = \frac{m_2 - m_4}{1 + m_2m_4}$$

$$\text{i.e., } \frac{-1 - \frac{1 + x_1}{1 - x_1}}{1 + (-1)\frac{x_1 + 1}{1 - x_1}} = \frac{\frac{x_1 - \frac{12}{5}}{-x_1 - \frac{2}{5}} + 1}{1 - \frac{x_1 - \frac{12}{5}}{-\frac{2}{5} - x_1}}$$

$\Rightarrow \dfrac{2}{2x_1} = \dfrac{-\frac{14}{5}}{2-2x_1}$ i.e., $5x_1 - 5 = 7x_1$

$\Rightarrow x_1 = -5/2$ and $y_1 = 9/2$. Hence, $B\left(-\dfrac{5}{2}, \dfrac{9}{2}\right)$

$\Rightarrow$ 5 answer Ans (c) option

$\Rightarrow$ Equation side BC is $\left(y-\dfrac{9}{2}\right) = -\dfrac{49}{21}\left(x+\dfrac{5}{2}\right)$

$\Rightarrow 7x + 3y + 4 = 0 \Rightarrow 4$ Ans (b) option

$\Rightarrow$ Equation of side AB is $y - 3 = \dfrac{-3}{7}(x-1)$

$\Rightarrow 3x + 7y = 24 \Rightarrow 6$ Ans (a) option

COMPREHENSION C

Important Note: The given transformation is not according to the standard system as discussed in chapter #1 (Point and Cartesian)

So proceed according to the given information in the comprehension

7. (c) According to the given $x = X\cos\theta + Y\sin\theta$ and $y = X\sin\theta - Y\cos\theta$

Now, $f(x, y) = (x + y)^2$

$\Rightarrow f(x, y) = \{X\cos\theta + Y\sin\theta + X\sin\theta - Y\cos\theta\}^2$

$= X^2\{1 + \sin2\theta\} + Y^2\{1 - \sin2\theta\} - 2XY\cos2\theta$

Comparing with $ax^2 + by^2$, we get $\cos2\theta = 0$ i.e., $\theta = \pi/4$

So $\sin2\theta = 1 \Rightarrow a = 2$

8. (a) When $\sin2\theta = 1$, then $b = 0$

$\therefore a + 2b = 2$

9. (d) when $\theta = \pi/4$, then $\sec2\theta$ is not defined

COMPREHENSION D

10. (a) The equation of line through (2, 3) is $\dfrac{x-2}{\cos\theta} = \dfrac{y-3}{\sin\theta} =$

Since it intersects $x + y - 7 = 0$ at a distance of $\sqrt{2}$, so 2

$+ r\cos\theta + 3 + r\sin\theta - 7 = 0$ or $2\sin\left(\theta + \dfrac{\pi}{4}\right) = 2$

$\Rightarrow \theta = \pi/4$

11. (a) The point at a distance of 4 units from P (2, 3) will be

$\left(2 - \dfrac{4\sqrt{3}}{2}, 3 - 4\left(\dfrac{1}{2}\right)\right) \Rightarrow (2 - 2\sqrt{3}, 1)$

12. (c) Given equation of AB is $3x + y = 4$ and equation of AC is $x + 3y = 4$. Mid point of BC is (2, 2)

Let $B(x_1, 4 - 3x_1)$, then $C = (4 - x_1, 3x_1)$

Since C satisfies equation of AC

$\Rightarrow 4 - x_1 + 9x_1 = 4 \Rightarrow x_1 = 0$

$\Rightarrow y_1 = 4$ i.e., B (0, 4)

Hence C (4, 0)

$\therefore$ Equation of BC is $x + y = 4$

COMPREHENSION E

13. (c) Line $y = mx$ intersects $L_1: x + y - 3 = 0$ at $A\left(\dfrac{3}{1+m}, \dfrac{3m}{m+1}\right)$ and

$L_2: x + y - 5 = 0$ at $B\left(\dfrac{5}{1+m}, \dfrac{5m}{1+m}\right) \Rightarrow AB = \dfrac{\sqrt{4+4m^2}}{|1+m|} = 2$

$\Rightarrow 4(1 + m^2) = 4(m^2 + 1 + 2m)$

$\Rightarrow m = 0, \infty$

Equation is $0m^2 + 8m + 0 = 0$ or $(4 - d^2)m^2 + 2md^2 + (4 - d^2) = 0$

14. (d) As worked out in Q #13

When $AB = d = 2$ units, then x-axis and y-axis both form intercepts of 2 units but these lines are at 90° to each other.

COMPREHENSION F

15. (b) Lines $L_1: m_1x - y + c_1 = 0$; $L_2: m_2x - y + c_2 = 0$ and $L_3: m_3x - y + c_3 = 0$ are concurrent

$\Rightarrow \begin{vmatrix} m_1 & -1 & c_1 \\ m_2 & -1 & c_2 \\ m_3 & -1 & c_3 \end{vmatrix} = 0$ i.e., $\begin{vmatrix} (m_1 - m_2) & 0 & (c_1 - c_2) \\ (m_2 - m_3) & 0 & (c_2 - c_3) \\ m_3 & -1 & c_3 \end{vmatrix} = 0$

$\Rightarrow (m_1 - m_2)(c_2 - c_3) = (m_2 - m_3)(c_1 - c_2)$ or

$\dfrac{m_2 - m_1}{m_3 - m_2} = \dfrac{c_2 - c_1}{c_3 - c_2}$

16. (c) Given $a \quad + bm + cn = 0$ or $\dfrac{a}{c}\ell + \dfrac{b}{c}m + n = 0$

So $\quad n + my + n = 0$ will match it for $x = \dfrac{a}{c}, y = \dfrac{b}{c}$

$\Rightarrow$ The point is $\left(\dfrac{a}{c}, \dfrac{b}{c}\right)$

17. (a) Lines $ax + by + c = 0$; L_2: $bx + cy + a = 0$ and L_3: $cx + ay + b = 0$ are concurrent

$\Rightarrow \begin{vmatrix} a & b & c \\ b & c & a \\ c & a & b \end{vmatrix} = a^3 + b^3 + c^3 - 3abc = 0$ or $(a + b + c)$

$\{(a + b + c)^2 - 3(ab + bc + ca)\} = 0$

Since a, b, c are the sides of a ΔABC

$\therefore a + b + c \neq 0$ but $a^2 + b^2 + c^2 = ab + bc + ca$

Using $\dfrac{a}{\sin A} = \dfrac{b}{\sin B} = \dfrac{C}{\sin C} = 2R$, we get $8R^3\{\sin^3 A +$

$\sin^3 B + \sin^3 C\} = 8R^3\{3\sin A \sin B \sin C\}$

i.e., $\sin^3 A + \sin^3 B + \sin^3 C = 3\sin A \sin B \sin C$

COMPREHENSION G

18. (b) Given $2y^2 - 5xy + 2x^2 = 0$ are the two lines

$\Rightarrow (2y - x)(y - 2x) = 0$

Let $L_1: x - 2y = 0$ and $L_2: y - 2x = 0$

Now diagonal $5x + 2y = 1$ intersects L_1 at $A\left(\dfrac{1}{6}, \dfrac{1}{12}\right)$

and L_2 at $B\left(\dfrac{1}{9}, \dfrac{2}{9}\right)$

Since O (0, 0) is also vertex

$\therefore$ Vertex $C\left(\dfrac{5}{18}, \dfrac{11}{36}\right)$

So, the other diagonal OC is $y = \dfrac{11x}{10}$ i.e., $11x - 10y = 0$

19. (c) The centroid of $||gm$ is mid point of diagonal, so $G=\left(\frac{5}{36},\frac{11}{72}\right)$

20. (a) The area of $||gm$ = (length of line segment AB) × (distance of origin from AB)

$$=\sqrt{\left(\frac{1}{18}\right)^2\left(\frac{5}{36}\right)^2}\times\frac{1}{\sqrt{29}}=\frac{\sqrt{29}}{36\times\sqrt{29}}=\frac{1}{36}\text{ square units}$$

21. (a) Length $OA=\frac{\sqrt{5}}{12}$ and length $OB=\frac{\sqrt{5}}{9}$. Now length OB is larger

$\therefore$ Required ratio is $\frac{1}{9}:\frac{1}{12}$ i.e., 4 : 3

COMPREHENSION H

22. (a) Given mid of AC is E (3, 4), mid point of AB is F (1, 2) and centroid G (3, 2).

Let A (x_1, y_1), B (x_2, y_2) and C (x_3, y_3)

Mid point of BC is D (5, 0), using $A(x_1, y_1) = 2\{F+E\} - 3G$, We get $x_1 = 2(3+1) - (3) = -1$ and $y_1 = 2(4+2) - 3(2) = 6$

$\Rightarrow A(-1, 6)$

Similarly $x_2 = 3$, $y_2 = -2$

$\Rightarrow B(3, -2)$ and $x_3 = 7$, $y_3 = 2 \Rightarrow C(7, 2)$

Equation of side AB is $y - 6 = (-2)(x + 1)$ or $2x + y = 4$

23. (b) As shown above the third mid point D is (5, 0)

24. (c) Equation of BC is $y + 2 = (x - 3)$

So altitude of ΔABC through A has length $=\frac{12}{\sqrt{2}}=6\sqrt{2}$ units

SECTION–VII (MATRIX-MATCH TYPE)

1. (i) → b; (ii) → (d); (iii) → a; (iv) → c

(i) Reflection of $x + y + 1 = 0$ in the line $2x + y + 1 = 0$ given by $L_R = L_I + (-2)\frac{(aa_1+bb_1)}{a^2+b^2}L_m = 0$

$\Rightarrow L_R:(x+y+1)+\frac{(-2)\{3\}(2x+y+1)}{5}=0$ i.e., $7x+y+1=0$

(ii) Reflection $x+y+1=0$ in $x-2y+1=0$ is L_R: $(x+y+1)$ $+\frac{(-2)(-1)(x-2y+1)}{5}=0$ i.e., $7x+y+7=0$

(iii) Reflection of $x + y + 1 = 0$ in $x + 2y - 1 = 0$ is L_R: $(x + y + 1) +\frac{(-2)(3)(x+2y-1)}{5}=0$ i.e., $x + 7y - 11 = 0$

(iv) Reflection of $x + y + 1 = 0$ in $2x + y - 1 = 0$ is L_R: $(x + y + 1) + \frac{(-2)(3)(2x+y-1)}{5}=0$ i.e., $7x + y - 11 = 0$

2. (i) → (a) & (d); (ii) → (a, c); (iii) → (a); (iv) → (a)

(i) Lines L_1: $x - 2y - 6 = 0$; L_2: $3x + y - 4 = 0$ and L_3: $\lambda x + 4y + \lambda^2 = 0$ are concurrent

$$\Rightarrow \Delta=\begin{vmatrix}1 & -2 & -6\\ 3 & 1 & -4\\ \lambda & 4 & \lambda^2\end{vmatrix}=0 \text{ or } \begin{vmatrix}1 & 0 & 0\\ 3 & 7 & 14\\ \lambda & (4+2\lambda) & (\lambda^2+6\lambda)\end{vmatrix}=0$$

$\Rightarrow 7\lambda^2 + 42\lambda - 28\lambda - 56 = 0$

$\Rightarrow 7(\lambda^2 + 2\lambda - 8) = 0$

$\Rightarrow \lambda = -4, 2$

(ii) Points $(\lambda + 1, 1)$, $(2\lambda + 1, 3)$ and $(2\lambda + 2, 2\lambda)$ are collinear

$\Rightarrow \frac{2}{\lambda}=\frac{2\lambda-3}{1} \quad \Rightarrow 2\lambda^2 - 3\lambda - 2 = 0$

$\Rightarrow \lambda = 2, -1/2$

(iii) Observe that line $x + y - (1 + \lambda) = 0$ is perpendicular to $x - y + 1 = 0$. The third concurrent line is $3x + y - 5 = 0$

$$\Rightarrow \begin{vmatrix}1 & 1 & (-1-\lambda)\\ 1 & -1 & 1\\ 3 & 1 & -5\end{vmatrix}=0$$

$\Rightarrow 8 - 4\lambda = 0 \Rightarrow \lambda = 2$

(iv) The line $x - y + (1 - \lambda) = 0$ is equally inclined to the axes.

Since (1, –2) and (3, 4) are equidistant

$\Rightarrow |4 - \lambda| = |-\lambda| \quad \Rightarrow \lambda = 2$

3. (i) → (c); (ii) → (d); (iii) → (a); (iv) → (b)

(i) Sides of the triangle are $x = 0$, $y = 0$ and L: $x + 2y = 3$

To find integral point $P(a, a^2)$ lying inside the triangle.

For $a \geq 1$ (where $a \in \mathbb{Z}$) the point will be in I[st] quadrant.

The point will be inside the triangle if $\frac{L(a,a^2)}{b}<0$, gives $a + 2a^2 - 3 < 0 \Rightarrow (2a + 3)(a - 1) < 0$

$\therefore a\in\left(-\frac{3}{2},1\right)$ but $a \geq 1$

Hence no such point. So number of points = 0

(ii) Consider two positions, when

(i) C (0, 0); A (5, 0) and B (0, 12)

(ii) B (0, 0), A (13, 0) and C (5 cos B, – 5sin B) = $\left(\frac{25}{13},-\frac{60}{13}\right)$

Hence the path locus is $y=-\frac{12}{5}x$ or $12x + 5y = 0$

(iii) Reflection of point $(t - 1, 2t + 2)$ in a line is $(2t + 1, t)$

$\Rightarrow$ Slope of perpendicular the line is $m_1=\frac{t+2}{-t-2}=-1$.

Hence the slope of line is $m_2 = 1$

(iv) Let B (b, b) and C (c, o), also AB and BC will be such that the product of slopes is 1.

Similarly slope of AC = – slope of BC

$\Rightarrow \frac{b-2}{b-1}.\frac{b}{b-c}=1 \quad \Rightarrow c=\frac{b}{b-1}$

Similarly $\frac{2}{1-c}=\frac{b}{c-b}$ gives {on putting $c=\frac{b}{b-1}$ }

$\Rightarrow \quad \dfrac{2}{1-c} = (-2)(b-1)$ and $\dfrac{b}{c-b} = \dfrac{-(b-1)}{(b-2)}$

$\Rightarrow \quad (-2)(b-1) = \dfrac{-(b-1)}{(b-2)}$

$\Rightarrow \quad b = 5/2$

$\therefore \quad B(5/2, 5/2)$ and $C(5/3, 0)$

$\Rightarrow$ Equation of BC is $y = 3(x - 5/3)$, i.e., $3x - y - 5 = 0$

Now, $A(1, 2)$

Hence $\sqrt{10}\, d(A, BC) = \sqrt{10}\, \dfrac{|3-2-5|}{\sqrt{10}} = 4$

4. (i) → (c) and (e); (ii) → (a) and (d); (iii) → (a)

(i) Given: $ax^2 - 6xy + y^2 = 0$.

Let the slopes be m, m^2

$\Rightarrow m + m^2 = 6 \quad \Rightarrow m = 2, -3$ and $m^3 = a$

$\therefore \quad a = 8, -27 \quad \Rightarrow$ (i) → (c) and (e)

(ii) Given $4y^3 - xy^2 - 9x^2y + ax^3 = 0$

Let $y = m_1x$, $y = m_2x$ and $y = m_3x$ be the roots, then $m_1 + m_2m_3 = \dfrac{1}{4}$ and $m_1m_2 + m_2m_3 + m_1m_3 = -9/4$ also $m_1m_2m_3 = -a/4$

Let $m_1m_2 = -1$ (for perpendicular lines)

$\Rightarrow m_3 = \dfrac{a}{4}$ and $m_3(m_1 + m_2) = -5/4$

$\Rightarrow m_1 + m_2 = -5/a$

$\Rightarrow m_1 + m_2 + m_3 = (-5/a) + (a/4) = 1/4$

$\Rightarrow a^2 - 20 = a$

$\Rightarrow (a-5)(a+4) = 0$ i.e., $a = 5, -4$

(iii) $x^4 + x^3y + cx^2y^2 - xy^3 + y^4 = 0$

Let $\dfrac{y}{x} = m$

$m^4 - m^3 + cm^2 + n + 1 = 0$

has 4 roots m_1, m_2, m_3, m_4 (say)

$\Rightarrow m_1 + m_2 + m_3 + m_4 = 1$ ...(i)

$m_1m_2 + m_1m_3 + m_1m_4 + m_2m_3 + m_2m_4 + m_3m_4 = c$ ...(ii)

$m_1m_2m_3 + m_1m_2m_4 + m_1m_3m_4 + m_2m_3m_4 = -1$ ....(iii)

$m_1m_2m_3m_4 = 1$ ...(iv)

Two lines of the set or angle bisector of other pair

$\Rightarrow$ Let $m_1m_2 = -1$

$\Rightarrow$ $m_3m_4 = -1$ (from equation (iv)

Substituting in (ii), we get $-(m_3 + m_4 + m_1 + m_2) = -1$

$-1 + m_1m_3 + m_1m_4 + m_2m_3 + m_2m_4 - 1 = c$

$-2 + m_1(m_3 + m_4) + m_2(m_3 + m_4) = c$

$c + 2 = (m_1 + m_2)(m_3 + m_4)$

$c + 2 = (m_1 + m_2)(1 - (m_1 + m_2))$

$c + 2 = m_1 + m_2 - (m_1 + m_2)^2$

$m_1 + m_2 = 2$

$\therefore \quad c + 2 = 2 - 4$

$\Rightarrow \quad c = -4$

SECTION–VIII (INTEGER-TYPE SECTION)

1. Given line $L: \dfrac{4x+3y}{24} = 1$ and curve $C: x^2 + y^2 - (6x + 8y) + (25 - c^2) = 0$

Homogenizing, we get $576x^2 + 576y^2 - 24(6x + 8y)(4x + 3y) + (25 - c^2)(4x + 3y)^2 = 0$

i.e., $16(25 - c^2)x^2 + 9(25 - c^2)y^2 + (-600 - 24c^2)xy = 0$

For 90° angle $a + b = 0$

$\Rightarrow (16 + 9)(25 - c^2) = 0$

$\Rightarrow c^2 = 25 \Rightarrow c = \pm 5 \Rightarrow |c| = 5$

2. Since the pair of lines are coincident the line $2x + 2y + k = 0$ touches the curve $\dfrac{x^2}{16} + \dfrac{y^2}{9} - 1 = 0$

Since the slope of tangent is -1

$\therefore \quad \left(\dfrac{16}{5}, \dfrac{9}{5}\right)$ or $\left(-\dfrac{16}{5}, -\dfrac{9}{5}\right)$ is the point of contact. The tangent is $x + y \pm 5 = 0$ or $2x + 2y \pm 10 = 0$

$\Rightarrow k = \pm 10$

3. $3y^2 + 2xy - 8y - x^2 + 4 = 0$ can be written as $(3y - x - 2)(y + x - 2) = 0$

The third line is passing through $(-5, 1)$

Let it be $p(x + 5) + q(y - 1) = 0$

Slope of third line $m = \dfrac{-p}{q}$

Now slope of the line joining $(-5, 1)$ and $(0, 0)$ is $-1/5$ and slopes of $x + y - 2 = 0$ is -1, these two are the extreme possibilities

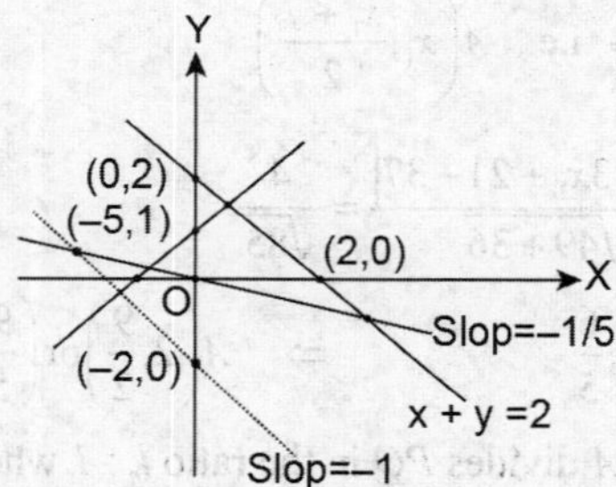

Hence $O(0, 0)$ will be interior point of the triangle when slope of the third line $m \in (-1, -1/5)$

Comparing with interval (a, b), we get $a = -1$ and $b = -1/5$

$\Rightarrow \quad a + \dfrac{1}{b^2} = -1 + 25 = 24$

4. $5x + y = 99$; where x and y are non-negative integers

$x = 0$;	$y = 99$
$x = 1$;	$y = 94$
$x = 2$;	$y = 84$
....	
....	
$x = 19$;	$y = 4$

So we have 20 such pairs which lie will on the line $5x + y = 99$.

Now distance origin (0, 0) from this line is $d = \dfrac{99}{\sqrt{26}}$

Let $P(x_1, y_1)$ and $Q(x_2, y_2)$ be two points on the line

$\Rightarrow \quad PQ = \sqrt{(x_2 - x_1)^2 + (y_2 - y_1)^2}$

Using $y_2 = 99 - 5x_2$ and $y_1 = 99 - 5x_1$ we get $PQ = |x_2 - x_1|\sqrt{26}$ and area of $\Delta OPQ = \dfrac{99}{2\sqrt{26}}|x_2 - x_1|\sqrt{26} = \dfrac{99}{2}|x_2 - x_1|$

For area to be non-zero integer $|x_2 - x_1|$ is non zero even integer.

⇒ Either $x_2, x_1 \in \{0, 2, 4, 6, \ldots, 18\}$ or $x_2, x_1 \in \{1, 2, 5, 7, \ldots, 19\}$

∴ Number of such possibilities = $2 \times {}^{10}C_2 = 2 \times 45 = 90$

5. Given A (1, 2), B (3, –1) and M lies on L: $x + y = 0$

Now, equation of AB is L_2: $3x + 2y = 7$

$|AM - BM|$ will be maximum = AB, when M is the point of intersection L and L_2

⇒ $M(7, -7)$

Hence distance between M (7, –7) and N (1, 1) is $MN = \sqrt{6^2 + 8^2} = 10$

6. Line $\frac{x}{6} + \frac{y}{8} = 1$ will intersect co-ordinate axes at M (6, 0) and N (0, 8) respectively so $MN = 10$ units

Now, ex-radius opposite to point N (0, 8) will $r_3 = \frac{\Delta}{s-c} = \frac{24}{12-6} = 4\,\text{units}$

7. Given: B (1, 5) and C (7, –2) ⇒ $BC = \sqrt{85}$ units and equation of BC is L_1 : $7x + 6y = 37$

Now equation P (–5, 1) and Q (3, 5) is L_2: $x - 2y + 7 = 0$

The area of ΔABC = 2 units, so distance of A from line BC is $h = \frac{4}{\sqrt{85}}$ i.e., $A\left(x_1, \frac{x_1+7}{2}\right)$

$$\Rightarrow \frac{|7x_1 + 3x_1 + 21 - 37|}{\sqrt{49+36}} = \frac{4}{\sqrt{85}}$$

$$\Rightarrow x_1 = 2, \frac{6}{5} \quad \Rightarrow \quad A\left(2, \frac{9}{2}\right) \text{or} \left(\frac{9}{5}, \frac{41}{10}\right)$$

Since A divides PQ in the ratio $k : 1$ where k is integer,

For $A\left(2, \frac{9}{2}\right)$, we get $\left(\frac{3k-5}{k+1}, \frac{5k+1}{k+1}\right) = (2, 9/2)$,

When $k = 7$ (which is an integer) for $A\left(\frac{6}{5}, \frac{41}{10}\right)$, we get it for $k = \frac{31}{9}$ (Not an integer). So value 7 is acceptable

8. Given B (12, 19) and C (23, 20)

⇒ $BC = \sqrt{11^2 + 1^2} = \sqrt{122}$ and equation of BC is L_1: $x - 11y - 197 = 0$

Area of ΔABC = 70 square units

∴ Altitude through A is $h = \frac{2 \times 70}{\sqrt{122}}$ units. Equation of median through A (p, q) is $y - \frac{39}{2} = (-5)\left\{x - \frac{35}{2}\right\}$

⇒ $5x + y - 107 = 0$ ⇒ A $(p, 107 - 5p)$ will be at distance h as $\left|\frac{p + 55p - 1177 + 197}{\sqrt{122}}\right| = \frac{140}{\sqrt{122}}$

⇒ $56p - 980 = \pm 140$ ⇒ $p = 20, 15$

⇒ (20, 7) or (15, 32) which gives $p + q = 27$ or 47

The larger value is 47

9. The line pairs are $(y - 3)(y - 1) = 0$ i.e., $y = 1, 3$

The other line pair is $4y^2 + 4xy + x^2 - 5x - 10y + 4 = 0$ i.e., $(2y + x - 4)(2y + x - 1) = 0$, gives $x + 2y = 4$ which will intersect $y = 1$ at (2, 1) and line $x + 2y = 1$ will intersect at B (1, 1)

For A (2, 1), B (–1, 1), we get $AB = 3$ units

∴ Area of polygon = 3 × 2 = 6 square units

10. $ax + 3y = 1$ and $ax + y = -1$ intercept at $A\left(\frac{-2}{a}, 1\right)$

⇒ slope of OA is $m_{OA} = \frac{-a}{2}$

Similarly $ax + 3y = 1$ and $x + 3y = 0$ intersect at $B\left(\frac{1}{a-1}, \frac{-1}{3a-3}\right)$, which will give a slope $m_{OB} = -1/3$

⇒ $m_{OA}\, m_{OB} = -1$ i.e., $\left(\frac{-a}{2}\right)\left(-\frac{1}{3}\right) = -1$

⇒ $a = -6$

11. given L_1: $-mx + y = 1$ ....(i)

And $x^2 + y^2 - 1 = 0$ ....(ii)

Homogenizing (ii) with (i), we get $x^2 + y^2 - (y - mx)^2 = 0$

⇒ $x^2(1 - m^2) + 0.\,y^2 + 2mxy = 0$

Now, $a + b = 0$ ⇒ $m^2 = 1$

⇒ $|m| = 1$

Circles and Family of Circles

CHAPTER 3

INTRODUCTION

In our day to day life, we come across numerous objects with circular shape, e.g., coins, dials of watches, flying discs, rings, wheels etc. Have you ever given a serious thought to the question what a 'circle' is?

A circle is perhaps the most regular object we know, as you know that each point on its circumference is equidistant from its centre. The shape and symmetry of circle has fascinated the thinkers of all fields throughout the ages. Its applications are rather more numerous. e.g., a king always wishes to have boundaries of his kingdom equally spread around his capital. Tantrics and priests used it in constructing different figurine, vedics and tantras. Ancient humans realized that smooth movement is only possible if the wheels of circular shape are used.

The circle has always been a mathematician delight. He has endlessly indulged himself in quest of different properties of circle, which are portrayed by the symmetry of circle.

In previous classes, you have learnt about properties, construction and different geometrical theorems on properties of circle, in Euclidian Geometry, but in this chapter, we will learn to study and analyse the circle and its properties taking the help of analytical geometry, i.e., Cartesian co-ordinate geometry.

We will also learn many interesting situations and tricks to solve complicated problems using concept of family of circles etc.

REVIEW OF BASIC GEOMETRY OF CIRCLES

- Equal chords subtend equal angles at the centre and vice versa.

  $CD = AB \Leftrightarrow \angle AOB = \angle COD = \theta.$

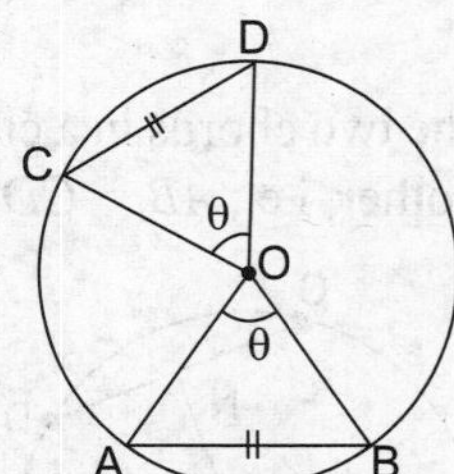

FIGURE 3.1

- Equal chords of a circle are equidistant from the centre and vice versa. (⊥ from centre, bisect the chords).

  $AB = CD \Rightarrow \frac{1}{2} AB = \frac{1}{2} CD$

  $BM = DN$ and $OB = OD$ (radius) $\Rightarrow OM = ON$

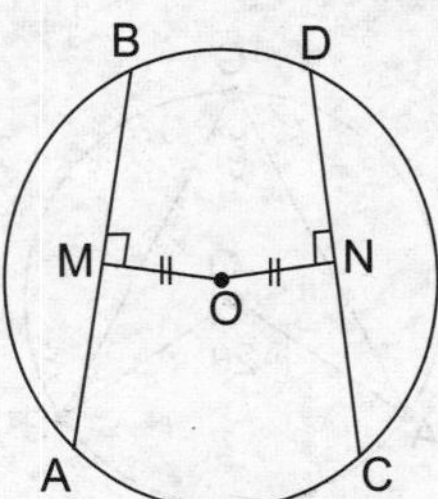

FIGURE 3.2

- A chord drawn across the circular region divides it into parts each of which is called a segment of the circle.

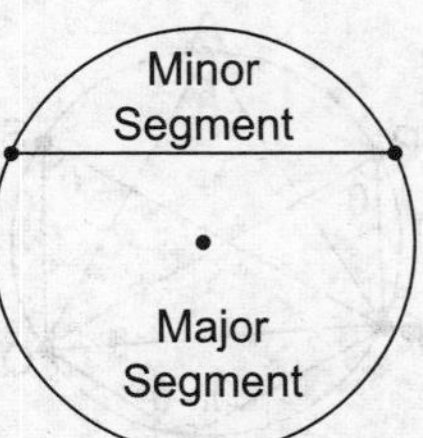

FIGURE 3.3

- The tangents drawn from external point P to the circle are equal, i.e., $PA = PB$. Also, the angle between the tangents is bisected by the straight line, which joins their point of intersection to the centre.

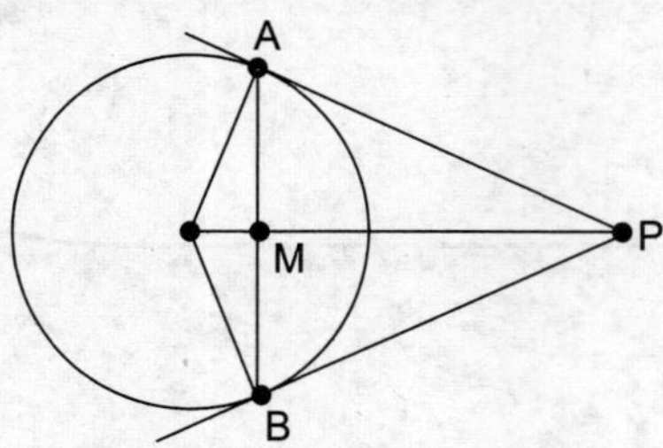

FIGURE 3.4

This straight line also bisects at right angles the chord, joining the points of contact, i.e., $\angle AMP = \pi/2$ and $AM = MB$.

- The greater of the two chords in a circle is nearer to the centre than the other, i.e., $AB > CD \Rightarrow OM < ON$.

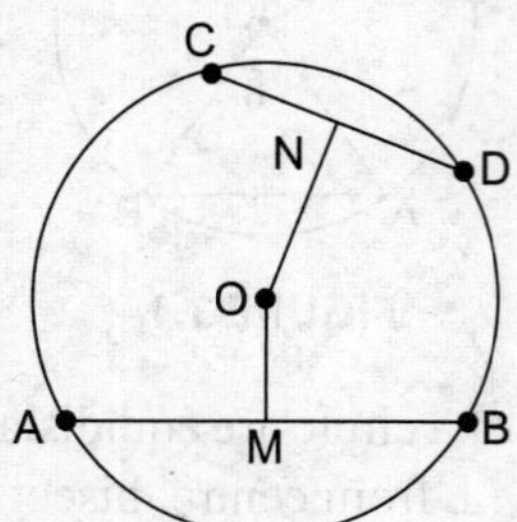

FIGURE 3.5

- Angle subtended by an arc at the centre is double the angle subtended at any point on the remaining part of the circle, i.e., $\angle ACB = \theta \;\Rightarrow\; \angle AOB = 2\theta$.

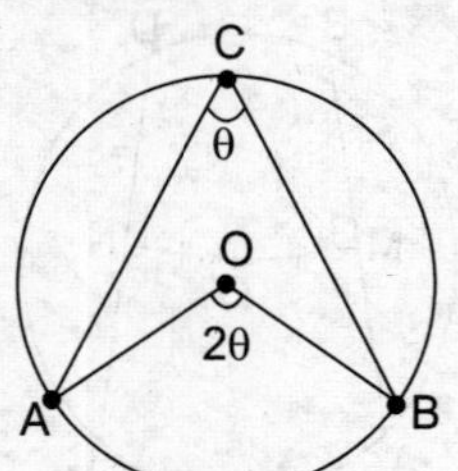

FIGURE 3.6

Angles in the same segment of a circle are equal. That is, $\angle PSQ = \angle PAQ = \angle PRQ = \theta$

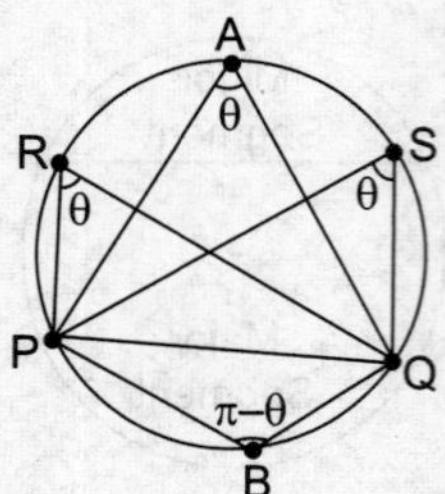

FIGURE 3.7

If θ is acute $\Rightarrow$ Arc $\overparen{PRASQ}$ is major and Arc $\overparen{PBQ}$ is minor.

If $\theta = \dfrac{\pi}{2} \Rightarrow$ Arc $\overparen{PRASQ}$ and Arc $\overparen{PBQ}$ are semicircles with PQ as diameter.

- The sum of the opposite angles of a cyclic quadrilateral is 180° and vice versa.

$\angle PAQ = \theta \;\Rightarrow\; \angle PBQ = \pi - \theta$

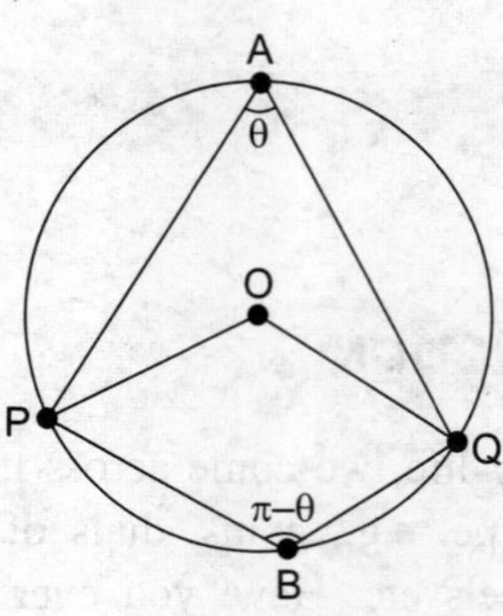

FIGURE 3.8

- If a line touches a circle and from the point of contact a chord is drawn, the angles which this chord makes with the given line are equal respectively to the angles formed in the corresponding alternate segments, i.e., $\angle ABT = \angle APB$ and $\angle ABS = \angle AQB$

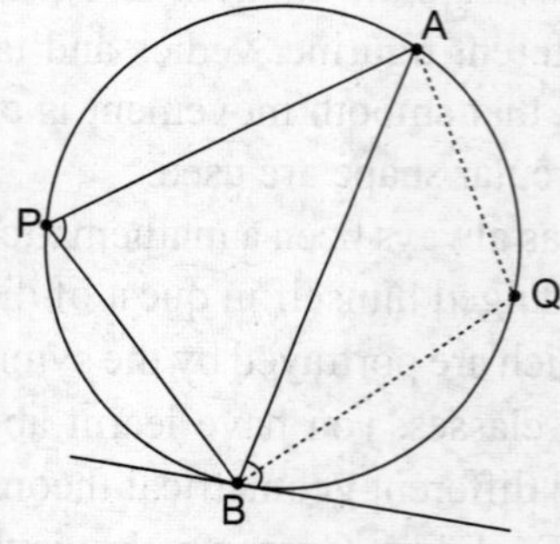

FIGURE 3.9

- If two chords of a circle intersect either inside or outside the circle, the rectangle constructed by the parts of one chord is equal in area to the rectangle constructed by the parts of the other, i.e., $AP \times PB = CP \times PD$

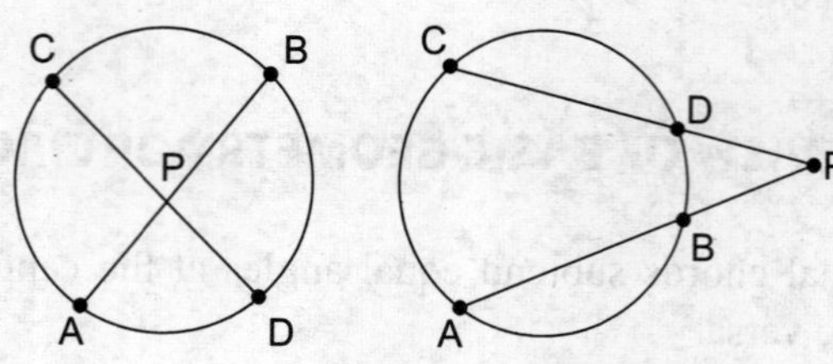

FIGURE 3.10

NOTE

To check whether four points A,B,C,D are concylic, we can also use the theorem that $AO \times OC = BO \times OD$, where O is the point of intersection of lines AC and BD.

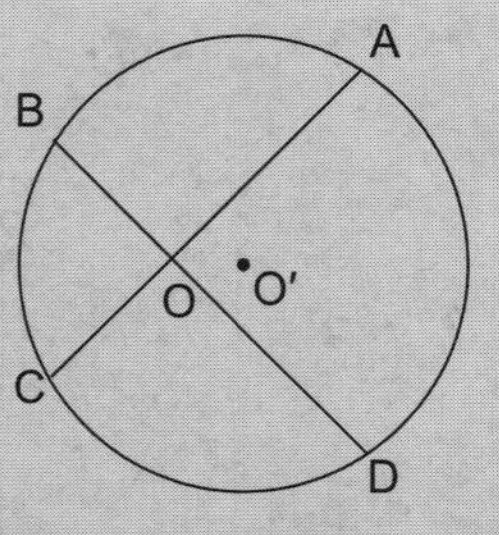

FIGURE 3.11

DEFINITION

A circle is the locus of a point moving in a plane so that its distance from a fixed point (in the same plane) remains constant. The fixed point is called ***centre*** of the circle and the constant distance is called the ***radius*** of the circle. The circle is the simplest non-trivial conic which possesses a lot of properties.

Equation of circle with centre at (α, β) and radius 'r' is

$$(x-\alpha)^2 + (y-\beta)^2 = r^2 \quad \text{....(i)}$$

$$\Rightarrow x^2 + y^2 - 2\alpha x - 2\beta y + \alpha^2 + \beta^2 - r^2 = 0 \quad \text{....(ii)}$$

Comparing with the general equation of conic section $ax^2 + by^2 + 2hxy + 2gx + 2fy + c = 0$, we conclude that, this equation will represent circle if and only if $a = b$ and $h = 0$ and $g^2 + f^2 - c \geq 0$. Therefore the equation $x^2 + y^2 + 2gx + 2fy + c = 0$ when $g^2 + f^2 - c \geq 0$ represents a circle and called as general form of a circle.

EQUATION OF A CIRCLE IN VARIOUS FORMS

Centre Radius Form/Point Circle Form

Equation of circle with centre at (α, β) and radius 'r' is given by $(x-\alpha)^2 + (y-\beta)^2 = r^2$. When centre is (0, 0) and radius is 'r' then the equation becomes $x^2 + y^2 = r^2$ and is called as standard form of circle.

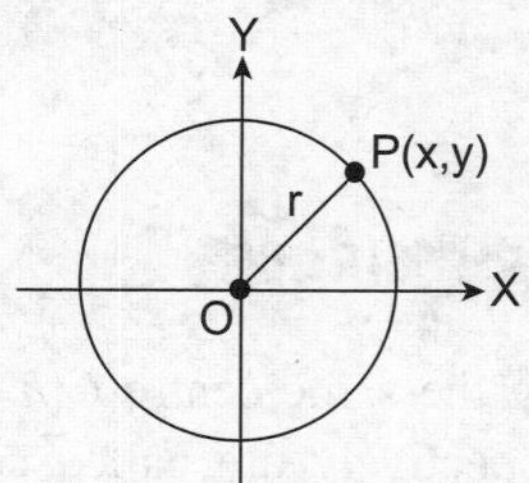

FIGURE 3.12

General Equation

General equation of circle $x^2 + y^2 + 2gx + 2fy + c = 0$ is also known as structural or canonical form of circle with co-ordinates of its centre as $(-g, -f)$ and radius $= \sqrt{g^2 + f^2 - c}$.

Rule to find the centre and radius of a circle whose equation is given:

(a) Make the co-efficients of x^2 and y^2 equal to 1 and right hand side equal to zero.

(b) Then the co-ordinates of centre will be (α, β), where $\alpha = -(1/2)$ (co-efficient of x) and $\beta = -(1/2)$ (co-efficient of y)

(c) radius $= \sqrt{\alpha^2 + \beta^2 - \text{constant term}}$

Notations: The expression $x^2 + y^2 + 2gx + 2fy + c$ is denoted by S. If we substitute the co-ordinates of a point $P(x_1, y_1)$ in the equation of the circle $S = 0$, then L.H.S . becomes a numerical quantity denoted by S_1 such that $S_1 = x_1^2 + y_1^2 + 2gx_1 + 2fy_1 + c$.

NOTES

1. *A general non-homogenous equation of second degree is $ax^2 + 2hxy + by^2 + 2gx + 2fy + c = 0$ in x, y represents a circle, if:*
   - *Co-efficients of x^2 = co-efficients of, y^2 i.e., $a = b \neq 0$*
   - *Co-efficient of xy is zero, i.e., $h = 0$.*
   - $g^2 + f^2 - c \geq 0$

- *The general equation may be of the form $Ax^2 + Ay^2 + 2Gx + 2Fy + C = 0$ represents an equation of circle.*

  Centre $\equiv \left(-\frac{G}{A}, -\frac{F}{A}\right)$ *and radius* $\equiv \frac{1}{A}\sqrt{G^2 + F^2 - AC}$.

2. *If $g^2 + f^2 - c < 0$, then the radius of circle will be an imaginary number. Hence in this case, circle is called a virtual circle or an imaginary circle, (with a real centre).*
3. *If $g^2 + f^2 - c = 0$, then the radius of circle will be real and zero, hence the circle is called a point circle.*
4. *If $g^2 + f^2 - c > 0$, then the radius of circle will be real. Hence in this case circle is called real non-trivial.*
5. *Clearly, the general equation of circle contains three independent parameters (effectively). Therefore to determine a circle uniquely, three independent conditions must be given.*

■ CONCENTRIC CIRCLE

Two circles having the same centre $C(h, k)$, but different, radii r_1 and r_2 respectively are called concentric circles. Thus, the circles $(x - h)^2 + (y - k)^2 = r_1^2$ and $(x - h)^2 + (y - k)^2 = r_2^2$, $r_1 \neq r_2$ are concentric circles.

Therefore, the equations of concentric circles differ in constant terms only.

ILUSTRATION 1: Find the equation of the circle whose centre is the point (–2, 3) and whose radius is 5.

SOLUTION: $(x - h)^2 + (y - k)^2 = r^2$

by putting the given values, we get $(x + 2)^2 + (y - 3)^2 = 5^2 \Rightarrow x^2 + y^2 + 4x - 6y = 12$.

ILLUSTRATION 2: Find the centre and radius of the circle $3x^2 + 3y^2 - 8x - 10y + 3 = 0$.

SOLUTION: The given circle is $3x^2 + 3y^2 - 8x - 10y + 3 = 0$

or $x^2 + y^2 - \frac{8}{3}x - \frac{10}{3}y + 1 = 0$

Here, $g = \frac{1}{2}(\text{coefficent of } x) = \frac{1}{2}\left(\frac{-8}{3}\right) = \frac{-4}{3}$

Similarly, $f = (1/2)$ (co-efficient of y) $= (1/2)(-10/3) = -5/3$ and $c = 1$.

Hence, centre of the circle is $(-g, -f)$ i.e., $(4/3, 5/3)$ and

radius of the circle $= \sqrt{g^2 + f^2 - c} = \sqrt{\frac{16}{9} + \frac{25}{9} - 1} = \sqrt{\frac{32}{9}} = \frac{4}{3}\sqrt{2}$

ILLUSTRATION 3: Find the equation of the circle concentric with the circle $x^2 + y^2 - 8x + 6y - 5 = 0$ and passing through the point (–2, –7).

SOLUTION: The given equation of circle is $x^2 + y^2 - 8x + 6y - 5 = 0$.

Therefore, the centre of the circle is at (4, –3). Since the required circle is concentric with this circle, therefore the centre of the required circle is also at (4, –3). Since the point (–2, –7) lies on the circle, the distance of the centre from this point is the radius of the circle. Therefore, we get $r = \sqrt{(4+2)^2 + (-3+7)^2} = \sqrt{52}$

Hence, the equation of the circle becomes $(x-4)^2 + (y+3)^2 = 52$

or $x^2 + y^2 - 8x + 6y - 27 = 0$

ILLUSTRATION 4: A circle has radius 3 units and its centre lies on the line $y = x - 1$. Find the equation of the circle if it passes through (7, 3).

SOLUTION: Let the centre of the circle be (h, k). Since the centre lies on $y = x - 1$, we get

$k = h - 1$ ....(i)

FIGURE 3.13

Since the circle passes through the point (7, 3), therefore the distance of the centre from this point is the radius r of the circle. We have $r = \sqrt{(h-7)^2 + (k-3)^2}$

$\Rightarrow \ 3 = \sqrt{(h-7)^2 + (h-1-3)^2}$ (from (i))

$\Rightarrow \ 9 = (h-7)^2 + (h-4)^2$ $\qquad \Rightarrow \ h^2 - 11h + 28 = 0$

$\Rightarrow \ (h-7)(h-4) = 0$ $\qquad \Rightarrow \ h = 7$ or $h = 4$

For $h = 7$, we get $k = 6$, and for $h = 4$, we get $k = 3$

Hence, there are two circles which satisfy the given conditions. They are

$(x-7)^2 + (y-6)^2 = 9$ or $x^2 + y^2 - 14x - 12y + 76 = 0$

and $(x-4)^2 + (y-3)^2 = 9$ or $x^2 + y^2 - 8x - 6y + 16 = 0$

ILLUSTRATION 5: Show that the four points (1, 0), (2, –7), (8, 1) and (9, –6) are concyclic.

SOLUTION: Since the given four points are to be shown concyclic, we are to show that they lie on a circle. Let the general equation of circle is $x^2 + y^2 + 2gx + 2fy + c = 0$ .... (i)

has three parameters. It is sufficient to obtain the equation of the circle passing through any three points. For concyclic, the fourth point should lie on this circle. Let three points $A(1, 0)$, $B(2, -7)$ and $C(8,1)$ lie on

(i), then $1 + 0 + 2g + 0 + c = 0$ or $1 + 2g + c = 0$ ...(ii)

$(2)^2 + (-7)^2 + 2g(2) + 2f(-7) + c = 0 \ \Rightarrow \ 53 + 4g - 14f + c = 0$ ...(iii)

and $(8)^2 + (1)^2 + 2g(8) + 2f(1) + c = 0 \ \Rightarrow \ 65 + 16g + 2f + c = 0$ ...(iv)

Now subtracting (ii) from (iii), we get $52 + 2g - 14f = 0$ or $26 + g - 7f = 0$ ...(v)

Subtracting (iii) from (iv),we get $12 + 12g + 16f = 0$ or $3 + 3g + 4f = 0$ ...(vi)

Solving (v) and (vi), we get $g = -5$ and $f = 3$

from (ii), $1 - 10 + c = 0 \Rightarrow c = 9$

therefore, equation of circle passing through three points is $x^2 + y^2 - 10x + 6y + 9 = 0$

Substituting the fourth point in the equation of this circle, we get

$(9)^2 + (-6)^2 - 10(9) + 6(-6) + 9 = 0$

Hence the point (9, –6) lies on the circle, that is the four points are concyclic.

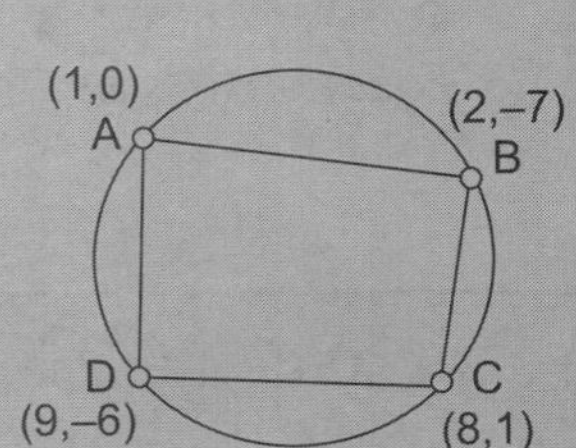

FIGURE 3.14

Aliter: Find equation of AC and BD, solve to get their point of intersection P and show that $PA.PC = PB.PD$.

Diameteric Form

Let $A(x_1, y_1)$ and $B(x_2, y_2)$ be the end points of a diameter and let $P(x, y)$ be any point on the circle.

Now slope of $AP = \dfrac{y-y_1}{x-x_1}$

and slope of $BP = \dfrac{y-y_2}{x-x_2}$

Since $\angle APB = 90°$

Slope of $AP \times$ Slope of $BP = -1$

$$\Rightarrow \left(\frac{y-y_1}{x-x_1}\right)\times\left(\frac{y-y_2}{x-x_2}\right) = -1$$

$$\Rightarrow (x-x_1)(x-x_2)+(y-y_1)(y-y_2)=0$$

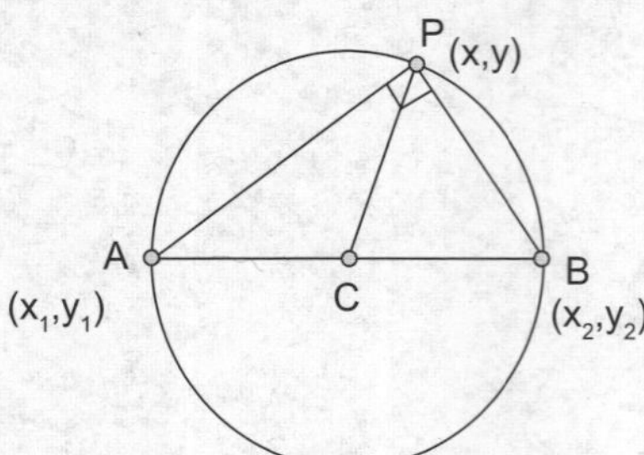

FIGURE 3.15

Aliter 1: Since diameter AB of the circle subtends right angle at any point P on the circumference, therefore in right angled triangle APB, $(AP)^2 + (PB)^2 = (AB)^2$

$$\Rightarrow (x-x_1)^2+(y-y_1)^2+(x-x_2)^2+(y-y_2)^2$$
$$=(x_1-x_2)^2+(y_1-y_2)^2$$

or $\quad 2(x^2 - xx_1 - xx_2 + x_1x_2) + 2(y^2 - yy_1 - yy_2 + y_1y_2) = 0$

$$\Rightarrow (x-x_1)(x-x_2)+(y-y_1)(y-y_2)=0$$

Aliter 2: Let the mid-point of AB is $C(h, k)$

$$\therefore \quad h=\frac{x_1+x_2}{2}, k=\frac{y_1+y_2}{2}$$

Here, C is the centre of the circle

Also, radius $r = CA = CB = CP$

$$\Rightarrow r=\frac{1}{2}AB=\frac{1}{2}\sqrt{(y_2-y_1)^2+(x_2-x_1)^2}$$

Equation of circle with centre $C(h, k)$ and radius r is $(x-h)^2 + (y-k)^2 = r^2$

$$\Rightarrow \left(x-\frac{x_1+x_2}{2}\right)^2+\left(y-\frac{y_1+y_2}{2}\right)^2$$
$$=\frac{1}{4}[(x_2-x_1)^2+(y_2-y_1)^2]$$

$$\Rightarrow x^2-(x_1+x_2)x+\frac{1}{4}(x_1+x_2)^2$$
$$+y^2-(y_1+y_2)y+\frac{1}{4}(y_1+y_2)^2$$
$$=\frac{(x_2-x_1)^2}{4}+\frac{(y_2-y_1)^2}{4}$$

$$\Rightarrow x^2+y^2-(x_1+x_2)x-(y_1+y_2)y+x_1x_2+y_1y_2=0$$

$$\Rightarrow [x^2-(x_1+x_2)x+x_1x_2]$$
$$+[y^2-(y_1+y_2)y+y_1y_2]=0$$

$$\Rightarrow (x-x_1)(x-x_2)+(y-y_1)(y-y_2)=0$$

NOTE

$(x-x_1)(x-x_2)+(y-y_1)(y-y_2)=0$ ..... (i)

Which can be written as $\underbrace{x^2-(x_1+x_2)x+x_1x_2}_{\text{quadratic in } x \text{ with } x_1x_2 \text{ as roots}}+\underbrace{y^2-(y_1+y_2)y+y_1y_2}_{\text{quadratic in } y \text{ with } y_1y_2 \text{ as roots}}=0$

- x_1, x_2 *are the roots of the equation* $x^2 - (x_1 + x_2)x + x_1x_2 = 0$ *and* y_1, y_2 *are the roots of the equation* $y^2 - (y_1 + y_2)y + y_1y_2 = 0$
- *Equation (i) is also the equation of the circle with least radius passing through* (x_1, y_1) *and* (x_2, y_2).

ILLUSTRATION 6: Find the equation of the circle, the extremities of one of the diameters of which are $A(6, 5)$, $B(0, -3)$.

SOLUTION: The centre C must be mid-point of AB

$\Rightarrow$ C must be $(3, 1)$

Radius $= AC = \sqrt{(6-3)^2+(5-1)^2} = 5$

Thus, equation of the circle must be $(x-3)^2+(y-1)^2=5^2$

or $x^2+y^2-6x-2y-15=0$

or equation is $(x-6)(x-0)+(y-5)(y+3)=0$

ILLUSTRATION 7: Find the equation of the circle the end points of whose diameter are the centres of the circles $x^2+y^2+6x-14y=1$ and $x^2+y^2-4x+10y=2$.

SOLUTION: The centres of $x^2+y^2+6x-14y=1$ and $x^2+y^2-4x+10y=2$. are $(-3, 7)$ and $(2, -5)$ respectively.

According to the question, the points $(-3, 7)$ and $(2, -5)$ are the extremities of the diameter of the required circle.

Hence equation of circle is $(x+3)(x-2)+(y-7)(y+5)=0$

$\Rightarrow x^2+y^2+x-2y-41=0$

ILLUSTRATION 8: The sides of a square are $x=2$, $x=3$, $y=1$ and $y=2$. Find the equation of the circle drawn on the diagonals of the square as its diameter.

SOLUTION: Let $ABCD$ be a square and the equation of its sides AB, BC, CD and DA are $y=1$, $x=3$, $y=2$ and $x=2$ respectively.

Then $A=(2, 1)$, $B=(3, 1)$, $C=(3, 2)$ and $D=(2, 2)$.

Since diagonals of squares are the diameters of the circle, then equation of circle is

$$(x-2)(x-3)+(y-1)(y-2)=0$$

$\Rightarrow x^2+y^2-5x-3y+8=0$ considering AC as diameter)

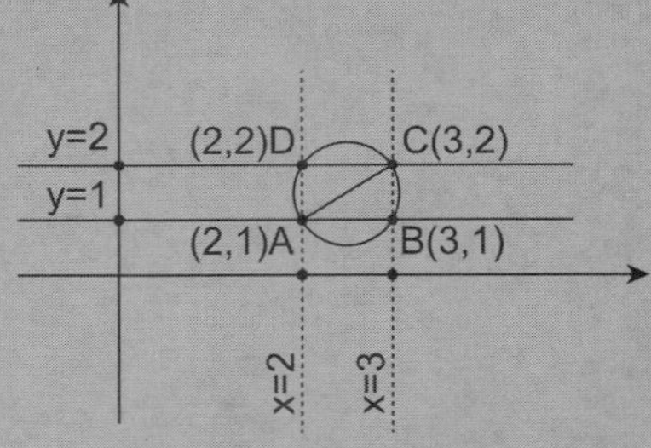

FIGURE 3.16

ILLUSTRATION 10: The abscissae of two points A and B are the roots of the equation $x^2+2ax-b^2=0$ and their ordinates are the roots of the equation $x^2+2px-q^2=0$. Find the equation and the radius of the circle with AB as diameter.

SOLUTION: Given equations are

$$x^2+2ax-b^2=0 \quad \text{....(i)}$$

$$x^2+2px-q^2=0 \quad \text{...(ii)}$$

Let the roots of equation (i) be α and β and those of equation (ii) be γ and δ. Then

$$\begin{cases}\alpha+\beta=-2a\\ \alpha\beta=-b^2\end{cases} \quad \text{and} \quad \begin{cases}\gamma+\delta=-2p\\ \gamma\delta=-q^2\end{cases}$$

Let $A=(\alpha, \gamma)$ and $B=(\beta, \delta)$

Now equation of circle with diameter AB will be $(x-\alpha)(x-\beta)+(y-\gamma)(y-\delta)=0$

$\Rightarrow x^2+y^2-(\alpha+\beta)x-(\gamma+\delta)y+\alpha\beta+\gamma\delta=0$

$\Rightarrow x^2+y^2+2ax+2py-b^2-q^2=0$ and radius $=\sqrt{a^2+p^2+b^2+q^2}$

ILLUSTRATION 11: Find the equation of a circle passing through origin whose centre is foot of perpendicular from origin to the line $4x+5y=20$

SOLUTION: Let C be the foot of perpendicular from O to AB. Therefore, 'C' becomes the centre of the circle. Therefore, if we take the image of O in AB (i.e., point D); then O and D should be the diametric ends of the circle.

Co-ordinates of D can be obtained by the formula

$$\frac{x-0}{4}=\frac{y-0}{5}=\frac{-2(4\times 0+5\times 0-20)}{4^2+5^2}$$

$$\Rightarrow \frac{x}{4}=\frac{y}{5}=\frac{40}{41}$$

$$\Rightarrow x=\frac{160}{41} \text{ and } y=\frac{200}{41}$$

FIGURE 3.17

$\therefore$ Equation of circle: $(x-0)\left(x-\frac{160}{41}\right)+(y-0)\left(y-\frac{200}{41}\right)=0$

$$\Rightarrow x\left(x-\frac{160}{41}\right)+y\left(y-\frac{200}{41}\right)=0 \Rightarrow 41x^2+41y^2-160x-200y=0$$

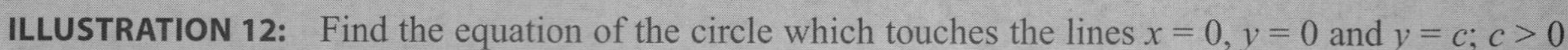

ILLUSTRATION 12: Find the equation of the circle which touches the lines $x = 0$, $y = 0$ and $y = c$; $c > 0$

SOLUTION: There can be two circles possible as shown in the figure.

Now since the circle touches x axis and y-axis both, therefore the equation of such a circle is

$(x-h)^2 + (y-h)^2 = h^2$ and $(x+h)^2 + (y-h)^2 = h^2$; $h > 0$

And since $OE = c$ and $OD = \frac{OE}{2} \Rightarrow OD = \frac{c}{2}$

$\Rightarrow$ Radius $= c/2 = h \Rightarrow$ equation of circle S_1 becomes $(x - c/2)^2 + (y - c/2)^2 = (c/2)^2$

And for circle S_2 : $(x + c/2)^2 + (y - c/2)^2 = (c/2)^2$

i.e., $4x^2 + 4y^2 - 4cx - 4cy + c^2 = 0$ and $4x^2 + 4y^2 + 4cx - 4cy + c^2 = 0$

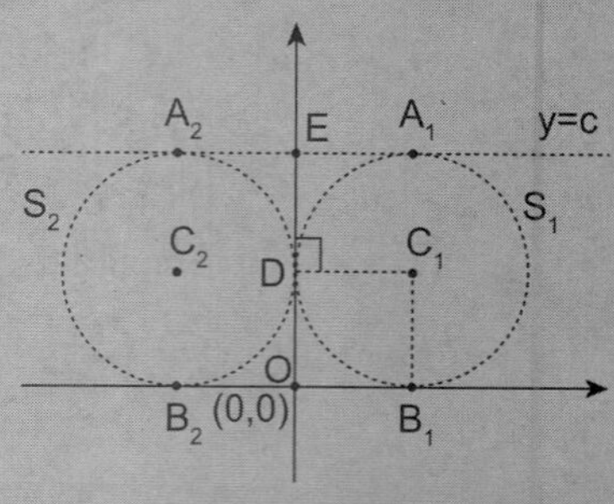

FIGURE 3.18

TEXTUAL EXERCISE–1 (SUBJECTIVE)

1. Find the equation of a circle whose diameter has the length 20 and the equations of two of its diameters are $2x + y = 6$ and $3x + 2y = 4$
2. Find the equation of the circle passing through the intersection of the lines $3x + y = 4$ and $x - 3y + 2 = 0$ and concentric with the circle $2(x^2 + y^2) - 3x + 8y - 1 = 0$.
3. Find the equation to the circle which passes through the points (1, –2) and (4, –3) and which has its centre on the straight line $3x + 4y = 7$.
4. If (4, 1) be an extremity of a diameter of the circle $x^2 + y^2 - 2x + 6y - 15 = 0$, find the co-ordinates of the other extremity of the diameter.

Answer Keys

1. $x^2 + y^2 - 16x + 20y + 64 = 0$
2. $2(x^2 + y^2) - 3x + 8y - 9 = 0$
3. $15x^2 + 15y^2 - 94x + 18y + 55 = 0$
4. (–2, –7)

TEXTUAL EXERCISE–1 (OBJECTIVE)

1. (i) The equation $x^2 + y^2 + 4x + 6y + 13 = 0$ represents
   (a) pair of straight lines
   (b) a circle with radius $r > 0$
   (c) pair of coincident lines
   (d) a point

   (ii) Precisely the condition that $x^2 + y^2 + 2gx + 2fy + c = 0$ represents a real circle with non-zero radius
   (a) for all real g, f and c
   (b) for all real g, f but $c < 0$
   (c) $g^2 + f^2 - c > 0$
   (d) None of these

   (iii) The equation of the circles having centre (a cos α, a sin α) and radius a is
   (a) $x^2 + y^2 + 2ax \cos\alpha + 2ay \sin\alpha = 0$
   (b) $x^2 + y^2 - 2ax \cos\alpha - 2ay \sin\alpha = 0$
   (c) $x^2 + y^2 - 2ax \cos\alpha + 2ay \sin\alpha = 0$
   (d) None of these

2. The centres of the three circles $x^2 + y^2 = 1$, $x^2 + y^2 + 6x - 2y = 1$ and $x^2 + y^2 + 6x - 2y = 4$
   (a) form a right angled triangle
   (b) form an isosceles triangle
   (c) are collinear
   (d) None of these

3. $ABCD$ is a square with side a, taking AB and AD as axes, then the equation to the circle circumscribing the square is
   (a) $x^2 + y^2 = a(x - y)$
   (b) $x^2 + y^2 = a(x + y)$.
   (c) $x^2 + y^2 + a(x + y) = 0$
   (d) None of these

4. If O is the origin and OP, OQ are distinct tangents to the circle $x^2 + y^2 + 2gx + 2fy + c = 0$, the circumcentre of the triangle OPQ is
   (a) $(-g/2, -f/2)$ (b) (g, f)
   (c) $(-f, -g)$ (d) None of these

5. If a circle of constant radius $3k$ passes through the origin and meets the axes at A and B, the locus of the centroid of ΔOAB is
   (a) $x^2 + y^2 = k^2$ (b) $x^2 + y^2 = 2k^2$
   (c) $x^2 + y^2 = 4k^2$ (d) None of these

6. The locus of the centre of a circle of radius 2 which rolls on the outside of the circle $x^2 + y^2 + 3x - 6y - 9 = 0$ is
   (a) $x^2 + y^2 + 3x - 6y + 5 = 0$
   (b) $x^2 + y^2 + 3x - 6y - 31 = 0$
   (c) $x^2 + y^2 + 3x - 6y + 19/4 = 0$
   (d) None of these

7. The radii of the circles $x^2 + y^2 - 1 = 0$, $x^2 + y^2 - 2x - 6y = 6$ and $x^2 + y^2 - 4x - 12y = 9$ are in
   (a) A.P. (b) G.P.
   (c) H.P. (d) None of these

8. If the equations of the two diameters of a circle are $x + y = 6$ and $x + 2y = 4$ and the radius of the circle is 10, then the equation of the circle is
   (a) $x^2 + y^2 + 16x + 4y - 32 = 0$
   (b) $x^2 + y^2 - 16x + 4y + 32 = 0$
   (c) $x^2 + y^2 - 16x + 4y - 32 = 0$
   (d) None of these

9. *A* square is inscribed in the circle $x^2 + y^2 - 2x + 4y + 3 = 0$, whose sides are parallel to the co-ordinates axes. One vertex of the square is
   (a) $(1+\sqrt{2}, -2)$ (b) $(1-\sqrt{2}, -2)$
   (c) $(1-\sqrt{2}, +2)$ (d) None of these

10. If a circle whose centre is $(1, -3)$ touches the line $3x - 4y - 5 = 0$, then the radius of the circle is
   (a) 2 (b) 4
   (c) 5/2 (d) 7/2

11. The equation of the circle whose radius is 5 and which touches the circle $x^2 + y^2 - 2x - 4y - 20 = 0$ externally at the point (5, 5), is
   (a) $x^2 + y^2 - 18x - 16y - 120 = 0$
   (b) $x^2 + y^2 - 18x - 16y + 120 = 0$
   (c) $x^2 + y^2 + 18x + 16y - 120 = 0$
   (d) $x^2 + y^2 + 18x - 16y + 120 = 0$

12. The lines $2x - 3y = 5$ and $3x - 4y = 7$ are the diameters of a circle of area 154 square units. The equation of the circle is
   (a) $x^2 + y^2 + 2x - 2y = 62$
   (b) $x^2 + y^2 - 2x + 2y = 47$
   (c) $x^2 + y^2 + 2x - 2y = 47$
   (d) $x^2 + y^2 - 2x + 2y = 62$

13. If the lines $3x - 4y + 4 = 0$ and $6x - 8y - 7 = 0$ are tangents to a circle, then the radius of the circle is
   (a) 3/2 (b) 3/4
   (c) 1/10 (d) 1/20

14. ABC is a triangle in which angle C is a right angle. If the co-ordinates of A and B be $(-3, 4)$ and $(3, -4)$ respectively, then the equation of the circumcircle of triangle ABC is
(a) $x^2 + y^2 - 6x + 8y = 0$
(b) $x^2 + y^2 = 25$
(c) $x^2 + y^2 - 3x + 4y + 5 = 0$
(d) None of these

15. The equation of the circle concentric with the circle $x^2 + y^2 + 8x + 10y - 7 = 0$ and passing through the centre of the circle $x^2 + y^2 - 4x - 6y = 0$ is
(a) $x^2 + y^2 + 8x + 10y + 59 = 0$
(b) $x^2 + y^2 + 8x + 10y - 59 = 0$
(c) $x^2 + y^2 - 4x - 6y + 87 = 0$
(d) $x^2 + y^2 - 4x - 6y - 87 = 0$

Answer Keys

1. (i) (d) (ii) (c) (iii) (b) **2.** (c) **3.** (b) **4.** (a) **5.** (c) **6.** (b) **7.** (a) **8.** (c)
9. (d) **10.** (a) **11.** (b) **12.** (b) **13.** (b) **14.** (b) **15.** (b)

EQUATION OF CIRCLE IN PARTICULAR FORMS

(a) Circle touching the x-axis: then absolute value of y-co-ordinate of the centre must be equal to the radius

(b) Circle touching y-axis: then absolute value of x-co-ordinate of the centre must be equal to the radius

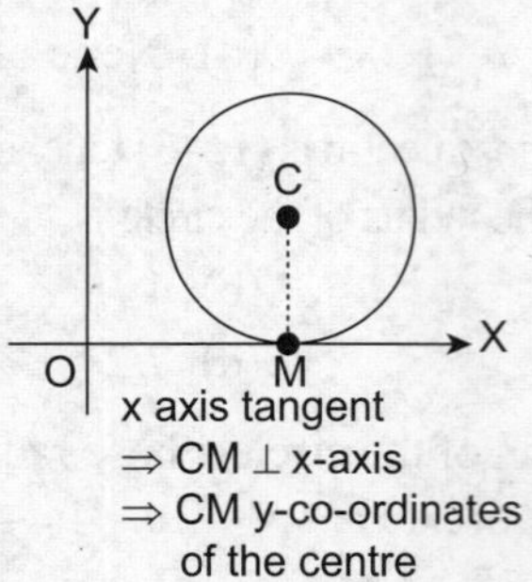

FIGURE 3.19

(c) Radius of circle is 'a' and touches x-axis at $(x_1, 0)$:
Centre of circle will be $(x_1, \pm a)$ and its equation will be

$$(x - x_1)^2 + (y \mp a)^2 = a^2$$

or, $x^2 + y^2 - 2xx_1 \mp 2ay + x_1^2 = 0$

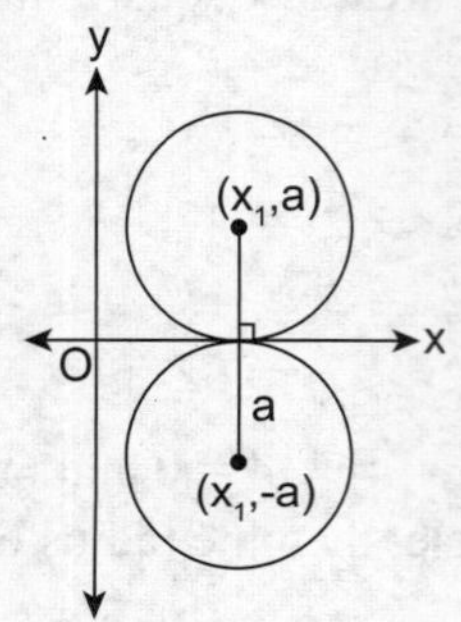

FIGURE 3.20

(d) Radius of circle is 'a' and touches y-axis at $(0, y_1)$:
Centre of circle will be $(\pm a, y_1)$ and its equation will be

$$(x \mp a)^2 + (y - y_1)^2 = a^2$$

or, $x^2 + y^2 \mp 2ax - 2yy_1 + y_1^2 = 0$

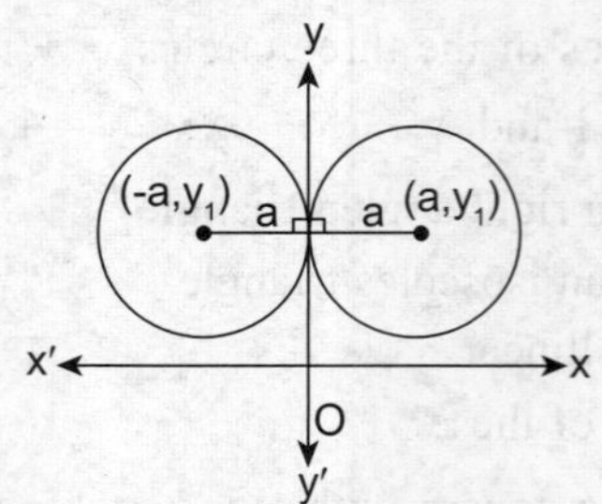

FIGURE 3.21

(e) Equation of circle touching both axes:
In this case, if 'a' is radius of circle then centre is $(\pm a, \pm a)$ and equation of circle will be $(x \pm a)^2 + (y \pm a)^2 = a^2$
or $x^2 + y^2 \pm 2ax \pm 2ay + a^2 = 0$

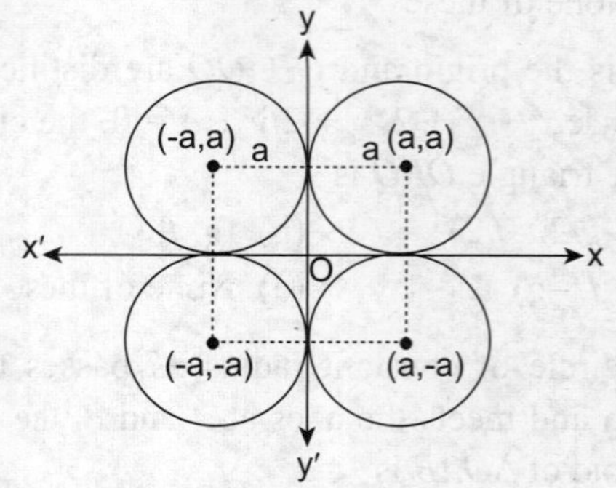

FIGURE 3.22

(f) Equation of circle when cicle pases through the origin and centre lies on x-axis:
In this case, if a is radius of circle, then centre is $(a, 0)$ and equation is $x^2 + y^2 - 2ax = 0$

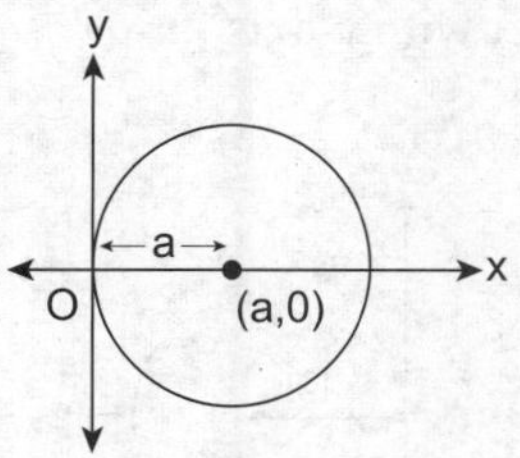

FIGURE 3.23

(g) **Equation of circle when circle passes through the origin and centre lies on y-axis:**

In this case, if a is radius of circle then centre is $(0, a)$ and equation is $x^2 + y^2 - 2ay = 0$

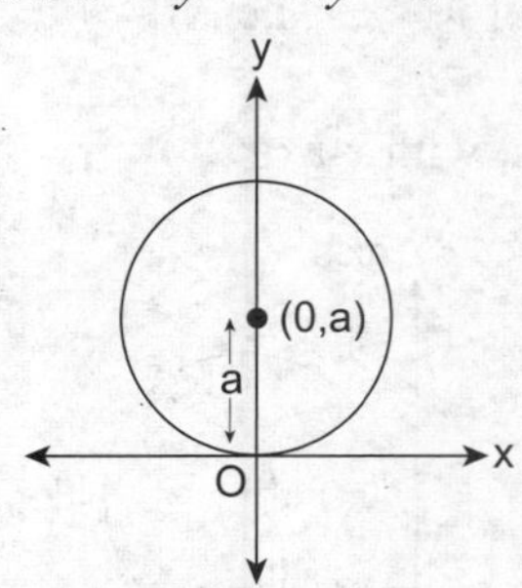

FIGURE 3.24

(h) **Equation of circle through origin and having x-intercept and y-intercept as 'a' and 'b' respectively:**

In this case, $(a/2, b/2)$ is centre and radius is $\sqrt{\dfrac{a^2+b^2}{4}}$.

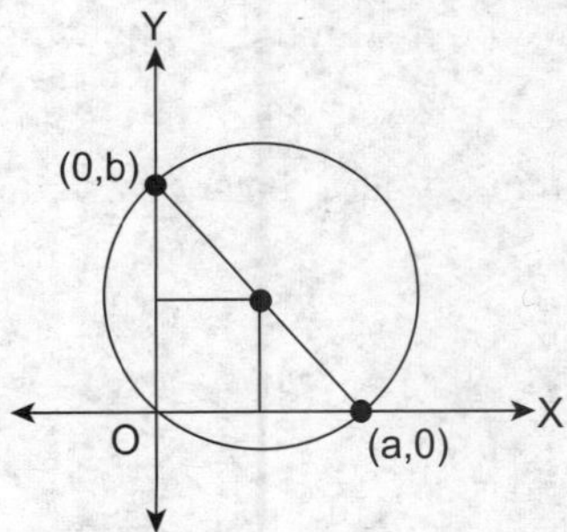

FIGURE 3.25

∴ The equation will be $x(x - a) + y(y - b) = 0$

or, $x^2 + y^2 - ax - by = 0$

■ LOCATION OF CIRCLE AND THEIR CENTRE

Case I:

Sign of co-efficient of *x*	Sign of co-efficient of *y*	Position of centre	
+	+	3 quadrant	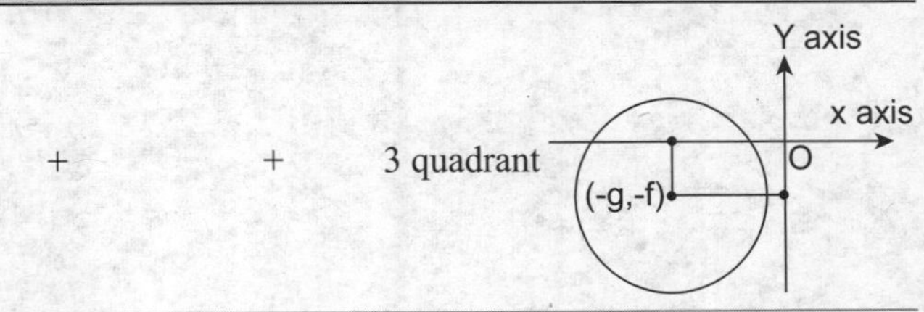
+	–	2 quadrant	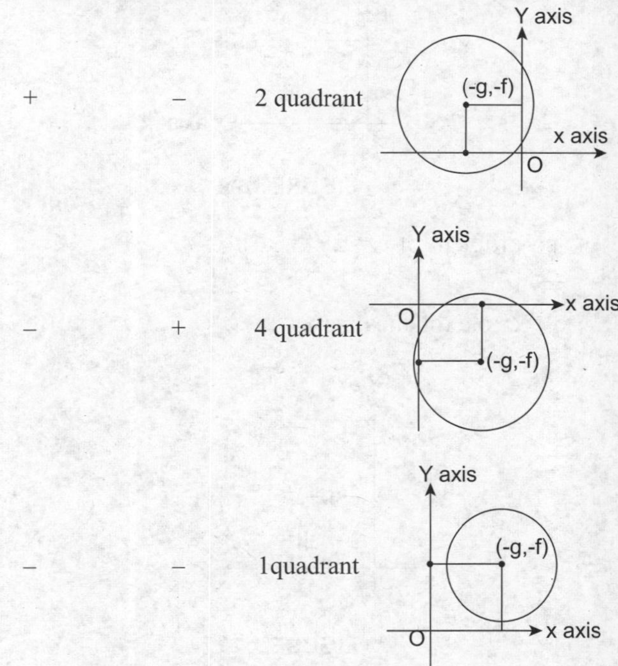
–	+	4 quadrant	
–	–	1quadrant	

Case II: For circle $S = x^2 + y^2 + 2gx + 2fy + c = 0$

- |co-efficient of x|< $2r$, i.e., $|g| < r \Rightarrow f^2 > c$ circle cuts y-axis

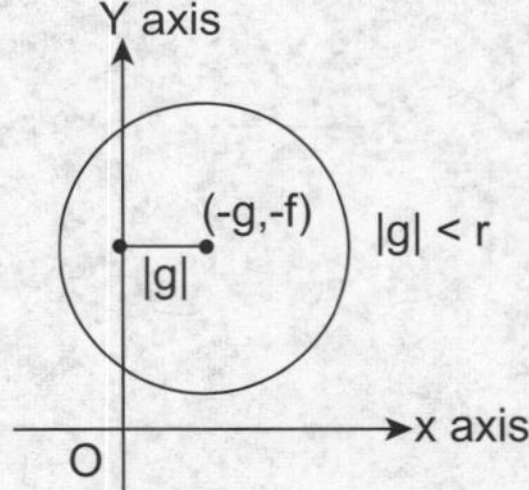

FIGURE 3.26

- |co-efficient of y|< $2r$ i.e., $|f| < r \Rightarrow g^2 > c$ circle cuts x-axis

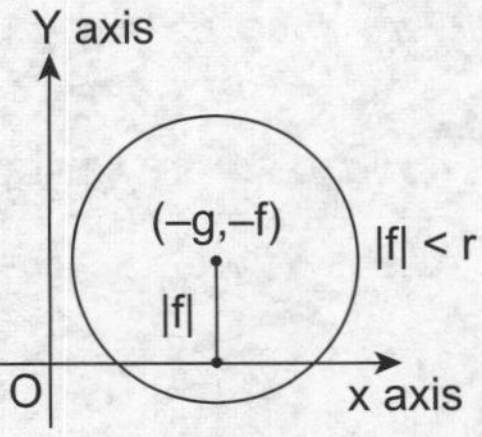

FIGURE 3.27

- |co-efficient of x| = $2r$, i.e., $|g| = r \Rightarrow f^2 = c$ circle touches y-axis

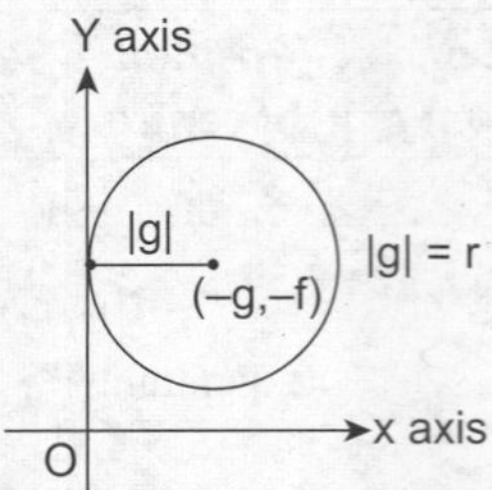

FIGURE 3.28

- |co-efficient of $y = 2r$,| i.e., $|f| = r$
  $\Rightarrow g^2 = c$
  circle touches x-axis

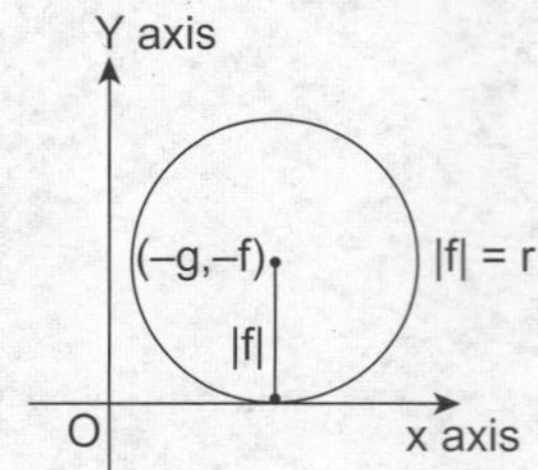

FIGURE 3.29

- |co-efficient of x| > $2r$, i.e., $|g| > r$
  $\Rightarrow f^2 < c$
  no contact with y-axis

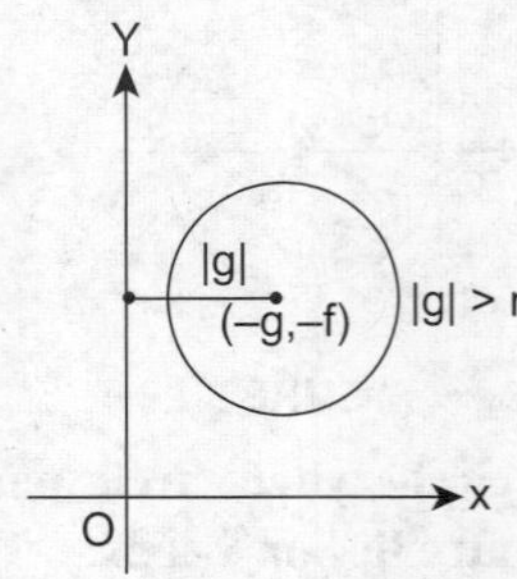

FIGURE 3.30

- |co-efficient of y| > $2r$, i.e., $|f| > r$
  $\Rightarrow g^2 < c$
  no contact with x-axis

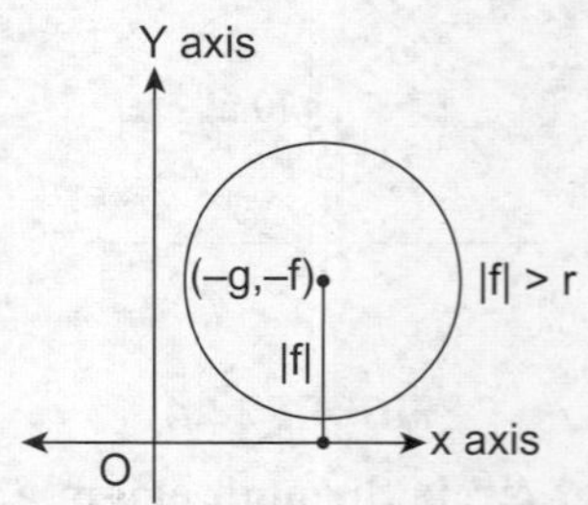

FIGURE 3.31

ILLUSTRATION 13: Find the equation of the circle which passes through two points on the x-axis which are at distances 4 unit from the origin and whose radius is 5.

SOLUTION: There are two circles which pass through two points A and A' at a distance of 4 units from the origin and whose radius is 5.

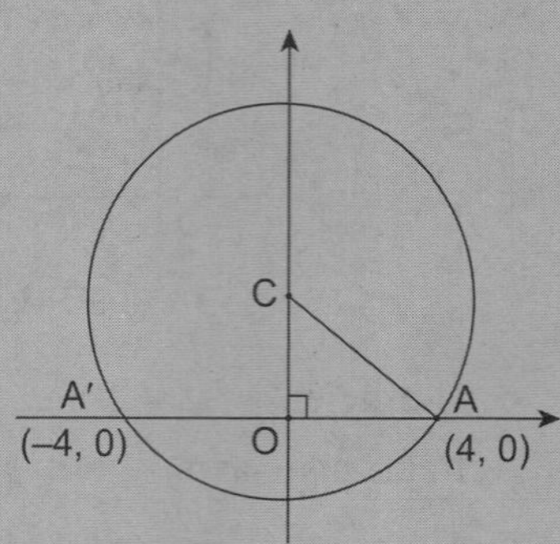

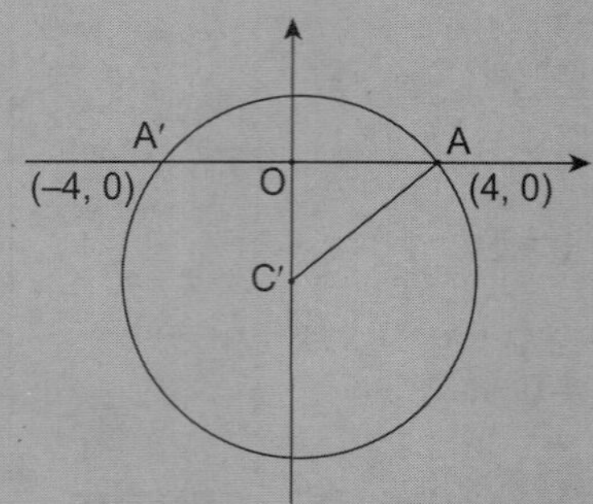

FIGURE 3.32

The centre of these circles lie on y-axis.

In ΔOAC, $AC^2 = OA^2 + OC^2$

$\Rightarrow \; 5^2 = 4^2 + OC^2 \quad \Rightarrow \quad OC = 3$

So the co-ordinates of the centres of the required circles are $C(0, 3)$ and C' $(0, -3)$.

Hence the equations of the required circles are $(x - 0)^2 + (y \mp 3)^2 = 5^2$

$\Rightarrow x^2 + y^2 \mp 6y - 16 = 0$

ILLUSTRATION 14: Find the equation of the circle which passes through the origin and cuts off intercept 3 and 4 from the positive parts of the axes respectively.

SOLUTION: Given that $OA = 3$ and $OB = 4$

$\therefore \quad OL = 3/2 \qquad \text{and} \qquad CL = 2$

In ΔOLC, $OC^2 = OL^2 + LC^2$

$\Rightarrow \; OC^2 = (3/2)^2 + 2^2$

$\Rightarrow \; OC = 5/2$

Thus the required circle has its centre at (3/2, 2) and radius 5/2.

Hence the required equation is $(x - 3/2)^2 + (y - 2)^2 = (5/2)^2$

Aliter: Using diametric equation of circle $x(x - 3) + y(y - 4) = 0$

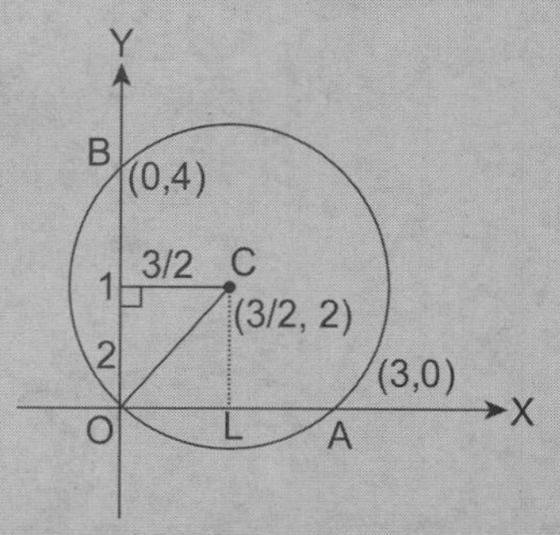

FIGURE 3.33

ILLUSTRATION 15: A circle of radius 2 lies in the first quadrant and touches both the axes of co-ordinates. Find the equation of the circle with the centre at (6, 5) and touching the above circle externally.

SOLUTION: Given $AC = 2$ units and $A = (2, 2)$, $B = (6, 5)$, then

$AB = \sqrt{(2-6)^2 + (2-5)^2} = \sqrt{16+9} = 5$

since $AC + CB = AB$

$\Rightarrow \; 2 + CB = 5$

$\Rightarrow CB = 3 \Rightarrow r = 3$

Hence, the equation of required circle with centre at (6, 5) and the radius 3 is

$(x-6)^2 + (y-5)^2 = 3^2$ or $x^2 + y^2 - 12x - 10y + 52 = 0$

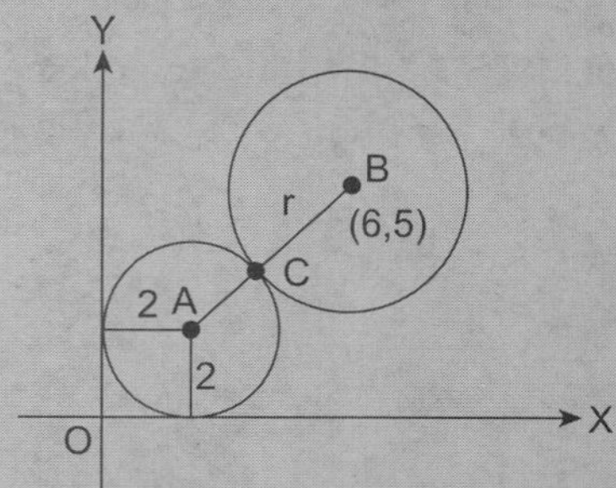

FIGURE 3.34

ILLUSTRATION 16: Find the equation to the circle which touches the axes of co-ordinates and also the line $\frac{x}{a} + \frac{y}{b} = 1$; the centre being in the positive quadrant.

SOLUTION: As the circle touches both the axes, centre may be taken as (h, h); $h > 0$ and its radius will also be h. The given line is $\frac{x}{a} + \frac{y}{b} = 1 \Rightarrow \qquad bx + ay - ab = 0$ ....(i)

As the circle touches (i), therefore length of the perpendicular from (h, h) upon (i) must be equal to h.

Hence $\frac{bh + ah - ab}{\sqrt{a^2 + b^2}} = h \qquad \text{or} \qquad bh + ah - ab = h\sqrt{a^2 + b^2}$

Squaring, we get $b^2h^2 + a^2h^2 + a^2b^2 + 2abh^2 - 2h(ab^2 + ba^2) = h^2(a^2 + b^2)$

$\Rightarrow \; 2h^2 - 2h(a+b) + ab = 0$

$$\Rightarrow \; h = \frac{2(a+b) \pm \sqrt{4(a+b)^2 - 8ab}}{2(2)}$$

$$\Rightarrow \; h = \frac{(a+b) \pm \sqrt{a^2 + b^2}}{2} \qquad \text{....(ii)}$$

Therefore the required equation of the circle will be $(x - h)^2 + (y - h)^2 = h^2$

or $x^2 + y^2 - 2hx - 2yh + h^2 = 0$; where $h = \frac{(a+b) \pm \sqrt{a^2 + b^2}}{2}$

ILLUSTRATION 17: Find the equation to the circle which touches the axis of y at a distance 4 from the origin and cuts off an intercept 6 units from the axis of x.

SOLUTION: Any circle is $x^2 + y^2 + 2gx + 2fy + c = 0$

This meets the axis of y in points given by $y^2 + 2fy + c = 0$

The roots of this equation must be and each equal to 4 or –4, so that it must be equivalent to $(y-4)^2 = 0$ or $(y+4)^2 = 0$.

Hence $2f = -8$ and $c = 16$ or $2f = 8$ and $c = 16$

The equation of the circle is then $x^2 + y^2 + 2gx \pm 8y + 16 = 0$

This meets the axis of x in points given by $x^2 + 2gx + 16 = 0$

i.e., at point given by $x = -g + \sqrt{g^2 - 16}$ and $x = -g - \sqrt{g^2 - 16}$

Hence $6 = 2\sqrt{g^2 - 16}$

Therefore $g = \pm 5$ and the required equation is $x^2 + y^2 \pm 10x \pm 8y + 16 = 0$

ILLUSTRATION 18: A circle of radius 5 units touches both the co-ordinate axes in the first quadrant. If the circle makes one complete roll on along the positive direction of x-axis. Find its equation in the new position.

SOLUTION: Let C be the centre of the circle in its initial position and D be its centre in the new position.

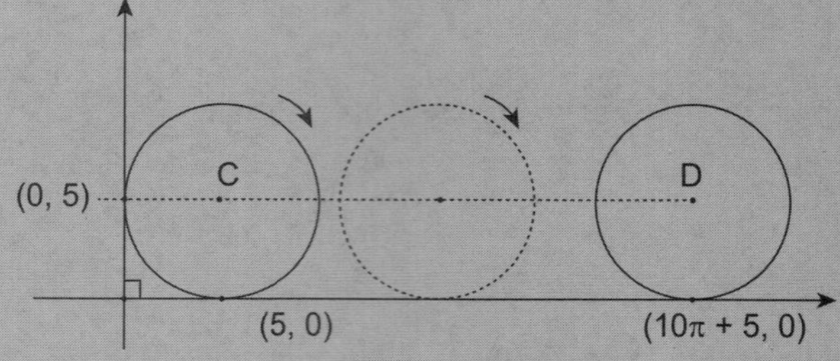

FIGURE 3.35

Since the circle touches the co-ordinate axes in first quadrant and the radius of circle be 5 units.

$\therefore$ Centre of circle is (5, 5)

Total distance by which centre of circle has moved = circumference of the circle

$= 2\pi r = 2\pi(5) = 10\pi$

Now centre of circle in new position is $(5 + 10\pi, 5)$ and radius is 5 units, therefore its equation will be $(x-5-10\pi)^2 + (y-5)^2 = 5^2$

or $x^2 + y^2 - 10(1+2\pi)x - 10y + 100\pi^2 + 100\pi + 25 = 0$

■ THE EQUATION OF CIRCLE THROUGH THREE NON-COLLINEAR POINTS

The general equation of circle $x^2 + y^2 + 2gx + 2fy + c = 0$ contains three independent parameters g, f and c. Therefore, equation of the circle through three non-collinear points $A(x_1, y_1)$, $B(x_2, y_2)$ and $C(x_3, y_3)$ can be obtained as follows:

Let the equation of circle be

$$x^2 + y^2 + 2gx + 2fy + c = 0 \quad \text{...(i)}$$

If three points (x_1, y_1), (x_2, y_2), (x_3, y_3) lie on the circle (i), their co-ordinates must satisfy its equation. Hence, we get

$$x_1^2 + y_1^2 + 2gx_1 + 2fy_1 + c = 0 \quad \text{...(ii)}$$

$$x_2^2 + y_2^2 + 2gx_2 + 2fy_2 + c = 0 \quad \text{...(iii)}$$

$$x_3^2 + y_3^2 + 2gx_3 + 2fy_3 + c = 0 \quad \text{...(iv)}$$

Eliminating g, f and c from the above four equations, we get the eliminant equation as

$$\begin{vmatrix} x^2+y^2 & x & y & 1 \\ x_1^2+y_1^2 & x_1 & y_1 & 1 \\ x_2^2+y_2^2 & x_2 & y_2 & 1 \\ x_3^2+y_3^2 & x_3 & y_3 & 1 \end{vmatrix} = 0$$

which is the equation of the circle through three non-collinear points $A(x_1, y_1)$, $B(x_2, y_2)$ and $C(x_3, y_3)$.

Cyclic quadrilateral: If all the four vertices $A(x_1, y_1)$, $B(x_2, y_2)$, $C(x_3, y_3)$ and $D(x_4, y_4)$ of a quadrilateral lie on a circle, then the quadrilateral is called a cyclic quadrilateral and the four vertices are said to be concyclic. The required condition for it can be given as

$$\begin{vmatrix} x_1^2+y_1^2 & x_1 & y_1 & 1 \\ x_2^2+y_2^2 & x_2 & y_2 & 1 \\ x_3^2+y_3^2 & x_3 & y_3 & 1 \\ x_4^2+y_4^2 & x_4 & y_4 & 1 \end{vmatrix} = 0$$

ILLUSTRATION 19: Find the equation of the circle passing through the three non-collinear points (1, 1), (2, –1) and (3, 2).

SOLUTION: Let the equation of circle be $x^2+y^2+2gx+2fy+c=0$ ...(i)

Since the three given points (1, 1), (2, –1) and (3, 2) lie on circle (i), we get

$2g+2f+c+2=0$ ....(ii)

$4g-2f+c+5=0$ ....(iii)

and $6g+4f+c+13=0$ ....(iv)

Subtracting (ii) from (iii) and subtracting (iii) from (iv), we get $2g-4f+3=0$ ...(v)

and $2g+6f+8=0$ ...(vi)

Solving (v) and (vi), we get $f=-1/2$ and $g=-5/2$

Now from (ii), $-5-1+c+2=0 \Rightarrow c=4$

Hence from (i), equation of circle is $x^2+y^2-5x-y+4=0$

Aliter:

$$\begin{vmatrix} x^2+y^2 & x & y & 1 \\ 1^2+1^2 & 1 & 1 & 1 \\ 2^2+1^2 & 2 & -1 & 1 \\ 3^2+2^2 & 3 & 2 & 1 \end{vmatrix} = 0$$

■ PARAMETRIC EQUATION OF A CIRCLE

For standard circle $x^2+y^2=a^2$, the parametric equation is given by

$x = a\cos\theta$,

$y = a\sin\theta$ for $0 \le \theta < 2\pi$

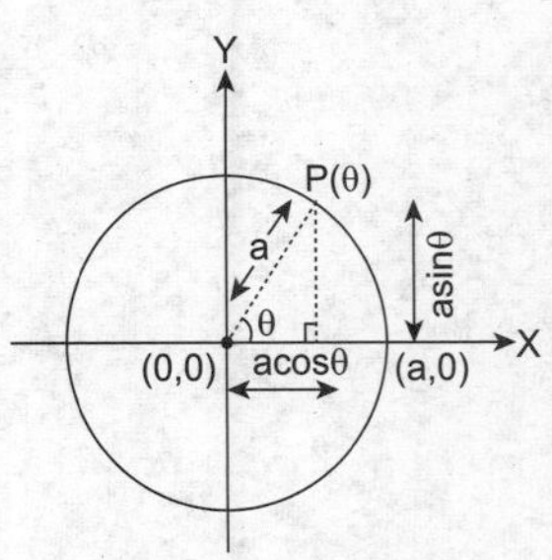

FIGURE 3.36

Therefore any point on the circumference of the circle can be represented as $P(\theta) = (a\cos\theta, a\sin\theta)$.

Similarly for a circle given by equation $x^2+y^2+2gx+2fy+c=0$; the parametric equation is given by $x=-g+r\cos\theta$, $y=-f+r\sin\theta$; where $r=\sqrt{g^2+f^2-c}$ is the radius of the circle.

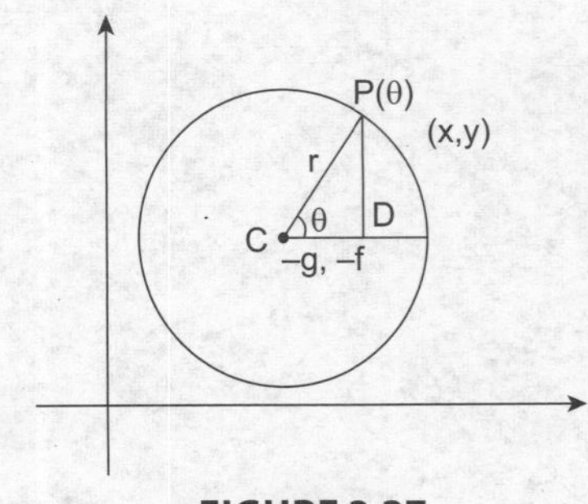

FIGURE 3.37

Proof: Let co-ordinates of $P(\theta)$ be (x, y), then in triangle CPD (with $CP = r$), $\cos\theta = \frac{CD}{r} \Rightarrow CD = r\cos\theta$ and similarly, $PD = r\sin\theta$

$\Rightarrow$ x co-ordinate of P = (x co-ordinate of C) + CD $= -g + r\cos\theta$

$\Rightarrow$ And y co-ordinate of P = (y co-ordinate of C) + $PD = -f + r\sin\theta$

$\therefore$ Co-ordinates of $P(\theta) = (-g + r\cos\theta, -f + r\sin\theta)$

NOTE

Parametric equation of a circle is very useful when we need to write the equation for a part of the circle only.

ILLUSTRATION 20: Find the parametric equations of the circle $x^2 + y^2 - 2x - 4y - 4 = 0$.

SOLUTION: Given that $x^2 + y^2 - 2x - 4y - 4 = 0$ is an equation of circle.

$\Rightarrow (x^2 - 2x + 1) + (y^2 - 4y + 4) = 9$

$\Rightarrow (x - 1)^2 + (y - 2)^2 = 9$

Hence the parametric equations of the given circles are

$x = 1 + 3\cos\theta, y = 2 + 3\sin\theta$; where θ is a variable lying between 0 and 2π.

ILLUSTRATION 21: Solve the following parametric equations:

(a) $x = 2 + 3\cos\alpha; y = 5 + 3\sin\alpha$

(b) $x = \frac{2rt}{(1+t^2)}; y = \frac{r(1-t^2)}{(1+t^2)}; t \in (-1, 1); r > 0$

SOLUTION: (a) $x = 2 + 3\cos\alpha; y = 5 + 3\sin\alpha$; comparing with $x = -g + r\cos\alpha; y = -f + r\sin\alpha$

$\Rightarrow g = -2, \quad f = -5$ and $r = 3$

$\Rightarrow 3 = \sqrt{g^2 + f^2 - c}$ $\quad (\because r = \sqrt{g^2 + f^2 - c})$

$\Rightarrow 3 = \sqrt{4 + 25 - c}$

$\Rightarrow c = 20$

Aliter: $\frac{x-2}{3} = \cos\alpha$

and $\frac{y-5}{3} = \sin\alpha$

Now using $\cos^2\alpha + \sin^2\alpha = 1$; we get $(x-2)^2 + (y-5)^2 = 9$

(b) $x = \frac{2rt}{(1+t^2)}; y = \frac{r(1-t^2)}{(1+t^2)}$; consider $t = \tan\theta$

$\because t \in (-1, 1) \Rightarrow \tan\theta \in (-1, 1) \Rightarrow \theta \in (-\pi/4, \pi/4)$

$\therefore \quad x = r\sin 2\theta; y = r\cos 2\theta$

Now, $2\theta \in (-\pi/2, \pi/2) \Rightarrow x \in (-r, r)$ and $y \in (0, r)$

$\therefore x^2 + y^2 = r^2 \Rightarrow y = \sqrt{r^2 - x^2}$, i.e., upper semi-circle with radis 'r' and centre at origin.

ILLUSTRATION 22: Find the parametric form of the equation of the circle $x^2 + y^2 + px + py = 0$

SOLUTION: Equation of the circle can be re-written in the form $\left(x+\dfrac{p}{2}\right)^2+\left(y+\dfrac{p}{2}\right)^2=\dfrac{p^2}{2}$

Therefore, the parametric form of the equation of the given circle is

$$x=-\frac{p}{2}+\frac{p}{\sqrt{2}}\cos\theta=\frac{p}{2}(-1+\sqrt{2}\cos\theta)\text{; and } y=-\frac{p}{2}+\frac{p}{\sqrt{2}}\sin\theta=\frac{p}{2}(-1+\sqrt{2}\sin\theta);$$

where $0 \le \theta < 2\pi$.

ILLUSTRATION 23: Find the equation of circle

(a) when the circle passes through the origin (0,0) and has intercepts 2α and 2β on the x-axis and y-axis respectively.

(b) when the circle touches x-axis and cuts off intercept of length $2l$ on y-axis.

(c) when the circle touches y-axis and cut off intercept of length $2k$ on x-axis.

(d) when the circle cut off intercepts of lengths $2l$ and $2k$ on x-axis and y-axis respectively and the circle does not pass through origin.

SOLUTION: (a) Centre of the circle is (α, β) and radius $OC=\sqrt{\alpha^2+\beta^2}$, then equation of circle is

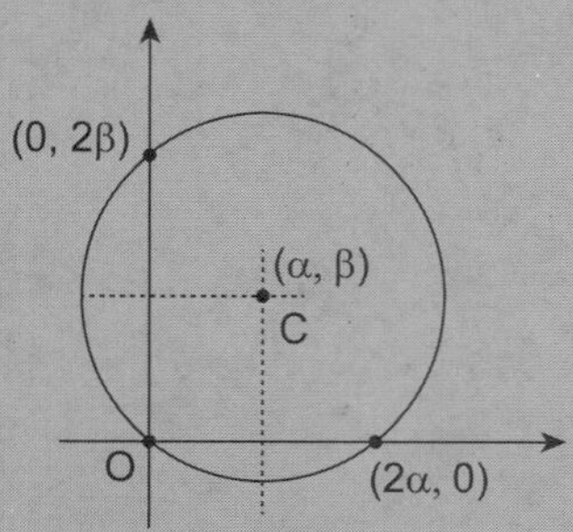

FIGURE 3.38

$$(x-\alpha)^2+(y-\beta)^2=\alpha^2+\beta^2 \qquad \text{or} \qquad x^2+y^2-2\alpha x-2\beta y=0$$

(b) Let the centre be $C(\alpha, \beta)$ and $MN = 2l$

$\therefore$ radius $= \beta$ and $CM = CN = \beta$

In ΔCMP, $CM^2 = CP^2 + PM^2$

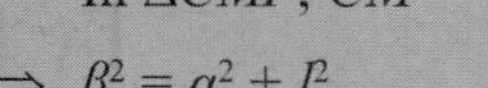

$\Rightarrow \beta^2 = \alpha^2 + l^2$

$\Rightarrow \alpha = \sqrt{\beta^2 - l^2}$ (for I quadrant); where β is a parameter.

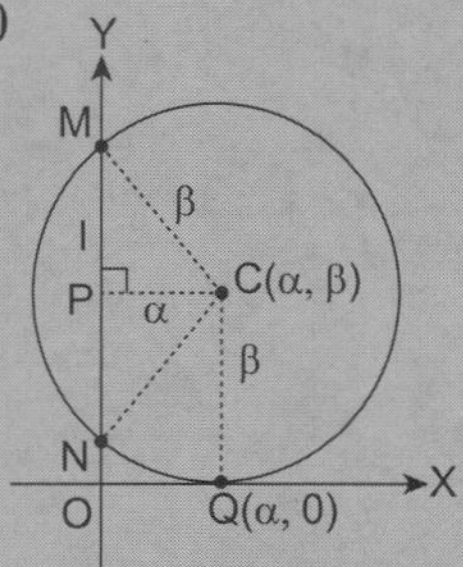

FIGURE 3.39

$\therefore$ Equation of circle is $[x-\sqrt{\beta^2-l^2}\,]^2+(y-\beta)^2=\beta^2$; where β is a parameter.

(c) Let the centre be $C(\alpha, \beta)$ and $MN = 2k$

$\therefore$ radius $= \alpha$ and $CM = CN = \alpha$

In ΔCMP, $CM^2 = CP^2 + PM^2 \Rightarrow \alpha^2 = \beta^2 + k^2$

$\beta = \sqrt{\alpha^2 - k^2}$ (for I quadrant)

$\therefore$ Equation of circle is $(x-\alpha)^2+(y-\sqrt{\alpha^2-k^2})^2=\alpha^2$, where α is a parameter.

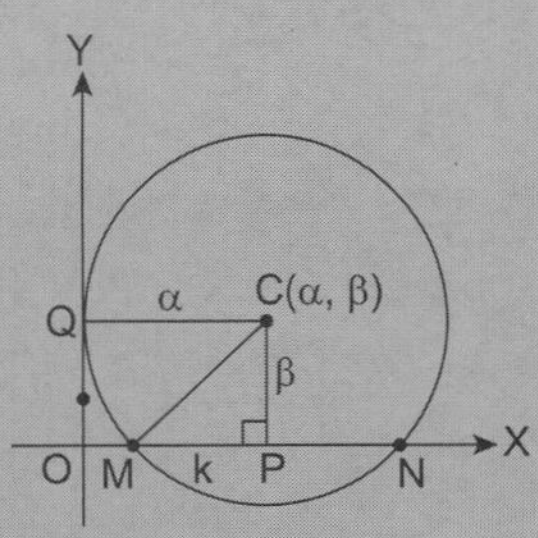

FIGURE 3.40

(d) Let the centre be $C(\alpha, \beta)$

$\therefore$ radius = λ $\qquad CP = CQ = \lambda$

$(CP)^2 = (CQ)^2 = \lambda^2$

$\Rightarrow \alpha^2 + l^2 = \beta^2 + k^2 = \lambda^2$

$\Rightarrow \alpha = \sqrt{\lambda^2 - l^2}$ and $\beta = \sqrt{\lambda^2 - k^2}$

$\therefore$ Equation of circle is: $(x - \sqrt{\lambda^2 - l^2})^2 + (y - \sqrt{\lambda^2 - k^2})^2 = \lambda^2$

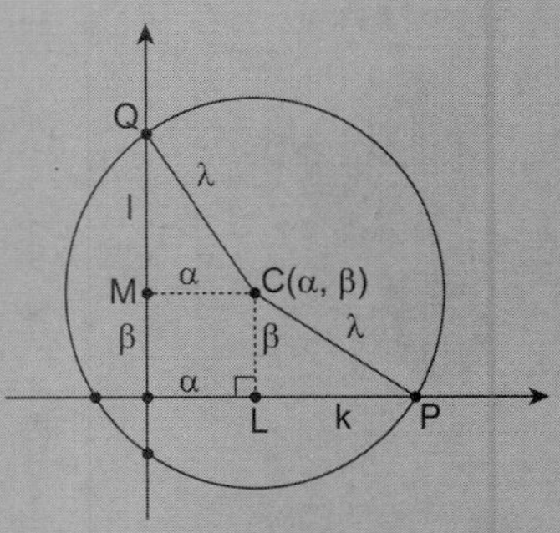

FIGURE 3.41

ILLUSTRATION 24: Find the parametric equation of the part of circle (lying in the quadrant IV) for the circle $x^2 + y^2 - 2x + 4y - 4 = 0$.

SOLUTION: Parametric equation of the circle $x^2 + y^2 - 2x + 4y - 4 = 0$ is given by $x = 1 + 3\cos\theta$ and $y = -2 + 3\sin\theta$ where θ is measured keeping CD as reference (CD is parallel to (+)ve x-axis)

Now maximum value of θ will be determined by the parametric co-ordinate of A

Cartesian co-ordinates of A are $\left(1+\sqrt{5}, 0\right)$

Let θ_1 be the parameter for A

$\Rightarrow 1 + 3\cos\theta_1 = 1 + \sqrt{5}$ and $-2 + 3\sin\theta_1 = 0$

$\Rightarrow \theta_1 = \sin^{-1}\frac{2}{3}$ *or* $\cos^{-1}\frac{\sqrt{5}}{3}$ ...(i)

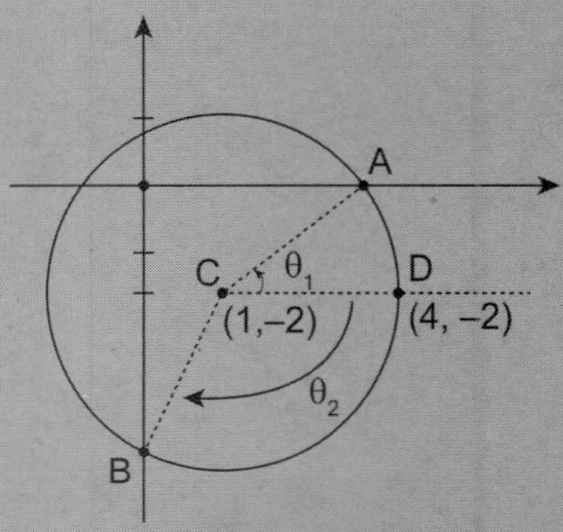

FIGURE 3.42

Similarly, minimum value of θ will be determined by the parametric co-ordinates of B

Cartesian co-ordinates of B are $\left(0, -2-2\sqrt{2}\right)$

And, if θ_2 be the parameter of B,

$\Rightarrow 1 + 3\cos\theta_2 = 0$ and $-2 + 3\sin\theta_2 = -2 - 2\sqrt{2}$

$\Rightarrow \theta_2 = \cos^{-1}\left(\frac{-1}{3}\right)$ or $\theta_2 = \sin^{-1}\left(\frac{-2\sqrt{2}}{3}\right)$ ...(ii)

Now from (i) and (ii), we can say that the complete portion of circle lying in the fourth quadrant is given by $x = (1 + 3\cos\theta, -2 + 3\sin\theta)$; $\theta \in \left(\sin^{-1}\left(\frac{-2\sqrt{2}}{3}\right), \sin^{-1}\left(\frac{2}{3}\right)\right)$

ILLUSTRATION 25: Given two circles $S_1 : x = -3 + 2\cos\theta, y = -1 + 2\sin\theta$ and $S_2 : y = \frac{6t}{1+t^2}$ and $x = \frac{3\left(1-t^2\right)}{\left(1+t^2\right)}$ for $\theta \in$ A and $t \in B$, then find the number of points of intersection of S_1 and S_2 under the following conditions:

(i) $A = \mathbb{R}, B = (-1, 1)$

(ii) $A = \left(0, \frac{3\pi}{4}\right); B = (-1, \infty)$

(iii) $A = \left(\frac{-\pi}{2}, \frac{\pi}{2}\right); B = (-\infty, -1) \cup (1, \infty)$

SOLUTION: For S_1: $(x+3)^2 + (y+1)^2 = 4$ i.e., $x^2 + y^2 + 6x + 2y + 6 = 0$

Centre : (–3, –1) and Radius = 2

For S_2: $\frac{y}{3} = \frac{2t}{1+t^2}$ and $\frac{x}{3} = \frac{1-t^2}{1+t^2}$

Substituting $t = \tan\alpha$; we get

$$\frac{y}{3} = \frac{2\tan\alpha}{1+\tan^2\alpha} \text{ and } \frac{x}{3} = \frac{1-\tan^2\alpha}{1+\tan^2\alpha}$$

$\Rightarrow$ $\frac{y}{3} = \sin 2\alpha$ and $\frac{x}{3} = \cos 2\alpha$

$\Rightarrow$ $y^2 + x^2 = 9$ is the equation of circle S_2

$\Rightarrow$ Centre: (0, 0); Radius = 3

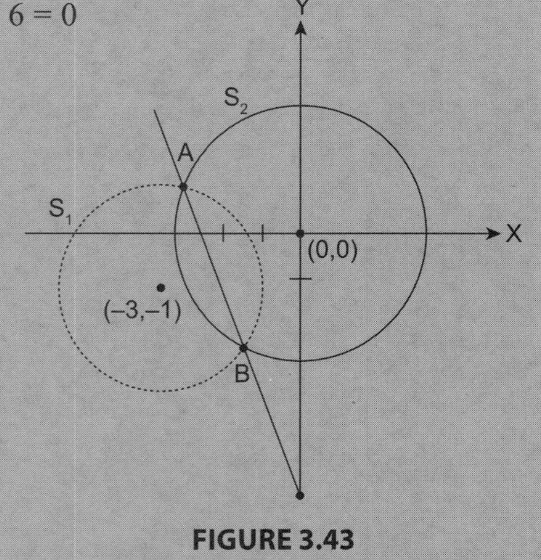

FIGURE 3.43

(i) For $t \in (-1,1)$; $\alpha \in (-\pi/4, \pi/4)$.

$\Rightarrow$ $2\alpha \in (-\pi/2, \pi/2)$

$\therefore$ S_2 will only have the part on right side of y-axis

Hence no point of intersection.

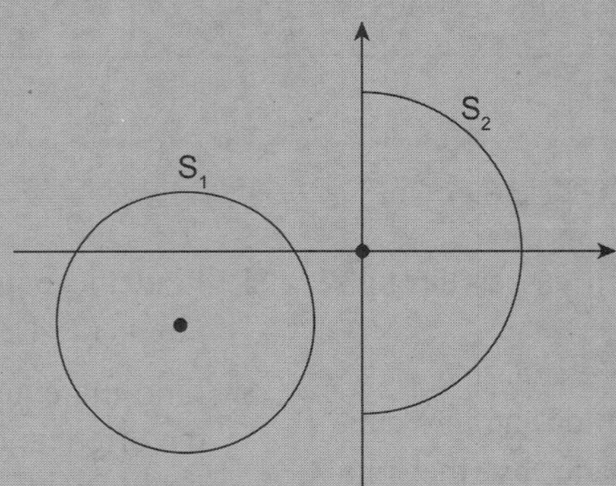

FIGURE 3.44

(ii) $t \in (-1,\infty)$

$\Rightarrow$ $\alpha \in (-\pi/4, \pi/2)$

$\Rightarrow$ $2\alpha \in (-\pi/2, \pi)$

Clearly, one point of intersection.

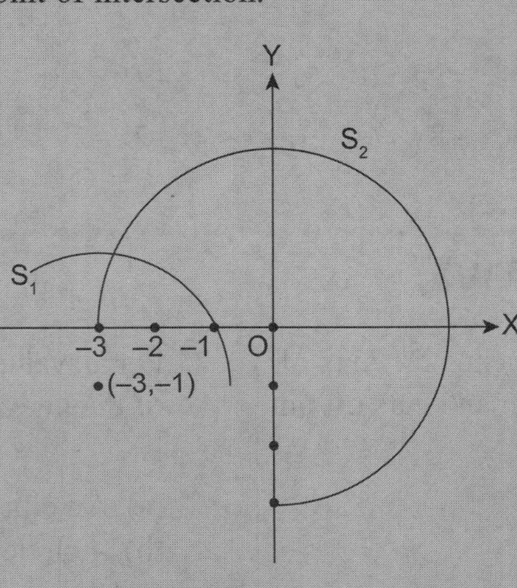

FIGURE 3.45

(iii) $t \in (-\infty,-1) \cup (1,\infty)$

$\Rightarrow \alpha \in (-\pi/2, -\pi/4) \cup (\pi/4, \pi/2)$,

$\Rightarrow 2a \in (-\pi, -\pi/2) \cup (\pi/2, \pi)$

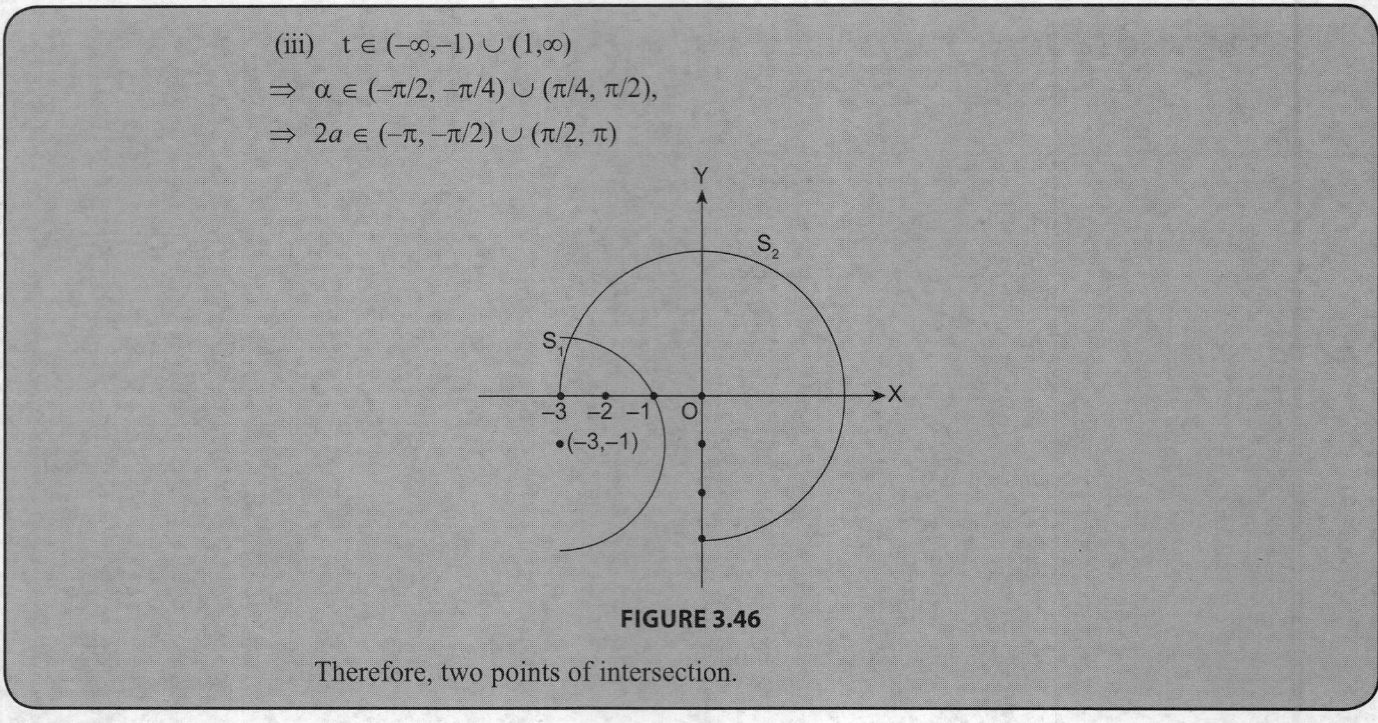

FIGURE 3.46

Therefore, two points of intersection.

TEXTUAL EXERCISE–2 (SUBJECTIVE)

1. Find the equation of the circle which passes through three points (0, 0), (a, 0) and (0, b).
2. Find the equation of the circle drawn on the intercept on diameter between the axes made by the line $3x + 4y = 12$.
3. Find the equation of the circle whose radius is 5 and the centre lies on the positive side of x-axis at a distance 5 from the origin.
4. Find the equation of the circle passing through the points (3, 4), (3, –6) and (–1,2).
5. Find the equation of the circle which passes through the origin and cuts off chords of length 4 and 6 on the positive side of the x-axis and y-axis respectively.
6. Find the equation of the circumcircle of the triangle formed by the lines $y = x$, $y = 2x$ and $y = 3x + 2$.
7. Find the equation of a circle touching lines $x = -1$, $x = 3$ and $y = 3$.

Answer Keys

1. $x^2 + y^2 - ax - by = 0$ **2.** $x^2 + y^2 - 4x - 3y = 0$ **3.** $x^2 + y^2 - 10x = 0$ **4.** $x^2 + y^2 - 6x + 2y - 15 = 0$

5. $x^2 + y^2 - 4x - 6y = 0$ **6.** $x^2 + y^2 - 6x + 8y = 0$ **7.** $x^2 + y^2 - 2x - 2y - 2 = 0$ or $x^2 + y^2 - 2x - 10y + 22 = 0$

TEXTUAL EXERCISE–2 (OBJECTIVE)

1. The equation to the circle which touches the axis of y at a distance 4 units from the origin and cuts off an intercept 6 units from the axis of x.
   (a) $x^2 + y^2 + 10x - 8y - 16 = 0$
   (b) $x^2 + y^2 - 10x - 8y - 16 = 0$
   (c) $x^2 + y^2 \pm 10x - 8y + 16 = 0$
   (d) None of these
2. For all values of θ, the locus of the point of intersection of the lines $x\cos\theta + y\sin\theta = a$ and $x\sin\theta - y\cos\theta = b$ is
   (a) An ellipse
   (b) *A* circle
   (c) *A* parabola
   (d) *A* hyperbola

3. If the circle $x^2 + y^2 + 2gx + 2fy + c = 0$ touches x-axis, then
(a) $g = f$ (b) $g^2 = c$
(c) $f^2 = c$ (d) $g^2 + f^2 = c$

4. The equation of the circle which touches both the axes and whose radius is a, is
(a) $x^2 + y^2 - 2ax - 2ay + a^2 = 0$
(b) $x^2 + y^2 + ax + ay - a^2 = 0$
(c) $x^2 + y^2 + 2ax + 2ay - a^2 = 0$
(d) $x^2 + y^2 - ax - ay + a^2 = 0$

5. The circle $x^2 + y^2 + 4x - 4y + 4 = 0$ touches
(a) x-axis
(b) y-axis
(c) x-axis and y-axis
(d) None of these

6. A circle touches the y-axis at the point (0, 4) and cuts the x-axis in a chord of length 6 units. The radius of the circle is
(a) 3 (b) 4
(c) 5 (d) 6

7. The number of circle having radius 5 and passing through the points (–2, 0) and (4,0) is
(a) One (b) Two
(c) Four (d) Infinite

8. The locus of the centre of the circle which cuts off intercepts of length $2a$ and $2b$ from x-axis and y-axis respectively, is
(a) $x + y = a + b$ (b) $x^2 + y^2 = a^2 + b^2$
(c) $x^2 - y^2 = a^2 - b^2$ (d) $x^2 + y^2 = a^2 - b^2$

9. The number of circle touching the line $y - x = 0$ and the y-axis is
(a) Zero (b) One
(c) Two (d) Infinite

10. If the vertices of a triangle be (2, –2),(–1, –1) and (5, 2), then the equation of its circumcircle is
(a) $x^2 + y^2 + 3x + 3y + 8 = 0$
(b) $x^2 + y^2 - 3x - 3y - 8 = 0$
(c) $x^2 + y^2 - 3x + 3y + 8 = 0$
(d) None of these

11. The equation of a circle which touches both axes and the line $3x - 4y + 8 = 0$ and whose centre lies in the third quadrant is
(a) $x^2 + y^2 - 4x + 4y - 4 = 0$
(b) $x^2 + y^2 - 4x + 4y + 4 = 0$
(c) $x^2 + y^2 + 4x + 4y + 4 = 0$
(d) $x^2 + y^2 - 4x - 4y - 4 = 0$

12. Circle $x^2 + y^2 + 6y = 0$ touches
(a) y-axis at the origin
(b) x-axis at the origin
(c) x-axis at the point (3, 0)
(d) the line $y + 3 = 0$

13. If the radius of the circle $x^2 + y^2 + 2gx + 2fy + c = 0$ be r, then it will touch both the axes, if
(a) $g = f = r$ (b) $g = f = c = r$
(c) $g = f = \sqrt{c} = r$ (d) $g = f$ and $c^2 = r$

14. The equation of the circle with centre on the x-axis, radius 4 and passing through the origin, is
(a) $x^2 + y^2 - 4x = 0$
(b) $x^2 + y^2 - 8y = 0$
(c) $x^2 + y^2 - 8x = 0$
(d) $x^2 + y^2 + 8y = 0$

15. For the circle $x^2 + y^2 + 3x + 3y = 0$, which of the following relations is true?
(a) Centre lies on x-axis
(b) Centre lies on y-axis
(c) Centre is at origin
(d) Circle passes through origin

16. The number of circles touching the lines $x = 0$, $y = a$ and $y = b$ is
(a) One (b) Two
(c) Four (d) Infinite

17. If the equation $x^2 + y^2 + 2gx + 2fy + c = 0$ represents a circle with x-axis as a diameter and radius a, then which of the following is possible
(a) $f = 2a, g = 0, c = 3a^2$
(b) $f = 0, g = a, c = 3a^2$
(c) $f = 0, g = -2a, c = 3a^2$
(d) None of these

Answer Keys

1. (c)	**2.** (b)	**3.** (b)	**4.** (a)	**5.** (c)	**6.** (c)	**7.** (b)	**8.** (c)	**9** (d)	**10.** (b)
11. (c)	**12.** (b)	**13.** (c)	**14.** (c)	**15.** (d)	**16.** (b)	**17.** (c)			

POSITION OF A POINT/LINE WITH RESPECT TO A CIRCLE

Position of Point with Respect to Circle

A circle divides *XY* plane in three regions (i) inside region (i.e., containing centre) (ii) outside region (i.e., not containing centre) (iii) points lying on the circle.

A point $P(x_1, y_1)$ lies inside, on or outside of the circle when the distance *CP* of the point *P* from the centre of circle (*C*) is less than, equal to or greater than the radius respectively. Given a point $P(x_1, y_1)$ and a circle $x^2 + y^2 + 2gx + 2fy + c = 0$.

Case I: Point *P* lies inside the circle.

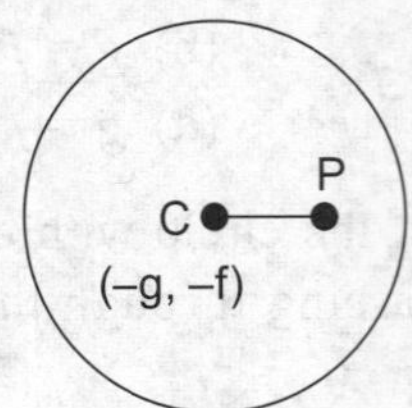

FIGURE 3.47

$CP < r$

$\Rightarrow CP^2 < r^2$

$\Rightarrow (x_1 + g)^2 + (y_1 + f)^2 < g^2 + f^2 - c$

$\Rightarrow S_1 = x_1^2 + y_1^2 + 2gx_1 + 2fy_1 + c < 0$

$\Rightarrow$ No tangent can be drawn from *P* to the given circle $(S_1 < 0)$

(Since the value of S_1 is square of the length of the tangent from $P(x_1, y_1)$ to the circle $S = 0$.)

Case II: Point *P* lies on the circle.

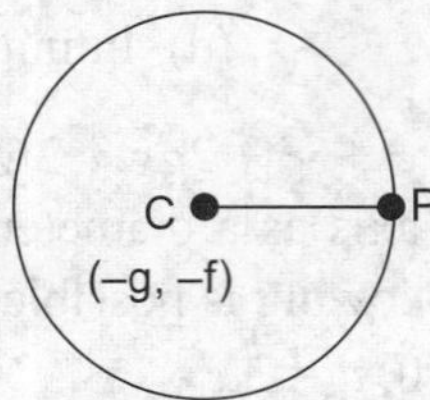

FIGURE 3.48

$CP = r \quad \Rightarrow \quad CP^2 = r^2$

$\Rightarrow (x_1 + g)^2 + (y_1 + f)^2 = g^2 + f^2 - c$

$\Rightarrow S_1 = x_1^2 + y_1^2 + 2gx_1 + 2fy_1 + c = 0$

$\Rightarrow$ Only one tangent can be drawn through P $(S_1 = 0)$

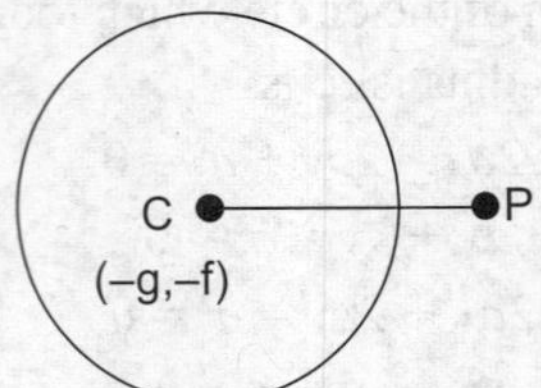

FIGURE 3.49

Case III: Point *P* lies outside the circle.

$CP > r \quad \Rightarrow \quad CP^2 > r^2$

$\Rightarrow (x_1 + g)^2 + (y_1 + f)^2 > g^2 + f^2 - c$

$\Rightarrow S_1 = x_1^2 + y_1^2 + 2gx_1 + 2fy_1 + c > 0$

$\Rightarrow$ Two tangents can be drawn from *P* to the given circle $(S_1 > 0)$

$L_T = PT = \sqrt{PC^2 - r^2}$

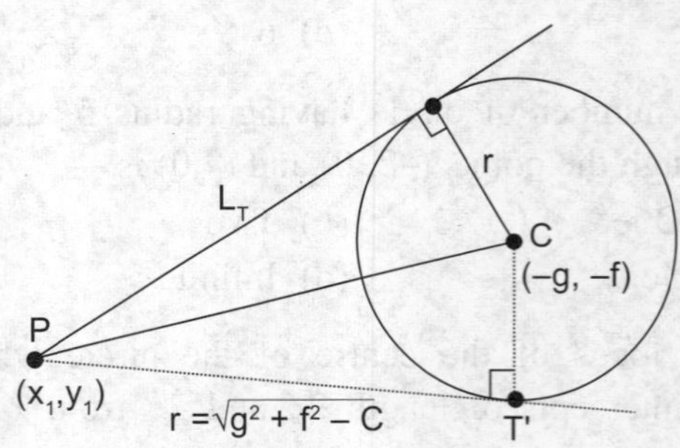

FIGURE 3.50

$= \sqrt{(x_1 + g)^2 + (y_1 + f)^2 - (g^2 + f^2 - c)}$

$= \sqrt{(x_1^2 + y_1^2 + 2gx_1 + 2fy_1 + c} = \sqrt{S_1}$

S_1 is called power of point *P* w.r.t circle $S = 0$

Thus, $\sqrt{S_1}$ = *l*ength of tangent drawn from *P* to circle.

If *P* lies outside; then S_1 is + ve $\Rightarrow$ two tangents drawn.

If *P* lies on circle; then $S_1 = 0 \Rightarrow$ only one tangent.

If *P* lies inside circle; then $S_1 < 0 \Rightarrow$ no (imaginary) tangent.

ILLUSTRATION 26: Discuss the position of the points (1, 2) and (6, 0) with respect to the circle $x^2 + y^2 - 4x + 2y - 11 = 0$.

SOLUTION: Let $S \equiv x^2 + y^2 - 4x + 2y - 11 = 0$ for the point (1, 2)

$S_1 = 1^2 + 2^2 - 4.1 + 2.2 - 11 = -6 \therefore S_1 < 0$

And for the point (6, 0); $S_2 = 6^2 + 0 - 4.6 + 2.0 - 11 = 36 - 24 - 11 = 36 - 35 = 1$

$\therefore\ S_2 > 0$

Hence the point (1, 2) lies inside the circle and the point (6, 0) lies outside the circle.

ILLUSTRATION 27: How are the following points situated with respect to the circle $x^2 + y^2 - 2x - 4y + 3 = 0$ and hence find the number of tangents drawn from these points on the circle. Also, find the length of the tangent if possible.

(i) (0,1) (ii) (3,1) (iii) (1,3)

SOLUTION: (i) P (0,1)

$S_p = 0^2 + 1^2 - 2 \times 0 - 4 \times 1 + 3 = 0$

$\therefore$ Point (0,1) lies on the circle

And hence exactly one tangent is possible from this point to the circle.

and length of tangent = 0

(ii) $P(3,1)$

$S_p = 3^2 + 1^2 - 2 \times 3 - 4 \times 1 + 3 = 3 > 0$

$\therefore$ point (3,1) lies outside the circle and length of each tangent = $\sqrt{3}$.

And hence, two tangents can be drawn from the point to the circle.

(iii) $P(1,3)$

$S_p = 1^2 + 3^2 - 2 \times 1 - 4 \times 3 + 3 = -1 < 0$

$\therefore$ point (1,3) lies inside the circle and hence no tangents are possible.

ILLUSTRATION 28: State how are the points (3,1) and (1,2) situated with respect to the circle $x^2 + y^2 - 2x - 4y + 5 = 0$ and hence find the number of tangents from these points on the circle. Also find the lengths of the tangents.

SOLUTION: (i) P (3,1)

$S_p = 3^2 + 1^2 - 2 \times 3 - 4 \times 1 + 5 = 5 > 0$

$\therefore$ Length of tangent = $\sqrt{5}$

$\therefore$ the point (3,1) lies outside the circle.

But, since the circle is of zero radius 'S' is a point circle.

Hence, only one tangent will be drawn from P to the circle.

(ii) $P(1,2)$

$S_p = 1^2 + 2^2 - 2 \times 1 - 4 \times 2 + 5 = 0$

$\therefore$ Point (1,2) satisfies the equation of the circle.

And since it is a point circle, therefore there are '∞' tangents possible from this point.

MAXIMUM AND MINIMUM DISTANCE OF A POINT FROM THE CIRCLE

Let any Point $P(x_1, y_1)$ and Circle

$S \equiv x^2 + y^2 + 2gx + 2fy + c = 0$

The centre and radius of the circle are

$C(-g,-f)$ and $\sqrt{(g^2 + f^2 - c)}$ respectively

The distance of point $P(x_1, y_1)$ from circle means distance of P from any arbitrary point on circle.

Case I: If P is inside the circle

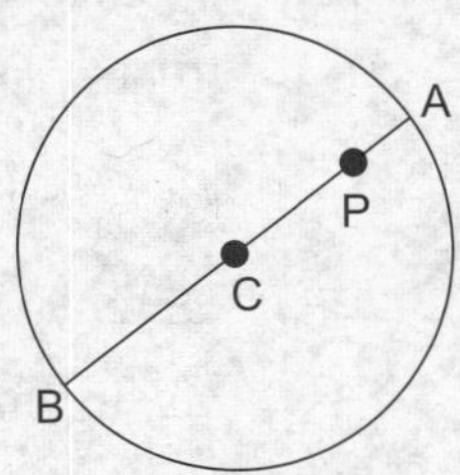

FIGURE 3.51

In this case, $S_1 < 0 \because r = \sqrt{(g^2 + f^2 - c)} = CA = CB$

The minimum distance of P from circle $= PA = CA - CP = |r - CP|$ and the maximum distance of P from circle $= PB = CB + CP = |r + CP|$

Case II: If P is outside the circle.

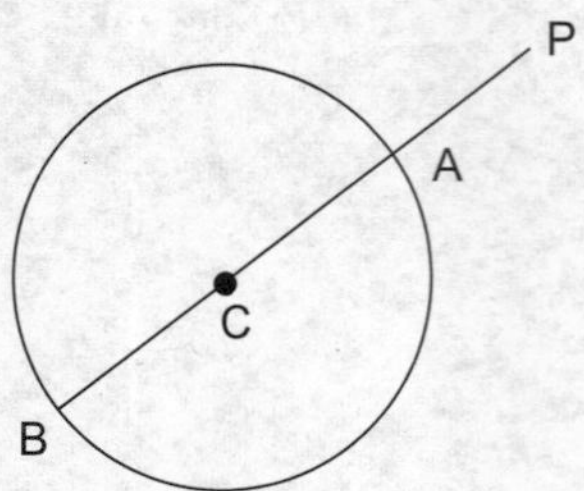

FIGURE 3.52

In this case, $S_1 > 0$, the minimum distance of P from circle $= PA = CP - CA = |CP - r|$ and the maximum distance of P from the circle $= PB = CP + CB = |r + CP|$.

Case III: If P is on the circle.
In this case, $S_1 = 0$, the minimum distance of P from the circle $= 0$ and the maximum distance of P from the circle $= PA = 2r$

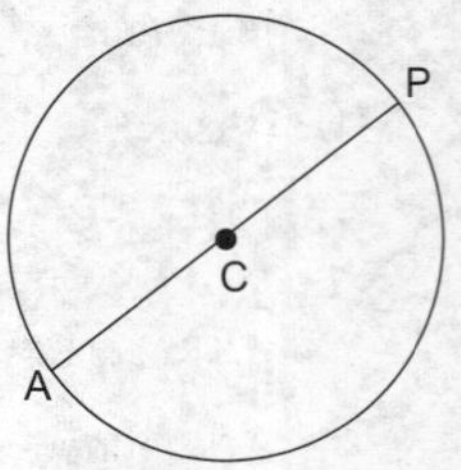

FIGURE 3.53

ILLUSTRATION 29: Given $P_1 : (2,3)$; $P_2 : (1,0)$; $P_3\ (6, -14)$; S_1: $x^2 + y^2 = 4$ and $S_2 : x^2 + y^2 - 10x + 24y + 144 = 0$
Then find the minimum and maximum distance between

(i) P_1 and S_1 (ii) P_2 and S_1
(iii) P_1 and S_2 (iv) P_3 and S_2

SOLUTION: $S_1 : x^2 + y^2 = 4$

Centre $C_1(0,0)$ and Radius $= r_1 = 2$

$S_2 : x^2 + y^2 - 10x + 24y + 144 = 0$

Centre $C_2\,(5, -12)$ and Radius $= r_2 = 5$

(i) P_1 and S_1

P_1 lies outside S_1

$P_1C_1 = \sqrt{2^2 + 3^2} = \sqrt{13}$

$\therefore$ Minimum distance: $\sqrt{13} - 2$

Maximum distance: $\sqrt{13} + 2$

(ii) P_2 and S_1

P_2 lies inside S_1

$P_2C_1 = \sqrt{0^2 + 1^2} = 1$

$\therefore$ Minimum distance $= |1 - 2| = 1$

Maximum distance $= |1 + 2| = 3$

(iii) P_1 and S_2

P_1 lies outside S_2

$P_1C_2 = \sqrt{(2-5)^2 + (3-(-12))^2} = \sqrt{234}$

$\therefore$ Minimum distance $= \sqrt{234} - 5$

Maximum distance $= \sqrt{234} + 5$

(iv) P_3 and S_2

P_3 lies inside S_2

$$P_3C_2 = \sqrt{1^2 + 2^2} = \sqrt{5}$$

$\therefore$ Minimum distance: $\left|\sqrt{5} - 5\right| = 5 - \sqrt{5}$

Maximum distance: $\sqrt{5} + 5$

ILLUSTRATION 30: An ant moves on the curves S: $x^2 + y^2 = 9$ and another ant moves on the curve $(x + 10)^2 + y^2 = 16$. Find the minimum and maximum distance between the two ants.

SOLUTION: Clearly, the distance will be minimum between the two ants, when they are at positions P_1 and P_1' respectively.

Hence minimum distance = 3

Also the distance between the two ants will be maximum, when they are at the positions P_2 and P_2' respectively.

Hence maximum distance = 17

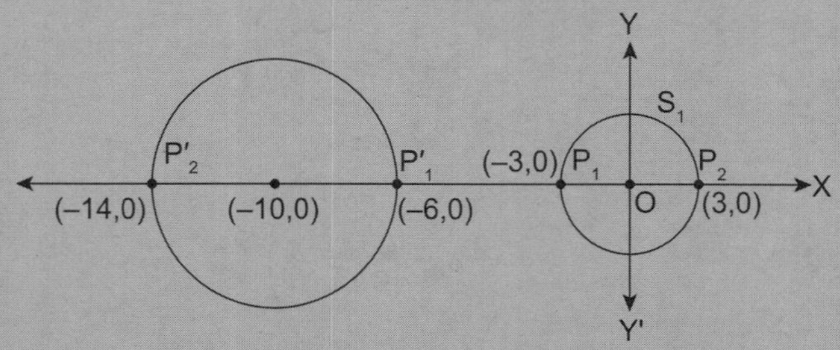

FIGURE 3.54

■ POSITION OF LINE WITH RESPECT TO CIRCLE

A line $L = y - mx - c = 0$ intersect a circle $(S = 0)$ with centre (α, β) and radius 'r' or touches or has no contact, depending on the condition that $CM < r$, $CM = r$ or $CM > r$, where CM is length of perpendicular dropped from the centre C to the line $y = mx + c$.

Analytically, any line will cut a circle either at two distinct points (intersection), two coincident points (touches) or at two imaginary points (no contact)

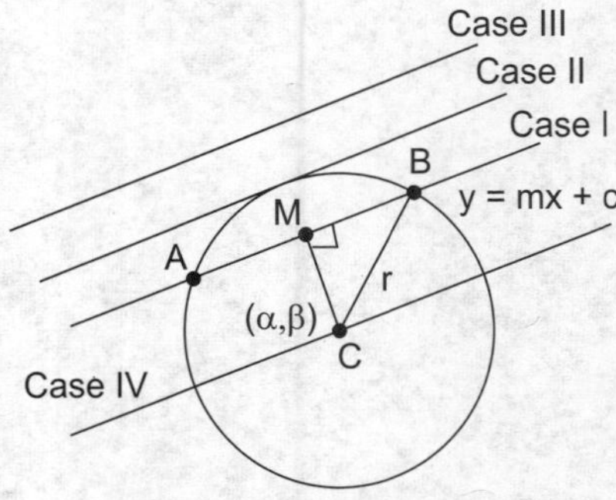

FIGURE 3.55

Case I: If line intersect the circle at two distinct points, i.e., Line is the secant of circle, i.e., has its segment AB as chord of circle, then

$$p < r \Rightarrow \frac{|\beta - m\alpha - c|}{\sqrt{1+m^2}} < r$$

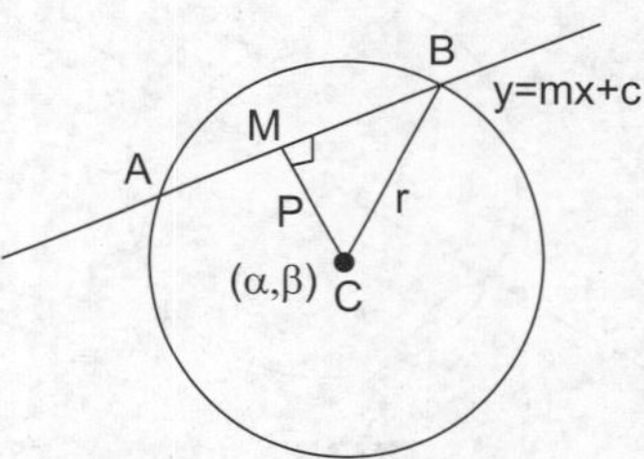

FIGURE 3.56

Using normal form of a line, we can also find the midpoint of the chord AB which is given by

$$\Rightarrow \frac{x-\alpha}{m} = \frac{y-\beta}{-1} = \frac{-(m\alpha + c - \beta)}{1+m^2}$$

Case II: If line intersect the circle at two coincident points i.e., Line is the tangent to the circle, then

$$p = r \Rightarrow \frac{|\beta - m\alpha - c|}{\sqrt{1+m^2}} = r$$

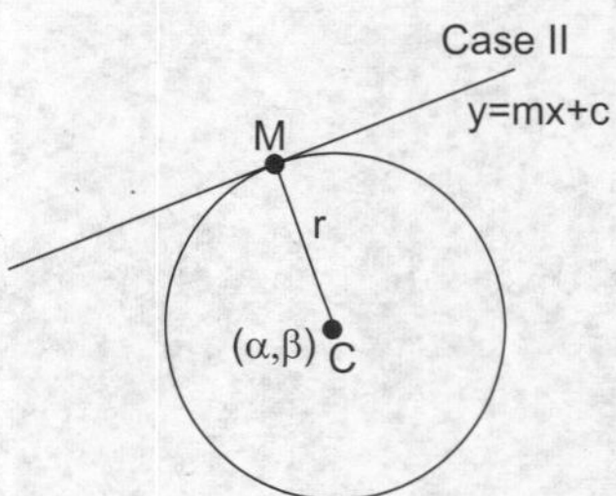

FIGURE 3.57

In this case, the chord becomes a tangent and hence the mid-point of the chord become the point of contact of the line and the circle (i.e., the point of tangency), which can be found using normal form of a line, which is given by

$$\Rightarrow \frac{x-\alpha}{m}=\frac{y-\beta}{-1}=\frac{-(m\alpha+c-\beta)}{1+m^2}$$

Case III: If line intersect the circle at two imaginary points, i.e., Line has no contact with the circle, then

$$p>r \Rightarrow \frac{|\beta-m\alpha-c|}{\sqrt{1+m^2}}>r$$

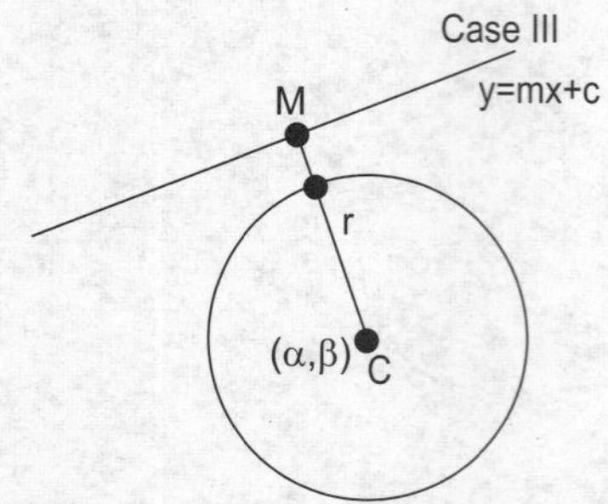

FIGURE 3.58

Case IV: $p = 0 \Rightarrow$ Line is a diameter of the circle

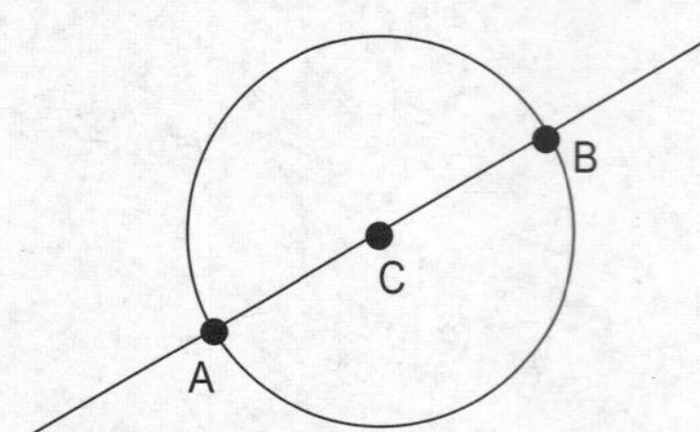

FIGURE 3.59

■ IMAGE OF THE CIRCLE IN THE LINE MIRROR

Let the circle be $x^2+y^2+2gx+2fy+c=0$ and line mirror is $lx + my + n = 0$ in this condition, radius of circle remains unchanged but centre changes. Let the centre of image circle be (x_1, y_1).

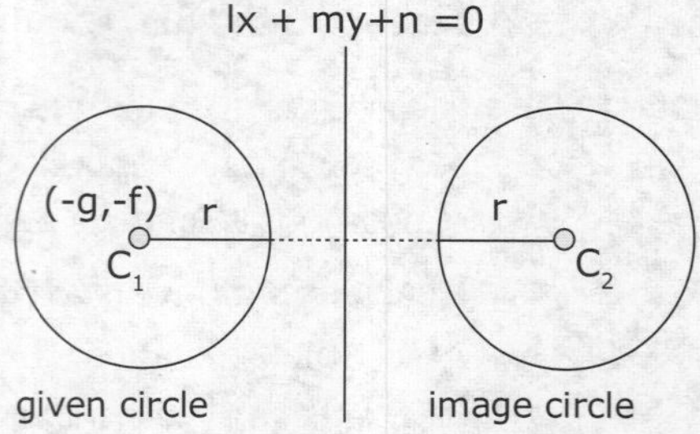

FIGURE 3.60

Slope $C_1C_2 \times$ slope of $lx + my + n = -1$ .....(i)

and mid-point of line segment joining C_1 $(-g, -f)$ and C_2 (x_1, y_1) lie on $lx + my + n = 0$

i.e., $l\left(\frac{x_1-g}{2}\right)+m\left(\frac{y_1-f}{2}\right)+n=0$ ...(ii)

Solving (i) and (ii), we get (x_1, y_1)

$\therefore$ Required image circle is $(x-x_1)^2+(y-y_1)^2=r^2$

where $r=\sqrt{g^2+f^2-c}$

ILLUSTRATION 31: Find the equation of the image of the circle $x^2+y^2+16x-24y+183=0$ by the line mirror $4x + 7y + 13 = 0$.

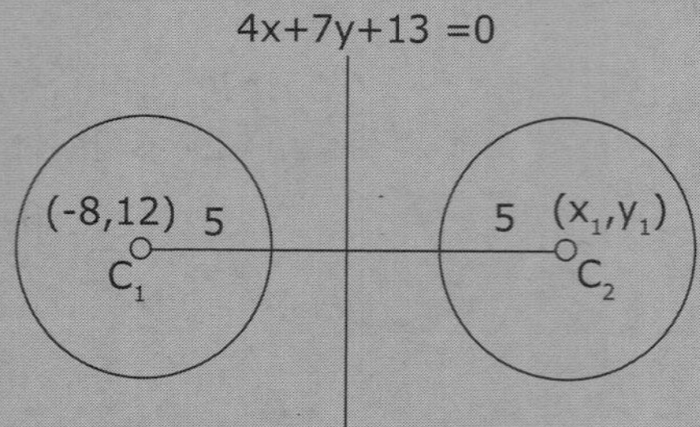

FIGURE 3.61

SOLUTION: The given circle and line are

$x^2+y^2+16x-24y+183=0$ ..(i)

and $4x + 7y + 13 = 0$ .. (ii)

Centre and radius of circle (i) are (–8, 12) and 5 respectively.

Let the centre of the image circle be (x_1, y_1).

Then slope of C_1C_2 × slope of $4x + 7y + 13 = -1$

$$\Rightarrow \frac{y_1-12}{x_1+8}\times -\frac{4}{7} = -1$$

or $4y_1 - 48 = 7x_1 + 56$

or $7x_1 - 4y_1 + 104 = 0$ ....(iii)

and mid-point of C_1C_2, i.e., $\left(\frac{x_1-8}{2}, \frac{y_1+12}{2}\right)$ lie on $4x + 7y + 13 = 0$,

then $4\left(\frac{x_1-8}{2}\right)+7\left(\frac{y_1+12}{2}\right)+13=0$ or $4x_1 + 7y_1 + 78 = 0$ ...(iv)

Solving (iii) and (iv), we get $(x_1,y_1) = (-16, -2)$

Equation of the image circle is $(x+16)^2+(y+2)^2=5^2$

or $x^2+y^2+32x+4y+235=0$

■ CONDITION FOR TANGENCY

In case of coincident points of intersection, the line will be tangent to the circle. Therefore,

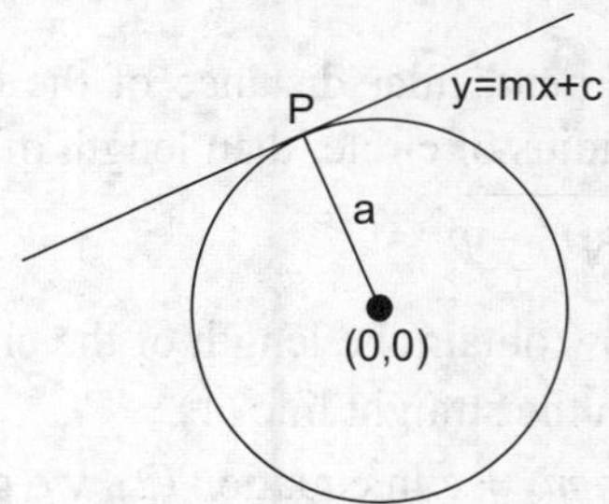

FIGURE 3.62

(i) the intersection of the line $y = mx + c$ is tangent to the circle $x^2 + y^2 = a^2$, if and only if, $c^2 = a^2(1 + m^2)$. If it is tangent, then the point of contact is given by:

$$\left(-\frac{a^2m}{c}, \frac{a^2}{c}\right).$$

Proof: The x-ordinates of points of intersection of the line and the circle are given by the equation

$x^2 + (mx + c)^2 = a^2$.

$$\Rightarrow (1+m^2)x^2+2mcx+c^2-a^2=0$$

$$\Rightarrow x=\frac{-2mc\pm\sqrt{(2mc)^2-4(c^2-a^2)(1+m^2)}}{2(1+m^2)}$$

$$=\frac{-mc\pm\sqrt{a^2+a^2m^2-c^2}}{(1+m^2)}$$

$D>0\Rightarrow$ line cuts the circle at two real and distinct points

$D<0\Rightarrow$ no contact with circle

$D=0\Rightarrow$ line touches the circle

The two values will be real and equal if $D = 0$, i.e., $a^2(1 + m^2) = c^2$.

In this case $x=\frac{-mc}{(1+m^2)}=\frac{mc}{c^2/a^2}=-\frac{a^2m}{c}$

The corresponding y-co-ordinate

$$= mx+c = m\left(-\frac{a^2m}{c}\right)+c=\frac{-a^2m^2+c^2}{c}=\frac{a^2}{c}$$

Thus the line $y = mx + c$, will be tangent to the circle $x^2+y^2=a^2$ if $c^2=a^2(1+m^2)$ and will touch it at $\left(-\frac{a^2m}{c}, \frac{a^2}{c}\right)$

or $\left(\frac{-am}{\sqrt{1+m^2}}, \frac{a}{\sqrt{1+m^2}}\right)$ and $\left(\frac{am}{\sqrt{1+m^2}}, \frac{-a}{\sqrt{1+m^2}}\right)$

and the line $y = mx + c$ will cut the circle $x^2 + y^2 = a^2$ at two distinct points if $a^2(1+m^2)>c^2$, will touch the circle if $a^2(1 + m^2) = c^2$ and will not meet it at all if $a^2(1 + m^2) < c^2$.

(ii) The line $lx + my + n = 0$ is tangent to the circle $x^2 + y^2 = a^2$ if and only if $n^2 = a^2(l^2 + m^2)$, and the point of contact is given by $\left(\frac{-la^2}{n}, \frac{-ma^2}{n}\right)$

Proof: $lx + my + n = 0 \Rightarrow y=\frac{-lx-n}{m}$

$\therefore\ x^2 + y^2 = a^2 \Rightarrow x^2 + \left(\dfrac{lx+n}{m}\right)^2 = a^2$

$\Rightarrow\ m^2x^2 + l^2x^2 + 2lnx + n^2 - a^2m^2 = 0$

$\Rightarrow\ (m^2 + l^2)x^2\ \ (2ln)x + (n^2 - a^2m^2) = 0$

$$\Rightarrow\ x = \frac{-2ln \pm \sqrt{(2ln)^2 - 4(m^2+l^2)(n^2-a^2m^2)}}{2(l^2+m^2)}$$

$$= \frac{-ln \pm \sqrt{l^2n^2 + a^2m^4 + a^2m^2l^2 - m^2n^2 - n^2l^2}}{l^2+m^2}$$

$$= \frac{-ln \pm m\sqrt{a^2m^2 + a^2l^2 - n^2}}{l^2+m^2}$$

For $D = 0 \Rightarrow a^2m^2 + a^2l^2 - n^2 = 0$

$\Rightarrow\ n^2 = a^2\ (m^2 + l^2)\ \therefore\ x = -\dfrac{ln}{n^2/a^2} = -\dfrac{a^2l}{n}$

$$\text{and } y = \frac{-l\left(\dfrac{-a^2l}{n}\right) - n}{m} = \frac{a^2l^2 - n^2}{nm}$$

$$= \frac{a^2l^2 - (a^2m^2 + a^2l^2)}{nm} = \frac{-a^2m}{n}\quad (\because n^2 = a^2(m^2+l^2))$$

$\therefore$ point of contact $= (x,y) = \left(\dfrac{-la^2}{n}, \dfrac{-ma^2}{n}\right)$

NOTES

1. *The line $y = mx \pm a\sqrt{1+m^2}$ will be a tangent to the circle $x^2 + y^2 = a^2$ for all values of m.*
2. *The condition of tangency can also be obtained by equating perpendicular distance of the centre (0, 0) to the line y-mx -c = 0 with the radius of the circle.*
3. *In case the circle is $x^2 + y^2 + 2gx + 2fy + c = 0$, we may shift the origin to the centre of the circle and apply the results described above.*

■ POINT OF INTERSECTION

Let a line $y = mx + c$ cut the circle $(x - \alpha)^2 + (y - \beta)^2 = r^2$

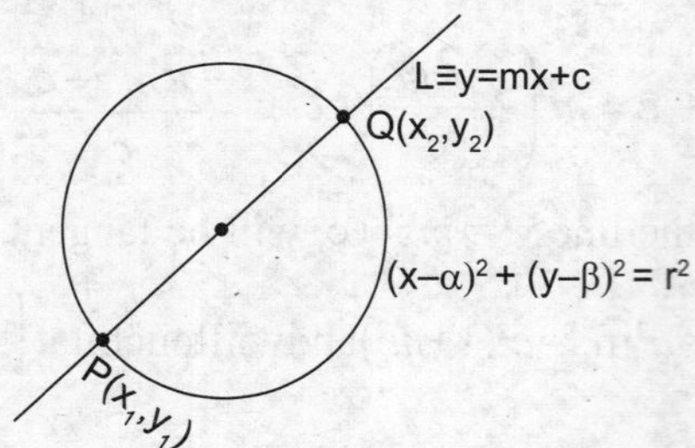

FIGURE 3.63

Putting $y = mx + c$ in $(x - \alpha)^2 + (y - \beta)^2 = r^2$, we get $(x - \alpha)^2 + (mx + c - \beta)^2 = r^2$

Solving this quadratic we will get x_1 and x_2

Substituting the above obtained values of x_1 and x_2; we will get y_1 and y_2

Hence; we get the points of intersection i.e., $P(x_1, y_1)$ and $Q(x_2, y_2)$

■ LENGTH OF INTERCEPT

Let a line $y = mx + c$ ... (1)

cuts the circle $x^2 + y^2 = a^2$ ....(2)

If p be perpendicular distance of the centre from the line and r be radius of circle, then length of intercept PQ is given as $L = 2\sqrt{r^2 - p^2}$

We can now obtain the length of the chord intercepted by the circle on the straight line (1)

Putting $y = mx + c$ in equation (2), we get

$$x^2 + (mx + c)^2 = a^2 \quad ...(3)$$

For if x_1 and x_2 be the roots of the equation (3), we have

Sum of roots $= x_1 + x_2 = -\dfrac{2mc}{1+m^2}$

and product of roots $= x_1x_2 = \dfrac{c^2 - a^2}{1+m^2}$

Hence, $|x_1 - x_2| = \sqrt{(x_1 + x_2)^2 - 4x_1x_2}$

$$= \frac{2}{1+m^2}\sqrt{m^2c^2 - (c^2 - a^2)(1+m^2)}$$

$$= \frac{2}{1+m^2}\sqrt{a^2(1+m^2) - c^2} \quad ......(4)$$

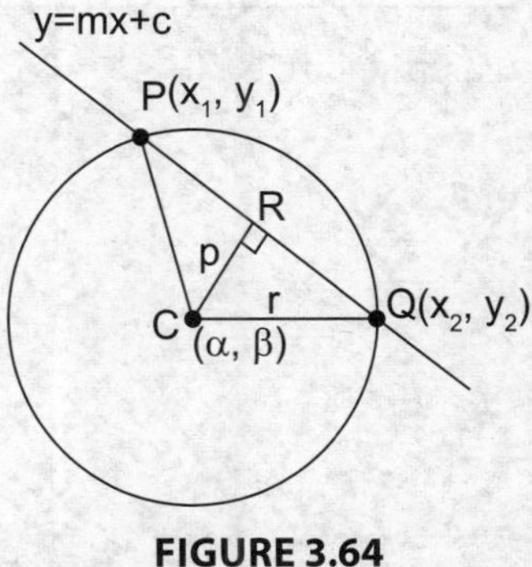

FIGURE 3.64

If y_1 and y_2 be the ordinates of P and Q we have

$|y_1 - y_2| = |(mx_1 + c) - (mx_2 + c)| = |m(x_1 - x_2)|$

Hence, $PQ = \sqrt{(y_1 - y_2)^2 + (x_1 - x_2)^2} = \sqrt{1+m^2}\,|x_1 - x_2|$

$$= 2\sqrt{\frac{a^2(1+m^2)-c^2}{1+m^2}} \qquad \text{(using (4))}$$

Hence, length of the intercept is given by

$$2\sqrt{\frac{a^2(1+m^2)-c^2}{1+m^2}}$$

NOTES

(i) *Length of the intercept made by the circle on the line is* $2\sqrt{r^2 - p^2}$.

(ii) *The length of the intercept made by the line* $y = mx + c$ *with the circle* $x^2 + y^2 = a^2$ *is* $2\sqrt{\frac{a^2(1+m^2)-c^2}{1+m^2}}$.

INTERCEPT MADE ON CO-ORDINATE AXES BY THE CIRCLE

General Method

Let the circle be $x^2 + y^2 + 2gx + 2fy + c = 0$ and length of intercepts on x-axis and y-axis are $|AB| = |x_2 - x_1|$ and $|CD| = |y_2 - y_1|$ respectively.

The circle intersect when $y = 0$, then $x^2 + 2gx + c = 0$. Therefore sum of roots $= x_1 + x_2 = -2g$ and product of roots $= x_1x_2 = c$

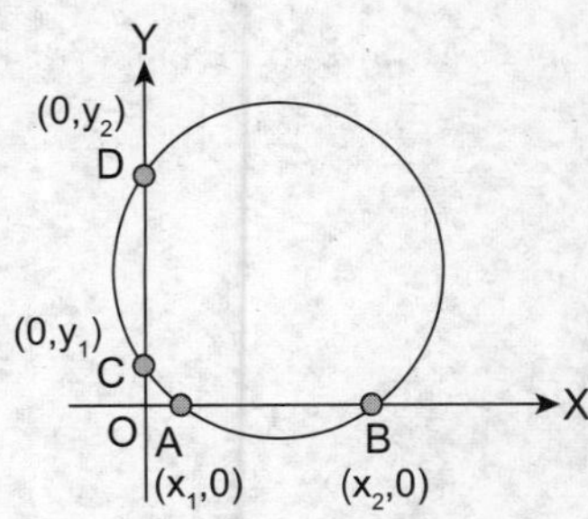

FIGURE 3.65

$\therefore |AB| = |x_2 - x_1| = \sqrt{(x_1 + x_2)^2 - 4x_1x_2} = 2\sqrt{g^2 - c}$

and the circle intersects the y-axis when $x = 0$, then $y^2 + 2fy + c = 0$. Now sum of roots $= y_1 + y_2 = -2f$ and product of roots $= y_1y_2 = c$

$\therefore |CD| = |y_2 - y_1| = \sqrt{(y_1 + y_2)^2 - 4y_1y_2} = 2\sqrt{f^2 - c}$'

Geometrical Method

$CM = |f|$ and $CB = \sqrt{g^2 + f^2 - c}$

$MB = \sqrt{CB^2 - CM^2} = \sqrt{g^2 - c}$

x intercept is $AB = 2\,BM = 2\sqrt{g^2 - c}$

Similarly, y-intercept is $PQ = 2\sqrt{f^2 - c}$

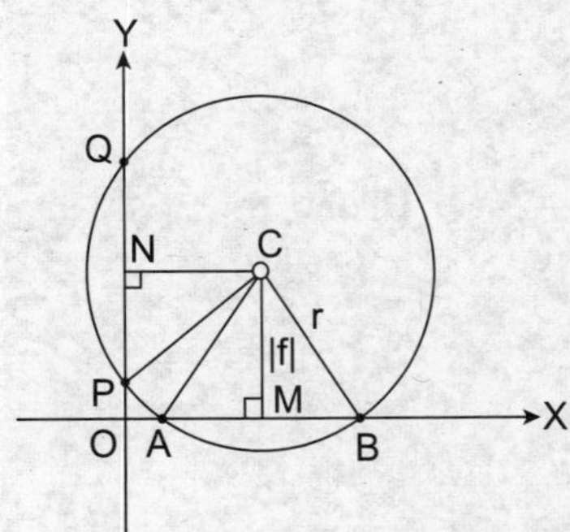

FIGURE 3.66

NOTES

1. *Intercepts are always positive.*
2. *If circle touches x-axis, then* $g^2 = c$ *and if circle touches y-axis, then* $f^2 = c$.
3. *If circle touches both the axes, then* $c = g^2 = f^2$
4. *If circle neither touches nor intersect x-axis, then* $g^2 < c$ *and similarly for y-axis;* $f^2 < c$.

ILLUSTRATION 32: Find the points of intersection of the line $2x + 3y = 18$ and the circle $x^2 + y^2 = 25$.

SOLUTION: We have $2x + 3y = 18$ ...(i)

and $x^2 + y^2 = 25$ ...(ii)

from (i), $y = \dfrac{18-2x}{3}$

Substituting in (ii), then $x^2 + \left(\dfrac{18-2x}{3}\right)^2 = 25$

$\Rightarrow\ 9x^2 + 4(9-x)^2 = 225 \Rightarrow 9x^2 + 4(81 - 18x + x^2) = 225 \Rightarrow 13x^2 - 72x + 324 - 225 = 0$

$\Rightarrow 13x^2 - 72x + 99 = 0 \Rightarrow (x - 3)(13x - 33) = 0$

$\Rightarrow x = 3$ or $x = \dfrac{33}{13}$ from (i) $x = 3 \Rightarrow y = 4$ and $x = \dfrac{33}{13} \Rightarrow y = \dfrac{56}{13}$

Hence the points of intersection of the given line and the given circle are (3, 4) and $\left(\dfrac{33}{13}, \dfrac{56}{13}\right)$

ILLUSTRATION 33: Find the equation of the circle whose diameter is the line segment joining the points (–4, 3) and (12, –1). Find also the intercept made by it on y-axis.

SOLUTION: Let $A \equiv (-4, 3)$ and $B \equiv (12, -1)$.

Now equation of the circle having A and B as the ends of a diameter is $(x + 4)(x - 12) + (y - 3)(y + 1) = 0$

$\Rightarrow\ x^2 - 8x - 48 + y^2 - 2y - 3 = 0 \qquad \Rightarrow\ x^2 + y^2 - 8x - 2y - 51 = 0$ ....(i)

Putting $x = 0$ in (i), we get $y^2 - 2y - 51 = 0$

Let y_1 and y_2 be its roots, then $y_1 + y_2 = 2$

and $y_1 y_2 = -51$

Now the intercept made by circle (i), on y-axis

$= |y_1 - y_2| = \sqrt{(y_1 + y_2)^2 - 4y_1y_2} = \sqrt{4 + 204} = \sqrt{208} = 4\sqrt{13}$

Aliter: Equation of the circle having A and B as the ends of a diameter is $(x + 4)(x - 12) + (y - 3)(y + 1) = 0$

$\Rightarrow\ x^2 - 8x - 48 + y^2 - 2y - 3 = 0$

$\Rightarrow\ x^2 + y^2 - 8x - 2y - 51 = 0$ .....(i)

Comparing (i) with standard equation of circle

$x^2 + y^2 + 2gx + 2fy + c = 0$, we have $g = -4, f = -1, c = -51$

$\therefore$ Intercept on y-axis $= 2\sqrt{f^2 - c} = 2\sqrt{(-1)^2 - (-51)} = 4\sqrt{13}$

ILLUSTRATION 34: Find the value of k for which the line $3x - 4y - k = 0$ is a tangent to the circle $x^2 + y^2 = 25$. Also, find the point of contact.

SOLUTION: The equation of the line can be written as $y = \dfrac{3}{4}x - \dfrac{k}{4}$

$\Rightarrow\ m = \dfrac{3}{4},\ c = -\dfrac{k}{4},\ a^2 = 25$

we must have $c^2 = a^2(1 + m^2)$

$\Rightarrow\ \dfrac{k^2}{16} = 25\left(1 + \dfrac{9}{16}\right) \Rightarrow k = \pm 25 \Rightarrow c = \mp\dfrac{25}{4}$

Thus there are two lines $3x - 4y - 25 = 0$ and $3x - 4y + 25 = 0$ which touch the circle $x^2 + y^2 = 25$. The first line will touch the circle $x^2 + y^2 = 25$ at $\left(-\frac{a^2m}{c}, \frac{a^2}{c}\right)$

$$= \left(-\frac{25\times\frac{3}{4}}{\frac{25}{4}}, \frac{25}{25/4}\right) = (-3, 4)$$

TEXTUAL EXERCISE–3 (SUBJECTIVE)

1. Find the co-ordinates of the points where the line $y = 2x + 1$ cuts the circle $x^2 + y^2 = 2$.
2. Show that the circle $x^2 + y^2 - 2ax - 2ay + a^2 = 0$ touches the axes of x and y.
3. Find the equation of the circle which touches the lines $x = 0$, $x = a$ and $3x + 4y + 5a = 0$.
4. Show that the line $y = m(x-a) + a\sqrt{1+m^2}$ touches the circle $x^2 + y^2 = 2ax$, whatever the value of m may be.
5. (i) Two lines are drawn through the points $(a, 0)$, $(-a, 0)$ respectively, and make an angle θ with one another, find the locus of their intersection.
   (ii) A line moves so that the sum of the perpendiculars drawn to it from the points $(a, 0)$, $(-a, 0)$ is constant, show that it always touches a circle.
6. Show that the circle of which the line joining the points $(am^2, 2am)(a/m^2, -2a/m)$ is a diameter touches $x + a = 0$ for all values of m.
7. Find the equation of the tangent to the circle $x^2 + y^2 = a^2$ which
   (i) is parallel to the straight line $y = mx + c$
   (ii) is perpendicular to the straight line $y = mx + c$
   (iii) passes through the point $(b, 0)$
   (iv) makes with the axes a triangle whose area is a^2.
8. A circle touches one given straight line and cuts off a constant length $(2l)$ from another straight line perpendicular to the former, find the equation of the locus of its centre.

Answer Keys

1. $(-1, -1)$ and $(1/5, 7/5)$
3. $x^2 + y^2 - ax + 2ay + a^2 = 0$ or $x^2 + y^2 - ax + \frac{9}{2}ay + \frac{81}{16}a^2 = 0$
5. (i) The circle $x^2 + y^2 - a^2 = \pm 2ay \cot\theta$
7. (i) $y = mx \pm a\sqrt{1+m^2}$ (ii) $my + x = \pm a\sqrt{1+m^2}$ (iii) $ax \pm y\sqrt{b^2 - a^2} = ab$ (iv) $x + y = \pm a\sqrt{2}$ or $x - y = \pm a\sqrt{2}$
8. $y^2 - x^2 = l^2$

TEXTUAL EXERCISE–3 (OBJECTIVE)

1. If the straight line $y = mx$ is outside the circle $x^2 + y^2 - 20y + 90 = 0$, then:
   (a) $m > 3$ (b) $m < -3$
   (c) $|m| > 3$ (d) $|m| < 3$
2. The equation $x^2 + y^2 - 6x + 8y - 11 = 0$ is a circle, then the points $(0, 0)$ and $(1, 8)$ lie,
   (a) both inside the circle
   (b) both outside the circle

(c) one outside the circle and one inside
(d) one on the circle and the other outside.

3. The circle $x^2 + y^2 + 2x - 4y + 1 = 0$ touches
(a) x-axis (b) y-axis
(c) both axes (d) None of the axes

4. The straight line $mx - y = 1 + 2x$ cuts the circle $x^2 + y^2 = 1$ at one point at least. Then the set of values of m is
(a) $\left[-\frac{4}{3}, 0\right]$ (b) $\left[-\frac{4}{3}, \frac{4}{3}\right]$
(c) $\left[0, \frac{4}{3}\right]$ (d) None of these

5. The straight line $y = mx + c$ cuts the circle $x^2 + y^2 = a^2$ at real points if
(a) $\sqrt{a^2(1+m^2)} \le |c|$ (b) $\sqrt{a^2(1-m^2)} \le |c|$
(c) $\sqrt{a^2(1+m^2)} \ge |c|$ (d) $\sqrt{a^2(1-m^2)} \ge |c|$

6. The length of the chord cut off by $y = 2x + 1$ from the circle $x^2 + y^2 = 2$ is
(a) 5/6 (b) 6/5
(c) $\frac{6}{\sqrt{5}}$ (d) $\frac{\sqrt{5}}{6}$

7. The maximum distance of the point (4, 4) from the circle $x^2 + y^2 - 2x - 15 = 0$ is
(a) 10 (b) 9
(c) 5 (d) None of these

8. The value of k for which two tangents can be drawn from (k, k) to the circle $x^2 + y^2 + 2x + 2y - 16 = 0$ is
(a) $k \in R^+$
(b) $k \in R^-$
(c) $k \in (-\infty, -4) \cup (2, \infty)$
(d) $k \in (0, 1]$

9. For the line $3x + 2y = 12$ and the circle $x^2 + y^2 - 4x - 6y + 3 = 0$, which of the following statements is true?
(a) Line is a tangent to the circle
(b) Line is a chord of the circle
(c) Line is a diameter of the circle
(d) None of these

10. The locus of the centre of the circle which cuts a chord of length $2a$ from the positive x-axis and passes through a point on positive y-axis distant b from the origin is
(a) $x^2 + 2by = b^2 + a^2$ (b) $x^2 - 2by = b^2 + a^2$
(c) $x^2 - 2by = a^2 - b^2$ (d) $x^2 - 2by = b^2 - a^2$

11. A circle touches x-axis and cuts off a chord of length $2l$ from y-axis. The locus of the centre of the circle is
(a) A straight line (b) A circle
(d) An ellipse (d) A hyperbola

12. The angle between the two tangents from the origin to the circle $(x - 7)^2 + (y + 1)^2 = 25$ is
(a) 0 (b) $\pi/3$
(c) $\pi/6$ (d) $\pi/2$

13. Equation of the pair of tangents drawn from the origin to the circle $x^2 + y^2 + 2gx + 2fy + c = 0$ is
(a) $gx + fy + c(x^2 + y^2) = 0$
(b) $(gx + fy)^2 = x^2 + y^2$
(c) $(gx + fy)^2 = c^2(x^2 + y^2)$
(d) $(gx + fy)^2 = c(x^2 + y^2)$

14. The line $(x - a)\cos\alpha + (y - b)\sin\alpha = r$ will be a tangent to the circle $(x - a)^2 + (y - b)^2 = r^2$
(a) If $\alpha = 30°$ (b) $\alpha = 60°$
(c) For all values of α (d) None of these

15. The equations of the tangents drawn from the origin to the circle $x^2 + y^2 - 2rx - 2hy + h^2 = 0$ are
(a) $x = 0, y = 0$
(b) $(h^2 - r^2)x - 2rhy = 0, x = 0$
(c) $y = 0, x = 4$
(d) $(h^2 - r^2)x + 2rhy = 0, x = 0$

16. An infinite number of tangents can be drawn from (1, 2) to the circle $x^2 + y^2 - 2x - 4y + \lambda = 0$, then $\lambda =$
(a) –20 (b) 0
(c) 5 (d) Cannot be determined

17. If the line $lx + my = 1$ be a tangent to the circle $x^2 + y^2 = a^2$, then the locus of the point (l, m) is
(a) a straight line (b) a circle
(c) a parabola (d) an ellipse

18. The equations of the tangents to the circle $x^2 + y^2 = a^2$ parallel to the line $\sqrt{3}x + y + 3 = 0$ are
(a) $\sqrt{3}x + y \pm 2a = 0$ (b) $\sqrt{3}x + y \pm a = 0$
(c) $\sqrt{3}x + y \pm 4a = 0$ (d) None of these

19. The number of tangents that can be drawn from (0, 0) to the circle $x^2 + y^2 + 2x + 6y - 15 = 0$ is
(a) None (b) One
(c) Two (d) Infinite

Answer Keys

1. (d)	**2.** (c)	**3.** (a)	**4.** (d)	**5.** (c)	**6.** (c)	**7.** (b)	**8.** (c)	**9.** (c)	**10.** (c)
11. (d)	**12.** (d)	**13.** (d)	**14.** (c)	**15.** (b)	**16.** (c)	**17.** (b)	**18.** (a)	**19.** (a)	

■ EQUATION OF TANGENT AND NORMAL

(a) Equation of tangent to circle $x^2 + y^2 = a^2$ at $P(x_1, x_2)$

Tangent line to a circle at a point $P(x_1, y_1)$ is defined as a limiting case of a chord PQ where Q is (x_2, y_2) such that $Q \to P$.

As $Q \to P$, i.e., $x_2 \to x_1$ and $y_2 \to y_1$

Then chord $PQ \to$ tangent at P.

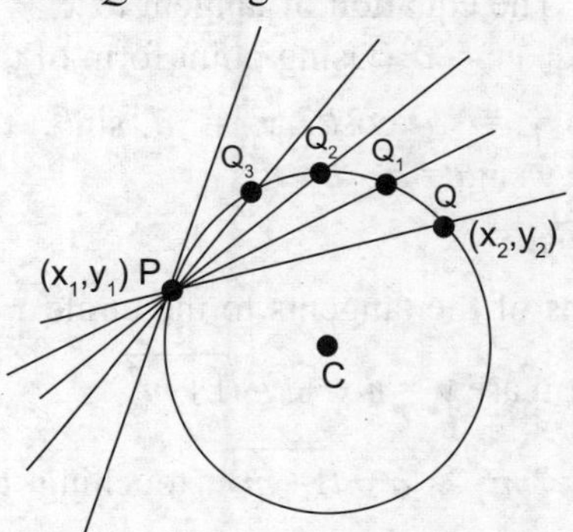

FIGURE 3.67

$\Rightarrow$ Slope of chord $PQ \to$ slope of tangent at P.

$$\Rightarrow m_t = \lim_{\substack{x_2 \to x_1 \\ y_2 \to y_1}} \left(\frac{y_2 - y_1}{x_2 - x_1}\right)$$

$$= \lim_{\substack{x_2 \to x_1 \\ y_2 \to y_1}} -\left(\frac{x_1 + x_2}{y_1 + y_2}\right) = -\frac{x_1}{y_1}$$

$$\left[\begin{array}{l} \because x_1^2 + y_1^2 = a^2 \quad ......(i) \\ \quad x_2^2 + y_2^2 = a^2 \quad ....(ii) \\ \Rightarrow (x_2^2 - x_1^2) = -(y_2^2 - y_1^2) \\ \Rightarrow \dfrac{y_2 - y_1}{x_2 - x_1} = -\left(\dfrac{x_1 + x_2}{y_1 + y_2}\right) \end{array}\right]$$

$$\therefore y - y_1 = -\frac{x_1}{y_1}(x - x_1)$$

$$\Rightarrow T = xx_1 + yy_1 - a^2 = 0$$

(b) Equation of Tangent to circle $x^2 + y^2 + 2gx + 2fy + c = 0$ at a point $P(x_1, y_1)$

The equation of tangent at any point on the circle $x^2 + y^2 = a^2$ can be easily obtained by using the fact that the tangent is perpendicular to the line joining point and centre. Similarly, result can also be derived as the equation of tangent at any point (x_1, y_1) of the circle $S \equiv x^2 + y^2 + 2gx + 2fy + c = 0$ is $xx_1 + yy_1 + g(x + x_1) + f(y + y_1) + c = 0$. This expression is denoted by T as

$$T \equiv xx_1 + yy_1 + g(x + x_1) + f(y + y_1) + c = 0$$

Mehod 1: Using slope point form a straight line:

Proof: Let P be (x_1, y_1) and C be the centre of the circle $S \equiv x^2 + y^2 + 2gx + 2fy + c = 0$, then C is $(-g, -f)$

and slope of $PC = \dfrac{y_1 + f}{x_1 + g}$

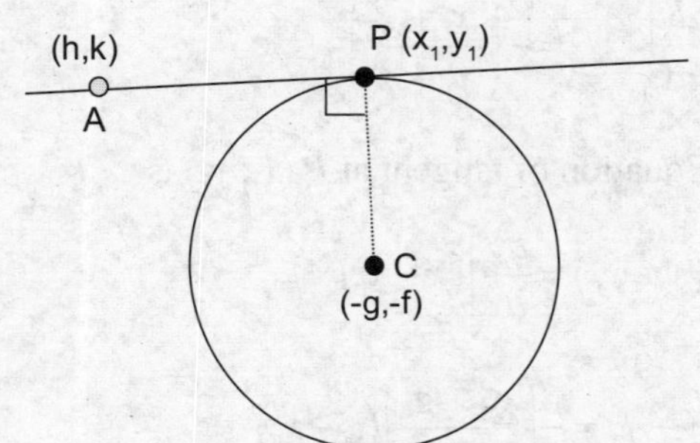

FIGURE 3.68

$\Rightarrow$ Slope of the tangent at $P = -\dfrac{x_1 + g}{y_1 + f}$

$\Rightarrow$ Equation of tangent at (x_1, y_1) must be

$$y - y_1 = -\frac{x_1 + g}{y_1 + f}(x - x_1)$$

$$\Rightarrow \quad yy_1 + fy - y_1^2 - fy_1 = -xx_1 + x_1^2 - gx + gx_1$$

$$\Rightarrow \quad xx_1 + yy_1 + gx + fy = x_1^2 + y_1^2 + fy_1 + gx_1 \quad ...(i)$$

But as (x_1, y_1) lies on the circle $S = 0$

$$\Rightarrow \quad x_1^2 + y_1^2 + 2fy_1 + 2gx_1 + c = 0$$

$$\Rightarrow \quad x_1^2 + y_1^2 + fy_1 + gx_1 = -c - gx_1 - fy_1$$

hence the equation (i) becomes

$$xx_1 + yy_1 + gx + fy = -c - gx_1 - fy_1$$

or $\quad xx_1 + yy_1 + g(x + x_1) + f(y + y_1) + c = 0$

Method 2: Using geometry : By Figure 3.68

$$AP^2 + PC^2 = AC^2$$

$\Rightarrow \quad (x_1 - h)^2 + (y_1 - k)^2 + (x_1 + g)^2 + (y_1 + f)^2 = (h + g)^2 + (k + f)^2$ and replacing h by x and k

by y; we get the equation of tangent as $xx_1 + yy_1 + g(x + x_1) + f(y + y_1) + c = 0$.

Method 3: using Calculus

Equation of circle is $x^2 + y^2 + 2gx + 2fy + c = 0$ .....(i)

Now, if $P(x_1, y_1)$ lie on circle, then

$$x_1^2 + y_1^2 + 2gx_1 + 2fy_1 + c = 0 \qquad ......(ii)$$

Differentiating (i) with respect to x, we get

$$2x + 2y\frac{dy}{dx} + 2g + 2f\frac{dy}{dx} + 0 = 0$$

$$\therefore \quad \frac{dy}{dx} = -\left(\frac{x+g}{y+f}\right)$$

$$\Rightarrow \quad \left.\frac{dy}{dx}\right|_{(x_1,y_1)} = -\left(\frac{x_1+g}{y_1+f}\right) \qquad ......(iii)$$

$\therefore$ Equation of tangent at $P(x_1, y_1)$ is

$$(y - y_1) = \left.\frac{dy}{dx}\right|(x - x_1)$$

$$\Rightarrow \quad y - y_1 = -\left(\frac{x_1+g}{y_1+f}\right)(x - x_1)$$

$$\Rightarrow \quad (y - y_1)(y_1 + f) + (x - x_1)(x_1 + g) = 0$$

$$\Rightarrow \quad xx_1 + yy_1 + gx + fy = x_1^2 + y_1^2 + gx_1 + fy_1$$

Adding $gx_1 + fy_1 + c$ to both sides, we get

$$xx_1 + yy_1 + g(x + x_1) + f(y + y_1) + c$$

$$= x_1^2 + y_1^2 + 2gx_1 + 2fy_1 + c$$

$$\Rightarrow \quad xx_1 + yy_1 + g(x + x_1) + f(y + y_1) + c = 0$$

This is the required equation of the tangent to the circle at the point $P(x_1, y_1)$.

(c) Parametric Form:

Theorem: The equation of tangent to the circle $x^2 + y^2 = a^2$ at the point $(a\cos\theta, a\sin\theta)$ is $x\cos\theta + y\sin\theta = a$

Proof: The equation of tangent to $x^2 + y^2 = a^2$ at (x_1, y_1) is $xx_1 + yy_1 = a^2$ (Using point form of the tangent)

Putting $x_1 = a\cos\theta$, $y_1 = a\sin\theta$, then we get $x\cos\theta + y\sin\theta = a$.

(d) Slope form:

Equations of the tangents to the circle $x^2 + y^2 = a^2$ in slope form are $y = mx + a\sqrt{1+m^2}$ and $y = mx - a\sqrt{1+m^2}$ touching the circle at

$$\left(\frac{-ma}{\sqrt{1+m^2}}, \frac{a}{\sqrt{1+m^2}}\right) \text{ and } \left(\frac{ma}{\sqrt{1+m^2}}, \frac{-a}{\sqrt{1+m^2}}\right)$$

respectively. i.e., the point of contact is also given as

$$\left(-\frac{a^2 m}{c}, \frac{a^2}{c}\right).$$

NOTES

1. *In case slope of PC = ∞, then slope of the tangent is zero and therefore, equation of the tangent to circle at point (x_1, y_1) is $y = y_1$. Also when slope PC = 0, then equation of such tangent is $x = x_1$.*
2. *The equation of tangent at any point (x_1, y_1) of the circle $x^2 + y^2 = a^2$ is $xx_1 + yy_1 = a^2$ $(g = 0, f = 0, c = -a^2)$*
3. *In order to write equation of tangent to any conic section at its point (x_1, y_1) replace in the equation*

   S = 0 (of conic) x^2 by xx_1, y^2 by yy_1, $2x$ by $x + x_1$, $2y$ by $y + y_1$, $2xy$ by $(xy_1 + x_1y)$

   The equation S = 0 transforms to equation of tangent, i.e., T = 0 (T symbolizes expression for tangent at (x_1, y_1))
4. *The point of intersection of the tangents at the point P(α) and Q (β) on the circle $x^2 + y^2 = a^2$ is*

$$\left(\frac{a\cos\left(\frac{\alpha+\beta}{2}\right)}{\cos\left(\frac{\alpha-\beta}{2}\right)}, \frac{a\sin\left(\frac{\alpha+\beta}{2}\right)}{\cos\left(\frac{\alpha-\beta}{2}\right)}\right)$$

Equation of Normal

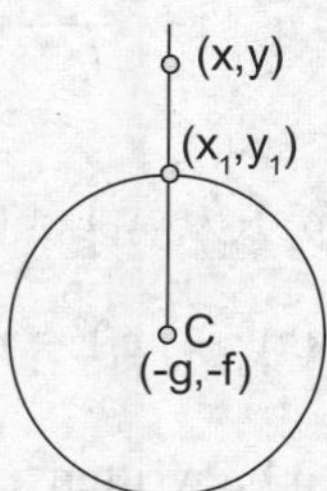

FIGURE 3.69

The normal at any point on a curve is a line which is perpendicular to the tangent to the curve at that point and in case of circle normal always passes through the centre. Equation of a normal at the point (x_1, y_1) lying on the circle $x^2 + y^2 + 2gx + 2fy + c = 0$ is $\dfrac{y - y_1}{y_1 + f} = \dfrac{x - x_1}{x_1 + g}$. Thus the condition for the line $y = mx + d$ to be normal to the circle $x^2 + y^2 + 2gx + 2fy + c = 0$ is $mg = f + d$ as line passes through the centre $(-g, f)$.

Equation of normal can also be given as

$$\begin{vmatrix} x & y & 1 \\ x_1 & y_1 & 1 \\ -g & -f & 1 \end{vmatrix} = 0 \text{ or } \begin{vmatrix} x & y \\ x_1 & y_1 \\ -g & -f \\ x & y \end{vmatrix} = 0$$

$\because$ Area of the triangle formed by 3 collinear points = 0

NOTES

1. *The equation of the normal to the circle $x^2 + y^2 = a^2$ at any point (x_1, y_1) is $xy_1 - x_1y = 0$*
2. *Parametric equation of normal to the circle $x^2 + y^2 = a^2$ at point $(a\cos\theta, a\sin\theta)$ is $\dfrac{x}{a\cos\theta} = \dfrac{y}{a\sin\theta}$*
3. *The equation of the normal to the circle $x^2 + y^2 + 2gx + 2fy + c = 0$ at point $(-g + r\cos\theta, -f + r\sin\theta)$ on it is $(\sin\theta)x - (\cos\theta)y + g(\sin\theta) - f(\cos\theta) = 0$*

■ LENGTH OF THE TANGENT

The length of tangent drawn from a point (x_1, y_1) to a circle

$S = x^2 + y^2 + 2gx + 2fy + c = 0$ is given by

$$PQ = PR = \sqrt{PC^2 - CQ^2} = \sqrt{x_1^2 + y_1^2 + 2gx_1 + 2fy_1 + c}$$

Therefore the length of tangent $= \sqrt{S_1}$, where the point (x_1, y_1) is exterior to the circle.

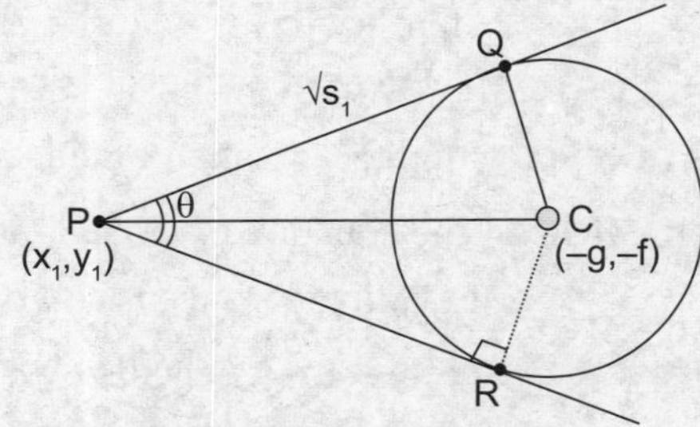

FIGURE 3.70

NOTES

1. *If PQ is a length of the tangent from a point P to a given circle, then $(PQ)^2$ is called the power of the point with respect to a given circle.*
2. *Area of quadrilateral PQCR $= r\sqrt{S_1}$ and angle between tangents PQ and PR is $\theta = 2\tan^{-1}\dfrac{r}{\sqrt{S_1}}$*

■ PAIR OF TANGENT

Let $P(x_1, y_1)$ be an exterior point of the circle $x^2 + y^2 - a^2 = 0$ and PT and PR are tangents from $P(x_1, y_1)$ to $x^2 + y^2 - a^2 = 0$. Let $Q(h, k)$ be any point on PT (or PR), then equation of PQ is $y - y_1 = \dfrac{k - y_1}{h - x_1}(x - x_1)$

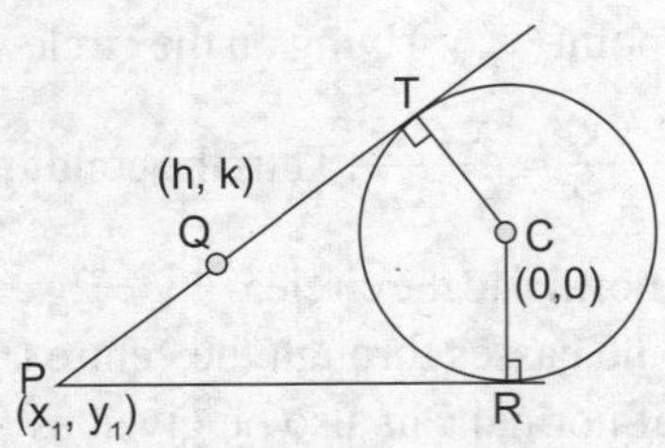

FIGURE 3.71

$$\Rightarrow \quad y = \frac{k - y_1}{h - x_1} x + \frac{hy_1 - kx_1}{h - x_1}$$

since PQ is tangent to $x^2 + y^2 = a^2$

we have $c^2 = a^2 (1 + m^2)$

$$\Rightarrow \quad \left(\frac{hy_1 - kx_1}{h - x_1}\right)^2 = a^2\left(1 + \left(\frac{k - y_1}{h - x_1}\right)^2\right)$$

$$\Rightarrow \quad (hy_1 - kx_1)^2 = a^2 [(h - x_1)^2 + (k - y_1)^2]$$

$$\Rightarrow \quad (h, k) \text{ satisfies } (xy_1 - yx_1)^2 = a^2[(x - x_1)^2 + (y - y_1)^2] \quad \text{....(i)}$$

The equation (i) can be written as

$$(x^2 + y^2 - a^2)(x_1^2 + y_1^2 - a^2) = (xx_1 + yy_1 - a^2)^2$$

which in notation becomes $\boldsymbol{SS_1 = T^2}$

NOTE

1. *Where circle is $x^2 + y^2 + 2gx + 2fy + c = 0$ and tangents are drawn from (x_1, y_1), then pair of tangents is*

   $(x^2 + y^2 + 2gx + 2fy + c)\left(x_1^2 + y_1^2 + 2gx_1 + 2fy_1 + c\right) = (xx_1 + yy_1 + g(x + x_1) + f(y + y_1) + c)^2$, *i.e.,* $SS_1 = T^2$

   where $S_1 \equiv x_1^2 + y_1^2 + 2gx_1 + 2fy_1 + c$

   and $T \equiv xx_1 + yy_1 + g(x + x_1) + f(y + y_1) + c$

ILLUSTRATION 35: Find the equation of tangent at the point (1, 1) of the circle $x^2 + y^2 + 4x + 6y - 12 = 0$.

SOLUTION: Given circle is $x^2 + y^2 + 4x + 6y - 12 = 0$

Here, $x_1 = 1, y_1 = 1, g = 2, f = 3$

The required tangent is $x + y + 2(x + 1) + 3(y + 1) - 12 = 0$ or $3x + 4y - 7 = 0$

ILLUSTRATION 36: Find the equation of tangent to circles $x^2 + y^2 - 2ax = 0$ at the point $[a(1 + \cos\alpha), a\sin\alpha]$.

SOLUTION: The equation of circle is $x^2 + y^2 - 2ax = 0$. At the point $[a(1 + \cos\alpha), a\sin\alpha]$.

Equation of tangent will be $x.a(1 + \cos\alpha) + y.a\sin\alpha - a[x + a(1 + \cos\alpha)] = 0$

$\Rightarrow \quad ax\cos\alpha + ay\sin\alpha - a^2(1 + \cos\alpha) = 0$ or $x\cos\alpha + y\sin\alpha = a(1 + \cos\alpha)$

ILLUSTRATION 37: Prove that the straight line $y = x + c\sqrt{2}$ touches the circle $x^2 + y^2 = c^2$, and find its point of contact.

SOLUTION: The given circle is, $x^2 + y^2 = c^2$, ...(1)

and the line is $y = x + c\sqrt{2}$ ...(2)

Putting the value of y from (2) in (1), we get

$\Rightarrow x^2 + x^2 + 2xc\sqrt{2} + 2c^2 - c^2 = 0$

$\Rightarrow 2x^2 + 2xc\sqrt{2} + c^2 = 0 \Rightarrow \left(\sqrt{2}.x + c\right)^2 = 0$ ...(3)

This being a perfect square gives two coincident values of x, hence (2) is a tangent to (1)

Again by (3): $\sqrt{2}.x + c = 0 \Rightarrow x = -\frac{c}{\sqrt{2}}$

Substituting the value of x in (2), we get $y = -\frac{c}{\sqrt{2}} + c\sqrt{2} = \frac{c}{\sqrt{2}}$

Hence, the point of contact is $\left(-\frac{c}{\sqrt{2}}, \frac{c}{\sqrt{2}}\right)$

ILLUSTRATION 38: Show that the circles $x^2 + y^2 - 4x + 6y + 8 = 0$ and $x^2 + y^2 - 10x - 6y + 14 = 0$ touch at (3, –1) and hence write the equation of common tangent.

SOLUTION: Equation of tangent at (3, –1) of the circle $x^2 + y^2 - 4x + 6y + 8 = 0$ is

$3x + (-1)y - 2(x+3) + 3(y-1) + 8 = 0 \Rightarrow x + 2y - 1 = 0$ ...(i)

and equation of tangent at (3, –1) of the circle $x^2 + y^2 - 10x - 6y + 14 = 0$ is

$3.x + (-1).y - 5(x+3) - 3(y-1) + 14 = 0$

$\Rightarrow -2x - 4y + 2 = 0 \Rightarrow x + 2y - 1 = 0$ ....(ii)

which is same as (i). Hence the given circles touch at (3, –1)

Clearly, common tangent is $x + 2y - 1 = 0$

ILLUSTRATION 39: Find the equation of the normal to the circle $x^2 + y^2 = 2x$ which is parallel to the line $x + 2y = 3$.

SOLUTION: Given circle is $x^2 + y^2 - 2x = 0$. Centre of given circle is (1, 0)

Since normal is parallel to $x + 2y = 3$

Let the equation of normal is $x + 2y = \lambda$

Since normal passes through the centre of the circle, i.e., (1, 0)

then $1 + 0 = \lambda \Rightarrow \lambda = 1$

Equation of normal is $x + 2y = 1$ or $x + 2y - 1 = 0$

Aliter: Let (x_1, y_1) be any arbitrary point on normal to circle $x^2 + y^2 - 2x = 0$

Then its equation is given by $\dfrac{x - x_1}{x_1 - 1} = \dfrac{y - y_1}{y_1 - 0}$ $\therefore$ slope $= \dfrac{y_1}{x_1 - 1} = m_1$

Now, since normal is parallel to $x + 2y = 3$

$\Rightarrow$ slope $= -\dfrac{1}{2} = m_2$ but given $m_1 = m_2$

$\Rightarrow \dfrac{y_1}{x_1 - 1} = -\dfrac{1}{2}$ or $x_1 + 2y_1 - 1 = 0$ Locus of (x_1, y_1) is $x + 2y - 1 = 0$

ILLUSTRATION 40: Find the equation of the normal to the circle $x^2 + y^2 - 5x + 2y - 48 = 0$ drawn at the point (5, 6).

SOLUTION: Equation of the normal at (5, 6) is $\dfrac{x-5}{5 - \frac{5}{2}} = \dfrac{y-6}{6+1} \Rightarrow \dfrac{x-5}{\frac{5}{2}} = \dfrac{y-6}{7} \Rightarrow \dfrac{2x-10}{5} = \dfrac{y-6}{7}$

$\Rightarrow 14x - 70 = 5y - 30 \Rightarrow 14x - 5y - 40 = 0$

Aliter 1: Since centre of the circle $x^2 + y^2 - 5x + 2y - 48 = 0$ is (5/2, –1), normal at (5, 6) is the equation of a line which passes through (5/ 2, –1) and (5, 6) is

$$y + 1 = \frac{6+1}{5 - 5/2}\left(x - \frac{5}{2}\right) \Rightarrow y + 1 = \frac{14}{5}\left(x - \frac{5}{2}\right)$$

$$\Rightarrow y + 1 = \frac{7}{5}(2x - 5) \Rightarrow 5y + 5 = 14x - 35 \Rightarrow 14x - 5y - 40 = 0$$

Aliter 2: Equation of tangent at (5, 6) is $5.x + 6.y - \dfrac{5}{2}(x+5) + (y+6) - 48 = 0$

$\Rightarrow 10x + 12y - 5x + 2y + 12 - 96 - 25 = 0$

$\Rightarrow 5x + 14y - 109 = 0$

$\Rightarrow$ Slope of tangent $= -\dfrac{5}{14}$ and hence slope of normal $= \dfrac{14}{5}$

Equation of normal: line through (5, 6) with slope $\frac{14}{5} \Rightarrow y-6=\frac{14}{5}(x-5)$

$\Rightarrow 5y - 30 = 14x - 70 \Rightarrow 14x - 5y - 40 = 0$

ILLUSTRATION 41: Find the equations of the tangents on the circle $x^2 + y^2 + 4x + 6y + 3 = 0$ which pass through the point (3, 2).

SOLUTION: Combined equation of the pair of tangents drawn from $A(3,2)$ to the given circle $x^2 + y^2 + 4x + 6y + 3 = 0$ can be written in the usual notation as

$T^2 = SS_1$

$\Rightarrow [3x + 2y + 2(x + 3) + 3(y + 2) + 3]^2 = [x^2 + y^2 + 4x + 6y + 3][9 + 4 + 12 + 12 + 3]$

$\Rightarrow (5x + 5y + 15)^2 = 40(x^2 + y^2 + 4x + 6y + 3)$

$\Rightarrow 5(x^2 + y^2 + 9 + 2xy + 6x + 6y) = 8(x^2 + y^2 + 4x + 6y + 3)$

$\Rightarrow 3x^2 + 3y^2 - 10xy + 2x + 18y - 21 = 0$

$\Rightarrow (3)x^2 + x(2 - 10y) + (3y^2 + 18y - 21) = 0$

$$\Rightarrow x=\frac{10y-2\pm 2\sqrt{(25y^2+1-10y)-(9y^2+54y-63)}}{6}=\frac{10y-2\pm 2\times 4\sqrt{y^2-4y+4}}{6}$$

$$=\frac{10y-2\pm 8(y-2)}{6}=\frac{10y-2+8y-16}{6},\frac{10y-2-8y+16}{6}=\frac{18y-18}{6},\frac{2y+14}{6}$$

$$=y-3,\ \frac{y+7}{3}$$

$\Rightarrow x = y - 3$ or $3x = y + 7$ are the required tangents to the circle from $A(3,2)$

Aliter: Let $S \equiv x^2+y^2+4x+6y+3=0$

Centre : $C \equiv (-2,-3)$ and radius $= \sqrt{2^2+3^2-3} = \sqrt{10}$

Let the slope of a tangent from A (3, 2) to $S = 0$ be 'm', then equation of tangent is $y - 2 = m(x - 3)$

Length of perpendicular from $C(-2, -3)$ on the tangent = radius of circle

$$\Rightarrow \frac{|-2m+3+2-3m|}{\sqrt{m^2+1}}=\sqrt{10}$$

$\Rightarrow 5(m-1)^2 = 2m^2 + 2 \Rightarrow 5m^2 + 5 - 10m = 2m^2 + 2$

$\Rightarrow 3m^2 - 10m + 3 = 0 \Rightarrow 3m^2 - 9m - m + 3 = 0 \Rightarrow (3m - 1)(m - 3) = 0$

$\Rightarrow m = \frac{1}{3}$ *or* $m = 3$

Substituting these values of m in equation of tangent; we get the equations as

$y - 2 = 3x - 9 \Rightarrow 3x - y - 7 = 0$

And $y - 2 = \frac{1}{3}(x-3) \Rightarrow x - y + 3 = 0$

ILLUSTRATION 42: Find the equation of the circle having the pair of lines $x^2+2xy+3x+6y=0$ as its normals and having the size just sufficient to contain the circle $x(x - 4) + y(y - 3) = 0$

SOLUTION: Given circle is $x(x - 4) + y(y - 3) = 0$ ...(i)

$$\Rightarrow x^2-4x+4+y^2-3y+\frac{9}{4}=4+\frac{9}{4} \Rightarrow (x-2)^2+\left(y-\frac{3}{2}\right)^2=\left(\frac{5}{2}\right)^2$$

Hence centre $B = (2, 3/2)$ and radius $= 5/2$

Given pair of lines is $x^2+2xy+3x+6y=0$

$\Rightarrow x(x+2y)+3(x+2y)=0 \Rightarrow (x+3)(x+2y)=0$

Hence $x+3=0$ and $x+2y=0$ are two straight lines

Solving $x+3=0$ and $x+2y=0$, we get $x=-3$,. $y=3/2$

Hence, centre of the required circle is $A(-3, 3/2)$

Now $AB=\sqrt{(2+3)^2+\left(\frac{3}{2}-\frac{3}{2}\right)^2}=5 \Rightarrow AB>5/2$

Hence point A lies outside the circle (i)

$\therefore$ Radius of the required circle $= AB + 5/2 = 5 + 5/2 = 15/2$

Hence equation of circle is $(x+3)^2+(y-3/2)^2=(15/2)^2$.

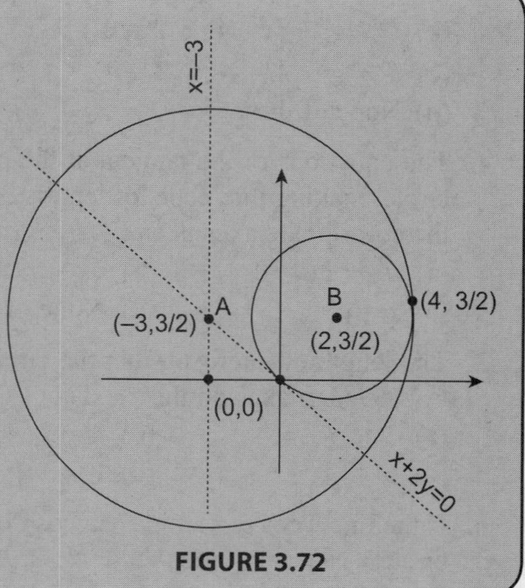

FIGURE 3.72

TEXTUAL EXERCISE–4 (SUBJECTIVE)

1. Find the equations of the tangents to the circle $x^2+y^2=4$ which are parallel to the line $x+2y+3=0$.

2. Find the condition that the straight line $ax - by + b^2 = 0$ may touch the circle $x^2+y^2=ax+by$ and find the point of contact.

3. Prove that the tangents to the circle $x^2+y^2=169$ at (5, 12) and (12, –5) are perpendicular to each other.

4. If the line $4x-3y=-12$ is tangent at the point (–3, 0) and the line $3x+4y=16$ is tangent at the point(4, 1) to a circle. Find the equation of the circle.

5. Find the value of p so that the straight line $x\cos\alpha + y\sin\alpha - p = 0$ may touch the circle $x^2+y^2-2ax\cos\alpha-2by\sin\alpha-a^2\sin^2\alpha=0$.

6. Prove that the tangent to the circle $x^2+y^2=5$ at the point (1, –2) also touches the circle $x^2+y^2-8x+6y+20=0$ and find its point of contact.

7. Find the condition that the straight line $Ax+By+C=0$ may touch the circle $(x-\alpha)^2+(y-\beta)^2=c^2$.

Answer Keys

1. $x+2y\pm 2\sqrt{5}=0$ **2.** (0, b) **4.** $(x-1)^2+(y+3)^2=25$ **5.** $p=a\cos^2\alpha+b\sin^2\alpha\pm\sqrt{a^2\cos^2\alpha+b^2\sin^2\alpha}$

6. $x^2+y^2-8x+6y+20=0$ at (3, –1) **7.** $A\alpha+B\beta+C=\pm c\sqrt{(A^2+B^2)}$

TEXTUAL EXERCISE–4 (OBJECTIVE)

1. If the line $ax+by+c=0$ touches the circle $x^2+y^2-2x=3/5$ and is normal to the circle $x^2+y^2+2x-4y+1=0$, then (a, b) are
   (a) 1, 3 (b) 3, 1
   (c) 1, 2 (d) 2, 1

2. The normal to the circle $(x-2)^2+y^2=4$ at the point (4,0) meets the circle again at:
   (a) (0, 4) (b) (0, 2)
   (c) (0, 0) (d) (2, 0)

3. If the length of tangent from (f, g) to the circle $x^2+y^2=6$ be twice the length of the tangent from (f, g) to circle $x^2+y^2+3x+3y=0$, then which of the following is true?

(a) $f^2 + g^2 + 4f + 4g - 2 = 0$
(b) $f^2 + g^2 + 4f + 4g + 2 = 0$
(c) $f^2 + g^2 - 4f - 4g + 2 = 0$
(d) None of these

4. The equations of the tangent at the point (0, 0) to the circle, making intercepts of length $2a$ and $2b$ units on the co-ordinate axes is /are

(a) $ax + by = 0$ (b) $ax - by = 0$
(c) $x = y$ (d) None of these

5. The length of intercept, that the circle $x^2 + y^2 + 10x - 6y + 9 = 0$ makes on the x-axis is

(a) 2 (b) 4
(c) 6 (d) 8

6. If the line $y \cos \alpha = x \sin\alpha + a \cos\alpha$ be a tangent to the circle $x^2 + y^2 = a^2$, then

(a) $\sin^2\alpha = 1$ (b) $\cos^2\alpha = 1$
(c) $\sin^2\alpha = a^2$ (d) $\cos^2\alpha = a^2$

7. If the ratio of the length of tangents drawn from the point (f, g) to the given circles $x^2 + y^2 = 6$ and $x^2 + y^2 + 3x + 3y = 0$ be 2 : 1, then

(a) $f^2 + g^2 + 2g + 2f + 2 = 0$
(b) $f^2 + g^2 + 4g + 4f + 4 = 0$
(c) $f^2 + g^2 + 4g + 4f + 2 = 0$
(d) None of these

8. The equation of the tangent at the point $\left(\frac{ab^2}{a^2+b^2}, \frac{a^2b}{a^2+b^2}\right)$ of the circle $x^2 + y^2 = \frac{a^2b^2}{a^2+b^2}$ is

(a) $\frac{x}{a}+\frac{y}{b}=1$ (b) $\frac{x}{a}+\frac{y}{b}+1=0$
(c) $\frac{x}{a}-\frac{y}{b}=1$ (d) $\frac{x}{a}-\frac{y}{b}+1=0$

9. If O is the origin and OP, OQ are tangents to the circle $x^2 + y^2 + 2gx + 2fy + c = 0$, the circumcentre of the triangle OPQ is

(a) $(-g, -f)$ (b) (g, f)
(c) $(-f, -g)$ (d) None of these

10. The angle between the tangents from (α, β) to the circle $x^2 + y^2 = a^2$, is

(a) $\tan^{-1}\left(\frac{a}{\sqrt{\alpha^2+\beta^2-a^2}}\right)$

(b) $\tan^{-1}\left(\frac{\sqrt{\alpha^2+\beta^2-a^2}}{a}\right)$

(c) $2\tan^{-1}\left(\frac{a}{\sqrt{\alpha^2+\beta^2-a^2}}\right)$

(d) None of these

11. The tangent at P, any point on the circle $x^2 + y^2 = 4$, meets the co-ordinate axes in A and B, then

(a) Length of AB is constant
(b) PA and PB are always equal
(c) The locus of the mid point of AB is $x^2 + y^2 = x^2y^2$
(d) None of these

12. The line $ax + by + c = 0$ is a normal to the circle $x^2 + y^2 = r^2$. The portion of the line $ax + by + c = 0$ intercepted by this circle is of length

(a) r (b) r^2
(c) $2r$ (d) $\sqrt{r}$

13. If a circle, whose centre is (–1, 1) touches the straight line $x + 2y + 12 = 0$, then the co-ordinates of the point of contact are

(a) $\left(\frac{-7}{2}, -4\right)$ (b) $\left(\frac{-18}{5}, \frac{-21}{5}\right)$
(c) (2, – 7) (d) (– 2, – 5)

14. The gradient of the normal at the point (–2, –3) on the circle $x^2 + y^2 + 2x + 4y + 3 = 0$ is

(a) 1 (b) –1
(c) 3/2 (d) 1/2

15. Equation of the tangent to the circle $x^2 + y^2 = a^2$ which is perpendicular to the straight line $y = mx + c$ is

(a) $y=-\frac{m}{x}\pm a\sqrt{1+m^2}$
(b) $x+my=\pm a\sqrt{1+m^2}$
(c) $x+my=\pm a\sqrt{1+(1/m^2)}$
(d) $x-my=\pm a\sqrt{1+1/m^2}$

16. The area of the triangle formed by the tangent at (3, 4) to the circle $x^2 + y^2 = 25$ and the co-ordinate axes is

(a) $\frac{24}{25}$ (b) 0
(c) $\frac{625}{24}$ (d) $-\left(\frac{24}{25}\right)$

17. If $\frac{x}{\alpha}+\frac{y}{\beta}=1$ touches the circle $x^2 + y^2 = a^2$, then point $(1/\alpha, 1/\beta)$ lies on a/an

(a) Straight line (b) Circle
(c) Parabola (d) Ellipse

18. If line $ax + by = 0$ touches $x^2 + y^2 + 2x + 4y = 0$ and is a normal to the circle $x^2 + y^2 - 4x + 2y - 3 = 0$ then value of (a, b) will be

(a) (2, 1) (b) (1, –2)
(c) (1, 2) (d) (–1, 2)

Answer Keys

1. (a) **2.** (c) **3.** (b) **4.** (a) **5.** (d) **6.** (b) **7.** (c) **8.** (a) **9.** (d) **10.** (c)
11. (c) **12.** (c) **13.** (b) **14.** (a) **15.** (b) **16.** (c) **17.** (b) **18.** (c)

■ DIRECTOR CIRCLE

The locus of point of intersection of two perpendicular tangents to a given circle is known as its *director circle.*

Method 1: Clearly, *PTCR* is always a square.

$\Rightarrow \quad PC = a\sqrt{2} \Rightarrow h^2 + k^2 = 2a^2$

$\Rightarrow \quad x^2 + y^2 = 2a^2$

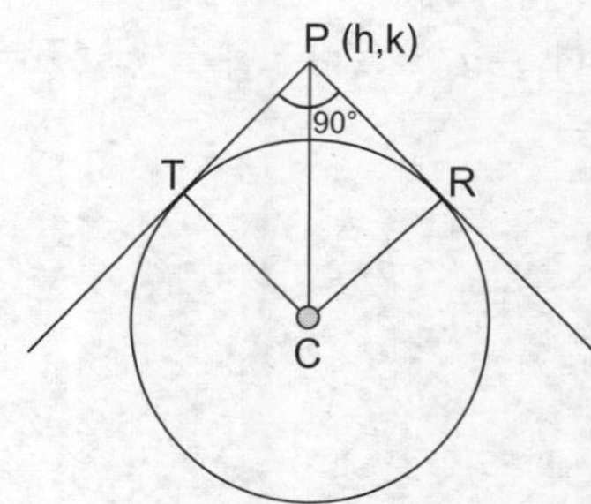

FIGURE 3.73

Method 2: The equation of any tangent to the circle $x^2 + y^2 = a^2$ is $y = mx \pm a\sqrt{1+m^2}$ ...(i)

Let $P(h, k)$ be the point of intersection of tangents, then $P(h, k)$ lies on (i).

$\therefore \quad k = mh \pm a\sqrt{1+m^2} \Rightarrow (k - mh)^2 = a^2(1 + m^2)$

$\Rightarrow \quad m^2(h^2 - a^2) - 2mkh + k^2 - a^2 = 0$

This is quadratic equation in m, let two roots are m_1 and m_2, but tangents are perpendiculars, then $m_1 m_2 = -1$

$\Rightarrow \quad \dfrac{k^2 - a^2}{h^2 - a^2} = -1$

$\Rightarrow \quad k^2 - a^2 = -h^2 + a^2 \Rightarrow h^2 + k^2 = 2a^2$

Hence locus of $P(h, k)$ is $x^2 + y^2 = 2a^2$

Method 3: The combined equation of the pair of tangents from (h,k) to $x^2 + y^2 = a^2$ is given $SS_1 = T^2$ where $S = x^2 + y^2 - a^2$; $S_1 = h^2 + k^2 - a^2$ and $T = hx + ky - a^2$

Now $(x^2 + y^2 - a^2)(h^2 + k^2 - a^2) = (hx + ky - a^2)^2$

$\Rightarrow \quad h^2x^2 + h^2y^2 - a^2h^2 + k^2x^2 + k^2y^2 - k^2a^2 - a^2x^2 - a^2y^2 + a^4 = h^2x^2 + k^2y^2 + a^4 - 2hxa^2 - 2kya^2 + 2hkxy$

Now the above equation of pair of straight line will represent two perpendicular lines if co-efficient of x^2 + co-efficient of $y^2 = 0$

$\Rightarrow \quad h^2 + k^2 - a^2 - h^2 + h^2 + k^2 - a^2 - k^2 = 0$

$\Rightarrow \quad h^2 + k^2 = 2a^2$

$\therefore \quad$ the locus of (h,k) is $x^2 + y^2 = 2a^2$

NOTES

1. *The director circle of the circle $(x - \alpha)^2 + (y - \beta)^2 = a^2$ is given by $(x - \alpha)^2 + (y - \beta)^2 = 2a^2$*
2. *The director circle of the circle $x^2 + y^2 + 2gx + 2fy + c = 0$ is given by $x^2 + y^2 + 2gx + 2fy + 2c - g^2 - f^2 = 0$.*

ILLUSTRATION 43: Prove that the tangents to the circle $x^2 + y^2 = 25$ at (3, 4) and (4, –3) are perpendicular to each other.

SOLUTION: The equation of tangents to the circle $x^2 + y^2 = 25$ at (3, 4) and (4, –3) are

$3x + 4y = 25$ ...(i)

and $4x - 3y = 25$ ...(ii)

Now slope of (i) is $\dfrac{-3}{4} = m_1$ and slope of (ii) is $\dfrac{4}{3} = m_2$

Clearly, $m_1 m_2 = -1$. Hence (i) and (ii) are perpendicular to each other.

CHORDS OF CIRCLE

A line segment joining any two points on the curve is called chord of the curve. We have already studied in Euclidian geometry that perpendicular drawn from centre to any chord always bisects the chord. So let us define two types of chords as shown in Figure 3.74.

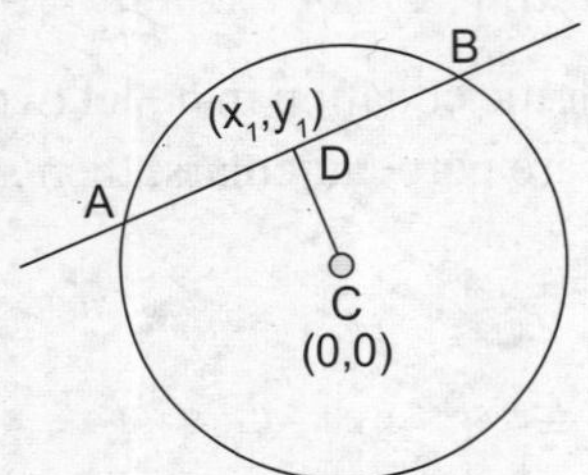

FIGURE 3.74

Chord with Mid-Point (x_1, y_1)

Let any chord AB of the circle $x^2 + y^2 = a^2$ be bisected at D (x_1, y_1). If centre of circle is representd by C. then slope of

$$DC = \frac{0 - y_1}{0 - x_1} = \frac{y_1}{x_1}$$

$\therefore$ Slope of the chord AB is $-\frac{x_1}{y_1}$,

then equation of AB is $y - y_1 = -\frac{x_1}{y_1}(x - x_1)$

$$\Rightarrow \quad yy_1 - y_1^2 = -xx_1 + x_1^2 \Rightarrow xx_1 + yy_1 = x_1^2 + y_1^2$$

$$\Rightarrow \quad xx_1 + yy_1 - a^2 = x_1^2 + y_1^2 - a^2 \Rightarrow \boldsymbol{T = S_1}$$

NOTES

1. *It must be noted that in above equation* $x_1^2 + y_1^2 - a^2 \neq 0$

   $\because$ *point* (x_1, y_1) *does not lie on the circle.*

2. *If* (x_1, y_1) *lies on the circle* $x^2 + y^2 = a^2$, *then* $x_1^2 + y_1^2 = a^2 \Rightarrow x_1^2 + y_1^2 - a^2 = 0 \Rightarrow xx_1 + yy_1 = a^2 \Rightarrow T = 0$

   $\Rightarrow$ *the chord becomes a tangent.*

3. *The equation of chord of the circle* $x^2 + y^2 + 2gx + 2fy + c = 0$, *which is bisected at* (x_1, y_1) *is* $T = S_1$. *Where* $T = xx_1 + yy_1 + g(x + x_1) + f(y + y_1) + c$ *and* $S_1 = x_1^2 + y_1^2 + 2gx_1 + 2fy_1 + c = 0$

4. *The chord obtained by* $T = S_1$ *is also*

   (a) the smallest chord passing through the point (x_1, y_1).

   (b) the chord passing through (x_1, y_1) *which is farthest from the centre.*

5. *Equation of the straight line joining two points* α *and* β *on the circle* $x^2 + y^2 = a^2$

   Required equation is $x\cos\left(\frac{\alpha+\beta}{2}\right) + y\sin\left(\frac{\alpha+\beta}{2}\right) = a\cos\left(\frac{\alpha-\beta}{2}\right)$

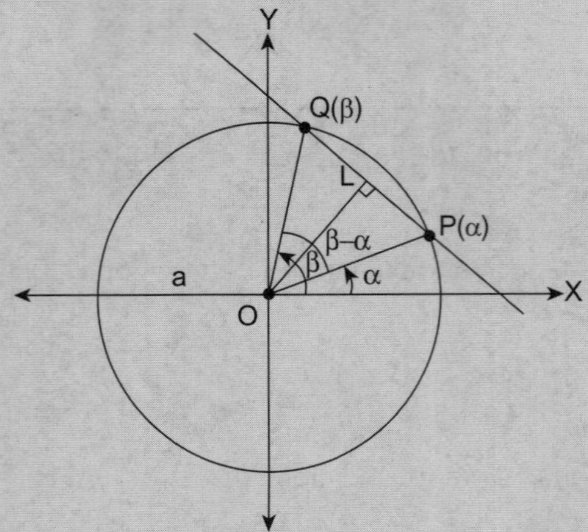

FIGURE 3.75

Length of chord of contact is $AB = \frac{2LR}{\sqrt{R^2+L^2}}$ *and area of the triangle formed by the pair of tangents and its chord of contact is* $\frac{RL^3}{R^2+L^2}$, *where R is the radius of the circle and L is the length of tangent from* $P(x_1, y_1)$ *on* $S = 0$.

Here $L = \sqrt{S_1}$

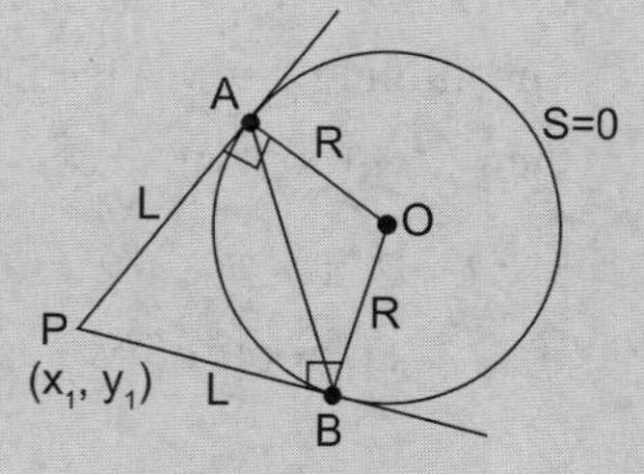

FIGURE 3.76

CHORD OF CONTACT

Let $P(x_1, y_1)$ be a point outside the circle $S = x^2 + y^2 + 2gx + 2fy + c = 0$. Let tangents from P to the circle meet the circle $S = 0$ at A and B then AB is called chord of contact whose equation is

$$T \equiv xx_1 + yy_1 + g(x + x_1) + f(y + y_1) + c = 0$$

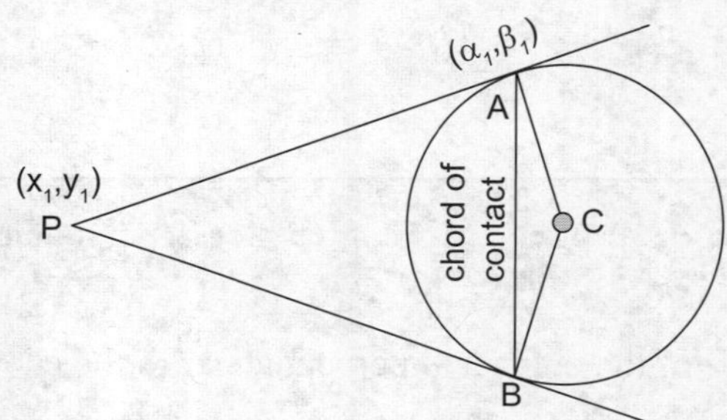

FIGURE 3.77

Proof: The equation $T = 0$ actually represents a line. If we could show that $T = 0$ is satisfied by co-ordinates of A and B then $T = 0$ must be the equation of line AB. Let A be (α, β) then in order to show that A lies on $T = 0$, we must show that $\alpha x_1 + \beta y_1 + g(\alpha + x_1) + f(\beta + y_1) + c = 0$ ...(i)

Equation of tangent at $A(\alpha, \beta)$ is

$\alpha x + \beta y + g(\alpha + x) + f(\beta + y) + c = 0$

Since this tangent passes through $P(x_1, y_1)$, we have $\alpha x_1 + \beta y_1 + g(\alpha + x_1) + f(\beta + y_1) + c = 0$ which is same as the relation (i).

Thus A lies on $T = 0$. In similar manner, it can be shown that B also satisfies the line $T = 0$.

Thus $T = 0$ represents the line AB (Chord of contact) in the figure.

Aliter: Equation of circle circumscribing ΔPAB

$S' = x(x - x_1) + y(y - y_1) = 0$

$\Rightarrow S' = x^2 + y^2 - xx_1 - yy_1 = 0$

Equation of locus through intersection of

$S = 0$ and $S' = 0$ is $S + \lambda S' = 0$

i.e., $(x^2 + y^2 - a^2) + \lambda(x^2 + y^2 - xx_1 - yy_1) = 0$

For $\lambda = -1$ the curve becomes $x x_1 + y y_1 = a^2$

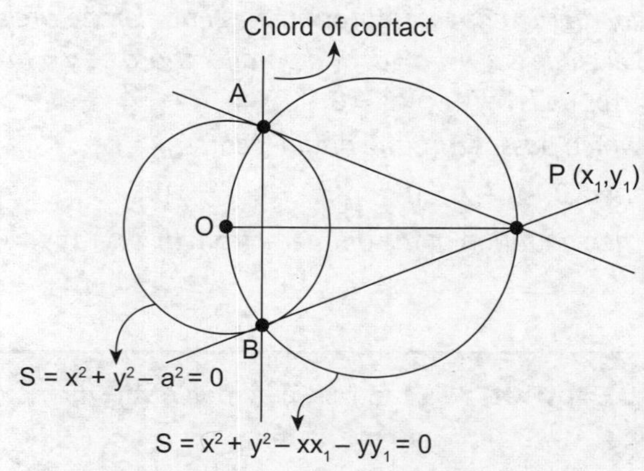

FIGURE 3.78

DIAMETER OF A CIRCLE

The locus of middle points of a system of parallel chords of a circle is called the diameter of circle. The diameter of the circle $x^2 + y^2 = r^2$ corresponding to the system of parallel chords $y = mx + c$ is $x + my = 0$. Let (h, k) be the middle point of the chord $y = mx + c$. Since P is the mid-point of segment joining $A(x_1, y_1)$ and $B(x_2, y_2)$, then $\frac{x_1 + x_2}{2} = h$ and $\frac{y_1 + y_2}{2} = k$

or $x_1 + x_2 = 2h$ and $y_1 + y_2 = 2k$ ...(i)

$\because$ $P(h, k)$ lie on $y = mx + c$

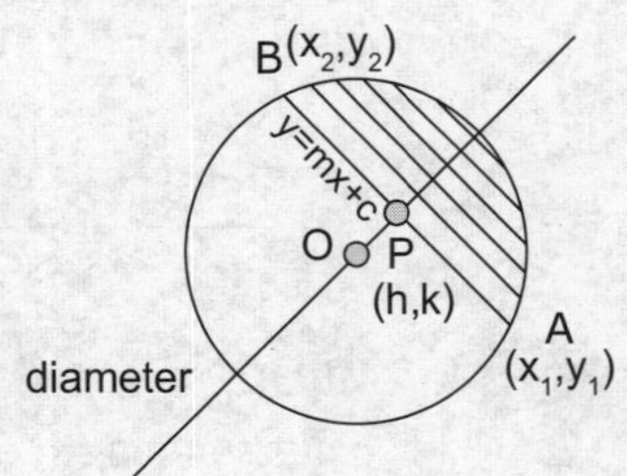

FIGURE 3.79

then $k = mh + c$ or $k - mh = c$ ...(ii)

Substituting $y = mx + c$ in $x^2 + y^2 = a^2$

then $x^2 + (mx + c)^2 = a^2$

$(1 + m^2)x^2 + 2mcx + c^2 - a^2 = 0$ ...(iii)

If x_1, x_2 are roots of (iii) then

$x_1 + x_2 = -\frac{mc}{} \Rightarrow 2h = -\frac{2m}{(1+m^2)}(k - mh)$

(from (i) and (ii))

$\Rightarrow \quad h + m^2h = -mk + m^2h \Rightarrow h + mk = 0,$

so locus of (h, k) is $x + my = 0$.

Aliter: Let (h, k) be the middle point of the chord $y = mx + c$ of the circle $x^2 + y^2 = a^2$, then equation of chord is given by $T = S_1 \Rightarrow xh + ky = h^2 + k^2$

$\Rightarrow$ slope $= -\frac{h}{k} = m \Rightarrow h + mk = 0$

Hence locus of mid-point is $x + my = 0$.

NOTE

Every diameter passes through the centre of the circle and perpendicular to the system of parallel chords.

Let circle is $x^2 + y^2 = a^2$ and parallel chord be $y = mx + c$ then equation of line perpendicular to $y = mx + c$ is $my + x + \lambda = 0$ ...(*i*)

which passes through origin (centre)

then $0 + 0 + \lambda = 0 \quad \therefore \quad \lambda = 0$

then equation of the diameter from (i) is $x + my = 0$.

ILLUSTRATION 44: Find the equation of the diameter of the circle $x^2 + y^2 + 2gx + 2fy + c = 0$ which is perpendicular to the chord $ax + by + d = 0$.

SOLUTION: The diameter of circle passes through the centre of the circle and perpendicular to the chord $ax + by + d = 0$ is $bx - ay + \lambda = 0$ ...(i)

which if passed through centre of circle i .e. $(-g, -f)$

gives $-bg + af + \lambda = 0 \ \therefore \lambda = bg - af$

from (i), the equation of the diameter is $bx - ay + bg - af = 0$

ILLUSTRATION 45: Tangents PQ and PR are drawn from $P(0, -2)$ to the circle $x^2 + y^2 + 2x - 4y = 0$. Find

(a) The equation of chord of contact QR

(b) The length QR

(c) Area of quadrilateral $PQCR$, where C is the centre of the circle

(d) The area of the triangle PQR

(e) The angle between PQ and PR.

SOLUTION: (a) The equation of chord of contact AB must be $x.0 + y(-2) + 1(x + 0) - 2(y - 2) + 0 = 0$

$\Rightarrow x - 4y + 4 = 0$ ....(i)

The perpendicular PM from $P(0, -2)$ to the chord given by (i)

$$= \left| \frac{0 - 4(-2) + 4}{\sqrt{1^2 + (-4)^2}} \right| = \frac{12}{\sqrt{17}}$$

(b) Length $PQ = \sqrt{S_1} = \sqrt{0^2 + (-2)^2 + 2\times 0 - 4\times(-2)} = \sqrt{12}$ hence the half of length $QR = QM$

$$\sqrt{(\sqrt{12})^2 - \left(\frac{12}{\sqrt{17}}\right)^2} = \sqrt{\frac{60}{17}} \qquad \left[\because (QM)^2 = (PQ) - (QM)^2\right]$$

Thus the length of the chord $QR = 2\sqrt{\frac{60}{17}}$

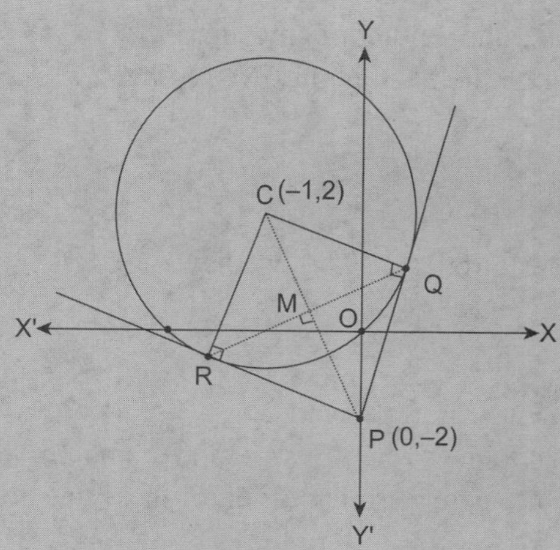

FIGURE 3.80

(c) The area of quadrilateral $PQCR = 2 \times$ Area of ΔPQC

$$= 2\times\frac{1}{2}(PQ)QC = \sqrt{12}\times\text{radius} = \sqrt{12}\times\sqrt{5} = \sqrt{60} = 2\sqrt{15} \text{ sq. units}$$

(d) The area of triangle $PQR = \frac{1}{2}\times(\text{length of } QR)\times PM = \frac{1}{2}\times 2\sqrt{\frac{60}{17}}\times\frac{12}{\sqrt{17}} = \frac{24\sqrt{15}}{17}$ sq. units

(e) If θ be the angle between PQ and PR, then $\tan\frac{\theta}{2} = \frac{QC}{PQ} = \frac{\sqrt{5}}{\sqrt{12}}$

$$\Rightarrow \tan\theta = \frac{2\tan\theta/2}{1-\tan^2\frac{\theta}{2}} = \frac{2\times\frac{\sqrt{5}}{\sqrt{12}}}{1-\frac{5}{12}} = \frac{2}{7}\sqrt{60} \Rightarrow \theta = \tan^{-1}\frac{2}{7}\sqrt{60}$$

ILLUSTRATION 46: Find the condition that chord of contact from any external points (h, k) to the circle $x^2 + y^2 = a^2$ should subtend right angle at the centre of the circle.

SOLUTION: Equation of the chord of contact AB is $hx + ky = a^2$ ...(i)

For equation of pair of lines OA and OB, make $x^2 + y^2 = a^2$ homogeneous with the help of line $hx + ky = a^2$

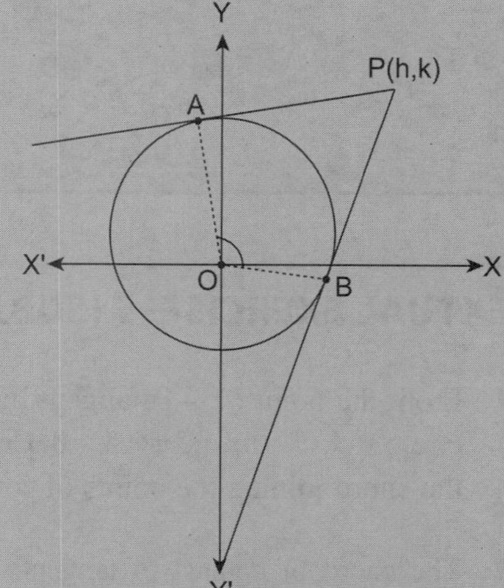

FIGURE 3.81

$$\Rightarrow \frac{hx+ky}{a^2} = 1$$

then $x^2 + y^2 = a^2\left(\frac{hx+ky}{a^2}\right)^2$

$\Rightarrow a^2(x^2 + y^2) = (hx + ky)^2$

$\Rightarrow x^2(a^2 - h^2) - 2hkxy + y^2(a^2 - k^2) = 0$

but $\angle AOB = \pi/2$

$\therefore$ co-efficient of x^2 + co-efficient of $y^2 = 0$

$\Rightarrow a^2 - h^2 + a^2 - k^2 = 0 \Rightarrow h^2 + k^2 = 2a^2$

ILLUSTRATION 47: Through a fixed point (h, k) secants are drawn to the circle $x^2 + y^2 = a^2$. Show that the locus of mid-point of the portions of secants intercepted by the circle is $x^2 + y^2 = hx + ky$.

SOLUTION: Let $P(x_1, y_1)$ be the middle point of any chord AB, which passes through the fixed point $C(h, k)$, then the equation of the chord AB is

$$T = S_1$$

$$\therefore \quad xx_1 + yy_1 - a^2 = x_1^2 + y_1^2 - a^2$$

But since AB passes through $C(h, k)$, we have

$$x_1^2 + y_1^2 = hx_1 + ky_1$$

$\Rightarrow$ locus of $P(x_1, y_1)$ is $x^2 + y^2 = hx + ky$

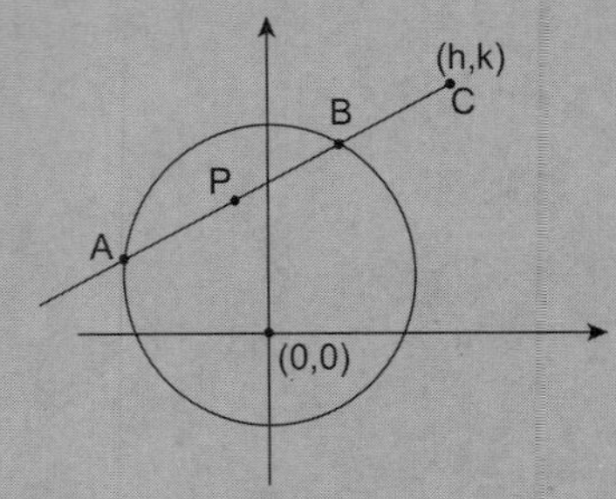

FIGURE 3.82

ILLUSTRATION 48: Tangents are drawn to the circle $x^2 + y^2 = 12$ at the points where it is met by the circle $x^2 + y^2 - 5x + 3y - 2 = 0$, find the point of intersection of these tangents.

SOLUTION: The given circles are $S_1 \equiv x^2 + y^2 = 12$ ...(i)

and $S_2 \equiv x^2 + y^2 - 5x + 3y - 2 = 0$ ..(ii)

If A and B are the points of intersection of (i) and (ii). Clearly AB will be the common chord whose equation will be $S_1 - S_2 = 0$

$$\Rightarrow \quad (x^2 + y^2 - 12) - (x^2 + y^2 - 5x + 3y - 2) = 0$$

$$\Rightarrow \quad 5x - 3y - 10 = 0 \qquad ...(iii)$$

If P be the point where the tangents at A and B with respect to (i), meet each other, AB will be the chord of contact of P. Let the co-ordinates of P be (α, β).

$\therefore$ Equation of the chord of contact of (α, β) with respect to (i) is

$$x\alpha + y\beta - 12 = 0 \qquad ...(iv)$$

As (iii) and (iv) represent the same equation, comparing the co-efficients, we get

$\dfrac{\alpha}{5} = \dfrac{\beta}{3} = \dfrac{-12}{-10}$ by which we get $\alpha = 6$ and $\beta = -18/5$. Hence the required point is $(6, -18/5)$.

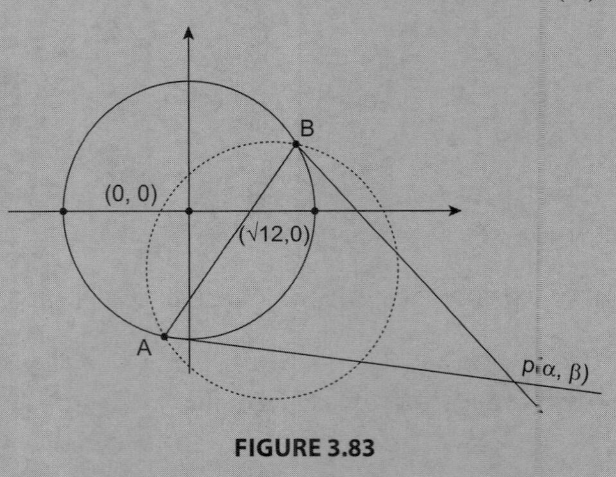

FIGURE 3.83

TEXTUAL EXERCISE–5 (SUBJECTIVE)

1. From the point (4, –4) tangent lines are drawn to the circle $x^2 + y^2 - 6x + 2y + 5 = 0$. Calculate the length of the chord joining the points of contact.

2. The chord of contact of tangents drawn from a point on the circle $x^2 + y^2 = a^2$ to the circle $x^2 + y^2 = b^2$ touches the circle $x^2 + y^2 = c^2$. Show that a, b, c are in G.P.

3. Tangents are drawn from the point (h, k) to the circle $x^2 + y^2 = a^2$. Prove that the area of the triangle formed by them and the straight line joining their point of contact is $\dfrac{a(h^2 + k^2 - a^2)^{3/2}}{h^2 + k^2}$.

4. From the origin, chords are drawn to the circle $(x - 1)^2 + y^2 = 1$. Find the locus of the middle points of the chords.

Answer Keys

1. $\sqrt{10}$ 4. $x^2 + y^2 = x$

TEXTUAL EXERCISE–5 (OBJECTIVE)

1. The chord of the circle $x^2 + y^2 = a^2$ subtends a right angle at the centre. The locus of middle point of the chord is:

(a) $x^2 + y^2 = a^2/2$ (b) $x^2 + y^2 = 4a^2$
(c) $x^2 + y^2 = 2a^2$ (d) $x^2 + y^2 = a^2/4$

2. From the origin, chords are drawn to the circle $x^2 + y^2 - 2y = 0$. The locus of the middle points of these chords is

(a) $x^2 + y^2 - y = 0$ (b) $x^2 + y^2 - x = 0$
(c) $x^2 + y^2 - 2x = 0$ (d) $x^2 + y^2 - x - y = 0$

3. Through a fixed point (h, k), secants are drawn to the circle $x^2 + y^2 = r^2$. The locus of mid-point of the portions of secants intercepted by the circle is

(a) $x^2 + y^2 = hx + ky$ (b) $x^2 + y^2 = hx - ky$
(c) $x^2 + y^2 = h + k$ (d) None of these

4. The common chord of the circle $x^2 + y^2 + 4x + 1 = 0$ and $x^2 + y^2 + 6x + 2y + 3 = 0$ is

(a) $x + y + 1 = 0$ (b) $5x + y + 2 = 0$
(c) $2x + 2y + 5 = 0$ (d) $3x + y + 3 = 0$

5. If the middle point of a chord of the circle $x^2 + y^2 + x - y - 1 = 0$ be (1, 1), then the length of the chord is

(a) 4 (b) 2
(c) 5 (d) None of these

6. $y = mx$ is a chord of a circle of radius 'a' and the diameter of the circle lies along x-axis and one end of this chord is at origin. The equation of the circle described on this chord as diameter is

(a) $(1 + m^2)(x^2 + y^2) - 2ax = 0$
(b) $(1 + m^2)(x^2 + y^2) - 2a(x + my) = 0$
(c) $(1 + m^2)(x^2 + y^2) + 2a(x + my) = 0$
(d) $(1 + m^2)(x^2 + y^2) - 2a(x - my) = 0$

7. The equation of the chord of the circle $x^2 + y^2 = a^2$ having (x_1, y_1) as its mid-point is

(a) $xy_1 + yx_1 = a^2$
(b) $x_1 + y_1 = a$
(c) $xx_1 + yy_1 = x_1^2 + y_1^2$
(d) $xx_1 + yy_1 = a^2$

8. The length of the chord intercepted by the circle $x^2 + y^2 = r^2$ on the line $\frac{x}{a} + \frac{y}{b} = 1$ is

(a) $\sqrt{\frac{r^2(a^2+b^2) - a^2b^2}{a^2+b^2}}$

(b) $2\sqrt{\frac{r^2(a^2+b^2) - a^2b^2}{a^2+b^2}}$

(c) $2\sqrt{\frac{r^2(a^2+b^2) + a^2b^2}{a^2+b^2}}$

(d) None of these

9. Middle point of the chord of the circle $x^2 + y^2 = 25$ intercepted on the line $x - 2y = 2$ is

(a) $\left(\frac{3}{5}, \frac{4}{5}\right)$ (b) $(-2, -2)$
(c) $\left(\frac{2}{5}, -\frac{4}{5}\right)$ (d) $\left(\frac{8}{3}, -\frac{1}{3}\right)$

10. The length of common chord of the circles $(x - a)^2 + y^2 = a^2$ and $x^2 + (y - b)^2 = b^2$ is

(a) $2\sqrt{a^2+b^2}$ (b) $\frac{|ab|}{\sqrt{a^2+b^2}}$
(c) $\frac{|2ab|}{\sqrt{a^2+b^2}}$ (d) None of these

11. The distance between the chord of contact of the tangent to the circle $x^2 + y^2 + 2gx + 2fy + c = 0$ from the origin and the point (g, f) is

(a) $\frac{1}{2}\left(\frac{|g^2+f^2-c|}{\sqrt{g^2+f^2}}\right)$ (b) $\left(\frac{|g^2+f^2-c|}{\sqrt{g^2\ f^2}}\right)$
(c) $-\left(\frac{|g^2+f^2-c|}{g^2\ f^2}\right)$ (d) None of these

12. The equation of the chord of contact, if the tangents are drawn from the point$(5, -3)$ to the circle $x^2 + y^2 = 10$, is

(a) $5x - 3y = 10$ (b) $5x + 3y = 10$
(c) $3x + 5y = 10$ (d) $3x - 5y = 10$

13. A line through (0, 0) cuts the circle $x^2 + y^2 - 2ax = 0$ at A and B, then locus of the centre of the circle drawn on AB as a diameter is

(a) $x^2 + y^2 - 2ay = 0$ (b) $x^2 + y^2 + ay = 0$
(c) $x^2 + y^2 + ax = 0$ (d) $x^2 + y^2 - ax = 0$

14. The radius of the circle, having centre at (2, 1) whose one of the chord is a diameter of the circle $x^2 + y^2 - 2x - 6y + 6 = 0$ is

(a) 1 (b) 2
(c) 3 (d) $\sqrt{3}$

15. The intercept on the line $y = x$ by the circle $x^2 + y^2 - 2x = 0$ is AB, equation of the circle on AB as diameter is

(a) $x^2 + y^2 + x - y = 0$
(b) $x^2 + y^2 - x + y = 0$
(c) $x^2 + y^2 + x + y = 0$
(d) $x^2 + y^2 - x - y = 0$

Answer Keys

1. (a)	**2.** (a)	**3.** (a)	**4.** (a)	**5.** (d)	**6.** (b)	**7.** (c)	**8.** (b)	**9.** (c)	**10.** (c)
11. (a)	**12.** (a)	**13.** (d)	**14.** (c)	**15.** (d)					

■ RELATIVE POSITION OF TWO CIRCLES

Given two circles of radius r_1 and r_2 and centres C_1 (x_1, y_1) and C_2 (x_2, y_2) respectively, then following conditions may arise.

Case 1: When circles lie outside of each other.

Case 2: When circles touch each other externally.

Case 3: When circles intersect each other at two distinct points.

Case 4: When circles touch each other internally.

Case 5: When smaller circle completely lies inside the bigger circle.

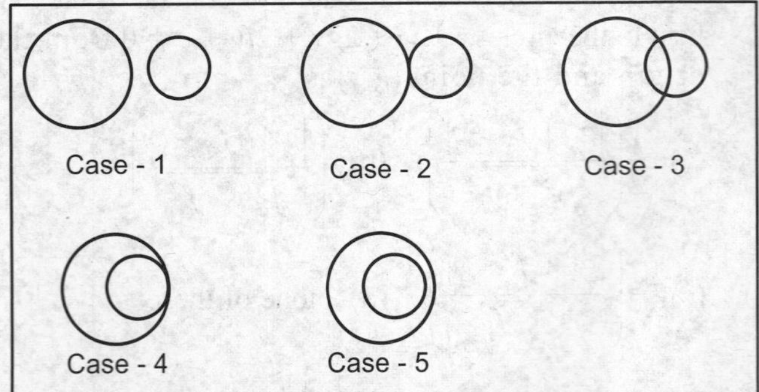

FIGURE 3.84

Case I: When Two Circles Lie Outside of Each Other

Let the two circles be $(x - x_1)^2 + (y - y_1)^2 = r_1^2$ ....(i)

and $(x - x_2)^2 + (y - y_2)^2 = r_2^2$ ....(ii)

with centres $C_1(x_1, y_1)$ and $C_2(x_2, y_2)$ and radii r_1 and r_2, respectively, then they will lie outside of each other when $|C_1 C_2| > r_1 + r_2$, i.e., the distance between the centres is greater than the sum of their radii.

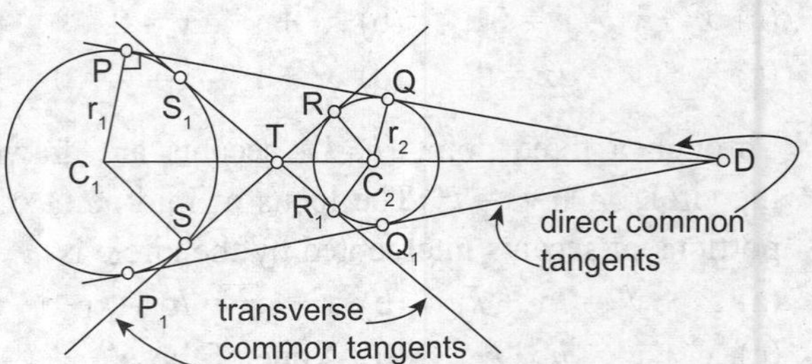

FIGURE 3.85

In this case, four common tangents can be drawn to two circles, in which two are direct common tangents (PQ and P_1Q_1) and the other two are transverse (or indirect) common tangents. (RS and R_1S_1).

Using similarity of triangles C_1ST and $C_2\,RT$, we get $\frac{C_1T}{C_2T} = \frac{C_1S}{C_2R} = \frac{r_1}{r_2}$, i.e., point T divides the line joining C_1 and C_2, internally, in the ratio $r_1 : r_2$.

Now, we can find the co-ordinates of T using the section formula.

$$\therefore\ T \equiv \left(\frac{r_1x_2 + r_2x_1}{r_1 + r_2}, \frac{r_1y_2 + r_2y_1}{r_1 + r_2}\right)$$

Using similarity of triangles C_1PD and $C_2\,QD$

$$\Rightarrow \frac{C_1D}{C_2D} = \frac{C_1P}{C_2Q} = \frac{r_1}{r_2}$$

That is, point D divides the line joining C_1 and C_2 externally in the ratio $r_1 : r_2$.

$\therefore$ Applying section formula, we get

$$D \equiv \left(\frac{r_1x_2 - r_2x_1}{r_1 - r_2}, \frac{r_1y_2 - r_2y_1}{r_1 - r_2}\right)$$

Algorithm: **To find direct common tangents:** Let equation of common tangent through $D(\alpha, \beta)$ is

$y - \beta = m\,(x - \alpha)$ ....(i)

Now, length of perpendicular from C_1 or C_2 on (i) $= r_1$ or r_2 respectively

Then, we get two values of m.

Substituting the values of m in (i); we get two direct common tangents.

Algorithm: **To find transverse common tangents:** Let equation of common tangent through $T(\gamma,\delta)$ is $y - \delta = m_o (x - \gamma)$ ......(ii)

Now, the length of $\perp$ from C_1 or C_2 on (ii) $= r_1$ or r_2 respectively; then we get two values of m_o.

Substituting the values of m_o in (ii); we get two transverse common tangents.

Length of Direct Common Tangent

Length of direct common tangent is defined as distance between the two points of contact, that is,

$$L_D = MN = \sqrt{d^2 - (r_1 - r_2)^2}\ ;\ d = |C_1\,C_2|$$

Angle between D.C.T. $= 2\theta = 2\sin^{-1}\left(\dfrac{|r_1 - r_2|}{d}\right)$

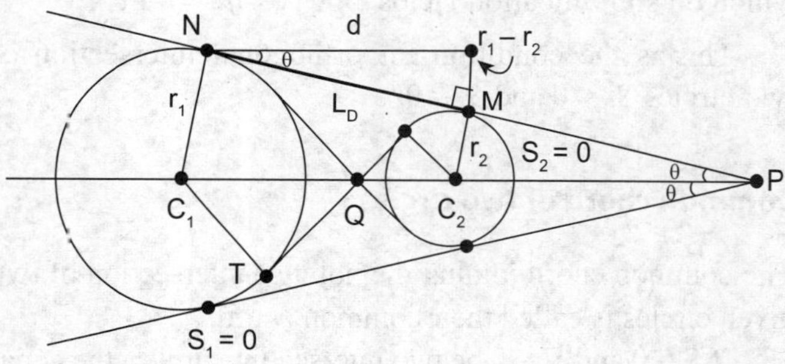

FIGURE 3.86

Length of Transverse Common Tangent

Length of transverse common tangent is defined as distance between point of contacts, i.e., S and T

$$L_T = ST = \sqrt{d^2 - (r_1 + r_2)^2}\ ;\ d = |C_1\,C_2|$$

Angle between $T.C.T. = 2\theta = 2\sin^{-1}\left(\dfrac{r_1 + r_2}{d}\right)$

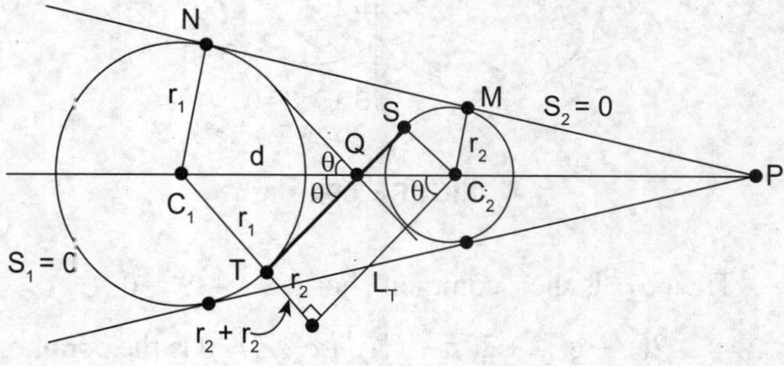

FIGURE 3.87

Case II: When Two Circles Touch Each Other Externally

Then distance between their centres is equal to the sum of their radii.

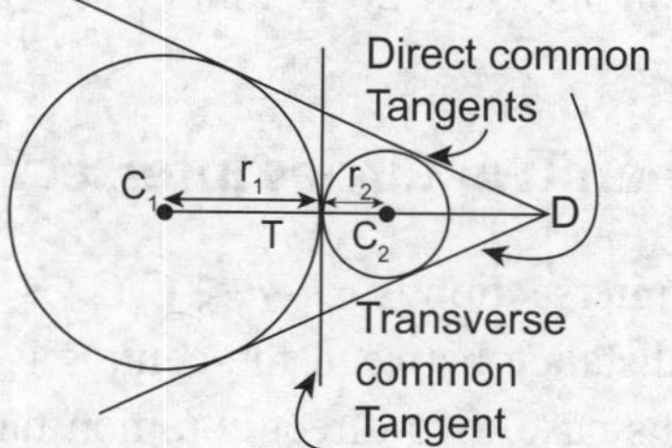

FIGURE 3.88

i.e., $|C_1C_2| = r_1 + r_2$

In such cases, the point of contact T divides the line joining C_1 and C_2 internally in the ratio $r_1 : r_2$

$$\Rightarrow \quad \frac{C_1T}{C_2T} = \frac{r_1}{r_2}$$

If $C_1 \equiv (x_1, y_1)$ and $C_2 \equiv (x_2, y_2)$ then co-ordinates of D are given by $D \equiv \left(\dfrac{r_1x_2 + r_2x_1}{r_1 + r_2}, \dfrac{r_1y_2 + r_2y_1}{r_1 + r_2}\right)$

In this case, two direct common tangents are real and distinct while the transverse tangents are coincident. The equation of transverse common tangent is given in many ways, the simplest of them is $S_1 - S_2 = 0$ where S_1 and S_2 are the expressions of the two circles and the equation of direct common tangent is obtained as discussed in the previous article.

Length of Direct Common Tangent

$$L_D = MN = \sqrt{d^2 - (r_1 - r_2)^2}$$

$$\Rightarrow \quad L_D = \sqrt{(r_1 + r_2)^2 - (r_1 - r_2)^2} = 2\sqrt{r_1\ r_2}$$

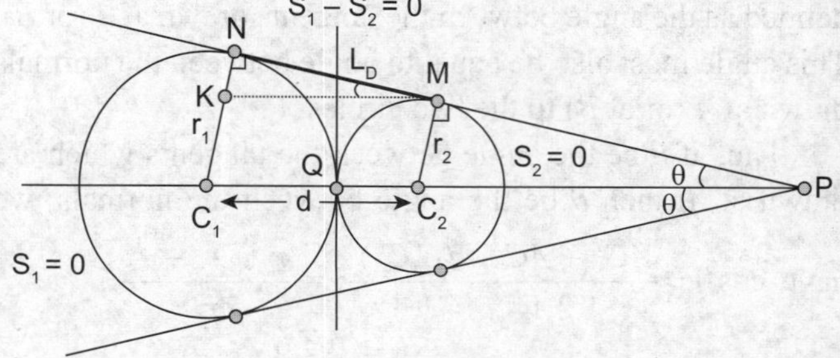

FIGURE 3.89

Angle between D.C.T. $= 2\theta = 2\sin^{-1}\left(\dfrac{r_1 - r_2}{r_1 + r_2}\right)$

Transverse common tangent

$$L_T = ST = \sqrt{d^2 - (r_1 + r_2)^2}$$

$$\Rightarrow \quad L_T = \sqrt{(r_1 + r_2)^2 - (r_1 + r_2)^2} = 0$$

Angle between T.C.T. $= 2\alpha = 2\sin^{-1}\left(\frac{r_1 + r_2}{r_1 + r_2}\right) = \pi$

Case III: When Two Circles Intersect Each Other

Condition of intersection is $r_1 - r_2 < | C_1C_2 | < r_1 + r_2$

i.e., the distance between the centres is less than sum of radii. In this case, two direct common tangents are real and distinct while the tranverse tangents are imaginary.The equations of *D.C.T's* is obtained as discussed earlier.

Length of direct common tangent:

$$L_D = MN = \sqrt{d^2 - (r_1 - r_2)^2}$$

Angle between D.C.T. $= 2\theta = 2\sin^{-1}\left(\frac{|r_1 - r_2|}{d}\right)$

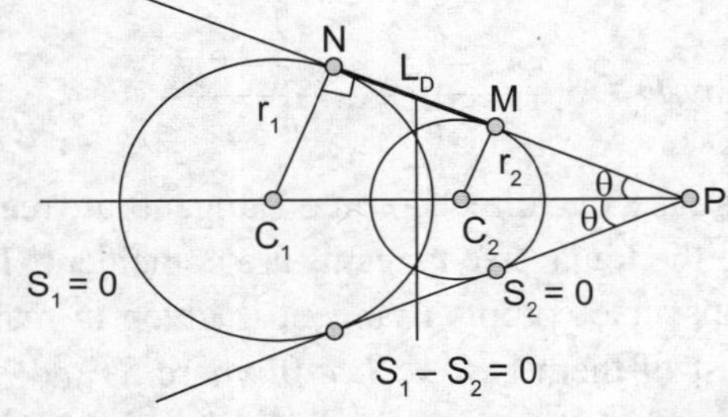

FIGURE 3.90

Angle of Intersection

Angle of intersection is defined as angle between the tangent to the circles at the point of intersection which is same as the angle between their normals (or its supplementary) drawn at the point of intersection. Therefore, if two cicles $S = 0$, $S' = 0$ intersect at A and B, then the angle of their intersection is defined as the angle between the tangents drawn at A (or B). This angle must also be equal to angle between the normals drawn at A (or at B) to the two circles.

Thus if θ be the angle between the tangents which are drawn at A, then θ be the angle between the normals, we have $\cos\theta = \frac{AC_1^2 + AC_2^2 - (C_1C_2)^2}{2AC_1.AC_2} = \frac{r_1^2 + r_2^2 - d^2}{2r_1.r_2}$.

Orthogonal intersection

Two circles are said to intersect orthogonally iff their tangents are perpendicular at the point of intersection. Therefore their normals are also perpendicular.

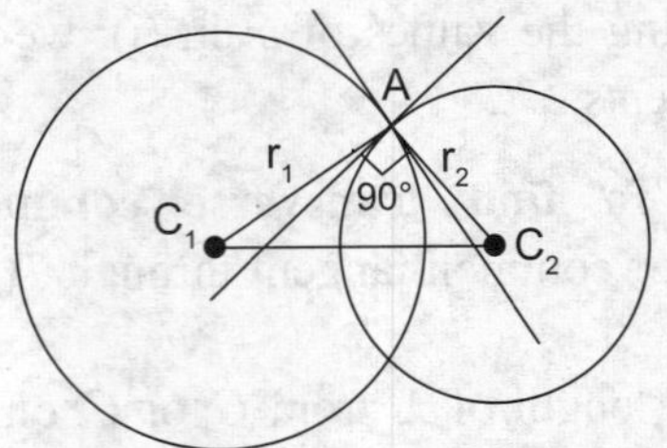

FIGURE 3.91

In case the tangents are perpendicular, the normals must also be perpendicular.

$$\Rightarrow \quad \cos\theta = 0 \Rightarrow r_1^2 + r_2^2 = d^2$$

If the two circles are

$S_1 : x^2 + y^2 + 2gx + 2fy + c = 0$

$S_2 : x^2 + y^2 + 2g'x + 2f'y + c' = 0,$

then $r_1^2 + r_2^2 = d^2$

$$\Rightarrow \quad g^2 + f^2 - c + g'^2 + f'^2 - c' = (-g + g')^2 + (-f + f')^2$$

which on simplification yields $2gg' + 2ff' = c + c'$.

This is the condition for orthogonal intersection of two circles $S_1 = 0$ and $S_2 = 0$.

Common chord of two circles

The common chord joining the point of intersection of two given circles is called their common chord.

If $S = 0$ and $S' = 0$ be two intersecting circles, the equation of their common chord is $S - S' = 0$.

Let $S \equiv x^2 + y^2 + 2gx + 2fy + c = 0$

and $S' \equiv x^2 + y^2 + 2g'x + 2f'y + c' = 0$

be two circles intersecting at P and Q.

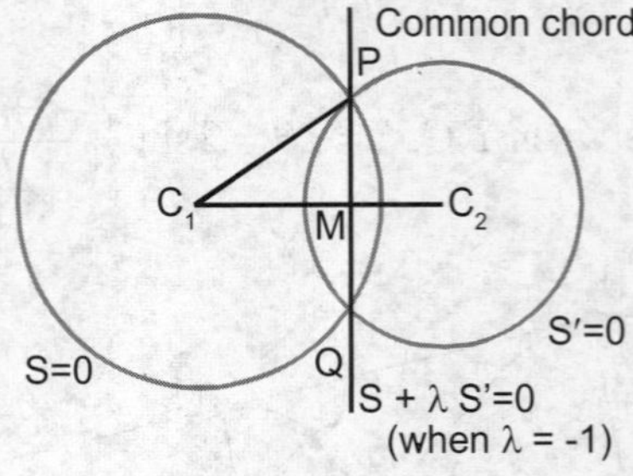

FIGURE 3.92

Then PQ is their common chord $\therefore S - S' = 0$

$\Rightarrow \quad 2(g - g')x + 2(f - f')y + c - c' = 0$ is the common chord of two circles $S = 0$ and $S' = 0$

Length of the common chord:

$PQ = 2(PM) = 2\sqrt{\{(C_1P)^2 - (C_1M)^2}$, where $C_1\ P$ = radius of the circle $S = 0$ and C_1M is the length of perpendicular from C_1 on common chord PQ.

Corollary

1. If the length of common chord is zero, then the two circles touch each other and the common chord becomes the common tangent to the two circles at the common point of contact.
2. The common chord PQ of two circles becomes of the maximum length when it is a diameter of the smaller circle. In this case, the common chord pases through the centre of the smaller circle.

Case IV: When Two Circles Touch Each Other Internally

Then distance between their centres is equal to the difference of their radii

i.e., $|C_1C_2| = |r_1 - r_2|$

In such cases, the point of contact P divides the line joining C_1 and C_2 externally in the ratio $r_1 : r_2$

$$\Rightarrow \quad \frac{C_1P}{C_2P} = \frac{r_1}{r_2}$$

If $C_1 \equiv (x_1, y_1)$ and $C_2 \equiv (x_2, y_2)$, then co-ordinate of P are given by $P \equiv \left(\frac{r_1x_2 - r_2x_1}{r_1 - r_2}, \frac{r_1y_2 - r_2y_1}{r_1 - r_2}\right)$.

In this case, two tangents are real and coincident while the other two tangents are imaginary.

If circles are represented by $S_1 = 0$ and $S_2 = 0$, then equation of common tangents is $S_1 - S_2 = 0$

Length of direct common tangent:

$$L_D = \sqrt{d^2 - (r_1 - r_2)^2} = \sqrt{(r_1 - r_2)^2 - (r_1 - r_2)^2} = 0$$

$$\text{Angle between D.C.T.} = 2\theta = 2\sin^{-1}\left(\frac{r_1 - r_2}{r_1 - r_2}\right) = \pi$$

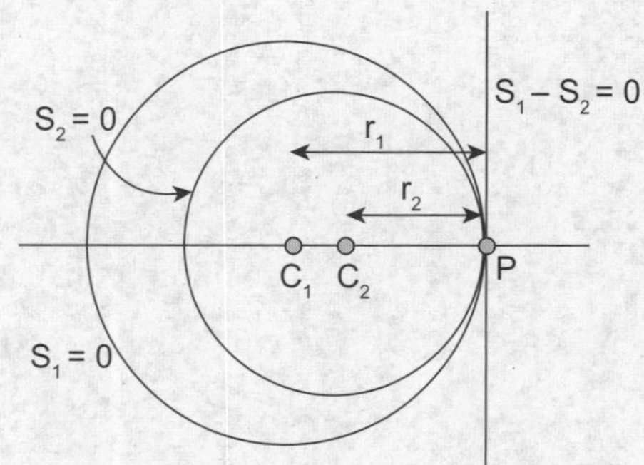

FIGURE 3.93

Case V: When Smaller Circle Completely Lies Inside the Bigger Circle

When $|C_1\ C_2| < r_1 - r_2$ i.e., the distance between the centre is less than the difference of their radii. In this case, all the four common tangents are imaginary.

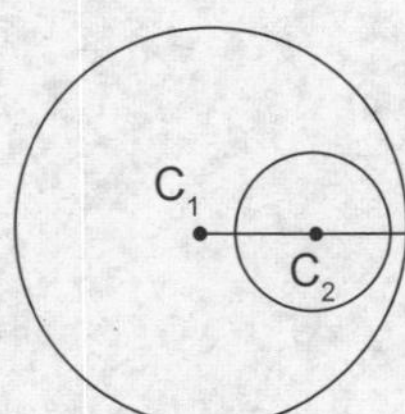

FIGURE 3.94

NOTE

If two circles $S = 0$ and $S' = 0$ touch each other, then their point of contact can be obtained by solving $S = 0$ and $S - S' = 0$ simultaneously.

ILLUSTRATION 49: Examine if the two circles $x^2 + y^2 - 2x - 4y = 0$ and $x^2 + y^2 - 8y - 4 = 0$ touch each other externally or internally.

SOLUTION: Given circles are $x^2 + y^2 - 2x - 4y = 0$ and $x^2 + y^2 - 8y - 4 = 0$.

Let centres and radii of circles (i) and (ii) are represented by C_1, r_1 and C_2, r_2 respectively.

$\therefore \quad C_1 \equiv (1, 2),\ r_1 = \sqrt{1+4}$

$\Rightarrow r_1 = \sqrt{5}$ and $C_2 \equiv (0, 4)$, $r_2 = \sqrt{0+16+4} \Rightarrow r_2 = 2\sqrt{5}$

Now $C_1C_2 = \sqrt{(1-0)^2 + (2-4)^2}$

$\Rightarrow C_1C_2 = \sqrt{5} = r_2 - r_1$

Hence the two circles touch each other internally.

ILLUSTRATION 50: Determine the number of common tangents to the two circles $C_1 : x^2 + y^2 = 25$, $C_2 : x^2 + y^2 - 4x - 6y + 4 = 0$ and find their lengths.

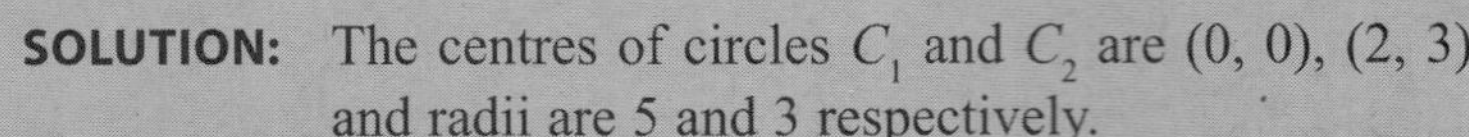

SOLUTION: The centres of circles C_1 and C_2 are (0, 0), (2, 3) and radii are 5 and 3 respectively.

$d = \sqrt{13}$, $r_1 = 5$, $r_2 = 3$

we can observe that $r_1 - r_2 < d < r_1 + r_2$

$\Rightarrow$ The circles intersect at two distinct real points.

$\Rightarrow$ There are two direct common tangents.

Their lengths $= \sqrt{d^2 - (r_1 - r_2)^2}$

$= \sqrt{13 - (5-3)^2}$

$= 3$ units

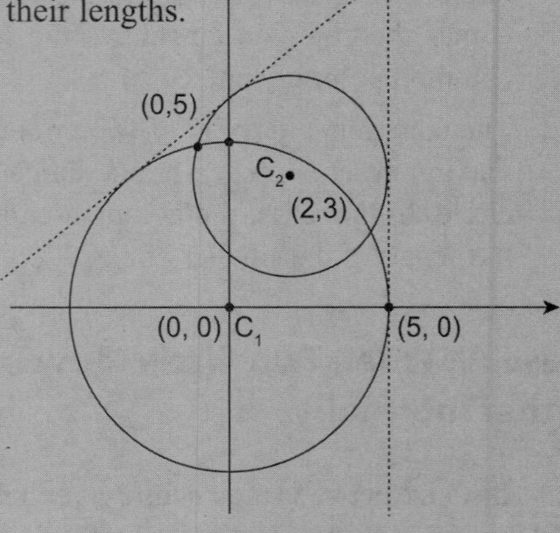

FIGURE 3.95

ILLUSTRATION 51: Find all the common tangents to the circles $x^2 + y^2 - 2x - 6y + 9 = 0$ and $x^2 + y^2 + 6x - 2y + 1 = 0$.

SOLUTION: Let $S_1 : x^2 + y^2 - 2x - 6y + 9 = 0$

$S_1 : (x-1)^2 + (y-3)^2 = 1$

$\therefore$ $C_1 : (1,3)$ and $r_1 = 1$ And $S_2 : x^2 + y^2 + 6x - 2y + 1 = 0$

$\Rightarrow S_2 : (x+3)^2 + (y-1)^2 = 9$

$\therefore$ $C_2 : (-3,1)$ and $r_2 = 3$

$\therefore$ Now, since C_1, C_2 are known, therefore we find $d = |C_1C_2| = \sqrt{16+4} = \sqrt{20}$.

And we can see that $|C_1C_2| > r_1 + r_2$.

Hence the two circles do not intersect.

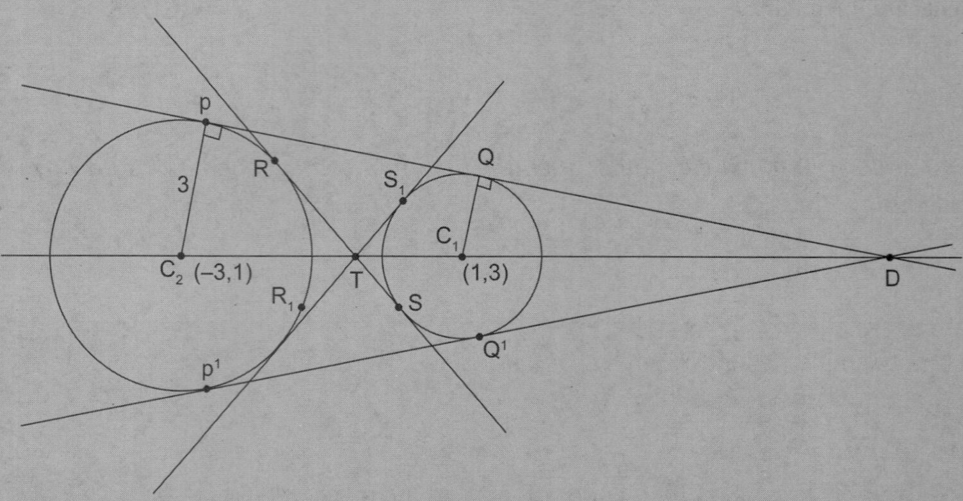

FIGURE 3.96

D divides $C_2\,C_1$ externally in the ratio of 3 : 1.

$\therefore$ Co-ordinates of D are $\left(\dfrac{3(1)-1(-3)}{3-1}, \dfrac{3(3)-1(1)}{3-1}\right)$ or (3, 4) and similarly point T divides C_2 C_1 in the ratio 3 : 1 (internally),

$\therefore$ co-ordinates of T are $\left(\dfrac{3(1)+1(-3)}{3+1}, \dfrac{3(3)+1(1)}{3+1}\right)$ or (0, 5/2)

Transverse* tan*gents:

Any line through T (0, 5/2) is $y - 5/2 = mx$

$\Rightarrow mx - y + 5/2 = 0$ ....(i)

Applying the condition of tangency to any of the circle; we get

$$\frac{m.1-3+5/2}{\sqrt{m^2+1}} = \pm 1 \Rightarrow m^2 + \frac{1}{4} - m = m^2 + 1$$

$\Rightarrow 0.\, m^2 - m - \dfrac{3}{4} = 0 \therefore m = \infty, m = -3/4$

Hence equations of trasverse tangents are

$x = 0$ and $3x + 4y - 10 = 0$

Direct* tan*gents:

Any line through D(3, 4) is

$y - 4 = m(x - 3)$ or $mx - y + 4 - 3m = 0$

Applying the condition of tangency to any of the circle; we get

$$\frac{m-3+4-3m}{\sqrt{m^2+1}} = \pm 1 \Rightarrow (-2m+1)^2 = m^2 + 1$$

$\Rightarrow 3m^2 - 4m = 0 \therefore m = 0, m = 4/3$

Equations of direct common tangents are $y = 4$ and $4x - 3y = 0$

ILLUSTRATION 52: Find the length of the common chord of the circles, whose equations are $(x-a)^2 + y^2 = a^2$ and $x^2 + (y-b)^2 = b^2$. Also prove that the equation to the circle whose diameter is the common chord is $(a^2+b^2)(x^2+y^2) = 2ab(bx+ay)$.

SOLUTION: The two given circles are given as $(x - a)^2 + y^2 = a^2$

or $x^2 + y^2 - 2ax = 0$ ...(i)

and $x^2 + (y - b)^2 = b^2$,

or $x^2 + y^2 - 2by = 0$ ..(ii)

Equation of the chord OA which is common to (i) and (ii) is (O is taken as (i) and (ii) are passing through origin) $(x^2 + y^2 - 2ax) - (x^2 + y^2 - 2by) = 0$

$by - ax = 0$, length $= \dfrac{2ab}{\sqrt{a^2+b^2}}$ ....(iii)

To get the co-ordinates of A, we solve equation (i) and (ii); so putting the values of y from (iii) in (i), we have $x^2 + \left(\dfrac{ax}{b}\right)^2 - 2ax = 0$ or $b^2x^2 + a^2x^2 - 2ab^2x = 0$ or $x[x(a^2+b^2) - 2ab^2] = 0$

so either $x = 0$ or $x = \dfrac{2ab^2}{(a^2+b^2)}$

the corresponding values of y are $y = 0$ or $y = \dfrac{2a^2b}{(a^2+b^2)}$

So the co-ordinates of O are (0, 0) and A are $\left(\dfrac{2ab^2}{a^2+b^2}, \dfrac{2a^2b}{a^2+b^2}\right)$

Hence the equation of the circle drawn on OA as diameter is

$$(x-0)\left(x-\frac{2ab^2}{a^2+b^2}\right)+(y-0)\left(y-\frac{2a^2b}{a^2+b^2}\right)=0$$

or $(a^2+b^2)x^2-2ab^2x+(a^2+b^2)y^2-2a^2by=0$

or $(x^2+y^2)(a^2+b^2)=2ab(bx+ay)$

ILLUSTRATION 53: Prove that the length of the common chord of the two circles whose equations are $(x-a)^2+(y-b)^2=c^2$ and $(x-b)^2+(y-a)^2=c^2$ is $\sqrt{\left[4c^2-2(a-b)^2\right]}$. Hence find the condition that the two circles may touch.

SOLUTION: Equations to the circles are given as $S_1 : (x-a)^2+(y-b)^2=c^2$

$\Rightarrow S_1 : x^2+y^2-2ax-2by+a^2+b^2-c^2=0$; $C \equiv (a, b)$ and $r_1 = c$ ...(i)

And $S_2 : (x-b)^2+(y-a)^2=c^2$

$\Rightarrow S_2 : x^2+y^2-2bx-2ay+a^2+b^2-c^2=0$; $C' \equiv (b, a)$ and $r_2 = c$ ..(ii)

Equation of the common chord is $S_1 - S_2 = 0$

or $2x(b-a) - 2y(b-a) = 0$ or $x - y = 0$ ......(iii)

Let AB be the common chord of the circle whose centre is C. If CN is perpendicular from the centre on the chord, then

$AB = 2BN = 2\sqrt{(CB^2 - CN^2)}$ or length of the chord

$= 2\sqrt{(\text{radius})^2 - (\text{length of the perpendicular from the centre on the chord})^2}$ ..(iv)

Perpendicular from (a, b) on the common chord (iii)

$= CN = \dfrac{a-b}{\sqrt{1+1}}$

Putting in (iv); length of the chord $= \sqrt{\left[c^2-\left\{\dfrac{|a-b|}{\sqrt{2}}\right\}^2\right]}$

$= \sqrt{\left[4c^2-2(a-b)^2\right]}$

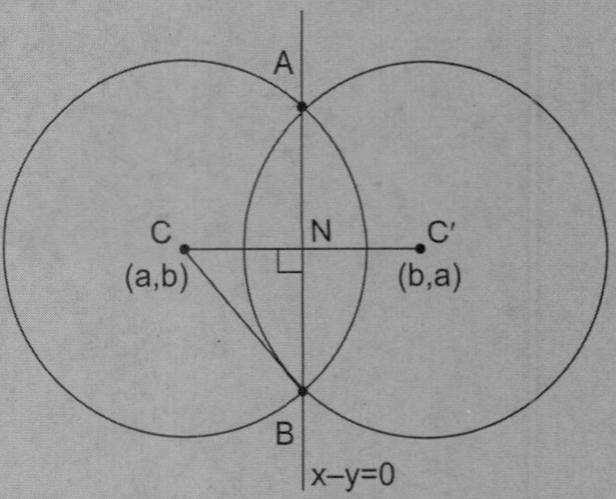

FIGURE 3.97

If the circles touch each other, the length of the common chord will be zero (as they simply touch each other); so $\sqrt{\left[4c^2-2(a-b)^2\right]} = 0$

or $(a-b)^2 = 2c^2$

ILLUSTRATION 54: Find the equation of the common tangents of the circles $x^2 + y^2 = c^2$ and $(x-a)^2 + y^2 = b^2$

SOLUTION: The circles are given as $x^2 + y^2 = c^2$ ...(i)

and $(x-a)^2 + y^2 = b^2$ .....(ii)

The equation of any to circle $x^2 + y^2 = c^2$ tangent may be given as $y = mx + c\sqrt{1+m^2}$ ...(iii)

If (iii) is a tangent to (ii), then the length of the perpendicular from the centre of (ii), i.e., $(a, 0)$ must be equal to its radius, i.e., b

$$\therefore \quad \frac{ma - 0 + c\sqrt{1+m^2}}{\sqrt{1+m^2}} = \pm b$$

Case I: $ma + c\sqrt{1+m^2} = b\sqrt{1+m^2}$ (taking $+ve$ sign)

or $ma = (b-c)\sqrt{1+m^2}$

Squaring we get, $(b-c)^2 m^2 - m^2 a^2 + (b-c)^2 = 0$

or $(b-c)^2 - m^2\,[\,a^2 - (b-c)^2] = 0$

whence $m = \pm \dfrac{b-c}{\sqrt{[\,a^2-(b-c)^2]}}$ ...(iv)

Case II: Again taking –ve in (iv), we get

$ma + c\sqrt{1+m^2} = -b\sqrt{1+m^2}$

Simplifying as before, we get, $m = \pm \dfrac{b+c}{\sqrt{[\,a^2-(b-c)^2]}}$ ....(v)

Hence the equation of the 4 common tangents can be obtained by substituting the values of 'm' form (iv) and (v) in equation (iii); we get $y = mx + c\sqrt{1+m^2}$ where

$$m = \pm \frac{b-c}{\sqrt{[\,a^2-(b-c)^2]}} \text{ or } m = \pm \frac{b+c}{\sqrt{[\,a^2-(b-c)^2]}}$$

ILLUSTRATION 55: Find the length of the common chord of the circles $x^2 + y^2 - 2ax - 4ay - 4a^2 = 0$ and $x^2 + y^2 - 3ax + 4ay = 0$. Find also the equations of the common tangents and show that the length of each is $4a$; $a > 0$.

SOLUTION: The circles are given as $S_1 \equiv x^2 + y^2 - 2ax - 4ay - 4a^2 = 0$ ...(i)

and $S_2 \equiv x^2 + y^2 - 3ax + 4ay = 0$ ...(ii)

The common chord is $S_1 - S_2 = 0$

or $ax - 8ya - 4a^2 = 0$ or $x - 8y - 4a = 0$ ...(iii)

Centre of (i) is $(a, 2a)$ and radius of (i) is $\sqrt{a^2 + 4a^2 + 4a^2} = 3a$

Length of the perpendicular from the centre $(a, 2a)$ upon the common chord (iii)

$$= \frac{|a - 8.2a - 4a|}{\sqrt{1^2 + 8^2}} = \frac{19a}{\sqrt{65}}$$

Now, length of the common chord

$$= 2\sqrt{(\text{radius})^2 - (\text{length of the perpendicular from the contre to the chord})^2}$$

Therefore; length of the common chord

$$= 2\sqrt{\left[(3a)^2 - \left(\frac{19a}{\sqrt{65}}\right)^2\right]} = 2\sqrt{\left[9a^2 - \frac{361a^2}{65}\right]} = 2\sqrt{\left[\frac{224}{65}.a^2\right]} = 8a\sqrt{\frac{14}{65}}$$

Centre of (ii) is $\left(\dfrac{3a}{2}, -2a\right)$ and its radius $\sqrt{\left(\dfrac{9a^2}{4} + 4a^2 + 0\right)} = \dfrac{5a}{2}$

Only direct common tangents are possible.

($\because |r_1 - r_2| < |C_1C_2| < r_1 + r_2$ Therefore the two circles S_1 and S_2 intersect each other in two distinct points and hence no transverse common tangent is possible)

Dividing the joins of $(a, 2a)$ and $\left(\frac{3a}{2}, -2a\right)$ in the ratio of $3a$ and — or 6 : 5 externally,

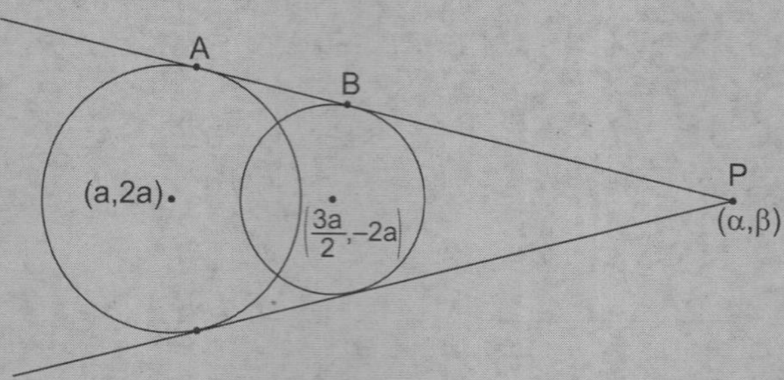

FIGURE 3.98

$\therefore$ Co-ordinates of point of intersection of direct common tangents will be $\alpha = \frac{9a-5a}{6-5}$,

$\beta = \frac{-12a-10a}{6-5}$ i.e., $(4a, -22a)$

Any line passing through $(4a, -22a)$ is $y + 22a = m(x - 4a)$ ....(iv)

If it is tangent to (i), then length of perpendicular from its centre will be equal to its radius.

So, $\frac{ma-4am-2a-22a}{\sqrt{1+m^2}} = 3a$ or $-3m-24 = 3\sqrt{1+m^2}$ or $-(m+8) = \sqrt{1+m^2}$

Squaring, we get $m^2+64+16m = 1+m^2$ or $0.m^2 + 16m + 63 = 0$

As it is a quadratic in m, and the co-efficient of m^2 is 0, so one root must be ∞.

i.e., $\frac{1}{m} = 0$ and other root is clearly $m = -\frac{63}{16}$

Putting $1/m = 0$ in (iv); the equation becomes $x - 4a = 0$

Putting $m = -\frac{63}{16}$ in (iv); the equation becomes $y+22a = -\frac{63}{16}(x-4a)$

or $63x + 16y + 100a = 0$

Hence the tangents are $x = 4a$ and $63x + 16y + 100a = 0$

Again length of tangents from $(4a, -22a)$ on S_1

$= \sqrt{\{(4a)^2+(-22a)^2-2a.4a-4a(-22a)-4a^2\}} = 24a \Rightarrow PA = 24a$

Similarly, length of the tangent from $(4a, -22a)$ on $S_2 = PB = 20a$

Hence, length of the common tangent $= AB = PA - PB = 24a - 20a = 4a$

TEXTUAL EXERCISE–6 (SUBJECTIVE)

1. Find the common tangents of the circles $x^2 + y^2 = 1$ and $(x - 1)^2 + (y - 3)^2 = 4$.
2. Find the equations of the straight lines which touch both the circles $x^2 + y^2 = 4$ and $(x - 4)^2 + y^2 = 1$.
3. Find the equations of the common tangents to the circles $x^2+y^2-2x-6y+9=0$, and $x^2+y^2+6x-2y+1=0$.

4. A line AB is divided at C such that $AC = 3CB$. Circles are described on AC and CB, as diameter and common tangent meets AB produced at D. Show that the radius of the smaller circle is equal to BD.

5. Two circles, each of radius 5, touch each other at (1, 2). If the equation of the common tangent is $4x + 3y = 10$, find the equation of the circles.

6. Prove that $x^2 + y^2 = a^2$ and $(x - 2a)^2 + y^2 = a^2$ are two equal circles touching each other. Find the equation of the circle(s) of equal radius touching both the circles.

7. Find the equation of the circle which touches the circle $x^2 + y^2 + 6x + 6y + 17 = 0$ externally and to which the lines $x^2 + 3xy + 3x + 9y = 0$ are normals.

Answer Keys

1. $4x - 3y - 5 = 0$, $3x + 4y - 5 = 0$, $x = -1$, $y = 1$ **2.** $3x \pm \sqrt{7}y - 8 = 0$ and $x \pm \sqrt{15}y - 8 = 0$

3. $x = 0$, $y = 4$, $3x + 4y = 10$, $3y = 4x + 5$ **5.** $x^2 + y^2 - 10x - 10y + 25 = 0$, $x^2 + y^2 + 6x + 2y - 15 = 0$

6. $x^2 + y^2 - 2ax \pm 2\sqrt{3}\,ay + 3a^2 = 0$ **7.** $x^2 + y^2 + 6x - 2y + 1 = 0$

TEXTUAL EXERCISE–6 (OBJECTIVE)

1. The number of common tangents that can be drawn to the circles $x^2 + y^2 - 4x - 6y - 3 = 0$ and $x^2 + y^2 + 2x + 2y + 1 = 0$ is
   (a) 1 (b) 2
   (c) 3 (d) 4

2. The circles $x^2 + y^2 - 10x + 16 = 0$ and $x^2 + y^2 = r^2$ intersect each other in two distinct points if
   (a) $r < 2$ (b) $r > 8$
   (c) $2 < r < 8$ (d) $2 \le r \le 8$

3. The number of common tangents to the circles $x^2 + y^2 = 4$ and $x^2 + y^2 - 6x - 8y = 24$ is
   (a) 0 (b) 1
   (c) 3 (d) 4

4. The common chord that can be drawn to the circle $x^2 + y^2 - 4x - 4y = 0$ and $x^2 + y^2 = 16$ subtends at the origin an angle equal to
   (a) $\pi/6$ (b) $\pi/4$
   (c) $\pi/3$ (d) $\pi/2$

5. The circles $x^2 + y^2 + x + y = 0$ and $x^2 + y^2 + x - y = 0$ intersect at an angle of
   (a) $\pi/6$ (b) $\pi/4$
   (c) $\pi/3$ (d) $\pi/2$

6. Equation of a circle with centre (4, 3) touching the circle $x^2 + y^2 = 1$ is
   (a) $x^2 + y^2 - 8x - 6y - 9 = 0$
   (b) $x^2 + y^2 - 8x - 6y + 11 = 0$
   (c) $x^2 + y^2 - 8x - 6y - 11 = 0$
   (d) $x^2 + y^2 - 8x - 6y + 9 = 0$

7. The circles $x^2 + y^2 - 4x + 6y + 8 = 0$ and $x^2 + y^2 - 10x - 6y + 14 = 0$ touch each other at the point
   (a) (3, –1) (b) (3, 1)
   (c) (–3, 1) (d) (1, –3)

8. The locus of the centre of a circle which touches externally the circle $x^2 + y^2 - 6x - 6y + 14 = 0$ and also touches the y-axis, is given by the equation
   (a) $x^2 - 6x - 10y + 14 = 0$
   (b) $x^2 - 10x - 6y + 14 = 0$
   (c) $y^2 - 6x - 10y + 14 = 0$
   (d) $y^2 - 10x - 6y + 14 = 0$

9. The locus of the point of intersection of the tangents at the extremities of a chord of the circle $x^2 + y^2 = a^2$ which touches the circle $x^2 + y^2 = 2ax$ is
   (a) $y^2 = a(a - 2x)$ (b) $x^2 = a(a - 2y)$
   (c) $x^2 + y^2 = (y - a)^2$ (d) None of these

10. The two circle which pass through (0, a) and (0, $-a$) and touch the line $y = mx + c$ will intersect each other at right angle, if
    (a) $a^2 = c^2 (2m + 1)$ (b) $a^2 = c^2 (2 + m^2)$
    (c) $c^2 = a^2 (2 + m^2)$ (d) $c^2 = a^2(2m + 1)$

11. The circles $x^2 + y^2 = 9$ and $x^2 + y^2 - 12y + 27 = 0$ touch each other. The equation of their common tangent is
    (a) $4y = 9$ (b) $y = 3$
    (c) $y = -3$ (d) $x = 3$

12. The two circles $x^2 + y^2 - 2x + 6y + 6 = 0$ and $x^2 + y^2 - 5x + 6y + 15 = 0$ touch each other. The equation of their common tangent is
    (a) $x = 3$ (b) $y = 6$
    (c) $7x - 12y - 21 = 0$ (d) $7x + 12y + 21 = 0$

Answer Keys

1. (c) **2.** (c) **3.** (b) **4.** (d) **5.** (d) **6.** (c, d) **7.** (a) **8.** (d) **9.** (a) **10.** (c)
11. (b) **12.** (a)

■ FAMILY OF CIRCLE

Since the general equation of a circle $x^2 + y^2 + 2gx + 2fy + c = 0$ contains three parameters (g, f, c); therefore three conditions are necessary to determine a circle uniquely.

If less than three conditions are given then only two parameters can be evaluated, resulting an equation of the circle having at least one unknown.

It represents infinite number of circles with common characteristics and referred as family of circle.

FIGURE 3.99

Depending upon the common characteristics, the family of circles can be classified in any of the following manner:

1. **Equation of a circle through the intersection of a circle and a line**
   Let $S \equiv x^2 + y^2 + 2gx + 2fy + c = 0$ be a circle
   and $L \equiv ax + by + c = 0$ be a straight line intersecting the circle $S = 0$. Then equation of any circle passing through the intersection of $S = 0$ and $L = 0$ is $S + \lambda L = 0$, where λ is a parameter.
   Remarks: If $L = 0$ is tangent to $S = 0$, then $S + \lambda L = 0$ and $S = 0$ touch each other and $L = 0$ is a common tangent.

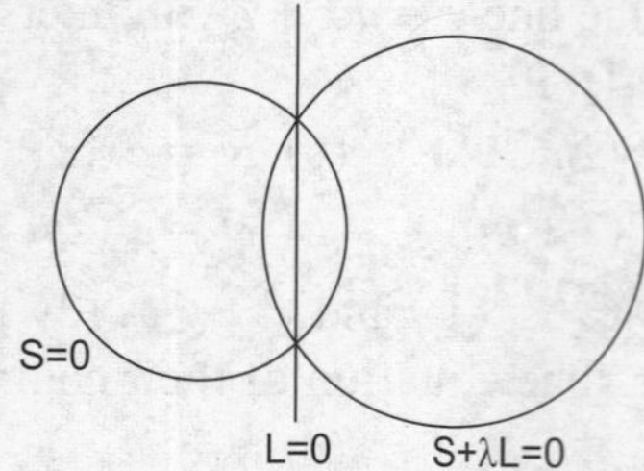

FIGURE 3.100

2. **Circle through the intersection of two circles**
   If $S = 0$ and $S' = 0$ are two intersecting circles, then equation of any circle passing through their points of intersection is $S + \lambda S' = 0$ for $\lambda \in \mathbb{R} \sim \{-1\}$. Also $S + \lambda(S - S') = 0$ represents any circle passing through the points of intersection of $S = 0$ and $S' = 0$, (for $\lambda \in \mathbb{R}$).

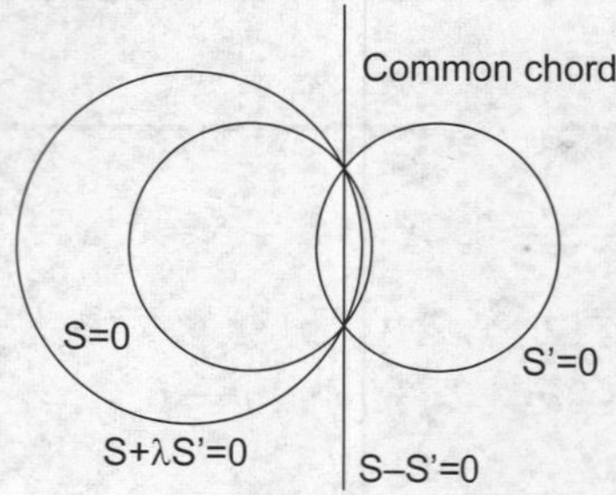

FIGURE 3.101

3. **Family of concentric circles**
   The family of circles with the same centre and different radii is called a family of concentric circles. The equation is given as $(x-\alpha)^2 + (y-\beta)^2 = r^2$ where (α, β) is the fixed point and r is a parameter.

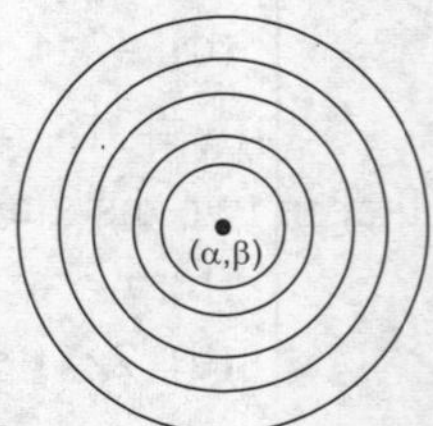

FIGURE 3.102

4. **Family of circles touching a given line at a given point**
   Equation of any circle that touches the line with slope m at the point (x_1, y_1) is
   $(x-x_1)^2 + (y-y_1)^2 + \lambda\{(y-y_1) - m(x-x_1)\} = 0$
   and if m is infinite, the family of circles is
   $(x-x_1)^2 + (y-y_1)^2 + \lambda(x-x_1) = 0$
   (where λ is a parameter)

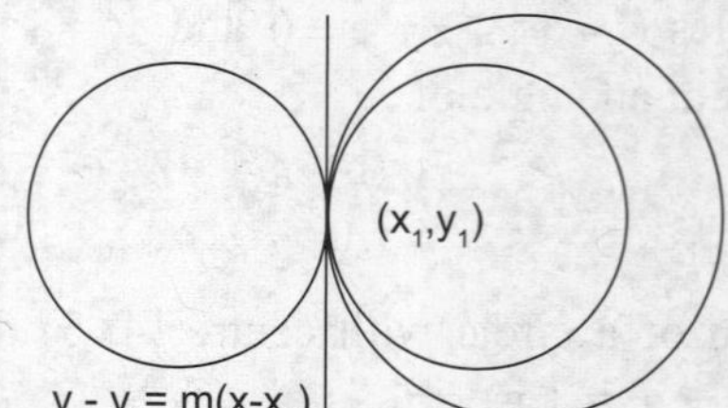

FIGURE 3.103

5. **Family of circle passing through two given points**
   Equation of any circle that passes through two given point $S(x_1, y_1)$ and (x_2, y_2) is

$$(x-x_1)(x-x_2)+(y-y_1)(y-y_2)+\lambda\begin{vmatrix} x & y & 1 \\ x_1 & y_1 & 1 \\ x_2 & y_2 & 1 \end{vmatrix}=0$$

where $S \equiv (x - x_1)(x - x_2) + (y - y_1)(y - y_2) = 0$ resperesents a circle passing through (x_1, y_1) and (x_2, y_2) and $L \equiv \begin{vmatrix} x & y & 1 \\ x_1 & y_1 & 1 \\ x_2 & y_2 & 1 \end{vmatrix} = 0$, represents a straight line passing through (x_1, y_1) and (x_2, y_2) and hence the required family of circles can be represented as $S + \lambda L = 0$, where λ is a parameter.

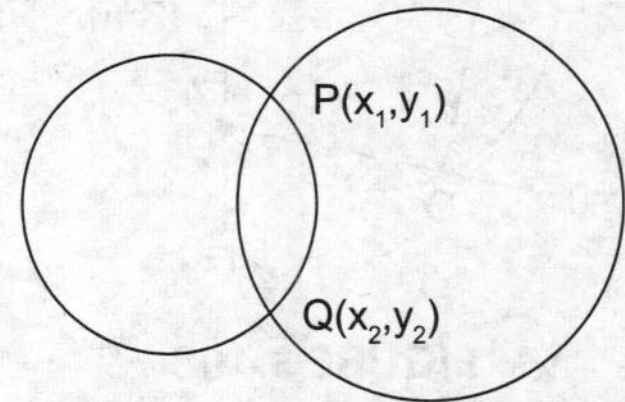

FIGURE 3.104

6. Family of pair of circles having a common chord (with fixed end points) and subtending equal angles on opposite segment.

Let the two fixed points be A (x_1, y_1) and $B(x_2, y_2)$, then the equation of the circle with diametric ends at these points can be written as $S \equiv (x - x_1)(x - x_2) + (y - y_1)(y - y_2) = 0$ and the equation of straight line thought (x_1, y_1) and (x_2, y_2) is $L \equiv \begin{vmatrix} x & y & 1 \\ x_1 & y_1 & 1 \\ x_2 & y_2 & 1 \end{vmatrix} = 0$

$$\Rightarrow \{(x-x_1)(y-y_2)-(x-x_2)(y-y_1)\}=0$$

Therefore the equation of the family of circles passing through the point of intersection of S and L is $S+\lambda L=0$

i.e., $\{(x-x_1)(x-x_2)+(y-y_1)(y-y_2)\}+$

$$\lambda\{(x-x_1)(y-y_2)+(x-x_2)(y-y_1)\}=0 \quad \text{.... (i)}$$

Now, the pair of circles on which AB subtends a fixed angle 'θ' on the opposite segments of the circle, will be the a members of the family of circles represented by (i)

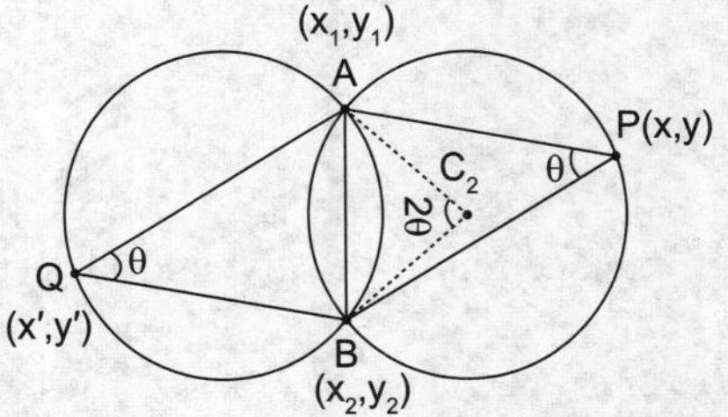

FIGURE 3.105

Now from Euclidean geometry, we know that a chord subtends a fixed angle on a particular segment of a circle. And if the arc changes, the angle is bound to change.

Therefore there can be only one circle on one side of the chord such that angle subtended is 'θ'.

And the other circle will be a mirror image of the first circle in the chord.

Now, having determined the number of circles for a particular value of 'θ'; we need to find the value of λ.

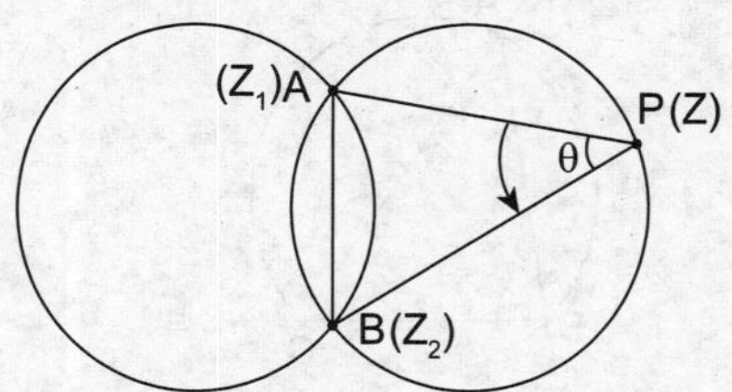

FIGURE 3.106

Now let $A \equiv x_1 + iy_1$ and $B \equiv x_2 + iy_2$ and a general point P on one of the circles be $P \equiv h + ik$

And $|\arg(Z_{BP}) - \arg(Z_{AP})| = \theta$

$$\Rightarrow |\arg(Z_2 - Z) - \arg(Z_1 - Z)| = \theta$$

$$\Rightarrow \left|\tan^{-1}\left(\frac{y_2-k}{x_2-h}\right)-\tan^{-1}\left(\frac{y_1-k}{x_1-h}\right)\right|=\theta$$

$$\Rightarrow \left|\tan^{-1}\frac{\left(\frac{y_2-k}{x_2-h}\right)-\left(\frac{y_1-k}{x_1-h}\right)}{1+\left(\frac{y_2-k}{x_2-h}\right)\left(\frac{y_1-k}{x_1-h}\right)}\right|=\theta$$

$$\Rightarrow \left|\tan^{-1}\left(\frac{(y_2-k)(x_1-h)-(y_1-k)(x_2-h)}{(x_1-h)(x_2-h)+(y_2-k)(y_1-k)}\right)\right|=\theta$$

$$\Rightarrow \frac{(k-y_2)(h-x_1)-(k-y_1)(h-x_2)}{(h-x_1)(h-x_2)+(k-y_1)(k-y_2)}=\pm\tan\theta$$

$$\Rightarrow ((h-x_1)(h-x_2)+(k-y_1)(k-y_2))\pm$$
$$\cot\theta((h-x_1)(k-y_2)-(h-x_2)(k-y_2))=0$$

Substituting (h, k) by (x, y); we get the required equations of the circle as

$$((x-x_1)(x-x_2)+(y-y_1)(y-y_2))\pm$$
$$\cot\theta((x-x_1)(y-y_2)-(x-x_2)(y-y_2))=0$$

8. Family of circles circumscribing a triangle whose sides are $L_1 = 0$; $L_2 = 0$ and $L_3 = 0$ is given by $L_1L_2 + \lambda L_2L_3 + \mu L_3L_1 = 0$ provided co-efficient of $xy = 0$ and co-efficient of x^2 = co-efficient of y^2.

If $L_r = a_r x + b_r y + c_r$; $r = 1, 2, 3$, then equation is

$$\begin{vmatrix} \frac{a_1^2+b_1^2}{a_1x+b_1y+c_1} & a_1 & b_1 \\ \frac{a_2^2+b_2^2}{a_2x+b_2y+c_2} & a_2 & b_2 \\ \frac{a_3^2+b_3^2}{a_3x+b_3y+c_3} & a_3 & b_3 \end{vmatrix}$$

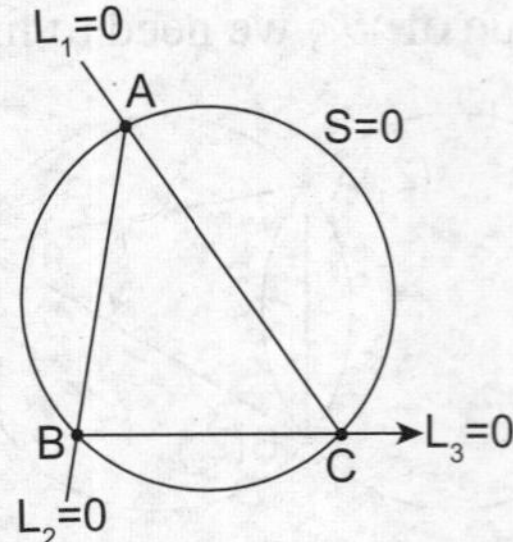

FIGURE 3.107

Proof: Let $S = 0$ be the circle which circumscribes the triangle ABC formed by the lines $L_1 = 0, L_2 = 0$ and $L_3 = 0$.

Also, we know that one and only one circle can be drawn passing through 3 non-collinear points.

And thereby, 3 points are the necessary and sufficient condition to determine a circle.

Let us assume that the equation of the circle be given by $L_1L_2 + \lambda L_2 L_3 + \mu L_1L_3 = 0$ ...(i)

where λ, μ are parameters.

circle passes through point A and point A lies on line L_1 and $L_2 = 0$

$\therefore$ $L_1 = 0$ and $L_3 = 0$

Substituting these values in (i); we get $0 + \lambda(0)L_3 + \mu\,(0)\,L_3 = 0$

Hence, equation (i) is satisfied by point A.

Similarly, it can also be proved that points B and C also lie on the circle.

$\therefore$ $S \equiv L_1L_2 + \lambda L_2L_3 + \mu L_1L_3 = 0$ is the required equation of the circle.

9. Equation of circle circumscribing a quadrilateral whose sides in order are represented by the lines $L_1 = 0, L_2 = 0, L_3 = 0$ and $L_4 = 0$ is given by $L_1L_3 + \lambda L_2L_4 = 0$ provided co-efficient of x^2 = co-efficient of y^2 and co-efficient of $xy = 0$.

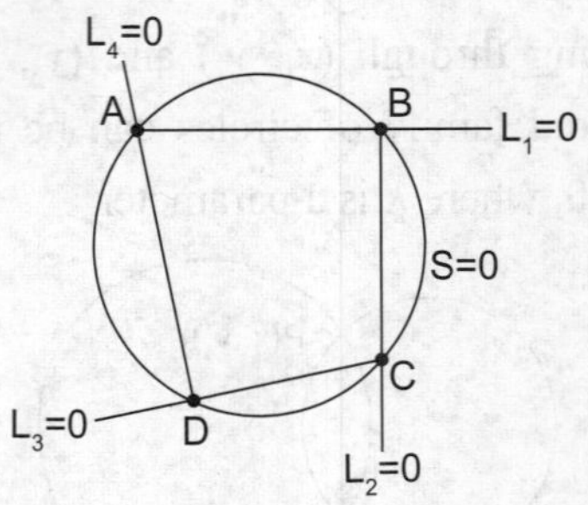

FIGURE 3.108

Let $ABCD$ be the quadrilateral formed by $L_1 = 0, L_2 = 0, L_3 = 0$ and $L_4 = 0$ and let us assume that the circle. circumscribing the quadrilateral $ABCD$ is given by $S = L_1L_3 + \lambda L_2L_4 = 0$. .... (i)

Now, it remains to be proved that A, B, C, D satisfy the equation (i)

For point A; $L_1 = 0$ and $L_4 = 0$

A is the point of intersection of these two lines.

$\therefore$ Putting the co-ordinates of A in equation (i)

$S = (0)\,L_3 + \lambda(0)\,L_2 = 0$

$\therefore$ Equation is satisfied.

Similarly, it can also be proved that the points B, C and D also lie on the circle.

Hence, our assumption is correct and $S = L_1L_3 + \lambda L_2L_4 = 0$ is the equation of the circle circumscribing the quadrilateral $ABCD$.

ILLUSTRATION 56: Find the equation of the circle passing through (1, 1) and the points of intersection of the circles $x^2 + y^2 + 13x - 3y = 0$ and $2x^2 + 2y^2 + 4x - 7y - 25 = 0$.

SOLUTION: The given circles are $x^2 + y^2 + 13x - 3y = 0$ ....(i)

and $2x^2 + 2y^2 + 4x - 7y - 25 = 0$

or $x^2 + y^2 + 2x - \frac{7}{2}y - \frac{25}{2} = 0$ ...(ii)

Equation of any circle passing through the point of intersection of the given circles (i) and (ii)

is $(x^2 + y^2 + 13x - 3y) + \lambda\left(x^2 + y^2 + 2x - \frac{7}{2}y - \frac{25}{2}\right) = 0$; $\lambda \neq -1$ ...(iii)

It passes through (1, 1) (given)

$$\Rightarrow \ (1+1+13-3)+\lambda\left(1+1+2-\frac{7}{2}-\frac{25}{2}\right)=0 \Rightarrow 12+\lambda(-12)=0 \therefore \ \lambda=1$$

Substituting the value of λ in (iii), the required equation is

$$x^2+y^2+13x-3y+x^2+y^2+2x-\frac{7}{2}y-\frac{25}{2}=0$$

$$\Rightarrow \ 2x^2+2y^2+15x-\frac{13}{2}y-\frac{25}{2}=0 \Rightarrow 4x^2+4y^2+30x-13y-25=0$$

ILLUSTRATION 57: Find the equation of the circle through points of intersection of the circle $x^2+y^2-2x-4y+4=0$ and the line $x + 2y = 4$ which touches the line $x + 2y = 0$.

SOLUTION: Equation of any circle through points of intersection of the given circle and the line is $x^2+y^2-2x-4y+4+\lambda(x+2y-4)=0$

or $x^2+y^2+(\lambda-2)x+(2\lambda-4)y+4(1-\lambda)=0$ ...(i)

It will touch the line $x + 2y = 0$ if solution of equation (i) and $x = -2y$ be unique. Hence, the roots of the equation

$(-2y)^2+y^2+(\lambda-2)(-2y)+(2\lambda-4)y+4(1-\lambda)=0$

or $5y^2 + 4(1 - \lambda) = 0$ must be equal, i.e., $D = 0$

$\Rightarrow \ 0 - 4.5.4(1 - \lambda) = 0$ or $1 - \lambda = 0 \Rightarrow \lambda = 1$

From (i), the required circle is $x^2+y^2-x-2y=0$.

ILLUSTRATION 58: Find the equation of a circle which passes through the intersection of the circles $x^2 + y^2 - 9 = 0$, $x^2+y^2+2x+4y+3=0$ and also passes through (0, 0).

SOLUTION: The equation of the required circle must be of the form

$x^2+y^2-9+\lambda(x^2+y^2-2x-4y+3)=0$

If this passes through (0, 0) we must have $-9 + 3\lambda = 0 \Rightarrow \lambda = 3$

The required circle is $x^2+y^2-9+3\,(x^2+y^2-2x-4y+3)=0$ or $2x^2 + 2y^2 - 3x - 6y = 0$

ILLUSTRATION 59: Find the equation of a circle which passes through the intersection of the circles $x^2+y^2-2x+4y-3=0$, $x^2+y^2-6x-8y+5=0$ and

(i) whose centre lies on y–axis
(ii) whose centre lies on the line $2x + y = 7$
(iii) which cuts the circle $x^2 + y^2 - 3x = 0$ orthogonally
(iv) whose diameter is the common chord of given circles

SOLUTION: The required circles must be of the form

$$x^2+y^2-2x+4y-3+\lambda(x^2+y^2-6x-8y+5)=0$$

which on rearranging becomes

$$(1+\lambda)x^2+(1+\lambda)y^2-2(1+3\lambda)x+2(2-4\lambda)y+5\lambda-3=0 \quad ...(i)$$

or $x^2+y^2-2\dfrac{(1+3\lambda)}{1+\lambda}x+2\dfrac{(2-4\lambda)}{1+\lambda}y+\dfrac{5\lambda-3}{1+\lambda}=0$

The centre $\left(\dfrac{(1+3\lambda)}{1+\lambda}, -\dfrac{(2-4\lambda)}{1+\lambda}\right)$

(i) We must have $\dfrac{(1+3\lambda)}{1+\lambda} = 0$ ($\because$ Centre lies on y-axis)

$\Rightarrow \lambda = -\dfrac{1}{3}$. On putting it in (i), we get $x^2 + y^2 + 10y - 7 = 0$

(ii) We must have $2\left(\dfrac{1+3\lambda}{1+\lambda}\right)+\left(\dfrac{4\lambda-2}{1+\lambda}\right)=7 \Rightarrow \lambda = \dfrac{7}{3}$

$\Rightarrow$ Required circle is $5x^2 + 5y^2 - 24x - 22y + 13 = 0$

(iii) Applying $2gg'+2ff'=c'+c$, we have $2\left(\dfrac{-1-3\lambda}{1+\lambda}\right)\left(-\dfrac{3}{2}\right)+2\left(\dfrac{2-4\lambda}{1+\lambda}\right)(0)=\left(\dfrac{5\lambda-3}{1+\lambda}\right)+0$

(for the second circle $x^2 + y^2 - 3x = 0$, $2g' = -3$, $2f' = 0$, $c' = 0$

$\Rightarrow \lambda = -3/2 \Rightarrow$ The required circle is $x^2 + y^2 - 14x - 32y + 21 = 0$

(iv) The common chord of the given circle is

$4x + 12y - 8 = 0$ or $x + 3y - 2 = 0$

Now the centre $\left(\dfrac{1+3\lambda}{1+\lambda}, \dfrac{4\lambda-2}{1+\lambda}\right)$ must lie on the common chord since common chord is the diameter.

$\Rightarrow \dfrac{1+3\lambda}{1+\lambda}+3\left(\dfrac{4\lambda-2}{1+\lambda}\right)-2=0 \Rightarrow \lambda = \dfrac{7}{13}$

$\Rightarrow$ The require circle is $5x^2 + 5y^2 - 17x - y - 1 = 0$

ILLUSTRATION 60: Find the equation of a circle passing through the points (2, 0), (3, –1) and (2, 5).

SOLUTION: The equation of circles passing through first two points (2, 0) and (3, –1) may be taken as

$$(x-2)(x-3)+(y-0)(y+1)+\lambda\begin{vmatrix} x & y & 1 \\ 2 & 0 & 1 \\ 3 & -1 & 1 \end{vmatrix}=0$$

$\Rightarrow x^2 - 5x + y^2 + y + 6 + \lambda(x + y - 2) = 0$

$\Rightarrow 4 - 10 + 25 + 5 + 6 + \lambda(2 + 5 - 2) = 0$ ($\because$ (2, 5) lies on it)

$\Rightarrow 30 + 5\lambda = 0$

$\Rightarrow \lambda = -6$

$\therefore$ Equation of required circle is $(x^2 - 5x + y^2 + y + 6) - 6(x + y - 2) = 0$

$\Rightarrow x^2 + y^2 - 11x - 5y + 18 = 0$

ILLUSTRATION 61: Find the equation of circle passing through (2, 0) and (3, –1) and cutting a chord of length 4 units on y-axis.

SOLUTION: From the equation $x^2+y^2-5x+y+6+\lambda(x+y-2)=0$ we conclude that length of the chord intercepted on y-axis $= 2\sqrt{\left(\dfrac{1+\lambda}{2}\right)^2-(6-2\lambda)}$ (applying $2\sqrt{f^2-c}$)

$\Rightarrow 2\sqrt{\left(\dfrac{1+\lambda}{2}\right)^2-(6-2\lambda)}=4 \Rightarrow \lambda = 3, \lambda = -13$

Thus one such circle is $x^2+y^2-5x+y+6+3(x+y-2)=0$

$\Rightarrow x^2 + y^2 - 2x + 4y = 0$

ILLUSTRATION 62: Show that any chord that arises as an intersection of a circle through $A(2, 0)$ and $B(3, -1)$ and the circle $x^2+y^2-2x+6y-11=0$ passes through a fixed point. Find the co-ordinates of that point.

SOLUTION: Any circles through $A(2, 0)$ and $B(3, -1)$ is

$$x^2+y^2-5x+y+6+\lambda(x+y-2)=0$$

This will cut the circle $x^2+y^2-2x+6y-11=0$ in a chord whose equation is

$$-3x-5y+17+\lambda(x+y-2)=0$$

It is clear that these chords are concurrent at a point whose co-ordinates are given by the point of intersection of $-3x-5y+17=0$ and $x+y-2=0$

We easily get the fixed point as $(-7/2, 11/2)$

ILLUSTRATION 63: Find the equation of the circle circumscribing the triangle formed by the line $x+y-3=0$; $2x=y$ and $x-y=5$.

SOLUTION: Let $L_1: x+y-3=0$

$L_2: 2x-y=0$

$L_3: x-y-5=0$

Method I: We find the points of intersection of these lines (i.e., the vertices of the triangle)

Solving $L_1=0$ and $L_2=0$; we get A (1, 2)

similarly solving $L_1=0$ and $L_3=0$; we get $B(4, -1)$

and from $L_2=0$ and $L_3=0$; we get $C(-5, -10)$

Now, since, we have three points lying on the circle, therefore circle can be formed.

FIGURE 3.109

Let equation of circle be S: $x^2+y^2+2gx+2fy+c=0$

Putting co-ordinates of A in $S=0$; we get $1^2+2^2+2g+4f+c=0$

$\Rightarrow 2g+4f+c=-5$ .....(i)

Putting co-ordinates of B in $S=0$; we get $4^2+1^2+8g-2f+c=0$

$\Rightarrow 8g-2f+c=-17$ ...(ii)

And putting co-ordinates of C in $S=0$; we get $25+100-10g-20f+c=0$

$\Rightarrow -10g-20f+c=-125$

$\Rightarrow 10g+20f-c=125$ ...(iii)

From (i) – (ii); we get $-6g+6f=12$

$\Rightarrow f-g=2$ ...(iv)

From (i) + (iii); we get

$12g+24f=120$

$\Rightarrow g+2f=10$ ...(v)

From (iv) + (v); we get

$3f=12 \Rightarrow f=4$

Putting this value in (iv); we get

$4-g=2 \Rightarrow g=2$

And putting $f=4$ and $g=2$ in (i); we get

$4+16+c=-5$

$\Rightarrow 20+c=-5 \Rightarrow c=-25$

$\therefore$ Equation of circle becomes

$$x^2+y^2+4x+8y-25=0$$

Method II: Since $S = 0$ circumscribes ΔABC

$\therefore$ $S \equiv L_1L_2 + \lambda L_2L_3 + \mu L_1L_3 = 0$

$\Rightarrow$ $S \equiv (x + y - 3)(2x - y) + \lambda (2x - y)(x - y - 5) + \mu (x + y - 3)(x - y - 5) = 0$

$\Rightarrow$ $(2x^2 + 2xy - 6x - xy - y^2 + 3y) + \lambda (2x^2 - xy - 2xy + y^2 - 10x + 5y) + \mu$
$(x^2 + xy - 3x - xy - y^2 + 3y - 5x - 5y + 15) = 0$

$\Rightarrow$ $S = x^2(2 + 2\lambda + \mu) + y^2(-1 + \lambda - \mu) + xy(1 - 3\lambda) + x(-6 - 10\lambda - 8\mu)$
$+ y(3 + 5\lambda - 2\mu) + 15\mu = 0$

Now for $S = 0$ to be the equation of a circle,

Co-efficient of x^2 = co-efficient of y^2

$\Rightarrow$ $2 + 2\lambda + \mu = -1 + \lambda - \mu$

$\Rightarrow$ $\lambda + 2\mu + 3 = 0$ ...(i)

Co-efficient of $xy = 0$

$\Rightarrow$ $1 - 3\lambda = 0 \Rightarrow \lambda = 1/3$

Putting this value of λ in equation (i); we get

$$\frac{1}{3} + 2\mu + 3 = 0 \Rightarrow 2\mu = \frac{-10}{3} \Rightarrow \mu = -5/3$$

$$\therefore\ S = x^2\left(2 + \frac{2}{3} - \frac{5}{3}\right) + y^2\left(-1 + \frac{1}{3} + \frac{5}{3}\right) + xy\left(1 - 3\times\frac{1}{3}\right) + x\left(-6 - 10\times\frac{1}{3} - 8\times\frac{-5}{3}\right)$$
$$+ y\left(3 + 5\times\frac{1}{3} - 2\times -\frac{5}{3}\right) + 15\times\frac{-5}{3} = 0$$

$x^2 + y^2 + 4x + 8y - 25 = 0$ is the equation of the required circle.

ILLUSTRATION 64: Find the equation of the circle whose diameter is the common chord of the circles $x^2 + y^2 + 4x + 3y + 4 = 0$ and $x^2 + y^2 + 6x + 3y + 10 = 0$.

SOLUTION: Let $S_1 \equiv x^2 + y^2 + 4x + 3y + 4 = 0$ and $S_2 \equiv x^2 + y^2 + 6x + 3y + 10 = 0$

Equation of common chord: $S_1 - S_2 = 0$, i.e., $-2x - 6 = 0 \Rightarrow x = -3$

Equation of family of circles passing though the point of intersection of $S_1 = 0$ and $S_2 = 0$ is

$S = S_1 + \lambda S_2 = 0$

i.e., $(x^2 + y^2 + 4x + 3y + 4) + \lambda (x^2 + y^2 + 6x + 3y + 10) = 0$

i.e., $x^2(1 + \lambda) + y^2(1 + \lambda) + x(4 + 6\lambda) + y(3 + 3\lambda) + (4 + 10\lambda) = 0$

Now, since the line $x = -3$ has to be a diameter to the above circle, therefore the centre of the circle must lie on $x = -3$.

Co-ordinates of centre of S are $\left(\frac{-(4+6\lambda)}{2}, \frac{-(3+3\lambda)}{2}\right)$

Putting in $x = -3$; we get

$$-\frac{(4+6\lambda)}{2} = -3 \Rightarrow 4 + 6\lambda = 6 \Rightarrow \lambda = 1/3$$

Putting this value of λ in $S = 0$; we get

$$S \equiv x^2\left(\frac{4}{3}\right) + y^2\left(\frac{4}{3}\right) + 6x + 4y + \frac{22}{3} = 0$$

i.e., $S \equiv x^2 + y^2 + \frac{9}{2}x + 3y + \frac{11}{2} = 0$ or $2x^2 + 2y^2 + 9x + 6y + 11 = 0$

TEXTUAL EXERCISE–7 (SUBJECTIVE)

1. Find the equation of the circle passing through the points (1, –2) and (4, –3) and whose centre lies on the line $3x + 4y = 7$.

2. Can the four lines $4x - 3y = 5$, $x - 2y = 10$, $7x + y = 40$, $x + 3y + 10 = 0$ form the sides of a cyclic quadrilateral? If no, Justify your answer.

3. Prove that the equation $x^2 + y^2 + 2(3 + p)x + (3 - p)y + 4 = 0$ represents a circle for all values of p, passing through two fixed points. Find the fixed points.

4. Prove that the circle $x^2 + y^2 - 6x - 4y + 9 = 0$ bisects the circumference of the circle $x^2 + y^2 - 8x - 6y + 23 = 0$.

5. Find the equation of the circle passing through the points of intersection of the circles $x^2 + y^2 = 4a^2$ and $x^2 + y^2 - 2x - 4y + 4 = 0$ and touching the line $x + 2y = 0$.

6. Find the equation of the circle which passes through the points of intersection of the circles $x^2 + y^2 - 6x + 2y + 4 = 0$ and $x^2 + y^2 + 2x - 4y - 6 = 0$ and whose centre lies on the line $y = x$.

7. Find the equation of the circle whose diameter is the common chord of the circles $x^2 + y^2 + 2x + 3y + 1 = 0$ and $x^2 + y^2 + 4x + 3y + 2 = 0$.

8. Prove that $x^2 + y^2 + \lambda(y - x - 2\sqrt{2}) = 4$ represents circles touching each other at a common point for all real λ. Also find the common point.

9. Prove that the length of the common chord of the two circles $x^2 + y^2 = a^2$ and $(x - c)^2 + y^2 = b^2$ is $\frac{1}{c}\sqrt{\{(a+b+c)(a-b+c)(a+b-c)(-a+b+c)\}}$.

10. Show that the equation of a straight line meeting the circle $x^2 + y^2 = a^2$ in two points at equal distances d from a point (x_1, y_1) on its circumference is $xx_1 + yy_1 - a^2 + \frac{d^2}{2} = 0$.

Answer Keys

1. $15(x^2 + y^2) - 94x + 18y + 55 = 0$
2. yes
3. $(-2, -4)$, $(-2/5, -4/5)$
5. $(1 + a^2)(x^2 + y^2) - 2a^2x - 4a^2y = 0$
6. $7(x^2 + y^2) - 10x - 10y - 12 = 0$
7. $2(x^2 + y^2) + 2x + 6y + 1 = 0$
8. $(-\sqrt{2}, \sqrt{2})$

TEXTUAL EXERCISE–7 (OBJECTIVE)

1. The intercept on the line $y = x$ by the circle $x^2 + y^2 - 2x = 0$ is AB. Equation of the circle with AB as a diameter is:
(a) $x^2 + y^2 + x + y = 0$ (b) $-x^2 + y^2 + x - y = 0$
(c) $x^2 + y^2 - x - y = 0$ (d) None of these

2. The equation of the circle described on the common chord of the circles $x^2 + y^2 + 2x = 0$ and $x^2 + y^2 + 2y = 0$ as diameters is:
(a) $x^2 + y^2 + x - y = 0$ (b) $x^2 + y^2 - x - y = 0$
(c) $x^2 + y^2 - x + y = 0$ (d) $x^2 + y^2 + x + y = 0$

3. If a circle $S(x, y) = 0$ touches at the point (2, 3) of the line $x + y = 5$ and $S(1, 2) = 0$, then radius of such circle
(a) 2 units (b) 4 units
(c) $\frac{1}{2}$ units (d) $\frac{1}{\sqrt{2}}$ units

4. A circle passing through the intersection of $x^2 + y^2 - 4 = 0$ and $x^2 + y^2 - 6x + 5 = 0$ which also passes through the point (2, 1).
(a) $x^2 + y^2 - 2y - 1 = 0$
(b) $x^2 + y^2 - 2x - 1 = 0$
(c) $x^2 + y^2 + 2x - 1 = 0$
(d) None of these

5. The equation of circle touching line $2x - y + 1 = 0$ at the point (1, 3) and passing through origin.

(a) $x^2 + y^2 + 22x + 4y = 0$
(b) $x^2 + y^2 - 4x + 22y = 0$
(c) $x^2 + y^2 - 22x + 4y = 0$
(d) None of these

6. The equation of the circle circumscribing the triangle formed by the lines

(i) $x + y = 6$, $2x + y = 4$ and $x + 2y = 5$
(a) $x^2 + y^2 - 19x - 17y + 50 = 0$
(b) $x^2 + y^2 - 17x + 19y + 50 = 0$
(c) $x^2 + y^2 - 17x - 19y + 50 = 0$
(d) None of these

(ii) $x + y = 6$, $2x + y = 4$ and $x - 2y = 5$
(a) $(x - 3/17)(x+2) + (y - 1/3)(y-8) = 0$
(b) $(x - 17/3)(x + 2) + (y - 1/3)(y-8) = 0$
(c) $(x - 17/3)(x + 2) + (y-3)(y-1/8) = 0$
(d) None of these

7. The equation of the circle circumscribing the quadrilateral formed by the lines $2x + 3y = 2$, $3x - 2y = 3$, $x + 2y - 3 = 0$ and $2x - y = 1$.
(a) $8x^2 + 8y^2 + 17x - 8y + 9 = 0$
(b) $x^2 + y^2 + 7x - y - 6 = 0$
(c) $x^2 + y^2 + 6x + y + 7 = 0$
(d) None of these

8. Length of the tangent drawn from any point on the circle $x^2 + y^2 + 2gx + 2fy + c_1 = 0$ to the circle $x^2 + y^2 + 2gx + 2fy + c = 0$
(a) $\sqrt{c_1 - c}$ (b) $\sqrt{c - c_1}$
(c) $\sqrt{c_1 + c}$ (d) None of these

9. If a circle passes through the points of intersection of the co-ordinate axis with the line $\lambda x - y + 1 = 0$ and $x - 2y + 3 = 0$, then the valus of λ is
(a) 1 (b) 2, 1/3
(c) 3 (d) 4

10. If $x^2 + y^2 + px + 3y - 5 = 0$ and $x^2 + y^2 + 5x + py + 7 = 0$ cut orthogonally, then p is
(a) 1/2 (b) 1
(c) 3/2 (d) 2

11. The point of contact of the given circles $x^2 + y^2 - 6x - 6y + 10 = 0$ and $x^2 + y^2 = 2$, is
(a) (0, 0) (b) (1, 1)
(c) (1, –1) (d) (–1, –1)

12. If a circle passes through the point (1, 2) and cuts the circle $x^2 + y^2 = 4$ orthogonally, then the equation of the locus of its centre is
(a) $x^2 + y^2 - 3x - 8y + 1 = 0$
(b) $x^2 + y^2 - 2x - 6y - 7 = 0$
(c) $2x + 4y - 9 = 0$
(d) $2x + 4y - 1 = 0$

13. The locus of centre of a circle passing through (a, b) and cuts orthogonally to circle $x^2 + y^2 = p^2$ is
(a) $2ax + 2by - (a^2 + b^2 + p^2) = 0$
(b) $2ax + 2by - (a^2 - b^2 + p^2) = 0$
(c) $x^2 + y^2 - 3ax - 4by + (a^2 + b^2 - p^2) = 0$
(d) $x^2 + y^2 - 2ax - 3by + (a^2 - b^2 - p^2) = 0$

14. The equation of circle which intersects circles $x^2 + y^2 + x + 2y + 3 = 0$, $x^2 + y^2 + 2x + 4y + 5 = 0$ and $x^2 + y^2 - 7x - 8y - 9 = 0$ at right angle, will be
(a) $x^2 + y^2 - 4x - 4y - 3 = 0$
(b) $3(x^2 + y^2) + 4x - 4y - 3 = 0$
(c) $x^2 + y^2 + 4x + 4y - 3 = 0$
(d) $3(x^2 + y^2) + 4(x + y) - 3 = 0$

15. From any point on the circle $x^2 + y^2 = a^2$ tangents are drawn to the circle $x^2 + y^2 = a^2 \sin^2 a$, the angle between them is
(a) $\frac{\alpha}{2}$ (b) a
(c) $2a$ (d) None of these

16. Any circle through the point of intersection of the lines $x + \sqrt{3}y = 1$ and $\sqrt{3}x - y = 2$ if intersects these lines at points P and Q, then the angle subtended by the arc PQ at its centre is
(a) 180°
(b) 90°
(c) 120°
(d) Depends on centre and radius

Answer Keys

1. (c) **2.** (d) **3.** (d) **4.** (b) **5.** (c) **6.** (i) (c) (ii) (b) **7.** (a) **8.** (b) **9.** (b)
10. (a) **11.** (b) **12.** (c) **13.** (a) **14.** (d) **15.** (c) **16.** (a)

RADICAL AXIS AND RADICAL CENTRE

The radical axis of two circles is the locus of a point which moves such that the lengths of the tangents drawn from it to the two circles are equal.

Consider $S = x^2 + y^2 + 2gx + 2fy + c = 0$ ...(i)

$S' = x^2 + y^2 + 2g_1x + 2f_1y + c_1 = 0$ ..(ii)

Let $P(x_1, y_1)$ be a point such that $|PA| = |PB|$

$\Rightarrow \sqrt{x_1^2 + y_1^2 + 2gx_1 + 2fy_1 + c}$

$= \sqrt{x_1^2 + y_1^2 + 2g_1x_1 + 2f_1y_1 + c_1}$

On squaring we get, $x_1^2 + y_1^2 + 2gx_1 + 2fy_1 + c$

$= x_1^2 + y_1^2 + 2g_1x_1 + 2f_1y_1 + c_1$

$\Rightarrow 2(g - g_1)x_1 + 2(f - f_1)y_1 + c - c_1 = 0$

$\therefore$ locus of $P(x_1, y_1)$ is

$2(g - g_1)x + 2(f - f_1)y + c - c_1 = 0$

which is the required equation of radical axis of the given circles. Clearly, this is a straight line.

Aliter: $PA = PB \Rightarrow \sqrt{S} = \sqrt{S_1}$ (where S_1 and S_2 are in canonical form)

$\Rightarrow S = S_1$

$\Rightarrow S - S_1 = 0$

$\Rightarrow 2(g - g_1)x + 2(f - f_1)y + c - c_1 = 0$

Properties of the Radical Axis

1. The radical axis and common chord are identical: Since the radical axis and common chord of two circles $S = 0$ and $S' = 0$ is the same straight line $S - S' = 0$, they are identical. The only difference is that the common chord exists only if the circles intersect in two real point, while the radical axis exists for all pair of circles irrespective of their position.

   The position of the radical axis of the two circles geometrically is shown here.

   From Euclidian geometry:

   $(PA)^2 = PR.\,PQ = (PB)^2$

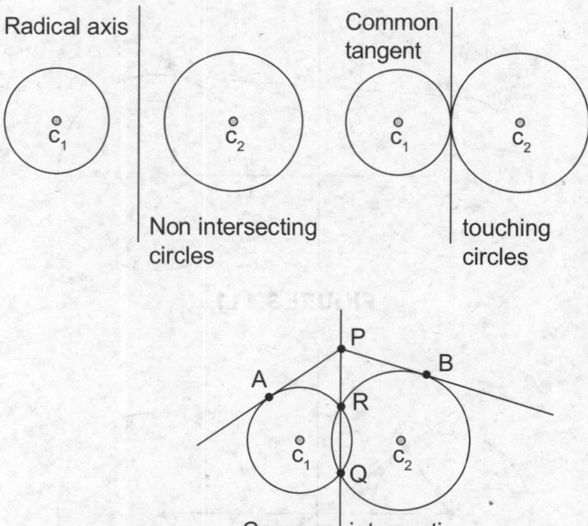

FIGURE 3.110

2. The radical axis is perpendicular to the straight line which joins the centres of the circles:

   Consider, $S \equiv x^2 + y^2 + 2gx + 2fy + c = 0$ ...(i)

   $S' \equiv x^2 + y^2 + 2g_1x + 2f_1y + c_1 = 0$ ....(ii)

   Since $C_1 \equiv (-g, -f)$ and $C_2 \equiv (-g_1, -f_1)$ are the centres of the circles (i) and (ii), then slope of

   $C_1C_2 = \dfrac{-f_1 + f}{-g_1 + g} = \dfrac{f - f_1}{g - g_1} = m_1$ (say)

   Equation of the radical axis is

   $2(g - g_1)x + 2(f - f_1)y + c - c_1 = 0$

   Slope of radical axis is $\dfrac{(g - g_1)}{(f - f_1)} = m_2$ (say)

   $\therefore\ m_1 m_2 = -1$

   Hence C_1C_2 and radical axis are perpendicular to each other.

3. The radical axis bisects common tangents of two circles:

   Let AB be the common tangent (direct or transverse). If it meets the radical axis LM in M, then MA and MB are two tangents to the circles. Hence $MA = MB$ since length of tangents are equal from any point on radical axis. Hence radical axis bisects the common tangent AB.

   If the two circles touch each other externally or internally, then A and B coincide. In this case, the common tangent itself becomes the radical axis

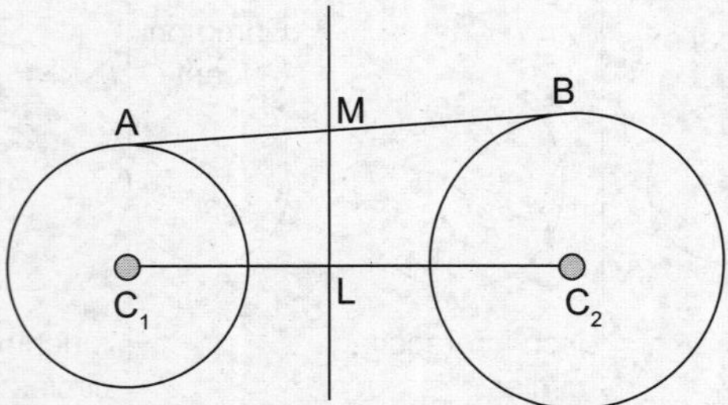

FIGURE 3.111

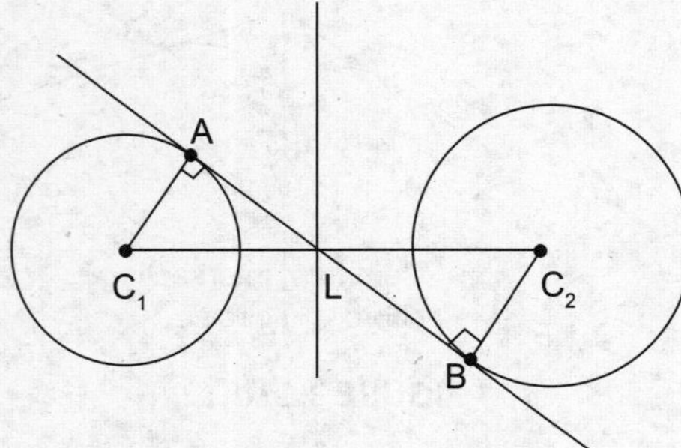

FIGURE 3.112

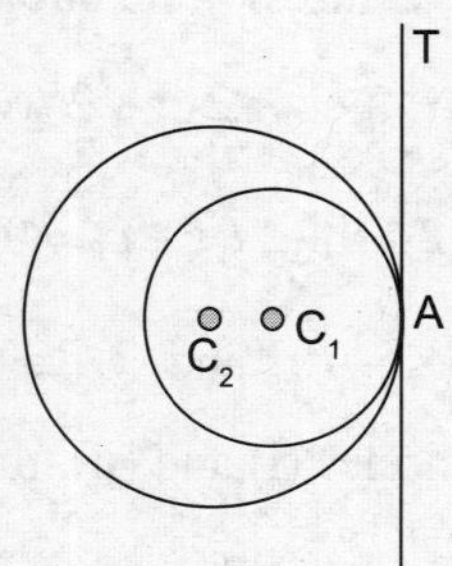

FIGURE 3.113

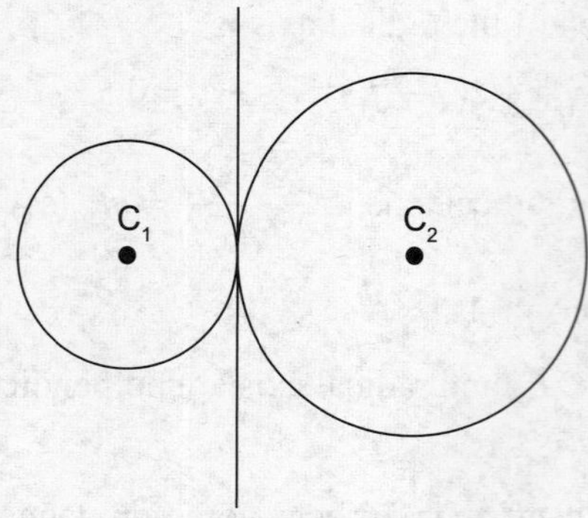

FIGURE 3.114

4. The radical axis of three circles taken in pairs are concurrent:

Let the equations of three circles be

$$S_1 \equiv x^2 + y^2 + 2g_1x + 2f_1y + c_1 = 0 \quad ....(i)$$

$$S_2 \equiv x^2 + y^2 + 2g_2x + 2f_2y + c_2 = 0 \quad ....(ii)$$

$$S_3 \equiv x^2 + y^2 + 2g_3x + 2f_3y + c_3 = 0 \quad ....(iii)$$

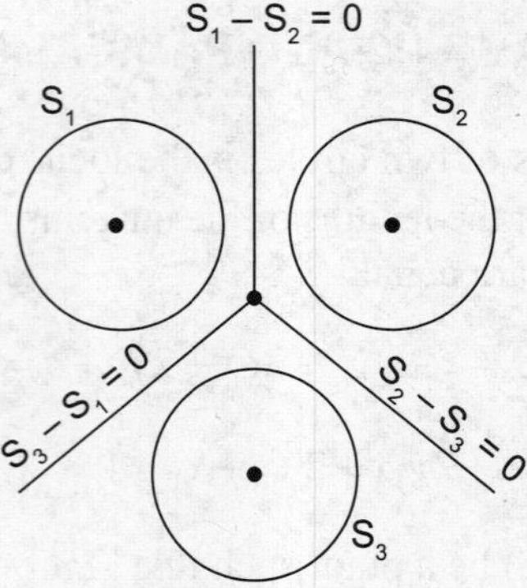

FIGURE 3.115

The radical axis of the above three circles taken in pairs are given by

$$S_1 - S_2 \equiv 2x(g_1 - g_2) + 2y(f_1 - f_2) + c_1 - c_2 = 0 \quad ....(iv)$$

$$S_2 - S_3 \equiv 2x(g_2 - g_3) + 2y(f_2 - f_3) + c_2 - c_3 = 0 \quad ....(v)$$

$$S_3 - S_1 \equiv 2x(g_3 - g_1) + 2y(f_3 - f_1) + c_3 - c_1 = 0 \quad ...(vi)$$

Adding (iv), (v) and (vi), we find L.H.S. vanished identically. Thus the three lines are concurrent.

5. If two circles cut a third circle orthogonally, the radical axis of the two circles will pass through the centre of the third circle

or,

The locus of the centre of a circle cutting two given circles orthogonally is the radical axis of the two circles.

Let $S_1 \equiv x^2 + y^2 + 2g_1x + 2f_1y + c_1 = 0$ ....(i)

$S_2 \equiv x^2 + y^2 + 2g_2x + 2f_2y + c_2 = 0$ ....(ii)

$S_3 \equiv x^2 + y^2 + 2g_3x + 2f_3y + c_3 = 0$ ....(iii)

Let (i) and (ii) both cut (iii) orthogonally

$\therefore \quad 2g_1g_3 + 2f_1f_3 = c_1 + c_3$

and $2g_2g_3 + 2f_2f_3 = c_2 + c_3$

Subtracting, we get

$2g_3(g_1 - g_2) + 2f_3(f_1 - f_2) = c_1 - c_2$ ...(iv)

Now radical axis of (i) and (ii) is

$S_1 - S_2 = 0$ or

$2x\,(g_1 - g_2) + 2y(f_1 - f_2) + c_1 - c_2 = 0$

Now, it will pass through the centre of circle (iii)

iff $-2g_3(g_1 - g_2) - 2f_3(f_1 - f_2) + c_1 - c_2 = 0$

or $\;2g_3(g_1 - g_2) + 2f_3(f_1 - f_2) = c_1 - c_2$ ....(v)

which is true by (iv)

NOTE

Radical axis need not always pass through the mid-point of the line joining the centres of the two circles.

ILLUSTRATION 65: Find the radical axis of the pairs of circles $x^2 + y^2 = 144$ and $x^2 + y^2 - 15x + 11y = 0$.

SOLUTION: Equation of the circles are given as

$$x^2 + y^2 = 144 \quad \text{....(i)}$$

and $x^2 + y^2 - 15x + 11y = 0$ ....(ii)

Subtracting (ii) from (i), equation of the radical axis is $15x - 11y = 144$

ILLUSTRATION 66: Find the radical axis of the pairs of circles $x^2 + y^2 - 3x - 4y + 5 = 0$ and $3x^2 + 3y^2 - 7x + 8y + 11 = 0$.

SOLUTION: Equations of the circles are given as

$$x^2 + y^2 - 3x - 4y + 5 = 0 \quad \text{.....(i)}$$

and $3x^2 + 3y^2 - 7x + 8y + 11 = 0$. ....(ii)

To make the co-efficients of x^2 same, multiplying (i) by 3, we get

$$3x^2 + 3y^2 - 9x - 12y + 15 = 0 \quad \text{....(iii)}$$

Subtracting (iii) from (ii), equation of the radical axis is $2x + 20y - 4 = 0$ or $x + 10y = 2$

ILLUSTRATION 67: Find the general equation of all circles, any pair of which have the same radical axis as the circles $x^2 + y^2 = 4$ and $x^2 + y^2 + 2x + 4y = 6$.

SOLUTION: The equations of the circles are

$$x^2 + y^2 = 4 \quad \text{...(i)}$$

and $x^2 + y^2 + 2x + 4y = 6$ ...(ii)

The general equation of all the circles having the same radical axis is given by

$$S + \lambda S' = 0; \lambda \neq -1;$$

where $S = 0$ and $S' = 0$ are the two circles and λ is a constant

Hence the required equation is $(x^2 + y^2 - 4) + \lambda(x^2 + y^2 + 2x + 4y - 6) = 0$

or $(x^2 + y^2)(\lambda + 1) + 2\lambda(x + 2y) = 4 + 6\lambda$

■ RADICAL CENTRE

Radical axes of three circles taken pairwise are always concurrent. The common point of intersection of the radical axes of three circles taken two at a time is called the radical centre of the three circles. In the given figure, O is the radical centre.

Let three circles be

$S_1 = 0$ ....(i)

$S_2 = 0$ ...(ii)

and $S_3 = 0$ ...(iii)

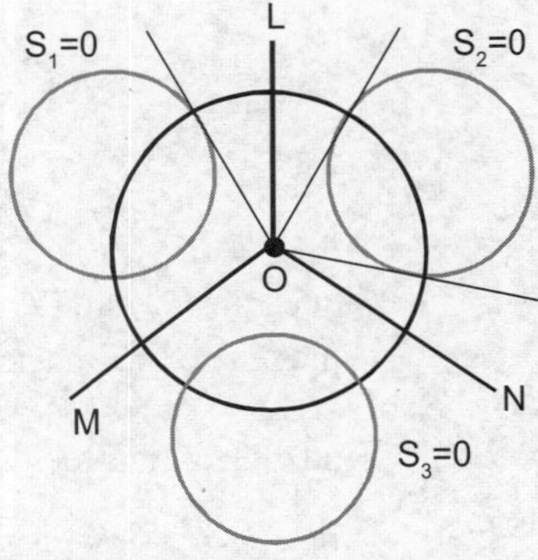

FIGURE 3.116

Let OL, OM and ON be radical axes of the pairs of the set of circles.

$\{S_1 = 0, S_2 = 0\}$, $\{S_3 = 0, S_1 = 0\}$ and $\{S_2 = 0, S_3 = 0\}$ respectively.

Equations of OL, OM and ON are respectively

$$S_1 - S_2 = 0 \quad \text{...(iv)}$$

$$S_3 - S_1 = 0 \quad \text{...(v)}$$

$$S_2 - S_3 = 0 \quad \text{...(vi)}$$

Let the straight lines (iv) and (v), i.e., OL and OM meet in O. The equation of any straight line passing through O is $(S_1 - S_2) + \lambda(S_3 - S_1) = 0$ where λ is any constant.

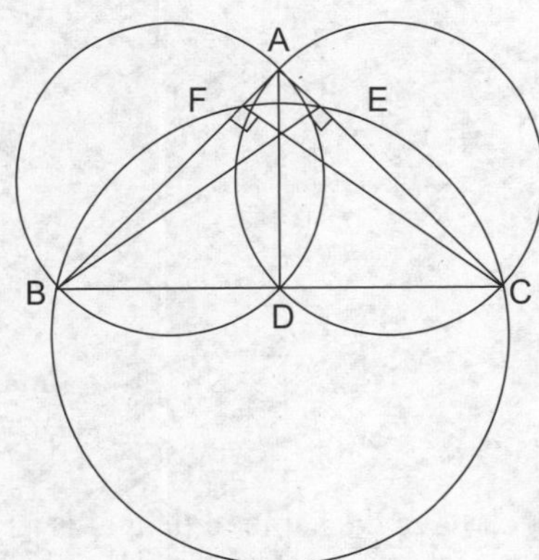

FIGURE 3.117

For $\lambda = 1$, this equation becomes $S_2 - S_3 = 0$. which is by (vi), equation of ON. Thus, the third radical axis also passes through the point where (iv) and (v) meet.

Properties of radical centre

1. Co-ordinates of radical centre can be found by solving the equation $S_1 - S_2 = 0$ and $S_2 - S_3 = 0$.
2. The radical centre of three circles described on the sides of a triangle as diameters is the orthocentre of the triangle.

Proof: Draw perpendicular from A on BC.

$$\therefore \angle ADB = \angle ADC = \frac{\pi}{2}$$

Therefore the circles whose diameters are AB and AC pass through D and A. Hence AD is their radical axis. Similarly the radical axis of the circles on AB and BC as diameter is the perpendicular line from B on CA and radical axis of the circles on BC and CA as diameter is the perpendicular line from C on AB. Hence the radical axis of three circles meet in a point. This point H is radical centre, but here radical centre is the point of intersection of altitudes. that is, AD, BE and CF. Hence radical centre = orthocentre.

3. The radical centre of the three given circles will be the centre of a fourth circle ($S_4 = 0$) which cuts all the three circles orthogonally and the radius of the fourth circle is the length of tangent drawn from radical centre of the three given circles to any of these circles.

Let the fourth circle be $(x - h)^2 + (y - k)^2 = r^2$ where (h, k) is centre of this circle and r be the radius. The centre of circle is the radical centre of the given circles and r is the length of tangent from (h, k) to any of the given three circles.

ILLUSTRATION 68: Find the radical centre of three circles described on the three sides $4x - 7y + 10 = 0$, $x + y - 5 = 0$ and $7x + 4y - 15 = 0$ of a triangle as diameters.

SOLUTION: Since the radical centre of three circles described on the sides of a triangle as diameters is the orthocentre of the triangle.

$\therefore$ Radical centre = orthocentre

Given sides are $4x - 7y + 10 = 0$ ...(i)

$x + y - 5 = 0$ ...(ii)

$7x + 4y - 15 = 0$ ...(iii)

Since lines (i) and (iii) are perpendicular, the point of intersection of (i) and (iii) is (1, 2) the orthocentre of the triangle. Hence radical centre is (1, 2)

ILLUSTRATION 69: Find the radical centre of the set of circles $x^2 + y^2 + x + 2y + 3 = 0$, $x^2 + y^2 + 2x + 4y + 5 = 0$ and $x^2 + y^2 - 7x - 8y - 9 = 0$

SOLUTION: The equations of the circles are

$$x^2+y^2+x+2y+3=0 \quad \text{....(i)}$$

$$x^2+y^2+2x+4y+5=0 \quad \text{....(ii)}$$

and $x^2+y^2-7x-8y-9=0$ ...(iii)

The radical axis of (ii) and (i) is

$$x^2+y^2+2x+4y+5-(x^2+y^2+x+2y+3)=0$$

or $x + 2y + 2 = 0$ ...(iv)

The radical axis of (ii) and (iii) is

$$x^2+y^2+2x+4y+5-(x^2+y^2-7x-8y-9)=0$$

or $9x + 12y + 14 = 0$ ....(v)

Solving equation (iv) and (v), we get

$x = -2/3$ and $y = -2/3$

Hence the radical centre is $(-2/3, -2/3)$

■ COAXIAL SYSTEM (FAMILY) OF CIRCLES

A family of circles is said to be co-axial if every pair of circles of this family has the same radical axis. Since the radical axis of any pair of the circles is perpendicular to the line joining their centres, therefore the centres of the circles of a coaxial system must be collinear.

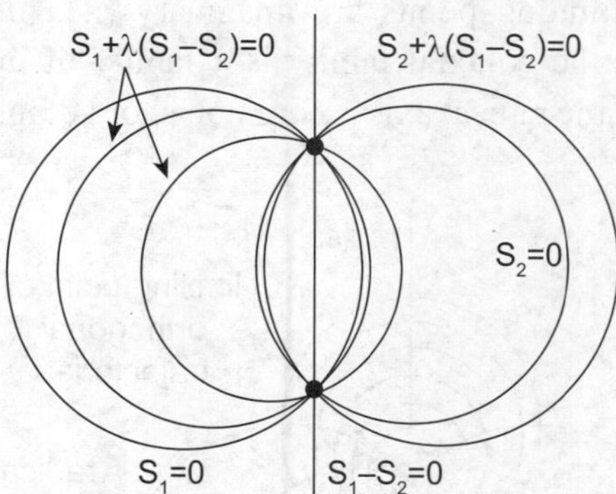

FIGURE 3.118

The equation of a co-axial system of circles where the equation of any two circles of the system are

$$S_1 \equiv x^2+y^2+2g_1x+2f_1y+c_1=0$$

and $S_2 \equiv x^2+y^2+2g_2x+2f_2y+c_2=0$ respectively is

$$S_1+\lambda(S_1-S_2)=0$$

or $\quad S_2+\lambda_1(S_1-S_2)=0$

Other form is $S_1+\lambda_1 S_2=0; \lambda_1 \neq -1$

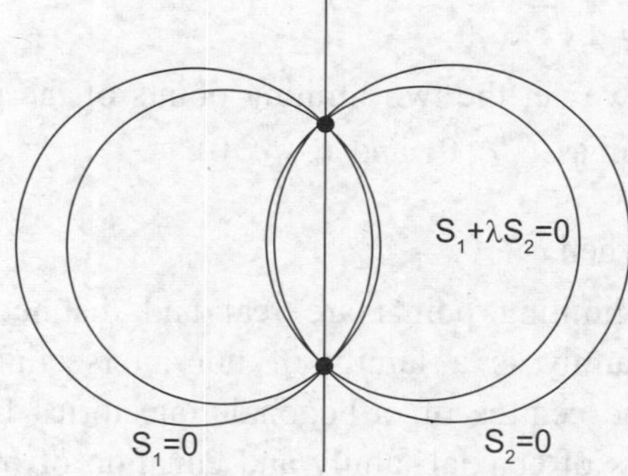

FIGURE 3.119

Because for any two different values of λ/λ_1, the two circles of the above family obtained always have radical axis as a fixed line given by $S_1 - S_2 = 0$. Treating as a line parallel to y-axis and line joining the centre as x-axis, we have

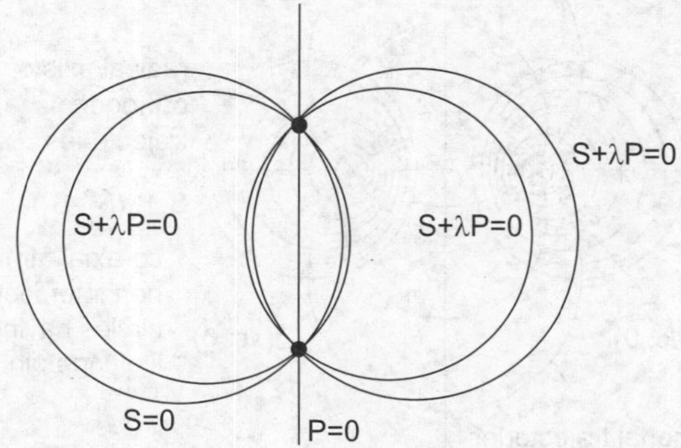

FIGURE 3.120

the equation of co-axial system is $x^2 + y^2 + 2gx + c = 0$. where g is a parameter and c is constant and the equation of other family of co-axial circle is $x^2 + y^2 + 2fy + c = 0$ where f is parameter and c is constant.

The equation of a system of co-axial circles, when the equation of the radical axis and of one circle of the system are

$$P \equiv lx + my + n = 0$$

and $S \equiv x^2 + y^2 + 2gx + 2fy + c = 0$ respectively is

$$S + \lambda P = 0 \text{ (where } \lambda \text{ is an arbitrary constant)}$$

Limiting Point of a Coaxial System

Limiting points of a system of co-axial circles are the centres of the point circles belonging to the family (circles whose radii are zero are called point circles)

Let the circle is $x^2 + y^2 + 2gx + c = 0$ where g is a variable and c is a constant. ......(i)

$\therefore$ centre and the radius of (i) are $(-g, 0)$ and $\sqrt{g^2 - c}$ respectively. Let $\sqrt{g^2 - c} = 0$

$$\Rightarrow g = \pm\sqrt{c}$$

Thus we get the two limiting points of the given co-axial system as $(\sqrt{c}, 0)$ and $(-\sqrt{c}, 0)$.

Case I: when $c > 0$:

Both the limiting points are real and distinct and the co-axial family is a family of non-intersecting circles. Clearly, the centres of orthogonal trajectories lie on the radical axis of coaxial family and equation of orthogonal trajectories is given as $x^2 + y^2 + 2\alpha y - c = 0$ (its radius is $\sqrt{\alpha^2 + c}$ which is never zero).

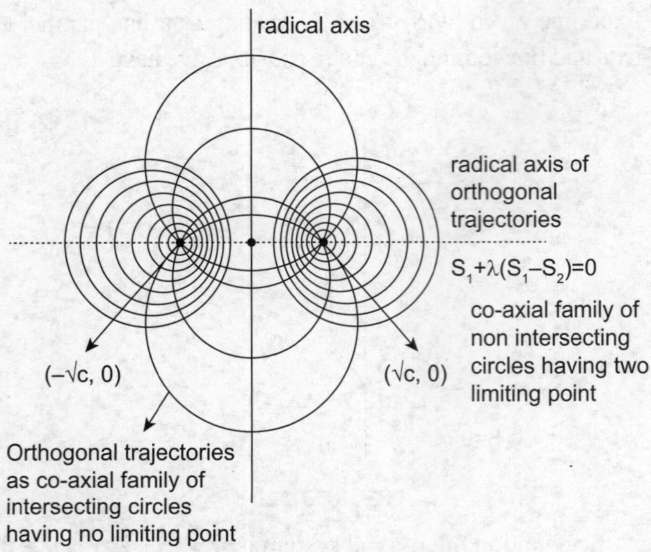

FIGURE 3.121

Case II: when $c = 0$:

both the limiting points are real and coincident and the co-axial family is a family of circles touching the radical axis at a point, radical axis is common tangent to them.

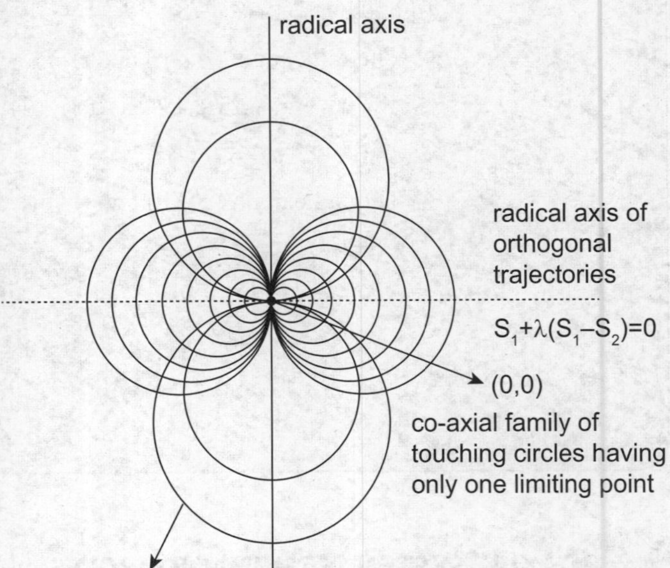

FIGURE 3.122

Case III: when $c < 0$:

Both the limiting points are imaginary (i.e., no limiting point) and the co-axial family is a family of intersecting circles and the radical axis is common chord of intersection.

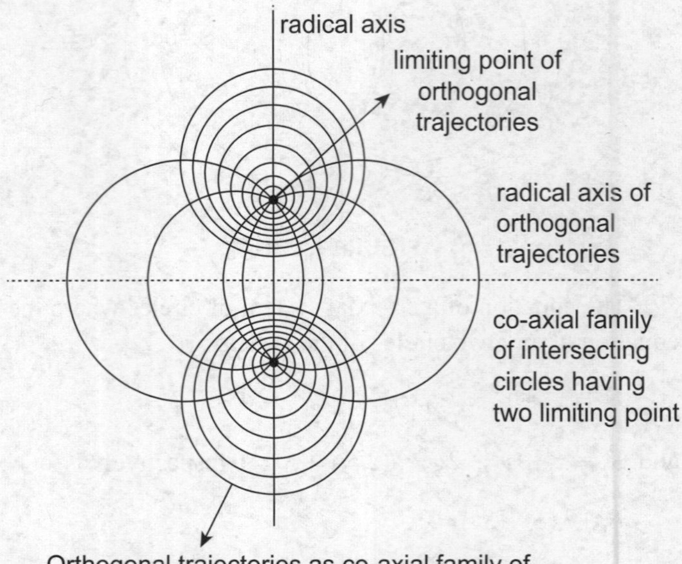

FIGURE 3.123

Orthogonal Circles of a Coaxial System

A set of coaxial circles can be cut orthogonally by another set of coaxial circles, the centres of each set lying on the radical axis of the other set, also one set is of the limiting point species and other set of the other species. The limiting points are therefore the intersection with the line of centres of any circle whose centre is on the common radical axis and whose radius is the tangent from it to any of the circles of the system.

Here limiting points are imaginary thus orthogonal circles do not meet the line of centres in real points. Hence they pass through limiting points L_1 and L_2

System of Coaxial Circles Whose Two Limiting Points are Given

Let (α, β) and (γ, δ) be the two given limiting points. Then the corresponding point circles with zero radii are

$(x - \alpha)^2 + (y - \beta)^2 = 0$ and $(x - \gamma)^2 + (y - \delta)^2 = 0$

or $x^2 + y^2 - 2\alpha x - 2\beta y + \alpha^2 + \beta^2 = 0$ and

$$x^2 + y^2 - 2\gamma x - 2\delta y + \gamma^2 + \delta^2 = 0$$

The equation of the co-axial system is

$$(x^2 + y^2 - 2\alpha x - 2\beta y + \alpha^2 + \beta^2)$$
$$+ \lambda(x^2 + y^2 - 2\gamma x - 2\delta y + \gamma^2 + \delta^2) = 0 \quad \text{.....(i)}$$

where $\lambda \neq -1$ is a variable parameter

$$\Rightarrow x^2(1+\lambda) + y^2(1+\lambda) - 2x(\alpha + \gamma\lambda)$$
$$-2y(\beta + \delta\lambda) + (\alpha^2 + \beta^2) + \lambda(\gamma^2 + \delta^2) = 0$$

or $$x^2 + y^2 - \frac{2x(\alpha+\gamma\lambda)}{(1+\lambda)}x - 2\frac{(\beta+\delta\lambda)}{(1+\lambda)}y + \frac{(\alpha^2+\beta^2)+\lambda(\gamma^2+\delta^2)}{(1+\lambda)} = 0$$

Centre of this circle is $\left(\frac{(\alpha+\gamma\lambda)}{(1+\lambda)}, \frac{(\beta+\delta\lambda)}{(1+\lambda)}\right)$ .....(ii)

For limiting point, radius

$$= \sqrt{\frac{(\alpha+\gamma\lambda)^2}{(1+\lambda)^2} + \frac{(\beta+\delta\lambda)^2}{(1+\lambda)^2} - \frac{(\alpha^2+\beta^2)+\lambda(\gamma^2+\delta^2)}{(1+\lambda)}} = 0$$

After solving, find λ. Substituting value of λ in (i), we get the co-axial system of circles with given limiting points.

Properties of Limiting Points

1. The limiting points of a system of co-axial circles are conjugate points with respect to any member of the system:

Let the equation of any circle be $x^2 + y^2 + 2gx + c = 0$ ...(i)

Limiting points of (i) are $(\sqrt{c}, 0)$ and $(-\sqrt{c}, 0)$.

The polar of the point $(\sqrt{c}, 0)$ with respect to (i) is

$$x\sqrt{c} + y.0 + g(x + \sqrt{c}) + c = 0$$
$$\Rightarrow x\sqrt{c} + g(x + \sqrt{c}) + c = 0$$
$$\Rightarrow (x + \sqrt{c}).(g + \sqrt{c}) = 0$$
$$\Rightarrow (x + \sqrt{c}) = 0$$

and it clearly passes through the other limiting point $(-\sqrt{c}, 0)$. Similarly, polar of the point $(-\sqrt{c}, 0)$ with respect to (i) also passes through $(\sqrt{c}, 0)$. Hence the limiting points of a system of co-axial circles are conjugate points.

2. Every circle through the limiting points of a co-axial system is orthogonal to all circles of the system:

Let the equation of any circle of the system be $x^2 + y^2 + 2gx + c = 0$ .....(i)

where g is a parameter and c is a constant. Limiting point of (i) are $(\sqrt{c}, 0)$ and $(-\sqrt{c}, 0)$

Now let $x^2 + y^2 + 2g'x + 2f'y + c' = 0$ ...(ii)

be the equation of any circle. If it passes through the limiting points of (i), then

$c + 2g'\sqrt{c} + c' = 0$ and $c - 2g'\sqrt{c} + c' = 0$

Solving, we get $c' = -c$ and $g' = 0$

From (ii) $x^2 + y^2 + 2f'y - c = 0$ ..(iii)

where c is constant and f' is variable. Applying the condition of orthogonality on (i) and (iii)

i.e., $2g_1 g_2 + 2f_1 f_2 = c_1 + c_2$, we find that

$$2 \times g \times 0 + 2 \times 0 \times f' = c - c$$

i.e., $0 = 0$

Hence, condition is satisfied for all values of g and f'.

ILLUSTRATION 70: Find the radical axis of co-axial system of circles whose limiting points are (–1,2) and (2, 3)

SOLUTION: Equations of circles with limiting points (–1, 2) and (2, 3) are

$(x + 1)^2 + (y - 2)^2 = 0$ or $x^2 + y^2 + 2x - 4y + 5 = 0$ ...(i)

and $(x - 2)^2 + (y - 3)^2 = 0$ or $x^2 + y^2 - 4x - 6y + 13 = 0$ ...(ii)

$\therefore$ Radical axis of circles (i) and (ii)

$(x^2 + y^2 + 2x - 4y + 5) - (x^2 + y^2 - 4x - 6y + 13) = 0$

or $6x + 2y - 8 = 0$ or $3x + y - 4 = 0$

ILLUSTRATION 71: Find the equation of the circle which passes through the origin and belongs to the co-axial system of circles whose limiting points are (1, 2) and (4, 3).

SOLUTION: Equations of circles whose limiting points are (1, 2) and (4, 3) are

$(x - 1)^2 + (y - 2)^2 = 0$ or $x^2 + y^2 - 2x - 4y + 5 = 0$ ...(i)

and $(x - 4)^2 + (y - 3)^3 = 0$ or $x^2 + y^2 - 8x - 6y + 25 = 0$ ...(ii)

Therefore the corresponding system of co-axial circles is

$(x^2 + y^2 - 2x - 4y + 5) + \lambda (x^2 + y^2 - 8x - 6y + 25) = 0$ ...(iii)

It passes through origin, then $5 + 25\lambda = 0$

$\Rightarrow \lambda = -1/5$

Substituting the value of λ in (iii), the required circle is

$5(x^2 + y^2 - 2x - 4y + 5) - (x^2 + y^2 - 8x - 6y + 25) = 0$

or $4x^2 + 4y^2 - 2x - 14y = 0$ or $2x^2 + 2y^2 - x - 7y = 0$

ILLUSTRATION 72: Prove that the tangents from any point of a fixed circle of co-axial system to two other fixed circles of the system are in a constant ratio.

SOLUTION: Let the equations of the circles be $x^2 + y^2 + 2g_i x + c = 0$, $i = 1, 2, 3$. Since all the three circles are fixed g_1, g_2 and g_3 are constants.

Let $P(h, k)$ be any point on the first circle, so that

$$h^2 + k^2 + 2g_1 h + c = 0 \qquad ....(i)$$

Let PQ and PR be the tangents from P on the other two circles

$$\therefore \quad PQ = \sqrt{h^2 + k^2 + 2g_2 h + c}$$

and $PR = \sqrt{h^2 + k^2 + 2g_3 h + c}$

$$\therefore \quad \frac{(PQ)^2}{(PR)^2} = \left(\frac{\sqrt{h^2 + k^2 + 2g_2 h + c}}{\sqrt{h^2 + k^2 + 2g_3 h + c}}\right)^2$$

$$= \frac{-2g_1 h + 2g_2 h}{-2g_1 h + 2g_3 h}$$

$$= \frac{g_2 - g_1}{g_3 - g_1} = \text{constant}$$

because g_1, g_2, g_3 are constants.

TEXTUAL EXERCISE–8 (SUBJECTIVE)

1. Find the equation of the family of circles described in the following problem and select the specified member
   (a) centre at $(2, -1)$; member which touches x-axis
   (b) centre at $(-4, 2)$; member tangent to $x - y = 3$

2. (a) Find a circle passing through the intersection of $x^2 + y^2 - 4 = 0$ and $x^2 + y^2 - 6x + 5 = 0$ which also passes through the point $(2, 1)$.
   (b) Find the equation of the circle described on the common chord of the circles $x^2 + y^2 - 4x - 5 = 0$ and $x^2 + y^2 + 8y + 7 = 0$ as diameter.
   (c) Find the equation of the circle through the intersection of the circles $x^2 + y^2 - 8x - 2y + 7 = 0$ and $x^2 + y^2 - 4x + 10y + 8 = 0$ and has its centre on the x-axis.

3. (a) Find the equation of the circle passing through $(1, 2)$ and $(3, 4)$ and touching the x-axis.
   (b) Find the equation of circle touching line $2x - y + 1 = 0$ at the point $(1, 3)$ and passing through origin.
   (c) Find the equation of the circle of radius 5 and touching the line $3x - 4y + 5 = 0$ at $(1, 2)$.
   (d) Find the equation of the circle passing through the point $A(4, 3)$, $B(2, 5)$ and touching y-axis. Also find the point P on the y-axis such that the angle APB has largest magnitude.

4. Find the equation of the system of circles co-axial with the circles $x^2 + y^2 + 4x + 2y + 1 = 0$ and $x^2 + y^2 - 2x + 6y - 6 = 0$. Also find the equation of that particular circle whose centre lies on the radical axis.

5. Find the co-ordinates of the limiting points of the system of circles determined by the two circles $x^2 + y^2 + 5x + y + 4 = 0$ and $x^2 + y^2 + 10x - 4y - 1 = 0$.

6. Find the equation of the circle co-axial with the circles $2x^2 + 2y^2 - 2x + 6y - 3 = 0$ and $x^2 + y^2 + 4x + 2y + 1 = 0$ and whose centre is on the radical axis of the circles.

7. If the origin be one limiting point of a system of co-axial circles of which $x^2 + y^2 + 3x + 4y + 25 = 0$ is a member, find the other limiting point.

Answer Keys

1. (a) $(x-2)^2 + (y+1)^2 = r^2$ specified member $(x - 2)^2 + (y + 1)^2 = 1$
   (b) $(x + 4)^2+(y-2)^2 = r^2$ member $r = 9/\sqrt{2}$
2. (a) $x^2 + y^2 - 2x - 1 = 0$ (b) $x^2 + y^2 - 2x + 4y + 1 = 0$ (c) $6(x^2 + y^2) - 44x + 43 = 0$
3. (a) $x^2 + y^2 - (\lambda - 4)x - (6 + \lambda) y + (11 + \lambda) = 0$ where $\lambda = 14, -2$ (b) $x^2 + y^2 - 22x + 4y = 0$
   (c) $x^2 + y^2 - 8x + 4y - 5 = 0$, $x^2 + y^2 + 4x - 12y + 15 = 0$ (d) $x^2 + y^2 - 4x - 6y + 9 = 0$, $P(0, 3)$
4. $26(x^2 + y^2) + 98x + 56y + 19 = 0$ 5. $(-2, -1)$ and $(0, -3)$
6. $4(x^2 + y^2) + 6x + 10y - 1 = 0$ 7. $(-6, -8)$

TEXTUAL EXERCISE–8 (OBJECTIVE)

1. The equation of a circle passing through origin and co-axial to circles $x^2 + y^2 = a^2$ and $x^2 + y^2 + 2ax = 2a^2$, is
   (a) $x^2 + y^2 = 1$ (b) $x^2 + y^2 + 2ax = 0$
   (c) $x^2 + y^2 - 2ax = 0$ (d) $x^2 + y^2 = 2a^2$

2. The radical centre of three circles described on the three side of a triangle as diameter is
   (a) The orthocentre
   (b) The circumcentre
   (c) The incentre of the triangle
   (d) The centroid

3. In the co-axial system of circle $x^2 + y^2 + 2gx + c = 0$, where g is a parameter, if $c > 0$ then the circles are
   (a) Orthogonal
   (b) Touching type
   (c) Intersecting
   (d) Non-intersecting type

4. The equation of radical axis of the circle $2x^2 + 2y^2 - 7x = 0$ and $x^2 + y^2 - 4y - 7 = 0$ is
(a) $7x + 8y + 14 = 0$
(b) $7x - 8y + 14 = 0$
(c) $7x - 8y - 14 = 0$
(d) None of these

5. The radical centre of the circles $x^2 + y^2 - 16x + 60 = 0$, $x^2 + y^2 - 12x + 27 = 0$, $x^2 + y^2 - 12y + 8 = 0$ is
(a) (13, 33/4) (b) (33/4, –13)
(c) (33/4, 13) (d) None of these

6. The radical axis of two circles and the line joining their centres are
(a) Parallel
(b) Perpendicular
(c) Neither parallel, nor perpendicular
(d) Intersecting, but at acute angle

7. The radical axis of the pair of circle $x^2 + y^2 = 144$ and $x^2 + y^2 - 15x + 12y = 0$ is
(a) $15x - 12y = 0$ (b) $3x - 2y = 12$
(c) $5x - 4y = 48$ (d) None of these

8. The circle $x^2 + y^2 = 4$ cuts the line joining the points $A(1, 0)$ and $B(3, 4)$ in two points P and Q. Let $\frac{BP}{PA} = \alpha$ and $\frac{BQ}{QA} = \beta$ (where the distances are of directed lines). Then α and β are roots of the quadratic equation
(a) $3x^2 + 2x - 21 = 0$ (b) $3x^2 + 2x + 21 = 0$
(c) $2x| + 3x - 21 = 0$ (d) None of these

9. A circle is inscribed in an equilateral triangle of side a, the area of any square inscribed in the circle is
(a) $\frac{a^3}{3}$ (b) $\frac{2a^3}{3}$
(c) $\frac{a^2}{6}$ (d) $\frac{a^2}{12}$

10. Let L_1 be a straight line passing through the origin and L_2 be the straight line $x + y = 1$. If the intercepts made by the circle $x^2 + y^2 - x + 3y = 0$ on L_1 and L_2 are equal, then which of the following equations can represent L_1?
(a) $x + y = 0$ (b) $x - y = 0$
(c) $x + 7y = 0$ (d) $x - 7y = 0$

11. The centre of the circle passing through (0, 0) and (1, 0) and touching the circle $x^2 + y^2 = 9$ is
(a) $\left(\frac{1}{2}, \frac{1}{2}\right)$ (b) $\left(\frac{1}{2}, -\sqrt{2}\right)$
(c) $\left(\frac{3}{2}, \frac{1}{2}\right)$ (d) $\left(\frac{1}{2}, \frac{3}{2}\right)$

12. If the tangent to the circle $x^2 + y^2 = r^2$ at the point (a, b) meets the co-ordinates axes at the point A and B, and O is the origin, then the area of the triangle OAB is
(a) $\frac{r^4}{2ab}$ (b) $\frac{r^4}{ab}$
(c) $\frac{r^2}{2ab}$ (d) $\frac{r^2}{ab}$

13. The tangents are drawn from the points (4, 5) to the circle $x^2 + y^2 - 4x - 2y - 11 = 0$. The area of quadrilateral formed by these tangents and radii, is
(a) 15 sq. units (b) 75 sq. units
(c) 8 sq. units (d) 4 sq. units

14. Let PQ and RS be tangents at the extremities of the diameter PR of a circle of radius r. If PS and RQ intersect at a point X on the circumference of the circle, then $2r$ equal
(a) $\sqrt{PQRS}$ (b) $\frac{PQ + RS}{2}$
(c) $\frac{2PQ.RS}{PQ + RS}$ (d) $\sqrt{\frac{PQ^2 + RS^2}{2}}$

15. P, Q and R are the centres and r_1, r_2, r_3 are the radiii respectively of the co-axial circles, then $QR\, r_1^2 + RP\, r_2^2 + PQ\, r_3^2$ is equal to
(a) $PQ \cdot QR \cdot RP$ (b) $-PQ \cdot QR \cdot RP$
(c) $PQ^2 \cdot QR \cdot RP^2$ (d) None of these

Answer Keys

1. (c)	**2.** (a)	**3.** (d)	**4.** (c)	**5.** (d)	**6.** (b)	**7.** (c)	**8.** (a)	**9.** (c)	**10.** (b, c)
11. (b)	**12.** (a)	**13.** (c)	**14.** (a)	**15.** (b)					

MULTIPLE-CHOICE QUESTIONS

SECTION–I

OBJECTIVE-TYPE SOLVED EXAMPLES

1. The point of which the line $9x + y - 28 = 0$, is the chord of contact of the circle $2x^2 + 2y^2 - 3x + 5y - 7 = 0$ is:

(a) $(3, -1)$ (b) $(3,1)$
(c) $(-3,1)$ (d) None of these

Solution: (d) Let the required point be (h, k). Then its chord of contact w.r.t the circle

$S \equiv x^2 + y^2 - (3/2)x + (5/2)y - (7/2) = 0$ is

$hx + ky - (3/4)(x + h) + (5/4)(y + k) - (7/2) = 0$

or $(4h - 3)x + (4k + 5)y - (3h - 5k + 14) = 0$ ....(1)

But given chord of contact is

$9x + y - 28 = 0$ ....(2)

Comparing co-efficients in (1) and (2), we get

$$\frac{4h-3}{9} = \frac{4k+5}{1} = \frac{3h-5k+14}{28}$$

i.e, $h - 9k = 12$ and $3h - 117k = 126$ ....(3)

Solving, we get $h = 3$, $k = -1$

Thus required point is $(3, -1)$. But this point must be outside the circle as any point inside or on the circle does not have chord of contact.

Now $S(3, -1) = -1 < 0$, i.e., $(3, -1)$ is inside the circle, so there is no point of which (2) is the chord of contact.

2. One of the diameters of the circle circumscribing the rectangle $ABCD$ is $4y = x + 7$. If A and B are the points $(-3, 4)$ and $(5, 4)$ respectively, then find the area of the rectangle.

(a) 6 sq units (b) 12 sq units
(c) 16 sq units (d) 32 sq units

Solution: (d) First, we note that none of the points $A(-3,4)$ $B(5, 4)$ lie on the diameter $4y = x + 7$

Let $E(\alpha, \beta)$ be the centre of the circle,

then $4\beta = \alpha + 7$ .... (i)

Since $ABCD$ is a rectangle. $|EA| = |EB|$

$\Rightarrow EA^2 = EB^2$

$\Rightarrow (\alpha + 3)^2 + (\beta - 4)^2 = (\alpha - 5)^2 + (\beta - 4)^2$

$\Rightarrow 6\alpha + 9 = -10\alpha + 25$

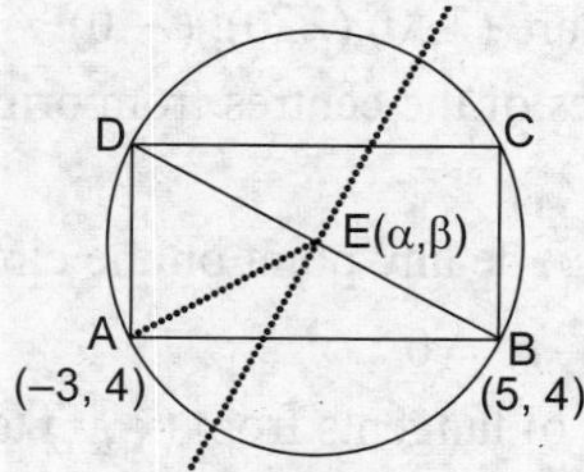

$\Rightarrow \alpha = 1$, and from (i) $\beta = 2$

Now $|AB| = \sqrt{(5+3)^2 + (4-4)^2} = 8$

and $|BD| = 2|EB|$

$= 2\sqrt{(5-1)^2 + (4-2)^2} = 4\sqrt{5}$

From right $\angle d \Delta ABD$,

$AD^2 = BD^2 - AB^2 = 80 - 64 = 16 \Rightarrow |AD| = 4$,

$\therefore$ area of the rectangle $ABCD$

$= |AB|.\,|AD| = 8.4 = 32$ sq. units

3. Find the equation of the circle which touches the circle $x^2 + y^2 - 6x + 6y + 17 = 0$ externally and to which the lines $x^2 - 3xy - 3x + 9y = 0$ are normals.

(a) $x^2 + y^2 + 6x - 2y + 1 = 0$
(b) $x^2 + y^2 - 6x + 2y + 1 = 0$
(c) $x^2 + y^2 - 6x - 2y + 1 = 0$
(d) None of these

Solution: (b) The centre of the given circle is $C_1(3, -3)$ and its radius $= \sqrt{9 + 9 - 17} = 1$

The given lines are $x^2 - 3xy - 3x + 9y = 0$

i.e., $(x - 3y)(x - 3) = 0$

i.e., $x - 3y = 0$ and $x - 3 = 0$

These lines intersect at the point $C_2(3, 1)$.

Since the given lines are normals to the required circle, therefore, the point $C_2(3, 1)$ is the centre of the required circle.

Let r_1, r_2 be the radii of the two circles.

As the circles touch externally, $r_1 + r_2 = |C_1C_2|$

$\Rightarrow 1 + r_2 = \sqrt{(3-3)^2 + (1+3)^2}$

$\Rightarrow 1 + r_2 = 4 \Rightarrow r_2 = 3$

$\therefore$ The equation of the required circle is

$(x - 3)^2 + (y - 1)^2 = 3^2$ i.e., $x^2 + y^2 - 6x - 2y + 1 = 0$

4. If the distance from the origin of the centres of the three circles $x^2 + y^2 - 2\lambda_i x = c^2$, $(i = 1,2,3)$ are in G.P., then the lengths of the tangents drawn to them from any point on the circle $x^2 + y^2 = c^2$ are in:

(a) AP (b) G.P

(c) H.P (d) None of these

Solution: Given circles are

$x^2 + y^2 - 2\lambda_i x - c^2 = 0$, $i = 1,2,3$

$\therefore$ Centres are $(\lambda_1,0),(\lambda_2,0),(\lambda_3,0)$

Distances of the centres from origin are in G.P.,

$\therefore$ $\lambda_2^2 = \lambda_1\lambda_3$.

Let (x_1,y_1) be any point on the circle $x^2 + y^2 = c^2$

$\therefore$ $x_1^2 + y_1^2 - c^2 = 0$

Lengths of tangents from (x_1,y_1) to the three given circles are $\ell_i = \sqrt{x_1^2 + y_1^2 - 2\lambda_i x_1 - c^2} = \sqrt{-2\lambda_i x_i}$

$\therefore$ $\ell_1^2\ell_3^2 = (-2\lambda_1 x_1)(-2\lambda_3 x_1) = 4\lambda_1\lambda_3 x_1^2$

$= 4\lambda_2^2 x_1^2 = l_2^4 \Rightarrow l_1 l_3 = l_2^2$

$\therefore$ l_1, l_2 and l_3 are in G.P

5. If P is a point on the circle $x^2 + y^2 = 9$, Q is a point on the line $7x + y + 3 = 0$ and the line $x - y + 1 = 0$ is the perpendicular bisector of PQ, the co-ordinates of P are:

(a) (3,0) (b) (–72/25,21/25)

(c) both (a) and (b) (d) None of these

Solution: Let $P(x_1,y_1)$ and $Q(x_2, y_2)$ be the points on the circle $x^2 + y^2 = 9$ and the line $7x + y + 3 = 0$ respectively.

$\therefore$ $y_2 = -(7x_2 + 3)$ and $x_1^2 + y_1^2 = 9$

Since $x - y + 1 = 0$ is perpendicular bisector of PQ

$\therefore$ $\frac{1}{2}(x_1 + x_2) - \frac{1}{2}[y_1 - (7x_2 + 3)] + 1 = 0$

or $x_1 - y_1 = -(8x_2 + 5)$ ....(1)

Further, $m_1 \times m_2 = -1$

$\Rightarrow$ $1.\frac{(y_1 - y_2)}{(x_1 - x_2)} = -1 \Rightarrow 1.\left(\frac{y_1 + 7x_2 + 3}{x_1 - x_2}\right) = -1$

i.e., $x_1 + y_1 = -(6x_2 + 3)$ ......(2)

from $(1)^2 + (2)^2$; we get

$2(x_1^2 + y_1^2) = 100x_2^2 + 116x_2 + 34$

or $100x_2^2 + 116x_2 + 16 = 0 \Rightarrow x_2 = -1$ or $-4/25$ ....(3)

Hence from (1) and (2), we get co-ordinates of P as (3,0) and (–72/25,21/25).

6. A region in the x - y plane is bounded by the curve $y = \sqrt{(25 - x^2)}$ and the line $y = 0$. If the point $(a, a+1)$ lies in the interior of the region, then

(a) $a \in (-4,3)$ (b) $a \in (-\infty, -1) \cup (3, \infty)$

(c) $a \in (-1,3)$ (d) None of these

Solution: (a),(c) $y = \sqrt{(25 - x^2)} \Rightarrow x^2 + y^2 = 25, y \geq 0$

$P(a, a + 1)$ is inside the region

$\therefore$ $a^2 + (a + 1)^2 - 25 < 0$ and $a + 1 > 0$

$\Rightarrow$ $2a^2 + 2a - 24 < 0$ and $a > -1$

$\Rightarrow$ $a^2 + a - 12 < 0$ and $a > -1$

$\Rightarrow$ $(a + 4) + (a - 3) < 0$ and $a > -1$

$\Rightarrow$ $-4 < a < 3$, $a > -1$

$\Rightarrow$ $-1 < a < 3$

$\Rightarrow$ $a \in (-1, 3) \subseteq (-4, 3)$

$\Rightarrow$ $a \in (-4, 3)$

7. If the equation of circle obtained by reflecting the circle $x^2 + y^2 - a^2 = 0$ in the line $y = mx + c$ is $x^2 + y^2 + 2gx + 2fy + c = 0$, then

(a) $g = \frac{2cm}{1+m^2}, a^2 + c = \frac{4c^2}{1+m^2}$

(b) $g = -\frac{2cm}{1+m^2}, a^2 + c = \frac{4c^2}{1+m^2}$

(c) $f = \frac{4c}{1+m^2}, a^2 + c = \frac{4c^2}{1+m^2}$

(d) None of these

Solution: (a) Centre of the circle $x^2 + y^2 = a^2$ is (0,0)

Let its reflection about the line $y = mx + c$ be (h, k).

Then $(h/2, k/2)$ lies on this line $m\frac{h}{2} + c = \frac{k}{2}$ ...(1)

and $-\frac{k}{h} \times m = -1 \Rightarrow mk = -h$ ....(2)

Solving (1) and (2), we get $h = -\frac{2cm}{1+m^2}, k = \frac{2c}{1+m^2}$

Also radius of reflected circle is 'a'

$\therefore$ Equation of reflected circle is

$$\left(x + \frac{2cm}{1+m^2}\right)^2 + \left(y - \frac{2c}{1+m^2}\right)^2 = a^2$$

But given equation of reflected circle is

$x^2 + y^2 + 2gx + 2fy + c = 0$

On comparing, we get

$$\frac{g}{\frac{2cm}{1+m^2}} = \frac{f}{\frac{-2c}{1+m^2}} = \frac{c}{\left(\frac{4c^2}{1+m^2} - a^2\right)} = 1$$

$\therefore$ $g = \frac{2cm}{1+m^2}, f = -\frac{2c}{1+m^2}, \frac{4c^2}{1+m^2} - a^2 = c$

8. If the circle touches the lines $x = \sqrt{3}y$, $y = x\sqrt{3}$ and has unit radius. If the centre of this circle lies in the first quadrant, then equation of this circle is

(a) $x^2 + y^2 - 2x\left(\sqrt{3}+1\right) - 2y\left(\sqrt{3}+1\right)+8+4\sqrt{3}=0$

(b) $x^2 + y^2 - 2x\left(1+\sqrt{3}\right)-2y\left(1+\sqrt{3}\right)+5+4\sqrt{3}=0$

(c) $x^2 + y^2 - 2x\left(1+\sqrt{3}\right)-2y\left(1+\sqrt{3}\right)+7+4\sqrt{3}=0$

(d) $x^2 + y^2 - 2x\left(1+\sqrt{3}\right)-2y\left(1+\sqrt{3}\right)+6+4\sqrt{3}=0$

Solution: (c) We observe that $y=\dfrac{1}{\sqrt{3}}x$ makes an angle of 30° with x-axis and $y=\sqrt{3}x$ makes an angle of 60° with x-axis. Clearly above two lines are tangents to two circles drawn from origin.

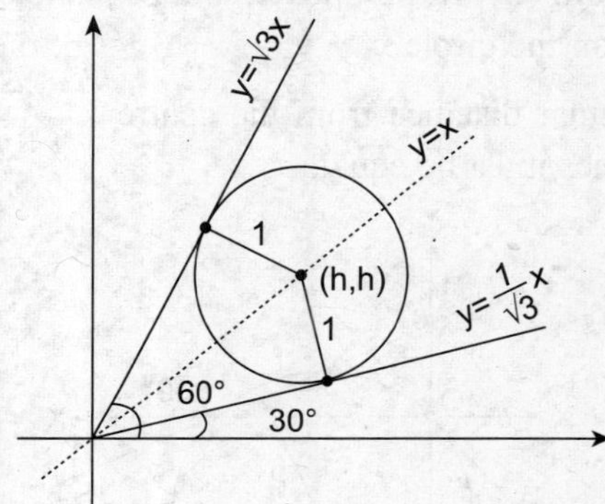

The centre of the required circle must lie on the line which makes an angle of 45° with x-axis, i.e., $y = x$

∴ Let the centre of the circle be (h, h)

$$\Rightarrow \frac{|h-\sqrt{3}h|}{\sqrt{1+(\sqrt{3})^2}}=1 \text{ and } \frac{\left|h-\frac{1}{\sqrt{3}}h\right|}{\sqrt{1+\left(\frac{1}{\sqrt{3}}\right)^2}}=1$$

Solving which, we get $\dfrac{(\sqrt{3}-1)h}{2}=\pm 1$

$$\Rightarrow h=\frac{\pm 2}{\sqrt{3}-1}$$

And since circle lies in I[st] quadrant therefore h > 0

$$\Rightarrow h=\frac{2}{\sqrt{3}-1}=\frac{2(\sqrt{3}+1)}{2}=\sqrt{3}+1$$

∴ Equation of required circle is

$$(x-(\sqrt{3}+1))^2+(y-(\sqrt{3}+1))^2=1^2$$

$$\Rightarrow x^2+y^2-2x(\sqrt{3}+1)-2y(\sqrt{3}+1)+2(\sqrt{3}+1)^2-1=0$$

$$\Rightarrow x^2+y^2-2x(\sqrt{3}+1)-2y(\sqrt{3}+1)+(7+4\sqrt{3})=0$$

9. The equation of largest circle passing through the points (1,1) and (2,2) and always in the first quadrant is

(a) $x^2 + y^2 - 4x - 2y + 4 = 0$

(b) $x^2 + y^2 - 2x - 4y + 4 = 0$

(c) $x^2 + y^2 - 3x - 3y + 4 = 0$

(d) None of these

Solution: (a), (b) Equation of the circle through (1, 1) and (2, 2) is $(x - 1)(x - 2) + (y - 1)(y - 2) + \lambda(x - y) = 0$

i.e., $x^2 + y^2 - (3 - \lambda)x - (3 + \lambda)y + 4 = 0$

$$\Rightarrow \text{Radius} = r = \sqrt{\left(\frac{3-\lambda}{2}\right)^2+\left(\frac{3+\lambda}{2}\right)^2-4}$$

Let (a, b) be the centre.Then we should have

(i) $|a| = r$, $|b| > r$ or (ii) $|b| = r, |a| > r$

(i) $|a| = r \Rightarrow a^2 = r^2$

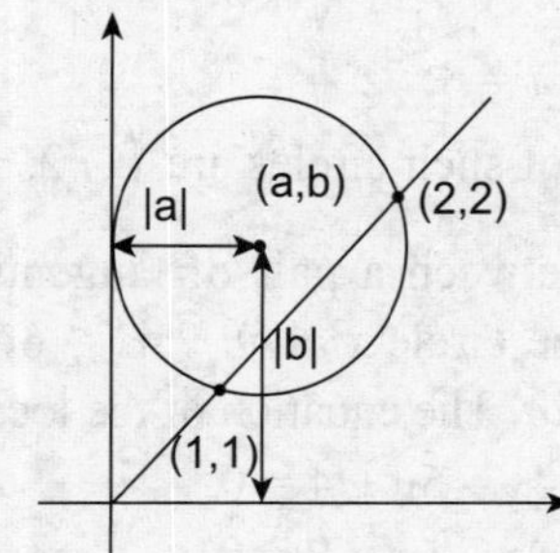

$$\Rightarrow \left(\frac{3-\lambda}{2}\right)^2=\left(\frac{3-\lambda}{2}\right)^2+\left(\frac{3+\lambda}{2}\right)^2-4$$

$\Rightarrow \lambda = 1$ or -7

$\lambda = 1 \Rightarrow |a| = 1$ and $|b| = 2 \Rightarrow r = 1$

$\lambda = -7 \Rightarrow |a| = 5$ and $|b| = 2 \Rightarrow r = 5$ (Rejected)

∴ $\lambda = 1$ and the required equation of circle is

$x^2 + y^2 - 2x - 4y + 4 = 0$ ....(1)

(ii) $|b| = r \Rightarrow \left(\dfrac{3-\lambda}{2}\right)^2 = 4 \Rightarrow \lambda = -1 \text{ or } 7$

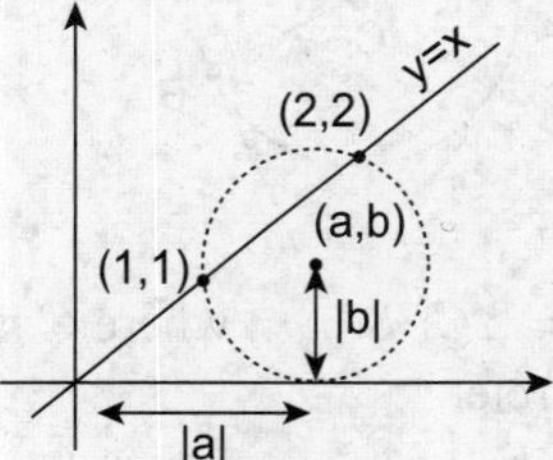

$\lambda = -1 \Rightarrow |a| = 2$, $r = 1$ so $|a| > r$

$\lambda = 7 \Rightarrow |a| = 2$, $r = 5$ so $|a| < r$

∴ $\lambda = -1$ is valid.

∴ Equation of circle is $x^2 + y^2 - 4x - 2y + 4 = 0$

10. The centre of a circle passing through the points (0,0), (1,0) and touching the circle $x^2 + y^2 = 9$ is:

(a) (3/2, 1/2)
(b) (1/2, 3/2)
(c) (1/2, 1/√2)
(d) (1/2, –√2)

Solution: (c),(d) Let the equation of the circle be

$x^2 + y^2 + 2gx + 2fy + c = 0$ .....(1)

It passes through the points (0,0) and (1,0)

$\therefore$ $c = 0$ and $1 + 2g + c = 0$; $g = -1/2$

Radius of circle (1) is

$$r_1 = \sqrt{(g^2 + f^2 - c)} = \sqrt{(1/4 + f^2)}$$

The centre of the circle $x^2 + y^2 = 9$ ....(2)

is (0,0) and radius $r_2 = 3$

Since the circle (1) passes through the centre (0,0) of circle (2) and it also touches the circle (2), so it will touch the circle (2) internally

$\therefore$ $C_1C_2 = \sqrt{(g^2 + f^2)} = |r_1 - r_2|$

$\Rightarrow$ $\sqrt{(1/4 + f^2)} = \left|3 - \sqrt{(1/4 + f^2)}\right|$

$\Rightarrow f^2 = 2 \Rightarrow f = \pm\sqrt{2}$

$\therefore$ Centres of such circles are $(1/2, \pm\sqrt{2})$.

11. The angle between a pair of tangents drawn from a point P to the circle $x^2 + y^2 + 4x - 6y + 9\sin^2\alpha + 13\cos^2\alpha = 0$ is 2α. The equation of the locus of point P is:

(a) $x^2 + y^2 + 4x - 6y + 4 = 0$
(b) $x^2 + y^2 + 4x - 6y - 9 = 0$
(c) $x^2 + y^2 + 4x - 6y - 4 = 0$
(d) $x^2 + y^2 + 4x - 6y + 9 = 0$

Solution: (d) Let PA and PB be the tangents from the point $P(x_1, y_1)$ to the circle

$x^2 + y^2 + 4x - 6y + 9\sin^2\alpha + 13\cos^2\alpha = 0$

Given $\angle APB = 2\alpha$

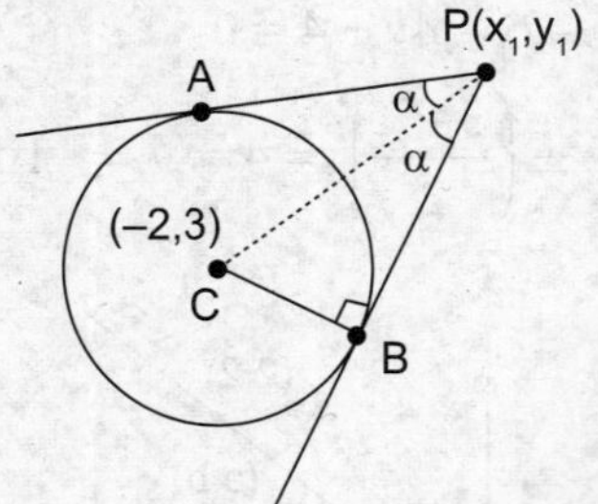

$\therefore$ $\angle CPA = \angle CPB = \alpha$; where C is the centre of the given circle,

Co-ordinates of centre C of the circle are $(-2,3)$

$$\therefore \quad \sin\alpha = \frac{CB}{CP} = \frac{\sqrt{4+9-\left[\left(9\sin^2\alpha + 13\cos^2\alpha\right)\right]}}{\sqrt{(x_1+2)^2 + (y_1-3)^2}}$$

$$= \frac{2\sin\alpha}{\sqrt{\left[(x_1+2)^2 + (y_1-3)^2\right]}}$$

or $(x_1 + 2)^2 + (y_1 - 3)^2 = 4$

or $x_1^2 + y_1^2 + 4x_1 - 6y_1 + 9 = 0$

$\therefore$ Locus of the point $P(x_1, y_1)$ is

$x^2 + y^2 + 4x - 6y + 9 = 0$

12. Let L_1 be a straight line passsing through the origin and L_2 be the straight line $x + y = 1$. If the intercepts made by the circle $x^2 + y^2 - x + 3y = 0$ on L_1 and L_2 are equal, then which of the following equations can represent L_1?

(a) $x + y = 0, x = 7y$
(b) $x - y = 0, x + 7y = 0$
(c) $x + 7y = 0, x + y = 0$
(d) None of these

Solution: (b) If the lines L_1 and L_2 make equal intercepts on the circle $x^2 + y^2 - x + 3y = 0$; then the perpendicular distance from the centre $\left(\frac{1}{2}, \frac{-3}{2}\right)$ on the two lines must be equal.

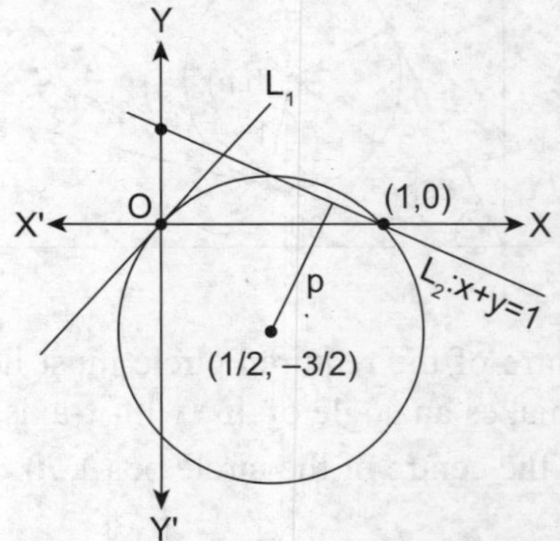

Let the equation of line L_1 be $y - mx = 0$.

By equating perpendicular distances from centre $\left(\frac{1}{2}, \frac{-3}{2}\right)$ on the lines $y - mx = 0$ and $x + y - 1 = 0$;

we get, $\dfrac{\left|\frac{-3}{2} - \frac{m}{2}\right|}{\sqrt{1+m^2}} = \dfrac{\left|\frac{1}{2} - \frac{3}{2} - 1\right|}{\sqrt{1^2+1^2}}$

Squaring both sides; we get $\dfrac{9 + m^2 + 6m}{4(1+m^2)} = 2$

$\Rightarrow$ $9 + m^2 + 6m = 8 + 8m^2$

$\Rightarrow$ $7m^2 - 6m - 1 = 0$

$\Rightarrow$ $m = \dfrac{-1}{7}, 1$

$\therefore$ Equations of required lines are $x - y = 0$ or $x + 7y = 0$

13. Tangents TP and TQ are drawn from a point T to the circle $x^2 + y^2 = a^2$. If the point T lies on the line $px + qy = r$, then locus of the centre of circumcircle of ΔTPQ is

(a) $2px + 2qy - r = 0$ (b) $2px + 2qy + r = 0$
(c) $px + 2qy - r = 0$ (d) None of these

Solution: (a) $T(x_1, y_1)$ lies on $px + qy = r$

$\therefore\ px_1 + qy_1 - r = 0$ ...(1)

Equation of (Chord of contact of T) is

$xx_1 + yy_1 - a^2 = 0$

Equation of circle through PQ is

$x^2 + y^2 - a^2 + \lambda(xx_1 + yy_1 - a^2) = 0$

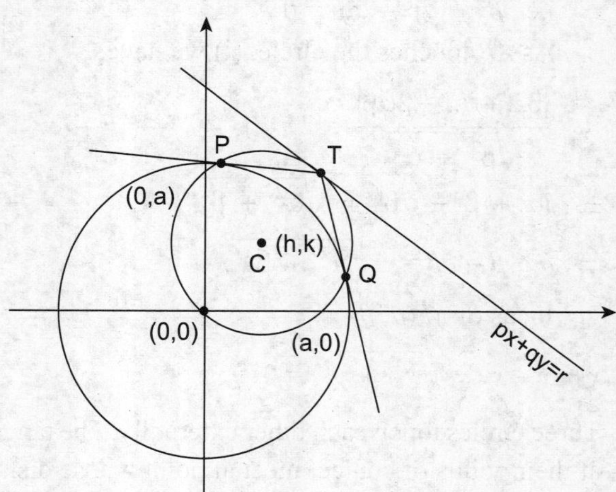

It passes through $T(x_1, y_1)$

$\therefore\ x_1^2 + y_1^2 - a^2 + \lambda(x_1^2 + y_1^2 - a^2) = 0 \Rightarrow \lambda = -1$

$\therefore$ circumcircle of ΔTPQ is $x^2 + y^2 - xx_1 - yy_1 = 0$

Its centre $C(h, k)$

$\therefore\ h = \frac{x_1}{2}, k = \frac{y_1}{2}$

From (1), we get $2ph + 2qk - r = 0$

Hence, locus of $C(h, k)$ is $2px + 2qy - r = 0$

14. Lines $5x + 12y - 10 = 0$ and $5x - 12y - 40 = 0$ touch circle C_1 of diameter 6. If the centre of C_1 lies in the first quadrant, find the equation of the circle C_2 which is concentric with C_1 and cuts intercepts of lengths 8 on these lines

(a) $x^2 + y^2 + 10x - 4y + 4 = 0$
(b) $x^2 + y^2 - 10x - 4y + 4 = 0$
(c) $x^2 + y^2 - 10x + 4y + 4 = 0$
(d) None of these

Solution: (b) Let $A(h, k)$ be the centre of circle C_1 where $h > 0$ and $k > 0$, as A lies in the first quadrant.

As diameter of C_1 is of length 6 units, its radius = 3.

Since the circle C_1 touches the lines $5x + 12y - 10 = 0$ and $5x - 12y - 40 = 0$;

we get $\frac{|5h+12k-10|}{\sqrt{5^2+12^2}} = 3$ and $\frac{|5h-12k-40|}{\sqrt{5^2+12^2}} = 3$

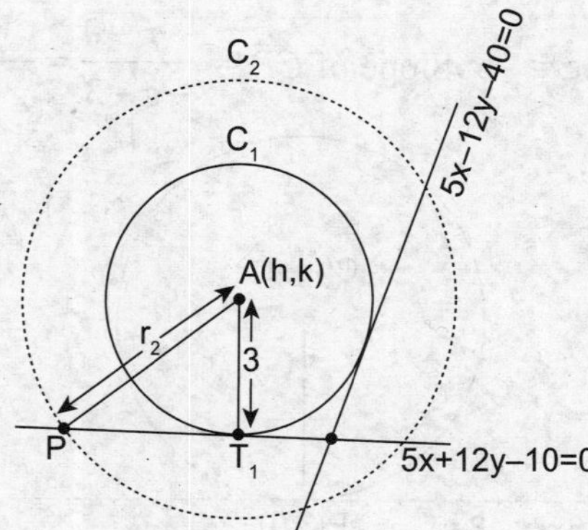

$\Rightarrow |5h + 12k - 10| = |5h - 12k - 40| = 39$

$\Rightarrow 5h + 12k - 10 = 5h - 12k - 40$

or $5h + 12k - 10 = -5h + 12k + 40$

$\Rightarrow 24k = -30$ or $10h = 50$

$\Rightarrow k = -\frac{5}{4}$ or $h = 5$ but $h > 0, k > 0 \Rightarrow h = 5$

Also $|5h + 12k - 10| = 39 \Rightarrow |25 + 12k - 10| = 39$

$\Rightarrow |12k + 15| = 39 \Rightarrow (12k + 15) = \pm 39$

$\Rightarrow 12k + 15 = 39$ or $12k + 15 = -39$

$\Rightarrow k = 2$ or $k = -\frac{9}{2}$ but $k > 0 \Rightarrow k = 2$

Hence centre of C_1 is (5, 2)

Since the circle C_2 cuts intercepts of length 8 on given lines.

$\therefore$ radius of $C_2 = \sqrt{AT_1^2 + T_1P^2} = \sqrt{3^2 + 4^2} = 5$

$\therefore$ The equation of the circle C_2 is

$(x - 5)^2 + (y - 2)^2 = 5^2$

i.e., $x^2 + y^2 - 10x - 4y + 4 = 0$

15. The equation of the circle which passes through the point (1, –1) and which touches the line $6x + y - 18 = 0$ at the point $P(3, 0)$.

(a) $13(x^2 + y^2) - 48x + 5y + 27 = 0$
(b) $13(x^2 + y^2) - 48x - 5y + 27 = 0$
(c) $11(x^2 + y^2) - 48x + 5y + 27 = 0$
(d) None of these

Solution: (a) Let the circle be $x^2 + y^2 + 2gx + 2fy + c = 0$ .... (i)

Its centre is $C(-g, -f)$ and radius = $\sqrt{g^2 + f^2 - c}$

Since points (1, –1), (3, 0) lie on (i), we have

$1 + 1 + 2g - 2f + c = 0$ and

$9 + 0 + 6g + 0 + c = 0$

i.e., $2g - 2f + c + 2 = 0$ .. .. (ii)

and $6g + c + 9 = 0$ .. .. (iii)

Given line is

$6x + y - 18 = 0$ .. .. (iv)

Its slope = –6; Slope of $CP = \frac{-f-0}{-g-3} = \frac{f}{g+3}$

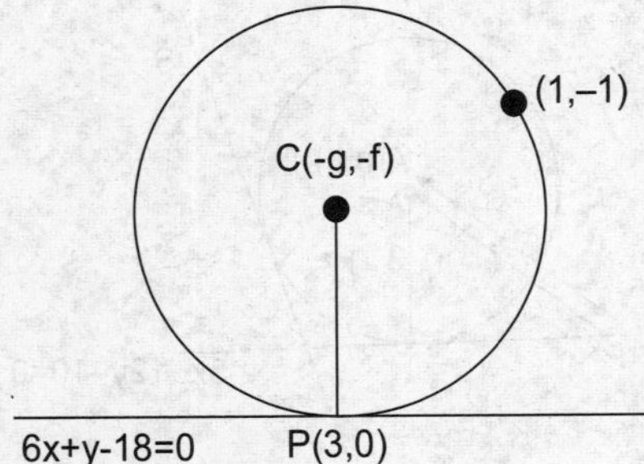

Since the line (iv) touches circle (i) at P,

$\therefore$ CP is perpendicular to (iv),

$\therefore \quad \frac{f}{g+3}(-6) = -1$

$\Rightarrow g + 3 = 6f$ .....(v)

Solving (ii), (iii) and (v), we get

$g = -\frac{24}{13}; f = \frac{5}{26}$ and $c = \frac{27}{13}$

Hence the equation of the circle is

$x^2 + y^2 - \frac{48}{13}x + \frac{10}{26}y + \frac{27}{13} = 0$

i.e., $13(x^2 + y^2) - 48x + 5y + 27 = 0$

16. The circle $x^2 + y^2 - 4x - 4y + 4 = 0$ is inscribed in a triangle which has two of its sides along the co-ordinates axes. The locus of the circumcentre of the triangle is $x + y - xy + k\sqrt{x^2 + y^2} = 0$, then $k =$

(a) 5 (b) – 1

(c) 2 (d) 1

Solution: (b), (d) Given circle is

$x^2 + y^2 - 4x - 4y + 4 = 0$ ... (i)

Its centre is (2, 2) and radius = 2

Let (i) be inscribed in ΔAOB (shown in the following figure).

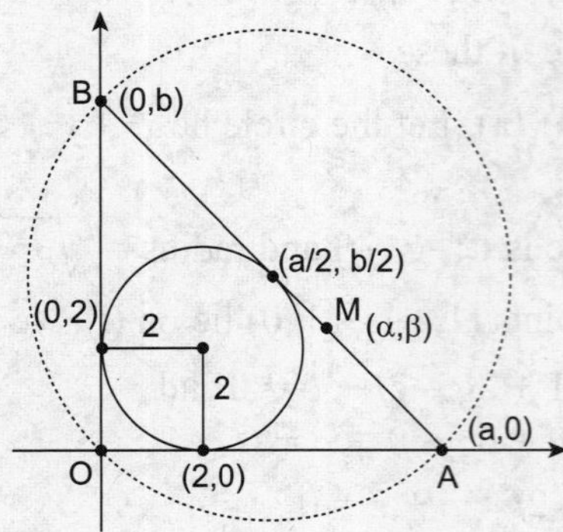

Let the equation of AB be $\frac{x}{a} + \frac{y}{b} = 1$,

so that A is $(a, 0)$ and B is $(0, b)$

Since $\angle AOB = 90^o$,

$\therefore$ AB is the diameter of the circumcircle of ΔOAB, and hence its centre, say $M(\alpha, \beta)$, is mid-point of AB, we have

$\frac{a+0}{2} = \alpha$ and $\frac{0+b}{2} = \beta$

$\Rightarrow a = 2\alpha$ and $b = 2\beta$

$\therefore$ Equation of AB becomes $\frac{x}{2\alpha} + \frac{y}{2\beta} = 1$

i.e., $\beta x + \alpha y - 2\alpha\beta = 0$ ... (ii)

As AB touches the circle, (i) we have

$\frac{|\beta.2 + \alpha.2 - 2\alpha\beta|}{\sqrt{\beta^2 + \alpha^2}} = 2$

$\Rightarrow |\alpha + \beta - \alpha\beta| = \sqrt{\alpha^2 + \beta^2}$

$\Rightarrow \alpha + \beta - \alpha\beta = \pm\sqrt{\alpha^2 + \beta^2}$

$\therefore$ locus of $M(\alpha, \beta)$ is

$x + y - xy \pm \sqrt{x^2 + y^2} = 0$

17. Three circles touch each other externally. The tangents at their points of contact meet at point whose distance from the point of contact is 4. The ratio of the product of the radii to the sum of the radii of the circles is.

(a) 16 : 1 (b) 16 : 3

(c) 16 : 5 (d) 16 : 7

Solution: (a) Let the centres of the three circles be A, B and C respectively and let their radii be R_1, R_2 and R_3 respectively.

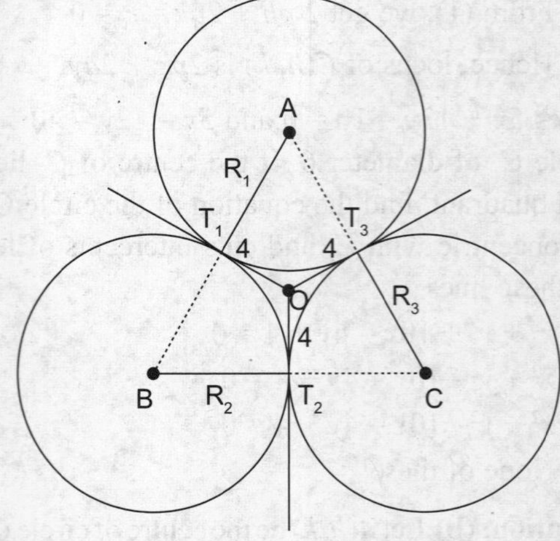

Let T_1, T_2 and T_3 be the points of contact of three circles and let the tangents at three points meet in O, then O is the radical centre of the three circles and hence $OT_1 = OT_2 = OT_3 = 4$ (given)

$\Rightarrow$ The circle with centre at O and radius 4 will touch the sides of the triangle ABC internally at T_1, T_2 and T_3, i.e., it is incircle of the ΔABC.

Now, for the ΔABC,

$a = BC = R_2 + R_3$

$b = CA = R_3 + R_1$

$c = AB = R_1 + R_2$

$\Rightarrow s = (1/2)(a + b + c) = R_1 + R_2 + R_3$

and $\Delta = \sqrt{s(s-a)(s-b)(s-c)}$

$= \sqrt{(R_1 + R_2 + R_3)R_1R_2R_3}$

Since, inradius of ΔABC is 4, therefore, we must have

$$4 = \frac{\Delta}{s} = \frac{\sqrt{(R_1 + R_2 + R_3)R_1 R_2 R_3}}{R_1 + R_2 + R_3}$$

$$\Rightarrow 16 = \frac{R_1 R_2 R_3}{R_1 + R_2 + R_3}$$

i.e., the required ratio is 16:1.

18. $P(\alpha)$ and $Q(\beta)$ are the two points on the circle having origin as its centre and radius 'a' and AB is the diameter along the axis of x. If $\alpha - \beta = 2\gamma$, then the locus of intersection of AP and BQ is.

(a) $x^2 + y^2 - 2ay \cot\gamma - a = 0$
(b) $x^2 + y^2 + 2ay \tan\gamma - a = 0$
(c) $x^2 + y^2 - 2ay \tan\gamma - a^2 = 0$
(d) None of these

Solution: (c) Equation of circle with centre (0, 0) and radius a is $x^2 + y^2 = a^2$

$\therefore\ P \equiv (a\cos\alpha, a\sin\alpha)$

and $Q \equiv (a\cos\beta, a\sin\beta)$

Co-ordinates of A and B are $A \equiv (-a, 0)$ and $B \equiv (a, 0)$

$\therefore$ Equation AP is $y - 0 = \dfrac{a\sin\alpha - 0}{a\cos\alpha + a}(x+a)$

or $y(1 + \cos\alpha) = \sin\alpha\,(x + a)$

or $y\left(2\cos^2\dfrac{\alpha}{2}\right) = 2\sin\left(\dfrac{\alpha}{2}\right)\cos\left(\dfrac{\alpha}{2}\right).(x+a)$

or $y\cos\left(\dfrac{\alpha}{2}\right) - x\sin\left(\dfrac{\alpha}{2}\right) = a\sin\left(\dfrac{\alpha}{2}\right)$

or $\tan\left(\dfrac{\alpha}{2}\right) = \dfrac{y}{x+a}$

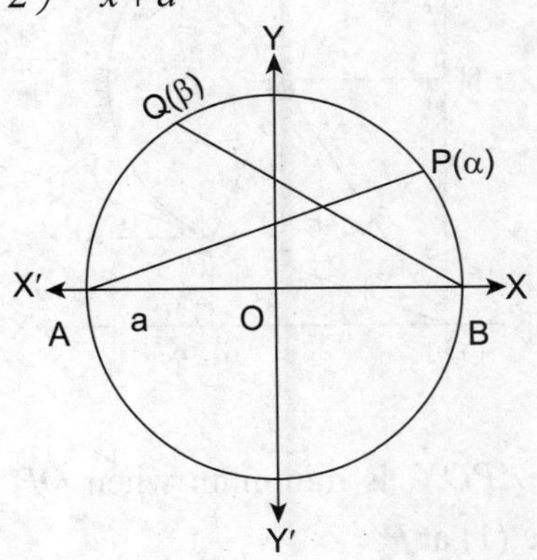

and equation of BQ is $y - 0 = \dfrac{a\sin\beta - 0}{a\cos\beta - a}(x-a)$

or $y(\cos\beta - 1) = \sin\beta(x-a)$

$\Rightarrow y\left(-2\sin^2\left(\dfrac{\beta}{2}\right)\right) = 2\sin\left(\dfrac{\beta}{2}\right)\cos\left(\dfrac{\beta}{2}\right)(x-a)$

$\Rightarrow y\sin\left(\dfrac{\beta}{2}\right) + x\cos\left(\dfrac{\beta}{2}\right) = a\cos\left(\dfrac{\beta}{2}\right)$

$\Rightarrow \tan\left(\dfrac{\beta}{2}\right) = \dfrac{a-x}{y};\ \because\ \gamma = \dfrac{\alpha-\beta}{2}$ (given)

$$\Rightarrow \tan\gamma = \frac{\tan(\alpha/2) - \tan(\beta/2)}{1 + \tan(\alpha/2)\tan(\beta/2)} = \frac{\dfrac{y}{x+a} - \dfrac{a-x}{y}}{1 + \dfrac{a-x}{a+x}}$$

or $x^2 + y^2 - 2ay\tan\gamma = a^2$

which is the locus of point of intersection of AP and BQ.

19. Consider a curve $ax^2 + 2hxy + by^2 = 1$ and a point P not on the curve. A line drawn from the point P intersects the curve at points Q and R. If the product $PQ.\ PR$ is independent of the slope of the line, then the curve is a

(a) circle (b) parabola
(c) hyperbola (d) ellipse

Solution: (a) Let P be (x_1, y_1) and line through $P(x_1, y_1)$ making an angle θ with x-axis, then $\dfrac{x - x_1}{\cos\theta} = \dfrac{y - y_1}{\sin\theta} = r$

Co-ordinates of any point on the curve is $(x_1 + r\cos\theta, y_1 + r\sin\theta)$. This point must lie on $ax^2 + 2hxy + by^2 = 1$

$\therefore\ a(x_1 + r\cos\theta)^2 + 2h(x_1 + r\cos\theta)(y_1 + r\sin\theta) + b(y_1 + r\sin\theta)^2 = 1$

$\therefore\ (a\cos^2\theta + h\sin 2\theta + b\sin^2\theta)r^2 + 2(ax_1\cos\theta + hx_1\sin\theta + hy_1\cos\theta)r + ax_1^2 + 2hx_1y_1 + by_1^2 = 0$

It is quadratic equation in r.

Let roots of this equation are r_1 and r_2, then

$$r_1 r_2 = \frac{ax_1^2 + 2hx_1y_1 + by_1^2}{(a\cos^2\theta + h\sin 2\theta + b\sin^2\theta)}$$

$$\therefore\ PQ.PR = \frac{ax_1^2 + 2hx_1y_1 + by_1^2}{a\cos^2\theta + h\sin 2\theta + b\sin^2\theta}$$

For $a = b$, $h = 0$

$$\therefore\ PQ.PR = \frac{ax_1^2 + 0 + ay_1^2}{a\cos^2\theta + 0 + a\sin^2\theta} = x_1^2 + y_1^2$$

which is independent of θ.

Then curve $ax^2 + 2hxy + by^2 = 1$ becomes $ax^2 + 0 + ay^2 = 1$

$\Rightarrow x^2 + y^2 = \dfrac{1}{a}$ is a circle, centered at (0,0) and radius $\dfrac{1}{\sqrt{a}}$

20. P is a variable point on the line $y = 4$. Tangents are drawn to the circle $x^2 + y^2 = 4$ from P to touch it at A and B. The parallelogram $PAQB$ is completed. The the equation of the locus of Q is

(a) $(x^2 + y^2)(y + 4) = 2x^2$
(b) $(x^2 + y^2)(x + 4) = 2y^2$
(c) $(x^2 + y^2)(x + 4) = 2x^2$
(d) $(x^2 + y^2)(y + 4) = 2y^2$

Solution: (d) Let $P(h,4)$ be a variable point. Given circle is $x^2 + y^2 = 4$ ....(1)

Draw tangents from $P(h,4)$ and complete parallelogram $PAQB$.

Equation of the diagonal AB which is chord of contact of $x^2 + y^2 = 4$

is $hx + 4y = 4$ ...(2)

Let co-ordinates of A and B are (x_a, y_a) and (x_b, y_b) respectively.

Since $A(x_a, y_a)$ and $B(x_b, y_b)$ lies on (2)

$\therefore \quad hx_a + 4y_a = 4$

and $hx_b + 4y_b = 4$

$\therefore \quad h(x_a + x_b) + 4(y_a + y_b) = 8$ ...(3)

Since $PAQB$ is parallelogram

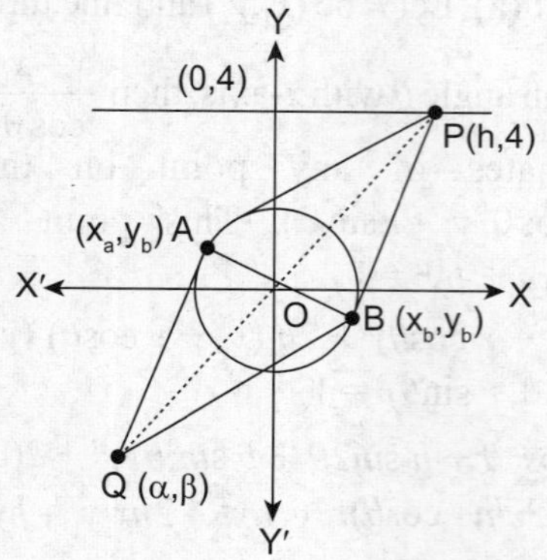

$\therefore$ mid point of AB = mid point of PQ

$$\Rightarrow \frac{x_a + x_b}{2} = \frac{\alpha + h}{2} \text{ and } \frac{y_a + y_b}{2} = \frac{\beta + 4}{2} \quad \text{.........(4)}$$

Eliminating x from (1) and (2), we get

$$\left(\frac{4 - 4y}{h}\right)^2 + y^2 = 4$$

$\Rightarrow 16 + 16y^2 - 32y + h^2y^2 = 4h^2$

$\Rightarrow (16 + h^2)y^2 - 32y + 16 - 4h^2 = 0$

$$\therefore \quad y_a + y_b = \frac{32}{16 + h^2} \quad \text{............(5)}$$

From (3) and (5), we get $x_a + x_b = \dfrac{8h}{16 + h^2}$ ......(6)

From (4) and (6), we get $\beta + 4 = \dfrac{32}{16 + h^2}$

or $(16 + h^2)(\beta + 4) = 32$ ....(7)

Also from (4) and (6), we get $\alpha + h = \dfrac{8h}{16 + h^2}$

$\Rightarrow (16 + h^2)(\alpha + h) = 8h$ ....(8)

Dividing (8) by (7), we get $\dfrac{\alpha + h}{\beta + 4} = \dfrac{h}{4}$ or $h = \dfrac{4\alpha}{\beta}$

Substituting the value h in (7), we get

$$\left(16 + \frac{16\alpha^2}{\beta^2}\right)(\beta + 4) = 32$$

$$\Rightarrow (\alpha^2 + \beta^2)(\beta + 4) = 2\beta^2$$

Hence locus of $Q(\alpha, \beta)$ is $(x^2 + y^2)(y + 4) = 2y^2$

21. The point P is on the circle $x^2 + y^2 - 4x - 6y + 9 = 0$

(i) If $\angle POX$ is minimum, then

(a) $P \equiv \left(\frac{36}{15}, \frac{15}{13}\right)$ (b) $P \equiv \left(\frac{36}{15}, \frac{17}{13}\right)$

(c) $P \equiv \left(\frac{36}{13}, \frac{15}{13}\right)$ (d) None of these

(ii) If OP is maximum, where O is the origin and OX is the x-axis, then

(a) $P \equiv \left(2 + \frac{4}{\sqrt{15}}, 3 + \frac{6}{\sqrt{15}}\right)$

(b) $P \equiv \left(2 + \frac{4}{\sqrt{13}}, 3 + \frac{6}{\sqrt{13}}\right)$

(c) $P \equiv \left(2 + \frac{4}{\sqrt{11}}, 3 + \frac{6}{\sqrt{11}}\right)$

(d) $P \equiv \left(2 + \frac{4}{\sqrt{17}}, 3 + \frac{6}{\sqrt{17}}\right)$

Solution: (i) → (c); (ii) → (b)

$x^2 + y^2 - 4x - 6y + 9 = 0$ or $(x - 2)^2 + (y - 3)^2 = 2^2$ ..(1)

Its centre is $C \equiv (2, 3)$ and radius $r = 2$

(i) Let OP and ON be the two tangents from O to the circle (1), then $OP = ON = 3$

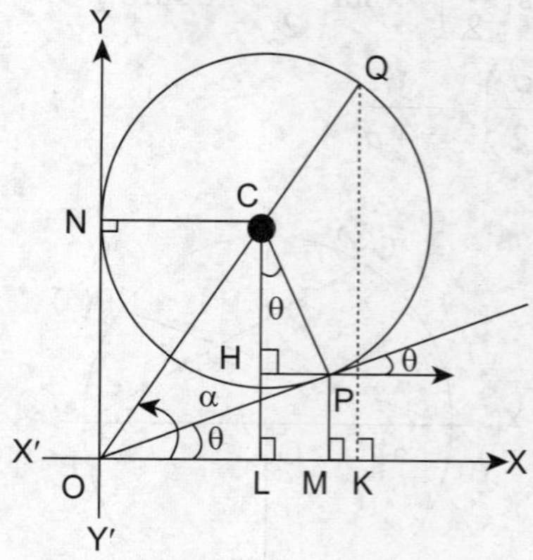

Further $\angle POX$ is minimum when OP is tangent to the circle (1) at P

Let $\angle POX = \theta$

$\therefore\ P \equiv (OP\cos\theta, OP\sin\theta)$

i.e., $P \equiv (3\cos\theta, 3\sin\theta)$ .....(2)

From the given figure $OM = OL + LM = NC + HP = NC + CP\sin\theta$

$\Rightarrow OP\cos\theta = NC + CP\sin\theta$

$\Rightarrow 3\cos\theta = 2 + 2\sin\theta\ [\because \sin\theta \neq 0]$

$\Rightarrow 9(1 - \sin^2\theta) = 4(1 + \sin\theta)^2$

$\Rightarrow 9(1 - \sin\theta) = 4(1 + \sin\theta)$

$\therefore\ \sin\theta = 5/13$ and $\cos\theta = 12/13$

From (2), $P \equiv \left(3\times\frac{12}{13}, 3\times\frac{5}{13}\right)$ i.e., $P \equiv \left(\frac{36}{13}, \frac{15}{13}\right)$

(ii) OP will be maximum, if P becomes the point where extended part of OC cuts the circle. Let this point be Q.

Then maximum value of $OP = OQ = OC + CQ = \left(\sqrt{13}+2\right)$. Let $\angle COX = \alpha$

then, $Q = (OQ\cos\alpha, OQ\sin\alpha)$

$\equiv\left(\left(2+\sqrt{13}\right)\cos\alpha, \left(2+\sqrt{13}\right)\sin\alpha\right)$ ....(3)

Now in ΔCOL, $\cos\alpha = \frac{OL}{OC} = \frac{NC}{OC} = \frac{2}{\sqrt{13}}$

$\sin\alpha = \frac{3}{\sqrt{13}}$

Now from (3), $Q \equiv \left(2+\frac{4}{\sqrt{13}}, 3+\frac{6}{\sqrt{13}}\right)$

22. The equation of the circle circumscribing the triangle formed by the lines: $x + y = 6$, $2x + y = 4$ and $x + 2y = 5$, is

(a) $x^2 + y^2 - 17x - 19y + 50 = 0$

(b) $x^2 + y^2 - 17x + 19y + 50 = 0$

(c) $x^2 + y^2 + 17x - 19y + 50 = 0$

(d) None of these

Solution: (a) Let the given lines represented by L_1, L_2 and L_3, then

$L_1 \equiv x + y - 6 = 0; L_2 \equiv 2x + y - 4 = 0$

and $L_3 \equiv x + 2y - 5 = 0$

Equation of conic of second degree passing through the point of intersection of lines $L_1 = 0$, $L_2 = 0$ and $L_3 = 0$ is given by $L_1L_2 + \lambda L_2L_3 + \mu L_3L_1 = 0$ ...(1)

$\Rightarrow (x + y - 6)(2x + y - 4) + \lambda(2x + y - 4)$

$(x + 2y - 5) + \mu(x + 2y - 5)(x + y - 6) = 0$ ...(2)

For circle, coeffcient of x^2 = co-efficient of y^2

$\therefore\ 2 + 2\lambda + \mu = 1 + 2\lambda + 2\mu$

$\therefore\ \mu = 1$ ......(3)

and co-efficient of $xy = 0$

$\therefore\ 3 + 5\lambda + 3\mu = 0$

or $3 + 5\lambda + 3 = 0$ (from 3))

$\therefore\ \lambda = -\frac{6}{5}$ ......(4)

Substituting the values of λ and μ from (3) and (4) in (2), we get $(x + y - 6)(2x + y - 4) - \frac{6}{5}(2x + y - 4)$

$(x + 2y - 5) + (x + 2y - 5)(x + y - 6) = 0$

or $5(2x^2 + y^2 + 3xy - 16x - 10y + 24) - 6(2x^2 + 2y^2 + 5xy - 14x - 13y + 20) + 5(x^2 + 2y^2 + 3xy - 11x - 17y + 30) = 0$

or $3x^2 + 3y^2 - 51x - 57y + 150 = 0$

or $x^2 + y^2 - 17x - 19y + 50 = 0$

23. Tangents are drawn from $P(6,8)$ to the circle $x^2 + y^2 = r^2$. The radius of the circle such that the area of the Δ formed by tangents and chord of contact is maximum is

(a) 4 (b) 5

(c) 7 (d) 8

Solution: (b) Equation of chord of contact (QR) is $6x + 8y - r^2 = 0$

$$PM = \frac{|6.6 + 8.8 - r^2|}{\sqrt{(6^2+8^2)}} = \frac{|100 - r^2|}{10}$$

and $OM = \frac{|0 + 0 - r^2|}{\sqrt{(6^2+8^2)}} = \frac{r^2}{10}$,

then $QR = 2.QM = 2\sqrt{\{(OQ)^2 - (OM)^2\}}$

$$= 2\sqrt{\left(r^2 - \frac{r^4}{100}\right)}$$

$\therefore$ Area of $\Delta QPR = 1/2\ QR.\ PM$

$$\Delta(\text{say}) = \frac{1}{2}\times 2\times\sqrt{\left(r^2 - \frac{r^4}{100}\right)}.\frac{|100 - r^2|}{10}$$

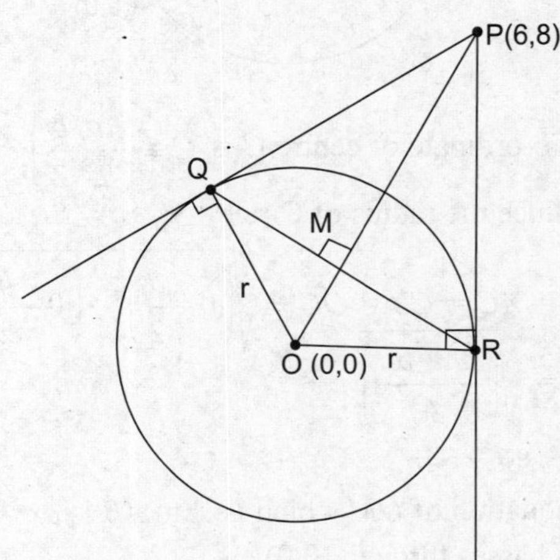

$\therefore\ \Delta^2 = \dfrac{r^2(100-r^2)^3}{10000} = z$ (say)

$\Rightarrow\ \dfrac{dz}{dr} = \dfrac{1}{10000}\left\{\begin{array}{l} r^2.3\left(100-r^2\right)^2.(-2r) \\ +\left(100-r^2\right)^3.2r \end{array}\right\}$

$= \dfrac{2r\left(100-r^2\right)^2}{10000}\left\{100-r^2-3r^2\right\}$

For maximum or minimum $\dfrac{dz}{dr} = 0$, then we get $r = 5$ ($r \neq 10$ otherwise P will be inside the circle) and

$\left.\dfrac{d^2 z}{dr^2}\right|_{r=5} = -ve$

$\therefore$ Δ is maximum at $r = 5$

24. A circle of constant radius r passes through the origin O, and cuts the axes at A and B. The locus of the foot of the perpendicular from O to AB is.

(a) $(x^2+y^2)^2(x^{-2}+y^{-2}) = 4r^2$

(b) $(x^2+y^2)^2(x^{-2}+y^{-2}) = 6r^2$

(c) $(x^2+y^2)^2(x^{-2}+y^{-2}) = 8r^2$

(d) None of these

Solution: (a) Let the co-ordinates of A and B are $(a, 0)$ and $(0,b)$

$\therefore$ Equation of AB is $\dfrac{x}{a}+\dfrac{y}{b} = 1$ .....(1)

Centre of circle lies on line AB, since AB is diameter of the circle $(\because \angle AOB = \pi/2)$

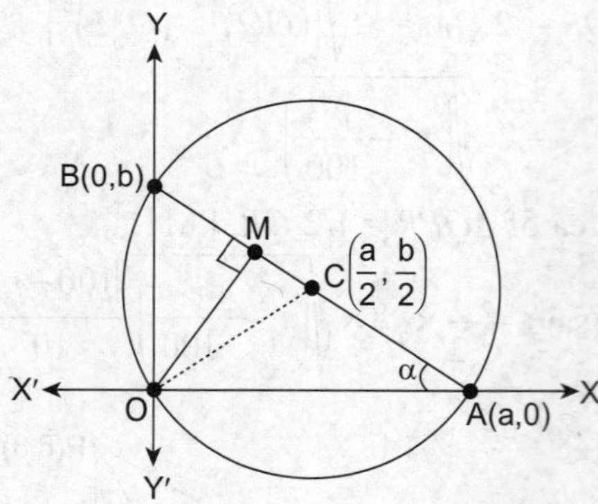

$\therefore$ Co-ordinate of centre C is $C \equiv \left(\dfrac{a}{2}, \dfrac{b}{2}\right)$

Since the radius of Circle $= r$

$r = AC = CB = OC = \sqrt{\left(0-\dfrac{a}{2}\right)^2+\left(0-\dfrac{b}{2}\right)^2}$

$= \sqrt{\left(\dfrac{a^2+b^2}{4}\right)}$

$\therefore\ a^2+b^2 = 4r^2$ .....(2)

Equation of OM which is $\perp$ to AB is $ax - by = \lambda$

It passes through $(0,0)$

$\therefore\ 0 = \lambda$

$\therefore$ Equation of OM is $ax - by = 0$ .....(2)

Solving (1) and (3), we get

$a = \dfrac{x^2+y^2}{x}$ and $b = \dfrac{x^2+y^2}{y}$

Substituting the values of a and b in (2), we get

$\left(x^2+y^2\right)^2\left(\dfrac{1}{x^2}+\dfrac{1}{y^2}\right) = 4r^2$

or $\left(x^2+y^2\right)^2\left(x^{-2}+y^{-2}\right) = 4r^2$

which is the required locus.

Alternative Method:

AB is the diameter of circle. If $\angle OAB = \alpha$, then $OA = 2r\cos\alpha$, $OB = 2r\sin\alpha$

Equation of AB is $\dfrac{x}{2r\cos\alpha}+\dfrac{y}{2r\sin\alpha} = 1$

$\Rightarrow\ \dfrac{x}{\cos\alpha}+\dfrac{y}{\sin\alpha} = 2r$ ......(1)

and equation of OM is $y = x\tan(90^0-\alpha)$

$\Rightarrow\ \cot\alpha = \dfrac{y}{x}$

$\therefore\ \sin\alpha = \dfrac{x}{\sqrt{(x^2+y^2)}}$ and $\cos\alpha = \dfrac{y}{\sqrt{(x^2+y^2)}}$

Then from (1)

$\dfrac{x}{y}\sqrt{(x^2+y^2)}+\dfrac{y}{x}\sqrt{(x^2+y^2)} = 2r$

$\Rightarrow\ \dfrac{(x^2+y^2)\sqrt{(x^2+y^2)}}{xy} = 2r^2$

On squaring, we have $(x^2+y^2)^2\dfrac{(x^2+y^2)}{x^2y^2} = 4r^2$

$\Rightarrow\ \left(x^2+y^2\right)^2\left(x^{-2}+y^{-2}\right) = 4r^2$

25. The intervals of the values of 'a' for which the line $y + x = 0$ bisects two chords drawn from a point $\left(\dfrac{1+\sqrt{2}a}{2}, \dfrac{1-\sqrt{2}a}{2}\right)$ to the circle $2x^2 + 2y^2 - (1+\sqrt{2}a)x-(1-\sqrt{2}a)y = 0$ is given by·

(a) $a \in (-\infty, -2)\cup(2, \infty)$

(b) $a \in (-\infty, -4)\cup(4, \infty)$

(c) $a \in (-\infty, -3)\cup(3, \infty)$

(d) None of these

Solution: The point $A\left(\dfrac{1+\sqrt{2}a}{2}, \dfrac{1-\sqrt{2}a}{2}\right)$ lies on the given circle as its co-ordinate satisfy the equation of the circle. Let AB and AC are two chords drawn from A. Let M and N be the mid-points of AB and AC respectively.

Let co-ordinates of M be $(h, -h)$ and co-ordinate of B be (α, β), then

$$h=\frac{\alpha+\frac{1+\sqrt{2}a}{2}}{2} \text{ and } -h=\frac{\beta+\frac{1-\sqrt{2}a}{2}}{2}$$

$$\therefore \quad \alpha = 2h-\left(\frac{1+\sqrt{2}a}{2}\right) \text{ and } \beta=-2h-\left(\frac{1-\sqrt{2}a}{2}\right)$$

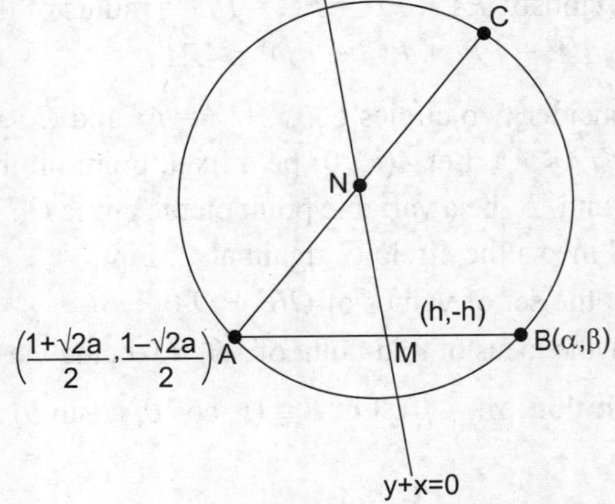

Since B (α,β) lies on the given circle, we have

$$\Rightarrow 2\left[2h-\frac{1+\sqrt{2}a}{2}\right]^2+2\left[-2h-\frac{1-\sqrt{2}a}{2}\right]^2$$

$$-\left(1+\sqrt{2}a\right)\left[2h-\frac{1+\sqrt{2}a}{2}\right]$$

$$-\left(1-\sqrt{2}a\right)\left[-2h-\frac{1-\sqrt{2}a}{2}\right]=0$$

$$\Rightarrow 16h^2-12\sqrt{2}ah+\left(1+\sqrt{2}a\right)^2+\left(1-\sqrt{2}a\right)^2=0$$

$$\Rightarrow 16h^2-12\sqrt{2}ah+2+4a^2=0$$

or $\quad 8h^2-6\sqrt{2}ah+1+2a^2=0$

Hence for two real and different values of h, we must have

$$\left(-6\sqrt{2}a\right)^2-4.8\left(1+2a^2\right)>0$$

or $\quad 72a^2-32\left(1+2a^2\right)>0 \Rightarrow 8a^2-32>0$

$\Rightarrow a^2-4>0 \qquad \Rightarrow (a+2)(a-2)>0$

Hence the required value of $a\in(-\infty,-2)\cup(2,\infty)$

Alternative Method:

Equation of chord AB whose mid-point is $(h,-h)$ is $T=S_1$

$$\Rightarrow 2xh-2yh-\left(1+\sqrt{2}a\right)\left(\frac{x+h}{2}\right)-\left(1-\sqrt{2}a\right)\left(\frac{y-h}{2}\right)$$

$$= 2h^2+2h^2-(1+\sqrt{2}a)h+(1-\sqrt{2}a)h$$

$$\Rightarrow 4xh-4yh-\left(1+\sqrt{2}a\right)(x+h)-\left(1-\sqrt{2}a\right)(y-h)$$

$$=8h^2-2\left(1+\sqrt{2}a\right)h+2\left(1-\sqrt{2}a\right)h$$

$$\Rightarrow x\left[4h-\left(1+\sqrt{2}a\right)\right]-y\left[4h+\left(1-\sqrt{2}a\right)\right]$$

$$-h\left(1+\sqrt{2}a\right)$$

$$+h\left(1-\sqrt{2}a\right)=8h^2-2\left(1+\sqrt{2}a\right)h+2\left(1-\sqrt{2}a\right)h$$

or $\quad 8h^2-\left(1+\sqrt{2}a\right)h+\left(1-\sqrt{2}a\right)h-x$

$$\left[4h-\left(1+\sqrt{2}a\right)+y\left[4h+\left(1-\sqrt{2}a\right)\right]\right]=0$$

It passes through $A\left(\frac{1+\sqrt{2}a}{2},\frac{1-\sqrt{2}a}{2}\right)$, then

$$8h^2-2\sqrt{2}ah-\left(\frac{1+\sqrt{2}a}{2}\right)\left[4h-\left(1+\sqrt{2}a\right)\right]$$

$$+\left(\frac{1-\sqrt{2}a}{2}\right)\left[4h+\left(1-\sqrt{2}a\right)\right]=0$$

or $8h^2-6\sqrt{2}ah+1+2a^2=0$

Hence for real and different values of h, we have

$$\left(-6\sqrt{2}a\right)^2-4.8.\left(1+2a^2\right)>0$$

$a^2-4>0$

$(a+2)(a-2)>0 \Rightarrow a\in(-\infty,-2)\cup(2,\infty)$

SECTION-II

SUBJECTIVE-TYPE SOLVED EXAMPLES

1. (a) A variable circle which always touches the line $x + y - 2 = 0$ at (1, 1) cuts the circle $x^2 + y^2 + 4x + 5y - 6 = 0$. Prove that all the common chords of intersection pass through a fixed point. Find that points.

(b) The circle $x^2 + y^2 - 6x - 6y + 9 = 0$ is inscribed in a triangle which has two of its sides along the co-ordinates axis. Find the locus of the circumcentre of the triangle.

Solution: (a) Any circle which touches the line $x + y - 2 = 0$ at (1, 1) will be of the form $(x-1)^2 + (y-1)^2 + \lambda(x + y - 2) = 0$

or, $x^2 + y^2 + (\lambda - 2)x + (\lambda - 2)y + 2 - 2\lambda = 0$

The common chord of this circle and $x^2 + y^2 + 4x + 5y - 6 = 0$ will be

$(\lambda - 6)x + (\lambda - 7)y + 8 - 2\lambda = 0$

or, $(-6x - 7y + 8) + \lambda(x + y - 2) = 0$

which is a family of lines, each member of which will be passing through a fixed point, which is the point of intersection of the lines $-6x - 7y + 8 = 0$ and $x + y - 2 = 0$ and that is $(6, -4)$.

(b) The equation of the incircle is $x^2 + y^2 - 6x - 6y + 9 = 0$; centre $\equiv (3, 3)$

Radius $= \sqrt{g^2 + f^2 - c} = \sqrt{9+9-9} = 3$

Let $BC = a$, $AB = c \Rightarrow AC = \sqrt{a^2 + c^2}$

Formula of incentre

$$= \left(\frac{ax_1 + bx_2 + cx_3}{a+b+c}, \frac{ay_1 + by_2 + cy_3}{a+b+c}\right)$$

$$\Rightarrow (3, 3) = \left(\frac{ca + a.0 + 0.\sqrt{a^2+c^2}}{a+c+\sqrt{a^2+c^2}}, \frac{ac+0+0}{a+c+\sqrt{a^2+c^2}}\right)$$

$$\Rightarrow \frac{ac}{a+c+\sqrt{a^2+c^2}} = 3$$

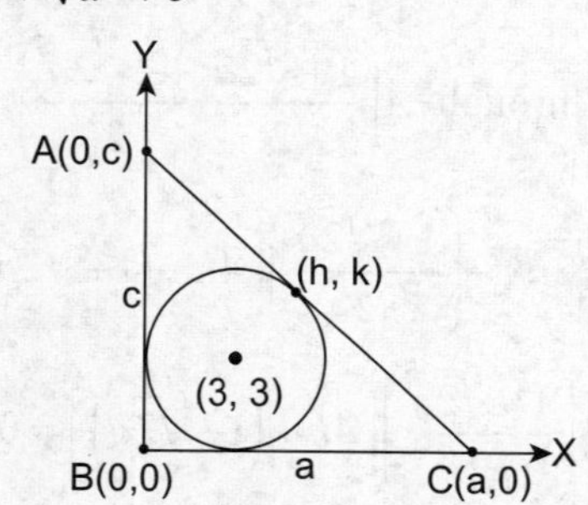

Now, let circumcentre $\equiv (h, k)$. Also $(h, k) = \left(\frac{a}{2}, \frac{c}{2}\right)$

$\Rightarrow a = 2h$, $c = 2k$

From (1) $\dfrac{4hk}{2h + 2k + 2\sqrt{h^2 + k^2}} = 3$

$\Rightarrow$ Locus of the circumcentre is $2xy - 3x - 3y - 3\sqrt{x^2 + y^2} = 0$

2. If the curves $ax^2 + 2hxy + by^2 - 2gx - 2fy + c = 0$ and $\ell x^2 - 2hxy + (a + \ell - b)^2 - 2mx - 2ny + k = 0$ intersect at four concyclic points A, B, C and D and P is the point $\left(\frac{g+m}{a+\ell}, \frac{f+n}{a+\ell}\right)$, then prove that $PA^2 + PB^2 + PC^2 + PD^2 = 4PA^2$.

Solution: The points of intersection of two curves $S_1 = 0$ and $S_2 = 0$ lie on $S_1 + \lambda S_2 = 0$

The given curves are $S_1 = ax^2 + 2hxy + by^2 - 2gx - 2fy + c = 0$

and $S_2 = \ell x^2 - 2hxy + (a + \ell - b)y^2 - 2mx - 2ny + k = 0$

As A, B, C and D are concyclic, equating the co-efficient of x^2 to the co-efficients of y^2 and the co-efficient xy be equal to 0; we get $\lambda = 1$.

Thus, the points A, B, C, D lie on $S_1 + S_2 = 0$

i.e., on $(a + \ell)x^2 + (a + \ell)y^2 - 2(g + m)x - 2(f + n)y + c + k = 0$

whose centre is $P \equiv \left(\frac{g+m}{a+\ell}, \frac{f+n}{a+\ell}\right)$

Obviously, $PA = PB = PC = PD =$ radius of the circle

$\Rightarrow PA^2 + PB^2 + PC^2 + PD^2 = 4PA^2$

3. Consider two circles C_1: $x^2 + y^2 = r_1^2$ and C_2: $x^2 + y^2 = r_2^2$ $(r_2 < r_1)$. Let $A(r_1, 0)$ be a fixed point on the circle C_1 and 'B' be a variable point on the circle C_2. The line AB meets the circle C_2 again at C. Find

(a) the set of values of $OB^2 + OA^2 + BC^2$

(b) the locus of mid-point of AB, 'O' being the origin.

Solution: $A(r_1, 0)$. Let $B \equiv (r_2 \cos\theta, r_2 \sin\theta)$

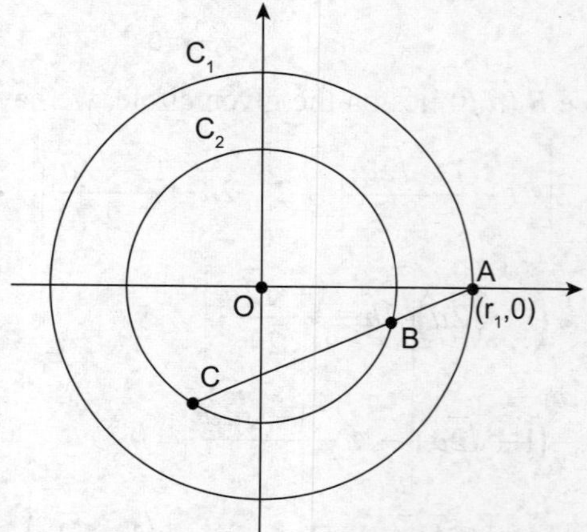

(a) It is clear that the length of BC is maximum when BC is the diameter of C_2 and minimum when BC is tangent to C_2 from A.

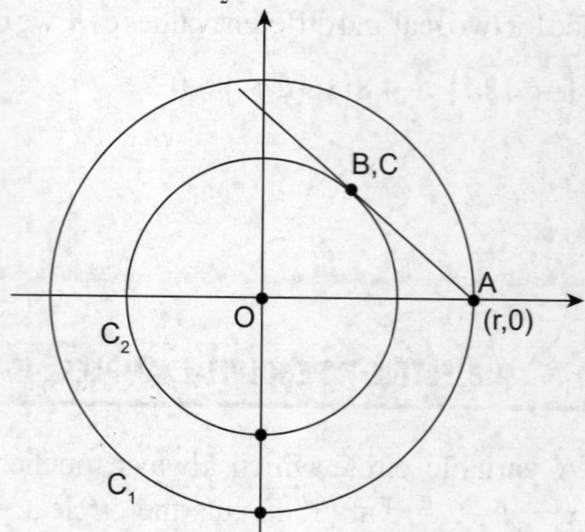

$\Rightarrow BC|_{max} = 2r_2$, $BC|_{min} = 0$

$\Rightarrow 5r_2^2 + r_1^2 \geq OA^2 + OB^2 + BC^2 \geq r_1^2 + r_2^2$

Thus the set of values of

$OA^2 + OB^2 + BC^2 \in [r_1^2 + r_2^2, 5r_2^2 + r_1^2]$

(b) Now let '*D*' (h, k) be the mid-point of AB.

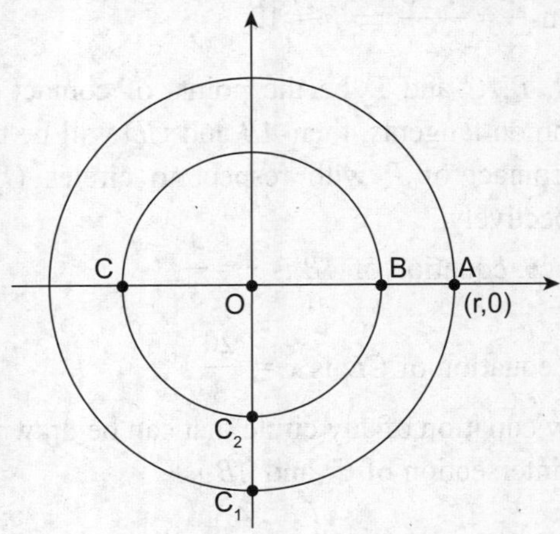

$$\Rightarrow D \equiv \left(\frac{r_1 + r_2\cos\theta}{2}, \frac{r_2\sin\theta}{2}\right)$$

$$\Rightarrow \sin\theta = \frac{2k}{r_2}, \cos\theta = \frac{2h - r_1}{r_2}$$

$$\Rightarrow 4k^2 + (2h - r_1)^2 = r_2^2$$

Locus of '*D*' is, $\left(x - \frac{r_1}{2}\right)^2 + y^2 = \frac{r_2^2}{4}$

4. Find the equation of the largest circle with centre at (1, 0) that can be inscribed in the ellipse $x^2 + 4y^2 = 16$.

Solution: The given ellipse is $\frac{x^2}{16} + \frac{y^2}{4} = 1$

Let the required circle touch the ellipse at $(4\cos\theta, 2\sin\theta)$ and $(4\cos\theta, -2\sin\theta)$

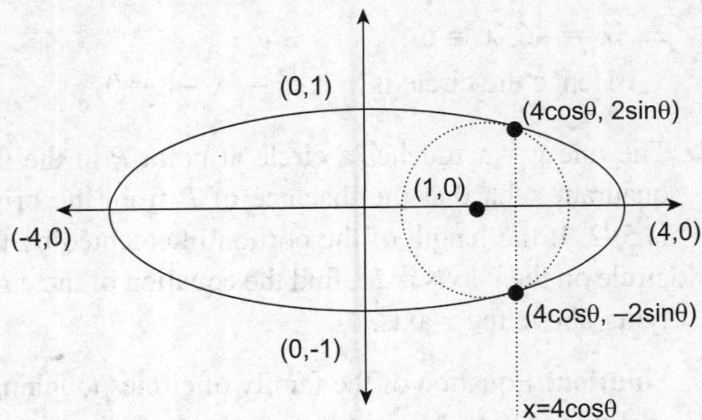

Then equation of this circle is

$$\frac{x^2}{16} + \frac{y^2}{4} - 1 + \lambda(x - 4\cos\theta)^2 = 0$$

$$or\ x^2\left(\frac{1}{16} + \lambda\right) + \frac{y^2}{4} - 8x\lambda\cos\theta + 16\lambda\cos^2\theta - 1 = 0$$

For this to represent a circle, we must have

$$\frac{1}{4} = \frac{1}{16} + \lambda \Rightarrow \lambda = \frac{3}{16}$$

$$\Rightarrow \text{circle is, } \frac{1}{4}(x^2 + y^2) - \frac{3}{2}x\cos\theta + 3\cos^2\theta - 1 = 0$$

or, $x^2 + y^2 - 6x\cos\theta + 12\cos^2\theta - 4 = 0$

It's centre is $(1, 0) \Rightarrow \cos\theta = 1/3$

$\Rightarrow$ Final equation of the circle is; $x^2 + y^2 - 2x - 8/3 = 0$

5. (a) Let $A = (-1, 0)$, $B = (3, 0)$ and PQ be any line passing through (4, 1) having slope m. Find the range of 'm' for which there exist two points on PQ at which AB subtends a right angle.

(b) The equation of radical axis of two circles is $x + y = 1$. One of the circles has the ends of a diameter at the points (1, –3) and (4, 1) and the other passes through the point (1, 2). Find the equations of these circles.

Solution: (a) Equation of line PQ is

$(y - 1) = m(x - 4)$ or; $y - mx + 4m - 1 = 0$

For the required 'm' we have to make sure that the line PQ meets the circle, with diameter AB, at real and distinct points.

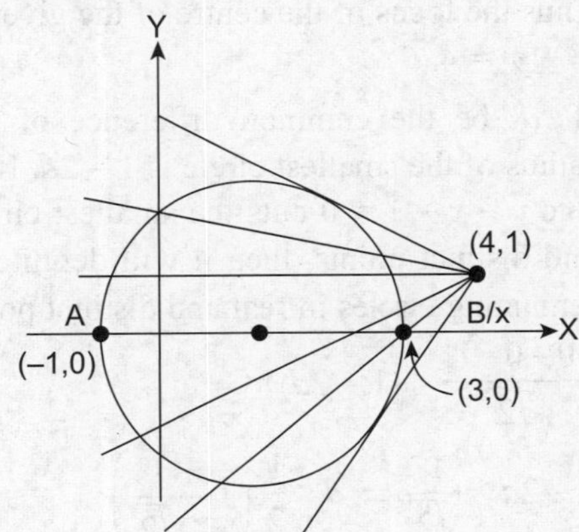

Equation of circle having AB as diameter is

$x^2 + y^2 - 2x - 3 = 0$

Thus we have to make sure that $\frac{|0 - m + 4m - 1|}{\sqrt{1 + m^2}} < 2$

$\Rightarrow 5m^2 - 6m - 3 > 0$

$$\Rightarrow m \in \left(\frac{3 - 2\sqrt{6}}{5}, \frac{3 + 2\sqrt{6}}{5}\right)$$

(b) Equation of the circle having the ends of diameter at (1, –3) and (4, 1) is

$(x - 1)(x - 4) + (y + 3)(y - 1) = 0$

$\Rightarrow x^2 + y^2 - 5x + 2y + 1 = 0$

Other circle will be

$x^2 + y^2 - 5x + 2y + 1 + \lambda(x + y - 1) = 0$

It passes through $(1, 2) \Rightarrow \lambda = -\frac{5}{2}$

$$\Rightarrow \text{circle is; } x^2 + y^2 - \frac{15x}{2} - \frac{y}{2} + \frac{7}{2} = 0$$

6. (a) Find the locus of the centres of the circles $x^2 + y^2 - 2ax - 2by + 2 = 0$, where '$a$' and '$b$' are parameters, if the tangents from the origin to each of the circles are orthogonal.

(b) Three concentric circles of which the biggest is $x^2 + y^2 = 1$, have their radii in A.P. If the line $y = x + 1$ cuts all the circles in real and distinct points, then find the interval in which the common difference of A.P. will lie.

Solution: (a) The given circle is: $x^2 + y^2 - 2ax - 2by + 2 = 0$

or $(x - a)^2 + (y - b)^2 = a^2 + b^2 - 2$ its director circle is $(x - a)^2 + (y - b)^2 = 2(a^2 + b^2 - 2)$

Given that tangents drawn from the origin to the circle are orthogonal, it implies that director circle of the circle must pass through the origin.

$\Rightarrow a^2 + b^2 = 2(a^2 + b^2 - 2) \Rightarrow a^2 + b^2 = 4$

Thus the locus of the centre of the given circle is, $x^2 + y^2 = 4$.

(b) If 'd' be the common difference of A.P., then radius of the smallest circle is $1 - 2d$. If the given line $y - x - 1 = 0$ cuts the smallest circle in real and distinct points, then it will definitely cut the remaining circles in real and distinct points.

$$\Rightarrow \frac{|0-0-1|}{\sqrt{2}} < (1 - 2d)$$

$$\Rightarrow 1 - 2d > \frac{1}{\sqrt{2}} \Rightarrow d < \frac{1}{2}\left(1 - \frac{1}{\sqrt{2}}\right)$$

Hence $d \in \left(0, \frac{1}{2}\left(1 - \frac{1}{\sqrt{2}}\right)\right)$

7. Find the equation of the circles passing through the points of contact of direct common tangents of $x^2 + y^2 = 16$ and $x^2 + y^2 - 12x + 32 = 0$.

Solution: The given circles are;

$C_1 : x^2 + y^2 = 16$; $C_2 : x^2 + y^2 - 12x + 32 = 0$

Let $P(h, 0)$ be the point of concurrency of direct tangents.

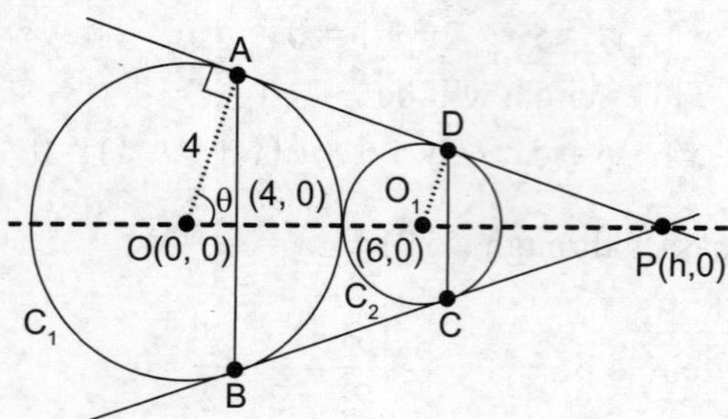

Then $\frac{h}{4} = \frac{h-6}{2} \Rightarrow h = 12$

If A, B, C and D be the points of contact of direct common tangents, then AB and CD will be the chord of contact of P with respect to circles C_1 and C_2 respectively.

Hence, equation of AB is $x = \frac{4}{3}$

and equation of CD is $x = \frac{20}{3}$

Now equation of any circle that can be drawn through the intersection of C_1 and AB is;

$$x^2 + y^2 - 16 + \lambda_1\left(x - \frac{4}{3}\right) = 0$$

$$\text{or } x^2 + y^2 + \lambda_1 x - \left(16 + \frac{4\lambda_1}{3}\right) = 0$$

and equation of circle passing through the intersection of CD and C_2 is

$$x^2 + y^2 - 12x + 32 + \lambda_2\left(x - \frac{20}{3}\right) = 0$$

$$\text{or, } x^2 + y^2 + x(\lambda_2 - 12) + \left(32 - \frac{20\lambda_2}{3}\right) = 0$$

These two circles should be same

$$\Rightarrow \frac{\lambda_2 - 12}{\lambda_1} = \frac{32 - \frac{20\lambda_2}{3}}{-\left(16 + \frac{4\lambda_1}{3}\right)} = 1$$

$\Rightarrow \lambda_1 = -6, \lambda_2 = 6$

Hence the circle is $x^2 + y^2 - 6x - 8 = 0$

8. The line $y = x$ touches a circle at point P in the first quadrant such that the distance of P from the origin is $5\sqrt{2}$. If the length of the portion intercepted by this circle on the y-axis is $2c$, find the equation of the circle lying above the x-axis.

Solution: Equation of the family of circles touching $y = x$ at (5, 5) can be written as

$(x - 5)^2 + (y - 5)^2 + \lambda(y - x) = 0$

Put $x = 0$, we get $y^2 - 10y + \lambda y + 50 = 0$

$\Rightarrow 4c^2 = (y_1 + y_2)^2 - 4y_1y_2$

$\Rightarrow \lambda^2 - 20\lambda - (4c^2 + 100) = 0$

$$\Rightarrow \lambda = \frac{20 \pm \sqrt{400 + 4(4c^2 + 100)}}{2}$$

Since y_1, y_2 are $< 10 \Rightarrow \lambda = 10 - 2\sqrt{c^2 + 50}$

$\therefore$ Equation of the required circle is

$$(x - 5)^2 + (y - 5)^2 + (10 - 2\sqrt{c^2 + 50})(y - x) = 0$$

9. Let S_1 be a circle passing through $A(0, 1)$, $B(-2, 2)$ and S_2 is a circle of radius $\sqrt{10}$ units such that AB is common chord of S_1 and S_2. Find the equation of S_2.

Solution: Equation of line AB is

$$y - 2 = \frac{2-1}{-2-0}(x+2) = -\frac{1}{2}(x+2)$$

$\Rightarrow x + 2y - 2 = 0$ ...(1)

Equation of circle whose diametrically opposite points are A and B: $(x - 0)(x + 2) + (y - 1)(y - 2) = 0$

$\Rightarrow x^2 + y^2 + 2x - 3y + 2 = 0$ ...(2)

Family of circles passing through the points of intersection of (1) and (2) $x^2 + y^2 + 2x - 3y + 2 + \lambda(x + 2y - 2) = 0$

$\Rightarrow x^2 + y^2 + (2 + \lambda)x + (2\lambda - 3)y + 2 - 2\lambda = 0$ ...(3)

Equation (3), represents a circle of radius $\sqrt{10}$ units

$$\Rightarrow \sqrt{\left(-\frac{2+\lambda}{2}\right)^2 + \left(-\frac{2\lambda-3}{2}\right)^2 - 2 + 2\lambda} = \sqrt{10}$$

$\Rightarrow (4 + 4\lambda + \lambda^2) + (4\lambda^2 + 9 - 12\lambda) + 8\lambda - 8 = 40$

$\Rightarrow \lambda = \pm\sqrt{7}$

Hence required circles are

$$x^2 + y^2 + 2x - 3y + 2 \pm \sqrt{7}(x + 2y - 2) = 0$$

There are two such circles possible.

10. *A* circle passing through the vertex C of a rectangle $ABCD$ and touches its sides AB and AD at M and N respectively. If the distance from C to the line-segment MN is equal to 5 units, find the area of the rectangle $ABCD$.

Solution: Let us take AB and AD as co-ordinate axes

If r be the radius of circle, then its centre is $P(r, r)$

Equation of circle is $(x - r)^2 + (y - r)^2 = r^2$

$\Rightarrow x^2 + y^2 - 2rx - 2ry + r^2 = 0$

Let the co-ordinates of $C \equiv (p, q)$

Equation of MN is $x + y = r$

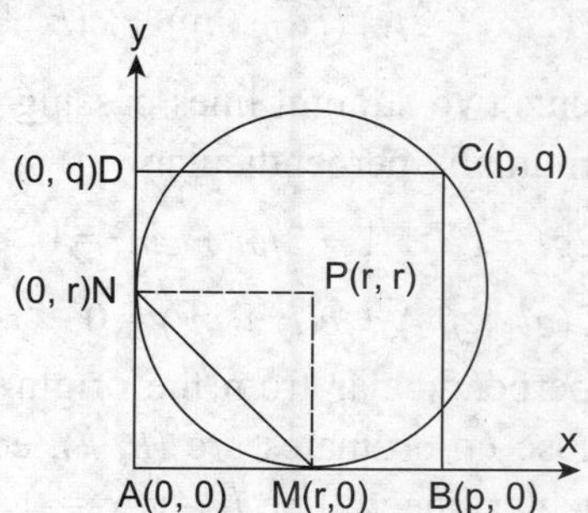

Its distance from C is 5 units $\Rightarrow \dfrac{|p+q-r|}{\sqrt{2}} = 5$

$\Rightarrow (p + q - r)^2 = 50$

Since (p, q) lies on the circle,

$p^2 + q^2 - 2rp - 2rq + r^2 = 0$

$\Rightarrow (p + q - r)^2 - 2pq = 0 \Rightarrow 50 - 2pq = 0 \Rightarrow pq = 25$

Area of rectangle = 25 sq. units.

11. Let ABC be a triangle right angled at A and S be its circumcircle. Let S_1 be the circle touching the lines AB, AC and the circle S internally. Further, let S_2 be the circle touching the lines AB and AC produce and the circle S externally. If r_1 and r_2 be the radii of the circles S_1 and S_2 respectively, such that $r_1r_2 = k$. Area (ΔABC). Find the value of k.

Solution: Take AB as the x-axis and AC as the y-axis

Let $BC = a$

Centre of $S = \left(\frac{c}{2}, \frac{b}{2}\right)$ and radius is $\frac{a}{2}$

If a circle S' touches the rays AB and AC, its centre must be (r, r) and its radius must be r, for some $r > 0$.

Circle S and S' touch each other if

$$\sqrt{\left(r - \frac{c}{2}\right)^2 + \left(r - \frac{b}{2}\right)^2} = \frac{a}{2} \pm r$$

Squaring both sides and using the fact that $a^2 = b^2 + c^2$, we get $r = b + c \pm a$

$\Rightarrow r_1 = b + c - a,$

$\Rightarrow r_2 = b + c + a$

$\Rightarrow r_1 . r_2 = (b + c)^2 - a^2 = 2bc$ (as $a^2 = b^2 + c^2$)

= 4 area (ΔABC)

$\therefore k = 4$

12. Find the range of parameter 'a' for which the variable line $y = 2x + a$ lies between the circles $x^2 + y^2 - 2x - 2y + 1 = 0$ and $x^2 + y^2 - 16x - 2y + 61 = 0$ without intersecting or touching either circle.

Solution: The given circles are $C_1 : (x - 1)^2 + (y - 1)^2 = 1$ and $C_2 : (x - 8)^2 + (y - 1)^2 = 4$

The line $y - 2x - a = 0$ will lie between these circle if centre of the circles lie on opposite sides of the line, i.e., $(1 - 2 - a)(1 - 16 - a) < 0 \Rightarrow a \in (-15, -1)$

Line wouldn't touch or intersect the circles if,

$$\frac{|1-2-a|}{\sqrt{5}} > 1, \frac{|1-16-a|}{\sqrt{5}} > 2$$

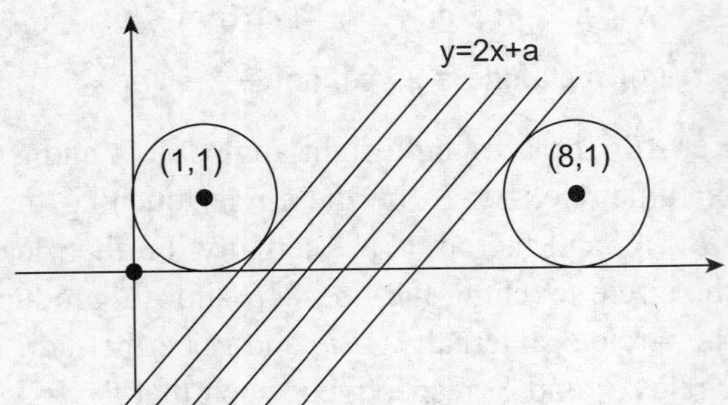

$\Rightarrow$ $|1 + a| > \sqrt{5}$, $|15 + a| > 2\sqrt{5}$

$\Rightarrow$ $a > \sqrt{5} - 1$ or $a < -\sqrt{5} - 1$, $a > 2\sqrt{5} - 15$ or

$a < -2\sqrt{5} - 15$

Hence common values of 'a' are $(2\sqrt{5}-15, -\sqrt{5}-1)$

13. If $al^2 - bm^2 + 2dl + 1 = 0$, a, b, d are fixed real numbers such that $a + b = d^2$, then prove that the line $lx + my + 1 = 0$ touches a fixed circle. Find its equation.

Solution: Let the equation of the circle be $x^2 + y^2 + 2gx + 2fy + c = 0$

If $lx + my + 1 = 0$ is a tangent to the circle

Then $\sqrt{g^2 + f^2 - c} = \dfrac{|-lg - mf + 1|}{\sqrt{l^2 + m^2}}$

$\Rightarrow$ $(g^2 + f^2 - c)(l^2 + m^2) = (lg + mf - 1)^2$

$\Rightarrow$ $(f^2 - c)l^2 + (g^2 - c)m^2 + 2l(g - mgf) + 2mf - 1 = 0$

Comparing with the given condition

$al^2 - bm^2 + 2dl + 1 = 0$; we get

$-(f^2 - c) = a, -(g^2 - c) = -b, -g(1 - mf) = d, f = 0$

$\Rightarrow$ $c = a, g = -d, g^2 - c = b, d^2 - a = b$

$\Rightarrow$ $a + b = d^2$ which is the required condition,

Hence the fixed circle is $x^2 + y^2 - 2dx + a = 0$

14. Show that the locus of a variable point, for which the ratio of length of tangents to two given concentric circles is inverse of ratio of their radii, is a concentric circle. Also show that the square of radius of this circle is equal to the sum of the square of the radii of the two given circles.

Solution: Taking the common centre as the origin and the radii of the two circle as a and b, the equation of the circle will be

$x^2 + y^2 = a^2$ ...(1)

and $x^2 + y^2 = b^2$ ...(2)

Let $P(h, k)$ be any point such that PT_1 and PT_2 are length of tangent from it to (1) and (2) respectively.

Now $PT_1 = \sqrt{h^2 + k^2 - a^2}$; $PT_2 = \sqrt{h^2 + k^2 - b^2}$

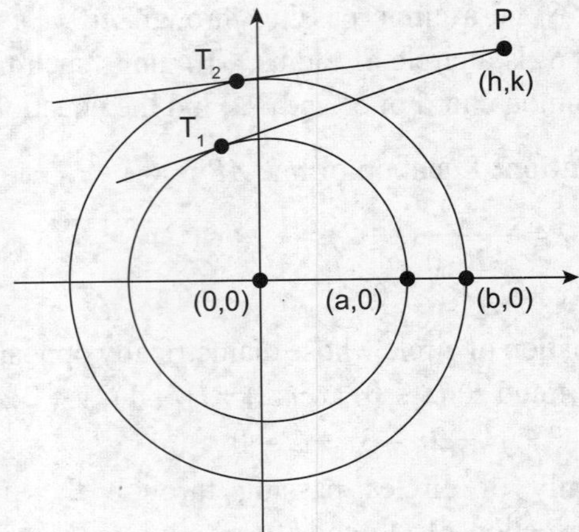

According to question: $\dfrac{PT_1}{PT_2} = \dfrac{b}{a}$ or $\dfrac{PT_1^2}{PT_2^2} = \dfrac{b^2}{a^2}$

Putting the values of PT_1 and PT_2; we get

$\dfrac{h^2 + k^2 - a^2}{h^2 + k^2 - b^2} = \dfrac{b^2}{a^2}$

or, $a^2h^2 + a^2k^2 - a^4 = b^2h^2 + b^2k^2 - b^4$

or, $h^2(a^2 - b^2) + k^2(a^2 - b^2) = a^4 - b^4$ or, $h^2 + k^2 = a^2 + b^2$

Therefore we get the required locus as $x^2 + y^2 = a^2 + b^2$ which is clearly a circle with centre (0, 0) and radius $\sqrt{a^2 + b^2}$.

15. Find the locus of the foot of the perpendicular from the origin upon any chord of the circle $x^2 + y^2 + 2gx + 2fy + c = 0$ which subtends a right angle at the origin.

Solution: Given circle is $x^2 + y^2 + 2gx + 2fy + c = 0$ ...(1)

let the equation of any chord be $lx + my = 1$ ...(2)

making (1) homogeneous with the help of (2) we get,

$x^2 + y^2 + (2gx + 2fy)(lx + my) + c(lx + my)^2 = 0$

It represents two straight lines passing through origin and mutually perpendicular

$\Rightarrow$ $1 + 2gl + cl^2 + 1 + 2fm + cm^2 = 0$

$\Rightarrow$ $2 + 2(gl + fm) + c(l^2 + m^2) = 0$ ...(3)

Let the perpendicular from the origin on (2) meet it at P, whose co-ordinates are (h, k), equation of this perpendicular line will be $ly - mx = 0$ ...(4)

as (h, k) is the point of intersection, it will satisfy (2) and (4)

Hence, $lh + mk - 1 = 0$ ...(5)

And $lk - mh = 0$ ...(6)

Eliminating l and m we get,

$l = \dfrac{h}{h^2+k^2}$ and $m = \dfrac{k}{h^2+k^2}$

putting these values in (3) we get required locus as $2x^2 + 2y^2 + 2gx + 2fy + c = 0$; which is a circle.

16. A straight line moves so that the product of length of the perpendiculars on it from two fixed points is constant. Prove that the locus of the feet of the perpendiculars from each of these points upon the straight-line is a unique circle.

Solution: Let $A\ (-a, 0)$ and $B(a, 0)$ be two fixed points. Taking AB as x-axis and its right bisector as y-axis. Let the equation of the given line be

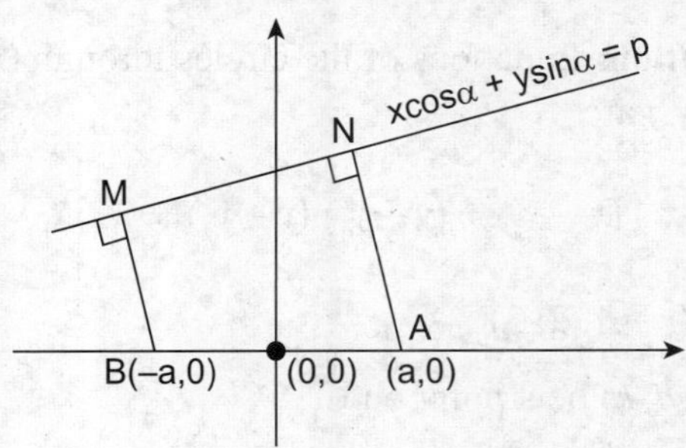

$$x \cos \alpha + y \sin \alpha = p \quad \text{...(1)}$$

and line perpendicular to it and passing through $(-a, 0)$ is given by

$$y \cos \alpha - x \sin \alpha = a \sin \alpha \quad \text{...(2)}$$

Let AN and BM be the perpendiculars from A and B, then

$$AN \times BM = \frac{-a\cos\alpha + 0\sin\alpha - p}{\sqrt{\sin^2\alpha + \cos^2\alpha}} \times \frac{a\cos\alpha + 0\sin\alpha - p}{\sqrt{\sin^2\alpha + \cos^2\alpha}} = \lambda^2 \text{ (const.)}$$

$$\Rightarrow p^2 = \lambda^2 + a^2 \cos^2 \alpha \quad \text{...(3)}$$

eliminating p and α from (1),(2) and (3) we get, $x^2 + y^2 = \lambda^2 + a^2$ which is the required locus, by changing a into $-a$, we get the same locus.

Hence proved.

17. Find the locus of the middle points of the chord of the circle $x^2 + y^2 = a^2$ which subtends a right angle at the point $(c, 0)$.

Solution: Let the equation of circle $x^2 + y^2 = a^2$ ...(1) taking any two points A and B on the circle as $(a \cos \alpha, a \sin \alpha)$ and $(a \cos \beta, a \sin\beta)$

Middle points of AB as (h, k) can be taken as

$$h = \frac{a}{2}(\cos\alpha + \cos\beta) \quad \text{.......(2)}$$

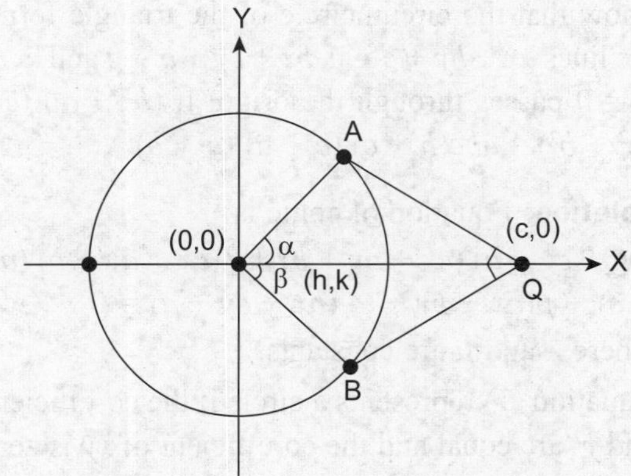

$$k = \frac{a}{2}(\sin\alpha + \sin\beta) \quad \text{.........(3)}$$

If Q be the given point $(c, 0)$, then since $\angle AQB = 90°$

$$\Rightarrow \frac{a\sin\alpha - 0}{a\cos\alpha - c} \times \frac{a\sin\beta - 0}{a\cos\beta - c} = -1$$

$$\Rightarrow a^2 (\sin \alpha \sin \beta + \cos \alpha \cos \beta) - ac(\cos \alpha + \cos \beta) + c^2 = 0 \quad \text{...(4)}$$

from (2) and (3); we get

$$h^2 + k^2 = \frac{a^2}{4}[2 + 2(\cos\alpha\cos\beta + \sin\alpha\sin\beta)] \quad \text{...(5)}$$

Substituting the values from (5) and (2) in (4), we get the required locus as

$2x^2 + 2y^2 - 2cx + c^2 - a^2 = 0$; which is a circle.

18. If the equation of circle having the pair of lines $x^2 + 2xy + 3x + 6y = 0$ as its normals and having the size just sufficient to contain the circle $x(x - 4) + y(y - 3) = 0$ is given by $x^2 + y^2 + 6x - 3y - k = 0$, then find the value of 'k'.

Solution: The given circle is $x(x - 4) + y(y - 3) = 0$ ...(1)

Clearly, centre is $C_1 = \left(2, \frac{3}{2}\right)$ and radius $= \frac{5}{2}$

Now, given lines are $x + 3 = 0$ and $x + 2y = 0$

And the intersection of these lines will give the centre $C_2 = \left(-3, \frac{3}{2}\right)$ of the required circle.

The required circle will just contain the circle (1), hence the circle (1) lies inside the required circle and touches internally.

$$\Rightarrow c_1c_2 = r_2 - r_1 \Rightarrow 5 = r_2 - \frac{15}{2} \Rightarrow r_2 = \frac{15}{2}$$

Hence the required circle is

$$(x + 3)^2 + \left(y - \frac{3}{2}\right)^2 = \left(\frac{15}{2}\right)^2$$

$$\Rightarrow x^2 + y^2 + 6x - 3y - 45 = 0$$

19. Show that the circumcircle of the triangle formed by the lines $ax + by + c = 0$, $bx + cy + a = 0$ and $cx + ay + b = 0$ passes through the origin if $(b^2 + c^2)(c^2 + a^2)(a^2 + b^2) = abc(b + c)(c + a)(a + b)$.

Solution: Equation of conic is

$(bx + cy + a)(cx + ay + b) + \lambda(cx + ay + b)(ax + by + c) + \mu(ax + by + c)(bx + cy + a) = 0$ ...(1)

where λ and μ are constants.

Equation (1) represents a circle if the co-efficient of x^2 and y^2 are equal and the co-efficient of xy is zero, i.e.,

$bc + \lambda ca + \mu ab = ca + \lambda ab + \mu bc$

or $(a - b)c + \lambda(b - c)a + \mu(c - a)b = 0$ ...(2)

and $(c^2 + ab) + \lambda(a^2 + bc) + \mu(b^2 + ac) = 0$ ...(3)

Solving (2) and (3) by cross-multiplication rule, we get

$$\therefore\ \lambda = \frac{(a^2 - bc)(b^2 + c^2)}{(c^2 - ab)(a^2 + b^2)} \text{ and } \mu = \frac{(b^2 - ac)(c^2 + a^2)}{(c^2 - ab)(a^2 + b^2)} \quad (4)$$

and given (1) passes through the origin, then

$ab + bc\,\lambda + ca\mu = 0$ ...(5)

From (4) and (5), we get

$abc^2(a^2 + b^2) + a^2bc(b^2 + c^2) + b^2ca(c^2 + a^2)$
$= a^2b^2(a^2 + b^2) + b^2c^2(b^2 + c^2) + c^2a^2(c^2 + a^2)$

$\Rightarrow abc\{c(a^2 + b^2) + a(b^2 + c^2) + b(c^2 + a^2)\} = a^2b^2(a^2 + b^2) + b^2c^2(b^2 + c^2) + c^2a^2(c^2 + a^2)$

$\Rightarrow abc\{(a + b)(b + c)(c + a) - 2abc\} = a^2b^2(a^2 + b^2) + b^2c^2(b^2 + c^2) + c^2a^2(c^2 + a^2)$

$\Rightarrow abc(a + b)(b + c)(c + a) = (a^2 + b^2)(b^2 + c^2)(c^2 + a^2)$

20. Find the radius of the largest circle, which passes through the focus of the parabola $y^2 = 4(x + y)$ and is contained in it.

Solution: $y^2 = 4(x + y) \Rightarrow (y - 2)^2 = 4(x + 1)$

$\Rightarrow Y^2 = 4X$

(shifting origin to (–1, 2), keeping direction of axes intact.)

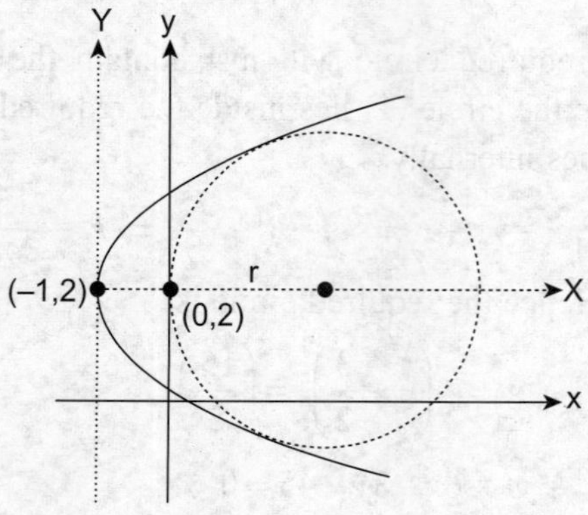

Circle referred to X, Y system will be

$(X - r - 1)^2 + Y^2 = r^2$, (where r is the radius of circle)

Eliminating Y, we get

$(X - r - 1)^2 + 4X = r^2$

$\Rightarrow X^2 + (2 - 2r)X + 2r + 1 = 0$

Discriminant $= 0 \Rightarrow r = 4$

[Note that the radius will be independent of the position of parabola]

21. If C_1, C_2 and C_3 belong to a family of circles through the points (x_1, y_1) and (x_2, y_2), prove that the ratio of the lengths of the tangent from any point on C_1 to the circles C_2 and C_3 is constant.

Solution: Equations of the circles through (x_1, y_1) and (x_2, y_2) are

$$(x - x_1)(x - x_2) + (y - y_1)(y - y_2) + \lambda_r \begin{vmatrix} x & y & 1 \\ x_1 & y_1 & 1 \\ x_2 & y_2 & 1 \end{vmatrix}$$

$= 0\ (r = 1, 2, 3)$

Let (h, k) be a point on C_1

$\Rightarrow \phi(h, k) + \lambda_1 \psi(h, k) = 0$

where $\phi(h, k) = (h - x_1)(h - x_2) + (k - y_1)(k - y_2)$

and $\psi(h, k) = \begin{vmatrix} h & k & 1 \\ x_1 & y_1 & 1 \\ x_2 & y_2 & 1 \end{vmatrix}$.

Let T_2 be the length of the tangent from (h, k) to C_2 and T_3 be the length of the tangent from (h, k) to C_3.

$\Rightarrow T_2 = \sqrt{\phi(h, k) + \lambda_2\psi(h, k)}$,

$T_3 = \sqrt{\phi(h, k) + \lambda_3\psi(h, k)}$

$$\Rightarrow \frac{T_2}{T_3} = \frac{\sqrt{\phi(h, k) + \lambda_2\psi(h, k)}}{\sqrt{\phi(h, k) + \lambda_3\psi(h, k)}}$$

$$= \frac{\sqrt{(\lambda_2 - \lambda_1)\psi(h, k)}}{\sqrt{(\lambda_3 - \lambda_1)\psi(h, k)}} = \sqrt{\frac{\lambda_2 - \lambda_1}{\lambda_3 - \lambda_1}}$$

which is independent of the choice of (h, k) and hence a constant.

22. C_1 and C_2 are two concentric circles of radii a and b respectively $(a < b)$ with centre at O. A tangent is drawn to circle C_2 at a given point P. This tangent is the diameter of a variable circle C_3, which touches C_1 externally. This tangent meets C_3 in Q_1 and Q_2. Prove that the line Q_1Q_2 subtends a constant non-zero angle at a fixed point T on the line joining O and P. Also find the locus of the point T for various positions of P on C_2.

Solution: Let O be the origin and the point P be on y-axis and T be $(0, k)$. Since the tangent at P is the diameter of the variable circle, we can take the centre of the variable circle to be $R(h, -b)$ with radius $r \Rightarrow Q_1$ is $(h - r, -b)$ and Q_2 is $(h + r, -b)$.

Then $(r + a)^2 = b^2 + h^2$

Also, slope of the line $TQ_1 = \tan\theta_1 = \dfrac{k+b}{r-h}$

Slope of line $TQ_2 = \tan\theta_2 = -\dfrac{k+b}{r+h}$

Also $\theta = \theta_2 - \theta_1 \Rightarrow \tan\theta = \dfrac{\tan\theta_2 - \tan\theta_1}{1+\tan\theta_1\tan\theta_2}$

$$= \frac{-\dfrac{k+b}{r+h} - \dfrac{k+b}{r-h}}{1 - \dfrac{(k+b)^2}{r^2-h^2}} = \frac{2(b+k)r}{h^2+(b+k)^2-r^2}$$

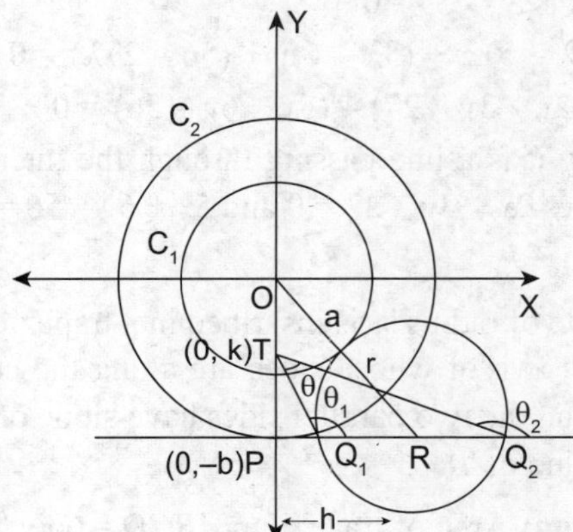

Put $h^2 = (r + a)^2 - b^2$ to

eliminate $h \Rightarrow \tan\theta = \dfrac{2(b+k)r}{(b+k)^2 + a^2 + 2ar - b^2}$

$$= \frac{2(b+k)}{2a + \dfrac{(b+k)^2 - (b^2 - a^2)}{r}}$$

In order to make $\tan\theta$ independent of r, we have to take $(b + k)^2 = b^2 - a^2$

$\Rightarrow \tan\theta = \dfrac{b+k}{a}$ = constant

Hence for $(b + k)^2 = b^2 - a^2$, $\angle Q_1TQ_2$ is constant for the fixed point T lying on the line joining O and P.

Now $(b + k)^2 = TP^2$

$\Rightarrow TP^2 = b^2 - a^2$

$\Rightarrow$ Distance between T and P is constant $= \sqrt{b^2 - a^2}$

$\Rightarrow$ Distance between O and T is constant

$= b \pm \sqrt{b^2 - a^2}$

$\Rightarrow$ Point T lies on either of the circles of radii

$b - \sqrt{b^2 - a^2}$ or $b + \sqrt{b^2 - a^2}$

with centre at the origin, i.e., $x^2 + y^2 = (b \pm \sqrt{b^2 - a^2})^2$

23. Consider the circles $x^2 + y^2 = a^2$. Let $A \equiv (a, 0)$ and D be a given interior point of the circle. If BC be any arbitrary chord of the circle through point D, prove that the locus of the centroid of triangle ABC is a circle whose radius is less than $a/3$.

Solution: Let $D \equiv (\alpha, \beta)$ where $\alpha^2 + \beta^2 - a^2 < 0$

$B \equiv (a\cos\theta_1, a\sin\theta_1)$, $C \equiv (a\cos\theta_2, a\sin\theta_2)$

Equation of line BC;

$$x\cos\left(\frac{\theta_1+\theta_2}{2}\right) + y\sin\left(\frac{\theta_1+\theta_2}{2}\right) = a\cos\left(\frac{\theta_1-\theta_2}{2}\right)$$

It is passing through $D(\alpha, \beta)$ is given

$$\Rightarrow \alpha\cos\left(\frac{\theta_1+\theta_2}{2}\right) + \beta\sin\left(\frac{\theta_1+\theta_2}{2}\right) = a\cos\left(\frac{\theta_1-\theta_2}{2}\right) \quad \text{.............(1)}$$

Let centroid of ΔABC be (h, k)

$\Rightarrow 3h = a\cos\theta_1 + a\cos\theta_2 + a$, $3k = a\sin\theta_1 + a\sin\theta_2$

$\Rightarrow \dfrac{3h}{a} = \cos\theta_1 + \cos\theta_2 + 1, \dfrac{3k}{a} = \sin\theta_1 + \sin\theta_2$

$$\Rightarrow 2\cos\left(\frac{\theta_1+\theta_2}{2}\right).\cos\left(\frac{\theta_1-\theta_2}{2}\right) = \frac{3h}{a} - 1 \ ..... \quad (2)$$

$$\text{and } 2\sin\left(\frac{\theta_1+\theta_2}{2}\right)\cos\left(\frac{\theta_1-\theta_2}{2}\right) = \frac{3k}{a} \ ...... \quad (3)$$

Multiplying (1) with $2\cos\left(\dfrac{\theta_1-\theta_2}{2}\right)$, we get

$$2\alpha\cos\left(\frac{\theta_1+\theta_2}{2}\right).\cos\left(\frac{\theta_1-\theta_2}{2}\right)$$

$$+2\beta\sin\left(\frac{\theta_1+\theta_2}{2}\right)\cos\left(\frac{\theta_1-\theta_2}{2}\right)$$

$$= 2a\cos^2\left(\frac{\theta_1-\theta_2}{2}\right)$$

$$\Rightarrow \alpha\left(\frac{3h-a}{a}\right) + \beta\left(\frac{3k}{a}\right) = 2a\cos^2\left(\frac{\theta_1-\theta_2}{2}\right)$$

Now, from (2) and (3), we get:

$$2\cos\left(\frac{\theta_1-\theta_2}{2}\right) = \frac{1}{2}\left(\frac{9k}{} + \left(\frac{3h \quad a}{}\right)\right)$$

$$\Rightarrow \alpha(3h - a) + 3\beta k = \frac{a^2}{2}\left(\frac{9k^2}{a^2} + \left(\frac{3h-a}{a}\right)^2\right)$$

$\Rightarrow 6\alpha h - 2a\alpha + 6\beta k = 9k^2 + 9h^2 + a^2 - 6ah$

$\Rightarrow$ locus of centroid is; $9x^2 + 9y^2 - 6x(a + \alpha) - 6y\beta + a^2 + 2a\alpha = 0$

which is clearly a circle of radius

$$\sqrt{\frac{(a+\alpha)^2}{9} + \frac{\beta^2}{9} - \frac{a^2+2a\alpha}{9}} = \frac{1}{3}\sqrt{\alpha^2+\beta^2} < \frac{a}{3}$$

24. A circle is inscribed in a trapezium $ABCD$ such that AD perpendicular DC. Prove that $\frac{1}{AD}+\frac{1}{BC}=\frac{2}{CD}$.

Solution: Let length of $AD = r + x$ and length of $BC = r + y$

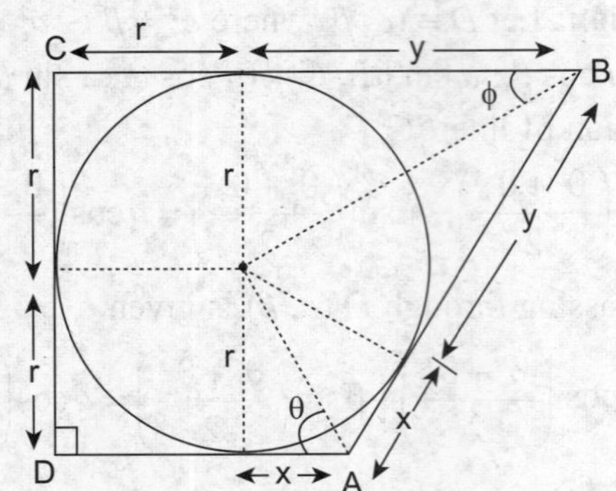

We have to prove that $\frac{1}{AD}+\frac{1}{BC}=\frac{2}{CD}$ or

$$\frac{1}{r+x}+\frac{1}{r+y}=\frac{2}{2r} \quad ...(1)$$

Let $\angle DAB = 2\theta$ and $\angle ABC = 2\phi$,

then $\angle DAB + \angle ABC = \pi$

or $2\theta + 2\phi = \pi$

or $\theta + \phi = \frac{\pi}{2}$

or $\tan(\theta + \phi) = \tan\frac{\pi}{2}$

or $\tan\theta \tan\phi = 1$

or $\left(\frac{r}{x}\right)\cdot\left(\frac{r}{y}\right) = 1$ {since $\tan\theta = \frac{r}{x}$ and $\tan\phi = \frac{r}{y}$}

or $r^2 = xy$ ...(2)

Now taking L.H.S. of equation (1)

$$\frac{1}{r+x}+\frac{1}{r+y}=\frac{r+y+r+x}{(r+x)(r+y)}$$

$$=\frac{2r+x+y}{r^2+r(x+y)+xy}$$

$$=\frac{2r+x+y}{r^2+r(x+y)+r^2} \text{ (from equation (2))}$$

$$=\frac{2r+x+y}{r(2r+x+y)}=\frac{1}{r}$$

$$=\frac{2}{2r}=\text{R.H.S}$$

25. Let S_1 be the family of circles passing through (3,7) and (6,5) and S_2 be the circle $x^2 + y^2 - 4x - 6y - 3 = 0$. Prove that the common chord of every two circles (i.e., one from family S_1 and other S_2) passes through fixed a point and hence find the co-ordinates of that fixed point.

Solution: The equation of the line joining two points A (3, 7) and $B(6, 5)$ is

$$y - 7 = \frac{5-7}{6-3}(x-3)$$

or $3y - 21 = -2x + 6$

or $2x + 3y - 27 = 0$ ...(1)

so the equation of the family of circles passing through $A(3, 7)$ and $B(6, 5)$ is

$(x - 3)(x - 6) + (y - 7)(y - 5) + \lambda(2x + 3y - 27) = 0$

or $S_1 \equiv x^2 + y^2 + x(2\lambda - 9) + y(3\lambda - 12) + (53 - 27\lambda) = 0$ ...(2)

The equation of the given circle is $S_2 = x^2 + y^2 - 4x - 6y - 3 = 0$

So the equation of the common chord of these two circles is $S_1 - S_2 = 0$

or $(2\lambda - 5)x + (3\lambda - 6)y + (56 - 27\lambda) = 0$

or $\lambda(2x + 3y - 27) - (5x + 6y - 56) = 0$

Clearly it is a line passing through the intersection of the lines $2x + 3y - 27 = 0$ and $5x + 6y - 56 = 0$ i.e., (2, 23/3).

26. A circle of radius 'r' is inscribed in a trapezium of area $900\sqrt{2}$; two of whose sides are along the co-ordinate axes and the two parallel sides have slope of '-1'. Find the value of 'r'.

Solution: Area of trapezium $ABCD = 900\sqrt{2}$

or $\frac{1}{2}\times(AD+BC)\times 2r = 900\sqrt{2}$

or $(AD+BC)r = 900\sqrt{2}$ ...(1)

Equation of parallel tangents with slope (-1) are $\Rightarrow y = -x + \lambda$

or $x + y = \lambda$ ...(2)

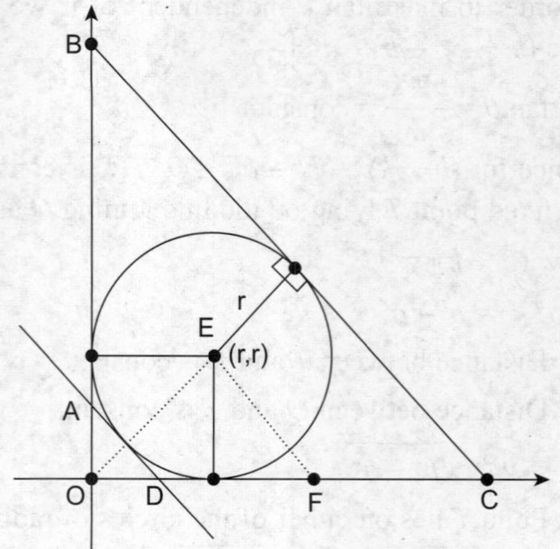

radius of inscribed circle = r

$\therefore$ centre : (r, r)

Now perpendicular distance from centre (r, r) to the tangent $x + y = \lambda$ is $\left|\frac{r+r-\lambda}{\sqrt{1^2+1^2}}\right| = r$ or $2r - \lambda = \pm r\sqrt{2}$

or $\lambda = (2 \pm \sqrt{2})r$ ...(3)

⇒ Equation of AD will be $x + y - (2 - \sqrt{2})r = 0$ and equation of BC will be $x + y - (2 + 2\sqrt{2})r = 0$

⇒ Co-ordinates of points A, B, C, D are

$A[0,(2-\sqrt{2})r]$; $B[0,(2+\sqrt{2})r]$;

$C[(2+\sqrt{2})r,0]$; $D[(2-\sqrt{2})r,0]$

distance, $BC = \sqrt{\{(2+\sqrt{2})r\}^2 + \{(2+\sqrt{2})r\}^2}$

$BC = \sqrt{2}(2+\sqrt{2})r$ ...(4)

Similarly distance, $AD = \sqrt{2}(2-\sqrt{2})r$ ...(5)

Now from equation (1), we get

$(AD + BC)\, r = 900\sqrt{2}$

$(2\sqrt{2}r - 2r + 2\sqrt{2}r + 2r)\, r = 900\sqrt{2}$

or $4\sqrt{2}r^2 = 900\sqrt{2} \Rightarrow r = 15$ units

27. (a) Find the equation of the circle which passes through the points of intersection of circles $x^2 + y^2 - 2x - 6y + 6 = 0$ and $x^2 + y^2 + 2x - 6y + 6 = 0$ and intersects the circle $x^2 + y^2 + 4x + 6y + 4 = 0$ orthogonally.

(b) Tangents TP and TQ are drawn from the point T to the circle $x^2 + y^2 = a^2$. If the point T lies on the line $px + qy = r$, find the locus of centre of the circumcircle of triangle TPQ.

Solution: (a) $S_1 + \lambda(S_1 - S_2) = 0$

$x^2 + y^2 - 2x - 6y + 6 + \lambda(-4x) = 0$

or $x^2 + y^2 - (2 + 4\lambda)x - 6y + 6 = 0$

Now, $2gg' + 2ff' = c + c'$ gives $2(-1 - 2\lambda)(2) + 2(-3)3 - 6 - 4 = 0$

⇒ $-2 - 4\lambda - 9 - 5 = 0$

⇒ $\lambda = -4$

∴ Required circle is $x^2 + y^2 + 14x - 6y + 6 = 0$

(b) Since $\angle P + \angle Q = 90^o + 90^o = 180^o$

∴ $\angle O + \angle T = 180^o$

Hence, points P, O, Q, T are concyclic. The circumcircle of ΔTPQ passes through the centre of given circle and the circumcenter of the ΔTPQ is the mid-point of point of line segment joining 0(0, 0) and $T(2h, 2k)$. Thus, the co-ordinates of the centre of circumcircle will be $S \equiv (h, k)$.

Since point $T\,(2h, 2k)$ lies on the given line, $px + qy = r$.

$T(2h, 2k)$ will satisfy this equation. Therefore, $p(2h) + q(2k) = r$.

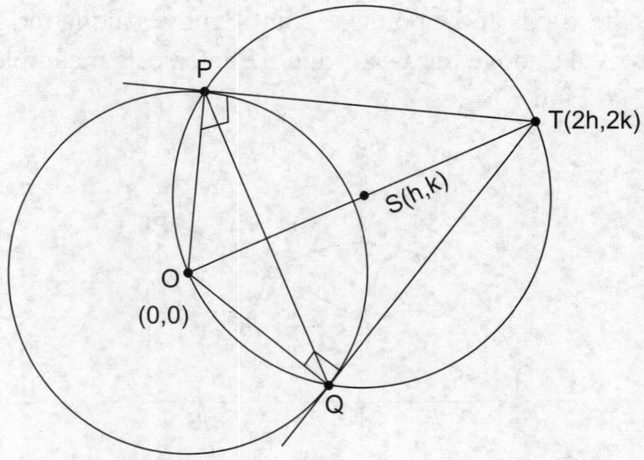

⇒ $2px + 2qy = r$

This is the locus of circumcentre of ΔTPQ.

28. A circle is given by $x^2 + (y - 1)^2 = 1$, another circle C touches it externally and also the x-axis, then the prove that locus of its centre is $\{(x, y): x^2 = 4y\} \cup \{(0, y): y \le 0\}$.

Solution: Let the locus of centre of circle be (h, k) touching $(y - 1)^2 + x^2 = 1$ and x-axis as shown in the figure, below.

⇒ $C_1C_2 = r_1 + r_2$ where $C_1 : (0, 1)$ and $C_2 : (h, k)$

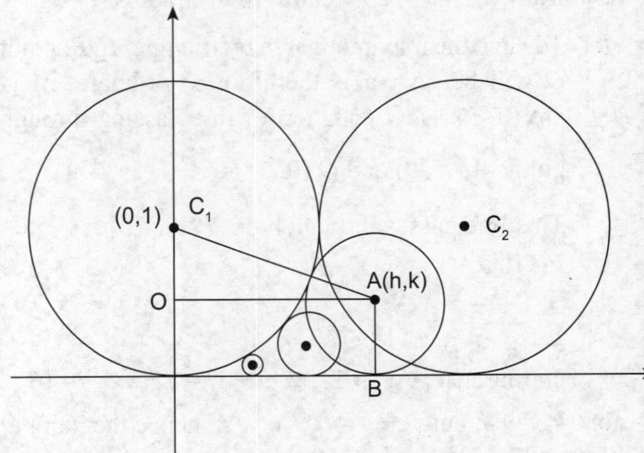

⇒ $r_1 = 1$ and $r_2 = |k|$ (since circle touches x-axis)

Now, $C_1C_2 = r_1 + r_2 = 1 + |k|$

or $h^2 + k^2 - 2k + 1 = 1 + k^2 + 2|k|$

⇒ $h^2 = 2|k| + 2k$

⇒ $x^2 = 2|y| + 2y$

where $|y| = \begin{cases} y, & y \ge 0 \\ -y, & y < 0 \end{cases}$

∴ $x^2 = 2y + 2y, y \ge 0$

and $x^2 = 0$ when $y \le 0$

∴ $\{(x, y): x^2 = 4y$, when $y \ge 0\} \cup \{(0, y); y \le 0\}$.

29. In the given figure, the circle $x^2 + y^2 = 25$ intersects the x-axis at the point A and B. The line $x = 11$ intersects the x-axis at the point C. Point P moves along the line $x = 11$ above the x-axis and AP intersects the circle at Q. Find

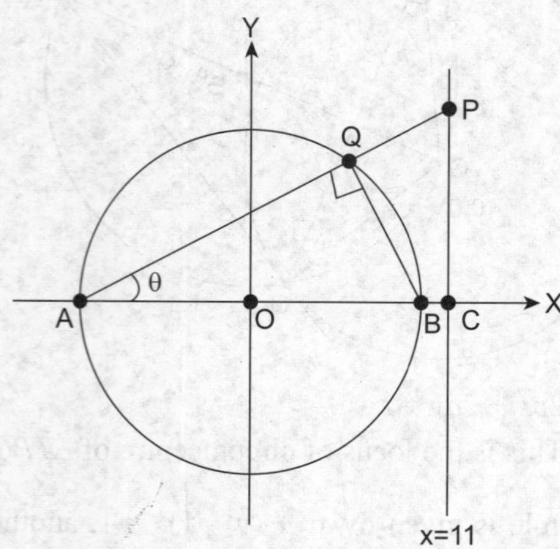

(i) The co-ordinates of the point P if the triangle AQB has the maximum area.

(ii) The co-ordinates of the point P if Q is the middle point of AP.

(iii) The co-ordinates of P if the area of the triangle AQB is $(1/4)^{th}$ of the area of the triangle APC.

Solution: $x^2 + y^2 = 25$; centre: (0, 0); radius (r) = 5

(i) To have the maximum area by triangle AQB, point Q will be on y-axis therefore co-ordinates of Q are (0, 5). Now equation of line passing through points $A(-5, 0)$ and $Q(0, 5)$ is $\frac{x}{-5}+\frac{y}{5}=1$

This line passes through point $P(11, 16)$

$\therefore$ $P(11, 16)$

(ii) $x_Q = \frac{11-5}{2} = 3$

putting in $x^2 + y^2 = 25 \Rightarrow 9 + y^2 = 25 \Rightarrow y^2 = 16$

or $y_Q = \pm 4$; but $y_Q \neq -4$ $\therefore$ $y_Q = 4$, hence the point Q will be (3, 4) or $y_P = 8$

$\Rightarrow x_P = 11$ and $y_P = 8$

$\therefore$ $P(11, 8)$

(iii) Let $\angle PAC = \theta$ and $P(11, \beta)$, then $\tan\theta = \frac{\beta}{16}$

$\Rightarrow \sin\theta = \frac{\beta}{\sqrt{\beta^2+16^2}}$ and $\cos\theta = \frac{16}{\sqrt{\beta^2+16^2}}$

Now, Area of $\Delta AQB = \frac{1}{4} \times$ area of ΔAPC (given)

$\Rightarrow \frac{1}{2} \times AQ \times QB = \frac{1}{4}\times\frac{1}{2}\times AC\times PC$

$\Rightarrow AQ \times QB = \frac{1}{4} \times AC \times PC$

$\Rightarrow AB\cos\theta \times AB\sin\theta = \frac{1}{4} \times 16 \times \beta$

$\Rightarrow 10\times\frac{16}{\sqrt{\beta^2+16^2}}\times 10\times\frac{\beta}{\sqrt{\beta^2+16^2}} = 4\beta$

$\Rightarrow 400 = \beta^2 + 16^2$

$\Rightarrow \beta^2 = 400 - 256$

$\Rightarrow \beta^2 = 144 \Rightarrow \beta = 12$

$\therefore$ P is (11, 12)

30. A point moving on circle $(x + 4)^2 + (y + 2)^2 = 25$ with centre C broke away from it either at the point A or point B on the circle and moved along a tangent to the circle passing through the point $D(3, -3)$. Find the following.

(i) Equation of the tangents at A and B

(ii) Co-ordinates of the points A and B

(iii) Angle ADB and the maximum and minimum distances of the point D from the circle.

(iv) Area of quadrilateral $ADBC$ and the ΔDAB.

(v) Equation of the circle circumscribing the ΔDAB and also the intercept made by this circle on the co-ordinate axes.

Solution: Given circle is $(x + 4)^2 + (y + 2)^2 = 25$ ..(1)

equation of $DA \Rightarrow y + 3 = m(x - 3)$

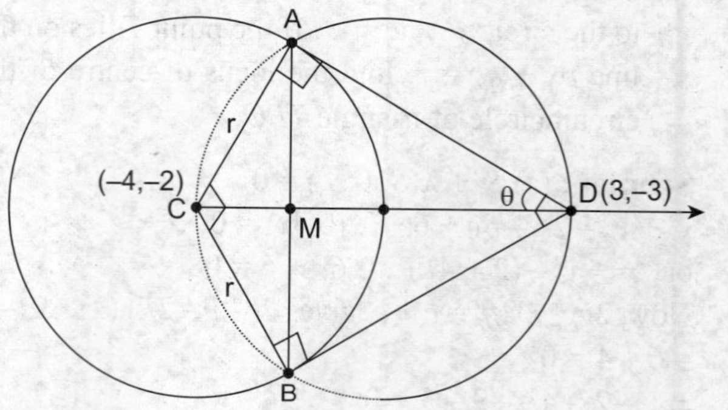

or $y = mx - 3m - 3$ ...(2)

now $p = r$

$\Rightarrow \left|\frac{-4m+2-3m-3}{\sqrt{1+m^2}}\right| = 5$

or $(-7m - 1)^2 = (5)^2 (\sqrt{1+m^2})^2$

or $49m^2 + 1 + 14m = 25(1 + m^2)$

or $24m^2 + 14m - 24 = 0$

or $12m^2 + 7m - 12 = 0$

or $12m^2 + 16m - 9m - 12 = 0$

or $4m(3m + 4) - 3(3m + 4) = 0$

or $m = \frac{3}{4}, -\frac{4}{3}$

(i) 1st tangent $\left(m = \frac{3}{4}\right)$; from equation ..(2)

$\Rightarrow y = \frac{3}{4}x - 3\left(\frac{3}{4}\right) - 3$

$\Rightarrow 3x - 4y - 21 = 0$ ...(3)

2nd tangent $\left(m = -\frac{4}{3}\right)$; from equation (2)

$\Rightarrow y = \left(\frac{-4}{3}\right)x - 3\left(\frac{-4}{3}\right) - 3$

or $3y = -4x + 12 - 9$

or $4x + 3y - 3 = 0$ ...(4)

(ii) perpendicular line to the line (3) be

$4x + 3y + \lambda = 0$ ...(5)

passing through $(-4, -2)$

$\Rightarrow -16 - 6 + \lambda = 0 \Rightarrow \lambda = 22$

putting in equation (5), we have $4x + 3y + 22 = 0$ ...(6)

intersection point of equation (3) and (6) is the point of tangency. Solving equation (3) and (6), we get $x = -1$ and $y = -6$

$\therefore$ A, the point of tangency $\equiv (-1, -6)$

Now perpendicular line to line (4) is

$3x - 4y + \mu = 0$ ...(7)

it passes through $(-4, -2)$ $-12 + 8 + \mu = 0$

$\Rightarrow \mu = 4$

putting in equation (7), we get

$\Rightarrow 3x - 4y + 4 = 0$ ...(8)

now solving equation (4) and (8), we get point of tangency $x = 0$ and $y = 1$

$\Rightarrow$ B, the point of tangency $\equiv (0, 1)$

(iii) maximum distance $= CD + r$

$= \sqrt{(3+4)^2 + (-3+2)^2} + 5 = \sqrt{49+1} + 5$

$= \sqrt{50} + 5 = 5\sqrt{2} + 5 = 5(\sqrt{2}+1)$

minimum distance $= CD - r$

$= 5\sqrt{2} - 5 = 5(\sqrt{2}-1)$

$\sin\theta = \frac{r}{CD} = \frac{5}{5\sqrt{2}} = \frac{1}{\sqrt{2}}$

or $\theta = \frac{\pi}{4}$

$\therefore \angle ADB = 2\theta = 2 \times \frac{\pi}{4} = \frac{\pi}{2}$

Aliter: $\because$ (slope of AD).(slope of BD) $= -1$

$\Rightarrow \angle ADB = \pi/2$

(iv) $AD = \sqrt{S_1} = \sqrt{(3+4)^2 + (-3+2)^2 - 25}$

$= \sqrt{49+1-25} = 5$

Area of quad. $ADBD = 2 \times \left(\frac{1}{2} \times AD \times AC\right)$

$= 2 \times \frac{1}{2} \times 5 \times 5 = 25$ sq. units

Area of $\Delta DAB = \frac{1}{2} \times AD \times BD$

$= \frac{1}{2}(AD)^2 = \frac{25}{2}$ square units

$\because \angle ADB = \pi/2$

$\Rightarrow$ circle through A, D and B would have AB as its diameter

$\therefore$ Equation of circumcircle of ΔABD will be $(x+1)(x-0) + (y+6)(y-1) = 0$ or $x^2 + y^2 + x + 5y - 6 = 0$

(v) $(x+4)(x-3) + (y+2)(y+3) = 0$

or $x^2 + x - 12 + y^2 + 5y + 6 = 0$

or $x^2 + y^2 + x + 5y - 6 = 0$

$g = \frac{1}{2}$, $f = \frac{5}{2}$, $c = -6$

x-intercept $= 2\sqrt{g^2 - c} = 2\sqrt{\frac{1}{4} + 6} = 5$ units

and y-intercept $= 2\sqrt{f^2 - c} = 2\sqrt{\frac{25}{4} + 6} = 7$ units

31. Find the equation of a line with gradient 1 such that the two circles $x^2 + y^2 = 4$ and $x^2 + y^2 - 10x - 14y + 65 = 0$ intercept equal length on it.

Solution: Given circles are $x^2 + y^2 = 4$ ...(1)

$x^2 + y^2 - 10x - 14y + 65 = 0$ ...(2)

$\Rightarrow C_1 \equiv (0, 0)$ and $C_2 \equiv (5, 7)$

and $r_1 = 2$ and $r_2 = 3$

Clearly, $|C_1C_2| = \sqrt{74} > r_1 + r_2$ (=5)

$\Rightarrow$ The two circle would not intersect each other as shown in the figure, below.

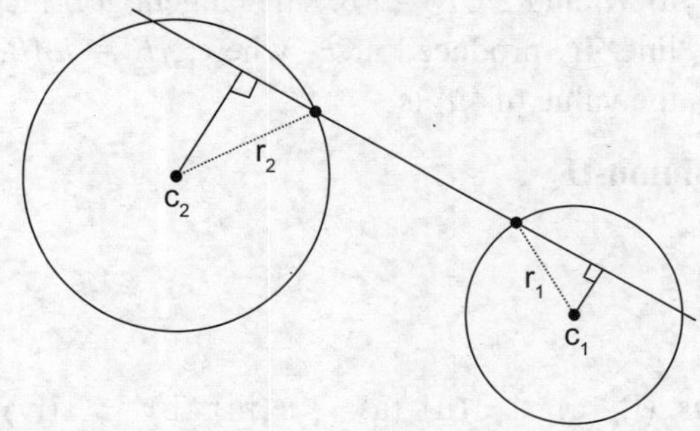

let the equation of line with 1 gradient be

$\Rightarrow\ y = x + \lambda$ ...(3)

now equal interception on line by both circles

$\Rightarrow\ 2\sqrt{r_1^2 - p_1^2} = 2\sqrt{r_2^2 - p_2^2}$

or $p_2^2 - p_1^2 = r_2^2 - r_1^2$

or $\left|\dfrac{5-7+\lambda}{\sqrt{1^2+1^2}}\right|^2 - \left|\dfrac{0-0+\lambda}{\sqrt{1^2+1^2}}\right|^2 = 3^2 - 2^2$

or $\dfrac{\lambda^2+4-4\lambda}{2} - \dfrac{\lambda^2}{2} = 9-4$

or $4 - 4\lambda = 10$

or $4\lambda = -6 \Rightarrow \lambda = -\dfrac{3}{2}$

putting $\lambda = -\dfrac{3}{2}$ in equation (3), we get $x - y + \lambda = 0$

$x - y - \dfrac{3}{2} = 0$

or $2x - 2y - 3 = 0$

Column Matching Type

32. Column-I

(i) If two distinct chords drawn from the point (p, q) on the circle $x^2 + y^2 = px + qy$ where $(pq \neq 0)$ are bisected by the x-axis, then $p^2 > kq^2$. Then the smallest positive integer such value of 'k' is

(ii) Let $ABCD$ be a quadrilateral having area 18 square units with side AB parallel to the side CD and AB = 2 CD. Let AD be perpendicular to AB and CD. If A circle is drawn inside the quadrilateral $ABCD$ touching all the sides, then its radius 'r' is

(iii) If a circle passes through the points of intersection of the co-ordinate axes with the lines $\lambda x - y + 1 = 0$ and $x - 2y + 3 = 0$, then the number of such values of 'λ' is

(iv) AB is a diameter of a circle. CD is a chord parallel to AB and $2\ CD = AB$. The tangent at B meets the line AC produced at E, where $AE = \mu AB$. Then the value of 'μ' is

Column-II

(a) 2

(b) 3

(c) 8

Ans. (i) (c) ; (ii) (a) ; (iii) (b) ; (iv) (a)

32. Solution:

(i) Suppose AB is a chord of the circle through $A(p, q)$ having $M(h,0)$ as its mid-point. Then co-ordinates of B are $(-p + 2h, -q)$

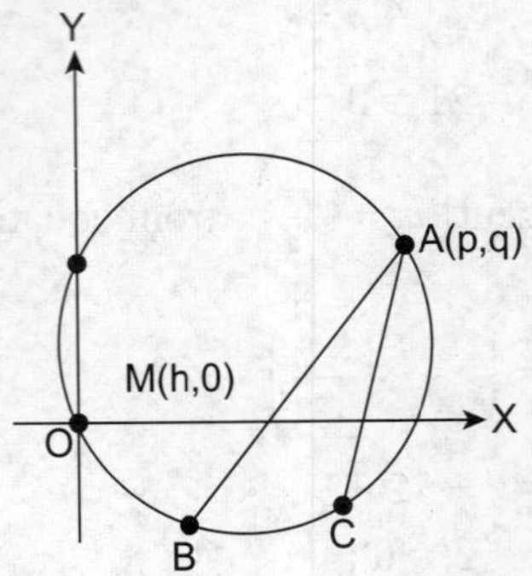

As B lies on the circle

$x^2 + y^2 = px + qy$, we have

$(-p + 2h)^2 + (-q)^2 = p(-p + 2h) + q(-q)$

$\Rightarrow 2p^2 + 2q^2 - 6ph + 4h^2 = 0$

$\Rightarrow 2h^2 - 3ph + p^2 + q^2 = 0$ ..(1)

As there are two distinct chords from $A(p, q)$, which are bisected on the x-axis, there must be two distinct values of h satisfying (1)

$\Rightarrow D = 9p^2 - (4)(2)(p^2 + q^2) > 0$

$\Rightarrow p^2 > 8q^2 \qquad \Rightarrow k = 8$

(ii) Let $CD = a$, $AB = 2a$. If r is the radius of the circle, then $AD = 2r$. Let us take A as the origin, AB as the x-axis and AD as the y-axis. Then co-ordinates $B \equiv (2a,0)$, $D \equiv (0,2r)$, $C \equiv (a,2r)$

Equations of BC is $y - 0 = \dfrac{2r-0}{a-2a}(x - 2a)$

$\Rightarrow 2rx + ay - 4ar = 0$ ....(1)

We are given that area $(ABCD)$ = 18 square units

$\Rightarrow \dfrac{1}{2}(a + 2a)(2r) = 18$

$\Rightarrow ar = 6$

$\therefore$ (1) can be written as $2rx + ay - 24 = 0$

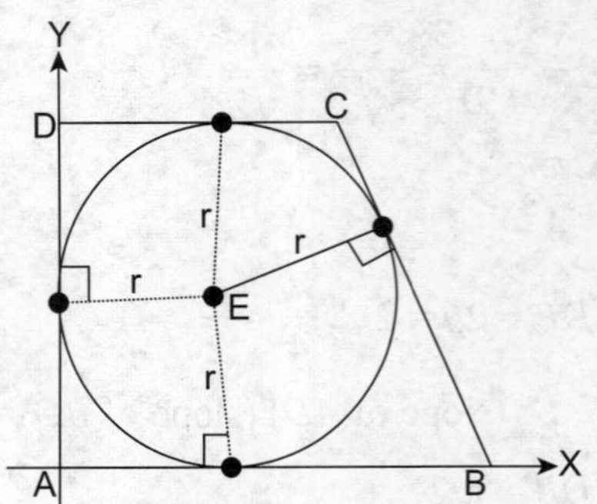

Co-ordinates of the centre of circle are $E(r, r)$,

Therefore $\dfrac{|2r(r)+ar-24|}{\sqrt{4r^2+a^2}}=r$

$\Rightarrow\ (2r^2-18)^2=r^2(4r^2+a^2)$

$\Rightarrow\ 4r^4-72r^2+324=4r^4+36$

$\Rightarrow\ 72r^2=288 \Rightarrow r^2=4 \Rightarrow r=2$

(iii) For $\lambda \neq 0$, the points of intersection of the lines $\lambda x-y+1=0$ and $x-2y+3=0$ with the axes are $(-1/\lambda,0)$, $(0,1)$, $(-3,0)$, $(0,3/2)$

Let the equation of the circle through (0,1), (–3,0) (0, 3/2) be

$x^2+y^2+2gx+2fy+c=0$

Then $1+2f+c=0$ (1)

$9-6g+c=0$ (2)

$9/4+3f+c=0$ (3)

Solving (1) and (2), we get $f=-5/4$, $c=3/2$. Putting $c=3/2$ in (2), we get $g=7/4$. Therefore, equation of the circle through (0,1), (–3,0) and (0,3/2) is

$$x^2+y^2+\frac{7}{2}x-\frac{5}{2}y+\frac{3}{2}=0$$

Now, $(-1/\lambda,0)$ will lie on (4) if

$$\frac{1}{\lambda^2}-\frac{7}{2\lambda}+\frac{3}{2}=0$$

$\Rightarrow\ 3\lambda^2-7\lambda+2=0$

$\Rightarrow\ \lambda=1/3, 2$

For $\lambda=0$ then points of intersections are (0,1), (–3,0), (0,3/2), which clearly lie on a circle. Therefore, the value of λ are 0, 1/3,2, which are 3 in counting.

(iv) Let us take the centre of circle as origin and AB along the x-axis.

Let equation of circle be $x^2+y^2=r^2$ ....(i)

As CD is parallel to AB and $CD=(1/2)AB$, and x-coordinate of C and D are $-r/2$ and $r/2$ respectively. Co-ordinates of C are

$\left(\dfrac{-r}{2}, \dfrac{\sqrt{3}}{2}r\right)$ [$\because$ C lies on (i)]

Equation of AC is

$$\frac{y-0}{\sqrt{3}r/2-0}=\frac{x+r}{-r/2+r}$$

or $y=\sqrt{3}(x+r)$ ....(ii)

Equation of tangent to the circle (1) at $B(r,0)$ is $rx+0y=r^2$ or $x=r$ ....(iii)

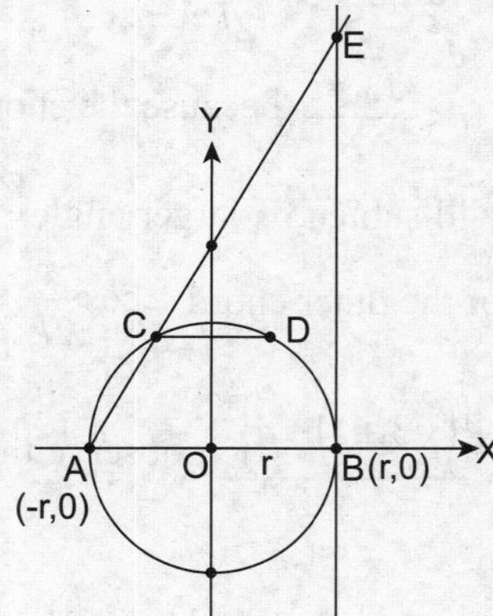

Solving (ii) and (iii) we obtain co-ordinates of E as $(r, 2\sqrt{3}r)$. We have

$$AE^2=(r+r)^2+(2\sqrt{3}r-0)^2=16r^2$$

$\Rightarrow\ AE=4r=2(AB)$

33. Column-I

(i) Two parallel chords of a circle of radius 2 are at a distance $\sqrt{3}+1$ apart. If the chords subtend at the centre, angles of π/k and $2\pi/k$ where $k>0$; then the value of $[k]$ is (where $[k]$ is the greatest integer function of k)

(ii) If $\left(m_i, \dfrac{1}{m_i}\right)$; $m_i>0$, $i=1,2,3,4$ are four distinct points on a circle, then $m_1m_2m_3m_4$ is equal to

(iii) The circle $x^2+y^2-4x-4y+4=0$ is inscribed in a triangle which has two of its sides along the co-ordinate axes. The locus of the circumcentre of the triangle is $x+y-xy+k(x^2+y^2)^{1/2}=0$. Then the value of '$k$' is

(iv) The equation of a circle C which touches the line $2x+3y+1=0$ at the point (1, –1) and is orthogonal to the circle C_1 which has (0, –1) and (–2,3) as the end points of a diameter is given by $\mu(x^2+y^2)-10x-5y+1=0$. Then the value of 'μ' is

Column-II

(a) 2

(b) 1

(c) 3

(d) None of these

Ans. (i) (d) (ii) (b) (iii) (b) (iv) (a)

33. Solution:

(i) Let the chord making larger angle (i.e., $2\pi/k$) be at distance a from the centre, so

$$2\cos\frac{\pi}{k} = a \text{ i.e., } \cos^2\frac{\pi}{2k} = \frac{a+2}{4}$$

Note: $a < \frac{\sqrt{3}+1}{2}$ because the chord nearer to the centre will subtend a larger angle.

Now, for the other chord $2\cos\frac{\pi}{2k} = \sqrt{3}+1-a$, so

$$\frac{a+2}{4} = \frac{\{(\sqrt{3}+1)-a\}^2}{4}$$, observe that $a = 1$ gives

$3 = (\sqrt{3})^2$

∴ the angle subtended by the chord which is at 1 unit distance from centre is 60° = $2\pi/6$ radians and the angle subtended by the chord which is at $\sqrt{3}$ unit distance from centre is 30° = $\pi/6$ radians

∴ $k = 6$

⇒ $[k] = 6$.

(ii) Suppose the four points $(m_i, 1/m_i)$, $i = 1,2,3,4$ lie on the circle

$x^2 + y^2 + 2gx + 2fy + c = 0$

$$\Rightarrow m_i^2 + \frac{1}{m_i^2} + 2gm_i + 2f\left(\frac{1}{m_i}\right) + c = 0\ (i = 1,2,3,4)$$

$$\Rightarrow m_i^4 + 2gm_i^3 + cm_i^2 + 2f\,m_i + 1 = 0\ \ (i = 1,2,3,4)$$

This shows that $m_1 m_2 m_3 m_4$ = product of the roots = 1

(iii) Equation of the given circle is

$x^2 + y^2 - 4x - 4y + 4 = 0$..... (1)

Centre and radius of the circle are respectively C (2,2) and 2.

Let the equation of third side of the triangle be

$\frac{x}{a} + \frac{y}{b} = 1$............. (2)

Since (2) touches the cirlce (1)

$$\left|\frac{\frac{2}{a}+\frac{2}{b}-1}{\sqrt{\left(\frac{1}{a^2}+\frac{1}{b^2}\right)}}\right| = 2 \text{ or } \frac{\frac{2}{a}+\frac{2}{b}-1}{\sqrt{\frac{1}{a^2}+\frac{1}{b^2}}} = -2 \quad (3)$$

[(0,0) and (2,2) lie on the same side of AB]

Since AOB is a right angled triangle, circumcentre of the triangle is the midpoint of the hypotenuse

i.e., $\left(\frac{a}{2}, \frac{b}{2}\right)$. From (3) locus of the circumcentre is

$$\frac{1}{x} + \frac{1}{y} - 1 = -2\sqrt{\left(\frac{1}{4x^2} + \frac{1}{4y^2}\right)}$$

or $x + y - xy + \sqrt{(x^2 + y^2)} = 0$ gives $k = 1$

(iv) Centre C lies on the line through $A(1,-1)$ and perpendicular to $2x + 3y + 1 = 0$, that is, on

$$y + 1 = \frac{3}{2}(x-1) \text{ or } \frac{x-1}{2} = \frac{y+1}{3} = \lambda(\text{say})$$

Let centre of C be $(2\lambda + 1, 3\lambda - 1)$

Equation of C is

$[x - (2\lambda + 1)]^2 + [y - (3\lambda - 1)]^2 = (2\lambda)^2 + (3\lambda)^2$

or $x^2 + y^2 - 2(2\lambda + 1)x - 2(3\lambda - 1)y - 2\lambda + 2 = 0$..(1)

Equation of circle C_1 is

$(x - 0)(x + 2) + (y + 1)(y - 3) = 0$

or $x^2 + y^2 + 2x - 2y - 3 = 0$.......(2)

As (1) and (2) intersect orthogonally,

⇒ $-2(2\lambda + 1)(1) - 2(3\lambda - 1)(-1) = -2\lambda + 2 - 3$

⇒ $l = 3/4$

Thus, equation of required circle is

$x^2 + y^2 - 5x - (5/2)y + 1/2 = 0$

or $2(x^2 + y^2) - 10x - 5y + 1 = 0$, gives $\mu = 2$

LINKED COMPREHENSION-TYPE

A: A circle C of radius 1 is inscribed in an equilateral triangle PQR. The points of contact of C with the side PQ, QR, RP are D, E, F respectively. The line PQ is given by the equation $\sqrt{3}x + y - 6 = 0$ and the point D is $\left(3\sqrt{3}/2, 3/2\right)$. Further it is given that the origin and the centre C are on the same side of PQ.

34. The equation of circle C is

(a) $\left(x - 2\sqrt{3}\right)^2 + (y-1)^2 = 1$

(b) $\left(x - 2\sqrt{3}\right)^2 + (y-1/2)^2 = 1$

(c) $\left(x - \sqrt{3}\right)^2 + (y+1)^2 = 1$

(d) $\left(x - \sqrt{3}\right)^2 + (y-1)^2 = 1$

35. Points E and F are given by

(a) $\left(\sqrt{3}/2, 3/2\right); \left(\sqrt{3}, 0\right)$

(b) $\left(\sqrt{3}/2, 1/2\right); \left(\sqrt{3}, 0\right)$

(c) $\left(\sqrt{3}/2, 3/2\right);\left(\sqrt{3}/2, 1/2\right)$

(d) $\left(3/2, \sqrt{3}/2\right);\left(\sqrt{3}/2, 1/2\right)$

36. Equation of the sides QR, RP are

(a) $y=\left(2/\sqrt{3}\right)x+1, =\left(-2\sqrt{3}\right)x-1$

(b) $y=\left(1/\sqrt{3}\right)x+1,\ y=0$

(c) $y=\left(\sqrt{3}/2\right)x+1,\ y=\left(-\sqrt{3}/2\right)x-1$

(d) $y=\sqrt{3}x,\ y=0$

Ans. (i) (d) (ii) (a) (iii) (d)

Solution:

34. Centre of C lies on the line perpendicular to PQ and passing through D. Thus centre of C lies on

$$y-\frac{3}{2}=\frac{1}{\sqrt{3}}\left(x-\frac{3\sqrt{3}}{2}\right) \text{ or } x=\sqrt{3}\,y$$

Let centre of the circle be $\left(\sqrt{3}y_1, y_1\right)$

Then $\left(\frac{3\sqrt{3}}{2}-\sqrt{3}y_1\right)^2+\left(\frac{3}{2}-y_1\right)^2=1$

$\Rightarrow 4\left(\frac{3}{2}-y_1\right)^2=1 \Rightarrow \frac{3}{2}-y_1=\pm\frac{1}{2}$

$\Rightarrow y_1 = 1,2$

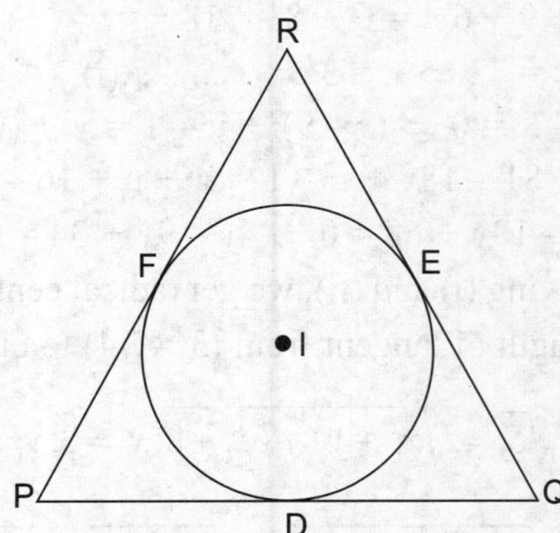

Thus, centre of the circle can be $\left(\sqrt{3},1\right)$ or $\left(2\sqrt{3},2\right)$. Since centre of the circle and the origin lie on the same side of $\sqrt{3}x+y-6=0$ and

$\sqrt{3}(0)+0-6<0 \Rightarrow \sqrt{3}\left(\sqrt{3}\right)+1-6>0$

We get centre of circle to be $I\left(\sqrt{3},1\right)$

$\therefore$ Equation of circle is $\left(x-\sqrt{3}\right)^2+\left(y-1\right)^2=1$

35. If z is affix of E or F then

$$\frac{z-\left(\sqrt{3}+i\right)}{\frac{3}{2}\left(\sqrt{3}+i\right)-\left(\sqrt{3}+i\right)}=\frac{IE}{ID}e^{\pm i2\pi/3}$$

$$\Rightarrow z-\left(\sqrt{3}+i\right)=\frac{1}{2}\left(\sqrt{3}+i\right)\left[\cos\frac{2\pi}{3}\pm i\sin\frac{2\pi}{3}\right]$$

$$=\frac{1}{2}\left(\sqrt{3}+i\right)\left(-\frac{1}{2}\pm\frac{\sqrt{3}}{2}i\right)$$

$$\Rightarrow z=\left(\sqrt{3}+i\right)\left[1-\frac{1}{4}\pm\frac{\sqrt{3}}{4}i\right]$$

$$=\frac{\sqrt{3}}{4}\left(\sqrt{3}+i\right)\left(\sqrt{3}\pm i\right)$$

$$=\sqrt{3}, \frac{\sqrt{3}}{2}\left(1+\sqrt{3}\,i\right)$$

Thus points E and F are given by $\left(\sqrt{3},0\right),\left(\sqrt{3}/2, 3/2\right)$

36. If m is slope of QR, then

$$\pm\tan\left(\frac{\pi}{3}\right)=\frac{m-\left(-\sqrt{3}\right)}{1+m\left(-\sqrt{3}\right)}$$

$$\Rightarrow \pm\sqrt{3}\left(1-m\sqrt{3}\right)=m+\sqrt{3} \Rightarrow m=0, \sqrt{3}$$

$\therefore$ Equations of QR and RP are

$$\Rightarrow y=0,\ y-3/2=\sqrt{3}\left(x-\frac{\sqrt{3}}{2}\right) \text{ or } y=\sqrt{3}x$$

B Three circles $S_1 : \{C_1 = (-3,0);\ r_1 = 1\}$
$S_2 : \{C_2(9,0);\ r_2 = 1\}$
$S_3 : \{C_3(0,4);\ r_3 = 2\}$
where C_i and r_i are the centre and the radius of the circle S_i; $\forall i\in\{1,2,3\}$.
On the basis of the above information, answer the questions that follow:

37. The equation of the circle with minimum radius which circumscribes S_1, S_2 and S_3 is

(a) $x^2 + y^2 - 10x + 72 = 0$

(b) $x^2 + y^2 - 4x + 49 = 0$

(c) $x^2 + y^2 - 6x - 40 = 0$\

(d) None of these

38. The equation of the circle with the shortest radius which touches at-least two of the three circle S_1, S_2 and S_3 will be

(a) $24x^2 + 24y^2 - 85x - 95y + 117 = 0$
(b) $25x^2 + 25y^2 - 90x - 80y + 108 = 0$
(c) $25x^2 + 25y^2 - 80x - 90y + 108 = 0$
(d) None of these

39. The equation of the circle which cuts all the three circles S_1, S_2 and S_3 orthogonally is
(a) $2x^2 + 2y^2 - 12x + 7y - 52 = 0$
(b) $2x^2 + 2y^2 - 12x + 7y + 24 = 0$
(c) $2x^2 + 2y^2 - 12x + 7y + 21 = 0$
(d) None of these

Ans. 1. (c) ; 2 (d) ; 3. (a)

Solution:

37. Mid-point of line segment joining $p_1(-4,0)$ and $P_2(10,0)$ is $M(3,0)$ and $|MP_1| = |MP_2| = 7$

$|MC_3| = \sqrt{3^2 + 4^2} = 5$

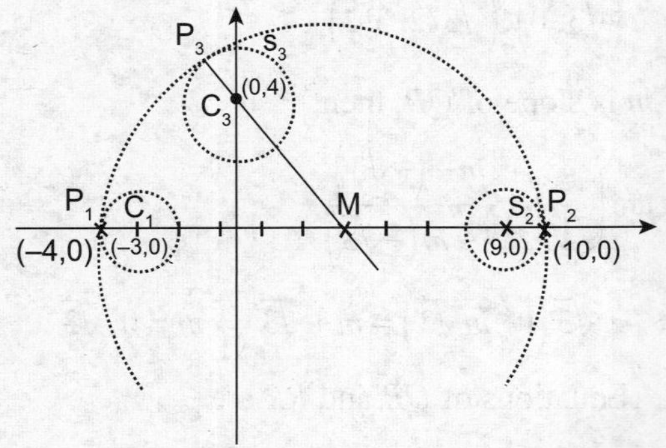

Extending MC_3 to meet S_3 at P_3

$\Rightarrow |MP_3| = 7 \Rightarrow |MP_1| = |MP_2| = |MP_3| = 7$

$\Rightarrow$ M must be the centre of the required circle with radius = 7 units

$\Rightarrow$ equations $(x - 3)^2 + y^2 = 7^2$

$\Rightarrow x^2 + 9 - 6x + y^2 = 49$

$\Rightarrow x^2 + y^2 - 6x - 40 = 0$

38. Obviously, the required such circle will only touch S_1 and S_3 but not S_2. Line segment joining C_1 and C_3 cuts circles S_1 and S_3 at point P_4 and P_5. Then the radius of the required circle will be the circle with diametric ends at P_4 and P_5.

Line joining C_1 C_3 is $\dfrac{x}{-3} + \dfrac{y}{4} = 1$

$\Rightarrow y = \dfrac{4(x+3)}{3}$ or $x = \dfrac{3(y-4)}{4}$ ......(i)

Solving ......(1) and S_1; we get $P_4\left(\dfrac{-12}{5}, \dfrac{4}{5}\right)$

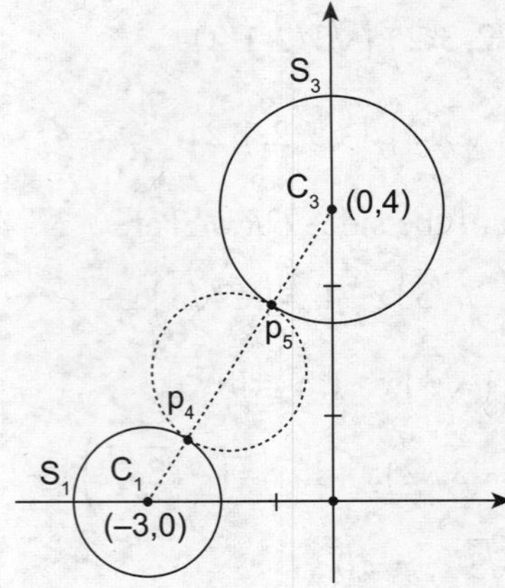

Solving .....(2) and S_3; we get P_5: $\left(\dfrac{-6}{5}, \dfrac{12}{5}\right)$

$\therefore$ Equations of required circle is

$$\left(x + \frac{12}{5}\right)\left(x + \frac{6}{5}\right) + \left(y - \frac{4}{5}\right)\left(y - \frac{12}{5}\right) = 0$$

$$\Rightarrow x^2 + \frac{18x}{5} + \frac{72}{25} + y^2 - \frac{16}{5}y + \frac{48}{25} = 0$$

$\Rightarrow 25x^2 + 90x + 25y^2 - 80y + 120 = 0$

39. The circle which cuts the three given circles orthogonally has its centre on the radical centre of the three given circles and whose radius is the length of tangent drawn from radical centre of the three given circles to any of these circles.

$S_1 - S_2 = 0 \Rightarrow (x + 3)^2 + y^2 - 1 = (x - 9)^2 - y^2 - 1$

$\Rightarrow x^2 + 9 + 6x = x^2 + 81 - 18x$

$\Rightarrow 24x = 72 \Rightarrow x = 3$ ....(i)

and $S_2 - S_3 = 0 \Rightarrow (x - 9)^2 + y^2 - 1 = x^2 + (y - 4)^2 - 4$

$\Rightarrow x^2 + 81 - 18x + y^2 - 1 = x^2 + y^2 + 16 - 8y - 4$

$\Rightarrow 8y - 18x + 68 = 0 \Rightarrow 4y - 9x + 34 = 0$ ....(ii)

Solving (i) and (ii), we get radical centre: $(3, -7/4)$

Length of tangent from $(3, -7/4)$ to circle S_1

$$= \sqrt{S_1} = \sqrt{x_1^2 + y_1^2 + 6x_1 + 8}; x_1 = 3, y_1 = -\frac{7}{4}$$

$$= \sqrt{3^2 + \frac{49}{16} + 18 + 8} = \sqrt{\frac{49}{16} + 35} = \frac{\sqrt{609}}{4}$$

$$\Rightarrow \text{Equations: } (x-3)^2 + \left(y + \frac{7}{4}\right)^2 = \left(\frac{\sqrt{609}}{4}\right)^2$$

$$\Rightarrow x^2 + 9 - 6x + y^2 + \frac{49}{16} + \frac{7}{2}y = \frac{609}{16}$$

$$\Rightarrow x^2 + y^2 - 6x + \frac{7}{2}y - 26 = 0$$

$\Rightarrow 2x^2 + 2y^2 - 12x + 7y - 52 = 0$

TUTORIAL EXERCISE

SECTION–III

OBJECTIVE-TYPE (ONLY ONE CORRECT ANSWER)

1. If a circle and a square have the same perimeter, then
(a) Their area are equal
(b) Area of circle is larger
(c) Area of square is larger
(d) None of these

2. Number of integral values of λ for which $x^2 + y^2 + 7x + (1 - \lambda) y + 5 = 0$ is the equation of circle whose radius cannot exceed 5, is
(a) 14 (b) 18
(c) 16 (d) None of these

3. Number of the points on the circle $2x^2 + 2y^2 - 3x = 0$ which are at a distance 2 from the point $(-2, 1)$ is
(a) 2 (b) 0
(c) 1 (d) None of these

4. The locus of the point of intersection of the tangents to the circle $x = r \cos \theta$, $y = r \sin \theta$ at the points whose parametric angles differ by a right angle is
(a) $x^2 + y^2 = \frac{r^2}{2}$ (b) $x^2 + y^2 = 2r^2$
(c) $x^2 + y^2 = 9r^2$ (d) None of these

5. The equations of the four circles are $(x \pm a)^2 + (y \pm a)^2 = a^2$ the radius of a circle touching all the four is
(a) $(\sqrt{2} \pm 1)a$ (b) $2\sqrt{2a}$
(c) $(2 + \sqrt{2})\, a$ (d) None of these

6. C_1 is the circle of radius 1 touching the x-axis and the y-axis. C_2 is another circle of radius > 1 and touching the axes as well as C_1, then the radius of C_2 is
(a) $3 - 2\sqrt{2}$ (b) $3 + 2\sqrt{2}$
(c) $3 + 2\sqrt{3}$ (d) None of these

7. The locus of the mid-points of the chords of the circle, $x^2 + y^2 + 4x - 6y - 12 = 0$ which subtends an angle of $\pi/3$ radians at its centre is
(a) $(x + 2)^2 + (y - 3)^2 = 6.25$
(b) $(x - 2)^2 + (y + 3)^2 = 6.25$
(c) $(x + 2)^2 + (y - 3)^2 = 18.75$
(d) $(x + 2)^2 + (y + 3)^2 = 18.75$

8. For a circle $x^2 + y^2 + 2gx + c = 0$, to have limiting points:
(a) $g = c$ (b) $g = c^2$
(c) $g = \pm \sqrt{c}$ (d) None of these

9. If AB is a diameter of circle and C is any point on the circumference then
(a) the area of ΔABC is maximum when it is isosceles
(b) the area of ΔABC is minimum when it is isosceles
(c) the perimeter of ΔABC is maximum when it is isosceles
(d) None of these

10. The equation of the image of the circle $(x - 3)^2 + (y - 2)^2 = 1$ by the line mirror $x + y = 19$, is
(a) $(x + 17)^2 + (y - 16)^2 = 1$
(b) $(x - 16)^2 + (y - 17)^2 = 1$
(c) $(x - 17)^2 + (y - 16)^2 = 1$
(d) $(x - 16)^2 + (y - 16)^2 = 1$

11. If a line segment $AM = a$ moves in the plane XOY remaining parallel to OX so that the left end point A slides along the circle $x^2 + y^2 = a^2$, the locus of M is
(a) $x^2 + y^2 = 4a^2$
(b) $x^2 + y^2 = 2ax$
(c) $x^2 + y^2 = 2ay$
(d) $x^2 + y^2 - 2ax - 2ay = 0$

12. A stick of length 10 units rests against the floor and a wall of a room. If the stick begins to slide on the floor then the locus of its middle point is
(a) $x^2 + y^2 = 2.5$ (b) $x^2 + y^2 = 25$
(c) $x^2 + y^2 = 100$ (d) None of these

13. The shortest distance from the point $(2, -7)$ to the circle $x^2 + y^2 - 14x - 10y - 151 = 0$ is.
(a) 1 (b) 2
(c) 3 (d) 4

14. The radical centre of three circles taken in pairs described on the sides of a triangle ABC as diameters is the:
(a) centroid of the $\Delta\ ABC$
(b) incentre of the $\Delta\ ABC$
(c) circumcentre of the $\Delta\ ABC$
(d) orthocentre of the $\Delta\ ABC$

15. The equation of a circle is $x^2 + y^2 = 25$. The equation of its chord whose middle point is (1, –2) is given by:
(a) $2x - y - 4 = 0$ (b) $5x + y - 3 = 0$
(c) $x - 2y - 5 = 0$ (d) $x + y + 1 = 0$

16. The locus of middle points of the chords of the circle, $x^2 + y^2 = a^2$, if length 'a' is:
(a) $x^2 + y^2 = \frac{a^2}{4}$ (b) $x^2 + y^2 = \frac{a^2}{2}$
(c) $x^2 + y^2 = \frac{3a^2}{4}$ (d) None of these

17. The range of values of 'a' such that the angle θ between the pair of tangents drawn from the point (a, 0) to the circle $x^2 + y^2 = 1$ satisfies $\frac{\pi}{2} < \theta < \pi$ is
(a) (1, 2) (b) $(1, \sqrt{2})$
(c) $(-\sqrt{2}, -1)$ (d) $(-\sqrt{2}, -1) \cup (1, \sqrt{2})$

18. Locus of the centre of the circle touching both the co-ordinates axes is
(a) $x^2 + y^2 = 0$
(b) $x^2 + y^2 = a$ non-zero constant
(c) $x^2 - y^2 = 0$
(d) $x^2 - y^2 = a$ non-zero constant

19. A line is drawn through a fixed point $P(\alpha, \beta)$ to cut the circle $x^2 + y^2 = r^2$ at A and B. then PA. PB is equal to
(a) $(\alpha + \beta)^2 - r^2$ (b) $\alpha^2 + \beta^2 - r^2$
(c) $(\alpha - \beta)^2 + r^2$ (d) None of these

20. The circles whose equations are $x^2 + y^2 - 2ax + c^2 = 0$ and $x^2 + y^2 - 2by + c^2 = 0$ will touch one another externally if
(a) $1/b^2 + 1/c^2 = 1/a^2$ (b) $1/c^2 + 1/a^2 = 1/b^2$
(c) $1/a^2 + 1/b^2 = 1/c^2$ (d) None of these

21. A line meets the co-ordinate axes in A and B. A circle is circumscribed about the ΔOAB. If m *and* n are the distances of the tangents to the circle at the origin from the point A and B respectively. The diameter of the circle is
(a) $m(m + n)$ (b) $m + n$
(c) $(m + n)n$ (d) $1/2\ (m + n)$

22. The locus of the centre of the circle which touches the circles $x^2 + y^2 = a^2$ and $x^2 + y^2 = 4ax$ externally is
(a) $12(x - a)^2 - 4y^2 = 3a^2$
(b) $9(x - a)^2 - 5y^2 = 2a^2$
(c) $8x^2 - 3(y - a)^2 = 9a^2$
(d) None of these

23. Equation of the circle which passes through origin, has its centre on the line $x + y = 4$ and cuts the circle $x^2 + y^2 - 4x + 2y + 4 = 0$ orthogonally, is
(a) $x^2 + y^2 - 2x - 6y = 0$
(b) $x^2 + y^2 - 6x - 3y = 0$
(c) $x^2 + y^2 - 4x - 4y = 0$
(d) None of these

24. The intercept on the line $y = x$ by the circle $x^2 + y^2 - 2x = 0$ is AB. Equation of circle with AB as diameter is
(a) $x^2 + y^2 + x + y = 0$ (b) $x^2 + y^2 - x - y = 0$
(c) $x^2 + y^2 + x - y = 0$ (d) $x^2 + y^2 - x + y = 0$

25. If OA and OB are the tangents from the origin to the circle $x^2 + y^2 + 2gx + 2fx + c = 0$ and C is the centre of the circle, the area of the quadrilateral $OACB$ is
(a) $\frac{1}{2}\sqrt{c(g^2 + f^2) - c}$ (b) $\sqrt{c(g^2 + f^2 - c)}$
(c) $c\sqrt{(g^2 + f^2 - c)}$ (d) $\frac{\sqrt{g^2 + f^2 - c}}{c}$

26. A tangent drawn from the point (4, 0) to the circle $x^2 + y^2 = 8$ touches it at a point A in the first quadrant. The co-ordinates of another point B on the circle such that $(AB) = 4$ are:
(a) (2, –2) (b) (–2, –2)
(c) $(-2\sqrt{2}, 0)$ (d) $(0, -2\sqrt{2})$

27. The equation of a circle which passes through ($2a$, 0) and whose radical axis in relation to the circle $x^2 + y^2 = a^2$ is $x = a/2$ is
(a) $x^2 + y^2 - ax = 0$ (b) $x^2 + y^2 + 2ax = 0$
(c) $x^2 + y^2 - 2ax = 0$ (d) $x^2 + y^2 + ax = 0$

28. The locus of the centre of a circle which touches externally the circle $x^2 + y^2 - 6x - 6y + 14 = 0$ and also touches the y-axis is given by the equation
(a) $x^2 - 6x - 10y + 14 = 0$
(b) $x^2 - 10x - 6y + 14 = 0$
(c) $y^2 - 6x - 10y + 14 = 0$
(d) $y^2 - 10x - 6y + 14 = 0$

29. *A* circle passes through the origin and has its centre on $y = x$. If it cuts $x^2 + y^2 - 4x - 6y + 10 = 0$ orthogonally, then the equation of circle is
(a) $x^2 + y^2 - x - y = 0$ (b) $x^2 + y^2 - 6x - 4y = 0$
(c) $x^2 + y^2 - 2x - 2y = 0$ (d) $x^2 + y^2 + 2x + 2y = 0$

30. The circle for which the line joining the points (am^2, $2am$) and (a/m^2, $-2a/m$) is a diameter, is touched, for all values of m, by the line

(a) $x = a$ (b) $x + a = 0$
(c) $x = 2a$ (d) $x + 2a = 0$

31. The equation of the circle having its centre on the line $x + 2y - 3 = 0$ and passing through the point of intersection of the circles $x^2 + y^2 - 2x - 4y + 1 = 0$ and $x^2 + y^2 - 4x - 2y + 1 = 0$ is
(a) $x^2 + y^2 - 6x + 1 = 0$
(b) $x^2 + y^2 - 3x + 4 = 0$
(c) $x^2 + y^2 - 2x - 2y + 1 = 0$
(d) $x^2 + y^2 + 2x - 4y + 7 = 0$

32. The equation of a circle is $x^2 + y^2 = 4$. The centre of the smallest circle touching this circle and the line $x + y = 5\sqrt{2}$ has the co-ordinates
(a) $\left(\frac{7}{2\sqrt{2}}, \frac{7}{2\sqrt{2}}\right)$ (b) $\left(\frac{3}{2}, \frac{3}{2}\right)$
(c) $\left(-\frac{7}{2\sqrt{2}}, -\frac{7}{2}\right)$ (d) None of these

33. The locus of the point of intersection of the lines $x = a\left(\frac{1-t^2}{1+t^2}\right)$ and $y = \frac{2at}{1+t^2}$ represents (t being a parameter)
(a) circle (b) parabola
(c) ellipse (d) hyperbola

34. The area of the triangle formed by the tangents from the points (h, k) to the circle $x^2 + y^2 = a^2$ and the line joining their points of contact is
(a) $a\frac{(h^2+k^2-a^2)^{3/2}}{h^2+k^2}$ (b) $a\frac{(h^2+k^2-a^2)^{1/2}}{h^2+k^2}$
(c) $\frac{(h^2+k^2-a^2)^{3/2}}{h^2+k^2}$ (d) $\frac{(h^2+k^2-a^2)^{1/2}}{h^2+k^2}$

35. From the point $A(0, 3)$ on the circle $x^2 + 4x + (y-3)^2 = 0$ a chord AB is drawn and extended to a point M such that $AM = 2AB$. The equation of the locus of M is:
(a) $x^2 + 8x + y^2 = 0$ (b) $x^2 + 8x + (y-3)^2 = 0$
(c) $(x-3)^2 + 8x + y^2 = 0$ (d) $x^2 + 8x + 8y^2 = 0$

36. If $a > 2b > 0$ then the positive value of m for which $y = mx - b\sqrt{1+m^2}$ is a common tangent to $x^2 + y^2 = b^2$ and $(x - a)^2 + y^2 = b^2$, is
(a) $\frac{2b}{\sqrt{a^2-4b^2}}$ (b) $\frac{\sqrt{a^2-4b^2}}{2b}$
(c) $\frac{2b}{a-2b}$ (d) $\frac{b}{a-2b}$

37. c_1 is a fixed circle and c_2 is a variable circle with fixed radius. The common transverse tangents to c_1 and c_2 are perpendicular to each other. The locus of the centre of variable circle is:
(a) circle (b) ellipse
(c) hyperbola (d) parabola

38. A circle passing throgh the origin and cutting off equal chords of length $\sqrt{2}$ from the curve $y = |x|$, is
(a) $x^2 + y^2 + 2y = 0$ (b) $x^2 + y^2 - 2x = 0$
(c) $x^2 + y^2 - 2y = 0$ (d) $x^2 + y^2 + 2x = 0$

39. Two distinct chords drawn from the point (p, q) on the circle $x^2 + y^2 = px + qy$ where $pq \neq 0$ are bisected by the x-axis then
(a) $|p| = |q|$ (b) $p^2 = 8q^2$
(c) $p^2 < 8q^2$ (d) $p^2 > 8q^2$

40. If θ and ϕ are the inclinations of the tangents drawn from point P to the circle $x^2 + y^2 = a^2$ such that $\tan\theta.\tan\phi$ = constant (K), then locus of P is:
(a) $K(x^2 - a^2) = 2xy$ (b) $K(x^2 - a^2) = y^2 - a^2$
(c) $K(y^2 - a^2) = 2xy$ (d) None of these

41. If from any point P on the circle $x^2 + y^2 + 2gx + 2fy + c = 0$, tangents are drawn to the circle, $x^2 + y^2 + 2gx + 2fy + c\sin^2\alpha + (g^2 + f^2)\cos^2\alpha = 0$, then the angle between the tangents is
(a) α (b) 2α
(c) $\alpha/2$ (d) None of these

42. Two equal circles with their centres on x and y axes will posses the radical axis in the following form:
(a) $ax - by - \frac{a^2+b^2}{4} = 0$
(b) $2gx - 2fy + f^2 - g^2 = 0$
(c) $g^2x + f^2y - g^4 - f^4 = 0$
(d) $2g^2x + 2f^2y - g^4 - f^4 = 0$

43. If one circle of a co-axal system is $x^2 + y^2 + 2gx + 2fy + c = 0$ and one limiting point is (a, b) then equation of the radical axis will be
(a) $(g + a)x + (f + b)y + c - a^2 - b^2 = 0$
(b) $2(g + a)x + 2(f + b)y + c - a^2 - b^2 = 0$
(c) $2gx + 2fy + c - a^2 - b^2 = 0$
(d) None of these

44. If two lines $a_1x + b_1y + c_1 = 0$ and $a_2x + b_2y + c_2 = 0$ cut the co-ordinate axes in concyclic points then:
(a) $a_1a_2 = b_1b_2$ (b) $\frac{a_1}{b_1} = \frac{a_2}{b_2}$
(c) $a_1b_1 = a_2b_2$ (d) None of these

45. Equation of the normal to the circle $x^2 + y^2 - 2ax = 0$ at the point $\{a(1 + \cos\theta), a \sin\theta\}$ is given by

(a) $y = (x - a) \tan\theta$

(b) $y = (x + a) \cot\theta$

(c) $y = x\tan\theta + a\cot\theta$

(d) None of these

46. The area of a circle is A_1 and the area of a regular pentagon inscribed in the circle is A_2. Then $A_1 : A_2$ is

(a) $\frac{\pi}{5}\text{cosec}\frac{\pi}{10}$ (b) $\frac{\pi}{5}\sec\frac{\pi}{10}$

(c) $\frac{2\pi}{5}\sec\frac{\pi}{10}$ (d) None of these

47. A variable circle always touches the line $y = x$ and passes through the point (0, 0). The common chords of above circle and $x^2 + y^2 + 6x + 8y - 7 = 0$ will pass through a fixed point, whose co-ordinate are

(a) (1/4, 1/4) (b) (1,1)

(c) (–1/2, –1/2) (d) (1/2, 1/2)

48. The locus of centres of all circle passing through (1, 2) and cutting $x^2 + y^2 = 16$ orthogonally is

(a) $4x - 2y + 21 = 0$

(b) $x + 2y + 11 = 0$

(c) $2x + 4y - 21 = 0$

(d) $2x - y + 11 = 0$

49. The equation of the circumcircle of an equilateral triangle is $x^2 + y^2 + 2gx + 2fy + c = 0$ and one vertex of the triangle is (1, 1). The equation of incircle of the triangle is

(a) $4(x^2 + y^2) = g^2 + f^2$

(b) $4(x^2 + y^2) + 8gx + 8fy$
$= (1 - g)(1 + 3g) + (1 + 3f)(1 - f)$

(c) $4(x^2 + y^2) + 8gx + 8fy = g^2 + f^2$

(d) None of these

50. Two circles are drawn through the points (1, 0) and (2, – 1) to touch the axis of *y*. They intersect at an angle:

(a) $\frac{\pi}{3}$ (b) $\cos^{-1}\frac{4}{5}$

(c) $\frac{\pi}{2}$ (d) $\tan^{-1} 1$

51. Equation of a circle through the origin and belonging to the co-axial system, of which the limiting points are (1, 2), (4, 3) is

(a) $x^2 + y^2 - 2x + 4y = 0$

(b) $x^2 + y^2 - 8x - 6y = 0$

(c) $2x^2 + 2y^2 - x - 7y = 0$

(d) $x^2 + y^2 - 6x - 10y = 0$

52. If (2, 1) is a limiting point of a co-axial system of circles containing $x^2 + y^2 - 6x - 4y - 3 = 0$, then the other limiting point is

(a) (2, 4) (b) (–5, –6)

(c) (3, 5) (d) (–2, 4)

53. If the circle $C_1 : x^2 + y^2 = 16$ intersects another circle C_2 of radius 5 in such a manner that the common chord is of maximum length and has a slope equal to 3/4, then the co-ordinates of the centre of C_2 are

(a) $\left(\pm\frac{9}{5}, \pm\frac{12}{5}\right)$ (b) $\left(\pm\frac{9}{5}, \mp\frac{12}{5}\right)$

(c) $\left(\pm\frac{12}{5}, \pm\frac{9}{5}\right)$ (d) $\left(\pm\frac{12}{5}, \mp\frac{9}{5}\right)$

54. A circle is inscribed into a rhombus *ABCD* with one angle 60°. The distance from the centre of the circle to the nearest vertex is equal to 1. If *P* is any point of the circle, then $|PA|^2 + |PB|^2 + |PC|^2 + |PD|^2$ is equal to

(a) 9 (b) 13

(c) 0 (d) 11

55. If two chords of the circle $x^2 + y^2 - ax - by = 0$, drawn from the point (*a*, *b*) is divided by the *x*-axis in the ratio 2 : 1 then

(a) $a^2 > 3b^2$ (b) $a^2 < 3b^2$

(c) $a^2 > 4b^2$ (d) $a^2 < 4b^2$

56. The locus of a point which moves such that the sum of the squares of its distance from the three vertices of a triangle is constant, is a circle whose centre is at the

(a) Incentre of the triangle

(b) Centroid of the triangle

(c) Orthocentre of the triangle

(d) None of these

57. For the equation $x^2 + y^2 + 2l\,x + 4 = 0$ which of the following statement is true?

(a) It represents a real circle for all l∈*R*.

(b) It represents a real circle for all $|\lambda| < 2$.

(c) The radical axis of any two circle of family is y-axis.

(d) The radical axis of any two circle of family is x-axis.

SECTION–IV

MORE THAN ONE CORRECT ANSWER

1. The locus of the point of intersection of the tangents at the extremities of the chord of the circle $x^2 + y^2 = a^2$ which touches the circle$x^2 + y^2 - 2ax = 0$ is

(a) $y^2 = a(a - 2x)$ (b) $x^2 = a(a - 2y)$

(c) $x^2 + y^2 = (x - a)^2$ (d) $x^2 + y^2 = (y - a)^2$

2. $\dfrac{x - x_1}{\cos\theta} = \dfrac{y - y_1}{\sin\theta} = r$, represents:

(a) equation of a straight line, if θ is constant and r is variable

(b) equation of a circle, if r is constant and θ is a variable

(c) a straight line passing through a fixed point and having a known slope

(d) a circle with a known centre and a given radius .

3. Line $\dfrac{x}{a} + \dfrac{y}{b} = 1$ cuts the co-ordinate axes at $A(a, 0)$ and $B(0, b)$ and the line $\dfrac{x}{a'} + \dfrac{y}{b'} = -1$ at $A'(-a', 0)$ and $B'(0, -b')$. If the points A, B, A', B' are concyclic then the orthocentre of the $\Delta ABA'$ is:

(a) (0, 0) (b) (0, –b')

(c) $\left(0, -\dfrac{aa'}{b}\right)$ (d) $\left(0, \dfrac{bb'}{a}\right)$

4. If $al^2 - bm^2 + 2dl + 1 = 0$, where a, b, d are fixed real numbers such that $a + b = d^2$ then the line $lx + my + 1 = 0$ touches a fixed circle

(a) which cuts the x-axis orthogonally

(b) with radius equal to b

(c) on which the length of the tangent from the origin is $\sqrt{d^2 - b}$

(d) None of these

5. The equation of circle which touches the line $2x - y = 1$ at (1, 1) and also touches the line $2x + y = 4$ is

(a) $x^2 + y^2 - 3y + 1 = 0$

(b) $x^2 + y^2 + 3y + 1 = 0$

(c) $4(x^2 + y^2) + 10x + 7y - 9 = 0$

(d) $4(x^2 + y^2) - 10x - 7y + 9 = 0$

6. The equation of a circle C_1 is $x^2 + y^2 = 4$. The locus of the intersection of orthogonal tangents to the circle is the curve C_2 and the locus of the intersection of perpendicular tangents to the curve C_2 is the curve C_3. Then

(a) C_3 is a circle

(b) The area enclosed by the curve C_3 is 8π

(c) C_2 and C_3 are circles with the same centre

(d) None of these

7. If $U_r = a_r x + b_r y + c_r = 0$, $r = 1, 2, 3$ are sides of a triangle ABC, then

$$\begin{vmatrix} a_1^2 + b_1^2 & a_1 u_1 & b_1 u_1 \\ a_2^2 + b_2^2 & a_2 u_2 & b_2 u_2 \\ a_3^2 + b_3^2 & a_3 u_3 & b_3 u_3 \end{vmatrix} = 0 \text{ represents}$$

(a) a pair of straight lines

(b) a circle

(c) a parabola circumscribing the ΔABC

(d) circumcircle of ΔABC

8. From the point $A(0, 3)$ on the circle $x^2 + 4x + (y - 3)^2 = 0$, a chord AB is drawn and extended to a point M, such that $AM = 2AB$. An equation of the locus of M is

(a) $x^2 + 6x + (y - 2)^2 = 0$

(b) $x^2 + 8x + (y - 3)^2 = 0$

(c) $x^2 + y^2 + 8x - 6y - 9 = 0$

(d) $x^2 + y^2 + 6x - 4y + 4 = 0$

9. Consider the circles $x^2 + y^2 = 1$ and $x^2 + y^2 - 2x - 6y + 6 = 0$. Then equation of a common tangent to the two circles is:

(a) $4x - 3y - 5 = 0$ (b) $x + 1 = 0$

(c) $3x + 4y - 5 = 0$ (d) $y - 1 = 0$

10. A circle passes through the points (– 1, 1), (0, 6) and (5, 5). The point(s) on this circle, the tangent(s) at which is/are parallel to the straight line joining the origin to its centre is/are:

(a) (1, – 5) (b) (5, 1)

(c) (–5, – 1) (d) (– 1, 5)

11. Equation of a chord joining two points ($a\cos\alpha$, $a\sin\alpha$) and ($a\cos\beta$, $a\sin\beta$) can be

(a) $(\cos\alpha + \cos\beta)x + (\sin\alpha + \sin\beta)y - a(1 + \cos(\alpha - \beta)) = 0$

(b) $(\sin\alpha - \sin\beta)x - (\cos\alpha - \cos\beta)y + a\sin(\alpha + \beta) = 0$

(c) $x\cos\dfrac{\alpha+\beta}{2} + y\sin\dfrac{\alpha+\beta}{2} - a\cos\dfrac{\alpha-\beta}{2} = 0$

(d) $\dfrac{x - a\cos\alpha}{y - a\sin\alpha} = \dfrac{\cos\alpha - \cos\beta}{\sin\alpha - \sin\beta}$

12. The circles $x^2 + y^2 + 2x + 4y - 20 = 0$ and $x^2 + y^2 + 6x - 8y + 10 = 0$
(a) are such that the number of common tangents on them is 2
(b) are orthogonal
(c) are such that the length of their common tangent is $5(12/5)^{1/4}$
(d) are such that the length of their common chord is $5\sqrt{\frac{3}{2}}$

13. AB is a diameter of a circle and 'C' is any point on the circumference of the circle then
(a) The area of ΔABC is maximum when it is isosceles
(b) The area of ΔABC is minimum when it is isosceles
(c) The perimeter of ΔABC is maximum when it is isosceles
(d) None of these

14. Let P be a point on the circle $x^2 + y^2 = 9$, Q a point on the line $7x + y + 3 = 0$, and the perpendicular bisector of PQ be the line $x - y + 1 = 0$. Then the co-ordinates of P are
(a) (3, 0) (b) (0, 3)
(c) $\left(\frac{72}{25}, -\frac{21}{25}\right)$ (d) $\left(-\frac{72}{25}, \frac{21}{25}\right)$

15. If the point $([P+1], [P])$ where [.] denotes the greatest integer function), lies inside the region bounded by the circle $x^2 + y^2 - 2x - 15 = 0$ and $x^2 + y^2 - 2x - 7 = 0$ then
(a) $P \in [-1, 2]$
(b) $P \in (-1, 2)$
(c) $P \in [-1, 3) - [2, 3)$
(d) $P \in [-1,0) \cup [0,1) \cup [1,2)$

16. Let x,y be real variable satisfying the $x^2 + y^2 + 8x - 10y - 40 = 0$.
Let $a = \sqrt{\max\{(x+2)^2 + (y-3)^2\}}$ and $b = \sqrt{\min.\{(x+2)^2 + (y-3)^2\}}$, then
(a) $a + b = 18$ (b) $a + b = 4\sqrt{2}$
(c) $a - b = 8\sqrt{2}$ (d) $a.b = 73$

17. Point M moved along the circle $(x - 4)^2 + (y - 8)^2 = 20$. Then it broke away from it and moving along a tangent to the circle, cuts the x-axis at the point $(-2,0)$. The co-ordinates of the point on the circle at which the moving point broke away can be
(a) $\left(-\frac{3}{5}, \frac{46}{5}\right)$ (b) $\left(-\frac{2}{5}, \frac{44}{5}\right)$
(c) (6, 4) (d) (3,5)

18. One tangent from the origin to a circle with centre at $(2, -1)$ is $3x + y = 0$. Then
(a) slope of the other tangent *is* + 1/3
(b) equation of the other tangent is $3y - x = 0$
(c) the radius of the circle is $\sqrt{\frac{5}{2}}$
(d) angle between the two tangents if $\frac{\pi}{2}$

19. If the straight line $3x - 4y - 5k = 0, \forall\, k \in I$ touches or lies inside the circle, $x^2 + y^2 - 4x - 8y - 5 = 0$, then the value of $|k + 2|$ can be
(a) 1 (b) 0
(c) 3 (d) 4

SECTION–V

ASSERTION AND REASON-TYPE

(a) Both A and R are individually true and R is the correct explanation of A
(b) Both A and R are individually true but R not the correct explanation of A
(c) A is true but R is false
(d) A is false but R is true

1. **A:** The line $(x - 3)\cos\theta + (y - 3)\sin\theta = 1$ touches circle $(x-3)^2 + (y-3)^2 = 1$ for all values of θ.
R: $x\cos\theta + y\sin\theta = a$ is a tangent of circle $x^2 + y^2 = a^2$ for all values of θ.

2. **A:** Two tangents are drawn from a point on the circle $x^2 + y^2 = 2a^2$ to the circle $x^2 + y^2 = a^2$, then tangents are always perpendicular.
R: $x^2 + y^2 = 2a^2$ is the director circle of $x^2 + y^2 = a^2$.

3. **A:** The line $2x + y = 5$ is a diameter of the circle $x^2 + y^2 - 6x + 2y = 0$.
R: Normal of a circle at any point always pass through centre of circle.

4. **A:** The number of common tangents to the circle $x^2 + y^2 = 4$ and $x^2 + y^2 - 6x - 8y = 24$ is 4.
R: Circles with centre C_1, C_2 and radii r_1, r_2 and if $|C_1C_2| > r_1 + r_2$, then circle have 4 common tangents.

5. A: Tangents cannot be drawn from the point $(1, \lambda)$ to the circle $x^2 + y^2 + 5x - 5y = 0$ if $2 < \lambda < 3$

R: $\left(1+\frac{5}{2}\right)^2 + \left(\lambda - \frac{5}{2}\right)^2 < \left(\frac{5}{2}\right)^2 + \left(\frac{5}{2}\right)^2$

6. A: Number of circles passing through $(-2,1)$, $(-1, 0)$, $(-4, 3)$ is 1.

R: Through three non collinear points in a plane only one circle can be drawn.

7. A: Circles $x^2 + y^2 = 4$ and $x^2 + y^2 - 8x + 7 = 0$ intersect each other in two distinct points.

R: Circles with centres C_1 and C_2 and radii r_1 and r_2 intersect at two distinct points, if $|C_1C_2| < r_1 + r_2$

8. A: Circle $x^2 + y^2 - 6x + 4y + 9 = 0$ bisects the circumference of the circle $x^2 + y^2 - 8x - 6y + 23 = 0$

R: Centre of first circle lie on the second circle.

9. A: The smallest possible radius of circle which pass through $(1, 0)$ and $(0,1)$ is $\frac{1}{\sqrt{2}}$.

R: Circle passes through origin.

10. A: Let C_1 and C_2 be two circles with C_2 lying inside C_1. A circle C lying inside C_1 touches C_1 internally and C_2 externally. Then the locus of the centre of C is an ellipse.

R: If A and B are foci and P be any point on the ellipse, then $AP + BP =$ Constant.

11. A: If the perpendicular tangents to the circle C_1 meet at P. Then the locus of P has a circle C_2. Then, $\frac{\text{radius of } C_1}{\text{radius of } C_2} = \frac{1}{\sqrt{2}}$.

R: Circles C_1 and C_2 are concentric.

12. A: A ray of light incident at the point $(-3, -1)$ gets reflected from the tangent at $(0, -1)$ to the circle $x^2 + y^2 = 1$. If the reflected ray touches the circle, then equation of the reflected ray is $4y - 3x = 5$.

R: The angle of incidence = angle of reflection, i.e., $\angle i = \angle r$.

13. A: If two circles $x^2 + y^2 + 2gx + 2fy = 0$ and $x^2 + y^2 + 2g'x + 2f'y = 0$ touches each other, then $f'g = fg'$.

R: Two circles touch each other, if line joining their centres is perpendicular to all possible common tangents.

14. A: The equations $S_1 : x^2 + y^2 + 7x - 9y + 5 = 0$; $S_2 : 2x^2 + 2y^2 + 5x + 7y + 11 = 0$ and $S_3 : 4x^2 + 4y^2 + x + 39y + 23 = 0$ belongs to a co-axial system of circles.

R: $2S_1 - S_2 - 0$ and $2S_2 - S_3 - 0$ represent the same line.

SECTION–VI

LINKED COMPREHENSION-TYPE

A: General equation of circle $x^2 + y^2 + 2gx + 2fy + c = 0$ contains 3 unknown parameters (effective) therefore three conditions are necessary in order to determine a circle uniquely and if only two conditions are given then the obtained equation contains a parameter and described as family of circle. Following are the ways of expressing some known family of circles:

Case I: Equation of circle through intersection of a circle $S = 0$ and a line $L = 0$; $S + \lambda L = 0$

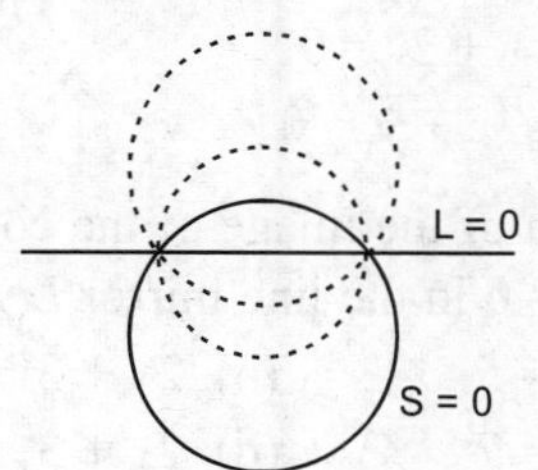

Case II: Equation of family of circle passing through intersection of two circles $S_1 = 0$ and $S_2 = 0$ is given as $S_1 + \lambda(S_1 - S_2) = 0$

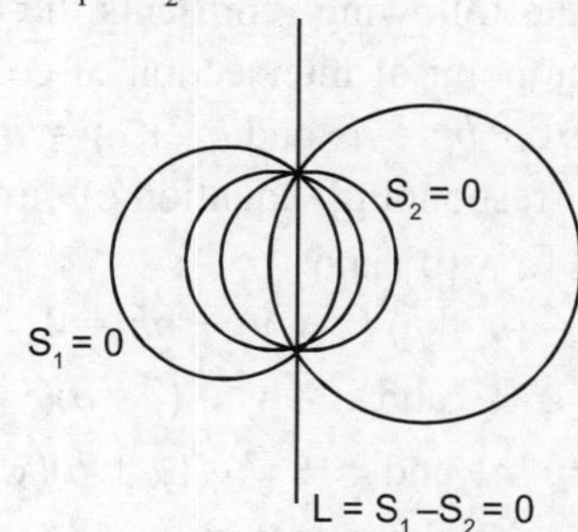

Case III: Equation of any circle touches the line with slope m at the point (x_1, y_1) is

$$(x - x_1)^2 + (y - y_1)^2 + \lambda\{(y - y_1) - m(x - x_1)\} = 0$$

and if m is infinite, the family of circle is

$(x - x_1)^2 + (y - y_1)^2 + \lambda(x - x_1) = 0$. (where λ is a parameter)

Case IV: Equation of any circle passes through two given point (x_1, y_1) and (x_2, y_2) is

$$(x-x_1)(x-x_2)+(y-y_1)(y-y_2)+\lambda\begin{vmatrix} x & y & 1 \\ x_1 & y_1 & 1 \\ x_2 & y_2 & 1 \end{vmatrix}=0$$

Case V: Equation of circle circumscribing a Δ with sides $L_1 = 0, L_2 = 0, L_3 = 0$ is $L_1L_2 + \lambda L_2L_3 \cdot +, \mu L_3L_1 - 0$ where λ, μ is obtained by applying the condition that co-efficient x^2 – co-efficient y^2 and co-efficient of $xy = 0$

Case VI: Family of conic circumscribing a quadrilateral with sides $L_1 = 0, L_2 = 0, L_3 = 0, L_4 = 0$ taken in order is $L_1L_3 + \lambda L_2L_4 = 0$ and condition of concyclicness and equation of possible circumcircle can be obtained by applying the condition that co-efficient of x^2 = co-efficient of y^2 and co-efficient $xy = 0$ and analysing the outcome mathematically.

1. Equation of circle circumscribing the triangle with vertices A (0,0), B (a, a), C (b, a) is
   (a) $x^2+y^2+(a+b)x+(a-b)y=0$
   (b) $x^2+y^2-(a+b)x-(a-b)y=0$
   (c) $x^2+y^2-(a-b)x-(a-b)y=0$
   (d) None of these

2. Which of the following is true about circle passing through intersection of circle $x^2 + y^2 - 2x - 4y + 4 = 0$ and the line $x + 2y = 4$ which also touches $x + 2y = 0$?
   (a) Centre of circle is $(\frac{1}{2}, 1)$
   (b) Equation of circle is $x^2+y^2-x-2y=0$
   (c) Radius of circle is $\frac{\sqrt{5}}{2}$
   (d) None of these

3. Which of the following represents the correct condition that the point of intersection of co-ordinate axes with lines $ax + by = ab$ and $bx + ay = ab$ are concylic and the (corresponding) equation of circle?
   (a) $a, b, \in \mathbb{R} - \{0\}$ and $x^2 + y^2 - (a + b)(x + y) + ab = 0$
   (b) $\forall\ a, b \in \mathbb{R}^+$ and $x^2+y^2+(a+b)(x+y)+ab=0$
   (c) $a^2 + b^2 - ab$ and $x^2 + y^2 - (a + b)(x + y) - ab = 0$
   (d) None of the above is true

4. The centre of circle circumscribing the triangle formed by lines $x + y - 6 = 0$, $2x + y - 4 = 0$ and $x + 2y - 5 = 0$
   (a) $\left(\frac{17}{2}, \frac{-19}{2}\right)$ (b) $\left(-\frac{17}{2}, \frac{-19}{2}\right)$
   (c) $\left(\frac{17}{2}, \frac{19}{2}\right)$ (d) None of these

5. For the circle passing through point of intersection of co-ordinate axis by line $\lambda x - y + 1 = 0$ and $x - 2y + 3 = 0$ which of the following statements are not true?
   (a) The value of λ is 2
   (b) Equation of circle is $2(x^2+y^2)+7x-5y=0$
   (c) The value of λ is 6
   (d) None of these

6. Find the equation of the circle on whcih line $2x + 3y - 5 = 0$ is tangent at (1, 1) and having radius $\sqrt{13}$.
   (a) $x^2+y^2+6x+8y-8=0$
   (b) $x^2+y^2+2x+4y-8=0$
   (c) $x^2+y^2-6x-8y+12=0$
   (d) $x^2+y^2-2x-4y+12=0$

B: The parametric equation of straight line is $\frac{x-x_1}{\cos\theta} = \frac{y-y_1}{\sin\theta} = r$ and it can be very conveniently used to obtain
   (i) Distance of a point along the line from fixed point on line (x_1, y_1)
   (ii) Co-ordinate of a point whose distance from (x_1, y_1) along the line is known.
   (iii) Foot of perpendicular of a point in a line
   (iv) Image of a point in a line etc.

In $x-y$ plane, a line mirror is place along the straight line L_M: $x - y - 1 = 0$ and a circular ring with equation S: $x^2 + y^2 - 6x + 8 = 0$ is lying in front of unsilvered face of the mirror and its reflection be the curve $S' = 0$ then

7. Equation of reflection of curve $S = 0$ in the mirror $L_M = 0$ is $S' = 0$ and it is given by
   (a) $x^2 + y^2 + 4x + 2y + 4 = 0$
   (b) $x^2 + y^2 - 2x - 4y + 4 = 0$
   (c) $x^2 + y^2 - 4x - 2y + 4 = 0$
   (d) $x^2 + y^2 - 4x + 2y + 4 = 0$

8. The equation of circle $S'' = 0$ passing through origin and point of contact of tangents drawn from origin to the circle $S' = 0$ is given as
   (a) $x^2 + y^2 - x - 2y = 0$
   (b) $x^2 + y^2 - 2x - y = 0$
   (c) $x^2 + y^2 + x + 2y = 0$
   (d) $x^2 + y^2 + 2x + y = 0$

9. The equation of the image of the common chord of $S = 0$ and $S'' = 0$ in the line mirror $L_M = 0$ is given as
   (a) $2x - 3y = 5$ (b) $2x + 3y = 7$
   (c) $2x - 3y = 7$ (d) $2x + 3y = 5$

10. The length intercept cut by $S = 0$ on the co-ordinate axes is given by
(a) 2 and 0 (b) 0 and 2
(c) 2 and 2 (d) None of these

11. If the curve $x^2 + xy - 2y^2 - 4x + 7y - 35 = 0$ is reflected in the mirror $L_M = 0$, then the equation of image will be
(a) $2x^2 - xy + y^2 + 7x + 4y + 5 = 0$
(b) $x^2 - xy - 2y^2 + 7x + 4y + 5 = 0$
(c) $2x^2 - xy - y^2 + 7x + 4y + 5 = 0$
(d) None of these

C: A ΔPQR is such that two of its vertices are variable, i.e., $P(-1 + \cos\theta, \sin\theta)$; $Q(-1 - \sin\theta, \cos\theta)$ where θ is a parameter and third vertex is $R(-1, -1)$ fixed.

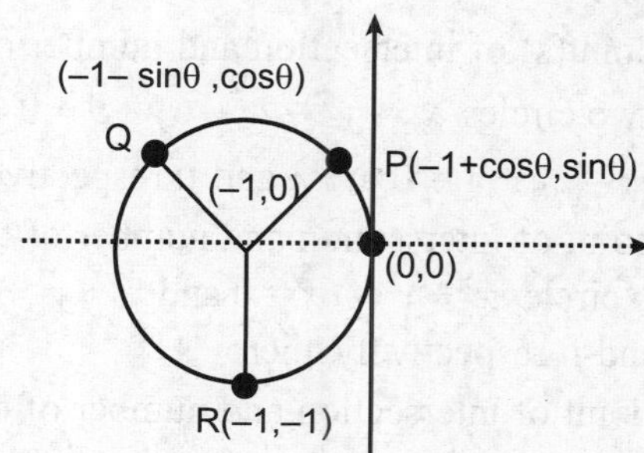

12. S_1 = be the locus of vertex P, then
(a) $(x - 1)^2 - y^2 = 1$ (b) $(x - 1)^2 + y^2 = 4$
(c) $(x + 1)^2 + y^2 = 1$ (d) $(x + 1)^2 - y^2 = 1$

13. $S_2 = 0$ be locus of its centroid and it is given by
(a) $(3x - 1)^2 + (3y - 3)^2 = 2$
(b) $(3x - 3)^2 - (3y + 1)^2 = 2$
(c) $(3x + 3)^2 + (3y + 1)^2 = 2$
(d) $(3x - 1)^2 - (3y - 3)^2 = 2$

14. The locus of its orthocentre ($S_3 = 0$) is given by
(a) $(x + 1)^2 + (y - 1)^2 = 2$
(b) $(x - 1)^2 + (y - 1)^2 = 2$
(c) $(x + 1)^2 + (y + 1)^2 = 2$
(d) None of these

15. The locus of its nine point centre is given as
(a) A parabola with vertex (–1, 0)
(b) A circle with radius 1 unit
(c) A circle with centre (–1, –1/2)
(d) An ellipse

16. If the origin is shifted to (–1, –1) and co-ordinate axes are rotated by angle $\pi/3$ then the transformed equation of $S_3 = 0$ will be
(a) $x^2 + y^2 = 4$ (b) $x^2 - y^2 = 4$
(c) $x^2 + y^2 = 2$ (d) $x^2 - y^2 = 2$

17. If S be any point other than P, Q on the locus of $S_1 = 0$ then the $\angle PSQ$ must be
(a) $\frac{3\pi}{4}$ (b) $\frac{\pi}{3}$
(c) $\frac{2\pi}{3}$ (d) $\frac{\pi}{4}$

18. Locus of mid point of chord drawn from origin to the curve $S_1 = 0$ be $S_4 = 0$ then equation of $S_4 = 0$ is
(a) $\left(x+\frac{1}{2}\right)^2 + y^2 = 1$ (b) $(2x+1)^2 + 2y^2 = 4$
(c) $(2x+1)^2 + 4y^2 = 1$ (d) None of these

19. A point is randomly selected within the locus $S_1 = 0$ what is the probability that selected point lies outside the smaller PQ arc
(a) $\frac{1}{2}$ (b) $\frac{3}{4}$
(c) $\frac{1}{3}$ (d) $\frac{1}{4}$

20. If $S_1 = 0$ takes one complete roll along + *ve* direction of *y*-axis, then $S'_1 = 0$ represents the new locus. The maximum number of possible rational points on the locus $S_1' = 0$ can be
(a) 0 (b) 4
(c) 3 (d) 2

D: Limiting points of a coaxial system of circles are the members of the system which are of zero radius. Let (a, b) and (α, β) be two limiting points of a coaxial system of circles, Then, the corresponding points circles are $S_1 \equiv (x-a)^2 + (y-b)^2 = 0$ and $S_2 \equiv (x-\alpha)^2 + (y-\beta)^2 = 0$. So, the coaxial system of circles is given by $S_1 + \lambda S_2 = 0, \lambda \neq -1$ or $\{(x-a)^2 + (y-b)^2\} + \lambda\{(x-\alpha)^2 + (y-\beta)^2\} = 0$, $\lambda \neq -1$.

The value of λ is determined from the given condition.

21. The co-ordinates of the limiting points of a system of coaxial circles determined by the circles

$S_1 \equiv x^2 + y^2 + 4x + 2y + 5 = 0$ and

$S_2 \equiv x^2 + y^2 + 2x + 4y + 7 = 0$ are

(a) (2, 1) and (0, 3)
(b) (–2, –1) and (0, –3)
(c) (–2, 1) ad (0, –3)
(d) None of these

22. The equation of the circle which passes through the origin and belongs to the coaxial system of which the limiting points are (1, 2) and (4, 3), is given by

(a) $2(x^2+y^2)-x-7y=0$

(b) $x^2+y^2-x-7y=0$

(c) $3(x^2+y^2)-x-7y=0$

(d) None of these

23. The equaiton of the radical axis of a coaxial system of circle whose limiting points are (2, –1) and (–3, 2) is given by

(a) $x-3y+4=0$

(b) $x+3y-4=0$

(c) $5x-3y+4=0$

(d) None of these

SECTION–VII

COLUMN MATCHING-TYPE

1. Column-I

(i) Locus of the point of intersection of the lines $x = at^2, y = 2at$ is

(ii) Locus of the point of intersection of perpendicular tangents to the circle $x^2 + y^2 - a^2 = 0$ is

(iii) Locus of the point intersection of the lines $x \cos \theta = y \cot \theta = a$ is

(iv) The locus of the mid-points of the chords of the circle $x^2+y^2-2ax=0$, passing through the origin is

Column-II

(a) $x^2 + y^2 = 2a^2$

(b) $y^2 = 4ax$

(c) $x^2 + y^2 - ax = 0$

(d) $x^2 = a^2 + y^2$

2. Column-I

(i) The circle $x^2+y^2+2x+c=0$ (where $c > 0$) and $x^2+y^2+2y+c=0$ touch each other

(ii) The circles $x^2+y^2+2x+3y+c=0$ and $x^2+y^2-x+2y+c=0$ intersect orthogonally

(iii) The circle $x^2 + y^2 = 9$ is contains the circle $x^2+y^2-2x+1-c^2=0$

(iv) The circle $x^2 + y^2 = 9$ is contains in the circle $x^2+y^2-2x+1-c/2=0$

Column-II

(a) if $c = 1$

(b) if $c < 2$

(c) if $c = \dfrac{1}{2}$

(d) if $c > 8$

3. Column-I

(i) If point(s) of intersection and number of tangents of two circles $x^2+y^2-2x-6y+9=0$ and $x^2+y^2+6x-2y+1=0$ are λ and μ respectively, then

(ii) If point of intersection and number of tangents of two circles $x^2+y^2-6x=0$ and $x^2+y^2+2x=0$ are λ and μ respectively, then

(iii) If point of intersection and number of tangents of two circles $x^2+y^2-2x-4y+1=0$ and $x^2+y^2-4x-2y+1=0$ are λ and μ respectively, then

Column-II

(a) $\mu - \lambda = 0$

(b) $\mu - \lambda = 2$

(c) $\mu - \lambda = 4$

(d) $\mu + \lambda = 4$

(e) $\lambda^{\mu} + \mu^{\lambda} = 4$

4. Column-I

(i) If the straight lines $y = a_1x + b$ and $y = a_2x + b$ $(a_1 \neq a_2)$ and $b \in R$ meet the co-ordinate axes in concyclic points, then

(ii) If the chord of contact of the tangents drawn to $x^2 + y^2 = b^2$ from any point on $x^2+y^2=a_1^2$, touches the circle $x^2 + y^2 = a_2^2 (a_1 \neq a_2)$. then immediate outcomes

(iii) If the circles $x^2 + y^2 + 2a_1x + b = 0$ and $x^2 + y^2 + 2a_2x + b = 0$ $(a_1 \neq a_2)$ and $b \in R$ cuts orthogonally, then

Column-II

(a) $a_1^2 + a_2^2 = 4$

(b) $a_1 + a_2 = 3$

(c) $a_1a_2 = b$

(d) $a_1a_2 = 1$

(e) $a_1a_2 = b^2$

SECTION–VIII

INTEGER-TYPE QUESTIONS

1. Two chords are drawn from the point $P(h, k)$ on the circle $x^2 + y^2 = hx + ky$. If the y-axis divides both the chords in the ratio 2: 3, such that $9k^2 > a\,h^2$, then find the value of a.

2. The circle $x^2 + y^2 - 4x - 8y + 16 = 0$ rolls up the tangent to it at $(2 + \sqrt{3}, 3)$ by k units. Assuming the x-axis as horizontal. If the equation of the circle in the new position is $x^2 + y^2 - 6x - 2(4+\sqrt{3})y + 24 + 8\sqrt{3} = 0$. Then find the value of 'k'.

3. The extremities of a diagonal of rectangle are $(-4, 4)$ and $(6, -1)$. A circle circumscribes the rectangle and cuts an intercept AB on y-axis. If the area of the triangle formed by AB and the tangents to the circle at A and B is $k/8$, then find the value of 'k'.

4. The point $(1, 4)$ lies inside the circle $x^2 + y^2 - 6x - 10y + k = 0$. Find the difference between the maximum and the minimum possible values of k if the circle neither touches nor cuts the axes.

5. If the circle on the chord $x \cos\alpha + y \sin\alpha - p = 0$ of the circle $x^2 + y^2 = a^2$ as diameter is $x^2 + y^2 - a^2 - kp\,(x \cos\alpha + y \sin\alpha - p) = 0$, then find the value of 'k'.

6. If the curves $ax^2 + 4xy + 2y^2 + x + y + 5 = 0$ and $ax^2 + 6xy + 5y^2 + 2x + 3y + 8 = 0$ intersect at four c oncyclic points, then find the value of 'a'.

7. If a circle $S(x, y) = 0$ touches at the point $(2, 3)$ of the line $x + y = 5$ and $S(1, 2) = 0$, then radius of such circle $\frac{1}{\sqrt{k}}$ units, then find the value of 'k'.

8. If the equation of the circle circumscribing the quadrilateral formed by the lines $2x + 3y = 2$, $3x - 2y = 3$, $x + 2y - 3 = 0$ and $2x - y = 1$ is given by $x^2 + y^2 + kx - y - 7 = 0$. Then find the value of 'k'.

9. The radical axes of the circles $x^2 + y^2 + 2gx + 2fy + c = 0$ and $2x^2 + 2y^2 + 3x + 8y + 2c = 0$ touches the circle $x^2 + y^2 + 2x - 2y + 1 = 0$; then find the value of 'f'.

10. Tangents are drawn from $P(6, 8)$ to the circle $x^2 + y^2 = r^2$. The radius of the circle such that the area of the triangle formed by tangents and chord of contact is maximum is

11. If $(1+\alpha x)^n = 1 + 8x + 24x^2 + \ldots$ and a line through $P(\alpha, n)$ cuts the circle $x^2 + y^2 = 4$ in A and B, then $PA.\ PB =$

12. The circle $x^2 + y^2 = 4$ cuts the line joining the points $A(1, 0)$ and $B(3, 4)$ in two points P and Q. Let $\frac{BP}{PA} = \alpha$ and $\frac{BQ}{QA} = \beta$ then α and β are roots of the quadratic equation is given by $3x^2 + 2x - k = 0$. Then find the value of 'k' (the distances are directed).

13. If the area bounded by the circles $x^2 + y^2 - r^2$, $r = 1, 2$ and the rays given by $2x^2 - 3xy - 2y^2 = 0$, $y > 0$ is $\frac{k\pi}{4}$ sq. units. Then find the value of 'k'.

14. If $a > 2b > 0$ then the positive value of m for which $y = mx - b\sqrt{1+m^2}$ is a common tangent to $x^2 + y^2 = b^2$ and $(x - a)^2 + y^2 = b^2$ is $\frac{2b}{\sqrt{a^2 - kb^2}}$, Then find the value of 'k'.

15. Find the maximum number of rational points (a point (a, b) is rational, if a and b both are rational numbers) on the circumference of a circle having centre (π, e).

16. Consider a circle S with centre at the origin and radius 4. Four circles A, B, C and D each with radius unity and centres $(-3, 0)$, $(-1, 0)$, $(1, 0)$ and $(3, 0)$ respectively are drawn. A chord PQ of the circle S touches the circle B and passes through the centre of the circle C. If the length of this chord can be expresses as $\sqrt{l}$, Find l.

Answer Keys

SECTION–III

1. (b) **2.** (c) **3.** (b) **4.** (b) **5.** (a) **6.** (b) **7.** (c) **8.** (c) **9.** (a) **10.** (c)
11. (b) **12.** (b) **13.** (b) **14.** (d) **15.** (c) **16.** (c) **17.** (d) **18.** (c) **19.** (b) **20.** (c)
21. (b) **22.** (a) **23.** (c) **24.** (b) **25.** (b) **26.** (a) **27.** (c) **28.** (d) **29.** (c) **30.** (b)
31. (a) **32.** (a) **33.** (a) **34.** (a) **35.** (b) **36.** (a) **37.** (a) **38.** (c) **39.** (d) **40.** (b)
41. (b) **42.** (b) **43.** (b) **44.** (a) **45.** (a) **46.** (c) **47.** (d) **48.** (c) **49.** (b) **50.** (a)
51. (c) **52.** (b) **53.** (b) **54.** (d) **55.** (a) **56.** (b) **57.** (c)

SECTION–IV

1. (a, c) **2.** (a, b, c, d) **3.** (b, c) **4.** (a, c) **5.** (a, d) **6.** (a, c) **7.** (a, d)
8. (b, c) **9.** (a, b, c, d) **10.** (b, d) **11.** (a, c, d) **12.** (a, b, c, d) **13.** (a, c) **14.** (a, d)
15. (a, c, d) **16.** (a, d) **17.** (b, c) **18.** (a, b, c, d) **19.** (a, b, c, d)

SECTION–V

1. (a) **2.** (a) **3.** (b) **4.** (d) **5.** (a) **6.** (d) **7.** (c) **8.** (c) **9.** (a) **10.** (a)
11. (b) **12.** (b) **13.** (c) **14.** (a)

SECTION–VI

1. (b) **2.** (a, b, c) **3.** (a) **4.** (c) **5.** (a) **6.** (b, c) **7.** (b) **8.** (a) **9.** (b) **10.** (a)
11. (d) **12.** (c) **13.** (c) **14.** (c) **15.** (d) **16.** (c) **17.** (d) **18.** (c) **19.** (b) **20.** (a)
21. (b) **22.** (a) **23.** (c)

SECTION–VII

1. (i) → (b) (ii) → (a) (iii) → (d) (iv) → (c)
2. (i) → (c) (ii) → (a) (iii) → (b) (iv) → (d)
3. (i) → (c, d) (ii) → (b, d, e) (iii) → (a, d)
4. (i) → (d) (ii) → (e) (iii) → (c)

SECTION–VIII

1. 40 **2.** 2. **3.** 1331 **4.** 4 **5.** 2 **6.** – 4 **7.** 2 **8.** 6 **9.** 2 **10.** 5
11. 16 **12.** 21 **13.** 3 **14.** 4 **15.** 2 **16.** 63

HINTS AND SOLUTIONS

TEXTUAL EXERCISE–1 (SUBJECTIVE)

1. Diameter $d = 20$ units (given) $\Rightarrow$ $r = 10$ units
The given diameters $2x + y = 6$ and $3x + 2y = 4$ intersect at $(8, -10)$
Hence the circle is $(x - 8)^2 + (y + 10)^2 = 10^2$ i.e., $x^2 + y^2 - 16x + 20y + 64 = 0$

2. The circle $(x^2 + y^2) - \frac{3}{2}x + 4y - \frac{1}{2} = 0$ is centered at $C\left(\frac{3}{4}, -2\right)$
The lines $3x + y = 4$ and $x - 3y + 2 = 0$ intersect at $P(1, 1)$
Now $CP = \sqrt{\left(\frac{1}{4}\right)^2 + 3^2} = \sqrt{\frac{145}{16}}$
The equation of the circle will be $\left(x - \frac{3}{4}\right)^2 + (y+2)^2 = \frac{145}{16}$
$\Rightarrow$ $16(x^2 + y^2) - 24x + 64y - 72 = 0$ or $2(x^2 + y^2) - 3x + 8y - 9 = 0$

3. Centre of the circle lies on $3x + 4y = 7$
Let it be $\left(x_1, \frac{7 - 3x_1}{4}\right)$, as $(1, -2)$, $(4, -3)$ lie on the circle.
$\therefore$ $(x_1 - 1)^2 + \left(\frac{7 - 3x_1}{4} + 2\right)^2 = (x_1 - 4)^2 + \left(\frac{7 - 3x_1}{4} + 3\right)^2$
$\Rightarrow$ $16x_1^2 + 16 - 32x_1 + (15 - 3x_1)^2$
$\Rightarrow$ $16x_1^2 + 256 - 128x_1 + (19 - 3x_1)^2$
$\Rightarrow$ $96x_1 = 240 + 136 - 24x_1$
$\Rightarrow$ $x_1 = \frac{47}{15}$ and $\frac{7 - 3x_1}{4} = -\frac{3}{5}$
$\Rightarrow$ Radius $= \sqrt{\left(\frac{32}{15}\right)^2 + \left(\frac{21}{15}\right)^2}$
So the circle is $\left(x - \frac{47}{15}\right)^2 + \left(y + \frac{9}{15}\right)^2 = \frac{(32)^2 + (21)^2}{(15)^2}$
$\Rightarrow$ $225(x^2 + y^2) - 1410x + 2209 + 81 + 270y = 1465$
$\Rightarrow$ $225(x^2 + y^2) - 1410x + 270y + 825 = 0$
$\Rightarrow$ $15(x^2 + y^2) - 94x + 18y + 55 = 0$

4. The given circle is $x^2 + y^2 - 2x + 6y - 15 = 0$ which has centre at $C(1, -3)$
Now, $A(4, 1)$ is one end of a diameter
$\Rightarrow$ other end is at $B(-2, -7)$

TEXTUAL EXERCISE–1 (OBJECTIVE)

1. **(i) (d)** Given: $x^2 + y^2 + 4x + 6y + 13 = 0$ as $a = b = 1$ and $h = 0$, also $g^2 + f^2 - c = 4 + 9 - 13 = 0$
The equation represents a point $P(-2, -3)$
(ii) (c) The equation will represent a circle with non-zero radius is $g^2 + f^2 - c > 0$
(iii) (b) The circle with radius $r = a$ and centre at $(a\cos\alpha, a\sin\alpha)$ will be $x^2 + y^2 - 2ax\cos\alpha - 2ay\sin\alpha = 0$

2. **(c)** Centre of C_1: $x^2 + y^2 = 1$ is $A(0, 0)$
Centre of C_2: $x^2 + y^2 + 6x - 2y = 1$ is $B(-3, 1)$
Centre of C_3: $x^2 + y^2 + 6x - 2y = 4$ is $C(-3, 1)$
Since point B and C are coincident
So, actually only two distinct points which are always collinear.

3. **(b)** Centre: $\left(\frac{a}{2}, \frac{a}{2}\right)$
One diameter end is at $A(0, 0)$
$\Rightarrow$ $C(a, a)$
$\therefore$ $x(x - a) + y(y - a) = 0$ or $x^2 - ax + y^2 - ay = 0$

Aliter: Centre of the circle is $\left(\frac{a}{2}, \frac{a}{2}\right)$
Radius of circumscribing circle $r = a/\sqrt{2}$
$\Rightarrow$ Equation of circle $x^2 + y^2 - ax - ay = 0$

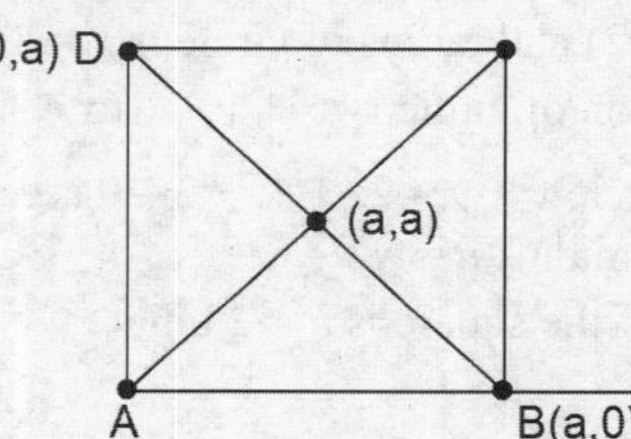

4. **(a)** Observe that ΔOPC and ΔOQC are right angled respectively at P and Q

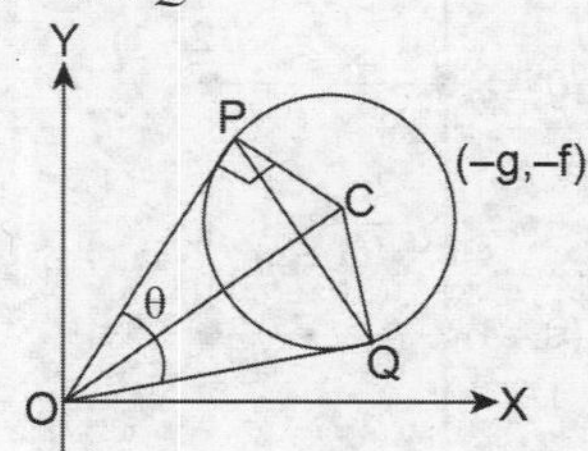

A circle that circumscribes O, P, C will also passes through Q and the centre of the circle will be at $\left(-\frac{g}{2}, \frac{f}{2}\right)$

5. **(c)** Let $C(x_1, y_1)$ be the centre of the circle, then $x_1^2 + y_1^2 = 9k^2$.
Observe that the circle will intersect the axes at $A(2x_1, 0)$ and $B(0, 2y_1)$, respectively.

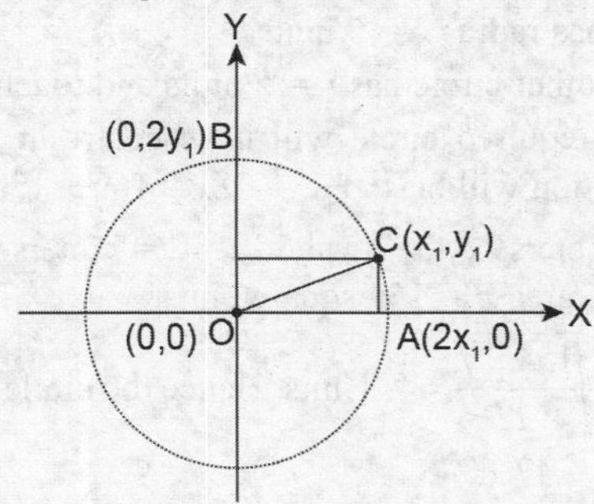

$\Rightarrow$ Centroid of $\Delta OAB = \left(\frac{2}{3}x_1, \frac{2}{3}y_1\right)$

$\Rightarrow \quad \frac{3}{2}x = x_1$ and $\frac{3}{2}y = y_1 \Rightarrow \quad \frac{9}{4}(x^2 + y^2) = 9k^2$

Hence, $x^2 + y^2 = 4k^2$ is the locus.

6. (b) The circle $x^2 + y^2 + 3x - 6y - 9 = 0$ is centered at $C\left(-\frac{3}{2}, 3\right)$ and has radius $r_1 = \frac{9}{2}$.

The circle that rolls outside has radius $r_2 = 2$ units, so the locus of centre has radius 13/2 units and its locus is $\left(x + \frac{3}{2}\right)^2 + (y - 3)^2 = \frac{169}{4}$ or $4(x^2 + y^2) + 12x - 24y - 124 = 0$

7. (a) Circle C_1: $x^2 + y^2 - 1 = 0 \quad \Rightarrow r_1 = 1$ units

Circle C_2: $x^2 + y^2 - 2x - 6y - 6 = 0$

$\Rightarrow r_2 = 4$ units.

Circle C_3: $x^2 + y^2 - 4x - 12y - 9 = 0$

$\Rightarrow r_3 = 7$ units

Observe that r_1, r_2, r_3 are in AP

8. (c) The two diameters $x + y = 6$ and $x + 2y = 4$ intersect at $C(8, -2)$ and radius of circle is $r = 10$ units.

$\Rightarrow$ Equation of circle is $x^2 + y^2 - 16x + 4y - 32 = 0$

9. (d) The circle $x^2 + y^2 - 2x + 4y + 3 = 0$ is centered at $C(1, -2)$ and its radius $r = \sqrt{2}$

$\Rightarrow$ Side of the square is a = 2 units.

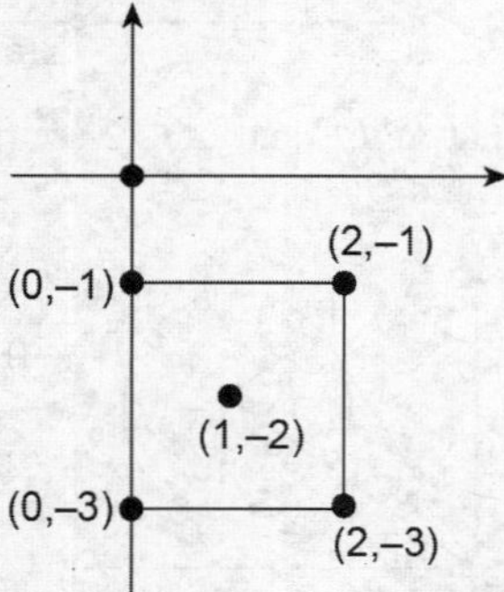

As sides are parallel to axes, so none of these points will form any vertex.

10. (a) Centre of circle is $C(1, -3)$ since line $L: 3x - 4y - 5 = 0$ touches the circle

$\Rightarrow \quad r = \frac{|L(1,-3)|}{5} = 2$ Units

11. (b) The circle $x^2 + y^2 - 2x - 4y - 20 = 0$ is centered at $C(1, 2)$ and has radius $r = 5$ units.

The other circle has $r = 5$ units and touches at (5, 5)

$\Rightarrow$ The required circle will have centre at (9, 8) and its equation will be $x^2 + y^2 - 18x - 16y + 120 = 0$

12. (b) Diameters $2x - 3y = 5$ and $3x - 4y = 7$ intersect at $C(1, -1)$.

Since area $A = 154$ square units.

$\Rightarrow \quad r = \sqrt{\frac{154 \times 7}{22}} = 7$ Units. Hence the circle is $x^2 + y^2 - 2x + 2y = 47$

13. (b) Given Line L_1: $3x - 4y + 4 = 0$ and L_2: $3x - 4y - \frac{7}{2} = 0$, are tangents to a circle.

Since these lines are parallel, so the distance between these lines is equal to the diameter of the circle.

$\Rightarrow \quad r = \frac{d}{2} = \left(\frac{1}{2}\right)\frac{\left|4 + \left(\frac{7}{2}\right)\right|}{5} = \frac{15}{20} = \frac{3}{4}$ Units

14. (b) In a right-angled triangle the hypotenuous is the diameter. Now, $A(-3, 4)$ and $B(3, -4)$. As $\angle C = 90°$.

$\Rightarrow (x + 3)(x - 3) + (y - 4)(y + 4) = 0$

$\Rightarrow x^2 + y^2 = 25$

15. (b) Centre of the second circle $x^2 + y^2 - 4x - 6y = 0$ is (2, 3)

Let the required equation be $x^2 + y^2 + 8x + 10y - c = 0$ since (2, 3) satisfies.

$\therefore \quad c = 59$

Aliter: Centre of first circle $x^2 + y^2 + 8x + 10y - 7 = 0$ is $C_1(-4, -5)$. The centre of second circle $x^2 + y^2 - 4x - 6y = 0$ is $C_2(2, 3)$.

$\Rightarrow \quad C_1C_2 = \sqrt{6^2 + 8^2} = 10$. Hence, the required circle is $(x + 4)^2 + (y + 5)^2 = 10$ i.e., $x^2 + y^2 + 8x + 10y - 59 = 0$.

TEXTUAL EXERCISE-2 (SUBJECTIVE)

1. Let the circle be $x^2 + y^2 + 2gx + 2fy + c = 0$.

Since $O(0, 0)$, $A(a, 0)$ and $B(0, b)$ lie on the circle and $\angle AOB = 90°$.

$\Rightarrow x(x - a) + y(y - b) = 0$ is the circle.

Aliter-1: Let the circle be $x^2 + y^2 + 2gx + 2fy + c = 0$. Since $O(0, 0)$ satisfies

$\therefore \quad c = 0$

For $A(a, 0)$, we get $a^2 + 2ga = 0$

$\Rightarrow 2g = -a$ and $B(0, b)$, gives $b^2 + 2fb = 0$, $2f = -b$

Hence the circle is $x^2 + y^2 - ax - by = a$

Aliter-2: Given point are $O(0, 0)$, $A(a, 0)$ and $B(0, b)$.

The circle passing through O, A, B will be

$$\begin{vmatrix} (x^2+y^2) & x & y & 1 \\ 0 & 0 & 0 & 1 \\ a^2 & a & 0 & 1 \\ b^2 & 0 & b & 1 \end{vmatrix} = 0 \Rightarrow \begin{vmatrix} (x^2+y^2) & x & y \\ a^2 & a & 0 \\ b^2 & 0 & b \end{vmatrix} = 0$$

$\Rightarrow ab\{x^2 + y^2 - ax - by\} = 0$

2. The line $3x + 4y = 12$ intersects the axes at $A(4, 0)$ and $B(0, 3)$

Since $\angle AOB = 90°$

$\Rightarrow AB$ will serve as the diameter with A and B as its ends

$\Rightarrow$ The circle is $x(x - 4) + y(y - 3) = 0$ or $x^2 + y^2 - 4x - 3y = 0$

3. As given $C(5, 0)$ and $r = 5$.

$\Rightarrow (x - 5)^2 + y^2 = 25 \quad \Rightarrow x^2 + y^2 - 10x = 0$

4. The given points are $A(3, 4)$, $B(3, -6)$ and $C(-1, 2)$

Let the circle be $x^2 + y^2 + 2gx + 2fy + c = 0$

$A(3, 4)$ gives $25 + 6g + 8f + c = 0$

$B(3, 6)$ gives $45 + 6g - 12f + c = 0$

$\Rightarrow f = 1 \quad \Rightarrow 6g = -c - 33$

$C(-1, 2)$ gives $5 - 2g + 4f + c = 0$

$\therefore\ 2g = 9 + c \qquad \Rightarrow 27 + 3c = -c - 33$

$\Rightarrow 4c = -60$ or $c = -15$ and $2g = -6$ and the circle is $x^2 + y^2 - 6x + 2y - 15 = 0$

5. From the given conclude that circle passes through $O(0, 0)$, A (4, 0) and B (0, 6).
Since $\angle AOB = 90°$ $\Rightarrow$ AB is a diameter
Hence $x(x - 4) + y(y - 6) = 0$
$\Rightarrow x^2 + y^2 - 4x - 6y = 0$

6. $y = x$ and $y = 2x$ intersect at $O(0, 0)$.
Now $y = 3x + 2$ intersects $y = x$ at A (–1, –1) and $y = 3x + 2$ intersect $y = 2x$ at A (–2, –4)
Since O (0, 0) lies on the circle so let the equation be $x^2 + y^2 + 2gx + 2fy = 0$
A (–1,–1) gives $2 - 2g - 2f = 0$
$\Rightarrow g + f = 1$
B (–2, –4) gives $20 - 4g - 8f = 0$
$\Rightarrow g + 2f = 5$
$\therefore$ $f = 4$ and $g = -3$ and the circle is $x^2 + y^2 - 6x + 8y = 0$

7. Distance between $x = -1$ and $x = 3$ is $d = 4$ units
$\Rightarrow r = 2$
Lines $y = 3$ touches the circles
$\Rightarrow C_1$ (1, 1) or C_2(1, 5) and the circle is $(x - 1)^2 + (y - 1) = 4$ or $(x - 1)^2 + (y - 5)^2 = 4$ i.e., $x^2 + y^2 - 2x - 2y - 2 = 0$ or $x^2 + y^2 - 2x - 10y + 22 = 0$

TEXTUAL EXERCISE–2 (OBJECTIVE)

1. **(c)** From the given considerations the centre of the circle is at (5, 4) and $r = 5$ units.
So $(x \pm 5)^2 + (y - 4)^2 = 25 \Rightarrow x^2 + y^2 \pm 10x - 8y + 16 = 0$

2. **(b)** Given $x \cos\theta + y \sin\theta = a$ and $x \sin\theta - y \cos\theta = b$
Squaring and adding, we get $x^2 + y^2 = a^2 + b^2$
$\Rightarrow$ Locus is a circle

3. **(b)** Since the circle $x^2 + y^2 + 2gx + c = 0$ touches x-axis
$\Rightarrow |\pm f| = r = \sqrt{g^2 + f^2 - c} \Rightarrow g^2 = c$

4. **(a)** The circle with radius $r = a$ that touches both the axes will have centre at $(a, \pm a)$ or $(-a, \pm a)$
$\Rightarrow (x \pm a)^2 + (y \pm a)^2 = a^2$
$\Rightarrow x^2 + y^2 \pm 2ax \pm 2ay + a^2 = 0$

5. **(c)** $x^2 + y^2 + 4x - 4x + 4 = 0$ has centre at (–2, 2) and radius $r = 2$, so it will touch both axes.

6. **(c)** From the given radius $r = |x$ co-ordinate of centre$)|$
$\Rightarrow r = \sqrt{4^2 + 3^2} = 5$

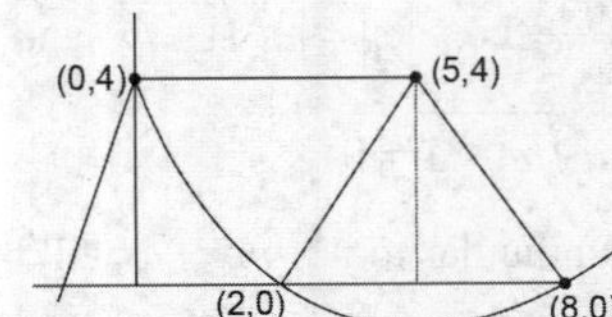

7. **(b)** We get two circles of radius 5 units having centers at $(1, \pm 4)$.

8. **(c)** From the given conditions the centre of the circle will be at $(x_1 + a, y_1 + b)$ and $(x_1 + a)^2 + b^2 = a^2 + (y_1 + b)^2$
$\Rightarrow (x_1 + a)^2 - (y_1 + b)^2 = a^2 - b^2$
or $x^2 - y^2 = a^2 - b^2$ is the locus of centre

9. **(d)** Observe that the circle to touch $y = x$ and y-axis
$\Rightarrow$ Centre of the circle will be on the angle bisectors which are at $\left(\frac{3\pi}{8} = 67\frac{1}{2}^\circ\right)$ and $\left(\frac{7\pi}{8} = 157\frac{1}{2}^\circ\right)$ as shown further the number of circles is infinite

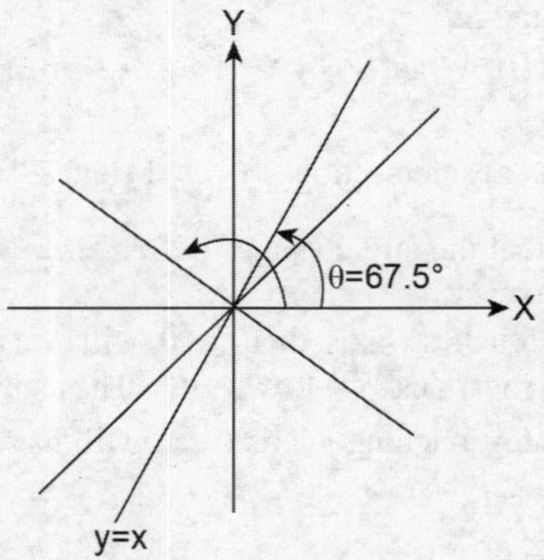

10. **(b)** Let A (2, –2), B (–1, –1) and C (5, 2) be the vertices of a triangle.
Circle $x^2 + y^2 + 2gx + c = 0$ will pass where A (2, –2)
$\Rightarrow 8 + 4g - 4f + c = 0$ and B (–1,–1)
$\Rightarrow 2 - 2g - 2f + c = 0 \quad \Rightarrow f = 3g + 3$ and C (5, 2)
$\Rightarrow 29 + 10g + 4f + c = 0$
From $f = 3g + 3$ and $27 + 12g + 6f = 0$ gives $2g = -3$ and $2f = -3$ and $c = -8$
$\Rightarrow x^2 + y^2 - 3x - 3y - 8 = 0$

11. **(c)** Let $(-a, -a)$ be the centre and radius $r = a$ (where $a > 0$)
Distance from $3x - 4y + 8 = 0$ is $\frac{|a + 8|}{5} = a$
$\Rightarrow$ $a = 2$, hence the circle is $(x + 2)^2 + (y + 2)^2 = 4$ or $x^2 + y^2 + 4x + 4y + 4 = 0$

12. **(b)** The circle is $x^2 + (y + 3)^2 = 3^2$
So its centre is C (0, –3) and $r = 3$ units. So it will touch x-axis at origin.

13. **(c)** The circle $x^2 + y^2 + 2gx + 2fy + c = 0$ will touch both the axes if $g = f = r$, so $c = r^2$
Hence $g = f = r = \sqrt{c}$

14. **(c)** Since the circle has radius $r = 4$ and centre on x-axis so for passing through origin $(x \pm 4)^2 + y^2 = 16$ or $x^2 + y^2 \pm 8x = 0$

15. **(d)** Circle $x^2 + y^2 + 3x + 3y = 0$ has centre at $\left(-\frac{3}{2}, -\frac{3}{2}\right)$ and it passes through origin

16. **(b)** Line $x = 0$ is y axis. The circle will touch these lines when the centre is at $\left(\pm\frac{b - a}{2}, \frac{a + b}{2}\right)$ and radius $r = \left|\frac{b - a}{2}\right|$
So two circles are possible

17. (c) The circle $x^2 + y^2 + 2gx + 2fy + c = 0$ has radius $r = a$ and x-axis as diameter

$\Rightarrow (x - x_1)^2 + y^2 = a^2$

$\Rightarrow f = 0$ and $-2x_1 = 2g$, also $x_1^2 - a^2 = c \Rightarrow g^2 = c + a^2$

So $g = a$, $c = 3a^2$ does not satisfy

Observe that $g = -2a$ and $c = 3a^2$ satisfies

TEXTUAL EXERCISE–3 (SUBJECTIVE)

1. Circle is $x^2 + y^2 - 2 = 0$ ....(i)

And the line $L: y = 2x + 1$

Putting in (i), we get $x^2 + 4x^2 + 1 + 4x - 2 = 0$ or $5x^2 + 4x - 1 = 0$

$\Rightarrow x = -1, 1/5$

$\Rightarrow$ Points of intersection $A(-1, -1)$ and $B\left(\frac{1}{5}, \frac{7}{5}\right)$

2. To show that the circle $x^2 + y^2 - 2ax - 2ay + a^2 = 0$ touches both the axes.

If a circle touches x-axis, then $(x_1, 0)$ will be a unique solution (repeated root) i.e., $x_1^2 - 2ax_1 + a^2 = 0$ has solution $x_1 = a, a$.

Similarly for touching y-axis $(0, y_1)$ will have

$y_1^2 - 2ay_1 + a^2 = 0$

$\Rightarrow y_1 = a, a$

The circle will touch both the axes.

Aliter: Circle $x^2 + y^2 - 2ax - 2ay + a^2 = 0$ can be written as $(x - a)^2 + (y - a)^2 + (y - a)^2 = a^2$

$\Rightarrow$ Centre of the circle is (a, a) and radius $|a|$, so it will touch both the axes as point (a, a) is at a distance $|a|$ from the axes.

3. $x = 0$, $x = a$ and $3x + 4y = -5a$

From the given considerations the centre of the circle will be on the line $x - a/2 = 0$ and radius $r = a/2$.

Let centre be $\left(\frac{a}{2}, y_1\right)$, then $\left|\frac{\frac{3a}{2} + 5a + 4y_1}{5}\right| = \frac{a}{2}$

$\therefore |13a + 8y_1| = 5a \Rightarrow y_1 = \frac{-9a}{4}, -a$

$\therefore$ The circles are $\left(x - \frac{a}{2}\right)^2 + (y + a)^2 = \left(\frac{a}{2}\right)^2$

$\Rightarrow x^2 - ax + y^2 + 2ay + a^2 = 0$

Or $\left(x - \frac{a}{2}\right)^2 + \left(y + \frac{9a}{4}\right)^2 = \left(\frac{a}{2}\right)^2$

$\Rightarrow x^2 + y^2 - ax + \frac{9}{2}ay + \frac{81}{16}a^2 = 0$ or $16(x^2 + y^2 - ax) + 72ay + 81a^2 = 0$

4. The circle $x^2 + y^2 - 2ax = 0$ may be written as $(x - a)^2 + y^2 = a^2$

$\Rightarrow$ Centre at $C(a, 0)$ and radius $r = |a|$ units.

Now the line $L: mx - y + a\left(\sqrt{1 + m^2} - m = 0\right)$ is at distance of $\frac{\left|am - 0 + a\sqrt{1+m^2} - am\right|}{\sqrt{1+m^2}} = |a|$ units. So, it will always touch.

5. (i) From the given considerations the circle will have centre at $C(0, y_1)$ where $y_1 = \pm a \cot\theta$ and radius $r = \sqrt{a^2 + a^2\cot^2\theta}$

$\Rightarrow r = |a| \operatorname{cosec}\theta$

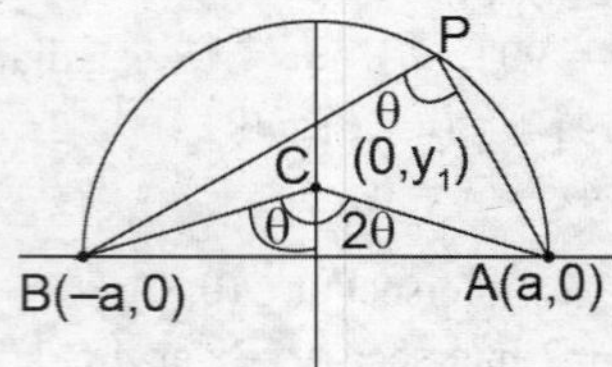

Hence the circle $x^2 + y^2 + a^2\cot^2\theta \pm 2ay\cot\theta = a^2\operatorname{cosec}^2\theta$ i.e., $x^2 + y^2 = a^2 \pm 2ay\cot\theta$

(ii) Let the line be $y = mx + c$

Let perpendiculars from $(-a, 0)$ and $(a, 0)$ be p_1 and p_2

$p_1 = \frac{ma + c}{\sqrt{1+m^2}}$ and $p_2 = \frac{-ma + c}{\sqrt{1+m^2}}$

$p_1 + p_2 =$ constant (say k)

$\frac{2c}{\sqrt{1+m^2}} = k \quad \Rightarrow \quad c = \frac{k}{2}\sqrt{1+m^2}$

$\Rightarrow$ The line is $y = mx + \frac{k}{2}\sqrt{1+m^2}$

Clearly it touches a fixed circle with centre $(0, 0)$ and radius $k/2$ i.e., $x^2 + y^2 = \frac{k^2}{4}$

6. The given points as ends of a diameter are $A(am^2, 2am)$ and $B\left(\frac{a}{m^2}, -\frac{2a}{m}\right)$.

Mid point of AB is $C\left(\frac{a(1+m^4)}{2m^2}, \frac{2a(m^2-1)}{2m}\right)$, distance of C from line $x + a = 0$ is $h = \frac{|a + am^4 + 2am^2|}{2m^2} = \frac{a(m^2+1)^2}{2m^2}$

$\Rightarrow AB^2 = \frac{a^2(m^4-1)^2}{m^4} + \frac{4a^2(m^2+1)^2}{m^2}$

$\Rightarrow AB = \left|\frac{a(m^2+1)}{m}\right|\left\{\sqrt{\frac{m^4 + 1 - 2m^2 + 4m^2}{m^2}}\right\}$

$= \left|\frac{a(m^2+1)^2}{m^2}\right|$, as $AB = 2h$ (Where AB = diameter)

$\therefore r = h$, so line will always touch the circle.

7. (i) The circle is $C: x^2 + y^2 = a^2$, which has centre at $(0, 0)$ and radius $r = |a|$

Line: $mx - y + c_1 = 0$ will be a tangent, when $|a| = \frac{|c_1|}{\sqrt{1+m^2}}$ i.e., $|c_1| = |a|\sqrt{1+m^2}$ and the line will be $mx - y \pm a\sqrt{m^2 - 1} = 0$

(ii) Line perpendicular to $y = mx + c$ will be $L: x + my + c_1 = 0$, this will be a tangent, when $|c_1| = |a|\sqrt{1+m^2}$ and the line will be $x + my \pm a\sqrt{1 + m^2} = 0$.

(iii) The line through $(b, 0)$ will be $y = mx - mb$

$\Rightarrow \quad \dfrac{|mb|}{\sqrt{1+m^2}} = |a|$ or $\dfrac{m^2}{m^2+1} = \dfrac{a^2}{b^2}$ (observe that $b^2 > a^2$)

$\Rightarrow \quad m = \dfrac{|a|}{\sqrt{b^2 - a^2}}$

Hence the line will be $y\sqrt{b^2-a^2} = |a|(x-b)$ or $ax \pm y\sqrt{b^2-a^2} = ab$

(iv) Let $y = mx + c$ be the line.

So $c = |a|\ \sqrt{1+m^2}$ or $c^2 = a^2(m^2+1)$. Now the area of

$\Delta OAB = \dfrac{1}{2}\left|(c)\dfrac{c}{m}\right| = a^2$

So $c^2 = 2a^2|m| \Rightarrow a^2(1+m^2) = 2a^2|m| \Rightarrow m = \pm 1$

Hence the lines are $y = x \pm a\sqrt{2}$ or $y = -x \pm a\sqrt{2}$

8. Without any loss of generality the lines be co-ordinate $(0, y_1 + 2\ \ )$ axes

Let $(x_1, y_1 + \ \ )$ be the centre.

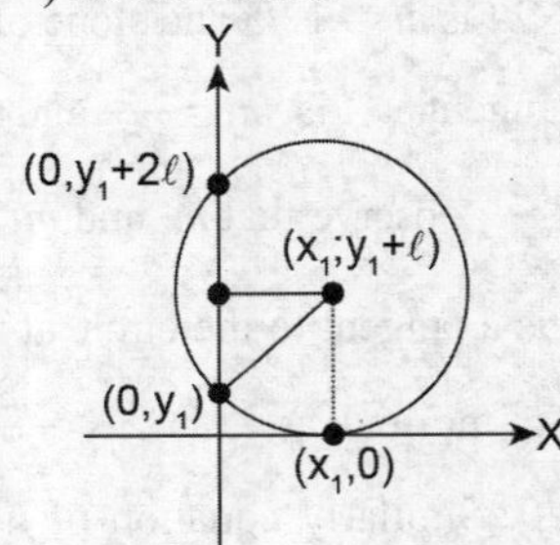

$\Rightarrow x_1^2 + \ell^2 = (y_1+\ell)^2$ i.e., $x^2 + \ \ ^2 = y^2$ or $y^2 - x^2 = \ \ ^2$

TEXTUAL EXERCISE-3 (OBJECTIVE)

1. (d) The circle $x^2 + y^2 - 20y + 90 = 0$ has centre at $C(0, 10)$ and radius $r = \sqrt{10}$ units, so line $mx - y = 0$ will not touch or intersect

when $\dfrac{|-10|}{\sqrt{1+m^2}} > \sqrt{10}$ i.e., $m^2 < 9$, so $|m| < 3$

2. (c) The points are $A(0, 0)$ and $B(1, 8)$ and the circle is $S: x^2 + y^2 - 6x + 8y - 11 = 0$. So $(0, 0) = -11$

$\Rightarrow A(0, 0)$ lies inside the circle and $S(1, 8) = 65 - 6 + 64 - 11 = 112 > 0$

$\Rightarrow B(1, 8)$ lies outside the circle

3. (a) For $y = 0$, we get $x^2 + 2x + 1 = 0$ is $(x+1)^2 = 0$

So $x = -1, -1$ (repeated roots)

$\therefore$ circle will touch x-axis

Aliter: The circle is $(x+1)^2 + (y-2)^2 = 2^2$

So centre $C(-1, 2)$ and radius $r = 2$ units

$\Rightarrow$ Circle will touch x-axis

4. (d) The line is L: $(m-2)x - y - 1 = 0$ and the circle $x^2 + y^2 = 1$ having $C(0, 0)$ and $r = 1$

For intersection (or contact) $\dfrac{|-1|}{\sqrt{1+(m-2)^2}} \le 1$

$\Rightarrow 1 + (m-2)^2 \ge 1$ which is always true so $m \in \mathbb{R}$.

5. (c) $mx - y + c = 0$ will intersect or touch the circle $x^2 + y^2 = a^2$ when $\dfrac{|c|}{\sqrt{m^2+1}} \le |a|$ i.e., $\sqrt{a^2(m^2+1)} \ge |c|$

6. (c) $L: 2x - y + 1 = 0$ and circle $x^2 + y^2 = 2$ has centre $C(0, 0)$ and radius $r = \sqrt{2}$

Length of chord $= 2\sqrt{r^2 - \dfrac{|1|}{5}} \Rightarrow 2\sqrt{\dfrac{9}{5}} = \dfrac{6}{\sqrt{5}}$

7. (b) The point $P(4, 4)$ and circle is $x^2 + y^2 - 2x - 15 = 0$ having centre $C(1, 0)$ and radius $= 4$

Now, $CP = 5$ units. So point lies outside the circle

$\therefore$ Maximum distance $= CP + r = 9$ units

8. (c) The point is $P(k, k)$ and circle $(x+1)^2 + (y+1)^2 = \left(\sqrt{18}\right)^2$ with $C(-1, -1)$ and $r = 3\sqrt{2}$ units

Two tangents will be possible, when $CP > 3\sqrt{2}$ i.e., $(k+1)^2 + (k+1)^2 > 18$

$\Rightarrow |k+1| > 3 \qquad \Rightarrow k \in (-\infty, -4) \cup (2, \infty)$

9. (c) The circle $(x-2)^2 + (y-3)^2 = \left(\sqrt{10}\right)^2$ has centre $C(2, 3)$ and radius $r = \sqrt{10}$ units.

The line Is $L: 3x + 2y - 12 = 0$. Now $L(2, 3) = 0$

$\Rightarrow C(2, 3)$ lies on the line so it is a diameter.

10. (c) From the given Let $(x_1 + a, y_1)$ be the centre

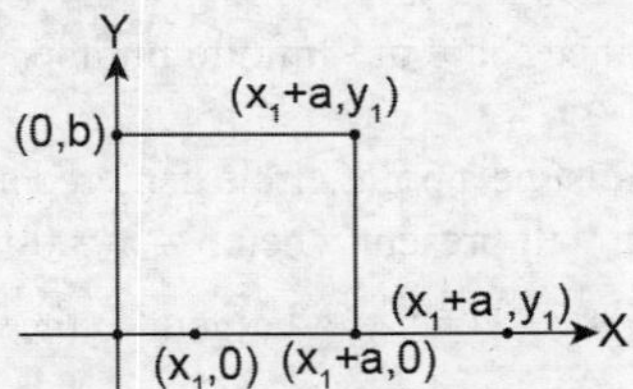

Now, $(x_1 + a)^2 + (y_1 - b)^2 = a^2 + y_1^{\,2}$ or $(x_1 + a)^2 + b^2 - 2by_1 = a^2$

$\Rightarrow x^2 - 2by = a^2 - b^2$

11. (d) Let the centre be $(x_1, y_1 + \ \ )$

Now $(y_1 + \ \ )^2 = x_1^2 + \ell^2$

$\Rightarrow (y_1+\ell)^2 - x_1^2 = \ell^2$ or $y^2 - x^2 = \ \ ^2$ which is a hyperbola

12. (d) The circle $(x-7)^2 + (y+1)^2 = 5^2$

$\Rightarrow C(7, -1)$ and $r = 5$

Now $O(0, 0) \qquad \therefore \quad OC = \sqrt{50}$

From $\sin\dfrac{\theta}{2} = \dfrac{r}{OC} = \dfrac{5}{5\sqrt{2}} = \dfrac{1}{2}$

$\therefore \quad \theta/2 = \pi/4$ i.e., $\theta = \pi/2$

13. (d) The circle is $S: x^2 + y^2 + 2gx + 2fy + c = 0$ and the point $P(0, 0)$

$\Rightarrow \quad S_1 = c$ and $T_1 \equiv gx + fy + c$

Equation of pair of tangents is $SS_1 = T_1^{\,2}$

$\Rightarrow c(x^2 + y^2 + 2gx + 2fy + c) = (gx + fy + c)^2$ or $c(x^2 + y^2) + (gx + fy)^2$

$\Rightarrow (gx + fy)^2 = c(x^2 + y^2)$

14. (c) The line $(x-a)\cos\alpha + (y-b)\sin\alpha - r = 0$ is $x\cos\alpha + y\sin\alpha - (a\cos\alpha + b\sin\alpha + \alpha + r) = 0$

The circle is $(x-a)^2 + (y-b)^2 = r^2$, where centre $C(a, b)$ and radius $= r$

Distance of C from the line is $= |r|$

$\Rightarrow$ Line will always touch the circle for all real value of α

15. **(b)** The circle is $S = x^2 + y^2 - 2rx - 2hy + h^2 = 0$ and the point $P(0, 0)$

$\Rightarrow$ $T_1 = (-rx - hy + h^2)$ and $S_1 = h^2$ from $SS_1 = T_1^2$, we get $h^2(x^2 + y^2) + h^2(h^2 - 2rx - 2hy)$
$= (rx + hy)^2 + (h^4 - 2h^2rx - 2h^3y)$.

So $(h^2 - r^2)x^2 - 2rhxy = 0$

$\Rightarrow$ $x = 0$ and $(h^2 - r^2)x - 2rhy = 0$ are the lines

Aliter: The circle is $(x - r)^2 + (y - h)^2 = r^2$ and $C(r, h)$, radius $= r$

$\Rightarrow$ The circle touches y-axis which passes through origin, so $x = 0$ is one tangent

Now slope of OC is $m = \dfrac{h}{r}$ and $\tan\dfrac{\theta}{2} = \left|\dfrac{r}{h}\right|$

Since $m_1 = \dfrac{m + \frac{r}{h}}{1 - m\frac{r}{h}}$ is not defined i.e., $x = 0$

So the other tangent has slope $m_2 = \dfrac{\frac{h}{r} - \frac{r}{h}}{1 + \frac{h}{r}\cdot\frac{r}{h}} = \dfrac{h^2 - r^2}{2rh}$

Hence the equation $y = \dfrac{h^2 - r^2}{2rh}x$ i.e., $(h^2 - r^2)x - 2rhy = 0$

16. **(c)** The circle is $(x - 1)^2 + (y - 2)^2 = 5 - \lambda$ and the point $P(1, 2)$, since there are infinite number of tangents from $P(1, 2)$

$\Rightarrow$ Circle must be a point circle centered at $P(1, 2)$ that the equation will present when $5 - \lambda = 0$ i.e., $\lambda = 5$

17. **(b)** if $x + my - 1 = 0$ is a tangent to the circle $x^2 + y^2 = a^2$

then $\dfrac{1}{\sqrt{\ell^2 + m^2}} = a$, so $\ ^2 + m^2 = \dfrac{1}{a^2}$

As a or $\dfrac{1}{a}$ are constants so $\ ^2 + m^2 =$ constant

$\therefore$ The locus is a circle

18. **(a)** The circle $x^2 + y^2 = a^2$ has centre at $C(0, 0)$ and radius $r = a$. Let the line be $\sqrt{3}x + y + c_1 = 0$

Which will become a tangent when $\dfrac{|c_1|}{2} = a$ gives $c_1 = \pm 2a$

Hence $\sqrt{3}\,x + y \pm 2a = 0$

19. **(a)** The circle $x^2 + 2x + 1 + y^2 + 6y + 9 = 25$ is centered at $(-1, -3)$ and $r = 5$.

Now $O(0, 0)$ gives $OC = \sqrt{10} < r$, so $O(0, 0)$ lies inside the circle

$\therefore$ no tangent is possible

Aliter: $x^2 + y^2 + 2x + 6y - 15 = 0$ and point $O(0, 0)$

$\Rightarrow$ $S(0, 0) = -15 < 0$

$\Rightarrow$ Point lies inside the circle so no tangent possible

TEXTUAL EXERCISE 4–(SUBJECTIVE)

1. The circle $x^2 + y^2 = 4$ has centre $C(0, 0)$ and radius $r = 2$.

The line $x + 2y + c_1 = 0$ will be a tangent when $\dfrac{|c_1|}{\sqrt{5}} = 2$

$\Rightarrow$ $c_1 = \pm 2\sqrt{5}$ and the tangents are $x + 2y \pm 2\sqrt{5} = 0$

2. The straight line $ax - by + b^2 = 0$ will be a tangent to the circle $\left(x - \dfrac{a}{2}\right)^2 + \left(y - \dfrac{b}{2}\right)^2 = \dfrac{a^2 + b^2}{4}$ when

$$\frac{\left[a\frac{a}{2} - b\frac{b}{2} + b^2\right]}{\sqrt{a^2 + b^2}} = \frac{\sqrt{a^2 + b^2}}{2}$$

$\Rightarrow$ $\dfrac{a^2 + b^2}{2} = \dfrac{a^2 + b^2}{2}$, which is always true. So the line is always a tangent for all real values of a and b.

The point of contact is the foot from $\left(\dfrac{a}{2}, \dfrac{b}{2}\right)$ given by

$$\frac{\frac{a}{2} - x}{a} = \frac{\frac{b}{2} - y}{-b} = \frac{\frac{a^2 + b^2}{2}}{a^2 + b^2}$$

$\Rightarrow$ $a - 2x = a$ $\Rightarrow$ $x = 0$ and $\dfrac{b - 2y}{-2b} = \dfrac{1}{2}$

$\Rightarrow$ $y = b$ $\quad\therefore$ Point of contact is $(0, b)$

3. We observe that $A(5, 12)$ and $B(12, -5)$ are two points on the circle $x^2 + y^2 = 169$ which has centre at $O(0, 0)$

Now, slope of OA is $m_1 = 12/5$ and slope of OB is $m_2 = -5/12$

$\Rightarrow$ Slope of tangent at A is $m_3 = \dfrac{-5}{12}$ and slope of tangent at B is $m_4 = \dfrac{12}{5}$, observe that m_3 and m_4 are perpendicular.

4. $4x - 3y = -12$ is a tangent to the circle at $A(-3, 0)$

$\Rightarrow$ Equation of normal at A is $y = -\dfrac{3}{4}(x + 3)$ i.e., $3x + 4y + 9 = 0$. Similarly equation of normal at $B(4, 1)$ is $y - 1 = \dfrac{4}{3}(x - 4)$ i.e., $4x - 3y - 13 = 0$. The normal intersect at $C(1, -3)$

$\therefore$ Radius $r = 5$ which gives the equation $x^2 - 2x + y^2 + 6y - 15 = 0$ or $(x - 1)^2\ (y + 3)^2 = 5^2$

5. The line $x\cos\alpha + y\sin\alpha - p = 0$ will be a tangent to the circle $(x - a\cos\alpha)^2 + (y - b\sin\alpha)^2 = a^2\cos^2\alpha + b^2\sin^2\alpha$

When $|a\cos^2\alpha + b\sin^2\alpha - p| = \sqrt{a^2\cos^2\alpha + b^2\sin^2\alpha}$ which gives $p = a\cos^2\alpha + b\sin^2\alpha \pm \sqrt{a^2\cos^2\alpha + b^2\sin^2\alpha}$

6 Tangent to the circle $x^2 + y^2 - 5 = 0$ at $A(1, -2)$ will be $x - 2y - 5 = 0$

Now, the other circle $(x - 4)^2 + (y + 3)^2 = (\sqrt{5})^2$ has centre at $B(4, -3)$ and radius $r_2 = \sqrt{5}$ units.

Now, distance of $B(4, -3)$ from the line $x - 2y - 5 = 0$ is $\dfrac{|4 + 6 - 5|}{\sqrt{5}} = \sqrt{5} = r_2$

The line is also tangent to the other circle. The point of contact is (x_1, y_1) given by $\dfrac{4 - x_1}{1} = \dfrac{-3 - y_1}{-2} = \dfrac{4 + 6 - 5}{5}$ gives $x_1 = 3$ and $y_1 = -1$

$\Rightarrow$ The point of contact is $(3, -1)$

7. The line $Ax + By + C = 0$ will touch the circle $(x - \alpha)^2 + (y - \beta)^2 = c^2$ when $\dfrac{|A\alpha + B\beta + C|}{\sqrt{A^2 + B^2}} = c$

$\Rightarrow$ $A\alpha + B\beta + C = \pm c\sqrt{A^2 + B^2}$

TEXTUAL EXERCISE–4 (OBJECTIVE)

1. (a) Since the line L: $ax + by + c = 0$ is a tangent to $x^2 + y^2 - 2x - 3/5 = 0$, centered at (1, 0) and radius $r = \frac{4}{\sqrt{10}}$

$\therefore \quad \frac{|a+c|}{\sqrt{a^2+b^2}} = \frac{4}{\sqrt{10}}$

Similarly the line $ax + by + c = 0$ will be a normal to circle $(x + 1)^2 + (y - 2)^2 = 2^2$ when $L(-1, 2) = -a + 2b + c = 0$ gives $c = a - 2b$,

so $\frac{|a+c|}{\sqrt{a^2+b^2}} = \frac{2|a-b|}{\sqrt{a^2+b^2}} = \frac{4}{\sqrt{10}}$

$\Rightarrow \frac{|a-b|}{\sqrt{a^2+b^2}} = \frac{2}{\sqrt{10}}$

Now, $a = 1$, $b = 3$ serves the purpose (for $c = -5$) completely.

2. (c) Circle $(x - 2)^2 + y^2 = 2^2$ has centre at C (2, 0) and radius $r = 2$

Normal at (4, 0) will meet the circle again at (0, 0), so that C becomes the mid point

3. (b) From the given equation $(f^2 + g^2 - 6) = 4\,(f^2 + g^2 + 3g + 3f)$

$\Rightarrow 3(f^2 + g^2) + 12g + 12f + 6 = 0$ i.e., $f^2 + g^2 + 4g + 4f + 2 = 0$

4. (a) Observe that the circle passing through O (0, 0), A (2a, 0) and B (0, 2b) is $x\,(x - 2a) + y\,(y - 2b) = 0$ i.e., $x^2 + y^2 - 2ax - 2by = 0$

So the tangent at (0, 0) will be $-ax - by = 0$ or $ax + by = 0$

5. (d) Length of intercept made by the circle $x^2 + y^2 + 10x - 6y + 9 = 0$ on x-axis is $I_x = 2\sqrt{g^2 - c} = 2\sqrt{25 - 9} = 8$ units

6. (b) The line $x \sin \alpha - y \cos \alpha + a \cos \alpha = 0$ will be a tangent to the circle $x^2 + y^2 - a^2 = 0$, when $|a \cos\alpha| = a$ i.e., $\cos^2\alpha = 1$

7. (c) From the given equation $f^2 + g^2 - 6 = 4\,(f^2 + g^2 + 3g + 3f)$, so $3(f^2 + g^2) + 3(4g + 4f) + 6 = 0$ or $f^2 + g^2 + 4g + 4f + 2 = 0$

8. (a) The equation of tangent at $\left(\frac{ab^2}{a^2+b^2}, \frac{a^2b}{a^2+b^2}\right)$ to the circle

$x^2 + y^2 - \frac{a^2b^2}{a^2+b^2} = 0$ is $\frac{ab^2x}{a^2+b^2} + \frac{a^2by}{a^2+b^2} - \frac{a^2b^2}{a^2+b^2} = 0$

or $bx - ay - ab = 0$

$\Rightarrow \frac{x}{a} + \frac{y}{b} = 1$

9. (d) The centre of circle $x^2 + y^2 + 2gx + 2fy + c = 0$ is at $(-g, -f)$ and radius $r = \sqrt{g^2 + f^2 - c}$.

The centre of the circum-circle of OPQ will be at $\left(-\frac{g}{2}, -\frac{f}{2}\right)$

10. (c) OQ = a and $PQ = \sqrt{\alpha^2 + \beta^2 - a^2}$

$\Rightarrow \tan\frac{\theta}{2} = \frac{a}{\sqrt{\alpha^2+\beta^2-a^2}}$

So angle between the tangents $\theta = 2\tan^{-1}\frac{a}{\sqrt{\alpha^2+\beta^2-a^2}}$

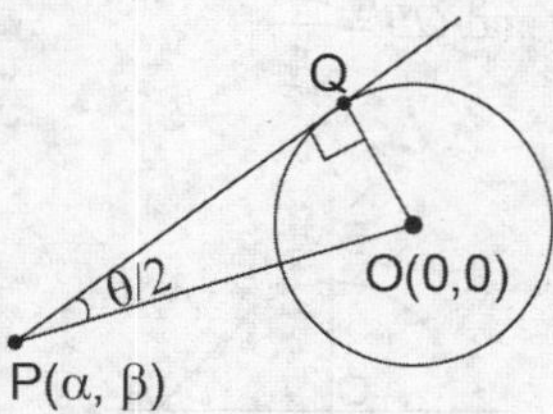

11. (c) Let (x_1, y_1) be a point on the circle $x^2 + y^2 - 4 = 0$, then the equation of tangent is $xx_1 + yy_1 - 4 = 0$ which gives

$A\left(\frac{4}{x_1}, 0\right)$ & $B\left(0, \frac{4}{y_1}\right)$

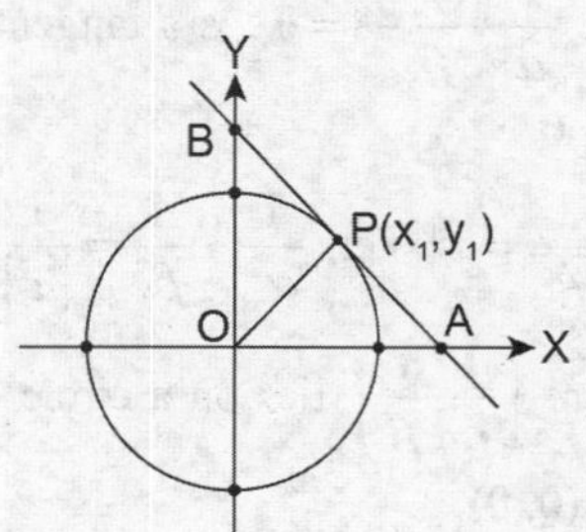

Mid point of AB is $M\left(\frac{2}{x_1}, \frac{2}{y_1}\right)$

$\Rightarrow \quad x = \frac{2}{x_1}$ and $y = \frac{2}{y_1} \Rightarrow x_1^2 + y_1^2 = \frac{4}{x^2} + \frac{4}{y^2}$

i.e., $\frac{4(x^2+y^2)}{x^2y^2} = 4 \ (as\, x_1^2 + y_1^2 = 4)$

$\therefore \quad x^2 + y^2 = x^2y^2$ is the locus mid-point of AB

12. (c) The line $ax + by + c = 0$ is a normal to the circle $x^2 + y^2 = r^2$

$\Rightarrow$ The line is a diameter $\therefore$ Length intercepted = 2r

13. (b) Centre of the circle is C (–1, 1). The line tangent o the circle is $x + 2y + 12 = 0$

$\therefore$ The point of contact is P (h, k) is given by $\frac{-1-h}{1} = \frac{1-k}{2} = \frac{13}{5}$

$\Rightarrow \quad h = -\frac{18}{5}$ and $k = \frac{-21}{5} \quad \Rightarrow P\left(-\frac{18}{5}, -\frac{21}{5}\right)$

14. (a) The circle $x^2 + y^2 + 2x + 4y + 3 = 0$ has centre at $C(-1, -2)$ and the point on the circle is P (–2, –3)

$\Rightarrow$ Slope of $CP = \frac{1}{1} = 1$

15. (b) A line perpendicular to $y = mx + c$ will be $x + my + c_1 = 0$

Now this line will be a tangent to the circle $x^2 + y^2 - a^2 = 0$ when $\frac{|c_1|}{\sqrt{1+m^2}} = |a|$

$\Rightarrow$ The equation of tangent line will be $x + my \pm a\sqrt{1+m^2} = 0$

16. **(c)** Equation of tangent at P (3, 4) to the circle $x^2 + y^2 - 25 = 0$ is $3x + 4y - 25 = 0$

$\Rightarrow \quad OA = \dfrac{25}{3}$ and $OB = \dfrac{25}{4}$

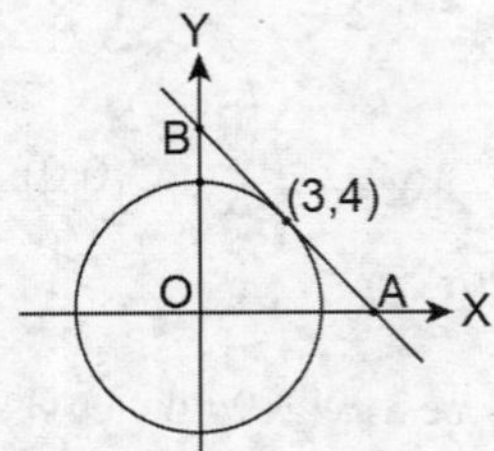

$\therefore$ Area of $\Delta OAB = \dfrac{625}{24}$ square units.

17. **(b)** The line $\dfrac{x}{\alpha}+\dfrac{y}{\beta}-1=0$, is a tangent to the circle $x^2 + y^2 - a^2 = 0$

$\Rightarrow \quad \dfrac{|\alpha\beta|}{\sqrt{\alpha^2+\beta^2}}=|a|$ i.e., $\dfrac{1}{\alpha^2}+\dfrac{1}{\beta^2}=\dfrac{1}{|a^2|}$

$\Rightarrow$ The point $\left(\dfrac{1}{\alpha},\dfrac{1}{\beta}\right)$ lies on a circle of radius $\dfrac{1}{|a|}$ centered at (0, 0)

18. **(c)** The line $ax + by = 0$ touches the circle $(x + 1)^2 (y + 2)^2 = \left(\sqrt{5}\right)^2$

So, $\dfrac{|-a-2b|}{\sqrt{a^2+b^2}}=\sqrt{5}$ or $\dfrac{(a+2b)^2}{a^2+b^2}=5$. The same lines will be a normal to the circle $x^2 - 4x + y^2 + 2y - 3 = 0$ when (2, –1) lies on the line, so $2a - b = 0$ i.e., $b = 2a$ solution gives $a = 1$ and $b = 2$

TEXTUAL EXERCISE–5 (SUBJECTIVE)

1. $x^2 + y^2 - 6x + 2y + 5 = 0$ circle has centered C (3, –1) and radius $r = \sqrt{5}$ units. The point P (4, –4) gives $PC = \sqrt{1^2+3^2} = \sqrt{10}$ units

Now, $\sin\dfrac{\theta}{2} = \dfrac{r}{PC}=\dfrac{\sqrt{5}}{\sqrt{10}}=\dfrac{1}{\sqrt{2}}$ and

$QM = \sqrt{S_1}\sin\dfrac{\theta}{2}=\dfrac{1}{\sqrt{2}}\times\sqrt{5}$

Length of chord of contact = 2 $QM = \sqrt{10}$ units.

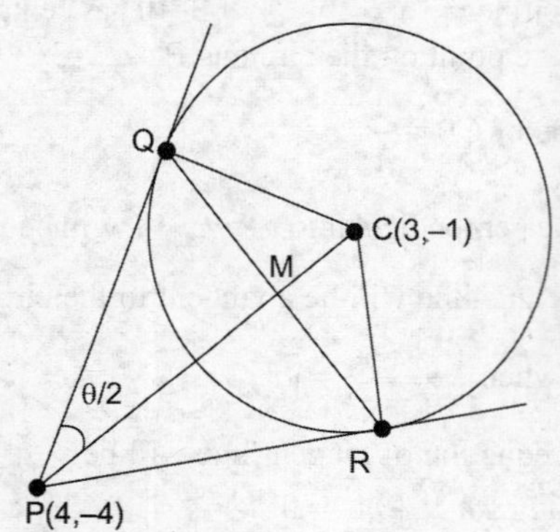

2. Let $P(x_1, y_1)$ be the point on $x^2 + y^2 = a^2$

$\Rightarrow \quad x_1^2 + y_1^2 = a^2$

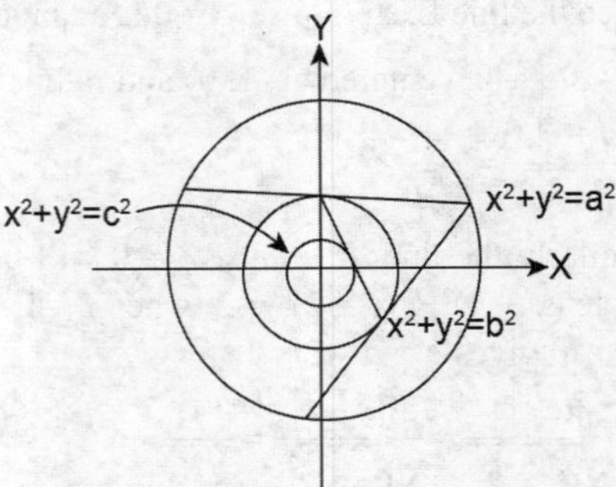

From where tangents are drawn to the circle $x^2 + y^2 - b^2 = 0$

$\Rightarrow \quad xx_1 + yy_1 + b^2 = 0$ is the chord of contact.

Since this chord of contact is a tangent to the circle $x^2 + y^2 = c^2$

$\therefore \quad \dfrac{|-b^2|}{\sqrt{x_1^2+y_1^2}}=|C| \qquad \Rightarrow \quad b^2 = ac$ i.e., a, b, c are in G.P.

3. Point P (h, k) and circle is $x^2 + y^2 = a^2$

$\therefore \quad \sin\dfrac{\theta}{2}=\dfrac{a}{\sqrt{h^2+k^2}}$ and $\cos\dfrac{\theta}{2}=\dfrac{\sqrt{h^2+k^2-a^2}}{\sqrt{h^2+k^2}}$

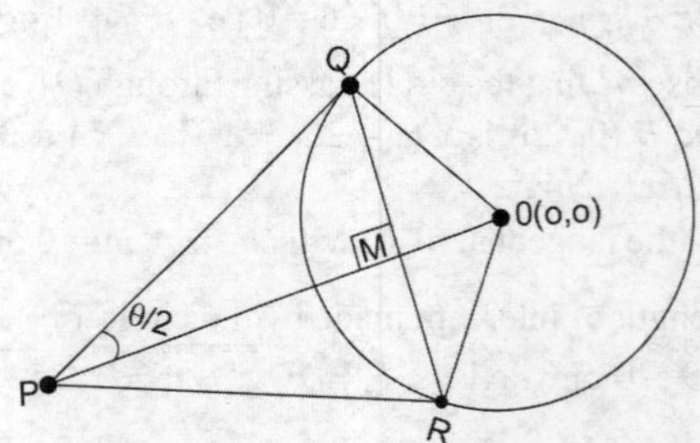

$\Rightarrow \quad PQ = \sqrt{h^2+k^2-a^2}$, now $QM = PQ \sin\dfrac{\theta}{2}$ and $PM = PQ\cos\dfrac{\theta}{2}$

$\therefore$ Area of $\Delta PQR = 2\times\dfrac{1}{2}$ $(PM \times QM) = PM \times QM$

$= PQ\cos\dfrac{\theta}{2}.PQ\sin\dfrac{\theta}{2} = (h^2 + k^2 - a^2)\dfrac{a\sqrt{h^2+k^2-a^2}}{(h^2+k^2)}$

$= \dfrac{a(h^2+k^2-a^2)^{3/2}}{(h^2+k^2)}$

4. Any line through O (0, 0) is $y = mx$, which will intersect the circle $(x - 1)^2 + y^2 = 1$, so $(x - 1)^2 + m^2x^2 = 1$ gives the point of intersection at $\left(\dfrac{2}{m^2+1},\dfrac{2m}{m^2+1}\right)$

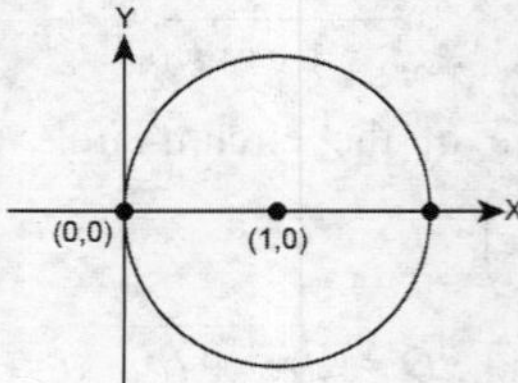

$\therefore$ Mid point of chord is $\left(\dfrac{1}{m^2+1},\dfrac{m}{m^2+1}\right)$

Hence, the locus is $x^2+y^2=\dfrac{m^2+1}{(m^2+1)^2}=\dfrac{1}{m^2+1}=x$ i.e., $x^2+y^2=x$

TEXTUAL EXERCISE–5 (OBJECTIVE)

1. (a) Since chord is subtending a right angle at the centre (origin)
$\Rightarrow$ Mid point of chord serves as centre of circle and ends points of chord serve as diameter.

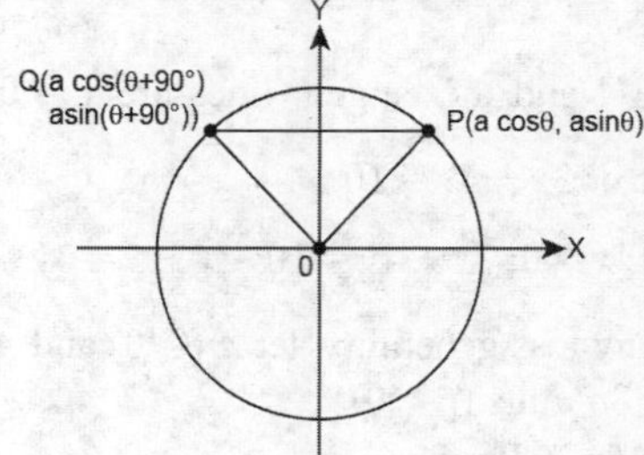

$\therefore$ Its distance of centre $=\dfrac{\sqrt{a^2+a^2}}{2}$. Hence the locus $x^2+y^2=\dfrac{a^2}{2}$

Aliter: A chord that subtends a right angle at the centre of $x^2+y^2=a^2$ will have ends as $P(a\cos\theta, a\sin\theta)$ and $Q(-a\sin\theta, a\cos\theta)$, so mid pint is $M\left(\dfrac{a(\cos\theta-\sin\theta)}{2}, \dfrac{a(\sin\theta+\cos\theta)}{2}\right)$.

By squaring and adding, we get $(2x^2)+(2y)^2=2a^2\{\cos^2\theta+\sin^2\theta\}$
$\Rightarrow x^2+y^2=\dfrac{a^2}{2}$ is the required locus

2. (a) The circle is $x^2+(y-1)^2=1$ any line through the origin is $y=mx$ which will intersect the circle at $\left(\dfrac{2m}{m^2+1}, \dfrac{2m^2}{m^2+1}\right)$ and the mid point of the chord is $M\left(\dfrac{m}{m^2+1}, \dfrac{m^2}{m^2+1}\right)$
$\Rightarrow x(m^2+1)=m$ and $y(m^2+1)=m^2$ on squaring and adding $(m^2+1)^2(x^2+y^2)=m^2(m^2+1)$
$\Rightarrow x^2+y^2=\dfrac{m^2}{m^2+1}=y$ or $x^2+y^2-y=0$

3. (a) Let PM be a secant where M is the mid point of the part of secant intercepted by the circle. Since at the mid point of the chord the perpendicular drawn always passes through the centre of the circle.
$\Rightarrow$ ΔOPM is always right angled at M.
$\Rightarrow$ M always lies on a circle with OP as its diameter. Hence the circle $x(x-h)+y(y-k)=0$ i.e., $x^2+y^2=xh+yk$

4. (a) The circle are S_1: $x^2+y^2+4x+1=0$ and S_2: $x^2+y^2+6x+2y+3=0$.
The common chord is $S_1-S_2=0$ i.e., $-2x-2y-2=0$ or $x+y+1=0$

5. (d) Let S: $x^2+y^2+x-y-1=0$ and the given point is $M(1, 1)$. Observe that $S(1, 1)=1>0$. So $M(1, 1)$ lies outside the circle
$\therefore$ No mid point

6. (b) Let the circle be $(x-a)^2+y^2=a^2$ and $y=mx$ be a chord
$\Rightarrow (m^2+1)x^2-2ax=0 \Rightarrow x=\dfrac{2a}{m^2+1}$ and $y=\dfrac{2am}{m^2+1}$
Now, $O(0, 0)$ and $A\left(\dfrac{2a}{m^2+1}, \dfrac{2am}{m^2+1}\right)$ are ends of a diameter
Hence the circle $x\left(x-\dfrac{2a}{m^2+1}\right)+y\left(y-\dfrac{2am}{m^2+1}\right)=0$
$\Rightarrow (x^2+y^2)(m^2+1)-2a(x+my)=0$

7. (c) $T=S_1$ is the equation, so for $P(x_1, y_1)$ and $S=x^2+y^2-a^2=0$, we get $x_1x+yy_1-a^2=x_1^2+y_1^2-a^2$
$\Rightarrow xx_1+yy_1=x_1^2+y_1^2$

8. (b) Length of the chord intercepted by $x^2+y^2-r^2=0$ on $\dfrac{x}{a}+\dfrac{y}{b}=1$ is
$$\ell=2\sqrt{r^2-\dfrac{1}{\dfrac{1}{a^2}+\dfrac{1}{b^2}}}=2\sqrt{\dfrac{(a^2+b^2)r^2-a^2b^2}{a^2+b^2}}$$

9. (c) The line $2y=x-2$ intersects the circle $x^2+y^2-25=0$
$\Rightarrow (2y+2)^2+y^2-25=0$
$\Rightarrow 5y^2+8y+4-25=0$, hence $y_1+y_2=-8/5$
i.e. $\dfrac{y_1+y_2}{2}=-\dfrac{4}{5}$ and then $\dfrac{x_1+x_2}{2}=(y_1+y_2)+2=\dfrac{2}{5}$
$\Rightarrow$ Mid point $\left(\dfrac{2}{5}, -\dfrac{4}{5}\right)$

10. (c) The circles are S_1: $x^2+y^2-2ax=0$ and S_2: $x^2+y^2-2by=0$, so the equation of common chord: $ax-by=0$
Length of the common chord $\ell=2\sqrt{a^2-\dfrac{a^4}{a^2+b^2}}$
$=|2a|\sqrt{\dfrac{a^2+b^2-a^2}{a^2+b^2}}=\dfrac{|2ab|}{\sqrt{a^2+b^2}}$

11. (a) The chord of contact form the origin $O(0, 0)$ to the circle $x^2+y^2+2gx+2fy+c=0$ is $T=0$
$\Rightarrow gx+fy+c=0$ or $2gx+2fy+2c=0$ is the equation of chord of contact and that from (g, f) is $2gx+2fy+g^2+f^2+c=0$
$\therefore$ Distance between above two chord of contact
$=\dfrac{|g^2+f^2+c-2c|}{\sqrt{4g^2+4f^2}}=\dfrac{|g^2+f^2-c|}{2\sqrt{g^2+f^2}}$

12. (a) The chord of contact form $P(5, -3)$ to the circle $x^2+y^2-10=0$ is $T=0$ i.e., $5x-3y-10=0$

13. (d) Let $y=mx$ be a line through $O(0, 0)$
Now, line $y=mx$ intersects circle $x^2+y^2-2ax=0$ locus of mid point of the chord intercepted lies on the circle with $O(0, 0)$ and $A(a, 0)$ as ends of the diameter
$\Rightarrow$ The locus (circle) is $x^2+y^2-ax=0$

14. (c) The given circle $(x-1)^2+(y-3)^2=2^2$ has centre at $C_1(1, 3)$ and radius $r_1=2$.
Since one the diameter is a chord to the other circle.
$\Rightarrow$ Centre $C_1(1, 3)$ is the mid point of the chord and line joining $C_1(1, 3)$ and $C_2(2, 1)$ is perpendicular to the chord
$\therefore r_2=\sqrt{r_1^2+(C_1C_2)^2}=\sqrt{4+1+4}=3$ units.

15. (d) Solving $x^2 + y^2 - 2x = 0$ and $y = x$, we get $2x(x - 1) = 0$, gives $A(0, 0)$ and $B(1, 1)$.
Hence the equation of the circle is $x^2 + y^2 - x - y = 0$

TEXTUAL EXERCISE-6 (SUBJECTIVE)

1. The circles are $S_1: x^2 + y^2 = 1$
$\Rightarrow$ $C_1(0, 0)$ and $r_1 = 1$ and $S_2\ (x - 1)^2 + y - 3)^2 = 2^2$
$\Rightarrow$ $C_2(1, 3)$ and $r_2 = 2$
Now $C_1C_2 = \sqrt{10} > 2 + 1 = 3$
$\Rightarrow$ Circles do not intersect for direct tangents the point of intersection is $D = \dfrac{C_1r_2 - C_2r_1}{r_2 - r_1}$
$\Rightarrow$ $D = (-1, -3)$.
Let $y + 3 = m(x + 1)$ be a tangent, so $\dfrac{|m-3|}{\sqrt{1+m^2}} = 1$
$\Rightarrow$ $m = \dfrac{4}{3}, \infty$
Hence the tangent lines are $x = -1$ and $4x - 3y - 5 = 0$. Similarly the point of intersection of transverse common tangents is
$$T = \frac{C_1r_2 + C_2r_1}{r_2 + r_1} = \left(\frac{1}{3}, 1\right)$$
Let $y - 1 = m(x - 1/3)$ be a tangent, so $\dfrac{|3-m|}{3\sqrt{1+m^2}} = 1$
$\Rightarrow$ $m = 0, -3/4$ and the tangent lines are $y = 1$ and $4y - 4 = (-3)(x - 1/3)$
$\Rightarrow$ $3x + 4y - 5 = 0$
The four tangents are $x = -1$; $y = 1$; $3x + 4y - 5 = 0$ and $4x - 3y - 5 = 0$

2. $S_1: x^2 + y^2 = 4$
$\Rightarrow$ $C_1(0, 0)$ and $r_1 = 2$ and $S_2: (x - 4)^2 + y^2 =$
$\Rightarrow$ $C_2 = (4, 0)$, $r_2 = 1$
Now, $C_1C_2 = 4 > r_1 + r_2$
$\Rightarrow$ Circles do not intersect of D.C.T is $(-8, 0)$
Let $y = mx + 8m$ be a tangent, so $\dfrac{|8m|}{\sqrt{1+m^2}} = 2$ gives m $= \pm 1/15$
The tangent lines are $y = \pm\dfrac{1}{\sqrt{15}}(x + 8)$ i.e., $x \pm y\sqrt{15} + 8 = 0$
Similarly the point of intersection of T.C.T is $T\left(\dfrac{8}{3}, 0\right)$
and $y = m(x - 8/3)$ is a tangent
So $\dfrac{|8m|}{3\sqrt{1+m^2}} = 2$ gives $m = \pm 3/\sqrt{7}$ and the tangents are
$y = \pm\dfrac{3}{\sqrt{7}}\left(x - \dfrac{8}{3}\right)$ gives $3x \pm y\sqrt{7} - 8 = 0$

3. The circles are $S_1: (x - 1)^2 + (y - 3)^2 = 1$
$\Rightarrow$ $C_1(1, 3)$ and $r_1 = 1$
$S_2: (x + 3)^2 + (y - 1)^2 = 3^2$
$\Rightarrow$ $C_2(-3, 1)$ and $r_2 = 3$
Now $C_1C_2 = \sqrt{4^2 + 2^2} = 2\sqrt{5} > 4$
$\Rightarrow$ Circles do not intersect
Now point of intersection of D.C.T is $D(3, 4)$. Let $y - 4 = m(x - 3)$ be a tangent
So $\left|\dfrac{1-2m}{\sqrt{1+m^2}}\right| = 1$ gives $m = 0, 4/3$ and the tangent lines are $y = 4$ and $y - 3 = \dfrac{4}{3}(x - 1)$ or $4x - 3y + 5 = 0$
Similarly point of intersection of TCT is $T\left(0, \dfrac{5}{2}\right)$. Let $y - \dfrac{5}{2} = mx$ be a tangent, so $\dfrac{|2m-1|}{2\sqrt{1+m^2}} = 1$ gives $m = \infty, -3/4$ and the tangent lines are $x = 0$ and $y - \dfrac{5}{2} = -\dfrac{3}{4}x$ or $3x + 4y = 10$
$\therefore$ $y = 4$; $x = 0$; $3x + 4y = 10$; $4x + 5 = 3y$

4. Without any loss generality, let $A(0, 0)$ and AB along x-axis and $AB = 8p$ units $(p > 0)$
$\Rightarrow$ $AC = 6p$, $CB = 2p$
$\therefore$ $C_1(3p, 0)$ and $C_2(7p, 0)$, $B(8p, 0)$, $r_1 = 3p$, $r_2 = p$

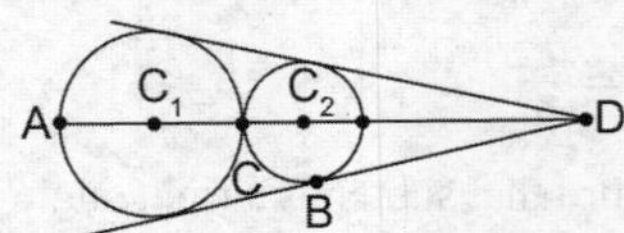

Since circles touch each other externally, so $C_1C_2 = 4p$
Point of intersection of D.C.T is $D = (9p, 0)$
Hence $BD = p = r_2$

5. Equation of the common tangent is $4x + 3y = 10$ which passes through $P(1, 2)$ where circles touch each other externally as circles are of equal radius
$\Rightarrow$ Centre of circles lie on $3x - 4y + 5 = 0$ at 5units distance on both side i.e., $C_1(1 + 4, 2 + 3)$ and $C_2(1 - 4, 2 - 3)$ gives $C_1(5, 5)$ and $C_2(-3, -1)$, both has radius 5 units. Hence $S_1: (x - 5)^2 + (y - 5)^2 = 5^2$ or $x^2 + y^2 - 10x - 10y + 25 = 0$ and $S_2: (x + 3)^2 + (y + 1)^2 = 5^2$ or $x^2 + y^2 + 6x + 2y - 15 = 0$

6. Given $S_1: x^2 + y^2 - a^2 = 0$
$\Rightarrow$ $C_1(0, 0)$, $r_1 = a$ units
$S_2: (x - 2a)^2 + y^2 = a^2 = 0$
$\Rightarrow$ $C_2(2a, 0)$, $r_2 = a$ units
As $|C_1C_2| = r_1 + r_2$
$\therefore$ Circles touch each other and these circles are of equal radius and without any loss of generality.
Let $a > 0$, then the centre of third Circe of equal radius will be at $(a, \pm\sqrt{3}a)$ and radius $r = a$.
Hence the circles will be $x^2 + y^2 - 2ax \pm 2\sqrt{3}\,ay + 3a^2 = 0$

7. The given pair of lines is $(x + 3)(x + 3y) = 0$ which intersect at $x = -3$ and $y = 1$ i.e., $C_2(-3, 1)$.
The given circle is $(x + 3)^2 + (y + 3)^2 = 1$, so $C_1(-3, -3)$ and $r_1 = 1$ units
Now, $C_1C_2 = 4$ units as circles touch each other externally so $r_2 = 3$ units
$\therefore$ $S_2: (x + 3)^2 + (y - 1)^2 = 3^2$ or $x^2 + y^2 + 6x - 2y + 1 = 0$

TEXTUAL EXERCISE-6 (OBJECTIVE)

1. (c) Given S_1: $(x-2)^2 + (y-3)^2 = 4^2$

$\Rightarrow$ C_1 (2, 3), $r_1 = 4$ units and S_2: $(x+1)^2 + (y+1)^2 = 1^2$

$\Rightarrow$ C_2 (–1,–1), $r_2 = 1$ unit

Hence $C_1C_2 = \sqrt{3^2+4^2} = 5$ units $= r_1 + r_2$

Hence circles touch each other externally

$\therefore$ 3 tangent can be drawn

2. (c) Given S_1: $(x-5)^2 + y^2 = 3^2$

$\Rightarrow$ $C_1(5, 0)$, $r_1 = 3$ units

S_2: $x^2 + y^2 = r^2$

$\Rightarrow$ C_2 (0, 0), $r_2 = r$

Now $C_1C_2 = 5$ units

The circles will intersect in two distinct points when $|r_1 - r_2| < C_1C_2 < r_1 + r_2$

Now, $r_1 + r_2 > C_1C_2$ $\Rightarrow$ $r_2 > 2$ and $|r_2 - r_1| < C_1C_2$

$\Rightarrow$ $r_2 < 8$

$\Rightarrow$ $2 < r_2 < 8$

3. (b) Given S_1: $x^2 + y^2 = 2^2$

$\Rightarrow$ C_1 (0, 0) and $r_1 = 2$ units and S_2: $(x-3)^2 + (y-4)^2 = 7^2$

$\Rightarrow$ C_2 (3, 4) and $r_2 = 7$ units. Now $C_1C_2 = 5$ units $= r_2 - r_1$

$\Rightarrow$ Circles touch each other internally.

So only one tangent is possible

4. (d) Given S_1: $x^2 + y^2 = 4^2$

$\Rightarrow$ $C_1(0, 0)$, $r_1 = 4$ units and S_2: $(x-2)^2 + (y-2)^2 = \left(2\sqrt{2}\right)^2$

$\Rightarrow$ C_2 (2, 2) and $r_2 = 2\sqrt{2}$ units.

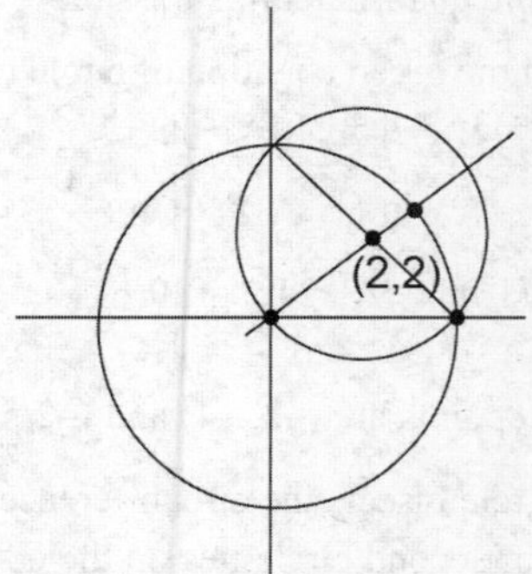

Now, $C_1C_2 = 2\sqrt{2}$ as $r_1 - r_2 < C_1C_2 < r_1 + r_2$

$\therefore$ The circles intersect at two distinct points which are (4, 0) and (0, 4) as and the common chord is $x + y = 4$

$\Rightarrow$ The angle subtended by the common chord at O (0, 0) is $\pi/2$.

5. (d) The circle $S_1: \left(x+\frac{1}{2}\right)^2 + \left(y+\frac{1}{2}\right)^2 = \left(\frac{1}{\sqrt{2}}\right)^2$

$\Rightarrow$ $C_1\left(-\frac{1}{2}, -\frac{1}{2}\right), r_1 = \frac{1}{\sqrt{2}}$ and

$S_2: \left(x+\frac{1}{2}\right)^2 + \left(y-\frac{1}{2}\right)^2 = \left(\frac{1}{\sqrt{2}}\right)^2$

$\Rightarrow$ $C_2\left(-\frac{1}{2}, \frac{1}{2}\right)$ and $r_2 = \frac{1}{\sqrt{2}}$

Now $C_1C_2 = 1$ also $r_1^2 + r_2^2 = 1 = C_1C_2^2 \Rightarrow$ Circles intersects at 90°

6. (c), (d) Centre of circle $x^2 + y^2 = 1$ is at C_1 (0, 0) and $r_1 = 1$ unit.

Now $C_2 = (4, 3) \Rightarrow C_1C_2 = 5$ units.

$\therefore$ Circle touching the circle at C_1 will have radius $r_2 = 4$ unit or 6 unit (external /internal touching respectively).

Hence $(x-4)^2 + (y-3)^2 = 4^2$

$\Rightarrow$ $x^2 + y^2 - 8x - 6y + 9 = 0$. The other circle is $x^2 + y^2 - 8x - 6y - 11 = 0$

7. (a) Since the circles touch each other

$\therefore$ Common chord has equation $6x + 12y - 6 = 0$ or $x + 2y = 1$. Now only (3, –1) satisfies it.

Aliter: Given S_1: $(x-2)^2 + (y+3)^2 = \left(\sqrt{5}\right)^2$

$\Rightarrow$ C_1 (2, –3) and $r_1 = \sqrt{5}$ and

S_2: $(x-5)^2 + (y-3)^2 = \left(2\sqrt{5}\right)^2 \Rightarrow 2\sqrt{5}$ and C_2 (5, 3)

Now $C_1C_2 = \sqrt{3^2+6^2} = 3\sqrt{5} = r_1 + r_2$

So circles touch externally and the point of contact lies on the line joining C_1C_2 and divides it in the ratio 1 : 2 internally

$\Rightarrow$ $C(3,-1)$

8. (d) The given circle S_1: $(x-3)^2 + (y-3)^2 = 2^2$

$\Rightarrow$ C_1 (3, 3) and $r_1 = 2$

Let C_2 (x_1, y_1) be the centre of circle S_2, then

$|x_1| = \sqrt{(x_1-3)^2 + (y_1-3)^2} - 2$

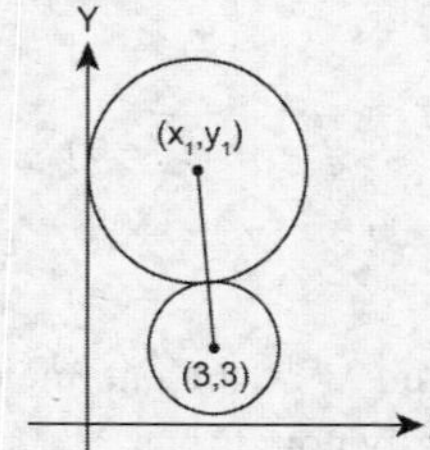

So $2 + |x_1| = \sqrt{(x_1-3)^2 + (y_1-3)^2}$

$\Rightarrow$ $4 + x_1^2 + 4|x_1| = x_1^2 - 6x + 9 + y_1^2 - 6y + 9$

Observe that $x_1 > 0$, so that S_2 is on RHS of y-axis so $y_1^2 - 10x_1 - 6y_1 + 14 = 0$ or $y^2 - 10x - 6y + 14 = 0$

9. (a) Given Circles are $x^2 + y^2 = a^2$ …. (i) and

$x^2 - 2ax + y^2 = 0$ or $x^2 - 2a + a^2 + y^2 = a^2$ i.e., $(x-a)^2 + y^2 = a^2$ ….(ii)

Lt P (h, k) be the point of intersection of tangents at the extremities of chord AB of circle $x^2 + y^2 = a^2$

$\Rightarrow$ AB is chord of contact of P (h, k) w.r.t. circle $x^2 + y^2 = a^2$

$\Rightarrow$ Equation of AB will be $hx + ky = a^2$ …..(iii)

Now (iii) is tangents to circle

$x^2 - 2ax + y^2 = 0$ …..(iv)

$\Rightarrow$ Perpendicular distance of (iii) form centre of circle (iv) i.e., $(a, 0)$ = radius of circle (iv)

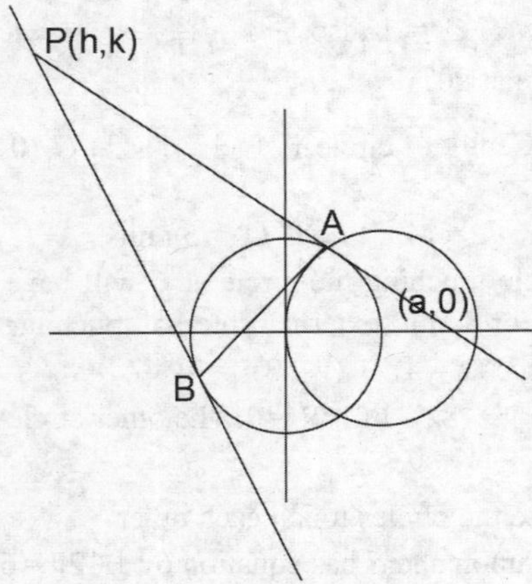

$\Rightarrow \quad \dfrac{|ha-a^2|}{\sqrt{h^2+k^2}} = a \quad \Rightarrow (ha-a^2) = \pm\sqrt{h^2+k^2}$

$\Rightarrow \quad a^2(h-a)^2 = a^2(h^2+k^2)$

$\Rightarrow \quad h^2 + a^2 - 2ah = h^2 + k^2$

$\Rightarrow \quad k^2 + 2ah - a^2 = 0$

$\therefore$ Required locus is $y^2 + 2ax - a^2 = 0$

10. (c) Equation of family of circle passing through A and B

i.e., $x^2(y-a)(y+a) + 2\lambda x = 0$

i.e., $x^2 + y^2 + 2\lambda x - a^2 = 0$ ...(i)

Now, let λ_1 and λ_2 be the parameters for two circles

$\therefore$ By orthogonally conditions $2\lambda_1\lambda_2 = -2a^2$

$\Rightarrow \lambda_1\lambda_2 = -a^2$ .....(ii)

Now $y = mx + c$ is a common tangent to two circles

$\Rightarrow$ Distance of common tangent from centre (by (i)) i.e., from $(-\lambda, 0)$ is radius $= \sqrt{\lambda^2+a^2}$ i.e., $\dfrac{|-\lambda m + c|}{\sqrt{1+m^2}} = \sqrt{\lambda^2+a^2}$

$\Rightarrow (\lambda m - c)^2 = (1+m^2)(\lambda^2+a^2)$

$\Rightarrow c^2 - 2mc\lambda = \lambda^2 + a^2 + a^2m^2$

$\Rightarrow \lambda^2 + 2mc\lambda + a^2 + a^2m^2 - c^2 = 0$

$\Rightarrow \lambda_1\lambda_2 = \dfrac{a^2 + a^2m^2 - c^2}{1}$ ...(iii)

$\therefore$ From (ii) and (iii), we have $a^2 + a^2m^2 - c^2 = -a^2$ or $c^2 = a^2m^2 + 2a^2$ or $c^2 = a^2(m^2+2)$

11. (b) The circles are $S_1: x^2 + y^2 - 9 = 0$ and $S_2: x^2 + y^2 - 12y + 27 = 0$.

The common chord is given by $y = 3$ which is at a distance 3 units $(= r_1)$ from $C_1(0, 0)$

$\Rightarrow$ The line is a tangent

12. (a) The circles are $S_1: x^2 + y^2 - 2x + 6y + 6 = 0$

$\Rightarrow \quad C_1(1, -3)$ and $r_1 = 2$ units and $S_2: x^2 + y^2 - 5x + 6y + 15 = 0$

The common chord is $3x = 9$ or $x = 3$ which is $1 - 3| = 2$ units $= r_1$ from (1, 3) .Hence it is a tangent

TEXTUAL EXERCISE–7 (SUBJECTIVE)

1. A circle passing though $A(1, -2)$ and $B(4, -3)$ wi l have equation $(x-1)(x-4) + (y+2)(y+3) + \lambda(x+3y+5) = 0$

Or $x^2 + y^2 + (\lambda - 5)x + (5 + 3\lambda)y + (10 + 5\lambda) = 0$

$\Rightarrow$ Centre $C_1 = \left(\dfrac{5-\lambda}{2}, -\dfrac{5+3\lambda}{2}\right)$

Since the centre lies on $3x + 4y = 7$

$\Rightarrow 15 - 3\lambda - 20 - 12\lambda = 14$

$\Rightarrow \lambda = \dfrac{19}{15}$

Hence the circle $15(x^2 + y^2) - 94x + 18y + 55 = 0$

2. Let $L_1: 4x - 3y - 5 = 0$ and $L_3: x - 2y - 10 = 0$; $L_2: 7x + y - 40 = 0$ and $L_4: x + 3y + 10 = 0$

The equation of the circle (if possible) is given by $S: L_1L_3 + \lambda L_2L_4 = 0$, provided coefficient of x^2 = coefficient of y^2 and coefficient of $xy = 0$

Solving S, we get: $4x^2 + 6y^2 - 11xy - 45x + 40y + 50 + \lambda(7x^2 + 3y^2 + 22xy + 30x - 110y - 400) = 0$

So $(4 + 7\lambda)x^2 + (6 + 3\lambda)y^2 + (22\lambda - 11)xy + (30\lambda - 45)x + (40 - 110\lambda)y + (50 - 400\lambda) = 0$

Now, $4 + 7\lambda = 6 + 3\lambda$

$\Rightarrow 4\lambda = 2$ i.e., $\lambda = 1/2$, also $22\lambda = 11$

$\Rightarrow \lambda = 1/2$

Hence a cyclic quadrilateral is possible

3. Observe that the given equation of circle can be rewritten as $x^2 + y^2 + 6x + 3y + 4 + p(2x - y) = 0$

So putting $y = 2x$ in $(x+3)^2 + \left(y + \dfrac{3}{2}\right)^2 = \dfrac{29}{4}$, we get $4(x^2 + 9 + 6x) + (16x^2 + 9 + 24x) = 29$

$\Rightarrow 4\{5x^2 + 12x + 4\} = 0 \Rightarrow x = -2, -2/5$

$\therefore$ The two points are $(-2, -4)$ and $(-2/5, -4/5)$

4. Since one circle bisects the circumference of the second

$\therefore$ Centre of second circle lies on the common chord.

Now $S_1: (x-3)^2 + (y-2)^2 = 2^2$ and $C_1(3, 2)$

$S_2: (x-4)^2 + (y-3)^2 = \left(\sqrt{2}\right)^2$

$\Rightarrow C_2(4, 3)$ and the common chord is $2x + 2y - 14 = 0$ or $x + y = 7$

Now (4, 3) lies on it. So circumference of S_2 is bisected by S_1

5. Let $x^2 + y^2 - 2x - 4y + 4 + \lambda(x^2 + y^2 - 4a^2) = 0$ be the circle so $(1+\lambda)(x^2+y^2) - 2x - 4y + (4 - 4\lambda a^2) = 0$ gives $C\left(\dfrac{1}{1+\lambda}, \dfrac{2}{1+\lambda}\right)$ and lines $x = -2y$ is a tangent, so

$\dfrac{1}{5}\left(\dfrac{5}{1+\lambda}\right)^2 = \dfrac{1+4-4+4\lambda a^2}{(1+\lambda)^2}$ gives $4\lambda a^2 = 4$

$\therefore \lambda = 1/a^2$ and the required circle is $(a^2+1)(x^2+y^2) - 2a^2x - 4a^2y = 0$

6. Let $(x^2 + y^2 - 6x + 2y + 4) + \lambda (x^2 + y^2 + 2x - 4y - 6) = 0$ be the required circle
So $(x^2 + y^2)(1 + \lambda) + 2(\lambda - 3)x + 2(1 - 2\lambda)y + (4 - 6\lambda) = 0$
and the centre is $C(3 - \lambda, 2\lambda - 1)$
If C lies on $y = x$, then $3 - \lambda = 2\lambda - 1$
$\Rightarrow \lambda = 4/3$
Hence the circle is $\frac{7}{3}(x^2+y^2) - \frac{10}{3}x - \frac{10}{3}y - 4 = 0$ or
$7(x^2+y^2) - 10x - 10y - 12 = 0$

7. Let the required circle be $(x^2 + y^2 + 2x + 3y + 1) + \lambda (x^2 + y^2 + 4x + 3y + 2) = 0$ or $(1 + \lambda)(x^2 + y^2) + 2(2\lambda + 1)x + 2\left(\frac{3}{2}+\frac{3\lambda}{2}\right)y + (2\lambda + 1) = 0$, which has centered at $C\left(-\frac{2\lambda+1}{\lambda+1}, -\frac{3\lambda+3}{2\lambda+2}\right)$, since it lies on the common chord i.e., $2x = -1$ (or $x = -1/2$)
$\Rightarrow (2\lambda+1) = \frac{\lambda+1}{2} \quad \Rightarrow \lambda = -1/3$
Hence the circle is $\frac{2}{3}(x^2+y^2) + \frac{2}{3}x + 2y + \frac{1}{3} = 0$
or $2(x^2 + y^2) + 2x + 6y + 1 = 0$

8. $(x^2 + y^2 - 4) + \lambda (y - x - 2\sqrt{2}) = 0$ represents a family of circles which will always pass through the intersection of circle $x^2 + y^2 = 4$ and line $L: y = x + 2\sqrt{2}$ which is given by
$x^2 + x^2 + 8 + 4\sqrt{2}x = 4$ i.e., $2\left(x+\sqrt{2}\right)^2 = 0$
$\Rightarrow x = -\sqrt{2}$ and $y = \sqrt{2}$
$\Rightarrow$ The line is a tangent to the circle at $P\left(-\sqrt{2}, \sqrt{2}\right)$, so all the circles will touch each other at $\left(-\sqrt{2}, \sqrt{2}\right)$

9. The circles are $S_1: x^2+y^2 = a^2$ and $S_2: x^2+y^2-2cx+(c^2-b^2)=0$
$\Rightarrow$ Equation of common chord $2cx + b^2 - c^2 - a^2 = 0$ or
$x + \frac{b^2-c^2-a^2}{2c} = 0$
Hence the length of common chord
$$\ell = 2\sqrt{a^2 - \left(\frac{b^2-c^2-a^2}{2c}\right)^2}$$
$$= \frac{1}{c}\sqrt{(2ac+b^2-c^2-a^2)(2ac-b^2+a^2+c^2)}$$
$$= \frac{1}{c}\sqrt{\{b^2-(a^2+c^2-2ac)\}\{(a^2+c^2+2ac)-b^2\}}$$
$$= \frac{1}{c}\sqrt{(b-a+c)(b+a-c)(a+c-b)(a+b+c)}$$

10. Let $O(0, 0)$, $P(x_1, y_1)$ such that $x_1^2 + y_1^2 = a^2$
Now $OP = OA = a$ and $AP = d$
$\therefore \quad \cos\theta = \frac{a^2+a^2-d^2}{2a^2} \Rightarrow OM = a\cos\theta = \frac{2a^2-d^2}{2a}$
$\therefore \quad M\left(\frac{2a^2-d^2}{2a^2}x_1, \frac{2a^2-d^2}{2a^2}y_1\right)$
Slope of $OP = \frac{y_1}{x_1} \quad \therefore$ slope of AM is $= \frac{-x_1}{y_1}$

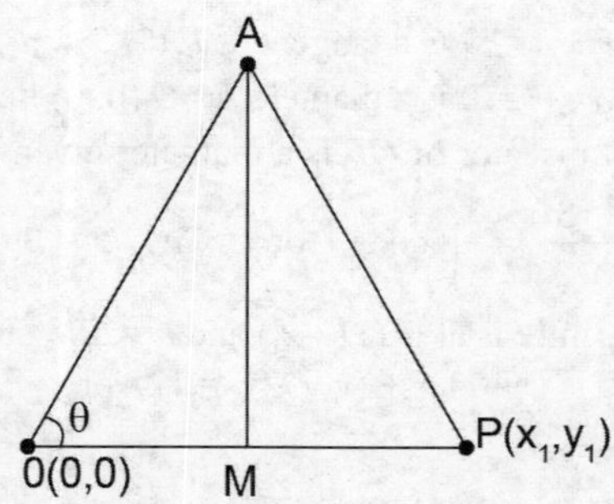

Hence equation the chord (of which AM is one-half part)
$$y - \frac{2a^2-d^2}{2a^2}y_1 = \frac{-x_1}{y_1}\left\{x - \frac{2a^2-d^2}{2a^2}x_1\right\}$$
$$\Rightarrow yy_1 - \frac{2a^2-d^2}{2a^2}y_1^2 = -x_1x - \frac{2a^2-d^2}{2a^2}x_1^2$$
or $yy_1 + xx_1 = \frac{2a^2-d^2}{2a^2}\left(x_1^2+y_1^2\right) = \frac{2a^2-d^2}{2}$
$(\text{as } x_1^2 + y_1^2 = a^2)$
$\Rightarrow xx_1 + yy_1 - a^2 + \frac{d^2}{2} = 0$, which is the required equation.

TEXTUAL EXERCISE-7 (OBJECTIVE)

1. (c) The circle is $S: (x - 1)^2 + y^2 = 1$ and the straight line $L: y = x$, so $2x^2 - 2x = 0$
Which will intersect the circle at $O(0, 0)$ and $A(1, 1)$
$\therefore$ The equation of required circle is with OA as diameter is $x^2 + y^2 - x - y = 0$

2. (d) The given circles are $S_1: x^2 + y^2 + 2x = 0$ and $S_2: x^2 + y^2 + 2y = 0$
$\Rightarrow$ The common chord is $x = y$
The circles will intersect at $O(0, 0)$ and $A(-1, -1)$. Hence the circle with OA as diameter will be $x^2 + y^2 + x + y = 0$

3. (d) $x + y = 5$ is a tangent at $P(2, 3)$
$\Rightarrow x - y + 1 = 0$ is a diameter as point $A(1, 2)$ also lies on the circle
$\therefore$ Right bisector of AP is also a diameter which is $x + y - 4 = 0$
Hence the centre is at $C\left(\frac{3}{2}, \frac{5}{2}\right)$ which is the mid point of $A.P.$
$\therefore$ Radius $r = \frac{\sqrt{1+1}}{2} = \frac{1}{\sqrt{2}}$ units

4. (b) Observe that (2, 1) satisfies only the circle under (b) option
Also $x^2 + y^2 - 4 + \lambda (x^2 + y^2 - 6x + 5) = 0$
$\Rightarrow (x^2 + y^2)(1 + \lambda) - 6\lambda x + (5\lambda - 4) = 0$
Which will be satisfied by (2, 1) when $5(1 + \lambda) - 12\lambda + 5\lambda - 4 = 0$, so $\lambda = 1/2$, then the circle is $\frac{3}{2}(x^2 + y^2) - 3x - 3/2 = 0$
Or $x^2 + y^2 - 2x - 1 = 0$

5. **(c)** $2x - y + 1 = 0$ is a tangent at $P(1, 3)$
$\Rightarrow$ $x + 2y - 7 = 0$ is a diameter as $O(0, 0)$ lies on the circle
$\Rightarrow$ Right bisector of OP is a diameter given by
$3\left(y-\frac{3}{2}\right) = -\left(x-\frac{1}{2}\right)$ or $x + 3y - 5 = 0$
The centre is at $C(11, -2)$ and $r = 5\sqrt{5}$ units. Hence the circle is $(x-11)^2 + (y+2)^2 = 125$ or $x^2 + y^2 - 22x + 4y = 0$

6. **(i)** **(c)** $x + y = 6$ and $2x + y = 4$ intersect at $A(-2, 8)$.
Further $2x + y = 4$ and $x + 2y = 5$ intersect at $B(1, 2)$.
Also $x + 2y = 5$ and $x + y = 6$ intersect at $C(7, -1)$ right bisector of AB is $x - 2y + \frac{21}{2} = 0$ or $2x - 4y + 21 = 0$.
Similarly right bisector of AC is $x - y + 1 = 0$
$\Rightarrow$ Circum centre is $H\left(\frac{17}{2}, \frac{19}{2}\right)$
$\therefore$ The circles is $\left(x-\frac{17}{2}\right)^2 + \left(y-\frac{19}{2}\right)^2 = \left(\frac{15}{2}\right)^2 + \left(\frac{15}{2}\right)^2$
or $x^2 + y^2 - 17x - 19y + 50 = 0$

(ii) **(b)** $x + y = 6$ and $2x + y = 4$ intersect at $A(-2, 8)$
Now $2x + y = 4$ and $x - 2y = 5$ intersect at $B\left(\frac{13}{5}, -\frac{6}{5}\right)$
and $x + y = 6$ and $x - 2y = 5$ intersect at $C\left(\frac{17}{3}, \frac{1}{3}\right)$
Observe that ΔABC is right angled at B as $2x + y = 4$ and $x - 2y - 5 = 0$ are at right angles.
Hence the circle is $\left(x-\frac{17}{3}\right)(x+2) + (y-8)\left(y-\frac{1}{3}\right) = 0$

Aliter: Let L_1: $x + y - 6 = 0$, L_2: $2x + y - 4 = 0$ and L_3: $x - 2y - 5 = 0$
The circle has equation $L_1L_2 + \lambda L_2L_3 + \mu L_1L_3 = 0$ (where $a = b$ and $h = 0$)
So $(2x^2 + y^2 + 3xy - 16x - 10y + 24) + \lambda(2x^2 - 2y^2 - 3xy + 20 - 14x + 3y) + \mu(x^2 - 2y^2 - xy - 11x + 7y + 30) = 0$
i.e., $(2 + 2\lambda + \mu)x^2 + (1 - 2\lambda - 2\mu)y^2 + (3 - 3\lambda - \mu)xy - (16 + 14\lambda + 11\mu)x + (3\lambda + 7\mu - 10)y + (24 + 20\lambda + 30\mu) = 0$
Now $3 = 3\lambda + \mu$ and $4\lambda + 3\mu = -1$
$\Rightarrow$ $\lambda = 2$, $\mu = -3$
$\Rightarrow$ The circle is $3x^2 + 3y^2 - 11x - 25y - 26 = 0$ which is
$\left(x-\frac{17}{3}\right)(x+2) + (y-8)\left(y-\frac{1}{3}\right) = 0$

7. **(a)** Let L_1: $2x + 3y - 2 = 0$; L_2: $3x - 2y - 3 = 0$; L_3: $x + 2y - 3 = 0$ and L_4: $2x - y - 1 = 0$
Now the circle circumscribing the quadrilateral is S: $L_1L_3 + \lambda L_2 L_4 = 0$ (where a = b and h = 0)
So $(2x + 3y - 2)(x + 2y - 3) + \lambda(3x - 2y - 3)(2x - y - 1) = 0$
i.e., $(2x^2 + 6y^2 + 7xy - 8x - 13y + 6) + \lambda(6x^2 + 2y^2 - 7xy - 9x + 5y + 3) = 0$
or $(2 + 6\lambda)x^2 + (6 + 2\lambda)y^2 + (7 - 7\lambda)xy - (8 + 9\lambda)x + (5\lambda - 13)y + (6 + 3\lambda) = 0$
Now $\lambda = 1$ gives $a = b$ and $h = 0$
Hence the circle $8x^2 + 8y^2 - 17x - 8y + 9 = 0$

8. **(b)** The circles are S_1: $(x + g)^2 + (y + f)^2 = g^2 + f^2 - c_1$ and S_2: $(x + g)^2 + (y + f)^2 = g^2 + f^2 - c$
Since tangents are drawn from S_1 to S_2
$\therefore$ $r_1 > r_2$ i.e., $c_1 < c$ and length of tangent $\sqrt{S_1 - S_2} = \sqrt{c - c_1}$

9. **(b)** From the given the points of intersection on x-axis are $A(-1/\lambda, 0)$ and $B(-3, 0)$. Similarly intersection on y-axis are $C(0, 1)$ and $D(0, 3/2)$. For the circle passing through $ABCD$
$(OA)(OB) = (OC)(OD)$
$\Rightarrow$ $\left|3\frac{1}{\lambda}\right| = \left|\frac{3}{2}\right|$. Hence $|\lambda| = 2$
$\Rightarrow$ $\lambda = 2$ for $\lambda > 0$ and point A and B would coincide for $\lambda = 1/3$ and there can always a circle passes through three non-collinear points.

10. **(a)** The circles S_1: $x^2 + y^2 + px + 3y - 5 = 0$ and S_2: $x^2 + y^2 + 5x + py + 7 = 0$ will cut orthogonally when $2gg' + 2ff' = c + c'$
i.e., $(2)\frac{p}{2}.\frac{5}{2} + (2)\left(\frac{3}{2}\right)\left(\frac{p}{2}\right) = 7 - 5 = 2$
$\Rightarrow$ $\frac{8p}{4} = 1$ $\Rightarrow$ $p = 1/2$

11. **(b)** The given circles are S_1: $x^2 + y^2 - 2 = 0$ and S_2: $x^2 + y^2 - 6x - 6y + 10 = 0$
$\Rightarrow$ Common chord is $6x + 6y = 12$ or $x + y = 2$,
Putting in $x^2 + y^2 = 2$, we get $x^2 + x^2 + 4 - 4x = 2$ or $2(x^2 - 2x + 1) = 0$
$\Rightarrow$ $x = 1$ $\Rightarrow$ $y = 1$.
Hence point of contact is $(1, 1)$

12. **(c)** Given S_1: $x^2 + y^2 - 4 = 0$
$\Rightarrow$ $C_1(0, 0)$ and $r_1 = 2$.
Let S_2 be centered at (x_1, y_1) and it passes through $(1, 2)$, then $(x_1 - 1)^2 + (y_1 - 2)^2 + 4 = x_1^2 + y_1^2$
$\therefore$ $-2x_1 - 4y_1 + 9 = 0$ $\Rightarrow$ $2x + 4y - 9 = 0$

13. **(a)** Given S_1: $x^2 + y^2 - p^2 = 0$
$\Rightarrow$ $C_1(0, 0)$ and $r_1 = p$
Let S_2 be centered at (x_1, y_1) and it passes through (a, b), then $(x_1 - a)^2 + (y_1 - b)^2 + p^2 = x_1^2 + y_1^2$
So $-2ax_1 - 2by_1 + p^2 + a^2 + b^2 = 0$ or $2ax + 2by - (p^2 + a^2 + b^2) = 0$

14. **(d)** Let S_1: $x^2 + y^2 + x + 2y + 3 = 0$ and S_2: $x^2 + y^2 + 2x + 4y + 5 = 0$
$\Rightarrow$ Radical axis is L_1: $S_2 - S_1 = 0$ i.e., $x + 2y + 2 = 0$ .....(i)
Similarly for S_3: $x^2 + y^2 - 7x - 8y - 9 = 0$
So radical axis L_2: $S_2 - S_3 = 0$ i.e., $9x + 12y + 14 = 0$ ....(ii)
The point of intersection of (i) and (ii) is $(-2/3, -2/3)$
and $r = \sqrt{S_1\left(-\frac{2}{3}, -\frac{2}{3}\right)}$
Hence the circle $(x + 2/3)^2 + (y + 2/3)^2 = S_1(-2/3, -2/3)$
i.e., $x^2 + \frac{4}{3}x + y^2 + \frac{4}{3}y + \frac{8}{9} = \frac{17}{9}$
$\Rightarrow$ $3(x^2 + y^2) + 4(x + y) - 1 = 0$

15. **(c)** Let P be a point on $x^2 + y^2 - a^2 = 0$, so OP = a and $OQ = a \sin\alpha$

$\Rightarrow \sin\theta = \dfrac{a\sin\alpha}{a} = \sin\alpha$

$\therefore \theta = \alpha$

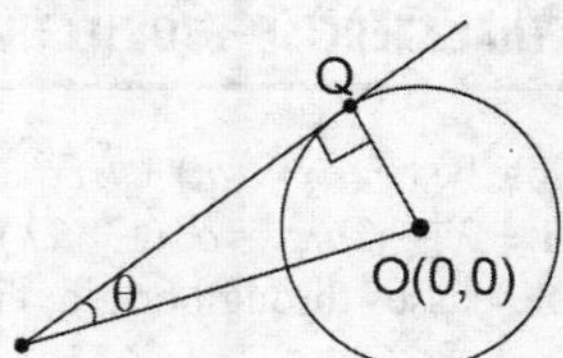

Now angle between tangents = $2\theta = 2\alpha$

16. **(a)** The lines are $L_1: x + \sqrt{3}y - 1$

$\Rightarrow m_1 = -\dfrac{1}{3}$ and $L_2: \sqrt{3}x - y - 2 = 0$

$\Rightarrow m_2 = \sqrt{3}$

Since $m_1 m_2 = -1 \quad \therefore L_1 \perp L_2$

Further since angle subtend at the centre of a circle is double that on the segment

$\therefore$ Angle formed is 180°.

TEXTUAL EXERCISE-8 (SUBJECTIVE)

1. **(a)** The family of circles with centre at (2, –1) is $(x - 2)^2 + (y + 1)^2 = r^2$ and the member that touches x-axis will have radius 1 unit

i.e., $(x - 2)^2 + (y + 1)^2 = 1$

(b) The family of circles cantered at (–4, 2) is $(x + 4)^2 + (y - 2)^2 = r^2$

Where $x - y - 3 = 0$ is a tangent, then $\left|\dfrac{-4-2-3}{\sqrt{2}}\right| = r$

$\Rightarrow r^2 = \dfrac{81}{2}$

Hence the circle $2(x + 4)^2 + 2(y - 2)^2 - 81 = 0$

2. **(a)** Any circle through the intersection can written as $(x^2 + y^2 - 4) + \lambda(x^2 + y^2 - 6x + 5) = 0$

Since this circle passes through (2, 1)

$\therefore 1 + \lambda(-2) = 0 \quad \Rightarrow \lambda = 1/2$

Hence $2(x^2 + y^2 - 4) + (x^2 + y^2 - 6x + 5) = 0$ or $3\{x^2 + y^2 - 2x - 1\} = 0$

(b) The circles are $S_1: x^2 + y^2 - 4x - 5 = 0$ and $S_2: x^2 + y^2 + 8y + 7 = 0$

$\Rightarrow$ Common chord is $4x + 8y + 12 = 0$ or $x + 2y + 3 = 0$

Now, any circle passing through the intersection of S_1 and S_2 can be $(x^2 + y^2 - 4x - 5) + \lambda(x^2 + y^2 + 8y + 7) = 0$

Or $(x^2 + y^2)(1+\lambda) + 2(-2)x + 2(4\lambda)y + (7\lambda - 5) = 0$,

which has centre at $\left(\dfrac{2}{1+\lambda}, \dfrac{-4\lambda}{1+\lambda}\right)$

Since it lies on $x + 2y + 3 = 0$

$\Rightarrow 2 - 8\lambda + 3\lambda + 3 = 0$ i.e., $\lambda = 1$

Hence the circle $2(x^2 + y^2) - 4x + 8y + 2 = 0$ or $x^2 + y^2 - 2x + 4y + 1 = 0$

(c) Any circle through the intersection of S_1 and S_2 can be $(x^2 + y^2 - 8x - 2y + 7) + \lambda(x^2 + y^2 - 4x + 10y + 8) = 0$

Or $(x^2 + y^2)(1 + \lambda) - 2(4 + 2\lambda)x - 2(1 - 5\lambda)y + (7 + 8\lambda) = 0$

Since the centre lies on x-axis

$\therefore \dfrac{1-5\lambda}{1+\lambda} = 0$ i.e., $\lambda = 1/5$.

Hence the circle $5(x^2 + y^2 - 8x + 2y + 7) + (x^2 + y^2 - 4x + 10y + 8) = 0$

$\Rightarrow 6(x^2 + y^2) - 44x + 43 = 0$

3. **(a)** Let any circle passing through A (1, 2) and B (3, 4) be $(x - 1)(x - 3) + (y - 2)(y - 4) + \lambda(x - y + 1) = 0$

Or $x^2 + y^2 + (\lambda - 4)x - (6 + \lambda)y + (11 + \lambda) = 0$

Since it touches x-axis

$\therefore x^2 + (\lambda - 4)x + (11 + \lambda) = 0$ has repeated roots (i.e. Disc. = 0), so $(\lambda - 4)^2 - 44 - 4\lambda = 0$

$\Rightarrow \lambda^2 - 12\lambda - 28 = 0$ or $(\lambda - 14)(\lambda + 2) = 0$

$\therefore \lambda = 14, -2$

The circles are $x^2 + y^2 + (\lambda - 4)x - (6 + \lambda)y + (11 + \lambda) = 0$ for $\lambda = 14, -2$

(b) Since the circle passes through O (0, 0), so let its equation be $x^2 + y^2 + 2gx + 2fy = 0$

Since line $2x - y + 1 = 0$ is a tangent at (1, 3)

$\therefore 2g + 6f + 10 = 0$ and also $x + 3y + g(x + 1) + f(y + 3) = 0$

Which is equation of tangent i.e., $2x - y + 1 = 0$, so $(g + 1)x + (f + 3)y + (g + 3f) = 0$ is identical to it.

$\therefore \dfrac{g+1}{2} = \dfrac{f+3}{-1} = \dfrac{g+3f}{1}$

$\Rightarrow g + 1 = -2f - 6$

$\Rightarrow g = -7 - 2f$ and $f + 3 = -g - 3f$

$\Rightarrow f + 3 = 7 + 2f - 3f$

$\Rightarrow 2f = 4$ i.e., f = 2 and, then g = –11

Hence the circle is $x^2 + y^2 - 22x + 4y = 0$

(c) From the given: line $3x - 4y + 5 = 0$ is a tangent at P (1, 2)

$\Rightarrow 4x + 3y - 10 = 0$ is a diameter

Hence centre at a distance 5 units form P (1, 2) is C_1 (1 + 3, 2 – 4) = (4, –2) or C_2 (1 – 3, 2 + 4) = (–2, 6)

Hence the circles are at $C_1: (x - 4)^2 + (y + 2)^2 = 25$ i.e., $x^2 + y^2 - 8x + 4y - 5 = 0$and

$C_2: (x + 2)^2 + (y - 6)^2 = 25$ i.e., $x^2 + y^2 + 4x - 12y + 15 = 0$

(d) Any circle passing through A (4, 3), B (2, 5) is $(x - 2)(x - 4) + (y - 3)(y - 5) + \lambda(x + y - 7) = 0$

Or $x^2 + y^2 - 6x - 8y + \lambda x + \lambda y + (23 - 7\lambda) = 0$

Since it touches y-axis

$\therefore y^2 + (\lambda - 8)y + (23 - 7\lambda) = 0$ has repeated roots (i.e., Disc. = 0)

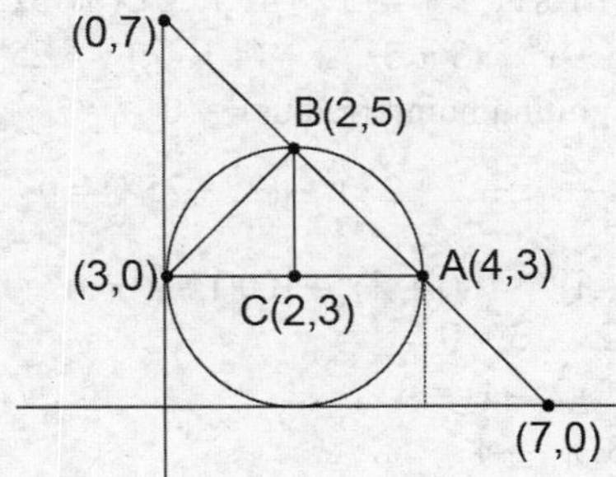

So $\lambda^2 + 64 - 16\lambda + 28\lambda - 92 = 0$ or $\lambda^2 + 12\lambda - 28 = 0$

$\Rightarrow \lambda = -14, 2$

Hence the circles are $x^2 + y^2 - 20x - 22y + 121 = 0$ or $(x-10)^2 + (y-1)^2 = 10^2$ and $x^2 + y^2 - 4x - 6y + 9 = 0$ or $(x-2)^2 + (y-3)^2 = 2^2$

Now consider the smaller circle:

Further observe that if a point is lying outside the circle, then angle subtended on AB will be smaller in comparison to the point lying on the circle. Since $P_1(0, 3)$ point lies on the circle (as well as on y-axis)

$\therefore$ $P_1(0, 3)$ is subtending the largest angle on AB than any other point on y-axis

$\therefore$ (0, 3)

4. Any circle passing through the intersection of S_1 and S_2 will be $(x^2 + y^2 + 4x + 2y + 1) + \lambda(x^2 + y^2 - 2x + 6y - 6) = 0$

Or $(x^2 + y^2)(1 + \lambda) + (4 - 2\lambda)x + (2 + 6\lambda)y + (1 - 6\lambda) = 0$

and its centre $C\left(\dfrac{\lambda-2}{\lambda+1}, \dfrac{-3\lambda-1}{\lambda+1}\right)$

Since C lies on the radical axis L: $6x - 4y + 7 = 0$

$\Rightarrow 6\lambda - 12 + 4 + 12\lambda + 7\lambda + 7 = 0$

$\Rightarrow \lambda = 1/25.$

Hence the circles is $26(x^2 + y^2) + 98x + 56y + 19 = 0$

5. The circle passing through the intersection of S_1 and S_2 will be $(x^2 + y^2 + 10x - 4y - 1) + \lambda(x^2 + y^2 + 5x + y + 4) = 0$

Or $(x^2 + y^2)(1 + \lambda) + (10 + 5\lambda)x + (\lambda - 4)y + (4\lambda - 1) = 0$

For it to represent the limiting points

$$\left(\frac{10+5\lambda}{2\lambda+2}\right)^2 + \left(\frac{\lambda-4}{2\lambda+2}\right)^2 - \left(\frac{4\lambda-1}{\lambda+1}\right) = 0$$

$\Rightarrow 25(\lambda^2 + 4 + 4\lambda) + (\lambda^2 + 16 - 8\lambda) - 16\lambda^2 - 12\lambda + 4 = 0$

Or $10(\lambda^2 + 8\lambda + 12) = 0 \Rightarrow \lambda = -6, -2$

Hence the limiting points are $C_1(-2, -1)$ and $C_2(0, -3)$

6. The circle are S_1: $x^2 + y^2 - x + 3y - 3/2 = 0$ and S_2: $x^2 + y^2 + 4x + 2y + 1 = 0$

Hence any circle of co axial system is $(x^2 + y^2)(1 + \lambda) + (4\lambda - 1)x + (2\lambda + 3)y + (\lambda - 3/2) = 0$ and the radical axis is L: $5x - y + 5/2 = 0$

$\Rightarrow \left(\dfrac{1-4\lambda}{2\lambda+2}, \dfrac{-3-2\lambda}{2\lambda+2}\right)$ lies on L

Hence $5 - 20\lambda + 3 + 2\lambda + 5\lambda + 5 = 0 \Rightarrow \lambda = 1$

Hence the required circle is $2(x^2 + y^2) + 3x + 5y - 1/2 = 0$ or $4(x^2 + y^2) + 6x + 10y - 1 = 0$

7. Equation of two circles of family are $x^2 + y^2 = 0$ ……(i)

and $x^2 + y^2 + 3x + 4y + 25 = 0$ …….(ii)

$\therefore$ Equation of system of coaxial family of circles of (i) and (ii) is $(x^2 + y^2 + 3x + 4y + 25) + \lambda(3x + 4y + 25) = 0$

Or $x^2 + y^2 + (3 + 3\lambda)x + (4 + 4\lambda)y + 25 + 25\lambda = 0$

For limiting point, $(\text{radius})^2 = 0$

$\Rightarrow \left(\dfrac{3+3\lambda}{4}\right)^2 + (2+2\lambda)^2 - (25+25\lambda) = 0$

$\Rightarrow 9(1+\lambda)^2 + 16(1+\lambda)^2 - 100(1+\lambda) = 0$

$\Rightarrow \lambda^2 - 2\lambda - 3 = 0$

$\Rightarrow (\lambda - 3)(\lambda - 1) = 0$

$\Rightarrow \lambda = 3, \lambda = -1$

But $\lambda = -1$ corresponds to origin i.e., $x^2 + y^2 = 0$ and $\lambda = 3$

$\Rightarrow x^2 + y^2 + 12x + 16y + 100 = 0$

$\Rightarrow (-6, -8)$

TEXTUAL EXERCISE-8 (OBJECTIVE)

1. **(c)** Let the circle be $(x^2 + y^2 - a^2) + \lambda(x^2 + y^2 + 2ax - 2a^2) = 0$

$\Rightarrow (x^2 + y^2)(1 + \lambda) + 2a\lambda x - a^2(1 + 2\lambda) = 0$

If the circle passes through origin, then $a^2(1 + 2\lambda) = 0$

$\Rightarrow \lambda = -1/2$ and the circle is $\dfrac{1}{2}(x^2 + y^2) + 2a\left(\dfrac{-1}{2}\right)x = 0$

Or $x^2 + y^2 - 2ax = 0$

2. **(a)** Since altitudes are at right angles to their respective side

$\therefore$ Ortho-center becomes the radical centre

3. **(d)** $x^2 + y^2 + 2gx + c = 0$ has centre at $C(-g, 0)$ and radius $r = \sqrt{g^2 - c}$ when $c > 0$ for $r = 0$, we get $g = \pm\sqrt{c}$

Hence $C_1(\sqrt{c}, 0)$ and $C_2(-\sqrt{c}, 0)$ which are distinct

$\therefore$ The circles are non-intersecting type (as $2\sqrt{c} > 0$)

4. **(c)** The circles are S_1: $x^2 + y^2 - \dfrac{7}{2}x = 0$ and S_2: $x^2 + y^2 - 4y - 7 = 0$

$\Rightarrow$ The radical axis is L: $S_1 - S_2 = 0$ i.e., $-\dfrac{7}{2}x + 4y + 7 = 0$

$\Rightarrow 7x - 8y - 14 = 0$

5. **(d)** Given circles are S_1: $x^2 + y^2 - 16x + 60 = 0$

$\Rightarrow C_1(8, 0)$ and $r_1 = 2$ units and

S_2: $x^2 + y^2 - 12x + 27 = 0$

$\Rightarrow C_2(6, 0)$ and $r_2 = 3$ units

Also S_3: $x^2 + y^2 - 12y + 8 = 0$

$\Rightarrow C_3(0, 6)$ and $r_3 = 2\sqrt{7}$ units

Observe that y-axis is the altitude form $C_3(0, 6)$, we can easily say that orthocenter is on y-axis.

Equation of C_1C_3 is $x + y = 6$

6. **(b)** We know that radical axis is always at 90° to the line joining the centers of the circles.

7. **(c)** The circles are S_1: $x^2 + y^2 - 144 = 0$ and S_2: $x^2 + y^2 - 15x + 12y = 0$

$\Rightarrow$ The radical axis is $15x - 12y - 144 = 0$ or $5x - 4y - 48 = 0$

8. **(a)** The circle is S: $x^2 + y^2 - 4 = 0$. Now the line joining $A(1, 0)$ and $B(3, 4)$ is L: $y = 2x - 2$ which will intersect S in $P(0, -2)$ and $Q\left(\dfrac{8}{5}, \dfrac{6}{5}\right)$. Now $\dfrac{BP}{PA} = \alpha = -3$ and $\dfrac{BQ}{QA} = \beta = \dfrac{7}{3}$

$\Rightarrow \alpha\beta = -7$ and $\alpha + \beta = -2/3$

$\therefore$ α, β are the roots of $3x^2 + 2x - 21 = 0$

9. **(c)** We know that for an equilateral ΔABC with side 'a', in radius $r = \dfrac{a}{2\sqrt{3}}$.

If a square is inscribed in the circle, then diameter = hypotenuous and area of square $= \left(\dfrac{a}{\sqrt{3}}\right)^2\left(\dfrac{1}{2}\right) = \dfrac{a^2}{6}$

10. **(b), (c)** The circle is $x^2 + y^2 - x + 3y = 0$

$\Rightarrow \quad C\left(\frac{1}{2}, \frac{-3}{2}\right)$ and radius $r = r = \sqrt{\frac{5}{2}}$

Distance of line L_2: $x + y - 1 = 0$ from C is $\sqrt{2}$. Now let L_1: $y - mx = 0$ and its distance from C is $\sqrt{2}$

$\therefore \quad \frac{\left|\frac{m}{2} + \frac{3}{2}\right|}{\sqrt{1+m^2}} = \sqrt{2} \Rightarrow m^2 + 6m + 9 = 8m^2 + 8$ i.e., $(m + 1)$ $(m - 1) = 0$

$\Rightarrow \quad m = 1, -1/7$ and L_1: $y = x$ or $7y + x = 0$

11. **(b)** Any circle passing though (0, 0) and (1, 0) will be $x^2 - x + y^2 + \lambda y = 0$ has center at $C_2\left(\frac{1}{2}, -\frac{\lambda}{2}\right)$ and $r_2 = \frac{\sqrt{\lambda^2+1}}{2}$.

Since it touches the circle $x^2 + y^2 = 9$ with C_1 (0, 0) and $r_1 = 3$ as $C_1C_2 = r_2 + r_1$ is not possible in this case

$\Rightarrow \quad C_1C_2 = |r_1 - r_2| \quad \Rightarrow \quad 3 - \frac{\sqrt{\lambda^2-1}}{2} = \frac{\sqrt{\lambda^2+1}}{2}$

$\Rightarrow \quad \sqrt{\lambda^2+1} = 3 \Rightarrow \lambda = \pm 2\sqrt{2}$

Hence the possible centre are $C_2\left(\frac{1}{2}, \pm\sqrt{2}\right)$

12. **(a)** Slope of $AB = -a/b$. Hence equation of AB is $y - b = \frac{-a}{b}(x - a)$

$\Rightarrow \quad ax + by = a^2 + b^2$

$\therefore \quad A\left(\frac{a^2+b^2}{a}, 0\right)$ and $B\left(0, \frac{a^2+b^2}{b}\right)$

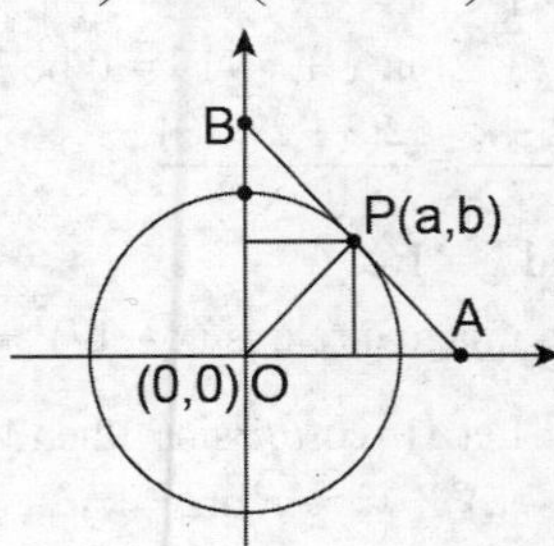

Hence area of $\Delta OAB = \frac{(a^2+b^2)^2}{2ab} = \frac{r^4}{2ab}$

13. **(c)** The circle S: $x^2 + y^2 - 4x - 2y - 11 = 0$, has centre at C (2, 1) and radius $r = 4$ units.

The given point P (4, 5). Area of quadrilateral formed = $r\sqrt{S(4,5)} = 4\sqrt{4} = 8$ square units

14. **(a)** Let C be the centre of circle as $\angle RPQ = \angle PRS = 90°$

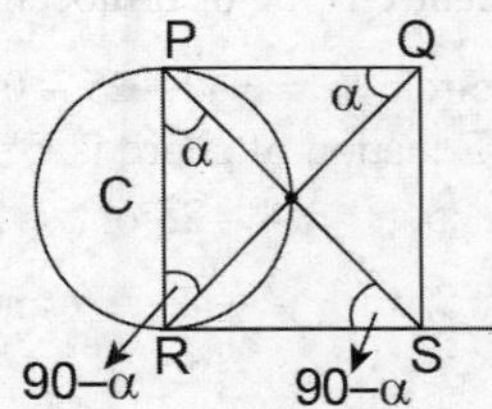

As shown $\tan\alpha = \frac{2r}{PQ} = \frac{RP}{PQ}$

Similarly $\tan(90° - \alpha) = \frac{PR}{RS} = \frac{2r}{RS}$

Hence $\frac{2r}{PQ} \cdot \frac{2r}{RS} = 1 \Rightarrow 2r = \sqrt{PQ.RS}$

15.

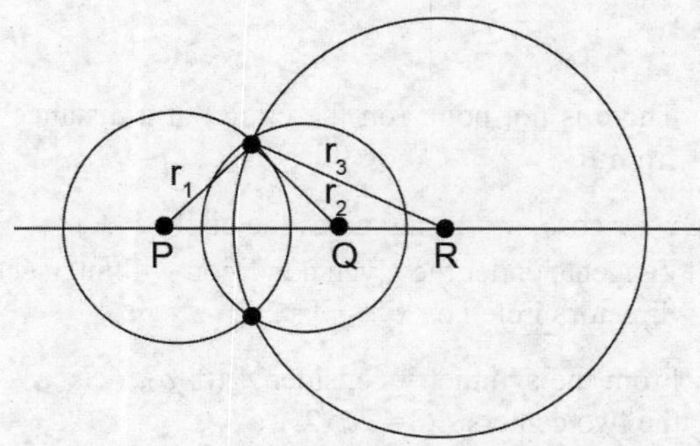

Let the equation of co-axial circle has equation $x^2 + y^2 - 2\lambda x + c = 0$

Where λ is a parameter $P(\lambda_1, 0)$, $Q(\lambda_2, 0)$, $R(\lambda_3, 0)$

$r_1\sqrt{\lambda_1^2 - C}$

$r_2\sqrt{\lambda_2^2 - c}$

$r_3 = \sqrt{\lambda_3^2 - C}$

$QR\,r_1^2 + R\,Pr_2^2 + PQr_3^2$

$= |\lambda_3 - \lambda_2|(\lambda_1^2 - c) + |\lambda_1 - \lambda_3|(\lambda_2^2 - c) + |\lambda_1 - \lambda_2|(\lambda_3^2 - c)$

$= (-c)(\lambda_3 - \lambda_2 + \lambda_1 - \lambda_3 + \lambda_2 - \lambda_1)$

$= (\lambda_1 - \lambda_2)(\lambda_2 - \lambda_3)(\lambda_3 - \lambda_1)$

$= -(\lambda_2 - \lambda_1)(\lambda_3 - \lambda_2)(\lambda_1 - \lambda_3)$

$= -PQ.\,QR.RP$

Hence option (b) is correct

TEXTUAL EXERCISE SECTION–III (OBJECTIVE)

1. **(b)** Let the perimeter be p units then $2\pi r = p$, so Area $A_c = \pi r^2 = \frac{p^2}{4\pi}$ square units

Similarly side of the square is $a = p/4$ and $A_s = a^2 = \frac{p^2}{16}$ square units

Since $4\pi < 16 \Rightarrow \frac{p^2}{16} < \frac{p^2}{4\pi}$. Hence circle has larger area

2. **(c)** The circle S: $x^2 + y^2 + 7x - (1 - \lambda)y + 5 = 0$ has radius

$0 < \frac{49}{4} + \frac{(1-\lambda)^2}{4} - 5 < 25$

$\Rightarrow \quad 0 < (\lambda - 1)^2 < 120 - 49 = 71$

Since λ is integer

$\therefore \quad 0 < (\lambda - 1)^2 < 71$ gives $\lambda = 2, 3, 4, \ldots, 9$ or $-7, -6, -5, \ldots, 0$ gives total 16 values

3. **(b)** The circle S: $x^2 + y^2 - \frac{3}{2}x = 0$ has centre at C (3/4, 0) and radius $r = 3/4$

Point P is $(-2, 1)$ gives $PC = \sqrt{\frac{121+16}{16}} = \frac{\sqrt{137}}{4} > \frac{\sqrt{121}}{4}$

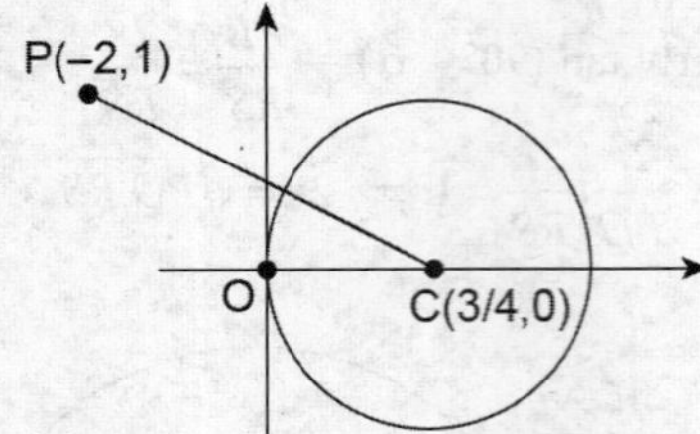

∴ There is not point (on the circle) at a distance 2 units from P.

4. (b) $x = r\cos\theta$, $y = r\sin\theta$ gives the circle $x^2 + y^2 = r^2$
Tangents under the given condition will intersect on the director circle i.e., $x^2 + y^2 = 2r^2$

5. (a) From the symmetry consider $A\ (a, a)$, $C\ (-a, -a)$ are the two centers $AC = 2a\sqrt{2}$.
The radius of a circle that will touch all the circles externally will have radius $r_1 = (\sqrt{2} - 1)a$
An other possibility where a circle will touch internally will have radius $r_2 = (\sqrt{2} + 1)a$.
Hence the radius $r = (\sqrt{2} \pm 1)a$

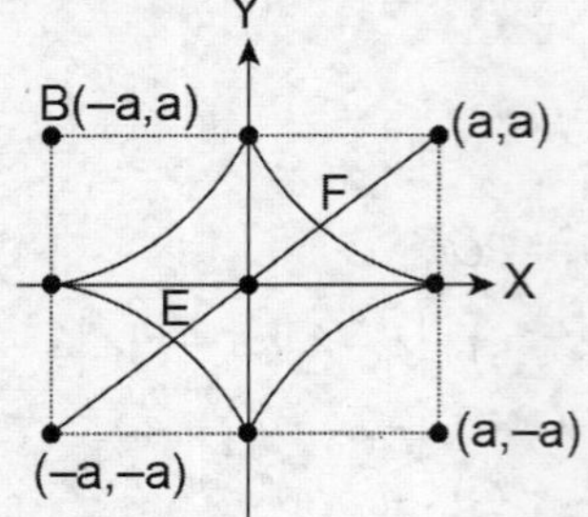

6. (b) Easily we can conclude that $|x_1| = |y_1| = |r_2|$

$$\Rightarrow \sqrt{2}r_2 - (\sqrt{2}+1) = r_2 \Rightarrow r_2 = \frac{\sqrt{2}+1}{\sqrt{2}-1} = (\sqrt{2}+1)^2 = 3+2\sqrt{2}$$

Aliter: Let (x_1, y_1) be the centre of a large circle $r_2 > 1$, then

$$y_1^2 = x_1^2 = \left(x_1 - 1 - \frac{1}{\sqrt{2}}\right)^2 + \left(y_1 - 1 - \frac{1}{\sqrt{2}}\right)^2 = r_2^2$$

$$\text{Hence } 2\left\{x_1^2 + \left(1+\frac{1}{\sqrt{2}}\right)^2 - 2x_1\left(1+\frac{1}{\sqrt{2}}\right)\right\} = x_1^2$$

$$\Rightarrow r_2 = (2+\sqrt{2})\left(1+\frac{1}{\sqrt{2}}\right) \text{ (as } r_2 > 1) \therefore r_2 = 3 + 2\sqrt{2}$$

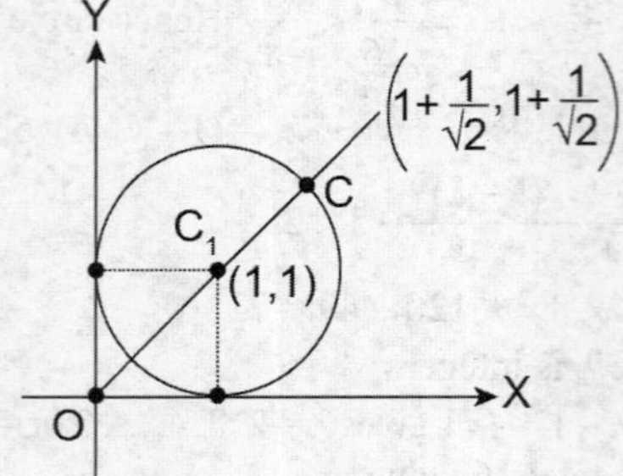

7. (c) As shown $CN = r\cos 30° = \frac{\sqrt{3}}{2}r$
Now the given circle is $(x + 2)^2 + (y - 3)^2 = 5^2$
Hence the locus of N is $(x + 2)^2 + (y - 3)^2 = (3/4) \times 25 = 18.75$

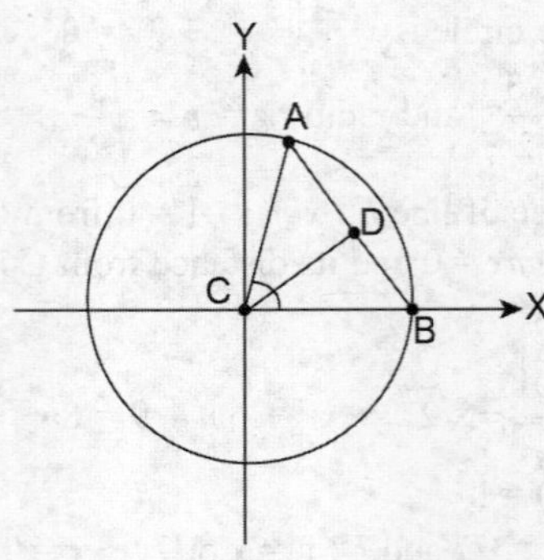

8. (c) The circle S: $x^2 + y^2 + 2gx + c = 0$ will have radius $r = \sqrt{g^2 - c}$ and centre at $(-g, 0)$ for limiting point $r = 0$
$\Rightarrow g = \pm\sqrt{C}$

9. (a) Consider a circle with AB as diameter. Let C be any variable point and $CN \perp AB$
Since area $\Delta ABC = (1/2)\ AB \times CN$ which will be maximum, when N coincides with the centre of the circle.
In that case ΔABC is right angled isosceles.

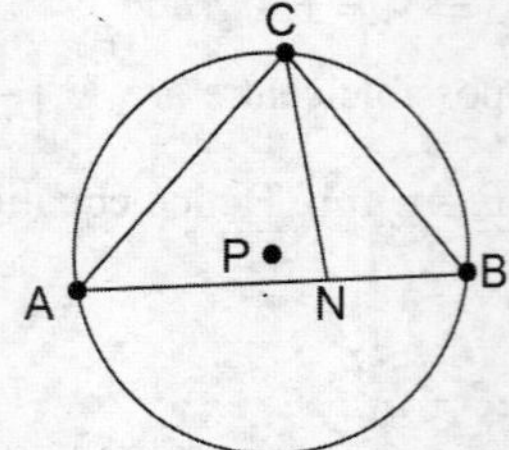

10. (c) The circle $(x - 3)^2 + (y - 2)^2 = 1$ has centre $C\ (3, 2)$ and radius $r = 1$.
Image of $(3, 2)$ in $x + y - 19 = 0$ be $M\ (h, k)$

$$\Rightarrow \frac{3-h}{1} = \frac{2-k}{1} = \frac{2(3+2-19)}{2}$$

$\Rightarrow h = 17$ and $k = 16$.
Hence the image circle is $(x - 17)^2 + (y - 16)^2 = 1$

11. (b) $x^2 + y^2 = a^2$. Let $A\ (a\cos\theta, a\sin\theta)$, then $M(a + a\cos\theta, a\sin\theta)$.
Hence $(x - a)^2 + y^2 = a^2$ or $x^2 + y^2 - 2ax = 0$

12. (b) $a^2 + b^2 = 100$, mid point of AB is $\left(\frac{a}{2}, \frac{b}{2}\right)$
$\Rightarrow (2x)^2 + (2y)^2 = 100$ or $x^2 + y^2 = 25$

13. (b) The circle S: $(x - 7)^2 + (y - 5)^2 = 15^2$
$\Rightarrow C(7, 5)$ and the point $P\ (2, -7)$
$\Rightarrow CP = \sqrt{5^2 + 12^2} = 13$. Observe that point lies inside the circle.
∴ The required shortest distance = 15 – 13 = 2 units

14. (d) The radical centre is the or orthocenter of the ΔABC

15. (c) Equation of circle S: $x^2 + y^2 - 25 = 0$, midpoint of chord is $P(1, -2)$. Equation of chord is $T = S_1$
i.e., $x - 2y - 25 = 1 + 4 - 25$ or $x - 2y - 5 = 0$

16. (c) The circle is S: $x^2 + y^2 - a^2 = 0$ and the length of the chord = a
∴ Distance of mid point of chord from the centre is $p = \frac{a\sqrt{3}}{2}$ and the locus of mid point is $x^2 + y^2 - \frac{3a^2}{4} = 0$

17. (d) The given circle is $x^2 + y^2 = 1$ and its director circle is $x^2 + y^2 = 2$ for angle between tangents to be $\pi/2 < \theta < \pi$

$\Rightarrow |a| \in (1, \sqrt{2}) \qquad \Rightarrow a \in (-\sqrt{2}, -1) \cup (1, \sqrt{2})$

18. (c) If the circle touch both the axes, then the centre of the circle will lie on lines $x + y = 0$ or $x - y = 0$.

Hence $x^2 - y^2 = 0$ is the locus the centre

19. (b) The circle is S: $x^2 + y^2 - r^2 = 0$. Now $P(\alpha, \beta)$ is the point

$\Rightarrow PA.PB = S(\alpha, \beta) = \alpha^2 + \beta^2 - r^2$

20. (c) The circles are S_1: $(x - a)^2 + y^2 = (a^2 - c^2)$

$\Rightarrow C_1(a, 0)$ and $r_1 = \sqrt{a^2 - c^2}$

And S_2: $x^2 + (y - b)^2 = \left(\sqrt{b^2 - c^2}\right)^2$

$\Rightarrow C_2(0, b)$ and $r_2 = \sqrt{b^2 - c^2}$

The circles will touch externally when $C_1C_2 = r_1 + r_2$

i.e., $\sqrt{a^2 + b^2} = \sqrt{a^2 - c^2} + \sqrt{b^2 - c^2}$

$\Rightarrow a^2 + b^2 = a^2 - c^2 + b^2 - c^2 + 2\sqrt{(a^2 - c^2)(b^2 - c^2)}$ or $(a^2 - c^2)(b^2 - c^2) = (c^2)^2$

$\Rightarrow a^2b^2 - a^2c^2 + c^4 - c^4 - b^4 - b^2c^2 = 0 \Rightarrow \dfrac{1}{c^2} = \dfrac{1}{b^2} + \dfrac{1}{a^2}$

21. (b) Let $A(a, 0)$ and $B(0, b)$

$\Rightarrow C\left(\dfrac{a}{2}, \dfrac{b}{2}\right) \qquad \therefore r = \dfrac{\sqrt{a^2 + b^2}}{2}$

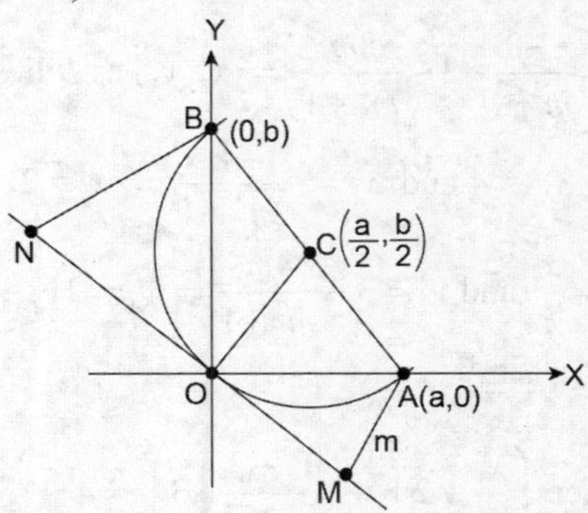

Now slope of OC is $m_1 = b/a$

$\Rightarrow$ slope of tangent at O is $m_2 = -a/b$ i.e., $ax + by = 0$

$\Rightarrow \dfrac{a^2}{\sqrt{a^2 + b^2}} = m$ and $\dfrac{b^2}{\sqrt{a^2 + b^2}} = n \Rightarrow r = \dfrac{a^2}{2m} = \dfrac{b^2}{2n}$

$\therefore 2r(m + n) = a^2 + b^2 = 4r^2 \Rightarrow 2r = (m + n)$

22. (a) The circles are S_1: $x^2 + y^2 = a^2$

$\Rightarrow C_1(0, 0)$ and $r_1 = a$ and S_2: $(x - 2a)^2 + y^2 = 4a^2$

$\Rightarrow C_2(2a, 0)$ and $r_2 = 2a$

Let (x_1, y_1) be the centre of the circle and radius r that touches S_1 and S_2 externally, then $\sqrt{x_1^2 + y_1^2} = a + r_3$ and $\sqrt{(x_1 - 2a)^2 + y_1^2} = 2a + r_3$

$\Rightarrow a + \sqrt{x_1^2 + y_1^2} = \sqrt{(x_1 - 2a)^2 + y_1^2}$

$\Rightarrow a^2 + x_1^2 + y_1^2 + 2a\sqrt{x_1^2 + y_1^2} = x_1^2 + y_1^2 + 4a^2 - 4ax_1$

$\Rightarrow a\left\{2\sqrt{x_1^2 + y_1^2}\right\} = a(3a - 4x_1)$

Squaring again, we get $4x_1^2 + 4y_1^2 = 16x_1^2 + 9a^2 - 24ax_1$

$\Rightarrow 12\left(x_1^2 + a^2 - 2ax_1\right) - 4y_1^2 = 3a^2 \Rightarrow 12(x - a)^2 - 4y^2 = 3a^2$

23. (c) The circle S_1: $(x - 2)^2 + (y + 1)^2 = 1^2$ has centre at $C(2, -1)$ and $r_1 = 1$.

The second circle S_2 has centre at $C_2(x_1, 4 - x_1)$ and radius $r_2 = \sqrt{x_1^2 + (4 - x_1)^2}$ for orthogonal intersection

$C_1C_2^2 = r_1^2 + r_2^2 \Rightarrow (x_1 - 2)^2 + (5 - x_1)^2 = 1 + x_1^2 + (4 - x_1)^2$

$\Rightarrow 3 - 4x_1 - 2x_1 - 9 \qquad \Rightarrow x_1 = 2$

The circle is S_2: $(x - 2)^2 + (y - 2)^2 = 2^2 + 2^2$ i.e., $x^2 + y^2 - 4x - 4y = 0$

24. (b) The circle is $x^2 + y^2 - 2x = 0$

The intersecting line is $y = x \Rightarrow 2x^2 - 2x = 0$

$\therefore x = 0, y = 0$ or $x = 1, y = 1$ i.e., $A(0, 0)$, $B(1, 1)$.

Hence circle with AB as diameter is $x^2 + y^2 - x - y = 0$

25. (b) From the given information

$OA = \sqrt{c}$ and $AC = r = \sqrt{g^2 + f^2 - c}$ and area of quadrilateral $OABC = OA \times AC = \sqrt{c(g^2 + f^2 - c)}$

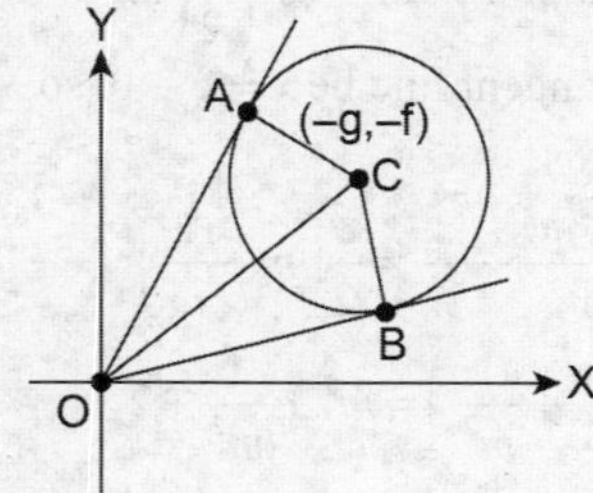

26. (a) From the given information $A(2, 2)$, now B lies on the circle and $AB = 4$ units

$\therefore B(2, -2)$ or $(-2, 2)$

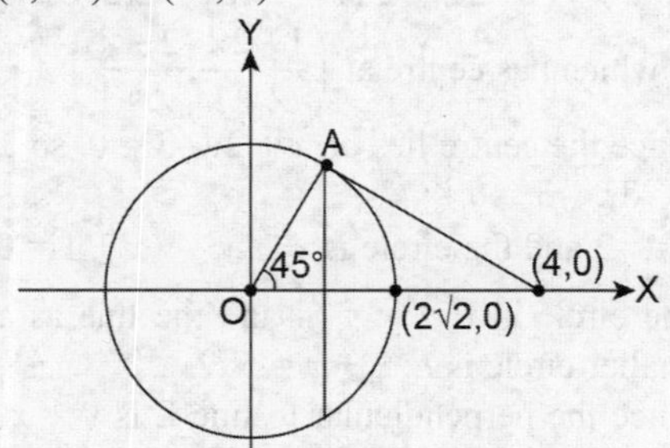

27. (c) Let S_2: $x^2 + y^2 - a^2 + \lambda\left(x - \dfrac{a}{2}\right) = 0$, be the required circle since $(2a, 0)$ lies on the circle.

Hence $4a^2 + 0 - a^2 + \lambda\left(2a - \dfrac{a}{2}\right) = 0 \Rightarrow \lambda = -2a$

$\therefore$ The circle S_2 is $x^2 + y^2 - a^2 - 2ax + a^2 = 0$ i.e., $x^2 + y^2 - 2ax = 0$

28. (d) The given circle S_1: $(x - 3)^2 + (y - 3)^2 = 2^2$

Let the other circle be S_2 with centre at (x_1, y_1), since it touches y-axis and S_1 externally

$\therefore |x_1| = r_2 \Rightarrow \sqrt{(x_1 - 3)^2 + (y_1 - 3)^2} = 2 + r_2 = 2 + |x_1|$

$\Rightarrow x_1^2 + 9 - 6x_1 + y_1^2 + 9 - 6y_1 = 4 + x_1^2 + 4|x_1|$

Since the circle S_1 is on R.H.S. of y-axis

$\therefore |x_1| = x_1$

Hence the locus of x_1 and y_1 is $y_1^2 - 10x_1 - 6y_1 + 14 = 0$

or $y^2 - 10x - 6y + 14 = 0$

29. (c) The given circle is S_1: $(x-2)^2+(y-3)^2=(\sqrt{3})^2$

$\Rightarrow$ C_1 (2, 3) and $r_1=\sqrt{3}$

The centre of the S_2 is $C_2(x_1, x_1)$ as it passes through origin

$\therefore$ $r_2=\sqrt{2}\,|x_1|$

Since circles intersect orthogonally, so $C_1C_2=r_1^2+r_2^2$

$\therefore$ $(x_1-2)^2+(x_1-3)^2 = 3+2x_1^2$

$\Rightarrow$ $-10x_1+13=3 \quad \Rightarrow x_1=1$

Hence S_2: $(x-1)^2+(y-1)^2=1^2+1^2$ or $x^2+y^2-2x-2y=0$

30. (b) Let $A\,(am^2, 2am)$ and $B\,(a/m^2, -2a/m)$

$$r=\frac{AB}{2}=\frac{a}{2}\sqrt{\left(m^2-\frac{1}{m^2}\right)^2+4\left(1+\frac{1}{m}\right)^2}$$

$$r=\frac{a}{2}\left(m+\frac{1}{m}\right)\sqrt{\left(m-\frac{1}{m}\right)^2+4}=\frac{a}{2}\left(m+\frac{1}{m}\right)^2$$

Now mid point of AB is $\left(\frac{a}{2}\left(m^2+\frac{1}{m^2}\right), a\left(m-\frac{1}{m}\right)\right)$

Let the tangent line be $x+c=0$, so

$$\frac{\frac{a}{2}\left(m^2+\frac{1}{m^2}\right)+c}{1}=\frac{a}{2}\left(m+\frac{1}{m}\right)^2$$

$$\Rightarrow \quad \frac{2c}{a}+\left(m^2+\frac{1}{m^2}\right)=m^2+\frac{1}{m^2}+2$$

$\Rightarrow$ $c=a$, hence line is $x+a=0$

31. (a) The circle through the intersection S_1 and S_2 is S: $(x^2+y^2-2x-4y+1)+\lambda\,(x^2+y^2-4x-2y+1)=0$

$\Rightarrow$ $(x^2+y^2)(1+\lambda)-2(1+2\lambda)x-2(2+\lambda)y+(\lambda+1)=0$, which has centre at $\left(\frac{1+2\lambda}{\lambda+1}, \frac{2+\lambda}{\lambda+1}\right)$

Since the centre lies on $x+2y-3=0$, so $2\lambda+1+2\lambda+4-3\lambda-3=0$

$\Rightarrow$ $\lambda=-2$ and the circle is $x^2+y^2-6x+1=0$

32. (a) The circle is $x^2+y^2=4$ and the line as a tangent for smaller circle is L: $x+y=5\sqrt{2}$.

Since the perpendicular to line L is $y=x$ and $ON=5$ units from symmetry the centre lies at the mid point of $L(\sqrt{2},\sqrt{2})$ and $N\left(\frac{5}{\sqrt{2}},\frac{5}{\sqrt{2}}\right)$, which is $M\left(\frac{7}{2\sqrt{2}},\frac{7}{2\sqrt{2}}\right)$

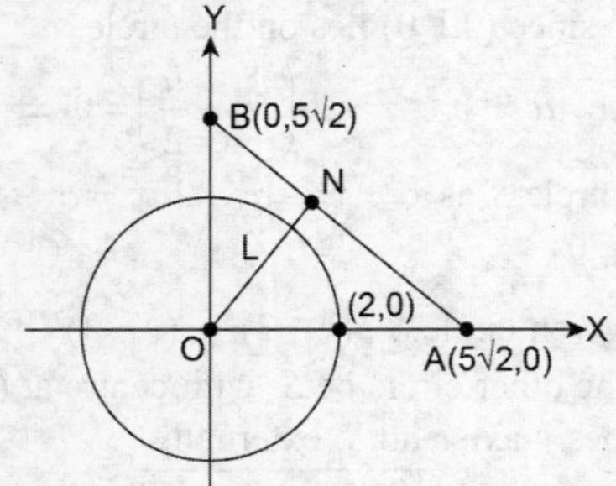

33. (a) $\frac{x}{a}=\frac{1-t^2}{1+t^2}$ and $\frac{y}{a}=\frac{2t}{1+t^2}$

On squaring and adding, we get

$$\frac{x^2}{a^2}+\frac{y^2}{a^2}=\frac{1+t^4-2t^2+4t^2}{(1+t^2)^2}=1$$

$\Rightarrow$ $x^2+y^2=a^2$ which is a circle

34. (a) $\sin\theta=\frac{a}{\sqrt{h^2+k^2}}, \cos\theta=\frac{AP}{\sqrt{h^2+k^2}}, AP=\sqrt{h^2+k^2-a^2}$

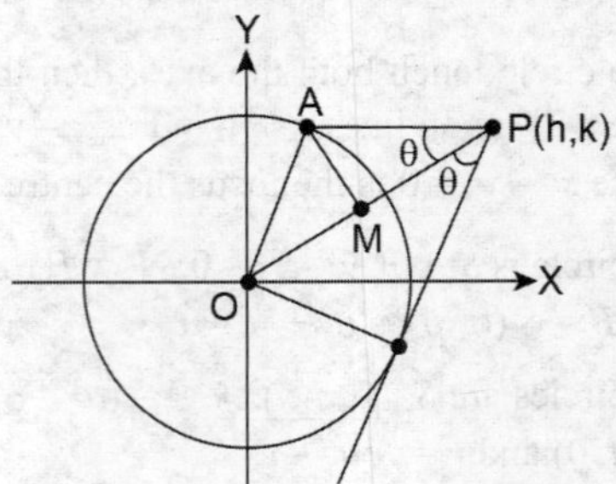

Required area $= 2\times\frac{1}{2}AP^2\sin\theta\cos\theta =$

$$(h^2+k^2-a^2)\frac{a}{\sqrt{h^2+k^2}}\cdot\frac{\sqrt{h^2+k^2-a^2}}{\sqrt{h^2+k^2}}=\frac{a\left(h^2+k^2-a^2\right)^{3/2}}{\left(h^2+k^2\right)}$$

35. (b) Any line through A (0, 3) will be $y-3=mx$, since it intersect the circle $(x+2)^2+(y-3)^2=4$

$\Rightarrow$ $x^2+4x+4+m^2x^2=4 \Rightarrow \{(m^2+1)\,x+4\}x=0$

$$\Rightarrow \quad x_1=0, x_2=\frac{-4}{m^2+1}$$

Hence A (0, 3) and $B\left(\frac{-4}{m^2+1}, 3+\frac{(-4m)}{m^2+1}\right)$

Since B is the mid point of A and M

$\therefore$ $M\left(\frac{-8}{m^2+1}, 3-\frac{8m}{m^2+1}\right)=(x_3, y_3)$ as B lies on $(x+2)^2+(y-3)^2=4$ and $x_2=-\frac{4}{m^2+1}\Rightarrow x_3=2x_2$

$\Rightarrow$ $x_2=\frac{x_3}{2}$ and $y_2=3-\frac{4m}{m^2+1}\Rightarrow y_3=2y_2-6+3=2y_2-3$

$$\Rightarrow \quad y_2=\frac{y_3+3}{2}$$

Hence $\left(\frac{x_3}{2}+2\right)^2+\left(\frac{y_3+3}{2}-3\right)^2=4$

i.e., $x_3^2+16+8x_3+y_3^2+9-6y_3=16$ or

$x^2+y^2+8x-6y+9=0$ i.e., $x^2+8x+(y-3)^2=0$

36. (a) $mx-y-b\sqrt{1+m^2}=0$ will be a common tangent to $x^2+y^2=b^2$ and $(x-a)^2+y^2=b^2$ when

$$\frac{\left|-b\sqrt{1+m^2}\right|}{\sqrt{1+m^2}}=\frac{\left|ma-b\sqrt{1+m^2}\right|}{\sqrt{1+m^2}}=b$$

$\Rightarrow$ $ma-b\sqrt{1+m^2}=\pm b\sqrt{1+m^2} \Rightarrow ma=0, 2b=\sqrt{1+m^2}$

Now $a>2b>0 \quad \Rightarrow \quad m=\frac{2b\sqrt{1+m^2}}{a}$

Squaring, we get $m^2a^2=4b^2(1+m^2)$

$\Rightarrow$ $m^2(a^2-4b^2)=4b^2 \quad \Rightarrow \quad m=\frac{2b}{\sqrt{a^2-4b^2}}$

37. (a) Let C_1 and C_2 be the centers of circles with r_1 and r_2 as their respective radii from the symmetry C_1C_2 line will intersect PQR at

45° (at Q) as shown $\therefore$ $C_1Q=\sqrt{2}\,r_1$ and $QC_2=\sqrt{2}\,r_2$

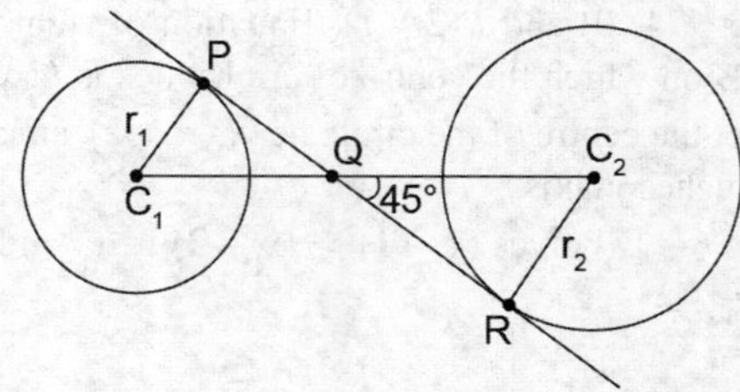

$\Rightarrow C_1C_2 = \sqrt{2}\ (r_1 + r_2)$ as r_1 and r_2 are fixed constants

$\therefore$ C_1C_2 is also a constant. Hence point C_2 will lie on a circle of radius $\sqrt{2}(r_1 + r_2)$ with C_1 as it centre.

38. (c) $y = |x|$ i.e., $y = \begin{cases} x; x \geq 0 \\ -x; x < 0 \end{cases}$. Let OA and OB be the chord with length $\sqrt{2}$.

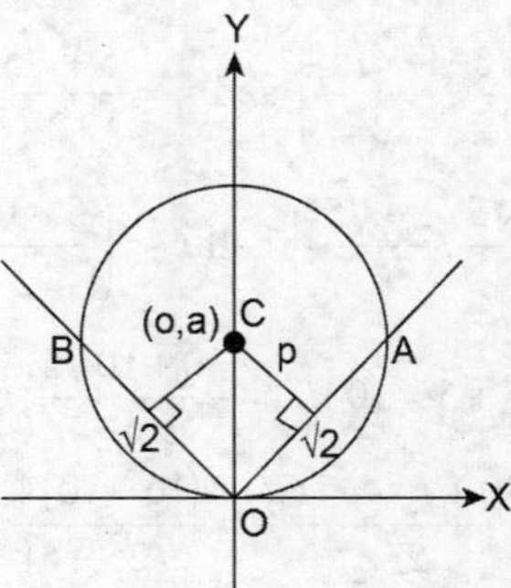

$\Rightarrow$ Centre will lie on positive y-axis. Let it be $(0, a)$ and radius $r = |a|$

Distance of chord from centre $C = \frac{|a|}{\sqrt{2}} = \frac{1}{\sqrt{2}} \Rightarrow |a| = 1$

$\Rightarrow a = 1\ (\because a > 0)$

$\therefore$ The centre will be at $(0, 1)$. Hence the circle $x^2 + y^2 - 2y = 0$

39. (d) The circle S_1: $x(x - p) + y(y - q) = 0$ has $A(0, 0)$ and $B(p, q)$ as ends of a diameter any chord through $B(p, q)$ is $y - q = m(x - p)$ or $mx - y + (q - mp) = 0$

Since chords are bisected on x-axis

$\therefore \quad x_1 = \frac{mp - q}{m}$

Hence the other end C is at $\left(\frac{mp - 2q}{m}, -q\right)$; which lies on the circle.

$$\Rightarrow \left(\frac{mp - 2q}{m}\right)^2 + q^2 = \frac{mp^2 - 2pq}{m} - q^2$$

$$\Rightarrow p^2 + \frac{4q^2}{m^2} - \frac{4pq}{m} + 2q^2 = p^2 - \frac{2pq}{m}$$

i.e., $2q^2\left(\frac{m^2 + 2}{m}\right) = 2pq$

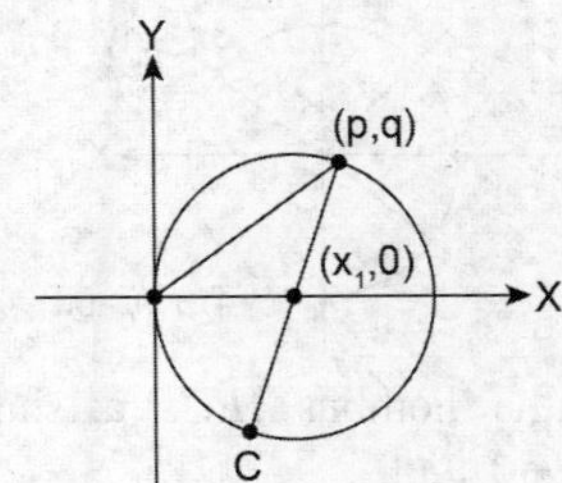

Rewriting as $qm^2 - mp + 2q = 0$, gives $m = \frac{p \pm \sqrt{p^2 - 8q^2}}{2q}$, since m has two real and distinct values

$\therefore \quad p^2 > 8q^2$

40. (b) Let $P(h, k)$, slope of $OP = \frac{k}{h} = \tan\beta$, $\tan\alpha = \frac{a}{\sqrt{h^2 + k^2 - a^2}}$, $\tan\theta$ = slope of $AP = \tan(\alpha + \beta)$ and $\tan\phi$ = slope of $BP = \tan(\beta - \alpha)$. Now, $\tan\theta . \tan\phi = K$

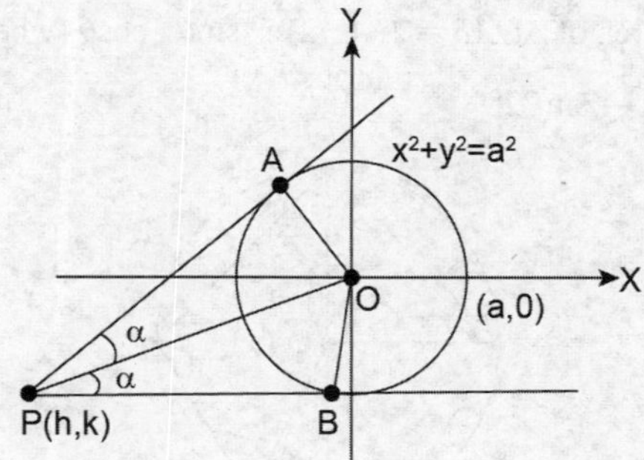

$$\Rightarrow \frac{(\tan\beta + \tan\alpha)(\tan\beta - \tan\alpha)}{(1 - \tan\alpha\tan\beta)(1 + \tan\alpha\tan\beta)} = K \text{ or } \tan^2\beta - \tan^2\alpha$$

$= K\ \{1 - \tan^2\alpha\ \tan^2\beta\}$

$$\Rightarrow \frac{k^2}{h^2} - \frac{a^2}{h^2 + k^2 - a^2} = k\left\{1 - \frac{k^2a^2}{h^2(h^2 + k^2 - a^2)}\right\}$$

$\Rightarrow (k^2 - a^2)(h^2 + k^2) = K(h^2 - a^2)(h^2 + k^2)$

$\Rightarrow K(x^2 - a^2) = y^2 - a^2$

41. (b) Let $S_1 : x^2 + y^2 + 2gx + 2fy + c = 0$

$\Rightarrow$ $C_1(-g, -f)$ and $r_1 = \sqrt{g^2 + f^2 - c}$ and S_2: $x^2 + y^2 + 2gx + 2fy + c\sin^2\alpha + (g^2 + f^2)\cos^2\alpha = 0$ has $C_2(-g, -f)$ and $r_2 = \sqrt{(g^2 + f^2 - c)\sin^2\alpha}$

Clearly $r_2 \leq r_1$ and observe that $\sin\theta = \frac{r_2}{r_1} = \sin\alpha \Rightarrow \theta = \alpha$

$\Rightarrow$ Tangents intersect at $2\theta = 2\alpha$.

42. (b) Let $A(a, 0)$ and $B(0, b)$ be the centre and r be the radius of the equal circles.

So S_1: $(x - a)^2 + y^2 = r^2$ and $S_2 : x^2 + (y - b)^2 = r^2$

Hence radical axis is $2by - 2ax + a^2 - b^2 = 0$

$\Rightarrow 2ax - 2by + b^2 - a^2 = 0$ or $2gx - 2fy + f^2 - g^2 = 0$

43. (b) Let L = 0 be the equation of radical axis. Since (a, b) is a limiting point.

$\therefore$ Limiting circle is $(x - a)^2 + (y - b)^2 = 0$. Further $x^2 + y^2 + 2gx + 2fy + c = 0$ is a member of the coaxial system

Hence $x^2 + y^2 + 2gx + 2fy + c = (x - a)^2 + (y - b)^2 + \lambda L$

$\Rightarrow \lambda L = (2g + 2a)x + (2f + 2b)y + c - a^2 - b^2$

$\Rightarrow (\lambda)L = 2(g + a)x + 2(f + b)y + c - (a^2 + b^2)$

Hence equation of radius axis is $2(g + a)x + 2(f + b)y + c - (a^2 + b^2) = 0$

44. (a) The lines L_1: $a_1x + x + b_1y + c_1 = 0$ and L_2: $a_2x + b_2y + c_2 = 0$ will intersect the co-ordinate axes respectively in

$$A\left(-\frac{c_1}{a_1}, 0\right); B\left(0, \frac{-c_1}{b_1}\right) \text{ and } C\left(-\frac{c_2}{a_2}, 0\right), D\left(0, -\frac{c_2}{b_2}\right)$$

Since point are concylic

$\therefore$ $OA.OC = OB.OD$, hence $\frac{c_1}{a_1}.\frac{c_2}{a_2} = \frac{c_1}{b_1}.\frac{c_2}{b_2}$ i.e., $a_1a_2 = b_1b_2$

45. (a) The circle is $S: (x - a)^2 + y^2 = a^2$

$\Rightarrow$ Centre C $(a, 0)$ and given point $P(a + a\cos\theta, a\sin\theta)$

Hence equation of CP is $y = \frac{a\sin\theta}{a\cos\theta}(x - a)$

$\Rightarrow$ $y = x\tan\theta - a\tan\theta$

46. (c) Consider ΔOAB, let $ON \perp AB$

$\therefore$ Area of $\Delta OAB = 2\times (1/2)\, r^2 \sin\theta \,.\, \cos\theta$ (where $\theta = 36°$)

$= \frac{r^2}{2}\sin 72°$

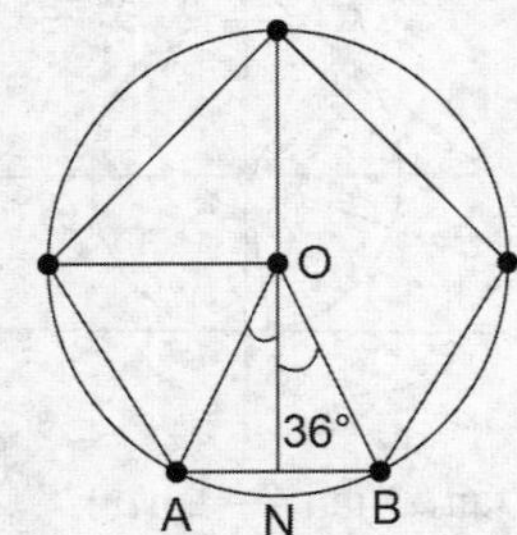

Hence area A_2 of the pentagon $= (5/2)\, r^2 \sin 72°$ square units

Area of circle $A_1 = \pi r^2$ square units.

$\Rightarrow$ $\frac{A_1}{A_2} = \frac{2\pi}{5\cos 18°} = \frac{2\pi}{5}\sec\frac{\pi}{10}$

47. (d) Since the circle passes through origin and touches the line $y = x$

$\Rightarrow$ Centre of the circle is on $y = -x$ say $(a, -a)$.

Hence the equation is $(x - a)^2 + (y + a)^2 = 2a^2$ i.e., $S_1: x^2 + y^2 - 2ax + 2ay = 0$.

The other circle is $S_2: x^2 + y^2 + 6x + 8y - 7 = 0$

Hence radical axis $(6 + 2a)\,x + (8 - 2a)\,y - 7 = 0$ or $(6x + 8y - 7) + 2a(x - y) = 0$

Now observe that $6x + 8y - 7 = 0$ when $x = y = 1/2$. Also $x - y = 0$ automatically

48. (c) Given: Circle $S_1: x^2 + y^2 = 16 \Rightarrow C_1\ (0, 0)$ and $r_1 = 4$

Let $C_2\ (x_2, y_2)$ and S_2 passes through $(1, 2)$

$\therefore$ $(x - x_2)^2 + (y - y_2)^2 = (x_2 - 1)^2 + (y_2 - 2)^2$

Since circles intersecting orthogonally, so $C_1C_2^2 = r_1^2 + r_2^2$

$\therefore$ $x_2^2 + y_2^2 = 16 + (x_2 - 1)^2 + (y_2 - 2)^2$

$\Rightarrow$ $2x_2 + 4y_2 = 21$ or $2x + 4y - 21 = 0$

49. (b) Equation of circum-circle of the equilateral triangle S_1: $x^2 + y^2 + 2gx + 2fy + c = 0$

$\Rightarrow$ $C_1\ (-g, -f)$ and $r_1 = \sqrt{g^2 + f^2 - c}$

The centre of incircle $C_2 = C_1 = (-g, -f)$ and radius $r_2 = \frac{r_1}{2}$

Hence the incircle $S_2: (x + g)^2 + (y + f)^2 = \frac{g^2 + f^2 - c}{4}$

$\Rightarrow$ $4(x^2 + y^2) + 8gx + 8fy + 3\,(g^2 + f^2) + c = 0$

Since one vertex is $(1, 1)$

$\Rightarrow$ $(g + 1)^2 + (f + 1)^2 = g^2 + f^2 - c \Rightarrow (-c) = 2 + 2g + 2f$

$\Rightarrow$ $4\,(x^2 + y^2) + 8gx + 8\,fy = (-c) - 3(g^2 + f^2) = 2 + 2g + 2f - 3\,(g^2 + f^2) = (1 - g)\,(1 + 3g) + (1 - f)\,(1 + 3f)$

50. (a) Let $A(1, 0)$ and $B(2, -1)$. Equation of right bisector of AB on which the centre of circles lies is $L: x - y = 2$

Let the centre of the circle be $(x_1, x_1 - 2)$, since the circle touches y-axis

So $r = |x_1|$ gives $(x_1 - 1)^2 + (x_1 - 2)^2 = x_1^2$ and $(x_1 - 2)^2 + (x_1 - 1)^2 = x_1^2$

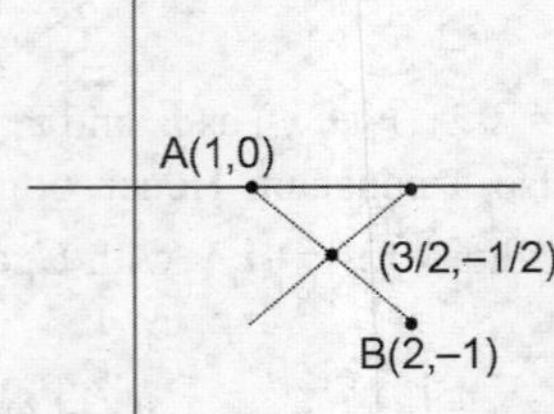

$\Rightarrow$ $x_1^2 - 5x_1 + 5 = 0$ and $x_1 = \frac{5 \pm \sqrt{5}}{2}$

$\Rightarrow$ $C_1\left(\frac{5+\sqrt{5}}{2}, \frac{1+\sqrt{5}}{2}\right)$ and $r_1 = \frac{5+\sqrt{5}}{2}$ and

$C_2\left(\frac{5-\sqrt{5}}{2}, \frac{1-\sqrt{5}}{2}\right)$ and $r_2 = \left(\frac{5-\sqrt{5}}{2}\right)$

$\Rightarrow$ $C_1C_2^2 = 5 + 5 = 10\,; r_1^2 = \frac{30 + 10\sqrt{5}}{4}; r_2^2 = \frac{30 - 10\sqrt{5}}{4}$

$\Rightarrow$ $\cos\theta = \frac{r_1^2 + r_2^2 - C_1C_2^2}{2r_1r_2} = \frac{(15-10)4}{2(25-5)} = \frac{1}{2} \Rightarrow \theta = \frac{\pi}{3}$

51. (c) Limiting point of the co-axial system are $A\,(1, 2)$, $B\,(4, 3)$ that gives line AB as $L: x - 3y + 5 = 0$

Observe that the centre of $2(x^2 + y^2) - x - 7y = 0$ is $\left(\frac{1}{4}, \frac{7}{4}\right)$ which lies on L. Option (b) circle $x^2 + y^2 - 8x - 6y = 0$ has centre on L but its radius is not zero (i.e. it should be a point circle)

52. (b) One limiting point is $A\ (2, 1)$. The given circle $S_1: x^2 + y^2 - 6x - 4y - 3 = 0$

$\Rightarrow$ $C_1\ (3, 2)$ and $r_1 = 4$ units

Line AC_1 is L: $x - y - 1 = 0$.

Now observe that only $(-5, -6)$ lies on this line

53. (b) From the given information $C_1\ (0, 0)$ and $r_1 = 4$. Slope of common chord $= 3/4$

Since length is maximum

$\therefore$ length $= 8$ and it will pass through the centre $C_1\ (0, 0)$.

Common chord is $L: 3x - 4y = 0$ and the centre C_2 will lie on $4x + 3y = 0$ at a distance of 3 units from C_1

Hence $C_2\left(\frac{9}{5}, -\frac{12}{5}\right)$ or $C_2\left(-\frac{9}{5}, \frac{12}{5}\right)$

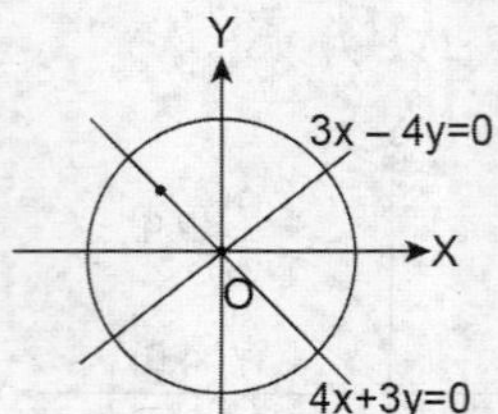

54. (d) Consider the rhombus $ABCD$ as shown $\angle ABO = 30°$.

Since $OA = 1$ unit

$\Rightarrow$ $AB = 2$ unit and $BO = \sqrt{3}$ units, hence $OQ = OP = r = \dfrac{\sqrt{3}}{2}$.

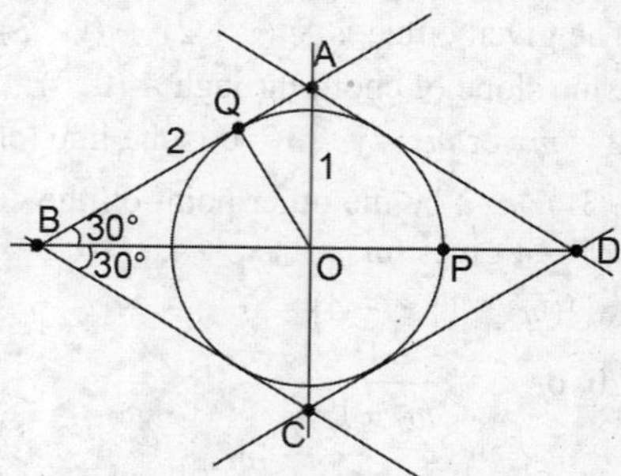

Consider point P on line BOD as shown P is the mid point of OD

$\Rightarrow$ $BP = \sqrt{3} + \dfrac{\sqrt{3}}{2} = \dfrac{3\sqrt{3}}{2}$ and $PD = \dfrac{\sqrt{3}}{2}$

Now, $CP^2 = AP^2 = 1^2 + \dfrac{3}{4} = \dfrac{7}{4}$

$\Rightarrow$ $AP^2 + PC^2 + PD^2 + BP^2 = \dfrac{7}{4} + \dfrac{7}{4} + \dfrac{3}{4} + \dfrac{27}{4} = 11$

55. **(a)** The given circle $x(x - a) + y(y - b) = 0$. Let the chord through $P(a, b)$ be $y - b = m(x - a)$ i.e., $mx - y + (b - am) = 0$

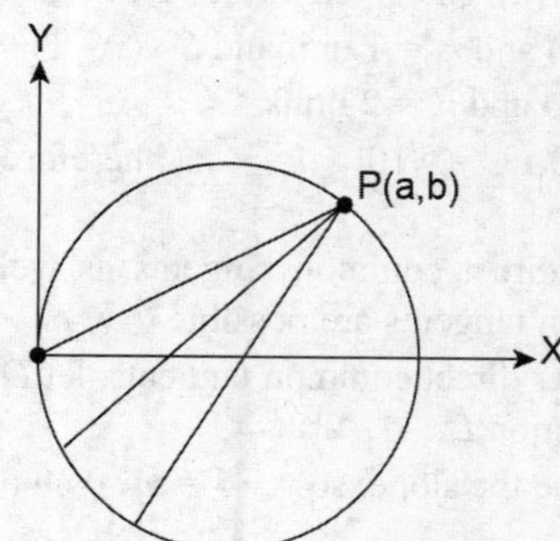

The point on x-axis is $M\left(a - \dfrac{b}{m}, 0\right)$. Since M divides the chord in the ratio 2: 1. The other point $Q(x_1, y_1)$ will be

$\dfrac{2x_1 + a}{3} = a - \dfrac{b}{m}$ and $\dfrac{2y_1 + b}{3} = 0 \Rightarrow y_1 = \dfrac{-b}{2}$ and x_1

$= \{3a - a - 3b/m\}\,(1/2) = a - \dfrac{-3b}{2m}$.

Since (x_1, y_1) lies on the circle

So $\left(a - \dfrac{3b}{2m}\right)\left(a - \dfrac{3b}{2m} - a\right) + \left(-\dfrac{b}{2}\right)\left\{-\dfrac{b}{2} - b\right\} = 0$

$\Rightarrow$ $\dfrac{9b^2}{4m^2} - \dfrac{3ab}{2m} + \dfrac{3b^2}{4} = 0$

$\Rightarrow$ $bm^2 - 2am + 3b = 0$

Since m has two real and distinct values

$\therefore$ $D = 4a^2 - 12b^2 > 0$ i.e., $a^2 > 3b^2$

56. **(b)** Let $A(x_1, y_1)$, $B(x_2, y_2)$ and $C(x_3, y_3)$ be the vertices of the triangle ABC and let $P(x, y)$ be a point of the circle

$\Rightarrow$ $\Sigma(x - x_1)^2 + (y - y_1)^2 =$ constant.

i.e., $3(x^2 + y^2) - 2(x_1 + x_2 + x_3)\,x - 2\,(y_1 + y_2 + y_3)\,y + \left\{\left(x_1^2 + x_2^2 + x_3^2\right) + \left(y_1^2 + y_2^2 + y_3^2\right)\right\} = const.$

Which gives a circle centered at $\left(\dfrac{x_1 + x_2 + x_3}{3}, \dfrac{y_1 + y_2 + y_3}{3}\right)$ which happens to be the Centroid of ΔABC

57. **(c)** $x^2 + y^2 + 2\ \ x + 4 = 0$ will present a real circle when $\ ^2 - 4 > 0$ i.e., $|\ \ | > 2$ consider $\ $ as a parameter

Radical axis of any two circles will be $2(\ _1 - \ _2)x = 0$ i.e., $x = 0$ which is the y-axis

SECTION–IV (MORE THAN ONE CORRECT)

1. **(a), (c)** Let the point of intersection of such a tangent be (p, q)

Chord of contact of such tangent $xp + 4q = a^2$

This chord of chord also touches the circle $x^2 + y^2 - 2ax = 0$

$\Rightarrow$ $\dfrac{|ap - a^2|}{\sqrt{p^2 + q^2}} = a$

$\Rightarrow$ $|ap - a^2| = a\sqrt{p^2 + q^2}$ replacing $p = x$ and $q = y$

$\Rightarrow$ $(ax - a^2) = a\sqrt{x^2 + y^2}$ $\Rightarrow$ Squaring $(x - a)^2 = x^2 + y^2$

$\Rightarrow$ **Option (c) is correct**

Also $y^2 = (x - a)^2 - x^2 = x^2 + a^2 - 2xa - x^2 = a^2 - 2xa = a(a - 2x)$

$\Rightarrow$ $y^2 = a(a - 2x)$ $\Rightarrow$ **Option (a) is also correct**

2. **(a), (b), (c), (d)** $x = x_1 + r\cos\theta$, $y = y_1 + r\sin\theta$

If θ is a constant but r varies, then it will represent a straight line.

If r is a constant but θ varies, then it will represent a circle of fixed radius 'r.

Since $P(x_1, y_1)$ is a point which is known. so option (c) and (d) also follow.

3. **(b), (c)** $L_1 : \dfrac{x}{a} + \dfrac{y}{b} = 1$ and $L_2 : \dfrac{x}{a'} + \dfrac{y}{b'} = 1$. Since $ABA'B'$ are concyclic

$\Rightarrow$ $-a'a = -bb'$ or $\dfrac{a}{b} = \dfrac{b'}{a'}$

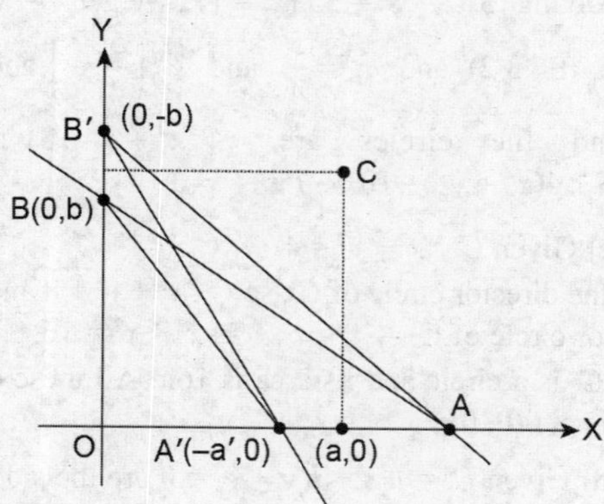

Now consider $\Delta A'B'A$, y-axis is the altitude through B'.

Slope of $A'B'$ $m_1 = -\dfrac{b'}{a'}$

Equation of altitude through A is $y = \dfrac{a'}{b'}(x - a)$ for $x = 0$, we get $y = \dfrac{-aa'}{b}$

Hence orthocenter of $\Delta AB'A'$ is $H\left(0, -\frac{aa'}{b}\right)$ using aa' = bb', $H = (0, -b')$

4. **(a), (c)** $a \quad -bm^2 + 2a \quad + 1 = 0$. If the fixed circle is $x^2 + y^2 + 2gx + 2fy + c = 0$. Since $\quad x + my + 1 = 0$ is a tangent to the circle.

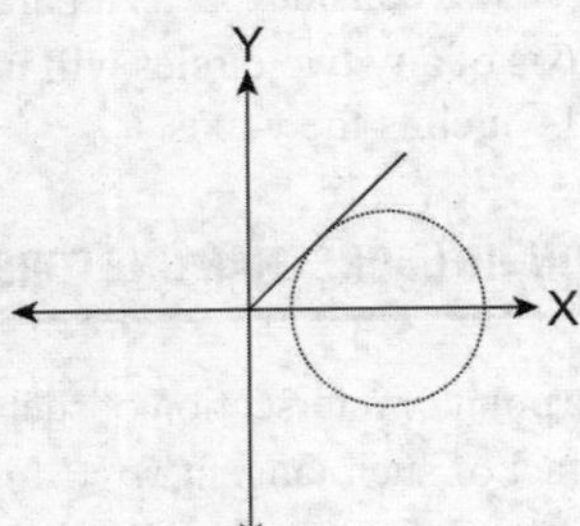

$\Rightarrow \sqrt{g^2 + f^2 - c} = \frac{|-g - mf + |}{\sqrt{\ \ ^2 \quad ^2}}$

$\Rightarrow (g^2 + f^2 - c)(\ \ ^2 + m^2) = (\ \ g + mf - 1)^2$

$\Rightarrow (g^2 + f^2 - c)(\ \ ^2 + m^2) = (\ \ g + mf - 1)^2$

$\Rightarrow$ Given $a \ \ ^2 - bm^2 + 2d \quad = 0$

$\Rightarrow C = a, g = -d, g^2 - c = b, d^2 = a = b$ also $a + b = d^2$

$\Rightarrow$ Circle reduces to $x^2 + y^2 - 2ddx + a = 0$

$\Rightarrow$ Clearly it cuts x-axis orthogonally

$\Rightarrow$ length of tangent from origin = using $a + b = a^2$

$\Rightarrow \sqrt{d^2 b}$

5. **(a), (d)** Given L_1: $2x - y = 1$ is a tangent to the circle at $P(1, 1)$
$\Rightarrow$ Centre lies on $x + 2y - 3 = 0$
Let $C(x_1, y_1) = \left(x_1, \frac{3 - x_1}{2}\right)$ be the centre of the circle,
then $CP^2 = (x_1 - 1)^2 + (y_1 - 1)^2 = r^2$,
Distance from L_2: $2x + y - 4 = 0$ is $\left|\frac{2x_1 + \frac{3 - x_1}{2} - 4}{\sqrt{5}}\right| = r$
$\Rightarrow (3x_1 - 5)^2 = 20r^2$
Also $(x_1 - 1)^2 + \frac{(x - 1)^2}{4} = r^2 \Rightarrow 5(x_1 - 1)^2 = 4r^2$
Solving $(3x_1 - 5)^2 = 25(x_1 - 1)^2$, we get $x_1 = 0, 5/4$
$\Rightarrow C_1(0, 3/2)$ and $r_1^2 = \frac{5}{4}$ and $C_2\left(\frac{5}{4}, \frac{7}{8}\right)$ and $r_2^2 = \frac{5}{64}$
and the circles are $S_1: x^2 + y^2 - 3y + 1 = 0$ and $S_2: 4(x^2 + y^2) - 10x - 7y + 9 = 0$

6. **(a), (c)** Given $C_1: x^2 + y^2 = 4$
C_2 is the director circle of C_1, so C_2: $x^2 + y^2 = 8$ and C_3 is the director circle of $C_2 \qquad \Rightarrow C_3$: $x^2 + y^2 = 16$
Now C_3 is a circle and its area is 16π. All these circles are centers at $0(0, 0)$

7. **(a), (d)** Given $U_r = a_r x + b_r y + c_r = 0$ are the sides of a triangle ABC for $r = 1, 2, 3$

Now $\begin{vmatrix} a_1^2 + b_1^2 & a_1u_1 & b_1u_1 \\ a_2^2 + b_2^2 & a_2u_2 & b_2u_2 \\ a_3^2 + b_3^2 & a_3u_3 & b_3u_3 \end{vmatrix} = 0$

Circum-circle ΔABC is given by $S_1 = L_1L_2 + \lambda(L_2L_3) + \mu L_1L_3 = 0$ where cofficient of x^2 = coefficient of y^2 and coefficient of $xy = 0$, opening the given determinant (using SARAS RULE) we will get the circumcircle

8. **(b), (c)** The given circle is S: $(x + 2)^2 + (y - 3)^2 = 2^2$
Let m be the slope of chord through $A(0, 3)$
$\Rightarrow y - 3 = mx$ or $mx - y + 3 = 0$ is the line (chord)
Let $B(x_1, 3 + mx_1)$ be the other point on the circle, so $(x_1 + 2)^2 + m^2x_1^2 = 4$ gives $(m^2 + 1)x_1^2 + 4x_1 = 0$
i.e., $x_1\{(m^2 + 1)x_1 + 4\} = 0$
$\Rightarrow x_1 = 0$, or $x_1 = \frac{-4}{m^2 + 1}$
Hence $B\left(-\frac{4}{m^2 + 1}, 3 - \frac{4m}{m^2 + 1}\right) = (x_1, y_1)$
Since B is the mid point of AM
$\Rightarrow M = \left(-\frac{8}{m^2 + 1}, 3 - \frac{8m}{m^2 + 1}\right) = (2x_1, 2y_1 - 3) = (x_2, y_2)$
Since B lies on the circle i.e., $\left(\frac{x_2}{2} + 2\right)^2 + \left(\frac{y_2 + 3}{2} - 3\right)^2 = 2^2$
$\Rightarrow x_2^2 + 16 + 8x_2 + y_2^2 + 9 - 6y_2 = 16$
or $x^2 + y^2 + 8x - 6y + 9 = 0$
Which can also be written as $x^2 + 8x + (y - 3)^2 = 0$

9. **(a), (b), (c), (d)** The given circles are S_1: $x^2 + y^2 = 1$
$\Rightarrow C_1(0, 0)$ and $r_1 = 1$ unit and S_2: $(x - 1)^2 + (y - 3)^2 = 2^2$
$\Rightarrow C_2(1, 3)$ and $r_2 = 2$ units
Since $C_1C_2 = \sqrt{10} > r_1 + r_2$, therefore circles do not intersect
Hence direct common tangents as well as transverse common tangents are possible
Consider direct common tangents, let D be the point of intersection $D(-1, -3)$
Let m be the slope, so $y + 3 = mx + m$ or $mx - y + (m - 3) = 0$
Solving $\frac{|m - 3|}{\sqrt{1 + m^2}} = 1$, we get $m = 4/3, \infty$
Hence tangents are $x + 1 = 0$, $4x - 3y - 5 = 0$.
The transverse common tangents intersect at T(1/3, 1)
For slope $\frac{|3 - m|}{\sqrt{9m^2 + 1}} = 1$
$\Rightarrow$ m = 0, –3/4 and the tangents are $y = 1$ and $3x + 4y - 5 = 0$

10. **(b), (d)** The circle passing through $A(-1, 1)$, $B(0, 6)$ and $C(5, 5)$ is centered at $P(2, 3)$ and $r = \sqrt{13}$ units, slope of OP is m = 3/2
$\Rightarrow$ Equation of diameter $y - 3 = (-2/3)(x - 2)$ i.e., $2x + 3y - 13 = 0$
Observe that (5, 1) and (–1, 5) lie on this diameter and also on the circle

11. **(a), (c), (d)** The two points on the chord are A ($a\cos\alpha$, $a\sin\alpha$) and B ($a\cos\beta$, $a\sin\beta$)

Slope of $AB = \frac{a(\sin\beta - \sin\alpha)}{a(\cos\beta - \cos\alpha)} = \frac{-\cos\left(\frac{\alpha + \beta}{2}\right)}{\sin\left(\frac{\alpha + \beta}{2}\right)}$

Equation of AB is $(y - a\sin\alpha)(\cos\beta - \cos\alpha) = (\sin\beta - \sin\alpha)(x - a\cos\alpha)$

∴ Equation of AB is $(y - a\sin\alpha)$

$$\sin\left(\frac{\alpha+\beta}{2}\right) = -\cos\left(\frac{\alpha+\beta}{2}\right)(x - a\cos\alpha)$$

$$\Rightarrow\ x\cos\left(\frac{\alpha+\beta}{2}\right) + \sin\left(\frac{\alpha+\beta}{2}\right)y -$$

$$a\sin\alpha\sin\left(\frac{\alpha+\beta}{2}\right) - a\cos\alpha.\cos\frac{\alpha+\beta}{2} = 0$$

i.e., $\cos\left(\frac{\alpha+\beta}{2}\right)x + \sin\left(\frac{\alpha+\beta}{2}\right)y - a\cos\left(\frac{\alpha-\beta}{2}\right) = 0$

Now $(y - a\sin\alpha)(\cos\beta - \cos\alpha) = (\sin\beta - \sin\alpha)(x - \cos\alpha)$

So $\frac{x - a\cos\alpha}{y - a\sin\alpha} = \frac{(\cos\beta - \cos\alpha)}{(\sin\beta - \sin\alpha)} = \frac{\cos\alpha - \cos\beta}{\sin\alpha - \sin\beta}$

Now consider

$$\cos\left(\frac{\alpha+\beta}{2}\right)x + \sin\left(\frac{\alpha+\beta}{2}\right)y = a\cos\left(\frac{\alpha-\beta}{2}\right)$$

On multiplying by $2\cos\left(\frac{\alpha-\beta}{2}\right)$, we get

$$2\cos\left(\frac{\alpha+\beta}{2}\right)\cos\left(\frac{\alpha-\beta}{2}\right)x + 2\sin\left(\frac{\alpha+\beta}{2}\right)\cos\left(\frac{\alpha-\beta}{2}\right)y$$

$$-2a\cos^2\left(\frac{\alpha-\beta}{2}\right) = 0$$

$\Rightarrow (\cos\alpha + \cos\beta)x + (\sin\alpha + \sin\beta)y - a$

$$\left\{1 + \cos\left(\frac{\alpha-\beta}{2}\right)\right\} = 0$$

12. (a), (b), (c), (d) The circle are $S_1: (x + 1)^2 + (y + 2)^2 = 5^2$

$\Rightarrow C_1(-1, -2)$ and $r_1 = 5$ and

$S_2: (x + 3)^2 + (y - 4)^2 = (\sqrt{15})^2$

$\Rightarrow C_2(-3, 4)$ and $r_2 = \sqrt{15}$

Now $C_1C_2 = \sqrt{2^2 + 6^2} = 2\sqrt{10} \Rightarrow C_1C_2^2 = r_1^2 + r_2^2$

Hence circles intersect orthogonally

∴ Only two common tangents are possible

Equation of common chord L: $2x - 6y + 15 = 0$

Length of common chord $= 2\sqrt{25 - \frac{625}{40}} = 2(5)$

$\sqrt{\frac{15}{40}} = 5\sqrt{\frac{3}{2}}$ units

Length of direct common tangents $= \sqrt{C_1C_2^2 - (r_1 - r_2)^2}$

$= \sqrt{40 - (25 + 15 - 2\times 5\sqrt{15})}$

$= \sqrt{10\sqrt{15}}$ units $= (1500)^{1/4} = \left(125\times 5\times\frac{12}{5}\right)^{1/4}$

$= 5\left(\frac{12}{5}\right)^{1/4}$ units

13. (a), (c) Given AB is a diameter and C is any point on the circle. Let $CN \perp AB$. The area of $\Delta ABC = (1/2)\, AB \times CN$, which will be

Maximum, when CN is maximum i.e., CN = radius and a right angled isosceles triangle is formed.

Let $\angle CAB = \theta$, so $AC = 2r\cos\theta$ and $BC = 2r\sin\theta$.

Observe that $AC + BC = 2r(\cos\theta + \sin\theta)$ will be maximum, when $\theta = 45°$ and then the perimeter of ΔABC will be maximum.

14. (a), (d) all the given points P lie on the circle. Image of point P in the line

$L_1: x - y + 1 = 0$ is point $Q(h, k)$ which must lie on $L_2: 7x + y + 3 = 0$

(a) $P(3,0) \Rightarrow \frac{3-h}{1} = \frac{0-k}{-1} = \frac{2\times 4}{2}$

$\Rightarrow Q(-1, 4)$ which lies on L_2

(b) $P(0, 3) \Rightarrow \frac{0-h}{1} = \frac{3-k}{-1} = \frac{2\times(-2)}{2}$

$\Rightarrow Q(2, 1)$ does not lie on L_2

(c) $P\left(\frac{72}{25}, \frac{-21}{25}\right) \Rightarrow \frac{\frac{72}{25} - h}{1} = \frac{-\frac{21}{25} - k}{-1} = \frac{2\times 118}{2\times 25}$

$\Rightarrow Q\left(-\frac{46}{25}, \frac{97}{25}\right)$ does not lie on L_2

(d) $P\left(\frac{-72}{25}, \frac{21}{25}\right) \Rightarrow \frac{\frac{-72}{25} - h}{1} = \frac{\frac{21}{25} - k}{-1} = \frac{2\times(-68)}{2\times 25}$

$\Rightarrow Q\left(-\frac{4}{25}, \frac{-47}{25}\right)$ which lies on L_2

15. (a), (c), (d) $C_1: x^2 + y^2 - 2x - 15 = 0$

$C_2: x^2 + y^2 - 2x - 7 = 0$

Clearly the two circles have the same centre. The region bounded implies of the circles itself.

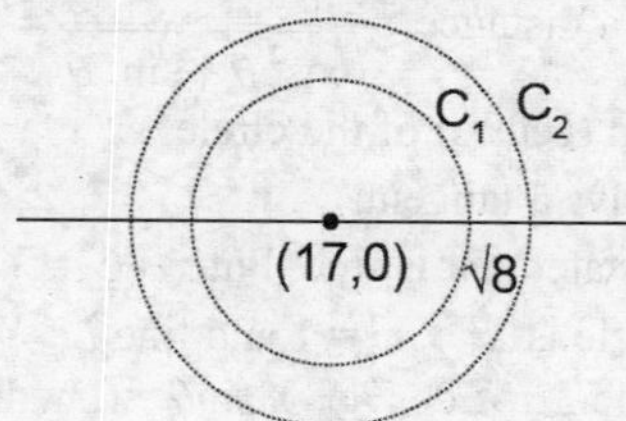

The points in the above region from the options lies in this region are

16. (a), (d) $(x + 4)^2 + (y - 5)^2 = 9^2$ to understand this question draw a diameter through $Q(-2, 3)$ and $P(-4, 5)$

$\Rightarrow a = \sqrt{AQ^2}$ and $b = \sqrt{BQ^2}$ i.e., $a = (9 + 2\sqrt{2})$ and $b = (9 - 2\sqrt{2})$

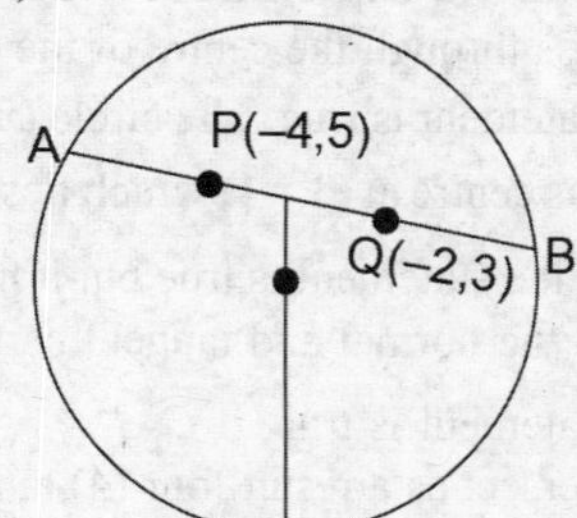

Hence $a + b = 18$, $a - b = 4\sqrt{2}$ and $a.b = 81 - 8 = 73$

17. (b), (c) Observe that indirectly we are to find the point of tangency from $P(-2, 0)$ to the circle $(x - 4)^2 + (y - 8)^2 = (2\sqrt{5})^2$.

Let $y = m(x + 2)$ be the tangent, so $mx - y + 2m = 0$

Now $\left|\frac{4m-8+2m}{\sqrt{1+m^2}}\right| = 2\sqrt{5}$ or $9m^2 + 16 - 24m = 5m^2 + 5$

$\Rightarrow$ $4m^2 - 24m + 11 = 0$ or $m = \frac{11}{2}, \frac{1}{2}$

The tangents are $L_1: 11x - 2y + 22 = 0$ and $L_2: x - 2y + 2 = 0$

Observe that $\left(-\frac{2}{5}, \frac{44}{5}\right)$ lies on L_1 and (6, 4) lies on L_2

18. (a), (b), (c), (d) Centre of circle $C(2,-1)$ one tangent from origin $O(0, 0)$ is $3x + y = 0$

$\Rightarrow$ Radius $r = \frac{5}{\sqrt{10}} = \sqrt{\frac{5}{2}}$ units.

Let $mx - y = 0$ be the other tangent

$\therefore$ $\frac{|2m+1|}{\sqrt{1+m^2}} = \sqrt{\frac{5}{2}} \Rightarrow 2(4m^2 + 1 + 4m) = 5m^2 + 5$

$\Rightarrow$ $3m^2 + 8m - 3 = 0$ $\Rightarrow$ $m = -3, 1/3$

$\Rightarrow$ Equation of the other tangent $x - 3y = 0$

Since $m_1 m_2 = (-3)(1/3) = -1 \therefore m_1 \perp m_2$

19. (a), (b), (c), (d) L: $3x - 4y - 5k = 0$ for $k \in \mathbb{Z}$ is a secant or tangent to the circle $(x-2)^2 + (y-4)^2 = 5^2$

$\therefore$ $0 \le \frac{|-10-5k|}{5} \le 5$ $\Rightarrow$ $0 \le |2 + k| \le 5$

Hence $|k + 2| = 0, 1, 3, 4$ are possible

SECTION-V (ASSERTION REASON)

1. (a) R: The statement is true. The line $L_1: x\cos\theta + y\sin\theta - a = 0$ is at a distance $\frac{|0+0-a|}{\sqrt{\cos^2\theta+\sin^2\theta}} = a$ = radius from the origin (centre) of the circle.

$\therefore$ It is always a tangent.

A: The statement is true. Put $x - 3 = X$ and $y - 3 = Y$

Now circle is $X^2 + Y^2 = 1$ and line L_2: $X\cos\theta + Y\sin\theta = 1$ and line L_2: $X\cos\theta + Y\sin\theta = 1$ will be a tangent to the given circle.

$\therefore$ Assertion is fully supported by reason.

2. (a) R: The statement is true by definition.

A: Tangents drawn from any point on circle $x^2 + y^2 = 2a^2$ to the circle $x^2 + y^2 = a^2$ will be perpendicular.

$\therefore$ Statement is true and it is implied by R.

3. (b) R: statement is true. Normal of a circle at any point will always pass through the centre of the circle

A: The statement is true. The circle $(x-3)^2 + (y+1)^2 = \left(\sqrt{10}\right)^2$ has centre at $(3, -1)$ which lies on $2x + y - 5 = 0$.

Although the statement is true but it has hardly any relevance to the normal and tangent concept of reason R.

4. (d) R: The statement is true, if $C_1C_2 > r_1 + r_2$, then circles do not intersect as a result four (4) tangent are possible

A: The given circles are S_1: $x^2 + y^2 - 4 = 0$

$\Rightarrow$ $C_1(0, 0)$ and $r_1 = 2$ and S_2: $(x-3)^2 + (y-4)^2 = 7^2$

$\Rightarrow$ $C_2(3, 4)$ and $r_2 = 7$. Now $C_1C_2 = 5 = r_2 - r_1$

$\Rightarrow$ circles touch the internally only one tangent is possible.

Hence A is false

5. (a) A: The statement is true, the circle is

$S: \left(x+\frac{5}{2}\right)^2 + \left(y-\frac{5}{2}\right)^2 = \frac{25}{2}$.

Now tangent from a point $P(1, \lambda)$ is possible only if $S(1, \lambda) \ge 0$ i.e., $\left(1+\frac{5}{2}\right)^2 \ge \left(\lambda-\frac{5}{2}\right)^2 \ge \frac{25}{2}$

$\Rightarrow$ $\left(\lambda-\frac{5}{2}\right)^2 \ge \left(\frac{1}{2}\right)^2$ $\Rightarrow$ $\lambda - \frac{5}{2} \ge \frac{1}{2}$ or $\lambda - \frac{5}{2} \le -\frac{1}{2}$

i.e., $1 \in (-\infty, 2] \cup [3, \infty)$

Obviously tangent is not possible from an internal point of a circle {where $\lambda \in (2, 3)$}

R: Reason $\left(1+\frac{5}{2}\right)^2 + \left(\lambda-\frac{5}{2}\right)^2 < \frac{25}{2}$ is true and it fully supports the Assertion A.

6. (a) R: The statement is true. Only on one circle is possible that will pass through three non-collinear points

A: The given point $A(-2, 1)$, $B(-1, 0)$ and $C(-4, 3)$ are collinear and lie on the line $x + y + 1 = 0$

$\Rightarrow$ A is false as no circle is possible.

7. (c) The statement is not completely true. Circles will intersect in two distinct points when $|r_2 - r_1| < C_1C_2 < r_2 + r_1$

A: The given circles are S_1: $x^2 + y^2 = 4$

$\Rightarrow$ $C_1(0, 0)$ and $r_1 = 2$ and S_2: $(x-4)^2 + y^2 = 3^2 \Rightarrow C_2(4, 0)$ and $r_2 = 3$. Now $1 < C_1C_2 = 4 < 5$

$\therefore$ Circles will intersect in two distinct points. $\Rightarrow$ A is true

8. (c) A: The given circles are S_1: $(x-3)^2 + (y+2)^2 = 2^2$

$\Rightarrow$ $C_1(3, -2)$ and $r_1 = 2$ and S_2: $(x-4)^2 + (y-3)^2 = \left(\sqrt{2}\right)^2$

$\Rightarrow$ $C_2(4, 3)$ and $r_2 = \sqrt{2}$

The common chord is $2x + 2y - 14 = 0$ which passes through $C_2 \Rightarrow$ Assertion is true

R: Reason is false. Centre of first circle does not lie on the second circle

9. (a) The smallest circle passing though $A(1, 0)$ and $B(0, 1)$ will be centered at $\left(\frac{1}{2}, \frac{1}{2}\right)$ and radius $= \frac{AB}{2} = \frac{1}{\sqrt{2}}$ units

$\Rightarrow$ A is true

R: Circle will passes through origin when AB is diameter since $\angle AOB = 90° \Rightarrow R$ is true

10. (a) R: The statement is true by the definition

A: Statement is true. Let C_1 circle has radius r_1 and C_2 with radius r_2.

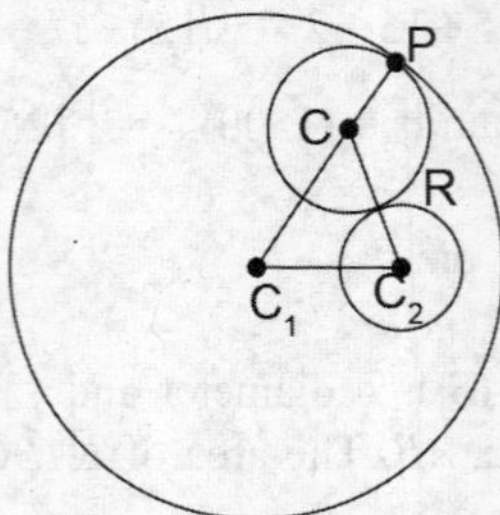

Further C circle has radius r. As shown observe (by joining C_1C) that $CP = r = CR$

$\therefore$ $C_1C + CC_2 = r_1 + r_2 =$ constant

11. **(b) A:** The statement is true as C_2 is the director circle to C_1 so $\frac{\text{radius of } C_1}{\text{radius of } C_2} = \frac{1}{\sqrt{2}}$

R: The statement is true on its own but it does not support statement A.

12. **(b) A:** statement is true, line $4y - 3x - 5 = 0$ passes though A (–3, –1) and its distance from O (0, 0) is 1 unit = radius of circle

The tangent at B (0, –1) is $y = -1$ and normal at A to line AB is $x = -3$

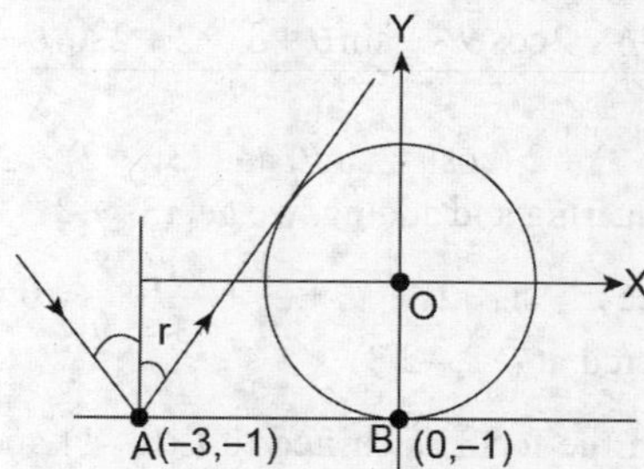

Equation of incident ray is $y + 1 = (-4/3)(x + 3)$

$\Rightarrow 4x + 3y + 15 = 0$

R: The statement is true on its own. There is no equation of incident ray for verification of $\angle i = \angle r$

13. **(c) A:** The statement is true. Observe that O (0, 0) satisfies both the circles.

Hence either circle touch internally or externally

$\Rightarrow C_1C_2 = r_1 + r_2$ or $C_1C_2 = |r_2 - r_1|$

Where $C_1(-g, -f)$, $r_1 = \sqrt{g^2 + f^2}$ and $C_2(-g', -f')$, $r = \sqrt{g'^2 + f'^2}$. Solving, we get $gf' = fg'$

R: The statement is false. This is possible only if the circles touch each other internally. When circles touch each other externally only the tangent at the common point is perpendicular to the line joining the centers. The direct common tangents can never be at 90° to the line joining the centers.

14. **(a) R:** The statement is true $2S_1 - S_2 = 0$ and $2S_2 - S_3 = 0$ represent the same line i.e., $9x - 25y - 1 = 0$

A: The statement is true as radical axis of a co-axial system is the same line. Hence Assertion is true and it follow completely from R.

SECTION–VI (LINKED COMPREHENSION-TYPE)

A:

1. **(b)** Let A (a, a,), B (b, a) and O (0, 0) right bisector of AB is $x = \frac{a+b}{2}$

Similarly right bisector of OA is $x + y - a = 0$ and the circumcentre is at $C\left(\frac{a+b}{2}, \frac{a-b}{2}\right)$ since the circle passes through origin

$\therefore$ Equation is $x^2 + y^2 - (a + b)x - (a - b)y = 0$

2. **(a), (b), (c)** Let the circle passing through the intersection of $S = 0$ and $L = 0$ be $S + \lambda L = 0$

So $x^2 + y^2 + (\lambda - 2)x + 2(\lambda - 2)y + (4 - 4\lambda) = 0$

Which has centre at $C\left(\frac{2-\lambda}{2}, 2-\lambda\right)$ and radius $= \sqrt{\frac{5}{4}(\lambda-2)^2 - (4-4\lambda)}$

Since circle touches L_2: $x + 2y = 0$

$\Rightarrow \frac{|2-\lambda+8-4\lambda|}{2\sqrt{5}} = r$.

On squaring, we get

$\frac{5}{4}(\lambda^2+4-4\lambda) - 4 + 4\lambda = \frac{25(\lambda^2+4-4\lambda)}{4\times 5}$

$\Rightarrow 4\lambda - 4 = 0 \Rightarrow \lambda = 1$

Hence $C\left(\frac{1}{2}, 1\right)$ and $r = \frac{\sqrt{5}}{2}$ unit and the circle is $x^2 + y^2 - x - 2y = 0$

3. **(a)** The lines are L_1: $bx + ay = ab$

$\Rightarrow A(a, 0)$, $B(0, b)$ and L_2 : $ax + by = ab$

$\Rightarrow C(b, 0)$, $D(0, a)$

The points will be concyclic when ab = ab which is always true but if a = 0, b = 0, then lines can not exist.

The circle will have centre at $\left(\frac{a+b}{2}, \frac{a+b}{2}\right)$ and radius

$r = \sqrt{\left(\frac{a+b}{2}\right)^2 + \left(\frac{a-b}{2}\right)^2} = \sqrt{\frac{a^2+b^2}{2}}$.

Hence the equation $x^2 + y^2 - (a + b)x - (a + b)y = a^2 + b^2 - \frac{(a^2+b^2)+2ab}{2}$

$\Rightarrow x^2 + y^2 - (a + b)(x + y) + ab = 0$

4 **(c)** Let $L_1: x + y - 6 = 0$; $L_2: 2x + y - 4 = 0$; $L_3: x + 2y - 5 = 0$

The equation circum circle is $S: L_1L_2 + \lambda L_2L_3 + \mu L_1L_3 = 0$

$\Rightarrow 2x^2 + y^2 + 3xy - 16x - 10y + 24 + \lambda\{2x^2 + 2y^2 + 5xy - 14x - 13y + 20\} + \mu\{x^2 + 2y^2 + 3xy - 11x - 17y + 30\} = 0$

$\Rightarrow x^2\{2 + 2\lambda + \mu) + y^2(1 + 2\lambda + 2\mu) + (3 + 5\lambda + 3\mu)xy - (16x + 14\lambda x + 11\mu x) - (10 + 13\lambda + 17\mu)y + 24 + 20\lambda + 30\mu = 0$

Now $2 + 2\lambda + \mu = 2\lambda + 2\mu + 1$

$\Rightarrow \mu = 1$ also $5\lambda + 3\mu + 3 = 0 \Rightarrow \lambda = -6/5$

We get the centre at $\left(\frac{8-\frac{42}{5}+\frac{11}{2}}{2-\frac{12}{5}+1}, \frac{5-\frac{39}{5}+\frac{17}{2}}{\frac{3}{5}}\right) = \left(\frac{17}{2}, \frac{19}{2}\right)$

5. **(a)** $L_1: x - 2y + 3 = 0$

$\Rightarrow A(-3, 0)$ and $B(0, 3/2)$ and

$L_2: \lambda x - y + 1 = 0 + C\left(-\frac{1}{\lambda}, 0\right)$ and $D(0, 1)$

Since A, B, C, D are concylic

$\Rightarrow (-3)\left(-\frac{1}{\lambda}\right) = \left(\frac{3}{2}\right)(1) \Rightarrow \lambda = 2$

Observe that the circle $2(x^2 + y^2) + 7x - 5y = 0$ does not pass through A or B

6. **(b), (c)** Since $2x + 3y - 5 = 0$ is a tangent at (1, 1)

$\Rightarrow$ The centre of circle lies on $3x - 2y - 1 = 0$

Since $r = \sqrt{13}$

$\Rightarrow C_1(1 + 2, 1 + 3)$ or $C_2(1 - 2, 1 - 3)$

Hence S_1: $(x - 3)^2 + (y - 4)^2 = 13$ i.e., $x^2 + y^2 - 6x - 8y + 12 = 0$ and S_2: $(x + 1)^2 + (y + 2)^2 = 13$ i.e., $x^2 + y^2 + 2x + 4y - 8 = 0$

7. (b) The circle is S: $(x - 3)^2 + y^2 = 1^2 \Rightarrow C(3, 0)$ and $r = 1$.

The line mirror is $L_m : x - y - 1 = 0$

Image of C is $M(h, k)$ given by

$$\frac{3-h}{1} = \frac{0-k}{-1} = \frac{(2)\{3-0-1\}}{2}$$

$\Rightarrow M(1, 2)$ hence the image circle is S: $(x - 1)^2 + (y - 2)^2 = 1$ i.e., $x^2 + y^2 - 2x - 4y + 4 = 0$

8. (a) The given circle S': $(x - 1)^2 + (y - 2)^2 = 1$ has centre $C(1, 2)$ and radius $r = 1$.

The required circle has $O(0,0)$ and $C(1,2)$ as its diameter.

$\therefore$ S'': $x^2 + y^2 - x - 2y = 0$

9. (b) Now S: $x^2 + y^2 - 6x + 8 = 0$ and S'': $x^2 + y^2 - x - 2y = 0$ Hence the common chord is $5x - 2y - 8 = 0$ this will intersect L_m: $x - y - 1 = 0$ at (2, 1) which lies on line $2x + 3y = 7$

10. (a) The circle S $(x - 3)^2 + y^2 = 1$ has centre on x-axis

$\Rightarrow$ Length of intercept on x-axis $= 2r = 2$ units.

Since centre lies at a distance of 3 unit which is larger than radius

$\Rightarrow$ Length of intercept on y-axis $= 0$

11. (d) Consider a point $P(x_1, y_1)$ on the curve S: $x^2 - 2y^2 + xy - 4x + 7y - 35 = 0$

Let $M(x_2, y_2)$ be its image in $L_m : x - y - 1 = 0$ .Using the equation for image $\dfrac{x_1 - x_2}{1} = \dfrac{y_1 - y_2}{-1} = \dfrac{(2)(x_1 - y_1 - 1)}{2}$

$\Rightarrow x_1 - x_2 = x_1 - y_1 - 1$

$\Rightarrow y_1 = x_2 - 1$ and $y_1 - y_2 = 1 + y_1 - x_1 \Rightarrow x_1 = y_2 + 1$

Putting in $S=0$, we get $(y_2+1)^2 - 2(x_2-1)^2 + (x_2-1)(y_2+1) - 4(y_2 + 1) + 7(x_2 - 1) - 35 = 0$

i.e., $y_2^2 + 1 + 2y_2 - 2x_2^2 - 2 + 4x_2 + x_2y_2 - 1 - y_2 + x_2 - 4y_2 - 4 + 7x_2 - 7 - 35 = 0$

$\Rightarrow 2x^2 - xy - y^2 - 12x + 3y + 48 = 0$

12. (c) As shown P, Q, R lie on the circle with centre.

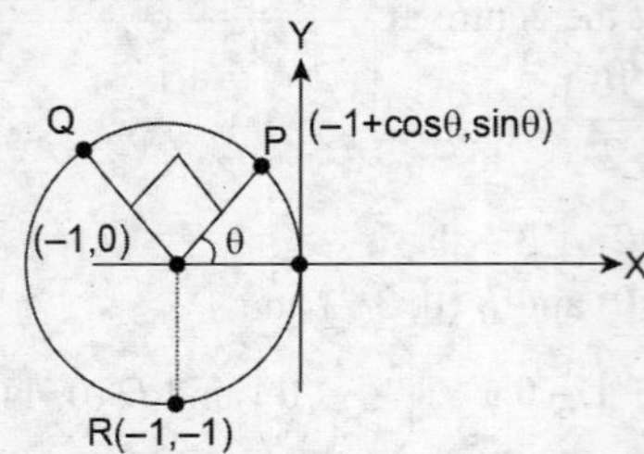

At (–1, 0) and radius $r = 1$ unit

$\Rightarrow$ Locus if P is S_1: $(x^2 + y^2) + 2x = 0$

13. (c) Centroid

$$G\left(\frac{-3+\cos\theta-\sin\theta}{3}, \frac{-1+\sin\theta+\cos\theta}{3}\right) = (x, y)$$

$\Rightarrow \cos\theta - \sin\theta = 3x + 3$ and $\sin\theta + \cos\theta = 3y + 1$ By squaring and adding, we get S_2: $(3x + 3)^2 + (3y + 1)^2 = 2$

14. (c) Circucenre $C(-1, 0)$ and

$$G\left(\frac{-3+\cos\theta-\sin\theta}{3}, \frac{-1+\sin\theta+\cos\theta}{3}\right)$$

From $ONGC$ rule $\dfrac{2C + 0}{3} = G$

$\Rightarrow O = (-3 + \cos\theta - \sin\theta + 2, -1 + \sin\theta + \cos\theta)$

$\Rightarrow x + 1 = \cos\theta - \sin\theta$ and $y + 1 = \sin\theta + \cos\theta$ by squaring and adding, we get $S_3 : (x + 1)^2 + (y + 1)^2 = 2$

15. (d) Similarly as in Q #14

$N = 2G - C$

$$= \left(\frac{-6+2\cos\theta-2\sin\theta+3}{3}, \frac{-2+2\sin\theta+2\cos\theta}{3}\right)$$

$\Rightarrow (3x + 3) = 2(\cos\theta - \sin\theta)$ and $(3y + 2) = 2(\sin\theta + \cos\theta)$ on squaring and adding, we get $(3x + 3)^2 + (3y + 2)^2 = 8$.

Hence $x^2 + 1 + 2x + y^2 + \dfrac{4}{9} + \dfrac{4y}{3} - \dfrac{8}{9} = 0$ which is circle centered at $(-1, -2/3)$

16. (c) When the locus is shifted to (–1, –1) and the axes are rotated by $\pi/3$ (anticlockwise)

Now the centre is at $(0, 0) \Rightarrow S_3$ becomes $x^2 + y^2 = 2$

17. (d) Arc PQ (P and Q) subtend 90° angle at the cenre of circumcircle

$\Rightarrow$ At any point S on the circle S_1 (other than P and Q), we get $\angle PSQ = \dfrac{\pi}{4}$

18. (c) Locus of S_1: $(x + 1)^2 + y^2 = 1$, any chord through $O(0, 0)$ is $y = mx$.

Let the other point be $B(x_1, y_1)$, so $x^2 + m^2x^2 + 2x = 0 \Rightarrow x\{(m^2 + 1)x + 2\} = 0$

$\Rightarrow O(0, 0)$ and $B\left(\dfrac{-2}{m^2+1}, \dfrac{-2m}{m^2+1}\right)$

Mid point of OB is $\left(\dfrac{-1}{m^2+1}, \dfrac{-m}{m^2+1}\right) = (x, y)$ by squaring and adding, we get $x^2 + y^2 = \dfrac{m^2+1}{(m^2+1)^2} = \dfrac{1}{m^2+1} = -x$

$\Rightarrow \left(x + \dfrac{1}{2}\right)^2 + y^2 = \left(\dfrac{1}{2}\right)^2 \Rightarrow (2x + 1)^2 + 4y^2 = 1$

19. (b) Since PQ arc subtends 90° at the centre of circle $S_1 = 0$

Probability of lying outside $= \dfrac{270}{360} = \dfrac{3}{4}$

20. (a) $S_1 = 0$ is the circle $(x + 1)^2 + y^2 = 1$

When it rolls once completely, it travels a distance equal to its own parameter on the positive y-axis

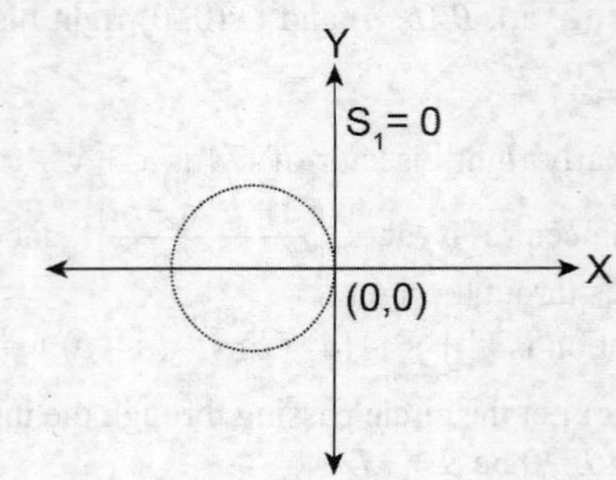

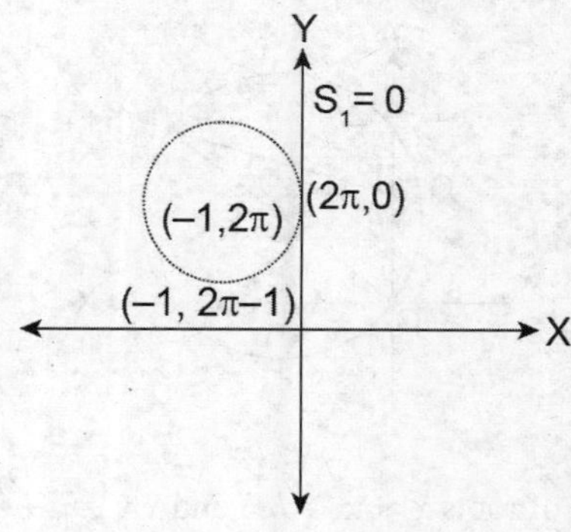

$S'_1: (x+1)^2 + (y-2\pi)^2 = 1$

$\Rightarrow$ There can be no rational points on the new locus

21. (b) Given $S_1: x^2 + y^2 + 4x + 2y + 5 = 0$ and $S_2: x^2 + y^2 + 2x + 4y + 7 = 0$

Let $S = S_1 + \lambda S_2 = 0$, be the system of co-axial system of circles then $S = (x^2 + y^2)(1 + \lambda) + 2(2 + \lambda)x + 2(1+2\lambda)y + (5 + 7\lambda) = 0$

The radius $r = \left(\frac{\lambda+2}{\lambda+1}\right)^2 + \left(\frac{2\lambda+1}{\lambda+1}\right)^2 - \left(\frac{7\lambda+5}{\lambda+1}\right) = 0$

$\Rightarrow \lambda^2 + 4 + 4\lambda + 4\lambda^2 + 1 + 4\lambda - 7\lambda^2 - 12\lambda - 5 - 2\lambda^2 - 4\lambda = 0$

$\Rightarrow (-2\lambda)\{\lambda + 2\} = 0$ i.e., $\lambda = 0, -2$

Now for $\lambda = 0$ the centre $C_1(-2, -1)$ and for $\lambda = -2$ then centre $C_2(0, -3)$

22. (a) The circles with the given limiting points are $S_1: (x-1)^2 + (y-2)^2 = 0$ and $S_2: (x-4)^2 + (y-3)^2 = 0$

Hence the coaxial system is $(x^2 + 1 - 2x + y^2 + 4 - 4y) + \lambda(x^2 + 16 - 8 + y^2 + 9 - 6y) = 0$

Since the circle passes through 0 (0, 0)

$\therefore$ $5 + 25\lambda = 0$, so $\lambda = -1/5$ and the required circle is $4(x^2 + y^2) - 2x - 14y = 0$ or $2(x^2 + y^2) - x - 7y = 0$

23. (c) Let $A(2, -1)$ and $B(-3, 2)$

Radical axis of the system is the right bisector of AB i.e., $5x - 3y + 4 = 0$

Aliter: The circles with the given limiting points are S_1: $(x-2)^2 + (y+1)^2 = 0$ and $S_2: (x+3)^2 + (y-2)^2 = 0$.

Let the circles of the coaxial system be $(x^2 + 4 - 4x + y^2 + 1 + 2y) + \lambda(x^2 + 9 + 6x + y^2 + 4 - 4y) = 0$

For radical axis $\lambda = -1$, gives $-10x + 6y + 5 - 13 = 0$ or $5x - 3y + 4 = 0$

SECTION– VII (COLUMN MATCHING)

1. (i) → (b); (ii) → (a); (iii) → (d); (iv) → (c)

(i) Given: $\frac{x}{a} = t^2$ and $\frac{y}{2a} = t \Rightarrow \frac{y^2}{4a^2} = \frac{x}{a}$ $i.e. y^2 = 4ax$

(ii) The director circle to the circle $x^2 + y^2 = a^2$ is $x^2 + y^2 = 2a^2$

(iii) Given $\frac{x}{a} = \sec\theta$ and $\frac{y}{a} = \tan\theta$, now $\sec^2\theta - \tan^2\theta = 1$

$\{\text{for } \theta \neq \frac{\pi}{2}(2n+1)\}$

$\Rightarrow x^2 - y^2 = a^2$

(iv) The circles $S: (x-a)^2 + y^2 = a^2$

The required locus is the circle with OA as its diameter i.e., $x^2 + y^2 - ax = 0$

2. Ans (i) → (c); (ii) → (a); (iii) → (b); (iv) → (d)

(i) The circles are $S_1: (x+1)^2 + y^2 = \left(\sqrt{\ -\ }\right)$

$\Rightarrow$ $C_1(-1, 0)$ and $r_1 = \sqrt{1-c}$ and $S_2: x^2 + (y+1)^2 = \left(\sqrt{1-c}\right)^2$

$\Rightarrow$ $C_2(0, -1)$ and $r_2 = r_1$; Now $C_1C_2 = \sqrt{2}$, as $r_1 = r_2$

$\therefore$ The circles will touch externally only and $C_1C_2 = r_1 + r_2$

$\Rightarrow$ $2\sqrt{1-c} = \sqrt{2}$; Hence $1 - c = \frac{1}{2} \Rightarrow c = \frac{1}{2}$

(ii) The circles are $S_1: (x+1)^2 + (y+3/2)^2 = \left(\sqrt{\frac{13}{4} - c}\right)^2$

$\Rightarrow$ $C_1\left(-1, -\frac{3}{2}\right)$ and $r_1^2 = \frac{13}{4} - c$

Now, $S_2: (x - 1/2)^2 + (y+1)^2 = \frac{5}{4} - c$

$\Rightarrow$ $C_2\left(\frac{1}{2}, -1\right)$ and $r_2^2 = \frac{5}{4} - c$

For orthogonal intersection $C_1C_2^2 = r_1^2 + r_2^2$

$\Rightarrow$ $\left(\frac{3}{2}\right)^2 + \left(\frac{1}{2}\right)^2 = \frac{18}{4} - 2c \Rightarrow c = 1$

(iii) The circle $S_1: x^2 + y^2 = 9$

$\Rightarrow$ $C_1(0, 0)$, $r_1 = 3$ and $S_2: (x-1)^2 + y^2 = c^2/4$

$\Rightarrow$ $C_2(1, 0)$, $r_2 = |c|/2$

The circle S_2 will be contained in S_1 when $C_1C_2 < r_1 - r_2$

$\Rightarrow$ $r_2 < 3 - 1 = 2$ $\Rightarrow$ $c < 2$

(iv) The circle $S_1: x^2 + y^2 = 9$

$\Rightarrow$ $C_1(0, 0)$ and $r_1 = 3$ and $S_2: (x-1)^2 + y^2 = \left(\sqrt{c}\right)^2$

$\Rightarrow$ $C_2(1, 0)$ and $r_2 = \sqrt{c}$

Now S_1 will be contained in S_2 when $r_2 - r_1 > C_1C_2$

$\Rightarrow$ $\sqrt{c} - 3 > 1$ $\Rightarrow$ $c > 16$

3. (i) → (c), (d); (ii) → (b) (d) (e); (iii) → (a), (d)

(i) The circles are $S_1: (x-1)^2 + (y-3)^2 = 1^2$

$\Rightarrow$ $C_1(1, 3)$, $r_1 = 1$ and $S_2: x^2 + y^2 + 6x - 2y + 1 = 0$

$\Rightarrow$ $C_2(-3, 1)$, $r_2 = 3$

Now $C_1C_2 = \sqrt{4^2 + 2^2} = 2\sqrt{5} > 4$ (as $r_1 + r_2 = 4$)

So circles do not intersect

$\Rightarrow$ Points of intersect of circles $\lambda = 0$

$\therefore$ Number of possible tangents $\mu = 4$

$\Rightarrow$ $\mu - \lambda = 4$, $\mu + \lambda = 4$

(ii) $S_1: (x-3)^2 + y^2 = 3^2 \Rightarrow C_1(3, 0), r_1 = 3$ and

$S_2: (x+1)^2 + y^2 = 1^2 \Rightarrow C_2(-1, 0), r_2 = 1$

Since $C_1C_2 = 4 = r_1 + r_2 \therefore \lambda = 1, \mu = 3$

Hence $\mu - \lambda = 2$, $\mu + \lambda = 4$

$\Rightarrow$ $\lambda^\mu + \lambda^1 + \mu^1 = 1^3 + 3^1 = 4$

(iii) The circles are $S_1: (x-1)^2 (y-2)^2 = 2^2$

$\Rightarrow$ $C_1(1, 2)$ and $r_1 = 2$ units and

$S_2: (x-2)^2 + (y-1)^2 = 2^2$

$\Rightarrow$ $C_2(2, 1)$ and $r_2 = 2$ units

Now $|r_1 - r_2| < C_1C_2 = \sqrt{2} < r_1 + r_2$

$\therefore$ The circles intersect at two points

$\Rightarrow$ $\lambda = 2$, $\mu = 2$; Hence $\mu - \lambda = 0$, $\mu + \lambda = 4$,

$\Rightarrow$ $\lambda^{\mu} + \mu^{1} = 2^2 + 2^1 = 6$

4. (i) → (d); (ii) → (e); (iii) → (c)

(i) $L_1: y = a_1x + b$ intersect the axes at and $A\left(-\frac{b}{a_1}, 0\right)$ and $B(0, b)$

$L_2: y = a_2x + b$ gives $C\left(-\frac{b}{a^2}, 0\right)$ and D (0, b). If these points are con-cyclic then $\frac{b^2}{a_1a_2} = b^2$

Hence $a_1a_2 = 1$ if $b \neq 0$

(ii) Let P be a point on $C_1: x^2 + y^2 = a_1^2$.

$P \equiv (a_1 \cos\theta, a_1 \sin\theta)$

Chord of contact of P to C_2 is $x(a_1\cos\theta) + y(a_1\sin\theta) = b^2$

If this chord of contact touches C_3

$\Rightarrow$ $\frac{|-b^2|}{a_1^2\cos^2\theta + a_1^2\sin^2\theta} = a_2 \Rightarrow b^2 = a_1.a_2$

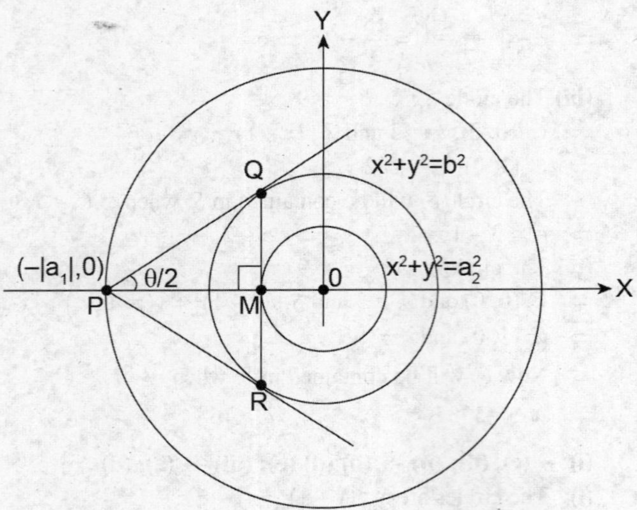

(iii) $S_1: (x + a_1)^2 + y^2 = \left(\sqrt{a_1^2 - b}\right)^2 \Rightarrow C_1(-a_1, 0)$ and $r_1 = \sqrt{a_1^2 - b}$ and S_2:

$(x + a_2)^2 + y^2 = \left(\sqrt{a_2^2 - b}\right)^2 \Rightarrow C_2(-a_2, 0)$ and $r_2 = \sqrt{a_2^2 - b}$

Circles will cut orthogonally, when $C_1C_2^2 = r_1^2 + r_2^2$

$\Rightarrow$ $(a_2 - a_1)^2 = a_1^2 - b + a_2^2 - b$

$2a_1a_2 = 2b$ i.e., $a_1a_2 = b$

SECTION–VIII (INTEGER-TYPE)

1. The circle is $S: x(x - h) + y(y - k) = 0$

Let $y - k = m(x - h)$ be a chord through $P(h, k)$, then $x^2 - xh + \{m(x - h) + k\}\{m(x - h)\} = 0$

$\Rightarrow$ $m^2\{x^2 + h^2 - 2xh\} + x^2 - xh + kmx - kh = 0$ i.e., $(m^2 + 1)x - (2hm^2 + h - km)x + m^2h^2 - kh = 0$

As chord is divided in the ratio 2: 3 at y-axis

$\therefore$ $Q(x_1, y_1)$ gives $x_1 = \frac{-2h}{3}$ and $x_2 = h$

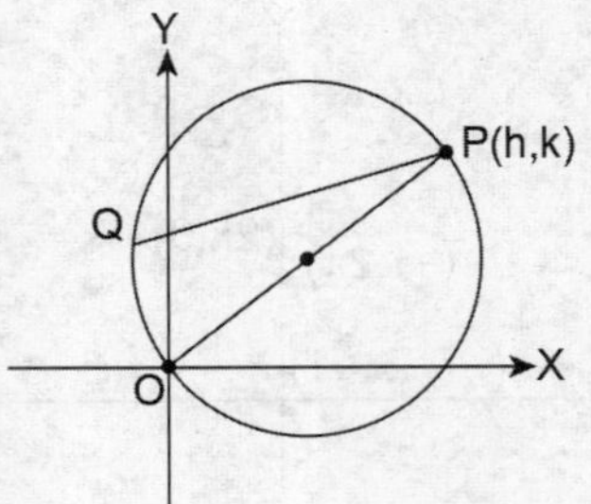

$\Rightarrow$ Sum of roots $x_1 + x_2 = h/3$ and $x_1x_2 = -\frac{2h^2}{3}$

$\Rightarrow$ $\frac{2hm^2 + h - km}{(m^2 + 1)} = \frac{h}{3}$

$\Rightarrow$ $5hm^2 - 3km + 2h = 0$

For real m, $9k^2 > 40h^2$

$\Rightarrow$ $a = 40$ $\therefore$ **40**

2. The circle is $S: (x - 2)^2 + (y - 4)^2 = 2^2$

$C_1 \to (2, 4)$ and $r = 2$ units

Now Point P $(2 + \sqrt{3}, 3)$ is at $-30°$ to the horizontal from the centre as shown

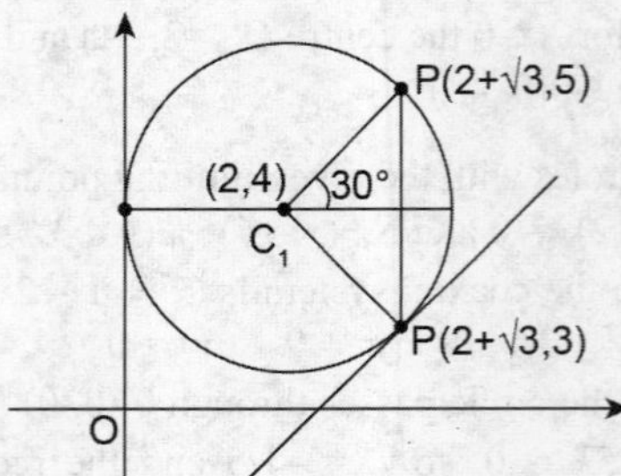

Circle in the new position has equation $(x - 3)^2 + \{y - (4 + \sqrt{3})\}^2 = 2^2$

New position of centre is C_2 $(3, 4+\sqrt{3})$.

Observe that the circle will move from P by a distance C_1C_2 along the tangent $k = c_1c_2 = \sqrt{1^2 + 3} = 2$ units.

$k = 2$ units

3. Let P (–4, 4) and Q (6, –1) be the extremities of a diagonal, then the circle is $x^2 + y^2 - 3y - 28 = 0$

This circles will intersect y-axis at A and B, so $y^2 - 3y - 28 = 0 \Rightarrow (y - 7)(y + 4) = 0$, i.e., A (0, 7) and B (0, –4).

Let the point of intersection of tangent from A and B be at $M(x_1, y_1)$, then chord of contact of tangents is $x_1x + y_1y - x - x_1 - \frac{3}{2}y - \frac{3}{2}y_1 - 28 = 0$ (i.e., $x = 0$)

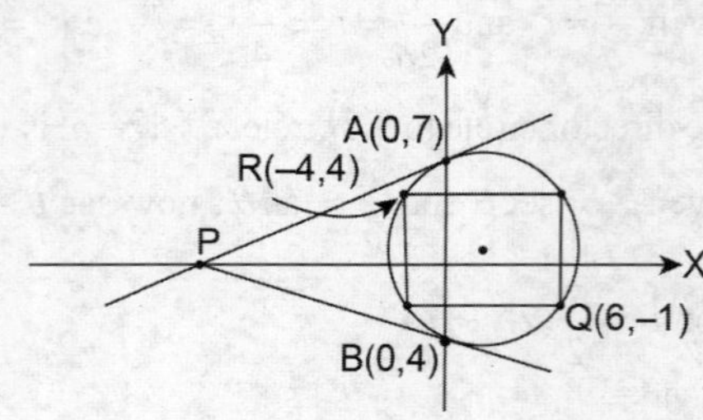

$\Rightarrow\ x(x_1 - 1) + y(y_1 - 3/2) - x_1 - \frac{3}{2}y_1 - 28 = 0$

$\Rightarrow\ y_1 = 3/2$ and $x_1 + \frac{9}{4} + 28 = 0 \Rightarrow x_1 = \frac{-121}{4}$

Hence $M = \left(-\frac{121}{4}, \frac{3}{2}\right)$

Now area ΔMAB is $A = \frac{121}{4} \times \frac{11}{2} = \frac{1331}{8}$ square units

$\Rightarrow\ \frac{k}{8} = \frac{1331}{8}$

$\Rightarrow\ k = 1331$

4. The given circle is S: $(x - 3)^2 + (y - 5)^2 = \left(\sqrt{34-k}\right)^2$
Since point P (1, 4) lies inside the circle
$\therefore\ OP < r$ or $CP^2 < r^2$
i.e., $5 < 34 - k$
$\Rightarrow\ k < 29$
Similarly $r < 3$, since circle does not touch (or cut) any axes so $34 - k < 9$ i.e., $k > 25$. Hence the difference $= 29 - 25 = 4$
$\therefore$ **4**

5. The circle is S: $x^2 + y^2 - a^2 = 0$ and the chord is L: $x \cos\theta + y \sin\theta - p = 0$.
Any circle through the intersection is $(x^2 + y^2 - a^2) - kp(x \cos\theta + y \sin\theta - p) = 0$
$\Rightarrow\ x^2 + y^2 - kp \cos\theta.\, x - kp \sin\theta.y + kp^2 - a^2 = 0$
The centre of this is at $\left(\frac{kp}{2}\cos\theta, \frac{kp\sin\theta}{2}\right)$
Since chord serves as diameter
$\therefore\ \frac{kp}{2}\cos^2\theta + \frac{kp}{2}\sin^2\theta = p \Rightarrow \frac{kp}{2} = p \Rightarrow k = 2$
$\therefore$ **2**

6. Let S: $(ax^2 + 4xy + 2y^2 + x + y + 5) + \lambda(ax^2 + 6xy + 5y^2 + 2x + 3y + 8) = 0$ be any curve through the points of intersection of given curves For $S = 0$ to represent a circle.
$\Rightarrow\ a(1 + \lambda) = (2 + 5\lambda)$ and $4 + 6\lambda = 0$
$\Rightarrow\ \lambda = -2/3$ and $\frac{a}{3} = 2 - \frac{10}{3} = -\frac{4}{3}$
$\Rightarrow\ a = -4$
$\therefore$ **–4**

7. Since $x + y - 5 = 0$ is a tangent to the circle at P (2, 3), so the centre of circle lies on $x - y + 1 = 0$, Now (1, 2) lies on the circle
Equation of right bisector of P (2, 3) and Q (1, 2) is $x + y - 4 = 0$.
The point of intersection of $x - y + 1 = 0$ and $x + y - 4 = 0$ is $\left(\frac{3}{2}, \frac{5}{2}\right)$. Hence the radius $r = 1/\sqrt{2} = 1/\sqrt{k}$
$\therefore$ **2**

8. L_1: $2x + 3y = 2$
L_2: $3x - 2y = 3$
L_3: $x + 2y - 3 = 0$
L_4: $2x - y = 1$

The lines intersect at four points and hence describe a quadrilateral.

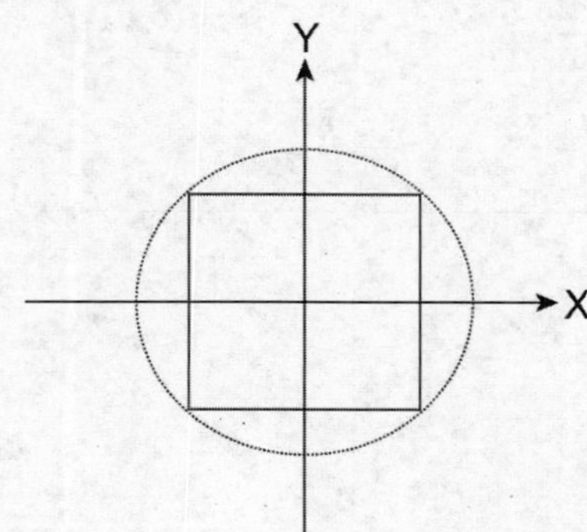

Let us find any one vertex such that. $2x + 3y = 2$ and $3x - 2y = 3$ intersect at (1,0)
The point (1,0) must lie on the circle $x^2 + y^2 + kx - y - 7 = 0$ putties (1,0) into it $\Rightarrow 1 + k - 7 = 0 \Rightarrow k = 6$
$\Rightarrow$ **6**

9. The circles are S_1: $x^2 + y^2 + 2gx + 2fy + c = 0$ and S_2: $x^2 + y^2 + \frac{3}{2}x + 4y + c = 0$
The radical axis is $\left(2g - \frac{3}{2}\right)x + (2f - 4)y = 0$. Since radical axis is a tangent to the circle is $(x + 1)^2 + (y - 1)^2 = 1^2$

$$\Rightarrow\ \frac{\left|\left(\frac{3}{2} - 2g\right) + (2f - 4)\right|}{\sqrt{\left(\frac{4g-3}{2}\right)^2 + (2f-4)^2}} = 1$$

$$\Rightarrow\ \left(2g - \frac{3}{2}\right)^2 + (2f-4)^2 - 2\left(2g - \frac{3}{2}\right)(2f-4)$$

$$= \left(2g - \frac{3}{2}\right)^2 + (2f-4)^2$$

$$\Rightarrow\ \left(\frac{4g-3}{2}\right)(2f-4) = 0$$

Either $4g - 3 = 0 \Rightarrow g = 3/4$ or $2f - 4 = 0$
$\Rightarrow f = 2$ $\therefore$ **2**

10. The circle is S: $x^2 + y^2 - r^2 = 0$ and the point P (6, 8)
$\Rightarrow\ OP = 10$ units
Length of tangent $PQ = PR = \sqrt{100 - r^2}$.

Now $PM = PQ\cos\frac{\theta}{2} = \sqrt{100-r^2}\cos\frac{\theta}{2}$ and QM

$= PQ\frac{\sin\theta}{2}\left(\text{where}\sin\frac{\theta}{2} = \frac{r}{10}\right)$, so $\cos\theta = \frac{\sqrt{100-r^2}}{10}$

$\Rightarrow$ Area of $\Delta PQR = QM \times PM$

$= \sqrt{100-r^2}\cos\frac{\theta}{2}.\sqrt{100-r^2}\frac{r}{10} = (100-r^2)^{3/2}.\frac{r}{100}$

Differentiate w.r.t 'r' we get

$\frac{r}{100}.\frac{3}{2}(100-r^2)^{1/2}(-2r) + \frac{(100-r^2)^{3/2}}{100} = 0$ or $\frac{(100-r^2)}{100}$

$\{100 - r^2 - 3r^2\} = 0$

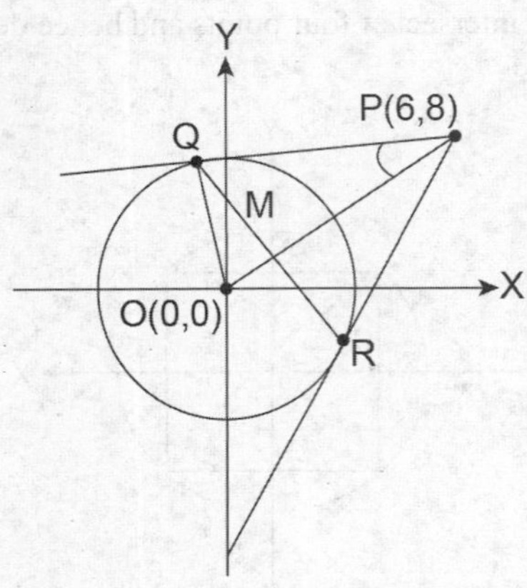

$\Rightarrow$ $100 = 4r^2 \Rightarrow r = 5$ ($r = 10$ is not possible)

$\therefore$ **r = 5 units**

11. $(1+\alpha x)^n = 1 + \frac{\alpha n}{1!}x + \frac{\alpha^2 x^2 n(n-1)}{2!} + ... = 1 + 8x + 24x^2 + ...$

Equating $\alpha n = 8$ and $\alpha^2 n(n-1) = 48$, we get $\alpha = 2$, $n = 4$. Now $P(\alpha, n) = P(2, 4)$. The circle is $S: x^2 + y^2 - 4 = 0$, any line through P intersects the circle in A and B

$\Rightarrow$ $PA.\ PB = S(2, 4) = 4 + 16 - 4 = 16$

$\therefore$ **16**

12. Line passing through A (1, 0) and B (3, 4) is $2x - y - 2 = 0$. This line will intersect the circle $x^2 + y^2 - 4 = 0$, give the points of intersection by $(5x - 8)x = 0$ i.e., $x = 0, 8/5$

The points are $P(0, -2)$ and $Q\left(\frac{8}{5}, \frac{6}{5}\right)$

$$\frac{BP}{PA} = \frac{\sqrt{3^2+6^2}}{\sqrt{1^2+2^2}} = -3 = \alpha \text{ and } \frac{BQ}{QA} = \frac{\sqrt{\left(\frac{7}{5}\right)^2 + \left(\frac{14}{5}\right)^2}}{\sqrt{\left(\frac{3}{5}\right)^2 + \left(\frac{6}{5}\right)^2}} = \frac{7}{3}$$

$\Rightarrow$ $\alpha = -3, \beta = 7/3$

Hence $\alpha + \beta = -2/3$ and $\alpha\beta = -7$

The quadratic equation will be $x^2 + \frac{2x}{3} - 7 = 0$ or $3x^2 + 2x - 21 = 0$ comparing, we get $k = 21$

$\therefore$ **21**

13. The circles are $S_1: x^2 + y^2 - 1 = 0$ and $S_2: x^2 + y^2 - 4 = 0$

The rays are given as $2x^2 - 3xy - 2y^2 = 0$ i.e., $(2x + y)(x - 2y) = 0$

$\Rightarrow$ $y = \frac{x}{2}$ or $y = -2x$

Observe that $L_1: y + 2x = 0$ and $L_2: x - 2y = 0$ are at right angles:

$\therefore$ required area $= \frac{4\pi - \pi}{4} = \frac{3\pi}{4}$ square unit.

Hence $\frac{k\pi}{4} = \frac{3\pi}{4}$

$\Rightarrow$ $k = 3$

14. Given $a > 2b > 0$ as $L: mx - y - b\sqrt{1+m^2} = 0$, is a common tangent to $S_1: x^2 + y^2 = b^2 \Rightarrow \frac{\left|-b\sqrt{1+m^2}\right|}{\sqrt{1+m^2}} = b$

Which is always true and L is also a common tangent to S_2:

$(x-a)^2 + y^2 = b^2 \Rightarrow \frac{\left|am - b\sqrt{1+m^2}\right|}{\sqrt{1+m^2}} = b$

$\Rightarrow$ $a^2m^2 + b^2(m^2+1) - 2abm\sqrt{1+m^2} = b^2(m^2+1)$

Hence $am\left\{am - 2b\sqrt{1+m^2}\right\} = 0$

$\Rightarrow$ $am = 2b\sqrt{1+m^2}$

On squaring, we get $a^2m^2 = 4b^2m^2 + 4b^2$ or $(a^2 - 4b^2)m^2 = 4b^2 \Rightarrow m = \pm\frac{2b}{\sqrt{a^2-4b^2}}$

For $m > 0$; $m = \frac{2b}{\sqrt{a^2-4b^2}} = \frac{2b}{\sqrt{a^2-kb^2}} \Rightarrow k = 4$

15. Centre of circle (π, e)

If there is a point (x, y) on the circle $(x-\pi)^2 + (y-e)^2 = r^2$

Clearly, there are two points on this circle.

For instance, let us assume that there are 3 such points say (P, Q, R). The perpendicular bisectors of PQ and QR would meet at the centre of the circle which must be a rational point.

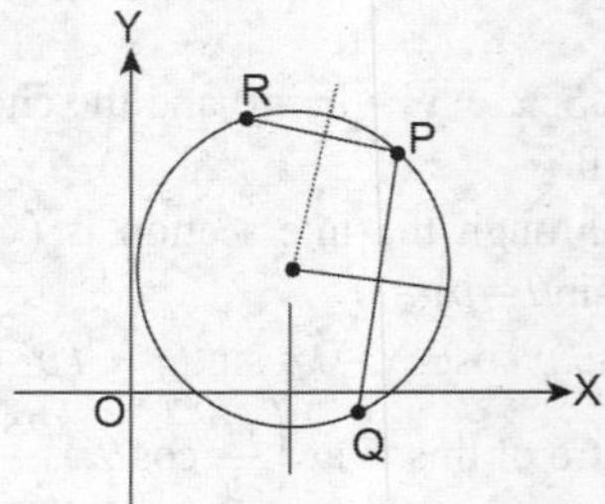

Hence the maximum number of such points is 2.

$\therefore$ **2**

16. From the given information observe that $BC = 2$ units and $BP = BQ = 1$unit, so $\angle PCB = \angle QCB = 30°$

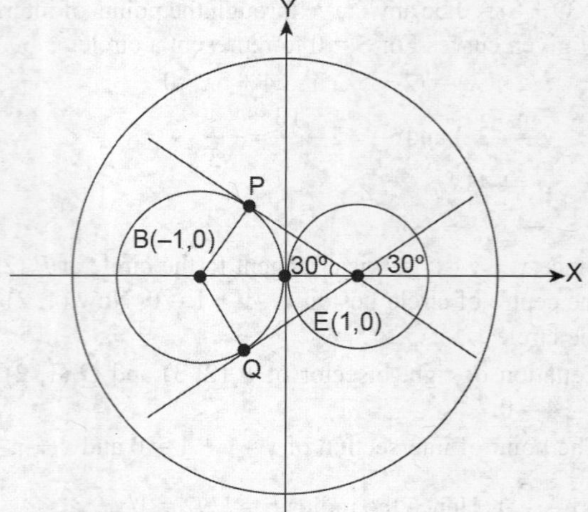

Equation QC line $\sqrt{3}y = (x-1)$ or $x - \sqrt{3}y - 1 = 0$

Similarly PC is $x + \sqrt{3}y - 1 = 0$

Length of the chord $= 2\sqrt{16 - \frac{1}{4}} = \sqrt{63}$ units $= \sqrt{\ell}$

$\Rightarrow$ $= 63$

Parabola

CHAPTER 4

■ INTRODUCTION

If we take a pen and randomly draw anything on a plane paper, we get some curve (locus). Is this a meaningful curve? Can we find out some mathematical representation of this? Does this curve follow certain rules? Is this curve useful to study some physical/practical phenomenon. All these queries naturally arise.

Now, let us start analysing the thought "a point always moves such that the ratio of its distance from a fixed point and a fixed line is constant". Can we get some meaningful locus? Yes, we get some curves which follow the above rule/thought. We can represent these curves mathematically using the co-ordinates, and this curve is useful in finding out many physical/practical phenomenon.

Interestingly, if we take a right circular cone and cut it by a horizontal plane, we get a cross-section which is circular. If we cut this cone by planes in different orientations, then different planes produce different type of curves. As all these curves are sections of a right circular cone, we call them 'conic sections'. These are named as circle, parabola, ellipse or hyperbola.The names 'parabola' and 'hyperbola' are given by Apollonious. Many important discoveries, both in mathematics and science have been associated to conic sections. The Greek mathematicians particularly Archimedes (287–212 BC) and Apollonious (260–190 BC) studied these curves for their own beauty. At present, these curves are very important tools for exploration of outer space and various researches into behaviour of atomic particles.

When we analyse these curves in detail, we find that when the ratio of the distances of a point on the curve from a fixed point to its distance from a fixed line is equal to 1 we have one type of curve; for ratio less than 1, we have second type of curve and for ratio greater than 1, we have third type of curve. With this basis, we can analyse these three curves.These curves have a very wide range of applications in the fields such as planetary motion, design of telescopes and antennas, reflectors in flashlights and automobile headlights, etc.

■ CONIC SECTION

Conic sections are sections obtained when a pair of two vertical cones with same vertex are intersected by a plane in various orientation as described here.

Let L_1 be a fixed vertical line and L_2 be another line intersecting it at a fixed point V and inclined to it at an angle θ as shown in the the following figure. Suppose we rotate the line L_2 around the line L_1 in such a way that the angle θ remains constant. Then the surface generated is a double-napped right-circular hollow cone. Herein after referred as cone and extending indefinitely far in both directions as shown in Figure 4.1.

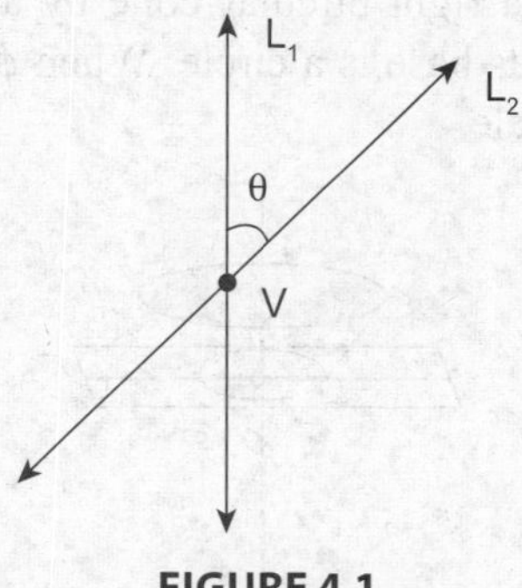

FIGURE 4.1

The point V is called the vertex; the line L_1 is the axis of the cone. The rotating line L_2 is called a **generator of the cone.** The vertex separates the cone into two parts called **nappes** and conic sections are the curves obtained by intersecting a right circular cone by a plane. We obtain different kinds of conic sections depending on the position of the intersecting plane.

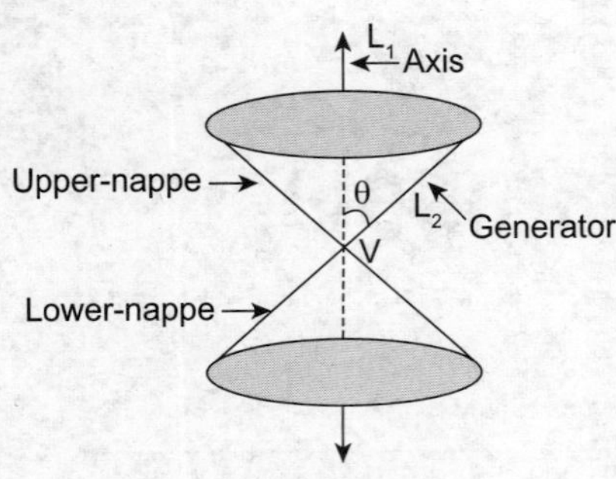

FIGURE 4.2

with respect to the cone and by the angle made by it with the vertical axis of the cone. Let ϕ be the angle made by the intersecting plane with the vertical axis of the cone.

The intersection of the plane with the cone can take place either at the vertex of the cone or at any other part of the nappe either below or above the vertex.

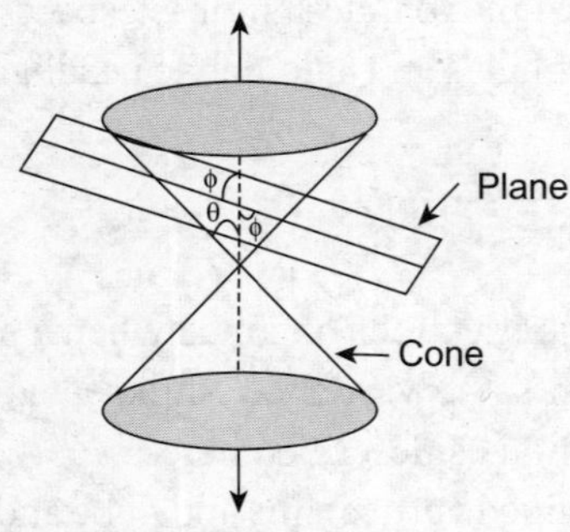

FIGURE 4.3

■ CIRCLE, ELLIPSE, PARABOLA AND HYPERBOLA

When the plane cuts the nappe (other than the vertex) of the cone, we have the following situations:

(a) Section of a right circular cone by a plane which is parallel to its base is a circle. When $\phi = 90°$, the section is a circle

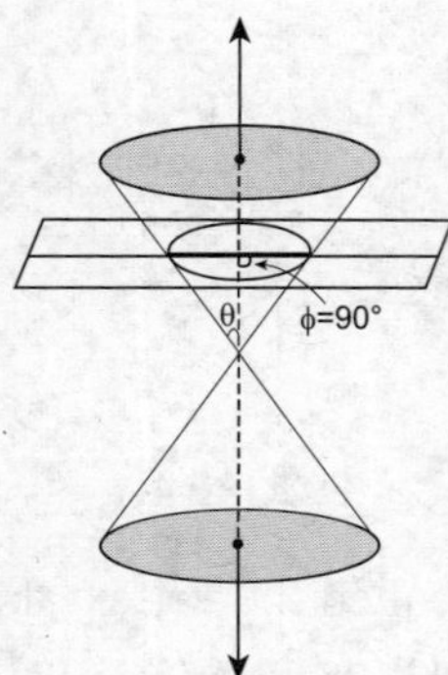

FIGURE 4.4

(b) Section of a right circular cone by a plane which is not parallel to any generator and not parallel or perpendicular to the axis of the cone is an ellipse. When $\theta < \phi < 90°$, the section is an ellipse.

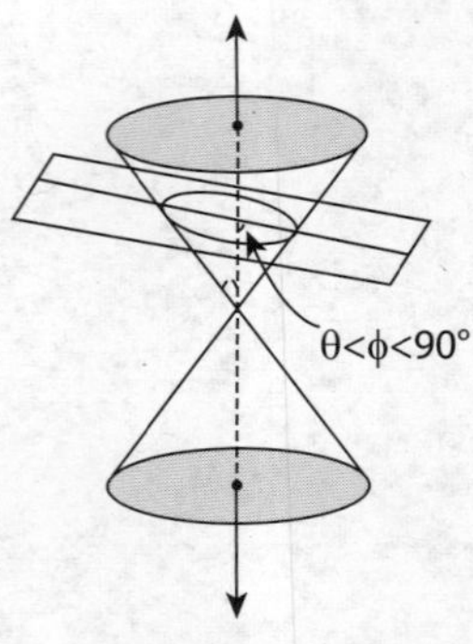

FIGURE 4.5

(c) Section of a right circular cone by a plane which is parallel to a generator of the cone is a parabola. When $\phi = \theta$; the section is a parabola shown in Figure 4.6. (In each of the above three situations, the plane cuts entirely across one nappe of the cone).

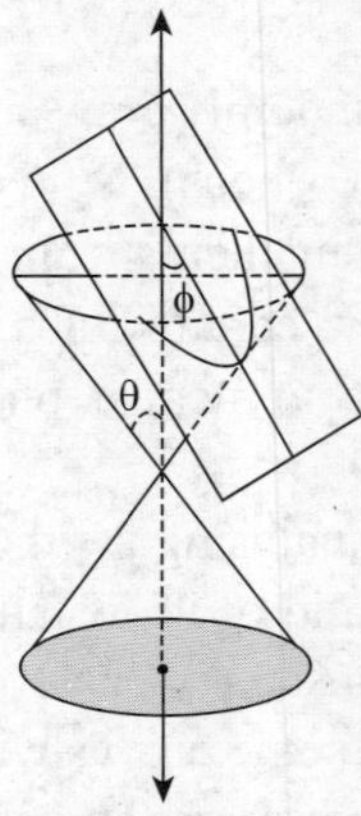

FIGURE 4.6

(d) Section of a right circular cone by a plane which cuts both the nappes such that the angle of intersection of plane and nappe with axis is greater than or equal to 0° but less than θ, is a hyperbola as shown in Figure 4.7.

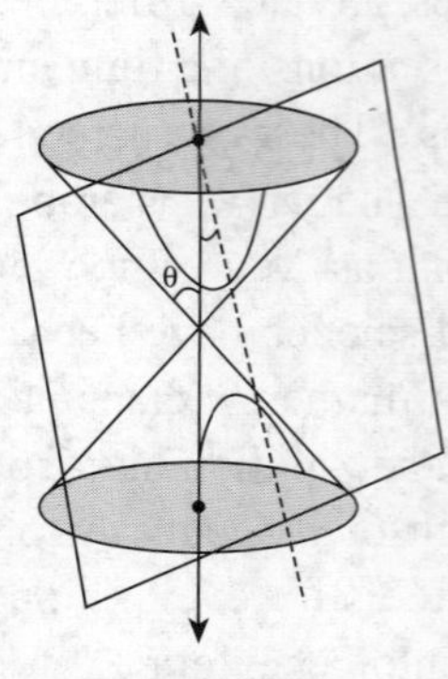

FIGURE 4.7

Degenerated Conic Sections

When the plane cuts at the vertex of the cone, the following cases are observed:

(a) When $\theta < \phi \leq 90°$, then the section is a point (Fig 4.8(a)).

(b) When $\phi = \theta$, the plane contains a generator of the cone and the section is a straight line (Fig. 4.8(b)). It is the degenerated case of a parabola.

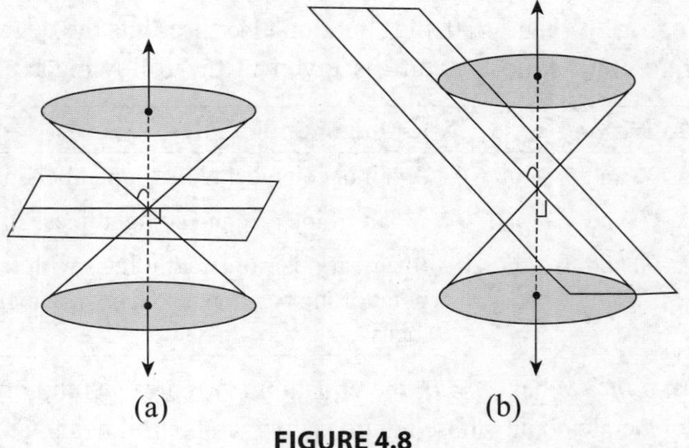

FIGURE 4.8

(c) Section of a right circular cone by a plane which is passing through its vertex is a pair of straight lines always passing through the vertex of the cone. When $0 \leq \phi < \theta$, the section is a pair of intersecting straight lines (Fig. 4.9). It is the degenerated case of a hyperbola.

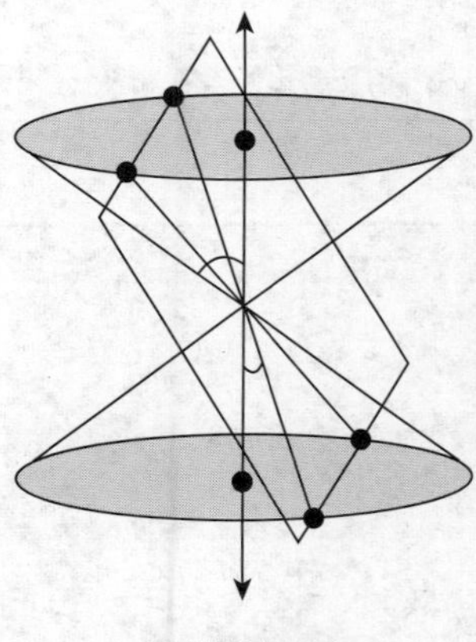

FIGURE 4.9

■ DEFINITION OF CONIC

The locus of a point which moves in a plane such that the ratio of its distance from a fixed point to its perpendicular distance from a fixed straight line is always constant, is known as a **conic section or a conic**.

The fixed point is called the focus of the conic and the fixed line known as the directrix of the conic. Also this constant ratio is called the eccentricity of the conic and is denoted by e.

If $e = 1$, the conic is called parabola
If $e < 1$, the conic is called ellipse
If $e > 1$, the conic is called hyperbola
If $e = 0$, the conic is called circle
If $e = \infty$, the conic is called pair of straight lines.

In Figure 4.10:

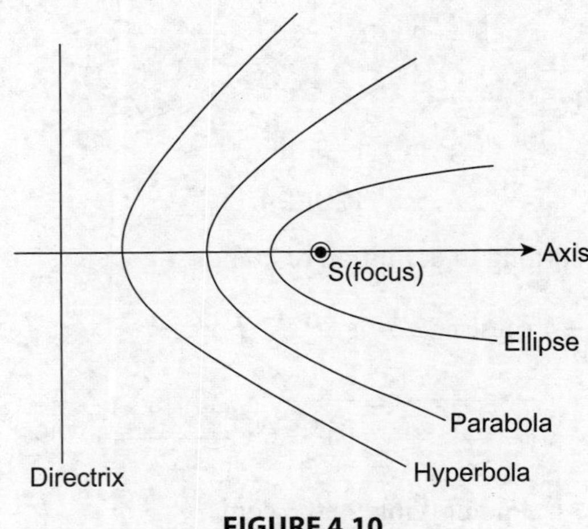

FIGURE 4.10

■ DEFINITION OF VARIOUS TERMS RELATED TO CONIC

Axis: The straight line passing through the focus and perpendicular to the directrix is called the axis of the conic section.

Vertex: The point of intersection of the conic section and the axis is (are) called vertex (vertices) of the conic section.

Focal Chord: Any chord passing through the focus is called focal chord of the conic section.

Focal Distance: The distance of any point on the conic from the focus.

Double Ordinate: A line segment drawn perpendicular to the principal axis and terminated at both ends on the curve is a double ordinate of the conic section provided the principal axis of conic is x-axis.

Tangent at Vertex: A line perpendicular to axis and passing through vertex.

Latus Rectum: The chord perpendicular to principal axis of conic and passing through the focus is called the latus rectum of the conic section.

Centre: The point which bisects every chord of the conic passing through it is called the centre of the conic section.

■ EQUATION OF CONIC SECTION

If the focus is $S(\alpha, \beta)$ and the directrix is $ax + by + c = 0$, then the equation of the conic section whose eccentricity is e, can be derived as follows:

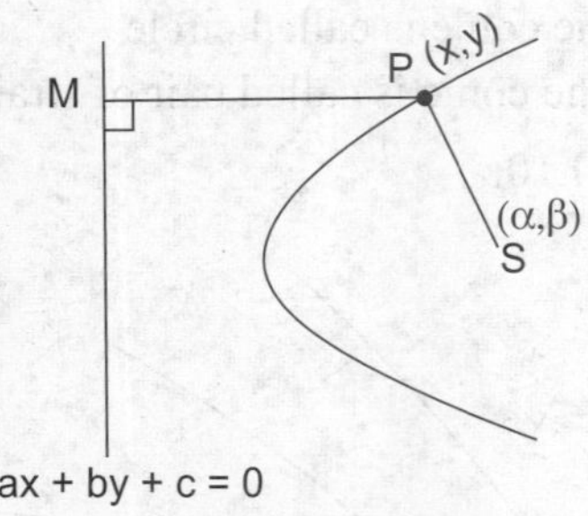

FIGURE 4.11

According to definition of conic

$$\frac{SP}{PM} = \text{constant} = e \text{ or } SP = e\,PM$$

$$\Rightarrow \sqrt{(x-\alpha)^2+(y-\beta)^2} = e\frac{|ax+by+c|}{\sqrt{(a^2+b^2)}}; \text{ where } P(x, y)$$

is a point lying on the conic.

$$\text{or } (x-\alpha)^2+(y-\beta)^2 = e^2.\frac{(ax+by+c)^2}{(a^2+b^2)}.$$

The equation of conics is represented by the general equation of second degree

$$ax^2+2hxy+by^2+2gx+2fy+c=0 \quad ....(1)$$

Conic can be recognized easily by the condition given in the tabular form. For this, first we have to find discriminant of the equation. We know that the discriminant of above equation is represented by Δ; where

$$\Delta = abc+2fgh-af^2-bg^2-ch^2 \text{ or } \Delta = \begin{vmatrix} a & h & g \\ h & b & f \\ g & f & c \end{vmatrix}$$

Case I: When $\Delta = 0$, the equation (1) represents the degenerate conic whose nature is given in the following table:

Condition	Nature of Conic
$\Delta = 0$ and $h^2 - ab = 0$	A pair of coincident lines or parallel lines.
$\Delta = 0$ and $h^2 - ab > 0$	A pair of intersecting straight lines.
$\Delta = 0$ and $h^2 - ab < 0$	Imaginary pair of straight lines with real point of intersection also known as point locus.

Case II: When $\Delta \neq 0$, the equation (1) represents the non-degenerate conic whose nature is given in the following table:

Condition	Nature of Conic
$\Delta \neq 0, h = 0, a = b$	a circle
$\Delta \neq 0, h^2 - ab = 0.$	a parabola
$\Delta \neq 0, h^2 - ab < 0$	an ellipse or empty set
$\Delta \neq 0, h^2 - ab > 0$	a hyperbola
$\Delta \neq 0, h^2 - ab > 0$ and $a + b = 0$	a rectangular hyperbola

ILLUSTRATION 1: What does the given equation $25[x^2 + y^2 - 2x + 1] = (4x - 3y + 1)^2$ represent?

SOLUTION: The given equation is $25[x^2 + y^2 - 2x + 1] = (4x - 3y + 1)^2$. ...(1)

Write the right hand side of this equation, so that it appears in perpendicular distance.

$$(4x - 3y + 1)^2 = 25\left(\frac{4x-3y+1}{\sqrt{(4^2+3^2)}}\right)^2 \text{ then equation (1) can be re-written as}$$

$$25[(x-1)^2 + (y-0)^2] = 25\left(\frac{4x-3y+1}{\sqrt{(4^2+3^2)}}\right)^2 \text{ or } \sqrt{(x-1)^2+(y-0)^2} = \frac{|4x-3y+1|}{\sqrt{4^2+3^2}}. \text{ Here } e = 1$$

Thus, the given equation represents a parabola. It may be noted that (1, 0) is the focus and $4x - 3y + 1 = 0$ is the directrix of the parabola.

ILLUSTRATION 2: What conic does the equation $13x^2 - 18xy + 37y^2 + 2x + 14y - 2 = 0$ represent?

SOLUTION: Comparing the given equation with

$ax^2 + 2hxy + by^2 + 2gx + 2fy + c = 0$; we get

$\therefore\ a = 13, h = -9, b = 37, g = 1, f = 7, c = -2$

Then, $\Delta = abc + 2fgh - af^2 - bg^2 - ch^2$

$= (13)(37)(-2) + 2(7)(1)(-9) - 13(7)^2 - 37(1)^2 + 2(-9)^2$

$= -962 - 126 - 637 - 37 + 162 = -1600 \neq 0$

and also $h^2 = (-9)^2 = 81$ and $ab = 13 \times 37 = 481$

Here, $h^2 - ab = -400 < 0$

So, we have $h^2 - ab < 0$ and $\Delta \neq 0$

Hence, the given equation represents an ellipse.

ILLUSTRATION 3: If the equation $x^2 - y^2 - 2x + 2y + \lambda = 0$ represents a degenerate conic, then find the value of λ.

SOLUTION: For degenerate conic $\Delta = 0$

Comparing the given equation of conic with $ax^2 + 2hxy + by^2 + 2gx + 2fy + c = 0$; we get

$\therefore\ a = 1, b = -1, h = 0, g = -1, f = 1, c = \lambda \therefore \Delta = abc + 2fgh - af^2 - bg^2 - ch^2 = 0$

$\Rightarrow (1)(-1)(\lambda) + 0 - 1 \times (1)^2 + 1 \times (-1)^2 - \lambda(0)^2 = 0$

or $-\lambda - 1 + 1 = 0$

$\therefore\ \lambda = 0$

ILLUSTRATION 4: For what value of λ the equation of conic $2xy + 4x - 6y + \lambda = 0$ represents two intersecting straight lines ? If $\lambda = 17$, then what does this equation represent?

SOLUTION: Comparing the given equation of conic with

$ax^2 + 2hxy + by^2 + 2gx + 2fy + c = 0$, we get

$\therefore\ a = 0, b = 0, h = 1, g = 2, f = -3, c = \lambda$

For two intersecting lines $h^2 - ab > 0, \Delta = 0$

$\because\ ab = 0, h = 1, \therefore h^2 - ab = 1 > 0$

and $\Delta = abc + 2fgh - af^2 - bg^2 - ch^2$

$\Rightarrow 0 + 2 \times (-3) \times 2 \times 1 - 0 - 0 - \lambda(1)^2 = 0 \Rightarrow \lambda = -12$

For $\lambda = 17$, the given equation of conic becomes $2xy + 4x - 6y + 17 = 0$

$\therefore\ a = 0, b = 0, h = 1, g = 2, f = -3, c = 17$

$\therefore\ \Delta = abc + 2fgh - af^2 - bg^2 - ch^2 = 0 + 2 \times -3 \times 2 \times 1 - 0 - 0 - 17 \times (1)^2 = -12 - 17 = -29 \neq 0$

$\therefore\ \Delta \neq 0$ and $h^2 - ab = 1 > 0$

$\therefore\ h^2 - ab > 0$

So we have $\Delta \neq 0$ and $h^2 - ab > 0$; and $a + b = 0$.

Hence, the given equation represents a rectangular hyperbola.

■ PARABOLA AND ITS RELATED TERMS AND PROPERTIES

Definition

A parabola is the locus of a point which moves in a plane so that its distance from a fixed point (*i.e., focus*) is equal to its distance from a fixed straight line (*i.e., directrix*) that is, conic with eccentricity 1.

Standard Parabola

Parabolas having their vertex at origin and axis as one of the co-ordinate axis are called 'standard parabolas'.

■ FOUR STANDARD TYPES OF PARABOLAS AND THEIR EQUATIONS AND RELATED TERMS

There are four standard forms of parabola, having their vertex at origin, focus lying on one of the 'axes called axis of parabola'. Four standard types of parabolae are shown in Figure 4.12.

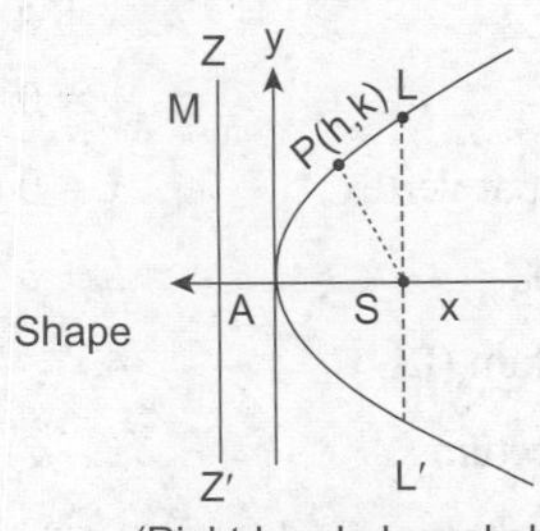

FIGURE 4.12

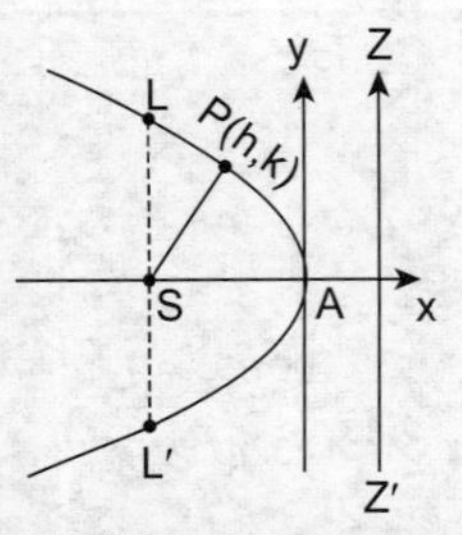

(Left-handed parabola)

FIGURE 4.13

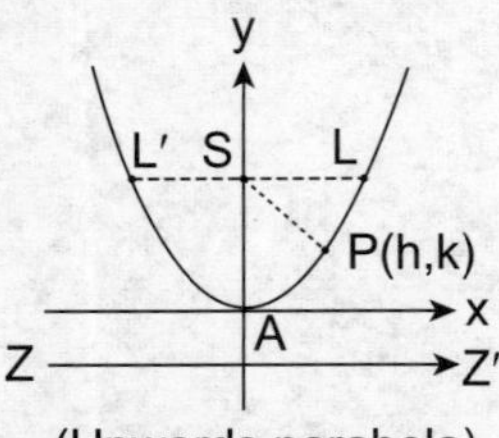

(Upwards parabola)

FIGURE 4.14

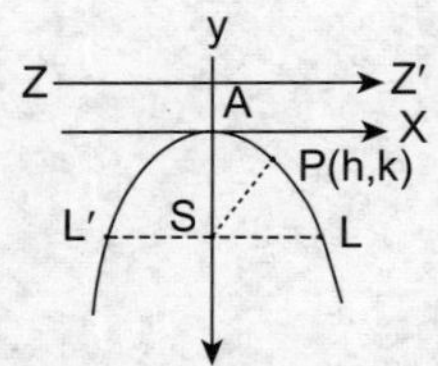

(Downwards parabola)

FIGURE 4.15

- ❑ Derivation of equation of standard parabola having its vertex at origin and axis as x-axis.

  Let $P(h, k)$ be any point on the parabola and PM be perpendicular drawn from P to directrix and S be the focus of parabola as shown in Figure 4.16.

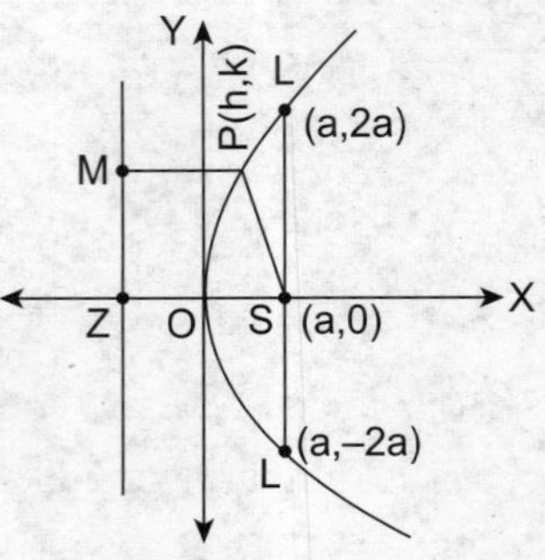

FIGURE 4.16

By definition of parabola

$$SP = PM = a + h \Rightarrow \sqrt{(h-a)^2 + k^2} = a + h$$

$$\Rightarrow a^2 + h^2 - 2ah + k^2 = a^2 + h^2 + 2ah$$

$\Rightarrow k^2 = 4ab$. The locus of point P is $y^2 = 4ax$; which is the required equation of parabola of this type.

Terms related to right handed parabola:

- ❑ Equation of parabola: $y^2 = 4ax$, $a > 0$
- ❑ Opening rightwards, passing through origin
- ❑ Symmetric about x-axis.
- ❑ Focus : $S(a, 0)$
- ❑ Vertex: $(0, 0)$
- ❑ Axis: $y = 0$
- ❑ Directrix: $x + a = 0$
- ❑ Tangent at vertex.: $x = 0$
- ❑ Focal distance $= a + h$
- ❑ **Latus rectum:** Equation $x - a = 0$ and length $4a$, extremities $(a, \pm 2a)$
- ❑ Parametric equation: $x = at^2$, $y = 2at$; where $t \in \mathbb{R}$

The following table shows the equation and related terms for standerd parabolas.

Diagram	Fig. 1	Fig. 2	Fig. 3	Fig. 4
Equation	$y^2 = 4ax$	$y^2 = -4ax$	$x^2 = 4ay$	$x^2 = -4ay$
Focus (S)	$(a, 0)$	$(-a, 0)$	$(0, a)$	$(0, -a)$
Vertex (A)	$(0, 0)$	$(0, 0)$	$(0, 0)$	$(0, 0)$
Eccentricity (e)	1	1	1	1
Directrix (ZZ′)	$x + a = 0$	$x - a = 0$	$y + a = 0$	$y - a = 0$
Axis (AS)	$y = 0$	$y = 0$	$x = 0$	$x = 0$
Equation of tangent at vertex	$x = 0$ (y-axis)	$x = 0$ (y-axis)	$y = 0$ (x-axis)	$y = 0$ (x-axis)
Focal distance (PS)	$h + a$	$a - h$	$k + a$	$a - k$
Length of latus rectum (LL′)	$4a$	$4a$	$4a$	$4a$
Equation of latus rectum	$x - a = 0$	$x + a = 0$	$y - a = 0$	$y + a = 0$
Extremities of latus rectum (L, L′)	$(a, \pm 2a)$	$(-a, \pm 2a)$	$(\pm 2a, a)$	$(\pm 2a, -a)$
Parametric equation	$(at^2, 2at)$	$(-at^2, 2at)$	$(2at, at^2)$	$(2at, -at^2)$

ILLUSTRATION 5: A circle is described with its centre at vertex and whose diameter is three-quarters of the latus rectum of a parabola $y^2 = 4ax$. Prove that the common chord of the circle and parabola bisects the distance between the vertex and the focus.

SOLUTION: Diameter of described circle $= \frac{3}{4}(4a) = 3a$ units

$\Rightarrow$ radius (r) $= \frac{3a}{2}$ and centre $V(0, 0)$

$\therefore$ equation of circle is given by $x^2 + y^2 = \frac{9a^2}{4}$ ...(i)

Now, equation of parabola is $y^2 = 4ax$

$\therefore$ At the point of intersections A and B; $x^2 + 4ax = \frac{9a^2}{4}$

$\Rightarrow 4x^2 + 16ax - 9a^2 = 0 \Rightarrow 4x^2 + 18ax - 2ax - 9a^2 = 0$

$\Rightarrow 2x(2x + 9a) - a(2x + 9a) = 0 \Rightarrow (2x - a)(2x + 9a) = 0$

$\Rightarrow x = \frac{a}{2}$ or $\frac{-9a}{2}$, but abscissae of A and B must be positive.

FIGURE 4.17

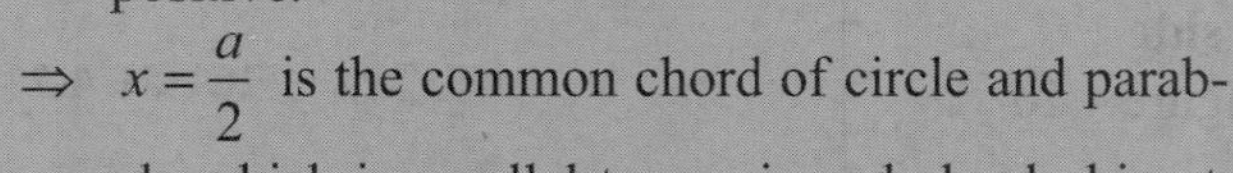

$\Rightarrow x = \frac{a}{2}$ is the common chord of circle and parabola which is parallel to y-axis and clearly bisects the distance between vertex $V(0, 0)$ and focus $S(a, 0)$.

ILLUSTRATION 6: O is the vertex of the parabola $y^2 = 4ax$ and L is the upper end of the latus rectum. If LH is drawn perpendicular to OL meeting OX in H, prove that the length of the double ordinate through H is $4a\sqrt{5}$ units.

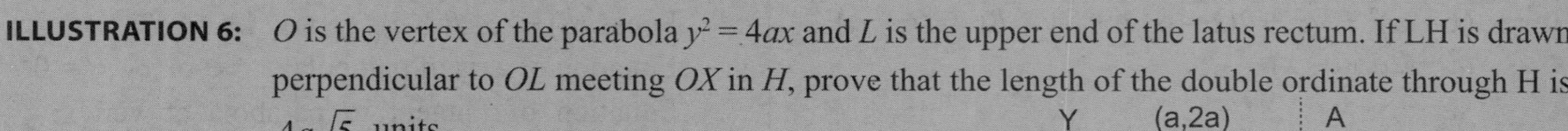

SOLUTION: Slope of OL is given by $S_{OL} = \frac{2a}{a} = 2$

Slope of LH is given by $S_{LH} = -\frac{1}{2}$

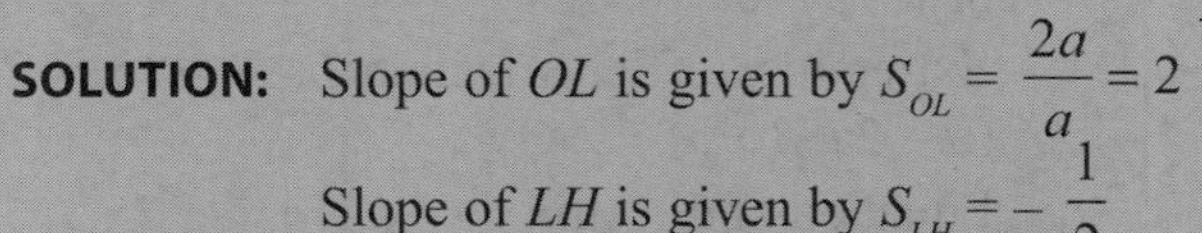

Equation of LH is given by $y - 2a = -\frac{1}{2}(x - a)$

$\therefore$ Co-ordinate of H are given by H:(5a, 0)

For double ordinate, put $x = 5a$ in

$y^2 = 4ax \Rightarrow y^2 = (4a)(5a) \Rightarrow y^2 = 20a^2 \Rightarrow y = \pm 2\sqrt{5}\,a$

$\Rightarrow$ length $= 4\sqrt{5}a$ units

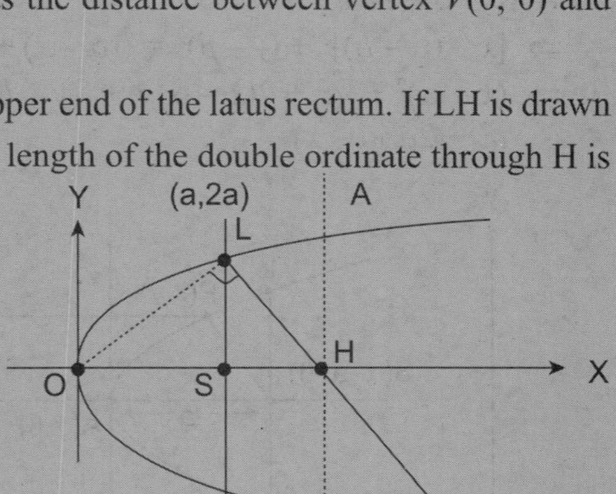

FIGURE 4.18

NON-STANDARD FORMS OF PARABOLA

1. Parabola having its vertex at (α, β) is not at origin, axis parallel to x-axis and length of latus rectum '4a'

These are of two types:

(a) Parabola opening towards right hand side

Let $P(x, y)$ be any point on parabola, with its vertex at $A(\alpha, \beta)$ and having its axis parallel to x-axis and opening towards right as shown in Figure 4.19.

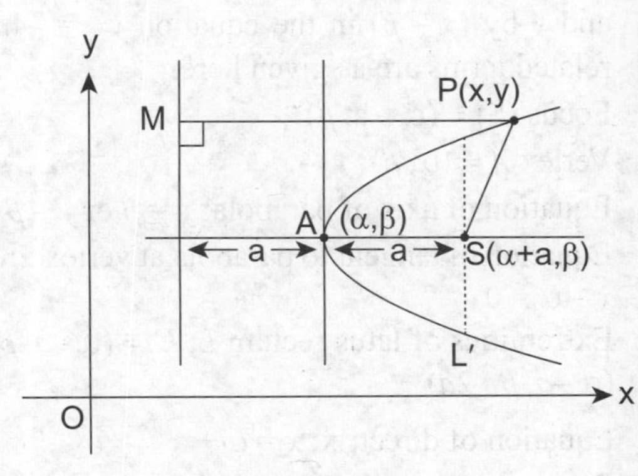

FIGURE 4.19

Now, $PS = PM$

$\Rightarrow (x - \alpha - a)^2 + (y - \beta)^2 = (x - \alpha + a)^2$

$\Rightarrow (x - \alpha)^2 + a^2 - 2a(x - \alpha) + (y - \beta)^2 = (x - \alpha)^2 + a^2 + 2a(x - \alpha) \Rightarrow (y - \beta)^2 = 4a(x - \alpha)$ ...(i)

(i) is the required equation of parabola and it can be obtained from the standard equation $y^2 = 4ax$, just by replacing y by $(y - \beta)$ and x by $(x - \alpha)$. The related terms are as given below.

Focus $(S) \equiv (\alpha + a, \beta)$; Vertex $(A) \equiv (\alpha, \beta)$

Equation of axis of parabola: $y = \beta$ or $y - \beta = 0$

Equation of tangent to parabola at vertex: $x = \alpha$ or $x - \alpha = 0$

Extremities of latus rectum $L, L' \equiv (\alpha + a, \beta + 2a)$; $(\alpha + a, \beta - 2a)$

Equation of directrix: $x = \alpha - a$

Parametric equation of parabola: $x = \alpha + at^2$; $y = \beta + 2at$; $t \in \mathbb{R}$

(b) Parabola opening towards left hand side

This type of parabola is as shown in Figure 4.20.

Here, $PS = PM$,

$\Rightarrow [x - (\alpha - a)]^2 + (y - \beta)^2 = ((\alpha - x) + a)^2$

$\Rightarrow (x - \alpha)^2 + a^2 + 2(x - \alpha)a + (y - \beta)^2 = (\alpha - x)^2 + a^2 + 2a(\alpha - x)$

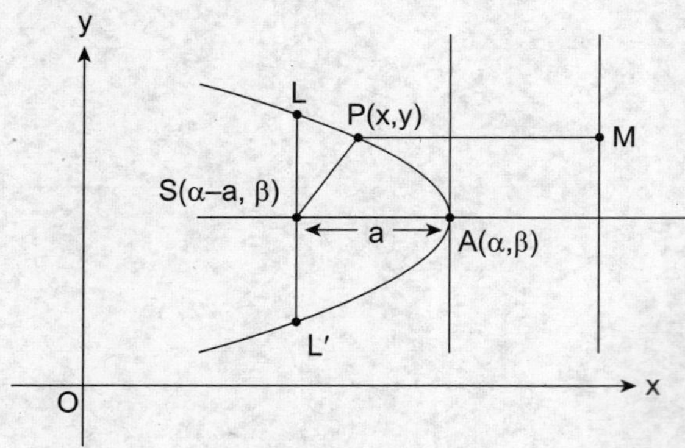

FIGURE 4.20

$\Rightarrow (y - \beta)^2 = 4a(\alpha - x)$

or $(y - \beta)^2 = -4a(x - \alpha)$ ...(ii)

(ii) can be easily obtained by replacing x by $(x - \alpha)$ and y by $(y - \beta)$ in the equation $y^2 = -4ax$. The related terms are as given here.

Focus $(S) \equiv (\alpha - a, \beta)$

Vertex $A \equiv (\alpha, \beta)$

Equation of axis of parabola: $y = \beta$ or $y - \beta = 0$

Equation of tangent to parabola at vertex : $x = \alpha$ or $x - \alpha = 0$

Extremities of latus rectum $L, L' \equiv (\alpha - a, \beta + 2a)$; $(\alpha - a, \beta - 2a)$

Equation of directrix: $x = \alpha + a$

Parametric equation of parabola: $x = \alpha - at^2, y = \beta + 2at$

2. Parabola having its vertex at (α, β), i.e., not at origin, axis parallel to y-axis and length of latus rectum '4a'

These are the following two types:

(a) Parabola opening upwards:

This type of parabola is as shown in Figure 4.21.

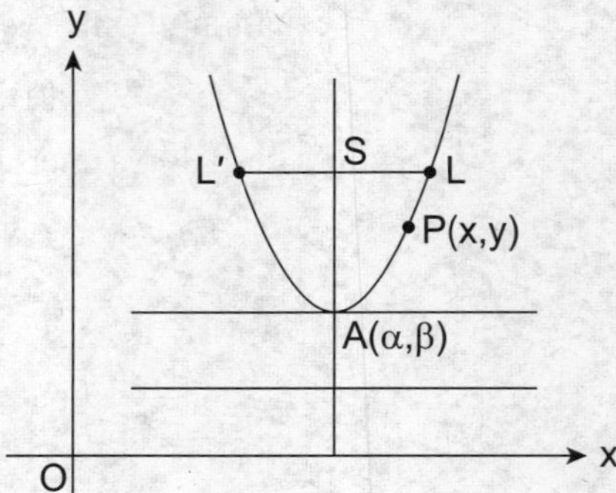

FIGURE 4.21

Its equation is given by $(x - \alpha)^2 = 4a(y - \beta)$ .....(iii)

and other related terms are as given here.

Focus $(S) \equiv (\alpha, \beta + a)$; Vertex $A \equiv (\alpha, \beta)$

Equation of axis of parabola: $x = \alpha$ or $x - \alpha = 0$

Equation of tangent to parabola at vertex: $y = \beta$ or $y - \beta = 0$

Extremities of latus rectum $L, L' \equiv (\alpha + 2a, \beta + a)$; $(\alpha - 2a, \beta + a)$

Equation of directrix : $y = \beta - a$

Parametric equation of parabola: $x = \alpha + 2at$, $y = \beta + at^2$

(b) Parabola opening downwards:

This type of parabola is as shown in Figure 4.22.

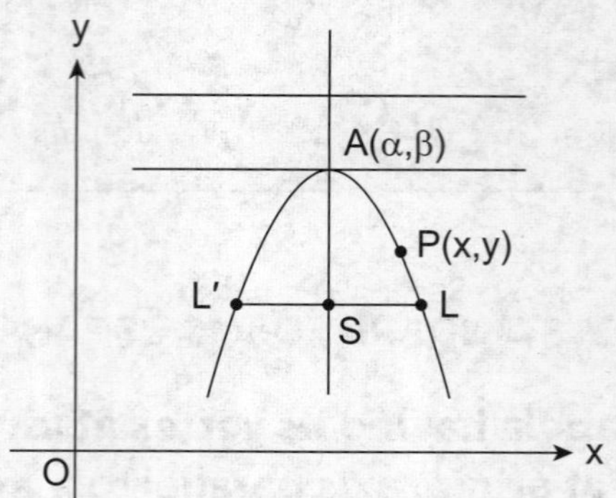

FIGURE 4.22

Its equation is given by $(x - \alpha)^2 = -4a(y - \beta)$ ...(iv)

And other related terms are as given here.

Focus $(S) \equiv (\alpha, \beta - a)$

Vertex $(A) \equiv (\alpha, \beta)$.

Equation of axis of parabola: $x = \alpha$ or $x - \alpha = 0$

Equation of tangent to parabola at vertex: $y = \beta$ or $y - \beta = 0$

Extremities of Latus rectum $L, L' \equiv (\alpha + 2a, \beta - a)$; $(\alpha - 2a, \beta - a)$

Equation of directrix : $y = \beta + a$

Parametric equation of parabola: $x = \alpha + 2at$, $y = \beta - at^2$

3. Parabola having its axis oblique, vertex at (α, β), and latus rectum 4a

Let the equations of axis of parabola and tangent at vertex be $lx + my + n = 0$ and $mx - ly + k = 0$ respectively. Let $A(\alpha, \beta)$ be the vertex and $P(x, y)$ be any point on the parabola as in Figure 4.23.

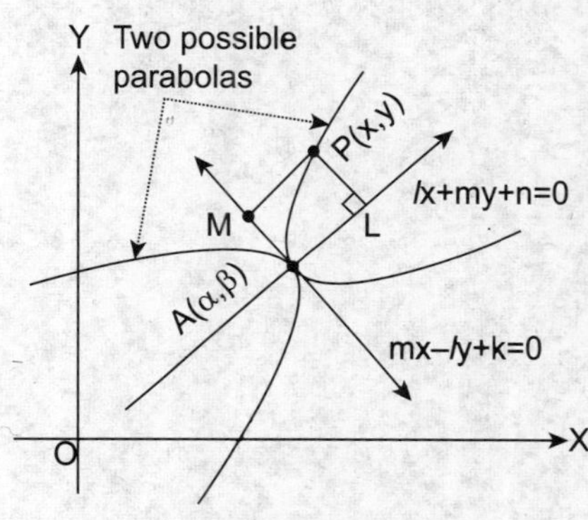

FIGURE 4.23

Let the length of latus rectum = 4 a

Then equation of parabola is given by

$(PL)^2 = 4a\,(PM)$

$$\Rightarrow \left(\frac{lx+my+n}{\sqrt{l^2+m^2}}\right)^2 = 4a\frac{|mx-ly+k|}{\sqrt{m^2+l^2}}$$

$$\Rightarrow (lx + my + n)^2 = \pm 4a\,(mx - ly + k)\sqrt{l^2+m^2}$$

which gives us two possible equations of parabola as shown in the above diagram.

■ GENERAL EQUATION OF PARABOLA

Let $S(x_0, y_0)$ be the focus of parabola, and let $lx + my + n = 0$ be the equation of directrix of parabola. Let $P(x, y)$ be any arbitrary point on parabola, as shown in Figure 4.24.

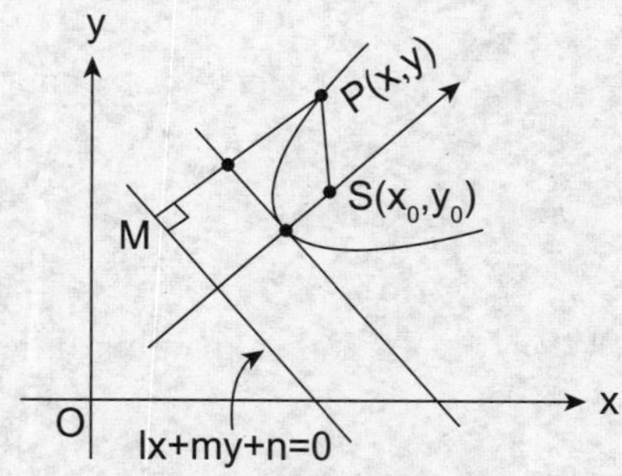

FIGURE 4.24

Then, $PS = PM$

$$\Rightarrow \sqrt{(x-x_0)^2+(y-y_0)^2} = \frac{|lx+my+n|}{\sqrt{l^2+m^2}}$$

$$\Rightarrow (x-x_0)^2 + (y-y_0)^2 = \frac{(lx+my+n)^2}{(l^2+m^2)}$$

$$\Rightarrow (x^2 + y^2 - 2xx_0 - 2yy_0 + x_0^2 + y_0^2)(l^2 + m^2) = l^2x^2 + m^2y^2 + n^2 + 2lmxy + 2mny + 2lnx$$

$$\Rightarrow m^2x^2 + l^2y^2 + x\,[-2x_0(l^2 + m^2) - 2ln] + y\,[-2y_0(l^2 + m^2) - 2mn] - 2lmxy + c = 0,$$ where c = constant term

$$\Rightarrow m^2x^2 + l^2y^2 + 2gx + 2fy - 2lmxy + c = 0$$

$$\Rightarrow (mx - ly)^2 + 2gx + 2fy + c = 0$$

Or simply $ax^2 + 2hxy + by^2 + 2gx + 2fy + c = 0$

Conversely, the second degree equation $ax^2 + 2hxy + by^2 + 2gx + 2fy + c = 0$ represents a parabola if $h^2 = ab$ and $abc + 2fgh - af^2 - bg^2 - ch^2 \neq 0$

i.e., $h^2 - ab = 0$ and $\Delta \neq 0$

REMARK

From the above, we note that second degree terms in the general equation of parabola form a perfect square.

■ PARAMETRIC EQUATION OF PARABOLA

Parametric form of the parabola $y^2 = 4ax$ is $x = at^2$, $y = 2at$.

∴ Parametric co-ordinates of any point on parabola are $(at^2, 2at)$.

ILLUSTRATION 7: Find the equation, axis, LR and vertex of the parabola whose focus is (0, 0) and directrix is $x + y + 1 = 0$.

SOLUTION: Let (x, y) be a point on the parabola. Then by definition $(x - 0)^2 + (y - 0)^2 = \left(\frac{x+y+1}{\sqrt{2}}\right)^2$

$\Rightarrow x^2 + y^2 - 2xy - 2x - 2y - 1 = 0$ which is the equation of the parabola.

Axis: It is a line passing through focus (0, 0) and perpendicular to the directrix $x + y + 1 = 0$.

Hence, its equation will be $x - y = 0$.

LR: It is a line passing through focus (0, 0) and parallel to the directrix $x + y + 1 = 0$.

Hence its equation will be $x + y = 0$

Length of LR: $2 \times$ (distance of S from $x + y + 1 = 0$) $= 2 \times 1/\sqrt{2} = \sqrt{2}$

Vertex: The point of intersection of $x - y = 0$ and $x + y + 1 = 0$ is $Z \equiv (-1/2, -1/2)$.

Now, vertex is the mid-point ZS, so it will be $(-1/4, -1/4)$.

ILLUSTRATION 8: Find the equation of the parabola whose focus is at $(-1, -2)$ and the directrix is the straight line $x - 2y + 3 = 0$.

SOLUTION: Let $P(x, y)$ be any point on the parabola whose focus is $S(-1, -2)$ and the directrix $x - 2y + 3 = 0$. Drawn PM perpendicular from $P(x, y)$ on the directrix.

$x - 2y + 3 = 0$. Then by definition $SP = PM$

$$\Rightarrow (SP)^2 = (PM)^2 \Rightarrow (x + 1)^2 + (y + 2)^2 = \left(\frac{|x-2y+3|}{\sqrt{(1)^2+(-2)^2}}\right)^2$$

$$\Rightarrow 5(x^2 + y^2 + 2x + 4y + 5) = (x^2 + 4y^2 - 4xy + 6x - 12y + 9)$$

$$\Rightarrow 4x^2 + y^2 + 4xy + 4x + 32y + 16 = 0$$

ILLUSTRATION 9: Find the equation of the parabola whose focus is $(4, -3)$ and vertex is $(4, -1)$.

SOLUTION: Let $A(4, -1)$ be the vertex and $S(4, -3)$ be the focus.

$\therefore$ Slope of $AS = \frac{-3+1}{4-4} = \infty \Rightarrow$ axis of parabola is parallel to y-axis

$\therefore$ Directrix will be parallel to x-axis.

Let $Z(x_1, y_1)$ be the point of intersection of axis of parabola and the directrix, then A is the mid-point of SZ

$\therefore$ $4 = \frac{x_1+4}{2} \Rightarrow x_1 = 4$ and $-1 = \frac{y_1-3}{2} \Rightarrow y_1 = 1 \therefore Z \equiv (4, 1)$

Also directrix is parallel to x-axis and passes through $Z(4, 1)$, so equation of directrix is $y = 1$ or $y - 1 = 0$

Now let $P(x, y)$ be any point on the parabola. Join SP and draw PM perpendicular to the directrix. Then by definition, $SP = PM \Rightarrow (SP)^2 = (PM)^2$

$$\Rightarrow (x - 4)^2 + (y + 3)^2 = \left(\frac{|y-1|}{\sqrt{1^2}}\right)^2 \Rightarrow (x - 4)^2 + (y + 3)^2 = (y - 1)^2 \text{ or } x^2 - 8x + 8y + 24 = 0$$

ILLUSTRATION 10: The focal distance of a point on a parabola $y^2 = 8x$ is 8. Find it.

SOLUTION: Comparing $y^2 = 8x$ with $y^2 = 4ax$

$\therefore$ $4a = 8 \Rightarrow a = 2$ $\therefore$ Equation of directrix is $x + 2 = 0$

Let $P(x_1, y_1)$ be on the parabola $y^2 = 8x$ whose focal distance is 8.

$\therefore$ $y_1^2 = 8x_1$ .....(i)

and $SP = 8$

$\Rightarrow PM = 8$ ($\because SP = PM$) $\Rightarrow x_1 + 2 = 8$ or $x_1 = 6$

From (1) $y^2_1 = 8 \times 6 \therefore y_1 = \pm 4\sqrt{3}$

$\therefore$ The required points are $(6, 4\sqrt{3})$ and $(6, -4\sqrt{3})$

ILLUSTRATION 11: QQ' is a double ordinate of a parabola $y^2 = 4ax$. Find the locus of its point of trisection.

SOLUTION: Let the double ordinate QQ' meet the axis of the parabola $y^2 = 4ax$ ...(1)

at L. Let the co-ordinates of Q be (x_1, y_1), then the co-ordinates of Q' will be $(x_1, -y_1)$.
Since Q and Q' lies on (1); $y^2_1 = 4ax_1$ ....(2)

Let R and T be the points of trisection of QQ'; then the co-ordinates of R and T are

$$\left(\frac{1.x_1 + 2.x_1}{1+2}, \frac{1(-y_1)+2.y_1}{1+2}\right) \text{ or } \left(x_1, \frac{y_1}{3}\right) \text{ and } \left(\frac{2.x_1 + 1.x_1}{2+1}, \frac{2.(-y_1)+1.y_1}{2+1}\right) \text{ or } \left(x_1, -\frac{y_1}{3}\right),$$

respectively.

Since R divide QQ' in 1: 2 (internally) and T divide QQ' in 2: 1 (internally)

for locus let (h, k) be the point of trisection, then $x_1 = h$ and $\pm\frac{y_1}{3} = k$ or $y_1 = \pm 3k$

Substituting the values of x_1 and y_1 in (2), we get

$(\pm 3k)^2 = 4a(h)$ or $9k^2 = 4ah$

Hence, the locus of point of trisection is $9y^2 = 4ax$.

Aliter: Let R and T be the points of trisection of double ordinates QQ'. Let (h, k) be the co-ordinates of R

then $AL = h$ and $RL = k$

$RT = RL + LT = k + k = 2k.$

Since $RQ = TR = Q'T = 2k.$

$\therefore \quad LQ = LR + RQ = k + 2k = 3k$

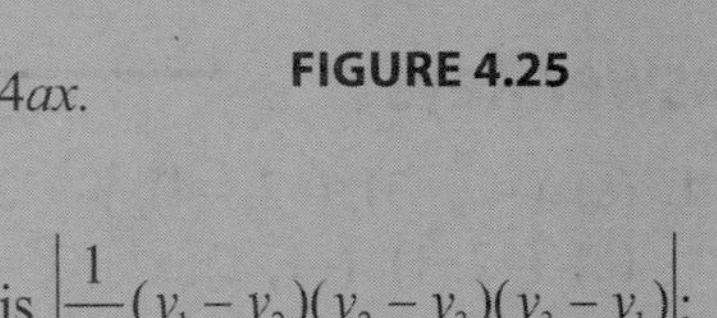

FIGURE 4.25

Thus, the co-ordinates of Q are $(h, 3k)$. Since $(h, 3k)$ lies on $y^2 = 4ax$.

$\Rightarrow 9k^2 = 4ah$. Hence, the locus of (h, k) is $9y^2 = 4ax$.

ILLUSTRATION 12: Prove that the area of the triangle inscribed in the parabola $y^2 = 4ax$ is $\left|\frac{1}{8a}(y_1 - y_2)(y_2 - y_3)(y_3 - y_1)\right|$; where y_1, y_2, y_3 are the ordinates of the vertices.

SOLUTION: Let the vertices of the triangle be (x_1, y_1); (x_2, y_2) and (x_3, y_3)

$\because (x_1, y_1)$ is a point on the parabola $y^2 = 4ax$

$\therefore y^2_1 = 4ax_1 \Rightarrow x_1 = \frac{y_1^2}{4a}$. Similarly, $x_2 = \frac{y_2^2}{4a}$ and $x_3 = \frac{y_3^2}{4a}$

Now, vertices of triangle are $\left(\frac{y_1^2}{4a}, y_1\right), \left(\frac{y_2^2}{4a}, y_2\right), \left(\frac{y_3^2}{4a}, y_3\right)$

$$\therefore \text{ Required area of the triangle} = \frac{1}{2}\left\|\begin{matrix} \frac{y_1^2}{4a} & y_1 & 1 \\ \frac{y_2^2}{4a} & y_2 & 1 \\ \frac{y_3^2}{4a} & y_3 & 1 \end{matrix}\right\| = \frac{1}{8a}\left\|\begin{matrix} y_1^2 & y_1 & 1 \\ y_2^2 & y_2 & 1 \\ y_3^2 & y_3 & 1 \end{matrix}\right\| = \left|\frac{1}{8a}(y_1 - y_2)(y_2 - y_3)(y_3 - y_1)\right|$$

TEXTUAL EXERCISE–1 (SUBJECTIVE)

1. Check whether the following equations represent parabola and if they do so, then find their vertex, focus, directrix, axis and latus-rectum:

(a) $x^2 + 4x + 4y + 16 = 0$

(b) $y^2 - 4x - 2y - 7 = 0$

(c) $y^2 - 8y - x + 19 = 0$

2. (a) Find the equation of parabola such that

(i) F: $(1, -1)$ and V: $(2, 1)$; F = focus, V = vertex

(ii) F: $(1, 1)$ and tangent at the vertex: $x + y = 1$

(b) Find the equation of a parabola whose focus is $(3, -4)$ and directrix is $x - y + 5 = 0$

(c) Find the equation of the parabola the extremities of whose latus rectum are $(1, 2)$ and $(1, -4)$.

(d) Prove that the equation to the parabola whose vertex and focus are on the axis of x at distances a and a' from the origin respectively is $y^2 = 4(a' - a)(x - a)$.

(e) Find the parabola of the form $Ax + By^2 + Cy + D = 0$ through the points $P_1(-1, 0)$, $P_2(2, 1)$, $P_3(1, -1)$. Find the vertex and axis of the parabola and sketch it.

(f) Find the equation of the parabola whose directrix makes an isosceles right angled triangle of area 4 square units with axes in 3rd quadrant and focus is on the line $y = x$, 2 units away from the origin.

3. Write the parametric equations of the parabolas:

(a) $y^2 = 4x$

(b) $(x + 1)^2 = 4(y - 1)$

(c) $3x^2 + 3x + 7y + 8 = 0$.

4. (a) Find the ordinate of a point on parabola $x^2 = 9y$ whose abscissa is its thrice

(b) A double ordinate of parabola $y^2 = 4ax$ is of length $8a$, prove that lines joining vertex to its two ends are at right angle.

5. For a parabola; directrix: $x + y = 4$ and F: $(6, 6)$. Find

(a) vertex

(b) axis of parabola

(c) equation of latus rectum

(d) length of latus rectum

(e) end points of latus rectum

(f) tangent at vertex

(g) equation of parabola

Answer Keys

1. (a) $(-2, -3)$, $(-2, -4)$, $y + 2 = 0$, $x + 2 = 0$ and $LR = 4a = 4$

(b) $(-2, 1)$, $(-1, 1)$, $x + 3 = 0$ $y = 1$, $x = -1$ and $LR = 4$

(c) $(3, 4)$, $(13/4, 4)$, $x = 11/4$, $y = 4$, $x = 13/4$ and 1

2. (a) (i) $4x^2 + y^2 - 4xy + 8x + 46y - 71 = 0$

(ii) $(x - y)^2 = 4(x + y - 1)$

(b) $x^2 + y^2 + 2xy - 22x + 26y + 25 = 0$

(c) $(y + 1)^2 = 3(2x + 1)$, $(y + 1)^2 = -3(2x - 5)$

(e) V: $\left(-\frac{41}{40}, -\frac{1}{10}\right)$ Eq. $5y^2 + y - 2 = 2x$

(f) $x^2 + y^2 - 8\sqrt{2}x - 8\sqrt{2}y - 2xy = 0$

3. (a) $(t^2, \pm 2t)$ (b) $x = 2t - 1$, $y = t^2 + 1$ (c) V: $(-1/2, -29/28)$

4. (a) $(0, 0)$; $(3, 1)$

5. (a) vertex $(4, 4)$;

(b) axis of parabola; $x - y = 0$

(c) equation of LR: $x + y = 12$

(d) length of $(LR) = 8\sqrt{2}$;

(e) end pts. of $LR \equiv (2, 10)$ and $(10, 2)$

(f) tangent at vertex: $x + y = 8$;

(g) equation of parabola: $x^2 + y^2 - 16x - 16y - 2xy + 128 = 0$

TEXTUAL EXERCISE–1 (OBJECTIVE)

1. $y^2 - 2x - 2y + 5 = 0$ is:
   (a) a circle with centre(1, 1)
   (b) a parabola with vertex (1, 2)
   (c) a parabola with directrix $x = 3/2$
   (d) a parabola with directrix $x = -1/2$

2. The centre of the conic represented by the equation $2x^2 - 72xy + 23y^2 - 4x - 28y - 48 = 0$ is
   (a) $\left(\frac{11}{15}, \frac{2}{25}\right)$ (b) $\left(\frac{2}{25}, \frac{11}{25}\right)$
   (c) $\left(\frac{11}{25}, -\frac{2}{25}\right)$ (d) $\left(-\frac{11}{25}, -\frac{2}{25}\right)$

3. The centre of $14x^2 - 4xy + 11y^2 - 44x - 58y + 71 = 0$ is
   (a) (2, 3) (b) (2, –3)
   (c) (– 2, 3) (d) (– 2, –3)

4. The equation of the conic with focus at (1, –1), directrix along $x - y + 1 = 0$ and with eccentricity $\sqrt{2}$ is
   (a) $x^2 - y^2 = 1$
   (b) $xy = 1$
   (c) $2xy - 4x + 4y + 1 = 0$
   (d) $2xy + 4x - 4y - 1 = 0$

5. The equation $x^2 - 2xy + y^2 + 3x + 2 = 0$ represents
   (a) A parabola (b) An ellipse
   (c) A hyperbola (d) A circle

6. The length of the latus rectum of the parabola, $y^2 - 6y + 5x = 0$ is
   (a) 1 (b) 5
   (c) 3 (d) 7

7. The focus of the parabola $(y - 1)^2 = 12(x - 2)$ is
   (a) (2, 1) (b) (1, –1)
   (c) (5, 1) (d) (3, 0)

8. The vertex of the parabola $y^2 + 6x - 2y + 13 = 0$ is
   (a) (1, –1) (b) (–2, 1)
   (c) (3/2, 1) (d) (–7/2, 1)

9. The equation of the directrix of the parabola $y^2 + 4y + 4x + 2 = 0$ is
   (a) $x = -1$ (b) $x = 1$
   (c) $x = -3/2$ (d) $x = 3/2$

10. The co-ordinates of a point on the parabola $y^2 = 8x$ whose focal distance is 4, are
    (a) (1/2, ± 2)
    (b) (1, $\pm 2\sqrt{2}$)
    (c) (2, ± 4)
    (d) None of these

11. The co-ordinates of the extremities of the latus rectum of the parabola $5y^2 = 4x$ are
    (a) (1/5, 2/5), (–1/5, 2/5)
    (b) (1/5, 2/5), (1/5, –2/5)
    (c) (1/5, 4/5), (1/5, –4/5)
    (d) None of these

12. The end points of latus rectum of the parabola $x^2 = 4ay$ are
    (a) $(a, 2a), (2a, -a)$
    (b) $(-a, 2a), (2a, a)$
    (c) $(a, -2a), (2a, a)$
    (d) $(-2a, a), (2a, a)$

13. $x - 2 = t^2$, $y = 2t$ are the parametric equations of the parabola
    (a) $y^2 = 4x$ (b) $y^2 = -4x$
    (c) $x^2 = -4y$ (d) $y^2 = 4(x - 2)$

14. The equation of the latus rectum of the parabola represented by equation $y^2 + 2Ax + 2By + C = 0$ is
    (a) $x = \frac{B^2 + A^2 - C}{2A}$ (b) $x = \frac{B^2 + A^2 + C}{2A}$
    (c) $x = \frac{B^2 - A^2 - C}{2A}$ (d) $x = \frac{A^2 - B^2 - C}{2A}$

15. The equation of the parabola with focus (a, b) and directrix $\frac{x}{a} + \frac{y}{b} = 1$ is given by
    (a) $(ax - by)^2 - 2a^3x - 2b^3y + a^4 - a^2 b^2 + b^4 = 0$
    (b) $(ax + by)^2 - 2a^3x - 2b^3y - a^4 + a^2 b^2 - b^4 = 0$
    (c) $(ax - by)^2 + a^4 + b^4 - 2a^3 x = 0$
    (d) $(ax - by)^2 - 2a^3 x = 0$

Answer Keys

1. (c) **2.** (d) **3.** (a) **4.** (c) **5.** (a) **6.** (b) **7.** (c) **8.** (b) **9.** (d) **10.** (c)
11. (b) **12.** (d) **13.** (d) **14.** (c) **15.** (a)

■ POSITION OF POINT AND LINE WITH RESPECT TO A PARABOLA

Position of Point with Respect to Parabola

The region towards focus is defined as the inside region of parabola and towards directrix is the outside region of parabola.

- ❑ Given a parabola $S : y^2 - 4ax = 0$ and a point $P(x_1, y_1)$
- ❑ Point P lies inside $\Leftrightarrow S_1 < 0$ (i.e., $y_1^2 - 4ax_1 < 0$)
- ❑ Point P lies on parabola $\Leftrightarrow S_1 = 0$ (i.e., $y_1^2 - 4ax_1 = 0$)
- ❑ Point P lies outside parabola $\Leftrightarrow S_1 > 0$ (i.e., $y_1^2 - 4ax_1 > 0$)

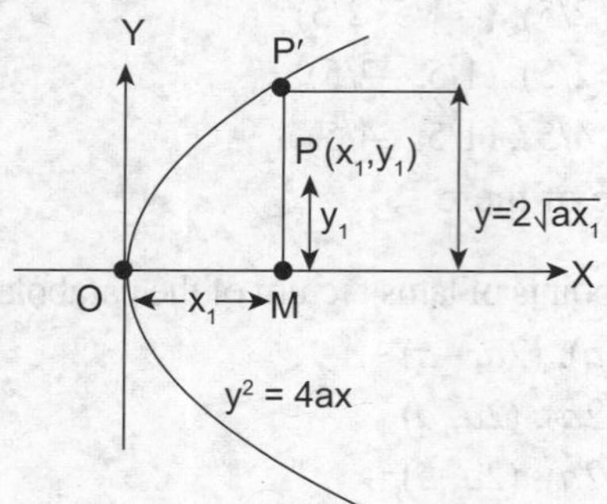

FIGURE 4.26

Proof: If point P lies inside the parabola, then $PM < P'M$

$\Rightarrow y_1 < 2\sqrt{ax_1} \Rightarrow y_1^2 < 4ax_1 \Rightarrow y_1^2 - 4ax_1 < 0 \Rightarrow S_1 < 0$

Similarly, if point $P(x_1, y_1)$ lies outside the parabola, then PM > P'M.

$\Rightarrow y_1 > 2\sqrt{ax_1} \Rightarrow y_1^2 - 4ax_1 > 0$

$\Rightarrow S_1 > 0$ and if $P(x_1, y_1)$ lies on parabola, then it would satisfy the equation $y^2 = 4ax$

$\Rightarrow y_1^2 - 4ax_1 = 0 \Rightarrow S_1 = 0$

■ POSITION OF LINE W.R.T. PARABOLA

Whether the straight line $y = mx + c$ cuts/touches/has no contact with the parabola $y^2 = 4ax$ can be determined by solving the parabola and straight line simultaneously by using two methods as follows:

Method I: Substituting $y = mx + c$ in the equation of parabola $y^2 = 4ax$, we get $(mx + c)^2 = 4ax$

$$\Rightarrow m^2x^2 + (2mc - 4a)x + c^2 = 0 \quad .....(i)$$

Now we have following three cases:

Case I: When the line $y = mx + c$ intersects the parabola $y^2 = 4ax$ at two points say $P(x_1, y_1)$ and $Q(x_2, y_2)$ as shown in the figure.

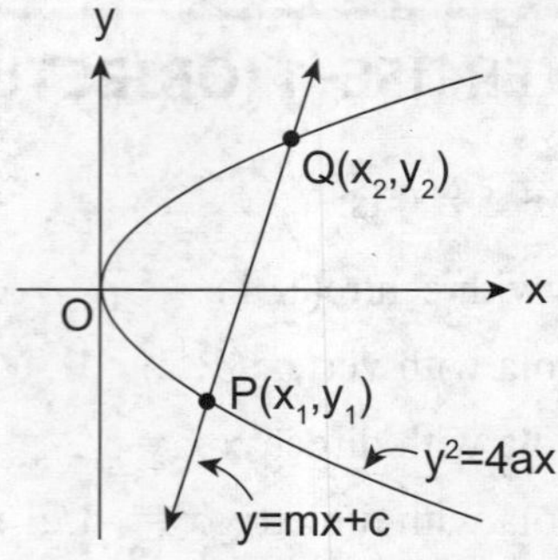

FIGURE 4.27

then, x_1 and x_2 will be the roots of equation (i) and would be real and distinct

$\Rightarrow$ Discriminant $(D) > 0$

$\Rightarrow (2mc - 4a)^2 - 4m^2c^2 > 0$

$\Rightarrow 4m^2c^2 + 16a^2 - 16mca - 4m^2c^2 > 0$

$\Rightarrow a^2 > amc$

$\Rightarrow a > mc$ ($a > 0$)

$\Rightarrow mc <$ a which is the required condition for the straight line $y = mx + c$ to intersect the parabola $y^2 = 4ax$ at two distinct points

Point of Intersection

From equation (i)

$$m^2x^2 + (2mc - 4a)x + c^2 = 0$$

Its two roots are x_1 and x_2

$$\Rightarrow x_1 + x_2 = \frac{4a - 2mc}{m^2} \text{ and } x_1x_2 = c^2/m^2$$

The above equation gives us the sum and product of abscissae of points of intersection. Solving the quadratic equation (i) we get roots x_1, x_2 and substituting these values in $y = mx + c$, we get corresponding values y_1 and y_2.

Case II: When the line $y = mx + c$ touches the parabola $y^2 = 4ax$ as shown below:

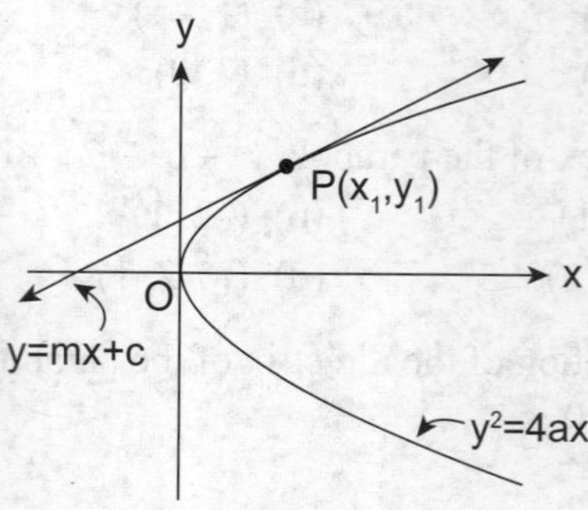

FIGURE 4.28

The two roots x_1, x_2 of equation (i) must coincide

$\Rightarrow D = 0 \Rightarrow a^2 = amc$

$\Rightarrow$ $a = mc \Rightarrow c = a/m$; which is the required condition of tangency of line $y = mx + c$ to the parabola $y^2 = 4ax$

Point of Contact

We have $x_1 + x_2 = \dfrac{4a - 2mc}{m^2}$, $x_1 x_2 = \dfrac{c^2}{m^2}$

$\therefore\ x_1 = x_2 \Rightarrow 2x_1 = \dfrac{4a - 2a}{m^2} = \dfrac{2a}{m^2} \Rightarrow x_1 = x_2 = a/m^2$

$\therefore\ y_1 = y_2 = mx_1 + c = m(a/m^2) + a/m\ (\because c = a/m)$

$\Rightarrow\ y_1 = y_2 = 2a/m \therefore$ point of contact is $(a/m^2, 2a/m)$

Case III: When the line $y = mx + c$ neither touches nor intersects the parabola i.e has no contact with parabola as shown below

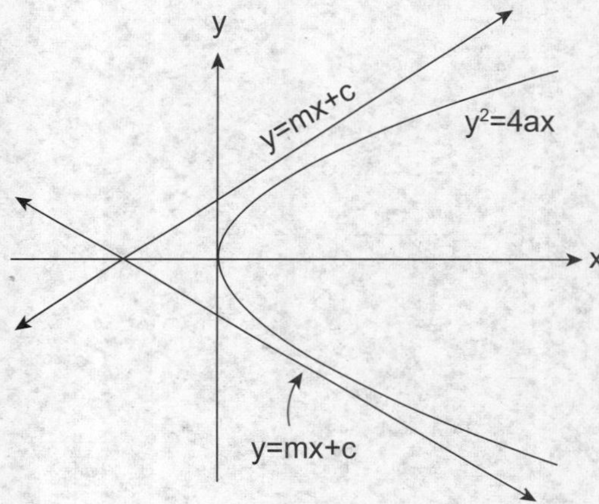

FIGURE 4.29

In this case, roots of equation (i) would be imaginary

$\Rightarrow\ D < 0 \Rightarrow a^2 - mca < 0$

$\Rightarrow\ mc > a\ (\quad a > 0)$

Method II Substituting $x = \dfrac{y - c}{m}$ in equation of parabola

$y^2 = 4ax$, we get $y^2 = 4a\left(\dfrac{y-c}{m}\right)$

$$\Rightarrow\ y^2 - \frac{4a}{m}y + \frac{4ac}{m} = 0 \qquad \text{....(ii)}$$

$\therefore$ For line $y = mx + c$ to intersect, to touch or having no contact with parabola $y^2 = 4ax$,

$D > 0$, $D = 0$ and $D < 0$ respectively

$$\Rightarrow\ \frac{16a^2}{m^2} - \frac{16ac}{m} >, =, < 0$$

$$\Rightarrow\ \frac{a}{m^2} - \frac{c}{m} >, =, < 0$$

$a - mc >, =, < 0$

i.e $mc < a$, $mc = a$ and $mc > a$

Point of Intersection

Let $P(x_1, y_1)$ and $Q(x_2, y_2)$ be the point of intersection of line $y = mx + c$ and parabola $y^2 = 4ax$, then y_1, y_2 would be the roots of equation (ii), then $y_1 + y_2 = 4a/m$ and $y_1 y_2 = 4ac/m$. Solving equation (ii), we get ordinates of points of intersection and substituting these values in $y = mx + c$, we get abscissae of points of intersection.

Point of Contact

When line $y = mx + c$ touches the parabola $y^2 = 4ax$, then $y_1 = y_2$

$\Rightarrow\ 2y_1 = 4a/m$

$\Rightarrow\ y_1 = 2a/m$

$$\text{and } x_1 = \frac{y_1 - c}{m} = \frac{y_1 - a/m}{m} = \frac{2a/m - a/m}{m} = a/m^2$$

$\Rightarrow$ point contact is $(a/m^2, 2a/m)$

$$\therefore\ y_1 = y_2 = mx_1 + c = m\left(\frac{a}{m^2}\right) + \frac{a}{m} (\because c = a/m)$$

$\Rightarrow\ y_1 = y_2 = 2a/m$

$\Rightarrow$ point of contact is $(a/m^2, 2a/m)$

Conclusions:

1. $y = mx + \dfrac{a}{m}\ \ \forall\ m \in \mathbb{R} \sim \{0\}$ represents the tangent of slope 'm' to the parabola $y^2 = 4ax$, also represents the family of tangents with parameter 'm'.

2. Point of contact $\left(\dfrac{a}{m^2}, \dfrac{2a}{m}\right)$

■ CHORDS OF PARABOLA

Given a parabola $y^2 = 4ax$, let AB be the chord joining $A(x_1, y_1)$ and $B(x_2, y_2)$:

$\Rightarrow\ y_1^2 = 4ax_1$ and $y_2^2 = 4ax_2$

$\Rightarrow y_2^2 - y_1^2 = 4a(x_2 - x_1)$

$$\Rightarrow \frac{y_2 - y_1}{x_2 - x_1} = \frac{4a}{y_1 + y_2}$$

$$\Rightarrow \text{Slope of chord } AB = \frac{4a}{y_1 + y_2} = \frac{2a}{\frac{y_1 + y_2}{2}}$$

$$\text{Equation of chord } AB:\ y - y_1 = \frac{4a}{y_1 + y_2}(x - x_1)$$

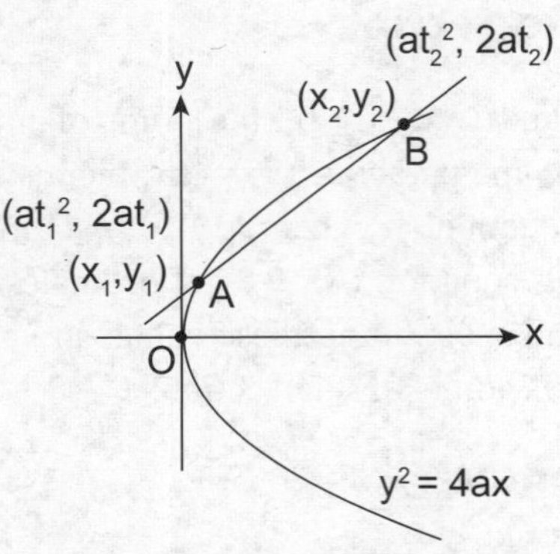

FIGURE 4.30

Chord of a Parabola in Parametric Form

Slope of chord $= \dfrac{4a}{y_1 + y_2} = \dfrac{4a}{2at_1 + 2at_2} = \dfrac{2}{t_1 + t_2}$

$\therefore$ Equation of chord : $y - 2at_1 = \dfrac{2}{t_1 + t_2}\left(x - at_1^2\right)$

$\Rightarrow \quad y\left(t_1 + t_2\right) - 2at_1^2 - 2at_1t_2 = 2x - 2at_1^2$

$\Rightarrow$ $y\ (t_1 + t_2) = 2\ (x + at_1\ t_2)$ which is the required equation of chord joining the points t_1 and t_2.

ILLUSTRATION 13: Let S be the focus of the parabola $y^2 = 4ax$ and R the point of intersection of axis and of the directrix, PP' is a double ordinate of the curve and PR meets the curve again at Q. Prove that $P'Q$ passes through focus.

SOLUTION: Co-ordinates of R are given by $R : (-a, 0)$

Equation of PR is given by $y = \dfrac{2at - 0}{at^2 + a}(x + a) \Rightarrow (1 + t^2)y = 2t\ (x + a)$ ......(1)

Now, (1) would intersect the parabola $y^2 = 4ax$, where $\dfrac{4t^2(x+a)^2}{(1+t^2)^2} = 4ax$

$\Rightarrow t^2(x + a)^2 = ax\ (1 + t^2)^2$

$\Rightarrow t^2[x^2 + a^2 + 2ax] = a[t^4 + 1 + 2t^2]x$

$\Rightarrow t^2x^2 + t^2a^2 = xat^4 + ax$

$\Rightarrow t^2x^2 - (a + at^4)x + a^2t^2 = 0$

$\Rightarrow xt^2(x - at^2) - a(x - at^2) = 0$

$\Rightarrow x = a/t^2$ or $x = at^2$, but $x = at^2$ corresponds to point P

$\Rightarrow x = a/t^2$ corresponds to Q.

$\Rightarrow Q \equiv \left(\dfrac{a}{t^2}, \dfrac{2a}{t}\right)$

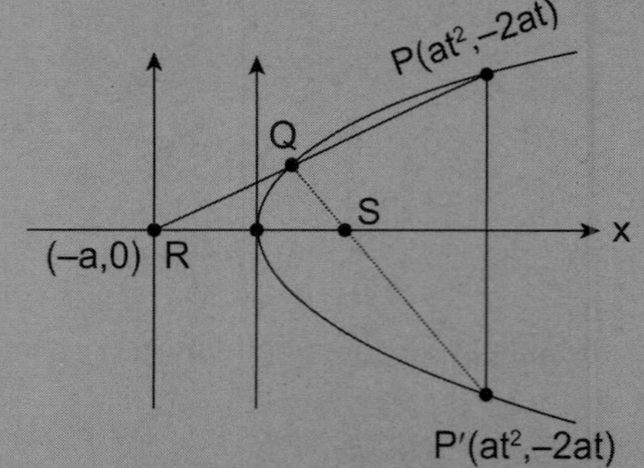

FIGURE 4.31

$\therefore$ Equation of $P'Q$ is given by $y + 2at = \dfrac{\dfrac{2a}{t} + 2at}{\dfrac{a}{t^2} - at^2}(x - at^2)$

$\Rightarrow (y + 2at)\ (1 - t^2) = 2t\ (x - at^2)$ ....(2)

which passes through focus as (a, 0) satisfies equation (2).

Condition for a Chord to be a Focal Chord

If the chord $y(t_1 + t_2) = 2(x + at_1t_2)$ is a focal chord, then point (a, 0) would lie on it.

Substituting, $y = 0, x = a \Rightarrow 0 = 2a\left(1 + t_1t_2\right)$

$\Rightarrow t_1t_2 = -1$

$\Rightarrow$ Condition to be focal chord: $t_1 \cdot t_2 = -1$

Since $t_1.t_2 = -1$

$\therefore$ if $t_1 = t \Rightarrow t_2 = -\dfrac{1}{t}$;

Also $y_1y_2 = 4a^2t_1t_2$ and $x_1x_2 = a^2(t_1t_2)^2$

$\therefore$ Condition to be a focal chord can also be written as $y_1y_2 = -4a^2$ and $x_1x_2 = a^2$

Properties of Focal Chord

- Extremities of focal chord PQ are given by: $P(at^2, 2at)$ and $Q\left(\frac{a}{t^2}, -\frac{2a}{t}\right)$
- Segments of focal chord PQ by focus S:
  $SP = l_1 = a + at^2$ and $SQ = l_2 = a + \frac{a}{t^2}$
- $H.M$ of segments of focal chord is semi latus-rectum:

$$\frac{2}{\frac{1}{l_1}+\frac{1}{l_2}} = \frac{2}{\frac{1}{a(1+t^2)}+\frac{t^2}{a(1+t^2)}} = \frac{2a(1+t^2)}{(1+t^2)} = 2a$$

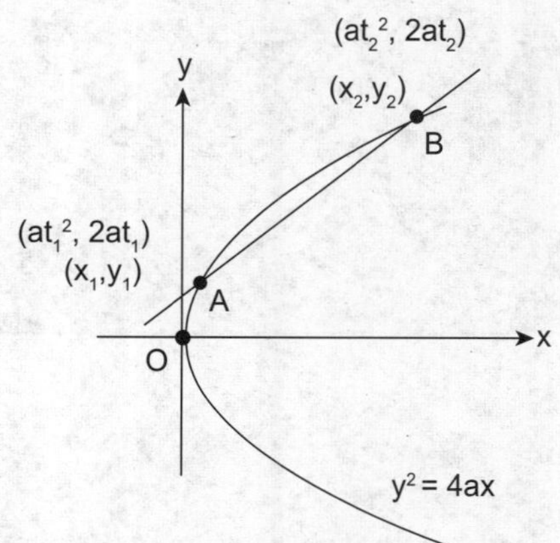

FIGURE 4.32

- Slope of focal chord at any point $(at^2, 2at)$: $m = \frac{2t}{t^2-1}$
- Equation of focal chord at point $(at^2, 2at)$: $y = \frac{2t}{t^2-1}(x-a)$
- If the line $y = mx + c$ intersects the parabola $y^2 = 4ax$, then the length of the chord intercepted is $\sqrt{1+m^2}\ |x_2 - x_1|$, where x_1 and x_2 are the abscissae of the points of intersection and are the roots of the equation $m^2x^2 + 2x(mc - 2a) + c^2 = 0$.

Equation of Chord Whose Mid-Point is (x_1, y_1)

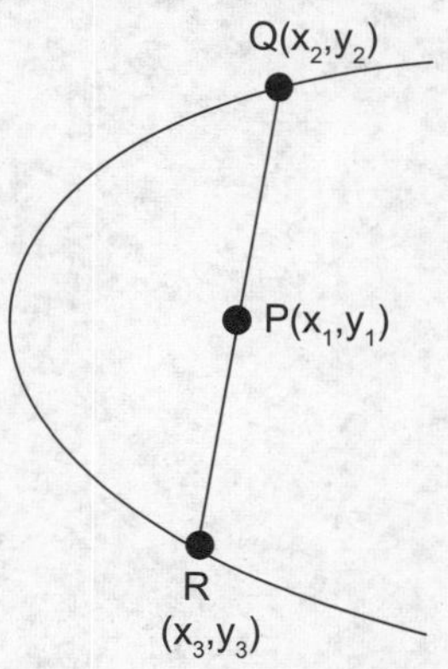

FIGURE 4.33

Equation of the parabola is $y^2 = 4ax$ (i)

Let QR be the chord of the parabola whose mid-point is $P(x_1, y_1)$

Since Q and R lie on parabola (i)

$\therefore\ y_2^2 = 4ax_2$ and $y_3^2 = 4ax_3$

$\Rightarrow\ y_3^2 - y_2^2 = 4a(x_3 - x_2)$

or $\frac{y_3 - y_2}{x_3 - x_2} = \frac{4a}{y_3 + y_2} = \frac{4a}{2y_1}$

$(\because P(x_1, y_1)$ is mid-point of $QR)$

$\therefore\ \frac{y_3 - y_2}{x_3 - x_2} = \frac{2a}{y_1} =$ slope of QR

$\Rightarrow$ Equation of QR is $y - y_1 = \frac{2a}{y_1}(x - x_1)$

$\Rightarrow\ yy_1 - y_1^2 = 2ax - 2ax_1$

$\Rightarrow\ yy_1 - 2a(x + x_1) = y_1^2 - 4ax_1$

(Subtracting $2ax_1$ from both sides)

$\Rightarrow\ T = S_1$;

where $T = yy_1 - 2a(x + x_1)$

and $S_1 = y_1^2 - 4ax_1$

ILLUSTRATION 14: Show that the point (2, 3) lies outside the parabola $y^2 = 3x$.

SOLUTION: Let the point $(h, k) = (2, 3)$, we have $k^2 - 3h = 3^2 - 3.2 = 9 - 6 = 3 > 0$

$k^2 - 3h > 0$

This shows that (2, 3) lies outside the parabola $y^2 = 3x$

ILLUSTRATION 15: Find the position of the point (–2, 2) with respect to the parabola $y^2 - 4y + 9x + 13 = 0$

SOLUTION: Let the point $(h.k) \equiv (-2, 2)$

We $k^2 - 4k + 9h + 13 = (2)^2 - 4(2) + 9(-2) + 13 = 4 - 8 - 18 + 13 = -9 < 0$

Hence $k^2 - 4k + 9h + 13 < 0$

Therefore the point (–2, 2) lies inside the parabola.

$y^2 - 4y + 9x + 13 = 0$

ILLUSTRATION 16: If the point $(at^2, 2at)$ be the extremity of a focal chord of parabola $y^2 = 4ax$, then show that the length of the focal chord is $a\left(t+\frac{1}{t}\right)^2$.

SOLUTION: Since one extremity of focal chord is $P(at^2, 2at)$, then the other extremity is $Q\left(\frac{a}{}\quad \frac{a}{}\right)$ (Replacing t by $- 1/t$)

$\therefore$ Length of focal chord = PQ

$= SP + SQ$ ($\because SP = PM$ and $SQ = QN$)

$= PM + QN$

$= at^2 + a + \frac{a}{t^2} + a$

$= a\left(t^2 + \frac{1}{t^2} + 2\right) = a\left(t \quad -\right)$

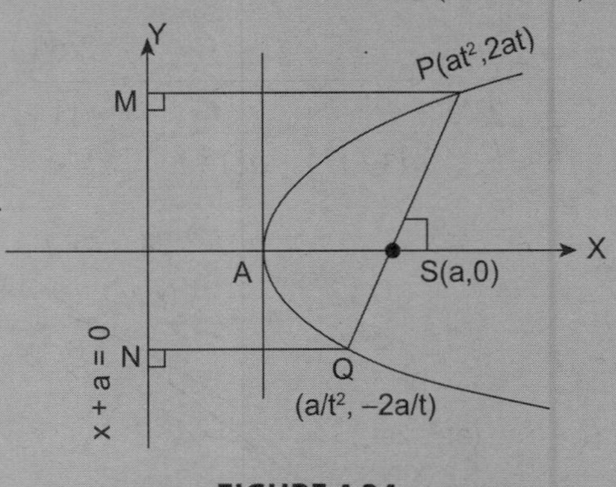

FIGURE 4.34

REMARK

$\because \left|t+\frac{1}{t}\right| \geq 2$ *for all* $t \neq 0$ ($\because AM \geq GM$)

$\therefore a\left(t+\frac{1}{t}\right)^2 \geq 4a$

$\Rightarrow$ *length of focal chord* $\geq$ *latus rectum*

i.e., The length of smallest focal chord of the parabola is 4a. Hence, the latus rectum of a parabola is the smallest focal chord.

ILLUSTRATION 17: Prove that the length of a focal chord of a parabola varies inversely as the square of its distance from the vertex.

SOLUTION: Let $(at^2, 2at)$ be one end of a focal chord of the parabola $y^2 = 4ax$. The focus of its parabola is $S(a, 0)$.

$\because$ Equation of focal chord is (i.e., equation of PS)

$y - 0 = \frac{2at - 0}{at^2 - a}(x - a) \qquad \Rightarrow y = \frac{2t}{(t^2 - 1)}(x - a)$

$\Rightarrow (t^2 - 1)y = 2tx - 2at \qquad \Rightarrow 2tx - (t^2 - 1)y - 2at = 0$

If d be the distance of this focal chord from the vertex (0, 0) of the parabola $y^2 = 4ax$, then

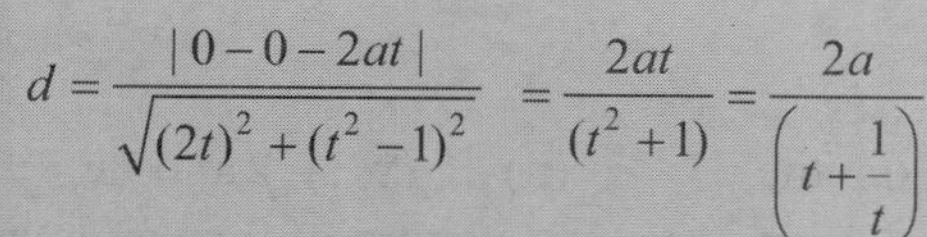

$$d = \frac{|0-0-2at|}{\sqrt{(2t)^2+(t^2-1)^2}} = \frac{2at}{(t^2+1)} = \frac{2a}{\left(t+\frac{1}{t}\right)}$$

$$d^2 = \frac{4a^2}{\left(t+\frac{1}{t}\right)^2} \qquad \text{......(1)}$$

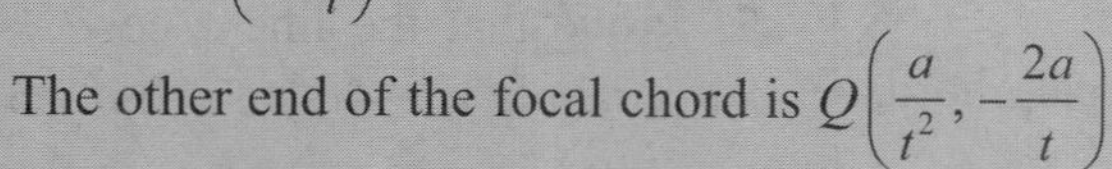

The other end of the focal chord is $Q\left(\frac{a}{t^2}, -\frac{2a}{t}\right)$

FIGURE 4.35

If length of focal chord = $PQ = \ell$ (say)

$\therefore \quad \ell = PQ = PS + SQ = PM + QN$

$\therefore \quad \ell = at^2 + a + \frac{a}{t^2} + a \Rightarrow \ell = a\left(t^2 + \frac{1}{t^2} + 2\right) \Rightarrow \ell = a\left(t + \frac{1}{t}\right)^2$

or $\quad \frac{\ell}{a} = \left(t + \frac{1}{t}\right)^2 \qquad \text{....(2)}$

From (1) and (2), $d^2 = \frac{4a^2}{(\ell/a)} = \frac{4a^3}{\ell}$

$\therefore \quad \ell = \frac{4a^3}{d^2}$ or $\ell \propto \frac{1}{d^2}$

i.e., the length of the focal chord varies inversely as the sqaure of its distance from vertex.

ILLUSTRATION 18: Find the equations of the chords of the parabola $y^2 = 4ax$ which pass through the point (–6a, 0) and which subtends an angle of 45° at the vertex.

SOLUTION: Equation of chord: $y = m(x + 6a)$

or $\quad \frac{y - mx}{6am} = 1$

Homogenising with the parabola, we have

$$y^2 - 4ax\left(\frac{y-mx}{6am}\right) = 0$$

i.e., $4amx^2 + 6am\,y^2 - 4axy = 0$; Angle between this pair of straight lines is 45°

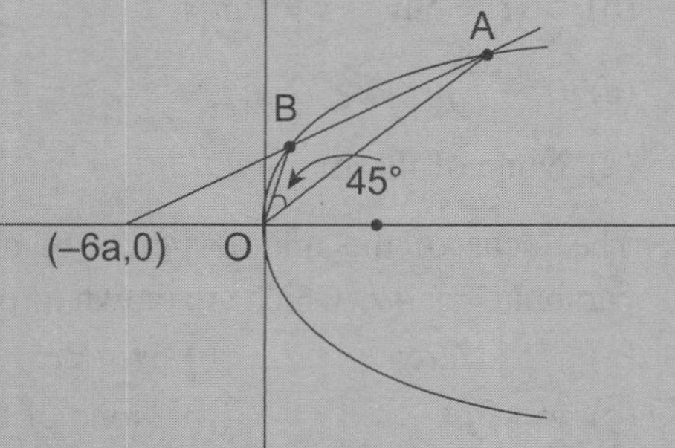

FIGURE 4.36

$\Rightarrow \tan 45° = \left|\frac{2\sqrt{4a^2 - 24a^2m^2}}{4am + 6am}\right| \Rightarrow 100\,a^2m^2 = 4(4a^2 - 24a^2m^2) \left(\because \tan\theta = \left|\frac{\sqrt{h^2 - ab}}{a+b}\right|\right)$

$\Rightarrow 25a^2m^2 + 24a^2m^2 = 4a^2$

$\Rightarrow 49m^2 = 4 \Rightarrow m = \pm\frac{2}{7}$

$\therefore$ Required chords are $y = \pm\frac{2}{7}(x + 6a)$ i.e., $2x - 7y + 12a = 0$ and $2x + 7y + 12a = 0$

TEXTUAL EXERCISE–2 (SUBJECTIVE)

1. If the focal chord passing through point t_1 on the parabola $y^2 = 4ax$ again meets the parabola at t_2, then prove that:
   (i) $t_1 t_2 = -1$
   (ii) length of such focal chord is $a(t_2 - t_1)^2$
2. If l_1, l_2 are the lengths of segments of a focal chord, then show that its latus rectum is $\dfrac{4\ell_1\ell_2}{\ell_1+\ell_2}$.
3. (a) Show that the focal chord of parabola $y^2 = 4ax$ making an angle α with the x-axis is of length $4a \operatorname{cosec}^2\alpha$.
   (b) If (x_1, y_1) and (x_2, y_2) are extremities of a focal chord of the parabola $y^2 = 4ax$, then find $x_1 x_2$.
4. Find the condition that $y = mx + c$ touches the parabola $y^2 = 4a(x + a)$.
5. Prove that the line $x \cos\alpha + y\sin\alpha = p$ touches the parabola $y^2 = 4a(x + a)$ if $p\cos\alpha + a = 0$.
6. Find the condition that the line $lx + my + n = 0$ touches the parabola $y^2 = 4a(x - b)$.
7. If the line $x + y = \lambda$ touches the parabola $y = x - x^2$ then find the value of λ.

Answer Keys

3. (b) a^2 **4.** $c = am + a/m$ **6.** $am^2 = bl^2 + nl$ **7.** $\lambda = 1$

TEXTUAL EXERCISE–2 (OBJECTIVE)

1. If t_1, t_2 are parametric points on the parabola $y^2 = 4ax$, then the equation of the chord through t_1 and t_2 is
   (a) $\frac{1}{2}(t_1 + t_2)y = x + at_1 t_2$
   (b) $\frac{1}{2}(t_1 + t_2)y = x + 2at_1 t_2$
   (c) $(t_1 + t_2)y = x + 2at_1 t_2$
   (d) None of these
2. The locus of the middle points of all chords of the parabola $y^2 = 4ax$ which are drawn through the vertex is
   (a) $y^2 = 1/2ax$ (b) $y^2 = 2ax$
   (c) $y^2 = 3ax$ (d) None of these
3. If the point $(at_1^2, 2at_1)$ is one extremity of a focal chord of the parabola $y^2 = 4ax$, then the length of the chord is.
   (a) $a\left(t_1 - \frac{1}{t_1}\right)^2$ (b) $a\left(\quad - \quad \right)$
   (c) $a\left(t_1 + \frac{1}{t_1}\right)^2$ (d) None of these
4. The equation of the chord of the parabola $y^2 = 4ax$ through the points (x_1, y_1) and (x_2, y_2) on it is
   (a) $(y - y_1)(y - y_2) = y^2 - 2ax$
   (b) $(y + y_1)(y + y_2) = y^2 - 4ax$
   (c) $(y - y_1)(y_1 + y_2) = y^2 - 4ax_1$
   (d) None of these
5. The equation to the chord of the parabola $y^2 = 8x$ which is bisected at the point $(1, -2)$ is
   (a) $y = -2x$ (b) $y = 2x - 4$
   (c) $y = x - 3$ (d) None of these
6. The locus of the middle points of chords of the parabola $y^2 = 4ax$ which pass through the focus is
   (a) $y^2 = ax - a^2$ (b) $y^2 = 2ax - a^2$
   (c) $y^2 = ax - 2a^2$ (d) $y^2 = 2ax - 2a^2$
7. The locus of the middle points of chords of the parabola $y^2 = 4x$ which pass through the fixed point (p, q) is
   (a) $y^2 - qx - x + 2p = 0$
   (b) $y^2 - qx - x - 2p = 0$
   (c) $y^2 - qx - 2x - 2p = 0$
   (d) $y^2 - qy - 2x + 2p = 0$
8. The length of the chord intercepted by the parabola $y^2 = 4x$ on the straight line $x + y = 1$ is
   (a) 4 (b) $4\sqrt{2}$
   (c) 8 (d) $8\sqrt{2}$

9. The length of the intercept on y-axis cut off by the parabola, $y^2 - 5y = 3x - 6$ is
(a) 1 (b) 2
(c) 3 (d) 5

10. If a line passing through a point $P(1, 0)$ and having slope 1 meets the parabola $y^2 = x$. at A and B, then $|PA| + |PB|$ is equal to
(a) $3\sqrt{2}$ (b) $\sqrt{8}$
(c) $\sqrt{10}$ (d) None of these

11. Length of the focal chord of the parabola $y^2 = 4ax$ at a distance p from the vertex is
(a) $\frac{2a^2}{p}$ (b) $\frac{a^3}{p^2}$
(c) $\frac{4a^3}{p^2}$ (d) $\frac{p^2}{a}$

12. AB is the chord of the parabola $y^2 = 4ax$ with vertex at A. BC is drawn perpendicular to AB meeting the axis at C. The projection of BC on the axis is
(a) a (b) $2a$
(c) $4a$ (d) $8a$

Answer Keys

1. (a) **2.** (b) **3.** (c) **4.** (c) **5.** (a) **6.** (d) **7.** (d) **8.** (c) **9.** (a) **10.** (c)
11. (c) **12.** (c)

■ EQUATION OF TANGENT IN DIFFERENT FORMS AND THEIR PROPERTIES

1. Point Form

(Equation of tangent at $P(x_1, y_1)$ lying on the parabola)

Slope of chord PQ: $m_{PQ} = \frac{4a}{y_1 + y_2}$

As $Q \to P \Rightarrow PQ \to$ tangent at P

$$\therefore\ m_t = \lim_{y_2 \to y_1}\left[\frac{4a}{y_1 + y_2}\right] \Rightarrow m_t = \frac{2a}{y_1}$$

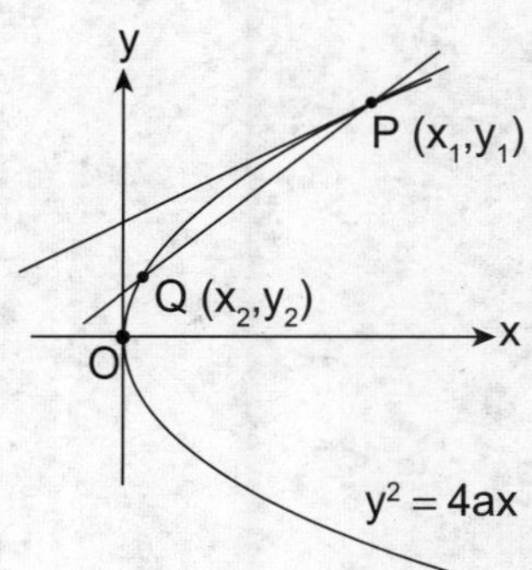

FIGURE 4.37

Equation of tangent at $P : y - y_1 = \frac{2a}{y_1}(x - x_1)$

$$\Rightarrow\ yy_1 - y_1^2 = 2ax - 2ax_1$$

$$\Rightarrow\ yy_1 - 2ax - 2ax_1 = y_1^2 - 4ax_1$$

$$\Rightarrow\ T : yy_1 - 2a(x + x_1) = 0\ (\because y_1^2 - 4ax_1 = 0)$$

2. Parametric Form

(Equation of tangent at $P(at^2, 2at)$)

Slope of chord PQ: $m_{PQ} = \frac{2}{t + t_1}$

As $Q \to P \Rightarrow PQ \to$ tangent at P

$$\therefore\ m_t = \lim_{t_1 \to t}\left[\frac{2}{t + t_1}\right] = \frac{1}{t} \Rightarrow m_t = \frac{1}{t}$$

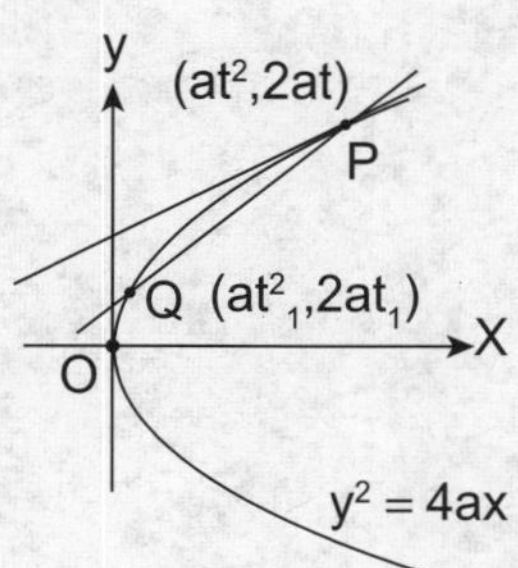

FIGURE 4.38

$\therefore$ Equation of tangent at $P : y - 2at = \frac{1}{t}(x - at^2)$

$$\Rightarrow\ yt - 2at^2 = x - at^2$$

$$\Rightarrow\ yt = x + at^2$$

ILLUSTRATION 19: Find the equations of tangents at the extremities of focal chord through point (3,6) of parabola $y^2 = 12x$ and hence find the area of triangle formed by these tangents and the tangent at vertex.

SOLUTION: Given point is $P(3,6)$ on parabola $y^2 = 12x = 4(3)x$

$\therefore \quad P(3,6) \equiv (at_1^2, 2at_1) \equiv (3t_1^2, 6t_1) \Rightarrow t_1 = 1$

$\therefore$ Other extremity of focal chord through P

will be $Q\left(\dfrac{a}{t_1^2}, \dfrac{-2a}{t_1}\right) \equiv \left(\dfrac{3}{1}, \dfrac{-2(3)}{1}\right) \equiv (3,-6)$

FIGURE 4.39

$\therefore$ equation of tangent at $P(t_1)$is given by $yt_1 = x + at_1^2$ i.e., $y = x + 3\,(1)^2$

i.e., $y = x + 3$ ...(i)

and equation of tangent at $Q(t_2)$ is given by $yt_2 = x + at_2^2$ i.e., $-y = x + 3(-1)^2$

i.e., $y = -x - 3$ ...(ii)

solving (i) and (ii), we get point of intersection $T(-3,0)$

Also point of intersection of tangents at P and Q with tangent at vertex are given by $A(0,3)$ and $B(0,-3)$

$\therefore$ Area of $\Delta TAB = \dfrac{1}{2}(VT)(AB) = \dfrac{1}{2}(3)(6) = 9$ sq.units

ILLUSTRATION 20: Find the equation of tangent to parabola $y^2 = 3x$ at point P(t) and hence find the locus of centroid of ΔTPS, where T is the point of intersection of tangent and axis of parabola and S the focus of parabola.

SOLUTION: Equation of tangent at P(t) is given by $yt = x + at^2$

i.e., $yt = x + \dfrac{3}{4}t^2$

It intersects x-axis where $y = 0$ i.e $x = -\dfrac{3}{4}t^2$

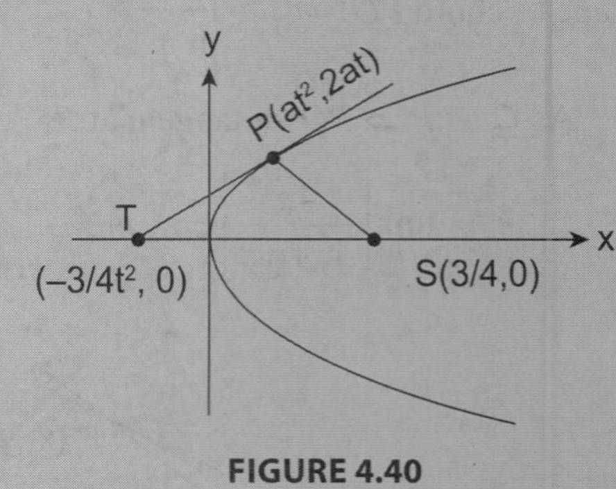

FIGURE 4.40

$\therefore$ Centroid of ΔPTS is given by (x, y)

$$\equiv \left(\frac{\frac{3}{4}t^2 - \frac{3}{4}t^2 + \frac{3}{4}}{3}, \frac{\frac{3}{2}t + 0 + 0}{3}\right)$$

i.e., $(x, y) \equiv \left(\dfrac{1}{4}, \dfrac{t}{2}\right)$

$\Rightarrow x = 1/4$ which is a straight line parallel to y-axis

3. Slope Form

The equation of the tangent to the parabola $y^2 = 4ax$ in slope form is given by $y = mx + a/m$.

Its point of contact is $(a/m^2, 2a/m)$, m being slope of the tangent.

ILLUSTRATION 21: Through the vertex V of the parabola $y^2 = 4ax$, a perpendicular is drawn to any tangent meeting it at P and the parabola at Q. Show that VP. VQ = constant.

SOLUTION: Any tangent to the parabola is $y = mx + \frac{a}{m}$ ...(1)

any line through vertex V and perpendicular to tangent is

$$y = -\frac{1}{m}x \quad ...(2)$$

VP is perpendicular distance of V(0, 0) from (1)

$$\therefore \quad VP = \frac{a}{\sqrt{1+m^2}}.\frac{1}{m}$$

The line (2) meets the parabola $y^2 = 4ax$ at Q

$$\therefore \quad \left(-\frac{1}{m}x\right)^2 = 4ax$$

$$\Rightarrow \quad \frac{1}{m^2}.x^2 = 4ax$$

$$\therefore \quad x = 4am^2 \text{ and } y = -\frac{1}{m}x$$

$$\Rightarrow y = -4am \Rightarrow VQ = \sqrt{(4am^2)^2 + (-4am)^2}$$

$$\therefore \quad VQ = 4am\sqrt{(m^2+1)}$$

$$\therefore \quad VP.\ VQ = \left(\frac{a}{\sqrt{1+m^2}}.\frac{1}{m}\right).4am\sqrt{1+m^2} = 4a^2$$

$$\Rightarrow PV\ .\ VQ = \text{constant}$$

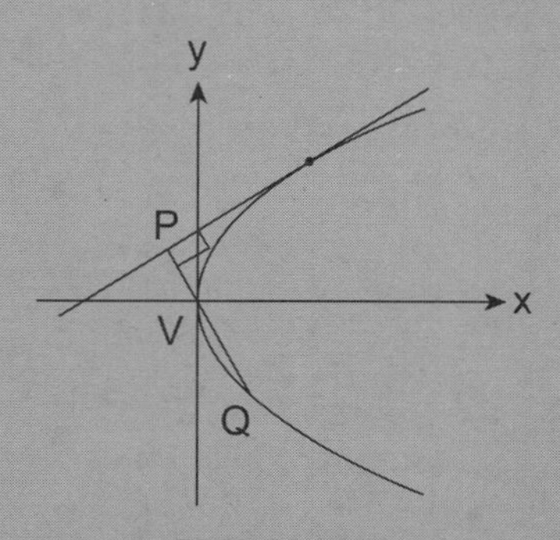

FIGURE 4.41

The following table represents the equations of tangents in different forms and related terms.

Equation	$y^2 = 4ax$	$y^2 = -4ax$	$x^2 = 4ay$	$x^2 = -4ay$
Tangent in point form	$yy_1 = 2a(x + x_1)$	$yy_1 = -2a(x + x_1)$	$xx_1 = 2a(y + y_1)$	$xx_1 = -2a(y + y_1)$
Parametric co-ordinate	$(at^2, 2at)$	$(-at^2, 2at)$	$(2at, at^2)$	$(2at, -at^2)$
Tangent in parametric form	$ty = x + at^2$	$ty = -x + at^2$	$tx = y + at^2$	$tx = -y + at^2$
Point of contact in terms of slope (m)	$\left(\frac{a}{m^2}, \frac{2a}{m}\right)$	$\left(-\frac{a}{m^2}, \frac{-2a}{m}\right)$	$(2am, am^2)$	$(-2am, -am^2)$
Condition of tangency	$c = \frac{a}{m}$	$c = -\frac{a}{m}$	$c = -am^2$	$c = am^2$
Tangent in slope form	$y = mx + \frac{a}{m}$	$y = mx - \frac{a}{m}$	$y = mx - am^2$	$y = mx + am^2$

ILLUSTRATION 22: Find the equations of the straight lines touching both $x^2 + y^2 = a^2$ and $y^2 = 4ax$.

SOLUTION: The given curves are $x^2 + y^2 = a^2$ ....(1)

and $y^2 = 4ax$ ....(2)

Equation of tangent to (2) is $y = mx + \frac{a}{m}$ or $m^2x - my + a = 0$ .....(3)

It is also tangent to (1), then the length of perpendicular from centre of (1) i.e., (0, 0) to (3) must be equal to the radius of (1) i.e., a

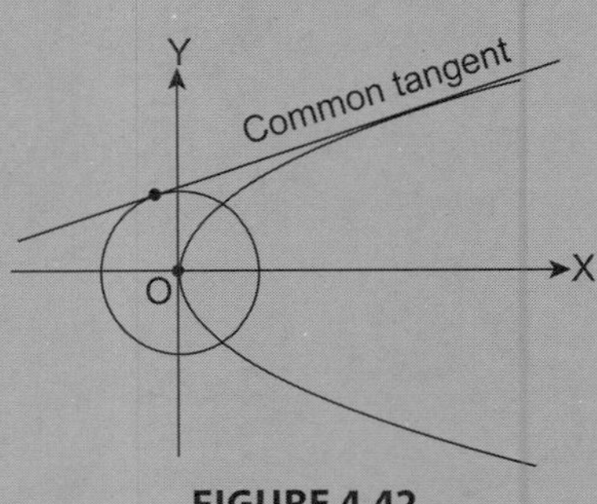

FIGURE 4.42

$$\therefore \quad \frac{|0-0+a|}{\sqrt{(m^2)^2+(-m)^2}} = a \text{ or } \frac{a^2}{m^4+m^2} = a^2$$

or $m^4 + m^2 - 1 = 0;\ m^2 = \frac{-1 \pm \sqrt{1+4}}{2} = \frac{-1 \pm \sqrt{5}}{2}$

But $m^2 \geq 0 \Rightarrow m^2 = \frac{\sqrt{5}-1}{2} \Rightarrow m = \pm\sqrt{\frac{\sqrt{5}-1}{2}}$

$\therefore$ Equation of required tangents are $y = mx + \frac{a}{m}$; where $m = \pm\sqrt{\frac{\sqrt{5}-1}{2}}$

ILLUSTRATION 23: Prove that the straight line $ax + by + c = 0$ touches the parabola $y^2 = 8x$ if $ac = 2b^2$.

SOLUTION: The given line is $ax + by + c = 0$

or $y = -\frac{a}{b}x - \frac{c}{b}$ ...(1)

Comparing this line with $y = mx + c'$, we get.

$\therefore \quad m = -\frac{a}{b}$ and $c' = -\frac{c}{b}$. The line (1) will touch the parabola $y^2 = 4ax$, if $c' = \frac{2}{m}$

or $c'm = 2$ or $\left(-\frac{c}{b}\right)\left(-\frac{a}{b}\right) = 2$ or $ac = 2b^2$

Alternative Method

The given line is $ax + by + c = 0$ .....(1)

and the parabola $y^2 = 8x$ .....(2)

Substituting the value of x from (1) i.e., $x = -\frac{c+by}{a}$ in (2), we get

$y^2 = 8\left(-\frac{c+by}{a}\right) \Rightarrow ay^2 + 8by + 8c = 0$ ....(3)

Since equation (1) touches the parabola (2), the roots of equation (3) must be coincident and condition for the same is Disc. = 0

That is, $64b^2 = 4a(8c)$ or $2b^2 = ac$.

ILLUSTRATION 24: Find the equation of the common tangents to the parabola $y^2 = 2ax$ and $x^2 = 2by$.

SOLUTION: The equation of any tangent in terms of slope (m) to the parabola $y^2 = 2ax$ is

$y = mx + \frac{a}{2m};\ m \neq 0$ ....(I)

If this line is also tangent to the parabola $x^2 = 2by$, then (1) meets $x^2 = 2\ by$ in two coincident points.

Substituting the value of y from (1), in $x^2 = 2by$ we get $x^2 = 2b\left(mx + \frac{a}{2m}\right)$

$\Rightarrow mx^2 - bmx - ab = 0.$

The roots of this quadratic are equal provided disc. = 0.

That is, $b^2m^2 = 4m\,(-ab)$

$\Rightarrow m = 0$ or $m = -4a/b$, but $m \neq 0 \Rightarrow m = -4a/b$

Substituting the value of m in (1) the required equation is $y = \dfrac{-4a}{b}x + \dfrac{ab}{(-8a)}$

or $y = \dfrac{-4ax}{b} - \dfrac{b}{8}$

$\Rightarrow 4ax + by + b^2/8 = 0$ or $32ax + 8by + b^2 = 0$

ILLUSTRATION 25: Two tangents to the parabola $y^2 = 4ax$ subtend angle α and β with x-axis. Find the locus of their point of intersection if $\cot\alpha + \cot\beta = 4$.

SOLUTION: Let the equation of any tangent to the parabola $y^2 = 4ax$ is

$y = mx + (a/m)$ ....(1)

Let (x_1, y_1) be the point of intersection of the tangents to $y^2 = 4ax$, then (1) passes through (x_1, y_1)

$\therefore\ y_1 = mx_1 + (a/m)$

or $m^2x_1 - my_1 + a = 0$

Let m_1 and m_2 be the roots of this quadratic equation, then $m_1 + m_2 = y_1/x_1$ and $m_1m_2 = a/x_1$

or $\tan\alpha + \tan\beta = y_1/x_1$

and $\tan\alpha\tan\beta = a/x_1$ ...(2)

Now $\cot\alpha + \cot\beta = 4$ (given)

$\Rightarrow \dfrac{1}{\tan\alpha} + \dfrac{1}{\tan\beta} = 4 \Rightarrow \dfrac{\tan\alpha + \tan\beta}{\tan\alpha\tan\beta} = 4$ [From (2)]

$\Rightarrow \dfrac{y_1/x_1}{a/x_1} = 4 \Rightarrow y_1 = 4a$

$\therefore$ The required locus is $y = 4a$ (which is a line parallel to x-axis).

The following table represents the equations of tangents in different forms and related terms to parabolas having vertex at (h, k) and axes parallel to the co-ordinate axes.

Equation	$(y-k)^2 = 4a(x-h)$	$(y-k)^2 = -4a(x-h)$	$(x-h)^2 = 4a(y-k)$	$(x-h)^2 = -4a(y-k)$
Tangent in point form	$(y-y_1)(y-k) = 2a(x-x_1)$	$(y-y_1)(y-k) = -2a\,(x-x_1)$	$(x-x_1)(x_1-h) = 2a\,(y-y_1)$	$(x-x_1)(x_1-h) = -2a(y-y_1)$
Parametric co-ordinate	$(h+at^2, k+2at)$	$(h-at^2, k+2at)$	$(h+2at, k+at^2)$	$(h+2at, k-at^2)$
Tangent in parametric form	$t(y-k) = (x-h) + at^2$	$t(y-k) = -(x-h) + at^2$	$t(x-h) = (y-k) + at^2$	$t(x-h) = -(y-k) + at^2$
Point of contact in terms of slope (m)	$\left(h + \dfrac{a}{m^2}, k + \dfrac{2a}{m}\right)$	$\left(h - \dfrac{a}{m^2}, k - \dfrac{2a}{m}\right)$	$(h+2am, k+am^2)$	$(h-2am, k-am^2)$
Condition of tangency	$c + mh = k + \dfrac{a}{m}$	$c + mh = k - \dfrac{a}{m}$	$c + mh = k - am^2$	$c + mh = k + am^2$
Tangent in slope form	$y = mx - mh + k + \dfrac{a}{m}$	$y = mx - mh + k - \dfrac{a}{m}$	$y = mx - mh + k - am^2$;	$y = mx - mh + k + am^2$

ILLUSTRATION 26: Prove that the line $\lambda x + \mu y = 1$ touches the parabola $y^2 = 4a(x-h)$ if $\lambda = \lambda^2 h - a\mu^2$

SOLUTION: The given parabola is $y^2 = 4a(x-h)$ .....(1)

Vertex of this parabola is $(h, 0)$

Now, shifting origin (0, 0) at $(h, 0)$, we have

$x = X + (h)$ and $y = Y + 0$

or $x - h = X$ and $y = Y$ ......(2)

$\therefore$ From (1), $Y^2 = 4aX$ ....(3)

and the line $\lambda x + \mu y = 1$

reduces to $\lambda(X+h) + \mu Y = 1 \Rightarrow \mu Y = 1 - \lambda(x + h)$

$$\Rightarrow Y = \frac{-\lambda}{\mu}x + \left(\frac{1}{\mu} - \frac{\lambda h}{\mu}\right) \quad ....(4)$$

The line (4) will touch the parabola (3), if $\frac{1}{\mu} - \frac{\lambda h}{\mu} = \frac{a}{-\lambda/\mu}$

$$\Rightarrow \frac{1-\lambda h}{\mu} = \frac{-a\mu}{\lambda}$$

$\Rightarrow \lambda - \lambda^2 h = -a\mu^2$

or $\lambda = \lambda^2 h - a\mu^2$

ILLUSTRATION 27: Let P be a point on the parabola $y^2 - 2y - 4x + 5 = 0$, such that the tangent on the parabola at P intersects the directrix at point Q. Let R be the point that divides the line segment QP externally in the ratio $\frac{1}{2}:1$. Find the locus of R.

SOLUTION: Given equation can be written as $(y - 1)^2 = 4(x - 1)$

whose parametric co-ordinates are $x - 1 = t^2$ and $y - 1 = 2t$

i.e., $P \equiv (1 + t^2, 1 + 2t)$

$\therefore$ equation of tangent at P is $t(y - 1) = x - 1 + t^2$, which meets the directrix $x = 0$ at Q.

$$\Rightarrow y = 1 + t - \frac{1}{t} \Rightarrow Q \equiv \left(0, 1 + t - \frac{1}{t}\right)$$

Let $R(h, k)$ divides QP externally in the ratio $\frac{1}{2}:1$.

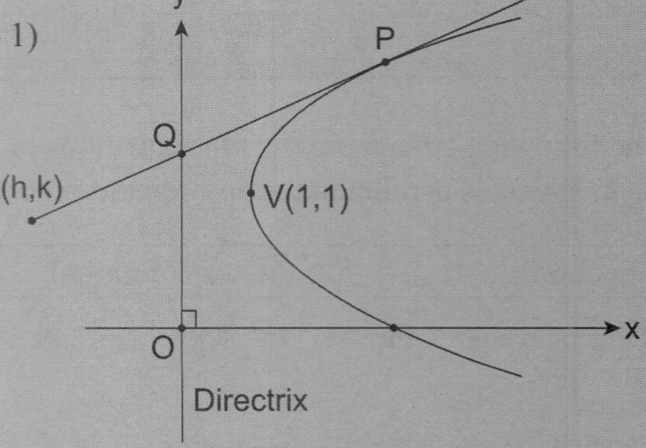

FIGURE 4.43

$\Rightarrow Q$ is mid-piont of $RP \Rightarrow 0 = \frac{h + t^2 + 1}{2}$ or $t^2 = -(h + 1)$ ...(1)

and $1 + t - \frac{1}{t} = \frac{k + 2t + 1}{2}$ or $t = \frac{2}{1-k}$ ...(2)

$\therefore$ from equations (1) and (2), we have $\frac{4}{(1-k)^2} + (h + 1) = 0$

or $(k - 1)^2 (h + 1) + 4 = 0$

$\therefore$ locus of point R is $(x + 1)(y - 1)^2 + 4 = 0$

PROPERTIES OF TANGENTS

P: 1. Point of intersection of tangents

The tangents to the parabola $S \equiv y^2 - 4ax = 0$ drawn at the points $P\,(at_1^2, 2at_1)$ and $Q\,(at_2^2, 2at_2)$ intersect at the point R $(at_1 t_2, a\,(t_1 + t_2))$

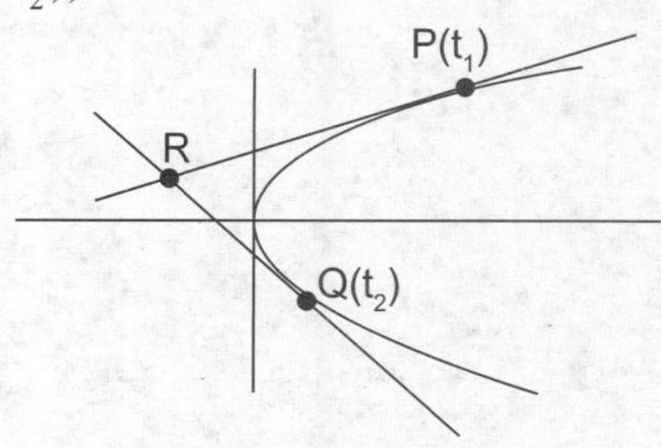

FIGURE 4.44

Proof: Equation of tangent to parabola $y^2 = 4ax$ at $P(at_1^2, 2at_1)$ is given by $t_1 y = x + at_1^2$ ... (i)

and at $Q(at_2^2, 2at_2)$ is given by $t_2 y = x + at_2^2$ ... (ii)

At the point of intersection $t_1 y - at_1^2 = t_2 y - at_2^2$

$\Rightarrow (t_1 - t_2)y = a(t_1^2 - t_2^2) \quad \Rightarrow \quad y = a(t_1 + t_2)$

$\therefore$ from (i) $x = t_1 y - at_1^2 = t_1(a(t_1 + t_2) - at_1^2$
$= at_1^2 + at_1 t_2 - at_1^2 = at_1 t_2$

$\therefore$ The point of intersection of tangents at P and Q is given by $(at_1 t_2, a(t_1 + t_2)) = \left(\sqrt{at_1^2 . at_2^2}, \dfrac{2at_1 + 2at_2}{2}\right)$

(G.M. of abscissae, A.M. of ordinates)

$= \left(\sqrt{x_P x_Q}, \dfrac{y_P + y_Q}{2}\right)$; if y_P, y_Q of same sign

$= \left(-\sqrt{x_P x_Q}, \dfrac{y_P + y_Q}{2}\right)$; if y_P, y_Q of same sign

ILLUSTRATION 28: Show that the points of intersection of the mutually perpendicular tangents to a parabola lie on the directix of parabola.

SOLUTION: Let the points be $P(at_1^2, 2at_1)$ and $Q(at_2^2, 2at_2)$ on the parabola $y^2 = 4ax$, tangents at P and Q are

$x = t_1 y - at_1^2$ ...(1)

and $x = t_2 y - at_2^2$ ...(2)

$\because$ point of intersection of tangents at $P(t_1)$ and $Q(t_2)$ is $(at_1 t_2, a(t_1 + t_2))$. Let this point is (h, k)

then $h = at_1 t_2$ ....(3)

and $k = a(t_1 + t_2)$ .....(4)

Now, slope of tangents (1) and (2) are $\dfrac{1}{t_1}$ and $\dfrac{1}{t_2}$ respectively

Since tangents are perpendiculars,

$\Rightarrow \dfrac{1}{t_1} \times \dfrac{1}{t_2} = -1$ or $t_1 t_2 = -1$ ....(5)

From (3) and (5), we get $h = -a$ or $h + a = 0$

$\therefore$ Locus of the point of intersection of tangents is $x + a = 0$ which is directrix of parabola $y^2 = 4ax$ i.e., points of intersection of perpendicular tangents lie on directix.

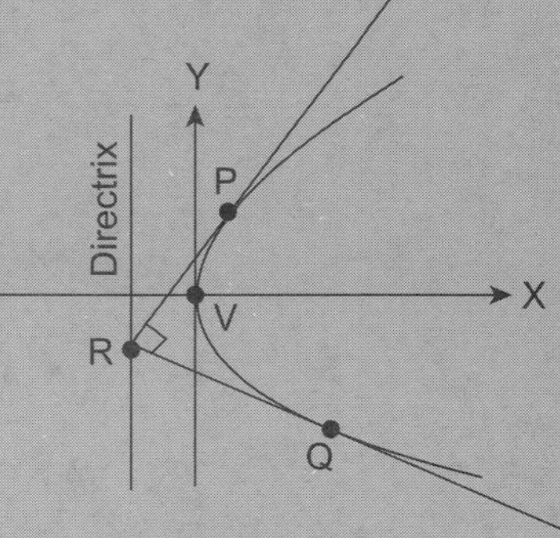

FIGURE 4.45

Alternative Method

Let the equation of any tangent to the parabola $y^2 = 4ax$ is $y = mx + a/m$ ..(i)

Let the point of intersection of the tangents to $y^2 = 4ax$ is (h, k), then (1) passes through (h, k)

$\therefore$ $k = mh + a/m$ or $m^2h - mk + a = 0$

Let m_1, m_2 be the roots of this quadratic equation.

$\Rightarrow m_1 m_2 = a/h = -1$(since tangents are perpendiculars) $\Rightarrow a + h = 0$

$\therefore$ Locus of the point of intersection of tangents is $x + a = 0$ which is directrix of $y^2 = 4ax$.

That is, points of intersection of tangents to parabola lie on directrix of the parabola.

ILLUSTRATION 29: If the tangents drawn at $P(3, 2\sqrt{3})$ and $Q(2, -2\sqrt{2})$ to the parabola $y^2 = 4x$. Intersects at T. Then find the co-ordinates of T and

(i) $ar\ \Delta PQT$: ar $\Delta\ VPQ$

(ii) $ar\ \Delta PQT$: ar ΔABT

(iii) $ar\ \Delta PVQ : ar\ \Delta ABT$; where V is the vertex of parabola, A and B are points where tangents at P and Q intersect y-axis, respectively.

SOLUTION: Tangents at $P(3, 2\sqrt{3})$ *and* $Q(2, -2\sqrt{2})$ intersects at T

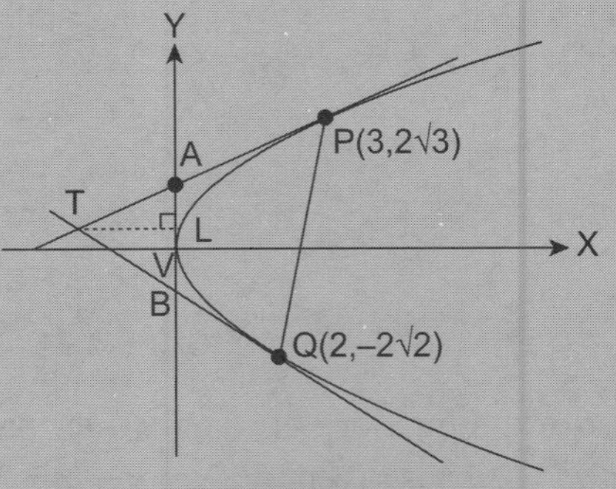

FIGURE 4.46

Given, parabola is $y^2 = 4x$

$\Rightarrow a = 1$

Let, $P \equiv (at_1^2, 2at_1) \equiv (3, 2\sqrt{3})$

$\Rightarrow t_1^2 = 3, t_1 = \sqrt{3}$

and $Q \equiv (at_2^2, 2at_2) \equiv (2, -2\sqrt{2})$

$\Rightarrow t_2 = -\sqrt{2}$

We know that point of intersection of tangents at t_1 and t_2 is given by $T(at_1t_2, a(t_1 + t_2))$

$= (-\sqrt{6}, (\sqrt{3} - \sqrt{2}))$

Also, equation of tangent at P is given by $yt = x + at^2$

$\therefore$ For A put $x = 0$, $t = \sqrt{3}, a = 1$

$\Rightarrow \sqrt{3}y = 3 \Rightarrow y = \sqrt{3}$

and for B, put $x = 0$, $t = -\sqrt{2}, a = 1$

$\Rightarrow -\sqrt{2}y = 2 \Rightarrow y = -\sqrt{2} \Rightarrow$ A$(0, \sqrt{3})$ and $B(0, -\sqrt{2})$ $ar\ \Delta PQT = \frac{1}{2}\begin{vmatrix} 3 & 2\sqrt{3} & 1 \\ 2 & -2\sqrt{2} & 1 \\ -\sqrt{6} & (\sqrt{3} - \sqrt{2}) & 1 \end{vmatrix}$

$= \frac{1}{2}\left|-9\sqrt{3} - 11\sqrt{2}\right| = \frac{9\sqrt{3} + 11\sqrt{2}}{2}$ square units $ar\ \Delta VPQ = \frac{1}{2}\begin{vmatrix} 0 & 0 & 1 \\ 3 & 2\sqrt{3} & 1 \\ 2 & -2\sqrt{2} & 1 \end{vmatrix}$

$= \frac{1}{2}\left|-6\sqrt{2} - 4\sqrt{3}\right| = \frac{4\sqrt{3} + 6\sqrt{2}}{2}$

and $ar\ \Delta ABT = \frac{1}{2}(TL) \times AB = \frac{1}{2}(\sqrt{6})(\sqrt{3} + \sqrt{2}) = \frac{1}{2}(3\sqrt{2} + 2\sqrt{3})$

$\therefore \frac{ar\ \Delta PQT}{ar \Delta VPQ} = \frac{9\sqrt{3} + 11\sqrt{2}}{4\sqrt{3} + 6\sqrt{2}}; \frac{ar\ \Delta PQT}{ar \Delta ABT} = \frac{9\sqrt{3} + 11\sqrt{2}}{2\sqrt{3} + 3\sqrt{2}}$

and $\frac{ar\ \Delta PVQ}{ar \Delta ABT} = \frac{4\sqrt{3} + 6\sqrt{2}}{3\sqrt{2} + 2\sqrt{3}} = \frac{2(2\sqrt{3} + 3\sqrt{2})}{(2\sqrt{3} + 3\sqrt{2})} = 2:1$

REMARK

It is a property of parabola that the area of Δ formed by joining three points on the parabola is twice the area formed by tangents at given three points.

ILLUSTRATION 30: Find the area enclosed by the triangle formed by tangents at the extremities of focal chord through a point $P\ (t = 4)$ on parabola $y^2 = 4x$ and the focal chord.

SOLUTION: We know for focal chord $t_1t_2 = -1$

$\therefore$ If $P \equiv (t = 4)$, then $Q \equiv \left(t = \dfrac{-1}{4}\right)$

$\therefore$ $P \equiv (16, 8)$ and $Q \equiv \left(\dfrac{1}{16}, \dfrac{-1}{2}\right)$

and also we know that tangents at the extremity of focal chord intersect at right angle on the directrix.

$\therefore$ Co-ordinates of point of intersection R will be $(at_1t_2, a(t_1 + t_2)) \equiv R\left(-1, \dfrac{15}{4}\right)$

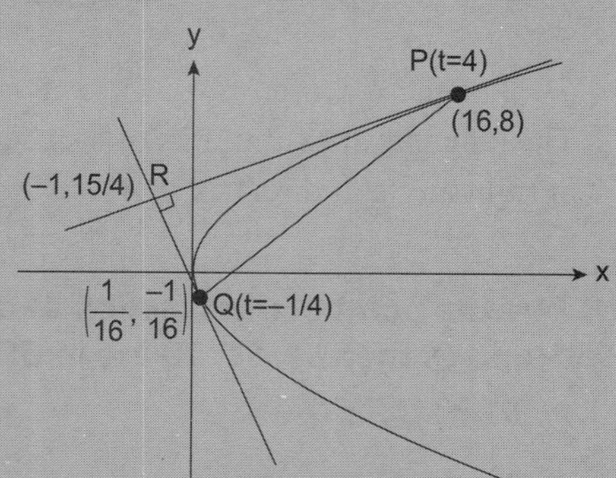

FIGURE 4.47

$$\therefore\ ar\ \Delta PQR = \frac{1}{2}\ RQ \times RP = \frac{1}{2}\sqrt{\left(\frac{1}{16}+1\right)^2 + \left(\frac{15}{4}+\frac{1}{2}\right)^2} \times \sqrt{(17)^2 + \left(8 - \frac{15}{4}\right)^2} = \frac{4913}{128} \text{ sq. units.}$$

ILLUSTRATION 31: Find the equations of the tangents to the parabola $y^2 = 12x$, which passes through the point (36, 21).

SOLUTION: Equation of parabola is $y^2 = 12x$

$\Rightarrow$ $a = 3$ and let $P(t_1)$ and $Q(t_2)$ be the points of contact of tangents drawn.

Let R is the intersection point of both tangents $\Rightarrow$ $R \equiv (36, 21)$

Also, the co-ordinates of R are given by

R: $(at_1t_2, a(t_1 + t_2)) \Rightarrow at_1t_2 = 36$ and $a(t_1 + t_2) = 21$ and $a = 3$

$\Rightarrow$ $t_1t_2 = 12$ ...(1)

and $a(t_1 + t_2) = 21 \Rightarrow (t_1 + t_2) = 7$ ...(2)

Solving equations (1) and (2), we get $t_1 = 4$, $t_2 = 3$ or $t_1 = 3$, $t_2 = 4$.

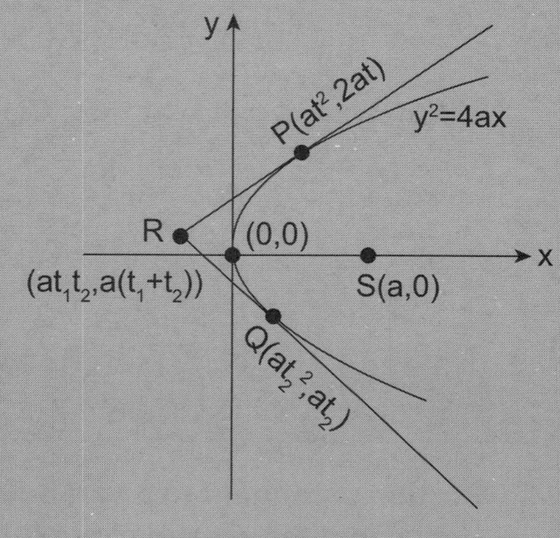

FIGURE 4.48

Now, the equation of tangent in parametric form is given by $ty = x + at^2$.

$\therefore$ For $t_1 = 4$, equation of tangent is $4y = x + 48$.

for $t_2 = 3$, equation of tangent is $3y = x + 27$.

P: 2. Tangents at extermities of any focal chord always intersect at directrix at right angle.

Proof: We know that point of intersection of tangents at two points $P(t_1)$ and $Q(t_2)$ is given by $R(at_1t_2, a(t_1 + t_2))$. If PQ is focal chord, then $t_1t_2 = -1$.

$\therefore$ $R \equiv (-a, a(t_1 + t_2))$.

$\Rightarrow$ R lies on the line $x = -a$ i.e., directrix. Thus the tangents at extremities of focal chord intersect at directrix.

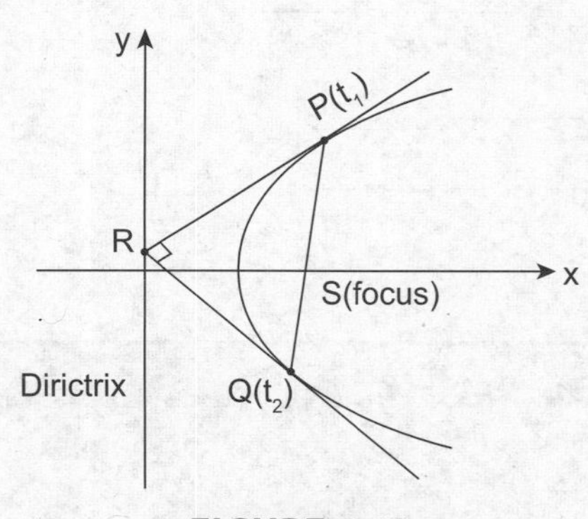

FIGURE 4.49

Further, tangents at $P(t_1)$ and $Q(t_2)$ have slopes $1/t_1$ and $1/t_2$.

$\therefore$ Product of slopes $= m_1.m_2 = \frac{1}{t_1t_2} = \frac{1}{-1} = -1$

($\because$ For focal chord $t_1t_2 = -1$).

$\therefore$ The tangents at extremities of focal chord intersect at right angle at directrix.

P: 3. If *T* be the point of intersection of tangent at *P* and *Q*, then *SP*, *ST*, *SQ* are in *GP*. i.e., $ST = \sqrt{SP.SQ}$

Proof: If co-ordinates of P and Q are $(at_1^2, 2at_1)$ and $(at_2^2, 2at_2)$ respectively, then the co-ordinates of their point of intersection are given by $R\ (at_1t_2, a(t_1 + t_2))$ and those of focus S are given by (a, 0)

$$\therefore \quad SP = \sqrt{(at_1^2 - a)^2 + (2at_1)^2} = a\sqrt{t_1^4 + 1 - 2t_1^2 + 4t_1^2}$$

$$= a\sqrt{(t_1^2+1)^2} = a(1+t_1^2)$$

$$SQ = a\sqrt{t_2^4 + 1 - 2t_2^2 + 4t_2^2} = a(1+t_2^2)$$

$$ST = \sqrt{(at_1t_2 - a)^2 + (a(t_1+t_2))^2}$$

$$= a\sqrt{t_1^2t_2^2 + 1 - 2t_1t_2 + t_1^2 + t_2^2 + 2t_1t_2}$$

$$= a\sqrt{t_1^2t_2^2 + 1 + t_1^2 + t_2^2}$$

$$= a\sqrt{t_1^2(1+t_2^2) + (1+t_2^2)} = a\sqrt{(1+t_1^2)(1+t_2^2)}$$

$$\therefore \quad ST^2 = a(1+t_1^2).a(1+t_2^2) = SP.SQ$$

Thus $(ST)^2 = (SP) . (SQ)$

$\Rightarrow$ SP, ST and SQ are in $G.P.$

ILLUSTRATION 32: Find the distance between the focus and the point of intersection of tangents at points $P(4,4\sqrt{3})$ and $Q(2, -2\sqrt{6})$ of parabola $y^2 = 12x$

SOLUTION: We know that SP, ST, SQ are in $G.P.$

$$\Rightarrow ST = \sqrt{SP.SQ}$$

$$\Rightarrow ST = \sqrt{\sqrt{(1)^2 + 48}.\sqrt{1+24}}$$

$$= \sqrt{7\times 5} = \sqrt{35} \text{ units}$$

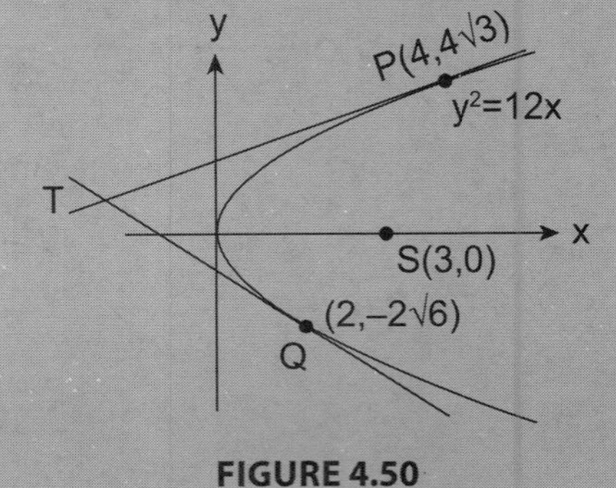

FIGURE 4.50

P: 4. Portion of tangent to a parabola intercepted between directrix and curve subtends right angle at focus.

Proof: Equation of tangent at $P(t)$ is $yt = x + at^2$

It meets directrix $x + a = 0$ at Q

$\therefore$ co-ordinates of Q are $\left\{-a, \frac{a(t^2-1)}{t}\right\}$

$$\therefore \quad M_{SP} = \frac{2at - 0}{at^2 - a} = \frac{2t}{t^2 - 1}$$

$$\text{and } M_{SQ} = \frac{a\left(\frac{t^2-1}{t}\right)}{-a-a} = -\frac{t^2-1}{2t}$$

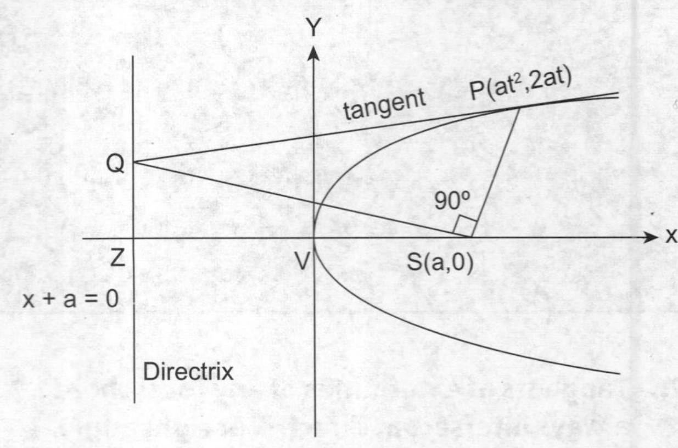

FIGURE 4.51

$\Rightarrow M_{SP} . M_{SQ} = -1 \Rightarrow \quad SP \perp r\ SQ.$

ILLUSTRATION 33: Find the equation of tangent at point $P(t = 3)$ to parabola $y^2 = 8x$ and find the area of ΔPQS, where Q is the point of intersection of tangent at P and directrix and S, focus of parabola.

SOLUTION: Equation of parabola is $y^2 = 8x \Rightarrow a = 2$

$\therefore$ focus $S \equiv (2,0)$

And $P \equiv (at^2, 2at) \equiv \left(2(3)^2, 2(2)(3)\right) \equiv (18,12)$

Equation of tangent at $P(t = 3)$ is given $yt = x + at^2$

i.e., $3y = x + 18$

Now, tangent $3y = x + 18$ intersects directrix at Q, where $x = -2 \Rightarrow y = 16/3$

$\Rightarrow$ Co-ordinates of Q are $(-2, 16/3)$

Also, we know that segment of tangent intercepted between the parabola and directrix subtend a right angle at focus.

$\Rightarrow \angle PSQ = \pi/2$

$\therefore \; ar\,\Delta PSQ = \frac{1}{2} PS.QS = \frac{1}{2}\sqrt{256+144}.\sqrt{16+256/9} = \frac{1}{2}(20).\frac{20}{3} = 200/3$ square units.

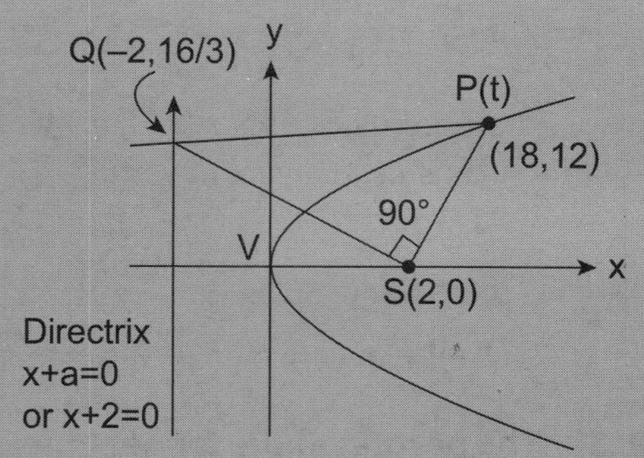

FIGURE 4.52

P: 5. Tangent at any point *P* of parabola bisects the angle between the focal chord through *P* and perpendicular from *P* to the directrix.

Proof: Let tangent at $P\ (at^2, 2at)$ meet x axis at T and equation of tangent at P is given by $yt = x + at^2$

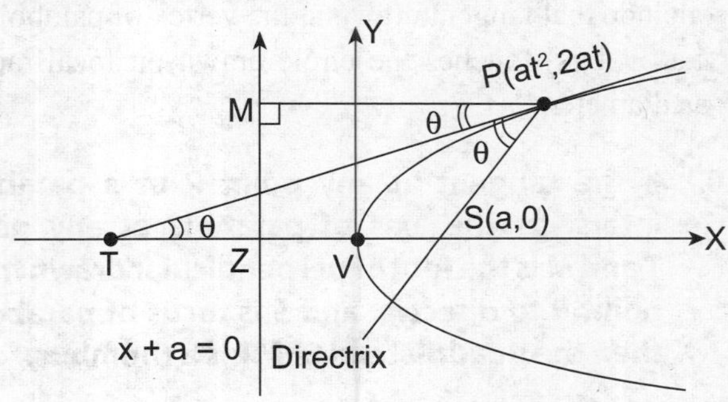

FIGURE 4.53

So point $T\ (-at^2, 0)$

Now, $ST = SV + VT = a + at^2 = a\,(1 + t^2)$

and $SP = PM = a + at^2 = a(1 + t^2) \Rightarrow ST = SP$

$\Rightarrow \angle SPT = \angle STP = \angle MPT$ alternate angle

P: 6. The foot of perpendicular from the focus on any tangent to a parabola lies on the tangent at vertex.

Proof: Tangent at 't' is given by $yt = x + at^2$ ...(i)

Equation of line through focus normal to (i) given by $y + xt = at$

$\Rightarrow yt + xt^2 = at^2$ ....(ii)

Subtracting (ii) form (i), we get $x(1 + t^2) = 0$

$\Rightarrow x = 0$, i.e., tangent at vertex.

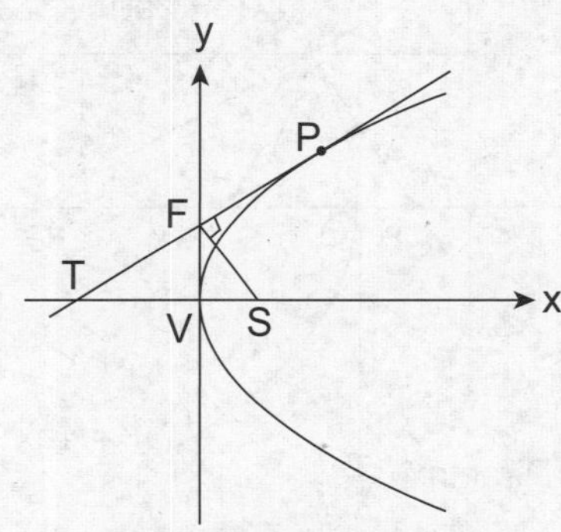

FIGURE 4.54

$\therefore$ Point of intersection of tangent and perpendicular from focus $(a, 0)$ to tangent lies on y-axis.

P: 7. If *N* is the foot of perpendicular, from focus to tangent at P, *A* (vertex) and *S*(focus), then $(SN)^2 = (SA)(SP)$.

Proof: Equation of tangent at $P(at^2, 2at)$ is given by $x - yt + at^2 = 0$. .... (i)

Foot of perpendicular drawn from focus $S(a, 0)$ to (i) is given by, $\frac{x-a}{1} = \frac{y-0}{-t} = -\frac{(a - 0 + at^2)}{(1)^2 + (-t)^2}$.

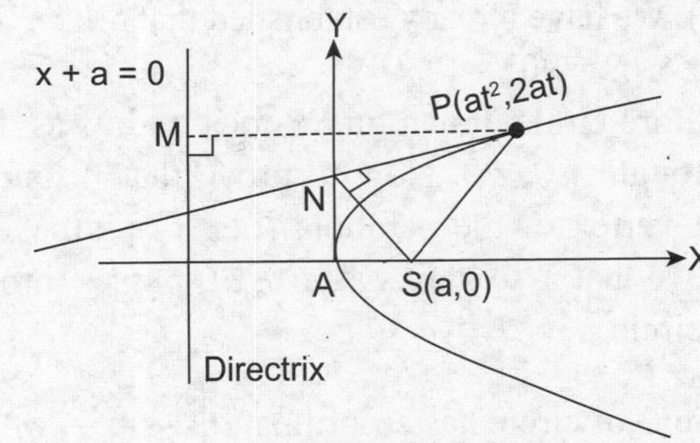

FIGURE 4.55

$\Rightarrow \quad x - a = \dfrac{y}{-t} = -a \Rightarrow x = 0; y = \text{at}$

$\therefore$ Co-ordinates of N are (0, at) which clearly lies on y-axis i.e., tangent at vertex.
Also $SP = PM = a(1 + t^2)$;

$SA = a$ and $SN = \sqrt{a^2 + a^2t^2} = a\sqrt{1+t^2}$

$\Rightarrow \quad (SN)^2 = a^2(1 + t^2) = a \,.\, a(1 + t^2) = (SA)\,.\,(SP)$

P: 8. The portion of the tangent intercepted between axis and point of contact is bisected by tangent at vertex.

Proof: ΔTPS is isosceles with $TS = PS$ and we know that foot of $\perp r$ drawn from focus to tangent lies on tangent at vertex.

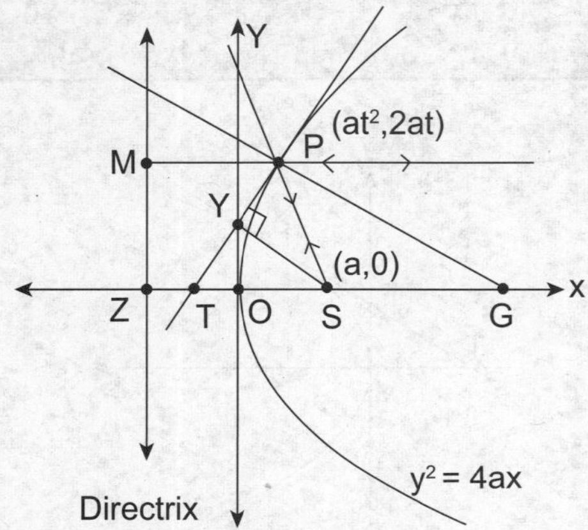

FIGURE 4.56

$\Rightarrow$ SY is an altitude of ΔTPS which is isosceles with $TS = PS$

$\Rightarrow$ SY is also median of ΔTPS through S

$\Rightarrow$ $TY = YP$

$\Rightarrow$ Tangent intercepted between the axis of parabola and point of contact is bisected by the tangent at vertex

P: 9. Circle drawn on focal radius as diameter touches tangent at vertex.

Given parabola $y^2 = 4ax$ with focus $S(a, 0)$. Let $P(at^2, 2at)$ be a point on it, tangent at which meets y-axis (tangent at vertex) at A (0, at). We have already established that $SA \perp PT$

$\Rightarrow \quad \angle SAP = 90°$

Therefore circle drawn on SP focal radius as diameter passes through $A(0, at)$. Now to prove that it touches the tangent at vertex of the parabola (i.e., y-axis). We would like to prove that $\perp$ to y-aixs at $A(0, at)$ passes through the centre of circle.

Centre C of circle has co-ordinate $\left(\dfrac{a + at^2}{2}, at\right)$

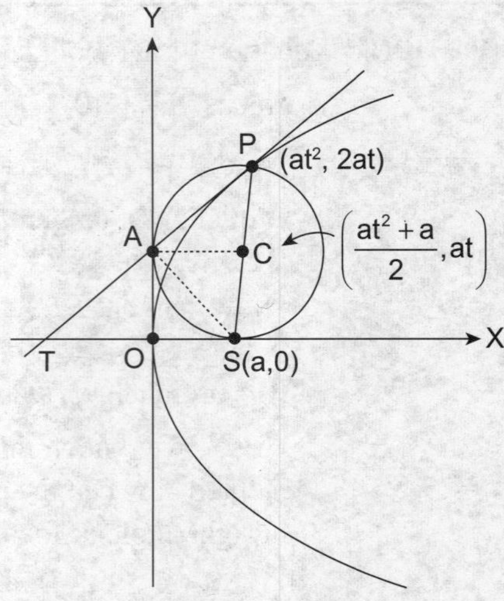

FIGURE 4.57

$\Rightarrow$ Slope of AC (M_{AC}) $= 0 \Rightarrow CA \perp y$-axis thus y-axis touches the circle.

Aliter: Equation of circle with SP as diameter is given by $(x - a)(x - at^2) + y(y - 2at) = 0$
Solving this equation with y-axis ($x = 0$) we get

$y^2 - 2aty + a^2t^2 = 0$

$\Rightarrow \quad (y - at)^2 = 0$

Thus we get two repeated roots $y = at$. Consequently we conclude that tangent drawn at the vertex of parabola.

(i.e., y-axis) touches the circle drawn on focal radius (SP) as diameter

P: 10. If the tangent at any point P of a parabola intersects the axis of parabola at any point T and M is the foot of perpendicular drawn from point P to directrix and S is focus of parabola, then the quadrilateral SPMT is a rhombus.

Proof: We proved $TS = SP = PM$
Also $MT =$

$$\sqrt{\left(-a + at^2\right)^2 + 4a^2t^2} = \sqrt{a^2 + a^2t^4 + 2a^2t^2} = a\left(1 + t^2\right)$$

FIGURE 4.58

$\left(\because M \equiv (-a, 2at) \, and \, T \equiv \left(-at^2, 0\right)\right)$

and $PM = at^2 + a = a(1 + t^2) \;\therefore\; TS = SP = PM = MT$

$\therefore$ $TSPM$ is a rhombus.

ILLUSTRATION 34: Find the area of quadrilateral *TSPM*, where *P*(*t*) is any point on parabola, $y^2 = 4ax$, where *S* focus, *M* foot of perpendicular drawn from point *P* to directrix and *T* point of intersection of tangent to parabola at *P* and axis of parabola. Hence find the area of rhombus due to point $P(t = 4)$ on the parabola $y^2 = 3x$

SOLUTION: We know that *TSPM* is a rhombus

$\therefore$ area of $\Delta YPS = \frac{1}{2}(SY)\times(PY)$ ....(i)

Also we know that portion of tangent between the parabola and its axis is bisected at tangent at vertex.

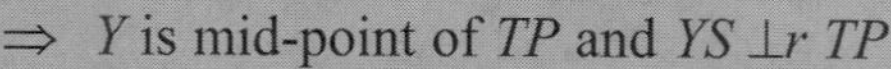

$\Rightarrow$ *Y* is mid-point of *TP* and $YS \perp r\ TP$

$\therefore$ Area of □ $TSPS = 2\left(\frac{1}{2}TP\times YS\right) = TP \times YS$ ...(ii)

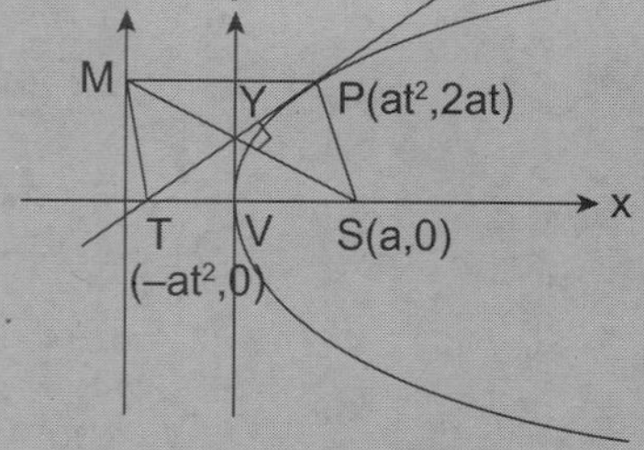

FIGURE 4.59

Also co-ordinates of *T* are $(-at^2,0) \Rightarrow$ co-ordinates of *Y* are $(0, at)$ (By mid-point formula)

$\therefore$ $TP = \sqrt{\left(2at^2\right)^2+\left(2at\right)^2} = 2at\sqrt{t^2+1}$ and $YS = \sqrt{a^2+a^2t^2} = a\sqrt{t^2+1}$

$\therefore$ From (ii), area of □ $TSPS = \left(2at\sqrt{t^2+1}\right)\left(a\sqrt{t^2+1}\right) = 2a^2t\left(t^2+1\right)$

For parabola $y^2 = 3x$; $4a = 3 \Rightarrow a = 3/4$ and at $t = 4$

area of rhombus $= 2a^2\, t(t^2+1) = 2\left(\frac{3}{4}\right)^2(4)\left((4)^2+1\right) = 153/2$ Sq.units

P: 11. Circle drawn on focal chord as diameter touches the directrix.

Proof: Equation of diametric circle is given by

$$(x-at^2)\left(x-\frac{a}{t^2}\right)+(y-2at)\left(y+\frac{2a}{t}\right)=0$$

$(\because t_1t_2 = -1$ for focal chord)

Solving with $x = -a$, we get

$$a^2(1+t^2)\left(1+\frac{1}{t^2}\right)+y^2-\left(2at-\frac{2a}{t}\right)y-4a^2=0$$

$$\Rightarrow\ y^2-2a\left(t-\frac{1}{t}\right)y+a^2\left(t^2+\frac{1}{t^2}\right)-2a^2=0$$

$$\therefore\ D = 4a^2\left(t^2+\frac{1}{t^2}-2\right)-4a^2\left(t^2+\frac{1}{t^2}\right)+8a^2=0$$

$\Rightarrow$ The circle touches the directrix $x = -a$

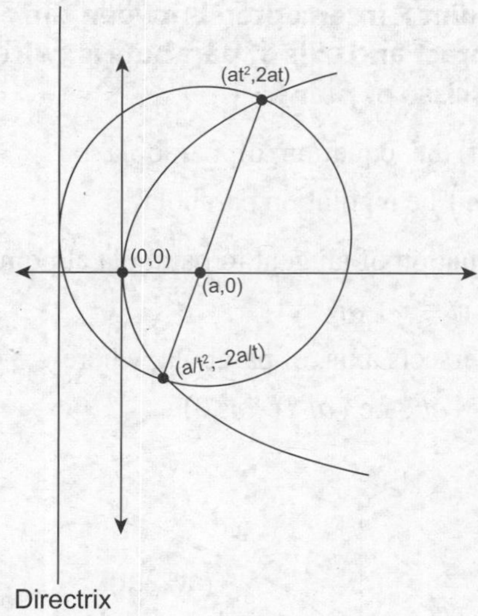

FIGURE 4.60

ILLUSTRATION 35: Find the equation(s) of circle having its centre at the point on directrix of parabola $y^2 = 8x$; where the circle drawn on focal chord through point $P(t = 4)$ as a diameter touches directrix and passing through the centre of given circle.

SOLUTION: $P(t=4) \equiv \left(at^2, 2at\right) \equiv \left(2(4)^2, 2(2)(4)\right) \equiv (32,16)$

Let $Q\left(t=\frac{-1}{4}\right)$ be the other extremity of focal chord through P

$$\therefore\quad Q\equiv\left(2\left(\frac{-1}{4}\right)^2,2(2)\left(\frac{-1}{4}\right)\right)\equiv\left(\frac{1}{8},-1\right)$$

$\therefore$ Centre of chord PQ (diameter of given circle) is

$$C\left(\frac{32+\frac{1}{8}}{2},\frac{16-1}{2}\right)\equiv\left(\frac{257}{16},\frac{15}{2}\right)$$

We know that circle drawn on the focal chord as diameter touches the directrix

$\therefore$ The required circle has its centre at $(-a,15/2)\equiv(-2,15/2)$ and radius $= TC = a + \frac{257}{16}=2+\frac{257}{16}=\frac{289}{16}$ (As required circle passes through C (given)).

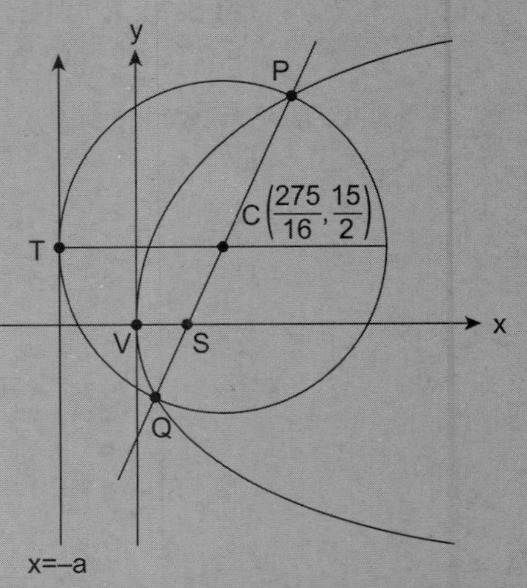

FIGURE 4.61

Thus, the equation of the required circle is $(x+2)^2+(y-15/2)^2=\left(\frac{289}{16}\right)^2$.

P: 12. The length of sub tangent to a parabola at a point P intercepted between the point of contact and axis of parabola is twice of the abscissa of point.

Proof: Let the equation of parabola be $y^2 = 4ax$. Let $P\equiv(at^2,2at)$ be a point on parabola

$\therefore$ Equation of tangent to parabola at point P is given by $yt = x + at^2$ ...(i)

(i) Intersects axis, of parabola, where $y = 0$

$\Rightarrow$ $x = -at^2$, i.e., at $T(-at^2,0)$

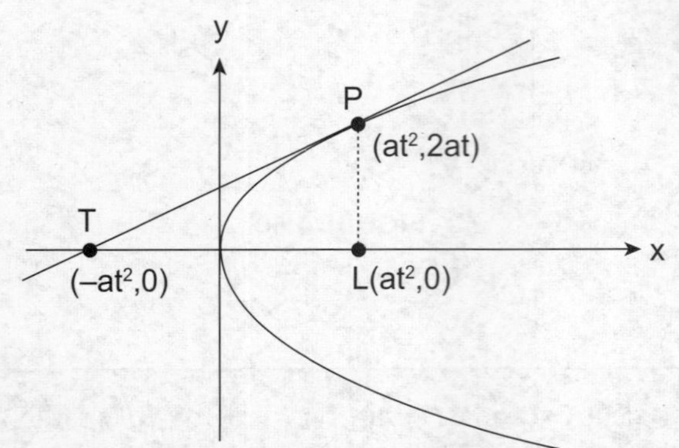

FIGURE 4.62

Clearly, length of the sub-tangent $= TL = 2at^2 = 2$ (Abscissa of point P)

P: 13. Tangents TP and TQ drawn from an external point T to a parabola subtend equal angles at focus.

Proof: Let $P(t_1)$ and $Q(t_2)$ be any two points on the parabola, such that tangents at P and Q intersect at T. Thus, co-ordinates of T are $(at_1t_2, a(t_1+t_2))$

Now equation of SP is $y=\frac{2at_1}{at_1^2-a}(x-a)$

or $y=\frac{2t_1}{t_1^2-1}(x-a)$

or $(t_1^2-1)y-2t_1x+2at_1=0$ ....(i)

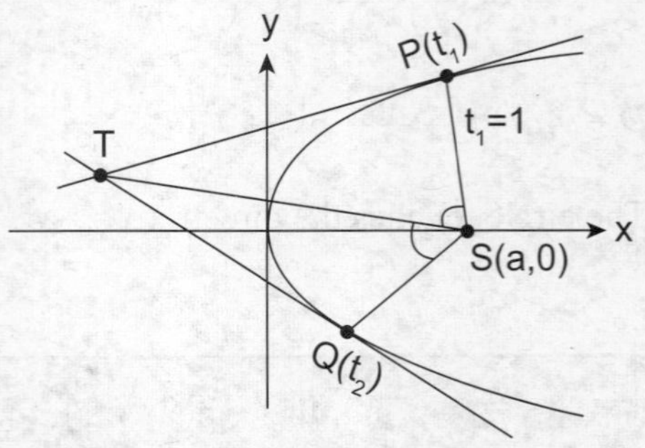

FIGURE 4.63

Similarly equation of SQ is $(t_2^2-1)\,y-2t_2x+2at_2=0$ ...(ii)

$\therefore$ Perpendicular distance of T from SP

$$= p_1 = \frac{\left|(t_1^2-1)a(t_1+t_2)-2t_1at_1t_2+2at_1\right|}{\sqrt{(t_1^2-1)^2+4t_1^2}}$$

$$= \frac{a\left|t_1^3+t_1^2t_2-t_1-t_2-2t_1^2t_2+2t_1\right|}{\sqrt{(t_1^2-1)^2+4t_1^2}}$$

$$= \frac{a\left|t_1^3-t_1^2t_2+t_1-t_2\right|}{\sqrt{(t_1^2+1)^2}} = \frac{a\left|t_1^2(t_1-t_2)+(t_1-t_2)\right|}{(t_1^2+1)}$$

$$= \frac{a\left|(t_1-t_2)(t_1^2+1)\right|}{t_1^2+1} = a\left|t_1-t_2\right|$$

Similarly, perpendicular distance of T from SQ

$$= p_2 = \frac{a\left|(t_1-t_2)(t_2^2+1)\right|}{(t_2^2+1)} = a\left|t_2-t_1\right|$$

Thus ⊥r distances of T from SP and SQ are same

$\Rightarrow$ T lies on angle bisector of $\angle PSQ$

$\Rightarrow$ $\angle TSP = \angle TSQ$

Hence, the result.

P: 14. The area of the triangle formed by three points on a parabola is twice the area of the triangle formed by the tangents at these points.

Proof: Let the three points on the parabola be $(at_1^2, 2at_1), (at_2^2, 2at_2)$, and $(at_3^2, 2at_3)$

The area of the triangle formed by these points is given by

$$A_1 = \frac{1}{2}\left|\begin{array}{l} at_1^2(2at_2-2at_3)+at_2^2 \\ (2at_3-2at_1)+at_3^2(2at_1-2at_2) \end{array}\right|$$

$$= \left|-a^2(t_2-t_3)(t_3-t_1)(t_1-t_2)\right|$$

The intersection of the tangents at these points are the points

$(at_2t_3, a(t_2+t_3)), (at_3t_1, a(t_3+t_1))$, and $(at_1t_2, a(t_1+t_2))$

The area of the triangle formed by these three points is given by

$$A_2 = \frac{1}{2}\left|\begin{array}{l} at_2t_3(at_3-at_2)+at_3t_1 \\ (at_1-at_3)+at_1t_2(at_2-at_1) \end{array}\right|$$

$$= \frac{1}{2}a^2\left|(t_2-t_3)(t_3-t_1)(t_1-t_2)\right|$$

$\therefore$ $A_1 = 2A_2$

P: 15. The orthocentre of any triangle formed by three tangents to a parabola lies on the directrix.

Proof: Let the equations to the three tangents be

$$y = m_1x + \frac{a}{m_1} \quad \text{...............(1)}$$

$$y = m_2x + \frac{a}{m_2} \quad \text{..................(2)}$$

$$y = m_3x + \frac{a}{m_3} \quad \text{...................(3)}$$

The point of intersection of (2) and (3) is found, by solving them to be

$$\left\{\frac{a}{m_2m_3}, a\left(\frac{1}{m_2}+\frac{1}{m_3}\right)\right\}$$

The equation to the straight line through this point perpendicular to (1) is

$$y - a\left(\frac{1}{m_2}+\frac{1}{m_3}\right) = -\frac{1}{m_1}\left[x - \frac{a}{m_2m_3}\right]$$

$$\text{i.e., } y + \frac{x}{m_1} = a\left[\frac{1}{m_2}+\frac{1}{m_3}+\frac{1}{m_1m_2m_3}\right] \quad \text{...................(4)}$$

Similarly, the equation to the straight line through the intersection of (3) and (1) perpendicular to (2) is

$$y + \frac{x}{m_2} = a\left(\frac{1}{m_3}+\frac{1}{m_1}+\frac{1}{m_1m_2m_3}\right) \quad \text{..................(5)}$$

and the equation to the straight line through the intersection of (1) and (2) perpendicular to (3) is

$$y + \frac{x}{m_3} = a\left(\frac{1}{m_1}+\frac{1}{m_2}+\frac{1}{m_1m_2m_3}\right) \quad \text{...............(6)}$$

The point which is common to the straight lines (4), (5) and (6)], i.e., the orthocentre of the triangle, is easily seen to be the point whose co-ordinates are

$$x = -a \;\; y = a\left(\frac{1}{m_1}+\frac{1}{m_2}+\frac{1}{m_3}+\frac{1}{m_1m_2m_3}\right),$$

and the point lies on the directrix.

■ EQUATION OF PAIR OF TANGENTS

To find the equation to the pair of tangents that can be drawn to the parabola from the point (x_1, y_1).

Let (h, k) be any point on either of the tangents drawn from (x_1, y_1). The equation to the line joining (x_1, y_1) to (h, k) is

$$y - y_1 = \frac{k - y_1}{h - x_1}(x - x_1)$$

i.e., $$y = \frac{k - y_1}{h - x_1}x + \frac{hy_1 - kx_1}{h - x_1}$$

If this be a tangent it must be of the form

$y = mx + \frac{a}{m}$ so that, $\frac{k - y_1}{h - x_1} = m$ and $\frac{hy_1 - kx_1}{h - x_1} = \frac{a}{m}$

Hence, by multiplication $a = \frac{k - y_1}{h - x_1}\frac{hy_1 - kx_1}{h - x_1}$

i.e., $a(h - x_1)^2 = (k - y_1)(hy_1 - yx_1)$

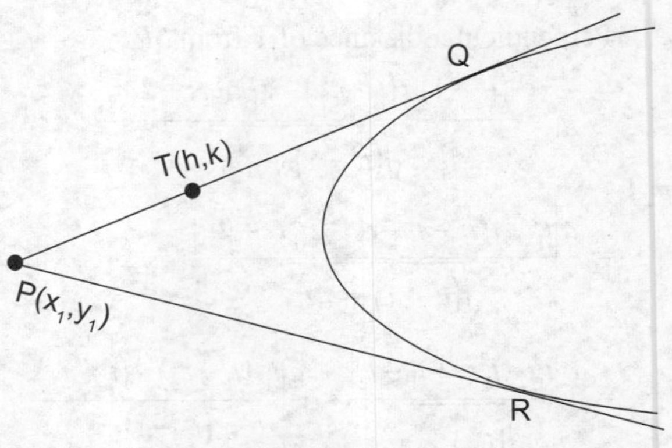

FIGURE 4.64

∴ The locus of the point (h, k), (i.e., the pair of tangents required) is therefore

$$(y^2 - 4ax)(y_1^2 - 4ax_1) = \{yy_1 - 2a(x + x_1)\}^2$$

i.e., $SS_1 = T^2$

ILLUSTRATION 36: Pair of tangents are drawn from point $P(-1, 1)$ to the parabola $y^2 = 8x$. Find their equations. Also find.

(i) Lengths of these tangents intercepted between the point $P(-1,1)$ and the line $x = 3$.

(ii) Length of line $x = 3$ intercepted by these two tangents.

SOLUTION: Equation of given parabola is $y^2 = 8x \Rightarrow a = 2$.

The equations of pair of tangents drawn from point (x_1, y_1) to parabola $y^2 = 4ax$ are given by $SS_1 = T^2$

i.e., $(y^2 - 8x)(y_1^2 - 8x_1) = (yy_1 - 4(x + x_1))^2$

$\Rightarrow (y^2 - 8x)(1 - 8(-1)) = (y - 4(x - 1))^2$

$\Rightarrow 9y^2 - 72x = [y - 4x + 4]^2$

$\Rightarrow 9y^2 - 72x = y^2 + 16x^2 + 16 - 8xy - 32x + 8y$

$\Rightarrow 8y^2 - 16x^2 - 40x + 8xy - 8y - 16 = 0$ ...(i)

which is the required equation of pair of tangents.

(i) would intersect the straight line $x = 3$

Where, $8y^2 - 144 - 120 + 24y - 8y - 16 = 0$

or $8y^2 + 16y - 280 = 0$

or $y^2 + 2y - 35 = 0$

$\Rightarrow (y + 7)(y - 5) = 0$

$\Rightarrow y = -7$ or $y = 5$

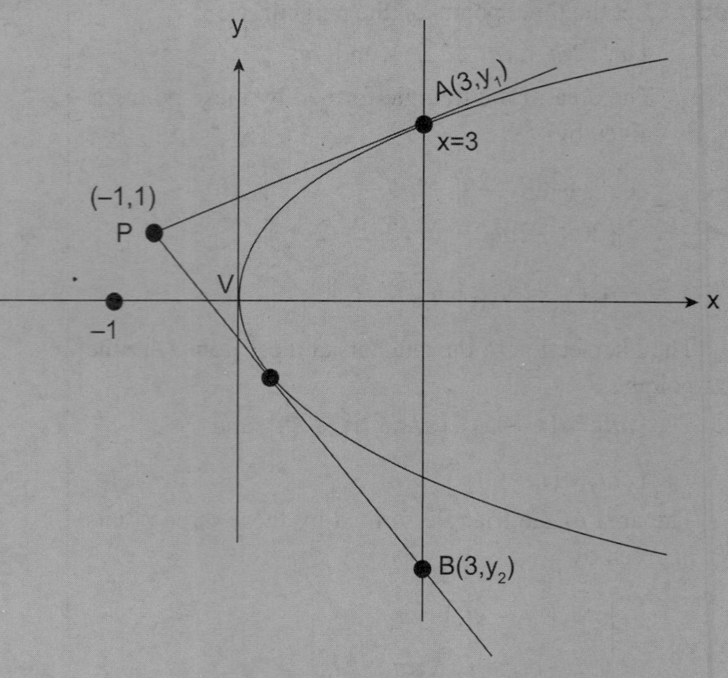

FIGURE 4.65

∴ Co-ordinates of A are (3, 5) and that of B are(3, –7).

∴ Length of tangents intercepted between the points (–1,1) and the line $x = 3$ are given by

$$PA = \sqrt{(3+1)^2+(5-1)^2} = 4\sqrt{2} \text{ and } PB = \sqrt{(3+1)^2+(-7-1)^2} = 4\sqrt{5}$$

Also, length of line $x = 3$ intercepted by pair of tangents $= AB = |y_1 - y_2| = |5-(-7)| = 12$ units.

ILLUSTRATION 37: Tangents are drawn to parabola $y^2 = 4x$ from the point (–1, 2). Find the area of triangle formed by these tangents and the line $x = 2$. Also find the centroid of that triangle.

SOLUTION : Equation of tangents drawn to parabola $y^2 = 4x$ from point $P(-1, 2)$ are given by $SS_1 = T^2$

i.e., $(y^2 - 4x)(y_1^2 - 4x_1) = (yy_1 - 2(x + x_1))^2$

$\Rightarrow (y^2 - 4x)(4 + 4) = [2y - 2(x-1)]^2$

$\Rightarrow 8y^2 - 32x = 4[y - x + 1]^2$

$\Rightarrow 2y^2 - 8x = (y^2 + x^2 + 1 - 2xy - 2x + 2y)$

$\Rightarrow y^2 - x^2 - 6x + 2xy - 2y - 1 = 0 \quad ...(i)$

Now, (i) would intersect line $x = 2$ where

$y^2 - 4 - 12 + 4y - 2y - 1 = 0$

i.e., $y^2 + 2y - 17 = 0$

$\Rightarrow y_1 + y_2 = -2;\ y_1y_2 = -17;$

where y_1 and y_2 are ordinates of points A and B respectively and $y_1 > y_2$

$\Rightarrow AB = |y_1 - y_2| = \sqrt{(y_1+y_2)^2 - 4y_1y_2}$

$= \sqrt{(-2)^2 - 4(-17)} = \sqrt{72} = 6\sqrt{2}$

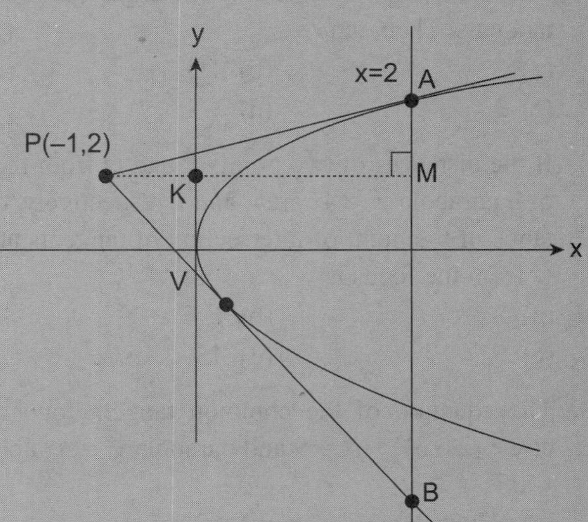

FIGURE 4.66

∴ Area of $\Delta PAB = \frac{1}{2}(PM)\times(AB) = \frac{1}{2}(PK + KM)\times(6\sqrt{2})$

$= \frac{1}{2}(1+2)\times 6\sqrt{2} = 9\sqrt{2}$ square units

Also centroid of $\Delta PAB \equiv \left(\frac{x_1+x_2+x_3}{3}, \frac{y_1+y_2+y_3}{3}\right)$

$\equiv \left(\frac{-1+2+2}{3}, \frac{2-2}{3}\right) \equiv (1,0)$. That is, focus of parabola

TEXTUAL EXERCISE–3 (SUBJECTIVE)

1. Prove that the tangents at the extremities of any focal chord intersect at right angles on the directrix.

2. A tangent to the parabola $y^2 = 8x$ makes an angle of 45° with the straight line $y = 3x + 5$. Find the equation of the tangent and its point of contact.

3. For what value of k does the line $x + y + 1 = 0$ touch the parabola $y^2 = kx$? Also find the point of contact.

4. Show that the line $x\cos\alpha + y\sin\alpha = p$ touches the parabola $y^2 = 4ax$ if $p\cos\alpha + a\sin^2\alpha = 0$ and that the point of contact is $(a\tan^2\alpha, -2a\tan\alpha)$

5. Prove that the line $lx + my + n = 0$ touches the parabola $y^2 = 4a(x - b)$ if $am^2 = bl^2 + nl$.

6. If the tangents at the points P and Q on the parabola $y^2 = 4ax$ meet at R and 'S' is its focus, prove that $SR^2 = SP.SQ$.

7. Find the common tangents of $x^2 + y^2 = 2a^2$ and $y^2 = 8ax$.

8. Prove that the locus of the point of intersection of tangents to the parabola $y^2 = 4ax$ which meet at an angle θ is $(x+a)^2\tan^2\theta = y^2 - 4ax$.

Answer Keys

2. $2x + y + 1 = 0$ at $(1/2, -2)$; $x - 2y + 8 = 0$ at $(8, 8)$ **3.** 4; $(1 - 2)$ **7.** $x \pm y + 2a = 0$

TEXTUAL EXERCISE–3 (OBJECTIVE)

1. Two tangents are drawn from the point $(-2, -1)$ to the parabola $y^2 = 4x$. If θ is the angle between these tangents. Then, $\tan\theta =$

(a) 3 (b) 1/3
(c) 2 (d) 1/2

2. If the distances of two points P and Q from the focus of a parabola $y^2 = 4x$ are 4 and 9 respectively, the distance of the point of intersection of tangents at P and Q from the focus is

(a) 8 (b) 6
(c) 5 (d) 13

3. The equation of the common tangent touching the circle $(x - 3)^2 + y^2 = 9$ and parabola $y^2 = 4x$ above the x-axis is

(a) $\sqrt{3}y = 3x + 1$ (b) $\sqrt{3}y = -(x + 3)$
(c) $\sqrt{3}y = (x + 3)$ (d) $\sqrt{3}y = -(3x + 1)$

4. The triangle formed by the tangent to the curve $f(x) = x^2 + bx - b$ at the point $(1, 1)$ and the co-ordinate axes lies in the first quadrant. If its area is 2, then the value of b is

(a) -1 (b) 3
(c) -3 (d) 1

5. The equation of the common tangent of the parabolas $x^2 = 108y$ and $y^2 = 32x$, is

(a) $2x + 3y = 35$ (b) $2x + 3y + 36 = 0$
(c) $3x + 2y = 36$ (d) $3x + 2y + 36 = 0$

6. The equation of a tangent to the parabola $y^2 = 4ax$ making an angle θ with x-axis is

(a) $y = x \cot\theta + a \tan\theta$
(b) $x = y \tan\theta + a \cot\theta$
(c) $y = x \tan\theta + a \cot\theta$
(d) None of these

7. The point of the contact of the tangent to the parabola $y^2 = 4ax$ which makes an angle of 60° with x-axis, is

(a) $\left(\frac{a}{3}, \frac{2a}{\sqrt{3}}\right)$ (b) $\left(\frac{2a}{\sqrt{3}}, \frac{a}{3}\right)$
(c) $\left(\frac{a}{\sqrt{3}}, \frac{2a}{3}\right)$ (d) None of these

8. The locus of a foot of perpendicular drawn to the tangent of parabola $y^2 = 4ax$ from focus, is

(a) $x = 0$ (b) $y = 0$
(c) $y^2 = 2a(x + a)$ (d) $x^2 + y^2 (x + a) = 0$

9. If the tangents to the parabola $y^2 = ax$ makes an angle of 45° with x-axis, then the point of contact is

(a) $\left(\frac{a}{2}, \frac{a}{2}\right)$ (b) $\left(\frac{a}{4}, \frac{a}{4}\right)$
(c) $\left(\frac{a}{2}, \frac{a}{4}\right)$ (d) $\left(\frac{a}{4}, \frac{a}{2}\right)$

10. If the tangent to the parabola $y^2 = 4ax$ meets the axis in T and tangent at the vertex A in y and the rectangle TAYG is completed, then the locus of G is equal to

(a) $y^2 + 2ax = 0$ (b) $y^2 + ax = 0$
(c) $x^2 + ay = 0$ (d) None of these

11. PT is the tangent at any point P on the parabola $y^2 = 4ax$ and SP is the focal distance and PM is the perpendicular from P to the directrix, then

(a) $\angle MPT = 2 \angle SPT$
(b) $\angle MPT = \angle SPT$
(c) $\angle MPT = \frac{1}{2} \angle SPT$
(d) None of these

12. The tangent drawn at any point P to the parabola $y^2 = 4ax$ meets the directrix at the point k, then the angle which KP subtends at its focus is

(a) 30^0 (b) 45^0
(c) 60^0 (d) 90^0

Answer Keys

1. (a) **2.** (b) **3.** (c) **4.** (c) **5.** (b) **6.** (c) **7.** (a) **8.** (a) **9.** (d) **10.** (b)
11. (b) **12.** (d)

EQUATIONS OF NORMALS IN DIFFERENT FORMS

Point Form of Normal

To find the equation of the normals to the parabola $y^2 = 4ax$ at the point (x_1, y_1).

Since the equation of the tangent to the parabola $y^2 = 4ax$ at (x_1, y_1) is

$$yy_1 = 2a(x + x_1) \qquad ....(1)$$

The slope of the tangent at $(x_1, y_1) = 2a/y_1$

Since the normal at (x_1, y_1) is perpendicular to the tangent at (x_1, y_1)

$\therefore$ Slope of normal at $(x_1, y_1) = -y_1/2a$

Hence the equation of normal at (x_1, y_1) is

$$y - y_1 = -\frac{y_1}{2a}(x - x_1)$$

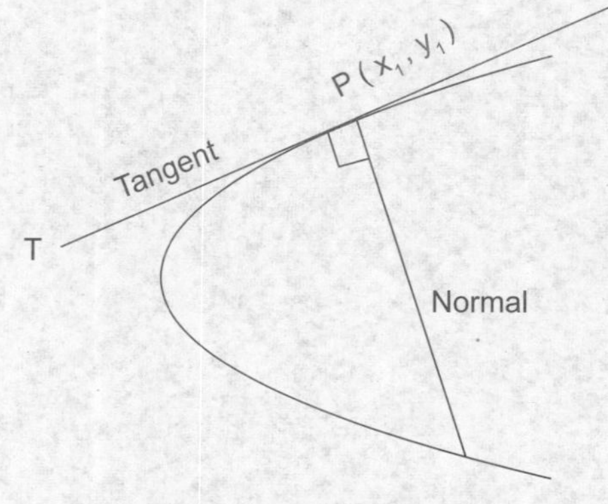

FIGURE 4.67

ILLUSTRATION 38: Show that normal to the parabola $y^2 = 4x$ at the point $P(1, 2)$ meets it again at $(9, -6)$. Find also the length of the normal chord (segment of normal intercepted by parabola).

SOLUTION: Comparing the given parabola (i.e., $y^2 = 4x$) with $y^2 = 4ax$, we have

$\therefore$ $4a = 4$

$\therefore$ $a = 1$

Since normal at (x_1, y_1) to the parabola $y^2 = 4ax$ is given by

$$y - y_1 = -\frac{y_1}{2a}(x - x_1)$$

Here, $x_1 = 1$ and $y_1 = 2$, i.e., $P \equiv (1, 2)$

$\therefore$ Equation of normal is $y - 2 = -\frac{2}{2}(x - 1)$

$\Rightarrow$ $y - 2 = -x + 1$

$\Rightarrow$ $x + y - 3 = 0 \qquad ....(1)$

Solving (1) and $y^2 = 4x$, we get $y^2 = 4(3 - y)$

$\Rightarrow$ $y^2 + 4y - 12 = 0 \Rightarrow (y + 6)(y - 2) = 0$

$\therefore$ $y = -6$ and $y = 2 \Rightarrow x = 9$ and $x = 1$

Hence point of intersection of normal and parabola are $Q(9, -6)$ and $P(1, 2)$ therefore normal meets the parabola again at $Q(9, -6)$ and length of normal chord is distance between these points $= PQ = \sqrt{(9-1)^2 + (-6-2)^2} = 8\sqrt{2}$.

ILLUSTRATION 39: Show that the normals at the points $(4a, 4a)$ and at the upper end of the latus ractum of the parabola $y^2 = 4ax$ intersect on the given parabola.

SOLUTION: Given, parabola is $y^2 = 4ax \qquad ...(1)$

Normal at (4a, 4a) to parabola (1) is given by

$y - y_1 = -\frac{y_1}{2a}(x - x_1)$; where $(x_1, y_1) \equiv (4a, 4a)$, i.e., $y - 4a = -\frac{4a}{2a}(x - 4a)$

or $y - 4a = -2x + 8a$

or $2x + y = 12a$ ...(2)

Now, upper end of latus rectum is $(a, 2a)$

and normal at (a, 2a) is given by $y - 2a = -\dfrac{2a}{2a}(x - a)$

or $y - 2a = -x + a$

or $x + y = 3a$ ...(3)

$\therefore$ intersection point of equation (2) and (3) will be $x = 9a$ and $y = -6a$

i.e., $(9a, -6a)$ which clearly satisfies the given parabola $y^2 = 4ax$, i.e., the point of intersection lies on the given parabola.

ILLUSTRATION 40: If the normal at $P(18, 12)$ to the parabola $y^2 = 8x$ cuts it again at Q, show that $9PQ = 80\sqrt{10}$.

SOLUTION: Equation of parabola is $y^2 = 8x \Rightarrow 4a = 8 \Rightarrow a = 2$

Now equation of normal to parabola at $P(x_1, y_1)$ is given by $y - y_1 = -\dfrac{y_1}{2a}(x - x_1)$

$\therefore$ normal at $P(18, 12)$ is given by $y - 12 = -\dfrac{12}{4}(x - 18)$

or $y = -3x + 66$. Solving with $y^2 = 8x$; we get $y^2 = \dfrac{8(y-66)}{-3}$

or $-3y^2 = 8y - 528$ or $3y^2 + 8y - 528 = 0$

or $3y^2 + 44y - 36y - 528 = 0$ or $y = 12, -\dfrac{44}{3}$ and $x = 18, \dfrac{242}{9}$

$\therefore$ Point Q is $\left(\dfrac{242}{9}, -\dfrac{44}{3}\right)$ and P is $(18, 12)$ $\therefore$ $PQ^2 = \left(18 - \dfrac{242}{9}\right)^2 + \left(12 + \dfrac{44}{3}\right)^2$

$\Rightarrow 9PQ = 80\sqrt{10}$.

Parametric Form of Normal

Given a parabola $y^2 = 4ax$, at point $P(t)$

Slope of normal: $m = -t$

Equation of normal: $y - 2at = -t(x - at^2)$

$\Rightarrow y + xt = 2at + at^3$ ......(i)

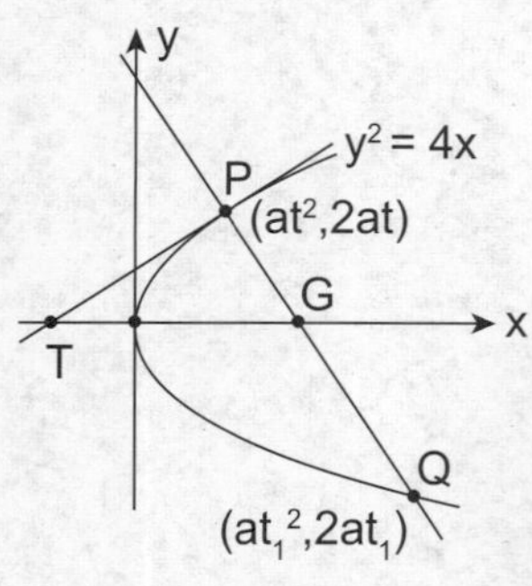

FIGURE 4.68

If (i) meets axis of parabola at G and the parabola again at $Q(t_1)$, then the

Co-ordinates of $G \equiv (2a + at^2, 0)$ and $t_1 = -t - \dfrac{2}{t}$.

Proof:

(i) meets axis of parabola at G, where $y = 0$

$\Rightarrow xt = 2at + at^3$

$\Rightarrow x = 2a + at^2$

$\therefore$ Co-ordinates of G are given by $G(2a + at^2, 0)$

$\because$ Chord PQ is normal to parabola at P

$\Rightarrow m_{PQ} = m_{\text{normal}}$ at $P \Rightarrow \dfrac{2}{t + t_1} = -t$

$\Rightarrow t + t_1 = -\dfrac{2}{t}$ $\Rightarrow t_1 = -t - \dfrac{2}{t}$

ILLUSTRATION 41: Prove that the locus of the middle point of segment of a normal $y^2 = 4ax$ intercepted between the curve and the axis is another parabola. Find the vertex and the latus rectum of the second parabola.

SOLUTION: Let point $M(h, k)$ is the mid-point of normal segment PR, where $P(at^2, 2at)$ and $R(b, 0)$, then $\Rightarrow k = \dfrac{2at+0}{2}$ and $h = \dfrac{at^2+b}{2}$.

$\Rightarrow 2k = 2at + 0$ and $2h = at^2 + b$

$\Rightarrow k = at$ and $h = (at^2 + b)/2$ ...(1)

Now, equation of normal at point $P(at^2, 2at)$ to the parabola $y^2 = 4ax$ is given by $y + tx = 2at + at^3$ .....(2)

Also, point $R(b, 0)$ satisfies this equation (2)

$\Rightarrow 0 + tb = 2at + at^3$

$\Rightarrow b = 2a + at^2$ ...(3)

$\therefore$ the co-ordinates of point $R \equiv (2a + at^2, 0)$

$\therefore$ from (1) we must have

or $h = a + at^2$ ...(4)

putting from equation (1), $t = k/a$ in equation (4)

we get, $h = a + a\dfrac{k^2}{a^2} \Rightarrow ah = a^2 + k^2 \Rightarrow k^2 = a(h - a)$

Hence, the locus of the mid-point (h, k) is $y^2 = a(x - a)$ which is a parabola with vertex at $(a, 0)$ and length of latus ractum is 'a', i.e., one-fourth of the latus rectum of the original parabola.

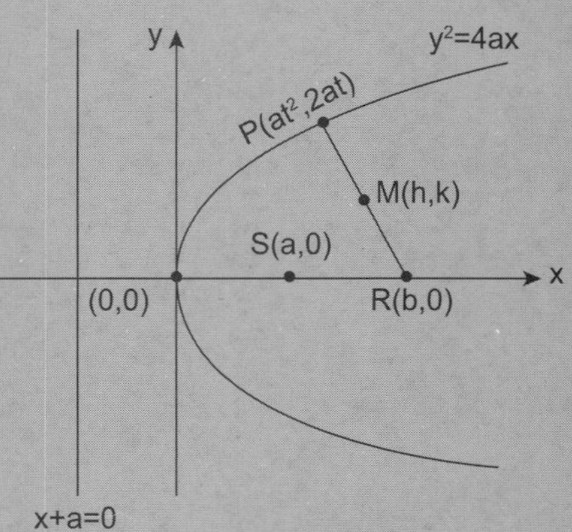

FIGURE 4.69

ILLUSTRATION 42: The tangent at any point $P(4a, y_1)$ to a parabola $y^2 = 4ax$ meets its axis at a point T and normal at P meets the curve again G, then show that $PT : PG = 4 : 5$.

SOLUTION: Let the co-ordinates of P be $(at^2, 2at)$ and that of G be $\left(at_1^2, 2at_1\right)$

Then $t_1 = -t - \dfrac{2}{t}$ ....(i)

Now $P \equiv \left(at^2, 2at\right) \equiv (4a, \pm 4a)$

$\Rightarrow t^2 = 4 \Rightarrow t = \pm 2$

Let us take $P \equiv \left(at^2, 2at\right)$ for $t = 2$

$\therefore$ From (i) $t_1 = -2 - 2/2 = -3$

$\therefore$ Co-ordinates of G are given by $(9a, -6a)$

Also, tangent at t is given by $yt = x + at^2$

i.e., $2y = x + 4a$

$\therefore$ Co-ordinates of T are given by $(-4a, 0)$

$\therefore PT = \sqrt{(4a+4a)^2 + (4a-0)^2} = 4\sqrt{5}\,a$ and

$PG = \sqrt{(4a-9a)^2 + (4a+6a)^2} = 5\sqrt{5}\,a$

$\therefore PT : PG = 4 : 5.$

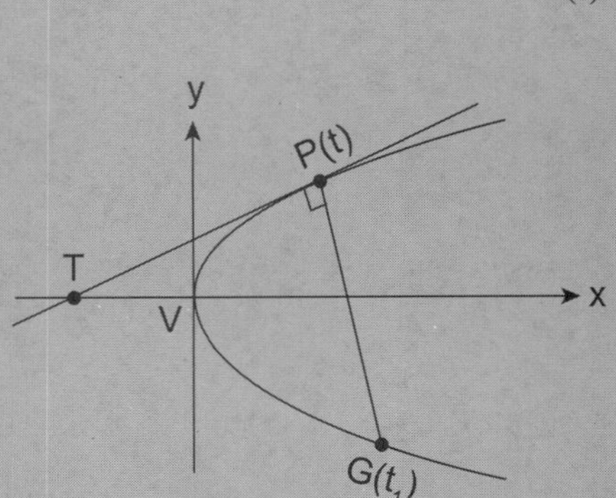

FIGURE 4.70

ILLUSTRATION 43: If a normal to a parabola $y^2 = 4ax$ at a point $P(t)$ meets it again at $Q(t')$, then prove that $t' \in \left(-\infty, -2\sqrt{2}\right] \cup \left[2\sqrt{2}, \infty\right)$

SOLUTION: We know that if the normal at a point $P(t)$ to the parabola $y^2 = 4ax$ meets it again at $Q(t')$, then $t' = -t - 2/t$, $t \in \mathbb{R} \sim \{0\}$

$\Rightarrow tt' = -t^2 - 2$

$\Rightarrow t^2 + t't + 2 = 0$ ...(i)

$t \in \mathbb{R} \sim \{0\}$, i.e., roots of (i) are non-zero real

$\Rightarrow$ Discriminant of (i) ≥ 0

$\Rightarrow t'^2 - 4(2) \geq 0 \Rightarrow t'^2 \geq 8 \Rightarrow t' \in (-\infty, -2\sqrt{2}] \cup [2\sqrt{2}, \infty)$

ILLUSTRATION 44: Find the length of normal intercepted by the parabola $y^2 = 4ax$, when the normal is subtending a right angle at the vertex of parabola.

SOLUTION: Let the co-ordinates of P be $(at^2, 2at)$ and that of G be $(at'^2, 2at')$

$$\Rightarrow t' = -t - \frac{2}{t} \quad \text{......(i)}$$

Now, slope of $AP = \dfrac{2at - 0}{at^2 - 0} = \dfrac{2}{t}$

And slope of $AG = \dfrac{2}{t'}$

But, $\angle PAG = \dfrac{\pi}{2} \Rightarrow \dfrac{4}{tt'} = -1$

$$\Rightarrow tt' = -4 \Rightarrow t' = \frac{-4}{t} \quad \text{.......(ii)}$$

Using (ii) in (i), we get $\dfrac{-4}{t} = -t - \dfrac{2}{t}$

$\Rightarrow \dfrac{-2}{t} = -t \Rightarrow t^2 = 2 \Rightarrow t = \pm\sqrt{2}$

without loss of generality let $t = \sqrt{2}$, $t' = -2\sqrt{2}$

$\therefore \quad P \equiv \left(2a, 2\sqrt{2}a\right)$ and $G \equiv \left(8a, -4\sqrt{2}a\right)$

$$\Rightarrow PG = \sqrt{(8a-2a)^2 + \left(-4\sqrt{2}a - 2\sqrt{2}a\right)^2} = \sqrt{36a^2 + 72a^2} = \sqrt{108a^2} = 6\sqrt{3}\,a \text{ units}$$

FIGURE 4.71

ILLUSTRATION 45: A quadrilateral is inscribed in a parabola $y^2 = 4ax$ and three of its sides pass through fixed points on the axis. Show that the fourth side also passes through fixed point on the axis of the parabola.

SOLUTION: Let $PQRS$ be a quadrilateral inscribed in parabola $y^2 = 4ax$ and sides RS, QP and PQ pass through fixed points $A(a_1, 0)$, $B(a_2, 0)$ and $D(a_4, 0)$ on the axis of parabola respectively. We have to prove that fourth side PS also passes through a fixed point on the axis of parabola i.e., point $C(a_3, 0)$. Equation of chord PQ is given by $y - 2at_1 = \dfrac{2at_1 - 2at_2}{at_1^2 - at_2^2}(x - at_1^2)$

$\Rightarrow y - 2at_1 = \dfrac{2}{t_1 + t_2}(x - at_1^2)$

$\Rightarrow y(t_1 + t_2) - 2at_1(t_1 + t_2) = 2(x - at_1^2)$

$\Rightarrow y(t_1 + t_2) - 2at_1^2 - 2at_1t_2 = 2(x - at_1^2)$

$\Rightarrow y(t_1 + t_2) = 2at_1^2 + 2at_1t_2 + 2x - 2at_1^2$

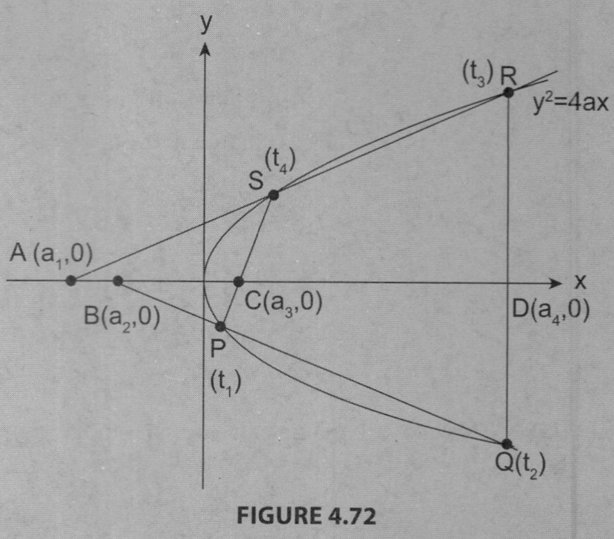

FIGURE 4.72

$\Rightarrow\ y(t_1 + t_2) = 2x + 2at_1t_2$

It passes through points $B(a_2, 0)$

$\Rightarrow\ 0 = 2a_2 + 2at_1t_2 \ \Rightarrow at_1t_2 + a_2 = 0$

$\Rightarrow\ t_1t_2 = -a_2/a$; similarly, $t_2t_3 = -a_4/a$; $t_3t_4 = -a_1/a$ ...(i)

Equation of chord PS is given by $y(t_4 + t_1) = 2x + 2a\,t_4t_1$

we have to prove that above line (iv) passes through fixed point $C(a_3, 0)$, i.e., $0 = 2a_3 + 2at_4t_1$

$\Rightarrow\ at_4t_1 + a_3 = 0 \Rightarrow t_4t_1 = -a_3/a$......... ...(ii)

Now, from equation (i), we have $\frac{(t_1t_2).(t_3t_4)}{(t_2t_3)} = \frac{\left(-\frac{a_2}{a}\right)\left(-\frac{a_1}{a}\right)}{\left(-\frac{a_4}{a}\right)} \Rightarrow t_1t_4 = -\frac{a_2a_1}{aa_4}$

Comparing equation (i) and (ii), we have $-\frac{}{a} = -\frac{a_2a_1}{a\,a} \Rightarrow a_3 = \frac{a_2a_1}{a_4}$

since a_2, a_1, a_4 are fixed so $\frac{a_2a_1}{a_4}$ is also fixed i.e., a_3 is fixed.

Hence, fourth side PS also passes through a fixed point $(a_3, 0) \equiv \left(\frac{a_1a_2}{a_4}, 0\right)$.

ILLUSTRATION 46: Find the locus of the points of intersection of tangents drawn at the ends of all chords normal to the parabola $y^2 = 8(x - 1)$.

SOLUTION: The given parabola is $y^2 = 8(x - 1)$ ...(1)

Let, $Y = y$ and $X = x - 1$. ...(2)

Then equation (1) becomes, $Y^2 = 8X$ ....(3)

Since if normal at 't_1' meets the parabola again at 't_2' then $t_2 = -t_1 - \frac{2}{t_1}$ ...(3)

(by property of normals), therefore the co-ordinates of the intersection point R is given by $R(h, k) = R(at_1t_2, a(t_1 + t_2))$ and here $a = 2$

$$\Rightarrow\ h = 2t_1t_2 = 2t_1\left(-t_1 - \frac{2}{t_1}\right) = -2t_1^2 - 4 \text{ or } h = -2t_1^2 - 4 \quad ...(4)$$

and $k = a(t_1 + t_2) = 2\left(t_1 - t_1 - \frac{2}{t_1}\right) = -\frac{4}{t_1} \Rightarrow t_1 = -\frac{4}{k}$

putting value of t_1 in equation (4), we get $h = -2\left(\frac{-4}{k}\right)^2 - 4$

Replace h by X and k by Y we have $X = -2\left(\frac{16}{Y^2}\right) - 4$

$\Rightarrow\ x - 1 = -\frac{32}{y^2} - 4$

($\because X = x - 1$ and $Y = y$)

$\Rightarrow\ xy^2 + 3y^2 + 32 = 0$

or $(x + 3)y^2 + 32 = 0$

which is the required locus of point of intersection.

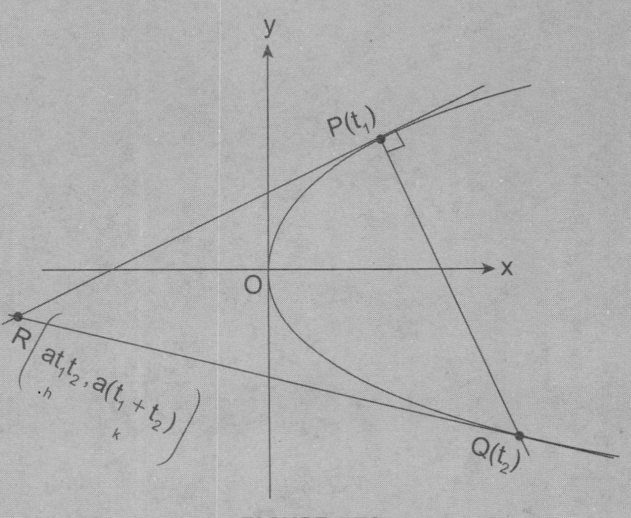

FIGURE 4.73

Slope Form of Normal

Since equation of normal: $y + xt = 2at + at^3$ at $(at^2, 2at)$ and slope of normal $= -t$

$\therefore$ Put $t = -m \Rightarrow y = mx - 2am - am^3$, where foot of normal is $(am^2, -2am)$

ILLUSTRATION 47: If $y = x + 4$ is normal to parabola $y^2 = -4ax$, then find the value of a

SOLUTION: Equation of line is $y = x + 4$ ...(1)

Equation of normal to parabola $y^2 = -4ax$ is

Given by $y = mx + 2am + am^3$ ...(2)

$\Rightarrow$ (1) and (2) are identical

$\Rightarrow m = 1;\ 2am + am^3 = 4$ $\quad\Rightarrow 2a + a = 4$

$\Rightarrow 3a = 4$ $\quad\Rightarrow a = 4/3$

ILLUSTRATION 48: Find the equation of normals to parabola $y^2 = 4x$ along the minimum distance between the parabola and the circle $x^2 + y^2 - 6x + 8 = 0$.

SOLUTION: Equation of parabola $y^2 = 4x$ ...(i)

and equation of circle is $x^2 + y^2 - 6x + 8 = 0$ ...(ii)

having its centre at (3, 0) and radius = 1

Shortest distance between the two curves are along the common normal. Equation of parabola $y^2 = 4x$. Slope of normal is $-t = \frac{-y}{2} \Rightarrow y = 2t$

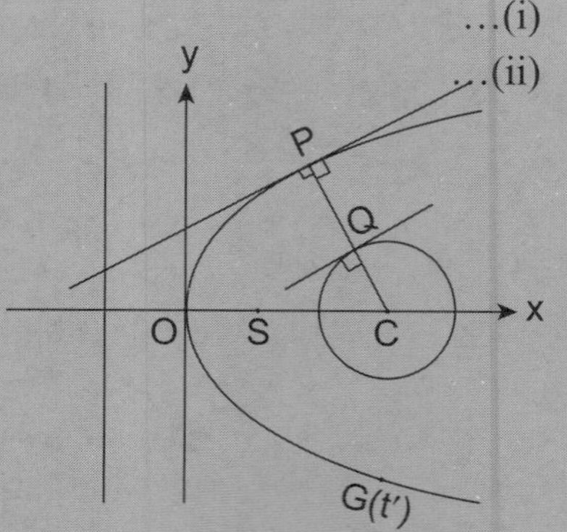

FIGURE 4.74

Let $P\ (x_1, y_1)$ be the point on parabola through which the minimum distance line passes.

Therefore, slope of shortest distance line $= -\frac{y_1}{2}$

Now, $P(x_1, y_1) \equiv P\left(\frac{y_1^2}{4}, y_1\right)$ and shortest distance line passes through the centre of circle $C(3,0)$

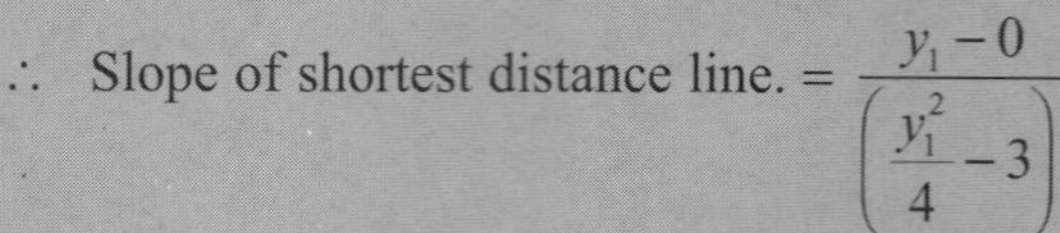

$\therefore$ Slope of shortest distance line. $= \frac{y_1 - 0}{\left(\frac{y_1^2}{4} - 3\right)}$

$\therefore \quad \frac{-y_1}{2} = \frac{4y_1}{y_1^2 - 12}$

$\Rightarrow -y_1^3 + 12y_1 = 8y_1 \Rightarrow -y_1^3 = -4y_1 \Rightarrow y_1^3 - 4y_1 = 0$

$\Rightarrow y_1 = 0$ or $y_1 = \pm 2$

$\therefore$ $P(x_1, y_1)$ can have co-ordinate (0, 0); (1, 2) or (1, –2)

Shortest distance $= \min\{PC - 1\} = \min\left\{2, 2\sqrt{2}-1, 2\sqrt{2}-1\right\} = 2\sqrt{2}-1$

$\therefore$ equation of normal along the shortest distance has slopes $= -\frac{y_1}{2} = -\left(\pm\frac{2}{2}\right) = \pm 1$

$\therefore$ equation of normals along shortest distance are given by $y = mx - 2am - am^3$ i.e., $y = x - 3$ and $y = -x + 3$

The following table represents the equations of normal and related terms to standard parabolas indifferent forms.

Equation of Parabola	$y^2 = 4ax$	$y^2 = -4ax$	$x^2 = 4ay$	$x^2 = -4ay$
Equation of normal in point form	$y - y_1 = \frac{-y_1}{2a}(x - x_1)$	$y - y_1 = \frac{y_1}{2a}(x - x_1)$	$y - y_1 = \frac{-2a}{x_1}(x - x_1)$	$y - y_1 = \frac{2a}{x_1}(x - x_1)$
Parametric co-ordinate	$(at^2, 2at)$	$(-at^2, 2at)$	$(2at, at^2)$	$(2at, -at^2)$
Normal in parametric form	$y + tx = 2at + at^3$	$y - tx = 2at + at^3$	$x + ty = 2at + at^3$	$x - ty = 2at + at^3$
Point of contact in terms of slope (m)	$(am^2, -2am)$	$(-am^2, 2am)$	$\left(-\frac{2a}{m}, \frac{a}{m^2}\right)$	$\left(\frac{2a}{m}, \frac{a}{m^2}\right)$
Condition of normality	$c = -2am - am^3$	$c = 2am + am^3$	$c = 2a + \frac{a}{m^2}$	$c = -2a - \frac{a}{m^2}$
Normal in slope form	$y = mx - 2am - am^3$	$y = mx + 2am + am^3$	$y = mx + 2a + \frac{a}{m^2}$	$y = mx - 2a - \frac{a}{m^2}$

The following table represents the equations of normals and related terms to parabolas having their vertex at (h, k) and axis parallel to co-ordinates axis.

Equation of Parabola	$(y - k)^2 = 4a(x - h)$	$(y - k)^2 = -4a(x - h)$	$(x - h)^2 = 4a(y - k)$	$(x - h)^2 = -4a(y - k)$
Equation of normal in point form	$y - y_1 = \frac{-(y_1 - k)}{2a}(x - x_1)$	$y - y_1 = \frac{(y_1 - k)}{2a}(x - x_1)$	$y - y_1 = \frac{-2a}{x_1 - h}(x - x_1)$	$y - y_1 = \frac{2a}{x_1 - h}(x - x_1)$
Parametric co-ordinate	$(at^2, 2at)$	$(-at^2, 2at)$	$(2at, at^2)$	$(2at, -at^2)$
Normal in parametric form	$(y - k) + t(x - h) = 2at + at^3$	$(y - k) - t(x - h) = 2at + at^3$	$(x - h) + t(y - k) = 2at + at^3$	$(x - h) - t(y - k) = 2at + at^3$
Point of contact in terms of slope (m)	$(h + am^2, k - 2am)$	$(h - am^2, k + 2am)$	$\left(h - \frac{2a}{m}, k + \frac{a}{m^2}\right)$	$\left(h + \frac{2a}{m}, k - \frac{a}{m^2}\right)$
Condition of normality	$c = k - mh - 2am - am^3$	$c = k - mh + 2am + am^3$	$c = k - mh + 2a + \frac{a}{m^2}$	$c = k - mh - 2a - \frac{a}{m^2}$
Normal in slope form	$(y - k) = m(x - h) - 2am - am^3$	$(y - k) = m(x - h) + 2am + am^3$	$(y - k) = m(x - h) + 2a + \frac{a}{m^2}$	$(y - k) = m(x - h) - 2a - \frac{a}{m^2}$

■ CO-NORMAL POINTS

From any point $P(h, k)$ in the plane of the parabola three normals can be drawn to the parabola. The foot of these normals are called co-normal points of the parabola.

Proof: If the normal $y = mx - 2am - am^3$ passes through (h, k), then

$\Rightarrow$ $k = mh - 2am - am^3$

$\Rightarrow$ $am^3 + (2a - h)m + k = 0$ (which is a cubic in slope m)

$\Rightarrow$ 3 slopes of normal are possible if it passes through (h, k)

ILLUSTRATION 49: Three normals to $y^2 = 4x$ pass through the point (15, 12). Find the equations of these normals..

SOLUTION: Given parabola is $y^2 = 4x$

$\Rightarrow a = 1$

Any normal to the parabola is given by $y = mx - 2am - am^3$

$a = 1$

$\Rightarrow y = mx - 2m - m^3$ ...(1)

If it passes through the point (15, 12), then $12 = 15m - 2m - m^3$

$\Rightarrow m^3 - 13m + 12 = 0$ ...(2)

Above equation (2) being a cubic in 'm' gives us three values of 'm' showing that there will be three normals to the parabola through the point (15, 12).

Clearly, $m = 1$ satisfies the equation (2) and hence it can be written as

$(m - 1)(m^2 + m - 12) = 0$ or $(m - 1)(m + 4)(m - 3) = 0$

$\Rightarrow m = 1, -4, 3$ and $a = 1$

putting these values in equations (1), we get $y = x - 3$

$y = -4x + 72;$

and $y = 3x - 33$; are the equation of three normals.

ILLUSTRATION 50: If $A(x_1, y_1)$, $B(x_2, y_2)$ and $C(x_3, y_3)$ be three points on the parabola $y^2 = 4ax$ and the normals at these points meet in a point, then prove that $\frac{x_1 - x_2}{y_3} + \frac{x_2 - x_3}{y_1} + \frac{x_3 - x_1}{y_2} = 0$.

SOLUTION: Equation of normal to parabola $y^2 = 4ax$ is given by $y = mx - 2am - am^3$.

If it passes through (h, k), then $am^3 + (2a - h)m + k = 0$

$\Rightarrow -at^3 - (2a - h)t + k = 0$ ($m = -t$)

$\Rightarrow at^3 + (2a - h)t - k = 0$ ...(1)

If t_1, t_2, t_3 are its three roots, then

$t_1 + t_2 + t_3 = 0,$ ...(2)

$t_1t_2 + t_2t_3 + t_3t_1 = \frac{2a - h}{a}$ ...(3)

and $t_1t_2t_3 = -\frac{k}{a}$ ...(4)

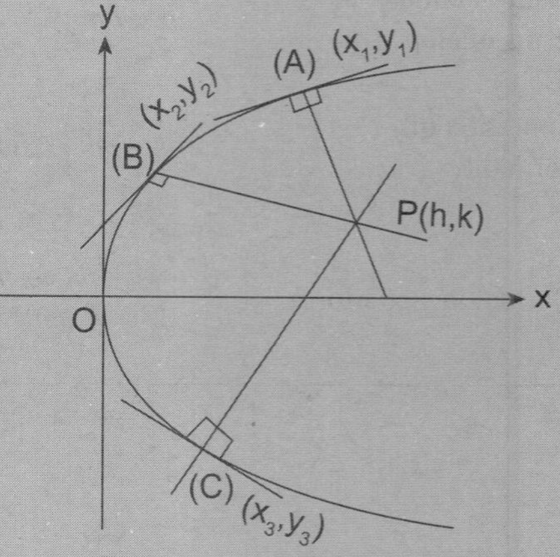

FIGURE 4.75

we have to prove that $\frac{x_1 - x_2}{y_3} + \frac{x_2 - x_3}{y_1} + \frac{x_3 - x_1}{y_2} = 0$

i.e., $\frac{at_1^2 - at_2^2}{2at_3} + \frac{at_2^2 - at_3^2}{2at_1} + \frac{at_3^2 - at_1^2}{2at_2} = 0$

or $\frac{t_1^2 - t_2^2}{t_3} + \frac{t_2^2 - t_3^2}{t_1} + \frac{t_3^2 - t_1^2}{t_2} = 0$

or $t_1t_2(t_1^2 - t_2^2) + t_2t_3(t_2^2 - t_3^2) + t_1t_3(t_3^2 - t_1^2) = 0$ ...(5)

Now $t_1t_2(t_1^2 - t_2^2) + t_2t_3(t_2^2 - t_3^2) + t_1t_3(t_3^2 - t_1^2)$

$= t_1t_2(t_1 - t_2)(-t_3) + t_2t_3(t_2 - t_3)(-t_1) + t_1t_3(t_3 - t_1)(-t_2)$ (using equation (2))

$= -t_1t_2t_3(t_1 - t_2 + t_2 - t_3 + t_3 - t_1) = 0$. Hence the result.

PROPERTIES OF NORMAL AND CO-NORMAL POINTS

P: 1. If normal at point P(t_1) meets the parabola $y^2 = 4ax$ again at Q(t_2), then $t_2 = -t_1 - \frac{2}{t_1}$

Proof: Equation of normal at $P(t_1)$ is given by $y + xt_1 = 2at_1 + at_1^3$ ....(i)

If it meets curve at $(at_2^2, 2at_2)$

Then $2at_2 + (at_2^2)t_1 = 2at_1 + at_1^3$

$\Rightarrow\ 2at_2 + at_1t_2^2 = 2at_1 + at_1^3$

$\Rightarrow\ 2a(t_2 - t_1) = at_1(t_1^2 - t_2^2) \Rightarrow -2 = t_1(t_1 + t_2)$

$\Rightarrow\ \frac{-2}{t_1} = t_1 + t_2$

$\Rightarrow\ t_2 = -t_1 \frac{-2}{t_1}$ which is real $\forall\, t_1 \in \mathbb{R} \sim \{0\}$

From above, clearly normal at vertex (0, 0) would never intersect the parabola again.

Thus all normals except for that at vertex would intersect the parabola again.

P: 2. If normal to parabola $y^2 = 4ax$ at point P(t_1) and Q(t_2) cuts the parabola at some point R (t_3), then (i) $t_1t_2 = 2$ (ii) $t_3 = -(t_1 + t_2)$

Proof: PR is normal to parabola $y^2 = 4ax$

$\Rightarrow\ t_3 = -t_1 - \frac{2}{t_1}$ ....(i)

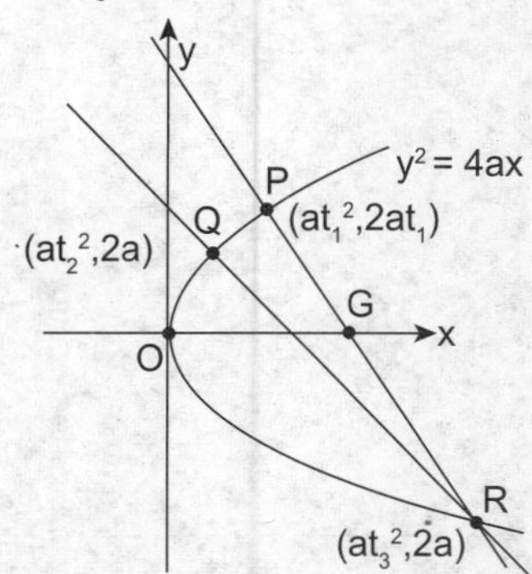

FIGURE 4.76

Also QR is normal to parabola $y^2 = 4ax$

$\Rightarrow t_3 = -t_2 - \frac{2}{t_2}$ ......(ii)

From (i) and (ii) we have $-t_1 - \frac{2}{t_1} = -t_2 - \frac{2}{t_2}$

$\Rightarrow\ t_1 - t_2 = \frac{2}{t_2} - \frac{2}{t_1} \quad \Rightarrow \quad (t_1 - t_2) = 2\left(\frac{t_1 - t_2}{t_1t_2}\right)$

$\Rightarrow\ t_1t_2 = 2\ (\because t_1 \neq t_2)$

Also from (i) $t_3 = -t_1 - \frac{2}{t_1} = -t_1 - 2\left(\frac{t_2}{2}\right) = -(t_1 + t_2)$

$\therefore\ t_3 = -(t_1 + t_2)$

P: 3. Point of intersection of normals at P(t_1) and Q(t_2) to parabola $y^2 = 4ax$ is given by $\left[2a + a(t_1^2 + t_2^2 + t_1t_2), -at_1t_2(t_1 + t_2)\right]$

Proof: Normals at $P(t_1)$ and $Q(t_2)$ are given by

$y + xt_1 = 2at_1 + at_1^3$ ...(i)

and $y + xt_2 = 2at_2 + at_2^3$ ...(ii)

Solving (i) and (ii), we get

$-xt_1 + 2at_1 + at_1^3 = -xt_2 + 2at_2 + at_2^3$

$\Rightarrow\ x(t_1 - t_2) = 2a(t_1 - t_2) + a(t_1^3 - t_2^3)$

$\Rightarrow\ x = 2a + a\left[t_1^2 + t_2^2 + t_1t_2\right]$

and $y = -xt_1 + 2at_1 + at_1^3$

$= -\left[2a + a(t_1^2 + t_2^2 + t_1t_2)\right]t_1 + 2at_1 + at_1^3$

$= -2at_1 - at_1^3 - at_1t_2^2 - at_1^2t_2 + 2at_1 + at_1^3$

$= -at_1t_2(t_1 + t_2)$

Thus normals (i) and (ii) intersect at

$\left[2a + a(t_1^2 + t_2^2 + t_1t_2), -at_1t_2(t_1 + t_2)\right]$

P: 5. The normal chord to a parabola $y^2 = 4ax$ at the point whose ordinate is equal to abscissa subtends a right angle at the focus

Proof: Let the normal at $P(at_1^2, 2at_1)$ meet the curve at $Q(at_2^2, 2at_2)$

$\therefore$ PQ is a normal chord.

and $t_2 = -t_1 - \frac{2}{t_1}$ ....(1)

By given condition $2at_1 = at_1^2$ .....(2)

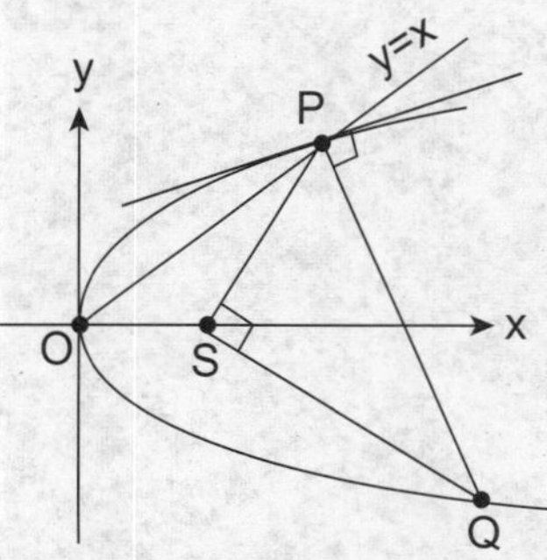

FIGURE 4.78

$\therefore$ $t_1 = 2$. From equation (1), $t_2 = -3$, then $P(4a, 4a)$ and $Q(9a, -6a)$ but focus $S(a, 0)$

$\therefore$ Slope of $SP = \frac{4a - 0}{4a - a} = \frac{4a}{3a} = \frac{4}{3}$

and slope of $SQ = -\frac{6a}{8a} = -\frac{3}{4}$

$\therefore$ Slope of $SP \times$ Slope of $SQ = \frac{4}{3} \times -\frac{3}{4} = -1$

$\Rightarrow \angle PSQ = \pi/2$

i.e., PQ subtends a right angle at the focus S.

P: 6. Chord joining points P(t_1) and Q(t_2) of parabola $y^2 = 4ax$ subtends a right angle at the vertex of parabola if $t_1t_2 = -4$.

Proof: The co-ordinates of point P are given by $(at_1^2, 2at_1)$ and that of Q are given by $(at_2^2, 2at_2)$.

Therefore slope of $OP = m_1 = 2/t_1$ and slope of $OQ = m_2 = 2/t_2$

Now, if chord PQ subtends a right angle at the vertex, then $m_1m_2 = -1$

$$\Rightarrow \frac{2}{t_1}.\frac{2}{t_2} = -1 \Rightarrow t_1t_2 = -4$$

P: 7. A normal chord through a point P(t) on parabola $y^2 = 4ax$ subtends a right angle at vertex iff $t^2 = 2$

Proof Let the normal at $P(t)$ to parabola $y^2 = 4ax$ meets the parabola again at $Q(t')$, then $t' = -t - \frac{2}{t}$ ...(i)

Now slope of $OP = \frac{2at - 0}{at^2 - 0} = \frac{2}{t}$

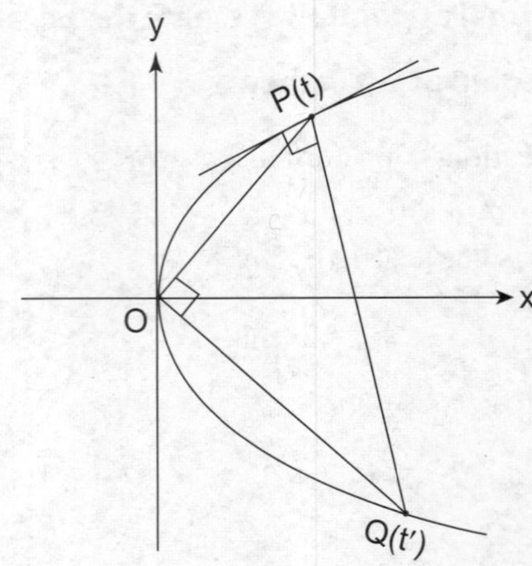

FIGURE 4.79

And slope of $OQ = \frac{2}{t'}$

$$OP \perp OQ \Rightarrow \frac{2}{t}.\frac{2}{t'} = -1$$

$\Rightarrow tt' = -4$

$$\Rightarrow t\left(-t - \frac{2}{t}\right) = -4 \Rightarrow \left(t^2 + 2\right) = 4 \Rightarrow t^2 = 2$$

ILLUSTRATION 51: Through the vertex O of a parabola $y^2 = 4x$, chords OP and OQ are drawn at right angles to each other. Show that for all positions of P, PQ cuts the axis of the parabola at a fixed point. Also find the locus of the middle point of PQ.

SOLUTION: Let $P(t_1)$ and $Q(t_2)$ be the points on parabola | $y^2 = 4ax$ such that $\angle POQ = \pi/2 \Rightarrow t_1t_2 = -4$.

Also, equation of chord PQ is

$$(y - 2t_1) = \frac{2}{t_1 + t_2}(x - t_1^2).$$

On x-axis, $-2t_1 = \frac{2}{t_1 + t_2}(x - t_1^2)$

$\Rightarrow -t_1t_2 = x \Rightarrow x = 4$

$\therefore$ PQ intersects x-axis at (4, 0) which is independent of t_1 i.e., independent of choice of point P.

Now, if (h, k) is the mid-point of PQ, then $2h = (t_1^2 + t_2^2)$ and $2k = 2(t_1 + t_2)$

$\Rightarrow 2h = [(t_1 + t_2)^2 - 2\,t_1t_2]$

$\Rightarrow 2x = \left[y^2 - 2t_1t_2\right] \Rightarrow 2x = y^2 + 8 \Rightarrow y^2 = 2(x - 4)$

($\because$ $t_1t_2 = -4$ as $\angle POQ = \pi/2$)

Which is the required locus of middle point of PQ.

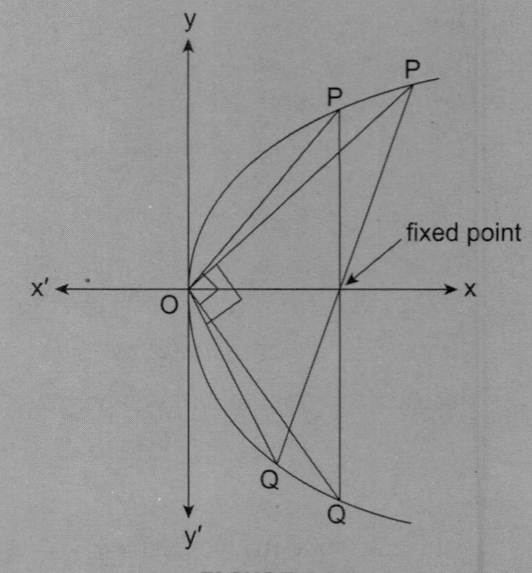

FIGURE 4.80

P: 8. No normal other than axis can pass through the focus of parabola.

Proof: Let the equation of normal to parabola $y^2 = 4ax$ be $y + xt = 2at + at^3$ ...(i)

If (i) passes through the focus (a, 0), then $0 + at = 2at + at^3$

$\Rightarrow$ $at^3 + at = 0$ $\Rightarrow$ $at(t^2 + 1) = 0$

$\Rightarrow$ $t = 0$ or $t^2 + 1 = 0$ (impossible)

$\therefore$ $t = 0$ is the only possibility

Thus axis is the only normal to parabola passing through focus.

P: 9. If *S* be focus of parabola and tangent and normal at any point *P* meet axis of parabola at *T* and *N* respectively, then *ST* = *SN* = *SP*

Proof: Equation of tangent is $yt = x + at^2$ and that of normal is $y = -xt + 2at + at^3$

Since tangent and normal meet *x*-axis (axis of parabola) at *T* and *N*

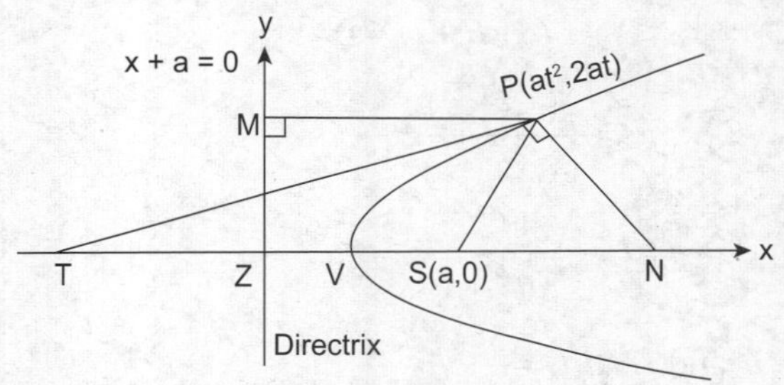

FIGURE 4.81

$\Rightarrow$ $T \equiv (-at^2, 0)$; $N \equiv (2a + at^2, 0)$
and $SP = PM = a(1 + t^2)$
$SN = VN - VS = a(1 + t^2)$
$ST = VS + VT = a + at^2 \Rightarrow$ clearly $SP = SN = ST$

P: 10. (Reflection property of parabola):

Tangent *PT* and normal *PN* are internal and external bisector of angle $\angle SPM$

i.e., $\angle QPN = \angle SPN$ and $\angle SPT = \angle MPT$

i.e., any ray of light coming parallel to axis of parabola after reflection through parabola passes through focus and conversely. i.e., ray after passing through focus becomes parallel to axis of parabola after reflecting through parabola.

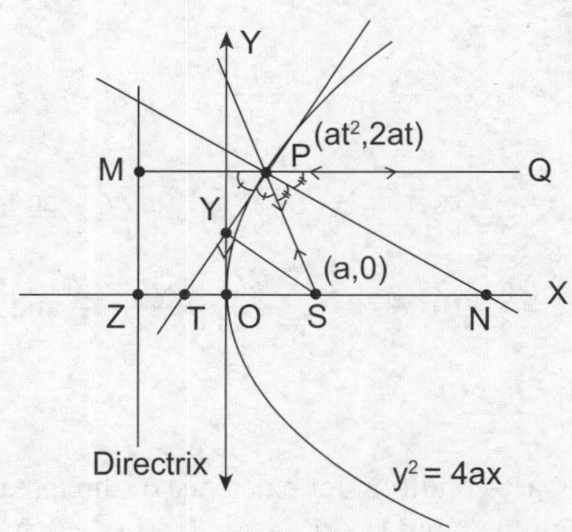

FIGURE 4.82

Proof: We already proved $ST = SP = SN$

$\Rightarrow$ $\angle SPT = \angle STP$ and $\angle SNP = \angle SPN$ (Angles opposite to equal sides of Δ are equal)

$\Rightarrow$ $\angle SPT = \angle MPT$ ($\because \angle STP = \angle MPT$, alternate angles)

and $\angle SPN = \angle QPN$ ($\because \angle SNP = \angle QPN$, alternate angle)

Hence the result.

REMARKS

1. *Light rays emerging from focus after reflection become parallel to the axis of parabolic mirror and all light rays coming parallel to axis of parabola converge at focus.*
2. *If Y is the foot of perpendicular from focus upon any tangent, then SY is altitude of ΔSPT which is isosceles, with ST = SP, hence SY is also median of Δ SPT. Thus PY = YT.*

P: 11. (Properties of co-normal points)

(i) Sum of ordinates of the feet of co-normal points vanishes.

(ii) Sum of the slopes of the normals drawn from a given point to a parabola is zero.

Proof: If the normal $y = mx - 2am - am^3$ passes through (h, k), then

$\Rightarrow$ $k = mh - 2am - am^3$

$\Rightarrow$ $am^3 + (2a - h)m + k = 0$ (which is a cubic in slope m)

$\Rightarrow$ 3 slopes of normal are possible if it passes through (h, k)
given as m_1, m_2, m_3 as the roots of above cubic equation.

$\Rightarrow$ $m_1 + m_2 + m_3 = 0$ ....(ii)

and $m_1m_2 + m_2m_3 + m_3m_1 = \frac{(2a-h)}{a}$ ....(iii)

and $m_1m_2m_3 = -\frac{k}{a}$ ....(iv)

⇒ Sum of ordinates of foot of conormal points $y_P + y_Q + y_R = -2a\,(m_1 + m_2 + m_3) = 0$.

(iii) Centroid of the Δ joining the co-normal point P, Q, R lies on axis of parabola.

Proof: $x_G = a\left(\frac{m_1^2 + m_2^2 + m_3^2}{3}\right)$ and

$$y_G = -2a\left(\frac{m_1 + m_2 + m_3}{3}\right) = 0$$

$$\Rightarrow x_G = \frac{a}{3}\left[(m_1 + m_2 + m_3)^2 - 2\sum m_1m_2\right] \text{ and } y_G = 0$$

$$\Rightarrow x_G = \frac{a}{3}\left[(0)^2 - 2\left(\frac{2a-h}{a}\right)\right] = \frac{2h-4a}{3} \text{ and } y_G = 0$$

$$\Rightarrow G\left(\frac{2h-4a}{3}, 0\right)$$

(iv) Necessary condition for existence of three real normal through the point (h, k), is $h > 2a$ if $a > 0$ and $h < 2a$ if $a < 0$.

(v) The converse of part (iv) is not true, i.e., if $h > 2a$ if $a > 0$ and $h < 2a$ if $a < 0$ does not necessarily implies that the three normals are real.

(vi) Sufficient condition for 3 real normals from (h, k) is $27ak^2 < 4(h - 2a)^3$:

Proof:

$f(m) = am^3 + (2a - h)m + k$, it has 3 real and distinct roots, then

$f'(m) = 3am^2 + 2a - h = 0$ has 2 real and distinct roots

i.e., $m = \pm\sqrt{\frac{h-2a}{3a}}$ (say) α, β are real roots

∴ Sufficient condition for 3 real slopes is $f(\alpha)\,.\,f(\beta) < 0$.

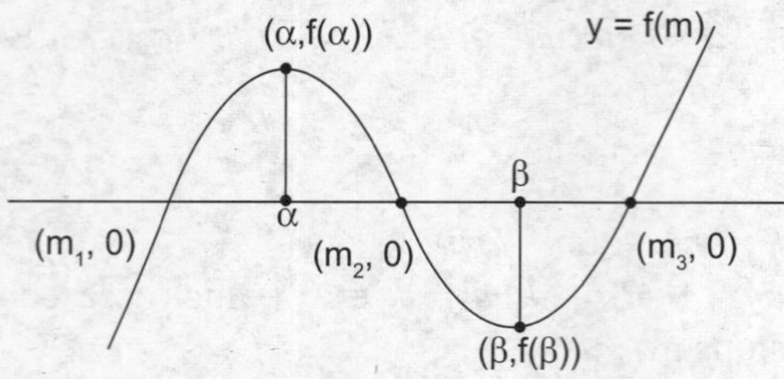

FIGURE 4.83

$\Rightarrow f(\alpha).\,f(\beta) < 0 \Rightarrow f(\alpha)\,.\,f(-\alpha) < 0$

$\Rightarrow [a\alpha^3 + k + \alpha(2a - h)]\,[-a\alpha^3 + k - \alpha(2a - h)] < 0$

$\Rightarrow k^2 - (a\alpha^3 + \alpha(2a - h))^2 < 0$

$$\Rightarrow k^2 - \left(a\left(\frac{h-2a}{3a}\right)^{\frac{3}{2}} - \sqrt{\frac{h-2a}{3a}}(h-2a)\right)^2 < 0$$

$$\Rightarrow k^2 - \left(\frac{(h-2a)^{\frac{3}{2}}}{3\sqrt{3a}} - \frac{(h-2a)^{\frac{3}{2}}}{\sqrt{3a}}\right)^2 < 0$$

$$\Rightarrow k^2 < (h-2a)^3.\left(\frac{2}{3\sqrt{3a}}\right)^2 \Rightarrow 27ak^2 < 4(h-2a)^3$$

(vii) Atmost there are four concylic points on parabola and sum of ordinates of these points vanishes.

Proof: If $(at^2, 2at)$ satisfies the circle $x^2 + y^2 + 2gx + 2fy + c = 0$

$\Rightarrow a^2t^4 + 4a^2t^2 + 2agt^2 + 4aft + c = 0$

$\Rightarrow a^2t^4 + 2a\,(2a + g)\,t^2 + 4aft + c = 0$

which being a biquadratic shows that atmost four points of parabola can be concyclic

Also, $t_1 + t_2 + t_3 + t_4 = 0$; ....(i)

$\sum t_1t_2 = \frac{2(2a+g)}{a}$ ....(ii)

$\sum t_1t_2t_3 = \frac{-4f}{a}$ ....(iii)

and $t_1t_2t_3t_4 = \frac{c}{a^2}$ ....(iv)

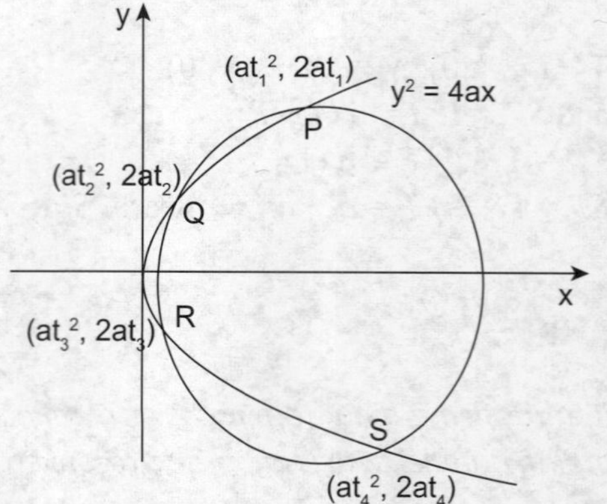

FIGURE 4.84

⇒ Sum of ordinates of four concyclic points on parabola $= 2a(t_1 + t_2 + t_3 + t_4) = 0\ (\because t_1 + t_2 + t_3 + t_4 = 0)$

(viii) Pair of chords obtained by joining four concyclic points are equally inclined to the axis of parabola.

Proof: $t_1 + t_2 + t_3 + t_4 = 0$ ...(i)

Slope of chord $PQ = m_1 = \frac{2at_1 - 2at_2}{at_1^2 - at_2^2} = \frac{2a(t_1 - t_2)}{a(t_1^2 - t_2^2)}$

$$= \frac{2}{t_1 + t_2}$$

and slope of chord

$$RS = m_2 = \frac{2at_3 - 2at_4}{at_3^2 - at_4^2} = \frac{2a(t_3 - t_4)}{a(t_3^2 - t_4^2)} = \frac{2}{t_3 + t_4}$$

For equally inclination to x-axis.

$|m_1| = |m_2| \Rightarrow m_1 = \pm m_2$

$$\Leftrightarrow \frac{2}{t_1 + t_2} = \frac{\pm 2}{t_3 + t_4}$$

$\Leftrightarrow t_1 + t_2 = \pm(t_3 + t_4)$

$\Rightarrow t_1 + t_2 + t_3 + t_4 = 0$ or $t_1 + t_2 - t_3 - t_4 = 0$

But $t_1 + t_2 + t_3 + t_4 = 0$ holds, so the result is true.

P: 12. The normal at one end of a focal chord is parallel to tangent at other end.

Proof: Let the equation of parabola be $y^2 = 4ax$ and let PQ be focal chord

By property of tangents, tangents drawn at P and Q are $\perp$ to each other and intersect at directrix at R.

$\Rightarrow RQ \perp RP$ ...(i)

Also PG is normal to parabola at P

$\Rightarrow$ PG $\perp$ RP ...(ii)

$\therefore$ from (i) and (ii) $RQ \parallel$ PG

i.e., normal at P is $\parallel$ to tangent at Q.

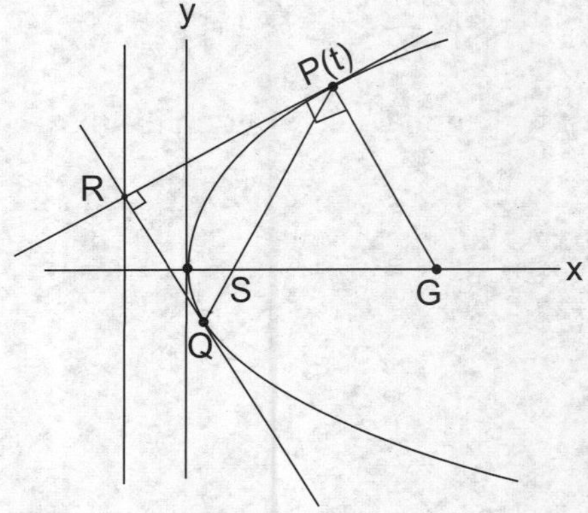

FIGURE 4.85

■ CIRCLES THROUGH CO-NORMAL POINTS

Three conormals are drawn from(α, β) to $y^2 = 4ax$ such that feet of normals be $(am_i^2, -2am_i)$; $i = 1, 2, 3$

$\Rightarrow am^3 + (2a - \alpha)m + \beta = 0$ ...(i)

$$\Rightarrow \sum_{i=1}^{3} m_i = 0;\ \sum m_i m_j = \frac{2a - \alpha}{a},\quad m_2 m_2 m_3 = \frac{-\beta}{a}$$

Let equation of circle through these three points be

$x^2 + y^2 + 2gx + 2fy + c = 0$ ...(ii)

$\Rightarrow a^2m^4 + (4a^2 + 2ag)m^2 - 4afm + c = 0$ ...(iii)

is biquadratic in m shows

$m_1 + m_2 + m_3 + m_4 = 0;$ ...(iv)

$\Rightarrow m_4 = 0\ [\ m_1 + m_2 + m_3 = 0]$

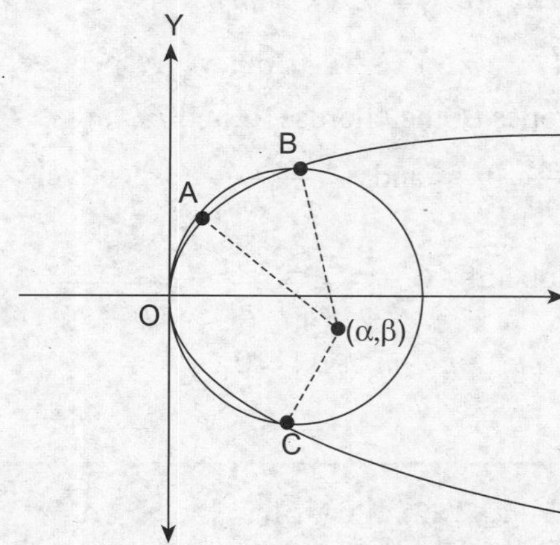

FIGURE 4.86

$\Rightarrow$ fourth point where circle meets this parabola is $(am_4^2, -2am_4)$ is (0, 0) that is through vertex.

$\Rightarrow$ $c = 0$ from equation of circle (ii)

$\Rightarrow$ $am^3 + (4a + 2g)m - 4f = 0$ ......(v)

$\Rightarrow$ Equation (i) and (v) are identical

$$\text{therefore, } 1 = \frac{4a + 2g}{2a - \alpha} = -\frac{4f}{\beta}$$

$$\Rightarrow g = \frac{-(2a + \alpha)}{2};\ 2f = -\beta/2$$

Hence, equation of the circle is

$x^2 + y^2 - (2a + \alpha)x - \beta/2\, y = 0$

Cor (1): Algebric sum of ordinate of four point of intersection of a circle and parabola = 0

Cor (2): Common chords of a circle and a parabola are pairwise equally inclined to the axis of parabola.

Proof:

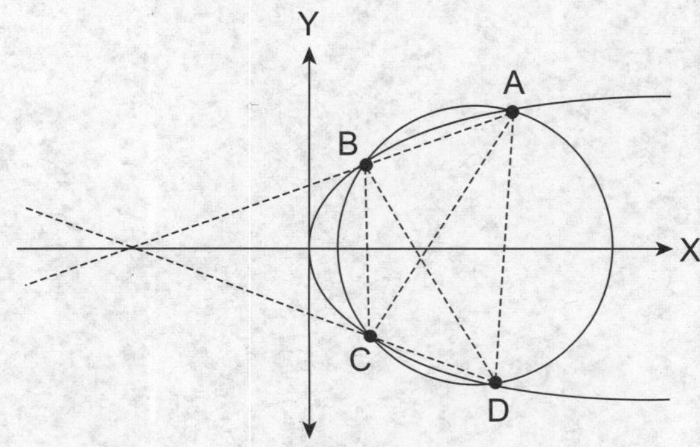

FIGURE 4.87

Let A, B, C, D be the points of intersection of the circle and the parabola with

$A(am_1^2, -2am_1)$; $B(am_2^2, -2am_2)$;

$C(am_3^2, -2am_3)$ and $D(am_4^2, -2am_4)$,

then equation of AC and BD are

$y(m_1 + m_3) = -2x - 2am_1m_3$

and $y(m_2 + m_4) = -2x - 2am_2m_4$ respectively

∴ Slopes of the chords AC and BD are

$-\dfrac{2}{m_1 + m_3}$ and $-\dfrac{2}{m_2 + m_4}$ respectively

∴ Slope of $AC = -\dfrac{2}{m_1 + m_3} = \dfrac{2}{m_2 + m_4}$

$[\because m_1 + m_2 + m_3 + m_4 = 0]$

$= -\left(-\dfrac{2}{m_2 + m_4}\right) = -$slope of BD

∴ Their slopes are equal in magnitude and opposite in sign.

∴ The chords of AC and BD are equally inclined to the axis.

Cor (3): Circle through co-normal points passes through vertex of parabola.

Cor (4): Centroid of four points in which circle intersect the parabola lie on axis $G \equiv \left(\dfrac{a}{4}\left(0 - \dfrac{2(4a^2 + 2ag)}{a^2}\right), 0\right)$

$\equiv (-2a - g, 0)$

ILLUSTRATION 52: Prove that the chord $y - x\sqrt{2} + 4a\sqrt{2} = 0$ is a normal chord of the parabola $y^2 = 4ax$. Also, find the point on the parabola where the given chord is normal to the parabola.

SOLUTION: $y - x\sqrt{2} + 4a\sqrt{2} = 0$

or $y = x\sqrt{2} - 4a\sqrt{2}$ ...(1)

Comparing the equation (1) with the equation $y = mx + c$, then

$m = \sqrt{2}$, $c = -4a\sqrt{2}$

Since $-2am - am^3 = -2a\sqrt{2} - a(\sqrt{2})^3$

$= -2a\sqrt{2} - 2a\sqrt{2} = -4a\sqrt{2} = c$

⇒ (1) is of the form $y = mx - 2am - am^3$

Hence the given chord is normal to the parabola $y^2 = 4ax$

The co-ordinate of the points are $(am^2, -2am)$ i.e., $(2a, -2\sqrt{2}a)$

ILLUSTRATION 53: Show that the locus of points such that two of the three normals drawn from them to the parabola $y^2 = 4ax$ coincide is $27ay^2 = 4(x - 2a)^3$.

SOLUTION: Let (h, k) be the point of intersection of three normals to the parabola $y^2 = 4ax$. The equation of any normal to $y^2 = 4ax$ is

$y = mx - 2am - am^3$

If it passes through (h, k), then

$k = mh - 2am - am^3$

$\Rightarrow am^3 + m(2a - h) + k = 0$ ......(i)

Let the roots of (i) be m_1, m_2 and m_3

Then from (i), $m_1 + m_2 + m_3 = 0$ ....(ii)

$m_1m_2 + m_2m_3 + m_3m_1 = \dfrac{(2a - h)}{a}$ .....(iii)

and $m_1m_2m_3 = -\dfrac{k}{a}$ .....(iv)

But here two of the three normals are given to be coincident i.e., $m_1 = m_2$

Putting $m_1 = m_2$ in (ii) and (iv), we get

$2m_1 + m_3 = 0$ .....(v)

and $m_1^2 m_3 = -\dfrac{k}{a}$ ....(vi)

Putting $m_3 = -2m_1$ from (v) in (vi), we get $-\ 2m_1^3 = -\frac{k}{a} \Rightarrow m_1^3 = \frac{k}{2a}$

Since m_1 is a root of equation (i)

$\therefore\ am_1^3 + m_1(2a - h) + k = 0$

$$\Rightarrow\ a\left(\frac{k}{2a}\right)+\left(\frac{k}{2a}\right)^{1/3}(2a-h)+k=0\ \left(\text{putting } m_1=\left(\frac{k}{2a}\right)^{1/3}\right)$$

$$\Rightarrow\ \left(\frac{k}{2a}\right)^{1/3}(2a-h)=-\frac{3k}{2}\ \text{ or }\ \frac{k}{2a}(2a-h)^3=-\frac{27k^3}{8}$$

or $27ak^2 = 4(h - 2a)^3$

Hence, the locus of (h, k) is $27ay^2 = 4(x - 2a)^3$.

ILLUSTRATION 54: Find the locus of the point through which three normals to the parabola $y^2 = 4ax$ pass such that two of them make angles α and β, respectively, with the axis such that $\tan\alpha \tan\beta = 2$.

SOLUTION: Let (h, k) be the point of intersection of three normals to the parabola $y^2 = 4ax$. The equation of any normal to $y^2 = 4ax$ is $y = mx - 2am - am^3$

If it passes through (h_1, k), then

$k = mh - 2am - am^3$

$\Rightarrow\ am^3 + m(2a - h) + k = 0$ ....(i)

Let roots of (i) be m_1, m_2, m_3, then from (i)

$m_1m_2m_3 = -\frac{k}{a}$ ....(ii)

also $m_1 = \tan\alpha$, $m_2 = \tan\beta$ and $\tan\alpha \tan\beta = 2$

$\therefore\ m_1m_2 = 2$ ....(iii)

From (ii) and (iii), $2m_3 = -\frac{k}{a}$ or $m_3 = -\frac{k}{2a}$

Which being a root of (i) must satisfy it i.e., $am_3^3 + m_3(2a - h) + k = 0$

or $a\left(-\frac{k}{2a}\right)^3-\frac{k}{2a}(2a-h)+k=0$ or $-\frac{k^3}{8a^2}-k+\frac{kh}{2a}+k=0$ or $k^2 - 4ah = 0$

$\therefore$ Required locus of (h, k) is $y^2 - 4ax = 0$

ILLUSTRATION 55: Find the point on the axis of the parabola $3y^2 + 4y - 6x + 8 = 0$ from where three distinct normals can be drawn.

SOLUTION: Given parabola is $3y^2 + 4y - 6x + 8 = 0$

$$\Rightarrow\ 3\left(y^2+\frac{4}{3}y\right)=6x-8 \Rightarrow 3\left\{\left(y+\frac{2}{3}\right)^2-\frac{4}{9}\right\}=6x-8$$

$$\Rightarrow\ 3\left(y+\frac{2}{3}\right)^2=\left(6x-8+\frac{4}{3}\right) \Rightarrow \left(y+\frac{2}{3}\right)^2=2\left(x-\frac{10}{9}\right)$$

Let $y + \frac{2}{3} = Y$, $x - \frac{10}{9} = X$. Then $Y^2 = 2X$

Comparing with $Y^2 = 4aX \therefore a = 1/2$

any point on the axis of parabola is $\left(X, -\frac{2}{3}\right)$

and $X > 2a \Rightarrow x - \frac{10}{9} > 1 \Rightarrow x > \frac{19}{9}$

CHORD OF CONTACT

Let PQ and PR be tangents to the parabola $y^2 = 4ax$ drawn from any external point $P(h, k)$, then QR is called chord of contact of the parabola $y^2 = 4ax$.

Let $Q \equiv (x_1, y_1)$ and $R \equiv (x_2, y_2)$

Equation of tangent PQ is

$yy_1 = 2a(x + x_1)$ (i)

and equation of the tangent PR is

$yy_2 = 2a(x + x_2)$ (ii)

Since (i) and (ii) pass through (h, k)

$\therefore\ ky_1 = 2a(h + x_1)$ (iii)

and $ky_2 = 2a(h + x_2)$ (iv)

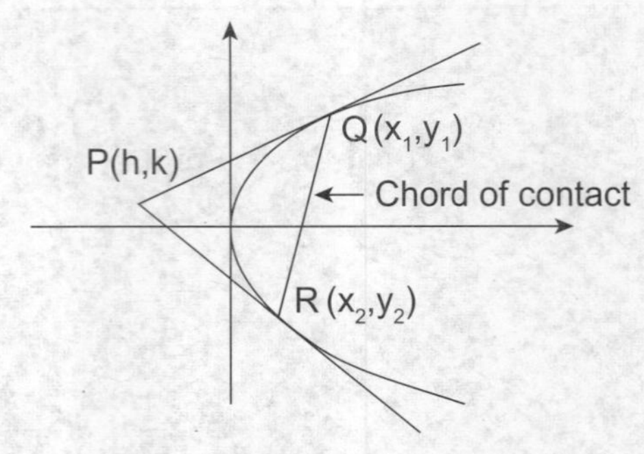

FIGURE 4.88

Hence, it is clear $Q(x_1, y_1)$ and $R(x_2, y_2)$ lie on $yk = 2a(x + h)$ which is chord of contact of QR.

ILLUSTRATION 56: Prove that the area of the triangle formed by the tangents drawn from (x_1, y_1) to to $y^2 = 4ax$ and their chord of contact is $(y^2_1 - 4ax_1)^{3/2}/2a$.

SOLUTION: Equation of QR chord of contact is

$yy_1 = 2a(x + x_1)$

or $yy_1 - 2a(x + x_1) = 0$

$\because\ PM$ = Length of perpendicular from $P(x_1, y_1)$ on QR

$$= \frac{|y_1y_1 - 2a(x_1 + x_1)|}{\sqrt{(y_1^2 + 4a^2)}} = \frac{|(y_1^2 - 4ax_1)|}{\sqrt{(y_1^2 + 4a^2)}}$$

[since $P(x_1, y_1)$ lies outside the parabola $\Rightarrow y_1{}^2 - 4ax_1 > 0$]

FIGURE 4.89

$\therefore$ area of $\Delta PQR = \frac{1}{2} QR.\ PM$

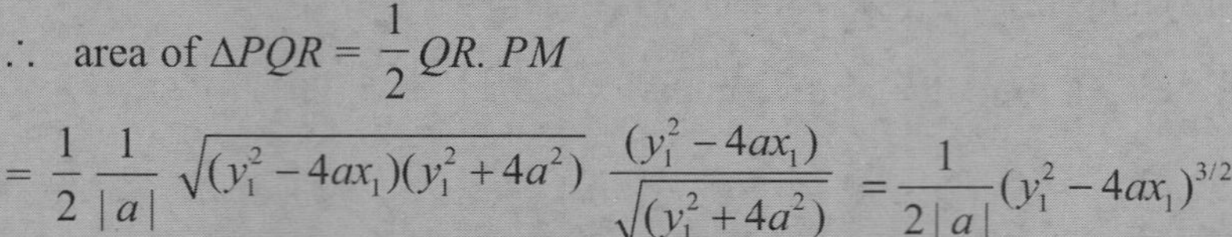

$$= \frac{1}{2}\frac{1}{|a|}\sqrt{(y_1^2 - 4ax_1)(y_1^2 + 4a^2)}\ \frac{(y_1^2 - 4ax_1)}{\sqrt{(y_1^2 + 4a^2)}} = \frac{1}{2|a|}(y_1^2 - 4ax_1)^{3/2}$$

ILLUSTRATION 57: Find the length of chord of contact of the tangents drawn from point (x_1, y_1) to the parabola $y^2 = 4ax$.

SOLUTION: Let tangent at $P(t_1)$ and $Q(t_2)$ meet at (x_1, y_1)

$\therefore\ at_1t_2 = x_1$ and $a(t_1 + t_2) = y_1$

$$\because\ PQ = \sqrt{(at_1^2 - at_2^2)^2 + (2a(t_1 - t_2))^2} = a\sqrt{((t_1 + t_2)^2 - 4t_1t_2)((t_1 + t_2)^2 + 4)}$$

$$= \sqrt{\frac{(y_1^2 - 4ax_1)(y_1^2 + 4a^2)}{a^2}}$$

ILLUSTRATION 58: If the line $x - y - 1 = 0$ intersect the parabola $y^2 = 8x$ at P and Q, then find the point of intersection of tangents at P and Q.

SOLUTION: Let (h, k) be point of intersection of tangents, then chord of contact is $yk = 4(x + h)$

$\Rightarrow 4x - yk + 4h = 0$ ....(i)

But $x - y - 1 = 0$ is same as (i)

$$\therefore\ \frac{4}{1} = \frac{-k}{-1} = \frac{4h}{-1} \Rightarrow h = -1, k = 4$$

$\therefore$ Intersection point $\equiv (-1, 4)$

TEXTUAL EXERCISE–4 (SUBJECTIVE)

1. Find the locus of point whose chord of contact w.r.t to the parabola $y^2 = 4bx$ is the tangents of the parabola $y^2 = 4ax$.

2. Prove that locus of a point whose chord of contact w.r.t parabola passes through focus is directrix.

3. If from a variable point 'p' on the line $x - 2y + 1 = 0$ pair of tangent's are drawn to the parabola $y^2 = 8x$, then prove that chord of contact passes through a fixed point, also find that point.

4. Find the points of the parabola $y^2 = 4ax$ at which the normal is inclined at 30° to the axis.

5. Prove that the chord of the parabola $y^2 = 4ax$, whose equation is $y - x\sqrt{2} + 4a\sqrt{2} = 0$, is a normal to the curve and that its length is $6\sqrt{3}a$.

6. Find the locus of the point N from which 3 normals are drawn to the parabola $y^2 = 4ax$ are such that
   (i) Two of them are equally inclined to x-axis
   (ii) Two of them are perpendicular to each other.

7. Find the equation and foot of normal to parabola $y^2 = 8x$ with following properties
   (i) parallel to line $y - 2x + 3 = 0$
   (ii) Perpendicular to line $x + 3y - 4 = 0$
   (iii) Passing through (6, 0)

8. Find the set of values of k if $P(2k^2 - 1, k)$ lie inside $y^2 = 4x$ and all three normals are real from P to $y^2 = 4x$.

9. Find the locus of the point $P(h, k)$ if three normals drawn from the point P to $y^2 = 4ax$, satisfying the following properties:
   (i) $m_1 + m_2 = 1$
   (ii) $m_1 . m_2 = 1$
   (iii) $m_1 . m_2 = -1$

10. Find the equations of the normals at the ends of the latus-rectum of the parabola $y^2 = 4ax$. Also prove that they intersect at right angles on the axis of the parabola.

Answer Keys

1. $y^2 = \dfrac{4b^2}{a}x$ **3.** (1, 8) **4.** $\left(\dfrac{a}{3}, -\dfrac{2a}{\sqrt{3}}\right), \left(\dfrac{a}{3}, \dfrac{2a}{\sqrt{3}}\right)$ **6.** (i) $y = 0$ (ii) $y^2 = a(x - 3a)$

7. (i) $y = 2x - 24$, (8, –8) (ii) $y = 3x - 66$, (18, –12)

(iii) $y = 0$ at (0, 0); $x + y - 6 = 0$ at (2, 4); $x - y - 6 = 0$ at (2, –4) **8.** $\left(-\infty, -\sqrt{\dfrac{3}{2}}\right] \cup \left[\sqrt{\dfrac{3}{2}}, \infty\right)$

9. (i) $x + y = 3a$ (ii) $ax - y^2 = a^2$ (iii) $ax - y^2 = 3a^2$ **10.** $x + y = 3a$; $x - y = 3a$

TEXTUAL EXERCISE–4 (OBJECTIVE)

1. The angle between the normals to the parabola $y^2 = 24x$ at points (6, 12) and (6, –12) is
   (a) 30° (b) 45°
   (c) 60° (d) 90°

2. The co-ordinates of the point on the parabola $y^2 = 8x$, which is at minimum distance from the circle $x^2 + (y + 6)^2 = 1$ are
   (a) (2, –4) (b) (18, –12)
   (c) (2, 4) (d) None of these

3. Three normals are drawn from a point (h, k) to the parabola $y^2 = 4ax$, the slopes of the normals are m_1, m_2 and m_3. The statements true is/are
   (a) $m_1 + m_2 + m_3 = 0$
   (b) $m_1m_2 + m_2m_3 + m_3m_1 = \dfrac{2a - h}{a}$
   (c) $m_1m_2m_3 = -\dfrac{k}{a}$
   (d) None of these

4. The point of intersection of normals drawn at the points t_1, t_2 on the parabola is

(a) $(a(2 + (t_1^2 + t_2^2 + t_1t_2)), at_1t_2(t_1 + t_2))$

(b) $a(2 + (t_1^2 + t_2^2 + t_1t_2)), -at_1t_2(t_1 + t_2))$

(c) $(a(2 - (t_1^2 + t_2^2 + t_1t_2)), -at_1t_2(t_1 + t_2))$

(d) None of these

5. P and Q are the points t_1, t_2 on the parabola $y^2 = 4ax$. If the normal at P, Q meet on parabola at $R(t_3)$, then $t_1t_2 = 2$ and $t_3 = -(t_1 + t_2)$ and locus of mid-point of PQ is

(a) $y^2 = -2ax + 4a^3$ (b) $y^2 = 2ax + 4a^2$

(c) $y^2 = -2ax + 4a^2$ (d) $y^2 = 2ax$

6. Normals are drawn from the extremities of the latus rectum of a parabola, then normals are

(a) parallel to each other

(b) perpendicular to each other

(c) intersect at the 45°

(d) None of these

7. A ray of light moving parallel to the x-axis gets reflected from a parabolic mirror whose equation is $(y-2)^2 = 4(x+1)$. After reflection, the ray must pass through the point

(a) (0, 2) (b) (2, 0)

(c) (0, –2) (d) (–1, 2)

8. If the normals at two points P and Q of a parabola $y^2 = 4ax$ intersect at a third point R on the curve, then the product of ordinates of P and Q is

(a) $4a^2$ (b) $2a^2$

(c) $-4a^2$ (d) $8a^2$

9. The normal chord to the parabola $y^2 = 4ax$ at the point whose ordinate is equal to abscissa subtends an angle θ at the focus, where θ is

(a) $\frac{\pi}{6}$ (b) $\frac{\pi}{4}$

(c) $\frac{\pi}{3}$ (d) $\frac{\pi}{2}$

10. If (h, k) is a point on the axis of the parabola $2(x - 1)^2 + 2(y-1)^2 = (x+y+2)^2$ from where three distinct normals may be drawn, then

(a) $h > 2$ (b) $h < 4$

(c) $h > 8$ (d) $h < 8$

11. The points on the axis of the parabola $3y^2 + 4y - 6x + 8 = 0$ from where 3 distinct normals can be drawn is given by

(a) $\left(a, \frac{4}{3}\right); a > \frac{19}{9}$ (b) $\left(a, -\frac{2}{3}\right); a > \frac{19}{9}$

(c) $\left(a, \frac{1}{3}\right); a > \frac{7}{9}$ (d) None of these

12. The set of points on the axis of the parabola $y^2 = 4x + 8$ from which the 3 normals to the parabola are all real and different is

(a) $\{ (k, 0) \mid k \le -2\}$

(b) $\{ (k, 0) \mid k > -2\}$

(c) $\{ (0, k) \mid k > -2\}$

(d) None of these

13. If the normals from any point to the parabola $x^2 = 4y$ cuts the line $y = 2$ in points whose abscissae are in A.P, then the slopes of the tangents at the three co-normal points are in

(a) A.P. (b) G.P

(c) H.P (d) None of there

14. The normals at three points P, Q, R of the parabola $y^2 = 4ax$ meet in (h, k). The centroid of triangle PQR lies on

(a) $x = 0$ (b) $y = 0$

(c) $x = -a$ (d) $y = a$

15. If a line $x + y = 1$ cut the parabola $y^2 = 4x$ in points A and B and normals drawn at A and B meet at C. The normal to the parabola from C other than above two meet the parabola in D, then the co-ordinates of D are

(a) (–4, 4) (b) (2, 1)

(c) (4, 4) (d) None of these

Answer Keys

1. (d)	**2.** (a)	**3.** (a, b, c)	**4.** (b)	**5.** (b)	**6.** (b)	**7.** (a)	**8.** (d)	**9.** (d)	**10.** (a)
11. (b)	**12.** (d)	**13.** (b)	**14.** (b)	**15.** (c)					

MULTIPLE-CHOICE QUESTIONS

SECTION–I

OBJECTIVE-TYPE SOLVED EXAMPLES

1. PQ is any focal chord of the parabola $y^2 = 32x$. The length of PQ can never be less than

(a) 8 unit (b) 16 unit
(c) 32 unit (d) 48 unit

Solution: (c) Let $P(at^2, 2at)$ be any point on parabola $y^2 = 4ax$ and $S(a, 0)$ be the foucs, then the other extremily of focal chord through P will be $Q\left(\frac{a}{t^2}, \frac{-2a}{t}\right)$. Thus the length of focal chord

$$PQ = \sqrt{a^2\left(t^2 - \frac{1}{t^2}\right)^2 + 4a^2\left(t + \frac{1}{t}\right)^2} = a\left(t + \frac{1}{t}\right)^2$$

$\because$ $t + \frac{1}{t} \geq 2 \Rightarrow a\left(t + \frac{1}{t}\right)^2 \geq 4a$. Here, $4a = 32$

$\therefore$ length of chord PQ can never be less than 32

2. AB is a chord of the parabola $y^2 = 4ax$ with vertex at A, BC is drawn perpendicular to AB meeting the axis at C. The projection of BC on the x-axis is

(a) a (b) $2a$
(c) $4a$ (d) $8a$

Solution: (c) Let B be $(at^2, 2at)$
slope of $AB = 2/t$

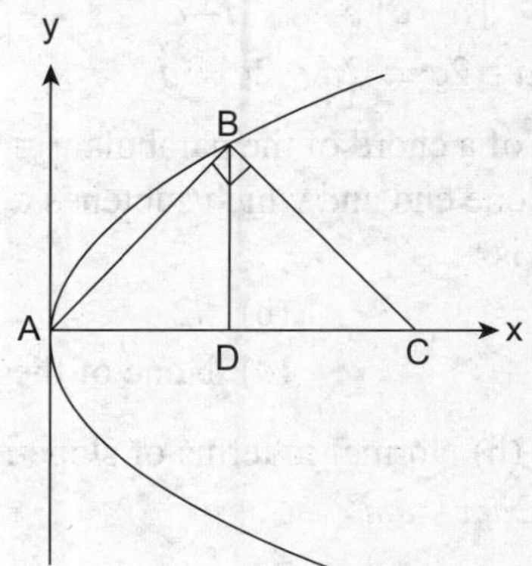

$\Rightarrow$ Slope of $BC = -t/2$

$\Rightarrow$ Equation of BC is $y - 2at = -\frac{t}{2}\left(x - at^2\right)$

This meets $y = 0$ at C whose x-co-ordinate $= 4a + at^2$ and $D \equiv (at^2, 0)$

$\therefore$ $DC = 4a + at^2 - at^2 = 4a$

3. The shortest distance between the parabola $y^2 = 4x$ and $y^2 = 2x - 6$ is

(a) 2 (b) $\sqrt{5}$
(c) 3 (d) None of these

Solution: (b) Shortest distance between two curves occured along the common normal.

Normal to $y^2 = 4x$ having slope m is $y = mx - 2m - m^3$ ....(i)

and normal to $y^2 = 2(x - 3)$ having slope m is

$$y = m(x - 3) - m - \frac{m^3}{2} \quad ......(ii)$$

$\because$ (i) and (ii) both are same

$\Rightarrow$ $-2m - m^3 = -4m - \frac{1}{2}m^3$ $\Rightarrow$ $m = 0, \pm 2$

$\Rightarrow$ Extremeties of common normal segment to paraolas will be and (5, 2) or (4, –4) and (5, –2) or (0, 0) and (3, 0)

Hence, shortest distance will be $\sqrt{(1+4)} = \sqrt{5}$.

4. The mirror image of the parabola $y^2 = 4x$ in the tangent to the parabola at the point (1, 2) is

(a) $(x - 1)^2 = 4(y + 1)$ (b) $(x + 1)^2 = 4(y + 1)$
(c) $(x + 1)^2 = 4(y - 1)$ (d) $(x - 1)^2 = 4(y - 1)$

Solution: (c) Any point on the given parabola is $(t^2, 2t)$.
The equation of the tangent at (1, 2) is $x - y + 1 = 0$.
The image (h, k) of the point $(t^2, 2t)$ in $x - y + 1 = 0$ is given by $\frac{h - t^2}{1} = \frac{k - 2t}{-1} = \frac{-2\left(t^2 - 2t + 1\right)}{1+1}$

$\therefore$ $h = t^2 - t^2 + 2t - 1 = 2t - 1$

and $k = 2t + t^2 - 2t + 1 = t^2 + 1$

Eliminating t from $h = 2t - 1$ and $k = t^2 + 1$ we get $(h + 1)^2 = 4(k - 1)$

The required equation of reflection is $(x + 1)^2 = 4(y - 1)$

5. If the 4th term in the expansion of $\left(px + \frac{1}{x}\right)^n$, $n \in \mathbb{N}$ is $\frac{5}{2}$ and three normals to the parabola $y^2 = x$ are drawn through a point $(q, 0)$, then

(a) $q = p$ (b) $q > p$
(c) $q < p$ (d) $pq = 1$

Solution: (b)

Given $\frac{5}{2} = {}^nC_3\left(px\right)^{n-3}\left(\frac{1}{x}\right)^3 = {}^nC_3 . p^{n-3}x^{n-6}$ .....(i)

Since LHS of equation (i) is independent of x

$\therefore\ n - 6 = 0 \Rightarrow n = 6$

From equation (i) $\frac{5}{2} = {}^6C_3\ p^3 = 20\ p^3$

$\Rightarrow\ p^3 = \left(\frac{1}{2}\right)^3 \Rightarrow p = \frac{1}{2}$

Given, parabola is $y^2 = x$

Here, $4a = 1 \Rightarrow a = 1/4$

Since, three normals are drawn from point $(q, 0)$

$\therefore\ q > 2a$ or $q > \frac{1}{2}$ or $q > p$ $\left[\because p = \frac{1}{2}\right]$

6. If the tangents at P and Q on a parabola meet at T, then SP, ST and SQ are in
(a) AP (b) GP
(c) HP (d) None of these

Solution: (b) Let the parabola be $y^2 = 4ax$

Let $P \equiv \left(at_1^2, 2at_1\right), Q \equiv \left(at_2^2, 2at_2\right)$

then, $T \equiv \left(at_1t_2, a\left(t_1 + t_2\right)\right)$

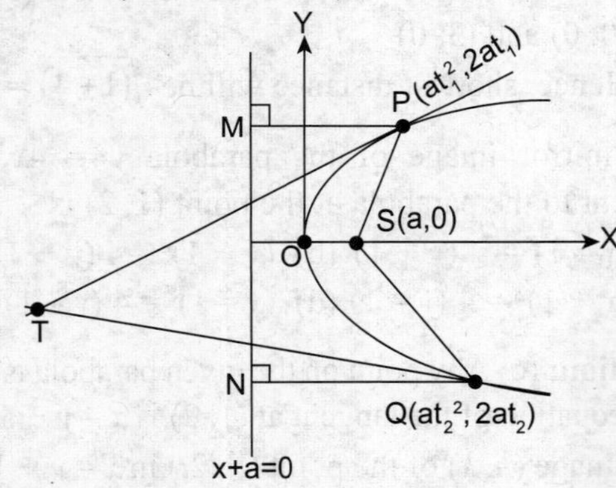

then, $SP = PM = a + at_1^2$

$SQ = QN = a + at_2^2$

and $ST = \sqrt{\left(a - at_1t_2\right)^2 + \left(0 - a\left(t_1 + t_2\right)\right)^2}$

$= a\sqrt{\left(1 - t_1t_2\right)^2 + \left(t_1 + t_2\right)^2} = a\sqrt{\left(1 + t_1^2t_2^2 + t_1^2 + t_2^2\right)}$

$= a\sqrt{\left(1 + t_1^2\right)\left(1 + t_2^2\right)} = \sqrt{a\left(1 + t_1^2\right)a\left(1 + t_2^2\right)}$

$= \sqrt{(SP)(SQ)}$

$\therefore\ SP, ST, SQ$ are in GP.

7. Radius of the largest circle which passes through the focus of the parabola $y^2 = 4x$ and contained in it, is
(a) 8 (b) 4
(c) 2 (d) 5

Solution: (b) Let the radius of circle be r and centre at $(r + a, 0) \equiv (r + 1, 0)$

$\Rightarrow$ equation of circle is $(x - r - 1)^2 + y^2 = r^2$

$\Rightarrow (x - r - 1)^2 + y^2 = r^2$

$\Rightarrow x^2 + \{4 - 2(r + 1)\}\ x + (2r + 1) = 0$ ($y^2 = 4x$)

It would have same roots due to symmetry

$\Rightarrow D = 0 \Rightarrow 4(1 - r)^2 - 4(2r + 1) = 0$

$\Rightarrow r = 0$ (rejected) $\Rightarrow r = 4$

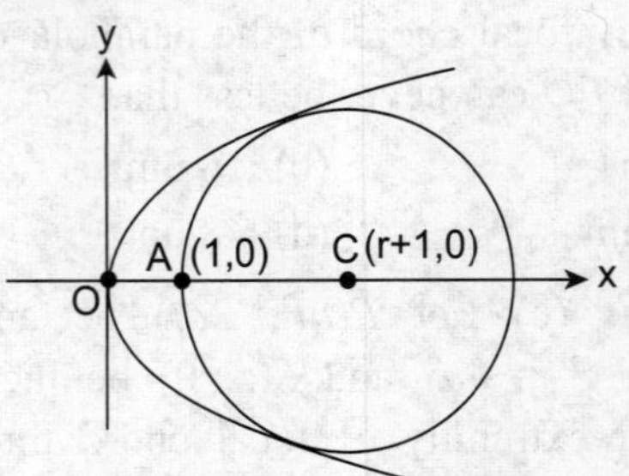

8. The condition that the parabolas $y^2 = 4c(x - d)$ and $y^2 = 4ax$ have a common normal other than x-axis $(a > 0, c > 0)$ is
(a) $2a < 2c + d$ (b) $2c < 2a + d$
(c) $2d < 2a + c$ (d) $2d < 2c + a$

Solution: (a) Normals of parabolas $y^2 = 4ax$ and $y^2 = 4c(x - d)$ in terms of slope are

$y = mx - 2am - am^3$ ....(i)

and $y = m(x - d) - 2cm - cm^3$ ......(ii)

Subtracting equation (ii) from (i), we get,

$md - 2am - am^3 + 2cm + cm^3 = 0;\ m \neq 0$

$\therefore\ d - 2a - am^2 + 2c + cm^2 = 0$

$\Rightarrow (a - c)m^2 = d - 2a + 2c$

$\Rightarrow\ m^2 = \frac{d - 2a + 2c}{(a - c)} \Rightarrow \frac{d}{a - c} - 2 > 0$

$\Rightarrow d > 2a - 2c \Rightarrow 2a < 2c + d$

9. The slope of a chord of the parabola $y^2 = 4ax$, which is normal at one end and which subtends a right angle at the origin is
(a) $1/\sqrt{2}$ (b) $\sqrt{2}$
(c) 2 (d) None of these

Solution: (b) Normal in terms of slope is given by

$y = mx - 2am - am^3$

$\Rightarrow\ \frac{mx - y}{2am + am^3} = 1$ .....(i)

and $y^2 = 4ax$

making homogenous equation (ii) with the help of equation (i)

We have, $4ax\left(\frac{mx - y}{2am + am^3}\right) = y^2$

$\Rightarrow \left(2am+am^3\right)y^2 = 4amx^2 - 4axy$

For right angle co-efficient of x^2 + co-efficient of $y^2 = 0$

$\Rightarrow -4am + 2am + am^3 = 0 \Rightarrow m = \pm\sqrt{2}$ ($\because$ m $\neq 0$)

10. Let α be the angle which a tangent to the parabola $y^2 = 4ax$ makes with its axis, the distance between the tangent and a parallel normal will be

(a) a $\sin^2\alpha\cos^2\alpha$ (b) $a \operatorname{cosec}\alpha \sec^2\alpha$

(c) $a^1 \tan^2\alpha$ (d) $a\cos^2\alpha$

Solution: (b) Tangent to parabola $y^2 = 4ax$ with slope m is given by $y = mx + a/m$ ....(i)

where $m = \tan\alpha$

and equation of normal parallel to equation (i) is $y = mx - 2am - am^3$

Distance between equations (i) and (ii) is

$$\frac{\left|\frac{a}{m}+2am+am^3\right|}{\sqrt{(1+m^2)}}$$

$$= \frac{a\left(1+m^2\right)^2}{m\left(\sqrt{1+m^2}\right)} = \frac{a\left(1+m^2\right)^{3/2}}{m} = \frac{a\left(1+\tan^2\alpha\right)^{3/2}}{\tan\alpha}$$

$$= \frac{a\sec^3\alpha}{\tan\alpha} = a\sec^2\alpha.\operatorname{cosec}\alpha$$

11. The locus of point of intersection of tangents to the parabolas $y^2 = 4(x + 1)$ and $y^2 = 8(x + 2)$ which are perpendicular to each other is

(a) $x + 7 = 0$ (b) $x - y = 4$

(c) $x + 3 = 0$ (d) $y - x = 12$

Solution: (c) $y = m(x + 1) + (1/m)$

or $y = mx + \left(m+\frac{1}{m}\right)$ ....(i)

is a tangent to the first parabola

and $y = m'(x + 2) + \frac{2}{m'} = m'x + 2\left(m'+\frac{1}{m'}\right)$ ....(ii)

is a tangent to the second parabola.

Now, $m.m' = -1$

Then from equation (ii)

$$y = -\frac{x}{m} + 2\left(-\frac{1}{m} - m\right)$$

$$\Rightarrow y = -\frac{x}{m} - 2\left(m+\frac{1}{m}\right) \quad ....(iii)$$

Subtracting equation (iii) from (i), we get,

$$x\left(m+\frac{1}{m}\right) + 3\left(m+\frac{1}{m}\right) = 0 \text{ or } x + 3 = 0$$

12. P is a point which moves in the $x - y$ plane such that the point P is nearer to the centre of a square than any of the sides. The four vertices of the square are $(\pm a, \pm a)$.The region in which P will move is bounded by parts of parabola/ parabolas having equations

(a) $y^2 = a^2 + 2ax$ (b) $x^2 = a^2 + 2ay$

(c) $y^2 + 2ax = a^2$ (d) None of these

Solution: (a, b, c) If $P(x, y)$ lie in region, then $OP < PM, RP, NP, PQ$

$$\Rightarrow \sqrt{x^2+y^2} < |a-x|, \sqrt{x^2+y^2} < |a+x|$$

$$\sqrt{x^2+y^2} < |a-y| \text{ and } \sqrt{x^2+y^2} < |a+y|$$

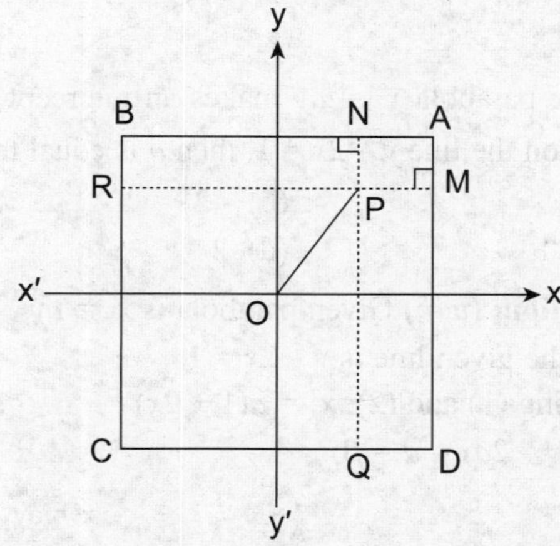

On squaring, we get

$\therefore$ The region bounded by the curves

$x^2 + y^2 = (a - x)^2$; $x^2 + y^2 = (a + x)^2$

$x^2 + y^2 = (a - y)^2$; $x^2 + y^2 = (a + y)^2$

13. The ends of a line segment are P (1, 3) and Q (1, 1). R is a point on the line segment PQ such that $PR: QR = 1: \lambda$. ($\lambda > 0$). If R is an interior point of the parabola $y^2 = 4x$, then

(a) $\lambda \in (0, 1)$ (b) $\lambda \in \left(-\frac{3}{5}, 1\right)$

(c) $\lambda \in \left(\frac{1}{2}, \frac{3}{5}\right)$ (d) None of these

Solution: (a) $R\left(1, \frac{1+3\lambda}{\lambda+1}\right)$. Since R is an interior point

Therefore $S_1 < 0$

$$\Rightarrow \left(\frac{1+3\lambda}{\lambda+1}\right)^2 - 4 < 0$$

$$\Rightarrow \left(\lambda+\frac{3}{5}\right)(\lambda - 1) < 0 \Rightarrow \lambda \in (0, 1) (\because \lambda > 0)$$

Aliter. As it is clear from figure

If $\lambda = 1$, then $R(1, 2)$ becomes mid-point of PQ.

$\Rightarrow$ R lies on the parabola

$\therefore$ $\lambda < 1$, then R lies inside the parabola. Also $\lambda > 0$

$\therefore$ $\lambda \in (0, 1)$

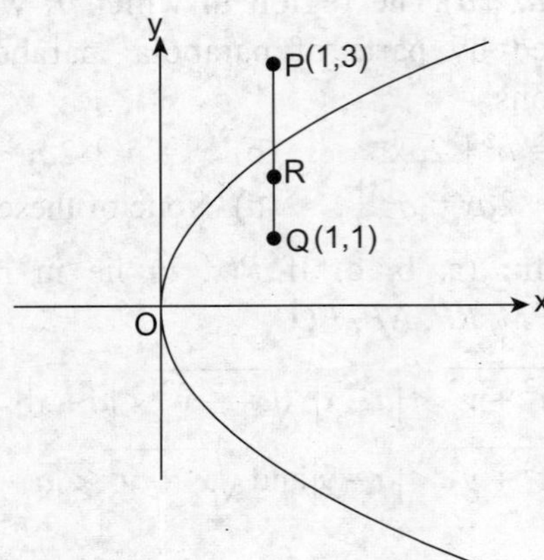

14. If the parabola $x^2 = ay$ makes an intercept of length $\sqrt{40}$ on the line $y - 2x = 1$, then a is equal to

(a) 1 (b) −2

(c) −1 (d) 2

Solution: (*a, b*) Given parabola is $x^2 = ay$ ... (1)

and the given line is $y - 2x = 1$ .... (2)

Solving (1) and (2); $x^2 = a(1 + 2x)$

$\Rightarrow x^2 - 2ax - a = 0$

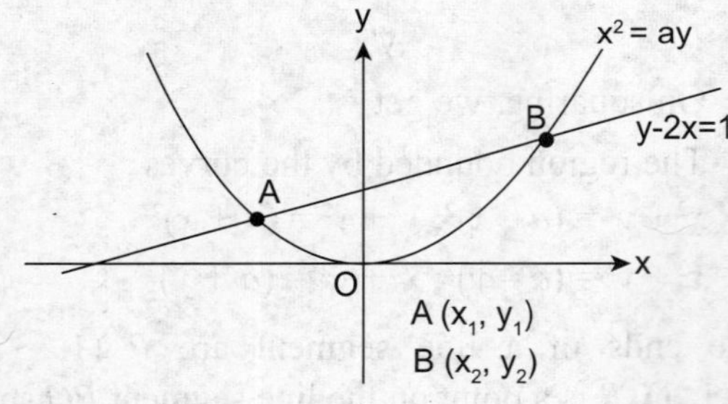

$$\Rightarrow |x_1 - x_2| = \frac{\sqrt{D}}{1} = \sqrt{4a^2 + 4a}$$

$$\Rightarrow (x_1 - x_2)^2 = 4a^2 + 4a$$

Since $y_1 = \dfrac{x_1^2}{a}, y_2 = \dfrac{x_2^2}{a}$

$$\therefore \quad |AB| = \sqrt{4a^2 + 4a + \left(\frac{x_1^2 - x_2^2}{a}\right)^2}$$

$$= \sqrt{4a^2 + 4a + \frac{1}{a^2} \times 4a^2(4a^2 + 4a)} = \sqrt{20a^2 + 20a}$$

$$\because \ |x_1^2 - x_2^2| = |x_1 - x_2|\,|x_1 + x_2| = \sqrt{4a^2 + 4a}\ |2a|$$

Given $\sqrt{20a^2 + 20a} = \sqrt{40}$

$\Rightarrow a^2 + a = 2 \Rightarrow a^2 + a - 2 = 0$

$\Rightarrow (a + 2)(a - 1) = 0 \Rightarrow a = -2, 1$

15. Let us define a region R in xy-plane as set of points (x, y) satisfying $[x^2] = [y]$ (where $[x]$ denotes greatest integer $\le x$), then the region R defines

(a) a parabola whose axis is horizontal

(b) a parabola whose axis is vertical

(c) integer point on the parabola $y = x^2$

(d) None of the above

Solution: (d) Given $[x^2] = [y]$

For $0 \le y < 1$, then $[y] = 0$

$\therefore$ $[x^2] = 0 \Rightarrow 0 \le x^2 < 1$

$\Rightarrow x \in (-1, 1)$

For $1 \le y < 2$, then $[y] = 1$

$\therefore$ $[x^2] = 1$

$\Rightarrow 1 \le x^2 < 2 \quad \Rightarrow \quad x \in \left(-\sqrt{2}, -1\right] \cup \left[1, \sqrt{2}\right)$

For $2 \le y < 3$, $[y] = 2$

then $[x^2] = 2 \Rightarrow 2 \le x^2 < 3$

$\therefore$ $x \in \left(-\sqrt{3}, -\sqrt{2}\right] \cup \left[\sqrt{2}, \sqrt{3}\right)$

The graph of the region will not only contain points of the parabola $y = x^2$ but $[x^2] = [y]$ contains points with in the rectangles of side 1, 2; 1, $\sqrt{2} - 1$; 1, $\sqrt{3} - \sqrt{2}$ etc. as shown in figure below.

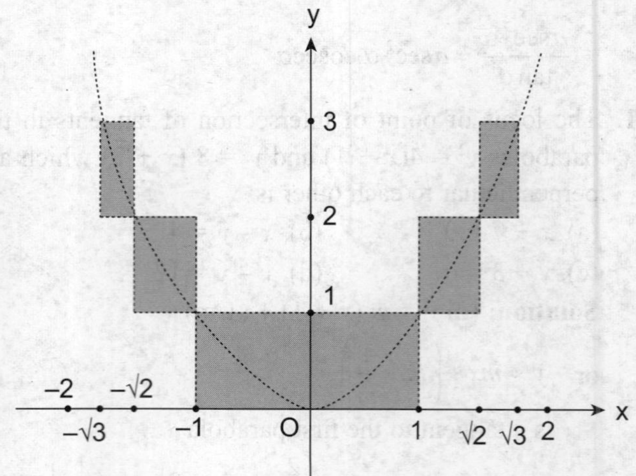

Hence, (a), (b), (c) are incorrect.

16. If the normals at three point P, Q, R of the parabola $y^2 = 4ax$ meet at a point O and S be its foucs, then $|SP|.|SQ|.\ |SR|$ is equal to

(a) a^2 (b) $a(SO)^3$

(c) $a(SO)^2$ (d) None of these

Solution: (c) Equation of normal in terms of slope is $y = mx - 2am - am^3$.

then, $am^3 - (h - 2a)\,m + k = 0$

or $m_1 + m_2 + m_3 = 0$;

$$m_1m_2 + m_2m_3 + m_3m_1 = -\frac{(h - 2a)}{a}$$

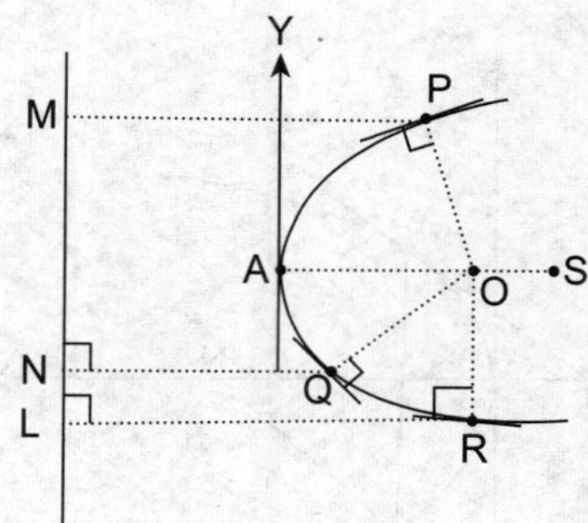

and $m_1m_2m_3 = -k/a$

Let $P \equiv (am_1^2, -2am_1)$; $Q \equiv (am_2^2, -2am_2)$ and $R \equiv (am_3^2, -2am_3)$ and $S \equiv (a, 0)$

$\therefore$ $|SP||SQ||SR| = |PM||QN||RL|$

$$= |a + am_1^2||a + am_2^2||a + am_3^2|$$

$$= a^3 \left|(1+m_1^2)(1+m_2^2)(1+m_3^2)\right|$$

$$= a^3 \left|1 + \Sigma m_1^2 + \Sigma m_1^2 m_2^2 + m_1^2 m_2^2 m_3^2\right|$$

$$= a^3 \left|\begin{matrix}1 + (\Sigma m_1)^2 - 2\Sigma m_1 m_2 + (\Sigma m_1 m_2)^2 \\ -2m_1m_2m_3\Sigma m_1 + (m_1m_2m_3)^2\end{matrix}\right|$$

$$= a^3 \left|1 + (0)^2 + \frac{2(h-2a)}{a} + \frac{(h-2a)^2}{a^2} - 0 + \frac{k^2}{a^2}\right|$$

$$= a\left|k^2 + 2a(h-2a) + a^2 + (h-2a)^2\right|$$

$$= a\left|k^2 + (h-2a+a)^2\right| = a\left|k^2 + (h-a)^2\right|$$

$$= a\left|(k-0)^2 + (h-a)^2\right| = a(SO)^2$$

17. The orthocentre of a triangle formed by any three tangents to a parabola lies on

(a) focus (b) directrix

(c) vertex (d) focal chord

Solution: (b) Let $P(t_1)$; $Q(t_2)$ and $R(t_3)$ be three points on the parabola $y^2 = 4ax$, the tangents at which meet at points A, B and C.

$\therefore$ Co-ordinates of A, B and C are

$(at_1t_3, a(t_1 + t_3)), (at_1t_2, a(t_1 + t_2))$

and $(at_2t_3, a(t_2 + t_3))$ respectively

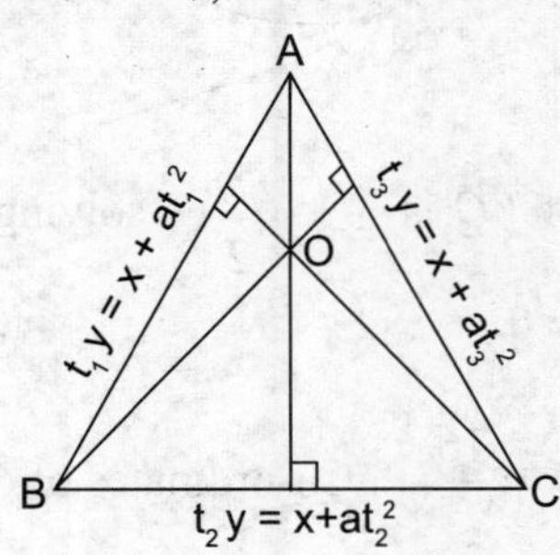

Let orthocentre is O (h, k) $\quad AO \perp BC$

$$\therefore \left\{\frac{k - a(t_1 + t_3)}{h - at_1t_3}\right\} \times \frac{1}{t_2} = -1$$

$\Rightarrow k - a(t_1 + t_3) = -ht_2 + at_1t_2t_3$ ....(i)

Also $BO \perp CA$

$\Rightarrow k - a(t_1 + t_2) = -ht_3 + at_1t_2t_3$ ....(ii)

Subtracting equation (ii) from (i), we get $h + a = 0$

$\therefore$ Locus of orthocentre is $x + a = 0$ which is directrix of parabola.

18. Let the line $lx + my = 1$ cut the parabola $y^2 = 4ax$ at the points A and B. Normals at A and B meet at point C. Normal from C other than these two meet the parabola at D, then co-ordinate of D are

(a) $(a, 2a)$ (b) $\left(\frac{4am}{l^2}, \frac{4a}{l}\right)$

(c) $\left(\frac{2am^2}{l^2}, \frac{2a}{l}\right)$ (d) $\left(\frac{4am^2}{l^2}, \frac{4am}{l}\right)$

Solution: (d) Let

$A \equiv (am_1^2, -2am_1)$ and $B(am_2^2, -2am_2)$

Now, A and B lie on $lx + my = 1$

$\Rightarrow l(am_1^2) + m(-2am_1) = 1$ ....(i)

and $l(am_2^2) + m(-2am_2) = 1$ ....(ii)

Subtracting equation (ii) from equation (i), we get

$$al(m_1^2 - m_2^2) - 2am(m_1 - m_2) = 0$$

$\Rightarrow l(m_1 + m_2) - 2m \neq 0$ $[\because a(m_1 - m_2) \neq 0]$

$\Rightarrow m_1 + m_2 = \frac{2m}{l}$ ....(iii)

Let $D \equiv (am_3^2, -2am_3)$ and $C \equiv (h, k)$

$\therefore$ Let the equation of normal in terms of slope M for m_1, m_2, m_3 be $y = Mx - 2aM - aM^3$ then $aM^3 - (h - 2a)M + k = 0$

$$\therefore m_1 + m_2 + m_3 = 0 \Rightarrow \frac{2m}{l} + m_3 = 0$$

$$\Rightarrow m_3 = \frac{-2m}{l}$$

$$\therefore D \equiv \left(a\left(\frac{-2m}{l}\right)^2, -2a\left(\frac{-2m}{l}\right)\right)$$

$$\Rightarrow D \equiv \left(\frac{4am^2}{l^2}, \frac{4am}{l}\right)$$

19. Two parabolas P_1 and P_2 intersect at two different points, where P_1 is $y = x^2 - 3$ and P_2 is $y = kx^2$. The abscissa which is positive is designated point A, and is 'a'. The tangent line l at A to the curve P_2 intersects curve P_1 at point B, other than A. If abscissa of point B is 1, then a is equal to

(a) 1 (b) 2

(c) 3 (d) 4

Solution: (c) $P_1: y = x^2 - 3$ and $P_2: y = kx^2$

Solving P_1 and P_2 we get $kx^2 = x^2 - 3$

$$\Rightarrow x^2 = \frac{3}{(1-k)}, \Rightarrow \frac{3k}{(1-k)} = y$$

$$\therefore \quad A \equiv \left(\sqrt{\left(\frac{3}{1-k}\right)}, \frac{3k}{(1-k)}\right)$$

($\because$ abscissa of A is positive)

and $a = \sqrt{\left(\frac{3}{(1-k)}\right)} \Rightarrow A \equiv (a, ka^2)$

$\equiv (a, a^2 - 3) \Rightarrow k = 1 - \frac{3}{a^2}$.

Now, tangent 'l' at A to the curve P_2 is given by

$$\frac{y + a^2 - 3}{2} = kx(a) \Rightarrow y + a^2 - 3 = 2ax\left(1 - \frac{3}{a^2}\right) \text{...(i)}$$

$\therefore \quad B \equiv (1, -2)$ (given abscissa of $B = 1$),

$\therefore$ from equation (i) $-2 + a^2 - 3 = 2a\left(1 - \frac{3}{a^2}\right)$

$= 2\frac{(a^2-3)}{a}$

$\Rightarrow a^3 - 2a^2 - 5a + 6 = 0 \Rightarrow (a-1)(a+2)(a-3) = 0$

$\Rightarrow a = 3 \ (\because a \neq 1, a \neq -2)$

20. P is the point 't' on the parabola $y^2 = 4ax$ and PQ is a focal chord, PT is the tangent at P and QN is the normal at Q. If the angle between PT and QN, be α and the distance between PT and QN be d, then

(a) $0 < \alpha < 90°$ (b) $\alpha = 0°$

(c) $d = 0$ (d) $d = \frac{a(t^2+1)^{3/2}}{t^2}$

Solution: (b, d) Let $Q \equiv (at_1^2, 2at_1)$ and $P(at^2, 2at)$ (given)

PQ is focal chord

$$\therefore \quad tt_1 = -1 \text{ or } t_1 = -\frac{1}{t} \Rightarrow Q \equiv \left(\frac{a}{t^2}, \frac{-2a}{t}\right)$$

Equation of PT is $ty = x + at^2$ having.

slope $= \frac{1}{t} = m_1$ (say)

Equation of normal at Q is $y + \frac{2a}{t} = \frac{1}{t}\left(x - \frac{a}{t^2}\right)$

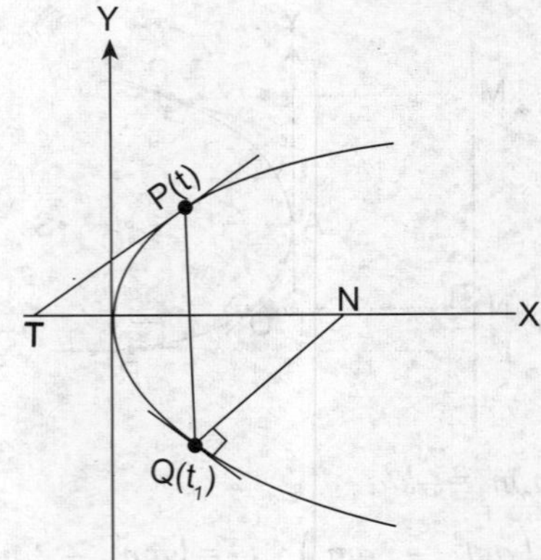

Its slope $= \frac{1}{t} = m_2$ (say) $\Rightarrow m_1 = m_2$

$\therefore$ PT and QN are parallel $\Rightarrow \alpha = 0^0$

$$\therefore \text{ Distance between them } d = \frac{at^2 + 2a + \frac{a}{t^2}}{\sqrt{(1+t^2)}}$$

$$= a\frac{(t^2+1)^2}{t^2\sqrt{(1+t^2)}} = \frac{a(t^2+1)^{3/2}}{t^2}$$

21. Let PQ be a chord of the parabola $y^2 = 4x$. A circle drawn with PQ as a diameter passes through the vertex V of the parabola. If area of $\Delta PVQ = 20$ sq unit, then co-ordinates of P are

(a) 4, –4) (b) (4, 4)

(c) (16, –8) (d) (16, 8)

Solution: (c, d) Slope of $VP = \frac{2t - 0}{t^2 - 0} = \frac{2}{t}$

$\therefore$ Slope of $VQ = -\frac{t}{2}$

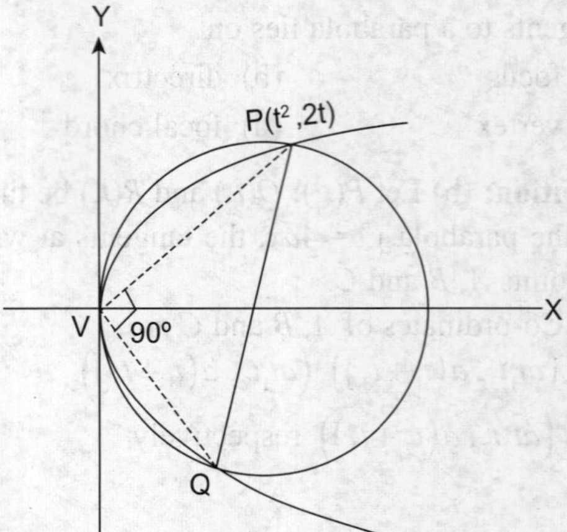

Equation of VQ is $y = -\frac{t}{2}x$, solving it with $y^2 = 4x$

$$\frac{t^2x^2}{4} - 4x = 0$$

$\therefore \quad x = 0, x = \frac{16}{t^2}$ $\therefore$ Co-ordinates of Q are $\left(\frac{16}{t^2}, -\frac{8}{t}\right)$

Area of $\Delta PVQ = (1/2)(PV)(VQ) = 20$

$\Rightarrow (PV)^2 (VQ)^2 = 1600$

$$\Rightarrow \left(t^4 + 4t^2\right)\left(\frac{256}{t^4} + \frac{64}{t^2}\right) = 1600$$

$$\Rightarrow t^2\left(t^2 + 4\right).\frac{64}{t^4}\left(t^2 + 4\right) = 1600$$

$$\Rightarrow \left(t^2 + 4\right)^2 = \frac{1600}{64}t^2 \Rightarrow t^2 + 4 = \pm\frac{40t}{8}$$

$\Rightarrow t^2 + 4 = \pm 5t$

we get $t = \pm 4, \pm 1$. Hence co-ordinates of P are $(16, \pm 8)$ and $(1, \pm 2)$

22. If $f(x + y) = f(x).f(y)$ for all $x, y \in \mathbb{R}$and $f(1) = 2$ and $a_r = f(r)$, for $r \in \mathbb{N}$, then the co-ordinates of a point on the parabola $y^2 = 8x$ whose focal distance is 4, may be

(a) (a_1, a_2) (b) $(a_1, -a_2)$

(c) (a_1, a_1) (d) None of these

Solution: (a, b) Given parabola is $y^2 = 8x$ ....(i)

Here, $a_1 = 2$. Let $P(2t^2, 4t)$ be a point on parabola (i), and S be the focus

Given, $SP = 4$

$\therefore\ a(1 + t^2) = 4 \Rightarrow 2(1 + t^2) = 4 \Rightarrow t = \pm 1$

$\therefore\ P \equiv (2, 4)$ or $(2, -4)$

Now, given $f(x+y) = f(x).f(y)$ for all x and $y \in \mathbb{R}$ ...(ii)

Given, $f(1) = 2$

From equation (ii), $f(2) = f(1 + 1) = f(1).f(1) = 2^2 = 4$

Similiary, $f(n) = 2^n$

$a_r = f(r) \therefore a_1 = f(1) = 2$

$a_2 = f(2) = 4$. Hence, $P \equiv (a_1, a_2)$ or $(a_1, -a_2)$

23. Minimum area of circle which touches the parabolas $y = x^2 + 1$ and $y^2 = x - 1$ is

(a) $\frac{9\pi}{16}$ sq. units (b) $\frac{9\pi}{32}$ sq. units

(c) $\frac{9\pi}{8}$ sq. units (d) $\frac{9\pi}{4}$ sq. units

Solution: (b) $y = x^2 + 1$ and $y^2 = x - 1$ are inverse relations of each other, their graphs are symmetric about $y = x$ and shortest distance between these occur along common normal i.e., a line $\perp$ to parallel tangent of both curves i.e., at the point where tangent is parallel to $y = x$

$\therefore\ PQ$ is $\perp$ to $y = x \Rightarrow$ slope of tangent at $P = 1$

$\therefore$ Diameter of circle $\Rightarrow P\left(\frac{5}{4}, \frac{1}{2}\right), Q\left(\frac{1}{2}, \frac{5}{4}\right)$

$$= 2r = \sqrt{\left(\frac{3}{4}\right)^2 + \left(\frac{3}{4}\right)^2} = \frac{3\sqrt{2}}{4}$$

$\Rightarrow r = \frac{3\sqrt{2}}{8}$ units

$\therefore$ Minimum area of circle touching both parabolas

$$= \pi\left(\frac{3\sqrt{2}}{8}\right)^2 = \frac{9\pi}{32} \text{ square units.}$$

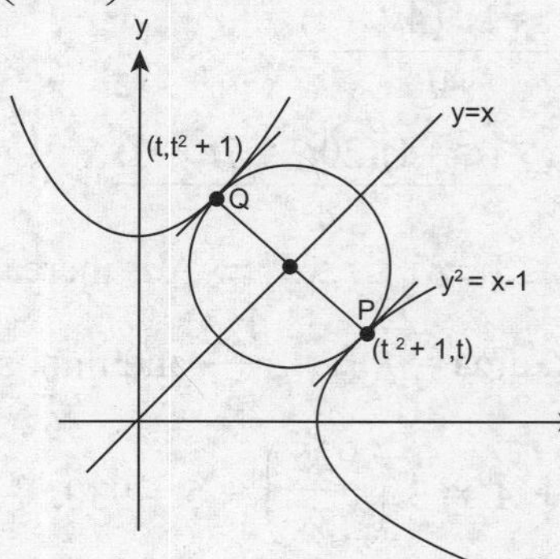

Aliter: Now, we shall find points P_0 and Q_0 on the parabolas C_1 and C_2 respectively such that,

$PQ \geq P_0Q_0$ for all pairs of points (P, Q) with P on C_1 and Q on C_2.

For this, we first find a point P_0 on the parabola C_1 such that P_0Q_0 is minimum, where Q_0 is the reflection of P_0 in $y = x$. Let $P_0(t, t^2 + 1)$ be a point on C_1.

Then $Q_0(t^2 + 1, t)$ is on C_2.

$$\text{Now, } P_0Q_0 = \sqrt{(t^2 + 1 - t)^2 + (t - t^2 - 1)^2}$$

$$= \sqrt{2(t^2 - t + 1)^2}\ .$$

$$P_0Q_0 = \sqrt{2\left\{\left(t - \frac{1}{2}\right)^2 + \frac{3}{4}\right\}}$$

Clearly, P_0Q_0 is minimum, when $t = \frac{1}{2}$

Hence, co-ordinates of P_0 and Q_0 are $P_0\left(\frac{1}{2}, \frac{5}{4}\right)$ and $Q_0\left(\frac{5}{4}, \frac{1}{2}\right)$ respectively.

24. If the maximum and minimum values of the area of the Δ formed by x-axis, tangent and normal at a point of the parabola $y = x^2 + 1$, $1 \leq x \leq 3$ be A_1 and A_2 respectively, then

(a) $3A_1 + A_2 = 930$ (b) $A_2 = 5$

(c) $A_1 = 925/6$ (d) None of these

Solution: (a, b) Any point on the parabola $y = x^2 + 1$ is $P(t, t^2 + 1)$ Equation of tangent at P. $y - t^2 - 1 = 2t(x - t)$.

Put $y = 0$, then $x = t - \frac{t^2 + 1}{2t}$. Hence, $T \equiv \left(t - \frac{t^2 + 1}{2t}, 0\right)$.

Equation of normal at P, $y - t^2 - 1 = -\frac{1}{2t}(x - t)$. Put $y = 0$, then $x = t + 2t(t^2 + 1)$.

Hence $G \equiv (2t(t^2+1)+t, 0)$

$\therefore \quad TG = 2t(t^2+1)+t-t+\dfrac{t^2+1}{2t} = 2t(t^2+1)+\dfrac{t^2+1}{2t}$

$\therefore \quad \text{Area } A = \dfrac{1}{2}(t^2+1)\left[2t(t^2+1)+\dfrac{t^2+1}{2t}\right]$

$= \dfrac{(t^2+1)^2(4t^2+1)}{4t}$

$\Rightarrow \quad \dfrac{dA}{dt} = \dfrac{(t^2+1)(20t^4+7t^2-1)}{4t^2} > 0$

$[\because 1 \le x \le 3 \Rightarrow 1 \le t \le 3] \Rightarrow$ A is increasing in $1 \le t \le 3$

$\therefore \quad$ Max. area $= f(3) = \dfrac{925}{3}$ and min. area $= f(1) = 5$

$\therefore \quad 3A_1 + A_2 = 3\left(\dfrac{925}{3}\right) + 5 = 930.$

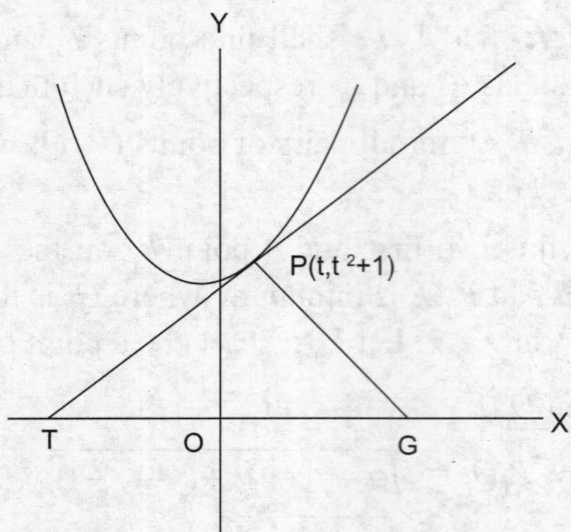

25. If $x + y = k$ is normal to $y^2 = 12x$, then 'k' is

(a) 3 (b) 9

(c) –9 (d) – 3

Solution: (b) We know that equation of normal to parabola $y^2 = 4ax$ is given by

$y = mx - 2am - am^3$

$\Rightarrow y = mx - 6m - 3m^3$ .....(1)

It is given to be $y = -x + k$

$\Rightarrow m = -1, -6m - 3m^3 = k \Rightarrow k = 6 + 3 = 9$

26. The locus of the mid-point of the line segment joining the focus and a moving point on the parabola $y^2 = 4ax$ is another parabola with directrix

(a) $x = -a$ (b) $x = -a/2$

(c) $x = 0$ (d) $x = a/2$

Solution: (c) Let $P(h, k)$ be the mid-point of the line segment joining the focus $(a, 0)$ and a general point $Q(x, y)$ on the parabola, then $h = \dfrac{x+a}{2}, k = \dfrac{y}{2}$

$\Rightarrow x = 2h - a, y = 2k$

putting these vlaues of x and y in $y^2 = 4ax$, we get $4k^2 = 4a(2h - a)$

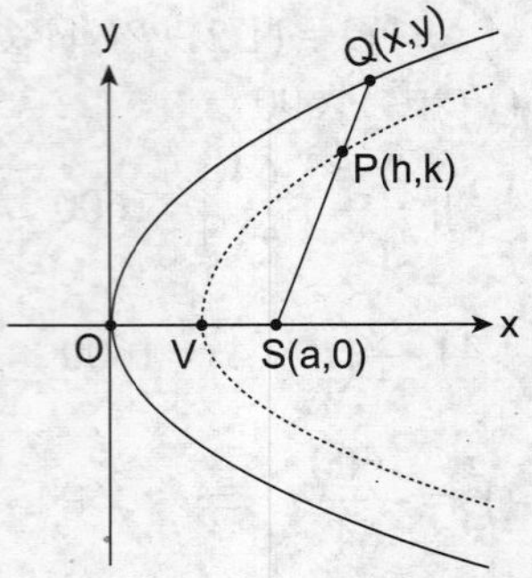

$\Rightarrow 4k^2 = 8ah - 4a^2 \Rightarrow k^2 = 2ah - a^2$

so locus of $P(h, k)$ is $y^2 = 2ax - a^2$

or $y^2 = 2a\left(x - \dfrac{a}{2}\right)$ which is a parabola and

its directrix is $x - \dfrac{a}{2} = -\dfrac{a}{2}$ i.e., $x = 0$

27. The axis of parabola is along the line $y = x$ and the distance of vertex from origin is $\sqrt{2}$ and that of origin from its focus is $2\sqrt{2}$. If vertex and focus both lies in the 1st quadrant, then the equation of the parabola is

(a) $(x + y)^2 = (x - y - 2)$

(b) $(x - y)^2 = (x + y - 2)$

(c) $(x - y)^2 = 4(x + y - 2)$

(d) $(x - y)^2 = 8(x + y - 2)$

Solution: (d) Since, distance of vertex from origin is $\sqrt{2}$ and focus is $2\sqrt{2}$.

$\therefore \quad V(1, 1)$ and $F(2, 2)$ (i.e., lying on $y = x$)

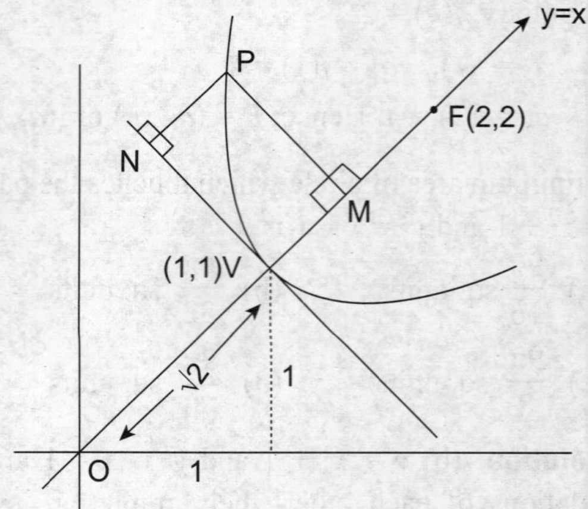

where, length of latusrectum $= 4a = 4\sqrt{2} \Rightarrow a = \sqrt{2}$

$\therefore$ By definition of parabola, $PM^2 = 4a(PN)$ where, PN is length of perpendicular upon $x + y - 2 = 0$ (i.e., tangent at vertex)

$\Rightarrow \quad \dfrac{(x-y)^2}{2} = 4\sqrt{2}\left(\dfrac{|x+y-2|}{\sqrt{2}}\right)$ ($\because$ (x, y) on parabola and origin lie on opposite sides of $x + y - 2 = 0$

$\Rightarrow x + y - 2 = 0) \Rightarrow (x - y)^2 = 8(x + y - 2)$

28. The equation of common tangents to the parabola $y = x^2$ and $y = -(x-2)^2$ is/are

(a) $y = 4(x-1)$ (b) $y = 0$

(c) $y = -4(x-1)$ (d) $y = -30x - 50$

Solution: (b, d) The equation of tangent to $y = x^2$, be

$$y = mx - \frac{m^2}{4}$$

putting in $y = -x^2 + 4x - 4$, we should get only one value of x, i.e., discriminant must be zero.

$$\Rightarrow mx - \frac{m^2}{4} = -x^2 + 4x - 4$$

$$\Rightarrow x^2 + x(m-4) + 4 - \frac{m^2}{4} = 0 \text{ and } D = 0$$

$$\Rightarrow (m-4)^2 - (16 - m^2) = 0$$

$$2m(m-4) = 0 \Rightarrow m = 0, 4$$

$\therefore$ $y = 0$ and $y = 4(x-1)$ are the required tangents.

29. The tangent PT and the normal PN to the parabola $y^2 = 4ax$ at a point P on it meet its axis at points T and N respectively. The locus of the centroid of the triangle PTN is a parabola whose

(a) vertex is $\left(\frac{2a}{3}, 0\right)$ (b) directrix is $x = 0$

(c) latus rectum is $\frac{2a}{3}$ (d) focus is (a, 0)

Solution: (a, d) Equation of tangent and normal at point $P(at^2, 2at)$ is $ty = x + at^2$ and $y = -tx + 2at + at^3$ respectively. Let the centroid of ΔPTN is $R(h, k)$

$$\Rightarrow h = \frac{at^2 + (-at^2) + 2a + at^2}{3} \text{ and } k = \frac{2at}{3}$$

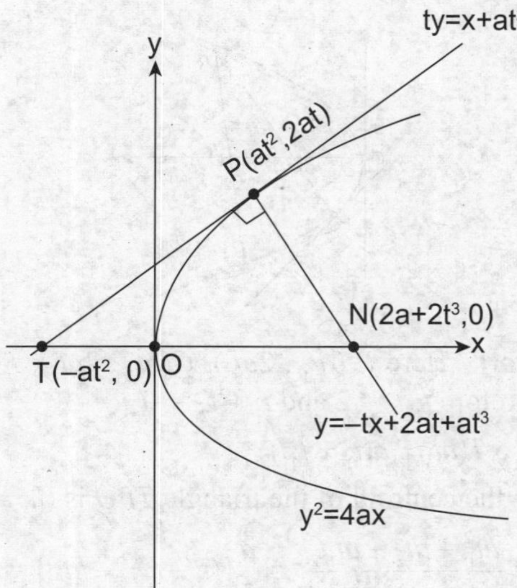

$$\Rightarrow 3h = 2a + a\left(\frac{3k}{2a}\right)^2 \Rightarrow 3h = 2a + \frac{9k^2}{4a}$$

$$\Rightarrow 9k^2 = 4a(3h - 2a)$$

$\therefore$ Locus of centroid is $y^2 = \frac{4a}{3}\left(x - \frac{2a}{3}\right)$ which is a parabola with its

$\therefore$ vertex at $\left(\frac{2a}{3}, 0\right)$; and directrix $x - \frac{2a}{3} = -\frac{a}{3}$

$\Rightarrow$ $x = \frac{a}{3}$; latus rectum $= \frac{4a}{3}$

$\therefore$ focus $\left(\frac{a}{3} + \frac{2a}{3}, 0\right)$, i.e., $(a, 0)$

SECTION-II

SUBJECTIVE-TYPE SOLVED EXAMPLES

1. The normals to the parabola $y^2 = 4ax$ at points P and Q meets the curve again at R. If T is the point of intersection of the tangents at P and Q to the parabola, then the locus of the centroid of the triangle TPQ is $y^2 = a(\lambda x + \mu a)$. Evaluate $\lambda + \mu$.

Solution: *Let* $P \equiv (at_1^2, 2at_1)$; $Q \equiv (at_2^2, 2at_2)$, then $T \equiv (at_1t_2, a(t_1 + t_2))$

$$\Rightarrow -t_1 - \frac{2}{t_1} = -t_2 - \frac{2}{t_2} \Rightarrow t_2 - t_1 = \frac{2}{t_1} - \frac{2}{t_2} = \frac{2(t_2 - t_1)}{t_1t_2}$$

$\therefore$ $t_1 \neq t_2 \Rightarrow t_1t_2 = 2$ .... (i)

Let the centroid of the triangle TPQ be (h, k), then

$$h = \frac{at_1^2 + at_2^2 + at_1t_2}{3}$$

$\therefore$ $3h = at_1^2 + at_2^2 + 2a$ (Using (i)) ....(ii)

and $k = \frac{2at_1 + 2at_2 + a(t_1 + t_2)}{3} = a(t_1 + t_2)$

therefore $k^2 = a^2t_1^2 + a^2t_2^2 + 2a^2t_1t_2$

$\therefore$ $k^2 = a^2t_1^2 + a^2t_2^2 + 4a^2$ (Using (i))

$\therefore$ $k^2 = a^2t_1^2 + a^2t_2^2 + 2a^2 + 2a^2 = 3ah + 2a^2$ (using (ii))

Therefore, required locus is $y^2 = a(3x + 2a)$

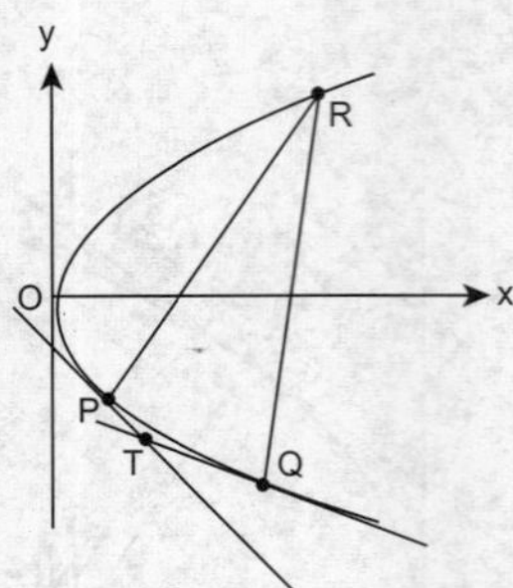

Alter: Here $P(at_1^2, 2at_1)$, $Q(at_2^2, 2at_2)$, $R(at_3^2, 2at_3)$ therefore $t_1 t_2 = 2$ and $t_1 + t_2 = -t_3$ ... (i)

Also $T(at_1t_2, a(t_1 + t_2))$

Let the centroid of the triangle TPQ is (h, k), then

$$h = \frac{at_1^2 + at_2^2 + at_1t_2}{3} = \frac{a}{3}\left((t_1+t_2)^2 - t_1t_2\right) = \frac{a}{3}\left(t_3^2 - 2\right)$$

(Using (i)) ...(ii)

$$\text{Also } k = \frac{2at_1 + 2at_2 + a(t_1+t_2)}{3} = \frac{a}{3}(2t_1 + 2t_2 - t_3)$$

(Using (i)) $\therefore\ k = \frac{a}{3}(2(t_1 + t_2) - t_3) = -at_3$

$\therefore\ t_3 = \frac{-k}{a}$. Put the value t_3 in equation (ii), we get

$$h = \frac{a}{3}\left(\frac{k^2}{a^2} - 2\right)$$

$\Rightarrow k^2 = 3ah + 2a^2$

$\therefore$ Therefore required locus is $y^2 = a(3x + 2a)$.

$\Rightarrow \lambda = 3, \mu = 2 \Rightarrow \lambda + \mu = 5$

2. Normals are drawn from the point 'P' with slopes m_1, m_2, m_3 to the parabola $y^2 = 4x$. If locus of P with $m_1 m_2 = \alpha$ is a part of the parabola itself then find α.

Solution: Equation of normal to $y^2 = 4x$ is

$y = mx - 2m - m^3$

Let it passes through (h, k)

$\Rightarrow k = mh - 2m - m^3$

$\Rightarrow m^3 + m(2 - h) + k = 0$ ...(1)

Here $m_1 + m_2 + m_3 = 0$,

$m_1 m_2 + m_2 m_3 + m_3m_1 = 2 - h$,

and $m_1m_2m_3 = -k$

where $m_1m_2 = \alpha$

$\Rightarrow m_3 = -\frac{k}{\alpha}$, it must satisfy equation (1)

$$\Rightarrow -\frac{k^3}{\alpha^3} - \frac{k}{\alpha}(2 - h) + k = 0$$

$\Rightarrow k^2 = \alpha^2h - 2\alpha^2 + \alpha^3$, thus locus of P is

$y^2 = \alpha^2x - 2\alpha^2 + \alpha^3$ which is same as $y^2 = 4x$ (given)

On comparing with $y^2 = 4x$, we have

$\Rightarrow \alpha^2 = 4$ and $-2\alpha^2 + \alpha^3 = 0 \Rightarrow \alpha = 2$

3. Find the locus of a point which is such that

(i) two of the normals drawn from it to the parabola are at right angles,

(ii) the three normals though it cut the axis in points whose distances from the vertex are in arithmetical progression.

Solution: Any normals to parabola $y^2 = 4ax$ is $y = mx - 2am - am^3$ and this passes through the point (h, k) if $am^3 + (2a - h)m + k = 0$ ....(1)

If m_1, m_2 and m_3 be the roots, we have

$m_1 + m_2 + m_3 = 0$ ......(2)

$m_2m_3 + m_3m_1 + m_1m_2 = \frac{2a-h}{a}$ ....(3)

and $m_1m_2m_3 = -k/a$ ....(4)

(i) If two of the normals, say m_1 and m_2 be at right angles, we have $m_1m_2 = -1$ and hence, from (4), $m_3 = k/a$

The quantity $\frac{k}{a}$ is therefore, a root of (1) and hence, by substitution, we have

$$\frac{k^3}{a^2} + (2a-h)\frac{k}{a} + k = 0$$

i.e., $k^2 = a(h - 3a)$

$\therefore$ The locus of the point (h, k) is therefore the parabola $y^2 = a(x - 3a)$ whose vertex is the point $(3a, 0)$ and whose latus rectum is one-quarter that of the given parabola

(ii) The normal $y = mx - 2am - am^3$ meets the axis of x at a point whose distance from the vertex is $2a + am^2$. The conditions of the question then gives

$$\left(2a + am_1^2\right) + \left(2a + am_3^2\right) = 2\left(2a + am_2^2\right)$$

i.e., $m_1^2 + m_3^2 = 2m_2^2$ ....(5)

If we eliminate m_1, m_2 and m_3 from the equations (2), (3), (4) and (5), we shall have relation between h and k.

$$\frac{2a-h}{a} = m_1m_3 + m_2\left(m_1 + m_3\right) = m_1m_2 - m_2^2 \quad ....(6)$$

Also (5) and (2) gives

$$2m_2^2 = \left(m_1 + m_3\right)^2 - 2m_1m_3 = m_2^2 - 2m_1m_3$$

i.e., $m_2^2 + 2m_1m_3 = 0$ ....(7)

Solving (6) and (7), we have

$$m_1m_3 = \frac{2a-h}{3a}, \text{ and } m_2^2 = -2 \times \frac{2a-h}{3a}$$

Substituting these values in (4), we have

$$\frac{2a-h}{3a}\sqrt{-2\frac{2a-h}{3a}} = -\frac{k}{a} \text{ i.e., } 27ak^2 = 2(h-2a)^3$$

So that the required locus is $27ay^2 = 2(x-2a)^3$

4. Find the equations of the tangents to the parabola $y^2 = 12x$, which passes through the point (2, 5).

Solution: Equation of parabola $y^2 = 12x \Rightarrow a = 3$

Let R is the intersection point of both tangents

$\therefore$ The co-ordinates of $R \equiv (at_1t_2, a(t_1 + t_2))$

But $R \equiv (2, 5)$................. (given)

and $a = 3$

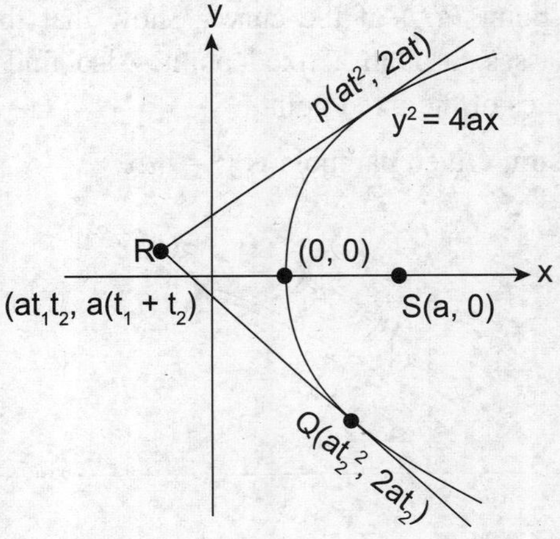

then $at_1t_2 = 2 \Rightarrow 3t_1t_2 = 2 \Rightarrow t_1t_2 = \frac{2}{3}$ ...(1)

and $a(t_1 + t_2) = 5 \Rightarrow 3(t_1 + t_2) = 5 \Rightarrow t_1 + t_2 = \frac{5}{3}$ ...(2)

Solving equation (1) and (2) we have $t_1 = 1$ or $t_1 = \frac{2}{3}$

when $t_1 = \frac{2}{3}$ then $t_2 = 1$ and when $t_1 = 1, t_2 = \frac{2}{3}$

without loss of generality. So, let $t_1 = 1$ and $t_2 = \frac{2}{3}$

Now the equation of tangent in parametric form is $ty = x + at^2$

$t_1 = 1 \Rightarrow$ equation of tangent is $y = x + 3$

and $t_2 = 2/3 \Rightarrow$ equation of tangent is $2y = 3x + 4$

5. The normal at a point P to the parabola $y^2 = 4ax$ meets its axis at G.Q is another point on the parabola such that QG is perpendicular to the axis of the parabola. Prove that $QG^2 - PG^2 =$ constant.

Solution: Co-ordinates of point P are given by $(at^2, 2at)$.

Also equation of normal at P is $y + xt = 2at + at^3$

$\Rightarrow$ G is $(2a + at^2, 0)$

$\therefore\ PG^2 = 4a^2 + 4a^2t^2$

Q is a point on the parabola such that QG is perpendicular to axis so that its ordinate is (QG) and abscissa is same as that of G.

Hence the point Q is $(2a + at^2, QG)$

But Q lies on the parabola $y^2 = 4ax$

$\therefore\ QG^2 = 4a(2a + at^2)$

$QG^2 = 8a^2 + 4a^2t^2$

$\therefore\ QG^2 - PG^2 = 4a^2 =$ constant

6. Find the equation of the circle which passes through the focus of the parabola $x^2 = 4y$ and touches it at the point (6, 9)

Solution: Equation of tangent at point (6, 9) is given by $x.6 = 2(y + 9)$; focus S:(0, 1)

$\Rightarrow y = 3x - 9$ .....(i)

Equation of required circle is given by

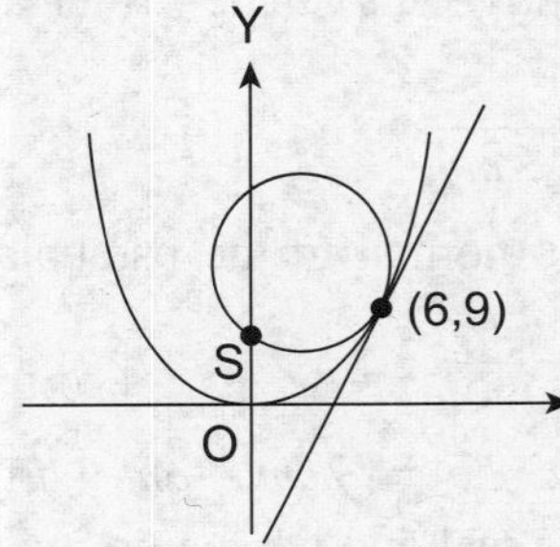

$(x - 6)^2 + (y - 9)^2 + \lambda(3x - y - 9) = 0$ ....(ii)

As it passes through (0, 1)

$\Rightarrow 36 + 64 + \lambda(-10) = 0$

$\Rightarrow \lambda = 10$

$\therefore$ equation of required circle: $(x - 6)^2 + (y - 9)^2 + 10(3x - y - 9) = 0$

i.e., $x^2 + y^2 + 18x - 28y + 27 = 0$.

7. Prove that the two parabolas $y^2 = 4ax$ and $y^2 = 4c(x - b)$ cannot have a common normal, other than the axis, unless $\frac{b}{(a-c)} > 2$.

Solution: The equation of any normal of slope m to the parabola $y^2 = 4c(x - b)$ is

$y = m(x - b) - 2cm - cm^3$

or $y = mx - mb - 2cm - cm^3$

For this to be normal to $y^2 = 4ax$, we must have

$-mb - 2cm - cm^3 = -2am - am^3$

$\Rightarrow b + 2c + cm^2 = 2a + am^2$

$\Rightarrow b + 2c - 2a = (a - c)m^2$

$\Rightarrow m^2 = \frac{b-2(a-c)}{a-c}$

or $m=\sqrt{\dfrac{b}{a-c}-2}$, since m is real, therefore

$\dfrac{b}{a-c}-2\ge 0$

$\Rightarrow \dfrac{b}{a-c}\ge 2$ for $\dfrac{b}{a-c}=2, m=0$

$\Rightarrow$ x-axis would be the common normal

$\therefore \dfrac{b}{a-c}>2$ is the required condition for two parabolas to have common normal other than axis.

8. Find the condition on 'a' and 'b' so that the two tangents drawn to the parabola $y^2 = 4ax$ from a point are normals to the parabola $x^2 = 4by$.

Solution: Let $y = mx + \dfrac{a}{m}$ be a tangent to the parabola $y^2 = 4ax$. This can be written as $x=\dfrac{y}{m}-\dfrac{a}{m^2}$

or $x=\left(\dfrac{1}{m}\right)y-\dfrac{a}{m^2}$

This will be a normal to the parabola $x^2 = 4$ by, if

$-\dfrac{a}{m^2}=-\dfrac{2b}{m}-\dfrac{b}{m^3}$

$\Rightarrow am = 2bm^2 + b \Rightarrow 2bm^2 - am + b = 0$

Since m is real and we require two different tangents/normals, therefore

discriminant $= a^2 - 8b^2 > 0$

$\Rightarrow a^2 > 8b^2$

9. Two tangents to parabola $y^2= 8x$ meet the tangent at its vertex in the points P and Q. If $PQ = 4$ units, prove that the locus of the point of the intersection of the two tangents is $y^2 = 8\,(x + 2)$

Solution: Let the two tangent be at points $R(t_1)$ and $S(t_2)$ given by $t_1 y = x + at_1^2$ ....(1)

and $t_2 y = x + at_2^2$ ....(2)

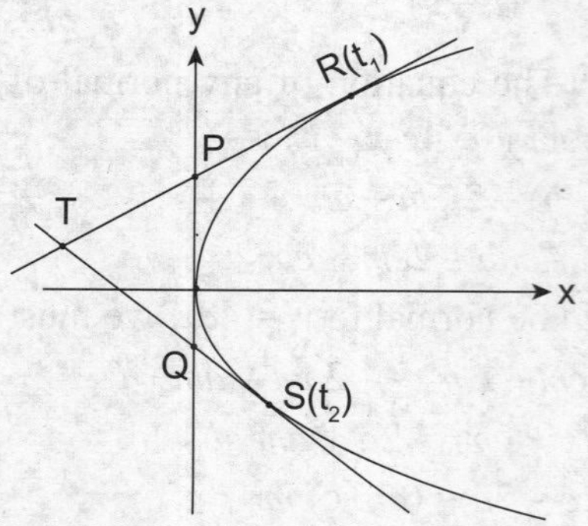

Here $4a = 8 \Rightarrow a = 2$; Let $T(h, k)$ be the point of intersection of (1) and (2)

$\Rightarrow h = at_1t_2;\ k = a(t_1 + t_2)$

$\Rightarrow h = 2t_1t_2;\ k = 2(t_1 + t_2)$ ....(3)

Point of intersection of tangents with tangent at the vertex $x = 0$ are $P\ (0, at_1)$ and $Q\ (0, at_2)$

Now, $PQ = 4 \Rightarrow a|t_1 - t_2| = 4$

$\Rightarrow |t_1 - t_2| = 2$ $(\because a = 2)$

$\Rightarrow (t_1 + t_2)^2 - 4t_1t_2 = 4$

$\Rightarrow \dfrac{k^2}{4}-\dfrac{4h}{2}=4$ from (3)

$\therefore$ locus of required point is $y^2 = 8(x + 2)$

10. A variable chord joining the points $Q(t_1)$ and $R(t_2)$ of the parabola $y^2 = 4ax$ subtends a right angle at a fixed point $P(t_0)$ of the curve. Show that the chord QR passes through a fixed point. Also find the co-ordinates of the fixed point.

Solution: Given parabola is $y^2 = 4ax$

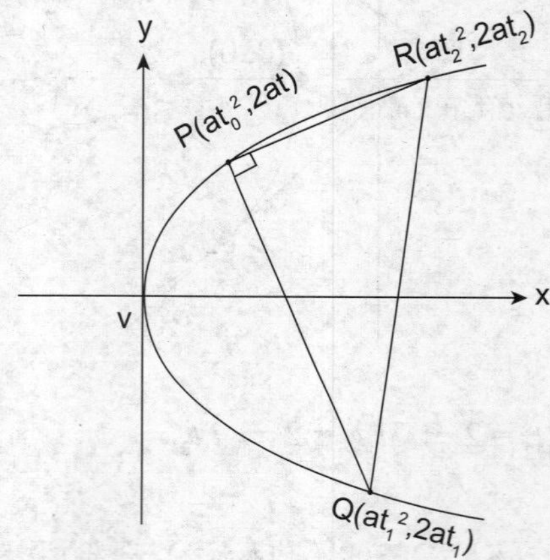

QR is a chord joining the point $Q(t_1)$ and $R(t_2)$ subtends a right angle at fixed point $P(t_0)$ of the parabola $y^2 = 4ax$ such that $PQ \perp PR$.

$\Rightarrow$ (slope of PQ) × (slope of PR) = – 1

$\Rightarrow \dfrac{2}{t_0+t_1}\times\dfrac{2}{t_0+t_2}=-1$

$\Rightarrow t_0^2 + t_0\,(t_1 + t_2) + t_1t_2 = -4$ ....(1)

The equation of chord QR is $(t_1 + t_2)y = 2x + 2at_1t_2$; putting the value of $t_1\, t_2$ from equation (1) we have

$y(t_1 + t_2) = 2x + 2a\,[-t_0^2 - t_0(t_1 + t_2) - 4]$

$\Rightarrow (y + 2a\,t_0)\,(t_1 + t_2) = 2x - 2at_0^2 - 8a$

$\Rightarrow (2x - 2at_0^2 - 8a) - (t_1 + t_2)\,(y + 2at_0) = 0$

It is in the form $L_1 + \lambda L_2 = 0$; i.e., a family of lines passing through the intersection of lines $L_1 = 0$

i.e., $2x - 2at_0^2 - 8a = 0$ and $L_2 = 0$ ie., $y + 2at_0 = 0$

Hence, QR passes through a fixed point $(at_0^2 + 4a, -2at_0)$

11. Two perpendicular straight lines through the focus of the parabola $y^2 = 4ax$ meets its directrix at T and T' respectively. Show that the tangents to the parabola parallel to the perpendicular lines intersect at the mid-point of TT'.

Solution: Any line through the focus (a, 0) is $y = m(x - a)$ ...(1)

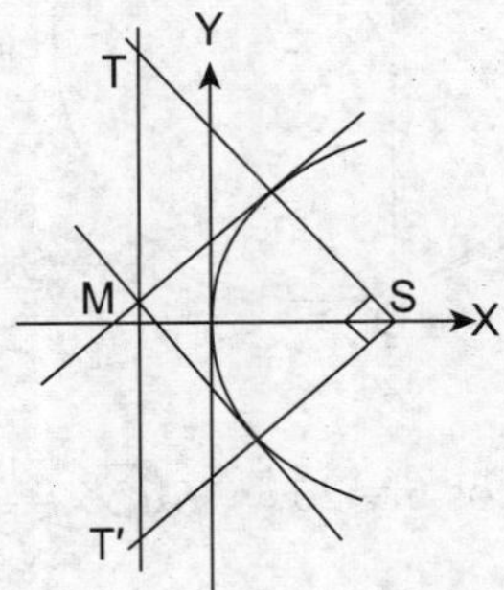

Any line perpendicular to line (1) and through (a, 0) is

$$y = -\frac{1}{m}(x - a) \quad ...(2)$$

Solving (1) and (2) with direction $x = -a$, we get the points T and T' as

$T(-a, -2am)$ and $T'\left(-a, \frac{2a}{m}\right)$

$\therefore$ mid-point of $TT' \equiv \left[-a, a\left(\frac{1}{m} - m\right)\right]$ ...(3)

Now tangents parallel to lines (1) and (2) are given by

$$y = mx + \frac{a}{m} \quad ...(4)$$

and $y = -\frac{1}{m}x - am$ ...(5)

(4) – (5) gives, $0 = x\left(m + \frac{1}{m}\right) + a\left(m + \frac{1}{m}\right)$

or $x = -a$ and hence on putting for x, we get $y = a\left(\frac{1}{m} - m\right)$

Thus, the point of intersection M is $\left[-a, a\left(\frac{1}{m} - m\right)\right]$ which is mid-point of TT' by (3).

12. A variable chord PQ of the parabola $y^2 = 4x$ is drawn parallel to the line $y = x$. If the parameters of the points P and Q on the parabola are p and q respectively, show that $p + q = 2$. Also show that the locus of the point of intersection of the normals at P and Q is $2x - y = 12$.

Solution: Equation of parabola is $y^2 = 4x \Rightarrow (a = 1)$

Also, equation of PQ is given by $(t_1 + t_2)y = 2x + 2at_1t_2$

or $y = \frac{2}{t_1 + t_2}x + \frac{2at_1t_2}{t_1 + t_2}$ ...(1)

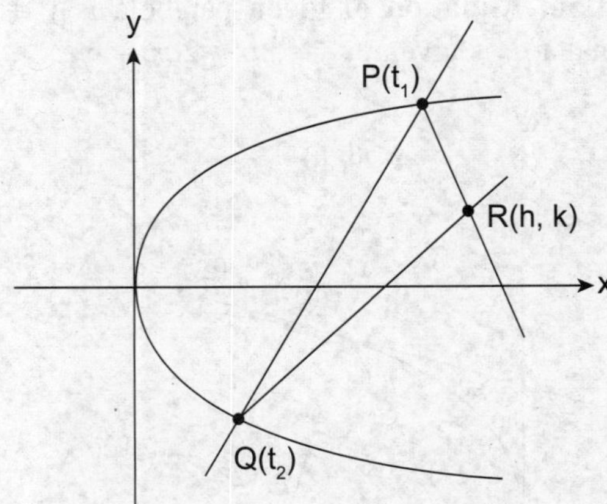

It is parallel to the line $y = x$

$\therefore$ slope of chord PQ is $= 1$

$\Rightarrow \frac{2}{t_1 + t_2} = 1$

$\Rightarrow t_1 + t_2 = 2$ ...(2)

Given that $t_1 = p$ and $t_2 = q$, from (2) we get $p + q = 2$ ...(3)

Normal at $P(p)$ is given by $y = -px + 2ap + ap^3$ ...(4)

Normal at $Q(q)$ is given by $y = -qx + 2aq + aq^3$ ...(5)

Let intersection point of (4) and (5) is $R(h, k)$

$\therefore$ $R(h, k)$ satisfies both equations

$\Rightarrow k = -ph + 2ap + ap^3$ ...(6)

and $k = -qh + 2aq + aq^3$ ...(7)

(6) + (7) gives $2k = -h(p + q) + 2a(p + q) + a(p^3 + q^3)$

Using equation (3), $p + q = 2$ we have

$2k = -2h + 4a + a(p + q)(p^2 + q^2 - pq)$

or $2k = -2h + 4a + 2a(p^2 + q^2 - pq)$

$\Rightarrow 2k = -2h + 4 + 2(p^2 + q^2 - pq)$ ($\because a = 1$)

$\Rightarrow k + h = 2 + (p^2 + q^2 - pq)$

$\Rightarrow k + h = 2 + 4 - 3pq$ ($\because p + q = 2 \Rightarrow p^2 + q^2 = 4 - 2pq$)

$\Rightarrow k + h = 6 - 3pq$ ...(8)

(6) – (7) gives $0 = h(q - p) - 2a(q - p) - a(q^3 - p^3)$

$\Rightarrow 0 = h(q - p) - 2(q - p) - (q^2 + p^2 + qp)(q - p)$

$\Rightarrow h = 2 + (q^2 + p^2 + pq)$

$\Rightarrow p^2 + q^2 + pq = h - 2$

or $4 - h + 2 = pq$ ($\because p^2 + q^2 = 4 - 2pq$)

or $pq = 6 - h$ ...(9)

putting vlaue of pq from (9) in (8) we have

$k + h = 6 - 3(6 - h)$

$\Rightarrow k + h = 6 - 18 + 3h$

The locus of point is $2x - y = 12$

13. PC is the normal at P to the parabola $y^2 = 4ax$, C being on the axis. CP is produced outwards to Q so that $PQ = CP$; show that the locus of Q is a parabola.

Solution: Equation of given parabola is $y^2 = 4ax$.
Normal at P is given by $y = mx - 2am - am^3$

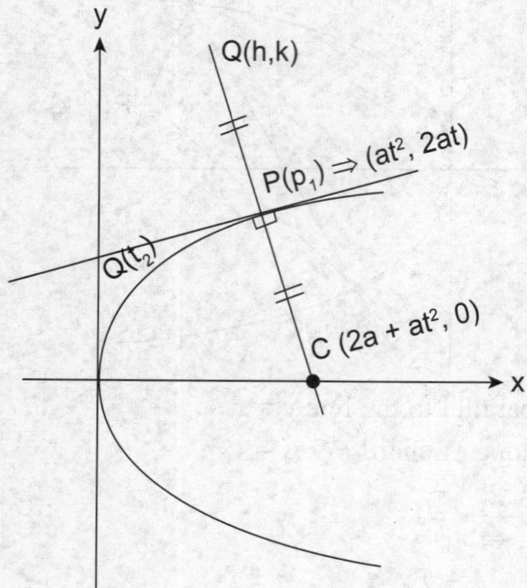

or $y = -tx + 2at + at^3$ ($\because m = -t$)

It would be intersect x-axis, where $y = 0$

i.e., $tx = 2at + at^3$

$\Rightarrow x = 2a + at^2 = a(2 + t^2)$

$\therefore$ C $\equiv (2a + at^2, 0)$

$\because$ $PQ = CP$, therefore P is the mid-point of QC

$\Rightarrow h + 2a + at^2 = 2at^2$ ...(1)

and $k = 4at \Rightarrow t = k/4a$ .....(2)

Substituting $t = k/4a$ from (2) in (1),

we get, $h + 2a = a\left(\dfrac{k}{4a}\right)^2$

$\Rightarrow h + 2a = \dfrac{k^2}{16a}$

$\therefore$ locus of point Q is $y^2 = 16a (x + 2a)$

which is a parabola having its vertex at (–2a, 0) and axis parallel to x-axis.

14. Prove that the parabola $y^2 = 16x$ and the circle $x^2 + y^2 - 40x - 16y - 48 = 0$ meet at the point $P(36, 24)$ and one other point Q, find that point. Prove that PQ is a diameter of the circle.

Solution: Equation of parabola is $y^2 = 16x$ ...(1)

and the equation of given circle is $x^2 + y^2 - 40x - 16y - 48 = 0$ ...(2)

Its centre: C (20, 8) and radius: $r = 16\sqrt{2}$

Also, point $P \equiv (36, 24)$ and $C \equiv (20, 8)$

$\therefore$ slope of $PC = \tan\theta = \dfrac{24-8}{36-20} = \dfrac{16}{16} = 1$

$\Rightarrow \theta = \dfrac{\pi}{4}$

Using parametric form, the co-ordinates of P and Q points are given by

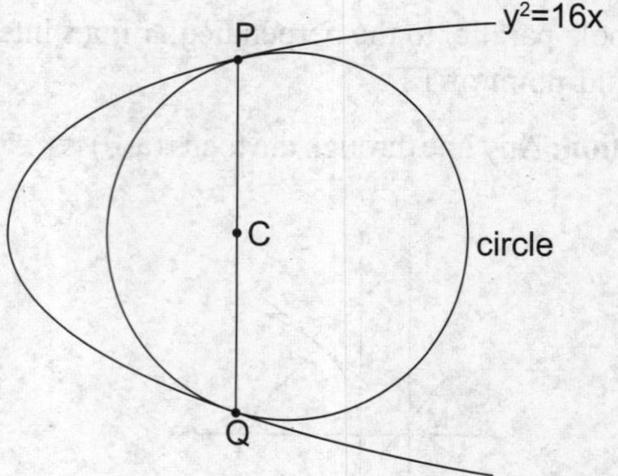

$\Rightarrow x = x_1 \pm r\cos\theta = 20 \pm 16\sqrt{2} \times \dfrac{1}{\sqrt{2}} = 36, 4$

and $y = y_1 \pm r\sin\theta = 8 \pm 16\sqrt{2} \times \dfrac{1}{\sqrt{2}} = 24, -8$

$\Rightarrow P \equiv (36, 24)$ and $Q \equiv (4, -8)$

$\therefore$ $Q(4, -8)$;

Now $PQ = \sqrt{(36-4)^2 + (24+8)^2}$

$= \sqrt{2(32)^2} = 32\sqrt{2} = 2r$

= diameter of circle

Thus PQ is diameter of given circle.

15. A variable tangent to the parabola $y^2 = 4ax$ meets the circle $x^2 + y^2 = r^2$ at P and Q. Prove that the locus of the mid-point of PQ is $x(x^2 + y^2) + ay^2 = 0$.

Solution: Equation of tangent to parabola $y^2 = 4ax$ at point (x_1, y_1) is given by $yy_1 = 2a(x + x_1)$ ......(1)

Also (x_1, y_1) lies on parabola

$\Rightarrow y_1^2 = 4ax_1$ ....(2)

Equation of given circle is $x^2 + y^2 = r^2$; Let (h, k) be the mid-point of PQ

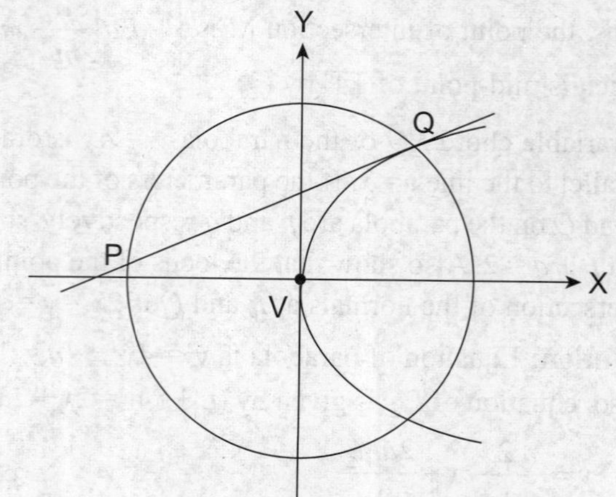

$\because$ equation of chord whose mid-point is (h, k) is given by $T = S_1$

i.e., $xh + yk = h^2 + k^2$ ...(3)

$\therefore$ equation (1) and (3) are identical. Therefore

$$\frac{2a}{h} = -\frac{y_1}{k} = \frac{2ax_1}{-(h^2+k^2)}$$

$$\Rightarrow y_1 = -\frac{2ak}{h} \quad ...(4)$$

$$\text{and } x_1 = -\frac{(h^2+k^2)}{h} \quad ...(5)$$

Putting the value of x_1 and y_1 from equation (4) and (5) in (2)

We get, $\left(\frac{-2ak}{h}\right) \quad \left(\frac{-(h^2+k^2)}{h}\right)$

$$\Rightarrow \frac{4a^2k^2}{h^2} = -\frac{4a}{h}(h^2+k^2)$$

$\Rightarrow 4ah\,(h^2 + k^2) + 4a^2k^2 = 0$

$\Rightarrow$ Locus of mid-point of PQ is given by $x(x^2 + y^2) + ay^2 = 0$

16. A fixed parabola $y^2 = 4\,ax$ touches a variable parabola. Find the equation to the locus of the vertex of the variable parabola. Assume that the two parabolas are equal and the axis of the variable parabola remains parallel to the x-axis.

Solution: We know that, two parabolas are equal, if their latus-rectums are equal. Also two parabolas having parallel axes can touch each other if they open in different directions.

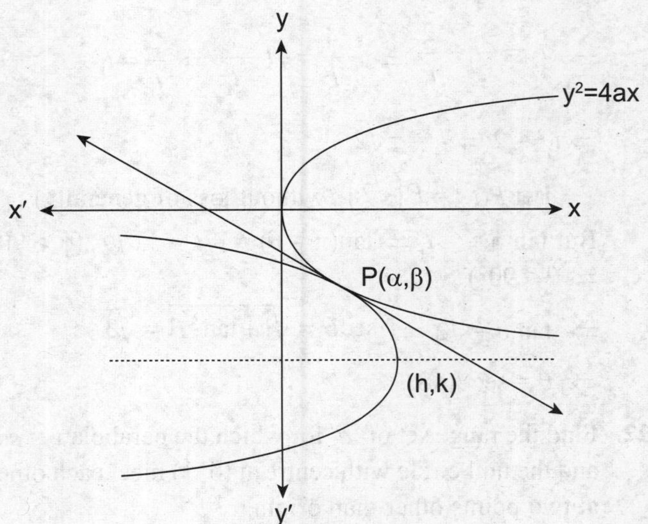

Let (h, k) be the vertex of the variable parabola. Let the fixed parabola be $y^2 = 4ax$ and having axis along OX. Then its equation will be $(y - k)^2 = -4a\,(x - h)$

Suppose these two parabolas touch each other at point $P(\alpha, \beta)$.

The equation of the tangent to $y^2 = 4ax$ at $P(\alpha, \beta)$ is $\beta y = 2a\,(x + \alpha)$

$\Rightarrow 2ax - \beta y + 2a\alpha = 0$ ...(1)

The equation of the tangent at $P(\alpha, \beta)$ to $(y - k)^2 = -4a\,(x - h)$ will be $(y - k)\,(\beta - k)$

$= -2a\,(x + \alpha - 2h)$

$\Rightarrow 2ax + (\beta - k)y + 2a\,(\alpha - 2h) - k(\beta - k) = 0$ ...(2)

Clearly, (1) and (2) represent the same line.

$$\text{Therefore } \frac{2a}{2a} = -\frac{\beta}{\beta - k} = \frac{2a\alpha}{2a(\alpha - 2h) - k(\beta - k)}$$

$$\Rightarrow \beta = \frac{k}{2} \text{ and } 2a\alpha = 2a(\alpha - 2h) - k\,(\beta - k)$$

$$\Rightarrow \beta = \frac{k}{2} \text{ and } 0 = -4ah - k\,(\beta - k)$$

$$\text{or } 4ah = -k\left(\frac{k}{2} - k\right)$$

$\Rightarrow k^2 = 8ah$, hence the locus of (h, k) is $y^2 = 8ax$

17. A tangent is drawn to the parabola $y^2 = 4x$ at the point 'P' whose abscissa lies in the interval [1,4]. Find the maximum possible area of the triangle formed by the tangent at 'P', ordinate of the point 'P' and the x-axis.

Solution: Equation of tangent to parabola at $P(t)$ is given by $ty = x + t^2$, $\tan\theta = \frac{1}{t}$

$\therefore$ Area of $\Delta APN = \Delta = \frac{1}{2}(AN)\,(PN) = \frac{1}{2}\,(2t^2)\,(2t)$

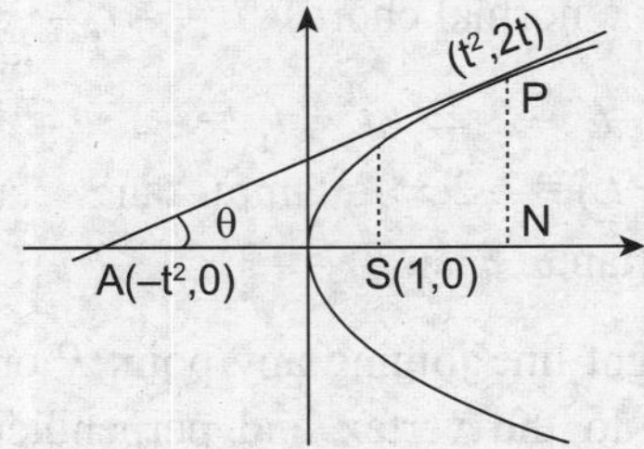

$\Rightarrow \Delta = 2t^3 = 2(t^2)^{3/2}$

$t^2 \in [1, 4] \Rightarrow \Delta_{max}$ occurs when $t^2 = 4$

$\Rightarrow \Delta_{max} = 16$ square units

$\Rightarrow$ The maximum area of Δ is 16 square units.

18. Find the length of the normal chord of the parabola, $y^2 = 4x$, which makes an angle of $\frac{\pi}{4}$ with the axis of x.

Solution: Let the chord whose length is to be found be normal at point $P(t_1)$.

$\therefore$ Equation of normal chord is $y + t_1x = 2t_1 + t_1^3$ ...(i)
Chord is subtending angle $\pi/4$ with positive direction of x-axis. Thus, the slope of chord (i) $= m = -t_1 = \tan \pi/4 = 1$.

$\Rightarrow t_1 = -1 \therefore$ co-ordinates of P will be $(1, -2)$

Hence parameter at $Q = t_2 = -t_1 - 2/t_1 = 1 + 2 = 3$

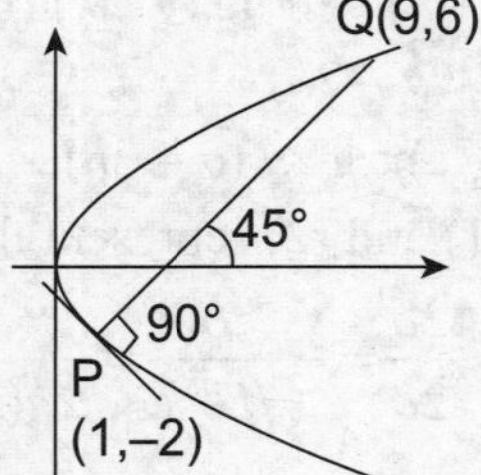

$\therefore$ Co-ordinates of Q are $(9, 6)$

$\therefore$ $l(PQ) = \sqrt{64+64} = 8\sqrt{2}$ units

19. If the normal to a parabola $y^2 = 4ax$ at P meets the curve again in Q and if the normal at P and the normal at Q, makes angles α and β respectively with the x-axis, then evaluate $\tan \alpha (\tan \alpha + \tan \beta)$.

Solution: Slope of normal at $P = -t_1$ and Slope of normal at $Q = -t_2$

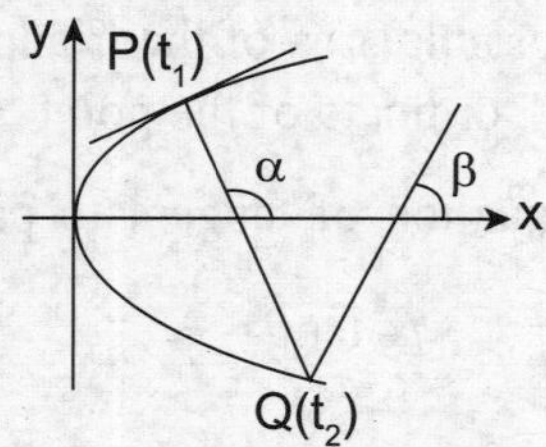

$\Rightarrow \tan \alpha = -t_1$ and $\tan \beta = -t_2$

Now, (for a normal chord) $t_2 = -t_1 - \dfrac{2}{t_1}$

$\Rightarrow t_1t_2 = -t_1^2 - 2 \Rightarrow t_1t_2 + t_1^2 = -2$

$\Rightarrow t_1(t_1 + t_2) = -2 \Rightarrow -\tan \alpha(-\tan\alpha - \tan\beta) = -2$

$\Rightarrow \tan \alpha(\tan \alpha + \tan \beta) = -2$.

20. The straight line joining any point P on the parabola $y^2 = 4ax$ to the vertex and perpendicular from the focus to the tangent at P, intersect at R, then find the locus of R.

Solution: Equation of tangent to parabola $y^2 = 4ax$ at point $P(t)$ is given by $ty = x + at^2$ ... (i)

Equation of line perpendicular to (i) is given by $y + tx = k$ .... (ii)

Now (ii) passes through focus $S(a, 0) \Rightarrow k = at$

$\therefore$ (ii) becomes $y + tx = at$ ... (iii)

and equation of OP is $y = \dfrac{2}{t}x$ .... (iv)

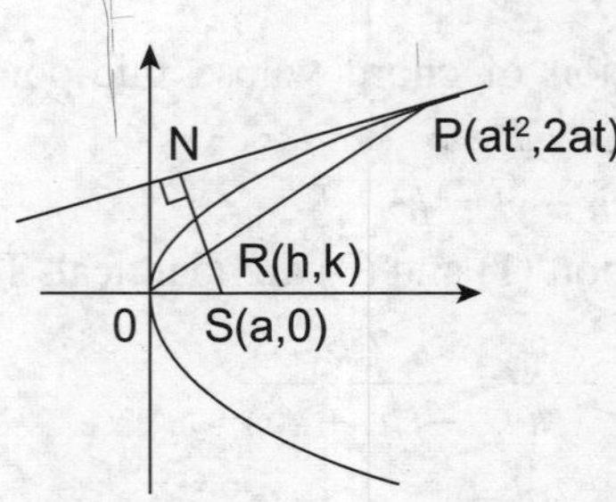

To find the locus of R. We are to eliminate 't' from (iii) and (iv).

$$\Rightarrow \quad y + \frac{2x}{y}.x = a\left(\frac{2x}{y}\right) \Rightarrow y^2 + 2x^2 = 2ax$$

which is the required locus of R.

21. A normal chord of the parabola $y^2 = 4x$ subtending a right angle at the vertex makes an acute angle θ with the x-axis, then find θ.

Solution: (slope of OP) . (Slope of OQ) = –1

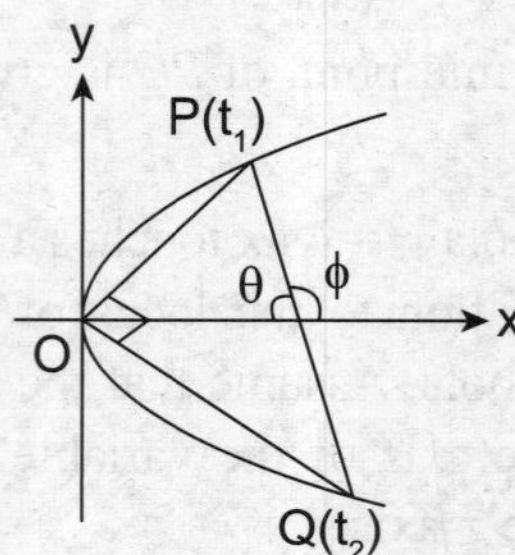

$\Rightarrow \quad \dfrac{2}{t_1}.\dfrac{2}{t_2} = -1 \Rightarrow t_1t_2 = -4$ .... (i)

Also, PQ is normal at P and meets the parabola again at t_2.

$$\Rightarrow \quad t_2 = -t_1 - \frac{2}{t_1} \Rightarrow -\frac{4}{t_1} = -t - \frac{2}{t_1} \Rightarrow \frac{2}{t_1} = t_1$$

$\Rightarrow t_1^2 = 2 \Rightarrow t_1 = \pm\sqrt{2}$

Let $P(t_1) = P\left(\sqrt{2}\right)$ (without loss of generally)

But $\tan \phi = -t_1 \Rightarrow \tan(\pi - \theta) = -t_1 = -\sqrt{2}$ ($\phi > 90°$ $\Rightarrow \theta < 90°$)

$\Rightarrow \tan \theta = \sqrt{2} \Rightarrow \sec\theta = \sqrt{1 + \tan^2\theta} = \sqrt{3}$

$\Rightarrow \theta = \sec^{-1}(\sqrt{3})$.

22. Find the range set of 'a' for which the parabola $y = ax^2$ and the unit circle with centre at (0, 1) meet each other at two points other than origin.

Solution: Equation of circle is $(x - 0)^2 + (y - 1)^2 = (1)^2$

i.e., $x^2 + y^2 - 2y = 0$ .... (i)

At the point of intersection of (i) and parabola $y = ax^2$

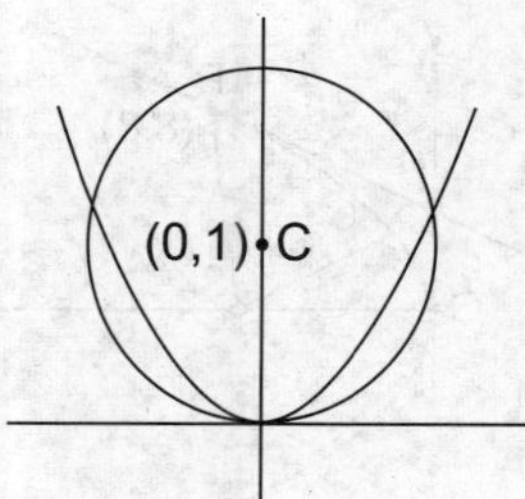

$\frac{y}{a} + y^2 - 2y = 0$ (substituting $x^2 = y/a$ in (i))

$\Rightarrow \; y\left(\frac{1}{a} + y - 2\right) = 0 \Rightarrow y = 0$ or $y = 2 - \frac{1}{a}$

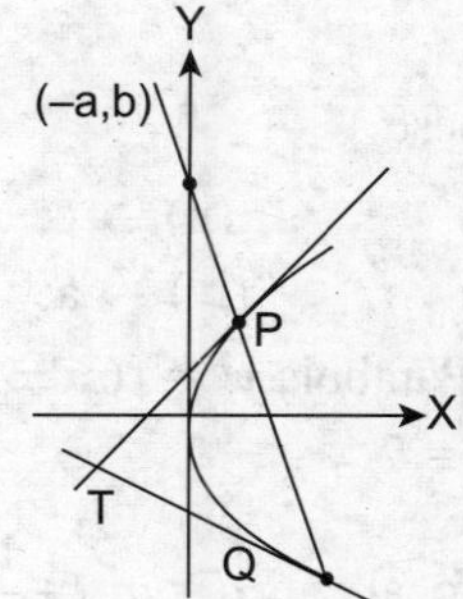

∴ For points of intersection other than origin,

$y = 2 - \frac{1}{a}$

$\Rightarrow \; x^2 = \frac{y}{a} = \frac{2}{a} - \frac{1}{a^2} > 0 \Rightarrow \frac{2a-1}{a^2} > 0 \Rightarrow 2a - 1 > 0$

$\Rightarrow \; a > \frac{1}{2} \Rightarrow a \in \left(\frac{1}{2}, \infty\right)$

23. Let TP and TQ are tangents to the parabola, $y^2 = 4ax$ at P and Q. If the chord PQ passes through a fixed point $(-a, b)$, then find the locus of T.

Solution: Let (h, k) be the co-ordinates of point T

∴ PQ will be the chord of contact of $T(h, k)$

∴ Its equation is given by $T = 0$

i.e., $ky = 2a(x + h)$ .... (i)

Now (i) passes through the fixed point $(-a, b)$

⇒ (i) would be satisfied by point $(-a, b)$

⇒ $kb = 2a(-a + h)$

∴ Locus of T is $2a^2 - 2ax + by = 0$

or $2ax - by = 2a^2$

24. Through the vertex O of the parabola, $y^2 = 4ax$ two chords OP and OQ are drawn and the circles on OP and OQ as diameters intersect in R. If θ_1, θ_2 and ϕ are the angles made with the axis by the tangents at P and Q on the parabola and by OR, then evaluate $\cot \theta_1 + \cot \theta_2$ in term of ϕ.

Solution: Since OP and OQ are diameters of circles, $\angle ORQ = \angle ORP = \pi/2$

⇒ PRQ is a straight line

⇒ $OR \perp$ to chord PQ of parabola

⇒ Slope of $OR = \frac{-1}{\text{slope of } PQ}$

$$\Rightarrow \; \tan\phi = \frac{-1}{\left(\frac{2}{t_1 + t_2}\right)} = -\frac{(t_1 + t_2)}{2} \quad ....(i)$$

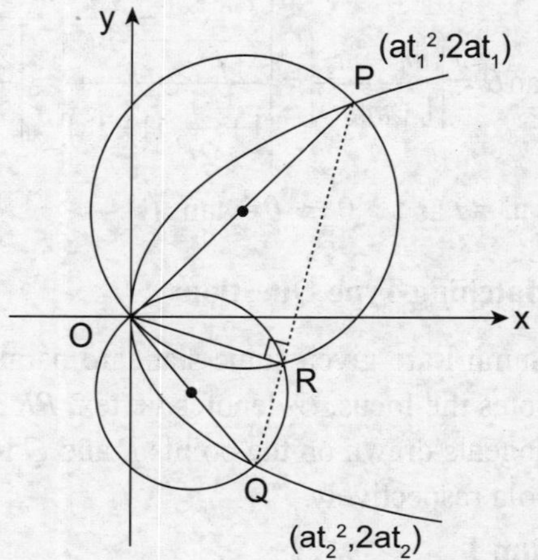

Also slope of tangent to parabola at point $P(t_1) = \frac{1}{t_1} = \tan\phi$

⇒ $t_1 = \cot\theta_1$ and slope of tangent to parabola at point $Q(t_2) = \frac{1}{t_2} = \tan\theta_2 \Rightarrow t_2 = \cot\theta_2$.

⇒ $\cot\theta_1 + \cot\theta_2 = t_1 + t_2 = -2\tan\phi$ (from (i))

25. The tangent and normal at $P(t)$, for all real positive t, to the parabola $y^2 = 4ax$ meet the axis of the parabola in T and G respectively, then find the acute angle at which the tangent at P to the parabola is inclined to the tangent at P to the circle passing through the points P, T and G.

Solution: Slope of tangent of parabola at point $P(t) = m_1 = 1/t$ ... (i)

We know that if tangent to parabola at a point P meets the axis of parabola at T and normal at P meets the axis at G, then $SP = ST = SG$.

∴ the circle passing through T, P and G must have its centre at S (focus)

∴ PS will be normal at P to the circle.

⇒ Slope of tangent to circle at point P.

$$P = \frac{-1}{\text{Slope of } PS} = \frac{-1}{\left(\frac{2t}{t^2 - 1}\right)} = \frac{1 - t^2}{2t} = m_2 \text{ (say)}$$

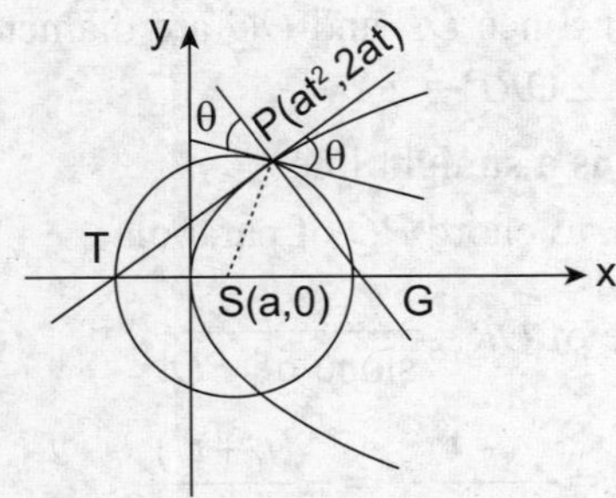

$\therefore$ Angle between tangent at P to circle and tangent at P to parabola is given by

$$\tan\theta = \left|\frac{m_1 - m_2}{1 + m_1 m_2}\right| = \left|\frac{\frac{1}{t} - \frac{1-t^2}{2t}}{1 + \frac{1}{2t^2}(1-t^2)}\right| = \left|\frac{2t - t + t^3}{(t^2+1)}\right|$$

$= |t| = t$ as $t > 0 \Rightarrow \theta = \tan^{-1}(t)$.

Column Matching Type Questions

26. In column I are given some standard parabolas. Here S denotes the focus, O denotes vertex, PR and QR are the tangents drawn on the points P and Q lying on the parabola respectively.

Column-I

(i)

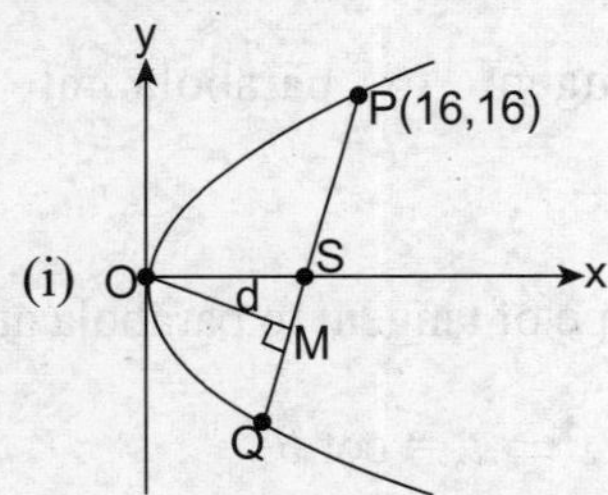

The value of $25d^2$ is (where $|OM| = d$)

(ii)

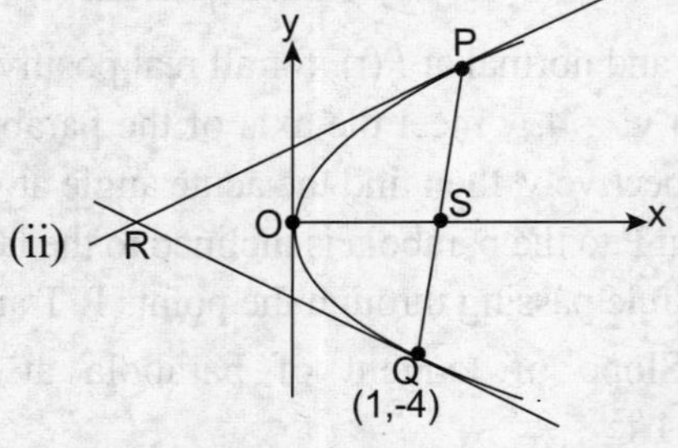

The value of $(SR)^2$ is

(iii)

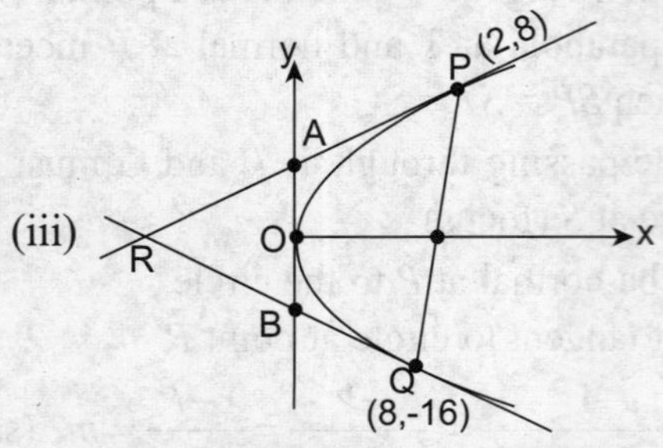

If area of $\Delta PQR = \lambda$, area of $\Delta OPQ = \mu$, area of $\Delta ABR = k$; then $\lambda + \mu + 2k$ equals

(iv)

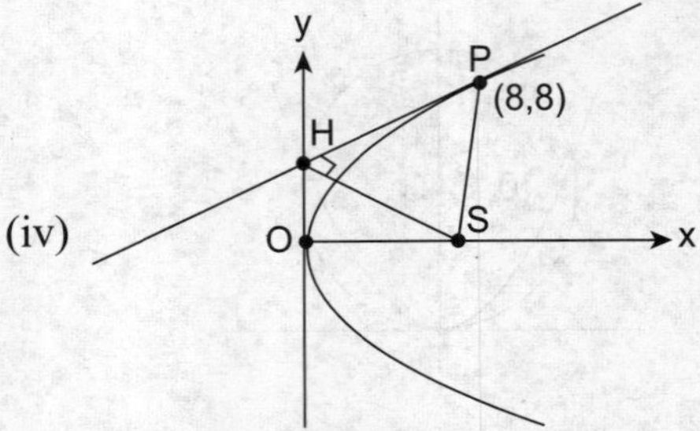

If $SH = \delta$, then $10\delta^2$ is

Column-II

(a) 200

(b) 256

(c) 100

(d) 204

(e) None of these

Ans. (i) → b; (ii) → c; (iii) → e; (iv) → a;

Solution: (i) Parabola $y^2 = 16x \Rightarrow a = 4$

$P(16, 16) \Rightarrow t = 2$

λ = length of focal chord $= a\left(t + \frac{1}{t}\right)^2 = 4.\ \frac{25}{4} = 25$

Aliter: $P(16, 16)$, $a = 4$, $SP = 20$

Also H.M. of SP and $SQ = 2a$

$\therefore \quad \frac{2}{\frac{1}{20} + \frac{1}{SQ}} = 2 \times 4 \Rightarrow SQ = 5$

$\therefore \quad \lambda$ = length of focal chord $= 20 + 5 = 25$

Also $d^2 = \frac{4a^3}{\lambda} = \frac{4 \times 64}{25} = \frac{256}{25}$

$\therefore \quad 25\, d^2 = 256$

(ii) $Q(1, -4)$; $P(16, 16) \Rightarrow SP = 20$, $SQ = 5$.

Since SP, SR and SQ are in GP. $\Rightarrow SR^2 = 100$.

(iii) Parabola is $y^2 = 32x$ and $P = (2, 8)$ and $Q \equiv (8, -16)$

$\Rightarrow t_1 = \frac{1}{2};\ t_2 = -1$

$\therefore \quad$ Area of $\Delta PQR = \frac{1}{2}\ a^2\, |t_1 - t_2|^3 = \frac{1}{2} \times 64 \times \left(\frac{1}{2} + 1\right)^3$

$= 32 \times \frac{27}{8} = 108 = \lambda$

Again area of $\Delta OPQ = |a^2\, t_1 t_2 (t_1 - t_2)|$

$= |64 \times 1/2 \times (-1) \times \frac{3}{2}| = 48, = \mu$

Also area of $\Delta ABR = 1/2 \times$ Area of ΔOPQ

$\Rightarrow 2k = 48 \Rightarrow \lambda + \mu + 2k = 204$

(iv) Parabola is $y^2 = 8x$, $OS = 2$, $SP = 10$

Since $(SH)^2 = SO \cdot SP$

$\Rightarrow (SH)^2 = 20 \Rightarrow \delta^2 = 20 \Rightarrow 10\delta^2 = 200$

27. Find the locus of the middle points of the chords for the following conditions in column I to the suitable answer in column II.

Column-I

(a) For parabola $y^2 = 4x$ chords which are of length $2b$

(b) For parabola $y^2 = 4x$; chord which touches the parabola $y^2 + 4bx = 0$; $(b > 0)$

(c) For parabola $y^2 = 4x$; chords which touch the circle $x^2 + y^2 = b^2$

(d) For parabola $y^2 = 4x$; chords which touch the circle $(x-1)^2 + y^2 = b^2$; $(b > 0)$

(e) For parabola $y^2 = 4bx$; chords which are normal to this parobola

Column-II

(p) $y^4 - 2b(x - 2b)y^2 + 8b^4 = 0$

(q) $y^2(2 + b) = 4x$

(r) $y^2 - 2x + 2 = b\sqrt{y^2+4}$

(s) $|y^2 - 2x| = b\sqrt{y^2+4}$

(t) $4b^2 = (4x - y^2)(y^2 + 4)$

(u) None of these

Ans. (a) → t; (b) → q; (c) → s; (d) → r; (e) → p;

Solution: Let $P(t_1)$ and $Q(t_2)$ be the point on parabola $y^2 = 4x$ and (h, k) be the mid-point of chord PQ.

$\Rightarrow t_1 + t_2 = k$

$t_1^2 + t_2^2 = 2h$

$\Rightarrow (t_1+t_2)^2 - 2t_1t_2 = 2h$

$\Rightarrow \dfrac{k^2 - 2h}{2} = t_1t_2$

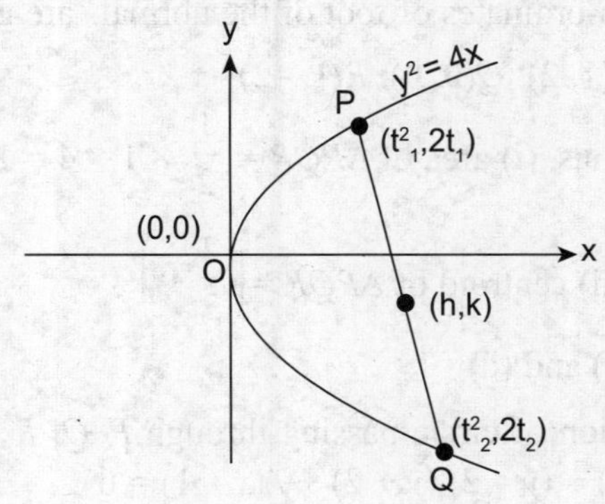

Similiary, $(t_1 - t_2)^2 = t_1^2 + t_2^2 - 2t_1t_2$

$= 2h - (k^2 - 2h) = 4h - k^2$

$\Rightarrow |t_1 - t_2| = \sqrt{4h - k^2}$

Equation of chord PQ: $y - 2t_1 = \dfrac{2}{t_1+t_2}(x - t_1^2)$

$$\Rightarrow y = \frac{2}{t_1+t_2}x + \frac{2t_1t_2}{t_1+t_2} \quad \text{......(i)}$$

(a) $2b = \sqrt{(t_1^2 - t_2^2)^2 + (2(t_1 - t_2))^2}$ = length of chord PQ

$\Rightarrow 4b^2 = (t_1 - t_2)^2\left[(t_1+t_2)^2 + 4\right]$

$\Rightarrow 4b^2 = (4h - k^2)(k^2 + 4)$

$\Rightarrow 4b^2 = (4x - y^2)(y^2 + 4)$ = *locus of mid-point of PQ*

(b) Equation of tangent to parabola $y^2 = -4bx$ is

$y = mx + \dfrac{-b}{m}$; $m = \dfrac{2}{t_1+t_2}$

$$\text{i.e., } y = \frac{2x}{t_1+t_2} - \frac{b}{2}(t_1+t_2) \quad \text{.....(ii)}$$

Now (h, k) satisfies (ii)

$\Rightarrow k = \dfrac{2h}{t_1+t_2} - \dfrac{b}{2}(t_1+t_2)$

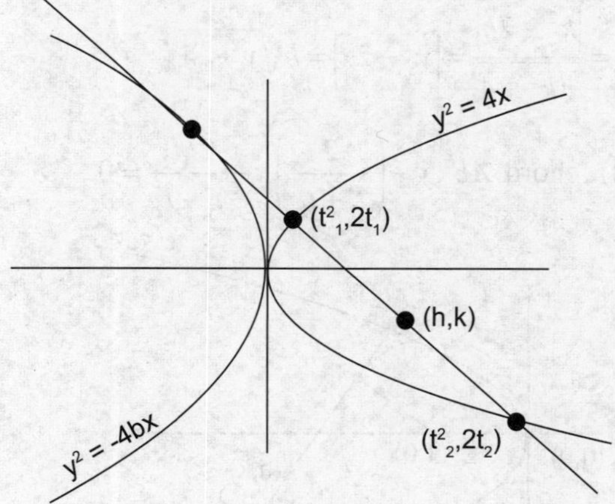

Also $(t_1+t_2) = k \Rightarrow$ we get $k = \dfrac{2h}{k} - \dfrac{b}{2}(k)$

$\Rightarrow 2k^2 = 4h - bk^2$

$\Rightarrow k^2(2 + b) = 4h$

$\Rightarrow y^2(2 + b) = 4x$

(c) $m(AB) = \dfrac{2}{t_1+t_2}$

$\Rightarrow |OT| = b$

$\Rightarrow$ distance of origin from $AB = b$

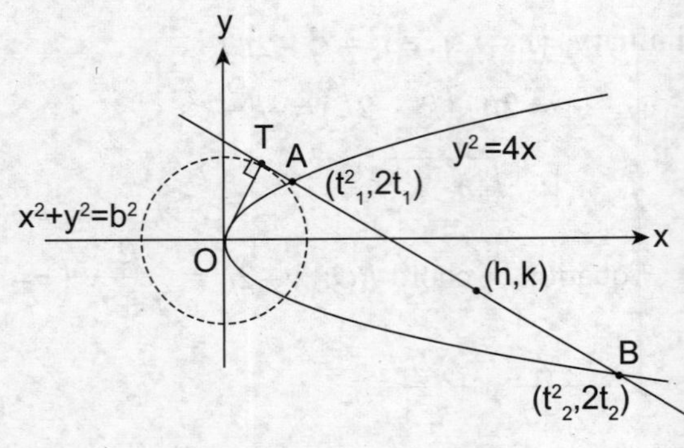

$$\Rightarrow OT = \frac{\left|O - \left(\frac{2}{t_1+t_2}\right)O - \frac{2t_1t_2}{t_1+t_2}\right|}{\sqrt{1+\left(\frac{2}{t_1+t_2}\right)^2}} = b$$

$$\Rightarrow \frac{\left|\frac{2t_1t_2}{t_1+t_2}\right|}{\frac{\sqrt{(t_1+t_2)^2+4}}{|t_1+t_2|}} = b \Rightarrow \frac{|k^2-2h|}{\sqrt{k^2+4}} = b$$

$$\Rightarrow |y^2-2x| = b\left(\sqrt{y^2+4}\right) \text{ is the required locus}$$

Aliter: S: $y^2 - 4x = 0$. Let (h, k) be the mid-point, so $T = S_1$ gives $ky - 2x - 2h = -k^2 - 4h$

i.e., $2x - ky + (k^2 - 2h) = 0$

Distance of origin from chord

$$b = \frac{|k^2-2h|}{\sqrt{k^2+4}} = |y^2-2x| = b\sqrt{y^2+4}$$

(d) chord $AB: y - \left(\frac{2}{t_1+t_2}\right)x - \frac{2t_1t_2}{t_1+t_2} = 0$

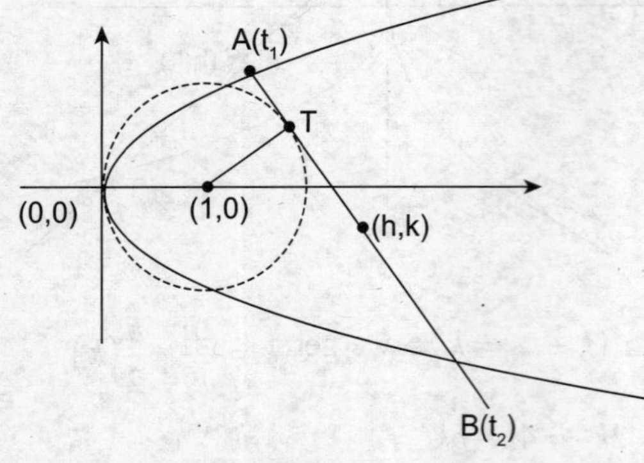

Distance of (1, 0) from $AB = b$

$$\Rightarrow \frac{\left|\left(\frac{2}{t_1+t_2}\right) + \frac{2t_1t_2}{t_1+t_2}\right|}{\sqrt{1+\left(\frac{2}{t_1+t_2}\right)^2}} = b$$

$$\Rightarrow \frac{\left|\left(\frac{2}{t_1+t_2}\right)\right| \times |1+t_1t_2|}{\frac{\sqrt{(t_1+t_2)^2+4}}{|t_1+t_2|}} = b \Rightarrow \frac{2\times\frac{|k^2-2h+2|}{2}}{\sqrt{k^2+4}} = b$$

$$\Rightarrow y^2 - 2x + 2 = b\sqrt{y^2+4} \text{ its the required locus}$$

(e) $y = mx - 2bm - bm^3$ is the equation of normal to parabola $y^2 = 4bx$ As it passes throgh (h, k)

$\Rightarrow bm^3 + m(2b - h) + k = 0$

$$\Rightarrow b\left(\frac{2}{t_1+t_2}\right)^3 + \frac{2}{t_1+t_2}(2b-h) + k = 0$$

$$\left(\because t_1 + t_2 = \frac{k}{b}\right)$$

$$\Rightarrow b \times \frac{8b^3}{k^3} + \frac{2b}{k}(2b-h) + k = 0$$

$$\Rightarrow 8b^4 + 4b^2k^2 - 2bhk^2 + k^4 = 0$$

$\Rightarrow y^4 - 2b(x - 2b)y^2 + 8b^4 = 0$ is the required locus.

28. Normals drawn at points P, Q and R lying on the parabola $y^2 = 4x$ intersect at (3, 0), then

Column I

(i) Area of ΔPQR

(ii) Radius of circumcircle of ΔPQR

(iii) Centroid of ΔPQR

(iv) Circumcentre of ΔPQR

Column II

(a) 2

(b) 5/2

(c) (5/2, 0)

(d) (2/3, 0)

Ans. (i) (a) (ii) (b) (iii) (d) (iv) (c)

Solution: Since equation of normal to parabola $y^2 = 4ax$ is $y + xt = 2at + at^3$ and it passes through (3, 0)

$\Rightarrow 3t = 2t + t^3$ ($a = 1$)

$\Rightarrow t = 0, 1, -1$

$\therefore$ co-ordinates of foot of the normals are given by $P(1, 2)$; $Q(0, 0)$: $R(1, -2)$

Thus, (i) area of $\Delta PQR = \frac{1}{2} \times 1 \times 4 = 2$ sq. units

(iii) centroid of $\Delta PQR = \left(\frac{2}{3}, 0\right)$

(b) and (d)

Equation of circle passing through P, Q, R is $(x - 1)(x - 1) + (y - 2)(y + 2) + \lambda(x - 1) = 0$

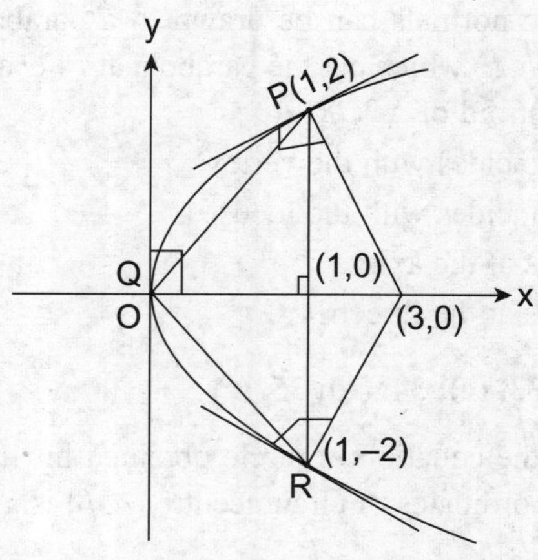

and is passing through (0, 0) $\Rightarrow 1 - 4 - \lambda = 0$

$\Rightarrow \lambda = -3$

therefore, required equation of circle is $x^2 + y^2 - 5x = 0$ having its

centre $\left(\frac{5}{2}, 0\right)$ and radius $= \frac{5}{2}$ units.

$\Rightarrow$ circumcentre $= \left(\frac{5}{2}, 0\right)$ and circumradius $\frac{5}{2}$

Assertion and Reason-type Questions

29. Assertion: The curve $y = \frac{-x^2}{2} + x + 1$ is symmetric with respect to the line $x = 1$.

Reason: A parabola is symmetric about its axis.

Solution: (a) Given curve is $y = -\frac{x^2}{2} + x + 1$

or $y = -\frac{1}{2}(x^2 - 2x - 2) = -\frac{1}{2}[(x^2 - 2x + 1) - 3]$

$y = -\frac{1}{2}[(x-1)^2 - 3]$

or $(x-1)^2 = -2\left(y - \frac{3}{2}\right)$

This is the equation of a parabola, whose axis is $x - 1 = 0$

or $x = 1$ about which the parabola is symmetric

Assertion and reason both are correct and reason is the correct explanation of assertion.

Comprehension Type Questions

A: Let C_1 and C_2 be respectively, the parabolas $x^2 = y - 1$ and $y^2 = x - 1$. Let P be any point on C_1 and Q be any point on C_2. Let P_1 and Q_1 be the reflections of P and Q respectively with respect to $y = x$, then

30. P_1 and Q_1 lie on:

(a) C_1 and C_2 respectively

(b) C_2 and C_1 respectively

(c) Cannot be discussed

(d) None of these

31. If the point $P(t_1, t_1^2 + 1)$ and $Q(t_2^2 + 1, t_2)$, then P_1 and Q_1 are:

(a) $(t_1^2 + 1, t_1)$ and $(t_2^2 + 1, t_2)$

(b) $(t_1^2 + 1, t_1)$ and $(t_2, t_2^2 + 1)$

(c) $(t_1, t_1^2 + 1)$ and $(t_2, t_2^2 + 1)$

(d) None of these

32. Arithmetic mean of PP_1 and QQ_1 is always less than:

(a) PQ (b) $\frac{1}{2}PQ$

(c) $2PQ$ (d) None of these

Solution: 30. (b); 31. (b); 32. (a)

We have,

$C_1: (x - 0)^2 = (y - 1)$ ....(i)

$C_2: (y - 0)^2 = (x - 1)$ ....(ii)

Let $P(t_1, t_1^2 + 1)$ and $Q(t_2^2 + 1, t_2)$ be points on C_1 and C_2 resepctively.

Since the reflection of a point (a, b) in $y = x$ is (b, a). So, the co-ordinates of P_1 and Q_1 *are* $P_1(t_1^2 + 1, t_1)$ and $Q_1(t_2, t_2^2 + 1)$ respectively

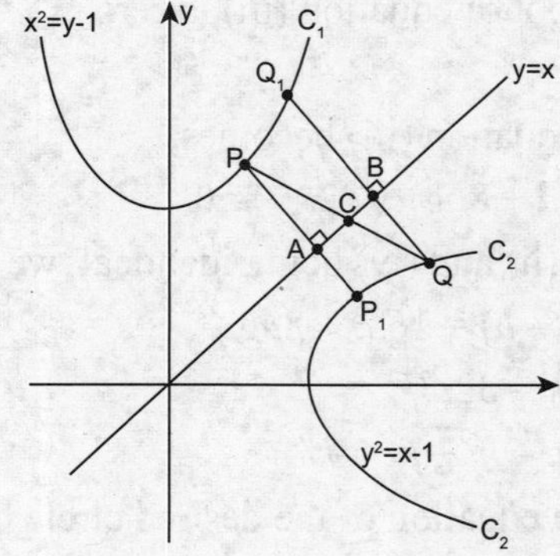

Clearly, P_1 satisfies the equation $y^2 = x - 1$ and Q_1 satisfies the equation $x^2 = y - 1$.

Thus P_1 lies on C_2 and Q_1 lies on C_1.

Since, P_1 is the reflection of P in $y = x$.

Therefore, PA $\perp$ $(y = x)$ and $PA = \frac{1}{2}PP_1$.

$\Rightarrow PC \geq PA$ ..... (iii)

Similarly, we have $QC \geq QB$ ...... (iv)

Adding (iii) and (iv), we get,

$$PC + QC \geq PA + QB \Rightarrow PQ \geq \left(\frac{1}{2}PP_1 + \frac{1}{2}QQ_1\right)$$

$$\Rightarrow PQ \geq \frac{1}{2}(PP_1 + QQ_1)$$

B: Normal to a parabola $y^2 = 4ax$ at the point $(at^2, 2at)$ is given by $xt + y = 2at + at^3$. If it passes through a point (h, k), then $at^3 + t(2a - h) - k = 0$ ....(i)

If t_1, t_2, t_3 be roots of (i), then three points P, Q, R are $(at_1^2, 2at_1)$, $(at_2^2, 2at_2)$, $(at_3^2, 2at_3)$ from which normals pass through the point (h, k). Points P, Q and R are called co-normal points. Putting $2at = y$, the ordinates of P, Q, R are the roots of $y^3 + 4a(2a - h)y - 8a^2k = 0$ .....(ii)

Let the circumcircle of ΔPQR be $x^2 + y^2 + 2gx + 2fy + c = 0$

Eliminating x from equation of the parabola and circle,

we have $\frac{y^4}{16a^2} + y^2 + 2g\frac{y^2}{4a} + 2fy + c = 0$, i.e.

$y^4 + y^2(16a^2 + 8ag) + 32a^2fy + 16a^2c = 0.$ ...(iii)

Equation (iii) gives the ordinates of the points of intersection of the parabola and the circle. Three roots of equation (iii) are the same as the roots of equation (ii). let these identical roots be y_1, y_2, y_3 and let the fourth root be y_4.

Then from (iii), $y_1 + y_2 + y_3 + y_4 = 0$

Also from (ii), $y_1 + y_2 + y_3 = 0$, so we get $y_4 = 0$

As, one root of equation (iii) is zero

$\Rightarrow c = 0$

And the equation (iii) becomes,

$y^3 + y(16a^2 + 8ag) + 32a^2f = 0$ .... (iv)

Equation (ii) and (iv) being identical, we have

$4a(2a - h) = 16a^2 + 8ag$

and $-8a^2k = 32a^2f$

$\Rightarrow g = -a - h/2, f = -k/4.$

Hence the equation of the desired circle can be found.

Based on above information answer the following:

33. If the normals at the point Q, R on parabola $y^2 - 4ax = 0$ meet the parabola at the same point P, then the locus of the circumcentre of the triangle PQR is

(a) a straight line (b) a circle
(c) another parabola (d) hyperbola

34. The centroid of four points in which the circle and parabola intersect is given by

(a) (2a – g, 0) (b) (–2a – g, 0)
(c) (2a – f, 0) (d) None of these

35. If three normals can be drawn to a parabola from a point (h, k) which cut the parabola at P, Q and R, then the centroid of ΔPQR.

(a) coincides with the vertex
(b) coincides with the focus
(c) lies at the axis
(d) lies at the directrix

Solutions: 33. (c); 34. (b); 35. (c)

33. From the equation of circle obtained in (ii), we have the co-ordinates of circumcentre (α, β) is given by

$\alpha = a + \frac{h}{2}$ and $\beta = \frac{k}{4}$

Now (h, k) lies on the parabola itself, so, $k^2 = 4ah$

Eliminating h and k, we get $(4\beta)^2 = 4a(2\alpha - 2a)$

$\Rightarrow 16\beta^2 - 8a\alpha + 8a^2 = 0 \Rightarrow 2\beta^2 - a\alpha + a^2 = 0$

So the locus of (α, β) is $2y^2 - ax + a^2 = 0$ which is clearly a parabola.

34. $at^3 + t(2a - h) - k = 0$

$\Rightarrow \Sigma t_1 = 0,\ \Sigma t_1t_2 = \frac{2a - h}{a}.$

Now centroid $\left(\sum_{i=1}^{4}\frac{at_i^2}{4}, \sum_{i=1}^{4}\frac{2at_i}{4}\right) = \left(\frac{a}{4}\Sigma t_i^2, 0\right)$

$$= \left(\frac{a}{4}\left(\left(\sum t_i\right)^2 - 2\sum_{i \neq j} t_it_j\right), 0\right)$$

$$\equiv \left(\frac{a}{4}\left(0 - 2\times\left[\frac{2a - h}{a}\right]\right), 0\right)$$

$= (-a + h/2, 0) = (-2a + a + h/2, 0) = (-2a - g, 0).$

35. In the comprehension, we found that $y_1 + y_2 + y_3 = 0$, So, the centroid lies on the axis of parabola

C: From the point $P(h, k)$ three normals are drawn to the parabola $x^2 = 8y$ and m_1, m_2 and m_3 are the slopes of three normals.

36. Find the algebraic sum of the slopes of these three normals.

(a) $(k - 2)/h$ (b) $(k + 2)/h$
(c) $(k - 4)/h$ (d) $(k + 4)/h$

37. If two of the three normals are at right angles, then the locus of point P is a conic, then the latus rectum of conic, is

(a) 1 (b) 2
(c) 3 (d) 4

38. If the two normals from P are such that they make complementary angles with the axis, then the locus of point P is a conic, then the directrix of conic is

(a) $y - 3 = 0$ (b) $2y + 3 = 0$
(c) $y + 3 = 0$ (d) $2y - 3 = 0$

Solutions: 36. (c); 37. (b); 38. (d)

36. Let normal to the parabola be

$$y = mx + 2a + \frac{a}{m^2};\ a = 2$$

$$\Rightarrow y = mx + 4 + \frac{2}{m^2}$$

which passes through (h, k)

$$\Rightarrow k - mh = 4 + \frac{2}{m^2}$$

$$\Rightarrow m^3h + (4 - k)\,m^2 + 2 = 0$$

$$\Rightarrow m_1 + m_2 + m_3 = \frac{k-4}{h}$$

37. The equation of any normal to $x^2 = 8y$ is

$y = mx + 4 + 2/m^2$ ...(1)

Let $P(h, k)$ be the point of the intersection of three normals drawn to $x^2 = 8y$.

If equation (1) passes through $P(h, k)$ then

$k = mh + 4 + (2/m^2)$ or $hm^3 + (4 - k)m^2 + 2 = 0$ ...(2)

This is a cubic equation in m so it gives three values of m such that corresponding to each value of m there is a normal passing through $P(h, k)$.

Let m_1, m_2, m_3 be the roots of the equation (2), then

$$m_1 + m_2 + m_3 = \frac{k-4}{h} \quad ...(3)$$

$$m_1m_2 + m_2m_3 + m_3m_1 = 0 \quad ...(4)$$

$$\text{and } m_1m_2m_3 = -\frac{2}{h} \quad ...(5)$$

If two of the three normals are at right angles. Then, $m_1m_2 = -1$

Substituting $m_1m_2 = -1$ in equation (5)

$$\text{we get } m_3 = \frac{2}{h}$$

since m_3 is a root of the equation (2), therefore

$$h\left(\frac{2}{h}\right)^3 + (4-k)\left(\frac{2}{h}\right)^2 + 2 = 0$$

$$\text{or } \frac{8}{h^2} + \frac{4(4-k)}{h^2} + 2 = 0$$

$$\Rightarrow h^2 + 2(4 - k) + 4 = 0$$

hence, the locus of (h, k) is $x^2 + 2(4 - y) + 4 = 0$

or $x^2 = 2y - 12$

or $x^2 = 2(y - 6)$; this is the equation of a parabola with $a = 1/2$ and length of latus rectum of parabola

$$= 4a = 4 \times \frac{1}{2} = 2$$

38. The equation of any normal to $x^2 = 8y$ is (here $a = 2$)

$y = mx + 4 + (2/m^2)$

If this normal passes through (h, k), then (h, k) will satisfy this equation, hence $k = mh + 4 + 2/m^2$

or $m^3h + (4 - k)\,m^2 + 2 = 0$ ...(1)

Let m_1, m_2 and m_3 be the roots of this equation, then

$$m_1 + m_2 + m_3 = \frac{k-4}{h} \quad ...(2)$$

$$m_1m_2 + m_2m_3 + m_3m_1 = 0 \quad ...(3)$$

$$\text{and } m_1m_2m_3 = -\frac{2}{h} \quad ...(4)$$

Let one of the normals make an angle θ with axis of x, then the other will make an angle of $(90° - \theta)$ by hypothesis. So $m_1 = \tan\theta$ and $m_2 = \tan(90° - \theta)$ $= \cot\theta$, so $m_1m_2 = 1$ ...(5)

Now we have to eliminate m_1, m_2 and m_3 from equation (2), (3), (4) and (5)

$$\text{Now from (4) and (5); } m_3 = -\frac{2}{h} \quad ...(6)$$

by equation (3); $m_1m_2 + m_3(m_2 + m_1) = 0$

$$\text{or } m_1m_2 + m_3\left(\frac{k-4}{h} + \frac{2}{h}\right) = 0 \quad .....(7)$$

putting the value of m_1m_2 from (5) and m_3 from (6) we

$$\text{have from (7); } 1 - \frac{2}{h} = \left(\frac{k-2}{h}\right) = 0 \text{ or } h^2 - 2k + 4 = 0$$

or $h^2 = 2k - 4$ or $x^2 = 2(y - 2)$

This is the equation of a parabola. The equation of directrix will be $(y - 2) = -a$

$$\text{or } y - 2 = -\frac{1}{2} \ \{\text{here } a = \frac{1}{2} \text{ for this parabola}\}$$

or $2y - 4 = -1$ or $2y - 3 = 0$

D: Consider the circle $x^2 + y^2 = 9$ and the parabola $y^2 = 8x$. They intersect at P and Q in the first and the fourth quadrants, respectively. Tangents to the circle at P and Q intersect the x-axis at R and tangents to the parabola at P and Q intersect the x-axis at S.

39. The ratio of the area of the triangle PQS and PQR is

(a) $1 : \sqrt{2}$ (b) $1 : 2$
(c) $1 : 4$ (d) $1 : 8$

40. The radius of the circumcircle of the triangle PRS is

(a) 5 (b) $3\sqrt{3}$

(c) $3\sqrt{2}$ (d) $2\sqrt{3}$

41. The radius of the incircle of the triangle PQR is

(a) 4 (b) 3

(c) 8/3 (d) 2

Solution: 39. (c); 40. (b); 41. (d)

39.

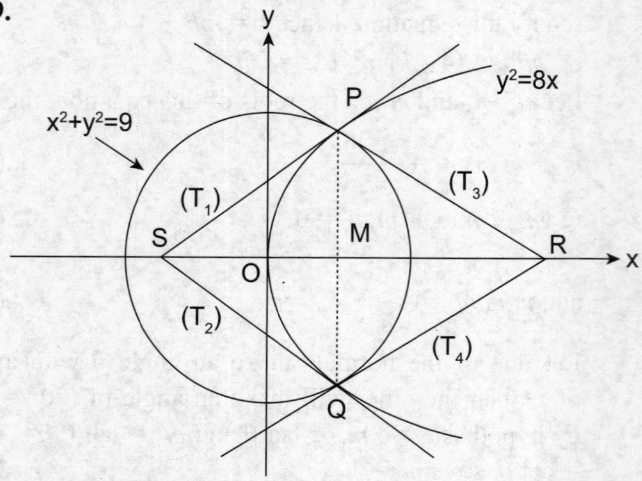

Given parabola is $y^2 = 8x$ ....(1)

and the circle is $x^2 + y^2 = 9$ ...(2)

on solving equations (1) and (2), we have $x^2 + (8x) = 9$

$\Rightarrow x = 1, -9$

since intersection points P and Q are in first and fourth quadrant respectively, therefore $x \neq -9$

$\Rightarrow x = 1$ putting in equation (2) we get $1 + y^2 = 9$

or $y = \pm 2\sqrt{2}$

$\therefore$ $P \equiv (1, 2\sqrt{2})$ and $Q \equiv (1, -2\sqrt{2})$

$\Rightarrow$ M ≡ (1, 0) (since M is the mid-point of P and Q)

Now PQ is a chord of parabola.

Therefore the intersection point of tangents drawn at end points of chord PQ

where $P(at_1^2, 2at_1)$ and $Q(at_2^2, 2at_2)$ are given by $S(at_1t_2, a(t_1 + t_2))$

$\Rightarrow 2at_1 = 2\sqrt{2} \Rightarrow t_1 = \frac{1}{\sqrt{2}}$

and $2at_2 = -2\sqrt{2} \Rightarrow t_2 = -\frac{1}{\sqrt{2}}$

$\therefore$ $S \equiv \left(2 \times \frac{1}{\sqrt{2}} \times \left(-\frac{1}{\sqrt{2}}\right), 0\right)$

or S ≡ (–1, 0)

Now tangent on circle at $P(1, 2\sqrt{2})$ is

$x.1 + 2\sqrt{2}\, y = 9$

It passes through x-axis where $y = 0$,

$\Rightarrow x + 0 - 9 = 0$

therefore, co-ordinates of point $R \equiv (9, 0)$

$$\frac{\text{Area of } \Delta PQS}{\text{Area of } \Delta PQR} = \frac{\frac{1}{2} \times PQ \times SM}{\frac{1}{2} \times PQ \times MR} = \frac{1}{4}$$

40. Now to find circum radius (r) of the triangle PRS, let circumcentre is $C(x, y)$, then $CP = CR =$ CS

Taking, $CP =$ CR we have

$(x - 1)^2 + (y - 2\sqrt{2})^2 = (x - 9)^2 + (y - 0)^2$

or $16x - 4\sqrt{2}y = 72$ ...(3)

Now taking $CR =$ CS, we have

$(x - 9)^2 + (y - 0)^2 = (x + 1)^2 + (y - 0)^2$

or $20x = 80 \Rightarrow x = 4$

putting in equation (3), we get $y = -\sqrt{2}$

hence, the co-ordinates of circumcentre is $(4, -\sqrt{2})$

Now circumradius = CR

$= \sqrt{(9-4)^2 + (-\sqrt{2} - 0)^2} = \sqrt{25 + 2} = 3\sqrt{3}$ units

41. Distance $PR = \sqrt{64 + 8} = \sqrt{72} = 6\sqrt{2}$

Distance $PQ = 4\sqrt{2}$; distance $QR = 6\sqrt{2}$

$$\text{Semi-perimeter (s)} = \frac{PR + PQ + QR}{2} = \frac{16\sqrt{2}}{2} = \frac{16}{\sqrt{2}}$$

$$\text{and radius of incircle} = r = \frac{\Delta}{s} = \frac{\frac{1}{2} \times 4\sqrt{2} \times (9-1)}{\frac{16}{\sqrt{2}}} = 2$$

TUTORIAL EXERCISE

SECTION–III

OBJECTIVE-TYPE (ONLY ONE CORRECT ANSWER)

1. Which one of the following equations parametrically, represents equation to a parabolic profile?
(a) $x = 3\cos t;\ y = 4\sin t$
(b) $x^2 - 2 = -2\cos t;\ y = 4\cos^2 t/2$
(c) $\sqrt{x} = \tan t;\ \sqrt{y} = \sec t$
(d) $x = \sqrt{1-\sin t}\ ;\ y = \sin t/2 + \cos t/2$

2. The number of distinct normals that can be drawn from (11/4, 1/4) to the parabola $y^2 = 4x$ is:
(a) 3 (b) 2
(c) 1 (d) 4

3. In a parabola, semi-latus rectum is the harmonic mean of the:
(a) segments of a focal chord
(b) segments of the directrix
(c) segments of a chord
(d) None of these

4. The radius of the circle whose centre is (–4, 0) and which cuts the parabola $y^2 = 8x$ at A and B such that its common chord AB subtends a right angle at the vertex of the parabola, is equal to
(a) $4\sqrt{13}$ (b) 3
(c) $\sqrt{18}$ (d) 5

5. A ray of light moving parallel to the x-axis gets reflected from a parabolic mirror whose equation is $(y-2)^2 = 4(x+1)$. After reflection, the ray must pass through the point
(a) (0, 2) (b) (2, 0)
(c) (0, –2) (d) (–1, 2)

6. If the normals at point P and Q of the parabola $y^2 = 4ax$ intersect at a point R on the parabola, then the product of the ordinates of P and Q is
(a) a^2 (b) $2a^2$
(c) $4a^2$ (d) $8a^2$

7. A circle with its centre at the focus of the parabola $y^2 = 4ax$ and touching its directrix intersects the parabola at points A, B Then length AB is equal to
(a) $4a$ (b) $2a$
(c) $a/2$ (d) None of these

8. The locus of a point which divides a chord of slope 2 of the parabola $y^2 = 4x$ internally in the ratio 1:2 is
(a) $\left(y+\frac{4}{9}\right)^2 = \frac{4}{9}\left(x-\frac{2}{9}\right)$
(b) $\left(y-\frac{8}{9}\right)^2 = \frac{4}{9}\left(x-\frac{2}{9}\right)$
(c) $\left(y-\frac{8}{9}\right)^2 = \frac{4}{9}\left(x+\frac{2}{9}\right)$
(d) $\left(y+\frac{8}{9}\right)^2 = \frac{4}{9}\left(x+\frac{2}{9}\right)$

9. The length of the chord of the parabola $y^2 = x$ which is bisected at (2, 1), is
(a) 3 (b) $\sqrt{14}$
(c) $\sqrt{6}$ (d) $2\sqrt{5}$

10. If (h, k) is a point on the axis of the parabola $2(x-1)^2 + 2(y-1)^2 = (x+y+2)^2$ from where three distinct normals may be drawn, then
(a) $h > 2$ (b) $h < 4$
(c) $h > 8$ (d) $h < 8$

11. A line touches the circle $x^2 + y^2 = 2a^2$ and also the parabola $y^2 = 8ax$. Its equation is:
(a) $y = \pm x$ (b) $y = \pm(x + c)$
(c) $y = \pm(x + 2a)$ (d) $y = \pm(x - 2a)$

12. A line L passing through the focus of the parabola $y^2 = 4(x-1)$ intersects the parabola in two distinct points. If m be the slope of the line L, then
(a) $m \in (-1, 1)$ (b) $m \in (-\infty, -1) \cup (1, \infty)$
(c) $m \in R$ (d) None of these

13. Two tangents are drawn from the point (–2, –1) to the parabola $y^2 = 4x$. If α is the angle between these tangents.Then $\tan\alpha =$
(a) 3 (b) 1/3
(c) 2 (d) 1/2

14. Equation of the line touching both the parabola $y^2 = 4x$ and $x^2 = -32y$ is
(a) $x + 2y + 4 = 0$ (b) $2x + y - 4 = 0$
(c) $x - 2y - 4 = 0$ (d) $x - 2y + 4 = 0$

15. The equation of the common tangent touching the circle $(x-3)^2+y^2=9$ and parabola $y^2 = 4x$ above the x-axis is

(a) $\sqrt{3}y=3x+1$ (b) $\sqrt{3}y=-(x+3)$

(c) $\sqrt{3}y=(x+3)$ (d) $\sqrt{3}y=-(3x+1)$

16. The circle drawn with variable chord $x + ay - 5 = 0$ (a being a parameter) of the parabola $y^2 = 20x$ as diameter will always touch the line

(a) $x + 5 = 0$ (b) $y + 5 = 0$

(c) $x + y + 5 = 0$ (d) $x - y + 5 = 0$

17. The line $x - b + \lambda y = 0$ cuts the parabola $y^2 = 4ax$ at $P(at_1^2, 2at_1)$ and $Q(at_2^2, 2at_2)$. If $b \in [2a, 4a]$ and $\lambda \in R$, then t_1t_2 belongs to

(a) $[-4, -2]$ (b) $[-4, -3]$

(c) $[-3, -2]$ (d) None of these

18. If the parabola $y = (a - b)x^2 + (b - c)x + (c - a)$ touches the x-axis in the intervals (0, 1], then the line $ax + by + c = 0$

(a) Always passes through a fixed point

(b) represents the family of parallel lines

(c) data is insufficient

(d) None of these

19. The co-ordinates of the point on the parabola $x^2 = 4y$ which is nearest to the circle $(x - 3)^2 + y^2 = 1$ are

(a) (0, 0) (b) (–2, 1)

(c) (2, 1) (d) (–4, 4)

20. The chord $x + y = 1$ cuts the parabola $y^2 = 4ax$ in points A, B. The normals at A and B intersect at C. A third line from C cuts the parabola normally at D whose co-ordinates are

(a) $(a, -2a)$ (b) $(4a, 4a)$

(c) (0, 0) (d) None of these

21. The number of points with integral co-ordinates $(2a, a - 1)$ that fall in the interior of the larger segment of the circle $x^2 + y^2 = 25$ cut off by the parabola $x^2 + 4y = 0$ is

(a) one (b) two

(c) three (d) None of these

22. If the line $y-\sqrt{3}\ x+3=0$ cuts the parabola $y^2 = x + 2$ at A and B, and if $P = \left(\sqrt{3},0\right)$, then PA. PB equals

(a) $\frac{2\left(\sqrt{3}+2\right)}{3}$ (b) $\frac{4\sqrt{3}}{2}$

(c) $\frac{4\left(2-\sqrt{3}\right)}{3}$ (d) $\frac{4\left(\sqrt{3}+2\right)}{3}$

23. If the distances of two points P and Q from the focus of a parabola $y^2 = 4x$ are 4 and 9 respectively, the distance of the point of intersection of tangents at P and Q from the focus is

(a) 8 (b) 6

(c) 5 (d) 13

24. The locus of point of intersection of tangents to the parabolas $y^2 = 4(x + 1)$ and $y^2 = 8(x + 2)$ which are perpendicular to each other, is

(a) $x + 7 = 0$ (b) $x - y = 4$

(c) $x + 3 = 0$ (d) $y - x = 12$

25. Let the line $lx + my = 1$ cut the parabola $y^2 = 4ax$ in the points A and B. Normals at A and B meet at point C. Normal from C other than these two meet the parabola at D then the co-ordinate of D are

(a) $(a, 2a)$ (b) $\left(\frac{4am}{l^2}, \frac{4a}{l}\right)$

(c) $\left(\frac{2am^2}{l^2}, \frac{2a}{l}\right)$ (d) $\left(\frac{4am^2}{l^2}, \frac{4am}{l}\right)$

26. The triangle formed by the tangent to the curve $f(x) = x^2 + bx - b$ at the point (1, 1) and the co-ordinate axes lies in the first quadrant. If its area is 2, then the value of b is

(a) –1 (b) 3

(c) –3 (d) 1

27. The co-ordinates of the point on the parabola $y^2 = 8x$, which is at minimum distance from the circle $x^2 + (y + 6)^2 = 1$ are

(a) (2, –4) (b) (18, –12)

(c) (2, 4) (d) None of these

28. The condition that the two tangents to the parabola $y^2 = 4ax$ become normal to the circle $x^2 + y^2 - 2ax - 2by + c = 0$ is given by

(a) $a^2 > 4b^2$ (b) $b^2 > 2a^2$

(c) $a^2 > 2b^2$ (d) $b^2 > 4a^2$

29. If the tangent to the parabola $y^2 = 4ax$ meets the axis at T and tangent at the vertex A at y and the rectangle $TAYG$ is completed, then the locus of G is

(a) $y^2+2ax=0$ (b) $y^2+ax=0$

(c) $x^2+ay=0$ (d) None of these

30. If O is vertex of parabola and the foot of perpendicular be H from the focus S on any tangent to a parabola at any point P, then OS, SH, SP are in

(a) A.P. (b) G.P

(c) H.P. (d) A.G.P

31. The locus of points of intersection of tangents drawn at the ends of all the chord normals to the parabola $y^2 = 8(x-1)$ is

(a) $y^2(x+3)-32=0$ (b) $y^2(x+3)+32=0$

(c) $y^2(3x+1)-32=0$ (d) None of these

32. If the angle between the tangents drawn from a point P to the parabola $y^2 = 4ax$ is 45°, then the locus of P is

(a) Parabola (b) Ellipse

(c) Hyperbola (d) Circle

33. The triangle formed by tangent to the parabola $y = x^2$ at the point whose abscissa is x_0 (where $x_0 \in [1, 2]$) the y-axis and the straight line $y = x_0^2$ has the greatest area if x_0 is equal to

(a) 0 (b) 1

(c) 2 (d) 3/2

34. The equation of the line of the shortest distance between the parabola $y^2 = 4x$ and the circle $x^2 + y^2 - 4x - 2y + 4 = 0$ is

(a) $x + y = 3$ (b) $x - y = 3$

(c) $2x + y = 5$ (d) None of these

35. Two parabolas $y^2 = 4ax$ and $x^2 = 4by$ intersect at two points. A circle passes through one of the intersection point of these parabolas and touches the directrix of first parabola, then the locus of the centre of the circle is

(a) straight line (b) ellipse

(c) circle (d) parabola

36. The triangle formed by the tangent to the parabola $y^2 = 4x$ at the point whose abscissa lies in the interval $[a^2, 4a^2]$, the ordinate and the x-axis, has greatest area given by

(a) $12\ a^3$ (b) $8\ a^3$

(c) $16\ a^3$ (d) None of these

37. A pair of tangents are drawn to parabola $y^2 = 4ax$, equally inclined to the straight line having slope $\tan\alpha$ then the locus of the intersection point of the tangent is

(a) $y = (x - a)\tan\alpha$ (b) $y = (x + a)\tan\alpha$

(c) $y = (x - a)\tan 2\alpha$ (d) $y = (x + a)\tan 2\alpha$

38. AB, AC are tangents to a parabola $y^2 = 4ax$, p_1, p_2, p_3 are the lengths of the perpendiculars from A, B, C on any tangent to the curve, then p_2, p_1, p_3 are in

(a) A.P. (b) G.P.

(c) H.P. (d) None of these

39. A parabola is drawn with its focus at (3, 4) and vertex at the focus of the parabola $y^2 - 12x - 4y + 4 = 0$. The equation of the parabola is

(a) $x^2 - 6x - 8y + 25 = 0$

(b) $y^2 - 8x - 6y + 25 = 0$

(c) $x^2 - 6x + 8y - 25 = 0$

(d) $x^2 + 6x - 8y - 25 = 0$

40. Length of the latus rectum of the parabola $25[(x-2)^2 + (y-3)^2] = (3x - 4y + 7)^2$ is

(a) 4 (b) 2

(c) 1/5 (d) 2/5

41. Let A be the vertex and L the length of the latus rectum of the parabola, $y^2 - 2y - 4x - 7 = 0$. The equation of the parabola with A as vertex 2 L the length of the latus rectum and the axis at right angles to that of the given curve is

(a) $x^2 + 4x + 8y - 4 = 0$

(b) $x^2 - 4x - 8y + 12 = 0$

(c) $x^2 + 4x + 8y + 12 = 0$

(d) $x^2 + 8x - 4y + 8 = 0$

42. The latus rectum of a parabola whose focal chord is PSQ such that $SP = 3$ and $SQ = 2$ is given by

(a) 24/5 (b) 12/5

(c) 6/5 (d) None of these

43. Consider a circle with its centre lying on the focus of the parabola, $y^2 = 2\,px$ such that it touches the directrix of the parabola. Then a point of intersection of the circle and the parabola is:

(a) $\left(\frac{p}{2}, p\right)$ (b) $\left(\frac{p}{2}, -\frac{p}{2}\right)$

(c) $\left(-\frac{p}{2}, p\right)$ (d) $\left(-\frac{p}{2}, -p\right)$

44. A normal chord of the parabola subtending a right angle at the vertex makes an acute angle θ with the x-axis, then $\theta =$

(a) arc tan 2 (b) arc sec$\sqrt{3}$

(c) arc cot$\sqrt{2}$ (d) $\frac{\pi}{2} + arc\cot\sqrt{2}$

45. Length of the focal chord of the parabola $y^2 = 4ax$ at a distance p from the vertex is:

(a) $\frac{2a^2}{p}$ (b) $\frac{a^3}{p^2}$

(c) $\frac{4a^3}{p^2}$ (d) $\frac{p^2}{a}$

46. T is a point on the tangent to a parabola $y^2 = 4ax$ at its point P. TL and TN are the perpendiculars on the focal radius SP and the directrix of the parabola respectively. Then,

(a) $SL = 2\ (TN)$ (b) $3\ (SL) = 2\ (TN)$

(c) $SL = TN$ (d) $2\ (SL) = 3\ (TN)$

47. PSQ is a focal chord of a parabola whose focus is S and vertex is A. PA and QA are produced to meet the directrix in R and T respectively. Then angle RST is equal to

(a) 90° (b) 60°
(c) 45° (d) 30°

48. Through the vertex O of the parabola, $y^2 = 4ax$ two chords OP and OQ are drawn and the circles on OP and OQ as diameters intersect in R. If θ_1, θ_2 and ϕ are the angles made with the axis by the tangents at P and Q on the parabola and by OR respectively, then the value of, $\cot\theta_1 + \cot\theta_2 =$

(a) $-2\tan\phi$ (b) $-2\tan(\pi - \phi)$
(c) 0 (d) $2\ \chi o\tau\ \phi$

49. The triangle PQR of area 'A' is inscribed in the parabola $y^2 = 4ax$ such that the vertex P lies at the vertex of the parabola and the base QR is a focal chord. The difference of the modulus of the ordinates of the points Q and R is:

(a) $\frac{A}{2a}$ (b) $\frac{A}{a}$
(c) $\frac{2A}{a}$ (d) $\frac{4A}{a}$

50. Through the vertex 'O' of the parabola $y^2 = 4ax$, variable chords OP and OQ are drawn at right angles. If the variable chord PQ intersects the axis of x at R, then distance OR

(a) varies with different positions of P and Q
(b) equals the semi latus rectum of the parabola
(c) equals latus rectum of the parabola
(d) equals double the latus rectum of the parabola

51. Let $y^2 = 4ax$ be a parabola and $x^2 + y^2 + 2bx = 0$ be a circle. If parabola and circle touch each other externally then:

(a) $a > 0, b > 0$ (b) $a > 0, b < 0$
(c) $a < 0, b > 0$ (d) None of these

52. From an external point P, pair of tangent lines are drawn to the parabola, $y^2 = 4x$. If θ_1 and θ_2 are the inclinations of these tangents with the axis of x such that, $\theta_1 + \theta_2 = \frac{\pi}{4}$, then the locus of P is:

(a) $x - y + 1 = 0$ (b) $x + y - 1 = 0$
(c) $x - y - 1 = 0$ (d) $x + y + 1 = 0$

53. If the chord of contact of tangents from a point P to the parabola $y^2 = 4ax$ touches the parabola $x^2 = 4by$, the locus of P is:

(a) circle (b) parabola
(c) ellipse (d) hyperbola

54. If the line, $y = mx$ – a is a common tangent to the parabolas, $y^2 = 4ax$ and $x^2 = 4ay$ then $m =$

(a) 0 (b) 1
(c) –1 (d) None of these

55. A tangent to the parabola $x^2 + 4ay = 0$ cuts the parabola $x^2 = 4by$, at A and B the locus of the mid-point of AB is

(a) $(a + 2b)\, x^2 = 4\, b^2y$ (b) $(b + 2a)\, x^2 = 4\, b^2y$
(c) $(a + 2b)\, y^2 = 4\, b^2x$ (d) $(b + 2x)\, x^2 = 4\, a^2y$

56. The equation of the other normal to the parabola $y^2 = 4ax$ which passes through the intersection of those at (4a, – 4a) and (9a, – 6a) is

(a) $5x - y + 115\, a = 0$ (b) $5x + y - 135\, a = 0$
(c) $5x - y - 115\, a = 0$ (d) $5x + y + 115 = 0$

57. The normal $y = mx - 2am - am^3$ to the parabola $y^2 = 4ax$ subtends a right angle at the vertex if

(a) $m = 1$ (b) $m = \sqrt{2}$
(c) $m = -2\sqrt{2}$ (d) $m = 1/\sqrt{2}$

58. If two normals to a parabola $y^2 = 4ax$ intersect at right angles, then the chord joining their feet passes through a fixed point whose co-ordinates are

(a) (–2a, 0) (b) (a, 0)
(c) (2a, 0) (d) None of these

59. If a normal to a parabola $y^2 = 4ax$ make an angle ϕ with its axis, then it will cut the curve again at an angle

(a) $\tan^{-1}\left(\frac{1}{2}\tan\phi\right)$ (b) $\frac{\phi}{2}$
(c) $\frac{\pi}{2} - \phi$ (d) $\cot^{-1}\left(\frac{1}{2}\tan\phi\right)$

60. If two normals to parabola $y^2 = 4x$ are inclined to its axis at angles a and 2 such that tan a . tan 2 = 2, then their point of intersection lies on

(a) $y = x$ (b) $x^2 = 4y$
(c) $y^2 = 4x$ (d) $y = -x$

61. If two distinct chords drawn from the point (4, 4) on the parabola $y^2 = 4x$ are bisected on the line y = mx, then the set of values of m is given by

(a) $\left(\frac{1-\sqrt{2}}{2}, \frac{1+\sqrt{2}}{2}\right)$ (b) R
(c) $(0, \infty)$ (d) $(-2, 2)$

62. Locus of a point P if the three normals drawn from it to the parabola $y^2 = 4ax$ are such that two of them make complementry angles with the axis of the parabola is
(a) $y^2 = a(x + a)$ (b) $y^2 = 2a(x - a)$
(c) $y^2 = a(x - 2a)$ (d) $y^2 = a(x - a)$

63. Two tangents to the parabola $y^2 = 4ax$ make angles α_1 and α_2 with the x-axis. The locus of their point of intersection if $\frac{\cot\alpha_1}{\cot\alpha_2} = 2$ is
(a) $2y^2 = 9ax$ (b) $4y^2 = 9ax$
(c) $y^2 = 9ax$ (d) None of these

64. Locus of the point of intersection of the normals at the ends of parallel chords of gradient m of the parabola $y^2 = 4ax$ is
(a) $2xm^2 - ym^3 = 4a(2 + m^2)$
(b) $2xm^2 + ym^3 = 4a(2 + m^2)$
(c) $2xm + ym^2 = 4a(2 + m)$
(d) $2xm^2 - ym^3 = 4a(2 - m^2)$

65. Locus of the point of intersection of the perpendicular tangents of the curve $y^2 + 4y - 6x - 2 = 0$ is
(a) $2x - 1 = 0$ (b) $2x + 3 = 0$
(c) $2y + 3 = 0$ (d) $2x + 5 = 0$

66. The equation to the locus of the points from which the parabola, $y = x^2$ can be seen at right angles is
(a) $4x + 1 = 0$ (b) $4y - 1 = 0$
(c) $4y + 1 = 0$ (d) $y + 4 = 0$

67. A point P moves such that the sum of the angles which the three normals makes with the axis drawn from P on the standard parabola, is constant. Then the locus of P is
(a) a straight line (b) a circle
(c) a parabola (d) a line pair

68. PN is an ordinate of the parabola $y^2 = 4ax$. A straight line is drawn parallel to the axis to bisect NP and meets the curve in Q. NQ meets the tangent at the vertex in apoint T such that $AT = kNP$, then the value of k is (where A is the vertex)
(a) $k = 2/3$ (b) $k = 1/3$
(c) $k = 4/3$ (d) None of these

69. If a circle be drawn so as always to touch a given straight line and also a given point. The locus of its centre is
(a) Parabola (b) Straight line
(c) Pair of line (d) Circle

70. Three tangents to a parabola, which are such that the tangents of their inclinations to the axis are in a given harmonical progression, form a triangle is
(a) which is isoceles (b) equilateral
(c) having contant area (d) None of these

71. A normal is drawn to a parabola $y^2 = 4ax$ at any point other than the vertex. Then it cuts the parabola again at a point whose distance from the vertex is
(a) $< 4\sqrt{6}a$ (b) $\geq 4\sqrt{6}a$
(c) $\geq 4\sqrt{3}a$ (d) None of these

72. The length of the intercept on the normal at the point $(at^2, 2at)$ made by the circle which is described on the focal distance of the given point as diameter is
(a) $a\sqrt{1+t}$ (b) $a^2\sqrt{1+t^2}$
(c) $a\sqrt{1+t^2}$. (d) None of these

73. If the normals at the three points P, Q, and R meet in a point and if PP', QQ' and RR' be chords parallel to QR, RP, and PQ respectively, then the normals at P', Q', and R' are
(a) concurrent (b) such that two intersect
(c) no two intersect (d) nothing can be said

74. AP and BP are tangents to the parabola, $y^2 = 4x$ at A and B. If the chord AB passes through a fixed point $(-1, 1)$ then the equation of locus of P is
(a) $y = 2(x - 1)$ (b) $y = 2(x + 1)$
(c) $y = 3(x - 2)$ (d) None of these

75. The normal at any point P meets the axis in G and the tangent at the vertex in G'; if A be the vertex and the rectangle $AGQG'$ be completed, prove that the equation to the locus of Q is
(a) $x^3 = 2ax^2 + ay^2$. (b) $x^2 = 2ax^2 - ay^2$
(c) $x^3 = 2ax^3 - ay^3$ (d) None of these

76. Two parabolas have a common axis and concavities in opposite direction. If any line parallel to the common axis meet the parabola in P and P' and the latus rectum of the given parabolas are unequal, then the locus of the middle point of PP' is
(a) parabola (b) circle
(c) straight line (d) None of thse

77. The locus of the centre of a circle, which intercepts a chord of given length '2a' on the axis of x and passes through a given point on the axis of y distant b from the origin, is
(a) $x^2 - 2yb + b^2 = a^2$ (b) $x^2 + 2yb + b^2 = a^2$
(c) $x^2 - 2yb - b^2 = a^2$ (d) None of these

78. Locus of the intersection of the tangents at the ends of the normal chords of the parabola $y^2 = 4ax$ is

(a) $(2a + x)\, y^2 + 4a^3 = 0$
(b) $(2a - x)\, y^2 + 4a^3 = 0$
(c) $(2a - x)\, y^2 - 4a^3 = 0$
(d) None of these

79. If a parabola whose length of latus rectum is $4a$ touches both the co-ordinate axis then find the locus of its focus.

(a) $x^2y^2 = a^2(x^2 + y^2)$ (b) $x^2y^2 = a^2(x^2 - y^2)$
(c) $x^2y^2 = a^3(x^3 + y^2)$ (d) None of these

80. A parabola is drawn to pass through A and B, the ends of a diameter of a given circle of radius a, and to have as directrix a tangent to a concentric circle of radius b; then axes being AB and a perpendicular diameter, then the locus of the focus of the parabola is

(a) $\frac{x^2}{b^2} + \frac{y^2}{b^2 - a^2} = 1$ (b) $\frac{x^2}{b^2} - \frac{y^2}{b^2 - a^2} = 1$
(c) $\frac{x^2}{b^2} + \frac{y^2}{b^2 + a^2} = 1$ (d) None of these

81. Find the equation to the locus of P when these tangents make angles θ_1 and θ_2 with the axis, such that $\tan^2\theta_1; + \tan^2\theta_2$ is constant $(=\lambda)$.

(a) $y^2 - \lambda x^2 = 2ax$ (b) $y^2 + \lambda x^2 = 2ax$
(c) $y^3 - \lambda x^3 = 2ax$ (d) None of these

82. The locus of the centre of a circle, which intercepts a chord of given length '$2a$' on the axis of x and passes through a given point on the axis of y distant b from the origin, is

(a) $x^2 \pm 2yb + b^2 = a^2$ (b) $x^2 + 2yb - b^2 = a^2$
(c) $x^2 \pm 2yb - b^2 = a^2$ (d) None of these

SECTION–IV

MORE THAN ONE CORRECT ANSWERS

1. Which of the following is true about the given conic section $b^2y^2 - 2ab(y+2x) + a^2 + 4b^2 = 0$?

(a) It is a parabola with vertex $\left(\frac{b}{a}, \frac{a}{b}\right)$
(b) It is a pair of straight lines with point of intersection $\left(\frac{a}{b}, \frac{b}{a}\right)$
(c) It is a parabola with focus at $\left(\frac{a^2+b^2}{ab}, \frac{a}{b}\right)$
(d) None of these

2. Let there be two parabolas with the same axis, focus of each being exterior to the other and the latus rectum being $4a$ and $4b$. The locus of the middle points of the intercepts between the parabolas made on the lines parallel to the common axis is a

(a) straight line if $a = b$
(b) parabola if $a \neq b$
(c) parabola for all a, b
(d) None of these

3. If the tangents drawn from the point (0, 2) to the parabola $y^2 = 4ax$ are inclined at an angle $3\pi/4$, then the value of a is

(a) 2 (b) –2
(c) 1 (d) None of these

4. Let the equations of a circle and a parabola be $x^2 + y^2 - 4x - 6 = 0$ and $y^2 = 9x$ respectively. Then

(a) (1, –1) is a point on the common chord of contact
(b) the equation of the common chord is $y + 1 = 0$
(c) the length of the common chord is 6
(d) None of these

5. A tangent to a parabola $y^2 = 4ax$ is inclined at $\pi/3$ with the axis of the parabola. The point of contact is:

(a) $(a/3, -2a/\sqrt{3})$ (b) $(3a, -2/\sqrt{3}a)$
(b) $(3a, 2\sqrt{3}a)$ (d) $(a/3, 2a/\sqrt{3})$

6. If two distinct chords of a parabola $y^2 = 4ax$, passing through $(a, 2a)$ are bisected on the line $x + y = 1$, then the length of the latus-rectum can be

(a) 2 (b) 1
(c) 4 (d) 5

7. A circle with centre lying on the focus of the parabola $y^2 = 2px$ it touches the directrix of the parabola. Then a point of intersection of the circle and the parabola is

(a) $\left(\frac{p}{2}, p\right)$ (b) $\left(\frac{p}{2}, -p\right)$
(c) $\left(-\frac{p}{2}, p\right)$ (d) $\left(-\frac{p}{2}, -p\right)$

8. Let P, Q and R are three co-normal points on the parabola $y^2 = 4ax$. Then the correct statement(s) is/are:

(a) algebraic sum of the slopes of the normals at P, Q and R vanishes

(b) algebraic sum of the ordinates of the points P, Q and R vanishes
(c) Centroid of the triangle PQR lies on the axis of the parabola
(d) circle circumscribing the ΔPQR passes through the vertex of the parabola

9. The locus of the mid-point of the focal radii of a variable point moving on the parabola, $y^2 = 4ax$ is a parabola whose
(a) Latus rectum is half the latus rectum of the original parabola
(b) Vertex is $(a/2, 0)$
(c) Directrix is y-axis
(d) Focus has the co-ordinates $(a, 0)$

10. Consider the parabolas $y^2 = 4ax$ and $x^2 = 4by$. The striaght line $b^{1/3}y + a^{1/3}x + a^{2/3}b^{2/3} = 0$
(a) touches $y^2 = 4ax$
(b) touches $x^2 = 4by$
(c) intersects both parabolas in real point
(d) touches first and intersects other.

11. Tangents are drawn from point $(-2, 0)$ to $y^2 = 8x$, radius of circle(s) that would touch these tangents and the corresponding chord of contact, can be equal to,
(a) $4(\sqrt{2} + 1)$ (b) $4(\sqrt{2} - 1)$
(c) $8\sqrt{2}$ (d) $4\sqrt{2}$

12. A quadrilateral is inscribed in a parabola, then
(a) quadrilateral may be cyclic
(b) diagonals of the quadrilateral may be equal
(c) all possible pairs of adjacent sides may be perpendicular
(d) None of these.

13. From a point $(\sin\theta, \cos\theta)$ if three normals can be drawn to the parabola $y^2 = 4ax$, then the value of 'a' belongs to
(a) $\left(\frac{1}{2}, 1\right)$ (b) $\left(-\frac{1}{2}, 0\right)$
(c) $\left(0, \frac{1}{2}\right)$ (d) $(1, \infty)$

14. Which of the following statements are true?
(a) Tangent at any point P of parabola bisects the angle between the focal chord through P and perpendicular from P to the directrix.
(b) The foot of perpendicular (H) from the focus (S) on any tangent to a parabola at any point P, lies on the tangent at vertex.
(c) If S be the focus of parabola and tangent and normal at any point P meet its axis in T and G respectively then $ST = SG = SP$.
(d) Light ray incident on the parabolic mirror parallel to the axis after reflection pass through focus of the mirror or they appear to be emerging out of focus of the mirror.

15. The locus of a point O when the three normals drawn from it are such that the line joining the feet of two of them is always in a given direction is
(a) a straight line (b) itself a normal
(c) circle (d) None of these

SECTION–V

ASSERTION AND REASON-TYPE

1. A: If the normals at two point P and Q of a parabola $y^2 = 4ax$ intersect at a third point R on the curve, then the product of the ordinates of P and Q is $8a^2$

R: If P and Q are the point $(at_1^2, 2at_1)$ and $\left(at_2^2, 2at_2\right)$ and normals at P and Q meet at one point on the parabola $y^2 = 4ax$; then $t_1t_2 = 2$.

2. A: The normals at three points P, Q, R of the parabola $y^2 = 4ax$ meet in (h, k). The centroid of triangle PQR lies on the axis of the parabola

R: $y = mx - 2am - am^3$ is the equation of the three normals from a point and hence, sum of ordinates of the feet of normals on the parabola is 0.

3. A: If $y + b = m_1 (x + a)$ and $y + b = m_2 (x + a)$; ($a \neq$ b, $m_1 \neq m_2$) are two tangents to the parabola $y^2 = 4ax$, then $m_1m_2 = -1$

R: Two distinct tangents cannot be drawn at any point on the parabola, therefore m_1 has to be equal to m_2

4. A: If the normals from any point to the parabola $x^2 = 4y$ cuts the line $y = 2$ in points whose abscissae are in A.P., then the slopes of the tangents at the three conormal points are in G.P.

R: If a^3, b^3, c^3 are in AP where $a + b + c = 0$; then a, b, c are in G.P.

5. A: If a and b be the segments of a focal chord and $2c$ the latus rectum of a parabola, then $a^3 + c^3 > 2b^3$

R: $A.M. > G.M. > H.M.$

6. A: If normal at the ends of double ordinate $x = 4$ of parabola $y^2 = 4a$ meet the curve again at P and P' respectively, then $PP' = 12$ units.

R: If normal at t_1 of $y^2 = 4ax$ meet the parabola again at t_2, then $t_2 = t_1 - \frac{2}{t_1}$.

7. A: If the parabola $y = (a - b)x^2 + (b - c)x + (c - a)$ touches the x-axis in the interval (0, 1], then the line $ax + by + c = 0$ always passes through a f ixed point.

R: The equation $L_1 + \lambda L_2 = 0$ or $\mu L_1 + \nu L_2 = 0$ represents a line passing through the intersection of the lines $L_1 = 0$ and $L_2 = 0$ which is a fixed point, when λ, μ, ν are constants.

8. A: The latus rectum of a parabola is 4 units, axis is the line $3x + 4y - 4 = 0$ and the tangent at the vertex is the line $4x - 3y + 7 = 0$, then the equation of directrix of the parabola is $4x - 3y + 8 = 0$.

R: If P be any point on the parabola and let PM and PN are perpendiculars from P on the axis and tangent at the vertex respectively, then $(PM)^2 = (\text{latus rectum})\ (PN)$

9. A: Tangent to the parabola $y^2 = 4x$ at (1, 2) touches the parabola $x^2 = 4y$ also.

R: If two equal parabolas have the same vertex and their axes are at right angles, then their common tangent touches each at the end of a latus rectum.

10. A: Circumcircle of a triangle formed by the lines $x = 0$, $x + y + 1 = 0$ and $x - y + 1 = 0$ also passes through the pont (1, 0).

R: Circumcircle of a triangle formed by three tangents of a parabola passes through its focus.

11. A: Length of focal chord of a parabola $y^2 = 8x$ making an angle of 60° with x-axis is 32/3.

R: Length of focal chord of a parabola $y^2 = 4ax$ making an angle a with x-axis is $4a\ \text{cosec}^2(a)$.

12. A: Area of triangle formed by pair of tangents drawn from a point (12, 8) to the parabola $y^2 = 4x$ and their corresponding chord of contact is 32 sq. units.

R: If from a point $P(x_1, y_1)$ tangents are drawn to a parabola $y^2 = 4ax$, then area of triangle formed by these tangents and their corresponding chord of contact is $\frac{(y_1^2 - 4ax_1)^{3/2}}{4|a|}$ sq.units.

13. A: The points of intersection of the tangents at three distinct points A, B, C on the parabola $y^2 = 4x$ can be collinear.

R: If a line L does not intersect the parabola $y^2 = 4x$, then from every point of the line two tangents can be drawn to the parabola.

14. A: A triangle ABC of area 6 square units is inscribed in the parabola $y^2 = 8x$ such that A lies on the vertex of parabola and side BC is focal chord, then the difference of the distances of B and C from the axis of the parabola is 3.

R: If a triangle ABC of area Δ is inscribed in the parabola $y^2 = 4ax$ such that the vertex A lies at the vertex of the parabola and the side BC is a focal chord, then the differences of distances of B and C from the axis of parabola is $\frac{2\Delta}{a^2}$.

15. A: In the given figure, $AS = 4$, $SP = 9$, then $SZ = 6$.

R: If SZ be perpendicular to the tangent at a point P of a parabola, then Z lies on the tangent at the vertex and $SZ^2 = AS.\ SP$, where A is the vertex of the parabola.

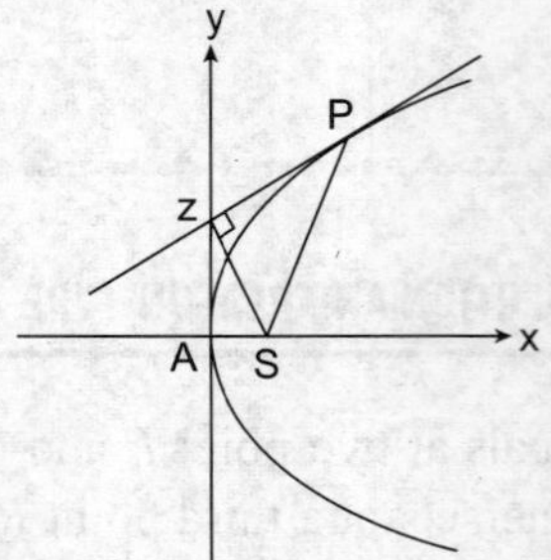

SECTION-VI

COLUMN MATCHING TYPE

1. Column-I

(i) Parabola $y^2 = 4x$ and the circle having its centre at (6, 5) intersects at right angle, at the point (a, a) then one value of a is equal to

(ii) The angle between the tangents drawn to $(y - 2)^2 = 4\ (x + 3)$ at the point where it is intersected by the line $3x - y + 8 = 0$ is $\frac{4\pi}{p}$, then p has the value equal to

(iii) If the line $x - 1 = 0$ is the directix of the parabola $y^2 - kx + 8 = 0$, then one of the value of k is

(iv) Length of the normal chord of the parabola $y^2 = 8x$ at the point where abscissa and ordinate are equal is

Column-II

(a) 13

(b) 8

(c) $10\sqrt{5}$

(d) 4

2. Column-I

(i) Normals AD, AA_1, AA_2 are drawn to parabola $y^2 = 8x$ from the point $A(h, 0)$. If triangle DA_1A_2 is equilateral, then possible value of 'h' is

(ii) A circle is drawn to pass through the extremities of the latus rectum of the parabola $y^2 = 8x$ and it also touches its directrix, then radius of the circle is

(iii) Area of quadrilateral formed by the tangents and normals at extremities of latus rectum of a parabola $y^2 = 4x + 16$, is

(iv) T is the point of intersection of tangents drawn at points $P(t_1)$ and $Q(t_2)$ of parabola $(y + 1)^2 = 8(x - 2)$ and given that $t_1 = 1/2$, and $t_2 = 1$, then value of $SP.\ SQ$, where 'S' is the focus, is

Column-II

(a) 8

(b) 10

(c) 4

(d) 28

3. Column-I

(i) Radius of the largest circle which passes through the focus of the parabola $y^2 = 4x$ and contained in it, is

(ii) Two perpendicular tangents PA and PB are drawn to parabola $y^2 = 16x$, then minimum value of AB is

(iii) The shortest distance between parabolas $y^2 = 4x$ and $y^2 = 2x - 6$ is d, then $d^2 =$

(iv) The harmonic mean of the segments of a focal chord of $y^2 = 8x$

Column-II

(a) 16

(b) 5

(c) 8

(d) 4

SECTION–VII

LINKED COMPREHENSION-TYPE

A: Standard equation of parabola $y^2 = 4x$ can be expressed as $x = \frac{y^2}{4a} = a\left(\frac{y}{2a}\right)^2$ and keeping $\frac{y}{2a} = t$ i.e., $y = 2at$, where t is some real parameter the parametric equation of parabola is generated as $\begin{cases} y = 2at \\ x = at^2 \end{cases}$ and $(at^2, 2at)$ and is called a parametric point 't' on the parabola and by suitably transforming we can write the parametric equation of parabola in other forms. Chord joining t_1 and t_2 on the parabola has slope $= \frac{2}{t_1 + t_2}$, so equation of chord $(y - 2at_1) = \frac{2}{(t_1 + t_2)}(x - at_1^2)$. Condition for this chord to be a focal chord is

$\Rightarrow\ t_1t_2 = -1$. (puting $x = a$, $y = 0$)

Length of focal chord: $L = (a + at_1^2 + a + at_2^2)$

$= a(t_1^2 + t_2^2 + 2) = a(t_1^2 + t_2^2 - 2t_1t_2) = a(t_1 - t_2)^2$

$\therefore\ L = a\left(t + \frac{1}{t}\right)^2$ $\{\because t_1t_2 = -1$ and Let $t_1 = t_1\}$

Harmonic mean of segment of focal chord is equal to semi latus rectum

i.e., if $SP = L_1$, $SQ = L_2$, $2a = \frac{2L_1L_2}{L_1 + L_2}$ as $L_1 = a + at_1^2$, $L_2 = a + at_2^2$ on the basis of above information answer the following questions:

1. Which of the following is true about the following two parametrically defined curves $C_1 : \begin{cases} x = at^2 \\ y = 2at \end{cases}$ and $C_2 : \begin{cases} x = \frac{a(1-t^2)}{1+t^2} \\ y = \frac{2at}{1+t^2} \end{cases}$; where $t \in [0, 1]$?

(a) C_1 and C_2 intersect at two distinct points

(b) they intersect at exactly one point

(c) The area enclosed between C_1, C_2 and x axis is less than $\frac{\pi a^2}{4}$.

(d) The area enclosed between C_1, C_2 and x-axis is greater than $\frac{\pi a^2}{4}$ but less than $\frac{\pi a^2}{2}$.

2. The length of focal chord of the parabola which makes angle α with +ve direction of x axis is $(a > 0)$

(a) $4a\,|\cot\alpha|$ (b) $4a\sec^2\alpha$
(c) $4a\,\mathrm{cosec}^2\alpha$ (d) $4a|\sin 2\alpha|$

3. Which of the following is true regarding the parabola $y^2 = 8x$. having focus S and a chord PQ such that $P(\alpha, 8)$ and $Q(\beta, 2)$?
(a) H.M. of $S.P.$ and $S.Q.$ is 4
(b) quatruple A.M. of $S.P.$ and $S.Q.$ is 25
(c) G.M. of $S.P.$ and $S.Q.$ is 5
(d) H.M. of $S.P.$ and $S.Q.$ is 8

4. Which one of the following equation parametrically represents equation of a parabolic arc?
(a) $x = 4\cos\alpha$ and $y = 3\sin\alpha$; $\alpha \in \left[0, \frac{3\pi}{4}\right]$
(b) $y^2 - 3 = 1 - 2\cos\theta$ and $x = 2\sin^2\frac{\theta}{2}$; $\theta \in \left(-\frac{\pi}{2}, \frac{\pi}{2}\right)$
(c) $x^2 = \sqrt{1+\sin 2t}$ and $y = \sin t + \cos t$; $t \in (0, 3\pi/4)$
(d) $x^2 = \sqrt{1-\sin 2t}$ and $y = \sin t + \cos t$; $t \in \left(0, \frac{\pi}{2}\right)$

5. If (α, β) and (γ, δ) be extremities of a focal chord of parabola $y^2 = 8x$, then which of the following is true.
(a) GM of α and γ is equal to 2
(b) Product of β and δ be equal to 16
(c) GM of α and γ is equal to 4
(d) None of these

6. The equation of chord joining focus to point t on parabola $y^2 = 4ax$ is given by
(a) $y = \frac{2t}{t-1}(x-a)$ (b) $y = \frac{2t}{t+1}(x-a)$
(c) $y = \frac{2t}{t^2-1}(x-a)$ (d) $y = \frac{2t}{t^2+1}(x-a)$

7. If chord joining focus of parabola $y^2 = 4x$ to a point 't' on it also passes through point (2, 1), then which of the following is not true?
(a) Product of possible values of t is – 1
(b) Product of possible values of t is 1
(c) sum of values of t is 2
(d) can't be said

B: If the locus of the circumcentre of a variable triangle having sides y-axis, $y = 2$ and $lx + my = 1$, where (l, m) lies on the parabola $y^2 = 4ax$ is a curve C, then

8. Co-ordinates of the vertex of this curve C is
(a) $\left(2a, \frac{3}{2}\right)$ (b) $\left(-2a, -\frac{3}{2}\right)$
(c) $\left(-2a, \frac{3}{2}\right)$ (d) $\left(2a, \frac{-3}{2}\right)$

9. The length of smallest focal chord of this curve C is
(a) $\frac{1}{12a}$ (b) $\frac{1}{4a}$
(c) $\frac{1}{16a}$ (d) $\frac{1}{8a}$

10. The curve C is symmetric about the line:
(a) $y = -\frac{3}{2}$ (b) $y = \frac{3}{2}$
(c) $x = -\frac{3}{2}$ (d) $x = \frac{3}{2}$

C: If $P(at^2, 2at)$ be any point on the parabola $y^2 = 4ax$, then slope of the tangent at P is

$$\frac{dy}{dx} = \frac{2a}{y} = \frac{2a}{2at} = \frac{1}{t}\left[\because \frac{dy}{dx} = \frac{2a}{y}\right]$$

Therefore, slope of the normal at P is $-$ t, and its equation is $y - 2at = -t(x - at^2)$ i.e., $y = -tx + 2at + at^3$. If we substitute m for $-t$ in the above equation, we have the follwoing result. The equation $y = mx - 2am - am^3$ is a normal to the parabola $y^2 = 4ax$ at the point $P(am^2, -2am)$.
Three normals real or imaginary can be drawn from any point to a given parabola and the algebraic sum of the ordinates of the feet of these three normals is zero. Let equation of a parabola be $y^2 = 4ax$ and that of a normal to it be $y = mx - 2am - am^3$. If this normal passes through the point (x_1, y_1), we have $y_1 = mx_1 - 2am - am^3$ i.e., $am^3 + (2a - x_1)m + y_1 = 0$. Which gives three values of m, real or imaginary . if m_1, m_2 and m_3 be the roots of the above equation, then we have $m_1 + m_2 + m_3 = 0$. Hence, the sum of the ordinates of the feet of these normals is $-2a(m_1 + m_2 + m_3) = 0$. If the normal at the point $P(at_1^2, 2at_1)$ of the parabola $y^2 = 4ax$ meets the parabola again at the point $Q(at_2^2, 2at_2)$, then $t_2 = -t_1 \quad -$.

11. From the point (15, 12) three normals are drawn to the parabola $y^2 = 4x$, then centroid of triangle formed by three co-normal points is
(a) $\left(\frac{16}{3}, 0\right)$ (b) (4, 0)
(c) $\left(\frac{26}{3}, 0\right)$ (d) (6, 0)

12. If $P(4, 8)$ and Q are points on the parabola $y^2 = 16x$ and the chord PQ subtends a right angle at the vertex of the parabola, then the co-ordinates of the point of intersection of normal at P and Q is
(a) (8, 20) (b) $\left(\frac{45}{4}, \frac{3}{4}\right)$
(c) (60, –48) (d) (4, –32)

13. Let P, Q and R are three co-normal points on the parabola $y^2 = 4ax$. Then the correct statement(s) is/are:

(a) Algebraic sum of the slopes of the normals at P, Q and R vanishes

(b) Algebraic sum of the ordinates of the points P, Q and R vanishes

(c) Centroid of the triangle PQR lies on the axis of the parabola

(d) Circle circumscribing the triangle PQR passes through the vertex of the parabola

14. Locus of the point of intersection of two normals to the parabola $x^2 = 8y$ which are at right angles to each other.

(a) $x^2 = 2(y - 6)$ (b) $x^2 = 2(y + 6)$

(c) $x^2 = (y - 6)$ (d) None of these

D: A slope of tangent at any point 't' is 1/t. Equation of tangent at 't' is $ty = x + at^2$

$\Rightarrow \quad \frac{x}{-at^2} + \frac{y}{at} = 1$. Also we have following properties of tangent

(i) x intercept of tang*ent* $= -at^2$, y intercept of tangent $= at$

(ii) The tangents to the parabola $S \equiv y^2 - 4ax = 0$ drawn at the points $(at_1^2, 2at_1)$ and $(at_2^2, 2at_2)$ intersect at the point $(at_1t_2, a(t_1 + t_2))$

(iii) Tangent at extermities of any focal chord always intersect at directrix.

(iv) Portion of tangent to a parabola intercepted between directrix and curve subtends right angle at focus.

(v) Tangent at any point P of parabola bisects the angle between the focal chord through P and perpendicular from P to the directrix. Based on these properties answer the following

15. AB, AC are tangents to a parabola $y^2 = 4ax$, p_1, p_2, p_3 are the lengths of the perpendiculars from A, B, C on any tangent to the curve, then p_2, p_1, p_3 are in

(a) AP (b) GP

(c) HP (d) None of these

16. If tangents at A and B on the parabola $y^2 = 4ax$ intersect at point C, then ordinates of A, C and B are

(a) always in AP (b) always in GP

(c) always in HP (d) None of these

17. The locus of point of intersection of tangents to the parabolas $y^2 = 8(x + 1)$ and $y^2 = 4(x + 2)$ which are perpendicular to each other is

(a) $x + 7 = 0$ (b) $x - y = 4$

(c) $x + 3 = 0$ (d) $y - x = 12$

18. Points A, B and C lie on the parabola $y^2 = 4ax$.The tangents to the parabola at A, B and C taken in pairs intersect at points P, Q and R respectively. then the ratio of the areas of the triangles ABC and PQR.

(a) 1: 2 (b) 2: 1

(c) 1: 1 (d) None of these

E: Three conormals are drawn from(α, β) to $y^2 = 4ax$ such that feet of normals be $(am_i^2, -2am_i)$ $i = 1, 2, 3$.

then $am^3 + (2a - \alpha)_m + \beta = 0$ ...(i)

$\Rightarrow \sum_{i=1}^{3} m_i = 0; \sum m_i m_j = \frac{2a - \alpha}{3}; m_1 m_2 m_3 = \frac{-\beta}{a}$.

Let the equation of circle through those three points be $x^2 + y^2 + 2gx + 2fy + c = 0$ ..(ii)

and $a^2m^4 + (4a^2 + 2ag)m^2 - 4afm + c = 0$ ...(iii)

being biquadratic in m shows that $m_1 + m_2 + m_3 + m_4 = 0$ ...(iv)

$m_4 = 0 \Rightarrow$ fourth point where circle meets this parabola is$(am_4^2, -2am_4)$ i.e., (0, 0) that is through vertex.

$\Rightarrow c = 0$ from, equation of circle (iii)

$\Rightarrow (am^3 + (4a + 2g)m - 4f) = 0$ ....(v)

$\Rightarrow$ Equation (i) and (v) are identical

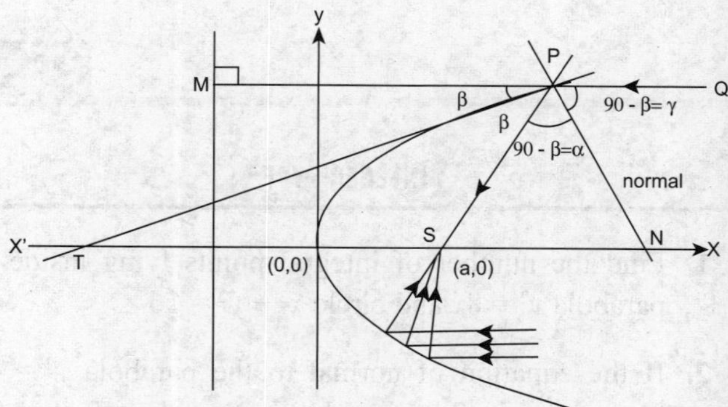

$$\Rightarrow 1 = \frac{4a + 2g}{2a - \alpha} = -\frac{4f}{\beta} \Rightarrow 2g = -(2a + \alpha); 2f = \frac{-\beta}{2}$$

$\therefore$ Equation of circle is $x^2 + y^2 - (2a + \alpha)x - \frac{\beta}{2}y = 0$

Cor (1): Algebric sum of ordinates of four points of intersection of a circle and parabola = 0

Cor (2): Common chords of a circle and a parabola are pairwise equally inclined to the axis of parabola.

Cor (3): Circle through conormal points passes thorough vertex of parabola

Cor (4): Centroid of four points in which circle intersect the parabola lie on axis

$$G \equiv \left(\frac{a}{4}\left(0 - \frac{2(4a^2 + 2ag)}{a^2}\right), 0\right) = (-2a - g, 0)$$

Reflection property of parabola: tangent PT and normal PN are internal and external bisectors of $\angle SPM$ and QP is parallel to Axis of parabola $\Rightarrow \angle QPN = \angle SPN$ as shown in figure. On the basis of above properties answer the following questions.

19. A ray of light travels along a line $y = 4$ and strikes the surface of a curve $y^2 = 4(x + y)$ then equation of the line along reflected ray travel

(a) $x = 0$ (b) $x = 2$
(c) $x + y = 4$ (d) $2x + y = 4$

20. If P be a point on the parabola $y^2 = 3(2x - 3)$ M is foot of the perpendicular drawn from P on the directrix of the parabola, then length of each side of an equilateral triangle SMP where 'S' is focus of the parabola is

(a) 2 (b) 4
(c) 6 (d) 8

21. If the locus of middle point of point of contact of tangent drawn to the parabola $y^2 = 8x$ and foot of perpendicular drawn from its focus to the tangent is a conic, then length of latusrecturm of this conic is

(a) 9/4 (b) 9
(c) 18 (d) 9/2

22. If the normal at three points P, Q, R of the parabola $y^2 = 4ax$ meet in a point O and S be its focus, then $|SP|.|SQ|.|SR|$ is equal to

(a) a^3 (b) $a^2(SO)$
(c) $a(SO^2)$ (d) None of these

23. A parabola $y = ax^2 + bx + c$ crosses the x-axis at $(\alpha, 0)$; $(\beta, 0)$ both to the right of the origin. A circle also passes through these two points. The length of a tangent from the origin to the circle is:

(a) $\sqrt{\dfrac{bc}{a}}$ (b) ac^2
(c) b/a (d) $\sqrt{\dfrac{c}{a}}$

24. The locus of the mid-point of the focal radii of a variable point moving on the parabola, $y^2 = 4ax$ is a parabola whose

(a) Latus rectum is half the latus rectum of the original parabola
(b) vertex is $(a/2, 0)$
(c) Directrix is y-axis
(d) Focus has the co-ordinates $(a, 0)$

SECTION–VIII

INTEGER-TYPE

1. Find the number of integral points lying inside the parabola $y^2 = 8x$ and circle $x^2 + y^2 = 16$

2. If the equation of normal to the parabola $y^2 = 4ax$ having slope -2 is given by $y + 2x + ka = 0$, then find the value of k.

3. The straight line $x + y = k$ touches the parabola $y = x - x^2$, then find the value of k.

4. The normals drawn at the end points of a variable chord PQ of the parabola $y^2 = 4\,ax$ intersect at parabola, and if the locus of the point of intersection of the tangent drawn at the points P and Q is given by $x - ka = 0$, then find the value of k.

5. Find the angle (in degrees) between the tangents drawn from the origin to the parabola $(x - a)^2 = -4a(y + a)$.

6. The normal at the ends of the latus rectum of the parabola $y^2 = 4x$ meet the parabola again at A and A', then find the length AA'.

7. If the line $4y - 3x - 8 = 0$ cuts the parabola $x^2 + y - 4 = 0$ at A and B, and $PA.PB$ is equal to $k/8$ (where $P = (0, 2)$), then find the value of k.

8. The length of the side of an equilateral triangle, inscribed in the parabola $y^2 = 8x$ so that one angular points is at the vertex is given by $k\sqrt{3}$, then find the value of k.

9. If one end point of latus rectum of parabola is L(1, 2) and slope of its axis is $y = x$. Given O is vertex and L' is other end of semi latus rectum and area of $\Delta OLL'$ is given by $1/k$, then find the value of k.

10. If the point P on the parabola $y^2 = 4ax$ for which $|PR - PQ|$ is maximum, where $R \equiv (-a, 0)$, $Q \equiv (0, a)$ is represented as $(k_1 a, k_2 a)$, then find the value of $k_1 + k_2$.

11. If the co-ordinates of the point on the parabola $y = x^2 + 7x + 2$, which is nearest to the straight line $y = 3x - 3$ is represented as (k_1, k_2), then find the value of $k_1 + k_2$.

12. The parabolas $y = x^2 - 9$ and $y = kx^2$ intersect at point A and B. if length AB is equal to $2a$, then the value of k is $\dfrac{a^2 - p}{a^2}$, then find the value of p.

13. If $y = 2x + 3$ is a tangent to the parabola $y^2 = 24x$, then it's distance from the parallel normal is given by $k\sqrt{5}$, then find the value of k.

14. If the distance of two points P and Q from the focus of a parabola $y^2 = 4ax$ are 4 and 9, respectively, then find the distance of the point of intersection of tangents at P and Q from the focus

15. If A and B are points on the parabola $y^2 = 4ax$ with vertex O such that OA is perpendicular to OB and having lengths r_1 and r_2, respectively, then find the value of $\frac{r_1^{4/3} r_2^{4/3}}{\left(r_1^{2/3} + r_2^{2/3}\right)a^2}$.

Answer Keys

SECTION–III

1. (b)	2. (b)	3. (a)	4. (a)	5. (a)	6. (d)	7. (a)	8. (b)	9. (d)	10. (a)
11. (c)	12. (d)	13. (a)	14. (d)	15. (a)	16. (a)	17. (a)	18. (a)	19. (c)	20. (b)
21. (b)	22. (d)	23. (b)	24. (c)	25. (d)	26. (c)	27. (a)	28. (d)	29. (b)	30. (b)
31. (b)	32. (c)	33. (c)	34. (a)	35. (a)	36. (c)	37. (c)	38. (b)	39. (a)	40. (d)
41. (a)	42. (a)	43 (a)	44. (b)	45. (c)	46. (c)	47. (a)	48. (a)	49. (c)	50. (c)
51. (a)	52. (c)	53. (d)	54. (c)	55. (a)	56. (b)	57. (b)	58. (b)	59. (a)	60. (c)
61. (a)	62. (d)	63. (a)	64. (a)	65. (d)	66. (c)	67. (a)	68. (a)	69. (a)	70. (c)
71. (b)	72. (c)	73. (a)	74. (b)	75. (a)	76. (a)	77. (a)	78. (a)	79. (a)	80. (a)
81. (a)	82. (a)								

SECTION–IV

1. (a, c)	2. (a, b)	3. (a, b)	4. (a, c)	5. (a, d)	6. (a, b)	7. (a, b)	8. all	9. all
10. (a, b)	11. (a, b)	12. (a, b)	13. (b, c)	14. all	15. (a, b)			

SECTION–V

1. (a)	2. (a)	3. (b)	4. (a)	5. (a)	6. (c)	7. (a)	8. (d)	9. (d)	10. (a)
11. (a)	12. (c)	13. (d)	14. (a)	15. (a)					

SECTION–VI

1. (i) → (d) (ii) → (b) (iii) → (d) (iv) → (c)
2. (i) → (d) (ii) → (c) (iii) → (a) (iv) → (b)
3. (i) → (d) (ii) → (a) (iii) → (b) (iv) → (d)

SECTION–VII

P-1	1. (b, c)	2. (c)	3. (a, b, c)	4. (b, c)	5. (a)	6. (c)	7. (a, c)
P-2	8. (c)	9. (d)	10. (b)				
P-3	11. (c)	12. (c)	13. (a, b, c, d)	14. (a)			
P-4	15. (b)	16. (a)	17. (c)	18. (b)			
P-5	19. (a)	20. (c)	21. (b)	22. (c)	23. (d)	24. (a, b, c, d)	

SECTION–VIII

1. 17	2. −12	3. 1	4. 2	5. 90	6. 12	7. 25	8. 16	9. 4	10. 3
11. −10	12. 9	13. 15	14. 6	15. 16					

HINTS AND SOLUTIONS

TEXTUAL EXERCISE 1–(SUBJECTIVE)

1. (a) Given $x^2 + 4x + 4y + 16 = 0$

$\Rightarrow a = 1, b = 0, h = 0, g = 0, f = 2, c = 16$

$$\Rightarrow \Delta = \begin{vmatrix} a & h & g \\ h & b & f \\ g & f & c \end{vmatrix} = \begin{vmatrix} 1 & 0 & 2 \\ 0 & 0 & 2 \\ 2 & 2 & 16 \end{vmatrix} = (-2)(2) = -4 \neq 0$$

Further $h^2 - ab = 0$ and $\Delta \neq 0$.

$\Rightarrow$ Equation represents a parabola

Now $(x + 2)^2 = -4y - 12$

So, $(x + 2)^2 = (-4)(y + 3)$

Which is a down wards parabola vertex $V = (-2, -3)$

Focus, $S = (-2, -4)$

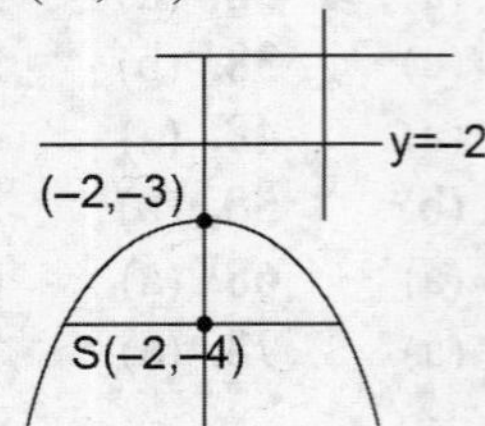

Directrix $y = -2$; axis $x = -2$ and equation of latus rectum $y = -4$, length of $LR = 4$ units

(b) Given: $y^2 - 4x - 2y - 7 = 0$ or $(y - 1)^2 = 4x + 8 = 4(1)(x + 2)$

Which is a parabola with vertex $V = (-2, 1)$

Focus $S = (-1, 1)$, Directrix $x = -3$ and axis $y = 1$, length of L.R. = 4 units

Equation of L.R. $x = -1$

(c) Given: $(y - 4)^2 = x - 3 = 4\left(\frac{1}{4}\right)(x - 3)$

Which is a parabola with Vertex $V = (3, 4)$, focus $\left(\frac{13}{4}, 4\right)$

Directrix $x = \frac{11}{4}$ and L.R. as $x = \frac{13}{4}$

Equation of axis $y = 4$, Length of L.R. = 1 unit

2. (a) (i) Given $V(2, 1)$, $F(1, -1)$

$\Rightarrow a = \sqrt{5}$, slope of axis = 2

Image of $(1, -1)$ in point $(2, 1)$ is $(3, 3)$

Equation of directrix $y - 3 = \left(-\frac{1}{2}\right)(x - 3)$

$\Rightarrow x + 2y - 9 = 0$

Now $\frac{|x + 2y - 9|}{\sqrt{5}} = \sqrt{(x-1)^2 + (y+1)^2}$

$\Rightarrow x^2 + 4y^2 + 81 + 4xy - 18x - 36y = 5x^2 + 5y^2 + 5 + 5 - 10x + 10y$ or $4x^2 + y^2 - 4xy + 8x + 46y - 71 = 0$

(ii) Given: $F(1, 1)$ tangent at the vertex $x + y = 1$

$\therefore$ $x = y$, vertex $= \left(\frac{1}{2}, \frac{1}{2}\right)$ and $a = \frac{1}{\sqrt{2}}$

Equation of directrix is $x + y = 0$. Hence the equation

$\frac{|x + y|}{\sqrt{2}} = \sqrt{(x-1)^2 + (y-1)^2}$

$\Rightarrow x^2 + y^2 + 2xy = 2x^2 + 2 + 2y^2 + 2 - 4x - 4y$ or $(x^2 + y^2 - 2xy) - 4x - 4y + 4 = 0$ and which can be show written as $(x - y)^2 = 4(x + y - 1)$

(b) Given: $F(3, -4)$ and directrix $x - y + 5 = 0$

Hence the equation of parabola

$\frac{|x - y + 5|}{\sqrt{2}} = \sqrt{(x-3)^2 + (y+4)^2}$

$\Rightarrow x^2 + y^2 + 25 - 2xy + 10x - 10y = 2[x^2 + y^2 + 9 + 16 - 6x + 8y]$

$\Rightarrow x^2 + y^2 + 2xy - 22x + 26y + 25 = 0$

(c) Ends of latus rectum are $(1, 2)$ and $(1, -4)$

Length of L.R. = 6 units $\Rightarrow a = \frac{3}{2}$ and $F(1, -1)$

Equation of axis $y = -1$, Vertex $V\left(\frac{5}{2}, -1\right)$ or $\left(-\frac{1}{2}, -1\right)$

Hence the parabola $(y+1)^2 = 4\left(\frac{3}{2}\right)\left(x + \frac{1}{2}\right)$

$\Rightarrow (y + 1)^2 = 6x + 3 = 3(x + 2)$

Or $(y + 1)^2 = -(6x - 15) = (-3)(2x - 5)$

(d) Let the vertex be at $V(a, 0)$ and $F(a', 0)$, there are two possibilities

(i) When $a' > a$, then it gives a parabola opening towards right then $y^2 = 4(a' - a)(x - a)$

(ii) When $a' < a$, then the parabola opens towards left, so $y^2 = -4(a - a')(x - a) = 4(a - a')(x - a)$

So in both the care the equation is $y^2 = 4(a - a')(x - a)$.

(e) Given the parabola is of the form $Ax + By^2 + Cy + D = 0$

We can easily observe that it is a horizontal parabola, let it be $(y - y_1)^2 = 4a(x - x_1)$ where (x_1, y_1) is the vertex and $a \in (R - \{0\})$

Putting the values of $P_1(-1, 0)$, $P_2(2, 1)$ and $P_3(1, -1)$, we get $y_1^2 = 4a(-1 - x_1)$ .... (i)

$(1 - y_1)^2 = 4a(2 - x_1)$ .... (ii)

and $(-1 - y_1)^2 = 4a(1 - x_1)$ .... (iii)

From (ii) and (iii), we have $-4y_1 = 4a$

$\Rightarrow y_1 = -a$

Putting in (i), we have $a^2 = 4a(-1 - x_1) \Rightarrow x_1 = -\frac{a}{4} - 1$

From (iii) and (ii), we have $2y_1 + 1 = 8a$

$\Rightarrow 10a = 1$ and $a = \frac{1}{10}$

$\therefore$ $x_1 = -\frac{41}{40}$ and $y_1 = -\frac{1}{10}$ and gives vertex $\left(-\frac{41}{40}, -\frac{1}{10}\right)$ and $\left(y + \frac{1}{10}\right)^2 = \frac{2}{5}\left(x + \frac{41}{40}\right)$ is the parabola i.e., $5y^2 + y - 2 = 2x$

(f) From the given information the equation of directrix is $x + y + 2\sqrt{2} = 0$, as it will pass through $A(-2\sqrt{2}, 0)$ and $B(0, -2\sqrt{2})$.

As the focus lies on $y = x$ at 2 units from of origin, so $F \leftrightarrow (\sqrt{2}, \sqrt{2})$ {and not $(-\sqrt{2}, -\sqrt{2})$ because this point is on directrix}

Hence vertex V (0, 0) and equation of parabola will be

$$(x-\sqrt{2})^2 + (y-\sqrt{2})^2 = \frac{(x+y+2\sqrt{2})^2}{2}$$

i.e., $2\{x^2 + y^2 - 2\sqrt{2}x - 2\sqrt{2}y + 4\}$

$= x^2 + y^2 + 4\sqrt{2}x + 4\sqrt{2}y + 8 + 2xy$

or $x^2 + y^2 - 8\sqrt{2}x - 8\sqrt{2}y - 2xy = 0$

3. (a) $y^2 = 4x$

$\Rightarrow a = 1$, hence $(x, y) = (at^2, 2at)$ becomes $(t^2, 2t)$ or $(t^2, \pm 2t)$

(b) $(x + 1)^2 = 4(y - 1) \Rightarrow a = 1$ and $x = 2t - 1, y = t^2 + 1, x = 2t - 1, y = t^2 + 1$

(c) $3(x^2 + x) = -7y - 8$

or $3\left(x^2 + x + \frac{1}{4}\right) = \frac{3}{4} - 8 - 7y \Rightarrow \left(x + \frac{1}{2}\right)^2 = -\frac{7y}{3} - \frac{29}{12}$

$= (-4)\left(\frac{7}{12}\right)\left\{y + \frac{29}{28}\right\}$

$\Rightarrow V\left(-\frac{1}{2}, -\frac{29}{28}\right)$ and $x + \frac{1}{2} = 2at = -\frac{7}{6}t \Rightarrow x = -\frac{7}{6}t - \frac{1}{2}$

Similarly $y = \left(-\frac{7}{12}\right)t^2 - \frac{29}{28}$

4. (a) Given $x^2 = 9y$ and abscissa is thrice the ordinate i.e., $x = 3y$

$\Rightarrow 9y^2 = 9y$

$\Rightarrow y = 0, 1$ (as $y \geq 0$) the points are (0, 0), (3, 1)

$\Rightarrow y = 0, 1$

(b) Given $y^2 = 4ax$

Now length of double ordinate is $8a$

$\Rightarrow y = \pm 4a$ for the required points

$\therefore (\pm 4a)^2 = 4ax \quad \Rightarrow x = 4a$

Hence the points are $(4a, \pm 4a)$ as a result slope of $V(0, 0)$ and A $(4a, 4a)$ is $m_1 = 1$ and slope of V (0, 0) and B $(4a, -4a)$ is $m_2 = -1$ as $m_1 m_2 = -1$

$\therefore \angle AOB = 90°$

5. (a) Given directrix is $x + y = 4$

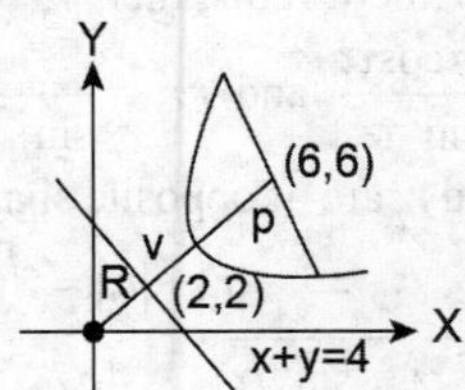

$\therefore R = (2, 2)$ and F (6, 6) $\Rightarrow V$ (4, 4)

(b) Axis of parabola is $x - y = 0$ or $x = y$

(c) Equation of L.R. is $x + y = c$ as F (6, 6) satisfies

$\Rightarrow x + y = 12$

(d) Length of L.R. = $2(RF) = 2\frac{|6+6-4|}{\sqrt{2}} = 8\sqrt{2}$ units

(e) End points of L.R. P (2, 10), P'(10, 2)

(f) Tangent at vertex is $x + y =$ constant since V (4, 4)

$\Rightarrow$ tangent is $x + y = 8$

(g) The equation of parabola

$$(x-6)^2 + (y-6)^2 = \frac{(x+y-4)^2}{2}$$

$\Rightarrow 2\{x^2 + y^2 - 12x - 12y + 72\} = x^2 + y^2 + 2xy + 16 - 8x - 8y$

or $x^2 + y^2 - 2xy - 16x - 16y + 128 = 0$

TEXTUAL EXERCISE 1–(OBJECTIVE)

1. (c) Given: $(y-1)^2 = 2x - 4 = 4\left(\frac{1}{2}\right)(x-2)$, which is parabola with vertex (2, 1) and directrix as $x = \frac{3}{2}$

2. (d) Given $2x^2 - 72xy + 23y^2 - 4x - 28y - 48 = 0$

Observe that $h^2 - ab = (-36)^2 - 46 > 0$

This represents a hyperbola as $\Delta = \begin{vmatrix} 2 & -36 & -2 \\ -36 & 23 & -14 \\ -2 & -14 & -18 \end{vmatrix} \neq 0$

Now $\frac{\partial S}{\partial x} = 4(x - 18y - 1) = 0$

and $\frac{\partial S}{\partial y} = (-2)(36x - 23y + 14) = 0$

Solving both we get centre at $\left(-\frac{11}{25}, -\frac{2}{25}\right)$

3. (a) Given $14x^2 - 4xy + 11y^2 - 44x - 58y + 71 = 0$

Observe that $h^2 - ab < 0$ {as $(-2)^2 - 14 \times 11 < 0$} and

$\Delta = \begin{vmatrix} 14 & -2 & -22 \\ -2 & 11 & -29 \\ -22 & -29 & 71 \end{vmatrix} \neq 0$

$\Rightarrow$ its represents an ellipse

Now $\frac{\partial S}{\partial x} = 4(7x - y - 11) = 0$ and

$\frac{\partial S}{\partial y} = (-2)\{2x - 11y + 29\} = 0$

Solving both we get $x = 2, y = 3$

$\therefore$ Center is at (2, 3)

4. (c) Given: F(1, 1) and directrix $x - y + 1 = 0$

For eccentricity $e = \sqrt{2}$, we get $\frac{SP}{SM} = \sqrt{2}$

$\Rightarrow (x-1)^2 + (y+1)^2 = 2\left\{\frac{(x-y+1)^2}{2}\right\}$

$\Rightarrow x^2 + 1 - 2x + y^2 + 1 + 2y = x^2 + y^2 + 1 - 2xy + 2x - 2y$

or $2xy - 4x + 4y + 1 = 0$

5. (a) The given equation is $x^2 - 2xy + y^2 + 3x + 2 = 0$.

Observe that $h^2 - ab = (-1)^2 - 1 = 0$

Now $\Delta = \begin{vmatrix} 1 & -1 & 3/2 \\ -1 & 1 & 0 \\ 3/2 & 0 & 2 \end{vmatrix} = \begin{vmatrix} 0 & -1 & 3/2 \\ 0 & 1 & 0 \\ 3/2 & 0 & 2 \end{vmatrix}$

$= \frac{3}{2}\left(-\frac{3}{2}\right) = -\frac{9}{4} \neq 0$

$\Rightarrow$ its will represent a parabola

6. (b) Given A parabola $(y-3)^2 = -5x + 9$ or

$$(y-3)^2 = (-4)\left(\frac{5}{4}\right)\left(x-\frac{9}{5}\right)$$

$\Rightarrow a = \frac{5}{4}$ $\Rightarrow$ L.R. $= 4a = 5$ units

7. (c) The parabola is $(y-1)^2 = 4(3)(x-2)$, which has vertex at $V(2, 1)$ and for $a = 3$ focus $F(5, 1)$

8. (b) The parabola is $(y-1)^2 = -6(x+2) = (-4)\left(\frac{3}{2}\right)(x+2)$

$\Rightarrow$ Vertex $V = (-2, 1)$

9. (d) The parabola is $(y+2)^2 = -4x + 2 = (-4)\left\{x-\frac{1}{2}\right\}$

Equation of directrix is $x = \frac{3}{2}$

10. (c) Given $y^2 = 8x = 4(2)(x)$

$\Rightarrow F(2, 0)$ and directrix $x = -2$. Now for a focal length of 4 units $x = 2$ $\Rightarrow P(2, \pm 4)$

11. (b) The parabola $y^2 = 4\left(\frac{1}{5}\right)x$ has $F\left(\frac{1}{5}, 0\right) = (a, 0)$ and points of L.R. are $(a, \pm 2a) = \left(\frac{1}{5}, \pm\frac{2}{5}\right)$

12. (d) End points of L.R. for parabola $x^2 = 4ay$ are $(\pm 2a, a)$

13. (d) $y = 2t$ and $x = t^2 + 2$ $\Rightarrow y^2 = 4(x-2)$

14. (c) Equation of parabola is $y^2 + 2By + B^2 = -2Ax - C + B^2$ or $(y+B)^2 = (-4)\left(\frac{A}{2}\right)\left\{x+\frac{C}{2A}-\frac{B^2}{2A}\right\}$; Vertex $V = \left(\frac{B^2-C}{2A}, -B\right)$

Equation of L.R. $x = \frac{B^2-C}{2A} - \frac{A}{2} = \frac{B^2-A^2-C}{2A}$

15. (a) Given $F(a, b)$ and directrix is $bx + ay - ab = 0$

Hence the parabola is $(x-a)^2 + (y-b)^2 = \frac{(bx+ay-ab)^2}{a^2+b^2}$

$\Rightarrow (a^2+b^2)\{x^2+y^2+a^2+b^2-2ax-2by\} = b^2x^2 + a^2y^2 + a^2b^2 - 2ab^2x - 2a^2by + 2abxy$

i.e., $(ax-by)^2 - 2a^3x - 2b^3y + a^4 + b^4 - a^2b^2 = 0$

TEXTUAL EXERCISE 2–(SUBJECTIVE)

1. (i) Let $P(t_1) = (at_1^2, 2at_1)$ be a point on the parabola $y^2 = 4ax$ $\Rightarrow$ focus $F(a, 0)$

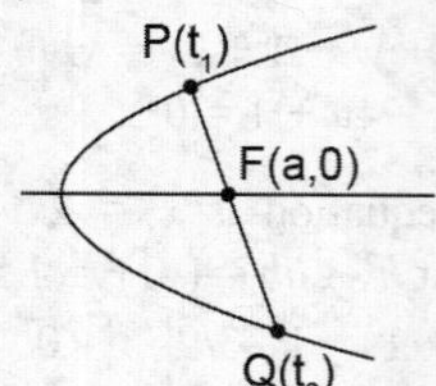

$\therefore$ Equation of PF is $y = \frac{2at_1}{a(t_1^2-1)}(x-a)$

$\Rightarrow y = \frac{2t_1}{(t_1^2-1)}(x-a)$ since $Q(t_2)$ lies on this focal chord

$\Rightarrow 2at_2 = \frac{2t_1}{(t_1^2-1)}\{at_2^2 - a\}$

$\Rightarrow 2a(t_1t_2^2 - t_2) = 2at_1\{t_2^2-1\}$ i.e., $t_2t_1(t_1 - t_2) = (t_2 - t_1)$ {since $t_1 \neq t_2$}

$\Rightarrow t_1t_2 = -1$

(ii) As $t_1t_2 = -1$

$\therefore PQ = \sqrt{a^2(t_1^2-t_2^2)^2 + 4a^2(t_1-t_2)^2}$

$\Rightarrow PQ = a|t_1 - t_2|\sqrt{t_1^2+t_2^2+2t_1t_2+4}$

$= a|(t_1-t_2)|\sqrt{t_1^2+t_2^2-2t_1t_2}$ (as $4 = -4t_1t_2$) $= a(t_1-t_2)^2$ or $a(t_2-t_1)^2$

2. Let $P(t_1)$ and $Q(t_2)$ be the end points of a focal chord

$\Rightarrow t_1t_2 = -1$

For $F(a, 0)$, we get $PF = \ell_1 = \sqrt{a^2(t_1^2-1)^2 + 4a^2t_1^2}$

$\Rightarrow \ell_1 = |a|(t_1^2+1)$

Similarly $FQ = \ell_2 = |a|(t_2^2+1)$

Now L.R. $= 4a$, we observe that

$$\frac{4\ell_1\ell_2}{\ell_1+\ell_2} = \frac{4a^2(t_1^2+1)(t_2^2+1)}{a\{(t_1^2+1)+(t_2^2+1)^2\}}$$

$$= \frac{4a(t_1^2+1)(t_1^2+1)}{t_1^2\{t_1^2+1\}+\frac{(t_1^2+1)}{t_1^2}} = \frac{4a(t_1^2+1)^2t_1^2}{t_1^2(t_1^2+1)^2} = 4a = LR$$

3. (a) for $y^2 = 4ax$, let $F(a, 0)$ be the focus and PFQ be the focal chord with angle α with the x-axis.

Any point on focal chord at a distance 'r' from focus will be $(a + r\cos\alpha, r\sin\alpha)$, since P and Q lie on the parabola

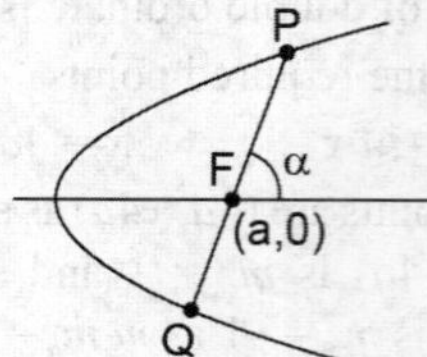

$\therefore r^2\sin^2\alpha = 4a\{a + r\cos\alpha\}$

$\Rightarrow \sin^2\alpha\, r^2 - 4a\cos\alpha r - 4a^2 = 0$

Let r_1 and r_2 be it roots, then $|r_1| + |r_2| = PF$. Now

$r_1 + r_2 = \frac{4a\cos\alpha}{\sin^2\alpha}$ and $r_1r_2 \frac{\quad}{\sin}$

Since r_1 and r_2 are of opposite sign (as these are directed distances) $\Rightarrow |r_1|+|r_2| = \sqrt{(r_1+r_2)^2 + 4|r_1r_2|}$

$$= 4a\sqrt{\frac{\cos^2\alpha}{\sin^4\alpha}+\frac{\sin^2\alpha}{\sin^4\alpha}} = \frac{4a}{\sin^2\alpha} = 4a\,\text{cosec}^2\,\alpha$$

(b) Given $P(x_1, y_1)$ and $Q(x_2, y_2)$ are the end points of a focal chord of parabola $y^2 = 4ax$

Let $P(t_1) \Rightarrow x_1 = at_1^2, y_1 = 2at_1$ and $Q(t_2)$

$\Rightarrow x_2 = at_2^2, y_2 = 2at_2$

$\Rightarrow x_1x_2 = a^2t_1^2t_2^2$, since $t_1t_2 = -1$

$\therefore x_1x_2 = a^2(-1)^2 = a^2$

4. Given $y = mx + c$ is a tangent to the parabola $y^2 = 4a(x+a)$

Let $x + a = X$ and $y = Y$

$\Rightarrow$ $Y^2 = 4ax$ which will have equation of tangent as $Y = mX + \dfrac{a}{m}$

i.e., $y = m(x+a) + \dfrac{a}{m} = mx + \left(am + \dfrac{a}{m}\right)$

$\Rightarrow \quad c = a\left(\dfrac{m^2+1}{m}\right)$ or $c = am + \dfrac{a}{m}$

5. The line is a $x \cos\alpha + y \sin\alpha - p = 0$

If it is a tangent to $y^2 = 4a(x + a)$, then $y = mx + \left(am + \dfrac{a}{m}\right)$ {as in Q # 4}

Comparing, we get $\dfrac{m}{\cos\alpha} = \dfrac{-1}{\sin\alpha} = \dfrac{\left(am + \dfrac{a}{m}\right)}{-p}$

$\Rightarrow \quad mp = -a\cos\alpha\left(\dfrac{m^2+1}{m}\right)$ or $mp = a(m^2 + 1)\sin\alpha$

From $m^2p + a\cos\alpha\, m^2 + a\cos\alpha = 0$, we get $\dfrac{\cos^2\alpha}{\sin^2\alpha}(p + a\cos\alpha) + \alpha\cos\alpha = 0$

$\left(\text{by putting } m = -\dfrac{\cos\alpha}{\sin\alpha}\right)$

$\Rightarrow$ $p\cos\alpha + a\cos^2\alpha + a\sin^2\alpha = 0$ i.e., $p\cos\alpha + a = 0$

6. Given $\quad x + my + n = 0$ is a tangent to $y^2 = 4a(x - b)$

$\Rightarrow \quad \left(\dfrac{\ell x + n}{m}\right)^2 = 4a(x - b)$

$\Rightarrow \quad {}^2x^2 + n^2 + 2\quad nx = 4am^2x - 4am^2b$ or $\quad {}^2x^2 + (2\quad n - 4am^2)x + n^2 + 4am^2b = 0$

For tangency x has a unique value, so $D = 0$ i.e., $(2\quad n - 4am^2)^2 - 4\quad {}^2(n^2 + 4am^2b) = 0$

$\Rightarrow$ $4\quad {}^2n^2 + 16a^2m^4 - 16a\quad nm^2 - 4\quad {}^2n^2 - 16a\quad {}^2m^2b$ or $am^2 = \quad n + b\quad {}^2$ which is the required condition.

7. Given $x + y = \lambda$ touches the parabola $x^2 - x = -y$

$\Rightarrow$ $(\lambda - y)^2 - (\lambda - y) = -y$

$\Rightarrow$ $\lambda^2 + y^2 - 2\lambda y - \lambda + y + y = 0$ i.e., $y^2 + (2 - 2\lambda)y + (\lambda^2 - \lambda) = 0$

Since the line is a tangent $\therefore D = 0$

$\Rightarrow$ $4(1 - \lambda)^2 - 4(\lambda^2 - \lambda) = 0$

$\Rightarrow$ $4 + 4\lambda^2 - 8\lambda - 4\lambda^2 + 4\lambda = 0$ i.e., $4\lambda = 4$

$\Rightarrow$ $\lambda = 1$

TEXTUAL EXERCISE 2–(OBJECTIVE)

1. (a) Given t_1 and t_2 are the parametric points

$\Rightarrow \quad p(t_1) = (at_1^2, 2at_1)$ and $Q(t_2) = (at_2^2, 2at_2)$

Hence the equation of PQ is

$$y - 2at_1 = \frac{2}{(t_1 + t_2)}(x - at_1^2)$$

$\Rightarrow$ $y(t_1 + t_2) = 2x + 2at_1t_2$

2. (b) Let (x_1, y_1) be the other end of chord through origin so mid points is $\left(\dfrac{x_1}{2}, \dfrac{y_1}{2}\right)$

$\Rightarrow$ $x_1 = 2x, y_1 = 2y$ as (x_1, y_1) lies on $y^2 = 4ax$ so $y_1^2 = 4ax_1$ putting values, we get $4y^2 = 4a(2x) \Rightarrow y^2 = 2ax$

3. (c) $P(at_1^2, 2at_1)$ is one end of the focal chord so the other end is $Q\left(\dfrac{a}{t_1^2}, \dfrac{-2a}{t_1}\right)$

$\Rightarrow$ Length of PQ is a $\left(t_1 + \dfrac{1}{t_1}\right)^2$

4. (c) $P(x_1, y_1)$ and $Q(x_2, y_2)$ are the ends of a chord.

Now $y_1^2 = 4ax_1$ and $y_2^2 = 4ax_2$

$\Rightarrow$ $y_2^2 - y_1^2 = 4a(x_2 - x_1)$ or $\dfrac{(y_2 - y_1)(y_2 + y_1)}{(x_2 - x_1)} = 4a$ .... (i)

The equation of line PQ is $y - y_1 = \dfrac{y_2 - y_1}{x_2 - x_1}(x - x_1)$

Substituting from (i), we get $y - y_1 = \dfrac{4a}{(y_2 + y_1)}(x - x_1)$

Hence $(y_2 + y_1)(y - y_1) = 4ax - 4ax_1 = y^2 - 4ax_1$ i.e., $(y - y_1)(t_1 + y_1) = y^2 - 4ax_1$

5. (a) The parabola is S: $y^2 - 8x = 0$, the mid - point of chord is $(1, -2)$ from $T = S_1$, we get $(-2)y - 4x - 4 = 4 - 8 = -4$

$\Rightarrow$ $2y = -4x$ i.e., $y = -2x$

6. (d) Let $P(t_1)$ and $Q(t_2)$ be the end points of the focal chords, then $t_1t_2 = -1$

The mid point $M\left(\dfrac{a}{2}(t_1^2 + t_2^2), a(t_1 + t_2)\right)$. Hence $2x = a(t_1^2 + t_2^2)$ and $y = a(t_1 + t_2)$

$\Rightarrow$ $y^2 = a^2\{t_1^2 + t_2^2 + 2t_1t_2\} = a^2(t_1^2 + t_2^2) - 2a^2 = 2ax - 2a^2$ i.e., $y^2 = 2ax - 2a^2$

7. (d) Let $P(t_1)$ and $Q(t_2)$ be the end points of the chord {passing trough (p, q)}, then $(t_1 + t_2)y = 2x + 2at_1t_2$. Since (p, q) lies on it.

$\Rightarrow$ $(t_1 + t_2)q = 2p + 2at_1t_2$

Hence $2at_1t_2 = (t_1 + t_2)q - 2p$

Now mid point of PQ is $M\left(\dfrac{a}{2}(t_1^2 + t_2^2), a(t_1 + t_2)\right)$

$\therefore \quad 2x = a(t_1^2 + t_2^2)$ and $y = a(t_1 + t_2)$ this gives

$$y^2 = a^2(t_1^2 + t_2^2 + 2t_1t_2) = 2ax + 2a^2t_1t_2$$

i.e., $y^2 = 2ax + a(t_1 + t_2)q - 2pa$ or $y^2 = 2ax + yq - 2pa$

Putting $a = 1$ for $y^2 = 4x$, we get $y^2 = qy - 2x + 2p = 0$

8. (c) S: $y^2 - 4x = 0$ and line L: $y = 1 - x$

So we get $x^2 + 1 - 2x - 4x = 0$

$\Rightarrow$ $x^2 - 6x + 1 = 0$

$\therefore \quad x = \dfrac{6 \pm \sqrt{36 - 4}}{2} = 3 + 2\sqrt{2}, 3 - 2\sqrt{2}$

$|x_2 - x_1| = 4\sqrt{2}$

$\Rightarrow$ Length of chord $\sqrt{2}|x_2 - x_1| = 8$ units

9. (a) S : $y^2 - 5y - 3x + 6 = 0$ and line $x = 0$

$\Rightarrow$ $y^2 - 5y + 6 = 0$

$\Rightarrow$ $y = 2, 3$

$\therefore$ Length of intercept = 1 unit

10. (c) $S: y^2 - x = 0$, $P(1, 0)$ and line is $y = x - 1$

$\Rightarrow x^2 + 1 - 3x = 0 \qquad \Rightarrow x_1 = \dfrac{3+\sqrt{5}}{2}, x_2 = \dfrac{3-\sqrt{5}}{2}$

$$\Rightarrow |PA| + |PB| = \sqrt{\frac{2(1+\sqrt{5})^2}{4}} + \sqrt{\frac{2(1-\sqrt{5})^2}{4}}$$

$$= \sqrt{2}\left\{\frac{\sqrt{5}+1}{2} + \frac{\sqrt{5}-1}{2}\right\} = \sqrt{2}\sqrt{5} = \sqrt{10}$$

11. (c) We know that the length of focal chord with one end at $P(t)$ is $a\left(t+\dfrac{1}{t}\right)^2$

Now $P(t) = (at^2, 2at)$ and $F(a, 0)$, hence the equation of focal chord will be $2tx + (1 - t^2)\,y - 2at = 0$

So its distance from origin is $\left|\dfrac{2at}{t^2+1}\right| = p$ (given) i.e.,

$$\left|\frac{2a}{t+\frac{1}{t}}\right| = p \Rightarrow \left(t+\frac{1}{t}\right)^2 = \frac{4a^2}{p^2}$$

Hence $a\left(t+\dfrac{1}{t}\right)^2 = \dfrac{4a^3}{p^2}$

12. (c) Consider B as $(at^2, 2at)$ and let $MB \perp AC$ and $\angle MAB = \theta$

$$\Rightarrow \tan\theta = \frac{MB}{AM} = \frac{2at}{at^2} = \frac{2}{t}$$

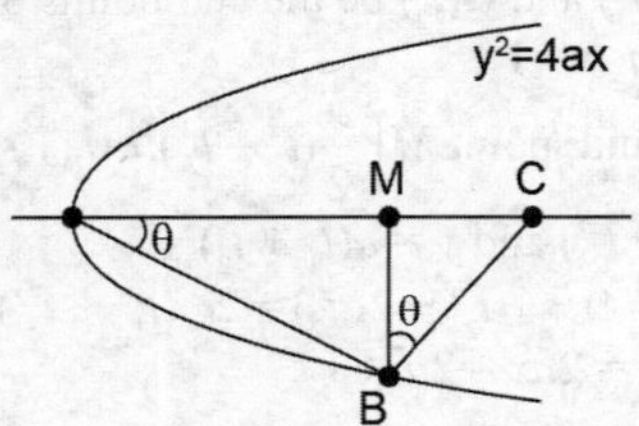

Similarly, projection $MC = MB\tan\theta = 2at\left(\dfrac{2}{t}\right) = 4a.$

TEXTUAL EXERCISE 3–(SUBJECTIVE)

1. Let PQ be a focal chord, so $P(t)$ and $Q\left(-\dfrac{1}{t}\right)$ are the points and the tangents are $ty = x + at^2$ and $-\dfrac{y}{t} = x + \dfrac{a}{t^2}$

i.e., $-ty = xt^2 + a$. Now observe that tangents are at 90° to each other as slopes are $\dfrac{1}{t}, -t$ respectively

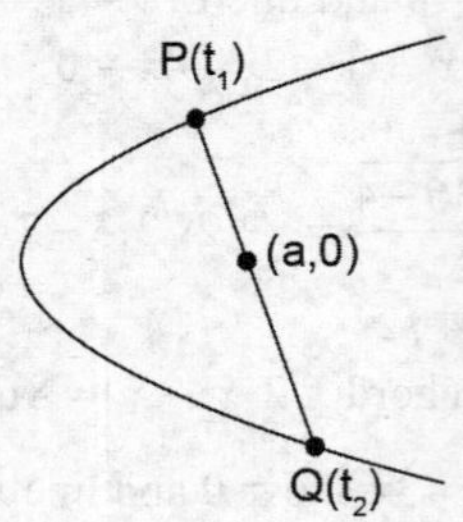

$\Rightarrow 0 = x(1 + t^2) + a(1 + t^2) = (x + a)(1 + t^2)$

Since $1 + t^2 \neq 0$

$\Rightarrow x = -a$ for all $t \in R$ which shows that tangents intersect at the directrix.

2. $y^2 = 8x = 4(2)\,x \qquad \Rightarrow a = 2$

As the tangent is at 45° to $y = 3x + 5$ (which has slope $m_1 = 3$)

$\Rightarrow m_2 = \dfrac{m_1+1}{1-m_1}$ or $\dfrac{m_1-1}{1+m_1}$ i.e., $m_2 = -2$ or $\dfrac{1}{2}$

Tangent with slope $m_2 = -2$ is $y = mx + \dfrac{a}{m} \Rightarrow y = -2x - 1$ or $2x + y + 1 = 0$ and its point of contact is $\left(\dfrac{a}{m^2}, \dfrac{2a}{m}\right) = \left(\dfrac{1}{2}, -2\right)$

Similarly for $m = \dfrac{1}{2}$, we get $x - 2y + 8 = 0$ at $(8, 8)$

3. $y^2 = kx \Rightarrow a = \dfrac{k}{4}$

Now the line $x + y + 1 = 0$ has slope $m = -1$

$\Rightarrow y = mx + \dfrac{a}{m} \qquad \Rightarrow mx - y + \dfrac{a}{m} = 0$ as $-x - y - a = 0$

$\Rightarrow a = 1$ or $k = 4$

Now the point of contact is $(1, -2)$

4. $y^2 = 4ax$ is the parabola so the line as tangent will be $mx - y + \dfrac{a}{m} = 0$, the given line is $x\cos\alpha + y\sin\alpha - p = 0$

$$\Rightarrow \frac{\cos\alpha}{m} = \frac{\sin\alpha}{-1} = \frac{-pm}{a}$$

Eliminating m, we get $m = -\cot\alpha = \dfrac{a\sin\alpha}{p}$

$\Rightarrow a\sin\alpha = -p\cot\alpha$ i.e., $p\cos\alpha + a\sin^2\alpha = 0$

Now the point of tangency is $\left(\dfrac{a}{m^2}, \dfrac{2a}{m}\right)$

$$= \left(\frac{a}{\cot^2\alpha}, \frac{2a}{-\cot\alpha}\right) = (a\tan^2\alpha, -2a\tan\alpha)$$

5. The parabola is $y^2 = 4a(x - b)$ shifting the origin to $(b, 0)$, we get $Y^2 = 4aX$ and the line $lx + my + n = 0$ becomes $(x + b) + my + n = 0$

i.e., $\quad X + mY + (n + b\quad) = 0$

Comparing with $Mx - Y + \dfrac{a}{M} = 0$,

we get $\dfrac{M}{\ell} = \dfrac{-1}{m} = \dfrac{a}{M(n+bl)}$

$\Rightarrow M = -\dfrac{l}{m} = \dfrac{-am}{(n+bl)} \Rightarrow am^2 = (n+b\quad) = b\quad{}^2 + n$

6. Let $P(t_1)$ and $Q(t_2)$ be two points on the parabola $y^2 = 4ax$ with focus at $S(a, 0)$.

Now R is the point of intersection of tangents at P and Q

$\Rightarrow R(at_1t_2, at_1 + at_2)$

Now $SR^2 = a^2\{t_1^2 + t_2^2 + 2t_1t_2 + t_1^2t^2 + 1 - 2t_1t_2\} = a^2\{(t_1 + 1)(t_2 + 1)\}$ .... (1)

Similarly, $SP.SQ = \sqrt{a^2(t_1^2-1)^2 + 4a^2t_1^2}\ \sqrt{a^2(t_2^2-1)^2 + 4a^2t_2^2}$

$= a\sqrt{(t_1+1)^2} \times a\sqrt{(t_2+1)^2} = a^2(t_1+1)(t_2+1)$

Clearly $SR^2 = SP.\,SQ$

7. The parabola is $y^2 = 8ax = 4(2a)\,x \Rightarrow A = 2a$

Now $y = mx + \frac{A}{m}$ is a tangent to the parabola .This line will also be a tangent to the circle $x^2 + y^2 = (\sqrt{2}a)^2$ when

$$\left|\frac{A}{m\sqrt{(1+m^2)}}\right| = \sqrt{2}a \Rightarrow \quad 2 = m^2(m^2+1)$$

Hence $m^2 = 1$

For $m = 1$, the tangent is $y = x + 2a$ or $x - y + 2a = 0$ and

For $m = -1$, the tangent is $y = -x - 2a$ or $x + y + 2a = 0$

i.e., $x \pm y + 2a = 0$

8. Let $P(t_1)$ and $Q(t_2)$ be the two points on the parabola $y^2 = 4ax$, so slope of tangent at $P(t_1)$ is $m_1 = \frac{1}{t_1}$ and slope of tangent at $Q(t_2)$ is $m_2 = \frac{1}{t_2}$ as given $\tan\theta = \left|\frac{m_1 - m_2}{1 + m_1m_2}\right| = \left|\frac{t_2 - t_1}{1 + t_1t_2}\right|$

$\Rightarrow (1 + t_1t_2)^2\tan^2\theta = (t_2 - t_1)^2$

Let $R(at_1t_2, at_1 + at_2) = (x, y)$ be the point of intersection of tangents $\Rightarrow \frac{x}{a} = t_1t_2$ and $\frac{y}{a} = (t_1 + t_2)$

Hence, $\left(1 + \frac{x}{a}\right)^2\tan^2\theta = (t_1 + t_2)^2 - 4t_1t_2$

$= \left(\frac{y}{a}\right)^2 - 4\left(\frac{x}{a}\right)$ i.e., $(a + x)^2\tan^2\theta = y^2 - 4ax$

TEXTUAL EXERCISE 3–(OBJECTIVE)

1. (a) $y^2 = 4x$ is the parabola $\Rightarrow a = 1$

$R(at_1t_2, at_1 + at_2)$ is the point of intersection

$\Rightarrow t_1t_2 = -2$ and $(t_2 + t_1) = -1$

Now $\tan\theta = \left|\frac{t_2 - t_1}{1 + t_1t_2}\right| = \frac{\sqrt{(1+8)}}{|1-2|} = 3$

$\{\ (t_2 - t_1)^2 = (t_1 + t_2)^2 - 4t_1t_2\}$

Aliter: Parabola is S: $y^2 - 4x = 0$ and the point is $(-2, -1)$ using $T^2 = SS_1$, we get $(-y - 2x + 4)^2 = (y^2 - 4x)(1 + 8) = 9y^2 - 36x$

i.e., $4x^2 + y^2 + 16 + 4xy - 16x - 8y = 9y^2 - 36x$ or $4x^2 - 8y^2 + 4xy + 16x - 8y + 16 = 0$

Now $\tan\theta = \frac{2\sqrt{h^2 - ab}}{|a+b|} = \frac{2\sqrt{4+32}}{|-4|} = 3$

2. (b) Using $SR^2 = SP.\,SQ$, we get $SR^2 = (4)(9) = 36$

$\Rightarrow SR = 6$

3. (c) Let $y = mx + \frac{a}{m}$ be the tangent to $y^2 = 4x$ and $(x - 3)^2 + y^2 = 3^2 \Rightarrow \left|\frac{3m + \frac{1}{m}}{\sqrt{1+m^2}}\right| = 3$

$\Rightarrow (3m^2 + 1)^2 = 9m^2(m^2 + 1)$, i.e., $3m^2 = 1$

For $m = \frac{1}{\sqrt{3}}$ we get the tangent $y = \frac{x}{\sqrt{3}} + \sqrt{3}$ or $\sqrt{3}y = x + 3$

4. (c) Let $y = f(x) = x^2 + bx - b$ be the curve

Now tangent at $(1, 1)$ with slope m is $y - 1 = m(x - 1)$ $(m < 0)$ for the Δ to be formed in 1st quadrant

Now the intercepts on axes by the tangent are $a = \frac{m-1}{m}, b = (1 - m)$

Since the area of Δ is $2sq$ units

$\Rightarrow \frac{1}{2}ab = 2$

Hence $-\frac{(m-1)^2}{m} = 4$

$\Rightarrow (m + 1)^2 = 0 \qquad \Rightarrow m = -1$

As a result $y + \frac{b^2}{4} + b = x^2 + \frac{b}{x} + \frac{b^2}{4} = \left(x + \frac{b}{2}\right)^2$

$\Rightarrow \left.\frac{dy}{dx}\right|_{(1,1)} = 2\left(x + \frac{b}{2}\right) = -1$

$\Rightarrow 1 + \frac{b}{2} = -\frac{1}{2}$ i.e., $\frac{b}{2} = -\frac{3}{2} \quad \therefore b = -3$

5. (b) Tangent to the parabola $y^2 = 32x = 4(8)\,x$ will be $y = mx + \frac{8}{m}(m \neq 0)$.

Similarly, tangent with slope m to the parabola $x^2 = 4(27)\,y$ will be $y = mx - 27\,m^2$.

Since the tangent is common so $\frac{8}{m} = -27m^2$

$\Rightarrow m^3 = -\frac{8}{27}$ i.e., $m = -\frac{2}{3}$

Hence, the tangent is $y = -\frac{2}{3}x - 12$ or $2x + 3y + 36 = 0$.

6. (c) Tangent to the parabola $y^2 - 4ax = 0$ will be $y_1y - 2ax - 2ax_1 = 0$ or $y = \frac{2a}{y_1}x + \frac{2ax_1}{y_1}$

Now, $\tan\theta = \frac{2a}{y_1}$ and $y_1^2 = 4ax_1$

Hence, $\frac{2ax_1}{y_1} = \frac{y_1}{2} = \frac{2a}{2\tan\theta} = a\cot\theta$

Hence. the equation becomes $y = x\tan\theta + a\cot\theta$.

7. (a) The parabola is $y^2 = 4ax$. As tangent makes 60° with x-axis

$\therefore \quad m = \sqrt{3}$

Hence, the point of contact is $\left(\frac{a}{m^2}, \frac{2a}{m}\right) = \left(\frac{a}{3}, \frac{2a}{\sqrt{3}}\right)$.

8. (a) Let $P(t)$ be a point on the parabola s: $y^2 - 4ax = 0$

Slope of tangent at $P(t)$ is $m = \frac{1}{t}$ and its equation is $ty = x + at^2$

$\Rightarrow$ Slope of perpendicular from focus $S(a, 0)$ will be $(-t)$ and the equation is $y = (-t)(x - a)$

Solving $ty = x + at^2$ and $ty = -t^2x + at^2 \Rightarrow x = 0$

9. (d) The parabola is $y^2 = ax$

$\Rightarrow A = \frac{a}{4}$. Now slope of tangent is $m = 1$

$\therefore$ The point of contact is $\left(\frac{A}{m^2}, \frac{2A}{m}\right) = \left(\frac{a}{4}, \frac{a}{2}\right)$

10. (b) As shown $A(0, 0)$, $Y(0, at)$, $T(-at^2, 0)$ for $TAYG$ to be a rectangle $G(-at^2, at)$

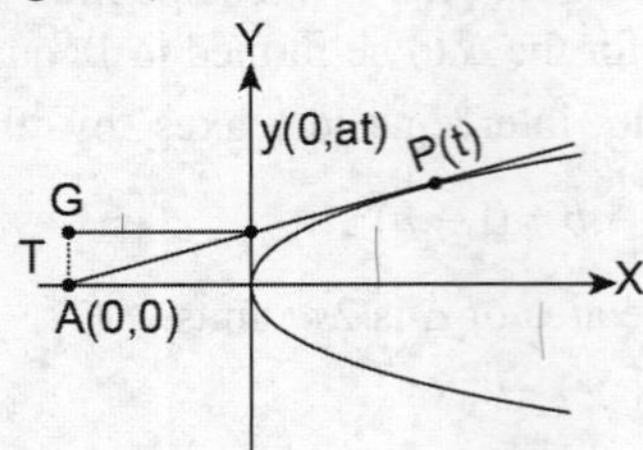

$\Rightarrow$ let $x = -at^2$ and $y = at$ i.e., $-\frac{x}{a} = t^2, \frac{y}{a} = t$

$\therefore \left(\frac{y}{a}\right)^2 = \left(-\frac{x}{a}\right)$, i.e., $y^2 + ax = 0$

11. (b) Equation of tangent at $P(t)$ is $ty = x + at^2$

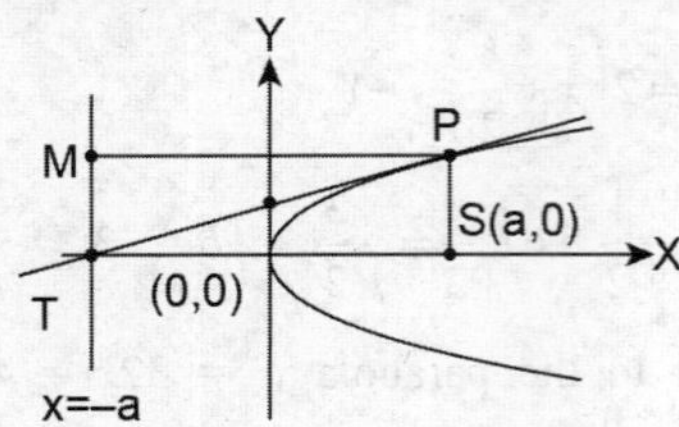

$SP = \sqrt{a^2(t^2-1)^2 + 4a^2t^2} = a(t^2+1)$ and $PM = at^2 + a = a(t^2 + 1)$

Now distance of M $(-a^2, at)$ from PT is $d_1 = \left|\frac{-a - 2at^2 + at^2}{\sqrt{1+t^2}}\right| = a\sqrt{t^2+1}$

Also distance of $S(a, 0)$ from PT is $d_2 = a\sqrt{(t^2+1)}$

Clearly for $SP = PM$, we get $d_1 = d_2 \Rightarrow \angle MPT = \angle SPT$

12. (d) Let $P(t) = (at^2, 2at)$, now $S(a, 0)$

$\Rightarrow$ Slope of SP is $m_1 = \frac{2at}{a(t^2-1)} = \frac{2t}{t^2-1}$

Now tangent at P is $x - ty + at^2 = 0$ which meets directrix at $x = -a$, $y = \frac{a(t^2-1)}{t}$

Hence slope of SK is $m_2 = \frac{a(t^2-1)}{t.2a} = \frac{t^2-1}{2t}$, observe that $m_1m_2 = -1 \Rightarrow \angle KSP = 90°$

TEXTUAL EXERCISE 4–(SUBJECTIVE)

1. Let (x_1, y_1) be the point then chord of contact for $S: y^2 - 4bx = 0$ will be $T = 0$

i.e., $y_1y - 2bx - 2bx_1 = 0$ or $2bx - y_1y + 2bx_1 = 0$

Since it is also a tangent to $y^2 = 4ax$

$\Rightarrow mx - y + \frac{a}{m} = 0$

$\Rightarrow \frac{2b}{m} = \frac{-y_1}{-1} = \frac{2bx_1(m)}{a}$ which gives $m = \frac{2b}{y_1}$ and $m^2 = \frac{a}{x_1}$

Hence $\left(\frac{2b}{y_1}\right)^2 = \frac{a}{x_1}$ or $y_1^2 = \frac{4b^2}{a}x_1$

$\Rightarrow$ The locus is $y^2 = \frac{4b^2}{a}x$

2. Let $P(x_1, y_1)$ be an external point and $y^2 - 4ax = 0$ be the parabola

$\Rightarrow$ The chord of contact is $yy_1 - 2ax - 2ax_1 = 0$

Since focus $S(a, 0)$ lies on the chord

$\Rightarrow -2a^2 - 2ax_1 = 0$

$\Rightarrow x_1 = -a$ which is independent of y_1

Hence, locus of $P(x_1, y_1)$ is $(-a, y)$ i.e., $x = -a$ which is the directrix

3. Let $P\left(x_1, \frac{x_1+1}{2}\right)$ be a variable point on $x - 2y + 1 = 0$ and the chord of contact from P to $y^2 = 4(2a)x$ will be $\frac{x_1+1}{2}y - 4ax - 4ax_1 = 0$ Rearranging, we get $(x_1 + 1)y = 8a(x + x_1)$. Observe that for $x = 1$, $y = 8a$ the equation is true for all real x_1, so the chord passes through $(1, 8a)$. For $a = 1$, we get $(1, 8)$ as the point.

4. The parabola is $y^2 = 4ax$. Slope of the normal is $m_1 = \pm\frac{1}{\sqrt{3}}$

$\therefore$ The slope of the tangent is $m_2 = \pm\sqrt{3}$ or $\frac{1}{t} = \pm\sqrt{3}$ and the points are $P(at^2, 2at) = \left(\frac{a}{3}, \pm\frac{2a}{\sqrt{3}}\right)$

5. The given chord is $y = \sqrt{2}x - 4a\sqrt{2}$ and slope of chord is $\sqrt{2}$. The parabola is $y^2 = 4ax$, we know that at $P(t)$ the slope of normal is $(-t)$ and the equation of normal is $y - 2at = (-t)(x - at)^2$

Putting $t = -\sqrt{2}$, we get $y = \sqrt{2}x + 2a(-\sqrt{2}) + 2a(-\sqrt{2})$ or $y = \sqrt{2}x - 4a\sqrt{2}$

$\Rightarrow$ The given chord is a normal chord at $P(t) = P(-\sqrt{2}) = (2a, -2a\sqrt{2})$ and the other end is $t_2 = -t - \frac{2}{t} = \sqrt{2} + \sqrt{2} = 2\sqrt{2}$, so the other end is $Q(2\sqrt{2}) = (8a, 4a\sqrt{2})$

$\Rightarrow$ The length $PQ = \sqrt{(6a)^2 + (6a\sqrt{2})^2} = 6a\sqrt{1+2} = 6\sqrt{3}a$

6. (a) The equation of co-normal is given by $y + 2am = mx - am^3$ for $y^2 = 4ax$

$\Rightarrow am^3 + 2am - mx + y = 0$

Which gives three slopes m_1, m_2, m_3. Now $m_1 = -m_2$ (given)

$\Rightarrow m_1 + m_2 + m_3 = 0 \Rightarrow m_3 = 0$

$\Rightarrow m_1m_2m_3 = -\frac{y}{a} = 0 \quad \Rightarrow y = 0$

Also $\sum m_1m_2 = m_1m_2 = -m^2 = \frac{2a - x}{a}$

$\therefore x = 2a - am^2 = a\{2 - m^2\}$. Since $x \geq 0 \Rightarrow |m| \leq \sqrt{2}$. Hence $y = 0$.

(b) When two normal are perpendicular to each other, then $m_1m_2 = -1$ (say)

$\Rightarrow m_1m_2m_3 = -m_3\frac{-y}{a} \quad \Rightarrow y = am_3$

Now $m_1 + m_2 + m_3 = m_1 - \frac{1}{m_1} + m_3 = 0$

$\Rightarrow m_1^2 - 1 = -m_1m_3$

Similarly $\sum m_1m_2 = -1 + m_1m_3 - \frac{m_3}{m_1} = \frac{2a - x}{a}$

$\Rightarrow \quad \frac{m_3}{m_1}(m_1^2 - 1) = \frac{3a - x}{a}$

or $\frac{m_3}{m_1}(-m_1m_3) = \frac{3a - x}{a} \Rightarrow -m_3^2 a = 3a - x$

or $-a^2m_3^2 = 3a^2 - ax$ i.e., $-y^2 = 3a^2 - ax$

$\Rightarrow y^2 = a(x - 3a)$

7. (i) The parabola is $y^2 = 8x \Rightarrow a = 2$

We know that the co-normal system has equation $am^3 + 2am - mx + y = 0$ (where $a = 2$)

Now let $m_1 = 2$, so the foot of normal is $(am^2, -2am)$ i.e., $(8, -8)$ and the equation will be $y = 2x - 24$

(ii) Slope of normal $m = 3$

$\therefore$ Foot of normal is $(18, -12)$ and the equation will be $y = 3x - 66$

(iii) Observe that (6, 0) is an interior point (on x-axis)

$\Rightarrow 2m^3 + 4m - 6m + 0 = 0$ i.e., $2m(m^2 - 1) = 0$

$\Rightarrow m = 0, \pm 1$

For $m = 0$, we get line $y = 0$; at $(0, 0)$

For $m = 1$, we get line $x - y - 6 =$ at $(2, -4)$

For $m = -1$, we get line $x + y - 6 = 0$ at $(2, 4)$

8. $S_1 < 0 \Rightarrow k^2 - 4(2k^2 - 1) < 0$

$\Rightarrow 4 - 7k^2 < 0$

$\Rightarrow |k| > \frac{2}{\sqrt{7}}$ ....(i)

Given three normal drawn are real $2k^2 - 1 \geq 2$

$\Rightarrow \quad k^2 \geq \frac{3}{2} \Rightarrow |k| \geq \sqrt{\frac{3}{2}}$ ...(ii)

Taking intersection of both sets

$k \in \left(-\infty, -\sqrt{\frac{3}{2}}\right] \cup \left[\sqrt{\frac{3}{2}}, \infty\right)$

9. (i) For the parabola $y^2 = 4ax$, we know that the co-normal system is $am^3 + 2am - mx + y = 0$

When $m_1 + m_2 = 1$, then $m_1 + m_2 + m_3 = 0 \Rightarrow m_3 = -1$

And from $m_1m_2m_3 = \frac{-y}{a}$, we get $m_1m_2 = \frac{y}{a}$

Similarly $\sum m_1m_2 = m_1m_2 + (m_1 + m_2)m_3 = \frac{2a - x}{a}$

$\Rightarrow \quad \frac{y}{a} + 1(-1) = \frac{2a - x}{a} \quad \Rightarrow \quad y - a = 2a - x$ or $x + y = 3a$

(ii) When $m_1m_2 = 1$, then $m_1m_2m_3 \equiv \frac{-y}{a} \Rightarrow m_3 \equiv \frac{-y}{a}$ and

$\sum m_1m_2 = 1 + (m_1 + m_2)m_3 = \frac{2a - x}{a}$

Also $m_1 + m_2 + m_3 = 0$

$\Rightarrow \quad m_1 + m_2 = \frac{y}{a}$

Hence $1 + \left(\frac{y}{a}\right)\left(-\frac{y}{a}\right) = \frac{2a - x}{a}$

$\Rightarrow a^2 - y^2 = 2a^2 - ax$ or $ax - y^2 = a^2$

(iii) $m_1m_2 = -1 \quad \Rightarrow \quad m_1m_2m_3 = -\frac{y}{a}$

$\Rightarrow \quad m_3 = \frac{y}{a}$ and $\Sigma m_1m_2 = -1 + (m_1 + m_2)\,m_3 = \frac{2a - x}{a}$

Hence $-1 + \left(-\frac{y}{a}\right)\left(\frac{y}{a}\right) = \frac{2a - x}{a}$

$\Rightarrow -a^2 - y^2 = 2a^2 - ax$

$\Rightarrow ax - y^2 = 3a^2$

10. The ends of L.R. for the parabola $y^2 = 4ax$ are $P(a, 2a)$, $Q(a, -2a)$ or $P(1)$, $Q(-1)$

Slope of normal at $P(t)$ is $(-t)$

$\therefore$ Normal at $(a, 2a)$ is $y - 2a = (-1)\{x - a\}$ i.e., $x + y = 3a$ and similarly normal at $(a, -2a)$ will be $y + 2a = (x - a)$ or $x - y = 3a$

Observe that the product of slope of normal is -1 and the point of intersection is $(3a, 0)$

TEXTUAL EXERCISE 4-(OBJECTIVE)

1. (d) $y^2 = 4(6)x \Rightarrow a = 6$

$\therefore$ $P(6, 12) = (at^2, 2at)$ gives $t = 1$ and $Q(6, -12)$ gives $t = -1$

Hence P and Q are ends of latus rectum (*L.R.*). Normal will have slopes -1 and 1 respectively

$\Rightarrow$ Angle = 90°

2. (a) $y^2 = 8x = 4ax \Rightarrow a = 2$

Slope of normal at (t) is '$-t$', so $y - 2at = (-t)(x - at^2)$ will be the equation of normal passing through the centre $(0, -6)$ of the circle

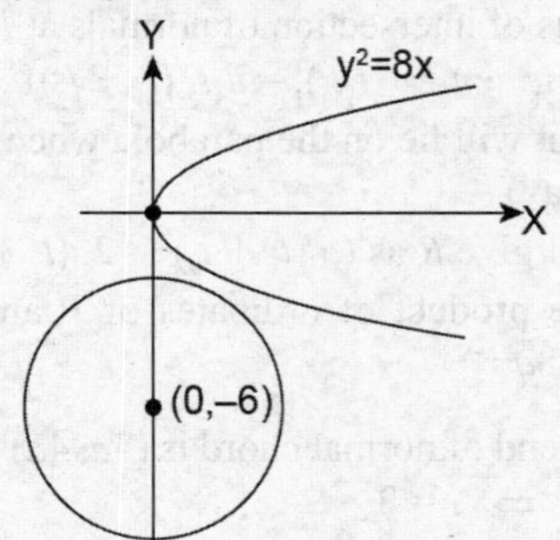

$\Rightarrow -6 - 4t = (-t)\{0 - 2t^2\}$

$\Rightarrow 2t^3 + 4t + 6 = 0$ observe that $t = -1$ satisfies

Hence the coordinates of the point on parabola are $(2, -4)$

3. (a), (b), (c) The equation of system of normals is $am^3 + 2am - mx + y = 0$.

Let $P(h, k)$ be the point and let m_1, m_2, m_3 be the slopes of normals, then $m_1 + m_2 + m_3 = 0$, $\sum m_1m_2 = \frac{2a - h}{a}$ and $m_1m_2m_3 = -\frac{k}{a}$

4. (b) The equation of normal at $P(t_1)$ is $y - 2at_1 = (-t_1)(x - at_1^2)$

Similarly $y - 2at_2 = (-t_2)(x - at_2^2)$ at $Q(t_2)$

The point of intersection will be $x = a(2 + t_1^2 + t_2^2 + t_1t_2)$ and $y = -at_1t_2(t_1 + t_2)$

5. (b) As worked above we know the point of intersection of normals at $P(t_1)$ and $Q(t_2)$ is $(a(2 + t_1^2 + t_2^2 + t_1t_2), -at_1t_2(t_1 + t_2)) = R(t_3)$

Which lies on the parabola when $t_3 = -(t_1 + t_2)$ and $t_1t_2 =$ 2, we get the mid point of PQ as $\left(\frac{a}{2}(t_1^2 + t_2^2), a(t_1 + t_2)\right)$

$\Rightarrow$ $2x = a\,(t_1^2 + t_2^2)$ and $y = a(t_1 + t_2)$

Hence $y^2 = a.a\,(t_1^2 + t_2^2) + 2a^2t_1t_2 = 2ax + 2a^2(2)$ i.e., $2ax + 4a^2$ is the locus

6. (b) Let $y^2 = 4ax$ be the parabola then ends of the latus rectum are $(a, \pm 2a)$ i.e., $t = \pm 1$

Now the slope of normal at $P(t)$ is $m = -t$, so we get slopes $m_1 = -1$ and $m_2 = 1$

Since $m_1m_2 = -1$

$\therefore$ Normal are perpendicular to each other

7. (a) The equation of parabola is $(y - 2)^2 = 4(x + 1)$

$\Rightarrow$ Vertex at $V(2,-1)$ and focus at $F\,(0, 2)$. Since axis of parabola is parallel to x-axis

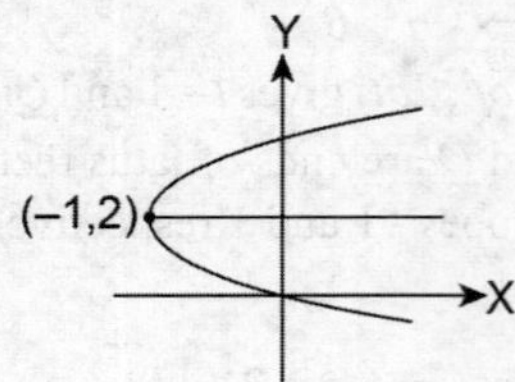

$\Rightarrow$ Any ray coming parallel to axis well pass through focus F at $(0, 2)$

8. (d) The locus of intersection of normals at $P(t_1)$ and $Q(t_2)$ is $R(a(2 + t_1^2 + t_2^2 + t_1t_2), -at_1t_2\,(t_1 + t_2))$

The point will lie on the parabola when $t_1t_2 = 2$ and $t_1 + t_2 = t_3$ (say)

This will give R as $(a\,(t_1 + t_2)^2, -2a(t_1 + t_2))$

Now the product of ordinates of P and Q $is = (2at_1)(2at_2) = 8a^2$

9. (d) The one end of normal chord is $(4a, 4a)$ for the parabola $y^2 = 4ax \Rightarrow t = 2$

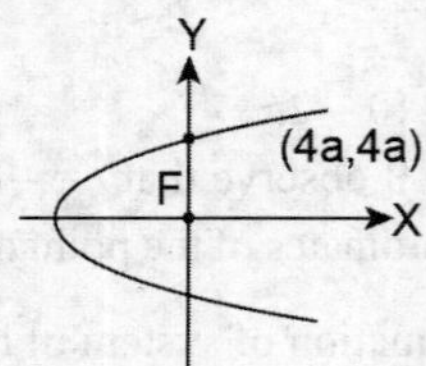

Slope of normal $= (-t) = -2$ from $t_2 = -t_1 - \frac{2}{4}$ for a normal chord, we get $t_2 = -3$ i.e., $Q\,(-3) = (9a, -6a)$

Slope of FP is $m_1 = \frac{4a}{3a} = \frac{4}{3}$ and slope of FQ is $m_2 = -\frac{3}{4}$

Since $m_1m_2 = -1$ $\quad\therefore\quad Q = 90°$

10. (a) The equation of parabola is $(x - 1)^2 + (y - 1)^2 = (x + y + 2)^2$

Observe that $y = x$ is the axis of parabola and $F(1, 1)$

$\Rightarrow$ $a = \sqrt{2}$ i.e., $Y^2 = 4(\sqrt{2})X$

If (h, k) is a point on the axis of the parabola, then $h = k$ let the point be $(X_1, 0)$

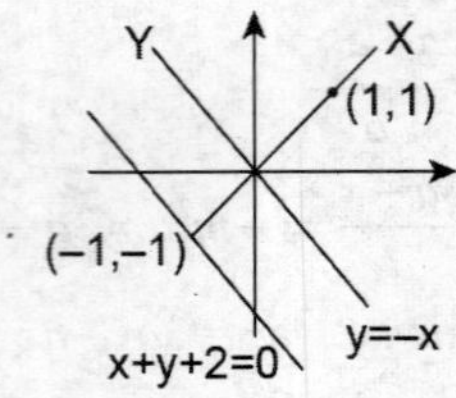

Now for three distinct normal $X_1 > 2a = 2\sqrt{2}$

Since $|X_1| = \sqrt{h^2 + k^2} = \sqrt{2}|h|$

Hence $2\sqrt{2} = h\sqrt{2} \quad\Rightarrow\quad h > 2$

11. (b) $3y^2 + 4y = 6x - 8$ or $3\left(y^2 + 2\frac{2}{3}y + \frac{4}{9}\right) = 6x - 8 + \frac{4}{3}$

$\Rightarrow$ $\left(y + \frac{2}{3}\right)^2 = 2x - \frac{20}{9} = 4\left(\frac{1}{2}\right)\left\{x - \frac{10}{9}\right\}$, gives vertex at $V\left(\frac{10}{9}, -\frac{2}{3}\right)$ and $a = \frac{1}{2}$

Axis is parallel to x-axis as $y = -\frac{2}{3}$ from $h > 2a$, we get $h > \frac{10}{9} + 1 = \frac{19}{9}$

Hence. the point gives $\left(a, -\frac{2}{3}\right)$ for $a > \frac{19}{9}$.

12. (d) $y^2 = 4x + 8 \quad\Rightarrow y^2 = 4(x + 2)$

$\Rightarrow$ Vertex is a $(-2, 0)$ and $a = 1$ for three distinct normals $h > 2a = 2$

$\Rightarrow$ $h > -2 + 2 = 0$. Since the point lies on the axis of parabola

$\Rightarrow$ $P(k, 0)$ where $k > 0$.

13. (b) Equation of normal at 't'; $y - t^2 = -1/t\,(x - 2t)$

$yt - t^3 = 2t - x$ i.e., $x + yt - 2t - t^3$ ....(i)

it passes through (α, β)

$t^3 + (2 - \beta)\,t - \alpha = 0$

$t_1 + t_2 + t_3 = 0$ ....(ii)

$t_1t_2 + t_2t_3 + t_1t_3 = 2 - \beta$ ...(iii)

$t_1t_2t_3 = \alpha$ ....(iv)

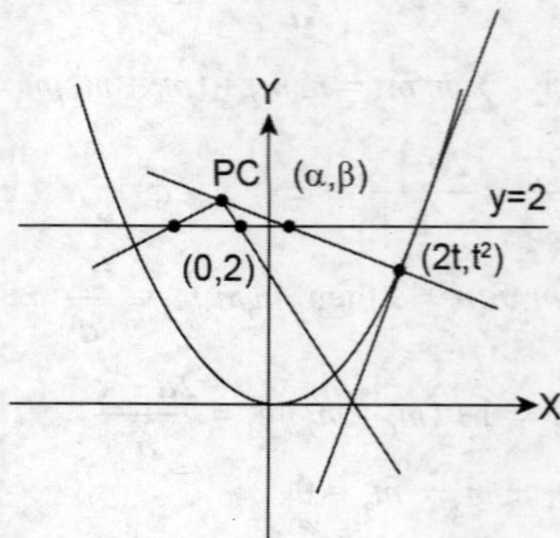

Normal at t_1 cuts $y = 2$ at $(t_1^3, 2)$

at t_2 cuts $y = 2$ at $(t_2^3, 2)$

at t_3 cuts $y = 2$ at $(t_3^3, 2)$ $\qquad t_1^3, t_2^3, t_3^3$ are in AP

$\Rightarrow$ $t_1^3 + t_3^3 = 2t_2^3$ $\qquad\Rightarrow$ $(t_1 + t_3)^3 - 3t_1t_3(t_1 + t_3) = 2t_2^3$

$\Rightarrow$ $-t_2^3 + 3t_1t_2t_3 = 2t_2^3$ $\qquad\Rightarrow$ $3t_2^3 = 3t_1t_2t_3$

$\Rightarrow$ $t_2^2 = t_1t_3$ $\qquad\Rightarrow$ t_1, t_2, t_3 are in G.P.

14. (b) Consider three points as $P(0, 0)$ $Q(at^2, 2at)$ and $R(at^2, -2at)$ as a result the normals will meet on x-axis. The centroid of ΔPQR (being an isosceles triangle) will be on the axis of the parabola i.e., $y = 0$

15. (c) The parabola is $y^2 = 4x \Rightarrow a = 1$
Now $x + y = 1$ will intersect the parabola at $A(t_1 = \sqrt{2} - 1)$ and $B(t_2 = -\sqrt{2} - 1)$ {using $x = 1 - y$ or $at^2 = 1 - 2at$}

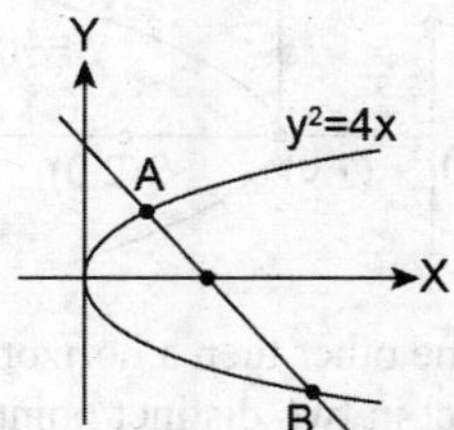

We know that the point of intersection of normals at t_1 and t_2 is $R(a(2 + t_1^2 + t_2^2 + t_1t_2), -at_1t_2(t_1 + t_2)) = C$ (given)
Now $t_1 + t_2 = -2$ and $t_1t_2 = -1$. Hence $C(7, -2)$
Equation of normal becomes $m^3 - 5m - 2 = 0$, which gives $m = -2$
Hence the point (4, 4)

TEXTUAL EXERCISE SECTION–III (OBJECTIVE)

1. (b) Observe the equation in (b) option

$x^2 - 2 = -2\cos t$ and $y = 4\cos^2\dfrac{t}{2}$

Now $y - 2 = 2\left(2\cos^2\dfrac{t}{2} - 1\right) = 2\cos t$

Hence $x^2 - 2 = 2 - y$ or $x^2 = -4\left(\dfrac{1}{4}\right)(y - 4)$ which is a parabola

2. (b) Equation of normals having slope 'm' is $y = mx - 2am - am^3$

$\Rightarrow y = mx - 2m - m^3$

As it passes through the point $P\left(\dfrac{11}{4}, \dfrac{1}{4}\right)$

$\Rightarrow \dfrac{1}{4} = \dfrac{11}{4}m - 2m - m^3 \Rightarrow 1 = 2m - 4m^3$

$\Rightarrow 4m^3 - 3m + 1 = 0 \quad \Rightarrow m = -1, 1/2, 1/2$

$\Rightarrow$ Two difference normals.

3. (a) Length of semi latus Rectum $= 2a$

We know $\dfrac{4L_1L_2}{L_1 + L_2} = 4a$

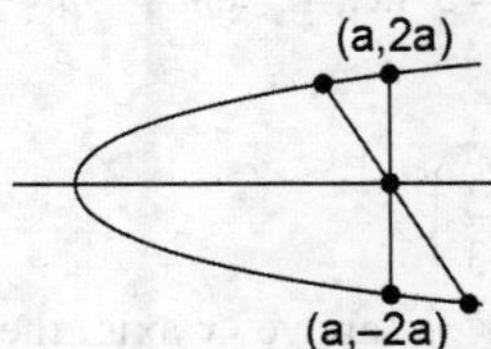

Where L_1 and L_2 are the segments of focal chord

$\Rightarrow \dfrac{2L_1L_2}{L_1 + L_2} = 2a \quad \Rightarrow 2a$ is the H.M. of L_1 and L_2

4. (a) The parabola is $y^2 = 8x \Rightarrow a = 2$
Observe that $(-4, 0)$ lies on the axis of the parabola

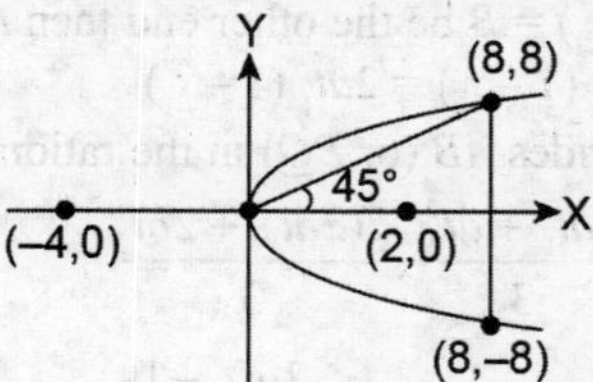

Obviously points A and B are A (t) and $B(-t)$ as $A(8, 8)$ and $B(8, -8)$

$\Rightarrow$ Radius of the circle $= \sqrt{12^2 + 8^2} = 4\sqrt{13}$ units

5. (a) The vertex of parabola is at $V(-1, 2)$ and its axis is parallel to x-axis .The focus will be at (0, 2)

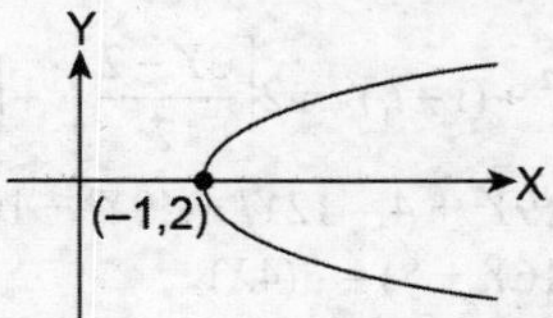

So. any ray coming parallel to the axis will pass through focus, i.e., (0, 2).

6. (d) The parabola is $y^2 = 4ax$, let $P(t_1)$ and $Q(t_2)$ be the two points on the parabola.
The normals at P and Q will intersect at R, $R(a(2 + t_1^2 + t_2^2 + t_1t_2), -at_1t_2(t_1 + t_2))$, when R lies on the parabola, then let it be
$R = (at_3^2, 2at_3)$
$\Rightarrow t_1t_2 = 2$ and $(t_1 + t_2) = -t_3$.
$\Rightarrow$ product of ordinates of P and Q is $(2at_1)(2at_2) = 4a^2(2) = 8a^2$

7. (a) The parabola $y^2 = 4ax$ has directrix as $x = -a$ and focus at $(a, 0)$
$\Rightarrow$ The equation of the circle will be $(x - a)^2 + y^2 = 4a^2$

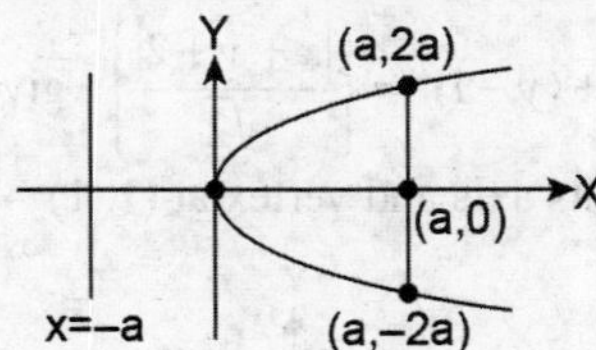

The intersection with $y^2 = 4ax$ will give $x^2 + a^2 - 2ax + 4ax = 4a^2$ i.e., $(x + a)^2 = (2a)^2$
$\Rightarrow x = a$, then $y = \pm 2a$ as a result we get latus rectum with length $AB = 4a$

8. (b) The given parabola is $y^2 = 4ax$ where $a = 1$
The chord system with slope 2 is $y = 2x + c$

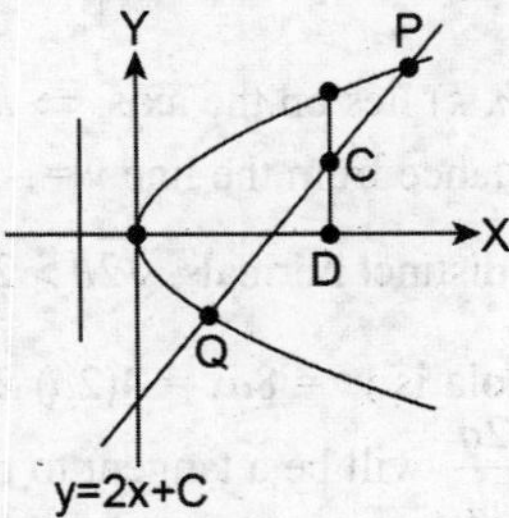

Let $P(t_1) = A$ be one end of such a chord, then $c = 2at_1(t - t_1)$

Let $Q(t_2) = B$ be the other end then $t_2 = 1 - t_1$ such that $c = 2at_2(1 - t_2) = 2at_1(1 - t_1)$

If C divides AB (or PQ) in the ration 1 : 2, then

$$C = \left(\frac{2at_1^2 + at_2^2}{3}, \frac{2(2at_1) + 2at_2}{3}\right)$$

$$= \left(\frac{a}{3}\{2t_1^2 + (1 - t_1)^2\}, \frac{2a(t_1 + 1)}{3}\right)$$

For $a = 1$, $C(X, Y) = \left(\frac{3t_1^2 + 1 - 2t_1}{3}, \frac{2(t_1 + 1)}{3}\right)$

Eliminating t_1, we get $\frac{3Y}{2} = t_1 + 1$

$$3X = 2t_1^2 + (1 - t_1)^2 = 2\left[\frac{3Y - 2}{2}\right]^2 + \left(\frac{2Y - 4}{2}\right)^2$$

$12X = 2\{9Y^2 + 4 - 12Y\} + \{9Y^2 + 16 - 24Y\}$

$\Rightarrow$ $3(9Y^2 - 16Y + 8) = 3(4X)$

i.e., $9\left\{y^2 - 2\left(\frac{8}{9}\right)y + \frac{64}{81}\right\} = 4x - \frac{8}{9}$

$\Rightarrow$ $\left(y - \frac{8}{9}\right)^2 = \frac{4}{9}\left(x - \frac{2}{9}\right)$

When $C = \left(\frac{2at_2^2 + at_1^2}{3}, \frac{4at_2 + 2at_1}{3}\right)$

$\Rightarrow$ $\left(y + \frac{8}{9}\right)^2 = \frac{4}{9}\left(x - \frac{2}{9}\right)$

9. (d) The equation of the chord which is bisected at $P(2, 1)$ in the parabola $y^2 = x$ is $(1)y - \frac{x}{2} - \frac{2}{2} = 1^2 - 2 \Rightarrow y = \frac{x}{2}$

Which has end points $O(0, 0)$ and $Q(4, 2)$.

Hence the length $OQ = \sqrt{20} = 2\sqrt{5}$ units

10. (a) $(x - 1)^2 + (y - 1)^2 = \left\{\frac{|x + y + 2|}{\sqrt{2}}\right\}^2$ gives a parabola with $y = x$ as its axis and vertex at (1, 1)

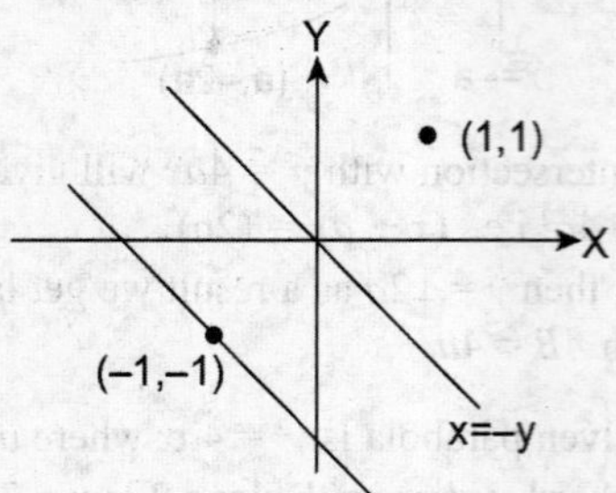

$\Rightarrow$ $a = \sqrt{2}$

Since $P(h, k)$ lies on the axis. $\Rightarrow h = k$

Hence distance from the line $y = -x$ is $\sqrt{h^2 + k^2} = \sqrt{2}h$

For three distinct normals $\sqrt{2}h > 2a$ i.e., $h > 2$

11. (c) The parabola is $y^2 = 8ax = 4(2a)x$, so $A = 2a$ any line $y = mx + \frac{2a}{m}$ will be a tangent to this parabola.

Now if this line is also a tangent to $x^2 + y^2 = (\sqrt{2}a)^2$, then $\left|\frac{2a}{m\sqrt{1 + m^2}}\right| = \sqrt{2}a$

$\Rightarrow$ $4 = 2m^2(1 + m^2) \Rightarrow m^2 = 1$ i.e., $m = \pm 1$.

Hence the lines possible are $y = x + 2a$ or $y = -x - 2a = -(x + 2a)$ i.e., $y = \pm(x + 2a)$

12. (d) The focus S (2, 0), for the parabola $y^2 = 4(x - 1)$

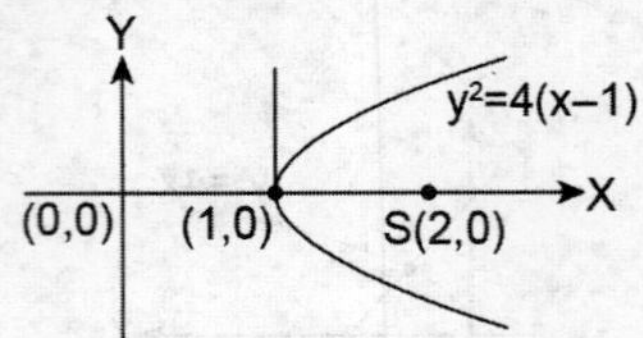

Now any line other then a horizontal line through focus will intersect in two distinct points.

It means slope $m \neq 0$, so $m \in R - \{0\}$ also line $x = 2$ is possible

13. (a) The given parabola is $y^2 = 4x \Rightarrow a = 1$

Let $P(t_1)$ and $Q(t_2)$ be the two points, then the point of intersection of tangents at P and Q is R given as $R(at_1t_2, at_1 + at_2) = (-2, -1)$

Hence $t_1t_2 = -2$ and $t_1 + t_2 = -1$

Now $\tan\alpha = \left|\frac{t_2 - t_1}{1 + t_1t_2}\right| = \left|\frac{\sqrt{(t_1 + t_2)^2 - 4t_1t_2}}{1 + t_1t_2}\right| = \left|\frac{3}{-1}\right| = 3$

14. (d) The given parabola $y^2 = 4x$ will have a line as its tangent when $y = mx + \frac{a}{m}$ where $a = 1$.

So a line with slope m will be a tangent to the parabola $x^2 = -32y$ when $y = mx + Am^2$ (where $A = 8$)

$\Rightarrow$ $Am^2 = \frac{a}{m}$ $\Rightarrow$ $8m^2 = \frac{1}{m} \Rightarrow m = \frac{1}{2}$

Hence $y = \frac{x}{2} + 2$ $\Rightarrow$ $x - 2y + 4 = 0$

15. (a) $y^2 = 4x$ will have $y = mx + \frac{1}{m}$ as its tangent

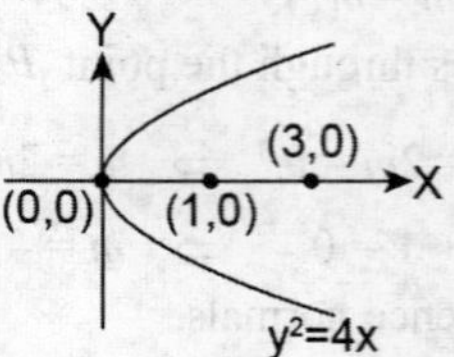

Since this line is also a tangent to $(x - 3)^2 + y^2 = 9$

$\Rightarrow$ $\left|\frac{3m + \frac{1}{m}}{\sqrt{1 + m^2}}\right| = 3$ i.e., $1 + 6m^2 + 9m^4 = 9m^4 + 9m^2$

$\Rightarrow$ $m^2 = \left(\pm\frac{1}{\sqrt{3}}\right)^2$

If tangent is above x-axis then $m = \frac{1}{\sqrt{3}}$. Hence $\sqrt{3}y = x + 3$

16. (a) The parabola is $y^2 = 4(5)x \Rightarrow x = -5$ is the directrix and $S(5, 0)$ is the focus any focal chord has equation

$y - 0 = \frac{1}{a}(x-5)$ where $m = \frac{1}{a}$ is the slope of the chord.

Since circle on a focal chord as diameter touches the directrix

∴ the circle will touch line $x + 5 = 0$

Aliter: The parabola is $y^2 = 20x$

⇒ $A = 5$

Now the line is $y = \frac{x-5}{a}$ where a is a parameter hence

$(x-5)^2 = 20a^2x \Rightarrow x^2 - 10x - 20a^2x + 25 = 0$

Let x_1 and x_2 be its roots, then $x_1 + x_2 = 10 + 20a^2$ and $x_1x_2 = 25$.

Now the mid point of $P(x_1, y_1)$ and $Q(x_2, y_2)$ is

$C\left(\frac{x_1+x_2}{2}, \frac{y_1+y_2}{2}\right)$

$= \left(5+10a^2, \frac{10+20a^2-10}{2a}\right) = (5+10a^2, 10a)$

The radius of the circle is $\frac{PQ}{2} = \frac{1}{2}\sqrt{(x_2-x_1)^2+(y_2-y_1)^2}$

Now $(x_2 - x_1)^2 = (x_1 + x_2)^2 - 4x_1x_2 = 400a^2(a^2 + 1)$

⇒ $(y_2 - y_1)^2 = \left(\frac{x_2-5-x_1+5}{a}\right)^2$

$= \left(\frac{x_2-x_1}{a}\right)^2 = 400(a^2+1)$

Hence $\frac{PQ}{2}$ = radius $= \frac{1}{2}\sqrt{400a^2(a^2+1)+400(a^2+1)}$

$= 10\sqrt{a^4+a^2+a^2+1} = 10(a^2+1)$

Now observe that $C(5 + 10a^2, 10a)$ is at a distance $10(a^2 + 1)$ from $x + 5 = 0$

17. (a) Given $P(t_1)$ and $Q(t_2)$ are the two points on the parabola $y^2 = 4ax$

⇒ Equation of chord PQ is $2x - (t_1 + t_2)y + 2at_1t_2 = 0$, also the chord is given as $x + \lambda y - b = 0$

⇒ $\frac{1}{2} = \frac{\lambda}{-(t_1+t_2)} = \frac{-b}{2at_1t_2}$

⇒ $2at_1t_2 = -2ab$ i.e., $t_1t_2 = -\frac{b}{a}$

Now $b \in [2a, 4a] \Rightarrow \frac{b}{a} \in [2, 4]$

Hence $-\frac{b}{a} \in [-4, -2]$

18. (a) The parabola is $y = (a - b)x^2 + (b - c)x + (c - a)$, observe that for $x = 1, y = 0$ as given the parabola touch x-axis in the interval (0, 1], clearly $y = (a - b)(x - 1)^2$ is the parabola.

Hence $y = (a - b)x^2 - 2(a - b)x + (a - b)$ is the same parabola

⇒ $b - c = -2a + 2b$ also $c - a = a - b$ both lead to $-2a + b + c = 0$

∴ The line $ax + by + c = 0$ is always true for $x = -2, y = 1$ i.e., it passes through $(-2, 1)$

19. (c) The parabola $x^2 = 4y$ (so $a = 1$).

$P(t) = (2t, t^2)$, let $P(t)$ be the point on the parabola at which the normal passes through center C (3, 0) of the circle, so (3, 0) lies on $y - t^2 = \left(-\frac{1}{t}\right)(x - 2t)$

⇒ $t^3 = 3 - 2t$ i.e., $t^3 + 2t - 3 = 0$ or $(t - 1)(t^2 - t + 3) = 0$ for $t = 1$ the point $P(2, 1)$

20. (b) The parabola is $y^2 = 4ax$

∴ Any point on it is $(at^2, 2at)$, since $x + y = 1$ intersects the parabola.

∴ $at^2 + 2at - 1 = 0$ ⇒ $t_1 + t_2 = -2$ and $t_1t_2 = -\frac{1}{a}$

Now the normals at t_1 and t_2 intersect at $R\,(a(2 + t_1^2 + t_2^2 + t_1t_2), -at_1t_2(t_1 + t_2)) = C \Rightarrow C = (6a + 1, 2)$

Now the system of co-normals is given by $am^3 + 2am - mx + y = 0 \Rightarrow C = (6a + 1, -2)$

We get $am^3 - m(4a + 1) - 2 = 0$ or $am(m^2 - 4) - (m + 2) = 0$

Observe that $m = -2$ satisfies this equation and $(m = -t)$

Hence the point (foot of normal) is $P(2) = (4a, 4a)$

21. (b) $x^2 + y^2 = 25$ is a circle and $x^2 = -4y$ is a downwards parabola

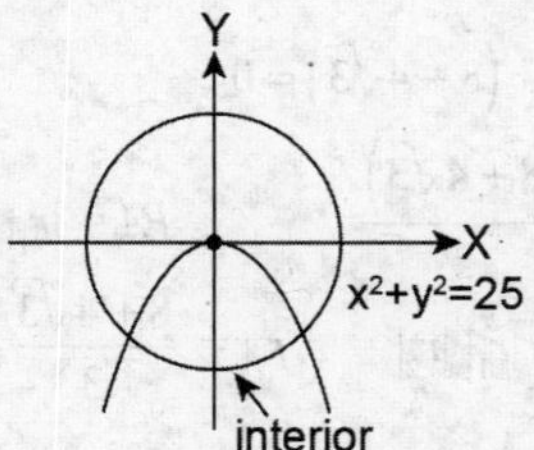

⇒ Larger segment of circle has the exterior of parabola

Hence the integral point $(2a, a - 1)$ will satisfy $S_1: x^2 + 4y > 0$ and $S_2: s^2 + y^2 - 25 < 0$

Hence $4a^2 + 4a - 4 > 0$ and $4a^2 + a^2 + 1 - 2a - 25 < 0$

Now $a^2 + a - 1 > 0$

⇒ $\left(a + \frac{\sqrt{5}+1}{2}\right)\left(a - \frac{\sqrt{5}-1}{2}\right) > 0$

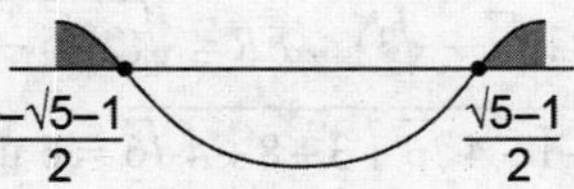

And $5a^2 - 2a - 24 < 0 \Rightarrow (5a - 12)(a + 2) < 0$

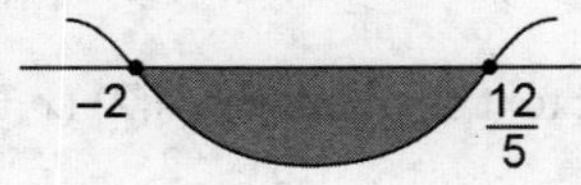

Finally we need to know the integral points $(2a, a - 1)$ where $a \in \left(-2, \frac{-\sqrt{5}-1}{2}\right) \cup \left(\frac{\sqrt{5}-1}{2}, \frac{12}{5}\right)$

Observe that $-2 < \frac{-\sqrt{5}-1}{2} < -1$ and $0 < \frac{\sqrt{5}-1}{2} < 1$

⇒ $a = 1, 2$ are the two values for which $(2a, a - 1)$ gives $(2, 0), (4, 1)$ as the two integral points.

22. (d) Equation of line $\dfrac{x-\sqrt{3}}{\cos 60°}=\dfrac{y-0}{\sin 60°}=r$

$\therefore$ Point A, B gives as $\left(\dfrac{r}{2}+\sqrt{3},\dfrac{r\sqrt{3}}{2}\right)$ where PA, PB be respective algebraic length given by values of 'r' solving line and $y^2 = x + 2$ together $\dfrac{3r^2}{4}=\dfrac{r}{2}+\sqrt{3}+2$

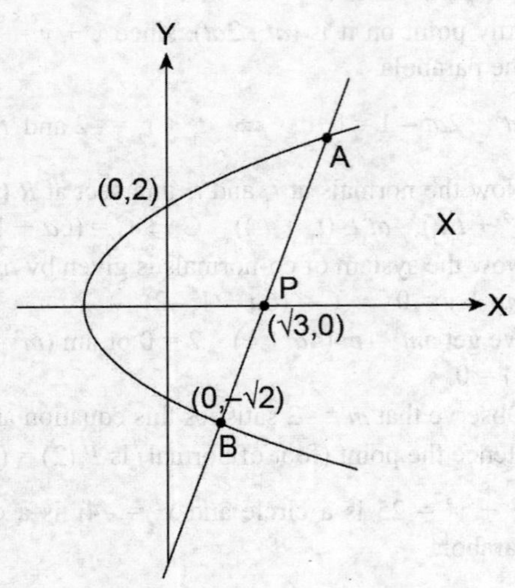

$\Rightarrow$ $3r^2 - 2r - (8+4\sqrt{3}) = 0$

$\Rightarrow$ $r_1r_2 = -\dfrac{(8+4\sqrt{3})}{3}$ $\quad PA = |r_1|$ and $PB = |r_2|$

$\Rightarrow$ $PA.\ PB = |r_1|\ |r_2| = |r_1 r_2| = \dfrac{8+4\sqrt{3}}{3}$

23. (b) The parabola is $y^2 = 4ax$ where $a = 1$.

Let $P(t_1)$ be a point on it, then $FP = a(t_1^2 + 1)$ (focal distance)

Now $FP = 4$ and $FQ = 9$

$\Rightarrow$ $t_1^2 + 1 = 4 \Rightarrow t_1 = \pm\sqrt{3}$ and similarly $t_2 = \pm 2\sqrt{2}$

Now the point of intersection of tangents is $R(at_1t_2, at_1 + at_2)$

Let $t_1 = \sqrt{3}$ and $t_2 = 2\sqrt{2}$

$\Rightarrow$ $R(2\sqrt{6}, 2\sqrt{2}+\sqrt{3})\ \therefore\ FR = \sqrt{(2\sqrt{6}-1)^2 + (2\sqrt{2}+\sqrt{3})^2}$

$= \sqrt{24+1-4\sqrt{6}+3+8+4\sqrt{6}} = 6$ units

24. (c) $y^2 = 4(x + 1) \quad \Rightarrow a = 1$ and $y = m_1(x+1)+\dfrac{1}{m_1}$ ... (i) is a tangent to it similarly $y^2 = 4(2)(x + 2)$ has tangent as $y = m_2(x+2)+\dfrac{2}{m_2}$ .... (ii)

Now $m_1 \perp m_2$ i.e., $m_1 m_2 = -1$

$\Rightarrow$ $m_2 = -\dfrac{1}{m_1}$

$\therefore$ (i) becomes $x + m_1 y + (2 + 2m_1^2) = 0$ and (ii) becomes $m_1^2 x - m_1 y + (1 + m_1^2) = 0$

On adding (i) + (ii), we have $(m_1^2 + 1)x + (3 + 3m_1^2) = 0$ i.e., $(x + 3) = 0$ (as $m_1^2 + 1 \neq 0$)

25. (d) Equation of normal form C will be $aM^3 + 2aM - Mx + y = 0$

$\Rightarrow$ $M_1 + M_2 + M_3 = 0$ i.e., $M_1 + M_2 = -M_3$

Now feet of normal are

$A(aM_1^2 - 2aM_1)$ and $B(aM_2^2 - 2aM_2)$ and $D(aM_3^2 - 2aM_3)$

Since A and B lie on $\quad x + my - 1 = 0$

$\therefore$ $a \quad M_1^2 - 2am\,M_1 - 1 = 0$ and $a \quad M_2^2 - 2amM_2 - 1 = 0$

Subtracting, we get $a(M_1 - M_2)\{\ell(M_1 + M_2) - 2m\} = 0$

Since $a \neq 0$, $M_1 \neq M_2 \quad \therefore\ M_1 + M_2 = \dfrac{2m}{\ell} = -M_3$

Hence $D = \left(\dfrac{4am^2}{\ell^2}, \dfrac{4am}{\ell}\right)$

26. (c) $y = x^2 + bx - b$

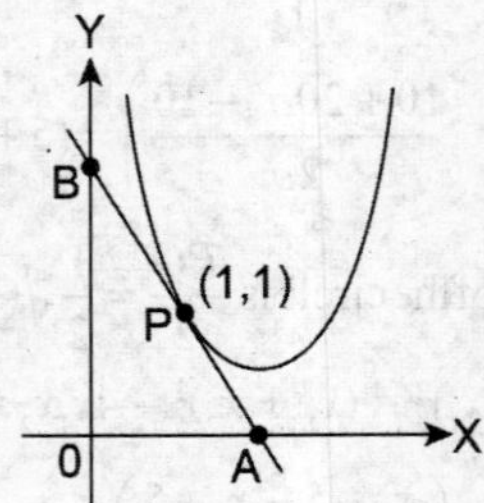

Any line through (1, 1) with slope m is $y - 1 = m(x - 1)$

$\Rightarrow$ Area of $\Delta OAB = \dfrac{1}{2}\times\dfrac{(m-1)}{m}(1-m) = 2$sq. units

$\Rightarrow$ $(m - 1)^2 = -4$ i.e., $(m + 1)^2 = 0 \Rightarrow m = -1$

Now $\left.\dfrac{dy}{dx}\right|_{(1,1)} = 2x + b \Rightarrow \quad m = 2 + b = -1 \Rightarrow b = -3$

27. (a) $y^2 = 4(2)x \quad \Rightarrow a = 2$

Let $P(t)$ be the point at which the normal will pass through (0, –6)

$\Rightarrow$ $y - 2at = (-t)(x - at^2)$

$\Rightarrow$ $-6 - 4t = 2t^3$ i.e., $2(t^3 + 2t + 3) = 0 \Rightarrow t = -1$

Hence the point is (2, –4)

28. (d) The circle $(x - a)^2 + (y - b)^2 = a^2 + b^2 - c$ has centre at $C(a, b)$ for $a^2 + b^2 \geq c$

Now tangents to the parabola $y^2 = 4ax$ at $P(t_1)$ and $Q(t_2)$ will intersect at $R(at_1t_2, at_1 + at_2) = C(a, b)$

$\Rightarrow$ $t_1 + t_2 = \dfrac{b}{a}$ and $t_1t_2 = 1$, from $(t_1 - t_2)^2 > 0$ (as $t_1 \neq t_2$), we get $(t_1 + t_2)^2 - 4t_1t_2 > 0$ i.e., $\dfrac{b^2}{a^2} > 4$ or $b^2 > 4a^2$

29. (b) From the given requirements $G(-at^2, at)$

$\Rightarrow$ $\dfrac{y}{a} = t$ and $-\dfrac{x}{a} = t^2$ i.e., $y^2 = -ax$

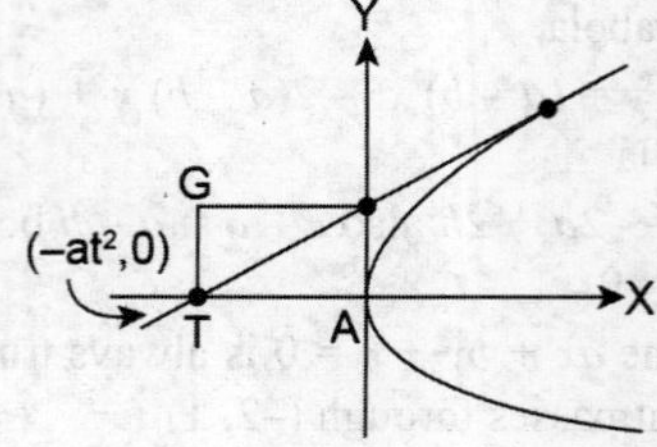

30. **(b)** Let $P(t_1)$ be any point on the parabola, so $OS = a$ and $SP = a(t^2 + 1)$ and $SH = \sqrt{a^2(t^2+1)}$

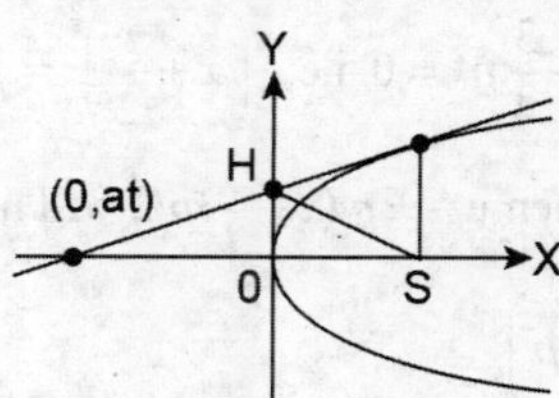

Observe that $SH^2 = OS.\ SP$

$\Rightarrow$ OS, SH, SP are in G.P.

31. **(b)** Let $Y^2 = 8(x-1)$ or $Y^2 = 8X = 4(a)X$

$\Rightarrow a = 2$

Now equation of normal at t_1 gives other point t_2 as $t_2 = -t_1 - \dfrac{2}{t_1}$

The point of intersection of $P(t_1)$ and $Q(t_2)$ tangents is $R(a\,t_1t_2,\ at_1 + at_2) = \left(-2t_1^2 - 4, -\dfrac{4}{t_1}\right)$

Let $X = -2t_1^2 - 4$ and $Y = -\dfrac{4}{t_1}$

$\Rightarrow t_1 = -\dfrac{4}{y}$ and $-(X+4) = 2t_1^2 = 2\times\dfrac{16}{Y^2}$

Hence $y^2(x+4) = -32$ i.e., $y^2(x+3) + 32 = 0$

32. **(c)** Let $P(t_1)$ and $Q(t_2)$ be two points on the parabola $y^2 = 4ax$ from where tangents intersect at $\angle Q$, then $\tan\theta = \left|\dfrac{t_2 - t_1}{1 + t_1t_2}\right| = 1$

$\Rightarrow (t_2 - t_1)^2 = (1 + t_1t_2)^2$

Now the point of intersection of tangent at $P(t_1)$ and $Q(t_2)$ is $R(at_1t_2,\ at_1 + at_2)$

Let $\dfrac{x}{a} = t_1t_2$ and $\dfrac{y}{a} = t_1 + t_2$

Now $(t_2 - t_1)^2 = (t_1 + t_2)^2 - 4t_1t_2$

$\Rightarrow \dfrac{y^2}{a^2} - \dfrac{4x}{a} = \left(1 + \dfrac{x}{a}\right)^2$ or $y^2 - 4ax = (x+a)^2$

$\Rightarrow x^2 + 6ax + 9a^2 - y^2 = 8a^2$ i.e., $\dfrac{(x+3a)^2}{(2\sqrt{2}a)^2} - \dfrac{y^2}{(2\sqrt{2}a)^2} = 1$

$\Rightarrow$ A rectangular hyperbola

33. **(c)** $y = x^2$ is the parabola at $P(x_0, x_0^2)$, the slope of tangent is $m = \left.\dfrac{dy}{dx}\right|_{(x_0, x_0^2)} = 2x_0$

Now the equation of tangent at $P(x_0, x_0^2)$ is $y - x_0^2 = 2x_0(x - x_0)$, gives point Q on y-axis as $Q(0, -x_0^2)$

$\therefore$ The area of $\Delta QPM = \dfrac{x_0}{2}.(2x_0^2) = x_0^3$, which will be maximum at $x_0 = 2$ where $x_0 \in [1, 2]$

34. **(a)** The parabola is $y^2 = 4x$, so $a = 1$ and the equation of normal is $am^3 + 2am - mx + y = 0$, which will pass through the centre C of circle $(x-2)^2 + (y-1)^2 = 1^2$ where C(2, 1)

$\Rightarrow m^3 + 2m - 2m + 1 = 0 \Rightarrow m = -1\ (t = -m)$

Now the point is $(am^2, -2am) = (1, 2)$, so the line is $y - 2 = (-1)(x - 1)$

$\Rightarrow x + y = 3$

35. **(a)** The parabolas $y^2 = 4ax$ and $x^2 = 4by$ will intersect as $\left(\dfrac{x^2}{4b}\right)^2 = 4ax$

$\Rightarrow x^4 = 64ab^2x$

$\Rightarrow x = 0$ or $x^3 = 64ab^2$ i.e., $x = 4a^{1/3}b^{2/3}$, when the circle touches $x = -a$ and (0, 0), then centre lies at $\left(-\dfrac{a}{2}, 0\right)$ which is a part of x-axis.

When the circle passes through $(4a^{1/3}b^{2/3}, 4a^{2/3}b^{1/3})$; considering centre of circle (h, k), then $(h - 4a^{1/3}b^{2/3})^2 + (k - 4a^{2/3}, b^{1/3})^2 = (h + a)^2$,

Required locus is $(x - 4a^{1/3}b^{2/3})^2 + (y - 4a^{2/3}b^{1/3})^2 = (x + a)^2$ which will form a parabola.

36. **(c)** $y^2 = 4x \quad \Rightarrow A = 1$

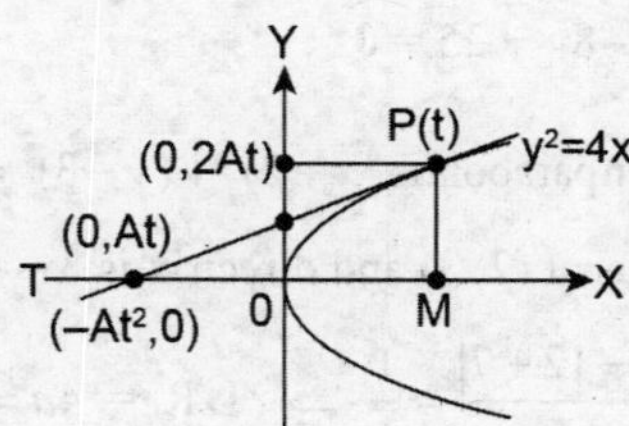

Now $x \in [a^2, 4a^2]$ i.e., $t \in [a, 2a]$

Consider area of $\Delta PTM = \dfrac{1}{2}(2At^2)\times 2At = 2A^2t^3 = 2t^3$

Now $\dfrac{d}{dt}(\text{area}) = 6t^2$, which shows that area will be maximum when t is maximum i.e., $2a$

$\Rightarrow$ The maximum possible $are = 2(2a)^3 = 16a^3$ square units.

37. **(c)** Let $\pm\theta$ be the angles with the line having slope $\tan\alpha$

$\Rightarrow m_1 = \dfrac{1}{t_1} = \tan(\alpha - \theta)$ and $m_2 = \dfrac{1}{t_2} = \tan(\alpha + \theta)$

Now the point of intersection of tangents at $P(t_1)$ and $Q(t_2)$ is $R(at_1t_2,\ at_1 + at_2)$

Now $\dfrac{x}{a} = \dfrac{1}{\tan(\alpha - \theta)\tan(\alpha + \theta)} = t_1t_2$ and

$\dfrac{y}{a} = t_1t_2\left\{\dfrac{1}{t_1} + \dfrac{1}{t_2}\right\} = t_1 + t_2$

Now $\frac{y}{a} = \frac{x}{a}\{\tan(\alpha-\theta)+\tan(\alpha+\theta)\}$

$= \frac{x}{\alpha}\left\{\frac{\tan(\alpha-\theta)+\tan(\alpha+\theta)}{1-\tan(\alpha-\theta)\tan(\alpha+\theta)}\right\}\times\{1-\tan(\alpha-\theta)\tan(\alpha+\theta)\}$

or $\frac{y}{a} = \frac{x}{a}\{\tan 2\alpha\}\left\{1-\frac{a}{x}\right\}$ i.e., $y = (x - a)\tan 2\alpha$

38. **(b)** Let $B(t_1)$ and $C(t_2)$ be the two points on parabola $y^2 = 4ax$

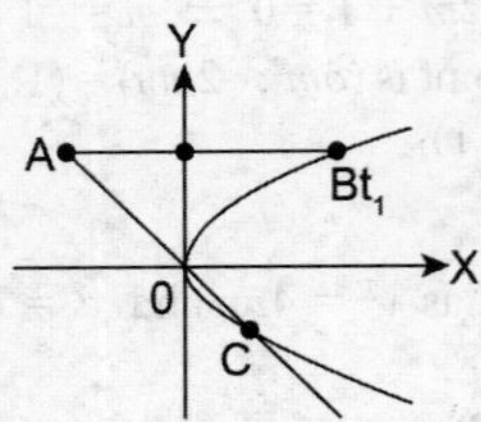

Let A $(at_1t_2, at_1 + at_2)$ be the point of intersection of tangents at B and C

Consider the tangent at vertex i.e., $x = 0$

Now the perpendicular are $p_1 = at_1t_2$, $p_2 = at_1^2$ and $p_3 = at_2^2$, observe that $p_1^2 = p_2p_3$.

So p_2, p_1 and p_3 are in G.P.

39. **(a)** The given parabola $y^2 - 4y + 4 = 12x$ is $(y-2)^2 = 4(3)x$ which will have vertex at (0, 2) and focus at (3, 2)

$\Rightarrow$ The required parabola will have vertex at (3, 2) and focus at (3, 4)

$\Rightarrow$ $a = 2$

Hence the parabola $(x-3)^2 = 8(y-2)$

$\Rightarrow$ $x^2 - 6x - 8y + 25 = 0$

40. **(d)** The given parabola is $(x-2)^2+(y-3)^2 = \frac{(3x-4y+7)^2}{25}$

i.e., focus at (2, 3) and directrix is $3x - 4y + 7 = 0$

$\Rightarrow$ $2a = \frac{|6-12+7|}{5} = \frac{1}{5}$ $\Rightarrow$ L.R. $= 4a = 0.4 = \frac{2}{5}$

41. **(a)** The parabola is $y^2 - 2y + 1 = 4x + 8 = 4(x + 2)$

$\Rightarrow$ $(y-1)^2 = 4(x+2)$, gives vertex at A (–2, 1) and L.R. $= L = 4$

The other curve has vertex at A (–2, 1) and L.R. $= 2L = 8$ units. Since the axis is at 90°

$\Rightarrow$ We have two possibilities $(x + 2)^2 = \pm 8(y - 1)$

$\Rightarrow$ $x^2 + 4x + 8y - 4 = 0$ or $x^2 + 4x - 8y + 12 = 0$

42. **(a)** Given the two parts of focal chord as $_1 = 3$ and $_2 = 2$

$\Rightarrow$ $L.R. = \frac{4\ell_1\ell_2}{\ell_1+\ell_2} = \frac{24}{5}$

43. **(a)** $y^2 = 2px$ $\Rightarrow$ $a = \frac{p}{2}$

The circle with centre at focus and that touches directrix has radius $r = 2a = p$

$\Rightarrow$ The circle is $\left(x-\frac{p}{2}\right)^2 + y^2 = p^2$

For $y^2 = 2px$, we get $x^2 + \frac{p^2}{4} - px + 2px = p^2$

$\Rightarrow$ $x^2 + px - \frac{3}{4}p^2 = 0$ i.e., $\left(x+\frac{p}{2}\right)^2 = p^2$

$\Rightarrow$ $x = \frac{p}{2}$ then $y = \pm p$ ($x = -3p/2$ will not give any real y),

so $\left(\frac{p}{2}, \pm p\right)$

44. **(b)** Let the normal chord be at $P(t_1)$, then $t_2 = -t_1 - \frac{2}{t_1}$.

Since the chord subtend a right angel at the vertex

$\Rightarrow t_1t_2 = -4$

Hence $t_1\left(-t_1 - \frac{2}{t_1}\right) = -4$

$\Rightarrow$ $t_1^2 + 2 = 4$ $\Rightarrow$ $t_1 = \pm\sqrt{2}$

Slope of normal at t_1 is $m = -t_1$

$\Rightarrow$ Slope of normal $= \pm\sqrt{2}$ i.e., $\tan\theta = \pm\sqrt{2}$ or $\theta = \tan^{-1}(\sqrt{2}) = \sec^{-1}\sqrt{3}$

45. **(c)** Let $P(t_1)$ be the point, so that focal chord is at a distance of p from the vertex.

Now the equation of focal chord through P (t_1) is $t_1 x - (t_1^2 - 1)\,y - 2at_1 = 0$

Now distance of vertex (0, 0) from the chord is $p = \frac{|2at_1|}{(t_1^2+1)} \Rightarrow \frac{t_1^2+1}{t_1} = \frac{2a}{p}$

Now focal distance of ends of the chord is $PF = a(t_1^2 + 1)$ and $QF = \frac{a}{t_1^2}(1+t_1^2)$

$\Rightarrow$ $PF + QF = a(t_1^2+1)\frac{(1+t_1^2)}{t_1^2} = \frac{a(t_1^2+1)^2}{t_1^2}$

$= a\left\{\frac{2a}{p}\right\}^2 = \frac{4a^3}{p^2}$

46. **(c)** Let P be a point on the parabola

$\Rightarrow$ $P \equiv (at^2, 2at)$

Obviously tours $S \equiv (a, 0)$

Tangent P = TP can be written as $TP \equiv ty = x + at^2$

Now since T is a point on TP

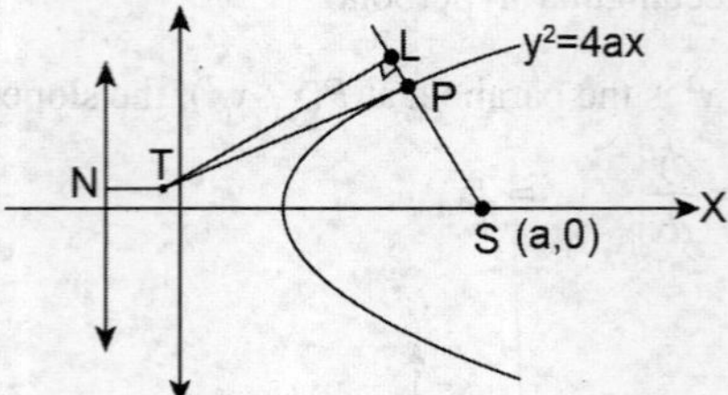

Let $T \equiv \left(p, \frac{p+at^2}{t}\right)$

TN = Distance of T from $x + a = 0$

$\Rightarrow$ $\frac{|p+a|}{\sqrt{1}} = |p+a|$ units .... (i)

TL is perpendicular to SP = slope of $sp = \frac{2at}{a(t^2-1)}$

Hence now can find the coordinates of L

$\Rightarrow \quad SL = \sqrt{\left[a-\left(-p^2+a-a^2-2ap\right)\right]}$

$\Rightarrow \quad \sqrt{(p+a)^2} = |p+a|$ units ....(ii)

$\Rightarrow \quad SL = TN$

47. (a) Let $P(t_1)$ and $Q(t_2)$ be the end points of a focal chord through focus S $(a, 0)$.

Also A $(0, 0)$ is the vertex, then $t_1t_2 = -1$ and the chord $P(at_1^2, 2at_1)$ and $Q\left(\frac{a}{t_1^2}, -\frac{2a}{t_1}\right)$

$\Rightarrow$ Equation of AP is $y = \frac{2}{t_1}x$ which will intersect directrix $x = -a$ at $R\left(-a, -\frac{2a}{t_1}\right)$

Similarly equation AQ is $y = -2t_1x$ and it will intersect the directrix at $(-a, 2at_1)$

Now slope of SR is $m_3 = \frac{2a}{t_1(2a)} = \frac{1}{t_1}$ and slope of ST is $m_4 = \frac{2at_1}{-2a} = -t_1$

Now $m_3m_4 = -1 \quad \Rightarrow \quad \angle RST = 90°$

48. (a) The circles will intersect at Q and R

Since OP is a diameter

$\Rightarrow \quad \angle ORP = 90°$

Similarly $\angle ORQ = 90°$

Hence PRQ is a straight line and $OR \perp PQ$

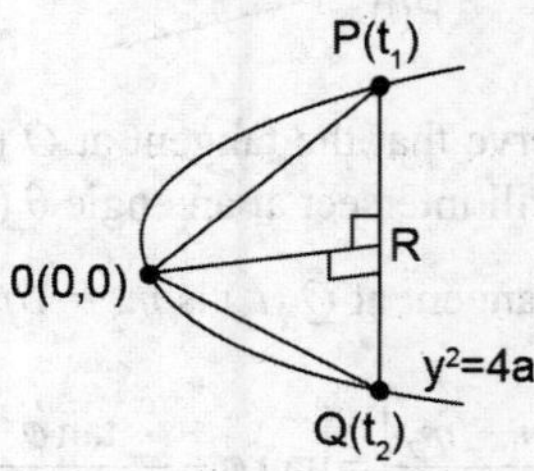

$\Rightarrow$ Slope of $PQ = -\frac{1}{\text{slope of } OR}$

$\Rightarrow$ Slope of $OR = -\frac{1}{\text{slope of } PQ} = \frac{-1}{\frac{2a(t_2-t_1)}{a(t_2^2-t_1^2)}}$

$= -\frac{(t_1+t_2)}{2}\tan\phi$

Now tangent at P (t_1) has slope $m_1 = \tan\theta_1 = \frac{1}{t_1}$ and

similarly tangent at Q (t_2) has slope $m_2 = \tan\theta_2 = \frac{1}{t_2}$

$\Rightarrow \quad \cot\theta_1 + \cot\theta_2 = t_1 + t_2 = -2\left\{\frac{-(t_1+t_2)}{2}\right\} = -2\tan\phi$

49. (c) Let Q (t_1) and R (t_2), $(P \equiv O)$ Since QR is a focal chord

$\Rightarrow \quad t_1t_2 = -1$

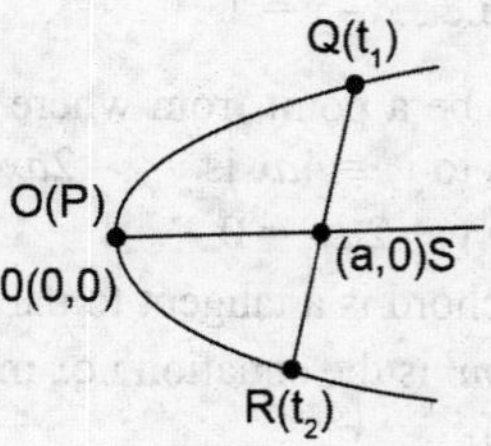

$\Rightarrow$ Area $\Delta PQR = \Delta PSQ + \Delta PSR = \frac{a}{2}(2at_1) + \frac{a}{2}(2at_2) = a^2(t_1+t_2) = \frac{a^2}{t_1}(t_1^2-1) = A$

Now the required difference of module is

$|y_1| - |y_2| = |2at_1| - \left|\frac{-2a}{t_1}\right| = \frac{2a}{t_1}(t_1^2-1) = \frac{2aA}{a^2} = \frac{2A}{a}$

50. (c) Let P (t_1) and Q (t_2) be the points on parabola $y^2 = 4ax$.

Since $\angle POQ = 90°$

$\therefore \quad t_1t_2 = -4 \qquad \Rightarrow P\ (t_1)$ and $Q\left(-\frac{4}{t_1}\right)$

Hence PQ has equation $y - 2at_1 = \frac{2t_1}{(t_1^2-4)}(x - at_1^2)$

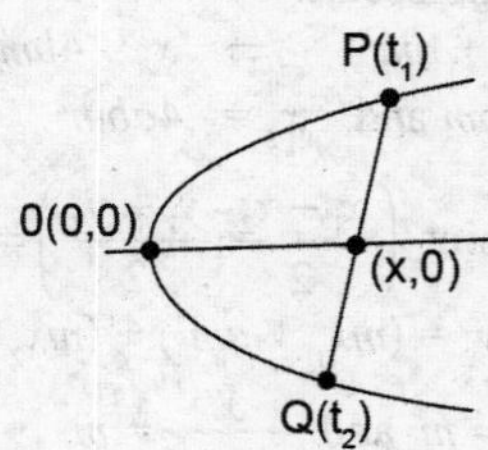

This will intersect x-axis at $y = 0$, so

$x = \frac{-2at_1(t_1^2-4)}{2t_1} + at_1^2 = 4a$, which is fixed point i.e., $(4a, 0)$ this is equal to L.R.

51. (a) Let $a > 0$ for parabola $y^2 = 4ax$

The circle is $(x + b)^2 + y^2 = b^2$ which has centre at $(-b, 0)$ and radius $r = |b|$

Since both the curves touch externally

$\therefore$ For $a > 0$, $b > 0$

52. (c) The parabola is $y^2 = 4x \qquad \Rightarrow \qquad a = 1$

Let R (t_1) and Q (t_2) be the two points on the parabola at which the slope will be $m_1 = \frac{1}{t_1}$ and $\frac{1}{t_2} = m_2$ respectively

Now $\theta_1 + \theta_2 = \frac{\pi}{4}$ (given)

$\Rightarrow \quad \frac{\tan\theta_1 + \tan\theta_2}{1 - \tan\theta_1\tan\theta_2} = 1$ i.e., $\frac{\frac{1}{t_1}+\frac{1}{t_2}}{1-\frac{1}{t_1t_2}} = 1$

$\Rightarrow \quad \frac{t_1+t_2}{t_1t_2-1} = 1$ or $t_1 + t_2 = t_1t_2 - 1$

Now the point of intersection of tangents is $P(at_1t_2, at_1 + at_2) = (t_1t_2, t_1t_2 - 1) = (x, y)$

$\Rightarrow x = y + 1$, i.e., $x - y = 1$

53. (d) Let (x_1, y_1) be a point from where the chord of contact of tangents to $y^2 = 4ax$ is $y_1y - 2ax - 2ax_1 = 0$

Or $2ax - y_1y + 2ax_1 = 0$ .... (i)

Since the chord is a tangent to parabola $x^2 = 4by$

$\Rightarrow y = mx - bm^2$ is the equation i.e., $mx - y - bm^2 = 0$ is the same as (i)

$$\Rightarrow \frac{m}{2a} = \frac{1}{y_1} = \frac{-bm^2}{2ax_1} \quad \Rightarrow x_1 = -bm$$

$$\text{and } y_1 = \frac{2a}{m} \quad \Rightarrow \frac{x_1}{-b} = \frac{2a}{y_1}$$

$\Rightarrow x_1y_1 + 2ab = 0$ or $xy = -2ab$ which is a hyperbola

54. (c) The equation of tangent to $y^2 = 4ax$ is $y = mx + \frac{a}{m}$ and for $x^2 = 4ay$ it is $y = mx - am^2$

Since the line is the same, so $\frac{a}{m} = -am^2 \Rightarrow m = -1$ and it gives $y = mx - a$

55. (a) The parabola $x^2 = -4ay$ will have a tangent as $y = mx + am^2$

Now line intersects $x^2 = 4by$

$\Rightarrow x^2 = 4b(mx + am^2) \quad \Rightarrow x^2 - 4bmx - 4abm^2 = 0$

$\Rightarrow x_1 + x_2 = 4bm$ and $x_1x_2 = -4abm^2$

$$\text{The mid point } \left(\frac{x_1 + x_2}{2}, \frac{y_1 + y_2}{2}\right) = (2bm, am^2 + 2bm^2)$$

using $y_1 + y_2 = (mx_1 + am^2) + (mx_2 + am^2)$

$$\text{Hence } \frac{x}{2b} = m \text{ and } \frac{y}{a + 2b} = m^2$$

$$\Rightarrow \left(\frac{x}{2b}\right)^2 = \frac{4}{a + 2b} \text{ or } (a + 2b)\, x^2 = 4b^2y$$

56. (b) For the parabola $y^2 = 4ax$ the point $P(4a, -4a)$

$\Rightarrow P(t_1 = -2)$ and $Q(9a, -6a)$

$\Rightarrow Q(t_2 = -3)$

$\Rightarrow m_1 = -t_1 = 2$ and $m_2 = -t_2 = 3$

Since $m_1 + m_2 + m_3 = 0$

$\Rightarrow m_3 = -5$

Hence $t_3 = 5$ and the third foot is $(25a, 10a)$ and the normal is $y - 10a = (-5)(x - 25a)$ i.e., $5x + y - 135a = 0$

57. (b) The normal with slope m is at $P(am_1^2, -2am_1)$ and its equation is $y = mx - 2am_1 - am_1^3$ where $m_1 = -t$

$$\text{The other end of normal chord is } t_2 = -t_1 - \frac{2}{t_1}$$

Now $P(at_1^2, +2at_1)$ and $Q(at_2^2, 2at_2)$ and vertex is $O(0, 0)$

$$\text{Slope of } OP = \frac{2at_1}{at_1^2} = \frac{2}{t_1} \text{ and slope of}$$

$$OQ = \frac{2}{t_2} = \frac{2}{-t_1 - \frac{2}{t_1}} = \frac{-2t_1}{t_1^2 + 2}$$

$$\text{Now, for } 90° \text{ angle } \left(\frac{2}{t_1}\right)\left(\frac{-2t_1}{t_1^2 + 2}\right) = -1$$

$\Rightarrow t_1^2 + 2 = 4 \quad \Rightarrow t_1 = \pm\sqrt{2} \Rightarrow m = \pm\sqrt{2}$

58. (b) The equation of normal system is $am^3 + 2am - mx + y = 0$

$$\Rightarrow m_1 + m_2 + m_3 = 0 \text{ and } m_1m_2 + m_2m_3 + m_1m_3 = \frac{2a - x}{a}$$

Also $m_1m_2m_2 = \frac{-y}{a}$ when two normals are at 90° (say m_1 and m_2), then $m_1m_2 = -1$.

$$\Rightarrow m_2 = \frac{-1}{m_1} \text{ and we get } P(am_1^2, -2am_1) \text{ and } Q\left(\frac{a}{m_1^2}, \frac{2a}{m_1}\right)$$

$\Rightarrow$ The equation of PQ is

$$y + 2am_1 = \frac{2a(1 + m_1^2)m_1^2}{am_1\{(1 + m_1^2)(1 - m_2^2)\}}(x - am_1^2),$$

Which on simplification gives $2m_1x + (m_1^2 - 1)y - 2am_1 = 0$, which becomes an identity for $x = a, y = 0$

Hence the chord PQ always passes through $(a, 0)$

59. (a) Let $P(t)$ be the point on parabola $y^2 = 4ax$, where the normal makes an angle and with the axis of parabola

$\Rightarrow m_1 = -t = \tan\phi$

Now the normal will intersect the parabola at $Q(t_2)$

$$\text{where } t_2 = -t - 2/t = \tan\phi + \frac{2}{\tan\phi}$$

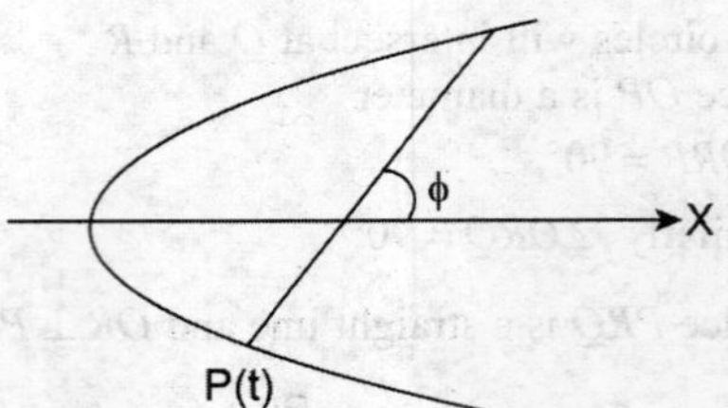

Now observe that the tangent at $Q(t_2)$ and the straight line PQ will intersect at an angle θ (say)

$$\Rightarrow \text{Slope of tangent at } Q(t_2) \text{ is } m_2 = 1/t_2 = \frac{\tan\phi}{2\tan^2\phi}$$

$$\therefore \quad \tan\theta = \left|\frac{m_1 - m_2}{1 + m_1m_2}\right| = \left|\tan\phi - \frac{\tan\phi}{2 + \tan^2\phi}\right| = \frac{\tan\phi}{2}$$

$$\text{Hence } \theta \tan^{-1}\left\{\frac{1}{2}\tan\phi\right\}$$

60. (c) The parabola is $y^2 = 4x \quad \Rightarrow a = 1$

The slope of first normal is $m_1 = \tan a = -t_1$ and slope of second normal is $m_2 = \tan 2 = -t_2$ (Also tan a. tan 2 = 2)

Now the point of intersect R of normals at $P(t_1)$ and $Q(t_2)$ is $(2 + \tan^2 a + \tan^2 2 + \tan 2 \tan a, 2\tan a + 2\tan 2)$

$$\Rightarrow x = (\tan a + \tan 2)^2 \text{ and } \frac{y}{2} = (\tan a + \tan 2)$$

$$\therefore \quad x = \left(\frac{y}{2}\right)^2 \text{ i.e., } y^2 = 4x$$

61. **(a)** Let (X_1, Y_1) be the mid points of chords for the parabola $y^2 = 4x = 0$

So $Y_1y - 2x - 2X_1 = Y_1^2 - 4X_1$

$\Rightarrow 2x - Y_1y + (Y_1^2 - 2X_1) = 0$

Since the chords pass through (4, 4)

$\Rightarrow 8 - 4Y_1 + (Y_1^2 - 2X_1) = 0$

As the mid point lies on $y = mx$, so let it be (X_1, mX_1)

$\Rightarrow 8 - 4mX_1 + m^2X_1^2 - 2X_1 = 0$ i.e., $m^2X_1^2 - (4m + 2)X_1 + 8 = 0$

Since we have two distinct chords, so X_1 will have two distinct real values hence $D > 0$

$\Rightarrow (4m + 2)^2 - 32m^2 > 0$ i.e., $16m^2 - 16m - 4 < 0$, which has roots $m = \frac{1 \pm \sqrt{2}}{2}$

Hence $m \in \left(\frac{1-\sqrt{2}}{2}, \frac{1+\sqrt{2}}{2}\right)$

62. **(d)** Let $P(x_1, y_1)$ be the point from where three normals exist $\Rightarrow am^3 + 2am - mx_1 + y_1 = 0$

Since two angles with axis are complementary, so let $m_1m_2 = 1$ i.e., $m_2 = \frac{1}{m_1}$

$\Rightarrow m_1 + m_2 + m_3 = 0$, gives $m_3 = -\left(m_1 + \frac{1}{m_1}\right)$ and

$m_1m_2m_3 = -\frac{y_1}{a}$, gives $m_3 = -\frac{y_1}{a}$

Now $m_1m_2 + m_3(m_1 + m_2) = \frac{2a - x_1}{a}$

$\Rightarrow 1 + m_3\left(m_1 + \frac{1}{m_1}\right) = \frac{2a - x_1}{a}$

i.e., $\left(-\frac{y_1}{a}\right)\left(\frac{y_1}{a}\right) = \frac{2a - x_1 - a}{a} = \frac{a - x_1}{a}$

$\Rightarrow \frac{y_1^2}{a^2} = \frac{x_1 - a}{a}$ i.e., $y^2 = a(x - a)$

63. **(a)** Let $P(t_1)$ and $Q(t_2)$ be the two points on the parabola $y^2 = 4ax$

Slope of tangent at P is $m_1 = \frac{1}{t_1} = \tan\alpha_1$

Similarly slope of tangent at Q is $m_2 = \frac{1}{t_2} = \tan\alpha_2$ as $\frac{\cot\alpha_1}{\cot\alpha_2} = 2$

$\Rightarrow \frac{t_1}{t_2} = 2$ or $t_1 = 2t_2$

Now the point of intersection of tangents at P and Q is $R(at_1t_2, at_1 + at_2)$

$\Rightarrow x = a(2t_2^2)$ and $y = a(3at_2)$

Hence $\frac{x}{2a} = t_2^2 = \left(\frac{y}{3a}\right)^2$

$\Rightarrow 2y^2 = 9ax$

64. **(a)** Let $P(t_1)$ and $Q(t_2)$ be the two end points of a chord with slope m, so $m = \frac{2a(t_1 - t_2)}{a(t_1^2 - t_2^2)} = \frac{2}{t_1 + t_2}$

$\Rightarrow t_1 + t_2 = \frac{2}{m}$, so the points will be t_1 and $t_2 = \frac{2}{m} - t_1$.

Now the point of intersection normals at $P(t_1)$ and $Q(t_2)$ is $R(a(t_1^2 + t_2^2 + t_1t_2 + 2), -at_1t_2(t_1 + t_2))$

So $x = a\{(t_1 + t_2)^2 + 2 - t_1t_2\}$

$\Rightarrow \frac{x}{a} = \frac{4}{m^2} + 2 - t_1t_2$ or $t_1t_2 = 2 + \frac{4}{m^2} - \frac{x}{a}$

and $y = -at_1t_2\left(\frac{2}{m}\right)$ or $t_1t_2 = \frac{-ym}{2a}$

$\Rightarrow 2 + \frac{4}{m^2} - \frac{x}{a} = \frac{-ym}{2a}$, which on rearrangement gives $2\{2am^2 + 4a - m^2x\}$

$= -m^3y$ or $4a(2 + m^2) = 2m^2x - m^3y = m^2(2x - my)$

65. **(d)** The curve is $(y + 2)^2 = 6x + 6 = 4\left(\frac{3}{2}\right)(x+1)$, which has vertex at $(-1, -2)$ and directrix as $x = -\frac{5}{2}$

Since the locus of perpendicular tangents intersection is directrix

$\Rightarrow x = -\frac{5}{2}$ or $2x + 5 = 0$

66. **(c)** The parabola is $x^2 = 4\left(\frac{1}{4}\right)y$, so directrix is $y = -\frac{1}{4}$.

The required locus is $4y + 1 = 0$

67. **(a)** Let θ_1, θ_2 and θ_3 be the three angles that the three normals make with the axis, so $\theta_1 + \theta_2 + \theta_3 =$ const. $= K$ (say)

$\Rightarrow \tan(\theta_1 + \theta_2 + \theta_3) =$ constant but $\tan\theta_1 + \tan\theta_2 + \tan\theta_3 = m_1 + m_2 + m_3 = 0$

Now $\frac{\tan(\theta_1 + \theta_2) + \tan\theta_3}{1 - \tan(\theta_1 + \theta_2).\tan\theta_3} =$ constant

$\Rightarrow \frac{(m_1 + m_2 + m_3) - m_1m_2m_3}{(1 - m_1m_2)\left\{1 - \frac{m_3(m_1 + m_2)}{1 - m_1m_2}\right\}} =$ constant $= k$(say)

So, $\frac{(m_1 + m_2 + m_3) - m_1m_2m_3}{1 - m_1m_2 - m_1m_3 - m_2m_3} = k =$ (constant)

Now the equation of normals from (x_1, y_1) is $am^3 + 2am - mx_1 + y_1 = 0$

Hence $\frac{0 + \frac{y_1}{a}}{1 - \frac{(2a - x_1)}{a}} = k$(constant)

$\Rightarrow y_1 = k\{x_1 - a\}$ which is a line i.e., $y = k(x - a)$ is the locus

68. **(a)** Let $P(t_1)$ be a point on parabola $y^2 = 4ax$ and N be the foot on axis from P, so $P(at_1^2, 2at_1)$; $N(at_1^2, 0)$ and $M(at_1^2, at_1)$

$\Rightarrow \quad Q\left(\frac{t_1}{2}\right)\left\{\text{as } 2a\left(\frac{t_1}{2}\right)=at_1\right\}$

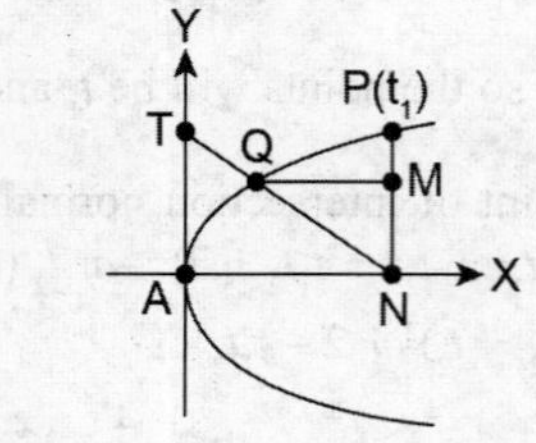

$\Rightarrow$ Equation of NQ will be $y-0=\dfrac{at_1}{\left(-\dfrac{3}{4}at_1^2\right)}(x-at_1^2)$

$= -\dfrac{4}{3t_1}(x-at_1^2)$

This line will intersect the tangent at vertex at $y=\dfrac{4at_1}{3}$

Hence $AT=\dfrac{4}{3}at_1$ and $NP = 2at_1$

Since $AT = kNP$

$\Rightarrow \quad k=\dfrac{4at_1}{3(2at_1)}=\dfrac{2}{3}$

69. (a) Since the circle is to always touch a straight line and a point then both distances are equal from the center of the circle

$\Rightarrow \quad e = 1$

Hence the locus of center will be a parabola.

70. (c) Let P (t_1), Q (t_2) and R (t_3) be the three points from where tan θ_1, tan θ_2, tan θ_3, (the slopes of tangents) are in H.P.

Since $\tan\theta_i = \dfrac{1}{t_i}$

$\Rightarrow$ $t_1 + t_2$, t_3 are in A.P. i.e., $t_2 - t_1 = t_3 - t_2 = d$ (say)

Now the points of intersection of the three tangents will be L $(at_1t_2, 2at_1 + at_2)$, M $(at_2t_3, at_2 + at_3)$, N $(at_1t_3, at_1 + at_3)$

$\therefore$ The ΔLMN will have area $= \dfrac{1}{2}\begin{vmatrix} at_1t_2 & at_1+at_2 & 1 \\ at_2t_3 & at_2+at_3 & 1 \\ at_1t_3 & at_1+at_3 & 1\end{vmatrix}$

$= \dfrac{1}{2}\left\|\begin{matrix} at_2(t_1-t_3) & a(t_1-t_3) & 0 \\ at_3(t_2-t_1) & a(t_2-t_1) & 0 \\ at_1t_3 & at_1+at_3 & 1\end{matrix}\right\|$

$= \left|\dfrac{a^2(t_1-t_3)(t_2-t_1)}{2}\right|\begin{vmatrix} t_2 & 1 & 0 \\ t_3 & 1 & 0 \\ at_1t_3 & (at_1+at_3) & 1\end{vmatrix}$

$= \left|\dfrac{a^2(2d)d}{2}\{t_2-t_3\}\right| = a^2d^3$, which is independent of individual values of t_1, t_2 and t_3 . For a fixed value of d and a the area is constant

71. (b) Let $P(t_1)$ b a point on parabola $y^2 = 4ax$

A normal at P (t_1) will meet the parabola again at Q (t_2) where $t_2 = -t_1 - \dfrac{2}{t_1}$

Now the minimum value of $|t_2| = \sqrt{8}$

$\left\{\text{using } AM \geq GM\text{, we get } \dfrac{1}{2}\left(t_1+\dfrac{2}{t_1}\right) \geq \sqrt{t_1.\dfrac{2}{t_1}} = \sqrt{2}\right\}$

$\Rightarrow$ The distance of Q from vertex V is $VQ = a\sqrt{(t_2^2+4)t_2^2}$ gives minimum value of gives ..a units.

72. (c) Let P (t) be a point on the parabola $y^2 = 4ax$, then the equation of the circle with SP (focal distance) as diameter is

$(x - at^2)(x - a) + (y - 0)(y - 2at) = 0$

The slope of normal at $P(t)$ is $m = (-t)$. Slope of SP is $\dfrac{2t}{t^2-1}$

Let θ be the angel between the normal and SP

$\Rightarrow \quad \tan\theta = t$

The required distance $d = SP \cos\theta = a(t^2 + 1)\cos\theta$

Since $\tan\theta = \dfrac{\sin\theta}{\cos\theta} = \dfrac{t}{1}$

$\Rightarrow \quad \cos\theta = \dfrac{1}{\sqrt{t^2+1}}$

$\therefore \quad$ Distance $d = a\sqrt{t^2+1}$

73. (a) Let P (t_1), Q (t_2) and R (t_3) be the three points on the parabola $y^2 = 4ax$ as PP' is parallel to QR

$\Rightarrow$ Slope of $PP' = \dfrac{2a(t_3-t_2)}{a(t_3^2-t_2^2)} = \dfrac{2}{t_3+t_2}$

Let P' be P' (t_4), then $\dfrac{2}{t_3+t_2} = \dfrac{2}{t_1+t_4}$, hence $t_4 = t_2 + t_3 - t_1 = -m_4$

Similarly from other slopes i.e., QQ' gives $t_5 = t_1 + t_3 - 2t_2 = -m_5$ and RR' gives $t_6 = t_2 + t_1 - t_3 = -m_6$.

Since normals at P, Q, R are concurrent. So $m = -t$, gives $t_1 + t_2 + t_3 = 0$

Hence $-m_4 = -2t_1$ or $m_4 = 2t_1$, $m_5 = 2t_2$ and $m_6 = 2t_3$

Clearly $m_4 + m_5 + m_6 = 2(t_1 + t_2 + t_3) = 0$

Similarly other factors, hence the chords will be concurrent

74. (b) Let $P(x_1, y_1)$ be the point outside the parabola $y^2 = 4ax$, so the chord of contact of tangents from will have equation

$y_1y - 2ax - 2ax_1 = 0$

Since it always passes through $(-1, 1)$ and $a = 1$

$\Rightarrow \quad y_1 - 2 - 2x_1 = 0$ i.e., $y_1 = 2(x_1 + 1)$ or $y = 2(x + 1)$

75. (a) Let P (t) be a point on the parabola $y^2 = 4ax$, then equation of normal at P (t) is $y - 2at = (-t)\{x - at^2\}$, so on axis G $(2a + at^2, 0)$ and G' $(0, at^3 + 2at)$

$\Rightarrow$ The point Q is $(2a + at^2, at^3 + 2at)$

So let $\dfrac{x}{a} = t^2 + 2$ i.e., $\dfrac{x-2a}{a} = t^2$ and $\dfrac{y}{a} = t(t^2+2)$

Hence $\dfrac{y}{a} = \sqrt{\dfrac{x-2a}{a}}\left(\dfrac{x}{a}\right)$

$\Rightarrow \quad ay^2 = x^2(x - 2a)$ i.e., $x^3 = 2ax^2 + ay^2$

76. (a) Without any loss of generality, let $y^2 = 4a_1x$ and $y^2 = -4a_2x$ be the two parabola and let the line parallel to the axis be $y = a$, so

$P = \left(\frac{a^2}{4a_1}, a\right)$ and $P' = \left(\frac{-a^2}{4a_2}, a\right)$ and the mid-point M is $\left(\frac{a^2}{4a_1} - \frac{a^2}{4a^2}, a\right)$

Let $x = a^2\left\{\frac{1}{4a_1} - \frac{1}{4a_2}\right\} = a^2k_1$ (say)

$\left(\text{as } \frac{1}{4a_1} - \frac{1}{4a_2} = \text{constant}\right)$ and $y = a$

$\Rightarrow$ $x = k_1y^2$ or $y^2 = kx$ which is a parabola

77. (a) Let $P(x_1, y_1)$ be the center of the circle and x_2, x_3 be the points of intersection on x-axis and $(0, b)$ be the point on y-axis so, $x_1^{\,2} + (y_1 - b)^2 = r^2$

Also $a^2 + y_1^{\,2} = r^2$, hence $x_1^{\,2} + y_1^{\,2} + b^2 - 2y_1b = a^2 + y_1^{\,2}$

$\Rightarrow$ $x^2 - 2yb + b^2 = a^2$

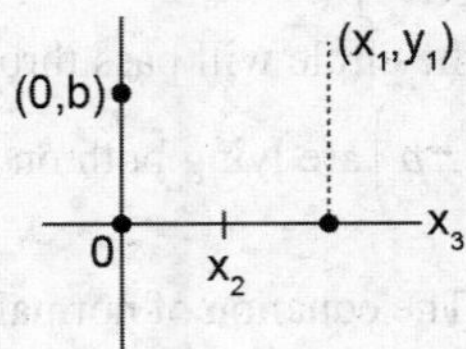

78. (a) Let (x_1, y_1) be an external point for a parabola $y^2 - 4x = 0$, then the chord of contact of tangents will be $y_1y - 2ax - 2ax_1 = 0$ or $2ax - y_1y + 2ax_1 = 0$

Putting $y = 2at$ and $x = at^2$, we get $2a^2t^2 - 2aty_1 + 2ax_1 = 0$ i.e., $2a\,\{at^2 - y_1t + x_1\} = 0$

Let t_1 and t_2 be its roots then $t_1 + t_2 = \frac{y_1}{a}$ and $t_1t_2 = \frac{x_1}{a}$

Since t_1 and t_2 are the end points of a normal chord, so

$t_2 = -t_1 - \frac{2}{t_1}$

$\Rightarrow$ $t_1 + t_2 = t_1 - t_1 - \frac{2}{t_1} = \frac{y_1}{a} \Rightarrow t_1 = \frac{2a}{y_1}$

and $t_1t_2 = t_1\left\{-t_1 - \frac{2}{t_1}\right\} = -t_1^2 - 2 = \frac{x_1}{a}$

Hence $\frac{x_1 + 2a}{a} = -t_1^2 = -\left(\frac{-2a}{y_1}\right)^2$

$\Rightarrow$ $(x_1 + 2a)\,y_1^{\,2} = -4a^3$

$\Rightarrow$ $(x + 2a)\,y^2 + 4a^3 = 0$ is the locus

79. (a) Since both the axes are its tangents (which are at 90°)

$\Rightarrow$ O (0, 0) lies on directrix. Let focus be $F(x_1, y_1)$, then slope of OF is $m = y_1/x_1$

$\Rightarrow$ Slope of focal chord is $m_2 = -1/m = -x_1/y_1 = \tan(90° + \theta)$.

Hence equation of focal chord is $y - y_1 = -\frac{x_1}{y_1}(x - x_1)$.

Which will intersect x-axis and y-axis at M and L respectively.

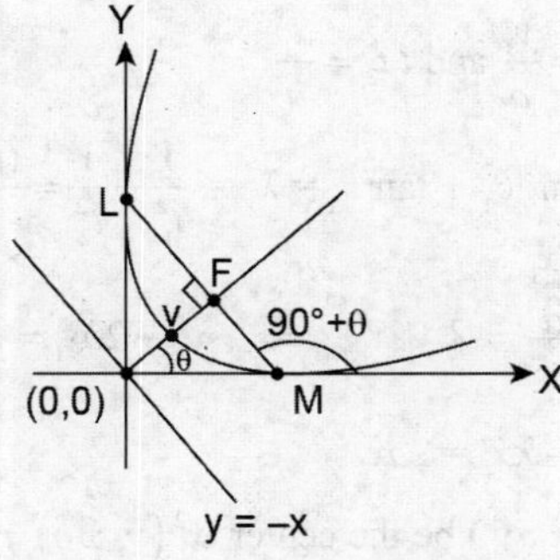

$\Rightarrow$ $M\left(\frac{x_1^2 + y_1^2}{x_1}, 0\right)$ and $L\left(0, \frac{x_1^2 + y_1^2}{y_1}\right)$

$\Rightarrow$ $FM = \ell_1 = \frac{y_1}{x_1}\sqrt{x_1^2 + y_1^2}$ and $FL = \ell_2 = \frac{x_1}{y_1}\sqrt{x_1^2 + y_1^2}$

Since H.M. of ${}_1$ and ${}_2$ will be $2a$ (semi L.R. length)

$\Rightarrow$ $\frac{1}{2a} = \frac{1}{2}\left\{\frac{1}{\ell_1} + \frac{1}{\ell_2}\right\}$

$\Rightarrow$ $a = \frac{\ell_1\ell_2}{\ell_1 + \ell_2} = \frac{(x_1^2 + y_1^2)\,x_1y_1}{\sqrt{x_1^2 + y_1^2}\,(x_1^2 + y_1^2)}$

i.e., $a^2(x_1^2 + y_1^2) = x_1^2y_1^2$.

Hence the local $a^2(x^2 + y^2) = x^2y^2$

80. (a) Consider the circle as shown in the diagram below without less of generality $b > a$

$C_1 \Rightarrow x^2 + y^2 = a^2$

$C_2 \Rightarrow x^2 + y^2 = b^2$ are two circles

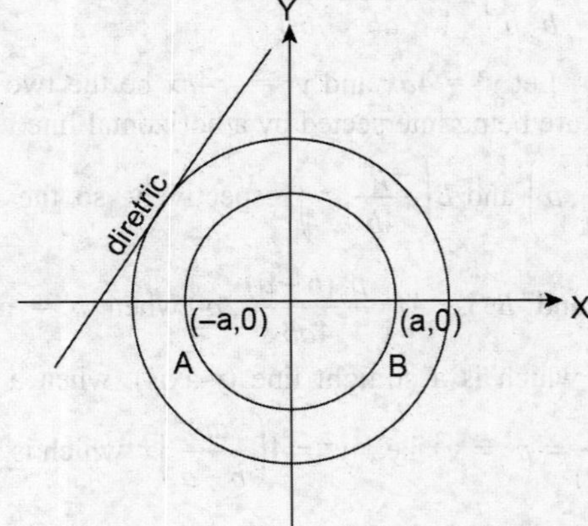

If $F(x_1, y_1)$ is the focus of parabola tangent to C_2

$\Rightarrow$ $x\cos\theta + y\sin\theta = b$ is the directrix.

$\Rightarrow$ $[-a\cos\theta - b]^2 = (x + a)^2 + y_1^2$

Similarly $[a\cos\theta - b]^2 = (x_1 - a)^2 + y_1^2$

$\Rightarrow$ $\cos\theta = \frac{x_1}{b}$ $\quad\Rightarrow$ $\frac{x^2}{b^2} + \frac{y^2}{b^2 - a^2} = 1$ is the locus

81. (a) Let $P\ (x_1, y_1)$ be an external point for a parabola $y^2 = 4ax$, then the chord of contact of tangents is $y_1y - 2ax - 2ax_1 = 0$

Putting $x = at^2$ and $y = 2at$, we get $(-2a)\,\{at^2 - y_1t + x_1\} = 0$

$\Rightarrow \quad t_1 + t_2 = \frac{y_1}{a}$ and $t_1t_2 = \frac{x_1}{a}$

Now $\tan^2\theta_1 + \tan^2\theta_2 = \lambda = \frac{1}{t_1^2} + \frac{1}{t_2^2} = \frac{(t_1+t_2)^2 - 2t_1t_2}{(t_1t_2)^2}$

$\Rightarrow \quad \frac{\frac{y_1^2}{a^2} - \frac{2x_1}{a}}{x_1^2/a^2} = \lambda \quad \Rightarrow \quad y_1^2 - 2ax_1 = \lambda x_1^2$

i.e., $y^2 - \lambda x^2 = 2ax$

82. **(a)** Let C (x_1, y_1) be the center $x_1^2 (y_1 \pm b)^2 r^2 = a^2 + y_1^2$

$\Rightarrow \quad x_1^2 + b^2 \pm 2by_1 = a^2$ i.e., $x^2 \pm 2by + b^2 = a^2$

(o,b) (x₁,y₁) X₂ X₃ (o,–b)

SECTION–IV (MORE THAN ONE CORRECT ANSWER)

1. **(a), (c)** The given curve is $y^2 + \frac{a^2}{b^2} - \frac{2ay}{b^2} = 4\frac{a}{b}x - 4$

Rewriting as $\left(y - \frac{a}{b}\right)^2 = 4\left(\frac{a}{b}\right)\left\{x - \frac{b}{a}\right\}$

Which is a parabola with vertex at $V\left(\frac{b}{a}, \frac{a}{b}\right)$ and its focus at is $\left(\frac{b}{a} + \frac{a}{b}, \frac{a}{b}\right) = \left(\frac{a^2+b^2}{ab}, \frac{a}{b}\right)$

2. **(a), (b)** Let $y^2 = 4ax$ and $y^2 = -4bx$ be the two parabolas which are being intersected by a horizontal line $y = p$ at $A\left(\frac{p^2}{4a}, p\right)$ and $B\left(\frac{-p^2}{4b}, p\right)$ respectively, so the mid point of A and B is $M\left(\frac{p^2(b-a)}{4ab}, p\right)$ when b = a, then M (o, p) which is a straight line (y-axis), when a ≠ b, then $\frac{4abx}{(b-a)} = p^2 = y^2$ i.e., $y^2 = 4\left(\frac{ab}{b-a}\right)x$ which is a parabola

3. **(a), (b)** Observe that one tangent is the y-axis, the other tangent is at $\theta = \pi/4$ and its equation is $ty = x + at^2$.
For $t = 1$, we get $y = x + a$
$\Rightarrow \quad a = 2$ from the symmetry $a = -2$ is also possible

4. **(a), (c)** The parabola is $y^2 = 9x$ and the circle is $(x-2)^2 + y^2 = 10$. Solving we get $x^2 + 4 - 4x + 9x = 10$
$\Rightarrow \quad x^2 + 5x - 6 = 0$
$\Rightarrow \quad x = 1, -6$ (but $x = -6$ is not possible) for $x = 1$, we get $y = \pm 3$
So the length of common chord is 6 units also (1, –1) lies on the common chord.

5. **(a), (d)** The parabola is $y^2 = 4ax$. The slope of tangent at $P(t)$ is $\tan\theta = 1/t = \pm\sqrt{3}$

So, $t = \pm 1/\sqrt{3}$ and the point $P(t) = (at^2, 2at)$
$\Rightarrow \quad \left(\frac{a}{3}, \frac{2a}{\sqrt{3}}\right), \left(\frac{a}{3}, \frac{-2a}{\sqrt{3}}\right)$

6. **(a), (b)** Let $y - 2a = m(x - a)$ be a chord through $(a, 2a)$ and the other end be at $(at_2^2, 2at_2)$
Now $t_1 = 1$, so mid point $x = \frac{a(t_1^2 + t_2^2)}{2}$ and $y = \frac{2a(t_1 + t_2)}{2}$ will satisfy $x + y = 1$.
Hence $\frac{a}{2}(t_1^2 + t_2^2) + \frac{2a}{2}(t_1 + t_2) = 1$.
Hence for $t_1 = 1$, $at_2^2 + 2at_2 + (3a - 2) = 0$.
Since two distinct real chords are possible, so $D > 0 \Rightarrow 4a^2 - 4a(3a - 2) > 0 \Rightarrow 8a(1 - a) > 0$
Hence $0 < a < 1$ $\qquad \therefore \; 0 < 4a < 4$
$\Rightarrow$ L.R. can be 1 or 2, from the given values

7. **(a), (b)** The given parabola is $y^2 = 2px$, so $a = p/2$ and the directrix is $x = -a = -p/2$.
Since the circle at focus S $(a, 0)$ touches the directrix
$\Rightarrow$ Radius $r = 2a = p$
Obviously the circle will pass through the ends of L.R.
Hence $\left(\frac{p}{2}, \pm p\right)$ are lying both on circle and parabola

8. **(a), (b), (c), (d)** The equation of normal with slope m (where $m = -t$) is $am^3 + 2am - ax + y = 0$
Which will give three values of $m = m_1, m_2, m_3$ as $m_1 + m_2 + m_3 = 0$
$\Rightarrow \quad -\{t_1 + t_2 + t_3\} = 0$
$\Rightarrow$ **Option (a) is true**
Obviously $2a(t_1 + t_2 + t_3) = 0$
$\Rightarrow$ **Option (b) is true**
Centroid of ΔPQR is $\left(\frac{x_1 + x_2 + x_3}{3}, \frac{y_1 + y_2 + y_3}{3}\right)$ as $\frac{y_1 + y_2 + y_3}{3} = 0$
$\Rightarrow$ Centroid lies on the axis of parabola
$\Rightarrow$ **Option (c) is true**
Let the circle passing through the three co-normal points be $x^2 + y^2 + 2gx + 2fy + c = 0$
$\Rightarrow \quad a^2m^4 + (4a^2 + 2ag)m^2 - 4amf + c = 0$
$\Rightarrow \quad m_1 + m_2 + m_3 + m_4 = 0$ but $m_1 + m_2 + m_3 = 0$
$\Rightarrow \quad m_4 = 0$
Hence (0, 0) lies on the circle.
$\Rightarrow$ **Option (d) is true**

9. **(a), (b), (c), (d)** Let $P(at^2, 2at)$ be a point on the parabola $y^2 = 4ax$ with focus $S(a, 0)$.
Now mid point of focal radii SP is $M\left(\frac{a}{2}(t^2 + 1), at\right) = (x, y)$ (say)
$\Rightarrow \quad \frac{2x}{a} - 1 = t^2$ and $\frac{y}{a} = t$
Hence $y^2 = 2ax - a^2$ or $y = 4\left(\frac{a}{2}\right)\left\{x - \frac{a}{2}\right\}$

Which is a parabola with vertex $V_2 = \left(\frac{a}{2}, 0\right)$ and $L.R = 2a$ and directrix is $x = 0$ and focus $S_2 = (a, 0)$

10. (a), (b) For the parabola $y^2 = 4ax$ the tangent will be $y = mx + a/m$ or $mx - y + a/m = 0$

The given line is $a^{1/3}x + b^{1/3}y + a^{2/3}b^{2/3} = 0$. On comparing we get $\frac{m}{a^{1/3}} = \frac{-1}{b^{1/3}} = \frac{a}{m.a^{2/3}b^{2/3}}$

$\Rightarrow m = \frac{a^{1/3}}{b^{1/3}}$ for all the parts

$\Rightarrow$ Line touches the parabola $y^2 = 4ax$.

Similarly for the parabola $x^2 = 4by$, the tangent will be $y = mx - bm^2$ or $mx - y - bm^2 = 0$

Comparing we get $\frac{m}{a^{1/3}} = \frac{-1}{b^{1/3}} = \frac{-bm^2}{a^{2/3}b^{2/3}}$

$\Rightarrow m = -\frac{a^{1/3}}{b^{1/3}}$ and $m^2 = a^{2/3}b^{2/3}$ which is true for $m = -\frac{a^{1/3}}{b^{1/3}}$

$\Rightarrow$ Line touches the parabola $x^2 = 4by$

11. (a), (b) The given parabola is $y^2 = 4(2)x$

$\Rightarrow a = 2.$

Since $(-2, 0)$ lies on the directrix and axis

$\Rightarrow$ The tangents will have slope $m = \pm 1$ and the equation is $y = \pm 1(x + 2)$ i.e., $x - y + 2 = 0$ or $x + y + 2 = 0$. The chord of contact of tangent is QR as $x = 2$ (*i.e.*, L.R)

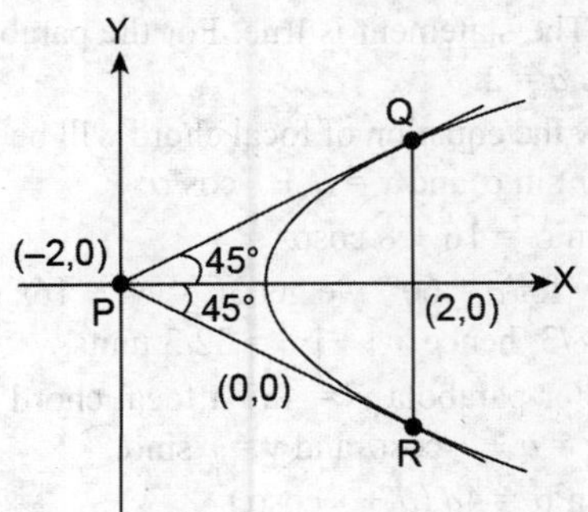

Now we are to find the in-radius and ex-radius of possible circles for the ΔPQR from $r_1 = \frac{\Delta}{s}$ we get, $r_1 = \frac{1\times 8\times 4}{2\times(4+4\sqrt{2})} = 4(\sqrt{2}-1)$ units. and from $r_2 = \frac{\Delta}{s-a}$,

we get $r_2 = \frac{16}{(4+4\sqrt{2}-8)} = \frac{16}{4(\sqrt{2}-1)} = 4\left(\sqrt{2}+1\right)$ Units

12. (a), (b) Consider a parabola $y^2 = 4ax$ and draw a circle with center at $(b, 0)$ $(b > 0)$ with radius r and let it intersect the parabola in four points A, B, C, D.

Now $ABCD$ is cyclic quadrilateral. Since the circle is centred on the axis of parabola

$\Rightarrow$ Diagonals will be equal because of symmetry.

Now it is not possible to form a rectangle (or a square) inside a parabola as the curve becomes wider and wider when we move away from the vertex.

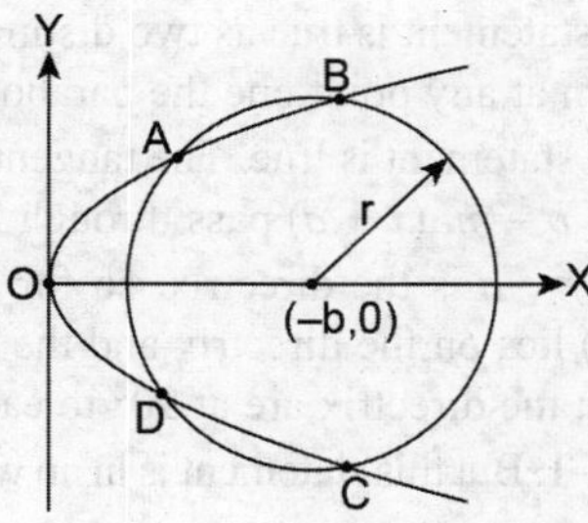

13. (b), (c) We know the system normals has equation $am^3 + 2am - mx + y = 0$

Three real (and distinct) normals are possible, when $h > 2a$ for $a > 0$, since $(h, k) = (\sin\theta, \cos\theta)$

$\Rightarrow 0 < a < 1/2$ and $h < 2a$ for $a < 0$

$\Rightarrow -1/2 < a < 0$

14. (a), (b), (c), (d) All the four statements are true

15. (a), (b) Consider the normal at $P(t_1)$ and $Q(t_2)$, then their point of intersection O is $(a(t_1^2 + t^2 + t_1t_2 + 2), -at_1t_2(t_1 + t_2))$

A third normal from O will have slope $m = -t_3 = t_1 + t_2$

Now the slope of the line joining P and Q is $M_{PQ} = \frac{2a(t_2 - t_1)}{a(t_2^2 - t_1^2)} = \frac{2}{t_1 + t_2}$

Since the line PQ is having a fixed direction

$\therefore \quad M_{PQ}$ = constant i.e., $\frac{2}{t_1 + t_2} = \frac{2}{k}$ (say)

$\Rightarrow t_1 + t_2 = k$ then, $O(x_1, y_1)$

$\Rightarrow x_1 = a(2 + k^2 - t_1t_2)$ and $y_1 = -t_1t_2k_a a$

$\Rightarrow \frac{x_1 - 2a - ak^2}{a} = -t_1t_2 = \frac{y_1}{ka}$

i.e., $y_1 = kx_1 - 2ak - ak^3$ or $ak^3 + 2ak - kx + y = 0$

Which is a straight line. Also it is the equation of a normal with slope $m = k$

SECTION–V (ASSERTION REASON)

1. **(a) R:** The statement is true for a parabola $y^2 = 4ax$. The normals at $P(t_1)$ and $Q(t_2)$ will intersect at $R(a(2 + t_1^2 + t_2^2 + t_1t_2), -at_1t_2(t_1 + t_2))$

   If point R lies on $y^2 = 4ax$, then $t_1 + t_2 = -t_3$ and $t_1t_2 = 2$

   A: Statement is true as the product of the ordinates is $4a^2t_1t_2 = 8a^2$ for $t_1t_2 = 2$

   Hence statement A is true and it is fully derivable from R.

2. **(a) R:** The statement is true as $y = mx - 2am - am^3$ is the equation of normals, so $am^3 + 2am - mx + y = 0$

   $\Rightarrow m_1 + m_2 + m_3 = 0$ i.e., $-\{t_1 + t_2 + t_3\} = 0$

   $\Rightarrow 2a(t_1 + t_2 + t_3) = 0$ where $P(t_1)$, $Q(t_2)$ and $R(t_3)$ are the three points (vertices of the ΔPQR) on parabola hence centroid G lies on the axis.

   A: Statement is true if three points be on the parabola and normals at these points intersect in (h, k), then the point will lie on the axis. Further the statement is fully supported by the reason R.

3. (b) **R:** The statement is true as two distinct tangents can not be drawn at any point one the parabola. $\Rightarrow m_1 = m_2$
A: The statement is true. The tangent $y + b = m_1(x + a)$ and $y + b = m_2(x + a)$ pass through $P(-a, -b)$.
Since $x = -a$ is the directrix, so for $a \neq b$, the point $P(-a, -b)$ lies on the directrix and the tangents from any point on the directrix are at 90° to each other.
$\Rightarrow m_1m_2 = -1$. But this statement is in no way supported by *R*.

4. (a) **R:** The statement is true when a^3, b^3, c^3 are in A.P. and $a + b + c = 0$, then $a^3 + c^3 = 2b^3$
$\Rightarrow (a + c)\{(a + c)^2 - 3ac\} = 2b^3$
$\Rightarrow (-b)\{b^2 - 3ac\} = 2b^3$ i.e., $3ac - b^2 = 2b^2$
Hence $3ac = 3b^2$
$\Rightarrow a, b, c$ are in G.P.
A: The statement is true. For the parabola $x^2 = 4ay = 4y$ $(a = 1)$, the equation of normal is $y = mx + 2a + \frac{a}{m^2}$
The (co-normal) points are $\left(-\frac{2a}{m}, \frac{a}{m^2}\right)$ and the slope of tangent at $P(t)$ is $m = t$, where $P(t)$ is $(2at, at^2)$
So the statement is true and it is fully supported by reason *R*.

5. (a) **R:** Statement is true as $AM \geq GM \geq HM$ for any two positive quantities (*n*umbers)
A: The statement is true when a, *b* are segments of focal chord, then semi-latus rectum $C = \frac{2ab}{a+b}$ {as L.R. = 2*c* (given)}
For a^3, b^3, c^3 in G.P., now $AM \geq GM$
$\Rightarrow \frac{a^3 + c^3}{2} \geq (ac)^{3/2} \quad \Rightarrow a^3 + c^3 \geq 2b^3$
The statement is true and it is support by *R*.

6. (c) **R:**Statement is false for $t = t_1$ the other end of the normal $t_2 = -t_1 - 2/t_1$ (*and* not $t_2 = t_1 - 2/t_1$)
A: The statement is true the normal at $x = 4$ for $y^2 = 4x$
$\Rightarrow t = \pm 2$ and $t_2 = \quad 3$
i.e., when $t_1 = 2$, then $t_2 = -3$
$\Rightarrow$ P$(-3) = (9, -6)$ and similarly $t_1 = -2 \Rightarrow t_2 = 3$
$\Rightarrow P'(3) = (9, 6) \quad \Rightarrow PP' = 6 + 6 = 12$ units

7. (a) **R:** The statement is true as the point of intersection of L_1 and L_2 will always satisfy any line in the form $L_1 + \lambda L_2 = 0$ or $\mu L_1 + \nu L_2 = 0$
A: The statements is true as $y = (a - b)x^2 + (b - c)x + (c - a)$ touches the x-axis in the interval (0, 1]
Observe that $x = 1$ satisfies $y = 0$
$\Rightarrow x_1 = 1$ and $x_2 = \frac{c-a}{a-b}$ but since the parabola touches x-axis
$\Rightarrow x_1 = x_2$ i.e., $\frac{c-a}{a-b} = 1$
$\therefore c - a = a - b$ or $-2a + b + c = 0$
Comparing with $ax + by + c = 0$, we get $x = -2$ and $y = 1$ i.e., the line always passes through a fixed point $(-2, 1)$

8. (d) **R:** The statement is true. Let $P(t)$ be a point on the parabola $y^2 = 4ax$, then $PM = 2at$ and $PN = at^2$ as $LR = 4a$
$\Rightarrow (PM)^2 = 4a^2t^2$ and $(PN)(LR) = 4a^2t^2$ i.e., $PM^2 = (PN)(LR)$
A: The statement is false. Observe that $4x - 3y + 7 = 0$ and $4x - 3y + 8 = 0$ are although parallel but distance between these lines is $1/5 \neq a$
Hence $4x - 3y + 8 = 0$ can not be the directrix where L.R. $= 4a = 4$ (*i.e.*, $a = 1$)

9. (d) **R:** The statement is true for the parabola $y^2 = 4x$ the tangent at $(1, -2)$ is $y + 2 = -(x - 1)$
i.e., $x + y + 1 = 0$ (*For* slope $m = -1$), we get $x^2 = 4y$ as $x^2 = 4(-1 - x) \Rightarrow x^2 + 4x + 4 = 0$ or $(x + 2)^2 = 0$, which gives $x = -2, -2$ (repeated root)
$\Rightarrow x + y + 1 = 0$ is also a tangent to $x^2 = 4y$ at $(-2, 1)$ an end of L.R.
A: The statement is false. The tangent $y = x + 1$ does not touch $x^2 = 4y$ at the end of latus rectum.

10. (a) **R:** The statement is true
A: For the parabola $y^2 = 4x$, the given lines L_1: $x = 0$, L_2: $x + y + 1 = 0$ and L_3: $x - y + 1 = 0$ are tangents at $P(0, 0)$, $Q(1, -2)$ and $R(1, 2)$ respectively. Now their circum-circle is centered at $(0, 0)$ and it is passing through $(-1, 0)$, $(0, 1)$, $(0, -1)$. Obviously it will pass through $F(1, 0)$. Hence the statement is true and fully supported by reason *R*.

11. (a) **A:** The statement is true. For the parabola $y^2 = 8x$, we have $a = 2$.
Now the equation of focal chord will be (for angle α) as $y = r \sin \alpha$ and $x = a + r \cos \alpha$
$\Rightarrow r^2 \sin^2\alpha = 16 + 8 \cos\alpha$
Now for $a = 60°$, we get $3r^2 = 64 + 16r$ which gives $r = 8, -8/3$, hence $|r_1| + |r_2| = 32/3$ units.
R: For parabola $y^2 = 4ax$ a focal chord at angle α will be $x = a + r\cos\alpha$ and $y = r\sin\alpha$
$\Rightarrow r^2 \sin^2\alpha = 4a(a + r\cos\alpha)$
$\Rightarrow r^2 \sin^2\alpha - 4ar\cos\alpha - 4a^2 = 0$
$\Rightarrow r = \frac{4a\cos\alpha \pm \sqrt{16a^2\cos^2\alpha + 16a^2\sin^2\alpha}}{2\sin^2\alpha}$
$= \frac{4a(\cos\alpha \pm 1)}{2\sin^2\alpha} = 2a\,\text{cosec}^2\alpha\,(\cos\alpha \pm 1)$
Since $|r_1|$, $|r_2|$ is the distance, so $|r_1| + |r_2|$ is the distance
$\Rightarrow |r_1| + |r_2| = 4a\text{cosec}^2\alpha$ {as r_1 and r_2 are of opposite sign}.
Hence reason *R* statement is true and it fully supports the assertion *A*.

12. (c) **R:** The statement is false. For the parabola $y^2 = 4ax$ the chord of contact for $P(x_1, y_1)$ is $yy_1 = 2ax + 2ax_1$.
Now distance of *P* from the chord is
$$d = \frac{|2ax_1 + 2ax_1 - y_1^2|}{\sqrt{4a^2 + y_1^2}} = \frac{|y_1^2 - 4ax_1|}{\sqrt{y_1^2 + 4a^2}}$$

Length of chord of contact from $P(x_1, y_1)$ to $y^2 = 4ax$ is $\frac{\sqrt{(y_1^2 - 4ax_1)(y_1^2 + 4a^2)}}{|a|}$

Hence area of Δ formed

$$= \frac{1}{2}\frac{(y_1^2 - 4ax_1)}{\sqrt{y_1^2 + 4a^2}}\frac{\sqrt{(y_1^2 - 4ax_1)(y_1^2 + 4a^2)}}{|a|}$$

$$= \frac{(y_1^2 - 4ax_1)^{3/2}}{|2a|} \neq \frac{(y_1^2 - 4ax_1)^{3/2}}{|4a|}$$

A: The statement is true for P (12, 8) and parabola $y^2 = 4x$, the chord of contact is $8y = 2x + 24$ or $x - 4y + 12 = 0$

Now distance of P from the chord of contact is $d = \frac{|12 - 32 + 12|}{\sqrt{17}} = \frac{8}{\sqrt{17}}$

The equation $y^2 = 4x$ becomes $\left(\frac{x+12}{4}\right)^2 = 4x$ i.e., $x^2 - 40x + 144 = 0$ or $(x - 36)(x - 4) = 0$

$\Rightarrow \Delta x = 32$ units

Hence Area $= \frac{d}{2}\Delta x.\sqrt{1+m^2} = \frac{8(32)}{2\sqrt{17}}.\frac{\sqrt{17}}{4} = 32$ sq units.

13. (d) R: The statement is true as two tangents are possible from every external point to the parabola.

A: The statement is false as it is not possible.

Let $P(t_1)$, $Q(t_2)$ and $R(t_3)$ be the three distinct points on parabola $y^2 = 4ax$.

Now the points of intersection of tangents are $I_1 = (at_1t_2, a(t_1+t_2))$, $I_2 = (at_2t_3, a(t_2 + t_3))$ and $I_3 = (at_1t_3, a(t_1 + t_3))$

If I_1, I_2, I_3 are collinear, then slope of I_2, I_1 is the same as of I_2I_3 joint i.e., $m_1 = \frac{a(t_3 - t_1)}{at_2(t_3 - t_1)} = \frac{1}{t_2} = \frac{a(t_1 - t_2)}{at_3(t_1 - t_2)} = \frac{1}{t_3}$

$\Rightarrow t_2 = t_3$ which is not true

14. (a) A: Parabola: $y^2 = 4ax = 8x$

$\Rightarrow a = 2$

Vertex = (0, 0) = A

Let B be $(at^2, 2at)$. sinθ BC is a focal chord

$\Rightarrow C = \left(\frac{2}{t^2}, -\frac{4}{t}\right)$.

Area of the $\Delta ABC = \frac{1}{2}\left|at^2\left(-\frac{4}{t}\right) + \frac{2}{t^2}(-2at)\right| = 6$

$\Rightarrow \frac{1}{2}\left|-8t - \frac{8}{t}\right| = 6$

Solving this $t = -2$ or 1/4

Now distance of B from $y = 0 = |2at| = |4t|$ and distance of C from $y = 0 = \left|\frac{4}{t}\right| = \left|\frac{4}{t}\right|$

$\Rightarrow \left|\frac{4}{t}\right| - |4t| = 3$ units

$\Rightarrow$ Assertion is correct

R: Let B be $(at^2, 2at)$ then C will be $\left(\frac{a}{t^2}, \frac{-2a}{t}\right)$ the vertex A is (0, 0)

$\Rightarrow \frac{1}{2}\left|at\left(\frac{2a}{t}\right) + \frac{a}{t^2}(2at)\right| = \Delta$

$\Rightarrow \left|\frac{2a}{t}\right| - |2at| = \frac{2\Delta}{a^2}$; Which means that difference of distances from axis is $2\Delta/a^2$ for $y^2 = 8x$

15. (a) R: The statement is true. We know that perpendicular from focus S on any tangent lies on the tangent at vertex.

Now consider the tangent at P $(at^2, 2at)$ which is $ty = x + at^2$ and it will intersect $x = 0$ line at z $(0, at)$

$\Rightarrow SP = \sqrt{a^2(t^2 - 1)^2 + 4t^2a^2} = a(t^2 + 1)$ and

$SZ = \sqrt{a^2 + a^2t^2} = a\sqrt{t^2 + 1}$ and $AS = a$

Clearly, $SZ^2 = a^2(t^2 + 1)$ and $AS.\,SP = a^2(t^2 + 1)$

$\Rightarrow SZ^2 = AS.SP$

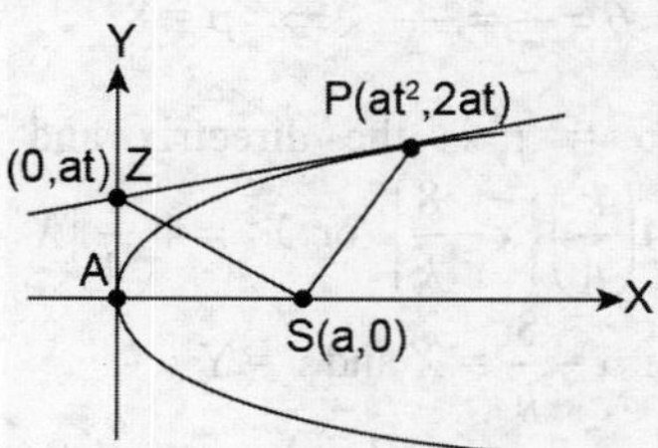

A: The statement is true for the given data and it is fully derivable from reason R.

SECTION– VI (COLUMN MATCHING)

1. (i) → (d); (ii) → (b); (iii) → (d); (iv) → (c)

(i) The parabola is $y^2 = 4Ax$ ($A = 1$), so any tangent at P (t) will be $ty = x + At^2$.

Since Circle (centered) at (6, 5) and the parabola intersect at 90°

∴ The tangent of parabola will be a normal to the circle hence passing through the centre (6, 5)

∴ $5t = t^2 + 6$ $\Rightarrow t = 2, 3$.

Hence the two possible points of intersection are (4, 4) and (9, 6)

Now (4, 4) has (a, a) form

$\Rightarrow a = 4$

(ii) Observe that indirectly we are told that the line $3x - y + 8 = 0$ is the chord of contact.

Shifting origin to (–3, 2) the parabola is $y^2 = 4x$ and the chord is $y = 3x - 3$ or $3x - y - 3 = 0$

Method I:

Now chord of contact from (X_1, Y_1) is $2X + 2X_1 - Y_1Y = 0$ or $3X - \frac{3}{2}Y_1Y + 3X_1 = 0$ $\Rightarrow X_1 = -1$ and $Y_1 = 2/3$

The equation of tangents from (–1, 2/3) to the parabola $Y^2 - 4X = 0$ will be $\left(\frac{6X - 2Y - 6}{3}\right)^2 = (Y^2 - 4X)\left(\frac{40}{9}\right)$

$\Rightarrow\ 9X^2 - 9Y^2 - 6XY + 22X + 6Y + 9 = 0$

$\Rightarrow\ m_1 + m_2 = -2/3,\ m_1m_2 = -1$, since $a + b = 0$

$\Rightarrow$ angle 90° i.e., $\dfrac{4\pi}{p} = \dfrac{\pi}{2}$ so $p = 8$

Method II:

The parabola $Y^2 = 4X$ will be intersected by $Y = 3X - 3$, so

$Y^2 = \dfrac{4(Y+3)}{3} \Rightarrow 3Y^2 - 4Y - 12 = 0$

Hence $Y = \dfrac{4 \pm 4\sqrt{10}}{6} = \dfrac{2 \pm 2\sqrt{10}}{3}$

Now slope of tangent is $m_1 = \dfrac{1}{t_1} = \dfrac{2}{Y_1}$ and

$m_2 = \dfrac{1}{t_2} = \dfrac{2}{Y_2} = \dfrac{3}{1-\sqrt{10}}$

Observe that $m_1m_2 = \dfrac{3}{1+\sqrt{10}} \cdot \dfrac{3}{1-\sqrt{10}} = -1$

$\therefore$ Angle $\theta = \dfrac{\pi}{2} = \dfrac{4\pi}{p} \Rightarrow p = 8$

(iii) Line $x = 1$ is the directrix and the parabola is

$y^2 = 4\left(\dfrac{k}{4}\right)\left\{x - \dfrac{8}{k}\right\}$ or $Y^2 = 4\left(\dfrac{k}{4}\right)X$

Where $x - \dfrac{8}{k} = X$ and $y = Y$

Now focus is at $\left(\dfrac{k}{4}, 0\right)$ and directrix $X = -\dfrac{k}{4} = 1 - \dfrac{8}{k}$

(as $x = 1$) $\Rightarrow k^2 = 32 - 4k \Rightarrow k = 4, -8$

(iv) The given parabola is $y^2 = 8x \Rightarrow a = 2$.

The point under consideration is $(p, p) \Rightarrow p = 8$

{As normal chord at (0, 0) is infinite} $(at^2, 2at) = A\ (8, 8)$

$\Rightarrow\ t_1 = 2$ for $a = 2 \Rightarrow t_2 = -t_1 - 2/t_1 = -3$

$\therefore$ The chord has end point A (8, 8) and B (18, –12). The length $AB = \sqrt{10^2 + 20^2} = 10\sqrt{5}$ units.

2. (i) → (d); (ii) → (c); (iii) → (a); (iv) → (b)

(i) Since A $(h, 0)$ lies on x-axis

$\therefore$ One foot of normal is the vertex and from the symmetry the two sides of ΔA_1A_2A are $y = \pm\dfrac{x}{\sqrt{3}}$

$\Rightarrow y^2 = 8x$, gives $x = 24$ and $y = \pm 8\sqrt{3}$. Now the normal at $x = 24$ for $a = 2$ has slope $m = -t = -2\sqrt{3}$

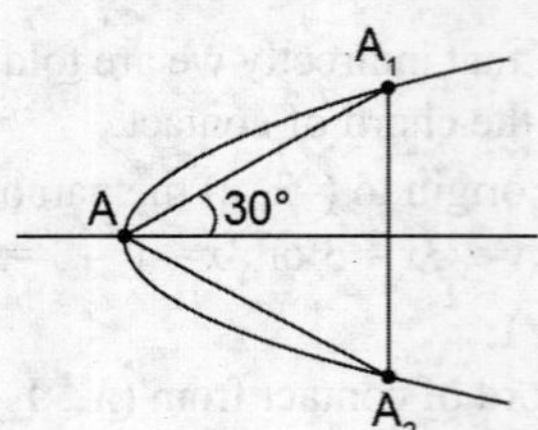

The normal will be $y - 8\sqrt{3} = \left(-2\sqrt{3}\right)\{x - 24\}$

Since normal passes through $(h, 0) \Rightarrow x = h = 28$

(ii) We know that the circle drawn with focal as diameter always touches the directrix

Now L.R is the smallest chord with centre at $(a, 0)$. Hence for parabola $y^2 = 8x$

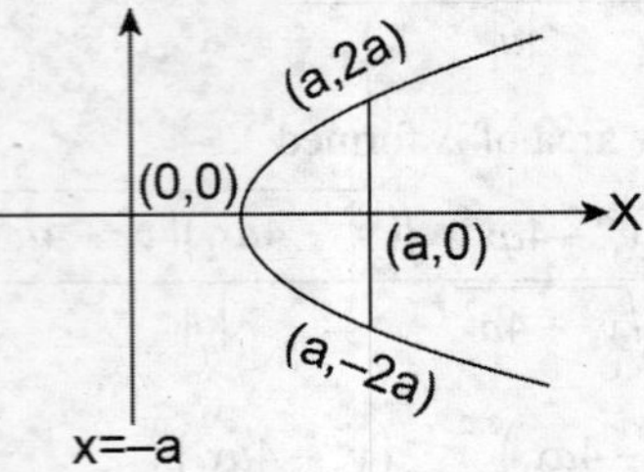

$a = 2 \Rightarrow$ Radius of the circle = 4 units

(iii) We know that tangents at the extremities of a focal chord intersect at directrix and in case of L.R. the tangents (as well normals) will have slope ± 1 (i.e. 45° with axis). Hence a square of side $2\sqrt{2}$ units is formed.

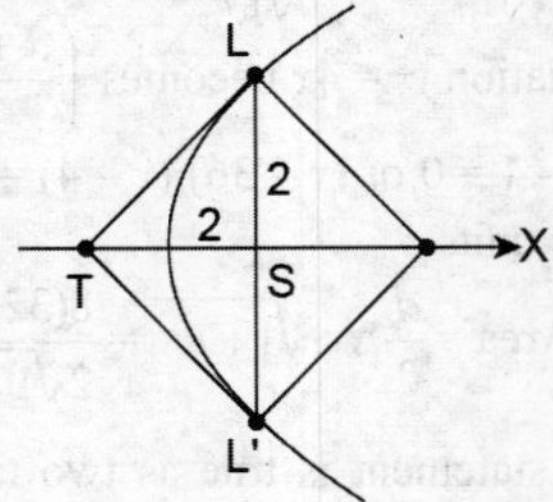

$\Rightarrow$ Area of the figure = 8 square units.

(iv) $(y + 1)^2 = 8\ (x - 2) = 4a\ (x - 2)$ where $a = 2$

We know that focal distance of a point M (t) is a $(t^2 + 1)$

$\Rightarrow\ P(t_1{=}1/2)$ and $Q(t_2 = 1)$, we get

SP. $SQ = 2\ (5/4).\ 2.(2) = 10$ units

3. (i) → (d); (ii) → (a); (iii) → (b); (iv) → (d)

(i) Let r be the radius of largest circle. Let the center of the circle be at $(1 + r, 0)$ (By symmetry considerations)

$\Rightarrow\ (x - 1 - r)^2 + y^2 = r^2$

Since the interior and boundary of the parabola is given by $y^2 - 4x \le 0$

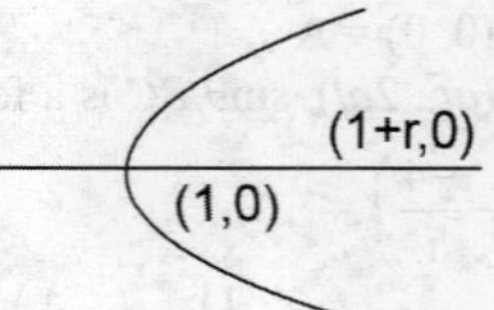

$\Rightarrow\ (x - 1 - r)^2 - r^2 + 4x \ge 0$ i.e., $(x + 1)^2 - 2r\ (x + 1) + 4r \ge 0$

Since r is real also $x + 1$ is real

$\Rightarrow$ Disc. ≤ 0

So r is real also $(x + 1)$ is real

$\Rightarrow$ Disc. ≤ 0

$\Rightarrow\ 4r^2 - 16r \le 0 \Rightarrow 0 \le r \le 4$

(ii) We know that tangents drawn at the ends of focal chord are, always at right angles.

The parabola is $y^2 = 16x$. Let A (t) be one point and then $B\left(\dfrac{-1}{t}\right)$ is the other end

$$\Rightarrow \quad AB = \sqrt{\left(at^2 - \frac{a}{t^2}\right)^2 + \left(2at + \frac{2a}{t}\right)^2}$$

$$= a\left|t + \frac{1}{t}\right|\sqrt{\left(t - \frac{1}{t}\right)^2 + 4} = a\left(t + \frac{1}{t}\right)^2$$

Which is minimum for $t = 1$, so $AB = 4a = 16$ units for $y^2 = 16x$

(iii) Let $P(t) = (at^2, 2at) = (t^2, 2t)$ be a point on $y^2 = 4x$ and $Q(t) = \left(\frac{t^2}{2} + 3, t\right)$ be the point on $y^2 = 4\left(\frac{1}{2}\right)(x - 3)$

Now slope of normal at $P(t) = -t$, so slope of $PQ = \dfrac{2t - t}{t^2 - \frac{t^2}{2} - 3} = -t$

$\Rightarrow \left(\frac{t^2}{2} - 3\right) = -1 \quad \Rightarrow \quad t^2 = 4$ i.e., $t = \pm 2$

So distance $PQ = d = \sqrt{1^2 + 2^2} = \sqrt{5}$. Hence $d^2 = 5$

(iv) Let $P(t)$ be one end of focal chord and $S(a, 0)$ be the focus, then $SP = a(t^2 + 1)$.

Now other end is $Q(1/t) \Rightarrow SQ = \dfrac{a(t^2 + 1)}{t^2}$

$HM = 2a \Rightarrow$ for $y^2 = 8x$ then $HM = 4$

SECTION– VII (COMPREHENSION)

Passage A:

1. (b), (c) Given C_1: $x = at^2$ and $y = 2at$ which is a parabola $y^2 = 4ax$ and $x = \dfrac{a(1 - t^2)}{1 + t^2}$ and $y = \dfrac{2at}{t^2 + 1}$ gives a circle in I[st] quadrant for $\in [0, 1]$

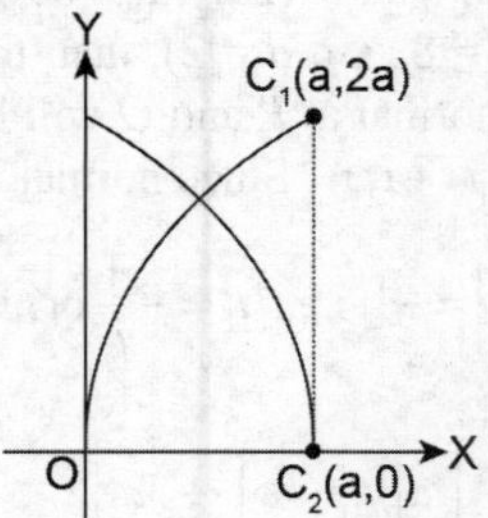

As shown these curves will intersect in exactly in one point in I[st] quadrant. Since area of curve C_2 in I[st] quadrant is $\dfrac{\pi a^2}{4}$, so the area common to C_1 and C_2 above x-axis is less than $\dfrac{\pi a^2}{4}$

2. (c) Let r be the length of the segment of the focal chord with angle α, so $x = a + r\cos\alpha$ and $y = r\sin\alpha$

$\Rightarrow$ At point A and B we get $y^2 = 4ax$ i.e., $r^2\sin^2\alpha = 4a^2 + 4ar\cos\alpha$

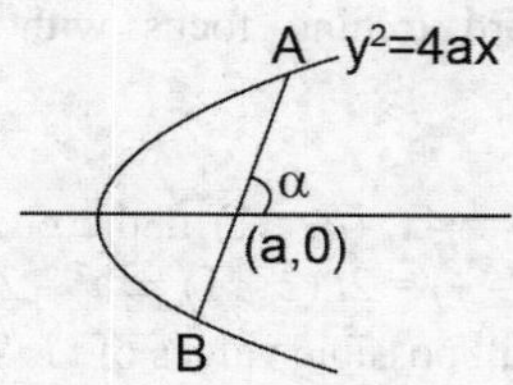

$\Rightarrow r^2\sin^2\alpha - 4ar\cos\alpha - 4a^2 = 0$ gives two values of r as

$$r = \frac{4a\cos\alpha \pm \sqrt{16a^2\cos^2\alpha + 16a^2\sin^2\alpha}}{2\sin^2\alpha}$$

Observe that r_1 and r_2 are of opposite sign

$$\Rightarrow \quad AB = |r_1| + |r_2| = \frac{8a}{2\sin^2\alpha} = 4a\,\text{cosec}^2\alpha$$

3. (a), (b), (c) Given $y^2 = 8x$ $\quad \Rightarrow a = 2$

$\Rightarrow P(\alpha, 8) = (at_1^2, 2at_1)$ gives $t_1 = 2$, hence $\alpha = 8$

$\therefore P(8, 8)$ and $Q(\beta, 2)$ gives $t_2 = 1/2$, hence $Q(1/2, 2)$

Now S (2, 0), so $SP = 10$ units and $SQ = 5/2$ units

Observe that H.M. of SP and SQ is 'H.M.' $= \dfrac{2 \times 10 \times \frac{5}{2}}{\left(10 + \frac{5}{2}\right)} = 4$

Now AM and SP and SQ is 'A.M.' $= \dfrac{1}{2}\left(10 + \dfrac{5}{2}\right) = \dfrac{25}{4}$

So quadruple of $AM = 4 \times \dfrac{25}{4} = 25$ and GM of SP and SQ is 'G.M.' $= \sqrt{10 \times \dfrac{5}{2}} = 5$

4. (b), (c) In (b) option observe that $x = 2\sin^2(\theta/2)$, so $1 - x = 1 - 2\sin^2(\theta/2) = \cos\theta$

Now $4 - y^2 = 2\cos\theta$

$\Rightarrow 4 - y^2 = 2(1 - x) = 2 - 2x$

$\therefore y^2 = 2x + 2 = 4(1/2)\{x + 1\}$, which is parabola arc for $\theta \in \left(-\frac{\pi}{2}, \frac{\pi}{2}\right)$

Similarly $x = |\sin t + \cos t|$ and $y = \sin t + \cos t$,

Observe that $|\sin t + \cos t| = \sin t + \cos t$ for $t \in \left(0, \frac{3\pi}{4}\right)$ (as $\sin t + \cos t \geq 0$)

5. (a) Let $P(\alpha, \beta)$ and $Q(\gamma, \delta)$ be the extremities of a focal chord of parabola $y^2 = 8x$ $\quad (\therefore a = 2)$

So $at^2 = \alpha$, $2at = \beta$ and $\dfrac{a}{t^2} = \gamma, \dfrac{-2a}{t} = \delta$. Observe that G.M. of α and γ is $\sqrt{a^2} = a = 2$

6. (c) Any point $P(t)$ on $y^2 = 4ax$ is $P(at^2, 2at)$ and focus $S(a, 0)$

$\Rightarrow$ Equation of SP is $y - 0 = \dfrac{2at}{a(t^2 - 1)}(x - a)$, i.e.,

$$y = \frac{2t(x - a)}{t^2 - 1}$$

7. (a), (c) A chord joining focus with a point P (t) is $y = \frac{2t}{t^2-1}(x-a)$

For parabola $y^2 = 4x$, ($a = 1$) if the chord passes through (2, 1), then $1(t^2 - 1) = 2t\,(2 - 1) \Rightarrow t^2 - 2t - 1 = 0$

$\Rightarrow$ Product of all possible values of t is 'p' = –1 and sum of all possible values of t is 's' = 2

Passage B:

8. (c) Since (, m) lies on parabola $y^2 = 4ax$

$\therefore$ $m^2 = 4a$. Now observe that ΔPQR is right angled at P (0, 2).

So its circum-centre is the mid point of Q and R, where

$Q\ (0, 1/m)$ and $R\left(\frac{(1-2m)4a}{m^2}, 0\right)$ as $x + my = 1$

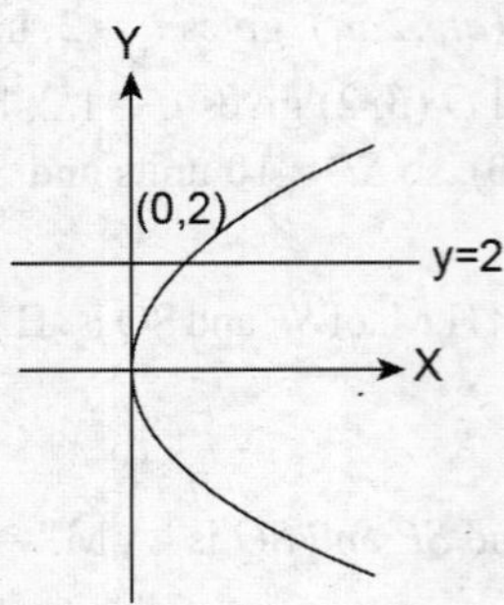

Hence $M = \left(\frac{2a(1-2m)}{m^2}, 1 + \frac{1}{2m}\right)$

Now, let $x = \frac{2a(1-2m)}{m^2}$ and $y - 1 = \frac{1}{2m}$

$\Rightarrow$ $x = \frac{2a\left\{1 - \frac{1}{y-1}\right\}}{\frac{1}{4}\left\{\frac{1}{y-1}\right\}^2} = \frac{4(2a)\{y-2\}(y-1)^2}{(y-1)}$, i.e., $x = 8a$

$(y-1)(y-2)$ or $8a\left(y - \frac{3}{2}\right)^2 = x + 18a - 16a$

Rewriting as $C: \left(y - \frac{3}{2}\right)^2 = 4\left\{\frac{1}{32a}\right\}\{x + 2a\}$

Hence the vertex of the parabola lies at (–2a, 3/2)

9. (d) Since L.R. is the smallest focal chord $\therefore$ L.R. $= \frac{1}{8a}$

10. (b) As C: $(y - 3/2)^2 = 4\left(\frac{1}{32a}\right)(x + 2a)$

$\therefore$ The curve is symmetric about $y = 3/2$

Passage: C

11. (c) The parabola is $y^2 = 4x$ (so $a = 1$)

Now the equation of normals is $am^3 + 2am - mx + y = 0$

$\Rightarrow$ The equation for M (15, 12) is $m^3 + 2m - 15m + 12 = 0$

$\Rightarrow$ $(m - 3)\ \{m^2 + 3m - 4\} = 0$

$\Rightarrow$ $m = 3, 1, -4$

Now the foot of normal (am^2, –2am) for m = 3, 1, –4

The centroid of ΔPQR will be

$$\left(\frac{a(m_1^2+m_2^2+m_3^2)}{3}, \frac{-2a(m_1+m_2+m_3)}{3}\right) = \left(\frac{26}{3}, 0\right)$$

12. (c) The parabola is $y^2 = 16x$ (so $a = 4$). Now, chord PQ subtends a right angle *at* the origin where P (4, 8)

$\Rightarrow$ Slope of $OP = \frac{8}{4} = 2$

$\Rightarrow$ Slope of $OQ = -1/2$

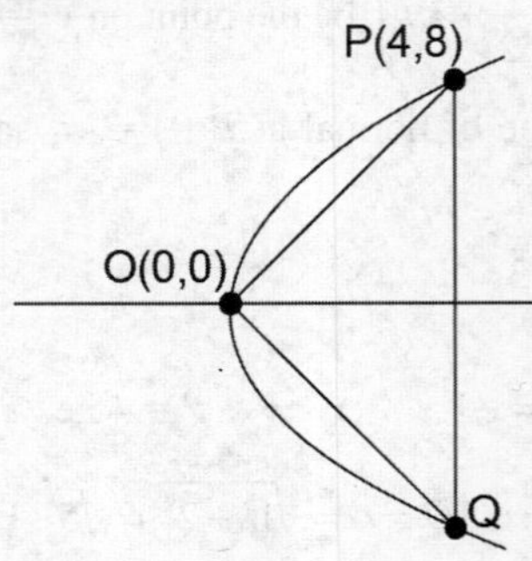

$\therefore$ Q (64, –32) {for $t_2 = -4$}

Now $t_1 = 1$ and $t_2 = -4$, $a = 4$ and the point of intersection of normals is $R(a(t_1^2 + t_2^2 + t_1t_2 + 2), -at_1t_2(t_1 - t_2))$

i.e., R (60, –48)

13. (a), (b), (c), (d) For $y^2 = 4ax$ the equation of normal with slope m is $am^3 + 2am - mx + y = 0$

$\Rightarrow$ $m_1 + m_2 + m_3 = 0$ $\Rightarrow$ $-2a\,(m_1 + m_2 + m_3) = 0$

$\Rightarrow$ $\frac{-2a(m_1+m_2+m_3)}{3} = 0$, so centroid lies on the axis (where $y = 0$)

Since the three slopes has sum zero so the fourth possible point is the origin where $m_4 = 0$

14. (a) Let $P(2at_1, at_1^2)$ and $(2at_2, at_2^2)$ be two points on the parabola $x^2 = 8y$ (so $a = 2$), then the point of intersection of the normal at P and Q will be $R(-at_1t_2(t_1 - t_2)$, $a(t_1^2 + t_2^2 + t_1t_2 + t_2))$. Since normal intersect at 90°

$\Rightarrow$ $\left(\frac{-1}{t_1}\right)\left(-\frac{1}{t_2}\right) = -1$ i.e., $t_1 = -\frac{1}{t_2}$ or $t_1t_2 = -1$

$\Rightarrow$ $R\left(a\left(t_1 - \frac{1}{t_1}\right), a\left(t_1 - \frac{1}{t_1}\right)^2 + 3a\right)$

Now, $\frac{x}{a} = \left(t_1 - \frac{1}{t_1}\right)$ and $\frac{y-3a}{a} = \left(t_1 - \frac{1}{t_1}\right)^2$

$\Rightarrow$ $\left(\frac{x}{a}\right)^2 = \frac{y-3a}{a}$ i.e., $x^2 = a\,(y - 3a)$

$\therefore$ For $a = 2$, we get $x^2 = 2\,(y - 6)$

Passage - D

15. (b) Let $B(at_1^2, 2at_1)$ and $C(at_2^2, 2at_2)$ be two points from where tangent intersect at A (at_1t_2, a (t_1+t_2)).

Now without any loss of generality consider the tangent at vertex i.e., $x = 0$

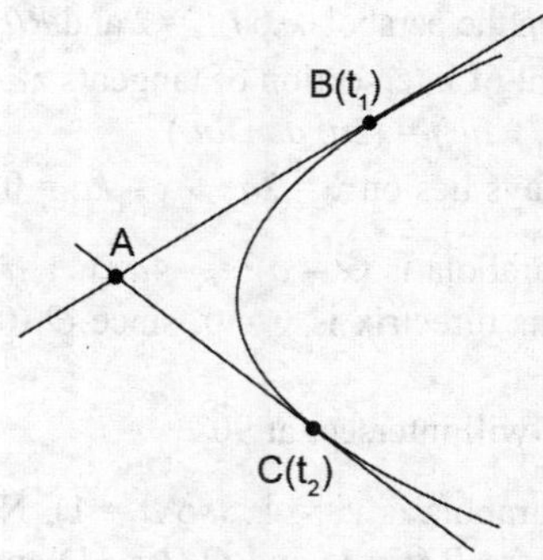

So $p_1 = at_1t_2$, $p_2 = at_1^2$ and $p_3 = at_2^2$

Now observe that $p_1^2 = p_2p_3 = a^2t_1^2t_2^2$

$\Rightarrow$ p_2, p_1, p_3 are in GP

16. (a) Let A $(at_1^2, 2at_1)$ and $B(at_2^2, 2at_2)$ be the two points on the parabola $y^2 = 4ax$, then their tangents will intersect at

$C(at_1t_2, at_1 + at_2)$. Since $a(t_1 + t_2) = \frac{1}{2}(2at_1 + 2at_2)$

$\Rightarrow$ Ordinates of A, C and B are in AP

17. (c) The parabola $y^2 = 8(x + 1)$ will have directrix as $x + 3 = 0$, similarly the parabola $y^2 = 4(x + 2)$ will have directrix as $x + 3 = 0$. And we know that tangents from the directrix are at right angles, so locus of intersection of perpendicular tangents is $x + 3 = 0$

18. (b) Let $A(t_1)$, $B(t_2)$ and $C(t_3)$ be the three points on the parabola $y^2 = 4ax$, then their point of intersection P (for tangent at t_1 and t_2) will be $(at_1t_2, at_1 + at_2)$. Similarly, we get Q and R.

Now area of

$$\Delta ABC = \frac{1}{2}\left\|\begin{matrix} at_1^2 & 2at_1 & 1 \\ at_2^2 & 2at_2 & 1 \\ at_3^2 & 2at_3 & 1 \end{matrix}\right\| = a^2|t_1 - t_2||t_2 - t_3||t_3 - t_1|$$

square units.

$$\text{Similarly area of } \Delta PQR = \frac{1}{2}\left\|\begin{matrix} at_1t_2 & at_1 + at_2 & 1 \\ at_2t_3 & at_2 + at_3 & 1 \\ at_3t_1 & at_3 + at_1 & 1 \end{matrix}\right\|$$

$$= \frac{a^2}{2}|t_1 - t_2||t_2 - t_3||t_3 - t_1|$$

So area $\Delta ABC = 2\ \Delta PQR$

19. (a) The parabola is $y^2 = 4(x + y)$ which can be rewritten as $(y - 2)^2 = 4(x + 1)$

$\Rightarrow$ The vertex is at $v(-1, 2)$ and focus is at $S(0, 2)$

A ray that is coming along $y = 4$

$\Rightarrow$ $2^2 = 4(x + 1)$ $\quad\Rightarrow x = 0$

Since a ray parallel to axes will pass through focus S

$\therefore$ The reflected ray passes through (0, 4) and (0, 2). Hence the equation $x = 0$

20. (c) The given parabola is $y^2 = 4\left(\frac{3}{2}\right)\left(x - \frac{3}{2}\right)$ which has vertex at $V(3/2, 0)$ and focus S (3, 0) and directrix is $x = 0$

Now let $P(x, y)$ be a point, then M $(0, y)$ and $\angle SMP = 60°$

$\Rightarrow$ $OM = y = 3\tan 60° = 3\sqrt{3}$

$\therefore$ $OM = 3\sqrt{3}$ $\quad\Rightarrow$ $SM = \sqrt{3^2 + 27} = 6$ units

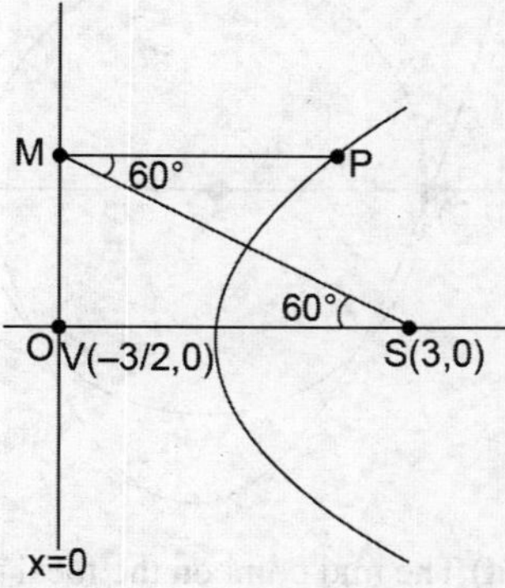

21. (b) Parabola is $y^2 = 8x$ (so $a = 2$). Observe that the point of contact $P(at^2, 2at)$ and $T(0, at)$ are the two points

$\Rightarrow$ Mid points $M\left(\frac{at^2}{2}, \frac{3}{2}at\right) = (t^2, 3t)$

Now, $x = t^2$ and $y = 3t$

$\Rightarrow$ $y^2 = 9t^2 = 9x$

Hence $L.R = 9$ units

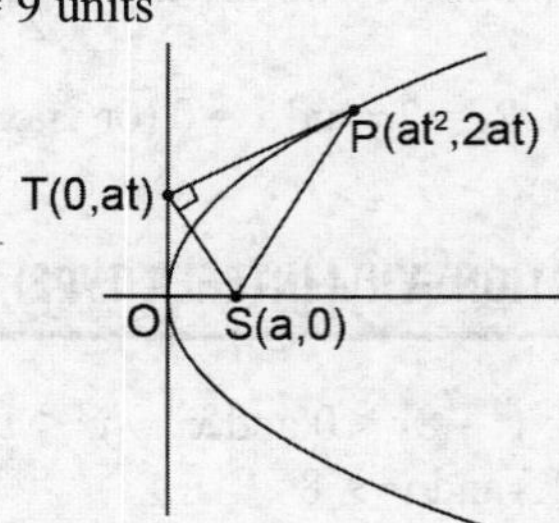

22. (c) Let $P(t_1)$, $Q(t_2)$ and $R(t_3)$ be the three points on parabola the normal at which intersect at $O(h, k)$

Now S $(a, 0)$ and $P(t_1) = (at_1^2, 2at)$

$\Rightarrow$ $|SP| = |a|\ (t_1^2 + 1)$

Similarly $SQ = |a|(t_2^2 + 1)$ and $SR = |a|(t_3^2 + 1)$

$\therefore$ $|SP.\ SQ.\ SR| = |a^3|\ (t_1^2 + 1)(t_2^2 + 1)(t_3^2 + 1) = |a^3|$ $\{1 + t_1^2 + t_2^2 + t_3^2 + t_1^2t_2^2 + t_1^2t_3^2 + t_2^2t_3^2 + t_1^2t_2^2t_3^2\}$

Now the point of intersection of normal at $P(t_1)$ and $Q(t_2)$ is $O = \left(a\left(t_1^2 + t_2^2 + t_1t_2 + 2\right)\right), -at_1t_2(t_1 + t_2))$

$$\Rightarrow SO^2 = a^2\left(1 + t_1^2 + t_2^2 + t_1t_2\right)^2 + a^2t_1^2t_2^2(t_1 + t_2)^2$$

$$= a^2\left(1 + t_2^2 + t_3^2 + t_2t_3\right)^2 + a^2t_2^2t_3^2(t_2 + t_3)^2$$

$$= a^2(1 + t_1^2 + t_3^2 + t_1t_3)^2 + a^2t_1^2t_3^2(t_1 + t_3)^2$$

Observe that $3|SP.\ SQ.\ SR| = 3a\,(SO^2)$

23. (d) Without any loss of generally, let $0 < \alpha < \beta$

Since α, β are the roots of $ax^2 + bx + c = 0$

$\Rightarrow$ $a(x-\alpha)(x-\beta) = ax^2 + bx + c$

Now the circle with $(\alpha, 0)$, $(\beta, 0)$ as a diameter is S: $(x-\alpha)(x-\beta) + y^2 = 0$

$\Rightarrow$ Length of tangent from $O(0, 0)$ is $d = \sqrt{\alpha\beta} = \sqrt{\dfrac{c}{a}}$

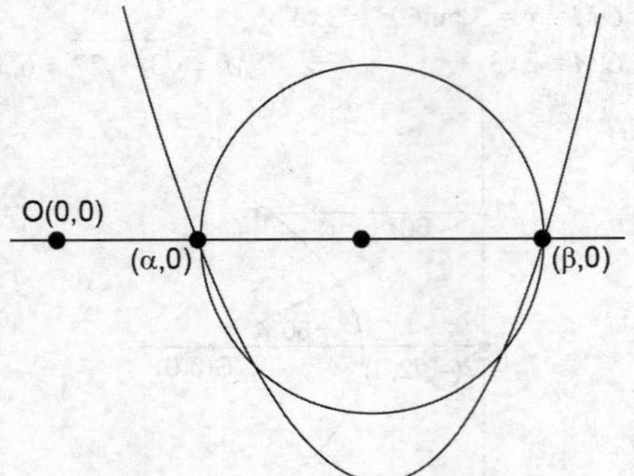

24. (a), (b), (c), (d) The mid point on the focal radius for a point $P(at^2, 2at)$ will be $M\left(\dfrac{a+at^2}{2}, at\right)$.

Hence $2x = a(t^2+1)$ and $y = at$

$\Rightarrow$ $a(2x-a) = a^2t^2 = y^2$ $\therefore$ $y^2 = 4\left(\dfrac{a}{2}\right)\left\{x - \dfrac{a}{2}\right\}$

Which is a parabola with vertex at $\left(\dfrac{a}{2}, 0\right)$ focus at $(a, 0)$;

Length of L.R. $= 2a$ and $x = 0$ (or y-axis) as its directrix

SECTION–VIII (INTEGER-TYPE)

1. From the given: $y^2 - 8x < 0$ and $x^2 + y^2 < 16$ where $x > 0$, so for $x = 1$, $y^2 < 15$ and $y^2 < 8$

$\Rightarrow$ $y = 0, \pm1, \pm2$ $\quad\therefore\quad$ 5 points

For $x = 2$, $y^2 < 12$ and $y^2 < 16$

$\Rightarrow$ $y = 0, \pm 1, \pm2, \pm3$ $\quad\therefore\quad$ 7 points

For $x = 3$, $y^2 < 7$ and $y^2 < 27$

$\Rightarrow$ $y = 0, \pm 1, \pm2$ $\quad\therefore\quad$ 5 points

Hence total 17 points

2. The equation of normal to the parabola $y^2 = 8x$ with slope m is $am^3 + 2am - mx + y = 0$ for $m = -2$, we get $2x + y - 12a = 0$.

Comparing with $2x + y + ka = 0$

$\Rightarrow$ $k = -12$

3. The given parabola $y = x - x^2$ can be rewritten as $\left(x-\dfrac{1}{2}\right)^2 = -\left(y-\dfrac{1}{4}\right)$ or $\left(x-\dfrac{1}{2}\right)^2 = -4\left(\dfrac{1}{4}\right)\left(y-\dfrac{1}{4}\right)$

And the equation of tangent with slope m will be $\left(y-\dfrac{1}{4}\right) = m\left(x-\dfrac{1}{2}\right) + am^2$

Now for $a = 1/4$ & $m = -1$, we get $x + y = 1$

$\Rightarrow x + y = $ k gives $\quad\Rightarrow\quad k = 1$

4. Let $P(t_1)$ and $Q(t_2)$ be the two points on the parabola to their normals will intersect at $R(at_1^2 + t_2^2 + t_1t_2 + 2), -at_1t_2(t_1+t_2))$

Since R lies on the parabola, so $t_1t_2 = 2$ and $R(t_3) = R(-(t_1+t_2))$

Now the point of intersection of tangents at $P(t_1)$ and $Q(t_2)$ is $T(at_1t_2, at_1 + at_2) = (2a, at_1 + at_2)$

Hence T always lies on $x = 2a \Rightarrow x - ka = 0 \Rightarrow k = 2$

5. The given parabola is $(x-a)^2 = -4a(y+a)$ with vertex at $(a, -a)$ and its directrix is $y = 0$, since $O(0, 0)$ lies on the directrix

$\therefore$ tangents will intersect at 90°

6. The given parabola is $y^2 = 4x$ (so $a = 1$). Now the ends of latus rectum are $P(t = 1)$ and $Q(t = -1)$ and we know that the other end of the normal is given by $t_2 = -t_1 - 2/t_1$

$\Rightarrow$ A (3) and A'(–3). Hence AA' $(2at_2) \times 2 = 12$ units

7. The parabola is $x^2 = -(y-4)$ and $P(0, 2)$. Now observe that the line $y = (3/4)x + 2$ passes though $P(0, 2)$

So a general point on AB (or APB) is given by $x = r\cos\theta$ and $y = 2 + r\sin\theta$ where $\cos\theta = 4/5$ and $\sin\theta = 3/5$

So $x^2 = -(y-4)$ becomes $r^2\cos^2\theta = -(r\sin\theta - 2)$ i.e., $r^2\cos^2\theta + r\sin\theta - 2 = 0$

Hence $|r_1.r_2| = \dfrac{2}{\cos^2\theta} = \dfrac{2\times5}{16} = \dfrac{25}{8} = \dfrac{k}{8}$

$\Rightarrow$ $k = 25$

8. From the symmetry $\tan 30° = \dfrac{2}{t} \Rightarrow t = 2\sqrt{3}$

Hence the *side* $= 4at = 16\sqrt{3} = k\sqrt{3}$

Hence $k = 16$

9. Observe that distance of L (1, 2) from $x - y = 0$ is $2a$

$\Rightarrow$ $2a = \dfrac{1}{\sqrt{2}}$ and $OS = a$

Now the area of ΔOLL' $= 2a^2 = \dfrac{1}{\sqrt{2}}.\dfrac{1}{2\sqrt{2}} = \dfrac{1}{4} = \dfrac{1}{k}$

$\therefore$ $k = 4$

10. The line through RQ is $x - y + a = 0$; If this line becomes a tangent to the parabola at P, then $|PR - PQ|$ will be maximum (equal to RQ)

Now for $m = 1$, the equation of tangent $y = mx + \dfrac{a}{m}$

$\Rightarrow y = x + a$ which is the same line as RQ

$\Rightarrow$ $m = 1$ is the slope of the tangent which touches the parabola at $(a, 2a)$. Hence $(k_1a, k_2a) = (a, 2a)$

$\Rightarrow$ $k_1 + k_2 = 3$

11. The parabola $x^2 + 7x + 2 = y$ can be rewritten as $\left(x+\dfrac{7}{2}\right)^2 = 4\left(\dfrac{1}{4}\right)\left(y+\dfrac{41}{4}\right)$

Now a tangent parallel to $y = 3x - 3$ will have slope $m = 3$

and its point of contact will be $x + 7/2 = 2am = 2\left(\dfrac{1}{4}\right)(3)$ (so $x = -2$) and $y + \dfrac{41}{4} = am^2 = \dfrac{9}{4}$ (so $y = -8$)

Hence the nearest point on the parabola for line $y = 3x - 3$ is $(-2, -8)$

$\Rightarrow$ $(-2, -8) = (k_1, k_2)$ $\quad\Rightarrow\quad k_1 + k_2 = -10$

12. The parabolas are $x^2 = y/k$ and $x^2 = y + 9$

$\Rightarrow \; x^2 = \dfrac{9}{1-k} \qquad \Rightarrow \; x = \pm\dfrac{3}{\sqrt{1-k}}$

Now from the symmetry about y-axis

$AB = 2a = \dfrac{6}{\sqrt{1-k}} \Rightarrow a = \dfrac{3}{\sqrt{1-k}}$ i.e., $a^2(1-k) = 0$

$\Rightarrow \; 1-k = \dfrac{9}{a^2}$ or $k = \dfrac{a^2-9}{a^2} = \dfrac{a^2-p}{a^2} \quad \therefore p = 9$

13. The parabola $y^2 = 24x$ (so $a = 6$) has a tangent as $y = 2x + 3$. Now the normal with slope $m = 2$ will be $am^3 + 2am - mx + y = 0$ as $y = 2x - 12a$ i.e., $2x - y - 72 = 0$
The tangent is $2x - y + 3 = 0$

The distance between the parallel lines is $d = \dfrac{75}{\sqrt{5}} = k\sqrt{5}$

$\Rightarrow \; k = 15$

14. Let $P(t_1)$ and $Q(t_2)$ be the two points. From the given $SP = a\left(t_1^2+1\right) = 4$ and $SQ = a\left(t_2^2+1\right) = 9$
Now the point of intersection of tangent at $P(t_1)$ and $Q(t_1)$ is $R = (at_1t_2, at_1 + at_2)$

Now distance $SR = \sqrt{a^2(t_1t_2-1)^2 + a^2(t_1+t_2)^2}$

$= \sqrt{a^2(t_1^2t_2^2 + 1 + t_1^2 + t_2^2} = \sqrt{a(t_1^2+1).a(t_2^2+1)}$

$= \sqrt{(4)(9)} = 6$

15. The parabola is $y^2 = 4ax$. Let A $(t_1) = \left(at_1^2, 2at_1\right)$ and $B(t_2) = \left(at_2^2, 2at_2\right)$

$\Rightarrow$ Slope of $OA = \dfrac{2}{t_1} = m_1$

$\Rightarrow$ Slope of $OB = \dfrac{2}{t_2} = m_2$

Now $m_1m_2 = \dfrac{4}{t_1t_2} = -1 \qquad \Rightarrow t_1t_2 = -4$

Let $OA^2 = r_1^2 = a^2(t_1^4 + 4t_1^2)$ and $OB = r_2^2 = a^2\left(t_2^4 + 4t_2^2\right)$

Since $t_2 = -4/t_1$

$\therefore \quad r_2^2 = a^2\left\{\dfrac{256}{t_1^4} + \dfrac{4(16)}{t_1^2}\right\} = \dfrac{64a^2(t_1^2+4)}{t_1^4}$

N

$r_1^{4/3}.r_2^{4/3} = (r_1^2r_2^2)^{2/3} = \left\{\dfrac{64a^4(t_1^2 + 4)^2}{t_1^2}\right\}^{2/3} = \dfrac{16a^{8/3}}{t_1^{4/3}}(t_1^2+4)^{4/3}$ W

Similarly $\left(r_1^{2/3} + r_2^{2/3}\right) = a^{2/3}t_1^{2/3}(t_1^2+4)^{1/3} + \dfrac{4a^{2/3}(t_1^2+4)^{1/3}}{t_1^{4/3}}$

$= a^{2/3}t_1^{2/3}(t_1^2+4)^{1/3}\left\{1 + \dfrac{4}{t_1^2}\right\} = \dfrac{a^{2/3}(t^2 \quad 4)^{1/3}}{t_1^{4/3}}$

$\Rightarrow \quad \dfrac{r_1^{4/3}r_2^{4/3}}{a^2(r_1^{2/3} + r_2^{2/3})} = 16$

CHAPTER 5

Ellipse

■ INTRODUCTION

In the previous chapter, we discussed the curve (parabola), which is defined as the locus of a point which moves such that the ratio (Eccentricity) of its distance from a fixed point (Focus) and a fixed line (Directrix) is equal to one. Now if this constant ratio (e) is less than one, i.e., the moving point remains nearer to the focus than the directrix, we get a curve looking like a circle but it is not exactly a circle, rather it is more like the edge of an egg. And if we plot the movement of the Earth and other planets around the Sun, we get the same curve with eccentricity less than one. This beautiful curve has been named as 'Ellipse'.

To understand the movement of Earth and other planets around Sun and other related properties, their behaviour and lot of many other physical phenomenon which in nature follow an elliptical path urged the scientists /mathematicians to analyse this curve in detail. The results of their tremendous work has resulted in the landing of human being on the beautiful Moon and many steps ahead towards the other planets and a better understanding of our Solar System and Galaxy.

■ DEFINITION OF ELLIPSE

An ellipse is the locus of a point which moves in a plane such that the ratio of its distance from a fixed point (Focus) to its distance from a fixed line (Directrix) is always constant 'e' less than one (called eccentricity of ellipse).

- Shape of ellipse: The shape of ellipse is oval as shown in figure 5.1.

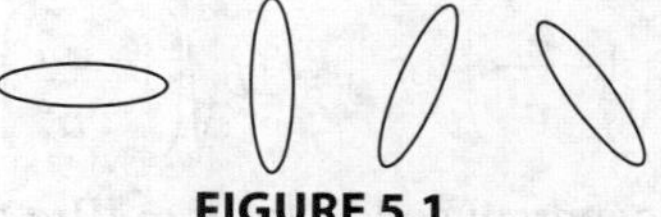

FIGURE 5.1

■ STANDARD ELLIPSE

The ellipses having their major axis (straight line along greatest chord) and minor axis (perpendicular bisector of greatest chord) along the co-ordinate axes are called standard ellipse. These are of two types:

(i) **Standard ellipse of first kind:** This type of ellipse has its major axis along x-axis and minor axis along y-axis as shown in figure 5.2.

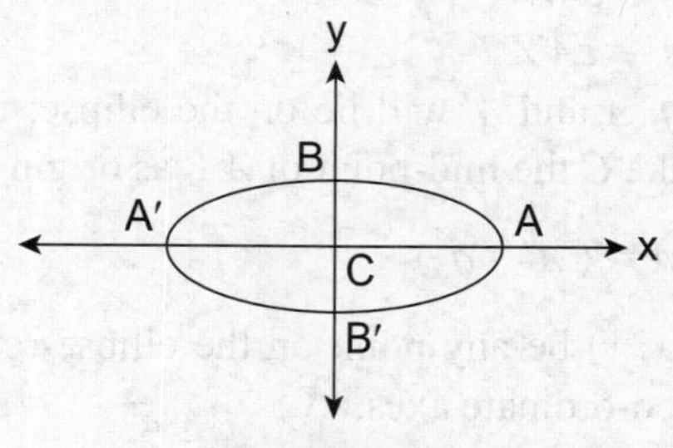

FIGURE 5.2

(ii) **Standard ellipse of second kind:** This type of ellipse has its major axis along y-axis and minor axis along x-axis as shown in figure 5.3.

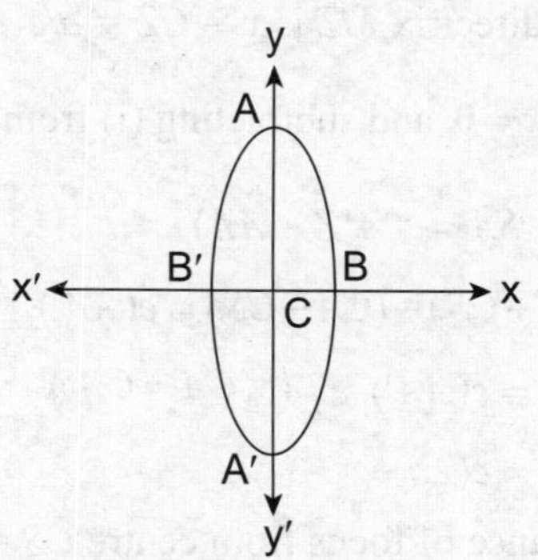

FIGURE 5.3

Equation of Standard Ellipse

Let us derive the equation of standard ellipse of first kind.

Let S be the focus and ZM the directrix of the ellipse and P any point on the ellipse such that PM be a perpendicular on directrix, then $\frac{SP}{PM} = e < 1$.

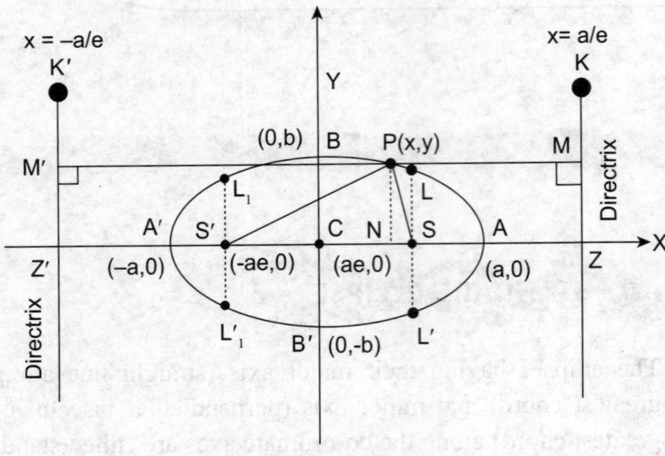

FIGURE 5.4

Draw $SZ \perp$ on directrix ZM and divide SZ internally and externally in the ratio $e : 1$ $(e < 1)$. Let A and A' be the internal and external point of division. Then we get

$SA = eAZ$ ....(i)

and $SA' = eA'Z$ ....(ii)

Clearly, A and A' will lie on the ellipse. Now, let $AA' = 2a$ and take C the mid-point of AA' as origin

$\therefore \quad CA = CA' = a$ ....(iii)

Let $P(x, y)$ be any point on the ellipse referred to CA and CB as co-ordinate axes.

Then, adding (i) and (ii), we get $SA + SA' = e(AZ + A'Z)$

$\Rightarrow \quad AA' = e(CZ - CA + CA' + CZ)$ (from figure)

$\Rightarrow \quad AA' = e(2CZ) \quad (\because CA = CA')$

$\Rightarrow \quad 2a = 2eCZ \qquad \Rightarrow CZ = a/e$

$\therefore$ The directrix MZ is $x = CZ = a/e$

or $\frac{a}{e} - x = 0$ and subtracting (i) from (ii), we get

$$SA' - SA = e(A'Z - AZ)$$

$\Rightarrow \quad (CA' + CS) - (CA - CS) = e(AA')$

$\Rightarrow \quad 2CS = e(AA') \quad (\because CA = CA')$

$\Rightarrow \quad 2CS = e(2a)$

$\therefore$ Distance of focus from centre $CS = ae$

$\therefore$ The focus S is $(CS, 0)$ i.e., $(ae, 0)$

Now draw $PM \perp MZ$.

From definition $\frac{SP}{PM} = e \Rightarrow (SP)^2 = e^2(PM)^2$

$\Rightarrow \quad (x - ae)^2 + (y - 0)^2 = e^2\left(\frac{a}{e} - x\right)^2$

$\Rightarrow \quad (x - ae)^2 + y^2 = (a - ex)^2$

$\Rightarrow \quad x^2 + a^2e^2 - 2aex + y^2 = a^2 - 2aex + e^2x^2$

$\Rightarrow \quad x^2(1 - e^2) + y^2 = a^2(1 - e^2)$

$\Rightarrow \quad \frac{x^2}{a^2} + \frac{y^2}{a^2(1 - e^2)} = 1$

i.e., $\frac{x^2}{a^2} + \frac{y^2}{b^2} = 1$ where $b^2 = a^2(1 - e^2) = (OB)^2$

This is the standard equation of an ellipse of first kind. Similarly, we can derive the equation of ellipse of second kind, given by

$\frac{x^2}{b^2} + \frac{y^2}{a^2} = 1$; $CA = a$ and $CB = b$

■ TERMS RELATED TO STANDARD ELLIPSE

(i) Major axis: AA' is called major axis and its length is taken to be $2a$ i.e., $AA' = 2a$.

(ii) Minor axis: BB' is called minor axis and its length is taken to be $2b$ i.e., $BB' = 2b$.

(iii) Centre of ellipse: The point of intersection of major and minor axes is called centre of ellipse and denoted by C. In standard ellipses centre is at origin.

(iv) Vertices of ellipse: The points of intersection of major axis with the curve (i.e., ellipse) are called vertices of ellipse i.e., A and A'.

(v) Foci: Due to symmetry of ellipse about the axes there is second focus S'. Thus S and S' are two foci of ellipse which lie on major axis at a distance of **'ae'** on either side of centre of ellipse.

(vi) Directrices: Due to symmetry of ellipse about the axes there is second directrix $Z'K'$. Thus ZK and $Z'K'$ are two directrices which are perpendicular to major axis and are at a distance of a/e on either side of centre of ellipse.

(vii) Double ordinates: If from a point P on the ellipse a perpendicular PN is drawn to the x-axis and produced to meet the curve at a point P', then |PP'| is called double ordinate of ellipse. If abscissa of P is x, then for standard ellipse of first kind ordinate of P is given by

$y^2 = b^2\left(1 - \frac{x^2}{a^2}\right)$ or $y = \pm b\sqrt{\left(1 - \frac{x^2}{a^2}\right)}$ ('+' for P in Ist or IInd quadrant and '–' for P in IIIrd or IVth quadrant)

and hence ordinate of P' are given by $y = \mp b\sqrt{\left(1-\frac{x^2}{a^2}\right)}$ such that $(y_P y_{P'}) < 0$.

$\therefore$ Double ordinate $= |PP'| = 2\frac{b}{a}\sqrt{a^2 - x^2}$

Similarly for second kind of ellipse double ordinate $= 2\frac{a}{b}\sqrt{b^2 - x^2}$

(viii) Latus rectum: The chord of ellipse through any one of foci and perpendicular to major axis of ellipse is called latus rectum. There are two latus rectum of ellipse LL' and L_1L_1'

Let $LL' = 2k$, then $LS = L'S = k$

Co-ordinates of L and L' are (ae, k) and $(ae, -k)$ and both lie on the ellipse $\frac{x^2}{a^2} + \frac{y^2}{b^2} = 1$

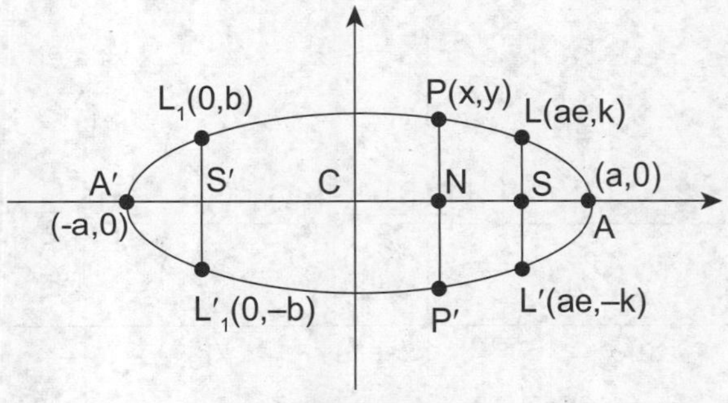

FIGURE 5.5

$$\Rightarrow \frac{a^2e^2}{a^2} + \frac{k^2}{b^2} = 1 \qquad \Rightarrow k^2 = b^2(1-e^2)$$

$$= b^2\left(\frac{b^2}{a^2}\right) \quad [\because b^2 = a^2(1-e^2)]$$

$$\therefore \quad k = \frac{b^2}{a} \quad (\because k > 0)$$

$$\Rightarrow 2k = \frac{2b^2}{a} = LL'$$

$\therefore$ Length of latus rectum $LL' = L_1L_1' = \frac{2b^2}{a}$ and end points of latus rectum are $L \equiv \left(ae, \frac{b^2}{a}\right); L' \equiv \left(ae, -\frac{b^2}{a}\right)$ and $L_1 \equiv \left(-ae, \frac{b^2}{a}\right); L_1' \equiv \left(-ae, -\frac{b^2}{a}\right)$ respectively for first kind of standard parabola.

(ix) Focal chord: A chord of the ellipse passing through its focus is called a focal chord. To explain this, the line segment joining any two points on the ellipse and passing through focus. Thus an ellipse has infinitely many focal chords, the longest one is major axis and the focal chord perpendicular to major axis is latus rectum.

REMARKS

1. *There exists a second focus and second directrix for the curve. On the negative side of the major axis take a point S', which is such that SC = S'C = ae and another point Z' such that ZC = CZ' = a/e. Draw Z'K' perpendicular to ZZ' and PM' perpendicular to Z'K'. The equation may also be written in the form* $(x + ae)^2 + y^2 = (a + ex)^2$

   $\Rightarrow (S'P)^2 = e^2 (PM')^2$

   Hence, any point P on the curve is such that its distance from S' is e times to its distance from Z'K', so we should have obtained the same curve, if we had started with S' as focus, Z'K' as directrix and the same eccentricity. We have considered a > b, now if we consider b > a, what will be the shape of the ellipse $\frac{x^2}{a^2} + \frac{y^2}{b^2} = 1$? *In this case, the major axis AA' of the ellipse is along the y-axis and is of length 2b. The minor axis along x-axis and of length of BB' = 2a. The foci S and S' are (0, be) and (0, –be) respectively. The directrices are MZ and M'Z' given by* $y \pm \frac{b}{e} = 0$, *respectively. Also here* $a^2 = b^2(1 - e^2)$.

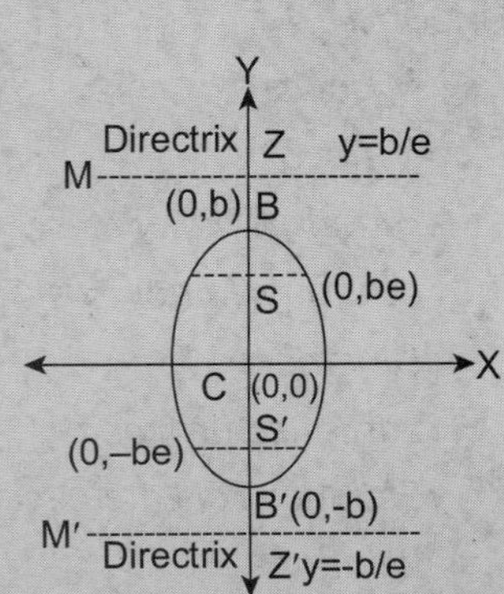

FIGURE 5.6

2. *If we always consider a > b, then the major axis is always of length = AA' = 2a and the minor axis is always having length = BB' = 2b and centred at origin, then the equation of this standard ellipse will be be* $\frac{x^2}{a^2}+\frac{y^2}{b^2}=1 (b<a)$ *and* $b^2 = a^2(1 - e^2)$*. If major axis is along y-axis and minor axis along x-axis, then it will be* $\frac{x^2}{b^2}+\frac{y^2}{a^2}=1$*, and* $b^2 = a^2(1 - e^2)$*; b < a. The above equations of ellipse are called standard ellipse of first kind and second kind respectively.*

3. *If P* (x_1, y_1) *be any point. This point lies outside, on or inside the ellipse according as* $\frac{x_1^2}{a^2}+\frac{y_1^2}{b^2}-1>0$ *or 0 or < 0.*

4. *Since for an ellipse,* $0<e<1$

$$\Rightarrow 0<e^2<1 \Rightarrow -1<-e^2<0 \Rightarrow 0<1-e^2<1 \text{ or } 0<a^2(1-e^2)<a^2 \Rightarrow b^2<a^2 \text{ i.e., } b<a$$

5. *If e = 0, then* $b^2 = a^2(1 - 0)$ $\quad\therefore b^2 = a^2$

then equation of ellipse $\frac{x^2}{a^2}+\frac{y^2}{b^2}=1$ *changes in circle, i.e.,* $x^2+y^2=a^2$

6. *Since* $\frac{x^2}{a^2}+\frac{y^2}{b^2}=1$

$$\Rightarrow \frac{y^2}{b^2}=1-\frac{x^2}{a^2}=\frac{a^2-x^2}{a^2}$$

$$\Rightarrow \frac{y^2}{b^2}=\frac{(a+x)(a-x)}{a^2}$$

$$\Rightarrow \frac{(PN)^2}{b^2}=\frac{A'N.AN}{a^2}$$

$$\Rightarrow \frac{(PN)^2}{AN.A'N}=\frac{b^2}{a^2}=\frac{(BC)^2}{(AC)^2}$$

i.e., $(PN)^2 : AN.A'N :: (BC)^2 : (AC)^2$.

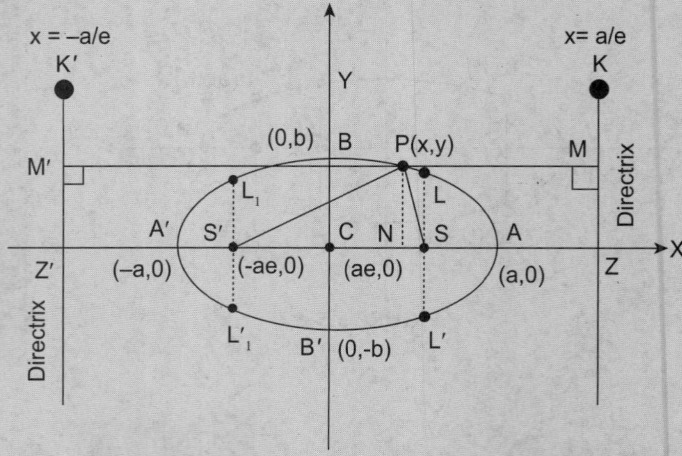

FIGURE 5.7

Table represents the related terms to standard ellipse.

	Standard Ellipse of first kind (a > b)	Standard ellipe of second kind (a > b)
Vertices	*A(a, 0); A'(–a, 0)*	*A(0, a); A'(0, –a)*
Foci	*S(ae, 0); S'(–ae, 0)*	*S(0, ae); S'(0, –ae)*
Major axis	*x-axis (y = 0)*	*y-axis (x = 0)*
Minor axis	*y-axis (x = 0)*	*x-axis (y = 0)*
Length of Major axis	*AA' = 2a*	*AA' = 2a*
Length of Minor axis	*BB' = 2b*	*BB' = 2b*
Dirictrices	*ZK : x = a/e; Z'K' : x = –a/e*	*ZK : y = a/e; Z'K' : y = –a/e*
Latus Rectum Length	$LL' = L_1L_1' = 2b^2/a$	$LL' = L_1L_1' = 2b^2/a$
Equations of Latera recta	$LL' : x = ae,\ L_1L_1' : x = -ae$	$LL' : y = ae,\ L_1L_1' : y = -ae$
Relation	$b^2 = a^2(1 - e^2)$	$b^2 = a^2(1 - e^2)$
Distance between foci	*SS' = 2ae*	*SS' = 2ae*
Distance between latera recta	*ZZ' = 2a/e*	*ZZ' = 2a/e*

FOCAL DISTANCE OF A POINT ON ELLIPSE

The focal distance of a point on ellipse is the distance of the point from one of the foci of the ellipse. i.e., SP and $S'P$ are focal distances of point P on the ellipse.

SECOND DEFINITION OF ELLIPSE

Let the equation of ellipse be $\frac{x^2}{a^2}+\frac{y^2}{b^2}=1$ ... (i)

Its foci S and S' are $(ae, 0)$ and $(-ae, 0)$. The equation of its directrices MZ and $M'Z'$ are $x = a/e$ and $x = -a/e$, respectively. Let $P(x, y)$ be any point on (i).

Now $SP = e\,PM = e\,NZ = e(CZ - CN)$

$$= e\left[\left(\frac{a}{e}\right)-x\right] = a - ex$$

and $S'P = e\,PM' = e(Z'N) = e\,(CZ' + CN)$

$$= e\left[\left(\frac{a}{e}\right)+x\right] = a + ex$$

$\therefore\quad SP + S'P = 2a = AA'$

So by this property an ellipse can also be defined as "the locus of a point which moves on a plane such that the sum of its distances from two fixed point is always constant." The fixed points are foci and the constant distance is length of major axis i.e., $2a$, length of greatest chord.

MECHANICAL CONSTRUCTION OF AN ELLIPSE

Let S and S' be two drawing pins and consider an inextensible string whose ends be tied at S and S' and having length equal to the sum of SP and $S'P$ i.e., $2a$ where P is point of pencil.

The point of pencil is being moved on paper keeping the string always tight. Satisfying these conditions point P of pencil will trace out a curve on the paper. This curve is an ellipse. Hence, the locus of the point of pencil is an ellipse as shown in figure 5.8.

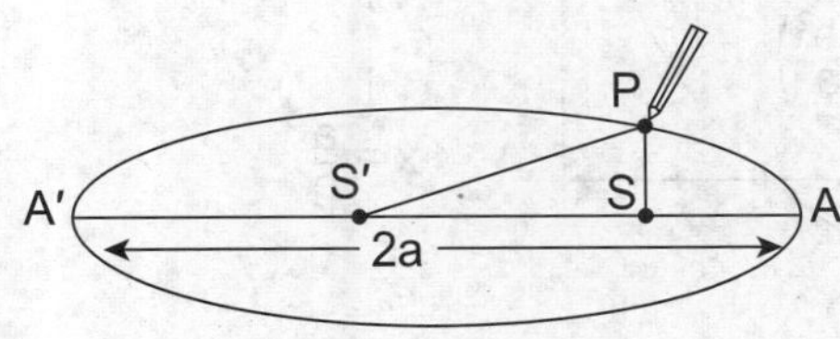

FIGURE 5.8

TRACING OF THE ELLIPSE

Equation of the ellipse is $\frac{x^2}{a^2}+\frac{y^2}{b^2}=1$ ..... (i)

1. **Intersection points of curve with co-ordinate axes:** The ellipse in figure 5.8 cuts x-axis at $A\,(a, 0)$ and $A'\,(-a, 0)$ and cuts y-axis at $B(0, b)$ and $B'\,(0, -b)$. AA' is major axis of the ellipse and BB' is called, minor axis of the ellipse.

2. **Symmetricity:** Since the equation does not change when y is replaced by $-y$ and x is replaced by $-x$, hence, ellipse is symmetrical about both axes that is why it has two foci, two directrices etc. from (i) $y=\pm b\sqrt{1-\frac{x^2}{a^2}}$ shows that for every value of x, y has two values one +ve and other –ve.

3. **Bounding region:** The equation may be written in either of the form

$$y=\pm b\sqrt{\left(1-\frac{x^2}{a^2}\right)} \quad \text{..... (ii)}$$

or $$x=\pm a\sqrt{\left(1-\frac{y^2}{b^2}\right)} \quad \text{..... (iii)}$$

From equations (ii) and (iii), we obtain that for y to be real,

$$1-\frac{x^2}{a^2}\ge 0 \Rightarrow a^2-x^2\ge 0$$

$$\Rightarrow\ x^2\le a^2 \Rightarrow -a\le x\le a$$

Similarly, for x to be real, $1-\frac{y^2}{b^2}\ge 0 \Rightarrow b^2-y^2\ge 0$.

$$\Rightarrow\ y^2\le b^2 \Rightarrow -b\le y\le b$$

$\Rightarrow$ Ellipse is a closed curve lying entirely within a rectangle having sides $x = a; x = -a; y = b$ and $y = -b$.

4. **Tangents parallel to co-ordinate axes:**

Here, $y^2=\frac{b^2}{a^2}(a^2-x^2)$

$$\Rightarrow\ 2y\frac{dy}{dx}=\frac{b^2}{a^2}(-2x)$$

$$\Rightarrow\ \frac{dy}{dx}=\frac{-x}{y}\frac{b^2}{a^2}$$

$$\therefore\ \frac{dy}{dx}=0\Rightarrow x=0$$

$$\Rightarrow y=\pm b$$

$\therefore$ $B(0, b)$ and B' $(0, -b)$ are the points on ellipse where the tangents are parallel to x-axis and $\frac{dy}{dx} \to \infty \Rightarrow y \to 0$

$\Rightarrow x = \pm a$

$\Rightarrow$ $A'(-a, 0)$ and $A(a, 0)$ are the points where the tangents are parallel to y-axis.

5. **Monotonicity:** If y increasses w.r.t. 'x', then

$$\frac{dy}{dx} > 0 \Rightarrow \frac{-x}{y}\frac{b^2}{a^2} > 0$$

$$\Rightarrow \frac{x}{y}\frac{b^2}{a^2} < 0$$

$\Rightarrow x/y < 0$

$\Rightarrow$ x and y must be of opposite signs.

Consequently, curve is an increasing function in 2nd and 4th quadrants, where as decreasing function in 1st and 3rd quadrant.

6. **Curvature:** $\frac{d^2y}{dx^2} = \frac{-b^2}{a^2}\left(\frac{y - x\frac{dy}{dx}}{y^2}\right)$

$$= \frac{b^2}{a^2y^2}\left(x\frac{dy}{dx} - y\right)$$

$$\therefore \frac{d^2y}{dx^2} > 0$$

$$\Rightarrow x\frac{dy}{dx} > y \Rightarrow \frac{-x^2}{y}\frac{b^2}{a^2} > y$$

$$\Rightarrow \frac{x^2}{y}\frac{b^2}{a^2} + y < 0 \quad \Rightarrow \frac{b^2x^2 + a^2y^2}{a^2y} < 0$$

$\Rightarrow y < 0$

$\Rightarrow$ Curve is concave upwards below x-axis and concave downwards above the x-axis.

$\therefore$ From above information regarding points of intersection of curve and co-ordinate axes, symmtricity, monotonicity, curvature, tangents parallel to co-ordinate axes and ploting some points (x, y) on $x - y$ plane, we can easily trace the ellipse and it will be as shown in figure 5.9.

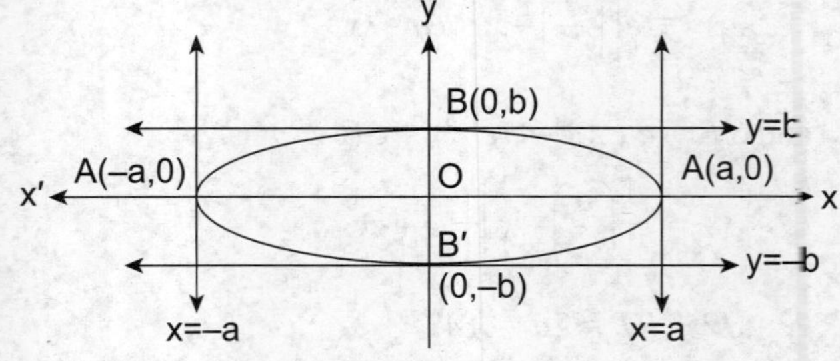

FIGURE 5.9

■ BASIC TERMINOLOGY AND ELLIPSE AT A GLANCE

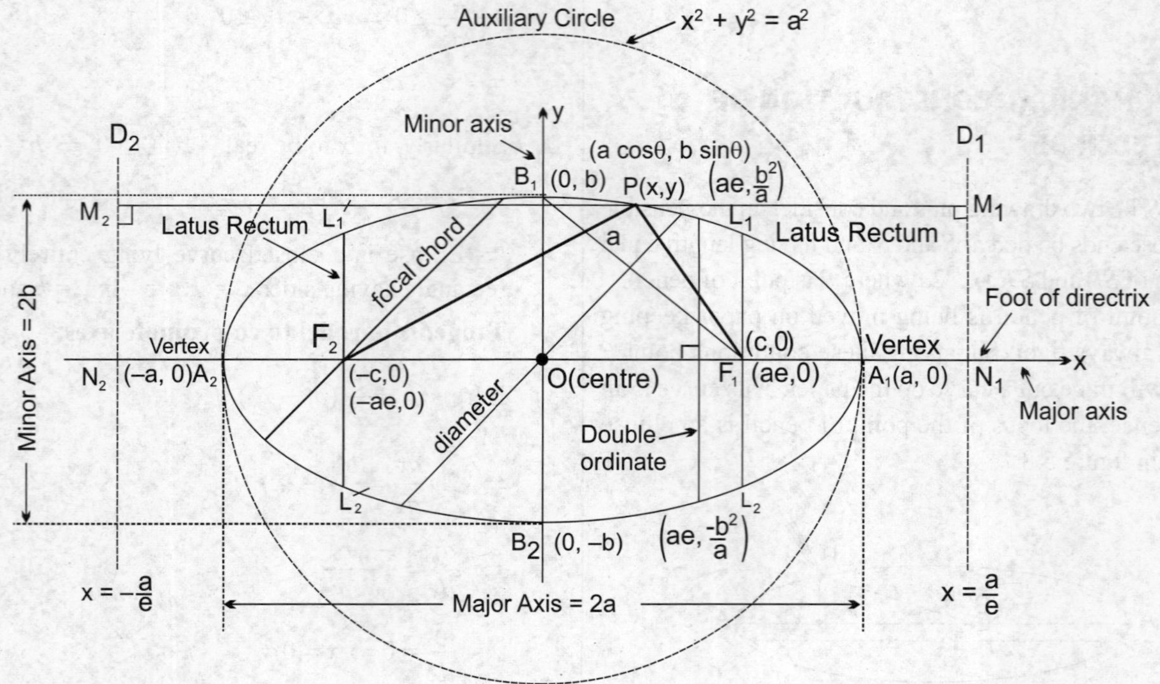

FIGURE 5.10

- The line containing the two fixed points (called 'foci') is called the focal axis and the points of intersection of the curve with focal axis are called the ***vertices*** of the ellipse A_1(a, 0) and A_2 (– a, 0). The distance between F_1 and F_2 is called the ***focal length***.
- Distance between the two vertices is $2a$ called the length of major axis. The distance between B_1B_2 is $2b$ called the minor axis.
- Point of intersection of the major and minor axis is called the centre of the ellipse. Any chord of the ellipse passing through it gets bisected by it and is called the diameter. Major and minor axes together are known as principal axes of the ellipse.
- Any chord through focus is called a focal chord and any chord perpendicular to the focal axis is called double ordinate, provided the focal axis is x-axis.
- Focal chord perpendicular to focal axis is called its ***latus rectum.***
- **Eccentricity:** Degree of flatness of ellipse defined as

$$e = \frac{c}{a} = \frac{OF_1}{OA_1} = \frac{\text{distance from centre to focus}}{\text{distance from centre to vertex}}$$

$$\therefore\ e2 = \frac{c^2}{a^2} = \frac{a^2 - b^2}{a^2} = 1 - \frac{b^2}{a^2} < 1;$$

Now if $c \to 0$ (i.e., the two foci comes closer and closer to form the centre $e \to 0 \Rightarrow b \to a$.

Hence ellipse gets thicker and tends to become a circle. Again if $c \to a$ (i.e., the two foci tends to coincide with the vertex of the ellipse), we have, $e \to 1$ $\Rightarrow b \to 0$

Hence ellipse gets thinner and thinner and tends to a line segment between the two foci.

Equation of ellipse in terms of eccentricity becomes

$$\frac{x^2}{a^2} + \frac{y^2}{a^2(1-e^2)} = 1$$

REMARKS

1. *Two ellipses are said to be similiar, if they have the same value of eccentricity.*
2. $b^2 = a^2(1 - e^2)$ *or* $a^2e^2 = a^2 - b^2$

   (i.e., distance of every focus from the extremity of minor axis is equal to a)

   The two foci are $(\pm ae, 0)$; *putting* $x = ae$ *in* $\frac{x^2}{a^2} + \frac{y^2}{b^2} = 1$, *we get*

   $$\frac{y^2}{b^2} = 1 - e^2 = 1 - \left(1 - \frac{b^2}{a^2}\right) = \frac{b^2}{a^2}; y = \pm\frac{b^2}{a};$$

   $\therefore$ *Co-ordinate of the extremities of LR* $= \pm ae, \pm$ —;

   and length of $LR = \frac{2b^2}{a} = \frac{4b^2}{2a} = \frac{(\text{minor axis})^2}{\text{major axis}}$
3. *The length of the latus rectum can alternatively be expressed as*

   $$L_1L_2 = \frac{2b^2}{a} = \frac{2a^2(1-e^2)}{a} = 2a(1 - e^2) = 2e\left(\frac{a}{e} - ae\right) = 2e\,(ON_1 - OF_1)$$

   =2(distance between focus and corresponding foot of the directrix)

ILLUSTRATION 1: Find the equation of the ellipse, which cuts intercepts of length 3 and 2 on positive x and y-axis. Centre of the ellipse is origin and major and minor axes are along the positive x-*axis* and along positive y-axis.

SOLUTION: Let the equation of the ellipse be $\frac{x^2}{a^2} + \frac{y^2}{b^2} = 1$ . . . (i)

According to the given condition the ellipse (i) passes through (3, 0) and (0,2), so we have

$$\frac{9}{a^2} = 1 \Rightarrow a^2 = 9$$

and $\frac{4}{b^2} = 1 \Rightarrow b^2 = 4$

Therefore, the equation of the ellipse is $\frac{x^2}{9} + \frac{y^2}{4} = 1$.

ILLUSTRATION 2: If minor-axis of ellipse subtend a right angle at its focus, then find the eccentricity of ellipse.

SOLUTION: Let the equation of ellipse be $\frac{x^2}{a^2} + \frac{y^2}{b^2} = 1$; $(a > b)$

Given $\angle BSB' = \frac{\pi}{2}$

And OB = OB' = ae

$\Rightarrow$ ΔBSB' is isosceles with $\angle S = \frac{\pi}{2}$

$\Rightarrow \angle OBS = \frac{\pi}{4} \Rightarrow \angle BSO = \frac{\pi}{4}$

$\Rightarrow OS = OB \Rightarrow ae = \text{b}$

$\Rightarrow e^2 = \frac{b^2}{a^2} = 1 - e^2 \Rightarrow e = \frac{1}{\sqrt{2}}$

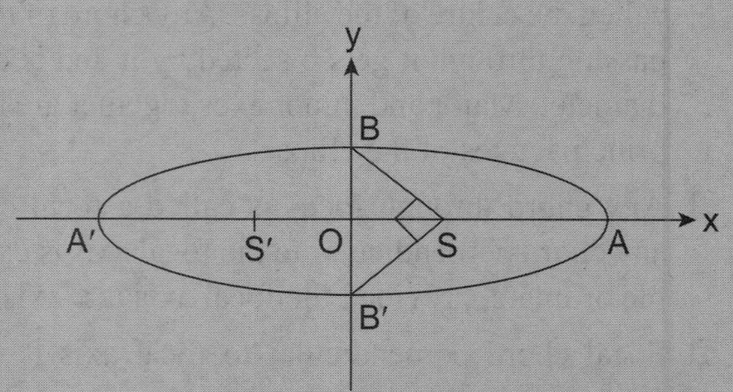

FIGURE 5.11

ILLUSTRATION 3: Find the equation of ellipse having its foci at (±4,0) and eccentricity 1/2.

SOLUTION: Clearly, foci lies on major axis i.e., on x-axis.

$\therefore$ $\text{foci} \equiv (\pm ae, 0) \equiv (\pm 4, 0)$

$\Rightarrow ae = 4$; $e = 1/2$ (given) $\Rightarrow a\frac{1}{2} = 4 \Rightarrow a = 8$

Also, we have relation $b^2 = a^2(1 - e^2)$

$\Rightarrow b^2 = 64\left(1 - \frac{1}{4}\right) = 64\left(\frac{3}{4}\right) = 48$ $\therefore$ $b^2 = 48$

$\therefore$ Equation of ellipse is $\frac{x^2}{64} + \frac{y^2}{48} = 1$

ILLUSTRATION 4: Find the equation of ellipse having one of its vertices at (–4,0); one of its foci at (–1,0) and centred at origin.

SOLUTION: Clearly, vertices and foci lie on x axis.

$\Rightarrow$ Major axis is x-axis and centre of ellipse is at origin

$\therefore$ $A'(-a, 0) \equiv (-4, 0)$ and $S'(-ae, 0) \equiv (-1, 0)$

$\Rightarrow a = 4$, $ae = 1$

$\Rightarrow 4e = 1 \Rightarrow e = 1/4$

Also, $b^2 = a^2(1 - e^2) = 16\left[1 - \frac{1}{16}\right] = 16 - 1 = 15$

$\Rightarrow a^2 = 16$; $b^2 = 15$

$\therefore$ Equation of ellipse will be $\frac{x^2}{16} + \frac{y^2}{15} = 1$.

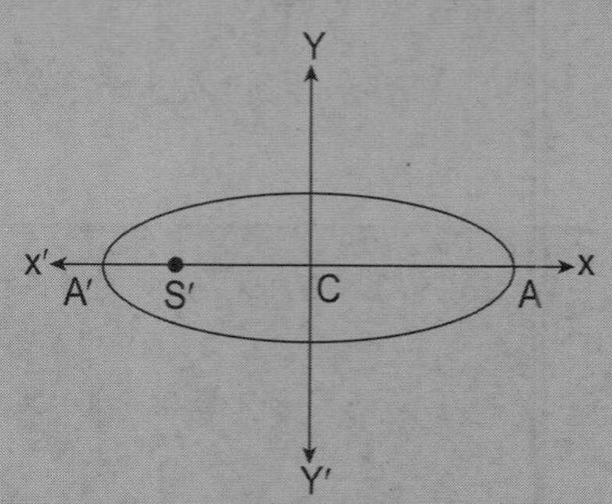

FIGURE 5.12

ILLUSTRATION 5: Find the equation of ellipse having its length of latus rectum 8 and axes along the co-ordinate axes and eccentricity = 1/2.

SOLUTION: Length of latus rectum = $\dfrac{2b^2}{a} = 8$

$\Rightarrow \; b^2 = 4a$

By relation $b^2 = a^2(1 - e^2)$, we have $4a = a^2\left[1-\dfrac{1}{4}\right]$

$\Rightarrow \; 4a = \dfrac{3}{4}\,a^2 \Rightarrow 3a^2 - 16a = 0$

$\Rightarrow \; a(3a - 16) = 0 \Rightarrow a = 0$ or $a = 16/3$

Rejecting zero value, we have $a = 16/3 \Rightarrow a^2 = \dfrac{256}{9}$.

$\therefore \; b^2 = 4\left(\dfrac{16}{3}\right) = 64/3$

$\therefore$ Equation of ellipse would be $\dfrac{x^2}{(256/9)} + \dfrac{y^2}{64/3} = 1$

$\Rightarrow \; \dfrac{9x^2}{256} + \dfrac{3y^2}{64} = 1 \;\Rightarrow\; 9x^2 + 12y^2 = 256$

ILLUSTRATION 6: Find the length of major and miror axis eccentricity, foci, equations of directrices, length of latus rectum, equations of latus rectum, centre of ellipse and end points of latera recta for the ellipse $25x^2 + 49y^2 = 1225$.

SOLUTION: Given equation is $25x^2 + 49y^2 = 1225$

$\Rightarrow \; \dfrac{25x^2}{1225} + \dfrac{49y^2}{1225} = 1$

$\Rightarrow \; \dfrac{x^2}{49} + \dfrac{y^2}{25} = 1$ ....(i)

which is standard ellipse of first kind. Compairing (i) with $\dfrac{x^2}{a^2} + \dfrac{y^2}{b^2} = 1$, we have $a^2 = 49$, $b^2 = 25$

$\Rightarrow \; a = 7, b = 5$

By relation $b^2 = a^2\,(1 - e^2)$, we have

$25 = 49\,(1 - e^2)$

$\Rightarrow \; \dfrac{25}{49} = 1 - e^2 \Rightarrow e^2 = 1 - \dfrac{25}{49} \Rightarrow e = \sqrt{\dfrac{24}{49}}$

$\Rightarrow \; e = \dfrac{2\sqrt{6}}{7}$

$\therefore$ Eccentricty $= \dfrac{2\sqrt{6}}{7}$

Length of major axis = $2a$ = 14 units

Length of minor axis = $2b$ =10 units

equation of major axis (x-axis) is $y = 0$

equation of minor axis (y-axis) is $x = 0$

Foci are given by $(\pm ae, 0) \equiv \left(\pm 7\left(\dfrac{2\sqrt{6}}{7}\right), 0\right) \equiv \left(\pm 2\sqrt{6}, 0\right)$

Equations of directrices are given by $x = \pm\frac{a}{e} \Rightarrow \frac{x = \pm 7}{\left(\frac{2\sqrt{6}}{7}\right)} = \pm\frac{49}{2\sqrt{6}}$

$\Rightarrow \quad x = \pm\frac{49}{2\sqrt{6}}$ or $2\sqrt{6}x \mp 49 = 0$

Length of latera recta $= \frac{2b^2}{a} = \frac{2(25)}{7} = \frac{50}{7}$ units

Centre of ellipse is (0,0).

End points of latus recta are $\left(\pm ae, \pm\frac{b^2}{a}\right) = \left(\pm 2\sqrt{6}, \pm\frac{25}{7}\right)$.

ILLUSTRATION 7: The tangent at any point P of a circle meets the tangent at a fixed point A in T and T is joined to B, the other end of the diameter through A, prove that the locus of the intersection AP and BT is an ellipse whose eccentricity is $\frac{1}{\sqrt{2}}$.

SOLUTION: Let the fixed point A be $(a, 0)$ and B be $(-a, 0)$ and P $(a \cos\theta, a \sin\theta)$ be any point on circle. Then, equation of tangent to circle at point P will be $ax \cos\theta + ay \sin\theta = a^2$

or $\quad x \cos\theta + y \sin\theta = a$ .... (1)

$\therefore$ Co-ordinates of T would be $\left(a, \frac{a - a\cos\theta}{\sin\theta}\right) = \left(a, a\tan\frac{\theta}{2}\right)$

Now equation of TB would be $(y - 0) = \frac{a\tan\frac{\theta}{2}}{2a}(x + a)$

$\Rightarrow \quad y = \frac{1}{2}(x + a)\tan\frac{\theta}{2}$ .... (2)

Also, equation of AP would be $(y - 0) = \frac{a\sin\theta}{a\cos\theta - a}(x - a)$

$\Rightarrow \quad y = \frac{\sin\theta}{(\cos\theta - 1)}(x - a)$ or $y = -(x - a)\cot\frac{\theta}{2}$ ....(3)

From (2) $\tan\frac{\theta}{2} = \frac{2y}{(x + a)}$

From (3) $\cot\frac{\theta}{2} = \frac{-(y)}{x - a}$

$\Rightarrow \quad \tan\frac{\theta}{2} = \frac{a - x}{y}$

$\therefore \quad \frac{2y}{x + a} = \frac{a - x}{y}$

$\Rightarrow \quad 2y^2 = a^2 - x^2 \Rightarrow x^2 + 2y^2 = a^2$

$\Rightarrow \quad \frac{x^2}{a^2} + \frac{y^2}{a^2/2} = 1,$

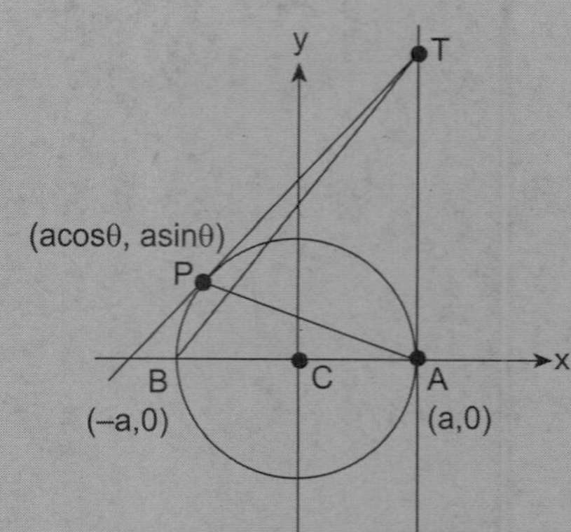

FIGRUE 5.13

which is an elllipse of the form $\frac{x^2}{A^2} + \frac{y^2}{B^2} = 1; A^2 = a^2, B^2 = \frac{a^2}{2}$

Now $B^2 = A^2(1-e^2)$

$\Rightarrow \quad \frac{a^2}{2} = a^2\left(1 - e^2\right) \Rightarrow \frac{1}{2} = 1 - e^2 \qquad \Rightarrow \quad e^2 = \frac{1}{2} \Rightarrow e = \frac{1}{\sqrt{2}};$

■ NON-STANDARD ELLIPSE WITH THEIR AXES PARALLEL TO CO-ORDINATES AXES AND CENTRE NOT AT ORIGIN

The ellipses of these types are classified in to two categories

1. **First kind:** This type of ellipse has its major axis parallel to x-axis and minor axis parallel to y-axis and centre at (h, k) other than origin and has its equation.

$\frac{(x-h)^2}{a^2}+\frac{(y-k)^2}{b^2}=1; a>b$ where (h, k) is the centre of ellipse.

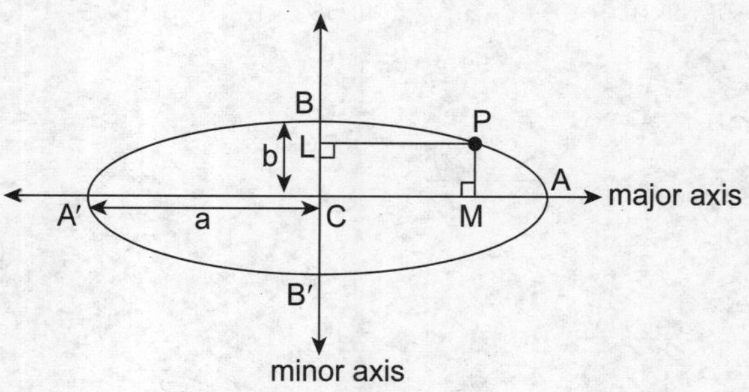

FIGURE 5.14

If P is any point on ellipse with AA' and BB' as major and minor axis, respectively, and PM and PL are perpendiculars on major and minor axis respectively, then we have identity.

$$\frac{(PL)^2}{a^2}+\frac{(PM)^2}{b^2}=1 \quad \text{.... (i)}$$

Let the centre be $C'(h, k)$ and the lengths of major and minor axes $2a$ and $2b$ respectively as shown in figure 5.15.

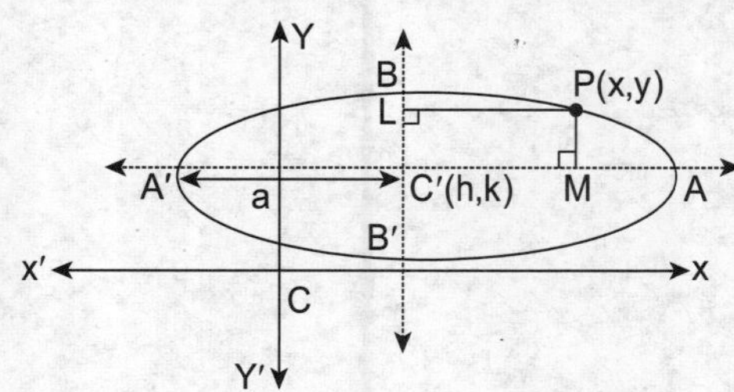

FIGURE 5.15

Now in the given case:

$PL = |x-h|$ and $PM = |y-k|$

$\therefore$ Equation (i) becomes $\frac{(x-h)^2}{a^2}+\frac{(y-k)^2}{b^2}=1$.

2. **Second kind:** This type of ellipse has its major axis parallel to y-axis and minor axis parallel to x-axis and centre at (h, k) other than origin and has its equation $\frac{(x-h)^2}{b^2}+\frac{(y-k)^2}{a^2}=1$; where (h, k) is the centre of ellipse and $a>b$.

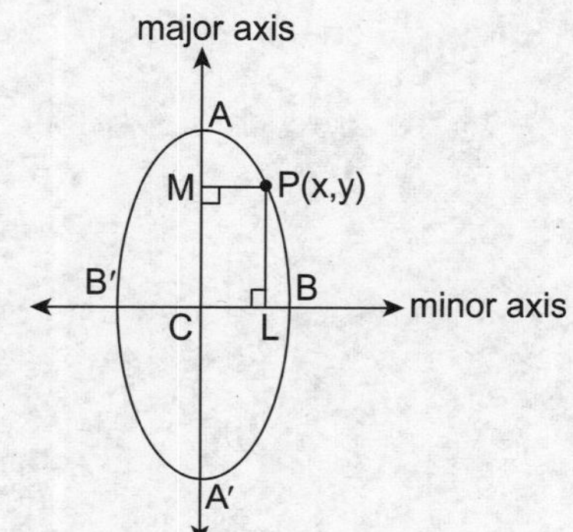

FIGURE 5.16

Here, $\frac{(PL)^2}{a^2}+\frac{(PM)^2}{b^2}=1$

$\therefore \quad \frac{|y-k|^2}{a^2}+\frac{|x-h|^2}{b^2}=1$

$\Rightarrow \quad \frac{(x-h)^2}{b^2}+\frac{(y-k)^2}{a^2}=1$

FIGURE 5.17

REMARK

The equations of the above types of ellipse are of second degree, not involving the terms containing (x y) i.e., of the form $ax^2+by^2+2gx+2fy+c=0$. *Whenever the axes of a conic are parallel to co-ordinate axis, it would not involve the term containing product (xy).*

ILLUSTRATION 8: Find the equation of ellipse having its centre at (2, 4) and axes parallel to co-ordinates axes and having their major axis and minor axis of length 4 and 3, respectively.

SOLUTION: Given $2a = 4$

$\Rightarrow a = 2;\ 2b = 3 \ \Rightarrow\ b = 3/2$, centre of ellipse is at $(2,4) \equiv (h, k)$

$\therefore$ Equation of ellipse would be $\dfrac{(x-h)^2}{a^2} + \dfrac{(y-k)^2}{b^2} = 1$

$$\Rightarrow \frac{(x-2)^2}{4} + \frac{(y-4)^2}{9/4} = 1 \Rightarrow \frac{(x-2)^2}{4} + \frac{4(y-4)^2}{9} = 1$$

$$\Rightarrow 9(x-2)^2 + 16(y-4)^2 = 36 \Rightarrow 9x^2 - 36x + 36 + 16y^2 - 128y + 256 - 36 = 0$$

$$\Rightarrow 9x^2 + 16y^2 - 36x - 128y + 256 = 0$$

ILLUSTRATION 9: Find the equation of axes, directrix, co-ordinates of foci, centre, vertices, length of latus rectum and eccentricity of an ellipse $\dfrac{(x-3)^2}{25} + \dfrac{(y-2)^2}{16} = 1$.

SOLUTION: Let $x - 3 = X$, $y - 2 = Y$, so equation of ellipse becomes $\dfrac{X^2}{5^2} + \dfrac{Y^2}{4^2} = 1$

$\therefore$ equation of major axis is $Y = 0$ $\Rightarrow y = 2$;

equation of minor axis is $X = 0$ $\Rightarrow x = 3$.

centre $\equiv (X = 0, Y = 0)$ $\Rightarrow x - 3 = 0, y - 2 = 0$

$\Rightarrow C \equiv (3, 2)$

Length of semi-major axis 'a' = 5 $\Rightarrow$ Length of major axis $= 2a = 10$

Length of semi-minor axis $= b = 4$ $\Rightarrow$ Length of minor axis $= 2b = 8$

Let 'e' be the eccentricity of given ellipse

$\therefore\ b^2 = a^2(1 - e^2)$

$$\Rightarrow e = \sqrt{\frac{a^2 - b^2}{a^2}} = \sqrt{\frac{25-16}{25}} = \frac{3}{5}$$

Now, Length of latus rectum $= LL' = \dfrac{2b^2}{a} = \dfrac{2 \times 16}{5} = \dfrac{32}{5}$ units

Co-ordinates foci are $X = \pm ae$, $Y = 0$

$\Rightarrow S \equiv (X = 3, Y = 0)$ and $S' \equiv (X = -3, Y = 0)$

$\Rightarrow S \equiv (6, 2)$ and $S' \equiv (0, 2)$

Vertices are given by $A, A' \equiv (X = \pm a, Y = 0) \equiv (x - 3 = \pm 5, y - 2 = 0) \equiv (8, 2); (-2, 2)$

Equation of directrices are given by $X = \pm a/e$, i.e., $x - 3 = \pm\dfrac{25}{3}$

$$\Rightarrow x = \frac{34}{3} \text{ and } x = -\frac{16}{3}$$

ILLUSTRATION 10: Find the length of major and minor axis, eccentricity, vertices, centre, foci, equations of directrices and length of latus rectum of ellipse $4x^2 + 9y^2 - 8x + 54y + 49 = 0$.

SOLUTION: Given equation is $4x^2 + 9y^2 - 8x + 54y + 49 = 0$

$\Rightarrow (4x^2 - 8x) + (9y^2 + 54y) + 49 = 0$ $\Rightarrow 4(x^2 - 2x) + 9(y^2 + 6y) + 49 = 0$

$\Rightarrow 4(x^2 - 2x + 1) + 9(y^2 + 6y + 9) + 49 - 4 - 81 = 0$

$\Rightarrow 4(x - 1)^2 + 9(y + 3)^2 + 49 - 85 = 0$ $\Rightarrow 4(x - 1)^2 + 9(y + 3)^2 = 36$

$$\Rightarrow \frac{(x-1)^2}{9} + \frac{(y+3)^2}{4} = 1 \qquad \text{... (i)}$$

Comparing (i) with $\frac{(x-h)^2}{a^2}+\frac{(y-k)^2}{b^2}=1$, we have $(h, k) \equiv (1, -3)$; $a = 3$; $b = 2$

$\therefore$ Length of major axis $= 2a = 6$ units; Length of minor axis $= 2b = 4$ units

By relation $b^2 = a^2(1 - e^2)$, we have $4 = 9(1 - e^2)$

$\Rightarrow 4/9 = 1 - e^2 \Rightarrow e^2 = 1 - \frac{4}{9} = \frac{5}{9} \Rightarrow e = \frac{\sqrt{5}}{3}$

$\therefore$ Eccentricity of ellipse $= \sqrt{5}/3$ and centre of ellipse $(h, k) \equiv (1, -3)$ vertices are on either sides of $C(1,-3)$ at a distance of $a = 3$ and on the line $y = -3$ is $(h \pm a, k) \equiv (1 \pm 3, -3)$ i.e., $A'(-2, -3)$ and $A(4, -3)$; foci are given by $(h \pm ae, k) \equiv \left(1 \pm 3\left(\frac{\sqrt{5}}{3}\right), -3\right) \equiv \left(1+\sqrt{5}, -3\right)$ and $\left(1-\sqrt{5}, -3\right)$

Directrices are at a distance of a/e from centre $C(h, k)$ and are parallel to y-axis i.e., $x = h \pm a/e$

$\Rightarrow x = 1 \pm \frac{3}{\sqrt{5}/3}$ i.e., $x = 1 \pm \frac{9}{\sqrt{5}}$.

$\Rightarrow \sqrt{5}x = \sqrt{5} \pm 9$ or $\sqrt{5}x - \sqrt{5} \mp 9 = 0$

and length of latus rectum of ellipse is given by $LL' = \frac{2b^2}{a} = \frac{2(4)}{3} = \frac{8}{3}$ units

ILLUSTRATION 11: Prove that the segment of direct common tangent between the point of contact to the ellipse $\frac{x^2}{a^2}+\frac{y^2}{b^2}=\frac{2x}{c}$ and $\frac{x^2}{b^2}+\frac{y^2}{a^2}+\frac{2x}{c}=0$, subtends a right angle at the origin; where $a^2 > b^2$.

SOLUTION: Given ellipses are $\frac{x^2}{a^2}-\frac{2x}{c}+\frac{y^2}{b^2}=0$ and $\frac{x^2}{b^2}+\frac{2x}{c}+\frac{y^2}{a^2}=0$

or $\frac{1}{a^2}\left[x^2-\frac{2a^2}{c}x\right]+\frac{y^2}{b^2}=0$ and $\frac{1}{b^2}\left[x^2+\frac{2b^2x}{c}\right]+\frac{y^2}{a^2}=0$

or $\frac{1}{a^2}\left(x-\frac{a^2}{c}\right)^2+\left(\frac{y^2}{b^2}-\frac{a^2}{c^2}\right)=0$ and $\frac{1}{b^2}\left(x+\frac{b^2}{c}\right)^2+\left(\frac{y^2}{a^2}-\frac{b^2}{c^2}\right)=0$

$$\Rightarrow \frac{\left[x-\left(\frac{a^2}{c}\right)\right]^2}{a^2}+\frac{y^2}{b^2}=\frac{a^2}{c^2};\quad \frac{\left[x-\left(\frac{-b^2}{c}\right)\right]^2}{b^2}+\frac{y^2}{a^2}=\frac{b^2}{c^2}$$

$$\Rightarrow \frac{\left[x-\frac{a^2}{c}\right]^2}{\left(\frac{a^4}{c^2}\right)}+\frac{y^2}{\left(\frac{a^2b^2}{c^2}\right)}=1 \qquad \text{...(1)}$$

and
$$\frac{\left[x-\left(-\frac{b^2}{c}\right)\right]^2}{\left(\frac{b^4}{c^2}\right)}+\frac{y^2}{\frac{a^2b^2}{c^2}}=1 \qquad \text{....(2)}$$

Centre of ellipse (1) is at $\left(\dfrac{a^2}{c}, 0\right)$ and that of ellipse (2) is at $\left(-\dfrac{b^2}{c}, 0\right)$

i.e., both lie on x-axis, and length of minor axis of (1) is same as that of major axis of (2)

$\left[\because \dfrac{b^4}{c^2} < \dfrac{a^2b^2}{c^2} < \dfrac{a^4}{c^2} \text{ as } b^2 < a^2\right]$

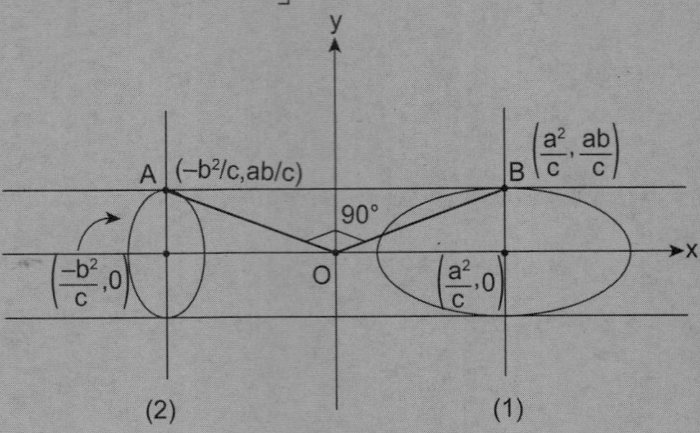

FIGURE 5.18

$\therefore$ Equation of common direct tangents will be $y = \pm \dfrac{ab}{c}$.

Let us consider $y = ab/c$ (Common direct tangent touching the ellipses above them)

Slope of $OA = \dfrac{ab/c}{-b^2/c} = m_1(say)$ and that of $OB = \dfrac{ab/c}{a^2/c} = m_2$ (say)

$$\therefore\ m_1 \cdot m_2 = \left(\left(\frac{ab}{c}\right)\cdot\frac{c}{-b^2}\left(\frac{ab}{c}\right)\frac{c}{a^2} = -1\right)$$

$\Rightarrow$ Required segment of common tangent subtents a right angle at the origin.

NON-STANDARD ELLLIPSE WITH AXES INCLINED TO CO-ORDINATE AXES

Let $l_1x + m_1y + n_1 = 0$ and $m_1x - l_1y + n_2 = 0$ be the equations of major axis and minor axis respectively as shown in figure 5.19. Equations of ellipse is given by

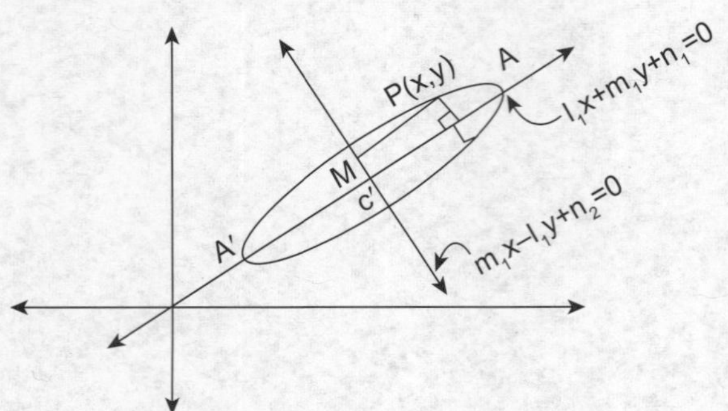

FIGURE 5.19

$$\frac{(PM)^2}{a^2} + \frac{(PL)^2}{b^2} = 1;$$ where a = length of semi-major axes and b = length of semi-minor axis.

$$\text{i.e., } \frac{\left(\dfrac{|m_1x - l_1y + n_2|}{\sqrt{m_1^2 + l_1^2}}\right)^2}{a^2} + \frac{\left(\dfrac{|l_1x + m_1y + n_1|}{\sqrt{l_1^2 + m_1^2}}\right)^2}{b^2} = 1$$

$$\text{or } \frac{(m_1x - l_1y + n_2)^2}{a^2\left(m_1^2 + l_1^2\right)} + \frac{(l_1x + m_1y + n_1)^2}{b^2\left(l_1^2 + m_1^2\right)} = 1$$

$$\text{or } \frac{(m_1x - l_1y + n_2)^2}{a^2} + \frac{(l_1x + m_1y + n_1)^2}{b^2} = l_1^2 + m_1^2$$

- Eccentricity of ellipse is given by $b^2 = a^2(1 - e^2)$
- Length of major axies = $2a$
- Length of minor axes = $2b$

- Centre of ellipse is the point of intersection of major axes and minor axis i.e., point of intersection of $l_1x + m_1y + n_1 = 0$ and $m_1x - l_1y + n_2 = 0$

i.e., $$\frac{x}{m_1n_2 + l_1n_1} = \frac{y}{m_1n_1 - n_2l_1} = \frac{1}{-(l_1^2 + m_1^2)}$$

i.e., $$C' \equiv \left(\frac{-(m_1n_2 + l_1n_1)}{l_1^2 + m_1^2}, \frac{n_2l_1 - m_1n_1}{l_1^2 + m_1^2}\right)$$

- **Foci:** Let (α, β) be the co-ordiantes of any focus, then its distance from minor axes is ae and from major axis is zero.

$$\Rightarrow \frac{(m_1\alpha - l_1\beta + n_2)}{\sqrt{l_1^2 + m_1^2}} = \pm ae, \frac{l_1\alpha + m_1\beta + n_1}{\sqrt{l_1^2 + m_1^2}} = 0$$

Solving these equations for α and β, we get co-ordinates of foci S and S'.

- **Directrices:** Let (x, y) be any point on any directrix, then its distance from minor axes would be a/e.

$\therefore \frac{(m_1x - l_1y + n_2)}{\sqrt{l_1^2 + m_1^2}} = \pm\frac{a}{e}$, gives us the equations of directrices.

REMARK

Equations of above type of ellipse involves terms containing xy i.e., their equations will be of the type $ax^2 + by^2 + 2hxy + 2gx + 2fy + c = 0$.

ILLUSTRATION 12: Find the equation of ellipse having one of its foci at (3,5); equation of a directix $2x - 3y + 7 = 0$ and eccentricity $= \frac{1}{\sqrt{2}}$.

SOLUTION: Clearly, the given ellipse is non-standard having its axes inclined to co-ordinate axes.

Let $P(x, y)$ be any point on the required ellipse, then $\frac{PS}{PM} = e$

$\Rightarrow (PS)^2 = e^2 (PM)^2$

$$\Rightarrow (x-3)^2 + (y-5)^2 = \frac{1}{2}\frac{|2x - 3y + 7|^2}{((4) + (9))}$$

$$\Rightarrow 26\left[x^2 + y^2 - 6x - 10y + 9 + 25\right]$$

$$= 4x^2 + 9y^2 + 49 - 12xy - 42y + 28x$$

i.e., $22x^2 + 17y^2 - 184x - 218y + 12xy + 835 = 0$

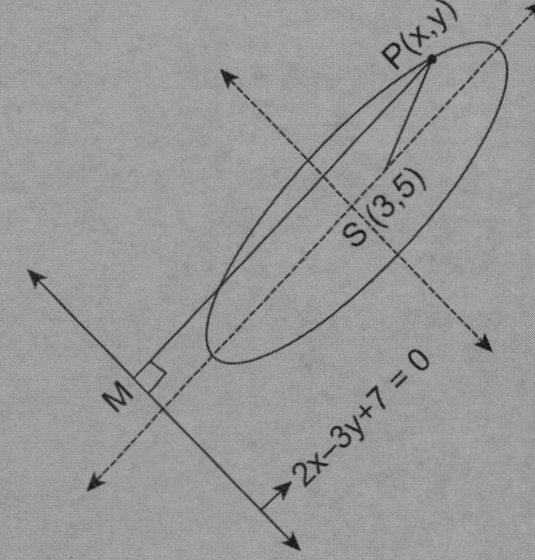

FIGURE 5.20

ILLUSTRATION 13: Obtain the equation of an ellipse of eccentricity 1/2, having one of the foci at (–1, 1); whose one directrix is the line passing through (2, 5) with unit gradient.

SOLUTION: Let $P(x, y)$ be any point on ellipse. Its focus is $S(-1, 1)$.

Let the directrix be $y = x + c$ ..(i) $\quad (\because$ gradient $m = 1)$

Line (i) passes through (2, 5) so,

$5 = 2 + c \Rightarrow c = 3$

$\therefore$ The directrix is $y = x + 3$

or $x - y + 3 = 0$ . . . (ii)

Now, let PM be the perpendicular from P, drawn to its directrix (ii). By definition of ellipse $SP = e\,PM$ or $SP^2 = e^2\,PM^2$

$$\Rightarrow (x+1)^2 + (y-1)^2 = \left(\frac{1}{2}\right)^2 \left[(x-y+3)/\sqrt{(1^2+1^2)}\right]^2$$

$$\Rightarrow 8[(x+1)^2+(y-1)^2] = (x-y+3)^2$$

$\Rightarrow 7x^2+7y^2+2xy+10x-10y+7=0$. Which is the required equation of ellipse.

ILLUSTRATION 14: If the focus, centre and eccentricity of an ellipse are respectively (3, 4); (2, 3) and 1/2, find its equation.

SOLUTION: The major axis is the line joining the points C (2, 3) and S(3, 4) i.e., centre and focus

i.e., $y - 3 = \dfrac{1}{1}(x-2)$ i.e., $x - y + 1 = 0$ ....(i)

The minor axis is the line through C(2, 3) perpendicular to the major axis (i)

$\therefore$ The equation of the minor axis is $x + y = 5$ ....(ii)

Now, $CS = ae$

i.e., $(CS)^2 = a^2e^2 = \dfrac{a^2}{4}\left(\because e = \dfrac{1}{2}, \text{given}\right)$

$\Rightarrow (3-2)^2 + (4-3)^2 = \dfrac{a^2}{4}$

$\Rightarrow a^2 = 8$

Now, $b^2 = a^2(1 - e^2)$

$\Rightarrow b^2 = 8\left(1 - \dfrac{1}{4}\right) = 6$

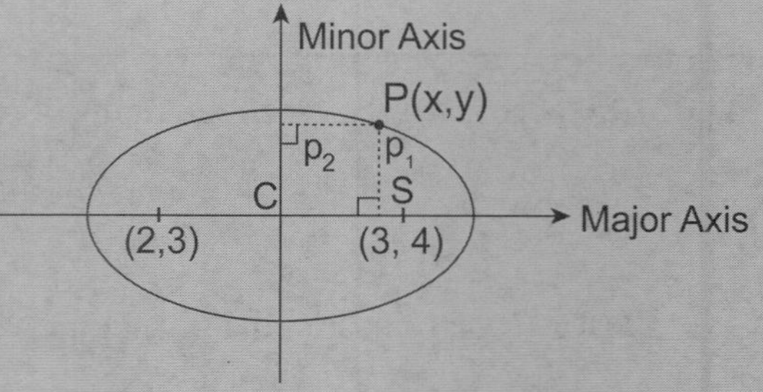

FIGURE 5.21

If p_1 and p_2 are the lengths of the perpendicular from any point (x, y) on the ellipse to $x - y + 1 = 0$ (major axis) and $x + y = 5$ (minor axis), then we must have $\dfrac{p_1^2}{b^2} + \dfrac{p_2^2}{a^2} = 1$;

i.e., $$\frac{\left(\frac{x-y+1}{\sqrt{2}}\right)^2}{6} + \frac{\left(\frac{x+y-5}{\sqrt{2}}\right)^2}{8} = 1 \Rightarrow \frac{(x-y+1)^2}{12} + \frac{(x+y-5)^2}{16} = 1$$

ILLUSTRATION 15: Determine the equations of major and minor axis, their lengths, centre of ellipse, length of latus rectum, eccentricity and foci of ellipse $4(x - 3y + 2)^2 + 9(3x + y + 1)^2 = 54$.

SOLUTION: The given equation can be written as

$$4\left(\frac{x-3y+2}{\sqrt{1+9}}.\sqrt{10}\right)^2 + 9\left(\frac{3x+y+1}{\sqrt{9+1}}.\sqrt{10}\right)^2 = 54$$

$$\Rightarrow 4\times10\left(\frac{x-3y+2}{\sqrt{10}}\right)^2 + 9\times10\left(\frac{3x+y+1}{\sqrt{10}}\right)^2 = 54 \Rightarrow \frac{40}{54}\left(\frac{x-3y+2}{\sqrt{10}}\right)^2 + \frac{90}{54}\left(\frac{3x+y+1}{\sqrt{10}}\right)^2 = 1$$

$$\Rightarrow \frac{\left(\frac{x-3y+2}{\sqrt{10}}\right)^2}{\left(\frac{27}{20}\right)} + \frac{\left(\frac{3x+y+1}{\sqrt{10}}\right)^2}{(3/5)} = 1. \Rightarrow \frac{\left(\frac{x-3y+2}{\sqrt{10}}\right)^2}{\left(\sqrt{\frac{27}{20}}\right)^2} + \frac{\left(\frac{3x+y+1}{\sqrt{10}}\right)^2}{\left(\sqrt{\frac{3}{5}}\right)^2} = 1$$

$\therefore$ Equation of minor axis is $x - 3y + 2 = 0$ ....(i)

and equation of major axis is $3x + y + 1 = 0$ ....(ii)

Length of major axis $= 2a = 2\sqrt{\frac{27}{20}} = \frac{2(3)}{2}\sqrt{\frac{3}{5}} = 3\sqrt{\frac{3}{5}}$ and length of minor axis $= 2b = 2\sqrt{\frac{3}{5}}$.

For eccentrcity, $b^2 = a^2(1 - e^2)$

$$\Rightarrow \frac{3}{5} = \frac{27}{20}(1-e^2) \Rightarrow \frac{3\times 20}{5\times 27} = 1-e^2 \Rightarrow e^2 = 1-\frac{4}{9} = \frac{5}{9} \Rightarrow e = \frac{\sqrt{5}}{3}$$

Now, centre of ellipse is the point of intersection (i) and (ii) i.e., $\left(-\frac{1}{2}, \frac{1}{2}\right)$

Length of latus rectum $= \frac{2b^2}{a} = \frac{2\left(\frac{3}{5}\right)}{\frac{3}{2}\sqrt{\frac{3}{5}}} = \frac{6}{5}\times\frac{2\sqrt{5}}{3\sqrt{3}} = \frac{4}{5}\frac{\sqrt{5}}{\sqrt{3}} = \frac{4}{\sqrt{15}}$

Foci: let (α, β) be the co-ordinates of foci

$$\Rightarrow \frac{\alpha - 3\beta + 2}{\sqrt{10}} = \pm ae = \pm\left(\left(\frac{3}{2}\right)\sqrt{\frac{3}{5}}\left(\frac{\sqrt{5}}{3}\right)\right) = \pm\frac{\sqrt{3}}{2}$$

$$\Rightarrow \alpha - 3\beta + 2 = \pm\frac{\sqrt{30}}{2} \text{ and } 3\alpha + \beta + 1 = 0$$

$$\Rightarrow (\alpha, \beta) \equiv \left(\frac{\sqrt{30}-10}{20}, \frac{10-3\sqrt{30}}{20}\right) \text{ and } \left(\frac{-\sqrt{30}-10}{200}, \frac{3\sqrt{30}+10}{20}\right)$$

Equations of directrices are given by: $\frac{x-3y+2}{\sqrt{1+9}} = \pm\frac{a}{e} = \pm\frac{\frac{3}{2}\sqrt{\frac{3}{5}}}{\sqrt{5}/3} = \pm\frac{9\sqrt{3}}{10}$

or $x - 3y + 2 = \pm 9\sqrt{\frac{3}{10}}$.

ILLUSTRATION 16: Find the equation of an ellipse from the definition that ellipse is the locus of a point which moves such that the sum of its distances from two fixed points is constant with the fixed foci.

SOLUTION: Let two fixed points be $S(ae, 0)$ and $S'(-ae, 0)$. Let $P(x, y)$ be a moving point such that

$SP + S'P = \text{constant} = 2a(say)$

i.e., $\sqrt{(x-ae)^2 + (y-0)^2} + \sqrt{(x+ae)^2 + (y-0)^2} = 2a$

or $\sqrt{(x^2 + y^2 - 2aex + a^2e^2} + \sqrt{(x^2 + y^2 + 2aex + a^2e^2)} = 2a$ ....(1)

Let $l = x^2 + y^2 + 2aex + a^2e^2$ ....(2)

and $m = x^2 + y^2 - 2aex + a^2e^2$ ....(3)

Equation (1) can be written as

$\sqrt{\ell} + \sqrt{m} = 2a$ ....(4)

From (2) and (3), $l - m = 4aex \Rightarrow (\sqrt{l} + \sqrt{m})(\sqrt{l} - \sqrt{m}) = 4aex$

$\Rightarrow 2a(\sqrt{l} - \sqrt{m}) = 4aex$ ....[from equation 4]

$\Rightarrow \sqrt{l} - \sqrt{m} = 2ex$ ....(5)

Adding (4) and (5), we get $2\sqrt{l} = 2a + 2ex$

$\Rightarrow \sqrt{l} = (a + ex)$ on squaring, we have $l = a^2 + 2aex + e^2x^2$

or $x^2 + y^2 + 2aex + a^2e^2 = a^2 + 2aex + e^2x^2$ (from(2))

$\Rightarrow x^2(1-e^2) + y^2 = a^2(1-e^2)$ or $\dfrac{x^2}{a^2} + \dfrac{y^2}{a^2(1-e^2)} = 1$

or $\dfrac{x^2}{a^2} + \dfrac{y^2}{b^2} = 1$ where $b^2 = a^2(1-e^2)$

∴ The equation of the ellipse, whose focus is the point (h, k) and directrix is $lx + my + n = 0$ and whose eccentricity is e, is $(x-h)^2 + (y-k)^2 = e^2 . \dfrac{(lx+my+n)^2}{(l^2+m^2)}$

TEXTUAL EXERCISE–1 (SUBJECTIVE)

1. Find the equations of major axis and minor axis, length of latus rectum, equations of directrices, foci and centre of following ellipses.

(i) $\dfrac{x^2}{9} + \dfrac{y^2}{4} = 1$

(ii) $\dfrac{x^2}{4} + \dfrac{y^2}{9} = 1$

(iii) $\dfrac{(x-1)^2}{16} + \dfrac{(y-2)^2}{9} = 1$

(iv) $\dfrac{(x+2)^2}{4} + \dfrac{(y-3)^2}{9} = 1$

2. Find the set of values of a for which the equation $\dfrac{x^2}{10-a} + \dfrac{y^2}{4-a} = 1$ represents an ellipse.

3. The distance between the foci of an ellipse is 10 and its latus rectum is 15; find its equation reffered to its axes as axes of co-ordinates.

4. Find the equation of the ellipse whose axes are parallel to the co-ordinate axes having its centre at the point (2, –3), one focus at (3, –3) and a vertex at (4, –3).

5. (a) Find the centre, the length of the axes, eccentricity and the foci of the ellipse.

(i) $12x^2 + 4y^2 + 24x - 16y + 25 = 0$

(ii) $2x^2 + 3y^2 - 4x - 12y + 13 = 0$

(b) Find the eccentricity of an ellipse, if its latus rectum is equal to one half of its major axis.

6. (a) Find the equation of the ellipse whose foci are (2, 3), (–2, 3) and whose semi minor axis is $\sqrt{5}$.

(b) Find the lengths of major and minor axes, the co-ordinates of foci, vertices and the eccentricity of the ellipse $3x^2 + 2y^2 = 6$. Also find the equations of the directrices.

(c) Find the equation of the ellipse with following specifications:

(i) centre at (1, 2), one of the foci at (6, 2) and passing through the point (4, 6)

(ii) a focus at (–1, 1), eccentricity 1/2 and a directrix $x - y + 3 = 0$.

7. Find the lengths and equations of the major and minor axes of the ellipse $5x^2 + 5y^2 + 6xy - 8 = 0$.

8. Determine the equations of major and minor axes of the ellipse $4(x - 2y + 1)^2 + 9(2x + y + 2)^2 = 25$ Also, find its centre, length of the latus rectum and eccentricity.

9. Show that the equation $(10x - 5)^2 + (10y - 5)^2 = (3x + 4y - 1)^2$ represents an ellipse, find the eccentricity of the ellipse.

10. Find the equation of ellipse such that the distance between its foci is 5 units and distance between its directrices is 10 units and axes parallel to co-ordinate axes.

11. Let P be a variable point on the ellipse $\dfrac{x^2}{49} + \dfrac{y^2}{25} = 1$; then find the maximum value of area of $\Delta PSS'$ and $\Delta PAA'$, where S and S' are foci and A and A' are vertices of ellipse.

Answer Keys

1. (i) $y = 0$; $x = 0$; 8/3 units; $x = \pm\frac{9}{\sqrt{5}}$; $(\pm\sqrt{5}, 0)$; $(0, 0)$ (ii) $x = 0$; $y = 0$; 8/3 units; $y = \pm\frac{9}{\sqrt{5}}$; $(0, \pm\sqrt{5})$; $(0, 0)$

(iii) $y = 2$; $x = 1$; 9/2 units; $x = 1\pm\frac{16}{\sqrt{7}}$; $\left(1\pm\sqrt{7},2\right)$; $(1, 2)$ (iv) $x = -2$; $y = 3$; 8/3 units; $y = 3\pm\frac{9}{\sqrt{5}}$;

$\left(-2,3\pm\sqrt{5}\right)$; $(-2, 3)$ **2.** $a \in (-\infty, 4)$ **3.** $3x^2 + 4y^2 = 300$ **4.** $\frac{(x-2)^2}{4}+\frac{(y+3)^2}{3} = 1$

5. (a) (i) $(-1, 2)$; $\sqrt{3}$; 1; $\sqrt{2/3}$; $(-1, 2 \pm 1/\sqrt{2})$ (ii) $(1, 2)$; $\sqrt{2}$; $2/\sqrt{3}$; $1/\sqrt{3}$; $(1\pm 1/\sqrt{6}, 2)$ (b) $1/\sqrt{2}$

6. (a) $5x^2 + 9y^2 - 54y + 36 = 0$ (b) $2\sqrt{3}$; $2\sqrt{2}$; $(0, \pm1)$; $(0, \pm\sqrt{3})$; $1/\sqrt{3}$; $y = \pm3$

(c) (i) $\frac{(x-1)^2}{45}+\frac{(y-2)^2}{20}=1$ (ii) $7x^2+7y^2+2xy + 10x - 10y + 7 = 0$

7. 4; 2; $y = -x$; $y = x$; **8.** $x - 2y + 1 = 0$; $2x + y + 2 = 0$; $(-1, 0)$; $(4\sqrt{5})/9$; $\sqrt{5}/3$ **9.** 1/2

10. $2x^2 + 4y^2 = 25$ **11.** $10\sqrt{6}$; 35 sq. units

TEXTUAL EXERCISE–1 (OBJECTIVE)

1. If the latus rectum of an ellipse is equal to half of its minor axis, then its eccentricity is

(a) 3/2 (b) $\sqrt{3}/2$
(c) 2/3 (d) $\sqrt{2}/3$

2. If the distance between the directrices is thrice the distance between the foci, then eccentricity of ellipse is

(a) 1/2 (b) 2/3
(c) $1/\sqrt{3}$ (d) 4/5

3. If the eccentricity of an ellipse is 5/8 and the distance between its foci is 10, then its latus rectum is

(a) 39/4 (b) 12
(c) 15 (d) 37/2

4. If the foci and vertices of an ellipse are respectively $(\pm 1, 0)$ and $(\pm 2, 0)$, then its minor axis length is

(a) $2\sqrt{5}$ (b) 2
(c) 4 (d) $2\sqrt{3}$

5. The eccentricity of an ellipse is 2/3, latus rectum is 5 and centre is (0, 0). The equation of the ellipse is

(a) $\frac{x^2}{81}+\frac{y^2}{45}=1$ (b) $\frac{4x^2}{81}+\frac{4y^2}{45}=1$
(c) $\frac{x^2}{9}+\frac{y^2}{5}=1$ (d) $\frac{x^2}{81}+\frac{y^2}{45}=5$

6. The latus rectum of an ellipse is 10 and the minor axis is equal to the distance between the foci .The equation of the ellipse is

(a) $x^2 + 2y^2 = 100$ (b) $x^2+\sqrt{2}y^2=10$
(c) $x^2 + 4y^2 = 100$ (d) None of these

7. The equation of the ellipse whose foci are at $(\pm5, 0)$ and one of its directrices is $5x = 36$, is

(a) $\frac{x^2}{36}+\frac{y^2}{11}=1$ (b) $\frac{x^2}{6}+\frac{y^2}{\sqrt{11}}=1$
(c) $\frac{x^2}{6}+\frac{y^2}{11}=1$ (d) None of these

8. If the eccentricity of an ellipse is $1/\sqrt{2}$, then its latus rectum is equal to its

(a) minor axis (b) semi-minor axis
(c) major axis (d) semi-major axis

9. If the distance between the foci of an ellipse is equal to its minor axis, then its eccentricity is

(a) 1/2 (b) $1/\sqrt{2}$
(c) 1/3 (d) $1/\sqrt{3}$

10. An ellipse passes through the point $(-3, 1)$ and its eccentricity is $\sqrt{\frac{2}{5}}$. The equation of the ellipse is

(a) $3x^2 + 5y^2 = 32$ (b) $3x^2 + 5y^2 = 25$
(c) $3x^2 + y^2 = 4$ (d) $3x^2 + y^2 = 9$

11. Eccentricity of the ellipse whose latus rectum is equal to the distance between foci is

(a) $\frac{\sqrt{5}+1}{2}$ (b) $\frac{\sqrt{5}-1}{2}$
(c) $\frac{\sqrt{5}}{4}$ (d) $\frac{\sqrt{3}}{2}$

12. An ellipse is described by using an endless string which is passed over two pins. If the axes are 8 cm and 4 cm, the required length of the string and the distance between the pins respectively in cm, are

(a) 16, $8\sqrt{3}$ (b) $(2+\sqrt{3})$, $8\sqrt{3}$
(c) 16, $4\sqrt{3}$ (d) None of these

13. The locus of the point of intersection of perpendicular tangents to the ellipse $\frac{x^2}{a^2}+\frac{y^2}{b^2}=1$, is

(a) $x^2+y^2=a^2-b^2$ (b) $x^2-y^2=a^2-b^2$
(c) $x^2+y^2=a^2+b^2$ (d) $x^2-y^2=a^2+b^2$

14. The equation of the ellipse whose one focus is at (4, 0) and whose eccentricity is 4/5, is

(a) $\frac{x^2}{3^2}+\frac{y^2}{5^2}=1$ (b) $\frac{x^2}{5^2}+\frac{y^2}{3^2}=1$
(c) $\frac{x^2}{5^2}+\frac{y^2}{4^2}=1$ (d) $\frac{x^2}{4^2}+\frac{y^2}{5^2}=1$

15. If $P\equiv(x, y)$, $F_1\equiv(3, 0)$, $F_2\equiv(-3, 0)$ and $16x^2+25y^2=400$, then PF_1+PF_2 equals

(a) 8 (b) 6
(c) 10 (d) 12

16. If the eccentricity of the two ellipses $\frac{x^2}{169}+\frac{y^2}{25}=1$ and $\frac{x^2}{a^2}+\frac{y^2}{b^2}=1$ are equal, then the value of a/b is

(a) 5/13 (b) 6/13
(c) 13/5 (d) 13/6

17. The equation of an ellipse whose eccentricity is 1/2 and the vertices are (4, 0) and (10, 0) is

(a) $3x^2+4y^2-42x+39=0$
(b) $3x^2+4y^2-42x+120=0$
(c) $3x^2+4y^2+42x-120=0$
(d) $3x^2+4y^2-42x-120=0$

18. The centre of the ellipse $\frac{(x+y-2)^2}{9}+\frac{(x-y)^2}{16}=1$ is

(a) (0, 0) (b) (1, 1)
(c) (1, 0) (d) (0, 1)

19. The eccentricity of the ellipse $9x^2+5y^2-30y=0$, is

(a) 1/3 (b) 2/3
(c) 3/4 (d) None of these

20. The eccentricity of the curve represented by the equation $x^2+2y^2-2x+3y+2=0$ is

(a) 0 (b) 1/2
(c) $1/\sqrt{2}$ (d) $\sqrt{2}$

21. The eccentricity of the ellipse $\frac{(x-1)^2}{9}+\frac{(y+1)^2}{25}=1$ is

(a) 4/5 (b) 3/5
(c) 5/4 (d) imaginary

22. The lengths of the axes of the conic $9x^2+4y^2-6x+4y+1=0$, are

(a) $\frac{1}{2}, 9$ (b) $3, \frac{2}{5}$
(c) $1, \frac{2}{3}$ (d) 3, 2

23. An ellipse has OB as semi minor axis, F and F' its foci and the angle FBF' is a right angle. Then the eccentricity of the ellipse is

(a) $\frac{1}{4}$ (b) $\frac{1}{\sqrt{3}}$
(c) $\frac{1}{\sqrt{2}}$ (d) $\frac{1}{2}$

Answer Keys

1. (b)	**2.** (c)	**3.** (a)	**4.** (d)	**5.** (b)	**6.** (a)	**7.** (a)	**8.** (d)	**9.** (b)	**10.** (a)
11. (b)	**12.** (d)	**13.** (c)	**14.** (b)	**15.** (c)	**16.** (c)	**17.** (b)	**18.** (b)	**19.** (b)	**20.** (c)
21. (a)	**22.** (c)	**23.** (c)							

■ POSITION OF A POINT WITH RESPECT TO AN ELLIPSE

The point $P(x_1, y_1)$ lies outside on or inside the ellipse $\frac{x^2}{a^2}+\frac{y^2}{b^2}=1$ according as $\frac{x_1^2}{a^2}+\frac{y_1^2}{b^2}-1>,=\text{or}<0$.

From point $P(x_1, y_1)$, draw perpendicular PM on AA' to meet the ellipse at $Q(x_1, y_2)$.

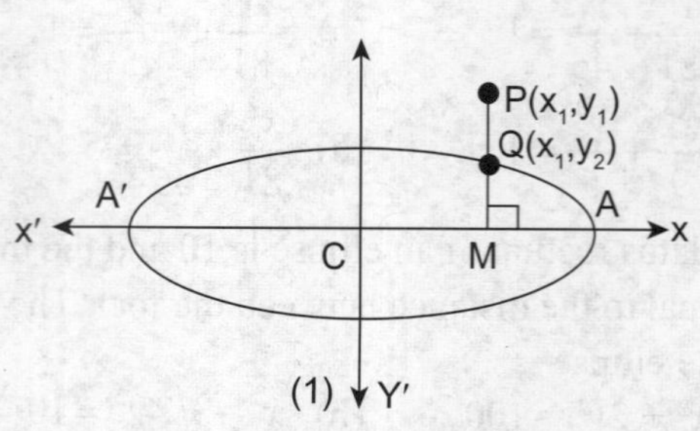

FIGURE 5.22

Since $Q(x_1, y_2)$ lies on the ellipse

$$\frac{x^2}{a^2}+\frac{y^2}{b^2}=1 \Rightarrow \frac{x_1^2}{a^2}+\frac{y_2^2}{b^2}=1 \Rightarrow \frac{y_2^2}{b^2}=1-\frac{x_1^2}{a^2} \quad (1)$$

Now, point P lies outside, on or inside the ellipse according as

$PM >, = \text{or} < QM$

$$\Rightarrow \; y_1 >, = \text{or} < y_2 \Rightarrow \frac{y_1^2}{b^2} >, = \text{or} < \frac{y_2^2}{b^2}$$

$$\Rightarrow \; \frac{y_1^2}{b^2} >, = \text{or} < 1-\frac{x_1^2}{a^2} \quad \text{[from (1)]}$$

$$\Rightarrow \; \frac{x_1^2}{a^2}+\frac{y_1^2}{b^2} >, = \text{or} < 1 \text{ or } \frac{x_1^2}{a^2}+\frac{y_1^2}{b^2}-1 >, = \text{or} < 0$$

Hence the point $P(x_1, y_1)$ lies outside, on or inside the ellipse $\frac{x^2}{a^2}+\frac{y^2}{b^2}=1$ according as $\frac{x_1^2}{a^2}+\frac{y_1^2}{b^2}-1 > 0, = \text{or} < 0$

NOTE

Let $S=\frac{x^2}{a^2}+\frac{y^2}{b^2}-1$, and $S_1=\frac{x_1^2}{a^2}+\frac{y_1^2}{b^2}-1$. Thus the point (x_1, y_1) lies outside, on or inside the ellipse $S=0$ according as $S_1 >, = \text{or} < 0$. If a point lies outside an ellipse, then we can draw two tangents from the point to ellipse. If point lies on ellipse, then only one tangent can be drawn to ellipse through the point. If point lies inside the ellipse, then no tangent can be drawn to ellipse through the point.

ILLUSTRATION 17: Find the position of the point (4, –3) relative to the ellipse $5x^2+7y^2=140$.

SOLUTION: Since $5(4)^2+7(-3)^2-140=80+63-140=3>0$

So, the point (4, –3) lies outside the ellipse $5x^2+7y^2=140$

LLUSTratioN 18: Find the set of value(s) of 'α' for which the point $P(\alpha, -\alpha)$ lies inside the ellipse $\frac{x^2}{16}+\frac{y^2}{9}=1$.

SOLUTION: If $P(\alpha, -\alpha)$ lies inside the ellipse $\frac{x^2}{16}+\frac{y^2}{9}-1=0$ i.e., $S=0$, then $S_1<0$

$$\Rightarrow \frac{\alpha^2}{16}+\frac{\alpha^2}{9}-1<0 \quad \Rightarrow \frac{25}{144}.\alpha^2<1$$

$$\Rightarrow \alpha^2<\frac{144}{25} \quad \Rightarrow \alpha \in \left(-\frac{12}{5}, \frac{12}{5}\right)$$

■ AUXILIARY CIRCLE AND ECCENTRIC ANGLE

The circle described on the major axis of an ellipse as diameter is called the **auxiliary circle** of the ellipse.

Let the equation of ellipse be $\frac{x^2}{a^2}+\frac{y^2}{b^2}=1$

∴ Equation of its auxiliary circle is

$x^2+y^2=a^2$ (∵ AA' is diameter of the circle)

Let Q be a point on the auxiliary circle $x^2+y^2=a^2$ such that QP produced is perpendicular to the major axis (here x-axis).

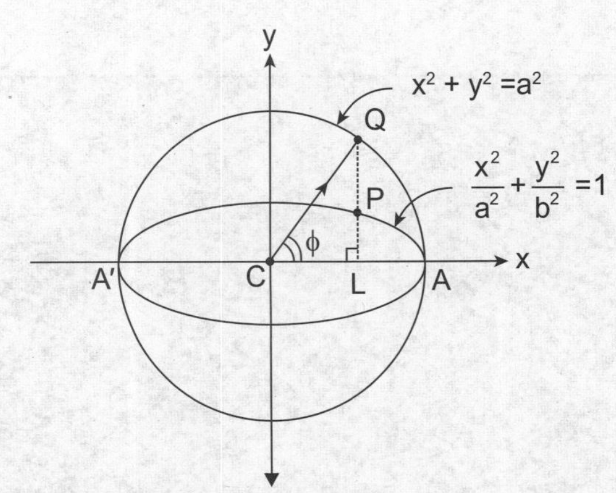

FIGURE 5.23

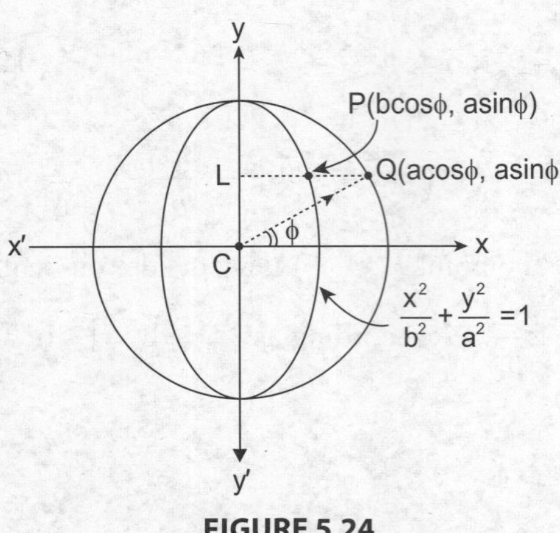

FIGURE 5.24

Then P and Q are the corresponding points on the ellipse and the auxiliary circle, respectively.

Let $\angle QCA = \phi$; $0 \le \phi < 2\pi$ and known as eccentric angle of point P.

Thus eccentric angle of P on an ellipse is the angle which the radius (or **radius vector**) through the corresponding point on the auxiliary circle makes with the major axis.

REMARK

we have, $\frac{PL}{PQ} = \frac{b\sin\phi}{a\sin\phi - b\sin\phi} = \frac{b}{a-b}$ *= constant, except when P coincides with A or A'.*

Hence, if from each point on a circle perpendiculars are drawn on a fixed diameter, then the locus of a point P dividing these perpendiculars internally in a constant ratio is an ellipse whose auxiliary circle is the original circle.

PARAMETRIC EQUATION OF THE ELLIPSE

Let $P(x, y)$ be any point on ellipse $\frac{x^2}{a^2}+\frac{y^2}{b^2}=1$; such that eccentric angle of point P is ϕ i.e., $\angle QCA = \phi$,

then $x = CL = CQ\cos\phi = a\cos\phi$

and $y = PL$ and $QL = a\sin\phi$

$\therefore$ $Q \equiv (a\cos\phi, a\sin\phi)$ and $P \equiv (a\cos\phi, y)$ and

P lies on the ellipse $\frac{x^2}{a^2}+\frac{y^2}{b^2}=1$

$\Rightarrow$ $\frac{a^2\cos^2\phi}{a^2}+\frac{y^2}{b^2}=1$

$\Rightarrow$ $y^2 = b^2\sin^2\phi \Rightarrow y = \pm b\sin\phi$

But ordinate y of point P on ellipse and $\sin\phi$ are of same sign and $b > 0$.

$\Rightarrow$ $y = b\sin\phi$

$\therefore$ co-ordinate of P are $(a\cos\phi, b\sin\phi)$. We have $x = a\cos\phi$, $y = b\sin\phi$ called parametric equations of the ellipse.

This point $(a\cos\phi, b\sin\phi)$ is also called the point 'ϕ'.

For the ellipse $\frac{x^2}{b^2}+\frac{y^2}{a^2}=1$, point '$\phi$' is given by $(b\cos\phi, a\sin\phi)$.

ILLUSTRATION 19: Distance of a point on the ellipse $\frac{x^2}{6}+\frac{y^2}{2}=1$ from its centre is 2. Find the eccentric angle of the point P.

SOLUTION: Let $P\left(\sqrt{6}\cos\theta, \sqrt{2}\sin\theta\right)$ be the given point on ellipse having its distance from centre C equal to 2.

$\therefore$ $CP = (\sqrt{6}\cos\theta)^2 + (\sqrt{2}\sin\theta)^2 = (2)^2$ $\Rightarrow 6\cos^2\theta + 2\sin^2\theta = 4$

$\Rightarrow 4\cos^2\theta = 2$ $\Rightarrow \cos^2\theta = 1/2$

$\Rightarrow$ $\cos\theta = \pm\frac{1}{\sqrt{2}}$ · As $\theta \in [0, 2\pi) \Rightarrow \theta = \frac{\pi}{4}, \frac{3\pi}{4}, \frac{5\pi}{4}, \frac{7\pi}{4}$ i.e. four positions of P are possible.

ILLUSTRATION 20: Find the eccentricity, foci, distance between foci of the ellipse having its parametric equation $x = 3\cos\theta, y = 2\sin\theta$.

SOLUTION: $x = 3\cos\theta$

$y = 2\sin\theta$

$\Rightarrow x^2 = 9\cos^2\theta; y^2 = 4\sin^2\theta$

$\Rightarrow \frac{x^2}{9}+\frac{y^2}{4}=1$ (First kind of ellipse) .... (i)

Comparing (i) with $\frac{x^2}{a^2}+\frac{y^2}{b^2}=1$, we have $a = 3, b = 2$

By relation, $b^2 = a^2(1 - e^2)$, we have $4 = 9(1 - e^2)$

$\Rightarrow 4 = 9 - 9e^2 \Rightarrow 9e^2 = 5 \Rightarrow e = \sqrt{5}/3$.

Foci are given by $(\pm ae, 0) \equiv (\pm\sqrt{5}, 0)$

Distance between foci $= 2ae = 6\left(\frac{\sqrt{5}}{3}\right) = 2\sqrt{5}$.

ILLUSTRATION 21: From a point Q on the circle $x^2 + y^2 = a^2$, perpendicular QM is drawn to x-axis, find the locus of point 'P' dividing QM in ratio 2:1.

SOLUTION: Let $Q \equiv (a\cos\theta, a\sin\theta)$,

$M \equiv (a\cos\theta, 0)$

Let (h, k) be the co-ordinates of point P.

$\Rightarrow h = a\cos\theta, k = \frac{a\sin\theta}{3}$

Now, $\cos^2\theta + \sin^2\theta = 1$

$\therefore \left(\frac{3k}{a}\right)^2 + \left(\frac{h}{a}\right)^2 = 1$

$\Rightarrow$ Locus of P is $\frac{x^2}{a^2}+\frac{y^2}{(a/3)^2}=1$ or $x^2 + 9y^2 = a^2$

FIGURE 5.25

ILLUSTRATION 22: Find the focal distances of point $P(\pi/3)$ on the ellipse $4x^2 + 9y^2 = 36$, hence show that the sum of focal distances is equal to the length of major axis.

SOLUTION: Given ellipse is $4x^2 + 9y^2 = 36 \Rightarrow \frac{x^2}{9}+\frac{y^2}{4}=1 \Rightarrow a = 3, b = 2$

$\therefore$ Parametric equation of ellipse is $x = 3\cos\phi, y = 2\sin\phi$.

$\therefore$ point $\left(3\cos\frac{\pi}{3}, 2\sin\frac{\pi}{3}\right)$, having its eccentric angle $= \pi/3$ lies on ellipse

$\therefore P \equiv \left(\frac{3}{2}, \sqrt{3}\right)$.

$\therefore$ foci are given by $S, S' \equiv (\pm ae, 0) \equiv \left(\pm 3\left(\frac{\sqrt{a^2-b^2}}{a}, 0\right)\right) \equiv \left(\pm 3\frac{\sqrt{9-4}}{3}, 0\right) \equiv (\pm\sqrt{5}, 0)$

$\therefore$ focal distances of P are given by SP and $S'P$

$\therefore SP = \sqrt{\left(\sqrt{5}-\frac{3}{2}\right)^2 + (0-\sqrt{3})^2} = \sqrt{5+\frac{9}{4}-3\sqrt{5}+3} = \sqrt{8+\frac{9}{4}-3\sqrt{5}}$

$= \sqrt{\frac{41-12\sqrt{5}}{4}} = \frac{\sqrt{41-12\sqrt{5}}}{2}$

$$\text{and } S'P = \sqrt{\left(-\sqrt{5}-\frac{3}{2}\right)^2+\left(0-\sqrt{3}\right)^2} = \sqrt{5+\frac{9}{4}+3\sqrt{5}+3} = \sqrt{8+\frac{9}{4}+3\sqrt{5}} = \frac{\sqrt{41+12\sqrt{5}}}{2}$$

$$\therefore\ SP + S'P = \frac{\sqrt{41-12\sqrt{5}}}{2}+\frac{\sqrt{41+12\sqrt{5}}}{2} = \frac{\sqrt{36+5-2(6)\sqrt{5}}+\sqrt{36+5+2(6)\sqrt{5}}}{2}$$

$$= \frac{6-\sqrt{5}+6+\sqrt{5}}{2} = 6 = 2a(\because a = 3)\ \textbf{Hence Proved.}$$

ILLUSTRATION 23: Find the distance from centre of the point P on the ellipse $\frac{x^2}{a^2}+\frac{y^2}{b^2}=1$ whose radius makes angle α with x-axis

SOLUTION: Let $P \equiv (a\cos\theta, b\sin\theta)$; θ is eccentric angle of point P

$\Rightarrow\ \angle XCQ = \theta$; Also $\angle PCX = \alpha$

$$\Rightarrow\ \frac{b}{a}\tan\theta = \tan\alpha \Rightarrow \tan\theta = \frac{a}{b}\tan\alpha$$

$$\Rightarrow\ CP = \sqrt{a^2\cos^2\theta + b^2\sin^2\theta} = \sqrt{\frac{a^2+b^2\tan^2\theta}{\sec^2\theta}}$$

$$= \sqrt{\frac{a^2+b^2\tan^2\theta}{1+\tan^2\theta}} = \sqrt{\frac{a^2+b^2\times\frac{a^2}{b^2}\tan^2\alpha}{1+\frac{a^2}{b^2}\tan^2\alpha}}$$

$$= \sqrt{\frac{a^2b^2(1+\tan^2\alpha)}{(b^2+a^2\tan^2\alpha)}} = \sqrt{\frac{a^2b^2\sec^2\alpha}{\left(\frac{a^2\sin^2 a+b^2\cos^2\alpha}{\cos^2\alpha}\right)}}$$

$$\Rightarrow\ CP = \frac{ab}{\sqrt{a^2\sin^2\alpha+b^2\cos^2\alpha}}$$

FIGURE 5.26

ILLUSTRATION 24: Let $P(\phi)$ be any point on ellipse $\frac{x^2}{a^2}+\frac{y^2}{b^2}=1$ and P' be the corresponding point on auxiliary circle. Find the ratio of area of ΔCBP and CPP', where B is one of the end points of minor axis and $a > b$.

SOLUTION: Given ellipse is $\frac{x^2}{a^2}+\frac{y^2}{b^2}=1$

Co-ordinates of P are $(a\cos\phi, b\sin\phi)$ and that of P' will be $(a\cos\phi, a\sin\phi)$

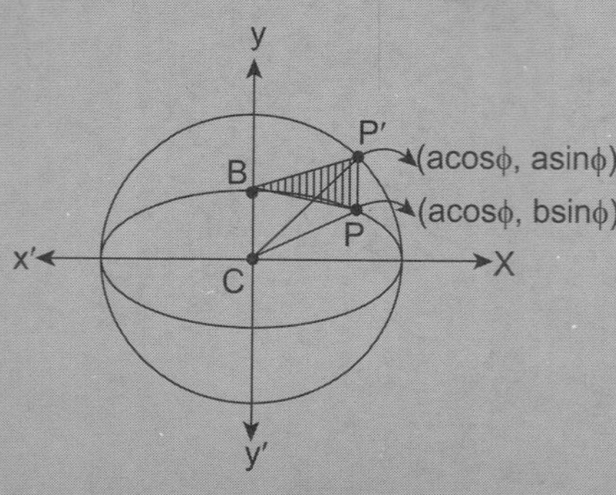

FIGURE 5.27

Area of ΔCBP' = $\frac{1}{2}CB \times (\perp$ distance of P' from minor axis)

$$= \Delta_1(\text{say}) = \frac{1}{2}\, b \times (|a\cos\phi|) = \frac{1}{2}\text{ab }|\cos\phi|$$

Area of $\Delta CPP'$ = $\frac{1}{2}$ (PP') × (⊥ distance of PP' from minor axis)

$$= \frac{1}{2}\,|a\cos\phi|\,|a\sin\phi - b\sin\phi| = \frac{1}{2}\,|(a-b)\sin\phi|\,|a\cos\phi|$$

$$= \Delta_2(\text{say}) = \frac{1}{2}(a-b)\ |\sin\phi|\,|\cos\phi|\left[\because a > b\right]$$

$$\therefore\quad \frac{\Delta_1}{\Delta_2} = \frac{\frac{1}{2}ab\,|\cos\phi|}{\frac{1}{2}(a-b)\,|\sin\phi|\,|\cos\phi|} = \frac{b}{(a-b)\,|\sin\phi|} = \frac{b}{(a-b)}\,|\operatorname{cosec}\phi|.$$

ILLUSTRATION 25: A variable point P on the ellipse of eccentricity 'e' is joined to the foci S and S'. Prove that the locus of the incentre of the triangle PSS' is an ellipse whose eccentricity is $\sqrt{\frac{2e}{1+e}}$.

SOLUTION: Let the ellipse be $\frac{x^2}{a^2}+\frac{y^2}{b^2}=1$, whose foci are $S(ae,0)$ and $S'(-ae,0)$. If $P(a\cos\theta, b\sin\theta)$ is any point on the ellipse, then $SP = a(1-e\cos\theta)$, $S'P = a(1+e\cos\theta)$ and $SS' = 2ae$. Let (h, k) be the incentre of the triangle PSS'.

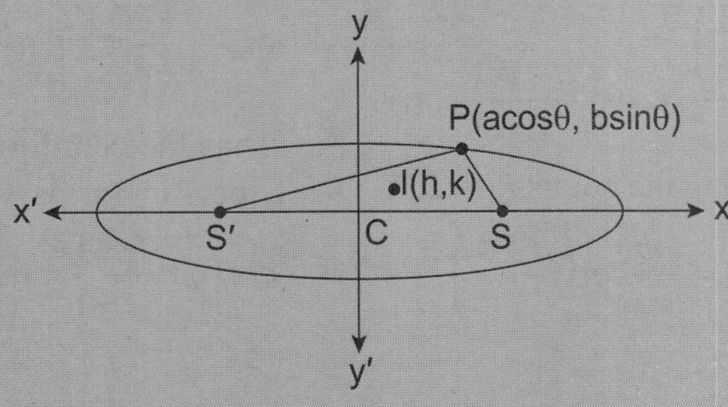

FIGURE 5.28

$$\text{Then } h = \frac{-ae.a(1-e\cos\theta)+ae.a(1+e\cos\theta)+a\cos\theta.2ae}{a(1-e\cos\theta)+a(1+e\cos\theta)+2ae} = ae\cos\theta$$

$$k = \frac{b\sin\theta.2ae}{a(1-e\cos\theta)+a(1+e\cos\theta)+2ae} = \frac{b\sin\theta e}{1+e}$$

$$\Rightarrow\quad \cos\theta = \frac{h}{ae}, \sin\theta = \frac{(1+e)k}{eb} \Rightarrow \frac{h^2}{a^2e^2}+\frac{(1+e)^2k^2}{e^2b^2}=1$$

Hence the locus of (h, k) is $\frac{x^2}{a^2e^2}+\frac{y^2}{e^2b^2/(1+e)^2}=1$

CHORD OF ELLIPSE

Let $R(\theta)$ and $Q(\phi)$ be two points on ellipse $\frac{x^2}{a^2}+\frac{y^2}{b^2}=1$. Let $P(x, y)$ be any arbitrary point on the line PQ. Then points P, Q, R will be collinear.

Thus $\begin{vmatrix} x & y & 1 \\ a\cos\theta & b\sin\theta & 1 \\ a\cos\phi & b\sin\phi & 1 \end{vmatrix}=0$

$\Rightarrow x(b\sin\theta - b\sin\phi) - y(a\cos\theta - a\cos\phi) + (ab\sin\phi\cos\theta - ab\cos\phi\sin\theta) = 0$

$\Rightarrow bx2\cos\left(\frac{\theta+\phi}{2}\right)\sin\left(\frac{\theta-\phi}{2}\right)$

$+ay.2\sin\left(\frac{\theta+\phi}{2}\right)\sin\left(\frac{\theta-\phi}{2}\right)+ab\sin(\phi-\theta)=0$

$\Rightarrow 2bx\cos\left(\frac{\theta+\phi}{2}\right)\sin\left(\frac{\theta-\phi}{2}\right)$

$+2ay\sin\left(\frac{\theta+\phi}{2}\right)\sin\left(\frac{\theta-\phi}{2}\right)=ab\sin(\theta-\phi)$

$\Rightarrow 2bx\cos\left(\frac{\theta+\phi}{2}\right)\sin\left(\frac{\theta-\phi}{2}\right)+2ay\sin\left(\frac{\theta+\phi}{2}\right)\sin\left(\frac{\theta-\phi}{2}\right)$

$=2ab\sin\left(\frac{\theta-\phi}{2}\right)\cos\left(\frac{\theta-\phi}{2}\right)$

$$\Rightarrow \frac{x\cos\left(\frac{\theta+\phi}{2}\right)}{a}+\frac{y\sin\left(\frac{\theta+\phi}{2}\right)}{b}=\cos\left(\frac{\theta-\phi}{2}\right) \quad \text{...(i)}$$

Thus (i) is the equation of chord joining the points $R(\theta)$ and $Q(\phi)$. Clearly, the slope of chord $RQ=-\frac{b}{a}\cot\left(\frac{\theta+\phi}{2}\right)$.

For ellipse $\frac{x^2}{b^2}+\frac{y^2}{a^2}=1$ equation of chord can be obtained by inter changing 'a' and 'b', i.e.,

$$\frac{x\cos\left(\frac{\theta+\phi}{2}\right)}{b}+\frac{y\sin\left(\frac{\theta+\phi}{2}\right)}{a}=\cos\left(\frac{\theta-\phi}{2}\right).$$

Focal Chord

If RQ is a focal chord, then it would pass through $S(ae, 0)$ or $S'(-ae, 0)$

$\Rightarrow \pm e\cos\left(\frac{\theta+\phi}{2}\right)=\cos\left(\frac{\theta-\phi}{2}\right)$

$\Rightarrow \cos\left(\frac{\theta-\phi}{2}\right)=e\cos\left(\frac{\theta+\phi}{2}\right)$

or $\cos\left(\frac{\theta-\phi}{2}\right)=-e\cos\left(\frac{\theta+\phi}{2}\right)$

Accordingly focal chord passes through S and S' respectively

$$\Rightarrow \frac{1}{e}=\frac{\cos\left(\frac{\theta+\phi}{2}\right)}{\cos\left(\frac{\theta-\phi}{2}\right)} \text{ or } \frac{1}{-e}=\frac{\cos\left(\frac{\theta+\phi}{2}\right)}{\cos\left(\frac{\theta-\phi}{2}\right)}$$

$\Rightarrow \frac{1-e}{1+e}=-\tan\frac{\theta}{2}\tan\frac{\phi}{2}$

or $\frac{1+e}{1-e}=-\tan\frac{\theta}{2}\tan\frac{\phi}{2}$

Thus the chord joining the points $R(\theta)$ and $Q(\phi)$ will be a focal chord passing through the focus $S(ae, 0)$ and $S'(-ae, 0)$ if $\frac{e-1}{e+1}=\tan\frac{\theta}{2}.\tan\frac{\phi}{2}$ and $\frac{e+1}{e-1}=\tan\frac{\theta}{2}.\tan\frac{\phi}{2}$.

ILLUSTRATION 26: If the chord joining the points $P\left(\frac{\pi}{12}\right)$ and $Q\left(\frac{\pi}{4}\right)$ on ellipse $\frac{x^2}{a^2}+\frac{y^2}{b^2}=1\ (a>b)$ is a focal chord, then find its eccentricity.

SOLUTION: We know that the equation of chord of ellipse $\frac{x^2}{a^2}+\frac{y^2}{b^2}=1$ joining $P(\theta)$ and $Q(\phi)$ is given by

$\frac{x}{a}\cos\left(\frac{\theta+\phi}{2}\right)+\frac{y}{b}\sin\left(\frac{\theta+\phi}{2}\right)=\cos\frac{\theta-\phi}{2}$.

If it is a focal chord, then it passes through $(ae, 0)$ or $(-ae, 0)$

$\Rightarrow \pm e\cos\left(\frac{\theta+\phi}{2}\right)=\cos\left(\frac{\theta-\phi}{2}\right)$

$$\therefore\ e=\left|\frac{\cos\left(\frac{\theta-\phi}{2}\right)}{\cos\left(\frac{\theta+\phi}{2}\right)}\right|$$

Let $\theta=\frac{\pi}{4};\ \phi=\frac{\pi}{12}$

$$\therefore\ e=\left|\frac{\cos\left(\frac{\frac{\pi}{4}-\frac{\pi}{12}}{2}\right)}{\cos\left(\frac{\frac{\pi}{4}+\frac{\pi}{12}}{2}\right)}\right|=\left|\frac{\cos\left(\frac{\pi}{12}\right)}{\cos\left(\frac{\pi}{6}\right)}\right|=\frac{\cos\left(\frac{\pi}{3}-\frac{\pi}{4}\right)}{\frac{\sqrt{3}}{2}}=\frac{\frac{1}{2}\cdot\frac{1}{\sqrt{2}}+\frac{\sqrt{3}}{2}\cdot\frac{1}{\sqrt{2}}}{\frac{\sqrt{3}}{2}}$$

$$=\frac{\left(\frac{\sqrt{3}+1}{2\sqrt{2}}\right)}{\frac{\sqrt{3}}{2}}=\frac{\sqrt{3}+1}{2\sqrt{2}}\times\frac{2}{\sqrt{3}}=\frac{\sqrt{3}+1}{\sqrt{6}}.$$

ILLUSTRATION 27: If PQ is focal chord of ellipse $\frac{x^2}{25}+\frac{y^2}{16}=1$ which passes through $S\equiv(3,0)$ and $PS=2$, then find the length of chord PQ.

SOLUTION: $a^2=25;\ b^2=16\Rightarrow 16=25(1-e^2)$

$$\Rightarrow\ \frac{16}{25}=1-e^2\Rightarrow e^2=1-\frac{16}{25}=\frac{9}{25}\Rightarrow e=\frac{3}{5}$$

$\therefore$ Foucs $S\equiv(ae,0)\equiv(3,0)$

Now, $PS=2$; Also $AS=2$.

$\Rightarrow$ P and A must coincide as circle with centre at S and passing through vertex A can't intersect the ellipse due to reflection property (to be discussed later).

$\therefore$ Q must coincide with A' $\therefore$ $PQ=AA'=2a=10$

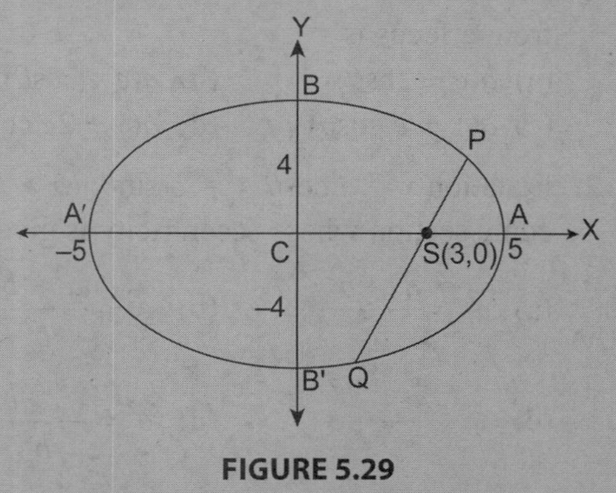

FIGURE 5.29

TEXTUAL EXERCISE–2 (SUBJECTIVE)

1. Find the position of the point (4, –3) relative to the ellipse $5x^2+7y^2=140$ and thus state the number of possible tangents which can be drawn from above point to the ellipse.

2. If a chord joining two points whose eccentric angles are α, β cut the major axis of an ellipse at a distance d from the centre. Show that $\tan\alpha/2\ \tan\beta/2=(d-a)/(d+a)$; where $2a$ is the length of major axis.

3. Find the condition on a and b for which two distinct chords of ellipse $\frac{x^2}{2a^2}+\frac{y^2}{2b^2}=1$ passing through $(a,-b)$ are bisected by the line $x+y=b$.

4. (a) If the eccentric angles of the ends of a focal chord of the ellipse $x^2/a^2 + y^2/b^2 = 1$ be θ and ψ, show that $\tan \theta/2 . \tan \psi/2 = \frac{e-1}{e+1}$.
   (b) Find the eccentricy of ellipse if chord joining θ and ϕ (where $\theta = 90^0$ and $\phi = -30^0$), is a focal chord.

5. Prove that the sum of the eccentric angles of the extremities of a chord which is drawn in a given direction is constant, and equal to twice the eccentric angles of the point, at which the tangent is parallel to the given direction.

6. Prove that the area of the triangle formed by point $P(\theta)$, $Q(\phi)$ and $R(\psi)$ on the ellipse $\frac{x^2}{a^2}+\frac{y^2}{b^2}=1; (a > b)$ is given by $2ab\left|\sin\frac{(\theta-\phi)}{2}\sin\frac{(\phi-\psi)}{2}\sin\frac{(\psi-\theta)}{2}\right|$

7. Find the area bounded between the auxiliary circle, and ellipse $\frac{x^2}{9}+\frac{y^2}{4}=1$ and the chord of auxiliary circle joining the corresponding points of points $P(\pi/4)$ and $Q(5\pi/4)$ on the ellipse (area of ellipse is given by πab).

Answer Keys

1. Point lies outside the ellipse, two tangent

3. $(a - b)(a + 7b) > 0$

4. (b) $1/\sqrt{3}$

7. $3\pi/2$

TEXTUAL EXERCISE–2 (OBJECTIVE)

1. The distance of the point 'θ' on the ellipse $\frac{x^2}{a^2}+\frac{y^2}{b^2}=1$ from a focus is
   (a) $a(e + \cos\theta)$
   (b) $a(e - \cos\theta)$
   (c) $a(1 + e \cos\theta)$
   (d) $a(1 + 2e \cos\theta)$

2. Equation $x = a \cos\theta, y = b \sin\theta; (a > b)$ represents a conic section whose eccentricity is given by
   (a) $e^2 = \frac{a^2+b^2}{a^2}$
   (b) $e^2 = \frac{a^2+b^2}{b^2}$
   (c) $\frac{a^2 \;\; b^2}{\;}$
   (d) $e^2 = \frac{a^2-b^2}{b^2}$

3. If $\tan\theta . \tan\phi = \frac{-a^2}{b^2}$, then the chord joining two points with eccentric angles θ and ϕ subtends a right angle at
   (a) focus
   (b) end of major axis
   (c) end of minor axis
   (d) centre

4. If a point P moves so that the sum of the squares of its distances from two intersecting straight lines is constant, then the locus of point is
   (a) circle
   (b) ellipse
   (c) parabola
   (d) hyperbola

5. The eccentric angles of the extremities of latus recta of the ellipse $\frac{x^2}{a^2}+\frac{y^2}{b^2}=1$ are given by
   (a) $\tan^{-1}\left(\pm\frac{ae}{b}\right)$
   (b) $\tan^{-1}\left(\pm\frac{ae}{a}\right)$
   (c) $\tan^{-1}\left(\pm\frac{b}{ae}\right)$
   (d) $\tan^{-1}\left(\pm\frac{a}{ae}\right)$

6. Eccentric angle of a point on the ellipse $x^2 + 3y^2 = 6$ at a distance 2 units from the centre of the ellipse is
   (a) $\frac{\pi}{4}$
   (b) $\frac{\pi}{3}$
   (c) $\frac{3\pi}{4}$
   (d) $\frac{2\pi}{4}$

7. If S_1 and S_2 be the foci (S_1 on left side and S_2 on right side of y-axis) of ellipse $\frac{x^2}{a^2}+\frac{y^2}{b^2}=1$; $(a > b)$. Let A be any point on the curve and the chords AS_2B, BS_1C, CS_2D, DS_1E and so on be drawn and eccentric angles of A, B, C, D, E....... be $\theta_1, \theta_2, \theta_3, \theta_4$....., then $\tan\frac{\theta_1}{2}\tan\frac{\theta_2}{2} = \cot\frac{\theta_2}{2}\cot\frac{\theta_3}{2} = \tan\frac{\theta_3}{2}\tan\frac{\theta_4}{2}$ equals
   (a) $\frac{e-1}{e+1}$
   (b) $\frac{e+1}{e-1}$
   (c) $\frac{e^2-1}{e^2+1}$
   (d) $\frac{e^2+1}{e^2-1}$

8. If S_1 and S_2 are the foci of an ellipse and P is any point on the curve, then $\tan\left(\frac{\angle PS_1S_2}{2}\right)\tan\left(\frac{\angle PS_2S_1}{2}\right) =$

(a) $\left(\frac{e-1}{e+1}\right)$ (b) $\frac{1-e}{1+e}$

(c) $\frac{e+1}{e-1}$ (d) $\frac{1+e}{1-e}$

9. Let P be any point on the ellipse $\frac{x^2}{9}+\frac{y^2}{4}=1$ and Q be its corresponding point on the auxiliary circle of ellipse respectively. The locus of point R such that $PR : RQ = 2 : 3$ is

(a) $9x^2 + 16y^2 = 144$ (b) $25x^2 + 16y^2 = 144$

(c) $16x^2 + 25y^2 = 144$ (d) None of these

Answer Keys

1. (c) **2.** (c) **3.** (d) **4.** (b) **5.** (c) **6.** (a) **7.** (a) **8.** (b) **9.** (c)

■ INTERSECTION OF A LINE AND AN ELLIPSE

Let the ellipse be $\frac{x^2}{a^2}+\frac{y^2}{b^2}=1$ ....(i)

and the given line be $y = mx + c$ ....(ii)

Eliminating y from equations (i) and (ii), we have

$$\frac{x^2}{a^2}+\frac{(mx+c)^2}{b^2}=1$$

$$\Rightarrow (a^2m^2+b^2)x^2+2mca^2x+c^2a^2-a^2b^2=0 \quad ....(iii)$$

The above equation being a quadratic in x gives two values of x, shows that every straight line will cut the ellipse in two points may be real and different, coincident or imaginary according as

Discriminant of (iii) >, =, <0

i.e., $4m^2c^2a^4 - 4(a^2m^2 + b^2)(c^2a^2 - a^2b^2) >, =, < 0$

or $- a^2b^2c^2 + a^4b^2m^2 + a^2b^4 >, =, < 0$

or $a^2m^2 + b^2 >, =, < c^2$ ....(iv)

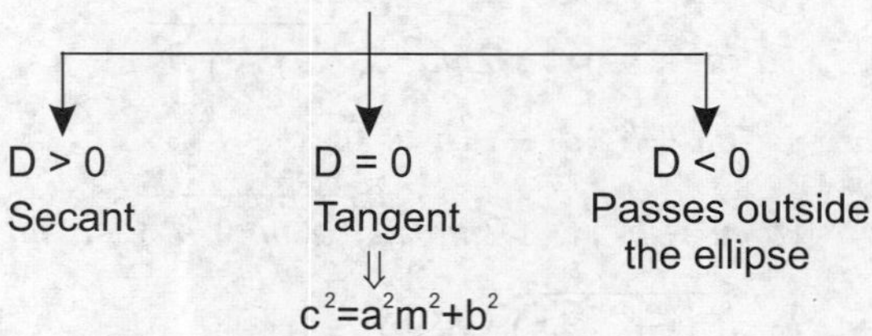

FIGURE 5.30

ILLUSTRATION 28: Find the set of value(s) of 'λ' for which the line $3x - 4y + \lambda = 0$ intersect the ellipse $\frac{x^2}{16}+\frac{y^2}{9}=1$ at two distinct points.

SOLUTION: Solving given line with ellipse, we get $\frac{(4y-\lambda)^2}{9\times16}+\frac{y^2}{9}=1$ or $\frac{16y^2+\lambda^2-8\lambda y}{9\times16}+\frac{y^2}{9}=1$

$$\frac{2y^2}{9}-\frac{y\lambda}{18}+\frac{\lambda^2}{144}-1=0$$

Since, line intersect the parabola at two distinct points,

∴ Roots of the above equation are real and distinct

∴ $D>0$

$$\Rightarrow \frac{\lambda^2}{(18)^2}-\frac{8}{9}\cdot\left(\frac{\lambda^2}{144}-1\right)>0 \qquad \Rightarrow -12\sqrt{2}<\lambda<12\sqrt{2}$$

CONDITION OF TANGENCY AND TANGENT TO ELLIPSE

If the line (ii) touches the ellipse (i), then equation (iii) has equal roots.

$\therefore$ Discriminant of (iii) = 0

$\Rightarrow \; c^2 = a^2m^2 + b^2$ or $c = \pm\sqrt{a^2m^2+b^2}$ ....(v)

so, the line $y = mx + c$ touches the ellipse

$$\frac{x^2}{a^2}+\frac{y^2}{b^2}=1 \text{ if } c^2 = a^2m^2+b^2$$

(which is condition of tangency)

Substituting the value of c from (v) in (ii), we have

$y = mx \pm \sqrt{a^2m^2+b^2}$

Hence the lines $\boldsymbol{y = mx \pm \sqrt{a^2m^2+b^2}}$ will always be tangents to the ellipse; $m \in \mathbb{R}$

REMARK

1. *$m \in \mathbb{R}$ and $c = \pm\sqrt{a^2m^2+b^2} \Rightarrow c \le -b$ or $\ge b$.*
2. *If slope of tangent m is known, then **c** can be determined and vice-versa by using the relation $c = \pm\sqrt{a^2m^2+b^2}$.*
3. *If slope of tangent is infinite, i.e., $m = \frac{1}{k}, k \to 0$, then equation of tangent becomes $y = \underset{k\to 0}{Lt}\left(\frac{x}{k} \pm \frac{1}{k}\sqrt{a^2+b^2k^2}\right)$*
   $\Rightarrow 0 = x \pm a \Rightarrow x = \pm a$ i.e., tangent at vertices A and A' of ellipse.

POINT OF CONTACT OF TANGENT

Substituting $c = \pm\sqrt{a^2m^2+b^2}$ in equation (iii), we have

$$(a^2m^2+b^2)x^2 \pm 2ma^2x\sqrt{(a^2m^2+b^2)} + (a^2m^2+b^2)a^2 - a^2b^2 = 0$$

or $\quad (a^2m^2+b^2)x^2 \pm 2ma^2x\sqrt{(a^2m^2+b^2)} + a^4m^2 = 0$

or $\quad (x\sqrt{(a^2m^2+b^2)} \pm a^2m)^2 = 0$

$\therefore \quad x = \mp\frac{a^2m}{\sqrt{(a^2m^2+b^2)}} = -\left(\frac{a^2m}{\pm\sqrt{(a^2m^2+b^2)}}\right) = \frac{-a^2m}{c}.$

Again, point of contanct lies on straight line $y = mx + c$

$\Rightarrow \; y = m\left(\frac{-a^2m}{c}\right) + c = \frac{-a^2m^2+c^2}{c} = \frac{b^2}{c}$

$(\quad c^2 = a^2m^2 + b^2)$

$\therefore$ Point of contact is given by $\left(\frac{-a^2m}{c}, \frac{b^2}{c}\right)$

REMARKS

1. *If we know one of slope (m) and intercept (c), then other can be determined by using the relation $c^2 = a^2m^2 + b^2$. After that point of contact can be determined by using the formula $\left(\frac{-a^2m}{c}, \frac{b^2}{c}\right)$.*
2. *For every value of 'm' there correspond two values of 'c' and corresponding to every value of 'c' $\in (-\infty, -b] \cup [b, \infty)$ there correspond two values of 'm'. Therefore, there are two tangents of given slope having y-intercept of same magnitude but opposite in sign. Similarly, for every possible value of c ($\neq b$) there exist two tangents having slopes which are same in magnitude but opposite in sign.*

ILLUSTRATION 29: Prove that the straight line $\alpha x + \beta y + \gamma = 0$ touches

(i) the ellipse $\frac{x^2}{a^2}+\frac{y^2}{b^2}=1$ if $a^2\alpha^2+b^2\beta^2=\gamma^2$

(ii) the ellipse $\frac{x^2}{b^2}+\frac{y^2}{a^2}=1$ if $\gamma^2 = a^2\beta^2+b^2\alpha^2$

SOLUTION: The given line is $\alpha x + \beta y + \gamma = 0$

or $y = -\dfrac{\alpha}{\beta}x - \dfrac{\gamma}{\beta}$ ......(1)

Comparing this line with $y = mx + c$, we have $m = -\dfrac{\alpha}{\beta}$ and $c = -\dfrac{\gamma}{\beta}$

(i) The line (1) will touch the ellipse $\dfrac{x^2}{a^2} + \dfrac{y^2}{b^2} = 1$ if $c^2 = a^2m^2 + b^2$

$\Rightarrow \dfrac{\gamma^2}{\beta^2} = a^2\dfrac{\alpha^2}{\beta^2} + b^2$ $\qquad \Rightarrow \gamma^2 = a^2\alpha^2 + b^2\beta^2$

(ii) The line (1) will touch the ellipse $\dfrac{x^2}{b^2} + \dfrac{y^2}{a^2} = 1$ if $c^2 = a^2 + b^2m^2$

$\Rightarrow \dfrac{\gamma^2}{\beta^2} = a^2 + b^2\left(\dfrac{\alpha^2}{\beta^2}\right)$ $\qquad \Rightarrow \gamma^2 = a^2\beta^2 + b^2\alpha^2$

ILLUSTRATION 30: Show that the line $x\sec\alpha + y\tan\alpha = p$ touches the ellipse $\dfrac{x^2}{a^2} + \dfrac{y^2}{b^2} = 1$ if $a^2\sec^2\alpha + b^2\tan^2\alpha = p^2$ and that point of contact is $\left(\dfrac{a^2}{p}\sec\alpha, \dfrac{b^2}{p}\tan\alpha\right)$

SOLUTION: The given line is $x\sec\alpha + y\tan\alpha = p$

$y = -x\text{cosec }\alpha + p\cot\alpha$

Comparing this line with $y = mx + c$, we have

$m = -\text{cosec}\,\alpha$ and $c = p\cot\alpha$

Hence the given line touches the ellipse if $c^2 = a^2m^2 + b^2$

$\Rightarrow p^2\cot^2\alpha = a^2\text{cosec}^2\alpha + b^2$

$\Rightarrow p^2 = a^2\sec^2\alpha + b^2\tan^2\alpha$

and the point of contact is given by $\left(-\dfrac{a^2m}{c}, \dfrac{b^2}{c}\right)$

$\equiv \left(\dfrac{-a^2(-\text{cosec }\alpha)}{p\cot\alpha}, \dfrac{b^2}{p\cot\alpha}\right) \equiv \left(\dfrac{a^2}{p}\sec\alpha, \dfrac{b^2}{p}\tan\alpha\right)$

ILLUSTRATION 31: For what value of k does the line $y = x + k$ touches the ellipse $4x^2 + 9y^2 = 36$?

SOLUTION: Equation of ellipse is $4x^2 + 9y^2 = 36$

or $\dfrac{x^2}{9} + \dfrac{y^2}{4} = 1$

Comparing this with $\dfrac{x^2}{a^2} + \dfrac{y^2}{b^2} = 1$, we have $a^2 = 9$ and $b^2 = 4$

and comparing the line $y = x + k$ with $y = mx + c$, we have $m = 1$ and $c = k$

If the line $y = x + k$ touches the ellipse $4x^2 + 9y^2 = 36$ then $c^2 = a^2m^2 + b^2$

$\Rightarrow k^2 = 9 \times 1^2 + 4$ $\qquad \Rightarrow k^2 = 13$

$\therefore \quad k = \pm\sqrt{13}$

■ EQUATIONS OF TANGENT TO ELLIPSE IN DIFFERENT FORMS

1. Point Form (Cartesian Form):

(i) Tangent at a point (x_1, y_1) on ellipse $\frac{x^2}{a^2}+\frac{y^2}{b^2}=1$ $(a > b)$.

Method 1: **(By first principle method)** Equation of ellipse is $\frac{x^2}{a^2}+\frac{y^2}{b^2}=1$ ....(1)

Let $P \equiv (x_1, y_1)$ and $Q \equiv (x_2, y_2)$ be any two points on (1), then $\frac{x_1^2}{a^2}+\frac{y_1^2}{b^2}=1$ ....(2)

and $\frac{x_2^2}{a^2}+\frac{y_2^2}{b^2}=1$ ....(3)

Subtracting (2) from (3), we have

$$\frac{1}{a^2}(x_2^2-x_1^2)+\frac{1}{b^2}(y_2^2-y_1^2)=0$$

$$\Rightarrow \frac{(x_2+x_1)(x_2-x_1)}{a^2}+\frac{(y_2+y_1)(y_2-y_1)}{b^2}=0$$

$$\Rightarrow \frac{y_2-y_1}{x_2-x_1}=-\frac{b^2(x_1+x_2)}{a^2(y_1+y_2)} \quad \text{....(4)}$$

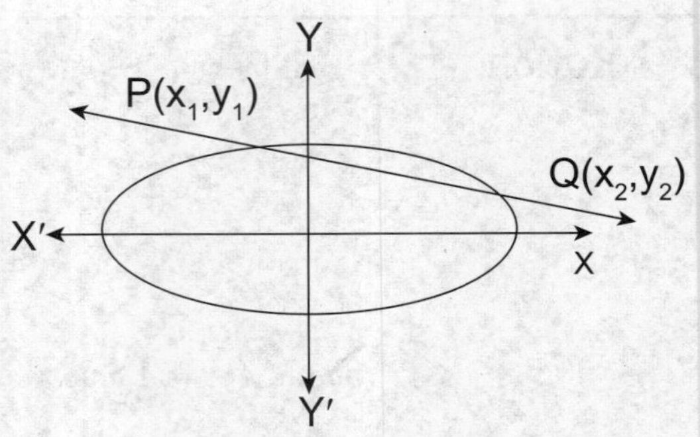

FIGURE 5.31

Equation of PQ is $y-y_1=\frac{y_2-y_1}{x_2-x_1}(x-x_1)$ ....(5)

$\therefore$ From (4) and (5), $y-y_1=-\frac{b^2(x_1+x_2)}{a^2(y_1+y_2)}(x-x_1)$ ...(6)

Now, for tangent at P, $Q \to P$ i.e., $x_2 \to x_1$ and $y_2 \to y_1$, then equation (6) becomes $y-y_1=-\frac{b^2(2x_1)}{a^2(2y_1)}(x-x_1)$

or $\frac{yy_1-y_1^2}{b^2}=-\left(\frac{xx_1-x_1^2}{a^2}\right)$ or $\frac{xx_1}{a^2}+\frac{yy_1}{b^2}=\frac{x_1^2}{a^2}+\frac{y_1^2}{b^2}$

or $\frac{xx_1}{a^2}+\frac{yy_1}{b^2}=1$ [from (2)]

which is the required equation of tangent at (x_1, y_1).

NOTE

The equation of tangent at (x_1, y_1) *to any conic of second degree can be obtained by replacing* x^2 *by* xx_1, y^2 by yy_1, x by $\frac{x+x_1}{2}$, y by $\frac{y+y_1}{2}$ and xy by $\frac{xy_1+x_1y}{2}$. *Thus equation of tangent to ellipse* $\frac{x^2}{a^2}+\frac{y^2}{b^2}=1$ *at point* (x_1, y_1) *is* $\frac{xx_1}{a^2}+\frac{yy_1}{b^2}=1$.

Method 2: **(By using calculus)**

The equation of given ellipse is $\frac{x^2}{a^2}+\frac{y^2}{b^2}=1$ ....(i)

Let (x_1, y_1) be any point on ellipse (i) at which we are to find the equation of tangent.

(x_1, y_1) lies on ellipse (i)

$\therefore$ $\frac{x_1^2}{a^2}+\frac{y_1^2}{b^2}=1$ ....(ii)

Differentiating both sides of (i) w.r.t x we get

$$\frac{2x}{a^2}+\frac{2y}{b^2}\cdot\frac{dy}{dx}=0 \Rightarrow \frac{dy}{dx}=\frac{-b^2x}{a^2y}$$

$\therefore$ Slope of tangent at (x_1, y_1) to ellipse (i) is

$$m=\frac{-b^2x_1}{a^2y_1}$$

$\therefore$ Equation of tangent at (x_1, y_1) to ellipse (i) will be

$$(y-y_1)=\frac{-b^2x_1}{a^2y_1}(x-x_1)$$

$$\Rightarrow (a^2yy_1)-(a^2y_1^2)=-b^2xx_1+b^2x_1^2$$

$$\Rightarrow a^2yy_1+b^2xx_1=a^2y_1^2+b^2x_1^2$$

$$\Rightarrow \frac{yy_1}{b^2}+\frac{xx_1}{a^2}=\frac{y_1^2}{b^2}+\frac{x_1^2}{a^2}=1(\quad \text{of (ii)})$$

$\Rightarrow \dfrac{xx_1}{a^2}+\dfrac{yy_1}{b^2}=1$, which is the required equation of tangent to given ellipse (i) at point (x_1, y_1)

2. (Parametric form): Tangent at a point $(a\cos\phi, b\sin\phi)$ on ellipse $\dfrac{x^2}{a^2}+\dfrac{y^2}{b^2}=1$; $(a>b)$

The equation of tangent to the ellipse $\dfrac{x^2}{a^2}+\dfrac{y^2}{b^2}=1$ at the point (x_1, y_1) is

$$\frac{xx_1}{a^2}+\frac{yy_1}{b^2}=1 \qquad \text{(by point form)}$$

Replacing x_1 by $a\cos\phi$ and y_1 by $b\sin\phi$, we get

$$\frac{x}{a}\cos\phi+\frac{y}{b}\sin\phi=1$$

ILLUSTRATION 32: The tangent at a point $P(a\cos\phi, b\sin\phi)$ to an ellipse $\dfrac{x^2}{a^2}+\dfrac{y^2}{b^2}=1$, meets its auxiliary circle in two points, the chord joining them subtends a right angle at the centre. Show that the eccentricity of the ellipse is $(1+\sin^2\phi)^{-1/2}$.

SOLUTION: Equation of the auxiliary circle is $x^2+y^2=a^2$ ....(1)

Equation of the tangent line at point $P(a\cos\phi, b\sin\phi)$ is $\dfrac{x}{a}\cos\phi+\dfrac{y}{b}\sin\phi=1$ ...(2)

$\therefore$ equation of pair of lines OA, OB is obtained by making equation (1) homogeneous with the help of (2)

$$\therefore\ x^2+y^2=a^2\left(\frac{x}{a}\cos\phi+\frac{y}{b}\sin\phi\right)^2$$

$$\Rightarrow (1-\cos^2\phi)x^2-\frac{2xya\sin\phi\cos\phi}{b}+y^2\left(1-\frac{a^2}{b^2}\sin^2\phi\right)=0$$

But $\angle AOB=90^o$

$\Rightarrow$ co-efficient of x^2 + co-efficient of $y^2=0$

$$\Rightarrow 1-\cos^2\phi+1-\frac{a^2}{b^2}\sin^2\phi=0 \Rightarrow 1=\frac{a^2-b^2}{b^2}\sin^2\phi$$

$$\Rightarrow 1=\frac{a^2e^2}{a^2(1-e^2)}\sin^2\phi \Rightarrow e^2=\frac{1}{1+\sin^2\phi}\quad \therefore\ e=(1+\sin^2\phi)^{-1/2}$$

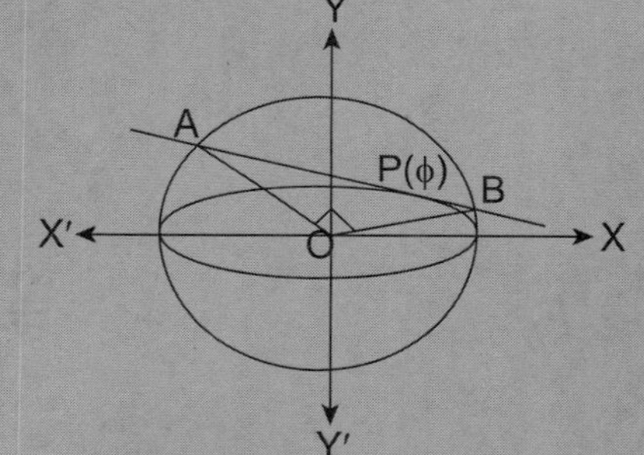

FIGURE 5.32

Point of intersection of tangents:

`$\theta$' and `ϕ' on the ellipse $\dfrac{x^2}{a^2}+\dfrac{y^2}{b^2}=1$

The point of intersection of tangent at $P(\theta)$ and $R(\phi)$ i.e., $\dfrac{x}{a}\cos\phi+\dfrac{y}{b}\sin\phi=1$ and $\dfrac{x}{a}\cos\theta+\dfrac{y}{b}\sin\theta=1$ is given by

$$\left(\frac{a\cos\left(\frac{\theta+\phi}{2}\right)}{\cos\left(\frac{\theta-\phi}{2}\right)}, \frac{b\sin\left(\frac{\theta+\phi}{2}\right)}{\cos\left(\frac{\theta-\phi}{2}\right)}\right)$$

Remembering Method:

$\because$ Equation of chord joining $(a\cos\theta, b\sin\theta)$ and $(a\cos\phi, b\sin\phi)$ is

$$\frac{x}{a}\cos\left(\frac{\theta+\phi}{2}\right)+\frac{y}{b}\sin\left(\frac{\theta+\phi}{2}\right)=\cos\left(\frac{\theta-\phi}{2}\right)$$

$$\Rightarrow \frac{x}{a}\left\{\frac{\cos\left(\frac{\theta+\phi}{2}\right)}{\cos\left(\frac{\theta-\phi}{2}\right)}\right\}+\frac{y}{b}\left\{\frac{\sin\left(\frac{\theta+\phi}{2}\right)}{\cos\left(\frac{\theta-\phi}{2}\right)}\right\}=1$$

$$\text{or}\ \frac{x}{a^2}\left\{\frac{a\cos\left(\frac{\theta+\phi}{2}\right)}{\cos\left(\frac{\theta-\phi}{2}\right)}\right\}+\frac{y}{b^2}\left\{\frac{b\sin\left(\frac{\theta+\phi}{2}\right)}{\cos\left(\frac{\theta-\phi}{2}\right)}\right\}=1$$

co-eficients of $\dfrac{x}{a^2}$ and $\dfrac{y}{b^2}$ gives us respectively the abscissa and ordinate of point of intersection.

ILLUSTRATION 33: Find the locus of the point of intersection of the pair of tangents to an ellipse if the sum of the ordinates of their point of contact is b.

SOLUTION: Let $P(x_1, y_1)$ be the point of intersection of pair of tangents having their points of contacts $A(a\cos\alpha, b\sin\alpha)$ and $B(a\cos\beta, b\sin\beta)$

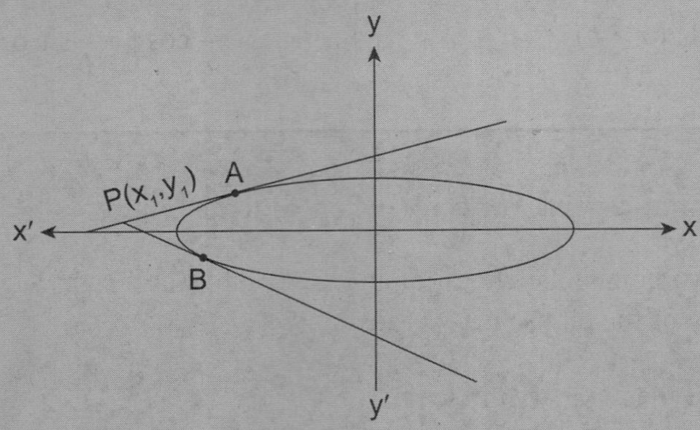

FIGURE 5.33

Given $b\sin\alpha + b\sin\beta = b$

$\Rightarrow \sin\alpha + \sin\beta = 1$ ...(1)

Also $x_1 = a\left(\dfrac{\cos\frac{\alpha+\beta}{2}}{\cos\frac{\alpha-\beta}{2}}\right)$; $y_1 = b\left(\dfrac{\sin\frac{\alpha+\beta}{2}}{\cos\frac{\alpha-\beta}{2}}\right)$

hence $\left(\dfrac{x_1^2}{a^2}+\dfrac{y_1^2}{b^2}\right)\cos^2\dfrac{\alpha-\beta}{2} = 1$ ....(2)

now (1) gives $2\sin\dfrac{\alpha+\beta}{2}\cos\dfrac{\alpha-\beta}{2} = 1$ ....(3)

$\Rightarrow \dfrac{2y_1}{b}\cos^2\dfrac{\alpha-\beta}{2} = 1$ $\left(\text{Using } \sin\dfrac{\alpha+\beta}{2} = \dfrac{y_1}{b}\cos\dfrac{\alpha-\beta}{2}\right)$

$\Rightarrow \cos^2\dfrac{\alpha-\beta}{2} = \dfrac{b}{2y_1}$; $\Rightarrow \left(\dfrac{x_1^2}{a^2}+\dfrac{y_1^2}{b^2}\right)\dfrac{b}{2y_1} = 1$ (from (2))

$\Rightarrow$ locus of point of intersection is $\dfrac{x^2}{a^2}+\dfrac{y^2}{b^2} = \dfrac{2y}{b}$

3. Slope form: given ellipse is $\dfrac{x^2}{a^2}+\dfrac{y^2}{b^2} = 1$ ....(i)

Let the equation of tangent be $y = mx + c$ ....(ii)

At the point of intersection of (i) and (ii), we have

$\dfrac{x^2}{a^2}+\dfrac{(mx+c)^2}{b^2} = 1$

$\Rightarrow b^2x^2 + a^2(m^2x^2 + c^2 + 2mcx) = a^2b^2$

$\Rightarrow (a^2m^2 + b^2)x^2 + 2mca^2x + a^2c^2 - a^2b^2 = 0$ ....(iii)

Now, (iii) being a quadratic in x gives two abscissae of points of intersection of (i) and (ii). But (ii) is a tangent to (i), therefore two roots of (iii) would coincide.

$\therefore$ Discriminant of (iii) $= 0 \Rightarrow 4m^2c^2a^4 - 4(a^2m^2 + b^2)(a^2c^2 - a^2b^2) = 0$

$\Rightarrow m^2c^2a^2 - (a^2m^2 + b^2)(c^2 - b^2) = 0$

$\Rightarrow m^2c^2a^2 - a^2m^2c^2 + a^2m^2b^2 - b^2c^2 + b^4 = 0$

$\Rightarrow c^2 = a^2m^2 + b^2$

$\Rightarrow c = \pm\sqrt{a^2m^2 + b^2}$

Therefore equations of tangents to ellipse (i) are

$y = mx \pm \sqrt{a^2m^2 + b^2}$; $m \in \mathbb{R}$

ILLUSTRATION 34: If two points are taken on the minor axis of an ellipse at the same distance from the centre as foci, prove that the sum of the square of the perpendicular from these points on any tangents to the ellipse is constant.

SOLUTION: Let the equation of the ellipse be $\frac{x^2}{a^2}+\frac{y^2}{b^2}=1$

Then, the distance of a focus from the centre $= ae$ $= \sqrt{a^2-b^2}$, so that the two points of minor axis are $A\left(0,\sqrt{a^2-b^2}\right)$ and $B\left(0,-\sqrt{a^2-b^2}\right)$

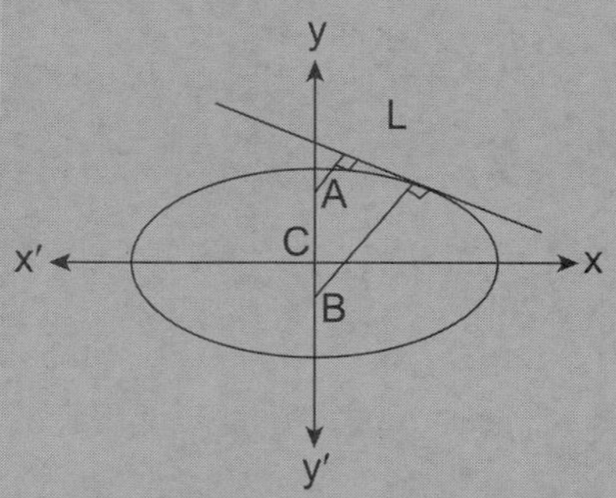

FIGURE 5.34

Now any tangent to the ellipse is $y=mx+\sqrt{a^2m^2+b^2}$; where m is a parameter.

The sum of the square of the perpendiculars on this tangent from the two points A and B

$$=\left(\frac{\sqrt{a^2-b^2}-\sqrt{a^2m^2+b^2}}{\sqrt{1+m^2}}\right)^2+\left(\frac{-\sqrt{a^2-b^2}-\sqrt{a^2m^2+b^2}}{\sqrt{1+m^2}}\right)^2$$

$= 2\,(a^2 - b^2 + a^2m^2 + b^2)/(1 + m^2) = 2a^2$ (constant)

ILLUSTRATION 35: Show that for all p, the line $2px+y\sqrt{1-p^2}=1$ touches a fixed ellipse. Find the eccentricity of the ellipse

SOLUTION: Let the ellipse be $\frac{x^2}{a^2}+\frac{y^2}{b^2}=1$. The line $y = \text{m}x \pm \sqrt{a^2m^2+b^2}$ touches this ellipse for all m

Hence it is identical with $y=\frac{-2px}{\sqrt{1-p^2}}+\frac{1}{\sqrt{1-p^2}}$

Hence $m=-\frac{2p}{\sqrt{1-p^2}}$ and $a^2\,m^2 + b^2 = \frac{1}{1-p^2}$

$\Rightarrow\ a^2.\frac{4p^2}{1-p^2}+b^2=\frac{1}{1-p^2}$ $\qquad\Rightarrow\ p^2\left(4a^2-b^2\right)+b^2-1=0$

This equation is true for all real p

$\Rightarrow\ b^2 = 1$ and $4a^2 = b^2$

$\Rightarrow\ b^2 = 1$ and $a^2=\frac{1}{4}$. Hence the ellipse is $\frac{x^2}{1/4}+\frac{y^2}{1}=1$

If e is its eccentricity, then $\frac{1}{4}=1-e^2$

$\Rightarrow\ e^2=\frac{3}{4}\Rightarrow e=\frac{\sqrt{3}}{2}$

ILLUSTRATION 36: A tangent to the ellipse $\frac{x^2}{27}+\frac{y^2}{48}=1$, having slope $\frac{-4}{3}$ cuts the x and y-axis at the points A and B respectively. If O is the orign, then find the area of ΔOAB

SOLUTION: Let $\left(\sqrt{27}\cos\theta,\sqrt{48}\sin\theta\right)$ be any point on the ellipse $\frac{x^2}{27}+\frac{y^2}{48}=1$, ...(1)

then equation of tangent at point (θ) to ellipse (1) is given by $\frac{x\sqrt{27}\cos\theta}{27}+\frac{y\sqrt{48}\sin\theta}{48}=1$

$\Rightarrow\ \frac{x\cos\theta}{\sqrt{27}}+\frac{y\sin\theta}{\sqrt{48}}=1$ ....(2)

Its slope $=-\left(\frac{\cos\theta}{\sqrt{27}}\right)\times\left(\frac{\sqrt{48}}{\sin\theta}\right)=\frac{-4}{3}\Rightarrow\cot\theta=\frac{4}{3}\times\sqrt{\frac{27}{48}}=\frac{4}{3}\times\frac{3\sqrt{3}}{4\sqrt{3}}=1$

$\therefore$ $\cot\theta = 1 \Rightarrow \theta = \pi/4$

$\therefore$ (2) becomes $\dfrac{x.1}{\sqrt{54}}+\dfrac{y.1}{\sqrt{96}}=1 \Rightarrow \dfrac{x}{3\sqrt{6}}+\dfrac{y}{4\sqrt{6}}=1$

$\therefore$ Area $\Delta OAB = \dfrac{1}{2}(3\sqrt{6})(4\sqrt{6}) = 6\times 6 = 36$ square units

ILLUSTRATION 37: Find the equation of tangents to the ellipse $\dfrac{x^2}{9}+\dfrac{y^2}{4}=1$ at the point of intersection or (point of contact) with the circle $(x-1)^2+y^2=4$. Also find the area of polygon made by points of intersection of ellipse and circle.

SOLUTION: Equation of ellipse is $\dfrac{x^2}{9}+\dfrac{y^2}{4}=1$ ....(1)

equation of circle is $(x-1)^2+y^2=4$ ....(2)

from (2) $y^2 = 4-(x-1)^2$ ....(3)

Putting in (1), we get $\dfrac{x^2}{9}+\dfrac{4-(x-1)^2}{4}=1$

$\Rightarrow 4x^2+36-9(x-1)^2=36$

$\Rightarrow 4x^2-9(x^2+1-2x)=0$ $\qquad \Rightarrow 4x^2-9x^2-9+18x=0$

$\Rightarrow -5x^2+18x-9=0$ $\qquad \Rightarrow 5x^2-18x+9=0$

$\Rightarrow x = \dfrac{18\pm\sqrt{(18)^2-4(5)(9)}}{2(5)}$ $\qquad \Rightarrow x = \dfrac{18\pm\sqrt{324-180}}{10}$

$\Rightarrow x = \dfrac{18\pm\sqrt{144}}{10}$ $\qquad \Rightarrow x = \dfrac{18\pm 12}{10} \Rightarrow x = 3, \dfrac{3}{5};$

$\Rightarrow$ At $x=3$; $(3-1)^2+y^2=4$ $\qquad \Rightarrow y = 0;$

$\Rightarrow$ At $x = 3/5$; $\left(\dfrac{3}{5}-1\right)^2+y^2=4$

$\Rightarrow y^2 = 4-\dfrac{4}{25}=\dfrac{96}{25}$ $\qquad \Rightarrow y = \pm\dfrac{\sqrt{96}}{5}=\pm\dfrac{4}{5}\sqrt{6}$

$\therefore$ $A(3,0)$ and $B\left(\dfrac{3}{5},\dfrac{4\sqrt{6}}{5}\right)$ and $C\left(\dfrac{3}{5},\dfrac{-4\sqrt{6}}{5}\right)$ are the points of intersection

$\therefore$ Equation of tangent at $A(3, 0)$ is given by $\dfrac{xx_1}{a^2}+\dfrac{yy_1}{b^2}=1$

i.e., $\dfrac{3x}{9}+\dfrac{y(0)}{4}=1 \Rightarrow x=3$

and equation of tangent at $B\left(\dfrac{3}{5},\dfrac{4\sqrt{6}}{5}\right)$ and $C\left(\dfrac{3}{5},\dfrac{-4\sqrt{6}}{5}\right)$ will be

$\dfrac{x\left(\frac{3}{5}\right)}{9}+\dfrac{y\left(\pm\frac{4\sqrt{6}}{5}\right)}{4}=1$ or $\dfrac{x}{15}\pm\dfrac{\sqrt{6}}{5}y=1$ or $x\pm 3\sqrt{6}y=15$

FIGURE 5.35

$\therefore$ $x=3, x+3\sqrt{6}y=15$ and $x-3\sqrt{6}y=15$

There are three points. of intersection A, B and C

$\therefore$ we are to find area of $\Delta ABC = \dfrac{1}{2}BC\times AL = \dfrac{1}{2}(2BL)\times AL = (BL).(AL)$

$=\left(\dfrac{4\sqrt{6}}{5}\right)\times\left(3-\dfrac{3}{5}\right)=\dfrac{4\sqrt{6}}{5}\times\dfrac{12}{5}=\dfrac{48\sqrt{6}}{25}$ sq. units.

ILLUSTRATION 38: Find the equations of tangents to ellipse $x^2 + 4y^2 = 4$ parallel to pair of straight lines $y^2 + xy - 6x^2 - 21x - 7y = 0$. Also find their points of contact.

SOLUTION: Given ellipse is $x^2 + 4y^2 = 4$

$\Rightarrow \frac{x^2}{4} + \frac{y^2}{1} = 1$

which is of the form $\frac{x^2}{a^2} + \frac{y^2}{b^2} = 1$

$\therefore$ $a^2 = 4$, $b^2 = 1$

Also equations of pairs of straight lines is $y^2 + xy - 6x^2 - 21x - 7y = 0$

$\Rightarrow (y^2 + 3xy - 2xy - 6x^2) - 7(3x + y) = 0$ $\Rightarrow y(y + 3x) - 2x(y + 3x) - 7(3x + y) = 0$

$\Rightarrow (y + 3x)(y - 2x - 7) = 0$ $\Rightarrow y = -3x$ and $y = 2x + 7$

$\therefore$ Slopes of pair of lines are $m_1 = -3$; $m_2 = 2$

Equations of tangents to ellipse $\frac{x^2}{a^2} + \frac{y^2}{b^2} = 1; (a > b)$

of slope m, are given by $y = mx \pm \sqrt{a^2m^2 + b^2}$

For $m = -3$,

Equations of tangents are $y = -3x \pm \sqrt{4(-3)^2 + 1}$ or $y = -3x \pm \sqrt{37}$

For $m = 2$

Equations of tangents are $y = 2x \pm \sqrt{4(2)^2 + 1}$ or $y = 2x \pm \sqrt{17}$

$\therefore$ Equation of tangents are $3x + y \pm \sqrt{37} = 0$ and $2x - y \pm \sqrt{17} = 0$

Also, points of contact of any tangent $y = mx + c$ with ellipse $\frac{x^2}{a^2} + \frac{y^2}{b^2} = 1$ are given by $\left(\frac{-a^2m}{c}, \frac{b^2}{c}\right)$

$\therefore$ Points of contact of tangent $y = 2x + \sqrt{17}$ are

$\left(-\frac{4(2)}{\sqrt{17}}, \frac{1}{\sqrt{17}}\right) \equiv \left(\frac{-8}{\sqrt{17}}, \frac{1}{\sqrt{17}}\right)$..........(in IInd quadrant)

that of $y = 2x - \sqrt{17}$ are

$\left(\frac{-4(2)}{-\sqrt{17}}, \frac{1}{-\sqrt{17}}\right) \equiv \left(\frac{8}{\sqrt{17}}, \frac{-1}{\sqrt{17}}\right)$..........(in IVth quadrant)

that of $y = -3x + \sqrt{37}$ are $\left(\frac{-4(-3)}{\sqrt{37}}, \frac{1}{\sqrt{37}}\right) \equiv \left(\frac{12}{\sqrt{37}}, \frac{1}{\sqrt{37}}\right)$..........(in Ist quadrant)

that of $y = -3x - \sqrt{37}$ are $\left(\frac{-4(-3)}{-\sqrt{37}}, \frac{1}{-\sqrt{37}}\right) \equiv \left(\frac{-12}{\sqrt{37}}, \frac{-1}{\sqrt{37}}\right)$..........(in IIIrd quadrant)

REMARKS

$$\begin{bmatrix} \text{for slope} > 0, c > 0, \text{point of contact lies in IInd quad.} \\ \text{for slope} > 0, c < 0, \text{point of contact lies in IVth quad.} \\ \text{for slope} < 0, c > 0, \text{point of contact lies in Ist quad.} \\ \text{for slope} < 0, c < 0, \text{point of contact lies in IIIrd quad.} \end{bmatrix}$$

$$\begin{bmatrix} \text{for slope} = 0, \text{there will be two tangents} \\ y = \pm b \text{ with point of contacts } (0, \pm b) \text{ respectively} \\ \text{for slope} = \infty, \text{there will be two tangents } x = \pm a \\ \text{with point of contact } (a, 0) \text{ and } (-a, 0) \text{ respectively} \end{bmatrix}$$

ILLUSTRATION 39: Find the equations of the tangents to the ellipse $81x^2+36y^2=2916$ which are ⊥r to the line $y+2x=6$.

SOLUTION: Let m be the slope of the tangent, since the tangent is perpendicular to the line $y + 2x = 6$.

$\Rightarrow\ m=\frac{1}{2}$; equation of ellipse is

$\therefore\ 81x^2+36y^2=2916$ or $\frac{x^2}{36}+\frac{y^2}{81}=1$

Comparing it with $\frac{x^2}{b^2}+\frac{y^2}{a^2}=1$, we have

$\therefore\ a^2=81$ and $b^2=36$

So the equations of the tangents are $y=mx\pm\sqrt{a^2+b^2m^2}$

$\Rightarrow\ y=\frac{1}{2}x\pm\sqrt{81+36\left(\frac{1}{4}\right)}$ $\qquad \Rightarrow\ y=\frac{1}{2}x\pm\sqrt{90}\Rightarrow 2y=x\pm 6\sqrt{10}$

ILLUSTRATION 40: Find the locus of the foot of the perpendicular drawn from centre upon any tangent to the ellipse $\frac{x^2}{a^2}+\frac{y^2}{b^2}=1$.

SOLUTION: Any tangent to $\frac{x^2}{a^2}+\frac{y^2}{b^2}=1$ is given by $y=mx+\sqrt{(a^2m^2+b^2)}$ ....(1)

Equation of the line perpendicular to (1) and passing through (0, 0) is

$y=-\frac{1}{m}x$ or $m=-\frac{x}{y}$ ....(2)

Substituting the value of m from (2) in (1), we get

$y=-\frac{x^2}{y}+\sqrt{\left(a^2\frac{x^2}{y^2}+b^2\right)}\Rightarrow(x^2+y^2)^2=a^2x^2+b^2y^2$ .....(3)

(3) is the required locus of point P in cartesian form. If we let $x = r\cos\theta$ and $y = r\sin\theta$

$\therefore$ (3) becomes $r^4 = a^2r^2\cos^2\theta + b^2r^2\sin^2\theta$, which is in the polar form.

$\Rightarrow r^2 = a^2\cos^2\theta + b^2\sin^2\theta$

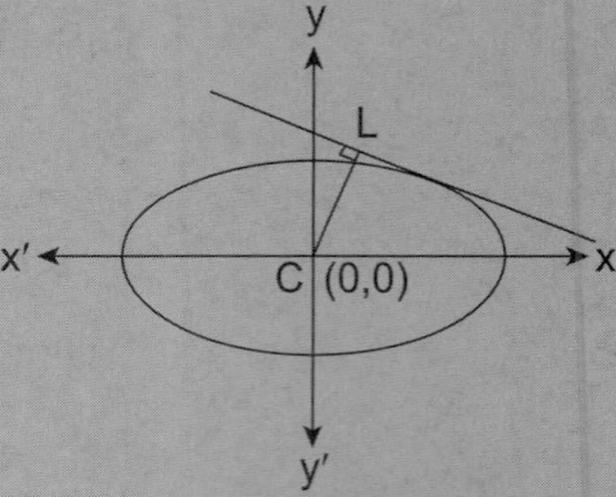

FIGURE 5.36

ILLUSTRATION 41: If a circle of radius 'r' concentric with an ellipse $\frac{x^2}{a^2}+\frac{y^2}{b^2}=1$ have common tangent(s) and θ is the angle subtended by common tangent with x-axis, then prove that $\theta=\tan^{-1}\sqrt{\frac{r^2-b^2}{a^2-r^2}}$ or $\pi-\tan^{-1}\sqrt{\frac{r^2-b^2}{a^2-r^2}}$ and hence find the range of 'r'.

SOLUTION: Equation of the circle of radius r and concentric with ellipse $\frac{x^2}{a^2}+\frac{y^2}{b^2}=1$ is $x^2+y^2=r^2$ ...(i)

Any tangent to ellipse $\frac{x^2}{a^2}+\frac{y^2}{b^2}=1$ is

$y=mx+\sqrt{(a^2m^2+b^2)}$ (where $m=\tan\theta$)

If it is a tangent to circle, then its perpendicular from (0, 0) is equal to radius r,

$$\therefore\ \frac{\sqrt{a^2m^2+b^2}}{\sqrt{m^2+1}}=r$$

$\Rightarrow\ a^2m^2+b^2=m^2r^2+r^2$

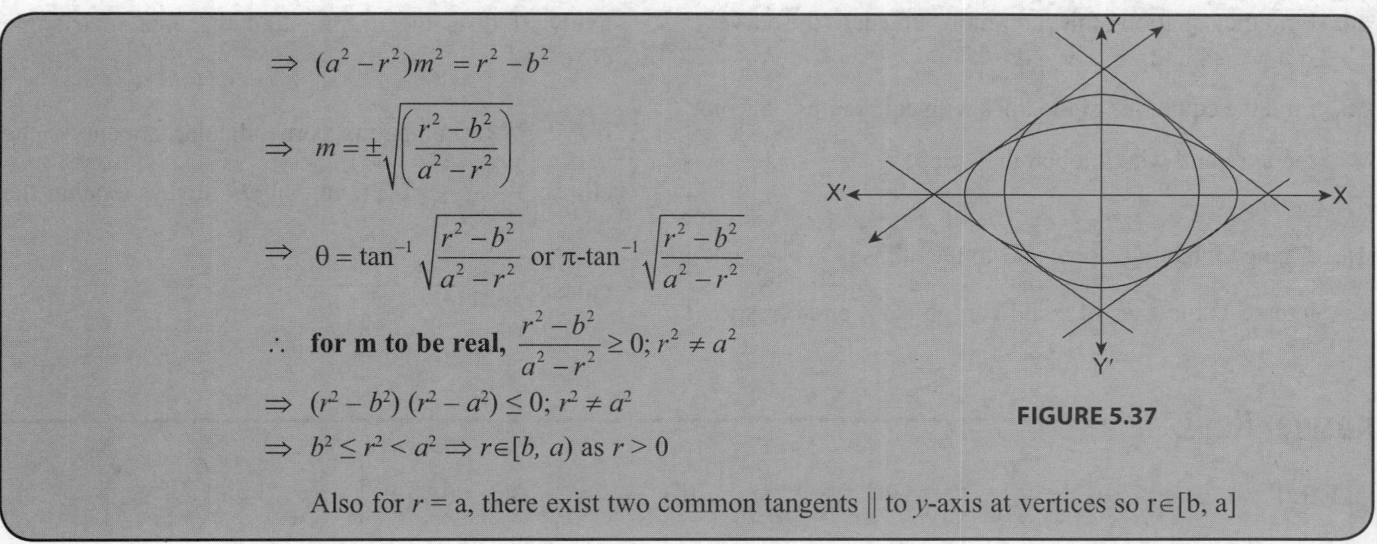

$\Rightarrow (a^2 - r^2)m^2 = r^2 - b^2$

$\Rightarrow m = \pm\sqrt{\left(\dfrac{r^2 - b^2}{a^2 - r^2}\right)}$

$\Rightarrow \theta = \tan^{-1}\sqrt{\dfrac{r^2 - b^2}{a^2 - r^2}}$ or $\pi\text{-}\tan^{-1}\sqrt{\dfrac{r^2 - b^2}{a^2 - r^2}}$

$\therefore$ **for m to be real,** $\dfrac{r^2 - b^2}{a^2 - r^2} \geq 0;\ r^2 \neq a^2$

$\Rightarrow (r^2 - b^2)(r^2 - a^2) \leq 0;\ r^2 \neq a^2$

$\Rightarrow b^2 \leq r^2 < a^2 \Rightarrow r \in [b, a)$ as $r > 0$

Also for r = a, there exist two common tangents || to y-axis at vertices so r∈[b, a]

FIGURE 5.37

TEXTUAL EXERCISE–3 (SUBJECTIVE)

1. Find the equation of tangent as well as point of contact of tangent to ellipse $\dfrac{x^2}{9} + \dfrac{y^2}{4} = 1$
   (a) If it is parallel to line $y = 2x + 1$
   (b) Parallel to the line that cuts intercepts of length 4 units on x-axis and 12 units on y-axis (+ve)
   (c) It passes through (0, 3)
   (d) It is drawn from point (4, 1)
   (e) It is drawn from point (–3, 1)

2. (a) For what value of λ does the line $y = x + \lambda$ touches the ellipse $9x^2 + 16y^2 = 144$?
   (b) Find the equations of the tangents to the ellipse $3x^2 + 4y^2 = 12$ which are perpendicular to the line $y + 2x = 4$. Also find the points of contacts.
   (c) Find the equation of the tangents to the ellipse $x^2 + 16y^2 = 16$ each one of which makes an angle of 60° with x-axis and corresponding points of contact.

3. Prove that the straight line $lx + my + n = 0$ touches the ellipse $x^2/a^2 + y^2/b^2 = 1$ if $a^2l^2 + b^2m^2 = n^2$

4. Find the points of intersection of the tangents at points α and β of the ellipse $x^2/a^2 + y^2/b^2 = 1$.

5. Show that the line $x \cos\alpha + y \sin\alpha = p$ touches the ellipse $x^2/a^2 + y^2/b^2 = 1$ if $a^2 \cos^2\alpha + b^2\sin^2\alpha = p^2$ and that the point of contact is $\left(\dfrac{a^2\cos\alpha}{p}, \dfrac{b^2\sin\alpha}{p}\right)$.

6. Find for the ellipse $x^2/a^2 + y^2/b^2 = 1$ the locus of mid-points of the portions of tangents included between the axes.

7. Find the points on the ellipse $\dfrac{x^2}{a^2} + \dfrac{y^2}{b^2} = 1$ such that the tangent at each of them makes equal angles with the axes. Prove also that the length of the perpendicular from the centre on either of these tangents is $\sqrt{\dfrac{a^2 + b^2}{2}}$.

8. Find the locus of foot of perpendicular drawn from centre upon any tangent to ellipse $\dfrac{x^2}{a^2} + \dfrac{y^2}{b^2} = 1$.

9. If a number of ellipses be described having the same major axes, but a variable minor axis, prove that the tangents at the ends of their latus rectum pass through one or other of two fixed points. Also find the point(s).

10. The tangent at point α on the ellipse $x^2/a^2 + y^2/b^2 = 1$ meets the auxiliary circle in two points which subtend a right angle at the centre. Show that the eccentricity of the ellipse is $\dfrac{1}{\sqrt{1 + \sin^2\alpha}}$.

11. Find the angle between the tangents to the ellipse $x^2/a^2 + y^2/b^2 = 1$ and the circle $x^2 + y^2 = ab$ at their points of intersection.

12. Find the equations of the tangents to the ellipse $3x^2 + y^2 = 3$ making equal intercepts on the axes.

13. Find the number of values of $\theta \in [0, 2\pi]$ for which the line $2x \cos\theta + 3y \sin\theta = 6$ touches the ellipse $4x^2 + 9y^2 = 36$.

14. Find the equation of the common tangent to the curves $x^2 + y^2 = 4$ and $2x^2 + y^2 = 2$.

15. Find the equations of the lines with equal intercepts on the axes and which touch the ellipse $\frac{x^2}{16}+\frac{y^2}{9}=1$.

16. A tangent having slope –4/3 to the ellipse $\frac{x^2}{18}+\frac{y^2}{32}=1$, intersects the axes of x and y in points A and B respectively. If O is the origin,. find the area of triangle OAB.

17. If PQ, PR form a pair of perpendicular tangents to the ellipse $\frac{x^2}{a^2}+\frac{y^2}{b^2}=1$. Prove that QR always touches the ellipse $\frac{x^2}{a^4}+\frac{y^2}{b^4}=\frac{1}{(a^2+b^2)}$.

Answer Keys

1. (a) $y = 2 \pm 2\sqrt{10}$; $\left(-\frac{9}{\sqrt{10}},\frac{2}{\sqrt{10}}\right)$ or $\left(\frac{9}{\sqrt{10}},\frac{-2}{\sqrt{10}}\right)$ (b) $3x + y \pm \sqrt{85} = 0$ (c) $\left(\pm\frac{\sqrt{5}}{3}x+3\right)$; $\left(\mp\sqrt{5},\frac{4}{3}\right)$

(d) $m=\frac{4\pm\sqrt{37}}{7}$ (e) $x - 2y + 5 = 0$; $\left(-\frac{9}{5},\frac{8}{5}\right)$

2. (a) $\lambda = \pm 5$ (b) $x - 2y \pm 4 = 0$ (c) $y = \sqrt{3}x \pm 7$

4. $\left(a\frac{\cos\frac{\alpha+\beta}{2}}{\cos\frac{\beta-\alpha}{2}}, b\frac{\sin\frac{\alpha+\beta}{2}}{\cos\frac{\beta-\alpha}{2}}\right)$

6. $a^2/x^2 + b^2/y^2 = 4$

7. $\left[\pm\frac{a^2}{\sqrt{a^2+b^2}},\pm\frac{b^2}{\sqrt{a^2+b^2}}\right]$

8. $(x^2 + y^2)^2 = a^2x^2 + b^2y^2$ 9. $(0, \pm a)$ 11. $\tan^{-1}\frac{a-b}{\sqrt{ab}}$

12. $y = \pm x \pm 2$ 13. ∞ 14. No tangent possible

15. $y = \pm x \pm 5$

16. 24 sq. unit.

TEXTUAL EXERCISE-3 (OBJECTIVE)

1. If the line $y = 2x + c$ be a tangent to the ellipse $\frac{x^2}{8}+\frac{y^2}{4}=1$, then c is equal to

(a) ± 4 (b) ± 6
(c) ± 1 (d) ± 8

2. The equation of the tangent to the ellipse $x^2 + 16y^2 = 16$ making an angle of 60° with x-axis is

(a) $\sqrt{3}x-y+7=0$ (b) $\sqrt{3}x-y-7=0$
(c) $\sqrt{3}x-y\pm7=0$ (d) None of these

3. The equations of the tangents to the ellipse $9x^2 + 16y^2 = 144$ which passes through the point (2, 3) is

(a) $y = 3, x + y = 5$ (b) $y = -3, x - y = 5$
(c) $y = 4, x + y = 3$ (d) $y = -4, x - y = 3$

4. The ellipse $\frac{x^2}{a^2}+\frac{y^2}{b^2}=1$ and the straight line $y = mx + c$ intersect at real points only if

(a) $a^2m^2 < c^2 - b^2$ (b) $a^2m^2 > c^2 - b^2$
(c) $a^2m^2 \geq c^2 - b^2$ (d) $c \geq b$

5. If any tangent to the ellipse $\frac{x^2}{a^2}+\frac{y^2}{b^2}=1$ cuts off intercepts of length h and k on the axes, then $\frac{a^2}{h^2}+\frac{b^2}{k^2}$ is equal to

(a) 0 (b) 1
(c) –1 (d) None of these

6. Minimum of the areas of the triangles formed by tangents to the ellipse $\frac{x^2}{a^2}+\frac{y^2}{b^2}=1$ with the co-ordinate axes is

(a) $\frac{a^2+b^2}{2}$ (b) $\frac{(a+b)^2}{2}$
(c) ab (d) $\frac{(a-b)^2}{2}$

7. The sum of the squares of the perpendiculars on any tangent to the ellipse $\frac{x^2}{a^2}+\frac{y^2}{b^2}=1$ from two points on the minor axis each at a distance $\sqrt{a^2-b^2}$ from the centre is

(a) $2a^2$ (b) $2b^2$
(c) $a^2 + b^2$ (d) $a^2 - b^2$

8. The product of the perpendicular drawn from the foci upon any tangent to an ellipse $x^2/a^2 + y^2/b^2 = 1$ is
(a) a^2 (b) b^2
(c) b^2/a^2 (d) None of these

9. The locus of the foot of perpendicular drawn from either focus upon any tangent to the ellipse is
(a) auxiliary circle
(b) director circle
(c) a circle concentric with the ellipse
(d) None of these

10. Tangents at extremities of the latus rectum of an ellipse always intersects at the
(a) major axis of the ellipse
(b) directrix
(c) minor axis of the ellipse
(d) None of these

11. In an ellipse $\frac{x^2}{48}+\frac{y^2}{36}=1$, if P is any point such that $\angle PS'S = \phi$ and $\angle PSS' = \theta$, then the value of $\frac{1-\cos\theta}{1+\cos\phi}:\frac{1+\cos\theta}{1-\cos\phi}=$
(a) 1 : 8 (b) 1 : 9
(c) 9 : 1 (d) 8 : 1

12. If a variable tangent to the circle $x^2 + y^2 = 1$, intersect the ellipse $x^2 + 2y^2 = 4$ at point P and Q, then the locus of the point of intersection of tangents to the ellipse at P and Q is a conic whose
(a) eccentricity $= \sqrt{3}/2$ (b) eccentricity $= \sqrt{5}/2$
(c) latus rectum $= 4$ (d) foci are $(\pm 2\sqrt{5}, 0)$

13. The locus of the point of intersection of tangents to an ellipse at 2 points, sum of whose eccentric angle is constant is/an
(a) ellipse (b) parabola
(c) hyperbola (d) straight line

14. The line $3x + 5y = k$ touches the ellipse $16x^2 + 25y^2 = 400$ if k is
(a) $\pm\sqrt{50}$ (b) $\pm\sqrt{15}$
(c) ± 5 (d) None of these

15. Which of the following is an equation of the ellipse with centre (–2, 1) major axis running from (–2, 6) to (–2, –4) and focus at (–2, 5)?
(a) $\frac{(x-2)^2}{25}+\frac{(y+1)^2}{16}=1$
(b) $\frac{(x+2)^2}{25}+\frac{(y-1)^2}{9}=1$
(c) $\frac{(x-2)^2}{9}+\frac{(y+1)^2}{25}=1$
(d) $\frac{(x+2)^2}{9}+\frac{(y-1)^2}{25}=1$

16. Extremities of the latera recta of the ellipse $\frac{x^2}{a^2}+\frac{y^2}{b^2}=1$ $(a > b)$ having a given major axis $2a$ lies on
(a) $x^2 = a(a - y)$ (b) $x^2 = a(a + y)$
(c) $y^2 = a(a + x)$ (d) $y^2 = a(a - x)$

17. The parametric angle θ, where $\theta \in (-\pi, \pi]$ of the point on the ellipse $\frac{x^2}{a^2}+\frac{y^2}{b^2}=1$, at which the tangent drawn cuts the intercept of minimum length on the co-ordinate axis, is/are
(a) $\tan^{-1}\sqrt{\frac{b}{a}}$ (b) $-\tan^{-1}\sqrt{\frac{b}{a}}$
(c) $\pi-\tan^{-1}\sqrt{\frac{b}{a}}$ (d) $\pi+\tan^{-1}\sqrt{\frac{b}{a}}$

Answer Keys

1. (b)	**2.** (c)	**3.** (a)	**4.** (c)	**5.** (b)	**6.** (c)	**7.** (a)	**8.** (b)	**9.** (a, c)	**10.** (b)
11. (b)	**12.** (a)	**13.** (d)	**14.** (c)	**15.** (d)	**16.** (a, b)	**17.** (a, b, c)			

■ PAIR OF TANGENTS

The combined equation of the pair of tangents drawn from a point (x_1, y_1) to the ellipse $\frac{x^2}{a^2}+\frac{y^2}{b^2}=1$ is $\left(\frac{x^2}{a^2}+\frac{y^2}{b^2}-1\right)\left(\frac{x_1^2}{a^2}+\frac{y_1^2}{b^2}-1\right)=\left(\frac{xx_1}{a^2}+\frac{yy_1}{b^2}-1\right)^2$

or $SS_1 = T^2$

where $S = \frac{x^2}{a^2} + \frac{y^2}{b^2} - 1$; $S_1 = \frac{x_1^2}{a^2} + \frac{y_1^2}{b^2} - 1$ and

$$T = \frac{xx_1}{a^2} + \frac{yy_1}{b^2} - 1$$

Let $T(h, k)$ be any point on the pair of tangents PQ or PR drawn from any external point $P(x_1, y_1)$ to the ellipse $\frac{x^2}{a^2} + \frac{y^2}{b^2} = 1$.

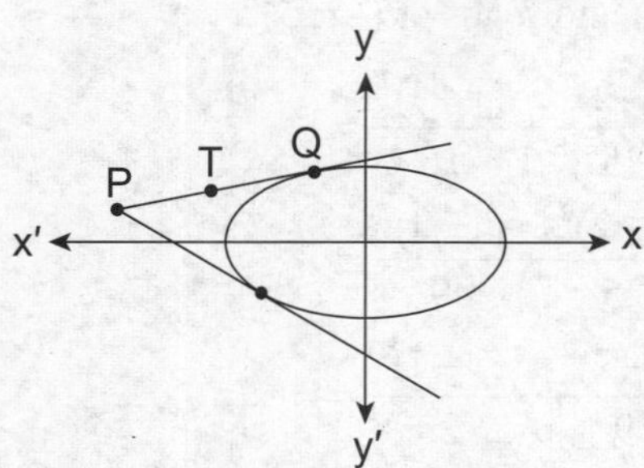

FIGURE 5.38

$\therefore$ Equation of PT is $y - y_1 = \frac{k - y_1}{h - x_1}(x - x_1)$ or $y = \left(\frac{k - y_1}{h - x_1}\right)x + \left(\frac{hy_1 - kx_1}{h - x_1}\right)$ which is the tangent to the ellipse $\frac{x^2}{a^2} + \frac{y^2}{b^2} = 1 \therefore c^2 = a^2m^2 + b^2$

$$\Rightarrow \left(\frac{hy_1 - kx_1}{h - x_1}\right)^2 = a^2\left(\frac{k - y_1}{h - x_1}\right)^2 + b^2$$

$$\Rightarrow (hy_1 - kx_1)^2 = a^2(k - y_1)^2 + b^2(h - x_1)^2$$

Hence locus of (h, k) is

$$(xy_1 - x_1y)^2 = a^2(y - y_1)^2 + b^2(x - x_1)^2$$

or $\left(\frac{xy_1 - x_1y}{ab}\right)^2 = \left(\frac{x^2}{a^2} + \frac{y^2}{b^2}\right) + \left(\frac{x_1^2}{a^2} + \frac{y_1^2}{b^2}\right) - 2\left(\frac{xx_1}{a^2} + \frac{yy_1}{b^2}\right)$

or $\left(\frac{xy_1 - x_1y}{ab}\right)^2 - \left(\frac{x^2}{a^2} + \frac{y^2}{b^2}\right) -$

$$\left(\frac{x_1^2}{a^2} + \frac{y_1^2}{b^2}\right) + \left(\frac{xx_1}{a^2} + \frac{yy_1}{b^2}\right)^2 + 1$$

$$= \left(\frac{xx_1}{a^2} + \frac{yy_1}{b^2}\right)^2 + 1 - 2\left(\frac{xx_1}{a^2} + \frac{yy_1}{b^2}\right)$$

or $\left(\frac{x^2}{a^2} + \frac{y^2}{b^2} - 1\right)\left(\frac{x_1^2}{a^2} + \frac{y_1^2}{b^2} - 1\right) = \left(\frac{xx_1}{a^2} + \frac{yy_1}{b^2} - 1\right)^2$

or $SS_1 = T^2$

Alternative Method

Let the ellipse be $\frac{x^2}{a^2} + \frac{y^2}{b^2} = 1$ .... (1)

Let $P(x_1, y_1)$ be any point outside the ellipse. Let a chord of the ellipse through the point $P(x_1, y_1)$ cut the ellipse at Q and let $R(h, k)$ be any arbitrary point on the line PQ. Let Q divides PR in the ratio $\lambda : 1$, then co-ordinates of Q will be

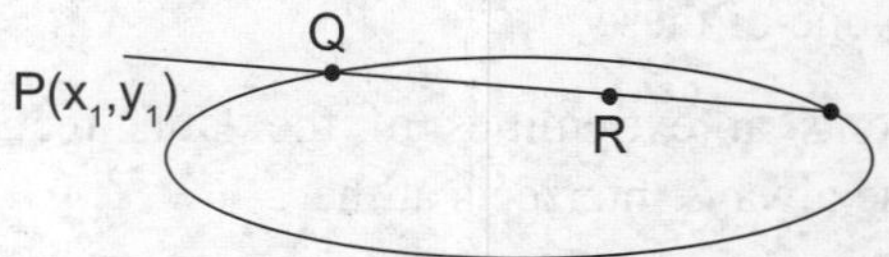

FIGURE 5.39

$$\left(\frac{\lambda h + x_1}{\lambda + 1}, \frac{\lambda k + y_1}{\lambda + 1}\right) \quad (\because PQ : QR = \lambda : 1)$$

Since Q lies on ellipse (1)

$$\Rightarrow \frac{1}{a^2}\left(\frac{\lambda h + x_1}{\lambda + 1}\right)^2 + \frac{1}{b^2}\left(\frac{\lambda k + y_1}{\lambda + 1}\right)^2 = 1$$

$$\Rightarrow b^2(\lambda h + x_1)^2 + a^2(\lambda k + y_1)^2 = a^2b^2(\lambda + 1)^2$$

$$\Rightarrow (a^2k^2 + b^2h^2 - a^2b^2)\lambda^2 + 2(hx_1b^2 + ky_1a^2 - a^2b^2)\lambda + (b^2x_1^2 + a^2y_1^2 - a^2b^2) = 0 \quad (2)$$

Line PR will become tangent to ellipse (1), then roots of equation (2) are equal

$$\therefore 4(hx_1b^2 + ky_1a^2 - a^2b^2)^2 - 4(a^2k^2 + b^2h^2 - a^2b^2)(b^2x_1^2 + a^2y_1^2 - a^2b^2) = 0$$

Dividing by $4a^4b^4$, we get

$$\left(\frac{hx_1}{a^2} + \frac{ky_1}{b^2} - 1\right)^2 = \left(\frac{k^2}{b^2} + \frac{h^2}{a^2} - 1\right)\left(\frac{x_1^2}{a^2} + \frac{y_1^2}{b^2} - 1\right)$$

Hence locus of $R(h, k)$ i.e., equation of pair of tangents from $P(x_1, y_1)$ to $S = 0$ is

$$T^2 = SS_1$$

or $SS_1 = T^2$

NOTE

$S = 0$ is the equation of the curve, S_1 is obtained from S by replacing x by x_1 and y by y_1 and $T = 0$ is the equation of tangent at (x_1, y_1) to $S = 0$.

ILLUSTRATION 42: Write down the equation of the pair of tangents drawn to the ellipse $3x^2 + 2y^2 = 5$ from the point (1, 2) and prove that the angle between them is $\tan^{-1}\left(\frac{12}{\sqrt{5}}\right)$.

SOLUTION: The equation of the given ellipse is $3x^2 + 2y^2 - 5 = 0$, and the co-ordinates of point are (1, 2)

$\therefore$ equation of pair of tangent are given by $SS_1 = T^2$.

$\Rightarrow (3x^2 + 2y^2 - 5)(3x_1^2 + 2y_1^2 - 5) = (3xx_1 + 2yy_1 - 5)^2$

$\Rightarrow (3x^2 + 2y^2 - 5)(3 + 8 - 5) = (3x(1) + 2y(2) - 5)^2$

$\Rightarrow (3x^2 + 2y^2 - 5)(6) = (3x + 4y - 5)^2$

$\Rightarrow 18x^2 + 12y^2 - 30 = 9x^2 + 16y^2 + 25 + 24xy - 40y - 30x.$

$\Rightarrow 9x^2 - 4y^2 - 24xy + 30x + 40y - 55 = 0$

$\Rightarrow 9x^2 - 4y^2 - 24xy + 30x + 40y - 55 = 0$

Comparing it with $ax^2 + by^2 + 2hxy + 2gx + 2fy + c = 0$

we get $a = 9, b = -4, h = -12, g = 15, f = 20, c = -55$.

Also, $\tan\theta = \dfrac{2\sqrt{h^2 - ab}}{|a+b|} = \dfrac{2\sqrt{144+36}}{|9-4|} = \dfrac{2\sqrt{180}}{5} = \dfrac{2\times 6\sqrt{5}}{5} = \dfrac{12}{\sqrt{5}}$.

$\Rightarrow \theta = \tan^{-1}\left(\frac{12}{\sqrt{5}}\right)$ Hence proved.

ILLUSTRATION 43: Tangents are drawn to the ellipse $\frac{x^2}{a^2} + \frac{y^2}{b^2} = 1$ from the point $\left(\frac{a^2}{\sqrt{a^2 - b^2}}, \sqrt{a^2 + b^2}\right)$; prove that they intersect the nearest latus rectum at two points at a distance equal to the length of major axis from each other.

SOLUTION: Pair of tangents from point $(x_1, y_1) = \left(\frac{a^2}{ae}, \sqrt{a^2 + b^2}\right) \equiv \left(\frac{a}{e}, \sqrt{a^2 + b^2}\right)$ are given by $SS_1 = T^2$

$$\Rightarrow \left(\frac{x^2}{a^2} + \frac{y^2}{b^2} - 1\right)\left(\frac{a^2/e^2}{a^2} + \frac{a^2 + b^2}{b^2} - 1\right) = \left(\frac{x}{ae} + y\frac{\sqrt{a^2 + b^2}}{b^2} - 1\right)^2$$

$$\Rightarrow \left(\frac{x^2}{a^2} + \frac{y^2}{b^2} - 1\right)\left(\frac{1}{e^2} + 1 + \frac{a^2}{b^2} - 1\right) = \left(\frac{x}{ae} + \frac{y\sqrt{a^2 + b^2}}{b^2} - 1\right)^2$$

$$\Rightarrow \left(\frac{x^2}{a^2} + \frac{y^2}{b^2} - 1\right)\left(\frac{1}{e^2} + \frac{a^2}{a^2(1-e^2)}\right) = \left(\frac{x}{ae} + \frac{y\sqrt{a^2 + b^2}}{b^2} - 1\right)^2$$

$$\Rightarrow \left(\frac{x^2}{a^2} + \frac{y^2}{b^2} - 1\right)\left(\frac{1}{e^2} + \frac{1}{(1-e^2)}\right) = \left(\frac{x}{ae} + \frac{y\sqrt{a^2 + b^2}}{b^2} - 1\right)^2$$

$$\Rightarrow \left(\frac{x^2}{a^2} + \frac{y^2}{b^2} - 1\right)\left(\frac{1}{e^2(1-e^2)}\right) = \left(\frac{x}{ae} + \frac{y\sqrt{a^2 + b^2}}{b^2} - 1\right)^2 \quad \text{.... (i)}$$

(i) would cut latus rectum, where $x = ae$

$$\therefore \left(e^2 + \frac{y^2}{b^2} - 1\right)\left(\frac{1}{e^2(1-e^2)}\right) = \left(1 + \frac{y\sqrt{a^2 + b^2}}{b^2} - 1\right)^2$$

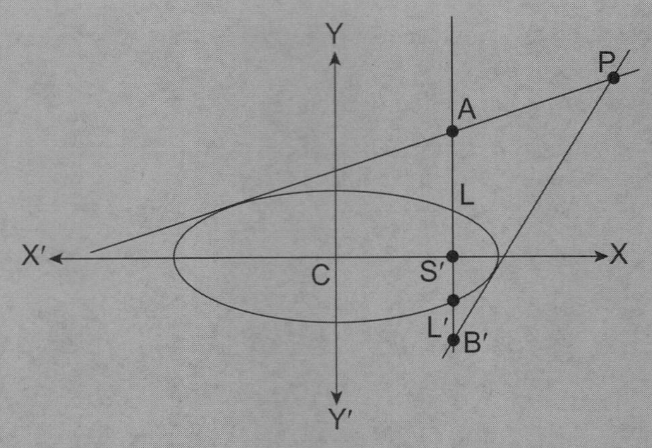

FIGURE 5.40

$$\Rightarrow \left(e^2-1+\frac{y^2}{b^2}\right)\left(\frac{1}{e^2\left(1-e^2\right)}\right)=\left(\frac{y^2\left(a^2+b^2\right)}{b^4}\right) \Rightarrow -\frac{1}{e^2}+\frac{y^2}{b^2e^2(1-e^2)}=\frac{y^2}{b^4}\left(a^2+b^2\right)$$

$$\Rightarrow y^2\left[\frac{a^2}{b^4}+\frac{1}{b^2}-\frac{1}{b^2e^2\left(1-e^2\right)}\right]=-\frac{1}{e^2}$$

$$\Rightarrow y^2\left[\frac{a^2}{a^4\left(1-e^2\right)^2}+\frac{1}{a^2\left(1-e^2\right)}-\frac{1}{a^2e^2\left(1-e^2\right)^2}\right]=-\frac{1}{e^2} \Rightarrow y^2\left[\frac{e^2+e^2\left(1-e^2\right)-1}{a^2e^2\left(1-e^2\right)^2}\right]=-\frac{1}{e^2}$$

$$\Rightarrow y^2\left[\frac{2e^2-e^4-1}{a^2e^2\left(1-e^2\right)^2}\right]=-\frac{1}{e^2} \qquad \Rightarrow -y^2\left[\frac{\left(e^2-1\right)^2}{a^2e^2\left(1-e^2\right)^2}\right]=-\frac{1}{e^2}$$

$\Rightarrow y^2 = a^2 \Rightarrow y = \pm a$

$\therefore$ Points of intersection of tangents with the nearest latus rectum are $A(ae, a)$ and $B(ae, -a)$

$\therefore$ $AB = 2a$ = length of major axis.

■ DIRECTOR CIRCLE

The locus of the point of intersection of the tangents to an ellipse $\frac{x^2}{a^2}+\frac{y^2}{b^2}=1$ which are perpendicular to each other is called director circle.

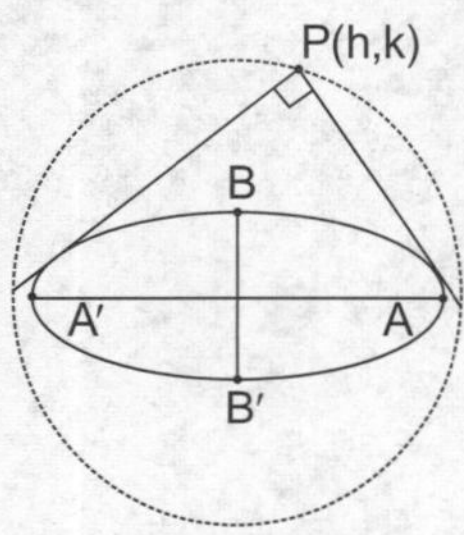

FIGURE 5.41

Let any tangent in terms of slope of ellipse $\frac{x^2}{a^2}+\frac{y^2}{b^2}=1$ is $y = mx + \sqrt{(a^2m^2+b^2)}$

If it passes through (h, k), then $k = mh+\sqrt{(a^2m^2+b^2)}$

or $\quad (k-mh)^2 = a^2m^2+b^2$

$\Rightarrow \quad k^2+m^2h^2-2mhk = a^2m^2+b^2$

$\Rightarrow \quad m^2(h^2-a^2)-2hkm+k^2-b^2=0$

It is a quadratic equation in m. Let the slopes of two tangents be m_1 and m_2

$$\therefore \quad m_1m_2 = \frac{k^2-b^2}{h^2-a^2}$$

$$\Rightarrow \quad -1 = \frac{k^2-b^2}{h^2-a^2} (\because \text{ tangents are perpendicular})$$

$\Rightarrow \quad -h^2 + a^2 = k^2 - b^2$ or $h^2 + k^2 = a^2 + b^2$

Hence locus of $P(h, k)$ is $x^2 + y^2 = a^2 + b^2$

Alternative method

If any tangent $y = mx + \sqrt{(a^2m^2 + b^2)}$ ....(i)

and $y = -\dfrac{x}{m} + \sqrt{\left\{a^2\left(-\dfrac{1}{m}\right)^2 + b^2\right\}}$ ....(ii)

touch the ellipse and intersect at right angles.

From equation (i), $y - mx = \sqrt{(a^2m^2 + b^2)}$ ....(iii)

(ii) can be re-written as $x + my = \sqrt{(a^2 + b^2m^2)}$ ....(iv)

squaring and adding (iii) and (iv) we have

$$(y - mx)^2 + (x + my)^2 = a^2m^2 + b^2 + a^2 + b^2m^2$$

$$\Rightarrow \quad (1 + m^2)(x^2 + y^2) = (1 + m^2)(a^2 + b^2)$$

Hence $x^2 + y^2 = a^2 + b^2$ is the required equation of director circle of the ellipse.

ILLUSTRATION 44: Find the locus of the points of the intersection of tangents to ellipse $\dfrac{x^2}{a^2} + \dfrac{y^2}{b^2} = 1$ including angle θ. Hence find the locus of point of intersection of perpendicular tangents.

SOLUTION: Given ellipse is $\dfrac{x^2}{a^2} + \dfrac{y^2}{b^2} = 1$ ....(i)

Equation of any tangent to ellipse (i) in terms of slope (m) is $y = mx + \sqrt{(a^2m^2 + b^2)}$

Since it passes through $P(\alpha, \beta)$, $\beta = m\alpha + \sqrt{(a^2m^2 + b^2)}$

$\Rightarrow \quad (\beta - m\alpha) = \sqrt{(a^2m^2 + b^2)} \qquad \Rightarrow \quad (\beta - m\alpha)^2 = a^2m^2 + b^2$

$\Rightarrow \quad m^2(a^2 - \alpha^2) + 2\alpha\beta m + (b^2 - \beta^2) = 0$ ....(ii)

(ii) being a quadratic equation in m. Let roots of equation (ii) are m_1 and m_2, then

$$m_1 + m_2 = -\frac{2\alpha\beta}{(a^2 - \alpha^2)}, \quad m_1m_2 = \frac{b^2 - \beta^2}{a^2 - \alpha^2}$$

$$\therefore \quad (m_1 - m_2) = \sqrt{(m_1 + m_2)^2 - 4m_1m_2} = \sqrt{\frac{4\alpha^2\beta^2}{(a^2 - \alpha^2)^2} - \frac{4(b^2 - \beta^2)}{(a^2 - \alpha^2)}}$$

$$= \sqrt{\frac{4\alpha^2\beta^2 - 4(b^2 - \beta^2)(a^2 - \alpha^2)}{(a^2 - \alpha^2)^2}}$$

$$= \sqrt{\frac{4\{\alpha^2\beta^2 - a^2b^2 + b^2\alpha^2 + a^2\beta^2 - \alpha^2\beta^2\}}{(a^2 - \alpha^2)^2}} = \frac{2}{|(a^2 - \alpha^2)|}\sqrt{(a^2\beta^2 + b^2\alpha^2 - a^2b^2)}.$$

$\because \quad \theta$ is the angle between these two tangents, $\tan\theta = \left|\dfrac{m_1 - m_2}{1 + m_1m_2}\right|$

$$= \left|\frac{\dfrac{2}{(a^2 - \alpha^2)}\sqrt{(a^2\beta^2 + b^2\alpha^2 - a^2b^2)}}{\left(1 + \dfrac{b^2 - \beta^2}{a^2 - \alpha^2}\right)}\right|$$

$\Rightarrow \quad \tan\theta = \left|\dfrac{2\sqrt{a^2\beta^2 + b^2\alpha^2 - a^2b^2}}{a^2 + b^2 - \alpha^2 - \beta^2}\right|$ or $(a^2 + b^2 - \alpha^2 - \beta^2)^2 \tan^2\theta = 4(a^2\beta^2 + b^2\alpha^2 - a^2b^2)$

$\Rightarrow \quad (\alpha^2 + \beta^2 - a^2 - b^2)^2 \tan^2\theta = 4(b^2\alpha^2 + a^2\beta^2 - a^2b^2)$

$\therefore$ Locus of $P(\alpha, \beta)$ is $(x^2 + y^2 - a^2 - b^2)^2 \tan^2\theta = 4(b^2x^2 + a^2y^2 - a^2b^2)$ ......(iii)

For perpendicular tangents $\theta = \dfrac{\pi}{2} \to \tan\theta \to \infty$, from (iii) we have $(x^2 + y^2 - a^2 - b^2) = 0$

or $x^2 + y^2 = a^2 + b^2$ i.e., the director circle of ellipse (i)

■ CHORD OF CONTACT

The equation of chord of contact of tangents drawn from a point (x_1, y_1) to the ellipse $\frac{x^2}{a^2}+\frac{y^2}{b^2}=1$ is $\frac{xx_1}{a^2}+\frac{yy_1}{b^2}=1$.

Let PQ and PR be the tangents drawn from a point $P(x_1, y_1)$ to the ellipse $\frac{x^2}{a^2}+\frac{y^2}{b^2}=1$, such that $Q\equiv(x',y')$ and $R\equiv(x'',y'')$ are the points of contacts of these tangents. The chord QR is called chord of contact of the ellipse $\frac{x^2}{a^2}+\frac{y^2}{b^2}=1$

Equations of tangents at $Q(x',y')$ and $R(x'',y'')$ are $\frac{xx'}{a^2}+\frac{yy'}{b^2}=1$ and $\frac{xx''}{a^2}+\frac{yy''}{b^2}=1$, respectively.

These tangents pass through $P(x_1, y_1)$ therefore, $\frac{x'x_1}{a^2}+\frac{y'y_1}{b^2}=1$ and $\frac{x''x_1}{a^2}+\frac{y''y_1}{b^2}=1$.

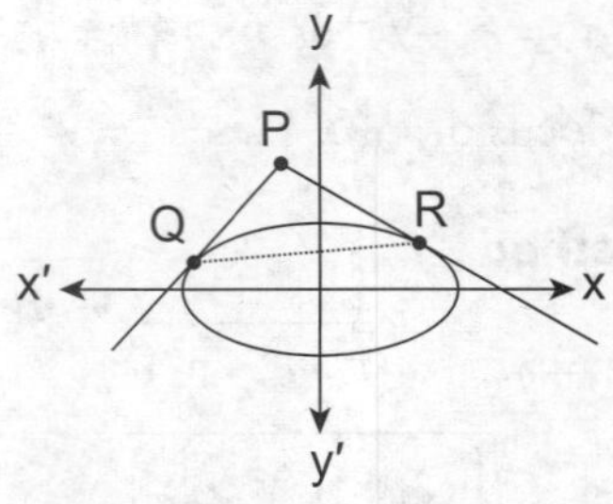

FIGURE 5.42

$\Rightarrow$ (x',y') and (x'',y'') lie on $\frac{xx_1}{a^2}+\frac{yy_1}{b^2}=1$

Hence the equation of QR is

$$\frac{xx_1}{a^2}+\frac{yy_1}{b^2}=1$$

which is same as the equation of tangent but position of point differs.

ILLUSTRATION 45: If a tangent to the parabola $y^2 = 4ax$ intersect the ellipse $\frac{x^2}{a^2}+\frac{y^2}{b^2}=1$ at A and B, then find the locus of point of intersection of tangents at A and B to ellipse.

SOLUTION: Let $P\equiv(h,k)$ be the point of intersection of tangents at A and B to ellipse.

$\therefore$ equation of chord of contact AB is $\frac{xh}{a^2}+\frac{yk}{b^2}=1$ ....(i)

which touches the parabola and hence is tangent to parabola $y^2 = 4ax$

$\Rightarrow$ (i) must be identical to $y = mx+\frac{a}{m}$

or $mx - y = -\frac{a}{m}$ ......(ii)

$\therefore$ from (i) on (ii), we have

$$\therefore \quad \frac{m}{\left(\frac{h}{a^2}\right)}=\frac{-1}{\left(\frac{k}{b^2}\right)}=\frac{-\frac{a}{m}}{1}$$

$\Rightarrow m=-\frac{h}{k}\frac{b^2}{a^2}$ and $m=\frac{ak}{b^2}$

$\Rightarrow -\frac{hb^2}{ka^2}=\frac{ak}{b^2}$ $\Rightarrow$ locus of P is $y^2 = -\frac{b^4}{a^3}.x$; which is again a parabola opening in opposite direction to that of given parabola:

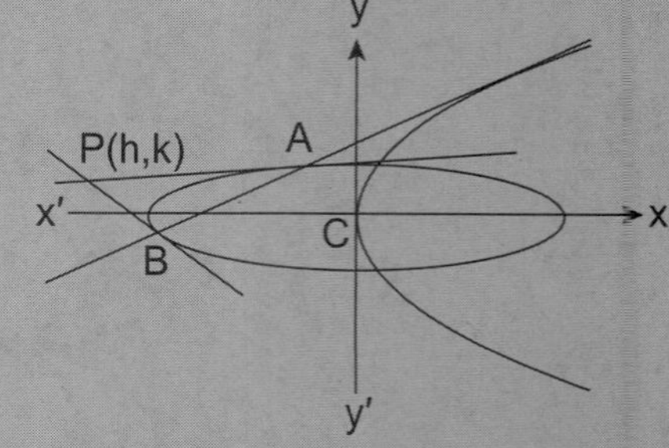

FIGURE 5.43

ILLUSTRATION 46: Tangents are drawn from the point (3, 2) to the ellipse $x^2 + 4y^2 = 9$. Find the equation to their chord of contact and the equation of the straight line joining (3, 2) to the middle point of this chord of contact.

SOLUTION: Equation of given ellipse is $x^2 + 4y^2 = 9$

Equation of the chord of contact of the pair of tangents from $P(3, 2)$ is given by

$3x + 8y = 9$ ....(1)

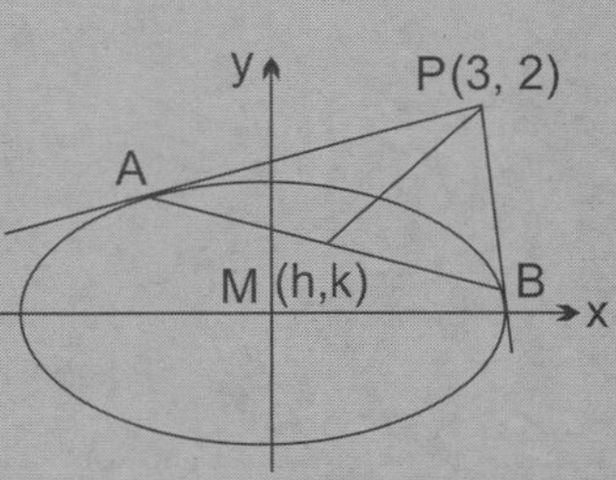

FIGURE 5.44

Let M (h, k) be the mid-point of chord of contact AB. It must be the same as chord whose middle point is (h, k). i.e., $T = S_1$

$\Rightarrow\ hx + 4ky = h^2 + 4k^2$ ....(2)

Comparing co-efficients of (1) and (2), we have

$$\frac{h}{3} = \frac{4k}{8} = \frac{h^2+4k^2}{9} \Rightarrow 2h = 3k \text{ and } 3h = h^2 + 4k^2$$

$$\Rightarrow 3h = h^2 + 4\cdot\frac{4h^2}{9} \Rightarrow \frac{25h^2}{9} = 3h \Rightarrow h = \frac{27}{25} \text{ and } k = \frac{18}{25}$$

$$\Rightarrow \text{Equation of PM will be } (y-2) = \frac{\frac{18}{25}-2}{\frac{27}{25}-3}(x-3) \Rightarrow (y-2) = \frac{-32}{-48}(x-3)$$

$\Rightarrow 3y - 6 = 2x - 6 \qquad \Rightarrow 3y - 2x = 0$

ILLUSTRATION 47: Prove that the chord of contact of tangents drawn from the point (h, k) to the ellipse $\frac{x^2}{a^2}+\frac{y^2}{b^2}=1$ will subtend a right angle at the centre, if $\frac{h^2}{a^4}+\frac{k^2}{b^4}=\frac{1}{a^2}+\frac{1}{b^2}$. Also, find the locus of (h, k). Hence find the area of circle through origin and passing through the points of contacts of tangents drawn from point $P\left(\frac{1}{\sqrt{2}},\frac{1}{4}\right)$ to ellipse $4x^2 + 8y^2 = 1$.

SOLUTION: The equation of chord of contact of tangents drawn from $P(h, k)$ to the ellipse $\frac{x^2}{a^2}+\frac{y^2}{b^2}=1$ is

$T = 0$ i.e., $\frac{xh}{a^2}+\frac{yk}{b^2}=1$ ....(i)

The equation of the straight lines CA and CB can be obtained by making

$\frac{x^2}{a^2}+\frac{y^2}{b^2}=1$ homogeneous with the help of (i)

i.e., $\frac{x^2}{a^2}+\frac{y^2}{b^2}=\left(\frac{hx}{a^2}+\frac{ky}{b^2}\right)^2$

$$\Rightarrow \left(\frac{h^2}{a^4}-\frac{1}{a^2}\right)x^2+\left(\frac{k^2}{b^4}-\frac{1}{b^2}\right)y^2+\frac{2hk}{a^2b^2}xy=0 \quad \text{.... (ii)}$$

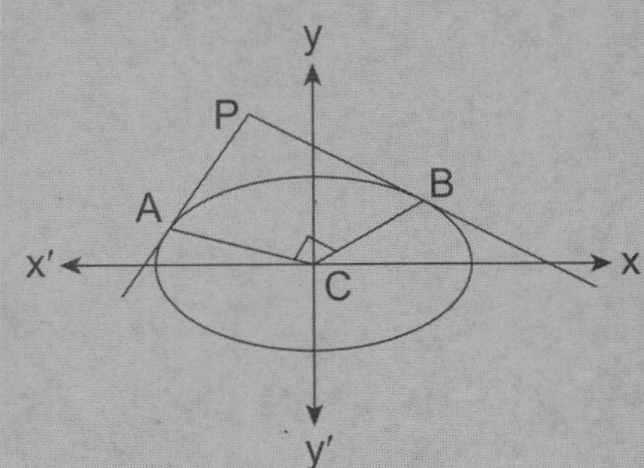

FIGURE 5.45

But given $\angle ACB = 90^o$

$\therefore$ co-efficient of x^2 + co-efficient of $y^2 = 0$

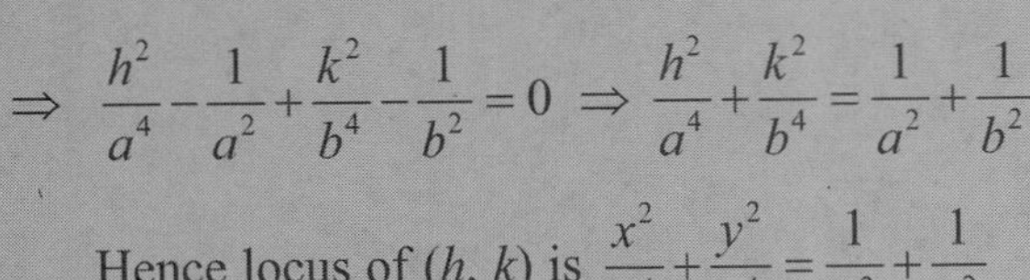

$$\Rightarrow \frac{h^2}{a^4}-\frac{1}{a^2}+\frac{k^2}{b^4}-\frac{1}{b^2}=0 \Rightarrow \frac{h^2}{a^4}+\frac{k^2}{b^4}=\frac{1}{a^2}+\frac{1}{b^2}$$

Hence locus of (h, k) is $\frac{x^2}{a^4}+\frac{y^2}{b^4}=\frac{1}{a^2}+\frac{1}{b^2}$

Let us find the points of contact of tangents and ellipse. Chord of contact QR is given by $4hx + 8ky - 1 = 0$.

i.e., $4\left(\frac{1}{\sqrt{2}}\right)x + 8\left(\frac{1}{4}\right)y - 1 = 0$

or $2\sqrt{2}x + 2y - 1 = 0$ .... (iii)

At the point of contact, $4x^2 + 8\left(\frac{-2\sqrt{2}x+1}{2}\right)^2 = 1$

$\Rightarrow 4x^2 + 2(8x^2 + 1 - 4\sqrt{2}x) - 1 = 0 = 20x^2 - 8\sqrt{2}x + 1 = 0$

$\Rightarrow x = \frac{8\sqrt{2} \pm \sqrt{128-80}}{40}$

$\Rightarrow x = \frac{8\sqrt{2} \pm 4\sqrt{3}}{40} = \frac{2\sqrt{2} \pm \sqrt{3}}{10}$

from (iii) $y = \frac{1-2\sqrt{2}x}{2} = \frac{1}{2} - \sqrt{2}\left(\frac{2\sqrt{2} \pm \sqrt{3}}{10}\right)$

$= \frac{1}{2} - \frac{2}{5} \mp \frac{\sqrt{6}}{10} = \frac{5-4}{10} \mp \frac{\sqrt{6}}{10} = \frac{1}{10} \mp \frac{\sqrt{6}}{10} = \frac{1 \mp \sqrt{6}}{10}$

Point of contact are $R\left(\frac{2\sqrt{2}+\sqrt{3}}{10}, \frac{1-\sqrt{6}}{10}\right)$ and $Q\left(\frac{2\sqrt{2}-\sqrt{3}}{10}, \frac{1+\sqrt{6}}{10}\right)$

$\therefore$ Length of chord of contact $QR = \sqrt{\left(\frac{\sqrt{3}}{5}\right)^2 + \left(\frac{\sqrt{6}}{5}\right)^2} = \sqrt{\frac{3}{25} + \frac{6}{25}} = \frac{3}{5}$

Now $\frac{h^2}{a^4} + \frac{k^2}{b^4} = \frac{1}{2}(4)^2 + \frac{1}{16}(64) = 12$ and $\frac{1}{a^2} + \frac{1}{b^2} = 4 + 8 = 12$ $\therefore$ $\frac{h^2}{a^4} + \frac{k^2}{b^4} = \frac{1}{a^2} + \frac{1}{b^2}$

$\Rightarrow$ Chord QR would subtend a right angle at the centre of ellipse.

As C lies on the circumference of circle passing through the points of contacts Q and R.

$\Rightarrow$ QR will be the diameter of that circle.

$\Rightarrow r = 3/10$

$\therefore$ Area of required circle $= \pi r^2 = \pi\left(\frac{3}{10}\right)^2 = \frac{9}{100}\pi$ sq. units.

FIGURE 5.46

ILLUSTRATION 48: If the tangent TP and TQ be drawn from a point T whose co-ordinates are (h, k) on the ellipse $\frac{x^2}{a^2} + \frac{y^2}{b^2} = 1$; then prove that the area of ΔTPQ is $\frac{ab\left(\frac{h^2}{a^2} + \frac{k^2}{b^2} - 1\right)^{3/2}}{\left(\frac{h^2}{a^2} + \frac{k^2}{b^2}\right)}$ and area of quadrilateral $TPCQ$ is $\sqrt{b^2h^2 + a^2k^2 - a^2b^2}$.

SOLUTION: Let ϕ_1 and ϕ_2 be eccentric angles of point P and Q respectively.

$\therefore$ Equation of PQ will be $\frac{x}{a}\cos\left(\frac{\phi_1 + \phi_2}{2}\right) + \frac{y}{b}\sin\left(\frac{\phi_1 + \phi_2}{2}\right) = \cos\left(\frac{\phi_1 - \phi_2}{2}\right)$ ...(i)

Also equation of chord of contant PQ will be $\frac{xh}{a^2} + \frac{yk}{b^2} = 1$ ... (ii)

Now (i) and (ii) are identical

$$\Rightarrow \frac{\cos\left(\dfrac{\phi_1+\phi_2}{2}\right)}{h/a}=\frac{\sin\left(\dfrac{\phi_1+\phi_2}{2}\right)}{k/b}=\frac{\cos\left(\dfrac{\phi_1-\phi_2}{2}\right)}{1}=\frac{1}{\sqrt{\left(\dfrac{h^2}{a^2}+\dfrac{k^2}{b^2}\right)}}=\frac{ab}{\sqrt{b^2h^2+a^2k^2}}.$$

$$\Rightarrow \cos\left(\frac{\phi_1+\phi_2}{2}\right)=\frac{h}{a}\frac{ab}{\sqrt{b^2h^2+a^2k^2}};$$

$$\Rightarrow \cos\left(\frac{\phi_1+\phi_2}{2}\right)=\frac{hb}{\sqrt{b^2h^2+a^2k^2}}$$

Similarly, $\sin\left(\dfrac{\phi_1+\phi_2}{2}\right)=\dfrac{ka}{\sqrt{b^2h^2+a^2k^2}}$;

$$\cos\left(\frac{\phi_1-\phi_2}{2}\right) \quad \frac{ab}{\sqrt{b^2h^2 \quad a^2k^2}}$$

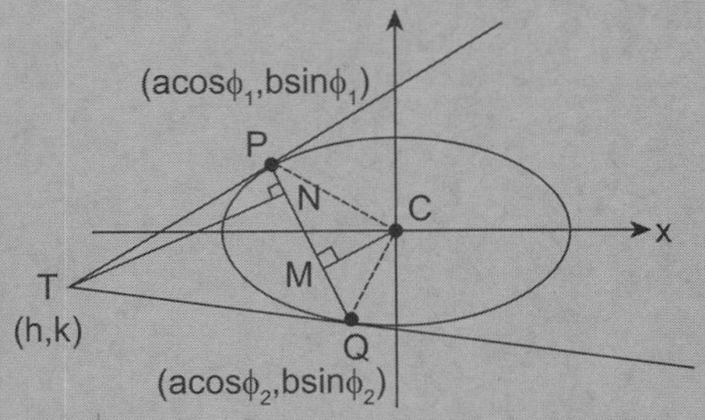

FIGURE 5.47

Let TN and CM be perpendiculars on PQ from T and centre C.

$$\therefore \quad \frac{\text{area } \Delta TPQ}{\text{area } \Delta CPQ}=\frac{\frac{1}{2}PQ\times TN}{\frac{1}{2}PQ\times CM}=\frac{TN}{CM}$$

$$\therefore \quad \text{area } \Delta TPQ=(\text{area } \Delta CPQ)\frac{TN}{CM}=(\text{area } \Delta CPQ)\frac{\left|\dfrac{h^2}{a^2}+\dfrac{k^2}{b^2}-1\right|}{\left(\sqrt{\dfrac{h^2}{a^4}+\dfrac{k^2}{b^4}}\right)(CM)}$$

$$=(\text{area } \Delta CPQ)\frac{\left|\dfrac{h^2}{a^2}+\dfrac{k^2}{b^2}-1\right|}{\left(\sqrt{\dfrac{h^2}{a^4}+\dfrac{k^2}{b^4}}\right)\cdot\dfrac{1}{\sqrt{\dfrac{h^2}{a^4}+\dfrac{k^2}{b^4}}}}=(\text{area of } \Delta CPQ)\left|\frac{h^2}{a^2}+\frac{k^2}{b^2}-1\right| \qquad \text{.....(iii)}$$

$$\therefore \quad \text{area } \Delta TPQ=(\text{area } \Delta CPQ)\left|\frac{h^2}{a^2}+\frac{k^2}{b^2}-1\right|=\frac{1}{2}|(x_1y_2-x_2y_1)|\left|\frac{h^2}{a^2}+\frac{k^2}{b^2}-1\right|$$

$$=\frac{1}{2}|a\cos\phi_1 b\sin\phi_2-a\cos\phi_2 b\sin\phi_1|\left|\frac{h^2}{a^2}+\frac{k^2}{b^2}-1\right|$$

$$=\frac{1}{2}ab|\sin\phi_2\cos\phi_1-\cos\phi_2\sin\phi_1|\left|\frac{h^2}{a^2}+\frac{k^2}{b^2}-1\right|=\frac{1}{2}ab|\sin(\phi_1-\phi_2)|\left(\frac{h^2}{a^2}+\frac{k^2}{b^2}-1\right)$$

[$\because$ (h, k) lies out side the ellipse, $\dfrac{h^2}{a^2}+\dfrac{k^2}{b^2}-1>0$]

$$=ab\left|\sin\left(\frac{\phi_1-\phi_2}{2}\right)\cos\left(\frac{\phi_1-\phi_2}{2}\right)\right|\left(\frac{h^2}{a^2}+\frac{k^2}{b^2}-1\right)$$

$$=ab\left|\cos\left(\frac{\phi_1-\phi_2}{2}\right)\sqrt{1-\cos^2\left(\frac{\phi_1-\phi_2}{2}\right)}\right|\left(\frac{h^2}{a^2}+\frac{k^2}{b^2}-1\right)$$

$$=ab\left|\frac{ab}{\sqrt{b^2h^2+a^2k^2}}\sqrt{1-\left(\frac{a^2b^2}{b^2h^2+a^2k^2}\right)}\right|\left(\frac{h^2}{a^2}+\frac{k^2}{b^2}-1\right)$$

$$= \left(\frac{a^2b^2}{b^2h^2+a^2k^2}\right)\frac{\sqrt{a^2k^2+b^2h^2-a^2b^2}}{1}\left(\frac{h^2}{a^2}+\frac{k^2}{b^2}-1\right)$$

$$= \frac{a^2b^2\sqrt{a^2k^2+b^2h^2-a^2b^2}}{(b^2h^2+a^2k^2)}\left(\frac{h^2}{a^2}+\frac{k^2}{b^2}-1\right)$$

$$= a^2b^2\frac{\sqrt{a^2b^2\left(\frac{k^2}{b^2}+\frac{h^2}{a^2}-1\right)}}{a^2b^2\left(\frac{h^2}{a^2}+\frac{k^2}{b^2}\right)}\left[\frac{h^2}{a^2}+\frac{k^2}{b^2}-1\right] = \frac{ab\left[\frac{h^2}{a^2}+\frac{k^2}{b^2}-1\right]^{3/2}}{\left(\frac{h^2}{a^2}+\frac{k^2}{b^2}\right)}.$$

$\therefore$ From (iii), ar $\Delta CPQ = \dfrac{ar\Delta TPQ}{\left|\frac{h^2}{a^2}+\frac{k^2}{b^2}-1\right|} = \dfrac{ab\left(\frac{h^2}{a^2}+\frac{k^2}{b^2}-1\right)^{1/2}}{\left(\frac{h^2}{a^2}+\frac{k^2}{b^2}\right)}$

Now, area of quadrilateral $TPCQ$ = area ΔTPQ + area ΔCPQ

$$= \frac{ab\left[\frac{h^2}{a^2}+\frac{k^2}{b^2}-1\right]^{3/2}}{\left(\frac{h^2}{a^2}+\frac{k^2}{b^2}\right)} + \frac{ab\sqrt{\left(\frac{h^2}{a^2}+\frac{k^2}{b^2}-1\right)}}{\left(\frac{h^2}{a^2}+\frac{k^2}{b^2}\right)} = \frac{ab}{\left(\frac{h^2}{a^2}+\frac{k^2}{b^2}\right)}\sqrt{\left(\frac{h^2}{a^2}+\frac{k^2}{b^2}-1\right)}\left[\frac{h^2}{a^2}+\frac{k^2}{b^2}-1+1\right]$$

$$= ab\left(\frac{h^2}{a^2}+\frac{k^2}{b^2}-1\right)^{1/2} = \sqrt{b^2h^2+a^2k^2-a^2b^2}.$$

■ CHORD WITH A GIVEN MID-POINT

The equation of a chord of the ellipse $\dfrac{x^2}{a^2}+\dfrac{y^2}{b^2}=1$ bisected at the point (x_1, y_1) is given by $\dfrac{xx_1}{a^2}+\dfrac{yy_1}{b^2}-1=\dfrac{x_1^2}{a^2}+\dfrac{y_1^2}{b^2}-1$ or $T = S_1$

Let $Q \equiv (x_2, y_2)$ and $R \equiv (x_3, y_3)$ be the end points of a chord QR of the ellipse $\dfrac{x^2}{a^2}+\dfrac{y^2}{b^2}=1$ and let $P \equiv (x_1, y_1)$ be its mid-point.

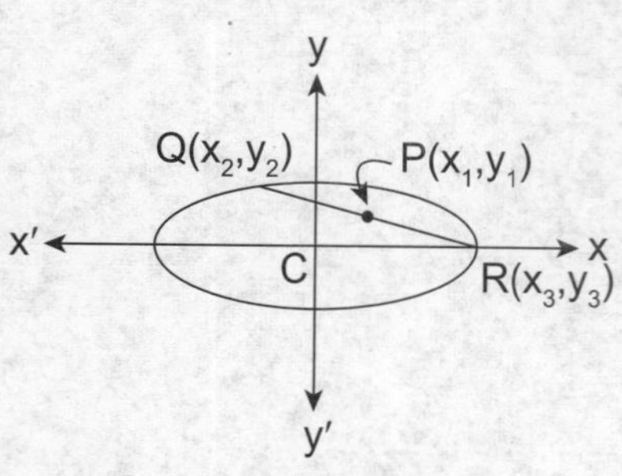

FIGURE 5.48

Now, $Q \equiv (x_2, y_2)$ and $R \equiv (x_3, y_3)$ lie on $\dfrac{x^2}{a^2}+\dfrac{y^2}{b^2}=1$

$\Rightarrow \quad \dfrac{x_2^2}{a^2}+\dfrac{y_2^2}{b^2}=1$ ....(i)

and $\dfrac{x_3^2}{a^2}+\dfrac{y_3^2}{b^2}=1$ ....(ii)

Subtracting (ii) from (i), we get

$$\frac{1}{a^2}(x_2^2-x_3^2)+\frac{1}{b^2}(y_2^2-y_3^2)=0$$

$$\Rightarrow \quad \frac{(x_2+x_3)(x_2-x_3)}{a^2}+\frac{(y_2+y_3)(y_2-y_3)}{b^2}=0$$

$$\Rightarrow \quad \frac{y_2-y_3}{x_2-x_3}=-\frac{b^2(x_2+x_3)}{a^2(y_2+y_3)}$$

$$= -\frac{b^2}{a^2}.\frac{2x_1}{2y_1}\left(\because x_1=\frac{x_2+x_3}{2} \text{ and } y_1=\frac{y_2+y_3}{2}\right)$$

$$= -\frac{b^2x_1}{a^2y_1} \qquad \text{...(iii)}$$

$\therefore$ Equation of QR is $y-y_1=\dfrac{y_2-y_3}{x_2-x_3}(x-x_1)$

$$\Rightarrow \quad y - y_1 = -\frac{b^2 x_1}{a^2 y_1}(x - x_1) \qquad \text{[from (iii)]}$$

$$\Rightarrow \quad \frac{yy_1}{b^2} - \frac{y_1^2}{b^2} = -\frac{xx_1}{a^2} + \frac{x_1^2}{a^2}$$

$$\Rightarrow \quad \frac{xx_1}{a^2} + \frac{yy_1}{b^2} - 1 = \frac{x_1^2}{a^2} + \frac{y_1^2}{b^2} - 1 \text{ or } T = S_1;$$

where $T = \frac{xx_1}{a^2} + \frac{yy_1}{b^2} - 1$ and $S_1 = \frac{x_1^2}{a^2} + \frac{y_1^2}{b^2} - 1$

ILLUSTRATION 49: Find the condition on 'a' and 'b' for which two distinct chords of the ellipse $\frac{x^2}{2a^2} + \frac{y^2}{2b^2} = 1$ passing through $(a, -b)$ are bisected by the line $x + y = b$.

SOLUTION: Let AB and CD be two chords of ellipse passing through point $P(a, -b)$ which are bisected by the line $x + y = b$

Let the line $x + y = b$ bisects the chord at $Q(\alpha, b - \alpha)$

$\Rightarrow$ Equations of chords would be

$$\therefore \quad \frac{x\alpha}{2a^2} + \frac{y(b-\alpha)}{2b^2} = \frac{\alpha^2}{2a^2} + \frac{(b-\alpha)^2}{2b^2} \quad (\because T = X_1)$$

Since it passes through $(a, -b)$

$$\therefore \quad \frac{\alpha}{2a} - \frac{(b-\alpha)}{2b} = \frac{\alpha^2}{2a^2} + \frac{(b-\alpha)^2}{2b^2}$$

$$\Rightarrow \quad \left(\frac{1}{a} + \frac{1}{b}\right)\alpha - 1 = \alpha^2\left(\frac{1}{a^2} + \frac{1}{b^2}\right) - \frac{2}{b}\alpha + 1$$

$$\Rightarrow \quad \alpha^2\left(\frac{1}{a^2} + \frac{1}{b^2}\right) - \left(\frac{3}{b} + \frac{1}{a}\right)\alpha + 2 = 0$$

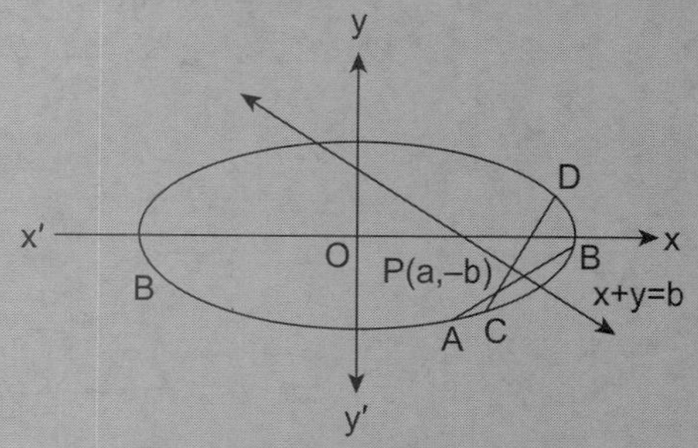

FIGURE 5.49

Since line bisects two chords

$\therefore$ The above quadratic equation in α must have two distinct real roots.

$$\therefore \quad \left(\frac{3}{b} + \frac{1}{a}\right)^2 - 4\left(\frac{1}{a^2} + \frac{1}{b^2}\right).2 > 0 \Rightarrow \frac{9}{b^2} + \frac{1}{a^2} + \frac{6}{ab} - \frac{8}{a^2} - \frac{8}{b^2} > 0 \Rightarrow \frac{1}{b^2} - \frac{7}{a^2} + \frac{6}{ab} > 0$$

$\Rightarrow a^2 - 7b^2 + 6ab > 0$

$\Rightarrow a^2 > 7b^2 - 6ab$, which is the required condition

ILLUSTRATION 50: Find the locus of the middle points of chords of an ellipse which pass through a fixed point.

SOLUTION: Let $P(h, k)$ be the middle point of any chord AB of the ellipse $\frac{x^2}{a^2} + \frac{y^2}{b^2} = 1$, then equation of chord AB will be

$$\Rightarrow \quad \frac{xh}{a^2} + \frac{yk}{b^2} - 1 = \frac{h^2}{a^2} + \frac{k^2}{b^2} - 1 \Rightarrow \frac{xh}{a^2} + \frac{yk}{b^2} = \frac{h^2}{a^2} + \frac{k^2}{b^2} \qquad \text{....(i)}$$

But it passes through a fixed point $R(x_1, y_1)$ (say); its co-ordinates must satisfy equation (i),

$$\therefore \quad \frac{hx_1}{a^2} + \frac{ky_1}{b^2} = \frac{h^2}{a^2} + \frac{k^2}{b^2}$$

$$\Rightarrow \quad \frac{\left(h - \frac{x_1}{2}\right)^2}{a^2} + \frac{\left(k - \frac{y_1}{2}\right)^2}{b^2} = -\left(\frac{x_1^2}{a^2} + \frac{k_1^2}{b^2}\right)$$; which is the required equation of locus and represents.

An ellipse with centre at $\left(\frac{x_1}{2}, \frac{y_1}{2}\right)$ and axes parallel to co-ordinate axes.

ILLUSTRATION 51: Find the locus of the middle points of normal chords of the ellipse $\frac{x^2}{a^2}+\frac{y^2}{b^2}=1$.

SOLUTION: Let (h, k) be the middle point of any chord of an ellipse, then its equation is $T = S_1$

or $\frac{xh}{a^2}+\frac{yk}{b^2}-1=\frac{h^2}{a^2}+\frac{k^2}{b^2}-1$

or $\frac{xh}{a^2}+\frac{yk}{b^2}=\frac{h^2}{a^2}+\frac{k^2}{b^2}$ ....(i)

If (i) is a normal chord, then it must be of the form

$ax\sec\phi - by\,\text{cosec}\phi = a^2-b^2$ ....(ii)

Thus the equations (i) and (ii) represent the same normal chord of the ellipse with its middle point (h, k). Hence they are identical and comparing their co-efficients, we get

$$\frac{h/a^2}{a\sec\phi}=\frac{k/b^2}{-b\,\text{cosec}\phi}=\frac{\frac{h^2}{a^2}+\frac{k^2}{b^2}}{(a^2-b^2)}$$

$$\Rightarrow \cos\phi=\frac{a^3}{h}\frac{\left(\frac{h^2}{a^2}+\frac{k^2}{b^2}\right)}{(a^2-b^2)} \quad \text{....(iii)}$$

$$\text{and } \sin\phi=-\frac{b^3}{k}\frac{\left(\frac{h^2}{a^2}+\frac{k^2}{b^2}\right)}{(a^2-b^2)} \quad \text{.... (iv)}$$

Squaring and adding (iii) and (iv), we have $\cos^2\phi+\sin^2\phi=\frac{\left(\frac{a^6}{h^2}+\frac{b^6}{k^2}\right)\left(\frac{h^2}{a^2}+\frac{k^2}{b^2}\right)^2}{(a^2-b^2)^2}$

$$\Rightarrow 1=\frac{\left(\frac{a^6}{h^2}+\frac{b^6}{k^2}\right)\left(\frac{h^2}{a^2}+\frac{k^2}{b^2}\right)^2}{(a^2-b^2)^2} \Rightarrow \left(\frac{h^2}{a^2}+\frac{k^2}{b^2}\right)^2\left(\frac{a^6}{h^2}+\frac{b^6}{k^2}\right)=(a^2-b^2)^2$$

Hence locus of (h, k) is $\left(\frac{x^2}{a^2}+\frac{y^2}{b^2}\right)^2\left(\frac{a^6}{x^2}+\frac{b^6}{y^2}\right)=(a^2-b^2)^2$

PROPERTIES OF TANGENTS TO ELLIPSE

P: 1. The product of the perpendiculars from the foci of any tangent to an ellipse is equal to the square of the semi-minor axis and the feet of these ⊥rs lie on the auxiliary circle.

Proof: Let the equation of ellipse be

$\frac{x^2}{a^2}+\frac{y^2}{b^2}=1$ .... (i)

Equation of any tangent to (i) in terms of slope (m) is given by $y=mx+\sqrt{(a^2m^2+b^2)}$

or $y-mx=\sqrt{a^2m^2+b^2}$ .... (ii)

Equation of a line perpendicular to (ii) and passing through $S\,(ae, 0)$ is

$$y-0=-\frac{1}{m}(x-ae)$$

or $x+my=ae$ .... (iii)

The lines (ii) and (iii) will meet at the foot of ⊥ whose locus is obtained by eliminating the variable m between (ii) and (iii), then squaring and adding (ii) and (iii) we get

$$(y-mx)^2+(x+my)^2=a^2m^2+b^2+a^2e^2$$

$$\Rightarrow (1+m^2)(x^2+y^2)=a^2m^2+b^2+a^2-b^2$$

$$\Rightarrow (1+m^2)(x^2+y^2)=a^2(1+m^2)$$

or $x^2+y^2=a^2$;

Which is auxiliary circle of ellipse. Similarly, we can show that the other foot drawn from second focus also lies on $x^2 + y^2 = a^2$.

Again if p_1 and p_2 be perpendiculars from foci $S(ae, 0)$ and $S'(-ae, 0)$ on (2), then $p_1 = \dfrac{|\sqrt{(a^2m^2+b^2)}+mae|}{\sqrt{(1+m^2)}}$

and $p_2 = \dfrac{|\sqrt{(a^2m^2+b^2)}-mae|}{\sqrt{(1+m^2)}}$

$$\Rightarrow \quad p_1p_2 = \frac{|a^2m^2+b^2-a^2e^2m^2|}{(1+m^2)}$$

$$= \frac{|a^2m^2+b^2-(a^2-b^2)m^2|}{(1+m^2)} = \frac{b^2(1+m^2)}{(1+m^2)} = b^2$$

= (semi minor axis)^2.

P: 2. The tangents at the extremities of latus rectum of an ellipse intersect on the corresponding directrix.

Let LSL' be a latus rectum of the ellipse $\dfrac{x^2}{a^2}+\dfrac{y^2}{b^2}=1$.

$\therefore$ The co-ordinates of L and L' are $\left(ae, \dfrac{b^2}{a}\right)$ and $\left(ae, -\dfrac{b^2}{a}\right)$ respectively.

$\therefore$ Equation of tangent at $L\left(ae, \dfrac{b^2}{a}\right)$ is

$$\Rightarrow \quad \frac{x(ae)}{a^2}+\frac{\left(\frac{b^2}{a}\right)y}{b^2}=1 \Rightarrow xe+y=a \qquad \text{.... (i)}$$

The equation of the tangent at $L'\left(ae, -\dfrac{b^2}{a}\right)$ is

$$\frac{x(ae)}{a^2}+\frac{y\left(-\frac{b^2}{a}\right)}{b^2}=1$$

$$\Rightarrow \; ex-y=a \qquad \text{.... (ii)}$$

Solving (i) and (ii) we get $x=\dfrac{a}{e}$ and $y=0$

Thus the tangents at L and L' intersect at $(a/e, 0)$ which is a point lying on the corresponding directrix i.e., $x = \dfrac{a}{e}$.

TEXTUAL EXERCISE-4 (SUBJECTIVE)

1. (a) Find the middle point of the chord intercepted on the line $2x - y + 3 = 0$ by the ellipse $3x^2 + 5y^2 = 30$.
   (b) Find the equation of the chord of $x^2 + 4y^2 = 36$ which is bisected at (2, 1)
2. (a) Find the locus of mid-points of chords drawn through the $+ve$ end of minor axis.
   (b) Find the locus of the middle points of chords of an ellipse which subtends a right angle at the centre.
3. A tangent to the ellipse $x^2 + 4y^2 = 4$ meets the ellipse $x^2 + 2y^2 = 6$ at P and Q. Prove that the tangents at P and Q of ellipse $x^2 + 2y^2 = 6$ are at right angles.
4. From a point O on the circle $x^2 + y^2 = d^2$, tangents OP and OQ are drawn to the ellipse $\dfrac{x^2}{a^2}+\dfrac{y^2}{b^2}=1$. Find the locus of the mid-point of the chord PQ.
5. Let d be perpendicular distance from centre of $x^2/a^2 + y^2/b^2 = 1$ to the tangent drawn at P on the ellipse. If F_1 and F_2 are the two foci of ellipse, then show that $(PF_1 - PF_2)^2 = 4a^2(1 - b^2/d^2)$.
6. If p is the length of perpendicular from the focus S of the ellipse $x^2/a^2 + y^2/b^2 = 1$, on the tangent at P, a point on the ellipse, then show that $\dfrac{b^2}{p^2}=\dfrac{2a}{SP}-1$.
7. A straight line AB touches the ellipse $x^2/a^2 + y^2/b^2 = 1$ and the circle $x^2 + y^2 = r^2$, where $b < r < a$. PQ is a focal chord of the ellipse. If PQ is parallel to AB and cuts the circle in P and Q, find the length of the perpendicular drawn from the centre of the ellipse to PQ. Hence show that $PQ = 2b$.
8. An ellipse of semi axes a, b slide between two perpendicular lines prove that the locus of its foci is $(x^2 + y^2)(x^2y^2 + b^4) = 4a^2x^2y^2$, the two lines being taken as co-ordinate axes.
9. Find the locus of the mid-point of chord of the ellipse $\dfrac{x^2}{a^2}+\dfrac{y^2}{b^2}=1$ $(a > b)$ passing through the point $(2a, 0)$.
10. Find the locus of the point the chord of contact of tangents from which to the ellipse $\dfrac{x^2}{a^2}+\dfrac{y^2}{b^2}=1$ touches the circle $x^2 + y^2 = c^2$.
11. Find the angle between pair of tangents drawn to the ellipse $3x^2 + 2y^2 = 5$ from the point (1, 2).

Answer Keys

1. (a) (–30/23, 9/23) (b) $x + 2y - 4 = 0$ **2.** (a) $x^2/a^2 + y^2/b^2 = y/b$ (b) $\frac{b^4x^2 + a^4y^2}{(a^2y^2 + b^2x^2)^2} = \frac{1}{a^2} + \frac{1}{b^2}$

4. $x^2 + y^2 = d^2 (x^2/a^2 + y^2/b^2)^2$ **7** $\sqrt{r^2 - b^2}$ **9.** $\frac{(x-a)^2}{a^2} + \frac{y^2}{b^2} = 1$

10. $b^4c^2x^2 + a^4c^2y^2 = a^4b^4$. **11.** $\tan^{-1}\left|\frac{12}{\sqrt{5}}\right|$

TEXTUAL EXERCISE–4 (OBJECTIVE)

1. If the ellipse $\frac{x^2}{4} + y^2 = 1$ meets the ellipse $x^2 + \frac{y^2}{a^2} = 1$ in four distinct points and $a = b^2 - 5b + 7$, then b belongs to

(a) (1, 4) (b) $(-\infty, 2) \cup (3, \infty)$
(c) (2, 3) (d) None of these

2. The locus of the point of intersection of ⊥ tangents to $\frac{x^2}{a^2} + \frac{y^2}{b^2} = 1$ and $\frac{x^2}{a^2 + \lambda} + \frac{y^2}{b^2 + \lambda} = 1$ is

(a) $x^2 + y^2 = a^2 + \lambda$ (b) $x^2 + y^2 = a^2 + b^2 + \lambda$
(c) $x^2 + y^2 = a^2 + b^2$ (d) $x^2 + y^2 = b^2 + \lambda$

3. $S(3,4)$ and $S'(9,12)$ are two foci of an ellipse. If the foot of the perpendicular from S on a tangent to the ellipse has the co-ordinates (1, –4), then the eccentricity of the ellipse is

(a) 8/13 (b) 9/23
(c) 5/13 (d) 7/9

4. Let d_1 and d_2 be the lengths of ⊥s drawn from foci S' and S of the ellipse $\frac{x^2}{a^2} + \frac{y^2}{b^2} = 1$ to the tangent at any point P on ellipse. Then $S'P : SP$ is equal to

(a) $d_1^2 : d_2^2$ (b) $d_2 : d_1$
(c) $d_1 : d_2$ (d) $\sqrt{d_1} : \sqrt{d_2}$

5. For an ellipse $\frac{x^2}{9} + \frac{y^2}{4} = 1$ with vertices A and A' tangent drawn at the point P in the Ist quadrant meets the y-axis in Q and the chord $A'P$ meet the y-axis in N. If C is the origin, then $CQ^2 - NQ^2$ is equal to

(a) 4 (b) 8
(c) 15 (d) 20

6. An ellipse is drawn with major and minor axis of lengths 10 and 8 respectively. Using one focus as centre, a circle is drawn that is tangent to the ellipse. The radius of the circle is

(a) 4 (b) $\sqrt{2}$
(c) 8 (d) 2

7. An ellipse slides between two lines at right angle to one another. Then the locus of its centre is

(a) a circle (b) a parabola
(c) an ellipse (d) a hyperbola

8. The locus of centroid of an equilateral triangle inscribed in the ellipse $\frac{x^2}{a^2} + \frac{y^2}{b^2} = 1$ is

(a) $\frac{(a^2 + 3b^2)^2}{a^2}x^2 + \frac{(b^2 + 3a^2)^2}{b^2}y^2 = (a^2 + b^2)^2$

(b) $\frac{(a^2 - 3b^2)^2}{a^2}x^2 + \frac{(b^2 - 3a^2)^2}{b^2}y^2 = (a^2 + b^2)^2$

(c) $\frac{(a^2 + 3b^2)^2}{a^2}x^2 - \frac{(b^2 + 3a^2)^2}{b^2}y^2 = (a^2 - b^2)^2$

(d) $\frac{(a^2 + 3b^2)^2}{a^2}x^2 + \frac{(b^2 + 3a^2)^2}{b^2}y^2 = (a^2 - b^2)^2$

9. If the tangent at a point $(a\cos\theta, b\sin\theta)$ on the ellipse $\frac{x^2}{a^2} + \frac{y^2}{b^2} = 1$ meets the auxillary circle in two points 1 If the chord joining them subtends a right angle at the centre, then the eccentricity of the ellipse is given by

(a) $(1 - \cos^2\theta)^{-1/2}$ (b) $(1 + \sin^2\theta)$
(c) $(1 + \sin^2\theta)^{-1/2}$ (d) $(1 + \cos^2\theta)$

10. Chord of an ellipse are drawn through the positive end of the minor axis. Then their mid-point lies on

(a) a circle (b) a parabola
(c) an ellipse (d) a hyperbola

11. A chord PQ of the ellipse $\frac{x^2}{9} + \frac{y^2}{4} = 1$ subtends a right angle at its centre. The locus of the point of intersection of tangents drawn at P and Q is

(a) a circle (b) a parabola
(c) an ellipse (d) a hyperbola

12. The point of the intersection of the tangents at the two points on the ellipse $\frac{x^2}{a^2}+\frac{y^2}{b^2}=1$ whose eccentric angles differ by a right angle lies on the ellipse

(a) $\frac{x^2}{a^2}+\frac{y^2}{b^2}=2$ (b) $\frac{x^2}{a}+\frac{y^2}{b}=2$

(c) $\frac{x^2}{a^2}+\frac{y^2}{b^2}=1$ (d) $\frac{x^2}{a}+\frac{y^2}{b}=1$

13. The line $lx + my + n = 0$; cut the ellipse $\frac{x^2}{a^2}+\frac{y^2}{b^2}=1$ in points whose eccentric angles differ by $\pi/2$. Then the value of $a^2l^2 + b^2m^2$ is

(a) $2n^2$ (b) $2m^2$

(c) $2m$ (d) $2n$

14. If the chords of contact of tangents drawn from two points (x_1, y_1) and (x_2, y_2) to the ellipse $\frac{x^2}{a^2}+\frac{y^2}{b^2}=1$ are at right angle, then $\frac{x_1x_2}{y_1y_2}$ is

(a) a^2/b^2 (b) b^2/a^2

(c) $-a^4/b^4$ (d) $-b^4/a^4$.

15. The locus of the point of intersection of two tangents if the sum of the eccentric angles of their points of contact be equal to a constant angle 2α is

(a) $ay = bx \tan \alpha$

(b) $ay = bx \cot \alpha$

(c) $ax = by \tan \alpha$

(d) None of these

16. Equation of chord of contact of pair of tangents drawn to ellipse $4x^2 + 9y^2 = 36$ from the point (m, n) where $m.n = m + n$; m, n being non-zero positive integers is

(a) $x + 9y = 15$

(b) $4x + 9y = 18$

(c) $x + 2y = 7$

(d) None of these

Answer Keys

1. (b)	**2.** (b)	**3.** (c)	**4.** (c)	**5.** (a)	**6.** (c, d)	**7.** (a)	**8.** (d)	**9.** (c)	**10.** (c)
11. (c)	**12.** (a)	**13.** (a)	**14.** (c)	**15.** (a)	**16.** (b).				

■ DIFFERENT FORMS OF NORMALS TO ELLIPSE

(a) Cartesian Form: $\frac{a^2x}{x_1}-\frac{b^2y}{y_1}=a^2-b^2 = a^2e^2$; is cartesian form of normal to ellipse $\frac{x^2}{a^2}+\frac{y^2}{b^2}=1$ at point (x_1, y_1).

Proof: Let $P(x_1, y_1)$ be any point on ellipse $\frac{x^2}{a^2}+\frac{y^2}{b^2}=1$ ....(i)

$\Rightarrow \frac{x_1^2}{a^2}+\frac{y_1^2}{b^2}=1$ ...(ii)

Now, slope of tangent to ellipse (i) at $P(x_1, y_1) = \frac{-b^2}{a^2}.\frac{x_1}{y_1}$

$\Rightarrow$ Slope of normal to curve at P is $m=\frac{a^2}{b^2}\frac{y_1}{x_1}$

$\therefore$ Equation of normal to curve at $P(x_1, y_1)$ is given by

$(y-y_1)=\frac{a^2}{b^2}.\frac{y_1}{x_1}(x-x_1)$

$\Rightarrow x_1b^2y - b^2x_1y_1 = a^2xy_1 - a^2x_1y_1$

$\Rightarrow x_1b^2y - a^2xy_1 = b^2x_1y_1 - a^2x_1y_1$

$\Rightarrow \frac{b^2y}{y_1}-\frac{a^2x}{x_1}=(b^2-a^2)$

$\Rightarrow \frac{a^2x}{x_1}-\frac{b^2y}{y_1}=a^2-b^2$

$\Rightarrow \frac{a^2x}{x_1}-\frac{b^2y}{y_1}=a^2e^2$ ....(iii)

$(\because b^2 = a^2(1 - e^2))$

which is the required equation of normal to ellipse at point $P(x_1, y_1)$.

(b) Paramatric Form of Normal: Replacing x_1 by a $\cos\phi$ and y_1 by $b \sin\phi$, (iii) becomes

$$\frac{a^2x}{a\cos\phi}-\frac{b^2y}{b\sin\phi}=a^2-b^2$$

or $ax\sec\phi - by\,\mathrm{cosec}\phi = a^2-b^2$.

is the equation of normal at $(a\cos\phi, b\sin\phi)$.

ILLUSTRATION 52: P and Q are corresponding points on the ellipse $\frac{x^2}{a^2}+\frac{y^2}{b^2}=1$ and the auxiliary circles, respectively. The normal at P to the ellipse meets CQ in R, where C is the centre of the ellipse. Prove that $CR = a + b$.

SOLUTION: Let $P \equiv (a\cos\theta, b\sin\theta)$

$\therefore\ Q \equiv (a\cos\theta, a\sin\theta)$

Equation of normal at P is

$(a\sec\theta)x - (b\,\text{cosec}\theta)\,y = a^2 - b^2$ ....(i)

Equation of CQ is $y = \tan\theta \,.\, x$

Solving equation (i) and (ii),

we get $(a - b)\,x = (a^2 - b^2)\cos\theta$

$\Rightarrow x = (a + b)\cos\theta$, and $y = (a + b)\sin\theta$

$\therefore\ R \equiv ((a+b)\cos\theta, (a+b)\sin\theta)$

$\Rightarrow CR = a + b$

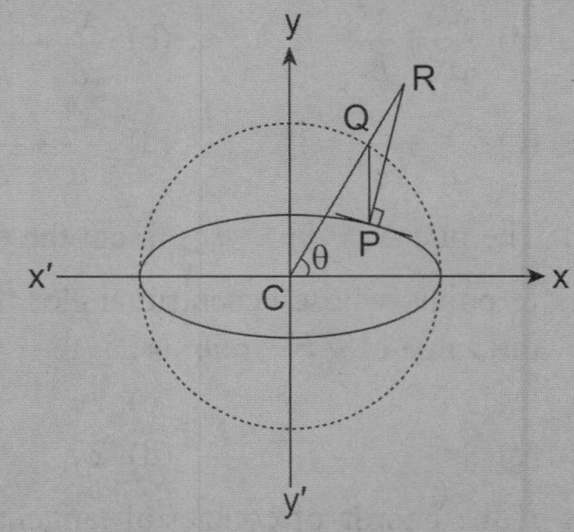

FIGURE 5.50

ILLUSTRATION 53: Prove that in ellipse, the distance between the centre and any normal does not exceed the difference between the semi-axes of the ellipse.

SOLUTION: Let the equation of ellipse is $\frac{x^2}{a^2}+\frac{y^2}{b^2}=1$

Equation of normal at $P(\theta)$ is $(a\sec\theta)x - (b\,\text{cosec}\theta)\,y - a^2 + b^2 = 0$

$\therefore$ Distance of normal from centre

$$= |CR| = \frac{|a^2-b^2|}{\sqrt{a^2+b^2+(a\tan\theta)^2+(b\cot\theta)^2}}$$

$$\Rightarrow |CR| = \frac{|a^2-b^2|}{\sqrt{(a+b)^2+(a\tan\theta-b\cot\theta)^2}}$$

$$\le \frac{|a^2-b^2|}{\sqrt{(a+b)^2}} = \frac{|a^2-b^2|}{|a+b|} = |a-b|$$

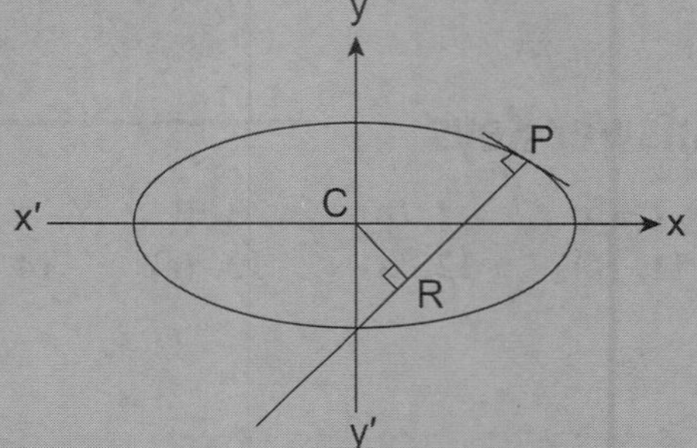

FIGURE 5.51

ILLUSTRATION 54: If the tangent drawn at a point $(t^2, 2t)$ on the parabola $y^2 = 4x$ is same as the normal drawn at a point $((\sqrt{5}\cos\phi, 2\sin\phi)$ on the ellipse $4x^2 + 5y^2 = 20$. Find the values of t and ϕ.

SOLUTION: Equation of the tangent at $(t^2, 2t)$ on the parabola $y^2 = 4x$ is $x - ty + t^2 = 0$ ....(1)

Equation of the normal at $(\sqrt{5}\cos\phi, 2\sin\phi)$ to the ellipse is $\sqrt{5}\sec\phi x - 2\text{cosec}\,\phi\, y = 1$ ...(2)

$\therefore$ (1) and (2) represent the same line

$$\Rightarrow \frac{\sqrt{5}\sec\phi}{1} = \frac{2\text{cosec}\phi}{t} = \frac{1}{-t^2}$$

$\Rightarrow \cos\phi = -\sqrt{5}t^2;\ \sin\phi = -2t$

$\Rightarrow 4t^2 + 5t^4 = 1 \Rightarrow t^2 = 1/5$

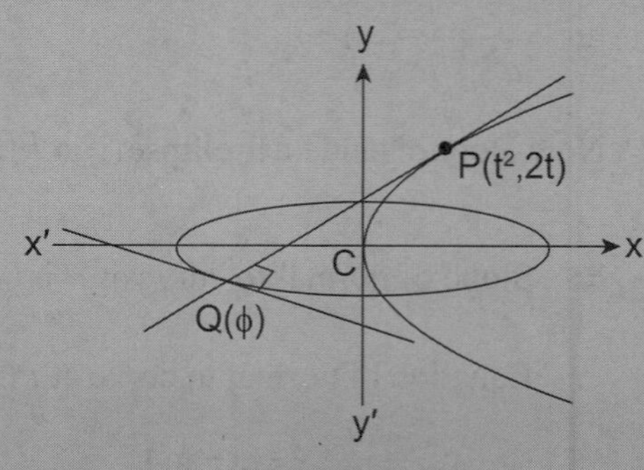

FIGURE 5.52

(c) Slope form of Normal: Also, the equations of the normals of slope m to the ellipse $\frac{x^2}{a^2}+\frac{y^2}{b^2}=1$ are given by $y = mx \pm \frac{m(a^2-b^2)}{\sqrt{(a^2+b^2m^2)}}$ and the points of contact are

$$\left(\frac{a^2}{\sqrt{a^2+b^2m^2}}, \frac{mb^2}{\sqrt{a^2+b^2m^2}}\right) \text{ and } \left(\frac{-a^2}{\sqrt{(a^2+b^2m^2)}}, \frac{-mb^2}{\sqrt{(a^2+b^2m^2)}}\right).$$

The equation of normal to the ellipse $\frac{x^2}{a^2}+\frac{y^2}{b^2}=1$ at (x_1, y_1) is

$$\frac{a^2x}{x_1}-\frac{b^2y}{y_1}=a^2-b^2 \quad \text{....(1)}$$

Since 'm' is the slope of the normal, $m=\frac{a^2y_1}{b^2x_1}$

$$\Rightarrow y_1=\frac{b^2x_1m}{a^2} \quad \text{....(2)}$$

Since (x_1, y_1) lies on $\frac{x^2}{a^2}+\frac{y^2}{b^2}=1$

$$\Rightarrow \frac{x_1^2}{a^2}+\frac{y_1^2}{b^2}=1 \quad \Rightarrow \frac{x_1^2}{a^2}+\frac{b^4x_1^2m^2}{a^4b^2}=1$$

$$\Rightarrow \frac{x_1^2}{a^2}+\frac{b^2x_1^2m^2}{a^4}=1 \Rightarrow x_1^2=\frac{a^4}{(a^2+b^2m^2)}$$

$$\therefore \quad x_1=\pm\frac{a^2}{\sqrt{a^2+b^2m^2}}$$

From equation (2), $y_1=\pm\frac{mb^2}{\sqrt{(a^2+b^2m^2)}}$

$\therefore$ Equation of normal in terms of slope is

$$y-\left(\pm\frac{mb^2}{\sqrt{a^2+b^2m^2}}\right)=m\left(x-\left(\pm\frac{a^2}{\sqrt{a^2+b^2m^2}}\right)\right)$$

$$\Rightarrow y=mx\mp\frac{m(a^2-b^2)}{\sqrt{(a^2+b^2m^2)}} \quad \text{.....(3)}$$

Thus $y=mx\mp\frac{m(a^2-b^2)}{\sqrt{(a^2+b^2m^2)}}$ is a normal to the ellipse $\frac{x^2}{a^2}+\frac{y^2}{b^2}=1$, where m is the slope of the normal.

The co-ordinates of the point of contact corresponding to normal of slope m are $\left(\frac{a^2}{\sqrt{a^2+b^2m^2}}, \frac{mb^2}{\sqrt{a^2+b^2m^2}}\right)$ and $\left(\frac{-a^2}{\sqrt{a^2+b^2m^2}}, \frac{-mb^2}{\sqrt{a^2+b^2m^2}}\right)$

Comparing (3) with, $y = mx + c$

$$\therefore \quad c=\mp\frac{m(a^2-b^2)}{\sqrt{(a^2+b^2m^2)}} \text{ or } c^2=\frac{m^2(a^2-b^2)^2}{(a^2+b^2m^2)} \quad \text{.....(iv)}$$

$\Rightarrow$ points of contact are also given by

$$\Rightarrow \left(\frac{a^2c}{m(a^2-b^2)}, \frac{b^2c}{(a^2-b^2)}\right)$$

Also (iv) is the condition of normality, when $y = mx + c$ is the normal to $\frac{x^2}{a^2}+\frac{y^2}{b^2}=1$.

ILLUSTRATION 55: If the normal at an end of a latus-rectum of an ellipse $\frac{x^2}{a^2}+\frac{y^2}{b^2}=1$ passes through one extremity of the minor axis, show that the eccentricity of the ellipse is given by $e^4+e^2-1=0$ or $e^2=\frac{\sqrt{5}-1}{2}$.

SOLUTION: The co-ordinates of an end of the latus-rectum are $(ae, b^2/a)$. The equation of normal at $P(ae, b^2/a)$ is $\frac{a^2x}{ae}-\frac{b^2(y)}{b^2/a}=a^2-b^2$

or $\quad \frac{ax}{e}-ay=a^2-b^2$

It passes through one extremity of the minor axis whose co-ordinate are $(0, -b)$

$\therefore \quad 0+ab=a^2-b^2$

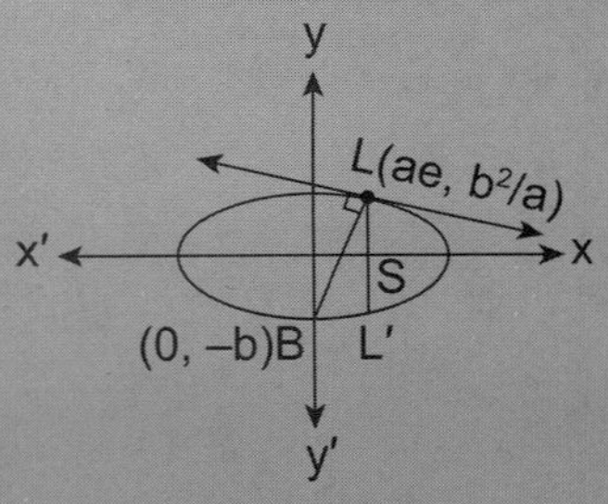

FIGURE 5.53

or $(a^2b^2) = (a^2 - b^2)^2$

or $a^2a^2(1-e^2) = (a^2e^2)^2$

or $1 - e^2 = e^4$ or $e^4 + e^2 - 1 = 0$

or $(e^2)^2 + e^2 - 1 = 0$ $\therefore\ e^2 = \dfrac{-1 \pm \sqrt{1+4}}{2}$

$\Rightarrow e^2 = \dfrac{\sqrt{5}-1}{2}$ [Taking +*ve* sign]

ILLUSTRATION 56: Prove that the straight line $lx + my + n = 0$ is a normal to the ellipse $\dfrac{x^2}{a^2} + \dfrac{y^2}{b^2} = 1$ if $\dfrac{a^2}{l^2} + \dfrac{b^2}{m^2} = \dfrac{(a^2-b^2)^2}{n^2}$.

SOLUTION: The equation of any normal to $\dfrac{x^2}{a^2} + \dfrac{y^2}{b^2} = 1$ is

$$ax\sec\phi - by\,\text{cosec}\,\phi = a^2 - b^2 \qquad \text{....(1)}$$

The straight line $lx + my + n = 0$ will be a normal to the ellipse $\dfrac{x^2}{a^2} + \dfrac{y^2}{b^2} = 1$

$\because$ (1) and $lx + my + n = 0$ represent the same line, $\dfrac{a\sec\phi}{l} = \dfrac{-b\,\text{cosec}\phi}{m} = \dfrac{a^2-b^2}{-n}$

$\Rightarrow \cos\phi = \dfrac{-na}{l(a^2-b^2)}$ and $\sin\phi = \dfrac{nb}{m(a^2-b^2)}$

$\because \sin^2\phi + \cos^2\phi = 1$

$\Rightarrow \dfrac{n^2b^2}{m^2(a^2-b^2)^2} + \dfrac{n^2a^2}{l^2(a^2-b^2)^2} = 1$ $\Rightarrow \dfrac{a^2}{l^2} + \dfrac{b^2}{m^2} = \dfrac{(a^2-b^2)^2}{n^2}$.

ILLUSTRATION 57: A normal inclined at a slope angle of 45° to the ellipse $\dfrac{x^2}{a^2} + \dfrac{y^2}{b^2} = 1$ is drawn. It meets the major and minor axes in P and Q respectively. If C is the centre of the ellipse, prove that area of ΔCPQ is $\dfrac{(a^2-b^2)^2}{2(a^2+b^2)}$ sq. units.

SOLUTION: Let R $(a\cos\phi, b\sin\phi)$ be any point on the ellipse, then equation of normal at R is

$$ax\sec\phi - by\,\text{cosec}\phi = a^2 - b^2$$

or $$\frac{x}{\dfrac{\cos\phi(a^2-b^2)}{a}} + \frac{y}{\dfrac{-\sin\phi(a^2-b^2)}{b}} = 1$$

It meets the major and minor axes at $P\left(\dfrac{(a^2-b^2)}{a}\cos\phi,\ 0\right)$ and $Q\left(0,\ -\dfrac{(a^2-b^2)}{b}\sin\phi\right)$ respectively.

$\therefore\ CP = \left(\dfrac{a^2-b^2}{a}\right)|\cos\phi|$ ......(i)

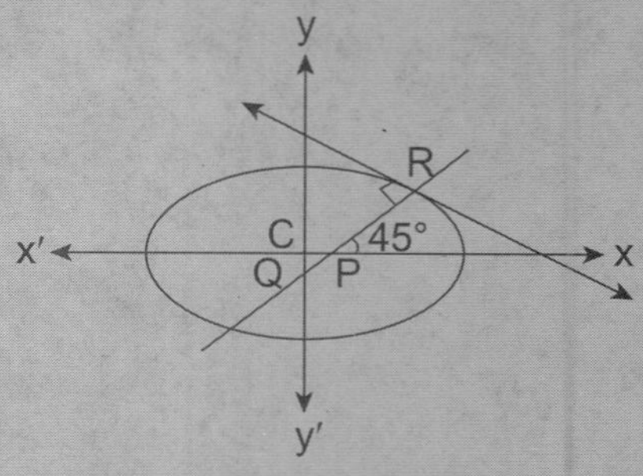

FIGURE 5.54

and $CQ = \left(\frac{a^2 - b^2}{b}\right)|\sin\phi|$ ......(ii)

$\because\ \sin 2\phi = \frac{2\tan\phi}{1+\tan^2\phi} = \frac{2ab}{a^2+b^2}$ ....(iii)

$\left(\because \text{ slope of normal} = \frac{a\sec\phi}{b\,\text{cosec}\,\phi} = \frac{a}{b}\tan\phi = \tan 45° = 1, \text{ given}\right)$

$\therefore$ Area of $\Delta CPQ = \frac{1}{2}|CP||CQ| = \frac{1}{2}\frac{(a^2-b^2)^2}{ab}\left|\frac{\sin 2\phi}{2}\right|$

$= \frac{(a^2-b^2)^2 \frac{ab}{(a^2+b^2)}}{2ab} = \frac{(a^2-b^2)^2}{2(a^2+b^2)}$ sq. units. (using (i), (ii) and (iii))

ILLUSTRATION 58: Any ordinate MP of an ellipse meets the auxiliary circle in Q. Prove that the locus of the point of intersection of the normals at P and Q is the circle $x^2 + y^2 = (a+b)^2$.

SOLUTION: Let the equation of ellipse be $\frac{x^2}{a^2} + \frac{y^2}{b^2} = 1$ Its auxiliary circle is $x^2 + y^2 = a^2$

Co-ordinate of P and Q are $(a\cos\phi, b\sin\phi)$ and $(a\cos\phi, a\sin\phi)$ respectively. Equation of normal at P to the ellipse $\frac{x^2}{a^2} + \frac{y^2}{b^2} = 1$ is

$ax\sec\phi - by\,\text{cosec}\,\phi = a^2 - b^2$ ....(1)

and equation of normal at Q to the circle $x^2 + y^2 = a^2$ is $y = x\tan\phi$ ....(2)

From equation (2), $\tan\phi = \frac{y}{x}$

$\therefore\ \sin\phi = \frac{y}{\sqrt{(x^2+y^2)}}$ *and* $\cos\phi = \frac{x}{\sqrt{(x^2+y^2)}}$

or $\text{cosec}\,\phi = \frac{\sqrt{(x^2+y^2)}}{y}$ and $\sec\phi = \frac{\sqrt{(x^2+y^2)}}{x}$ ....(3)

Substituting the values of $\sec\phi$ and $\text{cosec}\,\phi$ from (3) in (1), we get

$\therefore\ ax.\frac{\sqrt{(x^2+y^2)}}{x} - by.\frac{\sqrt{(x^2+y^2)}}{y} = a^2 - b^2$

or $(a-b)\sqrt{x^2+y^2} = (a+b)(a-b)$

or $\sqrt{x^2+y^2} = a+b$

$\therefore\ x^2 + y^2 = (a+b)^2$ which is the required locus.

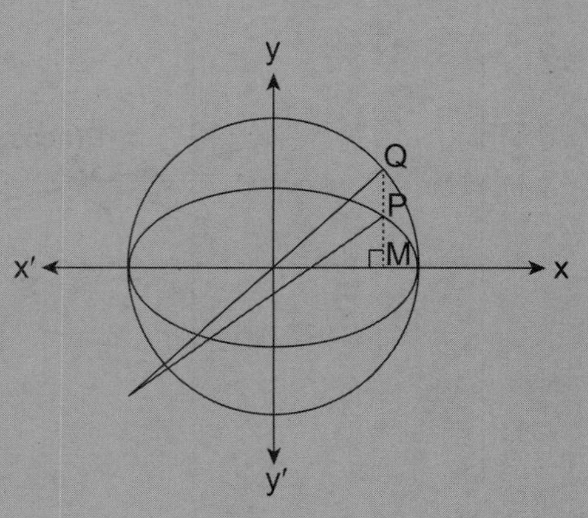

FIGURE 5.55

CO-NORMAL POINTS

In general, four normals can be drawn to any ellipse from any point in the plane of ellipse. The points of intersection of ellipse and four normals are called co-normal points.

Proof: Equation of normal to ellipse $\frac{x^2}{a^2}+\frac{y^2}{b^2}=1$ is given by $y=mx\pm\frac{m(a^2-b^2)}{\sqrt{a^2+b^2m^2}}$

or $(y-mx)^2=\frac{m^2(a^2-b^2)^2}{(a^2+b^2m^2)}$

or $(a^2+b^2m^2)(y-mx)^2=m^2(a^2-b^2)^2$ ....(i)

If (i) passes through a fixed point (h, k), then $(a^2+b^2m^2)(k-mh)^2=m^2(a^2-b^2)^2$ ....(ii)

Thus (ii) is a biquadratic in 'm' and each root of (ii) corresponds to slope of a normal

∴ In general, four normals can be drawn through a point to a given ellipse.

PROPERTIES OF NORMAL TO ELLIPSE AND CO-NORMAL POINTS

P: 1. Intersection with co-ordinate axes:

At point θ on $\frac{x^2}{a^2}+\frac{y^2}{b^2}=1$, the equation of normal is $ax\sec\theta-by\,\text{cosec}\theta=a^2-b^2=a^2e^2$

Intersection with x-axis $(ae^2\cos\theta, 0)$

Intersection with y axis $(0, \frac{-a^2e^2}{b}\sin\theta)$

P: 2. Tangent at any point bisects the external angle between focal distances whereas normal at same point bisects the internal angle between focal distances.

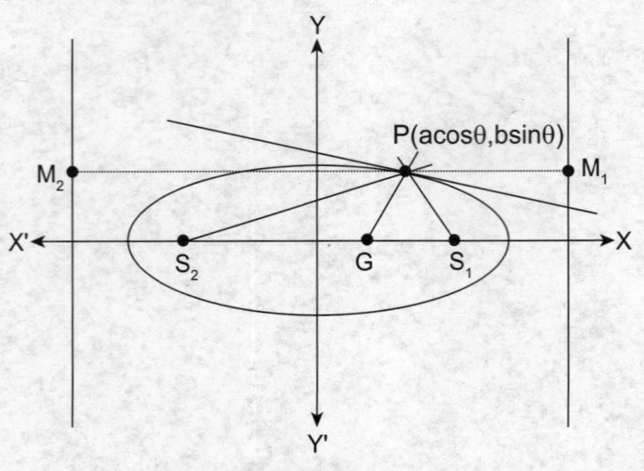

FIGURE 5.56

Proof: $\left.\begin{aligned}S_1G&=ae-ae^2\cos\theta\\S_2G&=ae+ae^2\cos\theta\end{aligned}\right\}$ ..... (1)

Also, $\left.\begin{aligned}S_1P&=a-ae\cos\theta\\S_2P&=a+ae\cos\theta\end{aligned}\right\}$ ....(2)

∴ $\frac{S_1G}{S_2G}=\frac{ae(1-e\cos\theta)}{ae(1+e\cos\theta)}=\frac{ae(1-e\cos\theta)}{ae(1+e\cos\theta)}=\frac{S_1P}{S_2P}$

∴ $\frac{S_1G}{S_2G}=\frac{S_1P}{S_2P}$

⇒ PG is angle bisector of $\angle S_2PS_1$ (By converse of angle bisector theorem).

Aliter: Let P be any point on the ellipse $\frac{x^2}{a^2}+\frac{y^2}{b^2}=1$.

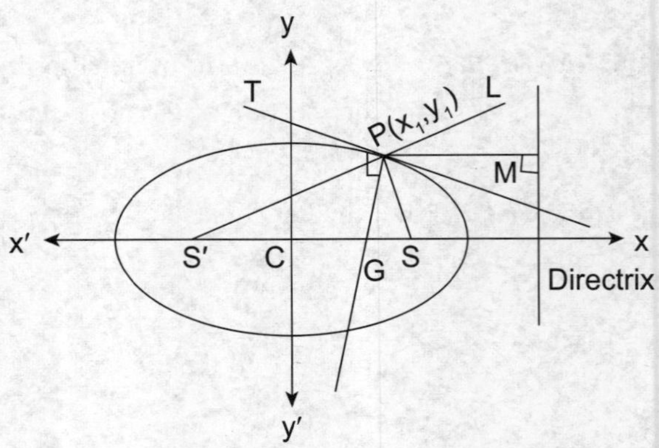

FIGURE 5.57

∴ Equation of normal PG is $(x-x_1)\frac{a^2}{x_1}=(y-y_1)\frac{b^2}{y_1}$

Putting $y=0$, for point G, we have $(x-x_1)\frac{a^2}{x_1}=-b^2$

∴ $x=CG=\left(\frac{a^2-b^2}{a^2}\right)x_1=\frac{a^2e^2}{a^2}x_1=e^2x_1$

∴ $SG=CS-CG=ae-e^2x_1=e(a-ex_1)$

$=eSP$ ∵ ($SP=ePM=e(a/e-x_1)=a-ex_1$)

Similarly, $S'G=eS'P$

∴ $\frac{SG}{S'G}=\frac{eSP}{eS'P}=\frac{SP}{S'P}$

∴ The normal PG bisects the internal $\angle SPS'$ between the focal distances but tangent and normal are at right angles, the tangent PT bisects the external angles SPL between them.

P: 3. In general, four normals can be drawn to an ellipse from any point and if α, β, γ, δ are the eccentric angles of these four co-normal points, then $\alpha+\beta+\gamma+\delta$ is an odd multiple of π.

Proof: Let $Q(h, k)$ be any given point and let $P(a\cos\phi, b\sin\phi)$ be any point on the ellipse $\frac{x^2}{a^2}+\frac{y^2}{b^2}=1$. ....(1)

Equation of normal at $P(a\cos\phi, b\sin\phi)$ is

$ax\sec\phi - by\,\text{cosec}\phi = a^2-b^2$. It passes through $Q(h, k)$

$\therefore\ ah\sec\phi - bk\,\text{cosec}\phi = a^2-b^2$

or $\frac{ah}{\cos\phi}-\frac{bk}{\sin\phi}=a^2-b^2$ ...(2)

or $\frac{ah}{\left(\frac{1-\tan^2(\phi/2)}{1+\tan^2(\phi/2)}\right)}-\frac{bk}{\left(\frac{2\tan(\phi/2)}{1+\tan^2(\phi/2)}\right)}=a^2-b^2$ ...(3)

Let $\tan\phi/2 = t$, then equation (3) reduces to

$$bkt^4+2\{ah+(a^2-b^2)\}t^3$$
$$+2\{ah-(a^2-b^2)\}t-bk=0 \quad ....(4)$$

which is a fourth degree equation in t, hence four normals can be drawn to an ellipse from any point. Consequently, it has four roots (say) t_1,t_2,t_3,t_4 and $t_1=\tan\alpha/2$, $t_2=\tan\beta/2$, $t_3=\tan\gamma/2$ and $t_4=\tan\delta/2$.

Now, $\tan\left(\frac{\alpha}{2}+\frac{\beta}{2}+\frac{\gamma}{2}+\frac{\delta}{2}\right)=\frac{S_1-S_3}{1-S_2+S_4}$

$$=\frac{\frac{-2\{ah+(a^2-b^2)\}}{bk}+\frac{2\{(ah-(a^2-b^2)\}}{bk}}{1-0-1}=\infty$$

(From trigonometry) ($\because a\neq b$)

or $\cot\left(\frac{\alpha}{2}+\frac{\beta}{2}+\frac{\gamma}{2}+\frac{\delta}{2}\right)=0$

or $\frac{\alpha}{2}+\frac{\beta}{2}+\frac{\gamma}{2}+\frac{\delta}{2}=$ an odd multiple of $\pi/2$

$\Rightarrow \alpha+\beta+\gamma+\delta=$ an odd multiple of π

Alternative Method: (Using complex number)

Let $z=e^{i\phi}=\cos\phi+i\sin\phi$ (By Euler's theorem)

$\therefore\ \frac{1}{z}=e^{-i\phi}=\cos\phi-i\sin\phi$

$\therefore\ \cos\phi=\frac{z+\frac{1}{z}}{2}=\frac{z^2+1}{2z}$ and $\sin\phi=\frac{z-\frac{1}{z}}{2i}=\frac{z^2-1}{2iz}$

Now, equation (2) reduces to

$$\frac{ah}{\left(\frac{z^2+1}{2z}\right)}-\frac{bk}{\left(\frac{z^2-1}{2iz}\right)}=a^2-b^2$$

$\Rightarrow\ (a^2-b^2)z^4-2(ah-ibk)z^3+$
$2(ah+ibk)z-(a^2-b^2)=0$ ....(5)

$\Rightarrow$ (4) has four roots z_1, z_2, z_3, z_4 and consequently gives four values of ϕ, say $\alpha, \beta, \gamma, \delta$ i.e., $z_1=e^{i\alpha}$, $z_2=e^{i\beta}$, $z_3=e^{i\gamma}$, $z_4=e^{i\delta}$

$\therefore\ z_1.z_2.z_3.z_4=-1$

$\Rightarrow\ e^{i\alpha}.e^{i\beta}.e^{i\gamma}.e^{i\delta}=-1$

$\Rightarrow\ e^{i(\alpha+\beta+\gamma+\delta)}=-1$

$\Rightarrow\ \cos(\alpha+\beta+\gamma+\delta)+i\sin(\alpha+\beta+\gamma+\delta)=-1$

$\Rightarrow\ \cos(\alpha+\beta+\gamma+\delta)=-1$ and $\sin(\alpha+\beta+\gamma+\delta)=0$

$\Rightarrow\ \alpha+\beta+\gamma+\delta=(2n+1)\pi$; where n, $\in\mathbb{Z}$

Hence, $\alpha+\beta+\gamma+\delta=$ odd multiple of π

P: 4. If α,β,γ are the eccentric angles of three points on the ellipse $\frac{x^2}{a^2}+\frac{y^2}{b^2}=1$, the normals at which are concurrent, then $\sin(\alpha+\beta)+\sin(\beta+\gamma)+\sin(\gamma+\alpha)=0$

Proof: From (5), $\sum z_1z_2=0$

$\Rightarrow\ z_1z_2+z_1z_3+z_1z_4+z_2z_3+z_2z_4+z_3z_4=0$

$\Rightarrow\ e^{i(\alpha+\beta)}e^{i(\alpha+\gamma)}e^{i(\alpha+\delta)}e^{i(\beta+\gamma)}e^{i(\beta+\delta)}+e^{i(\gamma+\delta)}=0$

$\Rightarrow\ [\cos(\alpha+\beta)+\cos(\alpha+\gamma)+\cos(\alpha+\delta)$
$+\cos(\beta+\gamma)+\cos(\beta+\delta)+\cos(\gamma+\delta)]$
$+i[\sin(\alpha+\beta)+\sin(\alpha+\gamma)+\sin(\alpha+\delta)+\sin(\beta+\gamma)$
$+\sin(\beta+\delta)+\sin(\gamma+\delta)]=0$

Comparing the imaginary parts, we get

$\sin(\alpha+\beta)+\sin(\alpha+\gamma)+\sin(\alpha+\delta)+\sin(\beta+\gamma)$
$+\sin(\beta+\delta)+\sin(\gamma+\delta)=0$ ...(6)

By property (3)

$\alpha+\beta+\gamma+\delta=$ odd multiple of π

$(\alpha+\delta)=$ (odd multiple of π) $-(\beta+\gamma)$

$(\beta+\delta)=$ (odd multiple of π) $-(\alpha+\gamma)$

$(\gamma+\delta)=$ (odd multiple of π) $-(\alpha+\beta)$

$$\left.\begin{array}{l}\sin(\alpha+\delta)=\sin(\beta+\gamma)\\ \sin(\beta+\delta)=\sin(\alpha+\gamma)\\ \sin(\gamma+\delta)=\sin(\alpha+\beta)\end{array}\right\}\left\{\begin{array}{l}\because \sin[(2n+1)\pi-\alpha]=\\ \quad\sin\alpha, \text{if } n \text{ is an integer})\end{array}\right\} \quad ...(7)$$

From (6) and (7), we get

$2\sin(\alpha+\beta)+2\sin(\beta+\gamma)+2\sin(\gamma+\alpha)=0$

$\Rightarrow\ \sin(\alpha+\beta)+\sin(\beta+\gamma)+\sin(\gamma+\alpha)=0$

Alternative Method:

From equation (4), $\sum t_1t_2 = 0$ ...(8)

and $t_1t_2t_3t_4 = -1$ ....(9)

Now $\sum t_1t_2 = 0$

$\Rightarrow t_1t_2 + t_2t_3 + t_3t_1 = -t_4(t_1 + t_2 + t_3)$

$\Rightarrow t_1t_2 + t_2t_3 + t_3t_1 = \dfrac{t_1 + t_2 + t_3}{t_1t_2t_3}$ {from (9)}

$\Rightarrow t_1t_2 + t_2t_3 + t_3t_1 = \dfrac{1}{t_2t_3} + \dfrac{1}{t_3t_1} + \dfrac{1}{t_1t_2}$

$\Rightarrow \tan\dfrac{\alpha}{2}\tan\dfrac{\beta}{2} + \tan\dfrac{\beta}{2}\tan\dfrac{\gamma}{2} + \tan\dfrac{\gamma}{2}\tan\dfrac{\alpha}{2}$

$= \cot\dfrac{\beta}{2}\cot\dfrac{\gamma}{2} + \cot\dfrac{\gamma}{2}\cot\dfrac{\alpha}{2} + \cot\dfrac{\alpha}{2}\cot\dfrac{\beta}{2}$

$\Rightarrow \sum\left(\tan\dfrac{\alpha}{2}\tan\dfrac{\beta}{2} - \cot\dfrac{\alpha}{2}\cot\dfrac{\beta}{2}\right) = 0$

$\Rightarrow \sum\left(\dfrac{\sin^2(\alpha/2)\sin^2(\beta/2) - \cos^2(\alpha/2)\cos^2(\beta/2)}{\sin(\alpha/2)\sin(\beta/2)\cos(\alpha/2)\cos(\beta/2)}\right) = 0$

$$\Rightarrow \sum -4\left(\frac{\{\cos(\alpha/2)\cos(\beta/2) + \sin(\alpha/2)\sin(\beta/2)\} \times \{\cos(\alpha/2)\cos(\beta/2) - \sin(\alpha/2)\sin(\beta/2)\}}{\sin\alpha\sin\beta}\right) = 0$$

$$\Rightarrow \sum -4\left(\frac{\cos\left(\dfrac{\alpha-\beta}{2}\right)\cos\left(\dfrac{\alpha+\beta}{2}\right)}{\sin\alpha\sin\beta}\right) = 0$$

$\Rightarrow \sum -2\dfrac{[\cos\alpha + \cos\beta]}{\sin\alpha\sin\beta} = 0$

$\Rightarrow \sum \dfrac{\sin\gamma(\cos\alpha + \cos\beta)}{\sin\alpha\sin\beta\sin\gamma} = 0$

$\Rightarrow \sum \sin\gamma(\cos\alpha + \cos\beta) = 0$

$\Rightarrow \sin\gamma(\cos\alpha + \cos\beta) + \sin\alpha(\cos\beta + \cos\gamma)$
$+ \sin\beta(\cos\gamma + \cos\alpha) = 0$

$\Rightarrow \sin(\alpha+\beta) + \sin(\beta+\gamma) + \sin(\gamma+\alpha) = 0$

P: 6. Co-normal Points lie on a fixed curve

Let $P\ (x_1, y_1), Q(x_2, y_2), R(x_3, y_3)$ and $S(x_4, y_4)$ be co-normal points so that normal drawn from them meet in $T(h, k)$.

Then, equation of normal at $P(x_1, y_1)$ is

$$\frac{a^2x}{x_1} - \frac{b^2y}{y_1} = a^2 - b^2$$

or $(a^2 - b^2)x_1y_1 + b^2yx_1 - a^2xy_1 = 0$

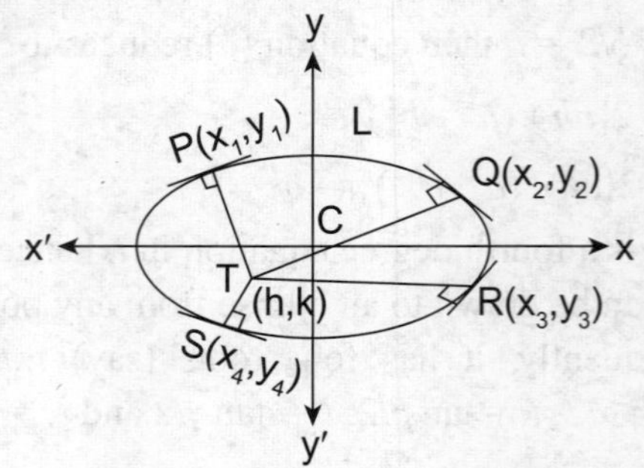

FIGURE 5.58

but the point $T(h, k)$ lies on it

$\Rightarrow (a^2 - b^2)x_1y_1 + b^2kx_1 - a^2hy_1 = 0$

similarly, $(a^2 - b^2)x_2y_2 + b^2kx_2 - a^2hy_2 = 0$

$(a^2 - b^2)x_3y_3 + b^2kx_3 - a^2hy_3 = 0$

and $(a^2 - b^2)x_4y_4 + b^2kx_4 - a^2hy_4 = 0$

Hence P, Q, R, S satisfy the curve

$(a^2 - b^2)xy + b^2kx - a^2hy = 0$ and

this curve is called **Apollonian rectangular hyperbola.**

NOTE

The feet of the normals from any fixed point to the ellipse lie at the intersections of the Apollonian rectangular hyperbola with the ellipse.

■ REFLECTION PROPERTY OF AN ELLIPSE

If an incoming light ray passes through one of the foci (say S') strike the concave side of the ellipse, then it will get reflected towards other focus (S), which in turn strikes the concave side of ellipse at a point Q and again it passes through S' and so on. and $\angle SPS' = \angle S'QS$.

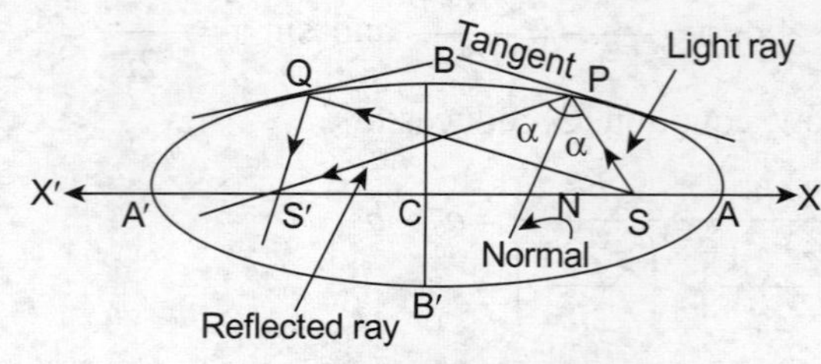

FIGURE 5.59

ILLUSTRATION 59: A ray emanating from the point (–3, 0) is incident on the ellipse $16x^2 + 25y^2 = 400$ at the point P with ordinate 4. Find the equation of the reflected ray after first reflection.

SOLUTION: For point P, y co-ordinate = 4

$\because$ given ellipse is; $16x^2 + 25y^2 = 400$

$$16x^2 + 25(4)^2 = 400$$

co-ordinates of P are (0, 4); $e^2 = 1 - \frac{16}{25} = \frac{9}{25}$

$\Rightarrow$ $e = 3/5$; foci $(\pm ae, 0)$ i.e., $(\pm 3, 0)$

$\therefore$ The ray S'P after reflection would pass through S (By reflection property)

$\therefore$ Equation of reflected ray (i.e., PS) will be $\frac{x}{3} + \frac{y}{4} = 1$ or $4x + 3y = 12$.

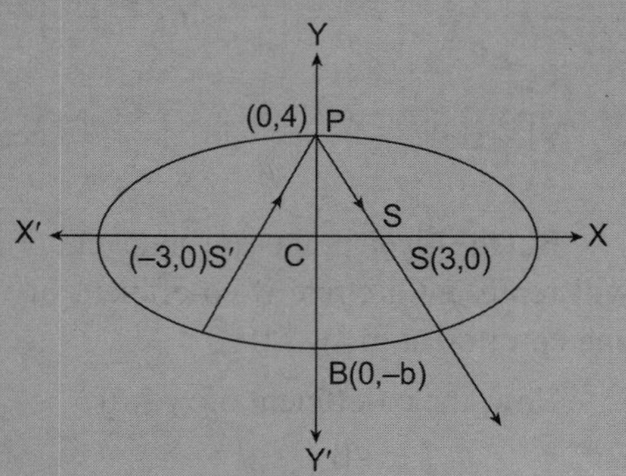

FIGURE 5.60

■ CONCYCLIC POINTS OF ELLIPSE

Any circle intersects an ellipse in two or four real points. These points are called concyclic points and the sum of their eccentric angles is an even multiple of π.

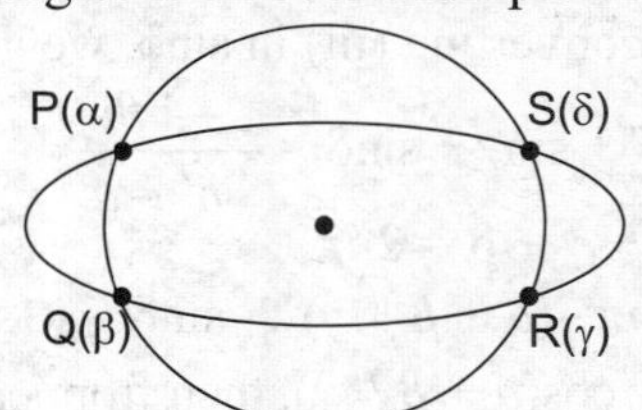

FIGURE 5.61

If $\alpha, \beta, \gamma, \delta$ be the eccentric angles of the four concyclic points P,Q,R, and S on an ellipse, then

$\alpha + \beta + \gamma + \delta = 2n\pi$, where n is any integer.

Let the given circle be

$$x^2 + y^2 + 2gx + 2fy + c = 0 \quad \text{.... (i)}$$

and the given ellipse be

$$\frac{x^2}{a^2} + \frac{y^2}{b^2} = 1 \quad \text{.... (ii)}$$

Let $(a \cos \phi, b \sin \phi)$ be a point of intersection of (i) and (ii).

As it lies on the circle (i).

$$\therefore \; a^2\cos^2\phi + b^2 \sin^2\phi + 2ga \cos\phi + 2fb \sin\phi + c = 0 \quad \text{.... (iii)}$$

$$\Rightarrow \; a^2\left(\frac{1-\tan^2(\phi/2)}{1+\tan^2(\phi/2)}\right)^2 + b^2\left(\frac{2\tan(\phi/2)}{1+\tan^2(\phi/2)}\right)^2 + 2ga\left(\frac{1-\tan^2(\phi/2)}{1+\tan^2(\phi/2)}\right) + 2fb\left(\frac{2\tan(\phi/2)}{1+\tan^2(\phi/2)}\right) + c = 0 \quad \text{...(iv)}$$

Put $\tan(\phi/2) = m$

$\therefore$ Equation (iv) reduces to

$$a^2\left(\frac{1-m^2}{1+m^2}\right)^2 + b^2\left(\frac{2m}{1+m^2}\right)^2 + 2ga\left(\frac{1-m^2}{1+m^2}\right) + 2fb\left(\frac{2m}{1+m^2}\right) + c = 0$$

$$\Rightarrow (a^2 - 2ga + c)m^4 + 4bfm^3 + (4b^2 - 2a^2 + 2c)m^2 + 4bfm + (a^2 + 2ga + c) = 0 \quad \text{... (v)}$$

which is a biquadratic equation in m.

i.e., it has four values of m; $m = \tan \phi/2$

Since four values of eccentric angles are $\alpha, \beta, \gamma, \delta$.

$$\therefore \quad \tan\left(\frac{\alpha}{2} + \frac{\beta}{2} + \frac{\gamma}{2} + \frac{\delta}{2}\right) = \frac{S_1 - S_3}{1 - S_2 + S_4}$$

$$= \frac{\sum m_1 - \sum m_1m_2m_3}{1 - \sum m_1m_2 + m_1m_2m_3m_4} = 0$$

$$\Rightarrow \quad \left(\frac{\alpha}{2} + \frac{\beta}{2} + \frac{\gamma}{2} + \frac{\delta}{2}\right) = n\pi.$$

$\therefore$ $\alpha + \beta + \gamma + \delta = 2n\pi$; $n \in \mathbb{Z}$ (set of integers)

Alternative Method: Let P, Q, R, S be four concylic points on ellipse $\frac{x^2}{a^2} + \frac{y^2}{b^2} = 1$, having the eccentric angles $\alpha, \beta, \gamma, \delta$ respectively.

Then equation of the chords PQ and RS will be given by $\frac{x}{a}\cos\left(\frac{\alpha+\beta}{2}\right) + \frac{y}{b}\sin\left(\frac{\alpha+\beta}{2}\right) - \cos\left(\frac{\alpha-\beta}{2}\right) = 0$

and $\frac{x}{a}\cos\left(\frac{\gamma+\delta}{2}\right) + \frac{y}{b}\sin\left(\frac{\gamma+\delta}{2}\right) - \cos\left(\frac{\gamma-\delta}{2}\right) = 0$

Now, the equation of any curve passing through P, Q, R and S is given by

$$\left(\frac{x^2}{a^2}+\frac{y^2}{b^2}-1\right)$$

$$+k\left[\frac{x}{a}\cos\left(\frac{\alpha+\beta}{2}\right)+\frac{y}{b}\sin\left(\frac{\alpha+\beta}{2}\right)-\cos\left(\frac{\alpha-\beta}{2}\right)\right]$$

$$\times\left[\frac{x}{a}\cos\left(\frac{\gamma+\delta}{2}\right)+\frac{y}{b}\sin\left(\frac{\gamma+\delta}{2}\right)-\cos\left(\frac{\gamma-\delta}{2}\right)\right]=0$$

But the given points are concyclic. Hence, this equation will represent a circle if co-effcient of x^2 = co-effcient y^2, and co-effcient of $xy = 0$.

Now, the co-effcient of $xy = 0$

$$\Rightarrow \cos\left(\frac{\alpha+\beta}{2}\right)\sin\left(\frac{\gamma+\delta}{2}\right)$$

$$+\sin\left(\frac{\alpha+\beta}{2}\right)\cos\left(\frac{\gamma+\delta}{2}\right)=0$$

$$\Rightarrow \sin\left(\frac{\alpha+\beta+\gamma+\delta}{2}\right)=0$$

$$\Rightarrow \frac{(\alpha+\beta+\gamma+\delta)}{2}=n\pi \Rightarrow \alpha+\beta+\gamma+\delta=2n\pi : n\in\mathbb{Z}$$

where n is any integer.

Hence the sum of eccentric angles of four concyclic points on an ellipse is always an even multiple of π.

Corollary 1: The common chords of circle and an ellipse are equally inclined to the axes of the ellipse.

If the point of intersection of chords PQ and RS is T, then the equation of chord PQ is

$$\frac{x}{a}\cos\left(\frac{\alpha+\beta}{2}\right)+\frac{y}{b}\sin\left(\frac{\alpha+\beta}{2}\right)=\cos\left(\frac{\alpha-\beta}{2}\right)$$

P(α) R(γ) T Q(β) S(δ)

FIGURE 5.62

$$\therefore \text{ Slope of } PQ = -\frac{b}{a}\cot\left(\frac{\alpha+\beta}{2}\right)$$

$$= -\frac{b}{a}\cot\left(n\pi-\frac{\gamma+\delta}{2}\right) \qquad (\because \alpha+\beta+\gamma+\delta=2n\pi)$$

$$= \frac{b}{a}\cot\left(\frac{\gamma+\delta}{2}\right)$$

= –(slope of RS); hence PQ and RS are equally inclined to axes of ellipse.

Corollary 2: The centre of the circle passing through the three points on an ellipse whose eccentric angles are α, β, γ is given by

$$\left(\left(\frac{a^2-b^2}{4a}\right)\left[\sum(\cos\alpha)+\cos\left(\sum\alpha\right)\right], \left(\frac{b^2-a^2}{4b}\right)\left[\sum(\sin\alpha)-\sin\left(\sum\alpha\right)\right]\right)$$

Let the point of intersection of ellipse $\frac{x^2}{a^2}+\frac{y^2}{b^2}=1$ ...(i)

and circle $x^2 + y^2 + 2gx + 2fy + c = 0$ ...(ii)

be $\alpha, \beta, \gamma, \delta$

$\therefore$ $\alpha+\beta+\gamma+\delta = 2n\pi$. Let θ be any point on (i)

$\therefore$ $x = a\cos\theta, y = b\sin\theta$. This point also lie on (ii)

$\therefore$ $a^2\cos^2\theta + b^2\sin^2\theta + 2ga\cos\theta + 2fb\sin\theta + c = 0$ ...(iii)

$\Rightarrow$ $\{(a^2-b^2)\cos^2\theta + 2ga\cos\theta + (b^2+c)\}^2 = 4f^2b^2(1-\cos^2\theta)$

$\Rightarrow$ $(a^2-b^2)^2\cos^4\theta + 4ga(a^2-b^2)\cos^3\theta + \{2(a^2-b^2)(b^2+c) + 4g^2a^2 + 4f^2b^2\}\cos^2\theta + 4ga(b^2+c)\cos\theta + \{b^2+c^2-4f^2b^2\} = 0$

It is a fourth degree equation in $\cos\theta$.

It has four roots (i.e., $\cos\alpha, \cos\beta, \cos\gamma, \cos\delta$)

$$\therefore \cos\alpha+\cos\beta+\cos\gamma+\cos\delta = -\frac{4ga}{(a^2-b^2)} \quad \text{...(iv)}$$

Similarly, converting (iii) in $\sin\phi$, we get

$$\sin\alpha+\sin\beta+\sin\gamma+\sin\delta = -\frac{4fb}{b^2-a^2} \quad \text{....(v)}$$

$\alpha+\beta+\gamma+\delta = 2n\pi$

$\therefore$ $\delta = 2n\pi-(\alpha+\beta+\gamma)$ $\therefore$ $\sin\delta = -\{\sin(\alpha+\beta+\gamma)\}$ and $\cos\delta = \cos(\alpha+\beta+\gamma)$, then from equatons (iv) and (v), we get

$$-=\left(\frac{a^2 \quad b^2}{}\right)[\cos\alpha+\cos\beta+\cos\gamma+\cos(\alpha+\beta+\gamma)]$$

$$\text{and } -f=\left(\frac{b^2-a^2}{4b}\right)[\sin\alpha+\sin\beta+\sin\gamma-\sin(\alpha+\beta+\gamma)]$$

which gives co-ordinates of centre of circle through $P(\alpha)$, $Q(\beta)$ and $R(\gamma)$.

Corollary 3: It $P'CP$ and $B'CB$ are conjugate diameters of an ellipse and α is the eccentric angle of P, then the eccentric angle of the point where the circle through P, P' and B again cuts the ellipse is $\left(\frac{\pi}{2}-3\alpha\right)$.

The eccentric angles of P, P' and D are $\alpha, \pi+\alpha, \frac{\pi}{2}+\alpha$ respectively. Let β be the eccentric angle of the fourth point.

As above,

$$\alpha+(\pi+\alpha)+\left(\frac{\pi}{2}+\alpha\right)+\beta = 2n\pi$$

$$\beta = 2n\pi-\left(\frac{3\pi}{2}+3\alpha\right)=\frac{\pi}{2}-3\alpha \text{ ; for } n = 1$$

Any other value of n gives the same point on the ellipse.

TEXTUAL EXERCISE–5 (SUBJECTIVE)

1. If the normal at an end of a latus rectum of an ellipse passes through an end of minor axis, then show that $e^4 + e^2 = 1$. (where e is the eccentricity of the ellipse)

2. Prove that four normals can be drawn to an ellipse from a point in its plane and hence show that sum of their eccentric angles is odd multiple of π.

3. (a) If the normal at any point P on the ellipse $\frac{x^2}{a^2}+\frac{y^2}{b^2}=1$, meets the axes in G and g respectively, then calculate the ratio $PG : Pg$.
   (b) Any ordinate NP of an ellipse meets the auxiliary circle in Q. Find the locus of the point of intersection of the normals at P and Q.

4. (a) Find the tangent of the angle between CP and the normal at P, and prove that its greatest value is $\frac{a^2-b^2}{2ab}$, where C is the centre of ellipse $\frac{x^2}{a^2}+\frac{y^2}{b^2}=1$, and P is a point on the ellipse.
   (b) P and Q are corresponding point of the ellipse $\frac{x^2}{a^2}+\frac{y^2}{b^2}=1$ and the circle $x^2 + y^2 = a^2$. The normal at P to the ellipse meets CQ in R, where C in the centre of the ellipse. Prove that $CR = a+b$.

5. The tangent and the normal at a point P on an ellipse meet the minor axis in T and G. Prove that TG subtends a right angle at each of the foci.

6. The tangents from (x_1, y_1) to the ellipse $\frac{x^2}{a^2}+\frac{y^2}{b^2}=1$ intersect at right angles, show that the normals at the points of contact meet on the line $\frac{y}{y_1}=\frac{x}{x_1}$.

7. The tangent and normal at any point P of an ellipse $\frac{x^2}{a^2}+\frac{y^2}{b^2}=1$ cut its major axis in point Q and R respectively. If $QR = a$, prove that the eccentric angle of the point P is given by $e^2 \cos^2 \phi + \cos \phi - 1 = 0$.

8. A ray emanating from the point $(0, -\sqrt{5})$ is incident on the ellipse $9x^2 + 4y^2 = 36$ at the point P with abscissa 2. Find the equation of the reflected ray after first reflection.

9. Prove that the straight line $lx + my + n = 0$ is a normal to the ellipse $x^2/a^2 + y^2/b^2 = 1$ if $\frac{a^2}{l^2}+\frac{b^2}{m^2}=\frac{(a^2-b^2)^2}{n^2}$.

10. If the tangent drawn at point $(\lambda^2, 2\lambda)$ on the parabola $y^2 = 4x$ is same as the normal drawn at a point $(\sqrt{5} \cos \theta, 2 \sin \theta)$ on the ellipse $4x^2 + 5y^2 = 20$. Find the values of λ and θ.

11. If the normal at the point $P(\theta)$ to the ellipse $5x^2 + 14y^2 = 70$ intersects it again at the point $Q(2\theta)$, show that $\cos\theta = 2/3$.

Answer Keys

2. (a) $(x^2 + y^2)^2 = a^2x^2 + b^2y^2$

3. (a) $b^2 : a^2$ (b) $x^2 + y^2 = (a + b)^2$

4. (a) $\frac{a^2-b^2}{2ab}\sin 2\theta$; where θ is the ecentric angle.

10. $L = x - 1/\sqrt{5}$, $\theta = \cos^{-1}\left(-\frac{1}{\sqrt{35}}\right)$

TEXTUAL EXERCISE–5 (OBJECTIVE)

1. The equation of the normal the ellipse $\frac{x^2}{a^2}+\frac{y^2}{b^2}=1$ at the point $(a \cos\theta, b \sin\theta)$ is

   (a) $\frac{ax}{\sin\theta}-\frac{by}{\cos\theta}=a^2-b^2$ (b) $\frac{ax}{\sin\theta}-\frac{by}{\cos\theta}=a^2+b^2$

   (c) $\frac{ax}{\cos\theta}-\frac{by}{\sin\theta}=a^2-b^2$ (d) $\frac{ax}{\cos\theta}-\frac{by}{\sin\theta}=a^2+b^2$

2. If the normal at the point $P(\theta)$ to the ellipse $\frac{x^2}{14}+\frac{y^2}{5}=1$ intersects it again at the point $Q(2\theta)$, then $\cos\theta$ is equal to

   (a) $\frac{2}{3}$ (b) $-\frac{2}{3}$

   (c) $\frac{3}{2}$ (d) $-\frac{3}{2}$

3. The line $y = mx + c$ is a normal to the ellipse $\frac{x^2}{a^2}+\frac{y^2}{b^2}=1$, if c is equal to

(a) $-(2am + bm^2)$ (b) $\frac{(a^2+b^2)m}{\sqrt{a^2+b^2m^2}}$

(c) $-\frac{(a^2-b^2)m}{\sqrt{a^2+b^2m^2}}$ (d) $\frac{(a^2-b^2)m}{\sqrt{a^2+b^2}}$

4. If the line $x \cos\alpha + y \sin\alpha = p$ be normal to the ellipse $\frac{x^2}{a^2}+\frac{y^2}{b^2}=1$, then

(a) $p^2 (a^2 \cos \alpha + b^2 \sin^2 \alpha) = a^2 - b^2$
(b) $p^2 (a^2 \cos^2 \alpha + b^2 \sin^2 \alpha) = (a^2 - b^2)^2$
(c) $p^2 (a^2 \sec^2 \alpha + b^2 \operatorname{cosec}^2 \alpha) = a^2 - b^2$
(d) $p^2 (a^2 \sec^2 \alpha + b^2 \operatorname{cosec}^2 \alpha) = (a^2 - b^2)^2$

5. The line $lx + my + n = 0$ is a normal to the ellipse $\frac{x^2}{a^2}+\frac{y^2}{b^2}=1$, if

(a) $\frac{a^2}{m^2}+\frac{b^2}{l^2}=\frac{(a^2-b^2)}{n^2}$ (b) $\frac{a^2}{l^2}+\frac{b^2}{m^2}=\frac{(a^2-b^2)^2}{n^2}$

(c) $\frac{a^2}{l^2}-\frac{b^2}{m^2}=\frac{(a^2-b^2)^2}{n^2}$ (d) None of these

6. The value of λ, for which the line $2x-\frac{8}{3}\lambda y=-3$ is a normal to the conic $x^2+\frac{y^2}{4}=1$ is

(a) $\frac{\sqrt{3}}{2}$ (b) $\frac{1}{2}$

(c) $-\frac{\sqrt{3}}{2}$ (d) $\frac{3}{8}$

7. Three normals are drawn to the curve $y^2 = x$ from a point $(c, 0)$. Out of three one is always on x-axis if two others normals are perpendicular to each other, then the value of c is

(a) 3/4 (b) 1/2
(c) 3/2 (d) 2

8. Let $\frac{x^2}{a^2}+\frac{y^2}{b^2}=1$ be an ellipse and *ABCD* are four points on the ellipse. Further, if normal drawn at these points are concurrent, then

(a) points *A, B, C, D* lie on a circle
(b) points *A, B, C, D* lie on an ellipse
(c) points *A, B, C, D* will lie on hyperbola
(d) cannot be determined

9. If *CF* is the perpendicular from the centre *C* of the ellipse $\frac{x^2}{a^2}+\frac{y^2}{b^2}=1$ on the tangent at any point *P*, and *G* is the point where the normal at *P* meets the major axis, then *CF. PG* =

(a) a^2 (b) ab
(c) b^2 (d) b^3

10. Let F_1, F_2 be two foci of the ellipse and *PT* and *PN* be the tangent and the normal respectively to the ellipse at point *P*, then

(a) *PN* bisects $\angle F_1PF_2$
(b) *PT* bisects $\angle F_1PF_2$
(c) *PT* bisects angle $(180° - \angle F_1PF_2)$
(d) None of these

11. The area of a triangle inscribed in an ellipse bears a constant ratio to the area of the triangle formed by joining points on the auxiliary circle correspponding to the vertices of the first triangle. This ratio is

(a) b/a (b) a^2/b^2
(c) $2a/b$ (d) None of these

12. If *S* be the focus and *G* be the point where the normal at *P* meets the axes of an ellipse, then

(a) *SG* = e. *SP*,
(b) tangent at *P* bisect the external angles between the focal distances of *P*.
(c) normal at *P* bisect the internal angles between the focal distances of *P*.
(d) All of the above

13. If α, β, γ, δ be the eccentric angles of the four concyclic points on an ellipse, then

(a) $\alpha+\beta+\gamma+\delta = (2n+1)\pi, n \in \mathbb{Z}$
(b) $\alpha+\beta+\gamma+\delta = n\pi, n \in \mathbb{Z}$
(c) $\alpha+\beta+\gamma+\delta = 2n\pi, n \in \mathbb{Z}$
(d) None of these

Answer Keys

1. (c) **2.** (b) **3.** (c) **4.** (d) **5.** (b) **6.** (d) **7.** (a) **8.** (c) **9.** (c)
10 (a, c) **11.** (a) **12.** (d) **13.** (c)

MULTIPLE-CHOICE QUESTIONS

SECTION–I

OBJECTIVE-TYPE SOLVED EXAMPLES

1. Let P be any point on a directrix of an ellipse of eccentricity e, S be the corresponding focus and C the centre of the ellipse. The line PC meets the ellipse at A. The angle between PS and tangent at A is α, then α, is equal to

(a) $\tan^{-1}e$ (b) $\pi/2$

(c) $\tan^{-1}(1-e^2)$ (d) None of these

Solution: (b)

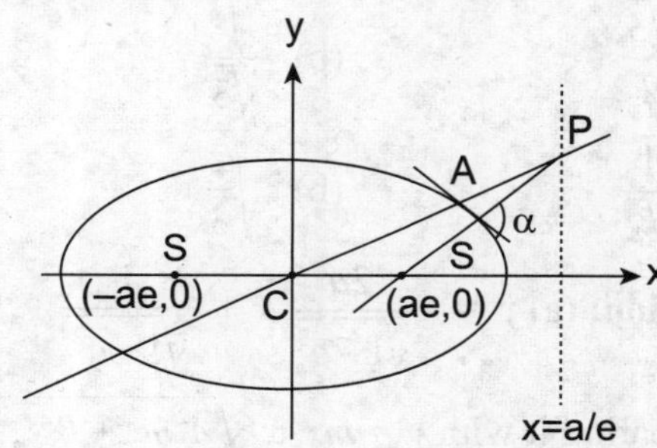

Let the ellipse be $\frac{x^2}{a^2}+\frac{y^2}{b^2}=1$

And let $A \equiv (a\cos\theta, b\sin\theta)$

$\therefore$ equation of line AC is $y=\left(\frac{b}{a}\tan\theta\right)x$

This line will intersect $x = a/e$ at $\left(\frac{a}{e},\frac{b}{e}\tan\theta\right)$

$\therefore$ Co-ordinates of $P \equiv \left(\frac{a}{e},\frac{b}{e}.\tan\theta\right)$

Slope of tangent at A : $\frac{-b}{a.\cot\theta}$

And slope of $PS = \frac{\frac{b}{e}.\tan\theta}{\frac{a}{e}-ae} = \frac{b\tan\theta}{a(1-e^2)} = \frac{a}{b}\tan\theta$

Since slope of tangent at A × slope of $PS = -1$

$\therefore \alpha = \pi/2$

2. P is a variable point on the ellipse $\frac{x^2}{a^2}+\frac{y^2}{b^2}=2$ whose foci are F_1 and F_2. The maximum area (in unit²) of the ΔPF_1F_2 is

(a) $2b\sqrt{a^2-b^2}$ (b) $\sqrt{2}b\sqrt{a^2-b^2}$

(c) $b\sqrt{a^2-b^2}$ (d) $2a\sqrt{a^2-b^2}$

Solution: (a) Let $P = (\sqrt{2}a\cos\phi, \sqrt{2}b\sin\phi)$.

F_1 and $F_2 = (\pm\sqrt{2}ae, 0)$.

The area of $\Delta PFF' = \left|\frac{1}{2}\begin{vmatrix}\sqrt{2}a\cos\phi & \sqrt{2}b\sin\phi & 1\\ \sqrt{2}ae & 0 & 1\\ -\sqrt{2}ae & 0 & 1\end{vmatrix}\right|$

$=\frac{2}{2}.\sqrt{2}b\sin\phi\,.\,\sqrt{2}\,ae$

$= 2abe\sin\phi$.

$\therefore$ Maximum area $= 2abe = 2ab.\frac{\sqrt{a^2-b^2}}{a}$

$= 2b\sqrt{a^2-b^2}$

3. If P and Q are the ends of a pair of conjugate diameters and C is the centre of the ellipse $4x^2 + 9y^2 = 36$, then the area of the ΔCPQ is.

(a) 6 unit² (b) 3 unit²

(c) 2 unit² (d) 12 unit²

Solution: (b) The ellipse is $\frac{x^2}{3^2}+\frac{y^2}{2^2}=1$. So $P =$ $(3\cos\phi, 2\sin\phi)$ and $Q = \left(3\cos\left(\phi+\frac{\pi}{2}\right), 2\sin\left(\phi+\frac{\pi}{2}\right)\right)$

$\therefore$ Area of the $\Delta CPQ = \frac{1}{2}\begin{vmatrix}0 & 0 & 1\\ 3\cos\phi & 2\sin\phi & 1\\ -3\sin\phi & 2\cos\phi & 1\end{vmatrix}$

$= \frac{1}{2}(6\cos^2\phi + 6\sin^2\phi) = \frac{1}{2}.\,6 = 3$ square units

4. If $y = x$ and $3y + 2x = 0$ are the equations of a pair of conjugate diameters of the ellipse $\frac{x^2}{a^2}+\frac{y^2}{b^2}=1$, then its eccentricity is

(a) $\frac{1}{2}$ (b) $\frac{1}{3}$

(c) $\frac{1}{\sqrt{3}}$ (d) $\frac{\sqrt{3}}{2}$

Solution: (c) We know that two diameters $y = m_1x$, $y = m_2x$ are conjugate diameters of $\frac{x^2}{a^2}+\frac{y^2}{b^2}=1$ if $m_1m_2 = \frac{-b^2}{a^2}$.

Hence, we have $(1)\left(-\frac{2}{3}\right)=\frac{-b^2}{a^2}$ or $b^2=\frac{2}{3}a^2$.

$\therefore\ b^2 = a^2(1-e^2) \Rightarrow \frac{2}{3}a^2 = (1-e^2)a^2$

$\Rightarrow\ \frac{2}{3} = 1-e^2 \Rightarrow e^2 = \frac{1}{3} \Rightarrow e = \frac{1}{\sqrt{3}}$

5. The line $y = mx + c$, will be a normal to the ellipse $\frac{x^2}{a^2}+\frac{y^2}{b^2}=1$, if

(a) $\frac{a^2}{m^2}+b^2=\frac{(b^2-a^2)^2}{c^2}$

(b) $\frac{c^2}{m^2}+b^2=\frac{(b^2-a^2)^2}{a^2}$

(c) $a^2+\frac{b^2}{m^2}=\frac{(b^2-a^2)^2}{c^2}$

(d) $c^2+\frac{b^2}{m^2}=\frac{(b^2-a^2)^2}{a^2}$

Solution: (a) Let the foot of normal is $(a\cos\theta, b\sin\theta)$, then $\frac{ax}{\cos\theta}-\frac{by}{\sin\theta}=a^2-b^2$ and the given lines will be identical. Thus, $\frac{m\cos\theta}{a}=\frac{\sin\theta}{b}=\frac{c}{(b^2-a^2)}$

$\Rightarrow \cos\theta=\frac{ac}{m(b^2-a^2)}, \sin\theta=\frac{bc}{(b^2-a^2)}$

$\Rightarrow\ \frac{a^2}{m^2}+b^2=\frac{(b^2-a^2)^2}{c^2}$

6. OA is the perpendicular drawn from centre O of the ellipse $\frac{x^2}{a^2}+\frac{y^2}{b^2}=1$, to the tangent at any point P on the ellipse. If the normal drawn to the ellipse at P cut axis at B, then $OA \times PB =$

(a) a^2 (b) $a\sqrt{a^2+b^2}$

(c) b^2 (b) $b\,a\sqrt{a^2+b^2}$

Solution: (c) Let $P \equiv (a\cos\theta, b\sin\theta)$.

Equation to tangent at 'P' is

$bx\cos\theta + ay\sin\theta - ab = 0$

$\Rightarrow OA = \frac{ab}{\sqrt{b^2\cos^2\theta + a^2\sin^2\theta}}$

Equation of normal at P is $ax\sec\theta - by\,\mathrm{cosec}\theta = (a^2 - b^2)$

$\Rightarrow B \equiv \left(\frac{(a^2-b^2)\cos\theta}{a}, 0\right)$

$\Rightarrow PB^2 = b^2\sin\theta + \left(a\cos\theta - \frac{(a^2-b^2)\cos\theta}{a}\right)^2$

$= b^2\sin^2\theta + \frac{b^4\cos^2\theta}{a^2} = \frac{b^2}{a^2}(a^2\sin^2\theta + b^2\cos^2\theta)$

$\Rightarrow PB = \frac{b\sqrt{a^2\sin^2\theta+b^2\cos^2\theta}}{a} \Rightarrow (OA)(PB) = b^2$

7. For all admissible values of the parameter 'a' the straight line $2ax + y\sqrt{1-a^2} = 1$ will touch an ellipse whose eccentricity is equal to

(a) $\frac{\sqrt{3}}{2}$ (b) $\frac{1}{\sqrt{3}}$

(c) $\frac{1}{\sqrt{2}}$ (b) $\sqrt{\frac{2}{3}}$

Solution: (a) $y = -\frac{2a}{\sqrt{1-a^2}}x + \frac{1}{\sqrt{1-a^2}}$.

Comparing it with $y = mx \pm \sqrt{A^2m^2+B^2}$, we get

$m = -\frac{2a}{\sqrt{1-a^2}}$,

$A^2m^2 + B^2 = \frac{1}{1-a^2}$

$\Rightarrow A^2.\frac{4a^2}{(1-a^2)} + B^2 = \frac{1}{1-a^2}$

$\Rightarrow \frac{a^2(4A^2-B^2)+B^2}{1-a^2} = \frac{1}{1-a^2}$

$\Rightarrow B^2 = 1,$

$4A^2 - B^2 = 0$

$\Rightarrow A^2 = \frac{1}{4} \Rightarrow A = \frac{1}{2}, B = 1$ i.e., $B > A$

$\therefore$ If 'e' be the eccentricity of the ellipse, then

$A^2 = B^2(1-e^2) \Rightarrow 1 - e^2 = \frac{1}{4} \Rightarrow e = \frac{\sqrt{3}}{2}$

8. The tangent and normal to the ellipse $x^2 + 4y^2 = 4$ at point $P(\theta)$ on it meets the major axes in Q and R respectively. If $QR = 2$, then $\cos\theta$ is equal to

(a) $-\frac{2}{3}$ (b) $\frac{2}{3}$

(c) $\frac{1}{3}$ (d) None of these

Solution: (a,b) Given ellipse is $\frac{x^2}{4}+y^2=1$

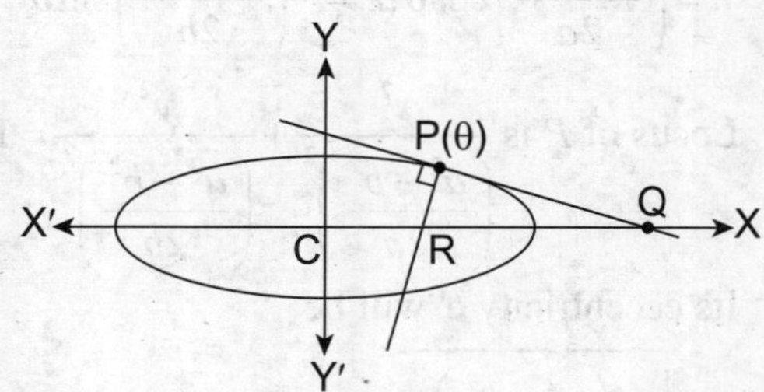

Let $P(\theta)\equiv(2\cos\theta,\sin\theta)$ be the given point on ellipse.

∴ Equation of the tangent at P is:

$$\frac{x\cos\theta}{2}+y\sin\theta=1 \quad ....(1)$$

$\Rightarrow Q\equiv(2\sec\theta,0)$

Equation of the normal at P is:

$2x\sec\theta - y\operatorname{cosec}\theta = 3$; on x-axis, $y = 0$

$\Rightarrow R\equiv\left(\frac{3}{2}\cos\theta,0\right)$

Therefore $QR=\left|2\sec\theta-\frac{3}{2}\cos\theta\right|=2$

$\Rightarrow \left|\frac{4-3\cos^2\theta}{2\cos\theta}\right|=2$

$\Rightarrow 16+9\cos^4\theta-24\cos^2\theta=16\cos^2\theta$

$\Rightarrow 9\cos^4\theta-40\cos^2\theta+16=0$

$\Rightarrow 9\cos^4\theta-36\cos^2\theta-4\cos^2\theta+16=0$

$\Rightarrow (9\cos^2\theta-4)(\cos^2\theta-4)=0$

$\Rightarrow \cos^2\theta=4/9 \Rightarrow \cos\theta=\pm 2/3$

($\because \cos^2\theta=4$ is impossible)

Hence (a), (b) are correct answers.

9. If $(\sqrt{3})\,bx+ay=2ab$ touches the ellipse $\frac{x^2}{a^2}+\frac{y^2}{b^2}=1$, then eccentric angle θ is

(a) $\frac{\pi}{6}$ (b) $\frac{\pi}{4}$

(c) $\frac{\pi}{3}$ (d) $\frac{\pi}{2}$

Solution: (a) Equation of tangent is $\frac{\sqrt{3}x}{2a}+\frac{y}{2b}=1$ and equation of tangent at the point $(a\cos\phi,b\sin\phi)$ is $\frac{x}{a}\cos\phi+\frac{y}{b}\sin\phi=1$.

But both are identical

$\Rightarrow \cos\phi=\frac{\sqrt{3}}{2}, \sin\phi=\frac{1}{2} \Rightarrow \phi=\frac{\pi}{6}$

10. If chord of contact of the tangents drawn from the point (α,β) to the ellipse $\frac{x^2}{a^2}+\frac{y^2}{b^2}=1$, touches the circle $x^2+y^2=c^2$, then the locus of the point (α,β) is

(a) $\frac{x^2}{a^2}+\frac{y^2}{b^2}=\frac{1}{c^2}$ (b) $\frac{x^2}{a^2}+\frac{y^2}{b^2}=\frac{1}{c^4}$

(c) $\frac{x^2}{a^4}+\frac{y^2}{b^4}=\frac{1}{c^2}$ (d) None of these

Solution: (c) Equation of chord of contact of tangents to ellipse $\frac{x^2}{a^2}+\frac{y^2}{b^2}=1$ from point (α,β) is given by $\frac{x\alpha}{a^2}+\frac{y\beta}{b^2}=1$, and it touches the circles $x^2+y^2=c^2$

⇒ perpendicular distance from centre of the circle to the line = radius of the circle

$$\Rightarrow \frac{|-1|}{\sqrt{\frac{\alpha^2}{a^4}+\frac{\beta^2}{b^4}}}=c \Rightarrow c^2\left(\frac{\alpha^2}{a^4}+\frac{\beta^2}{b^4}\right)=1$$

∴ Locus of (α,β) is $c^2\left(\frac{x^2}{a^4}+\frac{y^2}{b^4}\right)=1$

or $\frac{x^2}{a^4}+\frac{y^2}{b^4}=\frac{1}{c^2}$

Hence (c) is the correct answer.

11. A ladder of 12 units length slides in a vertical plane with its ends in contact with a vertical wall and a horizontal floor along x-axis. The locus of a point on the ladder 4 units from its foot has the equation.

(a) $\frac{x^2}{4}+y^2=1$ (b) $\frac{x^2}{16}+\frac{y^2}{64}=1$

(c) $\frac{x^2}{64}+\frac{y^2}{16}=1$ (d) $x^2+\frac{y^2}{4}=1$

Solution: (c) Let AB be the given ladder as shown in the figure below and let $P(h, k)$ be the given point on ladder such that $PB = 4$ units and $PA = 8$ units. Thus P divides AB in the ratio 2 : 1.

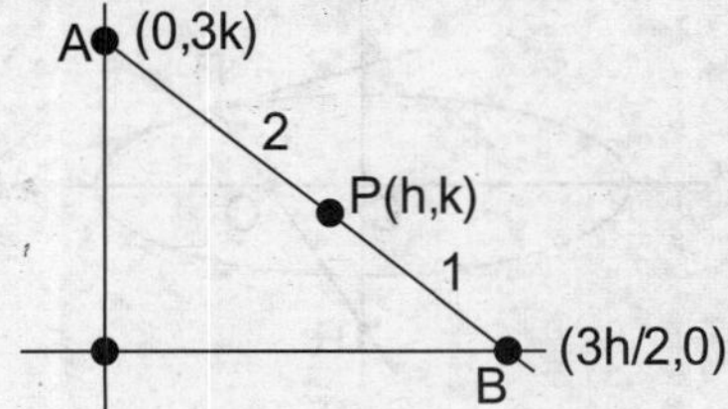

∴ Co-ordinates of A will be $(0, 3k)$ and that of B will be $(3h/2, 0)$

Now, $AB = 12 \Rightarrow \frac{9h^2}{4}+9k^2=144$

So, locus of P is $\frac{x^2}{64}+\frac{y^2}{16}=1$

12. The normal drawn to the ellipse $\frac{x^2}{a^2}+\frac{y^2}{b^2}=1$ at the extremity of the latus rectum passes through the extremity of the minor axis. Eccentricity of this ellipse is equal to

(a) $\sqrt{\frac{\sqrt{5}-1}{2}}$ (b) $\frac{\sqrt{5}-1}{2}$

(c) $\sqrt{\frac{\sqrt{3}-1}{2}}$ (d) $\frac{\sqrt{3}-1}{2}$

Solution: (a) Equation of normal at $\left(ae, \frac{b^2}{a}\right)$ will be $\frac{a^2x}{ae}-\frac{b^2y}{b^2/a}=a^2-b^2$. It passes through (0,– b), so $ab = a^2 - b^2$

$$\Rightarrow \left(\frac{b}{a}\right)^2+\left(\frac{b}{a}\right)-1=0 \Rightarrow \frac{b}{a}=\frac{-1\pm\sqrt{5}}{2}$$

$$\Rightarrow \frac{b}{a}=\frac{\sqrt{5}-1}{2} \text{ (rejecting } \frac{-1-\sqrt{5}}{2})$$

$$\Rightarrow e=\sqrt{1-\frac{b^2}{a^2}}=\sqrt{\frac{\sqrt{5}-1}{2}}$$

13. The normal at a variable point P on an ellipse $\frac{x^2}{a^2}+\frac{y^2}{b^2}=1$ of eccentricity 'e' meets the axes of the ellipse in Q and R, then the locus of the mid-point of QR is a conic with an eccentricity e' such that

(a) e' is independent of e

(b) e' = 1

(c) e' = e

(d) e' = 1/e

Solution: (c) Let point P be (a cosθ, b sinθ) be a variable point.

Equation of normal is given by $(a\sec\theta)\,x-(b\,\text{cosec}\,\theta)\,y=a^2-b^2$

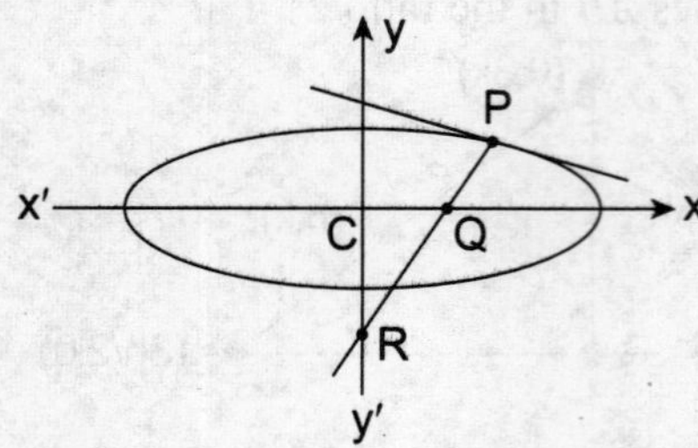

Now, on x axis $y=0 \Rightarrow \left(\left(\frac{a^2-b^2}{a}\right)\cos\theta, 0\right) \equiv Q$

On y axis, $x=0 \Rightarrow R\equiv\left(0, -\left(\frac{a^2-b^2}{b}\right)\sin\theta\right)$

Let P(h,k) be mid-point of QR

$$\Rightarrow h=\left(\frac{a^2-b^2}{2a}\right)\cos\theta, k=-\left(\frac{a^2-b^2}{2b}\right)\sin\theta$$

$\therefore$ Locus of P is $\frac{x^2}{\left(\frac{a^2-b^2}{2a}\right)^2}+\frac{y^2}{\left(\frac{a^2-b^2}{2b}\right)^2}=1$

Its eccentricity e' will be

$$=\sqrt{1-\frac{\left(\frac{a^2-b^2}{2a}\right)^2}{\left(\frac{a^2-b^2}{2b}\right)^2}}=\sqrt{1-\frac{b^2}{a^2}}=e$$

14. Let d be the perpendicular distance from the centre of the ellipse $\frac{x^2}{a^2}+\frac{y^2}{b^2}=1$ to the tangent drawn at a point P on the ellipse. If F_1 and F_2 are the two foci of the ellipse, and if $(PF_1-PF_2)^2=ka^2\left[1-\frac{b^2}{d^2}\right]$, then find the value of 'k'.

(a) 3 (b) 4

(c) 5 (d) None of these

Solution: (b) Equation of tangent to ellipse at $P(\theta)$

$$\frac{\cos\theta}{a}x+\frac{\sin\theta}{b}y=1$$

$$d=\frac{|0+0-1|}{\sqrt{\frac{\cos^2\theta}{a^2}+\frac{\sin^2\theta}{b^2}}} \Rightarrow \frac{1}{d^2}=\frac{\cos^2\theta}{a^2}+\frac{\sin^2\theta}{b^2}$$

$$\Rightarrow 1-\frac{b^2}{d^2}=1-\frac{b^2}{a^2}\cos^2\theta-\sin^2\theta$$

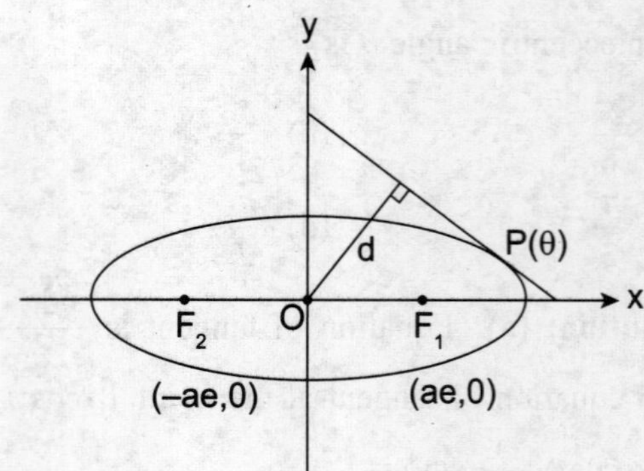

$$\Rightarrow 1-\frac{b^2}{d^2}=\cos^2\theta-\frac{b^2}{a^2}\cos^2\theta=\left(1-\frac{b^2}{a^2}\right)\cos^2\theta$$

$$\Rightarrow 1-\frac{b^2}{d^2}=e^2\cos^2\theta \qquad \text{....(i)}$$

Now, $PF_1 = a - ea\cos\theta$

and $PF_2 = a + ea\cos\theta$

$(PF_1 - PF_2)^2 = (a - ae\cos\theta - a - ea\cos\theta)^2$

$= (2ae\cos\theta)^2$

$\Rightarrow (PF_1 - PF_2)^2 = 4a^2e^2\cos^2\theta$ ...(ii)

from equation (i) and (ii)

$$(PF_1 - PF_2)^2 = 4a^2\left(1 - \frac{b^2}{d^2}\right)$$

15. Given the equation of the ellipse $\frac{(x-3)^2}{16} + \frac{(y+4)^2}{49} = 1$, a parabola is such that its vertex is the lowest point of the ellipse and it passes through the ends of the minor axis of the ellipse. The equation of the parabola is in the form $16y = a(x - h)^2 - k$. Determine the value of $(a + h + k)$.

(a) 152 (b) 173

(c) 186 (d) 204

Solution: (c) equation of parabola whose vertex is (3, –11) is given by

$\Rightarrow (x - 3)^2 = 4b(y + 11)$ ...(1)

Since, this parabola passes through (–1, –4)

$\therefore (-1 - 3)^2 = 4b(-4 + 11)$

or $16 = 4b(7)$ or $b = \frac{4}{7}$

Putting the value of 'b' in equation (1)

$\Rightarrow (x - 3)^2 = 4\left(\frac{4}{7}\right)(y + 11)$

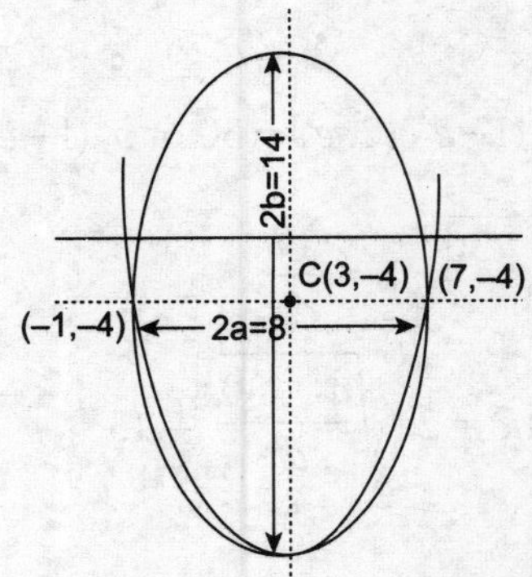

or $16y = 7(x - 3)^2 - 176$ .....(2)

Comparing this equation with $16y = a(x - h)^2 - k$

$\Rightarrow a = 7, h = 3, k = 176$

$\therefore$ Value of $(a + h + k) = 7 + 3 + 176 = 186$

16. Consider an ellipse $\frac{x^2}{25} + \frac{y^2}{9} = 1$ with centre C and a point P on it with eccentric angle $\frac{\pi}{4}$. Normal drawn at P intersects the major and minor axes in A and B respectively. N_1 and N_2 are the feet of the perpendiculars from the foci S_1 and S_2 respectively on the tangent at P and N is the foot of the perpendicular from the centre of the ellipse on the normal at P. Tangent at P intersects the axis of x at T. On the basis of the above information, identify the correct statements.

(a) (CA).(CT) is equal to 16

(b) (PN).(PB) is equal to 9

(c) $(S_1N_1).(S_2N_2)$ is equal to 25

(d) $(S_1P).(S_2P)$ is equal to 17

Solution: (a, d) Ellipse $\frac{x^2}{25} + \frac{y^2}{9} = 1$ $(a > b)$

major axis = $2a$ and minor axis = $2b$

$a^2 = 25$ and $b^2 = 9$

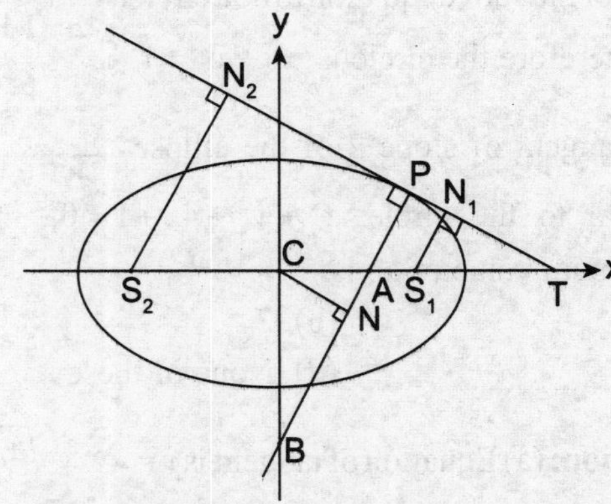

(a) (CA).(CT) = $(CS)^2$ (by property)

$= (ae)^2 = a^2e^2 = a^2 - b^2 = 25 - 9 = 16$

(b) (PN).(PB) = a^2 (by property)

= (semi major axis)^2 = 25

(c) $(S_1N_1).(S_2N_2) = b^2$ (by property) = 9

(d) $(S_1P).(S_2P)$ = (Focal distance of point P from S_1) × (focal distance of point P from S_2)

$= (a + ex_1)(a - ex_1) = a^2 - e^2x_1^2$; but

$x_1 = a\cos\theta$

$= 5 \times \cos\frac{\pi}{4} = \frac{5}{\sqrt{2}}$

$\therefore (S_1P).(S_2P) = a^2 - e^2x_1^2$

$$= 25 - \left(1 - \frac{b^2}{a^2}\right)\frac{25}{2} = 25 - \left(1 - \frac{9}{25}\right)\frac{25}{2}$$

$$= 25 - \left(\frac{16}{25}\right)\frac{25}{2} = 25 - 8 = 17$$

17. If the curves $\frac{x^2}{4} + y^2 = 1$ and $\frac{x^2}{a^2} + y^2 = 1$ for suitable value of 'a' cut on four concyclic points, the equation of circle passing through these points is

(a) $x^2 + y^2 = 2$ (b) $x^2 + y^2 = 1$

(c) $x^2 + y^2 = 4$ (d) None of these

Solution: (b) Let the equation of the circle be

$$\left(\frac{x^2}{4}+y^2-1\right)+\lambda\left(\frac{x^2}{a^2}+y^2-1\right)=0$$

$$\Rightarrow x^2\left(\frac{1}{4}+\frac{\lambda}{a^2}\right)+y^2(1+\lambda)=1+\lambda$$

$$\Rightarrow x^2\left(\frac{a^2+4\lambda}{4a^2}\right)+y^2(1+\lambda)=1+\lambda$$

$$\Rightarrow \quad x^2\left(\frac{a^2+4\lambda}{4a^2(1+\lambda)}\right)+y^2=1.$$

For the above equation to be a circle; $\frac{a^2+4\lambda}{4a^2(1+\lambda)}=1$

therefore the circle is $x^2+y^2=1$

18. If a tangent of slope 2 of the ellipse $\frac{x^2}{a^2}+\frac{y^2}{b^2}=1$ is normal to the circle $x^2+y^2+4x+1=0$, then the maximum value of ab is

(a) 4 (b) 2

(c) 1 (d) None of these

Solution: (a) Equation of tangent is $y=2x\pm\sqrt{4a^2+b^2}$

$\Rightarrow$ this is normal to the circle $x^2+y^2+4x+1=0$

$\Rightarrow$ this tangent passes through (–2, 0).

$\Rightarrow\ 0=-4\pm\sqrt{4a^2+b^2}\Rightarrow 4a^2+b^2=16$

$\Rightarrow$ using A.M. $\geq$ G.M.

we get, $\frac{4a^2+b^2}{2}\geq\sqrt{4a^2.b^2}\ \Rightarrow$ ab $\leq$ 4.

19. A series of ellipse are described with a given focus and corresponding directrix. The locus of the extremities of their minor axis is a

(a) circle (b) ellipse

(c) parabola (d) hyperbola

Solution: (c) Let focus S be (0,0) and ZM be a directrix, then ZS = constant, for all ellipses

i.e., $\frac{a}{e}-ae$ = constant or $\frac{b^2}{ae}=\lambda(say)$ ...(1)

But, if (x, y) be the end of minor axis, then $x = ae$, $y = b\ (or -b)$

$\therefore$ from (1), locus of the end of minor axis is $y^2=\lambda x$, which is a parabola.

20. A circle has the same centre as an ellipse and passes through the focii F_1 and F_2 of the ellipse, such that the two curves intersect in 4 points. Let 'P' be any one of their points of intersection. If the major axis of the ellipse is 17 and the area of the triangle PF_1F_2 is 30, then the distance between the focii is:

(a) 11 (b) 12

(c) 13 (d) 15

Solution: (c) Let ellipse $\frac{x^2}{a^2}+\frac{y^2}{b^2}=1$ and circle $x^2+y^2=a^2e^2$, radius of circle = ae. Point of intersection of circle $x^2+y^2=a^2e^2$ and ellipse $\frac{x^2}{a^2}+\frac{y^2}{a^2(1-e^2)}=1$ is given by

$\left(\frac{a}{e}\sqrt{2e^2-1},\frac{a}{e}(1-e^2)\right)$; where $a=\frac{17}{2}$

Now area of triangle

$$=\frac{1}{2}\begin{vmatrix}\frac{a}{e}\sqrt{2e^2-1} & \frac{a}{e}(1-e^2) & 1\\ ae & 0 & 1\\ -ae & 0 & 1\end{vmatrix}$$

$$=\frac{1}{2}\left|\frac{a}{e}(1-e^2)(2ae)\right|=30 \text{ or } a^2(1-e^2)=30$$

$$e=\sqrt{1-\frac{30}{a^2}};\ a=\frac{17}{2}$$

then, $2ae=2a\sqrt{\frac{a^2-30}{a^2}}=13$

SECTION–II

SUBJECTIVE-TYPE SOLVED EXAMPLES

1. Find the equation of the ellipse whose axes are of length 6 and 2√6 and their equations are $x-3y+3=0$ and $3x+y-1=0$ respectively.

Solution: Equation of ellipse would be

$$\frac{\left(\frac{3x+y-1}{\sqrt{9+1}}\right)^2}{\left(\frac{6}{2}\right)^2}+\frac{\left(\frac{x-3y+3}{\sqrt{(1)^2+(-3)^2}}\right)^2}{\left(\frac{2\sqrt{6}}{2}\right)^2}=1$$

$$\Rightarrow \frac{(3x+y-1)^2}{90}+\frac{(x-3y+3)^2}{60}=1$$

$\Rightarrow 2(3x+y-1)^2+3(x-3y+3)^2=180$

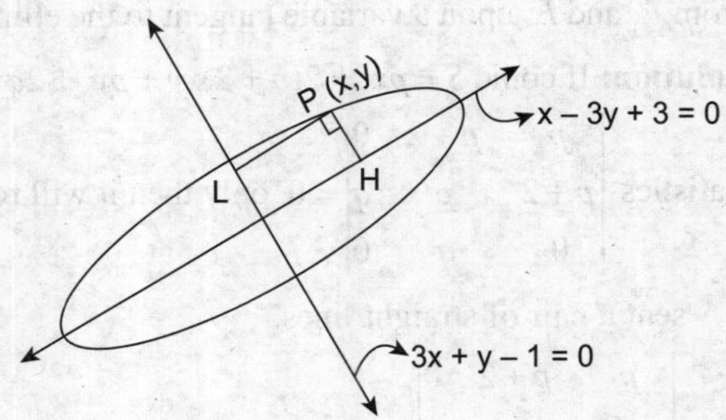

$\Rightarrow 2[9x^2+y^2+1+6xy-2y-6x]+3[x^2+9y^2+9-6xy-18y+6x]=180$

$\Rightarrow 21x^2+29y^2-6xy+6x-58y-151=0.$

2. If the focus of the parabola $y^2+8=4x$ coincides with one of the foci of the ellipse $3x^2+by^2-12x=0$, then find the eccentricity of the ellipse

Solution: Given parabola is $y^2=4x-8 \Rightarrow y^2=4(x-2)$
Its focus is given by $(x-2=a, y=0) \equiv (x=3, y=0)$ i.e., (3,0)

Given ellipse is $3x^2+by^2-12x=0$

or $3x^2-12x+by^2=0$

or $3(x^2-4x)+by^2=0$

or $3(x^2-4x+4)-12+by^2=0$

or $3(x-2)^2+by^2=12 \Rightarrow \frac{(x-2)^2}{4}+\frac{y^2}{(12/b)}=1.$

Case (i): If $4>\frac{12}{b}$, then ellipse $\frac{(x-2)^2}{4}+\frac{y^2}{12/b}=1$ is of first kind.

$$ae=2e=\sqrt{4-\frac{12}{b}}=\left[\because ae=\sqrt{a^2-b^2}\right]$$

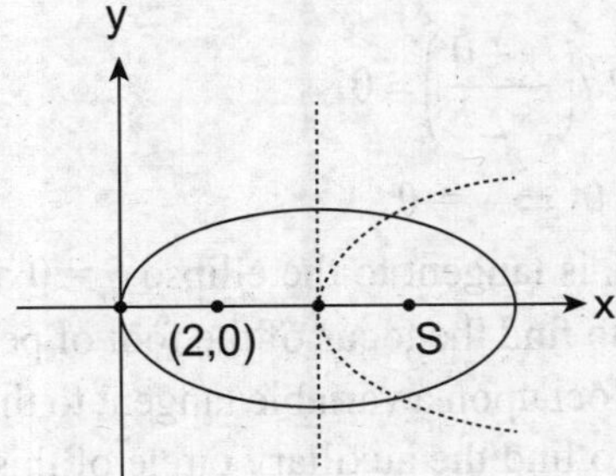

$\therefore$ foci are given by $(2\pm ae,0)\equiv\left(2\pm\sqrt{4-\frac{12}{b}},0\right)$
$\equiv (3, 0)$

$$\Rightarrow 2\pm\sqrt{4-\frac{12}{b}}=3$$

$$\Rightarrow \pm\sqrt{4-\frac{12}{b}}=1\Rightarrow 4-\frac{12}{b}=1$$

$$\Rightarrow \frac{12}{b}=3\Rightarrow b=4$$

$$\therefore\ 2e=\sqrt{4-\frac{12}{4}}=1\Rightarrow e=\frac{1}{2}$$

Case II: If $4<\frac{12}{b}$, then ellipse $\frac{(x-2)^2}{4}+\frac{y^2}{12/b}=1$ is of second kind.
But in this case, the foci will lie on the $x=2$ (which is not possible, since one focus is at (3, 0))

3. Let P be a point on the ellipse $\frac{x^2}{a^2}+\frac{y^2}{b^2}=1(a>b)$ in the Ist or 2nd quadrants whose foci are S_1 and S_2. Then find the least possible value of circumradius of ΔPS_1S_2.

Solution: For a triangle LMN, the circumradius is given by $R=\frac{lmn}{4\Delta}$

P(acosθ,bsinθ)
S_1 S_2

According to questions

$$R=\frac{(2ae)\times(a-ae\cos\theta)(a+ae\cos\theta)}{4\times\left[\frac{1}{2}\times(S_1S_2)\times b\sin\theta\right]}$$

$$=\frac{(2ae)(a^2)(1-e^2\cos^2\theta)}{2(2ae)(b\sin\theta)}=\frac{a^2(1-e^2\cos^2\theta)}{2b\sin\theta}$$

$$=\frac{a^2}{2b}\left[1-\left(1-\frac{b^2}{a^2}\right)\cos^2\theta\right]\frac{1}{\sin\theta}$$

$$=\frac{a^2}{2b}\left[\frac{1-\cos^2\theta}{\sin\theta}+\frac{b^2}{a^2}\frac{\cos^2\theta}{\sin\theta}\right]$$

$$=\frac{a^2}{2b}\left[\frac{\sin^2\theta}{\sin\theta}+\frac{b^2}{a^2}\frac{\cos^2\theta}{\sin\theta}\right]$$

$$=\frac{a^2}{2b}\left[\sin\theta+\frac{b^2}{a^2}\left(\frac{1-\sin^2\theta}{\sin\theta}\right)\right]$$

$$= \frac{a^2}{2b}\left[\sin\theta + \frac{b^2}{a^2}\text{cosec}\theta - \frac{b^2}{a^2}\sin\theta\right]$$

$$= \frac{a^2}{2b}\left[\left(1-\frac{b^2}{a^2}\right)\sin\theta + \frac{b^2}{a^2}\text{cosec}\theta\right]$$

$$= \frac{a^2}{2b}\left[e^2\sin\theta + \frac{b^2}{a^2}\text{cosec}\theta\right] \geq \frac{a^2}{2b}\times 2\sqrt{\frac{e^2b^2}{a^2}}$$

$$= \frac{a^2}{b}\times\frac{eb}{a} = ae$$

$\therefore$ Least value of circumradius $= ae$

4. From a point P, the chord of contact to the ellipse E_1: $\frac{x^2}{a}+\frac{y^2}{b}=(a+b)$ touches the ellipse $E_2:\frac{x^2}{a^2}+\frac{y^2}{b^2}=1$, then find the locus of the point P.

Solution: Let $P(h, k)$ be the point

$\therefore$ equation of chord of contact to ellipse

$$\frac{x^2}{a(a+b)}+\frac{y^2}{b(a+b)}=1 \text{ will be}$$

$$\frac{hx}{a(a+b)}+\frac{ky}{b(a+b)}=1 \qquad \text{.... (i)}$$

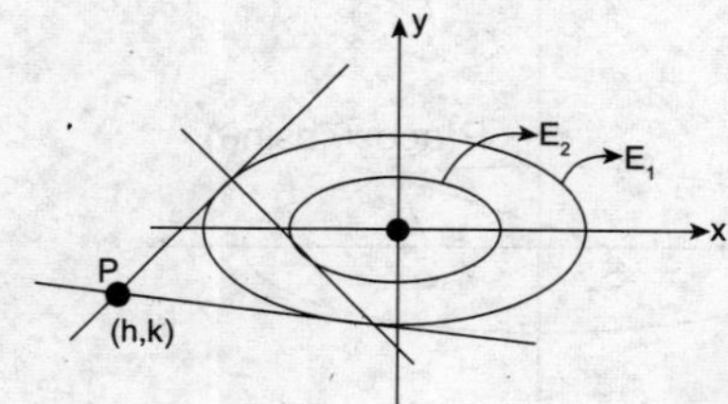

Now (i) is $y = \frac{-hbx}{ka}+\frac{b(a+b)}{k}$ ... (ii)

For (ii) to be tangent to ellipse $\frac{x^2}{a^2}+\frac{y^2}{b^2}=1$

$c^2 = a^2m^2 + b^2$

$$\Rightarrow \frac{b^2(a+b)^2}{k^2}=a^2\left(\frac{-hb}{ka}\right)^2+b^2$$

$$\Rightarrow \frac{(a+b)^2}{k^2}=\frac{h^2}{k^2}+1 \Rightarrow (a+b)^2=h^2+k^2$$

$\therefore$ locus is $x^2 + y^2 = (a + b)^2$;

Also, auxilliary circle of ellipse

$$\frac{x^2}{a(a+b)}+\frac{y^2}{b(a+b)}=1 \text{ is } x^2+y^2=a(a+b)+b(a+b)$$

or $x^2 + y^2 = (a + b)^2$

$\therefore$ locus of point P is

Auxillary circle of ellipse $E_1:\frac{x^2}{a}+\frac{y^2}{b}=(a+b)$.

5. Let foci of the conic represented by the equation $px^2 + 2(p + 2)xy + py^2 + 2qy = 0$, where $p < -1$ and $q \neq 0$ be $F_1 : (2, -3)$ and $F_2 : (8, -5)$.

Then, find the locus of the feet of the perpendicular from F_1 and F_2 upon a variable tangent to the ellipse.

Solution: If conic $S \equiv px^2 + 2(p+2)xy + py^2 + 2qy = 0$

Satisfies $\begin{vmatrix} p & p+2 & 0 \\ p+2 & p & q \\ 0 & q & 0 \end{vmatrix} = 0$ only then it will represent a pair of straight lines.

But $\begin{vmatrix} p & p+2 & 0 \\ p+2 & p & q \\ 0 & q & 0 \end{vmatrix} = -q^2p \neq 0$

Hence $S = 0$ does not represent a pair of straight lines

Now, we need to check if it represents circle or a parabola or an ellipse or a hyperbola.

$h^2 - ab = (p + 2)^2 - p^2 = 4(p + 1)$

And since $p < -1$ (given)

$\therefore$ $h^2 - ab < 0$

Hence $S = 0$ will represent an ellipse

Now, by observation, we can say that this ellipse passes through the origin.

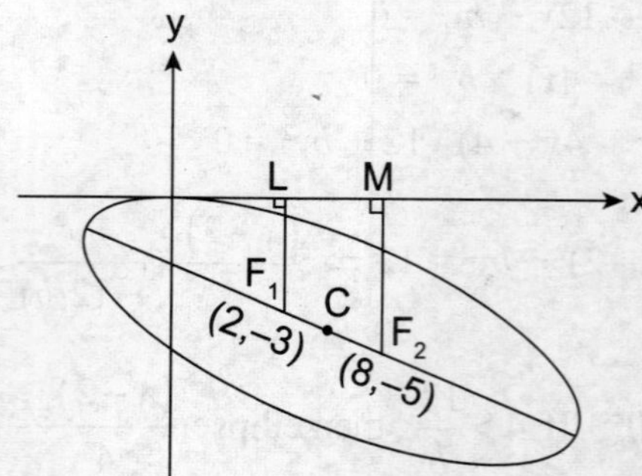

Now, we find the tangent to the ellipse $S = 0$ at (0,0)

$$T \equiv px(0)+2(p+2)\left(\frac{x(0)+y(0)}{2}\right)+$$

$$py(0)+2q\left(\frac{y+0}{2}\right)=0$$

$\Rightarrow qy = 0 \Rightarrow y = 0$

$\Rightarrow$ x-axis is tangent to the ellipse $S = 0$ at (0,0)

We need to find the locus of the foot of perpendiculars from the foci upon a variable tangent to the ellipse i.e., we need to find the auxiliary circle of this ellipse

Centre of auxiliary circle $\equiv$ (5, –4)

From the properties of ellipse, we know that the product of perpendiculars from foci to any tangent to the ellipse is constant and is equal to b^2 (square of semi-minor axis)

$\Rightarrow F_1L \times F_2M = b^2$

$\Rightarrow b^2 = 15$

Also $2ae = |F_1F_2| = 2\sqrt{10}$

$\Rightarrow ae = \sqrt{10} \Rightarrow a^2e^2 = 10$

Now, $b^2 = a^2(1 - e^2) \Rightarrow a^2 = b^2 + a^2e^2 = 15 + 10$

$\Rightarrow a = 5$

$\Rightarrow$ Equation of auxiliary circle will be

$(x - 5)^2 + (y + 4)^2 = 25$

6. *AB* is a common tangent to the ellipse $\frac{x^2}{a^2} + \frac{y^2}{b^2} = 1$ and the circle $x^2 + y^2 = r^2$ $(b < r < a)$. The chord *PQ* of the circle is a focal chord to the ellipse. If *PQ* is parallel to *AB*, then find its length.

Solution: Let $y = mx + \sqrt{a^2m^2 + b^2}$ be a tangent to the ellipse

It is also tangent to circle, and with slope m

It should be $y = mx + \sqrt{r^2m^2 + r^2}$

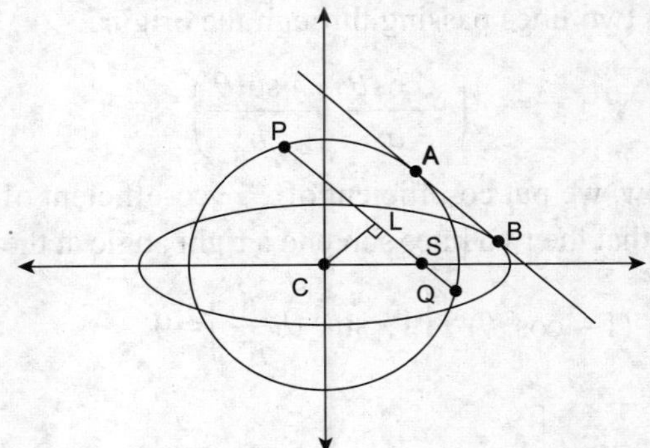

or $y = mx + r\sqrt{1 + m^2}$

$\therefore r\sqrt{1 + m^2} = \sqrt{a^2m^2 + b^2}$

$\Rightarrow r^2(1 + m^2) = a^2m^2 + b^2$

$\Rightarrow r^2m^2 - a^2m^2 = b^2 - r^2$

$\Rightarrow m^2 = \frac{b^2 - r^2}{r^2 - a^2} = \frac{r^2 - b^2}{a^2 - r^2}$

$\therefore$ Equation of focal chord parallel to this tangent will be

$(y - 0) = \sqrt{\frac{r^2 - b^2}{a^2 - r^2}}(x - ae)$ or $y = m(x - ae)$

$\therefore$ Length of chord $PQ = 2QL = 2\sqrt{CQ^2 - CL^2}$

$$= 2\sqrt{r^2 - \left(\frac{mae}{\sqrt{1 + m^2}}\right)^2} = 2\sqrt{r^2 - \frac{m^2a^2e^2}{1 + m^2}}$$

$$= 2\sqrt{\frac{r^2(1 + m^2) - m^2a^2e^2}{1 + m^2}}$$

$$= 2\sqrt{\frac{b^2 + a^2m^2 - m^2(a^2 - b^2)}{1 + m^2}}$$

$[\because r^2m^2 + r^2 = b^2 + a^2m^2]$

$$= 2\sqrt{\frac{b^2 + m^2b^2}{1 + m^2}} = 2\sqrt{b^2} = 2b$$

7. Consider an ellipse and a concentric circle. The circle passes though the foci of the ellipse and intersects the ellipse in four distinct points. The length of major axis of the ellipse is 15 units. If S_1 and S_2 are the foci of the ellipse. *P* be one of the points of intersections of ellipse and circle and area of triangle PS_1S_2 is 26 square units, then find the eccentricity of the elllipse.

Solution: Let $\frac{x^2}{a^2} + \frac{y^2}{b^2} = 1$ be the given ellipse

$\therefore$ Concentric circle passing through the foci would have radius $= ae$

$\therefore$ Its equation will be $x^2 + y^2 = a^2e^2$

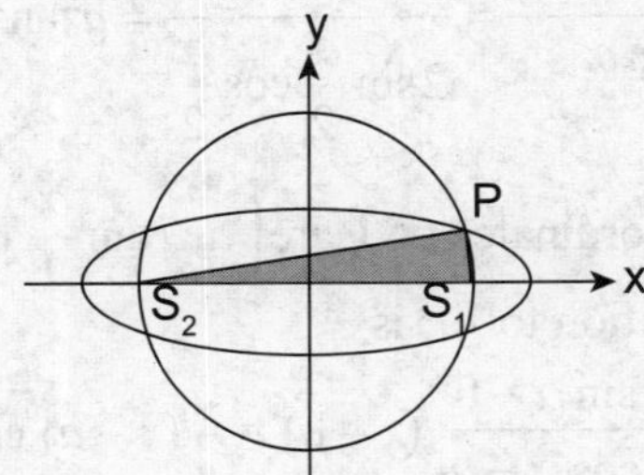

Let us eliminate x from above two curves

$a^2\left(1 - \frac{y^2}{b^2}\right) + y^2 = a^2e^2$

$\Rightarrow y^2\left(1 - \frac{a^2}{b^2}\right) = a^2e^2 - a^2 \Rightarrow y^2 = \frac{a^2b^2(e^2 - 1)}{(b^2 - a^2)}$;

$\Rightarrow y = ab\sqrt{\frac{e^2 - 1}{b^2 - a^2}}$ $[\because b^2 = a^2 - a^2e^2]$

$\therefore$ Height of point of intersection above x-axis would be

$$= ab\sqrt{\frac{1}{a^2e^2}} = \frac{\sqrt{b^2(1 - e^2)}}{e} = \frac{\sqrt{a^2(1 - e^2)}}{e}$$

$$= \frac{a(1 - e^2)}{e}$$

$\therefore$ Area of $\Delta PS_1S_2 = \frac{1}{2}(2ae)\left(\frac{a(1 - e^2)}{e}\right) = a^2(1 - e^2)$

A.T.Q. $a^2(1 - e^2) = 26$

Also length of major axis = $2a = 15 \Rightarrow a = 15/2$

$\therefore \quad \frac{225}{4}\left(1-e^2\right) = 26$

$\Rightarrow 1-e^2 = \frac{26\times 4}{225} \Rightarrow e^2 = 1-\frac{104}{225} = \frac{121}{225} \Rightarrow e = \frac{11}{15}$

$\therefore$ Eccentricity =11/15.

8. The tangent at any point P of a circle $x^2 + y^2 = a^2$ meets the tangent at a fixed point A(a, 0) in T and T is joined to B, the other end of the diameter through A, prove that the locus of the intersection of AP and BT is an ellipse whose eccentricity is $1/\sqrt{2}$.

Solution: The equation of circle is $x^2 + y^2 = a^2$ its radius is a and centre (0, 0).

Let (a cos α, a sin α) be any point on the circle say P and OA as x-axis, so A is (a, 0) and therefore B will be (–a, 0)

Equation of tangent at P will be $x \cos\alpha + y \sin\alpha = a$ ...(1)

and tangent at A is: $x = a$ ...(2)

solving (1) and (2)

$$y = \frac{a(1-\cos\alpha)}{\sin\alpha} = \frac{2a\sin^2\frac{\alpha}{2}}{2\sin\frac{\alpha}{2}\cos\frac{\alpha}{2}} = a\tan\frac{\alpha}{2} \quad \text{...(3)}$$

so the co-ordinates of T are $\left(a, a\tan\frac{\alpha}{2}\right)$

Now equation to AP is

$$y - 0 = \frac{a\sin\alpha - 0}{a\cos\alpha - a}(x-a) = -(x-a)\cot\frac{\alpha}{2}$$

$$\text{or } \cot\frac{\alpha}{2} = -\frac{y}{x-a} \quad \text{...(4)}$$

and equation of BT is

$$y - 0 = \frac{a\tan\frac{\alpha}{2} - 0}{a-(-a)}(x+a) = \frac{1}{2}(x+a)\tan\frac{\alpha}{2}$$

$$\text{or } \tan\frac{\alpha}{2} = \frac{2y}{x+a} \quad \text{...(5)}$$

to eliminate the variable α, we multiply (4) and (5), so

$\frac{-2y^2}{x^2-a^2} = 1$ or $2y^2 + x^2 = a^2$

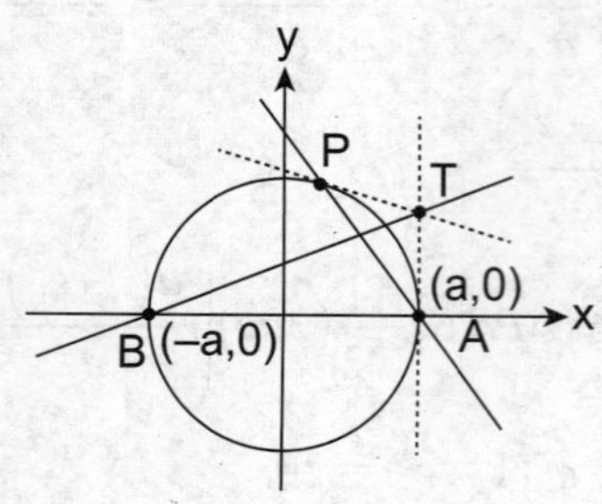

or $\frac{x^2}{a^2} + \frac{y^2}{\frac{a^2}{2}} = 1$

which is the locus of the point of intersection of AP and BT and this is an equation of ellipse.

Again $e = \sqrt{\left(\frac{A^2-B^2}{A^2}\right)} = \sqrt{\frac{a^2-\frac{a^2}{2}}{a^2}} = \frac{1}{\sqrt{2}}$ **Ans**

9. The tangent at the point α on a standard ellipse meets the auxiliary circle in two points which subtends a right angle at the centre. Show that the eccentricity of the ellipse is $(1 + \sin^2\alpha)^{-1/2}$.

Solution: Tangent at 'θ' is $\frac{x\cos\theta}{a} + \frac{y\sin\theta}{b} = 1$ ...(1)

Auxiliary circle is $x^2 + y^2 = a^2$ ....(2)

Putting the value of 1 form equation (1), in equation (2), we can get a homogenous equation in second degree. And since every homogenous equation in second degree represents a pair of straight lines, so we get two lines passing through the origin.

$$\therefore \quad x^2 + y^2 = a^2\left(\frac{x\cos\theta}{a} + \frac{y\sin\theta}{b}\right)^2$$

Now, we put co-efficient of x^2 + co-efficient of $y^2 = 0$, so that the two lines subtend a right angle at the origin.

$$\therefore \quad (1-\cos^2\theta) + \left(1-\sin^2\theta.\frac{a^2}{b^2}\right) = 0$$

$$\text{or } \sin^2\theta\left[1-\frac{a^2}{a^2(1-e^2)}\right] = -1$$

or $-e^2\sin^2\theta = -(1-e^2)$

$\therefore \quad e^2(1+\sin^2\theta) = 1$

$\therefore \quad e = (1+\sin^2\theta)^{-1/2}$

10. If on any arbitrary tangent to an ellipse $\frac{x^2}{a^2} + \frac{y^2}{b^2} = 1$, perpendiculars of length s, s', v, v' and c are drawn from the foci, verticles and centre respectively, then prove that $\frac{ss'-c^2}{vv'-c^2} = e^2$ where 'e' is the eccentricity of the ellipse.

Solution: The equation of given ellipse is $\frac{x^2}{a^2} + \frac{y^2}{b^2} = 1$,

the tangent at point (x_1, y_1) is $\frac{xx_1}{a^2} + \frac{yy_1}{b^2} = 1$

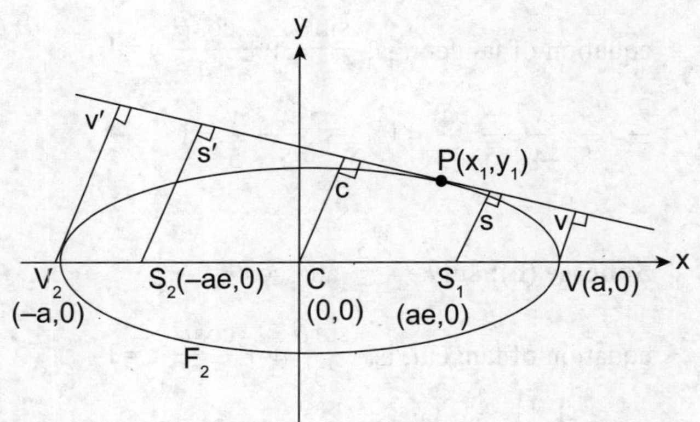

or $\left(\frac{x_1}{a^2}\right)x+\left(\frac{y_1}{b^2}\right)y-1=0$ ...(1)

Now

The length of perpendicular from focus S_1(ae, 0) to the line (1) is given by $s=\left|\dfrac{ae.\frac{x_1}{a^2}-1}{\sqrt{\left(\frac{x_1}{a^2}\right)^2+\left(\frac{y_1}{b^2}\right)^2}}\right|$ ...(2)

The length of perpendicular from focus S_2(–ae, 0) to the line (1) is given by $s'=\left|\dfrac{-ae.\frac{x_1}{a^2}-1}{\sqrt{\left(\frac{x_1}{a^2}\right)^2+\left(\frac{y_1}{b^2}\right)^2}}\right|$ ...(3)

The length of perpendicular from centre O(0, 0) to the line (1) is given by $c=\left|\dfrac{0+0-1}{\sqrt{\left(\frac{x_1}{a^2}\right)^2+\left(\frac{y_1}{b^2}\right)^2}}\right|$ ...(4)

The length of perpendicular from vertex V_1(a, 0) to the line (1) is given by $v=\left|\dfrac{a.\frac{x_1}{a^2}-1}{\sqrt{\left(\frac{x_1}{a^2}\right)^2+\left(\frac{y_1}{b^2}\right)^2}}\right|$ ...(5)

The length of perpendicular from vertex V_2(–a, 0) to the line (1) is given by $v'=\left|\dfrac{-a.\frac{x_1}{a^2}-1}{\sqrt{\left(\frac{x_1}{a^2}\right)^2+\left(\frac{y_1}{b^2}\right)^2}}\right|$ ..(6)

We need have to prove that $\dfrac{ss'-c^2}{vv'-c^2}=e^2$

Taking L.H.S $=\dfrac{ss'-c^2}{vv'-c^2}$

Substituting the values of s,s', c, v, v' from equations (2), (3), (4), (5) and (6), we get L.H.S. equals

$$\frac{\dfrac{\left|\left(\frac{ex_1}{a}\right)^2-1\right|}{\left(\frac{x_1}{a^2}\right)^2+\left(\frac{y_1}{b^2}\right)^2}-\dfrac{1}{\left(\frac{x_1}{a^2}\right)^2+\left(\frac{y_1}{b^2}\right)^2}}{\dfrac{\left|\left(\frac{x_1}{a}\right)^2-1\right|}{\left(\frac{x_1}{a^2}\right)^2+\left(\frac{y_1}{b^2}\right)^2}-\dfrac{1}{\left(\frac{x_1}{a^2}\right)^2+\left(\frac{y_1}{b^2}\right)^2}}=\left(\frac{1-\frac{e^2x_1^2}{a^2}-1}{1-\frac{x_1^2}{a^2}-1}\right)=e^2$$

$$\left[\begin{array}{l}\because e^2<1\Rightarrow\dfrac{e^2x_1^2}{a^2}<\dfrac{x_1^2}{a^2}\Rightarrow\dfrac{e^2x_1^2}{a^2}-1<\dfrac{x_1^2}{a^2}-1=-\dfrac{y_1^2}{b^2}\\ \Rightarrow\left(\dfrac{e^2x_1^2}{a^2}-1\right)<0,\text{Also}\dfrac{x_1^2}{a^2}-1=-\dfrac{y_1^2}{b^2}<0\end{array}\right]$$

11. Prove that the equation to the circle, having double contact with the ellipse $\frac{x^2}{a^2}+\frac{y^2}{b^2}=1$ (with eccentricity e) at the ends of a latus rectum, is $x^2+y^2-2ae^3x=a^2(1-e^2-e^4)$.

Solution: Equation of tangent at point $A\left(ae,\frac{b^2}{a}\right)$ on ellipse $\frac{x^2}{a^2}+\frac{y^2}{b^2}=1$ is given by

$\Rightarrow\ \frac{ae}{a^2}x+\frac{b^2}{a}\frac{y}{b^2}=1\Rightarrow ex+y=a$

It's also tangent on circle.

$\Rightarrow$ perpendicular from centre = r

$\Rightarrow\ \dfrac{|ex_1-a|}{\sqrt{1+e^2}}=\sqrt{(x_1-ae)^2+b^4/a^2}$

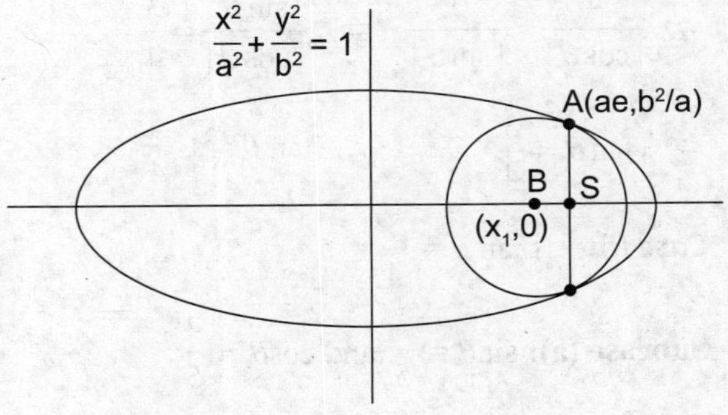

$\Rightarrow \dfrac{(ex_1 - a)^2}{1+e^2} = (x_1 - ae)^2 + b^4/a^2$

On solving, we get $x_1 = ae^3$

$\Rightarrow$ radius. $AB = \sqrt{(ae^3 - ae)^2 + b^4/a^2}$

$= a\sqrt{e^2(e^2-1)^2 + b^4/a^4}$

$= a\sqrt{e^2(e^2-1)^2 + (e^2-1)^2}$

$= a\sqrt{e^6 - e^4 - e^2 + 1}$

equation of circle

$(x - ae^3)^2 + y^2 = a^2(e^6 - e^4 - e^2 + 1)$

$\Rightarrow x^2 + y^2 - 2ae^3x = a^2(1 - e^2 - e^4)$

12. Find the equations of the lines with equal intercepts on the axes and which touch the ellipse $\dfrac{x^2}{16} + \dfrac{y^2}{9} = 1.$

Solution: Equation of ellipse $\dfrac{x^2}{16} + \dfrac{y^2}{9} = 1$

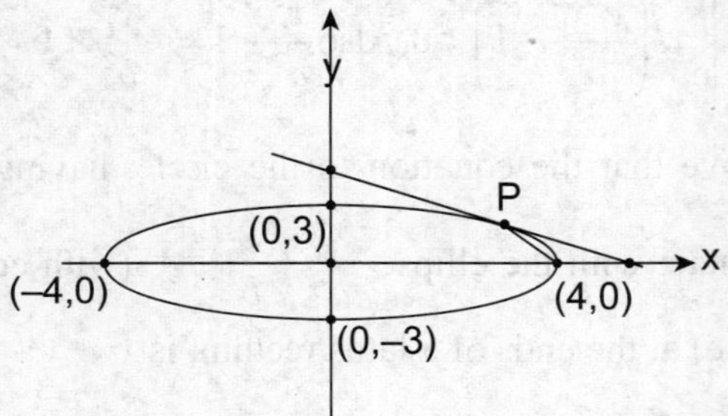

Equation of tangent on given ellipse at point $P(4\cos\theta, 3\sin\theta)$ is $\dfrac{4\cos\theta}{16}x + y\dfrac{3\sin\theta}{9} = 1$

$\Rightarrow \left(\dfrac{\cos\theta}{4}\right)x + \left(\dfrac{\sin\theta}{3}\right)y = 1$

$\Rightarrow \dfrac{x}{(4\sec\theta)} + \dfrac{y}{(3\operatorname{cosec}\theta)} = 1$

According to question,

Intercepts made by above tangent are equal in length.

$\therefore\ |4\sec\theta| = |3\operatorname{cosec}\theta|$

$\Rightarrow \dfrac{4}{|\cos\theta|} = \dfrac{3}{|\sin\theta|} \quad \Rightarrow \dfrac{|\sin\theta|}{|\cos\theta|} = \dfrac{3}{4}$

$\Rightarrow |\tan\theta| = \dfrac{3}{4} \qquad \therefore\ \tan\theta = \pm\dfrac{3}{4}$

Case (I): If $\tan\theta = \dfrac{3}{4}$

Subcase (a): $\sin\theta = \dfrac{3}{5}$ and $\cos\theta = \dfrac{4}{5}$

equation of tangent, is, $\dfrac{\sin\theta}{3}y + \dfrac{\cos\theta}{4}x = 1$

$\Rightarrow \dfrac{4}{5}\cdot\dfrac{x}{4} + \dfrac{3}{5}\cdot\dfrac{y}{3} = 1 \quad \Rightarrow \dfrac{x}{5} + \dfrac{y}{5} = 1$

$\therefore\ x + y = 5$

Subcase (b): $\sin\theta = -\dfrac{3}{5}$ and $\cos\theta = -\dfrac{4}{5}$

equation of tangent, is, $\dfrac{\sin\theta}{3}y + \dfrac{\cos\theta}{4}x = 1$

$\Rightarrow -\dfrac{4}{5}\cdot\dfrac{x}{4} - \dfrac{3}{5}\cdot\dfrac{y}{3} = 1$

$\Rightarrow \dfrac{x}{5} + \dfrac{y}{5} = -1 \qquad \therefore\ x + y = -5$

Case (ii): $\tan\theta = \dfrac{-3}{4}$

Subcase (a): $\sin\theta = \dfrac{3}{5}$ and $\cos\theta = \dfrac{4}{5}$

equation of tangent, is, $\dfrac{\sin\theta}{3}y + \dfrac{\cos\theta}{4}x = 1$

$\Rightarrow \dfrac{4}{5}\cdot\dfrac{x}{4} - \dfrac{3}{5}\cdot\dfrac{y}{3} = 1 \quad \Rightarrow \dfrac{x}{5} - \dfrac{y}{5} = 1$

$\therefore\ x - y = 5$

Subcase (b): $\sin\theta = \dfrac{3}{5}$ and $\cos\theta = -\dfrac{4}{5}$

equation of tangent, is, $\dfrac{\sin\theta}{3}y + \dfrac{\cos\theta}{4}x = 1$

$\Rightarrow -\dfrac{4}{5}\cdot\dfrac{x}{4} + \dfrac{3}{5}\cdot\dfrac{y}{3} = 1 \Rightarrow \dfrac{x}{5} - \dfrac{y}{5} = -1$

$\therefore\ x - y = -5$

13. Suppose x and y are real numbers and that $x^2 + 9y^2 - 4x + 6y + 4 = 0$, then find the maximum value of $(4x - 9y)$.

Solution: $x^2 + 9y^2 - 4x + 6y + 4 = 0$

$\Rightarrow x^2 - 4x + 9y^2 + 6y + 4 = 0$

$\Rightarrow x^2 - 2.2.x + 4 + (3y)^2 + 2(3y)\,1 + 1 - 1 = 0$

$\Rightarrow (x-2)^2 + (3y+1)^2 = 1$

$\Rightarrow (x-2)^2 + \dfrac{\left(y+\dfrac{1}{3}\right)^2}{\dfrac{1}{9}} = 1$

which is an equation of ellipse, with centre at $\left(2, -\dfrac{1}{3}\right)$

General point on ellipse is.

$P(x, y) = (2 + a\cos\theta, -1/3 + b\sin\theta)$

$= (2 + \cos\theta, -1/3 + 1/3\sin\theta);\ x = 2 + \cos\theta$

$y = -1/3 + 1/3 \sin\theta$

$\therefore\ 4x - 9y = 4(2 + \cos\theta) - 9\left(-\frac{1}{3}+\frac{1}{3}\sin\theta\right)$

$\Rightarrow f(\theta) = 8 + 4\cos\theta + 3 - 3\sin\theta$

$= 11 + 4\cos\theta - 3\sin\theta$

$= 11 + 5\left(\frac{4}{5}\cos\theta - \frac{3}{5}\sin\theta\right)$

$= 11 + 5\cos(\theta + \alpha)$ where $\left[\cos\alpha = \frac{4}{5}\right]$

$\therefore\ f(\theta)_{max} = 11 + 5 = 16$

14. 'O' is the origin and also the centre of two concentric circles having radii of the inner and the outer circle as 'a' and 'b' respectively. A line OPQ is drawn to cut the inner circle in P and the outer circle in Q. PR is drawn parallel to the y-axis and QR is drawn parallel to the x-axis. Prove that the locus of R is an ellipse touching the two circles. If the focii of this ellipse lie on the inner circle, find the ratio of inner: outer radii and find also the eccentricity of the ellipse.

Solution: $x^2 + y^2 = a^2$ (inner) ...(1)

$x^2 + y^2 = b^2$ (outer) ...(2)

$\therefore$ P is $(a\cos\theta, a\sin\theta)$

Q is $(b\cos\theta, b\sin\theta)$

$\therefore$ The point R has x-corodinate of P and y-co-ordinate of Q.

$\therefore$ R is $x = a\cos\theta$, $y = b\sin\theta$

$\therefore$ locus of R is $\frac{x^2}{a^2}+\frac{y^2}{b^2}=1$, $(a < b)$ ...(3)

$a < b$ therefore OX is minor axis

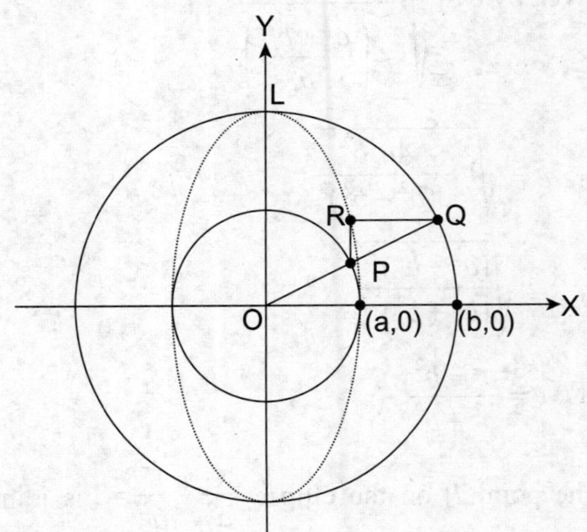

OY is major axis

$\therefore$ eccentricity $e = \sqrt{1-\frac{a^2}{b^2}}$ ...(4)

The equation no. (3) is an ellipse whose foci lie y-axis at (0, be) and (0, $-be$). If the foci lie on inner circle $x^2 + y^2 = a^2$ then $b^2e^2 = a^2$

but $a^2 = b^2(1 - e^2)$ (from equation (4))

$\Rightarrow b^2e^2 = b^2(1 - e^2) \quad \Rightarrow 2e^2 = 1$

$\Rightarrow e^2 = 1/2 \quad \Rightarrow e = \frac{1}{\sqrt{2}}$

Again we have to prove that the ellipse (3) touches both the circles. The point of intersection of (1) and (3) are given by $y^2\left(\frac{1}{a^2}-\frac{1}{b^2}\right)=0$

$\therefore\ y = 0$ (as $a \neq b$)

i.e., (a, 0), (–a, 0).

both equation (2) and (3) have the same tangents x = a, $x = -a$ at these points and hence they touches. Similarly (2) and (3) also touches.

Now, we have to find the ratio of inner and outer rad*ii* $= \frac{a}{b}$.

i.e., the ratio of minor axis to major axis of ellip*se* $= \frac{a}{b}$

$\therefore\ e = \sqrt{1-\left(\frac{a}{b}\right)^2}$

$\Rightarrow \left(-\right) = 1 - e^2 = 1 - \frac{1}{2} = \frac{1}{2}$

or $\frac{a}{b} = \frac{1}{\sqrt{2}}$

15. Find the equation of the largest circle with centre (1, 0) that can be inscribed in the ellipse $x^2 + 4y^2 = 16$.

Solution: Ellipse is $\frac{x^2}{16}+\frac{y^2}{4}=1$

so $a^2 = 16$, $b^2 = 4$

let the equation of the circle with centre $C(1, 0)$ be $(x - 1)^2 + y^2 = r^2$

since the circle is to be largest so it will touch the ellipse at some point $P\ (4\cos\theta, 2\sin\theta)$ where tangent is $\frac{x\cos\theta}{4}+\frac{y\sin\theta}{2}=1$

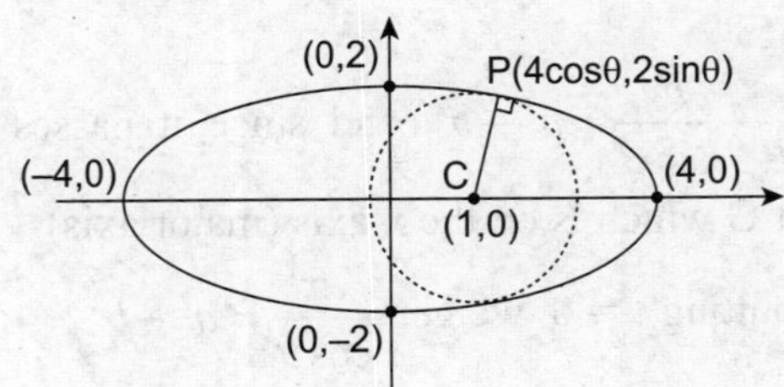

where slope is $m_1 = -\dfrac{1}{2}\dfrac{\cos\theta}{\sin\theta}$ ...(1)

Also m_2 = slope of $CP = \dfrac{2\sin\theta}{4\cos\theta - 1}$ ...(2)

but $m_1 m_2 = -1$

$\therefore \; -\dfrac{1}{2}\dfrac{\cos\theta}{\sin\theta} \times \dfrac{2\sin\theta}{(4\cos\theta - 1)} = -1$

$\therefore$ $\cos\theta = 4\cos\theta - 1$ or $\cos\theta = \dfrac{1}{3}$

$\therefore$ Radius of circle is

$\{(4\cos\theta - 1)^2 + 4\sin^2\theta\}^{1/2}$

or $\left\{\left(\dfrac{4}{3} - 1\right)^2 + 4\left(1 - \dfrac{1}{9}\right)\right\}^{1/2} = \sqrt{\dfrac{33}{9}} = \sqrt{\dfrac{11}{3}}$

$\therefore$ circle is $(x-1)^2 + y^2 = \dfrac{11}{3}$

16. PG is a normal to an ellipse $\dfrac{x^2}{a^2} + \dfrac{y^2}{b^2} = 1$ at P, G being on the major axis. GP is produced outwards to Q so that $PQ = GP$. Show that the locus of Q is an ellipse whose eccentricity is $\dfrac{a^2 - b^2}{a^2 + b^2}$.

Solution: Let $P(a\cos\theta, b\sin\theta)$ be any general point on the ellipse $\dfrac{x^2}{a^2} + \dfrac{y^2}{b^2} = 1$

$\therefore$ The equation of normal to ellipse at point P is given by

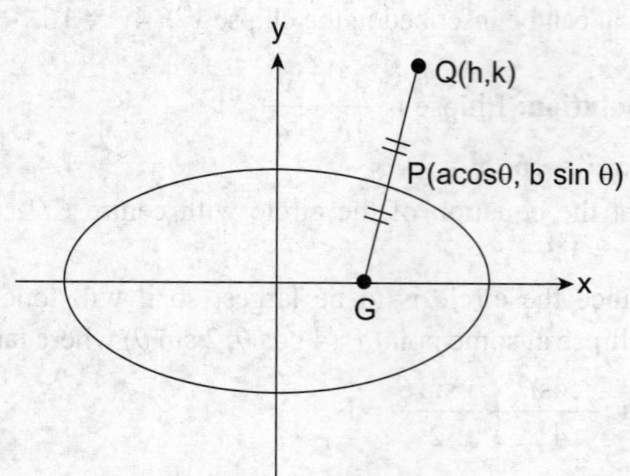

$\dfrac{ax}{\cos\theta} - \dfrac{by}{\sin\theta} = a^2 - b^2$ and since it passes through point G which is on the x-axis (major axis)

$\therefore$ putting $y = 0$, we get $\dfrac{ax}{\cos\theta} = a^2 - b^2$

or $x = \left(\dfrac{a^2 - b^2}{a}\right)\cos\theta$

$\therefore$ co-ordinates of G are $\left[\dfrac{(a^2 - b^2)}{a}\cos\theta, 0\right]$

Let co-ordinates of Q is (h, k)

since P is mid-point of QG.

Therefore. $a\cos\theta = \dfrac{h + \left(\dfrac{a^2 - b^2}{a}\right)\cos\theta}{2}$ ...(1)

and $b\sin\theta = \dfrac{k}{2}$ ...(2)

from equation (1); we get

$2a^2\cos\theta = ah + (a^2 - b^2)\cos\theta$

or $(a^2 - b^2 - 2a^2)\cos\theta = -ah$

or $\cos\theta = \dfrac{ah}{a^2 + b^2}$ ....(3)

And from (2), we get $\sin\theta = \dfrac{k}{2b}$ ...(4)

now $(3)^2 + (4)^2 \Rightarrow 1 = \dfrac{a^2h^2}{(a^2 + b^2)^2} + \dfrac{k^2}{4b^2}$

or $\dfrac{x^2}{\left(\dfrac{a^2 + b^2}{a}\right)^2} + \dfrac{y^2}{(2b)^2} = 1$

This locus represents an ellipse, whose eccentricity is given by $e = \sqrt{1 - \dfrac{4b^2}{\left(\dfrac{a^2 + b^2}{a}\right)^2}}$

or $e = \sqrt{1 - \dfrac{4a^2b^2}{(a^2 + b^2)^2}}$

or $e = \sqrt{\dfrac{(a^2 - b^2)^2}{(a^2 + b^2)^2}}$

or $e = \dfrac{a^2 - b^2}{a^2 + b^2}$

17. The point P on the ellipse $\dfrac{x^2}{a^2} + \dfrac{y^2}{b^2} = 1$ is joined to the ends A, A' of the major axis. If the lines through P perpendicular to PA, PA' meet the major axis in Q and R, then prove that (QR) = length of latus rectum.

Solution: Let P be any point (x_1, y_1) on the ellipse.

$\therefore$ Slope of $PA = \dfrac{y_1 - 0}{x_1 - a}$

$\therefore$ $(PQ)_{\text{slope}} = \dfrac{a - x_1}{y_1}$

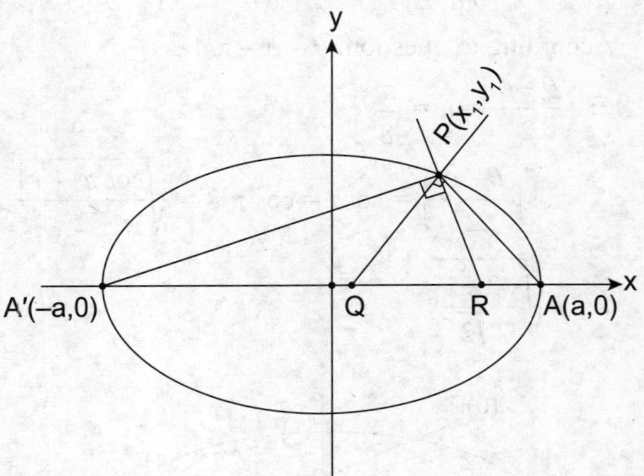

so, equation of PQ is $y - y_1 = \dfrac{a - x_1}{y_1}(x - x_1)$

Now PQ passes through Q point which is on the x-axis, therefore putting

$y = 0 \Rightarrow 0 - y_1 = \dfrac{a - x_1}{y_1}(x - x_1)$

$\Rightarrow x = \dfrac{-y_1^2}{a - x_1} + x_1$

$\therefore$ co-ordinates of Q are $\left[\dfrac{-y_1^2 - x_1^2 + ax_1}{a - x_1}, 0\right]$

Similarly, the co-ordinates of point R are $\left[\dfrac{y_1^2 + x_1^2 + ax_1}{x_1 + a}, 0\right]$

Now distance

$$QR = \sqrt{\left[\frac{y_1^2 + x_1^2 + ax_1}{x_1 + a} - \left(\frac{ax_1 - x_1^2 - y_1^2}{a - x_1}\right)\right]^2 + 0}$$

$$\text{or } QR = \frac{2ay_1^2}{a^2 - x_1^2} \qquad \text{...(1)}$$

Since point $P(x_1, y_1)$ satisfies the ellipse equation.

$\therefore$ $\dfrac{x_1^2}{a^2} + \dfrac{y_1^2}{b^2} = 1 \Rightarrow a^2 - x_1^2 = \dfrac{a^2 y_1^2}{b^2}$

putting this value of $a^2 - x_1^2$ in (i)

$\Rightarrow QR = 2ay_1^2 \times \dfrac{b^2}{a^2 y_1^2}$

$\Rightarrow QR = \dfrac{2b^2}{a}$ = length of the latus-rectum.

18. A tangent to the ellipse $x^2 + 4y^2 = 4$ meets the ellipse $x^2 + 2y^2 = 6$ at P and Q. Prove that the tangents at P and Q of the ellipse $x^2 + 2y^2 = 6$ are at right angles.

Solution: The equation of two ellipses are $\dfrac{x^2}{4} + \dfrac{y^2}{1} = 1$ ...(1)

and $\dfrac{x^2}{6} + \dfrac{y^2}{3} = 1$ ...(2)

Suppose the tangents at P and Q to the ellipse (2) intersect at R (h, k), then PQ is chord of contact of tangents drawn from $R(h, k)$ to the ellipse (2).
so the equation of PQ is

$$\frac{hx}{6} + \frac{ky}{3} = 1 \qquad \text{...(3)}$$

$\Rightarrow y = -\dfrac{hx}{2k} + \dfrac{3}{k}$

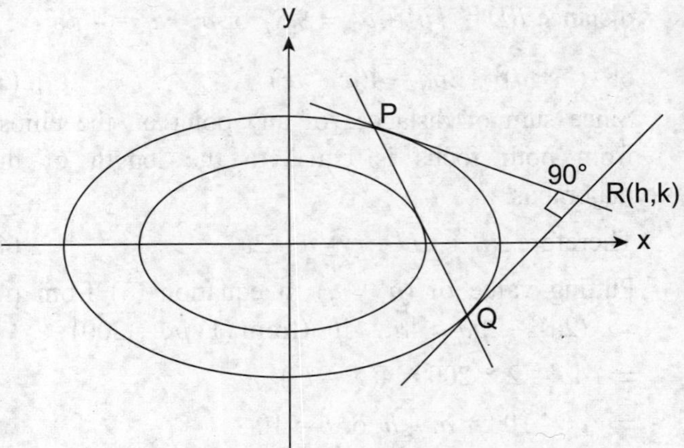

This touches the ellipse given in (1),
Therefore

$\dfrac{9}{k^2} = 4\left(\dfrac{-h}{2k}\right)^2 + 1$ (using $c^2 = a^2 m^2 + b^2$)

$\Rightarrow h^2 + k^2 = 9$

(h, k) lies on the circle $x^2 + y^2 = 9$
which is the director circle of ellipse (2).
And therefore, PR and QR intersect at 90°

19. Rectangle $ABCD$ has area 200 square units. An ellipse with area 200π square units passes through A and C and has foci at B and D. Find the perimeter of the rectangle.

Solution: Let the sides of rectangle be p and q

$\therefore$ Area of rectangle $= pq = 200$ ...(1)

Let the equation of the ellipse be $\frac{x^2}{a^2}+\frac{y^2}{b^2}=1$ with eccentricty 'e'

Now, area of ellipse $= \pi ab = 200\pi$

$\therefore$ $ab = 200$ ...(2)

We have to find out perimeter of rectangle i.e., $2(p + q)$. From triangle ABD

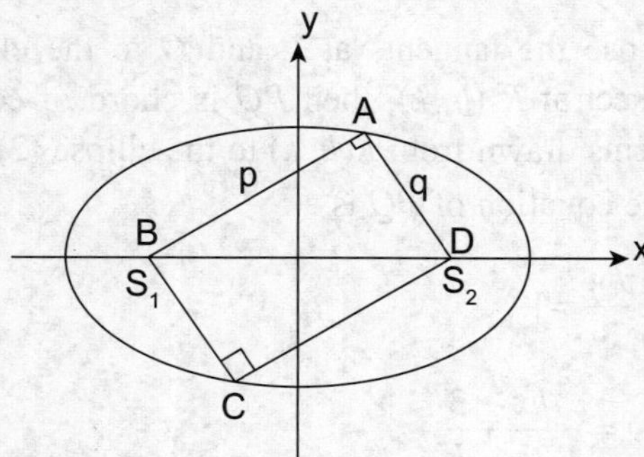

distance $BD = \sqrt{p^2+q^2} = S_1S_2$ or $p^2 + q^2 = 4a^2e^2$

or $(p + q)^2 - 2pq = 4(a^2 - b^2)$ ...(3)

Since sum of distances of any point on the ellipse from both focus is equal to the length of the major axis.

Therefore $AB + AD = p + q = 2a$ ...(4)

Putting value of $(p + q)$ in equation (3) from (4)

$\Rightarrow$ $(2a)^2 - 2pq = 4a^2 - 4b^2$ (from (1) $pq = 200$)

$\Rightarrow$ $4a^2 - 2 \times 200 = 4(a^2 - b^2)$

$\Rightarrow$ $a^2 - 100 = a^2 - b^2$ or $b = 10$

from equation (2) $\Rightarrow ab = 200$

or $10a = 200$ or $a = 20$

since $p + q = 2a$ (from equation (4))

therefore perimeter $= 2(p + q) = 4a = 4 \times 20 = 80$

20. Find the condition so that the line $px + qy = r$ intersects the ellipse $\frac{x^2}{a^2}+\frac{y^2}{b^2}=1$ in points whose eccentric angles differ by π/4.

Solution: Let the points of intersection of the line $px + qy = r$ intersect the ellipse $\frac{x^2}{a^2}+\frac{y^2}{b^2}=1$ at $P(\theta)$ and $Q(\phi)$

Equation of chord PQ is.

$$\cos\left(\frac{\phi+\theta}{2}\right)\frac{x}{a}+\sin\left(\frac{\phi+\theta}{2}\right)\frac{y}{b}=\cos\left(\frac{\phi-\theta}{2}\right)$$

According to question: $\theta - \phi = \pi/4$

$$\Rightarrow \frac{\theta-\phi}{2}=\alpha=\frac{\pi}{8}$$

$$\cos\left(\frac{\theta-\phi}{2}\right)=\cos\alpha=\cos\pi/8=\sqrt{\frac{\cos\pi/4+1}{2}}$$

$$=\sqrt{\frac{\sqrt{2}+1}{2\sqrt{2}}}$$

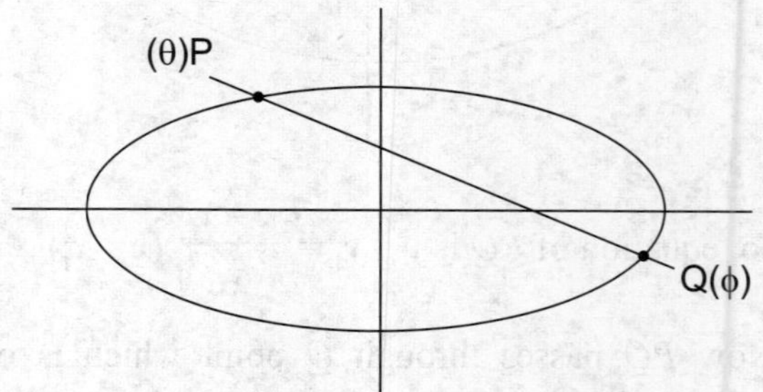

Equation of PQ,

$$\left(\frac{\cos\left(\frac{\theta+\phi}{2}\right)}{a}\right)x+\left(\frac{\sin\left(\frac{\theta+\phi}{2}\right)}{b}\right)y=\sqrt{\frac{\sqrt{2}+1}{2\sqrt{2}}} \quad \text{...(i)}$$

$px + qy = r$ ...(ii)

Equation (i) and (ii) are similar

$$\frac{\frac{\cos\left(\frac{\theta+\phi}{2}\right)}{a}}{p}=\frac{\frac{\sin\left(\frac{\theta+\phi}{2}\right)}{b}}{q}=\frac{\sqrt{\frac{\sqrt{2}+1}{2\sqrt{2}}}}{r}$$

$$\Rightarrow \cos\left(\frac{\theta+\phi}{2}\right)=\frac{pa}{r}\sqrt{\frac{\sqrt{2}+1}{2\sqrt{2}}};$$

$$\sin\left(\frac{\theta+\phi}{2}\right)=\frac{bq}{}\sqrt{\frac{\sqrt{2}\ \ 1}{2\sqrt{2}}}$$

$$\cos^2\left(\frac{\theta+\phi}{2}\right)+\sin^2\left(\frac{\theta+\phi}{2}\right)=1$$

$$\Rightarrow a^2p^2 + b^2q^2 = (4-2\sqrt{2})r^2$$

21. An archway is in the form of a semi ellipse, the major axis of which coincides with the road level. If the width of the road is 10 metres and a man 2 m high just reaches the top when 1 metre from a side of road, find the greatest height of the arch.

Solution: Let the arch way be a segment of ellipse $\frac{x^2}{a^2}+\frac{y^2}{b^2}=1$; $2a = 10$ and let origin be the centre of ellipse

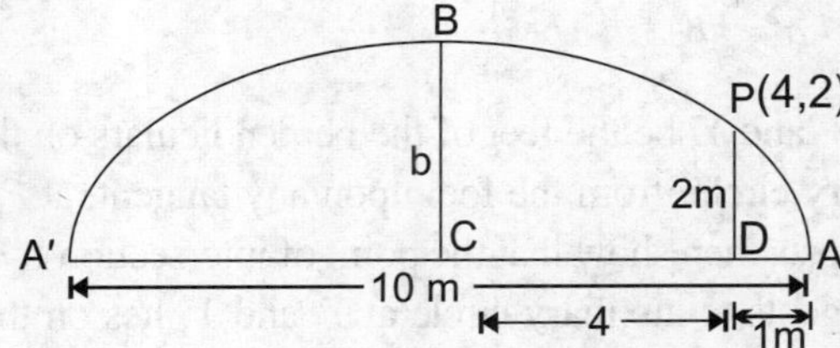

$\because$ Man of height 2 m just reaches the top of arch when at a point D 1 m from side A of road.

$\therefore$ point $P(4,2)$ lies on the elliptical arch

$\Rightarrow \frac{16}{25}+\frac{4}{b^2}=1 \Rightarrow \frac{4}{b^2}=1-\frac{16}{25}=\frac{9}{25}$

$\Rightarrow \frac{b^2}{4}=\frac{25}{9} \Rightarrow b^2=\frac{100}{9} \Rightarrow b=\frac{10}{3}$

The greatest height of arch $= BC = b = 10/3$ m.

22. Find the equation of the straight lines joining the foci of the ellipse $\frac{x^2}{25}+\frac{y^2}{16}=1$ to the foci of the ellipse $\frac{x^2}{24}+\frac{y^2}{49}=1$. Also find the area of the figure formed by the foci of these two ellipse.

Solution: Equation of ellipses are $\frac{x^2}{25}+\frac{y^2}{16}=1$ ...(i)

and $\frac{x^2}{24}+\frac{y^2}{49}=1$ ....(ii)

$\therefore$ for ellipse (i) $a^2 = 25$, $b^2 = 16$

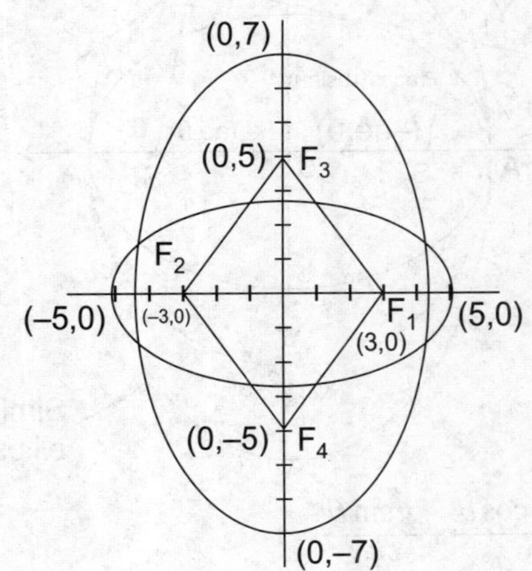

$\Rightarrow a^2e_1^2 = a^2 - b^2 \Rightarrow ae_1 = \pm 3$

and for ellipse (ii) $a^2 = 49$, $b^2 = 24$

$\Rightarrow a^2e_2^2 = a^2 - b^2 = 49 - 24$

$\Rightarrow a^2e_2^2 = 25 \Rightarrow ae_2 = \pm 5$

$\therefore$ foci of first ellipse are given by; $F_1 \equiv (3,0); F_2 \equiv (-3,0)$

and foci of second ellipse are given by $F_3 \equiv (0,5)$ and $F_4 \equiv (0,-5)$

$\therefore$ Area $F_1F_3F_2F_4$ (rhombus) $= \frac{6\times 10}{2} = 30$ sq. units]

23. Two tangents to an ellipse intersect at right angles. Prove that the sum of the square of the chords which the auxiliary circle intercepts on them is constant and is equal to the square of the line joining the foci.

Solution: Let the tangents be T and T'. Let Q be their point of intersection and AB, CD be the segments of T' and T intercepted by auxiliary circles as shown below

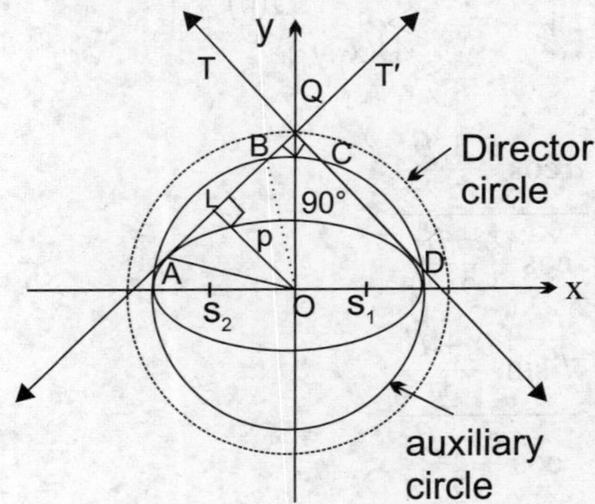

$\therefore$ We are to prove that $(AB)^2 + (CD)^2 = (S_1S_2)^2$

Let the equation of tangent T' be $y = mx + \sqrt{a^2m^2+b^2}$

or $mx - y + \sqrt{a^2m^2+b^2} = 0$

$\therefore$ perpendicular distance of T' from origin is given by $OL = \frac{\sqrt{a^2m^2+b^2}}{\sqrt{1+m^2}}$

$\therefore LB = \sqrt{(OB)^2-(OL)^2}$

$\Rightarrow AB = 2LB = 2\sqrt{a^2-\left(\frac{a^2m^2+b^2}{1+m^2}\right)}$

$\Rightarrow (AB)^2 = 4\left(\frac{a^2-b^2}{1+m^2}\right)$. Now, slope of $T = -1/m$

$\Rightarrow (CD)^2 = 4\left(\frac{a^2-b^2}{1+1/m^2}\right) = \frac{4(a^2-b^2)m^2}{1+m^2}$

$\therefore (AB)^2 + (CD)^2 = 4(a^2 - b^2) = 4a^2e^2 = (2ae)^2 = (S_1S_2)^2$

24. A parallelogram circumscribes the ellipse and two of its opposite angular points lie on the straight lines $x^2 = h^2$; prove that the locus of the other two is the conic $\frac{x^2}{a^2}+\frac{y^2}{b^2}\left(1-\frac{a^2}{h^2}\right)=1$

Solution: Let ABCD be the parallelogram circumscribing the ellipse $\frac{x^2}{a^2}+\frac{y^2}{b^2}=1$. Let $P(\alpha)$ and $Q(\beta)$ be the points of contact of tangents BC and CD respectively. Then the co-ordinates of point of intersection C are given by (h,k) as we assumed that A lies on line $x = -h$ and C lies on line $x = h$

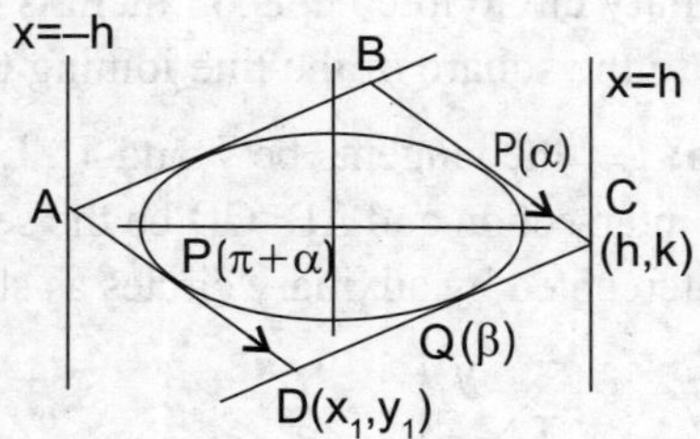

$$\therefore\; h=\frac{a\cos\frac{\alpha+\beta}{2}}{\cos\frac{\alpha-\beta}{2}} \quad ....(1)$$

$$\text{and } k=\frac{b\sin\frac{\alpha+\beta}{2}}{\cos\frac{\alpha-\beta}{2}} \quad ...(2)$$

On replacing α by $\pi + \alpha$, we will get the point of intersection of tangent at $Q(\beta)$ and $R(\pi + \alpha)$. Let it be (x_1,y_1)

$$\Rightarrow x_1=\frac{a\cos\frac{\pi+\alpha+\beta}{2}}{\cos\frac{\pi+\alpha-\beta}{2}};\; y_1=\frac{b\sin\frac{\pi+\alpha+\beta}{2}}{\cos\frac{\pi+\alpha-\beta}{2}}$$

$$\Rightarrow x_1=\frac{a\sin\frac{\alpha+\beta}{2}}{\sin\frac{\alpha-\beta}{2}} \quad ....(3)$$

$$y_1=-\frac{b\cos\frac{\alpha+\beta}{2}}{\sin\frac{\alpha-\beta}{2}} \quad ....(4)$$

$$\text{from (3) and (4), } \left(\frac{x_1^2}{a^2}+\frac{y_1^2}{b^2}\right)\sin^2\left(\frac{\alpha-\beta}{2}\right)=1 \quad ....(5)$$

$$\text{Also (1) and (4) gives, } \frac{h}{y_1}=-\frac{a}{b}\tan\left(\frac{\alpha-\beta}{2}\right)$$

$$\text{hence } \cot\left(\frac{\alpha-\beta}{2}\right)=-\frac{ay_1}{bh} \quad ...(6)$$

$$\therefore \text{ from (4), } \left(\frac{x_1^2}{a^2}+\frac{y_1^2}{b^2}\right)=1+\cot^2\left(\frac{\alpha-\beta}{2}\right)$$

$$=1+\frac{a^2y_1^2}{b^2h^2} \text{ (using (6))}$$

$\therefore$ Locus of other point of intersection $D(x_1,y_1)$, will be

$$\frac{x^2}{a^2}+\frac{y^2}{b^2}\left(1-\frac{a^2}{b^2}\right)=1$$

25. If Y_1 and Y_2 be the feet of the perpendiculars on the auxiliary circle from the foci upon any tangent, at P on the ellipse, then show that the point of intersection 'Q' of the tangents to auxiliary circle at Y_1 and Y_2 lies on the ordinate through P. If P varies i.e., θ varies, then show that the locus of Q is an ellipse having the same eccentricity as that of original ellipse.

Solution: Let the given ellipse be $\frac{x^2}{a^2}+\frac{y^2}{b^2}=1$; $(a > b)$ ....(i)

and auxiliary circle of ellipse (i) is given by $x^2 + y^2 = a^2$ ...(ii)

Let the tangents to auxiliary circle at points Y_1 and Y_2 intersect at $Q(h, k)$, then Y_1Y_2 will be the chord of contact of Q w.r.t auxiliary circle ...(ii)

Chord of contact w.r.t the circle $x^2 + y^2 = a^2$ is $hx + ky = a^2$ ....(1)

This must be the same as tangent at $P(\theta)$ on ellipse $\frac{x^2}{a^2}+\frac{y^2}{b^2}=1$

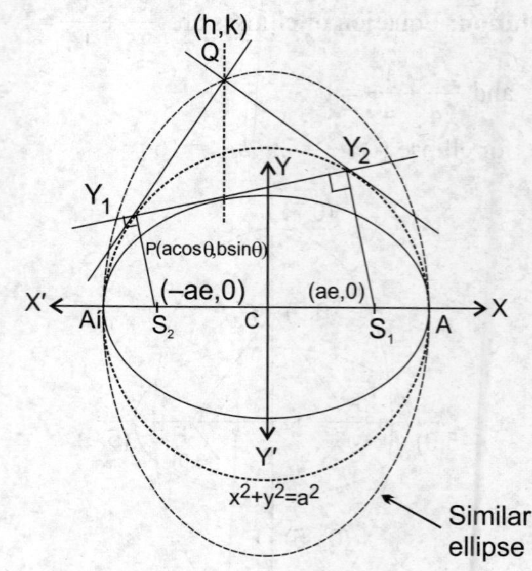

$$\text{i.e., } \frac{x\cos\theta}{a}+\frac{y\sin\theta}{b}=1 \quad ....(2)$$

Comparing (1) and (2), we have

$$h = a\cos\theta \quad \Rightarrow \quad k = \frac{a^2 \sin\theta}{b}$$

$$\therefore \left(\frac{h}{a}\right)^2 + \left(\frac{kb}{a^2}\right)^2 = \cos^2\theta + \sin^2\theta = 1;$$

Thus the locus of $Q(h,,k)$ will be given by $\frac{x^2}{a^2} + \frac{y^2}{\left(\frac{a^2}{b}\right)^2} = 1$

Which is an ellipse whose eccentricity e' is given by $e'^2 = 1 - \frac{a^2b^2}{a^4} = 1 - \frac{b^2}{a^2} = e^2$

Thus the locus of required point is also an ellipse with same eccentricity as that of given ellipse.

26. Lines joining centre to the feet of perpendicular from a focus on any tangent at P and the line joining other focus to the point of contant 'P' are parallel.

Solution: Let the given ellipse be $\frac{x^2}{a^2} + \frac{y^2}{b^2} = 1$ .Let $P(\theta)$ be the point on ellipse. Let N_2 be the foot of perpendicular drawn from focus S_2 to tangent at Point P. We are to prove CN_2 is parallel to S_1P.Now equation of tangent to ellipse at point $P(\theta)$ is given by

$$\frac{x\cos\theta}{a} + \frac{y\sin\theta}{b} = 1 \qquad .....(i)$$

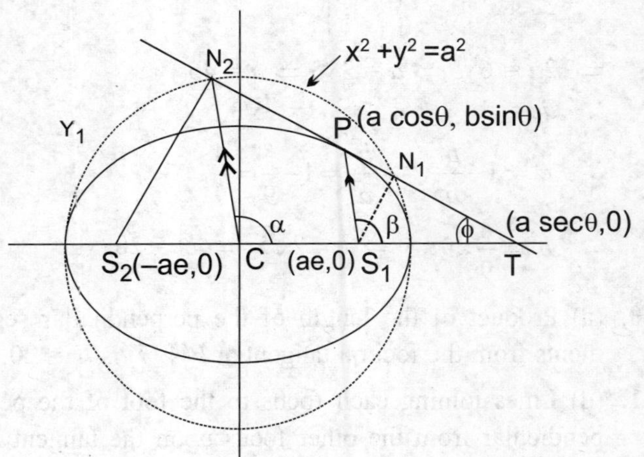

If T is the point of intersection of (i) and x-axis, then

$$\Rightarrow \frac{CT}{S_1T} = \frac{a\sec\theta}{a\sec\theta - ae} = \frac{a\sec\theta}{a(\sec\theta - e)}$$

$$\Rightarrow \frac{CT}{S_1T} = \frac{1}{1 - e\cos\theta}$$

$$\text{Again} \quad \frac{CN_2}{S_1P} = \frac{a}{e\left(\frac{a}{e} - a\cos\theta\right)}$$

$$= \frac{a}{a - ae\cos\theta} = \frac{1}{1 - e\cos\theta}$$

$$\Rightarrow \frac{CT}{S_1T} = \frac{CN_2}{S_1P}$$

$$\Rightarrow \Delta CN_2T \sim \Delta S_1PT$$

$$\Rightarrow \alpha = \beta$$

$$\Rightarrow CN_2 \| S_1P$$

Assertion and Reason Type

27. A : Locus of centre of a variable circle touching two circle $(x-1)^2 + (y-2)^2 = 25$ and $(x+2)^2 + (y-1)^2 = 16$ is an ellipse.

R : If a circle $S_1 = 0$ lies completely inside the circle $S_2 = 0$, then locus of centre of variable circle $S = 0$, which touches both the circles is an ellipse.

Solution: (d) Let C_1, C_2 the centres and R_1, R_2 be the radii of the two circles. Let $S_1 = 0$ lies completely inside the circle $S_2 = 0$. Let C and r be the centre and radius of the variable circle.

Then, $CC_2 = R_2 - r$ and $C_1C = R_1 + r$

$\therefore$ $C_1C + C_2C = R_1 + R_2$ (constant)

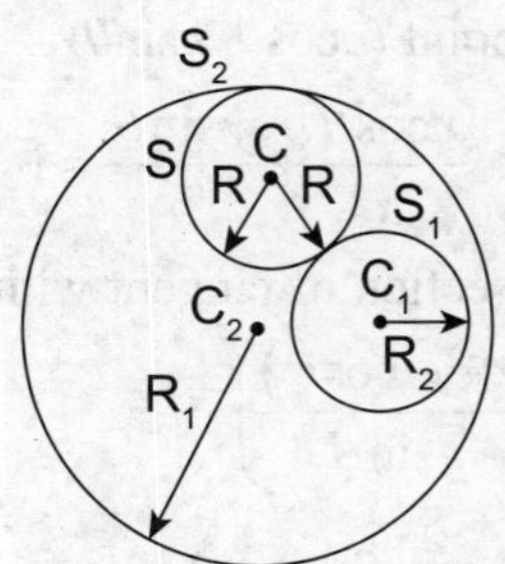

$\therefore$ locus of C is an ellipse

$\therefore$ Reason is true

Assertion is false. (Two circles are intersecting)

28. A: Feet of perpendiculars drawn from focii of an ellipse $4x^2 + y^2 = 16$ on the line $2\sqrt{3}x + y = 8$ lie on the circle $x^2 + y^2 = 16$.

R : If perpendiculars are drawn from focii of an ellipse to its any tangent, then feet of these perpendiculars lie on director circle of the ellipse.

Solution: (c) Simultaneously solving the equations of ellipse and the given line; we get

$4x^2 + (8 - 2\sqrt{3}x)^2 = 16$

$\Rightarrow x^2 - 2\sqrt{3}x + 3 = 0$

$\Rightarrow (x - \sqrt{3})^2 = 0$

$\therefore$ $2\sqrt{3}x + y = 8$ is a tangent to the ellipse, the auxiliary circle is $x^2 + y^2 = 16$.

$\therefore$ Assertion is true and reason is false.

29. A: Let tangent at a point P on the ellipse, which is not an extremity of major axis, meets a directrix at T. If circle drawn on PT as diameter cuts the directrix at Q, then $PQ = ePS$, where S is the focus corresponding to the directrix.

R : Let tangent at a point P on an ellipse, which is not an extremity of major axis, meets the directrix at T. Then, PT subtends a right angle at the focus corresponding to the directrix at which T lies.

Solution: (d) Since, $\angle PQT = \pi/2$ (Angle in a diameter)

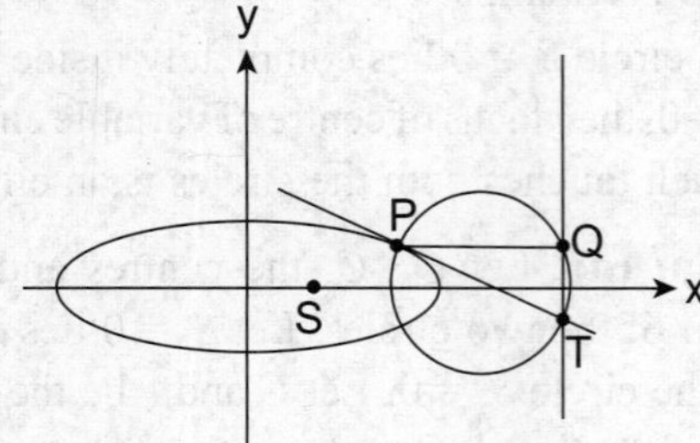

$\therefore$ PQ is perpendicular to the directrix

$\therefore$ By definition $PS = ePQ$

$\therefore$ Assertion is false.

Let P be the point $(a\cos\theta, b\sin\theta)$

Tangent at P : $\dfrac{x\cos\theta}{a} + \dfrac{y\sin\theta}{b} = 1$

Point of intersection of tangent with directrix

$$= T \equiv \left(a/e, \frac{b(e-\cos\theta)}{e\sin\theta}\right)$$

$$\therefore \quad M_{PS} = \frac{b\sin\theta}{a(\cos\theta - e)} \text{ and } M_{ST} = \frac{\dfrac{b(e-\cos\theta)}{e\sin\theta}}{\dfrac{a}{e} - ae}$$

$$m_{PS} \times m_{ST} = \frac{b\sin\theta}{a(\cos\theta - e)} \times \frac{b(e-\cos\theta)e}{e\sin\theta \times a(1-e^2)} = -1$$

$\therefore$ Reason is correct

Comprehension Type

A : Let the centre of an ellipse E be at (1, 3) and focus be at (6, 3) and passing through the point $P(4, 7)$, then

30. The product of the length of the perpendicular segments from the foci on tangent at point P is

(a) 20 (b) 55
(c) 30 (d) None of these

31. The point of intersection of the line joining each focus to the foot of the perpendicular from the other focus upon the tangent at point P, *is*.

(a) $\left(\dfrac{2}{3}, 4\right)$ (b) $\left(\dfrac{1}{3}, 3\right)$
(c) $\left(\dfrac{4}{7}, 3\right)$ (d) $\left(\dfrac{10}{3}, 5\right)$

32. If the normal at a variable point on the ellipse (E) meets its axes in Q and R, then the locus of the mid-point of QR is a conic with an eccentricity (e'), then

(a) $e' = \dfrac{3}{\sqrt{5}}$ (b) $e' = \dfrac{\sqrt{5}}{3}$
(c) $e' = \dfrac{1}{\sqrt{10}}$ (d) $e' = \dfrac{\sqrt{10}}{5}$

Solution: Equation of ellipse is $\dfrac{(x-1)^2}{45} + \dfrac{(y-3)^2}{20} = 1$ .... (i)

$\because$ $PS + PS' = 2a$

Now, $PS' = \sqrt{64+16} = 4\sqrt{5}$

and $PS = \sqrt{4+16} = 2\sqrt{5}$

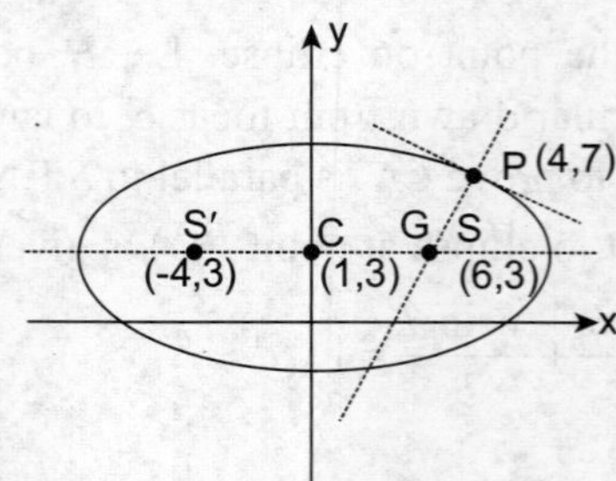

$\Rightarrow 2a = 6\sqrt{5} \Rightarrow a = 3\sqrt{5} \Rightarrow a = 45$

$\because$ $ae = 5$ $\quad e = \sqrt{5/3}$

$\therefore$ $e^2 = 1 - \dfrac{b^2}{a^2} \Rightarrow \dfrac{b^2}{a^2} = 1 - \dfrac{5}{9} = \dfrac{4}{9}$

$\therefore$ $b^2 = \dfrac{4}{9} \times 9 \times 5 \Rightarrow b = 2\sqrt{5} \Rightarrow b^2 = 20$

30. (a) Product of the length of the perpendicular segments from the foci on tangent at $P(4, 7)$ is $b^2 = 20$.

31. (d) Lines joining each focus to the foot of the perpendicular from the other focus upon the tangent at $P(4, 7)$ meet the normal PG and bisects it.

$\therefore$ Required point is mid-point of PG.

$\because$ Equation of normal at$P(4, 7)$ is given by

$3x - y - 5 = 0$

$\therefore$ $G\left(\dfrac{8}{3}, 3\right)$

$\therefore$ Required point is mid-point of PG, i.e., $\left(\dfrac{10}{3}, 5\right)$.

32. (b) For the ellipse $\frac{x^2}{a^2}+\frac{y^2}{b^2}=1$

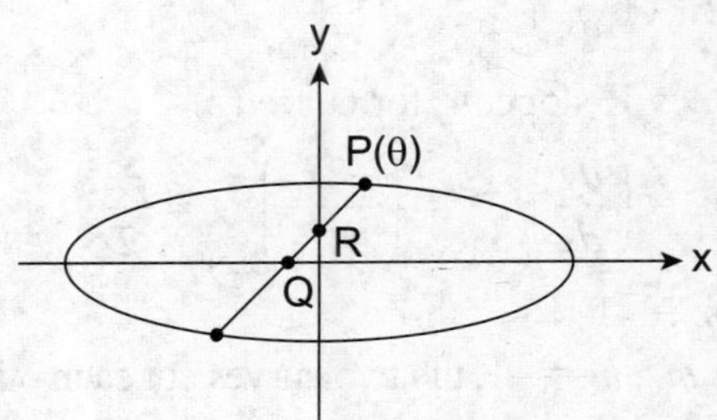

Let 'P' be any general point ($a\cos\theta$, $b\sin\theta$).

The normal at point P will be given by $(a\sec\theta)\,x-(b\cos\theta)\,y=a^2-b^2$

This normal will cut the major axis (i.e., x-axis) at $Q\left(\frac{a^2-b^2}{a\sec\theta},0\right)$

And minor axis at $R\left(0,\frac{a^2-b^2}{-b\operatorname{cosec}\theta}\right)$

Let the mid-point of QR be (h,k)

$$\therefore\quad h=\frac{\frac{a^2-b^2}{a\sec\theta}+0}{2}\text{ and } k=\frac{0+\frac{a^2-b^2}{-b\operatorname{cosec}\theta}}{2}$$

$\Rightarrow$ $2ha=(a^2-b^2)\cos\theta$ and $-2kb=(a^2-b^2)\sin\theta$

$$\Rightarrow \cos\theta=\frac{2ah}{a^2-b^2}\ \&\sin\theta=\frac{-2kb}{a^2-b^2}$$

Now; $\sin^2\theta+\cos^2\theta=1$

$$\Rightarrow\left(\frac{2ah}{a^2-b^2}\right)^2+\left(\frac{-2bk}{a^2-b^2}\right)^2=1$$

$$\Rightarrow\frac{h^2}{\frac{(a^2-b^2)^2}{4a^2}}+\frac{k^2}{\frac{(a^2-b^2)^2}{4b^2}}=1$$

$\therefore$ Locus of mid-point of QR will be given by

$$\frac{x^2}{\frac{(a^2-b^2)^2}{4a^2}}+\frac{y^2}{\frac{(a^2-b^2)^2}{4b^2}}=1$$

The above locus represents an ellipse whose eccentricity can be determined by

$$\frac{(a^2-b^2)^2}{4a^2}=\frac{(a^2-b^2)}{4b^2}(1-e^2)$$

$$\Rightarrow\frac{b^2}{a^2}=\left(1-(e')^2\right)\quad\Rightarrow (e')^2=1-\frac{b^2}{a^2}$$

$$\Rightarrow e'=\sqrt{1-\frac{b^2}{a^2}}$$

Which is same as the the eccentricity of the ellipse $\frac{x^2}{a^2}+\frac{y^2}{b^2}=1$

$\therefore$ Using the above derivation as a standard result, we can say that the eccentricity of the locus of the mid-point of the line QR will be the same as that of the given ellipse

i.e., $e=\sqrt{5}/3$

B: Let O be the centre of an ellipse (E): $\frac{x^2}{a^2}+\frac{y^2}{b^2}=1$ and having AB and CD as its major and minor axes respectively. If S_1 be one of the foci of the ellipse, radius of incircle of ΔOCS_1 be 1 unit and $OS_1=6$ unit, then.

33. The area of ellipse (E) is

(a) $\frac{65\pi}{4}$ (b) 16π

(c) $\frac{25\pi}{4}$ (d) 30π

34. Perimeter of ΔOCS_1 is

(a) 10 unit (b) 12 unit

(c) 15 unit (d) None of these

35. If the director circle of ellipse (E) be S, then the equation of director circle of S is

(a) $x^2+y^2=49$ (b) $x^2+y^2=\sqrt{194}$

(c) $x^2+y^2=97$ (d) $x^2+y^2=2\sqrt{97}$

Solution: $OS_1=ae=6$, $OC=b$ (suppose)

Also, Suppose $CS_1=a$

$\because$ Area of $\Delta OCS_1=\frac{1}{2}(OS_1)\times(OC)=3b$

$\because$ Semi-perimeter of $\Delta OCS_1=\frac{1}{2}(OS_1+OC+CS_1)$

$$=\frac{1}{2}(6+a+b)\qquad\text{.... (i)}$$

$\therefore$ Inradius of $\Delta OCS_1=1\Rightarrow\frac{3b}{\frac{1}{2}(6+a+b)}=1$

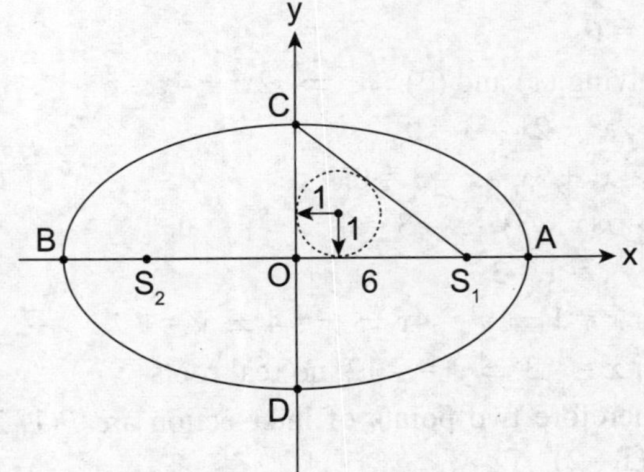

$\Rightarrow 5b = 6 + a$ ... (ii)

Also, $b^2 = a^2 - a^2e^2 = a^2 - 36$

$\therefore$ From *eq* (ii), $25b^2 = 36 + 12a + a^2$

$\therefore 25(a^2 - 36) = 36 + a^2 + 12a$

$\Rightarrow 2a^2 - a - 78 = 0 \therefore a = 13/2, -6$

When, $a = 13/2 \Rightarrow b = 5/2$

33. (a) Area of ellipse $= \pi ab = \dfrac{65\pi}{4}$ sq.units.

34. (c) Perimeter of $\Delta OCS_1 = 6 + a + b = 6 + 13/2 + 5/2 = 15$*unit*.

35. (c) S: $x^2 + y^2 = a^2 + b^2$

$\Rightarrow$ S: $x^2 + y^2 = 97/2 = r^2$

$\therefore$ equation of director circle of S is $x^2 + y^2 = 2r^2$.

$\therefore x^2 + y^2 = 97$.

C: Let P and Q be the points of intersection of the parabola $y^2 = 4x$ with the ellipse $2x^2 + y^2 = 6$. Then, on the basis of the information provided above, answer the questions that follow.

36. Find the angle at which the two curves intersect each other.

(a) $\dfrac{\pi}{6}$ (b) $\dfrac{\pi}{3}$

(c) $\dfrac{\pi}{4}$ (d) $\dfrac{\pi}{2}$

37. Find the area enclosed by the parabola and the common chord of the ellipse and parabola.

(a) 2 (b) 7/3

(c) 8/3 (d) None of these

38. If tangent and normal at the point P on the ellipse intersect the x-axis at T and G respectively, then find the area of ΔPTG.

(a) 4 (b) 5

(c) 9/2 (d) None of these

Solution: 36. (d) Parabola: $y^2 = 4x$ ...(1)

ellipse: $2x^2 + y^2 = 6$ ...(2)

$a^2 = 6/2 = 3$

$b^2 = 6$

solving (1) and (2) $\Rightarrow 2x^2 + 4x = 6$

$\Rightarrow x^2 + 2x - 3 = 0$

$\Rightarrow x^2 + 3x - x - 3 = 0$

$\Rightarrow x(x + 3) - x - 3 = 0$

$\Rightarrow x = 1, -3$

for $x = 1 \Rightarrow y^2 = 4x \Rightarrow y^2 = 4 \Rightarrow y = \pm 2$

for $x = -3 \Rightarrow y^2 = -12$, no real roots

Therefore two points of intersection are $P(1, 2)$ and $Q(1, -2)$

At point $P(1, 2) \Rightarrow$ slope m_1 for curve (1)

$$\Rightarrow m_1 = \left(\frac{dy}{dx}\right)_{(x_1, y_1)} = \left(\frac{2}{y}\right)_{(1,2)} = \frac{2}{2}$$

or $m_1 = 1$, slope m_2 for curve (2)

$$\Rightarrow m_2 = \left(\frac{dy}{dx}\right)_{(x_1, y_1)} = \left(\frac{-2x_1}{y_1}\right)_{(1,2)} = \frac{-2}{2}$$

or $m_2 = -1$

Since $m_1 . m_2 = -1$, the two curves cut each other at 90°.

37. (c) Area OPQ $= 2\int_0^1 ydx$

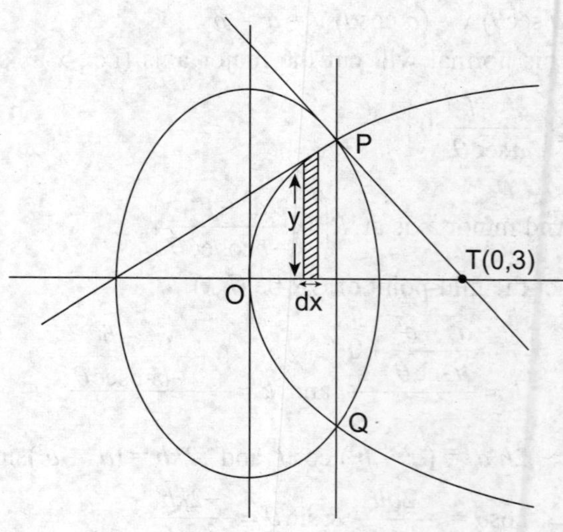

$$= 2\int_0^1 2\sqrt{x}\, dx = 4\left[\frac{2}{3}x\sqrt{x}\right]_0^1$$

$$= \frac{8}{3}[1 - 0] = \frac{8}{3} \text{ square units.}$$

38. Now tangent point $P(1,2)$ is given by $T = 0$

i.e., $2xx_1 + yy_1 - 6 = 0$

or $2x + 2y = 6$ or $x + y = 3$

$\therefore$ co-ordinates of point T $\Rightarrow (3, 0)$

Normal at point $P(1, 2)$ is given by

$x - y + k = 0$

Since it passes through $P(1, 2)$

$\therefore 1 - 2 + k = 0$

$\Rightarrow k = 1$

$\therefore$ normal $\Rightarrow x - y = -1$

co-ordinates of G $\Rightarrow (-1, 0)$

$$\text{Now area of triangle } PTG = \frac{1}{2}\begin{vmatrix} 1 & 2 & 1 \\ 3 & 0 & 1 \\ -1 & 0 & 1 \end{vmatrix}$$

$$= \frac{1}{2}[1(0) - 2(3 + 1) + 1(0)] = 4 \text{ sq. units}$$

Column matching Type

39. Column-I

(i) If P is any point on the ellipse $5x^2 + 4y^2 = 80$ whose foci are S and S', then sum of focal distances is

(ii) Eccentricity of the ellipse $2x^2 + 3y^2 - 4x - 12y + 13 = 0$ is

(iii) If from the points on the line $x - y - 5 = 0$ tangents are drawn to $x^2 + 4y^2 = 4$. Then, all the chords of contact pass through a fixed point having abscissa.

(iv) For the ellipse $\frac{x^2}{4}+\frac{y^2}{3}=1$, the minimum distance between the directrices is given by

Column-II

(a) 4/5

(b) $4\sqrt{5}$

(c) 8

(d) $1/\sqrt{3}$

Ans. (i) → (b); (ii) → (d); (iii) → (a); (iv) → (c)

Solution: (i) Here, $a^2 = 16,\ b^2 = 20$

$\Rightarrow a^2 < b^2$

$\therefore\ PS + PS' = 2b = 2\sqrt{20} = 4\sqrt{5}$

(ii) Given ellipse is $2x^2 + 3y^2 - 4x - 12y + 13 = 0$

$$\Rightarrow \frac{(x-1)^2}{1/2}+\frac{(y-2)^2}{1/3}=1$$

Here $a^2 = \frac{1}{2},\ b^2 = \frac{1}{3}$

$$\therefore e = \sqrt{1-\frac{1}{3}\times 2} = \frac{1}{\sqrt{3}}$$

(iii) Any point on the line $x - y - 5 = 0$ will be of the form $(t,\ t - 5)$. Chord of contact of this point w.r.t. curve $x^2 + 4y^2 = 4$ is

$tx + 4(t - 5)y - 4 = 0$

$\Rightarrow (-20y - 4) + t(x + 4y) = 0$

which is a family of straight lines. Each member of this family pass through point of intersection of straight lines.

$-20y - 4 = 0$ and $x + 4y = 0$

$\Rightarrow x = 4/5,\ y = -1/5$

(iv) $\frac{x^2}{4}+\frac{y^2}{3}=1$

$\Rightarrow a = 2,\ b = \sqrt{3}$ and $e = 1/2$

$\therefore$ Sum of distances $= \frac{2a}{e} = \frac{4}{1/2} = 8.$ units

TUTORIAL EXERCISE

SECTION–III

OBJECTIVE-TYPE (ONLY ONE COREECT ANSWER)

1. Equation $\sqrt{(x-2)^2+y^2}+\sqrt{(x+2)^2+y^2}>4$ represents
(a) Parabola (b) Ellipse
(c) Circle (d) Pair of straight lines

2. If the latus rectum of an ellipse be equal to half its minor axis, then its eccentricity is
(a) 3/2 (b) $\sqrt{3}/2$
(c) 2/3 (d) $\sqrt{2}/3$

3. If the eccentricity of an ellipse be 5/8 and the distance between its foci be 10, then its latus rectum is
(a) 39/4 (b) 12
(c) 15 (d) 37/2

4. The equation $\frac{x^2}{2-r}+\frac{y^2}{r-5}+1=0$ represents an ellipse, if
(a) $r>2$ (b) $2<r<5$
(c) $r>5$ (d) None of these

5. The equations of the tangents to the ellipse $9x^2+16y^2=144$ which passes through the point (2, 3) is
(a) $y=3,\ x+y=5$ (b) $y=-3,\ x-y=5$
(c) $y=4,\ x+y=3$ (d) $y=-4,\ x-y=3$

6. Find the equation of the chord of $x^2/36+y^2/9=1$ which is bisected at (2, 1)
(a) $x+2y+4=0$ (b) $x+2y-4=0$
(c) $x+4y-2=0$ (d) None of these

7. The equation of the tangents to the ellipse $4x^2+3y^2=5$, which are inclined at 60° to the axis of x are
(a) $y=\sqrt{3}x\pm\sqrt{\frac{65}{12}}$ (b) $y=\sqrt{3}x\pm\sqrt{\frac{12}{65}}$
(c) $y=\frac{x}{\sqrt{3}}\pm\sqrt{\frac{65}{12}}$ (d) None of these

8. The radius of the circle passing through the foci of the ellipse $\frac{x^2}{16}+\frac{y^2}{9}=1$, and having it centre at (0,3) is
(a) 4 (b) 3
(c) $\sqrt{12}$ (d) 7/2

9. The eccentricity of an ellipse with centre at the origin and axes along the co-ordinate axes, is 1/2. If one of the directrices is $x=4$, then the equation of the ellipse is
(a) $4x^2+3y^2=1$ (b) $3x^2+4y^2=12$
(c) $4x^2+3y^2=12$ (d) $3x^2+4y^2=1$

10. If the tangents are drawn to the ellipse $x^2+2y^2=2$, then the locus of the mid-point of the intercept made by the tangents between the co-ordinate axes is
(a) $\frac{1}{2x^2}+\frac{1}{4y^2}=1$ (b) $\frac{1}{4x^2}+\frac{1}{2y^2}=1$
(c) $\frac{x^2}{2}+\frac{y^2}{4}=1$ (d) $\frac{x^2}{4}+\frac{y^2}{2}=1$

11. Extremities of the latus rectum of the ellipse $\frac{x^2}{a^2}+\frac{y^2}{b^2}=1$ ($a>$b) having a given major axis $2a$ lies on
(a) $x^2=a(a-y)$ (b) $y^2=a(a+x)$
(c) $y^2=a(a-x)$ (d) None of these

12. P is a point on an ellipse $\frac{x^2}{a^2}+\frac{y^2}{b^2}=1$ having centre at point C. If M is the foot of perpendicular drawn from point P to the major axis and T is the intersection point of tangent at P with major axis, then geometric mean of CM and CT is?
(a) a (b) b
(c) $\sqrt{a^2-b^2}$ (d) $\sqrt{ab}$

13. C is the centre, BCB' the minor axis and S the focus (ae, 0) of the ellipse, $\frac{x^2}{a^2}+\frac{y^2}{b^2}=1$. B'S is produced to meet the ellipse again in P. If CP makes an angle ϕ with the x-axis, then tan ϕ equals
(a) $\frac{(1-e^2)^{3/2}}{e}$ (b) $\sqrt{1-e^2}$
(c) $\frac{(1-e^2)^{3/2}}{2e}$ (d) None of these

14. P and Q are corresponding points on the ellipse $\frac{x^2}{16}+\frac{y^2}{9}=1$, and the auxiliary circle respectively. The normal at P to the ellipse meets CQ in R where C is centre of the ellipse. Then $l(CR)$ equals

(a) 5 units (b) 7 units
(c) 10 units (d) None of these

15. The angle between the tangents to the ellipse $\frac{x^2}{a^2}+\frac{y^2}{b^2}=1$ and the circle $x^2 + y^2 = ab$ at their points of intersection is

(a) $\tan^{-1}\left(\frac{a+b}{\sqrt{ab}}\right)$ (b) $\tan^{-1}\frac{a-b}{\sqrt{ab}}$

(c) $\tan^{-1}\left(\frac{a-b}{ab}\right)$ (d) None of these

16. The eccentric angle of the point where the line, $5x - 3y = 8\sqrt{2}$ is a normal to the ellipse $\frac{x^2}{25}+\frac{y^2}{9}=1$ is

(a) $\frac{\pi}{3}$ (b) $\frac{\pi}{6}$

(c) $\frac{\pi}{4}$ (d) None of these

17. If the straight line $x = y\sqrt{3}$ cuts the ellipse $x^2 + y^2 + xy = 3$ at points P and Q, then the product of distance of P and Q from origin is

(a) $4 + \sqrt{3}$ (b) $\frac{12}{4-\sqrt{3}}$

(c) $\frac{12}{4+\sqrt{3}}$ (d) None of these

18. The locus of middle points of chords of an ellipse which are drawn through the positive end of the minor axis is

(a) $\frac{x^2}{a^2}+\frac{y^2}{b^2}=\frac{y}{b}$ (b) $\frac{x^2}{a}+\frac{y^2}{b}=by$

(c) $\frac{x^2}{a^2}+\frac{y^2}{b^2}=by$ (d) None of these

19. ABC is an isosceles triangle with its base BC twice its altitude. A point P moves within the triangle such that the square of its distance from BC is half the rectangle contained by its distances from the two sides. The locus of P is an ellipse with eccentricity

(a) $\sqrt{\frac{3}{4}}$ (b) $\sqrt{\frac{2}{3}}$

(c) $\sqrt{\frac{4}{5}}$ (d) None of these

20. A triangle ABC right angled at 'A' moves so that its sides touch the curve $\frac{x^2}{a^2}+\frac{y^2}{b^2}=1$ all the time. The locus of the point 'A' is

(a) $x^2+y^2=\sqrt{a^2+b^2}$ (b) $x^2 + y^2 = a^2b^2$
(c) $x^2 + y^2 = a^2 + b^2$ (d) None of these

21. The sum of the squares of the reciprocals of two perpendicular diameters of the ellipse $5x^2 + 4y^2 = 1$ is equal to

(a) $\frac{1}{4}$ (b) $\frac{3}{4}$

(c) $\frac{9}{4}$ (d) None of these

22. A conic passing through the point $A(1, 4)$ is such that the segment joining a point $P(x, y)$ on the conic and the point of intersection of the normal at P with the abscissa axis is bisected by the y-axis. The equation of the conic is

(a) $x^2 + 2y^2 = 18$ (b) $2x^2 + y^2 = 18$
(c) $3x^2 + 2y^2 = 32$ (d) None of these

23. The centre of the ellipse $\frac{(x+y-2)^2}{9}+\frac{(x-y)^2}{16}=1$ is

(a) (0, 0) (b) (1, 1)
(c) (1, 0) (d) (0, 1)

24. The curve represented by $x = 3(\cos t + \sin t)$, $y = 4(\cos t - \sin t)$ is

(a) Ellipse (b) Parabola
(c) Hyperbola (d) Circle

25. Minimum area of the triangle by any tangent to the ellipse $\frac{x^2}{a^2}+\frac{y^2}{b^2}=1$ with the co-ordinate axes is

(a) $\frac{a^2+b^2}{2}$ (b) $\frac{(a+b)^2}{2}$

(c) ab (d) $\frac{(a-b)^2}{2}$

26. The line $y = mx + c$ is a normal to the ellipse $\frac{x^2}{a^2}+\frac{y^2}{b^2}=1$, if c is equal to

(a) $-(2am + bm^2)$ (b) $\frac{(a^2+b^2)m}{\sqrt{a^2+b^2m^2}}$

(c) $-\frac{(a^2-b^2)m}{\sqrt{a^2+b^2m^2}}$ (d) $\frac{(a^2-b^2)m}{\sqrt{a^2+b^2}}$

27. The equation of the line passing through the centre and bisecting the chord $7x+ y -1 = 0$, of the ellipse $x^2+\frac{y^2}{7}=1$ is

(a) $x = y$ (b) $2x = y$
(c) $x = 2y$ (d) $x + y = 0$

28. P is a variable point on the ellipse $\frac{x^2}{a^2}+\frac{y^2}{b^2}=2$ whose foci are F_1 and F_2. The maximum area (in unit²) of the $\Delta\, PFF'$ is

(a) $2b\sqrt{a^2-b^2}$ (b) $\sqrt{2}b\sqrt{a^2-b^2}$

(c) $b\sqrt{a^2-b^2}$ (d) $2a\sqrt{a^2-b^2}$

29. If the tangents from the point $(\lambda,3)$ to the ellipse $\frac{x^2}{9}+\frac{y^2}{4}=1$ are at right $\angle s$ then λ is

(a) ±1 (b) ±3

(c) ±2 (d) None of these

30. The area of the quadrilateral formed by tangents at the end points of latus recta of the ellipse $\frac{x^2}{9}+\frac{y^2}{5}=1$ is

(a) 27/4 unit² (b) 9 unit²

(c) 27/2 unit² (d) 27 unit²

31. The maximum distance of the centre of the ellipse $\frac{x^2}{9}+\frac{y^2}{4}=1$ from the chord of contact of mutually perpendicular tangents of the ellipse is

(a) $\frac{9}{\sqrt{13}}$ (b) $\frac{3}{\sqrt{13}}$

(c) $\frac{6}{\sqrt{13}}$ (d) $\frac{36}{\sqrt{13}}$

32. For all admissible values of the parameter 'a' the straight line $2ax+y\sqrt{1-a^2}=1$ will touch an ellipse whose eccentricity is equal to

(a) $\frac{\sqrt{3}}{2}$ (b) $\frac{1}{\sqrt{3}}$

(c) $\frac{1}{\sqrt{2}}$ (b) $\sqrt{\frac{2}{3}}$

33. The locus of mid-points of a focal chord of the ellipse $\frac{x^2}{a^2}+\frac{y^2}{b^2}=1$,

(a) $\frac{x^2}{a^2}+\frac{y^2}{b^2}=\frac{ex}{a}$ (b) $\frac{x^2}{a^2}-\frac{y^2}{b^2}=\frac{ex}{a}$

(c) $x^2+y^2=a^2+b^2$ (d) None of these

34. Let $\frac{x^2}{a^2}+\frac{y^2}{b^2}=1$ be an ellipse and A, B, C, D are four points on the ellipse. Further if normal, drawn at these points are concurrent, then

(a) points A, B, C, D lie on a circle

(b) points A, B, C, D lie on an other ellipse

(c) points A, B, C, D will lie on hyperbola

(d) cannot be determined

35. If CF is the perpendicular from the centre C of the ellipse $\frac{x^2}{a^2}+\frac{y^2}{b^2}=1$ on the tangent at any point P, and G is the point when the normal at P meets the major axis, then $CF.\ PG=$

(a) a^2 (b) ab

(c) b^2 (d) b^3

36. If $\phi_1, \phi_2, \phi_3, \phi_4$ are the eccentric angles of four concyclic points on the ellipse $\frac{x^2}{a^2}+\frac{y^2}{b^2}=1$, then $\phi_1+\phi_2+\phi_3+\phi_4=\ldots\ (n\in N)$

(a) $n\pi$ (b) $n\frac{\pi}{2}$

(c) $(2n+1)\frac{\pi}{2}$ (d) $2n\pi$

37. Let P be any point on a directrix of an ellipse of eccentricity e. S be the corresponding focus and C the centre of the ellipse. The line PC meets the ellipse at A. The angle between PS and tangent at A is α then α, is equal to

(a) $\tan^{-1}e$ (b) $\pi/2$

(c) $\tan^{-1}(1-e^2)$ (d) None of these

38. The tangent at $(3\sqrt{3}\cos\theta, \sin\theta)$ is drawn to the ellipse $\frac{x^2}{27}+y^2=1$. Then the value of θ such that the sum of intercepts on axes made by the tangent is minimum is

(a) $\pi/3$ (b) $\pi/6$

(c) $\pi/8$ (d) $\pi/4$

39. P_1 and P_2 are the feet of altitudes drawn from the foci S_1 and S_2 respectively of the ellipse $\frac{x^2}{16}+\frac{y^2}{9}=1$ on one of its variable tangent. Maximum value of $(S_1P_1)(S_2P_2)$ is equal to

(a) 9 (b) 16

(c) 25 (d) 7

40. Maximum length of the chord of the ellipse $\frac{x^2}{a^2}+\frac{y^2}{b^2}=1$, such that eccentric angles of its extremities differ by $\pi/2$ is $(a>b)$

(a) $a\sqrt{2}$ (b) $b\sqrt{2}$

(c) $ab\sqrt{2}$ (b) None of these

41. If α, β are eccentric angles of the extremities of a focal chord of an ellipse, then eccentricity of the ellipse is

(a) $\frac{\cos\alpha+\cos\beta}{\cos(\alpha+\beta)}$ (b) $\frac{\sin\alpha-\sin\beta}{\sin(\alpha-\beta)}$

(c) $\sec\alpha+\sec\beta$ (d) $\frac{\sin\alpha+\sin\beta}{\sin(\alpha+\beta)}$

42. If $P(\theta)$, $Q\left(\theta+\frac{\pi}{2}\right)$ are points on ellipse and α is angle between normals at P and Q, Then

(a) $2\sqrt{1-e^2} = e\sin^2 2\theta.\tan\alpha$

(b) $2\sqrt{1-e^2} = e\sin^2\theta.\tan 2\alpha$

(c) $\sqrt{1-e^2} = 2e\sin^2 2\theta.\tan\alpha$

(d) $2\sqrt{1-e^2} = e^2\sin 2\theta.\tan\alpha$

43. If the focal distance of an end of the minor axis of an ellipse (refered to its axes as the axes of x and y respectively), is k and the distance between its foci is $2h$, then its equation is

(a) $\frac{x^2}{a^2}+\frac{y^2}{b^2}=1$ (b) $\frac{x^2}{k^2}+\frac{y^2}{k^2-h^2}=1$

(c) $\frac{x^2}{k^2}+\frac{y^2}{h^2-k^2}=1$ (d) $\frac{x^2}{k^2}+\frac{y^2}{k^2+h^2}=1$

44. The parametric representation of a point on the ellipse whose foci are $(-1, 0)$ and $(7, 0)$ and eccentricity $\frac{1}{2}$ is

(a) $(3+8\cos\theta, 4\sqrt{3}\sin\theta)$

(b) $(8\cos\theta, 4\sqrt{3}\sin\theta)$

(c) $(3+4\sqrt{3}\cos\theta, 8\sin\theta)$

(b) None of these

45. The semi-latus rectum of an ellipse is

(a) the arithmetic mean of the segments of its focal chord

(b) the geometric mean of the segments of its focal chord

(c) the harmonic mean of the segments of its focal chord

(d) None of these

46. Locus of all such points from where the tangents drawn to the ellipse $\frac{x^2}{a^2}+\frac{y^2}{b^2}=1$ are always inclined at 45°, is

(a) $(x^2+y^2-a^2-b^2)^2 = 4(a^2x^2+b^2y^1-1)$

(b) $(x^2+y^2-a^2-b^2)^2 = 4(a^2x^2+b^2y^2-1)$

(c) $(x^2+y^2-a^2-b^2)^2 = 4(b^2x^2+a^2y^2-1)$

(d) None of the above

47. The condition that the line $lx + my =$ n may be a normal to the ellipse $\frac{x^2}{a^2}+\frac{y^2}{b^2}=1$ is

(a) $\frac{n^2}{(a^2-b^2)^2}\left(\frac{a^2}{l^2}+\frac{b^2}{m^2}\right)=1$

(b) $\frac{n^2}{(a^2+b^2)^2}\left(\frac{a^2}{l^2}+\frac{b^2}{m^2}\right)=1$

(c) $\frac{n^2}{(a^2-b^2)^2}\left(\frac{a^2}{l^2}+\frac{b^2}{m^2}\right)=2$

(d) None of these

48. If the normal at θ on $\frac{x^2}{a^2}+\frac{y^2}{b^2}-1=0$ intersects the curve again at ϕ, then

(a) $\tan\theta\tan\phi=-1$

(b) $\tan\theta.\tan\left(\frac{\theta+\phi}{2}\right)=\frac{-b^2}{a^2}$

(c) $\tan\theta.\tan\left(\frac{\theta+\phi}{2}\right)=-\frac{a^2}{b^2}$

(d) None of these

49. If the chord of contact of the tangents drawn from the point (α, β) to the ellipse $\frac{x^2}{a^2}+\frac{y^2}{b^2}=1$ touches the circles $x^2+y^2=c^2$, then the locus of the point (α, β) is

(a) $\frac{x^2}{a^2}+\frac{y^2}{b^2}=\frac{1}{c^2}$ (b) $\frac{x^2}{a^2}+\frac{y^2}{b^2}=\frac{1}{c^4}$

(c) $\frac{x^2}{a^4}+\frac{y^2}{b^4}=\frac{1}{c^2}$ (d) None of these

50. S and S' are two foci of the ellipse $\frac{x^2}{a^2}+\frac{y^2}{b^2}=1$. The equation of the similar ellipse with S and S' as the ends of major axis is

(a) $\frac{x^2}{a^2e^2}+\frac{y^2}{b^2}=1$ (b) $\frac{x^2}{a^2}+\frac{y^2}{b^2e^2}=1$

(c) $\frac{x^2}{a^2}+\frac{y^2}{b^2}=e^2$ (d) $\frac{x^2}{1}+\frac{y^2}{(1-e^2)}=a^2e^2$

51. Number of distinct normal lines that can be drawn to ellipse $\frac{x^2}{169}+\frac{y^2}{25}=1$ from the point $P(0,6)$ is

(a) one (b) Two

(c) three (d) four

52. If $f(x)$ is a decreasing function, then the set of value of 'k', for which the major axis of the ellipse $\frac{x^2}{f(k^2+2k+5)}+\frac{y^2}{f(k+11)}=1$ is the x-axis, is:

(a) $k\in(-2, 3)$ (b) $k\in(-3, 2)$

(c) $k\in(-\infty,-3)\cup(2,\infty)$ (d) $k\in(-\infty,-2)\cup(3,\infty)$

53. The tangent at any point 'P' on the standard ellipse with foci as S and S' meets the tangents at the vertices A and A' in the points V and V' respectively, then which one does not holds good

(a) $(AV)(A'V')=b^2$

(b) $(AV)(A'V')=a^2$

(c) $\angle V'SV = 90°$

(d) $V'S'$ VS is a cyclic quadrilateral

54. x − 2y + 4 = 0 is a common tangent to $y^2 = 4x$ and $\frac{x^2}{4}+\frac{y^2}{b^2} = 1$. Then the value of '$b$' and the other common tangent are given by

(a) $b=\sqrt{3}$; $x + 2y + 4 = 0$

(b) $b = 3$; $x + 2y + 4 = 0$

(c) $b=\sqrt{3}$; $x + 2y - 4 = 0$

(d) $b=\sqrt{3}$; $x - 2y - 4 = 0$

55. An ellipse is such that the length of the latus rectum is equal to the sum of the lengths of its semi principal axes. Then:

(a) Ellipse bulges to a circle

(b) Ellipse becomes a line segment between the two foci

(c) Ellipse becomes a parabola

(d) None of these

56. A tangent is drawn to ellipse $x^2 + 2y^2 = 2$. Then the locus of mid-point of portion of the tangent intercepted between co-ordinate axes.

(a) $\frac{1}{2x^2}+\frac{1}{4y^2}=1$ (b) $\frac{1}{4x^2}+\frac{1}{2y^2}=1$

(c) $\frac{x^2}{2}+\frac{y^2}{4}=1$ (d) $\frac{x^2}{4}+\frac{y^2}{2}=1$

SECTION–IV

OBJECTIVE-TYPE (MORE THAN ONE COREECT ANSWER)

1. The locus of the foot of perpendicular drawn from either focus upon any tangent to the ellipse is

(a) auxiliary circle

(b) Director circle

(c) a circle concentric with the ellipse

(d) None of these

2. If P is any point on the ellipse $\frac{x^2}{a^2}+\frac{y^2}{b^2}=1$, whose foci are S and S'. Let $\angle PSS' = \alpha$ and $\angle PS'S = \beta$, then

(a) $PS + PS' = 2a$ if $a > b$

(b) $PS + PS' = 2b$ if $a < b$

(c) $\tan\frac{\alpha}{2}\tan\frac{\beta}{2}=\frac{1-e}{1+e}$

(d) $\tan\frac{\alpha}{2}\tan\frac{\beta}{2}=\frac{\sqrt{a^2-b^2}}{b^2}[a-\sqrt{a^2-b^2}\,]$

3. Let F_1, F_2 be two focii of the ellipse and PT and PN be the tangent and the normal respectively to the ellipse at point P then

(a) PN bisects $\angle F_1PF_2$

(b) PT bisects $\angle F_1PF_2$

(c) PT bisects angle $(180° - \angle F_1PF_2)$

(d) None of these

4. If S be the focus and G be the point where the normal at P meets the major axes of an ellipse, then

(a) $SG = e.\ SP$,

(b) tangent at P bisects the external angles between the focal distances of P.

(c) normal at P bisects the internal angles between the focal distances of P.

(d) None of these

5. The tangents from which of the following points to the ellipse $9x^2 + 4y^2 = 36$ are perpendicular

(a) $(1, 2\sqrt{3})$ (b) $(2\sqrt{3}, 1)$

(c) (2, 3) (d) (3, 2)

6. On the ellipse $4x^2+9y^2=1$, the points at which the tangents are parallel to the line $8x=9y$ are

(a) (2/5, 1/5) (b) (–2/5, 1/5)

(c) (–2/5, –1/5) (d) (2/5, –1/5)

7. Let A and B be two fixed (distinct) points and P be another point in the same plane, which moves in such a way that $\lambda_1 PA + \lambda_2 PB = \lambda_3$, where λ_1, λ_2, and λ_3 are real constants. The locus of P is

(a) a circle if $\lambda_1 = 0$ and $\lambda_2\ \lambda_3 > 0$

(b) a circle if $\lambda_1 > 0$, $\lambda_2 < 0$ and $\lambda_3 = 0$

(c) a ellipse if $\lambda_1 = \lambda_2 > 0$ and $\lambda_3\lambda_1 > AB$

(d) a hyperbola if $\lambda_2 = -1$ and $\lambda_1\ \lambda_3 > 0$

8. If ω is one of the angles between the normals to the ellipse $\frac{x^2}{a^2}+\frac{y^2}{b^2}=1$ at the points whose eccentric angles are θ and $\frac{\pi}{2}+\theta$, then $\frac{2\cot\omega}{\sin 2\theta}$ is

(a) $\frac{-e^2}{\sqrt{1-e^2}}$ (b) $\frac{e^2}{\sqrt{1+e^2}}$

(c) $\frac{e^2}{\sqrt{1-e^2}}$ (d) $\frac{e^2}{1+e^2}$

SECTION–V

ASSERTION AND REASON-TYPE

The questions, given below, consist of an assersion (A) and the reason (R). Use the following key to choose the appropriate answer.

(a) If both assertion and reason are correct and reason is the correct explanation of the assertion.
(b) If both assertion and reason are correct but reason is not correct explanation of the assertion.
(c) If assertion is correct, but reason is incorrect.
(d) If assertion is incorrect, but reason is correct.

Now, consider the following statements:

1. A: The equation $x^2 + 2\lambda xy + y^2 + 2x + 2y + 4 = 0$ represente an ellipse if $\lambda \in (-1, 1)$.
R: The general equation of second degree represents an ellipse if $\Delta \neq 0$, $h^2 < ab$.

2. A: In an ellipse the distance between foci is always less than the sum of focal distance of any point on it.
R: If e be the eccentricity of the ellipse then $0 < e < 1$.

3. A: The equation of the director circle to the ellipse $9x^2 + 16y^2 = 144$ is $x^2 + y^2 = 25$.
R: Director circle is the locus of point of intersection of perpendicular tangents to an ellipse.

4. A: The sum of focal distances of a point on the ellipse $9x^2 + 4y^2 - 18x - 24y + 9 = 0$ is 4.
R: The equation $9x^2 + 4y^2 - 18x - 24y + 9 = 0$ can be expressed as $9(x - 1)^2 + 4(y - 3)^2 = 36$.

5. A: The major and minor axes of the ellipse $5x^2 + 9y^2 - 54y + 36 = 0$ are 6 and 10 respectively.
R: The equation $5x^2 + 9y^2 - 54y + 36 = 0$ can be expressed as $5x^2 + 9(y - 3)^3 = 45$.

6. A: From the point $(\lambda, 3)$ tangents are drawn to $\frac{x^2}{9}+\frac{y^2}{4}=1$ are perpendicular to each other then $\lambda = \pm 2$.
R: The locus of point of intersection of perpendicular tangents to $\frac{x^2}{9}+\frac{y^2}{4}=1$ is $x^2 + 3y = 13$.

7. A: The condition on a and b for which two distinct chords of the ellipse $\frac{x^2}{2a^2}+\frac{y^2}{2b^2}=1$ passing through $(a, -b)$ are bisected by the line $x + y = b$ is $a^2 + 6ab - 7b^2 \geq 0$.
R: Equation of chord of the ellipse $\frac{x^2}{a^2}+\frac{y^2}{b^2}=1$ whose mid-point (x_1, y_1) is $T = S_1$.

8. A: Any chord of the ellilpse $x^2 + y^2 + xy = 1$ through $(0, 0)$ is bisected at $(0, 0)$.
R: The centre of an ellipse is a point through which every chord is bisected.

SECTION–VI

LINKED COMPREHENSION-TYPE

A: A sequence of ellipses $E_1, E_2, \ldots, E_n$ is constructed as follows: Ellipse E_n is drawn so as to touch ellipse E_{n-1} at the extremties of the major axis of E_{n-1} and to have its foci at the extremities of the minor axis of E_{n-1}.

1. If E_n is independent of n, then the eccentricity of ellipse E_{n-2} is

(a) $\left(\frac{3-\sqrt{5}}{2}\right)$ (b) $\left(\frac{\sqrt{5}-1}{2}\right)$
(c) $\frac{2-\sqrt{3}}{2}$ (d) $\frac{\sqrt{3}-1}{2}$

2. If eccentricity of ellipse E_n is e_n then the locus of (e_n^2, e_{n-1}^2) is
(a) a parabola
(b) an ellipse
(c) a non-rectangular hyperbola
(d) a rectangular hyperbola

3. If eccentricity of ellipse E_1 is $\sqrt{3}/2$, then the eccentricity of ellipse E_4 is
(a) $1/\sqrt{5}$ (b) $2/3$
(c) $3/\sqrt{23}$ (d) None of these

4. If equation of ellipse E_1 is $\frac{x^2}{9}+\frac{y^2}{16}=1$, then the equation of ellipse E_3 is

(a) $\frac{x^2}{9}+\frac{y^2}{16}=1$ (b) $\frac{x^2}{25}+\frac{y^2}{49}=1$

(c) $\frac{x^2}{25}+\frac{y^2}{41}=1$ (d) $\frac{x^2}{16}+\frac{y^2}{25}=1$

5. If eccentricity E_n is independent of n, then the locus of mid-point of chords of slope – 1 of E_n is (if axis of E_n along y-axis)

(a) $(\sqrt{5}-1)x=2y$ (b) $(\sqrt{5}+1)x=2y$

(c) $(3-\sqrt{5})x=2y$ (d) $(3+\sqrt{5})x=2y$

B: Consider the standard equation of an ellipse whose focus and corresponding foot of directrix are $(\sqrt{7}, 0)$ and $\left(\frac{16}{\sqrt{7}},0\right)$ and a circle with equation $x^2 + y^2 = r^2$. If in the first quadrant, the common tangent to a circle of this family and the above ellipse meets the co-ordinate axes at A and B.

6. The equation of the ellipse is

(a) $16x^2 + 9y^2 = 144$ (b) $9x^2 + 16y^2 = 144$

(c) $16x^2 + y^2 = 144$ (d) $x^2 + 9y^2 = 144$

7. Let P be a variable point on the ellipse with foci at S ad S' If Δ be the area of triangle PSS, then the maximum value of Δ is

(a) $\sqrt{7}$ sq unit (b) $2\sqrt{7}$ sq unit

(c) $3\sqrt{7}$ sq uint (d) $4\sqrt{7}$ sq uint

8. If mid-point of A and B is (x_1, y_1) and slope of common tangent be m, then

(a) $2mx_1 + y_1 = 0$ (b) $2my_1 + x_1 = 0$

(c) $my_1 + x_1 = 0$ (d) $mx_1 + y_1 = 0$

9. The locus of mid-point of A and B is

(a) $y=x\sqrt{\left(\frac{r^2-9}{16-r^2}\right)}$ (b) $y=x\sqrt{\left(\frac{16+r^2}{9-r^2}\right)}$

(c) $y=x\sqrt{\left(\frac{16-r^2}{r^2+9}\right)}$ (d) $y=x\sqrt{\left(\frac{r^2-9}{r^2-19}\right)}$

10. The domain of value of r $(r > 0)$ for which the above question is true is?

(a) $r \in (1, 2)$ (b) $r \in (3, 4)$

(c) $r \in (4, 5)$ (d) $r \in (5, 6)$

C: The standard equation of an ellipse and the general equation of a circle are respectively given by the equation.

$\frac{x^2}{a^2}+\frac{y^2}{b^2}-1=0$ and $C: x^2+y^2+2gx+2fy+c=0$,

then the equation $E + \lambda C = 0, \lambda \neq 0$

i.e., $\left(\frac{x^2}{a^2}+\frac{y^2}{b^2}-1\right)+\lambda(x^2+y^2+2gx+2fy+c)=0$,

represents a curve which passes through the common points of the ellipse and the circle.

11. If $\alpha, \beta, \gamma, \delta$ be the eccentric angles of the four concyclic points on the ellipse E, then $\alpha + \beta + \gamma + \delta$ is equal to

(a) $n\pi, n \in I$ (b) $(n + 1)\pi n \in I$

(c) $2n\pi, n \in I$, (d) $(2n + 1)\pi, n \in I$

12. If common chords of E and C are inclined at angles θ and ϕ with the axis of E, then

(a) $\theta = \phi$ (b) $\theta + \phi = 2\tan^{-1}\left(\frac{b}{a}\right)$

(c) $\theta + \phi = \frac{\pi}{2}$ (d) $\theta + \phi = \pi$

13. The radius of the circle passing through the points of intersection E and bisectors of the quadrants is

(a) $\frac{ab}{\sqrt{(a^2+b^2)}}$ (b) $\frac{ab\sqrt{2}}{\sqrt{(a^2+b^2)}}$

(c) $\frac{2ab}{\sqrt{(a^2+b^2)}}$ (d) $\frac{a^2-b^2}{\sqrt{(a^2+b^2)}}$

14. Let the eccentric angles of three points P, Q and R on the ellipse E are $\alpha, \frac{\pi}{2}+\alpha$ and $\pi + \alpha$. A circle C through P, Q and R cuts the ellipse E again at S, then the eccentric angle of S is

(a) $\frac{\pi}{2}-3\alpha$ (b) $\pi-3\alpha$

(c) $\frac{3\pi}{2}-3\alpha$ (d) $2\pi-3\alpha$

D: $9x^2 + 16y^2 = 144$, $y^2 - x + 4 = 0$ and $x^2 + y^2 - 12x + 32 = 0$ are the equations of an ellipse, a parabola and a circle respectively.

15. The parabola and the ellipse

(a) intersect in four points

(b) intersect in two points

(c) touch each other in one point

(d) They do not intersect

16. Maximum number of tangents that can be drawn from a point on the circle to the ellipse is

(a) 0 (b) 1

(c) 2 (d) None of these

17. The equation of the common tangent if any, to all the three above given curves is

(a) $x = 2$ (b) $x = 4$

(c) $y = 4$ (d) None of the above

SECTION–VII

MATRIX-MATCH TYPE

1. Column-I

(i) The minimum and maximum distance of a point (2, 6) from the ellipse $9x^2 + 8y^2 - 36x - 16y - 28 = 0$ are L and G, then

(ii) The minimum and maximum distance of a point (1, 2) from the ellipse $4x^2 + 9y^2 + 8x - 36y + 4 = 0$ are L and G, then

(iii) The minimum and maximum of a point $\left(\frac{9}{5}, \frac{12}{5}\right)$ from the ellipse $4(3x + 4y)^2 + 9(4x - 3y)^2 = 900$ are L and G, then

Column-II

(a) $L + G = 10$

(b) $L + G = 6$

(c) $G - L = 6$

(d) $G - L = 4$

(e) $L^G + G^L = 6$

2. Column-I

(i) A ray emanating from the point $(0, -\sqrt{5})$ is incident on the ellipse $9x^2 + 4y^2 = 36$ at the point P with abscissa 2. The equations of reflected ray after first reflection and second reflection are represented by F and S respectively, then

(ii) A ray emanating from the point $(\sqrt{5}, 0)$ is incident on the ellipse $4x^2 + 9y^2 = 36$ at the point P with ordinate -2. The equations of reflected rays after first reflection and second reflection are represented by F and S respectively, then

(iii) A ray emanating from the point $(-\sqrt{5}, 0)$ is incident on the ellipse $4x^2 + 5y^2 = 100$ at the point P with ordinate $2\sqrt{5}$. The equations of reflected rays after after first reflection and second reflection are represented by F and S respectively, then

Column-II

(a) $F: 2x + y - 2\sqrt{5} = 0$

(b) $S: x + 4\sqrt{5}y - \sqrt{5} = 0$

(c) $F: x\sqrt{5} + 2y - 2\sqrt{5} = 0$

(d) $S: 4\sqrt{5}x + y + \sqrt{5} = 0$

(e) $F: 2x + y\sqrt{5} + 2\sqrt{5} = 0$

(f) $S: x + 2y + \sqrt{5} = 0$

3. Column-I

(i) If $E: 2x^2 + y^2 = 2$ and director circle of E is C_1, director circle of C_1 is C_2, director circle of C_2 is C_3 and so on. If r_1, r_2, r_3.... are the radii of C_1, C_2, C_3 ... respectively. Then

(ii) IF $E: 3x^2 + 2y^2 = 6$ and director circle of E is C_1, director circle of C_1 is C_2, director circle of C_2 is C_3 and so on. If r_1, r_2, r_3..... are the radii of C_1, C_2, C_3...... respectively. Then

(iii) If $E: 3x^2 + 4y^2 = 12$ and director circle of E is C_1, director circle of C_1 is C_2, director circle of C_2 is C_3 and so on. If r_1, r_{2}, r_3........ are the radii of C_1, C_2, C_3...... respectively. Then

Column-II

(a) r_1^2, r_2^2, r_3^2.... are in GP

(b) GM of r_1^2, r_2^2, r_3^2 is 6

(c) $r_1^2 + r_2^2 + r_3^2 + ... + r_n^2 = r_1^2(2^n - 1)$

(d) GM of r_1^2, r_2^2, r_3^2 is 10

(e) GM of r_1^2, r_2^2, r_3^2 is 14

4. Column-I

(i) A point on the ellipse $4x^2 + 9y^2 = 36$ where the normal is parallel to the line $9x = 8y$ is

(ii) A point on the ellipse $4x^2 + 9y^2 = 36$; the tangent at which makes intercepts of equal length on the co-ordinate axes is

(iii) The point of contact of a tangent to the ellipse $9x^2 + 16y^2 = 144$ with a positive slope making an intercept of 5 units on the y-axis is

(iv) The mid-point of the chord of contact of tangents from (2,1) to the ellipse $x^2 + 2y^2 = 2$.

Column II

(a) $\left(-\frac{16}{5}, \frac{9}{5}\right)$

(b) $\left(\frac{2}{3}, \frac{1}{3}\right)$

(c) $\left(\frac{9}{\sqrt{13}}, \frac{4}{\sqrt{13}}\right)$

(d) $\left(\frac{12}{5}, \frac{6}{5}\right)$

5. **Column-I**

(i) A stick of length 10 meter slides on co-ordinate axes. The locus of a point, dividing this stick reckoning from x-axis in the ratio 6:4, is a curve whose eccentricity is e, then the $9e$ is equal to

(ii) AA' is axis of an ellipse $3x^2 + 2y^2 + 6x - 4y - 1 = 0$ and P is variable point on it, then greatest area of triangle APA' is

(iii) Distance between foci of the curve represented by the equation $x = 1 + 4\cos\theta$, $y = 2 + 3\sin\theta$ is

(iv) Tangents are drawn to the ellipse $\frac{x^2}{16}+\frac{y^2}{7}=1$ at end points of latus rectum. The area of quadrilateral so formed is

Column-II

(a) $\sqrt{6}$

(b) $2\sqrt{7}$

(c) 128/3

(d) $3\sqrt{5}$

SECTION–VIII

INTEGER-TYPE

1. Find the maximum value of $5h$ for which four normals can be drawn to ellipse $\frac{x^2}{25}+\frac{y^2}{16}=1$ through a point $(h, 0)$.

2. If 'e' be the eccentricity of the ellipse $(x - 3)^2 + (y - 4)^2 = y^2/9$; then evaluate '$9e$'.

3. Let E_1 and E_2 be two ellipse. The area of the ellipse E_2 is one-thrid the area of the quadrilateral formed by the tangents at the ends of the latusrectum of the ellipse $E_3 : 5x^2 + 9y^2 = 45$. The eccentricities of E_1, E_2 and E_3 are equal. E_1 is inscribed in E_2 in such a way that both E_1 and E_2 touch each other at one end of their common major axis. If the length of the major axis of E_1 is equal to the length of the minor axis of E_2, then find the aea of the ellipse E_2 outside the ellipse E_1.

4. If the normals at the four point (x_1,y_1) (x_2, y_2) (x_3, y_3) and (x_4, y_4) on the ellipse $\frac{x^2}{a^2}+\frac{y^2}{b^2}=1$ are concurrent, then find the value of $(x_1 + x_2 + x_3 + x_4)$. $\left(\frac{1}{x_1}+\frac{1}{x_2}+\frac{1}{x_3}+\frac{1}{x_4}\right)$.

5. A straight line PQ touches the ellipse $\frac{x^2}{16}+\frac{y^2}{9}=1$ and the circle $x^2+ y^2 = r^2 (3 < r < 4)$ RS is a focal chord of the ellipse which is parallel to PQ and meets the circle at points R and S. Then find the length of RS.

6. Let two concentic ellipses have their major axis equal in length and foci of each lie on other ellipse. Let their eccentricities be $1/\sqrt{2}$ and $\sqrt{3}/2$. If θ is the angle between their major axes, and it is given by $\cos^{-1}\left(\sqrt{\frac{2}{k}}\right)$, then evaluate k.

7. If two concentric elllipses having same length of major axis and foci of each lying on other have their eccentricities satisfying the relation $e_1^2 + e_2^2 = 1$, then find the numerical value of $\theta/10$; where θ is the angle between their major axis in degrees.

8. There are exactly two points on the ellipse $\frac{x^2}{a^2}+\frac{y^2}{b^2}=1$ whose distances from the centre of the ellipse are equal and equal to $\sqrt{\frac{a^2+3b^2}{3}}$. Then evaluate $27e^2$, where e is the eccentricity of the ellipse.

9. An ellipse with major and minor axis $6\sqrt{3}$ and 6 respectively slides along the co-ordinate axes and always remains confined in the first quadrant. If the length of arc desccribed by centre of ellipse is $\frac{k\pi}{6}$; then evaluate k.

10. A rod of fixed length $a + b$ slides between two $\perp r$ walls. Then the eccentricity of the conic described by the point on rod dividing the rod in the ratio $k : b$, when $a = 4$ and $b = 3$ is $\sqrt{7}/4$; then evaluate k.

11. If the area of the triangle formed by the three points on an ellipse, whose eccentric angle are θ, ϕ, and ψ, is Kab $\sin\frac{\varphi-\psi}{2}\sin\frac{\psi-\theta}{2}\sin\frac{\theta-\varphi}{2}$, then find the value of K.

12. Let P be a point on the ellipse $\frac{x^2}{a^2}+\frac{y^2}{b^2}=1, 0 < b < a$. Let the line parallel to y-axis passing through P meet the circle $x^2 + y^2 = a^2$ at the point Q such that P and Q are on

the same side of x-axis. For two positive real numbers r and s, and if the locus of the point R on PQ such that $PR : RQ = r : s$ as P varies over the ellipse is given by $\frac{x^2}{a^2}+\frac{y^2(r+s)^2}{(ra+sb)^2} = 1k$, then find the value of k.

13. Find the value of 'k' for which the length of the focal chord of the ellipse $\frac{x^2}{a^2}+\frac{y^2}{b^2}=1$ which makes an angle θ with the major axis is $\frac{kab^2}{a^2\sin^2\theta+b^2\cos^2\theta}$.

Answer Keys

SECTION–III

1. (b) **2.** (b) **3.** (a) **4.** (b) **5.** (a) **6.** (b) **7.** a) **8.** (a) **9.** (b) **10.** (a)
11. (a) **12.** (a) **13.** (c) **14.** (b) **15.** (b) **16.** (c) **17.** (c) **18.** (a) **19.** (b) **20.** (c)
21. (c) **22.** (b) **23.** (b) **24.** (a) **25.** (c) **26.** (c) **27.** (a) **28.** (a) **29.** (c) **30.** (d)
31. (a) **32.** (a) **33.** (a) **34.** (c) **35.** (c) **36.** (d) **37.** (b) **38.** (b) **39.** (a) **40.** (a)
41. (d) **42.** (d) **43.** (b) **44.** (a) **45.** (c) **46.** (d) **47.** (a) **48.** (b) **49.** (c) **50.** (c)
51. (c) **52.** (b) **53.** (b) **54.** (a) **55.** (a) **56.** (a)

SECTION–IV

1. (a, c) **2.** (a, b, c) **3.** (a, c) **4.** (a, b, c) **5.** (a, b, c, d) **6.** (b,d) **7.** (a, c) **8.** (a, c)

SECTION–V

1. (d) **2.** (a) **3.** (a) **4.** (d) **5.** (d) **6.** (c) **7.** (a) **8.** (a)

SECTION–VI

1. (b) **2.** (d) **3.** (d) **4.** (c) **5.** (b) **6.** (b) **7.** (c) **8.** (d) **9.** (a) **10.** (b)
11. (c) **12.** (d) **13.** (b) **14.** (a) **15.** (c) **16.** (c) **17.** (b)

SECTION–VII

1. (i) → (a, c) (ii) → (b, d, e) (iii) → (b, c),
2. (i) → (c, d) (ii) → (b, e) (iii) → (a), (f)
3. (i) → (a, b, c) (ii) → (a, c, d) (iii) → (a, c, e).
4. (i) → (d) (ii) → (c) (iii) → (a) (iv) → (b)
5. (i) → (d) (ii) → (a) (iii) → (b) (iv) → (c)

SECTION–VIII

1. 9 **2.** 3 **3.** 4 **4.** 4 **5.** 6 **6.** 3 **7.** 9 **8.** 9 **9.** 6 **10.** 4
11. 2 **12.** 1 **13.** 2

HINTS AND SOLUTIONS

TEXTUAL EXERCISE-1 (SUBJECTIVE)

1. (i) $\frac{x^2}{9}+\frac{y^2}{4}=1$; major axis → x-axis, minor axis → y-axis,

Length of $L.R = \frac{2b^2}{a}=\frac{8}{3}$ units,

$e=\sqrt{\frac{a^2-b^2}{a^2}}=\frac{\sqrt{5}}{3}$, so foci $(\pm\sqrt{5}, 0)$

Equations of directrice $x=\pm\frac{a}{e}=\pm\frac{9}{\sqrt{5}}$

(ii) $\frac{x^2}{4}+\frac{y^2}{9}=1$; major axis → y-axis, minor axis → x-axis.

Length of $L.R.$ = 8/3 units

$e=\frac{\sqrt{5}}{3}$, so foci $(0,\pm\sqrt{5})$ and equations of directrice $y=\pm\frac{9}{\sqrt{5}}$.

(iii) $\frac{(x-1)^2}{16}+\frac{(y-2)^2}{9}=1$; major axis $y=2$, minor axis $x=1$

Length of $L.R.$ = 9/2 units,

$e=\frac{\sqrt{7}}{4}$, so foci are $(1\pm\sqrt{7}, 2)$

Equations of directrice is $x=1+\frac{16}{7}\sqrt{7}$

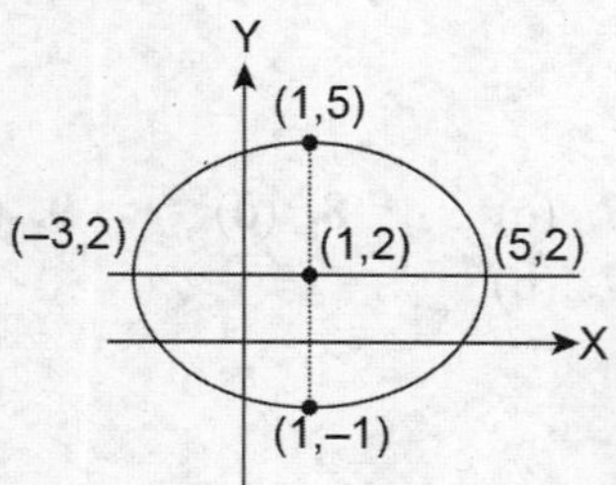

(iv) $\frac{(x+2)^2}{4}+\frac{(y-3)^2}{9}=1$; major axis $x=-2$, minor axis $y=3$.

Length of $L.R.$ = 8/3 units

$e=\frac{\sqrt{5}}{3}$, so foci $(-2, 3\pm\sqrt{5})$. Equations of directrice $y=3\pm\frac{9}{5}\sqrt{5}$

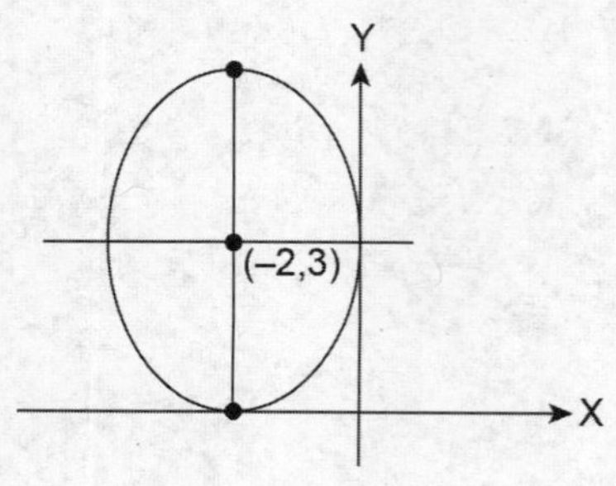

2. $\frac{x^2}{10-a}+\frac{y^2}{4-a}=1$ with represent an ellipse, when $(4-a)>0$ and $(10-a)>0$, i.e., $a<4$ $\Rightarrow a\in(-\infty, 4)$.

3. Given, $2ae=10$ $\Rightarrow ae=5$, also $\frac{2b^2}{a}=15$

$\Rightarrow \frac{2a^2(1-e^2)}{a}=15$

$\Rightarrow 2a^2-2a^2e^2=15a$ $\Rightarrow 2a^2-15a-50=0$

$\Rightarrow a=-\frac{5}{2},10$ as $a>0$

$\Rightarrow a=10$ $\Rightarrow e=1/2$

$\therefore b^2=75$

Hence the ellipse $\frac{x^2}{100}+\frac{y^2}{75}=1$ or $3x^2+4y^2=300$.

4. Given centre of ellipse is $(2,-3)$ focus $F_1(3,-3)$ and one vertex $A(4,-3)$.

Observe that $ae=1$ and $a=2$

$\Rightarrow e=1/2$ and $b^2=3$.

Hence the ellipse $\frac{(x-2)^2}{4}+\frac{(y+3)^2}{3}=1$

5. (a) (i) $12(x^2+2x+1)+4(y^2-4y+4)=3$

or $\frac{(x+1)^2}{\frac{1}{4}}+\frac{(y-2)^2}{\frac{3}{4}}=1$

Which gives an ellipse with centre at $(-1, 2)$. Length of major axis = $\sqrt{3}$ unit. Length of minor axis = 1 unit. Length of $L.R.$ = $1/\sqrt{3}$ units

$\Rightarrow e=\sqrt{\frac{2}{3}}=\frac{1}{3}\sqrt{6}$

$\Rightarrow$ Foci = $\left(-1,2\pm\frac{1}{\sqrt{2}}\right)$

(ii) The given equation is $\frac{(x-1)^2}{(1/2)}+\frac{(y-2)^2}{(1/3)}=1$

$\Rightarrow$ Centre of ellipse $(-1, 2)$

Major axis $=2a=\sqrt{2}$, minor axis $2b=2/\sqrt{3}$

Now $e^2=\frac{a^2-b^2}{a^2}=\frac{1}{3}$

$\Rightarrow e=\frac{1}{\sqrt{3}}$ $\Rightarrow ae=\frac{1}{\sqrt{6}}$

Hence foci $\left(1\pm\frac{1}{\sqrt{6}},2\right)$

(b) Given $LR=\frac{2b^2}{a}=a\Rightarrow a^2=2b^2$

$\Rightarrow a^2-b^2=b^2=\frac{a^2}{2}$

$\therefore e^2=\frac{a^2-b^2}{a^2}=\frac{1}{2}$ $\Rightarrow e=\frac{1}{\sqrt{2}}$

6. (a) Given foci (±2, 3)

$\Rightarrow$ centre of ellipse (0, 3) and the minor axis = $2b = 2\sqrt{5}$

Now $ae = 2$ and $b = \sqrt{5}$ $\Rightarrow a^2e^2 = a^2\left\{\frac{a^2-b^2}{a^2}\right\} = 4$

Hence $a^2 - b^2 = 4$ $\Rightarrow a^2 = 9$

$\therefore$ The equation of ellipse $\frac{x^2}{9} + \frac{(y-3)^2}{5} = 1$

(b) The given ellipse is $\frac{x^2}{2} + \frac{y^2}{3} = 1$; which has centre at (0, 0)

Length of major axis = $2\sqrt{3}$ units

Length of minor axis = $2\sqrt{2}$ units

Now $e^2 = \frac{3-2}{3} = \frac{1}{3}$ $\Rightarrow$ $\sqrt{\ }$

$\therefore$ $ae = 1$

Hence foci S (0, ±1) and vertices are (± $\sqrt{2}$, 0), (0, ± $\sqrt{3}$).

The diretrice are $y = \pm 3$

(c) (i) Given centre of ellipse (1, 2) and focus S_1 (6, 2)

$\Rightarrow$ $ae = 5$

From symmetry S_2 (–4, 2) as P (4, 6) lies on the ellipse

$\Rightarrow$ $S_1P + S_2P = 2a$

i.e., $\sqrt{20} + \sqrt{80} = 2$ $\Rightarrow a = 3\sqrt{5}$

Hence $e^2 = \frac{25}{45}$ $\Rightarrow e = \frac{\sqrt{5}}{3}$ gives

$\Rightarrow e^2 = \frac{a^2-b^2}{a^2} = \frac{5}{9}$ i.e., $\frac{45-b^2}{45} = \frac{25}{45}$

$\Rightarrow b^2 = 20$

Hence the ellipse is $\frac{(x-1)^2}{45} + \frac{(y-2)^2}{20} = 1$

(ii) Given focus of ellipse at (–1, 1) and the directrix is $x - y + 3 = 0$ for eccentricity $e = 1/2$, we get

$\frac{SP^2}{SM^2} = e^2$ i.e., $\frac{(2)\{(x+1)^2 + (y-1)^2\}}{(x-y+3)^2} = \frac{1}{4}$

$\Rightarrow 8\{x^2 + 2x + 1 + y^2 + 1 - 2y\} = x^2 + y^2 + 9 - 2xy + 6x - 6y$

or $7x^2 + 7y^2 + 2xy + 10x - 10y + 7 = 0$

7. The given ellipse is $5x^2 + 5y^2 + 6xy = 8$. Since no first degree term is existing.

$\Rightarrow y = m_1x$ and $y = -\frac{1}{m_1}x$ are its major and minor axis

$\Rightarrow \frac{(m_1x - y)^2}{(1+m_1^2)a^2} + \frac{(x + m_1y)^2}{(1+m_1^2)b^2} = 1$

$\Rightarrow \left(\frac{b^2m_1^2x^2 + b^2y^2 - 2b^2m_1xy + a^2x^2 + a^2m_1^2y^2 + 2a^2m_1xy}{a^2b^2(1+m_1^2)} = 1\right)$

On rearrangement, we get

$\frac{(a^2 + b^2m_1^2)x^2 + (a^2m_1^2 + b^2)y^2 + 2m_1(a^2 - b^2)xy}{(1+m^2)a^2b^2} = 1$

On comparing, we get $a^2 + b^2\ m_1^2 = 5,\ a^2m_1^2 + b^2 = 5$

$\Rightarrow m_1^2 = 1$

Let $m_1 = 1$

$\Rightarrow a^2 + b^2 = 5$ and $2\,(a^2 - b^2) = 6$

$\Rightarrow a^2b^2 = 4$ which is satisfied by $a^2 = 4$ and $b^2 = 1$

Hence, we get the ellipse as $\frac{(x-y)^2}{8} + \frac{(x+y)^2}{2} = 1$

Length of major axis = 4, Length of minor axis = 2

8. The given conic can be written as

$$\frac{4\left(\frac{x-2y+1}{\sqrt{5}}\right)^2(5)}{25} + \frac{9\left(\frac{2x+y+2}{\sqrt{5}}\right)^2(5)}{25} = 1 \text{ or}$$

$$\frac{\left(\frac{x-2y+1}{\sqrt{5}}\right)^2}{\frac{25}{20}} + \frac{\left(\frac{2x+y+2}{\sqrt{5}}\right)^2}{\frac{25}{45}} = 1 \text{ as } \frac{25}{20} = \frac{5}{4} > \frac{5}{9} = \frac{25}{45}$$

$\Rightarrow x - 2y + 1 = 0$ is minor axis and $2x + y + 2 = 0$ is major axis and centre at is at (–1, 0)

From $a^2 = 5/4$ and $b^2 = 5/9$, we get $L.R.$

$= \frac{2b^2}{a} = \frac{(2)(2)}{\sqrt{5}}\left(\frac{5}{9}\right) = \frac{4}{9}\sqrt{5}$ units

And $e^2 = \frac{a^2-b^2}{a^2} = \frac{4\left(\frac{5}{4} - \frac{5}{9}\right)}{5} = \frac{4\{9-4\}}{(4)(9)} = \frac{5}{9}$

$\Rightarrow e = \frac{\sqrt{5}}{3}$

9. We know that if (h, k) is a focus and $x + my + n = 0$ is a directrix, then the equation of ellipse for eccentricity e will be

$(x-h)^2 + (y-k)^2 = e^2\frac{(\ell x + my + n)^2}{(\ell^2 + m^2)}$

$\Rightarrow (10x - 5)^2 + (10y - 5)^2 = (3x + 4y - 1)^2$ can be rewritten

as $(2x - 1)^2 + (2y - 1)^2 = \left(\frac{3x + 4y - 1}{5}\right)^2$

i.e., $\left(x - \frac{1}{2}\right)^2 + \left(y - \frac{1}{2}\right)^2 = \frac{1}{4}\left(\frac{3x+4y-1}{5}\right)^2$, which

gives focus at $\left(\frac{1}{2}, \frac{1}{2}\right)$ directrix as $3x + 4y = 1$ and $e = 1/2$ for the ellipse

10. Given $2ae = 5$ and $\frac{2a}{e} = 10$ $\Rightarrow a^2 = \frac{25}{2}$ and $e^2 = \frac{1}{2}$

Hence $b^2 = a^2(1 - e^2) = \frac{25}{4}$

$\Rightarrow$ The equation of ellipse is $\frac{2x^2}{25} + \frac{4y^2}{25} = 1$ or $2x^2 + 4y^2 = 25$

11. The point P lies on $\frac{x^2}{49} + \frac{y^2}{25} = 1$

$\Rightarrow$ A (7, 0), A' (–7, 0) and B (0, 5) and B' (0, –5)

$\Rightarrow e^2 = \frac{49-25}{49} = \frac{24}{49}$

$\Rightarrow$ Foci are $S(2\sqrt{6}, 0)$ and $S' = \left(-2\sqrt{6}, 0\right)$

Now the $\Delta PAA'$ and $\Delta PSS'$ will have maximum area, when P lies at B or B'

Hence $max\ \Delta PSS' = \frac{1\times 4\sqrt{6}\times 5}{2} = 10\sqrt{6}$ square units

and $max\ \Delta PAA' = \frac{1}{2}\times 14\times 5 = 35$ sq.unit

TEXTUAL EXERCISE–1 (OBJECTIVE)

1. (b) Given $L.R. = \frac{2b^2}{a} = b \quad \Rightarrow a = 2b$

$\Rightarrow e^2 = \frac{a^2-b^2}{a^2} = \frac{4b^2-b^2}{4b^2} = \frac{3}{4}$

$\therefore e = \frac{\sqrt{3}}{2}$

2. (c) Given $\frac{2a}{e} \div 2ae = 3 \quad \Rightarrow \frac{1}{e^2} = 3$

$\Rightarrow e = \frac{1}{\sqrt{3}}$

3. (a) Given $2ae = 10$ and $e = 5/8 \ \Rightarrow a = 8$

Now $LR = \frac{2b^2}{a} = 2a(1-e^2) = 16\left\{\frac{64-25}{64}\right\} = \frac{39}{4}$ units

4. (d) From the given: $a = 2$ and $ae = 1$

$\Rightarrow e = \frac{1}{2} \Rightarrow 2b = 2\sqrt{a^2(1-e^2)} = 2\sqrt{4\left(\frac{3}{4}\right)} = 2\sqrt{3}$

5. (b) Given $\frac{2b^2}{a} = 5 \quad \Rightarrow b^2 = \frac{5a}{2}$, also $e^2 = \frac{4}{9} = \frac{a^2-b^2}{a^2}$

$\Rightarrow \frac{2a^2-5a}{2a^2} = \frac{4}{9} \quad \Rightarrow 10a^2 - 45a = 0$

$\Rightarrow a = 9/2$

$\therefore b^2 = \frac{5a}{2} = \frac{45}{4}$; hence the ellipse is $\frac{4x^2}{81} + \frac{4y^2}{45} = 1$

6. (a) $LR = \frac{2b^2}{a} = 10 \quad \Rightarrow b^2 = 5a$ and $2b = 2ae$

$\Rightarrow b^2 = a^2e^2$ or $a^2(1-e^2) = a^2e^2$

$\Rightarrow 2a^2e^2 = a^2$

$\Rightarrow e = \frac{1}{\sqrt{2}}$ and $b^2 = \frac{a^2}{2}$

$\Rightarrow a^2 = 10a \quad \therefore a = 10$

Hence the ellipse is $\frac{x^2}{100} + \frac{y^2}{50} = 1$ or $x^2 + 2y^2 = 100$

7. (a) Given foci $(\pm 5, 0)$

$\Rightarrow$ Centre is at (0, 0) the directrix is $x = \frac{36}{5}$

$\Rightarrow \frac{a}{e} = \frac{36}{5}$ and $ae = 5$

$\Rightarrow a^2 = 36$ and $e = 5/6 \quad \therefore b^2 = 36\left(1-\frac{25}{36}\right) = 11$

Hence the ellipse is $\frac{x^2}{36} + \frac{y}{11} = 1$

8. (d) Given $e = \frac{1}{\sqrt{2}} \quad \Rightarrow b^2 = a^2(1-e^2) = \frac{a^2}{2}$

Hence $L.R. = \frac{2b^2}{a} = \frac{a^2}{a} = a$ = semi major axis

9. (b) Given $2ae = 2b \Rightarrow b^2 = a^2e^2$ i.e., $a^2(1-e^2) = a^2e^2$

$\Rightarrow e = 1/\sqrt{2}$

10. (a) Observe that $(-3, 1)$ lies on $3x^2 + 5y^2 = 32$ only and not on any other given conic.

Further $\frac{x^2}{\left(\frac{32}{3}\right)} + \frac{y^2}{\left(\frac{32}{5}\right)} = 1$

$\Rightarrow \frac{32}{5} = \frac{32}{3}\{1-e^2\}$

$\Rightarrow 1 - e^2 = 3/5 \quad \Rightarrow e^2 = 2/5$ which is true

11. (b) Given $LR = \frac{2b^2}{a} = 2ae \Rightarrow b^2 = a^2e$ i.e., $a^2(1-e^2) = a^2e$

$\Rightarrow e^2 + e - 1 = 0$, hence $e = \frac{-1\pm\sqrt{5}}{2}$, since $e \geq 0$

$\therefore e = \frac{\sqrt{5}-1}{2}$

12. (d) Given $2a = 8$ cm and $2b = 4$cm

$\Rightarrow 4 = 16(1-e^2)$

$\Rightarrow e^2 = 3/4$. Hence $2ae = 2.\,4.\frac{\sqrt{3}}{2} = 4\sqrt{3}$

$\therefore$ The distance between the pins = $2ae = 4\sqrt{3}$ cms and required length of string $2a = 8$cm

13. (c) We knows that the locus of point of intersection of perpendicular tangents is the director circle.

$\therefore$ For ellipse $\frac{x^2}{a^2} + \frac{y^2}{b^2} = 1$ the locus will be $x^2 + y^2 = (a^2 + b^2)$

14. (b) Observe that the answers under given options indicate the ellipse to be in standard form, so $ae = 4$ and $e = 4/5$

$\Rightarrow a = 5$ and $b^2 = a^2(1-e^2)$

$\Rightarrow b^2 = 25\left\{\frac{25-16}{25}\right\} = 9$. Hence the ellipse $\frac{x^2}{25} + \frac{y^2}{9} = 1$

15. (c) Given F_1 (3, 0) and $F_2 = (-3, 0)$ and $\frac{x^2}{25} + \frac{y^2}{16} = 1$, then for $P(x, y)$, we get $PF_1 + PF_2 = 2a = 10$

16. (c) $a_1/b_1 = 13/5$ for the first ellipse

Since $\frac{a^2}{b^2} = \frac{1}{1-e^2}$ = constant

$\Rightarrow$ For the second ellipse ratio $a/b = 13/5$

Aliter: Eccentricity of $\frac{x^2}{169} + \frac{y^2}{25} = 1$ is $e^2 = \frac{a^2-b^2}{a^2} = \frac{144}{169}$

$\Rightarrow e = \frac{12}{13}$

Now the eccentricity of $\frac{x^2}{a^2} + \frac{y^2}{b^2} = 1$ is $e = 12/13$

$\Rightarrow \frac{a^2}{b^2} = \frac{1}{1-e^2} = \frac{169}{25}$

$\Rightarrow a/b = 13/5$

17. **(b)** Vertices are at (4, 0) and (10, 0)

$\Rightarrow$ Centre is at (7, 0) and $a = 3$.

Since $e = \frac{1}{2} \quad \Rightarrow \quad b^2 = 9\left(1 - \frac{1}{4}\right) = \frac{27}{4}$

Hence the ellipse $\frac{(x-7)^2}{9} + \frac{4y^2}{27} = 1$

$\Rightarrow 3\{x^2 + 49 - 14x\} + 4y^2 = 27$ or $3x^2 + 4y^2 - 42x + 120 = 0$

18. **(b)** The ellipse is $\frac{(x+y-2)^2}{9} + \frac{(x-y)^2}{16} = 1$

The lines $x + y - 2 = 0$ and $x - y = 0$ intersect at (1, 1) which will be the centre of the ellipse

19. **(b)** The ellipse is $9x^2 + 5(y^2 - 6y + 9) = 45$ i.e.,

$\frac{x^2}{5} + \frac{(y-3)^2}{9} = 1 \Rightarrow e^2 = \frac{9-5}{9} = \frac{4}{9}$ i.e., $e = 2/3$

20. **(c)** The ellipse is $(x^2 - 2x + 1) + 2\left(y^2 + \frac{(2)(3)}{(2)(2)}y + \frac{9}{16}\right) = \frac{1}{8}$

i.e., $\frac{(x-1)^2}{\frac{1}{8}} + \frac{\left(y + \frac{3}{4}\right)^2}{\left(\frac{1}{16}\right)} = 1$

Hence $e^2 = \frac{a^2 - b^2}{a^2} = \frac{\frac{1}{8} - \frac{1}{16}}{\frac{1}{8}} = \frac{1 \times 8}{16 \times 1} = \frac{1}{2}$

$\Rightarrow e = \frac{1}{\sqrt{2}}$

21. **(a)** $\frac{(x-1)^2}{9} + \frac{(y+1)^2}{25} = 1 \quad \Rightarrow e^2 = \frac{25-9}{25} = \frac{16}{25}$

$\Rightarrow e = \frac{4}{5}$

22. **(c)** $9\left\{x^2 - 2\left(\frac{1}{3}\right)x + \frac{1}{9}\right\} + 4\left\{y^2 + y + \frac{1}{4}\right\} = 1$

$\Rightarrow \frac{\left(x - \frac{1}{3}\right)^2}{\left(\frac{1}{9}\right)} + \frac{\left(y + \frac{1}{2}\right)^2}{\left(\frac{1}{4}\right)} =$

Hence $2a = 1$ and $2b = 2/3$.

23. **(c)** From the given information $b = ae$

Hence $b^2 = a^2(1 - e^2) = a^2e^2$

$\Rightarrow e^2 = 1/2 \quad \Rightarrow \quad e = \frac{1}{\sqrt{2}}$

TEXTUAL EXERCISE–2 (SUBJECTIVE)

1. The ellipse is $S: 5x^2 + 7y^2 - 140 = 0$ and the point is $P(4, -3)$.

Now, $S(4, -3) = 80 + 7 \times 9 - 140 > 0$ i.e., $3 > 0$.

Hence point lies outside the ellipse, so two tangents are possible from the ellipse.

2. Let $P(\alpha)$ and $Q(\beta)$ be the two ends of a chord for the ellipse $\frac{x^2}{a^2} + \frac{y^2}{b^2} = 1$; where $a > b$

$\Rightarrow$ Equation of PQ is

$\frac{x}{a}\cos\left(\frac{\alpha+\beta}{2}\right) + \frac{y}{b}\sin\left(\frac{\alpha+\beta}{2}\right) = \cos\left(\frac{\alpha-\beta}{z}\right)$

Since the chord intersect the major axis at $(\pm d, 0)$

$\therefore \pm\frac{d}{a}\cos\left(\frac{\alpha+\beta}{2}\right) = \cos\left(\frac{\alpha-\beta}{2}\right)$

Now from $\frac{d}{a} = \frac{\cos\left(\frac{\alpha-\beta}{2}\right)}{\cos\left(\frac{\alpha+\beta}{2}\right)}$, we get

$\frac{d+a}{d-a} = \frac{\cos\left(\frac{\alpha-\beta}{2}\right) + \cos\left(\frac{\alpha+\beta}{2}\right)}{\cos\left(\frac{\alpha-\beta}{2}\right) - \cos\left(\frac{\alpha+\beta}{2}\right)}$ or

$\frac{d+a}{d-a} = \frac{2\cos\frac{\alpha}{2}.\cos\frac{\beta}{2}}{2\sin\frac{\alpha}{2}.\sin\frac{\beta}{2}} \Rightarrow \frac{d-a}{d+a} = \tan\frac{\alpha}{2}.\tan\frac{\beta}{2}$

3. Let (x_1, y_1) be the mid point of the chord(s) for the ellipse $\frac{x^2}{2a^2} + \frac{y^2}{2b^2} = 1$

Then $T = S_1$ gives $\frac{x_1x}{2a^2} + \frac{y_1y}{2b^2} - 1 = \frac{x_1^2}{2a^2} + \frac{y_1^2}{2b^2} - 1 \qquad \ldots(i)$

Since the mid point lies on $x + y = b$

$\Rightarrow x_1 + y_1 = b$ or $y_1 = b - x_1$

$T = S_1$ chord passes through $(a, -b)$, so (i) becomes

$\frac{x_1}{2a} - \frac{y_1}{2b} - \left(\frac{x_1^2}{2a^2} + \frac{y_1^2}{2b^2}\right) = 0$

Putting $y_1 = b - x_1$, we get

$\frac{x_1}{2a} - \frac{b}{2b} + \frac{x_1}{2b} - \frac{x_1^2}{2a^2} - \frac{b^2 + x_1^2 - 2bx_1}{2b^2} = 0$

Rearrangement gives

$\frac{x_1(ab^2 + a^2b + 2a^2b)}{2a^2b^2} - \frac{(a^2 + b^2)x_1^2}{2a^2b^2} - 1 = 0$

i.e., $x_1^2(a^2 + b^2) - (ab^2 + 3a^2b)x_1 + 2a^2b^2 = 0$

Since x_1 is real $\quad \therefore \quad D > 0$

Hence $a^2b^2(3a + b)^2 - 8a^2b^2(a^2 + b^2) > 0$ or $(9a^2 + b^2 + 6ab - 8a^2 - 8b^2) > 0$

Which can be rearranged as $(a^2 - 7b^2 + 6ab) > 0$ i.e., $(a - b)(a + 7b) > 0$; which is the required condition.

4. **(a)** For ellipse $\frac{x^2}{a^2} + \frac{y^2}{b^2} = 1$ (for $a > b$) a chord with end points having eccentric angles θ and ψ will be

$\frac{x}{a}\cos\left(\frac{\theta+\psi}{2}\right) + \frac{y}{b}\sin\left(\frac{\theta+\psi}{2}\right) = \cos\left(\frac{\theta-\psi}{2}\right)$

If the chord happens to be a focal chord, then it passes through ($\pm ae$, 0) so from (ae, 0) we get

$$e\cos\left(\frac{\theta+\psi}{2}\right)=\cos\left(\frac{\theta-\psi}{2}\right) \text{ or } \frac{e}{1}=\frac{\cos\left(\frac{\theta-\psi}{2}\right)}{\cos\left(\frac{\theta+\psi}{2}\right)}$$

$$\Rightarrow \frac{e-1}{e+1}=\frac{\cos\left(\frac{\theta-\psi}{2}\right)-\cos\left(\frac{\theta+\psi}{2}\right)}{\cos\left(\frac{\theta-\psi}{2}\right)+\cos\left(\frac{\theta+\psi}{2}\right)}=\frac{2\sin\frac{\theta}{2}.\sin\frac{\psi}{2}}{2\cos\frac{\theta}{2}.\cos\frac{\psi}{2}}$$

i.e., $\tan\frac{\theta}{2}.\tan\frac{\psi}{2}=\frac{e-1}{e+1}$

Hence proved (for $\psi = \phi$)

(b) From $\tan\frac{\theta}{2}.\tan\frac{\phi}{2}=\frac{e-1}{e+1}$, we put $\theta = 90°$ and $\phi = -30°$

So, $\tan 45° \tan(-15°)=\frac{e-1}{e+1}$

$$\Rightarrow \frac{e-1}{e+1}=\frac{-(2-\sqrt{3})}{1} \text{ or } \frac{2e}{2}=\frac{\sqrt{3}-1}{\sqrt{3}\left(\sqrt{3}-1\right)} \text{ i.e., } e=\frac{1}{\sqrt{3}}$$

5 For the ellipse $\frac{x^2}{a^2}+\frac{y^2}{b^2}=1$. Consider a chord with extremities having eccentric angles α & β so the chord is

$$\frac{x}{a}\cos\left(\frac{\alpha+\beta}{2}\right)+\frac{y}{b}\sin\left(\frac{\alpha+\beta}{2}\right)=\cos\left(\frac{\alpha-\beta}{2}\right)$$

$\Rightarrow$ Slope of the chord $m=-\frac{b}{a}\cot\left(\frac{\alpha+\beta}{2}\right)$. Since m is constant, so ($\alpha+\beta$) is constant.

Now consider a point $P(\theta)=(x_1, y_1)$ where the slope of tangent will be m

$$\Rightarrow \frac{x_1x}{a^2}+\frac{y_1y}{b^2}=1 \Rightarrow m=\frac{-x_1b^2}{a^2y_1}$$

Putting $x_1 = a\cos\theta$ and $y_1 = b\sin\theta$, we get $m=-\frac{b}{a}\cot\theta$.

Hence from $m = -\frac{b}{a}\cot\left(\frac{\alpha+\beta}{2}\right)=-\frac{b}{a}\cot\theta$, we get

$$\frac{\alpha+\beta}{2}=\theta \text{ or } 2\theta=(\alpha+\beta)$$

6. Let $P(\theta)$, $Q(\phi)$ and $R(\psi)$ be the three points on the ellipse $\frac{x^2}{a^2}+\frac{y^2}{b^2}=1$, then the area of $\Delta PQR=\frac{ab}{2}\left\|\begin{matrix}\cos\theta & \sin\theta & 1\\ \cos\phi & \sin\phi & 1\\ \cos\psi & \sin\psi & 1\end{matrix}\right\|$

$$=\frac{ab}{2}\left|(\cos\theta-\cos\phi)(\sin\phi-\sin\psi)-(\cos\phi-\cos\psi)(\sin\theta-\sin\phi)\right|$$

$$=\frac{ab}{2}\left|\begin{matrix}-2\sin\frac{\theta+\phi}{2}\sin\frac{\theta-\phi}{2}.2\sin\frac{\phi-\psi}{2}.\cos\frac{\phi+\psi}{2}\\ +2\sin\frac{\phi+\psi}{2}\sin\frac{\phi-\psi}{2}.2\sin\frac{\theta-\phi}{2}.\cos\frac{\theta+\phi}{2}\end{matrix}\right|$$

Rearranging, we get $\Delta=2ab\left|\sin\frac{\theta-\phi}{2}.\sin\frac{\phi-\psi}{2}\right.$

$$\left(\cos\frac{\theta+\phi}{2}\sin\frac{\phi+\psi}{2}-\sin\frac{\theta+\phi}{2}\cos\frac{\phi+\psi}{2}\right)$$

$$=2ab\left|\sin\frac{\theta-\phi}{2}\sin\frac{\phi-\psi}{2}.\sin\frac{\psi-\theta}{2}\right|$$

7. Clearly chord PQ is diameter of auxiliary circle.

$\therefore$ we are to find the area between ellipse and auxiliary circle and bounded by the chord PQ as shown above by shaded region and equals.

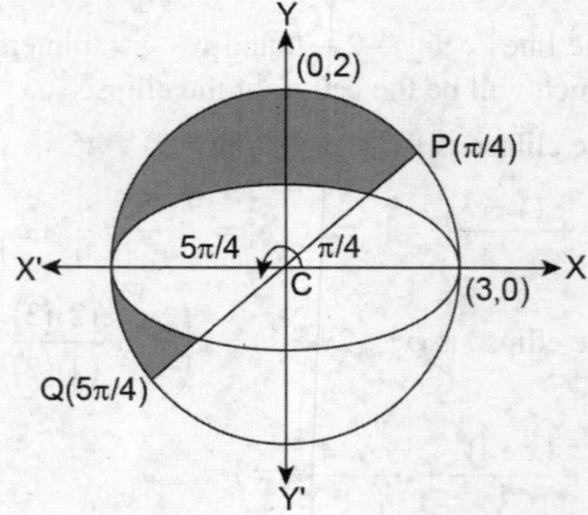

$$=\frac{1}{2}\left[\text{Area of auxiliary circle} - \text{Area of ellipse}\right]$$

$$=\frac{1}{2}\left[\pi(3)^2-\pi(3)(2)\right]=\frac{3\pi}{2} \text{ sq. units.}$$

TEXTUAL EXERCISE-2 (OBJECTIVE)

1. (c) A point $P(\theta)$ on the ellipse $\frac{x^2}{a^2}+\frac{y^2}{b^2}=1$ will be ($a\cos\theta$, $b\sin\theta$) and the focus will at $S(\pm ae, 0)$ for $a > b$

$\Rightarrow$ Distance $S'P=\sqrt{a^2\{\cos\theta+e\}^2+b^2\sin^2\theta}$ {where S' ($-ae$, 0)}

$\Rightarrow$ $S'P=\sqrt{a^2\cos^2\theta+a^2e^2+2a^2e\cos\theta+b^2\sin^2\theta}$

Substituting $b^2=a^2-a^2e^2$, we get $S'P=\sqrt{a^2(1+e\cos\theta)^2}$ $= a(1+e\cos\theta)$

2. (c) The equation $x = a\cos\theta$ and $y = b\sin\theta$ (where $a > b$) gives $\frac{x^2}{a^2}+\frac{y^2}{b^2}=\cos^2\theta+\sin^2\theta=1$

Which is a conic (ellipse) having eccentricity given by $e^2=\frac{a^2-b^2}{a^2}$

3. (d) Let $(x_1, y_1) = (a\cos\theta, b\sin\theta)$ and $(x_2, y_2) = (a\cos\phi, b\sin\phi)$.

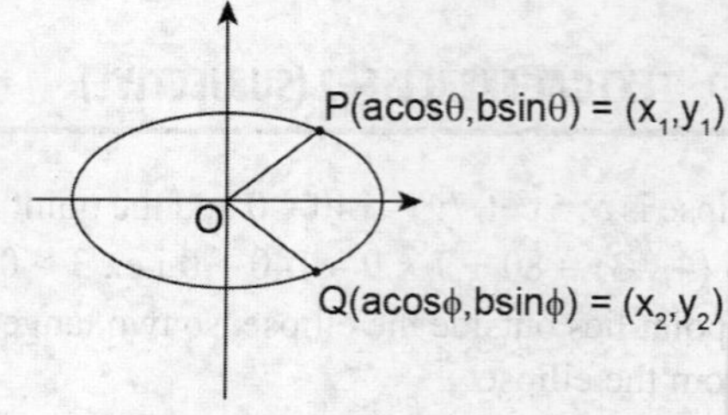

Now $\tan\theta . \tan\phi = \dfrac{-a^2}{b^2}$ (given)

i.e., $\dfrac{\sin\theta}{\cos\theta} \cdot \dfrac{\sin\phi}{\cos\phi} = \dfrac{(b\sin\theta)(b\sin\phi)a^2}{b^2(\cos\theta)(a\cos\phi)} = \dfrac{-a^2}{b^2}$ (from given)

$\Rightarrow \left(\dfrac{y_1}{x_1}\right)\left(\dfrac{y_2}{x_2}\right) = -1$; Which means product of {(slope of OP) × (slope of OQ)} = –1

$\therefore$ Chord PQ subtends a right angle at the centre

4. (b) Use $\Delta \neq 0$, $h^2 - ab < 0$

$\Rightarrow$ Locus of point is ellipse

5. (c) The ends of latus rectra for the ellipse $\dfrac{x^2}{a^2} + \dfrac{y^2}{b^2} = 1$ are $(ae, \pm y) = (a\cos\theta, \pm b\sin\theta) \Rightarrow e = \cos\theta$

$\Rightarrow \tan\theta = \dfrac{\sin\theta}{\cos\theta} = \dfrac{\sqrt{1-e^2}}{e}$

$= \pm\dfrac{1}{e}\sqrt{1-\left(\dfrac{a^2-b^2}{a^2}\right)} = \pm\dfrac{\sqrt{a^2-a^2+b^2}}{ae} = \pm\dfrac{b}{ae}$

$\Rightarrow \theta = \tan^{-1}\left(\pm\dfrac{b}{ae}\right)$

6. (a), (c) The ellipse is $\dfrac{x^2}{6} + \dfrac{y^2}{2} = 1$

Let $\left(\sqrt{6}\cos\theta, \sqrt{2}\sin\theta\right)$ be a point on the ellipse, so 6 $\cos^2\theta + 2\sin^2\theta = 4$

$\Rightarrow 6\cos^2\theta + 2(1 - \cos^2\theta) = 4$

Hence $4\cos^2\theta = 2$ i.e., $\cos^2\theta = 1/2$

$\Rightarrow \theta = 2n\pi \pm \dfrac{\pi}{4}, 2n\pi \pm \dfrac{5\pi}{4}\left(as \dfrac{3\pi}{4} = 2\pi - \dfrac{\pi}{4}\right)$

7. (a) As "AS_2B" chord passes through focus S_2 $(ae, 0)$

$\Rightarrow \tan\left(\dfrac{\theta_1}{2}\right).\tan\left(\dfrac{\theta_2}{2}\right) = \dfrac{e-1}{e+1}$

Since BS_1C is passing though S_1 $(-ae, 0)$

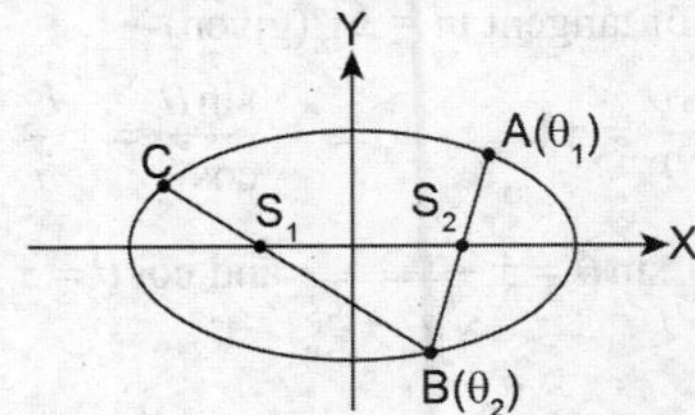

$\Rightarrow \tan\left(\dfrac{\theta_2}{2}\right).\tan\left(\dfrac{\theta_3}{2}\right) = \dfrac{e+1}{e-1}$

$\Rightarrow \cot\left(\dfrac{\theta_2}{2}\right).\cot\left(\dfrac{\theta_3}{2}\right) = \dfrac{e-1}{e+1}$ and so on

8. (b) Let P $(a\cos\theta, b\sin\theta)$ be a point on ellipse $\dfrac{x^2}{a^2} + \dfrac{y^2}{b^2} = 1$.

Now $S_1P = a(1 + e\cos\theta)$ and $S_2P = a(1 - e\cos\theta)$ as $S_1S_2 = 2ae$

$\Rightarrow$ Perimeter of $\Delta PS_1S_2 = 2a + 2ae$

Let $\angle PS_1S_2 = A$ and $\angle PS_2S_1 = B$

$\Rightarrow S_1S_2 = c$, $PS_1 = b$ and $PS_2 = a$

Now $\tan\dfrac{A}{2} = \dfrac{\sqrt{(s-b)(s-c)}}{s(s-a)}$ and $\tan\dfrac{B}{2}\sqrt{\dfrac{(s-a)(s-c)}{s(s-b)}}$

$\tan\dfrac{A}{2}.\tan\dfrac{B}{2} = \dfrac{(s-c)}{s} = \dfrac{a+ae-2ae}{a+ae} = \dfrac{1-e}{1+e}$

9. (c) The given ellipse is $\dfrac{x^2}{9} + \dfrac{y^2}{4} = 1$

$\Rightarrow P$ $(3\cos\theta, 2\sin\theta)$ and Q $(3\cos\theta, 3\sin\theta)$.

Now R is a point such that $PR : RQ = 2 : 3$

Let $R = (h, k)$, then $h = a\cos\theta = 3\cos\theta$ and $k = \dfrac{12}{5}\sin\theta$

$\Rightarrow \left(\dfrac{h}{3}\right)^2 + \left(\dfrac{5k}{12}\right)^2 = 1$ or $16x^2 + 25y^2 = 144$

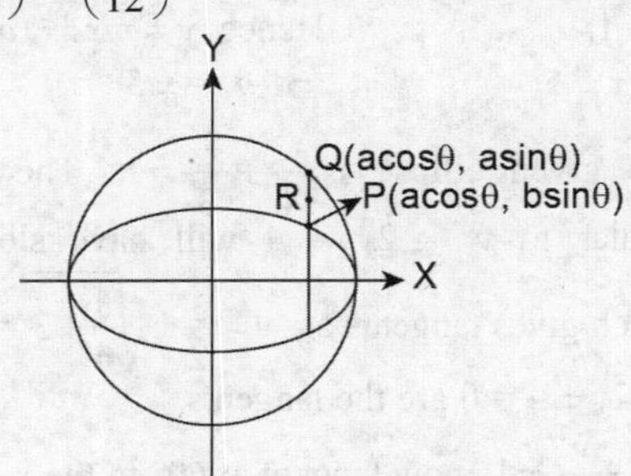

TEXTUAL EXERCISE–3 (SUBJECTIVE)

1. (a) The given ellipse is $\dfrac{x^2}{9} + \dfrac{y^2}{4} = 1$, line parallel to $y = 2x + 1$ will have slope $m = 2$

Hence $y = mx \pm \sqrt{a^2m^2 + b^2}$

$\Rightarrow y = 2x \pm \sqrt{36+4}$

$\Rightarrow y = 2x \pm 2\sqrt{10}$ are the possible straight lines.

The point of contact is given as $\left(-\dfrac{am}{c}, \dfrac{b^2}{c}\right)$

$\Rightarrow \left(-\dfrac{9}{\sqrt{10}}, \dfrac{2}{\sqrt{10}}\right)$ or $\left(\dfrac{9}{\sqrt{10}}, \dfrac{-2}{\sqrt{10}}\right)$

(b) The given ellipse is $\dfrac{x^2}{9} + \dfrac{y^2}{4} = 1$, the given line is $\dfrac{x}{4} + \dfrac{y}{12} = 1$ i.e., $3x + y - 12 = 0$ for $m = -3$, we get

$y = -3x \pm \sqrt{81+4} = -3x \pm \sqrt{85}$

$\Rightarrow 3x + y \pm \sqrt{85} = 0$

(c) The line passing through (0, 3) with slope m is $y - 3 = mx$ or $y = mx + 3 = mx + \sqrt{a^2m^2 + b^2}$

$\Rightarrow \sqrt{9m^2 + 4} = 3 \quad \Rightarrow m^2 = \dfrac{5}{9}$ i.e., $m = \pm\dfrac{\sqrt{5}}{3}$

Hence the lines are $y = \pm\dfrac{\sqrt{5}}{3}x + 3$ and the point of contact is $\left(\mp\sqrt{5}, \dfrac{4}{3}\right)$ respectively

(d) The line passing through (4, 1) is $y - 1 = m(x - 4)$ or $y = mx + (1 - 4m) = mx \pm \sqrt{a^2m^2 + b^2}$

$\Rightarrow 9m^2 + 4 = 16m^2 + 1 - 8m$

$\Rightarrow 7m^2 - 8m - 3 = 0$ i.e., $m = \dfrac{4 \pm \sqrt{37}}{7}$

(e) Line passing through (–3, 1) is $y - 1 = m(x + 3)$

$\Rightarrow y = mx + (3m + 1) = mx \pm \sqrt{m^2a^2 + b^2}$

$\Rightarrow 9m^2 + 4 = 9m^2 + 1 + 6m$

$\Rightarrow m = 1/2$ the other tangent is vertical at (–3, 0)

The tangent will slope $m = 1/2$ is $y = \dfrac{x}{2} + \dfrac{5}{2}$

i.e., $x - 2y + 5 = 0$ and its point of contact is $\left(-\dfrac{9}{5}, \dfrac{8}{5}\right)$

2 (a) The given ellipse is $\dfrac{x^2}{16} + \dfrac{y^2}{9} = 1$. Now the tangent line is $y = x + \lambda$

$\Rightarrow m = 1.$ Hence $y = mx \pm \sqrt{a^2m^2 + b^2}$

$\Rightarrow y = x \pm 5$ $\Rightarrow \lambda = \pm 5$

(b) The given ellipse is $\dfrac{x^2}{4} + \dfrac{y^2}{3} = 1$. The line perpendicular to $y + 2x = 4$ will have slope $m = 1/2$

which gives tangents as $y = \dfrac{x}{2} \pm \sqrt{\dfrac{4}{4} + 3} = \dfrac{x}{2} \pm 2$

$\Rightarrow x - 2y \pm 4 = 0$ are the tangents.

(c) $\dfrac{x^2}{16} + \dfrac{y^2}{1} = 1$, now tangent at 60° to the x-axis

$\Rightarrow m = \sqrt{3}$

Hence $y = mx \pm \sqrt{a^2m^2 + b^2}$

$\Rightarrow y = \sqrt{3}x \pm \sqrt{48 + 1} = \sqrt{3}x \pm 7$ i.e., $y = \sqrt{3}x \pm 7$

3. The straight line $x + my + n = 0$ can be rewritten as $y = \dfrac{-\ell}{m}x - \dfrac{n}{m}$

Comparing with $y = Mx \pm \sqrt{a^2m^2 + b^2}$, we get

$a^2M^2 + b^2 = \dfrac{n^2}{m^2}$, where $M = -\quad /m$

$\Rightarrow \dfrac{a^2\ell^2}{m^2} + b^2 = \dfrac{n^2}{m^2}$ i.e., $a^2 \quad ^2 + b^2m^2 = n^2$

4. $P(\alpha)$ and $Q(\beta)$ are the two points on the ellipse $\dfrac{x^2}{a^2} + \dfrac{y^2}{b^2} = 1$, then the chord PQ is

$$\frac{x}{a}\cos\left(\frac{\alpha+\beta}{2}\right) + \frac{y}{b}\sin\left(\frac{\alpha+\beta}{2}\right) = \cos\left(\frac{\alpha-\beta}{2}\right)$$

$$\Rightarrow \frac{x}{a}\frac{\cos\left(\frac{\alpha+\beta}{2}\right)}{\cos\left(\frac{\alpha-\beta}{2}\right)} + \frac{y}{b}\frac{\sin\left(\frac{\alpha+\beta}{2}\right)}{\cos\left(\frac{\alpha-\beta}{2}\right)} = 1$$

Comparing with $\dfrac{xx_1}{a^2} + \dfrac{yy_1}{b^2} = 1$ for chord of contact of tangent, we get $x_1 = \dfrac{a\cos\left(\frac{\alpha+\beta}{2}\right)}{\cos\left(\frac{\alpha-\beta}{2}\right)}$ and $y_1 = \dfrac{b\sin\left(\frac{\alpha+\beta}{2}\right)}{\cos\left(\frac{\alpha-\beta}{2}\right)}$

Hence the point of intersection of tangents is

$$(x_1, y_1) = \left(\frac{a\cos\left(\frac{\alpha+\beta}{2}\right)}{\cos\left(\frac{\alpha-\beta}{2}\right)}, \frac{b\sin\left(\frac{\alpha+\beta}{2}\right)}{\cos\left(\frac{\alpha-\beta}{2}\right)}\right)$$

5. The ellipse $\dfrac{x^2}{a^2} + \dfrac{y^2}{b^2} = 1$ and the straight line $y = (-\cot\alpha)x + p\,\text{cosec}\alpha$ will be a tangent to the ellipse, when $\sqrt{a^2m^2 + b^2} = p\,\text{cosec}\,\alpha$ for

$m = -\cot\alpha$

$\Rightarrow \dfrac{a^2}{\tan^2\alpha} + b^2 = \dfrac{p^2}{\sin^2\alpha}$ i.e., $a^2\cos^2\alpha + b^2\sin^2\alpha = p^2$; which is true.

Similarly the point of contact is $(x_1, y_1) = \left(\dfrac{-a^2m}{c}, \dfrac{b^2}{c}\right)$

$\Rightarrow x_1 = \dfrac{a^2.\sin\alpha}{\tan\alpha.p} = \dfrac{a^2}{p}\cos\alpha$ and $y_1 = \dfrac{b^2\sin\alpha}{p}$

Hence the point of contact is $\left(\dfrac{a^2\cos\alpha}{p}, \dfrac{b^2\sin\alpha}{p}\right)$

6. Let $P(x_1, y_1) = (a\cos\theta, b\sin\theta)$ be a point on the ellipse $\dfrac{x^2}{a^2} + \dfrac{y^2}{b^2} = 1$

$\Rightarrow$ Equation of tangent at P will be $bx\cos\theta + ay\sin\theta = ab$ which will intersect the axes at $X(a\sec\theta, 0)$ and $Y(0, b\,\text{cosec}\theta)$

Hence mid point of X and Y is $(h, k) = \left(\dfrac{a}{2}\sec\theta, \dfrac{b}{2}\text{cosec}\,\theta\right)$

Eliminating θ, we get $\dfrac{a}{2h} = \cos\theta$ and $\dfrac{b}{2k} = \sin\theta$ as

$\dfrac{a^2}{4h^2} + \dfrac{b^2}{4k^2} = 1$ or $\dfrac{a^2}{x^2} + \dfrac{b^2}{x^2} = 4$

7. Let $P(a\cos\theta, b\sin\theta) = (x_1, y_1)$ be a point on the ellipse $\dfrac{x^2}{a^2} + \dfrac{y^2}{b^2} = 1$

$\Rightarrow$ Equation of tangent at P is $bx\cos\theta + ay\sin\theta = ab$

Slope of tangent $m = \pm 1$ (given)

$\Rightarrow \dfrac{-b\cos\theta}{a\sin\theta} = \pm 1 \quad \Rightarrow \dfrac{\sin\theta}{\cos\theta} = \pm\dfrac{b}{a}$

Hence $\sin\theta = \pm\dfrac{b}{\sqrt{a^2 + b^2}}$ and $\cos\theta = \pm\dfrac{a}{\sqrt{a^2 + b^2}}$

$\Rightarrow (x_1, y_1) = (a\cos\theta, b\sin\theta) = \left(\pm\dfrac{a^2}{\sqrt{a^2 + b^2}}, \pm\dfrac{b^2}{\sqrt{a^2 + b^2}}\right)$; so we get four such points.

Now the distance of $O(0, 0)$ i.e., the centre of ellipse from the tangent $d = \dfrac{|ab|}{\sqrt{b^2\cos^2\theta + a^2\sin^2\theta}}$

Putting the values of $\sin\theta$ and $\cos\theta$, we get

$$d = \frac{ab}{\sqrt{\frac{a^2b^2}{a^2 + b^2} + \frac{a^2b^2}{a^2 + b^2}}} = \frac{|ab|}{|ab|\sqrt{\frac{2}{a^2 + b^2}}} = \sqrt{\frac{a^2 + b^2}{2}}$$

8. Let $y = mx \pm \sqrt{a^2m^2 + b^2}$ be a tangent to the ellipse $\frac{x^2}{a^2} + \frac{y^2}{b^2} = 1$

Now a line through (0, 0) and perpendicular to tangent will have slope $\frac{-1}{m} \Rightarrow y = -\frac{x}{m}$

$\Rightarrow m = -\frac{x}{y}$

Hence $y = -\frac{x^2}{y} \pm \sqrt{\frac{a^2x^2}{y^2} + b^2}$

$\Rightarrow \frac{y^2 - x^2}{y} = \pm\sqrt{\frac{a^2x^2 + b^2y^2}{y^2}}$ i.e., $(x^2 - y^2)^2 = (a^2x^2 + b^2y^2)$

9. Consider an ellipse $\frac{x^2}{a^2} + \frac{y^2}{b^2} = 1$; where a is constant $(a > b)$ and b is variable i.e., $0 < b < a$

Now the ends of latus rectum are $\left(ae, \pm\frac{b^2}{a}\right)$; where $e^2 = \frac{a^2 - b^2}{a^2}$ {as $0 < b < a \Rightarrow 0 < e < 1$}

Now tangent at $L(x_1, y_1) = \left(ae, \frac{b^2}{a}\right)$ will be $\frac{aex}{a^2} + \frac{b^2}{a}\frac{y}{b^2} - 1 = 0 \Rightarrow ex + y - a = 0$

Now observe that this line will always passes through $(0, a)$ for all values of e.

Similarly for $L'\left(ae, \frac{-b^2}{a}\right)$, we get $(0, -a)$ as the fixed point.

Note that $(0, \pm a)$ are also formed for the other latus rectram.

10. Let P $(a \cos\alpha, b \sin\alpha)$ be the point on the ellipse $\frac{x^2}{a^2} + \frac{b^2}{b^2} = 1$

Now the equation of tangent at P (α) will be $\frac{x}{a}\cos\alpha + \frac{y}{b}\sin\alpha = 1$

Further the auxiliary circle is $x^2 + y^2 = a^2$

Since the tangent (forming a chord for the auxiliary circle) subtends a right angle, so on homogenizing.

$\Rightarrow x^2 + y^2 = a^2 \left\{\frac{x}{a}\cos\alpha + \frac{y}{b}\sin\alpha\right\}^2$

$\Rightarrow x^2 (1 - \cos^2\alpha) + y^2 \left(1 - \frac{a^2}{b^2}\sin^2\alpha\right) - 2xy$

$\cos\alpha.\sin\alpha\left(\frac{a}{b}\right) = 0$ for 90° angle, we get $(1 - \cos^2\alpha)$

$+ 1 - \frac{a^2}{b^2}\sin^2\alpha = 0$

$\Rightarrow \sin^2\alpha\left(1 - \frac{a^2}{b^2}\right) = -1$ or $\frac{a^2 - b^2}{b^2} = \frac{1}{\sin^2\alpha}$

$\Rightarrow \frac{a^2e^2}{a^2(1 - e^2)} = \frac{1}{\sin^2\alpha}$

$\therefore$ $e^2 \sin^2\alpha = 1 - e^2$ or $e^2 (1 + \sin^2\alpha) = 1$

Hence $e = \frac{1}{\sqrt{1 + \sin^2\alpha}}$

11. The given ellipse is $\frac{x^2}{a^2} + \frac{y^2}{b^2} = 1$ and the circle $x^2 + y^2 = ab$

$\Rightarrow y^2 = ab - x^2 \Rightarrow \frac{x^2}{a^2} + \frac{ab - x^2}{b^2} = 1$

Hence $(b^2 - a^2) x^2 = a^2b (b - a)$

$\Rightarrow x^2 = \frac{a^2b}{a + b}$ in Ist quadrant. Hence $x = \pm a\sqrt{\frac{b}{a + b}}$

From $x = a \cos\theta$, we get $\cos\theta = \sqrt{\frac{b}{a + b}} \Rightarrow \sin\theta = \sqrt{\frac{a}{a + b}}$

Now slope of tangent for ellipse at (θ) is $m_1 = -\left(\frac{b}{a}\right)\frac{\cos\theta}{\sin\theta}$

and slope of tangent for the circle is $m_2 = \frac{(-1)\, a\cos\theta}{b\sin\theta}$.

This gives the angle ϕ between the tangents as

$\tan\phi = \frac{\left(\frac{a}{b} - \frac{b}{a}\right)\frac{\cos\theta}{\sin\theta}}{1 + \frac{\cos^2\theta}{\sin^2\theta}}$ i.e., $\tan\phi = \frac{a^2 - b^2}{ab}.\sin\theta\cos\theta$

Putting the value of $\sin\theta$ and $\cos\theta$, we get

$\tan\phi = \frac{a^2 - b^2}{ab}\frac{\sqrt{ab}}{a + b} = \frac{a - b}{\sqrt{ab}}$

Hence $\phi = \tan^{-1}\left(\frac{a - b}{\sqrt{ab}}\right)$

12. The ellipse is $\frac{x^2}{1} + \frac{y^2}{3} = 1$, which is of vertical type

The tangent that makes equal intercepts on axes has slope $m = \pm 1 \Rightarrow y = mx \pm \sqrt{a^2m^2 + b^2} \Rightarrow y = \pm x \pm \sqrt{1 + 3}$

i.e., $y = \pm x \pm 2$ (all four possibilities).

13. The given ellipse is $\frac{x^2}{9} + \frac{y^2}{4} = 1$

Hence $\frac{x}{3}\cos\theta + \frac{y}{2}\sin\theta = 1$ will be a tangent at any point θ since $\theta \in [0, 2\pi]$ (given), so there are infinite numbers of points

14. The given ellipse is $\frac{x^2}{1} + \frac{y^2}{2} = 1$; straight line $y = mx \pm \sqrt{a^2m^2 + b^2}$ will be a tangent to the ellipse

$\Rightarrow$ $mx - y \pm \sqrt{m^2 + 2} = 0$ will be a tangent. If this line is also a tangent to $x^2 + y^2 = 4$, then distance of (0, 0) from the tangent line must be 2 units i.e., $\frac{\sqrt{m^2 + 2}}{\sqrt{1 + m^2}} = 2$

$\Rightarrow$ $m^2 + 2 = 4m^2 + 4$ $\Rightarrow 3m^2 = -2$

$\Rightarrow$ No tangent is possible

15. The given ellipse is $\frac{x^2}{16} + \frac{y^2}{9} = 1$

$\Rightarrow$ $a = 4$ and $b = 3$

So a line $\frac{x}{4}\cos\theta+\frac{y}{3}\sin\theta=1$ will be a tangent at P $(a\cos\theta, b\sin\theta)$

Since the intercepts are equal

$\therefore\ m=\pm 1$

Hence $\frac{-3\cos\theta}{4\sin\theta}=\pm 1 \Rightarrow \frac{\sin\theta}{\cos\theta}=\pm\frac{3}{4}$

$\Rightarrow\ \sin\theta=\frac{\pm 3}{4}$ and $\cos\theta=\pm\frac{4}{5}$. Hence $y=\pm x\pm 5$

16. The given ellipse $\frac{x^2}{(3\sqrt{2})^2}+\frac{y^2}{(4\sqrt{2})^2}=1$

The equation of a tangent with slope $m=-\frac{4}{3}$ will be $y=-\frac{4}{3}x\pm\sqrt{\frac{18\times 16}{9}+32}=-\frac{4}{3}x\pm 8$

$\Rightarrow$ $4x+3y\pm 24=0$ are the tangent lines. Without any loss of generality, let the tangent be in Ist quadrant

$\Rightarrow$ $A=6$ and $B=8$. Hence $\Delta OAB=\frac{6\times 8}{2}=24$ square units.

17. Let PQ and PR be a pair of perpendicular tangents from P (h, k) to the ellipse $\frac{x^2}{a^2}+\frac{y^2}{b^2}=1$

$\Rightarrow$ $h^2+k^2=a^2+b^2$ and the chord contact of tangent will be $T=0$ i.e., $\frac{hx}{a^2}+\frac{ky}{b^2}=1$ or $y=-\frac{hb^2}{a^2k}x+\frac{b^2}{k}$ .....(i)

Now for the ellipse $\frac{x^2}{\left(\frac{a^4}{a^2+b^2}\right)}+\frac{y^2}{\left(\frac{b^4}{a^2+b^2}\right)}=1$

A line with slope m will be tangent, when $y=mx\pm\sqrt{A^2m^2+B^2}$ ......(ii)

From (i) and (ii), we get $A^2m^2+B^2=\left(\frac{b^2}{k}\right)^2$ for $A^2=\frac{a^4}{a^2+b^2}, B^2=\frac{b^4}{a^2+b^2}$ and $m=-\frac{hb^2}{ka^2}$ as shown below

$$A^2m^2+B^2=\frac{a^4}{(a^2+b^2)}\cdot\frac{b^4h^2}{k^2a^4}+\frac{b^4}{(a^2+b^2)}=\left(\frac{h^2}{k^2}+1\right)\frac{b^4}{a^2+b^2}$$

$\because\ h^2+k^2=a^2+b^2$

$\therefore\ \left(\frac{h^2+k^2}{k^2}\right)\left(\frac{b^4}{a^2+b^2}\right)=\frac{b^4}{k^2}=\left(\frac{b^2}{k}\right)^2$; which prove the condition.

TEXTUAL EXERCISE-3 (OBJECTIVE)

1. (b) The line $y=2x+c$ has slope $m=2$

The given ellipse is $\frac{x^2}{8}+\frac{y^2}{4}=1$, so $y=mx\pm\sqrt{a^2m^2+b^2}$

For $m=2$ $\Rightarrow \pm\sqrt{a^2m^2+b^2}=\pm 6$

2. (c) The given ellipse is $\frac{x^2}{16}+\frac{y^2}{1}=1$. The tangent with slope of $m=\sqrt{3}$ will be $y=\sqrt{3}x\pm\sqrt{16\times 3+1}$

$\Rightarrow$ $y=\sqrt{3}x\pm 7$ or $\sqrt{3}x-y\pm 7=0$

3. (a) The given ellipse $\frac{x^2}{16}+\frac{y^2}{9}=1$. A line with slope m will have equation $y=mx\pm\sqrt{16m^2+9}$.

Since (2, 3) lies on this line, so $3-2m=\pm\sqrt{16m^2+9}$.

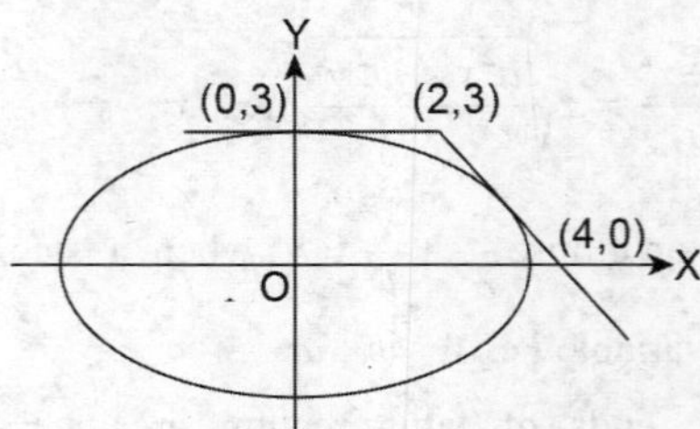

On squaring, we get $9+4m^2-12m=16m^2+9$

$\Rightarrow$ $12m^2+12m=0 \Rightarrow m=0, -1$

For $m=0$, we get $y=3$ as tangent (or $x+y=5$) and for $m=-1$, we get $y=-x+5$ (or $x+y=5$) as tangent

4. (c) A line $y=mx+c$ will intersect the ellipse $\frac{x^2}{a^2}+\frac{y^2}{b^2}=1$ will have real roots i.e., $b^2x^2+a^2(mx+c)^2-a^2b^2=0$ has real roots

$\Rightarrow$ $(a^2m^2+b^2)x^2+2mca^2x+a^2(c^2-b^2)=0$

$\therefore$ $D\geq 0$

$\Rightarrow$ $(a^2m^2+b^2)\geq c^2$ i.e., $a^2m^2\geq c^2-b^2$

5. (b) The given tangent will pass through $(h, 0)$ and $(0, k)$. Hence its slope $m=-\frac{k}{h}$ and equation is $y=-\frac{k}{h}x+k$.

Now any tangent with slope m for $\frac{x^2}{a^2}+\frac{y^2}{b^2}=1$ will be $y=mx\pm\sqrt{a^2m^2+b^2}$ which for $m=-\frac{k}{h}$ gives $\frac{a^2k^2}{h^2}+b^2=k^2$

$\Rightarrow$ $k^2\left(\frac{a^2}{h^2}+\frac{b^2}{k^2}\right)=k^2 \Rightarrow \frac{a^2}{h^2}+\frac{b^2}{k^2}=1$

6. (c) Let m be the slope of any tangent to the ellpse $\frac{x^2}{a^2}+\frac{y^2}{b^2}=1$, then $y=mx\pm\sqrt{a^2m^2+b^2}$ gives the area of $\Delta OAB=\frac{a^2m^2+b^2}{2m}$

The minimum value of ΔOAB area is given by $A.M.\geq G.M.$ i.e., $\frac{1}{2}\left\{\frac{a^2}{2}m+\frac{b^2}{2}\left(\frac{1}{m}\right)\right\}\geq\sqrt{\frac{a^2}{2}m,\frac{b^2}{2}\frac{1}{m}}=\frac{ab}{2}$

$\Rightarrow$ $\frac{a^2}{2}m+\frac{b^2}{2}\left(\frac{1}{m}\right)\geq ab$ i.e., Minimum value of area of $\Delta OAB=ab$

7. (a) Let $y = mx + \sqrt{a^2m^2 + b^2}$ be any tangent to the ellipse $\frac{x^2}{a^2} + \frac{y^2}{b^2} = 1$

Now the two points on minor axis are $\left(0, \pm\sqrt{a^2 - b^2}\right)$

So the square of distances of these points from the tangents is

$$\left(\frac{\sqrt{a^2 - b^2} + \sqrt{a^2m^2 + b^2}}{\sqrt{1+m^2}}\right)^2 + \left(\frac{-\sqrt{a^2b^2} + \sqrt{a^2m^2 + b^2}}{\sqrt{1+m^2}}\right)^2$$

$$= \frac{2(a^2m^2 + b^2 + a^2 - b^2)}{(1+m^2)} = 2a^2$$

8. (b) Let $y = mx + \sqrt{a^2m^2 + b^2}$ be any tangent to the ellipse $\frac{x^2}{a^2} + \frac{y^2}{b^2} = 1$ having foci at $(\pm ae, 0)$

∴ Product of distances of foci from the tangent

$$P = \frac{\left|aem + \sqrt{a^2m^2 + b^2}\right|}{\sqrt{1+m^2}} \cdot \frac{\left|-aem + \sqrt{a^2m^2 + b^2}\right|}{\sqrt{1+m^2}}$$

$$= \frac{a^2m^2 + b^2 - a^2e^2m^2}{(1+m^2)} = \frac{m^2(a^2 - a^2e^2) + b^2}{(1+m^2)}$$

$$= \frac{b^2(1+m^2)}{(1+m^2)} = b^2$$

9. (a), (c) Let $y = mx + \sqrt{a^2m^2 + b^2}$ be a tangent to the ellipse $\frac{x^2}{a^2} + \frac{y^2}{b^2} = 1$

Now consider a line perpendicular to the tangent and passing though foci.

Now foot of perpendicular from $(ae, 0)$ on $mx - y + \sqrt{a^2m^2 + b^2} = 0$ is $\frac{ae - h}{m} = \frac{0 - k}{-1} = \frac{mae + \sqrt{a^2m^2 + b^2}}{m^2 + 1}$

$\Rightarrow h^2 + k^2$

$$= \left\{\frac{m^2ae + m\sqrt{a^2m^2 + b^2} - aem^2 - ae}{m^2 + 1}\right\}^2 + \left(\frac{mae + \sqrt{a^2m^2 + b^2}}{m^2 + 1}\right)^2$$

$$= \frac{m^2(a^2m^2 + b^2) + a^2e^2 + (a^2m^2 + b^2) + m^2a^2e^2}{(m^2 + 1)^2}$$

$$= \frac{(1+m^2)\{a^2m^2 + b^2 + a^2e^2\}}{(1+m^2)^2} = \frac{a^2m^2 + a^2 - a^2e^2 + a^2e^2}{(1+m^2)} = a^2$$

10. (b) Let $\frac{x^2}{a^2} + \frac{y^2}{b^2} = 1$ be an ellipse wih $a > b$,

Now the ends of *L.R.* are $\left(ae, \pm\frac{b^2}{a}\right)$, because of symmetry the tangents will intersect on *x*-axis.

Now $\frac{x_1x}{a^2} + \frac{y_1y}{b^2} = 1$ will given for $\left(ae, \frac{b^2}{a}\right)$ *x*-axis at $x = a/e$ which happens to be on the corresponding directrix.

Similarly the tangent at $\left(ae, -\frac{b^2}{a}\right)$ will also intersect at the same point

11. (b) The given ellipse is $\frac{x^2}{48} + \frac{y^2}{36} = 1$

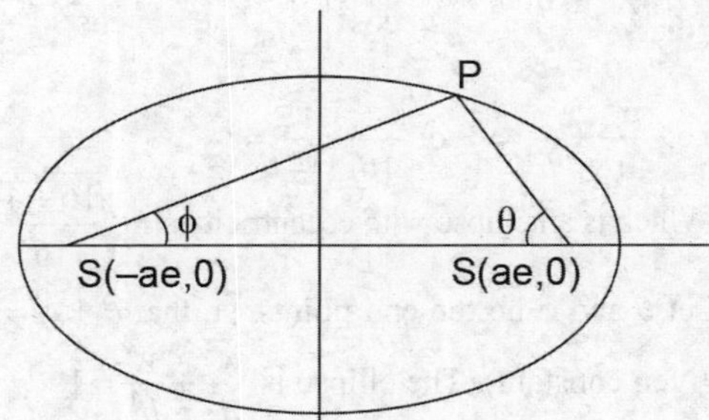

$\Rightarrow a = 4\sqrt{3}$, $b = 6$ and $e = \sqrt{\frac{48 - 36}{48}} = \frac{1}{2}$

Hence $ae = 2\sqrt{3}$

Now ratio $R = \frac{1 - \cos\theta}{1 + \cos\phi} : \frac{1 + \cos\theta}{1 - \cos\phi}$

$$\Rightarrow R = \frac{2\sin^2\left(\frac{\theta}{2}\right)}{2\cos^2\left(\frac{\phi}{2}\right)} : \frac{2\cos^2\left(\frac{\theta}{2}\right)}{2\sin^2\frac{\phi}{2}}$$

$$\Rightarrow \tan^2\left(\frac{\theta}{2}\right) : \cot^2\left(\frac{\phi}{2}\right) \textit{ i.e.} \left\{\tan\left(\frac{\theta}{2}\right) . \tan\frac{\phi}{2}\right\}^2 : 1$$

We know that $\tan\frac{\theta}{2} . \tan\frac{\phi}{2} = \frac{1 - e}{1 + e} \Rightarrow R = \left(\frac{1 - \frac{1}{2}}{1 + \frac{1}{2}}\right)^2 : 1$ i.e., 1 : 9

12. (a) The given ellipse is $\frac{x^2}{4} + \frac{y^2}{2} = 1$ and the circle is $x^2 + y^2 = 1$ to the circle

Let $y = mx + (1)\sqrt{1+m^2}$ be a tangent to the circle that intersects the ellipse $\frac{x^2}{4} + \frac{y^2}{2} = 1$ at two points with eccentric angle θ and ϕ (where $a = 2$ and $b = \sqrt{2}$)

∴ The tangent is a chord for the ellipse given as

$$\frac{x}{a}\cos\left(\frac{\theta + \phi}{2}\right) + \frac{y}{b}\left(\frac{\theta + \phi}{2}\right) = \cos\left(\frac{\theta - \phi}{2}\right)$$

Also distance of (0, 0) from the chord will be $r = 1$

$$\text{i.e., } \frac{\left|-\cos\left(\frac{\theta - \phi}{2}\right)\right|}{\sqrt{\frac{\cos^2\left(\frac{\theta + \phi}{2}\right)}{a^2} + \frac{\sin^2\left(\frac{\theta + \phi}{2}\right)}{b^2}}} = 1$$

$$\text{i.e., } \left(\frac{1}{a^2}\right)\cos^2\left(\frac{\theta + \phi}{2}\right) + \left(\frac{1}{b^2}\right)\sin^2\left(\frac{\theta + \phi}{2}\right) = \cos^2\left(\frac{\theta - \phi}{2}\right)$$

Now the tangents to the ellipse at $P(\theta)$ and $Q(\phi)$ will intersect in R given by $R\left(\frac{a\cos\left(\frac{\theta + \phi}{2}\right)}{\cos\left(\frac{\theta - \phi}{2}\right)}, \frac{b\sin\left(\frac{\theta + \phi}{2}\right)}{\cos\left(\frac{\theta - \phi}{2}\right)}\right) = (h, k)$

$$\Rightarrow \left(\frac{h}{a^2}\right)^2+\left(\frac{k}{b^2}\right)^2=\frac{\cos^2\left(\frac{\theta+\phi}{2}\right)}{a^2\cos^2\left(\frac{\theta-\phi}{2}\right)}+\frac{\sin^2\left(\frac{\theta+\phi}{2}\right)}{b^2\cos^2\left(\frac{\theta-\phi}{2}\right)}=1$$

i.e., $\frac{x^2}{a^4}+\frac{y^2}{b^4}=1$ or $\frac{x^2}{16}+\frac{y^2}{4}=1$

Which is an ellipse with eccentricity $e=\sqrt{\frac{16-4}{16}}=\sqrt{3}/2$

13. (d) Let θ and ϕ be the end points, so that $\theta+\phi=k$ (some given constant). The ellipse is $\frac{x^2}{a^2}+\frac{y^2}{b^2}=1$

Now the point intersection of tangents is given by

$$\frac{x}{a^2}\left\{\frac{a\cos\left(\frac{\theta+\phi}{2}\right)}{\cos\left(\frac{\theta-\phi}{2}\right)}\right\}+\frac{y}{b^2}\left\{\frac{b\sin\left(\frac{\theta+\phi}{2}\right)}{\cos\left(\frac{\theta-\phi}{2}\right)}\right\}=1$$

$$\text{As } (x_1, y_1)=\left(\frac{a\cos\left(\frac{\theta+\phi}{2}\right)}{\cos\left(\frac{\theta-\phi}{2}\right)},\frac{b\sin\left(\frac{\theta+\phi}{2}\right)}{\cos\left(\frac{\theta-\phi}{2}\right)}\right)$$

$$\Rightarrow \frac{x_1^2}{a^2}\cos^2\left(\frac{\theta-\phi}{2}\right)+\frac{y_1^2}{b^2}\cos^2\left(\frac{\theta-\phi}{2}\right)=1$$

From

$$\frac{x_1}{a}\cos\left(\frac{\theta-\phi}{2}\right)=\cos\left(\frac{\theta+\phi}{2}\right) \text{ and } \frac{y_1}{b}\cos\left(\frac{\theta-\phi}{2}\right)=\sin\left(\frac{\theta+\phi}{2}\right),$$

we get

$$\frac{x_1y_1}{ab}\cos^2\left(\frac{\theta-\phi}{2}\right)=\sin\left(\frac{\theta+\phi}{2}\right).\cos\left(\frac{\theta+\phi}{2}\right)=\frac{1}{2}\sin(\theta+\phi)$$

Since $\theta+\phi=k \quad \Rightarrow \sin(\theta+\phi)=$ constant c(say)

Hence the locus of (x_1, y_1) becomes $\left(\frac{x_1^2}{a^2}+\frac{y_1^2}{b^2}\right)\left(\frac{c\,ab}{2x_1y_1}\right)=1$

or $\frac{x_1^2}{a^2}+\frac{y_1^2}{b^2}=\frac{2x_1y_1}{abc}$ i.e., $b^2cx_1^2+a^2cy_1^2-2abx_1y_1=0$

Which is homogeneous in x_1 and y_1, so this is a combined equation of two straight lines (as $\Delta=0$)

14. (c) The line is $y=-\frac{3x}{5}+\frac{k}{5}$ and the elipse $\frac{x^2}{25}+\frac{y^2}{16}=1$

A line with slope $m=-3/5$ will be tangent, when $\frac{k}{5}=\pm\sqrt{25\times\frac{9}{25}+16}$ i.e., $k=\pm 5$

15. (d) From the given information $a=5$, $ae=4$

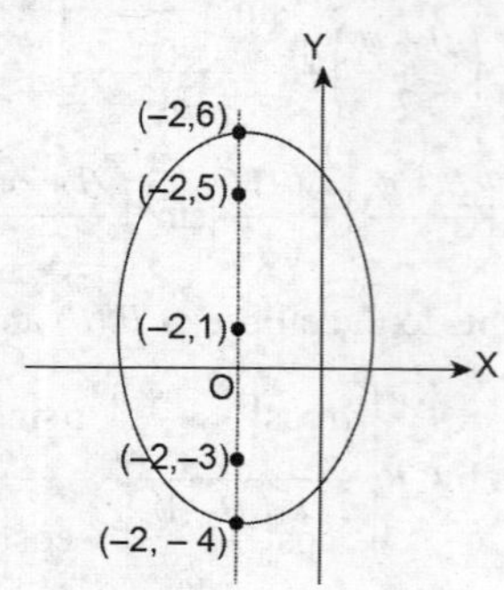

$\Rightarrow b^2=a^2-a^2e^2=9.$

Hence the ellipse is $\frac{(x+2)^2}{9}+\frac{(y-1)^2}{25}=1$

16. (a), (b) The ellipse is $\frac{x^2}{a^2}+\frac{y^2}{b^2}=1$. Let its ends of latus rectra be $\left(ae,\pm\frac{b^2}{a}\right)$

Putting $x=ae$ and $y=\pm b^2/a$, we get $x^2=a^2e^2$ and $b^2=\pm ay$

$\because \; b^2=a^2-a^2e^2$

$\therefore \; a^2-x^2=\pm ay$ i.e., $x^2=a(a\pm y)$

17. (a), (b), (c) Let θ be the parametric angle of a point P on the ellipse $\frac{x^2}{a^2}+\frac{y^2}{b^2}=1$.

The equation of tangent will be $\frac{x}{a}\cos\theta+\frac{y}{b}\sin\theta=1$ and the intercepts on the axes will be a $\sec\theta$ and b $\text{cosec}\theta$

$\Rightarrow$ Length of tangent (L) intercepted between axes is given by $L^2=a^2\sec^2\theta+b^2\text{cosec}^2\theta$

For smallest length L, we need $\frac{dL}{d\theta}=0$ and $\frac{d^2L}{d\theta^2}>0$

$\Rightarrow 2L\frac{dL}{d\theta}=2a^2\sec^2\theta.\tan\theta-2b^2\text{cosec}^2\theta.\cot\theta=0$ i.e.,

$$\left\{\frac{a^2\sin\theta}{\cos^3\theta}-\frac{b^2\cos\theta}{\sin^3\theta}\right\}\left(\frac{1}{L}\right)=0$$

Now $a^2\sin^4\theta-b^2\cos^4\theta=0$

$\Rightarrow a\sin^2\theta-b\cos^2\theta=0$ (as $a\sin^2\theta+b\cos^2\theta\neq 0$)

$\Rightarrow \frac{\sin\theta}{\cos\theta}=\pm\sqrt{\frac{b}{a}}$

Hence $\theta=\pm\tan^{-1}\sqrt{\frac{b}{a}}\Rightarrow\theta=n\pi\pm\tan^{-1}\sqrt{\frac{b}{a}}$

For $\theta\in(-\pi,\pi]$, we get $\theta=\pm\tan^{-1}\sqrt{\frac{b}{a}},\pi-\tan^{-1}\sqrt{\frac{b}{a}}$

TEXTUAL EXERCISE-4 (SUBJECTIVE)

1. (a) The given ellipse is $3x^2+5y^2-30=0$ and the line is $y=2x+3$

$\Rightarrow 3x^2+5(4x^2+9+12x)-30=0$

$\Rightarrow 23x^2+60x+15=0$

$\Rightarrow \frac{x_1+x_2}{2}=\frac{-60}{2\times 23}=-\frac{30}{23}$

$\therefore \frac{y_1+y_2}{2}=\frac{2(x_1+x_2)+6}{2}=(x_1+x_2)+3=\frac{-60}{23}+3=\frac{9}{23}$

Hence the mid point will be $\left(-\frac{30}{2},\frac{9}{23}\right)$

(b) The chord bisected at (2, 1) on the ellipse $x^2+4y^2=36$ is $T=S_1$

i.e., $\frac{2x}{36}+\frac{1y}{9}-1=\frac{4}{36}+\frac{4}{36}-1$

$\Rightarrow 2x+4y-8=0 \qquad \Rightarrow x+2y-4=0$

2. **(a)** Let (h, k) be the mid point of any chord on ellipse $\frac{x^2}{a^2}+\frac{y^2}{b^2}=1$, then $\frac{hx}{a^2}+\frac{ky}{b^2}-1=\frac{h^2}{a^2}+\frac{k^2}{b^2}-1$

Since the chords are passing through $(0, b)$

$\therefore \quad \frac{k}{b}=\frac{h^2}{a^2}+\frac{k^2}{b^2}$

Hence the locus will be $\frac{x^2}{a^2}+\frac{y^2}{b^2}=\frac{y}{b}$

(b) Let $P(h, k)$ be the mid-points of the chord that subtends a right angle at the centre of the ellipse $\frac{x^2}{a^2}+\frac{y^2}{b^2}=1$

From $T = S_1$, the chord become $\frac{hx}{a^2}+\frac{ky}{b^2}=\frac{h^2}{a^2}+\frac{k^2}{b^2}$ homogenizing the ellipse equation we get

$$\frac{\frac{x^2}{a^2}+\frac{y^2}{b^2}-\left(\frac{hx}{a^2}+\frac{ky}{b^2}\right)^2}{\left(\frac{h^2}{a^2}+\frac{k^2}{b^2}\right)^2}=0$$

For 90° angle at the origin $A + B = 0$

$$\Rightarrow \quad \frac{1}{a^2}\left(\frac{h^2}{a^2}+\frac{k^2}{b^2}\right)^2-\frac{h^2}{a^4}+\frac{1}{b^2}\left(\frac{h^2}{a^2}+\frac{k^2}{b^2}\right)^2-\frac{k^2}{b^4}=0$$

$$\Rightarrow \quad \left(\frac{h^2}{a^2}+\frac{k^2}{b^2}\right)^2\left(\frac{1}{a^2}+\frac{1}{b^2}\right)=\frac{h^2}{a^4}+\frac{k^2}{b^4}$$

Rearrangement, we get $(b^4h^2 + a^4k^2)\, a^4b^4$

$$\left\{\frac{b^2h^2+a^2k^2}{a^2b^2}\right\}^2=\frac{1}{a^2}+\frac{1}{b^2} \text{ i.e., } \frac{a^4y^2+b^4x^2}{(b^2x^2+a^2y^2)^2}=\frac{1}{a^2}+\frac{1}{b^2}$$

3. The tangent to the ellipse $\frac{x^2}{6}+\frac{y^2}{3}=1$ at P and Q intersect at 90°.

Let the point of intersect be (h, k) which will lie on the director circle hence $h^2 + k^2 = 9$

Now the chord of contact from (h, k) will be $\frac{h}{6}x+\frac{k}{3}y-1=0$

i.e., $y=-\frac{h}{2k}x+\frac{3}{k}$ which has slope $m = -\frac{h}{2k}$

Now a tangent to the ellipse $\frac{x^2}{4}+\frac{y^2}{1}=1$ with slope m will be $y=mx+\sqrt{a^2m^2+b^2}$

$$\Rightarrow \quad \sqrt{4\left(\frac{h^2}{4k^2}\right)+1}=\frac{3}{k} \text{ i.e., } \sqrt{\frac{h^2+k^2}{k^2}}=\frac{3}{k}$$

Which is true as $h^2 + k^2 = 9$

Hence a tangent to the ellipse $\frac{x^2}{4}+\frac{y^2}{1}=1$ is proved

4. Let $P(\theta)$ and $Q(\phi)$ be end the point with eccentric angles of the chord for the ellipse $\frac{x^2}{a^2}+\frac{y^2}{b^2}=1$

$\Rightarrow$ $P(a\cos\theta, b\sin\theta)$ and $Q(a\cos\phi, b\sin\phi)$

Now the point of intersection of tangent at P & Q is

$$R=\left(\frac{a\cos\left(\frac{\theta+\phi}{2}\right)}{\cos\left(\frac{\theta-\phi}{2}\right)}, \frac{b\sin\left(\frac{\theta+\phi}{2}\right)}{\cos\left(\frac{\theta-\phi}{2}\right)}\right)$$

Since R lies on $x^2 + y^2 = d^2$

$$\therefore \quad a^2\cos^2\left(\frac{\theta+\phi}{2}\right)+b^2\sin^2\left(\frac{\theta+\phi}{2}\right)=d^2\cos^2\left(\frac{\theta-\phi}{2}\right)$$

Now consider the mid-point M of P & Q; $M(x_1, y_1)$

$$=\left(\frac{a}{2}(\cos\theta+\cos\phi), \frac{b}{2}(\sin\theta+\sin\phi)\right)$$

$$\Rightarrow \quad x_1=\frac{a}{2}\left\{2\cos\left(\frac{\theta+\phi}{2}\right)\cos\left(\frac{\theta-\phi}{2}\right)\right\} \text{ and}$$

$$y_1=\frac{b}{2}\left\{2\sin\left(\frac{\theta+\phi}{2}\right)\cos\left(\frac{\theta-\phi}{2}\right)\right\}$$

$$\therefore \quad \cos^2\left(\frac{\theta+\phi}{2}\right)+\sin^2\left(\frac{\theta+\phi}{2}\right)=1=\left(\frac{x_1^2}{a^2}+\frac{y_1^2}{b^2}\right)\frac{1}{\cos^2\left(\frac{\theta-\phi}{2}\right)}$$

$$\text{i.e., } \cos^2\left(\frac{\theta-\phi}{2}\right)=\frac{x_1^2}{a^2}+\frac{y_1^2}{b^2}$$

$$\text{Also } a^2\cos^2\left(\frac{\theta+\phi}{2}\right)+b^2\sin^2\left(\frac{\theta+\phi}{2}\right)=d^2\cos^2\left(\frac{\theta-\phi}{2}\right)$$

$$=\frac{(x_1^2+y_1^2)}{\cos^2\left(\frac{\theta-\phi}{2}\right)}$$

$$\text{Clearly } (x_1^2+y_1^2)=d^2\cos^4\left(\frac{\theta-\phi}{2}\right)$$

Putting $\cos^2\left(\frac{\theta-\phi}{2}\right)=\frac{x_1^2}{a^2}+\frac{y_1^2}{b^2}$, we get

$$d^2\left(\frac{x_1^2}{a^2}+\frac{y_1^2}{b^2}\right)^2=(x_1^2+y_1^2) \text{ or } d^2\left\{\frac{x^2}{a^2}+\frac{y^2}{b^2}\right\}^2=x^2+y^2$$

5. Let $P(\theta) = P(\cos\theta, b\sin\theta)$ be a point on the ellipse $\frac{x^2}{a^2}+\frac{y^2}{b^2}=1$, then distance of origin from the tangent at P will be

$$d=\frac{|-ab|}{\sqrt{a^2\sin^2\theta+b^2\cos^2\theta}} \text{ or } \frac{a^2\sin^2\theta+b^2\cos^2\theta}{a^2}=\frac{b^2}{d^2}$$

Hence

$$4a^2\left(1-\frac{b^2}{d^2}\right)=4a^2\frac{\{a^2(a^2\sin^2\theta+b^2\cos^2\theta)\}}{a^2}$$

$$=4a^2\cos^2\theta-4b^2\cos^2\theta$$

$$\text{or } 4a^2\left(1-\frac{b^2}{d^2}\right)=4\cos^2\theta(a^2-b^2)=4a^2e^2\cos^2\theta$$

Now $F_1(ae, 0)$ and $F_2(-ae, 0)$, so $PF_1 = a(1 - e\cos\theta)$ and $PF_2 = a(1 + e\cos\theta)$

$\Rightarrow$ $(PF_1 - PF_2)^2 = 4a^2 e^2\cos^2\theta$

Since $4a^2e^2 \cos^2\theta = 4a^2 \left(1-\dfrac{b^2}{d^2}\right)$

$\Rightarrow$ $(PF_1 - PF_2)^2 = 4a^2 (1 - b^2/d^2)$

6. Le $P(\theta) = (a \cos\theta, b \sin\theta)$ be a point on the ellipse $\dfrac{x^2}{a^2}+\dfrac{y^2}{b^2}=1$ and $S(ae, 0)$ be a focus.

Then equation of tangent at P is $\dfrac{x}{a}\cos\theta+\dfrac{y}{b}\sin\theta=1$ and the distance (P) of focus (S) from the tangent is

$$p=\frac{|-ab+ab\cos\theta e|}{\sqrt{a^2\sin^2\theta+b^2\cos^2\theta}}=\frac{ab|1-e\cos\theta|}{a^2\sin^2\theta+b^2\cos^2\theta}$$

$$\Rightarrow \left(\frac{b}{p}\right)^2=\frac{a^2\sin^2\theta+b^2\cos^2\theta}{a^2(1-e\cos\theta)^2}$$

$$=\frac{a^2\sin^2\theta+(a^2\cos^2\theta-a^2e^2\cos^2\theta)}{a^2(1-e\cos\theta)^2}$$

i.e., $\dfrac{b^2}{p^2}=\dfrac{a^2(1-e^2\cos^2\theta)}{a^2(1-e\cos\theta)^2}=\dfrac{a(1+e\cos\theta)}{a(1-e\cos\theta)}=\dfrac{2a-a(1-e\cos\theta)}{a(1-e\cos\theta)}$

Now distance $SP=\sqrt{(a\cos\theta-ae)^2+\sin^2\theta}=a(1-e\cos\theta)$

$$\therefore \frac{b^2}{p^2}=\frac{2a}{a(1-e\cos\theta)}-\frac{a(1-e\cos\theta)}{a(1-e\cos\theta)}=\frac{2a}{SP}-1$$

7. Let $y=mx+\sqrt{a^2m^2+b^2}$ be a tangent to the ellipse $\dfrac{x^2}{a^2}+\dfrac{y^2}{b^2}=1$

This will be a tangent to the circle $x^2 + y^2 = r^2$ (for $b < r < a$), when $\dfrac{a^2m^2+b^2}{(1+m^2)}=r^2$

Now distance of origin (0, 0) from the tangent is $p=\dfrac{aem}{\sqrt{1+m^2}}$

$\because$ $a^2e^2 = a^2 - b^2$

$$\Rightarrow p=\frac{\sqrt{a^2m^2-b^2m^2}}{\sqrt{1+m^2}}=\frac{\sqrt{a^2m^2+b^2-b^2m^2-b^2}}{\sqrt{1+m^2}}$$

Putting $r^2(1+m^2) = a^2m^2 + b^2$, we get

$$p=\frac{\sqrt{r^2(1+m^2)-b^2(1+m^2)}}{\sqrt{1+m^2}}=\sqrt{r^2-b^2}$$

Now the chord $PQ = 2\sqrt{r^2-b^2}$

$$\Rightarrow PQ=2\sqrt{r^2-(r^2-b^2)}=2b$$

8. Without any loss generality, let $a > b$ and at any stage the major axis of the ellipse makes an angle θ with the horizontal.

Consider focus $S(h, k)$.

Now the centre (C) of the ellipse will be $C(h + ae\cos\theta, k + ae\sin\theta)$ from the focus $S(h, k)$ be perpendicular on the tangent will be at $P(h, 0)$ and $Q(0, k)$ {as tangents are along the axes}.

Since tangents are at at 90°

$\therefore$ $OC^2 = a^2 + b^2$ i.e., $(h + ae\cos\theta)^2 + (k + ae\sin\theta)^2 = a^2 + b^2$

$\Rightarrow$ $h^2 + k^2 + 2ae(h\cos\theta + k\sin\theta) = 2b^2$.

Further foot of perpendiculars from the focus on the tangent (s) is at the auxiliary circle $QC^2 = PC^2 = a^2$

$\Rightarrow$ $a^2e^2\cos^2\theta + k^2 + a^2e^2\sin^2\theta + 2ae\,k\sin\theta = b^2$ or $k^2 + 2aek\sin\theta = b^2$ .....(i)

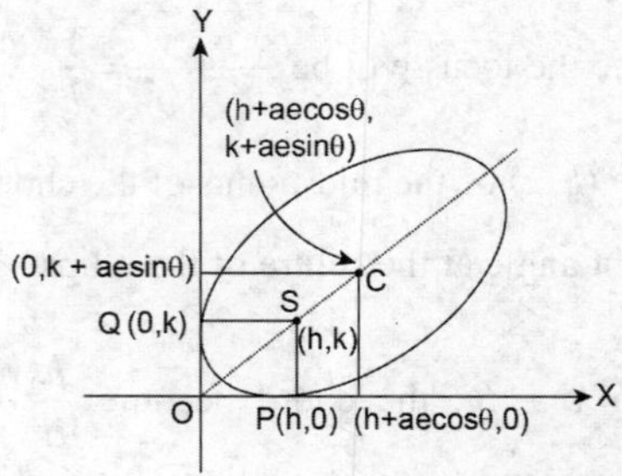

Similarly $h^2 + 2ae\,h\cos\theta = b^2$ ...(ii)

Eliminating, we get $\left(\dfrac{b^2-h^2}{2aeh}\right)^2+\left(\dfrac{b^2-k^2}{2aek}\right)^2=1$ i.e.,

$$\frac{b^4+h^4-2b^2h^2}{h^2}+\frac{b^4+k^4-2b^2k^2}{k^2}=4a^2e^2=4a^2-4b^2$$

$$\Rightarrow b^4\left\{\frac{1}{h^2}+\frac{1}{k^2}\right\}+h^2-2b^2+k^2-2b^2=4a^2-4b^2$$

$$\Rightarrow b^4\left\{\frac{h^2+k^2}{h^2k^2}\right\}+(h^2+k^2)=4a^2$$

$$\Rightarrow (h^2+k^2)\left\{\frac{h^2k^2+b^4}{h^2k^2}\right\}=4a^2$$

i.e., $(x^2 + y^2)(x^2y^2 + b^4) = 4a^2x^2y^2$

9. Let $P(h, k)$ be the mid point of the chords for ellipse $\dfrac{x^2}{a^2}+\dfrac{y^2}{b^2}=1; (a>b)$

$\therefore$ From $T = S_1$, we get $\dfrac{hx}{a^2}+\dfrac{ky}{b^2}-1=\dfrac{h^2}{a^2}+\dfrac{k^2}{b^2}-1$

Since (2, 0) lie on it

$$\Rightarrow \frac{2ah}{a^2}=\frac{h^2}{a^2}+\frac{k^2}{b^2} \text{ or } \frac{h^2}{a^2}-\frac{2ah}{a^2}+\frac{a^2}{a^2}-\frac{a^2}{a^2}+\frac{k^2}{a^2}=0$$

$$\Rightarrow \frac{(h-a)^2}{a^2}+\frac{k^2}{b^2}=1 \text{ or } \frac{(x-a)^2}{a^2}+\frac{y^2}{b^2}=1$$

10. Let $P(h, k)$ be an external point for the ellipse $\dfrac{x^2}{a^2}+\dfrac{y^2}{b^2}=1$, then chord of contact of tangents form P is $T = 0$

i.e., $\dfrac{hx}{a^2}+\dfrac{ky}{b^2}-1=0$. Since it is a tangent to the circle $x^2 + y^2 = c^2$

$$\therefore \frac{1}{\left(\dfrac{h^2}{a^4}+\dfrac{k^2}{b^4}\right)}=c^2$$

$$\Rightarrow c^2\left\{\frac{h^2b^4+k^2a^4}{a^4b^4}\right\}=1 \text{ i.e., } c^2\{b^4x^2+a^4y^2\}=a^4b^4$$

TEXTUAL EXERCISE–4 (OBJECTIVE)

1. **(b)** The ellipse $\frac{x^2}{1}+\frac{y^2}{a^2}=1$ will intersect in four distinct points with ellipse $\frac{x^2}{4}+\frac{y^2}{1}=1$, when $a^2 > 1$

$\Rightarrow a > 1$

Now $a = b^2 - 5b + 7 \Rightarrow b^2 - 5b + 6 > 0$

$\Rightarrow (b-2)(b-3) > 0 \quad \Rightarrow b \in (-\infty, 2) \cup (3, \infty)$

2. **(b)** Let $y = mx \pm \sqrt{(a^2+\lambda)m^2+(b^2+\lambda)}$ be a tangent of slope m to the ellipse $\frac{x^2}{a^2+\lambda}+\frac{y^2}{b^2+\lambda}=1$, then a tangent of slope $-1/m$ will be a tangent to the ellipse $\frac{x^2}{a^2}+\frac{y^2}{b^2}=1$, when $y=-\frac{x}{m}\pm\sqrt{\frac{a^2}{m^2}+b^2}$

$\Rightarrow (y - mx)^2 = (a^2 + \lambda) m^2 + (b^2 + \lambda)$, also $(x + my)^2 = a^2 + b^2m^2$

On adding, we get $(m^2 + 1)(x^2 + y^2) = (m^2 + 1)(a^2 + b^2 + \lambda)$

Hence $x^2 + y^2 = a^2 + b^2 + \lambda$

3. **(c)** Given foci $S(3, 4)$ and $S'(9, 12)$

$\Rightarrow$ Centre of the ellipse is $(6, 8)$ and $2ae = 10 \Rightarrow ae = 5$

Now foot of perpendicular from focus on a tangent is at $(1, -4)$

$\Rightarrow$ Radius of auxiliary circle $r = a = \sqrt{5^2+12^2}=13$.

Hence $e = 5/13$

4. **(c)** Let $P(a\cos\theta, b\sin\theta)$ be a point on the ellipse $\frac{x^2}{a^2}+\frac{y^2}{b^2}=1$ and $S(ae, 0)$ and $S'(-ae, 0)$ be the foci.

Now $SP = a(1 - e\cos\theta)$ and $SP' = a(1 + e\cos\theta)$. The tangent at P is $bx\cos\theta + ay\sin\theta - ab = 0$

$\Rightarrow d_1$ = distance form $S' = \frac{ab(1+e\cos\theta)}{\sqrt{b^2\cos^2\theta+a^2\sin^2\theta}}$ and

d_2 = distance from $S = \frac{ab(1-e\cos\theta)}{\sqrt{b^2\cos^2\theta+a^2\sin^2\theta}}$

$\Rightarrow \frac{S'P}{SP}=\frac{d_1}{d_2}$

5. **(a)** Let $P(3\cos\theta, 2\sin\theta)$ be a point in the first quadrant on ellipse $\frac{x^2}{9}+\frac{y^2}{4}=1$ and $A' = (-3, 0)$.

The equation of tangent at P is $2x\cos\theta + 3y\sin\theta = 6$ which intersect y-axis at $Q\left(0,\frac{2}{\sin\theta}\right)$.

The equation of $A'P$ is $y=\frac{2\sin\theta(x+3)}{3(1+\cos\theta)}$ which will intersect y-axis at $\frac{x^2}{4}+\frac{y^2}{b^4}=1$.

Now $CQ^2 - NQ^2 = CQ^2 - (CQ - CN)^2 = CN\{2CQ - CN\}$

$$=\frac{2\sin\theta}{(1+\cos\theta)}\left\{\frac{4}{\sin\theta}-\frac{2\sin\theta}{1+\cos\theta}\right\}$$

$$=\frac{4\sin\theta\{2+2\cos\theta-\sin^2\theta\}}{(1+\cos\theta)^2\sin\theta}=\frac{4\{1+\cos^2\theta+2\cos\theta\}}{(1+\cos\theta)^2}=4$$

6. **(c), (d)** Consider the ellipse with $a = 5$ and $b = 4$

$\Rightarrow ae=\sqrt{a^2-b^2}=\sqrt{25-16}$

$\therefore ae = 3$

Hence a circle of radius $r = 2, 8$ will touch the ellipse respectively at $(5, 0)$ and $(-5, 0)$

7. **(a)** **Refer to 8 subjective textual exercise-4:** Since the tangent are at 90°

$\Rightarrow$ The point of intersection of tangent (origin in this case) lies on the director circle

So either way if the ellipse slides then distance of its centre from the origin is fixed

$\Rightarrow$ locus will be circle

8. **(d)** Since there are two lines of symmetry

$\therefore$ Four such equilateral triangles are possible.

Without any loss of generality, let $a > b$

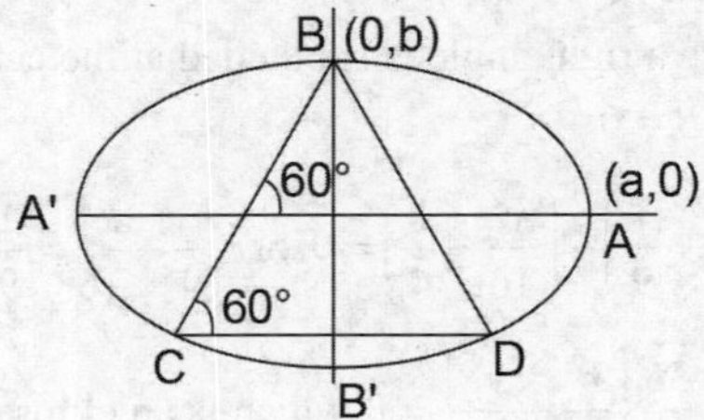

Consider one triangle with vertex $(0, b)$, so equation of BC is $y-b=\sqrt{3}x$. Solving with the ellipse $\frac{x^2}{a^2}+\frac{y^2}{b^2}=1$

$\Rightarrow x\{(b^2 + 3a^2)x + 2\sqrt{3}\, a^2b\} = 0$

$\Rightarrow x=\frac{-2\sqrt{3}\,ba^2}{3a^2+b^2}$

$\therefore y=b+\sqrt{3}x$

$\Rightarrow y=b-\frac{6a^2b}{3a^2+b^2}=\frac{b(b^2-3a^2)}{(3a^2+b^2)}$ from symmetry

$y_D=\frac{b(b^2-3a^2)}{(b^2+3a^2)}$

As centroid is 2: 1 from the vertex

$\therefore$ Centroid of ΔACD is $G_1=\left(0,\frac{b(b^2-a^2)}{(3a^2+b^2)}\right)$. From symmetry G_1 and G_2 are $\left(0,\pm\frac{b(b^2-a^2)}{(3a^2+b^2)}\right)$.

Similarly consider a side as $y=\frac{(x-a)}{\sqrt{3}}$, solving with ellipse, we get $x=\frac{a(a^2-3b^2)}{a^2+3b^2}$ and the centroid G_3 and G_4 are $\left(\pm\frac{a(a^2-b^2)}{(a^2+3b^2)},0\right)$.

9. **(c)** Let $\frac{x^2}{a^2}+\frac{y^2}{b^2}=1$ be an ellipse, so the chord with midpoint (h, k) will be $\frac{hx}{a^2}+\frac{ky}{b^2}-1=\frac{h^2}{a^2}+\frac{k^2}{b^2}-1$

Since the chord passes though (0, b)

$\Rightarrow \quad \frac{h^2}{a^2}+\frac{k^2}{b^2}=\frac{k}{b}$ or $\frac{h^2}{a^2}+\frac{k^2-bk+b^2/4}{b^2}=\frac{b^2}{4b^2}$

i.e., $4\left\{\frac{x^2}{a^2}+\frac{(y-b/2)^2}{b^2}\right\}=1$, which is an ellipse

10. (c) Let R (h, k) be an external point where tangents at P and Q to the ellipse $\frac{x^2}{9}+\frac{y^2}{4}=1$ intersect, so $T = 0$ is the chord of contact of tangents i.e., $\frac{hx}{9}+\frac{ky}{4}=1$

Now homogenizing the ellipse, we get

$\left(\frac{hx}{9}+\frac{k}{4}y\right)^2=\frac{x^2}{9}+\frac{y^2}{4}$ which can be rewritten as

$\frac{h^2x^2}{81}-\frac{x^2}{9}+\frac{k^2y^2}{16}-\frac{y^2}{4}+\frac{2hkxy}{36}=0$

Since a right angle is subtended at the centre

$\therefore \quad A + B = 0$

$\Rightarrow \quad \left(\frac{h^2}{81}-\frac{1}{9}\right)+\left(\frac{k^2}{16}-\frac{1}{4}\right)=0$ or $\frac{h^2}{81}+\frac{k^2}{16}=\frac{1}{9}+\frac{1}{4}$

i.e., $\frac{36}{13}\left\{\frac{x^2}{81}+\frac{y^2}{16}\right\}=1$ which is an ellipse

11. (a) Let P (θ) and Q (ϕ) be two points on the ellipse $\frac{x^2}{a^2}+\frac{y^2}{b^2}=1$

So the point of intersection of tangent is

$$R=(x,y)=\left\{\frac{a\cos\left(\frac{\theta+\phi}{2}\right)}{\cos\left(\frac{\theta-\phi}{2}\right)},\frac{b\sin\left(\frac{\theta+\phi}{2}\right)}{\cos\left(\frac{\theta-\phi}{2}\right)}\right\}$$

$\Rightarrow \quad \frac{x^2}{a^2}\cos^2\left(\frac{\theta-\phi}{2}\right)+\frac{y^2}{b^2}\cos^2\left(\frac{\theta-\phi}{2}\right)=1$

Since θ and ϕ differ by $\frac{\pi}{2}$

$\therefore \quad \theta-\phi=\pm\pi/2$ i.e., $\cos^2\left(\frac{\theta-\phi}{2}\right)=\frac{1}{2} \Rightarrow \frac{x^2}{a^2}+\frac{y^2}{b^2}=2$

12. (a) Consider the ellipse $\frac{x^2}{a^2}+\frac{y^2}{b^2}=1$ and the two points P (θ) and Q (ϕ), so the chord PQ is given as

$\frac{x}{a}\cos\left(\frac{\theta+\phi}{2}\right)+\frac{y}{b}\sin\left(\frac{\theta+\phi}{2}\right)-\cos\left(\frac{\theta-\phi}{2}\right)=0$

Comparing with $\quad x + my + n = 0$, we get

$$\frac{\ell.a\cos\left(\frac{\theta-\phi}{2}\right)}{\cos\left(\frac{\theta+\phi}{2}\right)}=\frac{m.b\sin\left(\frac{\theta-\phi}{2}\right)}{\sin\left(\frac{\theta+\phi}{2}\right)}=-\frac{n}{1}$$

$\Rightarrow \quad a^2\ell^2\cos^2\left(\frac{\theta-\phi}{2}\right)+m^2b^2\cos^2\left(\frac{\theta-\phi}{2}\right)=n^2$

13. (c) The given ellipse is $\frac{x^2}{a^2}+\frac{y^2}{b^2}=1$.

The chord of contact of tangents from P (x_1, y_1) is $T = 0$

$\Rightarrow \quad \frac{x_1}{a^2}x+\frac{y_1}{b^2}-1=0$

Similarly the chord for $Q(x_2, y_2)$ is $\frac{x_2}{a^2}x+\frac{y_2}{b^2}y-1=0$

Since the chords are at 90°

$m_1m_2=-1$ i.e., $\left(\frac{-x_1}{a^2}.\frac{b^2}{y_1}\right).\left(\frac{-x_2}{a^2}.\frac{b_2}{y_2}\right)=-1$

i.e., $\frac{x_1x_2}{y_1y_2}=-\frac{a^4}{b^4}$

14. (a) Let $\frac{x^2}{a^2}+\frac{y^2}{b^2}=1$ be an ellipse on which two points P (θ) and Q (ϕ) are considered.

So the point of intersection of tangents at P and Q is given by R (x, y) $=\left\{\frac{a\cos\left(\frac{\theta+\phi}{2}\right)}{\cos\left(\frac{\theta-\phi}{2}\right)},\frac{b\sin\left(\frac{\theta+\phi}{2}\right)}{\cos\left(\frac{\theta-\phi}{2}\right)}\right\}$

$\Rightarrow \quad \frac{y}{x}=\frac{b\sin\left(\frac{\theta+\phi}{2}\right)}{a\cos\left(\frac{\theta+\phi}{2}\right)}$; Since $\theta + \phi = 2\alpha$

$\therefore \quad \frac{y}{x}=\frac{b}{a}\tan\alpha$ i.e., $ay = bx\tan\alpha$

15. (b) The equation of chord of contact (m, n) to $\frac{x^2}{9}+\frac{y^2}{4}=1$ is $T = 0$ i.e., $\frac{m}{9}x+\frac{n}{4}y-1=0$

Since $m.n = m + n$ where $m, n \in \mathbb{N}$

$\Rightarrow \quad m = n = 2$ is the only possibility

$\Rightarrow \quad 8x + 18y = 36$ or $4x + 9y = 18$

TEXTUAL EXERCISE–5 (SUBJECTIVE)

1. Consider end of $L.R.$ at $(-ae, b^2/a)$ for the ellipse $\frac{x^2}{a^2}+\frac{y^2}{b^2}=1$

Now the equation of normal at $P(x_1, y_1)$ is $\frac{a^2}{x_1}x-\frac{b^2}{y_1}y=a^2-b^2$

i.e., $\frac{a^2x}{(-ae)}-\frac{b^2y}{\left(-\frac{b^2}{a}\right)}=a^2-b^2$

Since the normal passes though $(0, b)$

$\therefore \quad a^2 - b^2 = ab$ or $a^2e^2 = a\sqrt{a^2-a^2e^2}$

$\Rightarrow e^4 = 1 - e^2$; Hence $e^4 + e^2 = 1$

2. Let normal at P $(a\cos\theta, b\sin\theta)$ on an ellipse $\frac{x^2}{a^2}+\frac{y^2}{b^2}=1$ be given as $\frac{a^2x}{a\cos\theta}-\frac{b^2y}{b\sin\theta}=a^2-b^2$

Let it pass through some internal point (h, k)

$$\therefore \quad \frac{ah\left\{1+\tan^2\left(\frac{\theta}{2}\right)\right\}}{1-\tan^2\left(\frac{\theta}{2}\right)} - \frac{bk\left\{1+\tan^2\left(\frac{\theta}{2}\right)\right\}}{2\tan\left(\frac{\theta}{2}\right)} = a^2 - b^2$$

Putting $\tan\left(\frac{\theta}{2}\right) = t$, we get $bkt^4 + 2\{ah + (a^2 - b^2)\}t^3 + 2\{ah - (a^2 - b^2)\}\, t - bk = 0$; which is a fourth degree equation in 't' so four normals are possible.

Let, t_1, t_2, t_3, t_4 be the roots of above biquadrate and $\theta_1, \theta_2, \theta_3$ and θ_4 be the corresponding eccentric angle.

$$\therefore \quad \tan\frac{\theta_1}{2} + \tan\frac{\theta_2}{2} + \tan\frac{\theta_3}{2} + \tan\frac{\theta_4}{4}$$

$$= \frac{-2\{ah + (a^2 - b^2)\}}{bk} = s_1 \text{ and } \sum \tan\frac{\theta_1}{2}.\tan\frac{\theta_2}{2} = 0 = s_2$$

$$\sum \tan\frac{\theta_1}{2}.\tan\frac{\theta_2}{2}.\tan\frac{\theta_3}{2} = -\frac{2\{ah - (a^2 - b^2)\}}{bk} = S_3$$

$$\sum \tan\frac{\theta_1}{2}.\tan\frac{\theta_2}{2}.\tan\frac{\theta_3}{2}.\tan\frac{\theta_4}{2} = -1 = S_4$$

$$\therefore \quad \tan\left(\frac{\theta_1}{2} + \frac{\theta_2}{2} + \frac{\theta_3}{2} + \frac{\theta_4}{2}\right) = \frac{S_1 - S_3}{1 - S_2 + S_4}$$

$$= \frac{-2\{ah + (a^2 - b^2)\} + 2\{ah - (a^2 - b^2)\}}{(1 - 0 - 1)}$$

$$= \frac{-4(a^2 - b^2)}{0} = \infty$$

$$\Rightarrow \quad \frac{\theta_1}{2} + \frac{\theta_2}{2} + \frac{\theta_3}{2} + \frac{\theta_4}{2} = (2n+1)\frac{\pi}{4},\ n \in \mathrm{Z}$$

$$\Rightarrow \quad \sum \theta_1 (2n+1)\pi, n \in \mathbb{Z}$$

3. (a) Let P ($a\cos\theta$, $b\sin\theta$) be a general point on the ellipse $\frac{x^2}{a^2} + \frac{y^2}{b^2} = 1$ and the normal at P is $ax \sec\theta - by \operatorname{cosec}\theta = a^2 - b^2$

This will interest x-axis at $G = \left(\frac{(a^2 - b^2)}{a\sec\theta}, 0\right)$ and y-axis at $g = \left(0, \frac{b^2 - a^2}{b\operatorname{cosec}\theta}\right)$

Now

$$Pg = \sqrt{\left(\frac{b^2\cos\theta}{a}\right)^2 + b^2\sin^2\theta} = \frac{b}{a}\sqrt{b^2\cos^2\theta + a^2\sin^2\theta}$$

and

$$pg = \sqrt{a^2\cos^2\theta + \frac{a^4\sin^2\theta}{b^2}} = \frac{a}{b}\sqrt{b^2\cos^2\theta + a^2\sin^2\theta}$$

$$PG: Pg = \frac{b}{a}\frac{b}{a} = \frac{b^2}{a^2}$$

(b) Without any loss generality, let $a > b$ and ($a\cos\theta$, $b\sin\theta$) be a point on the ellipse $\frac{x^2}{a^2} + \frac{y^2}{b^2} = 1$

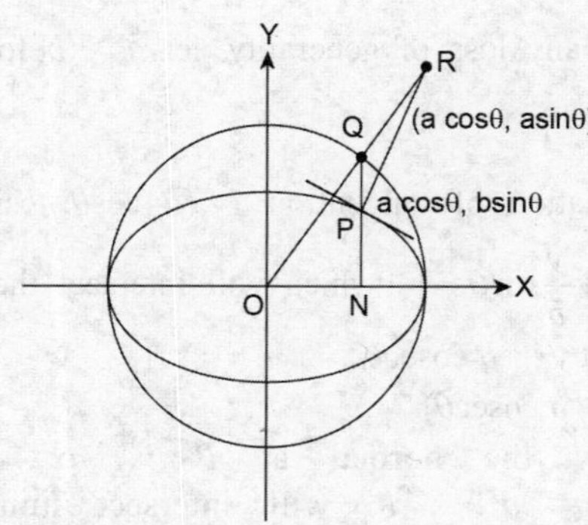

Now equation of normals at P is $ax\sec\theta - by\operatorname{cosec}\theta = a^2 - b^2$ ...(i)

And the equation of normal at Q is $t = x\tan\theta$ ...(ii)

Solving (i) and (ii), we get $y = (a + b)\sin\theta$ and $x = (a + b)\cos\theta$

$\therefore$ The locus the point of intersection is $x^2 + y^2 = (a + b)^2$

4. (a) Let P ($a\cos\theta$, $b\sin\theta$) be a point on the ellipse $\frac{x^2}{a^2} + \frac{y^2}{b^2} = 1$ where the equation of normal PN is $ax\sec\theta - by\operatorname{cosec}\theta = a^2 - b^2$.

Hence its slope is $m_1 = \frac{a\tan\theta}{b}$. Now the slope of CP is $m_2 = b/a \tan\theta$

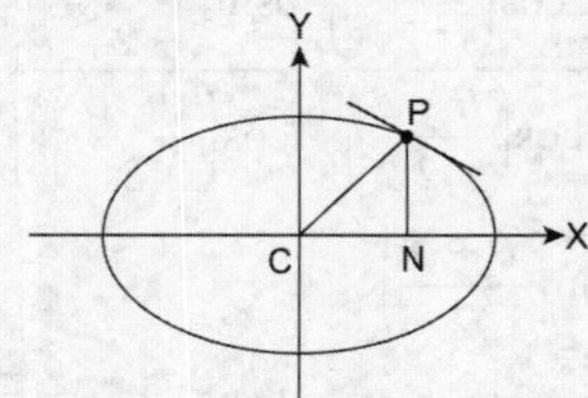

$$\Rightarrow \quad \tan\angle CPN = \frac{|m_1 - m_2|}{1 + m_1 m_2} = \frac{\left|\frac{a}{b} - \frac{b}{a}\right|\tan\theta}{1 + \tan^2\theta}$$

$$= \left(\frac{a^2 - b^2}{ab}\right)\sin\theta\cos\theta$$

$= \left(\frac{a^2 - b^2}{2ab}\right)\sin 2\theta$ {Where θ is the eccentric angle of point P}

(b) The normal at P ($\cos\theta$, $b\sin\theta$) for the ellipse $\frac{x^2}{a^2} + \frac{y^2}{b^2} = 1$ is $ax\sec\theta - by\operatorname{cosec}\theta = a^2 - b^2$.

The equation of CQ is $y = x\tan\theta$

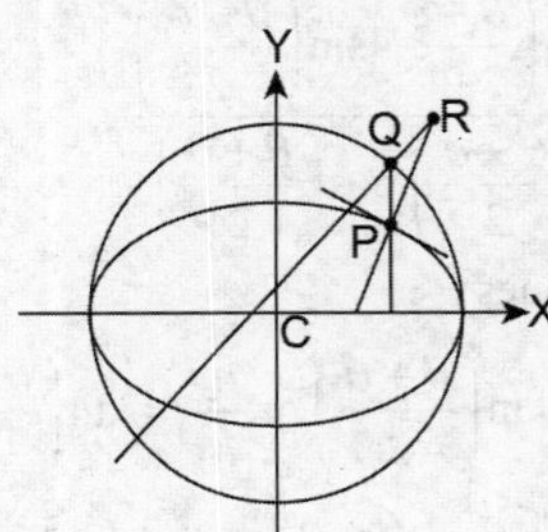

Solving, we get $x = (a + b)\cos\theta$ and $y = (a + b)\sin\theta$

$$\Rightarrow \quad CR = \sqrt{(a+b)^2(\sin^2\theta + \cos^2\theta)} = a + b$$

5. Without any loss of generality, let $a > b$ for the ellipse $\frac{x^2}{a^2}+\frac{y^2}{b^2}=1$

The equation of tangent at P ($a\cos\theta$, $b\sin\theta$) will be $\frac{x}{a}\cos\theta+\frac{y}{b}\sin\theta=1$; which will intersect the minor axis ($x = 0$) at $y_1 = b\,\text{cosec}\theta$

$\Rightarrow$ T (0, $b\,\text{cosec}\theta$).

Similarly the normal at P is $ax\sec\theta - by\,\text{cosec}\theta = a^2 - b^2$ will intersect minor axis at

$$y = \frac{(b^2-a^2)\sin}{b} \Rightarrow N\left(0, \frac{b^2-a^2}{b}\text{cosec}\right)$$

Let the focus be S ($\pm ae$, 0). For S_1 (ae, 0), we get slope of $ST = m_1 = \frac{b\,\text{cosec}\theta}{-ae}$ and slope of $SN = m_2 = \frac{(b^2-a^2)\sin\theta}{-aeb}$

Observe that $m_1m_2 = \frac{b^2-a^2}{a^2e^2} = \frac{b^2-a^2}{a^2-b^2} = -1$

Hence $SN \perp ST$. Simiarly, it is true for S_2 ($-ae$, 0)

6. Let $P(\theta)$ and $Q(\phi)$ be two points on the ellipse $\frac{x^2}{a^2}+\frac{y^2}{b^2}=1$, then the point intersection of tangents is

$$R(x_1, y_1) = \left(\frac{a\cos\left(\frac{\theta+\phi}{2}\right)}{\cos\left(\frac{\theta-\phi}{2}\right)}, \frac{b\sin\left(\frac{\theta+\phi}{2}\right)}{\cos\left(\frac{\theta-\phi}{2}\right)}\right)$$

$$\Rightarrow \frac{y_1}{x_1} = \frac{b}{a}\tan\left(\frac{\theta+\phi}{2}\right)$$

Since tangents at P and Q intersect at 90°; $m_1m_2 = -1$

i.e., $\frac{b^2}{a^2}\cot\theta.\cot\phi = -1$

So $\tan\theta\tan\phi = -b^2/a^2$

Normals at P and Q are respectively $\frac{ax}{\cos\theta}-\frac{by}{\sin\theta}=a^2-b^2$ and $\frac{ax}{\cos\phi}-\frac{by}{\sin\phi}=a^2-b^2$

Solving simultaneously, we get $\frac{by\sin(\theta-\phi)}{\sin\theta\sin\phi}=(a^2-b^2)$ $\{\cos\theta - \cos\phi\}$and $\frac{ax\sin(\theta-\phi)}{\cos\theta.\cos\phi}=(a^2-b^2)\{\sin\theta-\sin\phi\}$

$$\Rightarrow \frac{by\cos\theta.\cos\phi}{ax\sin\theta\sin\phi}=\frac{\cos\theta-\cos\phi}{\sin\theta-\sin\phi}$$

$$= \frac{(-2)\sin\left(\frac{\theta+\phi}{2}\right)\sin\left(\frac{\theta-\phi}{2}\right)}{2\sin\left(\frac{\theta-\phi}{2}\right).\cos\left(\frac{\theta+\phi}{2}\right)}$$

Putting the values

$$\frac{by}{ax}\left(\frac{a^2}{-b^2}\right) = -\tan\left(\frac{\theta+\phi}{2}\right) \text{ or } \frac{ay}{bx}=\tan\left(\frac{\theta+\phi}{2}\right),$$

we have $\frac{y}{x}=\frac{b}{a}\tan\left(\frac{\theta+\phi}{2}\right)=\frac{y_1}{x_1} \Rightarrow \frac{y}{x}=\frac{y_1}{x_1}$

Hence the point of intersection will satisfy $\frac{y}{x}=\frac{y_1}{x_1}$

7. Let P ($a\cos\phi$, $b\sin\phi$) be a point on $\frac{x^2}{a^2}+\frac{y^2}{b^2}=1$, then the tangent at P is $\frac{x}{a}\cos\phi+\frac{y}{b}\sin\phi=1$ and its intersection on major axes is Q ($a\sec\phi$, 0)

Similarly the normal is $ax\sec\phi - by\,\text{cosec}\phi = a^2 - b^2$

$$\Rightarrow R\left(\frac{a^2-b^2}{a\sec\phi}, 0\right)$$

Now $QR = \frac{a^2\sec^2\phi-(a^2-b^2)}{a\sec\phi} = a(\text{given})$

$\Rightarrow a^2\sec^2\phi - a^2e^2 = a^2\sec\phi$

$\Rightarrow 1 = \cos\phi + e^2\cos^2\phi$ i.e., $e^2\cos^2\phi + \cos\phi - 1 = 0$

8. $\frac{x^2}{4}+\frac{y^2}{9}=1$ .Observe that this is a vertical ellipse having foci $(0, \pm\sqrt{5})$, so a ray emanating from S_1 when stikes at (2, 0) will get reflected towards the other focus S_2 $(0, \sqrt{5})$

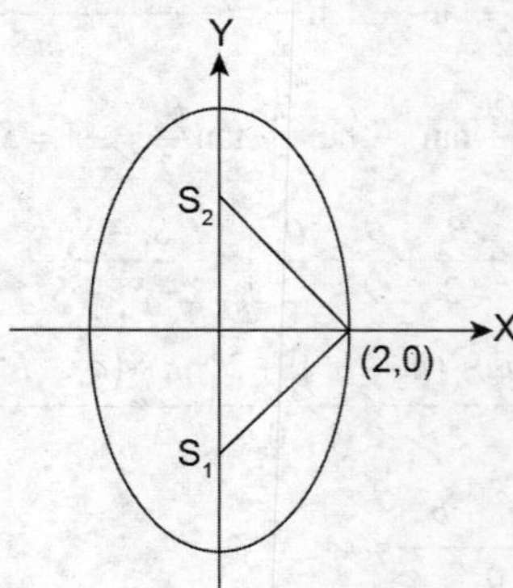

$\therefore$ The reflected ray is along the line $y-0=\frac{\sqrt{5}}{(-2)}(x-2)$

$\Rightarrow \sqrt{5}x + 2y - 2\sqrt{5} = 0$

9. Let P ($a\cos\theta$, $b\sin\theta$) be a point on the ellipse $\frac{x^2}{a^2}+\frac{y^2}{b^2}=1$, where the equation of normal is $\frac{ax}{\cos\theta}-\frac{by}{\sin\theta}-(a^2-b^2)=0$

Comparing with $x + my + n = 0$, we get $\frac{\ell\cos\theta}{a}=\frac{-m\sin\theta}{b}=\frac{n}{b^2-a^2}$

$$\Rightarrow \cos\theta=\frac{an}{\ell(b^2-a^2)} \text{ and } \sin\theta=\frac{bn}{m(a^2-b^2)}$$

$$\Rightarrow \left\{\frac{an}{\ell(b^2-a^2)}\right\}^2+\left\{\frac{bn}{m(a^2-b^2)}\right\}^2=1$$

i.e., $\frac{a^2}{\ell^2}+\frac{b^2}{m^2}=\frac{(a^2-b^2)}{n^2}$

10. Equation of tangent at $(\lambda^2, 2\lambda)$ to the parabola $y^2 = 4x$ is $\lambda y = x + \lambda^2$ or $x - \lambda y + \lambda^2 = 0$

Now the normal at $\left(\sqrt{5}\cos\theta, 2\sin\theta\right)$ to the ellipse $\frac{x^2}{5}+\frac{y^2}{4}=1$ is $\frac{a^2x}{a\cos\theta}-\frac{b^2y}{b\sin\theta}=a^2-b^2$ i.e.,

$$\frac{\sqrt{5}}{\cos\theta}x-\frac{2}{\sin\theta}y-1=0 \quad .....(ii)$$

Compairing, we get $\dfrac{\cos\theta}{\sqrt{5}} = \dfrac{\lambda\sin\theta}{2} = -\lambda^2$ i.e., $\cos\theta = -\sqrt{5}\,\lambda^2$ and $\sin\theta = -2\lambda$

$\Rightarrow\ 5\lambda^4 + 4\lambda^2 = 1$

Let $\lambda^2 = t > 0$, then $5t^2 + 4t - 1 = 0$

$\Rightarrow\ t = 1/5, -1$ rejecting $t = -1$; from $\lambda^2 = 1/5$, we get $\lambda = \pm 1/\sqrt{5}$

$\Rightarrow\ \cos\theta = -\dfrac{1}{\sqrt{5}}$

$\therefore\ \ \theta = \cos^{-1}\left(\dfrac{-1}{\sqrt{5}}\right)$ for $\lambda^2 = \dfrac{1}{5}$ and $\lambda = \pm 1/\sqrt{5}$

11. Let $P(\sqrt{14}\cos\theta, \sqrt{5}\sin\theta)$ be a point on the ellipse $\dfrac{x^2}{14} + \dfrac{y^2}{5} = 1$ where the equation of normal is $\dfrac{ax}{\cos\theta} - \dfrac{by}{\sin\theta} = a^2 - b^2$

i.e., $\dfrac{\sqrt{14}x}{\cos\theta} - \dfrac{\sqrt{5}\,y}{\sin\theta} - 9 = 0$

Since the normal chord ends at $Q\ (2\theta)$

$\therefore\ \ Q\left(\sqrt{14}\cos 2\theta, \sqrt{5}\sin 2\theta\right)$ satisfies it

$\Rightarrow\ \dfrac{14\cos 2\theta}{\cos\theta} - \dfrac{5\sin 2\theta}{\sin\theta} = 9 \Rightarrow \dfrac{14\{2\cos^2\theta - 1\}}{\cos\theta} - 10\cos\theta = 9$

i.e., $18\cos^2\theta - 9\cos\theta - 14 = 0$

$\Rightarrow\ \cos\theta = \dfrac{9 \pm 33}{36} = \dfrac{42}{36}, -\dfrac{24}{36}$

Rejecting $\dfrac{42}{36}$ we got $\cos\theta = -2/3$

TEXTUAL EXERCISE–5 (OBJECTIVE)

1. (c) Let $P\,(a\cos\theta, b\sin\theta)$ be a point on the ellipse $\dfrac{x^2}{a^2} + \dfrac{y^2}{b^2} = 1$, where the tangent is $\dfrac{x}{a}\cos\theta + \dfrac{y}{b}\sin\theta = 1$

$\therefore\ $ slope of tangent $m_1 = -\dfrac{b}{a}\cot\theta$

$\Rightarrow$ slope of normal $m_2 = \dfrac{a}{b}\tan\theta$

$\Rightarrow$ The equation of normal is

$(y - b\sin\theta) = \dfrac{a}{b}\tan\theta\{x - a\cos\theta\}$

$\Rightarrow\ \dfrac{ax\sin\theta}{b\cos\theta} - \dfrac{a^2}{b} = y - b\sin\theta$ or

$\dfrac{ax\sin\theta}{\cos\theta} - a^2 = by - b^2\sin\theta$

Dividing by $\sin\theta$, we get $\dfrac{ax}{\cos\theta} - \dfrac{by}{\sin\theta} = a^2 - b^2$

2. (b) Equation of normal at $P\ (\theta)$ for the ellipse $\dfrac{x^2}{14} + \dfrac{y^2}{5} = 1$ will be $\dfrac{\sqrt{14}x}{\cos\theta} - \dfrac{\sqrt{5}y}{\sin\theta} - 9 = 0$

Since the other end of the normal chord is at $Q\ (2\theta)$

$\therefore\ \ \left(\sqrt{14}\cos 2\theta, \sqrt{5}\sin 2\theta\right)$ lies on it

$\Rightarrow\ \dfrac{14\cos 2\theta}{\cos\theta} - \dfrac{5\sin 2\theta}{\sin\theta} = 9$

$\therefore\ \ \dfrac{14\{2\cos^2\theta - 1\}}{\cos\theta} - \dfrac{5(2\sin\theta\cos\theta)}{\sin\theta} = 9$

$\Rightarrow\ 28\cos^2\theta - 14 - 10\cos^2\theta = 9\cos\theta$

Rearrangement, we have $18\cos^2\theta - 9\cos\theta - 14 = 0$

Now $\cos\theta = \dfrac{9 \pm 33}{36} = \dfrac{42}{3}, -\dfrac{24}{36}$ rejecting $\dfrac{42}{36}$, we get $\cos\theta\ -2/3$

3. (c) Let $P\,(a\cos\theta, b\sin\theta)$ be a point on the ellipse $\dfrac{x^2}{a^2} + \dfrac{y^2}{b^2} = 1$; where the normal will be $\dfrac{ax}{\cos\theta} - \dfrac{by}{\sin\theta} - (a^2 - b^2) = 0$

Comparing with $mx - y + c = 0$, we get $\dfrac{m\cos\theta}{a} = \dfrac{\sin\theta}{b} = \dfrac{c}{b^2 - a^2}$

$\Rightarrow\ \cos^2\theta + \sin^2\theta = \dfrac{c^2 a^2}{m^2(b^2 - a^2)^2} + \dfrac{b^2c^2}{(b^2 - a^2)^2} = 1$

$\Rightarrow\ c^2 = \dfrac{(a^2 - b^2)^2 m^2}{a^2 + b^2m^2}$ i.e., $c = \pm\dfrac{(a^2 - b^2)m}{\sqrt{a^2 + b^2m^2}}$

4. (d) Comparing $\dfrac{ax}{\cos\theta} - \dfrac{by}{\sin\theta} - (a^2 - b^2) = 0$ with $x\cos\alpha + y\sin\alpha - p = 0$, we get $\dfrac{\cos\alpha\cos\theta}{a} = \dfrac{\sin\alpha\sin\theta}{(-b)} = \dfrac{p}{a^2 - b^2}$

$\Rightarrow \cos^2\theta + \sin\theta^2 = \dfrac{p^2a^2}{(a^2 - b^2)^2\cos^2\alpha} + \dfrac{p^2b^2}{(a^2 - b^2)^2\sin^2\alpha} = 1$

i.e., $p^2\ \{a^2\sin^2\alpha + b^2\cos^2\alpha\} = (a^2 - b^2)\cos^2\alpha\sin^2\alpha$ or $p^2(a^2\sec^2\alpha + b^2\text{cosec}^2\alpha\} = (a^2 - b^2)^2$

5. (b) Comparing $\dfrac{ax}{\cos\theta} - \dfrac{by}{\sin\theta} - (a^2 - b^2) = 0$ with $x + my + n = 0$, we get $\dfrac{\ell\cos\theta}{a} = \dfrac{m\sin\theta}{(-b)} = \dfrac{n}{b^2 - a^2}$

$\Rightarrow\ \cos^2\theta + \sin^2\theta = \left\{\dfrac{a^2}{\ell^2} + \dfrac{b^2}{m^2}\right\}\dfrac{n^2}{(a^2 - b^2)^2} = 1$

$\Rightarrow\ \left(\dfrac{a^2}{\ell^2} + \dfrac{b^2}{m^2}\right) = \dfrac{(a^2 - b^2)^2}{n^2}$

6. (a), (c) Normal to the ellipse $\dfrac{x^2}{1} + \dfrac{y^2}{4} = 1$ at $P\ (\cos\theta, 2\sin\theta)$ will be $\dfrac{x}{\cos\theta} - \dfrac{2y}{\sin\theta} = -3$

Comparing with $2x - \dfrac{8}{3}\lambda y = -3$, we get $2\cos\theta = \dfrac{8\lambda\sin\theta}{3(2)} = 1$

$\Rightarrow\ \cos\theta = 1/2 \qquad \therefore\ \ \sin\theta = \pm\dfrac{\sqrt{3}}{2}$

Hence $\dfrac{4\lambda}{3}\sin\theta \qquad \Rightarrow\ \lambda = \pm\dfrac{\sqrt{3}}{2}$

7. (a) The given conic (parabola) is $y^2 = 4\left(\frac{1}{4}\right)x$

$\therefore$ $a = 1/4$

The equation of normal with slope m is

$am^3 + 2am - mx + y = 0$, since $(c, 0)$ lies on it.

$\Rightarrow$ $am^3 + 2am - mc = 0$ or $m\ \{am^2 + 2a - c\} = 0$

$\Rightarrow$ $m_3 = 0$

Now $am^2 + 0m + (2a - c) = 0$ has $m_1 m_2 = -1$ as $m_1 \perp m_2$ but $m_1 + m_2 = 0 \Rightarrow m_1 = m_2 = -1$

$\therefore$ $\frac{2a-c}{a} = -1 \Rightarrow 2a - c = -a$ i.e., $c = 3a = 3/4$

8. (c) The four co normal points will lie on the Apollonian rectangular by hyperbola.

9. (c) At point P $(a\cos\theta, b\sin\theta)$ on an ellipse $\frac{x^2}{a^2}+\frac{y^2}{b^2}=1$, the tangent is PT given as $\frac{x}{a}\cos\theta+\frac{y}{b}\sin\theta=1$ (for $a > b$)

$\Rightarrow$ Distance of origin from the tangent is

$$CF = \frac{ab}{\sqrt{b^2\cos^2\theta + a^2\sin^2\theta}}.$$

Similarly the normal at P is $ax\sec\theta - by\ \mathrm{cosec}\theta = a^2 - b^2$, which will intersect x-axis (major axis) at

$$x = \frac{(a^2-b^2)}{a}\cos\theta$$

$$\Rightarrow G\left(\frac{(a^2-b^2)}{a}\cos\theta, 0\right)$$

$$\Rightarrow PG^2 = \left\{\frac{(a^2-b^2)\cos\theta}{a} - a\cos\theta\right\}^2 + b^2\sin^2\theta$$

or $$PG^2 = \frac{b^4\cos^2\theta}{a^2} + b^2\sin^2\theta$$

i.e., $$PG = \frac{b}{a}\sqrt{a^2\sin^2\theta + b^2\cos^2\theta}$$

Hence $CF.\ PG = (ab)\left(\frac{b}{a}\right) = b^2$

10. (a) Without any loss of generality let the three points be P $(a\cos\theta_1, b\sin\theta_1)$, Q $(a\cos\theta_2, b\sin\theta_2)$ and R $(a\cos\theta_3, b\sin\theta_3)$

The corresponding points on the auxiliary circle are A $(a\cos\theta_1, a\sin\theta_1)$, B $(a\cos\theta_2, a\sin\theta_2)$ and C $(a\cos\theta_3, a\sin\theta_3)$

Now area of $\Delta PQR = \left|\frac{1}{2}ab\begin{vmatrix}\cos\theta_1 & \sin\theta_2 & 1\\ \cos\theta_2 & \sin\theta_2 & 1\\ \cos\theta_3 & \sin\theta_3 & 1\end{vmatrix}\right|$ and area of $\Delta ABC = \left|\frac{a^2}{2}\begin{vmatrix}\cos\theta_1 & \sin\theta_1 & 1\\ \cos\theta_2 & \sin\theta_2 & 1\\ \cos\theta_3 & sin\theta_3 & 1\end{vmatrix}\right|$

$\therefore$ $$\frac{\text{area } \Delta\, PQR}{\text{area of } \Delta\, ABC} = \frac{b}{a}$$

11. (d) Consider the ellipse $\frac{x^2}{a^2}+\frac{y^2}{b^2}=1$. Let point P be at $(0, b)$.

Now SP = a and SG = ae as normal is along y-axis, so clearly $SG = eSP$.

Other options follow from the properties of reflection.

12. (c) If the four point on ellipse $\frac{x^2}{a^2}+\frac{y^2}{b^2}=1$ as $P(\alpha)$, $Q(\beta)$, $R(\gamma)$, $S(\delta)$ are concentric, then $\alpha + \beta + \gamma + \delta = 2n\pi$ (where $n \in \mathbb{Z}$ the set of integers)

TEXTUAL EXERCISE SECTION–III (OBJECTIVE)

1. (b) The equation $\sqrt{(x-2)^2+y^2}+\sqrt{(x+2)^2+y^2} > 4$ means sum of distances

From B (2, 0) and A (–2, 0) is greater than 4

$\Rightarrow$ it is an ellipse

2. (b) Let $a > b$

Given: $L.R = b$ i.e., $\frac{2b^2}{a} = b$

$\Rightarrow$ $2b = a$

$\Rightarrow$ $e = \sqrt{\frac{a^2-b^2}{a^2}} = \sqrt{\frac{3b^2}{4b^2}} = \frac{\sqrt{3}}{2}$

3. (a) Given $2ae = 10$ (distance between foci) and $e = 5/8$

$\Rightarrow$ $a = 8$

$$L.R. = \frac{2b^2}{a} = \frac{2(a)^2\{1-e^2\}}{a} = 16\times\frac{39}{64} = \frac{39}{4}$$

4. (b) The given equation of ellipse is $\frac{x^2}{r-2}+\frac{y^2}{5-r}=1$ will be true where $r - 2 > 0$ and $5 - r > 0$ i.e., $r > 2$ and $r < 5$.

$\Rightarrow$ $2 < r < 5$ satisfies

5. (a) For ellipse $\frac{x^2}{16}+\frac{y^2}{9}=1$ the combined equation of tangent from (2, 3) i.e., $T^2 = SS_1$ i.e., $\left(\frac{x}{8}+\frac{y}{3}-1\right)^2 = \frac{1}{4}\left(\frac{x^2}{16}-\frac{y^2}{9}-1\right)$

$\Rightarrow$ $9x^2 + 64y^2 + 576 - 144x - 384y + 48xy = 9x^2 + 16y^2 - 144$

or $48y^2 + 48xy - 384y - 144x + 720 = 0$

i.e., $y^2 + xy - 3x - 8y + 15 = 0$

$\Rightarrow$ $(y - 3)(x + y - 5) = 0$

$\Rightarrow$ $y = 3$ and $x + y = 5$ are the tangents

6. (b) The equation of chord with mid-point at (2, 1) for the ellipse $\frac{x^2}{36}+\frac{y^2}{9}=1$ is $T = S_1$ i.e., $\frac{x}{18}+\frac{y}{9}-1=\frac{1}{9}+\frac{1}{9}-1$

$\Rightarrow$ $x + 2y = 4$

7. (a) The given ellipse is $\frac{x^2}{\left(\frac{5}{4}\right)}+\frac{y^2}{\left(\frac{5}{3}\right)}=1$

The equation of tangent with slope $m = \sqrt{3}$ is

$$y = \sqrt{3}x \pm \sqrt{\frac{5}{4}(3)+\frac{5}{3}} \text{ i.e., } y = \sqrt{3}\,x \pm \sqrt{\frac{65}{12}}$$

8. (a) For the ellipse $\frac{x^2}{16}+\frac{y^2}{9}=1$, the foci are $(\pm\sqrt{7},0)$.

So radius of the centre circle at (0, 3) is $r=\sqrt{7+9}=\sqrt{16}$ = 4 units

9. (b) Given $e = 1/2$ and $x = a/e = 4 \Rightarrow a = 2$

$\Rightarrow b^2 = a^2\{1-e^2\} = 4\left(1-\frac{1}{4}\right)=3$

Hence the ellipse is $\frac{x^2}{4}+\frac{y^2}{3}=1$

10. (a) The given ellipse is $\frac{x^2}{2}+\frac{y^2}{1}=1$ at $P(\sqrt{2}\cos\theta,\sin\theta)$

The equation of tangent is $\frac{x}{\sqrt{2}}\cos\theta+\frac{y}{1}\sin\theta=1$

$\Rightarrow x=\sqrt{2}\sec\theta$ and $y=\operatorname{cosec}\theta$

The mid point is $\left(\frac{1}{\sqrt{2}\cos\theta},\frac{1}{2\sin\theta}\right)=(x_1,y_1)$ (say)

$\Rightarrow \left(\frac{1}{\sqrt{2}\,x_1}\right)^2+\left(\frac{1}{2y_1}\right)^2=1$ i.e., $\frac{1}{2x_1^2}+\frac{1}{4y_1^2}=1$

11. (a) Let $\frac{x^2}{a^2}+\frac{y^2}{b^2}=1$ be an ellipse with eccentricity e, then ends of $L.R$ are $\left(ae,\pm\frac{b^2}{a}\right)$ and $\left(-ae\pm-\right)$

Let $x_1 = ae$ and $y_1 = b^2/a$

$\Rightarrow y_1=\frac{a^2}{a}(1-e^2)\Rightarrow\frac{y_1}{a}=1-\left(\frac{x_1}{a}\right)^2\Rightarrow x^2=a(a-y)$

12. (a) Let $P(a\cos\theta, b\sin\theta)$ be a point on ellipse $\frac{x^2}{a^2}+\frac{y^2}{b^2}=1$ (for $a > b$)

$\Rightarrow M(a\cos\theta, 0)$ and $T = (a\sec\theta, 0)$

Hence $\sqrt{CM.CT}=\sqrt{a\cos\theta.\frac{a}{\cos\theta}}=a$

13. (c) Let $P(\alpha)$ and $B'(\beta)$ be the two points on the ellipse $\frac{x^2}{a^2}+\frac{y^2}{b^2}=1$ where $(ae, 0)$ is the focus.

The equation of focal PSB' gives $\frac{1-e}{1+e}=-\tan\frac{\alpha}{2}.\tan\frac{\beta}{2}$

Since $\beta = 270°\left(\text{or}-\frac{\pi}{2}\right)$

$\Rightarrow \tan\beta/2 = -1$

Hence $\tan\frac{\alpha}{2}=\frac{1-e}{1+e}$

Now $\tan\phi=\frac{b}{a}\tan\alpha=\frac{\sqrt{(1-e^2)}.2\left\{\frac{1-e}{1+e}\right\}}{1-\left\{\frac{1-e}{1+e}\right\}^2}$

$=\frac{2\sqrt{1-e^2}(1-e)(1+e)}{4e}=\frac{(1-e)^{3/2}}{2e}$

14. (b) Let $P(4\cos\theta, 3\sin\theta)$ be a point on the ellipse $\frac{x^2}{16}+\frac{y^2}{9}=1$, then $Q(4\cos\theta, 4\sin\theta)$

$\therefore$ CQ is $y = x\tan\theta$ ...(i)

The equation of normal at P is $\frac{4x}{\cos\theta}-\frac{3y}{\sin\theta}=7$ ...(ii)

Solving (i) and (ii), we get the point of intersection as R $(7\cos\theta, 7\sin\theta)$

$\Rightarrow CR = 7$ units

15. (b) Without any loss of generality consider the Ist quad.

Solving $\frac{x^2}{a^2}+\frac{y^2}{b^2}=1$ and $x^2+y^2=ab$, we get $\frac{x^2}{a^2}+\frac{ab-x^2}{b^2}=1$

$\Rightarrow (b^2-a^2)x^2=a^2b^2-a^3b=a^2b(b-a)$

$\therefore x^2=\frac{a^2b}{a+b}$ and $y^2=\frac{ab^2}{a+b}$

$\Rightarrow (x_1,y_1)=\left(\sqrt{\frac{a^2b}{a+b}},\sqrt{\frac{ab^2}{a+b}}\right)$

Slope of tangent to the ellipse at the point of intersection $m_1=-\frac{b^2}{a^2}.\frac{x_1}{y_1}=-\sqrt{\frac{b^3}{a^3}}$

Similarly slope of tangent to the circle is $m_2=-\frac{x_1}{y_1}=-\sqrt{\frac{a}{b}}$

Hence $\tan\alpha=\left|\frac{m_1-m_2}{1+m_1m_2}\right|=\left|\frac{\sqrt{\frac{a}{b}}-\sqrt{\frac{b^3}{a^3}}}{1+\frac{b}{a}}\right|=\frac{(a^2-b^2).a}{a\sqrt{ab}\,(a+b)}$

$\Rightarrow \tan\alpha=\frac{a-b}{\sqrt{ab}}\Rightarrow \alpha=\tan^{-1}\left(\frac{a-b}{\sqrt{ab}}\right)$

16. (c) The equation of normal at $P(5\cos\theta, 3\sin\theta)$ for the ellipse $\frac{x^2}{25}+\frac{y^2}{9}=1$ will be $\frac{ax}{\cos\theta}-\frac{by}{\sin\theta}=a^2-b^2$ is

$\frac{5x}{\cos\theta}-\frac{3}{\sin\theta}=16$

Comparing with $5x-3y=8\sqrt{2}$, we get $\sin\theta=\cos\theta=1/\sqrt{2}$

17. (c) The given ellipse is $C: x^2+y^2+xy-3=0$.

Observe that there is no term in x and y, since xy term is there, so it is rotation about origin

Now the line $L: x-\sqrt{3}\,y=0$ passes through O (0, 0)

Let $(x_1, y_1) = (r\cos30°, r\sin30°)$ be the point of intersection of line L and the ellipse C.

$\Rightarrow \frac{3r^2}{4}+\frac{r^2}{4}+\frac{\sqrt{3}r^2}{4}=3\Rightarrow(4+\sqrt{3})r^2-12=0$

$\Rightarrow |r_1.r_2|=\frac{12}{4+\sqrt{3}}=\frac{12(4-\sqrt{3})}{13}$

18. (a) Let (h, k) be the mid point of chord for ellipse $\frac{x^2}{a^2}+\frac{y^2}{b^2}=1$ $(a > b)$, then $\frac{hx}{a^2}+\frac{ky}{b^2}-1=\frac{h^2}{a^2}+\frac{k^2}{b^2}-1$.

If this chord passes though B (0, b), then $\frac{k}{b}=\frac{h^2}{a^2}+\frac{k^2}{b^2}$

Hence the locus is $\frac{x^2}{a^2}+\frac{y^2}{b^2}=\frac{y}{b}$

19. (b) As shown in the figure ΔABC is an isosceles Δ

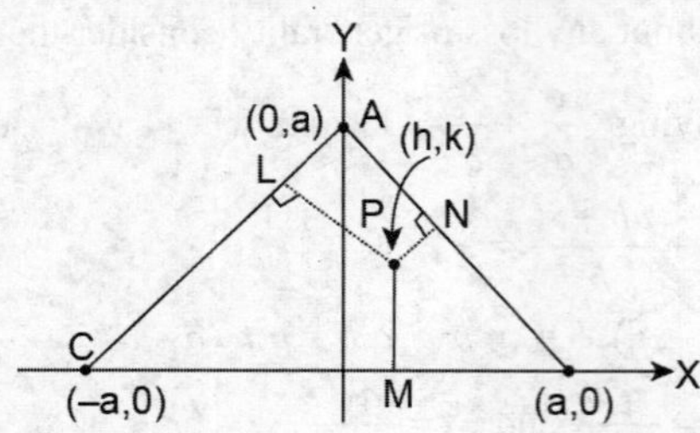

Given $PM^2 = 1/2\ PL.\ PN$ ...(i)

Equation of lines AB, AC are respectively

AB: $x + y = a$

AC: $y - x = a$

$$PL=\frac{|k-h-a|}{\sqrt{2}};\ PN=\frac{|k+h-a|}{\sqrt{2}}$$

Substituting in equation (i), we get $k^2=\frac{1}{2}\frac{\left|(k-a)^2-h^2\right|}{2}$

$\Rightarrow$ $4k^2 = |(k-a)^2 - h^2|$

$$\Rightarrow \begin{cases} 4k^2=k^2+a^2-2ak-h^2 \text{ for } (k-a)^2\ge h^2 \\ \text{or} \\ 4k^2=h^2-k^2-a^2+2ak \text{ for } (k-a)^2<h^2 \end{cases}$$

$$\Rightarrow \begin{cases} h^2+3k^2+2ak-a^2=0 \text{ for } a-k\le h\le k-a \\ \text{or} \\ h^2-5k^2+2ak-a^2=0 \text{ for } h\in(-\infty,a-k)\cup(k-a,\infty) \end{cases}$$

$$\Rightarrow \text{Required locus is } \begin{cases} x^2+3y^2+2ay=a^2 \\ \text{or} \\ x^2-5y^2+2ay=a^2 \end{cases} \text{ i.e.,}$$

$$\begin{cases} x^2+\left(y^2+\frac{2}{3}ay+\frac{1}{9}a^2\right)=a^2+\frac{1}{3}a^2 \\ \text{or} \\ x^2-\left(y^2-\frac{2}{5}ay+\frac{2}{25}a^2\right)=a^2=\frac{1}{5}a^2 \end{cases}$$

$$\text{i.e., } \begin{cases} \frac{x^2}{\frac{4}{3}a^2}+\frac{\left(y+\frac{a}{3}\right)^2}{\left(\frac{4}{9}a^2\right)}=1 \\ \text{or} \\ \frac{x^2}{\frac{4}{5}a^2}-\frac{\left(y-\frac{a}{5}\right)^2}{\left(\frac{4a^2}{25}\right)}=1 \end{cases}$$

Given that the locus is an ellipse

$$\Rightarrow e^2=1-\frac{B^2}{A^2}=1-\left(\frac{\frac{4}{9}a^2}{\frac{4}{3}a^2}\right)=1-\frac{1}{3}=\frac{2}{3}\Rightarrow e=\sqrt{\frac{2}{3}}$$

20. (c) The given ellipse is $\frac{x^2}{a^2}+\frac{y^2}{b^2}=1$ .Since tangent intersect at 90° at point A

$\Rightarrow$ A is on the director circle i.e., $x^2 + y^2 = a^2 + b^2$

21. (c) The ellipse $\frac{x^2}{\left(\frac{1}{\sqrt{5}}\right)^2}+\frac{y^2}{\left(\frac{1}{2}\right)^2}=1$

Consider two perpendicular diameters one along x-axis and the other along y-axis as shown

$\Rightarrow$ $D_x = 2/\sqrt{5}$ and $D_y = 1$

$$\Rightarrow \left(\frac{1}{D_x}\right)^2+\left(\frac{1}{D_y}\right)^2=\frac{5}{4}+1=\frac{9}{4}$$

22. (b) Observe that A (1, 4) lies on $2x^2 + y^2 = 18$ only.

Now a point P (x, y) = $(3\cos\theta, 3\sqrt{2}\sin\theta)$ will have equation of normal as $\frac{3x}{\cos\theta}-\frac{3\sqrt{2}y}{\sin\theta}=-9$

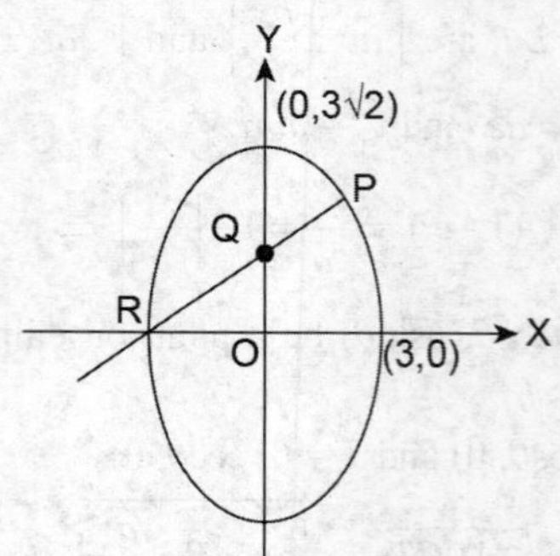

For $y = 0$, $x = -3\cos\theta$

$\Rightarrow$ R $(-3\cos\theta, 0)$ and $x = 0$,

$y=\frac{3}{\sqrt{2}}\sin\theta$ + $Q\left(0,\frac{3}{\sqrt{2}}\sin\theta\right)$

Mid-point of PR is $Q\left(0,\frac{3}{\sqrt{2}}\sin\theta\right)$ which is Q

Hence $2x^2 + y^2 = 18$ is the ellipse

23. (b) The centre of the ellipse $\frac{(x+y-2)^2}{9}+\frac{(x-y)^2}{16}=1$ will be given by the intersection of L_1: $x + y - 2 = 0$ and L_2: $x - y = 0$ i.e., $x = y = 1$

$\Rightarrow$ C (1, 1)

24. (a) Given $\frac{x}{3}=\cos t+\sin t$ and $\frac{y}{4}$ = cost – sint

Squaring addig, we get $\frac{x^2}{9}+\frac{y^2}{16}=2$

$\Rightarrow$ $\frac{x^2}{18}+\frac{y^2}{32}=1$ which is an ellipse

25. (c) Let P $(a\cos\theta, b\sin\theta)$ be a general point on the ellipse $\frac{x^2}{a^2}+\frac{y^2}{b^2}=1$, where the tangent is given by $\frac{x}{a}\cos\theta+\frac{y}{b}\sin\theta=1$

$\Rightarrow$ $OA = a\sec\theta$ and $OB = b\text{cosec}\theta$

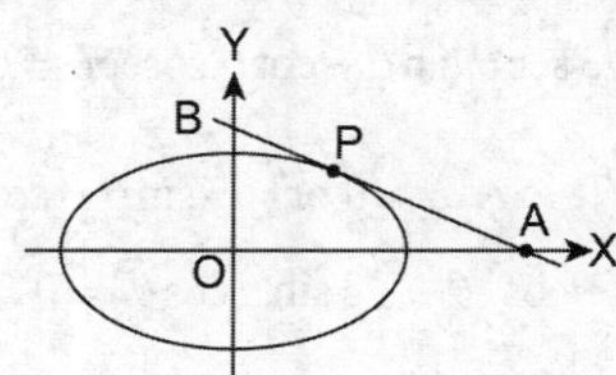

$\Rightarrow$ Area $\Delta OAB = \left|\frac{ab}{2\sin\theta\cos\theta}\right| = \left|\frac{ab}{\sin 2\theta}\right|$ which will have smallest value when $\sin 2\theta = 1$, then area $= ab$

26. (c) The normal to the ellipse $\frac{x^2}{a^2}+\frac{y^2}{b^2}=1$ at $P(a\cos\theta, b\sin\theta)$ is $\frac{ax}{\cos\theta}-\frac{by}{\sin\theta}=a^2-b^2$ or $ax\sin\theta - by\cos\theta = (a^2-b^2)\sin\theta\cos\theta$

Comparing with $mx - y + c = 0$, we get

$$\frac{m}{a\sin\theta}=\frac{1}{b\cos\theta}=\frac{c}{(a^2-b^2)\sin\theta\cos\theta}$$

$$\Rightarrow \left(\frac{a}{m}\right)^2+b^2=\frac{(a^2-b^2)^2\{\sin^2\theta+\cos^2\theta\}}{c^2}$$

Hence $c^2=\frac{(a^2-b^2)^2m^2}{a^2+b^2m^2}$

$$\Rightarrow c=\pm\frac{(a^2-b^2)m}{\sqrt{a^2+b^2m^2}}$$

27. (a) The ellipse is $\frac{x^2}{1}+\frac{y^2}{7}=1$

The chord bisected is $y = 1 - 7x$

$\Rightarrow$ $m = -7$

Equation of diameter (for a system of parallel chords with slope $m = -7$) is $y=-\frac{b^2x}{a^2m}$ i.e., $y=\frac{-7}{-7}x=x$

28. (a) Given ellipse is $\frac{x^2}{2a^2}+\frac{y^2}{2b^2}=$ (for $a > b$)

$\Rightarrow$ $Ae=\sqrt{A^2-B^2}=\sqrt{2\left(a^2-b^2\right)}$

$\therefore$ $F_1F_2=2\sqrt{2}\sqrt{a^2-b^2}$

Now ΔF_1PF_2 will have maximum area, when P is at maximum distance from the major axis which is at $(0, \pm b\sqrt{2})$

Hence the area $=\frac{1}{2}\times F_1F_2\times b\sqrt{2}$

$=\left(\sqrt{2}\sqrt{a^2-b^2}\right)\left(b\sqrt{2}\right)=2b\sqrt{a^2-b^2}$ square units.

29. (c) Since tangents from P $(\lambda, 3)$ to the ellipse $\frac{x^2}{9}+\frac{y^2}{4}=1$ are at right angles, so P lies on director circle $x^2 + y^2 = a^2 + b^2$

$\Rightarrow$ $\lambda^2 + 9 = 9 + 4$. Hence $\lambda = \pm 2$

30. (d) We know that the tangents at ends of LR will intersect on directrix.

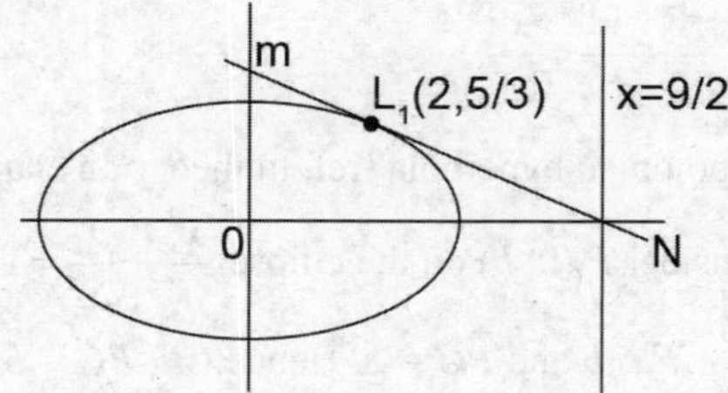

So for ellipse $\frac{x^2}{9}+\frac{y^2}{5}=1$ in the first quadrant L_1 (2, 5/3), N (9/2, 0) and M (0, 3)

Hence the area of quadrant = 4 × Area ΔOMN = $4\times\frac{1}{2}\times\frac{9}{2}\times 3$ = 27 squre units

31. (a) Le P (h, k) be the point on director circle, so the at tangents are at right angles to each other for the ellipse $\frac{x^2}{9}+\frac{y^2}{4}=1$

$\Rightarrow$ $h^2 + k^2 = 13$

Now the chord of contact of tangents is

$T=0$ $\Rightarrow$ $\frac{hx}{a^2}+\frac{ky}{b^2}-1=0$ i.e., $\frac{hx}{9}+\frac{k}{4}-1=0$

Distance of origin (0, 0) from the chord is $d=\frac{|a^2b^2|}{\sqrt{b^4h^2+a^4k^2}}$

$$\Rightarrow d=\frac{36}{\sqrt{16h^2+81k^2}}.$$

Now d will maximum when $16h^2 + 81k^2$ is minimum. Observe that $16h^2 + 81k^2 = 16\{13 - k^2\} + 81k^2 = 208 + 65k^2 \Rightarrow 16h^2 + 18k^2 \geq 208$ (when $k = 0$)

Hence max $d=\frac{36}{\sqrt{208}}=\frac{9}{\sqrt{13}}$

32. (a) Let $\frac{x^2}{A^2}+\frac{y^2}{B^2}=1$ be an ellipse. Now the tangent $y=\frac{-2a}{\sqrt{1-a^2}}x+\frac{1}{\sqrt{1-a^2}}$ becomes $y = -2\tan\theta\ .x + |\sec\theta|$, when a $\in (-1, 1)$ is substituted as $a = \sin\theta$ {where $\theta\in\left(-\frac{\pi}{2},\frac{\pi}{2}\right)$}

Hence for tangency $C^2 = A^2m^2 + B^2$

$\Rightarrow$ $\sec^2\theta = 4A^2\tan^2\theta + B^2$, which will hold when $A = 1/2$ and $B = 1 \Rightarrow B = 2A$

$\therefore$ $e^2=\frac{B^2-A^2}{B^2}=\frac{3A^2}{4A^2}$ i.e., $e=\frac{\sqrt{3}}{2}$

33. **(a)** Let (h, k) be the mid point of a chord, then $T = S_1$ for ellipse $\frac{x^2}{a^2}+\frac{y^2}{b^2}=1$ (where $a > b$), gives $\frac{hx}{a^2}+\frac{ky}{b^2}-1=\frac{h^2}{a^2}+\frac{k^2}{b^2}-1$

Since it is a focal chord

$\therefore$ $(\pm ae, 0)$ will lie on it for $(ae, 0)$, we get $\frac{hae}{a^2}=\frac{h^2}{a^2}+\frac{k^2}{b^2}$

or $\frac{x^2}{a^2}+\frac{y^2}{b^2}=\frac{xe}{a}$

34. **(c)** Appolonian hyperbola (ref. in theory on chapter ellipse)

35. **(c)** Consider $P(0, b)$ on the ellipse $\frac{x^2}{a^2}+\frac{y^2}{b^2}=1$ (for $a > b$)

$\Rightarrow$ $CF = b$ and $PG = b$. Hence $CF.\ PG = b^2$

36. **(d)** Since points are concylic

$\therefore$ $\phi_1 + \phi_2 + \phi_3 + \phi_3 = 2n\pi$ where $n \in \mathbb{N}$

37. **(b)** Consider the point P as $\left(\frac{a}{e}, 0\right)$ for the ellipse $\frac{x^2}{a^2}+\frac{y^2}{b^2}=1$ (where $a > b$)

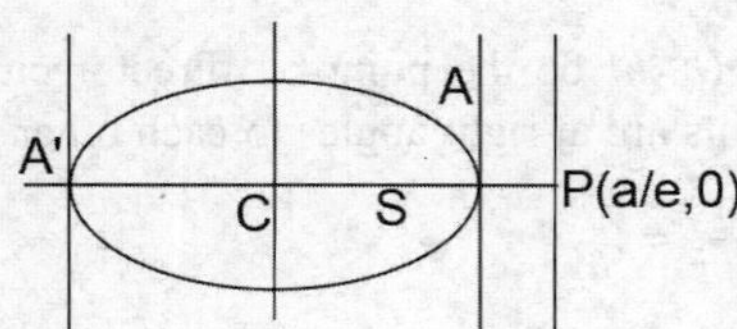

Clearly the tangent at A (or A') is vertical where as PS is horizontal $\therefore$ $\alpha = \pi/2$

Aliter: Let $P\left(\frac{a}{e}, k\right)$ and equation of PC is $y=\frac{ke}{a}x$.

This will intersect the ellipse $\frac{x^2}{a^2}+\frac{y^2}{b^2}=1$ (for $a > b$) as $\frac{x_1^2}{a^2}+\frac{k^2e^2}{a^2}x_1^2=1$

$\therefore$ $x_1^2=\frac{a^2}{1+k^2e^2}$ i.e., $x_1=\pm\frac{a}{\sqrt{1+k^2e^2}}$

These are points with eccentric angle θ and $\pi + \theta$, so tangents will have the same slope at both the points

$\Rightarrow$ Slope of tangent $m_1=-\frac{b^2}{a^2}\cdot\frac{a}{ke}=-\frac{b^2}{ake}=\frac{a^2e^2-a^2}{ake}$

Now slope of PS is $m_2=\frac{ke}{a(1-e^2)}$

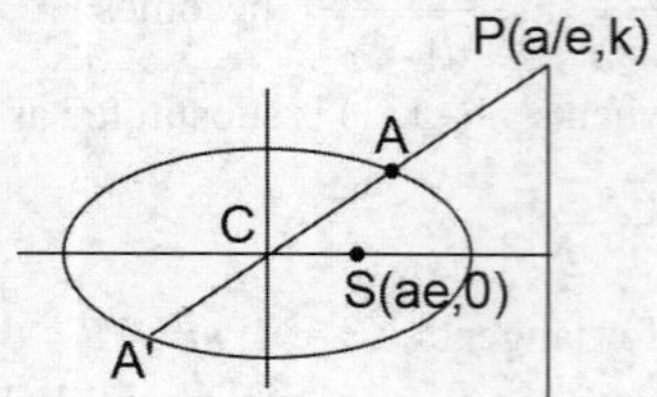

Observe that $m_1m_2=\frac{ke(a^2)(1-e^2)}{a(1-e^2)ake}=-1$

Hence $\alpha = \pi/2$

38. **(b)** Equation of tangent at $P(3\sqrt{3}\cos\theta, \sin\theta)$ for the ellipse $\frac{x^2}{27}+\frac{y^2}{1}=1$ is $\frac{x\cos\theta}{3\sqrt{3}}+\sin y-1=0$

$\Rightarrow$ x-intercept $= |3\sqrt{3}\ \sec\theta|$ and y-intersect $= |\operatorname{cosec}\theta|$

Let θ be in Ist quadrant, so the sum of intercepts $S = 3\sqrt{3}\ \sec\theta + \operatorname{cosec}\theta$

$\Rightarrow$ $\frac{dS}{d\theta}=3\sqrt{3}\sec\theta\tan\theta-\cot\theta.\operatorname{cosec}\theta=0$

$\Rightarrow$ $3\sqrt{3}\sin^3\theta-\cos^3\theta=0$ or $\left(\sqrt{3}\sin\theta-\cos\theta\right)\left(3\sin^2\theta+\cos^2\theta+\sqrt{3}\sin\theta\cos\theta\right)=0$

Hence $\frac{\sin\theta}{\cos\theta}=\frac{1}{\sqrt{3}}$ $\therefore$ $\theta=\frac{\pi}{6}$

39. **(a)** we know that the product of distances of a tangent from the foci is b^2, so for $\frac{x^2}{16}+\frac{y^2}{9}=1$

$\Rightarrow$ $(S_1P_1)(S_2P_2) = 9$

Aliter: For $\frac{x^2}{16}+\frac{y^2}{9}=1$ the tangent at $(4\cos\theta, 3\sin\theta)$

$\Rightarrow$ $\frac{x}{4}\cos\theta+\frac{y}{3}\sin\theta=1$, The foci are $\left(\pm\sqrt{7},0\right)$

Let $(-\sqrt{7}, 0)$ and $S_2(\sqrt{7}, 0)$

$\therefore$ $S_1P_1=\frac{\left|-\frac{\sqrt{7}\cos\theta}{4}-1\right|(12)}{\sqrt{9\cos^2\theta+16\sin^2\theta}}$ and $S_2\frac{\left|\frac{\sqrt{7}}{4}\cos\theta-1\right|(12)}{\sqrt{9\cos^2\theta+16\sin^2\theta}}$

$\Rightarrow$ $(S_1P_1)(S_2P_2)=\frac{144\left(1-\frac{7\cos^2\theta}{16}\right)}{16\sin^2\theta+9\cos^2\theta}=\frac{9\{16-7\cos^2\theta\}}{(16-7\cos^2\theta)}=9$

40. **(a)** Let $P(Q)$ and $Q(\phi)$ be the end points of a chord for the ellipse $\frac{x^2}{a^2}+\frac{y^2}{b^2}=1$ (where $a > b$)

Hence chord PQ is

$\frac{x}{a}\cos\left(\frac{\theta+\phi}{2}\right)+\frac{y}{b}\sin\left(\frac{\theta+\phi}{2}\right)=\cos\left(\frac{\theta-\phi}{2}\right)$

Now length of PQ is

$\ell=\sqrt{a^2(\cos\theta-\cos\phi)^2+b^2(\sin\theta-\sin\phi)^2}$

Since $\phi=\frac{\pi}{2}+\theta$

$\Rightarrow$ $\ell=\sqrt{a^2(\cos\theta+\sin\theta)^2+b^2(\cos\theta-\sin\theta)^2}$

Now will be maximum when ^{2} is maximum since $^2=(a^2+b^2)+(a^2-b^2)\sin 2\theta$ is maximum, when $\theta=\pi/4$

$\Rightarrow$ Maximum $= a\sqrt{2}$

41. **(d)** For the ellipse $\frac{x^2}{a^2}+\frac{y^2}{b^2}=1$, the chord joining $P(\alpha)$ and $Q(\beta)$ is $\frac{x}{a}\cos\left(\frac{\alpha+\beta}{2}\right)+\frac{y}{b}\sin\left(\frac{\alpha+\beta}{2}\right)=\cos\left(\frac{\alpha-\beta}{2}\right)$

Since it passes though focus $S(ae, 0)$

$\Rightarrow$ $e\cos\left(\frac{\alpha+\beta}{2}\right)=\cos\left(\frac{\alpha-\beta}{2}\right)$

$$\Rightarrow e=\frac{\cos\left(\frac{\alpha-\beta}{2}\right)}{\cos\left(\frac{\alpha+\beta}{2}\right)}=\frac{2\cos\left(\frac{\alpha-\beta}{2}\right)\sin\left(\frac{\alpha+\beta}{2}\right)}{2\sin\left(\frac{\alpha+\beta}{2}\right)\cos\left(\frac{\alpha+\beta}{2}\right)}$$

$$\Rightarrow e=\frac{\sin\alpha+\sin\beta}{\sin(\alpha+\beta)}$$

42. **(d)** For the ellipse $\frac{x^2}{a^2}+\frac{y^2}{b^2}=1$ (for $a>b$)

The equation of normal at $P(\theta)$ is $\frac{ax}{\cos\theta}-\frac{by}{\sin\theta}=a^2-b^2$

$\Rightarrow$ slope $m_1=\frac{a\sin\theta}{b\cos\theta}$

Similarly slope of normal at Q is $m_2=\frac{-a\cos\theta}{b\sin\theta}$

$$\Rightarrow \tan\alpha=\left|\frac{\frac{a\sin\theta}{b\cos\theta}+\frac{a\cos\theta}{b\sin\theta}}{1-\frac{a^2}{b^2}}\right|=\frac{\frac{a}{b}\left\{\frac{1}{\sin\theta\cos\theta}\right\}b^2}{(a^2-b^2)}$$

$$\Rightarrow \frac{a^2-b^2}{a^2}=\frac{2b}{a\sin2\theta\tan\alpha}$$

$$\Rightarrow e^2=\frac{2a\sqrt{1-e^2}}{a\sin2\theta\tan\alpha}$$ i.e., $2\sqrt{1-e^2}=e^2\sin2\theta\,.\tan\alpha$

43. **(b)** Distance between focus and an end of minor axis = k = semi-major axis

Distance between foci = $2h$

$\therefore$ semi minor axis = $\sqrt{k^2-h^2}$

Hence the equation of ellipse is $\frac{x^2}{k^2}+\frac{y^2}{k^2-h^2}=1$

44. **(a)** Given foci are (–1, 0), (7, 0)

$\Rightarrow$ centre at (3, 0) and $2ae = 8$

$\Rightarrow$ for $e = 1/2$, $a = 8$ and $b^2 = 48$

$\therefore$ Ellipse is $\frac{(x-3)^2}{64}+\frac{y^2}{48}=1$

The parametric representation $P(3+8\cos\theta+4\sqrt{3}\sin\theta)$

45. **(c)** For the ellipse $\frac{x^2}{a^2}+\frac{y^2}{b^2}=1(a>b)$

Consider the point P at $(a, 0)$

$\Rightarrow S_1P=L_1=a(1+e)$ and $S_2P=a(1-e)=L_2$

Now $H.M$ of S_1P and S_2P is

$$L=\frac{2L_1L_2}{L_1+L_2}=\frac{2a^2(1-e)^2}{2a}=\frac{b^2}{a}=\text{semi. } L.R$$

46. **(d)** For the ellipse $\frac{x^2}{a^2}+\frac{y^2}{a^2}-1=0$, the combined equation of tangnets from $P(h, k)$ will be $T^2 = SS_1$

$$\Rightarrow \left(\frac{hx}{a^2}+\frac{ky}{b^2}-1\right)^2=\left(\frac{x^2}{a^2}+\frac{y^2}{b^2}-1\right)\left(\frac{h^2}{a^2}+\frac{k^2}{b^2}-1\right)$$

$$\Rightarrow \left(\frac{h^2}{a^2}-\frac{h^2}{a^2}-\frac{k^2}{b^2}+1\right)\frac{x^2}{a^2}+\left(\frac{k^2}{b^2}-\frac{k^2}{b^2}-\frac{h^2}{a^2}+1\right)\frac{y^2}{b^2}+\frac{2hkxy}{a^2b^2}+\text{Other linear terms}=0$$

Since tangets intersect at 45°

$$\therefore \tan\theta=2\frac{\sqrt{H^2-AB}}{|A+B|}=1$$

$\Rightarrow 4(H^2-AB)=(A+B)^2$

$$\Rightarrow 4\left\{\frac{h^2k^2}{a^4b^4}-\frac{(b^2-k^2)(a^2-h^2)}{a^4b^4}\right\}=\left\{\frac{(a^2+b^2)-(h^2+k^2)}{a^2b^2}\right\}^2$$

Hence

$$4\left\{\frac{a^2k^2+b^2h^2-a^2b^2}{a^4b^4}\right\}=\frac{1}{a^4b^4}\{(a^2+b^2)-(h^2+k^2)\}^2$$

i.e., $4(a^2y^2+b^2x^2-a^2b^2)=\{(x^2+y^2)-(a^2+b^2)\}^2$

47. **(a)** For the ellipse $\frac{x^2}{a^2}+\frac{y^2}{b^2}=1$ (for $a>b$)

The equation of normal at $P(\theta)$ is $\frac{ax}{\cos\theta}-\frac{by}{\sin\theta}=a^2-b^2$

The line $x+my=n$ will be a normal, when

$$\frac{\ell\cos\theta}{a}=-\frac{m\sin\theta}{b}=\frac{n}{a^2-b^2}$$

$$\Rightarrow \frac{n^2}{(a^2-b^2)^2}\left\{\frac{a^2}{\ell^2}+\frac{b^2}{m^2}\right\}=1$$

48. **(b)** Let the normal at $P(\theta)$ on the ellipse $\frac{x^2}{a^2}+\frac{y^2}{b^2}=1$ (where $a>b$) intersect the ellipse again at $Q(\phi)$

$\Rightarrow$ Slope of normal at $P(\theta)$ = Slope of line PQ

i.e., $\frac{a}{b}\tan\theta=\frac{b(\sin\theta-\sin\phi)}{a(\cos\theta-\cos\phi)}$

$$\Rightarrow \tan\theta\left\{\frac{-2\sin\left(\frac{\theta+\phi}{2}\right).\sin\left(\frac{\theta-\phi}{2}\right)}{2\cos\left(\frac{\theta+\phi}{2}\right).\sin\left(\frac{\theta-\phi}{2}\right)}\right\}=\frac{b^2}{a^2}$$

i.e., $\tan\theta.\tan\left(\frac{\theta+\phi}{2}\right)=-\frac{b^2}{a^2}$

49. **(c)** Let (h, k) be the point then the equation of chord of contact for the ellipse $\frac{x^2}{a^2}+\frac{y^2}{b^2}=1$ is $\frac{hx}{a^2}+\frac{k}{b^2y}-1=0$

Since it touch the circle $x^2+y^2=c^2$.

$$\therefore \frac{(-1)^2}{\frac{h^2}{a^4}+\frac{k^2}{b^4}}=c^2 \quad\Rightarrow\quad \frac{x^2}{a^4}+\frac{y^2}{b^4}=\frac{1}{c^2}$$

50. **(c)** Let $\frac{x^2}{a^2}+\frac{y^2}{b^2}=1$ be the similar ellipse (with eccentricity e) under the given conditions.

$\Rightarrow A^2=a^2e^2$ and $\frac{A^2-B^2}{A^2}=E^2$ as $E=e$ for similarly

$$\therefore \frac{a^2e^2-B^2}{a^2e^2}=\frac{a^2-b^2}{a^2}$$

$\Rightarrow B^2=b^2e^2$.

51. **(c)** For the ellipse $\frac{x^2}{169}+\frac{y^2}{25}=1$ the equation of normal at $P(\theta)$ is $ax \sin\theta - by \cos\theta = (a^2 - b^2) \sin\theta\cos\theta$
Since the normal passes through (0, 6)
$\Rightarrow$ $-30 \cos\theta = 144 \cos\theta \sin\theta$; when $\cos\theta = 0$, then $\theta = \frac{\pi}{2}, \frac{3\pi}{2}$, which given the normal as y-axis, when $\cos\theta \neq 0$, then $\sin\theta = -5/24$ which will give two points one each in IIIrd quadrant and IVth quadrant.
Hence three distinct normals are possible.

52. **(b)** The major axis of $\frac{x^2}{f(k^2+2k+5)}+\frac{y^2}{f(k+11)}=1$ is along x-axis
$\Rightarrow$ $f(k^2 + 32k + 5) > f(k + 11)$
$\because$ f is a decreasing function
$\therefore$ $k^2 + 2k + 5 < k + 11$ or $k^2 + k - 6 < 0 \Rightarrow k \in (-3, 2)$.

53. **(b)** Consider point P at $(0, b)$ on the ellipse $\frac{x^2}{a^2}+\frac{y^2}{b^2}=1$

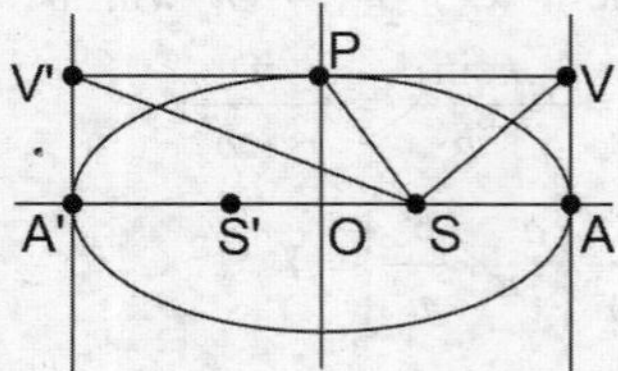

Obviously $(AV)(A'V') = b^2$
Also $PV = PV' = PS = a = PS'$
Hence $VS\,V'S'$ are con-cyclic points with centre at P.
Since V, P, V' are collinear
$\therefore$ $\angle VSV' = 90°$. Conclude that $(AV)(A'V') = a^2$ is incorrect.

54. **(a)** Straight line $y=\frac{x}{2}+2$ is a tangent to the parabola $y^2 = 4x$.
This straight line will be a tangent to the ellipse $\frac{x^2}{4}+\frac{y^2}{b^4}=1$, when $c^2 = a^2m^2 + b^2$ i.e., $4\left(\frac{1}{4}\right)+b^2=4$
$\Rightarrow b = \sqrt{3}$.
Now from the symmetry about x-axis a line with slope $m = -1/2$ passing through $(-4, 0)$ will also be a common tangent. i.e., $y=-\frac{1}{2}(x+4)$ or $x + 2y + 4 = 0$

55. **(a)** $2b^2 = a^2 + ab$ is $a^2 + ab - 2b^2 = 0$.
$\Rightarrow$ $(a - b)(a + 2b) = 0$
$\Rightarrow$ $a = b$ {as $a = -2b$ is meaningless}.
Hence ellipse becomes a circle.

Aliter: Let $\frac{x^2}{a^2}+\frac{y^2}{b^2}=1$ be an ellipse (with $a > b$)
$\Rightarrow$ $LR=\frac{2b^2}{a}=a+b$ (given)
$\Rightarrow$ $2b^2 = a^2 + ab$ or $2a^2(1 - e^2) = a^2 + a^2\sqrt{1-e^2}$
$\Rightarrow$ $2(1 - e^2) = 1+\sqrt{1-e^2}$. Let $t=\sqrt{1-e^2}$
$\Rightarrow$ $2t^2 - t - 1 = 0$
$\Rightarrow$ $t = 1, -1/2$, Now $t = 1$ gives $e = 0$ i.e., ellipse becomes a circle.

56. **(a)** The given ellipse is $\frac{x^2}{2}+y^2=1$ equation of tangent at P $(\sqrt{2}\cos\theta, \sin\theta)$ is $\frac{x}{\sqrt{2}}\cos\theta+y\sin\theta=1$ gives mid-point at $\left(\frac{\sec\theta}{\sqrt{2}},\frac{\operatorname{cosec}\theta}{2}\right)=(x_1,y_1)$ (say)
$\Rightarrow$ $\cos^2\theta + \sin^2\theta = \frac{1}{2x_1^2}+\frac{1}{4y_1^2}=1$
Hence the locus is $\frac{1}{2x^2}+\frac{1}{4y^2}=1$.

SECTION–IV (MORE THAN ONE CORRECT)

1. **(a), (c)** We know that foot of perpendicular from focus onto any tangent falls on the auxiliary circle which always has its centre at the centre of ellipse.

2. **(a), (b), (c)** Let $P(a \cos\theta, b \sin\theta)$ be a general point on $\frac{x^2}{a^2}+\frac{y^2}{b^2}=1$, then for $a > b$ foci are $(\pm ae, 0)$

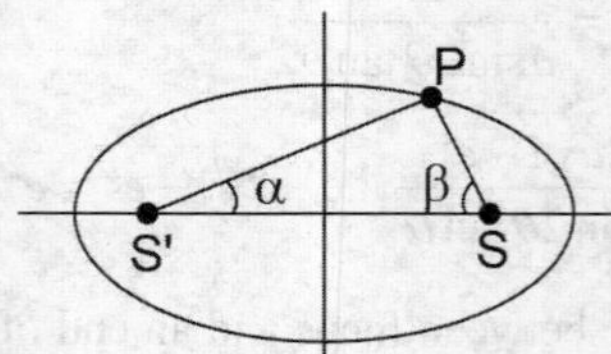

And for $a < b$ foci are $(0, \pm be)$. So in both the cases $SP + S'P$ = major axis.
So option (a) and (b) are true.
Now $\tan\frac{\alpha}{2}.\tan\frac{\beta}{2}=\frac{(s-c)}{s}$; where s is the semi perimeter and $c = 2ae$ {distance between the foci}
$\Rightarrow$ $\tan\frac{\alpha}{2}.\tan\frac{\beta}{2}=\frac{a(1-e)}{a(1+e)}=\frac{1-e}{1+e}$
$\Rightarrow$ (c) is correct
Observe that for $a > b$ we have
$$\sqrt{\frac{a^2-b^2}{b^2}}\left\{a-\sqrt{a^2-b^2}\right\}=\frac{a^2e^2}{a^2(1-e^2)}\{a-ae\}$$
$$=\frac{ae^2}{1+e}\neq\frac{1-e}{1+e}$$

3. **(a), (c)** We know that PT bisects the external angle i.e., $\angle F_1PF_2 = \pi - \angle F_1PF_2$, also PN bisects the internal angle i.e., $\angle F_1PF_2'$

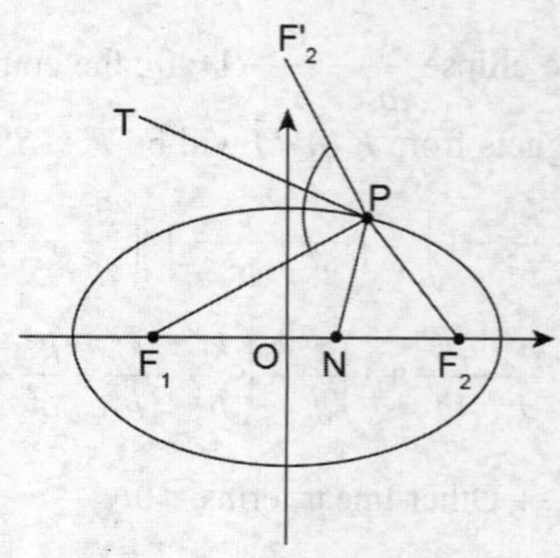

4. **(a), (b), (c)** Let P ($a\cos\theta$, $b\sin\theta$) for the ellipse $\frac{x^2}{a^2}+\frac{y^2}{b^2}=1$

As PG is the bisector of $\angle S'PS$

$\Rightarrow \frac{SG}{GS'}=\frac{PS}{PS'} \Rightarrow \frac{SG}{SG+GS'}=\frac{PS}{PS'+PS}$

Hence $SG=\frac{2ae.PS}{2a}=ePS$

Also PT bisects the external angle i.e., $\angle TPS''$. {By the reflection properties of ellipse}.

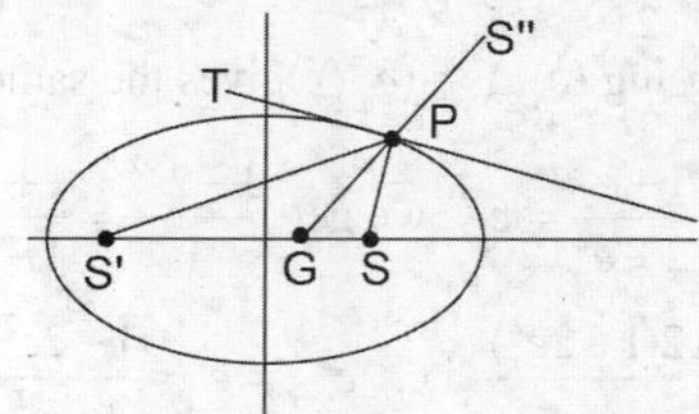

5. **(a), (b), (c), (d)** The locus of a point from which the tangents to the ellipse $\frac{x^2}{a^2}+\frac{y^2}{b^2}=1$ are at 90° will be director circle i.e., $x^2+y^2=a^2+b^2$. So for ellipse $\frac{x^2}{4}+\frac{y^2}{9}=1$ the director circle is $x^2+y^2=13$.

Observe that $(1, 2\sqrt{3})$, $(2\sqrt{3}, 1)$, $(2, 3)$, $(3, 2)$ all line on the director circle from where tangents will be at 90°.

6. **(b), (d)** Given ellipse is $\frac{x^2}{1/4}+\frac{y^2}{1/9}=1 \Rightarrow a=\frac{1}{2}, b=\frac{1}{3}$

Let P ($a\cos\theta$, $b\sin\theta$) $\equiv \left(\frac{1}{2}\cos\theta, \frac{1}{3}\sin\theta\right)$ be the point of contact

Then equation of tangent is $\frac{x}{a}\cos\theta+\frac{y}{b}\sin\theta=1$

$\Rightarrow 2\cos\theta+3y\sin\theta=1$

$\Rightarrow$ Slope $=-\frac{2\cos\theta}{3\sin\theta}=\frac{-2}{3}\cot\theta=\frac{8}{9}$

$\Rightarrow \cot\theta=\frac{-4}{3} \Rightarrow \theta$ is in II or IVth quad.

$\Rightarrow \text{cosec}^2\theta=1+\frac{16}{9}=\frac{25}{9}$

$\Rightarrow \text{cosec}\theta=\pm 5/3$

$\Rightarrow \sin\theta=\pm 3/5$; $\cos\theta=\mp\frac{4}{5}$

$\therefore$ The required. points are $\left(\frac{1}{2}\left(\mp\frac{4}{5}\right)\frac{1}{3}\left(\pm\frac{3}{5}\right)\right)$

$\equiv\left(\mp\frac{2}{5}, \pm\frac{1}{5}\right)$ i.e., $\left(\frac{-2}{5}, \frac{1}{5}\right)$ and $\left(\frac{2}{5}, \frac{-1}{5}\right)$

7. **(a), (c)** **(a)** Observe that when $\lambda_1=0$, and $\lambda_2\lambda_3>0$, then $\lambda_1 PA+\lambda_2 PB=\lambda_3$ becomes $PB=\frac{\lambda_3}{\lambda_2}=k>0$

Which will be a circle.

(b) When $\lambda_1>0$ and $\lambda_2<0$ and $\lambda_3=0$, we get $\lambda_1 PA-|\lambda_2| PB=0$ i.e., $\lambda_1^2 PA^2=\lambda_2^2 PB^2$

Now if $\lambda_1=|\lambda_2|$, then it will be the right bisector of AB

(c) When $\lambda_1=\lambda_2>0$, and $\lambda_3>0$, then clearly $PA+PB=\frac{\lambda_3}{\lambda_1}=k(say)>0$

Subject to the requirements i.e., $AB<k$, it will represent an ellipse

(d) Consider the case were $\lambda_1=-1$ and $\lambda_2\lambda_3>0$

If $\lambda_2<0$ and $\lambda_3<0$ we can not get a hyperbla, then the equation becomes $PA+|\lambda_2| PB=|\lambda_3|$

8. **(a), (c)** For the ellipse $\frac{x^2}{a^2}+\frac{y^2}{b^2}=1$ (where $a>b$)

The slope of normal at $P(\theta)$ is $m_1=\frac{a}{b}\tan\theta$.

Smilarly slope of normal at $Q(\theta+\pi/2)$ is $m_2=-\frac{a}{b}\cot\theta$

Now $\tan\omega=\pm\left|\frac{m_1-m_2}{1+m_1m_2}\right|=\pm\left|\frac{\frac{a}{b}\left(\tan\theta+\frac{1}{\tan\theta}\right)}{1-\frac{a^2}{b^2}}\right|$

$\Rightarrow \cot\theta=\pm\frac{-(a^2-b^2)\sin\theta\cos\theta}{ab}$

$\Rightarrow \frac{\cot w}{\sin 2\theta}=\pm\frac{-(a^2-b^2)}{2ab}$

or $\frac{2\cot\theta}{\sin 2\theta}=\pm\frac{-(a^2-b^2)}{a^2}\left(\frac{a}{b}\right)=(\pm e^2)\frac{1}{\sqrt{1-e^2}}$

SECTION–V (ASSERTION AND REASON)

1. **(d)** **R:** The statement is true

A: For ellipse $x^2+y^2+2\lambda xy+2x+2y+4=0$

$$\Delta=\begin{vmatrix}1 & \lambda & 1\\ \lambda & 1 & 1\\ 1 & 1 & \lambda\end{vmatrix}=3\{\lambda-4\lambda^2\}+\lambda-1=2+2\lambda-4\lambda^2 \text{ and}$$

$h^2-ab<0$

$\Rightarrow \lambda^2<1 \qquad \Rightarrow \lambda\in(-1, 1)$

Now, $(-1)(4\lambda^2-2\lambda-2)=0 \Rightarrow \lambda=\frac{2\pm 6}{8}=1, -\frac{1}{2}$

Since $-1/2$ lies in the interval $(-1, 1)$. So assertion is not completely true.

2. **(a)** **R:** The statement is true for an ellipse $0<e<1$.

A: The statement is true. For $a>b$ the distance between foci S and S' is $2ae$; where as the distance of any point P is $SP+S'P=$

$2a$, (the length of the major axis). Hence assertion is true and it is fully supported by reason.

3. **(a)** **R:** The statement is true (By definition of the director circle)

A: The statement is true for $\frac{x^2}{16}+\frac{y^2}{9}=1$ the director circle is $x^2+y^2=25$. The statement fully supported by reason.

4. **(d)** **R:** The statement is true. The given equation $9x^2+4y^2-18x-24y+9=0$ can be expressed as $\frac{(x-1)^2}{4}+\frac{(y-3)^2}{9}=1$

A: The statement is false sum of focal distances of a point = Length of major axis = 6 units ≠ 4 units.

5. **(d) R :** The statement is true $5x^2 + 9(y^2 - 6y + 9) + 36 - 81 = 0$ can be rewritten as $\frac{x^2}{9} + \frac{(y-3)^2}{5} = 1$

A: The statement is false for $\frac{x^2}{9} + \frac{(y-3)^2}{5} = 1$ the length of major axis is 6 and minor axis is $2\sqrt{5}$.

6. **(c) R:** The statement is false for the ellipse $\frac{x^2}{9} + \frac{y^2}{4} = 1$, the director circle i.e., the locus of point of intersection perpendicular tangents) is $x^2 + y^2 = 13$.

A: $(\lambda, 3)$ will be on the director circle of ellipse $\frac{x^2}{9} + \frac{y^2}{4} = 13$; when $\lambda^2 + 9 = 13$ i.e., $\lambda = \pm 2$

$\Rightarrow$ The statement is true.

7. **(a) R:** The statement is true. The equation of chord of ellipse with mid-point (x_1, y_1) is $T = S_1$.

A: The statement is true. Since chords are bisected on line $x + y = b$, so let the mid-point be $(h, b - h)$, then $T = S_1$

For ellipse $\frac{x^2}{a^2} + \frac{y^2}{b^2} = 1$ }

$$\Rightarrow \frac{hx}{a^2} + \frac{(b-h)y}{b^2} - 1 = \frac{h^2}{a^2} + \frac{(b\ \ h)^2}{b^2} -$$

Since the chords are passing through $(a, -b)$

$$\Rightarrow \frac{h}{a} - \frac{b-h}{b} = \frac{h^2}{a^2} + \frac{(b-h)^2}{b^2} \text{ i.e., } ab^2h - a^2b^2 + a^2bh = h^2b^2$$
$+ a^2b^2 + a^2h^2 - 2a^2bh$

Rewriting as $h^2(a^2 + b^2) - 2bh\{3a + b\} + 2a^2b^2 = 0$

Since h is real so $D \geq 0$

$\Rightarrow a^2b^2(3a + b)^2 - 8a^2b^2(a^2 + b^2) \geq 0$ or $9a^2 + b^2 + 6ab - 8a^2 - 8b^2 \geq 0$ i.e., $a^2 + 6ab - 7b^2 \geq 0$

Hence the statement is true and derivable from Reason R:

8. **(a) R:** The statement is true. Every chord passing through the centre of ellipse will be bisected at it.

A: The statement is true $x^2 + y^2 + xy = 1$ indicates that it is rotation of axis about the centre (origin) and the equation becomes

$\frac{(x-y)^2}{4} + \frac{(x+y)^2}{4/3} = 1$ as shown below $16(x^2 + y^2 + xy) = 16 \Rightarrow 4(x^2 + y^2 - 2xy) + 12(x^2 + y^2 + 2xy) = 16$

or $\frac{(x-y)^2}{4} + \frac{(x+y)^2}{4/3} = 1$. Hence the statement is true and it is fully supported by reason.

SECTION–VI (COMPREHENSION-TYPE)

Comprehension A:

From the given considerations, let E_{n-1} be a horizontal ellipse $\frac{x^2}{a^2} + \frac{y^2}{b^2} = 1$ (for $a > b$) having eccentricity $e_{n-1} = e$ where $e^2 = \frac{a^2 - b^2}{a^2}$, then E_n is $\frac{x^2}{a^2} + \frac{y^2}{a^2 + b^2} = 1$ (vertical) with

$$e_n^2 = \frac{b^2}{a^2 + b^2} = \frac{1 - e^2}{2 - e^2}$$

Also E_{n-2} is $\frac{x^2}{a^2 - b^2} + \frac{y^2}{b^2} = 1$ (vertical) with

$$e_{n-2}^2 = \frac{2b^2 - a^2}{a^2} = \frac{1 - 2e^2}{1 - e^2}$$

1. **(b)** Since E_n is independent of n, so $e_n^2 = e_{n-2}^2$

$\Rightarrow \frac{1 - 2e^2}{1 - e^2} = \frac{1 - e^2}{2 - e^2}$ i.e., $e^4 - 3e^2 + 1 = 0$

$\Rightarrow e^2 = \frac{3 \pm \sqrt{5}}{2}$ (rejecting $\frac{3 + \sqrt{5}}{2} > 1$), we get $e = \frac{\sqrt{5} - 1}{2}$

Also using $(e_{n-2})^2 = (e_{n-1})^2$ gives the same results.

2. **(d)** From $\frac{1 - e^2}{2 - e^2} = e_n^2$, we get $\frac{3 - 2e^2}{1} = \frac{1 + e_n^2}{1 - e_n^2}$

$\Rightarrow 2e^2 = \frac{2(1 - 2e_n^2)}{(1 - e_n^2)}$ i.e., $e_{n-1}^2 = e^2 = \frac{1 - 2e_n^2}{1 - e_n^2}$

Hence $(x_1, y_1) = (e_n^2, e_{n-1}^2) = \left\{e_n^2, \frac{1 - 2e_n^2}{1 - e_n^2}\right\}$

$\Rightarrow y_1 = \frac{1 - 2x_1}{1 - x_1} = 2 + \frac{1}{x_1 - 1}$ i.e., $(y_1 - 2) = \frac{1}{(x_1 - 1)}$ or $(x_1 - 1)$ $(y_1 - 2) = 1$ {equivalent to $XY = 1$, when origin is shifted to (1, 2)} which is a rectangular hyperbola.

3. **(d)** As calculated for $e_{n-1} = e$, we have $e_{n-2}^2 = \frac{1 - 2e^2}{1 - e^2}$ and $e_n^2 = \frac{1 - e^2}{2 - e^2}$. Now gives $e_1^2 = 3/4$

So for $n = 3$, we get $e_1^2 = \frac{1 - e_2^2}{1 - e_2^2} = \frac{3}{4}$

$\Rightarrow e_2^2 = \frac{1}{5}$ and $e_3^2 = \frac{1 - e_2^2}{2 - e_2^2} = \frac{(4/5)}{(9/5)} = \frac{4}{9}$

Putting this in $e_3^2 = \frac{1 - 2e_4^2}{1 - e_4^2} = \frac{4}{9}$, we get $e_4^2 = 5/14$

$\Rightarrow e_4 = \sqrt{\frac{5}{14}}$

4. **(c)** As calculated we know for $E_1 : \frac{x^2}{a^2 - b^2} + \frac{y^2}{b^2} = 1$ (vertical ellipse)

We will get E_3 as $\frac{x^2}{a^2} + \frac{y^2}{a^2 + b^2} = 1$ (vertical ellipse)

So, for $\frac{x^2}{9} + \frac{y^2}{16} = 1$ (Vertical ellipse), we get $\frac{x^2}{25} + \frac{y^2}{41} = 1$

5. **(b)** As calculated earlier for the independent situation $e^2 = \frac{3 - \sqrt{5}}{2}$

Now consider E_n as $\frac{x^2}{a^2} + \frac{y^2}{a^2 + b^2} = 1$

Since $e^2 = \frac{3 - \sqrt{5}}{2}$ or $e = \frac{\sqrt{5} - 1}{2}$

$\Rightarrow a^2 : a^2 + b^2 = \frac{\sqrt{5} - 1}{2} : 1 = e$

$\Rightarrow \frac{x^2}{ek^2} + \frac{y^2}{k^2} = 1$ (where $k^2 = a^2 + b^2$)

$\Rightarrow$ Locus of the mid-point (h, k) of chords with slope $m = -1$ is $h = ek$ or $y = \frac{x}{e} = \frac{x}{\left(\frac{\sqrt{5}-1}{2}\right)}$

Or $(\sqrt{5} - 1)\,y = 2x$ or $2y = (\sqrt{5} + 1)x$.

Comprehension B

6. (b) From the given considerations $ae = \sqrt{7}$ and $\frac{a}{e} = \frac{16}{\sqrt{7}}$

$\Rightarrow a^2 = 16$ and $b^2 = 9 \Rightarrow \left(e = \frac{\sqrt{7}}{4}\right)$

Hence $\frac{x^2}{16} + \frac{y^2}{9} = 1$ is the ellipse.

7. (c) Area of $\Delta PSS'$ will be maximum, when $P\ (0, \pm b)$ i.e.,

area $= \frac{1}{2} \times 2ae \times b = 3\sqrt{7}$ square units.

8. (d) In the first quadrant the tangent with slope m (where $m < 0$) for the ellipse is $y = mx + \sqrt{16m^2 + 9}$

$\Rightarrow (x_1, y_1) = \left(\frac{-\sqrt{16m^2+9}}{2m}, \frac{\sqrt{16m^2+9}}{2}\right)$

$\Rightarrow x_1 = \frac{y_1}{-m}$ or $mx_1 + y_1 = 0$.

9. (a) Since $mx - y + \sqrt{16m^2 + 9} = 0$ is a tangent to $x^2 + y^2 = r^2$

$\Rightarrow \frac{16m^2 + 9}{1 + m^2} = r^2 \qquad \Rightarrow r^2 = 16 - \frac{7}{1 + m^2}$

$\Rightarrow (m^2 + 1) = \frac{7}{16 - r^2} \qquad \therefore m = -\sqrt{\frac{r^2 - 9}{16 - r^2}}$ (as $m < 0$)

Hence $y = -mx = x\sqrt{\frac{r^2 - 9}{16 - r^2}}$ is locus.

10. (b) Since the locus is real

$\therefore \frac{r^2 - 9}{16 - r^2} > 0$ i.e., $r^2 \in (9, 16) \Rightarrow r \in (3, 4)$

Comprehension C

11. (c) Consider the intersection of $\frac{x^2}{a^2} + \frac{y^2}{b^2} = 1$ (for $a > b$) and $x^2 + y^2 = r^2$ where $b < r < a$

So that four con-cyclic points of intersection are obtained.

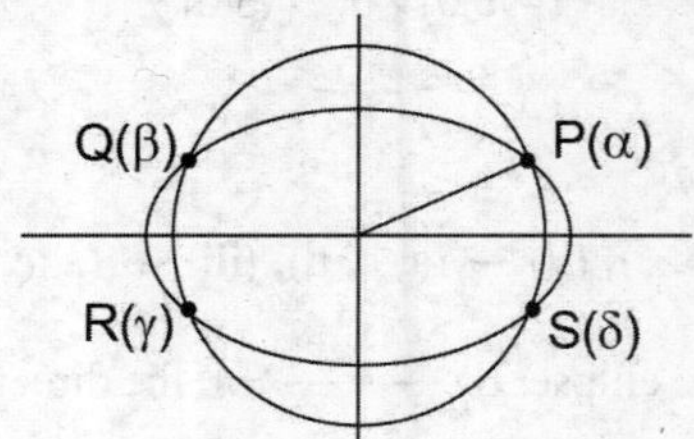

From symmetry, observe that $\alpha + \beta = \pi + 4k\pi$

Similarly $\gamma + \delta = 3\pi + 4k\pi$

$\Rightarrow \alpha + \beta + \delta = 4k\pi + 8k\pi = 2n\pi$ where $n \in \mathbb{Z}$.

12. (d) Since chords are symmetrically oriented with the major (or minor axis)

$\Rightarrow$ The angles will be α, $\pi - \alpha$ i.e., $\theta = \alpha$ and $\phi = \pi - \alpha$

$\Rightarrow \theta + \phi = \pi$

13. (b) Let r be the radius then $P\ (r\cos\theta, r\sin\theta)$; where $\theta = \frac{\pi}{4}$ (for 1st quadrant) will satisfy $\frac{x^2}{a^2} + \frac{y^2}{b^2} = 1$ i.e.,

$$\frac{r^2\cos^2\theta}{a^2} + \frac{r^2\sin^2\theta}{b^2} = 1 \Rightarrow \frac{r^2}{2}\left\{\frac{1}{a^2} + \frac{1}{b^2}\right\} = 1$$

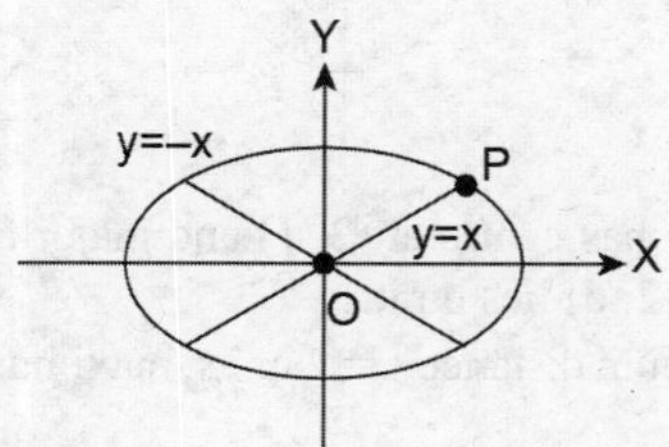

$\Rightarrow r = \sqrt{\frac{2a^2b^2}{a^2 + b^2}} = ab\sqrt{\frac{2}{a^2 + b^2}}$.

14. (a) As worked earlier the sum of four eccentric angles is $2n\pi$, $(n \in \mathbb{Z})$.

So for $S(\delta)$, we get $\alpha + \frac{\pi}{2} + \alpha + \pi + \alpha + \gamma = 2n\pi$

$\Rightarrow \delta = \frac{\pi}{2} - 3\alpha$ (let $n = 1$).

Comprehension D

15. (c) Rewriting as ellipse $E: \frac{x^2}{16} + \frac{y^2}{9} = 1$

Parabola $P: y^2 = 4\left(\frac{1}{4}\right)\{x - 4\}$ and circle C: $(x - 6)^2 + y^2 = 2^2$.

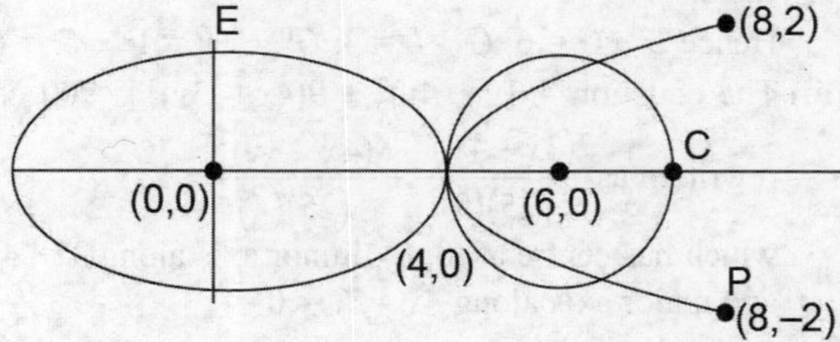

Now observe that parabola and ellipse touch each other at one point i.e., (4, 0)

16. (c) Since the circle touches the ellipse externally, so from a point on the circle maximum two tangents can be drawn to the ellipse.

17. (b) Clearly at (4, 0) a common tangent $x = 4$ is possible to all the three curves.

SECTION–VII (MATRIX-MATCH TYPE)

1. (i) → (a), (c), (ii) → (b), (d), (e), (iii) → (b), (c)

(i) The given ellipse is $9(x^2 - 4x + 4) + 8(y^2 - 2y + 1) = 72$

or $\frac{(x-2)^2}{8} + \frac{(y-1)^2}{9} = 1$

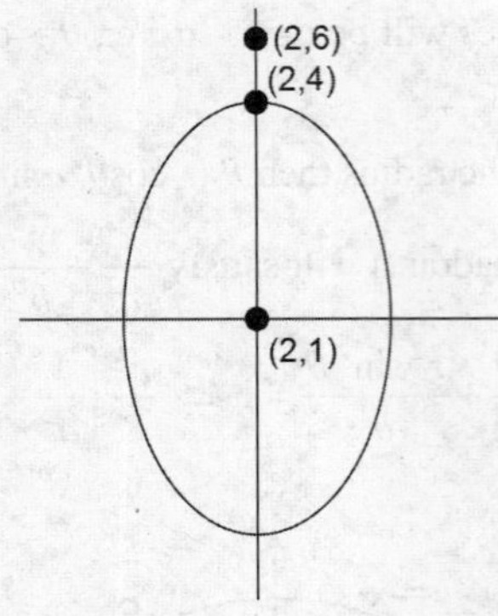

Which has centre at (2, 1) and major axis along $x = 2$. Now, (2, 6) lies on it.

$\therefore$ Minimum distance $L = 2$ units, maximum distance $G = 8$ units.

$\Rightarrow L + G = 10, G - L = 6, L^G + G^L = 2^8 + 8^2 = 256 + 64 = 320$

(ii) The given ellipse is $4(x^2 + 2x + 1) + 9(y^2 - 4y + 4) = 36$

or $\dfrac{(x+1)^2}{9} + \dfrac{(y-2)^2}{4} = 1$

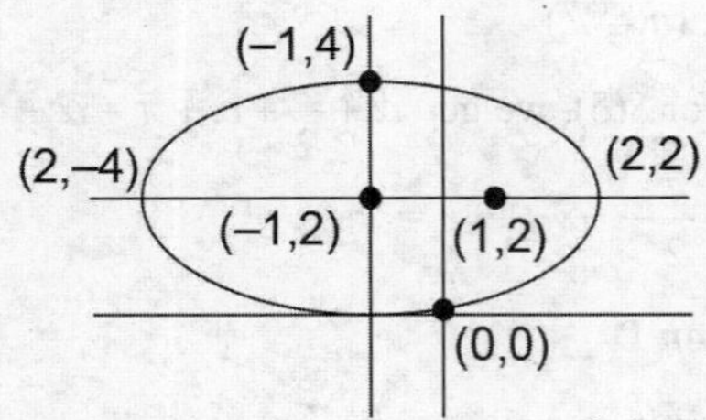

Which has centre at (–1, 2) and major axis along $y = 2$. Now (1, 2) lies on the major axis (inside the ellipse)

$\Rightarrow$ Minimum distance $L = 1$ unit, Maximum distance $G = 5$ units

Hence $L + G = 6, G - L = 4, G^L + L^G = 1^5 + 5^1 = 6$

(iii) The equation $4(3x + 4y)^2 + 9(4x - 3y)^2 = 900$ can be rewritten as $\dfrac{(3x+4y)^2}{(25)(9)} + \dfrac{(4x-3y)^2}{(25)(4)} = 1$;

which has centre at (0, 0), minor axis along $3x + 4y = 0$ and major axis along $4x - 3y = 0$

Now $\left(\dfrac{9}{5}, \dfrac{12}{5}\right)$ lies at the end of major axis (i.e., vertex)

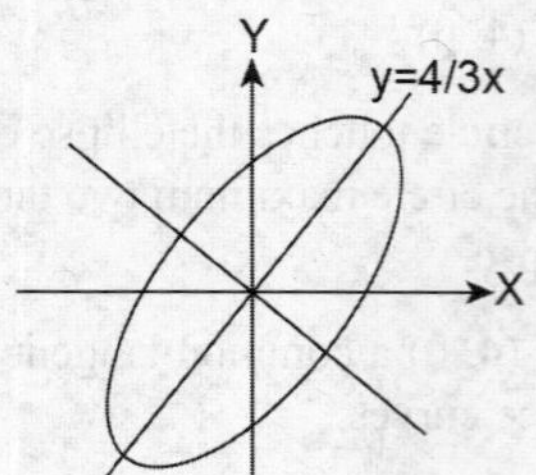

$\Rightarrow L = 0, G = 6$

$\therefore$ $G + L = 6, G - L = 6, G^L + L^G = 1$

2. (i) → (c, d), (ii) → (b, e), (iii) → (a, f)

(i) The given ellipse is $\dfrac{x^2}{4} + \dfrac{y^2}{9} = 1$ which has foci at $(0, \pm\sqrt{5})$

A ray emanating from $(0, -\sqrt{5})$ towards (2, 0) will get reflected from (2, 0) towards the other focus i.e., $(0, \sqrt{5})$.

So F: $\sqrt{5}\,x + 2y - 2\sqrt{5} = 0$ and this ray will strike the ellipse at $x = -\dfrac{4}{7}, y = \dfrac{9\sqrt{5}}{7}$ and it will get second reflection from $\left(-\dfrac{4}{7}, \dfrac{9\sqrt{5}}{7}\right)$ towards the first focus, i.e., $(0, -\sqrt{5})$

So $S: 4\sqrt{5}\ x + y + \sqrt{\ } = 0$

(ii) The given ellipse is $\dfrac{x^2}{9} + \dfrac{y^2}{4} = 1$.

After reflection from $P(0, -2)$ the ray will get reflected towards S' i.e., $(-\sqrt{5}, 0)$, so F: $2x + \sqrt{5}\ y + 2\sqrt{5} = 0$ and this ray will strike the ellipse at $\left(-\dfrac{9\sqrt{5}}{7}, \dfrac{4}{7}\right)$ and it will get reflected towards the first focus $(\sqrt{5}, 0)$.

$\Rightarrow$ S: $4\sqrt{5}\ y - \sqrt{5} = 0$

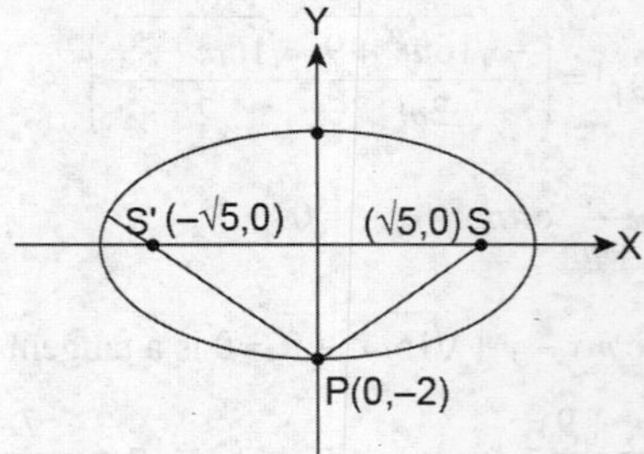

(iii) The given ellipse $\dfrac{x^2}{25} + \dfrac{y^2}{20} = 1$ After reflection at $P(0, 2\sqrt{5})$, the ray will be reflected towards $(\sqrt{5}, 0)$

So F: $y = (-2)(x - \sqrt{5})$ $\quad\Rightarrow 2x + y - 2\sqrt{5} = 0$

This ray will strike the ellipse at $\left(\dfrac{5}{3}\sqrt{5}, -\dfrac{4\sqrt{5}}{3}\right)$ and it will get second reflection towards the focus at $(-\sqrt{5}, 0)$,

$\Rightarrow$ S: $x + 2y + \sqrt{5} = 0$

3. (i) → (a, b, c), (ii) → (a, c, d), (iii) → (a, c, e)

(i) For the ellipse $E: \dfrac{x^2}{1} + \dfrac{y^2}{2} = 1$ the director circle of E is C_1: $x^2 + y^2 = 3$

$\Rightarrow$ Director circle of C_1 is C_2: $x^2 + y^2 = 6$. Hence C_3: $x^2 + y^2 = 12$ and so on ......

i.e., $r_1^2 = 3, r_2^2 = 6, r_3^2 = 12, r_4^2 = 24$ .....

$\Rightarrow$ $e_n^2 = 2^{n-1}(3)$ {for $n \geq 1, n \in \mathbb{N}$}, clearly, r_1^2, r_2^2, r_3^2 are in $GP \Rightarrow \sqrt[3]{r_1^2 r_2^2 r_3^2} = \sqrt[3]{(3)(6)(12)} = 6$

Now $r_1^2 + r_2^2 + r_3^2 + \ldots + r_n^2 = 3\{1 + 2 + 2^2 + 2^{n-1}\} = r_1^2\{2^n - 1\}$

(ii) For the ellipse $E: \frac{x^2}{2} + \frac{y^2}{3} = 1$

The director circle of E is $C_1: x^2 + y^2 = 5$

$\Rightarrow$ Director circle of C_1 is $C_2: x^2 + y^2 = 10$ and the director circle of C_2 is $C_3: x^2 + y^2 = 20$

$\Rightarrow r_1^2 = 5, r_2^2 = 10, r_3^2 = 20, r_4^2 = 40, \ldots\ldots, r_n^2 = (5)2^{n-1}$.

As worked for (i) r_1^2, r_2^2, r_3^2 are in $G.P.$

$\Rightarrow (r_1^2 r_2^2 r_3^2)^{1/3} = \{(5), (10), (20)\}^{1/3} = 10$ and $r_1^2 + r_2^2 + r_3^2 + \ldots\ldots + r_n^2 = 5(2^n - 1) = r_1^2(2^n - 1)$

(iii) For ellipse $E = \frac{x^2}{4} + \frac{y^2}{3} = 1$ the director circle of E is $C_1: x^2 + y^2 = 7$

The director circle of C_1 is $C_2: x^2 + y^2 = 14$ and $C_3: x^2 + y^2 = 28$

$\Rightarrow r_1^2 = 7, r_2^2 = 14, r_3^2 = 28, r_4^2 = 56, \ldots\ldots, r_n^2 = (7).2^{n-1}$.

Now r_1^2, r_2^2, r_3^2 are in $G.P.$

$\Rightarrow (r_1^2 r_2^2 r_3^2)^{1/3} = \{(7), (14), (28)\}^{1/3} = 14$ and $r_1^2 + r_2^2 + r_3^2 + \ldots\ldots + r_n^2 = 7(2^n - 1) = r_1^2(2^n - 1)$

4. (i) → (d), (ii) → (c), (iii) → (a), (iv) → (b)

(i) The given ellipse $E = \frac{x^2}{9} + \frac{y^2}{4} = 1$ the slope of normal

$$m = \frac{a}{b}\tan\theta = \frac{9}{8} \Rightarrow \frac{3}{2}\tan\theta = \frac{9}{8} \Rightarrow \tan\theta = \frac{3}{4}$$

$$\Rightarrow \sin\theta = \frac{3}{5}, \cos\theta = \frac{4}{5} \text{ or } \sin\theta = \frac{-3}{5}, \cos\theta = \frac{-4}{5}$$

Hence the point $(a\cos\theta, b\sin\theta) = \left(\frac{12}{5}, \frac{6}{5}\right)$ or $\left(-\frac{12}{5}, -\frac{6}{5}\right)$

(ii) For the ellipse $E = \frac{x^2}{9} + \frac{y^2}{4} = 1$, the slope of tangent at $P(\theta)$ is $m = \frac{-b}{a}\cot\theta$

Now for equal intercepts $m = \pm 1$

$$\Rightarrow -\frac{3\cos\theta}{2\sin\theta} = \pm 1 \Rightarrow \frac{\sin}{\cos} \pm \frac{2}{3}$$

Hence $\sin\theta = \frac{2}{\sqrt{13}}$ and $\cos\theta = \frac{3}{\sqrt{13}}$

$$\Rightarrow P\left(\frac{9}{\sqrt{13}}, \frac{4}{\sqrt{13}}\right)$$

(iii) A line passing through (0, 5) with slope $m > 0$ will be $y - 5 = mx \quad \Rightarrow \quad mx - y + 5 = 0$

For the ellipse $\frac{x^2}{16} + \frac{y^2}{9} = 1$ the tangent with slope m is

$$mx - y \pm \sqrt{a^2m^2 + b^2} = 0$$

$\Rightarrow 16m^2 + 9 = 25 \qquad \Rightarrow m = 1$

Now the point of contact (in 2nd quadrant) is

$$\left(\frac{-a^2m}{c}, \frac{b^2}{c}\right) = \left(-\frac{16}{5}, \frac{9}{5}\right)$$

(iv) For the ellipse $\frac{x^2}{2} + \frac{y^2}{1} = 1$, the equation of chord of contact of tangent is $T = 0$ i.e., $x + y - 1 = 0$

Now let $M(h, k)$ be the mid point then $T = S_1$

$$\Rightarrow \frac{h}{2}x + ky - \left(\frac{h^2}{2} + k^2\right) = 0$$

Comparing, we get $h = 2k$ and $\frac{1}{k}\left(\frac{h^2}{2} + k^2\right) = 1$

$\Rightarrow k = 1/3$. Hence $M\left(\frac{2}{3}, \frac{1}{3}\right)$

5. (i) → (d); (ii) → (a); (iii) → (b); (iv) → (c)

(i) Given that P divides AB in the ratio 6 : 4

$$\Rightarrow h = \frac{4a}{10}; k = \frac{6b}{10}$$

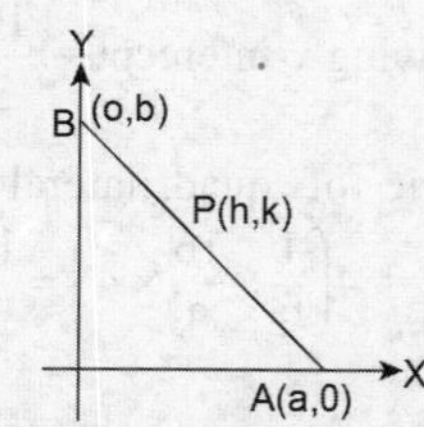

But $(AB) = 10$ (given)

$\Rightarrow a^2 + b^2 = 100 \qquad \Rightarrow \frac{100h^2}{16} + \frac{100k^2}{36} = 100$

$\Rightarrow \frac{h^2}{16} + \frac{k^2}{36} = 1$

$\Rightarrow$ Required locus is $\frac{x^2}{16} + \frac{y^2}{36} = 1$ having eccentricity

$$e = \sqrt{1 - \frac{16}{36}}$$

i.e., $e = \frac{\sqrt{5}}{3} \Rightarrow 9e = 3\sqrt{5}$

(ii) Given ellipse $3x^2 + 2y^2 + 6x - 4y - 1 = 0$

$\Rightarrow 3(x^2 + 2x) + 2(y^2 - 2y) = 1$

$\Rightarrow 3(x^2 + 2x + 1) + 2(y^2 - 2y + 1) = 6$

$$\Rightarrow \frac{(x+1)^2}{2} + \frac{(y-1)^2}{3} = 1$$

$\therefore$ Area of $\Delta AA'P$ will be greatest if P coincide with B and having value $= \frac{1}{2}(AA').(CB) = \frac{1}{2}(2\sqrt{3}).(\sqrt{2}) = \sqrt{6}$

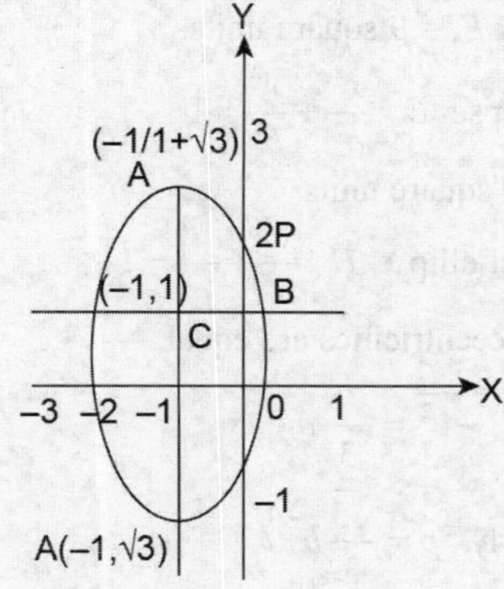

(iii) $x = 1 + 4\cos\theta$ and $y = 2 + 3\sin\theta$

$\Rightarrow \left(\frac{x-1}{4}\right)^2 + \left(\frac{y-2}{3}\right)^2 = 1$ is an ellipse with, $a = 4$, $b = 3$

and $a^2e^2 = a^2 - b^2 = 7 \Rightarrow ae = \sqrt{7}$

$\Rightarrow$ Distance between the foci $= 2ae = 2\sqrt{7}$

(iv) Given ellipse is $\frac{x^2}{16} + \frac{y^2}{7} = 1$

$\Rightarrow a = 4, b = \sqrt{7}, e = \sqrt{1 - \frac{b^2}{a^2}} = \sqrt{1 - \frac{7}{16}} = \frac{3}{4}$

$\therefore$ One of the extremities of latus a recta is $L\,(ae, b^2/a) \equiv L\left(3, \frac{7}{4}\right)$

$\therefore$ Equation of tangent to ellipse at $L\left(3, \frac{7}{4}\right)$ will be $\frac{3x}{16} + \frac{y}{4} = 1$ having x-intercept $= \frac{16}{3}$ and y-intercept = y

$\therefore$ By symmetric of quadrilateral in four quadrants, required area $= 4\left(\frac{1}{2} \times \frac{16}{3} \times 4\right) = \frac{128}{3}$ sq units.

SECTION–VIII (INTEGER-TYPE)

1. For the ellipse $\frac{x^2}{25} + \frac{y^2}{16} = 1$, a normal at $P\,(\theta)$ is $ax\sin\theta - by\cos\theta = (a^2 - b^2)\sin\theta \,.\, \cos\theta$ (where $a = 5$ and $b = 4$)

Since the normal is to pass through $(h, 0) \Rightarrow ah\sin\theta = (a^2 - b^2)\sin\theta\cos\theta$

For $\sin\theta \neq 0$ (i.e., $\theta \neq 0, \pi$), we get $ah = (a^2 - b^2)\cos\theta$

Since $-1 < \cos\theta < 1$ (as $\theta \neq 0, \sqrt{\pi}$)

$\Rightarrow 5h < 25 - 16 = 9$

2. Using the property definition of ellipse $PS^2 = e^2 PM^2$, we get focus (3, 4) and directrix $y = 0$.

$\Rightarrow (x-3)^2 + (y-4)^2 = e^2.y^2 = \frac{y^2}{9} \Rightarrow e = 1/3$

$\therefore$ $9e = 3$

3. The ellipse E_3 is $\frac{x^2}{9} + \frac{y^2}{5} = 1$ (gives $Ae = 2$)

$\Rightarrow e = 2/3$

Now the area of quadrant formed by tangents at the ends of

$L.R. = \frac{4 \times a \times a}{2 \times e} = \frac{2a^2}{e} = 27$ square units

$\Rightarrow$ Area of E_2 = 9 square units.

Let ellipse $E_2 : \frac{x^2}{a^2} + \frac{y^2}{b^2} = 1$

$\Rightarrow \pi ab$= 9 square units.

Now let ellipse $E_1 : \frac{x^2}{b^2} + \frac{y^2}{p^2} = 1$

$\Rightarrow$ since eccentricities are equal

$\therefore$ $b = a\sqrt{1 - e^2} = \frac{\sqrt{5}}{3}a$

Similarly, $p = \frac{\sqrt{5}}{3}b$

$\Rightarrow \pi bp = \pi\left(\frac{\sqrt{5}}{3}a\right)\left(\frac{\sqrt{5}}{3}b\right) = \frac{5}{9}(\pi ab) = 5$ square units.

$\therefore$ Area of E_2 lying out side $E_1 = 9 - 5 = 4$

4. For the ellipse $\frac{x^2}{a^2} + \frac{y^2}{b^2} = 1$ (where $a > b$)

Consider a point other than origin on major axis as $P\left(\frac{(a^2 - b^2)\cos\theta}{a}, 0\right)(0 \neq 0°, 90°)$ as a result the points as feet of normals will be.

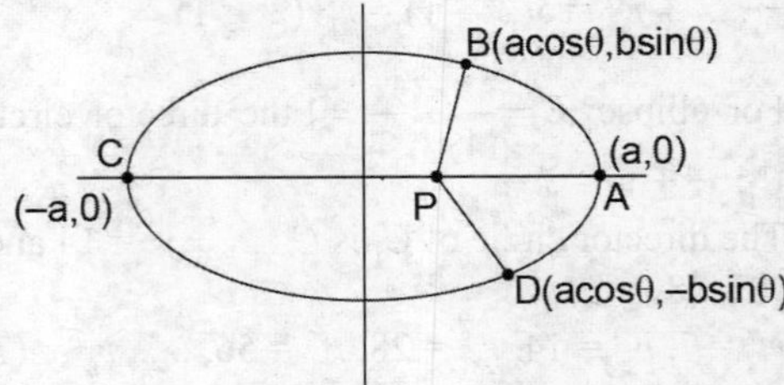

$A\,(a, 0)$, $B\,(a\cos\theta, b\sin\theta)$, $C\,(-a, 0)$ and $D\,(a\cos\theta, -b\sin\theta)$

{Because normal at $B\,(\theta)$ will be $\frac{ax}{\cos\theta} - \frac{by}{\sin\theta} = a^2 - b^2$

and it will intersect x-axis at $x = \frac{(a^2 - b^2)}{a}\cos\theta$ }

Now $\sum x_1 = 2a\cos\theta$ and $\sum \frac{1}{x_1} = \frac{2}{a\cos\theta}$

$\Rightarrow (x_1 + x_2 + x_3 + x_4)\left(\frac{1}{x_1} + \frac{1}{x_2} + \frac{1}{x_3} + \frac{1}{x_4}\right) = 4$

5 For the ellipse $\frac{x^2}{16} + \frac{y^2}{9} = 1$ a line with slope m will be a tangent when $y = mx \pm \sqrt{16m^2 + 9}$

This line will also be a tangent to the circle $x^2 + y^2 = r^2$ when $\frac{16m^2 + 9}{1 + m^2} = r^2$

Now a line through focus $S\,(\sqrt{7}, 0)$ with slope m will be $mx - y - \sqrt{7}\,m = 0$

Let p be its distance from centre (0, 0)

$\Rightarrow \quad p^2 = \frac{7m^2}{1 + m^2}$

Length of chord $RS = 2\sqrt{r^2 - p^2}$

$= 2\sqrt{\frac{16m^2 9}{1 + m^2} - \frac{7m^2}{1 + m^2}} = 2\sqrt{9} = 6$ units

6. Without any loss of generally, let $\frac{x^2}{a^2} + \frac{y^2}{b^2} = 1$ be the ellipse with $e_1 = \frac{1}{\sqrt{2}} \Rightarrow 1 - \frac{b^2}{a^2} = e_1^2 = \frac{1}{2}$

$\therefore$ $b = a/\sqrt{2}$. Let θ be the angle. Now consider the intersection of line $y = x\tan\theta$ with this ellipse $\frac{x^2}{a^2} + \frac{y^2}{b^2} = 1$

$\Rightarrow \frac{x^2}{a^2} + \frac{x^2\tan^2\theta}{b^2} = 1 \quad \Rightarrow x^2\{b^2 + a^2\tan^2\theta\} = a^2b^2$

Hence $x_1 = \pm\dfrac{ab}{\sqrt{a^2\tan^2\theta + b^2}}$

Now $F(x_1, y_1) = OF = x_1 \sec\theta = ae_2 = \dfrac{a\sqrt{3}}{2}$

$\therefore\ \dfrac{a\sqrt{3}}{2} = \dfrac{ab\sec\theta}{\sqrt{a^2\tan^2\theta + b^2}}$, squaring and putting $a^2 = 2b^2$,

we get $\dfrac{3}{4} = \dfrac{b^2\sec^2\theta}{b^2\{2\tan^2\theta + 1\}}$

$\Rightarrow\ \dfrac{3}{4} = \dfrac{\sec^2\theta}{(2\sec^2\theta - 1)}$

$\Rightarrow\ 6\sec^2\theta - 4\sec^2\theta = 3 \ \Rightarrow\ \sec^2\theta = 3/2$

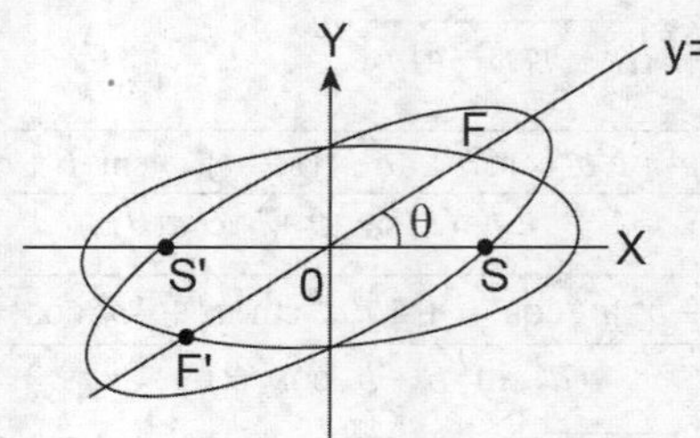

Now $\cos^2\theta = 2/k \ \Rightarrow\ \dfrac{2}{k} = \dfrac{2}{3}$

$\Rightarrow\ k = 3$

7. Without any loss of generality, let $\dfrac{x^2}{a^2} + \dfrac{y^2}{b^2} = 1$ (where $a > b$)

be the ellipse with eccentricity e_1 and θ be the angle between the major axes

$\Rightarrow\ 1 - \dfrac{b^2}{a^2} = e_1^2$ i.e., $\text{—} = -e_1^2 = e_2^2$ (as $e_1^2 + e_2^2 = 1$)

As worked for 6, $OF = x_1\sec\theta = ae_2 = b$

$\Rightarrow\ \dfrac{ab\sec\theta}{\sqrt{a^2\tan^2\theta + b^2}} = b \ \Rightarrow\ a^2\sec^2\theta = a^2\tan^2\theta + b^2$

Since $a \neq b$ $\Rightarrow\ \sec\theta + \tan\theta$ both $\to \infty$

Hence $\theta = 90°$ so $\theta/10 = 9°$

8. Indirectly we are told that $d = \dfrac{\sqrt{a^2 + 3b^2}}{3} = a$ (where $a > b$)

$\Rightarrow a^2 + 3b^2 = 3a^2$

Hence $3b^2 = 2a^2 \ \Rightarrow\ \dfrac{a^2 - b^2}{a^2} = \left(\dfrac{1}{3}\right)\dfrac{a^2}{a^2} = \dfrac{1}{3}$

i.e., $e^2 = 1/3 \ \Rightarrow\ 27e^2 = 9$

9. $a = 3\sqrt{3}$, $b = 3$ (given). Since axes are at 90° to each other

$\therefore\ O(0, 0)$ being the centre of arc and radius of acc is OC where $OC^2 = a^2 + b^2$

$\Rightarrow\ OC = \sqrt{27 + 9} = 6$ Units

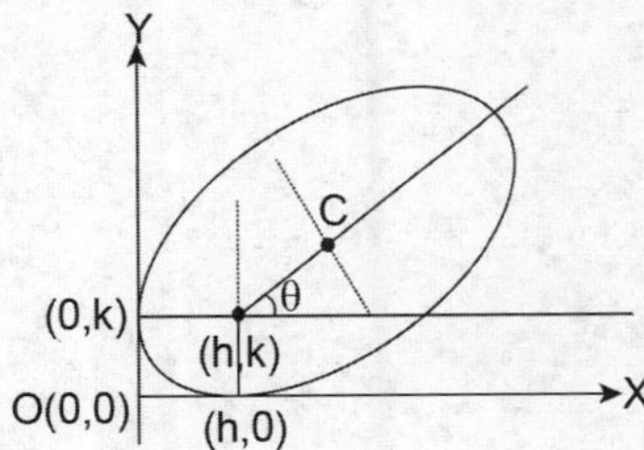

The two extreme positions are horizontal and vertical orientation

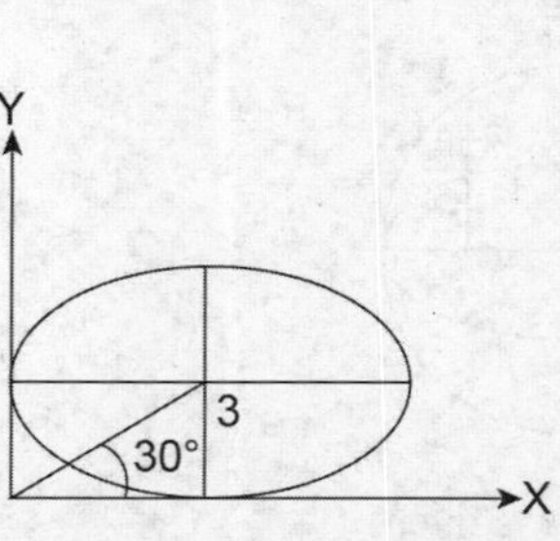

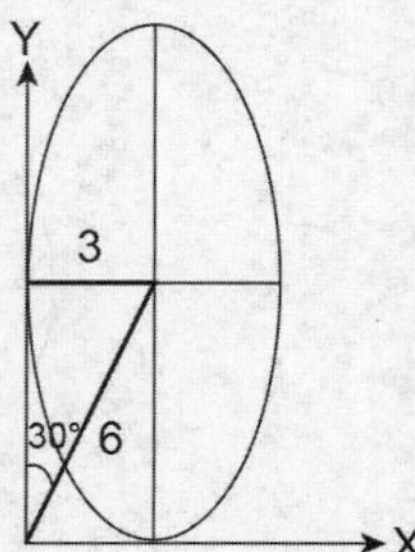

Hence rotation 30° takes place in totality

$\Rightarrow$ Length of *arc* $= \dfrac{2\pi r}{12}\left(\text{as } 30° = \dfrac{2\pi}{12}\right) = \pi = \dfrac{k\pi}{6}$ (given)

Hence $k = \mathbf{6}$

10. Given $a = 4$, $b = 3$. Let AB be the nod, so $A(0, y)$ and $B(x, 0)$

Length $AB = 7$ units (Now $x^2 + y^2 = 49$)

Let $P(x_1, y_1)$ be the point the divides AB in the $k : b$

$\Rightarrow\ x_1 = \dfrac{kx}{k+b}$ and $y_1 = \dfrac{by}{k+b}$

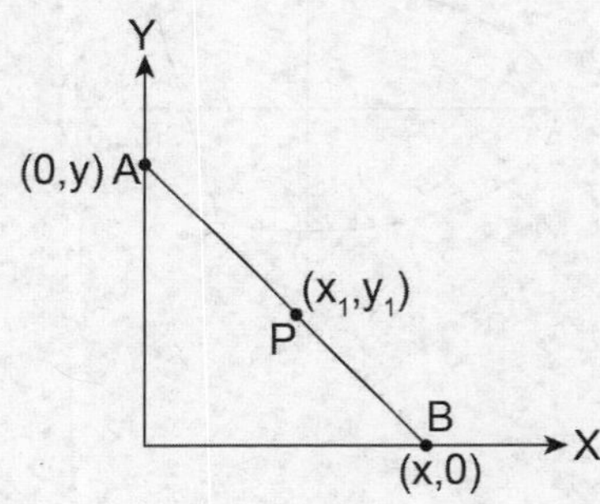

$\Rightarrow\ \dfrac{(k+b)^2 x_1^2}{k^2} + \dfrac{(k+b)^2 y_1^2}{b^2} = 49$

Now, eccentricity $e = \dfrac{\sqrt{7}}{4}$

$\Rightarrow\ \dfrac{A^2 - B^2}{A^2} = \dfrac{7}{16}$ i.e., $1 - \dfrac{B^2}{A^2} = \dfrac{7}{16}$

$\Rightarrow\ \dfrac{b^2}{k^2} = \dfrac{9}{16}$ as $b = 3 \ \Rightarrow\ k = 4$

11. Let $P(\theta)$, $Q(\phi)$ and $R(\psi)$ be the three points on the ellipse,

then area $\Delta PQR = \dfrac{1}{2}\left\|\begin{matrix} a\cos\theta & b\sin\theta & 1 \\ a\cos\phi & b\sin\phi & 1 \\ a\cos\psi & b\sin\psi & 1 \end{matrix}\right\|$

$= \dfrac{4ab}{2}\sin\left(\dfrac{\theta-\phi}{2}\right)\sin\left(\dfrac{\phi-\psi}{2}\right)\sin\left(\dfrac{\psi-\theta}{2}\right)$

$= k\,ab\,\sin\left(\dfrac{\theta-\phi}{2}\right)\sin\left(\dfrac{\phi-\psi}{2}\right)\sin\left(\dfrac{\psi-\theta}{2}\right)$

$\Rightarrow\ k = 2$

12. Let $P(a\cos\theta, b\sin\theta)$ be a point on the ellipse $\dfrac{x^2}{a^2} + \dfrac{y^2}{b^2} = 1$.

Now $Q(a\cos\theta, a\sin\theta)$ as $PR : RQ = r : s$

$\Rightarrow\ R = (x_1, y_1) = \left(a\cos\theta, \dfrac{(ar + bs)\sin\theta}{r+s}\right)$

$$\Rightarrow \frac{x_1}{a} = \cos\theta \text{ and } \frac{y_1(r+s)}{(ar+bs)} = \sin\theta$$

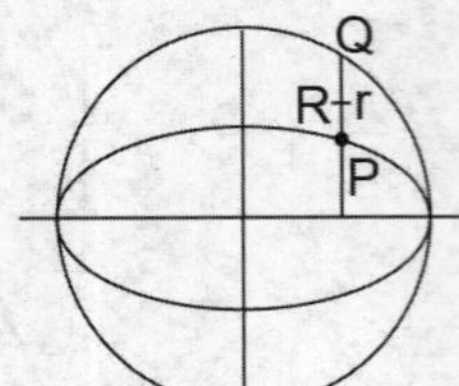

$$\therefore \frac{x_1^2}{a^2} + \frac{y_1^2(r+s)^2}{(ar+bs)^2} = 1 = 1k \text{ (given)}$$

$\Rightarrow k = 1$

13. we know that for $\theta = 90°$ the focal chord will be *L.R.* with length $\frac{2b^2}{a}$

$$\Rightarrow \frac{kab^2}{a^2\sin^2\theta + b^2\cos^2\theta} = \frac{kab^2}{a} = \frac{2b^2}{a}$$

$\Rightarrow k = 2$

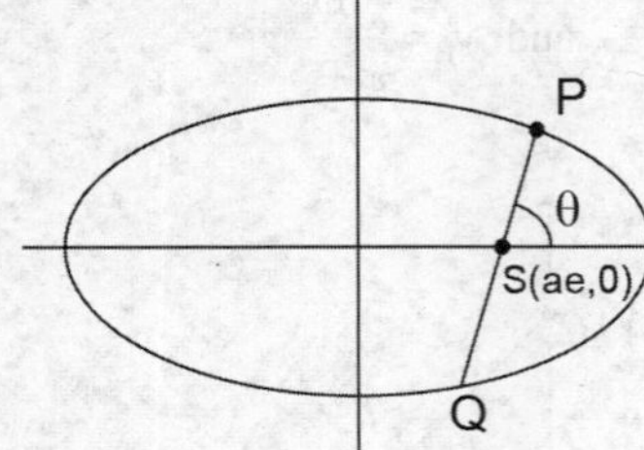

Aliter-1: Similarly for $\theta = 0°$, we get major axis as the focal chord so for $\theta = 0°$

$$\Rightarrow \frac{kab^2}{a^2\sin^2\theta + b^2\cos^2\theta} = \frac{kab^2}{b^2} = ka = 2a$$

Hence $k = 2$

Aliter-2: Let PQ a focal chord with angle θ, then $x = ae + r\cos\theta$ and $y = r\sin\theta$

Putting in $\frac{x^2}{a^2} + \frac{y^2}{b^2} = 1$, we get $\frac{(ae + r\cos\theta)^2}{a^2} + \frac{r^2\sin\theta^2}{b^2} = 1$

$$\Rightarrow b^2\left\{a^2e^2 + r^2\cos^2\theta + 2ae\cos\theta r\right\} + a^2r^2\sin^2\theta = a^2b^2$$

Rewritting as

$$r^2\{a^2\sin^2\theta + b^2\cos^2) + 2aeb^2\cos\theta r + a^2b^2(e^2 - 1) = 0$$

$$\therefore |r_1 - r_2| = \sqrt{(r_1 - r_2)^2 - 4r_1r_2}$$

$$\Rightarrow |r_1 - r_2| = \frac{\sqrt{4a^2e^2b^4\cos^2\theta + 4a^2b^2(1-e^2)(a^2\sin^2\theta + b^2\cos^2\theta)}}{(a^2\sin^2\theta + b^2\cos^2\theta)}$$

$$= \frac{\sqrt{4(a^2 - b^2)b^4\cos^4\theta + 4b^4a^2\sin^2\theta + 4b^6\cos^2\theta}}{(a^2\sin^2\theta + b^2\cos^2\theta)}$$

$$= \frac{\sqrt{4a^2b^4}}{(a^2\sin^2\theta + b^2\cos^2\theta)} = \frac{2ab^2}{a^2\sin^2\theta + b^2\cos^2\theta}$$

$$= \frac{kab}{(a^2\sin^2\theta + b^2\cos^2\theta)}$$

$\Rightarrow k = 2$

Hyperbola

CHAPTER 6

■ INTRODUCTION

We have studied earlier about parabola and ellipse as two conic sections. In this chapter, we will study another conic section called hyperbola, which can be obtained by cutting a right circular cone at both the nappes by a plane. Thus it has two branches, one on each nappe.

Hyperbola is the locus of a point which moves in a plane such that its distance from a fixed point is $e(e > 1)$ times its distance from a fixed straight line. It is symmetrical about two axes and one branch is the reflection of other about one of the axes. Since you have studied ellipse in the previous chapter, it will be easier to understand this curve. Also we can devices a transformation by which you can get the formulae for hyperbola if you know the corresponding formulae for ellipse.

Although most of the properties of hyperbola are similar to those of the ellipse, but few properties will be definitely new. We will introduce the concept of asymptotes, and learn about rectangular hyperbolas and conjugate hyperbolas.

To understand this curve well one should give stress on understanding the various concepts and solving good number of problems. Mostly, you need to plan the solution of the problem before starting, otherwise you may get lost in mathematical traps.

■ BASIC FEATURES OF HYPERBOLA

Definition

The locus of a point which moves in a plane such that its distance from a fixed point (i.e., focus) bears a constant ratio 'e' ($e > 1$) to its distance from a fixed straight line (directrix) is known as hyperbola.

Standard Equation of Hyperbola

Let S be the focus and ZM the directrix of the hyperbola. Draw $SZ \perp ZM$. Divide SZ internally and externally in the ratio $e : 1$ $(e > 1)$ and let A and A' be their internal and external points of division respectively. Then

$$SA = eAZ \qquad \text{....(1)}$$

$$\text{and } SA' = eA'Z \qquad \text{....(2)}$$

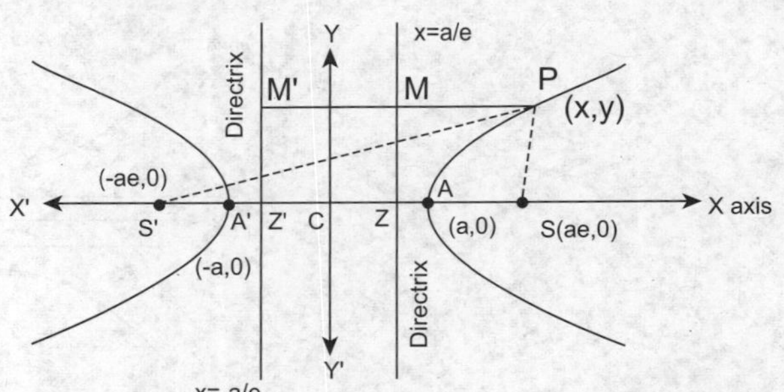

FIGURE 6.1

Clearly, A and A' will lie on the hyperbola. Let $AA' = 2a$ and take C the mid-point of AA' as origin.

$\therefore \quad CA = CA' = a$

Let $P(x, y)$ be any point on the hyperbola and CA be x-axis, the line through C perpendicular to CA be y-axis. Then, adding (1) and (2)

$\therefore \quad SA + SA' = e(AZ + A'Z)$

$\Rightarrow \quad CS - CA + CS + CA' = e(AA') \qquad (\because CA = CA')$

$\Rightarrow \quad 2CS = e(2a)$

$\therefore \quad CS = ae$

$\therefore \quad$ The focus S is $(ae, 0)$.

Subtracting (1) from (2), we get

$SA' - SA = e(A'Z - AZ)$

$AA' = e[(CA' + CZ) - (CA - CZ)]$

$\Rightarrow$ $AA' = e(2CZ)$ i.e., $2a = e(2CZ)$

$\therefore$ $CZ = a/e$

$\therefore$ The equation directrix MZ is $x = a/e$ or $x - a/e = 0$

$$\left(\because e > 1, \therefore \frac{a}{e} < a\right)$$

Now draw $PM \perp MZ$, $\therefore \frac{SP}{PM} = e$

or $(SP)^2 = e^2 (PM)^2$

or $(x - ae)^2 + (y - 0)^2 = e^2\left(x - \frac{a}{e}\right)^2$

or $(x - ae)^2 + y^2 = (ex - a)^2$

$\Rightarrow$ $x^2 + a^2e^2 - 2aex + y^2 = e^2x^2 - 2aex + a^2$

$\Rightarrow$ $x^2(e^2 - 1) - y^2 = a^2(e^2 - 1)$

$\Rightarrow$ $\frac{x^2}{a^2} - \frac{y^2}{a^2(e^2-1)} = 1$

or $\frac{x^2}{a^2} - \frac{y^2}{b^2} = 1$ where $b^2 = a^2(e^2 - 1)$ ....(i)

This is the standard equation of hyperbola.

REMARKS

1. *The above equation of hyperbola is called standard equation of hyperbola of first kind.*
2. *x-axis is called transverse axis of first kind of hyperbola and y-axis is called its conjugate axis. Thus foci S, S' and vertices A, A' and centre C of hyperbola lie on tranverse axis. For the hyperbola* $\frac{x^2}{a^2} - \frac{y^2}{b^2} = 1;$ *the length of transverse axis is given as '2a' and the length of conjngate axis is given as '2b'.*
3. *If y-axis is tranverse axis and x-axis is conjugate axis, then its eq. will be* $\frac{y^2}{a^2} - \frac{x^2}{b^2} = 1$ *and is called standard equation of second kind and its shape would be as shown in figure 6.2.*

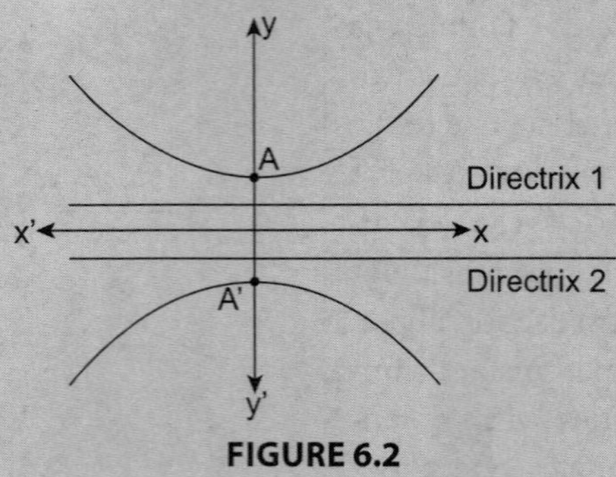

FIGURE 6.2

Tracing of the Hyperbola

For the hperbola

1. Since only even powers of x and y occurs in this equation therefore the hyperbola is symmetric about both the axes i.e., if (h, k) is a print on the gyperbola, then (h, k), $(-h, k)$ and $(-h, -k)$ will also lie on the hyperbola.
2. The hyperbola (1) does not cut y-axis in real points where as it cuts x-axis at $(a, 0)$ and $(-a, 0)$.
3. The equation (1) may be written as $y = \pm\frac{b}{a}\sqrt{(x^2 - a^2)}$

   It follows that $x^2 - a^2 \geq 0 \therefore x^2 \geq a^2$

   $\Rightarrow x \in (-\infty, -a] \cup [a, \infty)$

   Hence $x \notin (-a, a)$

   The curve does not exist in the region if $x \in (-a, a)$.
4. As $x \to \pm\infty$ then $y \to \pm\infty$ i.e., the curve extends to infinity.

Some Terms Related to Hyperbola

Let the equation of hyperbola is $\frac{x^2}{a^2} - \frac{y^2}{b^2} = 1$

1. **Centre:** It is a point on the plane of hyperbola which bisects all the chords of hyperbola passing through it. For the given hyperbola it is $C(0, 0)$ i.e., origin.
2. **Eccentricity:** For the hyperbola $\frac{x^2}{a^2} - \frac{y^2}{b^2} = 1$ we have $b^2 = a^2(e^2 - 1)$.

   $\Rightarrow e^2 = \frac{a^2 + b^2}{a^2} \Rightarrow e = \sqrt{1 + \left(\frac{b^2}{a^2}\right)}$

$$\Rightarrow e=\sqrt{\left\{1+\frac{(2b)^2}{(2a)^2}\right\}}$$

$$\Rightarrow e=\sqrt{\left\{1+\frac{(\text{conjugate axis})^2}{(\text{transverse axis})^2}\right\}}$$

3. **Foci and directrices:** S and S' are the foci of the ellipse and their co-ordinates are $(ae, 0)$ and $(-ae, 0)$ respectively and ZM and $Z'M'$ are two directrices of the hyperbola with equations $x = \frac{a}{e}$ and $x = -\frac{a}{e}$ respectively.

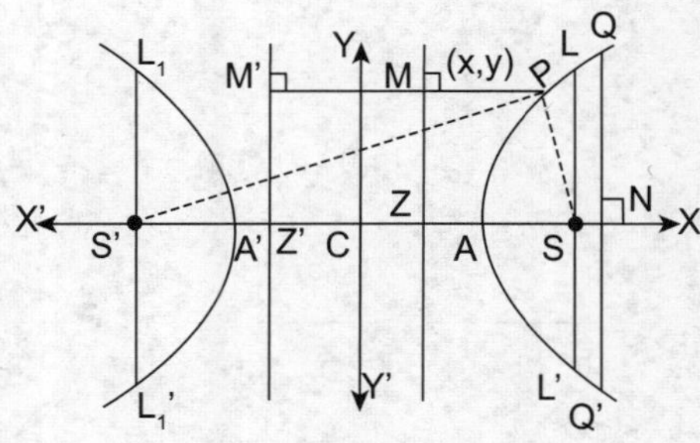

FIGURE 6.3

4. **Vertices and Axis:** The points $A(a, 0)$ and $A'(-a, 0)$ are called the vertices of the hyperbola and line AA' is called **transverse axis.** The line perpendicular to it through the centre (0, 0) of the hyperbola is called **conjugate axis** (i.e., $x = 0$, y-axis).
5. **Double ordinates:** If Q be a point on the hyperbola draw QN perpendicular to the transverse axis of the hyperbola and produce it to meet the curve again at Q'. Then QQ' is called a double ordinate of hyperbola.

   If abscissa of Q is h then ordinates of Q are

   $$\frac{y^2}{b^2}=\frac{h^2}{a^2}-1$$

   $$\therefore\ y=\pm\frac{b}{a}\sqrt{(h^2-a^2)}\ \therefore\ y=\frac{b}{a}\sqrt{(h^2-a^2)}$$

   (for 1 quadrant)

   and ordinate of Q' is $y=-\frac{b}{a}\sqrt{(h^2-a^2)}$

   (for IV quadrant).

   Hence co-ordinates of Q and Q' are $\left(h, \frac{b}{a}\sqrt{(h^2-a^2)}\right)$ and $\left(h, -\frac{b}{a}\sqrt{(h^2-a^2)}\right)$ respectively.
6. **Focal chord:** A chord of hyperbola passing through its focus is called a focal chord.
7. **Latera Recta: (Plural of Latus-rectum):** The double ordinates LL' and L_1L_1' are the latera-recta of the hyperbola. These lines are perpendicular to transverse axis AA' and through the foci S and S' respectively.

   Length of each latus rectum. Now let $LL' = 2k$ then $LS = L'S = k$

   Co-ordinates of L and L' (ae, k) and $(ae, -k)$

   lie on the hyperbola $\frac{x^2}{a^2}-\frac{y^2}{b^2}=1$

   $$\therefore\ \frac{a^2e^2}{a^2}-\frac{k^2}{b^2}=1 \text{ or } k^2=b^2(e^2-1)=b^2\left(\frac{b^2}{a^2}\right)$$

   $[\because\ b^2=a^2(e^2-1)]$

   $$\therefore\ k=\frac{b^2}{a} \qquad (\because k>0)$$

   $$\therefore\ 2k=\frac{2b^2}{a}=LL'$$

   $\therefore$ Length of latus rectum $= LL' = L_1L'_1 = \frac{2b^2}{a}$ and end points of latus rectums are

   $$L\equiv\left(ae,\frac{b^2}{a}\right);L'\equiv\left(ae,-\frac{b^2}{a}\right),$$

   $$L_1\equiv\left(-ae,\frac{b^2}{a}\right);L_1'\equiv\left(-ae,-\frac{b^2}{a}\right) \text{ respectively.}$$

   Hence, we observe that latus rectum are a special case of focal chord which are also the double ordinates.

ILLUSTRATION 1: For the hyperbola $\frac{x^2}{100}-\frac{y^2}{25}=1$, Prove that eccentricity $= \sqrt{5}/2$ and $SA.\ S'A = 25$, where S and S' are the foci and A is vertex of hyperbola.

SOLUTION: $\frac{x^2}{100}-\frac{y^2}{25}=1$, clearly $a^2 = 100$ and $b^2 = 25$

$\therefore$ eccentricity $e = \sqrt{1+\frac{b^2}{a^2}} = \sqrt{1+\frac{25}{100}} = \sqrt{\frac{5}{4}} = \frac{\sqrt{5}}{2}$

Now, foci are $S_1(ae, 0) = \left(10\times\frac{\sqrt{5}}{2}, 0\right) = (5\sqrt{5}, 0)$

and S_2 $(-ae, 0) = (-5\sqrt{5}, 0)$; vertices A, A' are (10, 0), (–10, 0).

Now, $S_1A = ae - a = a(e-1) = 10.\left(\frac{\sqrt{5}}{2}-1\right)$

and $S_2A = ae + a = a(e+1) = 10\left(\frac{\sqrt{5}}{2}+1\right)$

$\therefore$ $S_1A.\ S_2A = 25((\sqrt{5})^2 - 4) = 25.$

$\Rightarrow$ $S_1A.S_2A = 25$, hence proved.

FIGURE 6.4

As hyperbola $\frac{x^2}{A^2}-\frac{y^2}{B^2}=1$ passes through point (3, 0)

$\Rightarrow$ $\frac{9}{A^2} - 0 = 1 \Rightarrow A^2 = 9$

$e_2^2 = 1+\frac{B^2}{A^2} \Rightarrow \frac{25}{9} = 1+\frac{B^2}{9}$

$\Rightarrow$ $\frac{25}{9} = \frac{9+B^2}{9} \Rightarrow 25 = 9 + B^2$

FIGURE 6.5

$\therefore$ $B^2 = 16$; Hence, equation of the hyperbola is, $\frac{x^2}{9}-\frac{y^2}{16}=1$

Foci of hyperbola = $(\pm ae_2, 0) = (\pm 3 \times (5/3), 0) = (\pm 5, 0)$.

ILLUSTRATION 2: If a latusrectum LL' of a hyperbola forms an equilateral Δ with third vertex at the centre of hyperbola, then find the eccentricity of the hyperbola.

SOLUTION: $\because$ $\Delta CLL'$ is an equilateral Δ.(as shown in figure 6.6)

$\Rightarrow$ $\angle LCS = 30° \Rightarrow \tan 30° = \frac{b^2/a}{ae}$

$\Rightarrow$ $\frac{1}{\sqrt{3}} = \frac{b^2}{a^2e}$ and $b^2 = a^2(e^2-1)$

$\Rightarrow$ $\frac{1}{\sqrt{3}} = \frac{e^2-1}{e} \Rightarrow \sqrt{3}\,e^2 - e - \sqrt{3} = 0$

$\Rightarrow$ $e = \frac{1+\sqrt{13}}{2\sqrt{3}}$ (as $e > 1 > 0$)

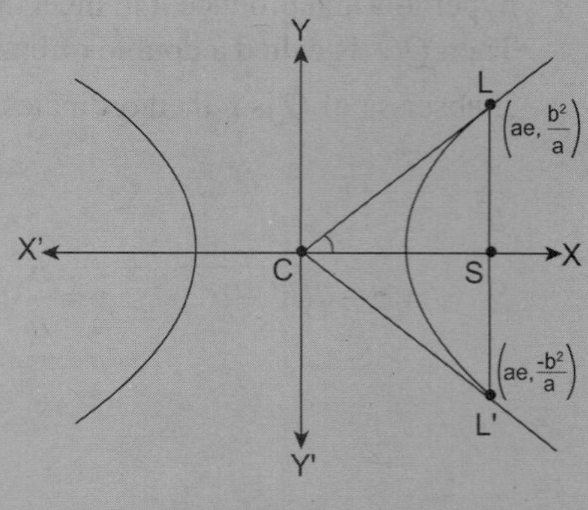

FIGURE 6.6

ILLUSTRATION 3: If a latus rectum LL' of a hyperbola forms an equilateral Δ with third vertex at the other focus, then find the eccentricity of the hyperbola.

SOLUTION: $\because$ $\Delta S'LL'$ is an equilateral Δ.

$\Rightarrow$ $\angle LS'S = 30° \Rightarrow \tan 30° = \frac{b^2/a}{2ae}$

$\Rightarrow \dfrac{1}{\sqrt{3}} = \dfrac{b^2}{2a^2 e}$

$\Rightarrow \dfrac{1}{\sqrt{3}} = \dfrac{1}{2}\left(\dfrac{e^2-1}{e}\right)$

$\Rightarrow 2e = \sqrt{3}e^2 - \sqrt{3}$

$\Rightarrow \sqrt{3}e^2 - 2e - \sqrt{3} = 0$

$\Rightarrow e = = \dfrac{2 \pm 4}{2\sqrt{3}} \Rightarrow e = \sqrt{3}.$

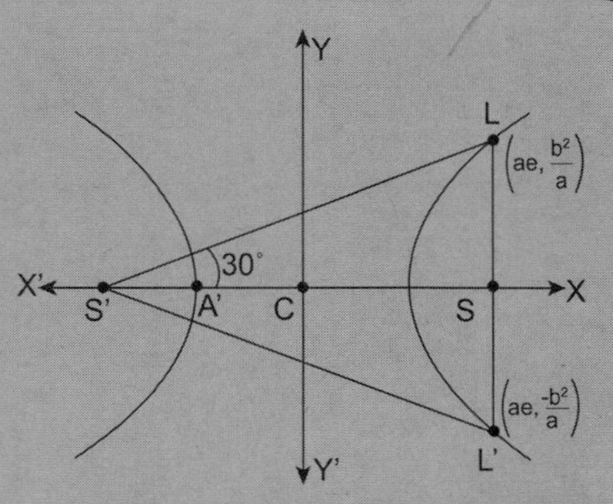

FIGURE 6.7

ILLUSTRATION 4: Find the equation of the hyperbola in standard form of given transverse axis $2a$ with a vertex mid-way between the centre and a focus.

SOLUTION: Let the equation of hyperbola be $\dfrac{x^2}{a^2} - \dfrac{y^2}{b^2} = 1$ ... (i)

According to the questions: A is mid-point of CS

$\Rightarrow a = \dfrac{ae}{2}.$

$\Rightarrow e = 2$. Also $\dfrac{b^2}{a^2} = e^2 - 1 \Rightarrow \dfrac{b^2}{a^2} = 3$

$\Rightarrow b^2 = 3a^2$

$\therefore$ Required hyperbola is $\dfrac{x^2}{a^2} - \dfrac{y^2}{3a^2} = 1$

$\Rightarrow 3x^2 - y^2 = 3a^2.$

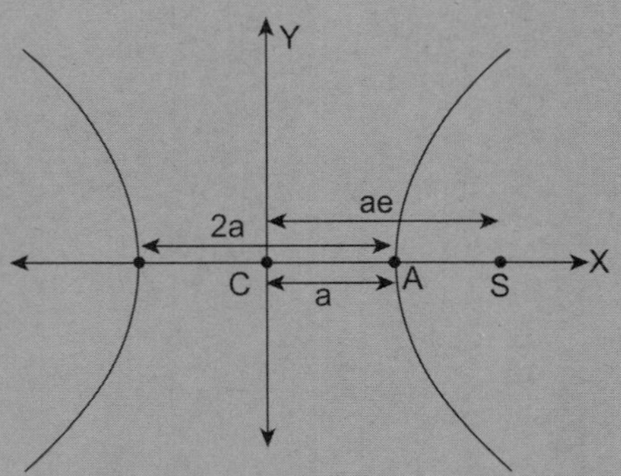

FIGURE 6.8

ILLUSTRATION 5: If foci of a hyperbola are foci of the ellipse $\dfrac{x^2}{25} + \dfrac{y^2}{9} = 1$ and if the eccentricity of the hyperbola be 2, then find its equation.

SOLUTION: Given ellipse: $\dfrac{x^2}{25} + \dfrac{y^2}{9} = 1$; where e = $\sqrt{\dfrac{a^2 - b^2}{a^2}} = \sqrt{\dfrac{25-9}{25}} = \dfrac{4}{5}$

Foci of ellipse are given by (± ae,0)

Foci of ellipse: (± 4,0)

Now, according to the question; focii of hyperbola : $(\pm 4, 0)$

Let the equation of hyperbola ≡ $\dfrac{x^2}{A^2} - \dfrac{y^2}{B^2} = 1$

Focii of hyperbola ≡ $(\pm Ae, 0)$, where e = 2

$\Rightarrow$ Ae = 4 $\Rightarrow A = 2$ $\qquad \because\ B^2 = A^2(e^2 - 1)$

$\Rightarrow 4\,(2^2 - 1) = 12$

$\therefore$ Equation of hyperbola is $\dfrac{x^2}{4} - \dfrac{y^2}{12} =$

ILLUSTRATION 6: The hyperbola $\dfrac{x^2}{a^2} - \dfrac{y^2}{b^2} = 1$ passes through the point of intersection of the lines, $7x + 13y - 87 = 0$ and $5x - 8y + 7 = 0$ and the length of the latus rectum is $\dfrac{32\sqrt{2}}{5}$. Find 'a' and 'b'.

SOLUTION: Given hyperbola is $\frac{x^2}{a^2}-\frac{y^2}{b^2}=1$, and the point of intersection of given lines is (5, 4), through which hyperbola passes.

$\Rightarrow \frac{25}{a^2}-\frac{16}{b^2}=1$ or $25b^2 - 16a^2 = a^2b^2$ ...(1)

Now, length of the latus rectum = $\frac{2b^2}{a}=\frac{32\sqrt{2}}{5} \Rightarrow \frac{b^2}{a}=\frac{16\sqrt{2}}{5}$ or $b^2=\frac{16\sqrt{2}a}{5}$ ...(2)

Substituting in equation (1), we get $25.\frac{16\sqrt{2}}{5}a-16a^2=a^2.\frac{16\sqrt{2}}{5}a$; as $a \neq 0$

or $80\sqrt{2}-16a=\frac{16\sqrt{2}}{5}a^2 \Rightarrow 5\sqrt{2}-a=\frac{\sqrt{2}}{5}a^2$

$\Rightarrow 2a^2+5\sqrt{2}\,a-50=0$

$\Rightarrow a=\frac{-5\sqrt{2}\pm\sqrt{450}}{4}=\frac{-5\sqrt{2}\pm 15\sqrt{2}}{4}=-5\sqrt{2},\frac{5}{\sqrt{2}}$

Now, 'a' cannot be $-5\sqrt{2}$ since if $a=-5\sqrt{2}$ then $b^2=\frac{32a}{5\sqrt{2}} \Rightarrow b^2=\frac{32(-5\sqrt{2})}{5\sqrt{2}}$

which gives $b^2 < 0$ which is impossible.

$\therefore a=\frac{5}{\sqrt{2}} \Rightarrow a^2=\frac{25}{2}$

using equation (3); $b^2=\frac{32a}{5\sqrt{2}} \Rightarrow b^2=16$.

Equation of hyperbola is $\frac{x^2}{25/2}-\frac{y^2}{16}=1 \Rightarrow 32x^2-25y^2=400$. **Ans.**

ILLUSTRATION 7: If PQ is a double ordinate of a hyperbola $\frac{x^2}{a^2}-\frac{y^2}{b^2}=1$, such that OPQ is an equilateral triangle 'O', being the centre of hyperbola. Then find the range of eccentricity 'e' of hyperbola.

SOLUTION: Let the co-ordinates of P be (h, k)

$\therefore PQ = 2k$ and $OP = OQ = PQ$

$\Rightarrow \sqrt{h^2+k^2}=2k \Rightarrow h^2+k^2=4k^2 \Rightarrow h^2=3k^2$ or $h = k\tan 60° = k\sqrt{3}$

$\therefore (h, k)$ lies on hyperbola $\Rightarrow \frac{h^2}{a^2}-\frac{k^2}{b^2}=1 \Rightarrow \frac{3k^2}{a^2}-\frac{k^2}{b^2}=1$

$\Rightarrow \frac{3b^2-a^2}{a^2b^2}=\frac{1}{k^2}>0 \Rightarrow \frac{3}{a^2}-\frac{1}{b^2}> \quad \Rightarrow \frac{3}{a^2}>\frac{1}{b^2}$

$\Rightarrow \frac{b^2}{a^2}>\frac{1}{3} \Rightarrow e^2-1>1/3 \Rightarrow e^2>4/3$

$\Rightarrow e>2/\sqrt{3} \Rightarrow e \in \left(\frac{2}{\sqrt{3}},\infty\right)$.

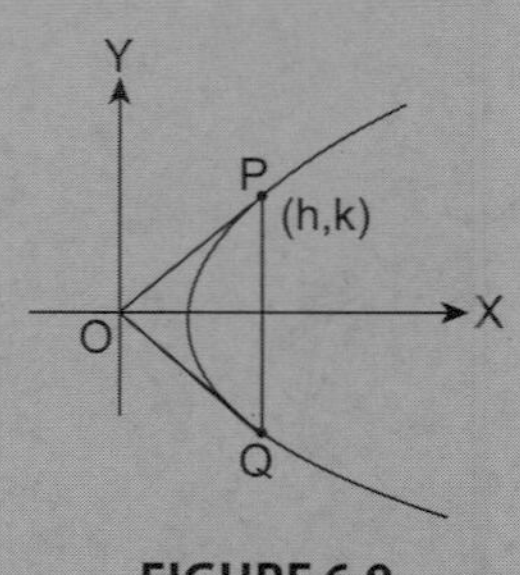

FIGURE 6.9

ILLUSTRATION 8: Find the set of values of 'α' for which the eccentricity of hyperbola $\frac{x^2}{5}-\frac{y^2}{5\cos^2\alpha}=1$ is $\sqrt{3}$ times the eccentricity of the ellipse $\frac{x^2}{25\cos^2\alpha}+\frac{y^2}{25}=1$.

SOLUTION: Let the eccentricities of hyperbola and ellipse be e_1 and e_2 respectively

$\therefore\ 5\cos^2\alpha = 5(e_1^2-1)$ and $25\cos^2\alpha = 25(1-e_2^2)$

$\therefore\ e_1^2 = 1+\cos^2\alpha$ and $e_2^2 = 1-\cos^2\alpha = \sin^2\alpha$

$A.T.Q.\ e_1 = \sqrt{3}\,e_2 \Rightarrow e_1^2 = 3e_2^2$.

$\Rightarrow\ 1+\cos^2\alpha = 3\sin^2\alpha$

$\Rightarrow\ 2 = 4\sin^2\alpha \Rightarrow \sin^2\alpha = 1/2 = \sin^2\pi/4$

$\Rightarrow \alpha = n\pi \pm \pi/4;\ n \in \mathbb{Z}$

$\therefore\ \alpha \in \left\{n\pi \pm \frac{\pi}{4};\ n \in \mathbb{Z}\right\}$.

ILLUSTRATION 9: If a hyperbola passes through the focus of the ellipse $\frac{x^2}{25}+\frac{y^2}{16}=1$ and its transverse and conjugate axis coincides with the major and minor axis of the ellipse, and product of their eccentricities is 1, then which of the following is/are correct?

(a) Equation of hyperbola is $\frac{x^2}{9}-\frac{y^2}{16}=1$

(b) Equation of hyperbola is $\frac{x^2}{9}-\frac{y^2}{25}=1$

(c) One focus of hyperbola is (5, 0)

(d) One focus of hyperbola is $(5\sqrt{3}, 0)$

SOLUTION: Equation of ellipse is $\frac{x^2}{25}+\frac{y^2}{16}=1$; $a^2 = 25$

$\Rightarrow\ a = 5;\ b^2 = 16 \Rightarrow b = 4$

Let the equation of hyperbola is $\frac{x^2}{A^2}-\frac{y^2}{B^2}=1$

Let e_1 is the eccentricity of ellipse and e_2 is the eccentricity of hyperbola

$$e_1 = \sqrt{1-\frac{b^2}{a^2}} = \sqrt{1-\frac{16}{25}} = \sqrt{\frac{9}{25}} \Rightarrow e_1 = \frac{3}{5}$$

Foci of ellipse $= (\pm\, ae_1, 0) = \left(\pm\frac{5\times 3}{5}, 0\right) = (\pm\, 3, 0)$

According to the question: Hyperbola passes through the foci of ellipse and its transverse and conjugate axis coincide with the major and mirnor axis of ellipse respectively.

$\because\ e_1.e_2 = 1 \Rightarrow (3/5)\,e_2 = 1$

$\Rightarrow\ e_2 = 5/3$.

ILLUSTRATION 10: If C is the centre of a hyperbola $\frac{x^2}{a^2}-\frac{y^2}{b^2}=1$ and S, S' its foci and P a point on it. Prove that

$SP.S'P = CP^2 - a^2 + b^2$.

SOLUTION: $SP = ea\sec\theta - a$

$S'P = ea\sec\theta + a$

$SP \,.\, S'P = e^2a^2 \sec^2\theta - a^2$

$= \dfrac{a^2(e^2 - \cos^2\theta)}{\cos^2\theta}$

$\because\ CP^2 = a^2\sec^2\theta + b^2\tan^2\theta$

$CP^2 - a^2 + b^2$

$= a^2\sec^2\theta - a^2 + b^2 + b^2\tan^2\theta$

$= a^2(\sec^2\theta - 1) + b^2(\sec^2\theta)$

$= a^2(\sec^2\theta - 1) + (a^2e^2 - a^2)\sec^2\theta$

$= a^2[\sec^2\theta - 1 + e^2\sec^2\theta - \sec^2\theta]$

$= a^2\left(\dfrac{e^2 - \cos^2\theta}{\cos^2\theta}\right) = SP \,.\, S'P$

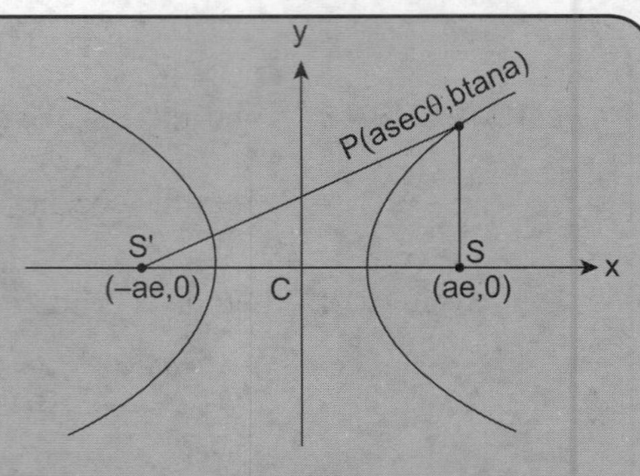

FIGURE 6.10

Aliter: $\dfrac{x^2}{a^2} - \dfrac{y^2}{b^2} = 1$ ...(1)

Let the point 'P' is (x_1, y_1)

$SP = \sqrt{(x_1 - ae)^2 + y_1^2}$ ...(2)

$S'P = \sqrt{(x_1 + ae)^2 + y_1^2}$ ...(3)

$CP = \sqrt{x_1^2 + y_1^2}$ ...(4)

From equation (1), point $P\,(x_1, y_1)$ satisfies it, therefore

$\dfrac{x_1^2}{a^2} - \dfrac{y_1^2}{b^2} = 1$ ...(5)

$\because\ SP - S'P = 2a$

Squring both sides, we get $SP^2 + (S'P)^2 - 2SP.\,S'P = 4a^2$

or $\dfrac{SP^2 + (S'P)^2 - 4a^2}{2} = SP.S'P$

using equation (2) and (3)

or $SP.\,S'P = \dfrac{1}{2} \times [(x_1 - ae)^2 + y_1^2 + (x_1 + ae)^2 + y_1^2 - 4a^2] = x_1^2 + y_1^2 + a^2(e^2 - 2)$

$= x_1^2 + y_1^2 + a^2\left[\dfrac{b^2 - a^2}{a^2}\right] = x_1^2 + y_1^2 - a^2 + b^2$

Now taking R.H.S. $\Rightarrow CP^2 - a^2 + b^2$

$= x_1^2 + y_1^2 - a^2 + b^2$ = L.H.S.

■ EQUACTION OF HYPERBOLA

The equation of the hyperbola whose focus is the point (h, k) and directrix is $lx + my + n = 0$ and whose eccentricity is e, is given by $(x-h)^2 + (y-k)^2 = e^2\dfrac{(lx+my+n)^2}{(l^2+m^2)}$.

■ SECOND DEFINITION OF HYPERBOLA

Hyperbola is a locus of the point which moves so that "the absolute difference of its distances from two fixed points S_1 and S_2 (foci of hyperbola) remains constant $(2a)$.

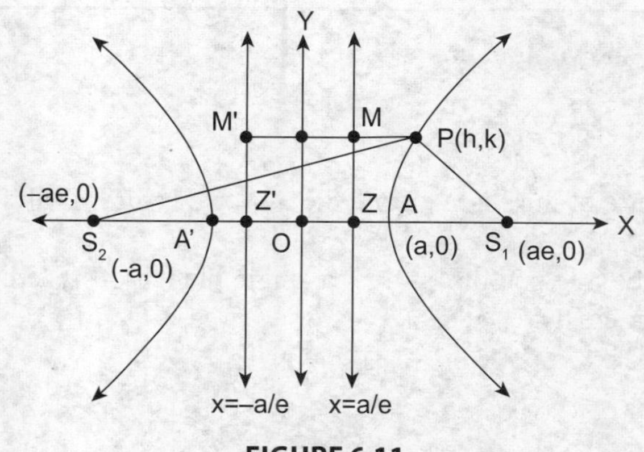

FIGURE 6.11

$\because \; S_1P = ePM = e(h - a/e) = eh - a;$

$$S_2P = ePM' = e\left(\frac{a}{e} + h\right) = eh + a$$

❑ $|S_2P - S_1P| = 2a$, where $2a$ is length of transverse axis.

Case I: If $2a < S_1S_2 = 2ae \Rightarrow$ hyperbola.

Case II: If $|S_2P - S_1P| = S_1S_2 \Rightarrow$ union of two rays (are towards left of S_2 and towards right of S_1)

$\longleftarrow \bullet S_2 \quad S_1 \bullet \longrightarrow$

FIGURE 6.12

Case III: If $|S_2P - S_1P| > S_1S_2 \Rightarrow$ No locus as it violates triangle inequality.

ILLUSTRATION 11: Identify the loci of point (x, y) satisfying the condition

$\left|\sqrt{(x-1)^2+(y-1)^2} - \sqrt{(x+1)^2+(y+1)^2}\right| = \lambda$. If (i) $\lambda = 2\sqrt{2}$ (ii) $\lambda = 1$ (ii) $\lambda = 6$

SOLUTION: Hyperbola satisfies the equation $|PS_1 - PS_2| = 2a;\ 2a < S_1S_2$.

(i) Observe the distance formula at R.H.S. consider $P(x, y)$ and $S_1(1, 1)$, $S_2(-1, -1)$

For $\lambda = 2\sqrt{2};\ S_1S_2 = \sqrt{(2)^2 + (2)^2} = 2\sqrt{2}$

$|PS_1 - PS_2| = 2\sqrt{2} = S_1S_2$

$\therefore$ Locus is a union of of two rays.

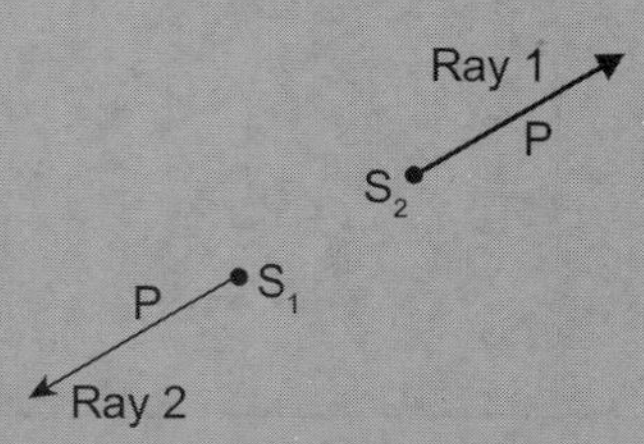

FIGURE 6.13

(ii) For $\lambda = 1$

$|PS_1 - PS_2| = 1 < S_1S_2 = 2\sqrt{2} \Rightarrow$ Locus is a hyperbola.

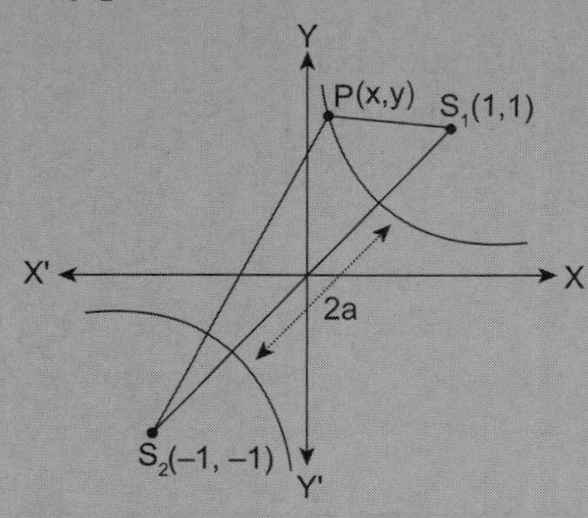

FIGURE 6.14

(iii) For $\lambda = 6;\ |PS_1 - PS_2| = 6 > 2\sqrt{2} = S_1S_2$

$\Rightarrow$ No locus is possible, because triangle inequality is violated.

ILLUSTRATION 12: Consider two circles with their centres at (–4, – 5) and (3, 4) and with their radii 3 and 2 respectively. Then prove that the locus of centre of a third circle touching both circles externally is a hyperbola. Also find

(i) Eccentricity of hyperbola

(ii) Equation of hyperbola

(iii) Length of tranverse and conjugate axes.

SOLUTION: When the circles are outside each other as shown in figure 6.15.

Clearly, $CC_1 = r_1 + r$, $CC_2 = r_2 + r$

$\Rightarrow |CC_1 - CC_2| = |r_1 - r_2|$

$\Rightarrow$ Thus difference of distances of the centre of third circle from the first two centres is always constant $= |r_1 - r_2| < |C_1 C_2|$ $\quad \because \; C_1C_2 \geq r_1 + r_2$

$\therefore$ Locus of centre of third circle would be a hyperbola (in fact a single branch of hyperbola oriented towards the circle of smaller radius)

If $2a$ = length of tranverse axis then

$2a = |r_1 - r_2| = |3 - 2| = 1 \Rightarrow a = 1/2$

and $2ae = S_1S_2$ (distance between foci) $= C_1C_2$

$\Rightarrow (1)(e) = \sqrt{(-4-3)^2 + (-5-4)^2} = \sqrt{49+81} = \sqrt{130}$

$\Rightarrow$ eccentricity $e = \sqrt{130}$.

Now, as for hyperbola, $b^2 = a^2(e^2 - 1)$

$\Rightarrow b^2 = (1/4)(130 - 1) = \dfrac{129}{4} \Rightarrow b = \dfrac{\sqrt{129}}{2}$.

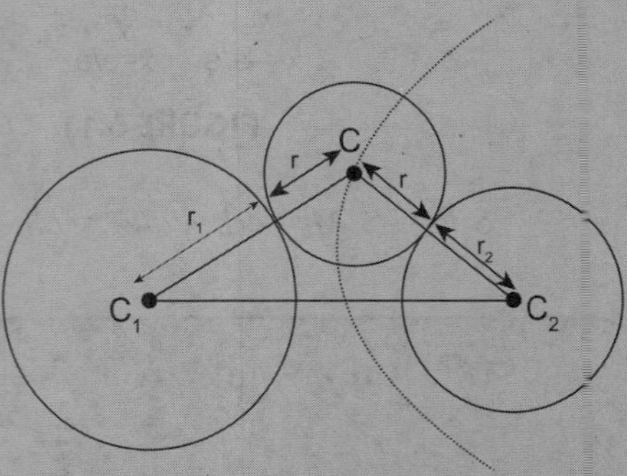

FIGURE 6.15

$\therefore$ equation of hyperbola is $\dfrac{x^2}{a^2} - \dfrac{y^2}{b^2} = 1$.

$\Rightarrow \dfrac{x^2}{1/4} - \dfrac{y^2}{129/4} = 1 \Rightarrow 516x^2 - 4y^2 = 129$

Length of tranverse axis $= 2a = 1$ and length of conjugate axis $= 2b = \sqrt{129}$.

■ RECTANGULAR HYPERBOLA

If the lengths of transverse and conjugate axes of hyperbola are equal i.e., $a = b$; hyperbola is said to be equilateral or rectangular and has the equation, $x^2 - y^2 = a^2$. Eccentricity for such a hyperbola is $\sqrt{2}$.

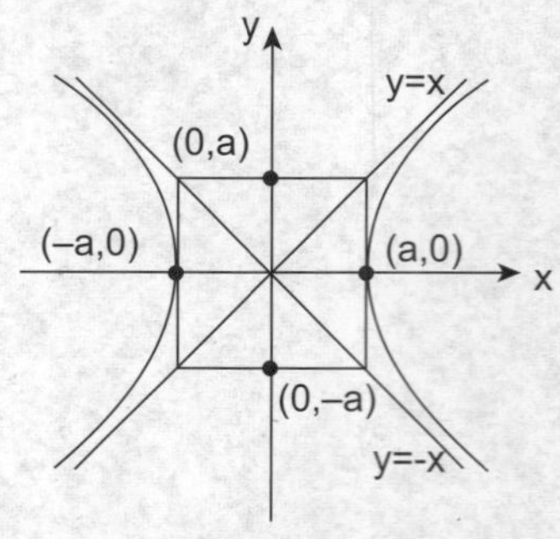

FIGURE 6.16

ILLUSTRATION 13: Prove that the perpendicular focal chords of a rectangular hyperbola are equal.

SOLUTION: Let the rectangular hyperbola is $x^2 - y^2 = a^2$

Let PQ and DE having equations $y = mx + c$ ... (i)

and $y = m_1x + c_1$ ... (ii)

respectively, be any two focal chords of rectangular hyperbola $x^2 - y^2 = a^2$.

We have to prove $PQ = DE$. Since $PQ \perp DE$,

$\Rightarrow mm_1 = -1$ ... (iii)

Also PQ passes through $S(a\sqrt{2}, 0)$ thus from (i), $0 = ma\sqrt{2} + c$

or $c^2 = 2a^2m^2$ ... (iv)

Let (x_1, y_1) and (x_2, y_2) be the co-ordinates of P and Q then

$(PQ)^2 = (x_1 - x_2)^2 + (y_1 - y_2)^2$ ... (v)

Since (x_1, y_1) and (x_2, y_2) lie on (i)

$\therefore\ y_1 = mx_1 + c$ and $y_2 = mx_2 + c$

$\therefore\ (y_1 - y_2) = m(x_1 - x_2)$ ... (vi)

From (v) and (vi)

$(PQ)^2 = (x_1 - x_2)^2 (1 + m^2)$ .. (vii)

Now solving $y = mx + c$ and $x^2 - y^2 = a^2$ then $x^2 - (mx + c)^2 = a^2$

or $(m^2 - 1)x^2 + 2mcx + (a^2 + c^2) = 0$

$\therefore\ x_1 + x_2 = -\dfrac{2mc}{m^2 - 1}$ and $x_1x_2 = \dfrac{a^2 + c^2}{(m^2 - 1)}$

$\Rightarrow (x_1 - x_2)^2 = (x_1 + x_2)^2 - 4x_1x_2$

$= \dfrac{4m^2c^2}{(m^2 - 1)^2} - \dfrac{4(a^2 + c^2)}{(m^2 - 1)}$

$= \dfrac{4\{a^2 + c^2 - a^2m^2\}}{(m^2 - 1)^2} = \dfrac{4a^2(m^2 + 1)}{(m^2 - 1)^2}$

$\{\because c^2 = 2a^2m^2\}$

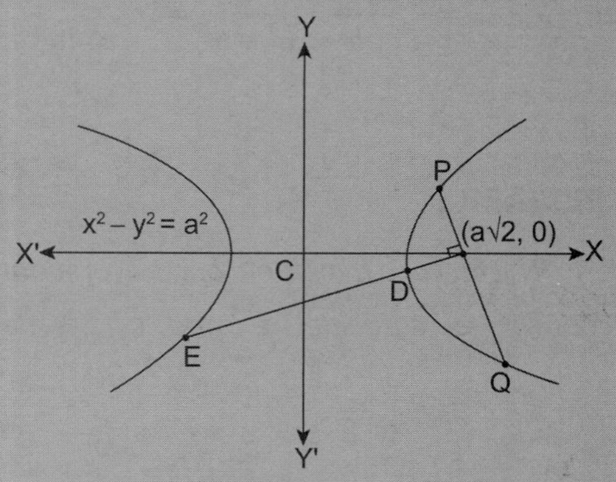

FIGURE 6.17

Form (vii), $(PQ)^2 = 4a^2\left(\dfrac{m^2 + 1}{m^2 - 1}\right)^2$

Similarly, $(DE)^2 = 4a^2\left(\dfrac{m_1^2 + 1}{m_1^2 - 1}\right)^2$

$$= 4a^2\left(\frac{\left(-\frac{1}{m}\right)^2 + 1}{\left(-\frac{1}{m}\right)^2 - 1}\right)^2 = 4a^2\left(\frac{m^2 + 1}{m^2 - 1}\right)^2 = (PQ)^2$$

$(\because mm_1 = -1)$

Thus $(PQ)^2 = (DE)^2$

$\Rightarrow PQ = DE$

Hence perpendicular focal chords of a rectangular hyperbola are equal in length.

■ CONJUGATE HYPERBOLA

For hyperbola $H = 0$, a hyperbola $C = 0$, whose transverse axis is conjugate axis of $H = 0$ and conjugate axis is transverse axis of $H = 0$, both in the sense of length and equation is called conjugate hyperbola of $H = 0$.

$\therefore$ If equation of hyperbola is $H \equiv \dfrac{x^2}{a^2} - \dfrac{y^2}{b^2} - 1 = 0$, then equation of conjugate hyperbola will be $C \equiv \dfrac{x^2}{a^2} - \dfrac{y^2}{b^2} + 1 = 0$.

- Eccentricity of conjugate hyperbola:
  $e_2 = \sqrt{1+\left(a^2/b^2\right)}$
- Foci of conjugate hyperbola: $(0, \pm be_2)$.
- Transverse axis of conjugate hyperbola: $x = 0$, Length $= 2b$.
- Latera recta of conjugate hyperbola: $y = \pm be_2$. Co-ordinates of extremities of latera recta are $\left(\pm\frac{a^2}{b}, \pm be_2\right)$ and their lengths $= \frac{2a^2}{b}$.
- $\frac{1}{e_2^2}+\frac{1}{e_1^2}=\frac{b^2}{a^2+b^2}+\frac{a^2}{a^2+b^2}=1$

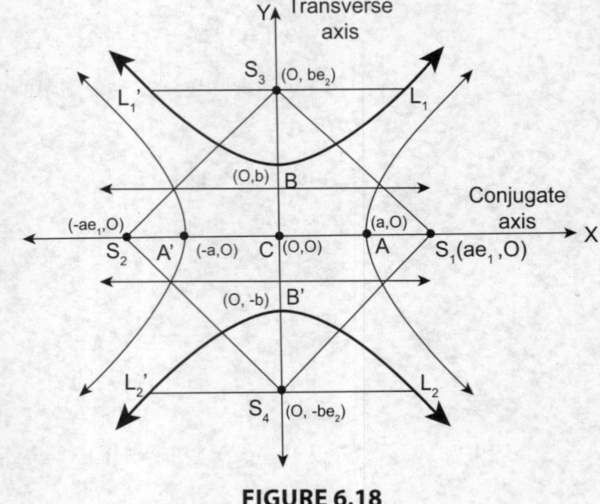

FIGURE 6.18

REMARKS

1. *The foci of a hyperbola and its conjugate are concylic and form the vertices of a square.*
2. *The eccentricities e_1 and e_2 of the hyperbola and its conjugate hyperbola respectively, satisfies the equation* $\frac{1}{e_1^2}+\frac{1}{e_2^2}=1$.

ILLUSTRATION 14: Prove that $\frac{1}{e_1^2}+\frac{1}{e_2^2}=1$; where e_1 and e_2 are ecccentricities of hyperbola and its conjugate hyperbola. Also show that the foci of a hyperbola and its conjugates hyperbola form the vertices of a square.

SOLUTION: Let $\frac{x^2}{a^2}-\frac{y^2}{b^2}=1$ and $\frac{y^2}{b^2}-\frac{x^2}{a^2}=1$ be the given hyperbola and its conjugate hyperbola.

Foci of given hyperbola is $(\pm ae_1, 0)$ and that of conjugate hyperbola is $(0, \pm be_2)$.

Also for given hyperbola, $b^2 = a^2(e_1^2 - 1)$ ... (i)

and for conjugate hyperbola, $a^2 = b^2(e_2^2 - 1)$ ... (ii)

$\therefore \quad \frac{b^2}{a^2}=e_1^2-1=\frac{1}{e_2^2-1}$ ... (iii)

$\Rightarrow (e_1^2 - 1)(e_2^2 - 1) = 1 \Rightarrow e_1^2e_2^2 - e_1^2 - e_2^2 + 1 = 1$

$\Rightarrow e_1^2 + e_2^2 = e_1^2e_2^2. \Rightarrow \frac{1}{e_1^2}+\frac{1}{e_2^2}=1$

As diagonal intersect at right angle, thus *SFS'F'* is a Rhombus.

Also, $e_1^2+e_2^2=e_1^2e_2^2 \Rightarrow e_1^2=e_2^2(e_1^2-1)$

$\Rightarrow \quad a^2e_1^2=e_2^2a^2(e_1^2-1)=b^2e_2^2 \Rightarrow ae_1=be_2$

$\Rightarrow$ Diagonals are equal, thus *SFS'F'* is square. Also being square is concylic.

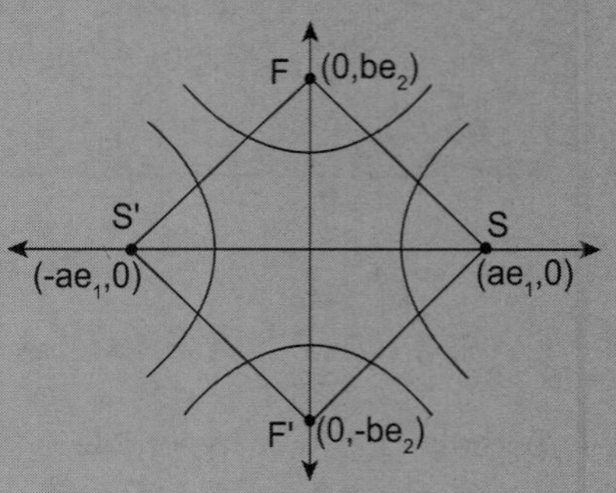

FIGURE 6.19

Hyperbola with Their Axes || to Co-ordinate Axis and Centre Shifted to (α, β)

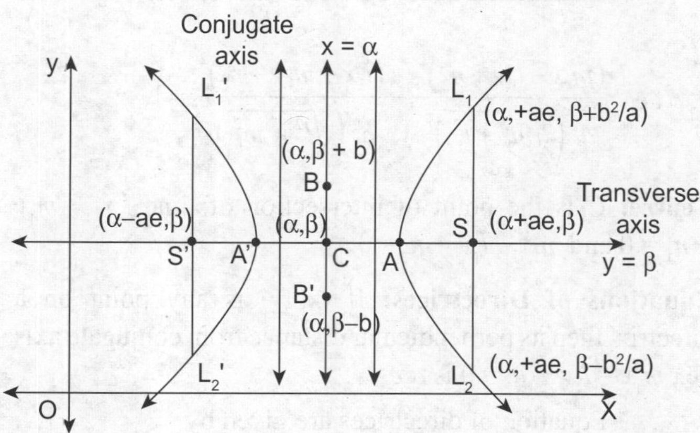

FIGURE 6.20

Equation of hyperbola with centre at (α, β), transverse axis parallel to x-axis and of length $2a$ and conjugate axis parallel to y-axis, of length $2b$.

- Transverse axis: $y = \beta$, Length $= 2a$
- Conjugate axis: $x = \alpha$, Length $= 2b$
- Equation of hyperbola: $\left(\dfrac{\text{Equation of } C.\text{Axis}}{\text{length of semi } T.\text{ Axis}}\right)^2 - \left(\dfrac{\text{Equation of } T.\text{ Axis}}{\text{length of semi } C.\text{ Axis}}\right)^2 = 1$

  i.e., $\dfrac{(x-\alpha)^2}{a^2} - \dfrac{(y-\beta)^2}{b^2} = 1$
- Foci: $S = (\alpha + ae, \beta)$, $S' = (\alpha - ae, \beta)$
- Directrices: $D_1: x = \alpha + a/e$, $D_2: x = \alpha - a/e$.

  Similarly, we can get equation if transverse axis is parallel to y-axis

ILLUSTRATION 15: Find the centre, the foci, the directrices, the length of the latus rectum, the length and the equations of the axes of the hyperbola $16x^2 - 9y^2 + 32x + 36y - 164 = 0$.

SOLUTION: Rearranging the expression and completing the square, we get,

$$6x^2 + 32x + 16 - (9y^2 - 36y + 36) - 144 = 0$$

or $16(x + 1)^2 - 9(y - 2)^2 = 144$ or $\dfrac{(x+1)^2}{9} - \dfrac{(y-2)^2}{16} = 1$ ... (i)

(i) has its centre at $C(-1, 2)$ and having its transverse axis parallel to x-axis.

Equations of axis of hyperbola: Equation of transverse axis is $y - 2 = 0$.

and equation of conjugate axis is $x + 1 = 0$.

Length of the axes: Length of transverse axis $= 2a = 2 \times 3 = 6$ units.

Length of conjugate axis $= 2b = 2(4) = 8$ units.

Eccentricity: Here $a^2 = 9$ and $b^2 = 16$; $b^2 = a^2(e^2 - 1)$

$$16 = 9(e^2 - 1) \text{ or } e^2 = \frac{25}{9} \quad \therefore\ e = \frac{5}{3}$$

Here, $a = 3$ and $b = 4$

Foci: $x + 1 = \pm ae$, $y - 2 = 0$

Foci: $(-1 + 5, 2)$ and $(-1 - 5, 2) = (4. 2)$ and $(-6, 2)$

Directrices: $(x + 1) = \pm\dfrac{a}{e} \Rightarrow x = -1 \pm \dfrac{3}{5/3} = -\dfrac{14}{5}$ or $\dfrac{4}{5}$.

$\therefore\ x = -\dfrac{14}{5}$ and $x = \dfrac{4}{5}$ are the equations of directrices

Length of each latus rectum $\dfrac{2b^2}{a} = 2 \times \dfrac{16}{3} = \dfrac{32}{3}$.

Equation of Hyperbola Referred to Two Perpendicular Straight Lines as Their Axes, but not Parallel to Co-ordinate Axes

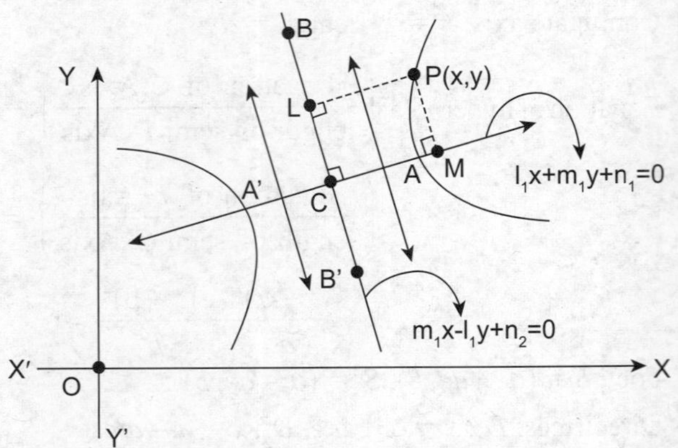

FIGURE 6.21

Let $l_1x + m_1y + n_1 = 0$ ...(1)

and $m_1x - l_1y + n_2 = 0$ ...(2)

be two $\perp r$ straight lines;

Let (1) be the equation of transverse axis and (2) the equation of conjugate axis, and let $2a$ and $2b$ be the lengths of transerve axis and conjugate axis respectively i.e., $AA' = 2a$ and $BB' = 2b$;

Then, equation of hyperbola is given by $\frac{(PL)^2}{a^2}-\frac{(PM)^2}{b^2}=1$.

$$\Rightarrow \frac{\left(\frac{m_1x-l_1y+n_2}{\sqrt{m_1^2+l_1^2}}\right)^2}{a^2}-\frac{\left(\frac{(l_1x+m_1y+n_1)}{\sqrt{l_1^2+m_1^2}}\right)^2}{b^2}=1$$

i.e., $$\frac{(m_1x-l_1y+n_2)^2}{a^2\left(\sqrt{m_1^2+l_1^2}\right)^2}-\frac{(l_1x+m_1y+n_1)^2}{b^2\left(\sqrt{l_1^2+m_1^2}\right)^2}=1$$

Centre: C is the point of intersection of; line $l_1x + m_1y + n_1 = 0$ and $m_1x - l_1y + n_2 = 0$.

Equations of Directrices: If (x, y) is any point on a directrix, then its perpendicular distance from conjugate axis i.e., $m_1x - l_1y + n_2 = 0$ is a/e

$\therefore$ Equation of directrices are given by

$$\frac{m_1x-l_1y+n_2}{\sqrt{m_1^2+l_1^2}}=\pm\frac{a}{e}.$$

Foci are the points of intersection of normal focal chords (Latera recta) and the transverse axis of hyperbola,

$\therefore$ Foci can be obtained by solving the equation $l_1x + m_1y + n_1 = 0$ and the pair of normal chords

(Latera Recta) $$\frac{m_1x-l_1y+n_2}{\sqrt{m_1^2+l_1^2}}=\pm ae$$

Length of each Latera recta $=\frac{2b^2}{a}$.

Equations of Latera recta are given by;

$$\frac{m_1x-l_1y+n_2}{\sqrt{m_1^2+l_1^2}}=\pm ae$$

REMARKS

Equation (1) is a second degree polynomial equation of the form $ax^2+by^2+2hxy+2gx+2fy+c=0$; $h\neq 0$ and $h^2 - ab > 0$ *i.e., term containing xy is present. If term containing xy is absent i.e., h = 0, then the axes of hyperbola would be parallel to co-ordinate axes.*

ILLUSTRATION 16: Show that equation of one of the latera recta of the hyperbola $(10x - 5)^2 + (10y - 2)^2 = 9(3x + 4y - 7)^2$ is $30x + 40y - 23 = 0$.

SOLUTION: $(10x - 5)^2 + (10y - 2)^2 = 9(3x + 4y - 7)^2$

$$\Rightarrow \left(x-\frac{1}{2}\right)^2+\left(y-\frac{1}{5}\right)^2=\frac{9}{4}\left(\frac{3x+4y-7}{5}\right)^2$$

Thus focus is $\left(\frac{1}{2},\frac{1}{5}\right)$, $e=\frac{3}{2}$, directrix is line $3x + 4y - 7 = 0$. Since latus rectum is line parallel directrix passing through focus. Thus equation of corresponding latus rectum will be in the form $3x + 4y + \lambda = 0$ and it should pass through $\left(\frac{1}{2},\frac{1}{5}\right)$.

$\Rightarrow \frac{3}{2}+\frac{4}{5}+\lambda=0 \Rightarrow \lambda=-\frac{23}{10}$

$\Rightarrow$ equation is $30x + 40y - 23 = 0$.

ILLUSTRATION 17: Find the equation to the hyperbola whose one directrix is $2x + y = 1$ and the corresponding focus (1, 1) and eccentricity $\sqrt{3}$. Also find the length of its latera recta.

SOLUTION: Let $S(1, 1)$ be the focus and $P(x, y)$ be a point on the hyperbola. Draw PM perpendicular from P on the directrix. Then by definition, we have $SP = ePM$; $e=\sqrt{3}$ (given)

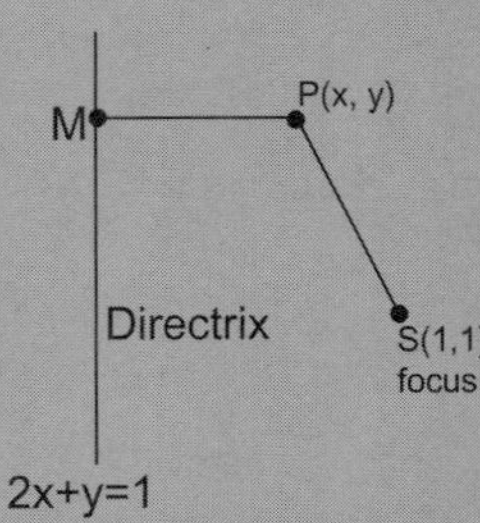

FIGURE 6.22

$\Rightarrow \sqrt{(x-1)^2+(y-1)^2}=\sqrt{3}\left|\frac{2x+y-1}{\sqrt{2^2+1^2}}\right|$

$\Rightarrow (x-1)^2 + (y-1)^2 = 3\frac{(2x+y-1)^2}{5}$; on solving, we get

$7x^2+12xy-2y^2-2x+4y-7=0$

Now $OS - OM = SM \Rightarrow ae-\left(\frac{a}{e}\right)=p$ = distance between focus and directrix.

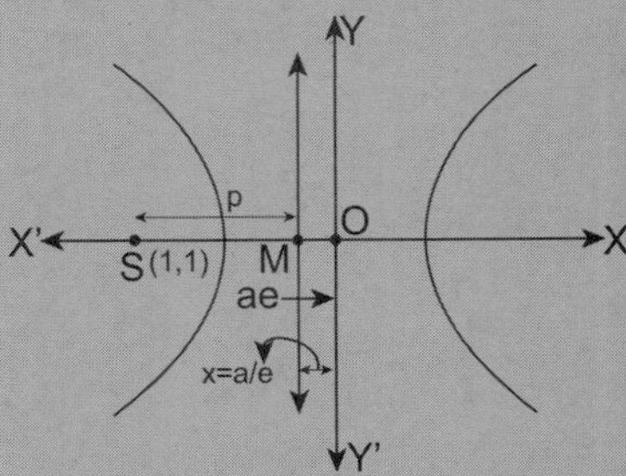

FIGURE 6.23

$\Rightarrow \frac{a}{e}(e^2-1)=\left|\frac{2+1-1}{\sqrt{2^2+1^2}}\right| \Rightarrow a=\sqrt{3/5}$.

Since $e^2 = 1 + \frac{b^2}{a^2}$ $\therefore$ $b^2 = 6/5$.

Length of each latera recta $=\frac{2b^2}{a}=2.\frac{6}{5}.\sqrt{\frac{5}{3}}=\sqrt{\frac{48}{5}}$.

ILLUSTRATION 18: Find the vertices, centre, foci, latus rectum, length of major and minor axis, and directrices for the hyperbola $x^2 - 3xy + y^2 + 10x - 10y + 21 = 0$.

SOLUTION: Given equation of hyperbola is $x^2 - 3xy + y^2 + 10x - 10y + 21 = 0$

Centre of hyperbola is given by the simultaneous linear equations $\frac{\partial F}{\partial x}=2x-3y+10=0$ ... (i)

and $\dfrac{\partial F}{\partial y} = -3x + 2y - 10 = 0$ ... (ii)

Solving (i) and (ii) simultaneously we get; $x = -2, y = 2$

∴ centre of hyperbola is given by (–2, 2)

Let us shift the origin to centre (–2, 2) by substituting $x = x' + (-2)$ and $y = y' + 2$ or simply replacing x by $x - 2$. and y by $y + 2$.

∴ Equation of hyperbola with centre at origin will be

∴ $(x-2)^2 - 3(x-2)(y+2) + (y+2)^2 + 10(x-2) - 10(y+2) + 21 = 0$

∴ $x^2 + y^2 - 3xy + 1 = 0$ ... (iii)

Let the axes be rotated through an angle θ.

So that transverse and conjugate axes coincide with co-ordinate axes, then

$\tan 2\theta = \dfrac{2h}{a-b}. \Rightarrow = \dfrac{-3}{1-1} \Rightarrow 2\theta = \dfrac{\pi}{2} \Rightarrow \theta = \dfrac{\pi}{4}$ i.e., 45°

Applying the concept of rotation of axes by an angle θ

	x	y
x'	$\cos\theta$	$\sin\theta$
y'	$-\sin\theta$	$\cos\theta$

where $\theta = 45°$

Representing new coordintes interms of old $x' = \dfrac{x-y}{\sqrt{2}}, y' = \dfrac{x+y}{\sqrt{2}}$

∴ The equation (iii) becomes $\left(\dfrac{x-y}{\sqrt{2}}\right)^2 + \left(\dfrac{x+y}{\sqrt{2}}\right)^2 - 3\left(\dfrac{y}{\sqrt{2}} + \dfrac{x}{\sqrt{2}}\right)\left(\dfrac{x}{\sqrt{2}} - \dfrac{y}{\sqrt{2}}\right) + 1 = 0$

or $x^2 + y^2 - 2xy + y^2 + x^2 + 2xy + 3y^2 - 3x^2 + 2 = 0$

$\Rightarrow x^2 - 5y^2 = 2$ or $\dfrac{x^2}{2} - \dfrac{y^2}{2/5} = 1$

or $\dfrac{x^2}{(\sqrt{2})^2} - \dfrac{y^2}{(\sqrt{2/5})^2} = 1$

∴ Length of major axis $= 2a = 2\sqrt{2}$

Length of minor axis $= 2b = 2\sqrt{\dfrac{2}{5}}$,

Eccentricity is given by $b^2 = a^2(e^2 - 1)$

$\Rightarrow \dfrac{2}{5} = 2(e^2 - 1) \Rightarrow e^2 = 6/5 \Rightarrow e = \sqrt{\dfrac{6}{5}}$

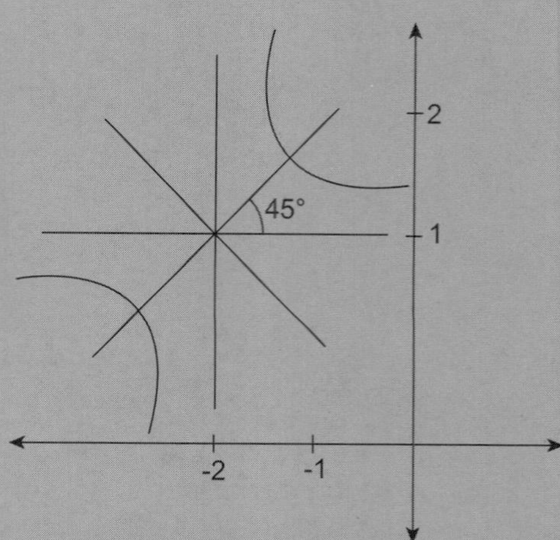

FIGURE 6.24

Equation of transverse axis: $(y - 2) = 1(x + 2)$ or $x - y + 4 = 0$

Equation of conjugate axis: $(y - 2) = -1(x + 2)$ or $x + y = 0$

Vertices are given by; $\dfrac{x+2}{\cos 45°} = \dfrac{y-2}{\sin 45°} = \pm\sqrt{2}$

$\Rightarrow x = -1, y = 3$ and $x = -3, y = 1$ i.e., (–1, 3) and (–3, 1)

$$\textbf{Foci are given by; } \frac{x+2}{\cos 45°} = \frac{y-2}{\sin 45°} = \pm ae = \pm\sqrt{2}\left(\frac{\sqrt{6}}{\sqrt{5}}\right) = \pm\sqrt{\frac{12}{5}}$$

$$\therefore \text{ foci will be } \left(-2+\sqrt{\frac{6}{5}}, 2+\sqrt{\frac{6}{5}}\right) \text{ and } \left(-2-\sqrt{\frac{6}{5}}, 2-\sqrt{\frac{6}{5}}\right)$$

Equation of directrices are given by: $\left|\frac{x+y}{\sqrt{2}}\right| = \frac{a}{e}$

$$\Rightarrow x+y = \pm\sqrt{2}\frac{a}{e} \Rightarrow x+y = \pm\sqrt{2}\left(\sqrt{2}\right)\times\sqrt{\frac{5}{6}} \text{ or } x+y = \pm 2\sqrt{\frac{5}{6}}$$

Length of each latera recta = $4a = 4(\sqrt{2})$ units.

Hyperbola and Basic Definitions at a Glance

❑ $\frac{1}{e'^2} + \frac{1}{e^2} = \frac{b^2}{a^2+b^2} + \frac{a^2}{a^2+b^2} = 1,$

where e and e' are eccentricities of hyperbola and its conjugate respectively.

❑ If $a = b$, hyperbola is called equilateral or rectangular.

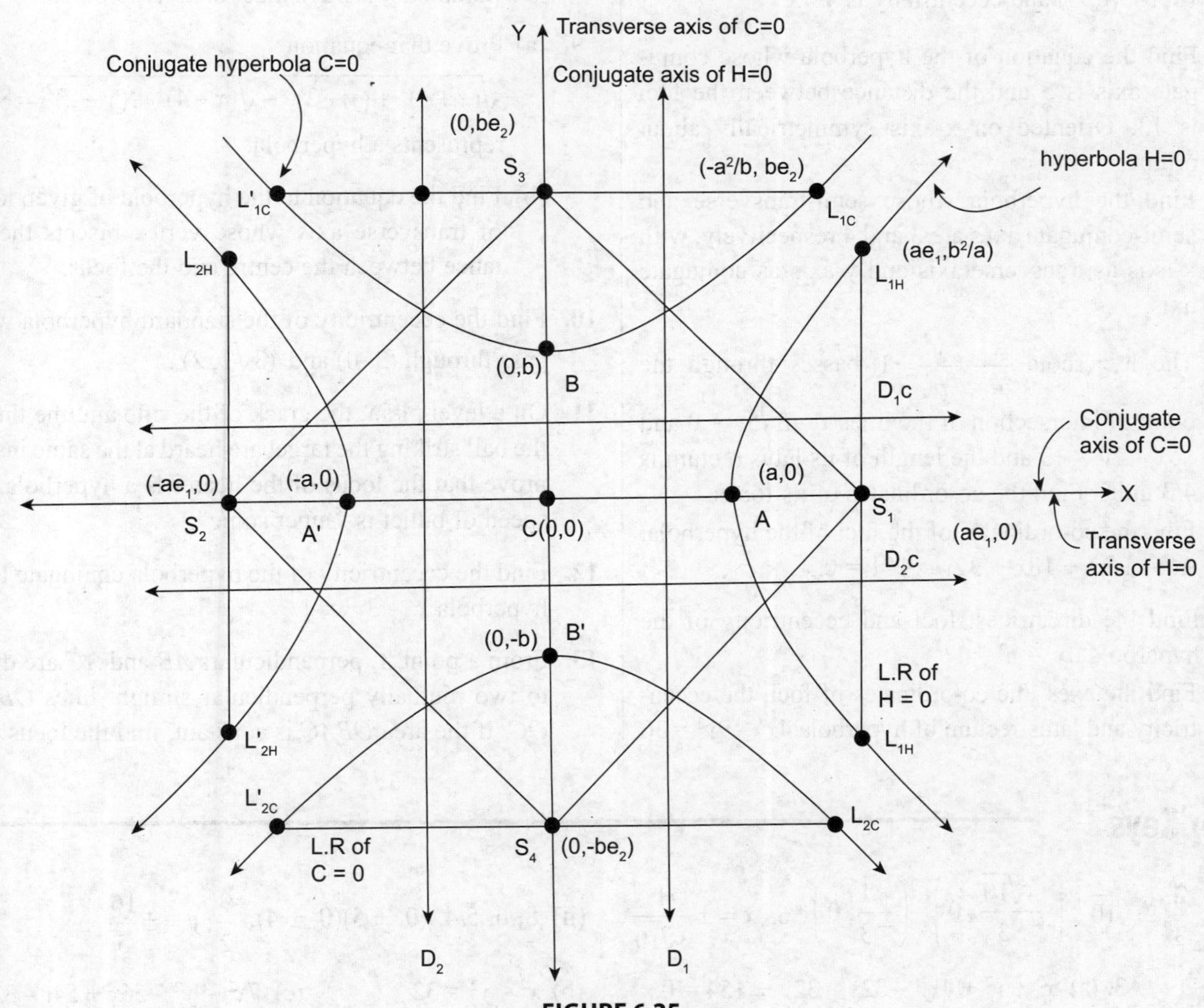

FIGURE 6.25

TEXTUAL EXERCISE–1 (SUBJECTIVE)

1. Find the lengths of transverse axis and conjugate axis, eccentricity, the co-ordinates of foci, vertices, lengths of each latera recta and equations of the directrices of the following hyperbolas.
   (i) $9x^2 - y^2 = 1$
   (ii) $16x^2 - 9y^2 = -144$.

2. Find the eccentricity of the hyperbola whose latus rectum is half of its transverse axis.

3. (a) Find the equation of the hyperbola one of whose foci is (2, 2), eccentricity = 2 and the equation of corresponding directrix is $x + y = 9$.
   (b) Find the equation of the hyperbola; the distance between whose foci is 16, whose eccentricity is $\sqrt{2}$ and whose transverse axis is along the x-axis with the origin as its centre.
   (c) Find the equation of the hyperbola whose foci are (8, 3); (0, 3) and eccentricity is 4/3.

4. (a) Find the equation of the hyperbola whose conjugate axis is 5 and the distance between the foci is 13. Oriented on x-axis symmetrically about origin.
   (b) Find the hyperbola whose semi-transverse and semi-conjugate axes are 3 and 4 respectively, with x-axis as transverse axis and y-axis as conjugate axis.
   (c) The hyperbola $\frac{x^2}{a^2} - \frac{y^2}{b^2} = 1$ passes through the point of intersection of the lines $x - 3\sqrt{5}y = 0$ and $\sqrt{5}x - 2y = 13$ and the length of its latus rectum is 4/3 units. Find the co-ordinates of its focus.
   (d) Find the co-ordinates of the foci of the hyperbola, $9x^2 - 16y^2 + 18x + 32y - 151 = 0$.

5. (a) Find the directrices, foci and eccentricity of the hyperbola $ax^2 - y^2 = 1$
   (b) Find the axes, the co-ordinates of foci, the eccentricity and latus-rectum of hyperbola $4x^2 - 9y^2 = 36$
   (c) Show that the equation $x^2 - 2y^2 - 2x + 8y - 1 = 0$ represents a hyperbola. Find the co-ordinates of the centre, lengths of the axes, eccentricity, latus-rectum, co-ordinates of foci and vertices and equations of directrices of the hyperbola.

6. Find the equation of the hyperbola, whose eccentricity is 5/4, a focus is at $(a, 0)$ and the corresponding directrix is $4x - 3y = a$. Also find the co-ordinates of the centre.

7. Prove that the point $\left\{\frac{a}{2}\left(t+\frac{1}{t}\right), \frac{b}{2}\left(t-\frac{1}{t}\right)\right\}$ lies on the hyperbola for all values of $t (t \neq 0)$.

8. Prove that the straight lines $\frac{x}{a} - \frac{y}{b} = m$ and $\frac{x}{a} + \frac{y}{b} = \frac{1}{m}$, where a and b are given positive real numbers and m is a parameter, always meet on a hyperbola.

9. (a) Prove that equation
   $$\sqrt{(x+4)^2 + (y+2)^2} - \sqrt{(x-4)^2 + (y-2)^2} = 8$$
   represents a hyperbola.
   (b) Find the equation to the hyperbola of given length of transverse axis whose vertex bisects the distance between the centre and the focus.

10. Find the eccentricity of the standard hyperbola which pass through (3, 0) and $(3\sqrt{2}, 2)$.

11. On a level plain, the crack of the rifle and the thud of the ball striking the target are heard at the same instant, prove that the locus of the hearer is a hyperbola, (the speed of bullet is "super sonic".)

12. Find the eccentricity of the hyperbola conjugate to the hyperbola $x^2 - 3y^2 = 1$.

13. From a point A, perpendiculars AB and AC are drawn to two mutually perpendicular straight lines OB and OC. If the area $OBAC$ is constant, find the locus of A.

Answer Keys

1. (i) $\frac{2}{3}, 2, \sqrt{10}$, $\left(\pm\frac{\sqrt{10}}{3}, 0\right)$, $\left(\pm\frac{1}{3}, 0\right)$, 6, $x = \pm\frac{1}{3\sqrt{10}}$ (ii) 8, 6, 5/4, (0, ± 5)(0, ± 4), $\frac{9}{2}, y = \pm\frac{16}{5}$

2. $\sqrt{\frac{3}{2}}$ 3. (a) $x^2 + y^2 + 4xy - 32x - 32y + 154 = 0$ (b) $x^2 - y^2 = 32$ (c) $7x^2 - 9y^2 - 56x + 54y - 32 = 0$

4. (a) $25x^2 - 144y^2 = 900$ (b) $16x^2 - 9y^2 = 144$ (c) $(\pm 2\sqrt{10}, 0)$ (d) (4, 1) and (– 6, 1)

5. (a) $x = \pm \dfrac{1}{\sqrt{a(a+1)}}$, $\left(\pm\sqrt{\dfrac{a+1}{a}}, 0\right)$, $\sqrt{a+1}$ (b) 6, 4, $(\pm\sqrt{13}, 0)$, $\dfrac{\sqrt{13}}{3}$ $\dfrac{8}{3}$ units

(c) (1, 2); $2\sqrt{3}$, $2\sqrt{6}$; $\sqrt{3}$; $4\sqrt{3}$; (1, –1), (1, 5); $(1, 2 \pm \sqrt{3})$; y = 1, y = 3

6. $7y^2 + 24xy - 24ax - 6ay + 15a^2 = 0$; $(-a/3, a)$ **9.** (b)$3x^2 - y^2 = 3a^2$ **10.** $\dfrac{\sqrt{13}}{3}$ **12.** 2

13. $x^2 - y^2$ = constant

TEXTUAL EXERCISE–1 (OBJECTIVE)

1. Co-ordinates of the foci of the hyperbola $\dfrac{(x-1)^2}{9} - \dfrac{(y-2)^2}{16} = 1$ are

(a) (4, –2) and (6, 2) (b) (–4, 2) and (6, –2)
(c) (–4, 2) and (6, 2) (d) (–4, 2) and (–6, 2)

2. Which of the following expressions ('t' being the parameter) can't represent a hyperbola?

(a) $\dfrac{tx}{a} - \dfrac{y}{b} + t = 0, \dfrac{x}{a} + \dfrac{ty}{b} - 1 = 0$

(b) $x = \dfrac{a}{2}\left(t + \dfrac{1}{t}\right), y = \dfrac{b}{2}\left(t - \dfrac{1}{t}\right)$

(c) $x = e^t + e^{-t}, y = e^t - e^{-t}$

(d) $x^2 = 2(\cos t + 3), y^2 = 2\left(2\cos^2\dfrac{t}{2} - 1\right)$

3. The equation

$$\left|\sqrt{(x-2)^2 + (y-1)^2} - \sqrt{(x+2)^2 + y^2}\right| = c$$

will represent a hyperbola if

(a) $c \in (0, 6)$ (b) $c \in (0, 5)$
(c) $c \in (0, \sqrt{17})$ (d) None of these

4. The co-ordinates of the centre of the hyperbola $x^2 + 2y^2 + 3xy + 2x + 3y + 2 = 0$ is

(a) (1, 0) (b) (–1, 0)
(c) (–1, 1) (d) (1, –1)

5. The equation of the hyperbola, having its axes along the co-ordinate axes, can be (it is given that distances of one of its vertex from its foci are 1 and 3 units respectively)

(a) $x^2 - 3y^2 = 3$ (b) $x^2 - 3y^2 + 3 = 0$
(c) $3x^2 - y^2 = 3$ (d) $3x^2 - y^2 + 3 = 0$

6. The locus of point of intersection of the lines $\sqrt{3}x - y - 4\sqrt{3}\,t = 0$ and $\sqrt{3}tx + ty - 4\sqrt{3} = 0$, '$t$' being the parameter, is a hyperbola, whose eccentricity is equal to

(a) $\sqrt{2}$ (b) 2
(c) $2/\sqrt{3}$ (d) None of these

7. If the eccentricity of the hyperbola $x^2 - y^2 . \sec^2\theta = 5$ is $\sqrt{3}$ times the eccentricity of the ellipse $x^2 \operatorname{cosec}^2\theta + y^2 = 25$, then smallest positive value of 'θ' is

(a) $\dfrac{\pi}{6}$ (b) $\dfrac{\pi}{4}$

(c) $\dfrac{\pi}{3}$ (d) None of these

8. Locus of the centre of a circle touching two given unequal circles externally, will be a hyperbola if

(a) The circles are orthogonal
(b) The circles are touching
(c) The circles are non-intersecting
(d) All of these

9. Find the equation of the corresponding conjugate hyperbola for the hyperbola represented by the equation $hx + ky = xy$.

(a) $yh + xk - xy + hk = 0$
(b) $xh + yk - xy + hk = 0$
(c) $yh + xk - xy - 2hk = 0$
(d) $xh + yk - xy - 2hk = 0$

10. The curves $\dfrac{x^2}{a^2} + \dfrac{y^2}{b^2} = 4$ and $\dfrac{x^2}{a^2} - \dfrac{y^2}{b^2} = 1$ are given to be confocal. If the eccentricities of the given curves be e_1 and e_2 respectively, then the value of $e_1^2 + e_2^2$ is equal to

(a) 4 (b) 2
(c) 3 (d) None of these

11. Product of the distances of a moving point P from the lines $x + y - 3 = 0$ and $x - y + 6 = 0$ is equal to 12. P will lie on a hyperbola, whose eccentricity is equal to

(a) $\sqrt{2}$ (b) $2\sqrt{2}$
(c) $\sqrt{3}$ (d) $2\sqrt{5}$

12. The ellipse $\frac{x^2}{a^2}+\frac{y^2}{b^2}=1$ and the hyperbola $\frac{x^2}{A^2}-\frac{y^2}{B^2}=1$ are given to be confocal and length of minor axis of ellipse is same as the conjugate axis of the hyperbola. If e_e and e_h represents the eccentricity of ellipse and hyperbola respectively, then the value of $\frac{1}{e_a^2}+\frac{1}{e_h^2}$ is equal to

(a) 1 (b) 2
(c) 4 (d) 6

13. P is any point on the hyperbola $x^2 - y^2 = a^2$. If F_1 and F_2 are the foci of the hyperbola and $PF_1 . PF_2 = \lambda . OP^2$, where O is the origin, then 'λ' is equal to

(a) 1 (b) $\sqrt{2}$
(c) 2 (d) None of these

14. The equation $2x^2 - 3y^2 - 12x + 12y = 0$ represents

(a) a parabola
(b) an ellipse
(c) a hyperbola
(d) a rectangular hyperbola

15. The equation of a hyperbola with a focus (2, 1), eccentricity $\sqrt{2}$ and corresponding directrix $2x + 3y - 1 = 0$ is

(a) $5x^2 - 5y^2 - 24xy - 44x - 14y + 63 = 0$
(b) $5x^2 - 5y^2 - 24xy - 44x + 14y - 63 = 0$
(c) $5x^2 - 5y^2 - xy + x + 14y + 63 = 0$
(d) None of these

16. The distance between the foci of a hyperbola is 16 and its eccentricity is $\sqrt{2}$, then the equation of hyperbola is

(a) $x^2 - y^2 = 32$ (b) $x^2 - y^2 = 16$
(c) $x^2 - y^2 = 4$ (d) $x^2 - y^2 = 64$

17. The equation $(x - \alpha)^2 + (y - \beta)^2 = k(lx + my + n)^2$ represents a hyperbola if

(a) $k > (l^2 + m^2)^{-1}$ (b) $k < (l^2 + m^2)^{-1}$
(c) $k = (l^2 + m^2)^{-1}$ (d) None of these

18. The centre of the hyperbola $7x^2 - 5y^2 - 28x - 30y - 35 = 0$ is

(a) (2, 3) (b) (–2, 3)
(c) (–2, – 3) (d) (2, – 3)

19. The equation of the conic with focus at (1, – 1), directrix along $x - y + 1 = 0$ and with eccentricity $\sqrt{2}$ is

(a) $x^2 - y^2 = 1$
(b) $xy = 1$
(c) $2xy - 4x + 4y + 1 = 0$
(d) $2xy + 4x - 4y - 1 = 0$

20. The eccentricity of the hyperbola whose latus rectum is 8 and length of conjugate axis is equal to half the distance between the foci, is

(a) $\frac{4}{3}$ (b) $\frac{4}{\sqrt{3}}$
(c) $\frac{2}{\sqrt{3}}$ (d) None of these

21. The curve represented by $x = ae^{\theta}, y = be^{-\theta}, \theta\in\mathbb{R}$ is

(a) a hyperbola (b) an ellipse
(c) a parabola (d) a circle

22. The hyperbola $\frac{x^2}{a^2}-\frac{y^2}{b^2}=1$ passes through (5, 4) and its latus rectum is $\frac{32\sqrt{2}}{5}$. Then b =

(a) 4 (b) 1
(c) 8 (d) 2

23. If the foci of the ellipse $\frac{x^2}{16}+\frac{y^2}{b^2}=1$ and $\frac{x^2}{144}-\frac{y^2}{81}=\frac{1}{25}$ coincide, then the value b^2 is equal to

(a) 9 (b) 1
(c) 5 (d) 7

24. For the hyperbola $\frac{x^2}{\cos^2\alpha}-\frac{y^2}{\sin^2\alpha}=1, 0<\alpha<\frac{\pi}{2}$, which one is independent of α?

(a) Eccentricity (b) Abscissae of the foci
(c) Directrices (d) Ends of transverse axis

25. The equation $\frac{x^2}{1-r}+\frac{y^2}{1+r}=1, r>1$ represents

(a) a parabola (b) an ellipse
(c) a hyperbola (d) a circle

Answer Keys

1. (c)	**2.** (a)	**3.** (c)	**4.** (b)	**5.** (c)	**6.** (b)	**7.** (b)	**8.** (d)	**9.** (d)	**10.** (b)
11. (a)	**12.** (b)	**13.** (a)	**14.** (c)	**15.** (a)	**16.** (a)	**17.** (a)	**18.** (d)	**19.** (c)	**20.** (c)
21. (a)	**22.** (a)	**23.** (d)	**24.** (b)	**25.** (c)					

AUXILIARY CIRCLE OF HYPERBOLA AND ECCENTRIC ANGLE OF POINT ON HYPERBOLA

Let $\frac{x^2}{a^2}-\frac{y^2}{b^2}=1$, be the hyperbola with centre C and transverse axis $A'A$. Therefore circle drawn with centre C and segment $A'A$ as a diameter is called 'auxiliary circle' of the hyperbola $\frac{x^2}{a^2}-\frac{y^2}{b^2}=1$.

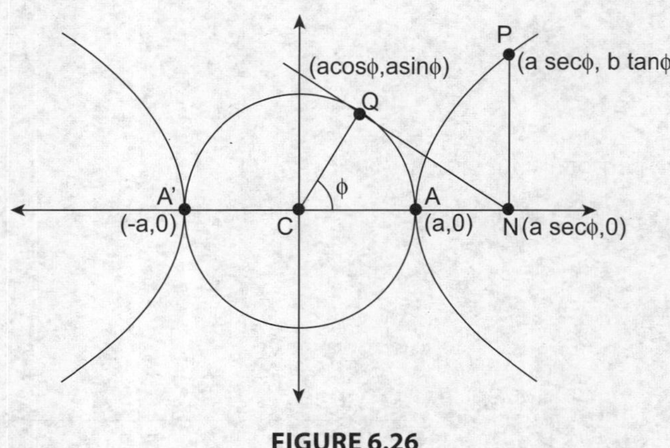

FIGURE 6.26

$\therefore$ Equation of the auxiliary circle is $x^2 + y^2 = a^2$

Let $P(x, y)$ be any point on the hyperbola $\frac{x^2}{a^2}-\frac{y^2}{b^2}=1$. Draw PN perpendicular to x-axis.

Let NQ be a tangent to the auxiliary circle $x^2 + y^2 = a^2$. Join C and Q. Let $\angle QCN = \phi$

Here P and Q are the corresponding points of the hyperbola and the auxiliary circle respectively and ϕ is called the eccentric angle of P. $(0 \le \phi < 2\pi)$. Clearly $Q \equiv (a \cos \phi, a \sin\phi)$

In ΔCQN, applying trigonometry, we get

$x = CN = a \sec \phi$

$\therefore$ $P(x, y) \equiv (a \sec \phi, y)$ as it lies on hyperbola

$$\frac{x^2}{a^2}-\frac{y^2}{b^2}=1$$

$$\therefore \quad \frac{a^2\sec^2\phi}{a^2}-\frac{y^2}{b^2}=1$$

or $\frac{y^2}{b^2}=\sec^2\phi-1=\tan^2\phi$

$\therefore$ $y = \pm b \tan \phi$

$\therefore$ $y = b \tan \phi$ ($\because$ P lies in 1st quadrant)

ECCENTRIC ANGLE

Of any point P on the hyperbola is the angle (ϕ) made by CQ with positive direction of transverse axis in anti-clockwise sense. (where C is centre and Q is point of contact of tangent drawn from foot of ordinate of P to the auxiliary circle).

$\therefore$ Coordinates of such point P (a secf, b tanf)

PARAMETRIC EQUATION

The equations $x = a \sec \phi$ and $y = b \tan\phi$, $\phi \in [0, 2\pi) \sim \{\pi/2, 3\pi/2\}$ are known as the parametric equations of the hyperbola $\frac{x^2}{a^2}-\frac{y^2}{b^2}=1$.

For every position of points Q on auxiliary circle and the corresponding point P which describes the hyperbola, $0 \le \phi < 2\pi$.

ϕ varies from	$Q(a \cos\phi, a \sin\phi)$	$P(a \sec \phi, b \tan \phi)$
0 to $\frac{\pi}{2}$	I quad.	I quad.
$\frac{\pi}{2}$ to π	II quad.	III quad.
π to $\frac{3\pi}{2}$	III quad.	II quad.
$\frac{3\pi}{2}$ to 2π	IV quad.	IV quad.

REMARK

The ratio of ordinate of point P on hyperbola and length of tangent from the foot of ordinate(N) to the auxilary circle is constant (b/a). $\frac{PN}{QN}=\frac{b\tan\phi}{a\tan\phi}=\frac{b}{a}$.

ILLUSTRATION 19: For hyperbola $4x^2 - 9y^2 = 36$. Sketch the curve when

(a) $\theta \in \left[0, \frac{\pi}{2}\right)$ (b) $\theta \in \left(\frac{\pi}{2}, \pi\right]$

(c) $\theta \in \left[\pi, \frac{3\pi}{2}\right)$ (d) $\theta \in \left(\frac{3\pi}{2}, 2\pi\right]$

SOLUTION: The parametric equation is very useful when it comes to tracing a part of hyperbola. For a hyperbola $\frac{x^2}{9} - \frac{y^2}{4} = 1$; the parametric equation is given by $x = 3\sec\theta$, $y = 2\tan\theta$.

❑ For $\theta \in \left[0, \frac{\pi}{2}\right)$

e.g., for hyperbola $\frac{x^2}{9} - \frac{y^2}{4} = 1$ and for $\theta = \frac{\pi}{3}$

$$(x, y) = \left(3\sec\frac{\pi}{3}, 2\tan\frac{\pi}{3}\right) = \left(2\sqrt{3}, 2\sqrt{3}\right)$$

$\therefore$ $\theta \in \left[0, \frac{\pi}{2}\right)$, we get only the part of hyperbola lying in first quadrant becomes each point has both co-ordinates positive.

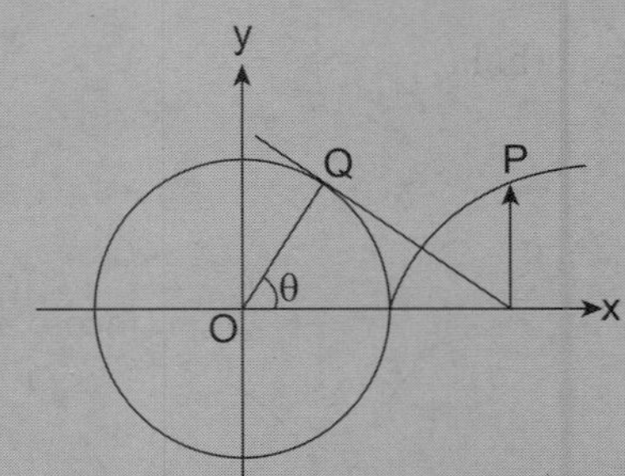

FIGURE 6.27

❑ For $\theta \in \left(\frac{\pi}{2}, \pi\right]$

E.g., for hyperbola $\frac{x^2}{9} - \frac{y^2}{4} = 1$ and $\theta = \frac{2\pi}{3}$

$$(x, y) = \left(3\sec\frac{2\pi}{3}, 2\tan\frac{2\pi}{3}\right) = \left(-2\sqrt{3}, -2\sqrt{3}\right)$$

$\therefore$ For $\theta \in \left(\frac{\pi}{2}, \pi\right]$ and for positive a, b; both $a\sec\theta$ and $b\tan\theta$ must be negative and hence we get only part of the hyperbola lying in third quadrant.

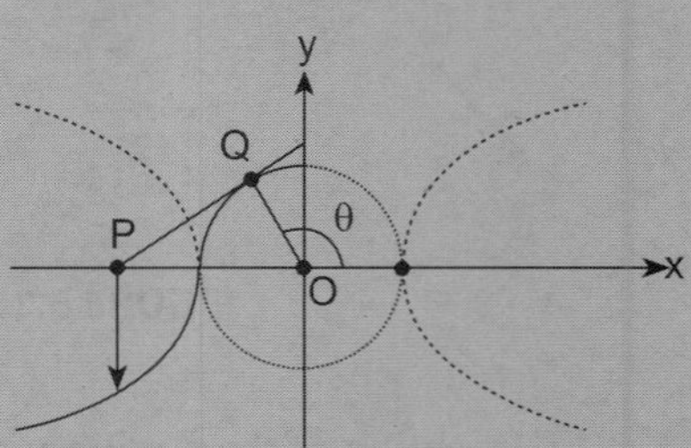

FIGURE 6.28

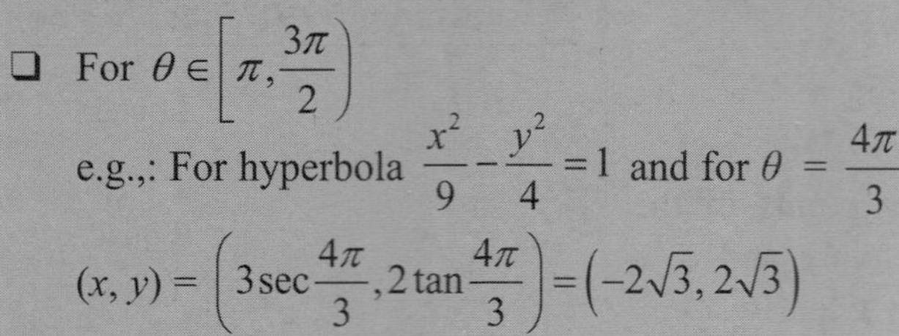

❑ For $\theta \in \left[\pi, \frac{3\pi}{2}\right)$

e.g.,: For hyperbola $\frac{x^2}{9} - \frac{y^2}{4} = 1$ and for $\theta = \frac{4\pi}{3}$

$$(x, y) = \left(3\sec\frac{4\pi}{3}, 2\tan\frac{4\pi}{3}\right) = \left(-2\sqrt{3}, 2\sqrt{3}\right)$$

$\therefore$ For $\theta \in \left[\pi, \frac{3\pi}{2}\right)$ and for positive a, b; $a\sec\theta$ must be negative and $b\tan\theta$ must be positive, and hence we get only the part of the hyperbola lying in second quadrant.

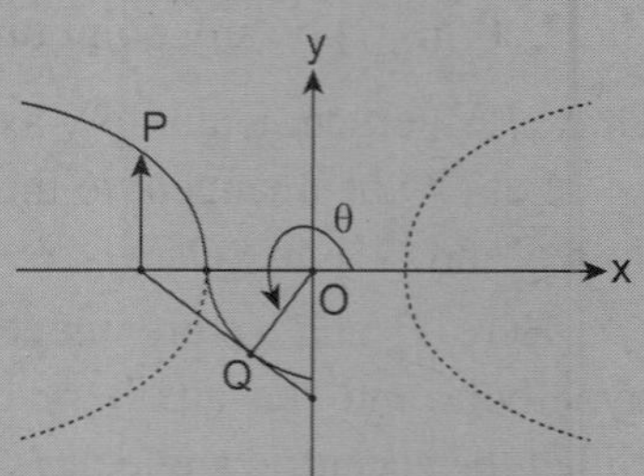

FIGURE 6.29

❑ For $\theta \in \left(\frac{3\pi}{2}, 2\pi\right)$

For hyperbola $\frac{x^2}{9} - \frac{y^2}{4} = 1$ and for $\theta = \frac{5\pi}{3}$

$$(x, y) = \left(3\sec\frac{5\pi}{3}, 2\tan\frac{5\pi}{3}\right) = \left(2\sqrt{3}, -2\sqrt{3}\right)$$

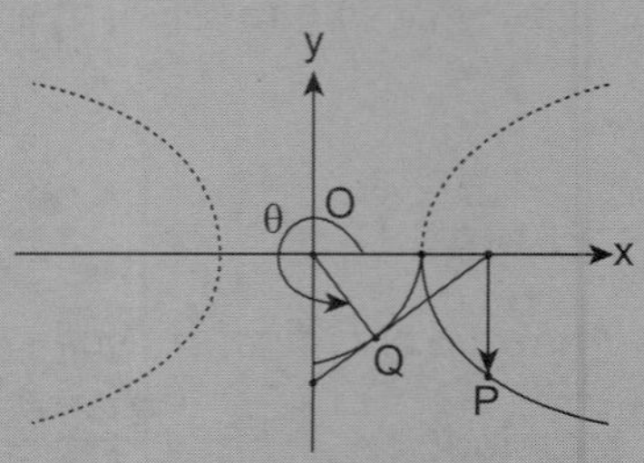

FIGURE 6.30

$\therefore$ For $\theta \in \left(\frac{3\pi}{2}, 2\pi\right]$; and positive a, b; $a \sec \theta$ must be positive and $b \tan \theta$ must be negative and hence, we get only the part of the hyperbola lying in the quadrant IV.

ILLUSTRATION 20: Let a hyperbola H be defined by $x = 8\sec\theta$, $y = 6 \tan\theta$ and a circle C be defined as $x = 12 \cos\alpha$, $y = 12\ r \sin \alpha$. Then find the points of intersection of H and C under the conditions:

(i) $\theta \in \left(-\frac{\pi}{2}, \frac{\pi}{2}\right)$ and $\alpha \in \left[-\frac{\pi}{2}, \pi\right]$

(ii) $\theta \in \left(\frac{\pi}{4}, \frac{\pi}{2}\right)$ and $\alpha \in [-\pi, \pi)$

(iii) $\theta \in \left(\frac{2\pi}{3}, \pi\right)$ and $\alpha \in \left[\frac{\pi}{2}, \pi\right)$

SOLUTION: **(i)** Clearly, there are two point of intersection of three quarter circles and branch of hyperbola opening rightward. (see Figure 6.31)

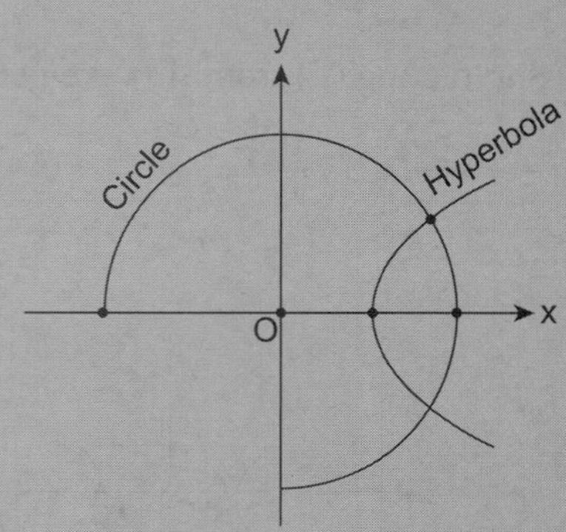

FIGURE 6.31

(ii) For $\theta = \left(\frac{\pi}{4}, \frac{\pi}{2}\right)$; the hyperbola will lie in the first quadrant with minimum abscissa equal to $8\sec(\pi/4) = 8\sqrt{2}$ (see Figure 6.32)

Now $8\sqrt{2} < 12$

Figure will be as shown in the diagram.

For hyperboa: at $\theta = \frac{\pi}{4}$

$\Rightarrow 8\sqrt{2}, y = 4$

For circle: at $x = 8\sqrt{2}$, $y = 4$ as $x^2 + y^2 = 144$ i.e., $128 + 16 = 144$

So, no point of intersection in hyperbola and circle

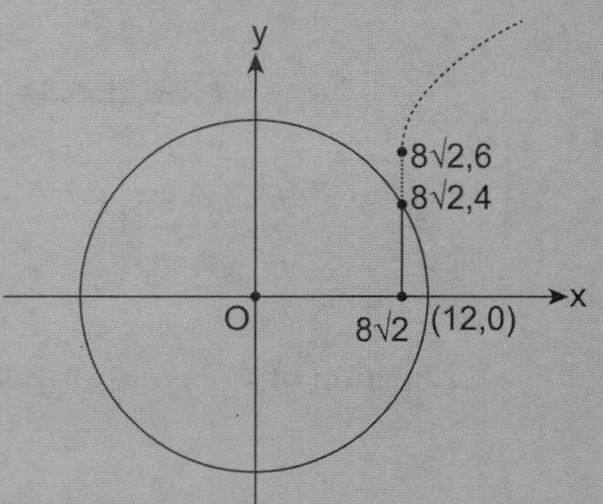

FIGURE 6.32

(iii) For $\theta \in \left(\frac{2\pi}{3}, \pi\right]$; the hyperbola will lie in third quadrant. (see Figure 6.33)

And for $\alpha \in \left[\frac{\pi}{2}, \pi\right]$; the circle will lie in quadrant II.

Clearly, there shall be no point of intersection.

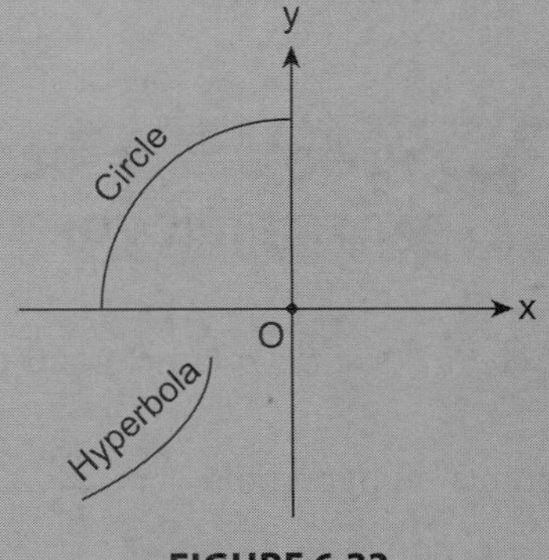

FIGURE 6.33

CHORDS OF HYPERBOLA IN CARTESIAN FORM

Given the equation of hyperbola $\frac{x^2}{a^2}-\frac{y^2}{b^2}=1$ .....(i)

Let $P\equiv(x_1,y_1)$ and $Q\equiv(x_2,y_2)$ be any two points on it then

$$\frac{x_1^2}{a^2}-\frac{y_1^2}{b^2}=1 \quad ......(ii)$$

and $$\frac{x_2^2}{a^2}-\frac{y_2^2}{b^2}=1 \quad ....(iii)$$

Subtracting (ii) from (iii), we get

$$\frac{1}{a^2}(x_2^2-x_1^2)-\frac{1}{b^2}(y_2^2-y_1^2)=0$$

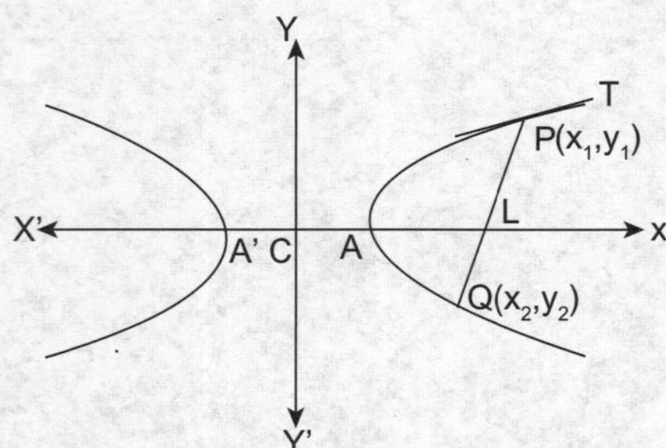

FIGURE 6.34

$$\Rightarrow \quad \frac{y_2-y_1}{x_2-x_1}=\frac{b^2}{a^2}\frac{(x_1+x_2)}{(y_1+y_2)} \quad ....(iv)$$

Equation of PQ is $y-y_1=\frac{y_2-y_1}{x_2-x_1}(x-x_1)$ ....(v)

From (iv) and (v), $y-y_1=\frac{b^2}{a^2}.\frac{(x_1+x_2)}{(y_1+y_2)}(x-x_1)$

which is equation of chord PQ in cartesian form.

CHORD OF HYPERBOLA IN PARAMETRIC FORM

Let P (a sec θ_1, b tan θ_1) and Q(a sec θ_2, b tan θ_2) be two points on hyperbola $\frac{x^2}{a^2}-\frac{y^2}{b^2}=1$, and R (x, y) be any point on chord, then point (x, y), P and Q are collinear.

$\Rightarrow$ Equation of chord is $\begin{vmatrix} x & y & 1 \\ a\sec\theta_1 & b\tan\theta_1 & 1 \\ a\sec\theta_2 & b\tan\theta_2 & 1 \end{vmatrix}=0$

(Multiply elements of 2*nd* and third row by cos θ_1 and $cos\theta_2$ respectively.

i.e., $R_2\to R_2\ cos\theta_1$, $R_3\to R_3\ cos\theta_2$.

$$\Rightarrow \begin{vmatrix} x & y & 1 \\ a & b\sin\theta_1 & \cos\theta_1 \\ a & b\sin\theta_2 & \cos\theta_2 \end{vmatrix}=0 \text{ ,(expand along 1st row)}$$

$\Rightarrow$ bx $\sin(\theta_1-\theta_2)-ay\,(cos\theta_2-cos\theta_1)+ab(sin\theta_2-sin\theta_1)=0$

Finally, we get,

$$\frac{x}{a}\cos\frac{\theta_1-\theta_2}{2}-\frac{y}{b}\sin\frac{\theta_1+\theta_2}{2}=\cos\frac{\theta_1+\theta_2}{2}.$$

FOCAL CHORD OF HYPERBOLA AND ITS PROPERTY

Since the chord of hyperbola in parametric form is

$\frac{x}{a}\cos\frac{\theta_1-\theta_2}{2}-\frac{y}{b}\sin\frac{\theta_1+\theta_2}{2}=\cos\frac{\theta_1+\theta_2}{2}$, thus

If it is a focal chord, then it passes through (ae, 0) or ($-ae$, 0), respctively

$$\Rightarrow \pm e\cos\left(\frac{\theta_1-\theta_2}{2}\right)-0=\cos\left(\frac{\theta_1+\theta_2}{2}\right)$$

$$\Rightarrow \frac{\cos\left(\frac{\theta_1-\theta_2}{2}\right)}{\cos\left(\frac{\theta_1+\theta_2}{2}\right)}=\pm\frac{1}{e}$$

$$\Rightarrow \frac{\cos\left(\frac{\theta_1-\theta_2}{2}\right)-\cos\left(\frac{\theta_1+\theta_2}{2}\right)}{\cos\left(\frac{\theta_1-\theta_2}{2}\right)+\cos\left(\frac{\theta_1+\theta_2}{2}\right)}=\frac{1-e}{1+e} \text{ or } \frac{1+e}{1-e}$$

$$\Rightarrow \tan\left(\frac{\theta_1}{2}\right)\tan\left(\frac{\theta_2}{2}\right)=\frac{1-e}{1+e} \text{ or } \frac{1+e}{1-e}$$

Hence if θ_1 and θ_2 are the eccentric angles of extremities of a focal chord of the hyperbola $\frac{x^2}{a^2}-\frac{y^2}{b^2}=1$ then $\tan\left(\frac{\theta_1}{2}\right)\tan\left(\frac{\theta_2}{2}\right)=\frac{1-e}{1+e}$ or $\frac{1+e}{1-e}$ according as focus is (ae, 0) or ($-ae$, 0).

ILLUSTRATION 21: If θ_1 and θ_2 are the parameters of the extremities of a chord through (5, 0) of a hyperbola $\frac{x^2}{16}-\frac{y^2}{9}=1$, then show that $9\tan\frac{\theta_1}{2}.\tan\frac{\theta_2}{2}+1=0$.

SOLUTION: The equation of chord joining points $P(\theta_1)$ and $Q(\theta_2)$ on the hyperbola is

$\frac{x}{4}\cos\left(\frac{\theta_1-\theta_2}{2}\right)-\frac{y}{3}\sin\left(\frac{\theta_1+\theta_2}{2}\right)=\cos\left(\frac{\theta_1+\theta_2}{2}\right)$. It passes through (5, 0)

$$\Rightarrow \frac{5}{4}.\cos\left(\frac{\theta_1-\theta_2}{2}\right)-0=\cos\left(\frac{\theta_1+\theta_2}{2}\right) \text{ or } \frac{5}{4}=\frac{\cos\left(\frac{\theta_1+\theta_2}{2}\right)}{\cos\left(\frac{\theta_1-\theta_2}{2}\right)}$$

Applying compenendo and dividendo, we get $\frac{\cos\left(\frac{\theta_1-\theta_2}{2}\right)-\cos\left(\frac{\theta_1+\theta_2}{2}\right)}{\cos\left(\frac{\theta_1-\theta_2}{2}\right)+\cos\left(\frac{\theta_1+\theta_2}{2}\right)}=-\frac{1}{9}$

$$\Rightarrow \tan\frac{\theta_1}{2}.\tan\frac{\theta_2}{2}=-\frac{1}{9}. \qquad \Rightarrow 9\tan\frac{\theta_1}{2}\tan\frac{\theta_2}{2}+1=0$$

ILLUSTRATION 22: The chord joining the points having eccentric angles α and β on the hyperbola $\frac{x^2}{a^2}-\frac{y^2}{b^2}=1$ passes through (given $\alpha+\beta=3\pi$).

(a) focus
(b) centre
(c) vertex
(d) None of these

SOLUTION: Equation of chord joining α and β is $\frac{x}{a}\cos\left(\frac{\alpha-\beta}{2}\right)-\frac{y}{b}\sin\left(\frac{\alpha+\beta}{2}\right)=\cos\left(\frac{\alpha+\beta}{2}\right)$

$$\because \alpha+\beta=3\pi \Rightarrow \frac{x}{a}\cos\left(\frac{3\alpha-2\beta}{2}\right)-\frac{y}{b}\sin\frac{3\pi}{2}=0 \Rightarrow y=\frac{b}{a}\sin\beta x$$

So, the above chord passes through centre. ⇒ Ans. (b)

TEXTUAL EXERCISE–2 (SUBJECTIVE)

1. Find the equation of the auxiliary circle of the hyperbola $\frac{x^2}{49}-\frac{y^2}{25}=1$. What it would be for its conjugate hyperbola?

2. Find the length of chord of auxiliary circle of hyperbola $\frac{x^2}{49}-\frac{y^2}{25}=1$, which is tangent to auxiliary circle of conjugate hyperbola of $\frac{x^2}{49}-\frac{y^2}{25}=1$.

3. Find the equation of tangent to auxiliary circle of conjugate hyperbola of hyperbola $4x^2-9y^2=36$ having its slope parallel to line $3x-4y=10$ and having no portion of it passing through fourth quadrant.

4. Find the eccentric angle (θ) of the point $\left(3\sqrt{2},2\right)$ on the hyperbola $4x^2-9y^2=36$. Find the co-ordinates of points which are at a distance of 4 units from the point (2, 3) and lying on the line containing the point (2, 3) and having its slope as $\tan\theta$.

5. Find the parametric equation of hyperbola $x^2-y^2-2x-6y-12=0$.

6. Find the equation of chord joining the point P and Q with eccentric angles α and β respectively for the hyperbola $\frac{x^2}{a^2}-\frac{y^2}{b^2}=1$ and that of joining points with eccentric angles α and β for its conjugate hyperbola.

Answer Keys

1. $x^2 + y^2 = 49$, $x^2 + y^2 = 25$ **2.** $4\sqrt{6}$ units **3.** $3x - 4y + 10 = 0$

4. $\pi/4$, $(2 + 2\sqrt{2}, 3 + 2\sqrt{2})$, $(2 - 2\sqrt{2}, 3 - 2\sqrt{2})$ **5.** $x = 1 + 2\sec\theta$; $y = -3 + 2\tan\theta$

6. $\frac{x}{a}\cos\left(\frac{\alpha-\beta}{2}\right) - \frac{y}{b}\sin\left(\frac{\alpha+\beta}{2}\right) = \cos\left(\frac{\alpha+\beta}{2}\right)$; $\frac{y}{b}\cos\left(\frac{\alpha-\beta}{2}\right) - \frac{x}{a}\sin\left(\frac{\alpha+\beta}{2}\right) = \cos\left(\frac{\alpha+\beta}{2}\right)$.

TEXTUAL EXERCISE–2 (OBJECTIVE)

1. If $P(\theta_1)$ and $Q(\theta_2)$ are the extremities of any focal chord of the hyperbola $\frac{x^2}{a^2} - \frac{y^2}{b^2} = 1$, then $\cos^2\frac{\theta_1+\theta_2}{2} = \lambda\cos^2\frac{\theta_1-\theta_2}{2}$, where 'λ' is equal to

(a) $\frac{a^2+b^2}{a^2}$ (b) $\frac{a^2+b^2}{b^2}$

(c) $\frac{a^2+b^2}{ab}$ (d) $\frac{a^2+b^2}{2ab}$

2. Locus of the mid-point of the chords of the hyperbola $x^2 - y^2 = a^2$, that touch the parabola $y^2 = 4ax$ is

(a) $x^2(x - a) = y^3$

(b) $y^2(x - a) = x^3$

(c) $x^3(x - a) = y^2$

(d) $y^3(x - a) = x^2$

3. If focal chord of hyperbola $\frac{x^2}{a^2} - \frac{y^2}{b^2} = 1$ joins the points $P\left(\frac{\pi}{3}\right)$ and $Q\left(-\frac{\pi}{6}\right)$ on it, then its eccentricity is

(a) $\frac{\sqrt{3}+2}{2}$ (b) $\frac{\sqrt{3}+1}{2}$

(c) $\sqrt{3}$ (d) None of these

4. Auxiliary circle of hyperbola $7x^2 - 5y^2 = 35$ is

(a) $x^2 + y^2 = 7$

(b) $x^2 + y^2 = 5$

(c) $x^2 + y^2 = 35$

(d) None of these

5. Area of auxiliary circle of hyperbola $x^2 - 10y^2 = 20$ is given by

(a) 20π

(b) 10π

(c) 5π

(d) None of these

6. The length of tangent to the auxiliary circle of hyperbola $16x^2 - 9y^2 = 144$ which is a chord of auxiliary circle of conjugate hyperbola is

(a) 7 (b) $\sqrt{7}$

(c) $2\sqrt{7}$ (d) None of these

7. Eccentric angle of point $\left(2\sqrt{\frac{7}{3}}, 1\right)$ on hyperbola $3x^2 - 7y^2 = 21$ is

(a) $\pi/3$ (b) $\pi/6$

(c) $\pi/4$ (d) None of these

8. Parametric equation of hyperbola $2x^2 - y^2 + 8x + 6y = 11$ is

(a) $x = -2 + \sqrt{5}\tan\theta$, $y = 3 + \sqrt{10}\sec\theta$

(b) $x = -2 - \sqrt{10}\sec\theta$, $y = 3 + \sqrt{5}\tan\theta$

(c) $x = -2 + \sqrt{5}\sec\theta$, $y = 3 + \sqrt{10}\tan\theta$

(d) None of these

9. The co-ordinates of the point where the chord joining the points $P(\pi/6)$, $Q(\pi/3)$ of hyperbola $9x^2 - 16y^2 = 144$ cuts x-axis are

(a) $(4(\sqrt{3} + 1), 0)$

(b) $(4\sqrt{3}, 0)$

(c) $(4(\sqrt{3} - 1), 0)$

(d) None of these

Answer Keys

1. (a) **2.** (b) **3.** (b) **4.** (b) **5.** (a) **6.** (c) **7.** (b) **8.** (c) **9.** (c)

■ POSITION OF A POINT WITH RESPECT TO HYPERBOLA

The point $P(x_1, y_1)$ lies outside (toward focus), on or inside the (towards centre) hyperbola $\frac{x^2}{a^2}-\frac{y^2}{b^2}=1$ according as $\frac{x_1^2}{a^2}-\frac{y_1^2}{b^2}-1 >, = \text{ or } < 0$

Let a point $Q \equiv (x_1, y_2)$ on the hyperbola.

Draw PL perpendicular to x-axis. If P lies outsides

Clearly, $PL > QL \Rightarrow y_1 > y_2$

$$\Rightarrow \frac{y_1^2}{b^2} > \frac{y_2^2}{b^2}$$

$$\Rightarrow -\frac{y_1^2}{b^2} < -\frac{y_2^2}{b^2}$$

$$\Rightarrow \frac{x_1^2}{a^2}-\frac{y_1^2}{b^2} < \frac{x_1^2}{a^2}-\frac{y_2^2}{b^2} \Rightarrow \frac{x_1^2}{a^2}-\frac{y_1^2}{b^2} < 1$$

$$\left(\because Q(x_1, y_2) \text{ lies on } \frac{x^2}{a^2}-\frac{y^2}{b^2}=1\right)$$

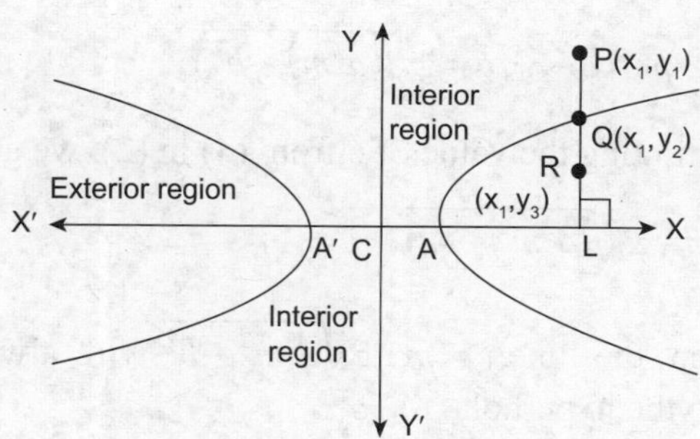

FIGURE 6.35

$$\Rightarrow \frac{x_1^2}{a^2}-\frac{y_1^2}{b^2}-1<0$$

$\Rightarrow$ P point lies inside the hyperbola if $S_1 < 0$. Similarly for point R lying outside the hyperbola; $RL < QL$.

$$\Rightarrow \frac{x_1^2}{a^2}-\frac{y_3^2}{b^2}-1>0$$

$\Rightarrow$ R point lies in the exterior region the hyperbola.

Thus a point (x_1, y_1) lies outside, on or inside the hyperbola $\frac{x^2}{a^2}-\frac{y^2}{b^2}=1$ according as $\frac{x_1^2}{a^2}-\frac{y_1^2}{b^2}-1 >, = \text{ or } < 0$.

ILLUSTRATION 23: Find the position of the point (5, – 4) relative to the hyperbola $9x^2 - y^2 = 1$.

SOLUTION: Since $9(5)^2-(-4)^2-1=225-16-1=208>0$,

So the point (5, –4) lies outside the hyperbola $9x^2 - y^2 = 1$.

■ POSITION OF A STRAIGHT LINE WITH RESPECT TO HYPERBOLA

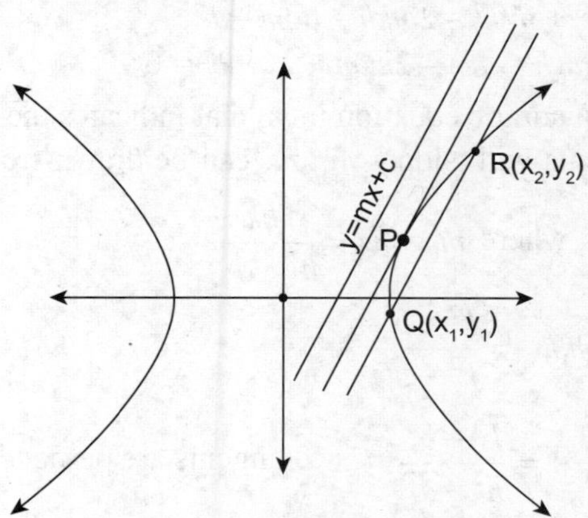

FIGURE 6.36

Let the hyperbola be $\frac{x^2}{a^2}-\frac{y^2}{b^2}=1$ ....(1)

and the given line be $y = mx + c$ ....(2)

Eliminating y from (1) and (2),

$$\frac{x^2}{a^2}-\frac{(mx+c)^2}{b^2}=1$$

$$\Rightarrow b^2x^2 - a^2(mx + c)^2 = a^2b^2$$

$$\Rightarrow (a^2m^2 - b^2)x^2 + 2mca^2x + a^2(b^2 + c^2) = 0 \quad ....(3)$$

Above equation being a quadratic in x gives two values of x. It shows that every straight line will cut the hyperbola in two points, may be real and distinct or coincident or imaginary according as discriminant of equation (3) >, =, < 0.

If $c^2 > a^2m^2 - b^2 \Rightarrow$ Line is secant to hyperbola

$c^2 = a^2m^2 - b^2$ ...$\Rightarrow$ Line is tangent to hyperbola

$c^2 > a^2m^2 - b^2$ $\Rightarrow$ Line has no contact with hyperbola

■ CONDITION OF TANGENCY OF A STRAIGHT LINE WITH RESPECT TO HYPERBOLA

Clearly, the condition of tangency for line $y = mx + c$ and hyperbola given in equation (i) is

$c^2 = a^2m^2 - b^2$ or $c = \pm\sqrt{(a^2m^2 - b^2)}$ ....(4)

Substituting the value of c from (4) in (2), we get

$y = mx \pm \sqrt{(a^2m^2 - b^2)}$

Hence the line $y = mx \pm \sqrt{a^2m^2 - b^2}$ will always be tangent to the hyperbola.

■ EQUATION OF TANGENT IN SLOPE FORM

The equations of tangents of slope m to the hyperbola $\frac{x^2}{a^2} - \frac{y^2}{b^2} = 1$ are given by $y = mx \pm \sqrt{(a^2m^2 - b^2)}$.

The co-ordinates of the points of contact are $\left(-\frac{a^2m}{c}, -\frac{b^2}{c}\right)$, where $c = \pm\sqrt{a^2m^2 - b^2}$

Put $c^2 = a^2m^2 - b^2$ and $c^2 + b^2 = a^2m^2$ in (3) we get

$c^2x^2 + 2mca^2x + a^4m^2 = 0$

$\Rightarrow$ $(cx + a^2m)^2 = 0 \Rightarrow x = -\frac{a^2m}{c}$

from $y = mx + c$; $y = -\frac{a^2m^2}{c} +$

$\Rightarrow$ $y = \frac{-a^2m^2 + c^2}{c}$

$\Rightarrow$ $y = -b^2/c$

Thus the co-ordinate of points of contact are $\left(\frac{-a^2m}{c}, \frac{-b^2}{c}\right)$ where $c = \pm\sqrt{a^2m^2 - b^2}$.

NOTE

1. *The equations of tongents of slope 'm' to the hyperbola* $\frac{x^2}{a^2} - \frac{y^2}{b^2} = 1$ *are given by* $y = mx + c$ *where* $c^2 = a^2m^2 - b^2$ *and the points of contact are given by* $\left(\frac{-a^2m}{c}, \frac{-b^2}{c}\right)$
2. *The equations of tangents of slope 'm' to hyperbola* $\frac{y^2}{a^2} - \frac{x^2}{b^2} = 1$ *are given by* $y = mx + c$; $c^2 = a^2 - b^2m^2$ *and the points of contact are* $\left(\frac{b^2m}{c}, \frac{a^2}{c}\right)$.

■ DIRECTOR CIRCLE

The locus of the point of intersection of the tangents to the hyperbola $\frac{x^2}{a^2} - \frac{y^2}{b^2} = 1$, which are perpendicular to each other is called director circle.

Consider the tangent with slope 'm' of above hyperbola as $y = mx + \sqrt{(a^2m^2 - b^2)}$

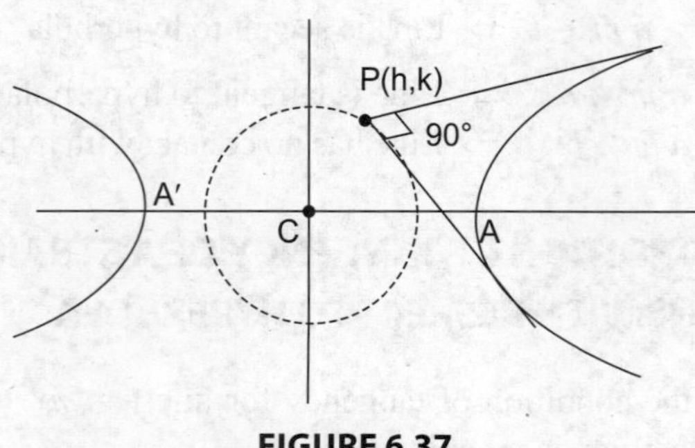

FIGURE 6.37

If it passes through (h, k), then $k = mh + \sqrt{(a^2m^2 - b^2)}$

or $(k - mh)^2 = a^2m^2 - b^2$

$\Rightarrow$ $k^3 + m^2h^2 - 2mkh = a^2m^2 - b^2$

$\Rightarrow$ $m^2(h^2 - a^2) - 2hkm + k^2 + b^2 = 0$

It is quadratic equation in m, that indicates the fact that two tangents with slopes m_1, m_2 can be drawn from (h, k) hyperbola, where $m_1 + m_2 = \frac{hk}{h^2 \quad a^2}$ .....(i)

$\therefore$ $m_1m_2 = \frac{k^2 + b^2}{h^2 - a^2}$ .....(ii)

$-1 = \frac{k^2 + b^2}{h^2 - a^2}$ ($\because$ tangents are perpendicular)

$\Rightarrow$ $-h^2 + a^2 = k^2 + b^2$ or $h^2 + k^2 = a^2 - b^2$

Hence locus of $P(h, k)$ is $x^2 + y^2 = a^2 - b^2 (a > b)$

Alternative Method: If tangents

$y = mx + \sqrt{(a^2m^2 - b^2)}$ ....(1)

and $$y=-\frac{x}{m}+\sqrt{\left\{a^2\left(-\frac{1}{m}\right)^2-b^2\right\}} \qquad ....(2)$$

touch the hyperbola and intersects at right angles

$\therefore$ from (1), $y-mx=\sqrt{(a^2m^2-b^2)}$ ....(3)

(2) can be rewritten as

$$x+my=\sqrt{(a^2-b^2m^2)} \qquad ....(4)$$

Squaring and adding (3) and (4), we get,

$(y-mx)^2+(x+my)^2=a^2m^2-b^2+a^2-b^2m^2$

$(1+m^2)(x^2+y^2)=(1+m^2)(a^2-b^2)$

Hence $x^2+y^2=a^2-b^2$ is the director circle of the hyperbola.

NOTE

1. *For director circle of* $\frac{x^2}{a^2}-\frac{y^2}{b^2}=1$, *to be a real circle of non-zero radius a must be greater than b.*

   If a < b then director circle $x^2+y^2=a^2-b^2$ *is an imaginary circle.*

2. *The equation of director circle of* $-\frac{x^2}{a^2}+\frac{y^2}{b^2}=1$, is $x^2+y^2=b^2-a^2 (b>a)$.

   If b < a, then director circle is an imaginary circle.

ILLUSTRATION 24: Prove that the straight line $lx+my+n=0$ touches the hyperbola $\frac{x^2}{a^2}-\frac{y^2}{b^2}=1$ if $a^2l^2-b^2m^2=n^2$.

SOLUTION: The given line is $lx+my+n=0$ or $y=-\frac{l}{m}x-\frac{n}{m}$

Comparing this line with $y=Mx+c$ ....(1)

$\therefore$ $M=-\frac{l}{m}$ *and* $c=-\frac{n}{m}$

This line (1) will touch the hyperbola

$\frac{x^2}{a^2}-\frac{y^2}{b^2}=1$ *if* $c^2=a^2M^2-b^2$

$\Rightarrow$ $\frac{n^2}{m^2}=\frac{a^2l^2}{m^2}-b^2$ or $a^2l^2-b^2m^2=n^2$

ILLUSTRATION 25: Show that the line $x\cos\alpha+y\sin\alpha=p$ touches the hyperbola $\frac{x^2}{a^2}-\frac{y^2}{b^2}=1$ if $a^2\cos^2\alpha-b^2\sin^2\alpha=p^2$.

SOLUTION: The given line is $x\cos\alpha+y\sin\alpha=p$

$\Rightarrow$ $y\sin\alpha=-x\cos\alpha+p \Rightarrow y=-x\cot\alpha+p\,\text{cosec}\,\alpha$

Comparing this line with $y=mx+c$

$\Rightarrow$ $m=-\cot\alpha$, $c=p\,\text{cosec}\,\alpha$; substituting in the condition of tangency $c^2=a^2m^2-b^2$, we get $p^2\text{cosec}^2\alpha=a^2\cot^2\alpha-b^2$ or $p^2=a^2\cos^2\alpha-b^2\sin^2\alpha$

ILLUSTRATION 26: For what value of λ does the line $y=2x+\lambda$ touches the hyperbola $16x^2-9y^2=144$?

SOLUTION: Equation of hyperbola is $16x^2-9y^2=144$ or $\frac{x^2}{9}-\frac{y^2}{16}=1$

Comparing this with standard equation, we get $a^2=9$, $b^2=16$ and comparing the line $y=2x+\lambda$ with $y=mx+c$, we find $m=2$ and $c=\lambda$

$\therefore$ If the line $y = 2x + \lambda$ touches the hyperbola $16x^2 - 9y^2 = 144$, then $c^2 = a^2m^2 - b^2$

$\Rightarrow \lambda^2 = 9(2)^2 - 16 = 20 \therefore \lambda = \pm 2\sqrt{5}$.

ILLUSTRATION 27: Find the equation of the tangent to the hyperbola $x^2 - 4y^2 = 36$ which is perpendicular to the line $x - y + 4 = 0$.

SOLUTION: Let m be the slope of the tangent. Since the tangent is perpendicular to the line $x - y + 4 = 0$, therefore its slope m must be -1

since equation of hyperbola is $\frac{x^2}{36} - \frac{y^2}{9} = 1$

Comparing this with standard equation, we get $a^2 = 36$ and $b^2 = 9$

Substituting the values in $y = mx \pm \sqrt{a^2m^2 - b^2}$, we get the equation of tangents

$$y = (-1)x \pm \sqrt{36 \times (-1)^2 - 9}$$

$$\Rightarrow y = -x \pm \sqrt{27} \text{ or } x + y \pm 3\sqrt{3} = 0$$

ILLUSTRATION 28: If the line $y = mx + \sqrt{a^2m^2 - b^2}$ touches the hyperbola $\frac{x^2}{a^2} - \frac{y^2}{b^2} = 1$ at the point $(a \sec\theta, b \tan\theta)$ then show that $\theta = \sin^{-1}\left(\frac{b}{am}\right)$.

SOLUTION: Since $(a \sec\theta, b \tan\theta)$ lies on $y = mx + \sqrt{a^2m^2 - b^2}$

$$\therefore \quad b\tan\theta = am\sec\theta + \sqrt{a^2m^2 - b^2}$$

$$\Rightarrow (b\tan\theta - am\sec\theta)^2 = a^2m^2 - b^2 \Rightarrow b^2\tan^2\theta + a^2m^2\sec^2\theta - 2abm\tan\theta\sec\theta = a^2m^2 - b^2$$

$$\Rightarrow a^2m^2\tan^2\theta - 2abm\tan\theta\sec\theta + b^2\sec^2\theta = 0$$

or $a^2m^2\sin^2\theta - 2abm\sin\theta + b^2 = 0$; which is a perfect square $(am \sin\theta - \text{b})^2 = 0$

$$\therefore \quad \sin\theta = \left(\frac{b}{am}\right); \therefore \theta = \sin^{-1}\left(\frac{b}{am}\right)$$

Aliter: tangent at $(\text{a} \sec\theta, \text{b} \tan\theta)$ is $\frac{x}{a}\sec\theta - \frac{y}{b}\tan\theta = 1$ which has slope.

$$\Rightarrow m = \frac{b}{a}\sec\theta\cot\theta = \frac{b}{a\sin\theta} \qquad \Rightarrow \sin\theta = \frac{b}{am} \quad \Rightarrow \quad \theta = \sin^{-1}\left(\frac{b}{am}\right)$$

ILLUSTRATION 29: Tangents are drawn from the point (α, β) to the hyperbola $3x^2 - 2y^2 = 6$ and are inclined at angles θ and ϕ to the x-axis. If $\tan\phi . \tan\theta = 2$, prove that $\beta^2 = 2\alpha^2 - 7$.

SOLUTION: Given hyperbola is $3x^2 - 2y^2 = 6$ i.e., $\frac{x^2}{2} - \frac{y^2}{3} = 1$

Let tangents from the point (α, β) be $y = mx \pm \sqrt{a^2m^2 - b^2}$ i.e., $(y - mx)^2 = a^2m^2 - b^2$

$\Rightarrow (\beta - m\alpha)^2 = 2m^2 - 3$ $\qquad (\because a^2 = 2 \text{ and } b^2 = 3)$

$\Rightarrow m^2\alpha^2 + \beta^2 - 2m\alpha\beta - 2m^2 + 3 = 0$

$\Rightarrow m^2(\alpha^2 - 2) - 2\alpha\beta m + \beta^2 + 3 = 0$; let m_1, m_2 be the roots of the equation

$\Rightarrow m_1.m_2 = \frac{\beta^2 + 3}{\alpha^2 - 2} = \tan\theta . \tan\phi = 2$ $\qquad (\because \tan\phi . \tan\theta = 2 \text{ given})$

$\therefore \beta^2 + 3 = 2(\alpha^2 - 2)$

$\Rightarrow \beta^2 = 2\alpha^2 - 7$

EQUATION OF TANGENT TO HYPERBOLA IN CARTESIAN FORM

Equation of tangent at point (x_1, y_1)

The equation of the tangent to the hyperbola $\frac{x^2}{a^2}-\frac{y^2}{b^2}=1$ at (x_1, y_1) is given as $\frac{xx_1}{a^2}-\frac{yy_1}{b^2}=1$.

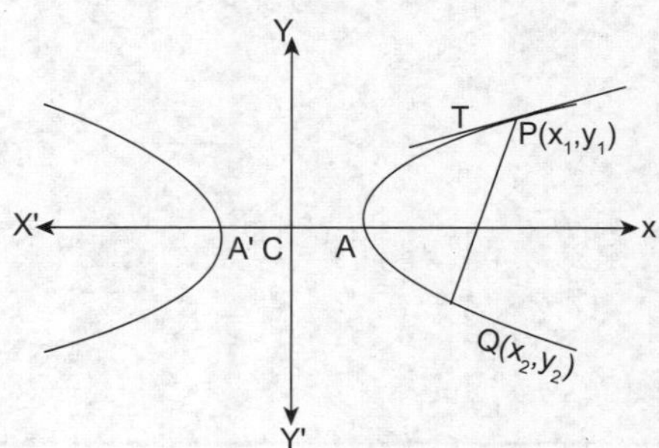

FIGURE 6.38

Proof: Let $P \equiv (x_1, y_1)$ and $Q \equiv (x_2, y_2)$ be any two points on hyperbola.

In the article "chord of hyperbola" we have obtained the equation of chord PQ as

$$y-y_1=\frac{b^2}{a^2}\cdot\frac{(x_1+x_2)}{(y_1+y_2)}(x-x_1) \qquad \text{....(i)}$$

Now as $Q \to P$ chord PQ $\to$ tangent at P

i.e., $x_2 \to x_1$ and $y_2 \to y_1$ then equation (i) becomes

$$y-y_1=\frac{b^2}{a^2}\cdot\frac{(2x_1)}{(2y_1)}(x-x_1) \text{ or } \frac{yy_1-y_1^2}{b^2}=\frac{xx_1-x_1^2}{a^2}$$

or $\frac{xx_1}{a^2}-\frac{yy_1}{b^2}=\frac{x_1^2}{a^2}-\frac{y_1^2}{b^2}$

Which is required equation of tangent at (x_1, y_1).

NOTE

The equation of tangent at (x_1, y_1) can be obtained by replacing x^2 by xx_1, y^2 by yy_1, x by $\frac{x+x_1}{2}$, y by $\frac{y+y_1}{2}$ and xy by $\frac{xy_1+x_1y}{2}$.
This method is applied to all conics whose equation is in form of polynomial of second degree in x and y.

EQUATION OF TANGENT TO HYPERBOLA IN PARAMETRIC FORM

The equation of tangent to the hyperbola $\frac{x^2}{a^2}-\frac{y^2}{b^2}=1$ at $(a\sec\phi, b\tan\phi)$ is $\frac{x}{a}\sec\phi-\frac{y}{b}\tan\phi=1$

Since equation of tangent at (x_1, y_1) is $\frac{xx_1}{a^2}-\frac{yy_1}{b^2}=1$

Now replacing x_1 by a sec ϕ and y_1 by b tan ϕ, we get

$\frac{x}{a}\sec\phi-\frac{y}{b}\tan\phi=1$.

ILLUSTRATION 30: Prove that the point of intersection of tangents at 'θ' and 'ϕ' on the hyperbola $\frac{x^2}{a^2}-\frac{y^2}{b^2}=1$ is

$$\left(\frac{a\cos\left(\frac{\theta-\phi}{2}\right)}{\cos\left(\frac{\theta+\phi}{2}\right)}, \frac{b\sin\left(\frac{\theta+\phi}{2}\right)}{\cos\left(\frac{\theta+\phi}{2}\right)}\right).$$

SOLUTION : To find the point of intersection of tangents. Let the point of intersection of tangents at θ and ϕ be (x, y)

Chord joining θ and ϕ is chord of contact of hyperbola w.r.t., $P(x, y)$, thus writing equation of chord QR as a chord and chord of contact generated from tangents drawn from $P(x_1, y_1)$ and comparing the two equations.

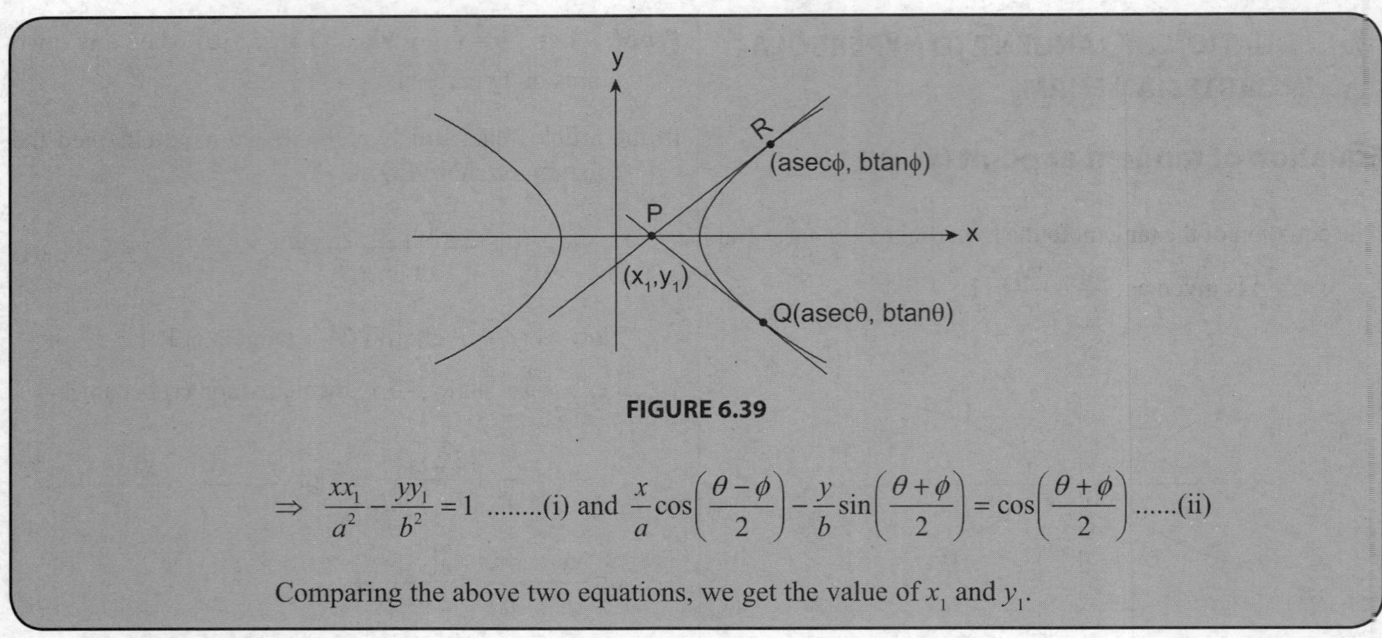

FIGURE 6.39

$$\Rightarrow \frac{xx_1}{a^2} - \frac{yy_1}{b^2} = 1 \text{ ....(i) and } \frac{x}{a}\cos\left(\frac{\theta-\phi}{2}\right) - \frac{y}{b}\sin\left(\frac{\theta+\phi}{2}\right) = \cos\left(\frac{\theta+\phi}{2}\right) \text{ ......(ii)}$$

Comparing the above two equations, we get the value of x_1 and y_1.

Memorising method:

$\because$ Equations of chord joining 'θ' and 'ϕ' is

$$\frac{x}{a}\cos\left(\frac{\theta-\phi}{2}\right) - \frac{y}{b}\sin\left(\frac{\theta+\phi}{2}\right) = \cos\left(\frac{\theta+\phi}{2}\right)$$

or $$\frac{x}{a}\left\{\frac{\cos\left(\frac{\theta-\phi}{2}\right)}{\cos\left(\frac{\theta+\phi}{2}\right)}\right\} - \frac{y}{b}\left\{\frac{\sin\left(\frac{\theta+\phi}{2}\right)}{\cos\left(\frac{\theta+\phi}{2}\right)}\right\} = 1$$

or $$\frac{x}{a^2}\left\{\frac{a\cos\left(\frac{\theta-\phi}{2}\right)}{\cos\left(\frac{\theta+\phi}{2}\right)}\right\} - \frac{y}{b^2}\left\{\frac{b\sin\left(\frac{\theta+\phi}{2}\right)}{\cos\left(\frac{\theta+\phi}{2}\right)}\right\} = 1$$

i.e., $$\left\{\frac{a\cos\left(\frac{\theta-\phi}{2}\right)}{\cos\left(\frac{\theta+\phi}{2}\right)}, \frac{b\sin\left(\frac{\theta+\phi}{2}\right)}{\cos\left(\frac{\theta+\phi}{2}\right)}\right\}.$$

■ EQUATION OF NORMAL TO HYPERBOLA IN CARTESIAN FORM

The equation of the normal to the hyperbola $\frac{x^2}{a^2} - \frac{y^2}{b^2} = 1$ at (x_1, y_1) is $\frac{a^2x}{x_1} + \frac{b^2y}{y_1} = a^2 + b^2$ where (x_1, y_1) lies on the hyperbola.

Since the equation of tangent to the hyperbola $\frac{x^2}{a^2} - \frac{y^2}{b^2} = 1$ at (x_1, y_1) is $\frac{xx_1}{a^2} - \frac{yy_1}{b^2} = 1$.

The slope of tangent at $(x_1, y_1) = \frac{b^2x_1}{a^2y_1}$.

$\therefore$ Slope of normal at $(x_1, y_1) = -\frac{a^2y_1}{b^2x_1}$

Hence the equation of normal at (x_1, y_1) is

$$y - y_1 = -\frac{a^2y_1}{b^2x_1}(x - x_1) \text{ or } \frac{a^2x}{x_1} + \frac{b^2y}{y_1} = \underbrace{a^2 + b^2}_{a^2e^2}$$

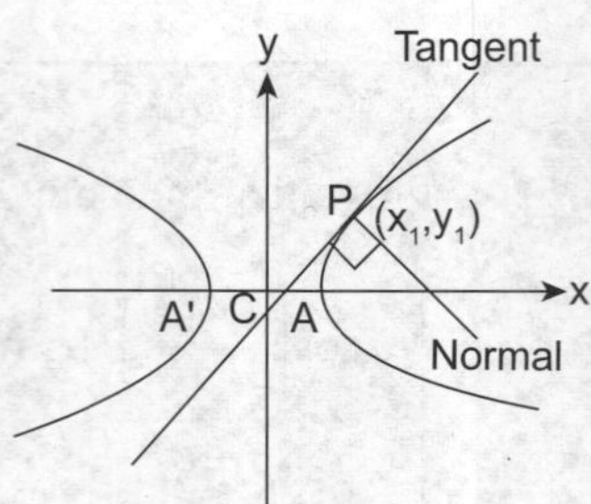

FIGURE 6.40

NOTE

The equation of normal at (x_1, y_1) *can also be obtained from the equation* $\frac{x - x_1}{a'x_1 + hy_1 + g} = \frac{y - y_1}{hx_1 + b'y_1 + f}$ ....(1)

a', b', g, f, h are obtained by comparing the given hyperbola with $a'x^2 + 2hxy + b'y^2 + 2gx + 2fy + c = 0$ ....(2)

Tip to Remember: *The denominator of (1) can easily be remembered by the first two rows of this determinant*

i.e., $\begin{vmatrix} a' & h & g \\ h & b' & f \\ g & f & c \end{vmatrix}$

Since first row is $a'(x_1)+h(y_1)+g(1)$ *and second row is* $h(x_1)+b'(y_1)+f(1)$

Here hyperbola is $\frac{x^2}{a^2}-\frac{y^2}{b^2}=1$ *or* $\frac{x^2}{a^2}-\frac{y^2}{b^2}-1=0$ ....(3)

Comparing (2) and (3), we get $a'=\frac{1}{a^2}, b'=-\frac{1}{b^2}, g=0, f=0, h=0$

From (1), equation of normal of (3) at (x_1, y_1) *is*

$$\frac{x-x_1}{\frac{1}{a^2}x_1+0+0}=\frac{y-y_1}{0-\frac{y_1}{b^2}+0} \text{ or } \frac{a^2x}{x_1}-a^2=-\frac{b^2y}{y_1}+b^2 \text{ or } \frac{a^2x}{x_1}+\frac{b^2y}{y_1}=a^2+b^2.$$

EQUATION OF NORMAL TO HYPERBOLA IN PARAMETRIC FORM

The equation of normal at $(a\sec\phi, b\tan\phi)$ to the hyperbola $\frac{x^2}{a^2}-\frac{y^2}{b^2}=1$ is $ax\cos\phi+by\cot\phi=a^2+b^2=a^2e^2$

Since the equation of normals of the hyperbola $\frac{x^2}{a^2}-\frac{y^2}{b^2}=1$ at (x_1, y_1) is

$$\frac{a^2x}{x_1}+\frac{b^2y}{y_1}=a^2+b^2 \quad ....(1)$$

Replacing x_1 by $a\sec\phi$ and y_1 by $b\tan\phi$ then (1) becomes

$$\frac{a^2x}{a\sec\phi}+\frac{b^2y}{b\tan\phi}=a^2+b^2$$

$$\Rightarrow \quad ax\cos\phi+by\cot\phi=a^2+b^2$$

is equation of normal at $(a\sec\phi, b\tan\phi)$.

ILLUSTRATION 31: Find the equation and the length of the common tangents to hyperbola $\frac{x^2}{a^2}-\frac{y^2}{b^2}=1$ and $\frac{y^2}{a^2}-\frac{x^2}{b^2}=1$ (for $a>b$).

SOLUTION: Tangent to $\frac{x^2}{a^2}-\frac{y^2}{b^2}=1$ is $y=mx\pm\sqrt{a^2m^2-b^2}$ ....(i)

(i) should also be tangent to other hyperbola $\frac{x^2}{(-b^2)}-\frac{y^2}{(-a^2)}=1$ ....(ii)

$$a^2m^2-b^2=(-b^2)m^2-(-a^2)$$

$$\Rightarrow (a^2+b^2)m^2=a^2+b^2 \Rightarrow m=\pm 1$$

Hence equations are $y=\pm x\pm\sqrt{a^2-b^2}$ and thus we can get the point of contact as well as length of common tangent $l=(a^2+b^2)\sqrt{\frac{2}{a^2-b^2}}$

Aliter: Tangents at $(a\sec\phi, b\tan\phi)$ on the 1*st* hyperbola is

$$\frac{x}{a}\sec\phi-\frac{y}{b}\tan\phi=1 \quad ....(1)$$

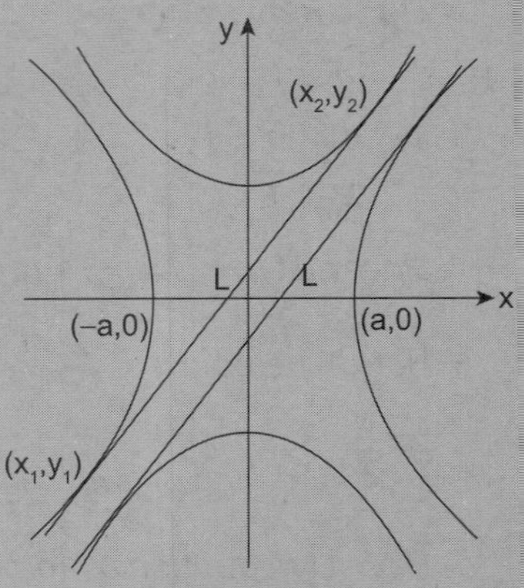

Similarly, tangents at any point (b tanθ, a secθ) on

2*nd* hyperbola is $\frac{y}{a}\sec\theta - \frac{x}{b}\tan\theta = 1$ ....(2)

If (1) and (2) are common tangents then they should be identical. Comparing the co-efficient of x and y

$$\Rightarrow \frac{\sec\theta}{a} = -\frac{\tan\phi}{b} \text{ or } \sec\theta = -\frac{a}{b}\tan\phi \quad ....(3)$$

$$\text{and } -\frac{\tan\theta}{b} = \frac{\sec\phi}{a} \text{ or } \tan\theta = -\frac{b}{a}\sec\phi \quad ....(4)$$

$$\because \sec^2\theta - \tan^2\theta = 1$$

$$\Rightarrow \frac{a^2}{b^2}\tan^2\phi - \frac{b^2}{a^2}\sec^2\phi = 1 \quad \{\text{from (3) and (4)}\}$$

$$\text{or } \frac{a^2}{b^2}\tan^2\phi - \frac{b^2}{a^2}(1+\tan^2\phi) = 1 \text{ or } \left(\frac{a^2}{b^2} - \frac{b^2}{a^2}\right)\tan^2\phi = 1 + \frac{b^2}{a^2}$$

$$\therefore \tan^2\phi = \frac{b^2}{a^2-b^2} \text{ and } \sec^2\phi = 1+\tan^2\phi = \frac{a^2}{a^2-b^2}$$

FIGURE 6.41

Hence the points of contact are $\left\{\pm\frac{a^2}{\sqrt{a^2-b^2}}, \pm\frac{b^2}{\sqrt{a^2-b^2}}\right\}$ and $\left\{\pm\frac{b^2}{\sqrt{a^2-b^2}}, \pm\frac{a^2}{\sqrt{a^2-b^2}}\right\}$.

Length of common tangent i.e., the distance between the above points is $\sqrt{2}\frac{(a^2+b^2)}{\sqrt{(a^2-b^2)}}$

and equation of common tangent on putting the values of sec ϕ and tanϕ in (1) is

$$\pm\frac{x}{\sqrt{a^2-b^2}} \mp \frac{y}{\sqrt{a^2-b^2}} = 1 \text{ or } x \mp y = \pm\sqrt{a^2-b^2}$$

ILLUSTRATION 32: Let PQ be chord joining the points ϕ_1 and ϕ_2 on the hyperbola $\frac{x^2}{a^2} - \frac{y^2}{b^2} = 1$. If $\phi_1 - \phi_2 = 2\alpha$, where α is constant, prove that PQ touches the hyperbola $\frac{x^2}{a^2}\cos^2\alpha - \frac{y^2}{b^2} = 1$.

SOLUTION: Given hyperbola is $\frac{x^2}{a^2} - \frac{y^2}{b^2} = 1$ ....(1)

Equation of the chord PQ to the hyperbola (1) is

$$\frac{x}{a}\cos\left(\frac{\phi_1-\phi_2}{2}\right) - \frac{y}{b}\sin\left(\frac{\phi_1+\phi_2}{2}\right) = \cos\left(\frac{\phi_1+\phi_2}{2}\right)$$

$$\Rightarrow \frac{x}{a}\cos\alpha - \frac{y}{b}\sin\left(\frac{\phi_1+\phi_2}{2}\right) = \cos\left(\frac{\phi_1+\phi_2}{2}\right) \quad (\text{Given } \phi_1 - \phi_2 = 2\alpha)$$

$$\text{i.e., } y = \frac{b}{a}\frac{\cos\alpha}{\sin\left(\frac{\phi_1+\phi_2}{2}\right)}x - \frac{b\cos\left(\frac{\phi_1+\phi_2}{2}\right)}{\sin\left(\frac{\phi_1+\phi_2}{2}\right)} \quad ....(2)$$

Comparing this line with $y = mx + c$

$$\therefore m = \frac{b}{a}.\frac{\cos\alpha}{\sin\left(\frac{\phi_1+\phi_2}{2}\right)} \text{ and } c = \frac{-b\cos\left(\frac{\phi_1+\phi_2}{2}\right)}{\sin\left(\frac{\phi_1+\phi_2}{2}\right)}$$

For line $y = mx + c$ to be a tangent on $\frac{x^2}{a^2}\cos^2\alpha - \frac{y^2}{b^2} = 1$, we have $c^2 = \frac{a^2}{\cos^2\alpha}m^2 - b^2$

Now consider R.H.S. $= \frac{a^2}{\cos^2\alpha}m^2 - b^2$

$$= \frac{a^2}{\cos^2\alpha} \times \frac{b^2\cos^2\alpha}{a^2\sin^2\left(\frac{\phi_1+\phi_2}{2}\right)} - b^2 = \frac{b^2}{\sin^2\left(\frac{\phi_1+\phi_2}{2}\right)} - b^2 = \frac{b^2\cos^2\left(\frac{\phi_1+\phi_2}{2}\right)}{\sin^2\left(\frac{\phi_1+\phi_2}{2}\right)} = \text{L.H.S.}$$

ILLUSTRATION 33: If SY and $S'Y'$ be perpendiculars drawn from foci to any tangent to a hyperbola, prove that Y and Y' lie on the auxiliary circle and that product of these perpendiculars is constant.

SOLUTION: Let hyperbola be $\frac{x^2}{a^2} - \frac{y^2}{b^2} = 1$ .....(1)

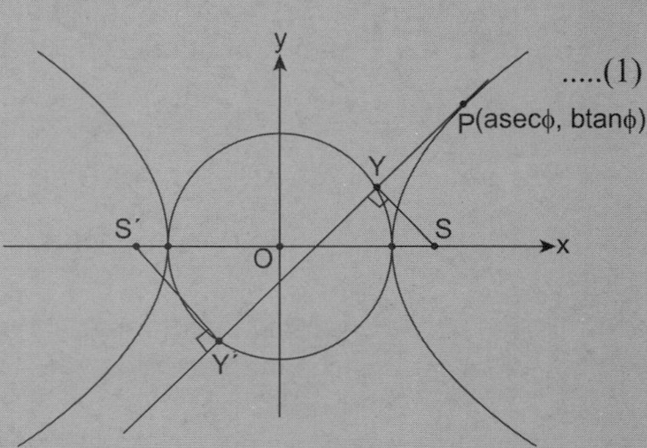

FIGURE 6.42

Tangent at P $(a\sec\phi, b\tan\phi)$ on (1) is

$$\frac{x}{a}\sec\phi - \frac{y}{b}\tan\phi = 1 \quad ....(2)$$

Equation of SY which is perpendicular to (1) and passes through focus S, i.e., $(ae, 0)$ is

$$y - 0 = -\frac{a\tan\phi}{b\sec\phi}(x - ae) \quad \text{i.e.,}$$

$$\frac{x}{b}\tan\phi + \frac{y}{a}\sec\phi = \frac{ae}{b}\tan\phi \quad ....(3)$$

∵ Lines (2) and (3) intersect at Y and in order to find its locus we have to eliminate ϕ between (2) and (3), for which squaring and adding (2) and (3), then

$$(x^2+y^2)\left(\frac{\sec^2\phi}{a^2} + \frac{\tan^2\phi}{b^2}\right) = 1 + \frac{a^2e^2}{b^2}\tan^2\phi$$

$$= 1 + \frac{a^2+b^2}{b^2}\tan^2\phi = (1+\tan^2\phi) + \frac{a^2}{b^2}\tan^2\phi$$

$$= \left(\sec^2\phi + \frac{a^2}{b^2}\tan^2\phi\right) = a^2\left(\frac{\sec^2\phi}{a^2} + \frac{\tan^2\phi}{b^2}\right)$$

∴ $x^2 + y^2 = a^2$ is the required locus auxiliary circle of hyperbola.

Similarly, the point Y' also lies on it. Again if p_1 and p_2 be the length of perpendiculars from S $(ae, 0)$ and $S'(-ae, 0)$ on the tangent (2), then

$$p_1p_2 = \frac{(e\sec\phi - 1).(e\sec\phi + 1)}{\sqrt{\left(\frac{\sec^2\phi}{a^2} + \frac{\tan^2\phi}{b^2}\right)}\sqrt{\left(\frac{\sec^2\phi}{a^2} + \frac{\tan^2\phi}{b^2}\right)}}$$

$$= \frac{a^2b^2(e^2\sec^2\phi - 1)}{b^2\sec^2\phi + a^2\tan^2\phi} = \frac{a^2b^2(e^2\sec^2\phi - 1)}{a^2(e^2-1)\sec^2\phi + a^2\tan^2\phi}$$

$$= \frac{b^2(e^2\sec^2\phi - 1)}{(e^2\sec^2\phi - 1)} = b^2 = \text{constant } [(\because b^2 = a^2(e^2-1))].$$

ILLUSTRATION 34: A normal to the hyperbola $\frac{x^2}{a^2}-\frac{y^2}{b^2}=1$ meets the axes in M and N and lines MP and NP are drawn perpendiculars to the axes meeting at P. Prove that the locus of P is the hyperbola $a^2x^2-b^2y^2=(a^2+b^2)^2$.

SOLUTION: The equation of normal at the point $(a\sec\phi, b\tan\phi)$ to the hyperbola $\frac{x^2}{a^2}-\frac{y^2}{b^2}=1$ is

$$ax\cos\phi+by\cot\phi=a^2+b^2 \quad \text{....(1)}$$

Let it intersect x-axis at M and y-axis at N as shown in figure

$$\Rightarrow\ M\left(\frac{a^2+b^2}{a}\sec\phi, 0\right) \text{ and } N\left(0, \frac{a^2+b^2}{b}\tan\phi\right).$$

FIGURE 6.43

Let the line through M and N perpendicular to axis meet at $P(h, k)$

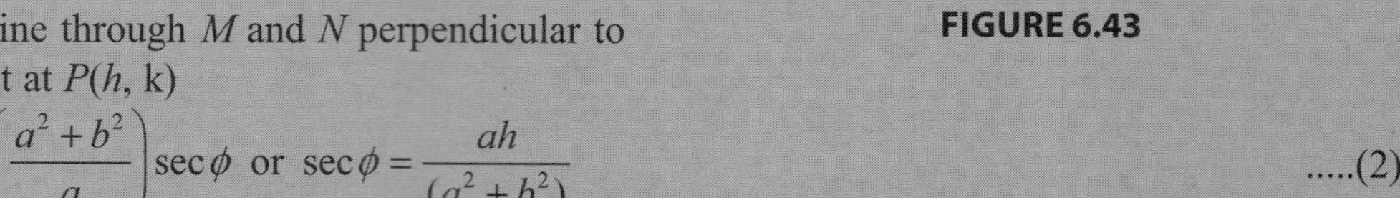

$$h=\left(\frac{a^2+b^2}{a}\right)\sec\phi \text{ or } \sec\phi=\frac{ah}{(a^2+b^2)} \quad \text{.....(2)}$$

$$\text{and } k=\left(\frac{a^2+b^2}{b}\right)\tan\phi \text{ or } \tan\phi=\frac{bk}{(a^2+b^2)} \quad \text{.... (3)}$$

The locus of the point of intersection of MP and NP will be obtained by eliminating ϕ from (2) and (3), As we have $\sec^2\phi-\tan^2\phi=1$

$$\Rightarrow \frac{a^2h^2}{(a^2+b^2)^2}-\frac{b^2k^2}{(a^2+b^2)^2}=1 \Rightarrow a^2x^2-b^2y^2=(a^2+b^2)^2\text{; is the required locus of } P.$$

ILLUSTRATION 35: Find the locus of the foot of perpendicular from the centre upon any normal to the hyperbola $\frac{x^2}{a^2}-\frac{y^2}{b^2}=1$.

SOLUTION: Normal at $P(a\sec\phi, b\tan\phi)$ is $ax\cos\phi+by\cot\phi=a^2+b^2$ ....(1)

and equation of line perpendicular to (1) and passing through origin is

$bx-ay\sin\phi=0$ ....(2)

Eliminating ϕ from (1) and (2), we will get the equation of locus of Q.

As from equation (2); we get $\sin\phi=\frac{bx}{ay}$

$$\therefore\ \cos\phi=\frac{\sqrt{(a^2y^2-b^2x^2)}}{ay} \text{ and}$$

$$\cot\phi=\frac{\sqrt{(a^2y^2-b^2x^2)}}{bx}$$

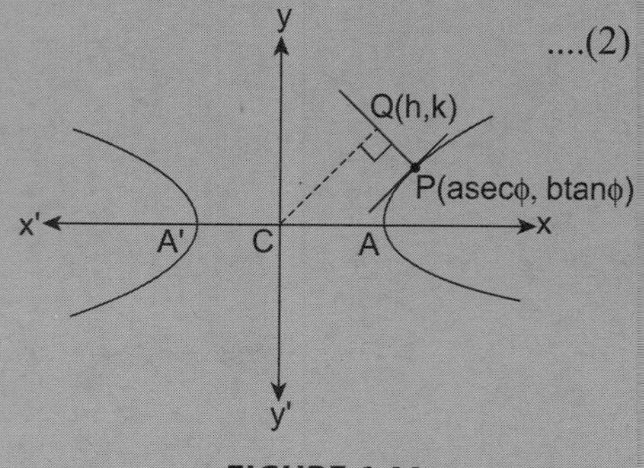

FIGURE 6.44

from (1),

$$ax\left(\frac{\sqrt{a^2y^2-b^2x^2}}{ay}\right)+by\left(\frac{\sqrt{a^2y^2-b^2x^2}}{bx}\right)=a^2+b^2$$

$$\Rightarrow (x^2+y^2)\sqrt{(a^2y^2-b^2x^2)}=(a^2+b^2)xy$$

which is the required locus.

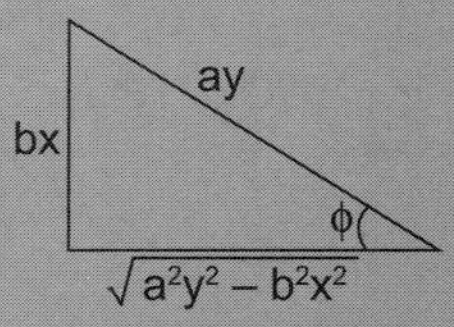

FIGURE 6.45

ILLUSTRATION 36: If the tangent at point $P(h, k)$ on the hyperbola $\frac{x^2}{a^2}-\frac{y^2}{b^2}=1$ cuts the circle $x^2+y^2=a^2$ at the point $P_1(x_1, y_1)$ and $P_2(x_2, y_2)$ then show that $\frac{1}{y_1}+\frac{1}{y_2}$ is always equal to $\frac{2}{k}$.

SOLUTION: Equation of tangent at $P(h, k)$ is $\frac{hx}{a^2}-\frac{ky}{b^2}=1$ where $\frac{h^2}{a^2}-\frac{k^2}{b^2}=1$

i.e., $b^2h^2-a^2k^2=a^2b^2$

Solving it with circle, we get $y^2+\left(1+\frac{yk}{b^2}\right)^2.\frac{a^4}{h^2}=a^2$

$\Rightarrow y^2(b^4h^2+a^4k^2)+2ya^4b^2k+a^2b^2(a^2b^2-b^2h^2)=0$

$\because$ Its roots are y_1 and y_2 thus

$$\frac{1}{y_1}+\frac{1}{y_2}=\frac{2a^4b^2k}{a^2b^2(b^2h^2-a^2b^2)}=\frac{2a^4b^2k}{a^2b^2a^2k^2}=\frac{2}{k}.$$

ILLUSTRATION 37: An ellipse has eccentricity 1/2 and one focus at the point $P(1/2, 1)$. Its one directrix is the common tangent, (nearer to the point P), to the circle $x^2+y^2=1$ and the hyperbola $x^2-y^2=1$. Find the equation of the ellipse in the standard form.

SOLUTION: Any point on the hyperbola $x^2-y^2=1$ is $(\sec\theta, \tan\theta)$.

Then tangent at $(\sec\theta, \tan\theta)$ to $x^2-y^2=1$ is $x\sec\theta-y\tan\theta=1$ ...(1)

This will also be a tangent to $x^2+y^2=1$ if

Length of perpendicular from (0, 0) to the equation (1) is equal to radius

$$\Rightarrow 1=\frac{1}{\sqrt{\sec^2\theta+\tan^2\theta}} \quad \Rightarrow \sec^2\theta+\tan^2\theta=1$$

$\therefore$ Putting for θ in (1), the common tangents are $x=1$ and $x+1=0$.

$x=1$ is nearer to $F\left(\frac{1}{2},1\right)$, thus the equation of directrix is $x-1=0$.

$\therefore$ The ellipse has the focus at $=\left(\frac{1}{2},1\right)$.

Corresponding directrix is $x-1=0$ and $e=\frac{1}{2}$.

$\therefore$ By focus – directrix property, the equation of the ellipse is $\left(x-\frac{1}{2}\right)^2+(y-1)^2=\left(\frac{1}{2}\right)^2\left(\frac{x-1}{\sqrt{1}}\right)^2$

i.e., $3x^2+4y^2-2x-8y+4=0$.

ILLUSTRATION 38: Chords of the hyperbola $\frac{x^2}{a^2}-\frac{y^2}{b^2}=1$ are tangents to the circle drawn on the line joining the foci as diameter. Find the locus of the point of intersection of tangents at the extremities of the chords.

SOLUTION: Let the end points of chord of hyperbola $\frac{x^2}{a^2}-\frac{y^2}{b^2}=1$ is $P(a \sec \theta, b \tan \theta)$ and $(\alpha \sec \phi, b \tan \phi)$

Equation of chord joining points P and Q is $\frac{x}{a}\cos\left(\frac{\theta-\phi}{2}\right)-\frac{y}{b}\sin\left(\frac{\theta+\phi}{2}\right)=\cos\left(\frac{\theta+\phi}{2}\right)$ ...(i)

According to the question, equation of line in (i) is tangent on the given circle with centre (0, 0).

$$\Rightarrow \frac{\left|0-0-\cos\left(\frac{\theta+\phi}{2}\right)\right|}{\sqrt{\frac{\cos^2}{a^2}\left(\frac{\theta-\phi}{2}\right)+\frac{1}{b^2}\sin^2\left(\frac{\theta+\phi}{2}\right)}}=ae$$

$$\Rightarrow \cos^2\left(\frac{\theta+\phi}{2}\right)=a^2e^2$$

$$\left[\frac{1}{a^2}\cos^2\left(\frac{\theta-\phi}{2}\right)+\frac{1}{b^2}\sin^2\left(\frac{\theta+\phi}{2}\right)\right] \quad ...(ii)$$

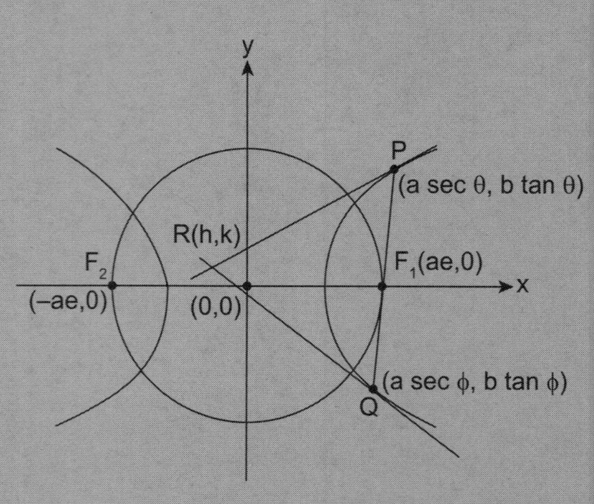

FIGURE 6.46

Point of intersection of tangents (*h*, k) at point $P(\theta)$ and $Q(\phi)$ is $\frac{h}{a}=\frac{\cos\left(\frac{\theta-\phi}{2}\right)}{\cos\left(\frac{\theta+\phi}{2}\right)}$ and

$$\frac{k}{b}=\frac{\sin\left(\frac{\theta+\phi}{2}\right)}{\cos\left(\frac{\theta+\phi}{2}\right)}$$

Considering equation (ii), $cos^2\left(\frac{\theta+\phi}{2}\right)=a^2e^2\left[\frac{1}{a^2}\cos^2\left(\frac{\theta-\phi}{2}\right)+\frac{1}{b^2}\sin^2\left(\frac{\theta+\phi}{2}\right)\right]$

$$\Rightarrow 1=a^2e^2\left[\frac{1}{a^2}\frac{\cos^2\left(\frac{\theta-\phi}{2}\right)}{\cos^2\left(\frac{\theta+\phi}{2}\right)}+\frac{1}{b^2}\frac{\sin^2\left(\frac{\theta+\phi}{2}\right)}{\cos^2\left(\frac{\theta+\phi}{2}\right)}\right]$$

$$\Rightarrow \frac{h^2}{a^4}+\frac{k^2}{b^4}=\frac{1}{a^2+b^2} \quad \therefore \frac{x^2}{a^4}+\frac{y^2}{b^4}=\frac{1}{a^2+b^2}$$

Chord of contact of hyperbola generated By point (*hk*) is given as

$T=0$

i.e., $\frac{hx}{a^2}-\frac{ky}{b^2}-1=0$ (i)

of it touches the circle with diameter F_2 $(-al, 0)$ and F_1 $(-al, 0)$

⇒ perpendicular distance form centre (0, 0) to above to sadius of circle i.e., al

$$\frac{1}{\sqrt{\left(\frac{h}{a^2}\right)^2+\left(\frac{k}{b^2}\right)}}=al$$

$$\Rightarrow \frac{h^2}{a^4}+\frac{k^2}{b^4}=\frac{1}{a^2e^2}=\frac{1}{a^2+b^2}$$

$\therefore$ the equation of low of R in $\frac{x^2}{a^4}+\frac{y^2}{b^4}=\frac{1}{a^2+b^2}$, which is an ellipse, concentric with both circle and hyperbola given.

ILLUSTRATION 39: Let 'p' be the perpendicular distance from the centre C of the hyperbola $\frac{x^2}{a^2}-\frac{y^2}{b^2}=1$ to the tangent drawn at a point R on the hyperbola. If S and S' are the two foci of the hyperbola, then show that $(RS + RS') = 4a^2\left(1+\frac{b^2}{p^2}\right)$.

SOLUTION: Equation of tangent at point R is, $\frac{x}{a}\sec\theta+\frac{y}{b}\tan\theta-1=0$

The perpendicular distance P from centre

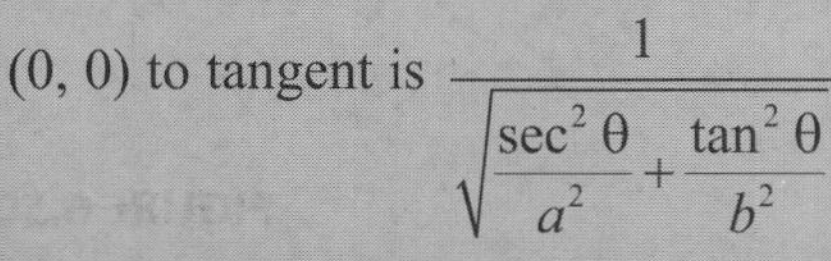

(0, 0) to tangent is $\frac{1}{\sqrt{\frac{\sec^2\theta}{a^2}+\frac{\tan^2\theta}{b^2}}}$

$$\Rightarrow \frac{1}{p^2}=\frac{\sec^2\theta}{a^2}+\frac{\tan^2\theta}{b^2}$$

$$\Rightarrow \frac{b^2}{p^2}=\frac{b^2}{a^2}\sec^2\theta+\tan^2\theta$$

$$\Rightarrow \frac{b^2}{p^2}+1=\left(\frac{b^2}{a^2}+1\right)\sec^2\theta$$

$$= e^2\sec^2\theta$$

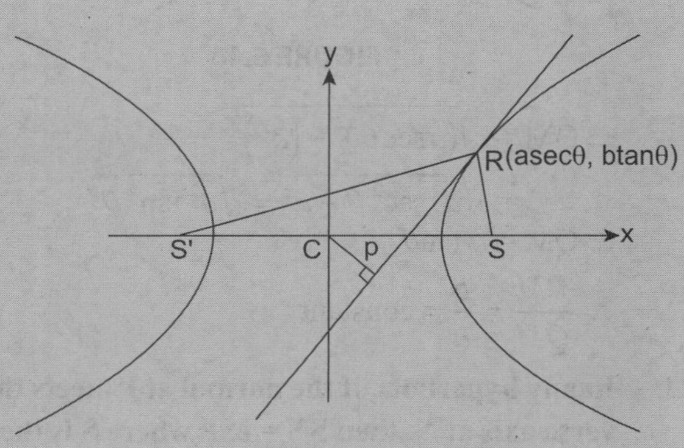

FIGURE 6.47

$$\Rightarrow 4a^2\left(\frac{b^2}{p^2}+1\right)=4a^2e^2\sec^2\theta=(2ae\sec\theta)^2=(RS+RS')^2$$

As we know that $RS = ex_1 - a = ea\sec\theta - a$ and $RS' = ex_1 + a = ea\sec\theta + a$

ILLUSTRATION 40: Let $P(a\sec\theta, b\tan\theta)$ and $Q(a\sec\phi, b\tan\phi)$, where $\theta+\phi=\frac{\pi}{2}$ be two points on the hyperbola $\frac{x^2}{a^2}-\frac{y^2}{b^2}=1$. If (h, k) is the point of intersection of the normals at P and Q, then find k.

SOLUTION: Given $\theta+\phi=\frac{\pi}{2}$ i.e., $\phi=\frac{\pi}{2}-\theta$ As normals at point $P(\theta)$ and $Q(\phi)$ passes through (h, k)

$\Rightarrow ah\cos\theta + bk\cot\theta = a^2 + b^2$ ..(1)

and $ah\cos\phi + bk\cot\phi = a^2 + b^2$ ..(2)

putting $\phi = \frac{\pi}{2}-\theta$, the equation (2) convert to $ah\sin\theta + bk\tan\theta = a^2 + b^2$ ...(3)

Now eliminating h we get $bk(\cot\theta\sin\theta - \tan\theta\cos\theta) = (a^2+b^2)(\sin\theta-\cos\theta)$

$\Rightarrow k = -(a^2+b^2)/b$.

PROPERTIES OF TANGENT AND NORMAL

P1: The ordinates of the hyperbola is in a constant ratio to the length of the tangent drawn from its foot to the auxiliary circle.

Proof: PM = $b\tan\theta$

To find QM = length of tangent from m(a secθ,0) on the circle $x^2 + y^2 = a^2$

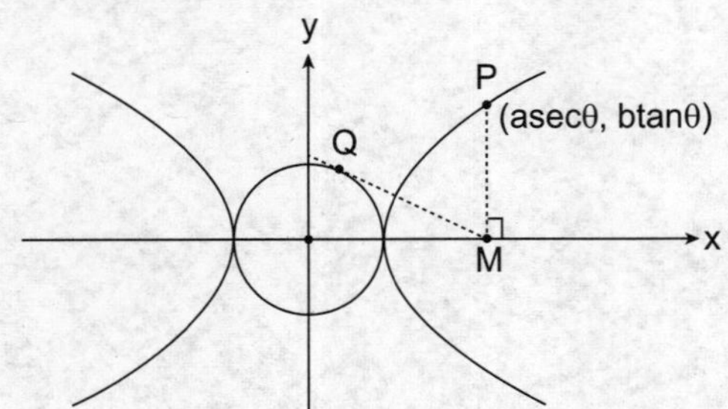

FIGURE 6.48

$$\Rightarrow QM = \sqrt{(a\sec\theta)^2 - (a^2)}$$

$$= \sqrt{a^2\sec^2\theta - a^2} = \sqrt{a^2\tan^2\theta}$$

$\Rightarrow$QM = $a\tan\theta$

$$\therefore \frac{PM}{QM} = \frac{b}{a} = \text{constant}$$

P2: In any hyperbola, if the normal at P meets the transverse axis at N, then SN = eSP where S is the focus.

Proof: Let the co-ordinates of P be (x_1, y_1)

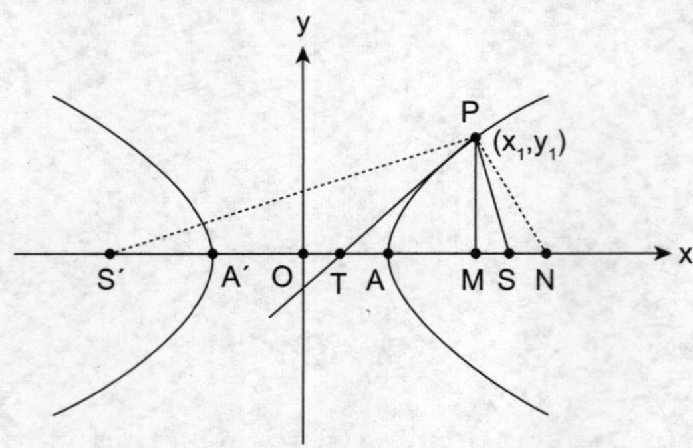

FIGURE 6.49

Then the equation of normal at P on the hyperbola

$$\frac{x^2}{a^2} - \frac{y^2}{b^2} = 1 \text{ is } \frac{a^2x}{x_1} + \frac{b^2y}{y_1} = a^2 + b^2 = a^2e^2 \quad \text{....(i)}$$

The normal (1) meets the x-axis i.e., y = 0 at 'N'; then the co-ordinates of N is $(e^2x_1, 0)$

$\therefore$ ON = e^2x_1

Now SN = ON – OS = e^2x_1 – ae

= e $(ex_1 - a)$ = eSP

Similarly S'N = eS'P

P3: In any hyperbola, the normal at the point P bisects the external angles between the two focal distances and the tangent at the point P bisects the interior angle between the two focal distances.

Proof: In the previous property (P2), we had established that SN = eSP and also S'N = eS'P therefore,

we get $\frac{SN}{S'N} = \frac{SP}{S'P}$

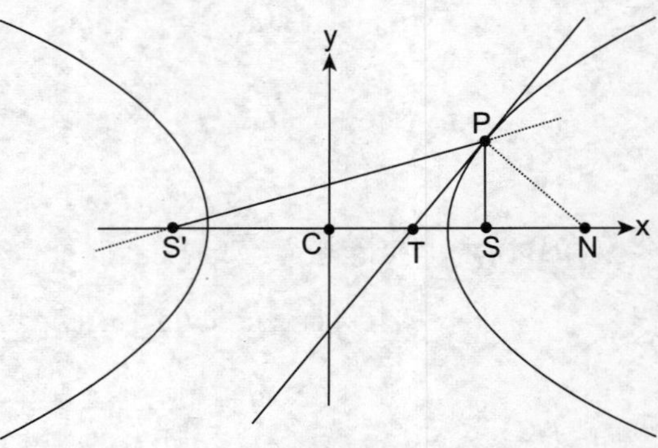

FIGURE 6.50

This relation shows that the normal PN is the external bisector of the angle SPS' therefore the tangent PT is the interior angle bisector of the angle SPS'.

P4: If an incoming light ray passing through one focus (*S*) strikes the convex side of the hyperbolic mirror then after reflection it will get reflected towards the other focus.

Proof: From the previous property, we have already established that PT is the internal angle bisector of ∠SPS'; and PN is the exterior angle bisector ∠SPS' And hence any ray emanating from *S* will pass through S' after reflection through the hyperbolic mirror and similarly, we can say that any ray emanating from S' will pass though *S* after reflection through hyperbolic mirror.

Aliter: SP = focal distance = ex_1 – a = ea sec$\theta - a$

And S'P = ex_1 + a = $ea\sec\theta + a$

Now $\frac{SP}{S'P} = \frac{e\sec\theta - 1}{e\sec\theta + 1}$

Also $ST = ae - a\cos\theta$ and S'T = $ae + \cos\theta$

$$\frac{ST}{S'T} = \frac{e - \cos\theta}{e + \cos\theta} = \frac{e\sec\theta - 1}{e\sec\theta + 1} = \frac{SP}{S'P}$$

P5: If an ellipse and hyperbola have the same foci *S* and *S'* then they cut orthogonally i.e., 90°.

Proof: Let the two curves hyperbola $\frac{x^2}{a^2} - \frac{y^2}{b^2} = 1$ and ellipse $\frac{x^2}{a^2} + \frac{y^2}{b^2} = 1$ cut each other at a point 'p'.

Now the ray emanating from *S* and getting reflected through the hyperbolic mirror at point P passes through S' then PT is the interior angle bisector of $\angle SPS'$ where T is the point of intersection of the tangent at P to the hyperbola with the x-axis.

And for the ellipse, the angle bisector of the interior angle $\angle SPS'$ is the normal to the ellipse therefore PT is the normal to the ellipse. Now, Since PT is normal to ellipse, it implies PN is tangent to ellipse (PN ⊥ PT)

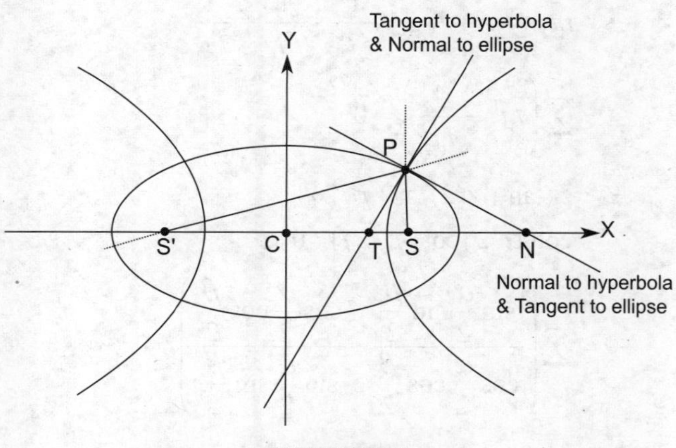

FIGURE 6.51

Conclusively, we can say that the tangent to ellipse at point 'P' i.e., PN is perpendicular to the tangent to hyperbola at point P i.e., PT

∴ The two curves cut orthogonally at P.
And by the virtue of symmetry, they will also cut at right angles at the point P', Q and Q'.

■ CO-NORMAL POINTS

In general, four normals can be drawn on a hyperbola each passing through a common point. The foot of perpendicular of these four normals lying on the hyperbola are called 'co-normal points'.

■ PROPERTIES OF CO-NORMAL POINTS

In general, four normals can be drawn to a hyperbola from any point and if $\alpha, \beta, \gamma, \delta$ be the eccentric angles of these four co-normal points, then $\alpha+\beta+\gamma+\delta$ is an odd multiple of π.

Proof: Let Q (h, k) be any given point and let $P(a \sec \phi, b \tan \phi)$ be any point on the hyperbola.

∵ Normal at P $(a\sec\phi, b\tan\phi)$,

If passes through $Q(h, k)$

$$\therefore \quad ah\cos\phi + bk\cot\phi = a^2 + b^2 = a^2e^2 \quad ....(1)$$

Here we can use the formula that

$$\cos\phi = \frac{1-\tan^2(\phi/2)}{1+\tan^2(\phi/2)} \text{ and } \sin\phi = \frac{2\tan(\phi/2)}{1+\tan^2(\phi/2)}$$

and since for a hyperbola; $\phi \neq [0,2\pi] - \{\pi/2, 3\pi/2\}$

$\therefore \quad \phi/2 \neq [0,\pi] \sim \{\pi/4, 3\pi/4\}$

Now substituting $\tan x = t$

$$\text{The equation ah } \cos\phi + \text{bk}\left(\frac{\cos\phi}{\sin\phi}\right) = a^2e^2 \quad ....(2)$$

Becomes $bkt^4 + 2(a^2e^2 + ah)t^3 + 2(a^2e^2 - ah)t - bk = 0$ ...(3)

And since $\tan x$ is monotonic for $\left(-\frac{\pi}{2}, \frac{\pi}{2}\right)$

Therefore the number of roots of the above equation will be the number of distinct value of the angle ϕ and hence we will get four value of ϕ say $\alpha, \beta, \gamma, \delta$.

$$\text{Now, } \tan\left(\frac{\alpha}{2}+\frac{\beta}{2}+\frac{\gamma}{2}+\frac{\delta}{2}\right) = \frac{S_1 - S_3}{1 - S_2 + S_4}$$

where Sk is sum of product of roots of equation (3), taken k at a time.

$$= \frac{-\frac{2(a^2e^2+ah)}{bk} + \frac{2(a^2e^2-ah)}{bk}}{1 - 0 + \left(\frac{-bk}{bk}\right)} \to \infty$$

or $\frac{\alpha}{2}+\frac{\beta}{2}+\frac{\gamma}{2}+\frac{\delta}{2} =$ an odd multiple of $\pi/2$

$\Rightarrow \quad \alpha, \beta, \gamma, \delta =$ an odd multiple of π.

Alternative Method:

Let $z = e^{i\phi} = \cos\phi + i\sin\phi$

$$\therefore \quad \frac{1}{z} = e^{-i\phi} = \cos\phi - i\sin\phi \Rightarrow \cos\phi = \frac{z+\frac{1}{z}}{2} = \frac{z^2+1}{2z}$$

$$\text{and } \sin\phi = \frac{z-\frac{1}{z}}{2i} = \frac{z^2-1}{2iz}.$$

Now equation (1) reduces to

$$ah\left(\frac{z^2+1}{2z}\right) + bk\left(i\left(\frac{z^2+1}{z^2-1}\right)\right) = a^2 + b^2$$

$$ah(z^4-1) + 2ibkz(z^2+1)$$

$$= (a^2+b^2)2z(z^2-1)$$

$$\Rightarrow \quad ahz^4 + [(2ibk - 2(a^2+b^2)]z^3 +$$

$$(2ibk + 2(a^2+b^2)]z - ah = 0 \quad ....(4)$$

Consequently, $z = e^{i\phi}$ gives four values of ϕ, say $\alpha, \beta, \gamma, \delta$ (*here* sum of four angles)

$\therefore\ z_1z_2z_3z_4 = -1$ or $e^{i\alpha}.e^{i\beta}.e^{i\gamma}.e^{i\delta} = -1$

$\Rightarrow\ e^{i(\alpha+\beta+\gamma+\delta)} = -1$

$\cos(\alpha+\beta+\gamma+\delta) = -1$ *and* $\sin(\alpha+\beta+\gamma+\delta) = 0$

$\Rightarrow\ \alpha+\beta+\gamma+\delta$ = odd integral multiplie of π.

Corollary: If α, β, γ are the eccentric angles of three points on the hyperbola $\frac{x^2}{a^2}-\frac{y^2}{b^2}=1$, the normals at which are concurrent, then $\sin(\alpha+\beta)+\sin(\alpha+\gamma)+\sin(\gamma+\alpha)=0$ Here each term contains sum of two eccentric angles.

$\therefore$ from (4), $\sum z_1z_2 = 0$

or $z_1z_2 + z_1z_3 + z_1z_4 + z_2z_3 + z_2z_4 + z_3z_4 = 0$

$\Rightarrow\ e^{i(\alpha+\beta)} + e^{i(\alpha+\gamma)} + e^{i(\alpha+\delta)} + e^{i(\alpha+\beta)} + e^{i(\beta+\gamma)} + e^{i(\gamma+\delta)} = 0$

$\Rightarrow\ [\cos(\alpha+\beta)+\cos(\alpha+\gamma)+\cos(\alpha+\delta)+\cos(\beta+\gamma)+\cos(\beta+\delta)+\cos(\gamma+\delta)]+$
$i[\sin(\alpha+\beta)+\sin(\alpha+\gamma)+\sin(\alpha+\delta)+\sin(\beta+\gamma)+\sin(\beta+\delta)+\sin(\gamma+\delta)]=0$

Comparing the imaginary part

then $\sin(\alpha+\beta)+\sin(\alpha+\gamma)+\sin(\alpha+\delta)+\sin(\beta+\gamma)+\sin(\beta+\delta)+\sin(\gamma+\delta)=0$ ....(5)

Since, $\alpha+\beta+\gamma+\delta$ odd multiple of π

$\therefore\ \delta+\alpha$ = odd multiple of $\pi-(\beta+\gamma)$

$\delta+\beta$ = odd multiple of $\pi-(\alpha+\gamma)$

$\delta+\gamma$ = odd multiple of $\pi-(\alpha+\beta)$

$\therefore\ \sin(\delta+\alpha) = \sin(\beta+\gamma)$

$\sin(\delta+\beta) = \sin(\alpha+\gamma)$

$\sin(\delta+\gamma) = \sin(\alpha+\beta)$ ....(6)

$\{\because \sin(n\pi-\alpha) = \sin\alpha$ if n is odd integer$\}$

From (5) and (6),

$2\sin(\alpha+\beta)+2\sin(\beta+\gamma)+2\sin(\gamma+\alpha)=0$

Hence, $\sin(\alpha+\beta)+\sin(\beta+\gamma)+\sin(\gamma+\alpha)=0$

Alternative Method:

From equation (3), $\sum t_1t_2 = 0$ ....(7)

and $t_1t_2t_3t_4 = -1$ ....(8)

Now $\sum t_1t_2 = 0$

$\Rightarrow\ t_1t_2 + t_1t_3 + t_2t_3 = -t_4(t_1+t_2+t_3)$

$\Rightarrow\ t_1t_2 + t_2t_3 + t_3t_1 = \frac{t_1+t_2+t_3}{t_1t_2t_3}$ {from(8)}

$\Rightarrow\ t_1t_2 + t_2t_3 + t_3t_1 = \frac{1}{t_2t_3}+\frac{1}{t_3t_1}+\frac{1}{t_1t_2}$

$\Rightarrow\ \sum\left(t_1t_2 - \frac{1}{t_1t_2}\right) = 0$

$\Rightarrow\ \Sigma(\tan(\alpha/2)\tan(\beta/2) - \cot(\alpha/2)\cot(\beta/2)) = 0$

$$\Rightarrow\ \sum\left(\frac{\sin\frac{\alpha}{2}\sin\frac{\beta}{2}}{\cos\frac{\alpha}{2}\cos\frac{\beta}{2}} - \frac{\cos\frac{\alpha}{2}\cos\frac{\beta}{2}}{\sin\frac{\alpha}{2}\sin\frac{\beta}{2}}\right) = 0$$

$$\Rightarrow\ \sum\frac{\sin^2\frac{\alpha}{2}\sin^2\frac{\beta}{2} - \cos^2\frac{\alpha}{2}\cos^2\frac{\beta}{2}}{\sin\frac{\alpha}{2}\sin\frac{\beta}{2}\cos\frac{\alpha}{2}\cos\frac{\beta}{2}} = 0$$

$$\Rightarrow\ \sum\frac{4\cos\left(\frac{\alpha+\beta}{2}\right)\cos\left(\frac{\alpha-\beta}{2}\right)}{4\sin(\alpha/2)\sin(\beta/2)\cos(\alpha/2)\cos(\beta/2)} = 0$$

$$\Rightarrow\ \sum\frac{2(\cos\alpha+\cos\beta)}{\sin\alpha\,\sin\beta} = 0$$

$$\Rightarrow\ 2\frac{\sum\sin\gamma\,(\cos\alpha+\cos\beta)}{\sin\alpha\,\sin\beta\,\sin\gamma} = 0$$

$\Rightarrow\ \sum\sin\gamma\,(\cos\alpha+\cos\beta) = 0$

$\Rightarrow\ \sin\gamma\,(\cos\alpha+\cos\beta)+\sin\alpha\,(\cos\beta+\cos\gamma)+\sin\beta(\cos\alpha+\cos\gamma) = 0$

$\Rightarrow\ \sin(\alpha+\beta)+\sin(\beta+\gamma)+\sin(\gamma+\alpha) = 0$.

ILLUSTRATION 41: Prove that the feet of the normals to $\frac{x^2}{a^2}-\frac{y^2}{b^2}=1$ from (h, k) lie on $a^2y(x-h)+b^2x(y-k)=0$.

SOLUTION: Let $P(x_1, y_1), Q(x_2, y_2), R(x_3y_3)$ and $S(x_4, y_4)$ be co-normal points so that normals drawn from them meet in $T(h, k)$. The equation of normal at $P(x_1, y_1)$ is $\frac{a^2x}{x_1}+\frac{b^2y}{y_1}=a^2+b^2$

or $a^2xy_1 + b^2x_1y = (a^2+b^2)x_1y_1$ or $(a^2+b^2)x_1y_1 - a^2xy_1 - b^2x_1y = 0$

but the point $T(h, k)$ lies on it.

$\therefore \quad (a^2+b^2)x_1y_1 - a^2hy_1 - b^2x_1k = 0$

Similarly, for points Q, R and S are

$(a^2+b^2)x_2y_2 - a^2hy_2 - b^2x_2k = 0$;

$(a^2+b^2)x_3y_3 - a^2hy_3 - b^2x_3k = 0$

and $(a^2+b^2)x_4y_4 - a^2hy_4 - b^2x_4k = 0$

Hence P, Q, R, S all lie on the curve

$a^2y(x-h) + b^2x(y-k) - 0$ or $a^2y(x-h)+b^2x(y-k)=0$.

TEXTUAL EXERCISE–3 (SUBJECTIVE)

1. (a) Find the condition for the line $lx + my + n = 0$ to touch the hyperbola $\frac{x^2}{a^2}-\frac{y^2}{b^2}=1$ and also find the same condition for line $x\cos\alpha + y\sin\alpha = p$.

(b) For what value of λ, does the line $y = 3x + \lambda$, touch the hyperbola $9x^2 - 5y^2 = 45$.

2. (a) Find the equations of the tangents to the hyperbola $3x^2 - y^2 = 3$ which are perpendicular to the line $x + 3y = 2$.

(b) Find the equation of the tangent to the hyperbola $4x^2 - 9y^2 = 1$ which is parallel to the line $4y = 5x + 7$. Also, find the points of contact.

3. Find the number of points on the plane of the hyperbola $\frac{x^2}{25}-\frac{y^2}{36}=1$ from where two perpendicular tangents can be drawn to the hyperbola.

4. The line $y = mx$ meets the circle $x^2 + y^2 = a^2$ at P and hyperbola $x^2 - y^2 = a^2$ at Q in the positive quadrant. Find the value of m if the tangent at P to the circle and the tangent at Q to the hyperbola intersect at the point $((\sqrt{5}+\sqrt{3})\,a/4), ((\sqrt{5}-\sqrt{3})\,a/2)$.

5. Find the locus of the middle points of the portion of tangents to $\frac{x^2}{a^2}-\frac{y^2}{b^2}=1$ included between the axes.

6. In any hyperbola, prove that the tangent at any point bisects the angle between the focal distances of the point.

7. Prove that the line $lx + my - n = 0$ will be a normal to the hyperbola $\frac{x^2}{a^2}-\frac{y^2}{b^2}=1$ if $\frac{a^2}{l^2}-\frac{b^2}{m^2}=\frac{\left(a^2+b^2\right)^2}{n^2}$.

8. A ray emanating from the point (5, 0) is incident on the hyperbola $9x^2 - 16y^2 = 144$ at the point P with abscissa 8. Find the equation of the reflected ray after first reflection when point P lies in first quadrant.

9. A normal to the hyperbola $x^2 - 4y^2 = 4$ meets the x and y axes at A and B respectively. Find the locus of the point of intersection of the straight lines drawn though A and B perpendicular to the x and y axes respectively.

10. Find the locus of the points of intersection of two tangents to a hyperbola $\frac{x^2}{25}-\frac{y^2}{16}=1$. If sum of their slopes is constant 'a'.

11. If the normal at 'ϕ' on the hyperbola $\frac{x^2}{a^2}-\frac{y^2}{b^2}=1$ meet transverse axis at G, prove that $AG.A'G = a^2(e^4\sec^2\phi - 1)$, where A and A' are the vertices of the hyperbola.

12. Normals are drawn to the hyperbola $\frac{x^2}{a^2}-\frac{y^2}{b^2}=1$ at the points P ($a\sec\theta_1$, $b\tan\theta_1$)and Q ($a\sec\theta_2$, $b\tan\theta_2$); where $\theta_1 + \theta_2 = \pi/2$, meeting the conjugate axis at G_1 and G_2 respectively. Prove that $CG_1.CG_2 = \frac{a^2e^4}{\left(e^2-1\right)}$. where C is the centre of the hyperbola and e is its eccentricity.

13. Find the equation of a circle touching the hyperbola $x^2 - y^2 = 8$ at the point $P(3, 1)$ and passing through its focus.

14. A straight line is drawn parallel to the conjugate axis of the hyperbola $\frac{x^2}{a^2}-\frac{y^2}{b^2}=1$, to meet it and conjugate hyperbola respectively in the points P and Q. Show that the normals at P and Q to the curves meet on the x-axis.

15. C is the centre of the hyperbola $\frac{x^2}{a^2}-\frac{y^2}{b^2}=1$. The tangent at any point P on this hyperbola meets the lines $bx - ay = 0$ and $bx + ay = 0$ in the points Q and R respectively. Show that $CQ.\ CR = a^2 + b^2$.

16. Prove that the part of the tangent at any point of the hyperbola $\frac{x^2}{a^2}-\frac{y^2}{b^2}=1$ intercepted between the point of contact and the transverse axis is a harmonic mean between the lengths of the perpendiculars drawn from the foci on the normal at the same point.

Answer Keys

1. (a) $n^2 = a^2 l^2 - b^2 m^2$; $a^2 \cos^2\alpha - b^2\sin^2\alpha = p^2$ (b) ± 6 **2.** (a) $y = 3x \pm \sqrt{6}$

(b) $5x-4y\pm\frac{\sqrt{161}}{6}=0;\left(\frac{\pm 15}{2\sqrt{161}},\frac{\pm 8}{3\sqrt{161}}\right)$ **3.** 0 **4.** $m = 1/2$ **5.** $a^2/x^2 - b^2/y^2 = 4$

8. $\left(3\sqrt{3}\right)x-13y+15\sqrt{3}=0$ **9.** $(2x)^2 - y^2 = 25$ **10.** $ax^2 - 2xy = 25a$

13. $2(x^2 + y^2) - 15x - 3y + 28 = 0$; $2(x^2 + y^2) + 3x - 9y - 20 = 0$.

TEXTUAL EXERCISE–3 (OBJECTIVE)

1. The line $x \cos\alpha + y \sin\alpha = p$ will be a tangent to the hyperbola $\frac{x^2}{a^2}-\frac{y^2}{b^2}=1$, provided
   (a) $a^2 \sin^2\alpha + b^2 \cos^2\alpha = p^2$
   (b) $a^2 \cos^2\alpha + b^2 \sin^2\alpha = p^2$
   (c) $a^2 \sin^2\alpha - b^2 \cos^2\alpha = p^2$
   (d) $a^2 \cos^2\alpha - b^2 \sin^2\alpha = p^2$

2. Shortest distance between the curves $\frac{x^2}{a^2}-\frac{y^2}{b^2}=$, $4x^2 + 4y^2 = a^2$ $(b > a)$ is
   (a) –   (b) $\frac{b}{\sqrt{2}}$
   (c) $\frac{a}{2}$ (d) $\frac{a}{\sqrt{2}}$

3. Total number of common tangents of the curves $\frac{x^2}{a^2}-\frac{y^2}{b^2}=1$ and $\frac{y^2}{a^2}-\frac{x^2}{b^2}=1$; $(a > b)$ is equal to
   (a) zero   (b) 2
   (c) 4   (d) None of these

4. The tangent at any arbitrary point 'P' on $\frac{x^2}{a^2}-\frac{y^2}{b^2}=1$ meets the line $bx - ay = 0$ at point 'Q', then locus of mid-point of PQ is
   (a) $\frac{x^2}{a^2}-\frac{y^2}{b^2}=\frac{1}{2}$ (b) $\frac{x^2}{a^2}-\frac{y^2}{b^2}=2$
   (c) $\frac{x^2}{a^2}-\frac{y^2}{b^2}=\frac{4}{3}$ (d) $\frac{x^2}{a^2}-\frac{y^2}{b^2}=\frac{3}{4}$

5. The locus of the foot of perpendicular drawn from the focus of the hyperbola $\frac{x^2}{9}-\frac{y^2}{4}=1$, to any arbitrary tangent of the hyperbola, is
   (a) $x^2 + y^2 = 4$ (b) $x^2 + y^2 = 9$
   (c) $x^2 + y^2 = 36$ (d) None of these

6. The co-ordinates of the point on the hyperbola $\frac{x^2}{24}-\frac{y^2}{18}=1$ that is nearest to the line $3x + 2y + 1 = 0$ are
   (a) (6, 3)   (b) (6, –3)
   (c) (–6, –3)   (d) (–6, 3)

7. The tangent at a point 'P' on the hyperbola $\frac{x^2}{a^2}-\frac{y^2}{b^2}=1$, meets one of it's directrices at the point Q. If the line segment PQ subtends an angle 'θ' at the corresponding focus, then 'θ' is always equal to

(a) $\frac{\pi}{4}$ (b) $\frac{\pi}{2}$

(c) $\frac{\pi}{3}$ (d) $\frac{\pi}{6}$

8. If m_1 and m_2 are the slopes of the tangents to the hyperbola $16x^2 - 25y^2 = 400$ which pass through the point (6, 2), then the harmonic mean of m_1 and m_2 is

(a) 3/5 (b) –3/5

(c) 2/3 (d) 5/3

9. If a tangent to the hyperbola $\frac{x^2}{a^2}-\frac{y^2}{b^2}=1$, whose centre is C, meets the transverse and conjugate axes at P and Q respectively, then $\frac{a^2}{CP^2}-\frac{b^2}{CQ^2}-1$ is

(a) $a^2 + b^2$ (b) $a^2 - b^2$

(c) 0 (d) None of these

10. A common tangent to $9x^2 - 16y^2 = 144$ and $x^2 + y^2 = 9$ is

(a) $y=\frac{3}{\sqrt{7}}x+\frac{15}{\sqrt{7}}$ (b) $y=3\sqrt{\frac{2}{7}}x+\frac{15}{\sqrt{7}}$

(c) $2\sqrt{\frac{3}{7}}x + 15\sqrt{7}$ (b) None of these

Answer Keys

1. (d)	2. (c)	3. (c)	4. (d)	5. (b)	6. (d)	7. (b)	8. (d)	9. (c)	10. (b)

■ EQUATION OF CHORD BISECTED AT A GIVEN POINT

The equation of the chord of the hyperbola $\frac{x^2}{a^2}-\frac{y^2}{b^2}=1$, bisected at a point (x_1, y_1) is $\frac{xx_1}{a^2}-\frac{yy_1}{b^2}-1=\frac{x_1^2}{a^2}-\frac{y_1^2}{b^2}-1$ or $T = S_1$, where $T=\frac{xx_1}{a^2}-\frac{yy_1}{b^2}-1$ and $S_1=\frac{x_1^2}{a^2}-\frac{y_1^2}{b^2}-1$

Since equation of the hyperbola is

$$\frac{x^2}{a^2}-\frac{y^2}{b^2}=1 \quad ....(1)$$

Proof: Let QR be the chord of the hyperbola whose mid-point is $P(x_1, y_1)$. Since Q and R lie on the hyperbola (1).

$$\therefore \frac{x_2^2}{a^2}-\frac{y_2^2}{b^2}=1 \quad ....(2)$$

$$\text{and } \frac{x_3^2}{a^2}-\frac{y_3^2}{b^2}=1 \quad ....(3)$$

Subtracting (3) from (2), $\frac{1}{a^2}(x_2^2-x_3^2)-\frac{1}{b^2}(y_2^2-y_3^2)=0$

$$\Rightarrow \frac{(x_2+x_3)(x_2-x_3)}{a^2}-\frac{(y_2+y_3)(y_2-y_3)}{b^2}=0$$

$$\Rightarrow \frac{y_2-y_3}{x_2-x_3}=\frac{b^2(x_2+x_3)}{a^2(y_2+y_3)}=\frac{b^2}{a^2}\frac{(2x_1)}{(2y_1)}=\frac{b^2x_1}{a^2y_1} \quad ....(4)$$

{P is the mid - point of QR}

$\therefore$ Equation of QR is $y-y_1=\frac{y_2-y_3}{x_2-x_3}(x-x_1)$

$$\Rightarrow y-y_1=\frac{b^2x_1}{a^2y_1}(x-x_1) \quad \text{from equation (4)}$$

$$\Rightarrow \frac{xx_1}{a^2}-\frac{yy_1}{b^2}=\frac{x_1^2}{a^2}-\frac{y_1^2}{b^2}$$

$$\Rightarrow \frac{xx_1}{a^2}-\frac{yy_1}{b^2}-1=\frac{x_1^2}{a^2}-\frac{y_1^2}{b^2}-1$$

$$\Rightarrow T = S_1$$

ILLUSTRATION 42: Find the locus of the mid-points of the chords of the hyperbola $\frac{x^2}{a^2}-\frac{y^2}{b^2}=1$ which subtend a right angle at the origin.

SOLUTION: Let (h, k) be the mid-point of the chord of the hyperbola. Then its equation is $T = S_1$ i.e.,

$$\frac{hx}{a^2}-\frac{ky}{b^2}=\frac{h^2}{a^2}-\frac{k^2}{b^2} \quad ....(1)$$

The equation of the pair of straight lines joining the origin to the points of intersection of the hyperbola and the chord (1) is obtained by equation of hyperbola. Making homogeneous hyperbola with the help of (1), as $\frac{x^2}{a^2}-\frac{y^2}{b^2}=\frac{\left(\frac{hx}{a^2}-\frac{ky}{b^2}\right)^2}{\left(\frac{h^2}{a^2}-\frac{k^2}{b^2}\right)^2}$

$$\Rightarrow \frac{1}{a^2}\left(\frac{h^2}{a^2}-\frac{k^2}{b^2}\right)^2 x^2-\frac{1}{b^2}\left(\frac{h^2}{a^2}-\frac{k^2}{b^2}\right)^2 y^2=\frac{h^2}{a^4}x^2+\frac{k^2}{b^4}y^2-\frac{2hk}{a^2b^2}xy \quad ....(2)$$

The lines represented by (2) will be at right angle if co-efficient of x^2 + co-efficient of $y^2 = 0$

$$\Rightarrow \frac{1}{a^2}\left(\frac{h^2}{a^2}-\frac{k^2}{b^2}\right)^2-\frac{h^2}{a^4}-\frac{1}{b^2}\left(\frac{h^2}{a^2}-\frac{k^2}{b^2}\right)^2-\frac{k^2}{b^4}=0$$

$$\Rightarrow \left(\frac{h^2}{a^2}-\frac{k^2}{b^2}\right)^2\left(\frac{1}{a^2}-\frac{1}{b^2}\right)=\frac{h^2}{a^4}+\frac{k^2}{b^4}$$

hence, the locus of (h, k) is

$$\left(\frac{x^2}{a^2}-\frac{y^2}{b^2}\right)^2\left(\frac{1}{a^2}-\frac{1}{b^2}\right)=\frac{x^2}{a^4}+\frac{y^2}{b^4}$$

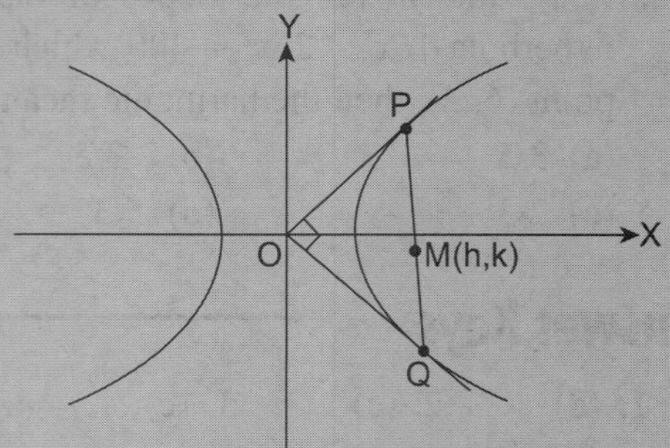

FIGURE 6.52

ILLUSTRATION 43: From the points on the circle $x^2+y^2=a^2$, tangents are drawn to the hyperbola $x^2-y^2=a^2$; prove that the locus of the middle points of the chords of contact is the curve $(x^2-y^2)^2=a^2(x^2+y^2)$.

SOLUTION: Since any point on the circle $x^2+y^2=a^2$ is $(a\cos\theta, a\sin\theta)$, the chord of contact of hyperbola $x^2-y^2=a^2$ w.r.to. this point is $x(a\cos\theta)-y(a\sin\theta)=a^2$

i.e., $x\cos\theta - y\sin\theta = a$ ....(1)

If its mid-point be (h, k) then it is same as $T=S_1$

i.e., $hx-ky=h^2-k^2$ ....(2)

Comparing (1) and (2), we get $\frac{\cos\theta}{h}=\frac{\sin\theta}{k}=\frac{a}{(h^2-k^2)} \Rightarrow (h^2-k^2)\cos\theta = ah$ ....(3)

and $(h^2-k^2)\sin\theta = ak$ ....(4)

Squaring and adding (3) and (4), we get $(h^2-k^2)^2=a^2(h^2+k^2)$

Hence the required locus is $(x^2-y^2)^2=a^2(x^2+y^2)$.

ILLUSTRATION 44: Prove that the locus of the middle points of the chords of the hyperbola $\frac{x^2}{a^2}-\frac{y^2}{b^2}=1$ which pass through a fixed point (α,β) is a hyperbola whose centre is $\left(\frac{\alpha}{2},\frac{\beta}{2}\right)$.

SOLUTION: Let the mid-point of the chord be (h, k). The equation of chord whose mid-point is (h, k) is $T = S_1$

$$\Rightarrow \frac{hx}{a^2}-\frac{ky}{b^2}=\frac{h^2}{a^2}-\frac{k^2}{b^2}$$

If it passes through (α,β) then $\frac{\alpha h}{a^2}-\frac{\beta k}{b^2}=\frac{h^2}{a^2}-\frac{k^2}{b^2}$

$\therefore$ locus of (h, k) is $\frac{x\alpha}{a^2} - \frac{y\beta}{b^2} = \frac{x^2}{a^2} - \frac{y^2}{b^2}$

$$\Rightarrow \frac{1}{a^2}(x^2 - \alpha x) - \frac{1}{b^2}(y^2 - \beta y) = 0$$

or $$\frac{\left(x - \frac{\alpha}{2}\right)^2}{a^2} - \frac{\left(y - \frac{\beta}{2}\right)^2}{b^2} = \frac{1}{4}\left\{\frac{\alpha^2}{a^2} - \frac{\beta^2}{b^2}\right\} = \lambda(\text{say})$$

$\therefore$ $\frac{\left(x - \frac{\alpha}{2}\right)^2}{a^2\lambda} - \frac{\left(y - \frac{\beta}{2}\right)^2}{b^2\lambda} = 1$; which is clearly a hyperbola with centre $\left(\frac{\alpha}{2}, \frac{\beta}{2}\right)$.

■ PAIR OF TANGENTS

The combined equation of the pair of tangents drawn from a point $P(x_1, y_1)$, lying outside the hyperbola $\frac{x^2}{a^2} - \frac{y^2}{b^2} = 1$ to the hyperbola $\frac{x^2}{a^2} - \frac{y^2}{b^2} = 1$ is $\boldsymbol{SS_1 = T^2}$

Where $S = \frac{x^2}{a^2} - \frac{y^2}{b^2} - 1$; $S_1 = \frac{x_1^2}{a^2} - \frac{y_1^2}{b^2} - 1$ and

$$T = \frac{xx_1}{a^2} - \frac{yy_1}{b^2} - 1$$

i.e., $$\left(\frac{x^2}{a^2} - \frac{y^2}{b^2} - 1\right)\left(\frac{x_1^2}{a^2} - \frac{y_1^2}{b^2} - 1\right) = \left(\frac{xx_1}{a^2} - \frac{yy_1}{b^2} - 1\right)^2$$

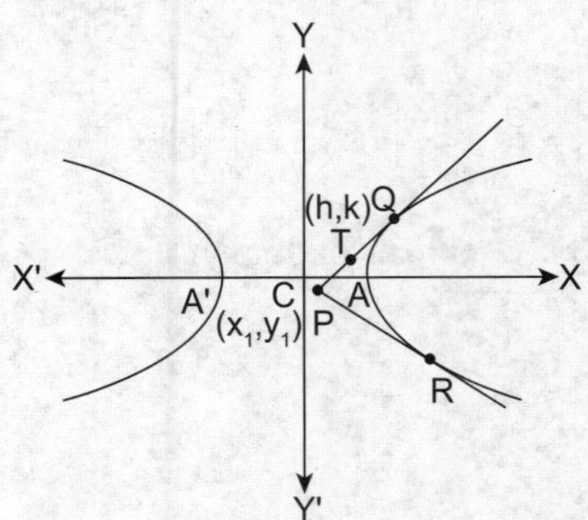

FIGURE 6.53

Let $T(h, k)$ be any point on the pair of tangents PQ or PR drawn from any external point $P(x_1, y_1)$ to the hyperbola $\frac{x^2}{a^2} - \frac{y^2}{b^2} = 1$

$\therefore$ Equation of PT is $y - y_1 = \frac{k - y_1}{h - x_1}(x - x_1)$

i.e., $$y = \left(\frac{k - y_1}{h - x_1}\right)x + \left(\frac{hy_1 - kx_1}{h - x_1}\right)$$

which is the tangent to the hyperbola $\frac{x^2}{a^2} - \frac{y^2}{b^2} = 1$

$\therefore$ $c^2 = a^2m^2 - b^2$.

or $$\left(\frac{hy_1 - kx_1}{h - x_1}\right)^2 = a^2\left(\frac{k - y_1}{h - x_1}\right) - b^2$$

$$\Rightarrow (hy_1 - kx_1)^2 = a^2(k - y_1)^2 - b^2(h - x_1)^2$$

Hence locus of (h, k) is;

$$(xy_1 - x_1y)^2 = a^2(y - y_1)^2 - b^2(x - x_1)^2$$

or $$(xy_1 - x_1y)^2 = -(b^2x^2 - a^2y^2) - (b^2x_1^2 - a^2y_1^2) - 2(a^2yy_1 - b^2xx_1)$$

or $$\left(\frac{xy_1 - x_1y}{ab}\right)^2 = -\left(\frac{x^2}{a^2} - \frac{y^2}{b^2}\right) - \left(\frac{x_1^2}{a^2} - \frac{y_1^2}{b^2}\right) + 2\left(\frac{xx_1}{a^2} - \frac{yy_1}{b^2}\right)$$

or $$\left(\frac{x^2}{a^2} - \frac{y^2}{b^2} - 1\right)\left(\frac{x_1^2}{a^2} - \frac{y_1^2}{b^2} - 1\right) = \left(\frac{xx_1}{a^2} - \frac{yy_1}{b^2} - 1\right)^2$$

or $SS_1 = T^2$

Alternative Method:

Let the hyperbola be

$$\frac{x^2}{a^2} - \frac{y^2}{b^2} = 1 \qquad .....(1)$$

Let $P(x_1, y_1)$ be any point outside the hyperbola.

Let a chord of the hyperbola through the point $P(x_1, y_1)$ cut the hyperbola at Q and R, Let $R\ (h, k)$ be any arbitrary point on the line PQ (R inside or outside).

Let Q divides PR in the ratio λ: 1 then co-ordinates of Q are

$$\left(\frac{\lambda h + x_1}{\lambda + 1}, \frac{\lambda k + y_1}{\lambda + 1}\right) \qquad (\because PQ : QR = \lambda : 1)$$

Since Q lies on hyperbola (1), then

$$\frac{1}{a^2}\left(\frac{\lambda h+x_1}{\lambda+1}\right)^2-\frac{1}{b^2}\left(\frac{\lambda k+y_1}{\lambda+1}\right)^2=1$$

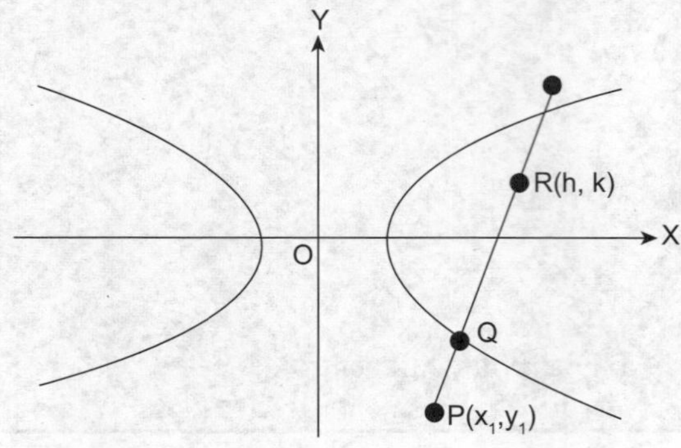

FIGURE 6.54

$$\Rightarrow\ b^2(\lambda h+x_1)^2-a^2(\lambda k+y_1)^2=a^2b^2(\lambda+1)^2$$

$$\Rightarrow\ (b^2h^2-a^2k^2-a^2b^2)\lambda^2+2(b^2hx_1-a^2ky_1-a^2b^2)\lambda$$

$$+(b^2x_1^2-a^2y_1^2-a^2b^2)=0 \qquad \text{.... (2)}$$

Let PR will become tangent to the hyperbola (1), then roots of equation (2) are equal.

$$4(b^2hx_1-a^2ky_1-a^2b^2)^2-4(b^2h^2-a^2k^2-a^2b^2)\times(b^2x_1^2-a^2y_1^2-a^2b^2)=0$$

Dividing by $4a^4b^4$.

$$\therefore\quad \left(\frac{hx_1}{a^2}-\frac{ky_1}{b^2}-1\right)^2=\left(\frac{h^2}{a^2}-\frac{k^2}{b^2}-1\right)\left(\frac{x_1^2}{a^2}-\frac{y_1^2}{b^2}-1\right)$$

Hence locus of $R(h, k)$, i.e., equation of pair of tangents from $P(x_1, y_1)$ is

$$\left(\frac{xx_1}{a^2}-\frac{yy_1}{b^2}-1\right)^2=\left(\frac{x^2}{a^2}-\frac{y^2}{b^2}-1\right)\left(\frac{x_1^2}{a_2}-\frac{y^2}{b^2}-1\right)$$

i.e., $T^2 = SS_1$ or $SS_1 = T^2$

NOTE

$S = 0$ is the equation of the curve, S_1 is obtained from S by replacing x by x_1 and y by y_1 and $T = 0$ is the equation of the tangent at (x_1, y_1) to $S = 0$.

ILLUSTRATION 45: Find the locus of the point of intersection of the tangent draws to the hyperbola $\frac{x^2}{a^2}-\frac{y^2}{b^2}=1$ which are at an angle of $\frac{\pi}{3}$ with each other.

SOLUTION: The equation of pair of tangents from P(h,k) is $\left(\frac{hx}{a^2}-\frac{yk}{b^2}-1\right)^2=\left(\frac{x^2}{a^2}-\frac{y^2}{b^2}-1\right)\left(\frac{h^2}{a^2}-\frac{k^2}{b^2}-1\right)$

$$\Rightarrow\ x^2\left(\frac{k^2}{a^2b^2}+\frac{1}{a^2}\right)+y^2\left(\frac{h^2}{a^2b^2}-\frac{1}{b^2}\right)+xy\left(\frac{-2hk}{a^2b^2}\right)+x\left(\frac{-2h}{a^2}\right)+y\left(\frac{2k}{b^2}\right)+\left(\frac{h^2}{a^2}-\frac{k^2}{b^2}\right)=0$$

Let θ be the angle, then using concept of angle between pair of straight lines

$$\tan\theta=\tan\frac{\pi}{3}=\frac{2\sqrt{\left(\frac{hk}{a^2b^2}\right)^2-\left(\frac{k^2}{a^2b^2}+\frac{1}{a^2}\right)\left(\frac{h^2}{a^2b^2}-\frac{1}{b^2}\right)}}{\left(\frac{k^2}{a^2b^2}+\frac{1}{a^2}\right)+\left(\frac{h^2}{a^2b^2}-\frac{1}{b^2}\right)}$$

$$\sqrt{3}=\frac{2\sqrt{\frac{h^2k^2}{a^4b^4}-\left(\frac{h^2k^2}{a^4b^4}+\frac{h^2}{a^4b^2}-\frac{k^2}{a^2b^4}-\frac{1}{a^2b^2}\right)}}{\left(\frac{k^2}{a^2b^2}+\frac{1}{a^2}\right)+\left(\frac{h^2}{a^2b^2}-\frac{1}{b^2}\right)}$$

$$\therefore\quad \left(\frac{y^2}{a^2b^2}+\frac{1}{a^2}\right)+\left(\frac{x^2}{a^2b^2}-\frac{1}{b^2}\right)=\frac{2}{\sqrt{3}}\sqrt{\frac{1}{a^2b^2}+\frac{y^2}{a^2b^4}-\frac{x^2}{a^4b^2}}$$

is the required locus.

CHORD OF CONTACT

If the tangents from a point $P(x_1, y_1)$ to the hyperbola $\frac{x^2}{a^2}-\frac{y^2}{b^2}=1$ touch the hyperbola at Q and R, then the equation of the chord of contact QR is $T=\frac{xx_1}{a^2}-\frac{yy_1}{b^2}-1=0$

Let $Q \equiv (x', y')$ and $R \equiv (x'', y'')$

Now equation of tangents PQ and PR are

$$\frac{xx'}{a^2}-\frac{yy'}{b^2}=1 \quad \text{....(1)}$$

and $$\frac{xx''}{a^2}-\frac{yy''}{b^2}=1 \quad \text{....(2)}$$

Since (1) and (2) pass through $P(x_1, y_1)$ then

$$\frac{x'x_1}{a^2}-\frac{y'y_1}{b^2}=1 \quad \text{....(3)}$$

and $$\frac{x''x_1}{a^2}-\frac{y''y_1}{b^2}=1 \quad \text{....(4)}$$

Hence it is clear that $Q(x', y')$ and $R(x'', y'')$ lie on $\frac{xx_1}{a^2}-\frac{yy_1}{b^2}=1$ or $T=0$

which is therefore the chord of contact QR.

ILLUSTRATION 46: Tangents are drawn from any point on the hyperbola $\frac{x^2}{9}-\frac{y^2}{4}=1$ to the circle $x^2+y^2=9$. Find the locus of mid-point of the chord of contact.

SOLUTION: Equation of hyperbola and circle are, $\frac{x^2}{9}-\frac{y^2}{4}=1$ or $x^2+y^2=9$

General point on hyperbola $P \equiv (3\sec\theta, 2\tan\theta)$

Equation chord of constant of circle w.r.t. point P is $T=0$, given as

$$(3\sec\theta)x+(2\tan\theta)y-9=0 \quad \text{....(i)}$$

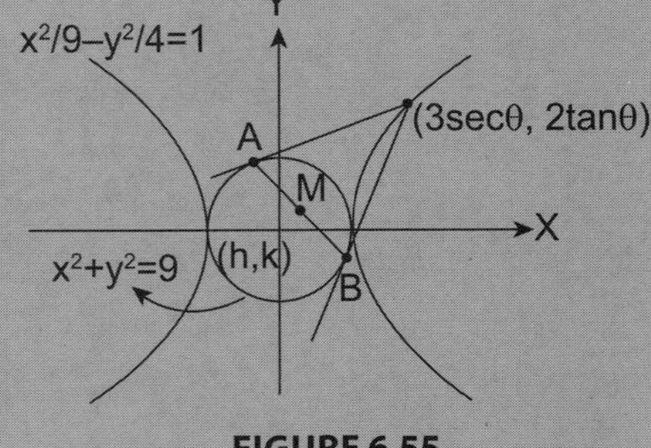

FIGURE 6.55

Also, the equation of chord AB of circle with mid-points $m(h, k)$ is $T=S_1$

$\Rightarrow hx+ky=h^2+k^2$

$\Rightarrow hx+ky-(h^2+k^2)=0 \quad \text{...(ii)}$

Since equation (i) and (ii) are identical

$$\frac{3\sec\theta}{h}+\frac{2\tan\theta}{k}=\frac{9}{h^2+k^2}$$

$$\Rightarrow \sec\theta=\frac{3h}{h^2+k^2} \text{ and } \tan\theta=\frac{9k}{2(h^2+k^2)}$$

Now, we know that $\sec^2\theta-\tan^2\theta=1$

$$\Rightarrow \frac{9h^2}{(h^2+k^2)^2}-\frac{81k^2}{4(h^2+k^2)^2}=1$$

$\Rightarrow 9(4x^2-9y^2)=4(x^2+y^2)^2$ is the desired locus.

ILLUSTRATION 47: How many real tangents can be drawn from the point (4, 3) to the hyperbola $\frac{x^2}{16}-\frac{y^2}{9}=1$. Find the equation of these tangents and angle between them.

SOLUTION: Given point $P \equiv (4, 3)$ and the equation of hyperbola $S \equiv \frac{x^2}{16}-\frac{y^2}{9}-1=0$

$$S_1 \equiv \frac{16}{16}-\frac{9}{9}-1=-1<0$$

$\Rightarrow$ Point $P \equiv (4, 3)$ lies towards centre of the hyperbola.

$\therefore$ Two tangents can be drawn from the point $P(4, 3)$ to the hyperbola.

Equation of pair of tangents is $SS_1 = T^2$

$$\Rightarrow \left(\frac{x^2}{16} - \frac{y^2}{9} - 1\right) . (-1) = \left(\frac{4x}{16} - \frac{3y}{9} - 1\right)^2$$

$$\Rightarrow -\frac{x^2}{16} + \frac{y^2}{9} + 1 = \frac{x^2}{16} + \frac{y^2}{9} + 1 - \frac{xy}{6} - \frac{x}{2} + \frac{2y}{3}$$

$\Rightarrow 3x^2 - 4xy - 12x + 16y = 0$ is the equation of pair of tangent

Clearly, the angle between them is $\theta = \tan^{-1}\left(\frac{4}{3}\right)$

ILLUSTRATION 48: Find the locus of point of intersection of perpendicular tangents to the hyperbola $\frac{x^2}{a^2} - \frac{y^2}{b^2} = 1$.

SOLUTION: Let $P(h, k)$ be the point of intersection of two perpendicular tangents equation of pair of tangents is $SS_1 = T^2$

$$\Rightarrow \left(\frac{x^2}{a^2} - \frac{y^2}{b^2} - 1\right)\left(\frac{h^2}{a^2} - \frac{k^2}{b^2} - 1\right) = \left(\frac{hx}{a^2} - \frac{ky}{b^2} - 1\right)^2$$

$$\Rightarrow \frac{x^2}{a^2}\left(-\frac{k^2}{b^2} - 1\right) - \frac{y^2}{b^2}\left(\frac{h^2}{a^2} - 1\right) + \text{the terms linear in } x \text{ and y} = 0 \qquad .....(i)$$

Since above equation represents two perpendicular lines.

$$\therefore \quad \frac{1}{a^2}\left(-\frac{k^2}{b^2} - 1\right) - \frac{1}{b^2}\left(\frac{h^2}{a^2} - 1\right) = 0$$

$\Rightarrow -k^2 - b^2 - h^2 + a^2 = 0 \Rightarrow$ the required locus is $x^2 + y^2 = a^2 - b^2$.

ILLUSTRATION 49: If tangents to the parabola $y^2 = 4ax$ intersect the hyperbola $\frac{x^2}{a^2} - \frac{y^2}{b^2} = 1$ at A and B, then find the locus of point of intersection of tangents at A and B.

SOLUTION: Let $P \equiv (h, k)$ be the point of intersection of tangents at A and B

$\therefore$ Equation of chord of contact AB is $\frac{xh}{a^2} - \frac{yk}{b^2} = 1$ i.e., $y = \frac{b^2h}{a^2k}x - \frac{b^2}{k}$ ....(i)

Which touches the parabola $y^2 = 4ax$

Equation of tangent to parabola $y^2 = 4ax$.

$\therefore$ y intercept c of the chord of contact satisfies c = a/m

$$\Rightarrow \frac{-b^2}{k} = \frac{a}{\frac{b^2h}{a^2k}} \Rightarrow -b^4h = a^3k^2$$

$$\Rightarrow k^2 = -\frac{b^4}{a^3}h \Rightarrow \text{locus of } P \text{ is } y^2 = -\frac{b^4}{a^3}x.$$

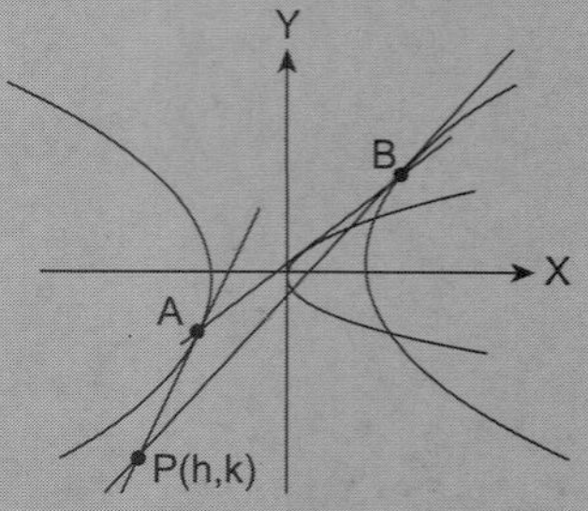

FIGURE 6.56

ILLUSTRATION 50: Find the locus of the mid-point of focal chords of the hyperbola $\frac{x^2}{a^2} - \frac{y^2}{b^2} = 1$.

SOLUTION: Let $P \equiv (h, k)$ be the mid-point of the chord.

$\therefore$ Equation of chord is is given by $\frac{xh}{a^2} - \frac{yk}{b^2} = \frac{h^2}{a^2} - \frac{k^2}{b^2}$

Since it is a focal chord,

$\therefore$ It passes through focus, either $(ae, 0)$ or $(-ae, 0)$

If it passes through (ae, 0) $\Rightarrow \dfrac{eh}{a} = \dfrac{h^2}{a^2} - \dfrac{k^2}{b^2}$

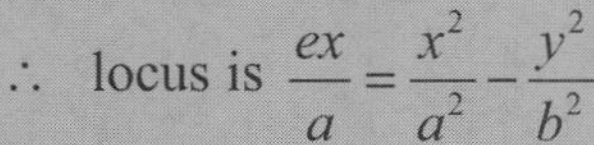

$\therefore$ locus is $\dfrac{ex}{a} = \dfrac{x^2}{a^2} - \dfrac{y^2}{b^2}$

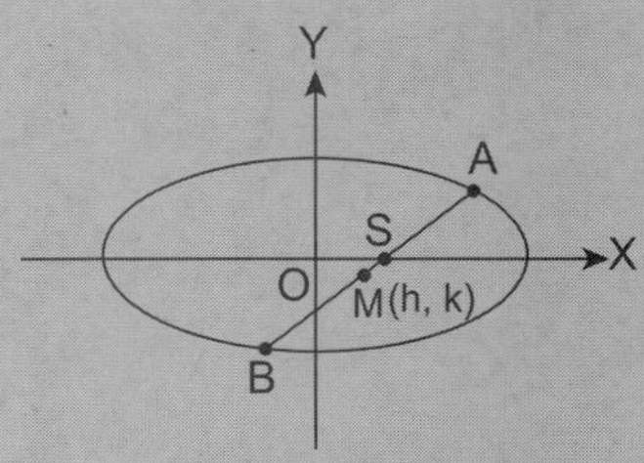

FIGURE 6.57

If it passes through $(-ae, 0)$

$\therefore$ locus is $-\dfrac{ex}{a} = \dfrac{x^2}{a^2} - \dfrac{y^2}{b^2}$.

ILLUSTRATION 51: Find the condition on 'a' and 'b' for which two distinct chords of the hyperbola $\dfrac{x^2}{2a^2} - \dfrac{y^2}{2b^2} = 1$ passing through (a, b) are bisected by the line $x + y = b$.

SOLUTION: Let the line $x + y = b$ bisect the chord at $P(\alpha, b - \alpha)$

$\therefore$ Equation of chord whose mid-point is $P(\alpha, b - \alpha)$

$$\frac{x\alpha}{2a^2} - \frac{y(b-\alpha)}{2b^2} = \frac{\alpha^2}{2a^2} - \frac{(b-\alpha)^2}{2b^2}$$

Since it passes through (a, b)

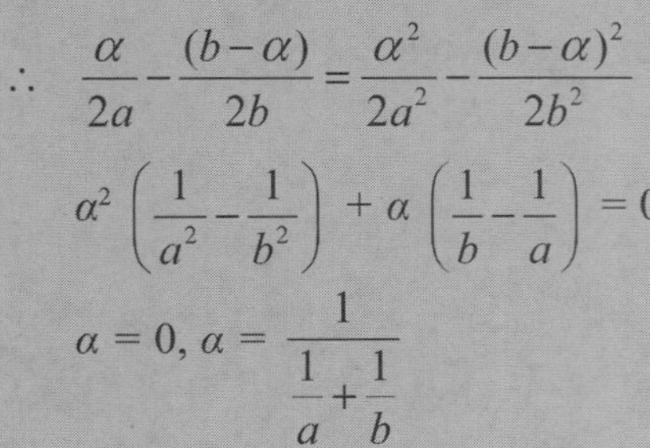

$$\therefore \quad \frac{\alpha}{2a} - \frac{(b-\alpha)}{2b} = \frac{\alpha^2}{2a^2} - \frac{(b-\alpha)^2}{2b^2}$$

$$\alpha^2\left(\frac{1}{a^2} - \frac{1}{b^2}\right) + \alpha\left(\frac{1}{b} - \frac{1}{a}\right) = 0$$

$$\alpha = 0,\ \alpha = \frac{1}{\dfrac{1}{a} + \dfrac{1}{b}}$$

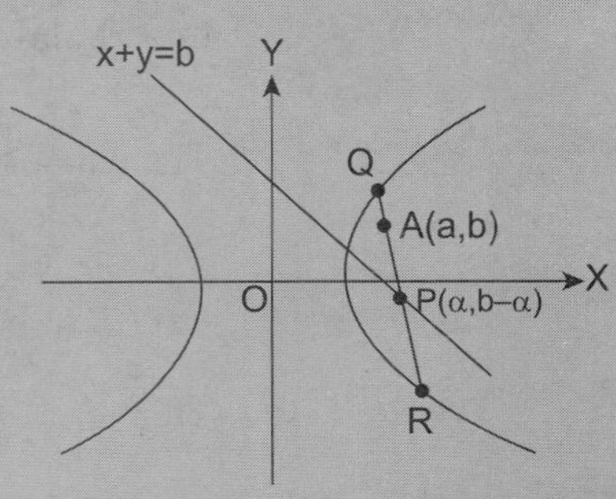

FIGURE 6.58

Therefore for two distinct chord we must get two distinct real values α for which $a \neq \pm b$

ILLUSTRATION 52: Find the equations of the tangents to the hyperbola $x^2 - 9y^2 = 9$ that are drawn from (3, 2). Find the area of the triangle that these tangents form with their chord of contact.

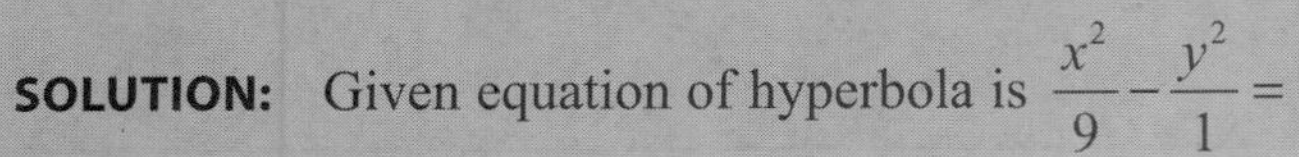

SOLUTION: Given equation of hyperbola is $\dfrac{x^2}{9} - \dfrac{y^2}{1} = 1$

Equation of tangent is $y = mx \pm \sqrt{a^2m^2 - b^2}$ ....(1)

It passes through (3, 2)

$$2 = 3m \pm \sqrt{9m^2 - 1} \text{ or } 4 + 9m^2 - 12m = 9m^2 - 1$$

or $m_1 = \dfrac{5}{12}$ and $m_2 = \infty$; equation of tangent (1) for

$$m_1 = \frac{5}{12}$$

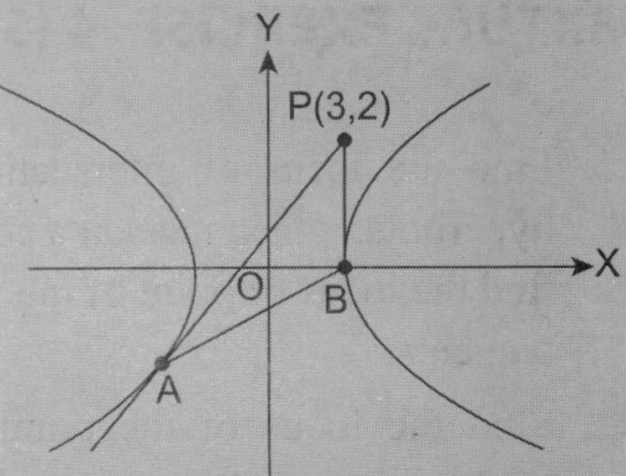

FIGURE 6.59

Given as $y - 2 = \dfrac{5}{12}(x - 3)$

$\Rightarrow 12y - 24 = 5x - 15$

$\Rightarrow 5x - 12y + 9 = 0$ .....(2)

Now, equation of tangent for $m_2 = \infty$ is a line passing through P(3, 2) perpendicular to x-axis, given as $x - 3 = 0$ .....(3)

Now equation of chord of contact w.r.t. point $P(3, 2)$ is $T = 0$

$\Rightarrow xx_1 - 9yy_1 = 9$ or $3x - 18y = 9$

$\Rightarrow x - 6y = 3$ ...(4)

Solving (2) and (4) $x = -5, y = -\dfrac{4}{3}$

Solving (3) and (4) $x = 3, y = 0$

Now vertices of triangle: P(3, 2), B(3, 0), A(–5, – 4/3)

$\therefore$ Area $= \dfrac{1}{2}$(PB)(perpendicular distance of A from PB) $= \dfrac{1}{2}(2)|3-(-5)| = 8$ sq. units.

ILLUSTRATION 53: A line through the origin meets the circle $x^2 + y^2 = a^2$ at P and the hyperbola $x^2 - y^2 = a^2$ at Q. Prove that the locus of the point of intersection of the tangent at P to the circle and the tangent at Q to the hyperbola is curve $a^4(x^2 - a^2) + 4x^2y^4 = 0$.

SOLUTION: Let $y = mx$ be a line passing through origin. This line intersects the circle $x^2 + y^2 = a^2$ and the hyperbola $x^2 - y^2 = a^2$ in $P\left(\dfrac{a}{\sqrt{1+m^2}}, \dfrac{+am}{\sqrt{1+m^2}}\right)$ and $Q\left(\dfrac{a}{\sqrt{1-m^2}}, \dfrac{am}{\sqrt{1-m^2}}\right)$ respectively.

The equation of the tangents to the circle $x^2 + y^2 = a^2$ at P and hyperbola $x^2 - y^2 = a^2$ at Q are

$x + my = a\sqrt{1+m^2}$ and $x - my = a\sqrt{}$ respectively.

Let (h, k) be the point of intersection of these two lines.

Then, $h + mk = a\sqrt{1+m^2}$ ...(1)

$h - mk = a\sqrt{1-m^2}$ ...(2)

$\Rightarrow (h + mk)^2 = a^2(1 + m^2)$ and $(h - mk)^2 = a^2(1 - m^2)$

$\Rightarrow m^2(k^2 - a^2) + 2mhk + h^2 - a^2 = 0$ and $m^2(k^2 + a^2) - 2mhk + h^2 - a^2 = 0$

Using cross multiplication method, we get $\dfrac{m^2}{4hk(h^2-a^2)} = \dfrac{m}{2a^2(h^2-a^2)} = \dfrac{1}{-4hk^3}$

$\Rightarrow m = \dfrac{2hk}{a^2}$ and $m = \dfrac{a^2(h^2-a^2)}{-2hk^3} \Rightarrow \dfrac{2hk}{a^2} = \dfrac{a^2(h^2-a^2)}{-2hk^3}$

$\Rightarrow -4h^2k^4 = a^4(h^2 - a^2) \Rightarrow a^4(x^2 - a^2) + 4x^2y^4 = 0$

TEXTUAL EXERCISE–4 (SUBJECTIVE)

1. Find the locus of intersection of the tangents to a hyperbola, which meet at a constant angle β and hence find the locus of point of intersection of perpendicular tangents.
2. Find the locus of the point of intersection of tangents drawn at the extremities of a latus rectum of hyperbola $\dfrac{x^2}{a^2} - \dfrac{y^2}{b^2} = 1$.
3. (a) Find the equation of the chord of the hyperbola $25x^2 - 16y^2 = 400$, which is bisected at the point (5, 3).

   (b) Find the locus of the mid-points of the chords of the circle $x^2 + y^2 = 16$ which are tangents to the hyperbola $9x^2 - 16y^2 = 144$..
4. Find the locus of the point whose chord of contact w.r.t hyperbola $\dfrac{x^2}{a^2} - \dfrac{y^2}{b^2} = 1$ is always at a constant distance d^2 from the centre of the hyperbola.
5. A variable chord of the hyperbola $\dfrac{x^2}{a^2} - \dfrac{y^2}{b^2} = 1$ is tangent to the circle $x^2 + y^2 = c^2$. Find the locus of its mid-point.

6. A tangent to the hyperbola $\frac{x^2}{a^2}-\frac{y^2}{b^2}=1$ cuts the ellipse $\frac{x^2}{a^2}+\frac{y^2}{b^2}=1$ in points P and Q. Find the locus of the mid-point of PQ.

7. Prove that the chords of the hyperbola $\frac{x^2}{a^2}-\frac{y^2}{b^2}=1$, which touch the conjugate hyperbola, are bisected at the point of contact.

8. From the points on the circle $x^2 + y^2 = a^2$, tangents are drawn to the hyperbola $x^2 - y^2 = a^2$, prove that the locus of the middle points of the chords of contact is the curve $(x^2 - y^2)^2 = a^2(x^2 + y^2)$.

Answer Keys

1. $(x^2 + y^2 + b^2 - a^2)^2 = 4\cot^2\beta\,(a^2y^2 - b^2x^2 + a^2b^2)^2$; $x^2 + y^2 = a^2 - b^2$; a > b **2.** $a^6y^2 - b^6x^2 = x^2y^2(a^2 + b^2)^2$

3. (a) $125x - 48y = 481$ (b) $(x^2 + y^2)^2 = 16x^2 - 9y^2$ **4.** $\frac{x^2}{a^4}+\frac{y^2}{b^4}=\frac{1}{d^4}$ **5.** $\left(\frac{x^2}{a^2}-\frac{y^2}{b^2}\right)^2 = c^2\left[\frac{x^2}{a^4}+\frac{y^2}{b^4}\right]$

6. $\left(\frac{x^2}{a^2}+\frac{y^2}{b^2}\right)^2=\frac{x^2}{a^2}-\frac{y^2}{b^2}$

TEXTUAL EXERCISE – 4 (OBJECTIVE)

1. Locus of mid-point of the parallel chord of the hyperbola $\frac{x^2}{a^2}-\frac{y^2}{b^2}=1$, giving slope '$m$'is
   (a) $y=\frac{b^2}{a^2m}x$ (b) $y=\frac{a^2}{b^2m}x$
   (c) $y=\frac{b^2m}{a^2}x$ (d) $y=\frac{a^2m}{b^2}x$

2. Angle between tangents drawn from the point $P(\sqrt{3}, 1)$ to the hyperbola $\frac{x^2}{9}-\frac{y^2}{5}=1$ is equal to
   (a) $\frac{\pi}{3}$ (b) $\frac{\pi}{6}$
   (c) $\frac{\pi}{4}$ (d) $\frac{\pi}{2}$

3. The equation of tangents drawn from the point (0, 4) to the hyperbola $x^2 - 4y^2 = 36$ are
   (a) $5x - 6y + 24 = 0$ and $5x + 6y - 24 = 0$
   (b) $x - 4y + 16 = 0$ and $x + 4y - 16 = 0$
   (c) $2x - y + 4 = 0$ and $2x + y - 16 = 0$
   (d) None of these

4. Equation of the chord of the hyperbola $25x^2 - 16y^2 = 400$ which is bisected at the point (6, 2) is
   (a) $15x - 2y = 86$ (b) $15x + 2y = 94$
   (c) $75x - 16y = 418$ (d) $75x + 16y = 418$

5. Locus of the point of intersection of two perpendicular tangents to the hyperbola $(x^2/a^2) - (y^2/b^2) = 1$; $a \geq b$ is
   (a) $x^2 + y^2 = a^2 - b^2$ (b) $x^2 + y^2 = a^2 + b^2$
   (c) $x^2 - y^2 = a^2 - b^2$ (d) $x^2 - y^2 = a^2 + b^2$

6. The tangents drawn from $(2\sqrt{2}, 1)$ to the hyperbola $16x^2 - 25y^2 = 400$ include between them an angle equal to
   (a) $\frac{\pi}{6}$ (b) $\frac{\pi}{3}$
   (c) $\frac{\pi}{2}$ (d) $\frac{\pi}{4}$

7. The angle between the tangents drawn from any point on the circle $x^2 + y^2 = 3$ to the hyperbola $(x^2/4) - y^2 = 1$ is
   (a) $\frac{\pi}{3}$ (b) $\frac{\pi}{4}$
   (c) $\frac{\pi}{2}$ (d) $\frac{\pi}{6}$

Answer Keys

1. (a) **2.** (d) **3.** (a) **4.** (c) **5.** (a) **6.** (c) **7.** (c)

■ ASYMPTOTES TO A HYPERBOLA

Asymptote to any curve is a straight line and touches the curve at infinity (∞).

The equations of two asymptotes of the hyperbola $\frac{x^2}{a^2}-\frac{y^2}{b^2}=1$ are $y=\pm\frac{b}{a}x$ or $\frac{x}{a}\pm\frac{y}{b}=0$.

Let $y = mx + c$ be an asymptote of the hyperbola

$$\frac{x^2}{a^2}-\frac{y^2}{b^2}=1 \quad ....(1)$$

Substituting the value of y in (1), $\frac{x^2}{a^2}-\frac{(mx+c)^2}{b^2}=1$

or $(a^2m^2 - b^2)x^2 + 2a^2mcx + a^2(b^2 + c^2) = 0$ ....(2)

If the line $y = mx + c$ is an asymptote to the given hyperbola, then it touches the hyperbola at infinity. So both roots of (2) must be infinite.

$\therefore$ $a^2m^2 - b^2 = 0$ and $-2a^2mc = 0 \Rightarrow m = \pm\frac{b}{a}$ and $c = 0$

Substituting the value of m and c in $y = mx + c$, we get

$y=\pm\frac{b}{a}x$ i.e., $\frac{x}{a}\pm\frac{y}{b}=0$

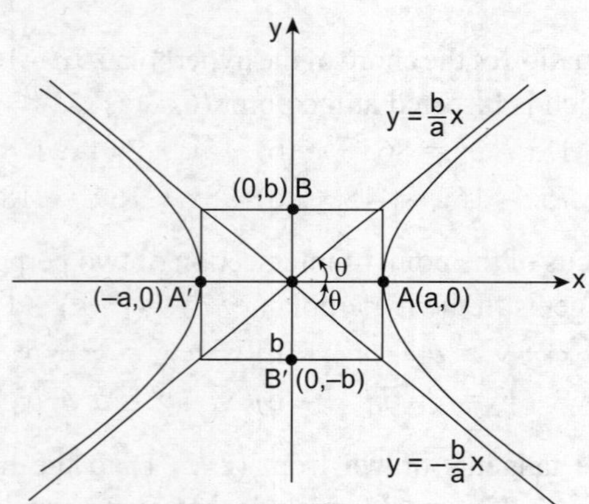

FIGURE 6.60

Alternative Method: The difference between the second degree curve and pair of asymptotes is constant.

$\because$ Given hyperbola is $\frac{x^2}{a^2}-\frac{y^2}{b^2}=1$

$\therefore$ Pair of asymptotes is $\frac{x^2}{a^2}-\frac{y^2}{b^2}+\lambda=0$ .....(1)

Equation (1) represents a pair of lines, then $\Delta = 0$

$\therefore$ $\frac{1}{a^2}\cdot\left(-\frac{1}{b^2}\right)\lambda + 0 - 0 - 0 - \lambda.0 = 0$

$\therefore$ $\lambda = 0$

From (1), pair of asymptotes is $\frac{x^2}{a^2}-\frac{y^2}{b^2}=0$

or $y=\pm\frac{b}{a}x$ or $\frac{x}{a}\pm\frac{y}{b}=0$

■ PROPERTIES OF ASYMPTOTES TO HYPERBOLA

1. The asymptotes form a pair of tangents to a hyperbola passing through its centre.($SS_1 = T^2$)
2. Both asymptotes pass through the centre of hyperbola and axes of hyperbola bisect the angles between the asymptotes.
3. Let the equation of hyperbola be $H\equiv\frac{x^2}{a^2}-\frac{y^2}{b^2}-1=0$; then equation of asymptotes will be $A\equiv\frac{x^2}{a^2}-\frac{y^2}{b^2}=0$ and the equation of corresponding conjugate hyperbola will be $C\equiv\frac{x^2}{a^2}-\frac{y^2}{b^2}+1=0$.

   Clearly, $H + C = 2A$ or $H - A = A - C$. i.e., expression for joint equation of asymptotes differ from that of hyperbola and conjugate hyperbola by same quantity.
4. **Relation between A, C and H:** From point (3), it is clear that $A=\frac{C+H}{2}$.
5. **Angle between asymptotes:** Included angle between two asymptotes of hyperbola $\frac{x^2}{a^2}-\frac{y^2}{b^2}=1$ is

   $\tan^{-1}\left(\frac{2ab}{a^2-b^2}\right)=2\tan^{-1}\left(\frac{b}{a}\right)$ *or* $2\,\text{Sec}^{-1}(e)$

   Proof: $\because 2\theta=2\tan^{-1}\left(\frac{b}{a}\right)$

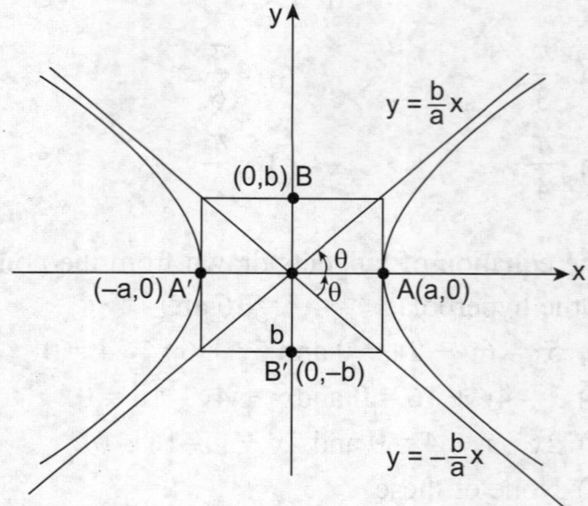

FIGURE 6.61

$\Rightarrow \tan 2\theta=\frac{2b/a}{1-b^2/a^2}$

$$\Rightarrow \tan 2\theta = \frac{2ab}{a^2-b^2};\ 2\theta = \tan^{-1}\left(\frac{2ab}{a^2-b^2}\right)$$

Also, $\cos\theta = \frac{a}{\sqrt{a^2+b^2}} = \frac{a}{\sqrt{a^2e^2}} = \frac{a}{ae} = \frac{1}{e}$.

$\Rightarrow\ \sec\theta = e \Rightarrow\ \theta = \sec^{-1} e$.

6. If the angle between the asymptotes is 90°, then $b = a$ and the hyperbola is called a **rectangular hyperbola.**

7. The asymptotes of a hyperbola meet the directrix on the auxilliary circle.

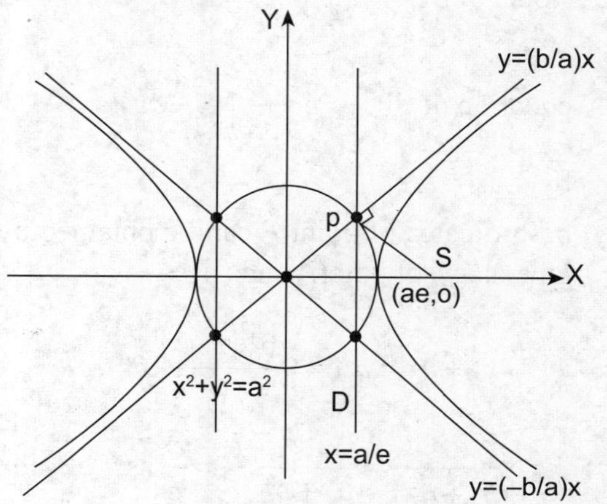

FIGURE 6.62

8. (i) The portion of any tangent to hyperbola intercepted between asymptotes is bisected at the point of contact.

(ii) Any tangent to the hyperbola makes with asymptote a triangle of constant area.

Proof: Let P ($a \sec \phi$, $b \tan \phi$) be any point on the hyperbola $\frac{x^2}{a^2} - \frac{y^2}{b^2} = 1$ .... (1)

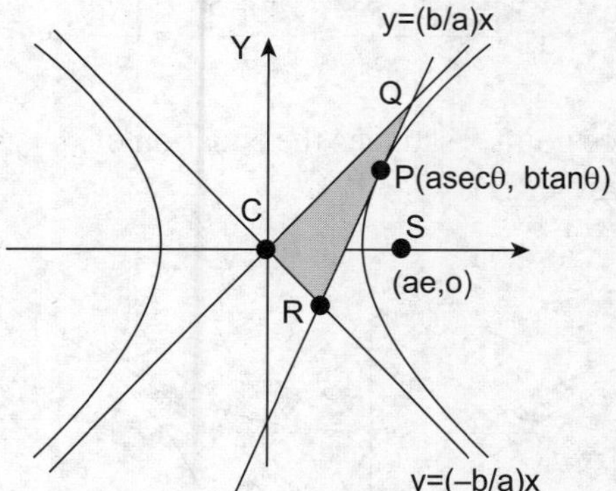

FIGURE 6.63

Asymptotes of (1) are $y = \pm\frac{b}{a}x$

Equation of tangent to (1) at P (a secϕ, b tanϕ) is

$$\frac{x}{a}\sec\phi - \frac{y}{b}\tan\phi = 1 \quad .... (2)$$

Solving (2) and $y = \frac{b}{a}x$

we get $\frac{x}{a}(\sec\phi - \tan\phi) = (\sec^2\phi - \tan^2\phi)$

$\therefore\ x = a(\sec\phi + \tan\phi)$

and $y = b(\sec\phi + \tan\phi)$

Let $Q \equiv [a(\sec\phi + \tan\phi), b(\sec\phi + \tan\phi)]$ and

solving (2) and $y = -\frac{b}{a}x$

We get $\frac{x}{a}(\sec\phi + \tan\phi) = (\sec^2\phi - \tan\phi)$

or $x = a(\sec\phi - \tan\phi)$ and $y = -b(\sec\phi - \tan\phi)$

Let $R \equiv [a(\sec\phi - \tan\phi), -b(\sec\phi - \tan\phi)]$

Mid-point of QR is $(a\sec\phi, b\tan\phi)$ which is co-ordinate of P.

Area of $\Delta CQR = \frac{1}{2}|x_1y_2 - x_2y_1|$

$= \frac{1}{2}|-ab - ab| = ab =$ constant.

9. The product of the perpendicular drawn from any point on a hyperbola to its asymptotes is constant.

Proof:

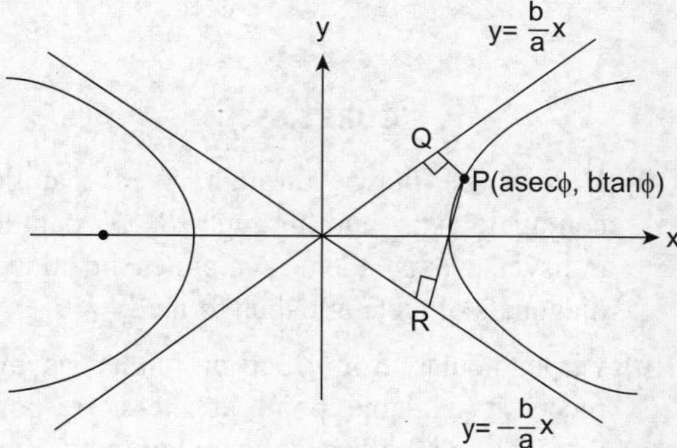

FIGURE 6.64

Let P (a secϕ, btanϕ,) be any point on the hyperbola.

Then the perpendicular from P on asymptote $y = \frac{b}{a}x$ is

$$|PQ| = \frac{\left|b\tan\phi - \frac{b}{a}(a\sec\phi)\right|}{\sqrt{1+(b/a)^2}}$$

Similarly, $|PR| = \frac{\left|b\tan\phi + \frac{b}{a}(a\sec\phi)\right|}{\sqrt{1+(b/a)^2}}$

$\Rightarrow$ Product of perpendiculars $= |PQ| \times |PR|$

$$= \frac{b^2 \tan^2 \phi - \left(\frac{b}{a}\right)^2 (a^2 \sec^2 \phi)}{1+\left(\frac{b}{a}\right)^2}$$

$$= \frac{b^2 \tan^2 \phi - b^2 \sec^2 \phi}{\frac{a^2+b^2}{a^2}} = \frac{a^2b^2}{a^2+b^2} = \text{constant}$$

10. (i) In case $m = \pm\frac{b}{a}$ but $c \neq 0$, then only one root of (2) will be infinite; then each of the lines $y = \pm\frac{b}{a}x + c$ will cut the hyperbola in one pt at infinity. Hence all the lines which are parallel to asymptote meet the hyperbola in one point at ∞ and in one finite point.

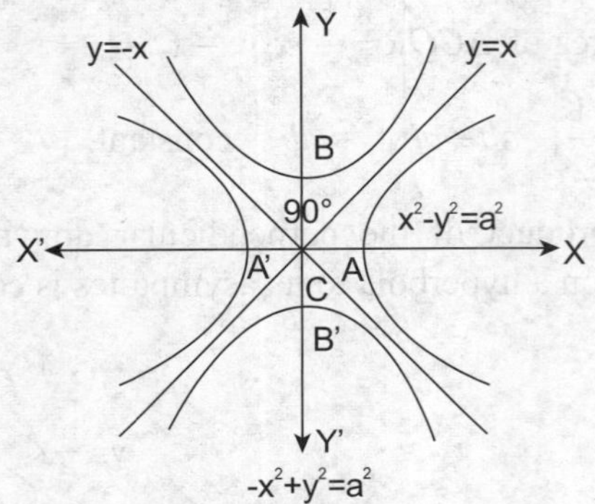

FIGURE 6.65

(ii) If lines be drawn through *A*, *A'* parallel to conjugate axis and through *B*, *B'* parallel to transverse axis, then asymptotes lie along the diagonals of rectangle thus formed.

(iii) Perpendicular from foci on either asymptote meets it in same point at the corresponding directrix and the common point of intersection lie on the auxiliary circle.

Proof:

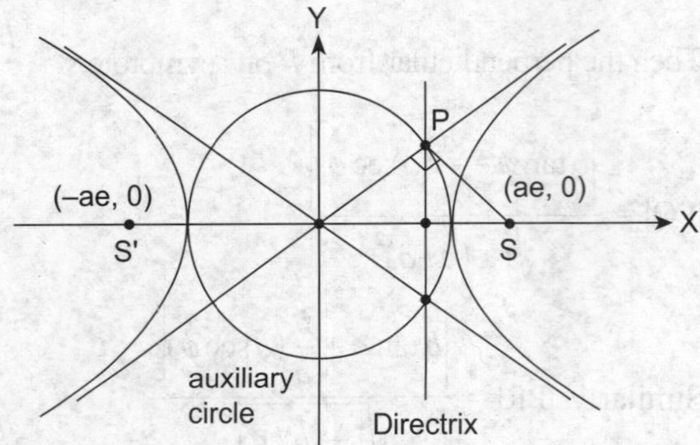

FIGURE 6.66

Let SP be the perpendicular from S(ae,0) to the asymptote $y = \frac{b}{a}x$. .... (i)

We need to prove that *P* point lies on the directrix (i.e., $x = \frac{a}{e}$) and auxiliary circle (i.e., $x^2 + y^2 = a^2$)

Slope of asymptote = $\frac{b}{a}$

$\therefore$ slope of SP = $-\frac{a}{b}$

$\therefore$ equation of SP $y = \left(-\frac{a}{b}\right)(x - ae)$

$\therefore$ by + ax = a^2e ...(ii)

$\therefore$ co-ordinates of point *P* can be obtained by simultaneously solving (i) and (ii)

$$\Rightarrow b\left(\frac{b}{a}x\right) + ax = a^2e \Rightarrow \frac{(b^2+a^2)}{a}x = a^2e$$

$$\Rightarrow x = \frac{a^2e \times a}{(b^2+a^2)} = \frac{a^2e^2}{a^2+b^2}.\frac{a}{e} = \frac{a}{e} \text{ as } a^2e^2 = a^2+b^2$$

Thus $x = \frac{a}{e}$ and hence the point *P* lies on corresponding directrix.

When $x = \frac{a}{e}$;

$$y = \frac{b}{a} \times \frac{a}{e} = \frac{b}{e}$$

$$\therefore x^2 + y^2 = \frac{a^2}{e^2} + \frac{b^2}{e^2} = \frac{a^2+b^2}{e^2} = \frac{a^2e^2}{e^2} = a^2$$

Hence point *P* satisfies the equation of $x^2 + y^2 = a^2$ i.e., the circle.

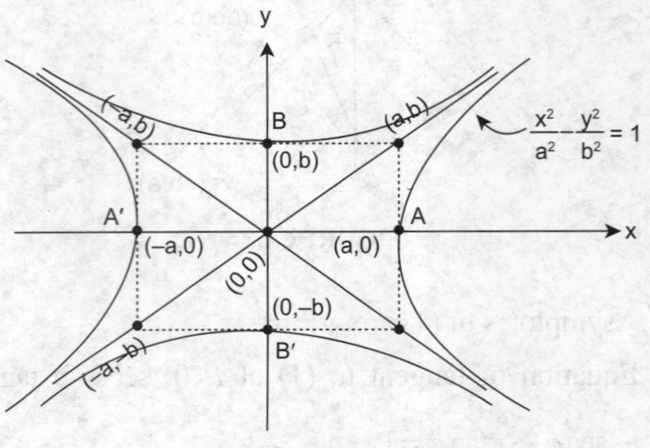

FIGURE 6.67

ILLUSTRATION 54: Find the asymptotes of the hyperbola $xy - 3y - 2x = 0$.

SOLUTION: Since equation of a hyperbola and its asymptotes differ in constant terms only,

$\therefore$ Pair of asymptotes is given by $xy - 3y - 2x + \lambda = 0$ .....(1)

Where λ is any constant such that it represents two straight lines.

$\therefore\ abc + 2fgh - af^2 - bg^2 - ch^2 = 0$

$\Rightarrow\ 0 + 2 \times -\dfrac{3}{2} \times -1 \times \dfrac{1}{2} - 0 - 0 - \lambda\left(\dfrac{1}{2}\right)^2 = 0$

$\therefore\ \lambda = 6$

From (1), the asymptotes of given hyperbola are given by $xy - 3y - 2x + 6 = 0$

or $(y - 2)(x - 3) = 0$

$\therefore$ Asymptotes are $x - 3 = 0$ and $y - 2 = 0$.

ILLUSTRATION 55: The asymptotes of a hyperbola having centre at the point (1, 2) are parallel to the lines $2x + 3y = 0$ and $3x + 2y = 0$. If the hyperbola passes through the point (5, 3). Show that its equation is $(2x + 3y - 8)(3x + 2y + 7) = 154$.

SOLUTION: Let the asymptotes be $2x + 3y + \lambda = 0$ and $3x + 2y + \mu = 0$.

Since asymptotes passes through (1, 2) therefore $\lambda = -8$ and $\mu = -7$

Thus the equation of asymptotes are $2x + 3y - 8 = 0$ and $3x + 2y - 7 = 0$

Let the equation of hyperbola be $(2x + 3y - 8)(3x + 2y - 7) + v = 0$ .....(1)

It passes through (5, 3), then $(10 + 9 - 8)(15 + 6 - 7) + v = 0$

$\Rightarrow\ 11 \times 14 + v = 0$

$\therefore\ v = -154$

Putting the value of v in (1), we obtain $(2x + 3y - 8)(3x + 2y - 7) - 154 = 0$

which is the equation of hyperbola.

ILLUSTRATION 56: The normal to the hyperbola $\dfrac{x^2}{a^2} - \dfrac{y^2}{b^2} = 1$ drawn at an extremity of one of its latera recta is parallel to an asymptote. Show that the eccentricity is equal to the square roots of $(1 + \sqrt{5})/2$.

SOLUTION: Given hyperbola is $\dfrac{x^2}{a^2} - \dfrac{y^2}{b^2} = 1$

Normal at $\left(ae, \dfrac{b^2}{a}\right)$ is $\dfrac{a^2x}{x_1} + \dfrac{b^2y}{y_1} = a^2 + b^2$ or $\dfrac{a^2x}{(ae)} + \dfrac{b^2y}{\left(\dfrac{b^2}{a}\right)} = a^2 + b^2$

or $y = -\dfrac{1}{e}x + \dfrac{a^2 + b^2}{a}$ ...(1)

The equation of asymptote to the hyperbola $\dfrac{x^2}{a^2} - \dfrac{y^2}{b^2} = 1$ is $y = -\dfrac{b}{a}x$ ...(2)

Slope of equation (1) $\Rightarrow m_1 = -\dfrac{1}{e}$

Slope of equation (2) $\Rightarrow m_2 = -\dfrac{b}{a}$

Since $b^2 = a^2(e^2 - 1) \Rightarrow \dfrac{b}{a} = \sqrt{e^2 - 1}$ ...(3)

It is given that $m_1 = m_2$

$\Rightarrow -\frac{1}{e} = -\frac{b}{a}$ or $\frac{1}{e} = \frac{b}{a}$, from equation (3) putting value of $\frac{b}{a}$

$\frac{1}{e} = \sqrt{e^2 - 1}$ or $e^4 - e^2 - 1 = 0$ or $e^2 = \frac{1 \pm \sqrt{5}}{2}$

or $e = \sqrt{\frac{1+\sqrt{5}}{2}}$ (we can't take (–)ve sign $\frac{1-\sqrt{5}}{2}$ is a negative quantity)

ILLUSTRATION 57: The parametric equation of a conic is given by $x = a \tan(\theta + \alpha)$ and $y = b \tan(\theta + \beta)$; then which conic is represented by given parametric equations?

(a) Parabola (b) Ellipse

(c) Hyperbola (d) Circle

SOLUTION: Given $x = a \tan(\theta + \alpha)$ and $y = b \tan(\theta + \beta)$

$$\Rightarrow \tan^{-1}\left(\frac{x}{a}\right) = \theta + \alpha \quad \text{....(1)}$$

$$\text{and } \tan^{-1}\left(\frac{y}{b}\right) = \theta + \beta \quad \text{....(2)}$$

Subtracting equation (2) from (1), we get $\tan^{-1}\left(\frac{x}{a}\right) - \tan^{-1}\left(\frac{y}{b}\right) = \alpha - \beta$

$$\Rightarrow \tan^{-1}\left[\frac{\frac{x}{a} - \frac{y}{b}}{1 + \frac{xy}{ab}}\right] = \alpha - \beta$$

$\Rightarrow \dfrac{\frac{x}{a} - \frac{y}{b}}{1 + \frac{xy}{ab}} = \tan(\alpha - \beta)$; Simplifying we get the required locus as

$xy + ab = (bx - ay) \cot(\alpha - \beta)$; which is a hyperbola.

ILLUSTRATION 58: Find the co-ordinates of the two points Q and R, where the tangent to the hyperbola $\frac{x^2}{45} - \frac{y^2}{20} = 1$ at the point $P(9, 4)$ intersects the two asymptotes. Prove that P is the middle point of QR. Also compute the area of the triangle CQR, where C is the centre of the hyperbola.

SOLUTION: Tangent at point $P(9, 4)$ is $\frac{xx_1}{a^2} - \frac{yy_1}{b^2} = 1$

$$\Rightarrow \frac{9x}{45} - \frac{4y}{20} = 1 \text{ or } x - y = 5 \quad \text{...(1)}$$

Asymptote is $\frac{x}{a} = \pm\frac{y}{b}$ i.e., $\frac{x}{\sqrt{45}} = \pm\frac{y}{\sqrt{20}}$;

taking (+) sign $y = \frac{2}{3}x$ or $2x - 3y = 0$ ...(2)

taking (–)ve sign $\Rightarrow y - \frac{2}{3}x$ or $2x + 3y = 0$ ...(3)

solving (1) and (2); we get Q : (15, 10)

solving (1) and (3); we get R : (3, –2)

$Q \equiv (15, 10)$ and $R \equiv (3, -2)$

It is clearly seen that the point $P(9, 4)$ is the mid-point of QR.

Now area of triangle whose vertices are $C(0, 0)$, $Q(15, 10)$, $R(3, -2)$ is given by

$$\Delta = \frac{1}{2} \times \begin{vmatrix} 0 & 0 & 1 \\ 15 & 10 & 1 \\ 3 & -2 & 1 \end{vmatrix} = \frac{1}{2}|1(-30-30)| = 30 \text{ sq. units (or } \Delta = ab = \sqrt{900} = 30 \text{ square units).}$$

ILLUSTRATION 59: A point P on the x-axis divides the focal length of the hyperbola $9x^2 - 16y^2 = 144$ in the ratio $S'P: PS = 2: 3$ where S and S' are the foci of the hyperbola. Through P a straight line is drawn at an angle of 135° to the axis OX. Find the points of intersection of this line with the asymptotes of the hyperbola.

SOLUTION: Equation of hyperbola is $\frac{x^2}{4^2} - \frac{y^2}{3^2} = 1$

$a^2 = 16 \Rightarrow a = 4$; $b^2 = 9 \Rightarrow b = 3$; $e = \sqrt{1 + \frac{b^2}{a^2}} = \sqrt{1 + \frac{9}{16}} = \frac{5}{4}$

Two foci are: $S(ae, 0) \Rightarrow S\left(4 \times \frac{5}{4}, 0\right) \Rightarrow S \equiv (5, 0)$ and $S' \equiv (-ae, 0) \equiv (-5, 0)$

point $P(x_1, y_1)$ divides length of SS' in the ratio of 2: 3 $\Rightarrow$ (given)

S' P(x_1,y_1) S

(–5,0) 2 3 (5, 0)

FIGURE 6.68

$$\Rightarrow x_1 = \frac{2(5) + 3(-5)}{5} = -1;\ y_1 = \frac{2(0) + 3(0)}{5} = 0 \Rightarrow P \equiv (-1, 0)$$

Now equation of line having slope $m = \tan 135° \equiv -1$ and passing through point $P(-1, 0)$ is

$y - (-1) = -1(x - 0)$ or $x + y + 1 = 0$ ...(1)

Equation of asymptotes: $y = \pm \frac{b}{a} x$,

taking (+) sign, $\frac{x}{4} - \frac{y}{3} = 0$ or, $3x - 4y = 0$ ...(2)

and similarly taking (–) sign, $\frac{x}{4} + \frac{y}{3} = 0$ $3x + 4y = 0$ ...(3)

Solving (1) and (2) $\Rightarrow$ point is $\left(\frac{-4}{7}, \frac{-3}{7}\right)$

Solving (1) and (3) $\Rightarrow$ point is $(-4, 3)$

ILLUSTRATION 60: The tangent at P on the hyperbola $\frac{x^2}{a^2} - \frac{y^2}{b^2} = 1$ meets one of the asymptote in Q. Show that the locus of the mid-point of PQ is a similar hyperbola.

SOLUTION: Tangent at point P ($a \sec \theta$, $b \tan \theta$) to the hyperbola $\frac{x^2}{a^2} - \frac{y^2}{b^2} = 1$ is $\frac{x \sec \theta}{a} - y\frac{\tan \theta}{b} = 1$ ...(1)

Equation of one of the asymptotes is $\Rightarrow \frac{x}{a} + \frac{y}{b} = 0$ ...(2)

$\Rightarrow x = a(\sec\theta - \tan\theta)$; $y = -b(\sec\theta - \tan\theta)$

Therefore the co-ordinates of Q on the asymptote are Q: $[a(\sec\theta - \tan\theta), -b(\sec\theta - \tan\theta)]$

Let co-ordinates of mid-point of PQ is $M(h, k)$

therefore, $h = \frac{a \sec \theta + (a \sec \theta - a \tan \theta)}{2}$ or $\frac{h}{a} = \sec \theta - \frac{\tan \theta}{2}$ ...(3)

Similarly, $k = \dfrac{b\tan\theta - b\sec\theta + b\tan\theta}{2}$ or, $\dfrac{k}{b} = \tan\theta - \dfrac{\sec\theta}{2}$ ...(4)

$(3)+(4) \Rightarrow \dfrac{h}{a}+\dfrac{k}{b}=\dfrac{\sec\theta+\tan\theta}{2}$ ...(5)

$(3)-(4) \Rightarrow \dfrac{h}{a}-\dfrac{k}{b}=\dfrac{3}{2}$ [$\sec\theta - \tan\theta$] ...(6)

now $(5)\times(6) \Rightarrow \dfrac{h^2}{a^2}-\dfrac{k^2}{b^2}=\dfrac{3}{4}$ or $4\left(\dfrac{x^2}{a^2}-\dfrac{y^2}{b^2}\right)=3$

Which is hyperbola of similar type.

ILLUSTRATION 61: Through any point P of $\dfrac{x^2}{a^2}-\dfrac{y^2}{b^2}=1$ a tangent line QPR is drawn with a fixed Gradient m, meeting the asymptotes in Q and R. Show that the product $(QP)(PR) = \dfrac{a^2b^2(1+m^2)}{a^2m^2-b^2}$.

SOLUTION: Equation of tangent at point $P(\theta)$ is. $\dfrac{\sec\theta}{a}x-\dfrac{\tan\theta}{b}y=1$ and asymptotes of hyperbola are $y=\pm\dfrac{bx}{a}$

Accordind to question, slope of tangent at point $P = m$

$\Rightarrow \dfrac{b\sec\theta}{a\tan\theta}=m \Rightarrow \sin\theta=\dfrac{b}{am}$

Calculating points Q and R; we have $R \equiv (a(\sec\theta + \tan\theta), b(\sec\theta+\tan\theta))$ and

$Q \equiv (a(\sec\theta - \tan\theta), -b(\sec\theta - \tan\theta))$

$\Rightarrow PQ^2 \cdot PR^2 = [a^2\tan^2\theta + b^2\sec^2\theta] \times [a^2\tan^2\theta + b^2\sec^2\theta]$

$= PQ.PR = (a^2\tan^2\theta + b^2\sec^2\theta)$

$$\frac{a^2b^2}{(a^2m^2-b^2)}+\frac{b^2-a^2m^2}{(a^2m^2-b^2)}=\frac{a^2b^2(1+m^2)}{a^2m^2-b^2}$$

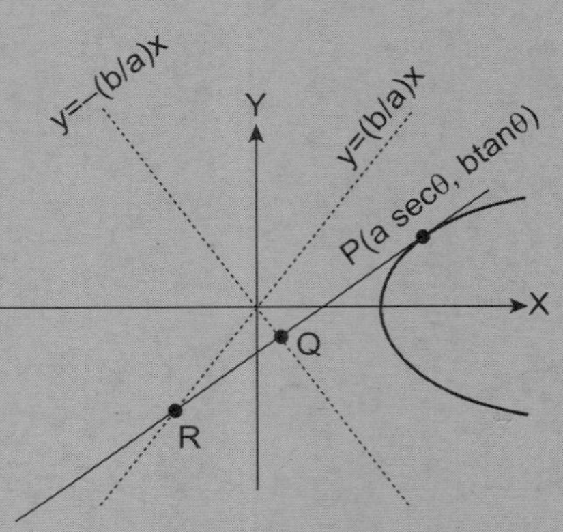

FIGURE 6.69

ILLUSTRATION 62: A transversal cuts the same branch of a hyperbola $x^2/a^2 - y^2/b^2 = 1$ in P, P' and the asymptotes in Q, Q'. Prove that:

(i) $PQ = P'Q'$ and

(ii) $PQ' = P'Q$

SOLUTION: To prove that: $PQ = P'Q'$ and $PQ' = P'Q$

equation of hyperbola is $b^2x^2 - a^2y^2 = a^2b^2$ ... (i)

Let the transversal be $y = mx + c$ .. (ii)

At the point of intersection of (i) and (ii) $b^2x^2 - a^2(mx + c)^2 = a^2b^2$

$\Rightarrow (b^2 - a^2m^2)x^2 - 2a^2mcx - a^2(c^2 + b^2) = 0$

$$\Rightarrow \frac{x_1+x_2}{2}=\frac{a^2mc}{b^2-a^2m^2} \quad \text{... (iii)}$$

where x_1, x_2 are the abscissae of point P and P' respectively.

Solving $y = mx + c$ with $b^2x^2 - a^2y^2 = 0$ (joint equation of asymptotes)

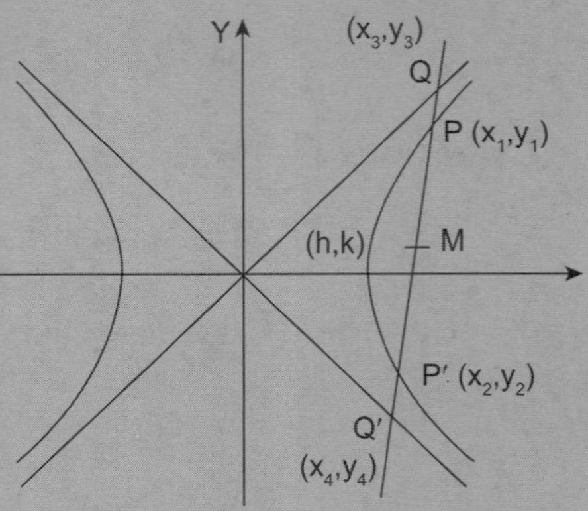

FIGURE 6.70

$(b^2 - a^2m^2)x^2 - 2a^2mcx - a^2c^2 = 0$ ....(iv)

$\Rightarrow \quad \dfrac{x_3 + x_4}{2} = \dfrac{a^2mc}{b^2 - a^2m^2}$ ... (v)

where x_3, x_4 are the abscissae of point Q and Q' respectively.

Now $y_1 = mx_1 + c$ and $y^2 = mx_2 + c$

$\Rightarrow \quad \dfrac{y_1 + y_2}{2} = m\left(\dfrac{x_1 + x_2}{2}\right) + c$

$\therefore$ The ordinatas of the mid-points of pp' and QQ' are same.
From equation (iii) and (v), we observe that mid-point M of PP' and QQ' are same.
Now $MQ = MQ'$ and $MP = MP'$

$\Rightarrow MQ - PM = MQ' - MP' \Rightarrow PQ = P'Q'$.

$\Rightarrow P'Q = P'P + PQ = P'P + P'Q' = PQ'$; Hence proved.

TEXTUAL EXERCISE–5 (SUBJECTIVE)

1. Find the equation of the hyperbola whose asymptotes are $x + 2y + 3 = 0$ and $3x + 4y + 5 = 0$ and which passes through the point (1, –1). Find also the equation of the conjugate hyperbola.

2. If ($a \sec\theta$, $b \tan\theta$) and ($a \sec\phi$, $b \tan \phi$) are the end points of a focal chord of the hyperbola $\dfrac{x^2}{a^2} - \dfrac{y^2}{b^2} = 1$ then prove that $\tan\dfrac{\theta}{2}\tan\dfrac{\phi}{2} + \left(\dfrac{e-1}{e+1}\right) = 0$.

3. Find the asymptotes of the hyperbola $2x^2 + 5xy + 2y^2 + 4x + 5y = 0$. Find also the general equation of all the hyperbolas having the same set of asymptotes

4. Show that the equations, $x = a \tan(\theta + \alpha)$ and $y = b \tan(\theta + \beta)$; where θ is a parameter represents a hyperbola; where $a - b \neq n\pi$, $n \in \mathbb{Z}$.

5. Find the hyperbola whose asymptotes are $2x - y = 3$ and $3x + y - 7 = 0$, which passes through the point (1, 2)

Answer Keys

1. $3x^2 + 10xy + 8y^2 + 14x + 22y + 23 = 0$
3. $x + 2y + 1 = 0$, $2x + y + 2 = 0$; $2x^2 + 5x + 2y^2 + 4x + 5y + 2 - k = 0$; $k \in \mathbb{R}$
5. $6x^2 - y^2 - 23x + 4y + 15 = 0$

TEXTUAL EXERCISE–5 (OBJECTIVE)

1. For the hyperbola $x^2 + 2y^2 + 3xy + 2x + 3y + 2 = 0$, combined equation of asymptotes is
   (a) $x^2 + 2y^2 + 3xy + 2x + 3y + 1 = 0$
   (b) $x^2 + 2y^2 + 3xy + 2x + 3y - 1 = 0$
   (c) $2x^2 + y^2 + 3xy + 2x + 3y = 0$
   (d) $2x^2 + y^2 + 3xy + 3x + 2y = 0$

2. The asymptotes of the hyperbola $hx + ky = xy$, are
   (a) $x - k = 0$, $y - h = 0$
   (b) $x + h = 0$, $y + k = 0$
   (c) $x - h = 0$, $y - k = 0$
   (d) $x + k = 0$, $y + h = 0$

3. For the hyperbola $\dfrac{x^2}{3} - y^2 = 3$, which of the following statements is wrong?
   (a) Its eccentricity is $\dfrac{2}{\sqrt{3}}$
   (b) Angle between its asymptotes is $\dfrac{\pi}{3}$

(c) Length of its latus rectum is 2 units

(d) Product of distances of any point of the curve from the asymptotes of the curve is less than 2

4. Consider the hyperbola $\frac{x^2}{a^2}-\frac{y^2}{b^2}=1$. Area of the triangle formed by the asymptotes and the tangent drawn to the hyperbola at the point $(a, 0)$ is equal to

(a) $\frac{ab}{2}$ (b) ab

(c) $2ab$ (d) $4ab$

5. From any point on the hyperbola $S_1 : \frac{x^2}{a^2}-\frac{y^2}{b^2}=1$, tangents are drawn to the hyperbola $S_2 : \frac{x^2}{a^2}-\frac{y^2}{b^2}=2$, locus of the mid-point of corresponding chord of contact is

(a) $\frac{a^2}{2x^2}-\frac{b^2}{2y^2}=\left(\frac{x^2}{a^2}-\frac{x^2}{b^2}\right)^2$

(b) $\frac{2x^2}{a^2}-\frac{2y^2}{b^2}=\left(\frac{x^2}{a^2}-\frac{x^2}{b^2}\right)^2$

(c) $\frac{a^2}{4x^2}-\frac{b^2}{4y^2}=\left(\frac{x^2}{a^2}-\frac{x^2}{b^2}\right)^2$

(d) $\frac{4x^2}{a^2}-\frac{4y^2}{b^2}=\left(\frac{x^2}{a^2}-\frac{x^2}{b^2}\right)^2$

6. Eccentricity of the hyperbola whose asymptotes are $4x + 5y = 5$ and $5x - 4y + 7 = 0$ is

(a) 2 (b) 1

(c) 1/2 (d) $\sqrt{2}$

7. If e is the eccentricity and θ is an angle between the asymptotes of the hyperbola $\frac{x^2}{a^2}-\frac{y^2}{b^2}=1$, then the value of $\sin(\theta/2)$ is

(a) $\frac{\sqrt{e^2-1}}{e}$ (b) $\frac{e}{\sqrt{e^2-1}}$

(c) $\sqrt{\frac{e^2-1}{e^2+1}}$ (d) $\sqrt{\frac{e^2+1}{e^2-1}}$

8. If m is the slope of an asymptote of the hyperbola $2x^2 + 5xy + 2y^2 + 4x + 5y = 0$, then $m =$

(a) 1/2 (b) 2

(c) –1/2 (d) –2.

Answer Keys

1. (a) **2.** (a) **3.** (d) **4.** (b) **5.** (d) **6.** (d) **7.** (a) **8.** (c, d)

■ RECTANGULAR HYPERBOLA AND THEIR PROPERTIES

A hyperbola whose asymptotes include a right angle is said to be a **rectangular hyperbola**. *OR*

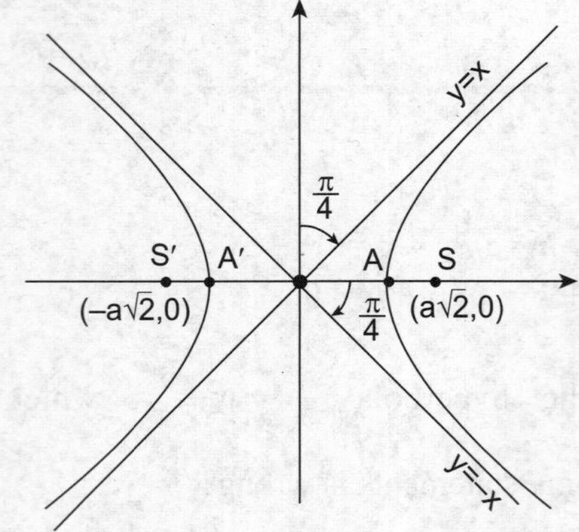

FIGURE 6.71

If the lengths of transverse and conjugate axes of any hyperbola be equal, it is called **rectangular or equilateral hyperbola.** According to the first definition, $2\tan^{-1}\left(\frac{b}{a}\right)=\frac{\pi}{2}$.

$\Rightarrow \tan^{-1}\left(\frac{b}{a}\right)=\frac{\pi}{4} \Rightarrow \frac{b}{a}=1 \Rightarrow a=b$

then, $\frac{x^2}{a^2}-\frac{y^2}{b^2}=1$ becomes $x^2-y^2=a^2$.

According to the second definition, when $a = b$, $\frac{x^2}{a^2}-\frac{y^2}{b^2}=1$ becomes $x^2-y^2=a^2$,

Eccentricity, $e=\sqrt{\left(1+\frac{b^2}{a^2}\right)}=\sqrt{2}$. Hence $x^2-y^2=a^2$ is the general form of the equation of the rectangular hyperbola.

■ FEATURES OF RECTANGULAR HYPERBOLA $x^2 - y^2 = a^2$

(i) **Transverse axis**: Equation: $y = 0$; Length: $2a = AA'$

(ii) **Conjugate axis**: Equation: $x = 0$; Length: $2b = BB'$

(iii) **Eccentricity** (e): √2
(iv) **Foci:** $(\pm a\sqrt{2}, 0)$
(v) **Directrices:** $x = \pm a/\sqrt{2}$
(vi) **Asymptotes:** $y = x$ and $y = -x$.
(vii) **Length of latus rectum:** $2a$
(viii) **Extremities of latus rectum:** $(\pm a\sqrt{2}, \pm a)$

■ THE RECTANGULAR HYPERBOLA WITH CO-ORDINATE AXES AS ASYMPTOTES

The equation of rectangular hyperbola is $x^2 - y^2 = a^2$ and its asymptotes are $x - y = 0$ and $x + y = 0$.

Since asymptotes are inclined at 45° and 135° to the positive x-axis respectively.

If we rotate the axes through $\theta = -45°$ without changing the origin, then the transformed equation is obtained by replacing (x, y) with

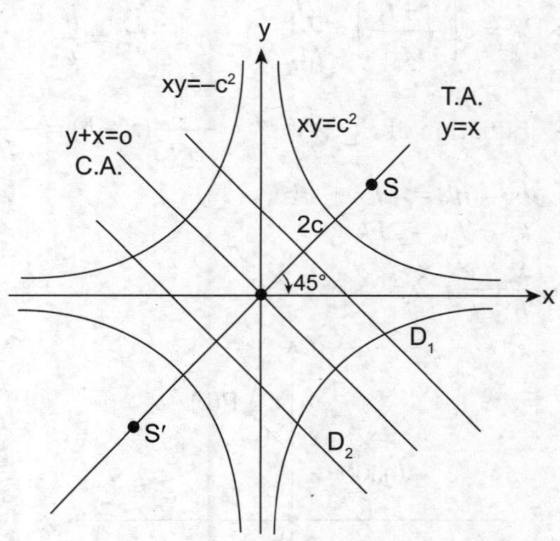

FIGURE 6.72

$$\left[x\cos(-45^o) - y\sin(-45^o), x\sin(-45^o) + y\cos(-45^o)\right].$$

as $$\left(\frac{x+y}{\sqrt{2}}\right)^2 - \left(\frac{-x+y}{\sqrt{2}}\right)^2 = a^2$$

or $$\frac{1}{2}\left\{(x+y)^2 - (-x+y)^2\right\} = a^2 \Rightarrow \frac{1}{2}(2y)(2x) = a^2$$

$$\Rightarrow \quad xy = \frac{a^2}{2} = \left(\frac{a}{\sqrt{2}}\right)^2 = c^2 \text{(say)}$$

■ PARAMETRIC FORM OF RECTANGULAR HYPERBOLA WITH *x*-AXIS AND *y*-AXIS AS THEIR ASYMPTOTES

For rectangular hyperbola $xy = c^2$ (figure 6.73) is given by $x = ct, y = \frac{c}{t}$; where $t \in \mathbb{R} \sim \{0\}$ and c is a constant.

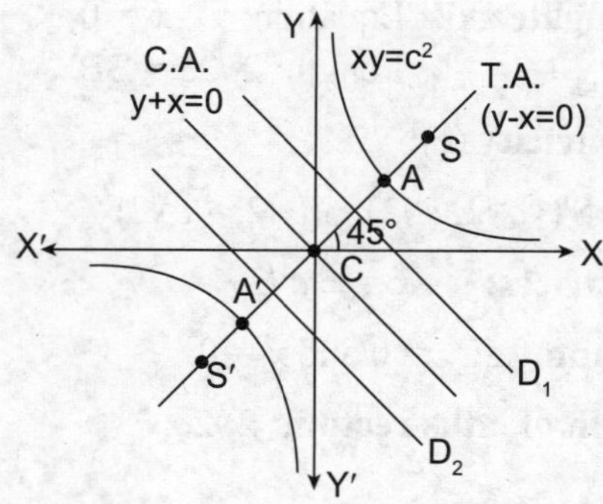

FIGURE 6.73

■ FEATURES OF ABOVE RECTANGULAR HYPERBOLA $xy = c^2$

(i) **Transverse axis**: Equation: $y - x = 0$;
Length: $2\sqrt{2}c = AA'$
(ii) **Conjugate axis**: Equation: $y + x = 0$;
(iii) **Vertices:** A, A': (c, c); (–c,–c)
Length: $2\sqrt{2}c$ = BB'
(iv) **Eccentricity** (e): √2
(v) **Foci:** $S'(-c\sqrt{2}, -c\sqrt{2})$; $S(c\sqrt{2}, c\sqrt{2})$
(vi) **Directrices:** $y + x = \pm c\sqrt{2}$
(vii) **Asymptotes:** $y = 0$ and $x = 0$.
(viii) **Length of latus rectum:** $2\sqrt{2}c$

■ CONJUGATE HYPERBOLA OF RECTANGULAR HYPERBOLA $xy = c^2$

For rectangular hyperbola $xy = -c^2$ (figure 6.74) is given by $x = ct, y = -c/t$ where $t \in R \sim \{0\}$ and c is a constant.

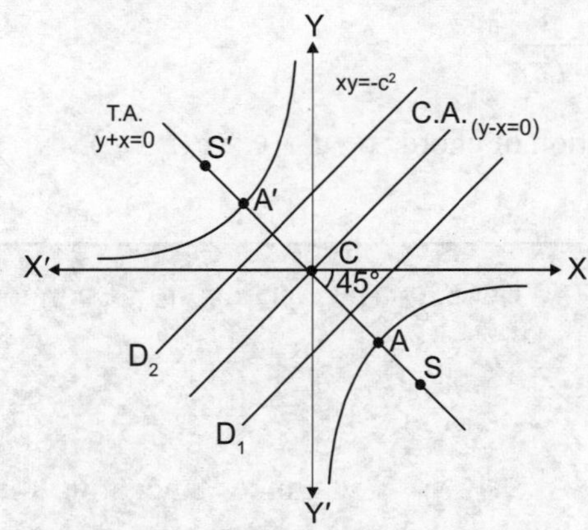

FIGURE 6.74

Features of above rectangular hyperbola

(i) **Transverse axis**: Equation: $y + x = 0$;
Length: $2\sqrt{2}c = AA'$

(ii) **Conjugate axis**: Equation: $y - x = 0$;
Length: $2\sqrt{2}c = BB'$

(iii) **Eccentricity** (e): $\sqrt{2}$

(iv) **Foci:** $S'(-c\sqrt{2}, c\sqrt{2})$; $S(c\sqrt{2}, -c\sqrt{2})$

(v) **Directrices:** $y - x = \pm c\sqrt{2}$

(vi) **Asymptotes:** $y = 0$ and $x = 0$.

(vii) **Length of latus rectum:** $2\sqrt{2}c$.

PARAMETRIC EQUATIONS OF CHORD/ TANGENT AND NORMAL TO RECTANGULAR HYPERBOLA $xy = c^2$

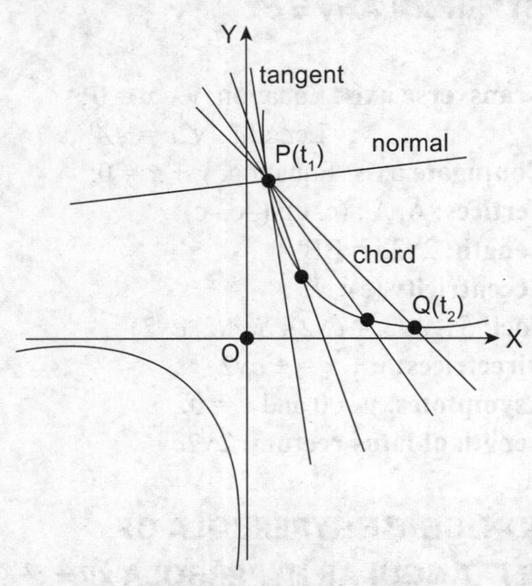

FIGURE 6.75

Slope of chord joining the points $P(t_1)$ and $Q(t_2)$: $m = -\dfrac{1}{t_1 t_2}$.

Equation of chord : $x + t_1 t_2 y = c(t_1 + t_2)$

Condition for focal chord: $\dfrac{t_1 + t_2}{1 - t_1 t_2} = \sqrt{2}$

Equation of the tangent at $P(x_1, y_1)$: $\dfrac{x}{x_1} + \dfrac{y}{y_1} = 2$

Equation of tangent at P(t): $x + yt^2 = 2ct$

Equation of normal at P (t): $y - \dfrac{c}{t} = t^2(x - ct)$
$xt^3 - yt = c(t^4 - 1)$.

CHORD OF RECTANGULAR HYPERBOLA $xy = c^2$ WITH A GIVEN MIDDLE POINT (h, k)

Proof: If the chord joining t_1 and t_2 has middle point (h, k)

$\Rightarrow 2h = c(t_1 + t_2)$ and $2k = c\left(\dfrac{1}{t_1} + \dfrac{1}{t_2}\right)$

$\Rightarrow 2k = c\left(\dfrac{t_1 + t_2}{t_1 t_2}\right) = \dfrac{2h}{t_1 t_2}$ $\quad\therefore\ t_1 t_2 = \dfrac{h}{k}$

$\Rightarrow$ Equation of PQ : $y - k = -\dfrac{1}{t_1 t_2}(x - h) = -\dfrac{k}{h}(x - h)$

$\Rightarrow hy - hk = -kx + hk$

$\Rightarrow kx + hy = 2hk$

$\Rightarrow \dfrac{x}{h} + \dfrac{y}{k} = 2$

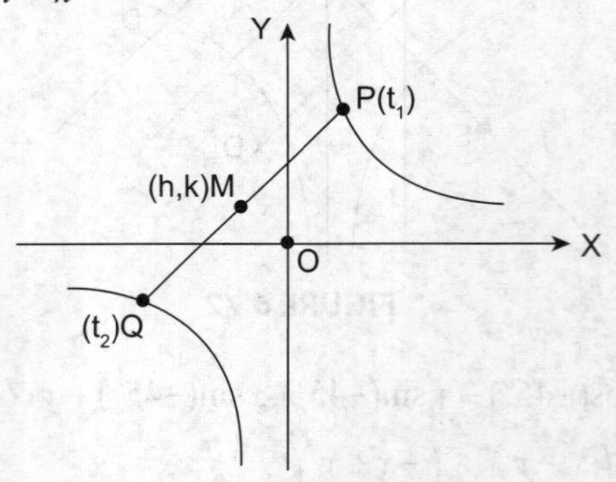

FIGURE 6.76

ILLUSTRATION 63: The equation of the common tangent to the curve $y^2 = 8x$ and $xy = -1$ is

(a) $3y = 9x + 2$
(b) $y = 2x + 1$
(c) $2y = x + 8$
(d) $y = x + 2$

SOLUTION: Equation of tangent for the curve, $y^2 = 8x$ is $y = mx + a/m$

$\Rightarrow y = mx + 2/m$ $\quad(\because a = 2)$

Since the above equation is also tangent on curve xy = – 1, thus solving together the quadratic obtained has equal roots $x\left(mx + \dfrac{2}{m}\right) = -1$

$\Rightarrow mx^2 + (2/m)\,x + 1 = 0$

$\Rightarrow D = \left(\frac{2}{m}\right)^2 - 4m = 0 \Rightarrow \frac{4}{m^2} - 4m = 0 \Rightarrow m^3 - 1 = 0$

$\therefore$ $m = 1$ hence, equation of common tangent $y = mx + 2/m$ is given as $y = x + 2$.

ILLUSTRATION 64: If normal to hyperbola $xy = c^2$ at the point $P(\mathrm{t})$ meet the hyperbola again at t', then prove that $t^3.t' = -1$.

SOLUTION: ***Proof:*** If normal at $P(t)$ meets hyperbola at $Q(t')$, then chord PQ is a normal to hyperbola at P

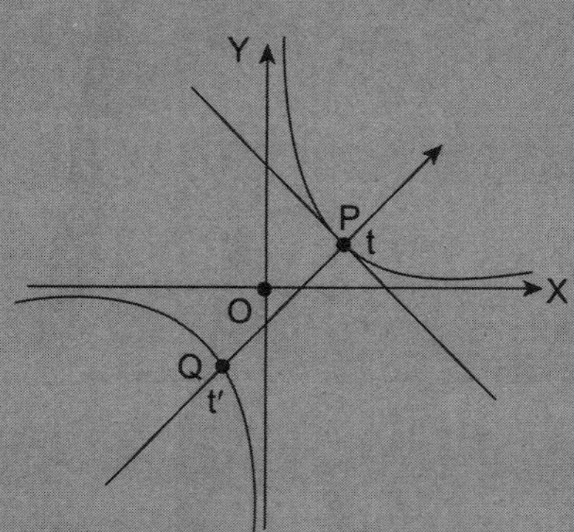

FIGURE 6.77

$\Rightarrow m_{PQ} = m_{\text{normal}}$ at $P \Rightarrow \frac{-1}{tt'} = t^2 \Rightarrow t^3t' = -1.$

ILLUSTRATION 65: Find the value of k for which the line $3x - 4y + k = 0$ is a tangent to the hyperbola $xy = 16$.

SOLUTION: Slope of the line $3x - 4y + k = 0$ is $\frac{3}{4} > 0$

$\therefore$ the given line cannot be a tangent to $xy = 16$ for any value of k. Ans: no value of k

ILLUSTRATION 66: A triangle has its vertices on a rectangular hyperbola. Prove that the orthocentre of the triangle also lies on the same hyperbola.

SOLUTION: Let 't_1, 't_2' and 't_3' are the vertices of the triangle ABC, described on the rectangular hyperbola $xy = c^2$.

$\therefore$ Co-ordinates of A, B and C are

$\left(ct_1, \frac{c}{t_1}\right), \left(ct_2, \frac{c}{t_2}\right)$ and $\left(ct_3, \frac{c}{t_3}\right)$ respectively.

Now slope of BC is $-\frac{1}{t_2t_3}$

$\Rightarrow$ Slope of AD is t_2t_3

$\Rightarrow$ Equation of altitude AD is $y - \frac{c}{t_1} = t_2t_3(x - ct_1)$ or $t_1y - c = x\,t_1t_2t_3 - ct_1^2t_2t_3$ .....(1)

Similarly, equation of altitude BE is $t_2y - c = x\,t_1t_2t_3 - ct_1t_2^2t_3$ ......(2)

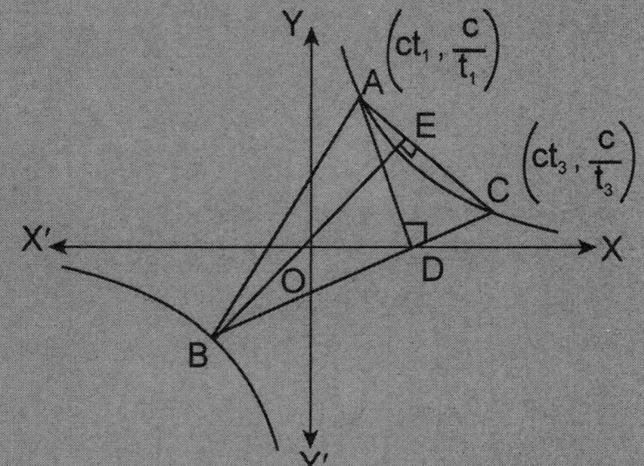

FIGURE 6.78

Solving (1) and (2), we get the orthocentre $\left(-\frac{c}{t_1t_2t_3}, -ct_1t_2t_3\right)$; which lies on $xy = c^2$.

ILLUSTRATION 67: A, B, C are three points on the rectangular hyperbola $xy = c^2$, find

(i) the area of the triangle ABC

(ii) the area of the triangle formed by the tangents at A, B and C.

SOLUTION: Let the co-ordinates of A, B and C on the hyperbola $xy = c^2$ be

$\left(ct_1, \frac{c}{t_1}\right), \left(ct_2, \frac{c}{t_2}\right)$ and $\left(ct_3, \frac{c}{t_3}\right)$ respectively.

(i) $\therefore$ Area of triangle $ABC = \left\| \begin{matrix} ct_1 & \frac{c}{t_1} & 1 \\ ct_2 & \frac{c}{t_2} & 1 \\ ct_3 & \frac{c}{t_3} & 1 \end{matrix} \right\| = \frac{c^2}{2}\left|\frac{t_1}{t_2} - \frac{t_2}{t_1} + \frac{t_2}{t_3} - \frac{t_3}{t_2} + \frac{t_3}{t_1} - \frac{t_1}{t_3}\right|$

$$= \frac{c^2}{2t_1t_2t_3}\left|t_1^2t_3 - t_2^2t_3 + t_1t_2^2 - t_3^2t_1 + t_2t_3^2 - t_1^2t_2\right| = \frac{c^2}{2t_1t_2t_3}\,|(t_1 - t_2)(t_2 - t_3)(t_3 - t_1)|$$

(ii) Equations of tangents at A, B, C are

$x + yt_1^2 - 2ct_1 = 0$

$x + yt_2^2 - 2ct_2 = 0$

and $x + yt_3^2 - 2ct_3 = 0$

$$\therefore \text{ Required area} = \frac{1}{2|C_1C_2C_3|}\begin{vmatrix} 1 & t_1^2 & -2ct_1 \\ 1 & t_2^2 & -2ct_2 \\ 1 & t_3^2 & -2ct_3 \end{vmatrix}^2 \qquad \text{.........(1)}$$

where $C_1 = \begin{vmatrix} 1 & t_2^2 \\ 1 & t_3^2 \end{vmatrix}, C_2 = -\begin{vmatrix} 1 & t_1^2 \\ 1 & t_3^2 \end{vmatrix}$ and $C_3 = \begin{vmatrix} 1 & t_1^2 \\ 1 & t_2^2 \end{vmatrix}$

$\therefore\ C_1 = t_3^2 - t_2^2,\ C_2 = t_1^2 - t_3^2$ and $C_3 = t_2^2 - t_1^2$

from (1), required area $= \dfrac{\{4c^2.(t_1 - t_2)^2\ (t_2 - t_3)^2\ (t_3 - t_1)^2\}}{2\left|(t_3^2 - t_2^2)\ (t_1^2 - t_3^2)(t_2^2 - t_1^2)\right|} = 2c^2\left|\dfrac{(t_1 - t_2)(t_2 - t_3)(t_3 - t_1)}{(t_1 + t_2)\ (t_2 + t_3)\ (t_3 + t_1)}\right|$

$\therefore$ Required area is, $2c^2\left|\dfrac{(t_1 - t_2)\ (t_2 - t_3)(t_3 - t_1)}{(t_1 + t_2)\ (t_2 + t_3)\ (t_3 + t_1)}\right|$.

ILLUSTRATION 68: If a rectangular hyperbola has the equation, $xy = c^2$, prove that the locus of the middle points of the chords of constant length $2d$ is $(x^2 + y^2)\ (xy - c^2) = d^2xy$.

SOLUTION: Let $P\left(ct_1, \frac{c}{t_1}\right), Q\left(ct_2, \frac{c}{t_2}\right)$ be the end point of a chord PQ of length $2d$ of the hyperbola $xy = c^2$.

Then, $2d = \sqrt{c^2(t_1 - t_2)^2 + c^2\left(\frac{1}{t_1} - \frac{1}{t_2}\right)^2} \Rightarrow 4d^2 = c^2(t_1 - t_2)^2\left\{1 + \frac{1}{t_1^2t_2^2}\right\}$ ...(1)

Let $R(h, k)$ be the mid-point of chord PQ. Then, $2h = c(t_1 + t_2)$ and $2k = c\left(\frac{1}{t_1} + \frac{1}{t_2}\right)$

$\Rightarrow t_1 + t_2 = \frac{2h}{c}$ and $t_1 t_2 = \frac{h}{k}$

Substituting these values in (1), $4d^2 = c^2\left(\frac{4h^2}{c^2} - \frac{4h}{k}\right)\left(1 + \frac{k^2}{h^2}\right)$

$\Rightarrow 4d^2hk = 4(hk - c^2)(h^2 + k^2)$

$\Rightarrow (h^2 + k^2)(hk - c^2) = d^2hk$

hence the locus of (h, k) is $(x^2 + y^2)(xy - c^2) = d^2xy$

ILLUSTRATION 69: If P, Q, R are the feet of normals drawn to a rectangular hyperbola from a point T on that rectangular hyperbola, then prove that the centroid of ΔPQR also lie on the same rectangular hyperbola.

SOLUTION: Let $xy = c^2$ be a rectangular hyperbola such that the normals at three points P, Q, R on the hyperbola intersect at point $T(h, k)$ on it.

The equation of the normal at a point $\left(ct, \frac{c}{t}\right)$ on $xy = c^2$ is $xt^3 - yt - ct^4 + c = 0$

It will pass through $T(h, k)$, if $ht^3 - kt - ct^4 + c = 0$

$\Rightarrow ht^3 - \frac{c^2}{h}t - ct^4 + c = 0$ [$T(h, k)$ lie on $xy = c^2 \therefore hk = c^2$]

$\Rightarrow h^2t^3 - c^2t - hct^4 + hc = 0 \Rightarrow ht^3(h - ct) + c(h - ct) = 0 \Rightarrow (h - ct)(ht^3 + c) = 0$

$\Rightarrow ht^3 + c = 0$ ($h \neq ct$)

Thus the parameters for three points other than T are given by $ht^3 + c = 0$ ...(1)

Let t_1, t_2, t_3 be the roots of this equation. Then

$t_1 + t_2 + t_3 = 0$ ...(2)

$t_1t_2 + t_2t_3 + t_3t_1 = 0$ ...(3)

$t_1t_2t_3 = -\frac{c}{h}$ ...(4).

The co-ordinates of P, Q and R are $\left(ct_1, \frac{c}{t_1}\right), \left(ct_2, \frac{c}{t_2}\right)$ and $\left(ct_3, \frac{c}{t_3}\right)$ respectively. Therefore, the co-ordinates of the centroid of ΔPQR are

$$\left(\frac{ct_1 + ct_2 + ct_3}{3}, \frac{1}{3}\left(\frac{c}{t_1} + \frac{c}{t_2} + \frac{c}{t_3}\right)\right) = \left[\frac{c}{3}(t_1 + t_2 + t_3), \frac{c}{3}\frac{\Sigma t_1t_2}{t_1t_2t_3}\right] = (0, 0) \text{ [using (2), (3), (4)]}$$

which is the centre of rectangular hyperbola $xy = c^2$.

ILLUSTRATION 70: Prove that the tangents to the ellipse $\frac{x^2}{a^2} + \frac{y^2}{b^2} = 1$ which are equally inclined to asymptotes of rectangular hyperbola $x^2 - y^2 = a^2 - b^2$ intersect on the rectangular hyperbola.

SOLUTION: We know in rectangular hyperbola $x^2 - y^2 = a^2 - b^2$, asymptotes are $x = \pm$ y. Hence any line which is equally inclined to both asymptotes is parallel to axes (x-axis, y-axis). So tangent on ellipse.

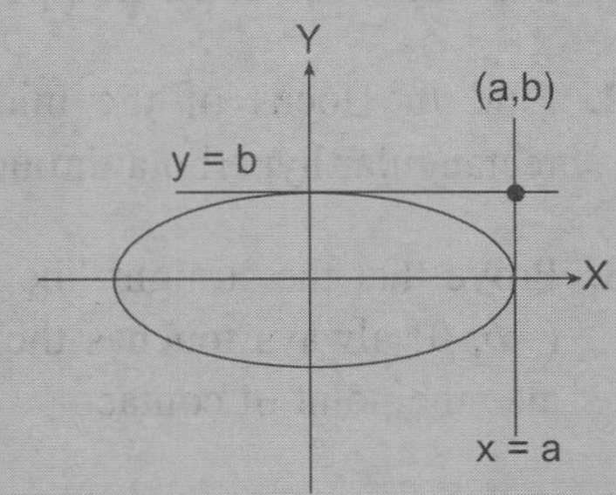

FIGURE 6.79

Intersection point of tangent $y = b$ and $x = a$ is (a, b) which lies on the rectangular hyperbola $x^2 - y^2 = a^2 - b^2$.

ILLUSTRATION 71: C is the centre of the hyperbola $\frac{x^2}{a^2}-\frac{y^2}{b^2}=1$ and the tangent at any point P meets the asymptotes in the point Q and R. Obtain the equation of the locus of centre of the circle circumscribing the triangle CQR.

SOLUTION: Equations of asymptotes be $L_1 : y=\frac{b}{a}x,\ L_2 : y=-\frac{b}{a}x$

And let L_3 be the tangent

$\therefore\ L_3 \equiv \frac{x\sec\phi}{a}-\frac{y\tan\phi}{b}-1=0$ at point P ($a\sec\phi$, $b\tan\phi$)

Point of intersection of L_1 and L_2 is C (0, 0)

Point of intersection of L_1 and L_3 is $Q\left(\frac{a\cos\phi}{1-\sin\phi},\frac{b\cos\phi}{1-\sin\phi}\right)$

and

Point of intersection of L_2 and L_3 is $R\left(\frac{a\cos\phi}{1+\sin\phi},\frac{-b\cos\phi}{1+\sin\phi}\right)$

FIGURE 6.80

Let the equation of desired circle be $x^2+y^2+2gx+2fy=0$ .....(1)

Putting co-ordinates of Q in (1); we get $\frac{(a^2+b^2)\cos\phi}{1-\sin\phi}+2ga+2fb=0$ ..... (2)

And putting co-ordinates of R in (i); we get $\frac{(a^2+b^2).\cos\phi}{1+\sin\phi}+2ga-2fb=0$ ...(3)

Solving (2) and (3), we get $f=\left(\frac{a^2+b^2}{2b}\right)\tan\phi$ and $g=\left(\frac{a^2+b^2}{2a}\right)\sec\phi$

$\because\ \sec^2\phi-\tan^2\phi=1$

$$\Rightarrow \left(\frac{2ag}{a^2+b^2}\right)^2-\left(\frac{2bg}{a^2+b^2}\right)^2=1$$

$\Rightarrow 4a^2g^2-4b^2f^2=(a^2+b^2)^2$

Substituting $(-g,-f)$ as (x, y), we have $4a^2x^2-4b^2y^2=(a^2+b^2)^2$ as the desired locus.

TEXTUAL EXERCISE–6 (SUBJECTIVE)

1. A circle and a rectangular hyperbola meet in four points A, B, C and D. If the line AB passes through the centre of the circle, prove that the centre of the hyperbola lies at the mid-point of CD.

2. Find the locus of the mid-points of chord of the rectangular hyperbola through the point $(-6c, -4c)$.

3. Prove that the straight line $y = mx + 2c\sqrt{-m}$, $m \in (-\infty, 0]$ always touches the hyperbola $xy = c^2$. Find also the point of contact.

4. A tangent to the parabola $x^2 = 4ay$, meets the hyperbola $xy = c^2$ in two points P and Q. Prove that the mid-point of PQ lies on a parabola and find its equation.

5. A variable straight line of slope 4 intersects the hyperbola $xy = 1$ at two points. Find the locus of the point which divides the line segment between these points in the ratio 1: 2.

6. A circle cuts the rectangular hyperbola $xy = 1$ in points (x_r, y_r), $r = 1, 2, 3, 4$ then prove that $x_1x_2x_3x_4 = y_1y_2y_3y_4 = 1$.

7. If the tangent and normal to a rectangular hyperbola cut off intercepts a_1 and a_2 on one axis and b_1 and b_2 on the other, show that $a_1a_2 + b_1b_2 = 0$.

Asnwer Key

2. $xy + 2cx + 3cy = 0$ **3.** $\left(\frac{c}{\sqrt{-m}}, c\sqrt{-m}\right)$

4. $x^2 = -(a/2)y$ **5.** $16x^2 + y^2 + 10xy = 2.$

TEXTUAL EXERCISE–6 (OBJECTIVE)

1. Total number of common tangents of $y^2 = 4ax$ and $xy = c^2$ is equal to
(a) 1 (b) 2
(c) 3 (d) 4

2. Tangent at any arbitrary point 'P' on the curve $xy = c^2$ meets the co-ordinate axes at A and B. Locus of the circumcenter of ΔAOB, 'O' being the origin: is
(a) $xy = c^2$ (b) $x = cy$
(c) $y = cx$ (d) None of these

3. Locus of the mid-point of the chords of the hyperbola $xy = c^2$ having slope m, is
(a) $y - mx = 0$ (b) $y + mx = 0$
(c) $my - x = 0$ (d) $my + x = 0$

4. Locus of the foot of perpendicular drawn from origin to any arbitrary tangent of the hyperbola $xy = c^2$, is
(a) $(x^2 + y^2)^2 = 4c^2$ (b) $x^2 + y^2 = 2c^2$
(c) $(x^2 + y^2)^2 = 2c^2xy$ (d) $(x^2 + y^2)^2 = 4c^2xy$

5. M and N are the feet of altitudes drawn from any point P on the hyperbola $xy = c^2$ on it's asymptotes. Locus of the mid-point of MN is
(a) $xy = 2c^2$ (b) $xy = 4c^2$
(c) $2xy = c^2$ (d) $4xy = c^2$

6. A circle cuts the rectangular hyperbola $xy = c^2$ at the points (x_i, y_i), $i = 1$ to 4. Then the values of $x_1.x_2.x_3.x_4$ and $y_1.y_2.y_3.y_4$ respectively
(a) $c^4, -c^4$ (b) $c^2, -c^2$
(c) c^4, c^4 (d) c^2, c^2

7. The equation of the chord PQ where $P(x_1, y_1)$ and $Q(x_2, y_2)$ are the points on the hyperbola $xy = c^2$, is
(a) $\frac{x}{x_1 - x_2} + \frac{y}{y_1 - y_2} = 1$
(b) $x(x_1 - x_2) + y(y_1 - y_2) = 1$
(c) $\frac{x}{x_1 + x_2} + \frac{y}{y_1 + y_2} = 1$
(d) $x(x_1 + x_2) + y(y_1 + y_2) = 1$

8. The chord PQ of the hyperbola $xy = c^2$, meets the x-axis at A. If 'R' is the mid-point of PQ and 'O' is the origin, then ΔARO is necessarily
(a) Equilateral (b) Isosceles
(c) Right angled (d) None of these

9. Consider any chord of the hyperbola $xy = c^2$, that is parallel to the line $y = x$. Circles are drawn having this chord as diameter. All these circles will pass through two fixed points, whose co-ordinates are
(a) $\left(c\sqrt{c}, c\sqrt{c}\right), \left(-c\sqrt{c}, -c\sqrt{c}\right)$
(b) $\left(-c\sqrt{c}, c\sqrt{c}\right), \left(c\sqrt{c}, -c\sqrt{c}\right)$
(c) (c, c), (–c, –c)
(d) (c, c), (–c, c)

10. A line drawn through the point $P(-1, 2)$ meets the hyperbola $xy = c^2$ at the points A and B. (Points A and B lie on the same side of P). A point 'Q' is chosen on this line such that PA, PQ and PB are in A.P., then locus of point Q is
(a) $x = y(1 + 2x)$ (b) $x = y(1 + x)$
(c) $2x = y(1 + 2x)$ (d) None of these

11. In $Q(10)$, if PA, PQ and PB are in G.P. Then locus of Q is
(a) $xy - y + 2x - c^2 = 0$
(b) $xy + y - 2x + c^2 = 0$
(c) $xy + y + 2x + c^2 = 0$
(d) $xy - y - 2x - c^2 = 0$

12. In $Q(10)$, if PA, PQ and PB are in H.P., then locus of Q is
(a) $2x - y = 2c^2$ (b) $x - 2y = 2c^2$
(c) $2x + y + 2c^2 = 0$ (d) $x + 2y = 2c^2$

13. AB and CD are any two perpendicular chords of the hyperbola $xy = c^2$. If m_{OA}, m_{OB}, m_{OC} and m_{OD} be the slopes of the line OA, OB, OC and OD respectively ('O' being the origin) then $m_{OA}\, m_{OB}\, m_{OC}\, m_{OD}$ is always equal to

(a) –1 (b) –2
(c) 1 (d) 2

14. If the normal at the point $P(x_i, y_i)$, $i \in \{1,2,3,4\}$ on the hyperbola $xy = c^2$ are concurrent at the point $Q(h, k)$, then $\frac{(x_1+x_2+x_3+x_4)(y_1+y_2+y_3+y_4)}{x_1x_2x_3x_4}$ is equal to

(a) $\frac{hk}{c^4}$ (b) $\frac{|hk|}{c^3}$
(c) $\frac{\sqrt{|hk|}}{c^3}$ (d) None of these

15. Locus of the point of intersection of tangent drawn to the hyperbola $xy = c^2$ at the extremities of any normal chord is
(a) $(x^2 - y^2)^2 - c^2xy = 0$
(b) $(x^2 - y^2)^2 + c^2xy = 0$
(c) $(x^2 - y^2)^2 - 4c^2xy = 0$
(d) $(x^2 - y^2)^2 + 4c^2xy = 0$

16. If the circle $x^2 + y^2 = a^2$ intersects the hyperbola $xy = c^2$ in four points $P(x_1, y_1)$, $Q(x_2, y_2)$, $R(x_3, y_3)$, $S(x_4, y_4)$ then
(a) $x_1 + x_2 + x_3 + x_4 = 0$
(b) $y_1 + y_2 + y_3 + y_4 = 0$
(c) $x_1x_2x_3x_4 = c^4$
(d) $y_1y_2y_3y_4 = c^4$

17. A straight line touching the rectangular hyperbola $9x^2 - 9y^2 = 8$ and the parabola $y^2 = 32x$ is (are)
(a) $9x + 3y - 8 = 0$ (b) $9x - 3y + 8 = 0$
(c) $9x + 3y + 8 = 0$ (d) $9x - 3y - 8 = 0$

18. If the line $ax + by + c = 0$ is a normal to the hyperbola $xy = 1$, then
(a) $a > 0, b > 0$ (b) $a > 0, b < 0$
(c) $a < 0, b < 0$ (d) $a < 0, b > 0$.

19. The transverse axis of a rectangular hyperbola is $2c$ and the asymptotes are the axes of co-ordinates. Then
(a) equation of the hyperbola is $xy = \frac{c^2}{2}$
(b) equation of the hyperbola is $xy = c^2$
(c) equation of the chord whose mid-point is $(2c, 3c)$ is $3x + 2y = 12c$.
(d) equation of the chord whose mid-point is $(2c, 3c)$ is $3x + 2y = 6c$.

20. If the sum of the slopes of the normals from a point P to the hyperbola $xy = c^2$ is equal to $\lambda(\lambda \in R^+)$, then locus of point P is
(a) $x^2 - y^2 = \lambda c^2$ (b) $y^2 = \lambda c^2$
(c) $xy = \lambda c^2$ (d) None of these

21. A, B, P are three points on $xy = c^2$ such that AB subtends a right angle at P. Then
(a) AB is parallel to the normal at P
(b) AB is parallel to the tangent at P
(c) the circumradius of the $\Delta APB = \frac{|c|}{2}\left|\frac{(t_1 - t_2)}{t_1t_2}\right| . \sqrt{t_1^2t_2^2 + 1}$
(d) the circumradius of the ΔAPB is $\frac{c}{2}\left|\frac{t_1 - t_2}{t_1t_2}\right|$.

22. Length of each latus rectum of the hyperbola $xy = c^2$. is equal to
(a) $2c$ (b) $\sqrt{2}\, c$
(c) $2\sqrt{2}\, c$ (d) $4c$

23. Area of the triangle formed by any arbitrary tangent of the hyperbola $xy = c^2$, with co-ordinate axes, is equal to
(a) $2c^2$ (b) $\sqrt{2}\, c^2$
(c) $2\sqrt{2}\, c^2$ (d) $4c^2$

24. Equation of a line passing through the center of a rectangular hyperbola is $x - y - 1 = 0$. If one of its asymptotes is $3x - 4y - 6 = 0$, then equation of its other asymptote is
(a) $4x + 3y - 17 = 0$ (b) $4x + 3y + 17 = 0$
(c) $4x + 3y - 15 = 0$ (d) $4x + 3y + 15 = 0$

Answer Keys

1. (a)	**2.** (a)	**3.** (b)	**4.** (d)	**5.** (d)	**6.** (c)	**7.** (c)	**8.** (b)	**9.** (c)	**10.** (c)
11. (b)	**12.** (a)	**13.** (c)	**14.** (d)	**15.** (d)	**16.** (a,b,c,d)	**17.** (b,c)	**18.** (b,d)	**19.** (a, c)	**20.** (d)
21. (a, c)	**22.** (c)	**23.** (a)	**24.** (a)						

MULTIPLE-CHOICE QUESTIONS

SECTION–I

OBJECTIVE-TYPE SOLVED EXAMPLES

1. If the foci of the ellipse $\frac{x^2}{16}+\frac{y^2}{b^2}=1$ and the hyperbola $\frac{x^2}{144}-\frac{y^2}{81}=\frac{1}{25}$ coincide, then length of each latera recta of ellipse is

(a) $\frac{5}{2}$ (b) $\frac{7}{2}$

(c) $\frac{9}{2}$ (d) None of these

Solution: (b) Let eccentricity of ellipse and hyperbola be e_1 and e_2 respectively, then

$b^2 = 16(1 - e_1^2)$ and $81 = 144(e_2^2 - 1)$

$\Rightarrow e_1^2 = 1-\frac{b^2}{16}$ and $e_2^2 = \frac{81}{144}+1=\frac{225}{144}$

$\Rightarrow e_2 = \frac{15}{12}$

Now given $4e_1 = \frac{12}{5}e_2$

$\Rightarrow 4\sqrt{1-\frac{b^2}{16}} = \frac{12}{5}\times\frac{15}{12}=3$

$\Rightarrow 1-\frac{b^2}{16}=\frac{9}{16}$

$\Rightarrow b^2 = \frac{7}{16}\times 16 \Rightarrow b^2 = 7$

Length of each latera recta of the ellipse $=\frac{2b^2}{a}=\frac{2\times 7}{4}=\frac{7}{2}$

2. The diameter of $16x^2 - 9y^2 = 144$ which is conjugate to $x = 2y$ is

(a) $y=\frac{16x}{9}$ (b) $y=\frac{32\,x}{9}$

(c) $x=\frac{16\,y}{9}$ (d) $x=\frac{32\,y}{9}$

Solution: (b) Diameters $y = m_1x$ and $y = m_2x$ are conjugate diameters of the hyperbola $\frac{x^2}{a^2}-\frac{y^2}{b^2}=1$ if $m_1m_2 = $ —

Here, $a^2 = 9$, $b^2 = 16$ and $m_1 = 1/2$

$\therefore\ m_1m_2 = \frac{b^2}{a^2} \Rightarrow \frac{1}{2}(m_2) = \frac{16}{9} \Rightarrow m_2 = \frac{32}{9}$

Thus, the required diameter is $y = \frac{32x}{9}$

3. The curve represented by $x = a(\cosh\theta + \sinh\theta)$, $y = b(\cosh\theta - \sinh\theta)$ is

(a) a hyperbola (b) an ellipse

(c) a parabola (d) a circle

Solution: (a) We have,

$$\frac{x}{a}.\frac{y}{b} = (\cosh\theta + \sinh\theta)(\cosh\theta - \sinh\theta) = \cosh^2\theta - \sinh^2\theta = 1$$

$\therefore\ xy, = ab$, which is a hyperbola.

4. Tangents at any point on the hyperbola $\frac{x^2}{a^2}-\frac{y^2}{b^2}=1$ cut the axes at A and B respectively. If the rectangle $OAPB$(where O is origin) is completed then locus of point P is given by

(a) $\frac{a^2}{x^2}-\frac{b^2}{y^2}=1$ (b) $\frac{a^2}{x^2}+\frac{b^2}{y^2}=1$

(c) $\frac{a^2}{y^2}-\frac{b^2}{x^2}=1$ (d) None of these

Solution: (a) Equation of the tangent at the point 'θ' is

$$\frac{x\sec\theta}{a}-\frac{y\tan\theta}{b}=1$$

$\Rightarrow A \equiv (a\cos\theta, 0)$, $B \equiv (0, -b\cot\theta)$

Let P be (h, k)

$\Rightarrow h = a\cos\theta$, $k = -b\cot\theta$

$\Rightarrow \frac{k}{h} = -\frac{b}{a\sin\theta} \Rightarrow \sin\theta = -\frac{bh}{ak}$

$\Rightarrow \frac{b^2h^2}{a^2k^2}+\frac{h^2}{a^2}=1$

$\Rightarrow \frac{b^2}{k^2}+1=\frac{a^2}{h^2} \Rightarrow \frac{a^2}{h^2}-\frac{b^2}{k^2}=1$

Thus locus of P is $\frac{a^2}{x^2}-\frac{b^2}{y^2}=1$

6. If PN is the perpendicular drawn from a point P on $xy = c^2$ to its asymptote, then locus of the mid-point of PN is

(a) circle (b) parabola

(c) ellipse (d) hyperbola

Solution: (d) Let $P \equiv (ct, c/t) \Rightarrow N \equiv (ct, 0)$ or $(0, c/t)$

If $Q(h, k)$ the mid-point of PN, then $h = ct/2$, $k = c/t$

$\Rightarrow hk = c^2/2$

$\Rightarrow$ Locus of Q is, $xy = c^2/2$ which is clearly a hyperbola.

6. If $x = 9$ is the chord of contact of the hyperbola $x^2 - y^2 = 9$, then the equation of the corresponding pair of tangents is:

(a) $9x^2 - 8y^2 + 18x - 9 = 0$

(b) $9x^2 - 8y^2 - 18x + 9 = 0$

(c) $9x^2 - 8y^2 - 18x - 9 = 0$

(d) $9x^2 - 8y^2 + 18x + 9 = 0$

Solution: (b) Let (h, k) be point whose chord of contact w.r.t. hyperbola $x^2 - y^2 = 9$ is $x = 9$

We know that chord of contact of (h, k) w.r.t. hyperbola $x^2 - y^2 = 9$

is $T = 0 \Rightarrow h.x + k(-y) - 9 = 0$

$\therefore$ $hx - ky - 9 = 0$ but it is the equation of the line $x = 9$.

This is possible when $h = 1, k = 0$ (by comparing both equations).

Again equation of pair of tangents is $T^2 = SS_1$

$\Rightarrow (x - 9)^2 = (x^2 - y^2 - 9)(1^2 - 0^2 - 9)$

$\Rightarrow x^2 - 18x + 81 = (x^2 - y^2 - 9)(-8)$

$\Rightarrow x^2 - 18x + 81 = -8x^2 + 8y^2 + 72$

$\Rightarrow 9x^2 - 8y^2 - 18x + 9 = 0$

7. For hyperbola $\frac{x^2}{\cos^2\alpha} - \frac{y^2}{\sin^2\alpha} = 1$, which of the following remains constant with change in 'α'?

(a) abscissae of vertices

(b) abscissae of foci

(c) eccentricity

(d) directrix

Solution: (b) Here, the hyperbola is given by,

$$\frac{x^2}{\cos^2\alpha} - \frac{y^2}{\sin^2\alpha} = 1$$

$\Rightarrow a^2 = \cos^2\alpha$ and $b^2 = \sin^2\alpha$

{i.e., comparing with standard equation $\frac{x^2}{a^2} - \frac{y^2}{b^2} = 1$}

Now, we know foci = $(\pm ae, 0)$

where $ae = \sqrt{a^2 + b^2} = \sqrt{\cos^2\alpha + \sin^2\alpha} = 1$

$\Rightarrow$ foci = $(\pm 1, 0)$

whereas vertices are $(\pm\cos\alpha, 0)$

$\Rightarrow$ foci remains constant with change in 'α'.

8. If $P(a \sec\alpha, b\tan\alpha)$ and $Q(a\sec\beta, b\tan\beta)$ are two points on the hyperbola $\frac{x^2}{a^2} - \frac{y^2}{b^2} = 1$ such that $\alpha - \beta = 2\theta$ (a constant), then PQ touches the hyperbola

(a) $\frac{x^2}{a^2\sec^2\theta} - \frac{y^2}{b^2} = 1$

(b) $\frac{x^2}{a^2} - \frac{y^2}{b^2\sec^2\theta} = 1$

(c) $\frac{x^2}{a^2} - \frac{y^2}{b^2} = \cos 2\theta$

(d) None of these

Solution: (a) The equation of chord PQ is

$$\frac{x}{a}\cos\left(\frac{\alpha-\beta}{2}\right) - \frac{y}{b}\sin\left(\frac{\alpha+\beta}{2}\right) = \cos\left(\frac{\alpha+\beta}{2}\right)$$

$$\Rightarrow \frac{x}{a}\cos\theta - \frac{y}{b}\sin\left(\frac{\alpha+\beta}{2}\right) = \cos\left(\frac{\alpha+\beta}{2}\right)$$

$$\Rightarrow \frac{x}{a\sec\theta}\sec\left(\frac{\alpha+\beta}{2}\right) - \frac{y}{b}\tan\left(\frac{\alpha+\beta}{2}\right) = 1$$

Clearly, it touches the hyperbola $\frac{x^2}{a^2\sec^2\theta} - \frac{y^2}{b^2} = 1$.

9. The locus of point of intersection of tangents at the ends of chord normal to the hyperbola $x^2 - y^2 = a^2$ is

(a) $y^4 - x^4 = 4a^2x^2y^2$ (b) $y^2 - x^2 = 4a^2x^2y^2$

(c) $a^2(y^2 - x^2) = 4x^2y^2$ (d) $y^2 + x^2 = 4a^2x^2y^2$.

Solution: (c) Let $P(h, k)$ be the point of intersection of tangents at the ends of a normal chord of the hyperbola $x^2 - y^2 = a^2$.

Then the equation of the chord is $hx - ky = a^2$....(i)

But, it is a chord normal to the hyperbola.

So, its equation must be of the form $x\cos\theta + y\cot\theta = 2a$ ......(ii)

Equation (i) and (ii), represent the same line.

$$\therefore \frac{\cos\theta}{h} = \frac{\cot\theta}{-k} = \frac{2a}{a^2}$$

$\Rightarrow \sec\theta = \frac{a}{2h}$ and $\tan\theta = -\frac{a}{2k}$

$\therefore \sec^2\theta - \tan^2\theta = 1 \Rightarrow \frac{a^2}{4h^2} - \frac{a^2}{4k^2} = 1$

$\Rightarrow a^2(k^2 - h^2) = 4h^2k^2$

$\Rightarrow$ locus of P is $a^2(y^2 - x^2) = 4x^2y^2$.

10. The equation of the common tangent to the curve $y^2 = 8x$ and $xy = -1$ is

(a) $3y = 9x + 2$ (B) $y = 2x + 1$

(c) $2y = x + 8$ (d) $y = x + 2$

Solution: (d) Equation of tangent for the curve, $y^2 = 8x$ is $y = mx + \frac{a}{m}$ i.e., $y = mx + \frac{2}{m}$.

Since above equation is also tangent to curve $xy = -1$

$\Rightarrow x\left(mx + \frac{2}{m}\right) = -1 \Rightarrow mx^2 + \frac{2}{m}x = -1$

$\Rightarrow mx^2 + \frac{2}{m}x + 1 = 0$ must have equal roots.

$\Rightarrow D = 0 \Rightarrow \left(\frac{2}{m}\right)^2 - 4m = 0 \Rightarrow \frac{4}{m^2} - 4m = 0$

$\Rightarrow \frac{4 - 4m^3}{m^2} = 0 = 0$

$\Rightarrow \frac{4(1-m^3)}{m^2} = 0$

$\Rightarrow m^3 - 1 = 0$

$\therefore$ $m = 1$; hence, equation of common tangent is $y = mx + \frac{2}{m} \Rightarrow y = x + 2$.

11. Consider a branch of the hyperbola, $x^2 - 2y^2 - 2\sqrt{2}x - 4\sqrt{2}y - 6 = 0$ with vertex at the point A. Let B be one of the end points of its latus rectum. If C is the focus of the hyperbola nearest to the point A, then the area of the triangle ABC is

(a) $1 - \sqrt{\frac{2}{3}}$ (b) $\sqrt{\frac{3}{2}} - 1$

(c) $1 + \sqrt{\frac{2}{3}}$ (d) $\sqrt{\frac{3}{2}} + 1$

Solution: (b) $x^2 - 2y^2 - 2\sqrt{2}x - 4\sqrt{2}y - 6 = 0$

$\Rightarrow x^2 - 2\sqrt{2}x - 2y^2 - 4\sqrt{2}y - 6 = 0$

$\Rightarrow (x^2 - 2\sqrt{2}x + 2) - 2(y^2 + 2\sqrt{2}y + 2) - 2 + 4 - 6 = 0$

$\Rightarrow (x-\sqrt{2})^2 - 2(y+\sqrt{2})^2 - 4 = 0$

$\Rightarrow (x-\sqrt{2})^2 - 2(y+\sqrt{2})^2 = 4$

$\Rightarrow \frac{(x-\sqrt{2})^2}{4} - \frac{2(y+\sqrt{2})^2}{4} = 1$

$\Rightarrow \frac{(x-\sqrt{2})^2}{4} - \frac{(y+\sqrt{2})^2}{2} = 1$

Centre $\equiv (\sqrt{2}, -\sqrt{2})$; $a^2 = 4$; $b^2 = 2$

$\Rightarrow e^2 = 1 + \frac{b^2}{a^2} = 1 + \frac{2}{4} = \frac{3}{2} \Rightarrow e = \sqrt{\frac{3}{2}}$

One end point of one of the latera recta is

$B(\sqrt{2} + ae, -\sqrt{2} + b^2/a) = \left(\sqrt{2} + 2.\sqrt{\frac{3}{2}}, -\sqrt{2} + \frac{2}{2}\right)$

$\equiv (\sqrt{2} + \sqrt{6}, -\sqrt{2} + 1)$; Vertex of hyperbola is

$A \equiv (\sqrt{2} + a, -\sqrt{2}) \equiv (\sqrt{2} + 2, -\sqrt{2})$

One of the foci of hyperbola is $C \equiv (\sqrt{2} + ae, -\sqrt{2})$

$\equiv (\sqrt{2} + 2.\sqrt{3/2}, -\sqrt{2}) \equiv (\sqrt{2} + \sqrt{6}, -\sqrt{2})$

$\therefore$ area $(\Delta ABC) = \frac{1}{2}$ AC.BC

$= \frac{1}{2}(\sqrt{6} + \sqrt{2} - \sqrt{2} - 2).(-\sqrt{2} + 1 + \sqrt{2})$

$= \frac{1}{2}(\sqrt{6} - 2).1$

$= \frac{\sqrt{6} - 2}{2} = \sqrt{\frac{6}{4}} - 1 = \sqrt{\frac{3}{2}} - 1$.

12. PQ is a double ordinate of the hyperbola $\frac{x^2}{a^2} - \frac{y^2}{b^2} = 1$ such that OPQ is an equilateral triangle, O being the centre of the hyperbola, then the eccentricity of the hyperbola satisfies.

(a) $1 < e < \frac{2}{\sqrt{3}}$ (b) $e = \frac{2}{\sqrt{3}}$

(c) $e = \frac{\sqrt{3}}{2}$ (d) $e > \frac{2}{\sqrt{3}}$

Solution: (d) Let the length of each side of the equilateral triangle OPQ be ℓ units ($l > a$). Then, the co-ordinates of p are $\left(\sqrt{3}\frac{\ell}{2}, \frac{\ell}{2}\right)$. Point P$\left(\frac{\sqrt{3}\ell}{2}, \frac{\ell}{2}\right)$ lies on the hyperbola $\frac{x^2}{a^2} - \frac{y^2}{b^2} = 1$

$\therefore \frac{3\ell^2}{4a^2} - \frac{\ell^2}{4b^2} = 1$

$\Rightarrow (3b^2 - a^2)\ell^2 = 4a^2b^2$

$\Rightarrow (3e^2 - 4)\ell^2 = 4a^2(e^2 - 1) \Rightarrow \ell = 2a\sqrt{\frac{e^2 - 1}{3e^2 - 4}}$

ℓ is real and $e > 1$

$\Rightarrow e^2 > \frac{4}{3} \Rightarrow e > \frac{2}{\sqrt{3}}$

13. A hyperbola, having the transverse axis of length $2\sin\theta$, is confocal with the ellipse $3x^2 + 4y^2 = 12$. Then its equation is

(a) $x^2 \operatorname{cosec}^2\theta - y^2\sec^2\theta = 1$

(b) $x^2\sec^2\theta - y^2\operatorname{cosec}^2\theta = 1$

(c) $x^2\sin^2\theta - y^2\cos^2\theta = 1$

(d) $x^2\cos^2\theta - y^2\sin^2\theta = 1$

Solution: Equation of ellipse is, $3x^2 + 4y^2 = 12$

$\Rightarrow \dfrac{x^2}{4} + \dfrac{y^2}{3} = 1$

$\Rightarrow a^2 = 4,\ b^2 = 3,\ e^2 = 1 - b^2/a^2 = 1/4$

Now, for hyperbola $A^2e_1^2 = A^2 + B^2$

$\Rightarrow (1)^2 = \sin^2\theta + B^2 \qquad (\because Ae_1 = ae)$

$\Rightarrow B^2 = \cos^2\theta$

Hence, equation of hyperbola is $\dfrac{x^2}{A^2} - \dfrac{y^2}{B^2} = 1$

i.e., $\dfrac{x^2}{\sin^2\theta} - \dfrac{y^2}{\cos^2\theta} = 1$ or $\operatorname{cosec}^2\theta\, x^2 - \sec^2\theta y^2 = 1$

SECTION–II

SUBJECTIVE-TYPE SOLVED EXAMPLES

1. If radii of director circles of $\dfrac{x^2}{a^2} + \dfrac{y^2}{b^2} = 1$ & $\dfrac{x^2}{a^2} - \dfrac{y^2}{b^2} = 1$ are $2r$ and r respectively and e_e and e_h be the eccentricities of the ellipse and hyperbola respectively, then show that $4e_h^2 - e_e^2 = 6$.

Solution: The equations of the director circles of ellipse and hyperbola are

$x^2 + y^2 = a^2 + b^2$ and $x^2 + y^2 = a^2 - b^2$ respectively.

$\therefore\ 2r = \sqrt{a^2+b^2}$ and $r = \sqrt{a^2-b^2}$

$\Rightarrow 2\sqrt{a^2-b^2} = \sqrt{a^2+b^2}$

$\Rightarrow 4(a^2-b^2) = a^2+b^2$

$\Rightarrow 4\left(1-\dfrac{b^2}{a^2}\right) = 1+\dfrac{b^2}{a^2}$

$\Rightarrow 4\{1-(e_h^2-1)\} = 1+1-e_e^2$

$\Rightarrow 8 - 4e_h^2 = 2 - e_e^2$

$\Rightarrow 4e_h^2 - e_e^2 = 6.$

2. A line drawn through the point $P(-1, 2)$ meets the hyperbola $xy = c^2$ at the point A and B. (point A and B lie on the same side of P). A point Q is chosen on the line such that PA, PQ and PB are in $GP.$, then find the possible value of 'c' such that locus of point Q is a hyperbola.

Solution: Any line drawn through $P(-1, 2)$ is

$\dfrac{x+1}{\cos\theta} = \dfrac{y-2}{\sin\theta} = r$

Any point on this line is $x = r\cos\theta - 1,\ y = 2 + r\sin\theta$.

Putting it in $xy = c^2$, we get

$(r\cos\theta - 1)(2 + r\sin\theta) = c^2$

$\Rightarrow r^2\sin\theta\cos\theta + r(2\cos\theta - \sin\theta) - 2 - c^2 = 0$

$\Rightarrow PA\,.\,PB = \dfrac{-(2+c^2)}{\sin\theta\cos\theta} = PQ^2$

$\Rightarrow 2 + c^2 + (PQ\sin\theta)(PQ\cos\theta) = 0$

$\Rightarrow 2 + c^2 + (y-2)(x+1) = 0$

$\Rightarrow xy + y - 2x + c^2 = 0$

For the above curve to represent hyperbola,

$c^2 \neq -2 \Rightarrow \forall\, c \in \mathbb{R}$, locus is hyperbola.

3. Find the equations of the tangents to the hyperbola $16x^2 - 25y^2 - 96x + 100y - 356 = 0$, which makes an angle of $\pi/4$ with conjugate axis.

Solution: Equation of given hyperbola is

$(16x^2 - 96x) - (25y^2 - 100y) = 356$

or $16(x^2 - 6x) - 25(y^2 - 4y) = 356$

or $16[x^2 - 6x + 9] - 25[y^2 - 4y + 4] =$

$356 + 144 - 100$

or $16(x-3)^2 - 25(y-2)^2 = 400$

or $\dfrac{(x-3)^2}{(20/4)^2} - \dfrac{(y-2)^2}{(20/5)^2} = 1$

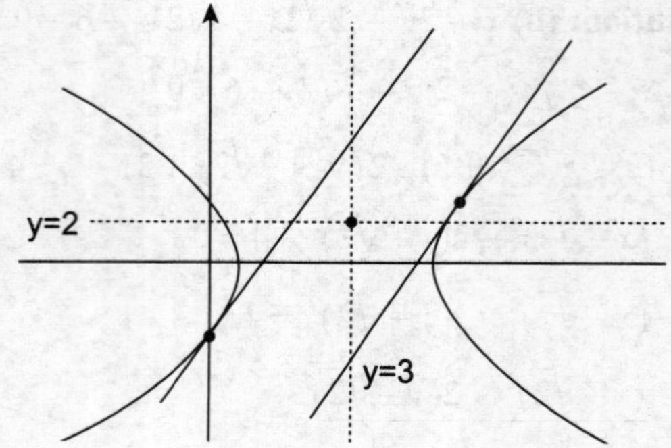

or $\frac{(x-3)^2}{(5)^2}-\frac{(y-2)^2}{(4)^2}=1$ i.e., $\frac{X^2}{(5)^2}-\frac{Y^2}{(4)^2}=1$

here $a^2 = 25$, $b^2 = 16$, equation of transverse axis is $Y = 0$ i.e., $y - 2 = 0$ and that of conjugate axis is $X = 0$ i.e., $x - 3 = 0$ and centre of hyperbola is at (3, 2)

If the tangent is forming an angle $\pi/4$ with conjugate axis, then it also forms an angle $\pi/4$ with transverse axis.

$\therefore$ $m = \tan \pi/4 = 1$

$\therefore$ Equation of tangents are given by

$Y = mX \pm \sqrt{a^2m^2 - b^2}$

$\Rightarrow$ $(y - 2) = m(x - 3) \pm \sqrt{25-16}$; m = 1

$\Rightarrow$ $y - 2 = (x - 3) \pm 3$

$\Rightarrow$ $y - 2 = x$ or $y - 2 = x - 6$

$\Rightarrow$ $x - y + 2 = 0$ or $x - y - 4 = 0$ are required tangents.

4. The equation of a line passing through the centre of a rectangular hyperbola is $x - y - 1 = 0$.
If one of its asymptotes is $3x - 4y - 6 = 0$, then find
(i) The equation of other asymptote.
(ii) The equation of hyperbola. If it passes through the point (4, 5).
(iii) Length of transverse axis of hyperbola.

Solution:

(i) Asymptotes pass through centre, therefore centre is the point of intersection of line $x - y - 1 = 0$ and asymptote $3x - 4y - 6 = 0$ i.e., $(-2, -3)$
In rectangular hyperbola, asymptotes are perpendicular to each other

$\therefore$ slope of other will be $-4/3$.

$\therefore$ equation of other asymptote is

$(y + 3) = \frac{-4}{3}(x+2)$

or $4x + 3y + 17 = 0$.

(ii) Further, we know that the equation of the hyperbola differs by a constant from the joint equation of the asymptotes.

$\therefore$ equation of hyperbola would be
$(3x - 4y - 6)(4x + 3y + 17) + \lambda = 0$
It passes through (4, 5)

$\Rightarrow$ $(12 - 20 - 6)(16 + 15 + 17) + \lambda = 0$

$\Rightarrow$ $(-14)(48) + \lambda = 0$

$\therefore$ Equation of hyperbola is
$(3x - 4y - 6)(4x + 3y + 17) + 672 = 0$ .. (1)

(iii) Axes of hyperbola will be the angle bisector of angle between the asymptotes; therefore axes are given by $\frac{3x-4y-6}{5} = \pm\frac{(4x+3y+17)}{5}$

or $3x - 4y - 6 = \pm(4x + 3y + 17)$

i.e., $x + 7y + 23 = 0$ ... (2)

and $7x - y + 11 = 0$ ... (3)

Solving (1) and (2); we do not get any real value of (x, y) and hence $x + 7y + 23 = 0$ cannot be the transverse axis of the hyperbola.

$\therefore$ $7x - y + 11 = 0$ is the transverse axis of the hyperbola.

Solving (1) and (3); we get the points of intersection as $\left(\frac{\sqrt{672}-50}{25}, \frac{7\sqrt{672}-75}{25}\right)$

and $\left(\frac{-\sqrt{672}-50}{25}, \frac{-7\sqrt{672}-75}{25}\right)$

And these points are the vertices of the hyperbola.

$\therefore$ Length of transverse axis $= 2a = (AA')$

$$= \sqrt{\left(\frac{2\sqrt{672}}{25}\right)^2 + \left(\frac{14\sqrt{672}}{25}\right)^2} = \frac{16}{5}\sqrt{21}$$

5. If radii of director circles of ellipse $\frac{x^2}{a^2}+\frac{y^2}{b^2}=1$ and hyperbola $\frac{x^2}{a^2}-\frac{y^2}{b_1^2}=1$ are in the ratio 3: 2, then prove that $9e_2^2 - 4e_1^2 = 10$; where e_1 = eccentricity of ellipse and e_2 = eccentricity of hyperbola.

Solution: Equation of director circle of ellipse $\frac{x^2}{a^2}+\frac{y^2}{b^2}=1$ is $x^2 + y^2 = a^2 + b^2$ ... (i)

and that of hyperbola $\frac{x^2}{a^2}-\frac{y^2}{b_1^2}=1$ is

$x^2 + y^2 = a^2 - b_1^2$ ... (ii)

$\therefore$ radius of (i) $= r_1 = \sqrt{a^2+b^2}$

and radius of (ii) $= r_2 = \sqrt{a^2-b_1^2}$

A.T.Q., $\frac{r_1}{r_2}=\frac{3}{2} \Rightarrow \frac{\sqrt{a^2+b^2}}{\sqrt{a^2-b_1^2}} = \frac{3}{2}$

$\Rightarrow$ $4(a^2 + b^2) = 9(a^2 - b_1^2)$

$\Rightarrow \; 4\left(1+\frac{}{a^2}\right)=9\left(1-\frac{}{a^2}\right)$

$\Rightarrow \; 4\left[1+(1-e_1^2)\right]=9\left[1-(e_2^2-1)\right]$

$\Rightarrow \; 4+4-4e_1^{\,2}=9-9e_2^{\,2}+9$

$\Rightarrow \; 9e_2^{\,2}-4e_1^{\,2}=10.$

6. Show that the locus of a point trisecting a variable line segment AB of slope 4, which intersects the rectangular hyperbola $xy = 1$ at points A and B is again a hyperbola. Find the equations of that hyperbola.

Solution: Let $A(x_1, y_1)$ and $B(x_2, y_2)$ be two points lying on the rectangular hyperbola.

Let $P(h, k)$ and $Q(h', k')$ be the points trisecting the line segment AB

i.e., $AP: PB = 1: 2$ and $BQ: QA = 1: 2$.

We shall find the equation of locus of point P

we have $h=\frac{2x_1+x_2}{3}$ $\quad k=\frac{2y_1+y_2}{3}$

now equation of AB is $(y - k) = 4(x - h)$ ... (i)

The abscissae of point of intersection of (i) and hyperbola $xy = 1$

are roots of equation $x[k + 4(x - h)] = 1$

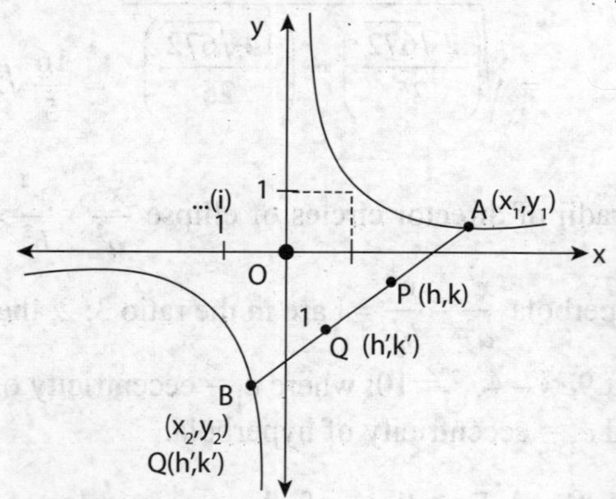

i.e., $kx + 4x^2 - 4hx - 1 = 0$

i.e., $4x^2 + (k - 4h)x - 1 = 0$

$\therefore \; x_1+x_2=\frac{h \quad k}{} ; \; x_1x_2=\frac{}{}$

$\Rightarrow \; 2x_1+x_2=\frac{h \quad k}{}+x_1;$

$\Rightarrow \; 3h=h-\_+x_1;$

$\Rightarrow \; x_1=2h+\_$

Also $x_2 = h - - - x = h-\frac{k}{4}-2h-\frac{k}{4}=-h-\frac{k}{2}$

$\therefore \; x_1x_2 = -1/4$

$\Rightarrow \; \left(2h+\frac{k}{4}\right)\left(-h-\frac{k}{2}\right)=\frac{-1}{4}$

$\Rightarrow \; \frac{(8h+k)}{4}\frac{(2h+k)}{2}=\frac{1}{4}$

$\Rightarrow \; (8h + k)(2h + k) = 2 \Rightarrow 16h^2 + 10kh + k^2 = 2$

$\therefore$ required locus is $16x^2 + 10xy + y^2 = 2$

Hence $h^2 - ab = 25 - 16 = 9 > 0$

Also $\Delta = abc + 2fgh - af^2 - bg^2 - ch^2$

$= (16)(1)(-2) - (-2)(5)^2 = -32 + 50 \neq 0.$

$\therefore$ The locus is a hyperbola.

Similarly, by symmetry locus of Q is also a hyperbola. Hence proved.

7. If $P(3\sec\alpha, 2\tan\alpha)$ and $Q(3\sec\beta, 2\tan\beta)$ are two points on a hyperbola $\frac{x^2}{9}-\frac{y^2}{4}=1$ such that $\alpha-\beta=\frac{2\pi}{3}$, then find the eccentricity of the hyperbola to which the line PQ is a tangent.

Solution: Equation of chord joining the points P and Q is given by

$$\frac{x}{a}\cos\left(\frac{\alpha-\beta}{2}\right)-\frac{y}{b}\sin\left(\frac{\alpha+\beta}{2}\right)=\cos\left(\frac{\alpha+\beta}{2}\right)$$

$$\Rightarrow \; \frac{x}{3}\cos\left(\frac{\pi}{3}\right)-\frac{y}{2}\sin\left(\frac{\alpha+\beta}{2}\right)=\cos\left(\frac{\alpha+\beta}{2}\right)$$

$$\Rightarrow \; \frac{x}{3(2)\cos\left(\frac{\alpha+\beta}{2}\right)}-\frac{y}{2}\frac{\sin\left(\frac{\alpha+\beta}{2}\right)}{\cos\left(\frac{\alpha+\beta}{2}\right)}=1$$

$$\Rightarrow \; \frac{x}{6}\sec\left(\frac{\alpha+\beta}{2}\right)-\frac{y}{2}\tan\left(\frac{\alpha+\beta}{2}\right)=1$$

Comparing it with $\frac{x\sec\theta}{a}-\frac{y\tan\theta}{b}=1$; $a = 6$, $b = 2$

which is tangent to hyperbola $\frac{x^2}{a^2}-\frac{y^2}{b^2}=1$,

i.e., $\frac{x^2}{36}-\frac{y^2}{4}=1$, having its eccentricity given by

$4 = 36(e^2 - 1) \Rightarrow 1/9 + 1 = e^2 \Rightarrow e=\frac{\sqrt{10}}{3}$.

8. If the tangents drawn from a point on $2x^2 - y^2 = 2a^2 - b^2$ to the ellipse $\frac{x^2}{a^2}+\frac{y^2}{b^2}=1$ make angle α and β with the positive x-axis, then find $\tan\alpha.\tan\beta$.

Solution: Any point on $2x^2 - y^2 = 2a^2 - b^2$ be (h, k)

$\Rightarrow 2h^2 - k^2 = 2a^2 - b^2$ ... (i)

Equation of any tangent to ellipse $\frac{x^2}{a^2}+\frac{y^2}{b^2}=1$ is given by

$y = mx \pm \sqrt{a^2m^2+b^2}$

If they passes through (h, k), then

$k - mh = \pm\sqrt{a^2m^2+b^2}$

squaring, $(k - mh)^2 = (a^2m^2 + b^2)$

$\Rightarrow k^2 + m^2h^2 - 2mkh = a^2m^2 + b^2$

$\Rightarrow m^2(h^2 - a^2) - 2kh(m) + (k^2 - b^2) = 0$ ... (ii)

Since m_1 and m_2 are roots of (ii) i.e., slopes of tangents to $\frac{x^2}{a^2}+\frac{y^2}{b^2}=1$ and passing through point (h, k).

$\Rightarrow \tan\alpha.\tan\beta = \frac{k^2-b^2}{h^2-a^2} = \frac{2(h^2-a^2)}{(h^2-a^2)} = 2$.

[$\because$ from (i) $2(h^2 - a^2) = k^2 - b^2$].

9. Find the equations of pair of tangents drawn from point (4, 5) to the locus of foot of perpendicular drawn from either of the foci to the variable tangent to hyperbola $16y^2 - 9x^2 = 1$.

Solution: Equation of hyperbola is

$16y^2 - 9x^2 = 1$ .. (i)

or $\frac{y^2}{1/16}-\frac{x^2}{1/9}=1 \Rightarrow a^2=\frac{1}{16};\ b^2=\frac{1}{9}$

We know that the locus of foot of perpendicular from either of foci to the variable tangent to hyperbola $\frac{x^2}{a^2}-\frac{y^2}{b^2}=1$ is its auxiliary circle $x^2 + y^2 = a^2$

$\therefore$ Auxiliary circle of (i) is $x^2 + y^2 = 1/16$ .... (ii)

$\therefore$ Equation of pair of tangents to (ii) from point $(x_1, y_1) = (4, 5)$ to second degree curve $S = 0$ is given by $SS_1 = T^2$.

i.e., $\left(x^2+y^2-\frac{1}{16}\right)\left((4)^2+(5)^2-\frac{1}{16}\right)$

$= \left(4x+5y-\frac{1}{16}\right)^2$

or $(16x^2 + 16y^2 - 1)(41 \times 16 - 1) = (64x + 80y - 1)^2$

or $(16x^2 + 16y^2 - 1)(655) = (64x + 80y - 1)^2$.

10. Find the foci of the conic, such that segment of tangent intercepted between the co-ordinate axes at any point of the conic is bisected at the point of contact, given that the conic passes through a point (2, 3).

Solution: Let P(h, k) be any point on the conic, then equation of tangent to conic at point P will be

$(y-k) = \frac{dk}{dh}(x-h)$

on x-axis $y = 0$

$\Rightarrow -k = \frac{dk}{dh}(x-h)$

$\Rightarrow -k\frac{dh}{dk} = (x-h)$

$\Rightarrow x = h - k\frac{dh}{dk}$

on y-axis; $x = 0$

$\Rightarrow y = k - h\frac{dk}{dh}$

$A.T.Q.\ h = \frac{x}{2};\ k = \frac{y}{2}$

$\therefore\ 2h = h - k\frac{dh}{dk};\ 2k = k - h\frac{dk}{dh}$

$\Rightarrow -\frac{h}{k} = \frac{dh}{dk}$ and $\frac{-k}{h} = \frac{dk}{dh}$

$\Rightarrow \frac{dk}{k} = -\frac{dh}{h}$

$\Rightarrow \ln k + \ln h = \ln c$

$\Rightarrow \ln xy = \ln c \Rightarrow xy = c$

$\because$ conic passes through the point $(2, 3) \Rightarrow c = 6$

$\therefore$ equation of conic is $xy = 6$ which is a rectangular hyperbola,

$\therefore$ Transverse axis of this hyperbola is $y = x$.

$\therefore$ vertices are given by $x^2 = 6$.

$\Rightarrow x = \pm\sqrt{6}$

$\therefore$ vertices are $(-\sqrt{6}, -\sqrt{6})$ and $(\sqrt{6}, \sqrt{6})$.

$\Rightarrow a = \sqrt{12} = 2\sqrt{3}$

for rectangular hyperbola $e = \sqrt{2}$

$\therefore$ foci are given by

$\left(-\sqrt{24},-\sqrt{24}\right)$ and $\left(\sqrt{24},\sqrt{24}\right)$

$\equiv\left(-2\sqrt{6},-2\sqrt{6}\right)$ and $\left(2\sqrt{6},2\sqrt{6}\right)$

11. If two points P and Q on the hyperbola $\frac{x^2}{a^2}-\frac{y^2}{b^2}=1$ whose centre is C be such that CP is perpendicular to CQ and $a < b$, then prove that $\frac{1}{CP^2}+\frac{1}{CQ^2}=\frac{1}{a^2}-\frac{1}{b^2}$.

Solution: Let the equation to CP be $y = mx$, its points of intersection with $\frac{x^2}{a^2}-\frac{y^2}{b^2}=1$

is given by $x^2 (b^2 - a^2m^2) = a^2b^2$ ...(1)

$\therefore\ CP^2 = x^2 + y^2 = x^2 + m^2 x^2$

$$= (1 + m^2) \frac{a^2b^2}{b^2 - a^2m^2}$$

Replacing m by $\left(-\frac{1}{m}\right)$, we get $(CQ)^2$

$$= \frac{1+m^2}{m^2} \cdot \frac{a^2b^2.m^2}{b^2m^2 - a^2} = \frac{(1+m^2)a^2b^2}{b^2m^2 - a^2}$$

$$\therefore\ \frac{1}{CP^2} + \frac{1}{CQ^2} = \frac{1}{(1+m^2)} \frac{b^2(1+m^2) - a^2(1+m^2)}{a^2b^2}$$

$$= \frac{1}{a^2} - \frac{1}{b^2}$$

12. The tangent and normal at a point on $\frac{x^2}{a^2} - \frac{y^2}{b^2} = 1$ cut the y-axis at A and B. Prove that the circle on AB as diameter passes through the foci of the hyperbola.

Solution: Equation of tangent at point at $P(\theta)$ is,

$$\frac{\sec\theta}{a} x - \frac{\tan\theta}{b} y = 1$$

$\therefore\ A(0, -b \cot \theta)$

equation of normal at point $P(\theta)$ is

$a \cos\theta\, x + b \cot\theta\, y = a^2 + b^2$

$\therefore\ B\left(0, \frac{a^2 + b^2}{b\cot\theta}\right)$.

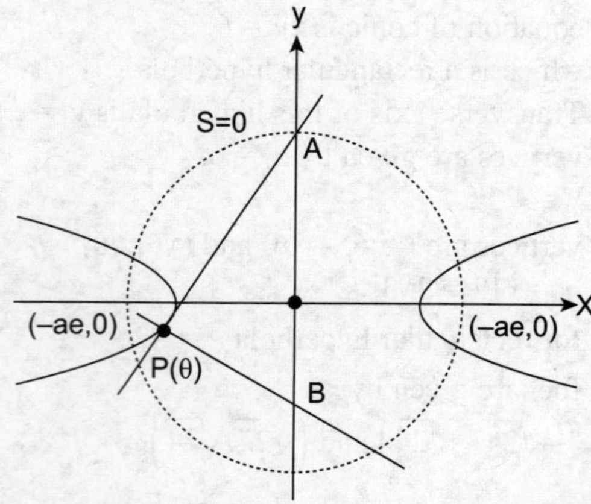

If A and B are the diametric ends of a circle, then equation of circle will be given by

$$S \equiv x^2 + (y + b\cot\theta)\left(y - \frac{a^2+b^2}{b\cot\theta}\right) = 0$$

$$\Rightarrow\ S \equiv x^2 + y^2 + y\left(b\cot\theta - \frac{a^2+b^2}{b\cot\theta}\right) - (a^2+b^2) = 0$$

substituting $x = \pm ae$, $y = 0$, we have $a^2e^2 - (a^2 + b^2) = 0$ which is true as $a^2e^2 = a^2 + b^2$ for given by hyperbola.

Thus the circle passes through the foci.

13. The perpendicular from the centre upon the normal on any point of the hyperbola $\frac{x^2}{a^2} - \frac{y^2}{b^2} = 1$ meets at R. Find the locus of R.

Solution: Let (x_1, y_1) be any point on the hyperbola so

$$\frac{x_1^2}{a^2} - \frac{y_1^2}{b^2} = 1 \quad ...(1)$$

The equation of the normal at (x_1, y_1) is $\frac{x - x_1}{\frac{x_1}{a^2}} = \frac{y - y_1}{-\frac{y_1}{b^2}}$

or $\frac{x_1}{a^2}(y - y_1) + \frac{(y_1)}{b^2}(x - x_1) = 0$ ...(2)

slope (m) of the normal $= -\frac{a^2 y_1}{b^2 x_1}$

$\therefore$ the equation of the perpendicular from the centre (0, 0) on (2) is $y = \frac{b^2 x_1}{a^2 y_1} x$ ...(3)

The intersection of (2) and (3) is R and the required locus is obtained by eliminating x_1, y_1 from (1), (2) and (3).

from (3), $\frac{x_1}{a^2 y} = \frac{y_1}{b^2 x} = t$ (say)

putting in equation (2), $yt\,(y - b^2xt) + xt\,(x - a^2yt) = 0$

or $(x^2 + y^2)t - (a^2 + b^2)xyt^2 = 0$ but t = 0 for (x_1, y_1) = (0, 0) which is not true.

$$\therefore\ t = \frac{x^2 + y^2}{xy(a^2 + b^2)} \text{ for } t \neq 0$$

$$\therefore\ x_1 = \frac{x^2 + y^2}{xy(a^2 + b^2)} a^2 y = \frac{a^2(x^2 + y^2)}{x(a^2 + b^2)} \text{ and}$$

$$y_1 = \frac{x^2 + y^2}{xy(a^2 + b^2)} b^2 x = \frac{b^2(x^2 + y^2)}{y(a^2 + b^2)}$$

$$\therefore \text{ from (1) } \frac{1}{a^2} \frac{a^4(x^2 + y^2)^2}{x^2(a^2 + b^2)^2} - \frac{1}{b^2} \frac{b^4(x^2 + y^2)^2}{y^2(a^2 + b^2)^2} = 1 \text{ or}$$

$$(x^2 + y^2)^2 \left(\frac{a^2}{x^2} - \frac{b^2}{y^2}\right) = (a^2 + b^2)^2$$

14. A conic C satisfies the differential equation, $(1 + y^2)\,dx - xydy = 0$ and passes through the point (1, 0). An ellipse E which is confocal with C has its eccentricity equal to $\sqrt{2/3}$.

(a) Find the length of the latus rectum of the conic C

(b) Find the equation of ellipse E.

(c) Find the locus of the point of intersection of the perpendicular tangents to the ellipse E.

Solution: $(1 + y^2)dx - xy\,dy = 0$

$\Rightarrow \dfrac{dx}{dy} - \dfrac{y}{(1+y^2)}x = 0$

$\Rightarrow \dfrac{dx}{x} = \dfrac{y}{1+y^2}dy$ Intergeting

$\Rightarrow \ell n x = \dfrac{1}{2}\ell n|1+y^2| + c$

$\Rightarrow \dfrac{x}{\sqrt{1+y^2}} = e^c$ = const. = k (say)

$\Rightarrow x^2 = k^2(1 + y^2) \Rightarrow \dfrac{x^2}{k^2} - y^2 = 1$

Above curve passes through point (1, 0)

$\dfrac{1}{k^2} = 1 \Rightarrow k^2 = 1$, thus the circle is

$C \equiv x^2 - y^2 = 1$, focus = $(\pm\sqrt{2}, 0) \Rightarrow E \equiv \dfrac{x^2}{a^2} + \dfrac{y^2}{b^2} = 1$

A/C to question; $ae = \sqrt{2} \Rightarrow a\sqrt{\dfrac{2}{3}} = \sqrt{2} \Rightarrow a = \sqrt{3}$

Now, for ellipse $e^2 = 1 - \dfrac{b^2}{a^2} \Rightarrow \dfrac{2}{3} = 1 - \dfrac{b^2}{3}$

$\Rightarrow b^2 = 1 \Rightarrow b = 1$

(a) length of latus rectum of conic = $\dfrac{2b^2}{a} = \dfrac{2}{\sqrt{3}}$

(b) Equation of ellipse E: $\dfrac{x^2}{3} + \dfrac{y^2}{1} = 1$

(c) Locus of perpendicular to E is its director circle i.e., $x^2 + y^2 = a^2 + b^2 = 4$

15. An ellipse and a hyperbola have their principale axes along the co-ordinate axes and have the foci common and separated by a distance $2\sqrt{13}$, the difference of their focal semi axes is equal to 4. If the ratio of their eccentricities is 3/7. Find the equation of these curves.

Solution: Let a be the semi major axis of the ellipse b be the semi minor axis of the ellipse and e be the eccentricity of the ellipse.

And let a' be the semi transverse axis of the hyperbola and b' be the semi conjugate axis of the hyperbola and e' be the eccentricity of the hyperbola. Given that

$a - a' = 4$ ..(1)

and $2ae = 2\sqrt{13} \Rightarrow ae = \sqrt{13}$ ...(2)

and $2a'e' = 2\sqrt{13} \Rightarrow a'e' = \sqrt{13}$ ...(3)

$\Rightarrow \dfrac{e}{e'} = \dfrac{3}{7} = \dfrac{a'}{a}$ ...(4)

$\Rightarrow a' = 3k, a = 7k$

Now, $a - a' = 4 \quad \Rightarrow k = 7$

$\Rightarrow a' = 3, a = 7 \quad \Rightarrow e = \dfrac{\sqrt{13}}{7}, e' = \dfrac{\sqrt{13}}{3}$ ...(5)

$\Rightarrow b'^2 = 4, b^2 = 36$ ...(6)

$\therefore$ Equation of ellipse E: $\dfrac{x^2}{49} + \dfrac{y^2}{36} = 1$ and

Equation of hyperbola H: $\dfrac{x^2}{9} - \dfrac{y^2}{4} = 1$.

16. If one axis of a varying hyperbola be fixed in magnitude and position, then prove that the locus of the point of contact of a tangent drawn to it from a fixed point on the other axis is a parabola.

Solution: Equation of a tangent at $P(\theta)$

i.e., at $h = a\sec\theta$; $k = b\tan\theta$ .... (i)

is $\dfrac{x\sec\theta}{a} - \dfrac{y\tan\theta}{b} = 1$ .... (ii)

(ii) passes through the point (0, c) (where 'c' any constant)

$\therefore \; -c\tan\theta = b$ ....(iii)

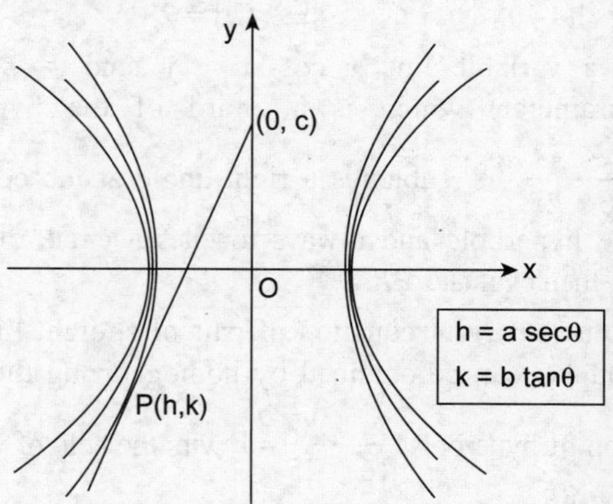

Now substituting the value of b in (i), we get $k/c = -\tan^2\theta$ and $h^2/a^2 = \sec^2\theta$

On adding $\dfrac{k}{c} + \dfrac{h^2}{a^2} = 1 \Rightarrow \dfrac{h^2}{a^2} = 1 - \dfrac{k}{c}$

$\Rightarrow h^2 = \dfrac{a^2}{c}(c - k)$

$\Rightarrow x^2 = -\dfrac{a^2}{c}(y - c)$ which is a parabola.

17. Find [λ]; where [.] is greatest integer function and λ is the minimum positive value of λ for which the portion of asymptote of hyperbola $\dfrac{x^2}{a^2} - \dfrac{y^2}{b^2} = 1$ (between centre and the tangent at vertex) in the first quadrant is cut by the line $y + \lambda\sqrt{\lambda^2 - 4}(x - a) = 0$ (λ is a parametre).

Solution: Given line is $y = -\lambda\sqrt{\lambda^2-4}(x-a)$ ... (i)
Clearly, line (i) passes through the vertex $(a, 0)$ and has its slope $= -\lambda\sqrt{\lambda^2-4}$, it would cut portion in first quadrant only if its slope is negative.

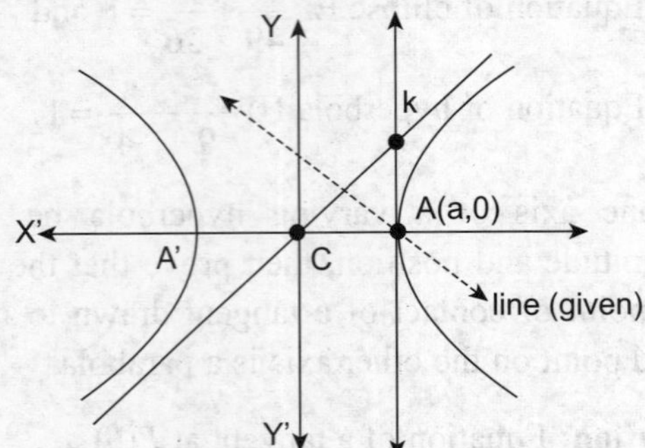

$\Rightarrow \; -\lambda\sqrt{\lambda^2-4} < 0$

$\Rightarrow \; \lambda\sqrt{\lambda^2-4} > 0$

$\Rightarrow \; \lambda > 0; \lambda^2 - 4 > 0 \Rightarrow \lambda > 0;$
$\lambda \in (-\infty, -2) \cup (2, \infty) \Rightarrow \lambda \in (2, \infty)$

$\Rightarrow$ Minimum positive value of $\lambda = 2 + h$,
$h \to 0^+$ $\quad \Rightarrow \; [\lambda] = 2$

18. If a variable line $x \cos\alpha + y \sin\alpha = p$ (α is parameter) which is a chord of the hyperbola $\frac{x^2}{16} - \frac{y^2}{25} = 1$, subtends a right angle at the centre of the hyperbola and always touches a circle of radius 'r', then evaluate '$3r$'.

Solution: Joint equation of pair of straight lines AC and BC can be obtained by homogenizing the equation of hyperbola $\frac{x^2}{16} - \frac{y^2}{25} = 1$ with the help of variable line AB.

$x \cos\alpha + y \sin\alpha = p$

$\therefore$ joint equation of AC and BC is given by

$$\frac{x^2}{16} - \frac{y^2}{25} = \left(\frac{x\cos\alpha + y\sin\alpha}{p}\right)^2$$

$$\Rightarrow x^2\left[\frac{1}{16} - \frac{\cos^2\alpha}{p^2}\right] + y^2\left[-\frac{1}{25} - \frac{\sin^2\alpha}{p^2}\right] - \frac{2xy\sin\alpha\cos\alpha}{p^2} = 0$$

$\therefore$ Angle between AC and BC is a right angle.

$\Rightarrow$ co-efficient of x^2 + co-efficeint of $y^2 = 0$

$$\Rightarrow \frac{1}{16} - \frac{\cos^2\alpha}{p^2} - \frac{1}{25} - \frac{\sin^2\alpha}{p^2} = 0$$

$$\Rightarrow \frac{1}{p^2} = \frac{1}{16} - \frac{1}{25} = \frac{9}{16\times 25} \Rightarrow p = \frac{20}{3}.$$

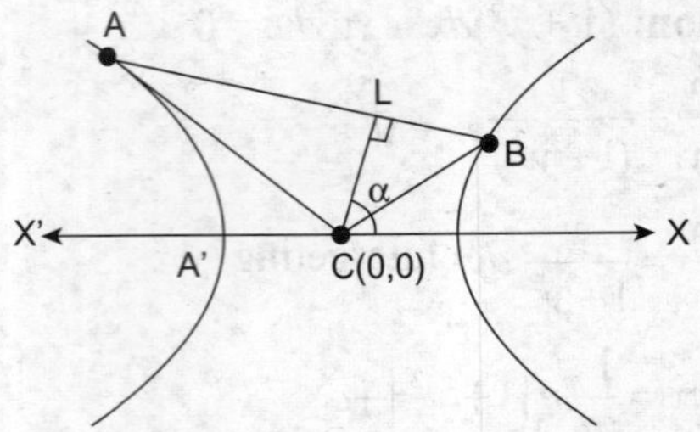

Also, $CL = p$ is always perpendicular to given variable line and hence variable line always touches a circle of radius $p = r = \frac{20}{3}$. $\Rightarrow 3r = 20$ units.

19. Let any double ordinate PNP' of hyperbola $\frac{x^2}{49} - \frac{y^2}{25} = 1$ be produced outwards to meet the asymptotes at Q and Q', then evaluate $PQ \cdot P'Q$.

Solution: Given equation of hyperbola is $\frac{x^2}{49} - \frac{y^2}{25} = 1$.

Its asymptotes are given by $\frac{x^2}{49} - \frac{y^2}{25} = 0$;

i.e., $\frac{y}{5} = \pm\frac{x}{7} \Rightarrow y = \pm\frac{5}{7}x$

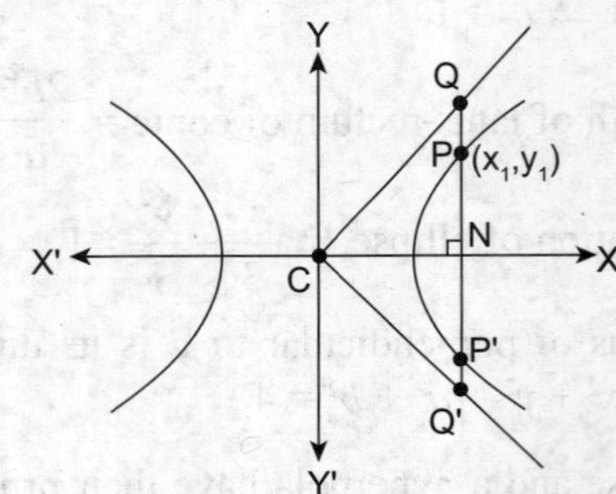

$\therefore$ Equation of asymptote CQ would be $y = \frac{5}{7}x$

Let the co-ordinates of P be (x_1, y_1)

$\therefore \; PN = y_1 = \frac{5}{7}\sqrt{x_1^2 - 49}$ and $QN = \frac{5}{7}x_1$

$\therefore \; PQ = QN - PN = \frac{5}{7}\left(x_1 - \sqrt{x_1^2 - 49}\right)$

Similarly, $P'Q = P'P + PQ = 2(NP) + PQ$

$$= 2\left(\frac{5}{7}\sqrt{x_1^2-49}\right) + \frac{5}{7}\left(x_1 - \sqrt{x_1^2-49}\right)$$

$$= \frac{5}{7}x_1 + \frac{5}{7}\sqrt{x_1^2-49} = \frac{5}{7}\left(x_1 + \sqrt{x_1^2-49}\right)$$

$$\therefore \; PQ \cdot P'Q = \left(\frac{5}{7}\right)^2\left(x_1 - \sqrt{x_1^2-49}\right)\left(x_1 + \sqrt{x_1^2-49}\right)$$

$$= \frac{25}{49}\left[x_1^2 - (x_1^2 - 49)\right] = \frac{25}{49}(49) = 25.$$

20. The equation of a latus rectum of rectangular hyperbola $xy = c^2$ is $lx + my = nc$, then evaluate $\left(\sqrt{l+m}\right)n$.

Solution: Co-ordinates of A are (c, c)

$\therefore\ OA = \sqrt{2}\,c$

$\Rightarrow OS = OA(e) = (OA)(\sqrt{2}) = (\sqrt{2}\,c)(\sqrt{2}) = 2c$

Let (*h*, h) be the co-ordinates of focus *S*. ($\because e = \sqrt{2}$)

$\Rightarrow 2h^2 = 4c^2 \Rightarrow h = \sqrt{2}\,c$

$\therefore$ co-ordinates of focus *S* are ($\sqrt{2}\,c$, $\sqrt{2}$ c)

Now, *LL'* (latus rectum) is straight line having slope equal to '–1' and passing through S($\sqrt{2}c$, $\sqrt{2}c$).

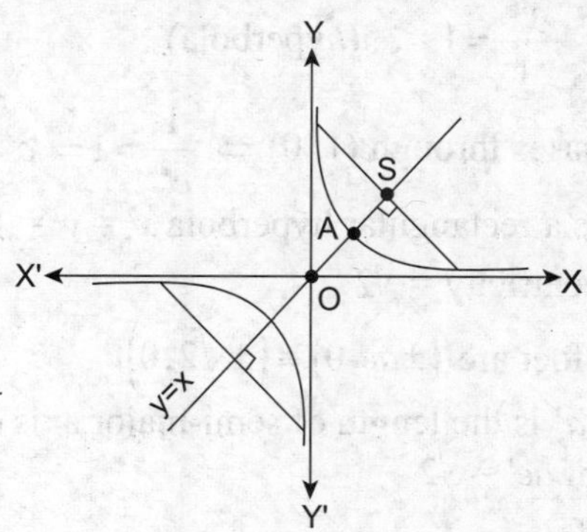

Its equation is given by $(y - \sqrt{2}c) = -1(x - \sqrt{2}c)$

$\Rightarrow x + y = -2\sqrt{2}\,c \Rightarrow l = 1, m = 1, n = 2\sqrt{2}$.

$\therefore\ \left(\sqrt{l+m}\right).n = \left(\sqrt{1+1}\right)(2\sqrt{2}) = (\sqrt{2})(2\sqrt{2}) = 4$.

21. The equations of the normals to the rectangular hyperbola $xy = 4$ which are parallel to the line $2x - y = 5$ are given by $l_1x + m_1y + n_1 = 0$ and $l_2x + m_2y + n_2 = 0$, then evaluate $|(l_1 + m_1 + n_1)(l_2 + m_2 + n_2)|$.

Solution: Parametric equation of rectangular hyperbola $xy = 4$ are given by $x = 2t, y = \frac{2}{t}$

$\therefore$ Slope of normal at a point 't' $= -\frac{dx}{dy}$

$$= -\left(\frac{dx/dt}{dy/dt}\right) = -\left(\frac{2}{-2/t^2}\right) = t^2$$

A.T.Q: $t^2 = 2 \Rightarrow t = \pm\sqrt{2}$

and the point are $(2\sqrt{2}, \sqrt{2})$ and $(-2\sqrt{2}, -\sqrt{2})$

Equation of normals are $(y - \sqrt{2}) = 2(x - 2\sqrt{2})$ and $(y + \sqrt{2}) = 2(x + 2\sqrt{2})$

$\Rightarrow 2x - y - 3\sqrt{2} = 0$ and $2x - y + 3\sqrt{2} = 0$

$$\therefore\ |(l_1 + m_1 + n_1)(l_2 + m_2 + n_2)|$$

$$= |(2 - 1 - 3\sqrt{2})(2 - 1 + 3\sqrt{2})|$$

$$= |1 - 18| = 17.$$

22. A tangent to the hyperbola $\frac{x^2}{4} - \frac{y^2}{1} = 1$ meets the ellipse $x^2 + 4y^2 = 4$ in two distinct points *P* and *Q*. Then evaluate the sum of abscissae of the points where locus of mid-point of chord *PQ* meets *x*-axis.

Solution: Let (2secθ, tanθ) be the point on hyperbola $\frac{x^2}{4} - \frac{y^2}{1} = 1$ at which the tangent is being drawn

$\therefore$ Its equation will be $\frac{2x\sec\theta}{4} - \frac{y\tan\theta}{1} = 1$

$$\text{or}\quad \frac{x\sec\theta}{2} - y\tan\theta = 1 \qquad ...(1)$$

Let (*h*, *k*) be the mid-point of chord *PQ* of ellipse $\frac{x^2}{4} + \frac{y^2}{1} = 1$

$\therefore$ Equation of chord *PQ* in terms of its mid-point (*h*, *k*) is $T = S_1$

$$\Rightarrow \frac{hx}{4} + \frac{ky}{1} = \frac{h^2}{4} + \frac{k^2}{1} \qquad ...(2)$$

Now (1) and (2) are identical

$$\Rightarrow \frac{\frac{\sec\theta}{2}}{\left(\frac{h}{4}\right)} = \frac{-\tan\theta}{k} = \frac{1}{\left(\frac{h^2}{4} + k^2\right)}$$

$$\Rightarrow \frac{2\sec\theta}{h} = \frac{-\tan\theta}{k} = \frac{1}{\left(\frac{h^2}{4} + k^2\right)}$$

$$\Rightarrow \sec\theta = \frac{2h}{(h^2 + 4k^2)}; \tan\theta = \frac{-4k}{(h^2 + 4k^2)}.$$

Now, $\sec^2\theta - \tan^2\theta = 1$.

$$\Rightarrow 4h^2 - 16k^2 = (h^2 + 4k^2)^2$$

$$\text{or}\quad (h^2 + 4k^2)^2 = 4(h^2 - 4k^2)$$

$\therefore$ Required locus of mid-point of chord *PQ* is $(x^2 + 4y^2)^2 = 4(x^2 - 4y^2)$

It cuts *x*-axis, where $y = 0$

$\Rightarrow x^4 = 4(x^2) \Rightarrow x^2(x^2 - 4) = 0 \Rightarrow x = 0, \pm 2$

$\therefore$ Sum of abscissae $0 + 2 + (-2) = 0$

23. The locus of a point from where tangents to the rectangular hyperbola $x^2 - y^2 = a^2$ contain an angle of 45° is $(x^2 + y^2)^2 + \lambda a^2(x^2 - y^2) = \mu a^4$, then evaluate '$\lambda + \mu$'.

Solution: Let the equation of pair of tangents is given by $SS_1 = T^2$

$\Rightarrow (x^2 - y^2 - a^2)(h^2 - k^2 - a^2) = (hx - ky - a^2)^2$

$\Rightarrow x^2(k^2 + a^2) + y^2(h^2 - a^2) + xy(-2hk) + x(-2a^2h) + y(2a^2k) + a^2(h^2 - k^2) = 0$

$$\Rightarrow \tan\frac{\pi}{4} = 1 = \frac{2\sqrt{(hk)^2 - (k^2 + a^2)(h^2 - a^2)}}{|k^2 + a^2 + h^2 - a^2|}$$

$\Rightarrow (k^2 + h^2)^2 = 4\,4\left[h^2k^2 + (a^2 + k^2)(a^2 - h^2)\right]$

$\Rightarrow (k^2 + h^2)^2 = 4\left(h^2k^2 + a^4 - a^2h^2 + a^2k^2 - h^2k^2\right)$

$\Rightarrow \left(k^2 + h^2\right)^2 = 4\left(a^4 + a^2k^2 - a^2h^2\right)$

$\Rightarrow \left(x^2 + y^2\right)^2 = 4a^4 + 4a^2\left(k^2 - h^2\right)$

$\Rightarrow \left(x^2 + y^2\right)^2 + 4a^2\left(x^2 - y^2\right) = 4a^4$ is the required locus.

$\Rightarrow \lambda = 4;\ \mu = 4.\ \therefore \lambda + \mu = 8.$

24. The differential equation $\frac{dx}{dy} = \frac{3y}{2x}$ represents a family of hyperbola (except when it represents a pair of lines) with possible eccentricities e_1 and e_2, then evaluate $6\left(e_1^2 + e_2^2\right)$.

Solution: Given different equation is $\frac{dx}{dy} = \frac{3y}{2x}$

$\Rightarrow 2x\,dy = 3y\,dy$

$\Rightarrow x^2 = \frac{3y^2}{2} + c$; for $c > O$,

the hyperbola will be $x^2 - \frac{3y^2}{2} = c$

or $\frac{x^2}{c} - \frac{y^2}{2c/3} = 1$

$\therefore\ e_1^2 = 1 + \frac{2c/3}{c} = 1 + \frac{2}{3} = \frac{5}{3}$;

Similarly for $c < 0$; hyperbola will be

$\frac{3y^2}{2} - x^2 = -c$

$\Rightarrow \frac{y^2}{(-2c/3)} - \frac{x^2}{(-c)} = 1$

$\therefore\ e_2^2 = 1 + \frac{-c}{-2c/3} = 1 + \frac{3}{2} = \frac{5}{2}$

$\therefore\ 6\left(e_1^2 + e_2^2\right) = 6\left[\frac{5}{3} + \frac{5}{2}\right] = 30\left[\frac{2+3}{6}\right] = 25.$

25. A conic C satisfies the differential equation $(1 + y^2)\,dx - xy\,dy = 0$ and passes through the point (1, 0). An ellipse E which is confocal with C having its eccentricity equal to $\frac{\sqrt{2}}{3}$. If A is the area of ellipse E, then evaluate $A^2 : \pi^2$.

Solution: Given differential equation is $(1 + y^2)\,dx = xydy$

$\Rightarrow \frac{dx}{x} = \frac{ydy}{1+y^2} \Rightarrow \ln x = (1/2)\ln(1 + y^2) + \ln c$

$\Rightarrow \ln x = \ln(1 + y^2)^{1/2} + \ln c \Rightarrow x = c\sqrt{1+y^2}$

$\Rightarrow x^2 = c^2(1 + y^2) \Rightarrow x^2 - c^2y^2 = c^2$

$\Rightarrow \frac{x^2}{c^2} - \frac{y^2}{1} = 1$ ......(*hyperbola*)

It pases through (1, 0) $\Rightarrow \frac{1}{c^2} = 1 \Rightarrow c^2 = 1$

$\therefore$ C is a rectangular hyperbola $x^2 - y^2 = 1$; having its eccentricity = $\sqrt{2}$

$\therefore$ Its foci are $(\pm ae, 0) \equiv (\pm\sqrt{2}, 0)$

If '*a*' is the length of semi-major axis of ellipse *E*, then $ae' = \sqrt{2}$

$\Rightarrow a\sqrt{\frac{2}{3}} = \sqrt{2} \Rightarrow a = \sqrt{3}$

Now, $b^2 = a^2(1 - e^2) = 3\left[1 - \frac{2}{3}\right] = 1 \Rightarrow b = 1$

Area of ellipse $E = \pi ab = \pi\left(\sqrt{3} \times 1\right) = \sqrt{3}\pi.$

$\therefore\ A^2 : \pi^2 = 3\pi^2 : \pi^2 = 3$

$\therefore$ **Ans. 3.**

26. Find the equation and the length of the common tangents to hyperbola $\frac{x^2}{a^2} - \frac{y^2}{b^2} = 1$ and $\frac{y^2}{a^2} - \frac{x^2}{b^2} = 1$.

Solution: Tangent at (a sec ϕ, b tan ϕ) on the 1st hyperbola is

$$\frac{x}{a}\sec\phi - \frac{y}{b}\tan\phi = 1 \quad \text{...(1)}$$

Similarly tangent at any point (b tan θ, a sec θ) on 2nd hyperbola is

$$\frac{y}{a}\sec\theta - \frac{x}{b}\tan\theta = 1 \quad \text{...(2)}$$

If (1) and (2) are common tangents, then they should be identical. Comparing the co-effecients of *x* and y

$$\Rightarrow \frac{\sec\theta}{a} = -\frac{\tan\phi}{b} \quad \text{...(3)}$$

and $-\frac{\tan\theta}{b} = \frac{\sec\phi}{a}$

$$\Rightarrow \sec\theta = -\frac{a}{b}\tan\phi \text{ and } \tan\theta = -\frac{b}{a}\sec\phi \quad \text{...(4)}$$

$\sec^2\theta - \tan^2\theta = 1$

$\Rightarrow \dfrac{a^2}{b^2}\tan^2\phi - \dfrac{b^2}{a^2}\sec^2\phi = 1$ (using (4))

or $\dfrac{a^2}{b^2}\tan^2\phi - \dfrac{b^2}{a^2}(1+\tan^2\phi) = 1$

or $\left(\dfrac{a^2}{b^2} - \dfrac{b^2}{a^2}\right)\tan^2\phi = 1 + \dfrac{b^2}{a^2}$

$\Rightarrow \tan^2\phi = \dfrac{b^2}{a^2-b^2}$

and $\sec^2\phi = 1 + \tan^2\phi = \dfrac{a^2}{a^2-b^2}$

Hence the point of contact are

$$\left\{\pm\frac{a^2}{\sqrt{(a^2-b^2)}}, \pm\frac{b^2}{\sqrt{(a^2-b^2)}}\right\} \text{ and}$$

$$\left\{\mp\frac{b^2}{\sqrt{(a^2-b^2)}}, \mp\frac{a^2}{\sqrt{(a^2-b^2)}}\right\} \quad \{\text{from (3) and (4)}\}$$

Length of common tangent i.e., the distance between the above point is $\sqrt{2}\dfrac{(a^2+b^2)}{\sqrt{(a^2-b^2)}}$ and equation of common tangent on putting the values of sec ϕ and tan ϕ in (1) is

$$\pm\frac{x}{\sqrt{(a^2-b^2)}} \mp \frac{y}{\sqrt{(a^2-b^2)}} = 1 \text{ or } x \mp y = \pm\sqrt{(a^2-b^2)}$$

27. How many real tangents can be drawn from the point (4,3) to the hyperbola $\dfrac{x^2}{16} - \dfrac{y^2}{9} = 1$? Find the equation of these tangents and the angle between them.

Solution: Given point $P \equiv (4,3)$

Hyperbola: $S \equiv \dfrac{x^2}{16} - \dfrac{y^2}{9} - 1 = 0$

$S_1 \equiv \dfrac{16}{16} - \dfrac{9}{9} - 1 = -1 < 0$

$\Rightarrow$ Point $P \equiv (4, 3)$ lies inside the hyperbola.

∴ Two tangents can be drawn from the point P(4,3). Equation of pair of tangents is $SS_1 = T^2$

$$\Rightarrow \left(\frac{x^2}{16} - \frac{y^2}{9} - 1\right).(-1) = \left(\frac{4x}{16} - \frac{3y}{9} - 1\right)^2$$

$$\Rightarrow -\frac{x^2}{16} + \frac{y^2}{9} + 1 = \frac{x^2}{16} + \frac{y^2}{9} + 1 - \frac{xy}{6} - \frac{x}{2} + \frac{2y}{3}$$

$\Rightarrow 3x^2 - 4xy - 12x + 16y = 0$

$\Rightarrow \theta = \tan^{-1}\left(\dfrac{4}{3}\right)$

28. From any point on the hyperbola $\dfrac{x^2}{a^2} - \dfrac{y^2}{b^2} = 1$ tangents are drawn to the hyperbola $\dfrac{x^2}{a^2} - \dfrac{y^2}{b^2} = 2$. Find the area cut off by the chord of contact on the asymptotes.

Solution: Let $P(x_1, y_1)$ be a point on the hyperbola $\dfrac{x^2}{a^2} - \dfrac{y^2}{b^2} = 1$. Then $\dfrac{x_1^2}{a^2} - \dfrac{y_1^2}{b^2} = 1$

The chord of contact of tangents from P to the hyperbola $\dfrac{x^2}{a^2} - \dfrac{y^2}{b^2} = 2$ is

$$\frac{xx_1}{a^2} - \frac{yy_1}{b^2} = 2 \quad \text{...(i)}$$

Then equation of asymptotes are

$\dfrac{x}{a} - \dfrac{y}{b} = 0$ and $\dfrac{x}{a} + \dfrac{y}{b} = 0$

The point of intersection of (i) with the two asymptotes are given by .

$$x_1 = \frac{2a}{\frac{x_1}{a} - \frac{y_1}{b}}, y_1 = \frac{2b}{\frac{x_1}{a} - \frac{y_1}{b}}$$

$$x_2 = \frac{2a}{\frac{x_1}{a} + \frac{y_1}{b}}, y_2 = \frac{-2b}{\frac{x_1}{a} + \frac{y_1}{b}}$$

∴ Area of $\Delta = (1/2)\,|(x_1y_2 - x_2y_1)|$

$$= \frac{1}{2}\left(\frac{4ab \times 2}{\frac{x_1^2}{a^2} - \frac{y_1^2}{b^2}}\right) = 4ab \text{ square units}$$

29. The normal at P to a hyperbola of eccentricity e, intersects its transverse and conjugate axes at L and M respectively. If locus of mid-point of LM is a hyperbola, then find the eccentricity of the hyperbola.

Solution: The equation of the normal at P(a secθ, b tanθ) to the hyperbola $\dfrac{x^2}{a^2} - \dfrac{y^2}{b^2} = 1$ is $ax\cos\theta + by\cot\theta = a^2 + b^2$

This intersects the transverse and conjugate axes at $L\left(\dfrac{a^2+b^2}{a}\sec\theta, 0\right)$ and $M\left(0, \dfrac{a^2+b^2}{b}\tan\theta\right)$ respectively

Let N(h,k) be the mid-point of LM. Then

$$h = \frac{a^2+b^2}{2a}\sec\theta \text{ and } k = \frac{a^2+b^2}{2b}\tan\theta$$

$\Rightarrow$ $\sec\theta = \frac{2ah}{a^2+b^2}$ and tan $\frac{2bk}{a^2+b^2}$

$\therefore$ $\sec^2\theta - \tan^2\theta = 1$

$\Rightarrow$ $4a^2h^2 - 4b^2k^2 = (a^2+b^2)^2$

Thus, the locus of (h,k) is $4a^2x^2 - 4b^2y^2 = (a^2+b^2)^2$

Let e_1 be, the eccentricity of this hyperbola, then

$$= \quad + \frac{a^2}{b^2} = \frac{a^2 \quad b^2}{b^2} = \frac{a^2e^2}{a^2\left(e^2 \quad\right)}$$

$\Rightarrow$ $\sqrt{\quad}$

30. Find the locus of the point of intersection of two tangents to the hyperbola $\frac{x^2}{a^2} - \frac{y^2}{b^2} = 1$. the product of whose slopes is c^2.

Solution: Let $P(h,k)$ be the point of intersection of two tangents to the hyperbola $\frac{x^2}{a^2} - \frac{y^2}{b^2} = 1$.

Now, $y = mx \pm \sqrt{a^2m^2 - b^2}$ are the equation of tangent and if it passes through $P(h,k)$, then

$k = mh + \sqrt{a^2m^2 - b^2}$

$\Rightarrow$ $(k - mh)^2 = a^2m^2 - b^2$

$\Rightarrow$ $m^2(h^2 - a^2) - 2m\,hk + k^2 + b^2 = 0$

Let m_1 and m_2 be the slopes of the tangents passing through P. Then m_1 and m_2 are the roots of the equation (i)

$\therefore$ $m_1m_2 = \frac{k^2+b^2}{h^2-a^2}$

$\Rightarrow$ $c^2 = \frac{k^2+b^2}{h^2-a^2} \Rightarrow (h^2-a^2)c^2 = k^2+b^2$

Hence $P(h,k)$ lies on $(x^2 - a^2)c^2 = y^2 + b^2$

32. If the tangent at point $P(h,k)$ on the hyperbola $\frac{x^2}{a^2} - \frac{y^2}{b^2} = 1$ cuts the circle $x^2 + y^2 = a^2$ at the point $P_1(x_1, y_1)$ and $P_2(x_2, y_2)$ then show that $\frac{1}{y_1} + \frac{1}{y_2}$ is always equal to $\frac{2}{k}$.

Solution: Equation of tangent at $P(h,k)$ is

$$\frac{xh}{a^2} - \frac{yk}{b^2} = 1$$

Where $\frac{h^2}{a^2} - \frac{k^2}{b^2} = 1$

i.e., $b^2h^2 - a^2k^2 = a^2b^2$

solving it with circle, we get

$$y^2 + \left(1 + \frac{yk}{b^2}\right)^2 \cdot \frac{a^4}{h^2} = a^2$$

$\Rightarrow$ $y^2(b^4h^2 + a^4k^2) + 2ya^4b^2k + a^2b^2(a^2b^2 - b^2h^2) = 0$

$$\frac{1}{y_1} + \frac{1}{y_2} = \frac{2a^4b^2k}{a^2b^2\left(b^2h^2 - a^2b^2\right)} = \frac{2}{k}$$

35. Find the equations of the tangents to the hyperbola $x^2 - 9y^2 = 9$ that are drawn from (3, 2). Find the area of the triangle that these tangents form with their chord of contact.

Solution: Hyperbola is $x^2 - 9y^2 = 9$

or $\frac{x^2}{9} - \frac{y^2}{1} = 1$

equation of tangent is $y = mx \pm \sqrt{a^2m^2 - b^2}$ ...(1)

it passes through (3, 2), so $2 = 3m \pm \sqrt{9m^2 - 1}$

or $4 + 9m^2 - 12m = 9m^2 - 1$

or $m_1 = \frac{5}{12}$ and $m_2 = \infty$

equation of tangent (1) for $m_1 = \frac{5}{12}$

$\Rightarrow$ $y = \frac{5}{12}x \pm \sqrt{9\left(\frac{5}{12}\right)^2 - 1}$

or $y = \frac{5}{12}x \pm \frac{9}{12}$ or $y = \frac{5}{12}x \pm \frac{3}{4}$

on taking (–)ve sign point P(3, 2) does not satisfy the equation of tangent, therefore rejecting (–)ve sign.

Now the first tangent is $5x - 12y = -9$

$\Rightarrow$ $y = \frac{5x}{12} + \frac{3}{4}$ .....(2)

now equation of tangent (1) for $m_2 = \infty = 1/0$ and passing through (3, 2) will be $x = 3$........(3)

Now equation of chord of contact w.r.t. point P(3, 2) is $T = 0$

i.e., $xx_1 - 9yy_1 = 9$

$\Rightarrow$ $3x - 18y = 9$

$\Rightarrow$ $x - 6y = 3$ ...(4)

solving (2) and (4), $x = -6$, $y = -3/2$

solving (3) and (4), $x = 3$, $y = 0$

Now vertices of triangle are (3, 2), (3, 0), (–6, –3/2)

$$\therefore \text{ Required are} = \frac{1}{2}\begin{vmatrix} 3 & 2 & 1 \\ 3 & 0 & 1 \\ -6 & -\frac{3}{2} & 1 \end{vmatrix}$$

$$= \frac{1}{2} \times \left|3\left(\frac{3}{2}\right) - 2(3+6) + 1\left(-\frac{9}{2}\right)\right| = 9 \text{ sq. units}$$

37. A line through the origin meets the circle $x^2 + y^2 = a^2$ at P and the hyperbola $x^2 - y^2 = a^2$ at Q in first quadrant. Prove that the locus of the point of intersection of the tangent at P to the circle and the tangent at Q to the hyperbola is curve $a^4(x^2 - a^2) + 4x^2y^4 = 0$.

Solution: Let $y = mx$ be a line passing through origin. This line intersects the circle $x^2 + y^2 = a^2$ and the hyperbola $x^2 - y^2 = a^2$ at $P\left(\frac{a}{\sqrt{1+m^2}}, \frac{+am}{\sqrt{1+m^2}}\right)$ and $Q\left(\frac{a}{\sqrt{1-m^2}}, \frac{am}{\sqrt{1-m^2}}\right)$ respectively in first quadrant.

The equation of the tangent to the circle $x^2 + y^2 = a^2$ at P and hyperbola $x^2 - y^2 = a^2$ at Q are $x + my = a\sqrt{1+m^2}$ and $x - my = a\sqrt{1-m^2}$ respectively.

Let (h, k) be the point of intersection of these two lines. Then, $h + mk = a\sqrt{1+m^2}$ and $h - mk = a\sqrt{1-m^2}$

$\Rightarrow (h + mk)^2 = a^2(1 + m^2)$ and $(h - mk)^2 = a^2(1 - m^2)$

$\Rightarrow m^2(k^2 - a^2) + 2mhk + h^2 - a^2 = 0$ and

$m^2(k^2 + a^2) - 2mhk + h^2 - a^2 = 0$

$$\Rightarrow \frac{m^2}{4hk(h^2 - a^2)} = \frac{m}{2a^2(h^2 - a^2)} = \frac{1}{-4hk^3}$$

$$\Rightarrow m = \frac{2hk}{a^2} \text{ and } m = \frac{a^2(h^2 - a^2)}{-2hk^3}$$

$$\Rightarrow \frac{2hk}{a^2} = \frac{a^2(h^2 - a^2)}{-2hk^3}$$

$\Rightarrow -4h^2k^4 = a^4(h^2 - a^2)$

$\Rightarrow a^4(x^2 - a^2) + 4x^2y^4 = 0$ is required locus.

Solved Assertion and Reason Type

38. A: A bullet is fired and hits a target. An observer in the same plane heard two sounds, the crack of rifle and the hitting of target at the same instant, then the locus of the observer is a hyperbola. (where speed of bullet is super sonic.)

R: If the difference of distances of a point P from two fixed points is constant and is less than the distance between the fixed points, then the locus of P is a hyperbola.

Solution: Reason is obviously true (second definition of hyperbola).

For assertion, let Q and R be the position of gun and the target and P be position of observer at any instant as shown above.

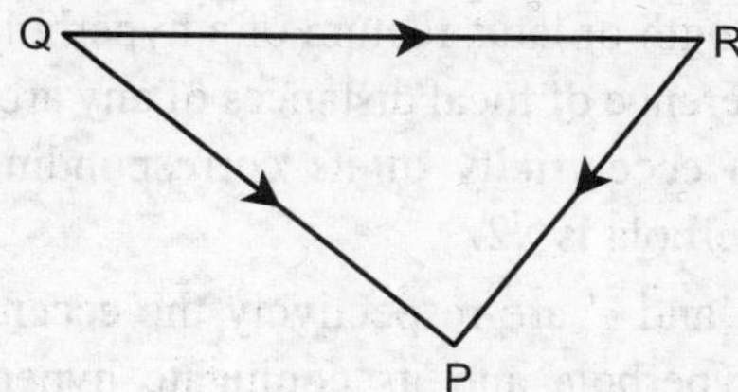

According to the question, $\frac{QP}{u} = \frac{QR}{v} + \frac{RP}{u}$; where u = velocity of sound and v = velocity of bullet.

$\Rightarrow (QP - RP) = \frac{u}{v}(QR) < QR$ as $u < v$.

i.e., $QP - RP = \frac{u}{v}(QR)$ = constant (independent of P)

$\therefore$ Locus of observer (P) would be a hyperbola with length of transverse axis $= \frac{u}{v}(QR)$ and distance between foci = QR.

$\therefore$ Assertion as well as reason both are correct and reason is the correct explanation of assertion.

$\Rightarrow$ **Ans.** (a)

39. A: Considering the hyperbola $\frac{x^2}{9} - \frac{y^2}{16} = 1$, we can conclude that perpendiuclars drawn from a point (5, 0) on the lines $3y \pm 4x = 0$, lie on circle $x^2 + y^2 = 16$.

R: If from any foci of a hyperbola, perpendiculars are drawn on the asymptotes of the hyperbola, then their feet lie on auxiliary circle.

Solution: Reason is correct as feet of perpendiculars drawn from foci on tangents lie on auxiliary circle and asymptote is also a tangent at infinity.

Now, equation of auxiliary circle of hyperbola $\frac{x^2}{9} - \frac{y^2}{16} = 1$ is $x^2 + y^2 = a^2$

i.e., $x^2 + y^2 = 9$; eccentricity is given by

$$\Rightarrow e^2 = 1 + \frac{b^2}{a^2} = 1 + \frac{16}{9} = \frac{25}{9} \Rightarrow e = 5/3.$$

$\therefore$ foci are $(\pm ae, 0) \equiv (\pm 5, 0)$;

Asymptotes of $\left(\frac{x^2}{9} - \frac{y^2}{16}\right) = 1$ are $y = \pm\frac{4}{3}x$.

i.e., $4x \pm 3y = 0$

Feet of perpendiculars drawn from (5, 0) to lines $3y \pm 4x = 0$ lie on circle $x^2 + y^2 = 9$ and not on $x^2 + y^2 = 16$.

$\therefore$ Assertion is incorrect $\Rightarrow$ **Ans.** (d)

40. **A:** If length of latus rectum of a hyperbola is equal to difference of focal distances of any arbitrary point, then eccentricity of its corresponding conjugate hyperbola is $\sqrt{2}$.

R: If e and e' are respectively the eccentricities of a hyperbola and its conjugate hyperbola, then $\frac{1}{e^2}+\frac{1}{e'^2}=1$.

Solution: Length of latera recta $= \frac{2b^2}{a} = 2a$

$\Rightarrow 2b^2 = 2a^2$

$\Rightarrow b^2 = a^2$

$\Rightarrow$ Given hyperbola is a rectangular hyperbola, whose eccentricity is $\sqrt{2}$.

Also by reason, $\frac{1}{e^2}+\frac{1}{e'^2}=1 \Rightarrow \frac{1}{e'^2}+\frac{1}{2}=1$

$\Rightarrow \frac{1}{e'^2}=\frac{1}{2} \Rightarrow e' = \sqrt{2}$.

$\therefore$ Assertion and reason both are correct and reason is the correct explanation of assertion.

$\therefore$ **Ans.** (a)

41. **A:** If a circle $S = 0$ intersect a rectangular hyperbola $xy = 4$ at four points, three of them are (2, 2); (4, 1) and (6, 2/3); then the co-ordinates of the fourth point are (1/6, 24).

R: If a circle $S = 0$ intersects a hyperbola $xy = c^2$ at t_1, t_2, t_3 and t_4, then $t_1 \cdot t_2 \cdot t_3 \cdot t_4 = 1$

Solution: Let $(x-h)^2 + (y-k)^2 = r^2$ intersects the hyperbola $xy = c^2$ at four points.

Let (ct, c/t) be any point of intersection, then

$(ct-h)^2 + (c/t-k)^2 = r^2$

or $t^2(ct-h)^2 + (c-kt)^2 = r^2t^2$

or $t^2[c^2t^2 + h^2 - 2cht] + [c^2 + k^2t^2 - 2ckt] - r^2t^2 = 0$

$\Rightarrow c^2t^4 - 2cht^3 + (h^2+k^2-r^2)t^2 - 2ckt + c^2 = 0$

$\therefore$ If t_1, t_2, t_3, t_4 are the points of intersection, then

$$t_1 \cdot t_2 \cdot t_3 \cdot t_4 = \frac{(-1)^4(c^2)}{c^2} = 1$$

$\therefore$ Reason is correct.

for point (2, 2); $2t_1 = 2 \Rightarrow t_1 = 1$

for point (4, 1); $2t_2 = 4 \Rightarrow t_2 = 2$

for point (6, 2/3); $2t_3 = 6 \Rightarrow t_3 = 3$

now $t_1 \cdot t_2 \cdot t_3 \cdot t_4 = (1)(2)(3)\, t_4 = 1$

$\Rightarrow t_4 = 1/6$

$\therefore$ Fourth point must be $(ct_4, c/t_4) = (1/3, 12)$

$\therefore$ Assertion is incorrect.

$\therefore$ **Ans.** (d).

Solved Comprehension Passages

A: The graph of the conic $x^2 - (y-1)^2 = 1$ has two tangents passing through the origin and touching the centre at points (a_1, b_1) and (a_2, b_2), then answer the questions that follows:

42. The points (a_1, b_1) and (a_2, b_2) are given by
(a) $(\sqrt{2}, 2); (\sqrt{2}, -2)$ (b) $(-\sqrt{2}, 2); (\sqrt{2}, 2)$
(c) $(-\sqrt{2}, 2); (-\sqrt{2}, -2)$ (d) None of these

43. If (a, b) is the point of contact of tangent with positive slope and passing through the origin, then $\sin^{-1}\left(\frac{a}{b}\right)$ equals
(a) $-\frac{\pi}{4}$ (b) $\frac{\pi}{3}$
(c) $\frac{\pi}{4}$ (d) $\frac{\pi}{6}$

44. The length of latera recta of conic is
(a) 2 (b) 4
(c) $2\sqrt{2}$ (d) None of these

Solution: Given conic is a hyperbola with centre at (0, 1) and vertices at $A'(-1, 1)$ and $A(1, 1)$ as shown in the following figure.

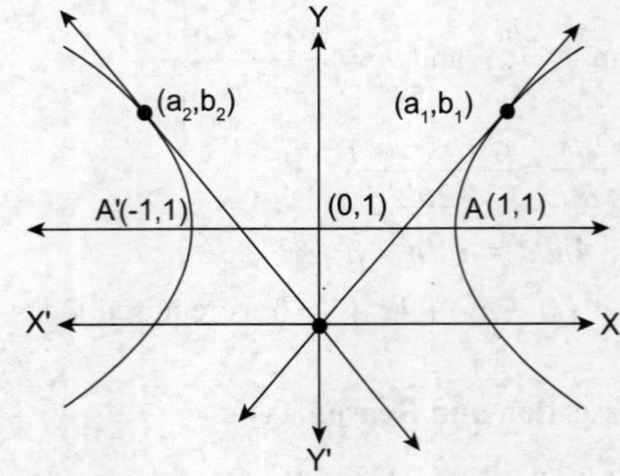

Equation of conic is $x^2 - (y-1)^2 = 1$

$\Rightarrow 2x - 2(y-1)\frac{dy}{dx} = 0 \Rightarrow \frac{dy}{dx} = \frac{x}{(y-1)}$

42. Slope of tangent at point (a, b) is given by

$$\left(\frac{dy}{dx}\right)_{(a,b)} = \frac{a}{b-1}$$

As the tangent passes through origin, then

$$\left(\frac{dy}{dx}\right)_{(a,b)} = \frac{b}{a}$$

$\therefore \quad \frac{a}{b-1} = \frac{b}{a} \Rightarrow a^2 = b^2 - b$ ... (i)

Also (a, b) lies on conic $\Rightarrow a^2 - (b-1)^2 = 1$

$\Rightarrow a^2 - [b^2 + 1 - 2b] = 1$

$\Rightarrow b^2 - b - b^2 - 1 + 2b = 1.$

$\Rightarrow b^2 - \text{b} - b^2 - 1 + 2b = 1$

$\Rightarrow b = 2 \therefore a^2 = b^2 - b = 4 - 2 = 2$

$\Rightarrow a = \pm\sqrt{2}.$

$\therefore$ The points are $(-\sqrt{2}, 2)$ and $(\sqrt{2}, 2)$. $\therefore$ Ans. (b).

43. For positive slope of tangent and passing through origin, a and b must be of same sign as slope = b/a

$\therefore$ $(a, b) \equiv (\sqrt{2}, 2)$

$$\therefore \sin^{-1}\left(\frac{a}{b}\right) = \sin^{-1}\left(\frac{a}{b}\right) = \sin^{-1}\left(\frac{\sqrt{2}}{2}\right)$$

$$= \sin^{-1}\left(\frac{1}{\sqrt{2}}\right) = \frac{\pi}{4}.$$

$\therefore$ **Ans.** (c)

44. Length of latus rectum $= \frac{2B^2}{A} = \frac{2A^2}{A} = 2A = 2(1) = 2$

[$\because$ conic is a rectangular hyperbola]. $\therefore$ Ans. (a)

B. Let S_1 and S_2 are two fixed points on a plane and $P(x, y)$ be a variable point on the plane of S_1 and S_2 such that $|PS_1 - PS_2| = 2a$; where $2a < S_1S_2$, then locus of P is a hyperbola. If S_1 and S_2 lie on x-axis and y-axis passes through mid-point of S_1S_2, then locus of P will represent hyperbola having its equation $\frac{x^2}{a^2} - \frac{y^2}{b^2} = 1$. Conjugate of this hyperbola is $\frac{x^2}{a^2} - \frac{y^2}{b^2} = -1$. Let $P(x, y)$ be a point such that $\left|\sqrt{(x-1)^2 + (y-2)^2} - \sqrt{(x-5)^2 + (y-5)^2}\right| = 4$. Now on the basis of above information answer the following questions.

45. If the locus of P is a hyperbola with eccentricity e, then e' the eccentricity of the corresponding hyperbola is

(a) 5/4 (b) 2

(c) 5/3 (d) 3/2

46. Locus of point of intersection of two mutually perpendicular tangents is

(a) $(x-3)^2 + \left(y - \frac{7}{2}\right)^2 = \frac{7}{4}$

(b) $(x-3)^2 + \left(y - \frac{7}{2}\right)^2 = \frac{1}{4}$

(c) $(x-3)^2 + \left(y - \frac{7}{2}\right)^2 = \frac{25}{4}$

(d) None of these

47. If origin is shifted to point (3, 7/2) and axes are rotated in anti-clockwise sense through an angle θ, so that the equation of hyperbola reduces to its standard form $\frac{x^2}{a^2} - \frac{y^2}{b^2} = 1$, then θ equals

(a) $\tan^{-1}\left(\frac{4}{3}\right)$ (b) $\tan^{-1}\left(\frac{3}{4}\right)$

(c) $\tan^{-1}\left(\frac{5}{3}\right)$ (d) None of these

Solution: Given locus is

$\left|\sqrt{(x-1)^2 + (y-2)^2} - \sqrt{(x-5)^2 + (y-5)^2}\right| = 4$

$= 2a$ (say)

Let $S_1(1, 2)$ and $S_2(5, 5)$, then $S_1S_2 = 5 > 4$.

$\therefore$ Locus is a hyperbola with its foci at $S_1(1, 2)$ and $S_2(5, 5)$ and length of transverse axis $= 2a = 4 \Rightarrow a = 2$. Also $2ae = S_1S_2 = 5$ $\Rightarrow 4e = 5 \Rightarrow e = 5/4$.

45. We know that for a hyperbola with eccentricity e and its conjugate hyperbola with eccentricity e';

$$\frac{1}{e^2} + \frac{1}{e'^2} = 1 \Rightarrow \frac{16}{25} + \frac{1}{e'^2} = 1 \Rightarrow e' = \frac{5}{3} \therefore \text{Ans. (c)}$$

46. Locus of point of intersection of two mutually perpendicular tangents is the director circle given by $(x-h)^2 + (y-k)^2 = a^2 - b^2$; where (h, k) is the centre of hyperbola given by $(h, k) \equiv \left(\frac{1+5}{2}, \frac{2+5}{2}\right) \equiv \left(3, \frac{7}{2}\right)$;

$a^2 = 4$; $b^2 = a^2(e^2 - 1) = 4\left[\frac{25}{16} - 1\right] = 4\left(\frac{9}{16}\right) = \frac{9}{4}$.

$\therefore$ Requried locus (director circle) is

$$(x-3)^2 + \left(y - \frac{7}{2}\right)^2 = 4 - \frac{9}{4} = \frac{7}{4}. \text{ Ans. (a)}$$

47. θ should be the angle between the transverse axis and x-axis, given by $\tan\theta = \frac{5-2}{5-1} = \frac{3}{4}$

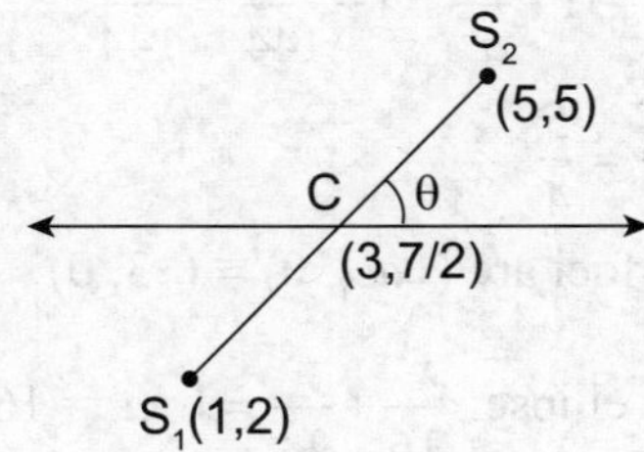

$\Rightarrow \quad \theta = \tan^{-1}\left(\frac{3}{4}\right)$

$\therefore$ Ans. (b)

Solved Column Matching Type

48. Column-I

(a) The number of point(s) inside the hyperbola $\frac{x^2}{16} - \frac{y^2}{25} = 1$ from where two perpendicular tangents can be drawn to hyperbola is/are

(b) The foci of the hyperbola $\frac{x^2}{144} - \frac{y^2}{81} = \frac{1}{25}$ and ellipse $\frac{x^2}{16} + \frac{y^2}{\lambda^2} = 1$ coincide, then λ^2 is

(c) The eccentricity of the conic $x^2 - y^2 - 4x + 4y + 16 = 0$ is

(d) The product of lengths of perpendiculars from any point of the hyperbola $x^2 - y^2 = 8$ to its asymptotes is

Column-II

(p) $\sqrt{2}$

(q) 4

(r) 0

(s) 7

Ans. (a) → (r); (b) → (s); (c) → (p); (d) → (q).

Solution:

(a) If from a point two perpendiculars tangents can be drawn, then it lies on the director circle.

Director circle for hyperbola $\frac{x^2}{16} - \frac{y^2}{25} = 1$.

is given by $x^2 + y^2 = a^2 - b^2 = 16 - 25 = -9$; which does not exist

$\therefore$ There is no such point. $\therefore$ **Ans. (a) → (r)**

(b) Foci of hyperbola: $\frac{x^2}{144} - \frac{y^2}{81} = \frac{1}{25}$;

$a^2 = \frac{144}{25}$; $b^2 = \frac{81}{25}$

$\therefore \quad e_1^{\,2} = 1 + \frac{b^2}{a^2} = 1 + \frac{81}{144} = \frac{225}{144}$.

$\Rightarrow \quad e_1 = \frac{15}{12} = \frac{5}{4}$

$\therefore$ Its foci are $(\pm ae_1, 0) \equiv (\pm 3, 0)$

Foci of ellipse: $\frac{x^2}{16} + \frac{y^2}{\lambda^2} = 1$; $a^2 = 16$; $b^2 = \lambda^2$

$\Rightarrow \quad e_2^{\,2} = 1 - \frac{b^2}{a^2} = 1 - \frac{\lambda^2}{16} = \frac{16 - \lambda^2}{16}$

$\Rightarrow e_2 = \frac{\sqrt{16-\lambda^2}}{4}$.

$\therefore$ Foci of ellipse are given by

$(\pm ae_2, 0) \equiv \left(\pm\sqrt{16-\lambda^2}, 0\right)$

$\therefore \quad \sqrt{16-\lambda^2} = 3 \Rightarrow 16 - \lambda^2 = 9. \Rightarrow \lambda^2 = 7$.

$\therefore$ **(b) → (s)**

(c) Given conic is $x^2 - 4x - y^2 + 4y + 16 = 0$

$\Rightarrow (x^2 - 4x) - (y^2 - 4y) + 16 = 0$

$\Rightarrow (x^2 - 4x + 4) - (y^2 - 4y + 4) = -16$

or $(y-2)^2 - (x-2)^2 = 16$ or $\frac{(y-2)^2}{16} - \frac{(x-2)^2}{16} = 1$

which represents a rectangular hyperbola and hence its eccentricity $= \sqrt{2}$.

$\therefore$ **(c) → (p)**

(d) Let us rotate positive x-axis by $-45°$

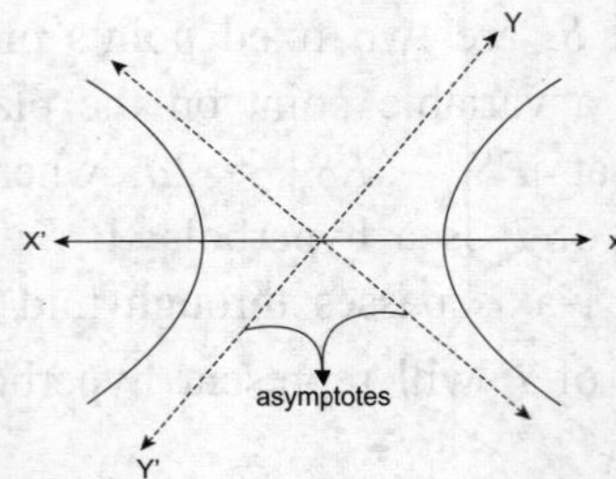

	x	y
X	$\cos(-45°)$	$\sin(-45°)$
Y	$-\sin(-45°)$	$\cos(-45°)$

$\therefore \quad x = X\cos 45° + Y\sin 45° = \frac{1}{\sqrt{2}}(X+Y)$

$y = -X\sin 45° + Y\cos 45° = \frac{1}{\sqrt{2}}(-X+Y)$

The hyperbola $x^2 - y^2 = 8$ reduces to

$\left(\frac{1}{\sqrt{2}}(X+Y)\right)^2 - \left(\frac{1}{\sqrt{2}}(-X+Y)\right)^2 = 8$

or $4XY = 16 \Rightarrow XY = 4$.

Now, asymptotes are x-axis and y-axis

$\therefore$ product of ⊥r distances of any point from hyperbola = product of co-ordinates of point $= xy = 4$

$\therefore$ (d) → **(q)**

49. Conic $3(x + y - 5)^2 - 2(x - y + 7)^2 = 6$

Column I		Column II
(a) Centre	(p)	$\left(\frac{-3}{2},\frac{13}{2}\right)$
(b) Vertices	(q)	$(-1, 6)$
(c) Foci	(r)	$\left(-1+\frac{1}{\sqrt{2}},6+\frac{1}{\sqrt{2}}\right)$
(d) Point of intersections of directrix and transverse axis	(s)	$\left(-1+\frac{\sqrt{5}}{2},6+\frac{\sqrt{5}}{2}\right)$
	(t)	$\left(-1+\frac{1}{\sqrt{5}},6+\frac{1}{\sqrt{5}}\right)$

Ans. (a) → (q), (b) → (r), (c) → (s), (d) → (t)

Solution:

(a) Given conic is $6\left(\frac{x+y-5}{\sqrt{2}}\right)^2-4\left(\frac{x-y+7}{\sqrt{2}}\right)^2=6$

or $$\frac{\left(\frac{x+y-5}{\sqrt{2}}\right)^2}{(1)^2}-\frac{\left(\frac{x-y+7}{\sqrt{2}}\right)^2}{\left(\sqrt{\frac{3}{2}}\right)^2}=1 \quad \text{... (i)}$$

It represents a hyperbola with conjugate axis

$x + y - 5 = 0$ .... (ii)

and transverse axis $x - y + 7 = 0$ ... (iii)

and length of conjugate axis $= 2b = 2\left(\sqrt{\frac{3}{2}}\right) = \sqrt{6}$

centre is the point of intersection of (ii) and (iii) i.e., $(-1, 6)$

∴ **Ans. (a) → (q)**

(b) Now vertices are the point of intersection of transverse axis $x - y + 7 = 0$ and the conic

$\Rightarrow 3(x + y - 5)^2 = 6$ and $(y = x + 7)$

$\Rightarrow (x + y - 7)^2 = 2$

$\Rightarrow (x + x + 7 - 5)^2 = 2$

$\Rightarrow (2x + 2)^2 = 2$

$\Rightarrow (x + 1)^2 = 1/2$

$\Rightarrow x + 1 = \pm 1/\sqrt{2}$

$\Rightarrow x = -1 \pm \frac{1}{\sqrt{2}}; y = -1 \pm \frac{1}{\sqrt{2}} + 7 = 6 \pm \frac{1}{\sqrt{2}}$.

∴ vertices are

$\left(-1+\frac{1}{\sqrt{2}},6+\frac{1}{\sqrt{2}}\right)$ and $\left(-1-\frac{1}{\sqrt{2}},6-\frac{1}{\sqrt{2}}\right)$

∴ **Ans. (b) → (r)**

(c) Now $b^2 = a^2(e^2 - 1) \Rightarrow \frac{3}{2} = 1(e^2 - 1)$

$\Rightarrow e^2 = 5/2 \Rightarrow e = \sqrt{\frac{5}{2}}$

foci are at distances of 'ae' from $(-1, 6)$ along the line $x - y + 7 = 0$

i.e., $\frac{x+1}{1/\sqrt{2}} = \frac{y-6}{1/\sqrt{2}} = \pm(1)\left(\sqrt{\frac{5}{2}}\right)$

$\Rightarrow x + 1 = y - 6 = \pm\frac{\sqrt{5}}{2}$

$\Rightarrow x = -1 \pm \frac{\sqrt{5}}{2}; y = 6 \pm \frac{\sqrt{5}}{2}$

∴ foci are $\left(-1+\frac{\sqrt{5}}{2},6+\frac{\sqrt{5}}{2}\right)$ and

$\left(-1-\frac{\sqrt{5}}{2},6-\frac{\sqrt{5}}{2}\right)$ ∴ **(c) → (s)**

(d) Directrices are parallel to conjugate axis i.e., $x + y - 5 = 0$.

$\Rightarrow \frac{|x+y-5|}{\sqrt{2}} = (a/e)$

$\Rightarrow \frac{|x+y-5|}{\sqrt{2}} = \frac{1}{\sqrt{5}/2} = \sqrt{\frac{2}{5}}$

$\Rightarrow x + y - 5 = \pm\frac{2}{\sqrt{5}}$.

∴ point of intersection of lines $x + y - 5 = \frac{2}{\sqrt{5}}$;

and $x + y - 5 = -\frac{2}{\sqrt{5}}$

with transverse axis $x - y + 7 = 0$ are

$\left(-1+\frac{1}{\sqrt{5}},6+\frac{1}{\sqrt{5}}\right)$ and $\left(-1-\frac{1}{\sqrt{5}},6-\frac{1}{\sqrt{5}}\right)$.

∴ **Ans. (d) → (t).**

TUTORIAL EXERCISE

SECTION–III

OBJECTIVE-TYPE (ONLY ONE CORRECT ANSWER)

1. The equation of the transverse and conjugate axes of a hyperbola $16x^2 - y^2 + 64x + 4y + 44 = 0$ are respectively
(a) $x = 2, y + 2 = 0$ (b) $x = 2, y = 2$
(c) $y = 2, x + 2 = 0$ (d) None of these

2. The locus of the point of intersection of the lines $ax \sec\theta + by \tan\theta = a$ and $ax \tan\theta + by \sec\theta = b$; where θ is the parameter and $a \neq b$, is
(a) A straight line (b) A circle
(c) An ellipse (d) A hyperbola

3. The locus of the point of intersection of the lines $\sqrt{3}x - y - 4\sqrt{3}k = 0$ and $\sqrt{3}kx + ky - 4\sqrt{3} = 0$ for different value of k ($\neq 0$) is
(a) Circle (b) Parabola
(c) Hyperbola (d) Ellipse

4. If the line $2x + \sqrt{6}y = 2$ touches the hyperbola $x^2 - 2y^2 = 4$, then the point of contact is
(a) $\left(-2, \sqrt{6}\right)$ (b) $\left(-5, 2\sqrt{6}\right)$
(c) $\left(\frac{1}{2}, \frac{1}{\sqrt{6}}\right)$ (d) $\left(4, -\sqrt{6}\right)$

5. The equation $\frac{x^2}{12-k} + \frac{y^2}{8-k} = 1$ represents
(a) a hyperbola if $k < 8$
(b) an ellipse if $k > 8$
(c) a hyperbola *if* $8 < k < 12$
(d) none of these.

6. The point of intersection of the curves whose parametric equations are $x = t^2 + 1$, $y = 2t$ and $x = 2s$ and $y = 2/s$ is given by
(a) $(1, -3)$ (b) $(2, 2)$
(c) $(-2, 4)$ (d) $(1, 2)$

7. Equation of a rectangular hyperbola whose aymptotes are $x = 3$ and $y = 5$ and passing through $(7, 8)$, is
(a) $xy - 3y + 5x + 3 = 0$
(b) $xy + 3y + 5x + 3 = 0$
(c) $xy - 3y + 5x - 3 = 0$
(d) $xy - 3y - 5x + 3 = 0$

8. Any tangent to the hyperbola makes with asymptotes a triangle of
(a) constant area $= ab/2$
(b) constant area $= 2ab$
(c) constant area $= ab$
(d) None of these

9. The product of the lengths of perpendiculars drawn from any point on the hyperbola $x^2 - 2y^2 - 2 = 0$ to its asymptotes is
(a) 1/2 (b) 2/3
(c) 3/2 (d) 2

10. The locus of mid-point of the portion of a line of constant slope m between two branches of a rectangular hyperbola $xy = 1$ is
(a) $y - mx = 0$ (b) $y + mx = 0$
(c) $my + x = 0$ (d) $y = x$

11. The equations to the common tangents to the two hyperbolas $\frac{x^2}{a^2} - \frac{y^2}{b^2} = 1$ and $\frac{y^2}{a^2} - \frac{x^2}{b^2} = 1$ are
(a) $y = \pm x \pm \sqrt{b^2 - a^2}$
(b) $y = \pm x \pm \sqrt{a^2 - b^2}$
(c) $y = \pm x \pm (a^2 - b^2)$
(d) $y = \pm x \pm \sqrt{a^2 + b^2}$

12. If a variable straight line $x \cos\alpha + y \sin\alpha = p$, which is a chord of the hyperbola $\frac{x^2}{a^2} - \frac{y^2}{b^2} = 1$ $(b > a)$, subtend a right angle at the centre of the hyperbola, then it always touches a fixed circle whose radius is
(a) $\frac{ab}{\sqrt{b-2a}}$ (b) $\frac{a}{\sqrt{b-a}}$
(c) $\frac{ab}{\sqrt{b^2-a^2}}$ (d) $\frac{ab}{\sqrt{b+a}}$

13. The point of intersection of tangents at t_1 and t_2 to the hyperbola $xy = c^2$ is
(a) $\left(\frac{c\,t_1t_2}{t_1+t_2}, \frac{c}{t_1+t_2}\right)$ (b) $\left(\frac{2c\,t_1t_2}{t_1+t_2}, \frac{2c}{t_1+t_2}\right)$
(c) $\left(\frac{t_1t_2}{c(t_1+t_2)}, \frac{t_1+t_2}{c}\right)$ (d) None of these

14. Asymptotes of the hyperbola $xy = 4x + 3y$ are
(a) $x = 3, y = 4$ (b) $x = 4, y = 3$
(c) $x = 2, y = 6$ (d) $x = 6, y = 2$

15. If e_1 is the eccentricity of the ellipse $\frac{x^2}{16}+\frac{y^2}{25}=1$ and e_2 is the eccentricity of the hyperbola passing through the foci of the ellipse and $e_1e_2 = 1$, then equation of the hyperbola is:

(a) $\frac{x^2}{9}-\frac{y^2}{16}=1$ (b) $\frac{x^2}{16}-\frac{y^2}{9}=-1$

(c) $\frac{x^2}{9}-\frac{y^2}{25}=1$ (d) None of these

16. P is a point on the hyperbola $\frac{x^2}{a^2}-\frac{y^2}{b^2}=1$. N is the foot of the perpendicular from P on the transverse axis. The tangent to the hyperbola at P meets the transverse axis at T. If O is the centre of the hyperbola, then OT.ON is equal to

(a) e^2 (b) a^2

(c) b^2 (d) $\frac{b^2}{a^2}$.

17. A series of hyperbola e is drawn having a common transverse axis of length $2a$. Then the locus of a point P on each hyperbola, such that its distance from the transverse axes is equal to its distance from an asymptote, is

(a) $(x^2 - y^2)^2 = 4x^2(x^2 - a^2)$

(b) $(x^2 - y^2)^2 = x^2(x^2 - a^2)$

(c) $(x^2 - y^2) = 4x^2(x^2 - a^2)$

(d) None of these

18. The number of triangle that can be inscribed in the rectangular hyperbola $xy = c^2$ whose all sides touch the parabola $y^2 = 4ax$ is

(a) infinite (b) 4

(c) 3 (d) 2

19. If $\frac{x^2}{a^2}+\frac{y^2}{b^2}=1$ $(a > b)$ and $x^2 - y^2 = c^2$ cut at right angles, then

(a) $a^2 + b^2 = 2c^2$ (b) $b^2 - a^2 = 2c^2$

(c) $a^2 - b^2 = 2c^2$ (d) $a^2 b^2 = 2c^2$

20. If $P(x_1, y_1)$, $Q(x_2, y_2)$, $R(x_3, y_3)$ and $S(x_4, y_4)$ are four concylic points on the rectangular hyperbola $xy = c^2$, the co-ordinates of orthocentre of the ΔPQR are

(a) $(x_4, -y_4)$ (b) (x_4, y_4)

(c) $(-x_4, -y_4)$ (d) $(-x_4, y_4)$

21. If e and e_1, are the eccentricities of hyperbola $xy = c^2$ and $x^2 - y^2 = c^2$, then $e^2 + e_1^2$ is equal to

(a) 1 (b) 4

(c) 6 (d) 8

22. If H $(x, y) = 0$ represents the equation of a hyperbola and $A(x, y) = 0$, $C(x, y) = 0$ the joint equation of its asymptotes and the conjugate hyperbola respectively, Then for any point (α, β) in the plane $H(\alpha, \beta)$, $A(\alpha, \beta)$, and $C(\alpha, \beta)$ are in

(a) A.P (b) G.P.

(c) H.P (d) None of these

23. Total number of tangents to the hyperbola $\frac{x^2}{9}-\frac{y^2}{4}=1$, that are perpendicular to the line $5x + 2y - 3 = 0$ is/are

(a) Zero (b) 2

(c) 4 (d) None of these

24. A tangent drawn to the hyperbola $\frac{x^2}{a^2}-\frac{y^2}{b^2}=1$ at $P\left(\frac{\pi}{6}\right)$ forms a triangle of area $3a^2$ sq. units, with co-ordinate axes. Eccentricity of the hyperbola is equal to

(a) $\sqrt{17}$ (b) $\sqrt{21}$

(c) 4 (d) $\sqrt{6}$

25. Total number of common tangents of the curves $\frac{x^2}{a^2}-\frac{y^2}{b^2}=1$ and $\frac{y^2}{a^2}-\frac{x^2}{b^2}=1$, (for $a > b$) is equal to

(a) zero (b) 2

(c) 4 (d) None of these

26. The equation of the conjugate hyperbola of hyperbola $hx + ky = xy$ is

(a) $yh + xk - xy + hk = 0$

(b) $xh + yk - xy + hk = 0$

(c) $yh + xk - xy - 2hk = 0$

(d) $xy + hk - xy - 2hk = 0$

27. Normal drawn to the hyperbola xy = 1 at any point 'P' on it, meets the co-ordinate axes at the point A and B respectively. If the rectangle OACB (O being the origin) is completed, then locus of C is

(a) $(x^2 - y^2)^2 - x^2y^2 = 0$ (b) $(x^2 + y^2)^2 - x^3y^3 = 0$

(c) $(x^2 - y^2)^2 + x^2y^2 = 0$ (d) $(x^2 - y^2)^2 + x^3y^3 = 0$

28. The chord PQ of the hyperbola xy = c², meets the x-axis at A. If 'R' is the mid-point of PQ and 'O' is the origin, then triangle ARO is necessarily

(a) Equilateral

(b) Isosceles and right angled

(c) Right angled but not isosceles

(d) None of these

29. Normal drawn to the hyperbola $xy = c^2$ at the point $P(t_1)$ meets the hyperbola against at $Q(t_2)$, then minimum distance between the point P and Q will be

(a) $2c$ (b) $\sqrt{2}c$

(c) $2\sqrt{2}c$ (d) None of these

30. A normal to the parabola $y^2 = 4ax$ with slope m touches the rectangular hyperbola $x^2 - y^2 = a^2$, if

(a) $m^6 + 4m^4 - 3m^2 + 1 = 0$
(b) $m^6 - 4m^4 - 3m^2 - 1 = 0$
(c) $m^6 + 4m^4 + 3m^2 + 1 = 0$
(d) $m^6 - 4m^4 - 3m^2 + 1 = 0$

31. If $xy = m^2 - 9$ be a rectangular hyperbola whose branches lie only in the second and fourth quadrant, then

(a) $|m| \geq 3$
(b) $|m| < 3$
(c) $\mathbb{R} - \{k: k \in \mathbb{R} \text{ and } |k| = 3\}$
(d) None of these

32. The tangents to the hyperbola drawn from the point (α, β) are inclined at angles 'θ' and 'ϕ' to the x axis. If $\tan\theta . \tan\phi = 2$ and hyperbola is $3x^2 - 2y^2 = 6$. Then

(a) $\beta^2 = \alpha^2 - 1$ (b) $\beta^2 = 2\alpha^2 - 7$
(c) $2\beta^2 = \alpha^2 - 1$ (d) $2\beta^2 + \alpha^2 = 2$

33. The two conics $bx^2 = y$ and $\frac{x^2}{a^2} - \frac{y^2}{b^2} = 1$ intersect iff

(a) $-\frac{1}{\sqrt{2}} \leq a \leq \frac{1}{\sqrt{2}}$ (b) $a < -\frac{1}{\sqrt{2}}$
(c) $a > \frac{1}{\sqrt{2}}$ (d) $a < b$

34. PM and PN are the perpendiculars from any point on a rectangular hyperbola to its asymptotes. If Q divides MN in the ratio 3:1, then the locus of Q is

(a) a circle (b) an ellipse
(c) a parabola (d) a rectangular hyperbola

35. Tangents PA and PB are drawn to circle $(x + 4)^2 + (y - 4)^2 = 1$ from a variable points P on $xy = 1$. The locus of circumcentre of the triangle PAB is:

(a) $(2x - 4)(2y - 4) = 1$
(b) $(2x + 4)(2y - 4) = 1$
(c) $(2x + 4)(2y + 4) = 1$
(d) None of these

36. From any point on a hyperbola $xy = c^2$ tangents are drawn to another hyperbola $xy = a^2$. Then the chord of contact cuts off a constant area from the asymptotes having magnitude:

(a) $\frac{2a^2}{c^2}$ (b) $\frac{a^2}{c^2}$
(c) $\frac{2c^2}{a^2}$ (d) $\frac{c^2}{a^2}$

37. If there are two points A and B on rectangular hyperbola $xy = c^2$ such that abscissa of A = ordinate of B, then locus of point of intersection of tangents at A and B is

(a) $y^2 - x^2 = 2c^2$ (b) $y^2 - x^2 = \frac{c^2}{2}$
(c) $y = x$ (d) None of these

38. At the point of intersection of the rectangular hyperbola $xy = c^2$ and the parabola $y^2 = 4ax$ tangents to the rectangular hyperbola and the parabola make an angle θ and ϕ respectively with x- axis, then–

(a) $\tan\theta = -2\tan\phi$ (b) $\tan 2\theta = -\tan\phi$
(c) $\tan\phi = -2\tan\theta$ (d) $\tan 2\phi = -\tan\theta$

39. If foci of hyperbola lie on $y = x$ and one of asymptote is $y = 2x$, then equation of hyperbola, given that it passes through (3, 4) is

(a) $x^2 - y^2 - \frac{5}{2}xy + 5 = 0$
(b) $2x^2 - 2y^2 + 5xy + 5 = 0$
(c) $2x^2 + 2y^2 + \frac{5}{2}xy + 5 = 0$
(d) None of these

40. The angle between the rectangular hyperbolas $(y - mx)(my + x) = a^2$ and $(m^2 - 1)(y^2 - x^2) + 4mxy = b^2$ is

(a) $\frac{\pi}{2}$ (b) $\frac{\pi}{3}$
(c) $\frac{\pi}{4}$ (d) $\frac{\pi}{6}$

SECTION–IV

OBJECTIVE-TYPE (MORE THAN ONE CORRECT ANSWER)

1. Equation of tangent to hyperbola $\frac{x^2}{3} - \frac{y^2}{2} = 1$ equally inclined to co-ordinate axis is

(a) $y = x + 1$ (b) $y = -x - 1$
(c) $y = x + 2$ (d) $y = x - 2$

2. In any hyperbola, if the normal at P meets the transverse axis at G, then

(a) The tangent at any point bisects the internal angle between the focal distances of the point.
(b) the normal bisect the external angle between the focal distances.
(c) $SG = e\, SP$.
(d) None of these

3. Let $\frac{x^2}{a^2}-\frac{y^2}{b^2}=1$ and $\frac{y^2}{a^2}-\frac{x^2}{b^2}=1$ be two hyperbolae, then the slope of their common tangent may be
(a) 1 (b) $\sqrt{2}$
(c) $-\sqrt{2}$ (d) –1

4. If (5, 12) and (24, 7) are the foci of a conic passing through the origin, then the eccentricity of conic is
(a) $\sqrt{386}/12$ (b) $\sqrt{386}/13$
(c) $\sqrt{386}/25$ (d) $\sqrt{386}/38$

5. For the hyperbola $9x^2 - 16y^2 - 18x + 32y - 151 = 0$
(a) one of the directrix is $x = \frac{21}{5}$
(b) Length of latus rectum $= \frac{9}{2}$
(c) foci are (6, 1) and (–4, 1)
(d) eccentricity is $\frac{5}{4}$

6. The point of contact of $5x + 12y = 9$ and $x^2 - 9y^2 = 9$ will lie on
(a) $4x + 15y = 0$ (b) $7x + 12y = 19$
(c) $4x + 15y + 1 = 0$ (d) $7x - 12y = 19$

7. The equation $16x^2 - 3y^2 - 32x + 12y - 44 = 0$, represents a hyperbola
(a) the length of whose transverse axis $4\sqrt{3}$.
(b) the length of whose conjugate axis is 4.
(c) whose centre is (1, 2)
(d) whose eccentricity is $\sqrt{\frac{19}{3}}$.

8. Let A and B be two fixed points and P, another point in the plane, moves such that $k_1PA + k_2PB = k_3$; where k_1, k_2 and k_3 being real constants. The locus of P is
(a) A circle if $k_1 = 0$ and $k_2.k_3 > 0$
(b) A circle if $k_1.k_2 < 0$ and $k_3 = 0$
(c) An ellipse if $k_1 = k_2 > 0$ and $k_3 > 0$, $k_3/k_1 > |AB|$
(d) A hyperbola if $k_2 = -1$ and $k_1 = 1$, $k_3 > 0$, $k_3 < |AB|$

9. If two tangents can be drawn to the different branches of hyperbola $\frac{x^2}{1}-\frac{y^2}{4}=1$ from the point (α, α^2), then
(a) $\alpha \in (-2, 0)$ (b) $\alpha \in (0, 2)$
(c) $\alpha \in (-\infty, -2)$ (d) $\alpha \in (2, \infty)$

10. If the ellipse $x^2 + k^2y^2 = k^2a^2$; $k^2 > 1$ is confocal with the hyperbola $x^2 - y^2 = a^2$, then -
(a) ratio of eccentricities of ellipse & hyperbola is $1/\sqrt{3}$
(b) ratio of major axis of ellipse and transverse axis of hyperbola is $\sqrt{3}$
(c) The ellipse and hyperbola cuts each other orthogonally.
(d) ratio of length of latus rectum of ellipse and hyperbola is 1/3.

SECTION–V

ASSERTION AND REASON-TYPE

The questions given below, consist of an assertion (A) and the reason (R). Use the following key to choose the appropriate answer.
(a) If both assertion and reason are correct and reason is the correct explanation of the assertion.
(b) If both assertion and reason are correct but reason is not correct explanation of the assertion.
(c) If assertion is correct, but reason is incorrect.
(d) If assertion is incorrect, but reason is correct.

Now consider the following statements:

1. **A:** In hyperbola the distance between foci is always greater than the difference of focal distance of any point on it.
R: If e be the eccentricity of the hyperbola, then $e > 1$.

2. **A:** The equation of the director circle to the hyperbola $4x^2 - 3y^2 = 12$ is $x^2 + y^2 = 1$.
R: Director circle is the locus of the point of intersection perpendicular tangents to a hyperbola.

3. **A:** The equation $7y^2 - 9x^2 + 54x - 28y - 116 = 0$ represents a hyperbola.
R: The square of the half the co-efficient of xy is greater than the product of co-efficient of x^2 and y^2 and $\Delta \neq 0$..

4. **A:** 5/3 and 5/4 are the eccentricites of two conjugate hyperbolas.
R: If e and e_1 are the eccentricites of two conjugate hyperbolas, then $ee_1 > 1$.

5. **A:** Director circle of hyperbola $\frac{x^2}{a^2}-\frac{y^2}{b^2}+1 = 0$ is defined only when $b \geq a$.
R: Director circle of hyperbola $\frac{x^2}{25}-\frac{y^2}{9}=1$ is $x^2 + y^2 = 16$.

6. A: A hyperbola whose asymptotes include a right angle will be an equilateral hyperbola.

R: If eccentricity of a hyperbola is √2, then it will be an equilateral hyperbola.

7. A: The point (5, – 4) lies inside the hyperbola $y^2 - 9x^2 + 1 = 0$.

R: If the point (x_1, y_1) is inside the hyperbola $\frac{x^2}{a^2} - \frac{y^2}{b^2} = 1$, then $\frac{x_1^2}{a^2} - \frac{y_1^2}{b^2} - 1 < 0$

8. A: A hyperbola and its conjugate hyperbola have the same asymptotes.

R: The difference between the second degree curve and pair of asymptotes is constant.

9. A: Let e_1 is eccentricity of $9x^2 + 4y^2 = 36$ and e_2 is the eccentricity of the conic $9x^2 - 4y^2 = 36$, then $[e_1^2 + e_2^2] = 3$; where [.] is greatest integer function.

R: For every ellipse eccentricity is less than one while for hyperbola eccentricity is greater than one.

10. A: The two conics $bx^2 = y$ and $\frac{x^2}{a^2} - \frac{y^2}{b^2} = 1$ intersect iff $a \in \left[-\frac{1}{\sqrt{2}}, \frac{1}{\sqrt{2}}\right] - \{0\}$.

R: On solving given two curves, real roots only exist for the $a \in \left[-\frac{1}{\sqrt{2}}, \frac{1}{\sqrt{2}}\right] - \{0\}$.

SECTION–VI

LINKED COMPREHENSION-TYPE

A: The difference between the second degree curve and pair of asymptotes is constant. If second degree curve represented by a hyperbola $S = 0$, then the equation of its asymptotes is $S + \lambda = 0$; where λ is constant, which will be a pair of straight lines, then we get λ. Then equation of asymptotes is $A \equiv S + \lambda = 0$ and if equation of conjugate hyperbola of S represented by S_1, then A is the arithmetic mean of S and S_1.

1. Pair of asymptotes of the hyperbola $xy - 3y - 2x = 0$ is

(a) $xy - 3y - 2x + 2 = 0$

(b) $xy - 3y - 2x + 4 = 0$

(c) $xy - 3y - 2x + 6 = 0$

(d) $xy - 3y - 2x + 12 = 0$

2. The asymptotes of a hyperbola having centre at the point (1, 2) are parallel to the lines $2x + 3y = 0$ and $3x + 2y = 0$. If the hyperbola passes through the point (5, 3) then its equation is

(a) $(2x + 3y - 3)(3x + 2y - 5) = 256$

(b) $(2x + 3y - 7)(3x + 2y - 8) = 156$

(c) $(2x + 3y - 5)(3x + 2y - 3) = 252$

(d) $(2x + 3y - 8)(3x + 2y - 7) = 154$

3. If angle between the asymptotes of hyperbola $\frac{x^2}{a^2} - \frac{y^2}{b^2} = 1$ is $\pi/3$ then the eccentricity of conjugate hyperbola is

(a) √2 (b) 2

(c) 2/√3 (d) 4/√3

4. A hyperbola passing thorugh origin has $3x - 4y - 1 = 0$ and $4x - 3y - 6 = 0$ as its asymptotes. Then the equation of its transverse and conjugate axes are

(a) $x - y - 5 = 0$ and $x + y + 1 = 0$

(b) $x - y = 0$ and $x + y + 5 = 0$

(c) $x + y - 5 = 0$ and $x - y - 1 = 0$

(d) $x + y - 1 = 0$ and $x - y - 5 = 0$

5. The tangent at any point of a hyperbola $16x^2 - 25y^2 = 400$ cuts off a triangle from the asymptotes, then the area of this triangle is

(a) 10 sq unit (b) 20 sq unit

(c) 30 sq unit (d) 40 sq unit

B: If we rotate the axes of the rectangular hyperbola $x^2 - y^2 = a^2$ through an angle $\pi/4$ in the clockwise direction, then the equation $x^2 - y^2 = a^2$ reduces to $xy = \frac{a^2}{2} = \left(\frac{a}{\sqrt{2}}\right)^2 = c^2$ (say). Since $x = ct$, $y = \frac{c}{t}$ satisfies $xy = c^2$.

$\therefore$ $(x, y) = \left(ct, \frac{c}{t}\right)$ $(t \neq 0)$ is called a 't' point on the rectangular hyperbola.

6. If t_1 and t_2 are the roots of the equation $x^2 - 4x + 2 = 0$, then the point of intersection of tangents at 't_1' and 't_2' on $xy = c^2$ is

(a) $\left(\frac{c}{2}, 2c\right)$ (b) $\left(2c, \frac{c}{2}\right)$

(c) $\left(\frac{c}{2}, c\right)$ (d) $\left(c, \frac{c}{2}\right)$

7. Let α, γ be the roots of the equation $t_1x^2 - 4x + 1 = 0$ and β, δ be the roots of $t_2x^2 - 6x + 1 = 0$ and $\alpha, \beta, \gamma, \delta$ are in HP, then the point of intersection of normals at 't_1' and 't_2' on $xy = c^2$ is

(a) $\left(\frac{327c}{264}, \frac{921c}{264}\right)$ (b) $\left(\frac{237c}{264}, \frac{291c}{264}\right)$

(c) $\left(\frac{723c}{264}, \frac{129c}{264}\right)$ (d) None of these

8. If e_1 and e_2 are the eccentricities of the hyperbolas $x + y = 9$ and $x^2 - y^2 = 25$, then (e_1, e_2) lie on a circle C_1 with centre origin then the (radius)^2 of the director circle of C_1 is

(a) 2 (b) 4

(c) 8 (d) 16

C: If the normals at (x_i, y_i); $i = 1, 2, 3, 4$ on the rectangular hyperbola $xy = c^2$, meet at the point (α, β).

On the basis of the above information, answer the following questions:

9. The value of $\sum x_i$ is

(a) c β (b) c α

(c) α (d) β

10. The value of $\sum y_i$ is

(a) c β (b) c α

(c) α (d) β

11. The value of $\sum x_i^2$ is

(a) c^2 (b) α^2

(c) $-c^2$ (d) $-\beta^2$

12. The value of $\sum y_i^2$ is

(a) β^2 (b) α^2

(c) $-c^2$ (d) c^2

13. The value of $\prod x_i = \prod y_i =$ is

(a) $-c$ (b) $-c^2$

(c) $-c^3$ (d) $-c^4$.

D: Cartesian graph of any equation $Ax^2 + Bxy + Cy^2 + Dx + Ey + F = 0$ in which A,B and C are not all zero, is nearly always a conic section. The exceptions are cases when there is a graph containing two parallel lines.

To eliminate xy terms in the equation of conic, we rotate the co-ordinate axis. The equations for rotation in new co-ordinate axes (X-axis and Y-axis) are

$x = X\cos\alpha - Y\sin\alpha$

$y = X\sin\alpha + Y\cos\alpha$

where α is the angle by which the old co-ordinate axes are rotated in anti-clockwise sense to obtain new axes (X –axis and Y-axis)

Applying above rotation, the equation of conic, we get $A'X^2 + B'XY + C'Y^2 + D'X + E'Y + F' = 0$

New and old coefficients are retated as

$A' = A\cos^2\alpha + B\cos\alpha\sin\alpha + C\sin^2\alpha$

$B' = B\cos 2\alpha + (C - A)\sin 2\alpha$

$C' = A\sin^2\alpha - B\sin\alpha\cos\alpha + C\cos^2\alpha$

$D' = D\cos\alpha + E\sin\alpha$

$E' = -D\sin\alpha + E\cos\alpha$

$F' = F$

Thus to eliminate the co-efficient of xy, we rotate the co-ordinate axis by angle α, such that

$B\cos 2\alpha + (C - A)\sin 2\alpha = 0$

$\Rightarrow 2\alpha = \cot^{-1}\left(\frac{A-C}{B}\right)$

It is then easy to check the nature of original conic by applying following test to the new equation

$A'X^2 + C'Y^2 + D'X + E'Y + F' = 0$

(i) if $A'C' = 0$ and A', C' are not both zeros, then graph is parabola

(ii) $A'C' > 0$ and $A' \neq C'$, then graph is ellipse

(iii) $A'C' < 0$, then graph is hyperbola

14. If x and y axis are rotated by an angle $\frac{\pi}{4}$ in counter-clockwise direction about origin, then the equation of hyperbola $2xy = 9$ transfers to

(a) $X^2 - Y^2 = 1$ (b) $X^2 - Y^2 = 9$

(c) $X^2 - Y^2 = 4$ (d) $X^2 - Y^2 = 9/2$

15. If axes be turned through, an angle 'α' such that the transformed equation of $xy - y^2 - 5y + 1 = 0$ does not contain any term of XY, then sinα is

(a) $\frac{1}{\sqrt{2}}$ (b) $\frac{1}{2}$

(c) $\frac{\sqrt{4-2\sqrt{2}}}{2\sqrt{2}}$ (d) $\frac{\sqrt{\sqrt{2}-1}}{2\sqrt{2}}$

16. If axes be rotated by positive obtuse angle 'α' such that the transformed equation of the curve $3x^2 - 6xy + 3y^2 + 2x - 7 = 0$ does not contain any term of XY, then co-efficient of X^2 in the transformed equation is

(a) –3 (b) 6

(c) 3 (d) 0

17. If co-ordinate axes are rotated through an angle 'α' in anti-clockwise sense such that the transformed equation of the curve $x^2 - 2xy + 3y^2 + 4x - 4y + 1 = 0$ does not contain linear term in Y, then α is

(a) $\frac{\pi}{4}$ (b) $\frac{\pi}{3}$

(c) $\frac{2\pi}{3}$ (d) $\frac{3\pi}{4}$

SECTION–VII

MATRIX-MATCH TYPE

1. Column-I

(i) Position of points (3, 4) and (5, 2) with respect to hyperbola $x^2 - 4y^2 - 2x + 24y - 37 = 0$ is

(ii) Position of points (3, 4) and (5, 2) with respect to hyperbola $xy + 2x + 3y - 12 = 0$ is

(iii) Position of points (3, 4) and (5, 2) with respect to hyperbola $xy = 12$ is

Column-II

(a) (3, 4) lie inside the hyperbola

(b) (5, 2) lie outside the hyperbola

(c) (3, 4) lie outside the hyperbola

(d) (5, 2) lie inside the hyperbola

(e) (3, 4) lie on the hyperbola

2. Column-I

(i) Director circles of $x^2 - 2y^2 = 2$ and $x^2 + 2y^2 = 2$ are respectively

(ii) Director circles of $3x^2 + 2y^2 = 6$ and $2x^2 - 3y^2 = 6$ respectively

(iii) Director circles of $5x^2 - 9y^2 = 45$ and $x^2 + y^2 = 1$ respectively

Column-II

(a) $x^2 + y^2 = 1$

(b) $x^2 + y^2 = 2$

(c) $x^2 + y^2 = 3$

(d) $x^2 + y^2 = 4$

(e) $x^2 + y^2 = 5$

3. Observe the following columns:

Column-I

(i) The foci of the hyperbola $12x^2 - 4y^2 - 24x + 32y - 127 = 0$ are

(ii) The foci of the hyperbola $8x^2 - y^2 - 64x + 10y + 71 = 0$ are

(iii) The foci of the hyperbola $9x^2 - 16y^2 - 36x + 96y + 36 = 0$ are

Column-II

(a) (10, 5)

(b) (2, 8)

(c) (–4, 4)

(d) (–2, 5)

(e) (6, 4)

4. Match the items of column I with that of column II.

Column-I

(i) Normals are drawn to the parabola $y^2 = 4x$ at $P(t^2, 2t)$; where $t < 0$, which meets the curve again at Q. If ordinate of Q is minimum then PQ^2 is

(ii) From point on circle $x^2 + y^2 = 36$, tangents are drawn to hyperbola $x^2 - y^2 = 36$, locus of mid points of chords of contact is $(x^2 - y^2)^2 = 4 \quad (x^2 + y^2)$, then equals

(iii) Maximum length of chord of ellipse $\frac{x^2}{2} + y^2 = 1$, such that eccentric angles of its extremities differ by $\pi/2$ is

(iv) Three distinct lines are drawn in a plane & there are exactly n circles in plane tangent to all three lines, then n can be

Column-II

(a) 9

(b) 0

(c) 108

(d) 2

(e) 4

5. Column-I

(i) The equation of the bisector of the angle between the lines $x - 7y + 5 = 0$ and $5x + 5y = 3$; which is the supplement of the angle containing the origin, is

(ii) The locus of a point such that tangents drawn from it to the curve $9x^2 + 4y^2 - 18x - 16y - 11 = 0$ are perpendicular to each other is

(iii) For all real values of m, the line $y = mx + \sqrt{3m^2 - 4}$ is a tangent to

(iv) The transverse axis of a hyperbola is of length $2a$ and a vertex divides the segment of the axis between the centre and the corresponding focus in the ratio 2 : 1. The equation of the hyperbola is

Column-II

(a) $5x^2 - 4y^2 = 5a^2$

(b) $x + 3y = 2$

(c) $4x^2 - 3y^2 = 12$

(d) $(x - 1)^2 + (y - 2)^2 = 13$

SECTION–VIII

INTEGER-TYPE

1. The normal at P to a hyperbola of eccentricity $\frac{3}{2\sqrt{2}}$ intersects the transverse and conjugate axis at M and N respectively. The locus of mid-point of MN is a hyperbola. Find its eccentricity.

2. The equation of tranverse axis of hyperbola (passing through origin) having asymptotes $3x - 4y - 1 = 0$ and $4x - 3y - 6 = 0$ is $ax + by - c = 0$; $a, b, c \in \mathbb{N}$ and g.c.d. $(a, b, c) = 1$, then evaluate $(a + b + c)$.

3. If abscissa of orthocentre of a triangle inscribed in a rectangular hyperbola $xy = 4$ is 1/2, then find the value of ordinate of orthocentre.

4. If the normal to the rectangular hyperbola $xy = c^2$ at any point 't' on it intersects the hyperbola at 't_1' then evaluate $8t^6t_1$.

5. The graph $xy = 1$ is reflected in $y = 2x$ to give the graph $12x^2 + rxy + sy^2 + t = 0$. Find rs

6. ABC is a triangle such that $\angle ABC = 2\angle BAC$. If now keeping AB fixed, C is moved so that the base angles satisfy the given relationship, showing that C describes a hyperbola, find the eccentricity of hyperbola.

7. Point P lie on hyperbola $2xy = 1$. A triangle is constructed by P, S and S' (where S and S' are foci). The locus of ex-centre opposite S (S and P lie in first quadrant) is $(x + my)^2 = (\sqrt{2} - 1)^2 (x - y)^2 + n$; then find $(m + n)$.

Answer Keys

SECTION–III

1. (c)	**2.** (d)	**3.** (c)	**4.** (d)	**5.** (c)	**6.** (b)	**7.** (d)	**8.** (c)	**9.** (b)	**10.** (b)
11. (b)	**12.** (c)	**13.** (b)	**14.** (a)	**15.** (b)	**16.** (b)	**17.** (a)	**18.** (a)	**19.** (c)	**20.** (c)
21. (b)	**22.** (a)	**23.** (a)	**24.** (a)	**25.** (c)	**26.** (d)	**27.** (d)	**28.** (b)	**29.** (c)	**30.** (c)
31. (b)	**32.** (b)	**33.** (a)	**34.** (d)	**35.** (b)	**36.** (a)	**37.** (c)	**38.** (a)	**39.** (d)	**40.** (a)

SECTION–IV

1. (a, b) **2.** (a, b, c) **3.** (a, d) **4.** (a, d) **5.** (a, b, c, d) **6.** (a, b) **7.** (c, d) **8.** (a, b, c)
9. (c, d) **10.** (a, b, c)

SECTION–V

1. (a)	**2.** (d)	**3.** (a)	**4.** (b)	**5.** (b)	**6.** (b)	**7.** (d)	**8.** (a)	**9.** (b)	**10.** (a)

SECTION–VI

1. (c)	**2.** (d)	**3.** (b)	**4.** (c)	**5.** (b)	**6.** (d)	**7.** (d)	**8.** (c)	**9.** (c)	**10.** (d)
11. (b)	**12.** (a)	**13.** (d)	**14.** (b)	**15.** (c)	**16.** (b)	**17.** (d)			

SECTION–VII

1. (i) → (a, b) (ii) → (b, c) (iii) → (d, e)
2. (i) → (a, c) (ii) → (a, e) (iii) → (b, d)
3. (i) → (c, e) (ii) → (a, d) (iii) → (b)
4. (i) → (c) (ii) → (a) (iii) → (d) (iv) → (b, d, e)
5. (i) → (b) (ii) → (d) (iii) → (c) (iv) → (a)

SECTION–VIII

1. 3 **2.** 7 **3.** 8 **4.** 8 **5.** 84 **6.** 2 **7.** 8

HINTS AND SOLUTIONS

TEXTUAL EXERCISE-1 (SUBJECTIVE)

1. (i) $\dfrac{x^2}{\left(\frac{1}{3}\right)^2} - \dfrac{y^2}{1} = 1$, clearly x-axis is its transverse axis with length $(2a) = 2/3$ units and y-axis is its conjugate axis with length $2b = 2$ units.

$\Rightarrow \quad e = \sqrt{\dfrac{a^2+b^2}{a^2}} = \sqrt{\dfrac{\left(\frac{1}{3}\right)^2+1}{\left(\frac{1}{3}\right)^2}} = \sqrt{10}$

$\Rightarrow$ Foci $(\pm ae, 0)$ are $\left(\pm\dfrac{\sqrt{10}}{3}, 0\right)$

The directrices are $x = \pm\dfrac{1}{3\sqrt{10}}$; vertices are $\left(\pm\dfrac{1}{3}, 0\right)$;

length of LR $= \dfrac{2b^2}{a} = \dfrac{2(1)^2}{\left(\frac{1}{3}\right)} = 6$ units

(ii) The hyperbola is $\dfrac{y^2}{16} - \dfrac{x^2}{9} = 1$ y-axis is its transverse axis which length 8 units, x-axis is its conjugate axis with length 6 units $e = \sqrt{\dfrac{a^2+b^2}{a^2}} = \sqrt{\dfrac{25}{16}} = \dfrac{5}{4}$

$\Rightarrow$ Foci are $(0, \pm 5)$

The directrices are $y = \pm\dfrac{16}{5}$; vertices are $(0, \pm 4)$

Length of LR = 9/2 units.

2. From the given $\dfrac{2b^2}{a} = a \Rightarrow 2b^2 = a^2$

$\therefore \quad e = \sqrt{\dfrac{a^2+b^2}{a^2}} = \sqrt{\dfrac{3b^2}{2b^2}} = \sqrt{\dfrac{3}{2}}$

3. (a) Given focus (2, 2), eccentricity $e = 2$, directrix L: $x + y - 9 = 0$. Let $P(x, y)$ be a point.

From the definition of conics $\dfrac{(PS)^2}{(PM)^2} = e^2$

$\Rightarrow \quad (x-2)^2 + (y-2)^2 = \dfrac{4}{2}\{x+y-9\}^2$

$\Rightarrow \ x^2 + y^2 + 4xy - 32x - 32y + 154 = 0$

(b) Given $e = \sqrt{2}$; $2ae = 16$. So $a = 4\sqrt{2}$, $b = \sqrt{a^2e^2 - a^2} = 4\sqrt{2}$ (rectangular hyperbola)

As centre is at the origin and transverse axis is along x-axis so equation of hyperbola is $x^2 - y^2 = 32$

(c) Given foci at (8, 3) and (0, 3) so centre is (4, 3) eccentricity $e = 4/3$

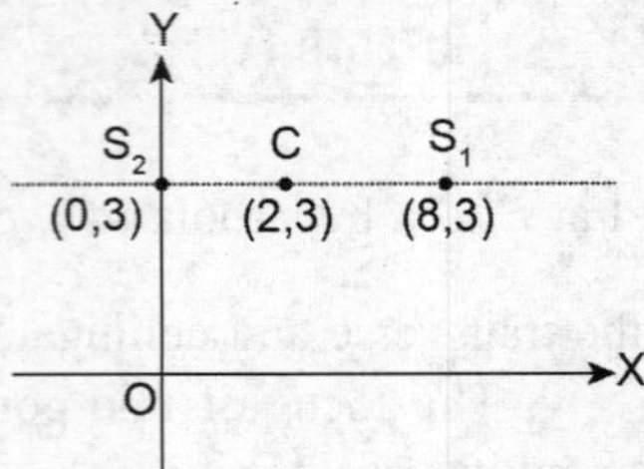

Now $ae = 4$

$\Rightarrow$ a = 3 and $a^2 + b^2 = a^2e^2 = 16 \Rightarrow b^2 = 7$

Hence the equation $\dfrac{(x-4)^2}{9} - \dfrac{(y-3)^2}{7} = 1$ i.e., $7x^2 - 9y^2 - 56x + 54y - 32 = 0$

4. (a) Length of conjugate axis is $2b = 5 \Rightarrow b = 5/2$. Distance between foci is $2ae = 13$, therefore $ae = \dfrac{13}{2}$

Now $a^2 + b^2 = a^2e^2 \quad \Rightarrow \quad a^2 + \dfrac{25}{4} = \dfrac{169}{4}$

$\Rightarrow \quad a^2 = 36$

Hence the equation is given as $\dfrac{x^2}{36} - \dfrac{4y^2}{25} = 1$ i.e., $25x^2 - 144y^2 = 900$

(b) Given $a = 3$, $b = 4$

Hence the parabola $\dfrac{x^2}{9} - \dfrac{y^2}{16} = 1$ i.e., $16x^2 - 9y^2 = 144$

(c) The points of intersection of lines L_1: $x = 3\sqrt{5}y$ and L_2: $\sqrt{5}x - 2y = 13$ is $(3\sqrt{5}, 1)$, which is satisfy the equation $\dfrac{x^2}{a^2} - \dfrac{y^2}{b^2} = 1$

$\Rightarrow \quad \dfrac{45}{a^2} - \dfrac{1}{b^2} = 1$ ....(*i*)

Since length of $LR = \dfrac{2b^2}{a} = \dfrac{4}{3}$

$\Rightarrow \quad b^2 = \dfrac{2a}{3}$ or $\dfrac{45}{a^2} - \dfrac{3}{2a} = 1$

$\Rightarrow \quad 2a^2 + 3a - 90 = 0$

i.e., $(2a + 15)(a - 6) = 0 \Rightarrow$ a = 6 and $b^2 = 4$

$\therefore \quad e = \dfrac{\sqrt{10}}{3}$ gives foci at $\left(\pm 2\sqrt{10}, 0\right)$

(d) The given hyperbola $9x^2 + 18x - 16y^2 + 32y = 151$ can be rewritten as $9(x + 1)^2 - 16(y - 1)^2 = 144$ i.e., $\dfrac{(x+1)^2}{16} - \dfrac{(y-1)^2}{9} = 1$

Now $ae = \sqrt{a^2+b^2} = 5$

$\Rightarrow$ foci are at (4, 1) and (–6, 1)

5. (a) The given hyperbola is $\dfrac{x^2}{\left(\frac{1}{a}\right)} - \dfrac{y^2}{1} = 1$

$\Rightarrow \quad A^2 = \frac{1}{a}, B^2 = 1$

Now, $Ae = \sqrt{1+\frac{1}{a}}$ so foci at $\left(\pm\sqrt{\frac{a+1}{a}}, 0\right)$; $e = \sqrt{a+1}$

Further $\frac{A}{e} = \frac{1}{\sqrt{a(a+1)}}$ so directrices are

$x = \pm\frac{1}{\sqrt{a(a+1)}}$

(b) The hyperbola is $\frac{x^2}{9} - \frac{y^2}{4} = 1$

Transverse axis along y-axis with length = 6 units, conjugate axis along y-axis with length = 4 units

Eccentricity = $\sqrt{\frac{9+4}{9}} = \frac{\sqrt{13}}{3}$

$ae = \sqrt{13}$ so, foci at $(\pm\sqrt{13}, 0)$

Length of $LR = \frac{2b^2}{a} = \frac{2(4)}{3} = \frac{8}{3}$ units

(c) The equation $x^2 - 2y^2 - 2x + 8y - 1 = 0$ can be rewritten as $(x-1)^2 - 2(y^2 - 4y + 4) = -6$ gives $\frac{(y-2)^2}{3} - \frac{(x-1)^2}{6} = 1$; which is *a* hyperbola with Centre at (1, 2). Transverse axis along $x = 1$ with length $2\sqrt{3}$ units.

Conjugate axis along $y = 2$ with length $2\sqrt{6}$ units.

Eccentricity $e = \sqrt{\frac{3+6}{3}} = \sqrt{3}$

Length of latus rectum = $\frac{2(6)}{\sqrt{3}} = 4\sqrt{3}$ units

Now $ae = 3$, so foci at (1, –1) and (1, 5)
Vertices at $(1, 2 \pm \sqrt{3})$ As a/e = 1. So directrices are y = 1, 3

6. Given focus S $(a, 0)$, $e = 5/4$, directrix $4x - 3y = a$. Let $P(x, y)$ be *a* point on hyperbola, the by definition

$\frac{PS^2}{PM^2} = e^2$ so $(x-a)^2 + (y-0)^2 = \frac{25}{16}\left\{\frac{4x-3y-a}{5}\right\}^2$ gives

$16\{x^2 + a^2 - 2ax + y^2\} = 16x^2 + 9y^2 + a^2 - 24xy - 8ax + 6ay$

∴ Equation of conic is $7y^2 + 24xy - 24ax - 6ay + 15a^2 = 0$

$\Rightarrow \quad \frac{\partial c}{\partial x} = 24y - 24a = 0$ and $\frac{\partial c}{\partial y} = 14y + 24x - 6a = 0$

$\Rightarrow$ $x = -a/3$, $y = a$, hence centre is at $\left(-\frac{a}{3}, a\right)$

7. Given $P(x_1, y_1) = \left\{\frac{a}{2}\left(t+\frac{1}{t}\right), \frac{b}{2}\left(t-\frac{1}{t}\right)\right\}$

So $\frac{2x_1}{a} = t + \frac{1}{t}$ and $\frac{2y_1}{b} = t - \frac{1}{t}$

$\Rightarrow \quad 2t = \left(\frac{2x_1}{a} + \frac{2y_1}{b}\right)$ and $\frac{2}{t} = \left(\frac{2x_1}{a} - \frac{2y_1}{b}\right)$

Multiplying, we get $\left(\frac{x_1}{a} + \frac{y_1}{b}\right)\left(\frac{x_1}{a} - \frac{y_1}{b}\right) = t.\frac{1}{t} = 1$

$\frac{x_1^2}{a^2} - \frac{y_1^2}{b^2} = 1$ i.e., $\frac{x^2}{a^2} - \frac{y^2}{b^2} = 1$; which is *a* hyperbola for all real $t \neq 0$

8. Given $\frac{x}{a} - \frac{y}{b} = m$ and $\frac{x}{a} + \frac{y}{b} = \frac{1}{m}$ multiplying, we get $\frac{x^2}{a^2} - \frac{y^2}{b^2} = 1$ which is *a* hyperbola as x and y will always satisfy it for all real $m \neq 0$,

9. **(a)** Rewriting as

$\sqrt{(x+4)^2 + (y+2)^2} = 8 + \sqrt{(x-4)^2 + (y-2)^2}$

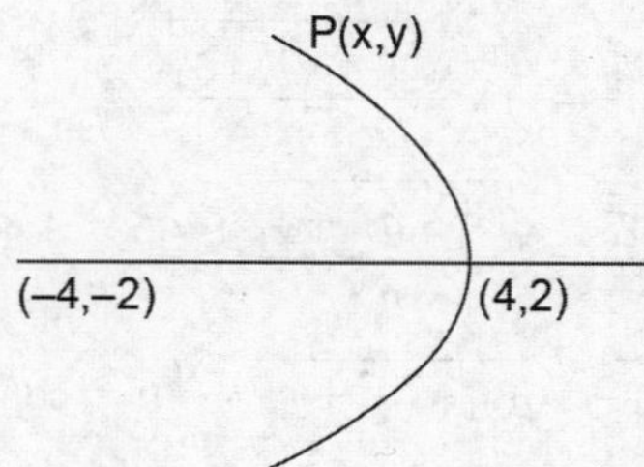

Now on squaring we get $(x+4)^2 + (y+2)^2 = 64 + (x-4)^2 + (y-2)^2 + 16\sqrt{(x-4)^2 + (y-2)^2}$ or $(x+4)^2 - (x-4)^2 + (y+2)^2 - (y-2)^2 = 64 + 16\sqrt{(x-4)^2 + (y-2)^2}$

$\Rightarrow \quad 16x + 8y - 64 = 16\sqrt{(x-y)^2 + (y-2)^2}$

Squaring again, we get $\{4x^2 + y^2 + 64 + 4xy - 32x - 16y\} = (4)\{x^2 + y^2 + 20 - 8x - 4y\}$ gives $3y^2 - 4xy + 16 = 0$; which is *a* hyperbola as $\Delta \neq 0$ where

$\Delta = \begin{vmatrix} 0 & -2 & 0 \\ -2 & 3 & 0 \\ 0 & 0 & 16 \end{vmatrix} = -64$ and $h^2 - ab = 4 - 0 > 0$.

(b) Without any loss of generality let O (0, 0) be the centre of the hyperbola, $(a, 0)$ as vertex and $(ae, 0)$ as its focus for the given length
$= 2a$ of the transverse axis.
Now according to the given $2a = ae$
So $e = 2$

$\Rightarrow \quad \frac{a^2 + b^2}{a^2} = 4$ gives $b^2 = 3a^2$, hence the equation

$\frac{x^2}{a^2} - \frac{y^2}{3a^2} = 1$ or $3x^2 - y^2 = 3a^2$

10. Let the hyperbola be $\frac{x^2}{a^2} - \frac{y^2}{b^2} = 1$. Since it passes through (3,0)

$\Rightarrow \quad a^2 = 9$

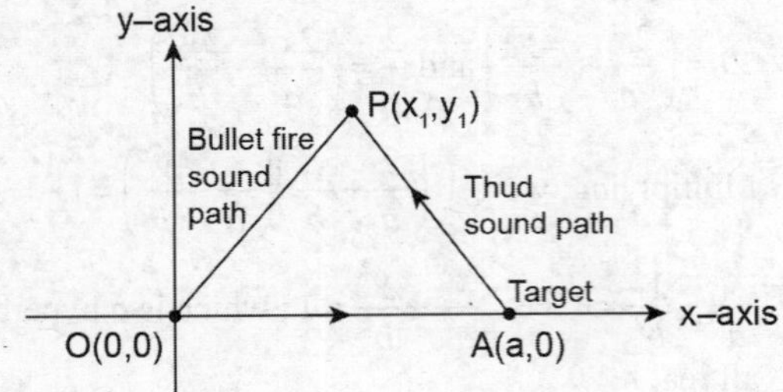

Further (3√2, 2) lies on it

$\Rightarrow \frac{18}{9} - \frac{4}{b^2} = 1 \qquad \Rightarrow b^2 = 4$

Hence its eccentricity $e = \sqrt{\frac{a^2+b^2}{a^2}} = \sqrt{\frac{9+4}{9}} = \frac{\sqrt{13}}{3}$

11. Without any loss of generality let the rifle be situated at O (0, 0) and the target at $(a, 0)$. Now consider the hearer at $P(x_1, y_1)$ the speed of bullet (super-sonic) ='s' m/sec. and the speed of sound in air (normal speed) = 'n' m/sec

Since sound is heard at the same time

$\therefore \frac{\sqrt{x_1^2+y_1^2}}{n} = \frac{a}{s} + \frac{\sqrt{(x_1-a)^2+y_1^2}}{n}$

$\Rightarrow \frac{\sqrt{x_1^2+y_1^2}}{n} - \frac{\sqrt{(x_1-a)^2+y_1^2}}{n} = \frac{a}{s}$

$\Rightarrow \sqrt{x_1^2+y_1^2} - \sqrt{(x_1-a)^2+y_1^2} = \frac{a}{s}n$ = constant {as a, s, n are all non-zero constants}; which is the locus of a point $P(x1, y1)$ such that $PO - PA = \frac{an}{S}$ i.e., a hyperbola with foci at O and A and length of transverse axis $= \frac{an}{S}$.

12. Given hyperbola $\frac{x^2}{1} - \frac{y^2}{(1/3)} = 1$

$\therefore \quad e_1^2 = \frac{a^2+b^2}{a^2} = \frac{1+\left(\frac{1}{3}\right)}{1} = \frac{4}{3}$

$\Rightarrow$ The conjugate hyperbola will have its eccentricity e_2 given by $\frac{1}{e_1^2} + \frac{1}{e_2^2} = 1$, so $\frac{1}{e_2^2} = 1 - \frac{3}{4} = \frac{1}{4} \Rightarrow e_2 = 2$

13. Without any loss of generality let OB be along x-axis and OC be along y-axis.

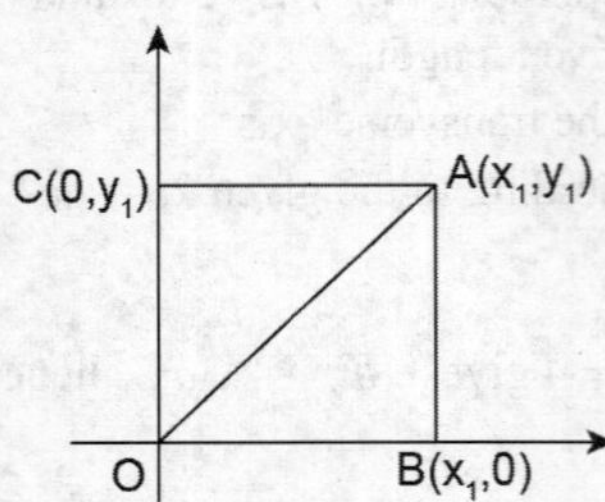

If $A(x_1, y_1)$, then $OB = x_1$ and $OC = y_1$. Now the area $OBAC$ $= x_1y_1$ = constant = C^2 (say)

i.e., $xy = C^2$ which is locus of a hyperbola. By rotating the axis by 45°, we get $X^2 - Y^2 = 2C^2 = k^2$.

TEXTUAL EXERCISE–1 (OBJECTIVE)

1. (c) The hyperbola is $\frac{(x-1)^2}{9} - \frac{(y-2)^2}{16} = 1$

$\Rightarrow$ Centre at (1, 2) and transverse axis along $y = 2$.

Now $a^2e^2 = a^2 + b^2 = 25$, so $ae = 5$

$\Rightarrow$ foci at (–4, 2) and (6, 2)

2. (a) $\frac{tx}{a} - \frac{y}{b} + t = 0$ can be rewritten as $t.\left(\frac{x+a}{a}\right) = \frac{y}{b}$

Similarly $\frac{-ty}{b} = \left(\frac{x-a}{a}\right)$

The product becomes $\frac{-ty^2}{b^2} = t\left(\frac{x^2-a^2}{a^2}\right)$ i.e., $\frac{x^2}{a^2} + \frac{y^2}{b^2} = 1$ which does not represents a hyperbola.

Other option can be shown to form a hyperbola.

3. (c) The points involved in the equation are P (2, 1) and Q (–2, 0).

Now $|PQ| = \sqrt{4^2+1} = \sqrt{17}$

Now the movable point must pairs in between P and Q So that $|PM| - |QM| < |PQ|$

$\Rightarrow c \in (0, \sqrt{17})$

4. (b) The given conic is S: $x^2 + 2y^2 + 3xy + 2x + 3y + 2 = 0$

Now $\frac{\partial S}{\partial x} = 2x + 3y + 2 = 0$ and $\frac{\partial S}{\partial y} = 4y + 3x + 3 = 0$

Which gives the centre of conic at their intersection at (–1, 0).

5. (c) Let $\frac{x^2}{a^2} - \frac{y^2}{b^2} = 1$ be a hyperbola.

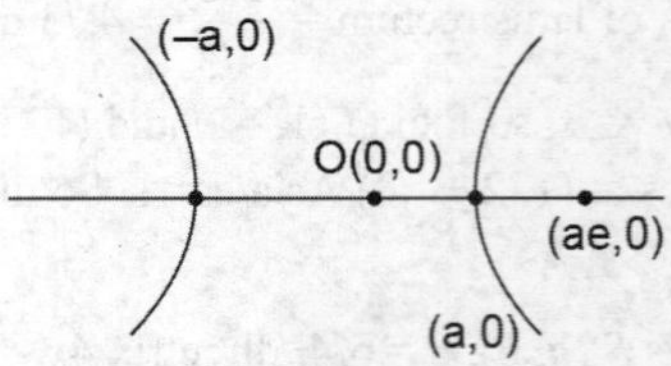

Now $a(e-1) = 1$ and $a(e+1) = 3$

$\Rightarrow a = 1, e = 2$

$\Rightarrow b^2 = 3$. Hence the hyperbola $\frac{x^2}{1} - \frac{y^2}{3} = 1$ or $3x^2 - y^2 = 3$

6. (b) Rewriting as $(\sqrt{3}x - y) = 4\sqrt{3}\, t$ and $(\sqrt{3}x + y) = \frac{4\sqrt{3}}{t}$

The product becomes $3x^2 - y^2 = 16 \times 3$ so $\frac{x^2}{16} - \frac{y^2}{48} = 1$

gives $e = \sqrt{\frac{16+48}{16}} = 2$

7. (b) Rewriting as $\frac{x^2}{5} - \frac{y^2}{5\cos^2\theta} = 1$ gives

$e_1^2 = \frac{5(1+\cos^2\theta)}{5} = 1 + \cos^2\theta$

Also, $\frac{x^2}{25\sin^2\theta} + \frac{y^2}{25} = 1$

$\Rightarrow\ e_2^2 = 1 - \dfrac{25\sin^2\theta}{25} = \cos^2\theta$

Now $e_1^{\,2} = 3e_2^{\,2}$ so $1 + \cos^2\theta = 3\cos^2\theta$ gives $\cos^2\theta = 1/2$

$\Rightarrow$ Smallest positive value of $\theta = 45° = \dfrac{\pi}{4}$

8. (d) Without any loss of generality let circle centered at C_1 be of radius r_1, circle without centre at C_2 be of r_2 and the circle touching both C_1 and C_2 externally be centred at C_3 and with radius r_3.

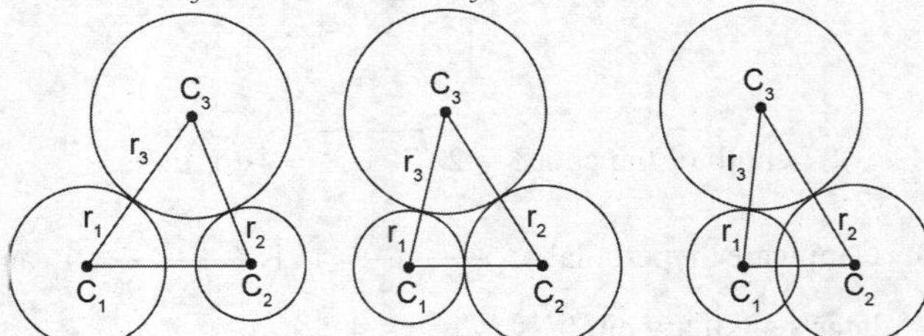

Given $r_1 \neq r_2$ i.e., $C_1C_3 = r_1 + r_3$ and $C_2C_3 = r_2 + r_3$.

$\therefore$ Locus of C_3 is such that $C_1C_3 - C_2C_3 = r_1 - r_2$ = constant $\in (0, C_1C_2)$ for $r_1 > r_2$ and $C_2C_3 - C_1C_3 = r_2 - r_1$ = constant $\in (0, C_1C_2)$ for $r_2 > r_1$

This locus will be *a* hyperbola in all condition when two given circles are disjoint, intersecting or touching. Also they may or may not be orthogonal, when intersecting.

Note that: when $r_1 = r_2$, then locus of C_3 will be along the right bisector of line segment C_1C_2.

9. (d) $hx + ky = xy$ is the given hyperbola which has centre at (k, h)

Shifting the origin to (k, h), we get $x = X + k$ and $y = Y + h$ and the equation becomes $hX + hk + kY + hk = (X + k)(Y + h)$ gives $hk = XY$ which is *a* rectangular hyperbola.

$\therefore$ Its conjugate hyperbola will be $XY = -hk$, so $(x - k)(y - h) = -hk$

$\Rightarrow hx + ky - xy - 2hk = 0$

10. (b) Let e_1 be the eccentricity of $\dfrac{x^2}{4a^2} + \dfrac{y^2}{4b^2} = 1$

$\Rightarrow\ e_1 = \sqrt{\dfrac{a^2 - b^2}{a^2}}$ and $2ae_1 = 2\sqrt{a^2 - b^2}$

Now let e_2 be the eccentricity of $\dfrac{x^2}{a^2} - \dfrac{y^2}{b^2} = 1$

$\Rightarrow\ e_2^2 = \dfrac{a^2 + b^2}{a^2}$ and $ae_2 = \sqrt{a^2 + b^2}$

Since curves are con-focal $\therefore\ 2ae_1 = ae_2$

So $4(a^2 - b^2) = a^2 + b^2$

$\Rightarrow 3a^2 = 5b^2$. As *a* result $= -\dfrac{3}{5} = \dfrac{2}{5}$ and $e_2^2 = 1 + \dfrac{3}{5} = \dfrac{8}{5}$

So $e_1^{\,2} + e_2^{\,2} = 2$.

11. (a) Let $P(x_1, y_1)$ be *a* moving point in *x-y* plane. According to the given condition $\dfrac{|x_1 + y_1 - 3||x_1 - y_1 + 6|}{2} = 12$

Can be written as $\left(x_1 + \dfrac{3}{2}\right)^2 - \left(y_1 - \dfrac{9}{2}\right)^2 = 24$ or $X^2 - Y^2 = 24$ {as $\left(-\dfrac{3}{2}, \dfrac{9}{2}\right)$ is centre}

This gives *a* rectangular hyperbola with $e = \sqrt{2}$.

12. (b) The ellipse with eccentricity e_e is $\dfrac{x^2}{a^2} + \dfrac{y^2}{b^2} = 1$ where $a > b$.

So $ae_e = \sqrt{a^2 - b^2}$ or $a^2e_e^{\,2} = a^2 - b^2$

Similarly, for the hyperbola $\dfrac{x^2}{A^2} - \dfrac{y^2}{b^2} = 1$ ($\because 2b = 2B$ (given))

$Ae_h = \sqrt{A^2 + b^2} \Rightarrow A^2e_h^2 = A^2 + b^2$

$\Rightarrow\ \dfrac{1}{e_e^2} + \dfrac{1}{e_h^2} = \dfrac{a^2}{a^2 - b^2} + \dfrac{A^2}{A^2 + b^2}$

Since both the curves are con-focal $\Rightarrow ae_e = Ae_h$.

Or $a^2e_e^2 = A^2e_h^2 \quad \Rightarrow a^2 - b^2 = A^2 + b^2$

So $\dfrac{1}{e_e^2} + \dfrac{1}{e_h^2} = \dfrac{a^2}{a^2 - b^2} + \dfrac{A^2}{a^2 - b^2} = \dfrac{A^2 + a^2}{(a^2 - b^2)}$

Since $a^2 - b^2 = A^2 + b^2$, so $A^2 = a^2 - 2b^2$

Hence $\dfrac{1}{e_e^2} + \dfrac{1}{e_h^2} = \dfrac{a^2 - 2b^2 + a^2}{a^2 - b^2} = 2$

13. (a) $PF_1.PF_2 = e\left(x_1 - \dfrac{a}{e}\right).e\left(x_1 + \dfrac{a}{e}\right) = \lambda\left(x_1^2 + y_1^2\right)$

$\Rightarrow\ \left(x_1^2e^2 + a^2\right) = \lambda\left(x_1^2 + x_1^2 - a^2\right)$

$\Rightarrow\ 2x_1^2 - a^2 = 2\lambda x_1^2 - \lambda a^2 \Rightarrow \lambda = 1$

14. (c) $2(x^2 - 6x + 9) - 3(y^2 - 4y + 4) = 6$ or $\dfrac{(x-3)^2}{3} - \dfrac{(y-2)^2}{2} = 1$ which is *a* hyperbola but not *a* rectangular one.

15. (a) Given focus $F = (2, 1)$, eccentricity $e == \sqrt{2}$ and directrix is $2x + 3y - 1 = 0$

So $(x-2)^2 + (y-1)^2 = \dfrac{2(2x + 3y - 1)^2}{13}$

$\Rightarrow\ 13\{x^2 + 4 - 4x + y^2 + 1 - 2y\} = 2\{4x^2 + 9y^2 + 1 - 4x - 6y + 12xy\}$ or $5x^2 - 5y^2 - 24xy - 44x - 14y + 63 = 0$

16. (a) Given $2ae = 16$, $e = \sqrt{2} \Rightarrow a = 4\sqrt{2} = b$, hence the hyperbola $\dfrac{x^2}{32} - \dfrac{y^2}{32} = 1$

17. (d) For equation to represent hyperbola, $h^2 - ab > 0$

$\Rightarrow\ (-k\ell m)^2 - (1 - k\ell^2)(1 - km^2) > 0$

$\Rightarrow\ k^2\ell^2m^2 - (1 - km^2 - k\ell^2 + k^2\ell^2m^2) > 0$

$\Rightarrow\ k(m^2 + \ell^2) > 1$

18. (d) Rewriting as $7(x^2 - 4x + 4) - 5(y^2 + 6y + 9) = 18$

$\Rightarrow\ \dfrac{(x-2)^2}{18/7} - \dfrac{(y+3)^2}{18/5} = 1$ which has centre at $(2, -3)$

19. (c) Given focus $(1, -1)$, eccentricity $e = \sqrt{2}$ and directrix $x - y + 1 = 0$

So $(x-1)^2 + (y+1)^2 = \dfrac{2(x - y + 1)^2}{2}$

$\Rightarrow\ x^2 + y^2 + 1 + 2y - 2x + 1 = x^2 + y^2 + 1 - 2xy + 2x - 2y$ or $2xy - 4x + 4y + 1 = 0$

20. (c) Given: $LR = \dfrac{2b^2}{a} = 8$ and $2b = ae$. So $4b^2 = a^2e^2 = a^2 + b^2$.

As $2b^2 = 8a$ so $4b^2 - b^2 = a^2$

$\Rightarrow\ e = \sqrt{\dfrac{a^2 + b^2}{a^2}} = \sqrt{\dfrac{4b^2}{3b^2}} = \sqrt{\dfrac{4}{3}} = \dfrac{2}{\sqrt{3}}$

21. (a) Given $x = ae^{\theta}$ and $y = be^{-\theta}$

$\Rightarrow$ $xy = ab$; which is a rectangular hyperbola.

22. (a) $\frac{x^2}{a^2} - \frac{y^2}{b^2} = 1$ passes through (5, 4) and its

$LR = \frac{2b^2}{a} = \frac{32\sqrt{2}}{5}$

$\Rightarrow$ $\frac{25}{a^2} - \frac{16}{b^2} = 1$ or $25b^2 - 16a^2 = a^2b^2$

Putting $2b^2 = \frac{32\sqrt{2}}{5}a$, we get $5 \times 16\sqrt{2}\,a - 16a^2 =$

$a^2\left\{\frac{16}{5}\sqrt{2}a\right\}$ or $\frac{16a}{5}\left\{\sqrt{2}a^2 + 5a - 25\sqrt{2}\right\} = 0$

$\Rightarrow$ $a = \frac{-5 \pm \sqrt{25+200}}{2\sqrt{2}} = \frac{10}{2\sqrt{2}}$ (rejecting the negative value)

$\Rightarrow$ $b^2 = \frac{16\sqrt{2}}{5} \cdot \frac{5}{\sqrt{2}} = 16 \Rightarrow b = 4$

23. (d) Focus of $\frac{x^2}{144} - \frac{y^2}{81} = \frac{1}{25}$ is at $(\pm\ ae,\ 0)$ =

$\left(\pm\sqrt{\frac{144}{25} + \frac{81}{25}}, 0\right) = (\pm 3, 0)$

Now focus of $\frac{x^2}{16} + \frac{y^2}{b^2} = 1$ is at $(\pm ae, 0) = \left(\pm\sqrt{16 - b^2}, 0\right)$

Since conics are con-focal

$\Rightarrow$ $16 - b^2 = 9 \Rightarrow b^2 = 7$.

24. (b) $\frac{x^2}{\cos^2\alpha} - \frac{y^2}{\sin^2\alpha} = 1$ the foci are at $(\pm\ ae,\ 0)$ =

$\left(\pm\sqrt{\cos^2\alpha + \sin^2\alpha}, 0\right) = (\pm 1,\ 0)$; $e = \sqrt{\frac{a^2+b^2}{a^2}} = \sec\alpha$;

equation of directrices:

$x = \pm\frac{a}{e} = \frac{\cos\alpha}{\sin\alpha} = \cos^2\alpha$ and ends of transverse axis are $(\pm a, 0) \equiv (\pm\cos\alpha, 0)$

So for all values of α the location of foci is fixed (i.e., abscissa of foci is fixed)

25. (c) For $r > 1$ the equation $\frac{x^2}{1-r} + \frac{y^2}{1+r} = 1$ becomes $\frac{y^2}{1+r} - \frac{x^2}{r-1} = 1$; which represents a hyperbola, $1 + r > 2$ and $r - 1 > 0$

TEXTUAL EXERCISE-2 (SUBJECTIVE)

1. For the hyperbola $\frac{x^2}{49} - \frac{y^2}{25} = 1$, $a^2 = 49$

So the auxiliary circle will be $x^2 + y^2 = 49$

Now the conjugate hyperbola for the given hyperbola will be $\frac{y^2}{25} - \frac{x^2}{49} = 1$ and $A^2 = 25$.

Hence its auxiliary circle will be $x^2 + y^2 = 25$

2. Auxiliary circle of $\frac{x^2}{49} - \frac{y^2}{25} = 1$ is $x^2 + y^2 = 49$

The auxiliary circle of $\frac{y^2}{25} - \frac{x^2}{49} = 1$ is $x^2 + y^2 = 25$

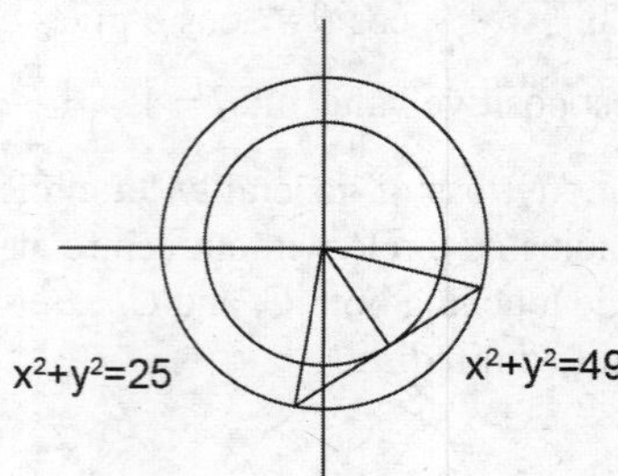

Length of the chord $= 2\sqrt{7^2 - 5^2} = 4\sqrt{6}$ units.

3. Conjugate hyperbola of $\frac{x^2}{9} - \frac{y^2}{4} = 1$ is $\frac{y^2}{4} - \frac{x^2}{9} = 1$ and later's auxiliary circle is $x^2 + y^2 = 4$.

Slope of tangent $m = 3/4$. Since no portion of tangent lies in the IVth quadrant, so the said tangent will be in 2nd quadrant from $y = mx + a\sqrt{1+m^2}$ $\Rightarrow$ $y = \frac{3}{4}x + 2\sqrt{1 + \frac{9}{16}}$ i.e., $3x - 4y + 10 = 0$

4. The hyperbola is $\frac{x^2}{9} - \frac{y^2}{4} = 1$ and the point P (a secθ, b tanθ) = $(3\sqrt{2}, 2)$. So $\theta = \pi/4$

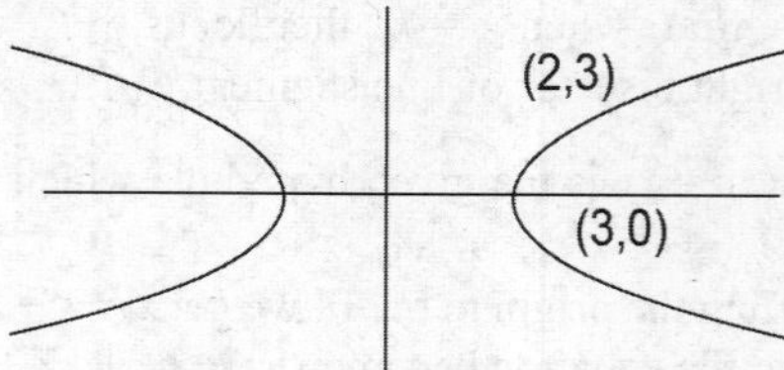

The line through (2, 3) with slope $m = \tan\theta = 1$ is $x - y + 1 = 0$

Any point on the line at a distance $|r|$ is $x = 2 + r\cos\theta = 2 + \frac{r}{\sqrt{2}}$ and $y = 3 + r\sin\theta = 3 + \frac{r}{\sqrt{2}}$ (given $|r| = 4$)

For $r = \pm 4$ units, we get $(2 + 2\sqrt{2}, 3 + 2\sqrt{2})$ or $(2 - 2\sqrt{2}, 3 - 2\sqrt{2})$

5. Given hyperbola is $(x^2 - 2x + 1) - (y^2 + 6y + 9) = 4$, so $\frac{(x-1)^2}{4} - \frac{(y+3)^2}{4} = 1$

$\Rightarrow$ $x = 1 + 2\sec\theta$ and $y = -3 + 2\tan\theta$ is the parametric form of given hyperbola.

6. For the ellipse $\frac{x^2}{a^2} - \frac{y^2}{b^2} = 1$, the equation of a chord with end points as P (α) and Q (β) will be $\frac{x}{a}\cos\left(\frac{\alpha-\beta}{2}\right) - \frac{y}{b}\sin\left(\frac{\alpha+\beta}{2}\right) = \cos\left(\frac{\alpha+\beta}{2}\right)$

Now the conjugate hyperbola will be $\frac{y^2}{b^2} - \frac{x^2}{a^2} = 1$

R (α) and S (β) will be (a tanα, b secα) and (a tanβ, b secβ)

Hence the chord will be $\begin{vmatrix} x & y & 1 \\ a\tan\alpha & b\sec\alpha & 1 \\ a\tan\beta & b\sec\beta & 1 \end{vmatrix} = 0$

$$\Rightarrow \begin{vmatrix} x & y & 1 \\ a\sin\alpha & b & \cos\alpha \\ a\sin\beta & b & \cos\beta \end{vmatrix} = 0$$

$$\Rightarrow \frac{y}{b}\cos\left(\frac{\alpha-\beta}{2}\right) - \frac{x}{a}\sin\left(\frac{\alpha+\beta}{2}\right) = \cos\left(\frac{\alpha+\beta}{2}\right)$$

TEXTUAL EXERCISE–2 (OBJECTIVE)

1. (a) For the hyperbola $\frac{x^2}{a^2}-\frac{y^2}{b^2}=1$, the equation of focal chord with end points $P(\theta_1)$ and $Q(\theta_2)$ will be $\pm e\cos\left(\frac{\theta_1-\theta_2}{2}\right)=\cos\left(\frac{\theta_1+\theta_2}{2}\right)$

$$\Rightarrow \lambda\cos^2\left(\frac{\theta_1-\theta_2}{2}\right)=\cos^2\left(\frac{\theta_1+\theta_2}{2}\right)$$

$$\Rightarrow \lambda = e^2 = \frac{a^2+b^2}{a^2}$$

2. (b) Let (h, k) be the mid-point of *a* chord to the hyperbola $\frac{x^2}{a^2}-\frac{y^2}{b^2}=1$

∴ Equation of chord is given by $T = S_1$

$$\Rightarrow \frac{hx}{a^2}-\frac{ky}{a^2}-1=\frac{h^2}{a^2}-\frac{k^2}{a^2}-1$$

i.e., $ky = hx + (k^2 - h^2) \Rightarrow y=\frac{hx}{k}+\left(\frac{k^2-h^2}{k}\right)$

Since it is *a* tangent to the parabola $y^2 = 4ax$. So $y = mx+\frac{a}{m}$ gives $\frac{k^2-h^2}{k}=\frac{ak}{h}$

i.e., $ak^2 = hk^2 - h^3$ or $h^3 = (h - a)\,k^2 \Rightarrow x^3 = (x - a)y^2$.

3. (b) On the hyperbola $\frac{x^2}{a^2}-\frac{y^2}{b^2}=1$, for focal chord with end points $P\left(\frac{\pi}{3}\right)$ and $Q\left(-\frac{\pi}{6}\right)$ will be $\frac{1-e}{1+e}=\tan\frac{\pi}{6}.\tan\left(-\frac{\pi}{12}\right)$

$$\Rightarrow \frac{1-e}{1+e}=\left(-\frac{1}{\sqrt3}\right)\left\{\frac{\sqrt3-1}{\sqrt3+1}\right\}=\frac{-(2-\sqrt3)}{\sqrt3}$$

$$\therefore \frac{2e}{2}=\frac{\sqrt3+2-\sqrt3}{\sqrt3+\sqrt3-2}=\frac{2}{2(\sqrt3-1)}$$

$$\Rightarrow e=\frac{\sqrt3+1}{2}$$

4. (b) Auxiliary circle of $\frac{x^2}{5}-\frac{y^2}{7}=1$ is $x^2 + y^2 = 5$

5. (a) Auxiliary circle of $\frac{x^2}{20}-\frac{y^2}{2}=1$ is $x^2 + y^2 = 20$ which has area $\pi r^2 = 20\pi$.

6. (c) Auxiliary circle to the hyperbola $\frac{x^2}{9}-\frac{y^2}{16}=1$ is $x^2+y^2=9$.

Now the conjugate hyperbola is $\frac{y^2}{16}-\frac{x^2}{9}=1$ which has auxiliary circle as $x^2 + y^2 = 16$

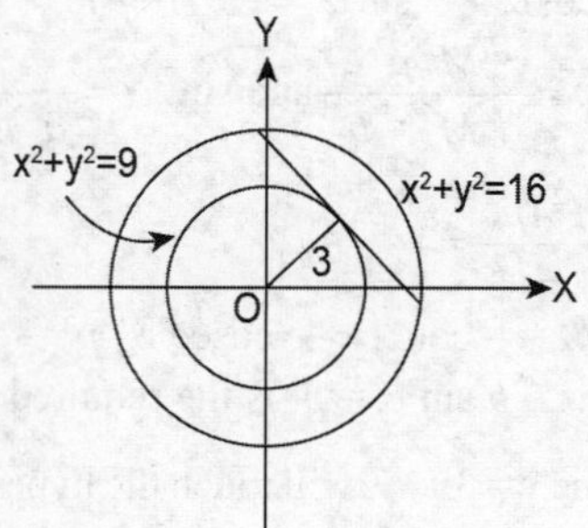

So the length of chord $= 2\sqrt{16-9}=2\sqrt7$ units

7. (b) Given hyperbola is $\frac{x^2}{7}-\frac{y^2}{3}=1$ having parametric form $\left(\sqrt7\sec\theta, \sqrt3\tan\theta\right)$

$$\Rightarrow \frac{2\sqrt7}{\sqrt3}=\sqrt7\sec\theta \text{ and } 1=\sqrt3\tan\theta$$

$$\Rightarrow \theta=\frac{\pi}{6}$$

8. (c) Given hyperbola is $2(x^2 + 4x + 4) - (y^2 - 6y + 9) = 10$ or $\frac{(x+2)^2}{5}-\frac{(y-3)^2}{10}=1$

⇒ $(x + 2) = \sqrt5\sec\theta$ and $(y - 3) = \sqrt{10}\tan\theta$

⇒ Parametric form of hyperbola is $(-2 + \sqrt5\sec\theta, 3 + \sqrt{10}\tan\theta)$

9. (c) For the hyperbola $\frac{x^2}{16}-\frac{y^2}{9}=1$ a chord with end points $P\left(\frac{\pi}{6}\right)$ and $Q\left(\frac{\pi}{3}\right)$ will be $\frac{x}{4}\cos\frac{\pi}{12}-\frac{y}{3}\sin\frac{\pi}{4}=\cos\frac{\pi}{4}$ which will intersect

x-axis where $x=\frac{4(2\sqrt2)}{\sqrt2\{\sqrt3+1\}}=4(\sqrt3-1)$

So the point of intersection with x-axis is $(4(\sqrt3 - 1), 0)$

TEXTUAL EXERCISE–3 (SUBJECTIVE)

1. (a) The line is $lx + my + n = 0$. *A* line with slope M will be *a* tangent to the hyperbola $\frac{x^2}{a^2}-\frac{y^2}{b^2}=1$ when $y = Mx \pm\sqrt{a^2M^2-b^2}$

i.e., $Mx - y \pm\sqrt{a^2M^2-b^2}=0$ and $x+my+n=0$ are identical

$$\Rightarrow \frac{M}{\ell}=\frac{-1}{m}=\frac{\pm\sqrt{a^2M^2-b^2}}{n}$$

$$\Rightarrow a^2M^2 - b^2 = \frac{n^2}{m^2} \text{ and } mM = -$$

$\Rightarrow$ $a^2\ {}^2 - m^2b^2 = n^2$ is the required condition

Similarly if the line $x \cos\alpha + y \sin\alpha - p = 0$ is tangent,

then $\dfrac{M}{\cos\alpha} = -\dfrac{1}{\sin\alpha} = \dfrac{\pm\sqrt{a^2M^2 - b^2}}{(-p)}$

$\Rightarrow$ $\cos^2\alpha = \dfrac{M^2p^2}{a^2M^2 - b^2}$ and $\sin^2\alpha = \dfrac{p^2}{a^2M^2 - b^2}$

$\Rightarrow$ $\dfrac{(1+M^2)p^2}{a^2M^2 - b^2} = 1$

Now $M = -\cot\alpha \Rightarrow$ $(\operatorname{cosec}^2\alpha)\, p^2 = a^2\cot^2\alpha - b^2$

$\Rightarrow$ $a^2\cos^2\alpha - b^2\sin^2\alpha = p^2$ is the required condition

(b) The line $y = 3x + \lambda$ will touch the hyperbola $\dfrac{x^2}{5} - \dfrac{y^2}{9} = 1$ when $\lambda = \pm\sqrt{5(9) - 9} = \pm 6$

2. (a) Slope of the line perpendicular to $x + 3y = 2$ is $m = 3$

Now, tangent to the hyperbola $\dfrac{x^2}{1} - \dfrac{y^2}{3} = 1$, with slope 3 is

$y = 3x \pm \sqrt{(3)^2 - 3} = 3x \pm \sqrt{6}$

(b) The hyperbola is $\dfrac{x^2}{\left(\frac{1}{4}\right)} - \dfrac{y^2}{\left(\frac{1}{9}\right)} = 1$

A tangent with slope $m = 5/4$ is

$y = \dfrac{5}{4}x \pm \sqrt{\left(\dfrac{1}{4}\right)\left(\dfrac{25}{16}\right) - \dfrac{1}{9}}$ or $y = \dfrac{5}{4}x \pm \dfrac{\sqrt{161}}{24}$

$\Rightarrow$ $5x - 4y \pm \dfrac{\sqrt{161}}{6} = 0$

The points of contact are $\left(\dfrac{-a^2m}{c}, \dfrac{-b^2}{c}\right)$

$= \left(\dfrac{\pm 15}{2\sqrt{161}}, \dfrac{\pm 8}{3\sqrt{161}}\right)$

3. Let (h, k) be the point on the plane of the hyperbola.

Let $y = mx \pm \sqrt{a^2m^2 - b^2}$ be tangent through (h, k)

$\Rightarrow$ $(k - mh)^2 = a^2m^2 - b^2$

$\Rightarrow$ $m^2(a^2 - h^2) + 2khm - k^2 - b^2 = 0$

$\Rightarrow$ $m_1.m_2 = -\dfrac{k^2 - b^2}{a^2 - h^2} = \dfrac{k^2 + b^2}{h^2 - a^2}$

Now, for perpendicular, $\dfrac{k^2 + b^2}{h^2 - a^2} = -1$

$\Rightarrow$ $k^2 + b^2 = a^2 - b^2 \Rightarrow$ $k^2 + h^2 = a^2 - b^2$

$\Rightarrow$ $k^2 + h^2 = 25 - 36 = -1$ $\quad(\because a^2 = 25, b^2 = 36)$

$\Rightarrow$ which is impossible.

Thus there is no such point (h, k). Note that except for above shown position of $P(h, k)$ two different tangents can't be drawn to hyperbola.

4. Equation of *a* tangent at $P(a\sec\theta, b\tan\theta)$ to the hyperbola $\dfrac{x^2}{a^2} - \dfrac{y^2}{b^2} = 1$ is $\dfrac{x}{a}\sec\theta - \dfrac{y}{b}\tan\theta = 1$ which has slope $= \dfrac{b}{a}\operatorname{cosec}\theta$

Now *a* line $y = mx + \sqrt{a^2m^2 - b^2}$ will be *a* tangent when $m = \dfrac{b}{a}\operatorname{cosec}\theta$ and $\sqrt{a^2m^2 - b^2} = -b\cot\theta$

i.e., $\left\{a^2\dfrac{b^2}{a^2}\operatorname{cosec}^2\theta - b^2\right\} = b^2\cot^2\theta$; which is true. Also

$m\sin\theta = \dfrac{b}{a} \Rightarrow \theta = \sin^{-1}\left(\dfrac{b}{am}\right)$

5. The line $y = mx$, solving with $\dfrac{x^2}{a^2} - \dfrac{y^2}{a^2} = 1$, gives $x^2 - m^2.x^2 = a^2$

$\Rightarrow$ $x = \pm\dfrac{a}{\sqrt{1-m^2}}$

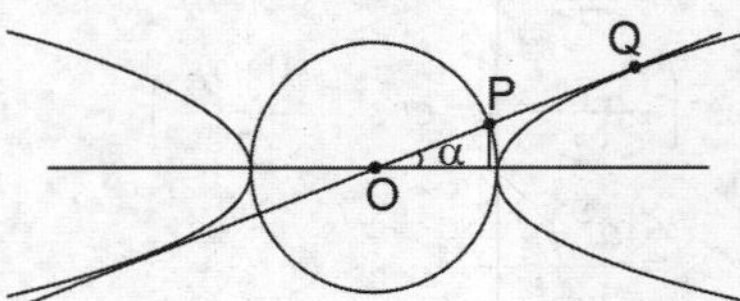

$\Rightarrow$ The point Q is $\left(\dfrac{a}{\sqrt{1-m^2}}, \dfrac{am}{\sqrt{1-m^2}}\right)$ and the equation of tangent at Q is $T = 0$ i.e., $\dfrac{a}{\sqrt{1-m^2}}x - \dfrac{amy}{\sqrt{1-m^2}} = a^2$ or

$x - my = a\sqrt{1-m^2}$ ...(i)

Similarly, solving with the circle, we get $P\left(\dfrac{a}{\sqrt{1+m^2}}, \dfrac{am}{\sqrt{1+m^2}}\right)$ and the equation of tangent at P is $ax + my = a^2\sqrt{1+m^2}$

i.e.. $x + my = a\sqrt{1+m^2}$ ...(ii)

From (i) and (ii), we get $x = \dfrac{a}{2}\left(\sqrt{1+m^2} + \sqrt{1-m^2}\right)$

and $y = \dfrac{a}{2m}\left(\sqrt{1+m^2} - \sqrt{1-m^2}\right)$

But the given point of intersection is $\left(\dfrac{\sqrt{5}+\sqrt{3}}{4}a, \dfrac{(\sqrt{5}-\sqrt{3})}{4}a\right)$

$\Rightarrow$ $\sqrt{1+m^2} + \sqrt{1-m^2} = \dfrac{\sqrt{5}+\sqrt{3}}{2}$ and

$\sqrt{1+m^2} - \sqrt{1-m^2} = \dfrac{\sqrt{5}-\sqrt{3}}{2}m$

Multiplying we get, $(2m^2) = m$

$\Rightarrow$ $m = 0$ or $m = 1/2$

$\Rightarrow$ $m = 1/2$ (rejecting $m = 0$ as otherwise P and Q coincide)

6. The tangent to the hyperbola $\dfrac{x^2}{a^2} - \dfrac{y^2}{b^2} = 1$ at $P(a\sec\theta, b\tan\theta)$ will be $\dfrac{\sec\theta}{a}x - \dfrac{\tan\theta}{b}y = 1$ gives the mid point between the coordinates axis as $\left(\dfrac{a}{2}\cos\theta, -\dfrac{b\cot\theta}{2}\right)$

So, let $x_1 = \frac{a}{2}\cos\theta \Rightarrow \frac{2x_1}{a} = \cos\theta$ and $y_1 = -\frac{b}{2}\cot\theta$

$$\Rightarrow \frac{-2y_1}{b} = \cot\theta = \frac{\cos\theta}{\sqrt{1-\cos^2\theta}}$$

$$\text{So, } \cot^2\theta = \frac{\cos^2\theta}{1-\cos^2\theta} = \frac{4y_1^2}{b^2}$$

$$\Rightarrow b^2\left(\frac{4x_1^2}{a^2}\right) = 4y_1^2\left\{1-\frac{4x_1^2}{a^2}\right\}$$

$$\Rightarrow x_1^2b^2 = y_1^2a^2 - 4x_1^2y_1^2$$

Rearranging, we get $\frac{a^2}{x^2} - \frac{b^2}{y_1^2} = 4$ or $\frac{a^2}{x^2} - \frac{b^2}{y^2} = 4$

7. Without any loss of generality let P (a secθ, $b$tanθ) be a point on the hyperbola $\frac{x^2}{a^2} - \frac{y^2}{b^2} = 1$ with focus at S (ae, 0) and S'($-ae$, 0); where $a^2e^2 = a^2 + b^2$

Now tangent at P is $\frac{x\sec\theta}{a} - \frac{y}{b}\tan\theta = 1$; which intersects the axis at T ($a$cosθ, 0)

Observe that $PS = a$ (e secθ – 1) and $PS' = a$ (e secθ + 1)

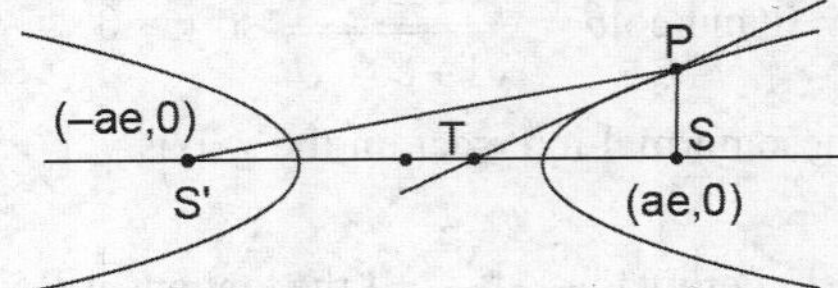

Now, $ST = ae - a\cos\theta = a\{e - \cos\theta\}$ and $S'T = a\cos\theta + ae = a\{e + \cos\theta\}$

$$\text{Since } \frac{PS'}{PS} = \frac{a(e\sec\theta+1)}{a(\sec\theta-1)} = \frac{a(e+\cos\theta)}{a(e-\cos\theta)} = \frac{S'T}{ST}$$

Hence PT is the internal angle bisector of $\angle SPS'$

8. At a point P (a secθ, b tanθ), the equation of normal to the hyperbola will be $\frac{a^2x}{a\sec\theta} + \frac{b^2y}{b\tan\theta} = a^2 + b^2$ or $ax\cos\theta + by\cot\theta - (a^2 + b^2) = 0$

Comparing with $x + my - n = 0$, we get

$$\frac{a\cos\theta}{\ell} = \frac{b\cot\theta}{m} = \frac{a^2+b^2}{n}$$

$$\Rightarrow \sec\theta = \frac{an}{\ell(a^2+b^2)} \text{ and } \tan\theta = \frac{bn}{m(a^2+b^2)}$$

$$\Rightarrow \frac{a^2n^2}{\ell^2(a^2+b^2)^2} - \frac{b^2n^2}{m^2(a^2+b^2)^2} = 1$$

$$\text{or } \frac{a^2}{\ell^2} - \frac{b^2}{m^2} = \frac{(a^2+b^2)^2}{n^2}$$

9. Hyperbola is $\frac{x^2}{16} - \frac{y^2}{9} = 1$; where foci are at (±5, 0).

Now for $x = 8$, we get $y = 3\sqrt{3}$ in the Ist quadrant. Since the ray will strike at $(8, 3\sqrt{3})$ on the concave side of hyperbola (after reflection)

So the ray will appear to be combing from the other focus (–5, 0)

Hence the equation of ray will be the line $y - 0 = \frac{3\sqrt{3}}{(8+5)}(x+5)$

i.e., $3\sqrt{3}x - 13y + 15\sqrt{3} = 0$

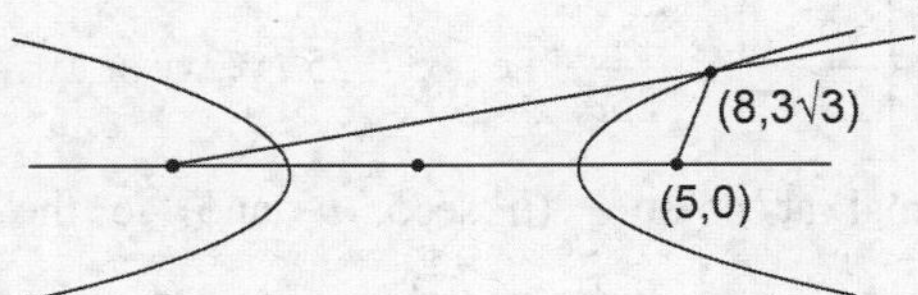

10. The normal at P (2secθ, tanθ) on the hyperbola $\frac{x^2}{4} - \frac{y^2}{1} = 1$ will be $2x\cos\theta + y\cot\theta = 2^2 + 1 = 5$

⇒ Point of intersection on x–axis = 5/2 secθ, i.e., $A\left(\frac{5}{2}\sec\theta, 0\right)$ and point of intersection y-axis is B = (0, 5tanθ)

Hence the point of intersection is $R\left(\frac{5}{2}\sec\theta, 5\tan\theta\right)$ = $R(x_1, y_1)$ (say)

∴ The locus of R is given by $\sec^2\theta - \tan^2\theta = \left(\frac{2x_1}{5}\right)^2 - \left(\frac{y_1}{5}\right)^2 = 1$, i.e., $(2x)^2 - y^2 = 25$.

11. Let P (θ) and Q (ϕ) be the two points on the hyperbola $\frac{x^2}{25} - \frac{y^2}{16} = 1$

Now, slope of tangent at P (θ) is $\frac{\quad}{5\sin}$.

Similarly, slope of tangent at Q (ϕ) is $m_2 = \frac{4}{5\sin\phi}$.

Since $m_1 + m_2 = a \Rightarrow \frac{5a}{4} = \frac{1}{\sin\theta} + \frac{1}{\sin\phi}$ gives

$$\frac{5a}{4} = \frac{2(\sin\theta+\sin\phi)}{2(\sin\theta.\sin\phi)} = \frac{4\sin\left(\frac{\theta+\phi}{2}\right).\cos\left(\frac{\theta-\phi}{2}\right)}{\{\cos(\theta-\phi)-\cos(\theta+\phi)\}}$$

$$\Rightarrow \frac{5a}{16} = \frac{\sin\left(\frac{\theta+\phi}{2}\right).\cos\left(\frac{\theta-\phi}{2}\right)}{\cos(\theta-\phi)-\cos(\theta+\phi)} \quad \text{....(i)}$$

Now the point of intersection of tangents is

$$R = (x_1, y_1) = \left(\frac{5\cos\left(\frac{\theta-\phi}{2}\right)}{\cos\left(\frac{\theta+\phi}{2}\right)}, \frac{4\sin\left(\frac{\theta+\phi}{2}\right)}{\cos\left(\frac{\theta+\phi}{2}\right)}\right)$$

$$\Rightarrow \frac{x_1}{5}.\frac{y_1}{4}.\cos^2\left(\frac{\theta+\phi}{2}\right) = \sin\left(\frac{\theta+\phi}{2}\right).\cos\left(\frac{\theta-\phi}{2}\right) \quad \text{.....(ii)}$$

So from (i) and (ii), we get

$$\frac{5a}{16}\{\cos(\theta-\phi)-\cos(\theta+\phi)\} = \frac{x_1y_1}{20}\cos^2\left(\frac{\theta+\phi}{2}\right) \text{ or}$$

$$\frac{5a}{16}\left\{2\cos^2\left(\frac{\theta-\phi}{2}\right) - 2\cos^2\left(\frac{\theta+\phi}{2}\right)\right\} = \frac{x_1y_1}{20}\cos^2\left(\frac{\theta+\phi}{2}\right)$$

$$\Rightarrow \frac{5a(2)}{16}\left\{\frac{\cos^2\left(\frac{\theta-\phi}{2}\right)}{\cos^2\left(\frac{\theta+\phi}{2}\right)}-1\right\}=\frac{x_1y_1}{20}$$

$$\Rightarrow \left\{\frac{x_1^2}{25}-1\right\}=\frac{2x_1y_1}{25a}$$ i.e. $ax^2 - 25a = 2xy$ or $ax^2 - 2xy = 25a$

12. Normal at point P (a secϕ, b tanϕ) to the hyperbola $\frac{x^2}{a^2}-\frac{y^2}{b^2}=1$ is $\frac{a^2x}{a\sec\phi}+\frac{b^2y}{b\tan\phi}=a^2+b^2$

Which will intersect transverse axis (x – axis) at $G\left(\frac{a^2+b^2}{a}\sec\phi,0\right)=(ae^2\sec\phi,0)$

If A' and A are vertices, then $A \equiv (a, 0)$ and $A' \equiv (a, 0)$

$\Rightarrow$ AG = a (e^2secϕ –1) and AG$'$ = a(e^2secϕ + 1)

$\Rightarrow$ AG . AG$'$ = a^2 {e^4sec$^2\phi$ –1}

13. Since $\theta_1 + \theta_2 = \pi/2 \Rightarrow \tan\theta_1 .\tan\theta_2 = 1$

Normal at P (a secθ_1, b tanθ_1) and θ (a secθ_2, b tanθ_2) respectively for the hyperbola $\frac{x^2}{a^2}-\frac{y^2}{b^2}=1$ are $ax\cos\theta_1 + by\cot\theta_1 = a^2 + b^2$ and $ax\cos\theta_2 + by\cot\theta_2 = a^2 + b^2$

Their points of intersection on conjugate axis are $G_1=\left(0,\frac{a^2+b^2}{b}\tan\theta_1\right)$ and $G_2=\left(0,\frac{a^2+b^2}{b}\tan\theta_2\right)$

$$\Rightarrow CG_1.CG_2=\frac{(a^2+b^2)^2}{b^2}\tan\theta_1\tan\theta_2=\frac{a^4e^4}{a^2(e^2-1)}$$

$$=\frac{a^2e^4}{(e^2-1)}$$

14. Hyperbola $\frac{x^2}{8}-\frac{y^2}{8}=1$ has its foci (±4, 0). Now the equation of normal at (3, 1) is $\frac{8x}{3}+\frac{8y}{1}=16$ gives $x + 3y = 6$.

Now P (3, 1) and S (4, 0)

$\Rightarrow$ PS is a chord of the circle so its right bisector passes through the centre of the circle right bisector of PS is y –1/2 = x –7/2 or $x - y - 3 = 0$ and its intersection with normal gives centre of the circle at $C_1\left(\frac{15}{4},\frac{3}{4}\right)$

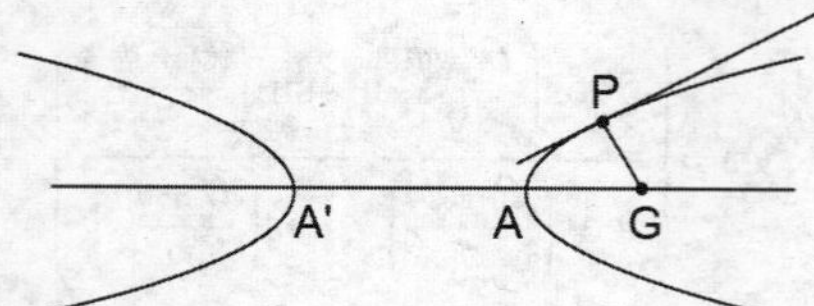

$$\Rightarrow r^2=\left(\frac{3}{4}\right)^2+\left(\frac{1}{4}\right)^2=\frac{10}{16}$$

Hence the circle is $(4x - 15)^2 + (4y - 3)^2 = 10$ gives $2(x^2 + y^2) - 15x - 3y + 28 = 0$

Similarly when focus is at $S'(-4, 0)$, we get the right bisector of PS' as $\left(y-\frac{1}{2}\right)=(-7)\left\{x+\frac{1}{2}\right\}$ i.e., $7x + y + 3 = 0$

Which will intersect the normal at $\left(-\frac{3}{4},\frac{9}{4}\right)$

$$\Rightarrow r^2=\left(3+\frac{3}{4}\right)^2+\left(1-\frac{9}{4}\right)^2=\left(\frac{15}{4}\right)^2+\left(\frac{5}{4}\right)^2=\frac{250}{16}$$

Hence the circle is $(4x + 3)^2 + (4y - 9)^2 = 250$ i.e., $2(x^2 + y^2) + 3x - 9y - 20 = 0$

15. Let $x = a\sec\theta$ be a vertical line intersecting the hyperbola $\frac{x^2}{a^2}-\frac{y^2}{b^2}=1$ at P (a secθ, $b$tanθ)

$\Rightarrow$ The normal at P will be $ax\cos\theta + by\cot\theta = a^2 + b^2$ .....(i)

Now this vertical line will intersect the conjugate hyperbola $\frac{y^2}{b^2}-\frac{x^2}{a^2}=$ at $y=\pm b\sqrt{1+\sec^2\theta}$

Now the normal at Q (a secθ, $b\sqrt{1+\sec^2\theta}$) will be

$$\frac{a^2x}{a\sec\theta}+\frac{b^2y}{b\sqrt{1+\sec^2\theta}}=a^2+b^2 \quad ...(ii)$$

From (i) and (ii), we get $\left\{\cot\theta-\frac{\cos\theta}{\sqrt{1+\cos^2\theta}}\right\}.y=0$ as $b \neq 0$ and $\cot\theta \neq \frac{\cos\theta}{\sqrt{1+\cos^2\theta}}$ so $y = 0$

$\therefore$ These normal intersect on the x-axis.

16. For the hyperbola $\frac{x^2}{a^2}-\frac{y^2}{b^2}=1$ the tangent at P ($a$secθ, $b$tanθ) $=(x_1,y_1)$ will be $\frac{x\sec\theta}{a}-\frac{y\tan\theta}{b}=1$ or $\frac{x_1x}{a^2}-\frac{y_1y}{b^2}=1$ which will intersect the line $bx - ay = 0$ (or $y = bx/a$)

$$\Rightarrow \frac{x_1}{a^2}x-\frac{by_1x}{ab^2}=1$$

$$\Rightarrow x=x_2=\frac{a^2b}{bx_1-ay_1} \text{ and } y=y_2=\frac{b^2a}{bx_1-ay_1}$$

$$\therefore Q\equiv(x_2,y_2)\equiv\left(\frac{a^2b}{bx_1-ay_1},\frac{ab^2}{bx_1-ay_1}\right)$$

$$\text{Now } CQ=x_2\sqrt{\frac{a^2+b^2}{a^2}}=\frac{ab\sqrt{a^2+b^2}}{(bx_1-ay_1)}$$

Similarly for the line $y=-\frac{b}{a}x$, we get $R(x_3, y_3)$; where

$$x_3=\frac{a^2b}{(bx_1+ay_1)} \text{ and } y_3=\frac{ab^2}{(bx_1+ay_1)}$$

$$\text{Hence } CR=\frac{ab\sqrt{a^2+b^2}}{(bx_1+ay_1)}$$

$$\text{Now } CQ.CR=\frac{a^2b^2\left(\sqrt{a^2+b^2}\right)^2}{(b^2x_1^2-a^2y_1^2)}$$

Putting x_1 = $a\sec\theta$ and y_1 = $b\ \tan\theta$, we get $b^2x_1^2 - a^2y_1^2 = a^2b^2$ $(\sec^2\theta - \tan^2\theta) = a^2b^2$

So $CQ.CR = (a^2 + b^2)$

17. Equation of the tangent at P ($a\sec\theta$, $b\tan\theta$) on the hyperbola $\frac{x^2}{a^2} - \frac{y^2}{b^2} = 1$ is $\frac{x}{a}\sec\theta - \frac{y}{b}\tan\theta = 1$, which will intersect the transverse axis at $T(a\cos\theta, 0)$

So PT = $\ell = \sqrt{a^2(\cos^2\theta + \sec^2\theta - 2) + b^2\tan^2\theta}$

$$= \sqrt{\frac{a^2\sin^4\theta + b^2\sin^2\theta}{\cos^2\theta}} = \sqrt{\frac{a^2\sin^2\theta + b^2}{\cot\theta}}$$

Now the equation of normal at P will be $ax\cos\theta + by\cot\theta - (a^2 + b^2) = 0$

The foci are S_1 $(ae, 0)$ and S_2 $(-ae, 0)$

So $\ell_1 = \dfrac{|a^2e\cos\theta - (a^2+b^2)|}{\sqrt{a^2\cos^2\theta + \dfrac{b^2\cos^2\theta}{\sin^2\theta}}}$ and

$$\ell_2 = \frac{|-a^2e\cos\theta - (a^2+b^2)|}{\sqrt{a^2\cos^2\theta + \dfrac{b^2\cos^2\theta}{\sin^2\theta}}}$$

$$\left(\begin{array}{ll} \because & a^2+b^2 = a^2e^2 \\ \Rightarrow & a^2e\cos\theta - (a^2+b^2) \\ & = a^2e\cos\theta - a^2e^2 \\ & = a^2e(\cos\theta - e) < 0 \end{array}\right)$$

$$\Rightarrow \ell_1 = \frac{(a^2+b^2-a^2e\cos\theta)|\tan\theta|}{\sqrt{a^2\sin^2\theta + b^2}};$$

$$\ell_2 = \frac{(a^2+b^2+a^2e\cos\theta)|\tan\theta|}{\sqrt{a^2\sin^2\theta + b^2}}$$

$(\because\ (a^2+b^2) - a^4e^2\cos^2\theta = (a^2+b^2)\{a^2+b^2-a^2\cos^2\theta)) = (a^2+b^2)\{b^2 + a^2\sin^2\theta\}$

$$\Rightarrow \text{H.M. of } \ell_1\ell_2 = \frac{2\ell_1\ell_2}{(\ell_1+\ell_2)} = \frac{2[(a^2+b^2)(b^2+a^2\sin\theta)]}{(a^2\sin^2\theta + b^2)\cot^2\theta}$$

$$= \frac{2(a^2+b^2)|\tan\theta|}{\sqrt{a^2\sin^2\theta+b^2}} = \frac{\sqrt{a^2\sin^2\theta+b^2}}{|\cot\theta|} = PT$$

TEXTUAL EXERCISE-3(OBJECTIVE)

1. (d) A line $y = mx \pm \sqrt{a^2m^2 - b^2}$ will be a tangent to the hyperbola $\frac{x^2}{a^2} - \frac{y^2}{b^2} = 1$ so $x\cos\alpha + y\sin\alpha = p$; slope ($m = -\cot\alpha$) will be a tangent when $\frac{p^2}{\sin^2\alpha} = a^2\cot^2\alpha - b^2$

i.e., $a^2\cos^2\alpha - b^2\sin^2\alpha = p^2$

2. (c) Observe that $x^2 + y^2 = \frac{a^2}{4}$ and $\frac{x^2}{a^2} - \frac{y^2}{b^2} = 1$ having origin as their common centre.

So vertex is the nearest point. Hence shortest distance $= a/2$

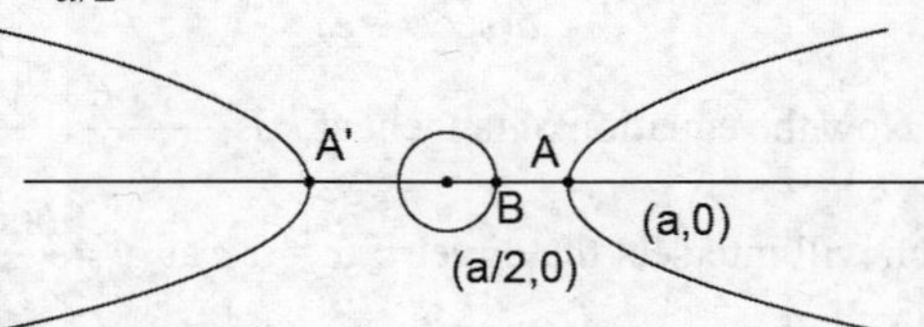

3. (c) Given hyperbolae are C_1: $\frac{x^2}{a^2} - \frac{y^2}{b^2} = 1$ and C_2: $\frac{y^2}{a^2} - \frac{x^2}{b^2} = 1$

For C_1 the line $y = mx \pm \sqrt{a^2m^2 - b^2}$ will be a tangent and for C_2 it will be $y = mx \pm \sqrt{a^2 - b^2m^2}$

$\Rightarrow a^2m^2 - b^2 = a^2 - b^2m^2 \Rightarrow m = \pm 1$ and $a > b$.

Thus the four common tangents are given by $y = \pm x \pm \sqrt{a^2 - b^2}$

4. (d) Equation of tangent at P($a\sec\theta$, $b\tan\theta$) to the hyperbola $\frac{x^2}{a^2} - \frac{y^2}{b^2} = 1$ is $\frac{x\sec\theta}{a} - \frac{y\tan\theta}{b} = 1$. It will intersect the line $y = \frac{b}{a}x$ at

$Q \equiv (a(\sec\theta + \tan\theta), b(\sec\theta + \tan\theta))$

Now mid-point of segment PQ is $M(x_1, y_1)$

$$\equiv \left(a\sec\theta + \frac{a}{2}\tan\theta, \frac{b}{2}\sec\theta + b\tan\theta\right)$$

$$\Rightarrow \frac{x_1}{a} - \frac{y_1}{b} = \frac{\sec\theta}{2} - \frac{\tan\theta}{2} \text{ and } \frac{x_1}{a} + \frac{y_1}{b} = \frac{3}{2}\sec\theta + \frac{3}{2}\tan\theta$$

$$\Rightarrow \frac{x_1^2}{a^2} - \frac{y_1^2}{b^2} = \frac{3}{4}\sec^2\theta - \frac{3}{4}\tan^2\theta = \frac{3}{4}$$

$\Rightarrow \frac{x^2}{a^2} - \frac{y^2}{b^2} = \frac{3}{4}$ is the locus of mid-point M.

5. (b) The foot of perpendicular drawn from focus onto any tangent will lie on the auxiliary circle for $\frac{x^2}{9} - \frac{y^2}{4} = 1$, thus the required locus is its auxiliary circle i.e., $x^2 + y^2 = 9$

6. (d) The equation of normal at P ($a\sec\theta$, $b\tan\theta$) to the hyperbola $\frac{x^2}{24} - \frac{y^2}{18} = 1$ is $\frac{24x}{2\sqrt{6}\sec\theta} + \frac{18y}{3\sqrt{2}\tan\theta} = 42$ which has slope

$m = -\frac{a}{b}\sin\theta$. Now this line will be perpendicular to $3x + 2y + 1 = 0$ if $\frac{-2}{\sqrt{3}}\sin\theta = 2/3$ i.e. $\sin\theta = -1/\sqrt{3}$

$\therefore\quad \tan\theta = \sqrt{\frac{1}{2}}$ and $\sec\theta = -\sqrt{\frac{3}{2}}$. Hence $P \equiv (-6, 3)$

7. (b) Let P ($a\sec\theta$, $b\tan\theta$) be a point on the hyperbola $\frac{x^2}{a^2} - \frac{y^2}{b^2} = 1$ having focus at S $(ae, 0)$ {where $a^2e^2 = a^2 + b^2$}

$\Rightarrow$ Slope of PS, $m_1 = \dfrac{b\tan\theta}{a(\sec\theta - e)}$ .....(i)

Now the equation of tangent at P is $\dfrac{x\sec\theta}{a} - \dfrac{y\tan\theta}{b} = 1$

which will intersect the directrix $x = a/e$ at $y = \dfrac{b(\sec\theta - e)}{e\tan\theta}$

So $Q \equiv \left(\dfrac{a}{e}; \dfrac{b(\sec\theta - e)}{e\tan\theta}\right)$

$\therefore$ Slope of QS is $m_2 = \dfrac{b(\sec\theta - e)}{e\tan\theta(1-e^2)a/e} = \dfrac{b(\sec\theta - e)}{a\tan\theta(1-e^2)}$

So $m_1m_2 = -1 \Rightarrow$ required angle $= \pi/2$

8. (d) For the hyperbola $\dfrac{x^2}{25} - \dfrac{y^2}{16} = 1$, the combined equation of tangents from (6, 2) will be $T^2 = SS_1$

So $\left(\dfrac{6x}{25} - \dfrac{2y}{16} - 1\right)^2 = \left(\dfrac{36}{25} - \dfrac{4}{16} - 1\right)\left(\dfrac{x^2}{25} - \dfrac{y^2}{16} - 1\right)$

$\Rightarrow 20x^2 + 11y^2 - 24xy - 192x + 100y + 476 = 0$

Now, shifting the origin to (6, 2) we get homogenous equation $20x^2 + 11y^2 - 24xy = 0$

$\Rightarrow 11m^2 - 24m + 20 = 0 \Rightarrow m_1 + m_2 = 24/11$ and $m_1m_2 = 20/11$

$\therefore$ H.M. $= \dfrac{2m_1m_2}{m_1 + m_2} = \dfrac{40}{24} = \dfrac{5}{3}$

Aliter: Equation of tangent with slope m to the hyperbola $\dfrac{x^2}{25} - \dfrac{y^2}{16} = 1$ is $y = mx \pm \sqrt{25m^2 - 16}$

Since (6, 2) lies on it so $(2 - 6m)^2 = 25m^2 - 16$ gives $11m^2 - 24m + 20 = 0$

So $m_1 + m_2 = \dfrac{24}{11}$ and $m_1m_2 = \dfrac{20}{11}$ gives H.M.

$m = \dfrac{2m_1m_2}{m_1 + m_2} = \dfrac{40}{24} = \dfrac{5}{3}$

9. (c) Equation of tangent to the hyperbola $\dfrac{x^2}{a^2} - \dfrac{y^2}{b^2} = 1$ at B (a sec θ, b tanθ) is $\dfrac{x}{a}\sec\theta - \dfrac{y}{b}\tan\theta = 1$ which intersect x-axis (transverse axis) at $x_1 = a\cos\theta$ and y-axis (conjugate axis) at $y_Q = -b\cot\theta$

So $\dfrac{a^2}{CP^2} - \dfrac{b^2}{CQ^2} - 1 = \dfrac{a^2}{a^2\cos^2\theta} - \dfrac{b^2}{b^2\cot^2\theta} - 1 = \sec^2\theta - \tan^2\theta - 1 = 0$

10. (b) The given hyperbola is $\dfrac{x^2}{16} - \dfrac{y^2}{9} = 1$ and the given circle is $x^2 + y^2 = 9$.

Let m be the slope of common tangent, then for hyperbola $y = mx \pm \sqrt{a^2m^2 - b^2}$ gives $y = mx \pm \sqrt{16m^2 - 9}$

Now its distance from O (0, 0) must be $r = 3$ i.e.

$\dfrac{|\sqrt{16m^2 - 9}|}{\sqrt{1 + m^2}} = 3$ so $m^2 = \dfrac{18}{7}$

$\therefore$ The tangents are $y = \pm\left(3\sqrt{\dfrac{2}{7}}\right)x \pm \sqrt{\dfrac{16\times 18}{7} - 9}$

$\Rightarrow y = \pm 3\sqrt{\dfrac{2}{7}}x \pm \dfrac{15}{\sqrt{7}}$

TEXTUAL EXERCISE–4 (SUBJECTIVE)

1. Equations of tangents from a point P (h, k) to the hyperbola $\dfrac{x^2}{a^2} - \dfrac{y^2}{b^2} = 1$ is $T^2 = SS_1$

i.e., $\left(\dfrac{hx}{a^2} - \dfrac{ky}{b^2} - 1\right)^2 = \left(\dfrac{h^2}{a^2} - \dfrac{k^2}{b^2} - 1\right)\left(\dfrac{x^2}{a^2} - \dfrac{y^2}{b^2} - 1\right)$

$\Rightarrow x^2\left\{\dfrac{h^2}{a^4} - \dfrac{h^2}{a^4} + \dfrac{k^2}{a^2b^2} + \dfrac{1}{a^2}\right\} + y^2\left\{\dfrac{k^2}{b^4} + \dfrac{h^2}{a^2b^2} - \dfrac{k^2}{b^4} - \dfrac{1}{b^2}\right\}$

$-\dfrac{2hk}{a^2b^2}xy$ + other terms = 0

$\Rightarrow \tan\beta = \dfrac{2\sqrt{\dfrac{h^2k^2}{a^4b^4} - \left(\dfrac{(k^2+b^2)(h^2-a^2)}{a^4.b^4}\right)}}{\left|\dfrac{k^2+b^2}{a^2b^2} + \dfrac{h^2-a^2}{a^2b^2}\right|}$

$= \dfrac{2}{a^2b^2} \times \dfrac{a^2b^2}{1}\left(\dfrac{\sqrt{a^2k^2 - b^2h^2 + a^2b^2}}{|h^2 + k^2 - (a^2 - b^2)|}\right)$

$\Rightarrow \tan^2\beta = \dfrac{4\{a^2k^2 - b^2h^2 + a^2b^2\}}{|h^2 + k^2 - (a^2 - b^2)|}$

$\Rightarrow \{x^2 + y^2 - (a^2 - b^2)\}^2 \tan^2\beta = 4\ \{a^2y^2 - b^2x^2 + a^2b^2\}$

If $\beta = \dfrac{\pi}{2}$, then $x^2 + y^2 + b^2 - a^2 = 0$

$\Rightarrow x^2 + y^2 = a^2 - b^2$; $a > b$

2. Let $T(h, k)$ be the point of intersection of tangents at the extremeties of normal chord PQ with $P(a\sec\theta, b\tan\theta)$

$\therefore$ Equation of PQ must be $T = 0$ i.e., $\dfrac{hx}{a^2} - \dfrac{ky}{b^2} = 1$ .....(i)

Or $\dfrac{ax}{\sec\theta} + \dfrac{by}{\tan\theta} = a^2 + b^2$ ....(ii)

$\Rightarrow \dfrac{h/a^2}{a/\sec\theta} + \dfrac{-k/b^2}{b/\tan\theta} = \dfrac{1}{a^2 + b^2}$

$\Rightarrow \dfrac{h\sec\theta}{a^3} = -\dfrac{-k\tan\theta}{b^3} = \dfrac{1}{a^2 + b^2}$

$\Rightarrow \sec\theta = \dfrac{a^3}{h(a^2 + b^2)}$ and $\tan\theta = \dfrac{-b^3}{k(a^2 + b^2)}$

Now, $\sec^2\theta + \tan^2\theta = 1$

$\Rightarrow \dfrac{a^6}{h^2(a^2 + b^2)^2} - \dfrac{b^6}{k^2(a^2 + b^2)^2} = 1$

Required locus will be $a^6y^2 - b^6x^2 = x^2y^2(a^2 + b^2)^2$

3. (a) The hyperbola is $\frac{x^2}{16}-\frac{y^2}{25}=1$ and the mid-point of chord is (5, 3) so $T=S_1$ gives $\frac{5x}{16}-\frac{3y}{25}=\frac{25}{16}-\frac{9}{25}$

$\Rightarrow$ $125x-48y=625-144 \Rightarrow 125x-48y=481$

(b) Equation of chord of circle $x^2+y^2=16$ having mid-point (h, k) is $T=S_1$

$\Rightarrow$ $hx+ky-16=h^2+k^2-16$

$\Rightarrow$ $hx+ky-(h^2+k^2)=0$ .....(i)

The equation of tangent to the hyperbola $\frac{x^2}{16}-\frac{y^2}{9}=1$ is

$y=mx\pm\sqrt{16m^2-9}$ .....(ii)

Comparing, (i) and (ii), we get $\frac{h}{m}=\frac{k}{-1}=\pm\frac{-(h^2+k^2)}{\sqrt{16m^2-9}}$

$\Rightarrow$ $\frac{h^2}{(h^2+k^2)^2}=\frac{m^2}{16m^2-9}$ and $\frac{k^2}{(h^2+k^2)^2}=\frac{1}{16m^2-9}$

$\Rightarrow$ $16x^2-9y^2=(x^2+y^2)^2$ is the required locus of mid point.

4. The chord of contact of tangents from $P(h, k)$ to the hyperbola $\frac{x^2}{a^2}-\frac{y^2}{b^2}-1=0$ is $T=0$

i.e., $\frac{hx}{a^2}-\frac{ky}{b^2}-1=0$. Its distance from origin (centre) is d^2

$\Rightarrow$ $d^2=\frac{|1|}{\sqrt{\frac{h^2}{a^4}+\frac{k^2}{b^4}}}$ gives $\frac{h^2}{a^4}+\frac{k^2}{b^4}=\frac{1}{d^4}$

$\Rightarrow$ $\frac{x^2}{a^4}+\frac{y^2}{b^4}=\frac{1}{d^4}$ is the required locus.

5. Chord with mid-point $P(h, k)$ to the given hyperbola is $\frac{hx}{a^2}-\frac{ky}{b^2}-\left(\frac{h^2}{a^2}-\frac{k^2}{b^2}\right)=0$

Since this chord is always a tangent to the circle $x^2+y^2=c^2$

$\Rightarrow$ $\frac{\left|\frac{h^2}{a^2}-\frac{k^2}{b^2}\right|}{\sqrt{\frac{h^2}{a^4}+\frac{k^2}{b^4}}}=c$

$\Rightarrow$ $\left(\frac{h^2}{a^2}-\frac{k^2}{b^2}\right)^2=c^2\left\{\frac{h^2}{a^4}+\frac{k^2}{b^4}\right\}$ i.e.,

$\left(\frac{x^2}{a^2}-\frac{y^2}{b^2}\right)^2=c^2\left\{\frac{x^2}{a^4}+\frac{y^2}{b^4}\right\}$

6. Let $M(h, k)$ be the mid point of chord PQ for the ellipse $\frac{x^2}{a^2}+\frac{y^2}{b^2}=1$ so $T=S_1$ gives $\frac{hx}{a^2}+\frac{ky}{b^2}-\left(\frac{h^2}{a^2}+\frac{k^2}{b^2}\right)=0$ ...(i)

Since this is also tangent to the hyperbola $\frac{x^2}{a^2}-\frac{y^2}{b^2}=1$

So $mx-y\pm\sqrt{a^2m^2-b^2}=0$ is the same as (i)

$\Rightarrow$ $\frac{h}{a^2m}=\frac{k}{b^2(-1)}=\frac{-\left(\frac{h^2}{a^2}+\frac{k^2}{b^2}\right)}{\pm\sqrt{a^2m^2-b^2}}$

$\Rightarrow$ $\frac{h^2}{\left\{\frac{h^2}{a^2}+\frac{k^2}{b^2}\right\}^2 a^2}=\frac{a^2m^2}{(a^2m^2-b^2)}$ and

$\frac{k^2}{\left\{\frac{h^2}{a^2}+\frac{k^2}{b^2}\right\}.b^2}=\frac{b^2}{(a^2m^2-b^2)}$

$\Rightarrow$ $\frac{h^2}{a^2}-\frac{k^2}{b^2}=\left(\frac{h^2}{a^2}+\frac{k^2}{b^2}\right)^2$ i.e., $\frac{x^2}{a^2}-\frac{y^2}{b^2}=\left(\frac{x^2}{a^2}+\frac{y^2}{b^2}\right)^2$

7. $H:\frac{x^2}{a^2}-\frac{y^2}{b^2}=1$ ...(i)

$C:\frac{y^2}{b^2}-\frac{x^2}{a^2}=1$ ...(ii)

Let $y=mx+c;\ c=\sqrt{b^2-a^2m^2}$ be the equation of tangent to conjugate hyperbola (ii) with point of contact $R\left(\frac{a^2m}{C},\frac{b^2}{C}\right)$

Now, $y=mx+c$ intersects (i), at $P(x_1, y_1)$ and $Q(x_2, y_2)$

$\Rightarrow$ $\frac{x^2}{a^2}-\frac{(mx+c)^2}{b^2}=1 \Rightarrow b^2x^2-a^2(mx+c)^2-a^2b^2=0$

$\Rightarrow$ $x^2(b^2-a^2m^2)-2a^2mcx-a^2c^2-a^2b^2=0$

$\Rightarrow$ $\frac{x_1+x_2}{2}=\frac{a^2mc}{b^2-a^2m^2}=\frac{a^2mc}{c^2}=\frac{a^2m}{c}$

$\therefore$ Abscissa of mid –point of PQ is $\frac{a^2m}{C}$. Also mid point of PQ on $y=mx+c$

$\Rightarrow$ Ordinate of mid point of PQ = $m\left(\frac{a^2m}{c}\right)+c$ = $\frac{a^2m^2+c^2}{c}=\frac{b^2}{c}$

$\Rightarrow$ $R\equiv\left(\frac{a^2m}{c},\frac{b^2}{c}\right)$ = mid point of PQ.

8. Let $M(h, k)$ be the mid point of a chord for the hyperbola $x^2-y^2=a^2$, then $T=S_1$ gives $hx-ky-(h^2-k^2)=0$ ...(i)

Now consider a point $P(x_1, y_1)$ on the given circle, then $x_1^2+y_1^2=a^2$ and the chord of contact of tangents from P to the hyperbola is

$T=0$ i.e., $x_1x-y_1y-a^2=0$ ....(ii)

Since (i) and (ii) are identical;

$\frac{h}{x_1}=\frac{k}{y_1}=\frac{h^2-k^2}{a^4}$ and $x_1^2+y_1^2=a^2$

$\Rightarrow$ $\frac{h^2+k^2}{x_1^2+y_1^2}=\frac{(h^2-k^2)^2}{a^4}$

$\Rightarrow$ $\frac{h^2+k^2}{a^2}=\frac{(h^2-k^2)^2}{a^4} \Rightarrow (x^2-y^2)^2=a^2(x^2+y^2)$

TEXTUAL EXERCISE 4–OBJECTIVE

1. (a) For the hyperbola $\frac{x^2}{a^2}-\frac{y^2}{b^2}-1=0$ the equation of a chord with mid point $M(h, k)$ is $T = S_1$

$\Rightarrow \frac{hx}{a^2}-\frac{ky}{b^2}-\left(\frac{h^2}{a^2}-\frac{k^2}{b^2}\right)=0$

$\Rightarrow$ Slope $m=\frac{hb^2}{a^2k}$ gives $k=\frac{b^2}{a^2m}h$ or $y=\frac{b^2}{a^2m}x$

2. (d) Let $y=mx\pm\sqrt{9m^2-5}$ be tangents to the hyperbola $\frac{x^2}{9}-\frac{y^2}{5}=1$. Since it passes through $P(\sqrt{3}, 1)$

So $\left(1-\sqrt{3}m\right)^2=\left(\pm\sqrt{9m^2-5}\right)^2$

$\Rightarrow 6m^2+2\sqrt{3}m-6=0$ as $m_1m_2=-1$

$\Rightarrow$ Angle $\pi/2$

3. (a) Let $y=mx\pm\sqrt{36m^2-9}$ be tangent to the hyperbola $\frac{x^2}{36}-\frac{y^2}{9}=1$

Since it passes through $P(0, 4)$ so $16 = 36m^2 - 9$ gives $m^2 = 25/36$

$\Rightarrow m = \pm 5/6$

Hence the tangents are $y - 4 = \pm 5/6\, x$ i.e., $5x - 6y + 24 = 0$ or $5x + 6y - 24 = 0$

4. (c) The equation of a chord with mid-point $M(6, 2)$ for the hyperbola $\frac{x^2}{16}-\frac{y^2}{25}=1$ is $T = S_1$ so

$\frac{6x}{16}-\frac{2y}{25}-\left(\frac{36}{16}-\frac{4}{25}\right)=0$

$\Rightarrow 75x - 16y = 418$

5. (a) The locus of perpendicular tangents for the hyperbola $\frac{x^2}{a^2}-\frac{y^2}{b^2}=1$ will be its director circle with equation $x^2 + y^2 = a^2 - b^2$

Aliter: Let $y=mx\pm\sqrt{a^2m^2-b^2}$ be a tangent to the hyperbola $\frac{x^2}{a^2}-\frac{y^2}{b^2}=1$ the perpendicular tangent will have equation

$-my=x\pm\sqrt{a^2-b^2m^2}\Rightarrow m^2x^2+y^2-2mxy=a^2m^2-b^2$

$\therefore$ $m^2y^2 + x^2 + 2mxy = a^2 - b^2m^2$, on adding, we get $(1 + m^2)(x^2 + y^2) = (a^2 - b^2)(1 + m^2)$ i.e., $x^2 + y^2 = a^2 - b^2$

6. (c) A tangent with slope m for the hyperbola $\frac{x^2}{25}-\frac{y^2}{16}=1$ will be $y=mx\pm\sqrt{25m^2-16}$. Since it passes through P $\left(2\sqrt{2},1\right)$

So $\left(1-2\sqrt{2}m\right)=\left(\pm\sqrt{25m^2-16}\right)^2$ gives $17m^2 + 4\sqrt{2}m - 17 = 0$

Since $m_1m_2 = -1$ $\Rightarrow$ Angle $= \pi/2$

7. (c) Since $x^2 + y^2 = 3$ is the director circle for the hyperbola $\frac{x^2}{4}-\frac{y^2}{1}=1$

$\therefore$ Angle between the tangents is $90° = \pi/2$

TEXTUAL EXERCISE–5 (SUBJECTIVE)

1. The asymptotes of the hyperbola are $L_1: x + 2y + 3 = 0$ and $L_2: 3x + 4y + 5 = 0$

$\Rightarrow$ The equation of hyperbola will be L_1: $(x + 2y + 3)(3x + 4y+5) + \lambda = 0$

Since the hyperbola passes through $(1, -1)$

$\therefore$ $(2)(4) + \lambda = 0$ gives the hyperbola as L_1 : $(x + 2y +3)(3x + 4y + 5) -8 = 0$ or $3x^2 + 8y^2 + 10xy + 14x + 22y + 7 = 0$

The conjugate hyperbola is given by C = 0; where H + $C = 2\lambda$

$\Rightarrow 2\lambda - H = 0$ or $H - 2\lambda = 0$

i.e., $3x^2 + 8y^2 + 10xy + 14x +22y + 23 = 0$

2. The equation of a chord with end points as $P(\theta)$ and $Q(\phi)$ on the hyperbola $\frac{x^2}{a^2}-\frac{y^2}{b^2}=1$ is

$\frac{x}{a}\cos\left(\frac{\theta-\phi}{2}\right)-\frac{y}{b}\sin\left(\frac{\theta+\phi}{2}\right)=\cos\left(\frac{\theta+\phi}{2}\right)$

For it to be a focal chord $S(ae, 0)$ will satisfy

$\Rightarrow e\cos\left(\frac{\theta-\phi}{2}\right)=\cos\left(\frac{\theta+\phi}{2}\right)$

$\Rightarrow \frac{\cos\left(\frac{\theta-\phi}{2}\right)-\cos\left(\frac{\theta+\phi}{2}\right)}{\cos\left(\frac{\theta-\phi}{2}\right)+\cos\left(\frac{\theta+\phi}{2}\right)}=\frac{1-e}{1+e}$

$\Rightarrow \tan\theta/2 \,.\, \tan\phi/2=\frac{1-e}{1+e}$ i.e., $\tan\frac{\theta}{2}\tan\frac{\theta}{2}+\frac{e-1}{e+1}=0$

3. For the hyperbola H: $2x^2 + 5xy + 2y^2 + 4x + 5y = 0$

Let the asymptotes be $A = H + \lambda = 0$. Now A being a pair of straight lines $\Delta = 0$

i.e., $\begin{vmatrix} 2 & 5/2 & 2 \\ 5/2 & 2 & 5/2 \\ 2 & 5/2 & \lambda \end{vmatrix}=0 \Rightarrow (2-\lambda)\left\{\frac{25}{4}-4\right\}=0$

$\Rightarrow \lambda = 2$, so A: $2x^2 + 5xy + 2y^2 + 4x + 5y + 2 = 0$ or $(x + 2y + 1)(2x + y + 2) = 0$

i.e., Asymptotes are $x + 2y + 1 = 0$ and $2x + y + 2 = 0$.

A general equation of family of hyperbola is $2x^2 + 5xy + 2y^2 + 4x + 5y + 2 + k = 0$ (where $k \neq 0$)

4. Given $\frac{x}{a}=\tan(\theta+\alpha)$ and $\Rightarrow \frac{y}{b}=\tan(\theta+\beta)$

Now, $\tan(\alpha-\beta)=\tan\left[(\theta+\alpha)-(\theta+\beta)\right]$

$=\frac{\tan(\theta+\alpha)-\tan(\theta+\beta)}{1+\tan(\theta+\alpha).\tan(\theta+\beta)}=\frac{\frac{x}{a}-\frac{y}{b}}{1+\frac{x}{a}.\frac{y}{b}}=\frac{bx-ay}{ab+xy}$

$\Rightarrow$ $bx - ay - \tan(\alpha - \beta).\ xy = ab\tan(\alpha - \beta)$

$\Rightarrow$ $bx - ay - kxy = abk$; where $k = \tan(\alpha - \beta)$ = constant $\neq$ 0 as $\alpha - \beta \neq n\pi$, $n \in \mathbb{Z}$

$\Rightarrow$ Clearly $\Delta = \dfrac{(k^2+1)\,k\,ab}{4} \neq 0$ and $h^2 - ab = \dfrac{k^2}{4} > 0$

$\Rightarrow$ locus of point is *a* hyperbola.

5. Let the equation of hyperbola be $(2x - y - 3)(3x + y - 7) + \lambda = 0$

It passes through (1, 2)

$\Rightarrow$ $(-3)(-2) + \lambda = 0$

$\Rightarrow$ $\lambda = -6$

$\therefore$ Equation of hyperbola is $(2x - y - 3)(3x + y - 7) - 6 = 0$

$\Rightarrow$ $6x^2 - y^2 - 23x + 4y + 15 = 0$

TEXTUAL EXERCISE–5 (OBJECTIVE)

1. (a) The equation of hyperbola H: $x^2 + 2y^2 + 3xy + 2x + 3y + 2 = 0$

$\Rightarrow$ Equation of asymptotes $A = H + \lambda = 0$

So $x^2 + 2y^2 + 3xy + 2x + 3y + (2 + \lambda) = 0$ where $\Delta = 0$

$$\Rightarrow \begin{vmatrix} 1 & \frac{3}{2} & 1 \\ \frac{3}{2} & 2 & \frac{3}{2} \\ 1 & \frac{3}{2} & \lambda \end{vmatrix} = 0 \text{ gives } \lambda = -1$$

Hence A: $x^2 + 2y^2 + 3xy + 2x + 3y + 1 = 0$

2. (a) The hyperbola is H: $xy - hx - ky = 0$

$\Rightarrow$ Asymptotes $A = H + \lambda = 0$, so $xy - hx - ky + \lambda = 0$

$$\text{Where } \Delta = 0, \text{ so } \begin{vmatrix} 0 & 1/2 & -h/2 \\ 1/2 & 0 & -k/2 \\ -h/2 & -k/2 & \lambda \end{vmatrix} = 0 \text{ gives}$$

$$\frac{hk}{4} - \frac{\lambda}{4} = 0$$

So $\lambda = hk$

$\therefore$ $A \equiv xy - hx - ky + hk = 0$ or $(x - k)(y - h) = 0$, so $x - k = 0$, $y - h = 0$

3. (d) The hyperbola is $\dfrac{x^2}{9} - \dfrac{y^2}{3} = 1$

$\therefore$ $e = \sqrt{\dfrac{9+3}{9}} = \dfrac{2}{\sqrt{3}}$

Angle between the asymptotes $2\theta = 2\tan^{-1}\left(\dfrac{b}{a}\right) = 2\tan^{-1}\left(\dfrac{1}{\sqrt{3}}\right) = \dfrac{\pi}{3}$

Length of L.R. $= \dfrac{2b^2}{a} = \dfrac{2(3)}{3} = 2$

The Asymptotes are: $y = \pm\dfrac{x}{\sqrt{3}}$

Let $P\,(a\sec\theta, b\tan\theta) \equiv \left(3\sec\theta, \sqrt{3}\sec\theta\right)$ be any point on hyperbola.

Then $d_1 d_2 = \dfrac{\left|\sqrt{3}\sec\theta - \sqrt{3}\tan\theta\right| . \left|\sqrt{3}\sec\theta - \sqrt{3}\tan\theta\right|}{\left(1+\frac{1}{3}\right)}$

$= \dfrac{3}{4/3} = \dfrac{9}{4} > 2$

4. (b) Equation of tangent drawn at *A* (*a*, 0) to hyperbola $\dfrac{x^2}{a^2} - \dfrac{y^2}{b^2} = 1$ is $x = a$, equation of asymptotes is $y = \pm\dfrac{b}{a}x$

$\Rightarrow$ Vertices of required Δ are $O\,(0, 0)$; $A\,(0, a)$ and $B\,(a, -b)$

$\Rightarrow$ Area of required $\Delta = \dfrac{1}{2}(a)(2b) = (ab)$ square units.

5. (d) Let $P\,(a\sec\theta, b\tan\theta)$ be *a* point on the hyperbola $S_1 : \dfrac{x^2}{a^2} - \dfrac{y^2}{b^2} = 1$

Now chord of contact of tangents from *P* to $S_2 : \dfrac{x^2}{2a^2} - \dfrac{y^2}{2b^2} - 1 = 0$ is $\dfrac{x\sec\theta}{2a} - \dfrac{y\tan\theta}{2b} - 1 = 0$ ...(i)

Let $M\,(h, k)$ be the mid-point of chord on H_2 then $T = S_1$ gives $\dfrac{h}{2a^2}x - \dfrac{ky}{2b^2} - \left(\dfrac{h^2}{2a^2} - \dfrac{k^2}{2b^2}\right) = 0$ .....(ii)

Since (i) and (ii) are identical

$$\Rightarrow \frac{a\sec\theta}{h} = \frac{b\tan\theta}{k} = \frac{1}{\left(\frac{h^2}{2a^2} - \frac{k^2}{2b^2}\right)}$$

$$\Rightarrow \frac{h^2}{a^2} - \frac{k^2}{b^2} = \left(\frac{h^2}{2a^2} - \frac{k^2}{2b^2}\right)^2 \text{ i.e., } \frac{x^2}{a^2} - \frac{y^2}{b^2} = \left(\frac{x^2}{2a^2} - \frac{y^2}{2b^2}\right)^2$$

6. (d) The asymptotes are $L_1 = 4x + 5y - 5 = 0$ and $L_2 = 5x - 4y + 7 = 0$

Observe that (slope of L_1) (slope L_2) = –1

$\Rightarrow$ Asymptotes are at = 90° so it is *a* rectangular hyperbola with $e = \sqrt{2}$

7. (a) We know that $\theta = 2\tan^{-1}\left(\dfrac{b}{a}\right)$

$\Rightarrow$ $\tan\theta/2 = \dfrac{\sin\frac{\theta}{2}}{\cos\frac{\theta}{2}} = \dfrac{b}{a}$ gives $\sin\theta/2 = \dfrac{b}{\sqrt{a^2+b^2}}$

$\Rightarrow$ $\sin\left(\dfrac{\theta}{2}\right) = \dfrac{b}{ae} = \dfrac{\sqrt{e^2-1}}{e}$

8. (c), (d) For the hyperbola *H*: $2x^2 + 5xy + 2y^2 + 4x + 5y = 0$. The equation of asymptotes is $A = H + \lambda = 0$

Shifting the origin at the centre of the conic we get $2x^2 + 5xy + 2y^2 = 0$

i.e., $(x + 2y)(2x + y) = 0 \Rightarrow m_1 = -\dfrac{1}{2}, m_2 = -2$

TEXTUAL EXERCISE–6 (SUBJECTIVE)

1. Let the rectangular hyperbola be $xy = c^2$ ....(i)
and the circle be $x^2 + y^2 + 2gx + 2fy + \lambda = 0$ ....(ii)
If $\left(ct, \frac{c}{t}\right)$ is any point on hyperbola (i) and (ii), then
$$c^2t^2 + \frac{c^2}{t^2} + 2gct + \frac{2fc}{t} + \lambda = 0$$
$\Rightarrow$ If $A(t_1), B(t_2), C(t_3), D(t_4)$ are points of intersection of (i) and (ii) then
$$t_1 + t_2 + t_3 + t_4 = \frac{(-1)^1(2gc)}{c^2} = \frac{-2g}{c} \quad ....(iv)$$
Now, we are given that AB passes through the centre of circle. But AB is *a* chord of circle. It means AB is the diameter of
$\therefore$ Centre of circle is $M \equiv \left(\frac{C(t_1+t_2)}{2}, \frac{C(t_1+t_2)}{2t_1t_2}\right) \equiv (-g, -f)$
$$\Rightarrow -g = \frac{c(t_1+t_2)}{2}, -f = \frac{c(t_1+t_2)}{t_1t_2}$$
$\therefore$ from (iv) $t_1 + t_2 + t_3 + t_4 = t_1 + t_2$
$\Rightarrow t_3 + t_4 = 0$
$\Rightarrow$ Mid point of $CD \equiv \left(\frac{c(t_3+t_4)}{2}, \frac{c(t_3+t_4)}{2t_3t_4}\right) \equiv (0,0) =$ centre of rectangular hyperbola

2. Let m (h, k) be the mid point of *a* chord for the hyperbola $xy = c^2$, then equation is $\frac{x}{h} + \frac{y}{k} = 2$. Since it passes through $(-6c, -4c)$
$\Rightarrow \frac{(-6c)}{h} + \frac{(-4c)}{k} = 2$ gives $3kc + 2hc = -hk$ i.e., $xy + 2cx + 3cy = 0$

3. Solving $xy = c^2$ with $y = mx + 2c\sqrt{-m}$, we get $mx^2 + 2c\sqrt{-m}x - c^2 = 0$ or $-\left\{\sqrt{-m}\,x - c\right\}^2 = 0$
$$\Rightarrow x = \frac{c}{\sqrt{-m}} \Rightarrow y = \frac{c^2}{x} = c\sqrt{-m}$$
$\therefore$ The point of contact is $\left(\frac{c}{\sqrt{-m}}, c\sqrt{-m}\right)$

4. Let m (h, k) be the mid point of chord PQ on the hyperbola $xy = c^2$ so $hx + hy = 2hk$ ....(i)
Which is also *a* tangent to *a* parabola $x^2 = 4ay$
So $y = mx - am^2$ is identical to (i)
$$\Rightarrow \frac{k}{m} = \frac{h}{-1} = \frac{2hk}{am^2} \Rightarrow m^2 = \left(\frac{-2k}{a}\right) \text{ and } m = \frac{2h}{a}$$
$\therefore \left(\frac{2h}{a}\right)^2 = \frac{-2k}{a}$ i.e., $2h^2 = -ak$ i.e., $2x^2 = -ay$
$\Rightarrow x^2 = \left(-\frac{a}{2}\right)y$; which is *a* parabola

5. Let $P\left(t_1, \frac{1}{t_1}\right)$ and $Q\left(t_2, \frac{1}{t_2}\right)$ be the points on hyperbola $xy = 1$, if h (h, k) be the point dividing PQ in 1:2, then $3h = 2t_1 + t_2$ and
$$3k = \frac{2}{t_1} + \frac{1}{t_2} = \frac{2t_2 + t_1}{t_1t_2}$$
Since $m = -\frac{1}{t_1t_2} = 4$ so $t_1t_2 = -1/4$
So $3h = 2t_1 + t_2$ and $3k = \frac{2t_2 + t_1}{\left(-\frac{1}{4}\right)}$
Observe that $(3h)\left(\frac{-3k}{4}\right) = 2t_1^2 + 2t_2^2 + 5t_1t_2$
$$= 2(t_2 - t_1)^2 + 9t_1t_2 = 2\left(-\frac{3k}{4} - 3h\right)^2 + 9\left(-\frac{1}{4}\right)$$
$$\Rightarrow \frac{-9hk}{4} = 18\frac{(k+4h)^2}{16} - \frac{9}{4}$$
$\Rightarrow -36hk = 18k^2 + 288h^2 + 144hk - 36$
$\Rightarrow 18k^2 + 288h^2 + 180hk = 36$
$\Rightarrow k^2 + 16h^2 + 10hk = 2 \Rightarrow 16x^2 + y^2 + 10xy = 2$

6. Let the circle be $x^2 + y^2 + 2gx + 2fy + c = 0$
Now the hyperbola is $xy = 1$ so $x = 1/y$
Solving, we get $x^2 + \frac{1}{x^2} + 2gx + \frac{2f}{x} + c = 0$ i.e., $x^4 + 2gx^3 + cx^2 + 2fx + 1 = 0$
Lets its roots be $x = x_1, x_2, x_3, x_4$ so $x_1x_2x_3x_4 = 1$
Since these points are lying on the hyperbola $xy = 1$
$\therefore \quad y_1y_2y_3y_4 = \frac{1}{1} = 1$
Hence $x_1x_2x_3x_4 = y_1y_2y_3y_4 = 1$

7. Let p (t) be *a* point the hyperbola $xy = c^2$, then equation of tangent at P is $x + t^2y = 2ct$ .... (i)
And equation of normal at P is $xt^3 - yt = c(t^4 - 1)$ ....(ii)
$$\Rightarrow x_1 = a_1 = 2ct \text{ and } a_2 = \frac{c(t^4-1)}{t^3}$$
$$\therefore \quad a_1a_2 = \frac{2c^2}{t^2}(t^4 - 1) \text{ and}$$
$$y_1 = b_1 = \frac{2c}{t} \text{ and } b_2 = \left(\frac{-c}{t}\right)\{t^4 - 1\}$$
So $b_1b_2 = \frac{-2c^2}{t^2}(t^4 - 1)$
Clearly $a_1a_2 = -b_1b_2$ so $a_1a_2 + b_1b_2 = 0$

TEXTUAL EXERCISE–6 (OBJECTIVE)

1. **(a)** For the parabola $y^2 = 4ax$ the equation of tangent is $y = mx + a/m$ and the point of contact is $\left(\frac{a}{m^2}, \frac{2a}{m}\right)$.

Equation of tangent to hyperbola $xy = c^2$ at $R(t)$ is $x + t^2y - 2ct = 0$

$\Rightarrow \dfrac{m}{1} = \dfrac{-1}{t^2} = \dfrac{-2ct}{a/m} = \dfrac{-2ctm}{a}$

$\Rightarrow t^2 = \dfrac{-1}{m}$; $t = \dfrac{-a}{2c}$ $\Rightarrow$ $m = -\dfrac{1}{t^2} = -\left(\dfrac{4c^2}{a^2}\right)$

Thus the only point of contact on rectangular hyperbola is $\left(ct, \dfrac{c}{t}\right) \equiv \left(-\dfrac{a}{2}, \dfrac{-2c^2}{a}\right)$ and $\left(\dfrac{a^5}{16c^4}, \dfrac{-a^3}{2c^2}\right)$ on parabola $y^2 = 4ax$

2. (a) Let $P(x_1, y_1)$ be *a* point on the hyperbola $xy = c^2$, then equation of tangent at P is $\dfrac{x}{x_1} + \dfrac{y}{y_1} = 2$

$\Rightarrow$ The points of intersection on co-ordinate axes are $A(2x_1, 0)$ and $B(0, 2y_1)$

Since ΔOAB is right angled at O

$\Rightarrow$ mid point of AB is the circumcentre i.e., $(h, k) = (x_1, y_1)$

Since $x_1y_1 = c^2$ so $xy = c^2$ is the locus of circumcentre of ΔOAB

3. (b) Equation of chord of hyperbola $xy = c^2$ having mid-point (h, k) is $\dfrac{x}{h} + \dfrac{y}{k} = 2$.

The slope $= m = \dfrac{-1/h}{1/k} = \dfrac{-k}{h}$

$\Rightarrow$ Required locus is $mx + y = 0$

4. (d) At *a* point $P(ct, c/t)$ on the hyperbola $xy = c^2$, the equation of tangent will be $x + t^2y - 2ct = 0$

So perpendicular to the tangent from origin will be $y = t^2x$ and intersects the tangent where $x\{1 + t^4\} = 2ct$

$\Rightarrow$ Foot (h, k) will be $h = \dfrac{2ct}{1+t^4}$ and $k = \dfrac{2ct^3}{t^4+1}$, hence $k = ht^2$

$\Rightarrow t^2 = \dfrac{k}{h}$

Now $hk = \dfrac{4c^2t^4}{(t^4+1)^2} = \dfrac{4c^2\left(\dfrac{k^2}{h^2}\right)}{\left[\dfrac{h^2+k^2}{h^2}\right]^2}$

$\Rightarrow \dfrac{hk(h^2+k^2)^2}{h^4} = \dfrac{4c^2k^2}{h^2} \Rightarrow (h^2 + k^2)^2 = 4c^2\,hk$

Hence the locus is $(x^2 + y^2)^2 = 4c^2xy$

5. (d) For the hyperbola $xy = c^2$, the asymptotes are x-axis and y-axis

$\therefore$ $M(ct, 0)$ and $N(0, c/t)$

$\Rightarrow$ Mid point of MN is $D \equiv \left(\dfrac{ct}{2}, \dfrac{c}{2t}\right)$

So $hk = \dfrac{c^2}{4}$ gives $4xy = c^2$

6. (c) Let the circle $x^2 + y^2 + 2gx + 2fy + k = 0$

Intersect the hyperbola $xy = c^2$, then

$y = c^2/x$ gives $x^2 + \dfrac{c^4}{x^2} + 2gx + \dfrac{2fc^2}{x} + k = 0$

or $x^4 + 2gx^3 + kx^2 + 2fc^2x + c^4 = 0$

$\Rightarrow x_1x_2x_3x_4 = c^4$ so $y_1y_2y_3y_4 = \dfrac{c^8}{x_1x_2x_3x_4} = \dfrac{c^8}{c^4} = c^4$

7. (c) The given points on the hyperbola $xy = c^2$ are $P(x_1, y_1)$ and $Q(x_2, y_2)$

$\Rightarrow$ Mid point of PQ is $M\left(\dfrac{x_1+x_2}{2}, \dfrac{y_1+y_2}{2}\right)$.

Now the equation of *a* chord with mid-point $M(h, k)$ will be $\dfrac{x}{h} + \dfrac{y}{k} = 2$ so $\dfrac{2x}{x_1+x_2} + \dfrac{2y}{y_1+y_2} = 2$ gives $\dfrac{x}{x_1+x_2} + \dfrac{y}{y_1+y_2} = 1$

8. (b) Let $R(x_1, y_1)$ be the mid-point of *a* chord PQ on the hyperbola $xy = c^2$, then equation of PQ is $\dfrac{x}{x_1} + \dfrac{y}{y_1} = 2$ which will intersect the x-axis at $A(2x_1, 0)$. Observe that $OR = \sqrt{x_1^2 + y_1^2} = AR$ and $OA = 2x_1$

So ΔOAR is isosceles

9. (c) Let $y = x + k$ be the chord for the hyperbola $xy = c^2$ {where k is *a* parameter}. At the point of intersection $x^2 + kx - c^2 = 0$ gives $x_1 + x_2 = -k$

$x_1x_2 = -c^2$, $|x_1 - x_2| = \sqrt{k^2 + 4c^2}$

and $y_1 + y_2 = (x_1 + k) + (x_2 + k) = x_1 + x_2 + 2k = k$

For $A(x_1, y_1)$ and $B(x_2, y_2)$, let M (*be* the mid point and the centre of the circle $M\left(-\dfrac{k}{2}, \dfrac{k}{2}\right)$ and diameter $= \sqrt{2}$ $|x_1 - x_2| = \sqrt{2}\sqrt{k^2 + 4c^2}$

Hence the family of circle is

$\left(x + \dfrac{k}{2}\right)^2 + \left(y - \dfrac{k}{2}\right)^2 = \dfrac{2(k^2 + 4c^2)}{4}$ i.e., $x^2 + kx + y^2 - ky = 2c^2$

Let $x^2 + y^2 + k_1x - k_1y - 2c^2 = 0$ and $x^2 + y^2 + k_2x - k_2y - 2c^2 = 0$ be two members

$\Rightarrow$ Common chord will be obtained by subtraction on i.e., $(k_1 - k_2)x - (k_1 - k_2)y = 0$; $x - y = 0$ i.e.; $y = x$

$\therefore$ The two fixed points will be given by $x^2 + kx + x^2 - kx = 2c^2$

$\Rightarrow x = \pm c$

$\therefore$ The two fixed points are (c, c) and $(-c, -c)$

10. (c) For the hyperbola $xy = c^2$, we have $PA + PB = 2PQ$

$\Rightarrow$ Q is the mid point of A and B, so let $Q \equiv (h, k)$

So the equation of chord with mid-point $Q(h, k)$ is $\dfrac{x}{h} + \dfrac{y}{k} = 2$. Since the chord passes through $P(-1, 2)$ so

$\dfrac{-1}{h} + \dfrac{2}{k} = 2$

$\Rightarrow 2hk + k - 2h = 0$ i.e., $2x = y(1 + 2x)$

11. (b) As given $PA.PB = (PQ)^2$ where the hyperbola is $xy = c^2$

Let a general point on line $PAQB$ be $x = -1 + r\cos\theta$ and $y = 2 + r\sin\theta$ and also let $r = r_1$ for $A(x_1, y_1)$ and $r = r_2$ for $Q(x_2, y_2)$ and $r = r_3$ or $B(x_3, y_3)$

Since A and B lie on the hyperbola so $x_1y_1 = x_3y_3 = c^2$

In general for any point lying on hyperbola as well as on line $(-1 + r\cos\theta)(2 + r\sin\theta) = c^2$

$\Rightarrow r^2\sin\theta\cos\theta + r(2\cos\theta - \sin\theta) - (c^2 + 2) = 0$

$$\therefore \quad r_1 + r_3 = \frac{\sin\theta - 2\cos\theta}{\sin\theta\cos\theta} \text{ and } r_1r_3 = \frac{-(c^2+2)}{\sin\theta\cos\theta} \qquad \ldots(i)$$

Now $Q = (x_2, y_2) = (h, k)$ (say) so $(h + 1) = r_2\cos\theta$ and $(k - 2) = r_2\sin\theta$ gives $(h + 1)(k - 2) = r_2^2\sin\theta\cos\theta$

Since $r_2^2 = r_1r_3$ so $(h + 1)(k - 2) = r_1r_3\cos\theta\sin\theta = -(c^2 + 2)$ (from (i))

$\Rightarrow hk - 2h + k - 2 = -2 - c^2$ gives $xy - 2x + y + c^2 = 0$

12. (a) Given $P(-1, 2)$ and hyperbola $xy = c^2$. As worked in Q.11

Let $x = -1 + r\cos\theta$ and $y = 2 + r\sin\theta$

Let $A(x_1, y_1)$ for $r = r_1$, $Q(x_2, y_2)$ for $r = r_2$ and $B(x_3, y_3)$ for $r = r_3$

Since A and B lie on the hyperbola so

$$r_1r_3 = \frac{-(c^2+2)}{\sin\theta\cos\theta} \text{ and } r_1 + r_3 = \frac{\sin\theta - 2\cos\theta}{\sin\theta\cos\theta}$$

Since PA, PQ and PB are in H.P. so $\dfrac{2PA.PB}{PA + PB} = PQ$

$$\text{i.e., } r_2 = \frac{2r_1r_3}{r_1 + r_3} = \frac{(-2)(c^2+2)}{\sin\theta - 2\cos\theta} \qquad \ldots(i)$$

Now $Q(x_2, y_2) = (h, k)$ (say) so $(h + 1) = r_2\cos\theta$ and $(k - 2) = r_2\sin\theta$

$\Rightarrow (k - 2) - 2(h + 1) = r_2(\sin\theta - 2\cos\theta) = (-2)(c^2 + 2)$ {from (i)}

So $k - 2 - 2h - 2 = -2c^2 - 4$ or $2h - k = 2c^2$ i.e., $2x - y = 2c^2$

13. (c) For the hyperbola $xy = c^2$, let the chord AB be $y = mx + k_1$ where $A(t_1) = (x_1, y_1)$ and $B(t_2) = (x_2, y_2)$

Similarly let the other perpendicular chord be CD as $y = \left(-\dfrac{1}{m}\right)x + k_2$ where $C(t_3) = (x_3, y_3)$ and $D(x_4, y_4) = D(t_4)$

Solving $y = mx + k_1$ and $xy = c^2$, we get $mx^2 + k_1x - c^2 = 0$

$\Rightarrow x_1x_2 = c^2 t_1t_2 = \dfrac{-c^2}{m}$ so $t_1t_2 = -1/m$

Similarly we will get (from CD and hyperbola) $t_3t_4 = m$

$\therefore \; t_1t_2t_3t_4 = -1$, now observe that $m_{OA} = \dfrac{y_1}{x_1} = \dfrac{c}{t_1(ct_1)} = \dfrac{1}{t_1^2}$

Similarly, other slopes, so $m_{OA}\; m_{OB}\; m_{OC}\; m_{OD} = \dfrac{1}{(t_1t_2t_3t_4)^2} = \dfrac{1}{(-1)^2} = 1$

14. (d) Equation of normal at $P(t)$ for the hyperbola $xy = c^2$ is $ct^4 - xt^3 + ty - c = 0$

Which gives four values of $t = t_1, t_2, t_3, t_4$

Let the feet normal be $A(t_1), B(t_2), C(t_3), D(t_4)$

Where $x_i = ct_i$ and $y_i = \dfrac{c}{t_i}$

Since the line passes through (h, k) so $ct^4 - t^3h + kt - c = 0$

$\therefore \quad ct^4 - t^3h + kt - c = 0$

$\therefore \quad t_1 + t_2 + t_3 + t_4 = \dfrac{h}{c}$ and $t_1t_2t_3t_4 = -1$ also $\sum t_1t_2t_3 = \dfrac{-k}{c}$

Now $x_1 + x_2 + x_3 + x_4 = c(t_1 + t_2 + t_3 + t_4) = h$ and $x_1x_2x_3x_4 = -c^4$

Similarly, $y_1 + y_2 + y_3 + y_4 = c$

$$\left\{\frac{1}{t_1} + \frac{1}{t_2} + \frac{1}{t_3} + \frac{1}{t_4}\right\} = c\left\{\frac{\sum t_1t_2t_3}{t_1t_2t_3t_4}\right\} = \frac{c(-k)}{c(-1)} = k$$

$$\therefore \quad \frac{(x_1 + x_2 + x_3 + x_4)(y_1 + y_2 + y_3 + y_4)}{x_1x_2x_3x_4} = \frac{-kh}{c^4}$$

15. (d) Let $T(h, k)$ be the point of intersection of tangents drawn at the extremities of the normal chord drawn at a point

$P(t)$ on the rectangular hyperbola $xy = c^2$, then equation of normal chord will be $ty - xt^3 + ct^4 - c = 0$ ....(i)

Also it is the chord of contact of point $T(t, k)$ given $hx + ky = 2c^2$ ..... (ii)

$\therefore$ (i) and (ii) are identical

$$\Rightarrow \quad \frac{t}{k} = \frac{-t^3}{h} = \frac{ct^4 - c}{-2c^2} \quad \Rightarrow \quad t^2 = -\frac{h}{k} \text{ and } \frac{t}{k} = \frac{1-t^4}{2c}$$

$\Rightarrow$ Required locus is $(k^2 - h^2)^2 + 4c^2kh = 0$ or $(x^2 - y^2)^2 + 4c^2xy = 0$

16. (a), (b), (c), (d) The given circle is $x^2 + y^2 = a^2$ and the hyperbola $xy = c^2$ so $y = c^2/x$ gives $x^2 + \dfrac{c^4}{x^2} = a^2$ or $x^4 - a^2x^2 + c^4 = 0$

$\therefore \quad x_1 + x_2 + x_3 + x_4 = 0$ and $x_1x_2x_3x_4 = c^4$

$$\Rightarrow \quad y_1 + y_2 + y_3 + y_4 = c^2\left\{\frac{1}{x_1} + \frac{1}{x_2} + \frac{1}{x_3} + \frac{1}{x_4}\right\} = c^2\left\{\frac{\sum x_1x_2x_3}{x_1x_2x_3x_4}\right\}$$

$$= c^2\left\{\frac{0}{c^4}\right\} = 0$$

$$\text{Similarly } y_1y_2y_3y_4 = c^8\left\{\frac{1}{x_1}\frac{1}{x_2}\frac{1}{x_3}\frac{1}{x_4}\right\} = c^4$$

17. (b), (c) The given parabola is $y^2 = 4(8)x$ and the hyperbola is $\dfrac{x^2}{\left(\frac{8}{9}\right)} - \dfrac{y^2}{\left(\frac{8}{9}\right)} = 1$

Now $y = mx \pm \sqrt{\dfrac{8}{9}(m^2 - 1)} = mx + \dfrac{8}{m}$ gives

$$\frac{64}{m^2} = \frac{8}{9}(m^2 - 1)$$

$\Rightarrow m^4 - m^2 - 72 = 0$ gives $m^2 = 9$ ($m^2 = -8$ is rejected)

$\therefore \; m = 3$ gives $y = 3x + 8/3$ i.e., $9x - 3y + 8 = 0$ and $m = -3$ gives $y = -3x - 8/3$ i.e., $9x + 3y + 8 = 0$

18. **(b), (d)** We know that the slope of normal $m = t^2$ for the hyperbola $xy = c^2 = 1$

The line $ax + by + c = 0$ has slope $m = -a/b$

$\Rightarrow \quad -\frac{a}{b} > 0$ gives $\frac{a}{b} < 0$ o opposite sign of a and b required i.e., $a > 0, b < 0$ or a, 0, $b > 0$

19. **(a), (c)** Given length of transverse axis = 2c and asymptotes are the coordinates axes. We know that the length of transverse axis of hyperbola $xy = k^2$ is $2\sqrt{2}k \Rightarrow k = \frac{c}{\sqrt{2}}$

$\therefore$ The equation of hyperbola is $xy = c^2/2$

Now equation of the chord with mid-point (2c, 3c) is

$\frac{x}{h} + \frac{y}{k}$ so $\frac{x}{2c} + \frac{y}{3c} = 2$ gives $3x + 2y = 12c$

20. Let the point of contact of normal from point $P(h, k)$ be $R(t)$ for the hyperbola $xy = c^2$

$\Rightarrow$ The equation of normal is $ct^4 - xt^3 + yt - c = 0$ which gives $ct^4 - ht^3 + kt - c = 0$ ($\because$ It passes through $P(h, k)$)

$\Rightarrow t_1 + t_2 + t_3 + t_4 = h/c, \Sigma t_1t_2 = 0, \Sigma t_1t_2t_3 = -k/c$ and $t_1t_2t_3t_4 = -1$

Now the slope of normal at $P(t)$ is $m = t^2$

$\Rightarrow t_1^2 + t_2^2 + t_3^2 + t_4^2 = \lambda > 0$ (as $\lambda \in \mathbb{R}^+$)

From $(t_1 + t_2 + t_3 + t_4)^2 = \frac{h^2}{c^2}$, we get

$t_1^2 + t_2^2 + t_3^2 + t_4^2 + 2\,\Sigma t_1t_2 = \frac{h^2}{c^2}$ or $\lambda = \frac{h^2}{c^2} \Rightarrow h^2 = \lambda c^2$

i.e., $x^2 = \lambda c^2$

21. **(a), (c)** Let $A(t_1) = B(t_2)$ and $P(t_3)$ be the three points on the hyperbola $xy = c^2$

Now slope of AP is $m_1 = -\frac{1}{t_1t_3}$ and slope of BP is $m_2 = -\frac{1}{t_2t_3}$

Since $m_1m_2 = -1$ so $\frac{1}{t_1t_2t_3^2} = -1$ i.e., $t_3^2 = \frac{-1}{t_1t_2}$

We know that the slope of normal at $R(t)$ is $m = t^2$ and the slope of chord joining $M(\alpha)$, N (β) is $m = -\frac{1}{\alpha\beta}$

$\Rightarrow$ Slope of normal at $P(t_3)$ = slope of chord AB

Further ΔABP is right angled at P

$\therefore$ AB is a diameter of the circum-circle of ΔABP

So circum radius $R = \frac{AB}{2} = \frac{1}{2}\sqrt{c^2(t_2 - t_1)^2 + c^2\left\{\frac{1}{t_1} - \frac{1}{t_2}\right\}^2}$

$= \frac{|c|}{2}\left|\frac{t_1 - t_2}{t_1t_2}\right|\sqrt{t_1^2t_2^2 + 1}$

22. **(c)** The hyperbola $xy = c^2$ can be rewritten as $\frac{X^2}{2c^2} - \frac{Y^2}{2c^2} = 1$ by rotating x-axis and y-axis such that where x-axis coincide with line $y = x$ and y-axis with line $y = -x$

Now length of $L.R. = \frac{2b^2}{a} = \frac{2(2c^2)}{\sqrt{2}c} = 2\sqrt{2}c$

Aliter: $xy = c^2$ is rectangle hyperbola where distance of vertex from centre is $a = \sqrt{2}c = b$

$\Rightarrow \quad L.R. = \frac{2b^2}{a} = \frac{4c^2}{\sqrt{2}c} = 2\sqrt{2}c$

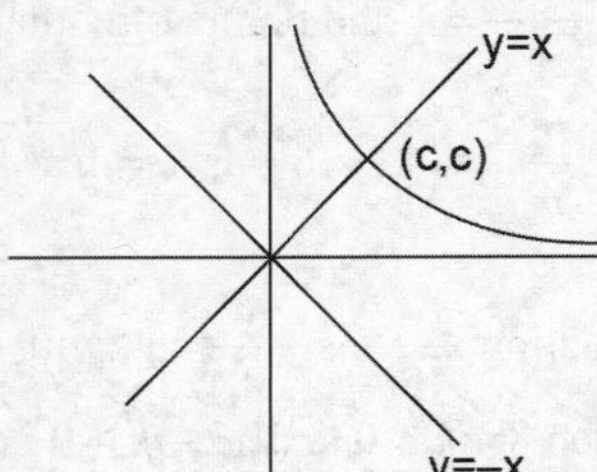

23. **(a)** Let the point be $P\left(ct, \frac{c}{t}\right)$ on the hyperbola $xy = c^2$

$\Rightarrow$ Equation of tangent at P will be $\left(y - \frac{c}{t}\right) = -\frac{1}{t^2}(x - ct)$.

Which interests x-axis, where $y = 0$

$\Rightarrow \quad -\frac{c}{t} - \frac{c}{t} = -\frac{x}{t^2}$

$\Rightarrow \quad x = 2ct$ and y-axis, where $x = 0$

$\Rightarrow \quad y = \frac{c}{t} + \frac{c}{t} = \frac{2c}{t}$

$\therefore$ required area $= \frac{1}{2}(2ct)\left(\frac{2c}{t}\right) = 2c^2$

24. **(a)** Since the hyperbola is rectangular.

$\therefore$ Asymptotes are 90° to each other and the point of intersection is given by L_1: $3x - 4y - 6 = 0$ and L_3: $x - y - 1 = 0$

$\Rightarrow$ (–2, –3)

$\therefore$ The other asymptotes is L_2 : $4x + 3y + 17 = 0$

TEXTUAL EXERCISE SECTION–III (OBJECTIVE)

1. **(c)** Equation of the hyperbola is $16x^2 + 64x - y^2 + 4y = -44$

i.e., $16(x^2 + 4x + 4) - (y^2 - 4y + 4) = 64 - 4 - 44 = 16$

$\therefore \quad \frac{(x+2)^2}{1} - \frac{(y-2)^2}{16} = 1$

$\Rightarrow$ Transverse axis is $x = -2$ and conjugate axis is $y = 2$

2. **(d)** Given L_1: $ax\sec\theta + by\tan\theta = $ a and L_2: $ax\tan\theta + by\sec\theta = b$

Squaring and subtracting, we get $a^2x^2(\sec^2\theta - \tan^2\theta) + b^2y^2(\tan^2\theta - \sec^2\theta) = a^2 - b^2$

$\frac{x^2}{\frac{a^2 - b^2}{a^2}} - \frac{y^2}{\frac{a^2 - b^2}{b^2}} = 1$ which is a hyperbola (for $a \neq b$)

3. **(c)** The lines are $L_1 : x\sqrt{3} - y = 4\sqrt{3}k$ and $L_2 : x\sqrt{3}k + ky = 4\sqrt{3}$ or $x\sqrt{3} + y = \frac{4\sqrt{3}}{k}$

Eliminating k by multiplication (for $k \neq 0$), we get $\left(x\sqrt{3}+y\right)\left(x\sqrt{3}-y\right)=48$ *or* $3x^2 - y^2 = 48$ or $\frac{x^2}{16}-\frac{y^2}{48}=1$, which is a hyperbola.

4. (d) Slope of line $2x+\sqrt{6}y=2$ is $m=-\sqrt{\frac{2}{3}}$. For the hyperbola $\frac{x^2}{4}-\frac{y^2}{2}=1$ the points of contact will be $\left(\frac{-a^2m}{c},\frac{b^2}{m}\right)\equiv\left(4,-\sqrt{6}\right)$ as $c=\sqrt{a^2m^2-b^2}=\sqrt{\frac{2}{3}}$

5. (c) The equation $\frac{x^2}{12-k}+\frac{y^2}{8}=1$ will represent a hyperbola when $8-k<0$ and $12-k>0$ i.e., $k>8$ and $k<12$

6. (b) Let parametric equation of C_1 be $x = t^2 + 1$ and $y = 2t$ i.e., $x-1\left(\frac{y}{2}\right)^2 \Rightarrow y^2 = 4(x-1)$

The other curve is C_2 as $x = 2s$ and $y = 2/s$, so $xy = 4$

Solving, we get $\frac{16}{x^2}=4x-4$ i.e., $4x^3 - 4x^2 - 16 = 0$ *or* $x^3 - x^2 - 4 = 0$

Observe that $x = 2$ satisfies, so $x^3 - x^2 - 4 = (x-2)\{x^2 + x + 2\} = 0$

$\Rightarrow$ $x = 2$ is the only real root

$\Rightarrow$ Only point of intersection is (2, 2)

7. (d) The Asymptotes are $x-3=0$ and $y-5=0$

$\therefore$ Equation of hyperbola $H = A + \lambda = 0$ i.e., $(x-3)(y-5) + \lambda = 0$

Since (7, 8) lies on it, $12 + \lambda = 0$ gives $\lambda = -12$ so $H = xy - 3y - 5x + 3 = 0$

8. (c) Let P (α sec, b tanθ) by any point on the hyperbola $\frac{x^2}{a^2}-\frac{y^2}{b^2}=1$

$\Rightarrow$ Equation of tangent at P: $\frac{x\sec\theta}{a^2}-\frac{y\tan\theta}{b}=1$ .... (i)

(i) intersects asymptotes $y = \pm\, b/a\; x$ where $\frac{x}{a}\sec\theta \mp x\frac{\tan\theta}{a}=1$

$\Rightarrow$ $x = a(\sec\theta \pm \tan\theta)$

$\Rightarrow$ Points of intersect area A (a (secθ + tanθ), b (secθ + tanθ)) and B (a(secθ – tanθ), –b (secθ – tanθ))

$\therefore$ Area of ΔOAB

$$=\frac{1}{2}\begin{vmatrix}0 & 0 & 1\\ a(\sec\theta+\tan\theta) & b(\sec\theta+\tan\theta) & 1\\ a(\sec\theta-\tan\theta), & -b(\sec\theta-\tan\theta) & 1\end{vmatrix}$$

$=\frac{1}{2}ab|-2| = ab =$ constant

9. (b) The hyperbola is $\frac{x^2}{2}-\frac{y^2}{1}=1$

So a general point P is $\left(\sqrt{2}\sec\theta,\tan\theta\right)$ and the equation of asymptotes are $y=\pm\frac{x}{\sqrt{2}}$.

Length of perpendicular from P on to $x - y\sqrt{2} = 0$ is $d_1=\frac{\left|\sqrt{2}\sec\theta-\sqrt{2}\tan\theta\right|}{\sqrt{3}}$, similarly

$$d_2=\frac{\left|\sqrt{2}\sec\theta+\sqrt{2}\tan\theta\right|}{\sqrt{3}}$$

$$\therefore\; d_1d_2=\frac{2\left(\sec^2\theta-\tan^2\theta\right)}{3}=\frac{2}{3}$$

10. (b) Let (h, k) be the mid point of a chord on hyperbola $xy = 1$, then its equation is $\frac{x}{h}+\frac{y}{k}=2$

Since it has a fixed slope 'm'

$\Rightarrow$ $-\frac{k}{h}=m$, so $k = -mh$ i.e., $y = -mx$

11. (b) Given $H_1:\frac{x^2}{a^2}-\frac{y^2}{b^2}=1$ and $H_2:\frac{x^2}{\left(-b^2\right)}-\frac{y^2}{\left(-a^2\right)}=1$

Equation of tangent with slope m for H_1 is $y=mx\pm\sqrt{a^2m^2-b^2}$ and that with slope m for H_2 is $y=mx\pm\sqrt{-b^2m^2+a^2}$

$\Rightarrow$ $a^2m^2 - b^2 = a^2 - b^2m^2$

$\Rightarrow$ $a^2m^2 - a^2 + b^2m^2 - b^2 = 0$ i.e., $(a^2 + b^2)(m^2 - 1) = 0$ gives $m = \pm 1$

Hence the tangents are $y=\pm x\pm\sqrt{a^2-b^2}$

12. (c) The hyperbola is $H:\frac{x^2}{a^2}-\frac{y^2}{b^2}-1=0$ ($b > a$) and the line $L:\left(\frac{x\cos\alpha+y\sin\alpha}{p}\right)=1$

Homogenizing, we get

$$\frac{x^2}{a^2}-\frac{y^2}{b^2}-\frac{x^2\cos^2\alpha+y^2\sin^2\alpha+2xy\sin\alpha\cos\alpha}{p^2}=0$$

Since angle subtended at the centre is 90° so $A + B = 0$

i.e., $\frac{1}{a^2}-\frac{\cos^2\alpha}{p^2}-\frac{1}{b^2}-\frac{\sin^2\alpha}{p^2}=0$

$\therefore$ $\frac{1}{a^2}-\frac{1}{b^2}=\frac{1}{p^2}\Rightarrow p^2=\frac{a^2b^2}{b^2-a^2}$

$\Rightarrow$ Distance of origin from line L is $|p|=\frac{ab}{\sqrt{b^2-a^2}}$, which is the radius of required circle.

13. (b) Equation of tangent at P (t_1) on the hyperbola $xy = c^2$ is $x+t_1^2y=2ct_1$.

Similarly tangent at Q (t_2) is $x+t_2^2y=2ct_2$ gives the point of intersection at R (h, k) as $h=\frac{2ct_1t_2}{t_1+t_2}$ and $k=\frac{2c}{t_1+t_2}$

$\Rightarrow$ $R\equiv\left(\frac{2ct_1.t_2}{t_1+t_2},\frac{2c}{t_1+t_2}\right)$

14. (a) The given hyperbola is H: $xy - 4x - 3y = 0$

$\Rightarrow$ $A = H + \lambda = 0$

i.e., $xy - 4x - 3y + \lambda = 0$ where $\Delta = 0$ (as pair of straight line is represented)

$$\Rightarrow \begin{vmatrix} 0 & \frac{1}{2} & -2 \\ \frac{1}{2} & 0 & -\frac{3}{2} \\ -2 & -\frac{3}{2} & \lambda \end{vmatrix} = 0 \Rightarrow \left(-\frac{1}{2}\right)\left\{\frac{\lambda}{2} - 3\right\} - 2\left(\frac{-3}{4}\right) = 0$$

$\Rightarrow$ $\lambda = 12$ and we get $(3 - x)(4 - y) = 0$ i.e., $y = 4$ and $x = 3$ are the asymptotes.

15. (b) Observe that the ellipse is a vertical type so it foci will be on y-axis

Now a hyperbola $\frac{x^2}{a^2} - \frac{y^2}{b^2} = 1$ never intersects y-axis

$\therefore$ The hyperbola must be of the type $\frac{y^2}{y^2} - \frac{x^2}{b^2} = 1$

Since it passes through the foci $(0, \pm 3)$, $\frac{9}{a^2} = 1 \Rightarrow a^2 = 9$

Further eccentricity of ellipse is $e_1 = \frac{3}{5}$ and that of hyperbola is $e_2 = \sqrt{1 + \frac{b^2}{9}}$.

It is given that $e_1 . e_2 = 1$

$\Rightarrow \frac{3}{5} . \frac{\sqrt{9 + b^2}}{3} = 1 \quad \Rightarrow \quad b^2 = 16$

$\Rightarrow$ equation of hyperbola is $\frac{y^2}{9} - \frac{x^2}{16} = 1$ *or* $\frac{x^2}{16} - \frac{y^2}{9} = 1$

16. (b) Let P $(a \sec \theta, b \tan \theta)$ be any point on hyperbola, then $N \equiv (a \sec\theta, 0)$. Equation of tangent at P is $\frac{x \sec\theta}{a} - \frac{y \tan\theta}{b} = 1$

$\Rightarrow \quad T \equiv (a \cos\theta, 0) \Rightarrow OT.ON = (a \cos\theta).(a \sec\theta) = a^2$

17. (a) Let P (h, k) be any point on any one member of hyperbola family, having equation $\frac{x^2}{a^2} - \frac{y^2}{b^2} = 1$; b is arbitrary constant then, its asymptotes are given by $y = \pm \frac{b}{a} x$,

then A.T.Q, $|k| = \frac{\left|\pm \frac{b}{a} h - k\right|}{\sqrt{\frac{b^2}{a^2} + 1}}$

$$\Rightarrow \quad k^2 = \left(\pm \frac{b}{a} h - k\right)^2 \frac{a^2}{(a^2 + b^2)}$$

$$\Rightarrow \quad k^2 = \frac{(\pm bh - ak)^2}{(a^2 + b^2)} \qquad \text{.....(i)}$$

Further (h, k) lies on hyperbola, $\frac{h^2}{a^2} - \frac{k^2}{b^2} = 1 \Rightarrow \frac{k^2}{b^2} = \frac{h^2 - a^2}{a^2}$

$$\Rightarrow \quad b = \sqrt{\frac{a^2k^2}{h^2 - k^2}} \qquad \text{.....(ii)}$$

$\therefore$ From (i) and (ii), we get $k^2 = \frac{[b^2h^2 + a^2k^2 \mp 2abhk]}{a^2 + \left(\frac{a^2k^2}{h^2 - a^2}\right)}$

$$\Rightarrow \quad k^2a^2\left(\frac{h^2 + k^2 - a^2}{h^2 - a^2}\right) = \left(b^2h^2 + a^2k^2 \mp 2abhk\right)$$

$$\Rightarrow \quad \frac{k^2a^2\left(h^2 + k^2 - a^2\right)}{h^2 - a^2} - a^2k^2 - b^2h^2 = \pm 2a\sqrt{\frac{a^2k^2}{h^2 - a^2}}.hk$$

$$\Rightarrow \quad \frac{k^2a^2\left(h^2 + k^2 - a^2\right)}{h^2 - a^2} - a^2k^2 - \left(\frac{a^2k^2}{h^2 - a^2}\right).h^2 = \pm 2a\sqrt{\frac{a^2k^2}{h^2 - a^2}}.hk$$

$$\Rightarrow \quad \frac{k^2a^2\left(k^2 - a^2\right) - a^2k^2\left(h^2 - a^2\right)}{h^2 - a^2} = \pm \frac{2a^2k^2h}{\sqrt{h^2 - a^2}}$$

$$\Rightarrow \quad \frac{a^2k^2\left(k^2 - h^2\right)}{\left(h^2 - a^2\right)} = \frac{\pm 2a^2k^2h}{\sqrt{h^2 - a^2}}$$

$\Rightarrow$ $(k^2 - h^2) = \pm 2h\sqrt{h^2 - a^2}$

$\Rightarrow$ $(k^2 - h^2) = 4h^2(h^2 - a^2)$

$\Rightarrow$ Required locus is $(x^2 - y^2)^2 = 4x^2(x^2 - a^2)$

18. (a) Let $P(t_1)$; $Q(t_2)$; $R(t_3)$ be the vertices ΔPQR inscribed in the rectangular hyperbola whose each side touches the parabola $y^2 = 4ax$.

Now equation of chord joining $P(t_1)$ and $Q(t_2)$ is

$$y = -\frac{1}{t_1t_2}x + c\frac{(t_1 + t_2)}{t_1t_2} \qquad \text{.....(i)}$$

Similarly that of QR: $y = -\frac{1}{t_2t_3}x + c\frac{(t_2 + t_3)}{t_2t_3}$ .....(ii)

And of PR: $y = -\frac{1}{t_1t_3}x + c\frac{(t_1 + t_3)}{t_1t_3}$ ....(iii)

Now (i), (ii) and (iii) are tangents to parabola $y^2 = 4ax$

$\Rightarrow$ They are identical to $y = mx + a/m$

$$\Rightarrow \quad \left(\frac{c(t_1 + t_2)}{t_1 + t_2}\right) = \frac{a}{\left(-\frac{1}{t_1 + t_2}\right)} \Rightarrow \frac{c(t_1 + t_2)}{t_1t_2} = -at_1t_2$$

$$\Rightarrow \quad c(t_1 + t_2) = -at_1^2t_2^2 \qquad \text{....(iv)}$$

$$c(t_2 + t_3) = -a_2^2t_3^2 \qquad \text{.....(v)}$$

$$\text{And } c(t_3 + t_1) = -at_3^2t_1^2 \qquad \text{....(vi)}$$

$$\Rightarrow \quad \frac{t_1 + t_2}{t_1^2 + t_2^2} = \frac{t_2 + t_3}{t_2^2.t_3^2} = \frac{t_3 + t_1}{t_3^2t_1^2} = -\frac{a}{c}$$

$\Rightarrow$ From first two members, we get $(t_3 - t_1)(t_1t_2 + t_2t_3 + t_3t_1) = 0$

$\Rightarrow$ $t_1t_2 + t_2t_3 + t_3t_1 = 0 \quad (\quad t_3 \neq t_1)$

$\Rightarrow$ $t_3 = -\frac{t_1t_2}{t_1 + t}$

$\Rightarrow$ For every two points $P(t_1)$ and $Q(t_2)$ for which $(t_1 + t_2 + 0)$ on hyperbola, we can get a point $R(t_3)$ given by $t_3 = -\frac{t_1 + t_2}{t_1 + t_2}$. Thus three will be infinitely many such triangles

19. (c) Let $P(x_1, y_1)$ be the points where the ellipse $E: \frac{x^2}{a^2} + \frac{y^2}{b^2} = 1$ and H: $x^2 - y^2 = c^2$ intersect

$\Rightarrow$ Slope of tangent at $P(x_1, y_1)$ to the ellipse (E) at is $m_1 = \frac{-b^2x_1}{a^2y_1}$ and slope of tangent at P to the hyperbola H is $m_2 = \frac{x_1}{y_1}$. Since $m_1m_2 = -1$

$\Rightarrow \frac{x_1^2}{y_1^2} = \frac{a^2}{b^2}$

Now for ellipse $P(x_1, y_1) = (a\cos\theta, b\sin\theta)$

$\Rightarrow \frac{a^2\cos^2\theta}{b^2\sin^2\theta} = \frac{a^2}{b^2}$ gives $\cot^2\theta = 1$ i.e., $\theta = \pm\frac{\pi}{4}, \pm\frac{3\pi}{4}$

$\Rightarrow P\left(\pm\frac{a}{\sqrt{2}}, \pm\frac{b}{\sqrt{2}}\right)$

Since P also lie on the hyperbola H so $\frac{a^2}{2} - \frac{b^2}{2} = c^2$ i.e., $a^2 - b^2 = 2c^2$

20. (c) Let the points be $P\left(ct_1, \frac{c}{t_1}\right); Q\left(ct_2, \frac{c}{t_2}\right); R\left(ct_3, \frac{c}{t_3}\right)$ and $S\left(ct_4, \frac{c}{t_4}\right)$ lie on a circle

Altitude though P: $\left(y - \frac{c}{t_1}\right) = t_2t_3(x - ct_1)$ ...(ii)

And Altitude through Q: $\left(y - \frac{c}{t_2}\right) = t_1t_3(x - ct_2)$ ...(iii)

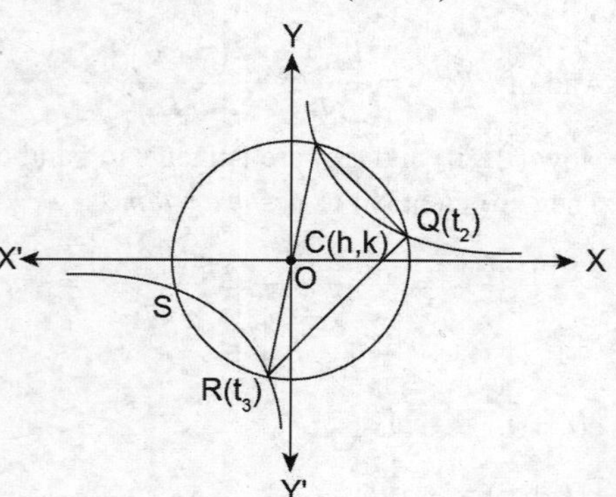

$\therefore$ Orthocenter of ΔPQR is the point of intersection (ii) and (iii)

$\Rightarrow \frac{c}{t_1} + t_2t_3(x - ct_1) = \frac{c}{t^2} + t_1t_3(x - ct_2)$

$\Rightarrow (t_1t_3 - t_2t_3)x = \frac{c}{t_1} - \frac{c}{t_2} + ct_1t_2t_3 - ct_1t_2t_3$

$\Rightarrow t_3(t_1 - t_2)x = -\frac{c(t_1 - t_2)}{t_1t_2} \Rightarrow x = \frac{-c}{t_1t_2t_3}$

$\Rightarrow y = \frac{c}{t_1} + xt_2t_3 - ct_1t_2t_3 = -ct_1t_2t_3$

$\therefore$ Orthocenter $\equiv \left(\frac{-c}{t_1t_2t_3}, -ct_1t_2t_3\right)$

We know that if $P(t_1)$, $Q(t_2)$, $R(t_3)$ and $S(t_4)$ one four concyclic points lying on rectangular Hyperbola $xy = c^2$, then $t_1t_2t_3t_4 = 1$

$\Rightarrow t_4 = \frac{1}{t_1t_2t_3} \quad \Rightarrow S(x_4, y_4) \equiv \left(\frac{c}{t_1t_2t_3}, ct_1t_2t_3\right)$

$\Rightarrow$ Orthocenter $\equiv (-x_4, -y_4)$

21. (b) We know that the eccentricity of a rectangle hyperbola is $\sqrt{2}$. Further $xy = c^2$ and $x^2 - y^2 = c^2$ are rectangular hyperbolas

$\Rightarrow e^2 + e_1^2 = 2 + 2 = 4$

22. (a) We know that $2A = H + C$ i.e., H, A, C are in AP, so $H(\alpha, \beta)$, $A(\alpha, \beta)$ and $C(\alpha, \beta)$ are in $A.P.$

23. (a) For the hyperbola $\frac{x^2}{9} - \frac{y^2}{4} = 1$ the asymptotes are $y = \pm\frac{2}{3}x$

The slope of the line perpendicular $5x + 2y - 3 = 0$ is $m = 2/5$.

Since $2/5 < 2/3$, so no tangent with such a slope is possible.

24. (a) For the hyperbola $\frac{x^2}{a^2} - \frac{y^2}{b^2} = 1$ equation of a tangent at $P(\theta)$ is $\frac{x}{a}\sec\theta - \frac{y}{b}\tan\theta = 1$

Which gives intersection with axes at $A(a\cos\theta, 0)$, $B(0, -b\cot\theta)$

Hence the area $\Delta OAB = \left|\frac{1}{2} \times a\cos\theta \times \frac{b\cos\theta}{\sin\theta}\right|$; when $\theta = \pi/6$

$\Rightarrow$ Area of $\Delta OAB = \frac{3}{4}ab$ square units

Since $\frac{3}{4}ab = 3a^2 \quad \Rightarrow b = 4a$

$\therefore e = \sqrt{\frac{a^2 + b^2}{a^2}} = \sqrt{17}$

25. (c) The tangent to the hyperbola of slope of m for $H_1 = \frac{x^2}{a^2} - \frac{y^2}{b^2} - 1 = 0$ (for $a > b$) is $y = mx \pm \sqrt{a^2m^2 - b^2}$ ....(i)

Also that of slope m for $H_2 = \frac{y^2}{a^2} - \frac{x^2}{b^2} = 1$ is $y = mx \pm \sqrt{a^2 - b^2m^2}$ .....(i)

Since (i) and (ii) are identical

$\therefore a^2m^2 - b^2 = a^2 - b^2m^2$ gives $(a^2 + b^2)(m^2 - 1) = 0 \Rightarrow m = \pm 1$

$\therefore$ Four tangents are possible given by $y = \pm x \pm \sqrt{a^2 - b^2}$

26. (d) The hyperbola is H: $xy - hx - ky = 0$

The asymptotes are $A = H + \lambda = 0$ where $\Delta = 0$, we get $\lambda = kh$ and the asymptotes as $y = h$, $x = k$

Now $C + H = 2A$ gives $C = 2A - H = 2A - (A - \lambda) \Rightarrow C = A + \lambda$ i.e. $xy - hx - ky + 2hk = 0$

Or $hx + ky - xy - 2hk = 0$

27. **(d)** The equation of normal at P (t) on the hyperbola $xy = 1$ is $xt^3 - yt = t^4 - 1$ gives $x = h = \frac{t^4 - 1}{t^3}$ and $y = k = \frac{t^4 - 1}{(-t)}$

Now $C = (h, k)$

$$\Rightarrow \frac{k}{h} = -t^2 \text{ and } kh = \frac{-(t^4 - 1)^2}{t^4}$$

$$\text{Hence } -kh = \frac{(k^2 - h^2)^2}{h^4\left\{\frac{k^2}{h^2}\right\}}$$

$\therefore$ $(x^2 - y^2)^2 + x^3y^3 = 0$ is the required locus

28. **(b)** Let P (t_1) and Q (t_2) be the two end points of a chord on $xy = c^2$.

Let R (h, k) be its midpoint, then $\frac{x}{h} + \frac{y}{k} = 2$ which will intersect x-axis at $x = 2h$

$\Rightarrow$ A $(2h, 0)$

$$\Rightarrow \text{Slope of } RA = \frac{K - 0}{h - 2h} = -\frac{k}{h} = m_1 \text{ (say)}$$

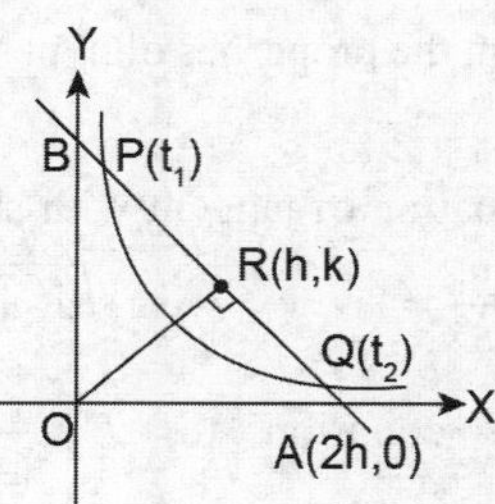

Now slope of OR is $m_2 = \frac{k}{h}$. Observe that $m_1m_2 = -1$

$\Rightarrow$ $\angle ORA = 90°$

Hence ΔOAR is right angled at R. Also $OR = \sqrt{h^2 + k^2} = AR$

$\therefore$ Δ is right angled

29. **(c)** Since the normal drawn at P (t_1) meets the hyperbola $xy = c^2$ again at Q (t_2),

$\Rightarrow$ $t_1t_2 = -$

$$\text{Now distance } PQ = \sqrt{(ct_1 - ct_2)^2 + \left(\frac{c}{t_1} - \frac{c}{t_2}\right)^2}$$

$$= c\sqrt{(t_1 - t_2)^2 + \left(\frac{t_1 - t_2}{t_1^2 t_2^2}\right)^2} = \frac{c|t_1 - t_2|}{t_1t_2}\sqrt{t_1^2t_2^2 + 1}$$

$$= \frac{c\left|t_1 + \frac{1}{t_1^3}\right|}{\left|-t_1 . \frac{1}{t_1^3}\right|}\sqrt{t_1^2 . \frac{1}{t_1^6} + 1} = \frac{c|t_1^4 + 1|}{|t_1^3| . \left(\frac{1}{t_1^2}\right)}\frac{\sqrt{1 + t_1^4}}{(t_1^2)}$$

$$= \frac{c|1 + t_1^4|^{3/2}}{|t_1|^3} = \frac{c|1 + t_1^4|^{3/2}}{|t_1^2|^{3/2}} = c\left|t_1^2 + \frac{1}{t_1^2}\right|^{3/2} = \geq c(2)^{3/2}$$

$= 2\sqrt{2}c$

30. **(c)** Equation of tangent with slope m to the hyperbola $\frac{x^2}{a^2} - \frac{y^2}{a^2} = 1$ is $y = mx \pm a\sqrt{m^2 - 1}$ so $mx - y = \pm a\sqrt{m^2 - 1}$

Now the normal with slope m to the parabola $y^2 = 4ax$ is $am^3 + 2am - mx + y = 0$

$$\Rightarrow a(m^3 + 2m) = mx - y = \pm a\sqrt{m^2 - 1}$$

On squaring, we get $m^2\{m^4 + 4 + 4m^2\} = m^2 - 1$ i.e., $m^6 + 4m^4 + 3m^2 + 1 = 0$ is the condition

31. **(b)** As branches lie in the second and fourth quadrant

$\therefore$ We have $xy < 0 \Rightarrow m^2 - 9 < 0 \Rightarrow |m| < 3$

32. **(b)** $S = 3x^2 - 2y^2 - 6 = 0$, joint equation of tangents drawn from point $P(\alpha, \beta)$ is given by $SS_1 = T^2$

$\Rightarrow$ Pair of tangents are given by $(3x^2 - 2y^2 - 6)$ $(3\alpha^2 - 2\beta^2 - 6) = (3\alpha x - 2\beta y - 6)^2$

$$\Rightarrow \text{Product of slopes of the tangents} = \frac{\text{Coefficient of } x^2}{\text{Coefficient of } y^2}$$

$$\text{i.e., } \tan\theta \tan\phi = \frac{3(3\alpha^2 - 2\beta^2 - 6) - 9\alpha^2}{-2(3\alpha^2 - 2\beta^2 - 6) - 4\beta^2}$$

$$\Rightarrow 2 = \frac{-6\beta^2 - 18}{-6\alpha^2 + 12} \Rightarrow \beta^2 = 2\alpha^2 - 7$$

33. **(a)** Substituting $x^2 = \frac{1}{b}$ y, the equation of the conic gives $a^2y^2 - by + a^2b^2 = 0$. This has real roots iff $b^2 - 4a^4b^2 \geq 0$

i.e., $a^4 \leq - \Rightarrow -\frac{1}{\sqrt{2}} \leq a \leq \frac{1}{\sqrt{2}}$

34. **(d)** If the rectangular hyperbola is $xy = c^2$, then P is $\left(ct, \frac{c}{t}\right)$

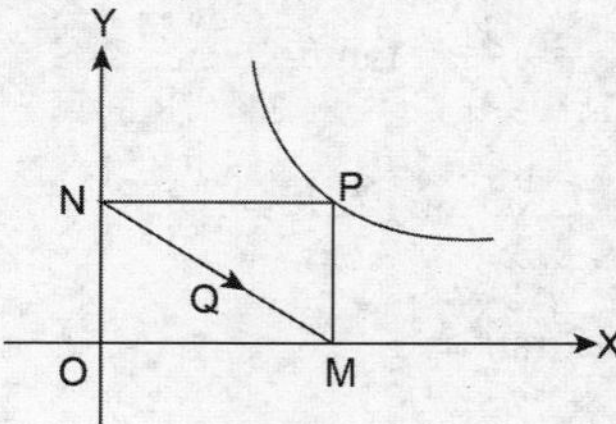

$$\Rightarrow M = (ct, 0); N = \left(0, \frac{c}{t}\right) \Rightarrow Q = \left(\frac{ct}{4}, \frac{3c}{4t}\right) = (x, y)$$

$\therefore$ $xy = \frac{3c^2}{16}$ which is another rectangular hyperbola.

35. **(b)** Locus is mid-point of line segment joining P and centre C of circle

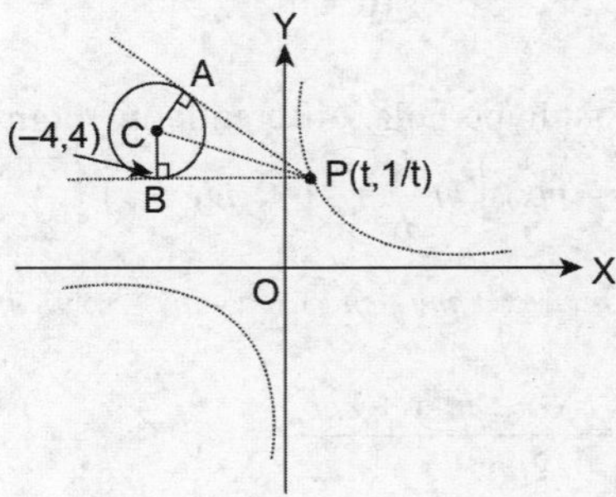

$\Rightarrow \quad x = \dfrac{-4+t}{2}; y = \dfrac{4+\frac{1}{t}}{2}$

$\Rightarrow$ $(2x + 4)(2y - 4) = 1$ is the required locus.

36. (a) Let $xy = c^2$ be the rectangular hyperbola referred to its asymptotes as the coordinate axis.

Let $P(h, k)$ be a point on $xy = c^2$

$\Rightarrow hk = c^2$ ... (i)

Tangents are drawn from $P(h, k)$ to the rectangular hyperbola $xy = a^2$

$\therefore$ Equation of chord of contact

$\Rightarrow kx + hy = 2a^2$

This cuts the coordinate axes at $A\left(\dfrac{2a^2}{k}, 0\right)$ and $B\left(0, \dfrac{2a^2}{h}\right)$

$\therefore$ Area of $\Delta OAB = \dfrac{1}{2}\left(\dfrac{2a^2}{k} \times \dfrac{2a^2}{h}\right) = \dfrac{2a^4}{kh} = \dfrac{2a^2}{c^2}$ (using (i))

37. (c) Let A is (α, β), then B is (β, α)

$\therefore$ A and B are symmetric about $y = x$,

So tangents at A and B will be mirror images of each other, about $y = x$. Thus point of intersection will lie on $y = x$.

38. (a) Let (x_1, y_1) be point of intersection

$\Rightarrow y_1^2 = 4ax_1$ and $x_1y_1 = c^2$

Now, slope of tangent to parabola at (x_1, y_1)

$= m_{T(x_1, y_1)} = \text{——} = \tan\phi$ and that of tangent to hyperbola at (x_1, y_1)

$= M_{T(x_1, y_1)} = -\dfrac{y_1}{x_1} = \tan\theta$

$\Rightarrow \dfrac{\tan\theta}{\tan\varphi} = \dfrac{-y_1^2}{2ax_1} = -\dfrac{4ax_1}{2ax_1} = -2$

$\Rightarrow \tan\theta = -2\tan\phi$

39. (d) Major axis of hyperbola bisects asymptote

$\therefore$ Equation of other asymptote is $y = \dfrac{x}{2}$

Equation of hyperbola is $(y - 2x)(x - 2y) + k = 0$ put (3, 4)

$\Rightarrow (4 - 6)(3 - 8) + k = 0$

$\Rightarrow k = -10$

$\Rightarrow (y - 2x)(x - 2y) - 10 = 0$

$\Rightarrow 2x^2 + 2y^2 - 5xy + 10 = 0$

40. (a) For first hyperbola, differentiating both sides w.r.t. x, we get $(y - mx)\left(m\dfrac{dy}{dx} + 1\right) + (my + x)\left(\dfrac{dy}{dx} - m\right) = 0$

$\Rightarrow \dfrac{dy}{dx}(my + x + my - m^2x) + y - mx - m^2y - mx = 0$

$\Rightarrow \dfrac{dy}{dx} = \dfrac{-y + m^2y + 2mx}{2my + x - m^2x} = m_1$

For second hyperbola, differentiating both sides w.r.t. x, we get $(m^2 - 1)\left(2y\dfrac{dy}{dx} - 2x\right) + 4m\left(x\dfrac{dy}{dx} + y\right) = 0$

$\Rightarrow \dfrac{dy}{dx}(2y(m^2 - 1) + 4mx) = -4my + 2x(m^2 - 1)$

$\Rightarrow \dfrac{dy}{dx} = \dfrac{-2my + m^2x - x}{m^2y - y + 2mx} = m_2. \quad m_1m_2 = -1$

$\Rightarrow$ Angle between the hyperbolas $= \dfrac{\pi}{2}$.

SECTION IV (MORE THAN ONE CORRECT)

1. (a), (b) The hyperbola is $\dfrac{x^2}{3} - \dfrac{y^2}{2} = 1$.

Equation of tangents with slope $m = 1$ are $y = x \pm \sqrt{3 - 2}$ gives $y = x \pm 1$ and similarly the tangents with slope $m = -1$ are

$y = -x \pm 1$

2. (a), (b), (c) From the properties of hyperbola options (a, b, c) are true

3. (a), (d) The equation of tangent with slope m to the hyperbola $H_1 : \dfrac{x^2}{a^2} - \dfrac{y^2}{b^2} = 1$ is $y = mx \pm \sqrt{a^2m^2 - b^2}$ .....(i)

Similarly, the equation to $H_2 : \dfrac{x^2}{(-b^2)} - \dfrac{y^2}{(-a^2)} = 1$ is $y = mx \pm \sqrt{a^2 - b^2m^2}$ ....(ii)

Since (i) and (ii) are identical so $a^2m^2 - b^2 = a^2 - b^2m^2$ gives $(a^2 + b^2)(m^2 - 1) = 0 \Rightarrow m = \pm 1$

4. (a), (d) The foci are A (5, 12) and B (24, 7) so $2ae = \sqrt{386}$.

Now the conic passes through O (0, 0)

If the conic is an ellipse, then $2a = 13 + 25 = 38$

$\Rightarrow$ Eccentricity $e_1 = \dfrac{\sqrt{386}}{38}$ and $2f$ the conic is a hyperbola, then $2a = 25 - 13 = 12$, so $e_2 = \dfrac{\sqrt{386}}{12}$

5. (a), (b), (c), (d) The given hyperbola can be rewritten as $9(x^2 - 2x + 1) - 16(y^2 - 2y + 1) = 151 + 9 - 16 = 144$

$\Rightarrow \dfrac{(x-1)^2}{16} - \dfrac{(y-1)^2}{9} = 1$

Centre at (1, 1) and eccentricity $e = 5/4$, directrix is $X = x - 1 = \dfrac{a}{e} = \dfrac{16}{5}$ gives $x = \dfrac{21}{5}$

Foci are (6, 1) and (–4, 1). Length of L.R. $= \dfrac{2b^2}{a} = \dfrac{2 \times 9}{4} = \dfrac{9}{2}$

6. (a), (b) $x^2 - 9y^2 = 9$ and $5x + 12y = 9$

$\Rightarrow \dfrac{x^2}{9} - \dfrac{y^2}{1} = 1$ and $y = -\dfrac{5}{12}x + \dfrac{3}{4}$

$\Rightarrow a^2 = 9, b^2 = 1, m = -\dfrac{5}{12}, c = \dfrac{3}{4}$

We know that point of contact are given by $\left(\frac{-a^2m}{C}, \frac{-b^2}{C}\right)$

$\equiv \left(\frac{(-9)\left(-\frac{5}{12}\right)}{\frac{3}{4}}, \frac{-(1)}{\frac{3}{4}}\right) \equiv \left(5, -\frac{4}{3}\right)$; which clearly lie on line.

7. (c), (d) The given hyperbola can be rewritten as 16 $(x^2 - 2x + 1) - 3(y^2 - 4y + 4) = 44 + 16 - 12 = 48$, so

$\frac{(x-1)^2}{3} - \frac{(y-2)^2}{16} = 1$

Which is a hyperbola with centre at (1, –2) and eccentricity $e = \sqrt{\frac{19}{3}}$

The length of conjugate axes = $2b$ = 8 units and the length of transverse axis is $2a = 2\sqrt{3}$ Units

8 (a), (b), (c), (d) (i) If $k_1 = 0$, $k_2.k_3 > 0$, $PB = \frac{k_3}{k_2} > 0$ and P describes a circle with B as centre and radius $\frac{k_3}{k_2}$.

(ii) If $k_3 = 0$, $k_1.k_2 < 0$ $\frac{PA}{PB} = \frac{k_2}{-k_1} = k > 0$ and P describes a circle with P_1P_2 as its diameter, P_1 and P_2 being the points which divides AB internally and externally in the ratio k : 1.

(iii) If $k_1 = k_2 > 0$ and $k_3 > 0$, $PA + PB = \frac{k_3}{k_1} = k > 0$ and $k > |AB|$, then P describes an ellipse with A and B as its foci and the length of the major axis = k.

(iv) Iff $k_1 = -k_2 > 0$ and $k_3 > 0$, P describes a hyperbola with A and B as foci and length of the transverse axis $= \frac{k_3}{k_1}$,

Provided $k_3/k_1 < |AB| \Rightarrow k_3 < |AB|$ for $k_1 = 1$

9 (c), (d) (α, α^2) lie on the parabola $y = x^2$

(α, α^2) must lie between the asymptotes of hyperbola $\frac{x^2}{1} - \frac{y^2}{4} = 1$ in 1st and 2nd quadrant.

$\therefore$ asymptotes are $y = \pm 2x$

$\Rightarrow 2\alpha < \alpha^2$

$\Rightarrow \alpha < 0$ *or* $\alpha > 2$ and $-2\alpha < \alpha^2$; $\alpha < -2$ *or* $\alpha > 0$

$\therefore \alpha \in (-\infty, -2)$ *or* $\alpha \in (2, \infty)$

10 (a), (b), (c) Given ellipse is $\frac{x^2}{k^2a^2} + \frac{y^2}{k^2} = 1$; $k^2a^2 > a^2$

$\Rightarrow$ Focus of ellipse ≡ $(kae_1, 0)$ and focus of hyperbola ≡ $(ae_2, 0)$

Now $kae_1 = a\,e_2 \Rightarrow k_a\sqrt{1 - \frac{a^2}{k^2a^2}} = a\sqrt{2}$

$\Rightarrow k^2 - 1 = 2$ *or* $k = \pm\sqrt{3}$

(a) Thus $\frac{e_1}{e_2} = \frac{1}{k} = \frac{1}{\sqrt{3}}$

(b) Major axis of ellipse = $2ka$ and transverse axis of hyperbola = $2a$

$\therefore \frac{2ka}{2a} = k = \sqrt{3}$

(c) At point of intersection (x, y) slope of tangent to ellipse and hyperbola are given by

$m_1 = -\frac{x_1}{k^2y_1}$ and $m_2 = \frac{x_1}{y_1}$; where $y_1^2 = \left(\frac{k^2-1}{k^2+1}\right)a^2 = \frac{a^2}{2}$ ($k^2 = 3$) and $x_1^2 = a^2 + y_1^2 = a^2 + \frac{a^2}{2} = \frac{3a^2}{2}$

$\therefore m_1m_2 = -\frac{x_1^2}{k^2y_1^2} = -1$

Hence the curves are orthogonal

(d) Latus rectum of ellipse and hyperbola are $\frac{2a^2}{ak}, \frac{2a^2}{a}$ respectively

$\therefore \frac{2a/k}{2a} = \frac{1}{k} = \frac{1}{\sqrt{3}}$

SECTION V (ASSERTION REASON)

1. R: $e > 1$ is true

A: The statement is true for any point $P(h, k)$ the difference of focal distances = $2a$. The distance between foci = $2ae > 2a$ (as $e > 1$) and it is fully supported by R

2. R: The statement is true from the definition director circle

A: The statement is false for $H: \frac{x^2}{a^2} - \frac{y^2}{b^2} = 1$ the director circles is $x^2 + y^2 = a^2 - b^2$ so for $H: \frac{x^2}{3} - \frac{y^2}{4} = 1$ the director circle is an imaginarily circle $(x2 + y = -1)$

3. R: The statement is true if $ax^2 + by^2 + 2hxy + 2gx + 2fy + c = 0$ is the equation representing a hyperbola, then $h^2 - ab > 0$ and $\Delta \neq 0$

A: The given hyperbola can be rewritten as $9x^2 - 7y^2 + 0\,xy - 54x + 28y + 116 = 0$, so $h^2 = 0$ and $ab = -63 < 0$ i.e., $h^2 > ab$ also

$$\Delta = \begin{vmatrix} 9 & 0 & -27 \\ 0 & -7 & 14 \\ -27 & 14 & 116 \end{vmatrix} \neq 0$$

$\therefore$ A is true and it is fully supported by R

4. Since $e + e_1$ are the eccentricities of two conjugate hyperboles, so $e > 1$ and $e_1 > 1$.

$\Rightarrow ee_1 > 1$

$\therefore$ reason is true

As for e and e_1 for hyperbola and its conjugate, $\frac{1}{e^2} + \frac{1}{e_1^2} = 1$.

Let $e = \frac{5}{3}$ and $e_1 = \frac{5}{4}$

Now $\frac{1}{e^2}+\frac{1}{e_1^2}=\frac{9}{25}+\frac{16}{25}=1$

$\Rightarrow$ Assertion is true, but the statement is not supported is any way by *R*.

5. R: The statement is true for $H:\frac{x^2}{a^2}-\frac{y^2}{b^2}=1$ as the director circle is $x^2 + y^2 = a^2 - b^2$ so for $\frac{x^2}{25}-\frac{y^2}{9}=1$, the director circle is $x^2 + y^2 = 16$

A: The statement is true Hyperbola $\frac{x^2}{a^2}-\frac{y^2}{b^2}+1=0$ can be rewritten as $\frac{x^2}{(-a^2)}-\frac{y^2}{(-b^2)}=1$, so the director circle will be $x^2 + y^2 = (-a^2) - (-b^2) = b^2 - a^2$, which will be defined only when $b \geq a$ (i.e. $b^2 - a^2 \geq 0$). Hence statement is true but it cannot defined from R.

6. R: The statement is true for $\frac{x^2}{a^2}-\frac{y^2}{b^2}=1$, if e = √2, then

$2=1+\frac{b^2}{a^2}$ $\Rightarrow$ $a^2 = b^2$

$\Rightarrow$ hyperbola is equilateral

A: Since the angle between asymptotes is $2\theta = 2\tan^{-1}\left(\frac{b}{a}\right)$ so when a = *b* then $2\theta = \pi/2$

$\Rightarrow$ $\theta=\frac{\pi}{4}$ $\Rightarrow$ $\tan^{-1}\left(\frac{b}{a}\right)=\frac{\pi}{4}$

$\Rightarrow$ $\frac{b}{a}=1$ $\Rightarrow$ b = a

$\therefore$ A is true but it is not supported by (or derivable) *R* is determining the constant (a^2).

7. R: The statement is true for *H*: $\frac{x^2}{a^2}-\frac{y^2}{b^2}=1$. The point $P(x_1,y_1)$ will lie inside if $H(x_1, y_1) < 0$

A: Statement is false. The hyperbola is $\frac{x^2}{\left(\frac{1}{9}\right)}-\frac{y^2}{1}=1$ and *P* (5, –4) gives 225 –16 –1 = 208 > 0

The point lies outside the conic

8. R: The statement is true by definition if $H(x, y) = 0$, then $A_1(x, y) = H(x, y) + \lambda_1$

A: The statement is true, since $C(x, y) = H(x, y) + 2 = A_1(x, y) + (2 - \lambda_1)$

Since for $A_1(x, y)$, we have $\Delta = 0$ and also for $A_2(x, y)$, $\Delta = 0$ So $A_1 = A_2$. So A is derivable form *R*.

9. A : Assertion is true

R: is also true but Reason is not correct explanation.

$e_1^2 = \frac{9-4}{9} = 5/9$ and $e_2^2 = \frac{4 \;\; 9}{} = \frac{13}{4}$

$\Rightarrow$ $e_1^2 + e_2^2 = \frac{137}{36} \in \left(\frac{108}{36},\frac{144}{36}\right) = (3, 4)$

$\Rightarrow$ $[e_1^2 + e_2^2] = 3$

10. A: Assertion is true

R: Reason is true and reason is the correct explanation of Assertion.

Substituting $x^2 = \frac{1}{b}y$ in second curve, we get $a^2y^2 - by + a^2b^2 = 0$. This has real roots if $b^2 - 4a^4b^2 \geq 0$.

$\Rightarrow$ $1 - 4a^4 \geq 0$ $\Rightarrow$ $a^4 \leq \frac{1}{4}$

$\Rightarrow$ $a \in \left[-\frac{1}{\sqrt{2}},\frac{1}{\sqrt{2}}\right]-\{0\}$

SECTION VI (COMPREHENSION)

Comprehension A

1. (c) The hyperbola is H: $xy - 3y - 2x = 0$

$\Rightarrow$ $A = H + \lambda = 0$ where $\Delta = 0$

So $\begin{vmatrix} 0 & \frac{1}{2} & -1 \\ \frac{1}{2} & 0 & -\frac{3}{2} \\ -1 & -\frac{3}{2} & \lambda \end{vmatrix} = 0$

$\Rightarrow$ $\left(-\frac{1}{2}\right)\left\{\frac{\lambda}{2}-\frac{3}{2}\right\}-1\left(-\frac{3}{4}\right)=0$ i.e., $\frac{-\lambda}{4}=-\frac{6}{4}$

$\therefore$ $\lambda = 6$, hence $A = xy - 3y - 2x + 6 = 0$

Aliter: Rewritten $H + \lambda = 0$ as $(x - 3)(y - 2) = 0$

$\Rightarrow$ $\lambda = 6$

2. (d) Asymptotes parallel to $2x + 3y = 0$ and $3x + 2y = 0$ and passing through *C* (1, 2) are $2x + 3y - 8 = 0$ and $3x + 2y - 7 = 0$

$\Rightarrow$ $H = (2x + 3y - 8)(3x + 2y - 7) + \lambda = 0$

Since *P* (5, 3) lies on it so $11 \times 14 + \lambda = 0$

Hence H: $(2x + 3y - 8)(3x + 2y - 7) = 154$

3. (b) For the hyperbola $H:\frac{x^2}{a^2}-\frac{y^2}{b^2}=1$, the angle between the asymptotes is $2\theta = 2\tan^{-1}\left(\frac{b}{a}\right)=\frac{\pi}{3}$

$\therefore$ $\frac{b}{a}=\frac{1}{\sqrt{3}}$ or $a = \sqrt{3}b$

Hence $e^2 = \frac{3b^2+b^2}{3b^2}=\frac{4}{3}$

So from $\frac{1}{e^2}+\frac{1}{e_1^2}=1$, we get $\frac{1}{e_1^2}=1-\frac{3}{4}=\frac{1}{4}$ gives $e_1=2$, where e_1 is the eccentricity of conjugate hyperbola

4. (c) The asymptotes are A_1: $3x - 4y - 1 = 0$ and A_2: $4x - 3y - 6 = 0$ which intersect at *C* (3, 2), now the transverse axis and conjugate axis will be along the angle bisectors of A_1 and A_2 through *C* (3, 2) shifting the origin to *C* (3, 2), we get A_1: $3x - 4y = 0$ and A_2: $4x - 3y = 0$

Now solving, $\frac{(3X_1-4Y_1)}{5}=\frac{\pm(4X_1-3Y_1)}{5}$

$\Rightarrow$ $x_1 + y_1 = 0$ and $x_1 - y_1 = 0$

$\Rightarrow$ $(x-3)+(y-2)=0$ i.e., $x+y-5=0$ and $x-y=0$ *or* $(x-3)-(y-2)=0$ i.e., $x-y-1=0$

5. (b) The given hyperbola is $\frac{x^2}{25}-\frac{y^2}{16}=1$. The asymptotes are $y=\pm\frac{4}{5}x$

We know that area of Δ formed by a tangent and asymptotes to a hyperbola $\frac{x^2}{a^2}-\frac{y^2}{b^2}=1$ is constant = ab

∴ Here the area of Δ will be = 5 × 4 = 20 square units.

Comprehension B

6. (d) The equations of tangents at $P(t_1)$ and $Q(t_2)$ on the hyperbola $xy=c^2$ are $x+t_1^2y=2c_1t_1$ and $x+t_2^2y=2ct_2$, which will intersect

at $R(h,k)=\left(\frac{2ct_1t_2}{t_1+t_2},\frac{2c}{t_1+t_2}\right)$

Since t_1 and t_2 are the roots of $x^2-4x+2=0$

$\Rightarrow$ $t_1+t_2=4$, $t_1t_2=2$

Hence $R=\left(c,\frac{c}{2}\right)$

7. (d) α,γ are roots of equation $t_1x^2-4x+1=0$

$\Rightarrow$ $\alpha+\gamma=\frac{4}{t_1}$ and $\alpha.\gamma=\frac{1}{t_1}$ ....(i)

Also β,δ are roots of equation $t_2x^2-6x+1=0$

$\Rightarrow$ $B+\delta=\frac{6}{t^2}$ and B.δ = 1/2 ....(ii)

α,β,γ,δ are in H.P.

$\Rightarrow$ $\alpha+\gamma=4\alpha\gamma$; $B+\delta=6B\delta$

$\Rightarrow$ $\frac{1}{\gamma}+\frac{1}{\alpha}=4$; $\frac{1}{8}+\frac{1}{\beta}=6$

$\Rightarrow$ $\frac{1}{\alpha}+\frac{1}{\beta}+\frac{1}{\gamma}+\frac{1}{\delta}=10$

Now $\frac{1}{\alpha},\frac{1}{\beta},\frac{1}{\gamma},\frac{1}{\delta}$ are in AP.

$\Rightarrow$ $\frac{1}{\alpha}=a-3d$, $\frac{1}{\beta}=a-d$, $\frac{1}{\gamma}=a+d$, $\frac{1}{\delta}=af+3d$

$\Rightarrow$ $\frac{1}{\alpha}=\frac{5}{2}-3d$, $\frac{1}{\beta}=\frac{5}{2}-d$, $\frac{1}{\gamma}=\frac{5}{2}+d$, $\frac{1}{8}=\frac{5}{2}+3d$

Now $\frac{1}{\gamma}+\frac{1}{\alpha}=4$

$\Rightarrow$ $\frac{5}{2}-3d+\frac{5}{2}+d=4 \Rightarrow 5-2d=4$

$\Rightarrow$ $2d=1+d=1/2$

$\Rightarrow$ $\frac{1}{\alpha}=1,\frac{1}{\beta}=2,\frac{1}{\gamma}=3,\frac{1}{\delta}=4$

$\Rightarrow$ $t_1=\frac{1}{\alpha}.\frac{1}{\gamma}=1\ .3=3$ and $t_2=\frac{1}{\beta}.\frac{1}{\delta}=2.4=8$

∴ Equation of normal at $P(t_1)$: $\left(y-\frac{c}{3}\right)=9(x-3c)$ and

equation of normal at $Q(t_2)$: $\left(y-\frac{c}{8}\right)=64(x-8c)$

Solving above two equations we get point of intersection $\equiv\left(\frac{2329}{264}C,\frac{13921}{264}C\right)$

8. (c) Electricity of $H_1: x^2-y^2=9$ (or $xy=9/2$) is $e_1=\sqrt{2}$ and eccentricity of $H_2: x^2-y^2=25$ is $e_2=\sqrt{2}$

Now (e_1, e_2) lies on C_1 $\Rightarrow C_1: x^2+y^2=2+2=4$

$\Rightarrow$ Director circle of C_1 is $C_2: x^2+y^2=8$

Comprehension C

9. (c) The equation of normal at $P(t)$ on the hyperbola $xy=c^2$ is $xt^3-yt=c(t^4-1)$

Since it passes through m(α,β), so $ct^4-t^3\alpha+0t^2+t\beta-c=0$

Which gives four value of t i.e., $t=t_1,t_2,t_3,t_4$

$\left\{\text{Now } x_i=ct_i \text{ and } y_i=\frac{c}{t_i}\right\}$

$\Rightarrow$ $t_1+t_2+t_3+t_4=\alpha/c$, $t_1t_2t_3t_4=-1$

Now $\sum x_i=x_1+x_2+x_3+x_4=c\{t_1+t_2+t_3+t_4\}=\alpha$

10 (d) The value of $\Sigma y_i=c\left\{\frac{1}{t_1}+\frac{1}{t_2}+\frac{1}{t_3}+\frac{1}{t_4}\right\}=c\left\{\frac{\sum t_1t_2t_3}{t_1t_2t_3t_4}\right\}=c\left\{\frac{-\beta}{c(-1)}\right\}=\beta$

11. (b) $\sum x_i^2=c^2\{t_1^2+t_2^2+t_3^2+t_4^2\}=c^2\{t_1+t_2+t_3+t_4\}^2-2\sum t_1t_2$

$=c^2\left\{\frac{\alpha^2}{c^2}-2(0)\right\}=\alpha^2$

12. (a) $\sum y_i^2=y_1^2+y_2^2+y_3^2+y_4^2=c^2\left\{\frac{1}{t_1^2}+\frac{1}{t_2^2}+\frac{1}{t_3^2}+\frac{1}{t_1^2}\right\}$

$=c^2\left\{\frac{\Sigma t_1^2t_2^2t_3^2}{t_1^2t_2^2t_3^2t_4^2}\right\}=c^2\{\Sigma t_1^2t_2^2t_3^2\}$

Observe that $\sum t_1^2t_2^2t_3^2=(\Sigma t_1t_2t_3)^2-2t_1t_2t_3t_4(\Sigma t_1t_2)$

$=\left(-\frac{\beta}{c}\right)^2-2(-1)(0)=\frac{\beta^2}{c^2}$

Hence $\Sigma y_i^2=c^2\left\{\frac{\beta^2}{c^2}\right\}=\beta^2$

13. (d) The value of $\Pi xi=x_1x_2x_3x_4=c^4\{t_1t_2t_3t_4\}=-c^4$ and the value of $\Pi yi=y_1y_2y_3y_4$

$=\frac{c^4}{t_1t_2t_3t_4}=-c^4 \Rightarrow$ $\Pi x_i=\Pi y_i=-c^4$

Comprehension D

14. (b) Hyperbola is $xy=\frac{9}{2}$. Since it is a rectangular hyperbola passing through $\left(\frac{3}{\sqrt{2}},\frac{3}{\sqrt{2}}\right) \Rightarrow OV=3$

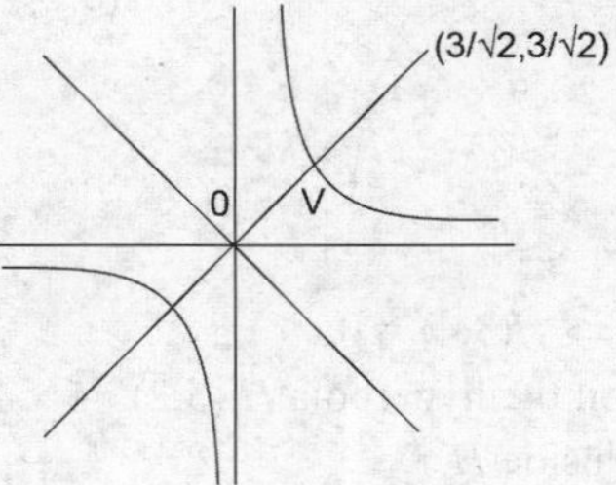

When axes are rotated by $\frac{\pi}{4}$ counter clock wise, then we will get $\frac{x^2}{(OV)^2}-\frac{y^2}{(OV)^2}=1$ i.e. $\frac{X^2}{9}-\frac{Y^2}{9}=1$

15. (c) The given hyperbola is $0x^2 + xy - y^2 - 5y + 1 = 0$

When axes are rotated anticlockwise the coefficient of XY in transformed equation vanishes.

So $\cot 2\alpha = \frac{A-C}{B} = \frac{0-(-1)}{1} = 1$; hence $2\alpha = \pi/4$

$\therefore$ $\sin\alpha = \sin \pi/8 = \frac{\sqrt{4-2\sqrt{2}}}{2\sqrt{2}}$

16. (b) Given hyperbola is $3x^2 - 6xy + 3y^2 + 2x - 7 = 0$

For the coefficient of XY in the transformed equation to vanish $\cot 2\alpha = \frac{A-C}{B} = \frac{3-3}{6} = 0$

So $2\alpha = 270° = 3\pi/2$ (since α is positive obtuse angle)

$\Rightarrow \alpha = 3\pi/4$

Now, $A' = A\cos^2\alpha + B\sin\alpha\cos\alpha + C\sin^2\alpha$

i.e., $A' = \frac{A}{2}\{1+\cos 2\alpha\} + \frac{B}{2}\sin 2\alpha + \frac{C}{2}\{1-\cos 2\alpha\}$

$= \frac{3}{2}\{1+0\} + \frac{(-6)}{2}(-1) + \frac{3}{2}\{1-0\} = 6$

17. (d) Given curve is $x^2 - 2xy + 3y^2 + 4x - 4y + 1 = 0$

Since after transformation the curve does not contain any term in Y

So $E' = -D\sin\alpha + E\cos\alpha = 0$ i.e., $-4\sin\alpha - 4\cos\alpha = 0$

$\Rightarrow$ $\sin\alpha + \cos\alpha = 0 \Rightarrow \tan\alpha = -1 \Rightarrow \alpha = \frac{3\pi}{4}$.

SECTION VII (MATRIX-MATCH TYPE)

1. (i) → (a, b); (ii) → (b, c); (iii) → (d, e)

(i) Hyperbola is $H(x, y) : x^2 - 4y^2 - 2x + 24y - 37 = 0$

Points are $P(3,4)$ and $Q(5,2)$ and $H(3,4) = 9 - 64 - 6 + 96 - 37 = -2 < 0$

and $H(5,2) = 25 - 16 - 10 + 48 - 37 = 10 > 0$

(ii) Hyperbola is $H(x,y) : xy + 2x + 3y - 12 = 0$ and $P(3,4)$

$\Rightarrow$ $H(3, 4) = 12 + 6 + 12 - 12 = 18 > 0$

$\Rightarrow$ P lies outside H. Similarly Q (5,2) gives $H(5,2) = 10 + 10 + 6 - 12 = 14 > 0$ Q lies outside H

(iii) The hyperbola is $H : xy - 12 = 0$

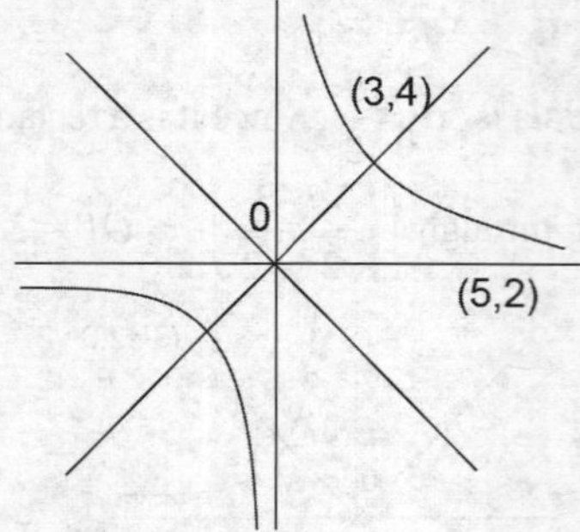

$\therefore$ $P(3,4) \Rightarrow H(3,4) = 0$

$\Rightarrow$ P lies on the hyperbola $H(5,2) : 5 \times 2 - 12 = -2 < 0$

Q lies inside H.

2. (i) → (a,c); (ii) → (e,a); (iii) → (d,b)

(i) Director circle of $\frac{x^2}{2}-\frac{y^2}{1}=1$ is $x^2 + y^2 = a^2 - b^2$ i.e., $x^2 + y^2 = 1$ and the director circle of $\frac{x^2}{2}+\frac{y^2}{1}=1$ is $x^2 + y^2 = a^2 + b^2$ i.e., $x^2 + y^2 = 3$

(ii) Director circle of $\frac{x^2}{2}+\frac{y^2}{3}=1$ is $x^2 + y^2 = 5$ and director circle of $\frac{x^2}{3}-\frac{y^2}{2}=1$ is $x^2 + y^2 = 1$.

(iii) Director circle of $\frac{x^2}{9}-\frac{y^2}{5}=1$ is $x^2 + y^2 = 4$ and director circle of $x^2 + y^2 = 1$ is $x^2 + y^2 = 2$.

3. (i) → (c,e); (ii) → (a,d); (iii) → (b)

(i) Rewriting as $12(x^2 - 2x + 1) - 4(y^2 - 8y + 16) = 75$

$\Rightarrow$ $\frac{(x-1)^2}{\left(\frac{5}{2}\right)^2} - \frac{(y-4)^2}{\left(\frac{5\sqrt{3}}{2}\right)^2} = 1$

Now $ae = \sqrt{a^2+b^2} = \sqrt{\frac{25}{4}+\frac{75}{4}} = 5$. Hence foci are at (6,4) and (–4,4)

(ii) Rewriting as $8(x^2 - 8x + 16) - (y^2 - 10y + 25) = 128 - 25 - 71 = 32$

$\Rightarrow$ $\frac{(x-4)^2}{4} - \frac{(y-5)^2}{32} = 1$, now $ae = 6$

Hence foci are (10,5) and (–2,5)

(iii) Rewriting as $9(x^2 - 4x + 4) - 16(y^2 - 6y + 9) = -144$

$\Rightarrow$ $\frac{(y-3)^2}{9} - \frac{(x-2)^2}{16} = 1 \Rightarrow ae = \sqrt{9+16} = 5$ gives foci at (2,8) and (2,–2)

4. (i) → c; (ii) → a; (iii) → d; (iv) → b,d,e

(i) Point P is $(t^2, 2t)$ and Q is $\left[\left(-t-\frac{2}{t}\right)^2, 2(-t-2/t)\right]$

$\therefore$ $-t - \frac{2}{t} \geq 2\sqrt{2}$ (by A.M. ≥ G.M. as $-t > 0$)

For minimum, $t = -\sqrt{2}$

$\therefore$ Points are $P(2, -2\sqrt{2})$ and $Q(8, 4\sqrt{2})$

$\therefore$ $PQ^2 = 108$

(ii) Equation of chord of contact is $T = 0$, therefore if P $(6\cos\theta, 6\sin\theta)$ is any point on circle $x^2 + y^2 = 36$, then

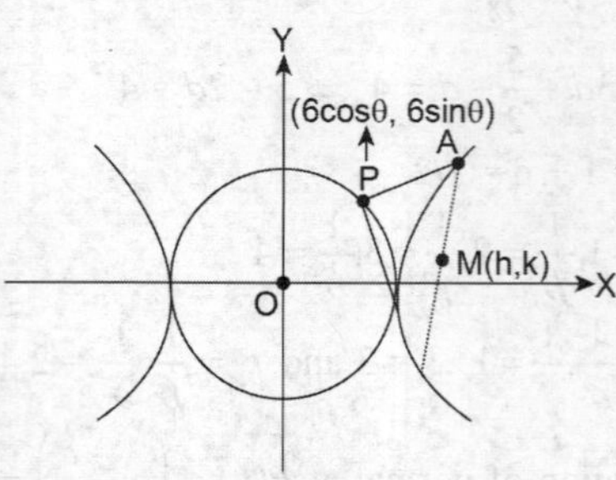

Equation of chord contact of hyperbola will be $x\cos\theta - y\sin\theta = 6$ .... (i)

Let $M(h, k)$ is mid point of AB

$\therefore$ Equation of AB is $T = S_1$

$\Rightarrow hx - ky = h^2 - k^2$ ....(ii)

$\therefore$ Comparing (i) and (ii), we get $\cos\theta = \dfrac{6h}{h^2 - k^2}$ and $\sin\theta = \dfrac{6k}{h^2 - k^2}$

Square and adding, we get $(x^2 - y^2)^2 = 36\,(x^2 + y^2)$, which is the required locus of mid point of chord of contact.

$\therefore$ 4 = 36 $\Rightarrow \lambda = 9$

(iii) Let $P\,(\sqrt{2}\cos\theta, \sin\theta)$ and $Q\,(-\sqrt{2}\sin\theta, \cos\theta)$ are extremities of chord

$\therefore$ Length $PQ = \sqrt{3+\sin 2\theta} \le 2$

$\therefore$ $(PQ)_{max.} = 2.$

(iv) Case – (i): If lines form a triangle then $n = 4$, i.e., 3 excircle and 1 incircle

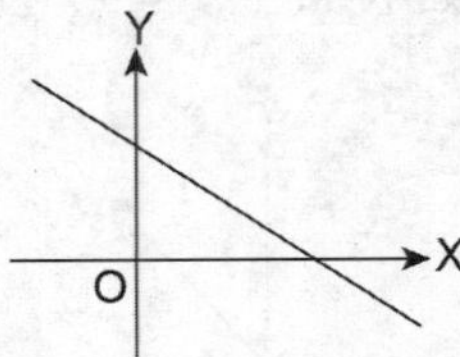

Case – (ii): Two lines are parallel and third intersect them, then n = 2

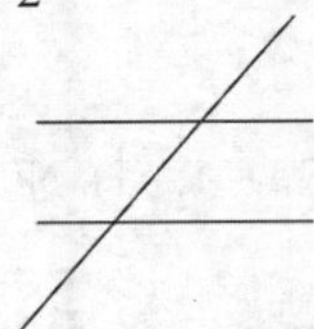

Case – (iii): All lines are parallel *or* lines are concurrent, then n = 0

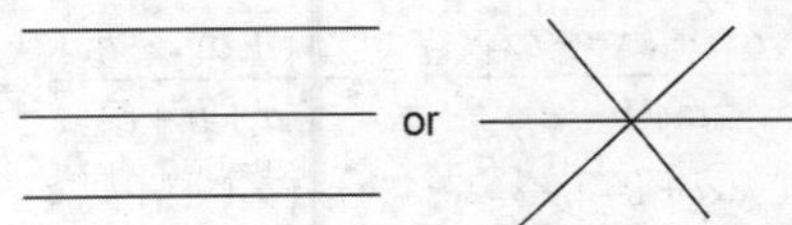

5. (i) → b, (ii) → d,(iii) → c, (iv) → a

(i) The required bisector is the one which bisects the angle supplementary to the one containing origin i.e., the bisector of angle not containing origin is required.

The two bisectors are: $\dfrac{(x-7y+5)}{\sqrt{50}} = \pm\dfrac{(-5x-5y+3)}{\sqrt{50}}$

(Here the sign of constant has to be same).

Now the required bisector is obtained by taking the negative sign. i.e., $x + 3y = 2$

(ii) The given curve is $\dfrac{(x-1)^2}{4} + \dfrac{(y-2)^2}{9} = 1$, which is an ellipse centred at (1, 2).

The required locus is the director circle of ellipse. Hence, the required locus is $(x - 1)^2 + (y - 2)^2 = 13$

(iii) $y = mx + c$ is a tangent to $\dfrac{x^2}{a^2} - \dfrac{y^2}{b^2} = 1$ if $c^2 = a^2m^2 - b^2$

Comparing the given equation with this condition, $a^2 = 3$, $b^2 = 4$

$\Rightarrow$ It is tangent to ellipse $\dfrac{x^2}{3} - \dfrac{y^4}{4} = 1$ i.e., $4x^2 - 3y^2 = 12$

(iv) In the standard form $\dfrac{x^2}{a^2} - \dfrac{y^2}{b^2} = 1$

Centre: (0, 0), vertex: (a, 0), focus: (ae, 0)

It is given that $a = \dfrac{2ae}{3}$

$\Rightarrow e = 3/2$ $\Rightarrow b^2 = \dfrac{5a^2}{4}$ $(\because b^2 = a^2(e^2 - 1))$

i.e., $\dfrac{x^2}{a^2} - \dfrac{4y^2}{5a^2} = 1$ or $5x^2 - 4y^2 = 5a^2$

However if we take $\dfrac{y^2}{a^2} - \dfrac{x^2}{b^2} = 1$; then we get equation of hyperbola as $\dfrac{y^2}{a^2} - \dfrac{4x^2}{5b^2} = 1 \Rightarrow 5y^2 - 4x^2 = 5a^2$

SECTION–VIII (INTEGER-TYPE)

1. Let $P\,(\theta\sec\theta, b\tan\theta)$ be a point on the hyperbola $\dfrac{x^2}{a^2} - \dfrac{y^2}{b^2} = 1$ with eccentricity $e = \dfrac{3}{2\sqrt{2}} = \dfrac{3}{4}\sqrt{2}$

Equation of normal at P is $\dfrac{a^2x}{a\sec\theta} + \dfrac{b^2y}{b\tan\theta} = a^2 + b^2$

Which will intersect x-axis at $L = \left\{\left(\dfrac{a^2+b^2}{a}\right)\sec\theta, 0\right\}$ and y-axis at $M = \left(0, \dfrac{a^2+b^2}{b}\tan\theta\right)$

$\Rightarrow$ Mid point of LM is $R\,(h,k)$

$\equiv \left\{\dfrac{(a^2+b^2)}{2a}\sec\theta, \dfrac{a^2+b^2}{2b}\tan\theta\right\}$

The locus of $R(h, k)$ will be $\dfrac{(2ah)^2}{(a^2+b^2)^2} - \dfrac{(2bk)}{(a^2+b^2)^2} = 1$

gives $4a^2x^2 - 4b^2y^2 = (a^2 + b^2)^2$

Which have eccentricity $e_2 = \sqrt{1+\dfrac{4a^2}{4b^2}}$ as $1 + \dfrac{b^2}{a^2} = e_1^2 = \dfrac{9}{8}$ so $\dfrac{b^2}{a^2} = \dfrac{1}{8}$ or $\dfrac{a^2}{b^2} = 8$

Hence $e_2 = \sqrt{1+8} = 3$

2. The equation of asymptotes are $A_1 = 3x - 4y - 1 = 0$ and $A_2 = 4x - 3y - 6 = 0$

Equation of transverse axis is given by

$\dfrac{3x-4y-1}{5} = \pm\dfrac{(4x-3y-6)}{5}$

$\Rightarrow$ $(3x - 4y - 1) = \pm\,(4x - 3y - 6) \Rightarrow x + y - 5 = 0$ and $7x - 7y - 7 = 0$ *or* $x - y - 1 = 0$

Equation of transverse axis is given to be $ax + by - c = 0$; $a, b, c \in \mathbb{N}$

$\Rightarrow$ $ax + by - c \equiv x + y - 5 = 0 \Rightarrow a + b + c = 1 + 1 + 5 = 7$

3. Without any loss of generality let $P(t_1)$, $Q(t_2)$ and $R(t_3)$ be the three points on the hyperbola $xy = c^2$

We know that the slope of chord with end points $P(t_1)$ and $Q(t_2)$ will be $m_{PQ} = -\dfrac{1}{t_1t_2}$

$\therefore$ Slope of perpendicular to PQ will be $m_1 = t_1t_2$

Hence equation of altitude though $R(t_3)$ will be $\left(y - \frac{c}{t_3}\right) =$ $t_1t_2(x - ct_3)$ gives $t_1t_2t_3x - t_3y - c\,t_1t_2t_3(t_3 - 1) = 0$ .....(i)

Similarly altitude through $P(t_1)$ will be $\left(y - \frac{c}{t_1}\right) = t_2t_3(x - ct_1)$

i.e., $t_1t_2t_3x - t_1y - c\,t_1t_2t_3(t_1 - 1) = 0$ ...(ii)

(ii) → (i) gives, $y(t_3 - t_1) = ct_1t_2t_3(t_1 - t_3)$

Hence $y = -ct_1t_2t_3$

Next we get $x = \frac{-c}{t_1t_2t_3}$ which lies on $xy = c^2$

Hence orthocenter will lie on the hyperbola it seft For the given hyperbola $xy = 4$

When $x = 1/2 \Rightarrow y = 8$

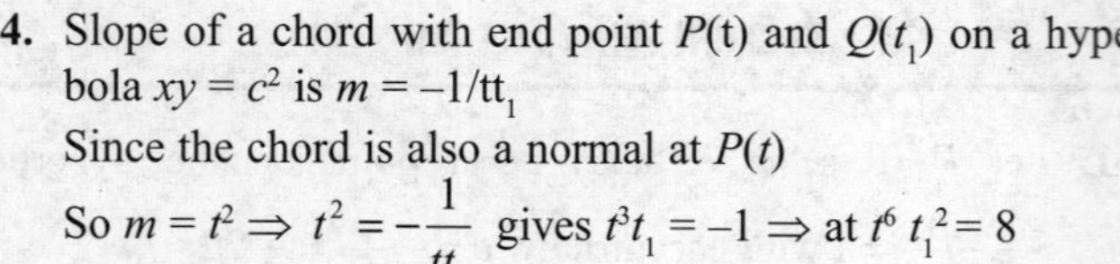

4. Slope of a chord with end point $P(t)$ and $Q(t_1)$ on a hyperbola $xy = c^2$ is $m = -1/tt_1$

Since the chord is also a normal at $P(t)$

So $m = t^2 \Rightarrow t^2 = -\frac{1}{tt_1}$ gives $t^3t_1 = -1 \Rightarrow$ at $t^6t_1^2 = 8$

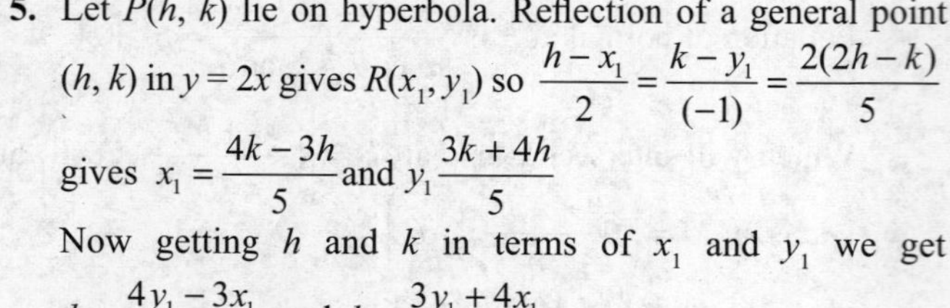

5. Let $P(h, k)$ lie on hyperbola. Reflection of a general point (h, k) in $y = 2x$ gives $R(x_1, y_1)$ so $\frac{h - x_1}{2} = \frac{k - y_1}{(-1)} = \frac{2(2h - k)}{5}$

gives $x_1 = \frac{4k - 3h}{5}$ and $y_1 \frac{3k + 4h}{5}$

Now getting h and k in terms of x_1 and y_1 we get $h = \frac{4y_1 - 3x_1}{5}$ and $k = \frac{3y_1 + 4x_1}{5}$

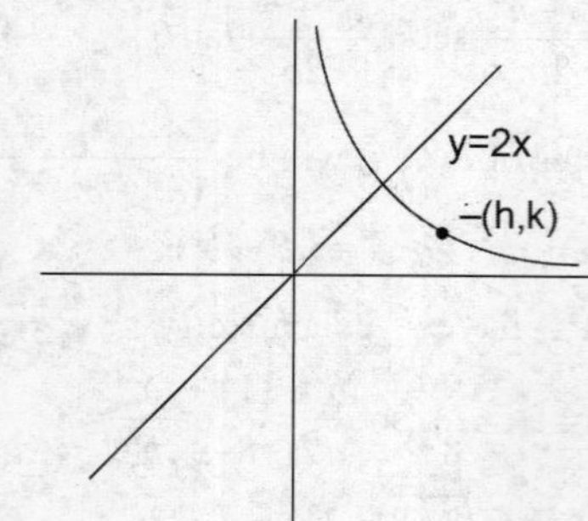

$\therefore$ $hk = 1$ gives $\left(\frac{4y_1 - 3x_1}{5}\right)\left(\frac{3y_1 + 4x_1}{5}\right) = 1$

i.e., $12x_1^2 - 7x_1y_1 - 12y_1^2 + 25 = 0 \Rightarrow rs = (-7)(-12) = 84$

6. Choose AB for x-axis and $A \equiv (-a, 0)$ and $B \equiv (a, 0)$. If $\angle A = \theta$ and $\angle B = \pi - \phi$, the given condition is $\pi - \phi = 2\theta$ *or* tan $\phi = -\tan 2\theta$, the locus of C is given by

$$\frac{y}{x - a} = -\frac{2\frac{y}{x + a}}{1 - \left(\frac{y}{x + a}\right)^2}$$

Which simplifies to $3x^2 + 2xa - y^2 = a^2$ and this can be written in the form $3\left(x + \frac{a}{3}\right)^2 - y^2 = \frac{4a^2}{3}$

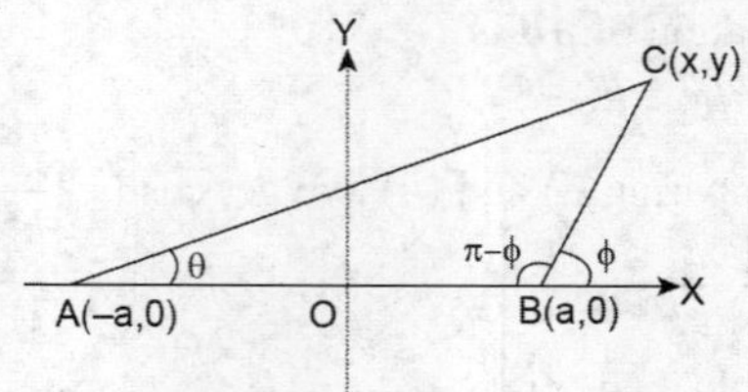

The locus is a hyperbola with centre at $\left(-\frac{a}{3}, 0\right)$ and transverse axis along the x-axis i.e., the base AB. As $(-a, 0)$ satisfies the equation, the point A is one of the vertices.

$$\text{Eccentricity} = \sqrt{\frac{\left(\frac{4a^2}{9} + \frac{4a^2}{3}\right)}{\frac{4a^2}{9}}} = 2, \text{ a (constant)}$$

7. Equation of directrix are $x + y = \pm 1$. Let $PS = a$, $SS' = b$ and $PS' = c$

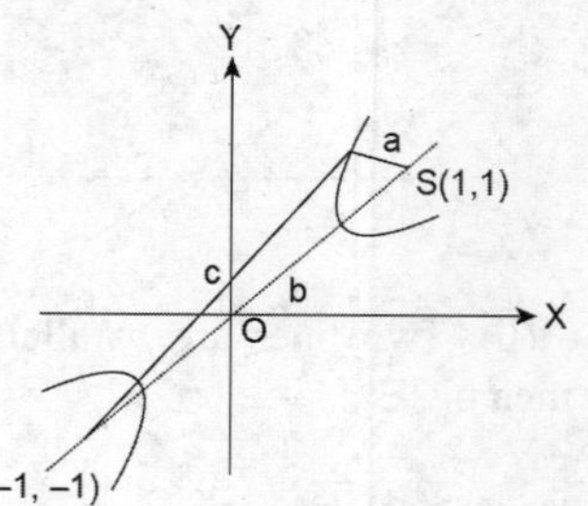

Foci are: $S(1, 1)$, S' $(-1, -1)$. Now, let $P \equiv \left(\frac{1}{\sqrt{2}}t, \frac{1}{\sqrt{2}t}\right)$;

$a = \left(\frac{1}{\sqrt{2}}t + \frac{1}{\sqrt{2}t} - 1\right)$, $b = 2\sqrt{2}$ and $c = \left(\frac{1}{\sqrt{2}}t + \frac{1}{\sqrt{2}t} - 1\right)$

Further let (h, k) be the excentre of $\Delta PSS'$ opposite to vertex S.

$$\Rightarrow h = \frac{ax_1 + bx_2 - cx_3}{a + b - c}, k = \frac{ay_1 + by_2 - cy_3}{a + b - c}$$

$$\Rightarrow h = \frac{-(a + c) + \frac{b}{\sqrt{2}}t}{2\sqrt{2} - 2}, k = \frac{-(a + c) + \frac{b}{\sqrt{2}t}}{2\sqrt{2} - 2}$$

$$\therefore h = \frac{-\sqrt{2}\left(t + \frac{1}{t}\right) + 2t}{2\sqrt{2} - 2} \quad \text{...(i)}$$

$$\text{And } k = \frac{-\sqrt{2}\left(t + \frac{1}{t}\right) + \frac{2}{t}}{2\sqrt{2} - 2} \quad \text{...(ii)}$$

$$\Rightarrow h + k = -\left(t + \frac{1}{t}\right) \quad \text{...(iii)}$$

$$\text{And } h - k = \frac{t - \frac{1}{t}}{(\sqrt{2} - 1)} \quad \text{...(iv)}$$

Eliminating t, we get $\left(t + \frac{1}{t}\right)^2 = \left(t - \frac{1}{t}\right)^2 + 4$

$\Rightarrow (h + k)^2 = (\sqrt{2} - 1)^2 (h - k)^2 + 4$

$\Rightarrow (x + y)^2 = (\sqrt{2} - 1)^2 (x - y)^2 + 4$

$\Rightarrow m = 1; n = 4$ $\Rightarrow m + n = 5$